HANDBOOK OF NEUROENDOCRINOLOGY

HANDBOOK OF NEUROENDOCRINOLOGY

Edited by

GEORGE FINK
DONALD W. PFAFF
JON E. LEVINE

ELSEVIER

AMSTERDAM • BOSTON • HEIDELBERG • LONDON
NEW YORK • OXFORD • PARIS • SAN DIEGO
SAN FRANCISCO • SINGAPORE • SYDNEY • TOKYO

Academic Press is an imprint of Elsevier

Academic Press is an imprint of Elsevier
32 Jamestown Road, London NW1 7BY, UK
225 Wyman Street, Waltham, MA 02451, USA
525 B Street, Suite 1800, San Diego, CA 92101-4495, USA

First edition 2012

British Library Cataloguing-in-Publication Data
A catalogue record for this book is available from the British Library

Library of Congress Cataloging-in-Publication Data
A catalog record for this book is available from the Library of Congress

ISBN : 978-0-12-3750976

For information on all Academic Press publications visit our website at elsevierdirect.com

Typeset by TNQ Books and Journals Pvt Ltd.

www.tnq.co.in

Printed and bound in China

12 13 14 15 16 10 9 8 7 6 5 4 3 2 1

Contents

IV
NEUROENDOCRINE PATHOLOGY

List of "How Do We Know?" Boxes

Preface

Reproduction, growth, stress, aggression, metabolism, birth, feeding and drinking, and blood pressure are some of the bodily functions that are triggered and controlled by neuroendocrine systems. Neuroendocrinology is the discipline concerned with the study of how the nervous system controls hormonal secretion by endocrine glands and, in turn, the ways in which hormones control the brain. The neural–endocrine nexus facilitates appropriate hormonal responses to (1) outside (so called exteroreceptive) influences such as day-length, stressful challenges, suckling and temperature, and (2) internal (interoreceptive) influences such as hunger, thirst, plasma osmolality, emotions, and afferent impulses from the uterus in parturition.

The main neuroendocrine nexus in vertebrates is at the base of the brain, where the pituitary gland is connected to the hypothalamus by way of the pituitary stalk – the "hypothalamic–pituitary system." The hypothalamus is connected by way of major nerve pathways with the rest of the forebrain and hindbrain, and with the peripheral and autonomic nervous systems through the spinal cord.

The human pituitary gland weighs no more than 1 gram, but nonetheless controls all the major endocrine systems and is indispensable for life. The gland is comprised of the neural and anterior lobes. Derived embryologically from a neural downgrowth, the neural lobe (neurohypophysis) is comprised of axons that project from perikarya in the hypothalamus to terminate on capillaries of the inferior pituitary artery. This is the site at which the neurohormones vasopressin and oxytocin are released into the systemic circulation, by means of which they are transported to their target organs – the kidneys, blood vessels, uterus and breast. Synthesized in hypothalamic nuclei, vasopressin controls the volume of body water, whereas oxytocin triggers milk ejection in response to suckling, or contraction of the uterus during parturition.

Derived from an outgrowth of the embryonic roof of the mouth, the anterior pituitary gland controls the adrenal and thyroid glands, the gonads, body growth and metabolism, and development of the breast and lactation. Anterior pituitary hormone secretion is under central nervous control, and is modulated by the feedback of hormones secreted by the pituitary target organs – i.e., the adrenal and thyroid glands, the gonads, and fat tissue. Neural control of anterior pituitary hormones is mediated by hypothalamic–pituitary regulatory neurohormones that are released from nerve terminals in the hypothalamus into the "hypophysial portal vessels." This set of small vessels (portal veins), in the pituitary stalk, transports the neurohormones to the anterior pituitary gland, where they either stimulate or inhibit the synthesis and release of the anterior pituitary hormones.

In addition to the hypothalamic–pituitary system, neurovascular organs (e.g., the pineal gland) that comprise neurohemal junctions surround the cerebral ventricles (the "circumventricular organs") and also satisfy the criteria of neuroendocrine systems in that they facilitate the transmission of chemical signals between soma and brain. The term "neuroendocrine" is also applied to junctional zones between nerve and endocrine cells in the periphery and especially the viscera, of which the adrenal medulla (the main source of the body's epinephrine) and the gastrointestinal system and its appendages are the most prominent.

Neuroendocrinology is pivotal to physiology and medicine in that this discipline has disclosed and underpins numerous fundamental molecular biological and genetic principles. Thus, for example, much of our knowledge of gene regulation, gene transcription and post-translational protein processing derives from neuroendocrine models. Similarly, much of our knowledge of the mechanisms of neurochemical transmitter synthesis, release (exocytosis) and stimulus–secretion coupling derives from studies of the neural lobe of the pituitary gland and the adrenal medulla.

Neuroendocrine systems have played a seminal role in enabling us to understand biological control, and especially positive and negative feedback mechanisms and biological rhythms (e.g., circhoral, circadian, seasonal, and reproductive rhythms such as estrous and menstrual cycles). Neuroendocrine mechanisms subserve many key functions of the body, and are pivotal to the body's ability to cope with stress. Consequently, neuroendocrine dysfunction due to genetic or other deficits leads to disorders such as infertility, impotence, precocious or delayed puberty, defective or excessive growth (dwarfism and acromegaly, respectively), obesity and anorexia, Cushing's syndrome, Addison's disease, hypertension, and hypothyroidism and hyperthyroidism. Neuroendocrine tumors are relatively common, and in turn neuroendocrine strategies are being used as adjunct therapy to control tumor growth.

Hormones exert powerful effects on brain and behavior. In birds and mammals, for example, development of the male brain ("sexual differentiation of the brain") depends, at least in part, on exposure to androgens secreted by the testes. Central nervous development also depends on the secretion of normal amounts of thyroid hormone in the fetus and early postnatal period. Absence of thyroid hormone during this critical period results in cretinism, due to failure of nerve cells to develop normal dendritic trees and synaptic connections. This defective brain development can be rectified by thyroid hormone administration before the end of the critical postnatal period (about 6 months of age in man).

Hormones exert powerful appetitive drives with respect to sex, water, food, brooding and nest-building, and parenting. Closely related to the need to find optimal conditions for reproduction and care of the young offspring, day-length dependent neuroendocrine mechanisms drive migration – perhaps seen most dramatically in birds, but also readily observed in other phyla. Aggression is a key behavior driven by hormones. In stags, for example, changes in day-length result in massive neuroendocrine changes that are manifested by androgen-induced growth of antlers in preparation for rutting – the annual fight for mating rights – that is also driven by androgen, and occurs each autumn. Similarly, changes in day-length switch the female into estrus, a state in which she is receptive to sexual advances of the male.

The stress response is closely related to aggressive behavior and the need to compete for territory, food, water, and appropriate mates. Orchestrated by neuroendocrine mechanisms that involve the forebrain, hypothalamus, pituitary gland, adrenal gland and autonomic nervous system, the stress response enables the individual to fight or flee, and to reset the metabolic and cardiovascular feedback control set-points to levels optimum for particular environmental situations or challenges.

This handbook is aimed at a level suitable for senior undergraduate students, graduate students, postdoctoral fellows and faculty in neuroscience, medicine, clinical psychology, neuropsychology, endocrinology and/or hormones and behavior. The handbook is also relevant for students and researchers in cognate disciplines such as psychiatry or neuropharmacology. In addition to covering neuroendocrine science, the handbook addresses clinically and socially topical areas of research, such as obesity, diabetes and the metabolic syndrome, stress and aggression, endocrine disrupters, sexual behavior and gender assignment, and neuroendocrine pathology.

We were fortunate to be able to attract internationally acknowledged experts in the field to prepare up-to-date reviews in most major areas of the field, and thereby satisfy what we believe to be a current unmet need in the discipline of neuroendocrinology. Key points have been underscored by the inclusion of "How do we know?" callout sections that highlight and demonstrate the experimental or technical foundation for a major concept, principle or methodological advance (both classic and modern). These callout sections will add to the educative value of this handbook for neuroscience and medicine.

We thank Mica Haley (Elsevier's Senior Neuroscience Editor) for her enthusiasm and encouragement, and invaluable assistance that enabled us to complete this work.

George Fink
Donald W. Pfaff
Jon E. Levine

About the Editors

George Fink

George Fink is a neuroendocrinologist and neuropharmacologist with a special interest in reproduction, stress and the pathogenesis of mental disorders. Currently, Fink is Honorary Professor in the University of Melbourne and Professorial Research Fellow of the Mental Health Research Institute, Australia, of which he was formerly Director (2004 – 2006). Following an appointment as senior lecturer in Anatomy at Monash University, Australia (1968–1971), Fink moved to Oxford University to serve as University Lecturer in Human Anatomy and Official Fellow and Tutor (Brasenose College) in Physiology and Medicine (1971 – 1980). This was followed by nearly 20 years as Director of the MRC Brain Metabolism Unit (BMU) Edinburgh. Under Fink's Directorship the BMU became renowned for integrated molecular and clinical neuroendocrinology and psychopharmacology. Between 1999 and 2003, Fink served as Vice President Research, Pharmos Corporation, an Israeli/USA biotechnology company focused on neuroinflammatory disorders.

Fink graduated MB BS (Hons and Prosector in Anatomy) and MD (awarded for a selection of his published neuroendocrine papers) from the University of Melbourne, Australia. An Oxford DPhil was gained under tenure of a Nuffield Dominions Demonstratorship in the laboratory of Geoffrey Harris FRS. Fink's recent research, focused on sex steroid control of central neurotransmission, is relevant to our understanding of gender differences in disorders of mood and mental state. His earlier research discoveries include (i) characterization of neurohormone release into hypophysial portal blood, (ii) the mechanism by which estrogen positive feedback triggers the ovulatory gonadotropin surge, (iii) the self-priming effect of gonadotropin releasing hormone, and (iv) aspects of adrenal and gonadal steroid control of pituitary hormone secretion by way of feedback and servomechanisms. Fink's research contributions are reported in more than 360 scientific publications including Elsevier's *Encyclopedia of Stress* of which Fink was founding Editor and which was awarded the 2001 BMA Medical Book Commendation in Mental Health.

Fink has served on numerous editorial boards and scientific committees including Chair of the Brain Research Panel of the EU Biomedical Programs 3 and 4, the Mental Health Panel of the Wellcome Trust, Council of the European Neuroscience Association and President of the European Neuroendocrine Association.

His distinctions include Fellow of the Royal Society of Edinburgh (elected 1989), Fellow of the Royal College of Physicians of Edinburgh (elected 1998), Honorary Professor in the University of Edinburgh (appointed 1984), Lifetime Achievement Award of the International Society of Psychoneuroendocrinology (2000), Honorary Member of the British Society for Neuroendocrinology (2005) and Distinguished Scholar of Melbourne High School (2000). Fink has given numerous invited lectures including the inaugural Geoffrey Harris Prize Lecture of the British Physiological Society and the Wolfson Lecture. His other appointments include: Visiting Professor at the Rockefeller University (Neurobiology and Behavior Laboratory) New York; Visiting Professor in Neurobiology, the Mayo Clinic, Rochester Minnesota; Arthur Fishberg Professor, Mount Sinai Medical Center, New York; Walter Cottman Visiting Professor at Monash University; Royal Society-Israel Academy Exchange Fellow at the Weizmann Institute, Rehovot, Israel.

Donald W. Pfaff

Donald W. Pfaff, Ph.D., professor and head of the Laboratory of Neurobiology and Behavior at The Rockefeller University, is a brain scientist who uses neuroanatomical, neurochemical and neurophysiological methods to study the cellular mechanisms by which the brain controls behavior. His laboratory's research has proceeded to demonstrate how steroid hormone effects on nerve cells can direct natural, instinctive behaviors.

In 2003, Pfaff received an NIH MERIT Award for the study of generalized arousal, responsible for activating all behavioral responses. His team formulated the first operational definition of nervous system arousal, enabling scientists to measure arousal quantitatively in laboratory animals, as well as in human beings. In humans, deficits in arousal contribute to such cognitive problems as attention deficit hyperactivity disorder, autism and Alzheimer's disease.

Pfaff has made fundamental contributions to our understanding of how the administration of sex hormones can affect health. Pfaff's lab recently showed that giving hormone doses in pulses, rather than as a steady exposure, may maximize the benefits and limit the side effects now associated with hormone therapies. By giving estrogen replacement to the rats, the scientists studied the actions of the hormone at the level of the brain cell's protective outer membrane, and inside the nucleus where the cell's DNA is housed. They found that both the membrane and the DNA pathways are crucial, with one facilitating the other, in triggering hormone-dependent gene expression and female mating behavior. By limiting the estrogen exposure of to short pulses, the total dose can be kept much smaller than with steady delivery, and therefore some of the negative effects will be reduced.

Born in Rochester, N.Y., on December 9, 1939, he received the A.B. degree magna cum laude from Harvard College in 1961 and a Ph.D. from the Massachusetts Institute of Technology in 1965. He held a National Merit Scholarship, Harvard National Scholarship, Woodrow Wilson Fellowship, MIT President's Award Fellowship, National Institutes of Health Predoctoral Fellowship and National Science Foundation Postdoctoral Fellowship.

Pfaff joined The Rockefeller University in 1966 as a postdoctoral fellow. He was named assistant professor in 1969, associate professor in 1971, granted tenure in 1973 and promoted to full professor in 1978. He is a member of the U.S. National Academy of Sciences and a fellow of the American Academy of Arts and Sciences. He also is a member of several scientific organizations related to studies of the central nervous system.

He is the author of "Estrogens and Brain Function" (Springer, 1980); "Drive: Neurobiological and Molecular Mechanisms of Sexual Motivation" (MIT Press, 1999); and "Brain Arousal and Information Theory" (Harvard University Press, 2005). He has edited 'The Physiological Bases of Motivation' (1982), 'Ethical Questions in Brain and Behavior' (1984), 'Genetic Influences on the Nervous System' (CRC Press, 1999) and 'Hormones, Brain and Behavior' (5 volumes, Academic Press, 2002, 2009). He also is on the editorial boards of several scientific journals.

Jon E. Levine

Dr. Jon E. Levine, Ph.D. completed his B.A. at Oberlin College in Oberlin, Ohio, and his Ph.D. from the University of Illinois, Champaign-Urbana. Dr. Levine completed postdoctoral training at the Oregon National Primate Research Center & Oregon Health Sciences University. Levine joined the faculty at Northwestern University in Evanston, Illinois in 1984, and remained there as Professor in the Department of Neurobiology and Physiology until 2010. While on the faculty at Northwestern, Dr. Levine served as Director of the Program in Biological Sciences (1999–2006), and as Director of a NIH-sponsored Training Program in Reproductive Biology (1991–2010). He is currently the Director of the Wisconsin National Primate Research Center and Professor in the Department of Neuroscience at the University of Wisconsin-Madison. For the past 30 years Dr. Levine has studied the neuroendocrine regulation of gonadotropin releasing hormone (GnRH) neurons. Dr. Levine's research has also focused on the molecular and cellular mechanisms by which ovarian steroids exert their physiological and behavioral effects in the brain, including the negative feedback mechanisms that maintain homeostatic control within the reproductive axis, as well as the positive feedback actions of steroids that trigger preovulatory gonadotropin surges. His recent work has made use of newly developed mutant mice to analyze the cell signaling mechanisms that mediate negative and positive feedback actions of estradiol, as well as the effects of estrogens on energy homeostasis and body weight. Dr. Levine is currently Editor-in-Chief of the journal Frontiers in Neuroendocrinology, and is a member of the Steering Council for the Office of Research on Women's Health at the NIH. He is an active member of numerous professional societies including the Endocrine Society, Society for Neuroscience, Society for the Study of Reproduction, American Neuroendocrine Society, and the Society for Behavioral Neuroendocrinology.

List of Contributors

Greti Aguilera Program on Developmental Endocrinology and Genetics, Eunice Kennedy Shriver Institute of Child Health and Human Development, NIH, Bethesda, Maryland, USA

Deanna M. Arble Center for Sleep and Circadian Biology, Department of Neurobiology and Physiology, Northwestern University, Evanston, IL, USA

Iñigo Azcoitia Departamento de Biología Celular, Facultad de Biología, Universidad Complutense de Madrid, Madrid, Spain

Sarah L. Berga Division of Reproductive Endocrinology and Infertility, Departments of Gynecology and Obstetrics and Psychiatry, Emory University School of Medicine, Atlanta, GA, USA

Wah Chin Boon Florey Neuroscience Institutes, Centre for Neuroscience, University of Melbourne, and Department of Anatomy and Development, Monash University, Melbourne, Australia

Pierre M.G. Bouloux Centre for Neuroendocrinology, University College London Medical School, Royal Free Hospital, Hampstead, London, UK

Cyril Y. Bowers Division of Endocrinology, Department of Internal Medicine, Tulane University Health Sciences Center, New Orleans, LA, USA

Ellen R. Busby University of Victoria, Department of Biology, Victoria, BC, Canada

Rebecca Campbell Department of Physiology, School of Medical Science, University of Otago, Dunedin, New Zealand

Frances A. Champagne Department of Psychology, Columbia University, New York, NY, USA

Evangelia Charmandari First Department of Pediatrics, University of Athens Medical School, "Aghia Sophia" Children's Hospital, Athens, Greece; and Division of Endocrinology and Metabolism, Biomedical Research Foundation of the Academy of Athens, Athens, Greece

Alon Chen Department of Neurobiology, Weizmann Institute of Science, Rehovot, Israel

George P. Chrousos First Department of Pediatrics, University of Athens Medical School, "Aghia Sophia" Children's Hospital, Athens, Greece; and [3]Division of Endocrinology and Metabolism, Biomedical Research Foundation of the Academy of Athens, Athens, Greece

Iain J. Clarke Department of Physiology, Monash University, Clayton, Victoria, Australia

Georges Copinschi Laboratory of Physiology, School of Medicine, Université Libre de Bruxelles, Brussels, Belgium

Virginia L. Crowder Department of Psychiatry, University of North Carolina at Chapel Hill, Chapel Hill, NC, USA

James P. Curley Department of Psychology, Columbia University, New York, NY, USA

Wouter W. de Herder Department of Internal Medicine, Section of Endocrinology, Erasmus Medical Center, Rotterdam, The Netherlands

E. Ronald de Kloet LACDR, Leiden University, Leiden, The Netherlands

Ines Donangelo Pituitary Center, Cedars-Sinai Medical Center, Los Angeles, CA, USA

Alison J. Douglas Centre for Integrative Physiology, University of Edinburgh, Edinburgh, UK

Dana Erickson Endocrine Research Unit, Mayo School of Graduate Medical Education, Clinical Translational Science Center, Mayo Clinic, Rochester, MN, USA

Richard A. Feelders Department of Internal Medicine, Section of Endocrinology, Erasmus Medical Center, Rotterdam, The Netherlands

George Fink Mental Health Research Institute, University of Melbourne, Melbourne, Victoria, Australia

Luis M. Garcia-Segura Instituto Cajal, Consejo Superior de Investigaciones Científicas, Madrid, Spain

Thang S. Han Centre for Neuroendocrinology, University College London Medical School, Royal Free Hospital, Hampstead, London, UK

Leo J. Hofland Department of Internal Medicine, Section of Endocrinology, Erasmus Medical Center, Rotterdam, The Netherlands

Ali Iranmanesh Endocrine Service, Medical Section, Salem Veterans Affairs Medical Center, Salem, VA, USA

Marian Joëls Rudolf Magnus Institute, UMC Utrecht, Utrecht, The Netherlands

Ilia N. Karatsoreos Harold and Margaret Milliken Hatch, Laboratory of Neuroendocrinology, The Rockefeller University, New York, NY, USA

Henk Karst Rudolf Magnus Institute, UMC Utrecht, Utrecht, The Netherlands

Tomoshige Kino Unit on Molecular Hormone Action, Program in Reproductive and Adult Endocrinology, *Eunice Kennedy Shriver* National Institute of Child Health and Human Development, National Institutes of Health, Bethesda, MD, USA

Joseph R. Kurian Wisconsin National Primate Research Center, University of Wisconsin, Madison, WI, USA

Dik J. Kwekkeboom Department of Nuclear Medicine, Erasmus Medical Center, Rotterdam, The Netherlands

Steven W. Lamberts Department of Internal Medicine, Section of Endocrinology, Erasmus Medical Center, Rotterdam, The Netherlands

Lawrence C. Layman Section of Reproductive Endocrinology, Infertility, & Genetics, Dept Obstetrics and Gynecology, Developmental Neurobiology Program, Institute of Molecular Medicine and Genetics, Neuroscience Program, Georgia Health Sciences University, Augusta, GA, USA

Gareth Leng Centre for Integrative Physiology, University of Edinburgh, Edinburgh, UK

Jon E. Levine Wisconsin National Primate Research Center, and Department of Neuroscience University of Wisconsin, Madison, WI, USA

Malcolm J. Low Department of Molecular & Integrative Physiology, Department of Internal Medicine, Division of Metabolism, Endocrinology and Diabetes, and Brehm Center for Diabetes Research, University of Michigan Medical School, Ann Arbor, MI, USA

Mike Ludwig Centre for Integrative Physiology, University of Edinburgh, Edinburgh, UK

Margaret M. McCarthy Departments of Physiology and Psychiatry, University of Maryland School of Medicine, Baltimore, MD, USA

Bruce S. McEwen Harold and Margaret Milliken Hatch, Laboratory of Neuroendocrinology, The Rockefeller University, New York, NY, USA

Marilyn Y. McGinnis Department of Pharmacology and Center for Biomedical Neuroscience, University of Texas Health Sciences Center at San Antonio, San Antonio, TX, USA

Shlomo Melmed Pituitary Center, Cedars-Sinai Medical Center, Los Angeles, CA, USA

Robert P. Millar Centre for Integrative Physiology, University of Edinburgh, School of Biomedical Sciences, Hugh Robson Building, Edinburgh, UK; UCT/MRC Group for Receptor Biology, University of Cape Town, Cape Town, South Africa; Mammal Research Institute, University of Pretoria, Hatfield, Pretoria, South Africa

Ali Mohamadi Division of Endocrinology, Johns Hopkins School of Medicine, Baltimore, MD, USA

Randy J. Nelson Departments of Neuroscience and Psychology, Ohio State University, Columbus, OH, USA

Claire L. Newton Centre for Integrative Physiology, University of Edinburgh, School of Biomedical Sciences, Hugh Robson Building, Edinburgh, UK

Donald W. Pfaff Neurobiology and Behavior, The Rockefeller University, New York, NY, USA

Vincent Prevot Inserm, Jean-Pierre Aubert Research Center, U837, Development and Plasticity of the Postnatal Brain, Lille, France; Université Nord de France, Lille, France; UDSL, Laboratory of Anatomy, School of Medicine, Place de Verdun, Lille, France; CHRU Lille, Department of Neurosurgery, Hôpital Roger Salengro, Lille, France

Graeme J. Roch University of Victoria, Department of Biology, Victoria, BC, Canada

Ferdinand Roelfsema Department of Endocrinology and Metabolic Diseases, Leiden University Medical Center, The Netherlands

Antonia K. Roseweir Centre for Integrative Physiology, University of Edinburgh, School of Biomedical Sciences, Hugh Robson Building, Edinburgh, UK

David R. Rubinow Department of Psychiatry, University of North Carolina at Chapel Hill, Chapel Hill, NC, USA

Katya B. Rubinow Division of Endocrinology in the Department of Medicine, University of Washington School of Medicine; Seattle, WA, USA

Randall R. Sakai University of Cincinnati College of Medicine, Department of Psychiatry and Behavioral Neurosciences, Cincinnati, OH, USA

Roberto Salvatori Division of Endocrinology, Johns Hopkins School of Medicine, Baltimore, MD, USA

Willis K. Samson Department of Pharmacological and Physiological Science, Saint Louis University School of Medicine, St Louis, MO, USA

Peter J. Schmidt Section on Behavioral Endocrinology, NIMH, Bethesda, MD, USA

Karen A. Scott University of Cincinnati College of Medicine, Department of Psychiatry and Behavioral Neurosciences, Cincinnati, OH, USA

Barbara B. Sherwin Department of Psychology, McGill University, Montreal, Quebec, Canada

Nancy M. Sherwood University of Victoria, Department of Biology, Victoria, BC, Canada

Aniket R. Sidhaye Division of Metabolism, Departments of Pediatrics and Medicine, Johns Hopkins University School of Medicine, Baltimore, MD, USA

Evan R. Simpson Prince Henry's Institute, Melbourne, Australia

Jeremy T. Smith Department of Physiology, Monash University, Clayton, Victoria, Australia

Yolanda R. Smith Division of Reproductive Endocrinology and Infertility, Department of Obstetrics and Gynecology, University of Michigan School of Medicine, Ann Arbor, MI, USA

Edward M. Stricker Department of Neuroscience, University of Pittsburgh, Pittsburgh, PA, USA

Yehezkel Sztainberg Department of Neurobiology, Weizmann Institute of Science, Rehovot, Israel

Kellie L.K. Tamashiro The Johns Hopkins University School of Medicine, Department of Psychiatry and Behavioral Sciences, Baltimore, MD, USA

Ei Terasawa Wisconsin National Primate Research Center, and Department of Pediatrics, University of Wisconsin, Madison, WI, USA

Brian C. Trainor Department of Psychology, University of California, Davis, CA, USA

Fred W. Turek Center for Sleep and Circadian Biology, Department of Neurobiology and Physiology, Northwestern University, Evanston, IL, USA

Eve Van Cauter Sleep, Metabolism and Health Center Department of Medicine, University of Chicago Pritzker School of Medicine, Chicago, IL, USA

Johannes D. Veldhuis Endocrine Research Unit, Mayo School of Graduate Medical Education, Clinical Translational Science Center, Mayo Clinic, Rochester, MN, USA

Joseph G. Verbalis Department of Medicine, Georgetown University Medical Center, Washington, DC, USA

Martha H. Vitaterna Center for Sleep and Circadian Biology, Department of Neurobiology and Physiology, Northwestern University, Evanston, IL, USA

Alan G. Watts Department of Biological Sciences, USC College of Letters, Arts, and Sciences, University of Southern California, Los Angeles, CA, USA

Fredric E. Wondisford Division of Metabolism, Departments of Pediatrics and Medicine, Johns Hopkins University School of Medicine, Baltimore, MD, USA

Susan Wray Cellular and Developmental Neurobiology Section, NINDS, NIH, Bethesda, Maryland, USA

Gina L.C. Yosten Department of Pharmacological and Physiological Science, Saint Louis University School of Medicine, St Louis, MO, USA

R. Thomas Zoeller Biology Department, University of Massachusetts, Amherst, MA, USA

Notes on Nomenclature

CATECHOLAMINES

Adrenaline and noradrenaline are catecholamines that play a pivotal role in the stress and other neuroendocrine responses. These terms are synonymous with epinephrine and norepinephrine, respectively. Both sets of terms are used interchangeably in the endocrine and neuroendocrine literature, and this principle has been adopted for the Handbook. Style has depended on author preference.

CORTICOTROPIN-RELEASING FACTOR/HORMONE

The central nervous regulation of the anterior pituitary gland is mediated by substances, mainly peptides, synthesized in the hypothalamus and transported to the gland by the hypophysial portal vessels. Because these compounds are transported in the blood stream, the term "neurohormone" gained acceptance in the neuroendocrine literature. This convention, endorsed by the Endocrine Society (USA), has been applied without dissension to all hypothalamic peptide neurohormones with the exception of the 41-amino acid residue corticotropin-releasing factor (CRF). Hauger and colleagues,[1] in liaison with the International Union of Pharmacology Committee on Receptor Nomenclature and Drug Classification, argue that the CRF's function extends well beyond the biology of a hormone, and that therefore it should be termed corticotropin-releasing factor (CRF) rather than hormone. Since the terminology of CRF versus CRH has yet to be resolved, the two terms and abbreviations are here used interchangeably, depending on author preference.

GONADOTROPIN-RELEASING HORMONE (GnRH)/LHRH

The anterior pituitary gland secretes two gonadotropins, luteinizing hormone (LH) and follicle-stimulating hormone (FSH). At the start of research into the nature of hypothalamic pituitary releasing (subsequently "regulatory") factors it was assumed that neural control of LH and FSH would be mediated by distinct LH-releasing and FSH-releasing factors. However, despite extensive studies over the past 55 years, current evidence points to the fact that neural control of the synthesis and release of both gonadotropins is mediated by one and the same decapeptide. This was at first termed LH-releasing factor (LRF), then LH-releasing hormone (LHRH), and then finally, because it mediates neural control of both gonadotropins, gonadotropin-releasing hormone (GnRH). The latter, adopted as the standard term and abbreviation by the Endocrine Society, is used by most authors in this Handbook. However, readers should be aware that GnRH, LHRH and LRF are synonymous.

Reference

1. Hauger RL, Grigoriadis DE, Dallman MF, Plotsky PM, Vale WW, Dautzenberg FM. International Union of Pharmacology. XXXVI. Current status of the nomenclature for receptors for corticotropin-releasing factor and their ligands. *Pharmacol Rev.* 2003;55(1):21–26.

PART I

BASIC PRINCIPLES OF NEUROENDOCRINOLOGY

CHAPTER

1

An Introduction to Neuroendocrine Systems

Jon E. Levine

Wisconsin National Primate Research Center, and Department of Neuroscience, University of Wisconsin, Madison, WI, USA

OUTLINE

Summary

The field of neuroendocrinology is concerned with the interrelationships between the neural and endocrine systems, and addresses regulatory mechanisms of two basic types: the neural control of hormone secretions, and the actions of peripheral hormones on neurophysiological processes and behavior. In most cases, neuroendocrine systems of both types are linked together in broader bidirectional control mechanisms which govern vitally important processes that include growth, reproduction, metabolism and energy homeostasis, electrolyte and water balance, and responses to stress. The editors and authors of this volume of work have endeavored to provide a comprehensive presentation of the "state-of-the-science" of neuroendocrinology, presented in a format that will be accessible for both generalist and specialist students and researchers who seek a current understanding of neuroendocrine systems. Chapter 1 introduces the basic anatomical and physiological components of neuroendocrine systems, and defines and describes their fundamental operating characteristics.

NEUROENDOCRINOLOGY DEFINED

Neuroendocrinology is a relatively new science that emerged in the mid-20th century as a branch of endocrinology, propelled in part by the realization that the brain produces neurohormones and thereby functions as an endocrine organ. During the same period, fundamental observations were made in the relatively new field of behavioral neuroscience, revealing effects of peripheral hormones on brain function, neural development, and behavior. We now consider that both processes — control of hormone secretions by the brain and effects of hormones on brain function — together define the scope of modern neuroendocrine science. *Neuroendocrine systems* can be defined as the sets of neurons, glands and non-endocrine tissues, and the neurochemicals,

Handbook of Neuroendocrinology, DOI: 10.1016/B978-0-12-375097-6.10001-0

hormones, and humoral signals they produce and receive, that function in an integrated manner to collectively regulate a physiological or behavioral state. *Neuroendocrine integration* is the process by which neuroendocrine systems register, transduce, and interpret important signals from the internal and external environment, and thereafter direct adaptive changes in prevailing physiological and behavioral states. In this introductory chapter, we define and describe different types of neuroendocrine systems, and review the basic integrative mechanisms that each employ to operate under normal physiological circumstances.

NEUROSECRETION

By the 1920s, the existence of the major hormones of the pituitary gland, and their effects on the gonads, adrenal glands and growth, had been established. In the years following, there was greater awareness that the functions of the pituitary gland are largely governed by neural influences. The concept of *neurosecretion* – the production and secretion of neurohormones by neurons, and the actions of these hormones at target tissues – was proposed by Ernst and Berta Scharrer[1] based on their work in fish, and subsequently confirmed in a variety of species in the succeeding decades. This seminal advance is regarded as the launching point for the study of neuroendocrine systems, and perhaps as the start of the field of neuroendocrinology as a whole. Recognition that specialized *neurosecretory cells* can release hormones at neurovascular junctions provided the conceptual framework for understanding the two major neuroendocrine systems that govern pituitary function. In one, neurosecretion of neurohormones (*vasopressin* and *oxytocin*) occurs at neurovascular junctions in the posterior pituitary gland, directly into the systemic circulation to act at distant target tissues. In the other, hypothalamic neurohormones that were later identified as hypothalamic releasing and inhibiting factors,[2,3] are released from neurovascular junctions in the median eminence of the hypothalamus, into a hypothalamo-hypophysial portal vessel system that conveys these factors to their target cells in the anterior pituitary gland. The basic structural and physiological features of these neuroendocrine systems are described below.

THE BASIC ANATOMY OF NEUROENDOCRINE SYSTEMS

The Hypothalamus

In 1859, the celebrated French physiologist Claude Bernard noted that the "constancy of the internal milieu" is essential for the survival and perpetuation of warm-blooded animals. The regulation of the internal state in the face of changing external and internal conditions is *homeostasis* – a process that requires coordinated control over endocrine, behavioral and autonomic nervous system responses to the environment. It is clear that the *hypothalamus* – and the neuroendocrine neuronal systems that reside in it – has evolved to assume a critically important role as a major integrative center for mediating these homeostatic functions. It is located at the base of the forebrain, where it can direct the endocrine functions of the pituitary gland, while also receiving hormonal signals from the periphery. The hypothalamus is also adjacent to – and highly interconnected with – limbic and cortical structures and brainstem autonomic centers. The hypothalamus is uniquely positioned to send and receive endocrine signals, as well as neural signals from sensory systems, emotion- and memory-processing circuitries, and autonomic centers. Neural and endocrine information is continuously transduced, integrated and interpreted in hypothalamic neurons, and appropriate homeostatic signals are conveyed back to these endocrine, autonomic and behavioral control systems to affect coordinated changes, when necessary, in hormone secretions, autonomic outflow and behavioral state. The hypothalamus is thus responsible for monitoring the internal and external environment and coordinating adaptive physiological responses among several systems.

The hypothalamus is defined anatomically as an area of gray matter that is located in the basal forebrain, consisting of two symmetrical halves divided medially by the third ventricle. It emerges in the developing diencephalon to be bounded rostrally by the optic chiasm, caudally by the mammillary bodies, laterally by the optic tracts, and dorsally by the thalamus. The preoptic area (POA) lies rostrally to the hypothalamus, and although it is considered telencephalic in origin, it is often regarded as a functional hypothalamic tissue. Groups of neuronal cell bodies and their neuropils constitute bilaterally symmetrical *hypothalamic nuclei*, as schematically depicted in Fig. 1.1. Although there are species differences in the topography of the major hypothalamic nuclei as well as the less distinct hypothalamic areas, in general the anterior region of the hypothalamus contains the *supraoptic nucleus* (*SON*), *paraventricular nucleus* (*PVN*), *suprachiasmatic nucleus* (*SCN*) and *anterior hypothalamic area* (*AHA*), and a *periventricular nucleus* that continues caudally; the middle region of the hypothalamus includes the *arcuate nucleus* (*AN*), *ventromedial nucleus* (*VMN*), *dorsomedial nucleus* (*DMN*) and *lateral hypothalamic area* (*LHA*); and within the posterior region of the hypothalamus lie the *posterior hypothalamic nucleus* (*PHN*) and the *premammillary nucleus* (*PMN*).

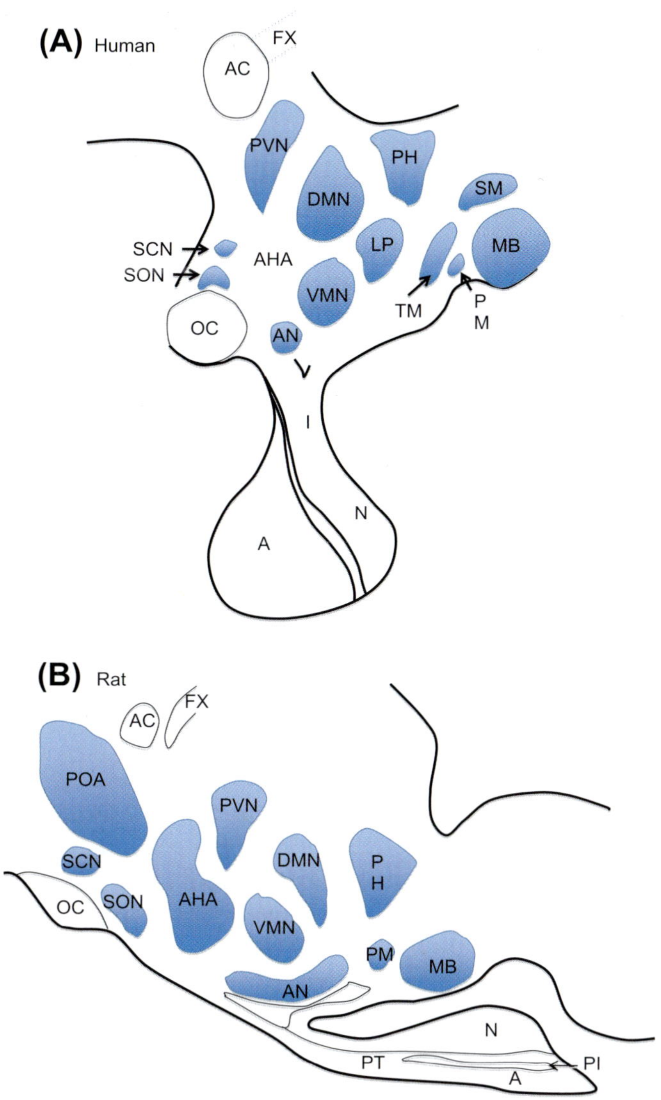

FIGURE 1.1 **Schematic diagrams depicting the anatomical locations of hypothalamic nuclei in parasagittal sections of (A) human and (B) rat basal forebrain.** Figures are oriented as anterior to posterior (rostral to caudal) from left to right. Abbreviations: AC, anterior commissure; A, adenohypophysis; AHA, anterior hypothalamic area; AN, arcuate nucleus; DMN, dorsomedial hypothalamic nucleus; FX, fornix; I, infundibulum; LP, lateral posterior nucleus; MB, mammillary body; N, neurohypophysis; OC, optic chiasm; POA, preoptic area; PH, posterior hypothalamic nucleus; PI, pars intermedia; PM, premammillary nucleus; PT, pars tuberalis; PVN, paraventricular nucleus; SCN, suprachiasmatic nucleus; SM, supramammillary nuceleus; SON, supraoptic nucleus; TM, tuberomammillary nucleus, VMN, ventromedial nucleus.

Classical studies utilized lesion and electrical stimulation techniques to ascribe integrative functions to specific nuclei as "centers" for the control of functions such as feeding, drinking, sexual behavior, stress responses, and electrolyte and water balance. The "center" concept has been somewhat outmoded with the recognition that these regulatory systems are distributed throughout many interconnected neuronal populations within and beyond the hypothalamus. Nevertheless, several of the hypothalamic nuclei do contain well-characterized neurohormone and neurotransmitter-producing cell groups that clearly serve specific integrative and effector functions that are essential components of these systems (Table 1.1). The PVN and SON, for example, are critically important in the regulation of electrolyte and water balance by virtue of the fact that magnocellular neurons within these nuclei produce the antidiuretic neurohypophysial hormone *vasopressin*, as described below; parvocellular neurons within the PVN express *corticotropin-releasing hormone* (CRH) and thereby regulate neural and hormonal responses to stress; the POA similarly contains neurons that produce the central neurohormonal effector of the reproductive axis, *gonadotropin-releasing hormone (GnRH)*, and is therefore essential for the maintenance of gonadal function in rodents; the AN includes neuronal groups that express orexigenic and anorexigenic peptides, as well as receptors for peripheral metabolic hormones, and is thus integrally important in the regulation of food intake and energy expenditure; and subsets of neurons in the VMN express receptors for ovarian hormones, *estrogens* and *progestins*, and mediate many of the actions of these hormones on female sexual behavior. The concept of a neural "center" might be best retained for the case of the SCN, which contains neurons that exhibit circadian pacemaking activity and hence function as a 24-hour clock that controls circadian rhythms in physiology and behavior. The roles played by all of the foregoing cell groups, and the nuclei within which they reside, are described in the context of each of the major neuroendocrine systems characterized in subsequent sections in this and succeeding chapters.

The homeostatic functions of the hypothalamus require that afferent neural signals, derived from brain regions concerned with sensory processing, memory and emotion, are processed and integrated with humoral signals; efferent neural signals that produce appropriate alterations in neurohormone secretions and transmission through efferent pathways are consequently initiated. Not surprisingly, the hypothalamus is known to receive and send neural signals through efferent and afferent pathways that connect it with a ring of subcortical limbic structures known to be critically important in emotional status, such as the *amygdala*; in learning and memory, such as the *hippocampus*; and in autonomic nervous system control, such as the *lower brainstem* nuclei. For example, the major connections between the amygdala and the hypothalamus include the *stria terminalis* and a *direct amygdalohypothalamic* tract; the hippocampus connects with the POA, AN and mammillary bodies via the *fornix*; the *medial forebrain bundle* connects the hypothalamus to the more anterior *septal area* of the basal forebrain; the

TABLE 1.1 Selected List of Chemical and Functional Characteristics of Several Hypothalamic Nuclei in the Rat

Nucleus	Cell bodies producing...	Functional roles
PVN, SON (magnocellular)	oxytocin, vasopressin	Electrolyte and water balance, blood pressure (vasopressin); milk ejection, uterine contractility (oxytocin)
PVN (parvicellular)	CRH, TRH, GR	Stress responses, neurosecretory control of HPA and HPT axis
SCN	vasopressin, VIP	Circadian rhythms
Periventricular nuclei	SST, kisspeptin, ERα, ERβ,	Inhibition of GH secretion, control of ovulatory cyclicity
POA	GnRH, TRH, ERα, ERβ, PR, AR	Neurosecretory regulation of HPG axis, HPT axis; male sexual behavior
AHA		Parasympathetic control, thermoregulation
VMN	GHRH, ERα, PR	Satiety, female sexual behavior
AN	POMC, NPY, AgRP, GHRH, DA, kisspeptin, ERα, PR, GR, leptin receptors	Food intake, energy expenditure, neurosecretory control of PRL and GH
DMN	NPY, GR	Behavioral rhythms, blood pressure, heart rate
PH		Sympathetic control, thermoregulation

Abbreviations not defined elsewhere in this chapter: GR, glucocorticoid receptor; VIP, vasoactive intestinal polypeptide; ERα, ERβ, estrogen receptors of the alpha and beta isoforms; PR, progesterone receptors; AR, androgen receptors; POMC, proopiomelanocortin; NPY, neuropeptide Y; AgRP, agouti-related peptide.

hypothalamus is highly interconnected with the midbrain and lower brainstem nuclei via the *dorsal longitudinal fasciculus*, *mammillotegmental tract* and *mammillopeduncular* tract; and the epithalamus (dorsal posterior segment of the diencephalon that includes the habenula and pineal gland) provides afferents to the POA via the *stria medullaris*.

Within the hypothalamus, well-characterized fiber tracts consist of bundles of axons that extend from the soma of neurosecretory cells, and terminate at sites of neurosecretion. Prominent *paraventricular hypophsyial tracts* and *supraoptic hypophysial tracts* project from the paraventricular and supraoptic nuclei, respectfully, to the posterior lobe. These projections consist of axons of the magnocellular neurons that transport vasopressin and oxytocin to their release sites at neurovascular junctions in the pars nervosa. *Tuberoinfundibular tracts* likewise extend from parvocellular (smaller soma) neurons that produce the hypothalamic releasing and inhibiting factors, to terminate in the median eminence where they release these neurosecretory products into the primary plexus of the hypothalamo-hypophysial portal vasculature.

The median eminence is one of several specialized structures that are located at sites about the cerebroventricular system, and are therefore called *circumventricular organs*. These structures lack the *blood–brain barrier*, in which the endothelium of brain capillaries normally restricts movement of compounds from blood to brain or brain to blood. The endothelia of these organs are typically fenestrated, revealing a morphological basis for the diffusion of substances secreted by neurosecretory neurons into the systemic or portal circulation. Conversely, some of the circumventricular organs clearly serve as targets of circulating factors that may activate neural circuitries. The area postrema, located in the caudal extremity of the fourth ventricle, monitors substances present in the circulation, and serves to trigger emesis as an appropriate response to certain blood-borne stimuli.

The Pituitary Gland

The pituitary gland was at one time considered the master endocrine gland of vertebrates, since it was known to control the activity of other major endocrine glands, such as the thyroid, adrenals and gonads. It is now known to be primarily controlled by hormonal stimuli delivered from the hypothalamus and other glands. The pituitary gland, also known as the *hypophysis*, is comprised of the *adenohypophysis*, alternatively referred to as the *anterior lobe*, and the *neurohypophysis*, also called the *posterior lobe*. The adenohypophysis is primarily glandular tissue, while the neurohypophysis consists of neuronal processes that originate from the soma of neurosecretory neurons in the PVN and SON. These axons pass through the *median eminence*, the mediobasal extremity of the hypothalamus that is continuous with the *infundibulum*, or *pituitary stalk*, and ultimately end in the posterior lobe, or pars nervosa.

The embryologic origins of the two lobes of the pituitary are distinct. The neural lobe arises from the neural

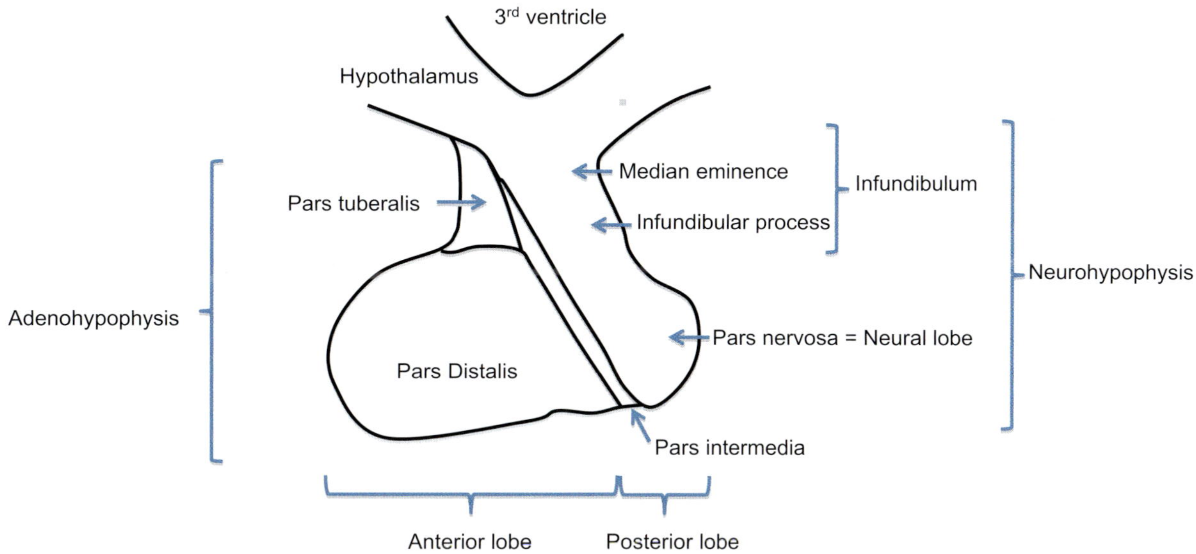

FIGURE 1.2 Anatomical subdivisions of the pituitary gland.

ectoderm of the floor of the developing forebrain, while the anterior lobe is derived from an inward invagination of the primitive mouth cavity (oral ectoderm) known as Rathke's pouch. Cells of the anterior wall of Rathke's pouch develop into the *pars distalis*, containing the majority of the hormone-producing cells of the adenohypophysis. The hormone-secreting cell types and their corresponding hormonal products include the corticotropes (adrenocorticotropic hormone; ACTH), somatotropes (growth hormone; GH), lactotropes (prolactin; PRL), gonadotropes (luteinizing hormone; LH, and follicle-stimulating hormone; FSH) and thyrotropes (thyroid-stimulating hormone; TSH). Dorsal extensions of the anterior lobe constitute a *pars tuberalis*, a non-secretory tissue that wraps around the infundibular stalk. An *intermediate lobe* develops between the two lobes that can vary greatly in size among different species; in humans, this regresses and disappears in adults. In many vertebrates the intermediate lobe produces hormones that include melanotropins, such as *melanocyte-stimulating hormone* (*MSH*). The anatomical components of the pituitary gland are given in Fig. 1.2.

The Hypothalamo-hypophysial Portal Vessel System

The hypophysial vasculature was initially studied in detail in the 1930s, when it was first demonstrated that blood flows downward in the hypophsyial portal vessels, rather than upward from pituitary to brain.[4] Detailed studies thereafter revealed that the superior hypophysial artery provides blood supply to the median eminence and pituitary stalk, from where blood passes via capillary loops through the long portal vessels to the sinusoids of the pars distalis.[5] These findings provided the conceptual basis for Harris' proposal that the hypothalamus exerts a neurohumoral control over anterior pituitary hormone secretions.[6] The pituitary transplantation studies of Harris (described in Chapter 5 of this volume), as well as elegant electrical stimulation experiments by Markee, Sawyer and Hollinshead, as well as Harris,[7,8] confirmed this hypothesis, and in turn revealed the existence of hypothalamic releasing factors that are conveyed by the hypothalamo-hypophysial portal vessel system to control release of anterior pituitary hormones. Two research groups, headed by Roger Guillemin and Andrew Schally, respectively, simultaneously determined the structure of thyrotropin-releasing hormone (TRH), as well as GnRH (or as it was known, LRF or LHRH) and somatostatin (SST), for which these two neuroendocrinologists were awarded the Nobel Prize in Physiology and Medicine in 1977.[2,3] Major releasing factors discovered, in the years following, included corticotropin-releasing hormone (CRH) in 1981,[9] and growth hormone-releasing hormone (GHRH) in 1982 (Fig. 1.3).[10,11]

NEUROSECRETORY CELLS

The major groups of neurosecretory cells in the hypothalamus include those in the neurohypophysial and the tuberoinfundibular systems. Neurons in both groups are predominantly peptidergic, although some may co-produce non-peptide products. An exception is the short-axon tuberoinfundibular dopaminergic (TIDA) neurons that primarily secrete the catecholamine, dopamine, into the portal vasculature. In most regards, neurosecretory neurons are similar in structure and function to neurons elsewhere in the brain. They have

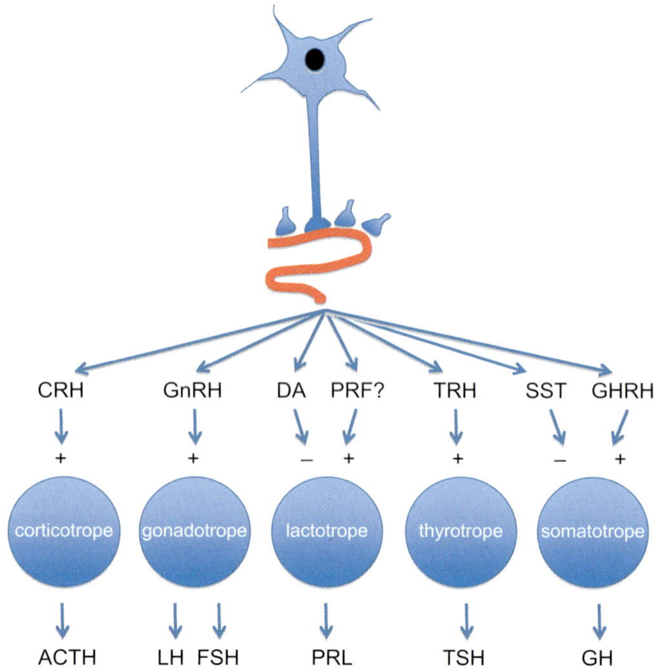

FIGURE 1.3 **Hypothalamic releasing- and inhibiting hormones and their major cellular targets in the anterior pituitary gland.** Abbreviations: ACTH, adrenocorticotropic hormone; CRH, corticotropin-releasing hormone; DA, dopamine; FSH, follicle-stimulating hormone; GH, growth hormone; GHRH, growth hormone-releasing hormone; GnRH, gonadotropin-releasing hormone; LH, luteinizing hormone; PRF, prolactin releasing factor; PRL, prolactin; SST, somatostatin; TRH, thyrotropin-releasing hormone; TSH, thyroid-stimulating hormone.

dendrites, perikarya and axons that resemble those in most central neurons. Furthermore, they exhibit relatively normal resting potentials, action potentials and synaptic potentials, and can display a normal range of intrinsic and synaptically-driven activity patterns that are relatively common among other brain cell types, regardless of their neurotransmitter phenotype. However, the morphology of some neurosecretory neurons, such as the magnocellular vasopressinergic and oxytocinergic neurons, differs in some respects due to the specialized neurosecretory functions. Because they release neurohormones that are delivered via the portal or peripheral circulation to distant target tissues, these neurosecretory cells are required to make copious amounts of neuropeptide hormone in their soma. Thus, these neurons have characteristics of active peptide-producing cells, including abundant rough endoplasmic reticulum, Golgi, and membrane bound granules.

Like other proteins, neuropeptide hormones are synthesized by a ribosomal mechanism. The process of neurohormone gene expression – the transcription of a neurohormone gene and translation of the corresponding mRNA – culminates in the ribosomal synthesis of a pre-prohormone protein that is longer, but inclusive of the specific neurohormone amino acid sequence. It is longer because it contains an amino terminal, hydrophobic signal peptide sequence that functions to permit movement of the protein across the ER membrane into the Golgi apparatus. There, enzymatic removal of the signal sequence, which yields the prohormone, is typically followed by the actions of proteolytic processing enzymes that additionally attack dibasic amino acid cleavage sites. This releases the mature protein hormone sequence from the remaining prohormone precursor molecule, and other peptide sequences with or without biological functions may be produced by these cleavages – for example, the neurophysins in magnocellular neurons. Within the Golgi elements the hormone may additionally be altered by processing enzymes that conjugate carbohydrate, C-terminal amide or other moieties, or create covalent linkages between two sulfur-containing cysteine residues, termed "disulfide bridges." These modifications typically endow the molecule with secondary or tertiary structures that are required for biological activity. Vesicles containing the mature neurohormone are pinched off at the terminal cisternae of the Golgi apparatus, and undergo axonal transport to neurovascular terminals. Release of the neurohormone from the neurosecretory cell thereafter occurs through calcium-dependent *exocytosis*, a process that involves fusion of the secretory vesicle with the plasma membrane and diffusion of its contents into the extracellular space within the neurovascular junction. The neurohormone diffuses through fenestrated capillary walls into the blood.

NEUROENDOCRINE TRANSDUCTION AND NEUROENDOCRINE SYSTEMS

The transformation of neural signals into the release of hormones is called *neuroendocrine transduction*. In general, secretion of neurohormones is a regulated process that reflects neuroendocrine transduction in its simplest form. Both synaptic and hormonal excitation of the neurosecretory cell body initiates action potentials that propagate along the axon and invade the axon terminal. The depolarization of the axon terminal, in turn, triggers an elevation of intracellular Ca^{2+}, which prompts vesicular fusion with the plasma membrane and exocytosis of the granule contents. The amplitude of neurohormone secretion is generally determined by the frequency of action potentials that depolarize the neurosecretory cell terminal. Notably, this stereotyped *stimulus-secretion coupling* process was first characterized by Douglas and Poisner[12] in their classical experiments on isolated posterior pituitary tissues.

Neuroendocrine transduction can take place in the context of several different types of neuroendocrine systems. In the case of the neurophypophysial system,

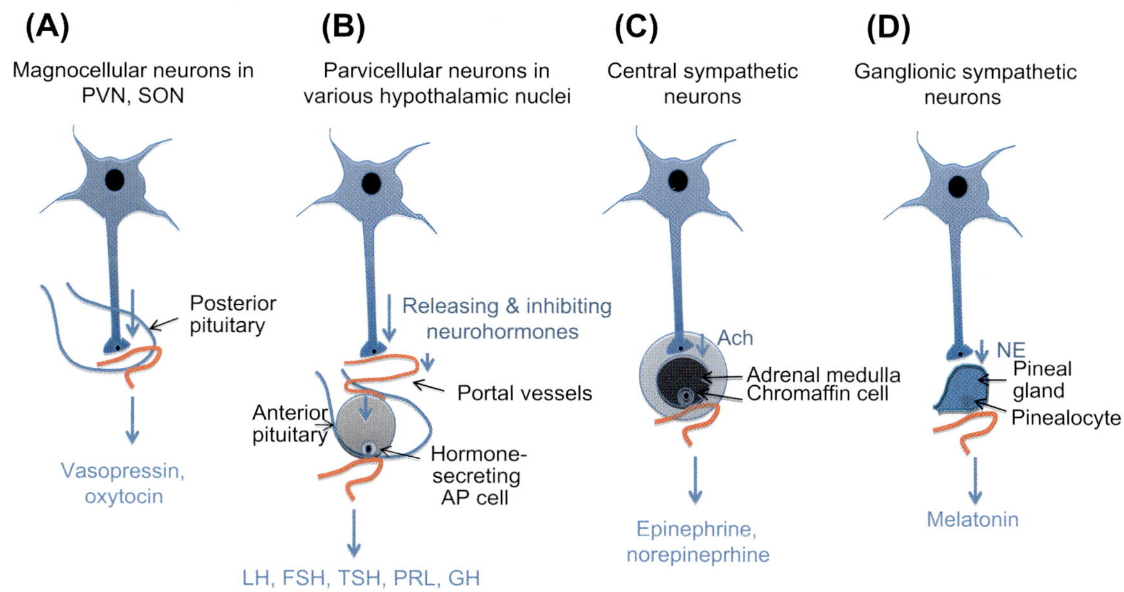

FIGURE 1.4 **Schematic representation of different types of neuroendocrine systems.** (**A**) Neurohypophysial systems releasing vasopressin and oxytocin into the peripheral circulation; (**B**) neurosecretion of hypothalamic releasing and inhibiting hormones into the hypothalamic–hypophysial portal vessel system, regulating anterior pituitary hormone secretions; (**C**) sympathetic innervation of the adrenal medulla, controlling secretion of epinephrine and norepinephrine; (**D**) sympathetic innervation of the pineal gland, controlling secretion of melatonin.

magnocellular neurons in the PVN and SON release vasopressin and oxytocin from neurovascular junctions in the posterior pituitary, into the systemic circulation, through which these hormones are delivered to target organs (Fig. 1.4A). Parvocellular neurons in various hypothalamic nuclei project axons into the median eminence, and release their neuropeptide hormones into the hypothalamo-hypophysial portal vessels, which conduct these neurohormones to the anterior pituitary gland; there, they bind cognate receptors on target pituitary cells and direct the synthesis and secretion of the major trophic hormones of the adenohypophysis (Fig. 1.4B). The sympathetic innervation of the adrenal medulla constitutes yet another variant of a neuroendocrine system; here, axons derived from central sympathetic neurons innervate the chromaffin cells of the adrenal medulla, and at these synapses the neurotransmitter acetylcholine is released, which in turn evokes secretion of epinephrine into the systemic circulation (Figure 1.4C). Similarly, postganglionic sympathetic neurons in the superior cervical ganglion innervate the pineal gland, releasing norepinephrine that in turn evokes release of the hormone melatonin into the bloodstream and cerebrospinal fluid (Figure 1.4D).

BASIC ASPECTS OF NEUROENDOCRINE INTEGRATION

Neuroendocrine regulatory systems can be described as predominantly reflexive or homeostatic control mechanisms. *Neuroendocrine reflexes* resemble their neuromuscular reflex counterparts, as they operate as transient, fixed-pattern or graded reactions to an applied sensory stimulus. The most important and well-studied of these involve the reflexive release of the neurohypophysial hormones, oxytocin and vasopressin. *Homeostatic systems* function to restrict a physiological or behavioral variable such that it is maintained about a set point, or within a restricted range of values. Nearly all neuroendocrine homeostatic systems use some form of *negative feedback* control as a servomechanism that maintains the output of the system, such as secretion of hormones, within a biologically predetermined state. The *hypothalamic–pituitary–end organ* axes, as well as hypothalamic systems that control food intake and other behaviors, are homeostatic systems that feature negative feedback control mechanisms. The basic components of these neuroendocrine systems are described and exemplified below.

Neuroendocrine Reflexes Mediated by Neurohypophysial Systems

Neuroendocrine reflexes mediate acute physiological responses to external and internal signals. A sensory stimulus, such as suckling by an infant at a mother's nipple, can depolarize and thereby activate afferent nerves that convey neural signals up the neuroaxis through multisynaptic pathways to the hypothalamus; these afferent signals activate an effector neuronal population, such as the oxytocinergic magnocellular neurons, and the net result is a physiological response in the form of neurosecretion and oxytocin-mediated milk ejection that is qualitatively and quantitatively appropriate for the physiological stimulus. Like the classical neuromotor reflexes, neuroendocrine reflexes operate via afferent and efferent loops that can be comprised of mono- or multi-order synaptic pathways. By definition, however, neuromotor reflexes are mediated by neural afferent and efferent pathways. The major neuroendocrine reflexes, by contrast, can be mediated by a humoral input and neural output, humoral input and neural output, humoral input and hormonal output, or neural input and hormonal output. Furthermore, some neuroendocrine reflexes are stereotyped, fixed-action responses to specific stimuli, while others mediate graded responses to stimuli of varying magnitude.

In the foregoing example, the suckling-induced neurosecretion of oxytocin can be considered a fixed-action neuroendocrine reflex mediated by a neural input and hormonal output. One of the major biological actions of oxytocin is to stimulate contractions of myoepithelial cells of the mammary glands, facilitating milk ejection. The reflexive secretion of oxytocin and its actions at the mammary gland represent a fixed-action response, since it is a relatively stereotyped output signal triggered by a specific input signal. The afferent limb of the reflex is comprised of the sensory endings in the mammary glands, primary sensory afferents, and multi-order afferents to the supraoptic and paraventricular nuclei. The suckling stimulus evokes transient increases in the action potentials conveyed along the processes in this afferent limb of the reflex, and thereafter in the firing rate of the oxytocinergic neurons, which is in turn transduced into an acute increase in the neurosecretion of oxytocin from neurovascular junctions in the posterior pituitary gland.[13] The oxytocinergic neurons and the circulation of oxytocin to the mammary gland thus constitute the efferent loop of the neuroendocrine reflex. Suckling-induced oxytocin secretion thus represents a relatively simple reflex circuit that consists of a neural input and hormonal output, and occurs as a fixed-action response to a specific stimulus. A similar neuroendocrine reflex governs the release of oxytocin during parturition, when it assumes an important role in stimulating myometrial contractility. In the latter stages of parturition, dilation of the cervix evokes activity in a neural afferent loop that conveys this mechanoreceptive information to the hypothalamus, culminating in the activation of oxytocinergic neurons and neurosecretion of oxytocin into the systemic circulation. The increase in circulating oxytocin prompts an increase in myometrial activity, and hence in the intensity of the labor. Similar to the milk-ejection reflex, the activation of oxytocin neurosecretion during labor is mediated by a reflex with a neural input and hormonal output, and reflects a fixed-action response to a specific sensory stimulus.

Vasopressinergic neurons also function as effectors of neuroendocrine reflexes. One major biological action of vasopressin is exerted in the distal tubules of the mammalian kidney, where it facilitates water resorption from the collecting ducts. These antidiuretic actions of vasopressin can thereby mediate restoration of blood volume in life-threatening situations such as hemorrhage or severe diarrhea. Vasopressin also induces contraction of vascular smooth muscle cells, stimulating increases in blood pressure. The two major stimuli for the reflexive release of vasopressin are thus decreases in blood pressure and volume, and increases in blood osmolality. Two anatomically distinct afferent loops mediate the transduction of these stimuli into the release of vasopressin. In the first, increased blood osmolality is registered and transduced by osmoreceptive cells in the hypothalamus, and these signals are then conveyed synaptically to magnocellular neurons, prompting activation of these cells and neurosecretion of vasopressin from neurovascular terminals in the posterior pituitary. Increased vasopressin in the circulation thereafter acts via vasopressin receptors in the renal tubules to promote water resorption, and thus dilution of body fluid osmolytes. The distinguishing features of this reflex is that it is comprised of a humoral input and endocrine output, and that it elaborates graded vasopressin secretory responses that are proportional to the magnitude of the initial increase in blood osmolarity. A second reflex involving magnocellular vasopressin neurons is triggered by low blood volume and/or low blood pressure. Baroreceptors located in the aortic arch, carotid sinus and right atrium register any suprathreshold drop in volume or pressure, and tranduce this information into neural signals conveyed up the neuroaxis and ultimately to the magnocellular vasopressinergic neurons; the electrophysiological and thus neurosecretory activity of these cells is accordingly increased, and thereafter the increased vasopressin in the circulation mediates both antidiuretic and pressor responses to the original depressor or hypovolumetric stimulus. The two modes of reflexive vasopressin neurosecretion are mediated by distinct neural input pathways and a common endocrine output (see also Chapter 6).[14]

The Basic Elements of Homeostatic Neuroendocrine Systems

Homeostatic regulation in physiological systems can be described in terms of control systems analysis, using basic terms and concepts borrowed from engineers. A neuroendocrine control system functions to control a physiological variable about a set point, or predetermined range of values, that is most adaptive to the

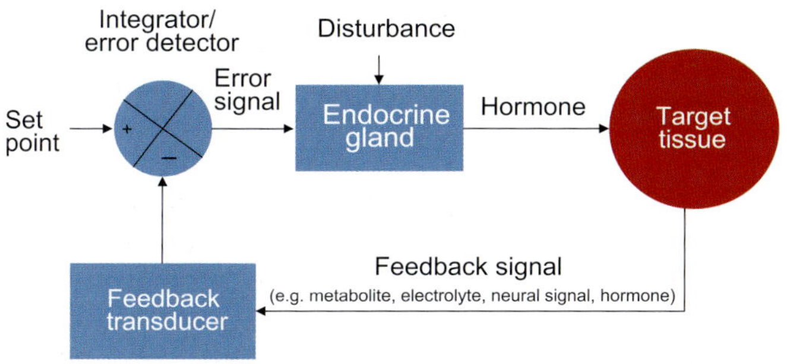

FIGURE 1.5 Basic components of a simple controlled endocrine system featuring negative feedback regulation of secretion (see text for explanation of terms).

animal in a given environment. The generic system features such a set point, an integrator (or error detector, or comparator), a controlling element, a controlled variable, a feedback signal and a feedback signal transducer (Fig. 1.5). This simplest version of a homeostatic control system makes use of negative feedback to maintain the controlled variable within a narrow range of values. The controlling element regulates the controlled variable, which in turn provides a feedback signal representing the momentary value of the variable. A feedback signal transducer registers the feedback signal, converts it to a readable signal, and conveys that information to an integrator. Here, a comparison is made between the ambient level of the variable and a desired set-point value. If a discrepancy is calculated between the real and preselected states, then the integrator delivers an error signal to the controlling elements. In almost all physiological systems, the error signal is inverted in sign to produce an adjustment of the controlling element activity in the opposite direction of the original deviation from the set point. Virtually all neuroendocrine homeostatic mechanisms use some form of negative feedback control that can be described in these terms. Whether there are in fact distinct preselected "set points" that are regulated by simple servomechanisms such as the foregoing control system has been debated, and more complex – and likely distributed – control mechanisms have been proposed to regulate hypothalamic functions, such as thermoregulation.[15] Nevertheless, negative feedback mechanisms, and integrative mechanisms that defend a given range of values along a variable such as body weight,[16] or hormone secretion, are readily demonstrable for most neuroendocrine systems, and thus models such as these prove instructive in understanding the physiological relationships within these systems.

Homeostatic Systems: Hypothalamic–Pituitary–End-Organ Axes

Neuroendocrine homeostatic systems can involve the hypothalamus, anterior pituitary, and an end-organ or target tissues functioning together as an axis. These hypothalamic–pituitary–end-organ axes are organized into distinct tiers of regulated activity, arranged so that hormone signals are conveyed from: (a) hypothalamus to anterior pituitary by hypothalamic releasing and/or inhibiting factors secreted into the portal vessels; and (b) anterior pituitary to end-organ by adenohypophysial hormones secreted into the peripheral circulation. In many of these axes, a third round of signaling proceeds to complete a feedback loop from (c) end-organ back to hypothalamic neurons and/or anterior pituitary gland. In all of these axes, experimental analyses have revealed that negative feedback regulation functions as the predominant control feature within the system. Feedback signals from the anterior pituitary gland to the hypothalamus constitute *short-loop feedback* mechanisms. *Long-loop feedback* mechanisms are those in which feedback signals are conveyed from the end-organ, such as the gonad, thyroid or adrenal cortex, to antecedent levels.

The three-tiered neuroendocrine systems include: (a) the hypothalamic–pituitary–thyroid (HPT) axis; (b) the hypothalamic–pituitary–gonadal (HPG) axis; and (c) the hypothalamic–pituitary–adrenocortical (HPA) axis (shown schematically in Figs 1.6A, B and C, respectively). In the HPT axis, a population of hypothalamic neurons primarily located in the PVN produces the releasing factor TRH, and secretes the tripeptide from neurovascular terminals in the median eminence. The TRH peptide is conveyed to the anterior pituitary gland in the portal vessels, where it diffuses through the fenestrated secondary capillary plexus into the interstitial spaces of the anterior pituitary gland; here, the peptide can bind to G-protein coupled TRH receptors on thyrotropes and cytoplasmic signaling cascades that in turn stimulate synthesis and secretion of thyrotropin (or thyroid-stimulating hormone, TSH). Thyrotropin is conveyed via the circulation to the thyroid, where it binds TSH receptors on follicular cells and promotes the production of T3 and thyroxine (T4). Thyroid hormones exert their widespread actions on target tissues via their cognate intracellular receptors, while also providing the major negative feedback signal within the axis. Thus, elevations in T3 and T4 exert long-loop feedback actions at the hypothalamic level to suppress TRH expression and neurosecretion, and at the pituitary level to suppress TRH-stimulated TSH secretion. Conversely, reduction in thyroid hormone, or complete removal thyroid hormones by thyroidectomy, is generally accompanied by elevations in TRH and TSH secretions.

Similar cascades of hormonal activity and feed back relationships prevail in other hypothalamic–pituitary–end-organ axes. As described fully in succeeding chapters in this book, the HPG axis and the HPA axis

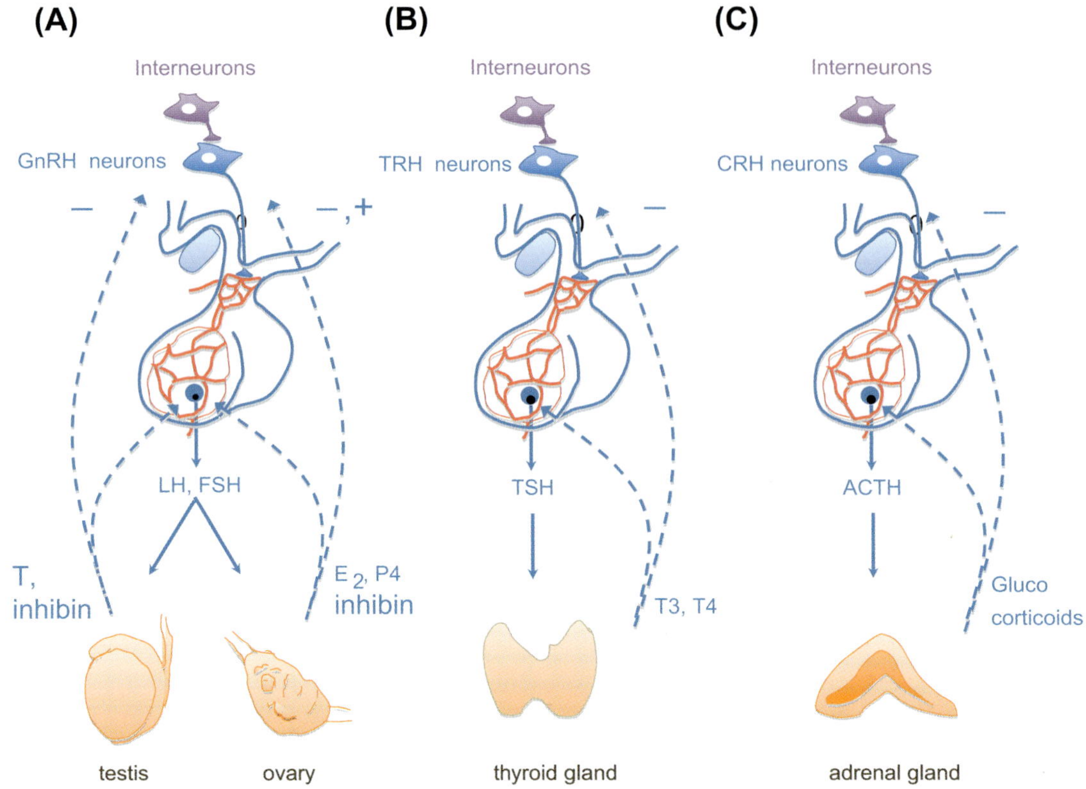

FIGURE 1.6 **"Three-tiered" (A–C) and "two-tiered" (D, E) homeostatic neuroendocrine systems.** (**A**) Hypothalamic–pituitary–gonadal (HPB) axis; (**B**) hypothalamic–pituitary–thyroidal (HPT) axis; (**C**) hypothalamic–pituitary–adrenocortical (HPA) axis; (**D**) hypothalamic control of GH secretion; (**E**) hypothalamic control of PRL secretion. Shown for each axis in (A–C) are the major hypothalamic releasing neurohormones, and their corresponding pituitary and end-organ hormone mediators. Homeostatic control in each axis is largely effected by long-loop negative feedback mechanisms, mediated by end-organ action in the hypothalamus and/or anterior pituitary gland. Hypothalamic feedback can be mediated by end-organ hormone actions on interneurons controlling the releasing factor neurons, or on the releasing factor neurons themselves. Feedback in the hypothalamic–pituitary–gonadal axis is exerted by gonadal steroids (primarily testosterone (T) in males, and estradiol (E_2) and progesterone (P4) in females), as well as the protein hormone inhibin, which selectively suppresses FSH secretion in both sexes. Triodothyronine (T3) and thyroxine (T4) exert feedback in the HPT axis, and glucocorticoids, principally cortisol in humans and corticosterone in rodents, exert major feedback effects in the HPA axis. Shown for the two-tiered systems in (D) and (E) are their hypothalamic releasing and inhibiting hormones, corresponding pituitary hormones, and major target tissues. Homeostatic controls in each of these systems are largely affected by short-loop feedback mechanisms mediated by GH and PRL actions in hypothalamic neurons controlling releasing of their corresponding releasing and inhibiting factors.

also function as three-tiered hormonal control systems in which long-loop feedback mechanisms predominate. Like the HPT axis, the HPG and HPA axes are organized so that the primary releasing factors (GnRH and CRH) stimulate their corresponding trophic pituitary hormones (gonadotropins and ACTH), which in turn stimulate end-organ hormones (gonadal steroid/peptide hormones, and glucocorticoids), which in turn exert long-loop feedback effects at preceding levels in the respective axis.

Secretions of GH and PRL are predominantly under the control of two-tier systems in which short-loop feedback functions are the major regulatory mechanism. The reduced importance of long-loop feedback control in these systems is probably a function of the distributed targets of GH and PRL actions. Growth hormone exerts actions in bone, cartilage, liver, muscle and other tissues, while PRL evokes responses in mammary tissue, gonads and accessory sex organs. Without a single endocrine end-organ to provide feedback control, GH and PRL have evolved the capacity to exert their own direct feedback control within the hypothalamus, ultimately influencing release of the hypothalamic releasing and inhibiting factors that control their own secretions. Furthermore, both stimulatory and inhibitory hypothalamic mechanisms have evolved to control GH and PRL secretion. Hypothalamic neurosecretion of GHRH stimulates, while SST inhibits, GH secretion; and dopamine inhibits, while one or more putative PRL-releasing factors stimulate, PRL release. The short-loop feedback mechanisms for both hormones appear to be mediated by both suppression of releasing factor (GHRH, PRL-RFs) release and stimulation of inhibitory factor (SST, dopamine) release. In the case of GH control, it should

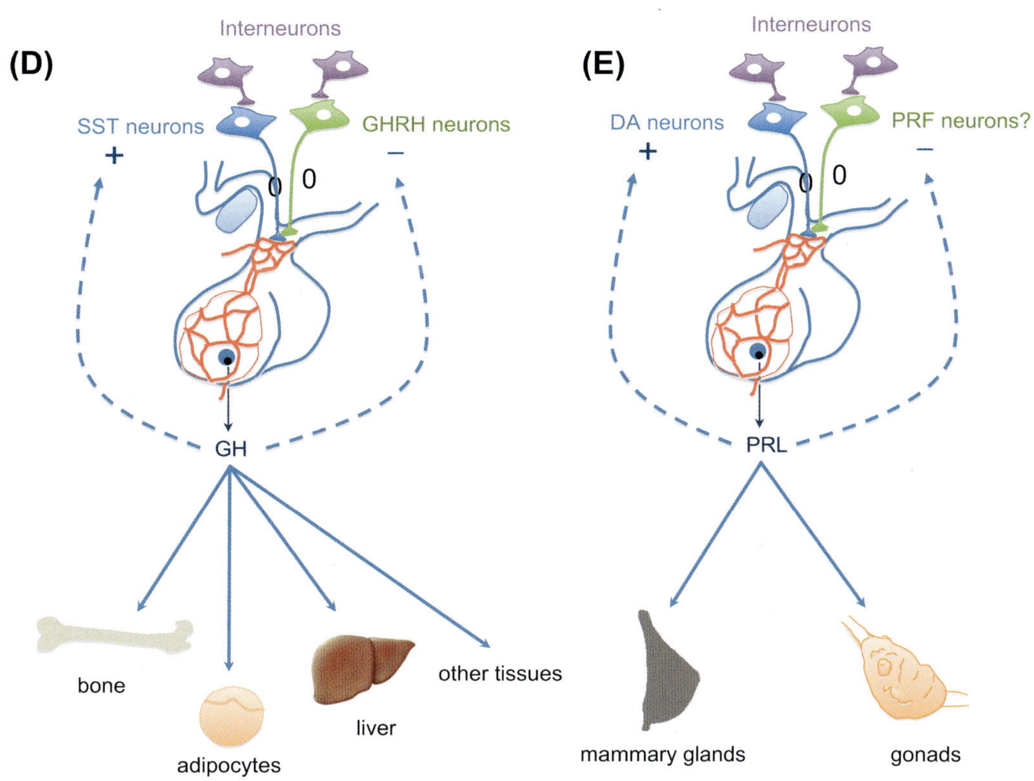

FIGURE 1.6—(*Continued*).

be noted that some long-loop feedback regulation has been shown to be superimposed upon the basic short-loop feedback control system – the actions of GH in promoting cartilage formation, leading to bone growth, are mediated in part by GH stimulation of somatomedins (insulin-like growth factors 1 and 2; IGF1, IGF2) from liver, and IGF1 may also exert negative feedback actions within the hypothalamus to suppress GHRH and stimulate SST release.

Homeostatic Neuroendocrine Systems Controlling Motivated Behaviors

Specific hypothalamic neuronal populations and circuitries have been shown to be intimately involved in the regulation of motivated behaviors, and to function as major targets of peripheral hormones and other humoral factors that regulate these behaviors. Many of the chapters in this book address in detail the neuroendocrine mechanisms that mediate the central control of motivational states that prompt reproductive, social and ingestive behaviors. Some of these mechanisms function as homeostatic neuroendocrine systems that feature stimulatory and inhibitory influences and both short-term and long-term feedback controls. This is perhaps best exemplified by the neuroendocrine systems that govern energy homeostasis.

Food intake and energy expenditure – and hence body weight – are controlled within narrow limits by homeostatic neuroendocrine systems. While these homeostatic control mechanisms have been known to exist for many decades, neuroendocrine research has only recently revealed several of their major endocrine and neurochemical components. Neurons located in the mediobasal hypothalamus are now known to be critically important in both food intake and the autonomic and behavioral systems through which energy expenditure is controlled. Many of these hypothalamic neurons produce neuropeptide neurotransmitters that are either orexigenic (stimulate food intake) or anorexigenic (inhibit food intake), while at the same time reduce or increase energy expenditure. The relative activities of these cell groups are believed to manifest the prevailing motivational state (i.e., satiety or hunger). These cell groups are influenced, in turn, by neural and hormonal signals that convey information about the availability of metabolic energy substrates. Some of these signals arise from the gut and provide short-term cues about the absence or presence of food in the stomach.[16,17] Others are hormonal signals that represent the levels of metabolic substrates in blood, and stored in fat cells and other tissues. The adipocyte-derived hormone, leptin, is one such hormonal modulator that has been shown to suppress food intake and increase energy expenditure.[18]

Insulin appears to exert similar feedback effects.[19] The targets of leptin and insulin feedback effects on food intake and energy expenditure are the same hypothalamic neuropeptidergic neurons that are crucial in inducing or inhibiting feeding behavior and energy expenditure. The long-term homeostatic regulation of body weight thus appears to be controlled by a neuroendocrine mechanism that controls energy intake and expenditure, and a negative feedback regulation of these same neuronal groups by leptin, insulin and other factors that are raised as a consequence of energy intake and storage. How the hypothetical set point for body-weight regulation may be established, and how it may be altered in morbidly obese individuals, remains the subject of intense study.

Experimental Characterization of Homeostatic Neuroendocrine Systems

Homeostatic neuroendocrine systems have been characterized through controlled experimental manipulation of the neuroendocrine axis, followed by careful analysis of the responses of the system. Selective blockade or stimulation of forward signals in the axis, or removal and replacement of feedback signals, produces responses of the system that are predicted by a control system theory. For example, in the HPG axis, the stimulatory actions of GnRH on LH secretion and LH on testosterone secretion are easily demonstrated *in vivo* and *in vitro*. Conversely, removal of these regulators *in vivo*, by immunoneutralization, pharmacological methods, hypothalamic lesions or hypophysectomy, results in a precipitous decline in circulating testosterone concentrations. Negative feedback control has also been characterized in virtually all of the neuroendocrine axes. The critical test requires removal of the end-organ, where possible, and measurement of the hypothalamic and pituitary secretory responses. In systems featuring negative feedback, removal of the end-organ (e.g., gonads) results in an acceleration of GnRH release and a large increase in LH secretion; similarly, removal of the adrenal glands results in a robust increase in CRH and ACTH secretion, and removal of the thyroid results in a substantial increase in TRH and thyrotropin secretion. In all cases, replacement with physiological concentrations of the appropriate end-organ hormone prevents these increases in hypothalamic and pituitary hormone secretions. Specific features of these control mechanisms can differ between the axes. For example, in some the principal feedback target may be the hypothalamus, and in others the pituitary may serve as the primary feedback target. Furthermore, in the systems governing GH and PRL secretion, short-loop feedback control may involve both suppression of the releasing factor and stimulation of the inhibiting factor.

Environmental Stimuli, Homeostatic Settings and Allostasis

While defending a preselected state, neuroendocrine homeostatic systems must also register and transduce acute stimuli, mount an appropriate response to the perturbation, and return the system to the basal operating state. In the HPA axis, for example, a stressful stimulus rapidly evokes neurosecretion of CRH into the portal vessels, raising ACTH and hence glucocorticoid release. The duration and pattern of the CRH, ACTH and glucocorticoid response will be limited to some extent by an increase in feedback suppression of CRH and ACTH; however, with unabated stress some elevated level of CRH and ACTH secretion would persist. Termination of the stress is then accompanied by a return to the original levels of CRH, ACTH and glucocorticoid secretions, and responsiveness to feedback in the system.

There are numerous examples of short-term responses by an axis to a physiological stimulus. These perturbations usually take the form of exteroceptive or interoceptive stimuli that are conveyed by sensory pathways, and they can be pheromonal, thermal, visual, auditory, tactile, olfactory or gustatory in nature. In reflex ovulators, for example, coitus stimulates neural pathways of virtually all of the sensory modalities, and these sensory signals converge in the hypothalamus. Here, integrative neurons produce a major ovulatory release of GnRH into the portal vessels, triggering a surge of LH into the circulation that in turn evokes ovulation.[20] This neuroendocrine mechanism ensures that the ovulation occurs in temporal register with the presence of viable sperm in the reproductive tract, increasing the chances of reproductive success. Following the perturbation of the HPG axis, it returns to its basal activity state under the control of steroid hormone feedback mechanisms. Another example is the response of the HPT axis to an acute cold stress, which produces a transient stimulation of TRH, and hence TSH and thyroid hormone secretions[21]; the latter secretions serve to elevate body metabolism and mount a thermoregulatory response to the external cold stress. Interoceptive sensory signals can likewise be conveyed from baroreceptors and proprioceptors in peripheral organs and tissues to the hypothalamus, and thereby effect changes in neurohormone output. A sudden drop in blood pressure, for example, is detected by baroreceptors in the atria, carotid bodies and elsewhere in the arterial system, and resulting neural signals are conveyed up the neuroaxis, into the hypothalamic centers that mediate stress responses. As a result, CRH and thus ACTH and glucocorticoid secretions (as well as vasopressin release) are acutely increased.[22] Restoration of blood pressure is accompanied by a return of the

HPA axis to its original basal activity state. Interoceptive signals can also be mediated by alterations in circulating concentrations of metabolic intermediates, osmolytes or growth factors. Specific neuronal populations function as sensory receptors, which monitor the blood levels of these physiological factors and relay appropriate commands to hypothalamic homeostatic regulatory systems. For example, elevated amino acid levels (e.g., arginine) in blood stimulate GHRH and GH secretion, while hypoglycemia induces CRH/ACTH/glucocorticoid secretion.

Homeostatic systems do not function in isolation from one another. Often a most adaptive state may entail responses that are integrated among the different homeostatic neuroendocrine systems. In some cases, these responses are synergistic; stressful stimuli activate both CRH and vasopressinergic neurosecretion in the median eminence (distinct from the magnocellular vasopressin release from posterior pituitary), and vasopressin can amplify the effects of CRH on ACTH secretion from corticotropes.[23] In other circumstances, the activation of one system is accompanied by the inhibition of another, providing an overall net adaptive benefit for the animal. Stressors of many types, as well as a prolonged state of negative energy balance that accompanies undernutrition or strenuous exercise training, raise stress hormone secretions while often suppressing reproductive hormone secretions.[24] These coordinated responses to environmental conditions and behavioral state are mediated in large part by activation of CRH neurons and suppression of GnRH neurons, through mechanisms that are largely unknown and are the subject of current studies. Clearly, the induction of "hypothalamic amenorrhea" in these circumstances is an adaptive response that conserves physiological resources that increase the chance of individual survival, at the expense of temporary suppression of reproductive activities that normally perpetuate the species.

As revealed by Bernard, homeostatic systems maintain the constancy of the internal milieu in the face of changing external conditions. However, homeostatic systems also confer a major adaptive advantage to the organism, because homeostatic settings can change during development, and can be altered in anticipation, in association and/or in response to changes in the internal and external environments. An example of the resetting of a homeostatic set point during development is the increase in the frequency and/or amplitude of GnRH secretion that occurs during the pubertal activation of the HPG axis.[25,26] The pubertal activation of GnRH release is sustained throughout the adult reproductive lifespan, as is the acquired equilibrium between hormone actions and feedback mechanisms in the HPG axis. The set point for neurohormone secretion can also vary sinusoidally over the circadian period, likely as a function of signals from the biological clock in the SCN; activity in the HPA axis, for example, varies with a 24-hour rhythm that peaks in the morning hours and decreases to a nadir in the evening. Sustained shifts in the set point of an axis can also occur over many days; pheromonal cues, for example, can manifest continued inhibition of GnRH neurosecretion and hence reproductive state, while social and psychological cues can alter food intake and energy homeostasis in humans. Seasonal alteration in reproductive status (seasonal breeding) is a reproductive strategy that represents major set-point adjustments in the HPG axis over periods of months. The reproductive axis receives photoperiodic cues through a signaling cascade that includes a retinohypothalamic pathway, synaptic pathways leading to the pineal gland, nocturnal melatonin secretions and, ultimately, neuronal circuitries that control GnRH neurosecretion. Information regarding day length is registered, transduced, and encoded in the duration of melatonin secretion as stimulatory or inhibitory signals for reproductive status. The resulting periods of reproductive activity and inactivity are thereby sustained throughout the months that are most adaptive for the reproductive success and survival of young of a particular species.

Some of the foregoing adaptive changes in set points can be considered as components of higher-level integrative and adaptive responses to changing environments and social, ecological, and physical variables. Superimposed upon the concept of homeostasis is the idea of *allostasis*, or achieving stability through change. Allostasis is thought of as a process that supports and maintains homeostatic systems to allow the organism to adapt to changes in the environment and life history stages.[27] Major mediators of allostasis include the HPA axis, catecholamines and cytokines, and allostatic state refers to altered and sustained activity of these mediators in response to environmental challenges. The adaptation-promoting responses of these mediators are recognized as having protective action in the short term, but may cause damage over the long term, including increased rates of brain and body aging, poor responses to stress, and increased susceptibility to metabolic, cardiovascular and psychiatric disease, as well as immune compromise. *Allostatic load* refers to the accumulated wear and tear caused by inefficiently operated adaptive responses over time.

CELLULAR MECHANISMS OF NEUROENDOCRINE INTEGRATION

Both neuroendocrine reflexes and homeostatic regulatory systems are critically dependent upon the functioning of the cellular signal transduction mechanisms in their constituent parts, especially those that occur within the hypothalamic neurons that control these systems. A given hypothalamic cell is endowed with certain complements of neurotransmitter and hormone receptors, second messenger systems, transcription factors and ion channels, all of which may determine the hypothalamic cell's role in receiving, integrating and transmitting neural signals that are vital to the operation of a given neuroendocrine system. A major challenge to neuroendocrine researchers is to elucidate mechanisms through which a cell registers neural and hormonal signals and integrates them in a process leading to the production of a physiological output signal. The difficulty of this task is greatly compounded by the tremendous complexity of cell–cell connections and interactions with the hypothalamus, and the likelihood that individual neurons are responsible for integrating many different types of signals conveyed simultaneously, or in an ordered temporal sequence. Signals to be integrated include: (a) synaptic activation via afferent neural pathways; (b) circulating hormones that bind their cognate receptors in the cell, thereby altering cytoplasmic signaling cascades and hence electrophysiological activity, secretory activity, and/or the expression of a variety of target genes; and (c) other humoral factors, such as metabolic intermediates, electrolytes, temperature, etc. Moreover, humoral or neurochemical actions may be stimulatory or inhibitory, or they may be permissive – that is, they render a cell more or less responsive to another (or the same) stimulus. Cellular integration occurs when a cell assesses the weight of each of these signals, computes their net effect and directs alterations in output signals. The final output signal is encoded in changes in the rate or pattern of neurotransmission or neurosecretion from that cell.

A fundamental property that is specific to many neuroendocrine cell groups is the propensity to release neurohormone in synchrony and at regular intervals. This coordinated, intermittent release pattern by a population of neurosecretory cells is referred to as *neuroendocrine pulsatility*. An example of pulsatile GnRH release from terminals in the median eminance and the corresponding pattern of pulsatile LH secretion in peripheral blood of an ovariectomized ewe is shown in Fig. 1.7.[28] Classic experiments by Ernst Knobil and colleagues[29] established the functional importance of pulsatile neurohormone secretion, as presentation of continuous neurohormonal stimuli is less effective, or even inhibitory, in releasing pituitary hormones. However, administration of neurohormone (e.g., GnRH) as a series of regular pulses can continue to evoke anterior pituitary hormone secretions for virtually unlimited periods. It appears likely that pulsatile hormone secretions serve to maintain responsiveness of the signal transduction events in pituitary cells, instead of inducing downregulation

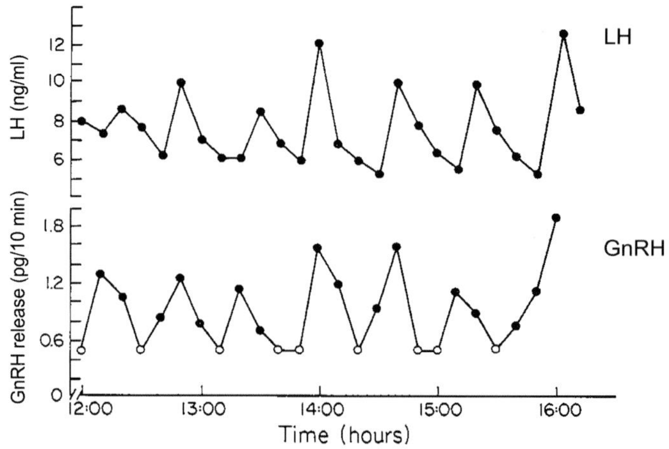

FIGURE 1.7 Pulsatile hypothalamic GnRH release and LH secretion in serum as simultaneously monitored in an ovariectomized ewe. *Modified from, Levine* et al.[28]

of receptors or their downstream signaling pathways. The pulsatile pattern of neurohormone secretion has been found to be of profound importance in clinical cases requiring neurohormone substitution therapies, such as the GnRH regimens used to induce or restore fertility in hypogonadotropic-hypogonadal patients.[30]

Neural and hormonal signals may also exert long-lasting, often permanent, effects on the function of a hypothalamic cell group by altering neurogenesis, cell growth, axon and dendrite morphogenesis, synaptic connectivity, and cell death during fetal and neonatal development, as well as neuroplasticity in adulthood. The well-known organizational effects of steroid hormones during fetal development are but one class of trophic signals that impact hypothalamic function. Metabolic hormones, such as leptin, have also been determined to exert trophic effects on developing hypothalamic circuitries.[31] Many of these developmental actions of hormones are discussed in subsequent chapters, particularly with relevance to sexual differentiation of the brain or prenatal or neonatal programming of neuroendocrine function and behavior.

CLINICAL NEUROENDOCRINOLOGY

Clinical consequences of neuroendocrine pathophysiologies are discussed in many of the chapters in this book, although in general they fall into categories of growth and developmental impairments, infertility, oligo- or galactorrhea, hypo- or hyperthyroidism, hypo- or hyperadrenocorticolism, diabetes insipidus, feeding disorders, metabolic diseases, sexual disorders, hormone-sensitive psychiatric and neurological diseases, and specific symptoms of diseases which present as a syndrome, such as McCune-Albright disease or polycystic ovarian syndrome. As with any diseases, the underlying causes of neuroendocrine dysregulation can be varied, and can be inherited (mono- or polygenic), acquired, associated with tumorigenesis and/or carcinogenesis, or otherwise associated with another medical condition. Diagnosis and treatment of diseases with neuroendocrine components must always take into account the multidimensional nature of the controlled neuroendocrine system. Diseases that cause dysregulation in the neuroendocrine homeostatic axes are considered *primary*, if they involve the major end-organ of the axis, or *secondary*, if defects occur at antecedent levels (pituitary or hypothalamus). Neuroendocrine diseases are usually diagnosed with the aid of X-rays, MRIs and other imaging diagnostic procedures, along with measurements of pituitary and end-organ hormones in serum. In some cases, provocative tests of a target-organ function are conducted with stimulatory hormone preparations.

THE STUDY OF NEUROENDOCRINOLOGY

The methods used to study neuroendocrine systems are as diverse as those employed by neuroscientists and endocrinologists across a full spectrum of investigative levels, from molecular biology to behavior. Classical *in vivo* and *in vitro* methods include surgical manipulations, hormone treatments, neuropharmacological approaches, immunohistochemical analysis of neuropeptide or hormone receptor expression, cell and tissue culture methods, radioligand receptor binding assays, *in situ* hybridization, quantitative real-time PCR, and immunoassays of hormones in body fluids and tissue extracts. Modern neuroendocrine research utilizes virtually any of the latest molecular, cellular, physiological, behavioral and genetic approaches – often in combination with the classical techniques – to further understanding of neurohormone synthesis, secretion, physiological roles and mechanisms of action, as well as the molecular and cellular actions of peripheral hormones on brain development and function.

There are inherent challenges in the study of neuroendocrine systems that derive from the heterogeneity of the cell groups in the hypothalamus, and the relative lack of stereotyped cell circuitries that are more easily identified and accessed in the hippocampus, cerebellum and other brain structures. Neuroendocrinologists have therefore often led the way in developing new methods to analyze the cellular and integrative properties of hypothalamic neurons, and their functional roles in physiological contexts. *Immortalized cell lines* producing the neurohormone GnRH were initially developed by Mellon *et al.*[32] and Radovick *et al.*[33] by targeted tumorigenesis, and these and subsequent cell lines have permitted sophisticated cell and molecular analyses of neuropeptide gene transcription, intra- and intercellular signaling mechanisms, and cellular substrates of neuronal migration. *Transgenic mouse lines* have been developed by neuroendocrinologists that have permitted targeted alteration or mapping of gene expression in specific subsets of hypothalamic cells. Transgenesis has been used to label cells of a certain phenotype, such that they may be visualized in the living state. For example, the promoter sequence for the GnRH gene was fused to a gene encoding the jellyfish green fluorescent protein (GFP) to create transgenic mice in which the transgene, and thus the fluorescent GFP molecule, is only expressed in GnRH neurons.[34,35] This animal model has been used to permit visualization of these few neurons in living brain tissues for electrophysiological experiments.

Gene knockout and *knockin* animals have provided a wealth of information about the physiological roles of hormones and their receptors in the central nervous system. The loss of function in gene-deletion mutant animals can confirm the essential roles of a protein in a specific neuroendocrine process, or it can even replicate a neuroendocrine disease state that is believed to arise from analogous gene mutations in humans. For example, the critical importance of the estrogen receptor α (ERα) in mediating estradiol feedback effects in the reproductive axis, sexual behavior and the neuroendocrine regulation of energy homeostasis has been firmly established using ERα null mutant mice.[36–38] Additional knowledge of the specific ERα signaling mechanisms mediating neural responses to estrogens has been provided by the study of ERα mutant gene knockin mice,[38,39] which can only exert effects through non-classical signaling mechanisms that do not utilize canonical geneotropic regulatory mechanisms. Similarly, an obligatory role of progesterone receptors in the stimulation of preovulatory gonadotropin surges was demonstrated through the study of progesterone receptor knockout (PRKO) mice.[40] Conditional gene targeting has likewise produced profound insights into the cellular and molecular physiology of neuroendocrine systems. In cell-specific gene targeting, the loxP–Cre recombination system is one such method involving the generation of mice bearing site-specific recombination sites, called loxP sites, in the intronic sequences that flank an essential exon of the target gene. A second line of mice is produced that harbor a transgene construct containing a promoter fused to the Cre recombinase gene; importantly, the promoter is known to be active only in the cells of interest. Mating of the two mouse lines results in the expression of Cre protein in the targeted cells, where it acts at the loxP sites to delete the exon of the target gene, thereby inactivating it. Thus, the specific gene deletion only occurs in the target cells. Many striking examples of the successful use of this system are found among studies that have produced cell-specific deletions of leptin-receptor or insulin-receptor genes in hypothalamic neurons, producing animals with altered energy homeostasis.[41,42] Targeted deletion of ERα in neurons that produce the reproductive neuropeptide, kisspeptin, was recently found to advance the onset and prevent the completion of puberty, unambiguously implicating these receptors in the neuroendocrine mechanisms that orchestrate reproductive maturation.[43]

SUMMARY

In this chapter, the science of neuroendocrinology has been introduced and general descriptions of neuroendocrine systems have been provided, with the intention of preparing the generalist for the detailed treatment of specific neuroendocrine topics in this book. The reader should now be acquainted with the functional anatomy and the basic operating principles of the major mammalian neuroendocrine systems. Intensive and up-to-date treatments of neuroendocrine signaling mechanisms, feedback mechanisms, neurohypophysial systems and hypothalamic–pituitary–end-organ axes, and the actions of hormones in the CNS, follow in subsequent chapters by renowned experts in the field.

References

1. Scharrer E, Scharrer B. Neurosecretion. *Physiol Rev.* 1945;25:171–181.
2. Schally AV. Aspects of hypothalamic regulation of the pituitary gland. *Science.* 1978;202(4363):18–28.
3. Guillemin R. Peptides in the brain: the new endocrinology of the neuron. *Science.* 1978;202(4366):390–402.
4. Houssay BA, Biasotti A, Sammartino R. Modifications fonctionelles de l'hypophyse après les lesions infundibulotuberiennes chez le crapaud. *C.R. Soc Biol Paris.* 1935;120:725–727.
5. Wislocki GB, King LS. The permeability of the hypophysis and the hypothalamus to vital dyes with a study of the hypophsyial vascular supply. *Am J Anat.* 1936;58:421–472.
6. Green JD, Harris GW. The neurovascular link between the neurohypophysis and adenohypophysis. *J Endocrinol.* 1947; 5:136–146.
7. Markee JE, Sawyer CH, Hollinshead WH. Activation of the anterior hypophysis by electrical stimulation in the rabbit. *Endocrinology.* 1946;38:345–357.
8. Harris GW. Electrical stimulation of the hypothalamus and the mechanism of neural control of the adenohypophysis. *J Physiol Lond.* 1948;107:418–429.
9. Vale W, Spiess J, Rivier C, Rivier J. Characterization of a 41-residue ovine hypothalamic peptide that stimulates secretion of corticotrophin and beta-endorphin. *Science.* 1981; 213(4514):1394–1397.
10. Guillemin R, Brazeau P, Bohlen P, Esch F, Ling N, Wehrenberg WB. Growth hormone releasing factor from a human pancreatic tumor that caused acromegaly. *Science.* 1982;218(4572):585–587.
11. Rivier J, Spiess J, Thorner M, Vale W. Characterization of a growth hormone-releasing factor from a human pancreatic islet tumour. *Nature.* 1982;300(5889):276–278.
12. Douglas WW, Poisner AM. Stimulus-secretion coupling in a neurosecretory organ: the role of calcium in the release of vasopressin from the neurohypophysis. *J Physiol.* 1964;172:1–18.
13. Lincoln DW, Wakerley JB. Electrophysiological evidence for the activation of supraoptic neurons during the release of oxytocin. *J Physiol.* 1974;242:533–554.
14. Yamashita H. Effect of baro- and chemoreceptor activation on supraoptic nuclei neurons in the hypothalamus. *Brain Res.* 1977;126:551–556.
15. Werner J. System properties, feedback control and effector coordination of human temperature regulation. *Eur J Appl Physiol.* 2010;109:13–25.
16. Levin BE. Developmental gene × environment interactions affecting systems regulating energy homeostasis and obesity. *Front Neuroendocrinol.* 2010;31(3):270–283.
17. Castaneda TR, Tong J, Datta R, Culler M, Tschop MH. Ghrelin in the regulation of body weight and metabolism. *Front Neuroendocrinol.* 31(1):44–60.
18. Sánchez-Lasheras C, Könner AC, Brüning JC. Integrative neurobiology of energy homeostasis-neurocircuits, signals, and mediators. *Front Neuroendocrinol.* 2010;31(1):4–15.
19. Niswender KD, Schwartz MW. Insulin and leptin revisited: adiposity signals with overlapping physiological and intracellular signaling capabilities. *Front Neuroendocrinol.* 24(1):1–10. Review.
20. Kaynard AH, Pau KY, Hess DL, Spies HG. Gonadotropin-releasing hormone and norepinephrine release from the rabbit mediobasal and anterior hypothalamus during the mating-induced luteinizing hormone surge. *Endocrinology* 1990; 27(3):1176–1185.
21. Arancibia S, Tapia-Arancibia L, Assenmacher I, Astier H. Direct evidence of short-term cold-induced TRH release in the median eminence of unanesthetized rats. *Neuroendocrinology.* 1983; 37(3):225–228.
22. Plotsky PM, Vale W. Hemorrhage-induced secretion of corticotropin-releasing factor-like immunoreactivity into the rat hypophysial portal circulation and its inhibition by glucocorticoids. *Endocrinology.* 114(1):164–169.
23. Rivier C, Rivier J, Mormede P, Vale W. Studies of the nature of the interaction between vasopressin and corticotrophin-releasing factor on adrenocorticotropin release in the rat. *Endocrinology.* 1984;115(3):882–886.
24. Schneider JE. Energy balance and reproduction. *Physiol Behav.* 2004;81(2):289–317.
25. Watanabe G, Terasawa E. *In vivo* release of luteinizing hormone releasing hormone increases with puberty in the female rhesus monkey. *Endocrinology.* 1989;125(1):92–99.
26. Harris GC, Levine JE. Pubertal acceleration of pulsatile gonadotropin releasing hormone release in male rats as revealed by microdialysis. *Endocrinology.* 2003;144(1):163–171.
27. McEwen BS, Wingfield JC. The concept of allostasis in biology and biomedicine. *Horm Behav.* 2003;43(1):2–15.
28. Levine JE, Pau KY, Ramirez VD, Jackson GL. Simultaneous measurement of luteinizing hormone-releasing hormone and luteinizing hormone release in unanesthetized, ovariectomized sheep. *Endocrinology.* 1982;111(5):1449–1455.
29. Belchetz PE, Plant TM, Nakai Y, Keogh EJ, Knobil E. Hypophysial responses to continuous and intermittent delivery of hypothalamic gonadotropin-releasing hormone. *Science.* 1978;202(4368):631–633.
30. Crowley Jr WF, Filicori M, Spratt DI, Santoro NF. The physiology of gonadotropin releasing hormone (GnRH) secretion in men and women. *Recent Prog Horm Res.* 1985;1985(41):473–531.

31. Bouret SG, Draper SJ, Simerly RB. Trophic action of leptin on hypothalamic neurons that regulate feeding. *Science.* 2004; 304(5667):108–110.
32. Mellon PL, Windle JJ, Goldsmith PC, Padula CA, Roberts JL, Weiner RI. Immortalization of hypothalamic GnRH neurons by genetically targeted tumorigenesis. *Neuron.* 1990;5(1):1–10.
33. Radovick S, Wray S, Lee E, Nicols DK, Nakayama Y, Weintraub BD, Westphal H, Cutler GB, Wondisford FE. Migratory arrest of gonadotropin-releasing hormone neurons in transgenic mice. *Proc Natl Acad Sci USA.* 1991;88(8):3402–3406.
34. Spergel DJ, Krüth U, Hanley DF, Sprengel R, Seeburg PH. GABA- and glutamate-activated channels in green fluorescent protein-tagged gonadotropin-releasing hormone neurons in transgenic mice. *J Neurosci.* 1999;19(6):2037–2050.
35. Suter KJ, Song WJ, Sampson TL, Wuarin JP, Saunders JT, Dudek FE, Moenter SM. Genetic targeting of green fluorescent protein to gonadotropin-releasing hormone neurons: characterization of whole-cell electrophysiological properties and morphology. *Endocrinology.* 2000;141(1):412–419.
36. Ogawa S, Eng V, Taylor J, Lubahn DB, Korach KS, Pfaff DW. Roles of estrogen receptor-alpha gene expression in reproduction-related behaviors in female mice. *Endocrinology.* 1998;139(12):5070–5081.
37. Park CJ, Zhao Z, Glidewell-Kenney C, Lazic M, Chambon P, Krust A, Weiss J, Clegg DJ, Dunaif A, Jameson JL, Levine JE. Genetic rescue of nonclassical ERα signaling normalizes energy balance in obese Erα-null mutant mice. *J Clin Invest.* 2011;121(2):604–612.
38. Glidewell-Kenney C, Hurley LA, Pfaff L, Weiss J, Levine JE, Jameson JL. Nonclassical estrogen receptor alpha signaling mediates negative feedback in the female mouse reproductive axis. *Proc Natl Acad Sci USA.* 2007;104(19):8173–8177.
39. McDevitt MA, Glidewell-Kenney C, Jimenez MA, Ahearn PC, Weiss J, Jameson JL, Levine JE. New insights into the classical and non-classical actions of estrogen:evidence from estrogen receptor knock-out and knock-in mice. *Mol Cell Endocrinol.* 2008;290(1-2):24–30.
40. Chappell PE, Schneider JS, Kim P, Xu M, Lydon JP, O'Malley BW, Levine JE. Absence of gonadotropin surges and gonadotropin-releasing hormone self-priming in ovariectomized (OVX), estrogen (E2)-treated, progesterone receptor knockout(PRKO) mice. *Endocrinology.* 1999;140(8):3653–3658.
41. Brüning JC, Gautam D, Burks DJ, Gillette J, Schubert M, Orban PC, Klein R, Krone W, Müller-Wieland D, Kahn CR. Role of brain insulin receptor in control of body weight and reproduction. *Science.* 2000;289(5487):2122–2125.
42. Ring LE, Zeltser LM. Disruption of hypothalamic leptin signaling in mice leads to early-onset obesity, but physiological adaptations in mature animals stabilize adiposity levels. *J Clin Invest.* 2010;120(8):2931–2941.
43. Mayer C, Acosta-Martinez M, Dubois SL, Wolfe A, Radovick S, Boehm U, Levine JE. Timing and completion of puberty in female mice depend on estrogen receptor alpha-signaling in kisspeptin neurons. *Proc Natl Acad Sci USA.* 2010;107(52):22693–22698.

CHAPTER

2

Neuroendocrine GPCR Signaling

Robert P. Millar[1,2,3], *Claire L. Newton*[1], *Antonia K. Roseweir*[1]

[1]Centre for Integrative Physiology, University of Edinburgh, School of Biomedical Sciences, Hugh Robson Building, Edinburgh, UK, [2]UCT/MRC Group for Receptor Biology, University of Cape Town, Cape Town, South Africa, [3]Mammal Research Institute, University of Pretoria, Hatfield, Pretoria, South Africa

OUTLINE

Summary

G-protein coupled receptors (GPCRs) mediate the majority of neuroendocrine signaling and are the major targets of current neuroendocrine therapeutics. Currently, the major neuroendocrine GPCR families are the rhodopsin, secretin and glutamate families. Most neuroendocrine ligands are neuropeptides, but lipid molecules and biogenic amines are also important regulators. Upon ligand binding of the cognate

Handbook of Neuroendocrinology, DOI: 10.1016/B978-0-12-375097-6.10002-2

GPCR, they undergo conformational change into an active state which facilitates binding of a heterotrimeric G protein comprised of α, β and γ subunits. This leads to activation of the G protein and the displacement of GDP bound to the α subunit by GTP. This results in dissociation of the α subunit from the βγ subunits, and activation or inhibition of intracellular effectors resulting in downstream signaling cascades which alter cellular activity and gene expression. There are four major classes of G proteins; G_s, which activates adenylate cyclase to generate cAMP, which then activates protein kinase A; $G_{i/o}$, which inhibits adenylate cyclase; $G_{q/11}$, which activates phospholipase Cβ to generate inositol trisphospate (which mobilizes intracellular Ca^{2+}) and diacyl glycerol (which activates protein kinase C); and $G_{12/13}$, whose targets are less well defined. There is a wide range of isoforms for each of the subunits, such that there is potentially a large number of combinations making up the heterotrimeric G proteins. While the Gα subunits are largely responsible for the activation or inhibition of the effector enzymes, the Gβγ is also able to alter cellular systems such as ion channels. Hydrolysis of GTP to GTP on the α subunit allows it to reassociate with the βγ subunits so that the heterotrimeric G protein is available for another cycle of GPCR activation. GPCRs have been shown recently to also activate or recruit non-G proteins such as β-arrestin to initiate cellular events. Selectivity of GPCRs for signaling pathways and/or desensitization may be modified by phosphorylation by kinases, homo- or hetero-dimerization or oligomerization, and by association with a host of intracellular proteins. In addition, these elements may modify the selectivity of the GPCR for ligands. GPCR activities may also be modulated by proteins that affect their expression and trafficking to the cell surface. Taken together, there is a vast array of mechanisms that can affect GPCR signaling, which is influenced by the cellular context and numerous inputs on cellular function. In addition to being major targets for development of therapeutics, dysfunction of GPCRs, G proteins and effectors through mutation leads to many disease states.

PRINCIPLES AND CLASSES OF RECEPTORS IN NEUROENDOCRINE SIGNALING

The detection and integration of diverse exogenous inputs (e.g., light, temperature, nutrient, visual, odorant and pheromone) and endogenous signals (e.g., hormones, growth factors, neurotransmitters, metabolites, ions and lipids) into the vertebrate hypothalamus is crucial for homeostasis and survival. Hypothalamic integration of, and responses to, these diverse inputs are mediated via four types of receptors. These are: (a) The cell surface enzyme-associated receptors such as the tyrosine kinase insulin receptors; (b) the ion-channel receptors such as the nicotinic acetylcholine receptors; (c) the G-protein coupled receptors; and (d) the intracellular transcription factor receptors such as the steroid hormone receptors. All of these receptors are involved in mediating signals to the hypothalamus, where they are integrated to culminate in the secretion of regulatory neuropeptides and biogenic amines. However, this chapter will focus exclusively on GPCRs, which are responsible for the majority of signaling in neuroendocrinology.

Two groups of GPCRs will be considered: (a) those that mediate neurohypophysial ligand regulation of pituitary (adenohypophysis) function (i.e., hypothalamic neurohormones); and (b) those that regulate the secretion of the neurohypophysial hormones. It is impractical to present a comprehensive review of the signaling of all of these GPCRs, so this chapter will review the spectrum of receptors involved and their coupling, and will then provide a few exemplar detailed descriptions of intracellular signaling of selected GPCRs. Established and putative ligands and their cognate receptors involved in neuroendocrine regulation are listed in Table 2.1, along with the G protein(s) they preferentially associate with. Although GPCRs may preferentially recruit a specific G protein for signaling, this can be highly modified by the intracellular protein milieu, which can also alter the ligand selectivity of the receptor. Table 2.1 lists the hypothalamic factors that are released into the hypophysial portal system which regulate pituitary hormone secretion, as well as the peptide, biogenic amine and lipid activated GPCRs that are proven or putative regulators of the secretion of the hypothalamic factors. It is evident that a wide diversity of ligands and cognate GPCRs modulate the neuroendocrine system, as might be expected of this major physiological integrator.

GPCR GENERAL STRUCTURE AND CLASSIFICATION

GPCRs convey approximately 80% of signal transduction across cell membranes, and are also the predominant signalers in neuroendocrinology. They are activated by diverse ligands, which vary from single light photons through cations, odorants, amino acids, lipids, fatty acids, neurotransmitters, peptides and polypeptides. GPCRs are located within the plasma membrane and have a common architecture consisting of seven transmembrane (TM) α-helical domains, connected by extracellular (ECL) and intracellular (ICL) loops (Fig. 2.1). One of the characteristics of GPCRs is that they are highly "druggable," and more than one-third of all current therapeutics are directed at them. Despite this, to date only a small percentage of the ~800 known and verified GPCRs have been targeted for therapeutics. There remains, therefore, an enormous scope for researchers to delineate the numerous roles of GPCRs in physiology and pathophysiology, and to thereby understand disease processes and develop new treatments.

GPCRs may be classified as five major families (Table 2.2). The largest family is the Rhodopsin family, which comprises 672 family members, including 388

TABLE 2.1 Established and Putative Neuroendocrine Ligands and their Cognate GPCRs

Ligand	GPCR family[1]	Receptor(s)	G-protein coupling predominance	Ref(s)
HYPOTHALAMIC SECRETED HORMONES				
Corticotropin-releasing factor (CRF) and urocortins	sec	CRF_1R, $CRF_{2a}R$, $CRF_{2b}R$ and $CRF_{2c}R$	G_s	212
Dopamine	rho	D_2R	$G_{i/o}$	165
Gonadotropin-releasing hormone (GnRH)	rho	GnRHR type I, GnRHR type II	$G_{q/11}$	5,78
Growth hormone-releasing hormone (GHRH)	sec	GHRHR	G_s and $G_{q/11}$	163
Oxytocin	rho	OTR	$G_{q/11}$	185
Pituitary adenylate cyclase-activating peptide (PACAP) and vasoactive intestinal peptide (VIP)	sec	$VPAC_1R$, $VPAC_2R$, PAC_1R	G_s	213
Somatostatin	rho	SST_1R, SST_2R, SST_3R, SST_4R, SST_5R	$G_{i/o}$	180
Thyrotropin-releasing hormone (TRH)	rho	TRH_1R, TRH_2R	$G_{q/11}$	161
Vasopressin	rho	$V_{1a}R$, $V_{1b}R$, V_2R	$G_{q/11}$ G_s	182
REGULATORS OF HYPOTHALAMIC HORMONES				
Peptides				
Apelin	rho	APJR	$G_{i/o}$	214
Cholecystokinin and gastrin	rho	CCK_1R, CCK_2R	$G_{q/11}$ and G_s	215
Galanin and galanin-like peptide (GALP)	rho	GAL_1R, GAL_2R, GAL_3R	$G_{i/o}$ and $G_{q/11}$ (GAL_2R only)	216
Ghrelin	rho	GHS-R1a, GHS-R1b	$G_{q/11}$	217
Gonadotropin-inhibitory hormone (GnIH)	rho	GPR147	$G_{i/o}$	218
Hypocretins (orexins)	rho	OX_1R, OX_2R	$G_{q/11}$	219
Kisspeptins	rho	GPR54	$G_{q/11}$	149
Melanin-concentrating hormone (MCH)	rho	MCH_1R, MCH_2R	$G_{i/o}$ and $G_{q/11}$	220
Melanocortins	rho	MC_1R, MC_2R, MC_3R, MC_4R, MC_5R	G_s	221
Melatonin	rho	MT_1R, MT_2R MT_3R	$G_{i/o}$ $G_{q/11}$	222
Neuromedin U	rho	NMU1R, NMU2R	$G_{q/11}$	223
Neuropeptide Y (NPY), peptide YY (PYY) and pancreatic polypeptiode (PP)	rho	Y_1R, Y_2R, Y_4R, Y_5R, Y_6R	$G_{i/o}$	224
Neurotensin	rho	NTS_1R, NTS_2R	$G_{q/11}$	225
Neurokinin B (NKB)	rho	NK_3R	$G_{q/11}$	226
Opioids	rho	μR, δR, κR	$G_{i/o}$	227
Urotensin-II	rho	UTR	$G_{q/11}$	228
BIOGENIC AMINES				
γ-amino butyric acid (GABA)	glu	$GABA_{B1}R$, $GABA_{B2}R$	$G_{i/o}$	229
Histamine	rho	H_1R H_2R H_3R, H_4R	$G_{q/11}$ G_s $G_{i/o}$	230
Metabotropic glutamate	glu	$mGlu_2R$, $mGlu_3R$, $mGlu_4R$, $mGlu_6R$, $mGlu_7R$, $mGlu_8R$ $mGlu_1R$, $mGlu_5R$	$G_{i/o}$ $G_{q/11}$	231

(*Continued*)

TABLE 2.1 Established and Putative Neuroendocrine Ligands and their Cognate GPCRs—cont'd

Ligand	GPCR family[1]	Receptor(s)	G-protein coupling predominance	Ref(s)
Noradrenaline	rho	α_{1A}R, α_{1B}R, α_{1D}R α_{2A}R, α_{2B}R, α_{2C}R β_1R, β_2R, β_3R	$G_{q/11}$ $G_{i/o}$ G_s	232
Serotonin (5-hydroxytrptamine/5-HT)	rho	5-HT$_{1A}$R, 5-HT$_{1B}$R, 5-HT$_{1D}$R, 5-HT$_{1E}$R, 5-HT$_{1F}$R 5-HT$_{2A}$R, 5-HT$_{2B}$R, 5-HT$_{2C}$R 5-HT$_4$R, 5-HT$_6$R, 5-HT$_7$R 5-HT$_{5A}$R, 5-HT$_{5B}$R	$G_{i/o}$ $G_{q/11}$ G_s ?	233
Lipids				
Cannabanoid	rho	CB$_1$R, CB$_1$R	$G_{i/o}$	234
Leukotriene and lipoxin	rho	BLT$_1$R, BLT$_2$R CysLT$_1$R, CysLT$_2$R Lipoxin ALX R	$G_{q/11}$ $G_{q/11}$ $G_{q/11}$	235
Lysophospholipid	rho	LPA$_1$R, LPA$_2$R, LPA$_3$R, LPA$_4$R, LPA$_5$R SIP$_1$R, SIP$_2$R, SIP$_3$R, SIP$_4$R, SIP$_5$R	$G_{q/11}$, $G_{i/o}$ and $G_{12/13}$ $G_{q/11}$ and $G_{i/o}$	236
Prostanoid	rho	DP$_1$R, EP$_2$R, EP$_4$R, IPR EP$_3$R, DP$_2$R EP$_1$R, FPR, TPR	G_s $G_{i/o}$ $G_{q/11}$	237

[1] *rho, Rhodopsin family; glu, Glutamate family; sec, Secretin family*

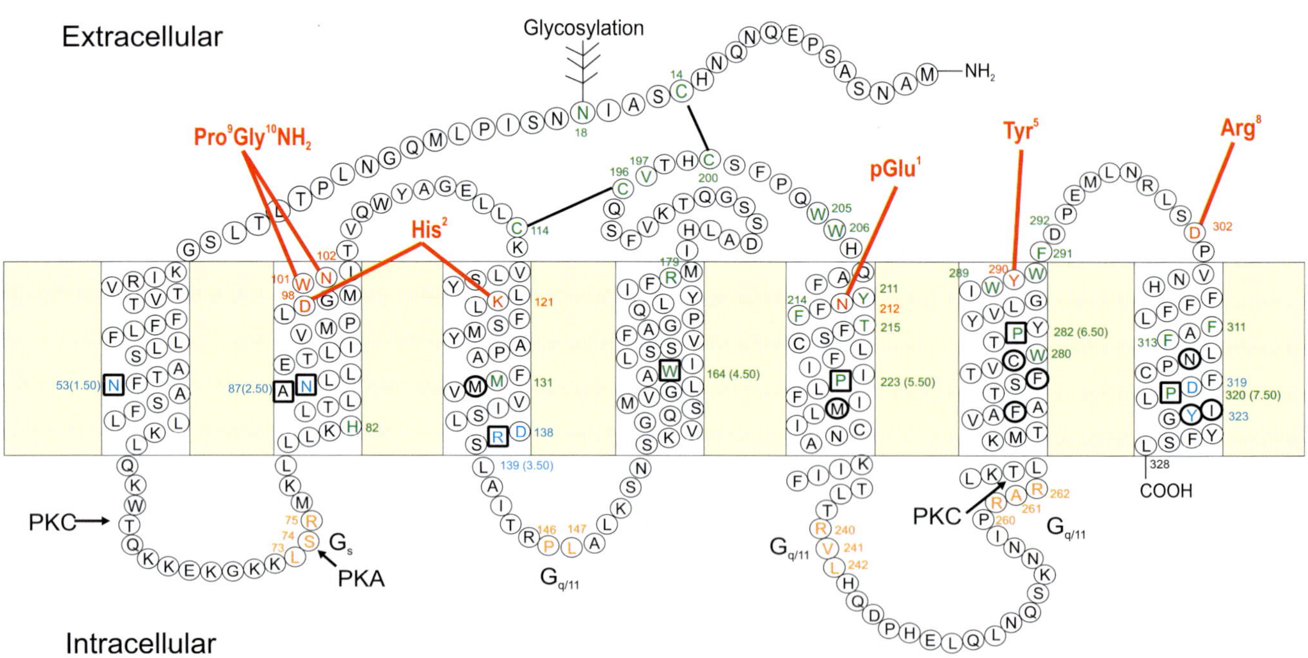

FIGURE 2.1 **Two-dimensional representation of the human GnRH receptor.** The 7-transmembrane (TM) α-helical domains (boxed) are connected by three extracellular loops (ECLs) and three intracellular loops (ICLs). Residues in squares are ones highly conserved throughout the Rhodopsin family of GPCRs. Ligand binding residues (red) and residues thought to be important in receptor structure or binding pocket configuration (green) are shown. These include disulfide bond formation (black lines) and a glycosylation site. Residues involved in receptor activation are shown in blue. Residues involved in coupling to G proteins are shown in orange. Putative protein kinase C (PKC) and protein kinase A (PKA) phosphorylation sites are indicated. The intermolecular interactions between GnRH I residues and the receptor are indicated with red lines. Residues shown to alter selectivity for relative binding affinities of GnRH I and GnRH II are shown in bold circles.

TABLE 2.2 GPCR Families

	Rhodopsin	Secretin	Adhesion	Glutamate	Frizzled & Taste	
Total number	672 (388 Ors[2])	15	33	22	11	25
Identified drug targets	40	4	0	3	0	0
Number of orphans	63 + Ors	0	30	7	0	21

[2] *Ors, odorant receptors.*

so-called odorant receptors. Despite this large number, and the fact that only about 60 of the non-odorant Rhodopsin family GPCRs are orphan receptors (i.e., ones for which ligands are unknown), relatively few have been identified as drug targets. Moreover, there are likely to be substantially more drug candidates in the family, as odorant receptors, which have been assumed to be involved exclusively in the detection of odorants, are now emerging as the target of ligands in tissues other than the nasal neuroepithelium. An example is an odorant receptor that was recently shown to be expressed in prostatic cancer cells (prostate-specific G-protein coupled receptor). Its ligand has been identified as a steroid hormone, androstenone, which has potent antiproliferative effects.[1]

The four other, smaller, families are the Secretin (15 members), Adhesion (33 members), Glutamate (22 members) and Frizzled/Taste (36 members) families. The family of GPCRs classified as Adhesion GPCRs is particularly fascinating. They have a very extended N-terminus thought to be involved in cell–cell contact. As yet, there are no drugs targeting them, and the majority remain orphan receptors. Nevertheless, it has become evident from knockout studies in *Caenorhabditis elegans* (*C. elegans*) and in rodents that they play a role in the early development of the embryo, and also appear to have a number of physiological roles.[2] There are no orphans amongst the 15 Secretin family and the 11 Frizzled GPCRs, but two-thirds of the Glutamate and most of the Taste GPCRs remain as orphans. Only seven of all these GPCRs have been identified as drug targets. To date, all of the established GPCRs involved in neuroendocrine regulation belong to the Rhodopsin, Secretin or Glutamate GPCR families (Table 2.1). As yet, no neuroendocrine regulators have been identified from the Adhesion and Frizzled/Taste families, but it is very likely that they will be found to play roles in neuroendocrinology.

ATOMIC-LEVEL STRUCTURE OF GPCRs

In the absence of structural knowledge derived from crystallographic studies, predictions of the atomic level structures of GPCRs were originally based on speculative molecular models, and later on low-resolution two-dimensional crystal structures of rhodopsin.[3,4] Nevertheless, this information provided insight regarding the relative positioning of the TM domains, and inspiration for testing predictions of molecular interactions derived from the molecular models.

A schematic of the two-dimensional structure of the gonadotropin-releasing hormone (GnRH) receptor is used to illustrate the salient features of Rhodopsin family GPCRs (Fig. 2.1). The GnRH receptor does have some unusual features that differ from the other members of this family. It lacks a C-terminal tail, and the conserved interacting TM2 Asn^{87} and TM7 Asp^{319} residues are more frequently in reciprocal locations. Nevertheless, the GnRH receptor does have the major hallmarks of a Rhodopsin family GPCR (see legend to Fig. 2.1). By developing molecular models of the GnRH receptor based on the early predicted structures of rhodopsin, residues were identified which were proposed to configure the receptor, bind the ligand, and be involved in receptor activation or engagement of intracellular signaling proteins (see Millar *et al.*[5] for review). These were then mutated to test the validity of the proposal and the model. For example, the putative interaction of TM residues mentioned above was tested by mutation and restoration of function by reciprocal mutation. Established interactions were then firmly fixed in the model, while negation of interactions led to refining the molecular model. As anticipated, residues conserved in all Rhodopsin family members (boxed in Fig. 2.1) were shown to be critical for function. Amongst these, the conserved Asp-Arg (DR) sequence in TM3 (sometimes the Asp is substituted with another acidic residue, Glu) was thought to be involved in receptor activation. Indeed, mutation of Asp to Asn, which would release Arg from the ionic bond, increases the coupling efficiency.[6] Confirmation that the Asp-Arg (DR) ionic bond is broken in the process of receptor activation has emerged from the recent crystal structures (see later). GnRH binding sites were also established by mutagenesis studies (red residues in Fig. 2.1), and the GnRH ligand NMR structure could be satisfactorily docked to the GnRH receptor structure.

Although considerable progress could be made with molecular models, physical structures of crystallized GPCRs were essential to transform experimental interpretation into fact. Crucial to an understanding of how any protein carries out its functions is the knowledge of the protein's three-dimensional structure, which provides the necessary framework for determining mechanisms of action of proteins. In 2000, Chris Palczewski and co-workers solved the structure of rhodopsin. This had a major impact in confirming and refuting previous predictions on the structure and function of rhodopsin.[7] The structure also provided more accurate information for sequence alignment and modeling of other GPCRs. In rapid succession the structures of the β2-adrenergic, β1-adrenergic and A_{2A}-adenosine receptors, as well as the structure of the unliganded active opsin (Ops) form of rhodopsin, and the active (Ops*) bound to an 11-amino acid C terminus sequence of the Gα transducin G protein, were solved.[8–13] These structures have been key to advancing our understanding of the molecular mechanisms underlying GPCR activation for transmission of signals into cells. Aside from the important biomedical implications of solving the structures of these receptors, the findings provide insight into understanding how proteins in general can take up different conformations in their associations with other molecules and in the signaling/activation process.

Solving the β2-adrenergic crystal structure presented a formidable challenge in producing sufficient pure protein, stabilizing flexible domains such as ICL3, defining appropriate detergent/lipidic environments and crystallization conditions, and developing microdiffraction technology to obtain X-ray data from very small crystals. Kobilka's group first used an antibody to stabilize ICL3 (which is highly mobile and compromises crystallization) and achieved a 3.4-to 3.7-Å resolution, which revealed the TM domains very clearly (Fig. 2.2).[10] This was followed up with another crystal structure in which ICL3 was substituted with lysozyme, a highly structured molecule, which improved the resolution to 2.4 Å so that extracellular loop structure could be visualized as well as elements of the intracellular loops.[13] The structure also revealed membrane cholesterol interaction with TM domains 2, 3 and 4. This is thought to be the site at which small molecules can act allosterically to potentially activate or inactivate GPCRs (Fig. 2.2).[14]

Comparison of the four solved GPCR structures reveals two important features: first, *structural convergence* – the similarities in structure; and second, *structural divergence* – the features that differ. The docked ligands in the β-adrenergic and adenosine receptors occupy much the same space in the binding pocket created by the TM cluster, but adenosine docks somewhat more superficially.[15] Thus, GPCRs tend to accommodate these small molecules with similar spatial arrangements, but the particular interactions with amino acid side-chains are quite different. The structures of the ECL2 are quite divergent. In rhodopsin, the top of the TM cluster seems to be occluded by the N-terminal and ECL2 domains, whereas the top of the TM cluster appears to be open in the other receptors, with ECL2 positioned to one side.[15] The occlusion of the TM cluster pocket by ECL2 in rhodopsin had presented a conundrum, as this would prevent access of

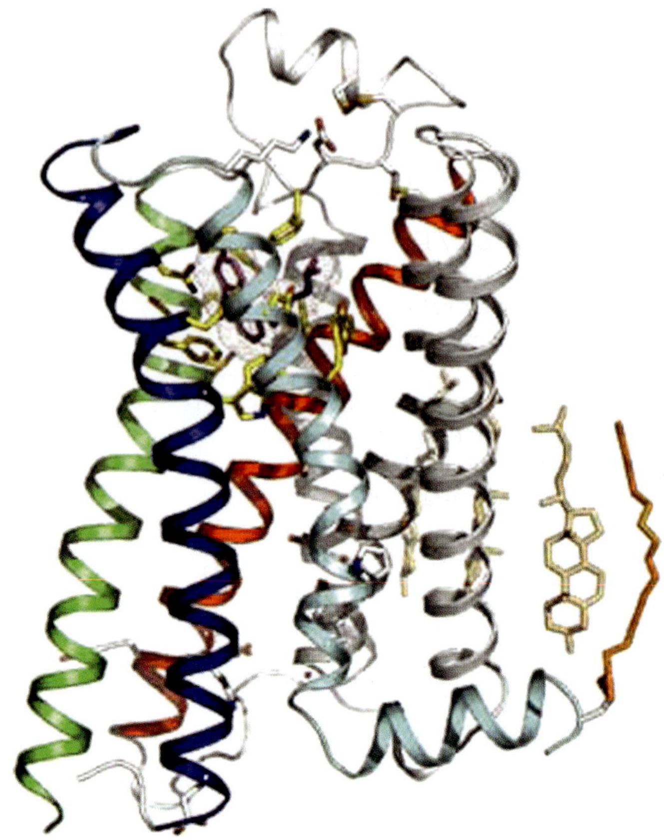

FIGURE 2.2 **Three-dimensional structure of the β2-adrenergic receptor.** The three-dimensional structure of the β2-adrenergic receptor stabilized by lysozyme substitution of intracellular loop 3 (ICL3), with a resolution of 2.4 Å. The inverse agonist carazolol co-crystallized with the receptor is shown in purple, bound to cognate residues (yellow). Extracellular loop 2 (ECL2) is shown as a helical structure above the transmembrane (TM) helices. Molecules of water are shown in red, membrane cholesterol molecules interacting with the receptor in beige, and a fourth intracellular loop, made up of the carboxyl terminus tethered to the membrane by a palmitoyl residue, in orange. *Figure kindly supplied by Brian Kobilka, Department of Molecular and Cellular Physiology, Stanford University School of Medicine, California, USA.*

small ligands known to bind in the pocket formed by the TM cluster. In forming this highly structured closure at the top of rhodopsin, retinol is retained within an environment where it will not be easily hydrolyzed, and can be activated by a photon of light. In contrast the new solved structures of the β2-adrenergic and adenosine A_{2A} receptors revealing the positioning of ECL2 to one side of the TM cluster provided a mechanism for small ligand access. Crystal structures of GPCRs from other families have yet to be solved, and are likely to differ significantly from the Rhodopsin family structure, as none of the amino acids conserved in the Rhodopsin family of GPCRs is conserved between the families. The only common feature is the disulfide bridge between ECL1 and ECL2, underlining its importance in configuring GPCRs. The overall structural similarities of the different GPCR families which are genetically unrelated are a remarkable molecular example of parallel or convergent evolution, whereby similar molecular or anatomical structures can arise independently through evolutionary selection drive.

The majority of GPCRs undergo post-translational modifications. N-linked glycosylation at the Asn of the Asn-x-x-x-Ser/Thr consensus site (see Fig. 2.1) occurs co-translationally in the endoplasmic reticulum, with further elaboration in the Golgi apparatus. Glycosylation is important for the stability of some neuroendocrine GPCRs at the cell surface, such as the GnRH and vasopressin V_{1a} receptors, and plays a role in membrane trafficking by facilitating correct folding of others, such as the follicle-stimulating hormone (FSH), thyrotropin-releasing hormone (TRH) and vasoactive intestinal peptide-1 (VIP-1) receptors. For some GPCRs, such as TRH and somatostatin (SST) receptors, glycosylation contributes to high-affinity binding. Many GPCRs undergo thio ester linkage to palmitate at a cysteine residue in the intracellular carboxyl terminal domain. This tethers the carboxyl terminal domain to the plasma membrane, thereby creating a fourth intracellular loop. Examples are the luteinizing hormone (LH), somatostatin 5, vasopressin V_2, dopamine, serotonin (5-HT) and endothelin receptors, in which elimination of the palmitoylation site compromises G-protein coupling. However, glycosylation and palmitoylation are not crucial for the function of some GPCRs.

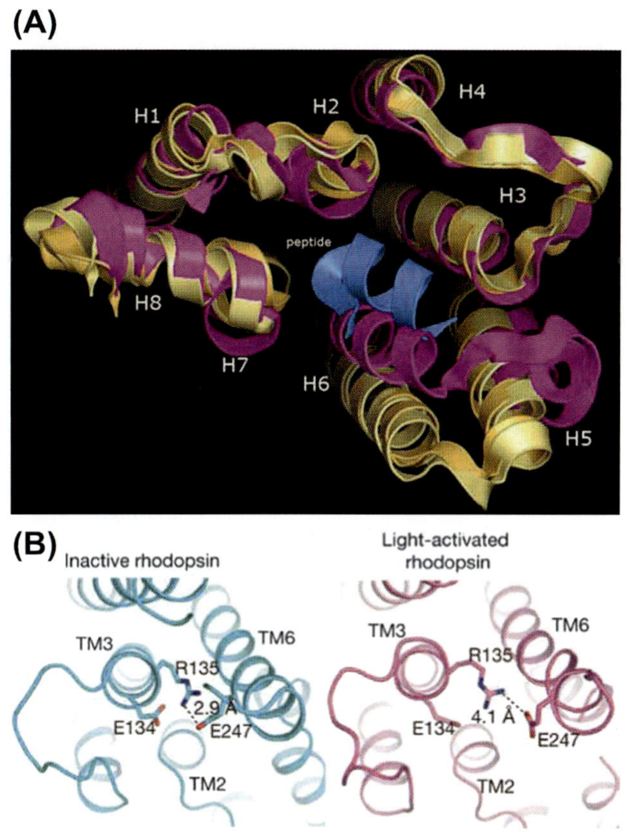

FIGURE 2.3 **Comparison of inactive and active rhodopsin conformations. (A)** Comparison of the structures of active opsin (Ops), active opsin bound to a C-terminal fragment of Gα transducin (Ops*) (both yellow) and rhodopsin (purple), depicting the change in position of TM6 (H6) upon activation.**(B)** Presence of the "ionic lock" between Arg^{135} and the adjacent Glu^{134} in TM3 and Glu^{247} in TM6. In the light-activated structure, the Arg^{135}/Glu^{134} ionic bond is broken and the distance of Glu^{247} from Arg^{135} has increased. *Figure 2.3 (B) kindly supplied by Richard Henderson, MRC Laboratory of Molecular Biology, Cambridge, UK.*

RECEPTOR MECHANISM OF ACTIVATION

A major question regarding GPCR function is, how does receptor activation occur? There have been many postulates over the years and a plethora of data and speculations presented on the basis of modeling, biophysical data and mutagenesis studies. The three-dimensional solved structures of GPCRs have contributed to advancing our understanding of the molecular mechanisms involved in receptor activation. By comparing the crystal structures of the inactive and active states of rhodopsin, differences in conformation of the TM domains could be identified. Fig. 2.3A demonstrates the structural changes that occur in the rhodopsin activation process. Opsin, which reflects the active state of rhodopsin, is depicted in yellow helices,[9] active state opsin bound to the C-terminus peptide of the G protein (transducin) in orange helices,[11] and inactive rhodopsin in mauve.[16] TM6, which is predicted to move, does so at around 3–4 Å in the active opsin crystal structures. This movement allows access for the G protein transducin to interact with TMs 5 and 6, and induction of an α-helical-type structure as it associates with TMs 5, 6 and 7. These molecular changes are brought about by the breaking of the "ionic lock" in rhodopsin comprising ionic interactions in the conserved Glu/Asp Arg motif between Glu^{134} and the adjacent Arg^{135}, and with a second Glu^{247} in TM6.[11] The disruption of the Arg ionic bond with its adjacent Asp acidic residue in TM3 had been implicated by Ballesteros and colleagues in the activation of the GnRH receptor a decade earlier, in 1998.[6] The recent crystal structures of active opsin and inactive rhodopsin clearly show that activation is accompanied by the disruption of the Arg^{135}/Glu^{134} ionic bond (Fig. 2.4B). This results in the formation of a new interaction of Arg^{135} with Tyr^{223} in TM5.[11]

Until recently, crystal structures obtained for the β-adrenergic and adenosine-A_{2A} receptors have all been of the inactive state of the GPCRs bound to inverse agonist or partial agonist. However, the structure of the active state of the β_2-adrenergic receptor tethered

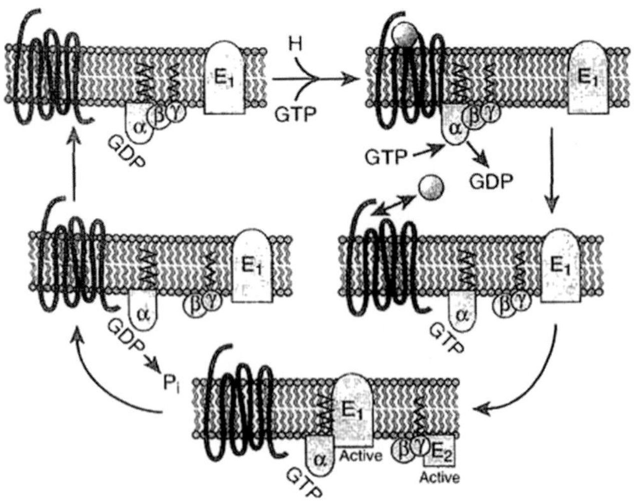

FIGURE 2.4 **GPCR activation of G proteins.** This schematic shows the G-protein cycle commencing with the inactive state of the GPCR (top left). Hormone (H) binding facilitates association of the GDP-bound heterotrimeric G protein and exchange of GDP by GTP. This is followed by dissociation of the Gα and Gβγ subunits from the receptor and their subsequent interaction with different effectors (E_2). The most common activation of effectors is through association with the Gα subunit. The intrinsic GTPase activity of the Gα subunit regenerates the GDP-bound Gα subunit, which can reassociate with Gβγ and is then ready to interact with further activated receptors. *Reproduced from Gonzalez-Maeso J, Sealfon SC. Chapter 5. In Jameson JL, De Groot LJ, eds.* Endocrinology*, Vol. I (6th ed.). Philadelphia: Saunders Elsevier; 2010: Ch. 5.*

irreversibly to an engineered agonist[17] and stabilized with a camelid antibody fragment[18] was recently solved. This reveals that there are subtle changes in the binding pocket associated with an 11-Å outward movement of the cytoplasmic end of TM6, and rearrangements of TM5 and TM7 which are very similar to those seen in opsin (Fig. 2.3). The studies reveal that binding events at both the extracellular and intracellular surfaces are required to stabilize an active conformation of the receptor.

BIOPHYSICAL APPROACHES TO GPCR STRUCTURE AND FUNCTION

Although crystal structures of GPCRs have revolutionized our understanding of the operation of these receptors, they have limitations. First, it is extremely difficult and labor-intensive to produce three-dimensional X-ray crystals; it is thus currently impractical to produce crystals of a large number of GPCRs, and accomplishment of this will require new and improved, methodology. Secondly, these structures represent a single conformational state, yet GPCRs are highly dynamic and assume different conformational states and go through various transitions when interacting with ligands and with intracellular signaling and modulating proteins. Thirdly, most of the current crystal structures (with the exception of opsin and the active β_2-adrenergic structure) appear to represent an inactive or partially active conformation, possibly because the stability of this state is greater and it is a preferred conformation for crystallization. Therefore, other biophysical techniques are being pursued to inform us about the dynamic changes in GPCR structures upon their activation.

The Hubbell laboratory has employed spin labeling of rhodopsin, which involves specifically labeling the receptor with spin labels (organic molecules able to interact with other molecules and which contain an unpaired electron) before analysis by electron paramagnetic resonance (EPR) spectroscopy. This dynamic new technology allowed monitoring of the real-time structural movements in rhodopsin and demonstrated that helix 6 moved several angstroms on activation.[19]

Another approach, utilized by the Kobilka group, involved substituting putative interacting receptor amino acids with cysteine and tryptophan. The cysteines were then labeled with the fluorophore bimane, and the quenching of the fluorescent signal by the tryptophan used to monitor the distance between the two residues when the receptor is in the active (agonist-bound) or inactive (inverse agonist-bound) states. Agonist activation induced an increase in signal which reflected the movement of the fluorescent cysteine residue in TM6 away from the tryptophan so that quenching no longer occurred.[20] This technique has now been used with other residues, facilitating the development of an overall picture of the movements and conformational changes in GPCRs on activation.

Changes in conformation of the β_2-adrenergic receptor have also been examined using two-dimensional nuclear magnetic resonance (NMR) spectroscopy after systematically labeling lysine (Lys) residues with ^{13}C. This enabled the identification of signals from specific lysines. The lysines at the intracellular face (distal TM regions and adjacent loops) were solvent exposed and mobile, while Lys^{235} in ECL3 was highly constrained in accordance with it forming a salt bridge with an aspartate residue in ECL2. The movement of this residue was then demonstrated by the spectral change after agonist, but not inverse agonist, binding. ^{13}C NMR spectroscopy has also been used to show the ECL2 is displaced from the retinal binding site upon activation of rhodopsin.[21]

These creative studies and ensuing insights into the molecular functioning of GPCRs represent a superb intellectual *tour de force*. But do these fundamental discoveries have relevance to solving medical problems? Do they provide the means to develop new drugs? In a recent study, Kolb and colleagues screened 972,608 molecules for their ability to occupy the carazolol binding site in the β_2-adrenergic receptor crystal structure. Since carazolol is an inverse agonist, the receptor is thought to be in the inactive conformation. From this screen, 25 of the best fitting molecules were selected and characterized pharmacologically. Six of them had binding affinities of $< 4\,\mu M$ (one with a K_i of 9 nM) and five were inverse agonists.[22] The finding is encouraging, as it demonstrates the potential of *in silico* screening based on crystal structure. In fact, the best selected compound is amongst the most effective inverse agonists at the β_2-adrenergic receptor. It is also intriguing that unprecedented new classes of molecules emerged from the screen, which were not expected to be ones that would occupy the binding site. This work is significant because it demonstrates the proof of the concept that GPCR crystal structures can be used for structure-based discovery of new ligands.

G-PROTEIN COUPLING OF GPCRs

GPCRs are named by virtue of their recruitment and activation of guanine nucleotide binding proteins (G proteins) at the cell membranes to initiate an intracellular signal cascade that culminates in altered activity of the target cell. However, more recent discoveries of a plethora of signaling elicited by GPCRs which do not involve G proteins (see "Non-G-protein Coupling," page 34) suggest that a more appropriate nomenclature for GPCRs is "Seven TM receptors." Nevertheless, signaling through G proteins remains the predominant mechanism of action. GPCRs may be highly selective for specific G proteins, or promiscuous in their coupling to a number of G proteins. They may also alter their coupling selectivity depending on the intracellular milieu of specific cell types, or dynamic protein changes in a single cell type brought about by physiological or pathological changes.

For example, GnRH signaling via the GnRH receptor on gonadotropes is likely to vary depending on gonadal steroid and peptide hormone effects on gene expression, translation and post-translational processing of numerous intracellular proteins involved in regulating the GnRH receptor, signaling proteins, gonadotropins and the exocytotic machinery. Thus, these clear limitations of studying neuropeptide signaling are relevant in cell lines, and also pertain to a "real" gonadotrope which represents a single state in a single physiological setting. As a prelude to neuroendocrinologists rising to the enormous challenge of delineating GPCR signaling in cells in relation to physiological dynamics, this section will describe the spectrum of G-protein signaling and cellular events that modulate the degree and flavor of the signaling.

Heterotrimeric G-protein Selectivity

GPCR activation results in conformational changes in the intracellular domains and the recruitment of G proteins which modulate activities of three major downstream effectors: adenylate cyclase,[23] phospholipase C-β (PLCβ)[24] and ion channels.[25] Heterotrimeric G proteins are a complex of α, β and γ subunits coded by separate genes. The G-protein complex dissociates from the GPCR upon ligand activation, and the Gα subunit dissociates from the Gβγ dimer, which remains associated. Although heterotrimeric G proteins are soluble, the Gα subunit may be myristoylated or palmitoylated and the Gγ farnesylated or geranylated, and thus both are docked to the plasma membrane via these fatty acids.

The Gα subunit comprises about 20 subtypes, which fall into $G\alpha_s$, $G\alpha_{q/11}$, $G\alpha_i$ and $G\alpha_{12}$ families. The major role of $G\alpha_s$ (three members) is the stimulation of adenylate cyclase, but it also stimulates Ca^{2+} channels and inhibits Na^+ channels. $G\alpha_i$ (nine members) inhibits adenylate cyclase and Ca^{2+} channels and activates K^+ channels. $G\alpha_{q/11}$ (five members) stimulates PLCβ, while $G\alpha_{12}$ stimulates PLC and the other member of the family ($G\alpha_{q/13}$) stimulates phospholipase D (PLD).

Solved crystal structures have shown that Gα has three domains. A GTPase domain, similar in structure to that found in monomeric G proteins, which binds nucleotide, Gβγ subunit and effectors. An associated helical domain is involved in effector binding, and enhancing guanosine triphospate (GTP) binding and GTPase activation. The amino terminus assumes an α-helix structure.

There are 6 subtypes of Gβ subunits and 12 subtypes of Gγ, each encoded by a different gene. The combinations of all the different α, β and γ subtype subunits can potentially create 72 different heterotrimeric G proteins, but the presence of these in biological systems and potential differences in function have yet to be elaborated.[26]

The Gβ subunit is characterized by the β-propeller structure consisting of seven β-sheets at the carboxy terminal domain and an α helix at the amino terminal domain. The γ subunit has two α helices which interact with the β-subunit helices, while the carboxy terminus makes intimate contact with the β-subunit surface.

A large number of studies have examined GPCR interactions with G proteins.[27] ICL2 and ICL3 are crucial determinants of G-protein selectivity, as demonstrated by chimeric GPCRs in which the recipient of ICL2 and ICL3 takes on the G-protein coupling characteristics of the donor GPCR.[28] ICL1 is highly conserved in length in GPCRs, and appears not to be involved in G-protein selectivity but is essential for GPCR/G-protein coupling. The conserved Asp/Glu-Arg sequence at the start of ICL2 is crucial for GPCR signal transduction, and mutation of N to neutral amino acids leads to a loss of binding to G proteins. This suggests that it interacts with negatively charged moieties of the G protein, and is core to the release of GTP.[6,11]

This event is an initiator in the cyclic activation and deactivation of G proteins (Fig. 2.4). Agonist binding leads to the disruption of the Asp/Glu-Arg salt bridge, and conformational change of the receptor into the active state. The liberated Arg is able to interact with the G protein, along with the intracellular loops, in which a cleft is opened up, exposing amino acid residue side-chains for interaction with the G protein. This results in a reduction in the affinity for guanosine diphosphate (GDP), which initiates its dissociation and replacement by guanosine triphosphate (GTP). GTP binding induces a dissociation of the Gα subunit from the G$\beta\gamma$ dimer and their activation of effectors. The intrinsic Mg^{2+}-dependent GTPase activity of the α subunit leads to hydrolysis of GTP to GDP, thereby restoring Gα association with G$\beta\gamma$ until the GPCR is again activated by agonists.

Bacterial toxins have been valuable tools for identifying G proteins involved in signaling by GPCRs. G_s is activated by cholera toxin (CTX), which ADP-ribosylates an Arg located in the nucleotide binding pocket on the α subunit. This blocks GTP hydrolysis, leading to prolonged activation. G_i is inhibited by pertussis toxin (PTX) through the ADP-ribosylation of a cysteine located in the GPCR-coupling region of the G protein, resulting in inhibition of signaling.

Modulating GPCR-coupling Selectivity

The dynamic regulation of the relative preponderance of molecules on the cell membrane and intracellularly will clearly modulate the flavor and magnitude of cellular signaling in response to GPCR activation. Similarly, variations in cellular proteins which influence the expression of GPCRs, their translation and trafficking to the cell membrane, and their internalization will have marked effects on ligand stimulation of cells. As summarized in Fig. 2.5, the first levels of GPCR regulation are gene transcription, translation and post-translational processing, which may be regulated by the ligand itself and by other hormones and factors that regulate the neuroendocrine system. The synthesized receptor may also be further regulated by other sets of proteins in its trafficking to the membrane of the cell. On arrival at the cell surface the GPCR may associate with numerous membrane and intracellular proteins, which will potentially alter ligand affinity, ligand selectivity, signaling, cytoskeletal and extracellular matrix interactions and internalization. In addition, GPCRs may undergo homo- or hetero-oligomerization to induce transactivation of other receptors or lead to signal modification. GPCR phosphorylation, acetylation, palmitoylation, ubiquitination and myristoylation also modify receptor functional properties. Clearly, the integrated effects of all these possibilities in regulating the neuroendocrine system are vast.

Differential Receptor Phosphorylation

The effects of differential receptor phosphorylation on signaling events have recently been reviewed.[29] Using the M_3 muscarinic receptor as an example, characteristic fingerprints of receptor phosphorylation were demonstrated in different cells, each with its own spectrum of kinases. Each fingerprint imparts both

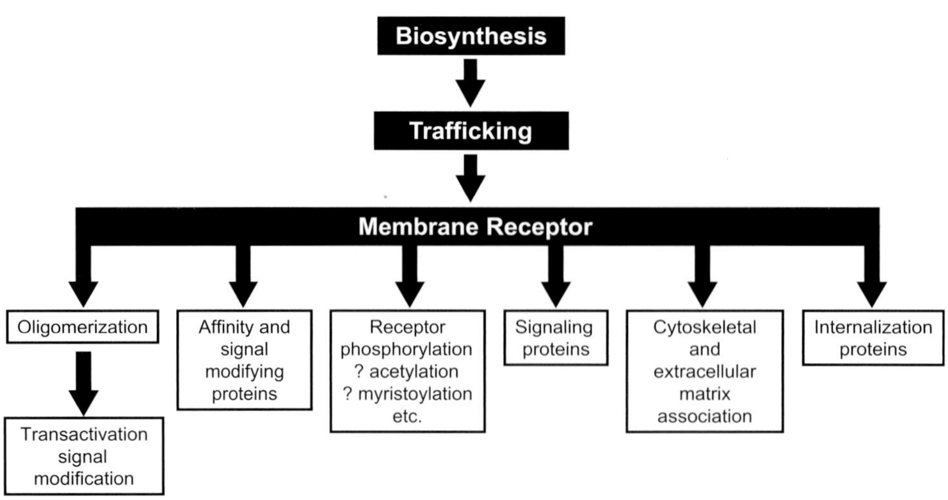

FIGURE 2.5 **Potential mechanisms for regulation of GPCRs.** Schematic describing how GPCRs can be regulated at many levels: from their biosynthesis (gene transcription, translation and post-translational processing) through their trafficking to the cell membrane and, once at the cell surface, through oligomerization and interactions with various other non-receptor proteins or through modifications such as differential phosphorylation.

a different flavor of signaling and a different phenotype of effects in cells. The M_3 receptor has also been mutated so that certain types of phosphorylation cannot take place, and when transgenically knocked into mice they have produced differential phenotypes, demonstrating the importance of differential phosphorylation in regulation of cellular responses to GPCR activation.[29] This level of regulation opens up many possibilities for differential phosphorylation of GPCRs, which will be affected by expression, activation, deactivation, and recruitment of kinases and phosphatases, all of which influence GPCR conformation and their ability to recruit and activate intracellular signaling pathways.

Receptor Oligomerization

Receptor oligomerization has recently attracted a great deal of attention. A conceptual framework was recently laid out to classify the various types of oligomerization and their resultant effects on receptor function. Receptors that are inactive in binding or signaling as monomers but that become active as oligomers are designated as "homomeric/heteromeric receptors," while receptors that are intrinsically active as monomers but have new activities as oligomers are designated as "receptor homomers/heteromers".[30]

Oligomerization of GPCRs can influence their signaling in several ways. For example, dopamine D_1 and D_2 receptors in monomer states signal through G_s and G_i, respectively, but D_1/D_2 heteromers recruit a different signaling pathway, G_q.[31] Interestingly, dopamine signaling in the brain is predominantly through G_q, suggesting that D_1/D_2 heteromers predominate *in vivo*.[32]

Another way in which receptor oligomerization can affect receptor function is through a process known as *transactivation*, in which oligomerization of two defective receptors is able to restore receptor functionality. A natural example of heterodimerization and transactivation is that of the $GABA_B$ receptors. The $GABA_{B1}$ has binding capacity but not signaling output, while the $GABA_{B2}$ has no binding capacity but has signaling capacity, such that the dimer has full high-affinity binding and signaling capacity.[33] Direct evidence for homodimerization has been obtained through the co-expression of two mutant LH receptors in cells; one that cannot signal but can bind LH, and a second that can signal but cannot bind LH. These non-functional receptors had full binding and signaling complementation when co-expressed.[34] Evidence for GPCR oligomerization *in vivo* was obtained by creating knock-in mice with each of these mutant receptors (both of which exhibit regressed reproductive tracts), and then demonstrating that crossing the two knock-in strains can produce offspring with one allele of each mutant receptor and restoration of reproductive activity.[35]

Receptor oligomerization is becoming an increasingly important area of study for understanding the biology and pathophysiology of disease, and the phenomenon offers novel therapeutic approaches. An example is that of rimonabant, a cannabinoid receptor (CB_1R) antagonist/inverse agonist, which was originally developed as an antagonist of nicotine addiction and therefore an aid to stop smoking. The drug was found to be effective in diminishing the urge to smoke, but a desirable side effect, of appetite suppression, was also observed. Research revealed that CB_1R dimerizes with the appetite-stimulating orexin-1 receptor (OX1R), and that rimonabant antagonizes orexin A stimulation of ERK 1/2 through OX1R.[36] Another example is δ-opioid/μ-opioid receptor heteromers: δ- and μ-opioid receptor interactions have been implicated in μ-opioid receptor mediated morphine tolerance and physical dependence, and a hybrid bivalent molecule consisting of μ-opioid receptor agonist coupled with a δ-opioid receptor antagonist, which targets these heteromers, has the potential to convey analgesia without generating these unwanted side effects.[37]

These are just a few examples of GPCR receptor heteromers. The formation of heteromers between neuroendocrine GPCRs and association with other receptors or non-receptor proteins suggest that these interactions create a huge spectrum of additional subtleties in the way that GPCR activities may be modulated and how they may be involved in cross-talk with other regulatory systems.

Regulation of GPCR Cell Surface Expression and Pharmacochaperones

Identifying the proteins involved in receptor internalization, downregulation and uncoupling from signaling proteins has been an area of intensive research for many years (see "Desensitization of GPCRs," below). An emerging area that is attracting great interest is the identification and delineation of the roles of proteins involved in escorting GPCRs through the various cellular compartments to the cell surface. GPCRs are dynamically regulated in order to adapt responses of cells to cognate ligands by regulation of receptor trafficking to the cell surface, or internalization and degradation/recycling of receptors. These processes therefore represent novel means of physiological regulation, and targets for potential therapeutic intervention.[38] It is not possible to describe the full spectrum of proteins involved, but of particular interest is a group of proteins, listed in Table 2.3, called *non-classical private GPCR chaperones*, which interact with GPCRs in intracellular compartments to facilitate their

TABLE 2.3 Non-classical Private GPCR Chaperones and Escort Proteins that Regulate GPCR Translocation to the Cell Surface

RAMPs (1–3)	receptor-activity modifying proteins
RTPs (1–4)	receptor-transporting proteins
REEP	receptor-expression enhancing protein
ODR4	odorant-response protein
M10s	MHC-like proteins
MRAP	MC2-receptor accessory protein
p11	calpactin 1
ATBP50	AT2-receptor-binding protein
RACK1	receptor-for-activated c kinase 1
GEC1	microtubule-associated protein
DRiP78	dopamine-receptor interacting protein 78
TcTex	tethering receptor to cytoskeleton

cell-surface expression.[39] A neuroendocrine example is the melanocortin receptor-2 (MC_2) accessory protein (MRAP). Genetically inherited hypocortisolism in patients with normal adrenocorticotropin hormone (ACTH) and melanocortin receptor-2 receptor genes presented a clinical conundrum until mutations in MRAP were identified which ablated its ability to shuttle the MC2 receptor to the surface of the cell.[40–42] Given the large number of intracellular proteins involved in regulating GPCR trafficking to the cell surface, there are likely to be many opportunities for understanding how cell surface of expression of GPCRs is regulated and how this may be modulated for therapeutic interventions. For example, inhibiting MRAP function could potentially be used as an adjunct therapy for Cushing's disease, as the overproduction of ACTH could be counteracted by the absence of MC2 receptor expression on the cell surface. The importance of MRAP, and its paralog MRAP2, in the regulation of melanocortin receptors has been recently reviewed in detail.[43]

Can the regulation of GPCR trafficking to the cell surface potentially be modulated by exogenous molecules that interact with GPCRs? A new arena of GPCR pharmacology has emerged which utilizes molecules called *pharmacochaperones*. These are hydrophobic small molecules that can penetrate the cell membrane, bind to the nascent GPCR, and "rescue" or shuttle mutant and under-expressing GPCRs to the cell membrane. Michael Conn and colleagues have pioneered the use of cell-permeant small-molecule GnRH receptor antagonists to rescue poorly expressed GnRH mutant receptors.[44] These kinds of GnRH antagonists are routinely employed to facilitate the characterization of GnRH receptor mutations directed at delineating receptor function.[45] As an example, a mutant human GnRH receptor, which exhibits no inositol phosphate response to GnRH, exhibits a robust response to GnRH by pre-incubation with a small-molecule GnRH antagonist from Neurocrine Bioscience (NBI42902) (J.A. Tello, unpublished data) (Fig. 2.6). Similar results have been observed using a mutant human LH receptor, which was identified in patients with impaired fertility and has been shown to be retained intracellularly.[46] When exposed

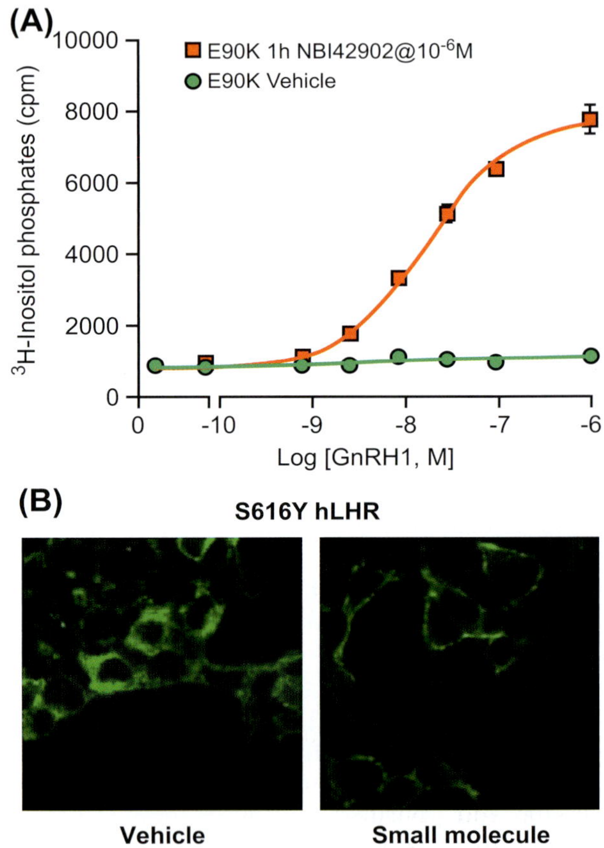

FIGURE 2.6 **"Rescue" of mutant GPCRs using pharmacochaperones.** **(A)** Rescue of mutant gonadotropin-releasing hormone (GnRH) receptors by the small-molecule pharmacochaperone NBI42902. Cells expressing mutant (E90 K) human GnRH R1 were pre-incubated in the presence/absence of NBI42902 (1 μM) for 1 h before measurement of inositol phosphate accumulation in response to stimulation with GnRH I. In the absence of the small molecule no inositol phosphate production in measured, but a robust response occurs after pre-incubation with the pharmacochaperone. **(B)** Rescue of mutant luteinizing hormone (LH) receptors by a small-molecule pharmacochaperone. Cells expressing mutant (S616Y) human LHR were pre-incubated in the presence/absence of a small molecule agonist for 24 h, before permeabilization of the cells, labeling of the receptors with fluorescent (green) antibodies and visualization by confocal microscopy. In the absence of the small-molecule receptor, localization is largely intracellular. However, pre-incubation with the pharmacochaperone causes translocation of the receptor to the cell surface, as seen in untreated wild-type human LH receptors.

to a cell-permeant small-molecule agonist, the mutant receptors were recruited to the cell surface (Fig. 2.6). In addition, treating the mutant receptor with LH elicits a very poor response compared with the wild type, whereas treatment with the small-molecule agonist restores functionality.[47] Other examples of the rescue of underexpressing mutant GPCR neuroendocrine receptors are the vasopressin V_2 receptor and the μ-opioid receptor.[48]

Regulators of G-protein Signaling (RGS) Proteins

Members of the RGS family of over 20 proteins accelerate GTP hydrolysis to increase the rate of the G protein/receptor/nucleotide cycle shown in Fig. 2.4, and attenuate or modulate GPCR-mediated responses.[49] They are thus GTPase activating proteins (GAPs), in common with PLCβ, RhoGEF, G-protein coupled receptor kinases (GRKs) and cyclic guanosine monophosphate (cGMP) phosphodiesterases. The RGS proteins are recruited to the plasma membrane by activated Gα, with which they interact to activate Gα GTPase. They have selectivity for G proteins and are variably expressed in cells, making them potential targets for drug development.

Activators of G-protein Signaling (AGS)

These proteins are capable of activating G proteins and inducing signaling in the absence of a GPCR. The precise role and regulation has yet to be fully elaborated, but they offer another potential dimension of regulation of signaling.

Receptor Activity of Modifying Proteins

Another class of proteins modulating receptor expression and pharmacology is the receptor activity modifying proteins (RAMPs) (Table 2.3). Three related proteins with extracellular, single-transmembrane and intracellular domains have been identified. RAMPs interact with certain secretin family GPCRs to affect trafficking to the membrane, and can also alter the pharmacology of ligand binding. Thus, the calcitonin receptor interacts with RAMPs to generate amylin binding. The calcitonin-receptor-like receptor binds the calcitonin-gene-related peptide when associated with RAMP1, but binds adrenomedulin when associated with RAMP2.[50]

A further group of modulators form the Homer family; these bind polyproline sequences in the large C-terminal tail of the metabotropic receptors and facilitate other interactions, such as with the inositol trisphosphate receptor.

G-PROTEIN EFFECTORS

As described earlier, G-protein subtypes predominantly interact with specific signaling effector pathways. G_s proteins activate adenylate cyclase to generate cAMP; $G_{q/11}$ proteins activate PLCβ to generate inositol trisphosphate, which mobilizes intracellular Ca^{2+}; and the Gβγ dimer can directly regulate ion channels (Fig. 2.7). In addition, an increasing number of signaling pathway recruitments by GPCRs have been found to be independent of G proteins. This aspect is covered below, in the section "Non-G-protein Coupling."

Adenylate Cyclase

Nine isoforms of adenylate cyclase have been identified. They all have the same structural architecture of two membrane-associated hydrophobic domains and two cytoplasmic domains that make up the catalytic site, which is regulated by the intracellular environment in an isoform-specific manner. Although $G\alpha_s$ is recognized as the classical activator and $G\alpha_i$ as the classical inhibitor of adenylate cyclase, Gβγ, protein kinase A (PKA), protein kinase C (PKC), calmodulin kinase, Ca^{2+} and phosphatases can all modulate activity of adenylate cyclase isoforms, thus providing integrative inputs from other signaling pathways.

Phospholipase C (PLC)

There are also multiple isoforms of PLC, comprising six family members: PLCβ (four subtypes), PLC-γ, PLC-δ, PLC-ε, PLC-δ and PLC-η. Although there is little homology between the isoforms, they each have

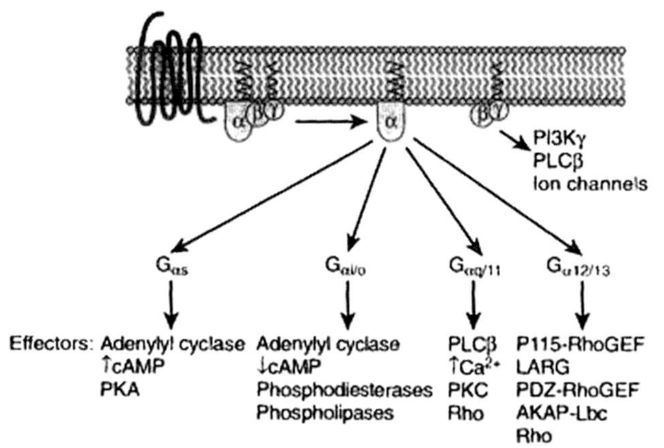

FIGURE 2.7 **GPCR recruitment of heterotrimeric G proteins and downstream effectors.** The spectrum of effectors regulated by the four Gα families are indicated. Effectors regulated by Gβγ, such as ion channels, are not shown. *Reproduced from Gonzalez-Maeso J, Sealfon SC. Chapter 5. In Jameson JL, De Groot LJ, eds.* Endocrinology, *Vol. I (6th ed.). Philadelphia: Saunders Elsevier; 2010: Ch. 5.*

a conserved catalytic region made up of the X and Y domains separated by a region containing Src homology 2 (SH2) domains. PLC-γ has an SH3 domain. All isoforms have a plekstrin homology (PH) domain which binds membrane phosphoinositides. PLC-β isoforms are activated differentially by the G_q family members (G_q, G_{11}, G_{14}, G_{15}, G_{16}), and often co-coordinately expressed in specific cell types. They are activated by Gα-GTP to hydrolyze inositol bisphosphate to inositol trisphosphate (IP_3) and diacyl glycerol (DAG), which mobilizes intracellular Ca^{2+} and activates PKC, respectively.

Ion Channels

Ion channels are ubiquitously expressed in neuroendocrine and other cells. There is a very large number of ion channels and complex associations. Most are regulated by GPCRs, either directly through Gβγ (e.g., voltage gated Ca^{2+} channel inhibition and inward rectifying K^+ channel activation), or indirectly through downstream second messenger molecules and protein kinases.

NON-G-PROTEIN COUPLING

It is becoming increasingly apparent that there are many ways in which GPCRs can signal independently of G proteins. The first convincing evidence for the existence of GPCR-independent signaling came from the Lefkowitz laboratory,[51] and has since been extensively reviewed (see, for example, Rajagopal *et al.*[52]). An example is angiotensin II at its AT_1 receptor activating both β-arrestin and G proteins. When antagonists such as angiotensin II-receptor blockers (losartan and valsartan) engage the binding site, neither signal is propagated. However, another type of antagonist (SII) does not activate the G-protein pathway but exclusively recruits β-arrestin and activates ERK.[53] Thus, it is imperative when developing drugs to target the signaling output appropriate for the clinical condition when screening small-molecule libraries. Furthermore, the use of generic high-throughput assays, such as monitoring β-arrestin recruitment, has limitations in small-molecule screening programs.

LIGAND-INDUCED SELECTIVE SIGNALING (LiSS)

The classical view of GPCR activation of signaling is conventionally viewed as the receptor occurring in two states: an inactive state, and an active state that is stabilized by agonist binding. In this active state, the receptor is able to co-opt the intracellular signaling machinery and activate a cascade of events culminating in the physiological function of the cell. This simple bimodal-switch model has been under scrutiny for some time. Recently, it has emerged that receptors can assume multiple conformations. Each of these conformations can potentially interact with a ligand in a highly selective manner. In turn, this specific receptor conformation selectively interacts with a specific intracellular signaling complex. Clearly, the preponderance of a specific signaling complex in a particular cell will tend to stabilize a certain receptor conformation, thereby inducing selectivity for a certain ligand (Fig. 2.8A). This concept has been called *ligand-induced-selective signaling (LiSS)*, and proposes that different ligands selectively recruit different intracellular signaling proteins to produce different phenotypic effects in cells.[54]

The LiSS concept was originally promulgated by Terry Kenakin,[55] and is increasingly becoming a generic theme for GPCR actions. This phenomenon is referred to by a variety of terms: *functional selectivity, biased agonism, ligand-selective agonism, agonist-directed trafficking of signaling* or *agonist-receptor trafficking*. It has important implications in specific drug development, and in minimizing side effects. A neuroendocrine example comes from the effects of the two naturally occurring GnRHs, GnRH I and GnRH II, operating through the single GnRH type I receptor to differentially recruit signaling pathways. GnRH I is much more potent in generating inositol phosphate (a gonadotrope pathway) than in its antiproliferative effects on certain cells, whereas GnRH II does not show much difference between these two effects. An extreme example is a GnRH antagonist (135–25) that has no intrinsic stimulation of inositol phosphate generation but has potent antiproliferative activity[56] (Fig. 2.8B). Of the three amino acid differences in GnRHII (His^5, Trp^7, Tyr^8), Tyr^8 is the main determinant of selective antiproliferative effects.[57] Residues in the TM domains[58] and extracellular loops[59] of the GnRH type I receptor which determine selectivity for ligand binding have also been identified.

The LiSS concept has now been demonstrated for many GPCRs (see, for example, references 60–64) and is creating a new level of sophistication which challenges the dogma that ligand engagement of a GPCR consistently elicits a specific intracellular signal. Instead, it has become increasingly evident that the nature of the ligand and the dynamically changing intracellular environment alter the flavor of the signaling. This concept is heralding a new era of drug development, in which screening for novel ligands will not simply involve receptor binding and/or the most convenient high-throughput functional signal output, but instead will screen for the appropriate intracellular signal, which targets the desired phenotypic response of a cell for a disease state. Part of this is the selection of cell lines that will ensure an appropriate intracellular context representative of the targeted cell type in the whole

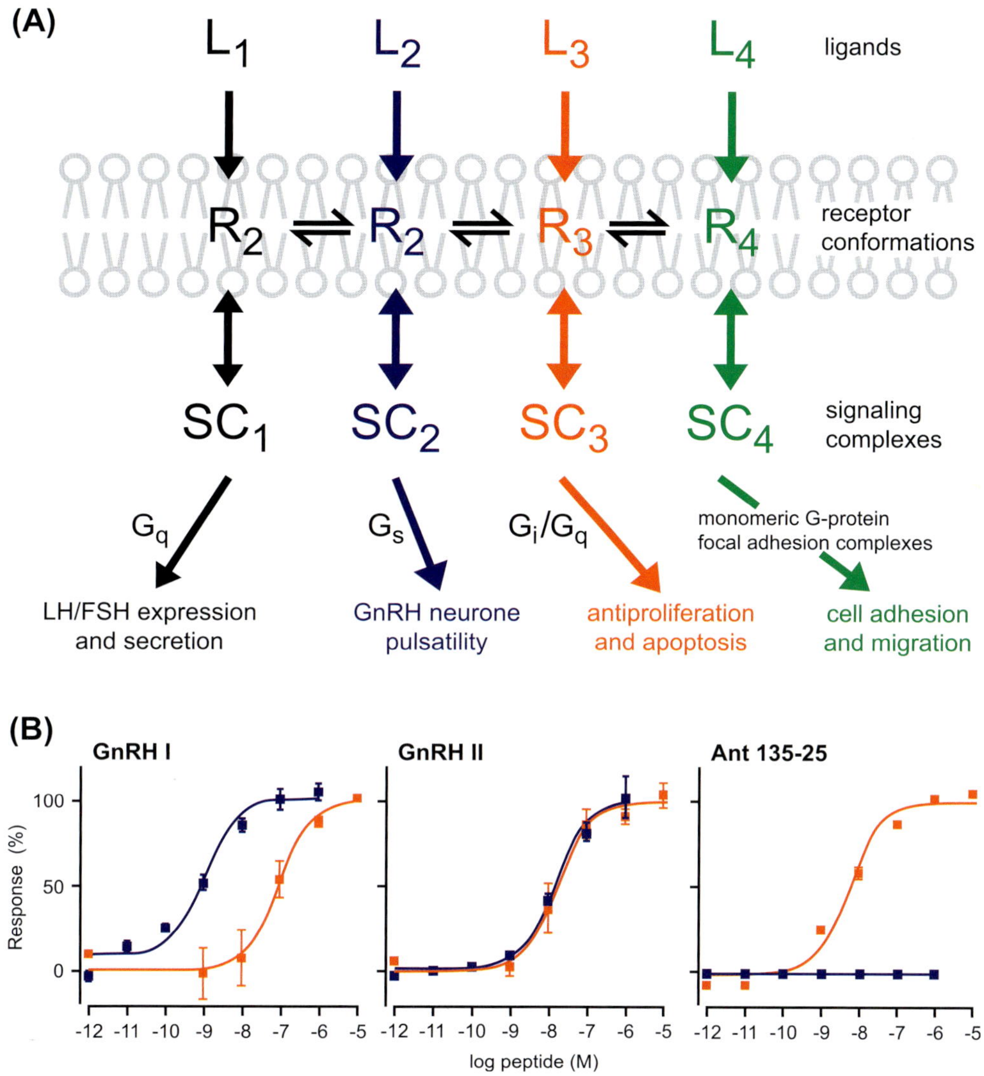

FIGURE 2.8 **GnRH receptor ligand-induced selective signaling.** (**A**) Schematic depicting the concept of multiple active states (R1–R4) of a single GPCR (GnRH receptor) that are selective for different agonist ligands (L1–L4). The different active receptor states are selectively coupled to different signaling complexes (SC1–SC4) that give rise to different effects in cells. (**B**) Effects of GnRH I, GnRH II and GnRH antagonist (135–25) on stimulation of G_q G proteins (inositol phosphate production) (blue) and inhibition of proliferation (red) in HEK 293 cells stably transfected with the rat GnRH receptor.

animal. Though these challenges are substantial, they are likely to have long-term benefits in GPCR drug discovery in the spin-off benefits of reduced failure in the clinic through lack of specificity and off-target effects.

DESENSITIZATION OF GPCRs

Systems for signal termination are a prerequisite for the maintenance of homeostasis, which is essential for a normally functioning neuroendocrine system. A great deal has been published on the mechanisms of GPCR desensitization involving receptor internalization and downregulation of surface expression, uncoupling from G proteins and other signal transduction systems, and downstream desensitization of effector systems and gene expression.

While rapid desensitization of GPCRs through uncoupling to G proteins is physiologically relevant, the slow mechanism of desensitization through receptor-endocytosis and degradation on prolonged exposure to ligand is probably not physiologically relevant, and the phenomenon may only be operative in pathological conditions of unregulated ligand secretion in tumors and in pharmacological settings of protracted exogoneous ligand supply. Such mechanisms of slow desensitization would seem unnecessary in the face of several mechanisms for ligand elimination. The pulsatile secretion of neuroendocrine hypothalamic factors ensures that the ligand signal is rapidly terminated. Moreover, some of these low molecular weight neuropeptides are massively diluted to ineffective concentrations when they enter the general circulation, and are also rapidly degraded in the circulation and tissues. Coupled with a very high renal clearance, the half-lives of hypothalamic neuroendocrine peptides are only a few minutes.

Uncoupling of GPCRs from G Proteins

GPCRs are rapidly desensitized through phosphorylation by two groups of intracellular kinases. These are the G-protein coupled receptor kinases (GRKs), consisting of seven family members with differences in membrane association and specificities for GPCRs;[65,66] and second messenger-dependent protein kinases; PKA and PKC, which are activated by DAG, and calmodulin kinase which is activated by Ca^{2+}. All of the kinases phosphorylate serine and threonine residues in ICL3 and the carboxyl-terminal tail of GPCRs, but the GRKs will only do so to the ligand-activated receptor. Thus, in principle GRKs can only induce homologous receptor uncoupling (mediated by agonist-dependent activation of the same receptors). In contrast, the second-messenger dependent kinases can phosphorylate GPCRs in the inactive as well as active conformations. This means that heterologous phosphorylation of a different receptor to the one activated can take place. Phosphorylation of ICL3 and the carboxyl-terminal tail promotes the binding of arrestin, which sterically uncouples GPCRs from G proteins. The GRKs and second-messenger activated kinases have different specificities for the amino acid sequence in which serine and threonine reside. While there are some general consensus motifs, these are rather poorly defined.[67] Clearly, there is considerable heterogeneity in the degree and sites of the phosphorylation fingerprint of ICL3 and the carboxyl-terminus, which appears relatively unimportant for arrestin recruitment but may alter the ligand binding specificity and coupling selectivity of the receptor (see "Ligand-induced Selective Signaling," above).[68]

While it is likely that many GPCRs are phosphorylated by GRKs, to date only somatostatin, prostaglandin E_1, substance P, olfactory receptors, and α_2-adrenergic, cholinergic, β-adrenergic, angiotensin-1_A, and muscarinic acetylcholine M_1 M_2 and M_3 receptors along with rhodopsin have been definitively shown to be phosphorylated by GRKs. There are considerable differences in the degree to which GPCRs are uncoupled by ligand stimulation. For example, the TRH receptor undergoes rapid desensitization by ligand exposure, whereas the mammalian GnRH type I receptor, which lacks the carboxyl-terminal tail, does not.[69] In contrast, the catfish GnRH receptor[70] and chicken GnRH type I receptor, which have carboxyl-terminal tails, undergo rapid desensitization in the generation of inositol phosphate. This is ablated by removal of the tail from the chicken receptor to mimic the rate of desensitization of the mammalian type I receptor.[70,71]

Arrestins bind to agonist-activated and GRK-phosphorylated GPCRs to sterically prevent G-protein coupling. They have a preference for binding GRK phosphorylated GPCRs over those phosphorylated by second-messenger dependent protein kinases.[67] There are four family members of the arrestin gene family,[65] two of which are mainly expressed in the retina. The other two are β-arrestin 1 and β-arrestin 2, which are ubiquitously expressed. Arrestins bind to the activated receptor via an amino terminal domain, with secondary binding via the carboxy terminus and a phosphate sensor located between the two domains.[72]

Persuasive evidence for a physiological role of the GRK/arrestin system in signal termination was provided by the demonstration that β-arrestin knockout mice exhibit an exaggerated response to β-adrenoreceptor agonists.

GPCR Internalization

In addition to inducing uncoupling, phosphorylation of GPCRs and binding of arrestins targets them for endocytosis.[65] This occurs predominantly through clathrin-coated pits which are major transporters of proteins. β-arrestin 1 or β-arrestin 2 bound to activated GPCRs interact with clathrin and the β2-adaptin of the tetrameric AP-2 adaptin complex to form an endocytic protein complex. This is then pinched off from the cell membrane by the GTPase, dynamin, to form vesicles. The internalized complex then enters acidic endosomes, where the GPCRs are degraded by lysosomes or dephosphorylated and recycled to the plasma membrane. There appear to be considerable differences in the degree of endocytosis and recycling of neuroendocrine GPCRs. For example, the GnRH receptors exhibit poor recycling, while the LH receptor demonstrates major recycling.

The ability of different ligands to stabilize different conformations of GPCRs, resulting in the recruitment of different intracellular signaling pathways (see "LiSS," above) is in principle extendable into differential rates of internalization. This appears to be the case for μ-opioid receptors, which are rapidly desensitized and internalized by the peptide opioid, DAMGO, whereas the unnatural ligand, morphine, does not induce rapid desensitization or endocytosis, suggesting explanations for tolerance and addiction.[73]

GPCR Ubiquitination

Ubiquitination of proteins and degradation in the proteasome has been described as a mechanism of degradation of many proteins. Recently, a number of GPCRs, including rhodopsin, the β_2-adrenergic opioid and vasopressin V_2 receptors, have been shown to be ubiquitinated and undergo internalization and degradation in the lysosomes.[74] Ubiquitin is covalently linked to the epsilon amino group of lysine in proteins through the COOH group of its carboxy terminal glycine. The first of three enzymes involved (E1) activates the glycine through thioester formation. The activated ubiquitin is then transferred to a carrier enzyme (E2) before being covalently linked to the protein by the third protein ligase enzyme (E3). For the β-adrenergic receptors, the process appears to be coordinated with phosphorylation and recruitment of β-arrestin to the receptor.[75]

INTRACELLULAR SIGNALING BY MAJOR NEUROENDOCRINE GPCR REGULATORS

A detailed and exhaustive description of intracellular signaling by every GPCR involved in neuroendocrine control is not practical within the confines of this chapter. Instead, only the major signaling pathways of well-established neuroendocrine GPCRs will be described, while a more comprehensive description of the spectrum of signaling of the GnRH receptor will be presented as an exemplar of the potentially vast array of pathways and their cross-talk. The G-protein preference for signaling by neuroendocrine GPCRs is summarized in Table 2.1. This section will be followed by a focused description of GPCRs involved in the neuroendocrine regulation of the GnRH neuron as an example of a network regulating a single neurosecretory neuron output.

GnRHR (Gonadotropin-releasing Hormone Receptor)

Receptor Coupling

GnRH agonist occupancy of GnRH receptors leads to activation of multiple signal transduction pathways (Fig. 2.9). In gonadotropes, GnRH activates PLCβ via $G_{q/11}$, resulting in the hydrolysis of membrane-bound phosphatidylinositol 4, 5-bisphosphate (PIP_2) to IP_3 and DAG, which mobilize intracellular calcium and activate PKC, respectively. These in turn stimulate the biosynthesis and secretion of the gonadotropins, LH and FSH. The mechanism of GnRH receptor activation and recruitment of signaling pathways has been comprehensively reviewed.[5]

$G_{q/11}$ is the predominant G protein coupled to the GnRH receptor in various cellular environments.[76–78] However, a number of studies have implicated other G proteins in mediating the actions of GnRH receptors. Pretreatment of rat pituitary cells with pertussis toxin decreased IP_3 production in response to GnRH, suggesting coupling to either G_i or G_o.[79] In addition, GnRH receptor coupling to G_i has been demonstrated in ovarian carcinomas,[80–82] uterine leiomyosarcomas,[80] uterine endometrial carcinomas[82,83] and human prostate cancer cells.[84] Pretreatment of rat pituitary cells with cholera toxin results in an increase in GnRH stimulation of LH, suggesting coupling to G_s.[85,86] In addition, G_s and G_i coupling has been revealed by the GnRH stimulation of cAMP in a number of experimental paradigms.[87–92]

What are the structural features of the GnRH receptor that facilitate the coupling to different heterotrimeric G proteins? The intracellular loops and carboxyl-terminal tail have been implicated in specific coupling of GPCRs to G proteins, but their degree of involvement varies amongst different receptors. Since all the mammalian GnRH type I receptors lack a carboxyl-terminal tail, effective receptor G-protein interactions must take place via one or more of the intracellular loops (see "G-protein Coupling of GPCRs," above). The conservation of the carboxyl-terminal sequence of ICL3 in vertebrate GnRH receptors suggested this region may be crucial for coupling to the primary mediator, $G_{q/11}$. A series of cassette substitutions covering the entire sequence of ICL3 confirmed this hypothesis (I.K. Wakefield, unpublished). Within this region, $Ala^{6.29}$(261) was identified as an important residue for coupling. When the equivalent Ala in ICL3 of biogenic amine receptors is mutated to large residues, the receptors are constitutively active.[93] However, mutation of $Ala^{6.29}$(261) to bulky amino acids resulted in an opposite effect in the GnRH receptor – namely, uncoupling of the receptor and failure to generate inositol phosphate.[94] Mutation of the evolutionarily conserved adjacent basic amino acid ($Arg^{6.30}$(262)) to Ala was also shown to result in uncoupling (Wakefield, unpublished), and natural mutations of $Arg^{6.30}$(262) have been shown to cause uncoupling in receptors of families with hypogonadotropic hypogonadism.[95–97] The effects of overexpression of rat GnRH receptor ICL3 peptides in GnRH receptor-expressing GGH3 cells on inositol phosphate production and cAMP accumulation demonstrated that this domain is involved in coupling to $G_{q/11}$ and G_s signal transduction pathways.[98]

The GnRH receptors have the conserved motif Asp-Arg-x-x-x-Iso/Val-x-x-Pro-Leu at the N-terminus of ICL2, which plays a role in coupling. The importance of the Asp-Arg-x-x-x-Iso element in receptor activation appears to extend to the $Pro^{3.57}$(146)$Leu^{3.58}$(147). Mutation of the preceding $Arg^{3.56}$(145) to Pro causes

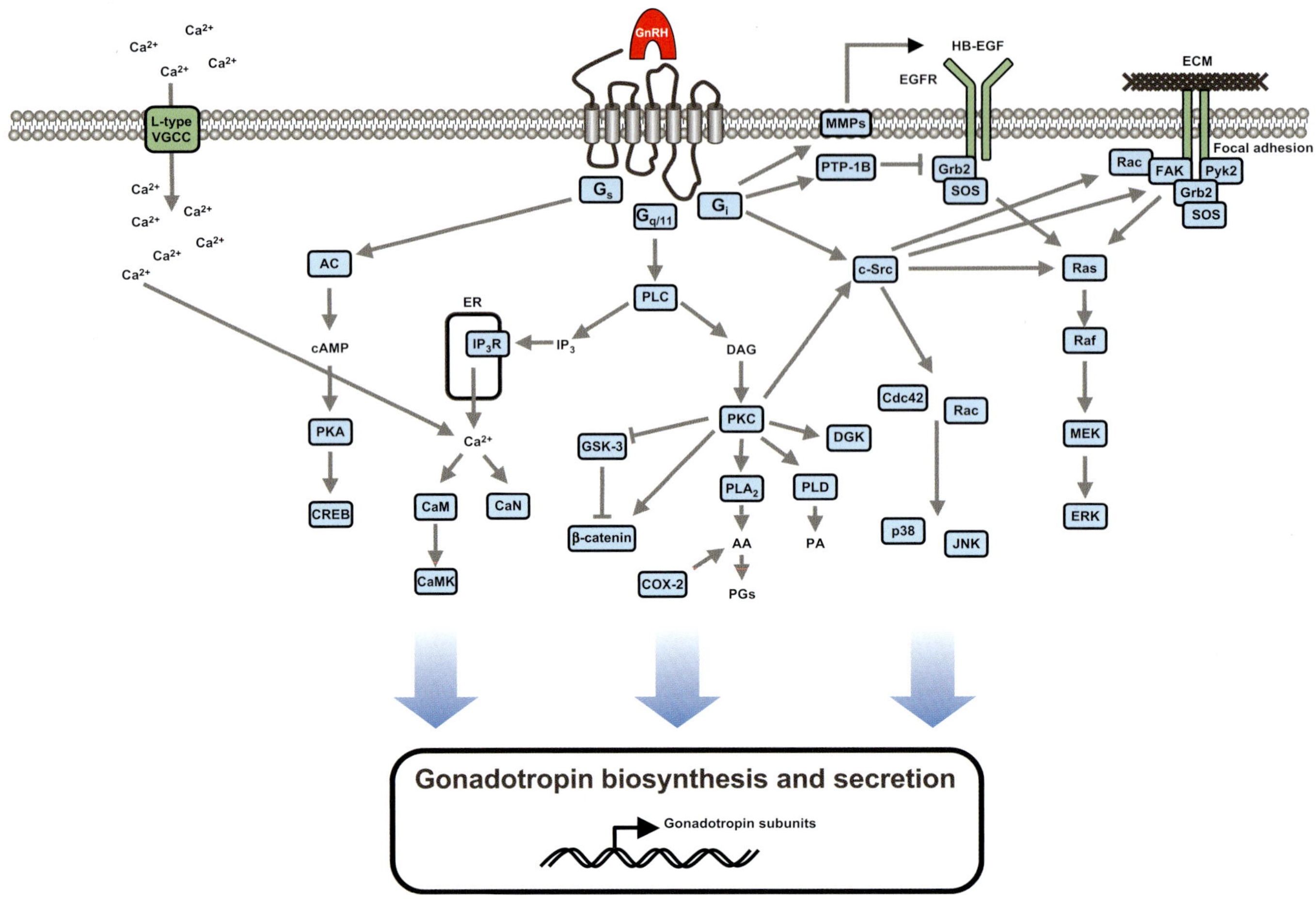

FIGURE 2.9 **Gonadotropin-releasing hormone (GnRH) receptor-activated pathways.** In the pituitary gonadotrope, the primary pathway involves the activation of phospholipase C (PLC) via $G_{q/11}$, resulting in the hydrolysis of membrane phosphatidylinositol 1,4-bisphosphate (PIP_2) to diacylglycerol (DAG), which activates protein kinase C (PKC), and to inositol 1,4,5-trisphosphate (IP_3), which triggers Ca^{2+} release from the endoplasmic reticulum (ER) by activating IP_3 receptors (IP3R). GnRH receptor occupancy by GnRH also leads to the activation of plasma-membrane L-type voltage-gated Ca^{2+} channels. PKC activation results in the activation of the numerous signaling cascades that lead the activation of extracellular signal-regulated kinase (ERK), p38 mitogen-activated protein kinase (p38) and c-Jun N-terminal kinase (JNK), and gonadotropin subunit gene transcription. The rise in intracellular Ca^{2+} leads to their exocytosis. The GnRH receptor can also couple to G_s, leading to the activation of adenyl cyclase (AC) and a rise in cAMP levels and the activation of protein kinase A (PKA), and cAMP response element-binding protein (CREB), which also plays a role in gonadotropin subunit gene transcription. In cancer cells, coupling to G_i has been demonstrated to enhance the activities of p38 and JNK, leading to cell apoptosis. In some cell types, GnRH receptor occupancy can lead to the transactivation of the EGF receptor (EGFR). Signaling via focal adhesion complexes in response to GnRH to promote cytoskeletal reorganization has also been demonstrated. (See text for in-depth description.)

uncoupling,[99] presumably because this mutation introduces a Pro-Pro motif known to disrupt secondary structure. Replacement of the conserved $Leu^{3.58}$(147) with Asp or Ala lead to defective $G_{q/11}$ coupling,[100] and mutation of $Arg^{3.50}$(139) of the mouse GnRH receptor to Gln produced a similar effect.[101] In ICL1 the sequence (Lys-Lys-Leu-Ser-Arg) is a G_s recognition motif (B-B-x-x-B, where B is a basic amino acid) mutation of certain of these residues leads to uncoupling of cAMP production but not of inositol phosphate production.[89]

Thus, the GnRH receptor is apparently able to couple to several G proteins and activate a number of effectors via different elements of the three intracellular loops. It appears that coupling to $G_{q/11}$ occurs through ICL2 and ICL3, and to G_s through ICL1. Coupling to G_i is less understood, but may be related to the cell type, stage of the cell cycle or availability of the G_i protein. It has been suggested that G_i activation underlies the antiproliferative effects of GnRH in many cancers.[81,102,103] Importantly, the intracellular loop elements required for G_i coupling have not been investigated.

The carboxyl-terminal tail of GPCRs has also been implicated in the regulation of signaling via receptor-coupled G proteins. In contrast to the mammalian GnRH receptors, but in common with other GPCRs, the cloned non-mammalian GnRH receptors all have a carboxyl-terminal tail.[104–108] The absence of a carboxyl-terminal tail in mammalian GnRH receptors is correlated with a lack of rapid desensitization,[69] in contrast to non-mammalian and Type II tailed receptors that exhibit rapid desensitization. A number of studies[70,71,109–111] involving truncation, site-directed mutagenesis and "carboxyl-terminal tail-swapping" have established the importance of this region for coupling, desensitization and receptor internalization (see below). Although mammalian type I GnRH receptors lack the carboxyl-terminal tail, the carboxyl-terminal residues of TM7 are important for $G_{q/11}$ effector coupling.[112] Since the last four residues (Tyr-Phe-Ser-Leu) of all mammalian GnRH receptors are conserved, these residues may be important for effective mammalian GnRH receptor signal transduction.

Intracellular Signaling

As mentioned above, GnRH receptor activation in gonadotropes mobilizes intracellular Ca^{2+} and activates PKC, which in turn stimulates the biosynthesis and secretion of LH and FSH. Through coupling to $G_{q/11}$ and PKC activation, GnRH activation of the GnRH receptor can lead to the activation of the four key mitogen-activated protein kinase (MAPK) subfamilies, including the extracellular signal regulated protein kinase (ERK), the c-Jun N-terminal protein kinase (JNK), the p38 MAPK, and the big MAPK (BMK) cascades; this has been the subject of numerous comprehensive reviews.[113–115] Recent studies have demonstrated that ERK activation induces c-fos and LHβ gene expression in LβT2 cells, and that this signaling process occurs by GnRH receptor activation of both $G_{q/11}$ and G_s,[91,116,117] further implicating coupling of the GnRH receptor to multiple G proteins and the activation of different MAPK signaling cascades to regulate selective gonadotropin subunit expression. Activation of ERK in LβT2 cells may regulate basal and GnRH-induced α-subunit and LHβ-subunit expression, while JNK activation regulates basal and GnRH-stimulated LHβ-subunit expression, both ERK and JNK involving c-Src participation (see Harris *et al.*[116] and references therein). The rise in intracellular calcium levels in response to GnRH is important in the regulation of LHβ-subunit expression. This was demonstrated by a study in the LβT2 gonadotrope cell line showing that GnRH induces the activation of calcium/calmodulin-dependent kinase type II (Ca/CaMK II), which is mediated by the elevation of intracellular calcium levels.[118] Ca/CaMK II probably acts at one or more GnRH-sensitive transcription factors at the LHβ-subunit promoter to regulate LH expression. Differential regulation of LH and FSH secretion in a teleost fish is achieved through divergent signaling via the GnRH receptor, with expression of α-subunit and LHβ-subunit mRNAs being increased by activation of the ERK cascade, while induction of FSHβ transcript expression is ERK-independent and is regulated by the cAMP-PKA signaling cascades.[119]

While it has been demonstrated that two activator protein-1 (AP-1) elements located in the proximal FSHβ promoter are not involved in GnRH regulation of FSHβ gene expression, GnRH activation of the ERK cascade in LβT2 cells is reported to regulate FSHβ gene expression.[120] GnRH activation of MAPK cascades not only regulates gonadotropin subunit gene expression, but also regulates the GnRH receptor gene. GnRH activation of the JNK pathway regulates the activation of AP-1 that acts via the AP-1 element in the GnRH receptor promoter to drive endogenous GnRH type I receptor gene expression in the αT3-1 gonadotrope cell line.[121]

The rise in intracellular calcium levels, derived in part from IP_3-activated intracellular storage pools and from the influx of extracellular calcium through L-type voltage-sensitive calcium channels, is of paramount importance in initiating the exocytotic events that lead to the release of the gonadotropins from the gonadotrope in the anterior pituitary.[78]

The roles of both heterotrimeric and monomeric G proteins in divergent GnRH receptor signaling have recently been described.[122] GnRH receptor occupancy leads to dramatic changes in cell adhesion and morphology, as the result of actin cytoskeleton remodeling, and these effects are mediated by FAK/c-Src/ERK activation at focal adhesion complexes, and are not dependent on PKC and intracellular calcium mobilization.[122] The importance of the focal adhesion complex, and in particular the focal adhesion kinase Pyk2, in signaling via the ERK pathway to LHβ gene transcription in the gonadotrope (LβT2) has now been established.[123]

c-Src plays a central role in many GnRH signaling pathways. GnRH mediates a functional interaction between c-Src and the lipid kinase diacylglycerol kinase-ζ (DGK-ζ) in both HEK 293 and LβT2 cells.[124] In addition, it has been demonstrated that the activation of DGK-ζ exerts a functional role in the LβT2 gonadotrope cell line, in which elevated expression of DGK-ζ resulted in a shortening of the timescale of ERK activation, suggesting a potential role of endogenous DGK-ζ in controlling the induction of gonadotropin subunit gene expression, which is stimulated by GnRH via activation of ERK1/2. This mechanism may be mediated by the depletion of DAG due to its hydrolysis by DGK-ζ to phosphatidic acid after translocation to the plasma membrane.[124]

Arachidonic acid (AA) is converted to prostanoids (PGs) via activated cyclooxygenase-2 (COX-2) and specific PG synthases.[125] GnRH activation of COX-2 results in PG synthesis. It has been shown that GnRH stimulates AA release from rat pituitary cells, followed by elevated expression of 12-lipoxygenase and formation of 5-lipoxygenase products, which participate in GnRH actions.[126,127] A novel GnRH signaling pathway is mediated by prostanoid-$F_{2\alpha}$ ($PGF_{2\alpha}$) via the prostanoid-F (FP) receptor, and prostanoid-I_2 (PGI_2) via the prostainoid-I (IP) receptor, which limit the homologous regulation of GnRH receptor, while $PGF_{2\alpha}$ also exerts selective inhibition of LH release.[128] This mechanism may underlie the cyclical responsiveness of pituitary gonadotropes to GnRH and the asynchronous LH and FSH release during the mammalian ovarian cycle.

GnRH stimulation of the LβT2 gonadotrope cell line results in an elevated mRNA level of the bone morphogenic protein (BMP) antagonist, differential screening-selected gene aberrative in neuroblastoma (DAN), as well as mature cytosolic and secreted DAN glycoprotein.[129] Overexpression of DAN in LβT2 cells inhibited the synergistic activation of GnRH receptor gene expression by GnRH and activin, but had no effect on GnRH- or activin-stimulated gonadotropin gene expression, thus revealing an additional signaling mechanism for the coordinated regulation of gonadotropin biosynthesis and secretion which is crucial for the asynchronous secretion of LH and FSH during the ovarian cycle.

The ability of GnRH analogs to inhibit proliferation of androgen-dependent tumor cells suggests that GnRH can to modulate androgen receptor (AR) activity. In fact, a GnRH-induced complex of the focal adhesion kinase Pyk2 with c-Src and the AR co-factor Hic-5 has been demonstrated to directly modulate AR subcellular location and nuclear activity, such that the AR is rendered transcriptionally inactive in response to GnRH.[130] This scaffold-dependent signaling of the GnRH receptor to the AR is potentially involved in the direct antiproliferative effects of GnRH analogs on prostatic neoplasms and cell lines.

A role for β-catenin as a member of a transcription factor complex that drives maximal activity of the LHβ promoter in response to GnRH has been demonstrated.[131] Co-localization of β-catenin with steroidogenic factor-1 (SF-1) and early response growth factor 1 (EGR-1) on the promoter of the LHβ subunit gene in response to GnRH suggested that endogenous SF-1 and β-catenin can physically associate in LβT2 cells. Stabilization and nuclear accumulation of β-catenin is a pre-eminent hallmark of the Wnt/β-catenin signaling pathway. The seminal demonstration of GPCR activation of β-catenin/T cell factor (TCF)-mediated transcriptional activity was reported for the prostanoid FP_B receptor, and suggested that this may be a property of other GPCRs.[132,133] In fact, β-catenin accumulation in the nucleus, activation of a TCF-Luciferase reporter and upregulation of Wnt target genes in response to GnRH in gonadotrope cells has been demonstrated.[134] Thus, β-catenin signaling in response to GnRH stimulation in LβT2 gonadotrope cells is not only important as a co-factor for TCF/lymphoid enhancer factor (LEF)-dependent transcriptional activity at Wnt target genes, but also has a signficant role in mediating gonadotropin gene expression.[131] Furthermore, the results from the study by Gardner *et al.*[134] demonstrate that GnRH stimulates β-catenin/TCF signaling in a heterologous cell system, which has implications for GnRH impacting on Wnt/β-catenin signaling processes in a variety of peripheral tissues and cancers that express the type I GnRH receptor. The activation of Wnt target genes by GnRH demonstrates the ability of GnRH to influence the output of Wnt signaling, and poses the question: why does GnRH activate the same target genes as Wnt ligands?

In addition to heterotrimeric and monomeric G proteins, the involvement in GnRH receptor-mediated signaling of RGS proteins has also been shown. RGS proteins interact directly with active Gα subunits to accelerate their intrinsic GTPase activity and limit their half-life.[135] RGS-3 and RGS-10 have been implicated in the regulation of GnRH receptor coupling,[136–138] and there is evidence that the carboxyl-terminal tail of non-mammalian GnRH receptors may contain elements for interactions with RGS-10, although the nature of this interaction is unclear.[137]

Absence of Rapid Mammalian GnRH Receptor Desensitization and Ligand-induced Internalization

Although chronic administration of GnRH agonists gives rise to a diminution in gonadotropin secretion and consequent decline in gonadal activity and sex steroid hormone production, this is a long-term pharmacological process that takes days to weeks, and is distinctly different from the classical rapid desensitization (minutes) and internalization of receptors that occurs in the majority of GPCRs. Indeed, it is evident that the unique absence of a cytoplasmic carboxyl-terminal tail conveys a resistance of the GnRH receptor to rapid desensitization and ligand-induced internalization, as there is no recruitment of β-arrestin.[5,69,109,139] Consequently, the receptor experiences prolonged activation, and the elimination of the cytoplasmic carboxyl-terminal tail during evolution may have arisen in order to allow a protracted duration of gonadotropin secretion required for oocyte stimulation and ovulation.

Despite there being no short-term rapid desensitization of mammalian GnRH receptors, there is a marked suppression of GnRH-mediated calcium response and uncoupling of IP_3 production in αT3-1 cells in response to chronic GnRH receptor activation by GnRH.[140,141] This is due to downregulation of IP_3 receptors at intracellular calcium storage pools[142] by a GnRH-induced rapid increase in IP_3 receptor polyubiquitination, thereby targeting them for degradation via the ubiquitin/proteosome pathway.[143] Chronic GnRH treatment has also been demonstrated to downregulate phospholipase A_2 (PLA_2) and PLD activity, suggesting that a host of second-messenger activities are affected.[77] Indeed, the desensitizing effect of sustained GnRH treatment on more distal downstream signaling pathways, including the ERK and p38 MAPK cascades, which are coupled to GnRH receptor activation in LβT2 cells, has now been shown.[144]

Nevertheless, there remains a relative lack of desensitization which has consequences in the ability of chronic GnRH exposure to inhibit proliferation and induce apoptosis in tumor cells expressing the GnRH receptor. Inhibition of gonadotropin synthesis and secretion by GnRH agonist treatment is frequently attributed to GnRH receptor "downregulation" and, loosely, as "desensitization." However, it is evident that this is not strictly correct, as gonadotropin α-subunit remains massively elevated even after years of treatment with GnRH agonist in prostate cancer patients. Thus, GnRH receptors and intracellular machinery are continuing to respond to GnRH agonist, but biologically active mature gonadotropins are no longer produced.

The internalization pathways utilized by GnRH receptors differ between receptor subtypes. The absence of a carboxyl-terminal domain in the mammalian type I GnRH receptors probably accounts for their β-arrestin independency for internalization,[109] as this region has been extensively reported to mediate receptor interactions with β-arrestin.[67] Murine GnRH receptors undergo a slow rate of internalization in αT3-1 and LβT2 cells, and the internalization of human type I GnRH receptor after expression in αT4 pituitary cells proceeds at a slow rate and is β-arrestin independent.[145] This is in contrast to the Xenopus type I GnRH receptor, which possesses a carboxyl-terminal tail and internalizes more rapidly in a β-arrestin dependent manner when expressed in αT4 cells.[145] It appears that both mammalian and nonmammalian GnRH receptors can be targeted for clathrin-mediated internalization, regardless of their β-arrestin dependence. The catfish and chicken GnRH receptors both exhibit rapid internalization kinetics, and were shown to be dependent on their carboxyl-terminal tail domains for this process.[70,71,146] The chicken GnRH receptor mediated internalization of ^{125}I-[His^5,D-Tyr^6]GnRH at a rate of 11.3%.min^{-1} to a maximal level of approximately 75%, compared with only 0.71%.min^{-1} and maximal level of 25% for the human GnRH receptor. To determine whether the presence of the cytoplasmic carboxyl-terminal tail was responsible for the more rapid internalization of the chicken GnRH receptor, the tail was truncated at Ser^{337}. Internalization of ^{125}I-[His^5,D-Tyr^6]GnRH by the S^{337}stop-chicken GnRH receptor was much slower than the wild-type chicken receptor, and displayed similar internalization kinetics to the human GnRH receptor, with a rate of 0.55%.min^{-1} and a maximal level of approximately 25%.[71] A threonine doublet located at the distal end of the chicken GnRH receptor cytoplasmic tail was shown to be critical, as was a membrane proximal cysteine residue.[111] The chicken GnRH receptor was shown to preferentially undergo rapid agonist-induced internalization in a dynamin- and caveolae-dependent manner, and palmitoylation of the membrane proximal cysteine residue may serve to target the chicken GnRH receptor to caveolae microdomains for signaling and internalization.[111] The internalization pathways of the three bullfrog GnRH receptor subtypes have been partially characterized.[147] The bullfrog type I GnRH receptor was shown to internalize via a β-arrestin- and dynamin-dependent pathway, whereas the bullfrog type II and III GnRH receptors internalize via a pathway that is β-arrestin independent but dynamin dependent, similar to the pathway utilized by the chicken GnRH receptor.[111,147] The above studies thus suggest that the carboxyl-terminal tail of the nonmammalian GnRH receptors plays a pivotal role in their function and subcellular trafficking to divergent internalization pathways. While there is clearly divergence in the internalization pathways utilized by the tailed GnRH receptors, the sequence motifs and structural elements within the cytoplasmic carboxyl-terminal domain that determine which internalization pathway will be utilized have yet to be fully elucidated.

By measuring the trafficking of radioactive GnRH agonists, the above studies concluded that mammalian type I GnRH receptors undergo slow ligand-dependent internalization. However, by the direct measurement of mammalian type I GnRH receptor trafficking, both in the presence and in the absence of unlabeled GnRH agonist, low basal levels of constitutive agonist-independent internalization have been reported.[148] Stimulation with GnRH agonist did not significantly enhance the level of mammalian type I receptor internalization above the basal level, in contrast to receptor chimeras with cytoplasmic tails, or the TRH receptor.[148] These data suggest that, as a result of the deletion of the cytoplasmic carboxyl-terminal tail during evolution, the mammalian type I GnRH receptors can be considered as natural internalization defective "mutants".[139,148]

GPR54 (Kisspeptin Receptor)

GPR54, the cognate receptor for kisspeptins, interacts with the $G_{q/11}$ subunit to activate PLCβ, which hydrolyzes PIP_2 to IP_3 and DAG, which in turn mobilize intracellular Ca^{2+} and activate PKC.[149] Mutations in the receptor have revealed that this interaction involves Leu^{148} of ICL2, since dissociation of the G protein subunits is disrupted by mutations at this residue.[150] ICL2 of GPR54 makes hydrophobic interactions with the Gα subunit that stabilizes the switch II region of the G protein. This activates the subunit into a conformation that facilitates GDP—GTP exchange, allowing dissociation and eliciting downstream signaling. Therefore, ICL2 of GPR54 acts as a guanine nucleotide exchange factor (GEF). Leu[148] has been shown to dock in close proximity to the GTPase domain of the $G_{q/11}$, where Pro[138] interaction causes a conformational change in ICL2 to allow Phe[139] to fit into the hydrophobic groove created by α2/β4 and α3/β5 loops of the G protein and interact with Phe[220], Val[223], Trp[263] and Phe[264] of ICL2.[150] Phe[139] can then interact with Phe[220], Val[223], Trp[263] and Phe[264] of the switch II region. This is now thought to be a mechanism common to all GPCRs in the Rhodopsin family.

In addition to mobilizing intracellular Ca^{2+} and activating PKC, there is downstream phosphorylation and activation of MAP kinases ERK1/2, reorganization of intracellular stress fibers, and induction of focal adhesion kinase to inhibit cell movement.[149,151] Since 2001, a limited number of tissue-specific pathways have been shown for GPR54. Most studies have focused on signaling associated with two areas of kisspeptin action: stimulation of GnRH secretion from the GnRH neurons (the current focus), and antimetastatic actions in cancer cells.

Kisspeptin neurons directly contact GnRH neurons expressing GPR54 to induce secretion of GnRH into the hypophysial portal blood and stimulate the pituitary to secrete gonadotropins. Kisspeptin causes depolarization of GFP-tagged GnRH neurons in brain slices in a dose-dependent manner. This occurs via activation of sodium-dependent, non-selective cationic channels, possibly transient receptor potential cation (TRPC)-like channels, as the depolarization is blocked by 2-aminoethoxydiphenyl borate (2-APB), an inhibitor of TRPC channels. This activation is also accompanied by inhibition of inwardly rectifying (Kir) potassium channels.[152,153] This inhibition may be due to a blockade of gamma-aminobutyric acid-B ($GABA_B$) receptors, as kisspeptin has been shown to inhibit $GABA_B$ and the subsequent Kir potassium channel activation.[154] This mechanism of cationic channel activation and potassium channel inhibition is dependent on PLC and the IP_3 receptor, and therefore influences both plasma membrane- and endoplasmic reticulum-driven increases in intracellular Ca^{2+}. Plasma membrane tetrodotoxin (TTX)-sensitive calcium channels have also been implicated in GnRH neuronal depolarization.[155] In the immortalized GnRH neuronal cell line, GT_{1-7}, mobilization of intracellular calcium also appears to be critical for GnRH secretion, and this is dependent on PKC activation.[156]

Since kisspeptin was originally discovered as the antimetastatic gene, metastin, there has been extensive study on signaling that inhibits cell invasion and migration. In brief, the predominant mechanisms appear to be through the kisspeptin inhibition of Akt phosphorylation by tyrosine kinase receptors. Kisspeptin abolishes epidermal growth factor receptor (EGFR) and insulin-receptor mediated phosphorylation of Akt in an ERK1/2-dependent manner. Kisspeptin also desensitizes the chemokine receptor, CXCR4, response to stromal-derived factor-1 (SDF-1), causing a decrease in calcium release and inhibiting phosphorylation of Akt/protein kinase B (PKB) via cross-talk between the two GPCRs. This in turn blocks chemotaxis mediated via CXCR4 in cancer cells.[157] As well as Akt inhibition, reduction of matrixmetaloproteinases (MMPs) has also been delineated as a mechanism for this inhibition of metastasis via kisspeptin. HT-1080 trophoblast cells stably expressing the KiSS-1 promoter have decreased MMP-9 and invasiveness. This is due to an upregulation of cytosolic IκBα, an inhibitor of NFκB, stopping its nuclear translocation. This causes a decrease in MMP-9 expression, as NFκB needs to bind to the promoter of MMP-9 for activation.[158]

GPR147 (Gonadotropin-inhibitory Hormone (GnIH) Receptor)

Gonadotropin-inhibitory hormone (GnIH) is a hypothalamic RF amide peptide first discovered in birds as an inhibitor of gonadotropin secretion from the quail pituitary (see "Novel Neuroendocrine GPCRs Regulating Reproduction," page 44). The homologs are the RF amide-related peptides (RFRPs) in mammals, which have also been shown to inhibit gonadotropin secretion. GPR147 activation by RFRP-3 potently inhibits GnRH-stimulated gonadotropin secretion from ovine pituitary cells.[159] In the same model of cultured ovine pituitary cells, RFRP-3 is a potent inhibitor of GnRH mobilization of intracellular Ca^{2+}. GPR147 is coupled to G_i, which inhibits adenylate cyclase and cAMP production. However, $G\alpha_i$ can also inhibit Ca^{2+} channels, and Gβγ has been shown to inhibit voltage-gated Ca^{2+} channels.[160] Thus, RFRP-3 inhibition of GnRH stimulation of intracellular Ca^{2+} may be due to either, or both, of its actions on extracellular Ca^{2+} entry.

TRHR (Thyrotropin-releasing Hormone Receptors)

The TRH-R receptor is a classical Rhodopsin family GPCR. It is activated by thyrotropin-releasing hormone (TRH), a hypothalamic hypophysiotropic tripeptide hormone, which stimulates the secretion of thyroid-stimulating hormone (TSH) and prolactin in the anterior pituitary. Two isoforms of the receptor have been identified to date; TRH_1R, which is found in thyrotropes and lactotropes of the anterior pituitary, and TRH_2R, which is located in mouse/rat brain and spinal cord. TRH_1R is the dominant isoform, and is primarily responsible for the secretion of TRH and prolactin. However, both isoforms bind TRHR with a similar affinity, and both signal via $G_{q/11}$ to activate PLCβ, which can then hydrolyze PIP_2 in the cell membrane to DAG and IP_3.[161] IP_3 then stimulates mobilization of intracellular calcium, which can couple to calmodulin and calmodulin kinase to regulate a diversity of cellular events including the cyclic adenosine monophosphate response element-binding (CREB) transcription factor. DAG activates PKC, which, in conjunction with intracellular calcium, can regulate activator protein-1 (AP-1) transcription and phosphorylate members of the MAPK family. This activation of MAPK is thought to be responsible for the stimulation of prolactin secretion by TRH. It has also been shown that TRH_1R can affect potassium (K^+) channels in the pituitary, possible through G_{13}, which inhibits rat ether a-go-go related gene (r-ERG) K^+ channels. K^+ channels may also increase GABA secretion in the hippocampus when activated by TRH, showing site-specific signaling pathways for this receptor.

GHRHR (Growth Hormone-releasing Hormone Receptors)

The GHRH receptor is a member of the Secretin family of GPCRs, which recruits both G_s and $G_{q/11}$.[162] G_s recruitment in somatotropes leads to adenylate cyclase activation, generation of cAMP and activation of PKA, which in turn has been reported to activate L-type voltage-sensitive Ca^{2+} channels (possibly involving Na^+ and K^+ channel modulation).[163] PKA also activates CREB, which modulates growth hormone (GH) gene expression. By recruiting $G_{q/11}$ GHRHR activates PLCβ to generate IP_3, which mobilizes intracellular Ca^{2+}, which, together with increased Ca^{2+} entry through Ca^{2+} channels, culminates in GH exocytosis. PLCβ activation also leads to the generation of DAG, which activates PKC and MAP kinases, which stimulate GH gene expression. Importantly, somatotropes also express Ghrelin receptors which are also $G_{q/11}$ coupled (Table 2.1), thus providing the potential for synergy in Ca^{2+} mobilization and GH exocytosis. Thus, hypothalamic secretion of GHRH and peripheral secretion of Ghrelin can integrate their stimulation of GH secretion through a common signaling pathway in somatotropes.

DR (Dopamine Receptors)

Dopamine (DA) is a catecholamine hormone synthesized primarily in the central nervous system from the amino acid tyrosine. It is widespread in the brain and affects aspects of motor function, cognition and behavior, but also has effects in the pituitary, where it is involved in the tonic inhibition of prolactin secretion.[164] This pituitary effect is mediated primarily through DA produced in tuberoinfundibular dopaminergic (TIDA) neurons, which originate in the arcuate nucleus of the hypothalamus and release DA into the portal vessels that carry it to the anterior lobe of the pituitary.

Dopamine receptors are members of the Rhodopsin family of GPCRs, and comprise five specific receptor subtypes (D_1-D_5).[165] The pituitary effects of DA are mediated primarily through D_2 receptors expressed on lactotropes of the anterior pituitary.[166,167] On these cells, a specific long splice-variant form of the D_2 receptor (D_{2L}) predominates.[168] Activation of the D_2 receptors on these cells by DA results in inhibition of prolactin secretion through a number of mechanisms (extensively reviewed Ben-Jonathan and Hnasko[164] and Missale *et al.*[165]) associated with their coupling to G_o and G_i.[169] In brief, adenylate cyclase activity is inhibited, which results in decreased production of cAMP, subsequent suppression of PKA activity, and decreases in prolactin gene expression.[170] As cAMP has mitogenic effects, these effects are also believed to result in reduced proliferation of lactotropes. Decreases in intracellular Ca^{2+} concentrations also result from inhibition of inward calcium channels.[171] DA activation of outward potassium currents causes hyperpolarization of the lactotrope membranes and further inhibition of inward calcium channels, which results in a decrease in the influx of extracellular Ca^{2+}, further reducing intracellular concentrations.[172] Indirect inhibition of phosphoinositol hydrolysis has also been implicated in decreasing intracellular Ca^{2+} concentrations.[173] The resultant low concentration of intracellular Ca^{2+} causes an inhibition in prolactin release from secretory granules within the cells by attenuation of calcium-regulated exocytosis.[174] DA has also been shown to indirectly suppress expression of the prolactin gene by inhibiting the expression of Pit-1, a pituitary-specific transcription factor that is involved in regulation of growth hormone and prolactin gene expression.[175]

SSTR (Somatostatin Receptors)

Somatostatin (SST) is an abundant neuropeptide that produces a wide range of physiological effects in the body, including inhibiting the secretion of many hormones, such as GH and thyroid stimulating hormone (TSH), from the pituitary.[176] The pituitary effects of SST are predominantly mediated through SST synthesized in the anterior periventricular region of the hypothalamus. The endogenous somatostatin ligands are two bioactive cyclic tetradecapeptide proteins produced by cleavage of a common precursor protein.[177] Somatostatin-14 is predominant, but somatostatin-28 is more potent.[176] Another endogenous SST receptor ligand, cortistatin, has also been identified.[178] The pituitary effects of SST are predominantly mediated by SST released from neurons projecting from the hypothalamus into the portal vessels, which then carry SST to the anterior lobe of the pituitary.[179]

Somatostatin receptors are members of the Rhodopsin family of GPCRs, and comprise five receptor subtypes ($SSTR_1$-$SSTR_5$).[176] The predominant subtypes expressed throughout the pituitary are $SSTR_2$ and $SSTR_5$, with SST_1 and SST_3 expressed at lower levels.[180] SST receptors elicit their effects through coupling to $G_{i/o}$.[176,179,180] In brief, SST receptor activation leads to decreases in intracellular cAMP through inhibition of adenylate cyclase, and decreases in intracellular Ca^{2+} concentration directly through blockade of calcium channels (through $SSTR_2$) and indirectly through hyperpolarization of cell membranes due to activation of potassium channels and subsequent inhibition of voltage-gated calcium channels. The reduction in intracellular cAMP, and particularly Ca^{2+}, results in reduced calcium-regulated exocytosis[174] of GH and TSH from cells of the pituitary. In addition, SST receptor stimulation causes activation of protein phosphatases, such as calcineurin, which inhibit exocytosis. SST activation of SST_2 and SST_5 receptors also inhibits proliferation of pituitary cells through attenuation of the MAPK signaling cascade.

VR (Vasopressin Receptors)

Vasopressin is a cyclic nonapeptide synthesized in the magnocellular neurons of the hypothalamus and secreted from the neural lobe of the pituitary to regulate the reabsorbtion of water in the tubules of the kidneys. It also acts, in synergy with corticotropin-releasing hormone (CRH), to stimulate ACTH secretion from the pituitary in response to stress. The renal effects of vasopressin are mediated by vasopressin released into the peripheral circulation, while the pituitary effects of vasopressin are mediated predominantly by vasopressin released from neurons originating in the paraventricular nucleus of the hypothalamus, which release vasopressin into the portal vessels, from where it passes to the anterior lobe of the pituitary.[181]

Vasopressin receptors are members of the Rhodopsin family of GPCRs, and comprise three receptor subtypes (V_{1a}, V_{1b} and V_2).[182] V_2 receptors are extensively expressed in the adult kidney, where they couple through G_s to stimulate adenylate cyclase and generate CAMP, which in turn activates PKA to regulate cellular permeability to water. V_{1a} receptors are widely expressed on blood vessels and in the central nervous system (CNS). The pituitary effects of vasopressin are mediated through V_{1b} receptors expressed on corticotrope cells[183] which couple through $G_{q/11}$.[184] Activation of these receptors results in activation of PLC, and subsequent production of IP_3 and DAG.[181] IP_3 interacts with receptors on the endoplasmic reticulum, causing opening of calcium channels and an increase in cytosolic Ca^{2+}. This increased intracellular Ca^{2+} concentration results in an increase in calcium-regulated exocytosis[174] of ACTH from these cells.

OTR (Oxytocin Receptor)

Like vasopressin, oxytocin is synthesized in magnocellular neurons of the hypothalamus and is transported to the neural lobe of the pituitary, from where it is secreted into the general circulation to induce uterine contraction during labor and milk ejection. It also is produced in smaller neurons, which project to CNS areas and affect reproductive behavior (see Chapter 6).

Oxytocin is a paralog of vasopressin with two amino acid substitutions. It targets a paralog receptor, OTR, which couples through $G_{q/11}$ to activate PLCβ, leading to hydrolysis of PIP_2 to IP_3 and DAG, and activation of PKC.[185]

NOVEL NEUROENDOCRINE GPCRs REGULATING REPRODUCTION: INTEGRATED NEURONAL REGULATION OF GNRH

For the past 40 years, GnRH secretion into the hypothalamic hypophysial portal system has been recognized as the final common output of diverse regulators (e.g., photoperiod, nutrition, behavior, stress, inflammation, gonadal hormones) of reproduction. However, it became increasingly evident that neurons upstream of GnRH neurons are the mediators of these diverse inputs, as GnRH neurons themselves lacked the recognition biochemical machinery (e.g., sex steroid receptors and leptin receptors) to respond to these signals. In 2004, the discovery that inactivating mutations of a novel GPCR (GPR54), which is the cognate receptor for kisspeptin, resulted in a failure to progress through puberty, and in hypogonadotropic hypogonadism in adults,[186–189] revolutionized our understanding of the neuroendocrine regulation of reproduction. A large body of research has now established that the activity of kisspeptin neurons is regulated by many external and internal inputs that influence reproduction, including steroid hormones, fat status, nutrition, stress, and inflammatory processes.

Following the seminal discovery that kisspeptin/GPR54 acts as a major "whole body sensor" mediating diverse effects on the GnRH neuron,[188–190] Topaloglu and colleagues described mutations in neurokinin B (NKB) and its receptor (NK_3R (also known as TAC3R)) that gave rise to hypogonadotropic hypogonadism and pubertal failure in four Turkish families.[191] The question arose as to how this relates to the hypogonadotropic hypogonadism which results from inactivating mutations of the kisspeptin/GPR54 system. Interestingly, NKB co-localizes to kisspeptin neurons in the arcuate nucleus, along with dynorphin (Dyn),[192] supplying a potential subtle interplay of these three neuropeptides in modulating the GnRH neuron through their cognate GPCRs. It is possible that they are co-secreted and interact at the level of the GnRH neuron, or that they operate by an autocrine feedback mechanism to modulate one anothers' secretion from kisspeptin neurons, or possibly a combination of both (Fig. 2.10). Experimental studies suggested that NKB acts in an autocrine mode on the kisspeptin neurons to amplify kisspeptin secretion. The relative biosynthesis, processing and secretion of the three peptides may also be differentially regulated by various neuronal inputs, providing a further level for subtlety of regulation. Kisspeptin neurons directly influence GnRH neurons expressing GPR54 to induce pulsatile secretion of GnRH. This occurs via GPR54 activation of sodium-dependent, non-selective cationic channels, possibly TRPC-like channels, as the kisspeptin-induced depolarization is blocked by 2-APB, an inhibitor of TRPC channels. This activation is also accompanied by an inhibition of Kir potassium channels. This mechanism of cationic channel activation and potassium channel inhibition is dependent on PLC and the IP_3 receptor, and therefore influences both plasma membrane- and endoplasmic reticulum-driven calcium release to drive calcium-dependent exocytosis of GnRH.[152,153] The secretion of kisspeptin appears to be partially regulated by NKB and Dyn acting on NK_3R and κ-opioid receptors, respectively, which are expressed in kisspeptin neurons. It is postulated that when estrogen levels decline kisspeptin neurons become spontaneously activated, and this is amplified by autosynaptic feedback where NKB stimulates a synchronized release of kisspeptin. As NKB acts via $G_{q/11}$, this release is probably via calcium-dependent exocytosis. Dyn is also released along with NKB, which acts with a small phase lag to inhibit kisspeptin release, therefore creating pulsatile kisspeptin secretion and downstream pulsatile GnRH secretion.[193] Recent studies in patients with loss-of-function NKB or NK_3R mutations found that continuous kisspeptin infusion restores pulsatile LH secretion. This supports the concept that NKB stimulates kisspeptin secretion, which is sufficient to elicit pulsatile GnRH secretion (R.P. Millar and J. Young, unpublished data).

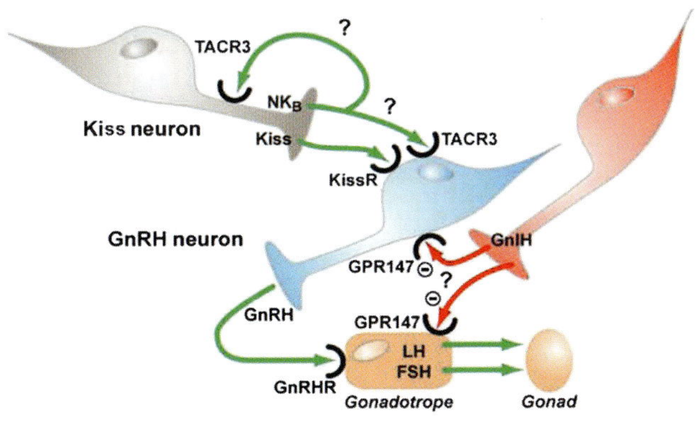

FIGURE 2.10 **Kisspeptin Neurokinin B and GnIH: novel regulators of gonadotropin in man.** Schematic describing the potential mechanisms for control of gonadotropin secretion by Kisspeptin, neurokinin B (NKB) and gonadotropin inhibitory hormone (GnIH). Kisspeptin (Kiss) released from kisspeptin neurons within the hypothalamus controls secretion of GnRH from GnRH neurons through interaction with its cognate receptor (GPR54). Secretion of GnRH from these neurons then leads to secretion of gonadotropins, luteinizing hormone (LH) and follicle-stimulating hormone (FSH) from pituitary gonadotropes. NKB is co-localized in kisspeptin neurons, and is postulated to be a novel regulator of GnRH secretion either through direct interaction with GnRH neurons or through autocrine interactions with kisspeptin neurons. The latter appears to be more likely, as kisspeptin infusion alone in patients with inactivating mutations of NKB restores GnRH and LH pulsatility. GnIH is a potent inhibitor of GnRH stimulation of gonadotropin secretion from cultured gonadotropes, but may also operate by inhibiting the activity of GnRH neurons.

Another discovery some years ago, that of a gonadotropin inhibitory hormone (GnIH) in birds,[194] has gathered momentum recently with the identification of a homologous gene in mammals and the purification and sequencing of two GnIHs (members of the RFRP family) from human hypothalamus.[195] GnIH is an RF-amide peptide like kisspeptin which binds to GPR147 to activate G_i (Table 2.1). It is a very potent inhibitor of LH pulsatility in ovariectomized sheep. It also potently inhibits GnRH stimulation of gonadotropin secretion from cultured ovine gonadotropes through an inhibition of Ca^{2+} mobilization.[159] GnIH stimulation of signaling via GPR147 is described in the previous section. GPR147 signals predominantly by activating G_i, which leads to the inhibition of cAMP generation. Inhibition of cAMP and G_i activation can diminish function of voltage-gated Ca^{2+} channels.[160] In addition, the liberated $G\beta\gamma$ from G_i activation can directly inhibit voltage-gated Ca^{2+} channels and activate Kir channels[25] to decrease the capacity of GnRH to mobilize intracellular Ca^{2+}. GnIH may also operate by directly or indirectly inhibiting the GnRH neurons themselves[196–198] (Fig. 2.10).

The discovery of kisspeptin, NKB, Dyn and GnIH as neuroendocrine regulators of gonadotropins has provided new opportunities for research on novel GPCRs in fine-tuning the hypothalamic–pituitary–gonadal axis, and provides new pathways in which to interrogate feedback mechanisms and metabolic, photoperiod and behavioral influences on the reproductive system (Fig. 2.10). As such, it presents one of the best elaborated neuroendocrine regulatory networks of a single output (gonadotropin secretion), and provides a model for delineating the diverse levels of regulation through neuroendocrine GPCRs.

DYSFUNCTION OF GPCR SIGNALING IN DISEASE

Loss-of-function mutations in GPCRs are responsible for many diseases.[199] These mutations disrupt ligand binding, receptor activation capacity, G-protein coupling or correct GPCR folding. The latter are the most common defects, as there are many more amino acids involved in configuring the receptor correctly than are involved in ligand binding, receptor activation and coupling, thus providing a greater chance of a mutation affecting this element of normal GPCR function. Misfolded GPCRs fail to traffic to the cell surface, and are directed into lysosomes and degraded (see "Regulation of GPCR Cell Surface Expression and Pharmacochaperones," above).

Amongst neuroendocrine GPCRs, loss-of-function mutations of the V_2 vasopressin receptor result in nephrogenic diabetes insipidus due to an inability of vasopressin to bind and activate the receptors. Many mutations in the GnRH type I receptor have been described that lead to loss of function and ensuing hypogonadotropic hypogonadism.[5,200,201] As expected, loss-of-function mutation in the GnRH ligand gene also results in the same phenotypes.[202] The majority of human GnRH-receptor inactivating mutations result from misfolding and poor surface expression of the receptor, but there are also examples of decreased ligand binding or a disruption of receptor activation or G-protein coupling. There are therapeutic possibilities in "rescuing" misfolded and poorly expressing GPCRs using cell-permeant pharmacological chaperone molecules that enter the cell and bind and stabilize the mutant GPCR as it is translated in the endoplasmic reticulum.[5,56,200,201] Mutations in two other GPCRs, GPR54 and NK_3R, also produce the same phenotype of hypogonadotropic hypogonadism. These are the cognate receptors for kisspeptin and NKB, which have now been shown to be upstream regulators of GnRH secretion. Mutations in the prokineticin GPCR also result in hypogonadotropic hypogonadism due to a failure of the GnRH neurons to migrate, presenting a similar phenotype to Kallman's syndrome but without anosmia.[203]

Inactivating mutations of LH and FSH receptors have been described that result in hypergonadotropic hypogonadism.[204] Mutations in the ACTH and TSH GPCRs, which give rise to familial glucocorticoid deficiency and hypothyroidism, respectively, have also been described. Inactivating mutations in GHRH give rise to familial GH deficiency and, in the melanocortin receptor, result in satiety disturbance and obesity.

Pathologies can also arise from activating mutations of neuroendocrine GPCRs. These mutations are usually in one or other amino acid in the TM domains, which, through their interactions, maintain the receptor in an inactive state. Precocious puberty has been reported for a mutation of the kisspeptin receptor (GPR54), which has prolonged signaling,[205] and for LH receptor mutations that signal constitutively in the absence of ligand.[204] Activating mutations of the TSH receptor produce thyroid adenomas and hyperthyroidism.[206] A mutation in the TSH receptor has been identified which increases its binding affinity for chorionic gonadotropin (hCG), resulting in hyperthyroidism that occurs only during the first trimester of pregnancy when hCG levels are high.[207] Continued spermatogenesis has been reported in a hypophysectomized man with an activating mutation in the FSH receptor.[208] Activating mutations causing inappropriate antidiuretic hormone (ADH) syndrome have also been described. One of these is in the TM3 DR activating switch (see "Receptor Mechanism of Activation," above).[209]

In addition to GPCRs producing disease states, mutations in the signaling machinery can also cause pathologies. A substantial percentage of somatotrope adenomas causing acromegaly result from Arg mutations in the GTPase domain of $G\alpha_s$ leading to a failure to hydrolyze GTP to GDP, and prolonged adenylate cyclase activation.[210] The same mutations in McCune-Albright syndrome originate in early development, giving rise to hyperfunction in diverse endocrine systems.[211] Interestingly, cholera toxin ADP-ribosylates the same Arg residue to cause prolonged cAMP production in mucous intestinal cells, resulting in diarrhea.

Inactivating mutations of $G\alpha_s$ have been described which result in pseudohypoparathyroidism, but similar mutations in G proteins in neuroendocrine signaling have not been described.

Acknowledgments

We are grateful to Brian Kobilka, Richard Henderson, Zhi-liang Lu, Adam Pawson and Stuart Sealfon for figure contributions. Adam Pawson also contributed in summarizing GnRH receptor signaling.

References

1. Neuhaus EM, Zhang W, Gelis L, Deng Y, Noldus J, Hatt H. Activation of an olfactory receptor inhibits proliferation of prostate cancer cells. *J Biol Chem.* 2009;284(24):16218–16225.
2. Yona S, Lin HH, Siu WO, Gordon S, Stacey M. Adhesion-GPCRs: emerging roles for novel receptors. *Trends Biochem Sci.* 2008;33(10):491–500.

3. Krebs A, Villa C, Edwards PC, Schertler GF. Characterisation of an improved two-dimensional p22121 crystal from bovine rhodopsin. *J Mol Biol.* 1998;282(5):991–1003.
4. Schertler GF. Structure of rhodopsin. *Eye (Lond)* 1998;12(Pt 3b):504–510.
5. Millar RP, Lu ZL, Pawson AJ, Flanagan CA, Morgan K, Maudsley SR. Gonadotropin-releasing hormone receptors. *Endocr Rev.* 2004;25(2):235–275.
6. Ballesteros J, Kitanovic S, Guarnieri F, Davies P, Fromme BJ, Konvicka K, et al. Functional microdomains in G-protein-coupled receptors. The conserved arginine-cage motif in the gonadotropin-releasing hormone receptor. *J Biol Chem.* 1998;273(17):10445–10453.
7. Palczewski K, Kumasaka T, Hori T, Behnke CA, Motoshima H, Fox BA, et al. Crystal structure of rhodopsin: A G protein-coupled receptor. *Science.* 2000;289(5480):739–745.
8. Jaakola VP, Griffith MT, Hanson MA, Cherezov V, Chien EY, Lane JR, et al. The 2.6 angstrom crystal structure of a human A2A adenosine receptor bound to an antagonist. *Science.* 2008;322(5905):1211–1217.
9. Park JH, Scheerer P, Hofmann KP, Choe HW, Ernst OP. Crystal structure of the ligand-free G-protein-coupled receptor opsin. *Nature.* 2008;454(7201):183–187.
10. Rasmussen SG, Choi HJ, Rosenbaum DM, Kobilka TS, Thian FS, Edwards PC, et al. Crystal structure of the human beta2 adrenergic G-protein-coupled receptor. *Nature.* 2007;450(7168): 383–387.
11. Scheerer P, Park JH, Hildebrand PW, Kim YJ, Krauss N, Choe HW, et al. Crystal structure of opsin in its G-protein-interacting conformation. *Nature.* 2008;455(7212):497–502.
12. Warne T, Serrano-Vega MJ, Baker JG, Moukhametzianov R, Edwards PC, Henderson R, et al. Structure of a beta1-adrenergic G-protein-coupled receptor. *Nature.* 2008;454(7203):486–491.
13. Cherezov V, Rosenbaum DM, Hanson MA, Rasmussen SG, Thian FS, Kobilka TS, et al. High-resolution crystal structure of an engineered human beta2-adrenergic G protein-coupled receptor. *Science.* 2007;318(5854):1258–1265.
14. Weis WI, Kobilka BK. Structural insights into G-protein-coupled receptor activation. *Curr Opin Struct Biol.* 2008;18(6):734–740.
15. Hanson MA, Stevens RC. Discovery of new GPCR biology: one receptor structure at a time. *Structure* 2009;17(1):8–14.
16. Li J, Edwards PC, Burghammer M, Villa C, Schertler GF. Structure of bovine rhodopsin in a trigonal crystal form. *J Mol Biol.* 2004;343(5):1409–1438.
17. Rosenbaum DM, Zhang C, Lyons JA, Holl R, Aragao D, Arlow DH, et al. Structure and function of an irreversible agonist-β_2 adrenoceptor complex. *Nature.* 2011;469(7329):236–240.
18. Rasmussen SG, Choi HJ, Fung JJ, Pardon E, Casarosa P, Chae PS, et al. Structure of a nanobody-stabilized active state of the β_2-adrenoceptor. *Nature.* 2011;469(7329):175–180.
19. Altenbach C, Kusnetzow AK, Ernst OP, Hofmann KP, Hubbell WL. High-resolution distance mapping in rhodopsin reveals the pattern of helix movement due to activation. *Proc Natl Acad Sci USA.* 2008;105(21):7439–7444.
20. Yao X, Parnot C, Deupi X, Ratnala VR, Swaminath G, Farrens D, Kobilka B. Coupling ligand structure to specific conformational switches in the beta2-adrenoceptor. *Nat Chem Biol.* 2006;2(8):417–422.
21. Ahuja S, Hornak V, Yan EC, Syrett N, Goncalves JA, Hirshfeld A, et al. Helix movement is coupled to displacement of the second extracellular loop in rhodopsin activation. *Nat Struct Mol Biol.* 2009;16(2):168–175.
22. Kolb P, Rosenbaum DM, Irwin JJ, Fung JJ, Kobilka BK, Shoichet BK. Structure-based discovery of beta2-adrenergic receptor ligands. *Proc Natl Acad Sci USA.* 2009;106(16):6843–6848.
23. Hanoune J, Defer N. Regulation and role of adenylyl cyclase isoforms. *Annu Rev Pharmacol Toxicol.* 2001;41:145–174.
24. Harden TK, Sondek J. Regulation of phospholipase C isozymes by ras superfamily GTPases. *Annu Rev Pharmacol Toxicol.* 2006;46:355–379.
25. Dascal N. Ion-channel regulation by G proteins. *Trends Endocrinol Metab.* 2001;12(9):391–398.
26. Krumins AM, Gilman AG. Targeted knockdown of G protein subunits selectively prevents receptor-mediated modulation of effectors and reveals complex changes in non-targeted signaling proteins. *J Biol Chem.* 2006;281(15):10250–10262.
27. Cabrera-Vera TM, Vanhauwe J, Thomas TO, Medkova M, Preininger A, Mazzoni MR, et al. Insights into G protein structure, function, and regulation. *Endocr Rev.* 2003;24(6): 765–781.
28. Yamashita T, Terakita A, Shichida Y. Distinct roles of the second and third cytoplasmic loops of bovine rhodopsin in G protein activation. *J Biol Chem.* 2000;275(44):34272–34279.
29. Tobin AB, Butcher AJ, Kong KC. Location, location, location ... site-specific GPCR phosphorylation offers a mechanism for cell-type-specific signalling. *Trends Pharmacol Sci.* 2008;29(8): 413–420.
30. Ferre S, Baler R, Bouvier M, Caron MG, Devi LA, Durroux T, et al. Building a new conceptual framework for receptor heteromers. *Nat Chem Biol.* 2009;5(3):131–134.
31. Lee SP, So CH, Rashid AJ, Varghese G, Cheng R, Lanca AJ, et al. Dopamine D_1 and D_2 receptor Co-activation generates a novel phospholipase C-mediated calcium signal. *J Biol Chem.* 2004;279(34):35671–35678.
32. Rashid AJ, So CH, Kong MM, Furtak T, El-Ghundi M, Cheng R, et al. D_1-D_2 dopamine receptor heterooligomers with unique pharmacology are coupled to rapid activation of Gq/11 in the striatum. *Proc Natl Acad Sci USA.* 2007;104(2):654–659.
33. Pin JP, Kniazeff J, Binet V, Liu J, Maurel D, Galvez T, et al. Activation mechanism of the heterodimeric $GABA_B$ receptor. *Biochem Pharmacol.* 2004;68(8):1565–1572.
34. Lee C, Ji I, Ryu K, Song Y, Conn PM, Ji TH. Two defective heterozygous luteinizing hormone receptors can rescue hormone action. *J Biol Chem.* 2002;277(18):15795–15800.
35. Rivero-Muller A, Chou YY, Ji I, Lajic S, Hanyaloglu AC, Jonas K, et al. Rescue of defective G protein-coupled receptor function in vivo by intermolecular cooperation. *Proc Natl Acad Sci USA.* 2010;107(5):2319–2324.
36. Ellis J, Pediani JD, Canals M, Milasta S, Milligan G. Orexin-1 receptor-cannabinoid CB1 receptor heterodimerization results in both ligand-dependent and -independent coordinated alterations of receptor localization and function. *J Biol Chem.* 2006;281(50):38812–38824.
37. Daniels DJ, Lenard NR, Etienne CL, Law PY, Roerig SC, Portoghese PS. Opioid-induced tolerance and dependence in mice is modulated by the distance between pharmacophores in a bivalent ligand series. *Proc Natl Acad Sci USA.* 2005;102(52): 19208–19213.
38. Cooray SN, Chan L, Webb TR, Metherell L, Clark AJ. Accessory proteins are vital for the functional expression of certain G protein-coupled receptors. *Mol Cell Endocrinol.* 2009;300(1-2):17–24.
39. Achour L, Labbe-Jullie C, Scott MG, Marullo S. An escort for GPCRs: implications for regulation of receptor density at the cell surface. *Trends Pharmacol Sci.* 2008;29(10):528–535.
40. Chung TT, Webb TR, Chan LF, Cooray SN, Metherell LA, King PJ, et al. The majority of adrenocorticotropin receptor (melanocortin 2 receptor) mutations found in familial glucocorticoid deficiency type 1 lead to defective trafficking of the receptor to the cell surface. *J Clin Endocrinol Metab.* 2008; 93(12):4948–4954.

41. Cooray SN, Almiro Do Vale I, Leung KY, Webb TR, Chapple JP, Egertova M, et al. The melanocortin 2 receptor accessory protein exists as a homodimer and is essential for the function of the melanocortin 2 receptor in the mouse y1 cell line. *Endocrinology.* 2008;149(4):1935–1941.
42. Metherell LA, Chapple JP, Cooray S, David A, Becker C, Ruschendorf F, et al. Mutations in MRAP, encoding a new interacting partner of the ACTH receptor, cause familial glucocorticoid deficiency type 2. *Nat Genet.* 2005;37(2):166–170.
43. Webb TR, Clark AJ. Minireview: The Melanocortin 2 Receptor Accessory Proteins. *Mol Endocrinol.* 2009;24(3):475–484.
44. Conn PM, Janovick JA. Trafficking and quality control of the gonadotropin releasing hormone receptor in health and disease. *Mol Cell Endocrinol.* 2009;299(2):137–145.
45. Lu Z-L, Gallagher R, Sellar R, Coetsee M, Millar RP. Mutations remote from the human gonadotropin-releasing hormone (GnRH) receptor-binding sites specifically increase binding affinity for GnRH II but not GnRH I. *J Biol Chem.* 2005; 280(33):29796–29803.
46. Mizrachi D, Segaloff DL. Intracellularly located misfolded glycoprotein hormone receptors associate with different chaperone proteins than their cognate wild-type receptors. *Mol Endocrinol.* 2004;18(7):1768–1777.
47. Newton CL, Whay AM, Millar RP. Rescue of human LH receptor mutant expression and function with an allosteric small molecule agonist. Proceedings of the Endocrine Society 92nd Annual Meeting and Expo, San Diego, California, USA, 2010.
48. Dunham JH, Hall RA. Enhancement of the surface expression of G protein-coupled receptors. *Trends Biotechnol.* 2009;27(9): 541–545.
49. Neubig RR, Siderovski DP. Regulators of G-protein signalling as new central nervous system drug targets. *Nat Rev Drug Discov.* 2002;1(3):187–197.
50. McLatchie LM, Fraser NJ, Main MJ, Wise A, Brown J, Thompson N, et al. RAMPs regulate the transport and ligand specificity of the calcitonin-receptor-like receptor. *Nature.* 1998;393(6683):333–339.
51. Wei H, Ahn S, Shenoy SK, Karnik SS, Hunyady L, Luttrell LM, Lefkowitz RJ. Independent beta-arrestin 2 and G protein-mediated pathways for angiotensin II activation of extracellular signal-regulated kinases 1 and 2. *Proc Natl Acad Sci USA.* 2003; 100(19):10782–10787.
52. Rajagopal K, Lefkowitz RJ, Rockman HA. When 7 transmembrane receptors are not G protein-coupled receptors. *J Clin Invest.* 2005;115(11):2971–2974.
53. Violin JD, Lefkowitz RJ. Beta-arrestin-biased ligands at seven-transmembrane receptors. *Trends Pharmacol Sci.* 2007;28(8): 416–422.
54. Millar RP, Pawson AJ. Outside-in and inside-out signaling: the new concept that selectivity of ligand binding at the gonadotropin-releasing hormone receptor is modulated by the intracellular environment. *Endocrinology.* 2004;145(8):3590–3593.
55. Kenakin T. Agonist-receptor efficacy. II. Agonist trafficking of receptor signals. *Trends Pharmacol Sci.* 1995;16(7):232–238.
56. Millar RP, Pawson AJ, Morgan K, Rissman EF, Lu ZL. Diversity of actions of GnRHs mediated by ligand-induced selective signaling. *Front Neuroendocrinol.* 2008;29(1):17–35.
57. Lopez de Maturana R, Pawson AJ, Lu ZL, Davidson L, Maudsley S, Morgan K, et al. Gonadotropin-releasing hormone analog structural determinants of selectivity for inhibition of cell growth: support for the concept of ligand-induced selective signaling. *Mol Endocrinol.* 2008;22(7):1711–1722.
58. Lu ZL, Coetsee M, White CD, Millar RP. Structural determinants for ligand-receptor conformational selection in a peptide G protein-coupled receptor. *J Biol Chem.* 2007;282(24):17921–17929.
59. Pfleger KD, Pawson AJ, Millar RP. Changes to gonadotropin-releasing hormone (GnRH) receptor extracellular loops differentially affect GnRH analog binding and activation: evidence for distinct ligand-stabilized receptor conformations. *Endocrinology.* 2008;149(6):3118–3129.
60. Leduc M, Breton B, Gales C, Le Gouill C, Bouvier M, Chemtob S, et al. Functional selectivity of natural and synthetic prostaglandin EP4 receptor ligands. *J Pharmacol Exp Ther.* 2009;331(1):297–307.
61. Cussac D, Boutet-Robinet E, Ailhaud MC, Newman-Tancredi A, Martel JC, Danty N, et al. Agonist-directed trafficking of signalling at serotonin 5-HT2A, 5-HT2B and 5-HT2C-VSV receptors mediated Gq/11 activation and calcium mobilisation in CHO cells. *Eur J Pharmacol.* 2008;594(1-3):32–38.
62. Woo AY, Wang TB, Zeng X, Zhu W, Abernethy DR, Wainer IW, et al. Stereochemistry of an agonist determines coupling preference of beta2-adrenoceptor to different G proteins in cardiomyocytes. *Mol Pharmacol.* 2009;75(1):158–165.
63. Georgieva T, Devanathan S, Stropova D, Park CK, Salamon Z, Tollin G, et al. Unique agonist-bound cannabinoid CB1 receptor conformations indicate agonist specificity in signaling. *Eur J Pharmacol.* 2008;581(1-2):19–29.
64. Jongsma M, van Unen J, van Loenen PB, Michel MC, Peters SL, Alewijnse AE. Different response patterns of several ligands at the sphingosine-1-phosphate receptor subtype 3 (S1P(3)). *Br J Pharmacol.* 2009;156(8):1305–1311.
65. Moore CA, Milano SK, Benovic JL. Regulation of receptor trafficking by GRKs and arrestins. *Annu Rev Physiol.* 2007;69: 451–482.
66. Premont RT, Gainetdinov RR. Physiological roles of G protein-coupled receptor kinases and arrestins. *Annu Rev Physiol.* 2007;69(1):511–534.
67. Ferguson SS. Evolving concepts in G protein-coupled receptor endocytosis: the role in receptor desensitization and signaling. *Pharmacol Rev.* 2001;53(1):1–24.
68. Lefkowitz RJ, Pierce KL, Luttrell LM. Dancing with different partners: Protein kinase A phosphorylation of seven membrane-spanning receptors regulates their G protein-coupling specificity. *Mol Pharmacol.* 2002;62(5):971–974.
69. Davidson JS, Wakefield IK, Millar RP. Absence of rapid desensitization of the mouse gonadotropin-releasing hormone receptor. *Biochem J.* 1994;300:299–302.
70. Blomenrohr M, Heding A, Sellar R, Leurs R, Bogerd J, Eidne KA, et al. Pivotal role for the cytoplasmic carboxyl-terminal tail of a nonmammalian gonadotropin-releasing hormone receptor in cell surface expression, ligand binding, and receptor phosphorylation and internalization. *Mol. Pharmacol.* 1999;56(6):1229–1237.
71. Pawson AJ, Katz A, Sun YM, Lopes J, Illing N, Millar RP, et al. Contrasting internalization kinetics of human and chicken gonadotropin-releasing hormone receptors mediated by C-terminal tail. *J Endocrinol.* 1998;156(3):R9–R12.
72. Luttrell LM, Lefkowitz RJ. The role of beta-arrestins in the termination and transduction of G-protein-coupled receptor signals. *J Cell Sci.* 2002;115(Pt 3):455–465.
73. Whistler JL, Chuang HH, Chu P, Jan LY, von Zastrow M. Functional dissociation of μ opioid receptor signaling and endocytosis: implications for the biology of opiate tolerance and addiction. *Neuron.* 1999;23(4):737–746.
74. Wojcikiewicz RJ. Regulated ubiquitination of proteins in GPCR-initiated signaling pathways. *Trends Pharmacol Sci.* 2004;25(1): 35–41.
75. Shenoy SK, McDonald PH, Kohout TA, Lefkowitz RJ. Regulation of receptor fate by ubiquitination of activated β_2-adrenergic receptor and β-arrestin. *Science.* 2001;294(5545):1307–1313.

76. Hsieh KP, Martin TF. Thyrotropin-releasing hormone and gonadotropin-releasing hormone receptors activate phospholipase C by coupling to the guanosine triphosphate-binding protens Gq and G11. *Mol Endocrinol.* 1992;10:1673–1681.
77. Poulin B, Rich N, Mas JL, Kordon C, Enjalbert A, Drouva SV. GnRH signalling pathways and GnRH-induced homologous desensitization in a gonadotrope cell line (alphaT3-1). *Mol Cell Endocrinol.* 1998;142(1-2):99–117.
78. Stojilkovic SS, Reinhart J, Catt KJ. Gonadotropin-releasing hormone receptors: structure and signal transduction pathways. *Endocr Rev.* 1994;15:462–499.
79. Hawes BE, Barnes S, Conn PM. Cholera toxin and pertussis toxin provoke differential effects on luteinizing hormone release, inositol phosphate production, and gonadotropin-releasing hormone (GnRH) receptor binding in the gonadotrope: evidence for multiple guanyl nucleotide binding proteins in GnRH action. *Endocrinology.* 1993;132(5):2124–2130.
80. Imai A, Takagi H, Furui T, Horibe S, Fuseya T, Tamaya T. Evidence for coupling of phosphotyrosine phosphatase to gonadotropin-releasing hormone receptor in ovarian carcinoma membrane. *Cancer* 1996;77(1):132–137.
81. Grundker C, Schulz K, Gunthert AR, Emons G. Luteinizing hormone-releasing hormone induces nuclear factor kappaB-activation and inhibits apoptosis in ovarian cancer cells. *J Clin Endocrinol Metab.* 2000;85(10):3815–3820.
82. Grundker C, Volker P, Emons G. Antiproliferative signaling of luteinizing hormone-releasing hormone in human endometrial and ovarian cancer cells through G protein alpha(I)-mediated activation of phosphotyrosine phosphatase. *Endocrinology.* 2001;142(6):2369–2380.
83. Imai A, Horibe S, Takagi A, Tamaya T. Gi protein activation of gonadotropin-releasing hormone-mediated protein dephosphorylation in human endometrial carcinoma. *Am J Obstet Gynecol.* 1997;176(2):371–376.
84. Limonta P, Moretti RM, Marelli MM, Dondi D, Parenti M, Motta M. The luteinizing hormone-releasing hormone receptor in human prostate cancer cells: messenger ribonucleic acid expression, molecular size, and signal transduction pathway. *Endocrinology.* 1999;140(11):5250–5256.
85. Young LS, Naik SI, Clayton RN. Increased gonadotrophin releasing hormone receptors on pituitary gonadotrophs: effect on subsequent LH secretion. *Mol Cell Endocrinol.* 1985;41(1):69–78.
86. Janovick JA, Conn PM. A cholera toxin-sensitive guanyl nucleotide binding protein mediates the movement of pituitary luteinizing hormone into a releasable pool: loss of this event is associated with the onset of homologous desensitization to gonadotropin-releasing hormone. *Endocrinology.* 1993;132(5): 2131–2135.
87. Kuphal D, Janovick JA, Kaiser UB, Chin WW, Conn PM. Stable transfection of GH3 cells with rat gonadotropin-releasing hormone receptor complementary deoxyribonucleic acid results in expression of a receptor coupled to cyclic adenosine 3′,5′-monophosphate-dependent prolactin release via a G-protein. *Endocrinology.* 1994;135(1):315–320.
88. Delahaye R, Manna PR, Berault A, Berreur-Bonnenfant J, Berreur P, Counis R. Rat gonadotropin-releasing hormone receptor expressed in insect cells induces activation of adenylyl cyclase. *Mol Cell Biol.* 1997;135(2):119–127.
89. Arora KK, Krsmanovic LZ, Mores N, O'Farrell H, Catt KJ. Mediation of cyclic AMP signaling by the first intracellular loop of the gonadotropin-releasing hormone receptor. *J Biol Chem.* 1998;273(40):25581–25586.
90. Nelson S, Horvat RD, Malvey J, Roess DA, Barisas BG, Clay CM. Characterization of an intrinsically fluorescent gonadotropin-releasing hormone receptor and effects of ligand binding on receptor lateral diffusion. *Endocrinology.* 1999; 140(2):950–957.
91. Liu F, Usui I, Evans LG, Austin DA, Mellon PL, Olefsky JM, et al. Involvement of both G(q/11) and G(s) proteins in gonadotropin-releasing hormone receptor-mediated signaling in L beta T2 cells. *J Biol Chem.* 2002;277(35):32099–32108.
92. Stanislaus D, Ponder S, Ji TH, Conn PM. Gonadotropin-releasing hormone receptor couples to multiple G proteins in rat gonadotrophs and in GGH3 cells: evidence from palmitoylation and overexpression of G proteins. *Biol Reprod.* 1998;59(3): 579–586.
93. Kjelsberg MA, Cotecchia S, Ostrowski J, Caron MG, Lefkowitz RJ. Constitutive activation of the alpha 1B-adrenergic receptor by all amino acid substitutions at a single site. Evidence for a region which constrains receptor activation. *J Biol Chem.* 1992;267(3):1430–1433.
94. Myburgh DB, Millar RP, Hapgood JP. Alanine-261 in intracellular loop III of the human gonadotropin- releasing hormone receptor is crucial for G-protein coupling and receptor internalization. *Biochem J.* 1998;331(Pt 3):893–896.
95. de Roux N, Young J, Misrahi M, Genet R, Chanson P, Schaison G, et al. A family with hypogonadotropic hypogonadism and mutations in the gonadotropin-releasing hormone receptor. *New Engl J Med.* 1997;337(22):1597–1602.
96. de Roux N, Young J, Brailly-Tabard S, Misrahi M, Milgrom E, Schaison G. The same molecular defects of the gonadotropin-releasing hormone receptor determine a variable degree of hypogonadism in affected kindred. *J Clin Endocrinol Metab.* 1999;84(2):567–572.
97. Bertherat J. Gonadotropin-releasing hormone receptor gene mutation: a new cause of hereditary hypogonadism and another mutated G-protein-coupled receptor. *Eur J Endocrinol.* 1998; 138(6):621–622.
98. Ulloa-Aguirre A, Stanislaus D, Arora V, Vaananen J, Brothers S, Janovick JA, et al. The third intracellular loop of the rat gonadotropin-releasing hormone receptor couples the receptor to Gs- and G(q/11)-mediated signal transduction pathways: evidence from loop fragment transfection in GGH3 cells. *Endocrinology.* 1998;139(5):2472–2478.
99. Chi L, Davidson JS, Zhou W, Millar RP, Sealfon SC. Mutations of the second intracellular loop domain of the GnRH receptor. Programe of the 76th Annual Meeting of the Endocrine Society, 1994, Anaheim, CA (Abstract 159). 1994.
100. Arora KK, Sakai A, Catt KJ. Effects of second intracellular loop mutations on signal transduction and internalization of the gonadotropin-releasing hormone receptor. *J Biol Chem.* 1995; 270(39):22820–22826.
101. Arora KK, Cheng Z, Catt KJ. Mutations of the conserved DRS motif in the second intracellular loop of the gonadotropin-releasing hormone receptor affect expression, activation, and internalization. *Mol Endocrinol.* 1997;11(9):1203–1212.
102. Everest HM, Hislop JN, Harding T, Uney JB, Flynn A, Millar RP, et al. Signaling and antiproliferative effects mediated by GnRH receptors after expression in breast cancer cells using recombinant adenovirus. *Endocrinology.* 2001;142(11):4663–4672.
103. Maudsley S, Davidson L, Pawson AJ, Chan R, de Maturana RL, Millar RP. Gonadotropin-releasing hormone (GnRH) antagonists promote proapoptotic signaling in peripheral reproductive tumor cells by activating a Galphai-coupling state of the type I GnRH receptor. *Cancer Res.* 2004;64(20):7533–7544.
104. Illing N, Troskie BE, Nahorniak CS, Hapgood JP, Peter RE, Millar RP. Two gonadotropin-releasing hormone receptor subtypes with distinct ligand selectivity and differential distribution in brain and pituitary in the goldfish (*Carassius auratus*). *Proc Natl Acad Sci USA.* 1999;96(5):2526–2531.

105. Sealfon SC, Weinstein H, Millar RP. Molecular mechanisms of ligand interaction with the gonadotropin-releasing hormone receptor. *Endocr Rev.* 1997;18(2):180–205.
106. Sun YM, Flanagan CA, Illing N, Ott TR, Sellar R, Fromme BJ, et al. A chicken gonadotropin-releasing hormone receptor that confers agonist activity to mammalian antagonists. Identification of D-Lys(6) in the ligand and extracellular loop two of the receptor as determinants. *J Biol Chem.* 2001;276(11): 7754–7761.
107. Tensen C, Okuzawa K, Blomenrohr M, Rebers F, Leurs R, Bogerd J, et al. Distinct efficacies for two endogenous ligands on a single cognate gonadoliberin receptor. *Eur J Biochem.* 1997;243:134–140.
108. Wang L, Bogerd J, Choi HS, Seong JY, Soh JM, Chun SY, et al. Three distinct types of gonadotropin-releasing hormone receptor characterized in the bullfrog. *Proc Natl Acad Sci USA.* 2001;98:361–366.
109. Vrecl M, Anderson L, Hanyaloglu A, McGregor AM, Groarke AD, Milligan G, et al. Agonist-induced endocytosis and recycling of the gonadotropin-releasing hormone receptor: effect of beta-arrestin on internalization kinetics. *Mol Endocrinol.* 1998;12(12):1818–1829.
110. Heding A, Vrecl M, Hanyaloglu AC, Sellar R, Taylor PL, Eidne KA. The rat gonadotropin-releasing hormone receptor internalizes via a beta-arrestin-independent, but dynamin-dependent, pathway: addition of a carboxyl-terminal tail confers beta-arrestin dependency. *Endocrinology.* 2000;141(1): 299–306.
111. Pawson AJ, Maudsley SR, Lopes J, Katz AA, Sun YM, Davidson JS, et al. Multiple determinants for rapid agonist-induced internalization of a nonmammalian gonadotropin-releasing hormone receptor: a putative palmitoylation site and threonine doublet within the carboxyl-terminal tail Are critical. *Endocrinology.* 2003;144(9):3860–3871.
112. Brothers SP, Janovick JA, Maya-Nunez G, Cornea A, Han XB, Conn PM. Conserved mammalian gonadotropin-releasing hormone receptor carboxyl terminal amino acids regulate ligand binding, effector coupling and internalization. *Mol Cell Endocrinol.* 2002;190(1-2):19–27.
113. Naor Z, Benard O, Seger R. Activation of MAPK cascades by G-protein-coupled receptors: the case of gonadotropin-releasing hormone receptor. *Trends Endocrinol Metab.* 2000;11(3):91–99.
114. Kraus S, Naor Z, Seger R. Intracellular signaling pathways mediated by the gonadotropin-releasing hormone (GnRH) receptor. *Arch Med Res.* 2001;32(6):499–509.
115. Naor Z. Signaling by G-protein-coupled receptor (GPCR): Studies on the GnRH receptor. *Front Neuroendocrinol.* 2009;30(1): 10–29.
116. Harris D, Bonfil D, Chuderland D, Kraus S, Seger R, Naor Z. Activation of MAPK cascades by GnRH: ERK and Jun N-terminal kinase are involved in basal and GnRH-stimulated activity of the glycoprotein hormone LHbeta-subunit promoter. *Endocrinology.* 2002;143(3):1018–1025.
117. Liu F, Austin DA, Mellon PL, Olefsky JM, Webster NJ. GnRH activates ERK1/2 leading to the induction of c-fos and LHbeta protein expression in LbetaT2 cells. *Mol Endocrinol.* 2002;16(3): 419–434.
118. Haisenleder DJ, Ferris HA, Shupnik MA. The calcium component of gonadotropin-releasing hormone-stimulated luteinizing hormone subunit gene transcription is mediated by calcium/calmodulin-dependent protein kinase type II. *Endocrinology.* 2003;144(6):2409–2416.
119. Gur G, Bonfil D, Safarian H, Naor Z, Yaron Z. GnRH signaling pathways regulate differentially the tilapia gonadotropin subunit genes. *Mol Cell Endocrinol.* 2002;189(1-2):125–134.
120. Vasilyev VV, Pernasetti F, Rosenberg SB, Barsoum MJ, Austin DA, Webster NJ, et al. Transcriptional activation of the ovine follicle-stimulating hormone-beta gene by gonadotropin-releasing hormone involves multiple signal transduction pathways. *Endocrinology.* 2002;143(5):1651–1659.
121. Ellsworth BS, White BR, Burns AT, Cherrington BD, Otis AM, Clay CM. c-Jun N-terminal kinase activation of activator protein-1 underlies homologous regulation of the gonadotropin-releasing hormone receptor gene in alpha T3-1 cells. *Endocrinology.* 2003;144(3):839–849.
122. Davidson L, Pawson AJ, Millar RP, Maudsley S. Cytoskeletal reorganization dependence of signaling by the gonadotropin-releasing hormone receptor. *J Biol Chem.* 2004;279(3):1980–1993.
123. Maudsley S, Naor Z, Bonfil D, Davidson L, Karali D, Pawson AJ, et al. Proline-rich tyrosine kinase 2 mediates gonadotropin-releasing hormone signaling to a specific extracellularly regulated kinase-sensitive transcriptional locus in the luteinizing hormone beta-subunit gene. *Mol Endocrinol.* 2007;21(5):1216–1233.
124. Davidson L, Pawson AJ, De Maturana RL, Freestone SH, Barran P, Millar RP, et al. Gonadotropin-releasing hormone-induced activation of diacylglycerol kinase-zeta and its association with active c-src. *J Biol Chem.* 2004;279(12): 11906–11916.
125. Hata K, Nakayama K, Fujiwaki R, Katabuchi H, Okamura H, Miyazaki K. Expression of the angopoietin-1, angopoietin-2, Tie2, and vascular endothelial growth factor gene in epithelial ovarian cancer. *Gynecol Oncol.* 2004;93(1):215–222.
126. Naor Z. Signal transduction mechanisms of Ca^{2+} mobilizing hormones: the case of gonadotropin-releasing hormone. *Endocr Rev.* 1990;11(2):326–353.
127. Limor R, Naor Z, Weisinger G, Naidich M, Knoll E, Sharon O, et al. Gonadotropin-releasing hormone activates the 12-lipoxygenase pathway in the LbetaT2 gonadotrope cell line. *Neuroendocrinology.* 2003;77(5):291–297.
128. Naor Z, Jabbour HN, Naidich M, Pawson AJ, Morgan K, Battersby S, et al. Reciprocal cross talk between gonadotropin-releasing hormone (GnRH) and prostaglandin receptors regulates GnRH receptor expression and differential gonadotropin secretion. *Mol Endocrinol.* 2007;21(2):524–537.
129. Lopez de Maturana R, Martin B, Millar RP, Brown P, Davidson L, Pawson AJ, et al. GnRH-mediated DAN production regulates the transcription of the GnRH receptor in gonadotrope cells. *Neuromol Med.* 2007;9(3):230–248.
130. Maudsley S, Davidson L, Pawson AJ, Freestone SH, Lopez de Maturana R, Thomson AA, et al. Gonadotropin-releasing hormone functionally antagonizes testosterone activation of the human androgen receptor in prostate cells through focal adhesion complexes involving Hic-5. *Neuroendocrinology* 2006;84(5): 285–300.
131. Salisbury TB, Binder AK, Grammer JC, Nilson JH. Maximal activity of the luteinizing hormone beta-subunit gene requires beta-catenin. *Mol Endocrinol.* 2007;21(4):963–971.
132. Fujino H, Regan JW. FP prostanoid receptor activation of a T-cell factor/beta-catenin signaling pathway. *J Biol Chem.* 2001; 276(16):12489–12492.
133. Fujino H, West KA, Regan JW. Phosphorylation of glycogen synthase kinase-3 and stimulation of T-cell factor signaling following activation of EP2 and EP4 prostanoid receptors by prostaglandin E2. *J Biol Chem.* 2002;277(4):2614–2619.
134. Gardner S, Maudsley S, Millar RP, Pawson AJ. Nuclear stabilization of beta-catenin and inactivation of glycogen synthase kinase-3beta by gonadotropin-releasing hormone: targeting Wnt signaling in the pituitary gonadotrope. *Mol Endocrinol.* 2007; 21(12):3028–3038.

135. Koelle MR. A new family of G-protein regulators – the RGS proteins. *Curr Opin Cell Biol.* 1997;9(2):143–147.
136. Neill JD, Duck LW, Sellers JC, Musgrove LC, Scheschonka A, Druey KM, et al. Potential role for a regulator of G protein signaling (RGS3) in gonadotropin-releasing hormone (GnRH) stimulated desensitization. *Endocrinology.* 1997;138(2):843–846.
137. Castro-Fernandez C, Conn PM. Regulation of the gonadotropin-releasing hormone receptor (GnRHR) by RGS proteins: role of the GnRHR carboxyl-terminus. *Mol Cell Endocrinol.* 2002;191(2): 149–156.
138. Castro-Fernandez C, Janovick JA, Brothers SP, Fisher RA, Ji TH, Conn PM. Regulation of RGS3 and RGS10 palmitoylation by GnRH. *Endocrinology.* 2002;143(4):1310–1317.
139. McArdle CA, Davidson JS, Willars GB. The tail of the gonadotrophin-releasing hormone receptor: desensitization at, and distal to, G protein-coupled receptors. *Mol Cell Endocrinol.* 1999;151(1-2):129–136.
140. McArdle CA, Forrest-Owen W, Willars G, Davidson J, Poch A, Kratzmeier M. Desensitization of gonadotropin-releasing hormone action in the gonadotrope-derived alpha T3-1 cell line. *Endocrinology.* 1995;136(11):4864–4871.
141. Willars GB, Challiss RA, Nahorski SR. Acute regulation of the receptor-mediated phosphoinositide signal transduction pathway. *J Lipid Mediat Cell Signal* 1996;14(1-3):157–168.
142. Willars GB, Royall JE, Nahorski SR, El-Gehani F, Everest H, McArdle CA. Rapid down-regulation of the type I inositol 1,4,5-trisphosphate receptor and desensitization of gonadotropin-releasing hormone-mediated Ca^{2+} responses in alpha T3-1 gonadotropes. *J Biol Chem.* 2001;276(5):3123–3129.
143. Wojcikiewicz RJ, Xu Q, Webster JM, Alzayady K, Gao C. Ubiquitination and proteasomal degradation of endogenous and exogenous inositol 1,4,5-trisphosphate receptors in alpha T3-1 anterior pituitary cells. *J Biol Chem.* 2003;278(2):940–947.
144. Liu F, Austin DA, Webster NJG. GnRH-desensitized LbetaT2 gonadotrope cells are refractory to acute PKC, cAMP, and calcium-dependent signaling. *Endocrinology.* 2003;144:4354–4365. p. DOI: 10.1210/en.2003-0204.
145. Hislop JN, Madziva MT, Everest HM, Harding T, Uney JB, Willars GB, et al. Desensitization and internalization of human and *Xenopus* gonadotropin-releasing hormone receptors expressed in alphaT4 pituitary cells using recombinant adenovirus. *Endocrinology.* 2000;141(12):4564–4575.
146. Blomenrohr M, Bogerd J, Leurs R, Schulz RW, Tensen CP, Zandbergen MA, et al. Differences in structure-function relations between nonmammalian and mammalian gonadotropin-releasing hormone receptors. *Biochem Biophys Res Comm.* 1997;238:517–522.
147. Acharjee S, Maiti K, Soh JM, Im WB, Seong JY, Kwon HB. Differential desensitization and internalization of three different bullfrog gonadotropin-releasing hormone receptors. *Mol Cells* 2002;14(1):101–107.
148. Pawson AJ, Faccenda E, Maudsley S, Lu ZL, Naor Z, Millar RP. Mammalian type I gonadotropin-releasing hormone receptors undergo slow, constitutive, agonist-independent internalization. *Endocrinology.* 2008;149(3):1415–1422.
149. Kotani M, Detheux M, Vandenbogaerde A, Communi D, Vanderwinden JM, Le Poul E, et al. The metastasis suppressor gene KiSS-1 encodes kisspeptins, the natural ligands of the orphan G protein-coupled receptor GPR54. *J Biol Chem.* 2001;276(37):34631–34636.
150. Wacker JL, Feller DB, Tang XB, Defino MC, Namkung Y, Lyssand JS, et al. Disease-causing mutation in GPR54 reveals the importance of the second intracellular loop for class A G-protein-coupled receptor function. *J Biol Chem.* 2008;283(45): 31068–31078.
151. Hori A, Honda S, Asada M, Ohtaki T, Oda K, Watanabe T, et al. Metastin suppresses the motility and growth of CHO cells transfected with its receptor. *Biochem Biophys Res Commun.* 2001;286(5):958–963.
152. Zhang C, Roepke TA, Kelly MJ, Ronnekleiv OK. Kisspeptin depolarizes gonadotropin-releasing hormone neurons through activation of TRPC-like cationic channels. *J Neurosci.* 2008; 28(17):4423–4434.
153. Liu X, Lee K, Herbison AE. Kisspeptin excites gonadotropin-releasing hormone neurons through a phospholipase C/calcium-dependent pathway regulating multiple ion channels. *Endocrinology.* 2008;149(9):4605–4614.
154. Zhang C, Bosch MA, Ronnekleiv OK, Kelly MJ. $GABA_B$ receptor mediated inhibition of gonadotropin-releasing hormone neurons is suppressed by kisspeptin-G protein-coupled receptor 54 signaling. *Endocrinology.* 2009;150(5):2388–2394.
155. Constantin S, Caligioni CS, Stojilkovic S, Wray S. Kisspeptin-10 facilitates a plasma membrane-driven calcium oscillator in gonadotropin-releasing hormone-1 neurons. *Endocrinology.* 2009;150(3):1400–1412.
156. Ozcan M, Alcin E, Ayar A, Yilmaz B, Sandal S, Kelestimur H. Kisspeptin-10 elicits triphasic cytosolic calcium responses in immortalized GT1-7 GnRH neurones. *Neurosci Lett.* 2011;492(1): 55–58.
157. Navenot JM, Wang Z, Chopin M, Fujii N, Peiper SC. Kisspeptin-10-induced signaling of GPR54 negatively regulates chemotactic responses mediated by CXCR4: a potential mechanism for the metastasis suppressor activity of kisspeptins. *Cancer Res.* 2005;65(22):10450–10456.
158. Yan C, Wang H, Boyd DD. KiSS-1 represses 92-kDa type IV collagenase expression by down-regulating NF-kappa B binding to the promoter as a consequence of Ikappa Balpha-induced block of p65/p50 nuclear translocation. *J Biol Chem.* 2001;276(2):1164–1172.
159. Clarke IJ, Sari IP, Qi Y, Smith JT, Parkington HC, Ubuka T, et al. Potent action of RFamide-related peptide-3 on pituitary gonadotropes indicative of a hypophysiotropic role in the negative regulation of gonadotropin secretion. *Endocrinology.* 2008; 149(11):5811–5821.
160. Carbone E, Carabelli V, Cesetti T, Baldelli P, Hernandez-Guijo JM, Giusta L. G-protein- and cAMP-dependent L-channel gating modulation: a manyfold system to control calcium entry in neurosecretory cells. *Pflügers Arch.* 2001;442(6):801–813.
161. Gershengorn MC, Osman R. Molecular and cellular biology of thyrotropin-releasing hormone receptors. *Physiol Rev.* 1996; 76(1):175–191.
162. Mayo KE, Miller LJ, Bataille D, Dalle S, Goke B, Thorens B, et al. International Union of Pharmacology. XXXV. The glucagon receptor family. *Pharmacol Rev.* 2003;55(1):167–194.
163. Chen C, Vincent JD, Clarke IJ. Ion channels and the signal transduction pathways in the regulation of growth hormone secretion. *Trends Endocrinol Metab.* 1994;5(6):227–233.
164. Ben-Jonathan N, Hnasko R. Dopamine as a Prolactin (PRL) Inhibitor. *Endocr Rev.* 2001;22(6):724–763.
165. Missale C, Nash SR, Robinson SW, Jaber M, Caron MG. Dopamine receptors: From structure to function. *Physiol Rev.* 1998;78(1):189–225.
166. Enjalbert A, Bockaert J. Pharmacological characterization of the D_2 dopamine receptor negatively coupled with adenylate cyclase in rat anterior pituitary. *Mol Pharmacol.* 1983;23(3): 576–584.
167. Kelly MA, Rubinstein M, Asa SL, Zhang G, Saez C, Bunzow JR, et al. Pituitary lactotroph hyperplasia and chronic hyperprolactinemia in dopamine D_2 receptor-deficient mice. *Neuron.* 1997;19(1):103–113.

168. Dal Toso R, Sommer B, Ewert M, Herb A, Pritchett DB, Bach A, et al. The dopamine D_2 receptor: two molecular forms generated by alternative splicing. *EMBO J.* 1989;8(13):4025–4034.
169. Lledo PM, Homburger V, Bockaert J, Vincent JD. Differential G protein-mediated coupling of D_2 dopamine receptors to K^+ and Ca^{2+} currents in rat anterior pituitary cells. *Neuron.* 1992;8(3): 455–463.
170. Diamond SE, Chiono M, Gutierrez-Hartmann A. Reconstitution of the protein kinase A response of the rat prolactin promoter: differential effects of distinct Pit-1 isoforms and functional interaction with Oct-1. *Mol Endocrinol.* 1999;13(2):228–238.
171. Lledo P-M, Legendre P, Israel J-M, Vincent J-D. Dopamine inhibits two characterized voltage-dependent calcium currents in identified rat lactotroph cells. *Endocrinology.* 1990;127(3): 990–1001.
172. Lledo PM, Legendre P, Zhang J, Israel JM, Vincent JD. Effects of dopamine on voltage-dependent potassium currents in identified rat lactotroph cells. *Neuroendocrinology.* 1990;52(6): 545–555.
173. Vallar L, Vicentini LM, Meldolesi J. Inhibition of inositol phosphate production is a late, Ca^{2+}-dependent effect of D_2 dopaminergic receptor activation in rat lactotroph cells. *J Biol Chem.* 1988;263(21):10127–10134.
174. Pang ZP, Südhof TC. Cell biology of Ca^{2+}-triggered exocytosis. *Curr Opin Cell Biol.* 2010;22(4):496–505.
175. Elsholtz HP, Lew AM, Albert PR, Sundmark VC. Inhibitory control of prolactin and Pit-1 gene promoters by dopamine. Dual signaling pathways required for D_2 receptor-regulated expression of the prolactin gene. *J Biol Chem.* 1991;266(34): 22919–22925.
176. Barnett P. Somatostatin and somatostatin receptor physiology. *Endocrine* 2003;20(3):255–264.
177. Pradayrol L, Jörnvall H, Mutt V, Ribet A. N-terminally extended somatostatin: The primary structure of somatostatin-28. *FEBS Letts* 1980;109(1):55–58.
178. de Lecea L, Criado JR, Prospero-Garcia O, Gautvik KM, Schweitzer P, Danielson PE, et al. A cortical neuropeptide with neuronal depressant and sleep-modulating properties. *Nature.* 1996;381(6579):242–245.
179. Patel YC. Somatostatin and its receptor family. *Front Neuroendocrinol.* 1999;20(3):157–198.
180. Ben-Shlomo A, Melmed S. Pituitary somatostatin receptor signaling. *Trends Endocrinol Metab.* 2010;21(3):123–133.
181. Scott LV, Dinan TG. Vasopressin and the regulation of hypothalamic-pituitary-adrenal axis function: implications for the pathophysiology of depression. *Life Sci.* 1998;62(22):1985–1998.
182. Birnbaumer M. Vasopressin receptors. *Trends Endocrinol Metab.* 2000;11(10):406–410.
183. Jard S, Gaillard RC, Guillon G, Marie J, Schoenenberg P, Muller AF, et al. Vasopressin antagonists allow demonstration of a novel type of vasopressin receptor in the rat adenohypophysis. *Mol Pharmacol.* 1986;30(2):171–177.
184. Todd K, Lightman SL. Vasopressin Activation of phosphatidylinositol metabolism in rat anterior pituitary *in vitro* and its modification by changes in the hypothalamo-pituitary-adrenal axis. *Neuroendocrinology* 1987;45(3):212–218.
185. Zingg HH, Laporte SA. The oxytocin receptor. *Trends Endocrinol Metab.* 2003;14(5):222–227.
186. de Roux N, Genin E, Carel J-C, Matsuda F, Chaussain J-L, Milgrom E. Hypogonadotropic hypogonadism due to loss of function of the KiSS1-derived peptide receptor GPR54. *Proc Natl Acad Sci USA* 2003;100(19):10972–10976.
187. Seminara SB, Messager S, Chatzidaki EE, Thresher RR, Acierno JS, Shagoury JK, et al. The GPR54 gene as a regulator of puberty. *New Engl J Med.* 2003;349(17):1614–1627.
188. Roseweir AK, Millar RP. The role of kisspeptin in the control of gonadotrophin secretion. *Hum Reprod Update* 2009;15(2): 203–212.
189. Oakley AE, Clifton DK, Steiner RA. Kisspeptin signaling in the brain. *Endocr Rev.* 2009;30(6):713–743.
190. Tena-Sempere M. KiSS-1 and reproduction: focus on its role in the metabolic regulation of fertility. *Neuroendocrinology* 2006;83(5-6):275–281.
191. Topaloglu AK, Reimann F, Guclu M, Yalin AS, Kotan LD, Porter KM, et al. TAC3 and TACR3 mutations in familial hypogonadotropic hypogonadism reveal a key role for Neurokinin B in the central control of reproduction. *Nat Genet.* 2009;41(3):354–358.
192. Goodman RL, Lehman MN, Smith JT, Coolen LM, de Oliveira CV, Jafarzadehshirazi MR, et al. Kisspeptin neurons in the arcuate nucleus of the ewe express both dynorphin A and neurokinin B. *Endocrinology.* 2007;148(12):5752–5760.
193. Navarro VM, Gottsch ML, Chavkin C, Okamura H, Clifton DK, Steiner RA. Regulation of gonadotropin-releasing hormone secretion by kisspeptin/dynorphin/neurokinin B neurons in the arcuate nucleus of the mouse. *J Neurosci.* 2009;29(38): 11859–11866.
194. Tsutsui K, Saigoh E, Ukena K, Teranishi H, Fujisawa Y, Kikuchi M, et al. A novel avian hypothalamic peptide inhibiting gonadotropin release. *Biochem Biophys Res Commun.* 2000; 275(2):661–667.
195. Ubuka T, Morgan K, Pawson AJ, Osugi T, Chowdhury VS, Minakata H, et al. Identification of human GnIH homologs, RFRP-1 and RFRP-3, and the cognate receptor, GPR147 in the human hypothalamic pituitary axis. *PLoS One* 2009;4(12):e8400.
196. Wu M, Dumalska I, Morozova E, van den Pol AN, Alreja M. Gonadotropin inhibitory hormone inhibits basal forebrain vGluT2-gonadotropin-releasing hormone neurons via a direct postsynaptic mechanism. *J Physiol.* 2009;587(Pt 7):1401–1411.
197. Ducret E, Anderson GM, Herbison AE. RFamide-related peptide-3, a mammalian gonadotropin-inhibitory hormone ortholog, regulates gonadotropin-releasing hormone neuron firing in the mouse. *Endocrinology.* 2009;150(6):2799–2804.
198. Anderson GM, Relf HL, Rizwan MZ, Evans JJ. Central and peripheral effects of RFamide-related peptide-3 on luteinizing hormone and prolactin secretion in rats. *Endocrinology.* 2009;150(4):1834–1840.
199. Thompson MD, Cole DE, Jose PA. Pharmacogenomics of G protein-coupled receptor signaling: insights from health and disease. *Methods Mol Biol.* 2008;448:77–107.
200. Conn PM, Ulloa-Aguirre A, Ito J, Janovick JA. G protein-coupled receptor trafficking in health and disease: lessons learned to prepare for therapeutic mutant rescue *in vivo*. *Pharmacol Rev.* 2007;59(3):225–250.
201. Millar RP, Newton CL. The year in G protein-coupled receptor research. *Mol Endocrinol.* 2010;24(1):261–274.
202. Bouligand J, Ghervan C, Tello JA, Brailly-Tabard S, Salenave S, Chanson P, et al. Isolated familial hypogonadotropic hypogonadism and a GNRH1 mutation. *N Engl J Med.* 2009; 360(26):2742–2748.
203. Balasubramanian R, Dwyer A, Seminara SB, Pitteloud N, Kaiser UB, Crowley Jr WF. Human GnRH deficiency: a unique disease model to unravel the ontogeny of GnRH neurons. *Neuroendocrinology* 2010;92(2):81–99.
204. Huhtaniemi IT, Themmen AP. Mutations in human gonadotropin and gonadotropin-receptor genes. *Endocrine* 2005;26(3): 207–217.
205. Teles MG, Bianco SD, Brito VN, Trarbach EB, Kuohung W, Xu S, et al. A GPR54-activating mutation in a patient with central precocious puberty. *N Engl J Med.* 2008;358(7):709–715.

206. Van Sande J, Parma J, Tonacchera M, Swillens S, Dumont J, Vassart G. Somatic and germline mutations of the TSH receptor gene in thyroid diseases. *J Clin Endocrinol Metab.* 1995;80(9): 2577–2585.
207. Rodien P, Bremont C, Sanson ML, Parma J, Van Sande J, Costagliola S, et al. Familial gestational hyperthyroidism caused by a mutant thyrotropin receptor hypersensitive to human chorionic gonadotropin. *N Engl J Med.* 1998;339(25):1823–1826.
208. Gromoll J, Simoni M, Nieschlag E. An activating mutation of the follicle-stimulating hormone receptor autonomously sustains spermatogenesis in a hypophysectomized man. *J Clin Endocrinol Metab.* 1996;81(4):1367–1370.
209. Feldman BJ, Rosenthal SM, Vargas GA, Fenwick RG, Huang EA, Matsuda-Abedini M, et al. Nephrogenic syndrome of inappropriate antidiuresis. *N Engl J Med.* 2005;352(18):1884–1890.
210. Vallar L, Spada A, Giannattasio G. Altered G_s and adenylate cyclase activity in human GH-secreting pituitary adenomas. *Nature.* 1987;330(6148):566–568.
211. Weinstein LS, Shenker A, Gejman PV, Merino MJ, Friedman E, Spiegel AM. Activating mutations of the stimulatory G protein in the McCune-Albright syndrome. *N Engl J Med.* 1991; 325(24):1688–1695.
212. Hauger RL, Grigoriadis DE, Dallman MF, Plotsky PM, Vale WW, Dautzenberg FM. International Union of Pharmacology. XXXVI. Current status of the nomenclature for receptors for corticotropin-releasing factor and their ligands. *Pharmacol Rev.* 2003;55(1):21–26.
213. Harmar AJ, Arimura A, Gozes I, Journot L, Laburthe M, Pisegna JR, et al. International Union of Pharmacology. XVIII. Nomenclature of receptors for vasoactive intestinal peptide and pituitary adenylate cyclase-activating polypeptide. *Pharmacol Rev.* 1998;50(2):265–270.
214. De Mota N, Lenkei Z, Llorens-Cortes C. Cloning, Pharmacological characterization and brain distribution of the rat apelin receptor. *Neuroendocrinology* 2000;72(6):400–407.
215. Noble F, Wank SA, Crawley JN, Bradwejn J, Seroogy KB, Hamon M, et al. International Union of Pharmacology. XXI. Structure, distribution, and functions of cholecytokinin receptors. *Pharmacol Rev.* 1999;51(4):745–781.
216. Branchek TA, Smith KE, Gerald C, Walker MW. Galanin receptor subtypes. *Trends Pharmacol Sci.* 2000;21(3):109–117.
217. Howard AD, Feighner SD, Cully DF, Arena JP, Liberator PA, Rosenblum CI, et al. A receptor in pituitary and hypothalamus that functions in growth hormone release. *Science.* 1996; 273(5277):974–977.
218. Gouarderes C, Mazarguil H, Mollereau C, Chartrel N, Leprince J, Vaudry H, et al. Functional differences between NPFF1 and NPFF2 receptor coupling: high intrinsic activities of RFamide-related peptides on stimulation of [^{35}S]GTPgammaS binding. *Neuropharmacology.* 2007;52(2):376–386.
219. Kukkonen JP, Akerman KE. Orexin receptors couple to Ca^{2+} channels different from store-operated Ca^{2+} channels. *Neuroreport.* 2001;12(9):2017–2020.
220. Kowalski TJ, McBriar MD. Therapeutic potential of melanin-concentrating hormone-1 receptor antagonists for the treatment of obesity. *Expert Opin Invest Drugs* 2004;13(9):1113–1122.
221. Adan RA, Gispen WH. Brain melanocortin receptors: from cloning to function. *Peptides* 1997;18(8):1279–1287.
222. Witt-Enderby PA, Bennett J, Jarzynka MJ, Firestine S, Melan MA. Melatonin receptors and their regulation: biochemical and structural mechanisms. *Life Sci.* 2003;72(20): 2183–2198.
223. Raddatz R, Wilson AE, Artymyshyn R, Bonini JA, Borowsky B, Boteju LW, et al. Identification and characterization of two neuromedin U receptors differentially expressed in peripheral tissues and the central nervous system. *J Biol Chem.* 2000; 275(42):32452–32459.
224. Michel MC, Beck-Sickinger A, Cox H, Doods HN, Herzog H, Larhammar D, et al. XVI. International Union of Pharmacology recommendations for the nomenclature of neuropeptide Y, peptide YY, and pancreatic polypeptide receptors. *Pharmacol Rev.* 1998;50(1):143–150.
225. Dobner PR. Multitasking with neurotensin in the central nervous system. *Cell Mol Life Sci.* 2005;62(17):1946–1963.
226. Khawaja AM, Rogers DF. Tachykinins: receptor to effector. *Intl J Biochem Cell Biol.* 1996;28(7):721–738.
227. Dhawan BN, Cesselin F, Raghubir R, Reisine T, Bradley PB, Portoghese PS, et al. International Union of Pharmacology. XII. Classification of opioid receptors. *Pharmacol Rev.* 1996;48(4): 567–592.
228. Aiyar N, Johns DG, Ao Z, Disa J, Behm DJ, Foley JJ, et al. Cloning and pharmacological characterization of the cat urotensin-II receptor (UT). *Biochem Pharmacol.* 2005;69(7): 1069–1079.
229. Bowery NG, Bettler B, Froestl W, Gallagher JP, Marshall F, Raiteri M, et al. International Union of Pharmacology. XXXIII. Mammalian gamma-aminobutyric acid(B) receptors: structure and function. *Pharmacol Rev.* 2002;54(2):247–264.
230. Hill SJ, Ganellin CR, Timmerman H, Schwartz JC, Shankley NP, Young JM, et al. International Union of Pharmacology. XIII. Classification of histamine receptors. *Pharmacol Rev.* 1997;49(3): 253–278.
231. Conn PJ, Pin JP. Pharmacology and functions of metabotropic glutamate receptors. *Annu Rev Pharmacol Toxicol.* 1997;37: 205–237.
232. Bylund DB, Eikenberg DC, Hieble JP, Langer SZ, Lefkowitz RJ, Minneman KP, et al. International Union of Pharmacology nomenclature of adrenoceptors. *Pharmacol Rev.* 1994;46(2):121–136.
233. Barnes NM, Sharp T. A review of central 5-HT receptors and their function. *Neuropharmacology* 1999;38(8):1083–1152.
234. Howlett AC, Barth F, Bonner TI, Cabral G, Casellas P, Devane WA, et al. International Union of Pharmacology. XXVII. Classification of cannabinoid receptors. *Pharmacol Rev.* 2002; 54(2):161–202.
235. Brink C, Dahlen SE, Drazen J, Evans JF, Hay DW, Nicosia S, et al. International Union of Pharmacology XXXVII. Nomenclature for leukotriene and lipoxin receptors. *Pharmacol Rev.* 2003;55(1):195–227.
236. Yu N, Lariosa-Willingham KD, Lin FF, Webb M, Rao TS. Characterization of lysophosphatidic acid and sphingosine-1-phosphate-mediated signal transduction in rat cortical oligodendrocytes. *Glia* 2004;45(1):17–27.
237. Coleman RA, Smith WL, Narumiya S. International Union of Pharmacology classification of prostanoid receptors: properties, distribution, and structure of the receptors and their subtypes. *Pharmacol Rev.* 1994;46(2):205–229.

CHAPTER

3

Neuroendocrine Feedback Control Systems: An Introduction

George Fink

Mental Health Research Institute, University of Melbourne, Parkville, Melbourne, Victoria, Australia

OUTLINE

Summary

Feedback control systems are fundamental for the normal physiological functioning and homeostasis of the body. There are two types of feedback, negative and positive, of which the former is the more common. Removal of the main pituitary target glands, the adrenal, gonads and thyroid, the most reproducible and reproduced experiment in classical endocrinology, demonstrates that the secretion of pituitary adrenocorticotropin (ACTH), the gonadotropins (luteinizing hormone, LH, and follicle-stimulating hormone, FSH), and thyrotropin (TSH) is controlled by negative feedback exerted by the adrenal corticosteroids, gonadal steroids and thyroid hormones, respectively.

Positive feedback, whereby the output of a system increases the output of the stimulator (gain in the system), is far less common than negative feedback, possibly because, taken to its logical conclusion, a positive feedback system, uncontrolled, will eventually self-destruct. Positive feedback is exemplified by (a) estrogen stimulation of gonadotropin secretion, which results in ovulation, and (b) the release of oxytocin induced during parturition by pressure of the fetal head on the uterine cervix. Estrogen stimulation of gonadotropin secretion is reinforced by the servomechanism of the "priming effect" of GnRH, whereby a small pulse of GnRH, by further amplifying pituitary responsiveness to itself, ensures the occurrence of a massive ovulatory gonadotropin surge.

Crucial for homeostasis, negative feedback control mechanisms comprise a system in which the output moderates the strength of the controller to a predetermined set-point level. Negative feedback control operates widely throughout the body

Handbook of Neuroendocrinology, DOI: 10.1016/B978-0-12-375097-6.10003-4

at the molecular (for example, end-product inhibition of enzyme activity), cellular and whole system/body levels. The mechanisms by which the set point is determined, functionally and anatomically, vary between systems and species.

Here, attention will focus on the principles of negative feedback control using the hypothalamic–pituitary–adrenal (HPA) system as example. The HPA system, together with the sympathetic–medullary system, plays a pivotal role in the neuroendocrine response to stress. Homeostasis within the hypothalamic HPA is maintained by a precise negative feedback system by which the adrenal glucocorticoids (afferent limb), cortisol or corticosterone, moderate ACTH synthesis and release (efferent limb). Allostasis – that is, change in HPA activity to cope with increased stress load – is thought to be brought about by change in feedback set-point.

INTRODUCTION

Feedback control systems are fundamental for the normal physiological functioning of the body. There are two types of feedback, negative and positive, of which the former is the more common. Removal of the main pituitary target glands, the adrenal, gonads and thyroid, the most reproducible and reproduced experiment in classical endocrinology, demonstrates that the secretion of pituitary adrenocorticotropin (ACTH), the gonadotropins (luteinizing hormone, LH, and follicle-stimulating hormone, FSH), and thyrotropin (TSH) is controlled by negative feedback exerted by the adrenal corticosteroids, gonadal steroids and thyroid hormones, respectively[1–3]; (see also Chapters 1 and 5 in this volume). Interruption of the negative feedback loop caused, for example, by surgical or pharmacological adrenalectomy, gonadectomy or thyroidectomy, or by an enzymatic defect in steroid or thyroid hormone biosynthesis, results in hypersecretion of pituitary ACTH, gonadotropins or TSH (see Chapters 5, 8–11, 19, 21, 30 and 31).

Not discussed here is the leptin feedback system. First characterized in 1994 by Jeffrey Friedman and his colleagues,[4] leptin is a hormone secreted from adipocytes, the body's fat depot. Its discovery galvanized the neuroendocrinology of feeding and metabolism by providing a new perspective for examining how hormones interact with the brain. For a detailed account of leptin and its feedback system, the reader is referred to Chapter 14, and the relatively massive literature in the field (for example, Fehm *et al.*[5]).

Crucial for homeostasis, negative feedback control mechanisms comprise a system in which the output moderates the strength of the controller to a predetermined set-point level[6] (Figs 3.1, 3.2). The set point–stimulator complex contains a "comparator" (error detector), which compares the strength of the feedback signal with a preset level.[7] An increase in the strength of the feedback signal above the preset level reduces the output of the stimulator, whereas a decrease in the strength of the feedback signal below the preset level results in an increase in output of the stimulator and corrects the "error".[7] Negative feedback control operates widely throughout the body at the molecular (for example, end-product inhibition of enzyme activity), cellular and whole system/body levels. The mechanisms by which the set point is determined, functionally and anatomically, vary between systems and

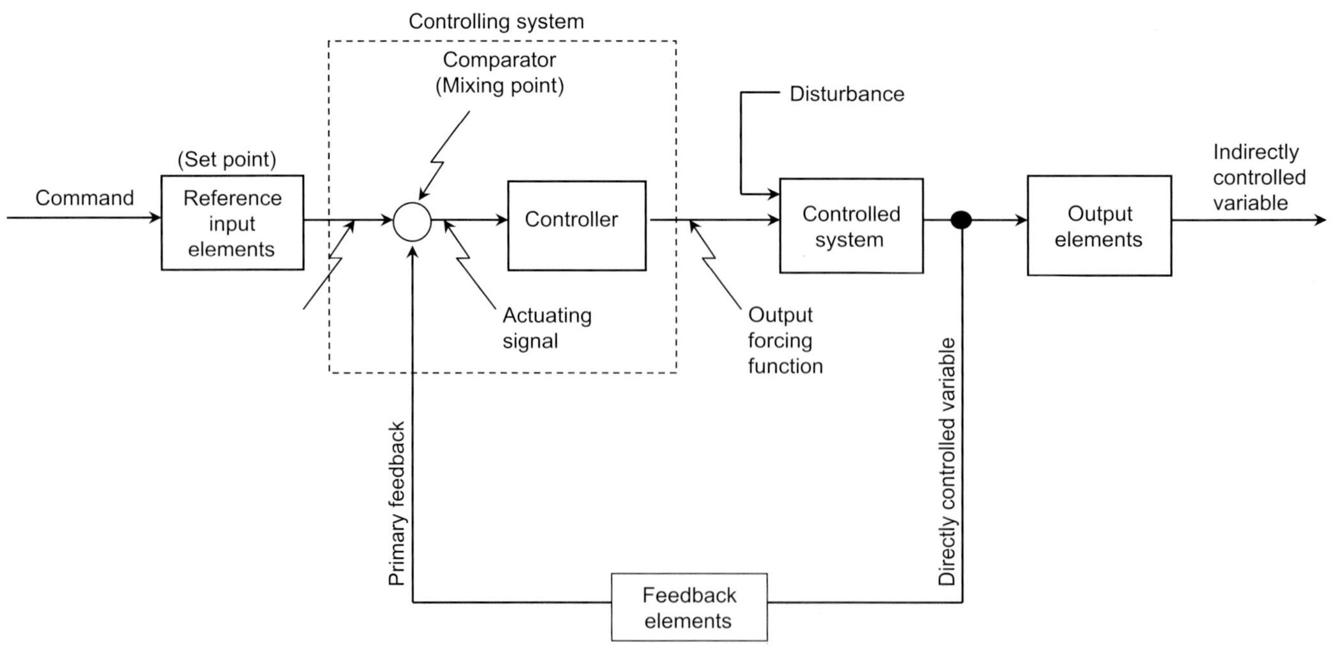

FIGURE 3.1 A generalized feedback control system. *Modified from Milhorn (1966).*[6]

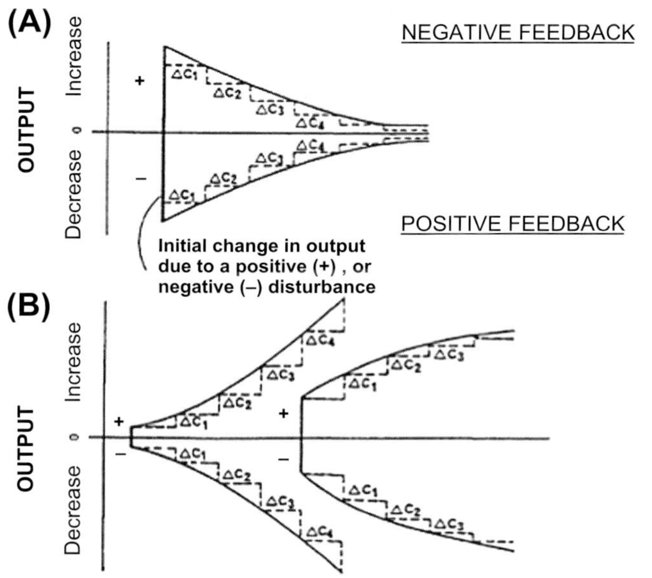

FIGURE 3.2 (**A**) Negative feedback minimizes the disturbance to a regulator, resulting in a system in which the output tends to remain constant. In this case, the ratios of the decrements of the controlled variable (C) (C_2/C_1, C_3/C_2) are less than unity. (**B**) In positive feedback (left-hand curves) an initial disturbance results in a continuous increase in output ("vicious cycle"). The increments of the controlled variable (C_2/C_1, C_3/C_2) are greater than unity. When the response does not result in a vicious cycle, the ratios of C_2/C_1, C_3/C_2, and so on are less than unity (see right-hand curves in B). *Reproduced from Milhorn HTJ.*[6] The Application of Control Theory to Physiological Systems. *Philadelphia: Saunders; 1966:386, with permission.*

species. In higher orders, the level of set point is thought to be regulated predominantly by the central nervous system. In the case of the hypothalamic–pituitary–adrenal axis (HPA) feedback system, for example, GABAergic, glutamatergic and monoamine neural projections from the limbic system and other brain regions to the paraventricular nucleus (PVN) of the hypothalamus may play a key role in set-point regulation[8,9]; (see below).

Positive feedback, whereby the output of a system increases the output of the stimulator (gain in the system) (Fig. 3.2B), is far less common than negative feedback, possibly because, taken to its logical conclusion, a positive feedback system, uncontrolled, will eventually self-destruct. Some systems loosely termed "positive feedback" are in fact "servomechanisms." A servomechanism is a closed-loop control system that increases significantly the power of a small signal. Positive feedback is exemplified by (a) estrogen stimulation of gonadotropin secretion, which results in ovulation[1–3]; (see also Chapter 5), and (b) the release of oxytocin induced during parturition by pressure of the fetal head on the uterine cervix (see Chapter 6). The latter triggers volleys of impulses that, by way of a multisynaptic pathway to the hypothalamus, stimulate oxytocin release. Oxytocin stimulates further contraction of the uterus, which results in a further increase in pressure on the uterine cervix. This "vicious cycle" is only broken when the fetus is expelled. Servomechanisms are illustrated by the way that just before ovulation in the human as well as in other spontaneously ovulating mammals, elevated plasma estrogen concentrations increase the responsiveness of the anterior pituitary gland to gonadotropin-releasing hormone (GnRH) by 20- to 50-fold. This estrogen effect is reinforced by the "priming effect" of GnRH, whereby a small pulse of GnRH, by further amplifying pituitary responsiveness to itself, ensures the occurrence of a massive ovulatory gonadotropin surge in response to a small surge or increased pulse frequency of GnRH.[1–3,10] For a detailed discussion of positive feedback and servomechanisms, see Chapter 5.

The present chapter focuses on the principles of negative feedback control using the hypothalamic–pituitary–adrenal (HPA) system as an example. While our discussion concentrates on negative feedback maintenance of homeostasis by glucocorticoids (cortisol in the human and corticosterone in rodents), the same mechanisms are involved in allostasis, which is brought about by a change in set point presumed to enable the organism to anticipate and deal with the physiological challenge or stress.[11–13] Allostatic regulation emphasizes feedforward regulatory systems, anticipatory and essential for adaptive social behaviors.[14]

THE NEUROENDOCRINE HPA CONTROL SYSTEM

Pattern of ACTH and Glucocorticoid Secretion

In the normal state, with the negative feedback loop intact, the set point in the brain–pituitary module maintains the secretion of ACTH within a relatively narrow bandwidth. Within this bandwidth, the basal secretion of ACTH is pulsatile, and is cleared rapidly from the blood by both metabolic degradation and distribution into several body compartments. Pulsatile, ultradian ACTH secretion is responsible for driving pulsatile glucocorticoid secretion,[15] and this is associated with parallel changes in steroidogenic acute regulatory protein (StAR) and P450scc heteronuclear RNA levels. The pulsatile pattern of StAR and P450scc is paralleled by pulsatile transcription of the melanocortin 2 receptor accessory protein.[15] Plasma ACTH concentrations show a circadian rhythm, with a peak in the morning of diurnal animals and a trough approaching a nadir around midnight. In nocturnal animals, such as rodents, the phase of this rhythm is reversed so that plasma ACTH concentrations reach a peak just before the onset of darkness. The circadian rhythm of ACTH results in a circadian rhythm in the plasma concentrations of adrenal corticosteroids: cortisol in humans and corticosterone in rodents. A key feature in both diurnal and nocturnal animals is that corticosteroid plasma concentrations reach a peak just before the animal is due to wake from sleep (see cortisol awakening response, below); the levels are lowest just before or during sleep.

The glucocorticoids are bound rapidly by albumin and a specific corticosteroid binding-protein, "transcortin," and metabolized by several organs, especially the liver and kidney (Fig. 3.3). Only unbound (free) glucocorticoids inhibit ACTH release; therefore, the degree of glucocorticoid binding and metabolism, as well as the magnitude of adrenal secretion (Fig. 3.3), determines the strength of the negative feedback signal. The HPA feedback system might also be influenced by a binding protein in plasma that has a high and selective affinity for human but not ovine corticotropin-releasing factor (CRF-41).[16] This CRF-binding protein may explain the brief biological action of hCRF as compared to ovine CRF in man, and why high concentrations of plasma immunoreactive hCRF in women during third-trimester pregnancy do not cause increased ACTH secretion.[16] The hCRF-binding protein was subsequently shown to be a 37-kDa protein.[17–19] Synthesized in the liver, the physiological role of the CRF binding (glyco) protein awaits determination. In the human, CRF binding protein concentrations in plasma increase during the third trimester of pregnancy.

Cortisol Awakening Response

In man, waking is closely associated with a cortisol awakening response (CAR) which typically involves an increase in salivary cortisol concentrations peaking around 30 minutes after waking.[20] The CAR has been increasingly studied in psychoneuroendocrinology in recent years, since the advent of sampling devices has allowed saliva samples to be collected by research participants at home. The CAR is a response to waking, rather than a reflection of any non-specific rise in cortisol levels in the early hours of the day.[21] The magnitude of the CAR has been associated with a number of psychosocial factors, including work stress and other forms of chronic life stress, psychological traits, depression, fatigue and post-traumatic stress disorder, with both increases and decreases being described (see, for example, Dockray *et al.*;[20] Chida and Steptoe[21]). Chronotype has a limited impact on the diurnal cortisol profile of healthy women, and may be somewhat impervious to individual preferences for morning or evening activity.[22] Cortisol, both total output and the awakening response, has consistently been shown to be lower among

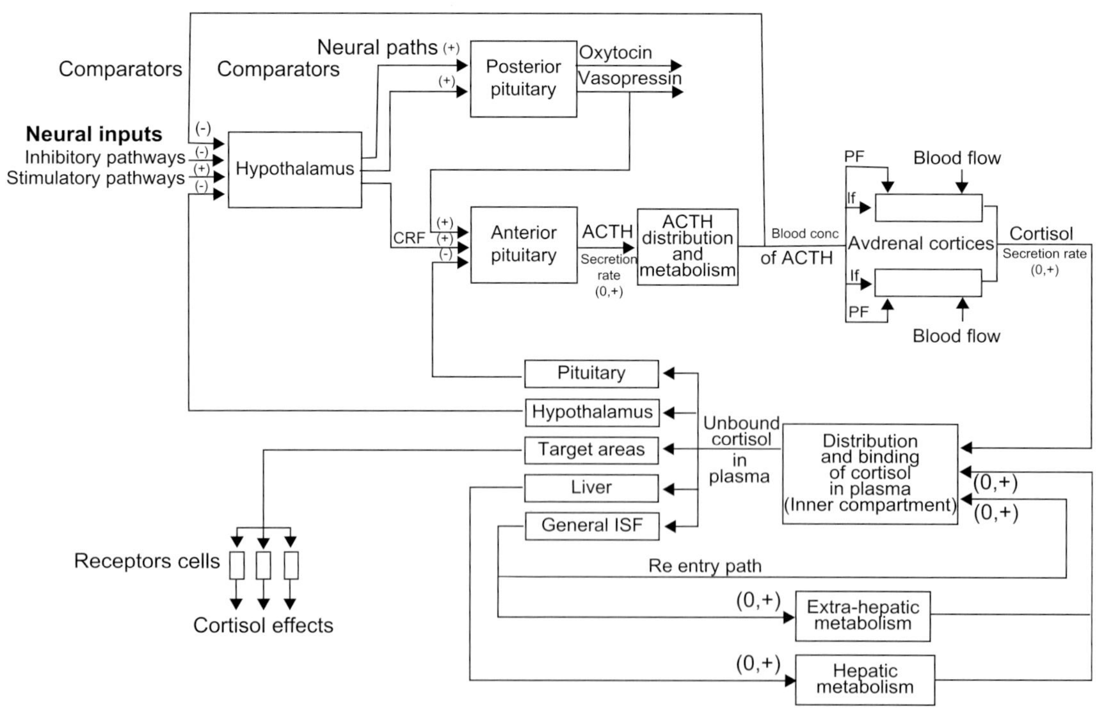

FIGURE 3.3 **Block diagram of the hypothalamic–pituitary–adrenal glucocorticoid control system.** If, input forcing of adrenal by ACTH; PF, parametric forcing of adrenal (hypertrophic effect) caused by ACTH over a longer time period. The parametric effect of changes in adrenal blood flow is also indicated. The designators 0, + and 0, − indicate that signals in pathways are restricted in values (e.g., there are no negative masses or frequencies, and removal processes or inhibitors are negative in effects). *Reproduced from Yates FE, Maran JW. In: Knobil E, Sawyer WH, eds.* Handbook of Physiology. *Washington, DC: American Physiological Society; 1975:367–404, with permission.*

individuals with higher levels of positive affect. The beneficial effects of positive mood on cardiovascular function, including heart rate and blood pressure, and the immune system have also been described. The influence of positive affect on these psychobiological processes is independent of negative affect, suggesting that positive affect may have characteristic biological correlates that may be partly responsible for the protective effects of positive affect on health outcomes.[22]

Glucocorticoid Negative Feedback Control in the HPA: Interactions with Neural Circuits

Glucocorticoids (cortisol in the human, corticosterone in rodents) secreted by the adrenal cortex exert their inhibitory action both on the brain and on the pituitary gland (Figs 3.3, 3.4). The paraventricular nuclei of the hypothalamus contain the final common pathway neurons that mediate the neural control of pituitary ACTH synthesis and release (see also Chapter 5). Many data point to PVN as the main component of the HPA set point. Thus, for example, our own data showed that glucocorticoid negative feedback in the rat is blocked by lesions of the PVN.[22] The PVN receives stimulatory and inhibitory neural inputs and an important projection from the suprachiasmatic nucleus (SCN), the key circadian oscillator in brain. PVN drive is mediated by the "stress neurohormones," CRF-41 and arginine vasopressin (AVP), which are released into hypophysial portal vessel blood and trigger pituitary ACTH synthesis and release.[24–28] ACTH, released into the systemic circulation, stimulates adrenal corticosteroid synthesis and release.

In addition to glucocorticoids, PVN activity is either moderated or stimulated by higher central nervous components. Thus, lesion studies of the hippocampus suggested that hippocampal efferents from the ventral subiculum and ventral CA1 exert tonic inhibition of the CRFmRNA and AVPmRNA synthesis in the PVN.[29] Furthermore, the posterior bed nucleus of the stria terminalis (BNST) is involved in inhibition of the HPA axis, whereas the anteroventral BNST activates the HPA axis. The BNST contains functional subdomains that play different roles in integrating and processing limbic information in response to stress.[30–32] The posterior medial BNST is likely to be a major component of the brain circuitry involved in both normal and pathological stress adaptation.[30,33,34]

In summary, the relationship between limbic structures and control of ACTH and glucocorticoid release suggests that: (a) in general, the hippocampus and anterior cingulate/prelimbic cortex inhibit stress-induced

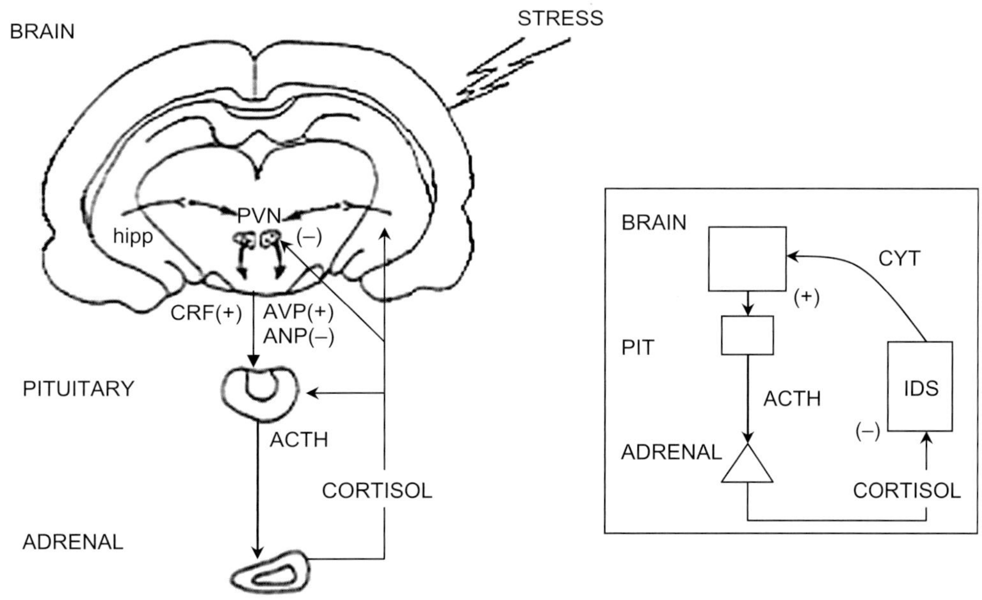

FIGURE 3.4 **Schematic diagram of the elements involved in glucocorticoid negative feedback.** The paraventricular nucleus (PVN) contains the main stress (final common pathway neurons) secreting both CRF and AVP. The negative feedback action of cortisol (corticosterone in rodents) is exerted mainly on the PVN and the pituitary corticotropes. However, long-term effects mediated through the hippocampus (Hipp) and amygdala (not shown) cannot be excluded. There are several possible indirect connections between the hippocampus and PVN. Neural control of ACTH secretion may also be mediated by a corticotropin inhibitory peptide, atrial natriuretic peptide (ANP). The inset shows the important inhibitory action of cortisol on the immune defense system (IDS). The IDS produces cytokines (CYT), which act on the brain to stimulate ACTH secretion. Cytokine secretion is inhibited by the negative feedback effect of glucocorticoids. *Reproduced with permission from Fink G Mechanisms of negative and positive feedback of steroids in the hypothalamic–pituitary system. In: Bittar EE, Bittar N, eds. Principle of medical Biology, Vol. 10A. New York: JAI Press; 1997:29–100.*

HPA activation, whereas the amygdala and perhaps the infralimbic cortex may enhance glucocorticoid secretion; (b) the role of limbic structures is both region- and stimulus-specific; (c) limbic sites have minimal direct projections to the HPA effector neurons of the paraventricular nucleus (PVN); (d) hippocampal, cortical and amygdala efferents apparently relay with neurons in the bed nucleus of the stria terminalis, hypothalamus and brainstem to access CRF neurons; (e) hippocampal, cortical and amygdala projection pathways show extensive overlap in regions such as the bed nucleus of the stria terminalis, hypothalamus and perhaps brainstem, implying that limbic information may be integrated at subcortical relay sites prior to accessing the PVN; and (f) all regions express both glucocorticoid and mineralocorticoid receptors, allowing for glucocorticoid modulation of limbic signaling patterns.[34,35] Our hypophysial portal blood studies showed that: (a) electrical stimulation of the PVN induced massive release of CRF-41; (b) stimulation of the amygdala inhibited CRF-41 release into portal vessel bood; and (c) hippocampal stimulation had no significant effect on CRF-41 release.[36]

INTERACTION BETWEEN NEGATIVE FEEDBACK AND CIRCADIAN RHYTHM IN THE HYPOTHALAMIC–PITUITARY–ADRENAL (HPA) SYSTEM

Already mentioned above, the circadian rhythm of ACTH and glucocorticoid secretion, which has a peak at the end and a nadir at the beginning of the sleep phase, is driven by a neural mechanism mediated mainly by the stress neurohormones. The twofold increase in the ACTH signal between the nadir and the peak of the circadian rhythm in the rat results in a nine-fold increase in corticosterone due to an increase in the responsiveness of the adrenal cortex.[3]

In the unstressed state the HPA system operates in an approximately linear domain, with all the loop variables (Fig. 3.3) showing circadian periodicity.[7] ACTH and cortisol are also released in pulsatile fashion, with circadian and ultradian rhythms governing secretion of these hormones.[37] ACTH and cortisol pulses are released approximately hourly, with ACTH pulses preceding cortisol pulses by approximately 10 minutes. In addition to these approximately hourly pulses, an ultradian rhythm of cortisol of 90–110 minutes has been found, which is linked to the basic rest–activity cycle, particularly the alternating arousal–sleep cycle, which continues throughout the day.[37] Van Cauter *et al.*[38] found a strong bidirectional link between cortisol secretory episodes and arousal.

The circadian rhythm of ACTH secretion is driven by the suprachismatic nucleus (SCN), the master circadian light entrainable oscillator (LEO).[39–41] In addition to the diurnal rhythm of corticosterone, the SCN, located in the hypothalamus, is involved in other key rhythms in the body, such as the sleep–wake cycle, motor activity, thermoregulation, pineal arylalkylamine *N*-acetyl transferase activity (rate-limiting enzyme for melatonin synthesis in the pineal gland), and the regular occurrence of ovulation (see Chapter 5). If any one of these functions is abnormal, then there is a high probability that the others will also be abnormal. The corticosterone rhythm is especially sensitive in that even partial lesions of the SCN which have no effect on any of the other circadian functions disrupt the adrenal rhythm.

A prominent exception to the concept that the SCN is dominant in the control of circadian rhythms in physiology and behavior is the fact that lesions of the SCN do not abolish the ability of rats to anticipate one meal per day. This anticipation is associated with an increase in motor activity, core body temperature and serum corticosterone, and would appear to be driven by another Zeitgeber termed the "food-entrainable oscillator" (FEO), the location of which has still to be determined.[42]

An approach to elucidating a possible link between the LEO and the FEO has recently been reported by Hayasaka *et al.*[42] and is based on G-protein signaling. Regulators of G-protein signaling (RGS) are a multifunctional protein family, which functions in part as GTPase-activating proteins (GAPs) of G-protein alpha-subunits to terminate G-protein signaling. Previous studies have demonstrated that the Rgs16 transcripts exhibit robust circadian rhythms both in the SCN and in the liver. To investigate the role of Rgs16 in the circadian clock *in vivo*, Hayasaka *et al.*[42] generated two independent transgenic mouse lines using lentiviral vectors expressing short hairpin RNA (shRNA) targeting the Rgs16 mRNA. The knockdown mice demonstrated significantly shorter free-running periods of locomotor activity rhythms and reduced total activity compared with wild-type siblings. Furthermore, when feeding was restricted during the daytime, FEO-driven elevated food-anticipatory activity (FAA) observed prior to the scheduled feeding time was significantly attenuated in the knockdown mice. Whereas the restricted feeding in wild-type animals phase advanced the rhythmic expression of the Per2 clock gene in liver and thalamus, this phase shift was not observed in the knockdown mice. The report by Hayasaka *et al.*[42] is the first *in vivo* demonstration that a common regulator of G-protein signaling is involved in the two separate, but interactive, circadian timing systems, LEO and FEO.

The SCN receives afferent projections from (a) the retina (direct as well as indirect after relay in the ventral lateral geniculate nucleus), (b) the 5-hydroxytryptamine

(5-HT) raphe neurons, and (c) the hippocampus by way of the medial corticohypothalamic tract.[43–50] Each of these inputs to the SCN is likely to affect the periodicity and amplitude of the SCN pacemaker. The retinal input affects the time of the light–dark cycle, while the raphe nuclei hippocampal input may be related to the sleep–wake cycle and other behaviors. The raphe input also plays a major role in determining the amplitude of diurnal ACTH oscillations.

Although the central action of glucocorticoids in moderating HPA activity is mainly on the PVN (see later), the noradrenergic locus coeruleus, the serotonergic dorsal raphe and the dopaminergic ventral tegmental area all express corticosteroid receptors and have been the focus of antidepressant research. All three of these nuclei express glucocorticoid receptor (GR), and the locus coeruleus and dorsal raphe have also been shown to express mineralocorticoid receptor (MR).[51] Monaminergic projections from these brainstem nuclei can also influence the activity of the HPA axis.[32,33] Locus coeruleus noradrenergic afferents to the hypothalamus stimulate HPA activity, while serotonergic projections from the raphe nuclei can both facilitate and inhibit HPA activity, depending on the limbic targets of these projections.[32] Dopaminergic projections from the ventral tegmental area affect HPA sensitivity to glucocorticoid feedback inhibition indirectly via the prefrontal cortex.[51]

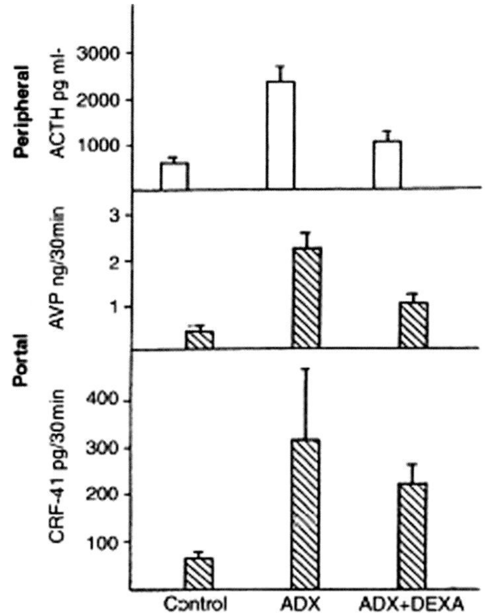

FIGURE 3.5 **The delayed negative feedback action of glucocorticoid in adult female Wistar rats anesthetized with sodium pentobarbitone.** ACTH was measured in plasma taken immediately before sectioning of the pituitary stalk for the collection of hypophysial portal vessel blood. Adrenalectomy (ADX) resulted in a fourfold increase in ACTH concentration and a similar increase in the output of AVP and CRF-41 release relative to that in intact, untreated animals (control). The administration of dexamethasone (DEXA) significantly reduced the levels of ACTH and AVP, but not CRF-41. *Modified from, Fink* et al. *(1988).*[28]

CORTICOSTEROID FEEDBACK ON THE HYPOTHALAMUS AND PITUITARY GLAND: PHASE DIFFERENCES

The seminal modeling studies of Eugene Yates and associates[7] suggested that corticosteroid negative feedback on ACTH release occurs in three phases: a fast and immediate rate-sensitive phase of about 30 minutes, followed by a level-sensitive phase that occurs 2–3 hours after the start of corticosteroid administration, followed by a long-term chronic phase. We investigated the effect of corticosteroids on the release of stress neurohormones into hypophysial portal blood in the intermediate and chronic phases.[28] Our findings were that adrenalectomy induced a three- to fourfold increase in the release of both CRF and AVP into hypophysial portal blood. Administration of the synthetic glucocorticoid dexamethasone 2.5 hours before portal blood collection significantly reduced the release of AVP but not CRF-41 (Fig. 3.5), and also blocked the ACTH response to CRF-41 (Fig. 3.6). These data suggest that "intermediate-delayed" (2- to 3-hour) glucocorticoid feedback is mediated by the blockade of pituitary responsiveness to CRF-41 and a reduction in AVP output into hypophysial portal blood. That is, in this

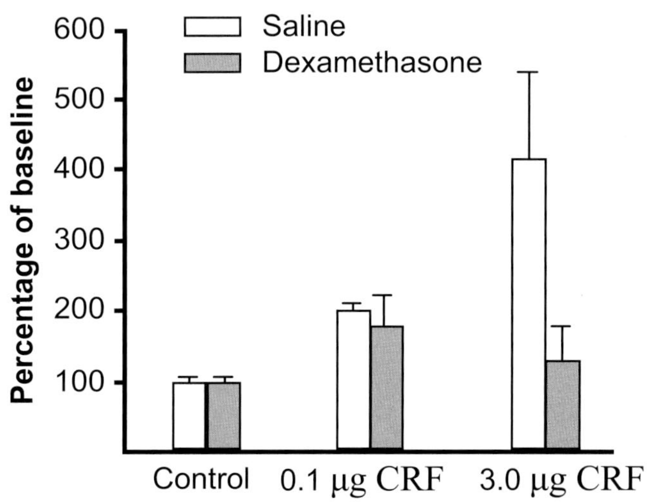

FIGURE 3.6 Mean (±SEM, $n = 5$) percentage increase over the basal concentration of ACTH after the injection of saline, 0.1 µg CRF, or 3 µg CRF. Female Wistar rats that had been adrenalectomized 3 weeks earlier were treated with either saline or dexamethasone 3 h before injection of CRF-41. Note that dexamethasone blocked the ACTH response to 3.0 µg of CRF. *Reproduced from Fink G, Robinson ICAF, Tannahill LA. Effects of adrenalectomy and glucocorticoids on the peptides, CRF-41, AVP and oxytocin in rat hypophysial portal blood.* J Physiol. *1988;401:329–345, with permission of the authors and Cambridge University Press.*

situation AVP is the regulatory or signaling neurohormone, whereas CRF-41 is the "permissive" neurohormone. In contrast, our studies on the long-term effects of corticosterone administered by a subcutaneous pellet suggested that long-term negative glucocorticoid feedback is due to a decreased release of CRF-41, as well as AVP, into portal blood.[52] That is, both stress neurohormones are sensitive to the long-term effects of corticosteroids.

Our findings agree broadly with those of Plotsky and associates, which showed that pharmacological "adrenalectomy" with metyrapone and aminoglutethimide (which block glucocorticoid biosynthesis) resulted, after 3 days, in a significant increase in the release of both CRF and AVP into hypophysial portal blood.[53] The intravenous infusion of corticosterone inhibited nitroprusside-evoked CRF release into portal blood, but had no effect on AVP release.[54]

Data in the rat are complemented by those obtained in the sheep. Thus, the concentrations of CRF-41 and AVP in hypophysial portal blood collected from conscious sheep were (a) similar to those in the rat, (b) increased by volume depletion, fear-associated audiovisual stimuli and by insulin-induced hypoglycemia, and (c) inhibited by dexamethasone.[55,56]

GLUCOCORTICOID FEEDBACK EFFECTS ON STRESS NEUROHORMONE BIOSYNTHESIS

Glucocorticoids have potent inhibitory effects on CRF-41 and ACTH biosynthesis and release.[57–66] Thus, adrenalectomy is followed by a significant increase in CRF-41 mRNA levels in the parvocellular PVN, and this increase can be reduced by either corticosterone or dexamethasone. As assessed by CRF-intron (CRFin) *in situ* hybridization, the stimulation of CRF gene transcription can be detected as early as 15–30 minutes after the injection of the glucocorticoid synthesis inhibitor metyrapone. This increase in CRF-41 gene transcription in the PVN was associated with a coincident increase in c-fos mRNA in the PVN. An increase in the levels of CRF-41 mRNA in the PVN after metyrapone injection was delayed by about 60 minutes, possibly a function of the high resting levels of CRF-41 mRNA and the time taken to assemble mRNA from the CRF-41 primary transcript.

The ACTH precursor pro-opiomelanocortin (POMC) has in its promoter one of the best characterized negative glucocorticoid response elements (nGRE).[58] POMC plays an important role in the regulation of the HPA axis (see above). The POMC promoter is stimulated by CRF and repressed by glucocorticoids. ACTH, which is the major POMC gene product in corticotropes, is generated by processing of the 266-amino acid POMC precursor.[58] The nGRE in the POMC promoter plays a key role in glucocorticoid negative feedback action in the HPA.

The effects of corticosterone[67] and dexamethasone are dose dependent, and can be demonstrated by systemic administration of the steroid, as well as by implantation of steroid pellets into the brain. This effect of glucocorticoids is cell-specific, in that the glucocorticoid-induced decrease in CRF-41 mRNA was localized to the dorsomedial parvocellular neurons of the PVN, the major source of CRF-41 fiber projections to the median eminence. In contrast, glucocorticoid increased the levels of CRF-41 mRNA in parvocellular neurons that project to the brainstem and the spinal cord. The implantation of dexamethasone micropellets in cerebral cortex, dorsal hippocampus, ventral subiculum, lateral septum or amygdala had no effect on CRF-41 mRNA levels in the PVN. The molecular mechanism by which glucocorticoids regulate CRF-41 gene expression remains unclear. Recent data show that the the repressor isoform of the cAMP response element modulator (CREM) is involved in CRF-41 gene regulation, but that stress-induced glucocorticoids do not limit CRF-41 gene transcription.[68] The latter may also be affected by GABAergic, glutamatergic and monoaminergic projections to the PVN from the forebrain and hindbrain limbic system (see also "Glucocorticoid Negative Feedback at the Pituitary Level," below).

The concentration of AVP mRNA in the dorsomedial parvocellular neurons of the PVN parallel those of CRF-41 mRNA – i.e., levels increased after adrenalectomy and decreased after treatment with glucocorticoids.[69] In unstressed intact rats, PVN CRF hnRNA, but not AVP hnRNA, showed a clear circadian rhythm that was correlated with plasma corticosterone concentrations. However, AVP hnRNA levels in the PVN did show a circadian rhythm in adrenalectomized rats that was moderated by corticosterone.[66] Corticosterone treatment also produced a modest reduction in the levels of enkephalin mRNA in all four parts of the PVN, but glucocorticoid manipulation had no significant effect on the PVN concentrations of the mRNAs for angiotensin, cholecystokinin, preprotachykinin and tyrosine hydroxylase.

Data on the release and synthesis of stress neurohormones suggest that even though corticosteroid receptors are present in high concentration in parts of the brain remote from the hypothalamus, including the hippocampus, amygdala and the monoaminergic nuclei of the hindbrain, central inhibition of CRF-41 and AVP synthesis and release may be due in large part to corticosteroid action mainly at the level of the PVN (but see "Possible Role of 11β-Hydroxysteroid Dehydrogenase," below).

ROLE OF HIPPOCAMPUS AND AMYGDALA IN GLUCOCORTICOID NEGATIVE FEEDBACK

It has long been assumed that the hippocampus and the amygdala, major components of the forebrain limbic system, play a key role in the stress response. The forebrain limbic system together with the thalamus and neocortex and brainstem structures forms an analyzer–integrator system involved in neuroendocrine control. The hypothalamus can be considered an intermediate relay station in the reciprocal circuits between the limbic forebrain and brainstem structures, receiving inputs from both. The forebrain limbic system is thought to be involved in stressors that require analysis by higher brain structures ("processive" stressors), whereas direct brainstem projections to the PVN subserve stressors that pose an immediate physiological threat ("systemic" stressors).[70–74]

Projections to the PVN from the hippocampus are mainly multisynaptic, involving (a) the hippocampal fimbria–fornix system including the lateral septum, bed nucleus of the stria terminalis (BNST) and anterior hypothalamus, which all project to the parvocellular PVN, and (b) the medial corticohypothalamic tract from the anteroventral subiculum to the ventromedial, arcuate and suprachiasmatic nuclei of the hypothalamus. The PVN also receive direct projections from the amygdala, as well as projections that relay in the BNST (Fig. 3.7).

Many but not all studies in which HPA activity was assessed by the assay of corticosterone, and, less frequently, ACTH, suggest that the hippocampus inhibits HPA activity. Hippocampal inhibition of the

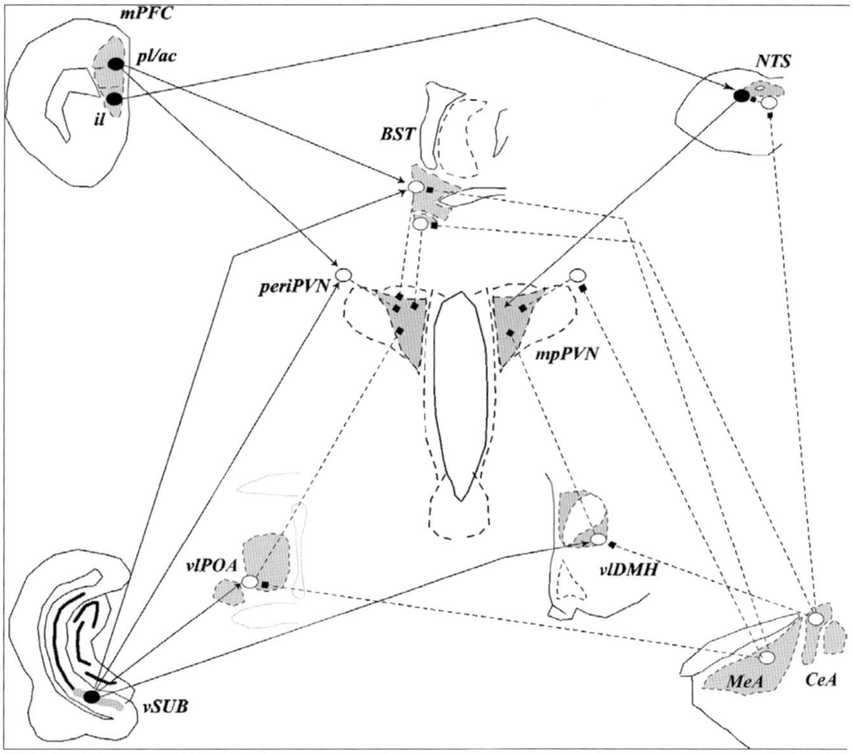

FIGURE 3.7 **Diagrammatic representations of limbic stress-integrative pathways from the prefrontal cortex, amygdala and hippocampus.** The medial prefrontal cortex (mPFC) subsumes neurons of the prelimbic (pl), anterior cingulate (ac) and infralimbic cortices (il), which appear to have different actions on the HPA axis stress response. The pl/ac send excitatory projections (designated as dark circles, filled line with arrows) to regions such as the peri-PVN zone and bed nucleus of the striaterminalis (BST), both of which send direct GABAergic projections to the medial parvocellular PVN (delineated as open circles, dotted lines ending in squares). This two-neuron chain is likely to be inhibitory in nature. In contrast, the infralimbic cortex projects to regions such as the nucleus of the solitary tract (NTS), which sends excitatory projections to the PVN, implying a means of PVN excitation from this cortical region. The ventral subiculum (vSUB) sends excitatory projections to numerous subcortical regions, including the posterior BST, peri-PVN region, ventrolateral region of the medial preoptic area (vlPOA) and ventrolateral region of the dorsomedial hypothalamic nucleus (vlDMH), all of which send GABAergic projections to the PVN and are likely to communicate transynaptic inhibition. The medial amygdaloid nucleus (MeA) sends inhibitory projections to GABAergic PVN-projecting populations, such as the BST, vlPOA and peri-PVN, eliciting a transynaptic disinhibition. A similar arrangement likely exists for the central amygdaloid nucleus (CeA), which sends GABAergic outflow to the ventrolateral BST and to a lesser extent, the vlDMH. The CeA also projects to GABAergic neurons in the NTS, which may disinhibit ascending projections to the PVN. *Reproduced from Herman JP, Ostrander MM, Mueller NK, Figueiredo H. Limbic system mechanisms of stress regulation: hypothalamo-pituitary–adrenocortical axis.* Prog Neuropsychopharmacol Biol Psychiatry *2005;29(8):1201–1213, with permission.*

HPA appears to be due mainly to corticosteroid negative feedback inhibition, although there is also evidence that the hippocampus may exert an inhibitory tone on the HPA independently of corticosteroid feedback. However, the hippocampus is neither the only nor necessarily the major site of corticosteroid negative feedback, as removal of its input to the hypothalamus reduces, but does not abolish, the efficacy of corticosteroid inhibition. Thus, for example: (a) although electrical stimulation of the PVN resulted in a two- to threefold increase in CRF-41 release into hypophysial portal blood, stimulation of the hippocampus had no effect on the release of CRF-41, AVP, or oxytocin;[36] (b) transection of the fornix, the major hippocampal–hypothalamic connection, had no significant effect on either basal or stress (nitroprusside-induced hypotension) evoked CRF-41 release into hypophysial portal blood; and (c) implantation of dexamethasone pellets in hippocampus had no effect on CRF-41 synthesis in the PVN. Nevertheless, fornix section did elevate AVP concentrations in portal blood and blocked corticosterone reduction of elevated CRF, but not AVP, levels in portal blood during hypotensive stress.[75]

Further evidence that the hippocampus normally moderates the synthesis of stress neurohormones is suggested by the finding that lateral fimbria–fornix lesions increased CRF-41 mRNA and AVP mRNA in medial parvocellular PVN and plasma ACTH concentrations.

Feldman and Weidenfeld[76] have shown that in freely moving male rats bearing cholesterol implants in the hippocampus, photic and acoustic stimuli depleted the CRF-41 content of the median eminence with a concomitant increase in plasma ACTH and corticosterone levels. This was inhibited by the systemic administration of dexamethasone, an effect inhibited by hippocampal implants of glucorticoid and, to a lesser extent, mineralocorticoid receptor antagonists. Thus, in freely moving conscious animals the hippocampus may play a role in corticosteroid feedback moderation of ACTH release in response to photic and acoustic stimuli.

Herman and associates have underscored the role of the BNST as a nucleus that integrates or processes hippocampal and amygdaloid modulation of the HPA. This hypothesis is based, first, on the anatomical connections of the BNST, in that the nucleus receives rich projections from the amygdala and hippocampus and projects to the parvocellular PVN. Second, lesion of the anterior BNST resulted in a 30% decrease in CRF-41 mRNA levels in the PVN, whereas lesion of the posterior BNST resulted in a 13% increase in the level of CRF mRNA in the PVN.[29,64] On the basis of these data, Herman and colleagues inferred that the anterior BNST integrates excitatory inputs mainly from the amygdala, whereas the posterior BNST integrates inhibitory inputs mainly from the hippocampus.[29,77] Lesion studies of the amygdala showed that glucocorticoid feedback was unaffected by lesions of the amygdala.[78]

In summary, the hippocampus plays an important role in moderating the HPA both by mediating corticosteroid negative feedback and by exerting an endogenous inhibitory tone on the HPA. Corticosteroid negative feedback is also exerted directly on the PVN and the pituitary gland. The hippocampus is a heterogeneous structure; thus, for example, Dunn and Orr found that electrical stimulation of the CA1 region increased corticosterone levels, whereas stimulation of the CA3, dentate and subiculum decreased plasma corticosterone concentrations.[79] This heterogeneity will need to be considered in the design of further studies on the role of the hippocampus in negative feedback control of the HPA.

GLUCOCORTICOID NEGATIVE FEEDBACK AT THE PITUITARY LEVEL

As well as exerting a central effect, corticosteroids inhibit ACTH synthesis and release by a profound action at the level of the anterior pituitary gland. This is illustrated by Fig. 3.6, in which dexamethasone completely blocked the ACTH response to a bolus injection of CRF-41.[28] As already mentioned in "Glucocorticoid Feedback Effects on Stress Neurohormone Biosynthesis," above, the ACTH precursor proopiomelanocortin (POMC) has in its promoter one of the best-characterized negative glucocorticoid response elements (nGRE).[58] POMC plays an important role in the regulation of the HPA axis (see above).

Studies on dispersed pituitary cells with inhibitors of mRNA and protein synthesis have shown that both the rapid and the delayed glucocorticoid inhibition of ACTH release depend upon mRNA and protein synthesis. Glucocorticoids exert potent inhibitory effects on the expression of POMC in the anterior lobe of the pituitary gland. In male Sprague-Dawley rats, transcription assays showed that dexamethasone inhibited POMC gene transcription by 10-fold within 30 minutes of a single injection of the glucocorticoid. Inhibition of POMC transcription was paralleled by a dramatic fall in plasma ACTH concentrations. The same study showed that CRF-41 stimulated POMC transcription by nearly twofold within 15 minutes, which coincided with a massive increase in ACTH release. The action of dexamethasone is cell-specific in that the steroid had no effect on the transcription rate of POMC in primary cultures of neurointermediate lobe cells.[80,81]

Studies in transgenic mice showed that no more than 769 base pairs of the rat POMC promoter are required for cell-specific expression and glucocorticoid inhibition of the POMC gene in the anterior pituitary gland.[63] A good deal is known about the glucocorticoid response elements in the POMC promoter and the transcription factors involved in POMC expression, but the precise molecular mechanism of glucocorticoid inhibition of POMC transcription remains to be elucidated.

POSSIBLE ROLE OF 11β-HYDROXYSTEROID DEHYDROGENASE

Two isozymes of 11β-hydroxysteroid dehydrogenase play a pivotal role in glucocorticoid synthesis and metabolism. 11beta-HSD type 1 (11β-HSD1) is an isozyme that predominantly catalyzes the reduction of inert cortisone to active cortisol in intact cells and organs. 11β-HSD type 2 (11β-HSD2) is an isozyme that catalyzes the rapid dehydrogenation of active cortisol to inert cortisone (11-dehydrocorticosterone). These two isozymes can play a key role in glucocorticoid feedback. For details, the reader is referred to the reviews by Seckl and colleagues.[82,83]

FUNCTIONAL IMPORTANCE OF GLUCOCORTICOID NEGATIVE FEEDBACK

Glucocorticoid feedback inhibition of ACTH release protects the organism against the deleterious effects of hypercortisolemia (excessive concentrations of cortisol in blood). Whether due to endocrine disorders such as Cushing's syndrome, or other causes such as trauma or chronic stress, hypercortisolemia is associated with at least three major deleterious effects. First, it suppresses the immune-inflammatory defense system, and so incapacitates the animal's ability to respond to infection by pathogenic microorganisms, or to chemical or physical insult. Second, persistent hypercortisolemia has major adverse effects on intermediary metabolism, resulting eventually in all the features of Cushing's syndrome: that is, android obesity, diabetes mellitus, hyperlipidemia, hypertension and osteoporosis. Third, hypercortisolemia and/or stress are thought to be associated with reduction in hippocampal volume ("atrophy") in several neuropsychiatric disorders, such as depression and post-traumatic stress disorder, as well as in Cushing's syndrome. In all three disorders, the hippocampal atrophy is associated with explicit memory deficits. For details regarding these stress-induced cognitive deficits and the mechanism of action of stress or hypercortisolemia on the hippocampus, the reader is referred to papers by Sapolsky.[84–86] Notwithstanding the technical excellence of the experimental work that has demonstrated the neurotoxicity of sustained high levels of glucocorticoids, especially in the hippocampus, translation of these findings to the human in terms of reduced hippocampal size as assessed by brain imaging might require further study.[87]

CLINICAL MANIFESTATIONS OF DISORDERED GLUCOCORTICOID FEEDBACK REGULATION OF THE HPA SYSTEM

No attempt is made here to give a detailed account of the clinical effects of disruption of feedback control within the HPA system (see Chapter 29). Rather, we consider two clinical examples that underscore the principles of negative feedback and illustrate the consequences of disruption of normal feedback control in the HPA system. The first example is of enzyme defect in the adrenal cortex that results in the absence or deficiency of the afferent glucocorticoid limb of the HPA negative feedback system, whereas the second is probably due to an alteration in the central set point of the negative feedback control system.

Congenital Adrenal Hyperplasia: Failure of Glucocorticoid Negative Feedback

There are several types of inherited enzymatic defects in cortisol synthesis known to result in congenital adrenal hyperplasia (CAH), also known as the adrenogenital syndrome. By far the most common form is due to a deficiency of $P450c_{21}$ (21-hydroxylase; see Fig. 3.8),[88] which leads to a deficiency in cortisol biosynthesis. Excessive androgen secretion results from a failure of glucocorticoid negative feedback and consequent, uncontrolled, high ACTH secretion. Excessive androgen levels may lead to virilization of females *in utero*. About two-thirds of patients also have mineralocorticoid deficiency, resulting in salt wasting. If not obvious during the neonatal period, androgen excess may appear in early infancy, resulting in sexual precocity in boys, and clitoral enlargement and pubic hair growth in girls. Excess androgen accelerates linear growth and epiphyseal closure, leading ultimately to diminished adult height. In adult women with untreated CAH, reproductive function is impaired due to (a) the disturbance of normal menstrual cycles as a consequence of the high plasma progesterone and androgen concentrations, and (b) labial fusion, which prevents successful coitus. The former can be corrected by glucocorticoid replacement therapy whereas the latter can be treated surgically.

The $P450c_{21}$ deficiency is transmitted as a single gene autosomal recessive trait linked to the major histocompatibility complex locus on the short arm of chromosome 6. An allelic variable of classical 21-hydroxylase

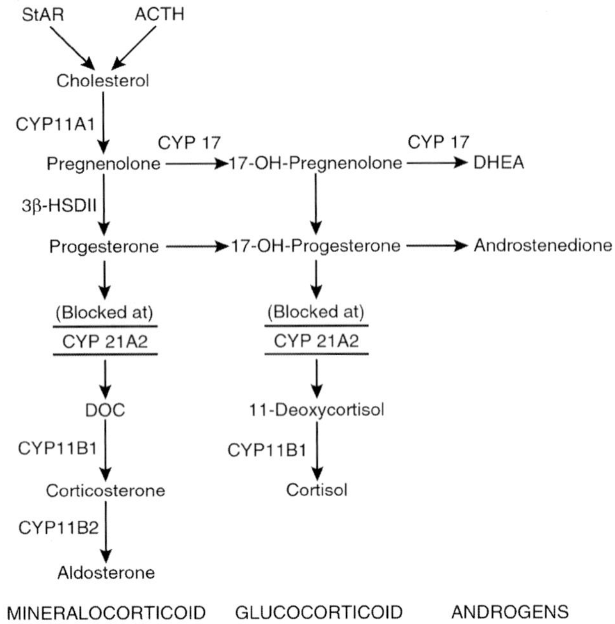

FIGURE 3.8 Congenital adrenal hyperplasia related to 21-hydroxylase deficiency. The normal synthesis of cortisol is impaired, and adrenocorticotropic hormone (ACTH) levels increase because of loss of normal negative feedback inhibition resulting in an increase in adrenal steroid precursors proximal to the block. The results are cortisol deficiency, mineralocorticoid excess related to excessive deoxycorticosterone (DOC) secretion, and excessive secretion of adrenal androgens. DHEA, dehydroepiandrosterone; StAR, steroidogenic acute regulatory protein. *Reproduced from Stewart PM. The adrenal cortex. In: Reed Larsen P, Kronenberg HM, Melmed S, Polonsky KS, eds.* Williams Textbook of Endocrinology, *10th edn. Philadelphia: Saunders; 2003:367–404, with permission.*

deficiency results in a late-onset type, which frequently presents with clinical features similar to those of polycystic ovarian disease.

Deficiency of P450c$_{11}$ (11β-hydroxylase) is a much less common cause of CAH. As in the case of P450c$_{21}$ deficiency, it is transmitted as an autosomal recessive disorder, but is not linked to the HLA locus. Also as in the case of P450c$_{21}$ deficiency, a deficiency in P450c$_{11}$ results in impaired glucocorticoid feedback and a consequent hypersecretion of ACTH and adrenal androgens. The condition is treated with glucocorticoid replacement therapy.

Much rarer forms of CAH are produced by deficiencies of 17α-hydroxylase and 3β-hydroxysteroid dehydrogenase, which result in defective adrenal androgen, as well as glucocorticoid secretion.

Hypercortisolemia in Major Depression: Possibly Due to an Altered Set Point in Glucocorticoid Negative Feedback

Major depressive disorder is characterized by a significant increase in plasma cortisol concentrations (hypercortisolemia), which is most prominent at the nadir of the circadian rhythm, toward midnight. It was first thought that hypercortisolemia and resistance to the suppression of endogenous cortisol secretion by dexamethasone were specific features of major depression, which led to the hope that the dexamethasone suppression test could be used as a specific biological marker of depression. However, extensive studies have shown that hypercortisolemia and resistance to dexamethasone suppression are also associated with other types of psychoses, such as schizoaffective disorder and organic dementia, including Alzheimer's disease.[89–91]

Hypercortisolemia in major depression is associated with a three-fold increase in the mean plasma concentration of β-endorphin.[92] In fact, resistance of β-endorphin to dexamethasone suppression appears to be a more robust marker of major depression than cortisol. Thus, in a study by Young and associates of 73 patients with major depressive disorder, 39 (53%) showed β-endorphin "non-suppression" to dexamethasone while only 8 (11%) showed cortisol "non-suppression".[93] These findings suggest that hypercortisolemia in major depression is due to a resistance of the brain–pituitary–ACTH module to glucocorticoid negative feedback: that is, an elevation of the set point for glucocorticoid feedback. The precise mechanism remains to be determined, but decreased responsiveness of the limbic system, PVN and/or pituitary gland to glucocorticoid negative feedback is a likely explanation. Reduced responsiveness of the PVN could be caused by transynaptic changes triggered by changes in function of the limbic system and frontal cortex. Because the serotonergic raphe neurons determine the amplitude of the circadian excursions of plasma ACTH and corticosterone, it is also conceivable that hypercortisolemia reflects dysregulation of serotonergic function, which seems to occur in major depression.

HPA FEEDBACK CONTROL: SUMMARIZED

The HPA system, together with the sympathetic–medullary system, plays a pivotal role in the neuroendocrine response to stress. Homeostasis within the HPA is maintained by a precise negative feedback system by which the adrenal glucocorticoids (the afferent limb) – cortisol in humans or corticosterone in rodents – moderate ACTH synthesis and release (efferent limb). Allostasis – that is, change in HPA activity to cope with increased stress load – is brought about by change in feedback set point. The major sites of negative feedback are the PVN, where glucocorticoids inhibit CRF and AVP synthesis and release, and the pituitary gland, where they block the ACTH response to CRF and inhibit

POMC/ACTH synthesis. The limbic system of the brain, especially the hippocampus and amygdala, plays an important role in glucocorticoid negative feedback.

Disruption of the HPA negative feedback system has serious deleterious effects, a point illustrated by the congenital adrenogenital syndrome and hypercortisolemia associated with serious mental illnesses. The adrenogenital syndrome, due to defective or absent cortisol secretion (loss of the afferent limb of the feedback system) consequent on a congenital enzyme defect in the adrenal cortex, results in massive uncontrolled pituitary ACTH release. The latter induces excessive androgen production, which in turn causes precocious puberty in males and masculinization of females. Hypercortisolemia, a prominent feature of major depressive disorder and other psychoses and organic dementias, is probably due to elevation of the set point of the glucocorticoid negative feedback in the HPA system. Elucidation of the precise cause of this change in feedback set point may provide insight into the central disorder in depression. Hypercortisolemia may exert adverse effects by (a) inhibiting immune-inflammatory defence mechanisms, (b) disrupting intermediary metabolism, (c) inducing effects akin to Cushing's syndrome that lead to obesity, type 2 diabetes and osteoporosis, and (d) compromising the viability of hippocampal structure, neurogenesis and function that might lead to cognitive impairment.

RELEVANCE FOR THE HYPOTHALAMIC–PITUITARY–THYROID AND –GONADAL AXES: SET POINTS

The principles set out in the HPA feedback system apply equally to the other two major hypothalamic pituitary axes. That is, for the major part all three systems are predominantly under negative feedback control. Negative feedback is mediated through the target hormones. As is the case for the glucocorticoids, a substantial proportion of the gonadal steroids and the thyroid hormones, tri-iodothyronin (T3) and thyroxin (T4), are bound to plasma proteins: it is only the small fraction of free target hormone that exerts a negative feedback action.

The key variant among the three hypothalamic–pituitary systems is the hypothalamic–pituitary–gonadal axis, in which positive feedback plays an essential role in triggering the ovulatory gonadotropin surge. Thus, in rodents, the arcuate nucleus and the anterior pituitary gland[94] tend to be the main site for estradiol negative feedback, while the preoptic area tends to be the main site of estradiol action in triggering positive feedback and the ovulatory GnRH/gonadotropin surge. For details, the reader is referred to Chapters 5, 9 and 19.

Thyroid negative feedback control is more complex than originally thought. It seems that most of thyroid negative feedback is actioned by an effect on TRH at the level of the paraventricular nucleus.[95] Furthermore, as reported by Ghamari-Langroudi *et al.*[96] TRH gene expression is influenced by leptin. Fasting-induced suppression of thyroid hormone levels is an adaptive response to reduce energy expenditure in both humans and mice. This suppression is mediated by the hypothalamic–pituitary–thyroid axis through a reduction in TRH levels expressed in neurons of the paraventricular nucleus of the hypothalamus. *TRH* gene expression is positively regulated by leptin. Whereas decreased leptin levels during fasting lead to a reduction in *TRH* gene expression, the mechanisms underlying this process are still unclear. Indeed, evidence exists that TRH neurons in the PVN are targeted by leptin indirectly via the arcuate nucleus, whereas correlative evidence for a direct action exists as well. The activity of the hypothalamic–pituitary–thyroid axis is regulated by both direct and indirect leptin regulation, with both leptin and α-MSH inducing significant neuronal activity mediated through a postsynaptic mechanism in TRH-expressing neurons of the PVN.[96]

For reasons outlined in Box 3.1, the nature of set points can be complex and defy precise definition. However, broadly, for the hypothalamic–pituitary adrenal system, the set point is comprised of the paraventricular nucleus and the anterior pituitary gland. In the case of the hypothalamic–pituitary–thyroid system, the set point seems to be comprised mainly of the paraventricular nucleus. The arcuate nucleus plus the anterior pituitary gland seem to comprise the set point for negative feedback in the hypothalamic–pituitary–gonadal system, and the preoptic area plus the anterior pituitary gland comprise the set point for estrogen positive feedback.

Glossary

allostasis Maintaining stability (or homeostasis) through change – term introduced by Sterling and Eyer in 1988 (see ref[11]) to describe cardiovascular adjustments to resting and active states.

homeostasis The maintenance of equilibrium, or constant conditions, in a biological system by means of automatic mechanisms (generally feedback systems) that counteract influences tending toward disequilibrium (term introduced by Walter Cannon in 1932[97]).

hypercortisolemia Excessively high concentrations of adrenal corticosteroids in plasma.

hypophysial portal vessels A small system of venules that connects a primary plexus of capillaries in the median eminence of the hypothalamus (base of the brain) with a secondary plexus of sinusoids in the pituitary gland. Blood flowing in these vessels, which run down the pituitary stalk, transports hypothalamic–pituitary regulatory neurohormones released at nerve

BOX 3.1

HOW DO WE KNOW THE VALIDITY OF THE CONCEPT OF THE "SET POINT"?

Unlike the classical example of the room thermostat, which is an all-in-one command and control center, most biological feedback set points and comparators are more complex.

By Analogy with Body Temperature Control

Consider, for example, body temperature control.[1] The concept of feedback of the thermal state of the body, conveniently represented by a high-weighted core temperature (T(c)) and low-weighted peripheral temperatures (T(s)) is equivalent to the control concept of "auxiliary feedback control," using a main (regulated) variable (T(c)), supported by an auxiliary variable (T(s)). This concept implies neither regulation of T(s) nor feedforward control. Steady-states result in the closed control-loop, when the open-loop properties of the (heat transfer) process are compatible with those of the thermoregulatory processors. No set-point and no comparison of signals (e.g., actual set value) are necessary. Metabolic heat production and sweat production, though receiving the same information about the thermal state of the body, are independent effectors with different thresholds and gains. Coordination between one of these effectors and the vasomotor effector is achieved by the fact that changes in the (heat transfer) process evoked by vasomotor control are taken into account by the metabolic/sweat processor.

Control theory deals primarily with the dynamic properties of control-loops (feedback systems). In contrast to such a feedback control, the simpler direct actuation, modification or triggering of a variable (without any feedback of such an action) is referred to as feedforward control. A synonymous term of feedback control is regulation. The term "regulation" is generally recommended in thermal physiology. Deep body temperatures are not as constant as possibly desired, and to a certain extent are dependent on the amount of environmental and internal disturbances, in spite of regulation or feedback control. This fact was interpreted as a consequence of additional aspects of biological complexity and plasticity. Only recently, it was proposed that the term "homeostasis" would be more adequate than "regulation," in view of the amount of integration and flexibility in biological systems. The sequence of controversies started with the erroneous assumption that every "regulator" needs as input a difference of signals, classically the difference between the actual and the reference value of the regulated variable, and that the aim and the consequence of regulation is a difference of zero, thus defining a "set point". Although it was hypothesized that temperature-insensitive neurons could play the role of the reference generators, there was ample evidence that the steady-states reached in temperature regulation were not due to stable reference generators. To solve this apparent discrepancy, a series of questionable theories, explaining the necessary difference of signals and the observed deviations of body temperature in spite of regulation, were developed, among them an "adjustable set point", a set point defined by the balance of warm and cold sensor signals or at least of signals representing information from sensors with different temperature coefficients, a concept of heat flow regulation, and of "comparators" instead of sensors. The term "set point" still is very popular in thermal physiology, although it was shown early that it is not necessary to explain the functional processes of temperature regulation, and indeed may be erroneous.

Romanovsky[2] argues that such a term is ambiguous and misleading in analyzing the system of temperature regulation. Romanovsky rejects the notion that deep body temperature (Tb) is regulated by a unified system with a single controller:

> It is proposed that Tb is regulated by independent thermoeffector loops, each having its own afferent and efferent branches. The activity of each thermoeffector is triggered by a unique combination of shell and core Tbs. Temperature-dependent phase transitions in thermosensory neurons cause sequential activation of all neurons of the corresponding thermoeffector loop and eventually a thermoeffector response. No computation of an integrated Tb or its comparison with an obvious or hidden set point of a unified system is necessary. Coordination between thermoeffectors is achieved through their common controlled variable, Tb.

Most of those controversies, though not in every detail, can be clarified now. Some physiological experimenters still tend to propose concepts which cannot be backed by the logics and physics of system and control theory, and, on the other hand, there are interested engineers still developing ideas and "models" without taking the physiological reality properly into account. None of the definitions of set point ("steady-state of body temperature," "central reference signal," "thermoneutral" or "null zone," or "thermoeffector threshold") have a biological basis.

Relevance for Neuroendocrine Control Systems

There are similarities in principle in the above examples and feedback control systems within the neuroendocrine systems. Specifically, the "set point" refers to a concept concerning a cluster of neuroanatomical

BOX 3.1 *(cont'd)*

structures whose function is modulated by a variety of neural and endocrine inputs. Within each "set point" complex there does seem to be a rough "benchmark" for determining the output of the main pituitary hormone within a known band width. Furthermore, internal and external environmental factors can alter the "set point." Nevertheless, the precise understanding of neuroendocrine "set points" still offers a major research challenge. In the case of the HPA and hypothalamic–pituitary–thyroid axis, for example, does each paraventricular neuron act independently of all other paraventricular neurons? How do neural or neuroendocrine inputs alter the sensitivity and responsiveness of individual paraventricular neurons, and do the latter respond coherently? In terms of the hypothalamic–pituitary–gonadal system, a large question remains as to how the alleged set points for negative (arcuate neurons in rodents) and positive (preoptic neurons) feedback sense and respond completely oppositely to relatively small differences in estrogen concentration?

References

1. Werner J. System properties, feedback control and effector coordination of human temperature regulation. *Eur J Appl Physiol.* 2010;109(1):13–25.
2. Romanovsky AA. Thermoregulation: some concepts have changed. Functional architecture of the thermoregulatory system. *Am J Physiol Regul Integr Comp Physiol.* 2007;292(1):R37–R46.

terminals in the median eminence of the hypothalamus to the anterior pituitary gland, where they stimulate or inhibit the release of anterior pituitary hormones.

limbic system An extensive brain region that includes the cingulate and parahippocampal cortex, the hippocampus, amygdala, hypothalamus, septal nuclei and other structures. The precise interaction of the several components of the system remains poorly understood, but since the proposal by Papez, the limbic system has been implicated in emotion. The connections between the limbic system and the neocortex, hypothalamus, and the olfactory bulb via the olfactory tract suggest that it serves as an important analyzer–integrator of signals, which it conveys from the neocortex to the hypothalamus. The input of olfactory information to the limbic system has led to its alternative name, the rhinencephalon (olfactory brain), and possibly reflects the strong emotive effects of smell in many animals. The structure and relative size of the limbic system have remained remarkably constant through evolution; in humans it is overshadowed by the massive development of the neocortex.

neurohormones Chemical neurotransmitters released from nerve terminals into the hypophysial portal vessels or the systemic circulation at neurohemal junctions and conveyed to their target cells by the blood stream. This contrasts with "classical neurotransmitters" that reach receptors on target cells by crossing synaptic clefts or neuromuscular junctions.

suprachiasmatic nuclei Two small hypothalamic nuclei located immediately above (dorsal to) the optic chiasm that are responsible for regulating most circadian rhythms of the body.

zeitgeber Generic term for the generator of bodily rhythms (from the German meaning time keeper or pacemaker).

References

1. Fink G. Feedback actions of target hormones on hypothalamus and pituitary with special reference to gonadal steroids. *Annu Rev Physiol.* 1979;41:571–585.
2. Fink G. The G.W. Harris Lecture: Steroid control of brain and pituitary function. *Q. Jf Exp Physiol.* 1988;73:257–293.
3. Fink G. Mechanisms of negative and positive feedback of steroids in the hypothalamic–pituitary system. In: Bittar EE, Bittar N, eds. *Principles of medical biology,* Vol. 10A. New York: JAI Press; 1997:29–100.
4. Zhang Y, Proenca R, Maffei M, Barone M, Leopold L, Friedman JM. Positional cloning of the mouse obese gene and its human homologue. *Nature.* 1994;372:425–432.
5. Fehm HL, Kern W, Peters A. The selfish brain: competition for energy resources. *Prog Brain Res.* 2006;153:129–140.
6. Milhorn HTJ. *The application of control theory to physiological systems.* Philadelphia: Saunders; 1966:386.
7. Yates FE, Maran JW. In: Knobil E, Sawyer WH, eds. *Handbook of Physiology.* Washington, DC: American Physiological Society; 1975:367–404.
8. Herman JP, Cullinan WE. Neurocircuitry of stress: Central control of the hypothalamo-pituitary–adrenocortical axis. *Trends Neurosci.* 1997;20:78–84.
9. Herman JP, Mueller NK. Figueiredo H. Role of GABA and glutamate circuitry in hypothalamo–pituitary–adrenocortical stress integration. *Ann NY Acad Sci.* 2004;1018:35–45.
10. Fink G. The self-priming effect of LHRH: A unique servomechanism and possible cellular model for memory. *Front Neuroendocrinol.* 1995;16:183–190.
11. Sterling P, Eyer J. Allostasis: a new paradigm to explain arousal pathology. In: Fisher S, Reason J, eds. *Handbook of Life Stress, Cognition, and Health.* New York: John Wiley and Sons; 1988:629–649.
12. McEwen BS, Stellar E. Stress and the individual: mechanisms leading to disease. *Arch Intl Med.* 1993;153:2093–3101.
13. Schulkin J. *Rethinking Homeostasis.* Cambridge: MIT Press; 2003.
14. Schulkin J. Social allostasis: anticipatory regulation of the internal milieu. *Front Evol Neurosci.* 2011;31(2):111.
15. Spiga F, Liu Y, Aguilera G, Lightman SL. Temporal effect of adrenocorticotrophic hormone on adrenal glucocorticoid steroidogenesis: involvement of the transducer of regulated cyclic AMP-response element-binding protein activity. *J Neuroendocrinol.* 2011;23(2):136–142.
16. Orth DN, Mount CD. Specific high-affinity binding protein for human corticotropin-releasing hormone in normal human plasma. *Biochem Biophys Res Commun.* 1987;143(2):411–417.

17. Potter E, Behan DP, Fischer WH, Linton EA, Lowry PJ, Vale WW. Cloning and characterization of the cDNAs for human and rat corticotropin releasing factor-binding proteins. *Nature.* 1991;349(6308):423–426.
18. Behan DP, Potter E, Lewis KA, Jenkins NA, Copeland N, Lowry PJ, Vale WW. Cloning and structure of the human corticotrophin releasing factor-binding protein gene (CRHBP). *Genomics.* 1993;16(1):63–68.
19. Behan DP, De Souza EB, Lowry PJ, Potter E, Sawchenko P, Vale WW. Corticotropin releasing factor (CRF) binding protein: a novel regulator of CRF and related peptides. *Front Neuroendocrinol.* 1995;16(4):362–382.
20. Dockray S, Bhattacharyya MR, Molloy GJ, Steptoe A. The cortisol awakening response in relation to objective and subjective measures of waking in the morning. *Psychoneuroendocrinology.* 2008;33:77–82.
21. Chida Y, Steptoe A. Cortisol awakening response and psychosocial factors: a systematic review and meta-analysis. *Biol Psychol.* 2009;80(3):265–278.
22. Dockray S, Steptoe A. Chronotype and diurnal cortisol profile in working women: Differences between work and leisure days. *Psychoneuroendocrinology.* 2011;36(5):649–655.
23. Antoni FA, Fink G, Sheward WJ. Corticotrophin-releasing peptides in rat hypophysial portal blood after paraventricular lesions: a marked reduction in the concentration of corticotrophin-releasing factor-41, but no change in vasopressin. *J Endocrinol.* 1990;125:175–183.
24. Vale W, Spiess J, Rivier C, Rivier J. Characterization of a 41 residue ovine hypothalamic peptide that stimulates the secretion of corticotropin and beta-endorphin. *Science.* 1981;213:1394–1397.
25. Bale TL, Vale WW. CRF and CRF receptors: role in stress responsivity and other behaviors. *Annu Rev Pharmacol Toxicol.* 2004;44:525–557.
26. Antoni FA. Hypothalamic control of adrenocorticotropin secretion: advances since the discovery of 41-residue corticotropin-releasing factor. *Endocr Rev.* 1986;7(4):351–378.
27. Antoni FA. Vasopressinergic control of pituitary adrenocorticotropin secretion comes of age. *Front Neuroendocrinol.* 1993;14(2):76–122.
28. Fink G, Robinson ICAF, Tannahill LA. Effects of adrenalectomy and glucocorticoids on the peptides, CRF-41, AVP and oxytocin in rat hypophysial portal blood. *J Physiol.* 1988;401:329–345.
29. Herman JP, Cullinan WE, Young EA, Akil H, Watson SJ. Selective forebrain fiber tract lesions implicate ventral hippocampal structures in tonic regulation of paraventricular nucleus corticotropin-releasing hormone (CRH) and arginine vasopressin (AVP) mRNA expression. *Brain Res.* 1992;592(1-2):228–238.
30. Herman JP, Mueller NK, Figueiredo H. Role of GABA and glutamate circuitry in hypothalamo-pituitary–adrenocortical stress integration. *Ann NY Acad Sci.* 2004;1018:35–45.
31. Choi DC, Furay AR, Evanson NK, Ostrander MM, Ulrich-Lai YM, Herman JP. Bed nucleus of the stria terminalis subregions differentially regulate hypothalamic-pituitary-adrenal axis activity: implications for the integration of limbic inputs. *J Neurosci.* 2007;27(8):2025–2034.
32. Lowry CA. Functional subsets of serotonergic neurones: implications for control of the hypothalamic–pituitary–adrenal axis. *J Neuroendocrinol.* 2002;14(11):911–923.
33. Herman JP, Figueiredo H, Mueller NK, Ulrich-Lai Y, Ostrander MM, Choi DC, Cullinan WE. Central mechanisms of stress integration: hierarchical circuitry controlling hypothalamo-pituitary–adrenocortical responsiveness. *Front Neuroendocrinol.* 2003;24(3):151–180.
34. Choi DC, Furay AR, Evanson NK, Ulrich-Lai YM, Nguyen MM, Ostrander MM, Herman JP. The role of the posterior medial bed nucleus of the stria terminalis in modulating hypothalamic–pituitary–adrenocortical axis responsiveness to acute and chronic stress. *Psychoneuroendocrinology.* 2008;33(5):659–669.
35. Herman JP, Ostrander MM, Mueller NK, Figueiredo H. Limbic system mechanisms of stress regulation: hypothalamo–pituitary–adrenocortical axis. *Prog Neuropsychopharmacol Biol Psychiatry.* 2005;29(8):1201–1213.
36. Tannahill LA, Sheward WJ, Robinson ICAF, Fink G. Corticotrophin-releasing factor-41, vasopressin and oxytocin release into hypophysial portal blood in the rat; effects of electrical stimulation of the hypothalamus, amygdala and hippocampus. *J Endocrinol.* 1991;129:99–107.
37. Young EA, Veldhuis JD. Disordered adrenocorticotropin secretion in women with major depression. *J Clin Endocrinol Metab.* 2006;91(5):1924–1928.
38. Van Cauter E, van Coevorden A, Blackman JD. Modulation of neuroendocrine release by sleep and circadian rhythmicity. In: Yen SSC, Vale WW, eds. *Neuroendocrine regulation of reproduction.* Norwell: Serono Symposia USA; 1990:113–122.
39. Moore RY, Eichler VB. Loss of a circadian adrenal corticosterone rhythm following suprachiasmatic lesions in the rat. *Brain Res.* 1972;42:201–206.
40. Hastings M, O'Neill JS, Maywood ES. Circadian clocks: regulators of endocrine and metabolic rhythms. *J Endocrinol.* 2007;195(2):187–198.
41. Karatsoreos IN, Silver R. Minireview: The neuroendocrinology of the suprachiasmatic nucleus as a conductor of body time in mammals. *Endocrinology.* 2007;148(12):5640–5647.
42. Hayasaka N, Aoki K, Kinoshita S, Yamaguchi S, Wakefield JK, Tsuji-Kawahara S, Horikawa K, Ikegami H, Wakana S, Murakami T, Ramabhadran R, Miyazawa M, Shibata S. Attenuated food anticipatory activity and abnormal circadian locomotor rhythms in Rgs16 knockdown mice. *PLoS One.* 2011;6(3):e17655.
43. Hay-Schmidt A, Vrang N, Larsen PJ, Mikkelsen JD. Projections from the raphe nuclei to the suprachiasmatic nucleus of the rat. *J Chem Neuroanat.* 2003;25(4):293–310.
44. Moore RY, Speh JC, Leak RK. Suprachiasmatic nucleus organization. *Cell Tissue Res.* 2002;309(1):89–98.
45. Krout KE, Kawano J, Mettenleiter TC, Loewy AD. CNS inputs to the suprachiasmatic nucleus of the rat. *Neuroscience.* 2002;110(1):73–92.
46. De La Iglesia HO, Blaustein JD, Bittman EL. Oestrogen receptor-alpha-immunoreactive neurones project to the suprachiasmatic nucleus of the female Syrian hamster. *J Neuroendocrinol.* 1999;11(7):481–490.
47. Pickard GE, Rea MA. Serotonergic innervation of the hypothalamic suprachiasmatic nucleus and photic regulation of circadian rhythms. *Biol Cell.* 1997;89(8):513–523.
48. Moga MM, Moore RY. Organization of neural inputs to the suprachiasmatic nucleus in the rat. *J Comp Neurol.* 1997;389(3):508–534.
49. Miller JD, Morin LP, Schwartz WJ, Moore RY. New insights into the mammalian circadian clock. *Sleep.* 1996;19(8):641–667.
50. Maxwell RC, Fink G. The connections between the suprachiasmatic, ventrolateral geniculate and raphe nuclei studied by uptake of [14C]2-deoxyglucose. *Neuroscience* 1988;24(1):265–274.
51. Heydendael W, Jacobson L. Glucocorticoid status affects antidepressant regulation of locus coeruleus tyrosine hydroxylase and dorsal raphé tryptophan hydroxylase gene expression. *Brain Res.* 2009;1288:69–78.
52. Sheward WJ, Fink G. Effects of corticosterone on the secretion of corticotrophin-releasing factor, arginine vasopressin and oxytocin into hypophysial portal blood in long-term hypophysectomized rats. *J Endocrinol.* 1991;129:91–98.

53. Plotsky PM, Sawchenko PE. Hypophysial portal plasma levels, median eminence content, and immunohistochemical staining of corticotropin-releasing factor, arginine vasopressin, and oxytocin after pharmacological adrenalectomy. *Endocrinology.* 1987; 120:1361–1369.
54. Plotsky PM, Otto S, Sapolsky RM. Inhibition of immunoreactive corticotropin-releasing factor secretion into the hypophysial-portal circulation by delayed glucocorticoid feedback. *Endocrinology.* 1986;119:1126–1130.
55. Canny BJ, Funder JW, Clarke IJ. Glucocorticoids regulate ovine hypophysial portal levels of corticotropin-releasing factor and arginine vasopressin in a stress-specific manner. *Endocrinology.* 1989;125(5):2532–2539.
56. Canny BJ, Clarke IJ, Funder JW. Adrenocorticotropin responses to endogenous secretagogues in the sheep: Specificity of glucocorticoid action. *Neuroendocrinology.* 1990; 51:181–189.
57. Drouin J, Sun YL, Chamberland M, et al. Novel glucocorticoid receptor complex with DNA element of the hormone-repressed POMC gene. *EMBO J.* 1993;12:145–156.
58. Dostert A, Heinzel T. Negative glucocorticoid receptor response elements and their role in glucocorticoid action. *Curr Pharm Des.* 2004;10(23):2807–2816.
59. Imaki T, Shibasaki T, Demura H. Regulation of gene expression in the central nervous system by stress: molecular pathways of stress responses. *Endocr J.* 1995;42(2):121–130.
60. Kovacs KJ, Mezey E. Dexamethasone inhibits corticotropin-releasing factor gene expression in the rat paraventricular nucleus. *Neuroendocrinology* 1987;46:365–368.
61. Swanson LW, Simmons DM. Differential steroid hormone and neural influences on peptide mRNA levels in CRH cells of the paraventricular nucleus: A hybridization histochemical study in the rat. *Journal Comp Neurol.* 1989;285:413–435.
62. Watts AG, Swanson LW. Diurnal variations in the content of preprocorticotropin-releasing hormone messenger ribonucleic acids in the hypothalamic paraventricular nucleus of rats of both sexes as measured by *in situ* hybridization. *Endocrinology.* 1989;125(3):1734–1738.
63. Tremblay Y, Tretjakoff L, Peterson A, Antakly T, Zhang CX, Drouin J. Pituitary-specific expression and glucocorticoid regulation of a proopiomelanocortin fusion gene in transgenic mice. *Proc Natl Acad Sci USA.* 1988;85:8890–8894.
64. Herman JP, Watson SJ. Glucocorticoid regulation of stress-induced mineralocorticoid receptor gene transcription *in vivo.* *Ann NY Acad Sci.* 1994;746:485–488.
65. Herman JP. Regulation of adrenocorticosteroid receptor mRNA expression in the central nervous system. *Cell Mol Neurobiol.* 1993;13(4):349–372.
66. Watts AG, Tanimura S, Sanchez-Watts G. Corticotropin-releasing hormone and arginine vasopressin gene transcription in the hypothalamic paraventricular nucleus of unstressed rats: daily rhythms and their interactions with corticosterone. *Endocrinology.* 2004;145(2):529–540.
67. Spiga F, Lightman SL. Dose-dependent effects of corticosterone on nuclear glucocorticoid receptors and their binding to DNA in the brain and pituitary of the rat. *Brain Res.* 2009;13(1293): 101–107. 69.
68. Shepard JD, Liu Y, Sassone-Corsi P, Aguilera G. Role of glucocorticoids and cAMP-mediated repression in limiting corticotropin-releasing hormone transcription during stress. *J Neurosci.* 2005;25(16):4073–4081.
69. Pace TW, Gaylord RI, Jarvis E, Girotti M, Spencer RL. Differential glucocorticoid effects on stress-induced gene expression in the paraventricular nucleus of the hypothalamus and ACTH secretion in the rat. *Stress.* 2009;12(5):400–411.
70. MacLean PD. The limbic system and its hippocampal formation: studies in animals and their possible application to man. *J Neurosurg.* 1954;11:29–44.
71. MacLean PD. Contrasting functions of limbic and neocortical systems of the brain and their relevance to psychophysiological aspects of medicine. *Am J Med.* 1958;25:611–626.
72. Nauta WJH. Central nervous organization and the endocrine motor system. In: Nalbandov AV, ed. *Advances in neuroendocrinology.* Urbana: University of Illinois Press; 1963:5–21.
73. Sawchenko PE, Swanson LW. The organization of forebrain afferents to the paraventricular and supraoptic nuclei of the rat. *J Comp Neurol.* 1983;218:121–144.
74. Smith OA, L DeVito J. Central neural integration for the control of autonomic responses associated with emotion. *Annu Rev Neurosci.* 1984;7:43–65.
75. Sapolsky RM, Armanini MP, Sutton SW, et al. Elevation of hypophysial portal concentrations of adrenocorticotropin secretagogues after fornix transection. *Endocrinology.* 1989;125: 2881–2887.
76. Feldman S, Weidenfeld J. Glucocorticoid receptor antagonists in the hippocampus modify the negative feedback following neural stimuli. *Brain Res.* 1999;821:33–37.
77. Herman JP, Cullinan WE, Watson SJ. Involvement of the bed nucleus of the stria terminalis in tonic regulation of paraventricular hypothalamic CRH and AVP mRNA expression. *J Neuroendocrinol.* 1994;6(4):433–442.
78. Prewitt CM, Herman JP. Hypothalamo-pituitary-adrenocortical regulation following lesions of the central nucleus of the amygdala. *Stress.* 1997;1(4):263–280.
79. Dunn JD, Orr SE. Differential plasma corticosterone responses to hippocampal stimulation. *Exp Brain Res.* 1984;54:1–6.
80. Gagner J-P, Drouin J. Opposite regulation of pro-opiomelanocortin gene transcription by glucocorticoids and CRH. *Mol Cell Endocrinol.* 1985;40:25–32.
81. Gagner JP, Drouin J. Tissue-specific regulation of pituitary proopiomelanocortin gene transcription by corticotropin-releasing hormone, 3′,5′-cyclic adenosine monophosphate, and glucocorticoids. *Mol Endocrinol.* 1987;1(10):677–682.
82. Seckl JR. 11β-Hydroxysteroid Dehydrogenases. Encylopedia of Stress. In: Fink G, ed. 2nd ed. Elsevier Inc; 2007:368–372.
83. Wyrwoll CS, Holmes MC, Seckl JR. 11β-Hydroxysteroid dehydrogenases and the brain: From zero to hero a decade of progress. *Front Neuroendocrinol.* 2010 Dec 7 [Epub ahead of print].
84. Sapolsky RM. Glucocorticoids, stress, and their adverse neurological effects: relevance to aging. *Exp Gerontol.* 1999;34(6): 721–732.
85. Sapolsky RM. Glucocorticoids and hippocampal atrophy in neuropsychiatric disorders. *Arch Gen Psychiatry.* 2000;57:925–935.
86. Sapolsky RM. Glucocorticoids – Adverse effects on the nervous system. In: Fink G, ed. *Encylopedia of Stress.* 2nd edn. Elsevier Inc; 2007.
87. Fink G. Stress controversies: Posttraumatic stress disorder, hippocampal volume, gastroduodenal ulceration. *J Neuroendocrinol.* 2011;23:107–117.
88. Stewart PM. The adrenal cortex. In: Reed Larsen P, Kronenberg HM, Melmed S, Polonsky KS, eds. *Williams Textbook of Endocrinology.* 10th ed. Philadelphia: Saunders; 2003: 491–551.
89. Christie JE, Whalley LJ, Dick H, Blackwood DH, Blackburn IM, Fink G. Raised plasma cortisol concentrations are a feature of drug-free psychotics and not specific for depression. *Br J Psychiatry.* 1986;148:58–65.
90. Copolov DL, Rubin RT, Stuart GW, Poland RE, Mander AJ, Sashidharan SP, Whitehouse AM, Blackburn IM, Freeman CP,

Blackwood DH. Specificity of the salivary cortisol dexamethasone suppression test across psychiatric diagnoses. *Biol Psychiatry.* 1989;25:879–893.
91. Christie JE, Whalley LJ, Bennie J, Dick H, Blackburn IM, Blackwood DH, Fink G. Characteristic plasma hormone changes in Alzheimer's disease. *Br J Psychiatry.* 1987;150: 674–681.
92. Goodwin GM, Austin M-P, Curran SM, Ross M, Murray C, Prentice N, Ebmeier KP, Bennie J, Carroll S, Dick H, Fink G. The elevation of plasma β-endorphin levels in major depression. *J Affect Disord.* 1993;29:281–289.
93. Young DA, Kotun J, Haskett RF, et al. Dissociation between pituitary and adrenal supression to dexamethasone in depression. *Arch Gen Psychiatry* 1993;50:395–403.
94. Iqbal J, Latchoumanin O, Sari IP, Lang RJ, Coleman HA, Parkington HC, Clarke IJ. Estradiol-17beta inhibits gonadotropin-releasing hormone-induced Ca^{2+} in gonadotropes to regulate negative feedback on luteinizing hormone release. *Endocrinology.* 2009;150(9):4213–4220.
95. Chiamolera MI, Wondisford FE. Minireview: Thyrotropin-releasing hormone and the thyroid hormone feedback mechanism. *Endocrinology.* 2009;150(3):1091–1096.
96. Ghamari-Langroudi M, Vella KR, Srisai D, Sugrue ML, Hollenberg AN, Cone RD. Regulation of thyrotropin-releasing hormone-expressing neurons in paraventricular nucleus of the hypothalamus by signals of adiposity. *Mol Endocrinol.* 2010;24(12):2366–2381.
97. Cannon WB. *The Wisdom of the Body.* New York: Norton; 1932.

CHAPTER

4

Evolution of Reproductive Neurohormones

Graeme J. Roch, Ellen R. Busby, Nancy M. Sherwood

University of Victoria, Department of Biology, Victoria, BC, Canada

OUTLINE

Summary

Gonadotropin-releasing hormone (GnRH) is a small neuropeptide secreted from the hypothalamus to stimulate the release of pituitary gonadotropins that regulate sexual development and function. Along with the four major types of GnRHs isolated from vertebrates, novel forms of the hormone have recently been characterized in a variety of invertebrates. Based on the structural homology of these peptides and their receptors, we now understand that GnRH forms a superfamily with two other invertebrate-specific neurohormones, adipokinetic hormone (AKH) and corazonin. Kisspeptin is another neurohormone, and it binds a receptor located on GnRH neurons, stimulating the release of GnRH. Kisspeptins are expressed throughout the vertebrates but not beyond; however, a homolog of their receptor is found in many invertebrate species, with unknown function. We discuss the evolution of reproductive neurohormones using these two peptides and others, and demonstrate that the pituitary was a vertebrate innovation.

INTRODUCTION

An array of neurohormones influences reproduction, but within the collection, gonadotropin-releasing hormone (GnRH) and kisspeptin (Kiss) are central in day-to-day control of the reproductive axis. Kisspeptin has considerable control over GnRH release at the hypothalamic level. The neurons containing either GnRH or Kiss integrate various inputs from the environment, steroids, and a myriad of nervous and hormonal signals to modulate GnRH release. This chapter considers the origin in evolution of these two hormones and their receptors. Oxytocin is another important reproductive neurohormone that is included along with its receptor in the chapter. The relatively new area of genomics makes available the DNA of entire genomes from select

Handbook of Neuroendocrinology, DOI: 10.1016/B978-0-12-375097-6.10004-6

vertebrate and invertebrate species. Now it is possible to systematically search genomes for genes, including those encoding reproductive hormones and their receptors.

Initially, the evolution of GnRH was considered to be a mammalian story. GnRH was characterized first as a decapeptide in pigs and sheep,[1,2] although we now know the identical structure (GnRH1) exists in humans and other mammals. Indeed, the same structure is also present in amphibians and some species of fish, such as sturgeon. Related forms of GnRH (GnRH2-4) are present throughout the vertebrates with the same length, although up to five amino acid substitutions occur. Also, vertebrates are not restricted to a single form of GnRH within an individual, but usually have two or three distinct forms within the brain. This chapter examines whether GnRH is restricted to vertebrates, or appears earlier in evolution within the invertebrates. The structures of some invertebrate GnRH-like peptides, including the nine tunicate GnRHs, are closely related to the vertebrate peptides, whereas other invertebrate GnRH-like peptides have one or two additional amino acids. In contrast, other neurohormones in invertebrates are related in structure to GnRH, but form individual families that together with GnRHs form a superfamily of neuropeptides. This concept changes our view of GnRH as a solitary family, and reveals a pattern of prolonged GnRH evolution.

A hormone can only effect a change after binding to a receptor; thus, receptors are essential throughout evolution. This expanded the search to finding the structure, location, specificity and action of receptors for each hormone. In truth, it is technically easier to follow the evolution of specific receptors than to follow short neurohormones such as GnRH using genomics. Clearly, the ligand and receptor can change in the course of evolution, but certain key domains for binding to each other and to the transducers of action are more highly conserved. The first GnRH receptor (GnRHR) was identified in a mouse,[3,4] but we now know that GnRH receptors and their putative relatives are found throughout the vertebrates and invertebrates, revealing the essence of these conserved domains.

Kisspeptin, discovered long after GnRH,[5–7] brought clarity to the mechanism of steroid feedback on the reproductive axis. The kisspeptin-expressing neurons in the brain were shown to be the target of steroid feedback, with subsequent control of GnRH release by kisspeptin. This added a new component in the cascade of events in the reproductive system. This chapter examines whether kisspeptin and its receptor emerged before or after GnRH, GnRHR and the sex-steroid receptors. We ask if the kisspeptin system is part of a superfamily, and if it also controls reproduction in invertebrates.

Evidence to date suggests that the classical vertebrate system of brain–pituitary–gonad is only partially based on an earlier reproductive system. We compare the vertebrate, one of its nearest invertebrate neighbors (amphioxus), and a more distantly related invertebrate (octopus) to illustrate the origin of the major neurohormonal and other hormonal components that serve as a foundation for the human reproductive system.

Finally, the origin of the pituitary gland is addressed based on a genomic search for pituitary hormones and their receptors. We consider whether the full cascade of reproductive hormonal components is present in invertebrates, or whether the pituitary arose as a vertebrate innovation. GnRH is well known in vertebrates to bind to its receptors in the pituitary to enhance the synthesis and release of luteinizing hormone (LH) and follicle-stimulating hormone (FSH) from the pituitary. The question of GnRH's role before the pituitary emerged is addressed.

GnRH AND ITS SUPERFAMILY OF PEPTIDES

Physiological Functions of Vertebrate GnRH

GnRH is well known as a hypothalamic neuropeptide involved in the regulation of the reproductive axis. After release from the hypothalamus, GnRH acts on the pituitary to stimulate the release of luteinizing hormone (LH) and follicle-stimulating hormone (FSH), which then induce steroid production and gametogenesis in the gonad. Lack of GnRH signaling results in immature gonads, decreased production of reproductive hormones, and other symptoms which together are defined as hypogonadism (Box 4.1). Studies on animals have also shown a role of GnRH in regulation of reproductive behaviors.[8] In addition to the hypothalamic regulation of reproduction in mammals, GnRH acts in a paracrine or autocrine fashion on reproductive tissues, including ovary, placenta, oviduct and testes, where GnRH binds locally expressed receptors to modulate reproduction.[9]

Skinner *et al.*[10] summarize evidence for the presence of GnRH target organs found outside of the reproductive axis in mammals, including the kidney, liver, heart, adrenal, skeletal muscle, retina, spinal cord, and regions of the brain such as the olfactory bulb and arcuate nucleus. GnRH peptides and receptors have been found in some of these same tissues, including the heart and neurons surrounding the cerebral spinal fluid. Some of the extra-hypothalamic brain tissues play peripheral roles in reproductive behavior, such as a neuromodulatory effect on the olfactory bulb by GnRH in response to

BOX 4.1

HOW DO WE KNOW THAT GnRH AND ITS RECEPTOR ARE ESSENTIAL FOR REPRODUCTION?

In 1977, a spontaneous mutation was reported in the gene encoding GnRH in the laboratory mouse that led to hypogonadism; thereafter, these animals were named hypogonadal (*hpg*) mice.[1] Similar to hypogonadism in humans, these mice possessed immature reproductive organs and greatly reduced reproductive hormone levels, including hypothalamic GnRH, pituitary luteinizing hormone (LH), follicle-stimulating hormone (FSH) and circulating steroid hormones. The *hpg* mouse provided the first opportunity to study the condition of familial hypogonadism in a model animal. Since 1977, many studies have been completed using this model to further describe the role of GnRH in regulating reproduction, such that the resulting phenotype has been more precisely defined as hypogonadotropic hypogonadism.

No one had done similar work with the GnRH receptor to demonstrate its effects on regulating reproduction until 2010, when we reported the effect of complete inactivation in a GnRH receptor knockout mouse.[2] This mouse exhibited very similar traits to the *hpg* mouse, with intact but totally immature reproductive organs (see figure). We also found very low levels of circulating LH, FSH, testosterone, estradiol and progesterone, resulting in reproductive immaturity and infertility. Unlike the *hpg* mouse, the GnRH receptor knockout mouse could be assessed for any effects of external sources of GnRH. Curiously, an intact receptor was not required for fetal testosterone secretion, as these mice showed typical sexual differentiation.

Mice have only one form of GnRH and a single GnRH receptor. Other mammals vary as to the number of GnRH peptides and receptors they express. For example, humans have GnRH1 and a small amount of GnRH2, but only one functional receptor. This knockout of the GnRH receptor gene in mice clarifies the reproductive processes that are blocked by disruption of the GnRH receptor, and allows for the determination of the stages and tissues in which the blockage occurs.

References

1. Cattanach BM, Iddon CA, Charlton HM, Chiappa SA, Fink G. Gonadotrophin-releasing hormone deficiency in a mutant mouse with hypogonadism. *Nature.* 1977;269(5626):338–340.
2. Wu S, Wilson MD, Busby ER, Isaac ER, Sherwood NM. Disruption of the single copy gonadotropin-releasing hormone receptor in mice by gene trap: severe reduction of reproductive organs and functions in developing and adult mice. *Endocrinology.* 2010;151(3):1142–1152.

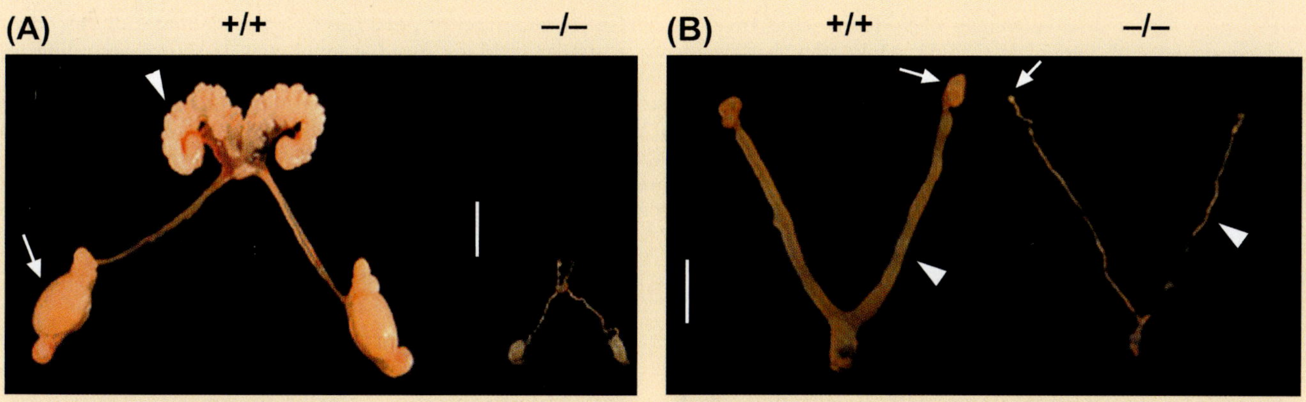

Reproductive organs of 2-month-old mice, wild type (+/+) and GnRH receptor null mice (–/–). **(A)** Testis (white arrow) and seminal vesicle (white arrowhead) from wild-type organs is indicated in the left pane. Null organs are very thin and small on the right. **(B)** Ovaries (arrows) and uterine horns (arrowheads) of null animals (on right) are significantly decreased in size compared with wild type (on left). Scale bars, 1 cm. *Modified from,*[2] *Wu* et al. *(2010), with permission from The Endocrine Society ©2010.*

scents relating to reproduction.[11] In the heart, activation of GnRH receptors may increase the risk of cardiovascular symptoms in patients treated with GnRH agonists.[10] Also, GnRH plays a role in the cardiac development of zebrafish.[12] These actions appear to be unrelated to reproduction, but may provide some insights into the role of GnRH in lineages where its effects are not solely reproductive.

		1	-	-	2	3	4	5	6	7	8	9	10		
GnRH2 Jawed	Vertebrates	pQ	-	-	H	W	S	H	G	W	Y	P	G	amide	V
GnRH1 Mammals		pQ	-	-	H	W	S	Y	G	L	R	P	G	amide	E
GnRH3 Teleost fish		pQ	-	-	H	W	S	Y	G	W	L	P	G	amide	R
GnRH4 Lamprey		pQ	-	-	H	W	S	H	D	W	K	P	G	amide	T
GnRH															
Sea squirt 3	Tunicate	pQ	-	-	H	W	S	Y	E	F	M	P	G	amide	
Sea squirt 6		pQ	-	-	H	W	S	K	G	Y	S	P	G	amide	
GnRH															
Sea urchin	Echinoderm	pQ	V	H	H	R	F	S	G	W	R	P	G	amide	
Octopus	Mollusk	pQ	N	Y	H	F	S	N	G	W	H	P	G	amide	
Aplysia		pQ	N	Y	H	F	S	N	G	W	Y	A	-	amide	
Limpet		pQ	H	Y	H	F	S	N	G	W	K	S	-	amide	
Marine worm	Annelid	pQ	A	Y	H	F	S	H	G	W	F	P	-	amide	I
															N
CORAZONIN															V
Locust	Insect	pQ	T	F	Q	Y	S	H	G	W	T	N	-	amide	E
Crane fly		pQ	T	F	Q	Y	S	R	G	W	Q	N	-	amide	R
Fruit fly/Crab/ *Daphnia*/Shrimp	Insect/Crust.	pQ	T	F	Q	Y	S	R	G	W	T	N	-	amide	T
															E
															B
ACP															R
Wasp	Insect	pQ	-	V	T	F	S	K	G	W	G	P	-	amide	A
Beetle		pQ	-	V	T	F	S	R	D	W	N	P	-	amide	T
Mosquito		pQ	-	V	T	F	S	R	D	W	N	A	-	amide	E
															S
AKH/RPCH															
Locust	Insect	pQ	-	L	N	F	T	P	W	W	-	-	-	amide	
Fruit fly		pQ	-	L	T	F	S	P	D	W	-	-	-	amide	
Mosquito		pQ	-	L	T	F	T	P	S	W	-	-	-	amide	
Daphnia	Crustacean	pQ	-	V	N	F	S	T	S	W	-	-	-	amide	
Stinkbug/Crab	Insect/Crust.	pQ	-	L	N	F	S	P	G	W	-	-	-	amide	
Aplysia	Mollusk	pQ	-	I	H	F	S	P	D	W	G	T	-	amide	
Limpet		pQ	-	I	H	F	S	P	T	W	G	S	-	amide	
C. elegans	Nematode	pQ	-	M	T	F	T	D	Q	W	T	-	-		

FIGURE 4.1 **Amino acid sequences of vertebrate and invertebrate GnRH superfamily peptides.** Colored shading indicates identical residues compared with vertebrate GnRH2, while gray shading identifies similar residues. Residue numbering is shown according to the vertebrate GnRH peptide sequence. Representative species from each invertebrate peptide group are listed below in order of appearance in figure: lamprey (*Petromyzon marinus*), tunicate (*Ciona intestinalis*), sea urchin (*Strongylocentrotus purpuratus*), octopus (*Octopus vulgaris*), sea hare (*Aplysia californica*), owl limpet (*Lottia gigantea*), annelid marine worm (*Capitella teleta*), locust (*Locusta migratoria*), crane fly (*Tipula* sp.), fruit fly (*Drosophila melanogaster*), crab (*Cancer borealis*), water flea (*Daphnia pulex*), shrimp (*Penaeus monodon*), wasp (*Nasonia vitripennis*), beetle (*Tribolium castaneum*), mosquito (*Anopheles gambiae*), stink bug (*Nezara viridula*), nematode (*Caenorhabditis elegans*). AKH (adipokinetic hormone), RPCH (red pigment-concentrating hormone), ACP (adipokinetic hormone/corazonin-related peptides).

Structure and Types of Vertebrate GnRHs

The first GnRH to be identified was mammalian GnRH, or GnRH1, with a primary sequence invariant across mammals, amphibians and bony fish that evolved before the teleost fishes. This GnRH sequence, like all other vertebrate GnRH peptides, is 10 amino acids in length, and is encoded as part of a prohormone precursor. In the precursor, GnRH peptide is flanked by an upstream signal sequence and the GnRH-associated peptide (GAP) downstream. Of this larger prohormone structure, only the 10 amino-acid mature GnRH sequence is highly conserved between different species. Characteristics that define vertebrate GnRH peptides include their length, an N-terminal pyroglutamate, high conservation of the first four and final two amino acids, and an amidated amino acid at the C-terminus (Fig. 4.1).

Once GnRHs were identified in non-mammalian vertebrates, sequence variations were discovered. GnRH peptides were classified into four types based on their sequence (Fig. 4.1) and location of expression. GnRH1 peptides are expressed primarily in the preoptic–hypothalamic area of vertebrates, whereas GnRH2 is consistently found in nuclei of the midbrain tegmentum, specifically the synencephalon/mesencephalon.[13–15] To date, 15 different vertebrate GnRH sequences have been identified; 9 are GnRH1s, 3 are GnRH2s, 1 is a single invariant GnRH3, and 2 are GnRH4 sequences. Most vertebrates express two or three of the different types of GnRHs. Mammals, birds, reptiles and amphibians express one of several distinct

GnRH1 peptides and a common GnRH2, whereas in teleosts (bony fish) most species express GnRH2 as well as a GnRH1 and GnRH3. In fish species that express both GnRH1 and GnRH3, the individual peptides are expressed in distinct populations of neurons in the forebrain, extending from the olfactory bulbs to the hypothalamic region.[14,15] Both identified GnRH4s have been found only in the lamprey, an ancient jawless fish.

Invertebrate GnRHs

Nine different GnRH sequences have been identified in tunicates (sea squirts), the closest invertebrate neighbor of the vertebrates, and all of them fulfill the vertebrate criteria for a GnRH.[16] These peptides most likely play a role in regulating reproduction, as they have been shown to activate tunicate GnRH receptors and to induce gamete release in tunicates (Box 4.2). A characteristic unique to the tunicate GnRHs is the organization of their genes: two tunicate genes each encode three distinct GnRH peptide sequences in tandem, suggesting exon duplication. At this point, tunicates are the only protochordates with characterized GnRHs, as an amphioxus gene has not yet been identified.

In addition to the invertebrate GnRHs described above for tunicates, a peptide was identified in octopus which shared some sequence similarity to the typical vertebrate GnRH, except it was 12 amino acids in length.[17] Other than the insertion of two amino acids after position one, this peptide still maintains the vertebrate GnRH-like characteristics of an N-terminal pyro-glutamate, sequence conservation in key positions, and an amidated C-terminal (Fig. 4.1). Through additional genome searching and the recent application of vigorous computational techniques (Box 4.3), more GnRH-like sequences have been identified in other invertebrates, including two more mollusks (the limpet *Lottia gigantea* and the sea hare *Aplysia californica*), an echinoderm (the sea urchin *Strongylocentrotus purpuratus*) and two annelids (the marine worm *Capitella teleta* and the leech *Helobdella robusta*).[18–20]

One major consideration is whether these invertebrate GnRH-like peptides are true orthologs (functionally related) to the vertebrate GnRH sequences. Experiments performed to determine the physiological role of these peptides in invertebrates have given contradicting conclusions. In *Aplysia*, Tsai *et al.*[19] found injection of its GnRH had little effect on the activation of reproduction. In contrast, Kanda *et al.*[21] showed that treatment with synthetic octopus GnRH *in vitro* resulted in steroidogenesis in octopus ovary and testes. The contradiction of these results demonstrates the diversity of possible GnRH roles, and the challenges of classifying peptides based solely on their identified biological functions.

BOX 4.2

HOW DO WE KNOW ABOUT ESTABLISHING A NEW GnRH RECEPTOR AND ITS SIGNALING PATH?

Identification of the structure of a novel receptor by genomics or cloning provides a hint about, but does not establish, its function based on similarity to known receptors. This is illustrated by the finding of two receptors with similar structures to the GnRH receptor in fruit fly (*Drosophila*). Careful testing showed that GnRH was not the true ligand, but rather adipokinetic hormone for one receptor and corazonin for the other.[1–3] For both vertebrates and invertebrates, one approach to defining receptor function is to transfect the novel receptor cDNA into a cell line and characterize the resulting expressed receptor *in vitro*. Although COS cells are a line developed from the monkey kidney, they will transcribe and translate foreign receptor cDNA from many species into protein. Also, COS cells are advantageous for GnRH work as they lack many endogenous hormone receptors, including GnRHR, avoiding interference. Meanwhile, the components needed for common receptor signaling pathways are present in COS cells, including G proteins, cAMP, inositol triphosphate (IP_3) and kinases. They provide an optimal environment to test GnRH receptors from any vertebrate or invertebrate source.

In 2005, the first invertebrate GnRH receptors were established using COS cells. We cloned four putative GnRH receptors from a tunicate (*Ciona intestinalis*), inserted them into a plasmid vector that also encoded green fluorescent protein, and transfected this into COS cells.[4] Once a receptor protein was expressed, as shown by fluorescence in the membrane, the cells were tested with the six different GnRH peptides known to exist in this species. As seen in the figure below, three of the four receptors responded with a measured increase in cAMP levels, implying coupling to the G_s protein, whereas the fourth receptor did not respond (not shown). Either the latter receptor did not bind GnRH, or the receptor activated a signaling pathway other than the ones tested (cAMP-dependent and IP_3).

(*Continued*)

BOX 4.2 *(cont'd)*

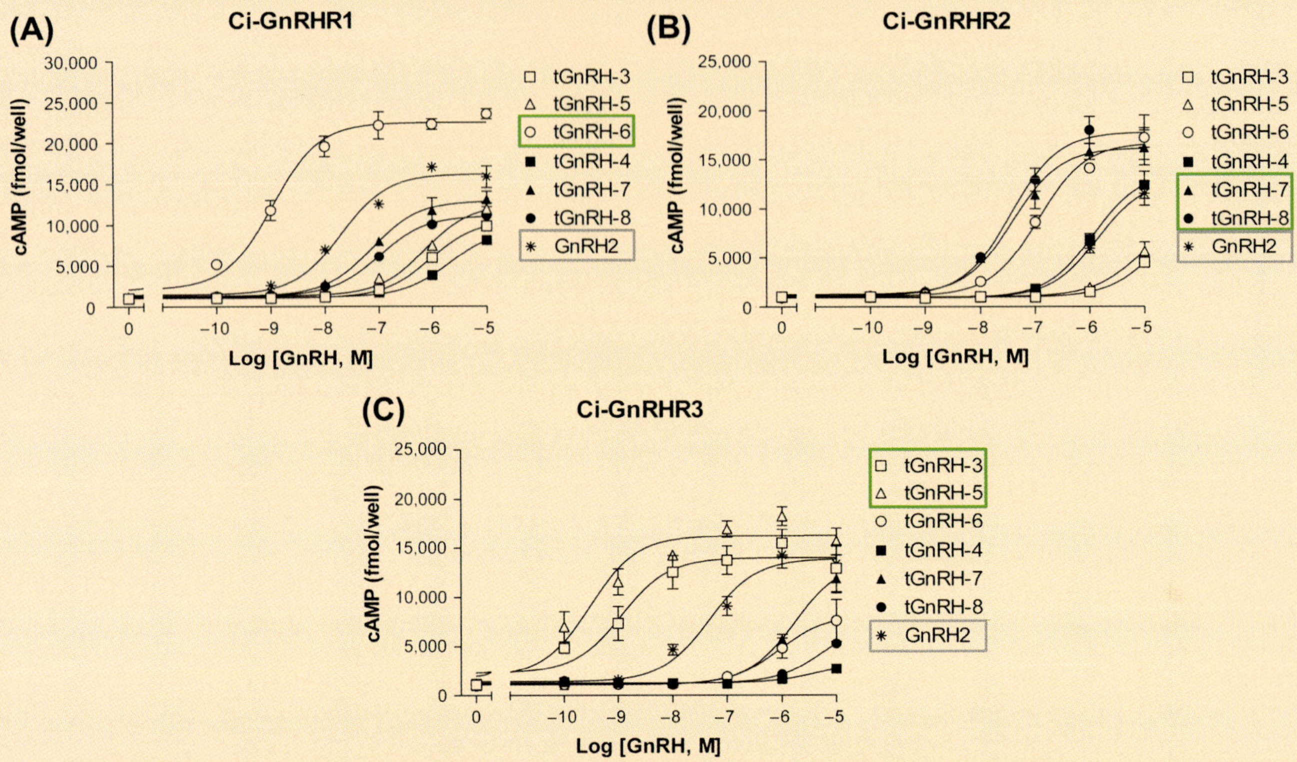

Cyclic AMP (cAMP) accumulation in COS7 cells expressing *Ciona intestinalis* GnRHR1 (A), GnRHR2 (B) or GnRHR3 (C) induced with different concentrations of peptide for 1 hour. Cells were transiently transfected with 0.8 μg GnRHR cDNA and stimulated with peptide after 48 hours. Error bars represent the mean SEM of at least three experiments performed in duplicate. *Modified from Tello* et al.,[4] *with permission from The Endocrine Society ©2005.*

The results indicated that each *Ciona* receptor had a different profile in terms of its sensitivity to the individual GnRHs. The most effective peptides were active in the physiological range, and were distinct for each receptor. In addition, vertebrate GnRH2 (not present in invertebrates) effectively activated the three *Ciona* receptors. In the future, it would be desirable to develop a tunicate cell line for expressing the receptors in a more native environment and to determine if alternative signaling pathways are used by the fourth receptor. In summary, each receptor must be fully characterized to determine its effective ligands and signaling pathways.

References

1. Cazzamali G, Saxild N, Grimmelikhuijzen C. Molecular cloning and functional expression of a *Drosophila* corazonin receptor. *Biochem Biophys Res Commun.* 2002;298(1):31–36.
2. Park Y, Kim YJ, Adams ME. Identification of G-protein coupled receptors for *Drosophila* PRXamide peptides, CCAP, corazonin, and AKH supports a theory of ligand–receptor coevolution. *Proc Natl Acad Sci USA.* 2002;99(17):11423–11428.
3. Staubli F, Jorgensen TJ, Cazzamali G, et al. Molecular identification of the insect adipokinetic hormone receptors. *Proc Natl Acad Sci USA.* 2002;99(6):3446–3451.
4. Tello JA, Rivier JE, Sherwood NM. Tunicate gonadotropin-releasing hormone (GnRH) peptides selectively activate *Ciona intestinalis* GnRH receptors and the green monkey type II GnRH receptor. *Endocrinology.* 2005;146(9):4061–4073.

More Invertebrate Sequences – The Beginning of a Superfamily

It has been proposed that the GnRH peptides belong to a superfamily containing several other invertebrate neuropeptides, based on their structural similarity and a possible shared ancestry.[20,22] Invertebrate corazonins (Crz), adipokinetic hormone/corazonin-related peptides (ACPs) and adipokinetic hormones (AKHs) all share some of the defining characteristics of vertebrate GnRHs: they possess the N-terminal pyroglutamate, sequence similarity in positions 3 (F/W/Y), 4 (S/T) and 7 (mostly W), and the C-terminal glycine residue that allows for peptide amidation (Fig. 4.1).

BOX 4.3

HOW DO WE KNOW ABOUT THE STRUCTURE OF NEUROPEPTIDES AND RECEPTORS BEYOND MAMMALS?

Throughout the history of endocrinology, techniques have expanded to identify ever more distantly related homologs to familiar mammalian peptide hormones and their receptors. Nowhere is this more evident than for GnRH, which was originally isolated in 1971, before the molecular biology revolution. GnRH was originally isolated as the peptide messenger synthesized in neurons by protein chemistry techniques.[1,2] The primary structure of GnRH (GnRH1) in several mammals, including humans, was determined to be an identical peptide of 10 amino acids, using a combination of Edman-dansylation sequencing and mass spectrometry. A confirmation of this structure was provided by synthetic GnRHs that induced the same effects *in vivo* as native hormones (see figure).

Protein sequencing required many brains (for example, 165,000 pig hypothalami), although improved techniques such as high performance liquid chromatography (HPLC) and the use of antibodies, rather than bioassays, reduced the amount of tissue due to greater yield during purification. More recently, molecular biology techniques have replaced protein chemistry for identification in most cases. The mRNA from a single brain can be used to determine GnRH peptide and receptor structure. This mRNA is converted to cDNA by reverse transcriptase for DNA sequencing. To date, isolation of the gene or mRNA transcript has been used to identify several novel forms of GnRH (guinea pig, frog, medaka, lamprey II, *Aplysia*) and to confirm known GnRH sequences in many species. One of the most important advantages of molecular biological techniques has been the identification of receptors for GnRH and other neurohormones. Molecular techniques also provide a method to transfect the receptors into cell lines to test the binding of hormones and determine their signaling pathways *in vitro*.

Invertebrate GnRHs are poorly conserved, compared with their vertebrate counterparts, and thus remained largely undiscovered until the advent of genomics. With the release of publicly available genomes and EST libraries from a variety of invertebrates, a wealth of data has appeared that can be mined for potential GnRH and receptor sequences using computational methods. The primary approach that has been applied to the discovery of most invertebrate GnRHs thus far is searching sequence databases using the popular BLAST (Basic Local Alignment Search Tool) program. This program compares either DNA or protein queries against a selected database looking for short, similar regions in other sequences. BLAST is a fast and accurate program, but one shortcoming is that the program relies on single sequences as queries, which may not provide enough homology to the targets being sought. Thus, to find divergent homologs with unknown primary structures in a variety of invertebrates, multiple queries using different GnRH sequences must be attempted,

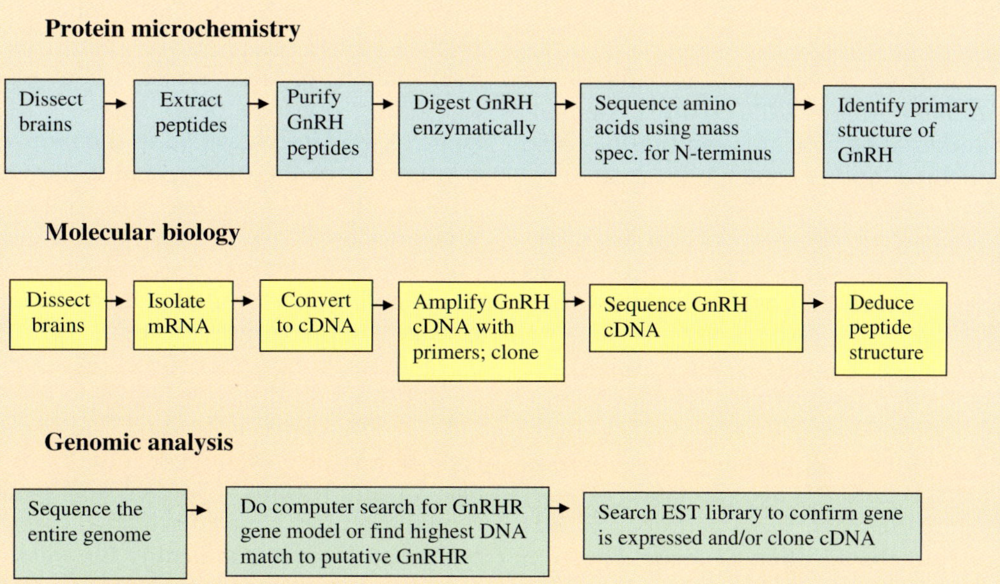

Techniques employed for identification of neuropeptides such as GnRH or kisspeptin.

(Continued)

BOX 4.3 *(cont'd)*

making the process more laborious. A more recent program, HMMER, overcomes this disadvantage by employing hidden Markov models to produce a weighted model from a selection of homologous sequences. With a high-quality model used as a query, HMMER searches of protein (or translated nucleotide) databases can be significantly more sensitive than BLAST searches for poorly conserved sequences like invertebrate GnRHs.

References

1. Burgus R, Butcher M, Amoss M, et al. Primary structure of the ovine hypothalamic luteinizing hormone-releasing factor (LRF). *Proc Natl Acad Sci USA.* 1972;69(1):278–282.
2. Matsuo H, Baba Y, Nair RM, Arimura A, Schally AV. Structure of the porcine LH- and FSH-releasing hormone. I. The proposed amino acid sequence. *Biochem Biophys Res Commun.* 1971;43(6):1334–1339.

Initial functional characterization of the Crz, ACP and AKH peptides does not immediately point to roles in reproduction, although in some species there are connections. Divergent functions are likely to have arisen as these peptides evolved from the ancestors they share with GnRH. Corazonins, neuropeptides found only in insects and crustaceans, have varied functions, including decreasing silk production in silk worms,[23] increasing cardioacceleration in cockroaches[24] and increasing pigmentation in locusts.[25,26] The varied effects of corazonins could be explained by the hypothesis that corazonin may be a stress-response peptide[22,27] that potentially modifies the reproductive readiness of the animal.

A similar regulatory connection to reproduction remains to be made for the adipokinetic hormones (AKHs), as the most common role found for these neuropeptides is the mobilization of lipids and carbohydrates for energy utilization during demanding activities, particularly flight in insects.[27] Besides insects, AKH is found in crustaceans, mollusks and nematodes. The AKH peptide found in crustaceans is called red pigment-concentration hormone (RPCH) due to demonstrated pigment modulation, but this peptide has also been shown to mobilize lipids in some crustaceans.[28,29] Additionally, one of the RPCH peptides in crustaceans has an identical sequence to the AKH of the stink bug, demonstrating the challenges of naming peptides based on identified function, without considering the structural context.

Based on this evidence, we suggest the classification of these invertebrate peptides into a GnRH superfamily based on a combination of structural similarity, functional connections, and the receptor-ligand co-evolution discussed in the next section. The determination of exact ancestral relationships between these neuropeptides is difficult at present, but the addition of sequences from a wider variety of invertebrate species and further physiological studies will shed more light on relationships and reproductive functions.

GnRH RECEPTOR SUPERFAMILY

GnRHR in the Vertebrates

The mammalian GnRH1 receptor was originally isolated by screening G protein-coupled receptors (GPCRs) from a mouse gonadotrope cell line, αT3-1.[3,4] After sequencing, it was determined to be a Rhodopsin family GPCR with a typical seven-transmembrane structure and an absent cytoplasmic C-terminal tail. Upon GnRH binding at a few key residues on the extracellular loops and juxtamembrane region, GnRHR undergoes a conformational shift, allowing it to activate G proteins within the cytoplasm. Its primary target is $G_{q/11}$, which activates the phospholipase C (PLC) pathway, resulting in the increase of second messengers like inositol triphosphate (IP_3), the release of intracellular calcium, and protein kinase C activation. The canonical effect of this is to promote the synthesis and release of gonadotropins within the pituitary, but, as mentioned in the previous section, GnRH receptors are found in a variety of other mammalian tissues, where they mediate an ever-expanding portfolio of functions.[10] To test the importance of the GnRH system in the mouse, we performed a knockout of the only receptor they express, GnRHR type I, and the resulting null mutants demonstrated severe reproductive deficits, illustrating the necessity of GnRH to sexually maturing mammals (Box 4.1).[30]

Following the initial characterization of mouse GnRHR, homologous receptors were isolated in a variety of mammals and other vertebrates. In addition to the identification of direct orthologs to mouse GnRHR type I in every mammal studied,[31] a paralogous GnRHR type II was also identified, initially in fish,[32] and then also in amphibians,[33] birds[34] and primates.[35,36] These type II receptors have an extended C-terminal tail critical for their function, confirming the missing tail on type I receptors is a derived condition.[37] GnRH type I receptors from every mammal tested bind GnRH1 with higher affinity than GnRH2 and are activated to a much greater extent.[38] GnRH type II receptors across the vertebrates are generally more sensitive to GnRH2, even though GnRH1 or GnRH3 is still the gonadotropin-inducing hormone. The mammals appear to have undergone a major change to their GnRH systems, as many do not express GnRH2 and/or the type II receptor, which is a transition to complete dependence on GnRH1 and the mammalian-specific type I receptor.[39]

Invertebrate GnRHRs

Following the discovery of tunicate GnRHs in *Ciona intestinalis*, we identified a group of four receptors in the tunicate genome and cloned them[40] (Box 4.2). All four receptors had extended C-terminal tails, like the type II vertebrate receptors, and three of the four receptors were activated by different tunicate ligands *in vitro*. In contrast to vertebrate GnRH receptors, however, only one tunicate receptor activated the PLC/IP_3 pathway, whereas they all stimulated an increase in the concentration of cyclic adenosine monophosphate (cAMP) through G_s. This is a minor pathway in vertebrate GnRHR signal transduction,[41–43] and may reflect disruptions in the tunicate genome leading to alternative signal pathways.[44] In a phylogenetic comparison with the vertebrate type I and II receptors, the tunicate GnRHRs grouped basally to all of the vertebrate receptors, as seen in Fig. 4.2.

We also identified four GnRH receptor homologs in the other model protochordate, amphioxus.[45] These four receptors formed two pairs structurally, with one group possessing a higher degree of conservation to vertebrate GnRH receptors, especially at those residues implicated in ligand binding. Similar to vertebrate receptors, the amphioxus GnRHRs stimulated the PLC/IP_3 pathway *in vitro*. As no native amphioxus GnRHs have been confirmed, vertebrate and invertebrate GnRH superfamily peptides were employed to test the ligand specificity of the receptors. In an interesting twist, the amphioxus receptors that had higher homology to the vertebrate receptors were preferentially activated by GnRH1 and GnRH2, whereas the other functional receptor with more homology to other putative invertebrate (besides tunicate) GnRH receptors was more strongly stimulated by octopus GnRH and silkworm AKH. This is reflected in their phylogenetic position in comparison with other vertebrate and invertebrate GnRH receptors (Fig. 4.2.)[45]

A putative GnRH receptor was isolated and characterized from the octopus, with its native GnRH.[21] It was

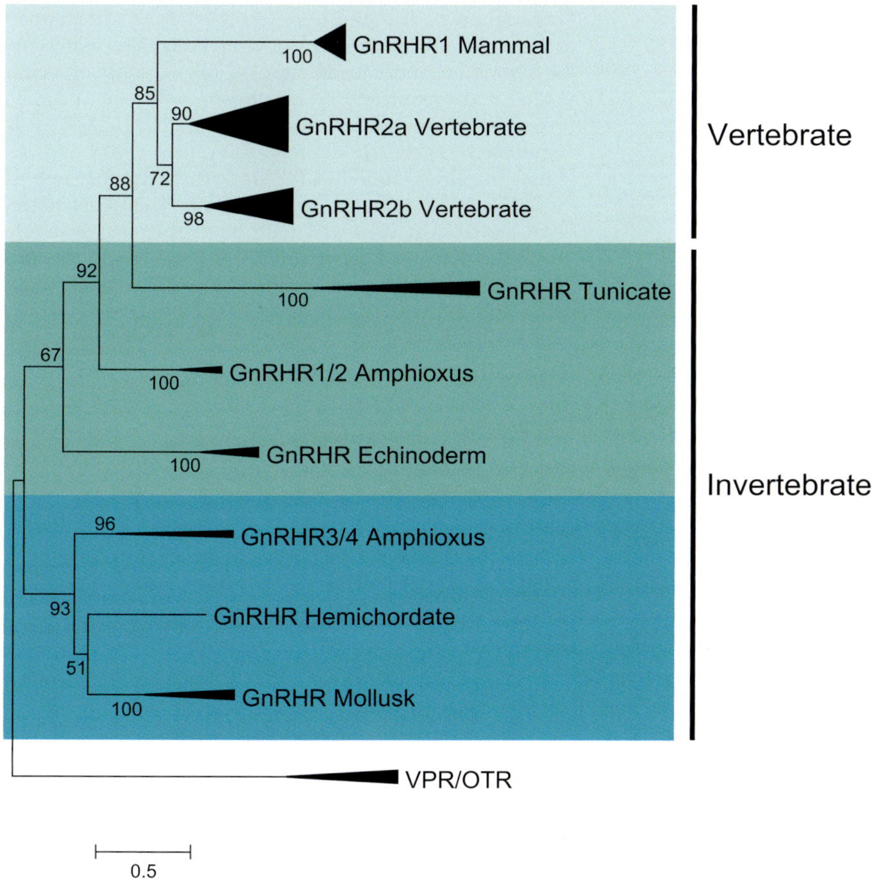

FIGURE 4.2 **Maximum likelihood tree of GnRH receptors from vertebrates and invertebrates.** The tree was generated from a degapped alignment of receptors, using conserved regions within the seven-transmembrane domain. Human vasopressin/oxytocin (VP/OT) receptors were used as an outgroup. PhyML 3.0[46] was employed to generate the tree, using the SH-like approximate likelihood ratio test for branchs (aLRT) for support (indicated in percentage at each node). The LG substitution matrix was used and both invariable sites and rate variation were estimated (LG+I+G), with four substitution rate categories. Tree searching was carried out by both NNI and SPR algorithms to find the phylogeny with the maximum likelihood.

expressed in a broad fashion throughout the octopus nervous system, and in several organs, including the digestive system, heart and gonads. In frog oocytes, the octopus receptor was activated specifically by its native ligand and stimulated the release of calcium. In one of the few physiological experiments conducted on invertebrates, octopus GnRH was able to stimulate steroidogenesis on gonadal cultures and produce *in vivo* contractions in muscle tissue where the receptor was expressed. Following the release of several invertebrate genomes, more putative GnRH receptor homologs have been identified in a wide range of animals, including echinoderms (sea urchin), hemichordates (*Saccoglossus kowalevskii*), annelids (marine worm) and another mollusk (limpet). Their phylogenetic comparison with vertebrate receptors is shown in Fig. 4.2.

GnRHR Superfamily: Vertebrates and Invertebrates

In addition to the GnRH receptors identified in protochordates, hemichordates, echinoderms, mollusks and annelids, receptors for other members of the GnRH superfamily have been identified and characterized in a variety of other invertebrates. Two groups identified the native receptor for fruit fly AKH,[47,48] and the corazonin receptor was also discovered at the same time.[48,49] This was followed by the identification of AKH and corazonin receptors in a swathe of insects,[20] as well as the identification of the related ACP receptors and peptides.[50] AKH peptides and receptors have also been isolated from a variety of nematodes, and in *C. elegans*, knockdowns on the hormone and/or receptor were found to delay egg laying.[51,52] An oyster (*Crassotrea gigas*) receptor, originally identified as a GnRH receptor but more homologous to the insect AKH receptors, was expressed exclusively in the gonads. This supports the possibility of the AKH and corazonin families playing a more prominent reproductive role than previously thought. Corresponding AKH receptors from the limpet, marine worm and leech were recently identified by the current authors,[20] and a phylogenetic analysis of the GnRH, AKH and corazonin receptors is presented in Fig. 4.3. As shown, the vertebrate GnRHRs, tunicate GnRHRs, amphioxus GnRHR1 and GnRHR2 and invertebrate AKHR/ACPRs group together, with the other invertebrate GnRHRs and the corazonin receptors grouping basally. It is clear that these receptors form

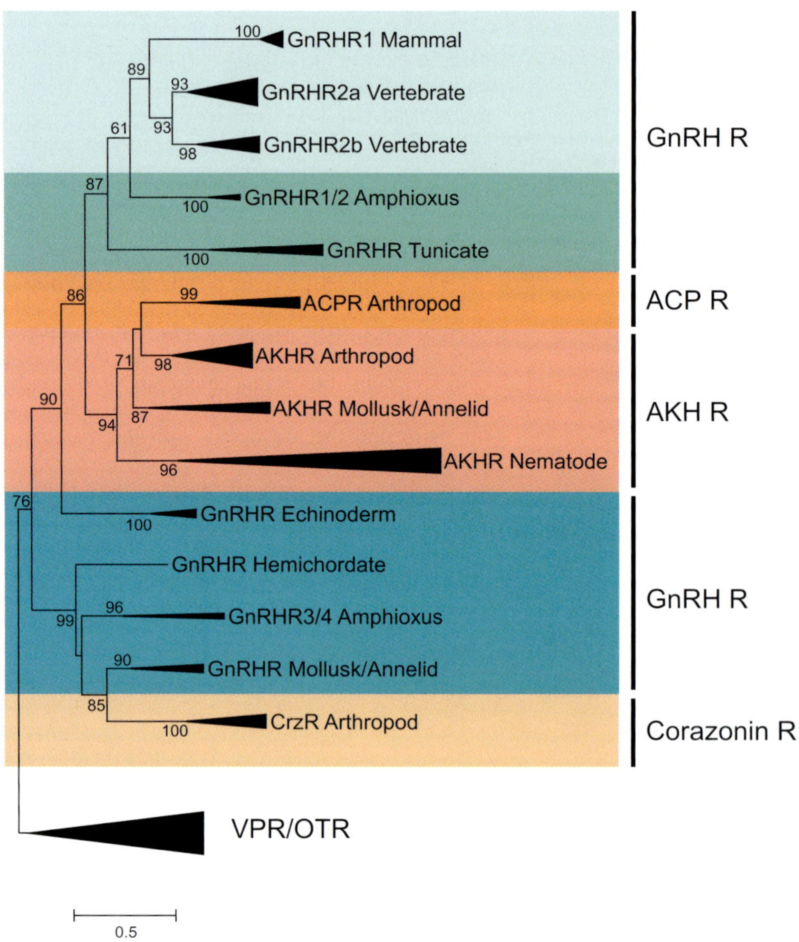

FIGURE 4.3 **Maximum likelihood tree of GnRH, corazonin and AKH/ACP receptors from vertebrates and invertebrates.** The tree was generated from a degapped alignment of receptors, using conserved regions within the seven-transmembrane domain. A large group of vertebrate and invertebrate vasopressin/oxytocin (VP/OT) receptors was used as an outgroup. PhyML 3.0[46] was employed to generate the tree, using the SH-like approximate likelihood ratio test for branchs (aLRT) for support (indicated in percentage at each node). The LG substitution matrix was used and both invariable sites and rate variation were estimated (LG+I+G), with four substitution rate categories. Tree searching was carried out by both NNI and SPR algorithms to find the phylogeny with the maximum likelihood.

a homologous superfamily with a shared common ancestor, and this phylogenetic perspective alters our previous classifications.

KISSPEPTIN AND ITS RECEPTOR

Physiology of Kisspeptin

The kisspeptins are the newest players on the field of neuropeptide hormones influencing vertebrate reproduction. Also known as metastatin for its role in the suppression of metastasis in tumors, in 2001 kisspeptin 1 (Kiss1) was found, by several groups to bind the previously orphaned GPR54 receptor.[53–55] Although the function of this hormone–receptor pairing was initially obscured, null mutations in the receptor (now referred to as Kiss1R) resulted in hypothalamic hypogonadism for both humans and mice,[5–7] indicating a crucial role in reproduction. Conversely, an activating mutation of Kiss1R results in precocious puberty.[56,57]

A connection between Kiss1 and GnRH was soon made, as administration of the peptide to rats[58,59] and monkeys[60] resulted in the release of LH and FSH. This stimulatory effect was ablated by the concurrent administration of a GnRH antagonist. Additionally, Kiss1R expression was localized to GnRH neurons in rodent brains,[59,61,62] and kisspeptin immunoreactive fibers were found in close proximity.[63] Kisspeptin can thus act directly to stimulate GnRH neurons, opposing the actions of another neuropeptide, GnRH inhibiting hormone (GnIH).

Kisspeptin also provides a link between GnRH-mediated stimulation of the sex steroids and their positive and negative feedback on the reproductive axis. Kiss1 neurons in the rodent brain are found in both the arcuate nucleus of the hypothalamus and the anteroventral periventricular nucleus (AVPV) of the third ventricle.[64] In both populations Kiss1 neurons express the estrogen receptor (ERα), and in the arcuate nucleus the androgen receptor (AR) is expressed as well. In castrated mice, Kiss1 expression increased in the arcuate nucleus, indicating negative feedback by sex steroids.[65,66] The opposite was true for Kiss1 expression in the AVPV, suggesting positive feedback. Other factors may also influence kisspeptin expression, including the metabolic regulator leptin, whose receptors can be found on Kiss1 neurons in the arcuate nucleus.[67] Additionally, both Kiss1 and its receptor are expressed in the pituitary, and there is debate as to whether they directly stimulate the gonadotropins.[64]

Kisspeptin Structure and Mechanism

Human Kiss1 is comprised of a 145-amino acid propeptide that is cleaved into a mature peptide of 54 amino acids, which may be further processed to shorter peptides of 14, 13 or 10 amino acids; they bind and activate the human Kiss1R equally.[53] Kisspeptins belong to the RFamide group of peptide neurohormones, indicating the amino acid motif at the C-terminus of the mature peptide. Other members of this group include GnIH, neuropeptide FF, prolactin-releasing peptide (PrRP) and QRFP in the vertebrates, and a host of unique peptides in the invertebrates, including the well-known molluskan FMRFamide. A comparison of putative mature kisspeptin sequences from a variety of vertebrate species is presented in Fig. 4.4A. As illustrated, the hormones are highly conserved within the C-terminal 10-amino acid residues (kisspeptin-10 region), except for position 5, which is a serine (S) in Kiss1 and a proline (P) in Kiss2. Positions 6 through 9 (FGLR) are invariant for all Kiss1 and Kiss2 homologs.

Similar to the other RFamides, kisspeptins bind specific GPCRs from the Rhodopsin (Family A) group. The kisspeptin receptor, formerly referred to as the orphan GPR54, is structurally most closely related to receptors of the unrelated neuropeptides galanin and the invertebrate-specific allatostatin (Fig. 4.4B). Regardless of this phylogenetic relationship, Kiss1R is specific for its cognate ligand and does not bind galanin.[68] Upon binding Kiss1, the activated receptor triggers the $G_{q/11}$-mediated phospholipase C pathway, resulting in the release of intracellular calcium stores and arachidonic acid, and the activation of ERK1/2 and p38 kinases.[53,69,70] This signaling cascade is hypothesized to mediate GnRH release through the depolarization of GnRH neurons via several types of cation channels.[71,72]

Evolution of Kisspeptin and Kiss1R/GPR54

Although initial work on kisspeptin biology focused on mammalian peptides, more recently homologs have been identified and characterized in fish and amphibians (see examples in Fig. 4.4B). Several idiosyncracies have emerged from the study of kisspeptins in these animals, including the expression of Kiss1R on GnRH3 neurons in the fathead minnow, which does not produce GnRH1.[73] Most importantly, the discovery of an orthologous peptide and receptor, Kiss2 and Kiss2R, has been reported in several non-mammalian species and platypus.[74,75] Although unique functions of this novel kisspeptin are not yet well understood, some preliminary observations have been made, including the release of gonadotropins in female zebrafish upon administration of Kiss2 but not Kiss1,[74] and estrogen insensitivity in medaka Kiss2 neurons.[76] In zebrafish, the Kiss2 neurons appear to be the more estrogen-sensitive, indicating the flexibility of these systems between different fish species.[77] Curiously, reptiles appear to lack Kiss1 but have Kiss2, and no kisspeptin has been identified

(A)

	1	2	3	4	5	6	7	8	9	10	
Kiss1											
Mouse	Y	N	W	N	S	F	G	L	R	Y	amide
Human	Y	N	W	N	S	F	G	L	R	F	amide
Platypus	Y	N	W	N	S	F	G	L	R	Y	amide
Frog a	Y	N	W	N	S	F	G	L	R	Y	amide
Frog b	Y	N	V	N	S	F	G	L	R	F	amide
Zebrafish	Y	N	L	N	S	F	G	L	R	Y	amide
Shark	Y	N	W	N	S	F	G	L	R	Y	amide
Lamprey	Y	N	W	N	S	F	G	L	R	F	amide
Kiss2											
Platypus	F	N	F	N	P	F	G	L	R	F	amide
Lizard	F	N	F	N	P	F	G	L	R	F	amide
Frog	F	N	F	N	P	F	G	L	R	F	amide
Zebrafish	F	N	Y	N	P	F	G	L	R	F	amide
Shark	F	N	F	N	P	F	G	L	R	F	amide
Lamprey	F	N	Y	N	P	F	G	L	R	F	amide

(B)

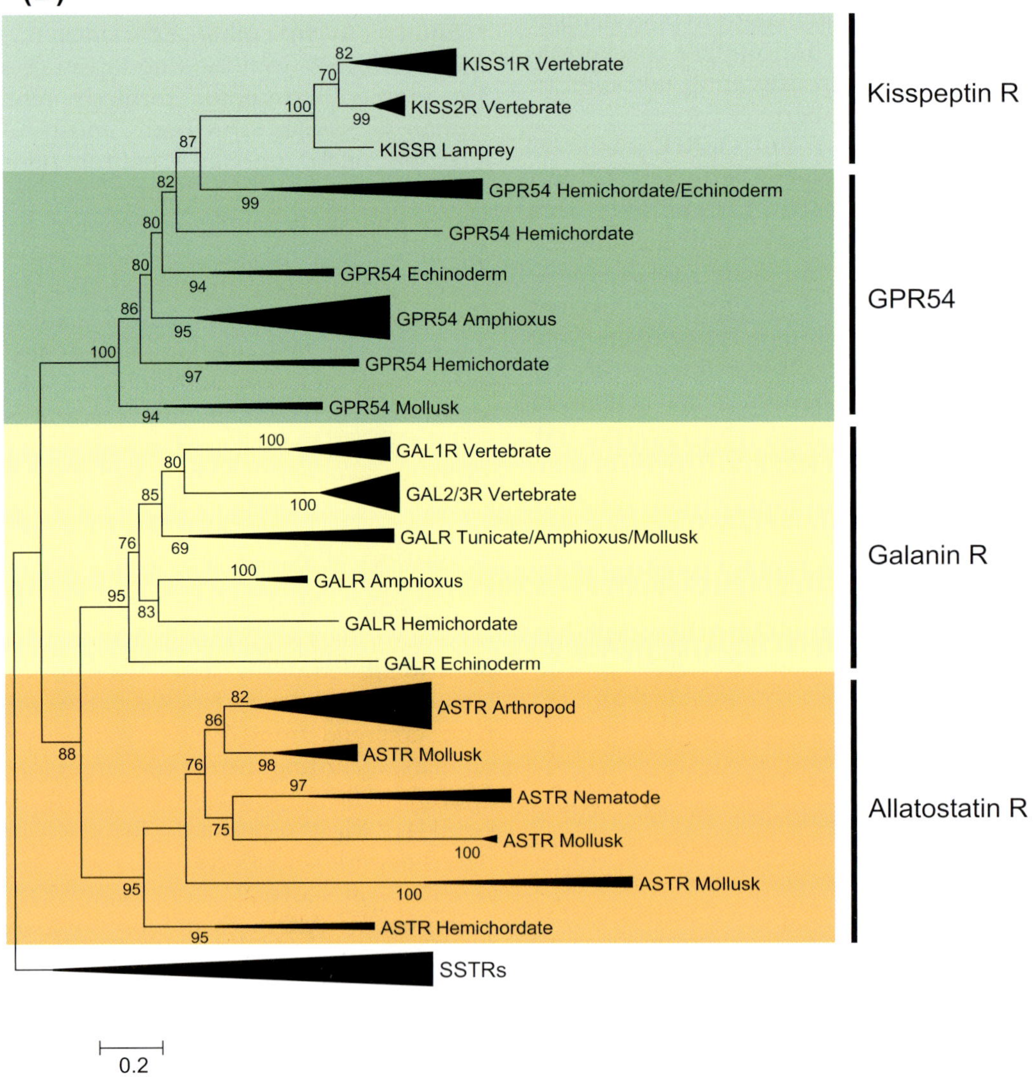

FIGURE 4.4 (**A**) Amino acid sequence of kisspeptin peptides. Colored shading indicates identical residues compared with mouse kisspeptin1, while gray shading identifies similar residues. Representative species are listed below in order of appearance in figure: mouse (*Mus musculus*), human (*Homo sapiens*), platypus (*Ornithorhynchus anatinus*), frog (*Xenopus tropicalis*), zebrafish (*Danio rerio*), shark (*Callorhinchus milii*), lamprey (*Petromyzon marinus*), lizard (*Takydromus tachydromoides*). (**B**) Maximum likelihood tree of kisspeptin, GPR54, galanin and allatostatin receptors from vertebrates and invertebrates. The tree was generated from a degapped alignment of receptors, using conserved regions within the seven-transmembrane domain. Somatostatin receptors were used as an outgroup. PhyML 3.0 [46] was employed to generate the tree, using SH-like approximate likelihood ratio test for branchs (aLRT) for support (indicated in percentage at each node). The LG substitution matrix was used and both invariable sites and rate variation were estimated (LG+I+G), with four substitution rate categories. Tree searching was carried out by SPR algorithms to find the phylogeny with the maximum likelihood.

from avian genomes, indicating that the kisspeptin system has been lost partially or completely within certain vertebrate lineages.[78]

Corresponding Kiss2Rs have been identified for many species with Kiss2 orthologs.[75] In an interesting development, we have identified kisspeptin receptor homologs from a wide variety of invertebrate species (Fig. 4.4B) (Box 4.3). During an investigation into the endocrine complement of the amphioxus *Branchiostoma floridae*, we found more than 10 putative receptor genes.[79] Although corresponding orthologs were not found in the tunicate genome, several are present in the genomes of the sea urchin and hemichordate, as well as the limpet and marine worm (within the "mollusk" branch). The function of these receptors is unknown, and as cognate kisspeptin genes have not been found, they have been labeled as orphan GPR54 receptors in Fig. 4.4B. The question of whether they might bind a related RFamide is pressing, and needs to be addressed to determine whether these receptors play a similar role to the vertebrate KissRs. The kisspeptins have performed a crucial part in the regulation of GnRH and the reproductive axis since the dawn of the vertebrates; it is possible a related neurohormone family extends this function to several invertebrate groups as well.

THE ORIGIN OF REPRODUCTIVE HORMONES

Origin of Reproductive Hormone Families

The evolution of GnRH, kisspeptin, oxytocin and their receptors is predicted to have occurred in a staggered fashion. The neurohormones of GnRH and oxytocin are ancient peptides extending back to early animals that first evolved bilateral symmetry of the body. GnRH-like peptides are found within the tunicates, mollusks and annelids, but related superfamily peptides such as corazonin and adipokinetic hormone (AKH) extend further into insects, crustaceans and nematodes, which together cover almost all major taxonomic groups of invertebrates (Fig. 4.1). Likewise, the oxytocin/vasopressin (OT/VP) superfamily is represented in all of these groups.[80] In contrast to the peptides, which are not structurally similar, the receptors for GnRH and oxytocin are closely related in structure (GPCRs in the Rhodopsin family) and both have roots extending to the origin of bilaterally symmetrical animals (Fig. 4.3).[20] It is likely that the GnRH and OT/VP receptors shared a common ancestral precursor before the rise of bilaterians. These deep evolutionary roots do not appear to be present for the kisspeptin neurohormone, which makes its appearance within the vertebrates, based on available data. The kisspeptin receptors that have a proven affinity for Kiss1/Kiss2 are currently confined to the vertebrates; there are structurally related GPR54 receptors in the invertebrates with no established ligand (Fig. 4.4B).

The appearance of the pituitary LH and FSH hormones and their specific receptors is also of interest, as they are a crucial part of the reproductive axis in vertebrates. Evidence from genomics suggests the pituitary, along with LH and FSH, did not arise until the evolution of jawless fishes (see below). The terminal components of the reproductive axis, sex steroids and their receptors, appeared earlier during the evolution of invertebrates. In annelids, estrogen binds its receptor, and together they activate transcription through the estrogen response element (ERE); reconstruction of the ancestral estrogen receptor suggests its origin from an ancient bilaterian.[81] Although we refer to all of these hormones as part of the reproductive system, as they are in vertebrates, the relationship remains primarily structural within the invertebrates. Nonetheless, estrogen receptors are usually found in invertebrate gonads, among other locations.

Reproductive Systems in Vertebrates and Invertebrates

The reproductive axis for a typical vertebrate, and the putative systems of a closely related invertebrate (amphioxus) and a distant invertebrate (octopus), are compared in Fig. 4.5, based on current evidence. The vertebrate axis in Fig. 4.5 begins with kisspeptin and its receptor; both are shown in color, as they have been identified in representative species throughout the vertebrates.[75] When kisspeptin is released from neurons in the hypothalamus, the Kiss1 receptors on GnRH neurons are activated, resulting in the release of GnRH and the downstream release of the gonadotropins LH and FSH.[64] In response to LH and FSH, gonadal cells synthesize and release sex steroids that bind intracellular (nuclear) receptors, which in turn bind to hormone response elements on steroid-sensitive genes. Vertebrates also have steroid receptors on the external membranes of cells. Intracellular sex-steroid receptors include estradiol receptors (ERα, ERβ), an androgen receptor (AR) and a progesterone receptor (PR). External membrane receptors (GPCRs) for the steroids include membrane estradiol receptor (mER), androgen receptor (mAR) and progesterone receptor (mPR).[82–84] Considerable evidence shows that fishes have similar reproductive hormonal components, although lampreys have only one ER, PR and CR (corticoid receptor, duplicated in jawed vertebrates to mineralocorticoid receptor (MR) and glucocorticoid receptor (GR)).[85] Another round of duplication resulted in a full repertoire of ERα and ERβ, PR, AR, MR and GR, as found in cartilaginous fish.[86]

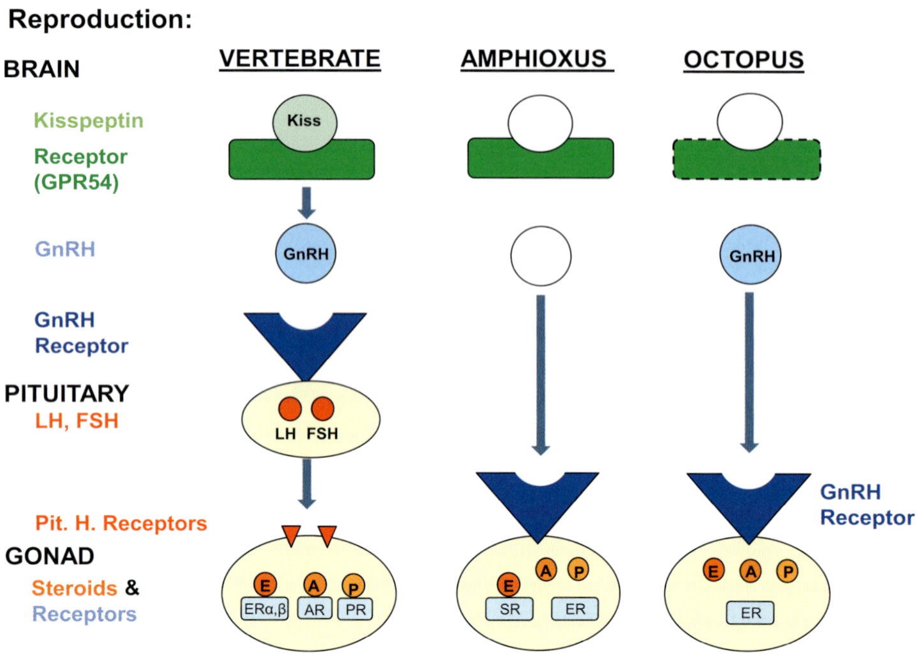

FIGURE 4.5 Origins of reproductive hormones: a schematic diagram of the evolution of the reproductive axis in a representative vertebrate, a closely related invertebrate (amphioxus, *Branchistoma floridae*) and a more distantly related invertebrate (octopus, *Octopus vulgaris*). Shapes representing hormones and their receptors are shown in color if they have been identified in representative species. A lack of color indicates the peptides have not yet been identified, and the broken line indicates the GPR54 was not identified in octopus but in a related mollusk, limpet (*Lottia gigantea*).

For comparison to vertebrates, we screened the amphioxus genome[87] for endocrine hormones and receptors.[79] Kisspeptin was not identified, but copies of GPCRs structurally related to its receptor were found and classified as GPR54s. A similar story is true for amphioxus GnRH; four receptors with homology to vertebrate GnRHRs were cloned and tested,[45] but the native ligand(s) has not yet been identified. An LH or FSH homolog was not detected in amphioxus, but the heterodimeric glycoprotein hormone thyrostimulin was identified.[79,88] As discussed in the next section, this hormone is thought to represent the ancestor of LH, FSH and TSH, which are only expressed in vertebrates.[44,87,89] Likewise, neither an LH nor an FSH receptor was present in the amphioxus genome.[79] Instead, the GnRH receptor is expressed in the gonad, where it is assumed that GnRH acts directly; in support, humans also have responsive gonadal GnRH receptors.[9,90]

Amphioxus has estradiol, testosterone and progesterone localized in the gonad, as detected by radioimmunoassay,[91] and this was confirmed by cloning the cDNAs for the biosynthetic enzymes of these steroids.[91,92] The steroidogenic pathway was determined using genomics to identify additional precursor enzymes,[79] and using ^{14}C-labeled intermediates in isolated ovarian tissue to show conversion to steroidal metabolites.[93] However, only estrogen activates a receptor (see below), so the role of testosterone and progesterone in amphioxus may only be as intermediates for estrogen synthesis.[85] Yet to be explored is a possible action on the membrane steroid receptors, if they exist outside of vertebrates.[94]

Two steroid receptors were identified in the amphioxus genome[79,95,96] and cloned.[95,97] Unexpectedly, estrogens activated the ambiguously classified steroid receptor (SR) and not the ER ortholog, as shown by transcription of reporter genes that contained the estrogen response element.[95,97,98] However, the amphioxus ER constitutively suppressed SR activity by competing for the ERE site,[95,97] and this may reflect a mutation in the ligand binding domain. Phylogenetic analysis shows that the amphioxus estrogen receptor is orthologous to the human ERα and ERβ, whereas the amphioxus SR is orthologous to the other group of steroid receptors including AR, PR, MR and GR.[95] The data indicate that the ancestral steroid receptor had already duplicated once in a common ancestor of the chordates, followed by further duplications within the vertebrates of ER to ERα and ERβ, and SR to AR, PR, GR and MR. It is proposed that after the ancestral steroid receptor duplicated, those in the amphioxus lineage developed different functions compared to the duplicates in the vertebrate lineage.[95] We know that amphioxus is not a model for all protochordates, as tunicates have GnRH peptides and receptors but do not have steroids or their respective biosynthetic enzymes and receptors. An explanation is that tunicates are highly derived and have likely lost their steroid components, although they continue to reproduce apace.

The octopus genome has not been released, but its GnRH was originally purified using a heart bioassay followed by protein chemistry; mass spectrometry and cloning confirmed the primary structure.[17] The octopus GnRH receptor was cloned and confirmed to be a functional GnRH receptor using the *Xenopus* oocyte expression system.[21] An octopus kisspeptin receptor-like (GPR54-like) gene has not yet been identified, but we have identified a putative GPR54 in another mollusk, the limpet (*Lottia gigantea*), with a sequenced genome. The relationship of mollusk GPR54s to other receptors is shown in Fig. 4.4B.

Octopus has estradiol, testosterone and progesterone present in the gonads, as measured by radioimmunoassay.[99] Only one receptor in octopus has been detected, and it is classified as an estrogen receptor that does not bind estrogen, but is able to constitutively activate gene transcription without a ligand.[94,100] This has led to the question of whether or not the ancestral ER was dependent on estrogen. We know that nuclear receptors are not found in either yeast or plants,[101] but are found in sponges and other multicellular animals.[102] Within this receptor family, steroid receptor genes are not found in the cnidarians – basal metazoans that include the jellyfish and sea anenomes.

The first steroid receptor is predicted to have been an estrogen-sensitive receptor present in an ancestral animal before the origin of the bilaterians.[81,103] It is curious that the ER is absent in fruit fly, nematode, sea urchin and sea squirt genomes, and that in mollusks (octopus, *Aplysia*) and amphioxus it does not bind estrogens or other steroids.[44,94,102,103] One hypothesis is supported by the ERs in two species of annelids. Both receptors activate gene transcription when bound to estradiol or estrone.[81] The reconstruction of the ancestral ER suggests it was estrogen-sensitive. Mutations in the ligand binding region of mollusk and amphioxus ER, and gene loss in other species, would explain the current status of ER in different lineages.[94,100] The vertebrate steroid receptors are considered to be the result of gene duplications and diversification of the ancestral chordate ER and SR genes. In conclusion, available evidence suggests that the reproductive molecules with the deepest evolutionary roots are the GnRH superfamily peptides and receptors, OT/VP peptides and receptors, GPR54 receptors, and the steroids and their receptors. To date, kisspeptin, GnIH and the pituitary hormones and receptors appear to be vertebrate innovations.

THE PITUITARY LINK BETWEEN BRAIN AND GONAD: A VERTEBRATE INNOVATION

Function of the Pituitary Gland

The origin of the pituitary gland has been of interest to biologists following early studies that showed removal resulted in cessation of reproduction, growth reduction, and secondary effects due to loss of function in the thyroid and adrenal cortex. Replacement of the rat pituitary by grafting restored the lost functions, provided the gland was revascularized by the portal blood.[104] Lactation was not restored as the neural part of the pituitary atrophied, resulting in the loss of oxytocin as a signal for milk release. These experiments and others in which a visible pituitary gland could be manipulated or sampled firmly established the pituitary gland's function throughout vertebrates.[105–107] Extending these results to invertebrates is difficult because a visible gland is not present, although it is possible that dispersed pituitary cells are present.

In vertebrates the pituitary gland lies below the brain, whether considering a jawless hagfish or a human. Mammals synthesize a full complement of hormones in the anterior, intermediate and posterior pituitary.[105–107] The six classical hormones in the anterior pituitary are follicle-stimulating hormone (FSH), luteinizing hormone (LH), thyroid-stimulating hormone (TSH), growth hormone (GH), prolactin (Prl) and adrenocorticotropic hormone (ACTH). In the intermediate pituitary, melanocyte-stimulating hormone (MSH) is synthesized. The posterior pituitary, unlike the other two parts, grows out from the brain during development, and acts as a release site for oxytocin (OT) and vasopressin (VP). The hormones are synthesized in hypothalamic neurons, but released from axon terminals in the posterior pituitary.

Pituitary Evolution in Vertebrates

The evolution of the pituitary in vertebrates has been deduced from both morphological changes and, more importantly, the expression of the pituitary hormones in the different classes of vertebrates (Fig. 4.6). In the most basal living vertebrate, hagfish,[89] three pituitary hormones are currently known. Hagfish gonadotropin (GTH) was recently cloned and found to be a glycoprotein hormone. The primary structures of hagfish GTHα and GTHβ subunits were homologous with both FSH and LH subunits in fishes, so it was not surprising that purified hagfish GTH released estradiol and testosterone from the testis *in vitro*.[108] The other two hagfish hormones, GH and ACTH, have only been identified with antisera,[109] and require final sequence identification in the genome.

In the jawless lamprey, four pituitary hormones, including GTH, GH, ACTH and MSH, are present.[110–113] Oxytocin is not present in the posterior pituitary; a vasopressin ortholog exists alone.[114] Gene duplications are thought to explain the expanded hormone complement in cartilaginous fish. ACTH and MSH remain, but the single gonadotropin gene appears to have duplicated and diversified to FSH and LH,[115] and the ancestral GH gene[116] is also predicted to have

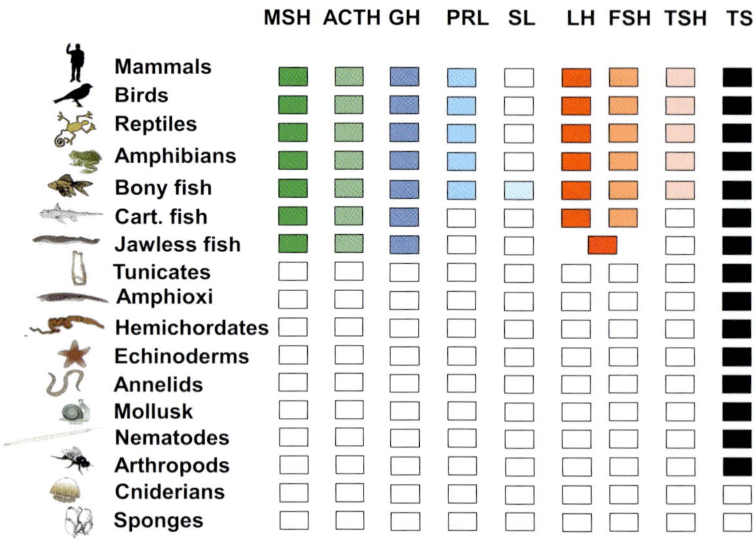

FIGURE 4.6 **The proposed evolution of eight pituitary hormones and thyrostimulin.** MSH (green), ACTH (light green) and GH (blue) are first detected in jawless fish, whereas LH (red) and FSH (pink) are not detected until cartilaginous fish. Gonadotropin-β, present in jawless fish, is considered to be the precursor after duplication to both LH and FSH. PRL (medium blue), SL (light blue) and TSH (light pink) appear first in the pituitary of bony fishes, although SL is only found in bony fish and PRL is found in cartilaginous fish but not in the pituitary. Thyrostimulin (black) is present in many invertebrates and vertebrates, and is likely the precursor of gonadotropins and TSH after duplication in early vertebrates. The genes of the eight pituitary hormones have not been found in the genomes of amphioxus or tunicates. ACTH, adrenocorticotropic hormone; FSH, follicle-stimulating hormone; GH, growth hormone; LH, luteinizing hormone; MSH, melanocyte-stimulating hormone; PRL, prolactin; SL, somatolactin; TSH, thyroid-stimulating hormone; TS, thyrostimulin.

duplicated, resulting in GH and Prl2. The latter is not present in the pituitary.[117] Both an OT and a VP family member are present in the posterior pituitary.

Within the bony fish the full set of pituitary hormones is present, including TSH and a pituitary-specific Prl1. This complete set includes all six anterior–intermediate pituitary hormones, the lineage-specific somatolactin (SL), and both members of the OT/VP family. This suite of hormones is maintained (except for somatolactin) throughout the amphibians, reptiles, birds and mammals (Fig. 4.6). The doubling of the genome twice in early vertebrates, followed by individual gene duplication and diversification, is thought to be the basis for the appearance and expansion of pituitary hormones.[118,119]

Origin of Pituitary Hormones

One strategy for examining the pituitary's origin is the utilization of genomics to determine whether the genes for any of the pituitary hormones or their receptors are present in invertebrates. Unlike neurohormones, some of the pituitary hormones are large (~200 amino acids for FSH, LH, TSH, GH and Prl), and their receptors are even larger, providing more opportunity for finding conserved regions. To date, the results reveal that the sequenced tunicate genomes and the amphioxus genome do not contain any pituitary hormone or receptor gene models homologous to those of vertebrates.[79,120] Likewise, the EST libraries did not yield any cDNAs representing typical pituitary hormones or receptors in these protochordates.

Another strategy to search for clues about the origin of the pituitary is to examine embryo development in the nearest invertebrate neighbors of vertebrates (protochordates and echinoderms) for morphological placodes or the presence of necessary transcription factors for induction of a pituitary. In vertebrate embryos, the adenohypophysial placode develops into the anterior and intermediate pituitary. Distinct from the nervous tissue and neural crest, the placode forms from cranial ectoderm at the anterior edge of the neural plate in the midline.[105] In contrast, the olfactory or nasal placode gives rise to GnRH1 cells, among many others.[105,121] Migrating cells from these two placodes have different destinations; the GnRH1 cells terminate dorsally in the preoptic-hypothalamus, whereas the pituitary cells migrate to an area ventral to the brain. Evidence to date does not support the presence of a functioning adenohypophysial placode in tunicates or amphioxus.[122,123] Support is also lacking for the presence of a full complement of transcription factors in tunicates and amphioxus necessary to specify pituitary organogenesis, or for differentiation and proliferation of the pituitary cells that synthesize hormones. A thorough examination of tunicates and amphioxus showed that homologs exist for transcription factors that play a central role in vertebrate placode formation, but a full regulatory network specifying placode formation is restricted to the vertebrate lineage.[124] The large-scale duplications of the genome in early vertebrates are credited with the expansion of the transcription factor families required for induction and regulation of pituitary formation.[79,120]

The ancestral origin of the pituitary hormones in jawless fish is of great interest, considering the lack of evidence for an invertebrate pituitary. The neural origin of the posterior pituitary hormones, oxytocin and vasopressin, puts them into a different classification. The ancestral origin of these neuropeptides has been clearly traced deep into the invertebrates.[80] The OT/VP structure, with a ring created by a disulfide bond in the 9-amino acid peptide, is very distinct and identified in most major invertebrate groups, including annelids, mollusks, insects, crustaceans, echinoderms and protochordates. Also, the classical anterior pituitary hormones (FSH, LH and TSH) have a suitable candidate for their ancestral origin. A glycoprotein with an ancient origin, thyrostimulin (TS), has α and β subunits (GPA2 and GPB5) homologous to pituitary hormones (Fig. 4.6). The pattern of cysteines in these glycoprotein hormones can be aligned, although FSHβ, LHβ and TSHβ have 12 residues, and thyrostimulin GPB5 has 10. Thyrostimulin has been identified from animals including all the major groups of vertebrates and invertebrates except for the basal cnidarians (hydra and sea anemone).[88,119,125] Both tunicates[120] and amphioxus[79] have TS, and in humans it is found in the pituitary, among other locations in the body.[119,126,127] Evidence strongly suggests that the FSHβ, LHβ, TSHβ and common α-subunit genes are the result of a duplication of the TSα/β genes that occurred during the genome duplications in early vertebrates.[44,87,89] Human chorionic gonadotropin (hCG) appeared more recently as a result of a LHβ gene duplication.[118,119,125] The receptors for vertebrate FSH, LH and TSH are likely to have appeared at a similar time in early vertebrates, resulting from duplications of ancestral leucine-rich repeat-containing G-protein coupled receptor (LGR) genes present in invertebrates.[118,119]

Although thyrostimulin appears to be the ancestral molecule for the vertebrate glycoprotein hormones in the pituitary, its functional role is not clear. In humans, TS binds to the TSH receptor with high affinity resulting in an increase in thyroxine (T4) in the thyroid gland.[126] In the human ovary, TS can stimulate an increase in cAMP.[128] Although TS is found within corticotropes in the human pituitary,[127] it is not specific to the pituitary and is expressed in many tissues.[119,127] In amphioxus, TS is expressed in the nerve cord, gills and oocytes, but not in the proposed pituitary homolog known as Hatschek's pit.[129]

Tracing the origin of the pituitary becomes more difficult when GH/Prl and MSH/ACTH are considered. These pituitary hormones are monomeric proteins, unlike FSH, LH and TSH. GH has a more ancient origin than prolactin, as GH has been identified in the jawless hagfish and lamprey,[109–111] whereas the oldest prolactin (Prl2) appears in elephant shark[117] and the pituitary-specific Prl1 is first detected in bony fish (Fig. 4.6). Invertebrate GH-like and prolactin-like proteins were reported, based on immunohistochemistry with vertebrate antibodies,[130,131] but this is not sufficient evidence for their presence; neither hormone is detected in the amphioxus[79] or tunicate genomes.[44,120] One hypothesis suggested the ancestral GH/Prl gene was the product of repeated internal duplications of a smaller primordial gene.[132] More recent evidence does not support this hypothesis, as growth hormone has four regions that lack sequence homology with each other.[133,134] Crystallization of GH showed it has four helices, but the conserved amino acids that are necessary to retain the 3D structure are not internal repeats.[135] The origin of the GH/Prl family remains murky.

Current evidence, based on several genome studies, does not provide any support for the presence of ACTH and MSH hormones from invertebrates. These two hormones are both encoded in a large precursor molecule (pro-opiomelanocortin, POMC), which also encodes endorphin and met-enkephalin. Two reports on invertebrate POMC cDNA and/or peptides from leech and mussel[136,137] were not supported by searches of the leech genome, or those of other invertebrates, including tunicates and amphioxus.

In conclusion, the anterior pituitary as a functioning organ appears to have its ancestral origin in early vertebrates after the large-scale genome duplications. Thus, invertebrate neurohormones including GnRH are likely to act in a direct or paracrine manner on tissues without the intermediary of a pituitary gland. The action of GnRH on the gonads in humans and other vertebrates[9,90] occurs in a paracrine fashion, and may be a residual pathway that existed before the pituitary evolved.

CONCLUSIONS

GnRH and kisspeptin are two neurohormones that are known to be central to the control of reproduction throughout the vertebrates, yet the evolutionary origin of these crucial peptides has remained obscure. The introduction of genomics and phylogenetic analysis has created the possibility of finding hormones in any animal with a sequenced genome. The result of these genomic searches has shown that some invertebrates also have GnRH-like peptides of 10–12 amino acids. Functional studies lag far behind the structural investigations, but in the few invertebrate species characterized there is evidence for a role in reproduction for some GnRH peptides. Based on genomic searches, a GnRH superfamily has been proposed. Invertebrates have several families of peptides with structures that are related to GnRH. These families include corazonin and adipokinetic hormone. Further, their receptors also form a superfamily, and appear to have shared a common ancestor in an ancient bilaterian.

Kisspeptin is a relatively "young" hormone compared with GnRH or oxytocin, but it is also essential for reproduction throughout the vertebrates. In contrast to the aforementioned neuropeptides, genomic searches have not established the presence of kisspeptin outside the vertebrates. The vertebrate receptor for kisspeptin, formerly known as GPR54, has homologs in several major invertebrate groups. The ligand for invertebrate GPR54 is not known, posing an important question regarding the evolution of these receptors.

Oxytocin and vasopressin were among the first neuropeptides to be sequenced. Today these peptide families have been identified throughout the vertebrates and invertebrates, suggesting they share a common ancestor in ancient bilaterians. Oxytocin has many important functions, including its role in the release of milk in mammals. Loss of oxytocin results in the inability of mammals to feed their young, and illustrates that a new function for an old peptide can occur. On the receptor side, genomics has shown us not only that the OT/VP receptors are present throughout the bilaterians, but also that the receptors are closely related in structure to GnRH receptors. Phylogenetic studies suggest the GnRHR and OTR/VPR superfamilies shared a common ancestor before the bilaterians evolved.

Genomics has also been useful for confirming species in which hormones and receptors do not exist, which is critical when specifying the origin of hormone families. The prime example is the classical pituitary hormones and receptors, which are not found outside the vertebrates. A plausible scenario for the evolution of FSH, LH, TSH and their receptors exists, based on the duplication and diversification of ancestral molecules such as thyrostimulin and the LGR receptors. Although the pituitary appears to be a vertebrate innovation, the origins of GH and POMC (ACTH and MSH) remain obscure in early vertebrates. The sex steroids and receptors are also ancient, with evidence that estrogen-sensitive receptors existed in annelids with a predicted ancestor in early bilaterians. In conclusion, the phylogenetic analysis of currently available genomic and expressed sequence tag (EST) databases has revealed new homologs for classical reproductive hormones and receptors from a diverse array of animals, including members of almost all major invertebrate groups.

FUTURE STUDIES

Technological advances in high-throughput sequencing point to a massive expansion in the number and range of genomes that will be available in the near future. The genomes that are currently sequenced have made pioneering studies possible in key organisms, but a fuller context is needed to make accurate deductions about the evolutionary origin and path of gene families. This anticipated wealth of sequenced genomes reminds us that physiological studies lag far behind. Functional experiments are often slow and expensive, but they must be performed to shed light on the roles of divergent hormones. In particular, the search for reproductive functions in the invertebrate peptides and non-reproductive functions for the vertebrate hormones discussed here is an important future endeavor. The goal of *in vivo* proof for hormone action remains paramount, and would be accelerated with gene knockout/knockdown techniques for more species.

One of the most important events that helped to define the GnRH superfamily was the discovery of the cognate ligands for the corazonin and AKH receptors, originally thought to be GnRH receptors based on structural homology. Instead, the invertebrate peptides were identified as corazonin and AKH by screening peptide libraries and purifying active fractions in conjunction with the receptors. In the future, identification of the ligand(s) for the invertebrate GPR54 receptors using similar techniques might revolutionize our knowledge of invertebrate reproduction or reveal novel functions.

Acknowledgments

We thank the Canadian Natural Sciences and Engineering Research Council (NSERC) for financial support.

References

1. Matsuo H, Baba Y, Nair RM, Arimura A, Schally AV. Structure of the porcine LH- and FSH-releasing hormone. I. The proposed amino acid sequence. *Biochem Biophys Res Commun.* 1971;43(6):1334–1339.
2. Burgus R, Butcher M, Amoss M, et al. Primary structure of the ovine hypothalamic luteinizing hormone-releasing factor (LRF). *Proc Natl Acad Sci USA.* 1972;69(1):278–282.
3. Tsutsumi M, Zhou W, Millar RP, et al. Cloning and functional expression of a mouse gonadotropin-releasing hormone receptor. *Mol Endocrinol.* 1992;6(7):1163–1169.
4. Reinhart J, Mertz LM, Catt KJ. Molecular cloning and expression of cDNA encoding the murine gonadotropin-releasing hormone receptor. *J Biol Chem.* 1992;267(30):21281–21284.
5. Seminara SB, Messager S, Chatzidaki EE, et al. The GPR54 gene as a regulator of puberty. *N Engl J Med.* 2003;349(17):1614–1627.

6. Funes S, Hedrick JA, Vassileva G, et al. The KiSS-1 receptor GPR54 is essential for the development of the murine reproductive system. *Biochem Biophys Res Commun.* 2003;312(4): 1357–1363.
7. de Roux N, Genin E, Carel JC, Matsuda F, Chaussain JL, Milgrom E. Hypogonadotropic hypogonadism due to loss of function of the KiSS1-derived peptide receptor GPR54. *Proc Natl Acad Sci USA.* 2003;100(19):10972–10976.
8. Okubo K, Nagahama Y. Structural and functional evolution of gonadotropin-releasing hormone in vertebrates. *Acta Physiol (Oxf).* 2008;193(1):3–15.
9. Ramakrishnappa N, Rajamahendran R, Lin YM, Leung PC. GnRH in non-hypothalamic reproductive tissues. *Anim Reprod Sci.* 2005;88(1-2):95–113.
10. Skinner DC, Albertson AJ, Navratil A, et al. Effects of gonadotrophin-releasing hormone outside the hypothalamic-pituitary-reproductive axis. *J Neuroendocrinol.* 2009;21(4):282–292.
11. Wirsig-Wiechmann CR. Function of gonadotropin-releasing hormone in olfaction. *Keio J Med.* 2001;50(2):81–85.
12. Sherwood NM, Wu S. Developmental role of GnRH and PACAP in a zebrafish model. *Gen Comp Endocrinol.* 2005;142(1-2):74–80.
13. Fernald RD, White RB. Gonadotropin-releasing hormone genes: phylogeny, structure, and functions. *Front Neuroendocrinol.* 1999;20(3):224–240.
14. Guilgur LG, Moncaut NP, Canario AV, Somoza GM. Evolution of GnRH ligands and receptors in gnathostomata. *Comp Biochem Physiol A Mol Integr Physiol.* 2006;144(3):272–283.
15. Zohar Y, Munoz-Cueto JA, Elizur A, Kah O. Neuroendocrinology of reproduction in teleost fish. *Gen Comp Endocrinol.* 2010;165(3):438–455.
16. Adams BA, Tello JA, Erchegyi J, et al. Six novel gonadotropin-releasing hormones are encoded as triplets on each of two genes in the protochordate, *Ciona intestinalis*. *Endocrinology.* 2003;144(5):1907–1919.
17. Iwakoshi E, Takuwa-Kuroda K, Fujisawa Y, et al. Isolation and characterization of a GnRH-like peptide from *Octopus vulgaris*. *Biochem Biophys Res Commun.* 2002;291(5):1187–1193.
18. Zhang L, Tello JA, Zhang W, Tsai PS. Molecular cloning, expression pattern, and immunocytochemical localization of a gonadotropin-releasing hormone-like molecule in the gastropod mollusk, *Aplysia californica*. *Gen Comp Endocrinol.* 2008;156(2):201–209.
19. Tsai PS, Sun B, Rochester JR, Wayne NL. Gonadotropin-releasing hormone-like molecule is not an acute reproductive activator in the gastropod, *Aplysia californica*. *Gen Comp Endocrinol.* 2010;166(2):280–288.
20. Roch GJ, Busby ER, Sherwood NM. Evolution of GnRH: diving deeper. *Gen Comp Endocrinol.* 2011;171(1):1–16.
21. Kanda A, Takahashi T, Satake H, Minakata H. Molecular and functional characterization of a novel gonadotropin-releasing-hormone receptor isolated from the common octopus (*Octopus vulgaris*). *Biochem J.* 2006;395(1):125–135.
22. Boerjan B, Verleyen P, Huybrechts J, Schoofs L, De Loof A. In search for a common denominator for the diverse functions of arthropod corazonin: a role in the physiology of stress? *Gen Comp Endocrinol.* 2010;166(2):222–233.
23. Tanaka Y, Hua Y, Roller L, Tanaka S. Corazonin reduces the spinning rate in the silkworm, *Bombyx mori*. *J Insect Physiol.* 2002;48(7):707–714.
24. Veenstra JA. Isolation and structure of corazonin, a cardioactive peptide from the American cockroach. *FEBS Lett.* 1989;250(2):231–234.
25. Tawfik AI, Tanaka S, De Loof A, et al. Identification of the gregarization-associated dark-pigmentotropin in locusts through an albino mutant. *Proc Natl Acad Sci USA.* 1999;96(12): 7083–7087.
26. Hua YJ, Ishibashi J, Saito H, et al. Identification of [Arg(7)] corazonin in the silkworm, *Bombyx mori* and the cricket, *Gryllus bimaculatus*, as a factor inducing dark color in an albino strain of the locust, *Locusta migratoria*. *J Insect Physiol.* 2000;46(6):853–859.
27. Veenstra JA. Does corazonin signal nutritional stress in insects? *Insect Biochem Molec.* 2009;39(11):755–762.
28. Marco HG, Gade G. Biological activity of the predicted red pigment-concentrating hormone of *Daphnia pulex* in a crustacean and an insect. *Gen Comp Endocrinol.* 2010;166(1):104–110.
29. Zrala J, Kodrik D, Zahradnickova H, Zemek R, Socha R. A novel function of red pigment-concentrating hormone in crustaceans: *Porcellio scaber* (Isopoda) as a model species. *Gen Comp Endocrinol.* 2010;166(2):330–336.
30. Wu S, Wilson MD, Busby ER, Isaac ER, Sherwood NM. Disruption of the single copy gonadotropin-releasing hormone receptor in mice by gene trap: severe reduction of reproductive organs and functions in developing and adult mice. *Endocrinology.* 2010;151(3):1142–1152.
31. Millar RP. GnRHs and GnRH receptors. *Anim Reprod Sci.* 2005; 88(1-2):5–28.
32. Tensen C, Okuzawa K, Blomenrohr M, et al. Distinct efficacies for two endogenous ligands on a single cognate gonadoliberin receptor. *Eur J Biochem.* 1997;243(1-2):134–140.
33. Troskie BE, Hapgood JP, Millar RP, Illing N. Complementary deoxyribonucleic acid cloning, gene expression, and ligand selectivity of a novel gonadotropin-releasing hormone receptor expressed in the pituitary and midbrain of Xenopus laevis. *Endocrinology.* 2000;141(5):1764–1771.
34. Sun YM, Flanagan CA, Illing N, et al. A chicken gonadotropin-releasing hormone receptor that confers agonist activity to mammalian antagonists. Identification of D-Lys(6) in the ligand and extracellular loop two of the receptor as determinants. *J Biol Chem.* 2001;276(11):7754–7761.
35. Millar R, Lowe S, Conklin D, et al. A novel mammalian receptor for the evolutionarily conserved type II GnRH. *Proc Natl Acad Sci USA.* 2001;98(17):9636–9641.
36. Neill JD, Duck LW, Sellers JC, Musgrove LC. A gonadotropin-releasing hormone (GnRH) receptor specific for GnRH II in primates. *Biochem Biophys Res Commun.* 2001;282(4): 1012–1018.
37. Blomenrohr M, Heding A, Sellar R, et al. Pivotal role for the cytoplasmic carboxyl-terminal tail of a nonmammalian gonadotropin-releasing hormone receptor in cell surface expression, ligand binding, and receptor phosphorylation and internalization. *Mol Pharmacol.* 1999;56(6):1229–1237.
38. Kah O, Lethimonier C, Somoza G, Guilgur LG, Vaillant C, Lareyre JJ. GnRH and GnRH receptors in metazoa: a historical, comparative, and evolutive perspective. *Gen Comp Endocrinol.* 2007;153(1-3):346–364.
39. Stewart AJ, Katz AA, Millar RP, Morgan K. Retention and silencing of prepro-GnRH-II and type II GnRH receptor genes in mammals. *Neuroendocrinology* 2009;90(4):416–432.
40. Tello JA, Rivier JE, Sherwood NM. Tunicate gonadotropin-releasing hormone (GnRH) peptides selectively activate *Ciona intestinalis* GnRH receptors and the green monkey type II GnRH receptor. *Endocrinology.* 2005;146(9):4061–4073.
41. Cheng KW, Leung PC. The expression, regulation and signal transduction pathways of the mammalian gonadotropin-releasing hormone receptor. *Can J Physiol Pharmacol.* 2000; 78(12):1029–1052.
42. Cheng CK, Leung PC. Molecular biology of gonadotropin-releasing hormone (GnRH)-I, GnRH-II, and their receptors in humans. *Endocr Rev.* 2005;26(2):283–306.
43. Arora KK, Krsmanovic LZ, Mores N, O'Farrell H, Catt KJ. Mediation of cyclic AMP signaling by the first intracellular loop

of the gonadotropin-releasing hormone receptor. *J Biol Chem.* 1998;273(40):25581–25586.
44. Dehal P, Satou Y, Campbell RK, et al. The draft genome of *Ciona intestinalis*: Insights into chordate and vertebrate origins. *Science* 2002;298(5601):2157–2167.
45. Tello JA, Sherwood NM. Amphioxus: Beginning of vertebrate and end of invertebrate type GnRH receptor lineage. *Endocrinology.* 2009;150(6):2847–2856.
46. Guindon S, Dufayard JF, Lefort V, Anisimova M, Hordijk W, Gascuel O. New algorithms and methods to estimate maximum-likelihood phylogenies: Assessing the performance of PhyML 3.0. *System Biol.* 2010;59(3):307–321.
47. Staubli F, Jorgensen TJ, Cazzamali G, et al. Molecular identification of the insect adipokinetic hormone receptors. *Proc Natl Acad Sci USA.* 2002;99(6):3446–3451.
48. Park Y, Kim YJ, Adams ME. Identification of G protein-coupled receptors for *Drosophila* PRXamide peptides, CCAP, corazonin, and AKH supports a theory of ligand-receptor coevolution. *Proc Natl Acad Sci USA.* 20 2002;99(17):11423–11428.
49. Cazzamali G, Saxild N, Grimmelikhuijzen C. Molecular cloning and functional expression of a *Drosophila* corazonin receptor. *Biochem Biophys Res Commun.* 2002;298(1):31–36.
50. Hansen KK, Stafflinger E, Schneider M, et al. Discovery of a novel insect neuropeptide signaling system closely related to the insect adipokinetic hormone and corazonin hormonal systems. *J Biol Chem.* 2010;285(14):10736–10747.
51. Vadakkadath Meethal S, Gallego MJ, Haasl RJ, Petras 3rd SJ, Sgro JY, Atwood CS. Identification of a gonadotropin-releasing hormone receptor orthologue in *Caenorhabditis elegans. BMC Evol Biol.* 2006;6:103.
52. Lindemans M, Liu F, Janssen T, et al. Adipokinetic hormone signaling through the gonadotropin-releasing hormone receptor modulates egg-laying in *Caenorhabditis elegans. Proc Natl Acad Sci USA.* 2009;106(5):1642–1647.
53. Kotani M, Detheux M, Vandenbogaerde A, et al. The metastasis suppressor gene KiSS-1 encodes kisspeptins, the natural ligands of the orphan G-protein coupled receptor GPR54. *J Biol Chem.* 2001;276(37):34631–34636.
54. Muir AI, Chamberlain L, Elshourbagy NA, et al. AXOR12, a novel human G-protein coupled receptor, activated by the peptide KiSS-1. *J Biol Chem.* 2001;276(31):28969–28975.
55. Ohtaki T, Shintani Y, Honda S, et al. Metastasis suppressor gene KiSS-1 encodes peptide ligand of a G-protein-coupled receptor. *Nature.* 2001;411(6837):613–617.
56. Navarro VM, Fernandez-Fernandez R, Castellano JM, et al. Advanced vaginal opening and precocious activation of the reproductive axis by KiSS-1 peptide, the endogenous ligand of GPR54. *J Physiol.* 2004;561(Pt 2):379–386.
57. Teles MG, Bianco SD, Brito VN, et al. A GPR54-activating mutation in a patient with central precocious puberty. *N Engl J Med.* 2008;358(7):709–715.
58. Thompson EL, Patterson M, Murphy KG, et al. Central and peripheral administration of kisspeptin-10 stimulates the hypothalamic-pituitary-gonadal axis. *J Neuroendocrinol.* 2004;16(10):850–858.
59. Irwig MS, Fraley GS, Smith JT, et al. Kisspeptin activation of gonadotropin releasing hormone neurons and regulation of KiSS-1 mRNA in the male rat. *Neuroendocrinology* 2004;80(4):264–272.
60. Plant TM, Ramaswamy S, Dipietro MJ. Repetitive activation of hypothalamic G protein-coupled receptor 54 with intravenous pulses of kisspeptin in the juvenile monkey (*Macaca mulatta*) elicits a sustained train of gonadotropin-releasing hormone discharges. *Endocrinology.* 2006;147(2):1007–1013.
61. Han SK, Gottsch ML, Lee KJ, et al. Activation of gonadotropin-releasing hormone neurons by kisspeptin as a neuroendocrine switch for the onset of puberty. *J Neurosci.* 2005;25(49):11349–11356.
62. Messager S, Chatzidaki EE, Ma D, et al. Kisspeptin directly stimulates gonadotropin-releasing hormone release via G-protein coupled receptor 54. *Proc Natl Acad Sci USA.* 2005;102(5):1761–1766.
63. Clarkson J, Herbison AE. Postnatal development of kisspeptin neurons in mouse hypothalamus; sexual dimorphism and projections to gonadotropin-releasing hormone neurons. *Endocrinology.* 2006;147(12):5817–5825.
64. Oakley AE, Clifton DK, Steiner RA. Kisspeptin signaling in the brain. *Endocr Rev.* 2009;30(6):713–743.
65. Smith JT, Cunningham MJ, Rissman EF, Clifton DK, Steiner RA. Regulation of Kiss1 gene expression in the brain of the female mouse. *Endocrinology.* 2005;146(9):3686–3692.
66. Smith JT, Dungan HM, Stoll EA, et al. Differential regulation of KiSS-1 mRNA expression by sex steroids in the brain of the male mouse. *Endocrinology.* 2005;146(7):2976–2984.
67. Smith JT, Acohido BV, Clifton DK, Steiner RA. KiSS-1 neurones are direct targets for leptin in the ob/ob mouse. *J Neuroendocrinol.* 2006;18(4):298–303.
68. Lee DK, Nguyen T, O'Neill GP, et al. Discovery of a receptor related to the galanin receptors. *FEBS Lett.* 1999;446(1):103–107.
69. Stafford LJ, Xia C, Ma W, Cai Y, Liu M. Identification and characterization of mouse metastasis-suppressor KiSS1 and its G-protein-coupled receptor. *Cancer Res.* 2002;62(19):5399–5404.
70. Castellano JM, Navarro VM, Fernandez-Fernandez R, et al. Ontogeny and mechanisms of action for the stimulatory effect of kisspeptin on gonadotropin-releasing hormone system of the rat. *Mol Cell Endocrinol.* 2006;257-258:75–83.
71. Zhang C, Roepke TA, Kelly MJ, Ronnekleiv OK. Kisspeptin depolarizes gonadotropin-releasing hormone neurons through activation of TRPC-like cationic channels. *J Neurosci.* 2008;28(17):4423–4434.
72. Liu X, Lee K, Herbison AE. Kisspeptin excites gonadotropin-releasing hormone neurons through a phospholipase C/calcium-dependent pathway regulating multiple ion channels. *Endocrinology.* 2008;149(9):4605–4614.
73. Filby AL, van Aerle R, Duitman J, Tyler CR. The kisspeptin/gonadotropin-releasing hormone pathway and molecular signaling of puberty in fish. *Biol Reprod.* 2008;78(2):278–289.
74. Kitahashi T, Ogawa S, Parhar IS. Cloning and expression of kiss2 in the zebrafish and medaka. *Endocrinology.* 2009;150(2):821–831.
75. Lee YR, Tsunekawa K, Moon MJ, et al. Molecular evolution of multiple forms of kisspeptins and GPR54 receptors in vertebrates. *Endocrinology.* 2009;150(6):2837–2846.
76. Mitani Y, Kanda S, Akazome Y, Zempo B, Oka Y. Hypothalamic Kiss1 but not Kiss2 neurons are involved in estrogen feedback in medaka (*Oryzias latipes*). *Endocrinology.* 2010;151(4):1751–1759.
77. Servili A, Le Page Y, Leprince J, et al. Organization of two independent kisspeptin systems derived from evolutionary-ancient kiss genes in the brain of zebrafish. *Endocrinology.* 2011;152(4):1527–1540.
78. Um HN, Han JM, Hwang JI, Hong SI, Vaudry H, Seong JY. Molecular coevolution of kisspeptins and their receptors from fish to mammals. *Ann NY Acad Sci.* 2010;1200:67–74.
79. Holland LZ, Albalat R, Azumi K, et al. The amphioxus genome illuminates vertebrate origins and cephalochordate biology. *Genome Res.* 2008;18(7):1100–1111.

80. Gwee PC, Tay BH, Brenner S, Venkatesh B. Characterization of the neurohypophysial hormone gene loci in elephant shark and the Japanese lamprey: origin of the vertebrate neurohypophysial hormone genes. *BMC Evol Biol.* 2009;9:47.
81. Keay J, Thornton JW. Hormone-activated estrogen receptors in annelid invertebrates: implications for evolution and endocrine disruption. *Endocrinology.* 2009;150(4):1731–1738.
82. Thomas P, Dressing G, Pang Y, et al. Progestin, estrogen and androgen G-protein coupled receptors in fish gonads. *Steroids* 2006;71(4):310–316.
83. Thomas P, Alyea R, Pang Y, Peyton C, Dong J, Berg AH. Conserved estrogen binding and signaling functions of the G protein-coupled estrogen receptor 1 (GPER) in mammals and fish. *Steroids* 2010;75(8-9):595–602.
84. Pi M, Parrill AL, Quarles LD. GPRC6A mediates the non-genomic effects of steroids. *J Biol Chem.* 2010;285(51): 39953–39964.
85. Thornton JW. Evolution of vertebrate steroid receptors from an ancestral estrogen receptor by ligand exploitation and serial genome expansions. *Proc Natl Acad Sci USA.* 2001;98(10):5671–5676.
86. Bridgham JT, Carroll SM, Thornton JW. Evolution of hormone–receptor complexity by molecular exploitation. *Science* 2006;312(5770):97–101.
87. Putnam NH, Butts T, Ferrier DE, et al. The amphioxus genome and the evolution of the chordate karyotype. *Nature.* 2008; 453(7198):1064–1071.
88. Dos Santos S, Bardet C, Bertrand S, Escriva H, Habert D, Querat B. Distinct expression patterns of glycoprotein hormone-alpha2 and -beta5 in a basal chordate suggest independent developmental functions. *Endocrinology.* 2009;150(8):3815–3822.
89. Kuraku S, Meyer A, Kuratani S. Timing of genome duplications relative to the origin of the vertebrates: did cyclostomes diverge before or after? *Mol Biol Evol.* 2009;26(1):47–59.
90. Hsueh AJ, Jones PB. Gonadotropin releasing hormone: extra-pituitary actions and paracrine control mechanisms. *Annu Rev Physiol.* 1983;45:83–94.
91. Mizuta T, Kubokawa K. Presence of sex steroids and cytochrome P450 genes in amphioxus. *Endocrinology.* 2007;148(8): 3554–3565.
92. Castro LF, Santos MM, Reis-Henriques MA. The genomic environment around the Aromatase gene: evolutionary insights. *BMC Evol Biol.* 2005;5:43.
93. Mizuta T, Asahina K, Suzuki M, Kubokawa K. *In vitro* conversion of sex steroids and expression of sex steroidogenic enzyme genes in amphioxus ovary. *J Exp Zool A Ecol Genet Physiol.* 2008;309(2):83–93.
94. Keay J, Bridgham JT, Thornton JW. The *Octopus vulgaris* estrogen receptor is a constitutive transcriptional activator: evolutionary and functional implications. *Endocrinology.* 2006;147(8):3861–3869.
95. Bridgham JT, Brown JE, Rodriguez-Mari A, Catchen JM, Thornton JW. Evolution of a new function by degenerative mutation in cephalochordate steroid receptors. *PLoS Genet.* 2008;4(9):e1000191.
96. Schubert M, Brunet F, Paris M, Bertrand S, Benoit G, Laudet V. Nuclear hormone receptor signaling in amphioxus. *Dev Genes Evol.* 2008;218(11-12):651–665.
97. Katsu Y, Kubokawa K, Urushitani H, Iguchi T. Estrogen-dependent transactivation of amphioxus steroid hormone receptor via both estrogen and androgen response elements. *Endocrinology.* 2010;151(2):639–648.
98. Paris M, Pettersson K, Schubert M, et al. An amphioxus orthologue of the estrogen receptor that does not bind estradiol: insights into estrogen receptor evolution. *BMC Evol Biol.* 2008;8:219.
99. D'Aniello A, Di Cosmo A, Di Cristo C, Assisi L, Botte V, Di Fiore MM. Occurrence of sex steroid hormones and their binding proteins in *Octopus vulgaris* lam. *Biochem Biophys Res Commun.* 1996;227(3):782–788.
100. Baker ME, Chandsawangbhuwana C. Analysis of 3D models of octopus estrogen receptor with estradiol: evidence for steric clashes that prevent estrogen binding. *Biochem Biophys Res Commun.* 2007;361(3):782–788.
101. Baker ME. Amphioxus, a primitive chordate, is on steroids: evidence for sex steroids and steroidogenic enzymes. *Endocrinology.* 2007;148(8):3551–3553.
102. Baker ME. Origin and diversification of steroids: co-evolution of enzymes and nuclear receptors. *Mol Cell Endocrinol.* 2011;334(1-2):14–20.
103. Thornton JW, Need E, Crews D. Resurrecting the ancestral steroid receptor: ancient origin of estrogen signaling. *Science* 2003;301(5640):1714–1717.
104. Harris GW, Jacobsohn D. Functional grafts of the anterior pituitary gland. *Proc R Soc Lond B Biol Sci.* 1952;139(895): 263–276.
105. Schlosser G. Induction and specification of cranial placodes. *Dev Biol.* 2006;294(2):303–351.
106. Scully KM, Rosenfeld MG. Pituitary development: regulatory codes in mammalian organogenesis. *Science* 2002;295(5563): 2231–2235.
107. Zhu X, Gleiberman AS, Rosenfeld MG. Molecular physiology of pituitary development: signaling and transcriptional networks. *Physiol Rev.* 2007;87(3):933–963.
108. Uchida K, Moriyama S, Chiba H, et al. Evolutionary origin of a functional gonadotropin in the pituitary of the most primitive vertebrate, hagfish. *Proc Natl Acad Sci USA.* 2010;107(36): 15832–15837.
109. Nozaki M, Oshima Y, Miki M, Shimotani T, Kawauchi H, Sower SA. Distribution of immunoreactive adenohypophysial cell types in the pituitaries of the Atlantic and the Pacific hagfish, *Myxine glutinosa* and *Eptatretus burgeri*. *Gen Comp Endocrinol.* 2005;143(2):142–150.
110. Kawauchi H, Sower SA. The dawn and evolution of hormones in the adenohypophysis. *Gen Comp Endocrinol.* 2006;148(1):3–14.
111. Kawauchi H, Suzuki K, Yamazaki T, et al. Identification of growth hormone in the sea lamprey, an extant representative of a group of the most ancient vertebrates. *Endocrinology.* 2002;143(12):4916–4921.
112. Sower SA, Moriyama S, Kasahara M, et al. Identification of sea lamprey GTHbeta-like cDNA and its evolutionary implications. *Gen Comp Endocrinol.* 2006;148(1):22–32.
113. Takahashi A, Kawauchi H. Evolution of melanocortin systems in fish. *Gen Comp Endocrinol.* 2006;148(1):85–94.
114. Lane TF, Sower SA, Kawauchi H. Arginine vasotocin from the pituitary gland of the lamprey (*Petromyzon marinus*): isolation and amino acid sequence. *Gen Comp Endocrinol.* 1988;70(1): 152–157.
115. Querat B, Tonnerre-Doncarli C, Genies F, Salmon C. Duality of gonadotropins in gnathostomes. *Gen Comp Endocrinol.* 2001; 124(3):308–314.
116. Moriyama S, Oda M, Yamazaki T, et al. Gene structure and functional characterization of growth hormone in dogfish, *Squalus acanthias*. *Zoolog Sci.* 2008;25(6):604–613.
117. Huang X, Hui MN, Liu Y, et al. Discovery of a novel prolactin in non-mammalian vertebrates: evolutionary perspectives and its involvement in teleost retina development. *PLoS One* 2009;4(7): e6163.

118. Sudo S, Kuwabara Y, Park JI, Hsu SY, Hsueh AJ. Heterodimeric fly glycoprotein hormone-alpha2 (GPA2) and glycoprotein hormone-beta5 (GPB5) activate fly leucine-rich repeat-containing G protein-coupled receptor-1 (DLGR1) and stimulation of human thyrotropin receptors by chimeric fly GPA2 and human GPB5. *Endocrinology.* 2005;146(8): 3596–3604.
119. Hsu SY, Nakabayashi K, Bhalla A. Evolution of glycoprotein hormone subunit genes in bilateral metazoa: identification of two novel human glycoprotein hormone subunit family genes, GPA2 and GPB5. *Mol Endocrinol.* 2002;16(7):1538–1551.
120. Campbell RK, Satoh N, Degnan BM. Piecing together evolution of the vertebrate endocrine system. *Trends Genet.* 2004;20(8): 359–366.
121. Metz H, Wray S. Use of mutant mouse lines to investigate origin of gonadotropin-releasing hormone-1 neurons: lineage independent of the adenohypophysis. *Endocrinology.* 2010;151(2): 766–773.
122. Schlosser G. Evolutionary origins of vertebrate placodes: insights from developmental studies and from comparisons with other deuterostomes. *J Exp Zool B Mol Dev Evol.* 2005; 304(4):347–399.
123. Schlosser G. Making senses development of vertebrate cranial placodes. *Intl Rev Cell Mol Biol.* 2010;283:129–234.
124. Schlosser G. Do vertebrate neural crest and cranial placodes have a common evolutionary origin? *Bioessays* 2008;30(7):659–672.
125. Park JI, Semyonov J, Chang CL, Hsu SY. Conservation of the heterodimeric glycoprotein hormone subunit family proteins and the LGR signaling system from nematodes to humans. *Endocrine* 2005;26(3):267–276.
126. Nakabayashi K, Matsumi H, Bhalla A, et al. Thyrostimulin, a heterodimer of two new human glycoprotein hormone subunits, activates the thyroid-stimulating hormone receptor. *J Clin Invest.* 2002;109(11):1445–1452.
127. Okada SL, Ellsworth JL, Durnam DM, et al. A glycoprotein hormone expressed in corticotrophs exhibits unique binding properties on thyroid-stimulating hormone receptor. *Mol Endocrinol.* 2006;20(2):414–425.
128. Sun SC, Hsu PJ, Wu FJ, Li SH, Lu CH, Luo CW. Thyrostimulin, but not thyroid-stimulating hormone (TSH), acts as a paracrine regulator to activate the TSH receptor in mammalian ovary. *J Biol Chem.* 2010;285(6):3758–3765.
129. Tando Y, Kubokawa K. Expression of the gene for ancestral glycoprotein hormone beta subunit in the nerve cord of amphioxus. *Gen Comp Endocrinol.* 2009;162(3):329–339.
130. Swinnen K, Broeck JV, Verhaert P, De Loof A. Immunocytochemical localization of human growth hormone- and prolactin-like antigenic determinants in the insects, *Locusta migratoria* and *Sarcophaga bullata. Comp Biochem Physiol A Comp Physiol.* 1990;95(3):373–378.
131. Terakado K, Ogawa M, Inoue K, Yamamoto K, Kikuyama S. Prolactin-like immunoreactivity in the granules of neural complex cells in the ascidian *Halocynthia roretzi. Cell Tissue Res.* 1997;289(1):63–71.
132. Niall HD, Hogan ML, Sauer R, Rosenblum IY, Greenwood FC. Sequences of pituitary and placental lactogenic and growth hormones: evolution from a primordial peptide by gene reduplication. *Proc Natl Acad Sci USA.* 1971;68(4):866–870.
133. Agellon LB, Davies SL, Chen TT, Powers DA. Structure of a fish (rainbow trout) growth hormone gene and its evolutionary implications. *Proc Natl Acad Sci USA.* 1988;85(14):5136–5140.
134. Nicoll CS, Mayer GL, Russell SM. Structural features of prolactins and growth hormones that can be related to their biological properties. *Endocr Rev.* 1986;7(2):169–203.
135. Abdel-Meguid SS, Shieh HS, Smith WW, Dayringer HE, Violand BN, Bentle LA. Three-dimensional structure of a genetically engineered variant of porcine growth hormone. *Proc Natl Acad Sci USA.* 1987;84(18):6434–6437.
136. Salzet M, Salzet-Raveillon B, Cocquerelle C, et al. Leech immunocytes contain proopiomelanocortin: nitric oxide mediates hemolymph proopiomelanocortin processing. *J Immunol.* 1997;159(11):5400–5411.
137. Stefano GB, Salzet-Raveillon B, Salzet M. Mytilus edulis hemolymph contains pro-opiomelanocortin: LPS and morphine stimulate differential processing. *Brain Res Mol Brain Res.* 1999;63(2):340–350.

PART II

NEURAL REGULATION OF ENDOCRINE SYSTEMS

CHAPTER

5

Neural Control of the Anterior Lobe of the Pituitary Gland (*Pars Distalis*)

George Fink

Mental Health Research Institute, University of Melbourne, Parkville, Melbourne, Victoria, Australia

OUTLINE

Summary

Neuroendocrinology is the study of how the nervous system controls hormonal secretion and, in turn, how hormones affect neural function and especially the brain. Neural–endocrine interactions occur at several sites in the body. Here, attention is focused on the neural control of the anterior lobe of the pituitary gland (*pars distalis*). The hormones secreted by the anterior lobe of the pituitary gland control reproduction (through gonadotropin stimulation of sex-steroid secretion, gametogenesis and ovulation), breast development and lactation (by way of prolactin), stress (through adrenocorticotropin stimulation of the adrenal glucocorticoid secretion), body temperature and metabolism (through thyrotropin stimulation of thyroid hormone secretion), and body growth and metabolism (through

DOI: 10.1016/B978-0-12-375097-6.10005-8

growth hormone secretion). The neural control of the synthesis and secretion of anterior pituitary hormones is mediated by neurohormones released from hypothalamic nerve terminals into hypophysial portal vessels in the pituitary stalk that transport them to the anterior pituitary gland. All but one (dopamine) of the hypothalamic neurohormones are peptides, ranging in size from the tripeptide thyrotropin-releasing hormone to the 44-amino acid residue growth hormone (GH)-releasing hormone. The function of the hypothalamic–pituitary axis is regulated by the negative feedback action of hormones secreted by the major target organs of the anterior pituitary gland – i.e., the gonads, the thyroid gland, the adrenal glands and adipose tissue. In reproduction, positive feedback also plays a pivotal role in triggering ovulation. Anterior pituitary target hormones, and especially thyroid hormones, sex steroids and the glucocorticoids, play key roles in neurodevelopment, sexual differentiation of the brain, central neurotransmission, behavior and cognition, mood and mental state. Neuroendocrine systems, and the adrenocorticotropin precursor proopiomelanocortin in particular, provided some of the first models for our understanding of gene transcription, translation and post-translational processing in vertebrates. This and other principles related to neurotransmitter/neurohormone synthesis, release, mode of action and control are highlighted as are neuroendocrine findings that have facilitated our understanding, diagnosis and management of disorders in man.

INTRODUCTION

The human pituitary gland weighs no more than 1 g, but nonetheless controls all the major endocrine systems and is indispensable for life. Located at the base of the brain and surrounded closely by protective dense bone and fibrous membranes, the gland is composed of the neurohypophysis (or neural lobe) and the adenohypophysis.[1,2] Derived embryologically from a neural downgrowth, the neural lobe is composed of axons that project from nerve cells in the hypothalamus and terminate on capillaries of the inferior hypophysial artery. This is the site at which nonapeptides, vasopressin and oxytocin are released into the systemic circulation. Synthesized in the supraoptic and paraventricular nuclei, vasopressin, also termed the antidiuretic hormone, controls the volume of body water, whereas oxytocin is concerned mainly with stimulating milk ejection during lactation, and contraction of the uterus during parturition (see Chapters 6 and 16). The adenohypophysis is derived from an invagination of the stomodeal ectoderm (Rathke's pouch), and is divided into the pars distalis and the pars intermedia.[1–4] The pars distalis is more commonly called the anterior lobe; the pars intermedia and the neural lobe or pars nervosa together form the posterior lobe. There is no distinct pars intermedia in the human. The anterior pituitary gland controls the adrenal and thyroid glands, the gonads, body growth, and development of the breast and lactation, by way of secreting adrenocorticotropin (ACTH), thryrotropin (TSH), the gonadotropins (luteinizing hormone, LH and follicle-stimulating hormone, FSH), growth hormone (GH), and prolactin, respectively. In addition to being growth promoting ("trophic") and stimulating immediate hormonal or cellular events ("tropic"), these hormones all affect metabolism. This is especially the case for ACTH, GH and TSH. The loss of ACTH with the consequent loss of adrenocortical hormone secretion, and, perhaps to a lesser degree, GH and TSH, makes removal of the gland ("hypophysectomy") lethal.[5–8]

Anterior pituitary hormone secretion is under central nervous control, and is modulated by the feedback of hormones secreted by the pituitary target organs[9] (see also Chapter 3) – that is, the gonads and the adrenal and thyroid glands, and, in addition, adipose tissue by way of adipokines such as leptin and adiponectin[10,11] (see also Chapter 14). Neural control of the anterior pituitary hormones is mediated by the hypothalamic–pituitary regulatory neurohormones (formerly termed "factors"), which are released into the hypophysial portal vessels that transport them to the anterior lobe, where they either stimulate or inhibit the release of the anterior pituitary hormones. The hypothalamic–pituitary regulatory factors are termed neurohormones because, instead of being released at synapses between nerve cells, they are released and transported to their target cells in the bloodstream.[8,12]

The hypothalamic–pituitary system is the interface between the central nervous and the endocrine systems by which external factors ("exteroreceptive"), such as day-length and stress, and internal factors, such as emotion, trigger endocrine responses.[1,8,13,14] It was therefore termed the neuroendocrine system, and the pituitary gland is said to be under neuroendocrine control. In addition to the hypothalamic–pituitary system, the circumventricular organs (see below) also satisfy the criteria of neuroendocrine systems. The term "neuroendocrine" is also applied to interactions between nerve and endocrine cells in the periphery and, especially, the viscera, of which the gastrointestinal system and its appendages (especially the pancreas) are the most prominent. Of special relevance to stress is the sympathetic component of the autonomic nervous system and the adrenal medulla. Derived from the embryonic neural crest, the cells of the adrenal medulla are modified postganglionic neurons that secrete their neurotransmitters, mainly epinephrine and norepinephrine, directly into the bloodstream.

This chapter focuses on the principles of neuroendocrine control of the anterior pituitary gland, and illustrates how the hypothalamic–pituitary system: (a) was used to demonstrate that peptides satisfy the criteria of neurotransmitters or neurohormones[8,14–17]; (b) played a significant role in the development of our understanding of gene transcription, translation and posttranslational processing; and (c) could be used as a "window" to study brain function ("neuroendocrine window of the brain"; see also Chapter 36). Neuroendocrinology is also concerned with the effects of pituitary target hormones on the brain–pituitary system – that is, the way that hormones secreted by the anterior pituitary target glands play an important role in brain differentiation and plasticity, affect central neurotransmission and thereby mood, mental state and cognition, and exert feedback effects on the brain–pituitary system.[9,18–22] Hormonal feedback systems constitute the afferent limb of homeostatic regulatory systems that ensure that the output of pituitary hormones is maintained at a preset and functionally optimal level (for details, see Chapter 3).

HYPOTHALAMIC–PITUITARY AXIS

Brief History

The importance of the anterior pituitary gland as "conductor of the endocrine orchestra" was not understood until the early 1930s, when P.E. Smith published his parapharyngeal approach for removing the gland ("hypophysectomy").[1,5,6,8,23,24] The effects of hypophysectomy proved to be so dramatic that for a short period most scientists in the field, including the distinguished neurosurgeon Harvey Cushing, thought that the pituitary gland was autonomous.[1,7,8] However, around the same time (late 1920s to early 1930s), William Rowan, working in Alberta on the annual migration of birds, showed that day length had a potent effect on the growth of the gonads. Rowan's experiments, together with those on seasonal breeding in animals, on the effects of stressful stimuli on endocrine organs, and of brain lesions on pituitary hormone secretion, led to the concept that the anterior pituitary gland must be under central nervous control – a view that Cushing soon adopted.[7,25] The observational and experimental evidence that supported this concept was summarized by Marshall in his 1936 Croonian lecture.[13]

It had long been known that the pituitary gland and brain were connected by the pituitary stalk, but several lines of evidence suggested that in mammals, neural control of the anterior pituitary gland was mediated not by nerve fibers but by chemical substances released into the hypophysial portal vessels. These vessels, first described by Popa and Fielding in the 1930s,[26,27] surround and run down the pituitary stalk, linking a primary plexus of capillaries at the base of the hypothalamus with a second capillary (sinusoidal) plexus in the anterior pituitary gland. Throughout the 1930s and 1940s, a debate raged about the direction of blood flow in the hypophysial portal vessels.[1] Based solely on histological evidence, this debate could have been avoided had researchers read the 1935 report by Houssay and associates, that in the living toad blood flowed from the hypothalamus down to the pituitary gland. However, Houssay's paper was published in French,[28] and so was missed until the late 1940s.

The neurohumoral hypothesis of the control of the anterior pituitary gland was first formally advanced by Friedgood in 1936 and Hinsey in 1937.[8,29,30] However, it was the elegant pituitary graft experiments of Harris and Jacobsohn that showed beyond doubt that the anterior pituitary gland was controlled by substances released at nerve terminals in the median eminence at the base of the hypothalamus, and transported to the pituitary gland by the hypophysial portal vessels.[31] The findings of Harris and Jacobsohn[31] were soon confirmed by the equally elegant pituitary grafting experiments of Nikitovitch-Winer and Everett.[32,33] The characterization of the first three of these substances, all peptides – thyrotropin-releasing factor (now "hormone"), luteinizing hormone-releasing factor (LRF, now termed "gonadotropin-releasing hormone", GnRH), and somatostatin – was to take a further 18 to 21 years of hard work in the laboratories of Roger Guillemin and Andrew Schally, for which they were awarded the 1977 Nobel Prize for Physiology and Medicine.[16,17,34] The sequence of corticotropin-releasing factor-41 (CRF-41) was determined by Wylie Vale and associates in 1981, a breakthrough that had eluded many workers for more than 25 years.[35,36] Evidence for the existence of CRF had been demonstrated by studies of CRF activity in extracts of hypophysial portal blood.[37] The discovery and characterization of GH-releasing hormone (GHRH) followed later with simultaneous publications by two separate groups, one led by Vale[38] and the other by his former mentor, Guillemin.[39] The concentration of GHRH in hypothalamus is exceedingly low, and so the simultaneous discoveries depended upon two pancreatic islet cell carcinomas, one from a patient in Virginia and the other in France, which secreted large amounts of the 44-amino acid residue GHRH, sufficient to enable peptide sequencing (see "A tale of two islets",[40]).

The concept that prolactin secretion was predominantly inhibited by the central nervous system was derived first from the pituitary graft experiments of Everett and Nikitovitch-Winer,[32,33,41] supported by findings that prolactin secretion was increased after hypothalamic lesion or stalk section.[42] Studies on the effects of reserpine and chlorpromazine by Barraclough and Sawyer[43] provided the first clue for the now well-established fact that dopamine is the prolactin inhibitory factor.[41,42] In spite of numerous investigations, no robust data for the existence of a prolactin-releasing factor have been adduced.

BOX 5.1

HOW DO WE KNOW THAT ANTERIOR PITUITARY HORMONE SECRETION IS UNDER NEUROHUMORAL CONTROL?

Between 1910 and 1930, workers were so dazzled by the effects of the newly discovered anterior pituitary hormones that many considered the gland to be autonomous – "the conductor of the endocrine orchestra".[1] However, earlier findings by Bramwell (1888) and Fröhlich (1901) regarding the effect of hypothalamic pituitary tumors on pituitary function had already suggested that the anterior pituitary gland was under CNS control.[2,3] These and the effects of hypothalamic lesions forced Harvey Cushing, a "pituitist" of long standing, to concede that the gland was not autonomous, but under neural control.[4] Inferences drawn from clinical observations were reinforced by the effects of exteroreceptive factors, such as day length and stress, on endocrine function,[5] as well as studies on the effects of experimental brain stimulation and lesions and pituitary stalk section on pituitary hormone secretion.[6,7]

The saga of the development of the neurohumoral hypothesis of anterior pituitary control is reviewed in detail by Geoffrey Wingfield Harris,[6,7] regarded as the "Father of Neuroendocrinology". Friedgood 1936[8] and Hinsey 1937[9] were the first to postulate that the anterior pituitary gland was controlled by substances liberated into the hypophysial portal vessels from nerve terminals in the median eminence. However, the hypothesis was not accepted until work from Harris' Laboratory in Cambridge[6,7] had established that: (a) the direction of blood-flow in the hypophysial portal vessels of a living mammal was from the hypothalamus to the pituitary[10]; (b) in experimental animals in which the portal vessels were permitted to regenerate after severance of the pituitary stalk, function of the anterior pituitary gland could be correlated with the degree of its revascularization by the portal vessels[11]; and (c) the morphological and functional integrity of pituitary grafts was maintained (or developed in the case of grafts from immature donor animals) when these grafts were vascularized by the hypophysial portal, but not the systemic, circulation.[12] The results of the latter study, which constituted the most important experimental evidence for the neurohumoral hypothesis, were confirmed by the equally elegant grafting experiments of Nikitovitch-Winer and Everett.[13,14]

Final proof for the neurohumoral hypothesis came from the isolation and characterization of peptide neurohormones TRH GnRH and SST by Andrew Schally and Roger Guillemin,[15–17] and the demonstration that these and other hypothalamic neurohormones were present in hypophysial portal blood at concentrations significantly greater than in systemic peripheral blood (see main text, and review by Fink and Sheward).[18]

References

1. Fink G. The development of the releasing factor concept. *Clin Endocrinol.* 1976;5(suppl):245–260.
2. Bramwell B. *Intracranial tumours.* Pentaland Edinburgh; 1888.
3. Fröhlich A. Ein Fall von Tumor der Hypophysis cerebri ohne Akromegalie. *Wiener klinische Rundschau.* 1901;15: 833–836.
4. Melmed S, Kleinberg DL. Anterior pituitary. In: Reed Larsen P, Kronenberg HM, Melmed S, Polonsky KS, eds. *Williams Textbook of Endocrinology.* 10th ed. Philadelphia: W.B. Saunders; 2003:177–279.
5. Marshall FHA. Sexual periodicity and the causes which determine it. The Croonian Lecture: *Philos Trans R Soc B.* 1936;226:423–456.
6. Harris GW. *Neural Control of the Pituitary Gland.* London: Edward Arnold; 1955.
7. Harris GW. Humours and hormones. *J Endocrinol* 1972; 53(2):2–23.
8. Friedgood HB. Studies on the sympathetic nervous control of the anterior hypophysis with special reference to a neuro-humoral mechanism. Symposium on Endocrine Glands: Harvard Tercentenary Celebrations. Cited by Friedgood in Textbook of Endocrinology. In: Williams RH, ed. Philadelphia: W.B. Saunders; 1936:635–698.
9. Hinsey JC. The relation of the nervous system to ovulation and other phenomena of the female reproductive tract. *Cold Spring Harb Symp Quant Biol.* 1937;5:269–279.
10. Green JD, Harris GW. Observation of the hypophysioportal vessels of the living rat. *J Physiol.* 1949;108:359–361.
11. Harris GW. Oestrous rhythm, pseudopregnancy and the pituitary stalk in the rat. *J Physiol.* 1950;111:347–360.
12. Harris GW, Jacobsohn D. Functional grafts of the anterior pituitary gland. *Proc R Soc. B* 1952;139:263–276.
13. Nikitovitch-Winer M, Everett JW. Functional Restitution of Pituitary Grafts Re-Transplanted From Kidney To Median Eminence. *Endocrinology.* 1958;63(6):916–930.
14. Nikitovitch-Winer M, Everett JW. Histocytologic Changes in grafts of rat pituitary on the kidney and upon retransplantation under the diencephalon. *Endocrinology.* 1959;65:357–368.
15. Guillemin R. Control of adenohypophysial functions by peptides of the central nervous system. *Harvey Lecture.* 1978;71:71–131.
16. Guillemin R. Hypothalamic hormones a.k.a. hypothalamic releasing factors. *J Endocrinol* 2005;184:11–28 (2005).
17. Fink G. Inadvertent collaboration. *Nature* 1977;269:747–748.
18. Fink G, Sheward WJ. Neuropeptide release *in vivo*: measurement in hypophysial portal blood. In: Fink G, Harmar AJ, eds. *Neuropeptides: a Methodology.* Chichester: John Wiley & Sons Ltd; 1989:157–188.

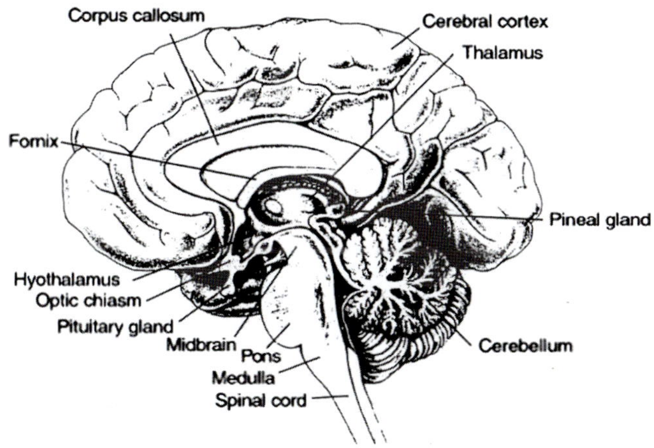

FIGURE 5.1 Diagram of a midsagittal section of the human brain showing the inside surface. Note the pituitary gland attached by way of the pituitary stalk to the base of the hypothalamus. The hypothalamus and the thalamus, which lies above it, form the wall of the third cerebral ventricle at the posterior end of which is the pineal gland.

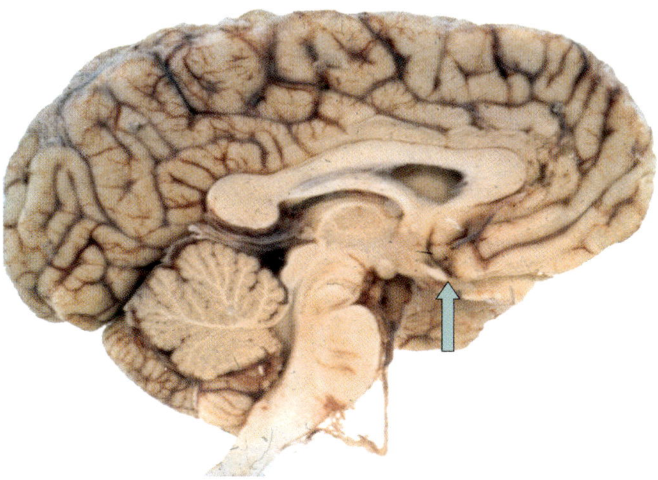

FIGURE 5.2 Medial aspect of the human brain postmortem, with the arrow pointing at the pituitary stalk (the pituitary gland was left in the skull on removal of the brain). Other pertinent regions and structures are shown diagrammatically in Figure 5.1.

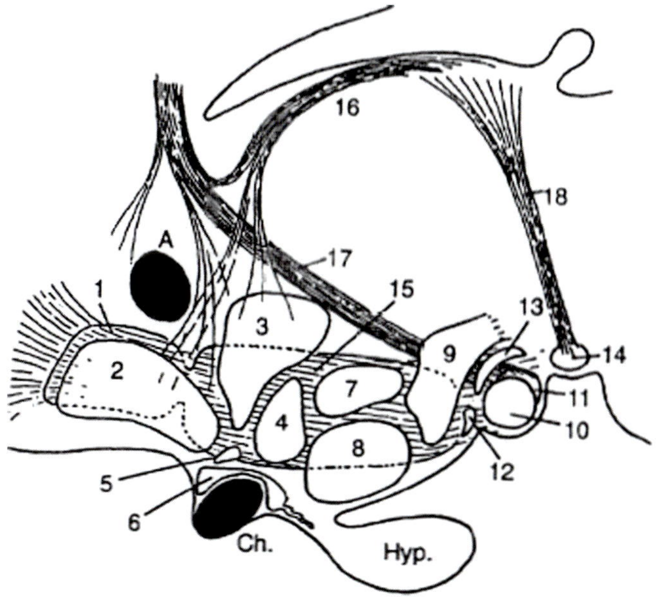

FIGURE 5.3 Diagram showing the relative positions in a sagittal plane of hypothalamic nuclei in a typical mammalian brain and their relation to the fornix, stria habenularis, and fasciculus retroflexus. A, anterior commissure; Ch., optic chiasma; Hyp., hypophysis (pituitary gland). (1) Lateral preoptic nucleus (permeated by the medial forebrain bundle). (2) Medial preoptic nucleus. (3) Paraventricular nucleus. (4) Anterior hypothalamic area. (5) Suprachiasmatic nucleus. (6) Supraoptic nucleus. (7) Dorsomedial hypothalamic nucleus. (8) Ventromedial hypothalamic nucleus. (9) Posterior hypothalamic nucleus. (10) Medial mamillary nucleus. (11) Lateral mamillary nucleus. (12) Premamillary nucleus. (13) Supramamillary nucleus. (14) Interpeduncular nucleus (a mesencephalic element in which the fasciculus retroflexus terminates). (15) Lateral hypothalamic nucleus (permeated by the medial forebrain bundle). (16) Stria habenularis. (17) Fornix. (18) Fasciculus retroflexus of Meynert (habenulopeduncular tract). *Reproduced from Le Gros Clark WE. Morphological aspects of the hypothalamus. In: Le Gros Clark WE, Beattie J, Riddoch G, Dott NM, eds.* The Hypothalamus. Morphological, Functional, Clinical and Surgical Aspects. *Edinburgh: Oliver and Boyd; 1938:1–68, with permission.*

Anatomy and Development

The pituitary gland is linked to the hypothalamus at the base of the brain (Figs 5.1, 5.2). The hypothalamus consists of a medial part adjacent to the third cerebral ventricle, in which are located the major hypothalamic nuclei; and a lateral part comprised mainly of the medial forebrain bundle, a large cable of nerve fibers that carries reciprocal fiber tracts between midbrain and forebrain (Fig. 5.3), in which are embedded a few aggregations of nerve cell bodies. Axons from nerve cell bodies located in the hypothalamic nuclei project to the median eminence, where they either terminate on the loops of primary capillaries of the hypophysial portal vessels in the external layer of the median eminence or form a cable that passes through the internal layer of the median eminence to form the bulk of the pituitary stalk and then the neural lobe (Fig. 5.4). The median eminence, so called because it protrudes as a small dome in the midline from the base of the hypothalamus, forms the floor of the third ventricle, and is delineated by the optic chiasm in front, the mammillary bodies behind, and a depression (hypothalamic sulcus) on either side. Arising from the median eminence is the neural stalk, which links the pituitary gland to the brain.

The pituitary gland is located in a fossa in the basisphenoid bone at the base of the skull, the "sella turcica," so called because its shape resembles a Turkish saddle. The close proximity of the hypothalamus and pituitary gland to the optic chiasm means that tumors in either the hypothalamus or the pituitary gland may press on,

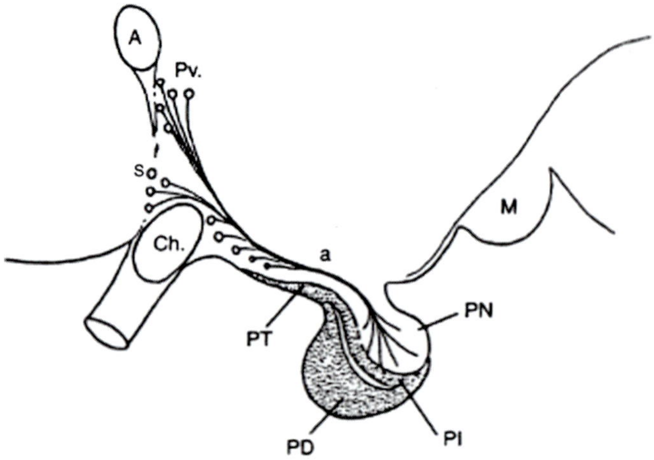

FIGURE 5.4 Schematic section of the mammalian hypothalamus and pituitary gland showing the neurohypophysial tract (labeled (a)) comprised mainly of fibers derived from the paraventricular (Pv) and supraoptic (So) nuclei. A, anterior commissure; Ch, optic chiasma; M, mamillary bodies; PT, pars tuberalus; PD, pars distalis; PI, pars intermedia; PN, pars nervosa.

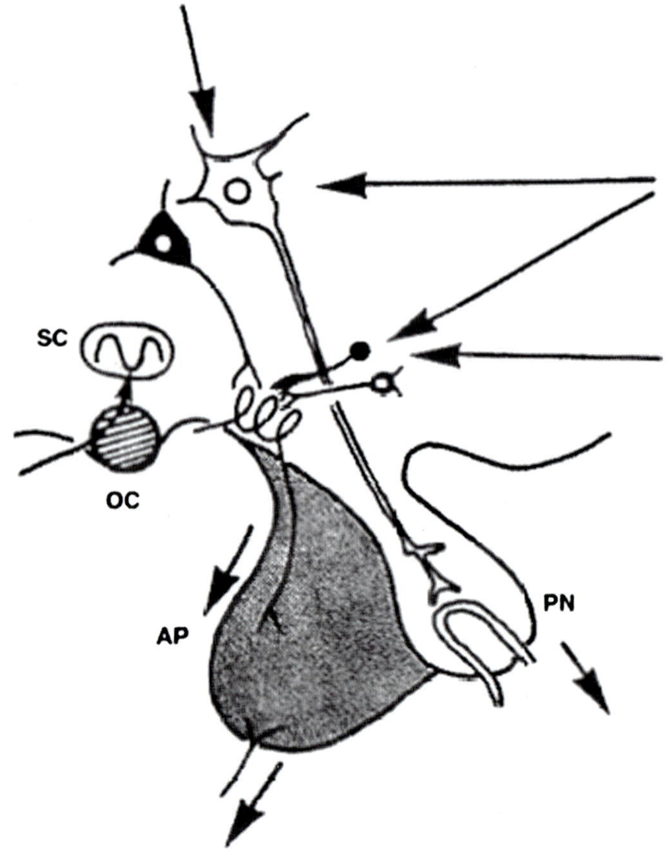

FIGURE 5.5 A schematic diagram of the hypothalamic–pituitary system showing the magnocellular (white) projections directly to the systemic capillaries of the pars nervosa (PN) and the parvocellular (black) projections to the primary plexus of the hypophysial portal vessels, which convey neurohormones to the pars distalis of the anterior pituitary gland (AP). Dorsal to the optic chiasm (OC) are the suprachiasmatic nuclei (SC), which receive direct projections from the retina and play a key role in the control of circadian rhythms (indicated by the sinusoidal curve). Activity of the intrinsic neurons of the hypothalamus is influenced greatly by projections (indicated by arrows) from numerous areas of the forebrain, midbrain and hindbrain, particularly the limbic system, as well as by hormones, mainly gonadal steroid hormones, adrenal glucocorticoids and thyroid hormone.

or more rarely invade, the optic chiasm or tracts, thereby leading to visual defects.

The hypothalamic–pituitary axis is divided functionally into two systems (Fig. 5.5). The hypothalamus, hypophysial portal vessels and adenohypophysis constitute the hypothalamo–adenohypophysial axis. The hypothalamus, neural stalk and neural lobe constitute the hypothalamo–neurohypophysial axis, described in detail in Chapter 6. The neural stalk is made up of numerous nerve fibers that project mainly from magnocellular neurons of the paraventricular and supraoptic nuclei (PVN and SON) of the hypothalamus to terminate on a capillary bed derived from the inferior hypophysial artery and located in the neural lobe of the pituitary gland. Nerve fibers in the neural lobe (or pars nervosa) are surrounded by pituicytes, the equivalent of glial cells. Also present in the neural stalk are nerve fibers of other types of chemical neurotransmitter (in addition to vasopressin and oxytocin), such as the endogenous opioids and dopamine, which are present in nerve fibers that project from the arcuate nucleus and innervate the pars intermedia. Because the stalk and median eminence are continuous, these dopaminergic neurons are a continuation of a dense palisade of dopaminergic fibers that also terminate on the primary plexus of the hypophysial portal vessels enmeshed with nerve fibers that contain other neurohormones.[44] The pituitary stalk and median eminence are covered by a single layer of cells termed the pars tuberalis, which is continuous with the pars distalis.

The adenohypophysis develops from an outgrowth of the ectodermal placode, which forms the roof of the embryonic mouth (or "stomadeum").[2,4] This ectodermal outgrowth forms Rathke's pouch and meets the neurohypophysis, which grows down from the floor of the embryonic third ventricle. Rathke's pouch closes and separates from the roof of the mouth. The caudal (rear) part of the pouch forms the pars intermedia, which becomes tightly juxtaposed to the rostral surface of the neurohypophysis (Fig. 5.6). The rostral part of the pouch develops into the pars distalis (Fig. 5.6). Vascularization of the median eminence and the pituitary gland in the rat begins at about day 15 of embryonic (E15) life, and the hypophysial portal vessels (see below) become defined by E18. Nerve terminals in the median eminence with granular vesicles (presumably neurohormones) are first evident on E16, and the first secretory

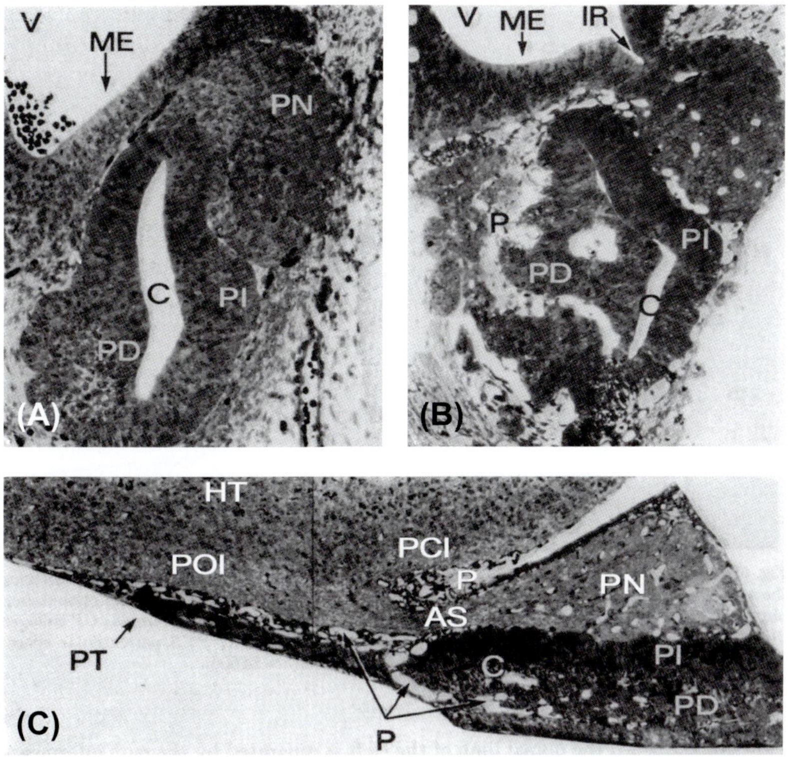

FIGURE 5.6 Development of the pituitary gland in the rat. Photomicrographs of midline sagittal sections through the hypothalamic–pituitary complex of rats at embryonic days 15 (**A**), 17 (**B**), and 20 (**C**). (**A**) The pituitary anlage shortly after closure of Rathke's pouch, which migrates dorsally to meet neurohypophysial downgrowth from the floor of the hypothalamus. Rotation of the pituitary gland caudally through 135° with respect to the base of the diencephalon (hypothalamus) is seen, as is the invasion of the pars distalis (PD) by the leash of portal vessels (P) at E17. AS, anatomical stem; C, hypophysial cleft; HT, hypothalamus; IR, infundibular recess; ME, median eminence; PCI, pars caudalis infundibuli; PI, pars intermedia; PN, pars nervosa; POI, pars oralis infundibuli; V, third ventricle. 1 micrometer Araldite sections, toluidine blue stain. A and B ×120; C ×80. *Reproduced from Fink G, Smith GC. Ultrastructural features of the developing hypothalamo-hypophysial axis in the rat: a correlative study.* Z. Zellforsch. mikrosk. Anat. *1971;119:208–226, with permission.*

granules appear in pars distalis cells on E17.[2] This sequence of embryonic development suggests that the development of secretory activity in cells in the pars distalis may depend in part on the development of nerve terminals in the median eminence and the anlage of the hypophysial portal vessels.[2] The factors that determine differentiation of hypothalamic neurohormonal and pituitary cells are not perfectly understood, but for accounts of genetic mechanisms involved the reader is referred to recent papers by Himes and Raetzman[45] and Auila *et al.*[46]

NEUROHEMAL JUNCTIONS AND CIRCUMVENTRICULAR ORGANS

Overview

Neurohemal junctions are the fundamental functional modules of the major central neuroendocrine system, the median eminence. They are composed of nerve terminals and capillaries that are closely juxtaposed and thereby facilitate the release of chemical messengers from nerve terminals into the bloodstream and *vice versa* (Fig. 5.7). Neurohemal junctions are also the fundamental units of the neurohypophysis and of the circumventricular organs (CVOs) that are located at various sites around the third and fourth cerebral

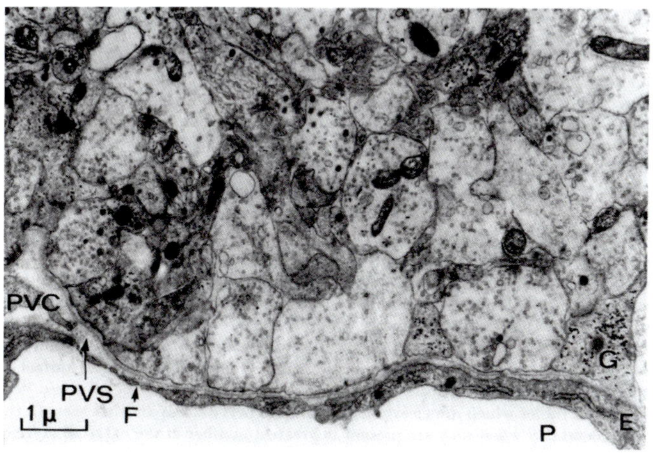

FIGURE 5.7 Electromicrograph of the external layer of the median eminence of a rat at the first postnatal day. Note the high density of nerve terminals around part of a primary portal capillary vessel (P), which is fenestrated (F). Note also the large number of agranular and granular vesicles in the nerve terminals. These vesicles contain the packets (quanta) of neurohormone or neurotransmitter that are released on nerve depolarization as a consequence of nerve action potentials. The neurohormones are released into the perivascular space, and from there move rapidly into portal vessel blood for transport to the pituitary gland. This arrangement is typical of neurohemal junctions found in the several circumventricular organs of the brain (see text). E, endothelial cell; F, fenestration; G, glial process; P, portal vessel; PVC, perivascular cell; PVS, perivascular space. ×13,200. *Reproduced from Fink G, Smith GC. Ultrastructural features of the developing hypothalamo-hypophysial axis in the rat: a correlative study.* Z. Zellforsch. mikrosk. Anat. *1971;119:208–226, with permission.*

ventricles[47–49] (see also Chapters 1,13). All CVOs are characterized by the fact that their vessels are fenestrated (Fig. 5.7), and that the blood–brain barrier (BBB) is less potent at these compared with other sites in brain.[47,48] Circumventricular organs play a crucial role as transducers of information between the blood, neurons and cerebral spinal fluid (CSF). They permit both the release and sensing of hormones without disrupting the blood–brain barrier. The CVOs thereby have essential regulatory actions in diverse physiological functions.

The neurohemal junctions in the median eminence, neurohypophysis and pineal gland facilitate the transport of neurohormones from the nerve terminals or nerve cell derivatives (pineal) into the bloodstream, whereas at the other CVOs, neurohemal junctions facilitate the sensing of blood hormones by nerve cells. The latter mechanism has been implicated in the "crosstalk" between peripheral organs and the brain, so that, for example, the hypertensive effect of the peptide angiotensin is due in part to its activation of neurons of the subfornical organ, which have a high density of angiotensin receptors.[48]

Classically, there are eight CVOs: three sensory, four secretory, and one that is of indiscriminate character.[47] The sensory CVOs include the subfornical organ (SFO), organum vasculosum of the lamina terminalis (OVLT), and the area postrema (AP). The secretory CVOs include the neurohypophysis, median eminence, intermediate lobe of the pituitary gland, and the pineal gland. The remaining indiscriminate CVO is the subcommissural organ (SCO). The SCO is not highly permeable and does not have fenestrated capillaries, and therefore is not considered among the standard categories of CVOs. However, the SCO still plays a significant neuroendocrine role.[47]

In addition to the lack of normal BBB potency and the presence of a dense vascular supply, the sensory CVOs (SFO, OVLT and AP) contain exceptionally dense aggregations of different receptors for peripheral signals, including regulatory peptides (e.g., angiotensin, cholecystokinin, ghrelin, leptin), steroids (e.g., estradiol) and specific ions (e.g., Ca^{2+}, Na^{+}). These specialized features uniquely position the sensory CVOs with the potential to directly monitor the constituents of peripheral circulation and send signals, via afferent projections, to autonomic control centers in the hypothalamus and medulla.[47,48] The sensory CVOs thus represent potential windows in the brain for autonomic feedback to the CNS. The sensory CVOs play a key role in cardiovascular regulation, and fluid and visceral control. The AP is a "vomiting center" that can detect noxious substances in the blood and trigger the vomiting reflex. Although the SFO has been viewed primarily as an angiotensin sensor, with roles in body fluid homeostasis and cardiovascular regulation, electrical activation studies have shown that the SFO also induces drinking in slaked and feeding in satiated rats.[50] Taken together with the fact that leptin may act at the SFO, the SFO may play a role in the integration of feeding control mechanisms involved in the hypothalamic control of energy homeostasis.[50]

Finally, recent data show that, in mice, CVO cells proliferate and undergo constitutive neurogenesis and gliogenesis.[51] These findings suggest that CVOs may constitute a previously unknown source of stem/progenitor cells, capable of giving rise to new neurons and/or glia in the adult brain.[51]

Pineal Gland

Philosophers (Descartes' "seat of the soul") and scientists have long been intrigued by the pineal, a CVO that deserves attention here because it reinforces the principles of neuroendocrine control. The pineal secretes melatonin into the circulation, and plays an important role in the photoperiodic control of reproduction in seasonal breeding animals.[52]

The secretion of melatonin is exquisitely sensitive to light. Pinealocytes in submammalian species are photoreceptors. The gland offers an excellent experimental model for studies of the transduction of light into nerve impulses and neurohormone secretion.[53,54] The outer segment (sensory pole) of the pinealocyte in fish, amphibia and reptiles has all the ultrastructural characteristics of a true photoreceptor. However, these features are only vestigial in mammals, and intermediate forms exist in birds. In fish, amphibia and reptiles, the effector pole of the pinealocyte "synapses" with secondary pineal neurons that give rise to the pineal tract, which propagates signals to the central nervous system. In birds and mammals, however, pinealocytes secrete melatonin directly into the circulation (or cerebrospinal fluid) in a neuroendocrine manner.

Melatonin, a derivative of serotonin, is synthesized within pinealocytes in two steps. First, serotonin is converted by the rate-limiting enzyme, arylalkylamine *N*-acetyltransferase (AANAT), to *N*-acetyl serotonin, which is then converted to melatonin by hydroxyindole-*0*-methyltransferase[53] (Fig. 5.8). The arylalkylamine *N*-acetyltransferase (AANAT) family is divided into structurally distinct vertebrate and non-vertebrate groups. Expression of vertebrate AANATs is limited primarily to the pineal gland and retina, where it plays a role in controlling the circadian rhythm in melatonin synthesis. Based on the role melatonin plays in biological timing, AANAT has been given the moniker "the Timezyme." Non-vertebrate AANATs, which occur in fungi and protists, are thought to play a role in

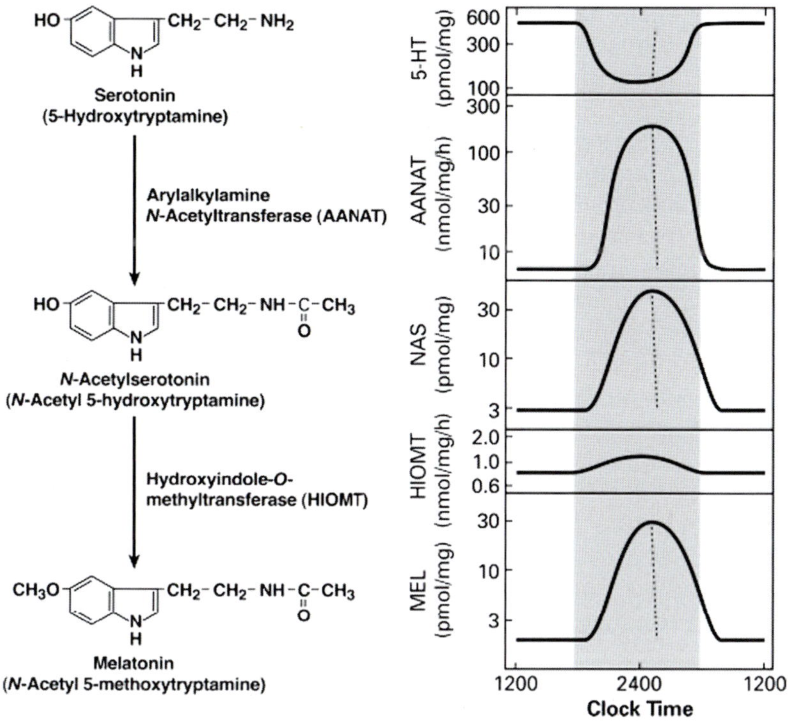

FIGURE 5.8 Daily rhythms in pineal indole metabolism. Shaded area represents darkness. The increase in AANAT activity and the resulting changes in indole metabolism normally occur at night in the dark. "Lights off" during the day does not cause these changes. The rhythmic pattern continues in constant darkness because AANAT is stimulated by an endogenous circadian clock. Under conditions of constant darkness, the rhythm is not entrained to the environmental lighting cycle and is "free-running" with a period of 23.5 to 24.5 h. In contrast, in a constant lighting regimen rhythmic changes in AANAT do not occur because circadian clock stimulation of AANAT is blocked in response to environmental lighting. The dotted lines represent the very rapid changes in each parameter, which occur following "lights on" at night. *Reproduced from Klein DC. Arylalkylamine N-acetyltransferase: "the Timezyme."* J Biol Chem. *2007;282(7):4233–4237, with permission.*

detoxification, and are not known to be associated with a specific tissue.[55]

Because little if any melatonin is stored, the rate of melatonin secretion is tightly linked to its synthesis, which depends on AANAT action, which in turn depends on noradrenaline release from the dense sympathetic innervation of the gland. In mammals, the control of melatonin secretion by light is mediated by a multisynaptic pathway that starts in the retina of the eye and successively involves synapses in the suprachiasmatic nucleus, the PVN, the intermediolateral column of the spinal cord, and the neurons of the superior cervical ganglion of the sympathetic nervous system. Light acts through the retinohypothalamic tract to entrain the SCN clock and to block SCN stimulation of the pineal.[53] At night the SCN induces sympathetic terminals in the pineal gland to release noradrenaline, which stimulates melatonin secretion by an action of beta1-adrenoreceptors on pinealocytes. Cyclic AMP is the main intracellular second messenger that mediates the action of noradrenaline in this system by promoting the formation of the AANAT complex.[53]

The secretion of melatonin starts with the onset of the dark period (night) and stops with the onset of the light period (day). The secretion of melatonin during the darkness is stopped abruptly by exposure to light. In blind persons, the secretion of melatonin takes on the typical 25-h free-running cycle. Taken together, these and other data suggest that the secretion of melatonin in mammals is predominantly controlled by light exposure superimposed on the intrinsic rhythm of the major neural clock, the suprachiasmatic nucleus. Over 600 genes that are important for immunity, cell cycle and death, intracellular signaling molecules, transcription factors and circadian rhythmicity are also under circadian regulation in the mammalian pineal gland.[56] In lower vertebrates, pinealocytes possess a photosensitive and autonomic circadian rhythm in melatonin secretion that persists in dissociated cell cultures.[57]

An interaction between melatonin and the hypothalamic–pituitary–adrenal stress response system has been suggested; indeed, melatonin has been implicated in glucocorticoid receptor-induced gene expression, exerts direct inhibitory effects upon several ACTH responses in the human adrenal gland,[58] and modulates the oscillation of clock genes in diurnal subhuman primates.[59] However, the precise role of melatonin in the circadian rhythm of ACTH and adrenal corticosteroids, and in the stress response, has yet to be established. Melatonin is a powerful antioxidant, which may explain some of its alleged beneficial effects with respect to aging.[60,61] So far, the most robust evidence regarding the function of melatonin in mammals relates to its inhibitory actions with respect to reproduction, especially in seasonal breeding animals such as the wallaby, hamster, vole and sheep.[62–68]

The suprachiasmatic nucleus constitutes the central generator of circadian rhythms of the body, with one notable exception – the daily anticipation of a meal (for details, see Chapters 12 and 14). That is, rats in which the suprachiasmatic nucleus has been lesioned are still able to anticipate one meal per day. This anticipation is associated with an increase in activity, core body temperature and plasma corticosterone concentrations, that occurs 2–4 hours before access to the meal. The precise nature and location of the "feeding entrained oscillator" (FEO) that presumably determines the daily rhythm of feeding anticipation have yet to be established. However, studies with c-Fos (early immediate gene) activation have implicated the nucleus accumbens and the limbic system. The relative importance of the pineal gland and its precise role in the photoperiodic control of circadian rhythms also await determination. However, recent studies have shown that exogenous melatonin can alter neuronal excitability in the majority of SCN neurons in the mouse, regardless of whether or not they overtly express the core clock gene Per1.[69] These effects of melatonin on neuronal excitability involve $GABA_A$ receptors.[69] The fact that melatonin may act mainly by modulating inhibitory GABAergic transmission within the SCN might explain why exogenous application of melatonin has heterogenous effects on individual SCN neurons. Although apparently robust, the definitive understanding of the molecular genetic mechanisms that control circadian rhythms remains to be elucidated.

NEUROHORMONAL CONTROL OF ANTERIOR PITUITARY HORMONE SECRETION

As outlined above, the transmission of signals between the brain and the anterior pituitary gland is mediated by chemical messengers, neurohormones, that are transported by the hypophysial portal vessels from the hypothalamus to the anterior pituitary gland (Figs 5.1–5.6), where they either stimulate or inhibit the synthesis and release of anterior pituitary hormones. Synthesized in nerve cells of the hypothalamic nuclei, the neurohormones are released from nerve terminals into the plexus of primary capillaries of the hypophysial portal vessel system. These primary capillaries are derived from the superior hypophysial arteries and coalesce to form the hypophysial portal veins, which run on the surface or through the pituitary stalk to the anterior pituitary gland, where they form a secondary plexus of vessels termed pituitary sinusoids. Vessels on the surface of the stalk are the long portal vessels, whereas those within the substance of the stalk are the short portal vessels.

The Hypophysial Portal Vessels

The hypophysial portal vessels (Fig. 5.9) are so called because they transport chemical messengers from one capillary bed (primary capillaries) to a second capillary bed before entering the general circulation. In principle this is identical to the hepatic portal system, which transports substances from the primary bed of capillaries in the intestine and its appendages (e.g., pancreas) to a second bed of capillaries or sinusoids in the liver. Both the primary and the secondary (sinusoids) plexus of capillaries are fenestrated (Fig. 5.7), which facilitates the transport of substances across the capillary wall. Hormones released from anterior pituitary cells are transported by pituitary veins into the systemic circulation, by which they are transported to their major target organs, the gonads and the adrenal and thyroid glands.

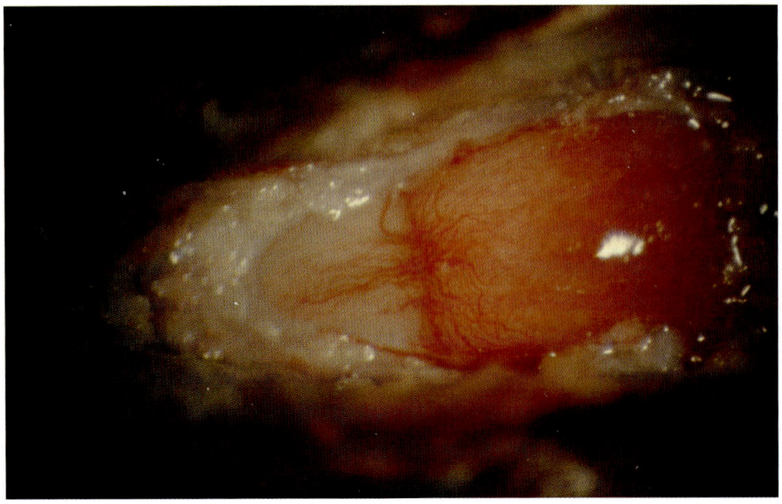

FIGURE 5.9 High-power view through a dissecting microscope of the hypophysial portal vessels on the anterior surface of the pituitary stalk (left) of an anesthetized rat. Note how the portal vessels arise from the primary capillary bed on the median eminence (pink area to the left) and fan out over the anterior pituitary gland at the pituitary stalk junction to the right. The tubero-infundibular artery, a branch of the superior hypophysial artery, can be seen arching across the top of the stalk–pituitary junction where it enters the anterior pituitary gland. This artery passes through the anterior pituitary gland to supply arterial blood to the neurohypophysis. *From G. Fink (unpublished).*

Criteria for Neurohormones and Neurotransmitters: The External Layer of the Median Eminence a Neurovascular Synapse

In 1972, Werman[70] summarized the criteria for a candidate compound to be classified as a neurotransmitter or neurohormone as follows:

> If it can be shown that a substance is released into the extracellular space from presynaptic nerves in quantities consistent with the amount and rate of stimulation and the physiology of transmitter release at that junction, and if it can be shown that the material released acts on postsynaptic membranes by using molecular mechanisms identical with those used by the physiologically evoked transmitter, then that substance is a transmitter. Thus, in order to identify a transmitter, one must satisfy the criteria of collectability and identity of action.

These criteria are difficult to satisfy in the central nervous system, where the study of neurotransmitter release and action under physiological conditions is difficult to achieve *in vivo*. The surgical accessibility of hypophysial portal vessels has made it possible to satisfy criteria of collectability and identity of action for the hypothalamic–pituitary neurohormones in the anesthetized rat,[14,15,37,71–74] anesthetized rhesus monkey[75–77] and conscious sheep.[78–84]

Comprised of nerve fibers that terminate on the primary plexus of the hypophysial portal vessels, the external layer of the median eminence is a unique neurovascular synapse which can be used as a window that allows the study of at least some characteristics of central neurotransmission. The anterior pituitary gland provides a readily accessible effector organ for studying the neurohormone–receptor interactions and "postsynaptic" events that underlie the action of the neurohormones/neurotransmitters. The neuropil that surrounds the primary plexus of the portal capillaries is heterogeneous (Fig. 5.6); this carries the advantage that interactions between different neurotransmitter classes can be studied, and the disadvantage that it is difficult to study one particular class of neurotransmitter neuron in isolation.

Most of the known hypothalamic neurohormones are peptides, but non-peptide transmitters such as dopamine are also released into the hypophysial portal vessels and have potent effects on pituitary function.[80,85] Information gained from the physiology and pharmacology of neurotransmitter/neurohormone release into hypophysial portal blood may be applicable to our understanding of how the same peptide and non-peptide transmitters are released in other, less accessible, parts of the central nervous system.

The technique of collecting hypophysial portal blood has also made it possible to clarify the physiological significance of post-translational processing. This is exemplified by the fact that somatostatin-14 as well as somatostatin-28, derived from the same precursor, are released into hypophysial portal blood.[86,87] The measurement of neurohormone release into hypophysial portal blood has also made it possible to ascertain whether newly discovered hypothalamic compounds could serve as hypothalamic–pituitary regulatory factors. Thus, for example, the concentrations of the cardiac peptide, atrial natriuretic peptide (ANP), were about four times greater in hypophysial portal than in systemic blood,[88,89] a finding that led to immunoneutralization studies which suggest that ANP is an ACTH-inhibiting factor[90,91] (see also below).

Studies of hypophysial portal blood can also exclude the neurohormonal role of a candidate neurotransmitter. Thus, for example, although immunohistochemistry suggests that angiotensin, cholecystokinin, substance P, neurotensin, galanine, neuropeptide Y and other peptides are present in nerve terminals of the external layer of the median eminence, their concentrations in portal blood are not greater than in peripheral blood, which makes it unlikely that they play a role as a neurohormone.[92] Projections into the external zone of the median eminence of neuronal systems containing peptides that are not released into hypophysial portal blood have been postulated to modulate the secretion of the well-established hypothalamic neurohormones by acting on their neurosecretory terminals.[92] This idea is possibly exemplified by the apparent modulatory effect of CCK on GnRH neurons, a finding that requires proof from studies of conditional CCK knockout mice.[93,94]

All of the known hypothalamic–pituitary regulatory neurohormones are present in regions of the nervous system outside the hypothalamus, where Werman's criteria for a neurotransmitter (above) need to be proved by determining whether the neurohormones are released (e.g., by push–pull cannulae or dialysis), activate cells (e.g., electrophysiologically), and are responsible for a behavioral effects. The functional significance of neurohormone action in areas of the CNS outside the hypothalamus is illustrated by CRF-41 and the urocortins, which are thought to be involved in coordinating the behavioral responses to stress,[95–97] as well as central control of energy and metabolism.[98,99]

Hypothalamic Neurohormones

Introduction

Most of the neurohormones that mediate the neural control of anterior pituitary hormone secretion are peptides that are synthesized in discrete hypothalamic nuclei. Hypothalamic neurohormone-secreting neurons are "the final common pathway" neurons for central nervous control of the anterior pituitary gland, a term borrowed by G.W. Harris[1] from Sherrington's description

BOX 5.2

HOW DO WE KNOW ABOUT THE TELEOLOGICAL ADVANTAGES OF NEUROHORMONAL CONTROL?

The hypothalamic–pituitary axis illustrates the remarkable economy of physiological systems. First, and perhaps most impressive, are the hypophysial portal vessels, which, by transporting the neurohormones from the hypothalamus to the pituitary gland, undiluted by mixture in the systemic circulation, ensure that hypothalamic neurohormones released in very small amounts from the hypothalamus will reach the pituitary gland at concentrations orders of magnitude greater than in the systemic circulation, and therefore sufficient to exert their effects. The corollary of this is that relatively little neurohormone needs to be released to exert its effect, and therefore only a small amount of new neurohormone needs to be synthesized. The "metabolic economy" of the hypophysial portal system is brought into sharp relief by the fact that the hypothalamic concentration and total content of the hypothalamic–anterior pituitary regulatory neurohormones (i.e., GnRH, TRH, CRF-41, SST, GHRH) are three or more orders of magnitude lower than those of the neurohypophysial nonapeptides, vasopressin and oxytocin, which reach their peripheral targets by the systemic circulation[1–3] (see also Chapter 6). Second, the transport of neurohormones at effective concentrations by the hypophysial portal vessels also protects the body from potential adverse effects of the high concentrations of the neurohormones necessary to stimulate or inhibit pituitary hormone secretion. Thus, for example, the high portal blood concentrations of somatostatin may, in the systemic circulation, inhibit insulin and glucagon secretion.[4] Similarly, the portal plasma concentrations of atrial natriuretic peptide that inhibit ACTH secretion would, in the systemic circulation, cause a potentially lethal drop in systemic blood pressure.[5]

Third, most of the neurohormones/neuropeptides of the hypothalamic–pituitary system have been implicated as neurotransmitters, neuromodulators or neurotropins elsewhere in the nervous system, although robust evidence for their precise function needs to be established. Thus, for example, somatostatin is secreted by cells of the pancreatic islets, and inhibits the secretion of insulin, glucagon and a range of gut peptides[4,6] (see also Chapter 35). And, as detailed by Bale and Vale,[7] and in Chapter 15, CRF-41 is present in higher brain centers and in the periphery, and has been implicated in stress-related behaviors, cardiovascular control, and the synchronization of the endocrine, autonomic and immunological components of the stress response.

References

1. Fink G. The external layer of the median eminence: a neurovascular synapse. *Neurochem Intl.* 1986;9:141–153.
2. Sherwood NM, Chiappa SA, Sarkar DK, Fink G. Gonadotropin-releasing hormone (GnRH) in pituitary stalk blood from proestrous rats: effects of anesthetics and relationship between stored and released GnRH and luteinizing hormone. *Endocrinology.* 1980;107:1410–1417.
3. Cross BA, Dyball RE, Dyer RG, Jones CW, Lincoln DW, Morris JF, Pickering BT. Endocrine neurons. *Recent Prog Horm Res.* 1975;31:243–294.
4. Strowski MZ, Parmar RM, Blake AD, Schaeffer JM. Somatostatin inhibits insulin and glucagon secretion via two receptors subtypes: an *in vitro* study of pancreatic islets from somatostatin receptor 2 knockout mice. *Endocrinology.* 2000; 141(1):111–117.
5. Fink G, Dow RC, Casley D, Johnson CI, Bennie J, Carroll S, Dick H. Atrial natriuretic peptide is involved in the ACTH response to stress and glucocorticoid negative feedback in the rat. *J Endocrinol.* 1992;135:37–43.
6. De Martino MC, Hofland LJ, Lamberts SW. Somatostatin and somatostatin receptors: from basic concepts to clinical applications. *Prog Brain Res.* 2010;182:255–280.
7. Bale TL, Vale WW. CRF and CRF receptors: role in stress responsivity and other behaviors. *Annu Rev Pharmacol Toxicol.* 2004;44:525–557.

of the alpha motor neurons of the spinal cord, which innervate and control the contraction of skeletal muscles. Like the α motor neurons, hypothalamic neurons are controlled by inputs to the hypothalamus from the brainstem, midbrain and higher brain centers. The hypothalamic neuroendocrine neurons are connected with many other regions of the nervous system, particularly the components of the limbic system, which is involved in several important higher brain functions, including emotions, olfaction and memory (see Chapter 14).

Neural control of the anterior pituitary hormones is covered in detail by several specialist chapters in this *Handbook*. Here we shall give a brief overview as a background to specialist reading, followed by outlines of individual neurohumoral control systems. Emphasis will be placed on the regulation of gonadotropin control, because it is crucial for propagation of species, and because the mechanisms involved illustrate key principles of neurohumoral control of anterior pituitary hormones.

Overview

The neural control of all established anterior pituitary hormones is mediated by at least one or more neurohormones. In some cases two neurohormones may act synergistically, as is the case for ACTH, the release of which is stimulated by both the 41-amino acid residue peptide, CRF-41, and the nonapeptide, arginine vasopressin (AVP). The control of ACTH secretion is further complicated by the fact that urocortins, members of the CRF-41 family of peptides, are involved in the stress response. In other cases two neurohormones may act antagonistically, as is the case for GH, the release of which is stimulated by the 44-amino acid residue peptide, GH-releasing hormone (GHRH-44), and inhibited by the 14- or 28-residue peptide, somatostatin-14 or -28. The neural regulation of pituitary GH secretion is even more complex than originally thought, in that ghrelin potentiates the stimulatory action of GHRH.

Prolactin seems to be the only anterior pituitary hormone that is predominantly under inhibitory control of the brain, mediated by dopamine. Neural control of thyrotropin release is mediated by only one neurohormone, thyrotropin-releasing hormone (TRH). This applies also to the gonadotropins, LH and FSH, the neural control of which is mediated by GnRH.

Neurohormones Induce Pituitary Hormone Synthesis as Well as Release; Exemplified by the Hypogonadal Mouse

The fact that neurohormones are crucial for pituitary hormone synthesis as well as release is poignantly illustrated by the hypogonadal (*hpg*) mouse, which has an autosomal recessive mutation in the GnRH gene that results in an isolated, massive deficiency in the pituitary gonadotropins, LH and FSH.[100] The hereditary autosomal recessive hypogonadotropic hypogonadism in hpg mice is caused by a deletional mutation of 33.5 kilobases encompassing the distal half of gnrh1.[101] This truncation leaves the region encoding the GnRH decapeptide intact, and so the reason underlying the profound phenotype of the hpg mouse remained elusive.[102] The apparent mystery seems to be explained by the fact that the genomic deletion of the two last exons results in the accumulation of the first intron, which inhibits the translational activity of the downstream open reading frame, resulting in the lack of functional GnRH and the consequent hypogonadism in the hpg mouse.[103]

Neural Control of Reproduction: Gonadotropin-Releasing Hormone (GnRH)

KEY PRINCIPLES ILLUSTRATED BY GNRH

1. In vertebrates, reproduction is dependent on GnRH which is required for the stimulation of gonadotropin synthesis and release which in turn stimulates sex-steroid hormone secretion and gamete production in the male and female, and ovulation in the female.
2. One and the same neurohormone can mediate the neural control of two pituitary hormones: the gonadotropins, LH and FSH.
3. The release of a neurohormone can be inhibited or triggered by feedback of a peripheral hormone – in this case, estradiol. Whether the action of estradiol is inhibitory or facilitatory depends on the amplitude and timing of the estradiol signal.
4. The action of estradiol is not direct on GnRH neurons; rather, it is mediated centrally by another peptide, kisspeptin, and also by non-peptide neurotransmitters.
5. The action of GnRH in triggering the ovulatory gonadotropin surge depends on a massive (20- to 50-fold) increase in pituitary responsiveness to GnRH, due in part to the actions of estrogen and progesterone on the anterior pituitary gland and in part to the self-priming effect of GnRH.
6. The self-priming effect of GnRH coordinates peak GnRH surge release with peak pituitary responsiveness, thereby ensuring an ovulatory gonadotropin surge.

Characteristics and Phylogeny

Gonadotropin-releasing hormone (GnRH) in vertebrates is a decapeptide hormone crucial for the initiation and maintenance of reproductive function. The amino acid residue sequence of the mammalian form of the decapeptide is: pGlu-His-Trp-Ser-Tyr-Gly-Leu-Arg-Pro-Gly-NH_2. GnRH mediates the neural control of the synthesis and release of the gonadotropins, LH and FSH.[8,104]

The evolution and phylogeny of GnRH and its receptors are described in detail in Chapter 4. Here we give a brief outline to set the scene. About 30 different different forms of the decapapetide and orthologs have been isolated and sequenced from a variety of different animal species.[105–107] The GnRH family of peptides is present in every single vertebrate class examined thus far, and has also been isolated and sequenced from invertebrate members of Phylum Chordata, the urochordates (105: 108). GnRH arose at least 540 million years ago in an ancestral chordate.[108] In vertebrates, GnRH is derived from a GnRH prohormone consisting of the GnRH decapeptide and a larger GnRH associated peptide (GAP).[109] The decapeptide and GAP are liberated from the prohormone by a series of proteolytic cleavages.[107] The only region of the GnRH prohormone that appears highly conserved is the GnRH decapeptide; the GAP sequences are highly variable across taxa.[105,108] As assessed by high performance liquid chromatography,

GnRH seems to be fully processed before being released into hypophysial portal blood.[110] The selective conservation of the GnRH decapeptide perhaps reflects its functional significance and, thus, low tolerance for structural alterations during evolution.

Genome synteny, combined with phylogenetic analyses of teleost fish and tetrapods, revealed that two rounds of genome duplication events may have generated three vertebrate lineages of the GnRH peptides, GnRH1, GnRH2 and GnRH3. The GnRH3 gene has been lost in tetrapods.[105,111] The presence of lamprey-specific GnRH-I and -III (GnRH4 group), however, reveals a complicated evolutionary scenario of the GnRH gene during early vertebrate evolution.[111,112]

The dominant function of GnRH in vertebrates, the activation of reproduction through the stimulation of pituitary gonadotropin secretion, is a function primarily of GnRH-I. GnRH3 is the prime mediator in some teleosts, such as salmon and zebrafish.[105] This neuroendocrine function is highly conserved in all organisms possessing a pituitary gland.[105,108] GnRH peptides may also serve as neurotransmitters/neuromodulators outside the hypothalamic–pituitary system. This view is underscored by the widespread distribution of GnRH2 in the nervous system (see below), and receives support from circumstantial evidence from protostomes such as the octopus and aplysia, which lack a pituitary homolog but in which protostomal GnRH seems to be involved in neural regulation.[108]

The genes encoding GnRH in vertebrates have duplicated, diversified and sometimes lost, suggesting changes in the roles of different peptide lineages.[105] Thus, in mammals, for example, many species have two GnRH genes, GnRH1 and GnRH2, whereas in others only GnRH1 is active.[105]

Chicken GnRH2 is ubiquitous in vertebrates, from primitive bony fish to man.[106] This complete conservation of structure over 500 million years suggests that GnRH2 has an important function and a discriminating receptor (or receptors) that has selected against any structural change in the ligand. The wide distribution of GnRH2 in the central and peripheral nervous systems suggests a neurotransmitter/neuromodulatory role, exemplified by the GnRH2 inhibition of M currents in bullfrog sympathetic ganglion, which sensitizes neurons to depolarization.[106,113,114] GnRH2 is present in the midbrain of most vertebrates. Although the function of GnRH2 is still poorly understood, its persistence in the midbrain suggests a possible role of GnRH2 as a neurotransmitter and/or neuromodulator.[108]

Residue substitutions have occurred during evolution in all locations of the decapeptide except those occupied by the pGlu at the N-terminal, the Ser4, the Pro, and the glycine-amide at the C-terminal (for details, see Chapter 4). These four residues therefore seem to be important for ligand binding and/or biological activity. Mammalian GnRH is active at low concentrations, whereas the remaining vertebrate forms – with the exception of chicken GnRH2 – show reduced binding affinity and gonadotropin release in mammals.[115] However, non-mammalian forms of GnRH are biologically potent in non-mammals.[106] The N- and C-terminal domains of GnRH are both involved in receptor binding, while the former also participates in receptor activation and contributes to the biological activity of the decapeptide.[107] Both the Pro and Gly-amide residues are critical for biological activity.

PHYSIOLOGY AND FUNCTION

EFFECTS OF ELECTRICAL BRAIN STIMULATION ON GnRH RELEASE The release of GnRH into hypophysial portal blood in anesthetized rats is increased severalfold by electrical stimulation of the median eminence (ME), preoptic area and suprachiasmatic nuclei[116–118] – areas known to contain the greatest concentrations of GnRH-containing cell bodies and fibers (preoptic-suprachiasmatic region) and nerve terminals (ME). Stimulation of either the hippocampus or amygdala has no effect on GnRH release.[118] The amount of GnRH released into portal blood is dependent on the amplitude and frequency of the stimulating pulses.[116,119,120]

SPONTANEOUS GnRH RELEASE

GnRH Surge Release In spontaneously ovulating mammals, such as rodents, sheep, non-human primates and the human, basal gonadotropin release (usually pulsatile) is interrupted by a massive ovulatory surge of luteinizing hormone (LH), which occurs once during each estrous or menstrual cycle ([9,18,104,121]; see also Chapters 9 and 19). The LH surge is accompanied by a surge of FSH, which tends to be smaller in amplitude but much longer in duration than the LH surge. In reflex ovulators, such as the rabbit, the ovulatory surge of LH is triggered by copulation.

In rodents, sheep and monkeys, the spontaneous gonadotropin surge is triggered by a surge of GnRH (Fig. 5.10) into hypophysial portal blood, which in turn is triggered by the surge of estradiol-17β that precedes the GnRH surge.[14,72,74,75,77–79,81–83,122,123] The positive feedback cascade that leads to the spontaneous LH surge is shown schematically in Fig 5.12. The spontaneous GnRH surge is relatively small in the rat, sheep and monkey; concentrations increase from basal values of about 20–30 pg GnRH/ml to 100–200 pg GnRH/ml of portal plasma. These concentrations are of the same order as those in hypophysial portal blood collected in sheep and rhesus monkeys. Intravenous infusion of GnRH into rats at diestrus and proestrus showed that the concentrations of GnRH reached at the peak of the

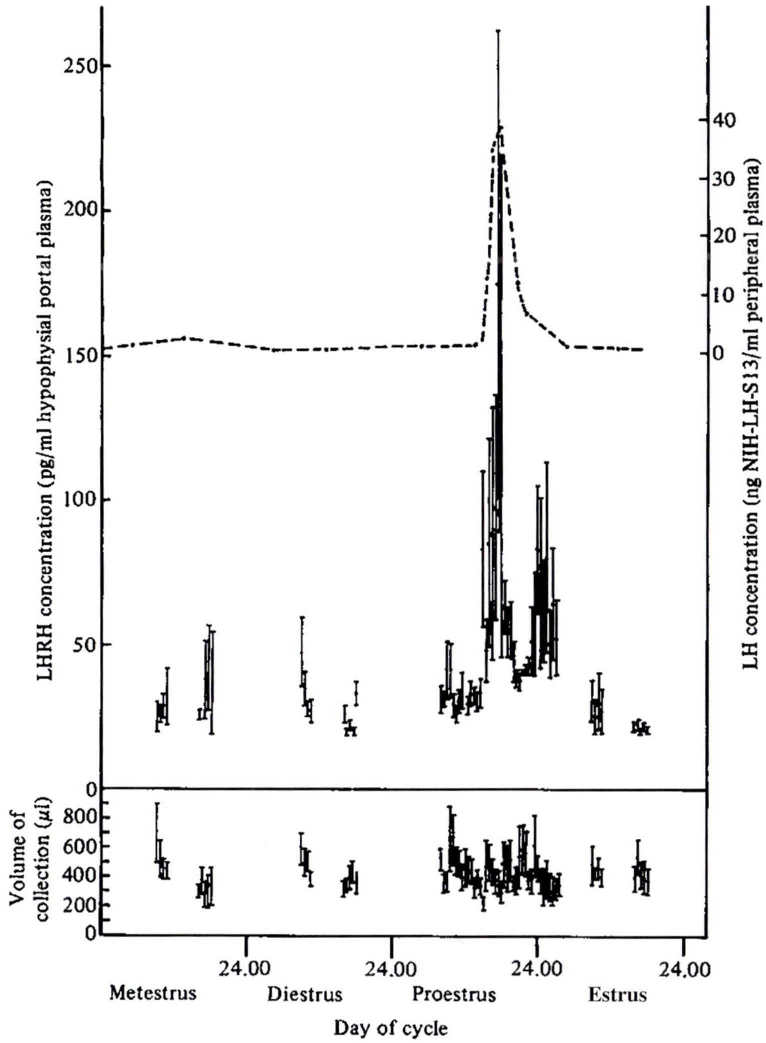

FIGURE 5.10 Mean (±S.E.M.) concentrations of GnRH (LHRH) in hypophysial portal plasma collected from female rats anesthetized with alphaxalone at various stages of the estrous cycle. For most of the cycle the concentrations of GnRH are low, but just before and during the surge of LH (dashed line) there is a surge of GnRH. The volumes of portal blood collected are shown in the lower panel. *Reproduced from Sarkar DK, Chiappa SA, Fink G, Sherwood NM. Gonadotropin-releasing hormone surge in pro-oestrous rats.* Nature 1976; *264:461–463, with permission from Macmillan Journals.*

spontaneous surge can only trigger the surge release of gonadotropin in proestrous animals in which the responsiveness of the pituitary gland to GnRH is 20- to 50-fold greater than at diestrus.[9,12,18,104,124,125] A 20- to 50-fold increase in pituitary responsiveness to GnRH also occurs between the early follicular and ovulatory phase of the human menstrual cycle.[121] While the GnRH surge is probably important for ensuring the precise timing of the spontaneous ovulatory LH surge, it must be stressed that exposure of the pituitary gland to small pulses of GnRH[126,127] or continuous infusions of small amounts of GnRH[128] can also trigger a massive surge of LH, probably by way of the priming effect of GnRH (see below). The mechanism and importance of increased pituitary responsiveness to GnRH is discussed below.

Pulsatile Release of GnRH For a detailed account, the reader is referred to Chapters 1, 9 and 19, and to reviews by Fink[18] and Clarke.[129] Briefly, pulsatile release of GnRH into hypophysial portal blood can most readily be detected in ovariectomized rats,[130] sheep[79,129] and rhesus monkeys.[123] The amplitude of GnRH pulses is related to the mean plasma LH concentrations,[130] and in the rat, but not the rhesus monkey, the pulsatile release of GnRH can be rapidly reduced by the intravenous injection of estradiol-17β. Pulsatile LH release also occurs in intact rats, monkeys and humans. Evidence from measurements of GnRH in portal blood (above) and the effects of injecting GnRH or anti-GnRH sera (immunoneutralization) shows that the pulses of LH in intact animals are dependent on pulses of GnRH.[9,131,132] The functional significance of the pulsatile release of GnRH and, as a consequence, LH appears to be as follows: (a) it provides the hypothalamic–pituitary–ovarian regulatory system with the capacity of control by both frequency and amplitude modulation; (b) as mentioned above, small frequent pulses of GnRH could, by way of the priming effect of GnRH, lead to a spontaneous ovulatory surge of LH[126,127]; and (c) pulsatile GnRH release prevents the downregulation of GnRH receptors that occurs during continuous exposure to high levels of GnRH and leads to pituitary refractoriness (tachyphylaxis) to GnRH.[131–133] In the female rat and human, changes in both pulse frequency and amplitude are important in the signalling system,[134,135] but in the male rhesus monkey changes in the pulse frequency alone[136] may play a major role in the negative feedback control of LH release by sex steroids. Thus, for example, testosterone exerts its inhibitory effects on the hypothalamic–pituitary–gonadotropin system by reducing LH pulse frequency.

Diurnal Critical Period for the Ovulatory Gonadotropin Surge Everett and Sawyer, using sodium pentobarbital (Nembutal) blockade, showed that in rodents "the neural apparatus controlling ovulation has a 24-hr periodicity":[137–139]

> Thus, if a 4-day cyclic rat was "blocked" with Nembutal injected shortly before 1400 hr on prestrus, ovulation was delayed for just 24 hr. There was again a critical period on the second afternoon, when further treatment with Nembutal would delay ovulation another 24 hr, provided that the dose was increased or supplemented later in the afternoon. Similar treatment on the third afternoon prevented ovulation on the third night.[138]

There may be some diurnal dependency in humans, as the LH surge in women tends to begin in the early morning and seems to be associated with the acrophase of the cortisol circadian rhythm.[140] Recently, Christian and Moenter[141] used the Everett/Sawyer model to demonstrate the role of fast synaptic transmission in mediating estradiol negative and positive feedback actions on GnRH neurons.

In earlier studies, we investigated the mechanism of the diurnal signal for the LH surge[138,139] using the powerful rodent models of Caligaris *et al.*[142,143] We found that ovariectomized rats treated with high doses of estradiol show diurnal surges of LH triggered by diurnal surges of GnRH.[130] However, as in the case of the spontaneous surge of LH, the diurnal surges of GnRH are relatively small, and only produce surges of LH because the responsiveness of the anterior pituitary gland to GnRH in ovariectomized rats treated with estrogen is more than two orders of magnitude greater than the level of pituitary responsiveness at diestrus.[130] Nevertheless, the diurnal surges of GnRH in long-term ovariectomized rats treated with estrogen provide strong support for the occurrence of a daily neural signal for LH release which is only expressed in the form of a GnRH/LH surge when the hypothalamic–pituitary system is exposed to high levels of estradiol.[9,104,130,137,139,141,144,145] The central origin and neural circuitory responsible for generating the daily neural signal for GnRH, which includes the suprachiasmatic nucleus,[146] remains to be elucidated. It seems probable that the Kisspeptin neurons (below) may be central to the daily "GnRH surge generator." However, estradiol has direct effects on GnRH neurons by way of other neurotransmitter neurons,[18,129,147,148] as well as by way of direct actions on voltage gated calcium channels on GnRH neurons.[149]

Mechanism of the Estradiol-Induced GnRH Surge Broadly, estrogen has two major effects on the GnRH/LH release system: low plasma concentrations of estrogen inhibit (negative feedback), while high plasma concentrations of estradiol stimulate (positive feedback), GnRH release.[9,18,104,129,141,145] The negative feedback action of estrogen is so fast that it may not necessarily involve nuclear receptors and genome-induced protein synthesis.[130] Conceivably, estrogen could inhibit GnRH by a direct action on the membranes and ion channels involved in GnRH release.[130,150–152] Specifically, estradiol can rapidly activate adenylate cyclase, increase intracellular [Ca^{2+}], activate phospholipase C to generate inositol 1,4,5-trisphosphate and diacylglycerol, stimulate the phosphatidylinositol 3 kinase pathway, stimulate nitric oxide synthase to liberate nitric oxide, increase intracellular cGMP to activate protein kinase G, and activate mitogen-activated protein (MAP) kinase pathways.[151,153] The ability of membrane-impermeable estradiol rapidly to activate ERK signaling in several different cell types indicates that rapid estrogenic effects are initiated at the plasma membrane. However, the precise nature of a membrane-associated estrogen receptor (mER) and the molecular mechanism through which the mER couples to the ERK signaling pathway remain to be established.[151,153] Recently, targeted mutagenesis studies in mice have addressed the role of estrogen response elements in estrogen negative-feedback control of gonadotropin secretion.[154]

In contrast, increased estrogen levels induce positive feedback stimulation of GnRH release (the GnRH surge) over a period of 26–28 hours.[9,18,104] This is of course more than sufficient time for receptor activation, transcription, translation, protein synthesis and structural changes in neuronal cytoskeleton, processes and synapses. The estradiol triggered GnRH surge is not due to a direct action of estrogen on GnRH neurons; rather, it must be mediated by intermediate neurons. We know this from three sequential seminal discoveries. First, Pfaff and colleagues, with the aid of a combined immunocytochemical and autoradiographic study, showed that in the rat only a few (1 in 435) hypothalamic GnRH neurons contain nuclear estrogen receptors.[155] Second, GnRH neurons were shown to express the beta but not the alpha isoform of the estradiol receptor (ERα and ERβ, respectively).[147,156] Third, a gene mutation experiment demonstrated that the preovulatory gonadotropin surge was normal in ERβ knockout mice: but estrogen positive feedback failed to occur in ERα knockout mice.[157] Therefore, it seemed reasonable to assume that the effects of estrogen that involve a genomic mechanism must be mediated by non-GnRH neurons that express ERα and are afferents to GnRH neurons.[18,155,157] Wintermantel *et al.*[157] used a GnRH neuron-specific Pseudorabies virus tracing approach to show that the ERα-expressing neurons that innervate GnRH neurons are located within the rostral periventricular region of the hypothalamus. In sum, ovulation is driven by estrogen actions upon ERα-expressing neuronal afferents to GnRH neurons,[147,157] and the cell bodies of these afferent neurons are located in the same brain region as kisspeptin neurons (see below).

BOX 5.3

HOW DO WE KNOW ABOUT THE APPARENT PARADOX OF NEGATIVE–POSITIVE FEEDBACK?

Whether estrogen and progesterone inhibit or stimulate gonadotropin output depends upon the duration, timing and level of exposure to the steroid.[1] Thus: Progesterone will stimulate LH release only when acting on a hypothalamo-hypophysial system which has been exposed for many hours to elevated plasma concentrations of estrogen. In the rat, the switch from an inhibitory to a facilitatory action of progesterone occurs relatively precisely during the early hours of pro-estrus, when anti-estrogens also cease to be effective in blocking ovulation.[2,3]

During most of the cycle, plasma estrogen concentrations are low. In the monkey, an LH surge will occur only if estradiol concentrations are maintained at about 200–400 pg/ml plasma for at least 36 hours.[4] Basal concentrations, or increments of less than 100 pg/ml, or of a duration shorter than 36 hours, reduce LH output. The magnitude of the plasma estradiol concentrations and the duration of exposure at which a switch from an inhibitory to a facilitatory effect occurs in the human is similar to that in the monkey.[5]

In the rat,[6,7] sheep[8] and human,[9] progesterone and estrogen act synergistically to inhibit gonadotropin release. Thus, in the presence of relatively high plasma progesterone concentrations, basal or elevated concentrations of estrogen will inhibit gonadotropin output. Progesterone by itself has little effect, but this may be because the abundance of progesterone receptors in the hypothalamus and pituitary depends upon estrogen.[10]

Estrogen induction of progesterone receptors plays a key role in the switch from negative to positive feedback.[11,12] Indeed, an interaction between between the priming effect of GnRH and active progesterone receptors seems to exist.[11–13] Progesterone receptor knockout blocks GnRH priming *in vivo* in mice[11] but not in rat pituitary gonadotropes *in vitro*.[13] Turgeon and Waring[13] suggest that this species difference may reflect differences in the balance of progesterone receptor isoforms (A: B) modulated by estradiol in gonadotropes.

References

1. Fink G. Feedback actions of target hormones on hypothalamus and pituitary with special reference to gonadal steroids. *Annu Rev Physiol.* 1979;41:571–585.
2. Aiyer MS, Fink G. The role of sex steroid hormones in modulating the responsiveness of the anterior pituitary gland to luteinizing hormone releasing factor in the female rat. *J Endocrinol.* 1974;62:553–572.
3. Brown-Grant K. The effects of progesterone and of pentobarbitone administered at the dioestrous stage on the ovarian cycle of the rat. *J Endocrinol.* 1969;43(4): 539–552.
4. Knobil E. On the control of gonadotropin secretion in the rhesus monkey. *Recent Prog Horm Res.* 1974;30:1–36.
5. Yen SSC, Lasley BL, Wang CF, Leblanc H, Siler TM. The operating characteristics of the hypothalamic pituitary system during the menstrual cycle and observations of biological action of somatostatin. Recent Prog. *Horm Res.* 1975;31:321–357.
6. McCann SM. Effect of progesterone on plasma luteinizing hormone activity. *Am J Physiol.* 1962;202:601–604.
7. Goodman RL. A quantitative analysis of the physiological role of estradiol and progesterone in the control of tonic and surge secretion of luteinizing hormone in the rat. *Endocrinology.* 1978;102:142–150.
8. Hauger RL, Karsch FJ, Foster DL. A new concept for control of the estrous cycle of the ewe based on temporal relationships between luteinizing hormone, estradiol and progesterone in peripheral serum and evidence that progesterone inhibits tonic LH secretion. *Endocrinology.* 1977;101:807–817.
9. Van Look PF. Failure of positive feedback. *Clin Obstet Gynaecol.* 1976 Dec;3(3):555–578.
10. MacLusky NJ, McEwen BS. Oestrogen modulates progestin receptor concentrations in some rat brain regions but not in others. *Nature.* 1978;274(5668):276–278.
11. Chappell PE, Schneider JS, Kim P, Xu M, Lydon JP, O'Malley BW, Levine JE. Absence of gonadotropin surges and gonadotropin-releasing hormone self-priming in ovariectomized (OVX), estrogen (E2)-treated, progesterone receptor knockout (PRKO) mice. *Endocrinology.* 1999;140: 3653–3658.
12. Attardi B, Scott R, Pfaff D, Fink G. Facilitation or inhibition of the oestradiol-induced gonadotrophin surge in the immature female rat by progesterone: effects on pituitary responsiveness to gonadotrophin-releasing hormone (GnRH), GnRH self-priming and pituitary mRNAs for the progesterone receptor A and B isoforms. *J Neuroendocrinol.* 2007;19(12):988–1000.
13. Turgeon JL, Waring DW. Differential expression and regulation of progesterone receptor isoforms in rat and mouse pituitary cells and LbetaT2 gonadotropes. *J Endocrinol.* 2006;190:837–846.

Possible candidates for facilitatory or inhibitory neurons involved in the estradiol-induced GnRH surge include noradrenergic, dopaminergic, serotonergic, opioid, GABAegic or glutamatergic.[14,129,147,148,158–161] While the role of these types of intermediate neuron remains important, recent attention has focused on Kisspeptin neurons.

KISSPEPTIN

Since the early 1970s, it had been assumed that GnRH was the key regulator – Grandmother neuron[162] – in the neural control of the gonadotropins. This view has recently been revised by the discovery that the positive and negative estradiol control of GnRH is mediated by Kisspeptin, which is expressed in the arcuate nucleus and anteroventral periventricular nucleus (AVPV) of the forebrain.[129,163–165] Kisspeptin was discovered as a consequence of studies on an orphan G-protein coupled receptor, GPR54, which has high affinity for the 54-amino acid residue kisspeptin (see Chapters 2 and 9). First implicated in cancer, two independent groups discovered that subjects with hypogonadotropic hypogonadism in consanguineous families possessed a mutation in the *GPR54* gene.[166–168] *Gpr54*-deficient mice created by site-directed mutagenesis also exhibited hypogonadotropic hypogonadism, despite the fact that they had normal brain levels of GnRH and were responsive to exogenous GnRH.[167] Several independent studies have shown that, in addition to playing a pivotal role in triggering puberty onset, kisspeptin neurons in the rodent arcuate nucleus mediate estradiol negative feedback control of GnRH, whereas AVPV kisspeptin neurons mediate estradiol-induced positive feedback regulation of GnRH secretion. While this also applies in principle to other mammals studied so far, there are species differences in the precise brain nuclei involved. Estradiol regulation of GnRH may involve interactions between kisspeptin and other neurotransmitters.[169] For a detailed review of the role of kisspeptin and other RF-amide peptides in gonadotropin control, see Chapters 2, 4, 9 and 19.

PITUITARY PESPONSIVENESS TO GnRH

Until the early 1970s, the anterior pituitary gland was thought to operate as a steady-state system which followed hypothalamic activity in a slave-like manner. Indeed, in her 1969 model for the regulation of ovulation in the rat, Nina Schwartz made no allowance for changes in pituitary responsiveness to GnRH.[170] However studies with synthetic GnRH demonstrated that, to the contrary, the responsiveness of anterior pituitary to GnRH changes quite dramatically under different physiological conditions.[9,12,18,104,124,125,171] As mentioned above, the responsiveness of the pituitary gland to GnRH increases 20 to 50-fold before and during the spontaneous surge of LH.[18,104,171] This increase in pituitary responsiveness to GnRH is initiated by the spontaneous, preovulatory surge of estradiol-17β, and is further increased by the priming effect of GnRH (Fig. 5.11). It seems that GnRH priming synchronizes the increasing concentrations of GnRH in portal blood with the increase in pituitary responsiveness, so that both events reach a peak at the same time and thus ensure the occurrence of a massive ovulatory surge of LH (Figs 5.12, 5.13).[126,127]

THE PRIMING EFFECT OF GNRH

The priming effect of GnRH is a mechanism whereby GnRH increases the responsiveness of gonadotropes to itself is a servomechanism apparently unique for this

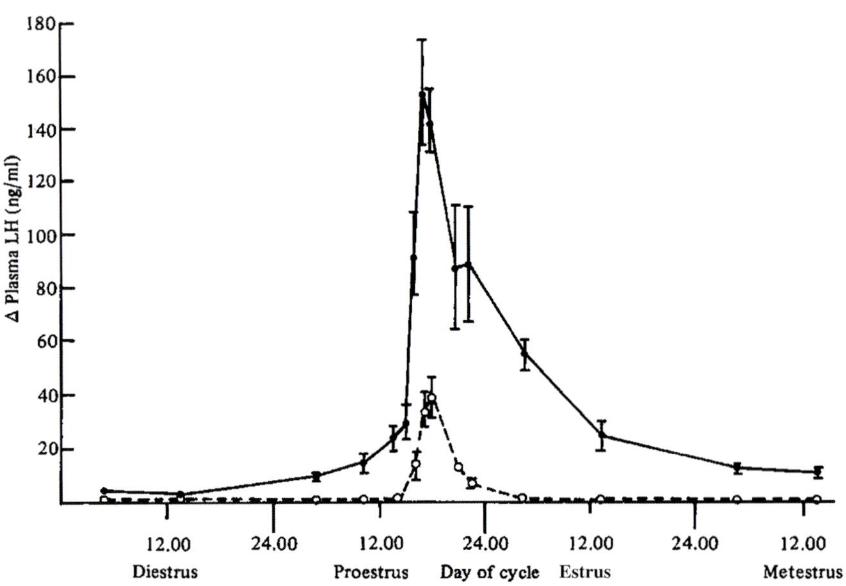

FIGURE 5.11 Changes in pituitary responsiveness to GnRH (LHRH) during the estrous cycle of the rat. The figure shows the mean (± S.E.M.) pre-injection concentrations (dashed line) and mean maximal increments (continuous line) in plasma LH concentrations (ng NIH-LH-S13/ml) in animals anesthetized with sodium pentobarbitone 30–60 minutes before the intravenous injection of 50 ng LHRH/100 g body weight at different stages of the estrous cycle. *Reproduced from Aiyer* et al., *1974 (see Ref 171) with permission from the* Journal of Endocrinology.

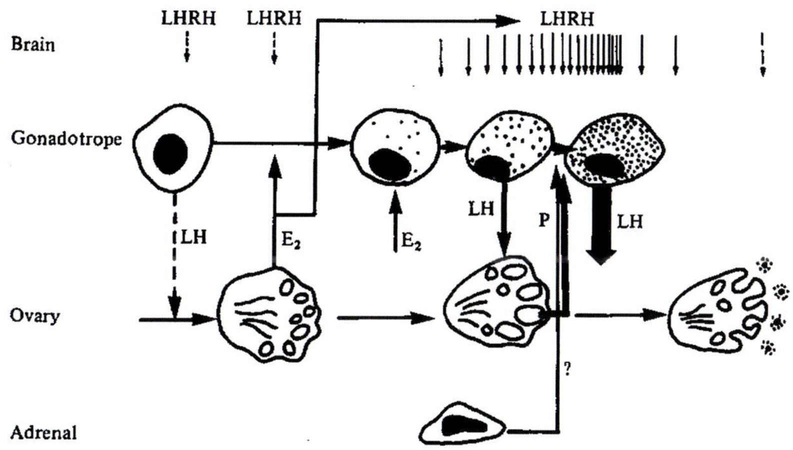

FIGURE 5.12 Schematic diagram showing the cascade of events that produce the spontaneous ovulatory LH surge in the rat. The increase in plasma concentrations of estradiol-17 beta (E2: the ovarian signal) increases the responsiveness of the pituitary gonadotropes (increased stippling) to LHRH (GnRH) and also triggers the surge of GnRH. Pituitary responsiveness to GnRH is further increased by progesterone (P) secreted from the ovary in response to the LH released during the early part of the LH surge, and by the priming effect of LHRH, the unique capacity of the decapeptide to increase pituitary responsiveness to itself. The priming effect of GnRH coordinates the surges of GnRH with increasing pituitary responsiveness so that the two events reach a peak at the same time. The conditions are thereby made optimal for a massive surge of LH. This positive feedback cascade is terminated by destruction of a major component of the system: that is, rupture of the ovarian follicles (ovulation). *Reproduced from Fink, 1979 (see Ref 104), with permission from the British Council.*

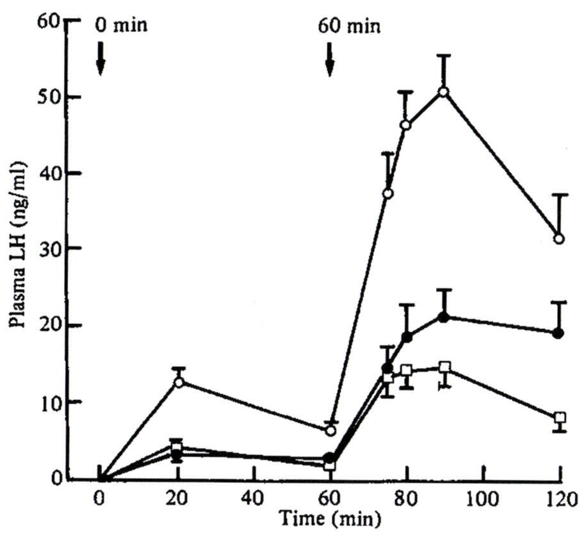

FIGURE 5.13 Priming effect of GnRH. Potentiation by estrogen of both the releasing action and priming effect of GnRH (LHRH). Mean (+S.E.M.) plasma luteinizing hormone (LH) concentrations (ng NIH-LH-S13/ml) after two successive i.v. injections of 50 ng/100 g body weight, 60 minutes apart (arrows). The animals were injected s.c. with 0.2 ml oil vehicle (closed circle), 10 μg estradiol benzoate (open circle) or 2–5 mg progesterone (square) at 1000 h of metestrus. The first dose of GnRH was injected 30–60 minutes after the administration of sodium pentobarbitone at 1330 h the next day, diestrus. *Reproduced from Aiyer Chiappa SA, Fink G, (1974),[172] with permission from the* Journal of Endocrinology.

peptide, possibly because, apart from the oxytocin uterine contraction system which operates during parturition, the ovulatory surge of LH is the only positive feedback endocrine system that operates under physiological conditions. The priming effect of GnRH can be demonstrated *in vivo* by different modes of administering exogenous GnRH, as well as by releasing endogenous GnRH by way of electrical stimulation of the preoptic area.[126,127,172] The priming effect of GnRH can also be elicited *in vitro*, and this has permitted a comparison to be made of the mechanisms of the priming effect and the releasing action of GnRH. The key differences between the releasing and priming actions of GnRH are: (a) GnRH priming, but not releasing, is dependent on protein synthesis; (b) in contrast to the GnRH-releasing action, GnRH priming cannot be mimicked by K^+ depolarization or Ca^{2+} ionophores; (c) priming involves potentiation of the IP_3 intracellular Ca^{2+} mechanisms and protein kinase C; and (d) priming involves activation of microtubule activated protein (MAP) kinase.[173–175]

The GnRH "priming protein" has a relative molecular mass (70 kDa) and similar electrophoretic properties to those of an estradiol-induced protein in the ventromedial nucleus of the hypothalamus which is associated with lordosis behavior in the female rat.[174,175] The amino acid residue sequence of the amino terminals of the GnRH and E_2-induced 70-kDa proteins (termed hormone induced protein-HIP-70) are identical to one another and to a protein also found in uterus.[175] The function of HIP-70 remains to be established. Full

GnRH priming *in vivo* in mice depends on active progesterone receptors, but the basis for this molecular cross-talk has yet to be resolved.[176–178]

Ultrastructural studies have shown that the GnRH priming effect involves an increase in length and a change in the angle of the microfilaments in gonadotropes, and a migration of secretory granules towards the plasmalemma of the gonadotrope.[179,180] This migration of granules ("margination") leads to an increase in the pool of LH available for release so that when the gonadotropes are exposed for a second time to a secretagog such as K+ depolarization, Ca^{2+} ionophores or GnRH itself, a massive second release of LH occurs.[127,173]

Neural Control of Lactation: Prolactin and Inhibitory Factor (PIF)

Prolactin seems to be the only anterior pituitary hormone that is predominantly under inhibitory control of the brain.[41] Evidence for this came first from the studies of Everett and Nikitowitch-Winer, who showed that prolactotropes were the only cell type that did not undergo atrophy in pituitary grafts under the kidney capsule, far removed from central neural control.[32,33] This histological observation was confirmed by the finding that prolactin concentrations in plasma are increased in animals bearing pituitary grafts under the kidney capsule. There is substantial evidence that dopamine, released into hypophysial portal blood from tuberoinfundibular dopaminergic neurons, is the prolactin inhibiting factor.[41] Indeed, dopamine agonists such as bromocriptine cabergoline and pergolide mesylate are highly effective in treating galactorrhea and hyperprolactinemia, a relatively common cause of infertility in women (see Chapter 34). Hyperprolactinemia is frequently due to benign tumors of the anterior pituitary gland, and these too can often be controlled or eradicated by treatment with dopamine agonists (see Chapter 34). Hyperprolactinemia is also caused by antipsychotic drugs that antagonize the actions of endogenous dopamine at dopamine 2 (D2) receptors in the brain. The propensity of antipsychotic agents to cause hyperprolactinemia is related to their potency in antagonizing D2 receptors on the anterior pituitary.[181] Symptoms of hyperprolactinemia include gynecomastia, galactorrhoea, sexual dysfunction, infertility, oligomenorrhea and amenorrhea.[182]

Prolactin is essential to the survival of most mammalian young after birth.[183] Indeed, the main physiological stimulus for prolactin release is suckling of the nipple by the young during lactation.[41,183] Suckling induces neural impulses that are relayed to the hypothalamus, where they affect neurohormonal release into hypophysial portal blood, which in turn results in prolactin release.[41,182,184] This is a classical neuroendocrine reflex, where the amount of prolactin released is proportional to the intensity and duration of suckling.[185] The neurohormonal mechanism involved in mediating the neural control of suckling-induced prolactin release remains uncertain.[41] Several neurohormones have been considered, with main contenders being inhibition of dopamine release and/or the release of TRH. However, TRH appears to have been ruled out by studies in the rat and sheep which found that suckling did not produce significant increases in TRH release into hypophysial portal blood.[41,184,186]

There is also a proestrous surge of prolactin in the rat which occurs coincidentally with the spontaneous LH surge, and which plays a major role in maintaining the function of the corpus luteum for about 1–2 days after ovulation.[18] Much less is known about the mechanisms that control the spontaneous prolactin compared with the spontaneous gonadotropin surge (see above). As already mentioned, a major deficiency in knowledge is in the precise nature of the neurohormones that mediate neural control of prolactin secretion.[18,41,183] Thyrotropin-releasing hormone (TRH) is one of the most potent prolactin-releasing factors when administered exogenously. Other peptides that release prolactin when administered exogenously include vasoactive intestinal peptide.[18,41] The control of prolactin release is further complicated by the fact that estrogen stimulates prolactin secretion by a direct action on pituitary cells, in which it also induces prolactin mRNA synthesis.[187] Estrogen stimulation of pituitary prolactin secretion overrides inhibition of prolactin secretion by dopamine and its agonists.[188] This could suggest that the spontaneous proestrous surge of prolactin is due mainly to the stimulatory action of estrogen on the pituitary gland. However, this explanation is unlikely to be correct for two reasons. First, elevated plasma concentrations of estrogen do not necessarily result in a sustained elevation of plasma prolactin concentrations; rather, elevated plasma estradiol concentrations produce prolactin surges which coincide with estrogen-induced LH surges.[189] Second, in rats treated with a single injection of EB, dopamine output into hypophysial portal vessel blood decreased by 50% coincident with the prolactin surge that occurs on the afternoon of the third day after EB injection.[188] In the same animals, TRH output increased by 240% at the time of the EB-induced prolactin surge.[188] The TRH findings of de Greef *et al.*[188] agree remarkably with those (100% increase of afternoon over morning values) obtained by Fink *et al.*[184] at the time of the endogenous prolactin surge on prestrus. The interactions between estrogen and progesterone in producing the LH and prolactin surge can also be investigated in several steroid models.[9,18] The existence of a daily neural signal for the LH surge[137] has already been discussed above:

prolactin release also exhibits a diurnal surge in long-term ovariectomized rats treated with high doses of estrogen. In animals so treated, diurnal surges of prolactin coincide with those of LH, which peak at 1700 h each day.[189] As explained above, the mechanisms of estradiol positive feedback that trigger the diurnal GnRH/gonadotropin surge and the prolactin surge remain to be established.

In addition to a prolactin-inhibiting factor (PIF), a search for a prolactin-releasing factor (PRF) has continued for many years. Potential candidates have included TRH, vasoactive intestinal peptide, GnRH associated peptide (GAP), and two novel, closely related hypothalamic "prolactin-releasing peptides" (termed PrRP31 and PrRP20, respectively). These two peptides, derived from the same precursor, were discovered in 1998 by Hinuma and associates when searching for a ligand for the orphan human G-protein coupled receptor GPR_{10} (also hGR3, ratUHR-1) with the aid of reverse pharmacology.[190–192] PrRP belongs to a family of RF-amidepeptides that contain an Arg=Phe-amide motif in their C-terminus.[192,193] The name of this peptide reflects the fact that PrRP was found to stimulate the secretion of prolactin (PRL) in primary pituitary cells from lactating rats, the rat cell lineRC-4B/C, and *in vivo* in rats.[192] However, the two PrRP s are not present in the external layer of the median eminence, and this, together with other findings, has questioned their physiological role in stimulating prolactin release *in vivo*.[41,192,193] The PrRPs and their receptors are widely distributed in the brain and may be involved in CNS function, including activation of the autonomic nervous system, and in the control of food intake and bodyweight.[192,193] However, the nature of the long-sought prolactin-releasing factor, if one exists, remains to be elucidated.

Hyperprolactinemia can cause infertility, and may do so by inhibiting the release of GnRH as well as through a complicated intracellular mechanism inhibiting the secretion of LH by gonadotropes.[41,194–196] The precise mechanism by which prolactin crosses the blood–brain barrier to reach prolactin receptors on GnRH neurons or intermediate afferent neurons (e.g., kisspeptin, GABA or POMC) that affect GnRH neuronal activity remains to be determined.[41,197] Kokay *et al.*[197] have recently shown that, in the rat, less than 5% of GnRH neurons express the prolactin receptor; prolactin effects on GnRH neurons would appear to be mediated mainly by hypothalamic kisspeptin and GABA neurons, a large proportion of which express prolactin receptors. The dopamine agonist bromocriptine has been shown to be a potent treatment for infertility due to hyperprolactinema caused mainly by pituitary tumors (prolactinomas) or dopamine antagonists, most commonly antipsychotics[198] (see also Chapter 34).

Prolactin has a variety of actions, which has led to its description as pleiotropic.[41,199] This is exemplified by prolactin actions peripartum. That is, the pleiotropic actions of prolactin in the brain result in a coordinated response that facilitates the neurobiological adaptation to pregnancy and lactation.[41] Elevated prolactin contributes to the establishment of maternal behavior, and the consequent suckling stimulus drives further prolactin secretion. Facilitated by increased prolactin receptor expression in the hypothalamus during pregnancy, prolactin stimulates food intake and suppresses the stress response during pregnancy and lactation.[41,200] Based on the effects of an antisense probe to the long form of the prolactin GPCR in rat brain on behavior in the elevated plus maze, it is thought that prolactin can reduce anxiety.[201] Intracerebral infusion of prolactin attenuated the ACTH response to novel environment or restraint, leading to the inference that prolactin also moderates the hypothalamic–pituitary–adrenal (HPA) stress response. However, this observation seems to contrast with the finding that prolactin potentiates gene transcription *in vitro*.[201] Clearly, further research is required to resolve this apparent paradox.

Stress Neurohormones: Corticotropin-Releasing Factor-41 and Arginine Vasopressin

The stress response is the subject of numerous reviews (see, for example, Refs 202–208). Adrenocorticotropin (ACTH) is the main anterior pituitary stress hormone. However, under certain conditions GH and prolactin are also released in response to stress. Thus, for example, insulin-induced hypoglycemia is a potent metabolic stressor that, in the human, results in a massive and rapid release of GH.[209]

The neural control of ACTH and its feedback regulation by glucocorticoids is covered in depth by several chapters in this *Handbook* (see Chapters 3, 8, 15, 21 and 29). Here, we shall give a brief outline to set the scene.

Briefly, ACTH release is stimulated by both the 41-amino acid residue peptide, CRF-41, and the nonapeptide, arginine vasopressin (AVP).[35,95,210,211] Since CRF-41 was first characterized, a family of CRF peptide ligands and receptors has been discovered. In mammals, the CRF family includes CRF, urocortin I (UcnI), UcnII and UcnIII, together with two G-protein coupled receptors, CRFR1 and CRFR2, and a CRF binding protein. Both CRFR1 and CRFR2 have isoforms, the characteristics and tissue distribution of which are described in detail in Chapter 15 and the review by Bale and Vale.[95]

CRF-41, the key stimulator of the stress response, is expressed in the hypothalamic paraventricular nucleus (PVN), central nucleus of the amygdala and the hindbrain, as well as in the gut, skin and adrenal gland.[95]

The urocortins are also expressed in brain regions as well as the periphery. CRF-41 has 10-fold higher affinity for CRFR1 than CRFR2; UcnI binds with equal affinity for both receptors, while UncII and -III appear to be selective for CRFR2. CRF-41 induces ACTH release by way of CRFR1 on anterior pituitary corticotropes, and is postulated to coordinate the central nervous response to stress by way of CRFR1 located in the cerebral cortex, hippocampus, amygdala, medial septum and other CNS regions. CRF-41 has been implicated in the causation of anxiety and mood disorders, and intense research is under way, currently focused on the discovery and development of small CRFR1 antagonist anxiolytics.[97] The actions of UcnII and UcnIII on CRFR2 may be important for dampening stress sensitivity[95] (see also Chapter 15). In addition to their central effects, CRF-41 and the urocortins exert actions on the gastrointestinal and cardiovascular systems,[95] and appear to be involved in the central control of metabolic and energy homeostasis.[98,99]

The action of CRF-41 in stimulating pituitary ACTH secretion is potentiated by AVP, which binds to the $V_{1\beta}$ receptor.[211] The AVP-stimulated $V_{1\beta}$ receptor simultaneously activates the Gq/11-inositol phosphate (IP) and, at higher AVP concentrations, G_S-cAMP pathways when transiently expressed in Chinese hamster ovary, human embryonic kidney (HEK) 293, and COS-7 cells.[212] The precise molecular mechanisms involved in the interactions between CRFR1 and the $V_{1\beta}$ that lead to enhanced ACTH secretion remain to be elucidated. However, vasopressin $V_{1\beta}R$ and CRFR1 are capable of forming homo- and heterodimers in a ligand-independent manner; an interaction that does not affect the binding properties of the receptors.[213]

One of several key questions is, which is more important for stimulating ACTH secretion, CRF-41 or AVP? The answer from CRF-41 gene and CRFR1 knockout and CRF-41 overexpression studies is CRF-41.[95,214,215] However, as assessed by studies of $V_{1\beta}$ knockout and the effect of $V_{1\beta}$ antagonists, a full ACTH response to stress depends also on the normal expression, release and action of AVP on functioning $V_{1\beta}$ receptors on pituitary corticotropes.[216,217]

The importance of AVP for a full ACTH response has been reinforced by concurrent measurement of AVP and CRF-41 in hypophysial portal blood in the rat and sheep (see, for example, Refs 218–223 Stress-challenge studies suggest that, in the sheep, AVP may be the more important secretagog,[220] and this seems to be especially the case when the stress is insulin-induced hypoglycemia.[221]

AVP plays dominant role in subacute glucocorticoid negative feedback regulation of the HPA. Thus, in adrenalectomized rats, dexamethasone significantly reduced the concentration of ACTH in peripheral plasma and the amount of AVP, but not CRF-41, released into hypophysial portal blood.[218] In contrast to these subacute (2.5-h) effects of dexamethasone, which suggested that AVP may dominate over CRF-41 in glucocorticoid negative feedback regulation, corticosterone administered continuously for 5 days to long-term hypophysectomized rats reduced CRF-41 to a greater extent than AVP.[222] The findings of Fink *et al.*[218] are perhaps consonant with the fact that glucocorticoid negative feedback selectively targets vasopressin transcription in the PVN parvocellular neurons[224]: direct inhibition of CRF-41 transcription appears to play a minor role in glucocorticoid negative feedback moderating HPA.[225] In fact, the major determinant of glucocorticoid negative feedback to be blockade of the pituitary response to CRF-41 [218,226] (see also below).

Overall, in keeping with studies in the AVP-deficient Brattelboro rat,[227] the dominance of either CRF-41 or AVP in mediating neural control of ACTH secretion seems to be species- and context-specific. This principle also seems to apply to the behavioral and inflammatory responses in animals subjected to genetic or pharmacological manipulation of CRF-41 and AVP (see, for example, Refs 95,214,215).

The neural control of ACTH release may be further complicated by the fact that hypophysial portal blood and immunoneutralization studies suggest that ACTH secretion may be inhibited by atrial natruretic peptide (ANP).[90–92,228–231] In addition to its production by atrial myocytes, ANP is expressed by neurons concentrated in the rostral hypothalamus (especially the AV3V region) that project to the median eminence.[228,230,232] ANP has been shown to inhibit CRF-41 induced ACTH in the human.[233,234] However, the investigation and use of ANP in the human is confounded by the fact that portal blood ANP concentrations when present in the systemic circulation induce hypotension.[91,233,234] Engler *et al.*[230] in their review of the literature, conclude that:

> Taken together, the studies with ANP represent the most compelling evidence to date that a neuropeptide may inhibit ACTH release. However, the inability of ANP to consistently decrease basal, or stimulated ACTH release from normal anterior pituitary cells and its inability to consistently affect the HPA axis in man suggest that ANP may not be the CRIF but it may be a subsidiary modulator of the HPA axis.

Not withstanding this caveat, high plasma ANP concentrations (and possibly other modulatory factors) may help to explain the low plasma ACTH concentrations in critically ill (multiple trauma and septic shock) patients.[235–237]

Finally, HPA activity is under the powerful inhibitory effects of glucocorticoids on the responsiveness of corticotropes to CRF-41 (see, for example, Refs 218,226).

Glucocorticoid negative feedback control is reviewed in Chapter 3, and in Chapters 8, 19 and 21.

Growth Hormone Control

In contrast to ACTH release, which is stimulated by the synergistic actions of CRF-41 and AVP, the neural control of growth hormone (GH) release is mediated by two neurohormones that are antagonistic to one another. That is, GH release is stimulated by the 44-amino acid residue peptide GH-releasing hormone (GHRH-44), and inhibited by the 14- or 28-residue peptide, somatostatin (SST)-14 or -28. GH release is further potentiated by the GH secretagog (GHS) ghrelin, which amplifies GH output by way of four complementary mechanisms: direct stimulation of somatotrope GH release; moderation of SST's inhibition of arcuate GHRH neurons; antagonism of SST inhibition of somatotrope GH release; and stimulation of GHRH secretion from arcuate neurons.[238–241]

For detailed reviews, the reader is referred to Chapters 2, 10 and 32, as well as several recent reviews.[240,241] Readers interested in how reverse pharmacology was used to develop small GH secretagogs are referred to the seminal review by Smith.[239]

The release of GH is pulsatile. The mechanism is thought to involve GH feedback via cognate receptors in the brain to regulate GHRH (inhibited), SST(stimulated), and GHS receptors (repressed).[240] Transgenic laboratory models and sporadic mutations in the mouse and human establish the importance of each of these four signals in directing GH secretion.[240] Pulsatile and continuous GH secretion patterns determine adult body size, inducible hepatic enzymes, lipoprotein metabolism, muscle IGF-I expression, and insulin sensitivity.[240] There are species-dependent sex differences in the pattern of GH secretion: the human and rat, for example, differ completely in sex-related GH pulsatility. Thus, GH pulse frequency is the same in women and men, but is significantly higher in female than in male rodents.[240] GH pulse amplitude is twofold greater in women than in men, but is 30-fold greater in male compared with female rodents. GH secretory patterns are less regular in the female than the male in both man and rodent. The physiological basis for such species- and sex-related distinctions remains to be established, but the role of sexual differentiation of the brain is under investigation.[242]

The action of GHRH is mediated by a G-protein coupled receptor that, in rat and human, is 423 amino acids in length.[243,244] The GHRH receptor is a member of the secretin/glucagon family of receptors.[243] When GHRH binds its receptor on the surface of pituitary somatotrope cells, G-protein coupling stimulates adenylate cyclase to produce cyclic AMP. Through the cAMP second messenger pathway, CREB is phosphorylated and stimulates the transcription of the pituitary-specific transcription factor Pit-1 gene, which in turn stimulates the transcription of the GH and GHRH receptor genes.[244] Binding of GHRH to its receptor also leads to an influx of calcium, which is involved in mediating GH secretion.[244] For a more detailed review of the GHRH receptor, see Chapter 2. Alternative splicing of the GHRH receptor and the pathophysiology of the GHRH receptor are reviewed by McElvaine and Mayo,[244] and Chapter 32.

SOMATOSTATIN ACTIONS BEYOND GH INHIBITION

SST is produced not only in the hypothalamus but also throughout the central nervous system and in most major peripheral organs.[245] SST-like immunoreactivity has been found in all vertebrates as well as in some invertebrate species and in the plant kingdom.

In addition to inhibiting growth hormone (GH) secretion, SST inhibits the secretion of many other hormones, including thyrotropin, CRF-41, ACTH, insulin, glucagon, secretin and vasoactive intestinal peptide.[245,246] SST also inhibits gastrointestinal motility, gastric acid production, pancreatic enzyme secretion, bile secretion and colonic fluid secretion, and exhibits potent antiproliferative activity (cell growth arrest and/or apoptosis) in both cultured cells and tumors in experimental animals.[246] SST also affects motor, sensory, cognitive and autonomic functions.

Along with its wide anatomical distribution, SST acts on multiple targets via a family of five receptors to produce a broad spectrum of biological effects.[245,246] The binding of SST to its receptors triggers a variety of pertussis-toxin sensitive ($G_{i/o}$) and insensitive (G_s, G_q, G_{12}, G_{14}, G_{16}) G proteins. Details of SST receptor interactions are reviewed in Chapter 2. The clinical use of SST, for example in the treatment of acromegaly and gastroenteropancreatic and other neuroendocrine tumors, is covered in Chapter 35.

Thyrotropin-Releasing Hormone

Thyrotropin-releasing hormone (TRH) is a tripeptide that mediates the neural control of thyrotropin (TSH) release. For a detailed review of TRH, its brain distribution and negative feedback control by thyroid hormones, see Chapters 11 and 31, and recent reviews such as Nillni.[247] Here, attention will focus on a brief outline, with special attention on studies on TRH in hypophysial portal blood.

Based on the finding that, when administered exogenously, TRH can stimulate prolactin release in the human and rodent (see, for example, Refs 248–250), TRH was considered as a possible candidate for the long-sought prolactin-releasing factor (PRF; see above). However, while TRH may influence prolactin secretion, the possible role of TRH as *the* physiological PRF has

BOX 5.4

HOW DO WE KNOW THAT SOMATOSTATIN 1–28 IS A PHYSIOLOGICAL NEUROHORMONE?

The cyclic peptide somatostatin (SST) or somatotropin release inhibiting factor (SRIF) is produced from a single gene that encodes preprosomatostatin of 116 amino acid residues including the signal peptide.[1,2] This is processed to prosomatostatin (96 amino acids), which is further cleaved to produce two bioactive products: SST14, the 14-amino acid peptide first isolated and characterized from ovine hypothalamus,[3] and SST28, isolated from porcine intestine, which has an N-terminal extension of 14 additional amino acids.[4] More detailed chromatographic studies showed that brain and other tissue extracts contained multiple forms of SST, of which the most prominent are SST14, SST28 and the N-terminal fragment of SST28, that is SST28 (1–12)[5–8] Measurements of extracts of hypophysial portal vessel blood using HPLC and radioimmunoassay showed that SST14 and SST28 are released into the portal vessels in about equimolar amounts[9], and that SST28 (1–12) is also released into portal blood[10] The concentrations in hypohysial portal blood of all three derivatives of the SST precursor are much greater than in peripheral blood, and can be increased six- to sevenfold by electrical stimulation of the median eminence. These results have been complemented by data on SST release from hypothalamus and ME in response to K^+ depolarization[7] Whereas both SST28 and SST14 were released from the ME, blocks of hypothalamic tissue from which ME had been removed released only SST14. Taken together, the results obtained by measuring SST14 and -28 release into portal blood and from hypothalamic and ME tissue suggest that the SST precursor is processed differently at different sites. A likely (but obviously not the only) explanation for the release of both SST28 and SST14 from the ME is that there are two different types of SST neurons: one in which processing stops at SST28, and another in which processing continues through to SST14. In the second type of neuron SST14 is derived directly from the precursor, an intermediate derivative of the precursor, or from SST28 itself. The SST14 type of neuron obviously predominates in the hypothalamus outside the area of the ME and also in other areas of the CNS, such as the amygdala.[7] The functions of SST28 are qualitatively similar to SST14, but the potency of the two forms of somatostatin in the pituitary gland is different, and the capacity for differential processing may point to a hypothalamic–pituitary regulatory function.[7,9]

References

1. Shen L-P, Pictet RL, Rutter WJ. Human somatostatin I: sequence of the cDNA. *Proc Natl Acad Sci USA.* 1982;79:4575–4579.
2. Goodman RH, Aron DC, Roos BA. Rat pre-prosomatostatin. Structure and processing by microsomal membranes. *J Biol Chem.* 1983;258:5570–5573.
3. Brazeau P, Vale W, Burgus R, Ling N, Butcher M, Rivier J, Guillemin R. Hypothalamic polypeptide that inhibits the secretion of immunoreactive pituitary growth hormone. *Science NY.* 1973;179:77–79.
4. Pradayrol L, Jornvall H, Mull V, Ribet A. N-terminally extended somatostatin: the primary structure of somatostatin-28. *FEBS Lett.* 1980;109:55–58.
5. Benoit R, Bohlen P, Ling N, Briskin A, Esch F, Brazeau P, Ying S-Y, Guillemin R. Presence of somatostatin-28 (1-12) in hypothalamus and pancreas. *Proc. Natl Acad Sci USA.* 1982;79:917–921.
6. Benoit R, Ling N, Alford B, Guillemin R. Seven peptides derived from pro-somatostatin in rat brain. *Biochem Biophys Res Commun.* 1982;107:944–950.
7. Pierotti AR, Harmar AJ. Multiple forms of somatostatin-like immunoreactivity in the hypothalamus and amygdala of the rat: selective localization of somatostatin-28 in the median eminence. *J Endocrinol.* 1985;105:383–389.
8. Charpenet G, Patel YC. Characterization of tissue and releasable molecular forms of somatostatin-28 [1-12] like immunoreactivity in rat median eminence. *Endocrinology* 1985;116:1863–1868.
9. Millar RP, Sheward WJ, Wegener I, Fink G. Somatostatin-28 is an hormonally active peptide secreted into hypophysial portal vessel blood. *Brain Res.* 1983;260:334–337.
10. Sheward WJ, Benoit R, Fink G. Somatostatin-28(1-12)-like immunoreactive substance is secreted into hypophysial portal vessel blood in the rat. *Neuroendocrinology* 1984;38:88–90.

been abandoned because: (a) TSH and prolactin surge release are not coincident (see, for example, Ref 184); (b) immunoneutralization with anti-TRH-serum does not block the proestrous or the suckling-induced prolactin surge in the rat[251,252]; and (c) copulsatile release of prolactin and TSH in the human occurs in the presence of constant TRH infusion.[253]

The amount of TRH released into hypophysial portal blood relative to TRH hypothalamic content is huge relative to that of GnRH and SST.[14] Thus, the amount of SST and GnRH released into portal blood per hour is much less than 1% of the total hypothalamic content (approximately 0.3% and 0.6%, respectively), while the amount of authentic TRH released into portal blood per hour can be as much as 34% relative to the total hypothalamic TRH content.[184,254] The uniquely high turnover of TRH suggests rapid synthesis, processing and release of the tripeptide. One factor that may contribute to the high "turnover" of TRH compared with GnRH and SST is that both in frog skin[255] and in the hypothalamus[247] the TRH precursor contains multiple copies of the tripeptide. In contrast, the precursors for SST and GnRH contain only one copy per molecule of the active peptide.

PITUITARY TARGET HORMONAL EFFECTS ON THE NERVOUS SYSTEM

Introduction

Gonadal ("sex") steroids, adrenal steroids and thyroid hormones, secreted by the three major pituitary target organs, exert powerful effects on the brain. These effects may be classified in terms of: (a) feedback actions; (b) brain differentiation and neural plasticity; (c) neurotransmission; (d) behavior; and (e) membranes and ion channels. Feedback actions of sex steroids on the brain and anterior pituitary gland are discussed above and in other chapters (see Chapters 3, 9, 19 and 20), as are the feedback actions of the adrenal corticosteroids (see Chapters 3, 8, 18, 21 and 29). For reviews on sexual differentiation of the brain, see Chapter 17 and the review by Fink.[18] The effects of steroids on behavior are reviewed in Chapter 20. The effect of steroids on membranes and ion channels is mentioned above, but is dealt with in detail in Chapter 18. Following a brief rehearsal of some of these aspects, attention here will focus on effects of sex steroids on central neurotransmission with reference to mood, mental state and cognition.

Effect of Sex Steroids on Sexual Behavior and Gender Assignment

Sex steroid hormones, estrogens and androgens, have long been known to exert powerful effects on sexual behavior.[18–20,256,257] The behavioral effects of these steroids are most clear-cut in lower species, where sexual behavior is driven by sex-steroid dependent reflexes. The term "estrus," meaning "gadfly," was used by the ancient Greeks to describe the frenzied behavior and mounting of cows during spring when, under the influence of increasing day length, there is an increased secretion of gonadotropin and, as a consequence, estrogen. Androgens, too, exert dramatic changes in the male. Thus, the annual growth of antlers and rutting of the stags each autumn is one of many examples of the powerful effect of testosterone and other androgens on musculoskeletal growth and instinctive and highly programmed aggressive behavior involved in fierce male competition for mating rights.

The influence of sex steroids on mating behavior in mammals has been investigated most extensively in the rat, where in females a spontaneous surge of estradiol-17β occurs cyclically and renders the female receptive to the advances of a male at estrus.[256] The neural circuits that subserve sexual reflexes in rodents have been established, as have the sites in the hypothalamus and other brain regions at which sex steroids exert their effects[256,257] (see also Chapter 20).

Cultural factors and gender assignment play a prominent role in sexual behavior in the human, and thereby blur sex-steroid effects on brain and behavior. However, the potent action of sex steroids is seen at puberty and other phases in the life cycle when significant changes occur in sex-steroid secretion. The importance of sex steroids in determining sexual outlook and behavior in man is illustrated by the androgen insensitivity syndrome, in which genetic (XY) males develop the outward appearance and self-perception of a female. This syndrome is due to one of several mutations of the androgen receptor gene which renders the androgen receptor defective or inactive (for details, see Chapter 17). That is, the testes, which remain undescended in the abdomen, secrete testosterone to which the body cannot respond because of the defective androgen receptors. The individual thus develops the outward appearance of a woman who is often voluptuous, but who is deficient in axillary and pubic hair and has no vagina and only vestigial remnants, if any, of the uterus or oviducts. The development of the vagina, uterus and the oviducts is inhibited by the Müllerian inhibitory factor secreted by the testes. The androgen insensitivity syndrome reinforces the dictum, derived first from work on rodents, that, irrespective of the genotype, the phenotype of the brain and body in mammals will develop as female unless exposed to testosterone at an early stage of development.[18,258] That is, the female phenotype may be regarded as the "default state."

Sex-steroid Effects on Central Serotonergic Mechanisms: Relevance for Mood, Mental State and Cognition

Clinical observations suggest that sex steroids may affect mood and mental state in the human.[259–265] The apparent increased incidence of mood disorder around the time of menstruation, puerperium and the menopause has been associated with the coincident fall of plasma estradiol concentrations at these times.[19,20,266] Estradiol has also been implicated in schizophrenia, in that the average age of onset of schizophrenia is later in women than in men; there is a qualitative sex difference in the symptoms of schizophrenia; and there is a second peak of schizophrenia onset in women, but not men, after the age of 40.[19,20,267] Furthermore, low estradiol levels appear to be a risk factor for psychotic disorders.[268,269]

The Serotonin 2A Receptor (5-HT_{2A} Receptor)

With these observations in mind, we were led into this field by investigations of the mechanisms by which estradiol-17β induces the spontaneous GnRH surge, and thereby pituitary gonadotropins, as discussed above.[18,270] Investigation of the classical neurotransmitter mechanisms involved in the estradiol-induced gonadotropin surge showed that estadiol-17β, in its positive feedback mode for GnRH/gonadotropin release in acutely ovariectomized female rats, induced a threefold increase in the amount of 5-HT_{2A} receptor (5-HT_{2A}R) mRNA in the dorsal raphe nucleus (DRN) with a concomitant increase in the density of 5-HT_{2A}R binding sites in the frontal, cingulate and primary olfactory cortex, the nucleus accumbens and caudate-putamen – brain regions that in the human are concerned with the control of mood, mental state and cognition.[271–273] The physiological significance of these experimental observations was underscored by our finding that in intact female rats the density of 5-HT_{2A}R binding sites in the frontal and cingulate cortex on the afternoon of proestrus was significantly greater than at diestrus[274], suggesting that the spontaneous preovulatory surge of estradiol-17β, which reaches a peak at about 1200 h of proestrus,[275] is able to increase significantly the density of 5-HT_{2A}R sites in forebrain. Acute estradiol treatment had no effect on any of the other receptor mRNAs studied (5-HT_{1A}, 5-HT_{2C}, dopamine 1, dopamine 2).[273] Long-term estradiol treatment of chronically ovariectomized rats or macaques has effects on the 5-HT_{2A}R that are similar to those of acute estradiol.[276–279] Furthermore, in the human, positron emission tomography (PET) has shown that estradiol treatment, administered long-term, also increases the density of 5-HT_{2A}R in the forebrain of women.[280,281]

Estrogen has similar effects on the central 5-HT_{2A}R in male rats. Our studies in the male were prompted by two factors. First, changes in mood and mental state occur in some men in association with an age-related decline in free plasma testosterone,[282,283] and testosterone replacement therapy improves mood in hypogonadal men[284] and in surgically menopausal women.[285] Second, we wished to determine whether there are sex differences in the effects of sex steroids on the central serotonergic mechanisms. Castration reduced, while treatment with testosterone propionate (TP) or estradiol benzoate (EB), but not 5α-dihydrotestosterone (5α-DHT) or oil, significantly increased the number of cells expressing 5-HT_{2A}R mRNA and the amount of 5-HT_{2A}R mRNA per cell in the DRN of male rats. Concomitantly, castration significantly reduced the density of 5-HT_{2A}R binding sites in frontal, cingulate and piriform cortex, olfactory tubercle and nucleus accumbens compared with that in intact animals.[20,286,287] Treatment of castrated rats with either TP or EB, but not the non-aromatizable androgen, 5α-DHT or oil (vehicle), increased the density of 5-HT_{2A}R binding sites in these same brain regions (as in the female) to levels similar to or greater than those in intact animals (Fig. 5.14).

In the caudate-putamen, the density of 5HT_{2A}R binding sites was increased by EB but not TP or 5α-DHT in the male, and by EB in the female. Aromatase is scarce in the caudate-putamen, and so this finding suggests that estradiol is the dominant steroid for increasing the density of 5-HT_{2A}R.[288] The apparent requirement of aromatase for the effect of testosterone on 5-HT_{2A}R density, suggesting that its action mediated by estrogen, resembles some of the other central actions of testosterone, such as sexual differentiation of the brain and the induction of arginine vasopressin in the bed nucleus of the stria terminalis.[18,288,289]

Concomitant measurement of LH release in the same male and female rats used to study effects on the 5-HT_{2A}R demonstrated that there is a major difference between the effects of testosterone and estradiol on higher brain centers compared with the neuroendocrine hypothalamus. In higher brain centers, estradiol and testosterone both increase the density of 5-HT_{2A}R. In the neuroendocrine hypothalamus, however, castration significantly increased, while TP and EB but not 5α-DHT significantly decreased, plasma LH concentrations. In contrast, EB significantly increased LH concentrations in ovariectomized female rats. The male phenotypic response can readily be produced in genetic females that have been masculinized by a single pulse of testosterone administered within the first 5 days postnatally.[9,258,290] Because of its importance, this issue is discussed in greater detail below.

The Serotonin Transporter

Because the serotonin transporter (SERT) has been implicated in mood disorders, and selective serotonin reuptake inhibitors are used as antidepressants, we also investigated the effects of acute sex-steroid manipulation on the SERT. Our results showed that, in female rats, estradiol-17β increased significantly the amount SERT mRNA in the DRN, with concomitant increases in SERT binding sites in the basolateral nucleus of the amygdala (Fig. 5.15), the lateral septum and the ventromedial hypothalamic nucleus – forebrain regions concerned with the control of emotions and behavior.[19,291] Similarly, in the male rat, castration decreased while testosterone or estradiol but not 5α-DHT increased

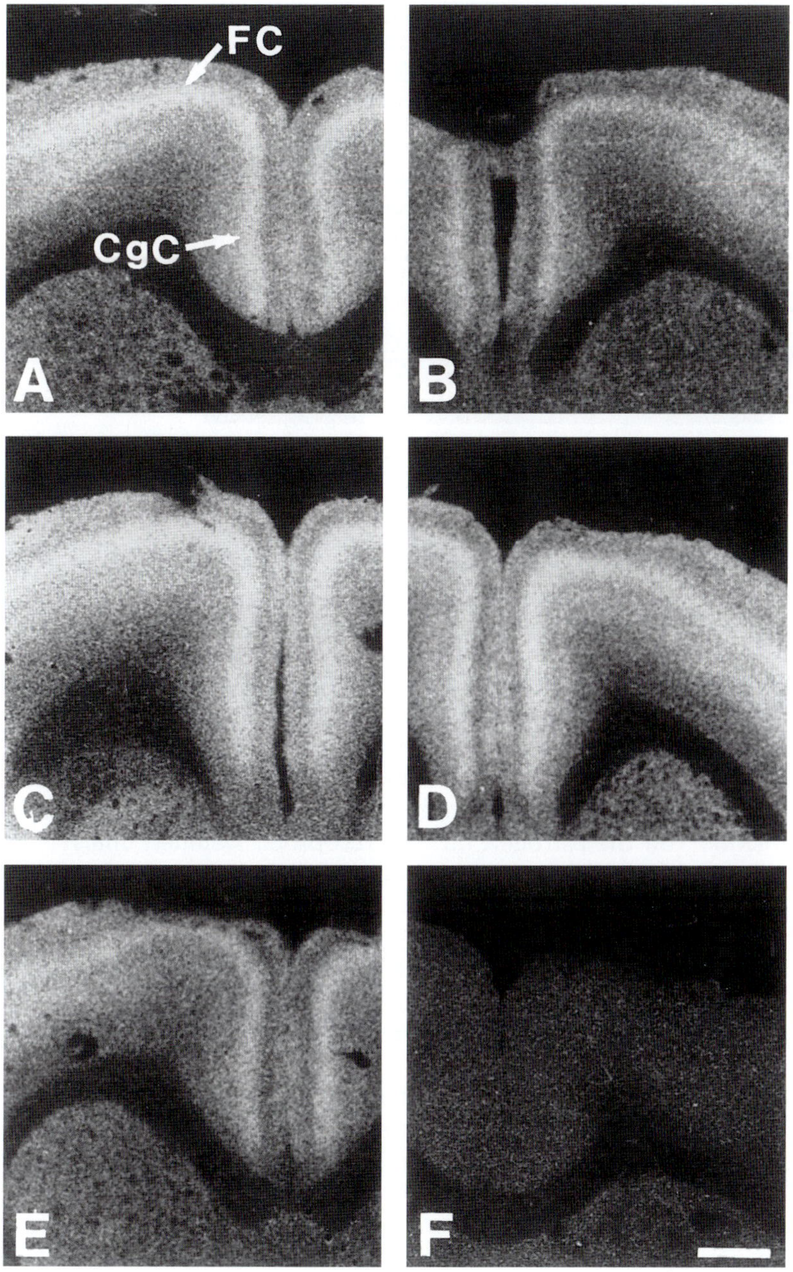

FIGURE 5.14 Density of 5-HT_{2A} receptors in the cerebral cortex of male rats as assessed by autoradiograph using the highly specific 5-HT_{2A}R ligand, [3H]RP62203. Note that the ligand binding is greatest in layer IV:Va of frontal and cingulate cortex (FC and CgC, respectively) in coronal sections of forebrain. The animals were either intact and untreated (**A**), or castrated and treated with oil vehicle (**B**), estradiol benzoate (**C**), testosterone propionate (**D**), or 5α-dihydrotestosterone (**E**). Non-specific binding is shown in (**F**). Note that the density of 5-HT_{2A}R is reduced by castration + oil treatment, and is increased to or above intact levels by either estradiol or testosterone, but not 5α-dihydrotestosterone. Darkfield photographs of autoradiographic images; scale, 1 mm. *Reproduced from Sumner BEH, Fink G, Testosterone as well as estrogen increases serotonin2A receptor mRNA and binding site densities in the male rat brain, Mol. Brain Res. 1998;59:205–214.*

significantly the content of SERT mRNA in the DRN and the density of SERT binding sites in the basolateral nucleus of the amygdala, the lateral septum and the ventromedial hypothalamic nucleus.[286]

The topochemical specificity of estrogen action is remarkable in that estradiol treatment had no effect on SERTmRNA levels in the median raphe nucleus (MRN). The reason for this is not clear, but Tao and Auerbach[292] demonstrated that there are striking differences in the inhibitory and excitatory afferents to the DRN compared with the MRN. GABA afferents were the predominant tonic influence on serotonergic neurons in the DRN. In contrast, glutamatergic but not GABAergic afferents had a strong tonic influence on serotonergic neurons in the MRN.[292] Other differences between the DRN and the MRN include the greater sensitivity of DRN neurons to the neurotoxic effects of (+/−)3,4-methylenedioxymethamphetamine ("ecstasy") in non-human primates.[293]

Mechanism of Estrogen Action on SERT and 5-HT_{2A}R Expression

There are several possible ways in which estrogen, whether secreted by the ovary or derived by enzymatic

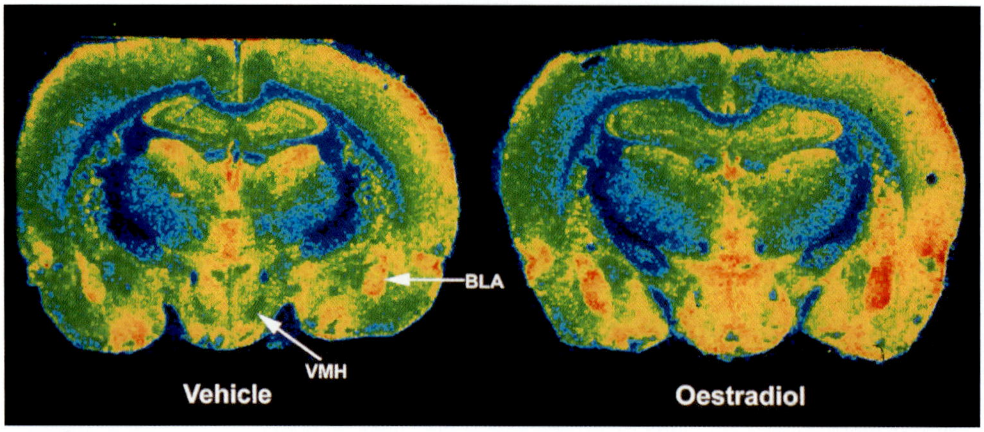

FIGURE 5.15 Estrogen increases density of SERT binding sites in rat forebrain. Pseudocolor representation of the density of SERT binding sites in the forebrain of female rats as assessed by autoradiography using [3H]paroxetine as ligand. The female rats had been ovariectomized and treated with either oil vehicle or estradiol benzoate. Note the estradiol-induced increase in [3H]paroxetine binding sites in the basolateral amygdala. *Based on data from,*[291] *McQueen JK, Wilson H, Fink G. Estradiol-17 beta increases serotonin transporter (SERT) mRNA levels and the density of SERT-binding sites in female rat brain Brain Res Mol Brain Res. 1997;45(1):13–23*

(aromatase) conversion from testosterone, could affect $5HT_{2A}R$ and SERT expression. These include: (a) a direct effect on the $5\text{-}HT_{2A}R$ or SERT gene by way of the classical genomic mechanism; (b) an indirect effect on the $5\text{-}HT_{2A}R$ or SERT gene by way of transynaptic mechanisms; and (c) a non-genomic mechanism.

At present, it is difficult to reach a firm conclusion about which of the three possible mechanisms is operative. This is because of species differences in the distribution and expression and co-localization of the estrogen receptors alpha and beta (ERα and ERβ, respectively) with the $5\text{-}HT_{2A}R$ and the SERT, and the lack of robust transcriptional data. Serotonergic neurons in the rat DRN express ERβ but not ERα,[261,294–298] but whether ERβ mediates estrogen induction of SERT and $5\text{-}HT_{2A}R$ expression in the DRN is not established. A transynaptic mechanism cannot be ruled out, since the 5-HT neurons of the DRN are activated by noradrenergic and glutamatergic neurons[299–302] and inhibited by GABA neurons[300]; one or more of these neuronal types, all of which express ERα,[292,295,303–306] might be involved in mediating the action of estradiol-17β.

The effect of EB on $5\text{-}HT_{2A}R$ mRNA and SERT mRNA in the DRN and the densities of 5-HT2AR and SERT binding sites in forebrain was completely blocked by the selective estrogen receptor modulators (SERMs) tamoxifen and raloxifene, which act as estradiol antagonists with respect to serotonergic mechanisms in brain.[21,22] The inhibitory effects of raloxifene on acute estrogen-induction of central serotonergic mechanisms were similar to those of tamoxifen, even though there are major differences between the two SERMs in their affinity for the two estrogen receptor subtypes.[307] The only difference between the actions of the two SERMS was that treatment of acutely ovariectomized rats with raloxifene alone increased the density of SERT sites in the mid-frontal cortex and decreased the density of $5\text{-}HT_{2A}R$ in the posterior olfactory tubercle. These results suggest that the action of estradiol on SERT and $5\text{-}HT_{2A}R$ gene expression in DRN may be dependent on the estrogen nuclear receptors. Detailed analysis of the effects of estradiol and tamoxifen on the DRN showed that constitutive SERT and $5\text{-}HT_{2A}R$ gene expression occurs only to a small extent in the caudal 25%–30% of DRN. SERT and $5\text{-}HT_{2A}R$ gene expression in the rostral 70% of the DRN depends upon estradiol induction. Indeed, a single injection of estradiol switches on $5\text{-}HT_{2A}R$ and SERT gene expression in the serotonergic neurons in the DRN [20,21] (Fig. 5.16).

The $5\text{-}HT_{2A}R$ in cortex would all appear to be postsynaptic with respect to projections from the DRN, and are located predominantly on the apical dendrites of pyramidal neurons and presynaptically on intrinsic glutamatergic neurons in frontal and cingulate cortex, and on intrinsic GABAergic neurons in the piriform cortex.[308] These conclusions, based on electrophysiological studies in the rat,[308] were confirmed by immunoelectronmicroscopical studies in the rhesus macaque which showed that in prefrontal cortex $5\text{-}HT_{2A}R$ are located in "hot spots" on apical dendrites of pyramidal neurons, and also in large to medium-sized interneurons.[309] The functional significance of this distribution of $5HT_{2A}R$ in cerebral cortex has been discussed in detail elsewhere.[19] Briefly, if the estrogen-induced increase in $5\text{-}HT_{2A}$ receptors in frontal, cingulate and piriform cortex results in an increase in receptor-mediated neuronal firing, as is the case for D2 receptors in striatum,[310–312] then estrogen could increase the activity of pyramidal neurons in the frontal and cingulate cortex and decrease pyramidal neuron activity in piriform cortex. The estrogen-induced increase in density of the 5-HT_{2A}R in cerebral cortex is not associated with an increase in 5-HT_{2A}R mRNA in cortex,[19,21,273,277,287] and is therefore likely to reflect a non-genomic action of estrogen or an action mediated by interneurons that express estrogen receptor. Given the close topographical relationship between the distribution of the 5-HT_{2A}R and the ERβ mRNA in rat cerebral cortex[19,21,313] and the paucity of ERα mRNA in cortex,[313] it seems possible that the effects of estradiol, which were blocked by tamoxifen, are mediated by the ERβ. The rationale for this assertion is based on the fact that studies on transfected cell lines showed that tamoxifen exerts a pure antagonist effect through ERβ.[314]

Clinical Implications

The clinical implications of the "estrogen—serotonin link" have been reviewed by Fink *et al.*,[20] where readers will find a detailed bibliography. Briefly, the 5-HT_{2A}R has been implicated in depression, and therefore the effects of estrogen on 5HT_{2A}R sites in higher brain centers offer a possible explanation for the association in some women between depression and the precipitous fall in plasma estrogen concentrations that occurs before menstruation, postnatally, or around the time of menopause. The hypothesis that estrogen or testosterone, by way of its conversion to estrogen, can improve mood by increasing the density of 5-HT_{2A}R in cerebral cortex is consistent with the PET findings that the density of 5-HT_{2A}R binding sites is decreased in the orbitofrontal and anterior insular cortex of drug-free depressed patients, and that there is a decrease with age in the density of 5-HT_{2A}R sites in cerebral cortex. Furthermore, repeated electroconvulsive shocks in rats, an experimental model for electroconvulsive antidepressant therapy in the human, increased the density of 5-HT_{2A}R sites in cerebral cortex.

Because it is a target of the "atypical" (second-generation) antipsychotics, such as clozapine and risperidone, and hallucinogens such as lysergic acid diethylamide, the 5-HT_{2A}R has also been implicated in schizophrenia. As mentioned above, sex differences in schizophrenia are characterized by a later first onset in women compared with men, a second peak of onset in women but not in men after the age of 40, and significant sex differences in symptomatology. The estrogen-induced increase in density of 5-HT_{2A}R sites in cortex and accumbens and D2 receptors in striatum are possible mechanisms which might be involved in the sex differences in schizophrenia. The estrogen-induced increase in 5-HT_{2A}R density appears at first sight to run contrary to the 5-HT_{2A}R antagonism of the atypical antipsychotics which are effective in schizophrenia. That is, if estradiol protects against schizophrenia, it might be expected to decrease, not increase, the density of 5-HT_{2A}R in cortex. The apparent paradox of a theoretically opposing action of estradiol and psychotic drugs also applies to the D2 receptor. The density of the D2 receptor is increased by estradiol, which nonetheless exerts neuroleptic effects and potentiates the rate of action of the typical antipsychotic, haloperidol. Resolution of these paradoxes awaits elucidation of the precise action of the antipsychotics and the role of 5-HT_{2A}R and estrogen in schizophrenia. Notwithstanding these apparent paradoxes, Kulkarni and associates have published preliminary data that suggest that estradiol and the selective estrogen receptor modulator, raloxifene, although not antipsychotic by themselves, potentiate the action of atypical antipsychotics in female and male patients with schizophrenia.[267,315,316]

Different Effects of Sex Steroids on Higher Brain Compared with Neuroendocrine Hypothalamus: Relevance of Sexual Differentiation of the Brain

The similar effect in both sexes of estrogen on the density of 5-HT_{2A}R in cerebral cortex and nucleus accumbens and the content of 5-HT_{2A}R mRNA in the DRN contrasts markedly with the striking sex difference in the neuroendocrine effects of estrogen and testosterone, as shown by the EB-induced release of LH in the ovariectomized female compared with EB and TP inhibition of LH release in the castrated male.[290] As mentioned above, the sex difference in estrogen action on the GnRH/LH release mechanism probably reflects a sex difference in the circuitry that mediates the effect of sex steroids on GnRH release, and which is established during the perinatal critical period of brain development, when in the male the brain is exposed to a burst of testosterone.[18,258,290] This point is exemplified by the fact that, in the female, 5-HT_{2A} receptors on interneurons are involved in the estrogen-induced GnRH surge,[317,318] whereas in the male, 5HT_{2A}R activation may inhibit GnRH synthesis.[319] The present data suggest that sexual differentiation of the brain does not appear to affect the responsiveness to sex steroids of the 5-HT_{2A}R gene in the DRN or 5-HT_{2A}R in forebrain outside the neuroendocrine circuits concerned with the control of GnRH release. What, then, of sex differences in mood disorders and schizophrenia?[19,266,320] The answer to this apparently simple question is probably complex. However, our comparison of intact male and female rats[274] shows that sex difference in the density of 5-HT_{2A}R binding sites in forebrain may be due to differences in the plasma concentrations of estradiol-17β. That is, sex differences in 5-HT_{2A}R binding sites in higher brain centers might be due to the concentrations of estradiol to which the brain is exposed, rather than to an irreversible sexual

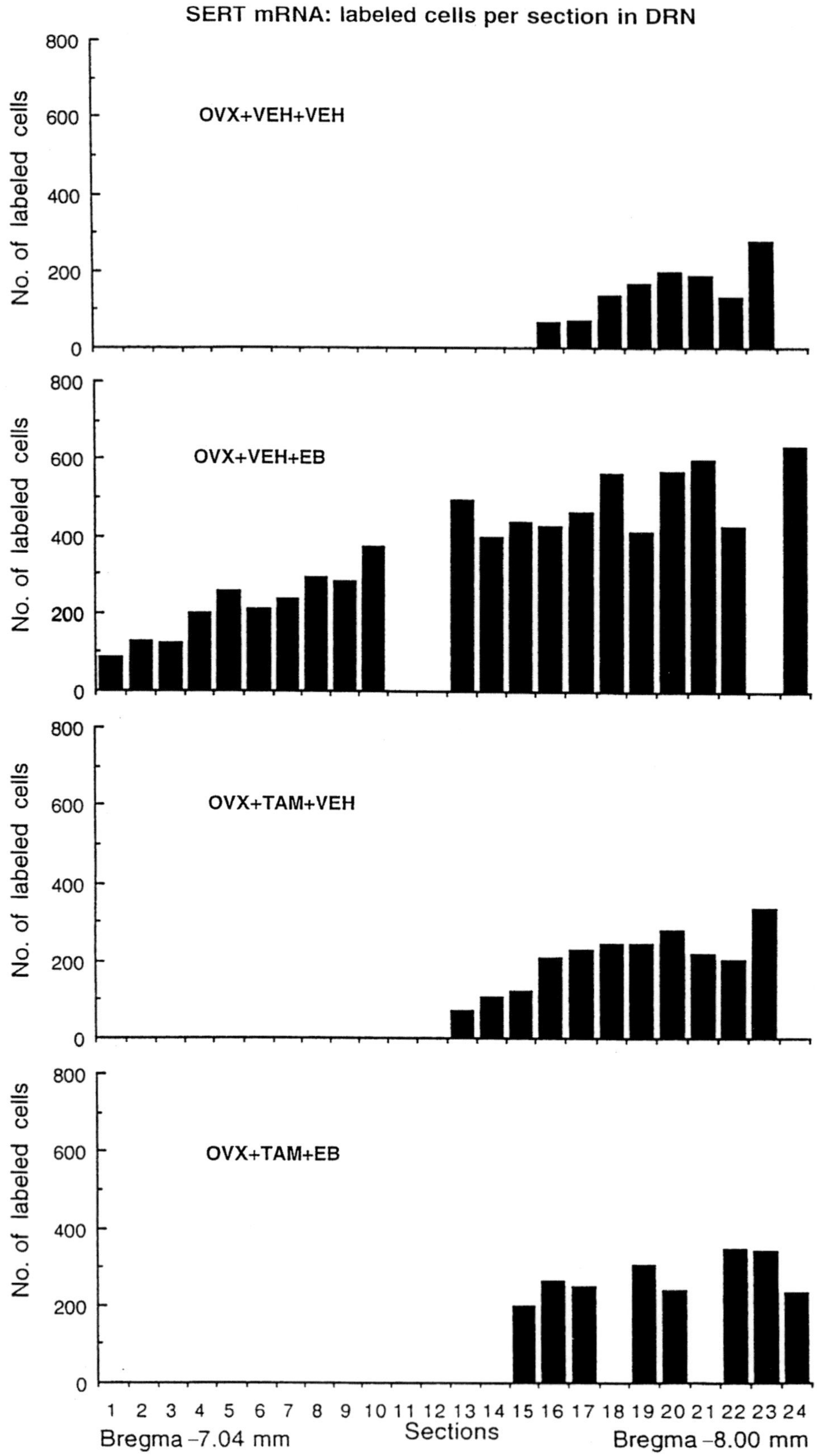
SERT mRNA: labeled cells per section in DRN
No. of labeled cells
OVX+VEH+VEH
OVX+VEH+EB
OVX+TAM+VEH
OVX+TAM+EB
0
200
400
600
800
1 2 3 4 5 6 7 8 9 10 11 12 13 14 15 16 17 18 19 20 21 22 23 24
Bregma -7.04 mm
Sections
Bregma -8.00 mm

dimorphism in neural circuitry or molecular mechanisms such as must underlie the sex difference in the GnRH/LH response to sex steroids.

REFLECTIONS ON NEUROENDOCRINE CONTRIBUTIONS TO SCIENCE AND MEDICINE

In addition to their intrinsic value, neuroendocrine systems have provided a robust basis and powerful generic models for neurophysiology, neurotransmission, synaptology, cybernetics, circadian control, molecular biology, behavior, genetics, gene × environment interactions, epigenetics, and endocrine, neuroendocrine and psychiatric disorders.

Thus, for example, the landmark paper by Shosaku Numa, Shigetada Nakanishi and their colleagues[321] which graced the cover page of the March 29, 1979 issue of *Nature* constituted a quantum leap in our understanding of the importance of post-translational processing. Entitled "Nucleotide sequence of cloned cDNA for bovine corticotropin-β-lipotropin precursor," the paper by Nakanishi *et al.*[321] showed how one gene, located on chromosome 2p23 in the human, could encode one protein precursor (proopiomelanocortin) that by post-translational enzymatic processing could generate three sets of biologically active proteins with diverse functions – that is, adrenocorticotropin (the key pituitary stress hormone); the melanocortins (involved in the regulation of food intake as well as skin pigmentation); and the endogenous opioids, enkephalin, and the endorphins (involved in the modulation of pain and other functions).

At the same time, Yuh Nung Jan, Lily Yeh Jan and Steve Kuffler discovered that the non-nicotinic substance responsible for the fourth type of excitatory postsynaptic potential (the "late slow epsp") in the sympathetic ganglia of the bullfrog was a peptide with characteristics similar to GnRH,[113] and subsequently shown to be GnRH2.

The suprachiasmatic nucleus is the circadian master clock in mammals: the clock genes have been identified and are described in detail in Chapter 12. Linked with the suprachiasmatic clock is the pineal gland, which provided an excellent model for the elegant work of David Klein and others that showed that arylalkylamine *N*-acetyltransferase, the "Timezyme," controls daily changes in melatonin production by the pineal gland, and thereby plays a unique role in biological timing in vertebrates.[53]

How the neurovascular synapse provided by the hypophysial portal vessels provides a unique investigative model has been described in detail above. Calcium-dependent stimulus–secretion coupling, a mechanism that is fundamental to virtually all known chemical neurotransmission, was discovered by W.W. (Bill) Douglas and associates by using as models the neurohypophysis (see Chapter 6) and adrenal medulla.[322]

Relatively new techniques include gene knockout or gene overexpression. Advances made in our understanding of genotype-to-phenotype relationships consequent on these techniques have occurred across the whole field of biology and medicine. Notable perhaps in neuroendocrinology are the way that gene knockout and overexpression have recently elucidated, for example, the function of the CRF family of peptides and their receptors (see Chapter 15).

The truism that evolution is dependent more on changes in receptor rather than ligand structure is widely seen in neuroendocrine systems, where in the case of somatostatin and melanocortin, for example, five different receptors enable each of these two neurohormones to have quite different functions in different cells and tissues (see Chapters 2, 7, 14 and 32).

All of the hypothalamic neurohormones are used extensively in diagnostic tests of neuroendocrine disorders such as Cushing's syndrome, hyogonadism, infertility, and disorders of growth, metabolism and the thyroid gland. In addition, neurohormone agonists and antagonists have been developed as therapeutic agents. Thus, as mentioned above, dopamine agonists such as bromocriptine, cabergoline and pergolide mesylate are effective in treating hyperprolactinemia and prolactin-secreting pituitary adenomata. GnRH superactive agonists and antagonists are used as an adjunct therapy for cancer of the breast, ovary and prostate gland; for fertility control; and for the treatment of infertility and precocious puberty. Somatostatin or its potent

FIGURE 5.16 Effect of estradiol on serotonin transporter (SERT) mRNA levels in the dorsal raphe nucleus (DRN) of female rats. The rats were ovariectomized (OVX) and treated with vehicle (VEH) for estradiol benzoate (EB) and Tamoxifen (TAM) [OVX+VEH+VEH], or EB + vehicle for TAM [OVX+EB+VEH], or [OVX + TAM + VEH], or [OVX + TAM + EB]. The histograms show the total number of labeled cells in the DRN, in each section of the midbrain sample in a typical set of four brains, each representing one treatment group. Note that EB induced SERT mRNA-expression in cells in the rostral region of the DRN, and also increased the numbers of SERT mRNA-expressing cells in caudal region. Tamoxifen blocked the action of estradiol. The data show that SERT gene expression in the caudal third of the DRN would appear to be constitutive: SERT gene expression in the rostral two-thirds of the DRN would appear to depend on estrogen. Abbreviations: OVX, ovariectomized; VEH, arachis oil vehicle; EB, estradiol benzoate; TAM, tamoxifen. The coordinates are according to Paxinos, G. and Watson, C. The rat brain in stereotaxic coordinates, 2nd edn. Sydney: Academic Press, 1986:264. *Reproduced from Sumner BE, Grant KE, Rosie R, Hegele-Hartung C, Fritzemeier KH, Fink G. Effects of tamoxifen on serotonin transporter and 5-hydroxytryptamine(2A) receptor binding sites and mRNA levels in the brain of ovariectomized rats with or without acute estradiol replacement.* Mol Brain Res. *1999;73:119–122, with permission.*

synthetic analogs, such as Octreotride, have been used in the treatment of GH-releasing tumors in the pituitary gland, which lead to acromegaly, and also ectopic hormone-secreting tumors, most commonly small cell carcinomas of the lung or tumors of the gastrointestinal system (see Chapter 35). Still under investigation is the possible use of CRF-41 antagonists for the treatment of anxiety and depression.

Finally, neuroendocrinology has spawned the discipline of psychoneuroendocrinology, which is based on two premises. The first is that pituitary target organ hormones, such as gonadal steroids, adrenal glucocorticoids and thyroid hormone, can affect mood, cognition and behavior. The second is that the secretion of pituitary hormones reflects the activity of hypothalamic neurons, which in turn reflects neurotransmitter activity that parallels normal or disordered neurotransmission in higher brain centers concerned with mood, mental state, cognition and behavior. That is, the premise assumes that disorders in central neurotransmission that accompany or underpin mental disorders, and especially the psychoses, will be reflected in altered pituitary hormone secretion. Robust data (some reviewed above) support the first premise. The second premise has some scientific support, but, for reasons exemplified by uncertainties related to the interactions between stress, genetic susceptibility to stress, post-traumatic stress disorder and changes in hippocampal volume associated with mental illness, more incisive studies are required before firm conclusions can be drawn.[207,323]

Chapter 36 provides a detailed discussion of psychoneuroendocrinology. Suffice to say that mental disorders are extremely common and costly to individuals and the community, and pose a daunting challenge to biomedical researchers. Nonetheless, several robust findings have been made, and these provide important clues for further neuroendocrine research that might optimally be carried out in conjunction with human genetics, human brain imaging, and sophisticated cognitive psychology and psychopharmacology.

References

1. Harris GW. *Neural Control of the Pituitary Gland*. London: Edward Arnold; 1955.
2. Fink G, Smith GC. Ultrastructural features of the developing hypothalamo–hypophysial axis in the rat: a correlative study. *Z Zellforsch mikrosk Anat*. 1971;119:208–226.
3. Cone RD, Low ML, Elmquist JK, Cameron JL. Neuroendocrinology. In: Reed Larsen P, Kronenberg HM, Melmed S, Polonsky KS, eds. *Williams Textbook of Endocrinology*. 10th ed. Philadelphia: W.B. Saunders; 2003:81–176.
4. Melmed S, Kleinberg DL. Anterior pituitary. In: Reed Larsen P, Kronenberg HM, Melmed S, Polonsky KS, eds. *Williams Textbook of Endocrinology*. 10th ed. Philadelphia: W.B. Saunders; 2003:177–279.
5. Smith PE. The disabilities caused by hypophysectomy and their repair. The tuberal (hypothalamic) syndrome in the rat. *J Am Med Assoc*. 1927;88:158–161.
6. Smith PE. The secretory capacity of the anterior hypophysis as evidenced by the effect of partial hypophysectomies in rats. *Anatomical Record*. 1932;52:191–207.
7. Cushing H. *Papers relating to the pituitary body. Hypothalamus and Parasympathetic Nervous System*. Bailliere: Tindall & Cox, London; 1932.
8. Fink G. The development of the releasing factor concept. *Clin Endocrinol*. 1976;5(suppl):245–260.
9. Fink G. Feedback actions of target hormones on hypothalamus and pituitary with special reference to gonadal steroids. *Annu Rev Physiol*. 1979;41:571–585.
10. Deng Y, Scherer PE. Adipokines as novel biomarkers and regulators of the metabolic syndrome. *Ann NY Acad Sci*. 2010 Nov;1212(1):E1–E19. doi: 10.1111/j.1749-6632.2010.05875.x.
11. Ouchi N, Parker JL, Lugus JJ, Walsh K. Adipokines in inflammation and metabolic disease. *Nat Rev Immunol*. 2011;11(2): 85–97.
12. Fink G. The hypothalamic pituitary ovarian axis. In: Stallworthy J, Bourne G, eds. *Recent Advances in Obstetrics and Gynaecology*. 12th ed. London: Churchill Livingstone; 1977:3–54.
13. Marshall FHA. Sexual periodicity and the causes which determine it. The Croonian Lecture: *Philos Trans R Soc. B* 1936;226: 423–456.
14. Fink G. The external layer of the median eminence: a neurovascular synapse. *Neurochem Intl*. 1986;9:141–153.
15. Fink G, Sheward WJ. Neuropeptide release *in vivo*: measurement in hypophysial portal blood. In: Fink G, Harmar AJ, eds. *Neuropeptides: a Methodology*. Chichester: John Wiley & Sons Ltd; 1989:157–188.
16. Guillemin R. Control of adenohypophysial functions by peptides of the central nervous system. *Harvey Lecture*. 1978; 71:71–131.
17. Guillemin R. Hypothalamic hormones a.k.a. hypothalamic releasing factors. *J Endocrinol*. 2005;184:11–28.
18. Fink G. The G W Harris lecture. Steroid control of brain and pituitary function. *QJ Exp Physiol*. 1988;73:257–293.
19. Fink G, Sumner BEH, McQueen JK, Wilson H, Rosie R. Sex steroid control of mood, mental state and memory. *Clin Exp Pharmacol Physiol*. 1998;25:764–775.
20. Fink G, Sumner B, Rosie R, Wilson H, McQueen J. Androgen actions on central serotonin neurotransmission: relevance for mood, mental state and memory. *Behav Brain Res*. 1999;105: 53–68.
21. Sumner BE, Grant KE, Rosie R, Hegele-Hartung C, Fritzemeier KH, Fink G. Effects of tamoxifen on serotonin transporter and 5-hydroxytryptamine(2A) receptor binding sites and mRNA levels in the brain of ovariectomized rats with or without acute estradiol replacement. *Mol Brain Res*. 1999; 73:119–122.
22 Sumner BE, Grant KE, Rosie R, Hegele-Hartung C, Fritzemeier KH, Fink G. Raloxifene blocks estradiol induction of the serotonin transporter and 5-hydroxytryptamine2A receptor in female rat brain. *Neurosci Lett*. 2007; 417:95–99.
23. Harris GW. Humours and hormones. *J Endocrinol*. 1972; 53(2):2–23.
24. Watts AG. Structure and function in the conceptual development of mammalian neuroendocrinology between 1920 and

1965. *Brain Res Rev.* 2011;66(1-2):174–204.
25. Rowan W. Experiments in Bird Migration. II. Reversed Migration. *Proc Natl Acad Sci USA.* 1930;16(7):520–525.
26. Popa GT, Fielding U. A portal circulation from the pituitary to the hypothalamic region. *J Anat.* 1930;65:88–91.
27. Popa GT, Fielding U. Hypophysio-portal vessels and their colloid accompaniment. *J Anat.* 1933;67:227–232.
28. Houssay BA, Biasotti A, Sammartino R. Modifications fonctionelles de l'hypophyse après les lesions infundibulotuberiennes chez le crapaud. *CR Soc Biol Paris.* 1935;120:725–727.
29. Friedgood HB. Studies on the sympathetic nervous control of the anterior hypophysis with special reference to a neurohumoral mechanism. Symposium on Endocrine Glands: Harvard Tercentenary Celebrations. Cited by Friedgood in Textbook of Endocrinology. In: Williams RH, ed. Philadelphia: W.B. Saunders; 1936:635–698.
30. Hinsey JC. The relation of the nervous system to ovulation and other phenomena of the female reproductive tract. *Cold Spring Harb Symp Quant Biol.* 1937;5:269–279.
31. Harris GW, Jacobsohn D. Functional grafts of the anterior pituitary gland. *Proc R Soc. B* 1952;139:263–276.
32. Nikitovitch-Winer M, Everett JW. Functional Restitution of Pituitary Grafts Re-Transplanted From Kidney To Median Eminence. *Endocrinology.* 1958;63(6):916–930.
33. Nikitovitch-Winer M, Everett JW. Histocytologic Changes In Grafts of Rat Pituitary on the Kidney and upon Retransplantation Under The Diencephalon. *Endocrinology.* 1959;65: 357–368.
34. Fink G. Inadvertent collaboration. *Nature.* 1977;269:747–748.
35. Vale W, Spiess J, Rivier C, Rivier J. Characterization of a 41 residue ovine hypothalamic peptide that stimulates the secretion of corticotropin and beta-endorphin. *Science.* 1981;213: 1394–1397.
36. Fink G. Has corticotropin-releasing factor finally been found. *Nature* 1981;294:270–272.
37. Fink G, Smith JR, Tibballs J. Corticotrophin releasing factor in hypophysial portal blood of rats. *Nature.* 1971;203: 467–468.
38. Rivier J, Spiess J, Thorner M, Vale W. Characterization of a growth hormone-releasing factor from a human pancreatic islet tumour. *Nature.* 1982;300(5889):276–278.
39. Guillemin R, Brazeau P, Böhlen P, Esch F, Ling N, Wehrenberg WB. Growth hormone-releasing factor from a human pancreatic tumor that caused acromegaly. *Science.* 1982;218(4572):585–587.
40. Fink G. Growth hormone releasing factor: a tale of two islets. *Nature.* 1983;301:562.
41. Grattan DR, Kokay IC. Prolactin: a pleiotropic neuroendocrine hormone. *J Neuroendocrinol.* 2008;20(6):752–763.
42. Tuomisto J, Männistö P. Neurotransmitter regulation of anterior pituitary hormones. *Pharmacol Rev.* 1985;37(3):249–332.
43. Barraclough CA, Sawyer CH. Induction of pseudopregnancy in the rat by reserpine and chlorpromazine. *Endocrinology.* 1959;65:563–571.
44. Smith GC, Fink G. Experimental studies on the origin of monoamine-containing fibres in the hypothalamo–hypophysial complex of the rat. *Brain Res.* 1972;43(1):37–51.
45. Himes AD, Raetzman LT. Premature differentiation and aberrant movement of pituitary cells lacking both Hes1 and Prop1. *Dev Biol.* 2009;325(1):151–161.
46. Aujla PK, Bora A, Monahan P, Sweedler JV. Raetzman LT The Notch effector gene Hes1 regulates migration of hypothalamic neurons, neuropeptide content and axon targeting to the pituitary. *Dev Biol.* 2011 Feb 23 [Epub ahead of print].
47. Cottrell GT, Ferguson AV. Sensory circumventricular organs: central roles in integrated autonomic regulation. *Regul Pept* 2004;117(1):11–23. Review.
48. Smith PM, Ferguson AV. Circulating signals as critical regulators of autonomic state–central roles for the subfornical organ. *Am J Physiol Regul Integr Comp Physiol.* 2010;299(2): R405–R415.
49. Duvernoy HM, Risold P-Y. The circumventricular organs: An atlas of comparative anatomy and vascularization. *Brain Res Rev.* 2007;56:119–147.
50. Smith PM, Rozanski G, Ferguson AV. Acute electrical stimulation of the subfornical organ induces feeding in satiated rats. *Physiol Behav.* 2010;99(4):534–537.
51. Bennett L, Yang M, Enikolopov G, Iacovitti L. Circumventricular organs: a novel site of neural stem cells in the adult brain. *Mol Cell Neurosci.* 2009;41(3):337–347.
52. Arendt J. *Melatonin and the Mammalian Pineal Gland.* London: Chapman and Hall; 1995.
53. Klein DC. Arylalkylamine N-acetyltransferase: "the Timezyme". *J Biol Chem.* 2007;282(7):4233–4237.
54. Klein DC. Evolution of the vertebrate pineal gland: the AANAT hypothesis. *Chronobiol Intl.* 2006;23(1-2):5–20.
55. Pavlicek J, Sauzet S, Besseau L, Coon SL, Weller JL, Boeuf G, Gaildrat P, Omelchenko MV, Koonin EV, Falcón J. Klein DC Evolution of AANAT: expansion of the gene family in the cephalochordate amphioxus. *BMC Evol Biol.* 2010;10:154.
56. Bailey MJ, Coon SL, Carter DA, Humphries A, Kim JS, Shi Q, Gaildrat P, Morin F, Ganguly S, Hogenesch JB, Weller JL, Rath MF, Møller M, Baler R, Sugden D, Rangel ZG, Munson PJ, Klein DC. Night/day changes in pineal expression of >600 genes: central role of adrenergic/cAMP signaling. *J Biol Chem.* 2009;284(12):7606–7622.
57. Ko GY, Shi L, Ko ML. Circadian regulation of ion channels and their functions. *J Neurochem.* 2009 Aug;110(4):1150–1169.
58. Campino C, Valenzuela FJ, Torres-Farfan C, Reynolds HE, Abarzua-Catalan L, Arteaga E, Trucco C, Guzmán S, Valenzuela GJ, Seron-Ferre M. Melatonin exerts direct inhibitory actions on ACTH responses in the human adrenal gland. *Horm Metab Res.* 2011 Feb 17 [Epub ahead of print].
59. Valenzuela FJ, Torres-Farfan C, Richter HG, Mendez N, Campino C, Torrealba F, Valenzuela GJ, Serón-Ferré M. Clock gene expression in adult primate suprachiasmatic nuclei and adrenal: is the adrenal a peripheral clock responsive to melatonin? *Endocrinology.* 2008;149(4):1454–1461.
60. Reiter RJ. The pineal gland and melatonin in relation to aging: a summary of the theories and of the data. *Exp Gerontol.* 1995;30(3-4):199–212.
61. Reiter RJ, Manchester LC, Tan DX. Neurotoxins: free radical mechanisms and melatonin protection. *Curr Neuropharmacol.* 2010;8(3):194–210.
62. Dardente H, Wyse CA, Birnie MJ, Dupré SM, Loudon AS, Lincoln GA, Hazlerigg DG. A molecular switch for photoperiod responsiveness in mammals. *Curr Biol.* 2010;20(24):2193–2198. Epub 2010 Dec 2.
63. Lincoln GA. Decoding the nightly melatonin signal through circadian clockwork. *Mol Cell Endocrinol.* 2006;252(1-2):69–73.
64. Reiter RJ. Pineal control of a seasonal reproductive rhythm in male golden hamsters exposed to natural daylight and temperature. *Endocrinology.* 1973;92(2):423–430.
65. Reiter RJ. Chronobiological aspects of the mammalian pineal gland. *Prog Clin Biol Res.* 1981;59C(00):223–233.
66. Renfree MB. Diapause, pregnancy, and parturition in Australian marsupials. *J Exp Zool.* 1993;266(5):450–462.
67. Schneider NY, Fletcher TP, Shaw G, Renfree MB. The effect of pregnant and oestrous females on male testosterone and

behaviour in the tammar wallaby. *Horm Behav.* 2010;58(3): 378–384.

68. Versi E, Chiappa SA, Fink G, Charlton HM. Pineal influences hypothalamic Gn-RH content in the vole, Microtus agrestis. *J Reprod Fertil.* 1983;67(2):365–368.
69. Scott FF, Belle MDC, Delagrange P, Piggins HD. Electrophysiological Effects of Melatonin on Mouse Per1 and non-Per1 Suprachiasmatic Nuclei Neurones *In Vitro*. *J Neuroendocrinol* 2010;22:1148–1156.
70. Werman RCNS. cellular level: membranes. *Annu Rev Physiol.* 1972;34:337–374.
71. Fink G. Nature of luteinizing hormone releasing factor in hypophysial portal blood. *Nature.* 1967;215:159–161.
72. Sarkar DK, Chiappa SA, Fink G, Sherwood NM. Gonadotropin-releasing hormone surge in pro-oestrous rats. *Nature.* 1976;264:461–463.
73. Sherwood NM, Fink G. Effect of ovariectomy and adrenalectomy on luteinizing-hormone releasing hormone in pituitary stalk blood from female rats. *Endocrinology.* 1980;106: 363–367.
74. Sherwood NM, Chiappa SA, Sarkar DK, Fink G. Gonadotropin-releasing hormone (GnRH) in pituitary stalk blood from proestrous rats: effects of anesthetics and relationship between stored and released GnRH and luteinizing hormone. *Endocrinology.* 1980;107:1410–1417.
75. Neill JD, Patton JM, Dailey RA, Tsou RC, Tindall GT. Luteinizing hormone releasing hormone (LHRH) in pituitary stalk blood of rhesus monkeys: Relationship to level of LH release. *Endocrinology.* 1977;101:430–434.
76. Wardlaw SL, Wehrenberg WB, Ferin M, Antunes JL, Frantz AG. Effect of sex steroids on beta-endorphin in hypophyseal portal blood. *J Clin Endocrinol Metab.* 1982;55(5):877–881.
77. Xia L, Van Vugt D, Alston EJ, Luckhaus J, Ferin M. A surge of gonadotropin-releasing hormone accompanies the estradiol-induced gonadotropin surge in the rhesus monkey. *Endocrinology.* 1992;131(6):2812–2820.
78. Clarke IJ, Cummins JT. The temporal relationship between gonadotropin releasing hormone (GnRH) and luteinizing hormone (LH) secretion in ovariectomized ewes. *Endocrinology.* 1982;111(5):1737–1739.
79. Clarke IJ, Cummins JT. Increased gonadotropin-releasing hormone pulse frequency associated with estrogen-induced luteinizing hormone surges in ovariectomized ewes. *Endocrinology* 1985;116(6):2376–2383.
80. Thomas GB, Cummins JT, Smythe GA, Gleeson RM, Dow RC, Fink G, Clarke IJ. Concentrations of dopamine and noradrenaline in hypophysial portal blood in the sheep and the rat. *J Endocrinol.* 1989;121(1):141–147.
81. Caraty A, Antoine C, Delaleu B, Locatelli A, Bouchard P, Gautron JP, Evans NP, Karsch FJ, Padmanabhan V. Nature and bioactivity of gonadotropin-releasing hormone (GnRH) secreted during the GnRH surge. *Endocrinology.* 1995;136(8): 3452–3460.
82. Moenter SM, Caraty A, Locatelli A, Karsch FJ. Pattern of gonadotropin-releasing hormone (GnRH) secretion leading up to ovulation in the ewe: existence of a preovulatory GnRH surge. *Endocrinology.* 1991;129(3):1175–1182.
83. Karsch FJ, Evans NP. Feedback actions of estradiol on GnRH secretion during the follicular phase of the estrous cycle. *Acta Neurobiol Exp (Wars).* 1996;56(3):715–725.
84. Evans NP, Dahl GE, Caraty A, Padmanabhan V, Thrun LA. Karsch FJ. How much of the gonadotropin-releasing hormone (GnRH) surge is required for generation of the luteinizing hormone surge in the ewe? Duration of the endogenous GnRH signal. *Endocrinology.* 1996;137(11):4730–4737.
85. Neill JD, Frawley LS, Plotsky PM, Tindall GT. Dopamine in hypophysial stalk blood of the rhesus monkey and its role in regulating prolactin secretion. *Endocrinology.* 1981;108(2): 489–494.
86. Millar RP, Sheward WJ, Wegener I, Fink G. Somatostatin-28 is an hormonally active peptide secreted into hypophysial portal vessel blood. *Brain Res.* 1983;260:334–337.
87. Sheward WJ, Benoit R, Fink G. Somatostatin-28(1-12)-like immunoreactive substance is secreted into hypophysial portal vessel blood in the rat. *Neuroendocrinology.* 1984;38: 88–90.
88. Lim AT, Sheward WJ, Copolov D, Windmill D, Fink G. Atrial natriuretic factor is released into hypophysial portal blood: Direct evidence that atrial natriuretic factor may be a neurohormone involved in hypothalamic pituitary control. *J Neuroendocrinol.* 1990;2(1):15–18.
89. Lim AT, Dow RC, Yang Z, Fink G. ANP(5-28) is the major molecular species in hypophysial portal blood of the rat. *Peptides.* 1994;15(8):1557–1559.
90. Antoni FA, Hunter EF, Lowry PJ, Noble JM, Seckl JR. Atriopeptin: an endogenous corticotropin-release inhibiting hormone. *Endocrinology.* 1992;130(3):1753–1755.
91. Fink G, Dow RC, Casley D, Johnston CI, Bennie J, Carroll S, Dick H. Atrial natriuretic peptide is involved in the ACTH response to stress and glucocorticoid negative feedback in the rat. *J Endocrinol.* 1992;135:37–43.
92. Clarke I, Jessop D, Millar R, Morris M, Bloom S, Lightman S, Coen CW, Lew R, Smith I. Many peptides that are present in the external zone of the median eminence are not secreted into the hypophysial portal blood of sheep. *Neuroendocrinology.* 1993;57(5):765–775.
93. Giacobini P, Wray S. Cholecystokinin directly inhibits neuronal activity of primary gonadotropin-releasing hormone cells through cholecystokinin-1 receptor. *Endocrinology.* 2007; 148(1):63–71.
94. Giacobini P, Wray S. Prenatal expression of cholecystokinin (CCK) in the central nervous system (CNS) of mouse. *Neurosci Lett.* 2008;438(1):96–101.
95. Bale TL, Vale WW. CRF and CRF receptors: role in stress responsivity and other behaviors. *Annu Rev Pharmacol Toxicol.* 2004;44:525–557.
96. Boorse GC, Denver RJ. Widespread tissue distribution and diverse functions of corticotropin-releasing factor and related peptides. *Gen Comp Endocrinol.* 2006;146(1):9–18.
97. Zorrilla EP. Koob GF.Progress in corticotropin-releasing factor-1 antagonist development. *Drug Discov Today.* 2010;15(9-10): 371–383.
98. Kuperman Y, Chen A. Urocortins: Emerging metabolic and energy homeostasis perspectives. *Trends Endocrinol Metab.* 2008;19:122–129.
99. Kuperman Y, Issler O, Regev L, Musseri I, Navon I, Neufeld-Cohen A, Gil S, Chen A. Perifornical Urocortin-3 mediates the link between stress-induced anxiety and energy homeostasis. *Proc Natl Acad Sci USA.* 2010;107:8393–8398.
100. Cattanach BM, Iddon CA, Charlton HM, Chiappa SA, Fink G. Gonadotrophin-releasing hormone deficiency in a mutant mouse with hypogonadism. *Nature.* 1977;269(5626):338–340.
101. Mason AJ, Hayflick JS, Zoeller RT, Young III WS, Phillips HS, Nikolics K, Seeburg PH. A deletion truncating the gonadotropin-releasing hormone gene is responsible for hypogonadism in the hpg mouse. *Science* 1986;1986(234):1366–1371.
102. Bouligand J, Ghervan C, Trabado S. Brailly-Tabard S Guiochon-Mantel A, Young J. Genetics defects in GNRH1: a paradigm of hypothalamic congenital gonadotropin deficiency. *Brain Res.* 2010;1364:3–9.

103. Son GH, Park E, Jung H, Han J, Lee KH, Seong JY, Kim K. GnRH pre-mRNA splicing: solving the mystery of a nature's knockout, hpg mouse. *Biochem Biophys Res Commun.* 2005;2005(326): 261–267.
104. Fink G. Neuroendocrine control of gonadotrophin secretion. *Br Med Bull.* 1979;35(2):155–160.
105. Roch GJ, Busby ER, Sherwood NM. Evolution of GnRH: Diving deeper. *Gen Comp Endocrinol.* 2011;171(1):1–16.
106. Millar RP, Lu ZL, Pawson AJ, Flanagan CA, Morgan K, Maudsley SR. Gonadotropin-releasing hormone receptors. *Endocr Rev.* 2004;25(2):235–275.
107. Wetsel WC, Srinivasan S. Pro-GnRH processing. *Prog Brain Res.* 2002;141:221–241.
108. Tsai P-S, Zhang L. The emergence and loss of gonadotropin-releasing hormone in protostomes:orthology, phylogeny, structure, and function. *Biol Reprod.* 2008;79:798–805.
109. Seeburg PH, Adelman JP. Characterization of cDNA for precursor of human luteinizing hormone releasing hormone. *Nature*; 1984. Lond. 311,666~668.
110. Sheward WJ, Harmar AJ, Fink G. LH-RH in the rat and mouse hypothalamus and rat hypophysial portal blood: confirmation of identity by high performance liquid chromatography. *Brain Res.* 1985;345:362–365.
111. Kim DK, Cho EB, Moon MJ, Park S, Hwang JI, Kah O, Sower SA, Vaudry H, Seong JY. Revisiting the evolution of gonadotropin-releasing hormones and their receptors in vertebrates: secrets hidden in genomes. *Gen Comp Endocrinol.* 2011;170(1):68–78.
112. Kavanaugh SI, Nozaki M, Sower SA. Origins of gonadotropin-releasing hormone (GnRH) in vertebrates: identification of a novel GnRH in a basal vertebrate, the sea lamprey. *Endocrinology.* 2008;149(8):3860–3869.
113. Jan YN, Jan LY, Kuffler SW. A peptide as possible transmitter in sympathetic ganglia of the frog. *Proc Natl Acad Sci USA.* 1979;76:1501–1505.
114. Jones SW. Chicken II luteinizing hormone-releasing hormone inhibits the M-current of bullfrog sympathetic neurons. *Neurosci Lett.* 1987;80(2):180–184.
115. Sealfon SC, Weinstein H, Millar RP. Molecular mechanisms of ligand interaction with the gonadotropin-releasing hormone receptor. *Endocr Rev.* 1997;18(2):180–205.
116. Fink G, Jamieson MG. Immunoreactive luteinizing hormone releasing factor in rat pituitary stalk blood: effects of electrical stimulation of the medial preoptic area. *J Endocrinol.* 1976; 1976(68):71–87.
117. Eskay RL, Mical RS, Porter JC. Relationship between luteinizing hormone releasing hormone concentration in hypophysial portal blood and luteinizing hormone release in intact, castrated, and electrochemically-stimulated rats. *Endocrinology.* 1977;100(2):263–270.
118. Chiappa SA, Fink G, Sherwood NM. Immunoreactive luteinizing hormone releasing factor (LRF) in pituitary stalk plasma from female rats: effects of stimulating diencephalon, hippocampus and amygdala. *J Physiol.* 1977;267:625–640.
119. Jamieson MG, Fink G. Parameters of electrical stimulation of the medial preoptic area for release of gonadotrophins in male rats. *J Endocrinol.* 1976;68:57–70.
120. Dyer RG, Mansfield S, Yates JO. Discharge of gonadotrophin-releasing hormone from the mediobasal part of the hypothalamus: effect of stimulation frequency and gonadal steroids. *Exp Brain Res.* 1980;39:453–460.
121. Yen SSC, Lasley BL, Wang CF, Leblanc H, Siler TM. The operating characteristics of the hypothalamic pituitary system during the menstrual cycle and observations of biological action of somatostatin. *Recent Prog Horm Res.* 1975;31:321–357.
122. Ching M. Correlative surges of LHRH, LH and FSH in pituitary stalk plasma and systemic plasma of rat during proestrus. *Neuroendocrinology.* 1982;34:279–285.
123. Carmel PW, Araki S, Ferin M. Pituitary stalk portal blood collection in rhesus monkeys: evidence for pulsatile release of gonadotropin-releasing hormone (GnRH). *Endocrinology.* 1976;99(1):243–248.
124. Aiyer MS, Fink G. The role of sex steroid hormones in modulating the responsiveness of the anterior pituitary gland to luteinizing hormone releasing factor in the female rat. *J Endocrinol.* 1974;62:553–572.
125. Fink G, Aiyer MS. Gonadotrophin secretion after electrical stimulation of the preoptic area during the oestrous cycle of the rat. *J Endocrinol.* 1974;62:589–604.
126. Fink G, Chiappa SA, Aiyer MS. Priming effect of luteinizing hormone releasing factor elicited by preoptic stimulation and by intravenous infusion and multiple injections of the synthetic decapeptide. *J Endocrinol.* 1976;69:359–372.
127. Fink G. The self-priming effect of LHRH: A unique servo-mechanism and possible cellular model for memory. *Front Neuroendocrinol.* 1995;16:183–190.
128. Speight A, Popkin R, Watts AG, Fink G. Oestradiol-17β increases pituitary responsiveness by a mechanism that involves the release and the priming effect of luteinizing hormone releasing factor. *J Endocrinol.* 1981;88:301–308.
129. Clarke IJ. Control of GnRH secretion: One step back. *Front Neuroendocrinol.* 2011 Jan 7 [Epub ahead of print].
130 Sarkar DK, Fink G. Luteinizing hormone releasing factor in pituitary stalk plasma from long-term ovariectomized rats: effects of steroids. *J Endocrinol.* 1980;86:511–524.
131. Knobil E. The neuroendocrine control of the menstrual cycle. *Recent Prog Horm Res* 1980;36:53–58.
132. Schuiling GA, Pols-Valkof N, van der Schaaf-Verdonk GCJ, Koiter TR. Blockade of LH and FSH secretion by LH-releasing hormone, by the LH-releasing hormone analogue, buserelin, and by combined treatment with LH-releasing hormone and oestradiol benzoate. *J Endocr.* 1984;103: 301–309.
133. Naor Z, Amsterdam A, Cart KJ. Binding and activation of gonadotropin-releasing hormone receptors in pituitary gonadotrophs. In: Saxena BB, Birnbaumber KJ, Lutz L, Martini L, eds. *Hormone Receptors in Growth and Reproduction.* New York: Raven Press; 1984:119.
134. Yen SSC, Tsou CC, Naftolin F, Vandenberg G, Ajabor L. Pulsatile patterns of gonadotropin release in subjects with and without ovarian function. *J Clin Endocrin Metab.* 1972;34: 671–675.
135. Fox SR, Smith MS. Changes in the pulsatile pattern of luteinizing hormone secretion during the rat estrous cycle. *Endocrinology.* 1985;116:1485–1492.
136. Plant TM, Dubey AK. Evidence from the rhesus monkey (*Macaca mulatta*) for the view that negative feedback control of luteinizing hormone secretion by the testis is mediated by a deceleration of hypothalamic gonadotropin-releasing hormone pulse frequency. *Endocrinology.* 1984;115: 2145–2153.
137. Everett JW, Sawyer CH. A 24-hour periodicity in the "LH-release apparatus" of female rats, disclosed by barbiturate sedation. *Endocrinology.* 1950;47:198–218.
138. Everett JW. The Third Annual Carl G. Hartman Lecture. Brain, pituitary gland, and the ovarian cycle. *Biol Reprod.* 1972; 6(1):3–12.
139. Everett JW. The timing of ovulation. *J Endocrinol.* 1977;75:3–13.
140. Kerdelhue B, Brown S, Lenoir V, Queenan JT, Jones GS, Scholler R, Jones HW. Timing of initiation of the preovulatory

luteinizing hormone surge and its relationship with the circadian cortisol rhythm in the human. *Neuroendocrinology* 2002;75:158–163.
141. Christian CA, Moenter SM. Critical roles for fast synaptic transmission in mediating estradiol negative and positive feedback in the neural control of ovulation. *Endocrinology.* 2008;149(11):5500–5508.
142. Caligaris L, Astrada JJ, Taleisnik S. Stimulating and inhibiting effects of progesterone on the release of luteinizing hormone. *Acta Endocrinol (Copenh).* 1968;59(2):177–185.
143. Caligaris L, Astrada JJ, Taleisnik S. Release of luteinizing hormone induced by estrogen injection into ovariectomized rats. *Endocrinology.* 1971;88(4):810–815.
144. Legan SJ, Karsh FJ. A daily signal for the LH surge in the rat. *Endocrinology.* 1975;96:57–62.
145. Christian CA, Moenter SM. The neurobiology of preovulatory and estradiol-induced gonadotropin-releasing hormone surges. *Endocr Rev.* 2010;31(4):544–577.
146. Raisman G, Brown-Grant K. The "suprachiasmatic syndrome": endocrine and behavioural abnormalities following lesions of the suprachiasmatic nuclei in the female rat. *Proc R Soc Lond B Biol Sci.* 1977;198(1132):297–314.
147. Moenter SM. Identified GnRH neuron electrophysiology: a decade of study. *Brain Res.* 2010;1364:10–24.
148. Rosie R, Sumner BE, Fink G. An alpha 1 adrenergic mechanism mediates estradiol stimulation of LHRH mRNA synthesis and estradiol inhibition of POMC mRNA synthesis in the hypothalamus of the prepubertal female rat. *J Steroid Biochem Mol Biol.* 1994;49(4-6):399–406.
149. Sun J, Chu Z, Moenter SM. Diurnal *in vivo* and rapid *in vitro* effects of estradiol on voltage-gated calcium channels in gonadotropin-releasing hormone neurons. *J Neurosci.* 2010;30(11): 3912–3923.
150. Kelly MJ, Rønnekleiv OK. Membrane-initiated estrogen signaling in hypothalamic neurons. *Mol Cell Endocrinol.* 2008; 290(1-2):14–23.
151. Kelly MJ, Qiu J. Estrogen signaling in hypothalamic circuits controlling reproduction. *Brain Res.* 2010;1364:44–52.
152. Hammes SR, Levin ER. Extranuclear steroid receptors: Nature and actions. *Endocr Rev.* 2007;28(7):726–741.
153. Belcher SM. Rapid signaling mechanisms of estrogens in the developing cerebellum. *Brain Res Rev.* 2008;57(2):481–492.
154. Glidewell-Kenney C, Weiss J, Hurley LA, Levine JE, Jameson JL. Estrogen receptor alpha signaling pathways differentially regulate gonadotropin subunit gene expression and serum follicle-stimulating hormone in the female mouse. *Endocrinology.* 2008;149(8):4168–4176.
155. Shivers BD, Harlan RE, Morell JI, Pfaff DW. Absence of oestradiol concentration in cell nuclei of LHRH-immunoreactive neurons. *Nature Lond.* 1983;304:345–347.
156. Hrabovszky E, Steinhauser A, Barabas K, Shughrue PJ, Petersen SL, Merchenthaler I, Liposits Z. Estrogen receptor--beta immunoreactivity in luteinizing hormone-releasing hormone neurons of the rat brain. *Endocrinology.* 2001;142: 3261–3264.
157. Wintermantel TM, Campbell RE, Porteous R, Bock D, Gröne HJ, Todman MG, Korach KS, Greiner E, Pérez CA, Schütz G, Herbison AE. Definition of estrogen receptor pathway critical for estrogen positive feedback to gonadotropin-releasing hormone neurons and fertility. *Neuron.* 2006;52(2):271–280.
158. Sarkar DK, Fink G. Gonadotropin-releasing hormone surge: possible modulation through postsynaptic alpha-adrenoreceptors and two pharmacologically distinct dopamine receptors. *Endocrinology.* 1981;108:862–867.
159. Fink G, Rosie R, Sheward WJ, Thomson E, Wilson H. Steroid control of central neuronal interactions and function. *J. Steroid Biochem Molec Biol.* 1991;40(1-3):123–132.
160. Hudgens ED, Ji L, Carpenter CD, Petersen SL. The gad2 promoter is a transcriptional target of estrogen receptor (ER) alpha and ER beta: a unifying hypothesis to explain diverse effects of estradiol. *J Neurosci.* 2009;29(27):8790–8797.
161. Iremonger KJ, Constantin S, Liu X, Herbison AE. Glutamate regulation of GnRH neuron excitability. *Brain Res.* 2010;1364: 35–43.
162 Gross CG. Genealogy of the "Grandmother Cell". *Neuroscientist* 2002;8(5):512–518.
163. Oakley AE, Clifton DK, Steiner RA. Kisspeptin signaling in the brain. *Endocr Rev.* 2009;30(6):713–743.
164. Clarkson J, Herbison AE. Oestrogen, kisspeptin, GPR54 and the pre-ovulatory luteinising hormone surge. *J Neuroendocrinol.* 2009;21(4):305–311.
165. Smith JT, Cunningham MJ, Rissman EF, Clifton DK, Steiner RA. Regulation of Kiss1 gene expression in the brain of the female mouse. *Endocrinology.* 2005;146(9):3686–3692.
166. de Roux N, Genin E, Carel JC, Matsuda F, Chaussain JL, Milgrom E. Hypogonadotropic hypogonadism due to loss of function of the KiSS1-derived peptide receptor GPR54. *Proc Natl Acad Sci USA.* 2003;100(19):10972–10976.
167. Seminara SB, Messager S, Chatzidaki EE, Thresher RR, Acierno Jr JS, Shagoury JK, Bo-Abbas Y, Kuohung W, Schwinof KM, Hendrick AG, Zahn D, Dixon J, Kaiser UB, Slaugenhaupt SA, Gusella JF, O'Rahilly S, Carlton MB, Crowley Jr WF, Aparicio SA, Colledge WH. The GPR54 gene as a regulator of puberty. *N Engl J Med.* 2003;349(17):1614–1627.
168. Kirby HR, Maguire JJ, Colledge WH, Davenport AP. International Union of Basic and Clinical Pharmacology. LXXVII. Kisspeptin receptor nomenclature, distribution, and function. *Pharmacol Rev.* 2010;62(4):565–578.
169. Zhang C, Bosch MA, Rønnekleiv OK, Kelly MJ. Gamma-aminobutyric acid B receptor mediated inhibition of gonadotropin-releasing hormone neurons is suppressed by kisspeptin-G protein-coupled receptor 54 signaling. *Endocrinology.* 2009;150(5):2388–2394.
170. Schwartz NB. A model for the regulation of ovulation in the rat. *Recent Prog Horm Res.* 1969;25:1–55.
171. Aiyer MS, Fink G, Greig F. Changes in sensitivity of the pituitary gland to luteinizing hormone releasing factor during the oestrous cycle of the rat. *J Endocrinol.* 1974;60:47–64.
172. Aiyer MS, Chiappa SA, Fink G. A priming effect of luteinizing hormone releasing factor on the anterior pituitary gland in the female rat. *J Endocrinol.* 1974;62:573–588.
173. Pickering AJMC, Fink G. Priming effect of luteinizing hormone releasing factor *in vitro*: role of protein synthesis, contractile elements, Ca^{2+} and cyclic AMP. *J Endocrinol.* 1979;81:223–234.
174. Curtis A, Lyons V, Fink G. The priming effect of LH-releasing hormone: effects of cold and involvement of new protein synthesis. *J Endocrinol.* 1985;105:163–167.
175. Mobbs CV, Fink G, Pfaff DW. HIP-70: An isoform of phosphoinositol-specific phospholipase C-? *Science.* 1990;249: 566–567.
176. Chappell PE, Schneider JS, Kim P, Xu M, Lydon JP, O'Malley BW, Levine JE. Absence of gonadotropin surges and gonadotropin-releasing hormone self-priming in ovariectomized (OVX), estrogen (E2)-treated, progesterone receptor knockout (PRKO) mice. *Endocrinology.* 1999;140:3653–3658.
177. Turgeon JL, Waring DW. Differential expression and regulation of progesterone receptor isoforms in rat and mouse pituitary cells and LbetaT2 gonadotropes. *J Endocrinol.* 2006;190: 837–846.

178. Attardi B, Scott R, Pfaff D, Fink G. Facilitation or inhibition of the oestradiol-induced gonadotrophin surge in the immature female rat by progesterone: effects on pituitary responsiveness to gonadotrophin-releasing hormone (GnRH), GnRH self-priming and pituitary mRNAs for the progesterone receptor A and B isoforms. *J Neuroendocrinol.* 2007;19(12):988–1000.
179. Lewis CE, Morris JF, Fink G. The role of microfilaments in the priming effect of LH-releasing hormone: an ultrastructural study using cytochalasin B. *J Endocrinol.* 1985;106: 211–218.
180. Lewis CE, Morris JF, Fink G, Johnson M. Changes in the granule population of gonadotrophs of hypogonadal (hpg) and normal female mice associated with the priming effect of LH-releasing hormone *in vitro*. *J Endocrinol.* 1986;109: 35–44.
181. O'Keane V. Antipsychotic-induced hyperprolactinaemia, hypogonadism and osteoporosis in the treatment of schizophrenia. *J Psychopharmacol.* 2008;22(2 Suppl):70–75.
182. Haddad PM, Weick A. Antipsychotic-induced hyperprolactinaemia: mechanisms, clinical features and management. *Drugs.* 2004;64:2291–2314.
183. Leong DA, Frawley LS, Neill JD. Neuroendocrine control of prolactin secretion. *Annu Rev Physiol.* 1983;45:109–127.
184. Fink G, Koch Y, Ben Aroya N. Release of thyrotropin releasing hormone into hypophysial portal blood is high relative to other neuropeptides and may be related to prolactin secretion. *Brain Res.* 1982;243:186–189.
185. Neill JD, Nagy GM. Prolactin secretion and its control. In: Knobil E, Neill JD, eds. *The Physiology of Reproduction.* 2nd ed. New York, NY: Raven Press; 1994:1833–1860.
186. Thomas GB, Cummins JT, Yao B, Gordon K, Clarke IJ. Release of prolactin is independent of the secretion of thyrotrophin-releasing hormone into hypophysial portal blood of sheep. *J Endocrinol.* 1988;117:115–122.
187. Stanley HF, Curtis A, Sheward WJ, Roberts JL, Fink G. Prolactin messenger ribonucleic acid levels in the normal and hypogonadal mouse pituitary gland. *Endocrinology.* 1986;119: 2422–2426.
188. de Greef WJ, Klootwijk W, Karels B, Visser TJ. Levels of dopamine and thyrotrophin-releasing hormone in hypophysial stalk blood during an oestrogen-stimulated surge of prolactin in the ovariectomized rat. *J Endocrinol.* 1985;105(1): 107–112.
189. Horn AM, Fink G. Effects of 5-hydroxytryptamine uptake blockers on the release of LH and prolactin in several different experimental steroid models in the rats. *J Endocrinol.* 1985; 104:397–406.
190. Hinuma S, Habata Y, Fujii R, Kawamata Y, Hosoya M, Fukusumi S, et al. A prolactin-releasing peptide in the brain. *Nature.* 1998;393:272–276.
191. Fukusumi S, Fujii R, Hinuma S. Recent advances in mammalian RFamide peptides: the discovery and functional analyses of PrRP, RFRPs and QRFP. *Peptides.* 2006;27(5):1073–1086.
192. Maixnerová J, Spolcová A, Pýchová M, Blechová M, Elbert T, Rezáčová M, Zelezná B, Maletínská L. Characterization of prolactin-releasing peptide: Binding, signaling and hormone secretion in rodent pituitary cell lines endogenously expressing its receptor. *Peptides.* 2011;32(4):811–817.
193. Taylor MM, Samson WK. The prolactin releasing peptides: RF-amide peptides. *Cell Mol Life Sci.* 2001;58(9):1206–1215.
194. Sarkar DK, Yen SS. Hyperprolactinemia decreases the luteinizing hormone-releasing hormone concentration in pituitary portal plasma: a possible role for beta-endorphin as a mediator. *Endocrinology.* 1985;116(5):2080–2084.
195. Hodson DJ, Townsend J, Tortonese DJ. Characterization of the effects of prolactin in gonadotroph target cells. *Biol Reprod.* 2010;83(6):1046–1055.
196. Anderson GM, Kieser DC, Steyn FJ, Grattan DR. Hypothalamic prolactin receptor messenger ribonucleic acid levels, prolactin signaling, and hyperprolactinemic inhibition of pulsatile luteinizing hormone secretion are dependent on estradiol. *Endocrinology.* 2008;149(4):1562–1570.
197. Kokay IC, Petersen SL, Grattan DR. Identification of prolactin-sensitive GABA and kisspeptin neurons in regions of the rat hypothalamus involved in the control of fertility. *Endocrinology.* 2011;152(2):526–535.
198. Eftekhari N, Mohammadalizadeh S. Pregnancy rate following bromocriptine treatment in infertile women with galactorrhea. *Gynecol Endocrinol.* 2009;25(2):122–124.
199. Brown RS, Kokay IC, Herbison AE, Grattan DR. Distribution of prolactin-responsive neurons in the mouse forebrain. *J Comp Neurol.* 2010;518(1):92–102.
200. Torner L, Toschi N, Nava G, Clapp C, Neumann ID. Increased hypothalamic expression of prolactin in lactation: involvement in behavioural and neuroendocrine stress responses. *Eur J Neurosci.* 2002;15(8):1381–1389.
201. van den Burg EH, Neumann ID. Bridging the Gap between GPCR activation and behaviour: oxytocin and prolactin signalling in the hypothalamus. *J Mol Neurosci.* 2011;43(2): 200–208.
202. McEwen BS. Physiology and neurobiology of stress and adaptation: central role of the brain. *Physiol Rev.* 2007;87:873–904.
203. de Kloet ER, Fitzsimons CP, Datson NA, Meijer OC, Vreugdenhil E. Glucocorticoid signaling and stress-related limbic susceptibility pathway: about receptors, transcription machinery and microRNA. *Brain Res.* 2009;1293:129–141.
204. Joels M, Baram TZ. The neuro-symphony of stress. *Nat Rev Neurosci.* 2009;10:459–466.
205. Fink G, ed. *Encyclopedia of Stress.* 2nd ed. Oxford: Academic Press; 2007.
206. Fink G. Stress: definition and history. In: Squire LR, ed. *Encyclopedia of Neuroscience*, Vol. 9. Oxford: Academic Press; 2009:549–555.
207. Fink G. Stress Controversies: Posttraumatic stress disorder, hippocampal volume, gastro-duodenal ulceration (2011). *J Neuroendocrinol.* 2011;23:107–117.
208. Palkovits M. Sympathoadrenal system: neural arm of the stress response. In: Squire LR, ed. *Encyclopedia of Neuroscience.* Oxford: Academic Press; 2009:679–684.
209. Greenwood FC, Landon J. Assessment of hypothalamic pituitary function in endocrine disease. *J Clin Pathol.* 1966;19(3): 284–292.
210. Antoni FA. Hypothalamic control of adrenocorticotropin secretion: advances since the discovery of 41-residue corticotropin-releasing factor. *Endocr Rev.* 1986;7(4):351–378.
211. Antoni FA. Vasopressinergic control of pituitary adrenocorticotropin secretion comes of age. *Front Neuroendocrinol.* 1993;14(2):76–122.
212. Orcel H, Albizu L, Perkovska S, Durroux T, Mendre C, Ansanay H, Mouillac B, Rabié A. Differential coupling of the vasopressin $V_{1\beta}$ receptor through compartmentalization within the plasma membrane. *Mol Pharmacol.* 2009;75(3): 637–647.
213. Young SF, Griffante C, Aguilera G. Dimerization between vasopressin $V_{1\beta}$ and corticotropin releasing hormone type 1 receptors. *Cell Mol Neurobiol.* 2007;27(4):439–461.
214. Muglia LJ, Bethin KE, Jacobson L, Vogt SK, Majzoub JA. Pituitary–adrenal axis regulation in CRH-deficient mice. *Endocr Res.* 2000;26(4):1057–1066.

215. Muglia LJ, Jacobson L, Weninger SC, Karalis KP, Jeong K, Majzoub JA. The physiology of corticotropin-releasing hormone deficiency in mice. *Peptides.* 2001;22(5):725–731.
216. Roper JA, Craighead M, O'Carroll AM, Lolait SJ. Attenuated stress response to acute restraint and forced swimming stress in arginine vasopressin 1b receptor subtype (Avpr1b) receptor knockout mice and wild-type mice treated with a novel Avpr1b receptor antagonist. *J Neuroendocrinol.* 2010;22(11): 1173–1180.
217. Roper J, O'Carroll AM, Young III W, Lolait S. The vasopressin Avpr1b receptor: molecular and pharmacological studies. *Stress.* 2011;14(1):98–115.
218. Fink G, Robinson ICAF, Tannahill LA. Effects of adrenalectomy and glucocorticoids on the peptides, CRF-41, AVP and oxytocin in rat hypophysial portal blood. *J Physiol.* 1988;401:329–345.
219. Tannahill LA, Sheward WJ, Robinson ICAF, Fink G. Corticotrophin-releasing factor-41, vasopressin and oxytocin release into hypophysial portal blood in the rat: effects of electrical stimulation of the hypothalamus, amygdala and hippocampus. *J Endocrinol.* 1991;129:99–107.
220. Engler D, Pham T, Fullerton MJ, Ooi G, Funder JW, Clarke IJ. Studies of the secretion of corticotropin-releasing factor and arginine vasopressin into the hypophysial-portal circulation of the conscious sheep. I. Effect of an audiovisual stimulus and insulin-induced hypoglycemia. *Neuroendocrinology.* 1989;49: 367–381.
221. Caraty A, Grino M, Locatelli A, Guillaume V, Boudouresque F, Conte-Devoix B, Oliver C. Insulin induced hypoglycenia stimulates corticotropin-releasing factor and arginine vasopressin secretion into hypophysial portal blood of conscious, unrestrained rams. *J Clin Invest.* 1990;85:1716–1721.
222. Sheward WJ, Fink G. Effects of corticosterone on the secretion of corticotrophin-releasing factor, arginine vasopressin and oxytocin into hypophysial portal blood in long-term hypophysectomized rats. *J Endocrinol.* 1991;129:91–98.
223. Antoni FA, Fink G, Sheward WJ. Corticotrophin-releasing peptides in rat hypophysial portal blood after paraventricular lesions: a marked reduction in the concentration of corticotrophin-releasing factor-41, but no change in vasopressin. *J Endocrinol.* 1990;125:175–183.
224. Kovács KJ, Földes A, Sawchenko PE. Glucocorticoid negative feedback selectively targets vasopressin transcription in parvocellular neurosecretory neurons. *J Neurosci.* 2000;20(10): 3843–3852.
225. Aguilera G, Kiss A, Liu Y, Kamitakahara A. Negative regulation of corticotropin releasing factor expression and limitation of stress response. *Stress.* 2007;10(2):153–161.
226. Rivier C, Brownstein M, Spiess J, Rivier J, Vale W. *In Vivo* Corticotropin releasing Factor-Induced secretion of adrenocorticotropin, /J-endorphin, and corticosterone. *Endocrinology.* 1982;110:272–278.
227. Zelena D, Domokos A, Jain SK, Jankord R, Filaretova L. The stimuli-specific role of vasopressin in the hypothalamus-pituitary-adrenal axis response to stress. *J Endocrinol.* 2009;202(2):263–278.
228. Porzionato A, Macchi V, Rucinski M, Malendowicz LK, De Caro R. Natriuretic peptides in the regulation of the hypothalamic-pituitary-adrenal axis. *Intl Rev Cell Mol Biol.* 2010;280: 1–39.
229. Perras B, Schultes B, Behn B, Dodt C, Born J, Fehm HL. Intranasal atrial natriuretic peptide acts as central nervous inhibitor of the hypothalamo–pituitary–adrenal stress system in humans. *J Clin Endocrinol Metab.* 2004;89(9):4642–4648.
230. Engler D, Redei E, Kola I. The corticotropin-release inhibitory factor hypothesis: a review of the evidence for the existence of inhibitory as well as stimulatory hypophysiotropic regulation of adrenocorticotropin secretion and biosynthesis. *Endocr Rev.* 1999;20:460–500.
231. Jessop DS. Review: central non-glucocorticoid inhibitors of the hypothalamo–pituitary–adrenal axis. *J Endocrinol.* 1999;160: 169–180.
232. Gutkowska J, Antunes-Rodrigues J, McCann SM. Atrial natriuretic peptide in brain and pituitary gland. *Physiol Rev.* 1997;77(2):465–515. Review.
233. Kellner M, Wiedemann K, Holsboer F. Atrial natriuretic factor inhibits the CRH-stimulated secretion of ACTH and cortisol in man. *Life Sci.* 1992;50:1835–1842.
234. Kellner M, Herzog L, Holsboer F, Wiedemann K. Circadian changes in the sensitivity of the corticotropin-releasing hormone-stimulated HPA system after arginine vasopressin and atrial natriuretic hormone in human male controls. *Psychoneuroendocrinology.* 1995;20(5):515–524.
235. Vermes I, Beishuizen A, Hampsink RM, Haanen C. Dissociation of plasma adrenocorticotropin and cortisol levels in critically ill patients: possible role of endothelin and atrial natriuretic hormone. *J Clin Endocrinol Metab.* 1995;80(4):1238–1242.
236. Vermes I, Beishuizen A. The hypothalamic–pituitary–adrenal response to critical illness. *Best Pract Res Clin Endocrinol Metab.* 2001;15(4):495–511.
237. Arafah BM. Hypothalamic pituitary adrenal function during critical illness: limitations of current assessment methods. *J Clin Endocrinol Metab.* 2006;91(10):3725–3745.
238. Luque RM, Kineman RD, Park S, Peng X-D, Gracia-Navarro F, Castaño JP, Malagon MM. Homologous and heterologous regulation of pituitary receptors for ghrelin and growth hormone-releasing hormone. *Endocrinology.* 2004;145(7): 3182–3189.
239. Smith RG. Development of growth hormone secretagogues. *Endocr Rev.* 2005;26(3):346–360.
240. Farhy LS, Bowers CY, Veldhuis JD. Model-projected mechanistic bases for sex differences in growth hormone regulation in humans. *Am J Physiol Regul Integr Comp Physiol.* 2007;292: R1577–R1593.
241. Osterstock G, Escobar P, Mitutsova V, Gouty-Colomer LA, Fontanaud P, Molino F, Fehrentz JA, Carmignac D, Martinez J, Guerineau NC, Robinson IC, Mollard P, Méry PF. Ghrelin stimulation of growth hormone-releasing hormone neurons is direct in the arcuate nucleus. *PLoS One.* 2010;5(2): e9159.
242. McArthur S, Robinson IC, Gillies GE. Novel ontogenetic patterns of sexual differentiation in arcuate nucleus GHRH neurons revealed in GHRH-enhanced green fluorescent protein transgenic mice. *Endocrinology.* 2011;152(2):607–617.
243. Mayo KE, Miller LJ, Bataille D, Dalle S, Göke B, Thorens B. Drucker DJ International Union of Pharmacology. XXXV. The glucagon receptor family. *Pharmacol Rev.* 2003;55(1):167–194.
244. McElvaine AT, Mayo KE. A dominant-negative human growth hormone-releasing hormone (GHRH) receptor splice variant inhibits GHRH binding. *Endocrinology.* 2006;147(4): 1884–1894.
245. Patel YC. Somatostatin and its receptor family. *Front Neuroendocrinol.* 1999;20:157–198.
246. Cervia D, Bagnoli P. An update on somatostatin receptor signaling in native systems and new insights on their pathophysiology. *Pharmacol Ther.* 2007;116(2):322–341.
247. Nillni EA. Regulation of the hypothalamic thyrotropin releasing hormone (TRH) neuron by neuronal and peripheral inputs. *Front Neuroendocrinol.* 2010;31(2):134–156.
248. Carlson HE, Jacob LS, Daughaday WH, Growth hormone. Thyrotropin and prolactin responses to thyrotropin-releasing

hormone following diethylstilbestrol pretreatment. *J Clin Endocrinol Metab.* 1973;37(3):488–490.
249. Donders SH, Pieters GF, Heevel JG, Ross HA, Smals AG, Kloppenborg PW. Disparity of thyrotropin (TSH) and prolactin responses to TSH-releasing hormone in obesity. *J Clin Endocrinol Metab.* 1985;61:56–59.
250. Ulloa-Aguirre A, Tarrasó J, Mendez JP, Garza-Flores J, Díaz-Sanchez V, Perez-Palacios G. Changes in the responsiveness of prolactin secretion to dopaminergic blockade and TRH stimulation throughout sexual maturation in men. *Psychoneuroendocrinology.* 1990;15(4):279–286.
251. Horn AM, Fraser HM, Fink G. Effects of antiserum to thyrotrophin-releasing hormone on the concentrations of plasma prolactin, thyrotrophin and LH in the pro-oestrous rat. *J Endocrinol.* 1985;104:205–209.
252. Sheward WJ, Fraser HM, Fink G. Effect of immunoneutralization of thyrotrophin-releasing hormone on the release of thyrotrophin and prolactin during suckling or in response to electrical stimulation of the hypothalamus in the anaesthetized rat. *J Endocrinol.* 1985;106:113–119.
253. Samuels MH, Veldhuis J, Ridgway EC. Copulsatile release of thyrotropin and prolactin in normal and hypothyroid subjects. *Thyroid.* 1995;5(5):369–372.
254. Sheward WJ, Harmar AJ, Fraser HM, Fink G. Thyrotropin-releasing hormone in rat pituitary stalk blood and hypothalamus: studies with high performance liquid chromatography. *Endocrinology.* 1983;113:1865–1869.
255. Richter K, Kawashima E, Egger R, Kreil G. Biosynthesis of thyrotropin releasing hormone in the skin of Xenopus laevis: partial sequence of the precursor deduced from cloned cDNA. *EMBO J.* 1984;3:617–621.
256. Pfaff DW. *Estrogens and Brain Function.* New York: Springer; 1980.
257. McEwen BS. Genomic regulation of sexual behavior. *J Steroid Biochem.* 1988;30:179–183.
258. Harris GW. Hormonal differentiation of the developing central nervous system with respect to patterns of endocrine function. *Phil Trans R Soc Lond.* 1970;259:165–177.
259. Rubinow DR, Schmidt PJ, Roca CA. Estrogen-serotonin interactions: implications for affective regulation. *Biol Psychiatry.* 1998;44(9):839–850.
260. Schmidt PJ, Rubinow DR. Sex hormones and mood in the perimenopause. *Ann NY Acad Sci.* 2009;1179:70–85.
261. Amin Z, Canli T, Epperson CN. Effect of estrogen–serotonin interactions on mood and cognition. *Behav Cogn Neurosci Rev.* 2005;4(1):43–58.
262. Kessler RC, McGonagle KA, Zhao S, Nelson CB, et al. Lifetime and 12-month prevalence of DSM-III-R psychiatric disorder in the United States. *Arch Gen Psychiatry.* 1994;51:8–19.
263. Wihlbäck A-C, et al. Influence of menstrual cycle on platelet serotonin uptake site and serotonin2A receptor binding. *Psychoneuroendocrinology.* 2004;29:757–766.
264. Almeida OP, Lautenschlager MD, Vasikaram S, Leedman P, Flicker L. Association between physiological serum concentration of estrogen and the mental health of community-dwelling postmenopausal women age 70 years and over. *Am J Geriatr Psychiatry.* 2005;13:142–149.
265. Schiff R, Bulpitt CJ, Wesnes KA, Rajkumar C. Short-term transdermal estradiol therapy, cognition and depressive symptoms in healthy older women. A randomized placebo controlled pilot cross-over study. *Psychoneuroendocrinology.* 2005;30(4): 309–315.
266. Fink G. The psychoprotective action of estrogen is mediated by central serotonergic as well as dopaminergic mechanisms. In: Takada A, Curzon G, eds. *Serotonin in the Central Nervous System and Periphery.* Amsterdam: Elsevier; 1995: 175–187.
267. Kulkarni J, de Castella A, Fitzgerald PB, Gurvich CT, Bailey M, Bartholomeusz C, Burger H. Estrogen in severe mental illness: a potential new treatment approach. *Arch Gen Psychiatry.* 2008;65(8):955–960.
268. Huber TJ, Borsutzky M, Schneider U, Emrich HM. Psychotic disorders and gonadal function: evidence supporting the oestrogen hypothesis. *Acta Psychiatr Scand.* 2004;109: 269–274.
269. Huber TJ, Tettenborn C, Leifke E, Emrich HM. Sex hormones in psychotic men. *Psychoneuroendocrinology.* 2005;30:111–114.
270. Herbison AE. Multimodal influence of estrogen upon gonadotropin-releasing hormone neurons. *Endocr Rev.* 1998;19: 302–330.
271. Fink G, Sumner BEH. Oestrogen and mental state. *Nature.* 1996;383:306.
272. Sumner BEH, Fink G. Effects of acute estradiol on 5-hydroxytryptamine and dopamine receptor subtype mRNA expression in female rat brain. *Mol Cell Neurosci.* 1993;4:83–92.
273. Sumner BEH, Fink G. Estrogen increases the density of 5hydroxytryptamine2A receptors in cerebral cortex and nucleus accumbens in the female rat. *J Steroid Biochem Mol Biol* 1995;54: 15–20.
274. Sumner BEH, Fink G. The density of 5-hydroxytryptamine2A receptors in forebrain is increased at pro-estrus in intact female rats. *Neurosci Lett.* 1997;234:7–10.
275. Henderson SR, Baker C, Fink G. Oestradiol-17 and pituitary responsiveness to luteinizing hormone releasing factor in the rat: a study using rectangular pulses of oestradiol-17 monitored by non-chromatographic radioimmunoassay. *J Endocrinol* 1977;73:441–453.
276. Biegon A, Reches A, Snyder L, McEwen BS. Serotonergic and noradrenergic receptors in the rat brain: modulation by chronic exposure to ovarian hormones. *Life Sci.* 1983;32: 2015–2021.
277. Cyr M, Bosse R, Di Paolo T. Gonadal hormones modulate 5-hydroxytryptamine2A receptors: emphasis on the rat frontal cortex. *Neuroscience.* 1998;83:829–836.
278. Cyr M, Landry M, Di Paolo T. Modulation by estrogen-receptor directed drugs of 5-hydroxytryptamine-2A receptors in rat brain. *Neuropsychopharmacology.* 2000;23(1):69–78.
279. Smith LJ, Henderson JA, Abell CW, Bethea CL. Effects of ovarian steroids and raloxifene on proteins that synthesize, transport, and degrade serotonin in the raphe region of Macaques. *Neuropsychopharmacology.* 2004;29: 2035–2045.
280. Kugaya A, Epperson CN, Zoghbi S, van Dyck CH, Hou Y, Fujita M, Staley JK, Garg PK, Seibyl JP, Innis RB. Increase in prefrontal cortex serotonin2A receptors following estrogen treatment in postmenopausal women, Am. *J Psychiatry.* 2003;160:1522–1524.
281. Moses-Kolko EL, Berga SL, Greer PJ. Widespread increases of cortical serotonin type 2A receptor availability after hormone therapy in euthymic postmenopausal women. *Fertil. Steril* 2003;80:554–559.
282. Mitchell R, Hollist S, Rothwell C, Robertson WR. Age related changes in the pituitary-testicular axis in normal men; lower serum testosterone results from decreased bioactive LH drive. *Clin Endocrinol.* 1995;42:501–507.
283. Vermeulen A. The male climacterium. *Ann Med.* 1993;25: 531–534.
284. Wang C, Alexander G, Berman N, Salehian B, Davidson T, McDonald V, Steiner B, Hull L, Callegari C, Swerdloff RS. Testosterone replacement therapy improves mood in

hypogonadal men—a clinical research center study. *J Clin Endocrinol Metab.* 1996;81:3578–3583.

285. Sherwin BB, Gelfand MM. Differential symptom response to parenteral estrogen and:or androgen administration in the surgical menopause. *Am J Obstet Gynecol.* 1985;151: 153–160.
286. McQueen JK, Wilson HM, Sumner EH, Fink G. Serotonin transporter (SERT)mRNA and binding site densities in male rat brain affected by sex steroids. *Mol Brain Res.* 1999;63: 241–247.
287. Sumner BEH, Fink G. Testosterone as well as estrogen increases serotonin2A receptor mRNA and binding site densities in the male rat brain. *Mol Brain Res.* 1998;59: 205–214.
288. Wagner CK, Morrell JI. Distribution and steroid hormone regulation of aromatase mRNA expression in the forebrain of adult male and female rats: a cellular-level analysis using *in situ* hybridization. *J Comp Neurol.* 1996;370:71–84.
289. Rosie R, Wilson H, Fink G. Testosterone induces an all-or-none, exponential increase in arginine vasopressin mRNA in the bed nucleus of stria terminalis of the hypogonadal mouse. *Mol Cell Neurosci.* 1993;4:121–126.
290. Fink G, Henderson SR. Steroids and pituitary responsiveness in female, androgenized female and male rats. *J Endocrinol.* 1977; 73:157–164.
291. McQueen JK, Wilson H, Fink G. Estradiol-17 beta increases serotonin transporter (SERT) mRNA levels and the density of SERT-binding sites in female rat brain. *Brain Res Mol Brain Res.* 1997;45(1):13–23.
292. Tao R, Auerbach SB. Influence of inhibitory and excitatory inputs on serotonin efflux differs in the dorsal and median raphe nuclei. *Brain Res.* 2003;961:109–120.
293. Ricaurte GA, Forno LS, Wilson MA, DeLanney LE, Irwin I, Molliver ME, Langston JW. (+/−)3,4-Methylenedioxy-methamphetamine selectively damages central serotonergic neurons in nonhuman primates. *J Am Med Assoc.* 1988;260(1): 51–55.
294. Lu H, Ozawa H, Nishi M, Ito T, Kawata M. Serotonergic neurones in the dorsal raphe nucleus that project into the medial preoptic area contain oestrogen receptor beta. *J Neuroendocrinol* 2001;13:839–845.
295. Donner N, Handa RJ. Estrogen receptor beta regulates the expression of tryptophan-hydroxylase 2 mRNA within serotonergic neurons of the rat dorsal raphe nuclei. *Neuroscience.* 2009;163(2):705–718.
296. Sheng Z, Kawano J, Yanai A, Fujinaga R, Tanaka M, Watanabe Y, Shinoda K. Expression of estrogen receptors (alpha, beta) and androgen receptor in serotonin neurons of the rat and mouse dorsal raphe nuclei; sex and species differences. *Neurosci Res.* 2004;49(2):185–196.
297. Merchenthaler I, Lane MV, Numan S, Dellovade TL. Distribution of estrogen receptor alpha and beta in the mouse central nervous system: *In vivo* autoradiographic and immunocytochemical analyses. *J Comp Neurol.* 2004;473: 270–291.
298. Alves SE, Weiland NG, Hayashi S, McEwen BS. Immunocytochemical localization of nuclear estrogen receptors and progestin receptors within the rat dorsal raphe nucleus. *J Comp Neurol.* 1998;391(3):322–334.
299. Aghajanian GK, Sprousse JS, Rasmussen K. Physiology of the midbrain serotonin system. In: Meltzer HY, ed. *Psychopharmacology: the third generation of progress.* New York: Raven Press; 1987:141–149.
300. Jolas T, Aghajanian GK. Opioids suppress spontaneous and NMDA-induced inhibitory postsynaptic currents in the dorsal raphe nucleus of the rat *in vitro*. *Brain Res.* 1997;755: 229–245.
301. Harsing Jr LG. The pharmacology of the neurochemical transmission in the midbrain raphe nuclei of the rat. *Curr Neuropharmacol.* 2006;4:313–339.
302. O'Leary OF, Bechtholt AJ, Crowley JJ, Valentino RJ, Lucki I. The role of noradrenergic tone in the dorsal raphe nucleus of the mouse in the acute behavioral effects of antidepressant drugs. *Eur Neuropsychopharmacol* 2007;17:215–226.
303. Curran-Rauhut MA, Petersen SL. Oestradiol-dependent and -independent modulation of tyrosine hydroxylase mRNA levels in subpopulations of A1 and A2 neurones with oestrogen receptor (ER)alpha and ER beta gene expression. *J Neuroendocrinol.* 2003;15(3):296–303.
304. Nomura M, Akama KT, Alves SE, Korach KS, Gustafsson JA, Pfaff DW, Ogawa S. Differential distribution of estrogen receptor (ER)-alpha and ER-beta in the midbrain raphe nuclei and periaqueductal gray in male mouse: Predominant role of ER-beta in midbrain serotonergic systems. *Neuroscience.* 2005;130(2):445–456.
305. Eyigor O, Lin W, Jennes L. Identification of neurones in the female rat hypothalamus that express oestrogen receptor-alpha and vesicular glutamate transporter-2. *J Neuroendocrinol.* 2004; 16(1):26–31.
306. Jelks KB, Wylie R, Floyd CL, McAllister AK, Wise P. Estradiol targets synaptic proteins to induce glutamatergic synapse formation in cultured hippocampal neurons: critical role of estrogen receptor-alpha. *J Neurosci.* 2007;27(26): 6903–6913.
307. Dutertre M, Smith CL. Molecular Mechanisms of Selective Estrogen Receptor Modulator (SERM) Action1. *J Pharmacol Exp Ther.* 2000;295:431–437.
308. Aghajanian GK, Marek GJ. Serotonin induces excitatory postsynaptic potentials in apical dendrites of neocortical pyramidal cells. *Neuropharmacology.* 1997;36:589–599.
309. Jakab RL, Goldman-Rakic PS. 5-Hydroxytryptamine2A serotonin receptors in the primate cerebral cortex: possible site of action of hallucinogenic and antipsychotic drugs in pyramidal cell apical dentrites. *Proc Natl Acad Sci USA.* 1998;95: 735–740.
310. Chiodo LA, Caggiula AR. Substantia nigra dopamine neurons: alterations in basal discharge rates and autoreceptor sensitivity induced by estrogen. *Neuropharmacology.* 1983;22:593–599.
311. Di Paolo T. Modulation of brain dopamine transmission by sex steroids. *Rev Neurosci.* 1994;5:27–42.
312. Tansey EM, Arbuthnott GW, Fink G, Whale D. Oestradiol-17. increases the firing rate of antidromically identified neurones of the rat neostriatum. *Neuroendocrinology.* 1983;37:106–110.
313. Shughrue PJ, Lane MV, Merchenthaler I. Comparative distribution of estrogen receptor-α and -β mRNA in the rat central nervous system. *J Comp Neurol.* 1997;388:507–525.
314. Barkhem T, Carlsson, Nilsson BY, Enmark E, Gustafsson J-A. Differential response of estrogen receptor α and estrogen receptor β to partial estrogen agonists/antagonists. *Mol Pharmacol.* 1998;54:105–112.
315. Kulkarni J, Gurvich C, Lee SJ, Gilbert H, Gavrilidis E, de Castella A, Berk M, Dodd S, Fitzgerald PB, Davis SR. Piloting the effective therapeutic dose of adjunctive selective estrogen receptor modulator treatment in postmenopausal women with schizophrenia. *Psychoneuroendocrinology.* 2010;35(8): 1142–1147.
316. Kulkarni J, de Castella A, Headey B, Marston N, Sinclair K, Lee S, Gurvich C, Fitzgerald PB, Burger H. Estrogens and men with schizophrenia: is there a case for adjunctive therapy? *Schizophr Res.* 2011;125(2-3):278–283.

317. Fink G, Dow RC, McQueen JK, Bennie JG, Carroll SM. Serotonergic 5–HT2A receptors important for the oestradiol-induced surge of luteinising hormone-releasing hormone in the rat. *J Neuroendocrinol.* 1999;11:63–69.
318. Wright DE, Jennes L. Lack of expression of serotonin receptor subtype-1a, -1c, and -2 mRNAs in gonadotropin-releasing hormone producing neurons of the rat. *Neurosci Lett.* 1993;163:1–4.
319. Li S, Pelletier G. Involvement of serotonin in the regulation of GnRH gene expression in the male rat brain. *Neuropeptides.* 1995;29:21–25.
320. Fink G. The psychoprotective action of oestrogen is mediated by central 5-hydroxytryptamine as well as dopamine receptors. In: Fuxe K, Hokfelt T, Olson L, Ottoson D, Dahlstrom A, Bjorklund A, eds. *Molecular Mechanisms of Neuronal Communication*. Oxford: Pergamon; 1996:177–204.
321. Nakanishi S, Inoue A, Kita T, Nakamura M, Chang AC, Cohen SN, Numa S. Nucleotide sequence of cloned cDNA for bovine corticotropin-beta-lipotropin precursor. *Nature.* 1979; 278(5703):423–427.
322. Douglas WW. Calcium-dependent links in stimulus-secretion coupling in the adrenal medulla and neurohypophysis. In: Von Euler US, Rosell S, Uvnxs B, eds. *Mechanism of Release of Biogenic Amines.* Oxford: Pergamon Press; 1966:267–290.
323. de Kloet ER, Joëls M, Holsboer F. Stress and the brain: from adaptation to disease. *Nat Rev Neurosci.* 2005;6(6):463–475.

CHAPTER

6

Neural Control of the Posterior Pituitary Gland (Neurohypophysis)

Gareth Leng, Mike Ludwig, Alison J. Douglas

Centre for Integrative Physiology, University of Edinburgh, Edinburgh, UK

OUTLINE

Summary

The peptide hormones vasopressin and oxytocin are synthesized by magnocellular neurosecretory neurons of the hypothalamus, and secreted into the systemic circulation from their axonal terminations in the posterior pituitary gland (neurohypophysis). Oxytocin secretion is essential for milk let-down in lactation, and plays an important role in parturition; in these circumstances, afferent stimuli from the suckling of young at the nipples or from the contracting uterus result in intermittent synchronized bursting activity of the oxytocin neurons. Vasopressin secretion is essential for electrolyte and body fluid homeostasis by its antidiuretic actions at the kidney, and occurs in response to increased extracellular osmolarity and in response to a fall in plasma volume. The vasopressin neurons also fire in bursts, but these are much longer and less intense than the bursts in oxytocin neurons, and are not synchronized. The magnocellular neurons also release vasopressin and oxytocin from their dendrites within the hypothalamus; this central release plays an important role in shaping the activity patterns of the magnocellular neurons, but also has behavioral effects at other brain sites.

INTRODUCTION

The endocrine cells of the anterior pituitary gland synthesize peptide hormones which are packaged in large stores of secretory vesicles; these are secreted in response to blood-borne factors that stimulate the release of Ca^{2+} from intracellular stores to trigger exocytosis. By contrast, although the two hormones of the posterior pituitary gland, vasopressin and oxytocin, are also peptides, they are made not in the pituitary gland itself but by magnocellular neuroendocrine neurons of the hypothalamus. In the cell bodies of these neurons, vasopressin and oxytocin are packaged into vesicles that are much like the vesicles that contain anterior pituitary hormones. These vesicles are transported along the axons to be stored in the posterior pituitary gland (also called the neurohypophysis or neural lobe), awaiting signals to trigger their release into the

Handbook of Neuroendocrinology, DOI: 10.1016/B978-0-12-375097-6.10006-X blood. As for the hormones of the anterior pituitary, secretion of oxytocin and vasopressin is by Ca^{2+}-dependent exocytosis, but the Ca^{2+} trigger for secretion comes as Ca^{2+} entry into the axons through channels that open in response to action potentials (spikes) that pass down the axons from the hypothalamus.[1]

Mammals have several thousand magnocellular neurons, aggregated in two main structures – the paraventricular nucleus (PVN) and the supraoptic nucleus (SON) – and in several smaller groups that are scattered throughout the medial hypothalamus. Each magnocellular neuron has one long axon that projects to the neurohypophysis, where it gives rise to numerous swellings and nerve endings, and usually two or three large dendrites that also contain many neurosecretory vesicles (Fig. 6.1). Nearly all of the neurons in the SON

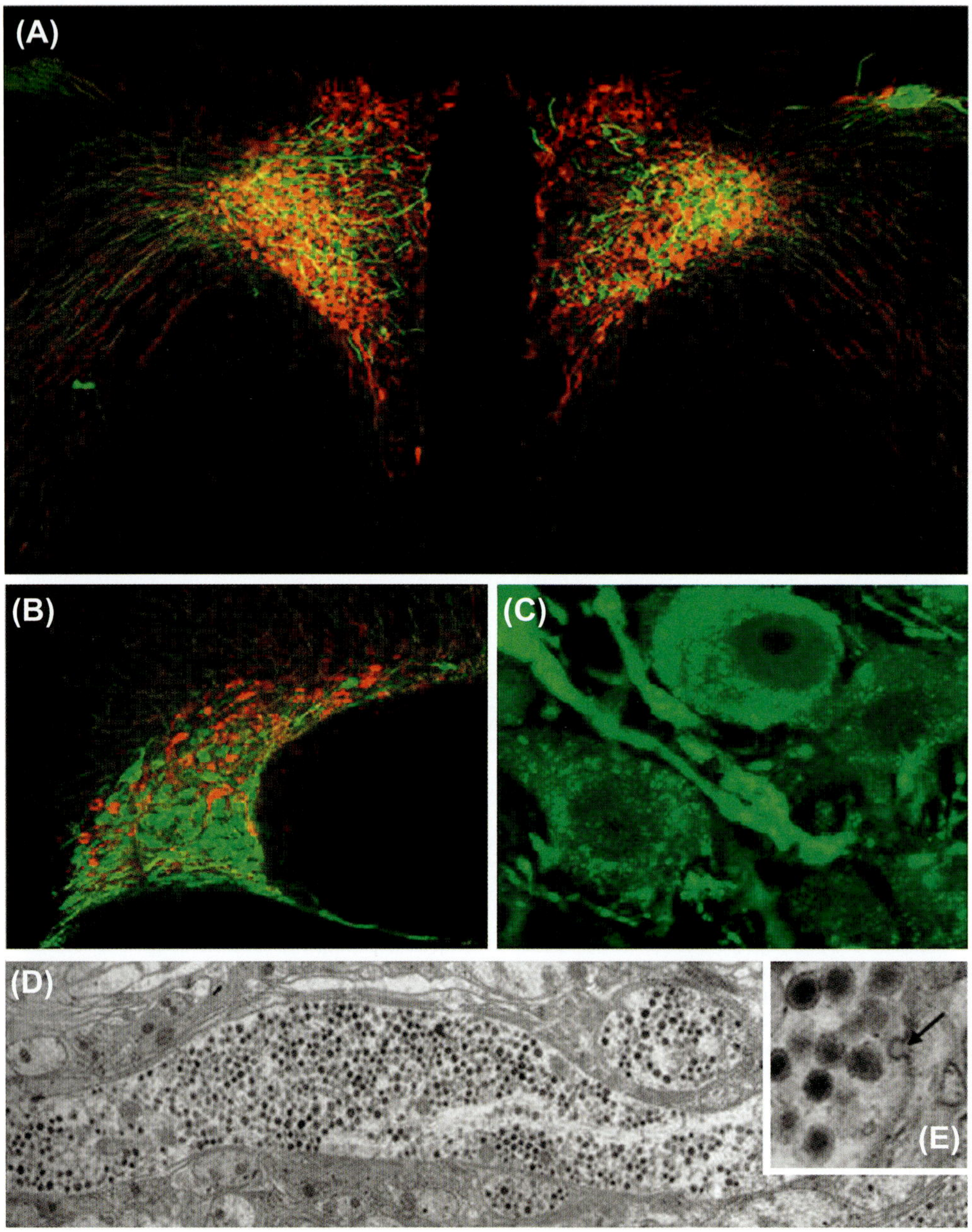

FIGURE 6.1 Vasopressin and oxytocin that circulate in the plasma are synthesized by a few thousand large (magnocellular) neurons (vasopressin cells are immunostained with fluorescent green and oxytocin cells with fluorescent red) whose cell bodies are located mainly in the (**A**) paraventricular and (**B**) supraoptic nuclei of the hypothalamus. (**C**) The vasopressin immunostaining is punctate and represents individual or aggregates of dense-cored vesicles. In the dendrite thickenings the vesicles are particularly abundant. (**D, E**) Large dense-core vesicles in an electron microscopic section of a dendrite appear as dark, round, membrane-bound organelles (black dots). (**E**) An "omega" fusion profile at the plasma membrane may show the pit in the dendritic membrane that remains after exocytosis of a large vesicle (arrow). *Adapted from Ludwig and Leng (2006);[26] immunofluorescent images courtesy of Dr V.A. Tobin.*

are magnocellular neurons, containing either oxytocin or vasopressin; the PVN is much more heterogeneous, containing neurons involved in many autonomic and neuroendocrine functions. As well as magnocellular neurons, the PVN contains smaller (parvocellular) oxytocin neurons that project not to the neurohypophysis but to the brainstem, spinal cord, and other parts of the hypothalamus; parvocellular vasopressin neuroendocrine neurons that project to the median eminence to regulate ACTH secretion; and centrally-projecting vasopressin neurons are involved in a variety of functions, including cardiovascular regulation and thermoregulation.

The magnocellular neurons receive afferent inputs from many hypothalamic, limbic and brainstem regions.[2] Most of these inputs use either the excitatory neurotransmitter glutamate or the inhibitory transmitter GABA, and many derive from the regions adjacent to the magnocellular nuclei. More distant inputs, including both glutamatergic and GABAergic projections, come from rostral structures of the anteroventral periventricular region, including from the subfornical organ (SFO) and the organum vasculosum of the lamina terminalis (OVLT); these circumventricular organs lack a blood–brain barrier and are sensitive to many blood-borne factors, including hormones such as relaxin and angiotensin. There are also prominent projections to the magnocellular neurons from the nucleus tractus solitarii (NTS) and ventrolateral medulla, including strong direct projections from the A1 and A2 noradrenergic cell groups in these regions. These neurons mediate signals from the periphery that act on the vagus, including uterine- and mammary-derived sensory afferents, signals from appetite-regulating peptides of the gastrointestinal tract such as cholecystokinin (CCK), and immune signals such as cytokines (e.g., interleukin IL-1β). From within the hypothalamus, magnocellular neurons are innervated by projections from the arcuate nucleus and the suprachiasmatic nucleus, and by a histaminergic innervation from the tuberomamillary nucleus; the magnocellular neurons express receptors for a huge diversity of peptides, as well as for all known transmitters, including serotonin, dopamine and acetylcholine. Many of the projections contain co-expressed peptides. For example, endogenous opioid peptides, including dynorphin, β-endorphin and enkephalin, are expressed in projections from the arcuate nucleus, caudal brainstem and other hypothalamic areas, and are inhibitory to oxytocin neurons in particular, mainly via κ- and μ-receptors.

Although magnocellular neurons are the only neurons in the SON, it (and the PVN) also contain many glial cells, including specialized astrocytes with long processes which envelop the magnocellular neurons; similar astrocytes are also found in the neurohypophysis, where they ensheathe the axons.[3] To some extent, these glial cells act as insulating sheaths, limiting cross-talk between neurons. They also regulate the environment of the magnocellular neurons, removing excess neurotransmitter resulting from spillover of synaptic release, and buffering the changes in extracellular electrolyte concentrations that accompany spike activity. These glial cells are dynamic components of the SON: their morphology changes in physiological circumstances, and they can release products (in particular taurine) that affect neuronal excitability.[4] Finally, the SON and PVN are amongst the most densely vascularized parts of the brain. As the cells are so metabolically active, this dense vascularity might simply reflect their high demand for metabolic sustenance. However, although these nuclei are within the blood–brain barrier and the blood vessels are unfenestrated, it is also possible that some blood-borne factors have privileged access to the magnocellular neurons through specific transport mechanisms.

HISTORY

Towards the end of the 19th century, Edward Albert Schafer (later known as Sharpey-Shafer, 1850–1935) first showed that extracts of the neurohypophysis could cause an increase in blood pressure; soon after, Oliver and Schafer reported that the extracts could affect urine flow, although they described a *diuretic* rather than an antidiuretic effect.[5] Confusion about the renal effects persisted, but by the 1950s it was clear that one of the factors present in these extracts must be *the* antidiuretic hormone. Ernest Verney (1894–1967) had shown that the isolated, perfused kidney of a dog would produce concentrated urine if the head was included in the preparation, but not if the pituitary was removed; he went on to show that extracts of the neurohypophysis could produce this antidiuretic effect, and that the secretion of this antidiuretic factor was regulated by changes in the osmotic pressure of the blood.[6] In elegant experiments, carefully ligating blood vessels, he presciently concluded that the "osmoreceptors" responsible for regulating secretion of the antidiuretic hormone must be located in or close to the SON, and that they must be in neural contact with the pituitary gland itself.

In 1906 Henry Dale (1875–1968) reported that extracts of the neurohypophysis could also promote uterine contractions,[7] and in 1941 James Ferguson proposed that distension of the birth canal by the fetus caused an increase in uterine contractility by stimulating the release of this second "oxytocic" of the neurohypophysis.[8] These extracts were also noted to increase milk let-down, and in 1913 Shafer reported the first effects in humans; the subject, a "young married lady" who was nursing her second baby, reported that after injections she "could feel the milk coming in, with a tingling sensation".[9] In

BOX 6.1

HOW DO WE KNOW THAT OXYTOCIN AND VASOPRESSIN ARE RELEASED FROM DENDRITES?

The ability of magnocellular neurons to release vasopressin and oxytocin from their whole membrane surface, and the fact that neuropeptides can reach their target sites via diffusion through the extracellular space, has predisposed these neuropeptides to specialized sampling techniques such as push–pull perfusion and *in vivo* microdialysis.

These techniques were first employed in the early 1960s, when push–pull cannulas and dialysis sacs were implanted into animal tissues, especially into rodent brains, to directly study the tissues' biochemistry.[1,2] Today, microdialysis is used for continuous measurement of molecules in the extracellular fluid of virtually any tissue. The probe is continuously perfused with a solution resembling the composition of the surrounding tissue fluid, and small molecules cross the semi-permeable membrane by diffusion. The direction of molecule flow is determined by the concentration gradient, and allows the usage of microdialysis probes for sampling as well as delivery of drugs. The solution leaving the probe (dialysate) can be repeatedly collected at defined time intervals for analysis.

Microdialysis was first used to measure vasopressin and oxytocin release in the supraoptic and paraventricular nuclei in the 1990s, proving that physiological conditions such as high osmolality and hemorrhage increase vasopressin concentration, and parturition and lactation increase oxytocin concentration in the extracellular fluid. That these peptides are released from dendrites was demonstrated by electron microscope studies showing fusion profiles of large dense-cored vesicles in dendrites.[3]

Many subsequent microdialysis studies in anesthetized or freely moving animals extended our understanding of the mechanisms and consequences of dendritic vasopressin and oxytocin release. Importantly, a striking dissociation between their central and peripheral release was discovered, proving independent regulation of the two neuronal compartments of somata/dendrites vs axon terminals. Later, the technique was significantly developed in studies combining microdialysis/retrodialysis with electrophysiological approaches. For instance, such an approach could identify that application of agents that mobilize intracellular Ca^{2+} induces oxytocin release from dendrites without affecting either the electrical activity of the cell bodies or oxytocin secretion from nerve terminals. Microdialysis has played a crucial role in the discovery and characterization of non-synaptic communication, and has substantially extended our knowledge about the communicatory capacities of peptidergic neurons.[4]

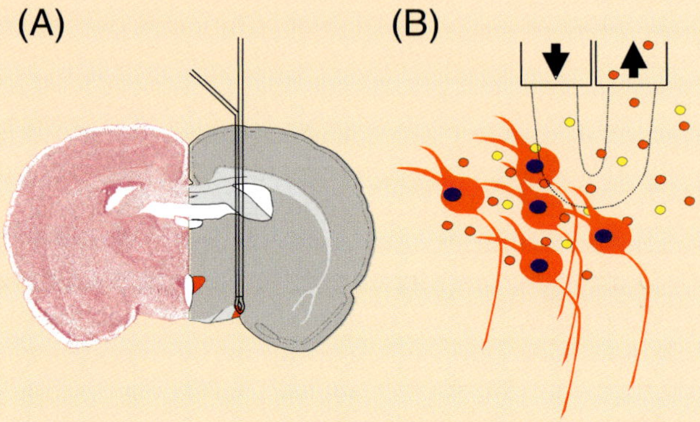

Microdialysis of the supraoptic nucleus. (**A**) Coronal section of a rat brain stained with cresyl fast violet (left) with a schematic drawing superimposed (right) to highlight the paraventricular and supraoptic nuclei (red). One microdialysis probe is depicted unilaterally in the supraoptic nucleus, as it might be located after insertion from the dorsal approach. (**B**). Schematic showing magnocellular neurons each with two large dendrites. Molecules of neurohypophysial peptide released from the dendrites and/or somata into the extracellular fluid diffuse down their concentration gradient across the dialysis membrane into the perfused buffer (indicated by arrows) inside the probe which is intermittently collected. (Not to scale.)

References

1. Gaddum JH. Push–pull cannulae. *J Physiol.* 1961;155:1–2.
2. Bito L, Davson H, Levin EM, Murray M, Snider N. The concentration of free amino acids and other electrolytes in cerebrospinal fluid, *in vivo* dialysate of brain and blood plasma of the dog. *J Neurochem.* 1966;13:1057–1067.
3. Pow DV, Morris JF. Dendrites of hypothalamic magnocellular neurons release neurohypophysial peptides by exocytosis. *Neuroscience.* 1989;32(2):435–439.
4. Ludwig M, Leng G. Dendritic peptide release and peptide-dependent behaviours. *Nat Rev Neurosci.* 2006;7(2):126–136.

1950, Barry Cross, then a student of Geoffrey Harris (the "father" of neuroendocrinology, 1913–1971), showed that electrical stimulation of the pituitary neural stalk could cause milk let-down in anesthetized lactating rabbits, and in 1951 Harris concluded that "Stimulation of the nipples by suckling appears to excite a reflex nervous release of oxytocin, and so contraction of the myoepithelial cells".[10] Confirmation soon came; relatively pure extracts of these factors were now available for physiological studies, and the structures of oxytocin and vasopressin were established. In 1955, Vincent du Vigneaud (1901–1978) won the Nobel Prize in Chemistry for his contributions to the isolation, structural identification, and synthesis of oxytocin.[11]

It was also becoming accepted that these factors were produced not in the pituitary but in the hypothalamus, and that they were packaged in "granules" that were transported along axons. In his PhD thesis in 1928, Ernst Scharrer suggested that neurons of the diencephalic brain structures are capable of releasing biologically active, chemical compounds into the systemic circulation.[12] During the following decades, he and his wife, Berta Scharrer, validated this concept of neurosecretion in many vertebrate and invertebrate species.[13] By 1951, Wolfgang Bargmann and Ernst Scharrer had shown that dehydration caused these granules to disappear from the neurohypophysis, and that when the axons of the hypothalamo-hypophysial tract were cut granules accumulated at the end of the neuronal stump.[14] This led doubters to accept that the neurohypophysial hormones were indeed manufactured in the hypothalamus. Then, in 1964, Douglas and Poisner showed that these vesicles were released by Ca^{2+}-dependent exocytosis, and drew attention to the similarities in the processes of "stimulus secretion coupling" in neurosecretory cells, chromaffin cells and neurons.[15] With this insight, it became apparent that the magnocellular neurons offered an opportunity not just for answering particular questions about neurohypophysial secretion, but also for establishing general principles of how the electrical activity of neurons relates to their physiological roles.

In the next 30 years the SON became a model system of profound importance to neuroscience, and it yielded a succession of fundamental advances. In the rat, the SON contains just two types of neurons – about 70% of these produce vasopressin, the remainder producing oxytocin. For the experimental neuroscientist, the SON is located very conveniently on the ventral surface of the brain abutting the optic chiasma, so it can be dissected away relatively easily and cleanly for biochemical studies. Indeed, it is relatively easy to keep the cells alive in slices or chunks of hypothalamus and study them *in vitro*. Finally, and importantly, the SON contains essentially just the cell bodies and dendrites of magnocellular neurons, as the axons of the magnocellular neurons leave the SON without giving rise to any collateral branches within its boundaries.

In 1973, Jon Wakerley and Dennis Lincoln published their observations of the behavior of oxytocin neurons during the milk-ejection reflex.[16] They found that during suckling the oxytocin neurons behaved in a way that was completely unanticipated: these cells did not simply increase their electrical activity in response to suckling – their average activity scarcely increased at all. Instead, their activity was *re-organized*: they fired in intermittent brief, high-frequency bursts, each lasting only a few seconds, which occurred synchronously in all the magnocellular oxytocin neurons in both the SON and PVN. The vasopressin neurons had surprises, too; many of them also showed *burst firing*, not in response to suckling, but in response to dehydration or hemorrhage. However, their bursts were much longer (tens of seconds) and slower, and, crucially, were not synchronized – a pattern known thereafter as *phasic firing*.

Synthesis of Vasopressin and Oxytocin

In the 1980s, the genes for both oxytocin[17] and vasopressin[18] were isolated and sequenced, beginning the modern era of molecular investigation. No longer was neuroendocrinology only about the regulation of secretion of these hormones; it was also about regulation of synthesis.

Oxytocin and vasopressin are each synthesized as part of a precursor protein that is packaged into the neurosecretory vesicles. While the vesicles are being transported to the neurohypophysis, the precursors are cleaved by enzymes to yield oxytocin and vasopressin. Each vesicle contains about 85,000 molecules of hormone,[19] and this cargo is very densely packed, giving an electron-dense appearance of the vesicle core when examined under the electron microscope (hence the term "large dense cored vesicles" commonly used to describe peptide containing vesicles, by contrast with the term "small synaptic vesicles" for vesicles that contain a conventional neurotransmitter). The vesicle core contains vasopressin (or oxytocin) and equimolar amounts of other fragments of the precursor, including the associated neurophysin. The neurophysins are large peptide molecules that have no known function after release, but which are important for the folding of the precursor molecule that allows it to be appropriately packaged in vesicles. Mutations in the neurophysin coding region of the peptide genes can give rise to deficiencies in hormone secretion – in particular, the rare disease of familial hypothalamic diabetes insipidus, where the kidney is unable to concentrate urine.

The Brattleboro rat was the first animal model of hypothalamic diabetes insipidus; this strain was derived from a litter of rats that fortuitously was noticed to be drinking excessively. The defect in the Brattleboro rat is the consequence of a single base deletion in exon B of the gene for the vasopressin precursor.[20] The mutation gives rise to an open reading frame, resulting in production of a greatly altered precursor protein. As the mutation is downstream of the vasopressin sequence, the aberrant precursor contains the normal vasopressin sequence, but cannot be properly packaged in vesicles, so rats homozygous for this mutation effectively have no vasopressin in the circulation. These rats can nevertheless maintain approximately normal electrolyte balance, but to achieve this they must drink (and urinate) about their own body weight in water each day. Virtually all cases of familial diabetes insipidus in humans similarly arise from a mutation in the precursor protein rather than in a mutation of the vasopressin sequence itself. Typically, this disease has a late, sudden onset, and appears to result because of a slow, cytotoxic accumulation in the cell body of aberrantly folded precursor.

The rate of synthesis of both oxytocin and vasopressin varies proportionally with the abundance of mRNA for these peptides, and mRNA expression in turn closely matches the level of secretory demand. Accordingly, there is a close link between the level of electrical activity in these cells and both the rate of secretion and the level of mRNA expression. However, the nature of the link between activity and gene expression is unclear. In both oxytocin neurons and vasopressin neurons, increases in spike activity and secretion are usually (though not invariably) accompanied by a rapid increase in expression of the immediate-early gene c-*fos*, which codes for Fos, a nuclear phospho-protein. Fos dimerizes with the protein product of another immediate-early gene, c-*jun*, to form the AP-1 transcription factor, which binds to specific DNA elements to regulate transcription of many genes. As the promoter region of the vasopressin gene includes a functional AP-1 binding element, this seems to be part of the mechanisms for linking activity and synthesis; for oxytocin, however, there is no evidence so far of a functional AP-1 element in the promoter region.

The propensity of both oxytocin and vasopressin neurons to express c-*fos* in response to activation has been extensively used in experimental studies. The Fos protein is readily visualized by immunocytochemistry in the nuclei of magnocellular neurons, and its presence is a generally reliable indicator of recent neuronal activation. In the brains of normonatremic rats in control conditions little if any Fos is found in magnocellular neurons, but within 90 minutes of, for example, an intraperitoneal injection of hypertonic saline, Fos is present in virtually every magnocellular neuron. However, spike activity *per se* is not a sufficient stimulus to induce Fos expression: some stimuli that mobilize intracellular Ca^{2+} stores are accompanied by induction of Fos expression even when electrical activity is inhibited.[21]

Molecular Evolution

Oxytocin and vasopressin are encoded by separate genes that appear to be descendants of a single gene presumed to have been present in the *Archaemetazoa*, a stem group from which both vertebrates and invertebrates diverged about 600 million years ago.[22,23] In mammals the genes for oxytocin and vasopressin are on the same chromosome, separated by just a few kilobases of sequence. Almost every vertebrate species has at least one vasopressin-like peptide and at least one oxytocin-like peptide. Exceptions include cyclostomes, which have just vasotocin (this differs from both oxytocin and vasopressin by just one amino acid), and teleost fish and amphibians, which express vasotocin and isotocin. Earthworms, leeches, octopus, snails and other multicellular invertebrates lack a pituitary gland, but all of these have at least one "neurohypophysial" peptide, produced by specialized nerve cells. For example, in the pond snail *Lymnaea stagnalis*, lys-conopressin has both oxytocin-like reproductive functions and vasopressin-like metabolic functions; in *Aplysia*, the same peptide is responsible for gill motility associated with feeding. However, *Octopus vulgaris* expresses two neurohypophysial peptides: cephalotocin, and octopressin (Fig. 6.2).

Most "new" genes arise by the duplication of an existing sequence, with subsequent mutations accumulating in a copy that is initially "redundant." The North American opossum and the northern brown bandicoot exhibit dual gene duplication; they have both arginine vasopressin and lysine vasopressin, and both oxytocin and mesotocin. We often think of oxytocin as a mammalian hormone, whose main functions are to control milk let-down and parturition, but the gene duplication that gave rise to oxytocin and vasopressin occurred early in evolution, and peptide family members have performed many different functions in their long evolutionary history. These include reproductive roles (notably in oviposition) that seem analogous to the roles of oxytocin in parturition, and roles in body fluid/electrolyte homeostasis that seem analogous to that of vasopressin in antidiuresis, but these peptides have also fulfilled other roles in different species.

Oxytocin and vasopressin family members all act via rhodopsin type (class I) G-protein coupled receptors. Four of these receptors have been identified in vertebrates, all of which are expressed in mammals.[24,25] The

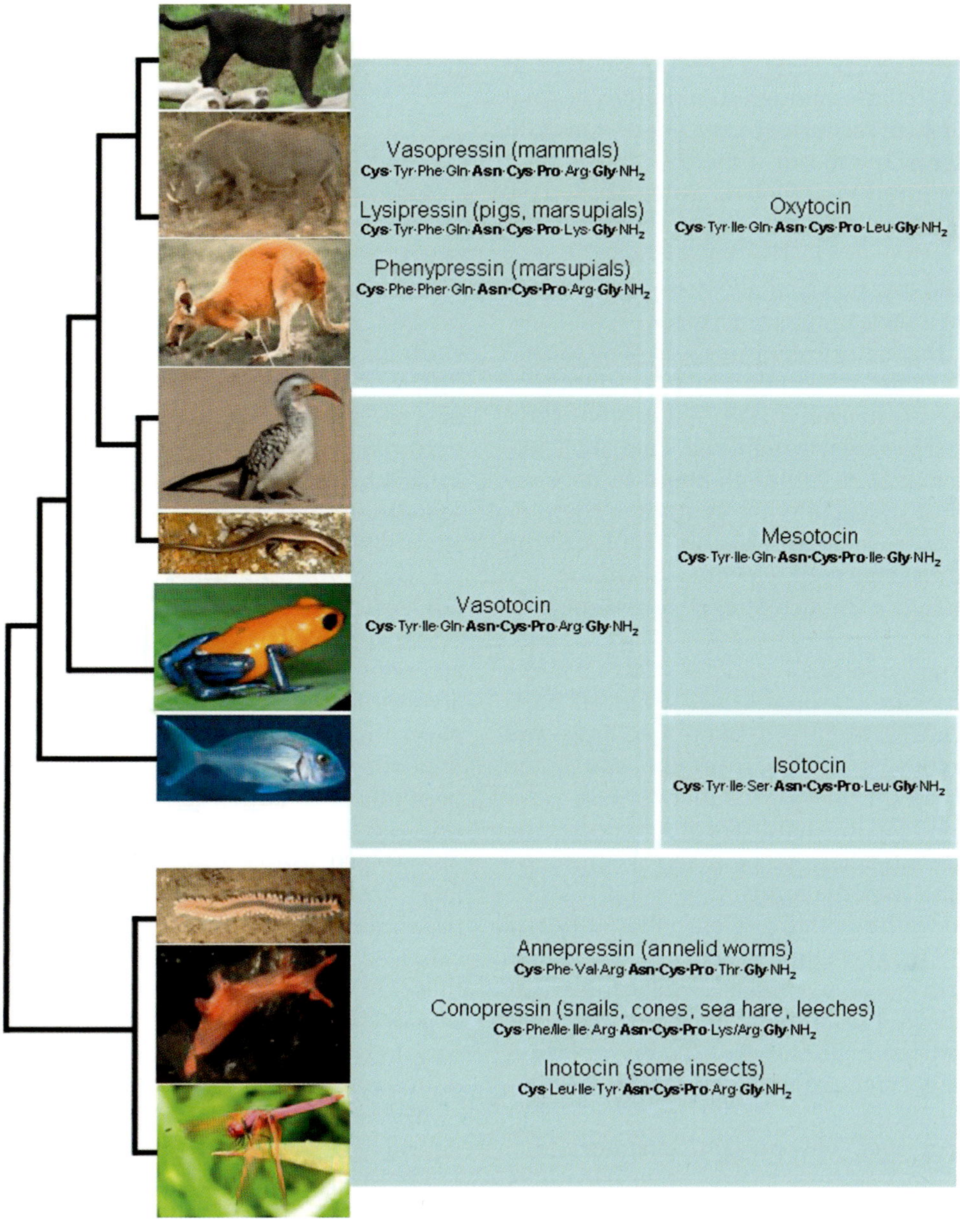

FIGURE 6.2 The mammalian oxytocin and vasopressin nonapeptides differ from each other at only two amino acid positions. Oxytocin, vasopressin, and their respective non-mammalian vertebrate lineages are thought to have arisen from a gene-duplication event before vertebrate divergence. Within these lineages, peptides vary by a single amino acid, and their genes are found near each other on the same chromosome. Invertebrates, with few exceptions, have only one oxytocin/vasopressin homolog, whereas vertebrates have two. Birds and amphibians generally express mesotocin and vasotocin. Mesotocin is involved in oviposition and vasotocin in electrolyte balance, and these are thus thought to be the evolutionary precursors of oxytocin and vasopressin in mammals. Most Eutherean marsupials express mesotocin rather than oxytocin, thought to be a neutral back-mutation. Some mammals (including the pig and hippopotamus) express lysine vasopressin – again a neutral mutation of arginine vasopressin. Some marsupials express phenypressin. *Adapted from Donaldson and Young (2008);[48] photographs courtesy of Dr S. Maciver.*

receptors are closely related, and again appear to derive from successive episodes of gene duplication. The oxytocin receptor and the vasopressin V2 receptor are most dissimilar, and appear therefore to be derived from an initial gene duplication. The vasopressin V1a and V1b receptors appear to have arisen from two subsequent duplications of the oxytocin receptor gene. Oxytocin receptors and both V1a and V1b receptors couple to Gq/11 and activate the inositol 1,4,5-trisphosphate–Ca^{2+} signal transduction cascade, whereas the V2 receptor is coupled to adenylate cyclase via G proteins followed by production of cyclic AMP.

Stimulus–Secretion Coupling

The neurohypophysis contains a very large store of vasopressin and oxytocin – enough to maintain normal basal concentrations of vasopressin for about 1 month even without replenishment, but if the demand is high the stores can be rapidly depleted. After 2–3 days without water, the neurohypophysial content is only about 10% of its usual level. The vasopressin that is secreted is replaced by newly-made vasopressin, but this replenishment takes time. In a steady state, new vesicles arrive at the neurohypophysis at a rate that closely matches the rate of secretion or degradation of old vesicles, and so the neurohypophysial content stays roughly constant. However, when the rate of vasopressin secretion increases, although there is a corresponding increase in the rate of synthesis there is a delay between activating synthesis and replenishing the stores, and thus states of elevated secretion are consistently associated with a depletion of neurohypophysial content.

In magnocellular neurons, vesicles that contain oxytocin or vasopressin can be released from all parts of the cell – the nerve terminals, axon swellings and even undilated axons, and also from the soma and dendrites. In the brain, most vesicles that contain vasopressin or oxytocin are in the dendrites of magnocellular neurons, and these are likely to be a major source of central release in the hypothalamus. Importantly, secretion is regulated differentially from the different compartments of magnocellular neurons. Secretion from the axons and nerve terminals is determined by spike activity, but in the soma and dendrites Ca^{2+} release from the intracellular stores of the endoplasmic reticulum can trigger release independently of spike activity.[26]

Besides oxytocin and vasopressin, magnocellular neurons produce and release many other peptides, including dynorphin, galanin, cholecystokinin, apelin and secretin. These "co-existing" peptides are all produced in very much lower quantities than vasopressin and oxytocin, but some are produced in quantities that are large compared to other peptidergic neurons. At least some of these peptides have important autocrine and paracrine actions subsequent to their central release, and some (notably dynorphin) have important autoregulatory effects on stimulus–secretion coupling in the neurohypophysis.[27] Both oxytocin and vasopressin neurons also express nitric oxide synthase in abundance, and release nitric oxide in response to intense activation. This gaseous transmitter inhibits the magnocellular neurons by a mixture of direct actions and indirect effects on afferent inputs, acting as a rapid negative feedback mechanism to limit over-activation. The magnocellular neurons additionally produce endocannabinoids and adenosine in response to activation, and these also function as negative feedback signals.[28]

Dendritic Peptide Release

Exocytosis of vasopressin and oxytocin from the axonal terminals in the neurohypophysis is linked to electrical activity. However, dendritic release does not necessarily parallel axonal release. The release of vasopressin and oxytocin from the dendrites and nerve endings of magnocellular neurons can be evoked on different time scales in response to the same stimulation, or can be regulated wholly independently.[26,29] For example, in oxytocin neurons, in response to suckling in lactating rats, oxytocin release in the SON precedes secretion into the blood, whereas after systemic osmotic stimulation it lags behind. The vasopressin and oxytocin released from dendrites have key autocrine and paracrine effects on their own neurons. Oxytocin and vasopressin can elicit dendritic peptide secretion without activating axonal release, while, conversely, electrical activity in the cell bodies can cause secretion from the nerve endings without evoking release from the dendrites. Part of the function of dendritically-released peptides is to autoregulate the electrical activity of the cells of origin. Vasopressin itself (and dynophin, an opioid peptide co-localized with vasopressin) modulates the phasic activity of vasopressin neurons by predominantly inhibitory actions. However, vasopressin can also activate quiescent vasopressin neurons, and these dual actions may reflect the fact that the neurons express both V1a and V1b receptors and these are coupled to different intracellular second messenger pathways.[30]

Normally, dendritic release from magnocellular neurons is not much influenced by electrical activity, but it can be evoked by certain peptides; most notably, oxytocin is a potent stimulator of oxytocin release, and vasopressin a potent stimulator of vasopressin release. Oxytocin mobilizes intracellular Ca^{2+} and triggers release from dendrites independently of electrical activity, and also primes dendritic stores of peptide, making them available for activity-dependent release.[26] Priming is an important phenomenon that underlies functional reorganization of the magnocellular system. Priming involves a relocation of peptide-containing vesicles close to the dendritic plasma membrane, and thus a dramatic facilitation in the size of the readily-releasable pool of vesicles available for activity-dependent dendritic release. Accordingly, priming determines the strength of dendro-dendritic communication, and hence the degree of coordination of magnocellular neuronal activity. As priming involves translocation of vesicles, its effects are delayed but long-lasting (> 90 min), which allows the neuropeptides to functionally reorganize neuronal networks through hormone-like or neuromodulatory actions on receptors which are widely expressed at diverse sites. This might result in peptide-dependent plasticity of the neuronal networks and gradually lead to coherent behavioral effects.

PHYSIOLOGICAL FUNCTIONS OF VASOPRESSIN

Antidiuretic Actions of Vasopressin

Vasopressin increases the permeability to water of the collecting duct cells in the kidney, and thus allows water reabsorption and excretion of concentrated urine – i.e., antidiuresis. This occurs through insertion of water channels (aquaporin-2) into the apical membrane of the collecting duct epithelial cells. Consequently, water moves out of the nephron, increasing the amount of water reabsorbed from the forming urine back into the bloodstream. Vasopressin increases transcription of the aquaporin-2 gene, increasing the number of aquaporin-2 molecules in collecting duct cells. Vasopressin also increases the permeability of the inner medullary portion of the collecting duct to urea, allowing increased reabsorption of urea into the medullary interstitium. Finally, vasopressin stimulates sodium and chloride reabsorption in the thick ascending limb of the loop of Henle.[31] This antidiuretic action is crucial for osmoregulation, and defects in vasopressin production, secretion or action result in diabetes insipidus, characterized by excessive thirst and frequent passing of large volumes of dilute urine. Most commonly this disorder arises from a deficiency in vasopressin production, which may be caused by damage to the hypothalamus or pituitary – for example, after an infection, operation or head injury – or may have a genetic cause, as described above. The rarer nephrogenic diabetes insipidus occurs when the kidneys fail to respond to vasopressin, either as a consequence of a defect in receptor mechanisms or as a result of kidney damage.[32] Administration of vasopressin analogs (such as Desmopressin) is effective in central but not in nephrogenic diabetes insipidus.

Vasopressin secretion is regulated by two key physiological variables: plasma osmotic pressure, and plasma volume. A rise in plasma osmotic pressure will follow from ingestion of salt, or from dehydration resulting from lack of access to drinking water. In both situations the increase in osmotic pressure can only be rectified by increased water intake or increased sodium excretion, and while sodium can be excreted in urine, this entails water loss and a reduction in plasma volume. By concentrating the urine, vasopressin can slow the rise in plasma osmotic pressure that results from continuing dehydration, but it cannot correct the raised osmotic pressure; only drinking can achieve this. In general, stimuli that cause thirst also cause vasopressin to be secreted, and many brain regions which control vasopressin secretion also control thirst. Sodium excretion (natriuresis) is also under hypothalamic control via the renal sympathetic innervation, via the aldosterone system, and via a cocktail of hormonal natriuretic factors. Minimizing water loss is also an important defense against low plasma volume, and hemorrhage strongly activates vasopressin secretion partly via relaxation of baroreceptors and, through hypoxia, via activation of chemoreceptors in the aortic arch and carotid body.[2]

Osmoreception

Magnocellular vasopressin neurons are the "command" neurons that regulate antidiuresis, and the rate of spike discharge increases as extracellular osmolality increases. As a result, vasopressin concentrations in the plasma are proportional to the plasma osmotic pressure above a threshold, known as the "set point," for osmoregulated vasopressin secretion. The set point can vary in different physiological circumstances: for example, in pregnancy the set point is lowered; as a result, more water is retained, allowing for an expansion of plasma volume, and this is accompanied by a lower plasma sodium concentration.[33]

Spikes in vasopressin neurons are discharged at a frequency that depends on: (a) the level of synaptic input to the neurons; and (b) their resting membrane potential. Vasopressin cells are themselves osmoreceptors – they are intrinsically responsive to small changes in extracellular osmotic pressure. When the extracellular osmotic pressure is raised, water leaves these cells, following the osmotic gradient, and thus they shrink. Because vasopressin neurons express specific stretch-sensitive cation channels, this shrinkage results in a depolarization of the membrane potential. In the absence of any synaptic input the response to osmolality changes that are within the normal physiological range would not result in any spike activity, as the resting membrane potential of vasopressin neurons is normally well below the threshold for spike initiation. However, graded depolarizations affect the probability that excitatory post-synaptic potentials (EPSPs) from randomly arriving synaptic input will trigger spikes. Accordingly, direct osmosensitivity interacts with the synaptic input rate to determine the *gain* of the neuronal spike response, which in turn determines the slope of the relationship between osmotic pressure and vasopressin secretion. Any stimulus that increases the level of synaptic input to vasopressin cells will increase the gain of the osmoregulatory response; conversely, any lesion that destroys a source of tonic excitatory input to the neurons will impair osmoregulation of vasopressin secretion.

Some sources of afferent input to the vasopressin neurons are also sensitive to changes in osmotic pressure. In particular, adjacent to the anterior wall of the third cerebral ventricle are two "circumventricular organs" – the SFO and the OVLT. These structures are densely vascularized and lack a blood–brain barrier, and they contain neurons that respond directly to many blood-borne factors that do not generally penetrate the brain, such as hormones like angiotensin, endothelin, relaxin and natriuretic hormones. Neurons from these structures project directly to the magnocellular oxytocin and vasopressin neurons, and also indirectly, via the nucleus medianus. Some neurons in both the SFO and OVLT are, like magnocellular neurons, intrinsically osmosensitive, via similar cellular mechanisms. Thus, damage to these anterior structures can result in severe impairment of vasopressin secretion, and even diabetes insipidus.

In addition, the magnocellular neurons receive inputs from oropharyngeal receptors, from the main olfactory bulb, and from visceral receptors, including especially from the gastrointestinal tract, and it is thought that these early signals of taste and ingestion trigger the anticipative changes in vasopressin secretion that precede the changes in osmotic pressure which arise from salt or water intake.

Phasic Firing

In response to dehydration, vasopressin neurons fire in a highly patterned manner (Fig. 6.3). Typically, bursts of spikes lasting for 20–60 s are separated by silent periods of 20–30 s. Within each burst, the neurons fire at between 4 and 10 spikes/s; this is relatively constant within a burst, and is consistent from one burst to the next in any one cell. With progressive dehydration, bursts become more frequent and more intense. When osmotic pressure rises rapidly, phasic cells tend to fire continuously at a progressively increasing frequency as the osmotic pressure is rising, settling down into a phasic pattern only once the osmotic pressure is again stable

This phasic firing is the consequence of activity-dependent changes in neuronal excitability. When a flurry of EPSPs depolarizes the neuron enough to exceed the spike threshold, voltage-dependent Na^+ channels open, depolarizing the cell membrane to produce an all-or-none action potential (a spike), followed by voltage-sensitive rectifying K^+ channels which repolarize the neuron. In addition, in both vasopressin and oxytocin neurons, voltage-sensitive Ca^{2+} channels open, resulting in a large Ca^{2+} influx. This opens slowly inactivating Ca^{2+}-dependent K^+ channels which hyperpolarize the cell for up to 40 ms, causing a hyperpolarizing afterpotential. In addition, in vasopressin neurons (but not oxytocin neurons) other even slower depolarizing channels open, producing, after this hyperpolarization, a rebound depolarizing afterpotential. These depolarizing events can summate to form a "plateau potential"; this brings the membrane potential of the vasopressin neuron to within just a few millivolts of the spike threshold, so that spikes are triggered by EPSPs relatively frequently. Eventually the plateau mechanism inactivates, in part due to dynorphin released from the dendrites of the vasopressin neurons, which acts back on the cell of origin to block the depolarizing mechanism that sustains the plateau. In part it is also because of yet another class of slowly inactivating Ca^{2+}-activated K^+ channels mediating a slowly accumulating hyperpolarization called an afterhyperpolarization, which differs from the hyperpolarizing afterpotential by its very slow prolonged time-course. Other signals generated by the cell (adenosine and endocannabinoids) act back upon the afferent nerve endings: vasopressin cells produce nitric oxide in response to a raised intracellular Ca^{2+} concentration, and this triggers GABA release from afferent nerve endings. Collectively, these mechanisms ensure that bursts will come to an end and be followed by a period of relative inexcitability. Thus, each vasopressin neuron has complex intrinsic properties which ensure that it alternates between periods of activity and silence.[27]

Although individual vasopressin neurons fire in bursts in response to dehydration, vasopressin secretion into the blood is *not* intermittent. The discharge activity of vasopressin neurons is asynchronous, and the activity of the population as a whole increases relatively linearly and smoothly as osmotic pressure rises. However, the cells, though asynchronous, do communicate with each other via vasopressin itself, released from the magnocellular dendrites, and this acts as a population negative-feedback signal.

PHYSIOLOGICAL SIGNIFICANCE OF PHASIC FIRING

If vasopressin cells do not fire phasically to produce pulsatile vasopressin secretion, then it is natural to ask why they fire phasically at all. This question was answered by studies that exploited the fact that the neurohypophysis contains just the axons and terminals of the magnocellular neurons. If it is kept in warm, oxygenated medium with an ionic composition similar to that of extracellular fluid, the isolated neurohypophysis will retain the ability to secrete vasopressin for many hours. This *in vitro* preparation became important for understanding how electrical activity is coupled to secretion, and some of the first experiments using this preparation addressed the question of why vasopressin neurons fire phasically.[34] They apparently do so because this pattern is the most efficient for

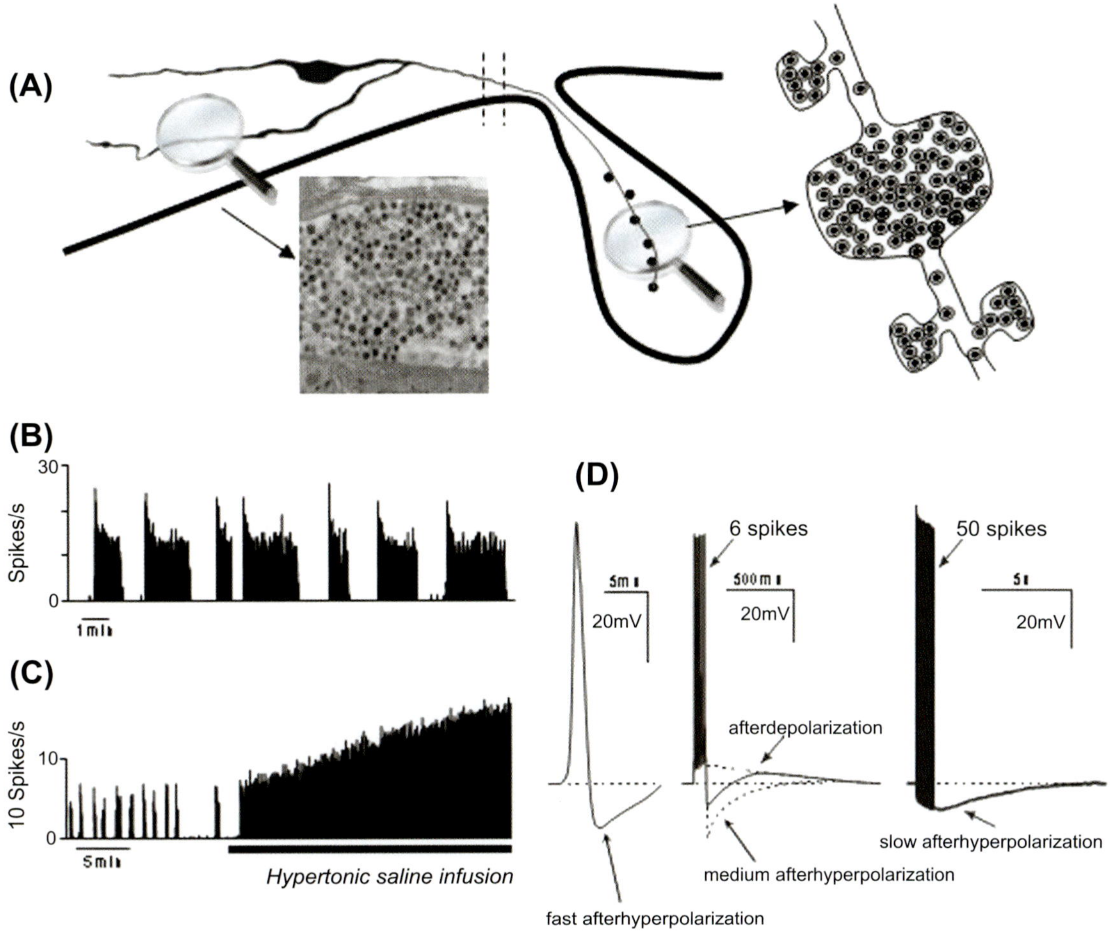

FIGURE 6.3 Phasic firing in vasopressin cells. (**A**) Each magnocellular vasopressin cell has a single axon, which projects to the posterior pituitary gland. There, axonal swellings and endings are packed with neurosecretory granules which can be secreted as a consequence of spike activity. In dendrites neurosecretory granules are particularly abundant and dendritic vasopressin release contributes to autoregulation vasopressin cells spike activity. (**B**) Phasic firing patterns are particularly important in this, as several hundred spikes are needed to release even a single vesicle from any given ending. Phasic firing comprises bursts of a few hundred spikes, separated by silent intervals. (**C**) If the plasma osmotic pressure is raised progressively, the cells increase their spike activity linearly with the change in osmotic pressure. (**D**) The phasic firing pattern arises because of intrinsic mechanisms that produce activity-dependent regulation of excitability. Importantly, every spike is followed by a fast afterhyperpolarization, which limits the maximum frequency at which a vasopressin cell can fire. However, this can be followed by a small afterdepolarization, which tends to make a vasopressin cell more likely to fire again once it has fired. These afterdepolarizations can summate to form a "plateau" of excitability that sustains a burst. This can, however, be terminated by yet other mechanisms – a medium afterhyperpolarization, which reduces firing rate during bursts (spike frequency adaptation), and a slow afterhyperpolarization that can summate to provide a hyperpolarization strong enough to stop a burst. *Adapted from Leng* et al. *(2008).*[56]

vasopressin secretion.[35] Vasopressin secretion is facilitated by an increasing spike frequency, mainly because this facilitates Ca^{2+} entry into the terminals. Facilitation is countered by a "fatigue" that reflects the temporary exhaustion of a readily releasable pool of vesicles. The most efficient spike frequency for vasopressin secretion is about 13 spikes/s – close to the maximum firing rate that vasopressin neurons can sustain. However, this high efficiency cannot be maintained for more than a few tens of seconds; and a few tens of seconds of silence is needed for recovery. As a consequence, the phasic discharge pattern appears to be optimally efficient, in the sense of achieving a given secretion with the fewest spikes.

Cardiovascular Actions of Vasopressin

Vasopressin increases peripheral vascular resistance and thus arterial blood pressure, and this can be important for restoring blood pressure in hypovolemic shock (during hemorrhage) and septic shock.[36] Afferent nerve impulses from stretch receptors in the left atrium, aortic arch and carotid sinus tonically inhibit vasopressin neurons, and a reduction in their discharge rate thus leads to increased vasopressin secretion. Whereas baroreceptors in the atrium and ventricles signal changes in blood volume, the receptors of the aortic arch and carotid sinuses signal changes in arterial blood pressure. Activation of chemoreceptors in the carotid bodies (by high P_{CO_2} or low P_{O_2} in the blood) excites vasopressin cells. Afferents from baro- and chemo-receptors are carried by the carotid sinus, aortic and vagus nerves, and terminate in the NTS, the area postrema and the nucleus ambiguous of the medulla oblongata. The NTS and the VLM are sources of major afferent innervation of the magnocellular nuclei, and are implicated in gastrointestinal, somatosensory and hemodynamic influences upon neurohypophysial hormone secretion. Both projections include a strong noradrenergic component, but several peptides have been identified in adrenergic and non-noradrenergic neurons projecting from these regions to the SON and PVN, including inhibin B, somatostatin, enkephalin, neuropeptide Y, galanin and substance P.

PHYSIOLOGICAL FUNCTIONS OF OXYTOCIN

Oxytocin classically stimulates smooth muscle contraction in the uterus and mammary glands in female mammals, but it is now also known to regulate natriuresis and to contract smooth muscle in the male reproductive tract.

Parturition

As indicated above, it was realized over 100 years ago that extracts from the neurohypophysis were uterotonic, and oxytocin is still one of the most potent endogenous uterotonic factors known today. Although there are significant species differences in mechanisms initiating parturition, the following explains the general mechanisms in women and rodents. In the last few hours before birth there is a marked change in the ratio of circulating estrogen and progesterone and/or their receptor subtypes in the uterus, as a result of which myometrial oxytocin receptor expression and binding increase, and the myoepithelial cells become highly sensitive to oxytocin. Soon after, myometrial contraction is initiated as a result of local production of prostaglandins. As the uterus starts to respond, signals from the contracting uterus are communicated via the vagal and hypogastric nerves to the caudal brainstem and from there to the hypothalamus, and to oxytocin neurons in particular. The resulting oxytocin secretion amplifies the uterotonic actions of prostaglandins by the direct actions of oxytocin on the myometrium and by actions on the endothelium to further increase the production of prostaglandins. As labor progresses, the contractions become stronger and more frequent and the neuronal afferents set up a positive-feedback loop known as the Ferguson Reflex.[37] During pregnancy, oxytocin action on the myometrium is in part prevented by degradation. The enzyme oxytocinase is primarily produced by the placenta, its activity being maximal around mid-gestation. Towards birth, oxytocin's role in myometrial contraction is then facilitated as oxytocinase expression and activity fall, resulting in slower degradation. Interestingly, the oxytocinase enzyme is also synthesized in the hypothalamus and other brain regions, so may control the availability of oxytocin in the brain as well as peripherally.

In rats, magnocellular oxytocin neurons fire in intermittent intense synchronized bursts during parturition, leading to large pulses of oxytocin that can precipitate the expulsion of a fetus or a placenta.[38] The pulsatile pattern of secretion is important because a sustained increase in oxytocin secretion can desensitize oxytocin receptors and can even result in delayed birth. The activation of oxytocin neurons is accompanied by increased dendritic release of oxytocin and induction of gene expression in the oxytocin neurons, including gene markers of activity (e.g., Fos) and other genes (including oxytocin itself), and over the course of parturition results in a substantial depletion of the pituitary stores of oxytocin.

The necessity for oxytocin in facilitating birth has been investigated using transgenic and pharmacological animal models. Targeting the oxytocin system pharmacologically (including with oxytocin antagonists) generally delays the onset and progress of birth in rodents. Transgenic mice lacking oxytocin or oxytocin receptor give birth apparently normally, indicating that there is redundancy in the mechanisms controlling parturition, but the onset of birth in rodents is normally very sensitive to changes in circadian cues and this sensitivity is lost in the absence of oxytocin. Furthermore, transgenic mice lacking the prostaglandin synthesizing enzyme cyclooxygenase 2 also do not enter labor, and this is apparently due to delayed induction of uterine oxytocin receptors, also in part because oxytocin facilitates luteolysis in this species. So although circulating oxytocin is not absolutely necessary for birth to occur, it has an important role in the timing of labor and birth in both women and rodents.[39]

Integrated Control of Parturition

In mid-pregnancy, magnocellular oxytocin neurons begin to adapt to prepare them for their imminent new function at birth. First, the secretion of oxytocin is suppressed while its synthesis is maintained, and as a result the neurohypophysial stores of oxytocin accumulate. In rats, this is achieved in part by upregulation of dynorphin expression in magnocellular neurons; dynorphin is co-packaged with oxytocin, and when released acts back at κ-opioid receptors on the axonal terminals to suppress activity-regulated secretion. Second, the electrical excitability of the oxytocin neurons is suppressed, reducing the risk of premature activation of oxytocin secretion. Under the control of progesterone and steroid metabolites such as allopregnanolone, oxytocin neuron responses to most stimuli are attenuated in the last third of gestation. Allopregnanolone interacts with the GABA-A receptor mechanisms on oxytocin neurons, enhancing their sensitivity to the inhibitory neurotransmitter GABA, and these inhibitory effects decline as the progesterone : estrogen ratio falls towards birth.[40] In addition, there is a powerful restraint of activity mediated by endogenous opioids. Oxytocin cells express the κ-opioid receptors by which dynorphin acts to inhibit secretion at the neurohypophysis, but they also abundantly express the μ-opioid receptors through which enkephalins and endorphins can act. Enkephalins are co-expressed by the noradrenergic neurons of the A2 group that are active in parturition – and the level of enkephalin expression is increased in these neurons in late pregnancy. Thus, in late pregnancy and during parturition, oxytocin secretion is extremely sensitive to both opioid agonists and antagonists.[41]

Labor begins when signals from the uterus and fetus trigger initial weak contractions of the myometrium. Sensory afferents from the contracting uterus activate A2 noradrenergic neurons of the NTS, which directly activate magnocellular oxytocin neurons. As a consequence, oxytocin is secreted from the neurohypophysis and also from the dendrites of oxytocin neurons. This dendritic oxytocin release is critical for the induction of synchronized bursting of the oxytocin neurons, and may also be involved in initiating maternal behaviors. Noradrenaline is a key neurotransmitter in this reflex, contributing to these mechanisms as its concentration increases in the SON at birth and blocking or mimicking adrenergic α1 receptor activation prevents or induces burst firing, peripheral oxytocin secretion and progress of birth, respectively.[42]

Lactation

Oxytocin has an essential role in lactation. Although not directly involved in milk synthesis or secretion, transgenic mice lacking oxytocin reveal that it is the only hormone able to drive the milk-ejection reflex. This is a defining mechanism in mammals, and indicates that oxytocin has a conserved function in the provision of milk to sucking young. Oxytocin receptor expression and binding in the alveolar myoepithelial cells increase in late gestation and their sensitivity to oxytocin becomes enhanced. Oxytocin causes their contraction, which exerts enough pressure within the mammary alveoli to expel milk.

Integrated Control of Lactation – A Neuroendocrine Reflex

In the lactating rat, pups suckle continuously, but are rewarded with milk only every 5–15 minutes, when they display a characteristic "stretching" response. About 10–15 s before each stretch response, all of the magnocellular neurons show a brief, high-frequency burst of activity, lasting just 1–3 s.[43] This intense burst of activity (Fig. 6.4), which is usually followed by a long silent period, triggers a large pulse of oxytocin to be secreted from the neurohypophysis. Pulsatile secretion is probably a feature of suckling in all mammals; most recently, imaging of milk-flow in humans has revealed that milk let-down in response to the continuous sucking of a baby occurs intermittently, even though the mother might not always be aware of this discrete response patterning.[44] The mammary gland responds to oxytocin only at relatively high concentrations, and it rapidly desensitizes in response to continued exposure, so pulsatile delivery of hormone is essential for efficient function. Moreover, bursting is efficient for releasing oxytocin; spikes release more oxytocin if they occur in a burst than if they occur at a lower frequency over a longer time. In the rat, a burst of activity in oxytocin cells releases about 1 mU of oxytocin – just right for a maximally effective contraction of the mammary gland.

The mechanisms that coordinate all the magnocellular oxytocin neurons in each of the SON and PVN nuclei involve excitatory, probably noradrenergic, inputs from the caudal brainstem and communication between the magnocellular nuclei. Communication between nuclei is probably mediated via oxytocin release from the long dendrites that extend well beyond the boundaries of the nuclei, but also may involve other limbic regions such as the bed nucleus of the stria terminalis. However, the most important element of control is the priming of somatodendritic oxytocin release during suckling, and its actions in the SON/PVN. A computational model of the milk-ejection reflex describes how synchronous burst firing can arise from these dendritic interactions.[45]

Natriuresis

While magnocellular oxytocin neurons respond to suckling and during parturition with intermittent bursts, in response to many other stimuli they show a graded, sustained, activation that leads to continuously elevated oxytocin secretion. In rodents, magnocellular oxytocin neurons, like vasopressin neurons, are osmoreceptors, and oxytocin concentrations in the blood rise in proportion to increasing blood osmolality. Oxytocin stimulates Na^+ excretion by the kidneys (natriuresis). In part this is mediated directly at the kidney (in rodents), but mainly it is mediated via stimulation of atrial natriuretic peptide secretion from atrial cardiac cells. However, this effect of oxytocin seems to vary considerably between species; in humans and dogs oxytocin appears to have no significant natriuretic effects, and only a weak antidiuretic effect that

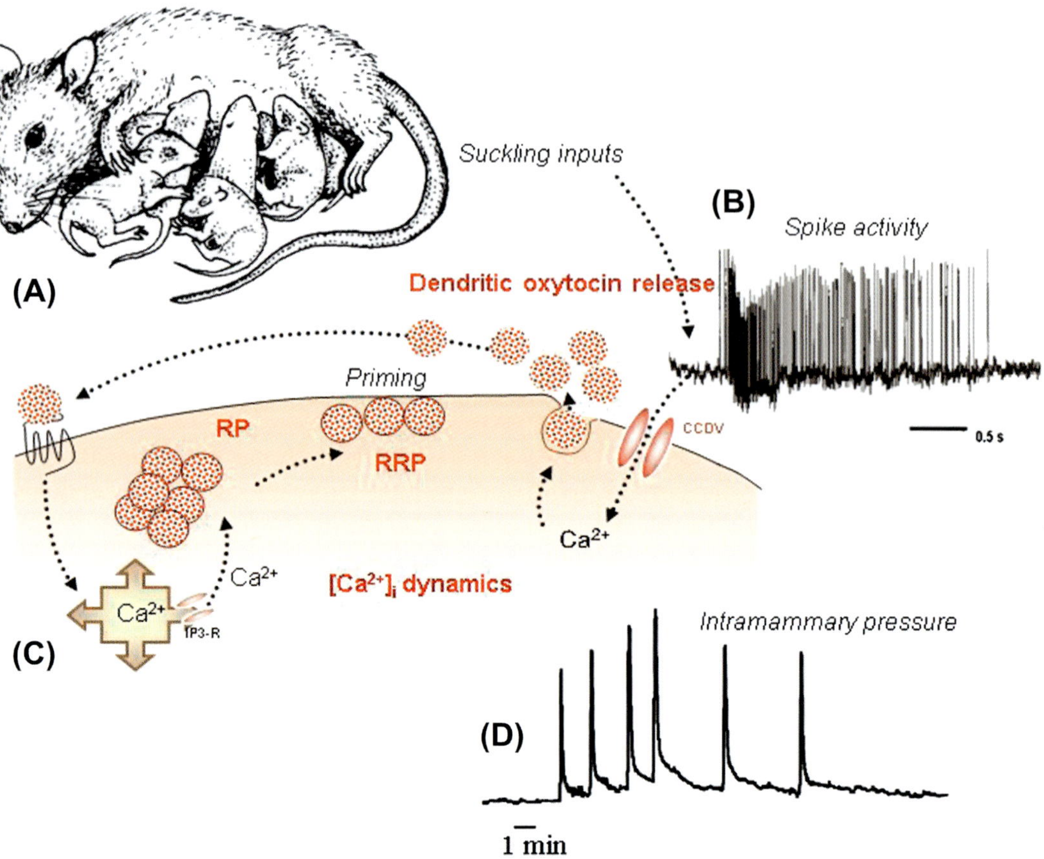

FIGURE 6.4 (**A**) During suckling, magnocellular oxytocin neurons in the SON and PVN display intense bursts of electrical activity (**B**) every few minutes. These bursts induce oxytocin secretion from both dendrites in the hypothalamus and axon terminals in the posterior pituitary gland. (**C**) The dendritic oxytocin release acts back on the oxytocin cells to mobilize intracellular calcium release ("priming"). Priming is an important phenomenon that underlies functional reorganization of the magnocellular system. Priming involves a relocation of peptide-containing vesicles close to the dendritic plasma membrane, and thus a dramatic facilitation in the size of the readily-releasable pool of vesicles available for activity-dependent dendritic release. Priming determines the strength of dendro-dendritic communication, and hence the degree of coordination of magnocellular neuronal activity. (**D**) Secretion from the nerve endings enters the blood and reaches the mammary gland, where pulses of oxytocin elicit abrupt rises in intramammary pressure, signaling milk let-down. The oxytocin that is released from dendrites subsequently diffuses to act at other brain sites to facilitate maternal behavior. *Adapted from Ludwig and Leng (2006).*[26]

might reflect a weak pharmacological activity at vasopressin receptors.

In rodents, oxytocin secretion is also stimulated by systemic injections of cholecystokinin (CCK), via CCK-A receptors on gastric vagal afferents, whereas vasopressin secretion is inhibited by the same stimulus. CCK is secreted from the duodenum in response to food ingestion, and its vagal actions mediate a satiety reflex. As central injections of oxytocin markedly suppress appetite in rodents, it is possible that the appetite-suppressing effects of CCK involve the central release of oxytocin; it is also possible that the stimulation of oxytocin secretion might be linked to the natriuretic actions of oxytocin.

Copulation

In both humans and rodents, oxytocin is secreted from the neurohypophysis during copulation, and particularly at ejaculation in males, and it stimulates transient smooth muscle contraction in reproductive organs such as the seminal vesicles and epididymis. Oxytocin also promotes penile erection through actions at the hypothalamus and spinal cord; the spinal effects are mediated by parvocellular oxytocin cells of the PVN, but the hypothalamic actions may reflect effects of dendritic release from magnocellular neurons.

BEHAVIORAL EFFECTS OF OXYTOCIN AND VASOPRESSIN

In mammals, vasopressin and oxytocin have a wide range of behavioral effects.

Vasopressin and oxytocin do not cross the blood–brain barrier in appreciable amounts, so the central actions of these hormones reflect the central release of these peptides, either from centrally projecting neurons or from the dendrites of magnocellular neurons.

Oxytocin is important for maternal behavior in several species, and in the female prairie vole it is instrumental in forging monogamous partner bonds. These roles seem to tally with behavioral observations that indicate more general effects of oxytocin on "social behaviours".[46] Oxytocin has been described as anti-anxiogenic for its ability to increase exploratory activity, and especially for olfactory investigations by mice of other mice. In humans, intranasal application of oxytocin has been reported to enhance "trust".[47] Oxytocin also facilitates sexual responsiveness in female rats and sexual behavior in male rats, and there have been reports of similar effects in humans. Vasopressin has effects on behavior that seem to complement those of oxytocin; in male prairie voles, vasopressin enhances the expression of aggression towards other males – behavior that seems analogous to territorial behavior designed to protect a mate from access by other males, or to defend a nest. This aggression can be facilitated by central injections of vasopressin, and can be attenuated by vasopressin antagonists. In the monogamous prairie vole, the vasopressin V1a receptor has a distribution in the brain quite unlike that of its close but promiscuous relative, the montane vole. Indeed, the distribution is variable even between different voles because of a high variability in the V1a receptor gene,[48] and these individual variations are predictive of differences in both receptor distribution and social behavior.[49] Moreover, experimental manipulation of the receptor gene, through either transgenic mutations or adenoviral gene transfer, show that behavioral changes accompany alterations in the distribution of V1a receptors.[50] Thus, it is not the distribution of vasopressin *fibers* in the brain that determines the behavior, but the distribution of vasopressin receptors, and this suggests that the key behavioral signal is not vasopressin released from conventional nerve endings with an effect localized at synapses, but a hormone-like release of vasopressin within the brain.

The behavioral effects of oxytocin and vasopressin do not consistently accompany peripheral secretion of these hormones, and concentrations measured in the CSF do not change in parallel with blood concentrations. The natural conclusion has been that the oxytocin and vasopressin measured in CSF or extracellular fluid in different brain areas do not derive from magnocellular neurons; this conclusion has, however, been thrown into doubt by the discovery that magnocellular neurons can regulate central release of their peptide independently of peripheral secretion.

PATHOPHYSIOLOGY OF THE NEUROHYPOPHYSIAL SYSTEMS

Oxytocin

Several conditions are associated with altered oxytocin activity or effects. The main clinical disorder directly associated with altered circulating (neurohypophysial) oxytocin is preterm labor, which typically arises due to stress or infection. This induces premature activation of the neurohypophysial oxytocin system, and resulting myometrial contraction rapidly leads to birth of a premature infant.[51] Preterm labor in women can be safely delayed by treating the mother with an oxytocin antagonist (Tractocile®). While treatment is only effective for a limited few days and needs to be repeated, this underlines the importance of neurohypophysial oxytocin in controlling birth.

Other conditions include autism spectrum disorders (comprising a range of defects in social behaviors) and sexual behavior dysfunction, which, although are primarily controlled by central oxytocin, are associated with altered circulating oxytocin secretion.[52,53]

Vasopressin

The most common diseases directly associated with dysregulation of the vasopressin system are the syndrome of inappropriate antidiuretic hormone secretion (SIADH), congestive heart failure, decompensated cirrhosis, diabetes insipidus and septic shock.[54,55] SIADH is characterized by hypotonic hyponatremia, increased urine osmolality, normal natriuresis, euvolemia or volume depletion, and normal renal and adrenal function. Elevated vasopressin secretion leads to renal water retention and extracellular fluid expansion, which is initially compensated for by increased urinary sodium excretion leading to hyponatremia. SIADH, where vasopressin secretion is regulated inadequately, is relatively common and can be present in a wide variety of clinical conditions, including tumors that produce ectopic vasopressin, human immunodeficiency virus/acquired immunodeficiency syndrome, pulmonary disease, endocrine disease, neurological disease or trauma, or administration of some drugs. Traditional management relies largely on restricting fluid intake.

Congestive heart failure is a complex syndrome in which the pathophysiology combines hemodynamic derangements and neurohormonal imbalances. In physiological conditions, the sympathetic nervous system and the renin–angiotensin–aldosterone system are key to the maintenance of normal blood pressure and circulation. Vasopressin plays a relatively small role in normal physiology. However, in congestive heart failure plasma concentrations of vasopressin are elevated despite the atrial distension, hyponatremia and low osmolality that would normally inhibit its secretion. Persistent stimulation of V_1 receptors by high concentrations of vasopressin causes vasoconstriction and an increase in systemic vascular resistance, and thus has a further adverse effect on cardiac function. V_2 receptor stimulation increases water retention, leading to edema, and can contribute to the hyponatremia commonly seen in patients with congestive heart failure. The mechanism underlying the persistently high vasopressin secretion in these patients is uncertain, but it is believed to be due to abnormal feedback control, such that hypoosmolality does not reduce vasopressin secretion, and to baroreceptor-stimulated secretion of vasopressin. The low cardiac output in patients with heart failure causes reduced arterial pressure, which unloads the carotid sinus and aortic arch baroreceptors, triggering vasopressin secretion in an effort to increase sodium and water retention.

Decompensated liver cirrhosis and ascites results in impaired renal water handling with water retention, associated with dilutional hyponatremia resulting in a persistently high level of vasopressin despite hypoosmolality and hyponatremia. High circulating concentrations of vasopressin result from decreased metabolism of vasopressin due to altered liver function, and nonosmotic secretion of vasopressin secondary to reduced effective arterial filling associated with arterial vasodilation characteristic of cirrhosis.

Arterial hypotension is typically associated with an appropriately high level of vasopressin secretion. In septic shock, the vasopressin response to hypotension seems to be blunted. The mechanism underlying these reduced vasopressin levels is unclear, with possible mechanisms including reduced production of vasopressin, depletion of vasopressin stores, inhibition of vasopressin secretion, impaired baroreflex-mediated vasopressin secretion, and increased vasopressin degradation.

References

1. Seward E, Chernevskaya N, Nowycky M. Exocytosis in peptidergic nerve terminals exhibits two sensitive phases during pulsatile calcium entry. *J Neurosci.* 1995;15:3390–3399.
2. Leng G, Brown CH, Russell JA. Physiological pathways regulating the activity of magnocellular neurosecretory cells. *Prog Neurobiol.* 1998;57:625–655.
3. Armstrong WE. Morphological and electrophysiological classification of hypothalamic supraoptic neurons. *Prog Neurobiol.* 1995;47:291–339.
4. Hussy N. Glial cells in the hypothalamo-neurohypophysial system: key elements of the regulation of neuronal electrical and secretory activity. *Prog Brain Res.* 2002;139:95–112.
5. Oliver G, Schäfer EA. On the physiological action of extracts of pituitary body and certain other glandular organs: preliminary communication. *J Physiol.* 1895;18:277–279.
6. Verney EB. The antidiuretic hormone and the factors which determine its release. *Proc R Soc Med B.* 1947;135:25–106.
7. Dale HH. On some physiological actions of ergot. *J Physiol.* 1906;34:163–206.
8. Ferguson JKW. A study of the motility of the intact uterus at term. *Surg Gynecol Obstet.* 1941;73:359–366.
9. Schäfer EA. The structure and functions of the pituitary body. *Proc R Soc Med.* 1913;6:xxxiv–liii.
10. Cross BA, Harris GW. The neurohypophysis and let-down of milk. *J Physiol.* 1951;113:35.
11. du Vigeaud V. Oxytocin, the principal oxytocic hormone of the posterior pituitary gland: its isolation, structure, and synthesis. *Experientia.* 1955;S2:9–26.
12. Scharrer E. Die Lichtempfindlichkeit blinder Elritzen: Untersuchungen über das Zwischenjirn der Fische I. *Z vergl Physiol.* 1928;7:1–38.
13. Scharrer E, Scharrer B. Über Drüsen-Nervenzellen und neurosekretorische Organe bei Wirbellosen und Wirbeltieren. *Biol Rev.* 1937;12:185–216.
14. Bargmann W, Scharrer E. The site of origin of the hormones of the posterior pituitary. *Am Sci.* 1951;39:255–259.
15. Douglas WW, Poisner AM. Stimulus–secretion coupling in a neurosecretory organ. The role of calcium in the release of vasopressin from the neurohypophysis. *J Physiol.* 1964;172:1–18.
16. Wakerley JB, Lincoln DW. The milk-ejection reflex of the rat: a 20- to 40-fold acceleration in the firing of paraventricular neurones during oxytocin release. *J Endocrinol.* 1973;57:477–493.
17. Ivell R, Richter D. Structure and comparison of the oxytocin and vasopressin genes from rat. *Proc Natl Acad Sci USA.* 1984;81:2006–2010.
18. Richter D, Schmale H. The structure of the precursor to arginine-vasopressin: a model preprohormone. *Prog Brain Res.* 1983;60:227–233.
19. Nordmann JJ, Morris JF. Method for quantitating the molecular content of a subcellular organelle: hormone and neurophysin content of newly formed and aged neurosecretory granules. *Proc Nat Acad Sci USA.* 1984;81:180–184.
20. Schmale H, Richter D. Single base deletion in the vasopressin gene is the cause of diabetes insipidus in Brattleboro rats. *Nature.* 1984;308:705–709.
21. Sabatier N, Caquineau C, Dayanithi G, et al. Alpha-melanocyte-stimulating hormone stimulates oxytocin release from the dendrites of hypothalamic neurons while inhibiting oxytocin release from their terminals in the neurohypophysis. *J Neurosci.* 2003;23:10351–10358.
22. Acher R, Chauvet J. The neurohypophysial endocrine regulatory cascade: precursors, mediators, receptors, and effectors. *Front Neuroendocrinol.* 1995;16:237–289.
23. Gwee PC, Amemiya CT, Brenner S, et al. Sequence and organization of coelacanth neurohypophysial hormone genes: Evolutionary history of the vertebrate neurohypophysial hormone gene locus. *BMC Evol Biol.* 2008;8:93.
24. Birnbaumer M. Vasopressin receptors. *Trends Endocrinol Metab.* 2000;11:406–410.
25. Zingg HH, Laporte SA. The oxytocin receptor. *Trends Endocrinol Metab.* 2003;14:222–227.
26. Ludwig M, Leng G. Dendritic peptide release and peptide-dependent behaviours. *Nat Rev Neurosci.* 2006;7:126–136.
27. Brown CH, Bourque CW. Mechanisms of rhythmogenesis: insights from hypothalamic vasopressin neurons. *Trends Neurosci.* 2006;29:108–115.
28. Brown CH, Ruan M, Scott V, et al. Multi-factorial somato-dendritic regulation of phasic spike discharge in vasopressin neurons. *Prog Brain Res.* 2008;170:219–228.

29. Ludwig M, Pittman QJ. Talking back: dendritic neurotransmitter release. *Trends Neurosci.* 2003;26:255–261.
30. Gouzenes L, Desarmenien MG, Hussy N, et al. Vasopressin regularizes the phasic firing pattern of rat hypothalamic magnocellular neurons. *J Neurosci.* 1998;18:1879–1885.
31. Schrier RW. Vasopressin and aquaporin 2 in clinical disorders of water homeostasis. *Semin Nephrol.* 2008;28:289–296.
32. Noda Y, Sohara E, Ohta E, et al. Aquaporins in kidney pathophysiology. *Nat Rev Nephrol.* 2010;6:168–178.
33. Bourque CW. Central mechanisms of osmosensation and systemic osmoregulation. *Nat Rev Neurosci.* 2008;9:519–531.
34. Leng G, Brown D. The origins and significance of pulsatility in hormone secretion from the pituitary. *J Neuroendocrinol.* 1997;9:493–513.
35. Dutton A, Dyball REJ. Phasic firing enhances vasopressin release from rat neurohypophysis. *J Physiol.* 1979;290:433–440.
36. Treschan TA, Peters J. The vasopressin system: physiology and clinical strategies. *Anesthesiology.* 2006;105:599–612.
37. Russell JA, Leng G, Douglas AJ. The magnocellular oxytocin system, the fount of maternity: adaptations in pregnancy. *Front Neuroendocrinol.* 2003;24:27–61.
38. Summerlee AJ. Extracellular recordings from oxytocin neurones during the expulsive phase of birth in unanaesthetized rats. *J Physiol.* 1981;321:1–9.
39. Ratajczak CK, Muglia LJ. Insights into parturition biology from genetically altered mice. *Pediatr Res.* 2008;64:581–589.
40. Brunton PJ, Russell JA. The expectant brain: adapting for motherhood. *Nat Rev Neurosci.* 2008;9:11–25.
41. Leng G, Meddle SL, Douglas AJ. Oxytocin and the maternal brain. *Curr Opin Pharmacol.* 2008;8:731–734.
42. Hatton GI, Wang YF. Neural mechanisms underlying the milk ejection burst and reflex. *Prog Brain Res.* 2008;170:155–166.
43. Lincoln DW, Wakerley JB. Electrophysiological evidence for the activation of supraoptic neurones during the release of oxytocin. *J Physiol.* 1974;242:533–554.
44. Ramsay DT, Kent JC, Owens RA, et al. Ultrasound imaging of milk ejection in the breast of lactating women. *Pediatrics.* 2004;113:361–367.
45. Rossoni E, Feng J, Tirozzi B, et al. Emergent synchronous bursting of oxytocin neuronal network. *PLoS Comput Biol.* 2008;4:e1000123.
46. Neumann ID. Brain oxytocin: a key regulator of emotional and social behaviours in both females and males. *J Neuroendocrinol.* 2008;20:858–865.
47. Baumgartner T, Heinrichs M, Vonlanthen A, et al. Oxytocin shapes the neural circuitry of trust and trust adaptation in humans. *Neuron.* 2008;58:639–650.
48. Donaldson ZR, Young LJ. Oxytocin, vasopressin, and the neurogenetics of sociality. *Science* 2008;322:900–904.
49. Hammock EA, Young LJ. Microsatellite instability generates diversity in brain and sociobehavioral traits. *Science* 2006;308:1630–1634.
50. Landgraf R, Frank E, Aldag JM, et al. Viral vector-mediated gene transfer of the vole V1a vasopressin receptor in the rat septum: improved social discrimination and active social behaviour. *Eur J Neurosci.* 2003;18:403–411.
51. Tara PN, Thornton S. Current medical therapy in the prevention and treatment of preterm labour. *Sem Fetal Neonatal Med.* 2004;9:481–489.
52. Hammock EA, Young LJ. Oxytocin, vasopressin and pair bonding: implications for autism. *Philos Trans R Soc Lond B Biol Sci.* 2006;361:2187–2198.
53. Gurrieri F, Neri G. Defective oxytocin function: a clue to understanding the cause of autism? *BMC Med.* 2009;7:63.
54. Verbalis JG. Disorders of body water homeostasis. *Best Pract Res Clin Endocrinol Metab.* 2003;17:471–503.
55. Vincent JL, Su F. Physiology and pathophysiology of the vasopressinergic system. *Best Pract Res Clin Anaesthesiol.* 2008;22:243–252.
56. Leng G, Brown CH, Sabatier N, et al. Population dynamics in vasopressin cells. *Neuroendocrinology.* 2008;88:160–172.

CHAPTER

7

Neural Control of the Intermediate Lobe of the Pituitary Gland (Pars Intermedia) and Proopiomelanocortin

Malcolm J. Low

Department of Molecular & Integrative Physiology, Department of Internal Medicine, Division of Metabolism, Endocrinology and Diabetes, and Brehm Center for Diabetes Research, University of Michigan Medical School, Ann Arbor, MI, USA

OUTLINE

Summary

The pituitary intermediate lobe (IL) magnificently illustrates the diversity of neuroendocrine communication and physiology of the pituitary gland across vertebrate phylogeny. Multiple CNS nuclei innervate the IL of amphibia and teleosts to coordinate the secretion of MSH in response to light and thereby alter skin pigmentation patterns by melanosome dispersal. This neuroendocrine system displays plasticity at the levels of synaptic input, receptor and ion channel abundance, POMC expression and processing, and control of MSH exocytosis. Mammalian IL melanotropes are controlled primarily by dopamine and GABA inhibition from a single source of periventricular hypothalamic neurons, although activity of the cells is also regulated by additional hormonal and paracrine factors. The physiological function of the mammalian IL is more obscure than that of its amphibian counterpart, and the lobe is vestigial in adult humans. Tumorigenesis of the IL in mutant mouse models has focused attention on cell-cycle control mechanisms in pituitary adenoma formation.

Handbook of Neuroendocrinology, DOI: 10.1016/B978-0-12-375097-6.10007-1

INTRODUCTION

The intermediate lobe (IL) of the pituitary gland came of age in 1981 when it was the focus of an influential Ciba Foundation Symposium, entitled "Peptides of the Pars Intermedia," that gathered together many luminaries of the time with interests in the newly discovered molecule proopiomelanocortin (POMC).[1–3] Seminal studies, including those of Eipper and Mains using a pituitary tumor cell line and primary cultures of rat IL melanotropes, had recently established the biochemical roadmap culminating in the synthesis of ACTH, MSH and beta-endorphin from a common POMC precursor.[4,5] These data intimately connected studies of regulatory processes in IL melanotropes to the broader field of pigment cell biology, with the latter's focus on skin melanocytes[6] and their stimulation by the peptide alpha-MSH, whose amino acid sequence had been determined two decades earlier.[7] (See Lerner[8] for a further historical review.)

DEVELOPMENT AND ANATOMY OF THE PARS INTERMEDIA

The adenohypophysis is derived from placodal ectoderm located at the anterior neural ridge in all vertebrates.[9] Except in teleosts, this primordium invaginates from the stomadeum due to cranial flexure into the structure known as Rathke's pouch. Development of the pituitary at this early stage is critically dependent on inductive signals, including bone morphogenic protein-4 (BMP4) and fibroblast growth factors-8/10/18 (FGF8/10/18) produced by the overlying ventral diencephalon. In mammals, including human embryos, the anterior wall of Rathke's pouch gives rise to the hormone secreting cells of the pars distalis and tuberalis and, due to its rotation, subsequently assumes a position ventral to the pars intermedia, which differentiates from the posterior wall.[10] The IL assumes its intimate apposition with the neurohypophysis, which arises from the infundibular portion of the hypothalamus. Birds, however, never develop a pituitary IL.

In zebrafish, FGF3 appears to play a specific role in establishing the lineage boundary between the two pituitary lobes.[11] Differential gradients of hedgehog and FGF3 are important for the demarcation between zebrafish anterior lobe (AL) and IL, with induction of the transcription factor pax7 in the IL.[12] The transcriptional repressor Hes1 is essential for melanotrope specification in mammals because cells of the posterior wall of Rathke's pouch differentiate into somatotropes instead of melanotropes in Hes1 knockout mice.[13] Similarly, all normal melanotropes express the T-box transcription factor Tpit, but Tpit knockout mice exhibit an IL with altered cell specification. POMC expression is absent, and many of the cells alternatively differentiate into gonadotropes or Pit1 transcription factor-independent thyrotropes.[14]

BOX 7.1

HOW DO WE KNOW THE ROLE OF IL MSH IN BACKGROUND-ADAPTED SKIN PIGMENTATION OF AMPHIBIANS?

Hypophysectomy (Hypox) results in a reduction of skin pigmentation in the tadpole due to melanosome or melanin granule translocation in dermal melanocytes, suggesting the importance of a pituitary-derived factor to maintain pigmentation.[1] Classical transplantation experiments subsequently demonstrated that IL but not AL tissue restored pigmentation in hypophysectomized tadpoles.[2] Similar findings were later confirmed in the dogfish.[3] The identical principle of melanosome dispersion in the skin of white-background adapted frogs was used effectively as a sensitive and relatively high throughput *in vitro* bioassay to purify α-MSH from tissue extracts.[4]

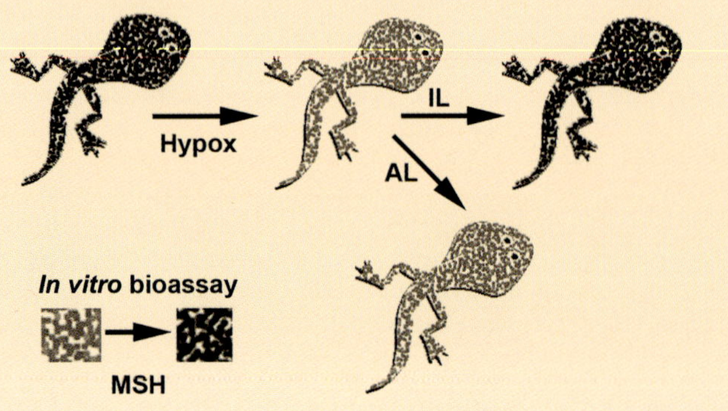

References

1. Smith PE. Experimental ablation of the hypophysis in the frog embryo. *Science.* 1916;44(1130):280–282.
2. Allen BM. Experiments in the transplantation of the hypophysis of adult *Rana pipiens* to tadpoles. *Science.* 1920; 52(1342):274–276.
3. Lundstrom HM, Bard P. Hypophysial control of cutaneous pigmentation in an elasmobranch fish. *Biol Bull.* 1931; 62(1):1–9.
4. Shizume K, Lerner AB, Fitzpatrick TB. *In vitro* bioassay for the melanocyte stimulating hormone. *Endocrinology.* 1954; 54(5):553–560.

POMC gene expression is first detectable in the IL of rat at embryonic age e15.5[15] and mouse at e14.5.[16,17] Glucocorticoid receptors are expressed in both walls of Rathke's pouch at E13–E14 in rat, but are largely restricted to the AL by E17–E18. At this same time the dopamine D2 receptor (D2R) can first be visualized in IL melanotropes by *in situ* hybridization. Tyrosine hydroxylase (TH)-positive fibers first reach the IL at postnatal day p1.5,[18] but brain derived neurotropic factor (BDNF) mRNA is detectable in IL 5 days earlier at e17.5, consistent with a possible axon guidance cue of BDNF for the hypothalamic dopaminergic neurons that innervate melanotropes.[19,20] Cultured adult rat IL melanotrope cells rapidly lose their BDNF immunoreactivity, and it can be restored by constant exposure of the cells to the dopamine receptor agonist apomorphine.[21] Dopamine innervation of the IL is also associated with a progressive decrease in peak Na^+ channel conductance.[22]

Rat melanotropes transiently respond to dexamethasone with an inhibition of corticotropin-releasing hormone (CRH) stimulated release of peptides for a period of 3 days from e19.5 to P2, indicating the temporary presence of functional glucocorticoid receptors (GR).[23] Krieger and colleagues were the first to demonstrate that GR expression could be reinduced in the adult rat IL by several manipulations.[24] Deafferentation of the NIL in adult rats leads to renewed expression of the GR after 3 weeks, but, unlike in the AL, glucocorticoids have a stimulatory effect on POMC gene expression.[25] GR expression can also be induced by ovariectomy and reversed by administration of estradiol in the adult female rat.[26] Similarly, primary cultures of disassociated melanotropes show the reappearance of GR expression and possible stimulatory effects on POMC expression. Syngenic NIL transplants into the anterior eye chamber also regain GR binding that is abolished by an excess of unlabeled corticosterone.[27]

The family leporidae (rabbit and hare) are unique among mammals in that they lack the otherwise ubiquitous dopaminergic innervation of the IL. Ontogeny of the rabbit IL is similar to other mammals, but the lobe is innervated by oxytocin- and neurophysin-immunoreactive fibers.[28] Consistent with the hypothesis that dopamine signaling represses GR expression, it was shown that rabbit melanotropes express the high affinity type-II GR and that dexamethasone inhibits MSH secretion in this species.[29]

Melanotropes are the principal hormone-secreting cell type of the IL. They are polyhedral cells of uniform size with smooth ovoid nuclei, and are characteristically grouped into lobules. There are several minority cell populations in the IL in addition to melanotropes. Some nestin-IR cells that co-stain for glial fibrillary associated protein (GFAP) and others of unidentified type border the pituitary cleft.[30] Glial-like cells, possibly analogous to the radial glial cells found in developing cerebral cortex, are found in the developing and adult rat IL. They have processes that span the entire thickness of the IL, and the cells initially express vimentin immunoreactivity, to be supplanted later by (GFAP) expression.[31] The GFAP-positive cells are a subpopulation of a larger group of S-100-expressing glial-like cells.

REGULATED EXPRESSION AND PROCESSING OF PROOPIOMELANOCORTIN IN THE PARS INTERMEDIA

Under basal conditions, the IL of rats and *Xenopus* contain two populations of melanotropes distinguished from each other by high or low POMC expression.[32] This heterogeneity has also been demonstrated by *in situ* hybridization for POMC mRNA using dispersed single cells and image quantification.[33] Dopamine signaling modulates this heterogeneity by direct actions on *Pomc* gene expression and indirect effects on D2R abundance.[34] The specific mechanisms by which D2R activation inhibits *Pomc* transcription remain obscure; however, promoter analyses performed in transgenic mice clearly implicate cis regulatory elements within the proximal 400 nucleotides upstream of exon 1 in the developmentally regulated, melanotrope-specific and dopamine-regulated expression of *Pomc*.[35,36]

The pituitary lobe-specific processing of POMC is largely determined by the prohormone convertases PC1/3 in the AL and PC2 and 7B2 in the IL.[37,38] 7B2 is a neuroendocrine-specific polypeptide contained in secretory granules that is important for both the transport and proteolytic maturation of proPC2 to PC2 in melanotropes.[39] Pituitary processing of POMC has been examined in detail in the fetal macaque. These studies demonstrated a progressive increase of mono- and diacetylated MSH in the IL during development, while the AL expressed almost entirely ACTH(1–39) and no acetylated-α-MSHs.[40] There were corresponding differences in processing of the carboxyterminal POMC peptides, including β-LPH and β-endorphins.[41] During lactation in rats, melanotropes synthesize ACTH in addition to MSH, and its release is regulated by dopamine.[42]

More recent data have also shown the importance of the cysteine protease cathepsin L for POMC processing in IL and AL.[43] Interestingly, gene knockouts of 7B2 and PC2 in mutant mice initially showed distinct phenotypes in contrast to the predictions based on the biochemical pathway which places 7B2 upstream of PC2 maturation.[44–46] These discrepancies now appear to have been resolved based on an analysis of background strain effects on the single gene mutations.[47]

BOX 7.2

HOW DO WE KNOW ABOUT REGULATION OF POMC TRANSGENES IN MOUSE IL?

A transgene was constructed that contained 770 bp from the rat *Pomc* gene between nucleotides −706 and +64 ligated to the coding sequences for *E. coli* β-galactosidase. Pituitary sections from the resulting transgenic mice were stained for beta-galactosidase activity using X-gal to develop a blue color.[1] Under basal conditions, there was strong activity throughout the IL and weaker activity in a subpopulation of cells in the AL. Adrenalectomy of the mice removed negative glucocorticoid feedback on AL corticotropes, causing a marked increase in their β-galactosidase activity due to increased expression of the transgene. In contrast, inhibition of dopamine D2R signaling in the IL by haloperidol treatment greatly increased β-galactosidase activity in melanotropes without affecting AL corticotropes. An identical pattern of results for endogenous *Pomc* mRNA detected by *in situ* hybridization is demonstrated in the right panels of the figure. Similar transgenic experiments with alternative reporter cassettes and truncations or internal deletions of the *Pomc* regulatory elements confirmed that the minimal contiguous sequences necessary for melanotrope cell-specific expression, developmental regulation and dopamine inhibition are contained within the proximal sequences between nucleotide positions −323 and −34 of the *Pomc* promoter.[2,3]

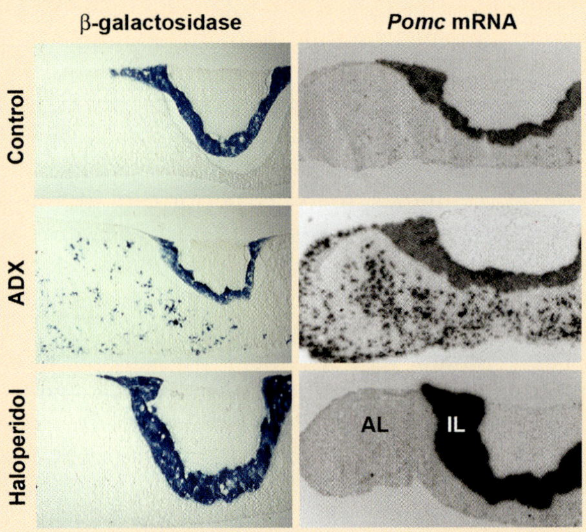

References

1. Hammer GD, Fairchild-Huntress V, Low MJ. Pituitary-specific and hormonally regulated gene expression directed by the rat proopiomelanocortin promoter in transgenic mice. *Mol Endocrinol.* 1990;4(11):1689–1697.
2. Liu B, Hammer GD, Rubinstein M, Mortrud M, Low MJ. Identification of DNA elements cooperatively activating proopiomelanocortin gene expression in the pituitary glands of transgenic mice. *Mol Cell Biol.* 1992;12(9): 3978–3990.
3. Rubinstein M, Mortrud M, Liu B, Low MJ. Rat and mouse proopiomelanocortin gene sequences target tissue-specific expression to the pituitary gland but not to the hypothalamus of transgenic mice. *Neuroendocrinology.* 1993;58(4): 373–380.

Finally, carboxypeptidase E (CPE)-deficient mice have elevated dopamine-insensitive POMC secretion from the IL accompanied by increased POMC mRNA and increased prohormone production.[48]

HYPOTHALAMIC REGULATORY FACTORS

Dopamine

There are three populations of hypothalamic neuroendocrine dopamine neurons in rodents: tuberoinfundibular dopaminergic (TIDA), tuberohypophysial dopaminergic (THDA) and periventricular-hypophysial dopaminergic (PHDA). The latter two groups were historically lumped together under the THDA rubric. Once thought not to contain the dopamine transporter (DAT), more recent studies indicate that all three populations do in fact express DAT and it is physiologically important.[49] Electrical simulation of the pituitary stalk causes synaptic release of dopamine from terminals in the IL that can be precisely measured by electrochemical techniques.[50] Tetrodotoxin prevents dopamine release by blocking action potentials. Dopamine overflow causes membrane hyperpolarization in melanotropes, and is augmented presynaptically by blockers of DAT and inhibition of D2 autoreceptors. Early patch-clamp studies of rat IL melanotropes demonstrated that dopamine application reversibly inhibited both low voltage activated (T-type) and high-voltage activated (N and L-type) Ca^{2+} currents, and therefore hyperpolarizes the resting membrane potential.[51]

BOX 7.3

HOW DO WE KNOW ABOUT NEUROANATOMIC MAPPING OF THE HYPOTHALAMIC INNERVATION OF IL MELANOTROPES?

Selective projections of A14 PHDA neurons to the rodent IL were mapped by anterograde tracing with *Phaseolus vulgaris* leukoagglutin (PHA-L) injected into the hypothalamic periventricular nucleus (PeV)[1] and retrograde tracing with the plant lectin wheat germ agglutinin (WGA) implanted in IL or lipophilic fluorescent DiI crystals applied to the IL.[2,3] Dopamine neurons were visualized by immunohistochemistry with an antibody directed against tyrosine hydroxylase. Arcuate dopamine neurons that project to the external zone of the median eminence were not double-labeled with tracer molecules under this paradigm. Similar studies were performed and identical results obtained in the cat, using horseradish peroxidase (HRP) as the retrograde tracer.[4] These results were also supported by surgical separation of the PeV from the medial basal hypothalamus followed by analysis of DOPA/DA ratios and MSH and prolactin secretion with or without electrical stimulation of the PeV.[5] It should be noted that the afferent projections to the amphibian IL, illuminated by the retrograde tracing technique with DiI, are much more extensive than the mammalian pattern, and include contributions from the preoptic nucleus, the supra- and retro-chiasmatic hypothalamus, and the brainstem, including the locus coeruleus.[6–8]

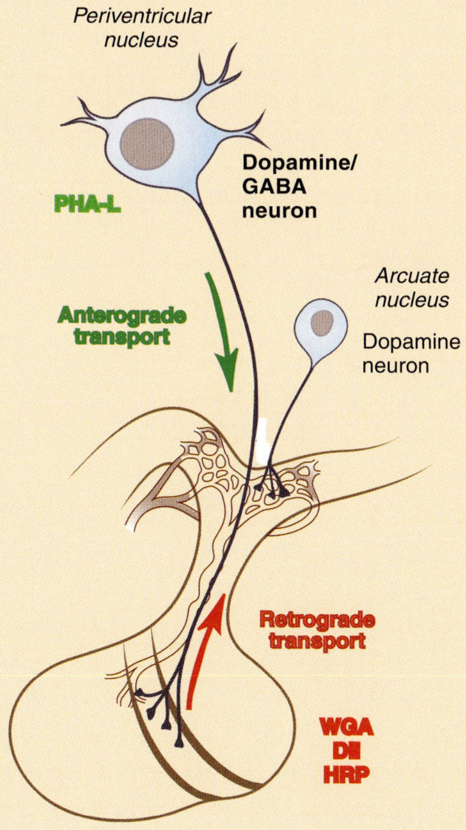

References

1. Goudreau JL, Falls WM, Lookingland KJ, Moore KE. Periventricular-hypophysial dopaminergic neurons innervate the intermediate but not the neural lobe of the rat pituitary gland. *Neuroendocrinology.* 1995;62(2):147–154.
2. Kawano H, Daikoku S. Functional topography of the rat hypothalamic dopamine neuron systems: retrograde tracing and immunohistochemical study. *J Comp Neurol.* 1987;265(2):242–253.
3. Makarenko IG, Ugrumov MV, Calas A. Axonal projections from the hypothalamus to the pituitary intermediate lobe in rats during ontogenesis: DiI tracing study. *Brain Res Dev Brain Res.* 2005;155(2):117–126.
4. Luppi PH, Sakai K, Salvert D, Berod A, Jouvet M. Periventricular dopaminergic neurons terminating in the neuro-intermediate lobe of the cat hypophysis. *J Comp Neurol.* 1986;244(2):204–212.
5. Goudreau JL, Lindley SE, Lookingland KJ, Moore KE. Evidence that hypothalamic periventricular dopamine neurons innervate the intermediate lobe of the rat pituitary. *Neuroendocrinology.* 1992;56(1):100–105.
6. Artero C, Fasolo A, Franzoni MF. Multiple sources of the pituitary pars intermedia innervation in amphibians: a DiI retrograde tract-tracing study. *Neurosci Lett.* 1994;169(1-2):163–166.
7. Tuinhof R, Artero C, Fasolo A, Franzoni MF, Ten Donkelaar HJ, Wismans PG, et al. Involvement of retinohypothalamic input, suprachiasmatic nucleus, magnocellular nucleus and locus coeruleus in control of melanotrope cells of *Xenopus laevis*: a retrograde and anterograde tracing study. *Neuroscience.* 1994;61(2):411–420.
8. Ubink R, Tuinhof R, Roubos EW. Identification of suprachiasmatic melanotrope-inhibiting neurons in *Xenopus laevis*: a confocal laser-scanning microscopy study. *J Comp Neurol.* 1998;397(1):60–68.

Of the five defined dopamine receptor subtypes, only the D2R long form is highly expressed on IL melanotropes. The classically defined signaling pathway for this G-protein coupled receptor involves $G_i\alpha$-mediated downregulation of adenylate cyclase and decreased production of the second messenger cAMP. In addition, $G_i\beta\gamma$-subunits activate inwardly rectifying K^+ currents that result in membrane hyperpolarization.[52] More recently, D2R activation in melanotropes has also been linked to the phosphoinositide pathway through diacylglycerol kinase beta, phospholipase C β4, and protein kinase C α.[53]

Dopamine Effects on Gene Expression in melanotropes

Dopamine regulates not only POMC gene expression in the IL but also the mRNAs for several of its important biosynthetic enzymes, including PC2, peptidylglycine α-amidating monooxynegase (PAM) and carboxypeptidase H.[54] Dopamine agonists downregulate and antagonists upregulate both PC1 and PC2 mRNAs, respectively, in rat melanotropes measured by *in situ* hybridization and Northern blots.[55] Dopamine and D2Rs also downregulate expression of high-threshold voltage-gated Ca^{2+} currents in IL melanotropes during postnatal development.[56] This effect can be mimicked *in vitro* and data showed that a signaling pathway involving the D2R and c-fos expression was involved in the depression of L-type Ca^{2+} channel activity and subsequent suppression of secretory activity.[57]

Dopamine Effects on Melanotrope Growth and Proliferation

Chronic treatment of rats with the dopamine agonist bromocriptine or the antagonist haloperidol results in decreases or increases, respectively, in the number of cell layers in the IL. These changes were paralleled by decreases or increases in the rate of proliferation measured by [^{3}H]-thymidine uptake.[58] In adult rats, bromocriptine also decreases the volume of the IL by a decrease in mean cell volume of melanotropes.[59]

Gary and Chronwall were the first to provide *in vivo* evidence for an antiproliferative role of dopamine on IL melanotropes.[60] They elegantly demonstrated a strong indirect correlation between DA innervation of the IL and melanotrope mitotic index during the early postnatal period, and, furthermore, destruction of the DA terminals by 6-hydroxydopamine in neonatal rats led to a persistently high mitotic index compared to control animals. Despite these strong data implicating dopamine in control of melanotrope growth, we have not observed melanotrope hyperplasia or adenomas in dopamine D2R knockout mice,[61] possibly due to compensatory adaptations.

Dopamine Effects on Melanotrope Peptide Secretion

Early studies focused on the role of cAMP as the sole mediator of the various stimulatory and inhibitory signals converging on melanotropes.[62,63] In particular, both CRH and catecholamines increase intracellular cAMP and stimulate POMC peptide secretion, while dopamine acting through the D2R decreases cAMP. CRH binding sites on IL melanotropes have identical pharmacological characteristics to those expressed on AL corticotropes.[64] We have found that a novel mouse IL melanotrope cell line generated in our lab[65] expressed mRNA for CRH-R1 receptors and high-affinity binding sites for [^{3}H]-nemonapride consistent with D2R (unpublished data). CRH dose-dependently stimulated cAMP accumulation, and the D2R agonist bromocriptine potently inhibited the CRH-induced stimulation in these cells (Fig. 7.1). However, the historical oversimplification of opposing receptor signaling limited to effects on intracellular cAMP levels has been revised by other lines of investigation.

Melanotropes exhibit discrete oscillations of $[Ca^{2+}]_i$ that are linked to their secretory capacity. These calcium waves are generated by the firing of Na^+- and Ca^{2+}-dependent action potentials,[66] and are also modulated by the extracellular Ca^{2+}-sensing receptor.[67] cAMP failed to increase $[Ca^{2+}]_i$ in melanotropes, but did stimulate the depolarization-induced secretion of storage vesicles.[68] The first wave of secretion probably represents exocytosis of the release-ready vesicles closely apposed to the plasma membrane, while the second wave requires activation of cAMP-dependent protein kinase A (PKA) and elevated $[Ca^{2+}]_i > 0.8\ \mu M$. Similar to other neuroendocrine cells, the Ca^{2+}-mediated exocytosis of secretory vesicles involves the Ca^{2+} sensor synaptotagmin, SNAP25, munc18-1, and the small GTPase Rab3A.[69–71]

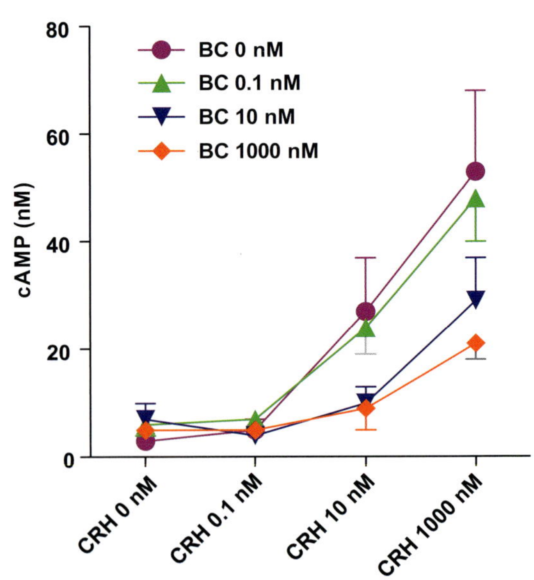

FIGURE 7.1 Competing actions of CRH1R and D2R activation on intracellular cAMP levels in an immortalized mouse IL melanotrope cell line. Triplicate wells of melanotrope cells were incubated with media containing the indicated concentrations of CRH and bromocriptine (BC) for 30 min and cAMP levels were quantified in the cell extracts by RIA. Data are means ± SD (K. Nusser and M.J. Low, unpublished data).

GABA

A central innervation of the rat IL by GABAergic neurons was firmly established in 1982.[72] GABA is systematically co-localized with TH in the PVDA fibers innervating the mammalian IL,[73,74] and GABA-immunoreactive terminals make direct contact with melanotropes. These contacts possess the ultrastructural features of true synapses, and have been recapitulated and studied *in vitro* with co-cultures of hypothalamic neurons and dissociated melanotropes from embryonic rats.[75,76] There are, however, many differences in the number and arrangement of synaptic contacts onto individual melanotropes and glia elements in the IL.[77]

Unlike DA, GABA has biphasic effects on MSH secretion from rat melanotropes mediated by the combined actions of ionotropic $GABA_A$ and inhibitory metabotropic $GABA_B$ receptors.[78] The initial stimulatory effects of $GABA_A$ receptor activation on MSH secretion from *Xenopus* melanotropes has been shown to result from Cl^- efflux (net inward current), due to the prevailing Cl^- ion equilibrium potential in melanotropes, that in turn triggers voltage-gated inward Ca^{2+} and Na^+ currents.[79] The neurosteroid allopregnanolone (3 alpha-hydroxy-5 α-pregnan-20-one) in physiological concentrations can modulate $GABA_A$ receptor evoked currents in patch-clamped rat melanotropes, but did not alter the GABA stimulation of MSH release from bath-cultured NILs.[80,81] Moreover, rat NIL tissue has the capacity to derivitize progesterone to allopregnanolone due to the presence of the enzyme 3 α-hydroxysteroid oxidoreductase.[82]

$GABA_B$ receptor expression is developmentally regulated and completed only after the IL is fully innervated.[83] mRNA transcripts corresponding to the two $GABA_B$ receptor subunits were detected at embryonic age e18, but functional activity was delayed until postnatal day 15, suggesting dual mechanisms controlling transcription and translation or post-translational maturation of subunits. Both subunits of the $GABA_B$ heterodimer are necessary for receptor function, as demonstrated by antisense knockdown in a mouse melanotrope cell line.[65] $GABA_B$ receptors reduce melanotrope membrane excitability by inhibiting voltage activated inward Ca^{2+} channels.[65]

ADDITIONAL REGULATORY TRANSMITTERS AND NEUROMODULATORY PEPTIDES

Although dopamine and GABA are the primary neurotransmitters released from terminals innervating the IL in most mammals, there are a large number of additional signaling molecules that regulate melanotrope function. Some of these factors are released from nerve terminals, others are secreted from pituitary cells, and many differ among species (Table 7.1). NPY coexists with DA and GABA in *Xenopus*, but not mammalian, nerve terminals, and has an independent slow but sustained inhibitory effect on MSH secretion mediated through NPY1 and NPY5 receptors, in contrast to the faster actions of DA and GABA.[84,85]

Catecholamines stimulate rat melanotropes through α2- and β-adrenergic receptors.[86] Although adult humans do not have a distinct IL, there is still a clear catecholaminergic control of MSH secretion from the pituitary that is independent of ACTH secretion. Isoproterenol stimulated and propranolol blocked MSH release.[87] There is limited evidence for noradrenergic innervation of the IL from a central brain origin based on stalk section experiments.[88] More likely, circulating catecholamines of adrenal medullary or other sympathetic system terminals are responsible for the rise in circulating β-endorphin and MSH in response to certain forms of stress.[89]

Serotonin stimulates melanotropes through 5-HT2 receptors, although it appears that much of the amine may not be directly synthesized in specific neurons, but rather recycled through the DAT back into dopamine terminals in the IL so that it acts as a false transmitter (reviewed in Saland[90]). Glutamate can stimulate action potential firing in neonatal rat IL melanotropes via AMPA receptors,[91] and less potently by NMDA receptors.[92] *Xenopus* melanotropes are stimulated by acetylcholine via a homolog of the M3 muscarinic receptor and a Ni^{2+}-sensitive Ca^{2+} current coupled to activation of the phospholipase C/PKC signaling pathway.[93] However, porcine melanotropes are stimulated by acetylcholine via a nicotinic receptor subtype and a depolarizing inward Na^+ current.[94]

CRH stimulates release of β-endorphin from rat IL melanotropes, and the effect can be blocked with the antagonist α-helical CRH, dopamine, or the benzodiazepine adinazolam that interacts with the $GABA_A$ receptor.[95] Urocortin 1 (UCN1) immunoreactivity occurs in the *Xenopus* magnocellular nucleus, with fiber projections in the internal zone of the ME to the neural lobe forming neurohemal axon terminals. *In vitro*, UCN1 stimulates MSH release from *Xenopus* melanotropes, suggesting that UCN1 released into the neural lobe can then flow through a portal plexus into IL and function as a stimulatory neurohormone.[96] CRH, urotensin I and sauvagine all stimulate the release of MSH from superfused teleost melanotropes.[97]

Amphibians and teleosts have a direct neural innervation of their IL by TRH-positive fibers, and in both vertebrate classes TRH is a potent secretagog for MSH acting through the receptor subtype xTRHR3.[98] PACAP acting on the VPAC1 receptor is also stimulatory for

TABLE 7.1 Regulatory Signals and Receptors on Vertebrate Melanotropes

Ligand	Source	Receptor	Action	Class
Dopamine	Neural	D2R	Inhibitory	Mamm[a], Amph[b], Tele[c]
GABA	Neural	$GABA_A$	Biphasic	Mamm, Amph, Tele
GABA	Neural	$GABA_B$	Inhibitory	Mamm, Amph, Tele
NPY	Neural	NPY1, NPY5	Inhibitory	Amph
5-HT	Neural	5-HT2	Stimulatory	Amph
	Neural (false transmitter)	5-HT2	Stimulatory	Mamm
TRH	Neural	xTRH3	Stimulatory	Amph, Tele
Glutamate	?	AMPA	Stimulatory	Mamm
Acetylcholine	?	Muscarinic	Stimulatory	Amph
		Nicotinic	Stimulatory	Mamm
NE	Neural	β2	Stimulatory	Amph
	Peripheral	β2, α2	Stimulatory	Mamm
CRH	Portal vessels	CRH1R	Stimulatory	Mamm
CRH, UCN, Sauvagine	Neural via posterior lobe	CRH1R	Stimulatory	Amph, Tele
PACAP/VIP	Neural	VPAC1R	Stimulatory	Amph
BDNF	Autocrine	TrkB	Trophic for DA axons?	Mamm, Amph

[a]*Mammalian.*
[b]*Amphibian.*
[c]*Teleost.*

Xenopus POMC expression, POMC biosynthesis and MSH secretion.

Several components of the endogenous opioid peptide system are expressed in mammalian ILs, including proenkephalin, prodynorphin, and kappa-opioid receptors.[99,100] Unlike POMC, prodynorphin expression in melanotropes is not regulated by dopamine, but prodynorphin mRNA levels are upregulated three- to four-fold in postpartum females.[101] Mu- and delta-opioid agonists both stimulated the *in vitro* release of POMC peptides from rat primary cultures, although kappa agonists had no effect.[102]

Xenopus melanotropes highly express the gene encoding the SS2 somatostatin prohormone variant and fully process the encoded prophormone to the mature SS2 variant peptide, but its functional relevance is not known.[103] SST-14 is also expressed in rat IL in vimentin-positive bipolar stellate-shaped glia, which are most abundant developmentally and then tend to regress in the adult.[104]

CENTRAL NEUROENDOCRINE REGULATION OF THE PARS INTERMEDIA

Enhanced secretory activity of rat IL melanotropes after unpredictable stress was noted by Pollard in 1983.[105] Stress causes the release of POMC peptides from both AL and IL of rats; however, the source can be differentiated by pretreatment with either dexamethasone to inhibit corticotropes, or bromocriptine to inhibit melanotropes. This approach showed that stress-induced ACTH is derived from the AL, MSH is from the IL, and β-endorphin comes from both lobes.[106] Furthermore, β-adrenergic blockade effectively inhibited stress-induced MSH and β-endorphin secretion from the IL. Swim stress-mediated secretion of β-endorphin into the circulation results from reduced DA inhibition of melanotropes.[107] Cold stress causes an increase in CRH receptor binding specifically on IL melanotropes and not AL corticotropes that is mimicked qualitatively by treatment with dopamine agonists.[108]

Restraint stress reduces the activity of THDA neurons, estimated by the ratios of DOPAC/DA following administration of the decarboxylase inhibitor NSD 1015, and simultaneously increases plasma MSH levels.[109] Central mechanisms of stress to reduce DA output from the hypothalamus include 5-HT effects mediated by central 5-HT2 receptors based on pharmacological studies.[110] Inhibition of endogenous brain histamine synthesis attenuated basal and ether-stress induced MSH responses[111] Centrally administered histamine or H1-R and H2-R agonists stimulate IL secretion of POMC peptides at least in part via a catecholaminergic pathway involving β-receptors because the effect is blocked peripherally by propranolol.[112] Bromocriptine can also block the release of MSH from the IL induced by ether stress or i.c.v. administration of histamine.[111]

Several other central modifiers of IL MSH secretion converge on the final common pathway of THDA neurons (Fig. 7.2). For example, i.c.v. administration of bombesin or gastrin-releasing peptide in rats increases THDA activity and decreases plasma concentrations of MSH.[113] In contrast, the κ-opioid receptor agonist U-50488 decreases the activity of DA projections to the IL and increases plasma MSH.[114] Similar central opioid effects were suggested in frogs by the action of the antagonist naloxone to prevent dark-background adaptation without any direct actions on the NIL.[115] These data are consistent with a mechanism whereby endogenous opioids inhibit the PVDA neurons that tonically inhibit MSH secretion from melanotropes. Exposure to light affects differentiation of the supraoptic DA neurons in frogs that receive a glutamatergic input via the retino-hypothalamic tract.[116] Finally, in rats the ovarian steroids estrogen and progesterone inhibit the activity of all three populations of hypothalamic neuroendocrine dopamine neurons.[117]

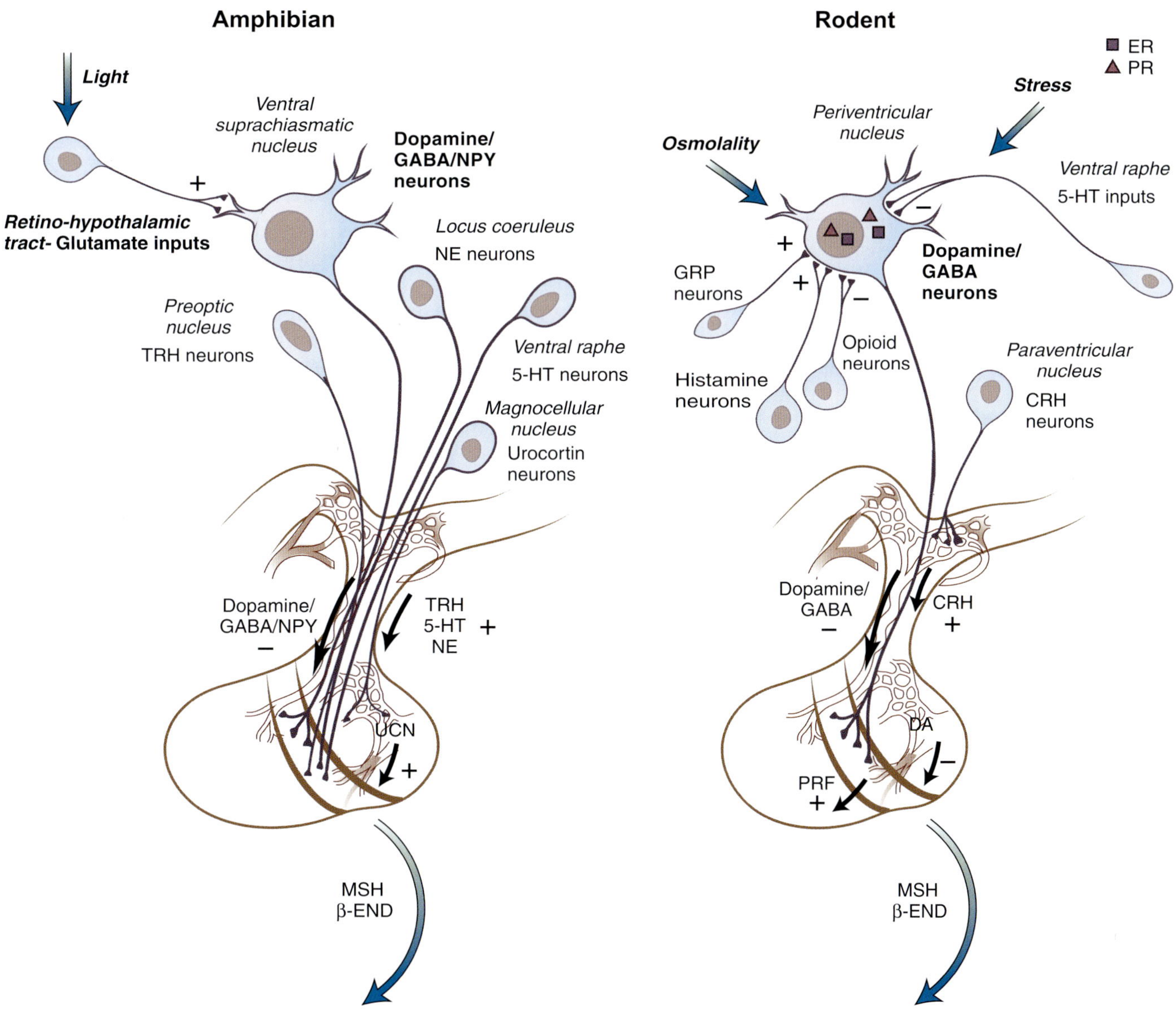

FIGURE 7.2 Comparison of the neuroendocrine regulation of IL melanotropes in amphibians and mammals. Multiple inhibitory and stimulatory factors modulate IL melanotrope activity in both amphibian and mammalian species, but there is a greater diversity of direct neural innervation to the IL from hypothalamic and midbrain nuclei in amphibians. Abbreviations: β-END, beta-endorphin; CRH, corticotropin-releasing hormone; DA, dopamine; GABA, gamma-amino butyric acid; GRP, gastrin-releasing peptide; 5-HT, 5-hydroxytryptamine (serotonin); NE, norepinephrine; NPY, neuropeptide Y; PRF, prolactin-releasing factor; TRH, thyrotropin-releasing hormone; UCN, urocortin.

PHYSIOLOGICAL FUNCTIONS OF THE PARS INTERMEDIA

Amphibians and Fish

Skin darkening during adaptation to a dark background occurs in both amphibians and teleosts. Dark-background adaptation-induced plasticity in the neuroendocrine system includes increased POMC expression, decreased completion of processing, increased expression of the processing enzyme PC1, increased expression of the p24 protein family, which functions at the endoplasmic reticulum–Golgi interface in the early secretory pathway to permit proper cargo transport and processing, and increased TRH receptors. There is hypertrophy, but not hyperplasia, of the frog melanotropes. Dark adaptation also leads to increased expression of the synaptic vesicle exocytosis protein homolog Xunc18,[69] and causes a distinct increase in the secretion of MSH from superfused NILS compared to secretion of the less bioactive des-acetylMSH from white-background adapted frogs.[118] *Xenopus* have two populations of melanotropes, highly active secretory cells and hormone storage cells.[119] There is a long-term balance between the two that is regulated by hypothalamic factors.[120] MSH causes pigment dispersion in dermal melanophores, and this response is counteracted by a second neuropeptide, melanin-concentrating hormone (MCH), in teleosts only.[121] Expression of the latter is upregulated in the neurohypophysis of white-background adapted eels.[122] However, the effects of MSH on pigmentation in mammals and birds is not dependent on IL melanotropes, but rather due to local autocrine/paracrine expression and secretion of MSH in skin and its follicular structures responsible for hair and feather growth.

Mammals

Lactation

Abundance evidence suggests that PHDA neuron-derived DA released into the IL has a role in regulating prolactin secretion from AL lactotropes.[123–125] Indeed, capillaries of the IL drain through a portal venous plexus into the AL, providing a route for dopamine overflow or paracrine secretion of IL peptides to modulate cells of the AL.[126]

Early physiological experiments suggested that IL-derived MSH mediated in part the acute suckling-induced stimulation of PRL secretion from AL lactotropes.[127] There are data in the mouse showing that estrogen upregulates both IL POMC expression and AL melanocortin 3 receptor (MC3R) expression, and that MSH acts on MC3R to stimulate prolactin release and lactotrope proliferation.[128]

However, there is controversy regarding whether MSH from the IL can access the AL to physiologically influence prolactin secretion from the suckling stimulus.[129] Similarly, it has been difficult to prove the existence of a distinct prolactin-releasing factor (PRF) produced and secreted from IL melanotropes.[130] We generated POMC-SV40 large T antigen transgenic mice that developed IL melanotrope tumors associated with hyperprolactinemia, and showed that extracts of these tumors exhibited PRF activity that was not related to products of the POMC gene.[131] However, an immortalized melanotrope cell line mIL39 derived from a T antigen-induced IL tumor did not secrete any PRF activity.[132] Other putative PRFs include heparin-binding growth factors,[133] the peptide intermedin (not to be confused with "intermedin" that was once used as a name for MSH)[134] and the endogenous tetrahydroisoquinoline salsolinol.[135,136]

Maintenance of Normal Blood Pressure and Responses to Osmotic Stimulation

An association has been demonstrated between IL POMC and D2R expression and hypertension through comparisons of the selectively bred, spontaneously hypertensive rat (SHR) and the normotensive Wistar Kyoto (WKY) rat.[137,138] This finding does not, however, establish a cause and effect relationship. Treatment of rats with the D2R agonist bromocriptine suppresses IL function and plasma gamma-MSH, leading to salt-sensitive hypertension. Blood pressure can be normalized by minipump administration of the stable melanocortin analog NDP-γ-MSH.[139] Furthermore, mutant mice with a deficiency in MC3R or targeted disruption of the prohormone convertase PC2 both develop salt-sensitive hypertension. Infusion of γ-MSH either peripherally or centrally via an i.c.v. cannula ameliorated the hypertension in PC2, but not MC3R, knockout mice, suggesting that γ-MSH acting through the MC3R is responsible for the pathological alterations.[140]

Salt-loading in Sprague-Dawley rats leads to increased expression of CRH and UCN1 in hypothalamic magnocellular neurons, and a corresponding increase in CRHR1 receptors and binding in the IL.[141] Salt loading with 2% saline causes a decrease in rat IL POMC mRNA levels via a dopamine-dependent mechanism, but the effect appeared to be primarily post-transcriptional and therefore likely due to altered mRNA stability.[142] In contrast, excessive drinking of a 5% glucose solution by mice causes an increase in POMC synthesis, with an opposite effect observed in response to dehydration.[143] Possibly related to the previous observations, wild gerbils captured in Algeria showed distinct seasonal differences in the melanotropes of their ILs. In the drought months of summer, when only dry food is available, the secretory activity of the cells was depressed, while the cells were highly functional morphologically during the winter or from animals that were fed hydrated food *ad libitum*.[144]

PATHOPHYSIOLOGY OF THE PARS INTERMEDIA

Cushing's Disease in Horses and Dogs

Equine Cushing's disease or pituitary pars intermedia dysfunction (PPID) is a spontaneous and progressive disorder of horses that is caused by degeneration of the PHDA neurons.[145] In some respects it may be a model for the spontaneous degeneration of nigrostriatal neurons that causes Parkinson's disease in humans, but at a frequency less than one-tenth that of equine PPID. With increasing age, the histologic findings progress from focal hypertrophy or hyperplasia to diffuse adenomatous hyperplasia, microadenomas and finally macroadenomas with increased plasma ACTH concentrations in response to the mixed dopamine receptor antagonist domperidone. POMC processing is substantially altered, with increased proportions of full-length β-endorphin(1–31) and desacetyl forms of β-endorphin (1–27) and (1–26).[146]

Canine Cushing's disease is much more frequent than the disorder in humans. One study showed that of 21 spontaneous pituitary adenomas producing hyperadrenocorticism in dogs, 6 were of pars intermedia origin and 15 were located in the pars distalis.[147] Immunoreactive ACTH and β-endorphin were found in most tumors, regardless of location, while MSH staining was generally confined to cells of the IL tumors. The canine IL appears to be a mosaic of two cytologically and biochemically distinct cell types resembling more closely either AL corticotropes or IL melanotropes of other mammals.[148] Consistent with the premise that dog corticotropes and melanotropes may be less distinct from each other than in other mammalian species, a 7-day treatment with domperidone did not increase circulating plasma MSH levels, but did augment CRH stimulation of ACTH release and partially recovered CRH stimulation of ACTH release in the presence of dexamethasone suppression.[149] However, in another study, an acute injection of haloperidol raised plasma levels of both ACTH and MSH in dogs.[150] HPLC and RIA analysis showed that canine IL tissue extracts contained mostly monoacetyl-α-MSH while AL extracts mostly desacetyl-α-MSH. However, primary cultures of canine AL contained much higher proportions of monoacetyl- and diacetyl-α-MSH, suggesting a disinhibition of the responsible acetyltransferase *in vitro*.[151]

There remains controversy regarding the nature of IL-like melanotropes in human adults. While developing human embryos have a distinct IL, it becomes vestigial later in life. A distinct IL secretion of MSH in human infants declines rapidly in the third trimester and following birth.[152] However, there are still IL-like cells scattered throughout the adult human AL based on their co-expression of ACTH with MSH and acetylated β-endorphin.[153]

MRI imaging studies have identified non-contrast-enhancing mass lesions associated with pituitary enlargement and interposed between the AL and neurohypophysis of some patients with Prop1 gene mutations.[154] The position is suggestive of an IL origin for the masses. Lamberts and colleagues once hypothesized that human Cushing's adenomas could be subdivided into those of presumptive AL vs IL origin based on accompanying hyperprolactinemia and bromocriptine suppression of cortisol levels in the latter type,[155] but this theory has not been supported in later studies.[156–158] Immunoreactive MSH is common in human corticotrope adenomas, is usually desacetylated, and does not preferentially locate in tumors near the neural lobe.[159] Similarly, normal fetal and adult human AL corticotropes often contain desacetyl-MSH regardless of their location in the pituitary gland.[160] No neural axons have been found in human Cushing's disease adenomas, consistent with AL and not IL origins.[161]

The IL is a Useful Experimental Model for Neuroendocrine Tumorigenesis

Regardless of the nature or location of melanotropes in humans, the IL has proven to be a useful experimental model for pituitary tumorigenesis and has illuminated the importance of cell-cycle control mechanisms in adenoma formation.[162] Mutant mice with knockouts in the genes for the cyclin-dependent kinase inhibitors KIP1/p27 and INK4c/p18, or the tumor suppressor genes Rb and p53, develop IL hyperplasia and adenomas with extremely high penetrance. Acceleration of pituitary tumor development is evident in compound Rb;p53 knockout mice, Rb;ARF tumor suppressor knockout mice, Rb;p27 knockout mice, and p18;p27 knockout mice.[163–165] E2F transcription factor overexpression is necessary, but not sufficient, to account for tumor formation in Rb heterozygous mice.[166] Increased expression of cyclin E produced by loss of the transcription factor Brg1 can also act synergistically with p27-deficiency to increase the frequency and size of pituitary adenomas.[167] Interestingly, the peculiar propensity for mice to develop IL adenomas at high frequency may result in part from inherent background strain differences. The embryonic stem cells used to generate most gene knockout mice are derived from 129 substrain mice that already have hyperplastic ILs and altered melanotrope function compared to other commonly used strains.[47,168]

IL-specific rescue of Rb expression in Rb heterozygous mice prevents IL tumors, but results in delayed expression of AL tumors possibly related to a putative inhibitory effect of MSH on AL cell proliferation.[169] Moreover, AL tumors of thyrotrope origin are preferentially induced in compound mutant mice that are deficient in p18 and the glycoprotein α-subunit gene.[170] Spontaneous human ACTH-secreting adenomas showed a particularly marked decrease in p27 expression, without mutations in the gene, compared to other hormonal types of pituitary adenomas, suggesting that the cyclin-dependent signaling cascades are also relevant in human pituitary pathophysiology.[162,171]

SUMMARY AND FUTURE QUESTIONS

This chapter has emphasized the remarkable diversity in the neural innvervation, regulation and function of the IL across vertebrate phylogeny, and even between families within one class. Melanotropes in the IL of teleosts and amphibians receive direct synaptic contacts from multiple CNS nuclei, and their secretory activity is tightly regulated by ambient lighting and background luminosity to directly alter skin pigmentation by hormonal actions of MSH on melanocytes. Birds and mammals utilize similar signaling and biochemical pathways to alter the pigmentation of their skin and skin appendages, but the source of MSH is primarily paracrine and not hormonal from the pituitary gland. Thus, the essential physiological functions of IL melanotropes in mammals are relatively obscure.

Several important questions remain for future research in this area of neuroendocrinology. Foremost among them are to fully define the importance of γ-MSH and the MC3R in regulation of blood pressure and renal salt excretion, and determine the relative importance of pituitary vs CNS as the source and site of action of γ-MSH. Adult pituitary stem cells have only recently been identified, and the optimal markers for these cells remain controversial. It will be of great interest to discover how stem cells contribute to the development and possible remodeling of the adult IL. Finally, there are still many gaps in our knowledge concerning the differential use of specific transcription factors and co-activator/co-repressor complexes in the regulation of *POMC* gene expression in AL corticotropes and IL melanotropes.

Acknowledgments

Research in the author's laboratory is supported by NIH grants DK066604 and DK068400.

References

1. Lerner AB. The intermediate lobe of the pituitary gland: introduction and background. *Ciba Found Symp.* 1981;81:3–12.
2. Ratter SJ, McLoughlin L, Gillies G, Clement-Jones V, Hope J, Rees LH. Pars intermedia peptides: studies in adult humans. *Ciba Found Symp.* 1981;81:224–243.
3. Silman RE, Street C, Holland D, Chard T, Falconer J, Robinson JS. The pars intermedia and the fetal pituitary–adrenal axis. *Ciba Found Symp.* 1981;81:180–195.
4. Eipper BA, Mains RE. Existence of a common precursor to ACTH and endorphin in the anterior and intermediate lobes of the rat pituitary. *J Supramol Struct.* 1978;8(3):247–262.
5. Mains RE, Eipper BA. Synthesis and secretion of corticotropins, melanotropins, and endorphins by rat intermediate pituitary cells. *J Biol Chem.* 1979;254(16):7885–7894.
6. Moellmann G, McGuire J, Lerner AB. Ultrastructure and cell biology of pigment cells. Intracellular dynamics and the fine structure of melanocytes with special reference to the effects of MSH and cyclic AMP on microtubules and 10-nm filaments. *Yale J Biol Med.* 1973;46(5):337–360.
7. Harris JI, Lerner AB. Amino-acid sequence of the alpha-melanocyte-stimulating hormone. *Nature.* 1957;179(4574):1346–1347.
8. Lerner AB. The discovery of the melanotropins. A history of pituitary endocrinology. *Ann NY Acad Sci.* 1993;680:1–12.
9. Rizzoti K, Lovell-Badge R. Early development of the pituitary gland: induction and shaping of Rathke's pouch. *Rev Endocr Metab Disord.* 2005;6(3):161–172.
10. Bazina M, Stefanovic V, Bozanic D, Saraga-Babic M. Ultrastructural and immunohistochemical characteristics of developing human pituitary gland. *Acta Histochem.* 2007;109(5):366–376.
11. Liu NA, Ren M, Song J, Rios Y, Wawrowsky K, Ben-Shlomo A, et al. *In vivo* time-lapse imaging delineates the zebrafish pituitary proopiomelanocortin lineage boundary regulated by FGF3 signal. *Dev Biol.* 2008;319(2):192–200.
12. Guner B, Ozacar AT, Thomas JE, Karlstrom RO. Graded hedgehog and fibroblast growth factor signaling independently regulate pituitary cell fates and help establish the pars distalis and pars intermedia of the zebrafish adenohypophysis. *Endocrinology.* 2008;149(9):4435–4451.
13. Raetzman LT, Cai JX, Camper SA. Hes1 is required for pituitary growth and melanotrope specification. *Dev Biol.* 2007;304(2): 455–466.
14. Pulichino AM, Vallette-Kasic S, Tsai JP, Couture C, Gauthier Y, Drouin J. Tpit determines alternate fates during pituitary cell differentiation. *Genes Dev.* 2003;17(6):738–747.
15. Tong Y, Pelletier G. Ontogeny of proopiomelanocortin (POMC) gene expression in the intermediate lobe of the rat pituitary gland. *Neuropeptides.* 1990;16(4):173–179.
16. Elkabes S, Loh YP, Nieburgs A, Wray S. Prenatal ontogenesis of pro-opiomelanocortin in the mouse central nervous system and pituitary gland: an *in situ* hybridization and immunocytochemical study. *Brain Res Dev Brain Res.* 1989;46(1):85–95.
17. Japon MA, Rubinstein M, Low MJ. *In situ* hybridization analysis of anterior pituitary hormone gene expression during fetal mouse development. *J Histochem Cytochem.* 1994;42(8): 1117–1125.
18. Makarenko IG, Ugrumov MV, Calas A. Axonal projections from the hypothalamus to the pituitary intermediate lobe in rats

during ontogenesis: DiI tracing study. *Brain Res Dev Brain Res.* 2005;155(2):117–126.

19. Jenks BG, Kidane AH, Scheenen WJ, Roubos EW. Plasticity in the melanotrope neuroendocrine interface of *Xenopus laevis. Neuroendocrinology.* 2007;85(3):177–185.
20. Nakakura T, Suzuki M, Watanabe Y, Tanaka S. Possible involvement of brain-derived neurotrophic factor (BDNF) in the innervation of dopaminergic neurons from the rat periventricular nucleus to the pars intermedia. *Zoolog Sci.* 2007;24(11):1086–1093.
21. Hopker VH, Kjaer B, Varon S. Dopaminergic regulation of BDNF content in the pituitary intermediate lobe. *NeuroReport.* 1997;8(5):1089–1093.
22. Lopez-Santiago LF, Gomora JC, Cota G. Postnatal decrease of sodium current density in rat pituitary melanotropes following the onset of dopaminergic innervation. *Neurosci Lett.* 2001;315(3):137–140.
23. Lugo DI, Pintar JE. Ontogeny of basal and regulated proopiomelanocortin-derived peptide secretion from fetal and neonatal pituitary intermediate lobe cells: melanotrophs exhibit transient glucocorticoid responses during development. *Dev Biol.* 1996;173(1):110–118.
24. Antakly T, Sasaki A, Liotta AS, Palkovits M, Krieger DT. Induced expression of the glucocorticoid receptor in the rat intermediate pituitary lobe. *Science.* 1985;229(4710):277–279.
25. Seger MA, van Eekelen JA, Kiss JZ, Burbach JP, de Kloet ER. Stimulation of pro-opiomelanocortin gene expression by glucocorticoids in the denervated rat intermediate pituitary gland. *Neuroendocrinology.* 1988;47(4):350–357.
26. Pfeiffer A, Barden N. Glucocorticoid receptor gene expression in rat pituitary gland intermediate lobe following ovariectomy. *Mol Cell Endocrinol.* 1988;55(2-3):115–120.
27. Ruhle HJ, Schnabel C, Lausch A. Autoradiographic demonstration of glucocorticoid receptors in the intermediate lobe of the rat pituitary transplanted to the anterior eye chamber. *Exp Clin Endocrinol.* 1989;94(1-2):141–150.
28. Schimchowitsch S, Plante M, Klein MJ, Stoeckel ME. Ontogenesis of the rabbit pituitary intermediate lobe. An ultrastructural and immunocytochemical study. *Anat Embryol (Berl).* 1993;187(1):87–97.
29. Schimchowitsch S, Plante M, Kienlen P, Felix JM, Koch B, Stoeckel ME. Glucocorticoids, but not dopamine, negatively regulate the melanotrophic activity of the rabbit pituitary intermediate lobe. *J Neuroendocrinol.* 1994;6(4): 385–390.
30. Krylyshkina O, Chen J, Mebis L, Denef C, Vankelecom H. Nestin-immunoreactive cells in rat pituitary are neither hormonal nor typical folliculo-stellate cells. *Endocrinology.* 2005;146(5):2376–2387.
31. Gary KA, Sands SA, Chronwall BM. Glial-like cells of the rat pituitary intermediate lobe change morphology and shift from vimentin to GFAP expression during development. *Intl J Dev Neurosci.* 1995;13(6):555–565.
32. Beatty DM, Morris SJ, Chronwall BM. Heterogeneity in POMC expression among explanted melanotropes decreases with time in culture and bromocriptine treatment. *Peptides.* 1998;19(4): 659–665.
33. Hatfield JM, Daikh DI, Adelman JP, Douglass J, Bond CT, Allen RG. *In situ* hybridization detection of marked differences in pre-proopiomelanocortin messenger ribonucleic acid content of individual corticotropes and melanotropes. *Endocrinology.* 1989;124(3):1359–1364.
34. Chronwall BM, Dickerson DS, Huerter BS, Sibley DR, Millington WR. Regulation of heterogeneity in D2 dopamine receptor gene expression among individual melanotropes in the rat pituitary intermediate lobe. *Mol Cell Neurosci.* 1994;5(1): 35–45.
35. Hammer GD, Fairchild-Huntress V, Low MJ. Pituitary-specific and hormonally regulated gene expression directed by the rat proopiomelanocortin promoter in transgenic mice. *Mol Endocrinol.* 1990;4(11):1689–1697.
36. Liu B, Hammer GD, Rubinstein M, Mortrud M, Low MJ. Identification of DNA elements cooperatively activating proopiomelanocortin gene expression in the pituitary glands of transgenic mice. *Mol Cell Biol.* 1992;12(9):3978–3990.
37. Bicknell AB. The tissue-specific processing of pro-opiomelanocortin. *J Neuroendocrinol.* 2008;20(6):692–699.
38. Zhou A, Bloomquist BT, Mains RE. The prohormone convertases PC1 and PC2 mediate distinct endoproteolytic cleavages in a strict temporal order during proopiomelanocortin biosynthetic processing. *J Biol Chem.* 1993;268(3):1763–1769.
39. Zhu X, Lindberg I. 7B2 facilitates the maturation of proPC2 in neuroendocrine cells and is required for the expression of enzymatic activity. *J Cell Biol.* 1995;129(6):1641–1650.
40. Allen RG, Hatfield JM, Stack J. Post-translational processing of pro-opiomelanocortin (POMC)-derived peptides during fetal monkey pituitary development. I. Adrenocorticotropin (ACTH) and alpha-melanotropins (alpha-MSHs). *Dev Biol.* 1988;126(1): 156–163.
41. Hatfield JM, Allen RG, Stack J, Ronnekleiv O. Post-translational processing of pro-opiomelanocortin (POMC)-derived peptides during fetal monkey pituitary development. II. Beta-Lipotropin (beta-LPH)-related peptides. *Dev Biol.* 1988;126(1):164–172.
42. Olah M, Feher P, Ihm Z, Bacskay I, Kiss T, Freeman ME, et al. Dopamine-regulated adrenocorticotropic hormone secretion in lactating rats: functional plasticity of melanotropes. *Neuroendocrinology.* 2009;90(4):391–401.
43. Funkelstein L, Toneff T, Mosier C, Hwang SR, Beuschlein F, Lichtenauer UD, et al. Major role of cathepsin L for producing the peptide hormones ACTH, beta-endorphin, and alpha-MSH, illustrated by protease gene knockout and expression. *J Biol Chem.* 2008;283(51):35652–35659.
44. Laurent V, Kimble A, Peng B, Zhu P, Pintar JE, Steiner DF, et al. Mortality in 7B2 null mice can be rescued by adrenalectomy: involvement of dopamine in ACTH hypersecretion. *Proc Natl Acad Sci USA.* 2002;99(5):3087–3092.
45. Sarac MS, Zieske AW, Lindberg I. The lethal form of Cushing's in 7B2 null mice is caused by multiple metabolic and hormonal abnormalities. *Endocrinology.* 2002;143(6):2324–2332.
46. Westphal CH, Muller L, Zhou A, Zhu X, Bonner-Weir S, Schambelan M, et al. The neuroendocrine protein 7B2 is required for peptide hormone processing *in vivo* and provides a novel mechanism for pituitary Cushing's disease. *Cell.* 1999;96(5):689–700.
47. Lee SN, Peng B, Desjardins R, Pintar JE, Day R, Lindberg I. Strain-specific steroidal control of pituitary function. *J Endocrinol.* 2007;192(3):515–525.
48. Shen FS, Aguilera G, Loh YP. Altered biosynthesis and secretion of pro-opiomelanocortin in the intermediate and anterior pituitary of carboxypeptidase E-deficient, Cpe(fat)/Cpe(fat)mice. *Neuropeptides.* 1999;33(4):276–280.
49. DeMaria JE, Nagy GM, Lerant AA, Fekete MI, Levenson CW, Freeman ME. Dopamine transporters participate in the physiological regulation of prolactin. *Endocrinology.* 2000;141(1): 366–374.
50. Williams PJ, Dunwiddie TV, Gerhardt GA. Direct *in vitro* electrochemical measurement of dopamine overflow in intermediate pituitary: characterization and pharmacology. *J Pharmacol Exp Ther.* 1992;261(2):403–411.

51. Nussinovitch I, Kleinhaus AL. Dopamine inhibits voltage-activated calcium channel currents in rat pars intermedia pituitary cells. *Brain Res.* 1992;574(1-2):49–55.
52. Nagata T, Harayama N, Sasaki N, Inoue M, Tanaka K, Toyohira Y, et al. Mechanisms of cytosolic Ca^{2+} suppression by prostaglandin E2 receptors in rat melanotrophs. *J Neuroendocrinol.* 2003;15(1):33–41.
53. Hozumi Y, Watanabe M, Goto K. Signaling cascade of diacylglycerol kinase β in the pituitary intermediate lobe: dopamine D2 receptor/phospholipase Cβ4/diacylglycerol kinase β/protein kinase Cα. *J Histochem Cytochem.* 2010;58:119–129.
54. Oyarce AM, Hand TA, Mains RE, Eipper BA. Dopaminergic regulation of secretory granule-associated proteins in rat intermediate pituitary. *J Neurochem.* 1996;67(1):229–241.
55. Day R, Schafer MK, Watson SJ, Chretien M, Seidah NG. Distribution and regulation of the prohormone convertases PC1 and PC2 in the rat pituitary. *Mol Endocrinol.* 1992;6(3):485–497.
56. Gomora JC, Avila G, Cota G. Ca^{2+} current expression in pituitary melanotrophs of neonatal rats and its regulation by D2 dopamine receptors. *J Physiol.* 1996;492(Pt 3):763–773.
57. Chronwall BM, Beatty DM, Sharma P, Morris SJ. Dopamine D2 receptors regulate *in vitro* melanotrope L-type Ca^{2+} channel activity via c-fos. *Endocrinology.* 1995;136(2):614–621.
58. Chronwall BM, Millington WR, Griffin WS, Unnerstall JR, O'Donohue TL. Histological evaluation of the dopaminergic regulation of proopiomelanocortin gene expression in the intermediate lobe of the rat pituitary, involving *in situ* hybridization and [3H]thymidine uptake measurement. *Endocrinology.* 1987;120(3):1201–1211.
59. Back N. The effect of bromocriptine on the intermediate lobe of the rat pituitary: an electron-microscopic, morphometric study. *Cell Tissue Res.* 1989;255(2):405–410.
60. Gary KA, Chronwall BM. The onset of dopaminergic innervation during ontogeny decreases melanotrope proliferation in the intermediate lobe of the rat pituitary. *Intl J Dev Neurosci.* 1992;10(2):131–142.
61. Kelly MA, Rubinstein M, Asa SL, Zhang G, Saez C, Bunzow JR, et al. Pituitary lactotroph hyperplasia and chronic hyperprolactinemia in dopamine D2 receptor-deficient mice. *Neuron.* 1997;19(1):103–113.
62. Cote TE, Eskay RL, Frey EA, Grewe CW, Munemura M, Stoof JC, et al. Biochemical and physiological studies of the beta-adrenoceptor and the D-2 dopamine receptor in the intermediate lobe of the rat pituitary gland: a review. *Neuroendocrinology.* 1982;35(3):217–224.
63. Meunier H, Labrie F. beta-Adrenergic, CRF-ergic and dopaminergic mechanisms controlling alpha-MSH secretion in rat pars intermedia cells in primary culture. *Prog Neuropsychopharmacol Biol Psychiatry.* 1982;6(4-6):411–415.
64. Grigoriadis DE, De Souza EB. Corticotropin-releasing factor (CRF) receptors in intermediate lobe of the pituitary: biochemical characterization and autoradiographic localization. *Peptides.* 1989;10(1):179–188.
65. Chronwall BM, Davis TD, Severidt MW, Wolfe SE, McCarson KE, Beatty DM, et al. Constitutive expression of functional GABA(B) receptors in mIL-tsA58 cells requires both GABA(B(1)) and GABA(B(2)) genes. *J Neurochem.* 2001;77(5):1237–1247.
66. Schwab Y, Jahke R, Jover E. Expression of tetrodotoxin-sensitive and resistant sodium channels by rat melanotrophs. *NeuroReport.* 2004;15(7):1219–1223.
67. van den Hurk MJ, Cruijsen PM, Schoeber JP, Scheenen WJ, Roubos EW, Jenks BG. Intracellular signal transduction by the extracellular calcium-sensing receptor of *Xenopus* melanotrope cells. *Gen Comp Endocrinol.* 2008;157(2):156 164.
68. Sedej S, Rose T, Rupnik M. cAMP increases Ca^{2+}-dependent exocytosis through both PKA and Epac2 in mouse melanotrophs from pituitary tissue slices. *J Physiol.* 2005;567(Pt 3):799–813.
69. Kolk SM, Berghs CA, Vaudry H, Verhage M, Roubos EW. Physiological control of Xunc18 expression in neuroendocrine melanotrope cells of. *Xenopus laevis. Endocrinology.* 2001;142(5):1950–1957.
70. Kolk SM, Groffen AJ, Tuinhof R, Ouwens DT, Cools AR, Jenks BG, et al. Differential distribution and regulation of expression of synaptosomal-associated protein of 25-kDa isoforms in the *Xenopus* pituitary gland and brain. *Neuroscience.* 2004;128(3):531–543.
71. Rupnik M, Kreft M, Nothias F, Grilc S, Bobanovic LK, Johannes L, et al. Distinct role of Rab3A and Rab3B in secretory activity of rat melanotrophs. *Am J Physiol Cell Physiol.* 2007;292(1):C98–105.
72. Oertel WH, Mugnaini E, Tappaz ML, Weise VK, Dahl AL, Schmechel DE, et al. Central GABAergic innervation of neurointermediate pituitary lobe: biochemical and immunocytochemical study in the rat. *Proc Natl Acad Sci USA.* 1982;79(2):675–679.
73. Vuillez P, Perez SC, Stoeckel ME. Colocalization of GABA and tyrosine hydroxylase immunoreactivities in the axons innervating the neurointermediate lobe of the rat pituitary: an ultrastructural immunogold study. *Neurosci Lett.* 1987;79(1-2):53–58.
74. Schimchowitsch S, Vuillez P, Tappaz ML, Klein MJ, Stoeckel ME. Systematic presence of GABA-immunoreactivity in the tubero-infundibular and tubero-hypophyseal dopaminergic axonal systems: an ultrastructural immunogold study on several mammals. *Exp Brain Res.* 1991;83(3):575–586.
75. Poisbeau P, Rene F, Egles C, Felix JM, Feltz P, Schlichter R. Characterization of functional GABAergic synapses formed between rat hypothalamic neurons and pituitary intermediate lobe cells in coculture: Ca^{2+} dependence of spontaneous IPSCs. *J Neurosci.* 1996;16(16):4835–4845.
76. Rene F, Poisbeau P, Egles C, Schlichter R, Felix JM. Co-culture of hypothalamic neurons and melanotrope cells: a model to study synaptogenesis between central neurons and endocrine cells. *Neuroscience.* 1997;76(1):203–214.
77. Chronwall BM, Sands SA, Cummings III KC, Hagler KE, Norberg M, Morris SJ, et al. Differential innervation of individual melanotropes suggests a role for nonsynaptic inhibitory regulation of the developing and adult rat pituitary intermediate lobe. *Synapse.* 1998;28(3):227–243.
78. Demeneix BA, Desaulles E, Feltz P, Loeffler JP. Dual population of GABAA and GABAB receptors in rat pars intermedia demonstrated by release of alpha MSH caused by barium ions. *Br J Pharmacol.* 1984;82(1):183–190.
79. Desrues L, Castel H, Malagon MM, Vaudry H, Tonon MC. The regulation of alpha-MSH release by GABA is mediated by a chloride-dependent $[Ca^{2+}]c$ increase in frog melanotrope cells. *Peptides.* 2005;26(10):1936–1943.
80. Hansen SL, Fjalland B, Jackson MB. Modulation of $GABA_A$ receptors and neuropeptide secretion by the neurosteroid allopregnanolone in posterior and intermediate pituitary. *Pharmacol Toxicol.* 2003;93(2):91–97.
81. Poisbeau P, Feltz P, Schlichter R. Modulation of $GABA_A$ receptor-mediated IPSCs by neuroactive steroids in a rat hypothalamo-hypophyseal coculture model. *J Physiol.* 1997;500(Pt 2):475–485.
82. Berman JA, Wu TJ, Roberts JL. Derivatization of progesterone to a neurally active steroid by pituitary neurointermediate lobe. *Steroids.* 1998;63(11):579–586.

83. Purisai MG, Sands SA, Davis TD, Price JL, Chronwall BM. $GABA_B$ receptor subunit mRNAs are differentially regulated in pituitary melanotropes during development and detection of functioning receptors coincides with completion of innervation. *Intl J Dev Neurosci.* 2005;23(4):315–326.
84. Galas L, Tonon MC, Beaujean D, Fredriksson R, Larhammar D, Lihrmann I, et al. Neuropeptide Y inhibits spontaneous alpha-melanocyte-stimulating hormone (alpha-MSH) release via a Y(5) receptor and suppresses thyrotropin-releasing hormone-induced alpha-MSH secretion via a Y(1) receptor in frog melanotrope cells. *Endocrinology* 2002;143(5):1686–1694.
85. Leenders HJ, de Koning HP, Ponten SP, Jenks BG, Roubos EW. Differential effects of coexisting dopamine, GABA and NPY on alpha-MSH secretion from melanotrope cells of *Xenopus laevis. Life Sci.* 1993;52(24):1969–1975.
86. Nemethy Z, Horvath G, Makara GB, Acs Z, Barna I. Catecholaminergic control of intracellular free calcium and beta-endorphin secretion of rat pituitary intermediate lobe cells. *J Neuroendocrinol.* 1998;10(2):85–91.
87. Limone P, D'Alessandro V, Rainero I, Ambrogio C, Pinessi L, De Gennaro T, et al. Adrenergic regulation of alpha-MSH secretion in man: evidence for a stimulatory role of beta-receptors. *Acta Endocrinol (Copenh).* 1993;128(2):150–155.
88. Saavedra JM. Central and peripheral catecholamine innervation of the rat intermediate and posterior pituitary lobes. *Neuroendocrinology* 1985;40(4):281–284.
89. Berkenbosch F, Tilders FJ, Vermes I. Beta-adrenoceptor activation mediates stress-induced secretion of beta-endorphin-related peptides from intermediate but not anterior pituitary. *Nature* 1983;305(5931):237–239.
90. Saland LC. The mammalian pituitary intermediate lobe: an update on innervation and regulation. *Brain Res Bull* 2001;54(6): 587–593.
91. Poisbeau P, Jo YH, Feltz P, Schlichter R. Electrophysiological characterization of non-NMDA glutamate receptors on cultured intermediate lobe cells of the rat pituitary. *Neuroendocrinology* 1996;64(2):162–168.
92. Giovannucci DR, Stuenkel EL. Glutamate receptor agonists modulate $[Ca^{2+}]i$ in isolated rat melanotropes. *Neuroendocrinology* 1995;62(2):111–122.
93. Garnier M, Lamacz M, Galas L, Lenglet S, Tonon MC, Vaudry H. Pharmacological and functional characterization of muscarinic receptors in the frog pars intermedia. *Endocrinology.* 1998;139(8):3525–3533.
94. Zhang ZW, Feltz P. Nicotinic acetylcholine receptors in porcine hypophyseal intermediate lobe cells. *J Physiol.* 1990;422:83–101.
95. Saland LC, Carr JA, Samora A, Tejeda D. Benzodiazepine suppression of corticotropin-releasing factor (CRF)-induced beta-endorphin release from rat neurointermediate pituitary. *Peptides.* 1992;13(5):913–917.
96. Calle M, Corstens GJ, Wang L, Kozicz T, Denver RJ, Barendregt HP, et al. Evidence that urocortin I acts as a neurohormone to stimulate alpha MSH release in the toad. *Xenopus laevis. Brain Res.* 2005;1040(1-2):14–28.
97. Tran TN, Fryer JN, Lederis K, Vaudry H. CRF, urotensin I, and sauvagine stimulate the release of POMC-derived peptides from goldfish neurointermediate lobe cells. *Gen Comp Endocrinol.* 1990;78(3):351–360.
98. Galas L, Raoult E, Tonon MC, Okada R, Jenks BG, Castano JP, et al. TRH acts as a multifunctional hypophysiotropic factor in vertebrates. *Gen Comp Endocrinol.* 2009;164(1):40–50.
99. Boersma CJ, Pool CW, Van Heerikhuize JJ, Van Leeuwen FW. Characterization of opioid binding sites in the neural and intermediate lobe of the rat pituitary gland by quantitative receptor autoradiography. *J Neuroendocrinol.* 1994;6(1):47–56.
100. Matthews SG, Heavens RP, Sirinathsinghji DJ. Distribution and cellular localization of preproenkephalin mRNA in the ovine brain and pituitary. *Brain Res Mol Brain Res.* 1992; 12(4):349–355.
101. Day R, Schafer MK, Collard MW, Weihe E, Akil H. Prodynorphin gene expression in the rat intermediate pituitary lobe: gender differences and postpartum regulation. *Endocrinology* 1993;133(6):2652–2659.
102. Carr JA, Lovering AT. Mu and delta opioid receptor regulation of pro-opiomelanocortin peptide secretion from the rat neurointermediate pituitary *in vitro. Neuropeptides.* 2000;34(1):69–75.
103. Tostivint H, Vieau D, Chartrel N, Boutelet I, Galas L, Fournier A, et al. Expression and processing of the [Pro(2), Met(13)] somatostatin-14 precursor in the intermediate lobe of the frog pituitary. *Endocrinology.* 2002;143(9):3472–3481.
104. Chronwall BM, Sands SA, Cummings III KC, Schwartz JP. Glial somatostatin-14 expression in the rat pituitary intermediate lobe: a possible neurotrophic function during development? *Intl J Dev Neurosci.* 2000;18(7):685–692.
105. Pollard I. Participation of the intermediate lobe of the pituitary gland after prolonged exposure to unpredictable stress in the rat. *J Endocrinol.* 1983;99(2):239–243.
106. Kjaer A, Knigge U, Bach FW, Warberg J. Stress-induced secretion of pro-opiomelanocortin-derived peptides in rats: relative importance of the anterior and intermediate pituitary lobes. *Neuroendocrinology.* 1995;61(2):167–172.
107. Young EA, Bronstein D, Akil H. Dopamine regulation of swim stress induction of the pituitary intermediate lobe proopiomelanocortin system. *Neuroendocrinology.* 1993;58(3):294–302.
108. Shiver T, Familari M, Aguilera G. Regulation of intermediate pituitary corticotropin-releasing hormone receptors by dopamine. *Endocrinology.* 1992;130(4):2299–2304.
109. Lookingland KJ, Gunnet JW, Moore KE. Stress-induced secretion of alpha-melanocyte-stimulating hormone is accompanied by a decrease in the activity of tuberohypophysial dopaminergic neurons. *Neuroendocrinology.* 1991;53(1):91–96.
110. Goudreau JL, Manzanares J, Lookingland KJ, Moore KE. 5HT2 receptors mediate the effects of stress on the activity of periventricular hypophysial dopaminergic neurons and the secretion of alpha-melanocyte-stimulating hormone. *J Pharmacol Exp Ther.* 1993;265(1):303–307.
111. Knigge U, Matzen S, Hannibal T, Jorgensen H, Warberg J. Involvement of histamine in the mediation of the stress-induced release of alpha-melanocyte-stimulating hormone in male rats. *Neuroendocrinology.* 1991;54(6):646–652.
112. Kjaer A, Knigge U, Matzen S, Warberg J. Beta-adrenergic receptors are involved in histamine-induced secretion of proopiomelanocortin-derived peptides and prolactin in rats. *Eur J Endocrinol.* 1995;132(2):223–228.
113. Manzanares J, Toney TW, Lookingland KJ, Moore KE. Activation of tuberoinfundibular and tuberohypophysial dopamine neurons following intracerebroventricular administration of bombesin. *Brain Res.* 1991;565(1):142–147.
114. Manzanares J, Lookingland KJ, Moore KE. Kappa-opioid-receptor-mediated regulation of alpha-melanocyte-stimulating hormone secretion and tuberohypophyseal dopaminergic neuronal activity. *Neuroendocrinology.* 1990;52(2):200–205.
115. Mennin SP, Saland LC. Naloxone prevents dark-background adaptation in amphibians. *Neuroendocrinology.* 1980;31(6): 385–389.
116. Dulcis D, Spitzer NC. Illumination controls differentiation of dopamine neurons regulating behaviour. *Nature.* 2008;456(7219): 195–201.

117. DeMaria JE, Livingstone JD, Freeman ME. Ovarian steroids influence the activity of neuroendocrine dopaminergic neurons. *Brain Res.* 2000;879(1-2):139–147.
118. van Strien FJ, Galas L, Jenks BG, Roubos EW. Differential acetylation of pro-opiomelanocortin-derived peptides in the pituitary gland of *Xenopus laevis* in relation to background adaptation. *J Endocrinol.* 1995;146(1):159–167.
119. Vazquez-Martinez R, Peinado JR, Gonzalez De Aguilar JL, Desrues L, Tonon MC, Vaudry H, et al. Melanotrope cell plasticity: a key mechanism for the physiological adaptation to background color changes. *Endocrinology.* 2001;142(7):3060–3067.
120. Vazquez-Martinez R, Castano JP, Tonon MC, Vaudry H, Gracia-Navarro F, Malagon MM. Melanotrope secretory cycle is regulated by physiological inputs via the hypothalamus. *Am J Physiol Endocrinol Metab.* 2003;285(5):E1039–E1046.
121. Baker BI. Melanin-concentrating hormone: a general vertebrate neuropeptide. *Intl Rev Cytol.* 1991;126:1–47.
122. Powell KA, Baker BI. Structural studies of nerve terminals containing melanin-concentrating hormone in the eel. *Anguilla anguilla. Cell Tissue Res.* 1988;251(2):433–439.
123. DeMaria JE, Livingstone JD, Freeman ME. Characterization of the dopaminergic input to the pituitary gland throughout the estrous cycle of the rat. *Neuroendocrinology.* 1998;67(6):377–383.
124. DeMaria JE, Zelena D, Vecsernyes M, Nagy GM, Freeman ME. The effect of neurointermediate lobe denervation on hypothalamic neuroendocrine dopaminergic neurons. *Brain Res.* 1998;806(1):89–94.
125. Vecsernyes M, Krempels K, Toth BE, Julesz J, Makara GB, Nagy GM. Effect of posterior pituitary denervation (PPD) on prolactin (PRL) and alpha-melanocyte-stimulating hormone (alpha-MSH) secretion of lactating rats. *Brain Res Bull.* 1997; 43(3):313–319.
126. Murakami T, Ohtsuka A, Taguchi T, Kikuta A, Ohtani O. Blood vascular bed of the rat pituitary intermediate lobe, with special reference to its development and portal drainage into the anterior lobe. A scanning electron microscope study of vascular casts. *Arch Histol Jpn.* 1985;48(1):69–87.
127. Hill JB, Lacy ER, Nagy GM, Gorcs TJ, Frawley LS. Does alpha-melanocyte-stimulating hormone from the pars intermedia regulate suckling-induced prolactin release? Supportive evidence from morphological and functional studies. *Endocrinology.* 1993;133(6):2991–2997.
128. Matsumura R, Takeuchi S, Takahashi S. Effect of estrogen on melanocortin-3 receptor mRNA expression in mouse pituitary glands *in vivo* and *in vitro. Neuroendocrinology.* 2004;80(3):143–151.
129. Vecsernyes M, Nagy G, Meszaros L, Bodnar L, Ahmed KW, Toth R, et al. Suckling-induced change in oxytocin but not in alpha-MSH concentrations of the median eminence, the neural, intermediate and anterior lobes of the pituitary gland. *Endocr Res.* 2000;26(3):333–345.
130. Laudon M, Grossman DA, Ben-Jonathan N. Prolactin-releasing factor: cellular origin in the intermediate lobe of the pituitary. *Endocrinology.* 1990;126(6):3185–3192.
131. Allen DL, Low MJ, Allen RG, Ben-Jonathan N. Identification of two classes of prolactin-releasing factors in intermediate lobe tumors from transgenic mice. *Endocrinology.* 1995;136(7): 3093–3099.
132. Hnasko R, Khurana S, Shackleford N, Steinmetz R, Low MJ, Ben-Jonathan N. Two distinct pituitary cell lines from mouse intermediate lobe tumors: a cell that produces prolactin-regulating factor and a melanotroph [see comments]. *Endocrinology.* 1997;138(12):5589–5596.
133. Hnasko R, Ben-Jonathan N. Prolactin regulation by heparin-binding growth factors expressed in mouse pituitary cell lines. *Endocrine.* 2003;20(1-2):35–44.
134. Lin Chang C, Roh J, Park JI, Klein C, Cushman N, Haberberger RV, et al. Intermedin functions as a pituitary paracrine factor regulating prolactin release. *Mol Endocrinol.* 2005;19(11):2824–2838.
135. Homicsko KG, Kertesz I, Radnai B, Toth BE, Toth G, Fulop F, et al. Binding site of salsolinol: its properties in different regions of the brain and the pituitary gland of the rat. *Neurochem Int.* 2003;42(1):19–26.
136. Radnai B, Mravec B, Bodnar I, Kubovcakova L, Fulop F, Fekete MI, et al. Pivotal role of an endogenous tetrahydroisoquinoline, salsolinol, in stress- and suckling-induced release of prolactin. *Ann NY Acad Sci.* 2004;1018:183–191.
137. Autelitano DJ, van den Buuse M. Concomitant up-regulation of proopiomelanocortin and dopamine D2-receptor gene expression in the pituitary intermediate lobe of the spontaneously hypertensive rat. *J Neuroendocrinol.* 1997;9(4):255–262.
138. Felder RA, Garland DS. POMC biosynthesis in the intermediate lobe of the spontaneously hypertensive rat. *Am J Hypertens.* 1989;2(8):618–624.
139. Ni XP, Humphreys MH. Prevention of salt-induced hypertension by an analog of gamma-melanocyte-stimulating hormone in the rat. *Am J Hypertens.* 2007;20(8):862–865.
140. Ni XP, Pearce D, Butler AA, Cone RD, Humphreys MH. Genetic disruption of gamma-melanocyte-stimulating hormone signaling leads to salt-sensitive hypertension in the mouse. *J Clin Invest.* 2003;111(8):1251–1258.
141. Imaki T, Katsumata H, Miyata M, Naruse M, Imaki J, Minami S. Expression of corticotropin releasing factor (CRF), urocortin and CRF type 1 receptors in hypothalamic–hypophyseal systems under osmotic stimulation. *J Neuroendocrinol.* 2001;13(4):328–338.
142. Pardy K, Carter D, Murphy D. Dopaminergic mediation of physiological changes in proopiomelanocortin messenger ribonucleic acid expression in the neurointermediate lobe of the rat pituitary. *Endocrinology.* 1990;126(6):2960–2964.
143. Leenders HJ, de Vries TJ, van den Loop FT, Jenks BG. Biosynthetic response of mouse intermediate pituitary gland to induced drinking and dehydration. *Acta Endocrinol (Copenh)* 1990;122(4):527–534.
144. Lebaili N, Bensalem M, Stoeckel ME. The pituitary intermediate lobe of the hydrated and dehydrated gerbil. An immunohistochemical and electron-microscopic study. *Cells Tissues Organs.* 1999;164(3):155–166.
145. McFarlane D. Advantages and limitations of the equine disease, pituitary pars intermedia dysfunction as a model of spontaneous dopaminergic neurodegenerative disease. *Ageing Res Rev.* 2007;6(1):54–63.
146. Millington WR, Dybdal NO, Dawson Jr R, Manzini C, Mueller GP. Equine Cushing's disease: differential regulation of beta-endorphin processing in tumors of the intermediate pituitary. *Endocrinology.* 1988;123(3):1598–1604.
147. Peterson ME, Krieger DT, Drucker WD, Halmi NS. Immunocytochemical study of the hypophysis in 25 dogs with pituitary-dependent hyperadrenocorticism. *Acta Endocrinol (Copenh)* 1982;101(1):15–24.
148. Halmi NS, Peterson ME, Colurso GJ, Liotta AS, Krieger DT. Pituitary intermediate lobe in dog: two cell types and high bioactive adrenocorticotropin content. *Science.* 1981;211(4477): 72–74.
149. Zerbe CA, Clark TP, Sartin JL, Kemppainen RJ. Domperidone treatment enhances corticotropin-releasing hormone stimulated adrenocorticotropic hormone release from the dog pituitary. *Neuroendocrinology.* 1993;57(2):282–288.
150. Kemppainen RJ, Sartin JL. *In vivo* evidence for dopaminergic regulation of the canine pituitary intermediate lobe. *Acta Endocrinol (Copenh).* 1986;113(4):471–478.

151. Young DW, Zerbe CA, Kemppainen RJ. Molecular forms of alpha-melanocyte-stimulating hormone in the canine pituitary anterior and intermediate lobe. *Peptides.* 1992;13(6):1061–1066.
152. Mauri A, Volpe A, Martellotta MC, Barra V, Piu U, Angioni G, et al. alpha-Melanocyte-stimulating hormone during human perinatal life. *J Clin Endocrinol Metab.* 1993;77(1):113–117.
153. Evans VR, Manning AB, Bernard LH, Chronwall BM, Millington WR. Alpha-melanocyte-stimulating hormone and N-acetyl-beta-endorphin immunoreactivities are localized in the human pituitary but are not restricted to the zona intermedia. *Endocrinology.* 1994;134(1):97–106.
154. Voutetakis A, Argyropoulou M, Sertedaki A, Livadas S, Xekouki P, Maniati-Christidi M, et al. Pituitary magnetic resonance imaging in 15 patients with Prop1 gene mutations: pituitary enlargement may originate from the intermediate lobe. *J Clin Endocrinol Metab.* 2004;89(5):2200–2206.
155. Lamberts SW, de Lange SA, Stefanko SZ. Adrenocorticotropin-secreting pituitary adenomas originate from the anterior or the intermediate lobe in Cushing's disease: differences in the regulation of hormone secretion. *J Clin Endocrinol Metab.* 1982;54(2):286–291.
156. McNicol AM. A study of intermediate lobe differentiation in the human pituitary gland. *J Pathol.* 1986;150(3):169–173.
157. McNicol AM, Teasdale GM, Beastall GH. A study of corticotroph adenomas in Cushing's disease: no evidence of intermediate lobe origin. *Clin Endocrinol (Oxf)* 1986;24(6):715–722.
158. Nagaya T, Kuwayama A, Seo H, Tsukamoto N, Matsui N, Sugita K. Endocrinological evaluation of ACTH-secreting pituitary microadenomas: their location and alpha-melanocyte stimulating hormone immunoreactivity. *J Neurosurg.* 1992;76(6):944–947.
159. Coates PJ, Doniach I, Holly JM, Rees LH. Demonstration of desacetyl alpha-melanocyte-stimulating hormone in fetal and adult human anterior pituitary corticotrophs. *J Endocrinol.* 1989;120(3):525–530.
160. Coates PJ, Doniach I, Wells C, Hale AC, Rees LH, Besser GM. Peptides related to alpha-melanocyte-stimulating hormone are commonly produced by human pituitary corticotroph adenomas: no relationship with pars intermedia origin. *J Endocrinol.* 1989;120(3):531–536.
161. Raffel C, Boggan JE, Eng LF, Davis RL, Wilson CB. Pituitary adenomas in Cushing's disease: do they arise from the intermediate lobe? *Surg Neurol.* 1988;30(2):125–130.
162. Musat M, Vax VV, Borboli N, Gueorguiev M, Bonner S, Korbonits M, et al. Cell cycle dysregulation in pituitary oncogenesis. *Front Horm Res.* 2004;32:34–62.
163. Franklin DS, Godfrey VL, Lee H, Kovalev GI, Schoonhoven R, Chen-Kiang S, et al. CDK inhibitors p18(INK4c) and p27(Kip1) mediate two separate pathways to collaboratively suppress pituitary tumorigenesis. *Genes Dev.* 1998;12(18):2899–2911.
164. Park MS, Rosai J, Nguyen HT, Capodieci P, Cordon-Cardo C, Koff A. p27 and Rb are on overlapping pathways suppressing tumorigenesis in mice. *Proc Natl Acad Sci USA.* 1999;96(11):6382–6387.
165. Tsai KY, MacPherson D, Rubinson DA, Nikitin AY, Bronson R, Mercer KL, et al. ARF mutation accelerates pituitary tumor development in Rb+/− mice. *Proc Natl Acad Sci USA.* 2002;99(26):16865–16870.
166. Lazzerini Denchi E, Attwooll C, Pasini D, Helin K. Deregulated E2F activity induces hyperplasia and senescence-like features in the mouse pituitary gland. *Mol Cell Biol.* 2005;25(7):2660–2672.
167. Roussel-Gervais A, Bilodeau S, Vallette S, Berthelet F, Lacroix A, Figarella-Branger D, et al. Cooperation between Cyclin E and p27(Kip1) in pituitary tumorigenesis. *Mol Endocrinol.* 2010;24(9):1835–1845.
168. Leung SW, Wloga EH, Castro AF, Nguyen T, Bronson RT, Yamasaki L. A dynamic switch in Rb+/− mediated neuroendocrine tumorigenesis. *Oncogene.* 2004;23(19):3296–3307.
169. Zhou Z, Flesken-Nikitin A, Levine CG, Shmidt EN, Eng JP, Nikitina EY, et al. Suppression of melanotroph carcinogenesis leads to accelerated progression of pituitary anterior lobe tumors and medullary thyroid carcinomas in Rb+/− mice. *Cancer Res.* 2005;65(3):787–796.
170. Lloyd RV, Ruebel KH, Zhang S, Jin L. Pituitary hyperplasia in glycoprotein hormone alpha subunit-, p18(INK4C)-, and p27(kip-1)-null mice: analysis of proteins influencing p27(kip-1) ubiquitin degradation. *Am J Pathol.* 2002;160(3):1171–1179.
171. Komatsubara K, Tahara S, Umeoka K, Sanno N, Teramoto A, Osamura RY. Immunohistochemical analysis of p27 (Kip1) in human pituitary glands and in various types of pituitary adenomas. *Endocr Pathol.* 2001;12(2):181–188.

CHAPTER

8

The Hypothalamic–Pituitary–Adrenal Axis and Neuroendocrine Responses to Stress

Greti Aguilera

Program on Developmental Endocrinology and Genetics, *Eunice Kennedy Shiver* Institute of Child Health and Human Development, NIH, Bethesda, MD, USA

OUTLINE

Summary

External or internal stressful stimuli threatening homeostasis induce coordinated responses, including behavioral, autonomic nervous system and endocrine changes, aimed at restoring the internal equilibrium. The major neuroendocrine stress response occurs via activation of the hypothalamic pituitary adrenal axis, initiated by corticotropin-releasing hormone (CRH) secretion, leading ultimately to increased adrenal glucocorticoid production. Acting through receptors in the brain, CRH and glucocorticoids play a critical role in coordinating the stress response. This includes not only behavioral and autonomic responses but also alterations in the activity of other neuroendocrine systems, such as the hypothalamic–pituitary–gonadal, hypothalamic–pituitary–thyroid and hypothalamic–growth hormone axes. Acute stress responses are transient and beneficial, but chronic stress or inadequate stress responses often lead to pathology. This chapter describes the neuroendocrine mechanisms regulating the HPA axis responses to stress, and its contribution to the integrated adaptive responses.

INTRODUCTION

Living organisms are constantly challenged by external or internal challenges or stressors, which threaten homeostasis, defined as constancy of the internal environment.[1] Adaptation to stress and restoration of homeostasis requires the coordinated activation of complex neuroendocrine responses, including behavior, endocrine, immune and autonomic nervous systems[2] (Fig. 8.1). While behavioral responses, including arousal, increased awareness and improved cognition, are necessary for avoidance and defense reactions, autonomic and endocrine responses are critical for physiological adaptive changes. The major endocrine response to stress is activation of the hypothalamic pituitary adrenal (HPA) axis leading to increases in circulating glucocorticoids, which are

Handbook of Neuroendocrinology, DOI: 10.1016/B978-0-12-375097-6.10008-3

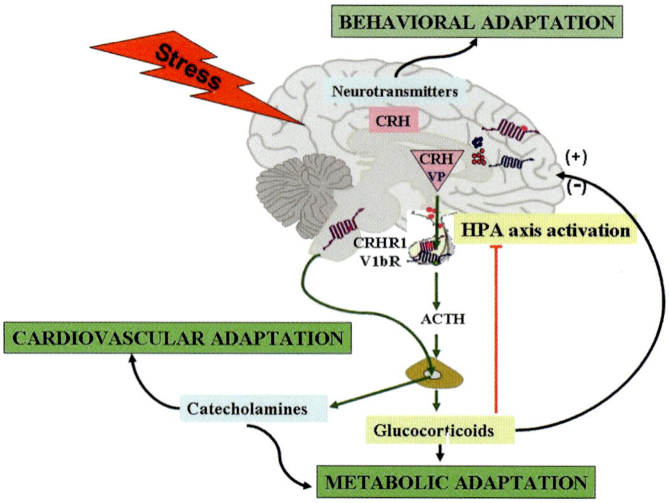

FIGURE 8.1 **Schematic representation of the interactions between activation of the hypothalamic–pituitary–adrenal (HPA) axis and major adaptive responses to stress.** Stress stimulates the release of corticotropin-releasing hormone (CRH) and vasopressin (VP) from the paraventricular nucleus of the hypothalamus (PVN) into the pituitary portal circulation. Binding of CRH to type 1 CRH receptors and VP to V1b receptors in the pituitary corticotrope stimulates ACTH secretion, which stimulates adrenal glucocorticoid secretion. Glucocorticoids are essential for metabolic adaptation, exert feedback inhibition on the HPA axis (red) and modulate brain function. In addition to stimulating the HPA axis, CRH released within the brain binds to CRHR1 in the brain and contributes to behavioral and autonomic adaptation to stress.

essential for the metabolic adaptation to stress. Other components of the HPA axis, especially the hypothalamic peptide corticotropin-releasing hormone (CRH) and, to a lesser extent, vasopressin (VP), are critical in this process. After describing the different components of the HPA axis, this chapter discusses the neuroendocrine mechanisms controlling HPA axis activity during stress, and the contribution of HPA axis activation to the overall stress response, including the impact of stress on control of appetite and metabolism, and the activity of other neuroendocrine axes, such as the hypothalamic–pituitary–gonadal, hypothalamic–pituitary–thyroid and hypothalamic–growth hormone axes.

HYPOTHALAMIC–PITUITARY–ADRENAL AXIS

Activation of the HPA axis involves stimulation of a specific group of neurons located in the dorsomedial parvocellular subdivision (see below) of the paraventricular nucleus (PVN) of the hypothalamus. These neurons synthesize corticotropin-releasing hormone (CRH) as well as vasopressin (VP), which are released into the pituitary portal circulation and reach corticotrope cells in the anterior pituitary gland, where they stimulate the secretion of adrenocorticotropic hormone (ACTH) into the peripheral circulation.[3,4] The main target of ACTH action is the zona fasciculata of the adrenal cortex, where it stimulates the production and secretion of glucocorticoids. While acute HPA axis responses to stress are rapid and transient, ACTH and corticosterone responses during chronic or repeated stress vary according to the intensity, frequency and the ability to control the stress

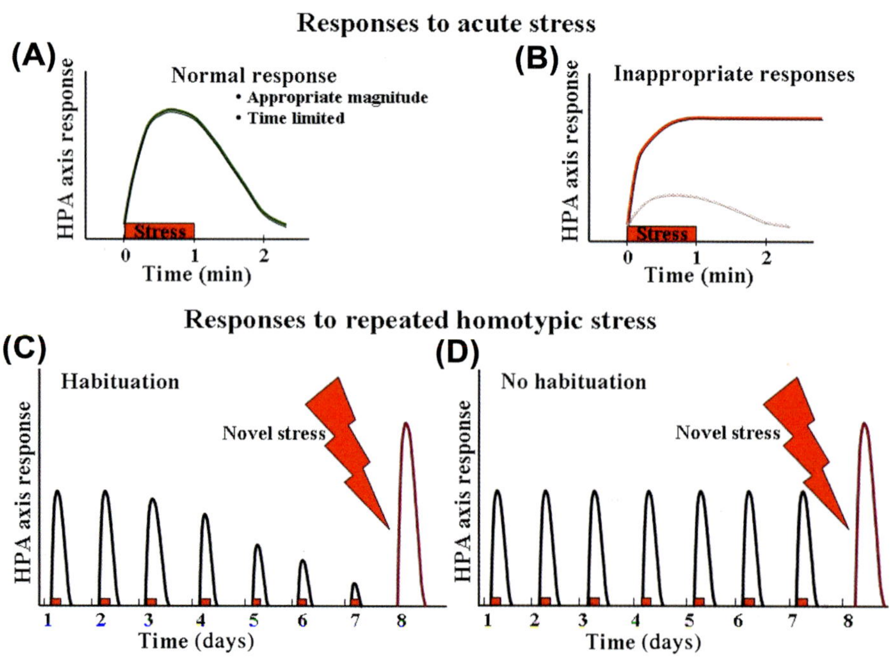

FIGURE 8.2 **Simplified representation of HPA axis responses to acute and repeated stress.** During acute stress, ACTH and glucocorticoid responses are rapid and marked but transient, returning to basal levels even in the presence of a persistent stimulus (**A**). Depending on the context of the stress and genetic background stress generation of either sustained or insufficient HPA axis responses can lead to pathology due to excessive or insufficient CRH and glucocorticoids (**B**). During repeated stress there is desensitization of the responses, depending on the nature and intensity of the repeated stress, but there is hyper-responsiveness to a novel stimulus, (indicated by the red arrow) (**C**). Physical stressors and chronic stressors (severe pain, blood loss, hypoglycemia) do not cause habituation, and also present with hyper-responsiveness to a novel stress (**D**). The duration of the stress is represented by the red bars.

(Fig. 8.2). In some cases responses are sustained, and in others there is habituation or desensitization in spite of the persistence of the stimulus.[3] In most chronic stress situations there is enhanced ACTH and glucocorticoid production in response to a heterotypical (or novel) stressor, even if there is habituation to the primary repeated stressor.[3,5] HPA axis activity can be modulated at each anatomical level – the hypothalamus, the pituitary and the adrenal – but the primary driver of HPA axis activation is the production and release of the hypothalamic peptides, CRH and VP. It is noteworthy that CRH and VP, as well as glucocorticoids, the end product of HPA axis stimulation, have important roles in regulating behavioral and autonomic stress responses.

The hypothalamic–pituitary–adrenal axis is a major mediator of the stress response.

Hypothalamic Regulation of the HPA Axis

The Paraventricular Nucleus of the Hypothalamus (PVN)

The paraventricular nucleus of the hypothalamus (PVN), site of production of hypophyseotropic CRH and VP, can be divided into three main functional regions controlling distinct physiological effectors of stress.[6] First, parvocellular neurons in the dorsomedial and anterior divisions of the PVN release peptides to the pituitary portal circulation from axonal terminals in the external zone of the median eminence. CRH-producing neurons, responsible for the stimulation of ACTH secretion, are located in this region. In all species studied, including rat, mouse and human, a proportion of CRH neurons also express VP, which is co-released with CRH and contributes to HPA axis regulation. The

BOX 8.1

HOW DO WE KNOW THE HYPOTHALAMIC–PITUITARY–ADRENAL AXIS IS A MAJOR MEDIATOR OF THE STRESS RESPONSE?

The first indications for the involvement of adrenal corticosteroids in the organism's response to threats to homeostasis were provided by the work of Selye in the late 1940s and early 1950s,[1] which showed that there was an association between responses to stress, increases in urinary 11-hydroxysteroids and plasma ACTH. Earlier, in 1932, the neurosurgeon Harvey Cushing had provided the first evidence for pituitary control of the adrenal cortex with his description of pituitary adenomas in autopsies of patients with clinical hypercortisolism.[2] Cushing postulated that the pituitary adenomas were the cause of adrenal hyperplasia and excess glucocorticoid secretion. Following the identification of adrenal stimulating activity in pituitary extracts, adrenocorticotropin (ACTH) was finally isolated and characterized in 1955 as the pituitary hormone responsible for stimulating the secretion of adrenal glucocorticoids.[3] At the same time, the seminal work of Geoffrey Harris demonstrated that electrical stimulation of the hypothalamus increased while hypothalamic lesions reduced the secretion of ACTH and other pituitary hormones, showing for the first time that the hypothalamus controlled pituitary ACTH production. Anatomical work by Harris, Bertha and Ernst Scharrer finally demonstrated the existence of neuronal secretion from the hypothalamus and the fact that the pituitary portal system was responsible for the transport of neurosecretion to the pituitary.[4–6] In the mid-1950s, McCann and Brobeck[7] and Martini and Morpurgo[8] described the ability of the antidiuretic hormone or vasopressin (VP) to stimulate ACTH secretion, and proposed VP as the putative CRH. However, major progress in our understanding of the mechanisms that regulate the HPA axis during stress was only possible after the isolation and characterization of corticotropin-releasing hormone (CRH) by Wylie Vale's laboratory in 1981.[9] The availability of synthetic agonists and antagonists of CRH and VP, as well as the production of antibodies against the two peptides, has been valuable for studying the physiological actions of these peptides in the regulation of ACTH secretion. In addition, studies in experimental animals showing that intracerebroventricular (i.c.v) administration of CRH reproduced the behavioral effects of stress and i.c.v. administration of CRH antagonists blocked these responses provided the first demonstration that CRH is a major mediator of the behavioral responses to stress.[10] The importance of CRH in mediating and orchestrating the HPA axis, autonomic and behavioral responses to stress was further confirmed by the creation of transgenic mice which either overexpressed or lacked the peptide. Thus, mice that overexpress CRH develop Cushing's-like symptoms due to excessive production of ACTH and corticosterone and increased anxiety, while CRH knockout mice die at birth of respiratory insufficiency due to the adverse effects of glucocorticoid

(*Continued*)

BOX 8.1 *(cont'd)*

deficiency on lung maturation: if rescued, CRH knockout mice lack the HPA axis, autonomic and behavioral responses to stress.[11,12]

References

1. Selye H. Stress and the general adaptation syndrome. *Br Med Journal.* 1950;1:1383–1392.
2. Cushing H. The basophil adenomas of the pituitary body and their clinical manifestations (pituitary basophilism). *Obesity Res.* 1932;2:486–508.
3. Li CH, Geschwind II, Dixon JS, Levy AL, Harris JI. Corticotropins (ACTH). *I. Isolation of alpha-corticotropin from sheep pituitary glands J Biol Chem.* 1955;213:171–185.
4. Harris GW. Neural control of the pituitary gland. *Physiol Rev.* 1948;28:139–179.
5. Lennette EH, Scharrer E. Neurosecretion; cytoplasmic inclusions in peripheral autonomic ganglion cells of the monkey. *Anat Rec.* 1946;94:85–92.
6. Scharrer B. Neuroendocrine communication (neurohormonal, neurohumoral, and intermediate). *Prog Brain Res.* 1972;38:7–18.
7. McCann SM, Brobeck JR. Evidence for a role of the supraopticohypophyseal system in regulation of adrenocorticotrophin secretion. *Proc Soc Exp Biol Med.* 1954;87: 318–324.
8. Martini L, Morpurgo C. Neurohumoral control of the release of adrenocorticotrophic hormone. *Nature.* 1955;175:1127–1128.
9. Spiess J, Rivier J, Rivier C, Vale W. Primary structure of corticotropin-releasing factor from ovine hypothalamus. *Proc Natl Acad Sci USA.* 1981;78:6517–6521.
10. Vale W, Rivier C, Brown MR, Spiess J, Koob G, Swanson L, Bilezikjian L, Bloom F, Rivier J. Chemical and biological characterization of corticotropin releasing factor. *Recent Prog Hormone Res.* 1983;39: 245–270.
11. Bale TL, Vale WW. CRF and CRF receptors: Role in stress responsivity and other behaviors. *Ann Rev Pharmacol Toxicol.* 2004;44:525–557.
12. Muller MB, Uhr M, Holsboer F, Keck ME. *Hypothalamic–pituitary–adrenocortical system and mood disorders: Highlights from mutant mice Neuroendocrinology.* 2004;79: 1–12.

second region consists of magnocellular neurons located in the anteromedial and dorsolateral divisions of the PVN. These neurons produce VP and oxytocin and project axons to the posterior pituitary (neurohypophysis) passing through the internal zone of the median eminence, and are responsible for releasing the peptides to the peripheral circulation.[7] The third region comprises a group of neurons located in the lateral parvocellular, dorsal parvocellular and ventromedial parvocellular divisions, which also express CRH. These neurons receive and send projections from and to nuclei in the brainstem, and are involved in the control of autonomic responses to stress.[4,6,8] Parvocellular CRH neurons are a major target of the neural pathways activated during stress. In addition to VP, these neurons co-express other peptides, such as angiotensin II, opiate peptides and neurotensin, all of which can modulate HPA axis responses to stress.[9]

CRH and VP controlling ACTH secretion are produced by the paraventricular nucleus of the hypothalamus.

CRH and CRH Receptors

The 41-amino acid peptide CRH is the principal member of a family of neuropeptides that also includes urocortins 1, 2 and 3.[10] The main site of production of CRH is the hypothalamic paraventricular nucleus, but the peptide is also expressed in limbic areas of the brain, such as the amygdala, bed nucleus stria terminalis, locus coeruleus and cortical areas.[11] This extrahypothalamic CRH expression and release within the brain plays a critical role in the behavioral and autonomic responses to stress, including arousal, decrease in feeding and sexual behavior, and increased sympathoadrenal activity. In addition, central CRH has prominent roles in the integration of the HPA axis with other neuroendocrine responses to stress. CRH mRNA and peptide have also been identified in a number of peripheral tissues, including the gonads, placenta and skin.[12–14] Urocortins are also expressed in the brain and peripheral organs. Urocortin 1 and 2 mediate behavioral effects in the brain, mostly counteracting the effects of stress and CRH[10]; urocortin 3 mediates metabolic-related effects of stress, such as inhibition of feeding behavior and increasing plasma glucose.[15] In the periphery, urocortins, especially urocortin 2, are important in cardiovascular regulation, having inotropic actions, independent of catecholamines.[16]

CRH exerts its effects in the pituitary corticotrope and other target cells by binding to plasma membrane receptors coupled to adenylate cyclase via guanyl nucleotide binding protein (G protein), specifically G_s. There are two major types of CRH receptors (Table 8.1): type 1 CRH receptors are located mainly in the brain and pituitary corticotropes, and also in peripheral tissues, including the reproductive organs and the immune system. They bind CRH and urocortin 1 with equal affinity, but do not recognize urocortins

BOX 8.2

HOW DO WE KNOW CRH AND VP CONTROLLING ACTH SECRETION ARE PRODUCED BY THE PARAVENTRICULAR NUCLEUS OF THE HYPOTHALAMUS?

Morphological studies, including immunohistochemistry and *in situ* hybridization, have been pivotal for the identification of the neuronal pathways responsible for the release of CRH and VP into the pituitary portal circulation. Immunohistochemical studies have made possible (a) the identification of neurons containing CRH and VP and mRNAs for the peptides, as well as visualization of their projections to the external zone of the median eminence from where they are released into the pituitary portal circulation; (b) the co-localization of VP in a proportion of parvocellular neurons of the PVN containing CRH; and (c) co-localization of CRH and other neuropeptides such as enkephalins and neurotensin.[1–3] These VP-containing parvocellular neurons are different from the large magnocellular neurons that project to the neural lobe of the pituitary. The combination of immunohistochemical techniques and retrograde and anterograde tracing has been invaluable for elucidating the afferent and efferent connections of CRH neurons.[4–6]

Molecular cloning of the precursor molecules for CRH and VP has allowed the characterization of the molecular structure of the genes encoding the peptides, and the design of antisense complementary probes to measure mRNA levels. *In situ* hybridization studies using ^{35}S-labeled oligonucleotide or riboprobes and cryostat sections of the hypothalamic region have been used to demonstrate the presence of CRH and VP mRNA in neurons located in the same region as the immunoactive peptides.

Since first used by Young and colleagues in 1986,[7] *in situ* hybridization has been a powerful tool widely used for studying the topographic localization and regulation of CRH gene expression.

References

1. Antoni FA, Palkovits M, Makara GB, Linton EA, Lowry PJ, Kiss JZ. Immunoreactive corticotropin-releasing hormone in the hypothalamo-infundibular tract. *Neuroendocrinology.* 1983;36:415–423.
2. Sawchenko PE, Swanson LW, Vale WW. Co-expression of corticotropin-releasing factor and vasopressin immunoreactivity in parvocellular neurosecretory neurons of the adrenalectomized rat. *Proc Natl Acad Sci USA.* 1984;81:1883–1887.
3. Swanson LW, Sawchenko PE, Rivier J, Vale WW. Organization of ovine corticotropin-releasing factor immunoreactive cells and fibers in the rat brain: an immunohistochemical study. *Neuroendocrinology.* 1983;36:165–186.
4. Cullinan WE, Herman JP, Watson SJ. Ventral subicular interaction with the hypothalamic paraventricular nucleus: evidence for a relay in the bed nucleus of the stria terminalis. *J Comp Neurol.* 1993;332:1–20.
5. Prewitt CMF, Herman JP. Anatomical interactions between the central amygdaloid nucleus and the hypothalamic paraventricular nucleus of the rat: a dual tract-tracing analysis. *J Chem Neuroanat.* 1998;15:173–186.
6. Swanson L, Sawchenko PE, Lind R, Rho JH. The CRH motoneuron: Differential peptide regulation in neurons with possible synaptic, paracrine, and endocrine outputs. *Ann NY Acad Sci.* 1987;512:12–23.
7. Young WS, Mezey E, Siegel RE. Quantitative *in situ* hybridization histochemistry reveals increased levels of corticotropin-releasing factor mRNA after adrenalectomy in rats. *Neurosci Lett.* 1986;70:198–203.

2 and 3. Type 2 CRH receptors, which have 70% homology with type 1 CRH receptors, bind urocortin 1, 2 and 3, and are located mostly in the periphery.[10] Being coupled to adenylyl cyclase, both type 1 and type 2 receptors activate cyclic AMP/protein kinase A dependent pathways upon binding to their ligand. Type 1 CRH receptor (CRHR1) is the major player in the stress response, and is essential for pituitary ACTH secretion and behavioral responses to stress.[10,17]

The distribution of CRH receptors in the brain is consistent with the pivotal role of CRH as a mediator of the integral stress responses.[18] Type 1 CRH receptors are located mainly in the limbic system, a number of hypothalamic nuclei, and the cerebral and cerebellar cortices. The highest limbic receptor concentration is found in the external plexiform layer of the olfactory bulb, with lower levels of expression in the amygdala; bed nucleus of the stria terminalis; lateral, intermediate and medial septal nucleus; nucleus accumbens; and caudate putamen. Diencephalic structures expressing abundant CRHR1 include the dorsomedial hypothalamic nucleus, dorsolateral thalamic nuclei, supramammillary nuclei, anterior paraventricular thalamic nuclei and arcuate nucleus. Low levels of CRHR1 are also present in the PVN, and stress induces dramatic increases in expression in different subdivisions, depending on the nature of the stimulus. Psychogenic and metabolic stressors (which stimulate CRH and VP in parvocellular neurons) induce type 1 CRH receptors in this region, while osmotic stress induces type 1 CRH receptors in

TABLE 8.1 CRH Receptor Subtypes

	CRHR1	CRHR2
Structure	7 TM domains 415 aa	7 TM domains CRHR2α:411 aa CRHR2β:431 aa
Signaling–transduction	cAMP-PKA PLC	cAMP-PKA
Binding properties	Ka nM CRH = Urocortin 1	Ka nM Urocortin 1 and 2
Tissue location	Pituitary Brain cortex Limbic structures Cerebellum Brain visceral and sensory structures Skin Gonads	Hypothalamus Lateral septum Olfactory bulb Heart Smooth muscle

the dorsolateral region of the PVN and in the SON – sites containing magnocellular VP neurons.[19] Type 1 CRH receptors are also expressed in a number of brainstem structures, including the interpeduncular and pontine nuclei, superior colliculus, inferior olive, dorsal tegmental nucleus, spinal trigeminal tract and locus coeruleus, and the nucleus of the solitary tract. Although this distribution is in general consistent with the distribution of CRHR1 mRNA in the brain and pituitary, there is not always a direct correlation between the two parameters (Table 8.2). The lateral septum, posterior colliculus and sensory trigeminal nucleus display solely type 2 CRH receptors. These extrapituitary CRH receptors in the brain are responsible for CRH-mediated neurotransmission in the complex circuitry regulating activation and termination of the stress response.

The biological function of CRH is further modified through interactions with a non-receptor CRH binding protein. This 37-kDa protein, first identified in blood of pregnant women and non-human primates,[24] is also widely distributed in the brain and pituitary. This protein binds both CRH and urocortin with an affinity equal to or greater than that of the receptors, and blocks CRH-mediated ACTH release *in vitro*. The CRH binding protein and the CRH receptors are co-located broadly in cortical regions, in limbic structures and certain sensory relay nuclei (Table 8.2), suggesting that this binding protein has a profound impact on the biological activity of CRH (or urocortin). Mouse models of CRH-BP overexpression or deficiency have shown changes in the HPA axis and in energy balance, and behavior consistent with the hypothesis that CRH-BP plays an important *in vivo* modulatory role by regulating levels of CRH and CRH-like peptides available to receptors in the pituitary and brain.[25]

TABLE 8.2 Distribution of CRF Receptor Binding, CRF Receptor 1 and 2 mRNA and CRF Binding Protein mRNA/Protein in the Rat Brain

Brain region	oCRH binding[1]	CRHR1 mRNA[3]	CRHR2 mRNA[3]	CRFBP mRNA/protein[4]
Cerebral cortex				
Prefrontal	2 to 4[2]	3		2
Anterior cingulated	2 to 3	3		
Frontoparietal (motor)	2 to 3	3		
Frontoparietal (sensory)	3	3		
Temporal, auditory area	4			
Subiculum	2			
Hippocampus	2	2	2	2
Entorhinal	2	2	2	
Basal telencephalon				
Olfactory bulb/tubercle	3			2
Lateral septal nucleus	2	1	4	
Medial septal nucleus		2	nd	
Nucleus accumbens	2			
Caudate putamen	1			1
Bed nucleus stria terminalis	3	2	2	2
Amygdaloid complex				
Anterior area	2			
Medial area	2	4	2	2
Basolateral area	3	4	nd	2
Claustrum	2			2
Diencephalon				
Paraventricular thalamic nuclei interior	1			1
Dorsolateral thalamic nuclei	1			
Supramamillary nuclei	2			2
Mamillary bodies				
Dorsomedial hypothalamic nuclei	2	3		2
Arcuate nucleus	1			2
Paraventricular nucleus	nd	nd	2	1
Periventricular nucleus	nd	1		2
Median eminence	nd			1

(Continued)

TABLE 8.2 Distribution of CRF Receptor Binding, CRF Receptor 1 and 2 mRNA and CRF Binding Protein mRNA/Protein in the Rat Brain—cont'd

Brain region	oCRH binding[1]	CRHR1 mRNA[3]	CRHR2 mRNA[3]	CRFBP mRNA/ protein[4]
Brainstem				
Superior colliculus	2	3	1	2
Inferior colliculus				1
Sensory trigeminal			3	2
Interpeduncular nucleus	3	4	3	2
Periaqueductal gray	1			
Locus coeruleus	1	nd	nd	
Inferior olive	2			
Pontine nucleus	3	4	1	2
Nucleus of the solitary tract	1			2
Dorsal tegmental nucleus	2	4	nd	2
Spinal/sensory trigeminal tract	2		3	
Cerebellum				
Granular layer	4	4		*

[1]*Data adapted from Wynn* et al.[20]. *125I-Tyr-oCRH was used for the binding assay binding.*
[2]*Binding intensity in cortical layers 1–3/layer 4.*
[3]*Data obtained from Chalmers* et al.[21] *and Smagin* et al.[22]
[4]*CRF binding protein expression data adapted from Potter* et al.[23] *CRFBP mRNA and protein were co-expressed at all sites. Levels of expression are estimated from hybridization histochemistry and are not quantitative.*
Signals for binding and hybridization to mRNA: nd, not detectable; 1, weak; 2, moderate; 3, strong; 4, very strong.

Vasopressin and VP Receptors

The nonapeptide, VP, is mainly produced by magnocellular neurons of the supraoptic nucleus and dorsolateral region of the PVN (Fig. 8.2). These neurons project axons to the neurohypophysis and release VP into the peripheral circulation, where it mediates its well-recognized effects on water conservation and vascular regulation.[7] The magnocellular neurons respond by producing VP in response to osmotic stimulation and to decreases in blood pressure, but they have limited input in the control of ACTH secretion. On the other hand, parvocellular VP, co-released into the pituitary portal circulation with CRH during stress, is involved in pituitary corticotrope function and ACTH secretion.[7,26] In addition, activation of VP neurons in the medial amygdala and the bed nucleus of the stria terminalis projecting to the lateral septum and ventral hippocampus modulate memory and behavior during stress.[27]

The actions of vasopressin are mediated by plasma membrane G-protein coupled receptors.[28,29] Three major receptor subtypes encoded by different genes have been identified and cloned; kidney V2 receptors, which are coupled to the G protein, G_S, and adenylate cyclase; V1a receptors, present in smooth muscle, liver and brain, coupled to phospholipase C (PLC); and the pituitary V1b receptor, also linked to PLC, which is responsible for the effects of VP on ACTH secretion.[30–32]

The main subtypes of VP receptors present in the brain are V1a and V1b receptors, with V1a being the most abundant.[33,34] In rats, VP binding and V1a receptor mRNA are widely distributed in the hypothalamic and limbic areas of the brain, including the olfactory bulb, frontal and piriform cortex; the CA1, CA2, CA3 and dentate gyrus of the hippocampus; the suprachiasmatic, paraventricular, ventromedial and arcuate nuclei in the hypothalamus; the lateral habenular nucleus; and the molecular and granular cell layers of the cerebellum. The anatomical distribution and levels of expression of V1aR mRNA are similar in male and female rats. A similar distribution has been described in non-human primates.[35] V1b receptors are less abundant, but are present at similar sites. The V1b receptor subtype is abundant in pituitary corticotropes, where it mediates the effects of VP on HPA axis regulation.[3]

CRH and VP act by interacting with plasma membrane receptors in the target tissue.

Regulation of CRH and VP during Stress

Acute stress causes rapid activation of parvocellular CRH neurons as shown by the rapid increases in c-fos and CRH primary transcripts, the latter indicating transcriptional activation of the CRH gene. Acute stress also induces a slightly delayed and more prolonged increase in VP transcription in CRH neurons.[36–38] In basal conditions about 50% of the parvicellular CRH neurons express VP, and this proportion increases markedly during adrenalectomy and chronic stress associated with hyper-responsiveness of the HPA axis.[3] Acute stress induces rapid release of CRH and VP into the pituitary portal circulation.[39,40] CRH antagonists or antibodies inhibit about 70% of the ACTH response to acute stress, indicating that CRH is largely responsible.[41] The release of CRH and VP is followed by rapid increases in gene transcription, and this is followed by elevation of steady-state mRNA levels and translation to the precursor peptide. In the stress models studied, CRH gene activation is rapid, reaching a peak within 5–30 minutes, while VP gene activation is delayed by 60–120 minutes.[36–38]

During chronic stress, changes in CRH expression in parvicellular neurons parallel the pattern of ACTH response to a repeated stress. In those types

BOX 8.3

HOW DO WE KNOW CRH AND VP ACT BY INTERACTING WITH PLASMA MEMBRANE RECEPTORS IN THE TARGET TISSUE?

Binding assays and pharmacological studies

Receptors for CRH were first identified and characterized in plasma membrane rich fractions from rat anterior pituitary using radioiodinated Tyr-ovine CRH. Subsequently, the development of autoradiographic techniques made it possible to identify receptors in areas of the brain involved in the responses to stress, as well as in peripheral tissues. The receptor binding assays involved incubation of membrane aliquots or cryostat tissue sections with the radio-iodinated ligand in a buffer containing a carrier albumin to reduce non-specific binding, and protease inhibitors to protect the receptor and ligand from degradation. The specificity of the binding is determined by the ability of the homologous but not heterologous ligands to compete and displace the radioactive ligand from the receptor. Analysis of competition curves of radioactive Tyr-oCRH by a range of concentrations of unlabeled ligand, using Scatchard plots, allows the calculation of the binding affinity and binding capacity of the receptor in different tissues. Study of the pituitary VP V1b receptor binding was more difficult because of the lack of biologically active iodinated VP. For this reason, VP binding assays required the use of tritiated vasopressin, which is bioactive and can interact with the receptor but has very low specific activity. These binding assays made it possible to determine the binding properties of CRH and VP receptors and to study their regulation during stress and other experimental manipulations.[1–3] Differences in the ability of different analog peptides to compete with the radioactive ligand provided the first indications of receptor subtypes. For example, the demonstration that the peptide sauvagine has higher affinity than CRH for the CRH receptor in the adrenal medulla suggested that this receptor was different from the pituitary and brain receptor.[4] Later studies after the cloning of receptor subtypes confirmed that the adrenal medulla contains CRH type 2 receptors. Similarly, unique pharmacological specificity of the pituitary VP receptor provided the first evidence for a receptor subtype different from the vascular or renal receptor.[2] The biological relevance of ligand binding has been demonstrated by the ability of CRH and VP to induce cell signaling (activation of adenylyl cyclase for CRH and inositol triphosphate formation for VP) and stimulate ACTH secretion in cultured pituicytes.[1–3]

Advances from molecular cloning of receptors and genetically modified mice

Major progress in the understanding of CRH and VP receptor subtypes, their properties and regulation was made following the molecular cloning of the receptors. Knowledge of the gene sequence and structure allowed: (a) the identification of CRH and VP receptor subtypes; (b) further characterization of receptor properties and signaling transduction mechanisms in transfected cells; (c) the investigation of the topographic distribution of mRNA expression in the brain, pituitary and some peripheral organs; (d) the regulation of mRNA levels in the target tissues; and (e) the development of genetically modified mice overexpressing or lacking one or more CRH or VP receptor subtype. The availability of these animals has been critical for understanding the physiological relevance of the different receptor subtypes, and the importance of CRH and VP during stress adaptation. For example, different groups have generated CRFR1 and CRHR2 knockout mice.[5,6] CRHR1-deficient mice display highly impaired HPA axis function under basal conditions and in response to stress, as well as increased exploratory activity and reduced anxiety behavior in a number of behavioral tests. In contrast, CRHR2 knockout mice show an opposite phenotype, with hyper-responsiveness to stress and increased anxiety behavior. For details on genetically modified animal models, see reviews by Bale and Vale[5] and Muller *et al.*[6] for CRH receptors, and Roper *et al.*[7] for VP receptors.

References

1. Aguilera G, Pham Q, Rabadan-Diehl C. Regulation of pituitary vasopressin receptors during chronic stress: relationship to corticotroph responsiveness. *J Neuroendocrinol.* 1994;6:299–304.
2. Antoni FA. Vasopressinergic control of pituitary adrenocorticotropin secretion comes of age. *Front Neuroendocrinol.* 1993;14:76–122.
3. Wynn PC, Aguilera G, Morell J, Catt KJ. Properties and regulation of high-affinity pituitary receptors for corticotropin-releasing factor. *Biochem Biophys Res Comm.* 1983;110:602–608.
4. Udelsman R, Harwood JP, Millan MA, Chrousos GP, Goldstein DS, Zimlichman R, Catt KJ, Aguilera G. Functional corticotropin releasing factor receptors in the primate peripheral sympathetic nervous system. *Nature.* 1986;319:147–150.
5. Bale TL, Vale WW. CRF AND CRF receptors: Role in stress responsivity and other behaviors. *Ann Rev Pharmacol Toxicol.* 2004;44:525–557.
6. Muller MB, Uhr M, Holsboer F, Keck ME. Hypothalamic–pituitary–adrenocortical system and mood disorders: Highlights from mutant mice. *Neuroendocrinology.* 2004; 79:1–12.
7. Roper JA, O'Carroll AM, Young WS, Lolait SJ. The vasopressin Avpr1b receptor: Molecular and pharmacological studies. *Stress.* 2011;14:98–115.

of stress associated with desensitization of ACTH responses to the repeated stimulus (restraint, cold exposure), CRH mRNA levels cease to respond after the first few exposures to the stimulus. Similarly, in physiological conditions associated with low stress responsiveness, CRH mRNA levels in the PVN are low.[42] However, in paradigms with no desensitization of ACTH responses (repeated i.p. hypertonic saline injection or foot shock), CRH mRNA in the PVN remains high for at least 24 hours after the last stress, suggesting that sustained increases in CRH are required to maintain ACTH responses during repeated homotypic stress.

In contrast to CRH, VP expression in the parvicellular system is increased in all chronic stress models studied, as shown by increases in VP mRNA and primary transcripts in parvicellular neurons, and the content of immunoreactive VP in the external zone of the median eminence.[43–45] This preferential increase in parvocellular VP expression compared with CRH suggests that VP becomes the main regulator of ACTH secretion during chronic stress.[42,45] The mechanism of the differential regulation of CRH and VP in the same neuron is not understood, and likely involves diverse responses of the two genes to transcription factors activated during chronic stress.

Regulation of CRH Transcription

Activation of CRH transcription depends on cAMP/protein kinase A (PKA)-dependent pathways, leading to recruitment of phosphorylated cAMP response element binding protein (pCREB) by the cAMP response element (CRE) of the CRH promoter.[46,47] However, CREB alone is not sufficient, and transcriptional activation of the CRH promoter requires activation and nuclear translocation of the CREB co-activator, Transducer of Regulated Cyclic AMP response element binding protein (TORC).[48] While a variety of signaling pathways lead to CREB phosphorylation, activation of TORC requires cyclic AMP.[49,50] The major receptors mediating the effect of neurotransmitters activating CRH neurons (α-adrenergic and glutamatergic) do not signal through cAMP, but induce CREB phosphorylation through calcium/phospholipid dependent pathways, as well as potentiating the action of small increases in cyclic AMP. Although the exact identity of the neurotransmitter(s) responsible for inducing cyclic AMP is unclear, it is likely to be a neuropeptide, such as PACAP,[51] and an autoregulatory action of CRH itself. The latter is supported by the existence of CRH synapses in CRH perikarya, as well as the increases in CRHR1 during stress.[19]

While activation of CRH transcription is important for HPA axis activation, behavioral and autonomic responses to stress, turning off the activation process is essential to prevent the direct and indirect deleterious effects of excessive CRH production. The increases in CRH transcription are usually transient not only during acute stress but also during sustained stimulation and even after adrenalectomy.[52,53] Mechanisms for limiting the stress response involve changes in stimulatory and inhibitory neurocircuitry, glucocorticoid feedback (see below), as well as intracellular feedback on the CRH neuron. *In vitro* and *in vivo* studies in rats have shown that the repressor, inducible cyclic AMP early repressor (ICER), a product of activation of the second promoter of the CREM gene,[54] is involved in terminating CRH transcription in the absence of glucocorticoids.[53,55] This is shown by the induction of ICER mRNA in CRH neurons and the recruitment of ICER protein by the CRH promoter during the declining phase of transcription. In addition, ICER knockout in primary cultures of hypothalamic neurons using siRNA attenuates the late decline in CRH transcription.[53,55]

There is new evidence that the CREB co-activator, TORC, could act as an on-and-off switch in the control of CRH transcription.[48] The parallel changes in nuclear translocation of TORC and activation of CRH transcription, and the requirement of the co-activator for activation of CRH transcription, suggest that TORC inactivation by phosphorylation serves as an intracellular mechanism limiting the transcriptional response. The relative contribution of these various intraneuronal feedback mechanisms in limiting CRH transcription in the context of a physiological stress remains to be elucidated.

The Pituitary Corticotrope

Hypothalamic CRH and VP released into the pituitary portal circulation gain access to receptors in the anterior pituitary corticotrope, where they stimulate adrenocorticotropic hormone (ACTH) secretion. ACTH is a 39-amino acid peptide derived from a larger precursor protein, pro-opiomelanocortin (POMC). POMC is expressed in a variety of tissues, but its main sites of expression are corticotrope cells in the anterior lobe of the pituitary, melanotropes in the intermediate lobe, and neurons in the arcuate nucleus of the hypothalamus. The POMC molecule undergoes post-translational processing to produce a variety of biologically active peptides according to the cell type, with ACTH, β-lipotropin and β-endorphin being major products in the corticotrope, and α-MSH, corticotropin-like intermediate lobe peptide and acetyl-β-endorphin in the melanotrope and arcuate neuron.[56]

CRH is a potent stimulant of ACTH secretion[57] and POMC transcription *in vitro* and *in vivo*.[58] Acting though cyclic AMP/PKA/CREB-dependent pathways, CRH mediates increases in POMC transcription during stress

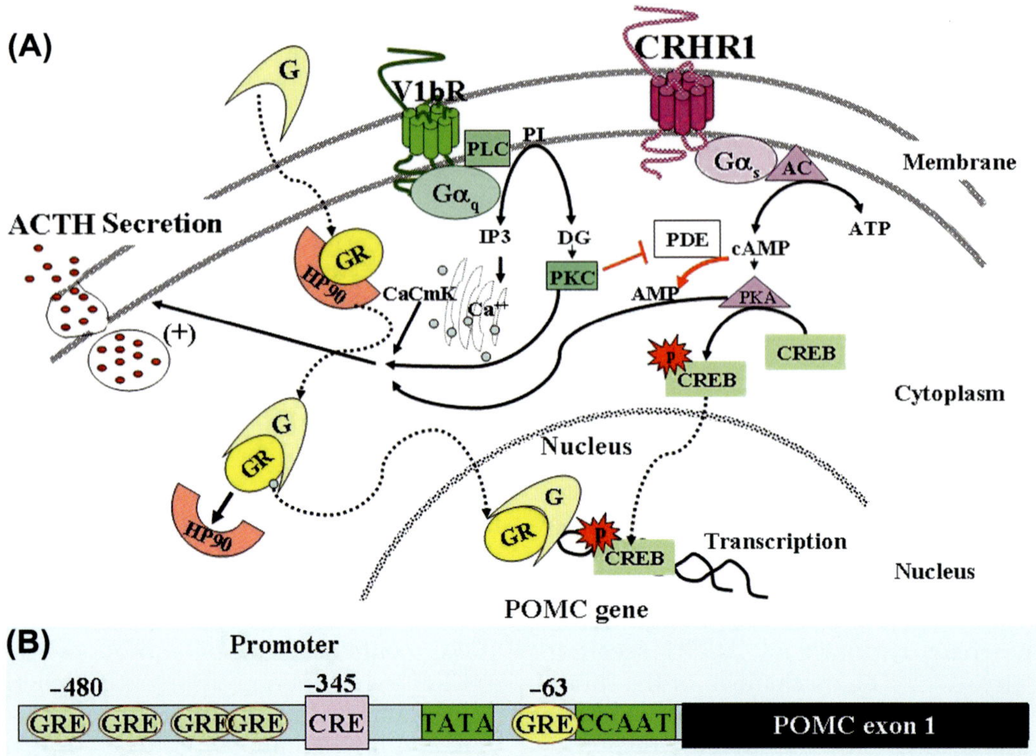

FIGURE 8.3 **Regulation of the anterior pituitary corticotrope.** CRH and VP released into the pituitary portal circulation bind to their specific plasma membrane receptors, type 1 CRH receptor (CRHR1) and type 1b vasopressin receptor (V1bR), in the pituitary corticotrope. Activation of CRHR1, coupled to the stimulatory guanyl nucleotide binding protein (G_s) and adenylate cyclase (AC), activates the cyclic AMP (cAMP)/protein kinase A (PKA) pathway and also increases calcium influx (not shown). Activation of the V1bR, coupled to the guanyl binding protein Gq/11 and phospholipase C (PLC), cleaves phosphatidyl inositol (PI) into inositol-3-phosphate (which releases calcium (Ca^{2+}) from intracellular stores) and diacyl glycerol, which activates protein kinase C (PKC). Calcium-activated calmodulin dependent protein kinase (CaCmK), PKC and PKA contribute to ACTH secretion from secretory granules. PKA phosphorylates cyclic AMP-responsive element binding protein (CREB) and phospho-CREB migrates into the nucleus, where it stimulates transcription of the ACTH precursor, pro-opiomelanocortin (POMC). Glucocorticoids inhibit POMC transcription and ACTH secretion. Glucocorticoids (G) enter the cell and bind to the receptor (GR), which dissociates from the scaffolding protein heat shock protein 90 (HSP90) and enters the nucleus, where it binds to the glucocorticoid responsive elements (GRE) in the POMC promoter and represses POMC transcription (**A**). The position of the major responsive elements in the POMC promoter is shown in (**B**).

and long-term adrenalectomy (Fig. 8.3). In contrast, VP is a weak stimulator of ACTH secretion on its own, but it markedly potentiates the stimulatory effect of CRH on ACTH release.[59,60] VP exerts this effect partly by potentiating CRH-stimulated cAMP production via protein kinase C-mediated inhibition of phosphodiesterases.[59] At variance with CRH, VP does not stimulate POMC transcription on its own or in combination with CRH. Moreover, *in vivo* and *in vitro* studies have shown inhibitory effects of VP on POMC transcription.[61] Basal levels of POMC transcription are not affected by PVN lesions, median eminence deafferentation or CRH antagonists, indicating that it is independent of CRH and other hypothalamic factors.

Type 1 CRH receptors are essential for regulating corticotrope function, since genetic ablation of CRHR1 in mice abolishes ACTH production in response to stress.[10] Although CRH and CRH receptors are required for pituitary ACTH response to stress, CRHR1 content in the pituitary undergoes downregulation during stress and adrenalectomy.[62,63] The decrease in receptors is likely the result of the higher exposure of the pituitary corticotrope to CRH and VP. A similar decrease is observed following exogenous CRH administration, and this effect is potentiated by VP.[64] The dissociation between CRH receptor downregulation and increased pituitary responsiveness during chronic stress indicates that the number of type 1 CRH receptors is not a determinant of corticotrope responsiveness *in vivo*.

On the other hand, there is a strong correlation between the pituitary content of V1b receptors and pituitary responsiveness to stress, with a high receptor content during psychogenic and metabolic stress, and a decrease during osmotic stimulation and lactation — conditions associated with low ACTH responsiveness.[65] While the pattern of regulation of hypothalamic CRH and VP, and pituitary type 1 CRH and V1b receptors, suggests that VP becomes the main regulator of ACTH secretion during chronic stress, pharmacological and genetic approaches do not support this view. It is clear that VP, acting through its pituitary V1b receptor, is required for the full ACTH response to acute stress, and to sustain responses during chronic stress. However, ACTH levels achieved in the absence of VP are sufficient to elicit a full adrenal glucocorticoid response.[26,66] Also, enhanced ACTH responsiveness during chronic stress does not require VP release during exposure to the novel stress. In addition to directly regulating ACTH secretion, there is evidence that VP exerts trophic actions in the pituitary.[67]

The number of corticotropes in the anterior pituitary increases after adrenalectomy and probably after chronic stress.[67,68] This effect is likely mediated by both CRH and VP. Administration of CRH increases the number of corticotropes *in vivo* and increases mitogenic activity in cultured pituitary cells *in vitro*. The mitogenic effect of CRH *in vitro* occurs with concentrations of the peptide submaximal for ACTH secretion, and appears to be a direct effect on corticotrope cells. Vasopressin acts as a mitogen in the pituitary, and V1 receptor antagonists prevent the increase in deoxybromouridine incorporation during adrenalectomy. However, the lack of labeling of differentiated pituitary cells suggests that newly-produced corticotropes following adrenalectomy and chronic stress originate from undifferentiated cells and not from division of existing corticotropes.[67] Thus, VP probably contributes to the production of precursor cells, while CRH mediates corticotrope differentiation.

The Adrenal Gland and Glucocorticoids

The adrenal gland, the end target of HPA axis activation, comprises two major zones: the cortex (site of production of glucocorticoids) and the medulla (source of catecholamines, a final product of autonomic nervous system stimulation). The adrenal cortex is composed of three zones (Fig. 8.4). The outer layer is the zona glomerulosa, which produces mineralocorticoids, the main one being aldosterone, which is responsible for sodium reabsorption. The adrenal zona glomerulosa function is under the control of the renin angiotensin system, but during acute stress also responds to ACTH. The intermediate layer is the zona fasciculata, which produces glucocorticoids under the control of ACTH. In several species, including humans, non-human primates and bovine species, cortisol is the dominant glucocorticoid, while in rodents such as rat and mice it is corticosterone. Glucocorticoid secretion by the adrenal follows a circadian rhythm, being high during the wake period – day in humans, and night in rats and mice (Fig. 8.5). This diurnal variation is controlled by the suprachiasmatic nucleus of the hypothalamus, which is responsible for circadian variations of CRH expression. In addition to the diurnal rhythm, glucocorticoid secretion displays an ultradian rhythm with hourly pulses.[69] The amplitude of the peaks varies according to the time of the day, and the rises during the wake period are determined by increases in pulse amplitude (Fig. 8.5). Preservation of both circadian and ultradian rhythms is essential for normal cell function and glucocorticoid-dependent gene transcription.[69]

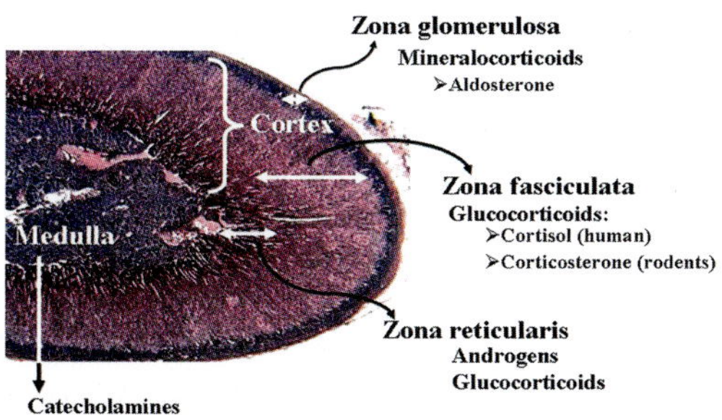

FIGURE 8.4 **Histological section of an adrenal gland showing the medulla in the cortex.** Each zone of the cortex is indicated by the two-way arrows.

Glucocorticoids are essential for the systemic adaptation to stress by controlling energy supply through stimulation of glycolysis and gluconeogenesis, as well as proteolysis to supply amino acids as substrates for gluconeogenesis. In addition, glucocorticoids modulate the synthesis and action of a number of hormones, and inhibit the immune system. The wide-ranging effects of glucocorticoids can explain some of the pathological consequences of prolonged exposure to stress levels of the steroid.[70,71]

Glucocorticoids act by interacting with cytoplasmic receptors, which bind the steroid and are translocated to the nucleus where they regulate gene transcription.[72,73] Glucocorticoids also have rapid non-genomic actions believed to occur by interaction with yet unidentified membrane-bound receptors. The genomic actions of glucocorticoids are mediated by two classes of cytoplasmic receptors, glucocorticoid (GR or type II) and mineralocorticoid (MR or type I) (Table 8.3). The type I receptor, initially identified in the kidney as the mediator of the effects of the mineralocorticoid, aldosterone, has high affinity for glucocorticoids and responds to low basal levels of circulating glucocorticoids. Type I receptors are discretely distributed in the brain, with high expression in the hippocampus and very weak expression in other areas.[74,75] Type II receptors are widely distributed in the brain,[76] bind glucocorticoids with much lower affinity than the type I receptors (about 10-fold), and recognize only stress levels of glucocorticoids.

Neural Regulation of HPA Axis Activity

When stressful stimuli occur, sensory information is integrated by the limbic system and conveyed to CRH neurons in the PVN through complex monoaminergic

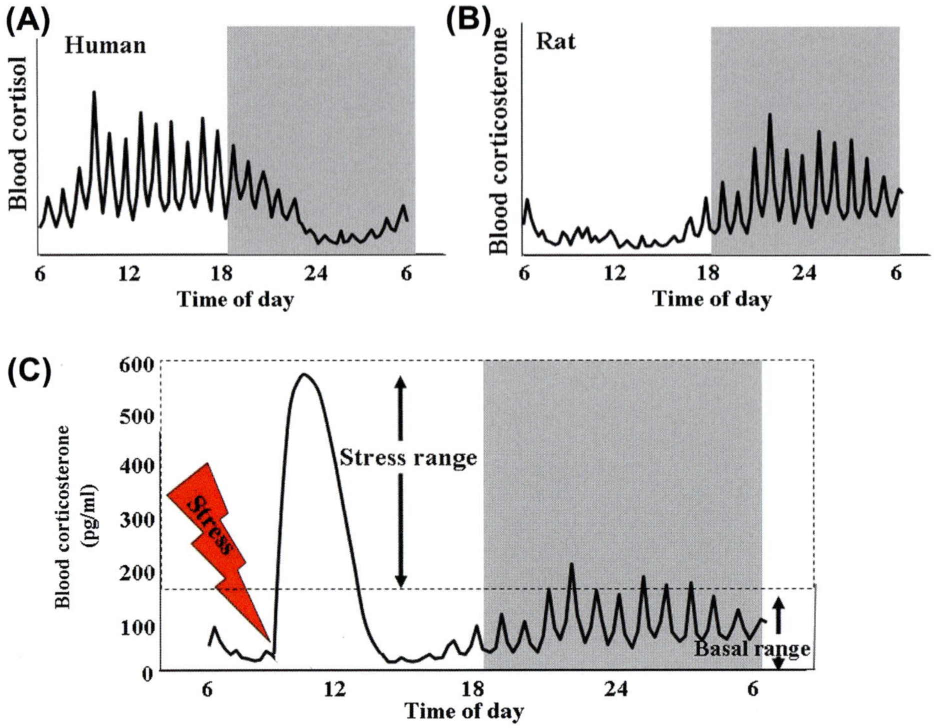

FIGURE 8.5 **Daily rhythms of glucocorticoid secretion.** Adrenal glucocorticoid secretion is higher during the wake period – day in humans, night in nocturnal rodents, (**A**) and (**B**), respectively. In addition, secretion is pulsatile, with one secretory pulse per hour. Increases during the wake period are due to increases in pulse amplitude. Stress overrides the pattern of secretion and causes a transient elevation above basal levels (**C**). The duration of the hypersecretory episode depends on the nature and duration of the stress.

and peptidergic neural pathways. The utilization of these various pathways and their interactions depend on the nature and intensity of the stressor. Several approaches have been used to localize and characterize brain structures and neuronal pathways that are involved in the organization of stress responses. Some of the techniques employed include immunohistochemistry and *in situ* hybridization for neurotransmitters, neuropeptides and protooncogenes; anterograde and retrograde tract-tracing techniques; specific brain lesions using electrocoagulation or neurotoxins; and transection of neuronal pathways.[9,77,78]

Acute Stressors

Physical and metabolic stressors such as loss of blood volume, immune challenge, pain and hypoglycemia constitute acute systemic stressors requiring immediate response. Such responses involve monosynaptic ascending pathways from the brainstem and spinal cord, with direct projections to the PVN.[79–81] These projections originate in the nucleus of the solitary tract (NTS) and C1 and C3 are mostly noradrenergic and adrenergic and stimulate CRH neurons by acting upon alpha adrenergic receptors in the CRH neuron (Fig. 8.6A, solid green lines). In addition to the direct projections to CRH neurons, brainstem pathways interact with other structures in the brain, including the dorsal raphe (which controls serotoninergic activity), the dorsomedial hypothalamic nucleus (DMH) (which modulates autonomic activity), and limbic structures in the forebrain. Catecholaminergic and non-catecholaminergic neurons in the NTS projecting to the PVN express other neuropeptides, such as NPY, glucagon-like peptide 1 (GLP-1), inhibin-β,

TABLE 8.3 Receptors Mediating Genomic Actions of Glucocorticoids in the Brain

Receptor type	Affinity (nM)	DNA responsive element	Location
Type I or mineralocorticoid receptor	0.1	AGAACANNNTGTTCT	High in hippocampus Low in other areas
Type II or glucocorticoid receptor (GR)	3	AGAACANNNTGTTCT	Widely distributed

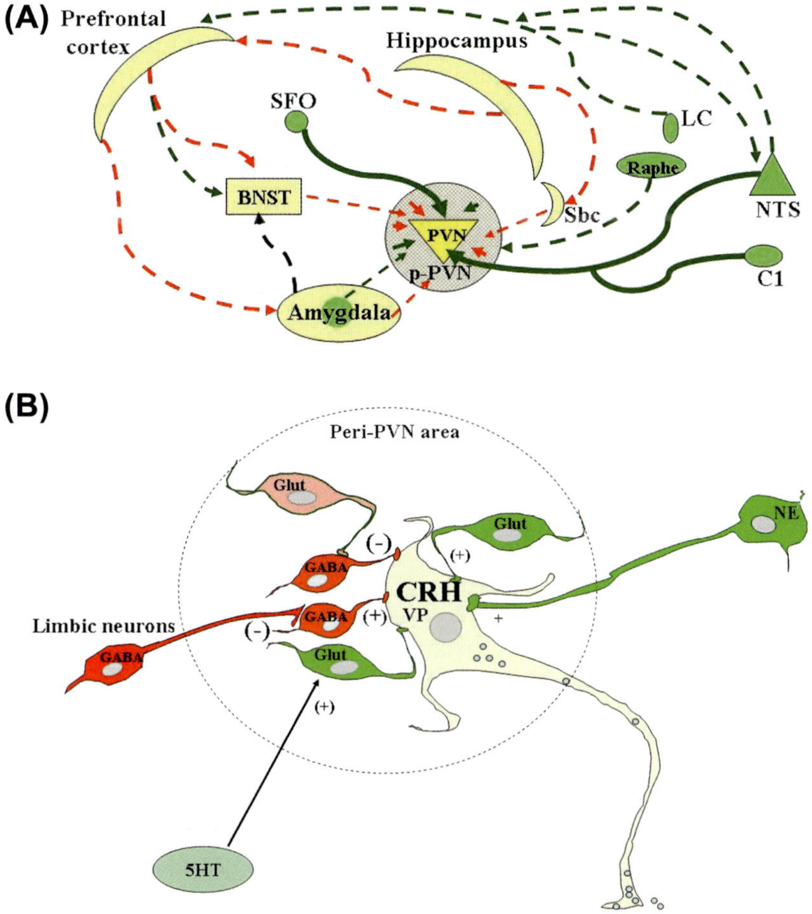

FIGURE 8.6 **(A) Neural pathways regulating the CRH neuron in the PVN during stress.** Homeostatic stressors (blood loss, pain, hypoglycemia) signal through direct (solid green lines) projections to the CRH neuron from the nucleus of the solitary tract (NTS) and C1 adrenergic neurons in the brainstem. Signals from psychogenic stressors utilize multi-synaptic pathways from sensory organs, the locus coeruleus (LC) and brainstem nuclei through limbic structures, including the prefrontal cortex, bed nucleus of the stria terminalis (BNST), and amygdala. These pathways interact with the CRH neuron indirectly (dotted lines) through glutamatergic and GABAergic neurons in the peri-PVN area (p-PVN), which make direct synapses with the CRH neuron. The hippocampus sends inhibitory signals through the prefrontal cortex and subiculum (Sbc). The dorsal raphe projects serotoninergic signals to the peri-PNV area, and directly to the CRH neuron. In addition, hormonal signals from the periphery interact with the subfornical organ (SFO), a circumventricular organ outside the blood–brain barrier, which has direct projections to the CRH neuron. **(B) Innervation of the CRH neuron.** CRH neurons in the PVN receive direct inputs from noradrenergic neurons of the brainstem (NA) and glutamergic and GABAergic interneurons in the peri-PVN area dotted circle. Other pathways regulate the CRH neuron by stimulating or inhibiting peri-PVN area interneurons. Limbic projections from the BNST and amygdala are mostly GABAergic, but stimulate CRH neurons by inhibiting GABA interneurons. Stimulatory glutamatergic inputs on GABAergic neurons (pink) are inhibitory for the CRH neuron by stimulating GABAergic activity.

somatostatin and enkephalin. Some of these peptides, such as GLP-1 and NPY, have been shown to influence HPA axis activity by acting on the CRH neuron.[82–84]

In contrast to systemic stress, responses to psychogenic stressors utilize complex polysynaptic pathways, including the participation of several limbic structures, the most important being the medial prefrontal cortex, hippocampus, amygdala, and bed nucleus of the stria terminalis (Fig. 8.6A, broken lines). The prefrontal cortex supplies inhibitory and stimulatory inputs, depending on the region. While the dorsomedial and prelimbic areas have inhibitory projections for the HPA axis and autonomic activity, the ventromedial region is stimulatory, initiating HPA axis and autonomic responses to psychogenic stress.[79] The prefrontal cortex inputs are mediated through interconnections with other limbic structures, such as the hippocampus, ventral subiculum and amygdala.

The hippocampus is predominantly inhibitory to the HPA axis and autonomic responses to psychogenic stressors. Hippocampal inputs reach the PVN indirectly through multi-synaptic pathways via the ventral subiculum for the HPA axis, and the prefrontal cortex for autonomic activity. These regions in turn innervate the periventricular region containing glutamatergic and GABAergic projections to CRH neurons[85] (Fig. 8.6B). The hippocampus acts to reduce the duration of HPA axis responses rather than the magnitude, as hippocampal or ventral subicular lesions prolong corticosterone responses to psychogenic stress.

The central nucleus of the amygdala, which expresses high levels of CRH, is essential for behavioral responses (especially fear) and for integrating autonomic responses during psychogenic stress. The basolateral and medial amygdala have positive modulatory actions on HPA axis activity, as lesions of these regions reduce responses to psychogenic stressors.

The bed nucleus of the stria terminalis (BNST) acts as an integrative center for limbic inputs to the PVN, and in fact most limbic inputs to the PVN relay in neurons of this nucleus. Lesion studies have shown that the anteroventral and anterolateral sub-regions relay excitatory pathways, while the posteromedial BNST conveys inhibitory signals. BNST inputs to the PVN are indirect, involving synapses with periventricular GABAergic and glutamatergic neurons. Most neurons in the BNST are GABAergic (inhibitory); thus, excitatory signals are likely to occur through inhibition of GABAergic periventricular interneurons.[85]

Chronic Stress

As indicated above, HPA axis responsiveness undergoes adaptation when the stress is repeated, and there is usually a hypersensitivity to a novel stress. This requires recruitment of new neural pathways not normally involved in the acute stress response. It is clear that exposure to many chronic stress paradigms (but not all) decreases the number of dendritic spines in the hippocampus and frontal cortex.[86,87] With repeated restraint stress (but not with variable inescapable stress) there are increases in dendritic projections and spines in the amygdala – changes converse to those in the hippocampus.[88] Neuronal plasticity at these sites and other brain regions controlling HPA and autonomic responses to stress is likely to mediate changes in HPA axis responsiveness typical of chronic stress. For example, increased inhibitory synaptic activity in the amygdala could mediate the habituation to homotypic repeated stressors. These changes are transient, and revert to normal 10–14 days after termination of the stress.[88] Another important nucleus involved in the regulation of HPA axis responsiveness during chronic stress is the paraventricular nucleus of the thalamus, serving as a relay center for both stimulatory and inhibitory inputs. While lesions of this nucleus have no effect on the acute stress response, they do prevent habituation to a homotypic repeated stress, as well as sensitization to a novel stimulus during repeated stress.[89,90]

Peripheral Influences on Stress Responses

Stress responses are modulated by peripheral hormonal signals, including a number of peptides, sex steroids and, most importantly adrenal glucocorticoids. Peripheral peptide hormones cannot cross the blood–brain barrier, but can signal in the brain through neural pathways from the circumventricular organs – structures which are outside the blood–brain barrier. These include the nucleus of the solitary tract (NTS); the subfornical organ (SFO); structures in the lamina terminalis, including the median preoptic nucleus (MPO); and the organum vasculosum of the lamina terminalis (OVLT). All of these structures express very high levels of angiotensin receptors which mediate the central actions of peripheral angiotensin II.[91,92] The peptide is the end product of the renin–angiotensin system, which is stimulated during stress as a consequence of adrenergic stimulation of renin production by the kidney. Elevations in circulating angiotensin II stimulate the HPA axis activity through direct neural angiotensinergic pathways from the SFO to the PVN. Angiotensin stimulates CRH release via type 1 receptors located in CRH neurons.[42] Angiotensin II receptors in the NTS and lamina terminalis are important in the regulation of cardiovascular homeostasis during alterations of blood volume, and water and electrolyte balance.[93] In addition, the arcuate nucleus is sensitive to peripheral signals, such as changes in blood glucose and leptin, and conveys metabolic signals to the PVN.[94,95] There is evidence that peripheral prolactin is taken up by receptors in the choroid plexus and acts in the brain.[96] Prolactin has inhibitory effects on HPA axis activity, and it is believed that the peptide partially mediates the hyper-responsiveness of the HPA axis during lactation.[97]

Glucocorticoid Feedback

Inhibitory feedback by glucocorticoids plays a major role in reducing HPA axis responses to stress, by acting at the pituitary and central levels. A major site of glucocorticoid feedback inhibition occurs at the level of the pituitary corticotrope, where glucocorticoids inhibit both ACTH secretion and the transcription of the precursor pro-opiomelanocortin (POMC).[98,99] These effects are mediated by type 2 glucocorticoid receptors in the pituitary corticotrope, and require circulating glucocorticoid levels in the stress range.

In addition to the pituitary, glucocorticoid feedback controls basal ACTH levels and HPA axis responsiveness to stress at the central level, ultimately by regulating the activity of hypothalamic CRH neurons. There is evidence that glucocorticoids inhibit CRH expression and release directly in the PVN, but the most important effects are mediated by receptors located in limbic structures.[100] The hippocampus expresses high levels of both glucocorticoid receptor subtypes, mineralocorticoid (MR) and glucocorticoid (GR). The high-affinity MRs are sensitive to very low levels of circulating glucocorticoids, and in the hippocampus are essential for maintaining adequate basal activity of the HPA axis. The lower-affinity receptors, GRs, respond to higher levels of glucocorticoids, and participate in restraining HPA axis responses to stress. In addition, evidence from targeted GR knockout and overexpression mouse models indicates that GRs located in the prefrontal cortex play an important role in limiting HPA axis responses to psychogenic stress.[101] The effect of glucocorticoids on these limbic pathways ultimately results in inhibition of CRH neuron activity by decreasing the release of stimulatory neurotransmitters, or by increasing inhibitory transmitters. For example, glucocorticoids inhibit stress-induced norepinephrine release from the PVN,[102] and reduce the content of alpha adrenergic receptors in the PVN.[103]

It is generally accepted that glucocorticoids have direct inhibitory actions in the PVN, including repression of CRH transcription. This is supported by the presence of glucocorticoid receptors in CRH neurons, the inhibitory effect of intra-PVN glucocorticoid injection on CRH mRNA, and the reduction in CRH promoter activity in reporter gene assays following incubation with glucocorticoids.[104–108] However, it is not clear that the inhibition occurs directly at the CRH neurons. There is evidence that glucocorticoids mediate rapid inhibition of CRH neurons through non-genomic mechanisms. According to these studies, glucocorticoids acting through a putative plasma membrane receptor stimulate synthesis and release of endocannabinoids from CRH neurons, which in turn inhibit presynaptic glutamatergic terminals, resulting in inhibition of CRH neuron activity.[109] Although it is clear that glucocorticoids have direct inhibitory actions in the PVN, the identity of the receptors mediating the rapid effects, as well as whether glucocorticoids affect transcriptional activity of the CRH gene, remains to be elucidated.

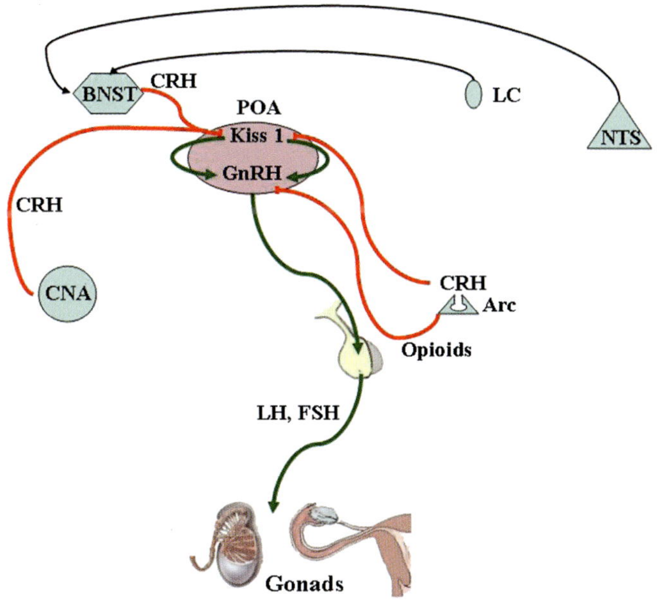

FIGURE 8.7 **Inhibitory effects of stress on the hypothalamic–pituitary–gonadal axis (HPG).** Gonadotropin-releasing hormone (GnRH) secreted into the pituitary portal circulation by neurons in the preoptic area stimulates the secretion of the gonadotropins, luteinizing hormone (LH) and follicle stimulating hormone (FSH), the regulators of gonadal function. Kisspeptin 1 (Kiss1) is an essential driver of GnRH pulse generation in the mPOA. Stimulation of CRH pathways from areas outside the PVN (bed nucleus stria terminalis (BNST), central nucleus of the amygdala, arcuate nucleus) inhibits the production of Kiss1 in the medial preoptic area (mPOA) and, consequently, GnRH secretion and gonadal function. Stress also increases opioid peptides in the arcuate nucleus which inhibit GnRH production. High glucocorticoid levels also affect gonadal function, but the effect depends on the species and sex-steroid status (not shown in the figure).

STRESS INFLUENCES ON OTHER NEUROENDOCRINE SYSTEMS

Stress and the Hypothalamic–Pituitary–Gonadal Axis

It is well established that stress suppresses the hypothalamic–pituitary–gonadal (HPG) axis in a variety of species, including non-human primates, sheep, and rats.[110–113] The primary controller of the HPG axis is a group of neurons located in the median preoptic area (mPOA), which release gonadotropin-releasing hormone (GnRH) in a pulsatile manner.[114,115] GnRH released into the pituitary portal circulation stimulates the secretion of the gonadotropins luteinizing hormone (LH) and follicle-stimulating hormone (FSH), the major regulators of ovarian and testicular function (Fig. 8.7). Stress inhibits the HPG axis at the central level by reducing GnRH production through different mechanisms, depending on the stress paradigm.[115] A major mediator in this inhibition is stress-induced stimulation of CRH. The negative effect of stress is suppressed by CRH antagonists but not by PVN lesions, indicating the involvement of extra-PVN sources of CRH, such as the amygdala, the BNST, the arcuate nucleus and the locus coeruleus.[116–118] Stimulation of adrenergic pathways during stress or norepinephrine injection leads to HPG axis inhibition by increasing CRH and by stimulating inhibitory GABAergic neurons in the mPOA.[115]

The mechanism by which CRH inhibits GnRH production involves different neural pathways and mediators. Both CRH receptor subtypes participate in the inhibitory process, depending on the stress paradigm. For example, the effect of restraint stress is prevented by CRHR1 antagonists, while the effect of hypoglycemia and immune stress is blocked by CRHR2 antagonists.[119] Type 2 CRH receptors have higher affinity for urocortin than for CRH, but the involvement of urocortin released during stress has not been studied. Although some reports argue for a direct inhibitory effect of CRH on the GnRH neuron, there is clear evidence that the effect is indirect, mainly by inhibiting the production of kisspeptin 1 in the mPOA and arcuate nucleus, and kisspeptin 1 receptors in the mPOA.[120] In addition, there is evidence that stress and CRH stimulate the secretion of other inhibitors of the GnRH neuron, including opiate peptides in the arcuate nucleus and the novel negative regulator of GnRH, RF-amide-related peptide.[111,121,122]

Whether elevations of glucocorticoids during stress have a direct role in inhibiting HPG axis activity is less clear, and the effects appear to be dependent on the species and experimental conditions. Chronic glucocorticoid administration decreases LH pulse frequency in monkeys and women[111,123,124] but not in rats.[125] Acute administration of high doses of glucocorticoids has no effect on LH pulses in women or in ewes in the follicular phase, but has a suppressing effect in ovariectomized ewes receiving estrogen and progesterone.[123] The precise site of action of glucocorticoids, at the pituitary or hypothalamic levels, and how glucocorticoids interact with gonadal steroids remains to be elucidated.

It should be noted that while it is clear that stress and especially CRH can inhibit HPG activity, it appears that activation of the HPA axis is not the only mediator of inhibition in LH secretion during stress. In this regard, CRH knockout and wild-type mice display equal suppression of the LH surge during pro-estrus in response to restraint stress or food withdrawal.[126] Although the development of compensatory mechanisms to the CRH deficiency cannot be ruled out, it is apparent that CRH is not required for the inhibition of the HPG axis induced by stress, and that factors other than CRH are able to suppress LH secretion during stress.

Stress and the Hypothalamic–Pituitary–Thyroid Axis

The hypothalamic–pituitary–thyroid axis (HPT axis) is under the control of neurons located in the medial region of the PVN that synthesize and release thyrotropin-releasing hormone (TRH) into the pituitary portal circulation. TRH stimulates the release of thyrotropin (TSH) from the anterior pituitary, which in turn stimulates the synthesis and release of thyroid hormones, 3,5,3′,5′-tetraiodothyronine (T_4) and 3,5,3′-triiodothyronine (T_3). T_3 is the main biologically active form because of its greater affinity for thyroid hormone receptors. T_4 is converted to T_3 by the activity of deiodinase enzymes located within most target tissues, including the central nervous system. Thyroid hormones are key regulators of metabolic activity, and are essential for adaptation to cold exposure or changes in energy supply (Fig. 8.8).

Acute stress causes transient activation of the HPT axis, which is mediated by increases in circulating TSH as a result of a direct stimulatory effect of glucocorticoids on the pituitary thyrotrope.[127] However, prolonged stress is invariably associated with decreased hypothalamic–pituitary–thyroid (HPT) activity in humans and experimental animals.[128–130] This leads to decreased metabolic activity, which contributes to energy conservation during chronic stress. Similarly, reduced HPT activity is observed following prolonged

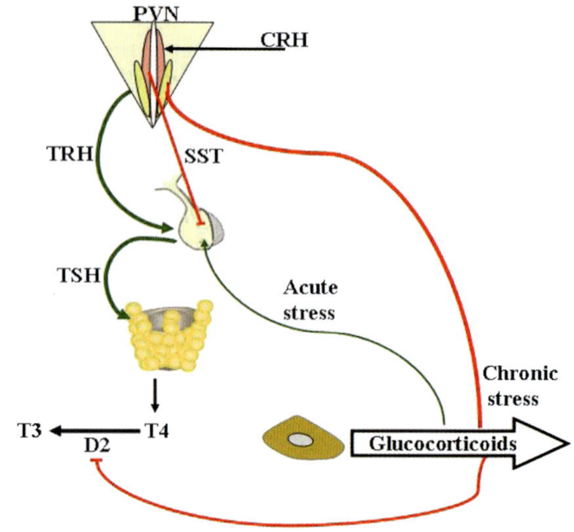

FIGURE 8.8 **Effects of stress on the hypothalamic–pituitary–thyroid (HPT) axis.** Thyrotropin-releasing hormone (TRH), released into the pituitary portal circulation from neurons in the medial region of the PVN (labeled in yellow), stimulates the secretion of thyrotropin (TSH) in the anterior pituitary, TSH stimulates the secretion of tetraiodothyronine (T4) from the thyroid gland. T4 is converted to the active hormone tri-iodo tyronine (T3) by deiodinase 2 (D2) in tissues, especially in the liver. During acute stress, glucocorticoids stimulate T4 production directly in the pituitary thyrotrope. Chronic stress inhibits the HPT axis through different mechanisms: (a) high levels of glucocorticoids inhibit THR production by the TRH neuron; (b) CRH produced outside the PVN stimulates somatostatin (SST) production by neurons located in the periventricular area (in orange), which inhibits TSH production; and (c) Glucocorticoids reduce the conversion of T4 to T3 by inhibiting D2.

exogenous administration or endogenous overproduction of glucocorticoids.[131] The inhibition is largely mediated by an inhibitory effect of glucocorticoids on TRH production in the hypothalamus. Enhanced somatostatin secretion as a result of increased intrahypothalamic CRH release may also contribute to decreasing TSH secretion during stress, though somatostatin is a weaker inhibitor of TSH secretion compared with growth hormone. CRH has no effect on pituitary TSH secretion in mammals, though it acts as a stimulant in amphibians.[132] Glucocorticoids do not affect the stimulatory effect of exogenous TSH on thyroid hormone production, thyroid hormone turnover, or levels of thyroxin-binding globulin.[131,133]

In addition to the inhibition in THR/TSH production, prolonged activation of the HPA axis also reduces the conversion of T4 to biologically active T3 in peripheral tissues – an effect that is mediated by glucocorticoids. Additional factors involved in HPT axis inhibition by stress include a number of cytokines. Inflammatory stress and, to a lesser extent, other stressors induce cytokine production.[134] TNF and other cytokines decrease THS bioactivity by preventing glycosylation and inhibiting deiodinase-2 in peripheral tissues.

Stress and the Hypothalamic-Growth Hormone Axis

Activation of the HPA axis has complex effects on the growth hormone (GH) axis (Fig. 8.9). The secretion of GH by the pituitary somatotrope is under the stimulatory control of growth hormone-releasing hormone (GHRH) produced in the arcuate nucleus, and the inhibitory effect of somatostatin, the 14-amino acid peptide produced in the periventricular nucleus. Somatostatin has dual inhibitory actions by suppressing GHRH secretion by the arcuate nucleus, and by directly restraining the pituitary somatotrope.[135] Growth hormone exerts its effects on growth and metabolism by stimulating the production and secretion of the mediator, insulin-like growth factor-1 (IGF-1), in the liver. Acute stress increases plasma GH levels in humans and experimental animals.[136,137] This effect is mediated by direct stimulatory effects of glucocorticoids in the pituitary somatotrope at the transcriptional and post-transcriptional levels.[138,139] However, prolonged activation of the HPA axis reduces circulating growth hormone at the central level – an effect that involves CRH. Intracerebroventricular injection of CRH markedly inhibits growth hormone secretion[140,141] by stimulating the production and secretion of the somatostatin.[140] As indicated above, somatostatin is a potent suppressor of GH, both by acting directly in the pituitary and by inhibiting the secretion of GHRH in the hypothalamus.[140] CRH has no direct effects on GH secretion in pituitary somatotropes.

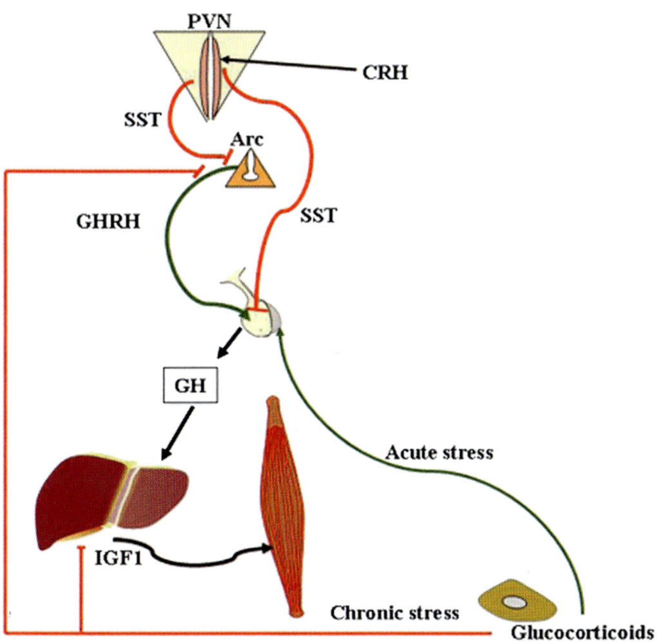

FIGURE 8.9 **Effects of stress on the hypothalamic–pituitary–growth hormone axis.** Growth hormone-releasing hormone (GHRH), produced in the arcuate nucleus (Arc), secreted into the pituitary portal circulation stimulates the secretion of growth hormone (GH) by the anterior pituitary somatotrope. Growth hormone exerts its anabolic actions through the production of insulin-like growth factor 1 (IGF1) in the liver. Acute stress increases GH secretion through a direct effect of glucocorticoids on GH transcription and translation. Chronic stress inhibits the HGH axis by different mechanisms: (a) increased CRH production by stress stimulates the secretion of somatostatin (SST), the potent inhibitor of GH; (b) high glucocorticoid levels inhibit GHRH in the arcuate nucleus; and (c) glucocorticoids impair GH signaling by inhibiting IGF1 production.

High levels of glucocorticoids also have direct effects on the growth hormone axis. These effects are complex, since dexamethasone injection decreases the expression of both GHRH in the arcuate nucleus, and somatostatin.[142] In addition to the effects of stress on growth hormone production, prolonged exposure to high glucocorticoid levels impairs growth hormone signaling in the target tissues by reducing IGF-1 production and IGF-1 receptors in target tissues. The negative effects of high glucocorticoid levels on growth are clearly evident in humans (especially in children) during glucocorticoid therapy or with Cushing's syndrome. Less pronounced increases in plasma glucocorticoid levels during chronic stress may also have direct effects on the GH axis activity, but the predominant factor during stress is CRH-induced inhibition of growth hormone production.

Stress and Energy Homeostasis

Stress leads to increased metabolic activity, which is necessary to sustain the increases in heart rate, blood pressure and skeletal muscle activity required for stress adaptation. This increase in energy expenditure is associated with protein catabolism and lipolysis, gluconeogenesis, glycogenolysis, and glucose release from the liver. The metabolic effects of stress are largely mediated by increases in glucocorticoids and catecholamines secondary to activation of the HPA axis and the sympathetic system. On the other hand, appetite is inhibited by stress, and this anorexigenic effect is largely mediated by CRH release at the hypothalamic level.

There is a close interaction between the HPA axis and the hypothalamic systems controlling energy homeostasis. Peripheral signals such as plasma glucose levels, insulin and leptin are detected by the arcuate nucleus.[95] CRH receptors are abundant in the arcuate nucleus, which is the site of production of the orexigenic peptides neuropeptide Y (NPY) and Agouti related protein (AGRP), and of the anorexigenic peptides α-MSH and cocaine, and amphetamine regulated transcript (CART). CRH receptors are also present in the ventromedial hypothalamic nucleus (VMH), an important site in appetite regulation. Localization of CRH receptors in these regions suggests an effect of CRH in regulating the expression of orexigenic and anorexigenic neuropeptides, but most evidence indicates that CRH exerts its anorexigenic actions in the PVN by inhibiting the action of orexigenic peptides. Peptidergic neurons of the arcuate nucleus have direct projections to the PVN, and arcuate peptides, including α-MSH and NPY, have been shown to activate the HPA axis by stimulating the secretion of CRH.[143–146] Microinjection of CRH antagonists in the PVN, but not at other sites, inhibits the orexigenic action of NPY.[147] There is also evidence that inhibitory effects of leptin on food intake are mediated by CRH, as leptin administration increases hypothalamic CRH content, and intracerebroventricular injection of a CRH antagonist prevents the anorexigenic action of leptin.[148] In addition to CRH, high levels of glucocorticoids have direct effects on neuropeptides in the arcuate nucleus.[149]

References

1. McEwen BS. Physiology and neurobiology of stress and adaptation: central role of the brain. *Physiol Rev.* 2007;87: 873–904.
2. Ulrich-Lai YM, Herman JP. Neural regulation of endocrine and autonomic stress responses. *Nat Rev Neurosci.* 2009;10: 397–409.
3. Aguilera G. *Regulation of Pituitary ACTH Secretion During Chronic Stress Frontiers in Neuroendocrinology.* 1994;15:321–350.
4. Antoni FA. Hypothalamic control of adrenocorticotropin secretion: advances since the discovery of 41-residue corticotropin-releasing factor. *Endocr Rev.* 1986;7:351–378.
5. Akana SF, Dallman MF. Chronic cold in adrenalectomized, corticosterone (B)-treated rats: facilitated corticotropin responses to acute restraint emerge as B increases. *Endocrinology.* 1997;138:3249–3258.
6. Swanson LW, Sawchenko PE. Hypothalamic integration: organization of the paraventricular and supraoptic nuclei. *Ann Rev NeuroSci.* 1983;6:269–324.
7. Antoni FA. Vasopressinergic control of pituitary adrenocorticotropin secretion comes of age. *Front Neuroendocrinol.* 1993;14: 76–122.
8. Palkovits M, Baffi J, Pacak K. The role of ascending neuronal pathways in stress-induced release of noradrenaline in the hypothalamic paraventricular nucleus of rats. *J Neuroendocrinol.* 1999;11:529–539.
9. Swanson L, Sawchenko P, Lind R, Rho JH. The CRH motoneuron: differential peptide regulation in neurons with possible synaptic, paracrine, and endocrine outputs. *Ann NY Acad Sci.* 1987;512:12–23.
10. Bale TL, Vale WW. CRF and CRF Receptors: Role in stress responsivity and other behaviors. *Ann Rev Pharmacol Toxicol.* 2004;44:525–557.
11. Swanson LW, Sawchenko PE, Rivier J, Vale WW. Organization of ovine corticotropin-releasing factor immunoreactive cells and fibers in the rat brain: an immunohistochemical study. *Neuroendocrinology.* 1983;36:165–186.
12. Murase M, Uemura T, Kondoh Y, Funabashi T, Hirahara F. Role of corticotropin-releasing hormone in ovarian steroidogenesis. *Endocrine.* 2002;18:255–260.
13. Perkins AV, Linton EA. Identification and isolation of corticotrophin-releasing hormone-positive cells from the human placenta. *Placenta.* 1995;16:233–243.
14. Slominsky A, Wortsman J, Pisarchik A, Zbytek B, Linton EA, Mazurkiewcz JE, Wei ET. Cutaneous expression of corticotropin-releasing hormone (CRH), urocortin, and CRH receptors. *FASEB J.* 2001;15:1678–1693.
15. Kuperman Y, Chen A. Urocortins: emerging metabolic and energy homeostasis perspectives. *Trends Endocrin Met.* 2005;19: 122–129.
16. Bale T, Hoshijima M, Gu Y, Dalton N, Anderson K, Lee K-F, Rivier J, Chien K, Vale W, Peterson K. The cardiovascular physiologic actions of urocortin II: acute effects in murine heart failure. *Proc Natl Acad Sci USA.* 2004;101:3697–3702.
17. Perrin MH, Vale WW. Corticotropin releasing factor receptors and their ligand family. *Ann NY Acad Sci.* 1999;885:312–328.
18. Aguilera G, Nikodemova M, Wynn PC, Catt KJ. Corticotropin releasing hormone receptors: two decades later. *Peptides.* 2004;25:319–329.
19. Luo X, Kiss A, Makara G, Lolait SJ, Aguilera G. Stress-specific regulation of corticotropin releasing hormone receptor expression in the paraventricular and supraoptic nuclei of the hypothalamus in the rat. *J Neuroendocrinol.* 1994;6:689–696.
20. Wynn PC, Hauger RL, Holmes MC, Millan MA, Catt KJ, Aguilera G. Brain and pituitary receptors for corticotropin releasing factor: localization and differential regulation after adrenalectomy. *Peptides.* 1984;5:1077–1084.
21. Chalmers DT, Lovenberg TW, De Souza EB. Localization of novel corticotropin-releasing factor receptor (CRF2) mRNA expression to specific subcortical nuclei in rat brain: comparison with CRF1 receptor mRNA expression. *J Neurosci.* 1995;15: 6340–6345.
22. Smagin GN, Heinrichs SC, Dunn AJ. The role of CRH in behavioral responses to stress. *Peptides.* 2001;22:713–724.
23. Potter E, Behan DP, Linton EA, Lowry PJ, Sawchenko PE, Vale WW. The central distribution of a corticotropin-releasing factor (CRF)-binding protein predicts multiple sites and modes of interaction with CRF. *Proc Natl Acad Sci USA.* 1992;89: 4192–4196.
24. Petraglia F, Potter E, Cameron VA, Sutton S, Behan DP, Woods RJ, Sawchenko PE, Lowry PJ, Vale W. Corticotropin-releasing factor-binding protein is produced by human placenta and intrauterine tissues. *J Clin Endocrinol Metab.* 1993;77: 919–924.
25. Karolyi IJ, Burrows HL, Ramesh TM, Nakajima M, Lesh JS, Seong E, Camper SA, Seasholtz AF. Altered anxiety and weight gain in corticotropin-releasing hormone-binding protein-deficient mice. *Proc Natl Acad Sci USA.* 1999;96: 11595–11600.
26. Aguilera G, Subburaju S, Young S, Chen J. The parvocellular vasopressinergic system and responsiveness of the hypothalamic–pituitary–adrenal axis during chronic stress. *Progr Brain Res.* 2008;170:29–39.
27. Alescio-Lautier B, Paban V, Soumireu-Mourat B. Neuromodulation of memory in the hippocampus by vasopressin. *Eur J Pharmacol.* 2000;405:63–72.
28. Jard S, Barberis C, Audigier S, Tribollet E. Neurohypophyseal hormone receptor systems in brain and periphery. *Prog Brain Res.* 1987;72:173–187.
29. Peter J, Burbach H, Adan RA, Lolait SJ, van Leeuwen FW, Mezey E, Palkovits M, Barberis C. Molecular neurobiology and pharmacology of the vasopressin/oxytocin receptor family. *Cell Mol Neurobiol.* 1995;15:573–595.
30. Lolait SJ, O'Carroll AM, Brownstein MJ. *Molecular biology of vasopressin receptors Ann NY Acad Sci.* 1995;771:273–292.
31. Seibold A, Brabet P, Rosenthal W, Birnbaumer M. Structure and chromosomal localization of the human antidiuretic hormone receptor gene. *Am J Hum Genet.* 1992;51:1078–1083.

32. Sugimoto T, Saito M, Mochizuki S, Watanabe Y, Hashimoto S, Kawashima H. Molecular cloning and functional expression of a cDNA encoding the human V1b vasopressin receptor. *J Biol Chem.* 1994;269:27088–27092.
33. Szot P, Bale TL, Dorsa DM. Distribution of messenger RNA for the vasopressin V1a receptor in the CNS of male and female rats. *Mol Brain Res.* 1994;24:1–10.
34. Vaccari C, Lolait SJ, Ostrowski NL. Comparative distribution of vasopressin V1b and oxytocin receptor messenger ribonucleic acids in brain. *Endocrinology.* 1998;139:5015–5033.
35. Young LJ, Toloczko D, Insel TR. Localization of vasopressin (V1a) receptor binding and mRNA in the rhesus monkey brain. *J Neuroendocrinol.* 1999;11:291–297.
36. Herman JP, Schäfer MK, Watson SJ, Sherman TG. *In situ* hybridization analysis of arginine vasopressin gene transcription using intron-specific probes. *Mol Endocrinol.* 1991;5:1447–1456.
37. Kovacs KJ, Sawchenko PE. Sequence of stress-induced alterations in indices of synaptic and transcriptional activation in parvocellular neurosecretory neurons. *J Neurosci.* 1996;16:262–273.
38. Ma XM, Aguilera G. Transcriptional responses of the vasopressin and corticotropin-releasing hormone genes to acute and repeated intraperitoneal hypertonic saline injection in rats. *Brain Res Mol Brain Res.* 1999;68:129–140.
39. Alexander SL, Irvine CH, Donald RA. Short-term secretion patterns of corticotropin-releasing hormone, arginine vasopressin and ACTH as shown by intensive sampling of pituitary venous blood from horses. *Neuroendocrinology.* 1994;60:225–236.
40. Plotsky PM. Hypophysiotropic regulation of stress-induced ACTH secretion. *Adv Exp Med Biol.* 1988;245:65–81.
41. Rivier CL, Plotsky PM. Mediation by Corticotropin Releasing Factor (CRF) of adenohypophysial hormone secretion. *Ann Rev Physiol.* 1986;48:475–494.
42. Aguilera G. Direct regulation of hypothalamic corticotropin-releasing-hormone neurons by angiotensin II. *Neuroendocrinology.* 1995;61:437–444.
43. Aubry JM, Bartanusz V, Jezova D, Belin D, Kiss JZ. Single stress induces long-lasting elevations in vasopressin mRNA levels in CRF hypophysiotrophic neurones, but repeated stress is required to modify AVP immunoreactivity. *J Neuroendocrinol.* 1999;11:377–384.
44. de Goeij DC, Kvetnansky R, Whitnall MH, Jezova D, Berkenbosch F, Tilders FJ. Repeated stress-induced activation of corticotropin-releasing factor neurons enhances vasopressin stores and colocalization with corticotropin-releasing factor in the median eminence of rats. *Neuroendocrinology.* 1991;53:150–159.
45. Ma XM, Lightman SL, Aguilera G. Vasopressin and corticotropin-releasing hormone gene responses to novel stress in rats adapted to repeated restraint. *Endocrinology.* 1999;140:3623–3632.
46. Guardiola-Diaz HM, Boswell C, Seasholtz AF. The cAMP-responsive element in the corticotropin-releasing hormone gene mediates transcriptional regulation by depolarization. *J Biol Chem.* 1994;269:14784–14791.
47. Nicholson RC, King BR, Smith R. Complex regulatory interactions control CRH gene expression. *Front Biosci.* 2004;9:32–39.
48. Liu Y, Coello AG, Grinevich V, Aguilera G. Involvement of transducer of regulated cAMP response element-binding protein activity on corticotropin releasing hormone transcription. *Endocrinology.* 2010;151:1109–1118.
49. Conkright MD, Canettieri G, Screaton R, Guzman E, Miraglia L, Hogenesch JB, Montminy M. TORCs: Transducers of Regulated CREB activity. *Mol Cell.* 2003;12:413–423.
50. Takemori H, Kanematsu M, Kajimura J, Hatano O, Katoh Y, Lin XZ, Min L, Yamazaki T, Doi J, Okamoto M. Dephosphorylation of TORC initiates expression of the StAR gene. *Mol Cell Endocrinol.* 2007;265–266:196–204.
51. Stroth N, Eiden LE. Stress hormone synthesis in mouse hypothalamus and adrenal gland triggered by restraint is dependent on pituitary adenylate cyclase-activating polypeptide signaling. *Neuroscience.* 2010;165:1025–1030.
52. Ma XM, Levy A, Lightman SL. Rapid changes of heteronuclear RNA for arginine vasopressin but not for corticotropin releasing hormone in response to acute corticosterone administration. *J Neuroendocrinol.* 1997;9:723–728.
53. Shepard JD, Liu Y, Sassone-Corsi P, Aguilera G. Role of glucocorticoids and cAMP-mediated repression in limiting corticotropin-releasing hormone transcription during stress. *J Neurosci.* 2005;25:4073–4081.
54. Molina CA, Foulkes NS, Lalli E, Sassone-Corsi P. Inducibility and negative autoregulation of CREM: an alternative promoter directs the expression of ICER, an early response repressor. *Cell.* 1993;75:875–886.
55. Liu Y, Aguilera G. Cyclic AMP inducible early repressor mediates the termination of corticotropin releasing hormone transcription in hypothalamic neurons. *Cell Mol Neurobiol.* 2009;29:1275–1281.
56. Eipper BA, Mains RE. Structure and biosynthesis of pro-adrenocorticotropin/endorphin and related peptides. *Endocr Rev.* 1980;1:1–27.
57. Vale W, Rivier C, Brown MR, Spiess J, Koob G, Swanson L, Bilezikjian L, Bloom F, Rivier J. Chemical and biological characterization of corticotropin releasing factor. *Recent Prog Horm Res.* 1983;39:245–270.
58. Autelitano DJ, Lundblud JR, Blum M, Roberts JL. Hormonal regulation of POMC gene expression. *Ann Rev Physiol.* 1989;51:715–726.
59. Abou-Samra AB, Harwood JP, Manganiello VC, Catt KJ, Aguilera G. Phorbol 12-myristate 13-acetate and vasopressin potentiate the effect of corticotropin-releasing factor on cyclic AMP production in rat anterior pituitary cells. Mechanisms of action. *J Biol Chem.* 1987;262:1129–1136.
60. Gillies GE, Linton EA, Lowry PJ. Corticotropin releasing activity of the new CRF is potentiated several times by vasopressin. *Nature.* 1982;299:355–357.
61. Levin N, Blum M, Roberts JL. Modulation of basal and corticotropin-releasing factor-stimulated proopiomelanocortin gene expression by vasopressin in rat anterior pituitary. *Endocrinology.* 1989;125:2957–2966.
62. Aguilera G. Corticotropin releasing hormone, receptor regulation and the stress response. *Trends Endocrinol Metab.* 1998;9:329–336.
63. Hauger RL, Aguilera G. Regulation of pituitary corticotropin releasing hormone (CRH) receptors by CRH: interaction with vasopressin. *Endocrinology.* 1993;133:1708–1714.
64. Hauger RL, Millan MA, Lorang M, Harwood JP, Aguilera G. Corticotropin-releasing factor receptors and pituitary adrenal responses during immobilization stress. *Endocrinology.* 1988;123:396–405.
65. Aguilera G, Pham Q, Rabadan-Diehl C. Regulation of pituitary vasopressin receptors during chronic stress: relationship to corticotroph responsiveness. *J Neuroendocrinol.* 1994;6:299–304.
66. Roper JA, O'Carroll AM, Young WS, Lolait SJ. The vasopressin Avpr1b receptor: molecular and pharmacological studies. *Stress.* 2011;14:98–115.
67. Subburaju S, Aguilera G. Vasopressin mediates mitogenic responses to adrenalectomy in the rat anterior pituitary. *Endocrinology.* 2007;148:3102–3110.

68. Childs GV, Lloyd J, Unabia G, Rougeau D. Growth and secretory responses of enriched populations of corticotropes. *Endocrinology.* 1989;125:2540–2549.
69. Lightman SL, Conway-Campbell BL. The crucial role of pulsatile activity of the HPA axis for continuous dynamic equilibration. *Nat Rev Neurosci.* 2010;11:710–718.
70. Buckingham JC. Glucocorticoids: exemplars of multi-tasking. *Br J Pharmacol.* 2006;147(S1):S258–S268.
71. Vegiopoulos A, Herzig S. Glucocorticoids, metabolism and metabolic diseases. *Mol Cell Endocrinol.* 2007;275:43–61.
72. Reul Jmhm, de Kloet ERD. Two receptor systems for corticosterone in rat brain: microdistribution and differential occupation. *Endocrinology.* 1985;117:2505–2511.
73. Evanson NK, Herman JP, Sakai RR, Krause. EG. Nongenomic actions of adrenal steroids in the central nervous system. *J Neuroendocrinol.* 2010;22:846–861.
74. Patel PD, Lopez JF, Lyons DM, Burke S, Wallace M, Schatzberg AF. Glucocorticoid and mineralocorticoid receptor mRNA expression in squirrel monkey brain. *J Psychiat Res.* 2000;34:383–392.
75. Reul JMHM, Gesing A, Droste S, Stec ISM, Weber A, Bachmann C, Bilang-Bleuel A, Holsboer F, Linthorst ACE. The brain mineralocorticoid receptor: greedy for ligand, mysterious in function. *Eur J Pharmacol.* 2000;405:235–249.
76. Aronsson M, Fuxe K, Dong Y, Agnati LF, Okret S, Gustafsson JA. Localization of glucocorticoid receptor mRNA in the male rat brain by *in situ* hybridization. *Proc Natl Acad Sci USA.* 1988;85:9331–9335.
77. Cullinan WE, Herman JP, Watson SJ. Ventral subicular interaction with the hypothalamic paraventricular nucleus: evidence for a relay in the bed nucleus of the stria terminalis. *J Comp Neurol.* 1993;332:1–20.
78. Prewitt CMF, Herman JP. Anatomical interactions between the central amygdaloid nucleus and the hypothalamic paraventricular nucleus of the rat: a dual tract-tracing analysis. *J Chem Neuroanat.* 1998;15:173–186.
79. Herman JP, Figueiredo H, Mueller NK, Ulrich-Lai Y, Ostrander MM, Choi DC, Cullinan WE. Central mechanisms of stress integration: hierarchical circuitry controlling hypothalamo-pituitary–adrenocortical responsiveness. *Front Neuroendocrinol.* 2003;24:151–180.
80. Pacak K, Palkovits M. Stressor specificity of central neuroendocrine responses: implications for stress-related disorders. *Endocr Rev.* 2001;22:502–548.
81. Palkovits M, Zaborszky L, Feminger A, Mezey E, Fekete MIK, Herman JP, Kanyicska B, Szabo D. Noradrenergic innervation of the rat hypothalamus: Experimental biochemical and electron microscopic studies. *Brain Res.* 1980;191:161–171.
82. Nakade Y, Tsukamoto K, Iwa M, Pappas TN, Takahashi T. Glucagon like peptide-1 accelerates colonic transit via central CRF and peripheral vagal pathways in conscious rats. *Autonom Neurosci.* 2007;131:50–56.
83. Suda T, Tozawa F, Iwai I, Sato Y, Sumitomo T, Nakano Y, Yamada M, Demura H. Neuropeptide Y increases the corticotropin-releasing factor messenger ribonucleic acid level in the rat hypothalamus. *Mol Brain Res.* 1993;18:311–315.
84. Wahlestedt C, Skagerberg G, Ekman R, Heilig M, Sundler F, Neuropeptide Y. (NPY) in the area of the hypothalamic paraventricular nucleus activates the pituitary–adrenocortical axis in the rat. *Brain Res.* 1987;417:33–38.
85. Ziegler DR, Cullinan WE, Herman JP. Organization and regulation of paraventricular nucleus glutamate signaling systems: N-methyl-D-aspartate receptors. *J Comp Neurol.* 2005;484:43–56.
86. Chen Y, Rex CS, Rice C, Dube CM, Gall CM, Lynch G, Baram TZ. Correlated memory defects and hippocampal dendritic spine loss after acute stress involve corticotropin-releasing hormone signaling. *Proc Natl Acad Sci USA.* 2010;107:13123–13128.
87. McEwen BS. Stress and hippocampal plasticity. *Ann Rev NeuroSci.* 1999;22:105–122.
88. Vyas A, Mitra R, Shankaranarayana Rao BS, Chattarji S. Chronic stress induces contrasting patterns of dendritic remodeling in hippocampal and amygdaloid neurons. *J Neurosci.* 2002;22:6810–6818.
89. Bhatnagar S, Dallman M. Neuroanatomical basis for facilitation of hypothalamic–pituitary–adrenal responses to a novel stressor after chronic stress. *Neuroscience.* 1998;84:1025–1039.
90. Bhatnagar S, Huber R, Nowak N, Trotter P. Lesions of the posterior paraventricular thalamus block habituation of hypothalamic–pituitary–adrenal responses to repeated restraint. *J Neuroendocrinol.* 2002;14:403–410.
91. Cottrell GT, Ferguson AV. Sensory circumventricular organs: central roles in integrated autonomic regulation. *Reg Peptides.* 2004;117:11–23.
92. McKinley MJ, McAllen RM, Davern P, Giles ME, Penschow J, Sunn N, Oldfield BJ. The sensory circumventricular organs of the mammalian brain. *Adv Anat Embryol Cell Biol.* 2003;172:1–122.
93. McKinley M, Allen A, May C, McAllen R, Oldfield B, Sly D, Mendelsohn F. Neural pathways from the lamina terminalis influencing cardiovascular and body fluid homeostasis. *Clin Exp Pharmacol Physiol.* 2001;28:990–992.
94. Banks WA, Obesity. *Blood–Brain Barrier and Energy Balance* 2006;14:234S–237S.
95. Konner AC, Klockener T, Bruning JC. Control of energy homeostasis by insulin and leptin: targeting the arcuate nucleus and beyond. *Physiol Behav.* 2009;97:632–638.
96. Mangurian LP, Walsh RJ, Posner BI. Prolactin enhancement of its own uptake at the choroid plexus. *Endocrinology.* 1992;131:698–702.
97. Torner L, Neumann ID. The brain prolactin system: involvement in stress response adaptations in lactation. *Stress.* 2002;5:249–257.
98. Antoni FA. Mortyn Jones Memorial Lecture – 1995. Calcium checks cyclic AMP–corticosteroid feedback in adenohypophysial corticotrophs. *J Neuroendocrinol.* 1996;8:659–672.
99. Drouin J, Charron J, Gagner JP, Jeannotte L, Nemer M, Plante RK, Wrange O. Pro-opiomelanocortin gene: a model for negative regulation of transcription by glucocorticoids. *J Cell Biochem.* 1987;35:293–304.
100. Dallman MF, Akana SF, Strack AM, Scribner KS, Pecoraro N, La Fleur SE, Houshyar H, Gomez F. Chronic stress-induced effects of corticosterone on brain: direct and indirect. *Ann NY Acad Sci.* 2004;1018:141–150.
101. Furay AR, Bruestle AE, Herman JP. The role of the forebrain glucocorticoid receptor in acute and chronic stress. *Endocrinology.* 2008;149:5482–5490.
102. Pacak K, Palkovits M, Kvetnansky R, Matern P, Hart C, Kopin I, Goldstein DS. Catecholaminergic inhibition by hypercortisolemia in the paraventricular nucleus of conscious rats. *Endocrinology.* 1995;136:4814–4819.
103. Day H, Campeau S, Watson SJ, Akil H. Expression of alpha(1b) adrenoceptor mRNA in corticotropin-releasing hormone-containing cells of the rat hypothalamus and its regulation by corticosterone. *J Neurosci.* 1999;19:10098–10106.
104. Fenoglio KA, Brunson KL, vishai-Eliner S, Chen Y, Baram TZ. Region-specific onset of handling-induced changes in corticotropin-releasing factor and glucocorticoid receptor expression. *Endocrinology.* 2004;145:2702–2706.
105. Harbuz MS, Lightman SL. Glucocorticoid inhibition of stress-induced changes in hypothalamic corticotrophin-releasing

factor messenger RNA and proenkephalin a messenger RNA. *Neuropeptides.* 1989;14:17–20.
106. Kovacs K, Kiss JZ, Makara GB. Glucocorticoid implants around the hypothalamic paraventricular nucleus prevent the increase of corticotropin-releasing factor and arginine vasopressin immunostaining induced by adrenalectomy. *Neuroendocrinology.* 1986;44:229–234.
107. Liposits Z, Uht RM, Harrison RW, Gibbs FP, Paull WK, Bohn MC. Ultrastructural localization of glucocorticoid receptor (GR) in hypothalamic paraventricular neurons synthesizing corticotropin releasing factor (CRF). *Histochemistry.* 1987;87:407–412.
108. Sawchenko PE. Evidence for a local site of action for glucocorticoids in inhibiting CRF and vasopressin expression in the paraventricular nucleus. *Brain Res.* 1987:213–224.
109. Di S, Malcher-Lopes R, Halmos KC, Tasker JG. Nongenomic glucocorticoid inhibition via endocannabinoid release in the hypothalamus: a fast feedback mechanism. *J Neurosci.* 2003;23:4850–4857.
110. Wagenmaker ER, Breen KM, Oakley AE, Tilbrook AJ, Karsch FJ. Psychosocial Stress Inhibits Amplitude of Gonadotropin-Releasing Hormone Pulses Independent of Cortisol Action on the Type II Glucocorticoid Receptor. *Endocrinology.* 2009;150:762–769.
111. Ferin M. Stress and the reproductive cycle. *J Clin Endocrinol Metab.* 1999;84:1768–1774.
112. Tilbrook AJ, Canny BJ, Serapiglia MD, Ambrose TJ, Clarke IJ. Suppression of the secretion of luteinizing hormone due to isolation/restraint stress in gonadectomised rams and ewes is influenced by sex steroids. *J Endocrinol.* 1999;160:469–481.
113. Rivier C, Rivest S. Effect of stress on the activity of the hypothalamic–pituitary–gonadal axis: peripheral and central mechanisms. *Biol Reprod.* 1991;45:523–532.
114. Charlton H. Hypothalamic control of anterior pituitary function: a history. *J Neuroendocrinol.* 2008;20:641–646.
115. Li XF, Knox AMI, O'Byrne KT. Corticotrophin-releasing factor and stress-induced inhibition of the gonadotrophin-releasing hormone pulse generator in the female. *Brain Res.* 2010;1364: 153–163.
116. Campbell RE, Herbison AE. Definition of brainstem afferents to gonadotropin-releasing hormone neurons in the mouse using conditional viral tract tracing. *Endocrinology.* 2007;148: 5884–5890.
117. Keen-Rhinehart E, Michopoulos V, Toufexis DJ, Martin EI, Nair H, Ressler KJ, Davis M, Owens MJ, Nemeroff CB, Wilson ME. Continuous expression of corticotropin-releasing factor in the central nucleus of the amygdala emulates the dysregulation of the stress and reproductive axes. *Mol Psychiatry.* 2009;14:37–50.
118. Weller KL, Smith DA. Afferent connections to the bed nucleus of the stria terminalis. *Brain Res.* 1982;232:255–270.
119. Li XF, Bowe JE, Kinsey-Jones JS, Brain SD, Lightman SL, O'Byrne KT. Differential role of corticotrophin-releasing factor receptor types 1 and 2 in stress-induced suppression of pulsatile luteinising hormone secretion in the female rat. *J Neuroendocrinol.* 2006;18:602–610.
120. Li XF, Kinsey-Jones JS, Cheng Y, Knox AMI, Lin Y, Petrou NA, Roseweir A, Lightman SL, Milligan SR, Millar RP, O'Byrne KT. Kisspeptin signalling in the hypothalamic arcuate nucleus regulates GnRH pulse generator frequency in the rat. *PLoS ONE* 2009;4:e8334.
121. Clarke IJ, Qi Y, Puspita SI, Smith JT. Evidence that RF-amide related peptides are inhibitors of reproduction in mammals. *Front Neuroendocrinol.* 2009;30:371–378.
122. Kirby ED, Geraghty AC, Ubuka T, Bentley GE, Kaufer D. Stress increases putative gonadotropin inhibitory hormone and decreases luteinizing hormone in male rats. *Proc Natl Acad Sci USA.* 2009;106:11324–11329.
123. Breen KM, Karsch FJ. New insights regarding glucocorticoids, stress and gonadotropin suppression. *Front Neuroendocrinol.* 2006;27:233–245.
124. Saketos M, Sharma N, Santoro NF. Suppression of the hypothalamic–pituitary–ovarian axis in normal women by glucocorticoids. *Biol Reproduct.* 1993;49:1270–1276.
125. Li XF, Mitchell JC, Wood S, Coen CW, Lightman SL, O'Byrne KT. The effect of oestradiol and progesterone on hypoglycaemic stress-induced suppression of pulsatile luteinizing hormone release and on corticotropin-releasing hormone mRNA expression in the rat. *J Neuroendocrinol.* 2003;15:468–476.
126. Jeong K-H, Jacobson L, Widmaier EP, Majzoub JA. Normal suppression of the reproductive axis following stress in corticotropin-releasing hormone-deficient mice. *Endocrinology.* 1999;140:1702–1708.
127. Morley JE. Neuroendocrine control of thyrotropin secretion. *Endocr Rev.* 1981;2:396–436.
128. Van den Berghe G. Dynamic neuroendocrine responses to critical illness. *Front Neuroendocrinol.* 2002;23:370–391.
129. Weinmann M. Stress-induced hormonal alterations. *Crit Care Clin.* 2001;17:1–10.
130. Helmreich DL, Parfitt DB, Lu XY, Akil H, Watson SJ. Relation between the Hypothalamic–Pituitary–Thyroid (HPT) axis and the Hypothalamic–Pituitary–Adrenal (HPA) axis during repeated stress. *Neuroendocrinology.* 2005;81:183–192.
131. Bruhn T, Huang S, Vaslet C, Nillni E. Glucocorticoids modulate the biosynthesis and processing of proThyrotropin-releasing hormone (proTRH). *Endocrine.* 1998;9:143–152.
132. De Groef B, Van der Geyten S, Darras VM, Knhn ER. Role of corticotropin-releasing hormone as a thyrotropin-releasing factor in non-mammalian vertebrates. *Gen Comp Endocrinol.* 2006;146:62–68.
133. Nicoloff JT, Fisher DA, Appleman MD. The role of glucocorticoids in the regulation of thyroid function in man. *J Clin Invest.* 1970;49:1922–1929.
134. Elenkov IJ, Wilder RL, Chrousos GP, Vizi ES. The sympathetic nerve – an integrative interface between two supersystems: the brain and the immune system. *Pharmacol Rev.* 2000;52:595–638.
135. García-Tornadu I, Risso G, Perez-Millan MI, Noain D, Diaz-Torga G, Low MJ, Rubinstein M, Becu-Villalobos D. Neurotransmitter modulation of the GHRH–GH axis. *Front Horm Res.* 2010;38:59–69.
136. Raza J, Massoud AF, Hindmarsh PC, Robinson IC, Brook CG. Direct effects of corticotrophin-releasing hormone on stimulated growth hormone secretion. *Clin Endocrinol.* 1998;48:217–222.
137. Casanueva FF, Burguera B, Muruais C, Dieguez C. Acute administration of corticoids: a new and peculiar stimulus of growth hormone secretion in man. *J Clin Endocrinol Metab.* 1990;70:234–237.
138. Casanueva FF. Depending on the time of administration, dexamethasone potentiates or blocks growth hormone-releasing hormone-induced growth hormone release in man. *Neuroendocrinology.* 1988;47:46–49.
139. Karin M, Castrillo JL, Theill LE. Growth hormone gene regulation: a paradigm for cell-type-specific gene activation. *Trends Genet.* 1990;6:92–96.
140. Katakami H, Arimura A, Frohman LA. Involvement of hypothalamic somatostatin in the suppression of growth hormone secretion by central corticotropin-releasing factor in conscious male rats. *Neuroendocrinology.* 1985;41:390–393.
141. Rivier C, Vale W. Involvement of corticotropin-releasing factor and somatostatin in stress-induced inhibition of growth hormone secretion in the rat. *Endocrinology.* 1985;117:2478–2482.
142. Senaris RM, Lago F, Coya R, Pineda J, Dieguez C. Regulation of hypothalamic somatostatin, growth hormone-releasing

hormone, and growth hormone receptor messenger ribonucleic acid by glucocorticoids. *Endocrinology.* 1996;137: 5236–5241.
143. Smith SM, Vaughan JM, Donaldson CJ, Rivier J, Li C, Chen A, Vale WW. Cocaine- and amphetamine-regulated transcript activates the hypothalamic–pituitary–adrenal axis through a corticotropin-releasing factor receptor-dependent mechanism. *Endocrinology.* 2004;145:5202–5209.
144. Dhillo WS, Small CJ, Seal LJ, Kim MS, Stanley SA, Murphy KG, Ghatei MA, Bloom SR. The hypothalamic melanocortin system stimulates the hypothalamo-pituitary–adrenal axis *in vitro* and *in vivo* in male rats. *Neuroendocrinology.* 2002;75:209–216.
145. Dimitrov EL, DeJoseph MR, Brownfield MS, Urban JH. Involvement of neuropeptide Y Y1 receptors in the regulation of neuroendocrine corticotropin-releasing hormone neuronal activity. *Endocrinology.* 2007;148:3666–3673.
146. Fekete C, Legradi G, Mihaly E, Tatro JB, Rand WM, Lechan RM. [alpha]-Melanocyte stimulating hormone prevents fasting-induced suppression of corticotropin-releasing hormone gene expression in the rat hypothalamic paraventricular nucleus. *NeuroSci Letters.* 2000;289:152–156.
147. Heinrichs SC, Menzaghi F, Pich EM, Hauger RL, Koob GF. Corticotropin-releasing factor in the paraventricular nucleus modulates feeding induced by neuropeptide Y. *Brain Res.* 1993;611:18–24.
148. Uehara Y, Shimizu H, Ohtani K, Sato N, Mori M. Hypothalamic corticotropin-releasing hormone is a mediator of the anorexigenic effect of leptin. *Diabetes.* 1998;47:890–893.
149. Vrang N, Larsen PJ, Tang-Christensen M, Larsen LK, Kristensen P. Hypothalamic cocaine-amphetamine regulated transcript (CART) is regulated by glucocorticoids. *Brain Res.* 2003;965:45–50.

CHAPTER

9

Neuroendocrine Control of Reproduction

Iain J. Clarke[1], *Rebecca Campbell*[2], *Jeremy T. Smith*[1], *Vincent Prevot*[3], *Susan Wray*[4]

[1] Department of Physiology, Monash University, Clayton, Victoria, Australia, [2] Department of Physiology, School of Medical Science, University of Otago, Dunedin, New Zealand, [3] Inserm, Jean-Pierre Aubert Research Center, U837, Development and Plasticity of the Postnatal Brain, Lille, France; Université Nord de France, Lille, France; UDSL, Laboratory of Anatomy, School of Medicine, Place de Verdun, Lille, France; CHRU Lille, Department of Neurosurgery, Hôpital Roger Salengro, Lille, France, [4] Cellular and Developmental Neurobiology Section, NINDS, NIH, Bethesda, Maryland, USA

OUTLINE

Handbook of Neuroendocrinology, DOI: 10.1016/B978-0-12-375097-6.10009-5

Summary

In this chapter the operation of the hypothalamic-pituitary gonadal axis is considered, with special reference to the way that gonadotropin-releasing hormone (GnRH) and gonadotropin secretion is controlled by gonadal hormones. The nature of GnRH neurons and how they function is considered, especially with respect to the way that GnRH secretion is controlled by the feedback effects of gonadal steroids. Feedback effects are also important at the level of the gonadotropes, and this too is covered in detail. With this background, a working model of the female estrous/menstrual cycle is presented.

INTRODUCTION

Reproduction is totally dependent upon the secretion of gonadotropin-releasing hormone (GnRH) from the brain. GnRH provides trophic support and acts as a secretagog for the release of gonadotropins from the pituitary gonadotropes. The gonadotropins, luteinizing hormone (LH) and follicle-stimulating hormone (FSH) circulate in the bloodstream to act upon the gonads, supporting the production of gametes as well as stimulating secretion of gonadal steroids and proteinaceous hormones. In turn, the gonadal hormones provide feedback signals to the brain and the pituitary gland. This closed-loop system is characteristic of all physiological systems, which operate as self-regulating units, ensuring maintenance of function within normal limits. In this chapter, the control of secretion and action of GnRH will be outlined, with special attention to regulation by gonadal hormone feedback.

Early studies that indicated the essential role of GnRH in the process of reproduction were as follows:

- GnRH was shown to be secreted into hypophysial portal blood.[1–3]
- Active or passive immunization against GnRH led to the loss of reproductive function, in terms of gonadotropin secretion as well as gonadal function.[4]
- Animals (hpg mice) that are genetically infertile were found not to produce GnRH, but this could be corrected by implantation of normal fetal preoptic area into the brain.[5,6] These hpg mice were subsequently found to have a large deletion in their GnRH gene that could be corrected by gene therapy.[7,8]
- Individuals with Kallman's syndrome have anosmia and do not have reproductive function. The olfactory bulbs also fail to form or have hypoplasia, leading to anosmia or hyposmia. As will be described below, GnRH cells originate from the nasal placode and migrate to the brain. In Kallman's syndrome, the GnRH neurons do not migrate properly from the olfactory placode to the hypothalamus during development[9] (see also below).

At the level of the GnRH cells and gonadotropes, function is regulated by the feedback effects of gonadal steroids, whereas factors such as inhibin act at the level of the gonadotropes only. Within the brain, steroids act on cells with the relevant receptors and relay information to GnRH cells. In the past decade, significant advances have been made in our understanding of the means by which steroid feedback control of GnRH cells and gonadotropes is effected – especially the role of kisspeptin. The kisspeptin cells appear to play a major role in the relay of feedback effects to GnRH cells within the brain. In addition, it is now recognized that gonadotropin inhibitory hormone (GnIH) plays a role as a negative regulator of GnRH cells and gonadotropes.[10]

Whereas gonadal feedback is a fundamental determinant of how GnRH cells function, it is also recognized that there is a range of other internal and external regulatory factors. Endogenous factors (within the body) also modulate the function of GnRH cells and gonadotropes; these include hormones of the stress axis (cortisol)[11] and from fat (leptin),[12] amongst others. In addition, environmental factors such as pheromones,[13] photoperiod,[14] stress[15] and immune response[16] modulate GnRH secretion and act in concert with gonadal steroid feedback.[17,18] A considerable understanding of the mechanisms underlying seasonality of reproduction has been obtained from studies in the female sheep, which have been reviewed recently.[19] The changes in the reproductive axis that occur during puberty (Chapter 10) and aging (Chapter 27), and the impact of developmental disorders (Chapter 28) and rhythms (Chapter 12), are found elsewhere in this book.

DEVELOPMENT OF THE GNRH SYSTEM AND LOCATION WITHIN THE BRAIN

Development of the GnRH System

Classical studies in two laboratories[20–22] produced the somewhat surprising finding that GnRH cells originate from the nose! The cells are found in the early embryo within the nasal placode and migrate into the brain during fetal development. The route of migration is through the cribriform plate into the olfactory lobe and then caudally to the preoptic area and the hypothalamus. As to how these neurons find their way to the correct place within the brain has been the subject of considerable investigation for two decades, but this remains unknown, in spite of

BOX 9.1

HOW DO WE KNOW THAT THE GNRH-1 NEURONS COME FROM THE NOSE?

Forebrain GnRH-1 cells are essential for reproductive function. During embryonic development, GnRH-1 cells are first detected in the nasal placode and thereafter migrate in association with a subset of vomeronasal olfactory-derived axons through the cribriform plate into the developing forebrain. This phenomenon has been documented by *in vivo* observations during normal prenatal development in a variety of vertebrates – from fish to primates. In mouse, the number of GnRH-1 cells found in the nose during prenatal development equals the number of GnRH-1 cells found in the brain postnatally. As development progresses, the number of GnRH-1 cells in the nose reduces as the number of GnRH-1 cells in brain concomitantly increases. In addition, GnRH-1 neurons are found in nasal explants, devoid of brain tissue, taken from prenatal mice, rat, sheep and primates. Other lines of research that support a nasal origin of forebrain GnRH-1 cells include: (a) removal of the anterior area of the nasal placode in chick eliminated GnRH-1 cells; (b) mutant mice with a loss of Fgf8 were found to lack the vomeronasal organ (which develops from the nasal placode) and GnRH-1 neurons; and (c) several mutant mice lines have been reported that exhibit aberrant migration of GnRH-1 cells in nasal areas, resulting in changes in GnRH-1 cells within the forebrain postnatally. Finally, developmental pathologies affecting functional GnRH-1 expression, neuronal development or migration translate into various forms of hypogonadism, with lack or delay of pubertal onset. In humans, the pathology associated with defects in olfactory system development and impaired migration of GnRH-1 neurons is Kallmann's syndrome, characterized by anosmia (lack of smell) and lack of pubertal onset. Examination of Kallmann's fetal material revealed lack of olfactory axons into the forebrain, and GnRH-1 cells "stuck" around the cribriform plate and in nasal regions. To date, many molecules have been found that alter migration of GnRH-1 neurons by disrupting normal growth of olfactory sensory axons. In addition, altering the GnRH-1 specific migratory route and/or the signaling of molecules directly on the GnRH-1 cells themselves can prevent entrance of GnRH-1 cells into the CNS or lead to abnormal distribution and/or numbers of GnRH-1 cells within the brain. Taken together, these data indicate that GnRH-1 neurons, essential for reproductive function, come from the nose.

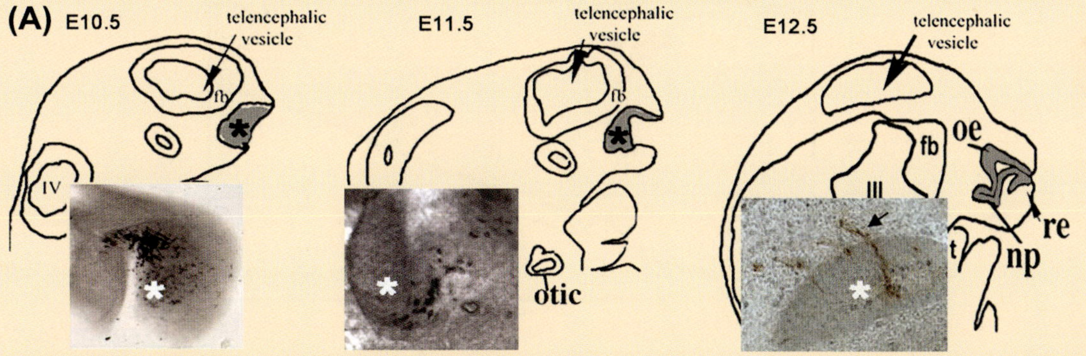

Panel (A): Development of the nasal placode/GnRH-1 neurons. Camera lucida drawing from E10.5–E12.5 mice embryos showing invagination of the nasal placode (asterisk) and the development of the respiratory epithelium (re), olfactory epithelium (oe) and presumptive VNO (pvno)/nasal pit (np). III, third ventricle; IV, fourth ventricle; t, tongue; fb, forebrain. E10.5: Parasagittal section immunocytochemically stained for Hu (early neuronal marker) shows many positive cells within the nasal placode at E10.5 (asterisk). E11.5: Parasagittal section from an E11.5 GnRH-GFP mouse immunocytochemically stained for GFP shows positive cells within the presumptive VNO (asterisk). E12.5: Parasagittal section immunocytochemically stained for GnRH-1 (brown, arrow) shows GnRH-1 cells leaving the VNO (asterisk). **Panel (B)**: Migration of GnRH-1 neurons. (**A**) Parasagittal section of an E14.5 mouse embryo head immunocytochemically stained for GnRH-1. GnRH-1 cells can be found migrating from the vomeronasal organ (VNO) through the nasal forebrain junction (NFJ) into the forebrain (asterisk). OE, olfactory epithelium; OB, olfactory bulbs. A1: Higher magnification of GnRH-1 cells in forebrain (asterisk, A). A2: Higher magnification of GnRH-1 cells migrating across nasal area (arrows, A). A3: Double labeling for GnRH-1 (brown) and peripherin (blue, marks olfactory axons). GnRH-1 cells (brown, arrows) migrate along olfactory axons (blue) in nasal areas, across the nasal forebrain junction (NFJ) and into the forebrain. (**B**) Schematic parasagittal section of a rodent brain indicating the location of GnRH-1 cells (blue dots). GnRH-1 cells are distributed in a continuum from the olfactory bulbs to the caudal hypothalamus. H, hippocampus; f, fornix; ob, olfactory bulb; mot, medial olfactory tract; ms, medial septum; dbb, diagonal band of Broca; mpoa, medial preoptic area; dh, dorsomedial hypothalamus; ot, optic tract; vh, ventromedial hypothalamus; a, arcuate nucleus; me, median eminence.

(Continued)

BOX 9.1 *(cont'd)*

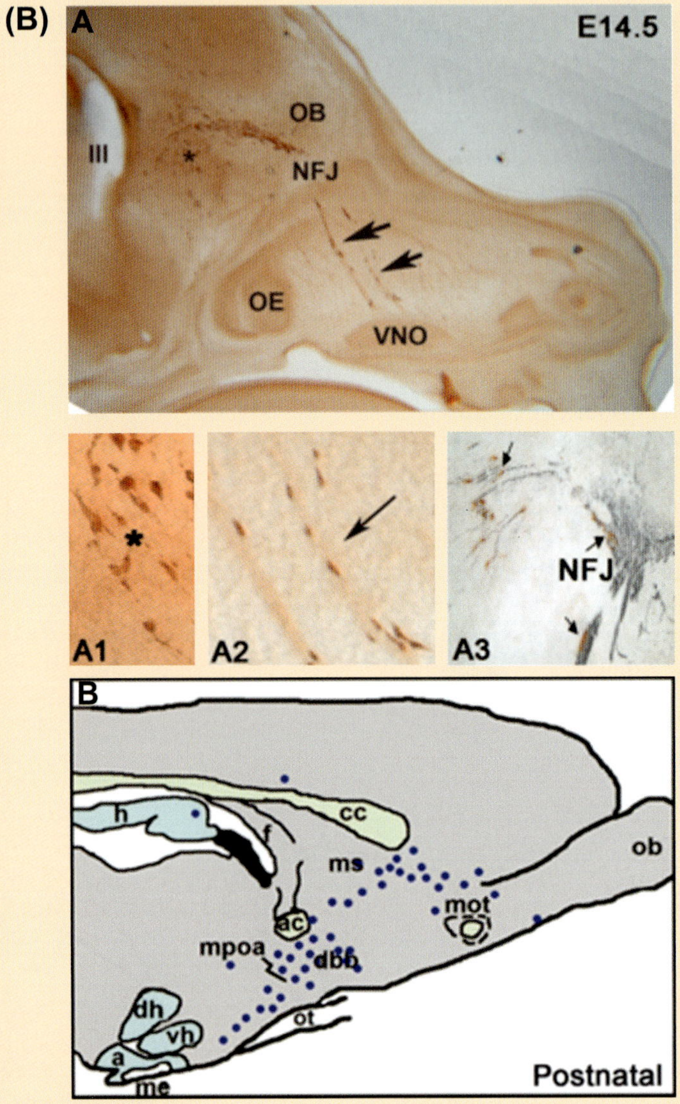

considerable effort (reviewed in Wray 2010[23]). Various factors involved in brain development are known to be involved in the progression of GnRH cells from the nasal placode to the brain and within the brain. These include guidance molecules (e.g. semaphorins), the chemokine SDF-1 and a gene called NELF, which was identified by screening single cell libraries and comparing cDNA from migrating and non-migrating GnRH cells.[24] Defects of nasal development are associated with loss of reproductive function, due to the lost of GnRH cells (Chapter 28). As to why GnRH neurons originate from the brain, why they migrate, why they stop in the right place and why they then project to the neurosecretory zone of the median eminence are questions that remain to be answered (see Box 9.1).

Location of the GnRH Neurons

Within the brain, the GnRH neurons are clustered in reasonably discrete groupings. There are only a few thousand GnRH cells, and they do not form a tight grouping such as is seen for other neuroendocrine cells.[25] Based on immunohistochemical studies, it has been estimated that only a few thousand GnRH cells exist in the brain.[25,26] For example, very tight groupings of vasopressin cells are seen in the supraoptic nucleus and the same is true for corticotropin-releasing hormone cells of the paraventricular nucleus.[27] In rodent species and sheep, the vast majority of GnRH cells are found in the ventral preoptic area of the brain, with a smaller number scattered throughout the hypothalamus, especially the mediobasal hypothalamus.[28] In other species such as rabbits and guinea pigs, and in primates, significant populations of GnRH cells are found in the mediobasal hypothalamus (reviewed in Ref 29, but more rostrally located populations are also abundant in the primate.[30] Given that GnRH cells exist within the preoptic area and the mediobasal hypothalamus, the question arises as to the relative importance of both of these populations in the control of reproduction. Original studies by two groups attempted to answer this question in the non-human primate by lesion/ablation studies, and these are detailed in Clarke and Cummins.[29] Electrolytic ablation of the neuronal population in the preoptic area caused loss of reproductive function that was subsequently restored.[31] Others favored the view that the GnRH cells in the medio-basal hypothalamus (arcuate nucleus) drive reproduction.[32] A series of studies by the Knobil group[33] consolidated the opinion that, in the primate, the GnRH cells of the arcuate nucleus drive the reproductive system. Whereas preoptic GnRH cells in primates may play a role in normal animals, ablation of these cells may lead to the function being taken over by the GnRH cells of the arcuate nucleus. The primate model in which the arcuate nucleus is ablated by electrolytic lesions provides a useful model in which pituitary function may be interrogated.[34] With pulsatile replacement of GnRH by intravenous administration, this model was used extensively to examine the pituitary action of GnRH and feedback effects of estrogen[35] (see below).

Characteristics of GnRH Cells

GnRH cells are relatively small (10–20 μm) and are either oval or fusiform in shape.[36] Dendrites extend from one or both poles and axons that emerge from the cell body or dentrites.[36] A range of shapes is found, from those with smooth contours to those that are "rough" or have spiny processes, with the possibility that the latter receive more dense innervations than those that are smooth.[36] These cells form a network, by contacting one another through axo-somatic and axo-dentritic connections,[36] which is presumably a means by which coordinated firing and secretion occurs. On the basis of studies in which agonists and antagonists have been delivered to animals, the receptor profiles, and electrophysiology (in brain slices), it is apparent that GnRH cells receive input from neurons with a range of different transmitter systems. The detail of all these networks is beyond the scope of the present chapter, but early work is reviewed by Kalra and Kalra,[37] input from ERα-responsive neurons by Kalra[38] and Herbison,[39] and tabulation of afferents to GnRH cells is found in Tilbrook *et al.*[40] In the past decade, two neuropeptides (kisspeptin and gonadotropin-releasing hormone – GnIH) have assumed significance in the regulation of GnRH/gonadotropin secretion, and these are considered below.

Dendrites of GnRH Neurons

Recent work has revealed the nature of the dendrites of GnRH cells and cast a new perspective on how these cells are regulated by other neuronal systems. Using transgenic mice in which green fluorescent protein (GFP) is expressed in GnRH cells, the nature of the dentrites was examined in detail.[41–43] Confocal examination of biocytin-filled GnRH cells showed extensive filopodia and spines on the cell bodies, as well as the dendrites. Most spines are found within 50 μm of the cell body but can be seen 1350 μm along the dentrites, and are indicative of synaptic input along the dendrites. Further analysis using electron microscopy showed that synapses between axons and the dendrites of GnRH cells do occur,[41] so tracing studies that analyzed the input to the perikarya of GnRH cells alone have provided an underestimate of the extent to which these cells receive input from other neuronal systems in the brain. In spite of this, the level of input (contact with GnRH cells) has been shown to vary with physiological state, such as season.[44] Another important point is that lack of demonstrable input to GnRH cells bodies, such as from the arcuate nucleus cells in the ovine brain,[45] does not preclude the possibility of input of afferents to the GnRH dendrites (see Box 9.2).

GnRH Terminals in the Median Eminence

The GnRH cells project to the external zone of the median eminence and to the organum vasculosum of the lamina terminalis. Presumably, there is some attractant molecule that guides the secretory axons to these circumventricular organs. The median

BOX 9.2

HOW DO WE KNOW THAT GNRH CELLS ARE INTERCONNECTED AND RECEIVE INPUT THROUGH THEIR DENDRITES?

Recent neuroanatomical findings have suggested a new hypothesis in which the long dendritic extensions of GnRH neurons play a significant role in their coordinated activity.[1] The full extent of the GnRH neuron dendritic tree can be visualized by using a patch-clamp electrode to fill individual GnRH neurons identified in slices of brain from GnRH-green fluorescent protein (GFP) transgenic mice.[2–4] Filled GnRH neurons reveal long dendrites decorated with spiny protrusions, which indicate sites of excitatory synaptic input 2. Observations of the relationship between lengthy, filled GnRH neuron dendrites and GFP-expressing GnRH neuron dendrites indicate dendro-dendritic interactions between different GnRH neurons. Using a confocal microscope, it was revealed that the great majority of GnRH neurons possess dendrites that bundle and intertwine with other GnRH neuron dendrites. The neurons wrap around each other and exhibit multiple points of close contact.[5] These new findings embellish the earlier findings with conventional microscopy,[6,7] but cell-filling has provided the advantage of revealing more distal regions of the dendrite to appreciate the full extent of the contact between the GnRH cells.

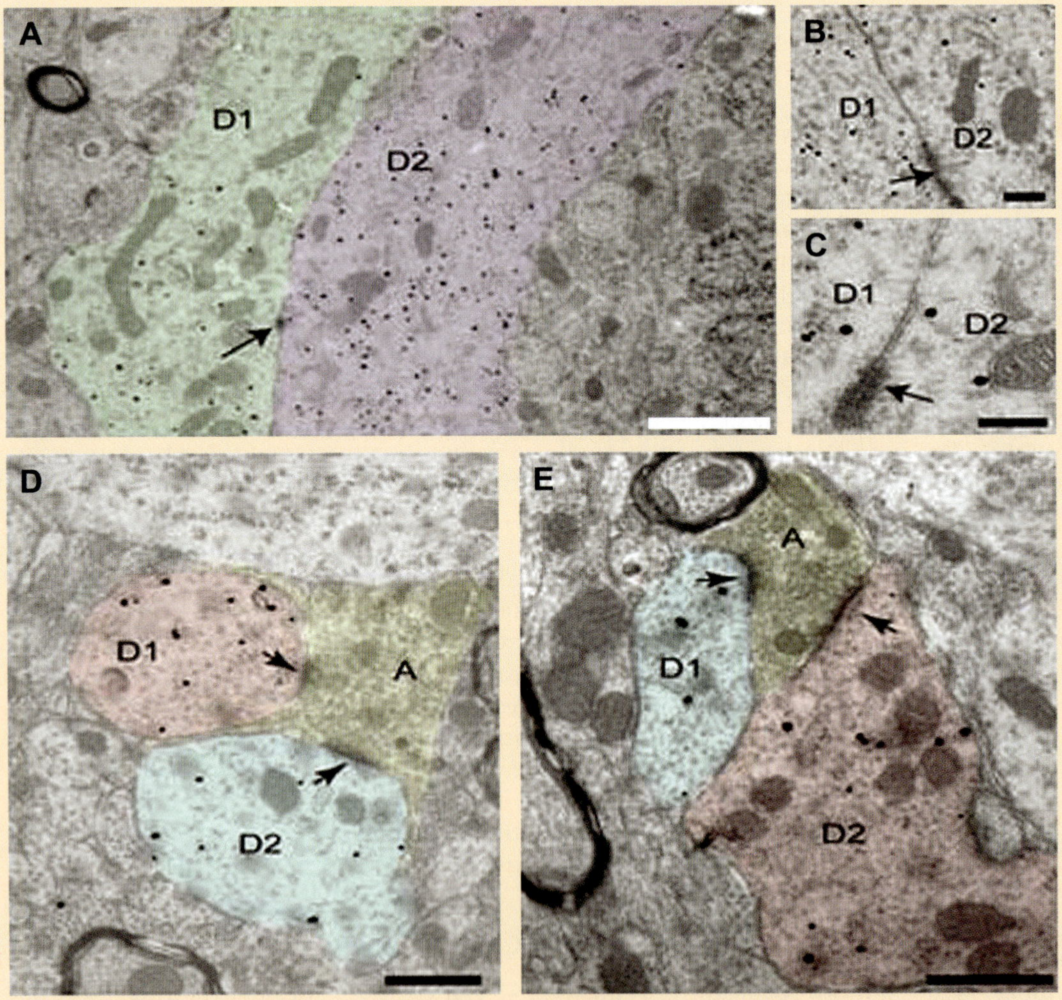

Electron microscopic images showing ultrastructural evidence of the bundling of GnRH neuronal dendrites and shared synapses (arrow in Panel A). Panels B and C show synaptic densities between two individual neurons (D1 and D2). Silver-enhanced immunogold labeled GnRH neuronal dendrites (D1 and D2 with pseudo-color for clarity) are shown in Panels D and E, with synaptic densities arising from afferent input neurons (A). Scale bar in A = 1 μm, B–E = 200 nm. *Modified from Campbell* et al. *(1999)*[5] *with permission.*

BOX 9.2 *(cont'd)*

Close contact and bundling was confirmed using electron microscopy, revealing immediately juxtaposed GnRH neuron dendrites that are "stuck" together in places by membrane specializations. These morphological findings have shown that GnRH neurons are not isolated from one another, as often inferred by their scattered distribution, but, rather, are highly interconnected via their long dendritic extensions. As yet, there is no evidence of electrical coupling between adult GnRH neuron dendrites, but ultra-structural examination of bundled dendrites has revealed a different structure that may mediate communication amongst scattered GnRH neurons. The phenomenon of shared synapses, originating from an individual afferent input, has been found on separate, bundling dendrites.[5] This neuroanatomical feature, coupled with recently described active properties of GnRH neuron dendrites,[8–10] indicates that an individual afferent input would be capable of communicating to at least two disparate GnRH neurons simultaneously via sites of dendritic bundling. The coordinated firing of many GnRH neurons is the most likely means of developing secretory bursts of GnRH, and regulation of the frequency of these bursts may be via afferents to the dendrites.

References

1. Campbell RE, Suter KJ. Redefining the gonadotrophin-releasing hormone neurone dendrite. *J Neuroendocrinol.* 2010;22(7):650–658.
2. Campbell RE, Han SK, Herbison AE. Biocytin filling of adult gonadotropin-releasing hormone neurons *in situ* reveals extensive, spiny, dendritic processes. *Endocrinology.* 2005;146(3):1163–1169.
3. Cottrell EC, Campbell RE, Han SK, Herbison AE. Postnatal remodeling of dendritic structure and spine density in gonadotropin-releasing hormone neurons. *Endocrinology.* 2006;147(8):3652–3661.
4. Roberts CB, Best JA, Suter KJ. Dendritic processing of excitatory synaptic input in hypothalamic gonadotropin-releasing hormone neurons. *Endocrinology.* 2006;147(3): 1545–1555.
5. Campbell RE, Gaidamaka G, Han SK, Herbison AE. Dendro-dendritic bundling and shared synapses between gonadotropin-releasing hormone neurons. *Proc Natl Acad Sci USA.* 2009;106(26):10835–10840.
6. Silverman AJ. The gonadotropin-releasing hormone (GnRH) neuronal system: Immunocytochemistry. In: Knobil E, Neill JD, eds. *The Physiology of Reproduction.* New York: Raven Press; 1988:1283–1304.
7. Witkin JW, Silverman AJ. Synaptology of luteinizing hormone-releasing hormone neurons in rat preoptic area. *Peptides.* 1985;6(2):263–271.
8. Roberts CB, Hemond P, Suter KJ. Synaptic integration in hypothalamic gonadotropin releasing hormone (GnRH) neurons. *Neuroscience.* 2008;154(4):1337–1351.
9. Roberts CB, O'Boyle MP, Suter KJ. Dendrites determine the contribution of after depolarization potentials (ADPs) to generation of repetitive action potentials in hypothalamic gonadotropin releasing-hormone (GnRH) neurons. *J Comput Neurosci.* 2009;26(1):39–53.
10. Roberts CB, Campbell RE, Herbison AE, Suter KJ. Dendritic action potential initiation in hypothalamic gonadotropin-releasing hormone neurons. *Endocrinology.* 2008;149(7): 3355–3360.

eminence is a highly specialized structure, providing a good example of how anatomy subserves function in a particular way. First, the portal vessels of the primary capillary bed of the hypophysial portal system are fenestrated, allowing transport of hypothalamic substances into the bloodstream, for delivery to the anterior pituitary. Secondly, all hypothalamic factors known to regulate the anterior pituitary gland are found in neuronal terminals within the neurosecretory zone of the median eminence, so discharge from these neurosecretory terminals produces "spikes" in the levels of the relevant factors in portal blood. Thirdly, sophisticated neuroanatomical rearrangement occurs in the median eminence across the estrous cycle of the female, facilitating GnRH secretion. Regarding the third of these features, recent work[46] highlights the importance of tanycytes, which project from the floor of the third ventricle into the neurosecretory region of the median eminence, between neurosecretory terminals. The projections of the tanycytes have "end-feet" specializations that make contact with the blood vessels of the primary capillary plexus of the hypophysial portal system. Neurosecretory GnRH terminals are virtually "encased" by these tanycyte processes during periods of low gonadotropin secretion, but retract at times of high gonadotropin secretion (Fig. 9.1).[47,48] Other remodeling events occur in the median eminence across the estrous cycle of the rat, including the "sprouting" of GnRH axon terminals, towards the portal capillary bed (see Box 9.3).

Other aspects of relevance within the neurosecretory bed of GnRH neurons in the median eminence are whether there is some form of axo-axonic communication and whether there is some form of control over secretion at this level by other neuronal systems.

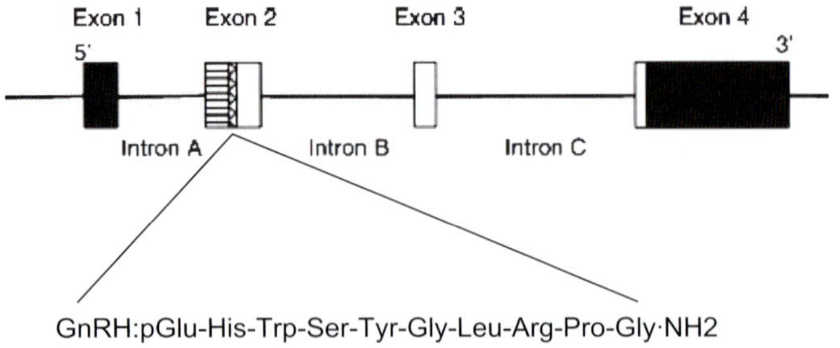

FIGURE 9.1 The GnRH-I gene consists of four relatively short exons, 1, 2, 3 and 4 (the rat gene has 138, 142, 96 and 186 bp, respectively). The exons are separated by three large introns A, B and C (the rate gene has 800, 1590 and 1350 bp respectively). Black bars indicate 5′- and 3′-untranslated regions in exons 1 and 4 respectively. The hatched bar indicates the signal sequence, the cross-hatched bar indicates the GnRH-I-coding region and the open bar indicates the GnRH-associated peptide (GAP) coding region. Note that the GnRH decapeptide is amidated post-translationally. The sequence of GnRH is conserved across most mammalian species. *Modified from Fernald and White (1999)[10] with permission.*

There is no good evidence for the former, at least between GnRH terminals but a number of studies have suggested that the latter pertains. For example, kisspeptin neurons project to the median eminence and varicose fibers come into close proximity to GnRH terminals[49] for possible axo-axonic regulation of GnRH secretion. Kisspeptin action at the level of the ME has been demonstrated in mice using an *in vitro* explants incubation.[50] Delivery of kisspeptin to the median eminence of non-human primates stimulates LH release,[51] but this may be due to diffusion of the stimulating peptide to GnRH cell bodies in the basal hypothalamus. It remains an open question as to how important any effect of a regulatory neurotransmitter/neuropeptide might be at the level of GnRH terminals.

BOX 9.3

HOW DO WE KNOW THAT THERE IS PLASTICITY WITHIN THE MEDIAN EMINENCE THAT INFLUENCES GNRH SECRETION?

GnRH neurons project to the median eminence, where the axons terminate in the neurosecretory zone, to release the peptide into the pituitary portal vasculature. It has long been appreciated that GnRH axons and terminals have a close relationship with a specific type of glia, named tanycytes.[1] Tanycytes are specialized unciliated ependymal cells, which line the floor of the third ventricle and send processes to the pericapillary space, where they make contact with the surface of the brain, inside the pia mater, and come into close proximity to the fenestrated capillaries of the primary bed of the hypophysial portal system through "endfeet" specialized structures.[2] The GnRH terminals are closely apposed to and ensheathed by the tanycyte membranes in all studied species: birds, rodents and primates, including humans.[3] Electron microscopic studies have revealed that reversible structural changes in the median eminence control the direct access of GnRH nerve terminals to the portal blood vasculature during the ovarian cycle.[4,5] Median eminence dynamics involve neurosecretory axons, tanycytes and the basal lamina of the brain, the last of which secreted peptides must cross to enter the bloodstream (see figure). Under conditions of low gonadotropin output such as in diestrus, GnRH-secreting nerve terminals are completely surrounded by tanycytes, which prevents direct access to the portal vessels and thus create a diffusion barrier impeding GnRH entry into the pituitary portal circulation (left panel of figure). During the preovulatory surge on the day of proestrus, a structural remodeling of tanycytes occurs, resulting in the release of the engulfed axons and the establishment of a direct neurohemal relationship between GnRH neuroendocrine neurons and the pituitary portal vessels (top right panel of figure). In parallel to tanycytic endfeet retraction, GnRH axon terminals are frequently seen to sprout new terminals towards the pericapillary space and thus appear to be attracted by the endothelial walls of the blood vessels, which they eventually contact (bottom right panel of figure).[4] Similarly, electron microscopic studies performed in gonadectomized rats, an experimental condition that results in increased GnRH release, showed that the distance of GnRH axon terminal from the pericapillary space was positively correlated to the plasma levels of LH.[6] Interestingly, microstructural changes in the median eminence have also been seen in the Japanese quail, a seasonal breeder, in response to the changing environmental context – e.g., changes in photoperiod. Thus, during long-day conditions (when GnRH secretion is induced) GnRH axon terminals were seen in close contact to the pericapillary space, whereas during short-day conditions (when GnRH secretion is inhibited) GnRH nerve terminals were enclosed in tanycytic processes and located far from the basal lamina delineating the vascular wall.[7] Treatment of median eminence explants with molecules that induce tanycyte movement and reconfiguration (e.g., nitric oxide, prostaglandin E2) has been shown to enhance GnRH release.[8,9] Conversely, local infusion of nitric oxide synthase (NOS) or cyclooxygenase (COX, an enzyme

BOX 9.3 *(cont'd)*

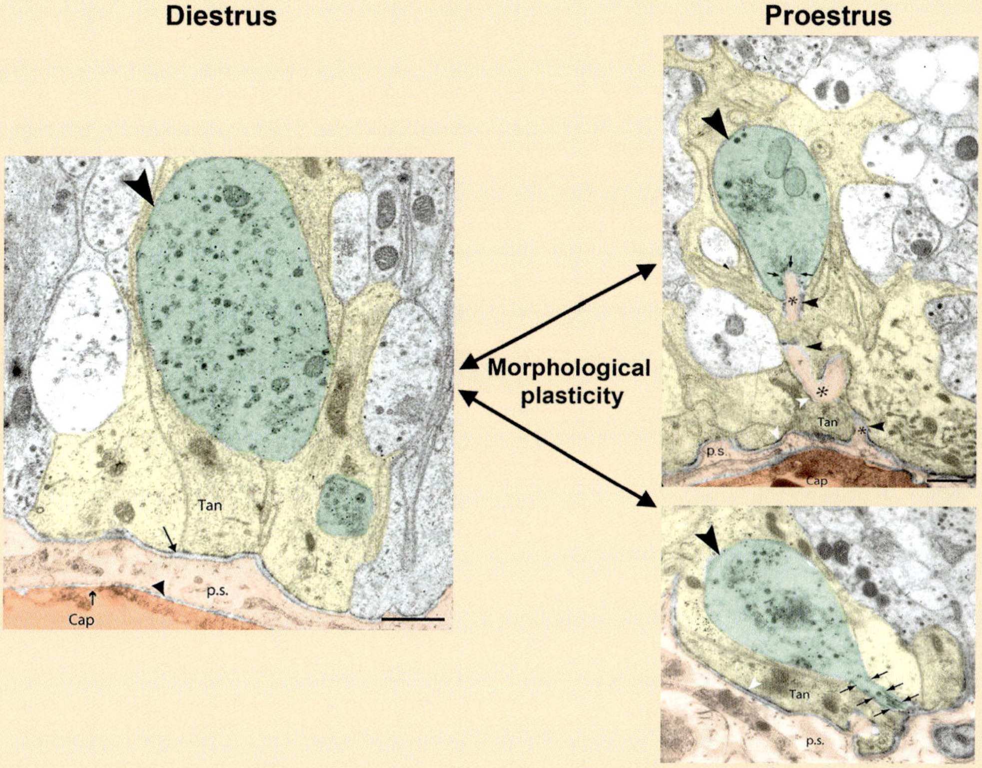

Electron micrographs illustrating the dynamic changes occurring in the external zone of the median eminence that control the direct access of GnRH nerve terminals to the pericapillary space during the reproductive cycle in the rat. **Left panel**: Electron micrograph of GnRH-immunoreactive terminals (large arrowhead) in the external zone of the median eminence in close proximity of the fenestrated capillaries (Cap) of the portal vasculature. At most stages of the reproductive cycle, GnRH nerve terminals (labeled with 15-nm gold particles) are entirely embedded in tanycytic endfeet (Tan), which prevent them from contacting the pericapillary space (p.s.) delineated by the parenchymatous basal lamina (arrow). Arrowhead, endothelial basal lamina; short arrows, fenestration of the endothelium. Scale bar: 0.5 mm. **Right panels**: On proestrus, the time of the occurrence of the preovulatory GnRH/LH surge, a significant fraction of GnRH nerve endings (large arrowhead) directly contact the pericapillary space (p.s.) either through filopodial extension of the nerve terminal (arrows) (bottom right panel) or (top right panel) by evaginations of the parenchymatous basal lamina (small black arrowheads) that allow the pericapillary space (p.s., asterisk) to penetrate into the nerve parenchyma. In the top right panel note the fusion of secretory granules (large-sized black vesicles) with the axo-plasmic membrane of the GnRH nerve terminal in direct apposition with the parenchymatous basal lamina (small arrows). The penetration of the pericapillary space into the nerve parenchyma on the day of proestrus may result from the morphological remodeling of tanycytic endfeet (tan) anchored to the parenchymatous basal lamina through hemidesmosomes seen as dark thickenings within the tanycytic processes in apposition with the basal lamina, small white arrowhead. Scale bar: 0.5 μm. *From Prevot* et al.[4,5] *with permission.*

involved in prostaglandin synthesis) inhibitors into the median eminence arrests the ovarian cycle in either the diestrus or the estrus phase[8] when GnRH release is low and GnRH neuroendocrine terminals are enclosed by tanycyte endfeet. Furthermore, recent studies of the effect of aging (a physiological condition where both GnRH release and the responsiveness of the GnRH neural network to estrogens are diminished[10] on the GnRH nerve terminal microenvironment suggest that alterations of the relationship between neuroendocrine terminals and tanycyte processes may contribute to the senescence of the hypothalamic–pituitary–gonadal axis.[11]

References

1. Kozlowski GP, Coates PW. Ependymoneuronal specializations between LHRH fibers and cells of the cerebroventricular system. *Cell Tissue Res.* 1985;242(2):301–311.
2. Mullier A, Bouret SG, Prevot V, Dehouck B. Differential distribution of tight junction proteins suggests a role for tanycytes in blood-hypothalamus barrier regulation in the adult mouse brain. *J Comp Neurol.* 518(7):943-962.
3. Prevot V, Hanchate NK, Bellefontaine N, Sharif A, Parkash J, Estrella C, et al. Function-related structural plasticity of the GnRH system: a role for neuronal-glial-endothelial interactions. *Front Neuroendocrinol.* 31(3):241-258.
4. Prevot V, Croix D, Bouret S, Dutoit S, Tramu G, Stefano GB, et al. Definitive evidence for the existence of morphological

(*Continued*)

BOX 9.3 *(cont'd)*

plasticity in the external zone of the median eminence during the rat estrous cycle: implication of neuro-glio-endothelial interactions in gonadotropin-releasing hormone release. *Neuroscience.* 1999;94(3):809–819.
5. Prevot V, Dutoit S, Croix D, Tramu G, Beauvillain JC. Semi-quantitative ultrastructural analysis of the localization and neuropeptide content of gonadotropin releasing hormone nerve terminals in the median eminence throughout the estrous cycle of the rat. *Neuroscience.* 1998;84(1):177–191.
6. King JC, Letourneau RJ. Luteinizing hormone-releasing hormone terminals in the median eminence of rats undergo dramatic changes after gonadectomy, as revealed by electron microscopic image analysis. *Endocrinology.* 1994;134(3): 1340–1351.
7. Yamamura T, Hirunagi K, Ebihara S, Yoshimura T. Seasonal morphological changes in the neuro-glial interaction between gonadotropin-releasing hormone nerve terminals and glial endfeet in Japanese quail. *Endocrinology.* 2004; 145(9):4264–4267.
8. Prevot V, Bellefontaine N, Baroncini M, Sharif A, Hanchate NK, Parkash J, et al. Gonadotrophin-releasing hormone nerve terminals, tanycytes and neurohaemal junction remodelling in the adult median eminence: functional consequences for reproduction and dynamic role of vascular endothelial cells. *J Neuroendocrinol.* 2010;22(7): 639–649.
9. Ojeda SR, Lomniczi A, Sandau US. Glial-gonadotrophin hormone (GnRH) neurone interactions in the median eminence and the control of GnRH secretion. *J Neuroendocrinol.* 2008;20(6):732–742.
10. Wise PM. Alterations in the prestrous pattern of median eminence LHRH, serum LH, FSH, estradiol and progesterone concentrations in middle-aged rats. *Life Sci.* 1982; 31(2):165–173.
11. Yin W, Wu D, Noel ML, Gore AC. Gonadotropin-releasing hormone neuroterminals and their microenvironment in the median eminence: effects of aging and estradiol treatment. *Endocrinology.* 2009;150(12):5498–5508.

GnRH, its Gene, the Peptide and Relevant Receptors Expressed by GnRH Neurons

The original purification of GnRH (Chapter 1) was achieved from mammalian hypothalamic tissues, but it has subsequently become apparent that a number of different forms of GnRH and GnRH receptors exist, to varying extents, in different species.[52,53] The predominant form of GnRH found in mammals is GnRH-1, which acts by binding to the GnRH-1 receptor. The outline of reproductive function given in this chapter is given in relation to GnRH-1 and its relevant receptor.

The GnRH gene (GnRH-1) was cloned by Seeburg and Adelman,[54] and has a structure as indicated in Fig. 9.1. The gene sequence and amino acid sequence for the decapeptide is highly conserved across a wide range of species, with a notable exception being the guinea pig, in which there are amino acid substitutions at positions 2 and 7. Following the cloning of this gene, there was an indication that the GnRH-associated peptide (GAP) may be a prolactin-releasing factor;[55] even though it is co-secreted with GnRH,[56] it does not appear to regulate the secretion of gonadotropins or prolactin.[57]

Steroid Receptor Expression in GnRH Cells

An original finding in 1983 reported that, using a combination of immunohistochemistry (for GnRH) and autoradiography (for estrogen binding), GnRH cells did not display estrogen binding properties.[58] This led to the general concept that estrogenic regulation of GnRH synthesis/secretion is effected via intermediary neurons that express estrogen receptors, which is discussed in more detail below. Surprisingly, in 1999 Skynner *et al.*[59] reported that polymerase chain reaction analysis of the contents of GnRH neurons did show the existence of ERα and ERβ in GnRH cells. This turned out to be a mistake, and a later publication by the same group confirmed that GnRH cells do not express ERα, but do express ERβ.[60]

Whereas sex-steroid feedback regulation of GnRH neurons involves intermediary cells that express ERα (an example being the kisspeptin cells; see below), it is now clearly apparent that estrogens can act directly on GnRH cells. This may be via membrane associated receptors and/or ERβ receptors, which add complexity to the action of estrogen, beyond that effected by classical liganded estrogen receptor regulation of genes.

Electrophysiological recordings of GnRH cells in slices of the guinea pig hypothalamus, identified *post hoc* by injecting dye into the cells and performing immunohistochemistry for GnRH, showed direct effects of estrogen.[61,62] These electrophysiological responses, involving the opening of potassium channels, heralded a new dimension in our understanding of how estrogen (and other steroids) acts to regulate GnRH cells. The exact nature of this putative membrane receptor remains unknown, but it could be GRPR30.[63,64] Various possible models are proposed for the membrane estrogen receptor.[65] The rapid effects of the estrogens on GnRH cells are most likely caused by "non-genomic" mechanisms, such that intracellular messengers are activated and these then have further effect. Thus, estrogen administration elicits a rapid activation of messenger signaling, such as phosphorylation of CREB.[66] Estrogen can also activate a number of other intracellular kinases,[67] which may then have genomic action. In neurons close to GnRH cells in the preoptic area of the female mouse brain the intracellular signaling molecule 21-activated kinase 1 (PAK1) is rapidly phosphorylated by estrogen injection, and a series of studies with ERα knockout animals show that this may be at least one way that rapid negative feedback occurs.[68]

With the discovery of ERβ, it was soon shown that this receptor is expressed in GnRH neurons.[69] The feedback effects of estradiol-17β are disrupted in mice with targeted disruption of ERα.[70,71] On the other hand, ERβ knockout animals do not show major deficit in feedback.[72,73] This strongly suggests that ERβ is not required for feedback regulation of the reproductive neuroendocrine axis, and unequivocal proof of this came from a study by Wintermantel *et al.*[74] which showed that neuron-specific knockout of ERα caused major deficit, whereas global knockout of ERβ had no effect.

GnRH Synthesis and Levels in Perikarya and Terminals

The levels of GnRH within the cells of the hypothalamus provide some index of the amount of stored neuropeptide available for release. This is a function of the levels of synthesis of the peptide and the amount of storage in neurosecretory terminals. There is an increase in the expression of the GnRH gene in mice as puberty occurs.[75] Even though there is a loss of reproductive function with aging, most studies show that this is not due to a deficit in the levels of GnRH in the hypothalamus.[76] In terms of the actual amount of GnRH present in the hypothalamus (cell bodies or terminals), short-term changes across the female estrous cycle or across seasons in seasonal breeding animals generally show that this is not a major factor in the regulation of reproductive function.[77,78]

The Inherent Function of GnRH Cells and the Concept of a "Pulse Generator"

Early electrophysiological studies indicated that magnocellular neuroendocrine cells display inherent phasic firing patterns.[79,80] With the advent of transgenic incorporation of green fluorescent protein genes into GnRH cells it was possible to record from single cells, and inherent phasic function of the cells was apparent.[81] Other work had shown that multi-unit electrical activity of the hypothalamus was correlated with pulsatile LH secretion in rats,[82] sheep[83] and non-human primates.[84] A pulsatile secretory pattern of secretion of GnRH is also observed in cells of the fetal nasal placode of rhesus monkeys when these are maintained in culture.[85] At least one model that would explain the generation of pulses by GnRH is one in which the cells themselves have phasic rhythm, which is modulated by a range of other neuronal elements, transducing the effects of steroid feedback, season, pheromones, stress, nutritional status, etc. Coordinated function of the cells, to allow regular pulses to be generated in terms of secretory events, could be achieved through dendro-dendritic connections.[41–43] In this model the neuronal elements that *modulate* GnRH neurons might be regarded as the "pulse generators," since these are the elements that dictate the pulse rate/amplitude. In the free-running situation, unrestrained by gonadal steroid feedback, the ultradian pulse rhythm of GnRH cells (reflected in LH pulses) is roughly 1 hour in non-human primates,[86] sheep[1] and other species.

GnRH Secretion and the Relationship to Gonadotropin Secretion

Original work in the non-human primate provided evidence that the pattern of secretion of luteinizing hormone (LH) was pulsatile.[86] Early studies from the laboratory of Fink[3] showed that GnRH was secreted into the hypophysial portal blood and that there was a pro-estrous (preovulatory) surge in secretion in the female rat, but technical constraints made it difficult to demonstrate circhoral patterns of secretion. Meanwhile, studies in the non-human primate showed that pulsatile GnRH input is mandatory for the maintenance of pituitary gonadotropin secretion, and that constant infusion of GnRH was not effective.[87] Definitive studies using a unique surgical approach[1] and push–pull perfusion of the basal hypothalamus[88] in the sheep showed that secretion of the neuropeptide in conscious animals was pulsatile, and that there was a 1:1 relationship between secretory episodes of GnRH in the portal blood and pulses of LH in the peripheral blood (Fig. 9.2).

Regarding FSH secretion, the relationship to pulsatile GnRH secretion is more complicated. A study in which the cavernous sinus was cannulated to obtain the pattern of secretion of FSH at the point of exit from the pituitary gland,[89] and the analysis of entropy of the pulse patterns, showed that FSH pulses occur in a random pattern and not in strict relationship to GnRH/LH secretion (Fig. 9.3).

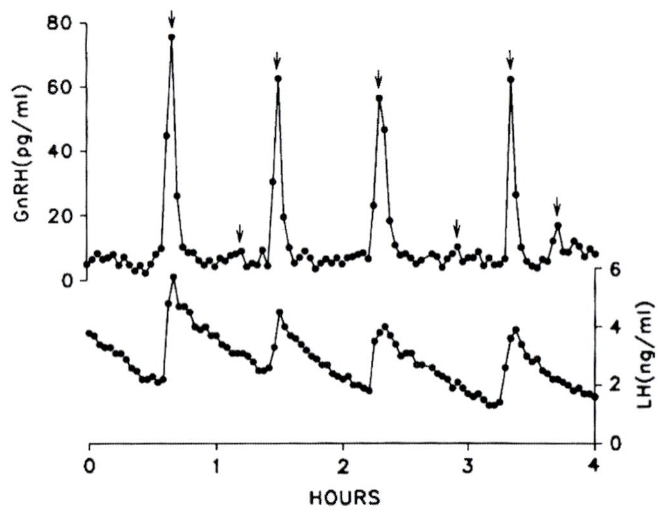

FIGURE 9.2 **The one-to-one relationship between the secretion of GnRH from the hypothalamus and LH from the anterior pituitary gland.** GnRH secretion was measured by serial sampling of hypophysial portal blood of an ovariectomized ewe, and LH levels were measured in jugular venous blood. Major pulses of GnRH drive the secretion of LH. Note that occasional small pulses of GnRH appear to occur without the secretion of LH. *Modified from Clarke (1993)*[17] *with permission.*

GnRH is required for the maintenance of FSH synthesis in the pituitary gonadotropes, but secretion of this gonadotropin does not require the pulsatile input of GnRH.[90] This is demonstrated in models of GnRH elimination (see Box 9.4).

The amplitude of the pulses of LH that are produced in response to GnRH is determined by the frequency of GnRH pulses (and feedback effects of gonadal steroids; see effect of estrogen in eliciting rapid feedback effects, below). This is an inverse relationship, whereby LH pulse amplitude increases with reduced frequency in primate and ovine models.[91,92] This effect of pulse frequency is seen in ovariectomized HPD ewes receiving fixed amplitude pulses of GnRH in the absence of gonadal steroids (Fig. 9.4).

In studies where ovariectomized, HPD ewes received varying frequencies of GnRH input, the releasable pool of LH in the pituitary gland was determined by a large-dose infusion of GnRH[93] (Table 9.1). These studies showed a 1:1 relationship between the LH pulse amplitude and the releasable pool of LH. This infers that slower pulse frequencies allow accumulation of LH in a releasable pool within the gonadotropes.

Interestingly, a study in the non-human primate showed that slower GnRH pulse frequencies favor the secretion of FSH over LH,[91] and subsequent studies in rats showed that slower pulse frequencies favor the transcription of FSHβ subunit rather than LHβ subunit in the gonadotropes.[94] Interestingly, in a sheep model of *in vivo* GnRH action, the effect of a change in frequency in increasing plasma FSH levels is virtually immediate, implying that it does not require a change at the transcriptional level of the gonadotropin subunit gene.[92]

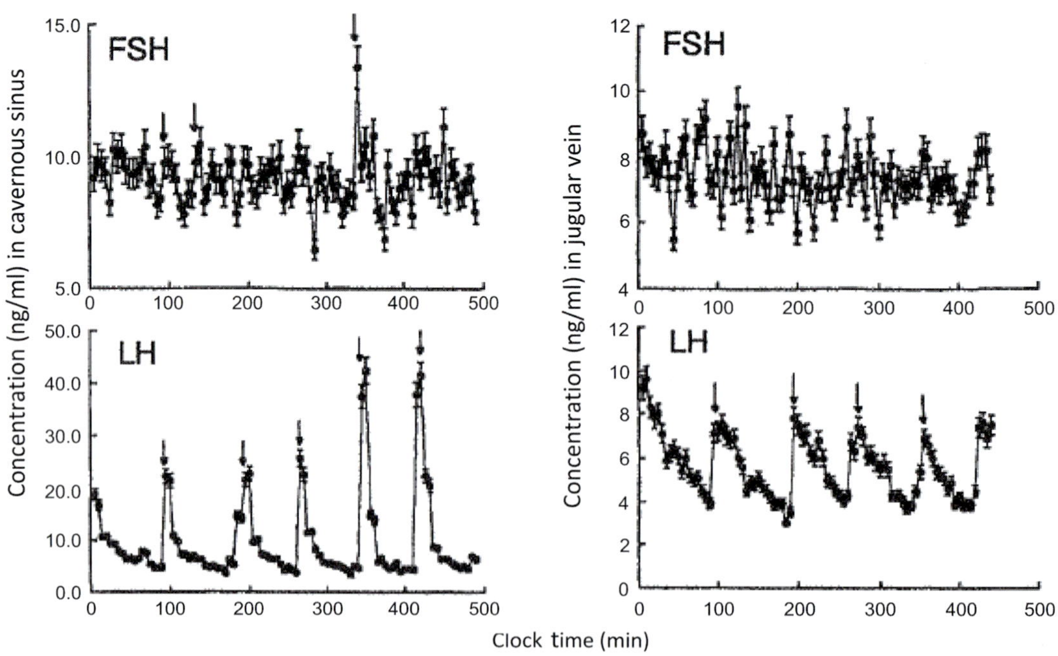

FIGURE 9.3 **The relationship between the secretion of LH and FSH, determined by cannulation of the cavernous sinus in a single sheep.** The data show assay values and standard errors for each value in cavernous sinus and jugular samples taken at 5-min intervals. Arrows indicate statistically significant pulses. Note that the concentrations are higher in the cavernous sinus than in the jugular vein, indicating that the samples from the former are close to the point of secretion from the pituitary gland. Whereas pulses of FSH occur and are sometimes contemporaneous with LH pulses, sophisticated pulse analysis showed that FSH secretion is random *from clarke et al.(2002)*[89] *with permission.*

BOX 9.4

HOW DO WE KNOW THAT THE SECRETION OF FSH DOES NOT REQUIRE THE PULSATILE INPUT OF GNRH TO THE PITUITARY GLAND?

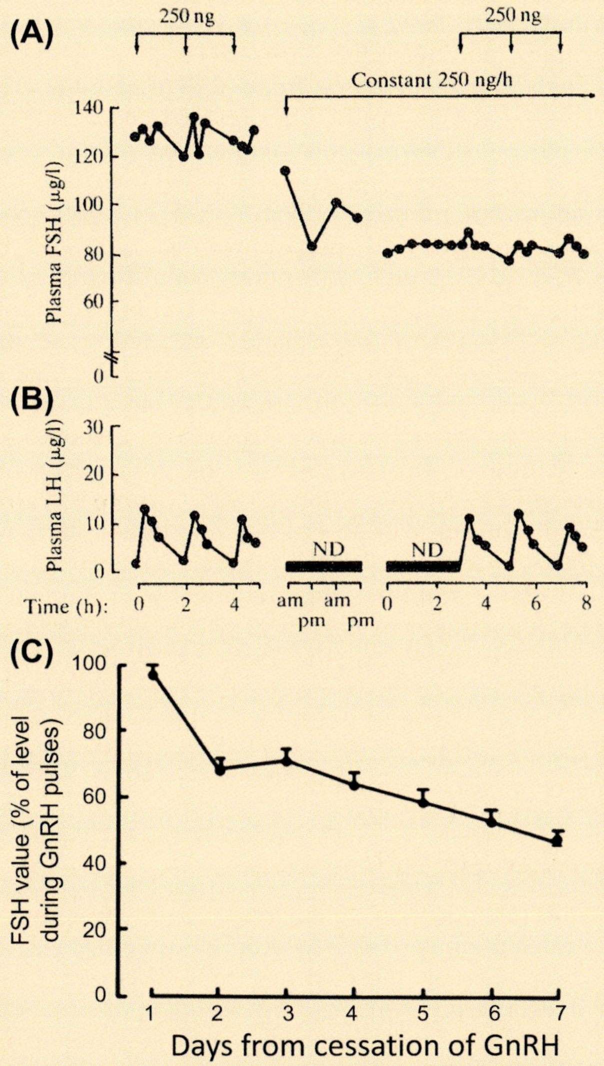

Effects of pulsatile vs constant infusion of GnRH on the secretion of LH and FSH and the secretion of FSH in the absence of GnRH input to the anterior pituitary gland. In (**A**), ovariectomized HPD ewes were given 250-ng pulses of GnRH, which elicited pulsatile secretion of LH and constant FSH secretion. With a change to a constant mode of GnRH input, LH pulses ceased but FSH secretion continued, albeit at a reduced rate. With reinstatement of LH pulses, over a constant infusion mode, pulsatile LH secretion was restored. (**B**) shows the effect of replacing GnRH input with a constant saline infusion in the ovariectomized, HPD model. Note that FSH secretion continues for up to 7 days, albeit at a declining level. *Adapted from Clarke* et al.[1] *with permission.*

This was determined using the model of ovariectomized ewes with hypothalamo-pituitary disconnection (HPD), in which endogenous GnRH input to the anterior pituitary gland was eliminated and gonadotropin secretion was restored by pulses of GnRH delivered intravenously.[1] Once stable and constant LH and FSH secretion was established, the input mode of GnRH was switched to constant infusion or the animals received an infusion of saline (no GnRH). With constant GnRH infusion, pulsatile LH secretion ceased but FSH levels in plasma decreased over time (see figure below, A). Even when the constant GnRH infusion was replaced with saline infusion, FSH secretion continued (see figure below, B). The declining level of secretion of FSH presumably reflects the passive secretion of stored FSH, which does not require the secretagog properties of GnRH. On the other hand, pulsatile input of GnRH to the pituitary gland is mandatory for the secretion of LH.

Reference

1. Clarke IJ, Burman KJ, Doughton BW, Cummins JT. Effects of constant infusion of gonadotrophin-releasing hormone in ovariectomized ewes with hypothalamo-pituitary disconnection: further evidence for differential control of LH and FSH secretion and the lack of a priming effect. *J Endocrinol.* 1986;111(1):43–49.

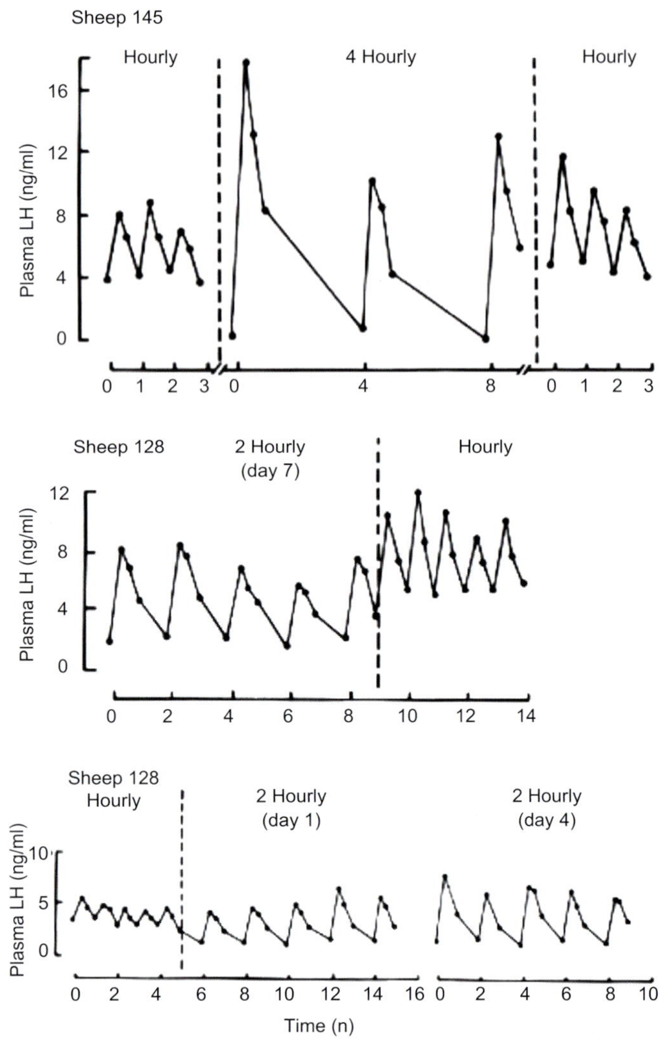

FIGURE 9.4 **Examples of plasma LH levels on ovariectomized HPD ewes given GnRH (i.v.) at varying frequencies.** The upper panel shows LH levels in an animal that received GnRH at frequencies of hourly for a week, then 4-hourly for a week, reverting to hourly in the third week. The middle panel shows the immediate effect of changing GnRH pulse frequency on LH pulse amplitude. The lower panel shows that the effect of GnRH pulse frequency on LH pulse amplitude increases with time. *Reproduced from Clarke et al.*[92] *with permission.*

PITUITARY GONADOTROPES AND GONADOTROPINS

Gonadotropin Subunits and Gonadotropin Synthesis

Pituitary gonadotropes are oval cells of either large or small size. Indeed, one study in the male rat found gonadotropes that ranged in size from 30 to 160 μm^2.[95] The gonadotropes produce luteinizing hormone (LH) and follicle-stimulating hormone (FSH). LH was originally named interstitial-cell stimulating hormone (ICSH), describing its effect in testes, and FSH was similarly named because of its ability to stimulate ovarian follicle growth in the ovaries. The realization that the pituitary produced gonadotropins that stimulate the gonads was made in a series of early studies between the 1920s and the 1950s, in which hypophysectomy and castration with various replacement treatments were used. For example, gonadectomy reduced comb size in male chickens, but testicular extracts did restore the comb. Pituitary extracts could also restore comb size, but only when testes were present (summarized in Nalbandov).[96] Whereas some studies have shown that some gonadotropes produce only LH or FSH, the majority of cells produce both gonadotropins.[97] LH and FSH are both glycoprotein hormones consisting of α and β subunits. The former is common to both gonadotropins, whereas the β subunits are encoded by separate genes. Although synthesis can be controlled by the rate of gene expression, there is also significant post-translational modification, including glycosylation. The synthesis and secretion of the gonadotropins is *absolutely* dependent upon GnRH, which is evident from the immunization studies mentioned earlier, as well as studies in which the pituitary gland was surgically isolated from the hypothalamus with the maintenance of the hypophysial portal system which provides vital blood supply to the former.[98] Without GnRH input, gonadotropin subunit gene expression is lost, and this is restored with pulsatile replacement of GnRH.[99] GnRH also regulates post-translational glycosylation of gonadotrophins.[100]

Gonadotropin Subunit Genes

GnRH-responsive elements exist within the promoter regions of gonadotropin subunit genes, and the response of these genes to GnRH stimulation involves a large number of intracellular proteins.[101,102] GnRH induces early response genes in gonadotropes, and these act to regulate transcription of the subunit genes.[103] Studies in castrated, testosterone-replaced rats (in which gonadotropin secretion is suppressed) showed that administration of GnRH at frequencies of 8–30 min increased the mRNA levels of α subunit and LHβ subunit, but the levels of FSHβ subunit were increased at lower pulse frequencies of 120–480 min.[104,105] Further regulation is through the feedback effects of gonadal steroids, but these differ both between species and in the way that the steroids regulate expression of the gonadotropin subunit genes. The means by which sex steroids may regulate gonadotropin subunit genes is complicated and appears to involve regulation of transcription factors, rather than the binding of liganded steroid receptors to consensus

TABLE 9.1 The Relationship Between the LH Pulse Amplitude and the Releasable Pool of LH.

	LH pulse amplitude	LH pulse area	Infusion area	First peak
Pulse area	0.974			
Infusion area	0.940	0.974		
1st peak	0.943	0.960	0.993	
2nd peak	0.928	0.986	0.999	0.992

In the absence of gonadal steroid feedback, the amplitude of LH pulses in response to GnRH is determined by the frequency of pulsatile GnRH input to the pituitary gonadotropes which, in turn, reflects the releasable pool of LH. These studies were carried out in ovariectomized HPD animals receiving pulsatile GnRH. The releasable pool of LH in the pituitary gland was determined by administering a large dose infusion (0.5 μg/minute) of GnRH (i.v.) and calculating the area under the response curve.[93] The large-dose infusion of GnRH was initiated at the time that the next GnRH pulse would have been administered on each particular pulse regimen. The data show correlation coefficients for LH pulse characteristics and the first and second phases of the response to high-dose GnRH infusion. The characteristics of the LH pulses were determined in animals given GnRH pulses at a frequency of one/h or one/2h, where the amplitudes of the pulses were lower at the higher pulse frequency. The data are correlation coefficients, all of which are highly significant ($P < 0.001$).

sequences on the promoters of the genes for the subunits.[106,107]

GONADAL STEROID FEEDBACK ON THE GNRH/GONADOTROPIN SYSTEM

General Concepts of Feedback

Feedback effects of sex steroids may occur at the level of the GnRH cell, to regulate secretion of the neuropeptide or at the level of the pituitary gonadotrope. To interrogate the former, the best model is the sentient sheep from which serial hypophysial portal and jugular venous samples may be collected over many hours.[1] The fact that maintenance of plasma levels of gonadal steroids (androgens, estrogens and progesterone) acts as a negative brake on the GnRH/gonadotropin axis is most simply demonstrated by gonadectomy in both sexes, in which case levels of GnRH and the gonadotropins rise. Thus, studies in rams showed that the main effect of castration was an increase in GnRH pulse frequency (Fig. 9.5), although it is difficult to determine changes in pulse amplitude (which may also occur) across days or weeks because of individual animal variation.[108] GnRH pulses are followed by LH pulses, which in turn cause pulses of testosterone secretion from the testes.[109]

Feedback effects of gonadal steroids on the GnRH/gonadotropin axis may involve changes in the frequency and/or amplitude of GnRH pulses. Whereas frequency of pulses is easily measured using the portal access model in the sheep, it is somewhat difficult to measure changes in amplitude of GnRH pulses. Nevertheless, such changes can be detected between the luteal and follicular phases of the ewe estrous cycle.[110] At the level of the pituitary gland, only pulse amplitude can be altered.

There is a distinct difference between males and females regarding the feedback effects of sex steroids on the brain, most notably the presence of a "positive feedback" mechanism in the female and the absence of the same in the male. This positive feedback mechanism drives the preovulatory surge in GnRH and LH secretion in the female and is programmed during fetal life or early postnatal life, depending upon species.[111] Androgenization of female fetuses, or newborn animals in the case of rats and mice, causes the loss of the positive feedback mechanism, demonstrating the programming function of the testes in the male.[111]

At least three types of feedback effect of steroids are known to regulate the GnRH/gonadotropin axis.[112] The three feedback types can be demonstrated in ovariectomized females. Thus, when estrogen is administered to such animals, a rapid negative effect occurs, followed some hours later by a transient positive feedback event (Fig. 9.6). If estrogen treatment is continued, a long-term negative feedback mechanism is operative.[18]

Steroid Feedback Regulation of GnRH and Gonadotropin Subunit Gene Expression

As mentioned above, the expression of the GnRH gene is not significantly affected by sex steroids or stage of estrous cycle in the female. The gonadotropes express ERα, ERβ and the androgen receptor,[113–115] and the expression of gonadotropin genes is regulated by steroid feedback. The expression of ERβ in gonadotropes is not a characteristic of all species, and is not present in the pituitary gland of the sheep (CJ Scott and IJ Clarke, unpublished data). In some instances steroid receptor binding to the promoter of the gene occurs, whereas in other cases indirect regulation occurs via second messengers within the gonadotropes. For example, in the case of the bovine LHβ subunit, estrogen regulation is conferred by indirect means, via second messenger pathways in the gonadotropes.[116] An imperfect binding

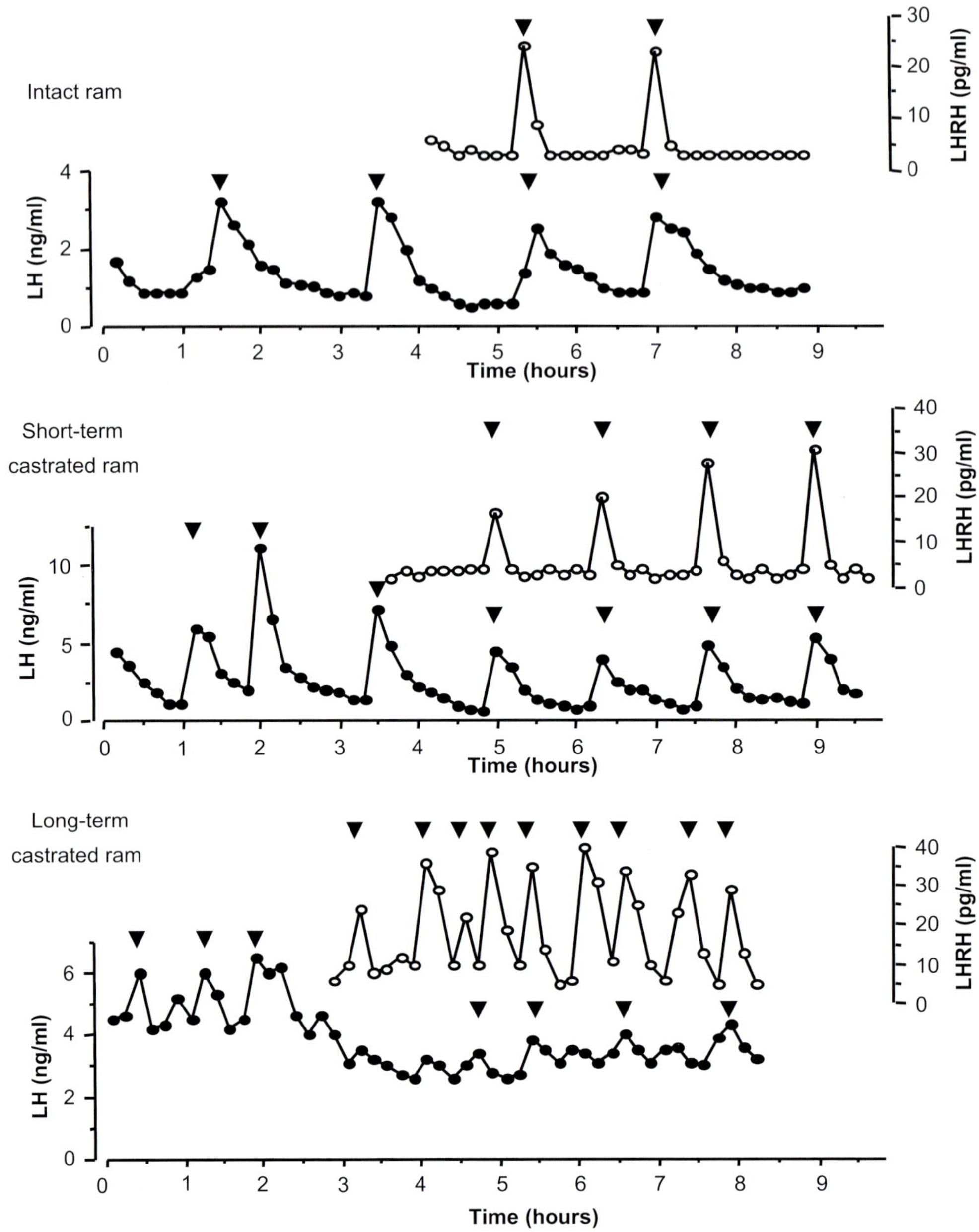

FIGURE 9.5 **The effect of castration of rams on GnRH levels.** Note that there is an increase in GnRH pulse frequency with time after castration. The "short-term" animals were sampled 2–15 days after castration, and the "long-term" animals were sampled 1–6 months after castration. The inter-pulse interval of intact animals was 180 minutes, and this was reduced to 70 minutes and 36 minutes in short- and long-term castrates respectively. *Adapted from Caraty and Locatelli (1988)*[108] *with permission.*

region for the estrogen receptor is found on the promoter of the LHβ subunit in the rat,[117] and estrogen stimulates LHβ-subunit transcription in the rat[118] and in some other species. An example of indirect regulation is that of estrogen suppression of FSHβ-subunit transcription via AP1 sites demonstrated *in vitro* in the sheep pituitary.[106] This subunit gene does not bind liganded estrogen receptors.[106]

In female mice, gene knock-in of a gene expressing a mutant form of ERα, with an ablated ERE-dependent activity, rescued negative feedback effects which were lost in ERα knockout animals.[119] This further demonstrates that feedback effects involve non-classical pathways. At the level of the pituitary in mice, this type of non-ERE dependent signaling negatively regulated α-subunit and FSHβ-subunit, but not LHβ-subunit, gene expression.[120] Using mice with a floxed ERα gene that were cross-bred to animals in which Cre recombinase expression is under the control of the mouse common glycoprotein α-subunit promoter, it was shown that

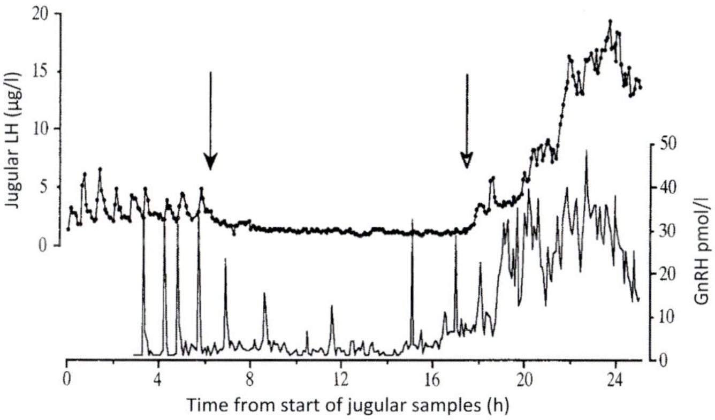

FIGURE 9.6 **Biphasic negative feedback followed by positive feedback when a single injection of estrogen is given to an ovariectomized ewe.** Note the initial negative feedback effect on GnRH (lower trace) and LH secretion (upper trace), followed by a positive feedback effect. The former indicates an hypothalamic effect of estrogen, and the latter is due to a combination of reduced GnRH secretion as well as a rapid feedback effect of estrogen on the gonadotropes (see below). The closed arrow indicates the injection of estradiol benzoate (25 μg i.m. and 25 μg i.v.), and the open arrow indicates the start of the surge approximately 12 h later. *From Caraty et al. (1989)*[124] *with permission.*

both negative and positive feedback effects of estrogen were impaired;[121] this demonstrates the importance of estrogen signaling at the level of the gonadotrope.

Studies in sheep[122] showed that estradiol-17β bound to BSA, so that it could not enter cells, had a rapid effect to cause negative feedback, without a change in GnRH secretion; this further substantiates the notion that estrogen can act on gonadotropes via non-classical means, probably via a membrane receptor. In other studies in sheep, rapid (within 15 min) phosphorylation of second messengers was seen in gonadotropes.[123] These effects could explain the rapid effect of estrogen in acting on gonadotropes, causing negative feedback, but how these events relate to positive feedback is not yet known.

In order to determine specific effects of sex steroids on the pituitary gonadotropes, two main models were used. Nakai *et al.*[35] employed the arcuate nucleus-lesioned monkey in which GnRH secretion was eliminated, and in the sheep a surgical extirpation of the neurosecretory zone of the median eminence achieved the same.[98] In both of these isolated-pituitary models, intravenous pulsatile administration to ovariectomized females (to remove fluctuating levels of ovarian steroids) restored pulsatile LH secretion as well as FSH secretion; any treatment that was imposed to alter gonadotropin levels in blood therefore indicated direct pituitary gonadotrope action.

In the ovine model of an isolated pituitary gland, in which gonadotropin secretion is restored with pulsatile GnRH treatment, estrogen treatment caused a rapid suppression of LH pulse amplitude and reduction in FSH levels).[124,125] This effect, which represents reduction in pulse amplitude of LH in response to GnRH, is due to the rapid signaling effect of estrogen in the gonadotropes,[126] with phosphorylation of second messengers such as CREB, ERK and Akt;[123,127] a similar rapid effect may reduce the secretion of GnRH, since there are rapid effects of estrogen on GnRH cells.[128] Also in the brain, non-GnRH cells that are responsive to estrogen display a rapid and transient phosphorylation of p21-activated kinase 1 (PAK1), and the inhibition of this phosphorylation prevents the rapid negative effect of estrogen on gonadotropin secretion.[68]

Positive Feedback

The Positive Feedback Effect of Estrogen on GnRH Secretion

The positive feedback effect of estrogen is the mechanism by which the GnRH cells of the brain and the pituitary gonadotropes produce surges in the secretion of GnRH and gonadotropins, respectively, that lead to ovulation.[129] Early studies in sheep showed that positive feedback involved a time-delayed mechanism, so the surge in LH secretion occurred many hours after i.m. injection in ovariectomized ewes.[130] Positive feedback is due to a surge in the secretion of GnRH, as well as an increase in the responsiveness of the pituitary to the action of GnRH. The estrogen-induced GnRH surge has been most thoroughly studied in the sheep because of the ease of sampling hypophysial portal blood. An early demonstration of this phenomenon was obtained in seasonally anestrous ewes, which are acyclic. This model is useful because GnRH secretion is held in check by the negative feedback effect of estrogen (see below). Administration of estradiol-benzoate to such animals, as a single i.m. injection, resulted in the surge secretion of GnRH[131] (Fig. 9.7). Later studies showed a time-delayed surge in secretion of GnRH in response to the same injection in ovariectomized ewes.[17,124] It is important to note that this surge secretion in GnRH is procured by nothing other than a single injection of estrogen, and that the event is transient. Early studies[132] also showed that the biphasic (negative and then positive) feedback effects of estrogen could be obtained in ovariectomized ewes with *intracarotid* injections of 1.25 μg of estradiol, but others have indicated that, with the use of subcutaneous implants, 6–8 hours of estrogen stimulation is required for surge induction.[133] In any event, it is clear that the effect of estrogen to cause the positive feedback is due to activation of subcellular systems in the brain and pituitary that require a time-delay for manifestation.[134] This involves a switch from the initial negative feedback effect of estrogen, that holds the GnRH responsiveness of the gonadotropes in check, even though there GnRH receptor levels increase

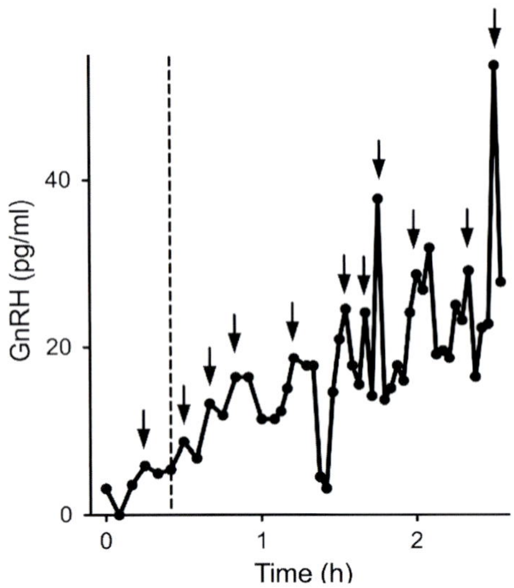

FIGURE 9.7 **Rapid sampling identifies GnRH pulses at the start of an estrogen-induced GnRH surge in the ovariectomized ewe.** Note that the levels for identified pulses (arrows) do not return to starting values prior to the next pulse occurring; this is a surge-generating mechanism at the level of GnRH secretion. *From Clarke (1993)*[17] *with permission.*

(induced by estrogen) at this time.[135] There is a marked increase in the responsiveness to GnRH at the onset of the GnRH/LH surge.[110] Thus, the surge in GnRH and the increased response of the pituitary gonadotropes to GnRH both contribute to the positive feedback event. The biphasic (negative then positive) feedback effect of injection of estradiol on GnRH and LH secretion in an ovariectomized ewe is seen in Fig. 9.6.

Further evidence that the molecular mechanisms which lead to positive feedback are set in train well before the onset of the GnRH/LH surge is found in studies where progesterone treatment prevents the same in ovine models. Early studies showed that progesterone prevents the induction of the surge by estrogen,[130] and this is due to the action of progesterone to prevent the surge in GnRH secretion.[136] In ovariectomized ewes the LH surge may be blocked by the administration of progesterone after the administration of estrogen, but before the onset of the surge.[133] Thus, the surge-inducing activational event that is caused by estrogen can be prevented by progesterone in the period of the "transmission" of the positive feedback signal.

Studies in the laboratory of Karsch used a different model, which recapitulated the events of the estrous/menstrual cycle by serial treatment with progesterone and estrogen. In this model, ewes are ovariectomized and given estrogen and progesterone implants. The progesterone treatment is withdrawn 1 week later, and additional estrogen implants are inserted 16 h after this. A surge in the secretion of GnRH and LH ensues approximately 35 h after progesterone withdrawal.[137] Interestingly, the GnRH surge continues well beyond the surge in LH secretion. The latter is limited by the pituitary stores of LH, which are almost totally depleted during the surge.[138] In at least one study where rapid (2.5 min) samples were taken of hypophysial portal blood, there was evidence that the pattern of secretion of GnRH during the estradiol-induced surge continued to be pulsatile in nature, suggesting that this surge is due to a marked increase in frequency of pulses.[17] This is based on evidence of GnRH pulses occurring at such a rapid rate that return to baseline does not occur between pulses (Fig. 9.7). Others have suggested that secretion is continuous at this time.[139] In reality, the generation of the GnRH surge probably involves increased firing of GnRH neurons with increased pulsatile release, as well as continuous release of the neuropeptide. Whatever the case, this unique positive feedback effect occurs in response to rising levels of estrogen in the preovulatory period of the female estrous/menstrual cycle, and causes ovulation.[140]

Using the arcuate nucleus-lesioned, GnRH replaced non-human primate (see "Location of the GnRH Neurons," above),[34,35] estrogen administration to animals receiving hourly intravenous pulses of GnRH was able to elicit a biphasic LH response similar to that seen in the hypothalamo-pituitary intact animal.[34,35] This prompted the suggestion that a GnRH surge is not required for the positive feedback effect on LH secretion in this species. Administration of invariant hourly GnRH pulses to gonad-intact animals caused a rise in estrogen values that led to positive feedback and ovulation.[33] In spite of this, a preovulatory surge in GnRH is indeed seen in the rhesus monkey,[141] as in other species, including lower phyla.[142]

Whilst the positive feedback of estadiol is generally regarded as the signal that precipitates the preovulatory surge in GnRH and LH, progesterone also has positive feedback-inducing action. Positive feedback in response to progesterone is seen in various species following estrogen-priming, and this is likely to be of physiological significance in species where a rise in plasma progesterone levels precedes the cyclic LH surge; such species include rats,[37] humans[143] and non-human primates.[144] Progesterone appears to facilitate or augment the estrogen-induced surge in women[145] and in rhesus monkeys.[146] This is not the case in sheep, where there is no progesterone surge prior to the ovulatory GnRH/LH surge, and continuous high levels of progesterone will block the surge in GnRH[146] and LH secretion.[130]

The estrogen-primed progesterone-treated rat model for the induction of LH surges was developed many years ago,[147] and is a commonly used laboratory tool for surge induction in rats and mice. The importance of progesterone in the positive feedback response in rats was indicated by data showing that the pharmacological blockade of progesterone receptors PR with anti-progestins[148–150] can prevent the LH surge in intact rats. When wild-type and progesterone receptor knockout mice were, treated with both 17β-estradiol and estradiol benzoate, and then killed at 1900 h on day 7 post-ovariectomy, the wild-type animals all showed LH surges: this was not the case in the knockout animals.[151] These data indicate the importance of progesterone signaling for the positive feedback event in rats and mice.

The Positive Feedback Effect of Estrogen on Pituitary Gonadotropes

As indicated above, there is an increase in the responsiveness of the pituitary gonadotropes to GnRH at the onset of the LH surge, which is part of the positive feedback mechanism. This involves multiple actions of estrogen. In addition to the response to a single pulse of GnRH, the "self-priming" effect of GnRH is produced by estrogen, being a phenomenon whereby the response to GnRH is amplified by a preceding GnRH pulse.[129] It is demonstrated during the positive feedback phase by the administration of two GnRH pulses, an hour apart, with the response to the second pulse being far greater than the response to the first. An increase in GnRH receptor levels, which may predispose to the increased response to GnRH caused by estrogen, occurs before the onset of the surge (see above), so this is not a surge-initiating event. Other effects of estrogen are also manifest before the onset of the surge, and a dramatic example is the mobilization of LH secretory granules to the periphery of the gonadotrope cells[152] (Fig. 9.8). Interestingly, this is not accompanied by the same migration of FSH-containing granules, reinforcing the notion that the two gonadotropins are in different secretory granules.

Other events that occur prior to the surge include an increase in the ERα levels in gonadotropes and the activation of a range of second messengers (by phosphorylation), but, once again, this is well prior to surge onset, consistent with the notion (see above) that the activational events that cause the surge occur in a critical period after estrogen action, but prior to surge onset. Interestingly, the rise in ERα is caused by an increased level of estrogen, but also by increased GnRH pulse frequency; the combined effects of both cause the greatest increase in ERα.[153] There is an increase in the expression of the genes for the α subunit and the LHβ subunit in the late part of the follicular phase of the estrous cycle in the ewe, with no change in the expression of the FSHβ subunit gene (Fig. 9.9).

An important mechanism for the induction of the LH surge is the alteration in the shape of LH pulses. As the surge commences, the LH pulses become longer. In other words, the decay phase of the surge is not as steep as in the absence of steroids. This was

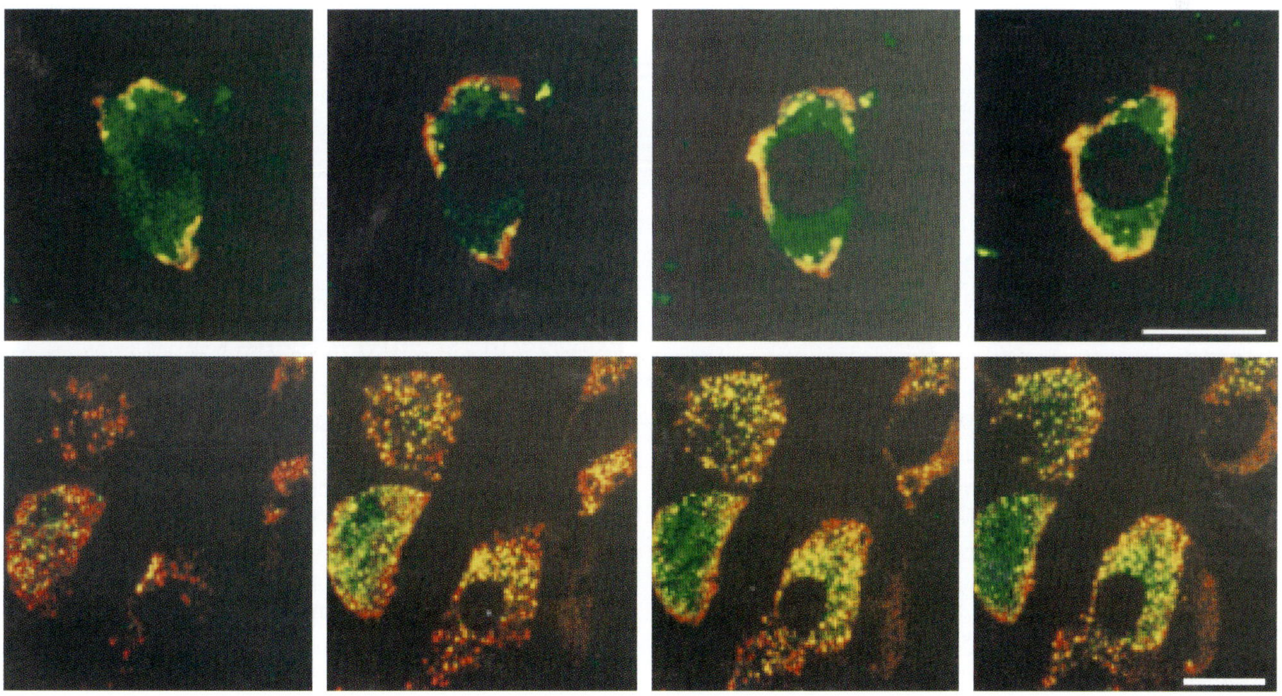

FIGURE 9.8 The positive feedback effect of estrogen on the pituitary gonadotropes involves the mobilization of LH secretory granules, but not FSH secretory granules. Movement of granules to the periphery of the cells occurs prior to the surge in GnRH and the increase in responsiveness to GnRH that initiates the LH surge. *Adapted from Thomas and Clarke (1997)*[152] *with permission.*

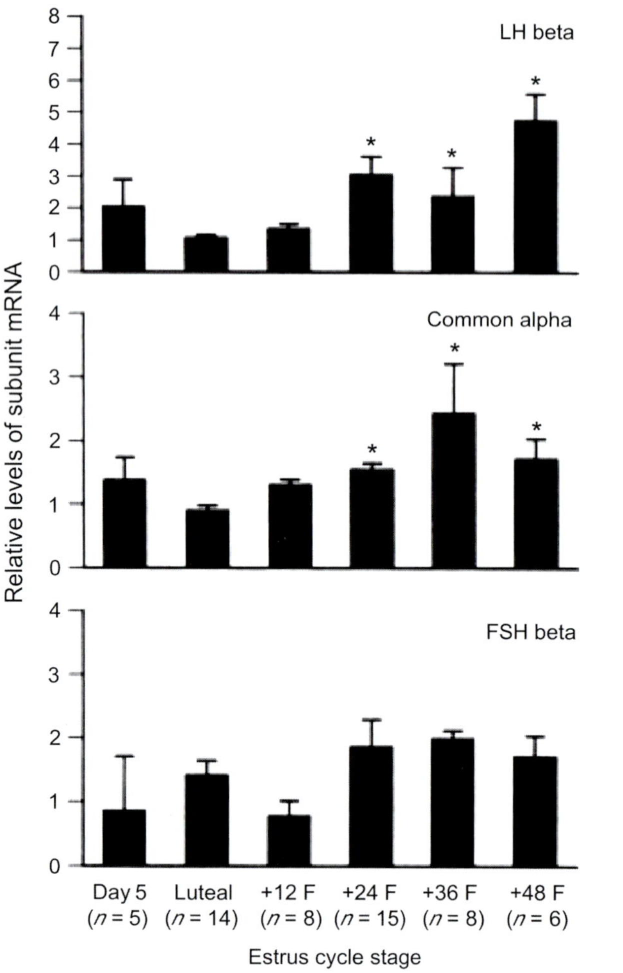

FIGURE 9.9 **Expression of genes for α subunit, LHβ subunit and FSHβ subunit in pituitaries of normally cycling ewes.** Expression of subunit gene expression was measured using polymerase chain reaction. *Reproduced from Cowley et al.*[54] *with permission.*

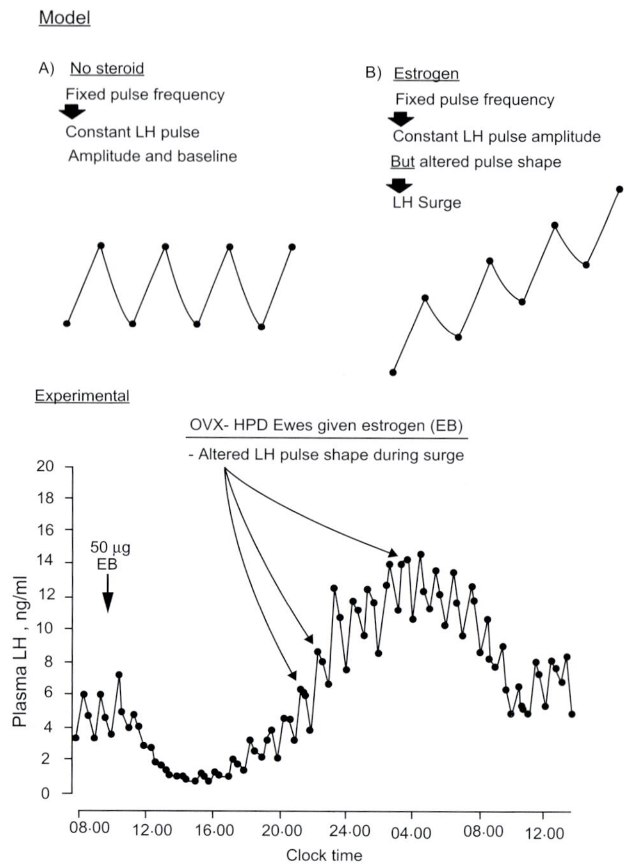

FIGURE 9.10 The estrogen-induced LH surge is partly due to altered LH pulse shape. The change in shape of the pulses, compared to those seen with no steroid effect, is a surge-generating mechanism at the level of the pituitary gonadotrope. *From Clarke and Cummins (1984)*[125] *with permission.*

demonstrated in the ovariectomized HPD ewe given hourly GnRH pulses i.v. and a single i.m. injection of estradiol benzoate (Fig. 9.10).[125] This lengthening of the LH pulses is not due to altered biological : immunological ratio of LH,[155] implying that the lengthening of the pulses occurs due to a longer secretory phase, rather than altered properties of the gonadotropin (which might change the half-life of the gonadotropin in plasma). In spite of this, under conditions of constant GnRH pulse frequency (1 pulse/h), the magnitude of the rise in plasma LH levels is only two-fold, so some other factor is required for full surge generation – the most important perhaps being the surge in GnRH secretion (see above). In particular, increasing the frequency of GnRH pulses at the time of the onset of the surge (without a change in amplitude) in the ovariectomized, HPD ewe leads to LH surges approaching the magnitude of those seen in hypothalamo-pituitary intact ovariectomized ewes given the same estrogen treatment (Fig. 9.11).[156] Thus, the generation of the preovulatory surge is due to the combined positive feedback effects of estrogen at the level of GnRH secretion as well as GnRH action on the gonadotropes.

Long-term Negative Feedback

This form of feedback is seen following acute phase responses in females, and in general in the male. When constant and sustained circulating levels of sex steroids prevail, negative feedback ensues. As indicated above, removal of the gonads invariably leads to increased GnRH and gonadotropin secretion, indicative of long-term negative feedback. In females, this form of feedback occurs during the luteal and follicular phases of the estrous cycle and (in seasonal mammals) during anestrus. In males, negative feedback is operative as a regulatory mechanism to counter the "feed-forward" effect of gonadotropins on the production of sex steroids from the testes.

In males, testosterone from the testes acts within the brain to limit the secretion of GnRH.[108,157] In the sheep at least, in studies where castrated HPD rams were given pulsatile replacement with GnRH, testosterone had minimal effect at the level of the pituitary gland.[157] On the other hand, testosterone treatment of castrated rams is seen to reduce GnRH secretion.[157] Thus, the negative feedback effect of gonadal steroids in the male is due to action on steroid-responsive neurons in the brain that relay information to GnRH cells. Testosterone can be aromatized in the brain to estradiol, so feedback effects of androgens on GnRH cells may involve cells that possess either estrogen or androgen receptors. These may be kisspeptin cells or other cells that express either or both sex-steroid receptors.

During the luteal phase of the estrous cycle of the female, progesterone (from the corpus luteum of the ovary), acts to reduce GnRH pulse frequency,[18,110,158,159] with the resultant reduction in the pulse frequency of LH. This negative brake on GnRH secretion is probably due also, in part, to the low level of estradiol secreted from small ovarian follicles. Although the fall in progesterone levels that occurs at the time of luteolysis causes an increase in GnRH/LH pulse frequency, levels of GnRH and the gonadotropins are still held in check by negative feedback action, until the onset of the preovulatory GnRH/LH surge. This may involve rapid negative feedback (*vide supra*). Whereas long-term negative feedback is due to a change in frequency of GnRH secretion, the action of constant estrogen or progesterone on the

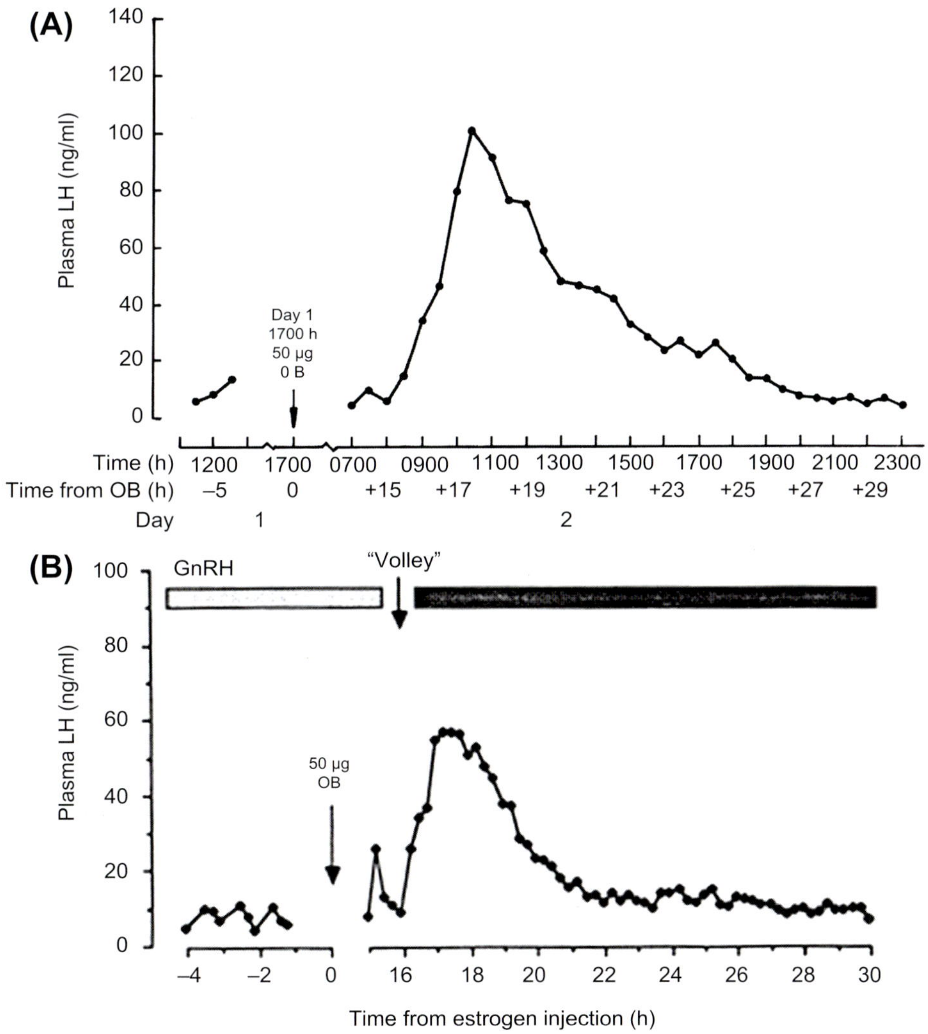

FIGURE 9.11 Increased GnRH pulse frequency is mandatory for full manifestation of the estrogen- induced LH surge. Panel (**A**) shows the response to an intramuscular injection of estradiol benzoate (OB), causing positive feedback and an LH surge. Panel (**B**) shows the effect of OB in an ovariectomized, HPD ewe that was given hourly GnRH pulses (shaded bar) and then a volley of five 15-min pulses of GnRH (arrow) at the time predicted to be the time of the onset of the LH surge, then half-hourly pulses (black bar). Note that the response depicted in (B) is lower than that depicted in (A), but greater than that seen in Fig. 9.10 (with invariant hourly pulses). This shows that the estrogen-induced LH surge is due, in part, to an increase in GnRH pulse frequency. *Modified from Clarke* et al. *(1989)*[156] *with permission.*

pituitary gland is minimal, at least in sheep models where this has been well examined.[125,160] The situation regarding control of FSH secretion, involving combined action of estrogen, inhibin and activin on the pituitary gland (*vide infra*).

Relaying the Feedback Effects of Steroids on GnRH Secretion: A Major Role for Kisspeptin

As indicated above, GnRH cells operate in a phasic manner, and the secretion of GnRH is pulsatile in nature. Such a system can be subject to alterations in frequency and amplitude with exquisite precision. The GnRH cells are modulated by neuronal afferents that provide direct and/or relayed input, which arises from brainstem noradrenergic cells[44,161] and the serotoninergic elements of the raphe nucleus,[162] as well as various cell types of the hypothalamus and the forebrain.[44,163–165] The input to GnRH neurons from various regions of the brain presumably allows a means of fine control of the reproductive system, incorporating information in relation to metabolic status, season, stress state, immune status, olfactory stimuli, etc. Many of these afferents, such as the A1 noradrenergic cells of the ventrolateral medulla, are estrogen-responsive and show increased activity at the time of the estrogen-induced surge.[161] In some cases, modulatory neurons may influence GnRH cells by a polysynaptic pathway. Thus, estrogen-responsive neurons of the brainstem project to the bed nucleus of the stria terminalis (BNST) or preoptic regions in close vicinity to the GnRH cells, and may then act via interneurons to affect GnRH secretion.[164,166] Anterograde and retrograde neuronal tracing between the arcuate nucleus of the hypothalamus (ARC) and the preoptic area of the ovine brain indicate that there is very limited direct input to GnRH cells from the former,[45,167] but this is probably an underestimate of the level of input because of likely input to dendrites that are not seen with conventional tracing methods and microscopy (see above).

Most recently, a range of studies in many species indicates the importance of the kisspeptin cells of the hypothalamus and preoptic area in the regulation of GnRH cells and positive feedback.[168,169] This became apparent when it was found that normal reproductive function required the kisspeptin receptor.[168] The product of the kisspeptin gene is a 145 residue peptide that is processed to shorter forms as small as 10 amino acids, all retaining the C-terminus, which is the active region.[168] Irrespective of size, the kisspeptin peptides stimulate GnRH secretion.[170, 171] Whereas this is not profound in itself, the kisspeptin cells express sex-steroid receptors at a high level and are therefore well placed to mediate the effects of sex steroids on GnRH cells. These cells are located in two regions, namely the arcuate nucleus and the rostral hypothalamus/POA. In rodents, the latter group of cells is located in the anteroventral periventricular nucleus (AVPV) and preoptic periventricular nucleus (PeV).[170,172–174] In the sheep, as in other non-rodent species, the two synonymous populations of kisspeptin cells are in the arcuate nucleus and in the lateral POA.[175] More than 60% of kisspeptin cells in the AVPV of the female mouse brain express ERα.[172,176] In the AVPV/PeV of the female rat brain, approximately 90% of kisspeptin cells were shown to express ERα, as assessed by immunohistochemical staining.[177] In the ovine brain, virtually all of the kisspeptin cells in the ARC express ERα, whereas only 50% of the population of cells in the lateral POA do so.[175] Kisspeptin cells also express progesterone receptors[172,178] and androgen receptors.[174] Accordingly, these cells have all the essential prerequisites to act as conduits of sex-steroid feedback to GNRH cells. The central role of kisspeptin cells in the feedback regulation of GnRH cells is seen in Box 9.5.

BOX 9.5

HOW DO WE KNOW THAT KISSPEPTIN RELAYS NEGATIVE AND POSITIVE FEEDBACK EFFECTS OF ESTROGEN TO GNRH NEURONS?

Extraordinarily low doses of kisspeptin injected into the cerebral ventricles (i.c.v.) robustly stimulate LH and FSH secretion in mice,[1] rats,[2] hamsters,[3] sheep,[4] monkeys[5] and humans.[6] It is now indisputable that kisspeptin directly stimulates GnRH secretion, and GnRH is secreted in response to kisspeptin.[7] The discrete localization of Kiss1 mRNA in the brain was first determined in the mouse, and immediately provided clues to the physiological role of kisspeptin. Cellular populations are located in the anteroventral periventricular nucleus (AVPV), extending to the periventricular nucleus (PEN) and the arcuate nucleus (ARC).[1] In sheep, Kiss1 mRNA expressing cells are located in the dorso-lateral region of the POA (perhaps a homologous population to the rodent AVPV/PEN) and the ARC.[8–10]

BOX 9.5 *(cont'd)*

Whereas cells projecting from the ARC to the preoptic area have long been thought to play a role in the negative feedback control of GnRH secretion by estrogen (E2) in rodent species,[11,12] the AVPV/PEN region is a sexually dimorphic nucleus thought to be involved in generating the positive feedback effect of E2 to cause the preovulatory surge of GnRH in females.[13] In rodent species, therefore, kisspeptin cells in the ARC and AVPV are ideally placed to mediate both negative and positive feedback effects. In accordance with this, E2 regulates the expression of the kisspeptin gene (*Kiss1*) in the ARC and AVPV of female mice in a manner consistent with negative and positive feedback, respectively[14] (see figure below).

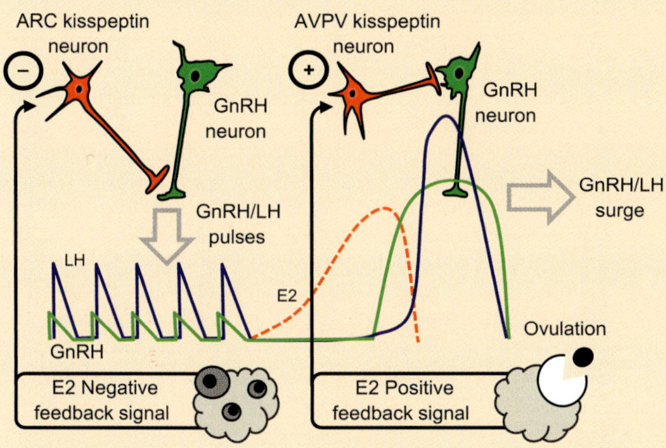

Proposed negative and positive feedback mechanism exerted by ovarian steroids on the regulation of kisspeptin neurons and, in turn, GnRH secretion, in female rodents. In the ARC, E2 inhibits ERα-positive kisspeptin neurons, leading to reduced secretion of GnRH, consistent with negative feedback regulation, resulting in reduced LH pulses. ARC kisspeptin neurons project directly to GnRH cell bodies by forming synapses on the dendrites of the GnRH cells, but this has not been shown. Kisspeptin may also regulate GnRH secretion at the level of the terminals in the median eminence. In the AVPV, rising concentrations of E2 (*) during the evening of proestrus may stimulate ERα-positive kisspeptin neurons, which leads to an increase in GnRH secretion, facilitating positive feedback regulatory control and the preovulatory GnRH/LH surge. AVPV kisspeptin neuron projections contact GnRH cell bodies.

Kiss1 expression in the AVPV increases at the time of estrogen-induced positive.[15] In female sheep, key differences from the rodent are apparent in feedback regulation of GnRH. Estrogen-sensitive cells in the ARC appear to be fundamental to both negative and positive feedback GnRH regulation in this species.[16,17] Thus, in the sheep, Kiss1 mRNA expression in the ARC is reduced by chronic E2 treatment,[9] indicating a role in negative feedback regulation (Fig. 9.12). At the time of the normal preovulatory LH surge Kiss1 expression in the caudal ARC is increased,[10] indicative of the role of these kisspeptin cells in acute E2 positive feedback (see figure below).

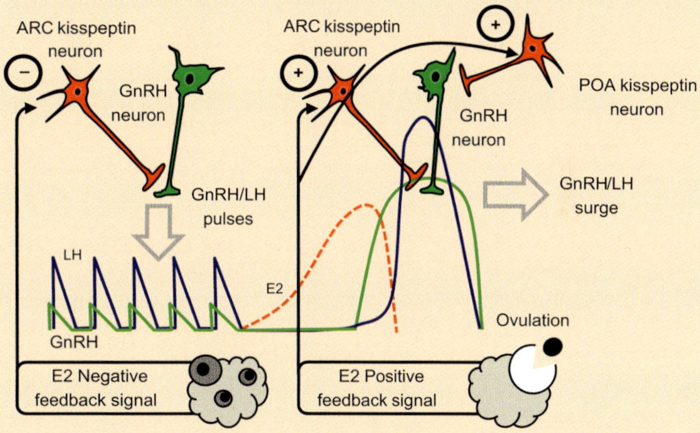

Proposed negative and positive feedback mechanism exerted by ovarian steroids via kisspeptin neurons and, in turn, on GnRH secretion in the ewe. E2 inhibits ERα-positive kisspeptin neurons in the ARC, consistent with negative feedback control, resulting in GnRH and LH pulses. During the late follicular phase of the estrous cycle, rising E2 (*) stimulates ERα-positive kisspeptin neurons in the ARC, and also kisspeptin neurons in the POA. It is unclear whether the effect in the latter is a direct effect of E2 or whether POA kisspeptin cells act as an interneuronal pathway linking ARC kisspeptin cells (or other ERα-expressing cells) to GnRH neurons. Data suggest POA kisspeptin neurons project directly to GnRH cell bodies.

E2 increases Kiss1 expression in the ovine POA.[18] Recent data also show that Kiss1 expression in the POA increases at the time of the preovulatory LH surge.[19,20] These data implicate kisspeptin cells, in both the ARC and the POA, as central processors of the feedback effects of estrogen that cause the GnRH/LH surge in the ovine species (see figure above). Overall, kisspeptin cells in the hypothalamus appear to provide the "missing link" in the steroid feedback control of GnRH secretion.

(Continued)

BOX 9.5 *(cont'd)*

References

1. Gottsch ML, Cunningham MJ, Smith JT, Popa SM, Acohido BV, Crowley WF, et al. A role for kisspeptins in the regulation of gonadotropin secretion in the mouse. *Endocrinology.* 2004;145(9):4073–4077.
2. Matsui H, Takatsu Y, Kumano S, Matsumoto H, Ohtaki T. Peripheral administration of metastin induces marked gonadotropin release and ovulation in the rat. *Biochem Biophys Res Commun.* 2004;320(2):383–388.
3. Greives TJ, Mason AO, Scotti MA, Levine J, Ketterson ED, Kriegsfeld LJ, et al. Environmental control of kisspeptin: implications for seasonal reproduction. *Endocrinology.* 2007; 148(3):1158–1166.
4. Messager S, Chatzidaki EE, Ma D, Hendrick AG, Zahn D, Dixon J, et al. Kisspeptin directly stimulates gonadotropin-releasing hormone release via G protein-coupled receptor 54. *Proc Natl Acad Sci USA.* 2005;102(5):1761–1766.
5. Shahab M, Mastronardi C, Seminara SB, Crowley WF, Ojeda SR, Plant TM. Increased hypothalamic GPR54 signaling: a potential mechanism for initiation of puberty in primates. *Proc Natl Acad Sci USA.* 2005;102(6):2129–2134.
6. Dhillo WS, Chaudhri OB, Patterson M, Thompson EL, Murphy KG, Badman MK, et al. Kisspeptin-54 stimulates the hypothalamic–pituitary–gonadal axis in human males. *J Clin Endocrinol Metab.* 2005;90(12):6609–6615.
7. Smith JT, Li Q, Yap KS, Shahab M, Roseweir AK, Millar RP, Clarke IJ. Kisspeptin is essential for the full preovulatory LH surge and stimulates GnRH release from the isolated ovine median eminence. *Endocrinology;* 2011. in press.
8. Franceschini I, Lomet D, Cateau M, Delsol G, Tillet Y, Caraty A. Kisspeptin immunoreactive cells of the ovine preoptic area and arcuate nucleus co-express estrogen receptor alpha. *Neurosci Lett.* 2006;401(3):225–230.
9. Smith JT, Clay CM, Caraty A, Clarke IJ. KiSS-1 messenger ribonucleic acid expression in the hypothalamus of the ewe is regulated by sex steroids and season. *Endocrinology.* 2007; 148(3):1150–1157.
10. Estrada KM, Clay CM, Pompolo S, Smith JT, Clarke IJ. Elevated KiSS-1 expression in the arcuate nucleus prior to the cyclic preovulatory gonadotrophin-releasing hormone/ lutenising hormone surge in the ewe suggests a stimulatory role for kisspeptin in oestrogen-positive feedback. *J Neuroendocrinol.* 2006;18(10):806–809.
11. Soper BD, Weick RF. Hypothalamic and extrahypothalamic mediation of pulsatile discharges of luteinizing hormone in the ovariectomized rat. *Endocrinology.* 1980;106(1):348–355.
12. Simerly RB. Wired for reproduction: organization and development of sexually dimorphic circuits in the mammalian forebrain. *Annu Rev Neurosci.* 2002;25:507–536.
13. Wintermantel TM, Campbell RE, Porteous R, Bock D, Grone HJ, Todman MG, et al. Definition of estrogen receptor pathway critical for estrogen positive feedback to gonadotropin-releasing hormone neurons and fertility. *Neuron.* 2006;52(2):271–280.
14. Smith JT, Cunningham MJ, Rissman EF, Clifton DK, Steiner RA. Regulation of Kiss1 gene expression in the brain of the female mouse. *Endocrinology.* 2005;146(9):3686–3692.
15. Smith JT, Popa SM, Clifton DK, Hoffman GE, Steiner RA. Kiss1 neurons in the forebrain as central processors for generating the preovulatory luteinizing hormone surge. *J Neurosci.* 2006;26(25):6687–6694.
16. Blache D, Fabre-Nys CJ, Venier G. Ventromedial hypothalamus as a target for oestradiol action on proceptivity, receptivity and luteinizing hormone surge of the ewe. *Brain Res.* 1991;546(2):241–249.
17. Caraty A, Fabre-Nys C, Delaleu B, Locatelli A, Bruneau G, Karsch FJ, et al. Evidence that the mediobasal hypothalamus is the primary site of action of estradiol in inducing the preovulatory gonadotropin releasing hormone surge in the ewe. *Endocrinology.* 1998;139(4):1752–1760.
18. Smith JT, Coolen LM, Kriegsfeld LJ, Sari LP, Jaafarzadehshirazi MR, Maltby M, Bateman K, Goodman RL, Tilbrook AJ, Ubuka T, Bentley GE, Clarke IJ, Lehman MN. Variation in kisspeptin and gonadotropin-inhibitory hormone expression and terminal connections to GnRH neurons in the brain: a novel medium for seasonal breeding in the sheep. *Endocrinology.* 2008;149(11): 5770–5782.
19. Hoffman GE, Le WW, Franceschini I, Caraty A, Advis JP. Expression of Fos and *in vivo* median eminence release of LHRH identifies an active role for preoptic area kisspeptin neurons in synchronized surges of LH and LHRH in the ewe. *Endocrinology.* 2011;152(1):214–222.
20. Smith JT, Li Q, Pereira A, Clarke IJ. Kisspeptin neurons in the ovine arcuate nucleus and preoptic area are involved in the preovulatory luteinizing hormone surge. *Endocrinology.* 2009;150(12):5530–5538.

Proof of the essential role of kisspeptin as a stimulator of GnRH secretion was obtained with a kisspeptin antagonist.[179] In ovariectomized ewes, the antagonist blocks pulsatile LH secretion; this is due to direct action on GnRH neurons, as demonstrated with electrophysiological studies of GnRH cells and the inhibition of GnRH secretion in prepubertal non-human primates.[179] The antagonist also blocked the rise in gonadotropin secretion due to castration, suggesting that restraint of the kisspeptin neurons is the mechanism of sex-steroid negative feedback. Finally, the kisspeptin antagonist blocks the estrogen-induced positive feedback effect,[180] indicating an essential role in surge generation.

The administration of kisspeptin to acyclic, seasonally anestrous ewes causes ovulation.[181] The mechanism for this appears to be stimulation of basal gonadotropic secretion and stimulation of estrogen secretion by ovarian follicles, which in turn act to cause positive feedback and an ovulatory LH surge.

An important observation that further consolidated the importance of kisspeptin cells as major regulators of reproduction was that which showed co-expression of dynorphin (DYN) and neurokinin B (NKB) with kisspeptin.[182] Virtually all kisspeptin cells in the ARC co-express these peptides. This has led to the naming of these cells as K (kisspeptin) N (neurokinin B) Dy (dynorphin) (KNDy) cells.[183] These KNDy cells may mediate both negative and positive feedback effects of sex steroids, since there is good evidence that dynorphin plays a role in mediating the negative feedback effect of progesterone,[184,185] in addition to the evidence that chronic estrogen treatment downregulates the KNDy cells in OVX ewes.[186] Further support for the notion that these cells participate in transmitting the negative feedback effect to GnRH cells is the observation that kisspeptin expression in the ARC is reduced in the luteal phase of the estrous cycle.[186]

GONADOTROPIN INHIBITORY HORMONE (GNIH)

Localization of GnIH Neurons and their Projections

Although reproductive function depends on the stimulatory action of gonadotropin-releasing hormone (GnRH), original work in birds identified and isolated a peptide that inhibits gonadotropin release, named gonadotropin inhibitory hormone (GnIH).[187,188] A significant body of evidence has now accumulated to show that a mammalian ortholog of GnIH also exists, and may act on GnRH neurons and the pituitary gonadotropes to negatively regulate reproductive function.[189,190] GnIH has also been named RFamide-Related Peptide (RFRP) because it belongs to the family of peptides that have a common Arg-Phe-NH_2 C-terminus, but there seems no reason why the mammalian forms of the peptide cannot be called GnIH. In the avian, *GnIH* gene encodes a precursor protein of 173 amino acids, from which GnIH and two other GnIH-related peptides (originally named GnIH-RP-1 and GnIH-RP-2) are derived.[191,192] In mammals, the *GnIH* (*RFRP*) gene encodes at least two peptides, named RFRP-1 and RFRP-3. RFRP-2 has also been identified in humans and in bovine species.[191–194] In the mammalian brain, GnIH-producing neurons are found in the paraventricular/dorsomedial hypothalamic nuclei.[189] Most of the work carried out in mammals has been with GnIH-3, as this appears to be the most genetically related and functionally similar to avian GnIH. Mammalian GnIH-1 however, also has significant homology with avian GnIH, as it shares the LPLRF-amide sequence at the C-terminus and the two peptides appear to be equipotent.[189] In the ovine brain, *in situ* hybridization identified GnIH-expressing cells in the ventral region of the paraventricular nucleus and the dorsomedial nucleus.[195–197] These cells project to GnRH neurons and other neurons in the hypothalamus that subserve other functions, such as food intake.[198,199] In fact, as a general rule, hypothalamic neuropeptides that inhibit reproductive function also stimulate food intake, and *vice versa*.[200] Based on data from various species, including primates, rat, sheep, hamster and mouse,[195,199,201–204] 40–80% of GnRH neurons are contacted by GnIH afferents, suggesting direct regulation of the former by the latter; this is substantiated by functional data (see below). Most recently, GnIH-expressing cells have been localized in the non-human primate hypothalamus. In this case the cells are found in the intermediate periventricular nucleus,[202] which is immediately medial to the dorsomedial nucleus.

Importantly, GnIH neurons project into the neurosecretory zone of the median eminence, as seen in hamsters, sheep and non-human primates,[196,198,202,203] so that the peptide may be delivered into the portal blood. Although it has not yet been shown that the peptide is actually secreted from the hypothalamus, there is evidence that GnIH acts on the anterior pituitary gonadotropes to counter the action of GnRH. This includes *in vitro* data and *in vivo* data (see below). Similar projections were not observed in the rat.[199,205] On this basis, a hypophysiotropic role for GnIH has been proposed for the sheep, but not the rat. This issue is unresolved, because direct pituitary effects of GnIH have been demonstrated in the rat, predicating a role for peptide secreted from the hypothalamus.[206]

Action of GnIH to Regulate GnRH Neurons in Mammals

Intracerebroventricular (i.c.v.) injection of GnIH-3 (presumably indicating an action on GnRH neurons) reduced plasma LH secretion in ovariectomized (OVX) hamsters[203] and gonad-intact rats,[199,207] but others have shown that similar treatments were not effective in reducing plasma LH levels in OVX, estrogen-treated rats[208] and OVX rats.[206] Using an estradiol/progesterone GnRH/LH surge-induction protocol in OVX rats, GnIH was seen to reduce the activation of GnRH neurons (assessed by detection of the immediate early gene *c-Fos*,) and also reduced neuronal activation in the anteroventral periventricular (AVPV) region, which provides stimulatory input to GnRH neurons,[209] but the LH surge was only slightly reduced in magnitude. Other evidence of a central role for GnIH in the rat is obtained from studies showing that GnIH antisense oligonucleotide administration (via i.c.v.) to prepubertal rats reduced immunochemical detection of GnIH cells in the brain and increased plasma LH levels,[207] but the antisense treatment did not advance puberty. Using electrophysiology to record GnRH cells in slices of mouse brain, GnIH treatment was seen to inhibit the firing rate of GnRH neurons.[201,208] This supports a role for GnIH in the regulation of GnRH cells.

In sheep, GnIH protein expression is higher during the non-breeding season than in the breeding season,[210] and terminal projections from GnIH cells to GnRH neurons are also increased during the non-breeding season. Thus, it appears that the activity of GnIH may be a contributing factor to the inhibition of the reproductive system during the non-breeding season. Similarly, *GnIH* gene expression was higher in Soay ewes on artificial long-day photoperiod (synonymous with the non-breeding season) than in ewes on a short-day protocol.[196] These data further support a role for GnIH in the regulation of GnRH cells. Contrary data have been found in other seasonally breeding species (hamsters),[211,212] so the role in control of breeding season may vary between species.

Action of GnIH to Regulate Gonadotropes in Mammals

In one study, measurement of plasma LH levels in ovariectomized rats showed minimal effect of intravenously (i.v.) administered GnIH-3,[205] which was taken as evidence of there being no major effect at the level of the gonadotrope; however, others have presented evidence of a pituitary effect.[206] GnIH-3 reduces pulsatile LH secretion in OVX ewes,[213] and the lack of effect on the plasma levels of other pituitary hormones (GH, prolactin) suggests specificity; similar results have recently been obtained in bovine species.[214] In support of a direct effect on the pituitary gonadotrope, dose-dependent reduction in GnRH-stimulated LH secretion is seen in cultures of rat, sheep and bovine pituitary cells.[206,213,214] These data strongly suggest that GnIH acts on gonadotropes. Various forms of the GnIH peptide are equipotent in ovine pituitary cell culture assays (Fig. 9.12).

In support of the pituitary gonadotrope action of GnIH-3, the peptide is seen to eliminate the GnRH-stimulated mobilization of intracellular calcium in these cells, which is considered mandatory for gonadotropin release.[213] GnRH-stimulated upregulation of *LHβ* mRNA levels is also counteracted by GnIH in ovine cells *in vitro*.[215] Moreover, GnRH-stimulated phosphorylation of extracellular signal-regulated kinase (ERK) was abolished by GnIH-3 treatment, indicating that GnIH-3 may inhibit gonadotropin synthesis through this second messenger pathway.[215] *In vitro* treatment of ovine

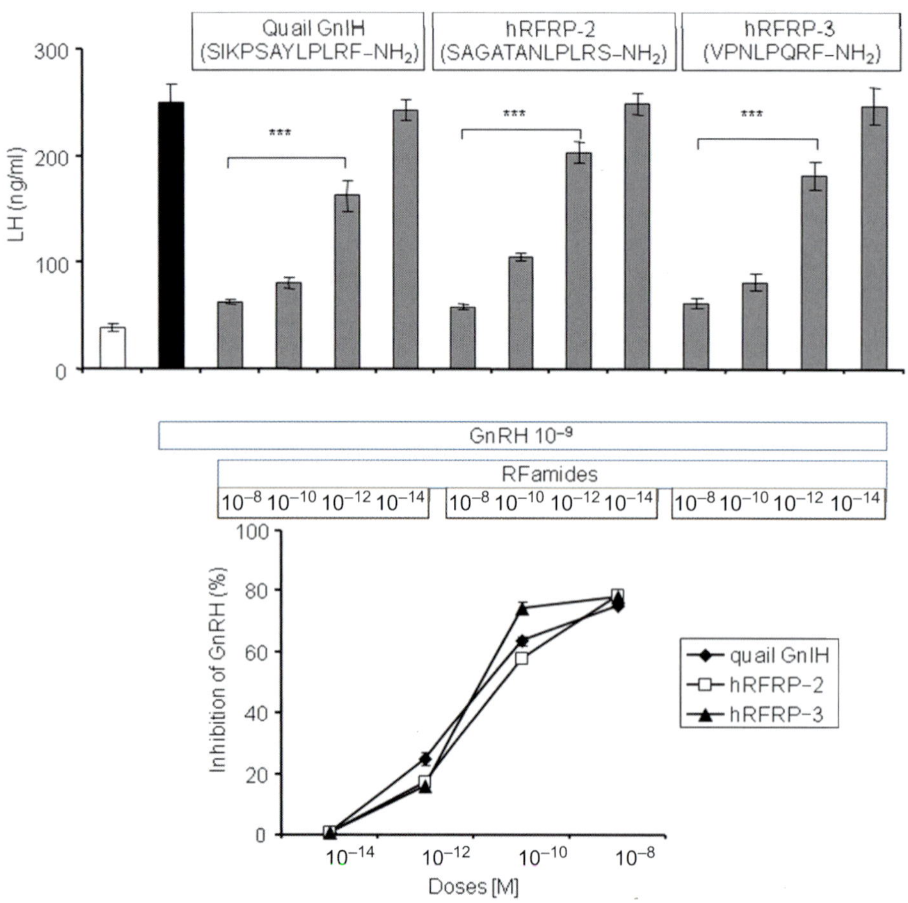

FIGURE 9.12 **Inhibitory effect of three GnIH (RFRP) peptides on secretion of LH from ovine pituitary cells in culture, using cells from the same pituitary of an ovariectomized ewe.** The upper panels shows mean LH levels in medium of cells receiving vehicle (open bar), GnRH (10^{-9} M: black bar) or with quail GnIH, human RFRP-2 or human RFRP-3 (10^{-8}–10^{-14} M) in combination with GnRH. The treatment sequences are indicated under the graph. The lower panel illustrates the potency of each peptide in reducing the secretory response to GnRH (% inhibition). Open bars show basal secretion, black bars show GnRH-stimulated secretion, and shaded bars show GnRH stimulated cells with various doses of GnIH, as indicated. ***$P < 0.001$ vs GnRH treatment alone. *Reproduced from Clarke et al.[189] with permission.*

pituitary cell cultures also showed that GnIH-3 reduced FSH secretion in response to GnRH[213] as well as *FSHβ* mRNA levels,[215] so it may inhibit the production/secretion of both gonadotropins. The evidence that GnIH is a *bona fide* regulatory peptide in mammals inhibiting the reproductive axis is summarized in Box 9.6.

BOX 9.6

HOW DO WE KNOW GNIH IS A PEPTIDE THAT NEGATIVELY REGULATES THE GNRH/GONADOTROPIN AXIS IN MAMMALS?

Following the identification of GnIH 10 years ago, a considerable body of data was forthcoming for avian species, but the significance of the peptide(s) in mammals has only gained acceptance in the last few years. Emerging data in a range of mammalian species now ascribe importance to GnIH as a negative regulator of reproduction. The data obtained in mammals show that GnIH negatively regulates reproduction in various experimental paradigms. It seems likely that GnIH cells transmit information to GnRH cells and/or gonadotropes, with some species variations.

References

1. Ubuka T, Kim S, Huang YC, Reid J, Jiang J, Osugi T, et al. Gonadotropin-inhibitory hormone neurons interact directly with gonadotropin-releasing hormone-I and -II neurons in European starling brain. *Endocrinology.* 2008;149(1):268–278.
2. Bentley GE, Ubuka T, McGuire NL, Chowdhury VS, Morita Y, Yano T, et al. Gonadotropin-inhibitory hormone and its receptor in the avian reproductive system. *Gen Comp Endocrinol.* 2008;156(1):34–43.
3. Yin H, Ukena K, Ubuka T, Tsutsui K. A novel G protein-coupled receptor for gonadotropin-inhibitory hormone in the Japanese quail (*Coturnix japonica*): identification, expression and binding activity. *J Endocrinol.* 2005;184(1):257–266.
4. Smith JT, Clarke IJ. Gonadotropin inhibitory hormone function in mammals. *Trends Endocrinol Metab.* 2010;21(4):255–260.

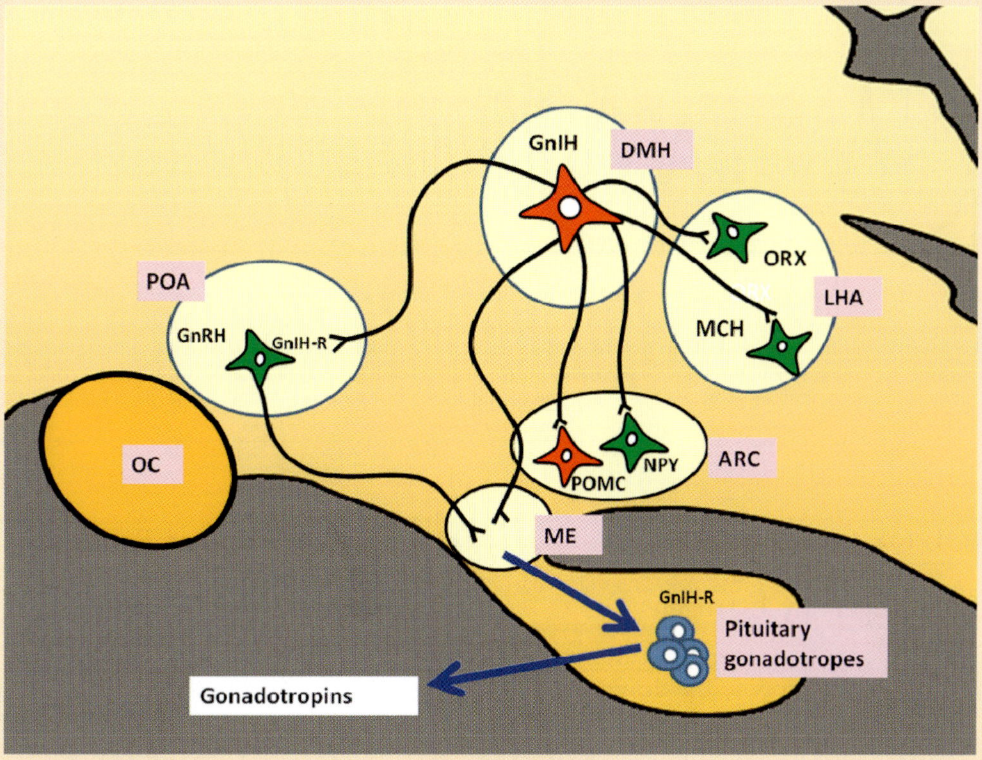

GnIH neurons (red) are localized to the dorsomedial nucleus (DMH) of the hypothalamus in mammalian species, with some cells also being found in the paraventricular nucleus and the intermediate periventricular nucleus. These cells project to GnRH neurons (green) in the preoptic area (POA) that are known to express the cognate receptor (GnIH-R).[1,2] GnIH neurons are also known to project to orexin (ORX, blue), and melanin concentrating hormone (MCH, purple) neurons in the lateral hypothalamic area (LHA) as well as pro-opiomelanocortin (POMC, yellow) and neuropeptide Y (NPY, orange) cells in the arcuate nucleus (ARC). Projections to the neurosecretory zone of the median eminence (ME) in the sheep brain suggest a hypophysiotropic role for the peptide and action on pituitary gonadotropes (which also express GnIH receptors[3]) to regulate the synthesis and secretion of gonadotropins. GnIH-R has not yet been described for "appetite-regulating" cells. OC, optic chiasm. *From Smith and Clarke[4] with permission.*

(*Continued*)

BOX 9.6 *(cont'd)*

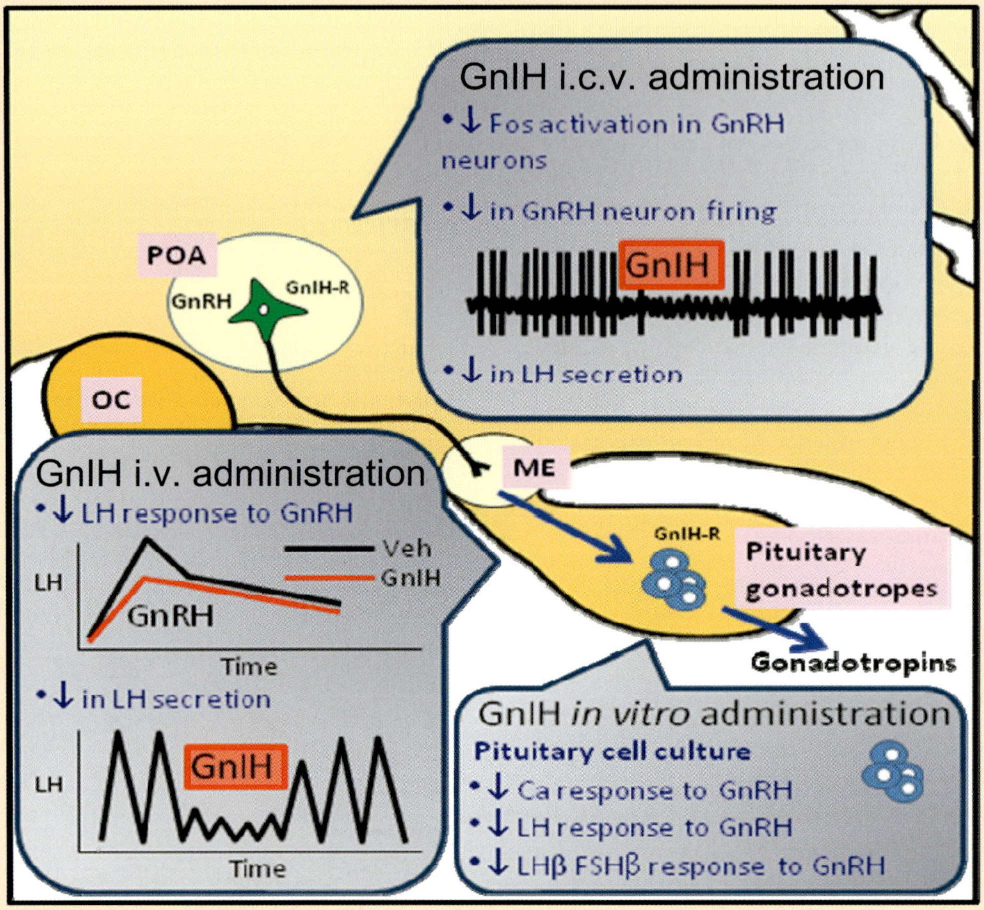

Inhibitory effects of GnIH on the hypothalamic pituitary gonadal axis. Effects are shown for both intracerebroventricular (i.c.v.) and intravenous (i.v.) administration *in vivo*, including effects on GnRH neurons (which express the GnIH-R in birds, at least) and LH secretion from the pituitary gonadotropes (which express GnIH-R). *In vitro* effects of GnIH on primary pituitary cell cultures are also summarized, including a decline in calcium (Ca^{2+}) induction, LH release, and LHβ-, FSHβ-subunit expression in response to GnRH treatment. POA, preoptic area; OC, optic chiasm; ME, median eminence. *From Smith and Clark*[4] *with permission.*

CONTROL OF FSH SECRETION

Regulation of FSH Levels by Estrogen

Estradiol negatively regulates the production of FSH by the pituitary gonadotropes, by downregulation of the FSHβ subunit gene.[216] This direct pituitary action of estradiol reduces FSH secretion, as seen in the *in vivo* isolated sheep pituitary model,[125] but estradiol regulation of FSHβ gene expression and FSH secretion is not seen in the mouse.[217] The mode of regulation of FSH secretion seen in the sheep pituitary conforms to the rule indicated above, whereby secretion of FSH is an index of the rate of synthesis of the gonadotropin.

Inhibins, Activins and Follistatin

FSH secretion by the gonadotropes is positively regulated by activin and negatively regulated by the combined action of estrogen and inhibin. Follistatin is an activin binding protein, and therefore counters the stimulatory action of activin. Inhibin and activin hormones are disulfide linked protein dimers, being various combinations of α and β subunits.[218–221] There are diverse forms of these hormones, depending upon whether they are heterodimeric or homodimeric combinations of the α and β subunits belonging to the transforming growth factor-β (TGF-β) superfamily, and they play a role in regulating growth and differentiation factors in different tissues. There are four mammalian activin-β subunits (β_A, β_B, β_C and β_E) that form homodimeric or heterodimeric proteins, such as activin A ($\beta_A\beta_A$), activin B ($\beta_B\beta_B$) and activin AB ($\beta_A\beta_B$). Most recently, β_C and β_E were discovered, and little is known about their function.[222,223] The activin-β_A and -β_B subunits heterodimerize with the inhibin-α subunit to form inhibin A (α-β_A) or inhibin B (α-β_B). Inhibins are produced in the gonads and circulate in the bloodstream, as hormones, acting on the pituitary gonadotropes to suppress FSH secretion. There is little difference in the actions of inhibin A or inhibin B, but inhibin A appears to be the major species in the female and inhibin B in the male.[224,225] Activins are produced by a range of tissues in the body, and inhibins oppose the actions of activins.[226] In addition to the opposing actions of activins and inhibins on the synthesis and secretion of FSH, follistatin is a binding protein for activins, negating their activity.[227,228] This is a single chain polypeptide hormone, originally purified from ovarian fluid,[229] that has a similar effect to inhibin in reducing FSH secretion,[230] albeit by a different mechanism (the neutralization of activin).

Feedback Effects of Inhibin

Inhibin A rapidly and powerfully suppresses FSHβ gene expression by direct action on the pituitary gland,[231] leading to reduced secretion of FSH,[232] again by direct pituitary action[232] (Fig. 9.13). This is due to

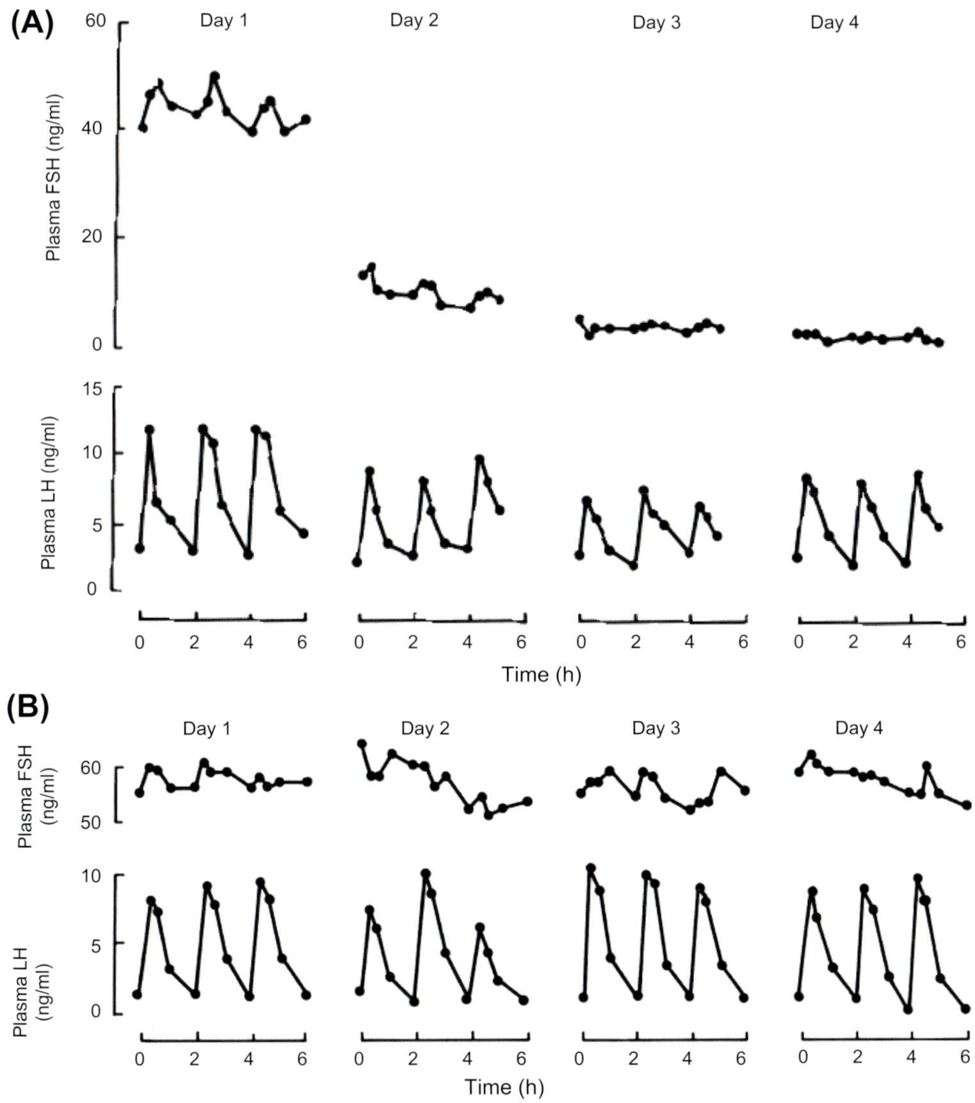

FIGURE 9.13 **Plasma FSH and LH levels in an ovariectomized HPD ewe given i.v. pulses of GnRH every 2 h.** Panel (**A**) shows data for an animal given twice daily treatment with charcoal-treated ovine follicular fluid (inhibin). Panel (**B**) shows data for an animal receiving control serum. Note the effect of inhibin on FSH levels but not LH levels. *Reproduced from Clarke et al. (1986)[240] with permission.*

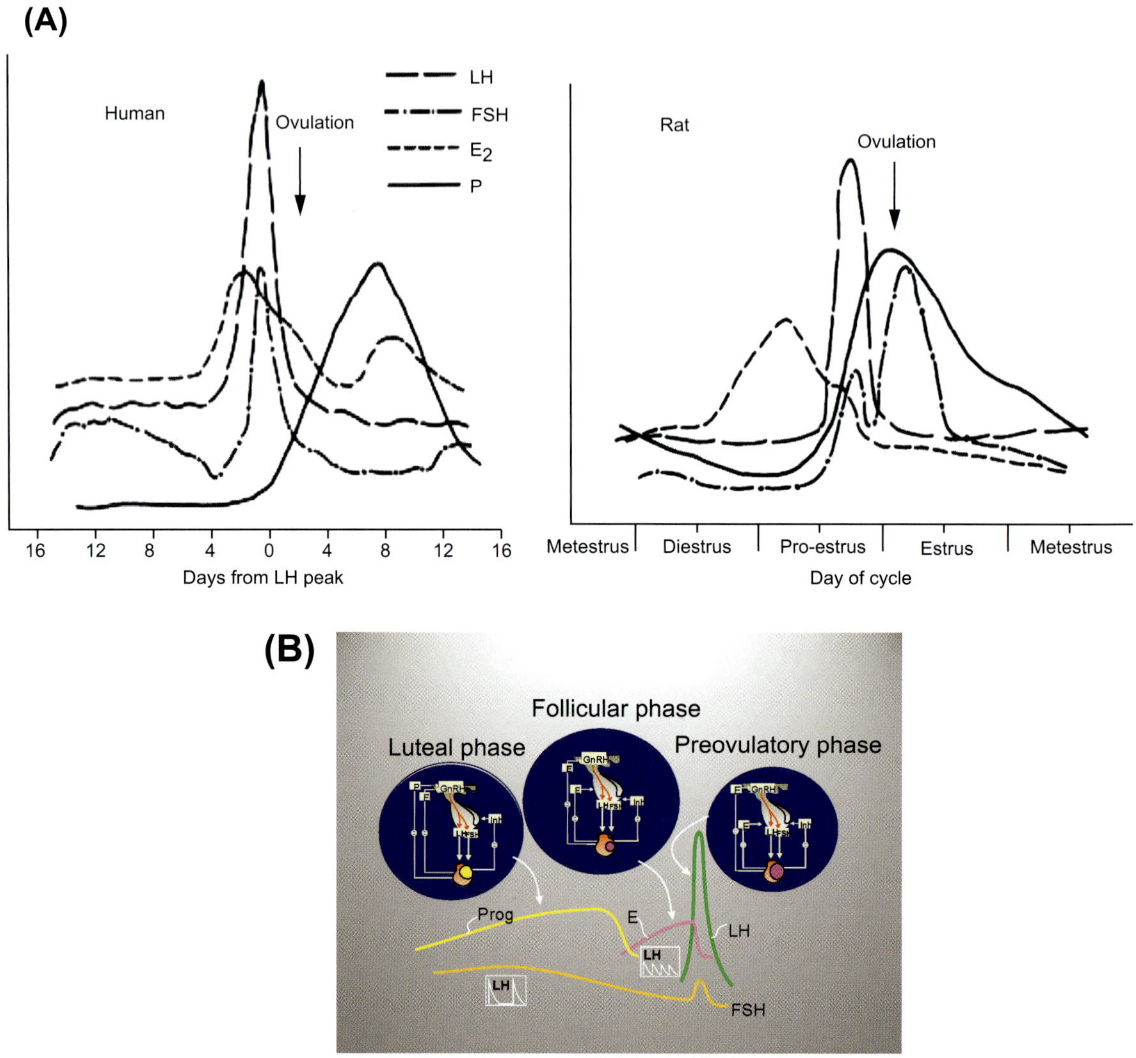

FIGURE 9.14 **The female estrous/menstrual cycle.** (**A**) The left-hand panel shows a schematic of hormonal changes during the human menstrual cycle; the right-hand left panel shows similar data for the rat estrous cycle. Ovulation is preceded by a surge of luteinizing hormone (LH) which is triggered by a surge of estradiol-17β (E2), and accompanied and followed by a surge of progesterone (P) secretion. Most of the progesterone is secreted by the ovary in response to LH, but in the rat a small amount of progesterone is secreted by the adrenal gland just before the onset of the LH surge. In the human, the timing and magnitude of the FSH surge is less consistent than for the LH surge, and in the rat a major peak of the FSH secretion occurs on the morning of estrus (after or about the time of ovulation). From Fink G. Gonadotropin secretion and its control. In Fink (1988)[129] with permission. (**B**) Detailed data for the ovine estrous cycle. From Clarke (1996)[140] with permission. In this species, the relationship between the gonadal steroids and protein hormones and the pulsatile secretion of GnRH and LH as well as the secretion of FSH has been very well characterized. During the luteal phase of the cycle, GnRH pulse frequency is reduced by progesterone (and estrogen acting in concert with progesterone), with an increase in pulse amplitude; this leads to a reduction in LH pulse frequency. The increase in amplitude of GnRH pulses is further amplified by the frequency : amplitude relationship at the pituitary level. Following regression of the corpus luteum and release of the negative brake of progesterone, GnRH pulse (and LH pulse) frequency increases, leading to stimulation of ovarian follicles and an increase in the production and secretion of estradiol. This in turn has an initial (rapid) negative feedback effect to hold down gonadotropin levels, but at a threshold level of estradiol a positive feedback event is initiated, leading to a surge in GnRH and LH secretion. Importantly, the positive feedback signal is initiated during the negative feedback phase of the follicular phase, but the process of the generation of the preovulatory LH surge is time-delayed. FSH levels fall during the follicular phase of the cycle due to increased secretion of inhibin from the ovarian follicles.

regulation at the transcriptional level.[233] There is no effect of inhibin on GnRH secretion, based on the lack of effect on pulsatile LH secretion,[232,234] although there are no published data to document such a lack of effect.

Role of Activin in control of FSH Secretion

Activins also were purified from follicular fluid, but because circulating activin levels generally are low, most actions of the hormones are likely to be paracrine in nature.[235] The activins selectively stimulate FSH production and secretion in the pituitary gonadotropes,[236] but also have a range of other non-reproductive functions.[237] This action in the gonadotropes involves promoter activation through Smad-associated binding proteins.[238,239]

FEEDBACK REGULATION OF GNRH AND THE GONADOTROPINS DURING THE ESTROUS/MENSTRUAL CYCLE

Whereas the feedback regulation of GnRH and gonadotropins in the male is a relatively straightforward case of negative control by gonadal hormones in a homeostatic model, that of the female is more complicated due to cyclic changes in gonadal function. The changes in plasma hormone levels that occur across the human menstrual cycle and the rat and sheep estrous cycles are shown in Fig. 9.14.

FUTURE DIRECTIONS

Considerable progress has been made in the past decade to improve our understanding of the way that the GnRH–gonadotropin axis operates, especially since the discovery of the role of kisspeptin and GnIH. In addition, the role of rapid signaling of steroids has gained appreciation. Future directions for research in this area are as follows:

- Increased understanding of how rapid steroid signaling occurs, either at the level of cell membranes or within the cells, via conventional receptors. Whereas a considerable amount is known about the rapid signaling of estrogen, similar effects of other steroids are less well studied.
- Information regarding the means by which sex steroids direct transcription and affect the synthesis of signaling intermediaries (kisspeptin) and the gonadotropins. In particular, even though we have identified some of the intracellular molecules that are involved in the transmission of sex-steroid signals to promoter regions of genes, this is a fruitful area of research that should yield a significantly greater understanding of the molecular mechanisms underlying feedback regulation of GnRH and the gonadotropins.
- The means by which inhibin acts to specifically regulate FSH synthesis and secretion; this is currently poorly understood.
- Factors that affect the synthesis and secretion of GnIH, which are not well documented at present.
- Species differences in the way that feedback effects of steroids (and inhibin, activin and follistatin) regulate the GnRH-gonadotropin system are known, but the actual points of difference are not well understood. In particular, it will be interesting to know which species provide the best models for the human.
- It is not known how the feedback effect of estrogen may "switch" from a negative mode to a positive mode. Presumably, this involves molecular mechanisms that have a specific time course.
- Regulation of gene transcription has been well studied in some systems, with the aid of immortalized cells or transgenic animals, but a greater appreciation of the events that occur in normal cells would provide an advance in the field. This could perhaps be achieved with better use of neuronal cultures and the study of pure populations of pituitary gonadotropes, which can be obtained from animals such as sheep that have large pituitary glands.
- Knowledge regarding the "wiring" of GnRH neurons would significantly improve our understanding of how these cells are regulated. Whereas a great body of work has been done with neuronal tracing techniques and confocal microscopy, a means of examining the synaptic input to the dendrites of GnRH cells is now available (in the mouse at least), and this could be utilized to improve our knowledge in this area. Greater use of electron microscopy would be an advantage.
- The neuronal plasticity of cells in the brain that regulate GnRH cells (including glia) will lead to a greater perspective on how various factors, including environmental factors, may impact on the GnRH system.

References

1. Clarke IJ, Cummins JT. The temporal relationship between gonadotropin releasing hormone (GnRH) and luteinizing hormone (LH) secretion in ovariectomized ewes. *Endocrinology.* 1982;111(5):1737–1739.
2. Fink G, Jamieson MG. Effect of electrical stimulation of the preoptic area on luteinizing hormone releasing factor in pituitary stalk blood. *J Physiol.* 1974;237(2):37P–38P.
3. Sarkar DK, Chiappa SA, Fink G, Sherwood NM. Gonadotropin-releasing hormone surge in pro-estrous rats. *Nature.* 1976; 264(5585):461–463.

4. Clarke IJ, Fraser HM, McNeilly AS. Active immunization of ewes against luteinizing hormone releasing hormone, and its effects on ovulation and gonadotrophin, prolactin and ovarian steroid secretion. *J Endocrinol.* 1978;78(1):39–47.
5. Cattanach BM, Iddon CA, Charlton HM, Chiappa SA, Fink G. Gonadotrophin-releasing hormone deficiency in a mutant mouse with hypogonadism. *Nature.* 1977;269(5626):338–340.
6. Silverman AJ, Zimmerman EA, Gibson MJ, Perlow MJ, Charlton HM, Kokoris GJ, et al. Implantation of normal fetal preoptic area into hypogonadal mutant mice: temporal relationships of the growth of gonadotropin-releasing hormone neurons and the development of the pituitary/testicular axis. *Neuroscience.* 1985;16(1):69–84.
7. Mason AJ, Hayflick JS, Zoeller RT, Young III WS, Phillips HS, Nikolics K, et al. A deletion truncating the gonadotropin-releasing hormone gene is responsible for hypogonadism in the hpg mouse. *Science.* 1986;234(4782):1366–1371.
8. Mason AJ, Pitts SL, Nikolics K, Szonyi E, Wilcox JN, Seeburg PH, et al. The hypogonadal mouse: reproductive functions restored by gene therapy. *Science.* 1986;234(4782): 1372–1378.
9. Schwanzel-Fukuda M, Bick D, Pfaff DW. Luteinizing hormone-releasing hormone (LHRH)-expressing cells do not migrate normally in an inherited hypogonadal (Kallmann) syndrome. *Brain Res Mol Brain Res.* 1989;6(4):311–326.
10. Fernald RD, White RB. Gonadotropin-releasing hormone genes: phylogeny, structure, and functions. *Front Neuroendocrinol.* 1999;20(3):224–240.
11. Oakley AE, Breen KM, Clarke IJ, Karsch FJ, Wagenmaker ER, Tilbrook AJ. Cortisol reduces gonadotropin-releasing hormone pulse frequency in follicular phase ewes: influence of ovarian steroids. *Endocrinology.* 2009;150(1):341–349.
12. Henry BA, Goding JW, Tilbrook AJ, Dunshea FR, Clarke IJ. Intracerebroventricular infusion of leptin elevates the secretion of luteinising hormone without affecting food intake in long-term food-restricted sheep, but increases growth hormone irrespective of bodyweight. *J Endocrinol.* 2001; 168(1):67–77.
13. Delgadillo JA, Gelez H, Ungerfeld R, Hawken PA, Martin GB. The "male effect" in sheep and goats – revisiting the dogmas. *Behav Brain Res.* 2009;200(2):304–314.
14. Karsch FJ, Dahl GE, Evans NP, Manning JM, Mayfield KP, Moenter SM, et al. Seasonal changes in gonadotropin-releasing hormone secretion in the ewe: alteration in response to the negative feedback action of estradiol. *Biol Reprod.* 1993;49(6): 1377–1383.
15. Pierce BN, Hemsworth PH, Rivalland ET, Wagenmaker ER, Morrissey AD, Papargiris MM, et al. Psychosocial stress suppresses attractivity, proceptivity and pulsatile LH secretion in the ewe. *Horm Behav.* 2008;54(3):424–434.
16. Karsch FJ, Battaglia DF. Mechanisms for endotoxin-induced disruption of ovarian cyclicity: observations in sheep. *Reprod Suppl.* 2002;59:101–113.
17. Clarke IJ. Variable patterns of gonadotropin-releasing hormone secretion during the estrogen-induced luteinizing hormone surge in ovariectomized ewes. *Endocrinology.* 1993;133(4): 1624–1632.
18. Karsch FJ, Cummins JT, Thomas GB, Clarke IJ. Steroid feedback inhibition of pulsatile secretion of gonadotropin-releasing hormone in the ewe. *Biol Reprod.* 1987;36(5):1207–1218.
19. Goodman RL, Jansen HT, Billings HJ, Coolen LM, Lehman MN. Neural systems mediating seasonal breeding in the ewe. *J Neuroendocrinol.* 2010;22(7):674–681.
20. Schwanzel-Fukuda M, Pfaff DW. Origin of luteinizing hormone-releasing hormone neurons. *Nature.* 1989;338(6211):161–164.
21. Wray S, Grant P, Gainer H. Evidence that cells expressing luteinizing hormone-releasing hormone mRNA in the mouse are derived from progenitor cells in the olfactory placode. *Proc Natl Acad Sci USA.* 1989;86(20):8132–8136.
22. Wray S, Nieburgs A, Elkabes S. Spatiotemporal cell expression of luteinizing hormone-releasing hormone in the prenatal mouse: evidence for an embryonic origin in the olfactory placode. *Brain Res Dev Brain Res.* 1989;46(2):309–318.
23. Wray S. From nose to brain: development of gonadotrophin-releasing hormone-1 neurones. *J Neuroendocrinol.* 2010;22(7): 743–753.
24. Kramer PR, Wray S. Novel gene expressed in nasal region influences outgrowth of olfactory axons and migration of luteinizing hormone-releasing hormone (LHRH) neurons. *Genes Dev.* 2000;14(14):1824–1834.
25. Silverman AJ, Kokoris GJ, Gibson MJ. Quantitative analysis of synaptic input to gonadotropin-releasing hormone neurons in normal mice and hpg mice with preoptic area grafts. *Brain Res.* 1988;443(1-2):367–372.
26. Wray S, Hoffman G. A developmental study of the quantitative distribution of LHRH neurons within the central nervous system of postnatal male and female rats. *J Comp Neurol.* 1986;252(4):522–531.
27. Watts AG. The impact of physiological stimuli on the expression of corticotropin-releasing hormone (CRH) and other neuropeptide genes. *Front Neuroendocrinol.* 1996;17(3): 281–326.
28. Clarke IJ. GnRH and ovarian hormone feedback. *Oxf Rev Reprod Biol.* 1987;9:54–95.
29. Clarke IJ, Cummins JT. Pulsatility of reproductive hormones: physiological basis and clinical implications. *Baillieres Clin Endocrinol Metab.* 1987;1(1):1–21.
30. Silverman AJ, Antunes JL, Abrams GM, Nilaver G, Thau R, Robinson JA, et al. The luteinizing hormone-releasing hormone pathways in rhesus (*Macaca mulatta*) and pigtailed (*Macaca nemestrina*) monkeys: new observations on thick, unembedded sections. *J Comp Neurol.* 1982;211(3):309–317.
31. Norman RL, Resko JA, Spies HG. The anterior hypothalamus: how it affects gonadotropin secretion in the rhesus monkey. *Endocrinology.* 1976;99(1):59–71.
32. Hess DL, Wilkins RH, Moossy J, Chang JL, Plant TM, McCormack JT, et al. Estrogen-induced gonadotropin surges in decerebrated female rhesus monkeys with medial basal hypothalamic peninsulae. *Endocrinology.* 1977;101(4):1264–1271.
33. Knobil E, Plant TM, Wildt L, Belchetz PE, Marshall G. Control of the rhesus monkey menstrual cycle: permissive role of hypothalamic gonadotropin-releasing hormone. *Science.* 1980; 207(4437):1371–1373.
34. Plant TM, Krey LC, Moossy J, McCormack JT, Hess DL, Knobil E. The arcuate nucleus and the control of gonadotropin and prolactin secretion in the female rhesus monkey (*Macaca mulatta*). *Endocrinology.* 1978;102(1):52–62.
35. Nakai Y, Plant TM, Hess DL, Keogh EJ, Knobil E. On the sites of the negative and positive feedback actions of estradiol in the control of gonadotropin secretion in the rhesus monkey. *Endocrinology.* 1978;102(4):1008–1014.
36. Silverman AJ. The gonadotropin-releasing hormone (GnRH) neuronal system: Immunocytochemistry. In: Knobil E, Neill JD, eds. *The Physiology of Reproduction.* New York: Raven Press; 1988:1283–1304.
37. Kalra SP, Kalra PS. Neural regulation of luteinizing hormone secretion in the rat. *Endocr Rev.* 1983;4(4):311–351.
38. Kalra SP. Mandatory neuropeptide-steroid signaling for the preovulatory luteinizing hormone-releasing hormone discharge. *Endocr Rev.* 1993;14(5):507–538.

39. Herbison AE. Multimodal influence of estrogen upon gonadotropin-releasing hormone neurons. *Endocr Rev.* 1998;19(3):302–330.
40. Tilbrook AJ, Turner AI, Clarke IJ. Stress and reproduction: central mechanisms and sex differences in non-rodent species. *Stress.* 2002;5(2):83–100.
41. Campbell RE, Gaidamaka G, Han SK, Herbison AE. Dendro-dendritic bundling and shared synapses between gonadotropin-releasing hormone neurons. *Proc Natl Acad Sci USA.* 2009;106(26):10835–10840.
42. Campbell RE, Han SK, Herbison AE. Biocytin filling of adult gonadotropin-releasing hormone neurons *in situ* reveals extensive, spiny, dendritic processes. *Endocrinology.* 2005; 146(3):1163–1169.
43. Cottrell EC, Campbell RE, Han SK, Herbison AE. Postnatal remodeling of dendritic structure and spine density in gonadotropin-releasing hormone neurons. *Endocrinology.* 2006; 147(8):3652–3661.
44. Pompolo S, Pereira A, Kaneko T, Clarke IJ. Seasonal changes in the inputs to gonadotropin-releasing hormone neurones in the ewe brain: an assessment by conventional fluorescence and confocal microscopy. *J Neuroendocrinol.* 2003;15(5): 538–545.
45. Pompolo S, Rawson JA, Clarke IJ. Projections from the arcuate/ventromedial region of the hypothalamus to the preoptic area and bed nucleus of stria terminalis in the brain of the ewe; lack of direct input to gonadotropin-releasing hormone neurons. *Brain Res.* 2001;904(1):1–12.
46. Prevot V, Bellefontaine N, Baroncini M, Sharif A, Hanchate NK, Parkash J, et al. Gonadotrophin-releasing hormone nerve terminals, tanycytes and neurohaemal junction remodelling in the adult median eminence: functional consequences for reproduction and dynamic role of vascular endothelial cells. *J Neuroendocrinol.* 2010;22(7):639–649.
47. Prevot V, Croix D, Bouret S, Dutoit S, Tramu G, Stefano GB, et al. Definitive evidence for the existence of morphological plasticity in the external zone of the median eminence during the rat estrous cycle: implication of neuro-glio-endothelial interactions in gonadotropin-releasing hormone release. *Neuroscience.* 1999;94(3):809–819.
48. Prevot V, Dutoit S, Croix D, Tramu G, Beauvillain JC. Semi-quantitative ultrastructural analysis of the localization and neuropeptide content of gonadotropin releasing hormone nerve terminals in the median eminence throughout the estrous cycle of the rat. *Neuroscience.* 1998;84(1):177–191.
49. Ramaswamy S, Guerriero KA, Gibbs RB, Plant TM. Structural interactions between kisspeptin and GnRH neurons in the mediobasal hypothalamus of the male rhesus monkey (*Macaca mulatta*) as revealed by double immunofluorescence and confocal microscopy. *Endocrinology.* 2008;149(9):4387–4395.
50. d'Anglemont de Tassigny X, Fagg LA, Carlton MB, Colledge WH. Kisspeptin can stimulate gonadotropin-releasing hormone (GnRH) release by a direct action at GnRH nerve terminals. *Endocrinology.* 2008;149(8):3926–3932.
51. Keen KL, Wegner FH, Bloom SR, Ghatei MA, Terasawa E. An increase in kisspeptin-54 release occurs with the pubertal increase in luteinizing hormone-releasing hormone-1 release in the stalk-median eminence of female rhesus monkeys *in vivo.* *Endocrinology.* 2008;149(8):4151–4157.
52. Okubo K, Nagahama Y. Structural and functional evolution of gonadotropin-releasing hormone in vertebrates. *Acta Physiol (Oxf).* 2008;193(1):3–15.
53. Stewart AJ, Katz AA, Millar RP, Morgan K. Retention and silencing of prepro-GnRH-II and type II GnRH receptor genes in mammals. *Neuroendocrinology.* 2009;90(4):416–432.
54. Seeburg PH, Adelman JP. Characterization of cDNA for precursor of human luteinizing hormone releasing hormone. *Nature.* 1984;311(5987):666–668.
55. Nikolics K, Mason AJ, Szonyi E, Ramachandran J, Seeburg PH. A prolactin-inhibiting factor within the precursor for human gonadotropin-releasing hormone. *Nature.* 1985;316(6028):511–517.
56. Clarke IJ, Cummins JT, Karsch FJ, Seeburg PH, Nikolics K. GnRH-associated peptide (GAP) is cosecreted with GnRH into the hypophyseal portal blood of ovariectomized sheep. *Biochem Biophys Res Commun.* 1987;143(2):665–671.
57. Van Look PF, Clarke IJ, Davidson WG, Scaramuzzi RJ. Ovulation and lambing rates in ewes actively immunized against androstenedione. *J Reprod Fertil.* 1978;53(1):129–130.
58. Shivers BD, Harlan RE, Morrell JI, Pfaff DW. Absence of oestradiol concentration in cell nuclei of LHRH-immunoreactive neurones. *Nature.* 1983;304(5924):345–347.
59. Skynner MJ, Sim JA, Herbison AE. Detection of estrogen receptor alpha and beta messenger ribonucleic acids in adult gonadotropin-releasing hormone neurons. *Endocrinology.* 1999;140(11):5195–5201.
60. Herbison AE, Skynner MJ, Sim JA. Erratum: lack of detection of estrogen receptor-alpha transcripts in mouse gonadotropin releasing-hormone neurons. *Endocrinology.* 2001;142.
61. Kelly MJ, Ronnekleiv OK, Eskay RL. Identification of estrogen-responsive LHRH neurons in the guinea pig hypothalamus. *Brain Res Bull.* 1984;12(4):399–407.
62. Lagrange AH, Ronnekleiv OK, Kelly MJ. Estradiol-17 beta and mu-opioid peptides rapidly hyperpolarize GnRH neurons: a cellular mechanism of negative feedback? *Endocrinology.* 1995;136(5):2341–2344.
63. Noel SD, Keen KL, Baumann DI, Filardo EJ, Terasawa E. Involvement of G protein-coupled receptor 30 (GPR30) in rapid action of estrogen in primate LHRH neurons. *Mol Endocrinol.* 2009;23(3):349–359.
64. Terasawa E, Noel SD, Keen KL. Rapid action of oestrogen in luteinising hormone-releasing hormone neurones: the role of GPR30. *J Neuroendocrinol.* 2009;21(4):316–321.
65. Kelly MJ, Ronnekleiv OK. Membrane-initiated estrogen signaling in hypothalamic neurons. *Mol Cell Endocrinol.* 2008;290(1-2):14–23.
66. Abraham IM, Herbison AE. Major sex differences in non-genomic estrogen actions on intracellular signaling in mouse brain *in vivo.* *Neuroscience.* 2005;131(4):945–951.
67. Roepke TA, Qiu J, Bosch MA, Ronnekleiv OK, Kelly MJ. Cross-talk between membrane-initiated and nuclear-initiated oestrogen signalling in the hypothalamus. *J Neuroendocrinol.* 2009;21(4):263–270.
68. Zhao Z, Park C, McDevitt MA, Glidewell-Kenney C, Chambon P, Weiss J, et al. p21-Activated kinase mediates rapid estradiol-negative feedback actions in the reproductive axis. *Proc Natl Acad Sci USA.* 2009;106(17):7221–7226.
69. Herbison AE, Pape JR. New evidence for estrogen receptors in gonadotropin-releasing hormone neurons. *Front Neuroendocrinol.* 2001;22(4):292–308.
70. Couse JF, Curtis SW, Washburn TF, Eddy EM, Schomberg DW, Korach KS. Disruption of the mouse oestrogen receptor gene: resulting phenotypes and experimental findings. *Biochem Soc Trans.* 1995;23(4):929–935.
71. Lubahn DB, Moyer JS, Golding TS, Couse JF, Korach KS, Smithies O. Alteration of reproductive function but not prenatal sexual development after insertional disruption of the mouse estrogen receptor gene. *Proc Natl Acad Sci USA.* 1993;90(23): 11162–11166.

72. Dupont S, Krust A, Gansmuller A, Dierich A, Chambon P, Mark M. Effect of single and compound knock-outs of estrogen receptors alpha (ERalpha) and beta (ERbeta) on mouse reproductive phenotypes. *Development.* 2000;127(19): 4277–4291.
73. Krege JH, Hodgin JB, Couse JF, Enmark E, Warner M, Mahler JF, et al. Generation and reproductive phenotypes of mice lacking estrogen receptor beta. *Proc Natl Acad Sci USA.* 1998;95(26): 15677–15682.
74. Wintermantel TM, Campbell RE, Porteous R, Bock D, Grone HJ, Todman MG, et al. Definition of estrogen receptor pathway critical for estrogen positive feedback to gonadotropin-releasing hormone neurons and fertility. *Neuron.* 2006;52(2):271–280.
75. Gore AC, Roberts JL, Gibson MJ. Mechanisms for the regulation of gonadotropin-releasing hormone gene expression in the developing mouse. *Endocrinology.* 1999;140(5):2280–2287.
76. Gore AC, Wu TJ, Oung T, Lee JB, Woller MJ. A novel mechanism for endocrine-disrupting effects of polychlorinated biphenyls: direct effects on gonadotropin-releasing hormone neurones. *J Neuroendocrinol.* 2002;14(10):814–823.
77. Clarke IJ. Control of GnRH secretion. *J Reprod Fertil Suppl.* 1987;34:1–8.
78. Clarke IJ, Pompolo S. Synthesis and secretion of GnRH. *Anim Reprod Sci.* 2005;88(1-2):29–55.
79. Andrew RD, Dudek FE. Burst discharge in mammalian neuroendocrine cells involves an intrinsic regenerative mechanism. *Science.* 1983;221(4615):1050–1052.
80. Andrew RD, Dudek FE. Analysis of intracellularly recorded phasic bursting by mammalian neuroendocrine cells. *J Neurophysiol.* 1984;51(3):552–566.
81. Suter KJ, Wuarin JP, Smith BN, Dudek FE, Moenter SM. Whole-cell recordings from preoptic/hypothalamic slices reveal burst firing in gonadotropin-releasing hormone neurons identified with green fluorescent protein in transgenic mice. *Endocrinology.* 2000;141(10):3731–3736.
82. Kawakami M, Uemura T, Hayashi R. Electrophysiological correlates of pulsatile gonadotropin release in rats. *Neuroendocrinology.* 1982;35(1):63–67.
83. Thiery JC, Pelletier J. Multiunit activity in the anterior median eminence and adjacent areas of the hypothalamus of the ewe in relation to LH secretion. *Neuroendocrinology.* 1981;32(4):217–224.
84. Wilson RC, Kesner JS, Kaufman JM, Uemura T, Akema T, Knobil E. Central electrophysiologic correlates of pulsatile luteinizing hormone secretion in the rhesus monkey. *Neuroendocrinology.* 1984;39(3):256–260.
85. Terasawa E, Keen KL, Mogi K, Claude P. Pulsatile release of luteinizing hormone-releasing hormone (LHRH) in cultured LHRH neurons derived from the embryonic olfactory placode of the rhesus monkey. *Endocrinology.* 1999;140(3):1432–1441.
86. Dierschke DJ, Bhattacharya AN, Atkinson LE, Knobil E. Circhoral oscillations of plasma LH levels in the ovariectomized rhesus monkey. *Endocrinology.* 1970;87(5):850–853.
87. Belchetz PE, Plant TM, Nakai Y, Keogh EJ, Knobil E. Hypophysial responses to continuous and intermittent delivery of hypopthalamic gonadotropin-releasing hormone. *Science.* 1978;202(4368):631–633.
88. Levine JE, Pau KY, Ramirez VD, Jackson GL. Simultaneous measurement of luteinizing hormone-releasing hormone and luteinizing hormone release in unanesthetized, ovariectomized sheep. *Endocrinology.* 1982;111(5):1449–1455.
89. Clarke I, Moore L, Veldhuis J. Intensive direct cavernous sinus sampling identifies high-frequency, nearly random patterns of FSH secretion in ovariectomized ewes: combined appraisal by RIA and bioassay. *Endocrinology.* 2002;143(1):117–129.
90. Clarke IJ, Burman KJ, Doughton BW, Cummins JT. Effects of constant infusion of gonadotrophin-releasing hormone in ovariectomized ewes with hypothalamo-pituitary disconnection: further evidence for differential control of LH and FSH secretion and the lack of a priming effect. *J Endocrinol.* 1986;111(1):43–49.
91. Wildt L, Hausler A, Marshall G, Hutchison JS, Plant TM, Belchetz PE, et al. Frequency and amplitude of gonadotropin-releasing hormone stimulation and gonadotropin secretion in the rhesus monkey. *Endocrinology.* 1981;109(2):376–385.
92. Clarke IJ, Cummins JT, Findlay JK, Burman KJ, Doughton BW. Effects on plasma luteinizing hormone and follicle-stimulating hormone of varying the frequency and amplitude of gonadotropin-releasing hormone pulses in ovariectomized ewes with hypothalamo-pituitary disconnection. *Neuroendocrinology.* 1984;39(3):214–221.
93. Clarke IJ, Cummins JT. GnRH pulse frequency determines LH pulse amplitude by altering the amount of releasable LH in the pituitary glands of ewes. *J Reprod Fertil.* 1985;73(2):425–431.
94. Ferris HA, Shupnik MA. Mechanisms for pulsatile regulation of the gonadotropin subunit genes by GNRH1. *Biol Reprod.* 2006;74(6):993–998.
95. Fellmann D, Bresson JL, Clavequin MC, Bugnon C. Quantitative immunocytochemical studies on the gonadotrophs isolated from the pituitary of the male rat. *Cell Tissue Res.* 1982;224(1): 137–144.
96. Nalbandov AV. *Reproductive Physiology.* 2nd ed. London: W.H. Freeman and Company; 1964.
97. Tougard C, Tixier-Vidal A. *Lactotropes and Gonadotropes. The Physiology of Reproduction.* New York: Raven Press; 1988. 1305–1334.
98. Clarke IJ, Cummins JT, de Kretser DM. Pituitary gland function after disconnection from direct hypothalamic influences in the sheep. *Neuroendocrinology.* 1983;36(5):376–384.
99. Mercer JE, Clements JA, Funder JW, Clarke IJ, eds. *Studies on Regulation of Gonadotropin Gene Expression in the Hypothalamo-pituitary Intact and Hypothalamo-pituitary Disconnected Ewe.* Massachusetts, USA: Serono Symposia; 1989.
100. Vogel DL, Magner JA, Sherins RJ, Weintraub BD. Biosynthesis, glycosylation, and secretion of rat luteinizing hormone alpha and beta subunits: differential effects of orchiectomy and gonadotropin-releasing hormone. *Endocrinology.* 1986;119(1): 202–213.
101. Ruf F, Sealfon SC. Genomics view of gonadotrope signaling circuits. *Trends Endocrinol Metab.* 2004;15(7):331–338.
102. Yuen T, Wurmbach E, Ebersole BJ, Ruf F, Pfeffer RL, Sealfon SC. Coupling of GnRH concentration and the GnRH receptor-activated gene program. *Mol Endocrinol.* 2002;16(6): 1145–1153.
103. Salisbury TB, Binder AK, Grammer JC, Nilson JH. GnRH-regulated expression of Jun and JUN target genes in gonadotropes requires a functional interaction between TCF?LEF family members and beta-catenin. *Mol Endocrinology.* 2009;23:402–411.
104. Dalkin AC, Haisenleder DJ, Ortolano GA, Ellis TR, Marshall JC. The frequency of gonadotropin-releasing-hormone stimulation differentially regulates gonadotropin subunit messenger ribonucleic acid expression. *Endocrinology.* 1989;125(2):917–924.
105. Haisenleder DJ, Katt JA, Ortolano GA, el-Gewely MR, Duncan JA, Dee C, et al. Influence of gonadotropin-releasing hormone pulse amplitude, frequency, and treatment duration on the regulation of luteinizing hormone (LH) subunit messenger ribonucleic acids and LH secretion. *Mol Endocrinol.* 1988;2(4):338–343.

106. Miller CD, Miller WL. Transcriptional repression of the ovine follicle-stimulating hormone-beta gene by 17 beta-estradiol. *Endocrinology.* 1996;137(8):3437–3446.
107. Kowase T, Walsh HE, Darling DS, Shupnik MA. Estrogen enhances gonadotropin-releasing hormone-stimulated transcription of the luteinizing hormone subunit promoters via altered expression of stimulatory and suppressive transcription factors. *Endocrinology.* 2007;148:6083–6091.
108. Caraty A, Locatelli A. Effect of time after castration on secretion of LHRH and LH in the ram. *J Reprod Fertil.* 1988;82(1): 263–269.
109. Caraty A, Bouchard P, Blanc MR, eds. *Studies of LHRH Secretion into the Hypophyseal Portal Blood of the Ram:Gonadal Regulation of LH Secretion is Exerted Mainly at the Hypothalamic Level.* New York: Springer-Verlag; 1992.
110. Clarke IJ. Evidence that the switch from negative to positive feedback at the level of the pituitary gland is an important timing event for the onset of the preovulatory surge in LH in the ewe. *J Endocrinol.* 1995;145(2):271–282.
111. Clarke IJ. Prenatal sexual development. In: Finn CA, ed. *Oxford Reviews of Reproductive Biology.* Oxford: Clarendon Press; 1982:100–148.
112. Clarke IJ. Ovarian feedback regulation of gonadotropin releasing hormone secretion and action. In: Leung PCK, ed. *Endocrinology and Physiology of Reproduction.* New York: Plenum Press; 1987:27–36.
113. Mitchner NA, Garlick C, Ben-Jonathan N. Cellular distribution and gene regulation of estrogen receptors alpha and beta in the rat pituitary gland. *Endocrinology.* 1998;139(9):3976–3983.
114. Schreihofer DA, Stoler MH, Shupnik MA. Differential expression and regulation of estrogen receptors (ERs) in rat pituitary and cell lines: estrogen decreases ERalpha protein and estrogen responsiveness. *Endocrinology.* 2000;141(6): 2174–2184.
115. Thieulant ML, Duval J. Differential distribution of androgen and estrogen receptors in rat pituitary cell populations separated by centrifugal elutriation. *Endocrinology.* 1985;116(4): 1299–1303.
116. Keri RA, Wolfe MW, Saunders TL, Anderson I, Kendall SK, Wagner T, et al. The proximal promoter of the bovine luteinizing hormone beta subunit gene confers gonadotrope-specific expression and regulation by gonadotropin-releasing hormone, testosterone, and 17 beta-estradiol in transgenic mice. *Mol Endocrinol.* 1994;8(12):1807–1816.
117. Shupnik MA. Gonadotropin gene modulation by steroids and gonadotropin-releasing hormone. *Biol Reprod.* 1996;54(2): 279–286.
118. Shupnik MA, Gharib SD, Chin WW. Divergent effects of estradiol on gonadotropin gene transcription in pituitary fragments. *Mol Endocrinol.* 1989;3(3):474–480.
119. Glidewell-Kenney C, Hurley LA, Pfaff L, Weiss J, Levine JE, Jameson JL. Nonclassical estrogen receptor alpha signaling mediates negative feedback in the female mouse reproductive axis. *Proc Natl Acad Sci USA.* 2007;104(19):8173–8177.
120. Glidewell-Kenney C, Weiss J, Hurley LA, Levine JE, Jameson JL. Estrogen receptor alpha signaling pathways differentially regulate gonadotropin subunit gene expression and serum follicle-stimulating hormone in the female mouse. *Endocrinology.* 2008;149(8):4168–4176.
121. Singh SP, Wolfe A, Ng Y, DiVall SA, Buggs C, Levine JE, et al. Impaired estrogen feedback and infertility in female mice with pituitary-specific deletion of estrogen receptor alpha (ESR1). *Biol Reprod.* 2009;81(3):488–496.
122. Arreguin-Arevalo JA, Ashley RL, Wagenmaker ER, Oakley AE, Karsch FJ, Nett TM. Membrane-initiated actions of estradiol (E2) in the regulation of LH secretion in ovariectomized (OVX) ewes. *Reprod Biol Endocrinol.* 2010;8:40.
123. Iqbal J, Latchoumanin O, Clarke IJ. Rapid *in vivo* effects of estradiol-17beta in ovine pituitary gonadotropes are displayed by phosphorylation of extracellularly regulated kinase, serine/threonine kinase, and 3′,5′-cyclic adenosine 5′-monophosphate-responsive element-binding protein. *Endocrinology.* 2007;148(12):5794–5802.
124. Caraty A, Locatelli A, Martin GB. Biphasic response in the secretion of gonadotrophin-releasing hormone in ovariectomized ewes injected with oestradiol. *J Endocrinol.* 1989;123(3):375–382.
125. Clarke IJ, Cummins JT. Direct pituitary effects of estrogen and progesterone on gonadotropin secretion in the ovariectomized ewe. *Neuroendocrinology.* 1984;39(3):267–274.
126. Arreguin-Arevalo JA, Nett TM. A nongenomic action of estradiol as the mechanism underlying the acute suppression of secretion of luteinizing hormone in ovariectomized ewes. *Biol Reprod.* 2006;74(1):202–208.
127. Iqbal J, Latchoumanin O, Sari IP, Lang RJ, Coleman HA, Parkington HC, et al. Estradiol-17beta inhibits gonadotropin-releasing hormone-induced Ca^{2+} in gonadotropes to regulate negative feedback on luteinizing hormone release. *Endocrinology.* 2009;150(9):4213–4220.
128. Herbison AE. Rapid actions of oestrogen on gonadotropin-releasing hormone neurons; from fantasy to physiology? *J Physiol.* 2009;587(Pt 21):5025–5030.
129. Fink G. Gonadotropin secretion and its control. In: Knobil E, Neill JD, eds. *The Physiology of Reproduction.* New York: Raven Press; 1988:1349–1378.
130. Scaramuzzi RJ, Tillson SA, Thorneycroft IH, Caldwell BV. Action of exogenous progesterone and estrogen on behavioral estrus and luteinizing hormone levels in the ovariectomized ewe. *Endocrinology.* 1971;88(5):1184–1189.
131. Clarke IJ. Gonadotrophin-releasing hormone secretion (GnRH) in anoestrous ewes and the induction of GnRH surges by oestrogen. *J Endocrinol.* 1988;117(3):355–360.
132. Clarke IJ, Wynne K, Funder JW, Findlay JK. Effects of 2-hydroxyoestradiol-17 beta on plasma luteinizing hormone, follicle-stimulating hormone and prolactin, and nuclear translocation of pituitary oestrogen receptors in ovariectomized ewes. *J Endocrinol.* 1982;95(1):19–27.
133. Harris TG, Dye S, Robinson JE, Skinner DC, Evans NP. Progesterone can block transmission of the estradiol-induced signal for luteinizing hormone surge generation during a specific period of time immediately after activation of the gonadotropin-releasing hormone surge-generating system. *Endocrinology.* 1999;140(2):827–834.
134. Clarke IJ. The preovulatory LH surge A case of a neuroendocrine switch. *Trends Endocrinol Metab.* 1995;6(7):241–247.
135. Clarke IJ, Cummins JT, Crowder ME, Nett TM. Pituitary receptors for gonadotropin-releasing hormone in relation to changes in pituitary and plasma gonadotropins in ovariectomized hypothalamo/pituitary-disconnected ewes. II. A marked rise in receptor number during the acute feedback effects of estradiol. *Biol Reprod.* 1988;39(2):349–354.
136. Kasa-Vubu JZ, Dahl GE, Evans NP, Thrun LA, Moenter SM, Padmanabhan V, et al. Progesterone blocks the estradiol-induced gonadotropin discharge in the ewe by inhibiting the surge of gonadotropin-releasing hormone. *Endocrinology.* 1992;131(1):208–212.
137. Moenter SM, Caraty A, Karsch FJ. The estradiol-induced surge of gonadotropin-releasing hormone in the ewe. *Endocrinology.* 1990;127(3):1375–1384.
138. Roche JF, Foster DL, Karsch FJ, Cook B, Dziuk PJ. Levels of luteinizing hormone in sera and pituitaries of ewes during

the estrous cycle and anestrus. *Endocrinology.* 1970;86(3):568–572.

139. Moenter SM, Brand RC, Karsch FJ. Dynamics of gonadotropin-releasing hormone (GnRH) secretion during the GnRH surge: insights into the mechanism of GnRH surge induction. *Endocrinology.* 1992;130(5):2978–2984.
140. Clarke I. The Hypothalamo-Pituitary Axis. In: Hillier SG, Kitchener HC, Neilson JP, eds. *Scientific Essentials of Reproductive Medicine.* London: WB Saunders; 1996:120–133.
141. Pau KY, Berria M, Hess DL, Spies HG. Preovulatory gonadotropin-releasing hormone surge in ovarian-intact rhesus macaques. *Endocrinology.* 1993;133(4):1650–1656.
142. Wibbels T, Owens DW, Licht P, Limpus C, Reed PC, Amoss Jr MS. Serum gonadotropins and gonadal steroids associated with ovulation and egg production in sea turtles. *Gen Comp Endocrinol.* 1992;87(1):71–78.
143. Hoff JD, Quigley ME, Yen SS. Hormonal dynamics at midcycle: a reevaluation. *J Clin Endocrinol Metab.* 1983;57(4):792–796.
144. Weick RF, Dierschke DJ, Karsch FJ, Butler WR, Hotchkiss J, Knobil E. Periovulatory time courses of circulating gonadotropic and ovarian hormones in the rhesus monkey. *Endocrinology.* 1973;93(5):1140–1147.
145. Liu JH, Yen SS. Induction of midcycle gonadotropin surge by ovarian steroids in women: a critical evaluation. *J Clin Endocrinol Metab.* 1983;57(4):797–802.
146. Woller MJ, Terasawa E. Changes in pulsatile release of neuropeptide-Y and luteinizing hormone (LH)-releasing hormone during the progesterone-induced LH surge in rhesus monkeys. *Endocrinology.* 1994;135(4):1679–1686.
147. Brown-Grant K. Steroid hormone administration and gonadotrophin secretion in the gonadectomized rat. *J Endocrinol.* 1974;62(2):319–332.
148. Rao IM, Mahesh VB. Role of progesterone in the modulation of the preovulatory surge of gonadotropins and ovulation in the pregnant mare's serum gonadotropin-primed immature rat and the adult rat. *Biol Reprod.* 1986;35(5):1154–1161.
149. Ringstrom SJ, Szabo M, Kilen SM, Saberi S, Knox KL, Schwartz NB. The antiprogestins RU486 and ZK98299 affect follicle-stimulating hormone secretion differentially on estrus, but not on proestrus. *Endocrinology.* 1997;138(6):2286–2290.
150. Szabo M, Knox KL, Ringstrom SJ, Perlyn CA, Sutandi S, Schwartz NB. Mechanism of the inhibitory action of RU486 on the secondary follicle-stimulating hormone surge. *Endocrinology.* 1996;137(1):85–89.
151. Chappell PE, Schneider JS, Kim P, Xu M, Lydon JP, O'Malley BW, et al. Absence of gonadotropin surges and gonadotropin-releasing hormone self-priming in ovariectomized (OVX), estrogen (E2)-treated, progesterone receptor knockout (PRKO) mice. *Endocrinology.* 1999;140(8):3653–3658.
152. Thomas SG, Clarke IJ. The positive feedback action of estrogen mobilizes LH-containing, but not FSH-containing secretory granules in ovine gonadotropes. *Endocrinology.* 1997;138(3):1347–1350.
153. Clarke IJ, Tobin VA, Pompolo S, Pereira A. Effects of changing gonadotropin-releasing hormone pulse frequency and estrogen treatment on levels of estradiol receptor-alpha and induction of Fos and phosphorylated cyclic adenosine monophosphate response element binding protein in pituitary gonadotropes: studies in hypothalamo-pituitary disconnected ewes. *Endocrinology.* 2005;146(3):1128–1137.
154. Cowley MA, Rao A, Wright PJ, Illing N, Millar RP, Clarke IJ. Evidence for differential regulation of multiple transcripts of the gonadotropin releasing hormone receptor in the ovine pituitary gland; effect of estrogen. *Mol Cell Endocrinol.* 1998;146(1–2):141–149.
155. Clarke IJ, Foulds LM, Hayward S, Cummins JT, Robertson DM. Analysis of the ratio of biological to immunological LH secreted during the oestrogen-induced LH surge in the ewe. *J Endocrinol.* 1990;127(2):217–222.
156. Clarke IJ, Cummins JT, Jenkin M, Phillips DJ. The oestrogen-induced surge of LH requires a "signal" pattern of gonadotrophin-releasing hormone input to the pituitary gland in the ewe. *J Endocrinol.* 1989;122(1):127–134.
157. Tilbrook AJ, de Kretser DM, Cummins JT, Clarke IJ. The negative feedback effects of testicular steroids are predominantly at the hypothalamus in the ram. *Endocrinology.* 1991;129(6):3080–3092.
158. Karsch FJ. Central actions of ovarian steroids in the feedback regulation of pulsatile secretion of luteinizing hormone. *Annu Rev Physiol.* 1987;49:365–382.
159. Moenter SM, Caraty A, Locatelli A, Karsch FJ. Pattern of gonadotropin-releasing hormone (GnRH) secretion leading up to ovulation in the ewe: existence of a preovulatory GnRH surge. *Endocrinology.* 1991;129(3):1175–1182.
160. Mercer JE, Clements JA, Funder JW, Clarke IJ. Luteinizing hormone-beta mRNA levels are regulated primarily by gonadotropin-releasing hormone and not by negative estrogen feedback on the pituitary. *Neuroendocrinology.* 1988;47(6):563–566.
161. Rawson JA, Scott CJ, Pereira A, Jakubowska A, Clarke IJ. Noradrenergic projections from the A1 field to the preoptic area in the brain of the ewe and Fos responses to oestrogen in the A1 cells. *J Neuroendocrinol.* 2001;13(2):129–138.
162. Kiss J, Halasz B. Demonstration of serotoninergic axons terminating on luteinizing hormone-releasing hormone neurons in the preoptic area of the rat using a combination of immunocytochemistry and high resolution autoradiography. *Neuroscience.* 1985;14(1):69–78.
163. Iqbal J, Pompolo S, Sakurai T, Clarke IJ. Evidence that orexin-containing neurones provide direct input to gonadotropin-releasing hormone neurones in the ovine hypothalamus. *J Neuroendocrinol.* 2001;13(12):1033–1041.
164. Pompolo S, Ischenko O, Pereira A, Iqbal J, Clarke IJ. Evidence that projections from the bed nucleus of the stria terminalis and from the lateral and medial regions of the preoptic area provide input to gonadotropin releasing hormone (GNRH) neurons in the female sheep brain. *Neuroscience.* 2005;132(2):421–436.
165. Pompolo S, Pereira A, Scott CJ, Fujiyma F, Clarke IJ. Evidence for estrogenic regulation of gonadotropin-releasing hormone neurons by glutamatergic neurons in the ewe brain: An immunohistochemical study using an antibody against vesicular glutamate transporter-2. *J Comp Neurol.* 2003;465(1):136–144.
166. Pereira A, Rawson J, Jakubowska A, Clarke IJ. Estradiol-17beta-responsive A1 and A2 noradrenergic cells of the brain stem project to the bed nucleus of the stria terminalis in the ewe brain: a possible route for regulation of gonadotropin releasing hormone cells. *Neuroscience.* 2010;165(3):758–773.
167. Backholer K, Smith JT, Rao A, Pereira A, Iqbal J, Ogawa S, et al. Kisspeptin cells in the ewe brain respond to leptin and communicate with neuropeptide Y and proopiomelanocortin cells. *Endocrinology.* 2010;151(5):2233–2243.
168. Oakley AE, Clifton DK, Steiner RA. Kisspeptin signaling in the brain. *Endocr Rev.* 2009;30(6):713–743.
169. Clarke IJ. Control of GnRH Secretion: one step back. *Frontiers in Neuroendocrinology.* 2011. in press.
170. Gottsch ML, Cunningham MJ, Smith JT, Popa SM, Acohido BV, Crowley WF, et al. A role for kisspeptins in the regulation of

gonadotropin secretion in the mouse. *Endocrinology.* 2004;145(9):4073–4077.

171. Messager S, Chatzidaki EE, Ma D, Hendrick AG, Zahn D, Dixon J, et al. Kisspeptin directly stimulates gonadotropin-releasing hormone release via G protein-coupled receptor 54. *Proc Natl Acad Sci USA.* 2005;102(5):1761–1766.
172. Clarkson J, d'Anglemont de Tassigny X, Moreno AS, Colledge WH, Herbison AE. Kisspeptin-GPR54 signaling is essential for preovulatory gonadotropin-releasing hormone neuron activation and the luteinizing hormone surge. *J Neurosci.* 2008;28(35):8691–8697.
173. Smith JT, Cunningham MJ, Rissman EF, Clifton DK, Steiner RA. Regulation of Kiss1 gene expression in the brain of the female mouse. *Endocrinology.* 2005;146(9):3686–3692.
174. Smith JT, Dungan HM, Stoll EA, Gottsch ML, Braun RE, Eacker SM, et al. Differential regulation of KiSS-1 mRNA expression by sex steroids in the brain of the male mouse. *Endocrinology.* 2005;146(7):2976–2984.
175. Franceschini I, Lomet D, Cateau M, Delsol G, Tillet Y, Caraty A. Kisspeptin immunoreactive cells of the ovine preoptic area and arcuate nucleus co-express estrogen receptor alpha. *Neurosci Lett.* 2006;401(3):225–230.
176. Smith JT, Acohido BV, Clifton DK, Steiner RA. KiSS-1 neurones are direct targets for leptin in the ob/ob mouse. *J Neuroendocrinol.* 2006;18(4):298–303.
177. Adachi S, Yamada S, Takatsu Y, Matsui H, Kinoshita M, Takase K, et al. Involvement of anteroventral periventricular metastin/kisspeptin neurons in estrogen positive feedback action on luteinizing hormone release in female rats. *J Reprod Dev.* 2007;53(2):367–378.
178. Smith JT, Clay CM, Caraty A, Clarke IJ. KiSS-1 messenger ribonucleic acid expression in the hypothalamus of the ewe is regulated by sex steroids and season. *Endocrinology.* 2007;148(3):1150–1157.
179. Roseweir AK, Kauffman AS, Smith JT, Guerriero KA, Morgan K, Pielecka-Fortuna J, et al. Discovery of potent kisspeptin antagonists delineate physiological mechanisms of gonadotropin regulation. *J Neurosci.* 2009;29(12):3920–3929.
180. Smith JT, Li Q, Yap KS, Shahab M, Roseweir AK, Millar RP, Clarke IJ. Kisspeptin is essential for the full preovulatory LH surge and stimulates GnRH release from the isolated ovine median eminence. *Endocrinology.* 2011. in press.
181. Caraty A, Smith JT, Lomet D, Ben Said S, Morrissey A, Cognie J, et al. Kisspeptin synchronizes preovulatory surges in cyclical ewes and causes ovulation in seasonally acyclic ewes. *Endocrinology.* 2007;148(11):5258–5267.
182. Goodman RL, Lehman MN, Smith JT, Coolen LM, de Oliveira CV, Jafarzadehshirazi MR, et al. Kisspeptin neurons in the arcuate nucleus of the ewe express both dynorphin A and neurokinin B. *Endocrinology.* 2007;148(12):5752–5760.
183. Cheng G, Coolen LM, Padmanabhan V, Goodman RL, Lehman MN. The kisspeptin/neurokinin B/dynorphin (KNDy) cell population of the arcuate nucleus: sex differences and effects of prenatal testosterone in sheep. *Endocrinology.* 2010;151(1):301–311.
184. Foradori CD, Goodman RL, Adams VL, Valent M, Lehman MN. Progesterone increases dynorphin A concentrations in cerebrospinal fluid and preprodynorphin messenger ribonucleic acid levels in a subset of dynorphin neurons in the sheep. *Endocrinology.* 2005;146(4):1835–1842.
185. Goodman RL, Coolen LM, Anderson GM, Hardy SL, Valent M, Connors JM, et al. Evidence that dynorphin plays a major role in mediating progesterone negative feedback on gonadotropin-releasing hormone neurons in sheep. *Endocrinology.* 2004;145(6):2959–2967.
186. Smith JT, Li Q, Pereira A, Clarke IJ. Kisspeptin neurons in the ovine arcuate nucleus and preoptic area are involved in the preovulatory luteinizing hormone surge. *Endocrinology.* 2009;150(12):5530–5538.
187. Tsutsui K, Saigoh E, Ukena K, Teranishi H, Fujisawa Y, Kikuchi M, et al. A novel avian hypothalamic peptide inhibiting gonadotropin release. *Biochem Biophys Res Commun.* 2000;275(2):661–667.
188. Tsutsui K, Bentley GE, Bedecarrats G, Osugi T, Ubuka T, Kriegsfeld LJ. Gonadotropin-inhibitory hormone (GnIH) and its control of central and peripheral reproductive function. *Front Neuroendocrinol.* 2010;31(3):284–295.
189. Clarke IJ, Qi Y, Puspita Sari I, Smith JT. Evidence that RF-amide related peptides are inhibitors of reproduction in mammals. *Front Neuroendocrinol.* 2009;30(3):371–378.
190. Smith JT, Clarke IJ. Gonadotropin inhibitory hormone function in mammals. *Trends Endocrinol Metab.* 2010;21(4):255–260.
191. Satake H, Hisada M, Kawada T, Minakata H, Ukena K, Tsutsui K. Characterization of a cDNA encoding a novel avian hypothalamic neuropeptide exerting an inhibitory effect on gonadotropin release. *Biochem J.* 2001;354(Pt 2):379–385.
192. Osugi T, Ukena K, Bentley GE, O'Brien S, Moore IT, Wingfield JC, et al. Gonadotropin-inhibitory hormone in Gambel's white-crowned sparrow (*Zonotrichia leucophrys gambelii*): cDNA identification, transcript localization and functional effects in laboratory and field experiments. *J Endocrinol.* 2004;182(1):33–42.
193. Hinuma S, Shintani Y, Fukusumi S, Iijima N, Matsumoto Y, Hosoya M, et al. New neuropeptides containing carboxy-terminal RFamide and their receptor in mammals. *Nat Cell Biol.* 2000;2(10):703–708.
194. Ubuka T, Kim S, Huang YC, Reid J, Jiang J, Osugi T, et al. Gonadotropin-inhibitory hormone neurons interact directly with gonadotropin-releasing hormone-I and -II neurons in European starling brain. *Endocrinology.* 2008;149(1):268–278.
195. Smith JT, Coolen LM, Kriegsfeld LJ, Sari LP, Jaafarzadehshirazi MR, Maltby M, Bateman K, Goodman RL, Tilbrook AJ, Ubuka T, Bentley GE, Clarke IJ, Lehman MN. Variation in kisspeptin and gonadotropin-inhibitory hormone expression and terminal connections to GnRH neurons in the brain: a novel medium for seasonal breeding in the sheep. *Endocrinology.* 2008;149:5770–5782.
196. Dardente H, Birnie M, Lincoln GA, Hazlerigg DG. RFamide-related peptide and its cognate receptor in the sheep: cDNA cloning, mRNA distribution in the hypothalamus and effect of photoperiod. *J Neuroendocrinol.* 2008;20(11):1252–1259.
197. Clarke IJ, Sari IP, Qi Y, Smith JT, Parkington HC, Ubuka T, et al. Potent action of RFamide-related peptide-3 on pituitary gonadotropes indicative of a hypophysiotropic role in the negative regulation of gonadotropin secretion. *Endocrinology.* 2008;149(11):5811–5821.
198. Qi Y, Oldfield BJ, Clarke IJ. Projections of RFamide-related peptide-3 neurones in the ovine hypothalamus, with special reference to regions regulating energy balance and reproduction. *J Neuroendocrinol.* 2009;21(8):690–697.
199. Johnson MA, Tsutsui K, Fraley GS. Rat RFamide-related peptide-3 stimulates GH secretion, inhibits LH secretion, and has variable effects on sex behavior in the adult male rat. *Horm Behav.* 2007;51(1):171–180.
200. Backholer K, Bowden M, Gamber K, Bjorbaek C, Iqbal J, Clarke IJ. Melanocortins mimic the effects of leptin to restore reproductive function in lean hypogonadotropic ewes. *Neuroendocrinology.* 2010;91(1):27–40.
201. Wu M, Dumalska I, Morozova E, van den Pol AN, Alreja M. Gonadotropin inhibitory hormone inhibits basal forebrain

vGluT2-gonadotropin-releasing hormone neurons via a direct postsynaptic mechanism. *J Physiol.* 2009;587(Pt 7):1401–1411.
202. Ubuka T, Lai H, Kitani M, Suzuuchi A, Pham V, Cadigan PA, et al. Gonadotropin-inhibitory hormone identification, cDNA cloning, and distribution in rhesus macaque brain. *J Comp Neurol.* 2009;517(6):841–855.
203. Kriegsfeld LJ, Mei DF, Bentley GE, Ubuka T, Mason AO, Inoue K, et al. Identification and characterization of a gonadotropin-inhibitory system in the brains of mammals. *Proc Natl Acad Sci USA.* 2006;103(7):2410–2415.
204. Qi Y, Iqbal J, Oldfield BJ, Clarke IJ. Neural connectivity in the mediobasal hypothalamus of the sheep brain. *Neuroendocrinology.* 2008;87(2):91–112.
205. Rizwan MZ, Porteous R, Herbison AE, Anderson GM. Cells expressing RFamide-related peptide-1/3, the mammalian gonadotropin-inhibitory hormone orthologs, are not hypophysiotropic neuroendocrine neurons in the rat. *Endocrinology.* 2009;150(3):1413–1420.
206. Murakami M, Matsuzaki T, Iwasa T, Yasui T, Irahara M, Osugi T, et al. Hypophysiotropic role of RFamide-related peptide-3 in the inhibition of LH secretion in female rats. *J Endocrinol.* 2008;199(1):105–112.
207. Johnson MA, Fraley GS. Rat RFRP-3 alters hypothalamic GHRH expression and growth hormone secretion but does not affect Kiss-1 gene expression or the onset of puberty in male rats. *Neuroendocrinology.* 2008;88(4):305–315.
208. Ducret E, Anderson GM, Herbison AE. RFamide-related peptide-3, a mammalian gonadotropin-inhibitory hormone ortholog, regulates gonadotropin-releasing hormone neuron firing in the mouse. *Endocrinology.* 2009;150(6):2799–2804.
209. Anderson GM, Relf HL, Rizwan MZ, Evans JJ. Central and peripheral effects of RFamide-related peptide-3 on LH and prolactin secretion in rats. *Endocrinology.* 1008;150(4): 1834–1840.
210. Smith JT, Coolen LM, Kriegsfeld LJ, Sari IP, Jaafarzadehshirazi MR, Maltby M, et al. Variation in kisspeptin and RFamide-related peptide (RFRP) expression and terminal connections to gonadotropin-releasing hormone neurons in the brain: a novel medium for seasonal breeding in the sheep. *Endocrinology.* 2008;149(11):5770–5782.
211. Paul MJ, Pyter LM, Freeman DA, Galang J, Prendergast BJ. Photic and nonphotic seasonal cues differentially engage hypothalamic kisspeptin and RFamide-related peptide mRNA expression in Siberian hamsters. *J Neuroendocrinol.* 2009; 21(12):1007–1014.
212. Revel FG, Saboureau M, Pevet P, Simonneaux V, Mikkelsen JD. RFamide-related peptide gene is a melatonin-driven photoperiodic gene. *Endocrinology.* 2008;149(3):902–912.
213. Clarke IJ, Sari IP, Qi Y, Smith JT, Parkington HC, Ubuka T, Iqbal J, Li Q, Tilbrook AJ, Morgan K, Pawson AJ, Tsutsui K, Millar RP, Bentley GE. Potent action of RFRP-3 on pituitary gonadotropes indicative of an hypophysiotropic role in the negative regulation of gonadotropin secretion. *Endocrinology.* 2008;149(11):5811–5821.
214. Kadokawa H, Shibata M, Tanaka Y, Kojima T, Matsumoto K, Oshima K, et al. Bovine C-terminal octapeptide of RFamide-related peptide-3 suppresses luteinizing hormone (LH) secretion from the pituitary as well as pulsatile LH secretion in bovines. *Domest Anim Endocrinol.* 2009;36(4):219–224.
215. Sari IP, Rao A, Smith JT, Tilbrook AJ, Clarke IJ. Effect of RF-amide-related peptide-3 on luteinizing hormone and follicle-stimulating hormone synthesis and secretion in ovine pituitary gonadotropes. *Endocrinology.* 2009;150(12):5549–5556.
216. Mercer JE, Phillips DJ, Clarke IJ. Short-term regulation of gonadotropin subunit mRNA levels by estrogen: studies in the hypothalamo-pituitary intact and hypothalamo-pituitary disconnected ewe. *J Neuroendocrinol.* 1993;5(5):591–596.
217. Huang HJ, Sebastian J, Strahl BD, Wu JC, Miller WL. The promoter for the ovine follicle-stimulating hormone-beta gene (FSHbeta) confers FSHbeta-like expression on luciferase in transgenic mice: regulatory studies *in vivo* and *in vitro*. *Endocrinology.* 2001;142(6):2260–2266.
218. Ling N, Ying SY, Ueno N, Esch F, Denoroy L, Guillemin R. Isolation and partial characterization of a Mr 32,000 protein with inhibin activity from porcine follicular fluid. *Proc Natl Acad Sci USA.* 1985;82(21):7217–7221.
219. Miyamoto K, Hasegawa Y, Fukuda M, Nomura M, Igarashi M, Kangawa K, et al. Isolation of porcine follicular fluid inhibin of 32K daltons. *Biochem Biophys Res Commun.* 1985;129(2): 396–403.
220. Rivier J, Spiess J, McClintock R, Vaughan J, Vale W. Purification and partial characterization of inhibin from porcine follicular fluid. *Biochem Biophys Res Commun.* 1985;133(1): 120–127.
221. Robertson DM, Foulds LM, Leversha L, Morgan FJ, Hearn MT, Burger HG, et al. Isolation of inhibin from bovine follicular fluid. *Biochem Biophys Res Commun.* 1985;126(1):220–226.
222. Fang J, Yin W, Smiley E, Wang SQ, Bonadio J. Molecular cloning of the mouse activin beta E subunit gene. *Biochem Biophys Res Commun.* 1996;228(3):669–674.
223. Lau AL, Nishimori K, Matzuk MM. Structural analysis of the mouse activin beta C gene. *Biochim Biophys Acta.* 1996; 1307(2):145–148.
224. Burger HG, Robertson DM, Cahir N, Mamers P, Healy DL, Jobling T, et al. Characterization of inhibin immunoreactivity in post-menopausal women with ovarian tumours. *Clin Endocrinol (Oxf).* 1996;44(4):413–418.
225. Illingworth PJ, Groome NP, Byrd W, Rainey WE, McNeilly AS, Mather JP, et al. Inhibin-B: a likely candidate for the physiologically important form of inhibin in men. *J Clin Endocrinol Metab.* 1996;81(4):1321–1325.
226. de Kretser DM, Meinhardt A, Meehan T, Phillips DJ, O'Bryan MK, Loveland KA. The roles of inhibin and related peptides in gonadal function. *Mol Cell Endocrinol.* 2000;161(1–2):43–46.
227. Carroll RS, Corrigan AZ, Gharib SD, Vale W, Chin WW. Inhibin, activin, and follistatin: regulation of follicle-stimulating hormone messenger ribonucleic acid levels. *Mol Endocrinol.* 1989;3(12):1969–1976.
228. Phillips CL, Lin LW, Wu JC, Guzman K, Milsted A, Miller WL. 17 Beta-estradiol and progesterone inhibit transcription of the genes encoding the subunits of ovine follicle-stimulating hormone. *Mol Endocrinol.* 1988;2(7):641–649.
229. Robertson DM, Klein R, de Vos FL, McLachlan RI, Wettenhall RE, Hearn MT, et al. The isolation of polypeptides with FSH suppressing activity from bovine follicular fluid which are structurally different to inhibin. *Biochem Biophys Res Commun.* 1987;149(2):744–749.
230. Tilbrook AJ, Clarke IJ, de Kretser DM. Human recombinant follistatin-288 suppresses plasma concentrations of follicle-stimulating hormone but is not a significant regulator of luteinizing hormone in castrated rams. *Biol Reprod.* 1995;53(6): 1353–1358.
231. Mercer JE, Clements JA, Funder JW, Clarke IJ. Rapid and specific lowering of pituitary FSH beta mRNA levels by inhibin. *Mol Cell Endocrinol.* 1987;53(3):251–254.
232. Tilbrook AJ, de Kretser DM, Clarke IJ. Human recombinant inhibin A and testosterone act directly at the pituitary to suppress plasma concentrations of FSH in castrated rams. *J Endocrinol.* 1993;138(2):181–189.

233. Clarke IJ, Rao A, Fallest PC, Shupnik MA. Transcription rate of the follicle stimulating hormone (FSH) beta subunit gene is reduced by inhibin in sheep but this does not fully explain the decrease in mRNA. *Mol Cell Endocrinol.* 1993;91(1-2):211–216.
234. Tilbrook AJ, Clarke IJ. Negative feedback regulation of the secretion and actions of gonadotropin-releasing hormone in males. *Biol Reprod.* 2001;64(3):735–742.
235. Lin SJ, Lerch TF, Cook RW, Jardetzky TS, Woodruff TK. The structural basis of TGF-β, bone morphogenetic protein, and activin ligand binding reproduction. 2006;132:179–190.
236. Huang HJ, Sebastian J, Strahl BD, Wu JC, Miller WL. Transcriptional regulation of the ovine follicle-stimulating hormone-beta gene by activin and gonadotropin-releasing hormone (GnRH): involvement of two proximal activator protein-1 sites for GnRH stimulation. *Endocrinology.* 2001;142(6):2267–2274.
237. Phillips DJ, Jones KL, Clarke IJ, Scheerlinck JP, de Kretser DM, Activin A. From sometime reproductive factor to genuine cytokine. *Vet Immunol Immunopathol.* 2005; 108(1–2):23–27.
238. Shafiee-Kermani F, Han SO, Miller WL. Chronic gonadotropin-releasing hormone inhibits activin induction of the ovine follicle-stimulating hormone beta-subunit: involvement of 3′,5′-cyclic adenosine monophosphate response element binding protein and nitric oxide synthase type I. *Endocrinology.* 2007;148(7):3346–3355.
239. Su P, Shafiee-Kermani F, Gore AJ, Jia J, Wu JC, Miller WL. Expression and regulation of the beta-subunit of ovine follicle-stimulating hormone relies heavily on a promoter sequence likely to bind Smad-associated proteins. *Endocrinology.* 2007;148(9):4500–4508.
240. Clarke IJ, Findlay JK, Cummins JT, Ewens WJ. Effects of ovine follicular fluid on plasma LH and FSH secretion in ovariectomized ewes to indicate the site of action of inhibin. *J Reprod Fertil.* 1986;77(2):575–585.

CHAPTER

10

Lifetime Regulation of Growth Hormone (GH) Secretion

Johannes D. Veldhuis [1], *Ali Iranmanesh* [2], *Dana Erickson* [1], *Ferdinand Roelfsema* [3], *Cyril Y. Bowers* [4]

[1] Endocrine Research Unit, Mayo School of Graduate Medical Education, Clinical Translational Science Center, Mayo Clinic, Rochester, MN, USA, [2] Endocrine Service, Medical Section, Salem Veterans Affairs Medical Center, Salem, VA, USA, [3] Department of Endocrinology and Metabolic Diseases, Leiden University Medical Center, The Netherlands, [4] Division of Endocrinology, Department of Internal Medicine, Tulane University Health Sciences Center, New Orleans, LA, USA

OUTLINE

Handbook of Neuroendocrinology, DOI: 10.1016/B978-0-12-375097-6.10010-1

Summary

Lifelong homeostasis of GH-IGF-I availability is controlled by multiple interacting pathways. Pivotal regulatory signals are sex steroids, GHRH, GHS/ghrelin and somatostatin. Feedback is enforced via GH and its tissue product, IGF-I. Numerous ancillary peptides (such as galanin, CRH, NPY and leptin) and neurotransmitter receptors (such as muscarinic receptors, dopamine receptors and norepinephrine) exert modulating effects, but human diseases involving collateral pathways with attendant short stature or gigantism are rare. A deficiency in the field is the lack of comprehensive ensemble models that incorporate all current knowledge, and permit analytical (quantitative) estimates of physiological adaptations and pathological disturbances in key regulatory pathways. Synergistic neurophysiological, gene-silencing and pharmacological investigations should aid progress to that end.

OVERVIEW OF GH AXIS

The growth-promoting pituitary-dependent GH axis mediates soft-tissue (organ) and skeletal (statural) growth in puberty,[1] and maintains body composition within relatively narrow bounds in puberty and adulthood.[2] There is an intimate collaboration between the GH axis and the reproductive axis, which is transduced primarily via IGF-I and sex steroids.[3,4] During the adolescent growth spurt, and to a lesser degree in aging individuals, aromatizable androgens (like testosterone, T) stimulate production of GH and its primary messenger, insulin-like growth factor type I [IGF-I], and synergize with GH and, in selected cases, with IGF-I.[5] Silencing estrogen receptor (ER) alpha reduces liver IGF-I production, whereas estradiol (E_2) can stimulate GH secretion, inhibit or potentiate IGF-I action and induce the IGF-I receptor gene. For example, both T and E_2 induce the IGF-I gene in bone. In combination, GH/IGF-I and androgens constitute powerful anabolic signals to muscle, bone, kidney, liver and brain, while concomitantly promoting lipolysis of visceral and subcutaneous fat depots, and sex-specific hepatic expression of metabolic enzymes, substrates and exported proteins.[6]

BOX 10.1

HOW DO WE KNOW THERE IS LIFETIME REGULATION OF GROWTH HORMONE (GH) SECRETION?

Estrogenic Stimulation of GH Release in the Human

GH production doubles in the estrogen-enriched preovulatory phase of the menstrual cycle, and is higher in pre- than postmenopausal women or comparably aged men.[1] Acute and long-term estrogen administration in girls with Turner's syndrome, postmenopausal women, male-to-female transsexual patients and men with prostatic cancer elevates GH concentrations by 1.8- to 3.3-fold. The route is not critical, since both oral and higher-dose transdermal E_2 will stimulate GH secretion. In men, administration of an anti-estrogen or aromatase inhibitor attenuates testosterone-stimulated GH secretion, indicating that estrogen derived from aromatizable androgen is the proximate effector.[2] Understanding such basic physiological mechanisms should foster new rational non-steroidal and steroidal strategies to maintain GH and IGF-I availability.

Investigation of Ensemble Regulatory Mechanisms

A consequence of interactive (multipathway) control is that modulation of any given regulatory locus will perforce influence the behavior of all interconnected signals. For example, the finding that postmenopausal age reduces (and estrogen amplifies) GH secretory-burst mass following a submaximal GHRH stimulus could in principle signify that aging (and conversely for estrogen): (a) downregulates somatotrope GHRH receptors; (b) reduces GHRH-releasable GH stores; (c) augments central SSergic inhibition; (d) blunts endogenous ghrelin action, which normally synergizes with GHRH; and/or (e) increases negative feedback by systemic GH or IGF-I.[1] Therefore, from an interpretative viewpoint, precisely how aging (or estrogen) directs multipeptide regulation cannot be inferred from the GH response to any single peptide. One approach to this impasse is an investigative platform comprising: (i) simultaneous delivery of complementary peptides; (ii) studies in a low vs high estradiol-clamped milieu; (iii) comparisons in young and older women with similar body composition; and (iv) ensemble model-assisted analyses.[2] This tetrapartite approach might permit inferences about how age, estradiol availability or body composition determine the outflow (release and actions) of endogenous regulatory peptides.

References

1. Giustina A, Veldhuis JD. Pathophysiology of the neuroregulation of growth hormone secretion in experimental animals and the human. *Endocrine Rev.* 1998;19:717–797.
2. Veldhuis JD, Roemmich JN, Richmond EJ, Bowers CY. Somatotropic and gonadotropic axes linkages in infancy, childhood, and the puberty–adult transition. *Endocrine Rev.* 2006;27:101–140.

Maximal GH secretion in the human occurs on the first day of postnatal life, and in Tanner stages III and IV of puberty. Childhood GH production is otherwise low.[3] Ultrasensitive GH assays reveal a nearly 20-fold diminution in GH secretion between mid-puberty and old age. Age-related hyposomatotropism is accompanied by a gradual fall in sex hormone concentrations, and a reciprocal rise in late-day cortisol availability. Concomitant changes in GH, androgens and cortisol may accentuate sarcopenia (deficiency of flesh), osteopenia (deficient bone), mild memory and cognitive decline, skin fragility (due to collagen loss) and intra-abdominal visceral adiposity. Visceral adiposity in turn heightens age-related insulin resistance, dyslipidemia, endothelial dysfunction, pro-inflammatory cytokine accumulation, atherogenesis and cardiovascular morbidity. Whereas this metabolic phenotype resembles that of organic pituitary GH deficiency in young adults, major unresolved clinical issues include the safety, indications, merits, adverse effects, dosimetry and perpetuity of GH or sex-steroid supplementation in older or obese individuals.

In many countries, GH treatment in adults is approved by drug-regulatory agencies for use in organic GH deficiency, HIV-associated lipodystrophy and the short-bowel syndrome. Indications in children remain multiple, including growth promotion in GH-deficient individuals with hypothalamo-pituitary disease and in patients with functional GH-insufficiency states, such as chronic renal failure, gonadal dysgenesis and Prader-Willi syndrome. Although GH is currently administered once daily, or less often in repository forms, normal physiology in humans is marked by 12–20 discrete pulses of GH daily, the amplitude of which is enhanced by deep sleep, exercise and fasting. In rodents, large infrequent GH pulses with low intervening nadirs are required in the male to drive maximal bone and muscle growth.[1] In addition, the time pattern (that is, pulsatile *vis-à-vis* continuous) of GH delivery defines which liver genes are repressed or induced. Future innovative therapies are likely to exploit knowledge of the basic mechanisms that maintain physiologically pulsatile (burst-like) and nycthemeral (day–night) patterns of GH secretion.[3] One notion is an annual injection of a recombinant DNA plasmid vector expressing the GH-releasing hormone, GHRH, constitutively, to maintain pulsatile GH secretion.

HYPOTHALAMIC DRIVE AND RESTRAINT

The hypothalamo-pituitary unit comprises a unique neuroendocrine interface, in which intermittent brain signals are delivered via a finite portal microvasculature to the multicellular anterior pituitary gland. Hypothalamic factors comprise classical neurotransmitters (e.g., norepinephrine, dopamine), peptidyl inhibitors (GH release-inhibiting factor or somatostatin, SS) and peptidyl stimulators (GH-releasing hormone/GHRH, and ghrelin, a GH-releasing peptide/GHRP)[7,8] (Fig. 10.1). A fundamental implication of multiple-signal exchange is that no given effector acts alone *in vivo*. Systemic hormones, metabolites and substrates reach hypothalamic and pituitary cells via segmental hypophysial arteries, which arborize into a precapillary network confluent with the hypothalamo-pituitary portal microcirculation. Blood-borne effectors can modulate certain pituitary cell types directly (e.g., ghrelin and E_2 act via corresponding receptors in somatotropes; triiodothyronine (T3) via cognate receptors expressed in thyrotropes; glucocorticoid via its type II receptor in corticotropes; and IGF-I via its type I receptor in somatotropes). Among these diverse inputs, salient effectors of somatotropes are somatostatin and GHRH, secreted by the hypothalamus, and ghrelin, secreted by the stomach, pituitary gland and hypothalamus. Other less well established factors, like direct neuronal GH secretion, will not be reviewed here.

Somatostatin

Somatostatin (SS) is a 14- and 28-amino acid peptide discovered in 1973, which is synthesized in neurons and in delta cells of the pancreas. Periventricular-nucleus neurons release SS within the brain (Fig. 10.1) and into portal blood in an episodic (irregularly pulsatile) fashion in the adult male rat, pig and sheep.[1] Indirect evidence for SS's episodic release in humans is that constant intravenous infusion of a maximally effective dose of GHRH and/or GHRP-2 (a synthetic ghrelin-like hexapeptide) stimulates pulsatile GH secretion for long intervals. Such observations point to intermittent restraint by a non-competitive inhibitor, such as SS, or intermittent co-stimulation by a synergistic factor as yet unknown. Physiologically timed GH pulses also persist in patients with loss-of-function mutations of the GHRH receptor, albeit in less orderly (more entropic) patterns and at approximately one-thirtieth the normal daily GH secretion rate. Neuronal SSergic circuits, non-GHRH products of GHRH neurons (e.g., galanin, neurotensin), cortistatin, neuropeptide Y (NPY), glutamate and bioactive (acylated) ghrelin, the principal endogenous GH-releasing peptide (GHS), might support GH pulsatility in such patients.

A major impediment to clarifying regulatory concepts in humans is the absence of specific (blood–brain barrier-permeant) SS, GHS and GHRH antagonists for safe clinical use. However, a linear hexapeptide SS receptor blocker paradoxically inhibited somatic growth in the young male rat. This initially unexpected

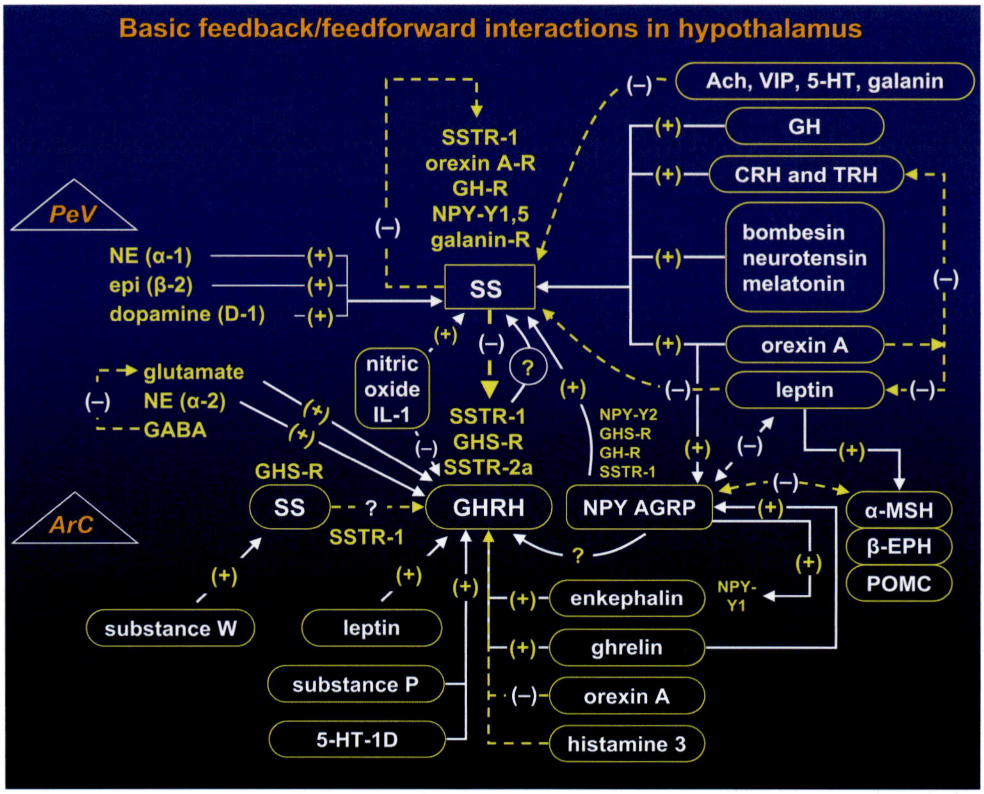

FIGURE 10.1 **Principal network-like connections among hypothalamic regulators of SS (somatostatin) and GHRH secretion in the rat.** Abbreviations: Ach, acetylcholine; Ach-M3, muscarinic-3 receptor; AGRP, Agouti-related peptide; α-MSH, alpha-melanocyte stimulating hormone; β-EPH, beta-endorphin; CRH, corticotropin-releasing hormone; GABA, gamma aminobutyric acid; GH-R, GH receptor; GHS, growth hormone secretagog; 5-HT, serotonin; 5-HT-1D, serotonin receptor 1D; IL-1, interleukin 1; NE, norepinephrine; NPY, neuropeptide Y; orexin A-R, orexin A receptor; POMC, proopio-melanocortin; SSTR, somatostatin receptor; TRH, thyrotropin-stimulating hormone; VIP, vasoactive intestinal peptide. Bombesin peptides include GRP (gastrin-releasing peptide) and NM-B (neuromedin B).

outcome suggested that intermittent SSergic input not only inhibits but also facilitates high-amplitude burst-like GH release. Investigations have delineated plausible *in vivo* mechanisms subserving SSergic facilitation of GH pulsatility; namely, SS withdrawal-induced release of neuronal GHRH, resensitization of somatotrope GHRH receptor signaling by phosphatase activation, and reversible inhibition by SS of exocytotic GH release but not *de novo* GH synthesis. Key factors that release periventricular SS are GH itself, anticholinergic (muscarinic-3) agents, β2- and α1-adrenoreceptor agonists, melatonin, neuropeptide Y (NPY), TSH-releasing hormone (TRH), nitric oxide, interleukins, neurotensin, GHRH (in some studies), dopamine-1 receptors, corticotropin-releasing hormone (CRH) and bombesins (gastrin-releasing peptide and neuromedin M), but not substance P, β-endorphin, metenkephalin or ghrelin. Conversely, muscarinic-3 cholinergic agonists, leptin, SS and both α1- and β2-adrenergic antagonists, galanin, serotonin and vasoactive intestinal peptide (VIP) suppress SS release.[1,3] Less well-defined pathways entail reciprocal connectivity between arcuate GHRH, NPY, SS or galanin neurons and periventricular SS neurons, and the cortistatin-14 neuropeptide, which, like SS, binds all SSTR subtypes.

Brain SS-28 may have somewhat greater efficacy in inhibiting GH secretion than SS-14, acting especially via the type 4 SS receptor. SS's effects are mediated via six subtypes of SS receptors, SSTRs 1–5, which include types 2a and 2b. Studies with SS-14 indicate that primarily SSTRs 2a and 5 transduce non-competitive inhibition of GHRH and GHS-stimulated exocytosis of GH. *In vitro* data suggest that SS receptor subtype 1 may repress basal (constitutive) GH secretion. These points are summarized in Table 10.1. Analyses with high-specificity subtype agonists and antagonists will be needed to clarify the precise roles of SS in health and disease, including in infancy, puberty and aging, and in supervising species-specific, gender-related and GH-mediated feedback effects. For example, somatotrope expression of SSTR-5 exceeds that of SSTR-2 in the rat, but the opposite is true in the human; pituitary SSTR-2a is stimulated by E_2 in the rat, but inhibited by E_2 in the mouse; and SS's actions may be either stimulatory (low-dose effect) or inhibitory (high-dose) with

TABLE 10.1 Principal SS Receptor-subtype Expression

PeV	SSTR-1	↓ by SS
	[SS neurons]	↑ by GH
		↑ by IGF-I
ArC	SSTR-1	
	[NPY neurons]	? Role in feedback
	[GHRH neurons]	♂ > ♀ expression
	SSTR-2a	↑ by SS
	[GHRH neurons]	↓ by T
		♀ > ♂ expression
Pituitary	SSTR-1	↓ by E_2 in rat and mouse
		Inhibit basal GH release
	SSTR-2a	↓ by E_2 in fish
	[Somatotropes]	↑ by E_2 in rat
		↓ by E_2 in mouse
	SSTR-4	Sensitive to SS-28
	SSTR-5	↓ by E_2 in rat

PeV, periventricular nucleus; ArC, arcuate nucleus.

TABLE 10.2 Comparisons Between GHRH and GHS

	GHRH	GHS/Ghrelin
Location	Arcuate nucleus	Brain, pituitary, stomach
Peptide	44 Amino acids	28 Amino acids (octanoylation of Ser^3)
Pituitary receptors	↑ Adenylyl cyclase	↑ Phospholipase C
GH gene	↑	_
Brain receptors	_	Arcuate NPY, GHRH neurons
GH feedback	↓ Arcuate GHRH	↓ Pituitary ghrelin, ↓ GHS receptor
CRH/AVP release	_	↑
NPY neurons	_	↑
Appetite	_	↑
Sleep	↑ Stage II	↑ Slow-wave sleep (III-IV)
Fat cells	_	↓ Fat oxidation, ↑ adipogenesis
Insulin secretion	_	↓
Blood pressure	_	↓
GHRH	↑ or ↓ GHRH-R	↑ GHS-R
Synergy	Threshold GHS dose	Maximal GHRH dose
Neuroprotection	No	Yes
Release GHRH	No	Yes
Antiapoptosis	No	Yes
Oppose CNS SS	No	Yes

respect to GH release by pig somatotropes, indicating a further level of regulatory complexity.

GH-Releasing Hormone

Isolation and cloning of GHRH were achieved in 1983 in two patients, one with clinical hypersomatotropism, by extraction and sequencing of peptides with GH-releasing activity from pancreatic neoplasms. GHRH is a 40- and 44-amino acid peptide synthesized by an interlaced network of homonymous neurons in the arcuate nucleus of the mediobasal hypothalamus (Fig. 10.1). In the human fetal hypothalamus, GHRH immunoreactivity is detected by 30 weeks of gestation. GHRH is also produced in gonads and the intestine. GHRH neurons are activated by opiates, norepinephrine, leptin, testosterone, serotonin agonists, ghrelin, glutamate, and alpha-2A agonists, and suppressed by somatostatin, orexin A, histamine-3 agonists, estrogen and GABA.[3] NPY appears to inhibit GHRH neurons by stimulating periventricular SS release. Regulation of GHRH is important, since this peptide is a potent agonist of both somatotrope-cell growth and GH-gene transcription (see Table 10.2). The dual effects are inferable in GHRH-transgenic mice, which develop pituitary somatotrope hypertrophy, hyperplasia and microadenomas. GHRH-transgenic animals also exhibit decreased VIP and increased galanin gene expression in the anterior pituitary gland. Conversely, GHRH receptor-inactivated (*lit*/*lit*) mice and GHRH-immunized rats are profoundly hyposomatotropic, stunted in growth, and exhibit pituitary hypoplasia.

The GHRH receptor is a GTP-dependent adenylyl cyclase-activating protein, with sequence homology to secretin and vasoactive intestinal polypeptide (VIP) receptors. In the testis and brain, high concentrations of GHRH may act via the VIP receptor. Intracellular cAMP accumulation and cellular calcium influx are key mediators of GHRH action. Unlike SS receptors, the GHRH receptor is expressed almost exclusively by somatotrope cells. Physiological amounts of thyroxine, cortisol and GHRH upregulate, while IGF-I downregulates, the GHRH receptor in animal models.[1,3] Developmental stage is important, since GHRH receptors are maximally expressed in the infant rodent, and minimally in senescence. In this regard, exogenous GHRH induces pituitary GHRH receptors in the young but not the aged rat. Species differences also exist, wherein

E_2 elevates pituitary GHRH receptor gene expression in mice, but inhibits the same in rats.

Direct sampling of hypothalamo-pituitary portal blood in sheep, rats and swine has revealed that GHRH pulses are more conspicuous and consistent in their temporal association with GH pulses than SS nadirs.[1,3] In addition, selective GHRH receptor antagonism or GHRH immunoneutralization essentially abolishes GH pulses in the rat and human, and antagonizes the effects of clonidine (an α2-adrenergic agonist) and L-dopa (a dopamine precursor). Accordingly, GHRH is the primary physiological agonist of GH secretory bursts. This is important because more than 85% of GH secretion in humans occurs in bursts.[1]

Somatostatin, when delivered briefly and withdrawn before a GHRH pulse, potentiates GH secretion (facilitative effect), but when delivered concurrently with a GHRH pulse inhibits GH release non-competitively by activating G_i, an inhibitory subunit of adenylyl cyclase, and by restricting cellular Ca^{2+} uptake necessary for exocytosis. Given the dual facilitative and inhibitory capabilities of SS, the relative timing of SS and GHRH signaling to somatotropes constitutes a key determinant of pulsatile GH secretion. Simple models of GHRH, GHS and SS actions yield predictions consistent with this thesis.[3]

GH-Releasing Peptides (GH Secretagogs, GHS)

Ghrelin Expression

The prototypical endogenous GHS is ghrelin, a 28-amino acid acylated peptide, which requires an octanoyl group esterified to serine in N-terminal position 3 for maximal receptor activation. Other fatty acyl groups can also confer bioactivity.[7] The N-terminal pentapeptide containing octanoylated serine is biologically active. The native gene and peptide are jointly expressed in stomach (X/A-like cells), pancreas (epsilon cells), placenta, B lymphocytes, gonads, adrenal, pituitary and brain. In the hypothalamus, ghrelin is produced by neurons within and dorsal to the arcuate nucleus and inferomedial to the periventricular nucleus, which extend axonal projections to suprachiasmatic, paraventricular, supraptic, dorsomedial and periventricular nuclei. Hypoglycemia, prolonged fasting and insulin repress hypothalamic ghrelin and preproghrelin-gene expression. Pituitary ghrelin is downregulated by excessive GH, thyroxine or glucocorticoid, but upregulated by GHRH. In the rat, the gastric ghrelin gene is reduced by obesity, feeding, hyperglycemia, acute vagal denervation or atropine administration, and induced by fasting, hypoglycemia and leptin. An acylating enzyme, ghrelin-octanoyl acyltransferase (GOAT), was cloned recently, which is expressed strongly in the stomach, pancreas, heart, liver and bone. Except for substrate autoinhibition and estrogenic stimulation, regulation of gastric GOAT activity has not yet been well delineated. Non-specific esterases of acylghrelin exist in the blood and stomach.

Ghrelin's Actions

Ghrelin (unlike GHRH) is a multifaceted agonist of appetite, gastric motility and acid secretion, GH secretion, vasodilation and cell survival, and an antagonist of insulin secretion and adipocyte fat oxidation (see Table 10.2). The gene encoding ghrelin is expressed in the stomach, hypothalamus, pituitary, gonads, placenta, thyroid and adrenal glands, breast, skeletal muscle, fat, heart, spleen, pancreas, lymph nodes and liver. The cognate receptor is a G-protein coupled heptahelical transmembrane protein cloned by Howard *et al.* in 1996,[9] three years before Kojima and colleagues identified rat and human ghrelin.[10] The existence of a native GHS and matching receptor had been proposed by Bowers in 1977,[11] after demonstrating potent direct pituitary GH-releasing effects of oligopeptide derivatives of enkephalins. Mu opiate receptor agonists, such as the alkaloid morphine and the peptide met-enkephalin, but not delta opiate receptor agonists, also evoke GH secretion *in vivo*. The fact that three decades elapsed between identification of synthetic GHS and cloning of native ghrelin reflects in part the unexpected fatty acid-ester requirement of the peptide (but not synthetic ligands) for full bioactivity. Unacylated ghrelin also appears to either mimic or antagonize selected effects of acylghrelin. Proof would require demonstrating that the administered peptide is neither esterified nor metabolized to other active peptides in target or neighboring cells. A second cleavage product of preproghrelin, obestatin, does not consistently affect gastric acidity, gastric motility, appetite, or GH or insulin secretion, but does suppress drinking behavior, and may activate CRH and urocortin-2 neurons. Obestatin may be better termed ghrelin-associated peptide, since knockout of its proposed receptor, GPR39, also does not affect food intake.

Ghrelin/GHS Receptor

Ghrelin and synthetic analogs, which include both peptides and non-peptidyl mimetics, activate phospholipase C-dependent, mitogen protein-kinase and biphasic intracellular Ca^{2+}-mediated signaling via the ghrelin type 1a receptor. Evidence also exists for mediation of certain ghrelin effects via nitric oxide and modulation by excitatory amino acids. Prominent sites of GHS receptor expression include somatotropes, GHRH neurons in both the arcuate and ventromedial nucleus, neuropeptide Y and orexin B neurons, gastric mucosa and myenteric plexus, vagal nodose ganglion, endothelium, cardiomyocytes, adrenal, thyroid, spleen and pancreas, among others. The type-1b receptor is also widely expressed as an unspliced non-functional protein. Activation of the functional GHS receptor on GHRH neurons upregulates GHRH gene expression. The receptor is distinguished by high constitutive activity, loss of which is associated with short stature. Transgenic reduction of GHS receptors on GHRH neurons in the female but not male mouse decreases GH pulse size, IGF-I concentrations, GHRH gene transcripts and somatic growth.

Contrasts between Ghrelin and GHRH

Ghrelin differs from GHRH not only in structure, topographic localization and receptorology, but also in the former's pleiotropic functions and distinctive mechanisms of driving GH secretion. Peak lifetime GHS receptor expression occurs in the neonatal and pubertal rat, but only in the neonate in the case of GHRH receptor. Intracerebroventricular instillation of ghrelin stimulates, and of GHRH, inhibits GH secretion in the male rat. GHS/ghrelin is unique mechanistically in partially antagonizing both hypothalamic and pituitary inhibition by SS. Analogously, GHS is less inhibited by glucose and free fatty acids than GHRH is in man. Although ghrelin may inhibit SS release by rat hypothalamic tissue *in vitro*, it does not block SS release into portal blood *in vivo* in the sheep or pig.[3] Sufficiently high concentrations of SS block Ca^{2+} entry into somatotropes, thereby non-competitively impeding both GHRH and GHS-induced GH secretion. Unlike GHRH, GHSs rapidly excite NPY and orexin neurons, and evoke GHRH release from arcuate nuclei in rats, sheep and goats *in vivo* and rat hypothalamic slices *in vitro* (Fig. 10.2). L-arginine potentiates GH release induced by a maximal GHRH stimulus, but only by a submaximal GHS stimulus. Ghrelin potentiates pulsatile GH secretion, but, unlike GHRH, does not promote GH gene transcription except in the prepubertal rat and embryonic fish *in vivo*, as well as ovine somatotropes *in vitro*. Furthermore, ghrelin/GHS but not GHRH briefly stimulates ACTH, cortisol and prolactin release, elevates blood glucose, and inhibits insulin secretion. The exact mechanisms of stimulating ACTH secretion are not certain, but ghrelin does not synergize with maximal CRH or AVP stimulation. GHS induces the CRH gene and AVP secretion *in vivo*, and stimulates both CRH and AVP release *in vitro* (Table 10.2).

Metabolic and Appetitive Effects of Ghrelin/GHS

The metabolic and appetitive effects of ghrelin are complex. Metabolic effects include enlarging adipocytes, blocking fat oxidation, increasing the respiratory quotient, stimulating osteoblast proliferation and differentation with attendant bone formation, promoting short-term weight gain, elevating serum glucose

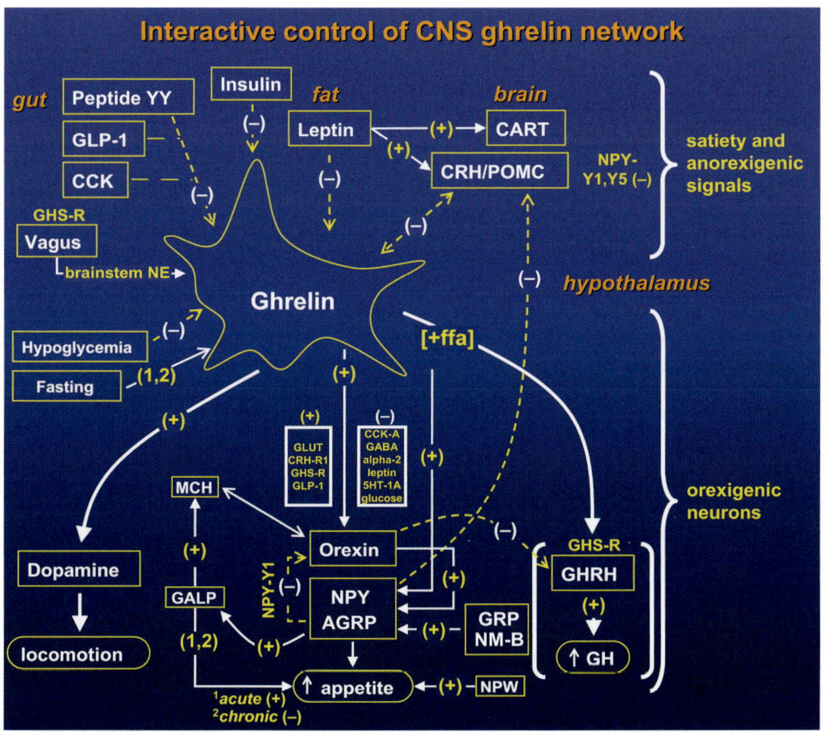

FIGURE 10.2 **Primary inputs to and output from ghrelin neurons proposed in the rat.** See abbreviations in Fig. 10.1. CART, cocaine and amphetamine-regulated transcript; CCK, cholecystokinin; ffa, free fatty acids; GALP, galanin-like peptide; GHS-R, GH-secretagog receptor; GLP-1, glucagon-like peptide 1; GLUT, glutamate; MCH, melanin-concentrating hormone; NPW, neuropeptide W.

concentrations, and inhibiting insulin secretion. A competitive antagonist of ghrelin exerts the opposite effects – namely, elevates insulin and reduces GH secretion by about 50%. Appetitive (orexogenic) effects reflect ghrelin-, hypoglycemia- and vagal-dependent activation of peripheral and central neural pathways, which ultimately impinge upon a rich plexus of arcuate, ventromedial and lateral hypothalamic nutritionally regulated neurons (Fig. 10.2). Critical intermediaries are NPY and AGRP neurons, in which ghrelin stimulates mitochondrial respiration by inducing uncoupling-protein-2. Ghrelin's effects are inhibited by satiety factors, such as certain gut peptides (peptide YY, glucagon-like peptide (GLP-1) and cholecytokinin (CCK)), the adipokine leptin, and stress-associated peptides (proopiomelanocortin (POMC), cocaine-amphetamine related transcript (CART), and corticotropin-releasing hormone (CRH)). Acylated ghrelin enters the brain, where it: (a) stimulates GHS receptor-bearing neurons expressing orexin, NPY and AGRP in the arcuate nucleus; and (b) inhibits neurons expressing anorexigenic peptides, such as leptin, the POMC derivative alpha-melanocyte stimulating hormone (α-MSH), CRH and possibly CART (Fig. 10.2).

Collaborating activators of NPY neurons are glucocorticoids, orexin and the bombesins, gastrin-releasing peptide (GRP) and neuromedin B (NM-B). Orexin, which stimulates NPY/AGRP neurons and inhibits GHRH outflow, is required for ghrelin's orexigenic effects. Orexin's fasting-associated inhibition of GHRH and stimulation of periventricular SS gene expression may contribute to suppression of pulsatile GH secretion in the calorically restricted rat. In turn, NPY, peptide YY (3–36), catecholamines and cannabinoids inhibit, whereas glucagon-like peptide 1 (a gut satiety factor) stimulates, orexin neurons, thus respectively curtailing and enhancing appetite. Orexin neurons stimulate and NPY neurons suppress POMC, a potent anorexigenic precursor, thereby repressing and inducing feeding, respectively. NPY may also exert autonegative feedback via NPY–Y2 receptors to quench prolonged feeding. Other signals such as galanin-like peptide (GALP) and NPW also modulate appetite, possibly by modifying orexin and melanin-concentrating hormone (MCH), illustrating the high complexity of ghrelin-associated feeding networks. Elements of the network interact manifoldly with leptin, a potent antagonist of ghrelin action in the arcuate nucleus (Fig. 10.2). A major need in the field is to clarify how nutritional networks operate distinctly in infancy, childhood, puberty, adulthood and senescence. However, in the rat, male sex and testosterone treatment are associated with increased NPY expression. In general, senescence impairs appetitive effects of key neuropeptides.

Circulating Ghrelin

The abundant expression and secretion of ghrelin by stomach X/A-like cells (now known to contain both ghrelin and motilin) contributes about 70% of systemic acylated peptide. Plasma total ghrelin concentrations are reduced in visceral obesity, after food intake, and by insulin, somatostatin, hyperthyroidism, cortistatin and urocortin-1 (a stress-induced neuropeptide), but increased by fasting, estradiol, chronic hepatic or renal failure, and other catabolic states. Acylghrelin levels are normal in endstage kidney disease, but fall during hemodialysis. Ghrlein does not consistently change with age, when its inverse relationships with BMI and hyperinsulinemia are considered. Food-related inhibition of ghrelin may be less in children than adults. The degree to which brain ghrelin neurons *per se* mediate central appetitive effects or release the cognate peptide into portal blood and modulate GH secretion has not been defined.[7] Indeed, prolonged fasting and glucoprivation repress (whereas acute fasting elevates) hypothalamic ghrelin gene transcripts in the rat (Fig. 10.2), while upregulating the same in the stomach and potentiating CNS effects of ghrelin. Neuron-selective silencing of arcuate and ventromedial-nuclear ghrelin genes and GHS receptors would be necessary to assess their *de facto* roles.

Ghrelin's Potentiation of GHRH-evoked GH Release

Ghrelin and analogs infused in submaximally stimulatory doses are capable of synergizing with a maximally effective dose of GHRH in evoking GH secretion in the human, rat, pig, cow and dog. Synergy does not occur *in vitro*. The precise mechanisms subserving this synergy are not established, but putatively include ghrelin-mediated: (a) opposition to hypothalamo-pituitary actions of SS; and/or (b) stimulation of an unknown ("U") synergy factor.[3] A substance like galanin or an endogenous opiatergic peptide might represent such a factor, since both peptides release and synergize with GHRH and their terminals innervate periventricular SS neurons. Reduced pituitary action of SS is unlikely to be the sole potentiating mechanism subserving GHS action, since GHS and GHRH synergize even after immunoneutralization of SS. The necessity of an intact hypothalamo-pituitary unit capable of secreting GHRH for maximal ghrelin drive is evident by marked impairment of GHS's stimulation of GH release in patients with pituitary-stalk interruption and in subjects given an antagonist of GHRH.

The trophic and anabolic effects of ghrelin have prompted preliminary studies in patients with cachexia associated with cancer, AIDS-related wasting syndrome, and endstage cardiac and pulmonary failure. On the other hand, ghrelin antagonists or vaccines may find utility in increasing energy expenditure, improving glucose homeostasis, and reducing excessive body fat.

Collateral Modulators of GHRH, GHS and SS

Concept of Neuronal Network-like Interactions

Neurons identified by transgenic expression of fluorescently tagged GHRH protein are assembled as a complex mediobasal hypothalamic network. They often co-express galanin, neurotensin and tyrosine hydroxylase (synthesizes L-dopa). By the same technique, hypothalamic ghrelin-producing neurons and pituitary GH-secreting cells manifest similarly remarkable network-like organization. In principle, GHRH and/or other products of GHRH neurons (such as L-dopa) might act by ultrashort-loop negative feedback (reversible inhibition after a time delay) to generate recurrent bursts of GHRH, and thereby GH secretion by somatotrope cells. Excitatory amino acids, like glutamate and aspartate, may potentiate GHRH release in the rat and pig. Whether a primary GHRH oscillator (pulse generator) exists is not known. GHRH secretion is both amplified and repressed by multiple collateral synaptic inputs, which might collectively initiate or stabilize GHRH oscillations. Internuncial neurons express glutamate, norepinephrine (NE), dopamine, enkephalins, nitric oxide, cortistatin, gamma-amino butyric acid (GABA), orexin, leptin, SS, ghrelin, neurotensin, bombesin, substance P, vasoactive intestinal peptide (VIP), NPY/AGRP, CRH, neuropeptide W, acetylcholine, galanin and β-endorphin (EPN): Fig. 10.1. For example, glutaminergic nerve terminals from periventricular neurons make multi- and monosynaptic excitatory contact with arcuate-nucleus GHRH neurons bearing SSTR-2, potentially effecting GHRH pulse generation. Conversely, GABA induces hypothalamic SS gene expression in the female rat, but inhibits the same in the male rat. Connections between SS and GHRHergic neurons suggest an indirect pathway by which GABA may sex-selectively modify GHRH pulses in the rat and mouse. Leptin, an adipokine, both decreases SS and increases GHRH secretion, thereby stimulating GH release. On the other hand, ghrelin induces the leptin gene in fat cells, thereby initiating negative feedback. Short-loop pituitary–hypothalamic feedback may also operate, since GH can cross the blood–brain barrier and rapidly excite arcuate NPY neurons and periventricular SS neurons.

Regulatory Roles of SS

Periventricular SS is a primary restraining signal to arcuate GHRH and pituitary GH secretion in the adult animal (Fig. 10.3). SS outflow simultaneously inhibits arcuate-nucleus GHRH neurons and GH exocytosis in the pituitary. In the hypothalamus, SS acts via at least SSTR-1 and SSTR-2a. SSTR-1 is an autoreceptor, which could mediate inhibition of periventricular SSergic neurons via ultrashort-loop feedback. In this regard, intracerebroventricular SS infusion stimulates GH release, possibly by disinhibiting SS-restrained GHRH secretion. Periventricular and arcuate GHRHneuronal SSTR-1 expression is greater in the adult male than female rodent. Conversely, SSTR-2a is expressed by

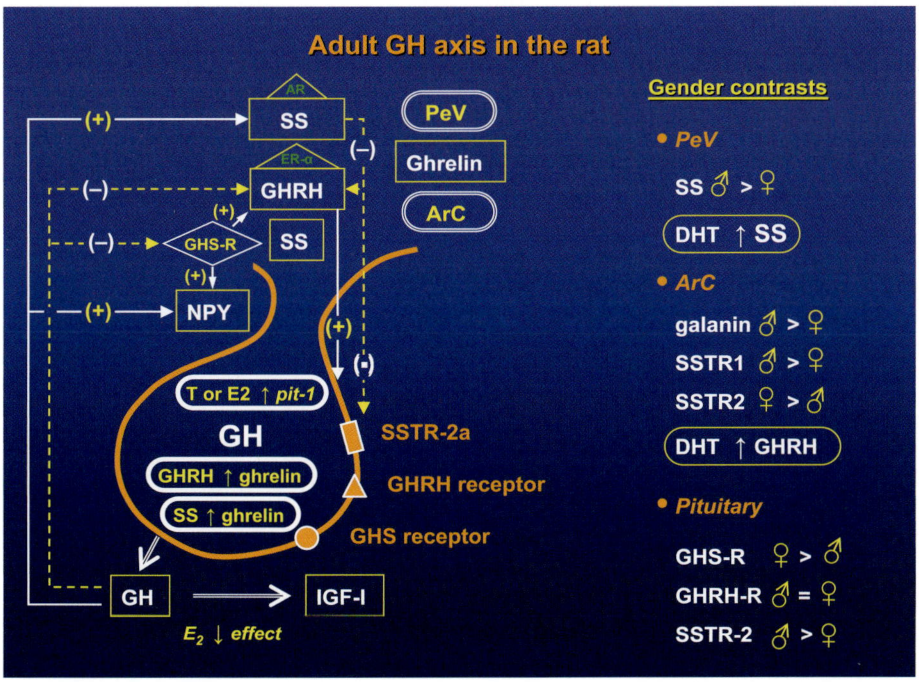

FIGURE 10.3 **Basic SS, GHRH and GH regulatory loci in the adult rat defined by anatomic site and gender.** AR, androgen receptor; ERα, estrogen receptor-alpha, 5α-DHT, 5-alpha-dihydrotestosterone; T, testosterone; PeV, periventricular nucleus; GHS, GH secretagog, of which ghrelin is the prototype.

about 75% of arcuate GHRH neurons in the adult female and 25% in the male rodent. Testosterone (T) represses whereas orchidectomy induces GHRH-neuronal SSTR-2a expression. In contrast, E_2 upregulates SSTR-2, and downregulates SSTR-5 and SSTR-1 in the pituitary gland of the rat. In mice, the effect of E_2 on pituitary SSTR-2 is opposite to that in rats. Other interactions are also possible, since nerve terminals containing SS, NPY, GHRH, galanin and β-endorphin of arcuate and ventromedial-nuclear origin project to periventricular SS neurons. Indeed, calcitonin injection suppresses GH secretion in men, ostensibly via mechanism(s) not reversed fully by SS antiserum in the rat, and immunoneutralization of both GHRH and SS in the rat does not abolish GH pulses, consistent with multifactorial regulation.[3]

Sex-steroidal Determinants of Network-like Interactions

Periventricular-nuclear neurons that co-express all three of androgen, GH receptors and SS peptide predominate in the adult male compared with female animal.[1,3] These neurons project terminals not only to the median-eminence portal microvascular system that delivers SS to pituitary somatotropes, but also directly to arcuate-nucleus neurons that secrete GHRH. This neuroanatomic framework confers an androgen-dependent, pubertally initiated and GH-activated negative feedback route to block GH exocytosis transportally and quench GHRH secretion transsynaptically. Periventricular SSergic neurons also express SSTR-1 and the NPY receptor, allowing for possible: (a) periventricular SS oscillations driven via ultrashort-loop feedback by SSTR-1 autoreceptors; and (b) GH-initiated feedback via arcuate-derived NPY-mediated stimulation of periventricular SS release. In the latter regard, GHS receptors expressed on NPY neurons could explain otherwise paradoxical inhibition of GH secretion induced by intracerebroventricular GHS injection. NPY–Y2 autoreceptors on homonymous arcuate neurons may in principle generate ultrashort-loop NPY oscillations, since intracerebroventricular injection of antiserum to NPY augments pulsatile GH secretion (Figs 10.1, 10.3). The roles of less-abundant arcuate SS, GHRH, galanin and β-endorphin neurons that project to the periventricular nucleus are less clear.

Modeling Neuroendocrine Network Control

Basic neuroanatomic connections within the hypothalamus undergird recent mathematical models of GH pulsatility driven by reciprocal GH–NPY, SS–GHRH or SS–SS interactions with amplification by GHS.[3] Innervation of GHRH neurons by glutamate and GABA-expressing terminals may provide excitatory and inhibitory neuromodulation. Indeed, glutamate receptor antagonists reduce pulsatile GH secretion in both the male and female rat, whereas GABA is stimulatory in the neonatal and inhibitory in the adult male rat. Current models do not yet include these basic neurotransmitters or other less wellstudied modulators such as enkephalins and galanin. The latter peptide, like ghrelin, acts via hypothalamic (and possibly also pituitary) receptors to disinhibit SSergic repression of GHRH drive in children and adults. Additional neurophysiologic, anatomic and genetic studies are needed to elucidate the precise interneuronal connections, feedback time delays, and dose–response relationships that collectively mediate GH pulse generation. Such models would be helpful in framing how a systemic stimulus, like endurance exercise training, augments pulsatile GH secretion, IGF-I concentrations and GHRH efficacy in men and women.

Pituitary–Hypothalamic Pathways

Anatomic evidence exists in seven mammalian species that the pituitary gland can secrete hormones retrograde to the brain via the valveless portal microvascular route. This pathway could mediate rapid direct GH-induced modulation of arcuate GHRH, SS and NPY neurons as well as periventricular SS neurons. Activated NPY neurons stimulate periventricular–nuclear SS output to inhibit arcuate GHRH neurons and pituitary somatotropes (and possibly quench NPY release). GH acting via CNS GH receptors also directly stimulates periventricular SS gene expression and SS release.[3] Circulating GH can also reach brain neurons via: (a) axon terminals in the median eminence outside the blood–brain barrier; (b) GH receptors in the organum–vasculosum complex containing the choroid plexus, area postrema and lamina terminalis; (c) olfactory and trigeminal perineural spaces; and (d) endothelial transendocytosis. What remains uncertain is what additional pituitary factors are secreted to the brain. For example, although cytokines (e.g., IL-1), nitric oxide, IGF-I and ghrelin (and their receptors) are expressed in both the hypothalamus and anterior pituitary gland, the extent of their antegrade and retrograde exchange is unknown.

Feedback Control Mechanisms

A hallmark of physiologically regulated neuroendocrine systems is an organized array of homotypic and heterotypic feedback (inhibitory), feedforward (stimulatory) and modulatory signals, which collectively preserve stable hormone availability[1,3] (Fig. 10.3). In mathematical and mechanical systems, negative feedback and an attendant time delay are sufficient to create oscillations. Adding stochastic (random) variability to the strength of the feedback signal or associated time delay generates amplitude-varying and irregularly timed events. Given that the amplitude and timing of GH pulses together specify end-organ responses (e.g., large infrequent GH pulses preferentially induce the muscle IGF-I gene), feedback modulation not only maintains homeostatic mean output but also determines sex-patterned pulse signaling to target organs.[1]

Hypothalamus

FEEDBACK VIA THE BRAIN

Autofeedback by GH or IGF-I is readily demonstrable in the adult animal and human as well as in childhood by transient suppression of fasting and secretagog-evoked GH secretion by injection of GH or IGF-I. Infant rats exhibit minimal GH or IGF-I-induced feedback on GHRH or SS gene expression. In part, this may reflect low pituitary SSTR-2a expression in neonatal animals. In humans, injection of GH or IGF-I suppresses fasting, hypoglycemia-, GHS- and GHRH-stimulated, as well as deep-sleep and exercise-induced, GH secretion after a time delay of 30–90 min. Although the precise reasons are not clear, GH responses to stimuli such as GHS and hypoglycemia are more difficult to suppress than the GH response to GHRH. Reduced SS action and/or outflow in response to GHS and hypoglycemia may be a factor.[1,3]

GH feedback is exerted via CNS pathways. Evidence that systemic GH feeds back on the brain includes: (a) reversible inhibition of GH secretion by either central or peripheral injection of GH; (b) the absence of direct GH effects on somatotrope cells *in vitro*; (c) GH-induced synthesis of periventricular neuronal SS, phosphorylation of STAT5b and release of SS into hypothalamo-pituitary portal blood; (d) impaired feedback when brain GH or SS receptors are silenced; (e) neutralization of GH feedback by systemic or intracerebroventricularly infused antibodies to SS; (f) rapid activation of *C-fos* gene expression in brain SS and NPYergic neurons by systemic GH; (g) feedback failure after inactivation of the SSTR-2 gene in the arcuate nucleus; and (h) *in vitro* release of SS from hypothalamic fragments after incubation with GH or GHRH.[1,3] In the rat, but not sheep, combined intracerebroventricular infusion of IGF-I and IGF-II suppresses GH secretion by inducing the SS and repressing the GHRH gene.

GENDER CONTRASTS IN FEEDBACK

Gender differences in feedback control emerge vividly in puberty, and are species-dependent[1,3] (Fig. 10.4). For example, percentage autoinhibition by GH is considerably more marked in the adult male than female rat, but more prominent in women than men. In the ovariectomized rat, a single dose of testosterone heightens GH feedback, as inferred by emergence of low interpulse GH nadirs and large post-nadir GH pulses. In the orchidectomized rat, E_2 rapidly enforces the adult female pattern of diminutive GH feedback defined by elevated interpulse nadirs and high-frequency low-amplitude GH pulses. In this species,

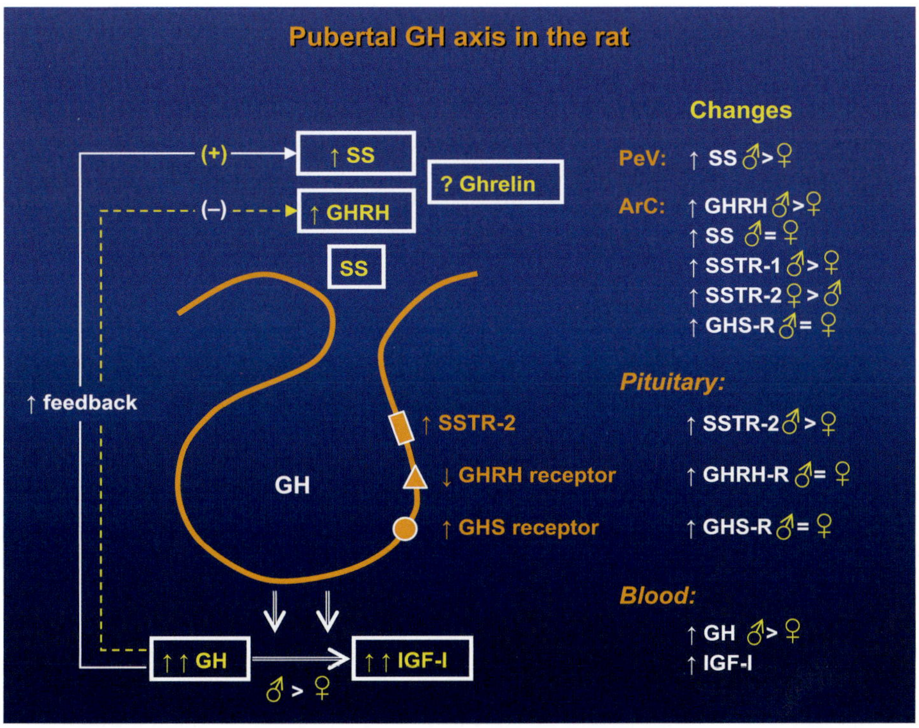

FIGURE 10.4 **Elements of pubertal GH axis.** ↑ increase or ↓ decrease with respect to prepubertal axis. Gender effects are shown by ♀ and ♂ symbols.

castration decreases arcuate GHRH and periventricular SS gene transcripts in both sexes.[1] Administration of testosterone or 5α-DHT normalizes both transcripts, whereas E_2 replacement restores SS expression inconsistently if at all in the adult rat, and in some contexts represses the GHRH gene. In one study in the rat, E_2 plus progesterone (but neither alone) increased periventricular SS immunoreactivity. Unlike the rat, sheep SS but not GHRH neurons and somatotropes express ERα and are activated by endogenous and exogenous E_2. A mathematical model supports, but does not prove, the thesis that SS-dependent feedback distinctions could adequately explicate differences between male and female patterns of GH pulsatility in the rat.

FEEDBACK IN HUMANS

Testosterone administration to prepubertal boys and older men attenuates GH and IGF-I feedback by elevating nadir GH concentrations and accelerating recovery of GH secretion after feedback suppression. E_2 treatment in older women restricts GH and IGF-I autofeedback in part by augmenting GHRP-2-stimulated, but not exercise- or maximal GHRH-driven, GH pulses.

MECHANISMS OF FEEDBACK

Transgenic silencing of germline SSTR-2 or the CNS GH receptor, mutation of STAT5b (a transcriptional mediator of GH action), and antisense RNA-directed attenuation of brain SSTR-1 signaling all reduce GH autofeedback and feminize the pattern of GH pulsatility in the male rodent, consistent with model-based predictions of sex-dependent GH-induced SS outflow. *In vitro* immunoneutralization of SS increases GHRH release by hypothalamic fragments. In humans, the feedback-mediatory role of hypothalamic SS release has been inferred indirectly by demonstrating reduced stimulatory efficacy of GHRH and GHS shortly after GH infusion, suggesting GH-induced SS outflow to the pituitary. Autofeedback is relieved by prior injection of L-arginine, which is a putative inhibitor of SS release as well as a precursor for nitric oxide. Methionine but not phenylalanine also potentiates GHRH action. In the ram, systemic delivery of a brain-permeant somatostatin analog (octreotide) suppresses portal-venous GHRH pulses, presumptively by directly inhibiting SSTR-2-expressing GHRH neurons. GHRH and GH pulses emerge again when octreotide is metabolized. By analogy, normal GHRH pulses may be initiated when endogenous GH feedback-suppressed interpulse GH concentrations are too low to evoke SS release from periventricular neurons, thereby relieving SSergic inhibition of GHRH neurons and pituitary somatotropes. What is uncertain is the degree to which such feedback mechanisms mediate *de facto* GH pulse generation in different species.

HYPOTHALAMIC ROLE IN PULSE GENERATION

Other theoretic GH pulse-generating models include an autonomous GHRH neuronal-bursting mechanism, GH–NPY–SS feedback, and primary intrahypothalamic SS–GHRH or SS–SSergic oscillatory mechanisms.[3] For example, electrical stimulation of periventricular neurons can evoke time-delayed GH and GHRH release. Critical limiting factors in distinguishing among various pulsatility constructs are difficulties in: (a) assembling a functional neuronal network *in vitro*; and (b) recording peptide release simultaneously from all four of periventricular SS, arcuate SS and arcuate GHRH neurons and pituitary somatotropes *in vivo*. In addition, delineating the complete physiological pulse-generating ensemble will ultimately require defining the roles of other putative internuncial neurons (Fig. 10.1).

Pituitary

The exact roles of systemic and intrapituitary factors in regulating GH secretion are not established. Although direct regulation of somatotropes by GH is not demonstrable *in vitro* except in fish, various systemic hormones, substrates and metabolites influence somatotrope function[1,3] (Fig. 10.5). Most inferences are based upon *in vitro* experiments using pituitary cells. Studies variously report: (a) direct stimulatory effects of β2- and α1- adrenergic agonists, lipoxygenase and cyclooxygenase products (like prostaglandin E_2), thymosin, thyroxine, T, E_2, nitric oxide, retinoic acid, pituitary adenylyl cyclase-activating peptide, VIP, galanin, resistin and NPY; and (b) direct inhibitory effects of IL-1β, IL-6, TNFα (cachexin), γ-interferon, corticostatin, dopamine, free fatty acids, adiponectin, leptin, insulin and IGF-I. An analytical model of the interrelated pituitary effects of GHRH, SS and GHS has been proposed. Definitive evidence of direct pituitary modulation by other factors would require somatotrope cell-specific silencing of the corresponding receptors in a conditional (exogenously or developmentally controlled) manner. Germ-cell gene deletions are less selective, strain-dependent, and confounded by developmental adaptations in the fetus, infant and pubertal animal.[3] Systemic immunoneutralization strategies also are less explicit, because antibodies gain access to axons and neuron terminals in the median eminence.

Foremost among potentially direct somatotrope-cell inhibitors are IGF-I and free fatty acids, as inferred by infusing or depleting these agents *in vivo*, and by culturing simian, rat, swine, bovine and human pituitary cells *in vitro*. Nonetheless, in clinical studies L-arginine, a putative hypothalamic effector, reverses

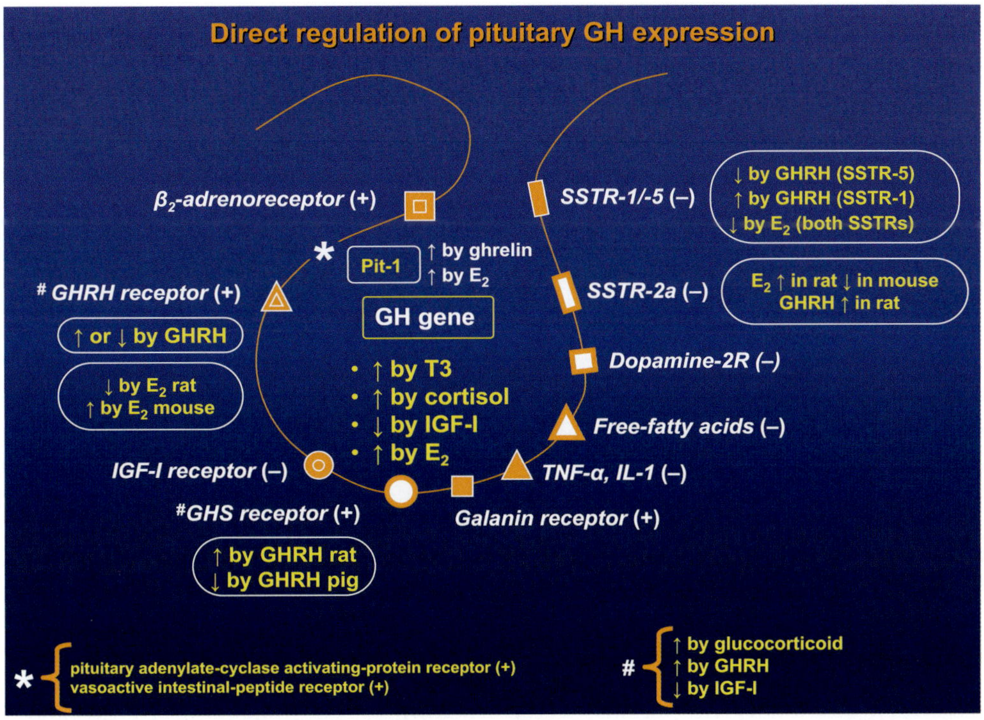

FIGURE 10.5 **Inferred direct regulators of pituitary somatotropes.** Pit-1 is a developmental transcription factor specifying GH, prolactin and TSH cell lineages. TNFα denotes tumor-necrosis factor alpha. Plus or minus signs in parentheses denote stimulation or inhibition, respectively.

the inhibitory effect of a single injection of IGF-I on a GHRH stimulus in healthy young adults. L-arginine also evokes comparable GH secretion in the presence and absence of exogenous IGF-I. These data suggest either an unrecognized direct effect of L-arginine on somatotropes, or a hypothalamic antagonism of SS secretion into portal blood. The most plausible hypothalamic factor repressed by L-arginine is periventricular SS release, but this has not been proven by hypothalamo-pituitary portal-venous sampling. Other human and animal data are consistent with either pituitary or hypothalamic sites of feedback by IGF-I. Evidence favors significant CNS feedback in the mouse, rat and human, possible pituitary feedback as well in mice, and primarily a pituitary locus in sheep. In mice, systemic IGF-I deficiency induces pituitary expression of the GH gene and receptors for both GHRH and GHS. This tripartite response is consistent with increased GHRH outflow from the hypothalamus, wherein low IGF-I induces GHRH secretion indirectly by reducing SSergic outflow. A similar mechanistic inference has been made for free fatty acids in the male rat, since SS antibodies impede their inhibition of GH secretion.

A plausible intrapituitary feedback concept in mammals is that IGF-I, which is synthesized by folliculostellate and somatotrope cells, feeds back to suppress GH synthesis and secretion via IGF-I receptors on somatotropes. Intrapituitary IGF-I-binding proteins (namely, IGFBP-2, -4 and -5) might modulate such a circuit. Systemic IGFBP-1 administration in mice induces GH secretion, possibly by reducing free IGF-I levels. In contradistinction, intravenous infusion of insulin in humans to achieve postprandial concentrations does not suppress fasting GH pulsatility. In addition, *in vitro* inhibition of GHRH-stimulated GH secretion requires pharmacological insulin ($>$3-nM) or IGF-II ($>$10-nM) concentrations, suggesting mediation of inhibition via the type I IGF-I receptor.

Fasting or Nutrient Deprivation

Inadequate caloric intake decreases pulsatile GH secretion profoundly in the male rat,[1] but stimulates the same in the turkey, mouse, guinea pig, dog, sheep, pig, cow and human. The mechanisms of inhibition in the rat entail fasting-dependent: (a) withdrawal of leptin's stimulation of GHRH and inhibition of SS release; (b) NPY–CRH–α-MSH-dependent decreases in both GHRH and SS gene expression; (c) systemic release of gut SS; and (d) orexin A-imposed suppression of GHS drive and GHRH secretion. Whether pituitary sensitivity to SS is increased in fasting remains to be seen.

Sex-steroidal Determinants of GH Secretion

Species Differences

The effects of gender and sex steroids on GH secretion differ by species.[1,3] Figs 10.3 and 10.4 highlight major regulatory distinctions by sex in the rat. Available data identify several prominent differences between the rat and human: (a) higher-amplitude GH pulses occur in women than men, but the opposite applies in the male and female rat; (b) GH enforces greater fractional feedback in women than men, and the reverse in male and female rats; (c) fasting and hypoglycemia stimulate pulsatile GH secretion in the human, but inhibit it in the rat; (d) illness and female gender increase hypothalamic GHRH neuron number in the human, but conversely in the rat[1,3]; and (e) E_2 increases the stimulatory potency of GHRH and ghrelin and decreases the inhibitory potency of SS, thus augmenting GH pulse size in the human, whereas E_2 downregulates pituitary GHRH receptors and upregulates pituitary SSTR-2, thereby diminishing GH-pulse size in the rat.[3] In contrast, E_2 increases pituitary GHRH receptors and decreases pituitary SSTR-2, thereby augmenting GH pulses in mice[1] (Fig. 10.5).

Sex-steroid Imprinting

Sex differences in the adult GH/IGF-I axis are putatively conferred primarily by E_2 and T acting *in utero* or neonatally (rat) and in puberty and adulthood (rat and human). An unproven postulate is that sex steroid-independent gonadal and extragonadal factors may also mediate some gender differences in neuroendocrine regulation. One such factor is intrauterine growth retardation secondary to placental insufficiency. However, fetal, neonatal and pubertal sex-hormone imprinting effects on neuroglial microarchitecture may explain many gender differences that persist in an adult castrate or hypogonadal milieu. Imprinting is not permanent, since gonadectomized adult rats remain responsive to sex-steroid induced reversal of genetically expected GH secretory patterns.

In the pubertal and adult rat, sex differences exist in relation to hypothalamic GH receptor, GHRH, galanin and SS gene expression, brain and pituitary SS receptor subtypes, and the number of hypophysial GHRH and GHS receptors[3] (Figs 10.3, 10.4). These contrasts are putatively mediated in part by sex-dependent effects of key neurotransmitters, such as GABA, which inhibits periventricular SS release in male but stimulates the same in female rats. Notable concepts are that non-aromatizable androgens, such as 5α-DHT, stimulate periventricular SS and arcuate GHRH gene expression; T treatment maintains GH responses to α2-adrenoreceptor agonism; T or E_2 stimulates pituitary *pit-1* (a somatotrope-specifying transcriptional factor); and T but not E_2 represses GHRH neuronal expression of SSTR-2a. Conversely, estrogens induce CNS genes encoding IGFBP-2, IGF-I receptor and neurotensin (an inducer of SS release) and ventromedial-nuclear GHS receptor in the adult rat; repress the arcuate GH receptor and GHRH peptide in the adult; elevate periventricular SS gene transcripts in the neonatal animal; and increase pituitary GH and SSTR-2 and decrease pituitary SSTR-1 and SSTR-5 expression[1] (Fig. 10.5).

Somatotropes and hypothalamic neurons express ERα and ERβ. Receptor isoforms may mediate distinct effects, such as inhibition and stimulation of NPY/AGRP gene expression by ERα and -β, respectively, and stimulation of neuronal secretion of IGF-I or IGFBP-2 by ERα and -β, respectively. GHRH neurons express ERα, which is putatively inhibitory in the rat and stimulatory in sheep; and periventricular SSergic neurons express AR, which is demonstrably stimulatory. In contrast, ERα is detectable only in 1–5% of arcuate nucleus SSergic neurons where, unlike the periventricular area, E_2 increases SS in the rat and sheep. AR is essentially undetectable in arcuate GHRH neurons. Confirmation of receptor functionality would require cell-selective silencing of AR, ERα and ERβ in periventricular SS neurons, arcuate nucleus GHRH neurons and somatotropes. Models have been developed in an effort to relate basic sex-steroid effects to gender differences in GH secretion. A simplified construct of T and E_2's supervision of GH autofeedback in the adult rat includes amplification by T and blunting by E_2 of arcuate GHRH gene expression and GH-evoked periventricular SS outflow to GHRH neurons and somatotropes.

Estrogenic Effects in Mice

E_2 replacement in aromatase-deficient mice upregulates pituitary genes for GH, *pit-1*, SSTR-2 and the GHRH receptor, but not the GHS receptor. The changes in SSTR-2 and GHRH receptors are opposite to those in the rat, and resemble those inferred indirectly in humans. E_2 action in mice may involve relieving inhibition of GHRH neurons by SS or IGF-I, because: (a) female (but not male) SS knockout animals exhibit increased pituitary expression of GH, GHRH and GHS receptors; (b) neonatal IGF-I depletion in this species activates the same three pituitary genes; and (c) selective knockdown of the GHRH neuronal GHS receptor diminishes GHRH gene expression, GH pulse amplitude and GH/IGF-I concentrations only in female animals.

Testosterone-dependent Regulation of GH Secretion

E_2 does not consistently alter SS gene expression in adult animals. However, T administration in the neonatal rat induces periventricular SS after aromatization to E_2 (Fig. 10.6). Neonatal presence of the testis is

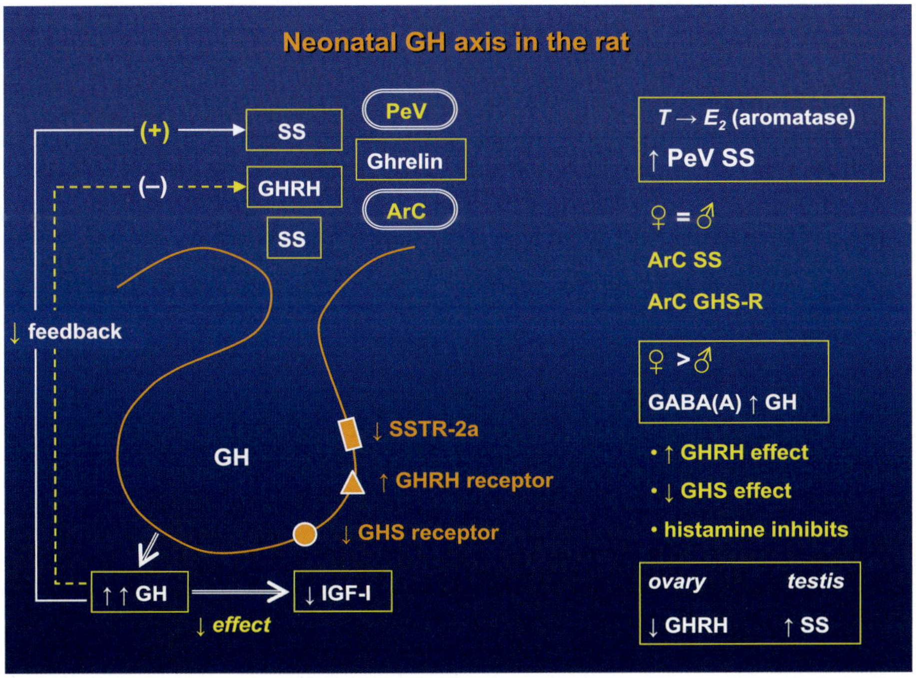

FIGURE 10.6 **Neonatal GH axis compared with adult axis.** Abbreviations: Ach, acetylcholine; Ach-M3, muscarinic-3 receptor; AGRP, Agouti-related peptide; α-MSH, alpha-melanocyte stimulating hormone; β-EPH, beta-endorphin; CRH, corticotropin-releasing hormone; GABA, gamma aminobutyric acid; GH-R, GH receptor; GHS, growth hormone secretagog; 5-HT, serotonin; 5-HT-1D, serotonin receptor 1D; IL-1, interleukin 1; NE, norepinephrine; NPY, neuropeptide Y; orexin A-R, orexin A receptor; POMC, proopiomelanocortin; SSTR, somatostatin receptor; TRH, thyrotropin-stimulating hormone; VIP, vasoactive intestinal peptide.

associated with greater periventricular SS, and neonatal presence of the ovary with less arcuate GHRH, gene expression in the adult.[1,3] In sexually mature rats, E_2 administration increases SS binding in the arcuate nucleus, augments SSTR-2a in somatotropes, and represses brain GHRH peptide and pituitary GHRH receptor gene expression. These effects predict less GHRHergic stimulation and greater constitutive SSergic restraint in the adult female than in the male animal. Reduced GHRH outflow and action in the female rat would result in smaller GH peaks, whereas high-nadir GH concentrations in this sex may reflect estrogenic repression of SSTR-1 and SSTR-5. Conversely, in the adult male the capabilities of AR to mediate both PeV SS synthesis and GH feedback-evoked SS release would explain low interpulse nadir GH concentrations (Fig. 10.4). Concomitantly, T's marked suppression of SSTR-2a expression by GHRH neurons could account for larger GHRH-driven GH pulses in the male rat.[3] Low nadirs and large GH pulses also develop in 5α-DHT-treated castrate female animals.

Human Gender Differences

In humans, puberty is marked by a greater efficacy of GHS in girls than boys. In tall adolescents, pubertal GH responses to GHRH are also greater in the female than in the male. E_2 supplementation potentiates the maximal GH response to GHS and L-dopa, but not to GHRH. Activation of an ERresponse element in the GHS receptor gene may provide a basis for these effects. T supplementation potentiates GHS drive in boys, and heightens recovery of GH secretion in boys given a bolus of GH, and older men administered a pulse of GH or IGF-I. Very low doses of T (25 mg) evoke pulsatile GH secretion in prepubertal boys. Estrogens rather than non-aromatizable androgens stimulate pulsatile GH secretion in humans, which is opposite to findings in the rat. In men, inhibition of aromatase activity reduces, whereas inhibition of 5α-reductase types I and II augments, GH secretion. High doses of injected 5α-DHT suppress GH concentrations in adolescent boys. Thus, it is possible that 5α-DHT or its ERβ-stimulating metabolite, 5α-androstane-3,17 β-diol, evokes SS or represses GHRH secretion. If more generally true, this hypothesis might explain why T administration in the adult rat and dog inhibited GH release by a synthetic GHS. The positive role of E_2 in humans is supported by the inhibitory effects of antiestrogens on GH pulses, and by the amplifying effects of exogenous and endogenous E_2 on GH pulses. A caveat is that earlier antiestrogens retained E_2-agonistic effects.

E_2 supplementation in postmenopausal women attenuates GH and IGF-I-mediated negative feedback. The mechanisms transducing these outcomes are not established. Animal data indicate that estrogens can repress hypothalamic GH receptor expression, and mute signal transducer and transcriptional regulator (STAT-5b) responses to GH in the liver and kidney via activation of a suppressor of cytokine synthesis (SOCS-3), which antagonizes STAT.[4] Induction of SOCS provides an explanation for attenuation of GH drive of hepatic IGF-I synthesis in estrogen-sufficient women. In contrast, none of aging, GH deficiency or obesity significantly diminishes hepatic IGF-I responses to small doses of GH.

Neonatal and Pubertal GH Axis

Neonate

Patterns of human newborn, infantile and prepubertal GH secretion differ principally by way of elevated fasting and GHRH- but attenuated GHS-stimulated GH secretion in the first days of life. In monkeys, the maximal response to GHRH is also on the first day of life, when SS's inhibition of GH release is diminished putatively due to limited SSTR-2 expression. In the cow, pig, sheep and rat, pulsatile GH secretion also peaks on the first postnatal day (Fig. 10.6). This pattern is consistent with higher concentrations of pituitary GHRH than GHS receptors at birth, and a maximal somatotrope-cell number just before birth. GH secretion in neonatal animals is paradoxically stimulated by TSH-releasing hormone, and sparingly inhibited by SS. However, SS and L-dopa are inhibitory in human infants. Indeed, maternal ghrelin or SS infusion by transplacental transfer stimulates or reduces cord-blood GH concentrations in the newborn.

Excitatory amino acids stimulate GH secretion in the immature rat. The immature pituitary gland differs from that in the adult, since in the former exogenous GHS or GHRH can upregulate the GH gene as well as the GHS receptor. The exact basis for this distinction is not established. On the other hand, GH's feedback onto hypothalamic expression of GHRH (negative effect) and SS (positive effect) is impaired in the immature rat. However, negative feedback is demonstrable in young children, indicating that hypothalamic SS peptide and pituitary SS receptors are functional at this time in humans.

The GH pattern is not sexually dimorphic in neonates, unlike LH secretion, which differs markedly in male and female newborns. Premature infants or those with intrauterine growth failure exhibit marked oversecretion of GH in association with lower IGF-I and IGFBP-3 and higher ghrelin concentrations. This pattern suggests GH resistance in liver, and decreased negative feedback. African–American men, but not boys, have higher E_2 levels and secrete more GH than their Caucasian counterparts. The racial difference is not evident in women.

Puberty

Normal puberty is associated with increased GH pulse size. Puberty also imposes an irregular pattern of GH secretion, and elevates IGF-I concentrations. At this time, E_2 concentrations correlate positively with basal (non-pulsatile) GH secretion, and T concentrations correlate directly with the size of GH pulses and IGF-I concentrations. In one study, the absolute increment in IGF-I concentrations induced by a single large dose of GH was also higher in adolescents than in prepubertal children. Unlike in the rodent, in children T and E_2, but not 5α-DHT, stimulate pulsatile GH secretion and elevate IGF-I concentrations. In contrast, 5α-DHT and T (but not E_2) repress plasma GH-binding protein. Childhood growth stunting due to psychosocial stress is marked by reduced pulsatile GH secretion and therefore low fasting IGF-I concentrations, whereas insulinopenic (type 1) diabetes mellitus, malabsorption, anorexia nervosa and untreated inflammatory bowel disease are associated with elevated pulsatile GH secretion and low IGF-I due to tissue resistance to GH.[1,3]

Both GHRH and GHS treatment can accelerate linear growth in children with idiopathic short stature, demonstrating responsivity of corresponding hypothalamo-pituitary receptor-effector pathways. Sequential SS infusion and withdrawal also evokes GH secretion in children, indicating that endogenous GHRH is releasable. During human puberty, responsivity to GHS but not GHRH increases markedly (Fig. 10.4). Nonetheless, the critical requirement for GHRH in growth is affirmed by profound hyposomatotropism in both mice and humans bearing inactivating mutations of the GHRH receptor. In young adults, women exhibit greater responsivity to a low dose of GHRH (reflecting increased potency) and to a high dose of GHS (denoting possible increased efficacy) than men. Enhanced GHRH potency and GHS efficacy may be estrogen-dependent, since each can be mimicked by administering E_2 in postmenopausal women. Gender differences persist in adulthood, wherein women exhibit larger GH pulses and more irregular GH release patterns than men. Less orderly GH secretion in part reflects the capability of E_2 to diminish the negative-feedback effects of GH and IGF-I, which otherwise impose regularity.

Impact of Aging on the GH Axis

In experimental animals and humans, aging, or processes associated with aging, diminish pulsatile GH secretion by 50–95% compared with young adult values. Major age-related inhibitory factors include increased visceral adiposity, decreased sex-steroid availability, and reduced physical fitness. However, the primary biochemical origins of aging are not known. Before mid-life, the age-associated decline in GH secretion over 24 hours is twofold greater in men than women, whether GH production is normalized per unit BMI or per unit maximal oxygen consumption. Ultrasensitive chemiluminescence-based GH assays were required to verify these inferences. Reduction in the size rather than the number of GH secretory bursts accounts for > 85% of the decrease in GH concentrations in aged individuals. In this regard, visceral-fat accumulation, low sex-steroid availability and reduced aerobic capacity all diminish pulsatile more than basal GH secretion.[1,3]

Reduced abundance of hypothalamic SS, galanin, neurotensin, IGF-I, GHRH (but not GHRH neuron number) and GHS receptor gene transcripts, and diminished pituitary SSTR-2/-5, GHRH- and GHS-receptor and GH gene expression with elevated SSTR-1/-3 characterize senescence in the rodent (Fig. 10.7). Possible explanations for pituitary changes are multiple. First, GHRH normally upregulates pituitary SSTR-2, ghrelin, galanin, and receptors for GHRH and GHS (Fig. 10.5), whereas age attenuates both the secretion and actions of GHRH. GHRH tends to downregulate its receptor *in vitro*, and in old (but not young) animals *in vivo*. Second, E_2 normally represses pituitary SSTR-1, but old male animals have low E_2. Third, diminished arcuate NPY and periventricular SS gene transcripts in aged rats could impair putatively feedback-induced renewal of GHRH and GH pulses.[3] Fourth, aging may heighten SS's dose-dependent inhibition of pituitary GH release, an inference requiring confirmation. Fifth, the number and size of immunoreactive GH-expressing anterior pituitary cells decline concurrently with the fall in GH secretion and IGF-I concentrations in aging. Finally, since IGF-I normally induces all 3 of SSTR-1/-2/-5 in somatotropes, reduced IGF-I availability may result in lower SSTR-2/-5. Because exogenous GH-induced hepatic IGF-I synthesis is preserved in older humans, impoverished IGF-I production must reflect decreased hypothalamo-pituitary drive in this setting. In fact, low IGF-I concentrations would be expected to unleash GHRH and GH secretion due to feedback withdrawal, as demonstrated in mice with transgenically silenced IGF-I receptor, GH receptor or IGF-I peptide, in men and women given a GH receptor antagonist to lower systemic IGF-I concentrations, and in rare patients with inactivating mutations of the GH receptor, IGF-I-peptide or IGF-I receptor.

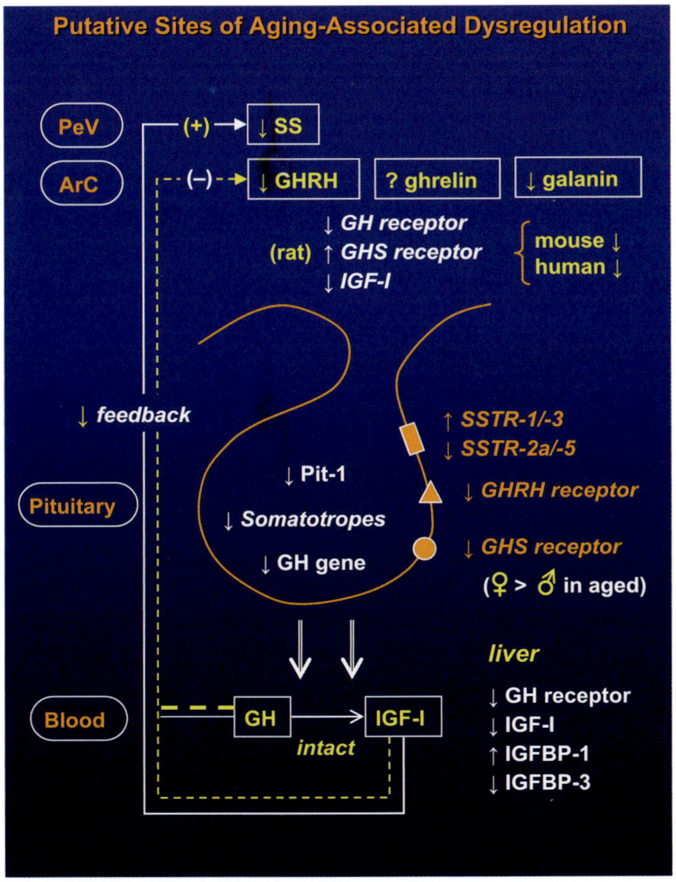

FIGURE 10.7 **Putative sites of aging-associated changes in GH axis compared with young adult system.**

Secretagog Actions in Aging

GH responses to all secretagogs, except insulin-induced hypoglycemia and a triple-secretagog infusion (L-arginine (to putatively limit hypothalamic SS release), GHRH and GHS), are reduced in aging, and overlap with hypopituitarism. Aging impairs GH responses to exercise, deep sleep, L-dopa, clonidine, galanin, morphine, fasting, L-arginine, GHRH, GHS/ghrelin, combined GHRH/GHS, propranolol, and glucagon stimulation.[1,3] Normal acute GH responses to hypoglycemia and a triple secretagog combination in older humans would indicate significant preservation of hypothalamo–pituitary reserve. Intermittent or continuous administration of GHRH for ≥24 h can stimulate pulsatile GH secretion and elevate IGF-I concentrations in elderly women and men, but considerably less markedly in older than young individuals. Ghrelin, synthetic GHS peptides and non-peptidyl GHS analogs administered for one or more days are also able to double pulsatile GH secretion and elevate IGF-I levels in older men and women. Ghrelin administration for 4 days in old mice also enhances appetite and, putatively via GH's stimulation of lipolysis, maintains low body adiposity. However, GH responses in young adults treated identically are not known. Moreover, although strongly effective over 24 hours, whether sustained delivery of combined GHRH and GHS could restore young adult-like GH production in aging individuals has not been assessed.

Negative Feedback Control in Aging

A single pulse of exogenous GH inhibits GH secretion in a dose-dependent fashion in the rat and human.[1,3] Neonatal and aged animals may have attenuated feedback (Figs 10.6, 10.7). Albeit not definitive, older humans may have attenuation of GH and IGF-I's feedback on fasting and GHRH-stimulated GH secretion.

Sex-steroid Deficiency in Aging

Short-term E_2 supplementation in women augments fasting and SS withdrawal-induced pulsatile GH secretion, reduces the inhibitory potency of exogenous SS, and amplifies the stimulatory potencies of GHRH and GHS, although to a lesser degree in post- than premenopausal individuals. Precisely comparable SS, GHRH and GHS dose–response data are not available in young and older men given T. Although T is a potent activator of GH secretion in boys and hypogonadal men, T does not augment maximal GHRH or GHS drive in the human or monkey. Nonetheless, T can relieve negative feedback by IGF-I, and (at least in children) augment GH stimulation by GHS. T concentrations positively predict pulsatile GH responses to submaximal GHRH stimuli, possibly explaining why either physiological or pharmacological T supplementation stimulates pulsatile GH secretion in older men. How increased visceral adiposity and diminished aerobic capacity in older adults interact with deficient sex-steroid drive of GH secretion is not established.

Model of Aging Effects

Decreased GH and, secondarily, IGF-I production in healthy older adults may be most parsimoniously explained by deficient hypothalamic GHRH outflow (release and action). In transgenic mice expressing fluorescently tagged GHRH neurons, GHRH axons contain large autophagic vesicles. Low GHRH would predict reduced pituitary galanin, ghrelin, GH, SSTR-2, GHRH- and GHS-receptor, and elevated SSTR-1 expression, as observed in aged rodents, and impaired stimulation of GH secretion by galanin, GHRH and ghrelin, as noted in elderly humans.[3] The basis for reportedly decreased pituitary SSTR-5 in aging is less clear. In fact, in the pig GHRH and GHS downregulate SSTR-5, which is the opposite of the change expected in aging when GHRH and GHS decline. The notion of impoverished GHRH outflow or action is supported experimentally by: (a) impaired rebound-like GH release after cessation of an i.v. SS infusion, since this paradigm putatively elicits GHRH release; (b) increased susceptibility to inhibition of fasting GH pulsatility by a GHRH receptor antagonist; and (c) fewer functional somatotropes. However, definitive proof would require documenting reduced hypothalamo-pituitary portal-venous GHRH secretion, inasmuch as GHRH action is itself decreased in aging. In the rat, prolonged caloric restriction may attenuate the age-related fall in GHRH receptor signaling. The precise roles of decreased norepinephrine, NPY, SS, galanin, IGF-I, GH and other signals in aging-associated hyposomatotropism are not known. Occasional *in vitro* studies also point to greater hypothalamic SS release and greater somatotrope inhibitory potency of SS-28 on GH release in aged rats.

Other Contributions to Hyposomatotropism in Aging

Low pulsatile GH secretion in aged individuals may be reversed acutely by simultaneous administration of GHRH, GHS and L-arginine. More chronically, low GH input is reversed partially (but not normalized) by aerobic exercise, short-term fasting or sex-steroid supplementation.[1,3] When the effects of sex-steroid milieu and visceral adiposity are controlled for experimentally in women, about 50% of the age-associated fall in pulsatile GH secretion remains unexplained. One factor may be reduced aerobic capacity, inasmuch as 18 months of endurance training doubles 24-hour GH secretion in young women. Among other considerations, the facilitative effect of endurance training may reflect increased insulin sensitivity, decreased insulin and IGF-I concentrations, reduced visceral adiposity, and greater lean body mass.

Influence of Obesity

Clinical Data

In multivariate analyses, adiposity emerges as a dominant negative correlate of 24-h pulsatile GH secretion. Excessive visceral and intermuscular fat may also explain some of the negative effects of aging, hypogonadism and poor physical fitness on the GH axis in adults. In children, subcutaneous rather than visceral fat is a strong negative correlate of GH secretion. Clinical studies indicate that age, sex-steroid deficiency and visceral adiposity are jointly negative covariates of GH responses to GHRH and GHRP-2. When E_2 concentrations are controlled experimentally in healthy women by combined administration of leuprolide (a GnRH agonist used to downregulate the gonadal axis) and E_2 or placebo, abdominal visceral fat is the principal negative covariate of pulsatile GH secretion induced by GHRH infusion. In the same sex-steroid clamp, age and E_2 status are primary determinants of pulsatile GH responses to GHRP-2. Comparable studies in healthy young and older men given leuprolide plus T or placebo addback show that visceral fat, IGF-I and IGFBP-3 together explain 60% of the interindividual variability in GHRH/GHRP-2 synergy, whereas abdominal visceral fat accounts for about 40% of the variability in individual GHRH and GHRP-2 efficacy. In morbidly obese subjects, a protocol of sequential SS infusion and withdrawal is less effective in initially suppressing and then evoking GH release, pointing to reduced GHRH release and/or diminished SS and GHRH action.[1,3] Massive weight loss is required to normalize pulsatile GH secretion, IGF-I, IGFBP-3 and sex-steroid concentrations.

GH Dynamics in Obesity

Obesity is associated with a shorter GH half-life, larger GH distribution volume, lower-amplitude GH secretory bursts, and less regular (less orderly) GH secretion patterns. This complex neurosecretory phenotype resembles that of aging, except that insulin-induced hypoglycemia evokes normal GH release in aging but not obesity.[3] The neuroendocrine pathophysiology in obesity seems to comprise attenuation of GHRH and ghrelin feedforward and GH/IGF-I feedback (inferred by more disorderly GH secretion patterns). In this regard, obese male rats exhibit reduced GHRH gene expression. Whereas increased SSergic restraint might otherwise explain impaired GH responses to GHRH and/or GHS in overweight adults, exogenous SS and its analogs enforce more regular (rather than less regular) patterns of GH secretion. Serum ghrelin concentrations decrease in proportion to total and abdominal visceral fat, and increase with weight loss. The mechanisms mediating these changes appear to be complex. In the rat, the insulin-antagonizing adipokine, resistin, can inhibit GHS-R1a expression in pituitary cells. GH-secretory responses to somatostatin withdrawal and to maximally stimulating doses of exogenous ghrelin/GHS are attenuated in obesity.[3] In murine models of obesity, pituitary GHRH and GHS receptors and GH gene transcripts are reduced. In the obese rat, impoverished responses to GHS could reflect GHRH deficiency, since GHRH normally represses pituitary SSTR-5 expression, maintains pituitary receptors for GHRH and GHS, induces the pituitary ghrelin gene, and synergizes with GHS in this species.

Implications for Clinical Strategies

Principles of neuroendocrinology have relevance to the understanding, diagnosis and treatment of pathophysiology. This precept can be readily illustrated.

Anabolic Support in Catabolic States

Prolonged catabolic states, such as starvation, metastatic cancer, severe thyrotoxicosis, hypercortisolism, poorly controlled type 1 diabetes mellitus, granulomatous processes, HIV-associated wasting, diffuse surface burns, multiple trauma and major surgery, are associated with reduced muscle and bone mass, decreased immune competence, hypogonadism, diminished pulsatile GH secretion, low IGF-I levels, and impaired insulin action. The catabolic state of ultra-endurance exercise also reduces IGF-I concentrations. Tissue breakdown reflects not only excessive catabolic factors – namely, glucocorticoids, cytokines, adipokines, glucagon and epinephrine – but also impoverished anabolic drive by GH, androgens, IGF-I and insulin. From a homeostatic perspective, tissue repair and organ growth require an optimal quotient of anabolic to catabolic factors. Accordingly, no single anabolic (or anticatabolic) agent is likely to suffice in treating marked protein wasting. In fact, neither GH nor IGF-I administration fully overcomes catabolism associated with prolonged fasting, glucocorticoid excess or gonadal-steroid deficiency. Together with hyperalimentation, combined GH and IGF-I infusion may maintain positive nitrogen balance in critically ill patients. A high dose of GH, however, increases mortality in critical-care settings. The latter issue has prompted pilot studies of the combined effects of lower doses of GH and androgen (men) and of GH and estrogen (women), as well as single and combined GHRH and GHS infusions. Definitive investigations are needed in larger cohorts.

Combination of GH/IGF-I with Sex Steroids

An important mechanistic concept is that prolonged exposure to endotoxins, immunological peptides and inflammatory cytokines (such as TNFα, interferon-γ, IL-1β and IL-6) suppress GH secretion and promote tissue resistance to the actions of GH and insulin. The degree of tissue resistance to IGF-I-driven anabolism is less clear. However, cytokines induce suppressors of their own synthesis (SOCS), which antagonize GH-induced STAT5b (signal transducer and transcriptional activator), a prominent mediator of the anabolic effects of GH. SOCS also mediates autoinhibitory effects of GH and ERα. New therapeutic avenues therefore could include tissue-selective inhibitors of SOCS and combinatorial anabolic agents such as androgen and IGF-I, with or without anticatabolic agents, such as glucocorticoid antagonists or insulin-sensitizing drugs. A foreseeable limitation is that local production of IGF-I in certain tissues, like skeletal muscle and bone, is more strongly anabolic than systemic delivery of IGF-I. One study used a triple infusion of TSH-releasing hormone, GnRH and GHRP-2 in men with extended critical illness. More comprehensive understanding is required of local tissue-specific *vis-à-vis* systemic anabolic and catabolic mechanisms, as well as of the risks of promoting neoplasia or metabolic abnormalities during long-term trophic-hormone treatment.

Tumoral GH Secretion

Excessive GH action causes gigantism prior to puberty, and enlargement of visceral organs and acral tissues (acromegaly) in adults. Three hallmarks of tumoral GH secretion are elevated IGF-I concentrations, increased interpulse nadir GH concentrations (due to increased non-pulsatile basal GH secretion), and greater irregularity (reduced orderliness) of GH secretory patterns. Successful surgical removal of the pituitary microadenoma normalizes not only mean GH concentrations but also basal GH secretion and pattern regularity. Basal secretion is quantifiable by deconvolution methods, and pattern orderliness by the approximate entropy (ApEn) statistic. GH ApEn, which is estimable from as few as 12 consecutive samples, achieves $> 90\%$ sensitivity and $> 90\%$ specificity in discriminating irregular GH-release patterns due to GH-secreting adenomas from normal. Irregular patterns are reflections of impaired physiological feedback control. Less orderly GH release patterns also typify organic GH deficiency, obesity, Turner syndrome, chronic renal failure, hepatic disease, starvation and aging. Orderly GH secretion is enforced by somatostatin administration in healthy fasting adults but not in patients with GH-secreting adenomas, consistent with tumoral autonomy. Although one patient was described with a GH-secreting pituitary tumor deficient in SSTR-5, tumoral expression of SSTR-2 best predicts suppression of GH secretion by the SSTR-2/SSTR-5 agonist octreotide. Two variants of SSTR-5 may exist in human tumors. GH-secreting tumors are also less sensitive to feedback by IGF-I, possibly in part reflecting the fact that IGF-I receptor expression is reduced in tumor tissue.

Glucocorticoid Use

Excessive glucocorticoid exposure, whether due to endogenous factors (e.g., Cushing's disease) or exogenous agents (e.g., high-dose prednisone use in chronic asthma), suppresses fasting GH secretion and GH responses to all known secretagogs, including insulin-induced hypoglycemia. Prevailing inferences are that pathological amounts of glucocorticoids induce periventricular SS release, decrease GH receptors in the periventricular and arcuate nuclei, and repress hypothalamo-pituitary galanin expression, which would reduce arcuate GHRH gene transcripts and impair GHRH action. How the ghrelin system is affected by high-dose glucocorticoids is less clear, except that serum ghrelin concentrations are increased in adrenalectomy and decreased in glucocorticoid-induced obesity.

Pharmacological doses of glucocorticoids also antagonize the synthesis of IGF-I in bone, and the anabolic actions of GH and IGF-I on bone, liver and muscle. Accordingly, consequences of chronic pharmacological cortisol excess include short stature in children, and sarcopenia, osteopenia and hyperinsulinemia in adults. GH therapy partially ameliorates these sequelae. Data in the rat suggest that GHS administration may be effective in partially antagonizing glucocorticoid-induced hyposomatotropism and protein catabolism. In contradistinction, physiological amounts of glucocorticoids stimulate and maintain normal GH secretion, inferably by blocking SS and IGF-I's inhibition of somatotrope GH secretion and maintaining expression of pituitary GH-peptide and GHRH receptor genes.

Thyroid Disease

The T3 receptor is expressed in somatotropes, where it induces GH gene expression. In addition, the pituitary gland can synthesize T3 from T4 via the type 2 seleno-deiodinase enzyme. Excessive tissue actions of T3 and T4 define clinical hyperthyroidism, which accelerates catabolism of muscle and bone proteins. Biochemical features include increased GH, IGFBP-1 and IGF-I concentrations in proportion to T3 concentrations in mild disease (but reduced GH and IGF-I in more severe illness); insulin resistance; decreased suppression of GH secretion by glucose; diminished actions of GHRH, GHS, pyridostigmine and galanin; lower ghrelin levels; and irregular patterns of GH secretion. Somatotrope secretory vesicles are depleted on electron microscopy, and GHRH/GHS synergy declines, but GH feedback is preserved. In the male rat, hyperthyroidism elevates GH concentrations and decreases hypothalamic GHRH transcripts, the latter putatively via negative feedback.

Hypothyroidism in neonatal and adult animals reduces GH secretion, pituitary GH and GHRH receptor and hypothalamic SS gene transcripts, while increasing GHRH transcripts due to decreased GH feedback. Similar changes unfold in the hypothyroid infant rat. GHRH responses are attenuated *in vivo* and *in vitro*. In hypothyroid humans, serum GH and IGF-I concentrations are decreased, and stimulation of GH release by GHRH, L-arginine and pyridostigmine is impaired. Thyroxine treatment restores IGF-I and GH production. In primary hypothyroidism, isosexual precocious puberty occurs in occasional children via unknown CNS mechanisms.

Pregnancy and Lactation

GH is important to maintain anabolism during pregnancy and lactation. In humans, a placental GH variant (chorionic somatomammotropin) termed GH-V is distinguishable from pituitary 20-kDa and 22-kDa GH-N. Non-pulsatile secretion of GH-V is detectable using highly specific two-site immunoradiometric assays. Glycosylated GH-V is a 25-kDa protein. Maternal concentrations of GH-V/IGF-I and ghrelin respectively peak at about 36 and 18 weeks' gestation, decline with maternal obesity, and in some studies correlate with maternal and umbilical-cord serum IGF-I concentrations as well as newborn weight. In contrast, maternal GH-N concentrations fall in gestation, putatively due to negative feedback by elevated GH-V and IGF-I levels. Maternal GH and IGF-I fall postpartum, reaching a nadir at 3 days. Stimulation of GH release by GHS is impaired in hyperprolactinemic nursing mothers.

In the rat, pituitary and serum GH concentrations and hypothalamo–pituitary ghrelin gene expression increase in pregnancy. GH remains elevated during lactation, more so in non-suckling than suckling dams, along with serum ghrelin concentrations and GHS receptor transcripts in the brain and pituitary gland. Blood IGF-I concentrations and liver GH receptor expression decline as GH concentrations rise in pregnancy, predicting GH resistance. Hypothalamic SS transcripts are said to be unchanged in pregnancy and lactation, but better *in situ* data are needed. Milk-related and unknown maternal factors may contribute to suckling-associated GH secretion in the neonatal rat. In the cow, maximal GHS stimulation of GH concentrations occurs in the lactating mother and suckling calf. Lactation in this species is associated with attenuation of GH's inhibition by exogenous SS and its stimulation by GHRH until 1 month postpartum. In the pig, ghrelin concentrations are maximal during gestation and decline in the lactating sow. Daily ghrelin injections in the nursing rat can augment food intake, GH secretion and milk yield, suggesting a role for endogenous GHS in these processes.

GH Receptor Variations

Deletion or truncation of the GH receptor gene results in profound reduction of IGF-I concentrations, short stature, visceral adiposity, and delayed onset of puberty (so-called Laron syndrome). Polymorphisms of the human GH receptor exist, which are said to enhance or restrict GH action at target tissues. However, relevance remains to be established in population-based epidemiologic studies. In a meta-analysis of an exon 3- deleted GH receptor polymorphism, this allelic variant was associated with greater pretreatment height in GH-deficient but not normal children ,and a 0.5-cm larger increment in height in the first year of GH treatment.

SUMMARY

Dynamic characteristics of the GH axis are determined in part by developmental age, including fetal, neonatal, childhood, pubertal, young-adult and older-adult stages of life.[1,3] Dynamics across the lifetime arise from multifold regulation of GH and IGF-I output by hypothalamic, pituitary and systemic factors discussed here. Additional incompletely delineated modulatory signals are associated with infancy, senescence, adiposity, nutritional status, lactate, physical fitness, inflammation, non-steroidal gonadal factors, and environmental temperature. In addition, control processes exist that stipulate organ and cell-selective GH signaling, IGF-I generation, IGF-I action, IGF-binding proteins and their proteases.

Acknowledgments

We thank Donna Scott for support in manuscript preparation, and Ashley Bryant for graphics assistance. This review was supported in part via the Center for Translational Science Activities (CTSA) Grant Number 1 UL 1 RR024150 to Mayo Clinic and Foundation from the National Center for Research Resources (Rockville, MD), and R01 NIA AG19695, and AG29362 from the National Institutes of Health (Bethesda, MD).

References

1. Giustina A, Veldhuis JD. Pathophysiology of the neuroregulation of growth hormone secretion in experimental animals and the human. *Endocrine Rev.* 1998;19:717–797.
2. Veldhuis JD, Roemmich JN, Richmond EJ, Rogol AD, Lovejoy JC, Sheffield-Moore M, Mauras N, Bowers CY. Endocrine control of body composition in infancy, childhood and puberty. *Endocrine Rev.* 2005;26:114–146.
3. Veldhuis JD, Roemmich JN, Richmond EJ, Bowers CY. Somatotropic and gonadotropic axes linkages in infancy, childhood, and the puberty–adult transition. *Endocrine Rev.* 2006;27:101–140.
4. Ho KK, Gibney J, Johannsson G, Wolthers T. Regulating of growth hormone sensitivity by sex steroids: implications for therapy. *Front Hormone Res.* 2006;35:115–128.
5. Moller N, Jorgensen JO. Effects of growth hormone on glucose, lipid, and protein metabolism in human subjects. *Endocrine Rev.* 2009;30:152–177.
6. Laursen T, Jorgensen JO, Christiansen JS. The management of adult growth hormone deficiency syndrome. *Expert Opin Pharmacother.* 2008;9:2435–2450.
7. Kojima M, Kangawa K. Ghrelin: structure and function. *Physiol Rev.* 2005;85:495–522.
8. Frohman LA. New insights into the regulation of somatotrope function using genetic and transgenic models. *Metabolism.* 1996;45:1–3.
9. Howard AD, Scott D, Feighner SD, et al. A receptor in pituitary and hypothalamus that functions in growth hormone release. *Science.* 1996;273:974–977.
10. Kojima M, Hosoda H, Date Y, et al. Ghrelin is a growth hormone-releasing acylated peptide from stomach. *Nature.* 1999;402: 656–660.
11. Bowers CY, Chang J, Momany F, et al. Effect of the enkephalins and enkaphlin analogs on release of pituitary hormones *in vitro.* In: MacIntyre I, Szelke M, eds. *Molecular Endocrinology.* Amsterdam, The Netherlands: Elsevier/North-Holland Biomedical Press; 1977:287–292.

CHAPTER

11

Neuroendocrine Regulation of Development, Growth and Metabolism – Thyroid

R. Thomas Zoeller

Biology Department, University of Massachusetts, Amherst, MA, USA

OUTLINE

Summary

Thyroid hormone is essential for normal development and for adult health. The regulation of blood levels of thyroid hormone is complex, requiring the activities of diverse proteins that are being uncovered by molecular and genetic studies. These studies are also revealing how the delivery of thyroid hormone to target tissues in the correct amount and at the correct time is controlled. Understanding these mechanisms will undoubtedly lead to new insights into the cause of disease. In addition, these studies are providing important new insight into the molecular mechanisms of thyroid hormone action. The goal of this chapter is to provide a brief overview of the thyroid system, and an entryway into this exciting field.

INTRODUCTION

The goal of this chapter is to provide an entryway into the modern field of thyroidology, highlighting new ideas and new approaches for its study. Classically, thyroid hormone is recognized for its role in metabolism.[1] In fact, this perspective was so pervasive at one time that the brain was not considered to be sensitive to thyroid hormone because it did not increase oxygen consumption in response to thyroid hormone administration.[2] We now know that thyroid hormone exerts very specific effects not only on metabolism, but also on specific aspects of energy regulation. In addition, thyroid hormone exerts specific actions on development and physiology of the brain and of many – perhaps all – organs and organ systems. We also know that thyroid hormone affects these developmental and physiological processes largely by acting on nuclear proteins that regulate various elements of gene expression, and by acting through potentially important non-genomic mechanisms. The classical view of the neuroendocrine control of thyroid hormone action has also improved, as it has become clear that the reference range for thyroid hormones in the human population is much broader than the range over which individual hormone levels normally vary,[3] and because of the recognition that the regulation of thyroid hormone action in tissues may be independent of circulating levels of thyroid hormone itself.[4] Finally, the use of transgenic rodent models has increased the toolkit for researchers to investigate the role of thyroid hormone in many physiological and developmental processes, as well as the mechanisms regulating thyroid function and thyroid hormone action. The combination of new information and new approaches makes the study of thyroid endocrinology more exciting and relevant than ever before.

Handbook of Neuroendocrinology, DOI: 10.1016/B978-0-12-375097-6.10011-3

THYROID PHYSIOLOGY

The anatomy of the thyroid gland itself is an important aspect of the function of the system, and requires some consideration (reviewed in detail by Zoeller *et al.*[5]). The mature human and rodent thyroid gland consists of two elongated oval lobes, one on each side of the trachea, joined near their posterior poles by a thin isthmus crossing the trachea ventrally (see Braverman and Utiger[6]). The lobes are richly vascularized and made up of groups of fluid ("colloid")-filled spheres, or follicles, often visible macroscopically. Microscopically, the gland is made up of follicles of varying sizes, surrounded by a fibrous connective tissue capsule and supported by vascular interfollicular connective tissues. Simple cuboidal cells with distinct outlines, large spherical nuclei and clear cytoplasm line the follicles. The height of the epithelial cells and the amount and staining quality of the internal fluid or colloid are sometimes used as indicators of secretory activity.[7] Because of developmentally associated dynamic changes in the histological structure of the mammalian thyroid gland, experiments in which thyroid histology is to be evaluated must be timed to ensure that normal variation in structure does not confound experimental findings.

Thyroid hormone is synthesized in a manner that is very different from that of other hormones that employ similar signaling pathways (e.g., steroid hormones)[8] (Fig. 11.1). Thyroxine (tetraiodothyronine, T_4) is a small bi-phenolic compound derived from separate tyrosine residues connected by an ether bond, on a large protein – thyroglobulin (Tg).[9] Thyroglobulin, in turn, is synthesized on ribosomes and transported (by exocytosis) to the colloid. Thyroglobulin is then iodinated at specific tyrosine residues as it is being exocytosed, before being stored in large quantities in the colloid until it is required for the synthesis of hormone. At that time, colloid droplets are ingested by thyroid follicle cells by endocytosis, and transported to the side of the cell that is bathed in interstitial fluid. On the way through the cell, the endocytotic vesicle fuses with a lysosome. Enzymes within the lysosome then digest the iodinated Tg and liberate T_4 and T_3.

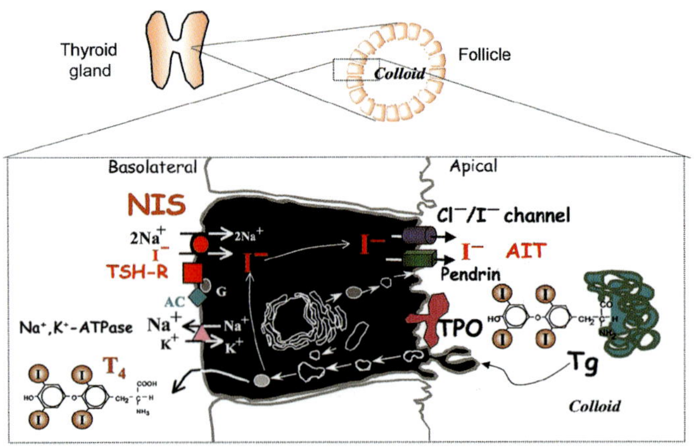

FIGURE 11.1 **Thyroid hormone synthesis.** Thyrocytes surround a fluid-filled "colloid," and this anatomical arrangement is required for thyroid hormone synthesis. Iodine is taken up into the thyrocyte, where it is oxidized and attached to thyroglobulin (TG) at the interface of the apical membrane and the colloid. Thyroid-stimulating hormone ("thyrotropin," TSH) binds to a G-protein coupled receptor, which activates adenylate cyclase (AC) and accounts for activating the cellular machinery responsible for thyroid hormone synthesis and release.[5] *Reprinted from Dohan O. The sodium/iodide symporter.* Endocrine Rev. *2003:24:48–77, with permission. Copyright 2003, The Endocrine Society.*

These synthetic events are uniformly stimulated by thyrotropin (thyroid-stimulating hormone, or TSH), including synthesis and release of thyroid hormones, uptake of iodine, and even cell hypertrophy and hyperplasia.[10] When TSH binds to its receptor on the thyroid cell, the intracellular domains of the receptor activate several guanine nucleotide-binding (G) proteins.[11] Cyclic AMP-dependent protein kinases (PKA) mediate many of the actions of the activated TSH receptor, including its mitogenic action.[12] The PKA signaling pathway activates CREM (cAMP-response element modulator) and CREB (cAMP-response element binding protein), which interact with specific regulatory regions on specific genes. The TSH receptor also activates protein kinase C (PKC) and diacylglycerol (DAG).[10] Thus, activation of the TSH receptor appears to fully control the actions of the thyroid gland.

There are several physiological implications of the anatomy of the thyroid gland and the chemistry of thyroid hormone. First, the observation that iodinated thyroglobulin is stored in large quantities in the adult thyroid gland indicates that there is a great deal of resilience in the system. That is, under conditions where thyroid hormone synthesis is impaired, symptoms would not manifest for a significant amount of time. However, thyroglobulin storage is age-dependent and newborns have little to no storage of thyroglobulin, indicating that there is little if any resilience to impairments in thyroid hormone synthesis at this age.[13,14] Second, because thyroid hormones (T_4 and T_3) are poorly soluble in aqueous media, binding proteins in serum must carry the majority of hormone in the blood. Moreover, these hormones require active transport across membranes. Finally, the long half-life of T_4 in serum (7–10 days in humans, a single day in rodents) results in normal T_4 levels that are higher than almost all other hormones.

THYROID HORMONE AND METABOLISM

Thyroid hormone is known to increase metabolic rate, as measured by oxygen consumption.[1] Moreover, the hyperthyroid state is known to produce symptoms, such as high blood pressure and tachycardia, that overlap with those of the hyperadrenergic state.[15] This overlap in symptoms associated with these two states is in part caused by a direct interaction between thyroid hormone and adrenalin action.

This interaction, especially in the context of metabolism, has been most extensively evaluated in brown fat. Golozoubova *et al.*[16] found that mice devoid of thyroid hormone receptors (TRα1 and TRβ isoforms) exhibited a slightly lower body temperature but were able to rapidly increase metabolic rate when exposed to cold temperatures. However, these animals were cold-intolerant because heat production did not result from the increased metabolic rate, despite significant recruitment of brown fat and the presence of a functional uncoupling protein-1. Isolated brown adipocytes from these animals were 10-fold less sensitive to adrenergic stimulation, indicating that an important role of thyroid hormone in heat production of brown fat is to maintain this sensitivity.

Importantly, brown fat can be activated through thyroid hormone signaling without changes in the levels of circulating thyroid hormones. Specifically, Watanabe *et al.*[4] showed that the bile acid cholic acid prevents weight gain in mice fed a high-fat diet, and that this effect is attributable to the activation of thyroid hormone signaling in brown fat. Moreover, they showed that cholic acid induced the expression and activation of the type 2 deiodinase in brown fat, which converts T_4 to the biologically active T_3, by activating a G-protein coupled receptor that increases cyclic AMP. This study not only demonstrates that thyroid hormone plays an important role in body weight regulation, but also shows that thyroid hormone action can be regulated by humoral factors in specific tissues in a manner that is independent of changes in circulating levels of thyroid hormone. The control of seasonal changes in fat deposition in birds may also represent an example whereby the induction of deiodinase in specific brain regions enhances thyroid hormone action in a specific brain region that controls metabolic functions in the absence of changes in serum thyroid hormone levels.[17]

These studies show that there are tissue- and cell-intrinsic mechanisms that control thyroid hormone action independent of changes in circulating levels of thyroid hormone itself. An important example is that of mammalian brain development. Thyroid hormone appears to control fate specification[18] and migration[19–21] of early neuroblasts in the cerebral cortex, but does not appear to influence their proliferation.[22] In contrast, TH clearly controls proliferation and migration of granule cells[23] in the cerebellum. Moreover, TH controls synaptogenesis in the cortex and hippocampus,[24,25] development of specific elements of the cochlea,[26] differentiation of specific photoreceptors in the retina,[27] the balance of production of oligodendrocytes and astrocytes in bridging white matter,[28] and many other specific features of brain development. However, TH does not produce all of these effects simultaneously; rather, these events occur in an orderly sequence. Thus, a key feature of TH action is the temporal sequence of events it supports, and the temporal nature of TH action must be controlled by tissue- and cell-intrinsic mechanisms.

There are two major reasons that it has been difficult to characterize the role of specific mechanisms controlling cells and tissues. First, there is a large number of potential mechanisms that may control tissue-level sensitivity to TH, and the relative importance of these may differ across tissues or developmental age[29] (Fig. 11.2). Second, it is not always possible to identify an unambiguous marker of TH action that can be used to test the efficacy of these mechanisms across tissues and

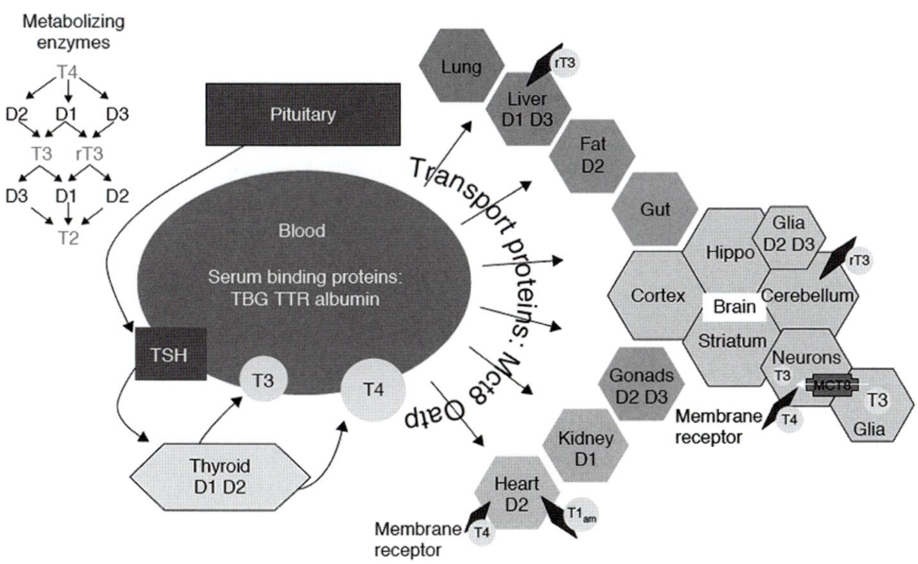

FIGURE 11.2 Relationship between serum levels of thyroid hormones and the regulation of thyroid hormone action in tissues. Serum thyroid hormone levels remain relative stable in individuals,[3] and it is well known that thyroid hormone transport and metabolism represent important mechanisms by which thyroid hormone action is controlled. Moreover, there are many studies indicating a tight relationship between serum thyroid hormone and downstream effects.[29] However, new studies are showing that metabolism of thyroid hormones and/or uptake into tissues can control thyroid hormone actions in tissues independent of serum hormone levels. *Reprinted from Gilbert ME, Zoeller RT. Thyroid hormones—impact on the developing brain: possible Mechanisms of neurotoxicity. In: Harry G, Tilson HA, eds.* Neurotoxicology. *New York: Informa Healthcare.*

developmental time. Specifically, we know that T_3 receptors (TRs) are ligand-dependent transcription factors that regulate gene expression,[30] but this regulation is highly complex. TRs do not regulate the same genes in all cells (see, for example, Iniguez *et al.*[31]), nor do they regulate the same gene in the same cell over developmental time.[32] Therefore, it would be useful to have a direct measure of intracellular "free" T_3 that is independent of the complexity of TH-regulated gene expression *per se*.

This issue has been addressed recently, and, somewhat surprisingly, type 3 deiodinase (D3) appears to be an important modulator of T_3 action in the developing mouse brain in a manner that is independent of serum TH levels.[33] D3 is a deiodinase enzyme that inactivates TH by converting T_4 to reverse T_3 (rT_3), or T_3 to T_2. This enzyme is expressed predominantly in neurons in the brain,[34] and is believed to protect the neuron from supraphysiological levels of T_3.[35] In addition, there is good evidence that the separate conversion of T_4 to T_3 and from T_3 to T_2 occurs in separate cells[36] because D2, which converts T_4 to the more biologically active T_3 in the brain, is expressed in glial cells. Thus, the picture emerging is that glial cells produce T_3, which must be transported to neurons for their use. Moreover, D3 is expressed in neurons to control their exposure to T_3. This concept has been elegantly evaluated both *in vitro* and *in vivo* by Freitas et al.[37] who provide strong support for this hypothesis. Previously, Hernandez and colleagues generated a mouse model with a targeted deletion of D3[38] and showed that D3 is critical for maturation and proper functioning of the hypothalamic–pituitary–thyroid axis, presumably by controlling T_3 availability in neurons.[39] However, to test this hypothesis empirically, they teamed up with another group of thyroid specialists who had generated a transgenic mouse model allowing them to track neurons that exhibit changes in bioavailable T_3.[40]

The strategy employed to test this required the use of a mouse model that was created by inserting a transgene that responds to a small amount of T_3 with the expression of β-galactosidase. The way this system works in general is as follows, as originally described by Mata de Urquiza *et al.*[41] First, the GAL4 DNA-binding domain was fused to the TRα1 ligand binding domain. This fusion protein is a T_3-inducible GAL4 transcription faction, and binds to an upstream activation site (UAShsp68) in response to T_3. The transgene inserted into this mouse line contains two UAS sites; one is in front of the GAL4/TRα1 coding region, while the other is in front of a lacZ (β-galactosidase) gene. There is a minimal (tk) promoter in front of both the GAL4/TRα1 and the lacZ to provide a small amount of fusion protein in the basal state. When T_3 binds to this basally-expressed GAL4/TRα1 fusion protein, it activates self-expression as well as expression of β-galactosidase. This system was aptly named "Feedback Inducible Nuclear-Driven reporter for T_3" (FINDT3).[41] When FINDT3 mice were made D3-deficient, the expression pattern of β-galactosidase in the brain revealed those cells that have too much T_3, and in this manner the investigators conclusively showed that D3 is an essential enzyme for normal thyroid hormone signaling in the developing brain.

These findings provide a very clear indication that D3 is an important regulator of TH action not only during development, but also throughout life. Moreover, they employ a novel and innovative approach to study one of several proteins that are likely to be key in regulating T_3 availability to individual cells in the developing and adult brain, and in other tissues. As the authors point out:

> Given that certain chemical compounds and endocrine disruptors might influence D3 activity by either inhibiting the enzyme or altering the epigenetic regulation of the D3 gene, our results raise the possibility that environmental influences could alter brain function in the absence of significant changes of serum TH levels.

Clearly, understanding the mechanisms by which TH influences development requires understanding the mechanisms by which cells and tissues regulate their own exposure to T_3. Likewise, understanding how disease or environmental factors can influence TH action will require an understanding of these mechanisms, and D3 will likely be an important enzyme to understand.

Thyroid Hormone Receptors

Thyroid hormone exerts its effect on development and physiology perhaps primarily by interacting with specific nuclear proteins, the thyroid hormone receptors (TRs).[42,43] TRs are members of the superfamily of ligand-dependent transcription factors,[44–46] which include receptors for steroids (estrogen, androgen, corticoids) and thyroid hormones, retinoids, and vitamin D.[44,45,47] Two separate genes encode the TRs, designated α-c-*erb*A and β-c-*erb*A.[48,49] Together, these two genes produce four known functional TRs: TR1α, TRβ1, TRβ2, and TRβ3.[46,50] The gene encoding TRα has 10 exons; TRα1 is composed of exons 1–9. A second major product, TRα2, is generated by the addition of a long c-terminal domain (exon 10) that disrupts the ligand-binding domain of the TR (see review by Flamant and Samarut.[51] Thus, TRα2 does not bind to thyroid hormone and is generally not considered to be a *bona fide* TR. Moreover, there is an internal promoter that drives the transcription of two additional short forms of the TRα gene.[52] These short forms, designated TRΔα1 and TRΔα2, are encoded by exons 8–9 and 8–10, respectively. These proteins are able to bind to thyroid hormone, but do not bind to DNA. In contrast, there are three promoters that drive the expression of the three functional TRs from the TRβ gene.[51] In addition, the TRβ3 transcript is differentially spliced to produce a TRΔβ3 isoform. Again, this small product of the TRβ gene binds to thyroid hormone, but not to DNA.

Triiodothyronine binds to these four *bona fide* TRs with equal, or nearly equal affinity,[53–55] although it binds with about 50-fold greater affinity than does T_4 for each receptor. Thus, T_3 is considered to be the physiologically relevant hormone on the receptor. Although T_3 binds to the various TRs with nearly equal affinity, there is good evidence that the different receptors vary enough to bind selectively to various TR analogs. For example, desethylamiodarone is a non-competitive inhibitor of T_3 binding to the TRβ1, but a competitive inhibitor of T_3 binding to the TRα1.[56,57] In addition, the T_3 analog, 3,5,3′-triodothyroacetic acid (triac) has a higher affinity for TRβ1 than does T_3 but the same affinity for TRα1 as does T_3.[58–60] A synthetic thyromimetic compound, GC-1, binds to TRβ-1 with the same affinity as T_3, but binds to TRα-1 with 10-fold lower affinity.[60–62] Finally, another synthetic chemical, NH-3, is a TR antagonist, but its isoform selectivity is not fully characterized.[63–65] An important emerging literature focuses on the differential ability of these TH analogs to activate the various receptors on native promoters. For example, Messier and Langlois[66] showed that triac exerts a more potent effect on TRβ1- and TRβ2-mediated transcription depending on the sequence of the TR response element (TRE). This issue is important because it is likely that xenobiotics will be found to exert these kinds of complex effects on TR activation.

Expression and Regulation of the Alpha TRs

Thyroid hormone exerts pleiotropic effects on development and physiology. One mechanism by which thyroid hormone can exert such different, but specific, effects is for these actions to be mediated by different receptor isoforms. The differential expression of TRα and TRβ products is especially robust in the developing brain. By far, the work of Bradley and colleagues remains the most comprehensive mapping to date of TR isoforms in the developing rat brain.[67] Subsequent studies have provided insight into the expression of TR isoforms discovered since Bradley's work (e.g., TRβ3). The earliest embryonic time evaluated for TRα1 expression in the fetal rat brain was E13.5.[67] At this time, TRα1 is relatively highly expressed in the neuroepithelium that gives rise to the cortex, hippocampus and basal telencephalon. However, it is also expressed in the trigeminal ganglion and striatum. TRα2 is expressed in these same regions, and appears to be more abundant than TRα1. As the brain expands over the next few days of embryonic development, TRα1 becomes more highly expressed in regions to which neurons are migrating and differentiating. For example, TRα1 becomes more intensely expressed in cells of the cortical plate compared to the ventricular zone. This pattern is also true for TRα2, and continues throughout fetal development. In addition, TRα1 and TRα2 are particularly abundant in the fetal thalamus, cerebellum and inferior colliculus. At all fetal time points evaluated (E13.5, 15.5, 17.5 and 19.5), TRα2 was more abundant than TRα1, and this became more obvious in the later stages.

TRα isoforms are expressed in nearly all tissues of the body,[68] but have been especially studied in bone, liver, heart and fat (in addition to brain and pituitary).[69] The cardiovascular effects of thyroid hormone have been realized for many years. Thyroid hormone decreases vascular resistance and increases heart rate, stroke volume and contractility, with an overall increase in cardiac output.[70] Thus, hyperthyroidism leads to cardiac arrhythmias and ultimately heart failure, whereas hypothyroidism leads to reduced cardiac function. This is likely to be attributed to the dominant expression of TRα1 in the heart.[68] In addition, TRα null mice exhibit decreased heart rate and contractility, but these parameters are unaffected in TRβ1 null mice.[71–73]

The liver is also a major site of action for thyroid hormone.[74–76] There are a large number of genes affected by thyroid hormone in the liver, as identified by cDNA microarray.[77,78] These genes may be regulated directly or indirectly by thyroid hormone. However, there are a number of genes that have been explored more extensively and are known to be regulated directly by thyroid hormone. These genes are under differential regulation by TRα and TRβ gene products. TRα1, TRα2, and TRβ1 are all present in liver.[68,76] Interestingly, TRα1 and TRα2 are expressed in specific cells, organized around the central vein.[79] Moreover, TRα2 exhibits a daily rhythm of expression, whereas TRα1 expression is stable throughout the day. However, the physiological relevance of TRα2 expression is difficult to interpret because it does not bind to thyroid hormone.

Expression and Regulation of the Beta TRs

Bradley *et al.*[67] also mapped the temporal and spatial distribution of TRβ1 and TRβ2 expression in the developing brain. The TRβ1 transcript is more widely expressed in the developing brain than the TRβ2 transcript. TR2 is expressed in the otic vesicle,[80] and in the upper tegmental nucleus and pituitary gland. In contrast, TRβ1 is widely expressed, though less so than TRα1 and TRα2. In early development (E13.5) TRβ1 is very low in abundance, or absent from most of the brain. Cortical expression occurs by E17.5, and this is restricted to cells of the ventricular zone. During postnatal development, TRβ1 expression in the brain increases considerably.[67,81] TRβ1 expression increases first in the striatum, then in the cortex, thalamus and olfactory bulb. TRβ2 expression is absent, or nearly so, in these studies. Recent studies indicate that TRα1 and TRβ1 are differentially expressed in the cerebellum,[82,83] with TRα1 expressed in cerebellar granule cells and both TRβ1 and TRα1 expressed in Purkinje cells. The regional distribution of the so-called "delta" forms of TR (TRΔα1, TRΔα2, TRΔβ3) has not been performed. TRβ1 is expressed in nearly all tissues.[68,81,84] However, like TRα1 and TRα2, TRβ1 is regionally expressed in the liver[85] and exhibits a diurnal rhythm of expression.

An important outgrowth of studies focused on the role of thyroid hormone in the control of metabolism and body weight is the search for thyroid hormone mimetics with potential applications in the control of atherosclerosis, obesity and type II diabetes.[86] This search has resulted in a number of thyroid hormone analogs, some with the ability to dramatically reduce serum LDL cholesterol and triglycerides, lower blood glucose and cause weight loss. For example, a recent 12-week clinical trial in people with elevated LDL-cholesterol taking statins, one of these analogs (Eprotirome) was shown to significantly reduce serum LDL cholesterol in a dose-dependent manner as well as to reduce serum triglycerides. This particular analog is reported to have a higher affinity for the TRβ than the TRα, and to be selectively taken up by the liver. This receptor isoform and tissue uptake selectivity may be an important reason that subjects did not exhibit effects of eprotirome on cardiovascular functions, since the cardiovascular system is particularly sensitive to thyroid hormone mediated by the TRα receptor. Another observation of this study was that serum total and free T_4 was reduced, although thyrotropin was not elevated. One possible explanation of this observation is that eprotirome could contribute to the negative feedback action of thyroid hormone on the hypothalamus and pituitary gland, leading to a reduction in circulating levels of T_4 without apparent effect. Another is that the modest reduction in serum T_4 could be the result of activated

BOX 11.1

HOW DO WE KNOW THAT THYROID HORMONE RECEPTOR REGULATES GENE EXPRESSION IN THE ABSENCE OF T_3?

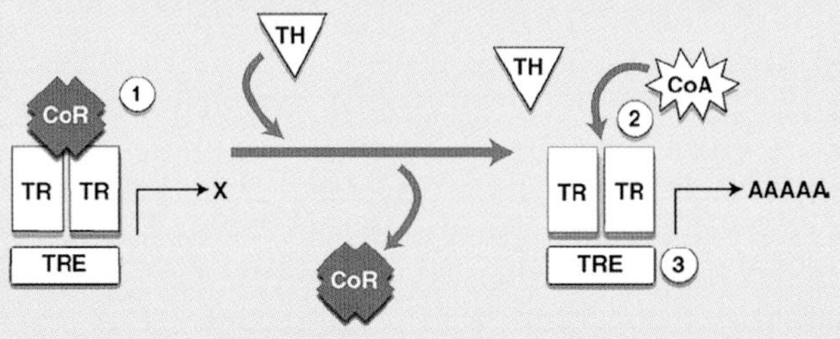

Thyroid hormone receptors are nuclear proteins that bind to T_3 and alter gene expression. These proteins are members of the Steroid/Thyroid superfamily of proteins,[1] whose members are ligand (hormone)-regulated transcription factors. In addition, thyroid hormone receptors are evolutionarily primitive proteins with a high degree of similarity across vertebrate taxa[2,3] indicating that important selective evolutionary pressures have maintained the genetic and functional stability of these receptors and the endocrine system they subserve. Therefore, it was surprising that the targeted deletion of TRα and/or TRβ in mice failed to reproduce the effects of low thyroid hormone.[4] Thus, although thyroid hormone is essential for normal development, and this is true across vertebrate taxa, thyroid hormone receptors are not essential for life. This conundrum was partially explained by the observation that the "knock-in" of a mutant thyroid hormone receptor that can bind to DNA but not to thyroid hormone produced the developmental damage of low thyroid hormone in the brain.[5] Thus, the unliganded thyroid hormone receptor can bind to DNA and cause developmental damage. The reciprocal experiment also supports this conclusion. That is, the "knock-in" of a mutant thyroid hormone receptor that can bind to T_3 but not to DNA fails to reproduce the effect of low thyroid hormone.[6] So, we know that thyroid hormone receptors bind to DNA and regulate gene expression because of these molecular studies that were characterized *in vitro*, but tested critically *in vivo* by the development of these transgenic lines.[7]

References

1. McEwan IJ. Nuclear receptors: one big family. *Methods Mol Biol.* 2009;505:3–18.
2. Paris M, Brunet F, Markov GV, Schubert M, Laudet V. The amphioxus genome enlightens the evolution of the thyroid hormone signaling pathway. *Dev Genes Evol.* 2008;218:667–680.
3. Bertrand S, Brunet FG, Escriva H, Parmentier G, Laudet V, Robinson-Rechavi M. Evolutionary genomics of nuclear receptors: from twenty-five ancestral genes to derived endocrine systems. *Mol Biol Evol.* 2004;21:1923–1937.
4. Gothe S, Wang Z, Ng L, Kindblom JM, Barros AC, Ohlsson C, Vennstrom B, Forrest D. Mice devoid of all known thyroid hormone receptors are viable but exhibit disorders of the pituitary–thyroid axis, growth, and bone maturation. *Genes Dev.* 1999;13:1329–1341.
5. Hashimoto K, Curty FH, Borges PP, Lee CE, Abel ED, Elmquist JK, Cohen RN, Wondisford FE. An unliganded thyroid hormone receptor causes severe neurological dysfunction. *Proc Natl Acad Sci USA.* 2001;98:3998–4003.
6. Shibusawa N, Hashimoto K, Nikrodhanond AA, Liberman MC, Applebury ML, Liao XH, Robbins JT, Refetoff S, Cohen RN, Wondisford FE. Thyroid hormone action in the absence of thyroid hormone receptor DNA-binding *in vivo*. *J Clin Invest.* 2003;112:588–597.
7. Wondisford FE. Thyroid hormone action: insight from transgenic mouse models. *J Investig Med.* 2003;51:215–220.

type 1 deiodinase in liver, as would be expected for a liver-selective T_4 analog.

Neuroendocrine Regulation of Thyroid Function

The thyroid gland is controlled principally and proximally by an interaction between iodine availability – a requirement for thyroid hormone synthesis – and thyrotropin (TSH) from the pituitary gland. This is true for humans and rodents,[87,88] as well as for other vertebrates.[89] In turn, TSH, a glycoprotein hormone, is regulated both by thyroid hormone itself (negative feedback) and by the releasing factor, thyrotropin-releasing hormone (TRH), from the hypothalamus. TRH release is controlled by thyroid hormone (the long loop of the negative feedback system) as well as by neural inputs that relay information about a variety of physiological states, including food availability, body temperature and, perhaps, cardiovascular functioning. Greer *et al.*[90] proposed that TRH controlled the set point around which thyroid hormone regulates TSH release, suggesting that TSH regulation is a pivotal point of regulation of the HPT axis.

Although these are the central features regulating thyroid function within the HPT axis, there are many additional processes that contribute to the overall regulation of the HPT axis, and of thyroid hormone action at target tissues. Among these are the metabolic enzymes that control changes in the iodination state of thyroid hormone. Specifically, three classes of deiodinase enzymes control the conversion of thyroid hormone to various active and inactive forms. It is becoming clear that the activity of these enzymes can contribute significantly to regulating tissue sensitivity to thyroid hormone, as highlighted above. In addition, enzymes in the liver target thyroid hormone for covalent modifications that lead to removal of thyroid hormones from the circulation. Finally, serum-binding proteins are important in regulating total hormone levels by increasing the carrying capacity of serum for the iodothyronines, which are only poorly soluble in aqueous media. Serum-binding proteins are themselves regulated by a variety of factors, including thyroid hormone and sex steroids; thus, the dynamics of the thyroid system can be altered in a variety of ways.

Recent studies of the HTP axis have revealed important new insights into this system. Therefore, we will selectively review these studies as an introduction to new concepts and new technologies that are applied to this system.

Hypophysiotropic TRH Neurons

Early work was unable to determine whether thyroid hormone exerted a negative feedback effect on the hypothalamus, or whether this action was directed only at the pituitary gland. It was not until 1981 that Aizawa and Greer[91] identified the specific region of the rodent brain (the hypothalamic paraventricular nucleus) that produced TRH and regulated pituitary thyrotropin secretion (i.e., the "hypophysiotropic TRH neurons"). The restricted distribution of these TRH neurons made it difficult to test the hypothesis that TRH release was regulated by thyroid hormone in a negative feedback manner. Then, in 1987, Koller *et al.*[92] and Segersen *et al.*[93] simultaneously reported that thyroid hormone reduces cellular levels of TRH mRNA specifically in the PVN. Following these reports, a number of papers were published showing that TRH peptide is reduced in the PVN[94] and in the pituitary-portal blood of rats.[95] Later reports demonstrated that both T_4 and T_3 can exert a negative feedback effect on TRH mRNA in the PVN, and that this is mediated by the TRβ receptor.[96–99]

The potential role of TSH in controlling TRH secretion is unclear, but is not likely to be dominant.[100] In contrast, there are TSH receptors expressed in human pituitary thyrotropes,[101] indicating that TSH may play a role in its own release. Although there are TRH immunoreactive terminals contacting TRH neurons of the PVN, revealed by immunocytochemistry at the electron microscopic level,[102] it is unclear whether TRH release exerts an ultra-short-loop feedback on hypophysiotropic TRH neurons.

The use of transgenic mouse models has led to important progress in clarifying the relative roles of the hypothalamus and pituitary in controlling thyroid function. This has been reviewed recently by Chiamolera and Wondisford.[103] An important outcome of these studies is the conclusion that the hypothalamus plays a dominant role both in establishing the set-point around which negative feedback operates and in controlling the baseline of TSH/TH secretion. This conclusion is based in part on the observation that serum T_4 levels are elevated in animals carrying a targeted deletion of the TRβ receptor, but reduced in animals carrying a targeted deletion of both the TRβ receptor and TRH.[104] In fact, TSH and T_4 in serum and TRH mRNA in the PVN are all elevated in TRβ-deficient mice, indicating that TRβ (TRβ2) is an important mediator of the negative feedback action of TH on the axis. Thus, it is important that if the TRH gene is deleted, serum T_4 levels decline to the same extent whether or not TRβ is present. However, serum TSH becomes elevated in response to hypothyroidism in both TRβ and TRH knockout animals, and exogenous T_3 can suppress TSH levels in these animals. In contrast, in TRβ/TRH double knockout animals TSH does not become elevated by hypothyroidism, nor does T_3 affect it. This combination of the use of transgenic animals with careful and thorough analysis of HPT axis functioning indicates that the hypothalamus plays a dominant role in the control

BOX 11.2

HOW DO WE KNOW THAT THYROID HORMONE FROM THE MOTHER ACTS ON THE FETAL BRAIN?

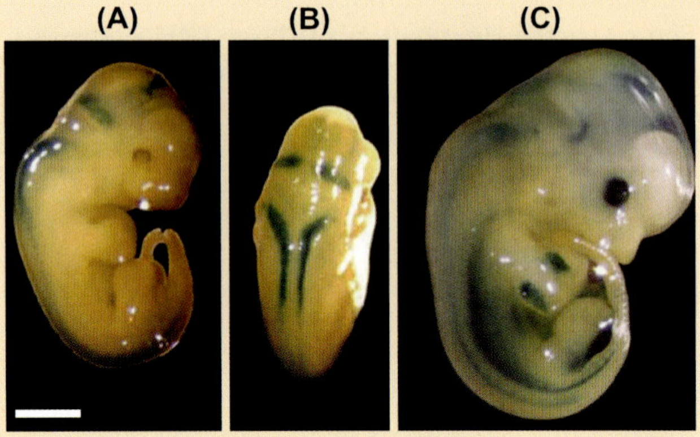

The human fetus does not begin to produce thyroid hormone until about 12–20 weeks' gestation.[1] If thyroid hormone is important before this time, then the only source of the hormone is the mother. An early indication that there is maternal–fetal transfer of thyroid hormone in the human came from the work of Evelyn Man,[2] who showed that maternal thyroid status was associated with measures of cognitive function in the offspring. Later, Vulsma *et al.*[3] showed that children born with a genetic inability to synthesize thyroid hormone still had significant serum T_4 levels in their umbilical cord at birth. Moreover, thyroid hormone receptors were found to be expressed in the fetal brain before the onset of thyroid function in both the rodent and the human.[4–6] These findings meant that thyroid hormone was traveling from the mother to the fetus, and that the machinery required for thyroid hormone signaling was present in the brain. However, these observations do not prove that thyroid hormone was having a direct effect on the brain. To address this, one group sought to identify thyroid hormone-responsive genes in the fetal cortex before the onset of fetal thyroid function.[7] It remained possible, though, that fetal genes affected by maternal thyroid status could be affected indirectly. Perhaps the most direct evidence that thyroid hormone of maternal origin can directly impact the fetus before the onset of fetal thyroid function was recently produced by Nucera *et al.*[8] This group developed a transgenic mouse carrying a β-galactosidase (β-Gal) gene that is driven by a thyroid hormone response element (TRE). Therefore, when T_3 becomes available in a cell and binds to the receptor, the receptor will drive the transcription of β-Gal, which will produce a blue color upon development. As can be seen in the figure, a blue color appears in the nervous system on day 11.5 (A and B), and also in the eyes and ears on day 12.5 (C). We still do not know what genes the TRs are directly regulating, and what developmental events these genes regulate. Reprinted with permission from Nucera *el al.*[8]

References

1. Smallridge RC, Glinoer D, Hollowell JG, Brent G. Thyroid function inside and outside of pregnancy: what do we know and what don't we know? *Thyroid.* 2005;15:54–59.
2. Man EB. Thyroid function in pregnancy and infancy. Maternal hypothyroxinemia and retardation of progeny. *Crit Rev Clin Lab Sci.* 1972;3:203–225.
3. Vulsma T, Gons MH, de Vijlder JJ. Maternal–fetal transfer of thyroxine in congenital hypothyroidism due to a total organification defect or thyroid agenesis. *New Engl J Med.* 1989;321:13–16.
4. Kilby MD, Gittoes N, McCabe C, Verhaeg J, Franklyn JA. Expression of thyroid receptor isoforms in the human fetal central nervous system and the effects of intrauterine growth restriction. *Clin Endocrinol (Oxf).* 2000;53: 469–477.
5. Ferreiro B, Bernal J, Goodyer CG, Branchard CL. Estimation of nuclear thyroid hormone receptor saturation in human fetal brain and lung during early gestation. *J Clin Endocrinol Metab.* 1988;67:853–856.
6. Perez-Castillo A, Bernal J, Ferreiro B, Pans T. The early ontogenesis of thyroid hormone receptor in the rat fetus. *Endocrinology.* 1985;117:2457–2461.
7. Dowling ALS, Martz GU, Leonard JL, Zoeller RT. Acute changes in maternal thyroid hormone induce rapid and transient changes in specific gene expression in fetal rat brain. *J Neurosci.* 2000;20:2255–2265.
8. Nucera C, Muzzi P, Tiveron C, Farsetti A, La Regina F, Foglio B, Shih SC, Moretti F, Della Pietra L, Mancini F, Sacchi A, Trimarchi F, Vercelli A, Pontecorvi A. Maternal thyroid hormones are transcriptionally active during embryo–foetal development: results from a novel transgenic mouse model. *J Cell Mol Med.* 2010;14:2417–2435.

of basal thyroid function, as well as the set-point around which negative feedback operates.

These mechanisms within the HPT axis that control thyroid function need to be able to respond to various environmental and physiological conditions. A potentially important modifier is 9-cis-retinoic acid. An early observation was that hyperthyroid patients could be treated with high doses of vitamin A to reduce circulating levels of thyroid hormone.[105] More recently, Sharma *et al.*[106] have demonstrated that rexinoids (9-cis-RA analogs) act directly on thyrotropes to inhibit TSH secretion. This has important practical implications, since these analogs are often used to treat a variety of medical conditions. However, it is not yet clear what role 9-cis-RA plays in the control of the HPT axis. Interestingly, PPARγ mRNA is expressed in the hypothalamus, and PPARγ agonists injected directly into the area of the PVN increased TRH transcription in the hypothalamus and increased serum T_4.[107] It is possible that these two observations are related; that is, perhaps the overexpression or activation of PPARγ competes for the availability of RXR that is required for TRβ action.

These observations also indicate a link between the control of elements of metabolism and control of the HPT axis. Early work indicated that acute starvation (caloric restriction) can cause a decrease in both serum TSH and T_4 in rats[108] and humans.[109] Moreover, food restriction could suppress TRH mRNA levels in the PVN even in the face of low thyroid hormone, indicating that the mechanism by which food restriction could suppress TRH mRNA is more powerful than the negative feedback effect of TH.[110]

CONCLUSIONS

The endocrinology of the thyroid is highly complex, but new techniques and reagents are becoming available to make important headway in understanding this complexity. Experimental studies are clearly demonstrating that the delivery of thyroid hormone to target cells is an important mechanism for controlling thyroid hormone action. However, while these observations clearly have clinical significance, measures of serum hormone concentrations remain the most important diagnostic tool for identifying thyroid disease. New areas of research have opened in nearly every level of organization of the thyroid system, from the molecular mechanisms of thyroid hormone action to the public health consequences of thyroid disruption by industrial chemicals. Students of endocrinology at all levels would be well-advised to begin to focus on this important system.

References

1. Aub JC, Bauer W, Heath C, Ropes M. Studies of calcium and phosphorus metabolism: III. The effects of the thyroid hormone and thyroid disease. *J Clin Invest* 1929;7:97–137.
2. Cocks JA, Balazs R, Johnson AL, Eayrs JT. Effect of thyroid hormone on the biochemical maturation of rat brain: conversion of glucose-carbon into amino acids. *J Neurochem* 1970; 17:1275–1285.
3. Andersen S, Pedersen KM, Bruun NH, Laurberg P. Narrow individual variations in serum T(4) and T(3) in normal subjects: a clue to the understanding of subclinical thyroid disease. *J Clin Endocrinol Metab.* 2002;87:1068–1072.
4. Watanabe M, Houten SM, Mataki C, Christoffolete MA, Kim BW, Sato H, Messaddeq N, Harney JW, Ezaki O, Kodama T, Schoonjans K, Bianco AC, Auwerx J. Bile acids induce energy expenditure by promoting intracellular thyroid hormone activation. *Nature.* 2006;439:484–489.
5. Zoeller RT, Tan SW, Tyl RW. General background on the hypothalamic–pituitary–thyroid (HPT) axis. *Crit Rev Toxicol.* 2007;37:11–53.
6. Braverman LE, Utiger RD, eds. *Ingbar's The Thyroid: A Fundamental and Clinical Text.* 9th ed. Philadelphia: Lippincott-Raven; 2004.
7. Finch JM, Osimitz TG, Gabriel KL, Martin T, Henderson WJ, Capen CC, Butler WH, Lake BG. A mode of action for induction of thyroid gland tumors by pyrethrins in the rat. *Toxicol Appl Pharmacol.* 2006;214:253–262.
8. Dohan O. The sodium/iodide symporter. *Endocrine Rev.* 2003;24:48–77.
9. Taurog A. Hormone Synthesis: Thyroid Iodine Metabolism. In: Braverman LE, Utiger RD, eds. *The Thyroid: A Fundamental and Clinical Text.* 9 ed. Philadelphia: Lippincott-Raven; 2004:61–85.
10. Spaulding SW. Biological actions of thyrotropin. In: Braverman LE, Utiger RD, eds. *The Thyroid: A Fundamental and Clinical Text.* 8 ed. Philadelphia: Lippincott Williams & Wilkins; 2000:227–233.
11. Wonerow P, Neumann S, Gudermann T, Paschke R. Thyrotropin receptor mutations as a tool to understand thyrotropin receptor action. *J Mol Med.* 2001;79:707–721.
12. Dremier S, Coulonval K, Perpete S, Vandeput F, Fortemaison N, Van Keymeulen A, Deleu S, Ledent C, Clement S, Schurmans S, Dumont JE, Lamy F, Roger PP, Maenhaut C. The role of cyclic AMP and its effect on protein kinase A in the mitogenic action of thyrotropin on the thyroid cell. *Ann NY Acad Sci.* 2002;968:106–121.
13. Savin S, Cvejic D, Nedic O, Radosavljevic R. Thyroid hormone synthesis and storage in the thyroid gland of human neonates. *J Pediatr Endocrinol Metab.* 2003;16:521–528.
14. van den Hove MF, Beckers C, Devlieger H, de Zegher F, De Nayer P. Hormone synthesis and storage in the thyroid of human preterm and term newborns: effect of thyroxine treatment. *Biochimie.* 1999;81:563–570.
15. Alkemade A. Central and peripheral effects of thyroid hormone signalling in the control of energy metabolism. *J Neuroendocrinol.* 2010;22:56–63.
16. Golozoubova V, Gullberg H, Matthias A, Cannon B, Vennstrom B, Nedergaard J. Depressed thermogenesis but competent brown adipose tissue recruitment in mice devoid of all hormone-binding thyroid hormone receptors. *Mol Endocrinol.* 2004;18:384–401.
17. Ebling FJ, Barrett P. The regulation of seasonal changes in food intake and body weight. *J Neuroendocrinol.* 2008;20:827–833.
18. Bansal R, You SH, Herzig CT, Zoeller RT. Maternal thyroid hormone increases HES expression in the fetal rat brain: An effect mimicked by exposure to a mixture of polychlorinated biphenyls (PCBs). *Brain Res.* 2005;156:13–22.
19. Tan XJ, Fan XT, Kim HJ, Butler R, Webb P, Warner M, Gustafsson JA. Liver X receptor beta and thyroid hormone receptor alpha in brain cortical layering. *Proc Natl Acad Sci USA.* 2010;107:12305–12310.
20. Nakamura K, Itoh K, Yaoi T, Fujiwara Y, Sugimoto T, Fushiki S. Murine neocortical histogenesis is perturbed by prenatal exposure to low doses of bisphenol A. *J Neurosci Res.* 2006;84:1197–1205.
21. Cuevas E, Auso E, Telefont M, Morreale de Escobar G, Sotelo C, Berbel P. Transient maternal hypothyroxinemia at onset of corticogenesis alters tangential migration of medial ganglionic eminence-derived neurons. *Eur J Neurosci.* 2005;22:541–551.
22. Zoeller RT, Dowling ALS, Herzig CTA, Iannacone EA, Gauger KJ, Bansal R. Thyroid hormone, brain development, and the environment. *Environ Health Persp.* 2002;110(Suppl 3):355–361.
23. Koibuchi N, Qiu CH, Miyazaki W, Iwasaki T, Shimokawa N. The role of thyroid hormone in developing cerebellum. *Cerebellum.* 2008;7:499–500.
24. Ahmed OM, El-Gareib AW, El-Bakry AM, Abd El-Tawab SM, Ahmed RG. Thyroid hormones states and brain development interactions. *Intl J Dev Neurosci.* 2008;26:147–209.
25. Bernal J. Thyroid hormone receptors in brain development and function. *Nat Clin Pract Endocrinol Metab.* 2007;3:249–259.

26. Jones I, Ng L, Liu H, Forrest D. An intron control region differentially regulates expression of thyroid hormone receptor β2 in the cochlea, pituitary, and cone photoreceptors. *Mol Endocrinol.* 2007;21:1108–1119.
27. Ng L, Hurley JB, Dierks B, Srinivas M, Salto C, Vennstrom B, Reh TA, Forrest D. A thyroid hormone receptor that is required for the development of green cone photoreceptors. *Nat Genets.* 2001;27:94–98.
28. Sharlin DS, Tighe D, Gilbert ME, Zoeller RT. The balance between oligodendrocyte and astrocyte production in major white matter tracts is linearly related to serum total thyroxine. *Endocrinology.* 2008;149:2527–2536.
29. Sharlin DS, Gilbert ME, Taylor MA, Ferguson DC, Zoeller RT. The nature of the compensatory response to low thyroid hormone in the developing brain. *J Neuroendocrinol.* 2010;22:153–165.
30. Chen Y, Young MA. Structure of a thyroid hormone receptor DNA-binding domain homodimer bound to an inverted palindrome DNA response element. *Mol Endocrinol.* 2010;24:1650–1664.
31. Iniguez MA, DeLecea L, Guadano-Ferraz A, Morte B, Gerendasy D, Sutcliffe JG, Bernal J. Cell-specific effects of thyroid hormone on RC3/neurogranin expression in rat brain. *Endocrinol.* 1996;137:1032–1041.
32. Shi YB. Dual functions of thyroid hormone receptors in vertebrate development: the roles of histone-modifying cofactor complexes. *Thyroid.* 2009;19:987–999.
33. Hernandez A, Quignodon L, Martinez ME, Flamant F, St Germain D. Type 3 deiodinase deficiency causes spatial and temporal alterations in brain T3 signaling that are dissociated from serum thyroid hormone levels. *Endocrinology;* 2010. in press.
34. Gereben B, Zeold A, Dentice M, Salvatore D, Bianco AC. Activation and inactivation of thyroid hormone by deiodinases: local action with general consequences. *Cell Mol Life Sci.* 2008;65:570–590.
35. Ng L, Hernandez A, He W, Ren T, Srinivas M, Ma M, Galton VA, St Germain DL, Forrest D. A protective role for type 3 deiodinase, a thyroid hormone-inactivating enzyme, in cochlear development and auditory function. *Endocrinology.* 2009;150:1952–1960.
36. Bernal J. The significance of thyroid hormone transporters in the brain. *Endocrinology.* 2005;146:1698–1700.
37. Freitas BC, Gereben B, Castillo M, Kallo I, Zeold A, Egri P, Liposits Z, Zavacki AM, Maciel RM, Jo S, Singru P, Sanchez E, Lechan RM, Bianco AC. Paracrine signaling by glial cell-derived triiodothyronine activates neuronal gene expression in the rodent brain and human cells. *J Clin Invest.* 2010;120:2206–2217.
38. Hernandez A, Martinez ME, Fiering S, Galton VA, St Germain D. Type 3 deiodinase is critical for the maturation and function of the thyroid axis. *J Clin Invest.* 2006;116:476–484.
39. Hernandez A, Martinez ME, Liao XH, Van Sande J, Refetoff S, Galton VA, St Germain DL. Type 3 deiodinase deficiency results in functional abnormalities at multiple levels of the thyroid axis. *Endocrinology.* 2007;148:5680–5687.
40. Quignodon L, Legrand C, Allioli N, Guadano-Ferraz A, Bernal J, Samarut J, Flamant F. Thyroid hormone signaling is highly heterogeneous during pre- and postnatal brain development. *J Mol Endocrinol.* 2004;33:467–476.
41. Mata De Urquiza A, Solomin L, Perlmann T. Feedback-inducible nuclear-receptor-driven reporter gene expression in transgenic mice. *Proc Natl Acad Sci USA* 1999;96:13270–13275.
42. Hu X, Lazar MA. Transcriptional repression by nuclear hormone receptors. *Trends Endocrinol Metab.* 2000;11:6–10.
43. Wu Y, Xu B, Koenig RJ. Thyroid hormone response element sequence and the recruitment of retinoid X receptors for thyroid hormone responsiveness. *J Biol Chem.* 2001;276:3929–3936.
44. Lazar MA. Thyroid hormone receptors: Update. *Endocr Rev Monogr.* 1994;3:280–283.
45. Mangelsdorf DJ, Evans RM. The RXR heterodimers and orphan receptors. *Cell.* 1995;83:841–850.
46. Zhang J, Lazar MA. The mechanism of action of thyroid hormones. *Annu Rev Physiol.* 2000;62:439–466.
47. Lazar MA. Thyroid hormone receptors: multiple forms, multiple possibilities. *Endocrine Rev.* 1993;14:184–193.
48. Sap J, Munoz A, Damm K, Goldberg Y, Ghysdael J, Lentz A, Beug H, Vennstrom B. The c-erbA protein is a high affinity receptor for thyroid hormone. *Nature.* 1986;324:635–640.
49. Weinberger C, Thompson CC, Ong ES, Lebo R, Gruol DJ, Evans RM. The c-erbA gene encodes a thyroid hormone receptor. *Nature.* 1986;324:641–646.
50. Williams GR. Cloning and characterization of two novel thyroid hormone receptor β isoforms. *Mol Cell Biol* 2000;20:8329–8342.
51. Flamant F, Samarut J. Thyroid hormone receptors: lessons from knockout and knock-in mutant mice. *Trends Endocrinol Metab.* 2003;14:85–90.
52. Chassande O, Fraichard A, Gauthier K, Flamant F, Legrand C, Savatier P, Laudet V, Samarut J. Identification of transcripts initiated from an internal promoter in the c-erbAα locus that encode inhibitors of retinoic acid receptor-α and triiodothyronine receptor activities. *Mol Endocrinology.* 1997;11:1278–1290.
53. Oppenheimer JH. The nuclear receptor-triiodothyronine complex: relationship to thyroid hormone distribution, metabolism, and biological action. In: Oppenheimer JH, Samuels HH, eds. *Molecular Basis of Thyroid Hormone Action.* New York: Academic Press; 1983:1–35.
54. Schwartz HL, Strait KA, Ling NC, Oppenheimer JH. Quantitation of rat tissue thyroid hormone binding receptor isoforms by immunoprecipitation of nuclear triiodothyronine binding capacity. *J Biol Chem.* 1992;267:11794–11799.
55. Oppenheimer JH, Schwartz HL, Strait KA. Thyroid hormone action 1994: the plot thickens. *Eur J Endocrinol.* 1994;130:15–24.
56. Bakker O, Beeren HCv, Wiersinga WM. Desethylamiodarone is a noncompetitive inhibitor of the binding of thyroid hormone to the thyroid hormone beta-1 receptor protein. *Endocrinology.* 1994;134:1665–1670.
57. Bakker O, Beeren HCv, Wiersinga WM. Desethylamiodarone is a competitive inhibitor of the binding of thyroid hormone to the alpha-1 receptor protein. *Mol Cell Endocrinol.* 1995;112:15–19.
58. Schueler PA, Schwartz HL, Strait KA, Mariash CN, Oppenheimer JH. Binding of 3,5,3′-triiodothyronine (T3) and its analogs to the *in vitro* translational products of c-erbA protooncogenes: differences in the affinity of the alpha and beta forms for the acetic acid analog and failure of the human testis and kidney alpha-2 products to bind T3. *Mol Endocrinol.* 1990;4:227–234.
59. Takeda T, Suzuki S, Liu R-T, DeGroot LJ. Triiodothyroacetic acid has unique potential for therapy of resistance to thyroid hormone. *J Clin Endocrinol Metab.* 1995;80:2033–2040.
60. Baxter JD, Dillmann WH, West BL, Huber R, Furlow JD, Fletterick RJ, Webb P, Apriletti JW, Scanlan TS. Selective modulation of thyroid hormone receptor action. *J Steroid Biochem Mol Biol.* 2001;76:31–42.
61. Kinugawa K, Yonekura K, Ribeiro RC, Eto Y, Aoyagi T, Baxter JD, Camacho SA, Bristow MR, Long CS, Simpson PC. Regulation of thyroid hormone receptor isoforms in physiological and pathological cardiac hypertrophy. *Circ Res.* 2001;89:591–598.

62. Chiellini G, Apriletti JW, al Yoshihara H, Baxter JD, Ribeiro RC, Scanlan TS. A high-affinity subtype-selective agonist ligand for the thyroid hormone receptor. *Chem Biol.* 1998;5:299–306.
63. Nguyen NH, Apriletti JW, Cunha Lima ST, Webb P, Baxter JD, Scanlan TS. Rational design and synthesis of a novel thyroid hormone antagonist that blocks coactivator recruitment. *J Med Chem.* 2002;45:3310–3320.
64. Webb P, Nguyen NH, Chiellini G, Yoshihara HA, Cunha Lima ST, Apriletti JW, Ribeiro RC, Marimuthu A, West BL, Goede P, Mellstrom K, Nilsson S, Kushner PJ, Fletterick RJ, Scanlan TS, Baxter JD. Design of thyroid hormone receptor antagonists from first principles. *J Steroid Biochem Mol Biol.* 2002;83:59–73.
65. Lim W, Nguyen NH, Yang HY, Scanlan TS, Furlow JD. A thyroid hormone antagonist that inhibits thyroid hormone action *in vivo*. *J Biol Chem.* 2002;277:35664–35670.
66. Messier N, Langlois MF. Triac regulation of transcription is T(3) receptor isoform- and response element-specific. *Mol Cell Endocrinol.* 2000;165:57–66.
67. Bradley DJ, Towle HC, Young WS. Spatial and temporal expression of alpha- and beta-thyroid hormone receptor mRNAs, including the beta-2 subtype, in the developing mammalian nervous system. *J Neurosci.* 1992;12:2288–2302.
68. Hodin RA, Lazar MA, Chin WW. Differential and tissue-specific regulation of the multiple rat c-erbA messenger RNA species by thyroid hormone. *J Clin Invest.* 1990;85:101–105.
69. Yen PM. Physiological and molecular basis of thyroid hormone action. *Physiol Rev.* 2001;81:1097–1142.
70. Klein I, Ojamaa K. Thyrotoxicosis and the heart. *Endocrinol Metab Clin N Am.* 1998;27:51–62.
71. Johansson C, Vennstrom B, Thoren P. Evidence that decreased heart rate in thyroid hormone receptor-a1-deficient mice is an intrinsic defect. *Am J Physiol Regul Integr Comp Physiol.* 1998;275:R640–R646.
72. Gloss B, Trost S, Bluhm W, Swanson E, Clark R, Winkfein R, Janzen K, Giles W, Chassande O, Samarut J, Dillmann W. Cardiac ion channel expression and contractile function in mice with deletion of thyroid hormone receptor alpha or beta. *Endocrinology.* 2001;142:544–550.
73. Weiss RE, Gehin M, Xu J, Sadow PM, O'Malley BW, Chambon P, Refetoff S. Thyroid function in mice with compound heterozygous and homozygous disruptions of SRC-1 and TIF-2 coactivators: evidence for haploinsufficiency. *Endocrinology.* 2002;143:1554.
74. McClain RM. The significance of hepatic microsomal enzyme induction and altered thyroid function in rats: implications for thyroid gland neoplasia. *Toxicol Pathol.* 1989;17:294–306.
75. Song MK, Grieco D, Rall JE, Nikodem VM. Thyroid hormone-mediated transcriptional activation of rat liver malic enzyme gene by dehydroepiandrosterone. *J Biol Chem.* 1989;264: 18981–18985.
76. Weiss RE, Murata Y, Cua K, Hayashi Y, Seo H, Refetoff S. Thyroid hormone action on liver, heart, and energy expenditure in thyroid hormone receptor beta-deficient mice. *Endocrinology.* 1998;139:4945–4952.
77. Feng X, Jiang Y, Meltzer P, Yen PM. Thyroid hormone regulation of hepatic genes *in vivo* detected by complementary DNA microarray. *Mol Endocrinol.* 2000;14:947–955.
78. Yen PM, Feng X, Flamant F, Chen Y, Walker RL, Weiss RE, Chassande O, Samarut J, Refetoff S, Meltzer PS. Effects of ligand and thyroid hormone receptor isoforms on hepatic gene expression profiles of thyroid hormone receptor knockout mice. *EMBO Rep.* 2003;4:581–587.
79. Zandieh-Doulabi B, Dop E, Schneiders M, Schiphorst MP, Mansen A, Vennstrom B, Dijkstra CD, Bakker O, Wiersinga WM. Zonal expression of the thyroid hormone receptor alpha isoforms in rodent liver. *J Endocrinol.* 2003;179:379–385.
80. Bradley DJ, Towle HC, Young WS. Alpha and beta thyroid hormone receptor (TR) gene expression during auditory neurogenesis: evidence for TR isoform-specific transcriptional regulation *in vivo*. *Proc Natl Acad Sci USA.* 1994;91:439–443.
81. Strait KA, Schwartz HL, Perez-Castillo A, Oppenheimer JH. Relationship of c-erbA content to tissue triiodothyronine nuclear binding capacity and function in developing and adult rats. *J Biol Chem.* 1990;265:10514–10521.
82. Manzano J, Morte B, Scanlan TS, Bernal J. Differential effects of triiodothyronine and the thyroid hormone receptor beta-specific agonist GC-1 on thyroid hormone target genes in the brain. *Endocrinology.* 2003;144:5480–5487.
83. Guadano-Ferraz A, Benavides-Piccione R, Venero C, Lancha C, Vennstrom B, Sandi C, DeFelipe J, Bernal J. Lack of thyroid hormone receptor alpha1 is associated with selective alterations in behavior and hippocampal circuits. *Mol Psychiatry.* 2003;8:30–38.
84. Falcone M, Miyamoto T, Fierro-Renoy F, Macchia E, DeGroot LJ. Antipeptide polyclonal antibodies specifically recognize each human thyroid hormone receptor isoform. *Endocrinology.* 1992;131:2419–2429.
85. Zandieh Doulabi B, Platvoet-ter Schiphorst M, van Beeren HC, Labruyere WT, Lamers WH, Fliers E, Bakker O, Wiersinga WM. TR(beta)1 protein is preferentially expressed in the pericentral zone of rat liver and exhibits marked diurnal variation. *Endocrinology.* 2002;143:979–984.
86. Baxter JD, Webb P. Thyroid hormone mimetics: potential applications in atherosclerosis, obesity and type 2 diabetes. *Nat Rev Drug Discov.* 2009;8:308–320.
87. Morreale de Escobar GM, Obregon MJ, Calvo R, Pedraza P, Escobar del Rey F. Iodine deficiency, the hidden scourge: The rat model of human neurological cretinism. In: Hendrich CE, ed. *Recent Research Developments in Neuroendocrinology – Thyroid Hormone and Brain Maturation.* Trivandrum: Research Signpost; 1997:55–70.
88. Dunn JT, Dunn AD. Thyroglobulin: chemistry, biosynthesis, and proteolysis. In: Braverman LE, Utiger RD, eds. *The Thyroid: A Fundamental and Clinical Text.* 8th ed. Philidelphia: Lippincott, Williams & Wilkins; 2000:91–104.
89. Norris DO. *Vertebrate Endocrinology.* 3rd ed. San Diego: Academic Press; 1997.
90 Greer MA, Sato N, Wang X, Greer SE, McAdams S. Evidence that the major physiological role of TRH in the hypothalamic paraventricular nuclei may be to regulate the set-point for thyroid hormone negative feedback on the pituitary thyrotroph. *Neuroendocrinology.* 1993;57:569–575.
91. Aizawa T, Greer MA. Delineation of the hypothalamic area controlling thyrotropin secretion in the rat. *Endocrinology.* 1981;109:1731–1738.
92. Koller KJ, Wolff RS, Warden MK, Zoeller RT. Thyroid hormones regulate levels of thyrotropin-releasing-hormone mRNA in the paraventricular nucleus. *Proc Natl Acad Sci USA.* 1987; 84:7329–7333.
93. Segersen TP, Kauer J, Wolfe HC, Mobtaker H, Wu P, Jackson IMD, Lechan RM. Thyroid hormone regulates TRH biosynthesis in the paraventricular nucleus of the rat hypothalamus. *Science.* 1987;238:78–80.
94. Yamada M, Rogers D, Wilber JF. Exogenous triiodothyronine lowers thyrotropin-releasing hormone concentrations in the specific hypothalamic nucleus (paraventricular) involved in thyrotropin regulation and also in posterior nucleus. *Neuroendocrinol.* 1989;50:560–563.

95. Rondeel JMM, deGreef WJ, van der Schoot P, Karels B, Klootwijk W, Visser TJ. Effect of thyroid status and paraventricular area lesions on the release of thyrotropin-releasing hormone and catecholamines into hypophysial portal blood. *Endocrinology.* 1988;123:523–527.
96. Kakucska I, Rand W, Lechan RM. Thyrotropin-releasing hormone gene expression in the hypothalamic paraventricular nucleus is dependent upon feedback regulation by both triiodothyronine and thyroxine. *Endocrinology.* 1992;130:2845–2850.
97. Dyess EM, Segerson TP, Liposits Z, Paull WK, Kaplan MM, Wu P, Jackson IMD, Lechan RM. Triiodothyronine exerts direct cell-specific regulation of thyrotropin-releasing hormone gene expression in the hypothalamic paraventricular nucleus. *Endocrinology.* 1988;123:2291–2297.
98. Lechan RM, Kakucska I. Feedback regulation of thyrotropin-releasing hormone gene expression by thyroid hormone in the hypothalamic paraventricular nucleus. In: Chadwick DJ, Marsh J, eds. *Functional anatomy of the neuroendocrine hypothalamus*. Budapest: John Wiley & Sons; 1992:144–164.
99. Lechan RM, Qi Y, Jackson IMD, Mahdavi V. Identification of thyroid hormone receptor isoforms in thyrotropin-releasing hormone neurons of the hypothalamic paraventricular nucleus. *Endocrinology.* 1994;135:92–100.
100. Zoeller RT, Wolff RS, Koller KJ. Thyroid hormone regulation of messenger ribonucleic acid encoding thyrotropin (TSH)-releasing hormone is independent of the pituitary gland and TSH. *Mol Endocrinol.* 1988;2:248–252.
101. Theodoropoulou M, Arzberger T, Gruebler Y, Korali Z, Mortini P, Joba W, Heufelder AE, Stalla GK, Schaaf L. Thyrotrophin receptor protein expression in normal and adenomatous human pituitary. *J Endocrinol.* 2000;167:7–13.
102. Merchenthaler I, Liposits Z. Mapping of thyrotropin-releasing hormone (TRH) neuronal systems of rat forebrain projecting to the median eminence and the OVLT. Immunocytochemistry combined with retrograde labeling at the light and electron microscopic levels. *Acta Biol Hung.* 1994;45:361–374.
103. Chiamolera MI, Wondisford FE. Minireview: Thyrotropin-releasing hormone and the thyroid hormone feedback mechanism. *Endocrinology.* 2009;150:1091–1096.
104. Nikrodhanond AA, Ortiga-Carvalho TM, Shibusawa N, Hashimoto K, Liao XH, Refetoff S, Yamada M, Mori M, Wondisford FE. Dominant Role of thyrotropin-releasing hormone in the hypothalamic–pituitary–thyroid axis. *J Biol Chem.* 2006;281:5000–5007.
105. Simpkins S. Use of massive doses of vitamin A in the treatment of hyperthyroidism. *J Clin Endocrinol Metab.* 1947;7:574–585.
106. Sharma V, Hays WR, Wood WM, Pugazhenthi U, St Germain DL, Bianco AC, Krezel W, Chambon P, Haugen BR. Effects of rexinoids on thyrotrope function and the hypothalamic–pituitary–thyroid axis. *Endocrinology.* 2006;147:1438–1451.
107. Kouidhi S, Seugnet I, Decherf S, Guissouma H, Elgaaied AB, Demeneix B, Clerget-Froidevaux MS. Peroxisome proliferator-activated receptor-gamma (PPARgamma) modulates hypothalamic Trh regulation *in vivo*. *Mol Cell Endocrinol.* 2010;317:44–52.
108. Harris AR, Fang SL, Azizi F, Lipworth L, Vagenakis AG, Barverman LE. Effect of starvation on hypothalamic–pituitary–thyroid function in the rat. *Metab Clin Exp.* 1978;27:1074–1083.
109. Douyon L, Schteingart DE. Effect of obesity and starvation on thyroid hormone, growth hormone, and cortisol secretion. *Endocrinol Metab Clin N Am.* 2002;31:173–189.
110. Blake NG, Johnson MR, Eckland DJ, Foster OJ, Lightman SL. Effect of food deprivation and altered thyroid status on the hypothalamic–pituitary–thyroid axis in the rat. *J Endocrinol.* 1992;133:183–188.

CHAPTER

12

Circadian Rhythms in Neuroendocrine Systems

Deanna M. Arble[1], *Georges Copinschi*[2], *Martha H. Vitaterna*[1], *Eve Van Cauter*[3], *Fred W. Turek*[1]

[1] Center for Sleep and Circadian Biology, Department of Neurobiology and Physiology, Northwestern University, Evanston, IL, USA, [2] Laboratory of Physiology, School of Medicine, Université Libre de Bruxelles, Brussels, Belgium, [3] Sleep, Metabolism and Health Center, Department of Medicine, University of Chicago Pritzker School of Medicine, Chicago, IL, USA

OUTLINE

Summary

A prominent feature of the neuroendocrine system is its high degree of temporal organization. This precise organization of hormonal concentrations during the 24-hour cycle results from the activity of two interacting time-keeping mechanisms in the central nervous system: endogenous circadian rhythmicity, and sleep-wake homeostasis. This chapter begins by reviewing basic circadian rhythm characteristics, including the sleep-wake and feeding-fasting rhythms, as well as circadian genetics as they pertain to neuroendocrine and endocrine rhythms. Further detail is then given specifically to rhythms within the corticotropic, somatotropic, lactotropic, gonadotropic and thyrotropic axes. The final portion of the chapter is dedicated to conditions of altered circadian rhythms and sleep, and the resulting effects on the neuroendocrine axes. As circadian and neuroendocrine disruptions become increasingly common in modern society, there is a growing need for therapeutic strategies which will focus on both circadian and neuroendocrine health.

MAJOR MECHANISMS CONTROLLING NEUROENDOCRINE AND ENDOCRINE RHYTHMS

A prominent feature of the neuroendocrine system is its high degree of temporal organization. Indeed, far from

Handbook of Neuroendocrinology, DOI: 10.1016/B978-0-12-375097-6.10012-5

obeying the concept of "constancy of the internal milieu," circulating hormonal levels undergo pronounced temporal oscillations ranging in period from a few minutes to a year. This intricate temporal organization provides the neuroendocrine system with remarkable flexibility. Not only can specific physiological processes be turned on and off depending on the presence or absence of a particular hormone, but the precise pattern of hormonal release may provide specific signaling information.

As schematically illustrated in Fig. 12.1, the temporal variability and organization of hormonal concentrations during the 24-hour cycle ultimately results from the activity of two interacting time-keeping mechanisms in the central nervous system: endogenous circadian rhythmicity; and sleep—wake homeostasis, a mechanism relating the timing and intensity of sleep to the duration of prior wakefulness. While this dual control was first demonstrated for neurohormones of the hypothalamo-pituitary axis, a similar regulation appears to apply for other endocrine subsystems. In mammals, endogenous circadian rhythmicity is generated by a "master circadian clock" located in the paired suprachiasmatic nucleus (SCN) of the hypothalamus.[1] The SCN controls the timing of most, if not all, circadian rhythms, including circadian behavior rhythms such as the sleep—wake and feeding—fasting cycles. These circadian behaviors in turn regulate the timing of many rhythms that depend on the presence or absence of sleep and wakefulness and/or feeding and fasting. Indeed, the timing and expression of many neuroendocrine and endocrine rhythms appear to depend upon a direct control from the SCN, as well as on the presence (or absence) and quality of sleep, with some 24-hour endocrine rhythms more influenced by the SCN (e.g., melatonin and cortisol) while others are more regulated by the sleep—wake state (e.g. growth hormone). A similar relationship is seen with the feeding—fasting cycle; a number of hormonal fluctuations (e.g., leptin, glucose) are influenced by both the endogenous circadian clock and the clock-controlled feeding—fasting behavior. Thus, the combined inputs of the master circadian clock and of the rhythmic behaviors of the sleep—wake or feeding—fasting state[2] control the overall temporal organization of the neuroendocrine and endocrine system, as well as many other behavioral and physiological systems, across the 24-hour day.

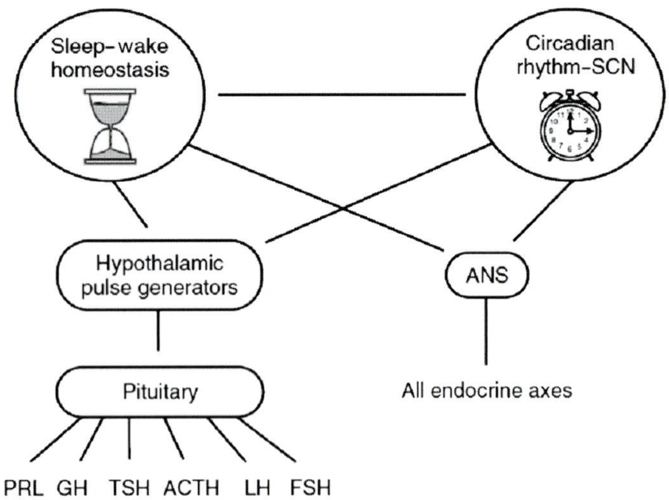

FIGURE 12.1 Schematic representation of the central mechanisms involved in the control of temporal variations in pituitary hormone secretions over the 24-hour cycle. ACTH, Adrenocorticotropic hormone; ANS, autonomous nervous system; FSH, follicle-stimulating hormone; GH, growth hormone; LH, luteinizing hormone; PRL, prolactin; SCN, suprachiasmatic nucleus of the hypothalamus; TSH, thyrotropin. *Reproduced from Copinschi G, Turek FW, Van Cauter E. Endocrine rhythms, the sleep—wake cycle, and biological clocks. In Jameson JL, De Groot LJ, eds.* Endocrinology, *6th edition. Philadelphia, PA: Saunders; 2010: Ch. 11, with permission.*

The first section of this chapter provides an overview of current concepts and recent advances in the understanding of circadian rhythmicity, behavioral rhythmicity, and the genetics that underlie the circadian expression of neuroendocrine rhythms. Specific examples of neuroendocrine systems dependent on circadian rhythms are described in the second section. The third section summarizes the present state of knowledge on conditions of altered or abnormal circadian and/or sleep or feeding regulation that have implications for the temporal organization of hormonal release.

Due to limitations on length and scope of this chapter, this review is limited to findings in adults.

Circadian Rhythmicity

General Characteristics

One of the most obvious characteristics of life on Earth is the ability of almost all species to change their behavior on a daily or 24-hour basis. A remarkable feature of these daily or diurnal rhythms is that they are not simply a response to the 24-hour changes in the physical environment imposed by the principles of celestial mechanics, but instead arise from an internal time-keeping system[1,3] that has the intrinsic capability to continuously generate rhythmic activity with a near 24-hour period. Thus, under laboratory conditions devoid of any external time-giving cues, it has been found that nearly all 24-hour rhythms continue to be expressed. However, under such constant conditions, the period of the rhythm rarely remains exactly 24 hours but instead is "about" 24 hours; this is why these rhythms are referred to as "circadian," from the Latin *circa diem*, meaning around a day. When a circadian rhythm is expressed in the absence of any 24-hour signals in the external environment, it is said to be "free-running." Under free-running conditions, the endogenous period is generally close to, but nearly never exactly, 24 hours. Strictly speaking, a diurnal rhythm should not be referred to as "circadian" until it has been demonstrated that such a rhythm persists under constant environmental conditions. The purpose of this distinction is to separate out those rhythms that are simply a response to 24-hour changes in the environment from those that are endogenous. However, for practical purposes there is little reason to make a distinction between "diurnal" and "circadian" rhythms, since an endogenous timing device underlies the generation of almost all diurnal rhythms. In this chapter, we will therefore extend the use of the term "circadian rhythm" to all diurnal variations recurring regularly at a time interval of approximately 24 hours.

An immense variety of circadian rhythms has been observed in man. Human circadian rhythms have been characterized for blood constituents, such as white blood cells, amino acids, and hormones; innumerable physiological variables, such as body temperature, heart rate, blood pressure and urinary volume; as well as behavioral parameters, such as food intake, sleep, mood, vigilance and cognitive performance. There are also rhythms in responsiveness to various challenges, such as drugs and stress. Circadian rhythmicity is maintained when subjects are sleep deprived, when they are starved, or when they receive equal amounts of food at short intervals over the day. The timing of single meals, however, can have effects on the temporal pattern of at least some variables, including hormones, and the timing, duration and quality of sleep and wake can alter the expression of many rhythms, especially those of the endocrine system.

The Suprachiasmatic Nucleus: A Master Circadian Pacemaker

In mammals, the suprachiasmatic nucleus (SCN) – i.e., two small bilaterally paired nuclei in the anterior hypothalamus, each containing about 10,000 cells in rodents – functions as the master circadian clock. Under both free-running and entrained conditions, destruction of the SCN in a variety of species leads to the abolishment or the severe disruption of many endocrine, behavioral and physiological rhythms.[1,4] The role of the SCN as the control center for the circadian system, first suggested by lesion studies, was confirmed by studies involving transplantation of the SCN from one animal to another. Indeed, circadian rhythmicity can be restored in adult arrhythmic SCN-lesioned rodents by transplanting fetal SCN tissue into the region of the SCN.[5,6] A number of SCN rhythms persist *in vitro*, including those of neural firing, vasopressin release, glucose metabolism and gene expression.[7–10] The ability of SCN cells to generate a circadian signal does not rely on some inherent network property of many cells acting together: single SCN cells in culture can generate circadian neural signals.[11] Neurons within the SCN appear to be organized in two groups: a "core" group of light responsive but non-rhythmic cells, and a "shell" of rhythmic cells.[12,13]

The generation and maintenance of circadian oscillations in the SCN involve a series of clock genes (including *per1, per2, per3, cry1, cry2, tim, clock, B-mal1, CKIε/δ*), often referred to as "canonical clock genes," which interact in a complex feedback loop of transcription/translation.[14,15] Indeed, improved tissue culture techniques and real-time monitoring of the expression of circadian clock genes has revealed that circadian oscillations in gene expression can persist in the SCN *in vitro* (as well as in other tissues, see below) for many weeks and months.[14] Importantly, mutations or deletions of canonical circadian clock genes have been found to profoundly affect endocrine rhythms and normal endocrine function in a variety of central and peripheral tissues.[16–18]

In recent years it has been recognized that circadian oscillations can be generated in areas of the brain other than the SCN, as well as in many peripheral tissues.[7,19] These "local" oscillators appear to be under the control of the SCN and of the canonical circadian clock genes in the SCN, through the synchronization or entrainment of the same circadian clock machinery at the local tissue level. The SCN controls or synchronizes these autonomous circadian clocks in non-SCN tissues, either directly via neural and/or endocrine signals, or indirectly via its control of behavioral rhythms such as the sleep–wake cycle and the rhythm of feeding. However, the precise mechanisms by which the SCN and its molecular core clock regulate the timing of other central, as well as peripheral, clocks that share common clock genes is not well understood.

Photic Entrainment of Circadian Rhythms

The fact that the endogenous circadian period observed under constant conditions is not exactly equal to 24 hours implies that changes in the physical environment must synchronize or entrain the internal clock. Otherwise, a clock with a period only a few minutes shorter or longer than 24 hours would soon be totally out of synchrony with the environmental day. Agents that are capable of entraining or synchronizing circadian rhythms are often called "zeitgebers," a German neologism meaning "time giver."

The light–dark (LD) cycle is the primary agent that synchronizes most circadian rhythms. Thus, in the presence of a 24-h LD cycle, the period of circadian rhythms exactly matches the period of the LD cycle. In addition to establishing "period control," an entraining LD cycle establishes "phase control" such that specific phases of the circadian rhythm occur at the same time in each cycle. Entrainment is restricted to cycles with periods that are "close" to 24 hours in duration, and, in general, is not possible for LD cycles that are more than a few hours shorter or longer than the endogenous circadian period. If the period of the LD cycle is too short or long for entrainment to occur, the circadian rhythm free runs. This rigidity of the circadian pacemaker has been used in so-called "forced desynchrony" studies, where the subjects are maintained on a light–dark and sleep–wake cycle with a period outside of the range of entrainment, such as 20 hours or 28 hours.[20,21] The fact that the human endogenous circadian period is probably very close to 24 hours is consistent with the findings of a study showing that a schedule of sleep–wake and dark–light cycles with very low light intensity during wakefulness is able to maintain entrainment to the 24-h day, but not to a 23.5- or 24.6-h day.[22]

The eyes are involved in relaying entraining information from the LD cycle to the circadian timing system in mammals via a unique pathway, separate from the visual system, referred to as the retinohypothalamic tract.[23] At the level of the optic chiasm, retinal projections first enter the brain in the region of the SCN and surrounding hypothalamic areas.[23] Thus, the integrity of the primary visual centers of the brain and/or the "perception" of light is not necessary for entrainment of circadian rhythms by the LD cycle. The independence of the photic input to the circadian clock from the visual system in mammals is actually not surprising from an evolutionary point of view. In all non-mammalian vertebrates, entrainment of circadian rhythms can occur in the absence of the eyes and relies on non-retinal photoreceptors in the brain.[24] The independence of the circadian light-sensing system from the visual system in mammals was also suggested in early studies showing that rodless/coneless mice could still entrain to the light–dark cycle, even though the eyes were necessary.[25] Early in this century, a number of laboratories discovered almost simultaneously that a special subset of retinal ganglion cells containing melanopsin could act as photoreceptors for relaying light information to the circadian clock in the SCN.[26–29] This breakthrough demonstrated not only a new light-sensing system in the retina, but also opened up new avenues of research regarding the impact of light on brain function independently of the visual system, including possible neuroendocrine effects.[30,31] The finding that visual light perception is not necessary for circadian light perception may explain why, in some totally blind humans, light exposure is capable of suppressing melatonin levels, indicating that visual blindness should not be equated with circadian blindness.[32] In addition to the retinohypothalamic tract, the SCN also receives retinal information indirectly from the lateral geniculate nucleus (LGN), which receives a direct projection from the retina.[23] For information on how human circadian rhythms can be entrained or phase-shifted by a light–dark cycle, see Copinschi *et al.*[33]

Behavioral cycles, such as the sleep–wake or feeding–fasting cycle, may also influence neuroendocrine rhythms independent of the direct control of the central circadian clock in the SCN. Under normal, well-aligned settings, both the sleep–wake and feeding–fasting cycles are dictated by the SCN. However, humans are unique in their ability to override the SCN, choosing to sleep or eat at whatever time they prefer, and not necessarily at times dictated by the circadian system. Recent research has begun to tease apart how each of these behavioral rhythms affects neuroendocrine rhythms independent of direct input from the SCN central clock.

Sleep–Wake Regulation

GENERAL CHARACTERISTICS

The sleep–wake cycle may be viewed as a 24-hour rhythm driven partly by the circadian pacemaker and partly by the homeostatic regulation of sleep pressure. Sleep itself is an ultradian rhythm, as it involves two states of distinct brain activity that are each generated in specific brain regions. The ultradian rhythm of normal sleep is an approximate 90-minute oscillation between non-REM (rapid eye movement) stages and REM stages. In young, healthy subjects, this pattern is usually repeated four to six times per night. REM sleep and non-REM sleep are characterized by distinct patterns of both cerebral and peripheral activity.

In the normal sequence, sleep onset corresponds to the appearance of the lighter stages of non-REM sleep (i.e., stages I and II), followed within 10–20 minutes by slow-wave sleep (SWS; stages III and IV). These deeper stages of sleep are maintained for nearly 60 minutes in normal young subjects, but are usually much shorter (5–10 minutes), if at all present, in older adults. Then, lighter stages of non-REM sleep reappear and the first REM period is initiated. As the night progresses, non-REM sleep becomes shallower, the duration of REM episodes becomes longer, and the number and duration of awakenings increase. In normal young subjects, approximately 50% of a normal night is spent in stage I and II sleep, 20% in SWS, 25% in REM, and 5% awake. In adults over 60 years of age, SWS is usually reduced to only 5–10% and REM sleep to 10–15%, while the proportion of time awake may reach 30% of the night.

During deep non-REM sleep (SWS) the electroencephalogram (EEG) is synchronized with low-frequency, high-amplitude waveforms, referred to as slow waves or delta waves. During REM sleep, eye movements are present, muscle tone is inhibited, and the EEG resembles that of active waking. During REM sleep, cerebral glucose utilization is similar to that of waking, while it is decreased during SWS.

An all-night recording of EEG, muscle tone and eye movements is called a polysomnogram, and is visually scored over 20- or 30-s periods in stages I, II, III, IV, REM and Wake using standardized criteria.[34] This procedure allows the determination of the duration of each sleep stage, but does not quantify the intensity of non-REM sleep. In contrast, the quantification of EEG recordings by power spectral analysis provides useful information regarding sleep depth or sleep intensity, because spectral analysis is sensitive to the amplitude of the delta waves. Higher-amplitude delta waves reflect more intense, deeper sleep, less sensitive to arousal stimuli. Slow-wave activity (SWA) is spectral EEG power in the low frequency range (also called delta range; 0.5–4.0 Hz), and is a marker of the intensity of non-REM sleep.

As detailed below, the timing, duration and architecture of sleep are under the dual control of a homeostatic mechanism relating sleep pressure to the duration of prior wakefulness and to central circadian rhythmicity.

NEUROANATOMICAL BASIS OF SLEEP REGULATION

Normal waking is associated with neuronal activity in regions of the so-called "ascending arousal system," including monoaminergic neurons in the brainstem and posterior hypothalamus, cholinergic neurons in the brainstem and basal forebrain, and orexin (hypocretin) neurons in the lateral hypothalamus.[35,36] The initiation of sleep therefore requires the inhibition of these multiple arousal systems. In recent years, the ventro-lateral preoptic area (VLPO) of the hypothalamus has been identified as being involved in the inhibition of arousal. The VLPO contains "sleep-active neurons" that use the inhibitory neurotransmitter GABA (gamma-aminobutyric acid), and have much higher firing rates during deep sleep than during wakefulness.[37–39] Lesions of the central cell cluster of the VLPO drastically reduce SW activity. Neurons of the VLPO provide GABAergic inhibitory innervation of the major monoamine arousal systems in the brainstem. Reciprocally, there are inhibitory pathways from the monoamine arousal nuclei to the VLPO.[40] The orexin neurons in the lateral hypothalamus project to all components of the ascending arousal system and stimulate the cortex.[41] REM sleep is primarily regulated by cholinergic nuclei in the pons.

INTERACTIONS BETWEEN CIRCADIAN RHYTHMICITY AND SLEEP–WAKE HOMEOSTASIS

There are several features of the interaction between sleep and circadian rhythmicity that appear to be fairly unique to the human species. First, human sleep is generally consolidated in a single 6- to 9-hour period, whereas fragmentation of the sleep period in several bouts is the rule in the majority of other mammals. Possibly as a result of this consolidation of the sleep period, the wake–sleep transition in man is associated with physiological changes that are usually more marked than those observed in animals. For example, the secretion of growth hormone (GH) in normal adults is tightly associated with the beginning of the sleep period, whereas the relationship between GH secretory pulses and sleep stages is much less evident in rodents, primates and dogs. Secondly, man is also unique in having the capacity to ignore circadian signals and to maintain wakefulness despite an increased pressure to go to sleep. Finally, approximately 25% of human subjects maintained for prolonged periods of time in temporal isolation have shown behavioral modifications that have not been observed in laboratory animals under constant conditions. These modifications consist of a desynchronization between the sleep–wake cycle and other rhythms, such as those of body temperature and cortisol secretion, which continue to free-run with a circadian period. Under conditions of so-called "internal desynchronization," the sleep–wake cycle may be suddenly lengthened to 30 hours or more while the rhythm of body temperature continues to free-run with a circadian period.[42] Wakefulness may last more than 30 hours. Remarkably, the subjects are not aware of these drastic changes in their way of living; instead, most of them believe they are living on a more or less regular 24-hour schedule. This can be explained by the observation that time perception is profoundly altered: subjective estimations of 1-hour intervals are positively correlated with the duration of wakefulness.[43] Of particular interest is that the subjects continue to have three meals per "day," irrespective of the actual number of hours they are awake.[44] The intervals between meals, as well as those between wake-up and breakfast, or between dinner and bedtime, are stretched or compressed in strong proportionality to the duration of wakefulness.[45] The mechanisms causing spontaneous internal desynchronization are not completely understood.

Detailed analyses of data obtained during temporal isolation and forced desynchrony protocols showed that the timing, duration and architecture of sleep are partially regulated by circadian rhythmicity.[21,46] Thus, the duration of sleep episodes is correlated with the phase of the circadian rhythm of body temperature, and not with the duration of prior wakefulness. Short (i.e., 7–8 hours) sleep episodes occur in free-running conditions when the subject goes to sleep around the minimum of body temperature, whereas long (i.e., 12–14 hours) sleep episodes occur when sleep starts around the maximum of body temperature. Moreover, the distributions of REM sleep and sleep spindle activity are also markedly modulated by circadian timing. In contrast, the hourglass-like mechanism of sleep–wake homeostasis was originally thought to be largely independent of the circadian system, and to involve one or several putative neural sleep factor(s) (factor(s) "S") that rise during waking and decay exponentially during sleep.[47] This homeostatic mechanism regulates the timing, amount and intensity of SWS and SWA. The VLPO has been proposed as a neuroanatomical locus for the interaction of the homeostatic process and central circadian rhythmicity because it receives dense projections from the dorsomedial hypothalamic nucleus, which itself receives direct and indirect projections from the SCN.[48] Based on human studies (described below), it is thought that the SCN generates a waking signal that promotes alertness during the active period. In support of this theory, studies have shown that rodents and monkeys with SCN lesions have increased sleep duration.[49,50] Furthermore, studies in rats have described an indirect neuronal circuit from the SCN to the locus coeruleus, an area of the midbrain involved in the control of arousal.[51] A role for the SCN in promoting sleep at other circadian times is suggested by the finding of decreased sleep in a mouse with a mutation of the *Clock* gene.[52] The finding that the mutation or deletion of a number of circadian clock genes affects not only the timing of sleep but also many other sleep–wake traits, including traits linked to the homeostatic drive to sleep,[13,53–55] indicates that the circadian and homeostatic processes underlying the regulation of the sleep–wake cycle may be linked at molecular, as well as anatomical levels of organization.

The dual control of sleep by circadian and homeostatic mechanisms extends to the control of objective and subjective measures of sleep tendency, mood and vigilance.[56–58] When wakefulness is extended beyond the usual 16–18 hours, maximum subjective sleepiness coincides with the minimum of body temperature, mood and performance. Remarkably, despite continued sleep deprivation, subjective fatigue then decreases, and mood and performance partially recover during the daytime hours, reflecting an interaction of circadian timing with the accumulation of waking time.[57–60] It is currently thought that the circadian clock generates a waking signal that increases from morning to evening and is maximally expressed in the early evening hours, one to two hours prior to the onset of nocturnal melatonin secretion.[58] This circadian waking signal counteracts the build-up of the putative factor "S" underlying the homeostatic process, allowing maintenance of a high level of alertness throughout the usual waking period. Current data from human studies are also compatible with the hypothesis that the SCN too, generates a "sleep" signal in the early evening hours.[61]

Circadian rhythmicity and sleep–wake homeostasis also interact to regulate hormonal secretion. These modulatory effects were long thought to be present only in hormones directly dependent of the hypothalamo-pituitary axis. However, it is now clear that modulation by circadian rhythmicity and sleep is also present in other endocrine systems, such as glucose regulation and the renin–angiotensin system.[62,63] The pathways by which circadian rhythmicity, sleep–wake homeostasis and their interaction modulate hormonal release are largely unknown. As illustrated in Fig. 12.1, humoral and/or neural signals originating from the hypothalamic circadian pacemaker and from brain regions involved in sleep regulation, affect the activity of the hypothalamic structures responsible for the pulsatile release of neuroendocrine factors which stimulate or inhibit intermittent secretion of pituitary hormones. The autonomic nervous system is another pathway linking the central control of sleep–wake homeostasis and circadian rhythmicity with peripheral endocrine organs. It appears that stimulatory or inhibitory effects of sleep on endocrine release are primarily associated with SW sleep, rather than REM sleep.[64–68] Theoretically, the modulation of neuroendocrine release by sleep and circadian rhythmicity could be achieved either by modulation of pulse amplitude or by modulation of pulse frequency, or by a combination of both. The data available so far seem to indicate that circadian rhythmicity of pituitary hormonal release is achieved primarily by modulation of pulse amplitude without changes in pulse frequency, whereas sleep–wake and REM–non-REM transitions affect pulse frequency. Pituitary hormones that influence endocrine systems not directly controlled by hypothalamic factors probably mediate, together with the autonomous nervous system, the modulatory effects of sleep and circadian rhythmicity on these systems (e.g., counterregulatory effects of GH and cortisol on glucose regulation).[63]

To delineate the relative roles of circadian and sleep effects in the temporal organization of hormonal secretion, strategies based on the fact that circadian rhythmicity needs several days to adapt to abrupt shifts of the sleep–wake cycle have been used. Thus, by shifting the sleep times by 8–12 hours, masking effects of sleep on circadian inputs are removed and the effects of sleep at an abnormal circadian time are revealed. Fig. 12.2 illustrates mean profiles of plasma cortisol, growth hormone (GH), prolactin and thyrotropin-stimulating hormone (TSH) observed in normal subjects who were studied before and during an abrupt 12-hour shift of the sleep–wake and dark–light cycles. The study period extended over a 53-hour span, and included an 8-hour period of nocturnal sleep, a 28-hour period of continuous wakefulness and a daytime period of recovery sleep. To eliminate the effects of feeding, fasting and postural changes, the subjects remained recumbent throughout the study and the normal meal

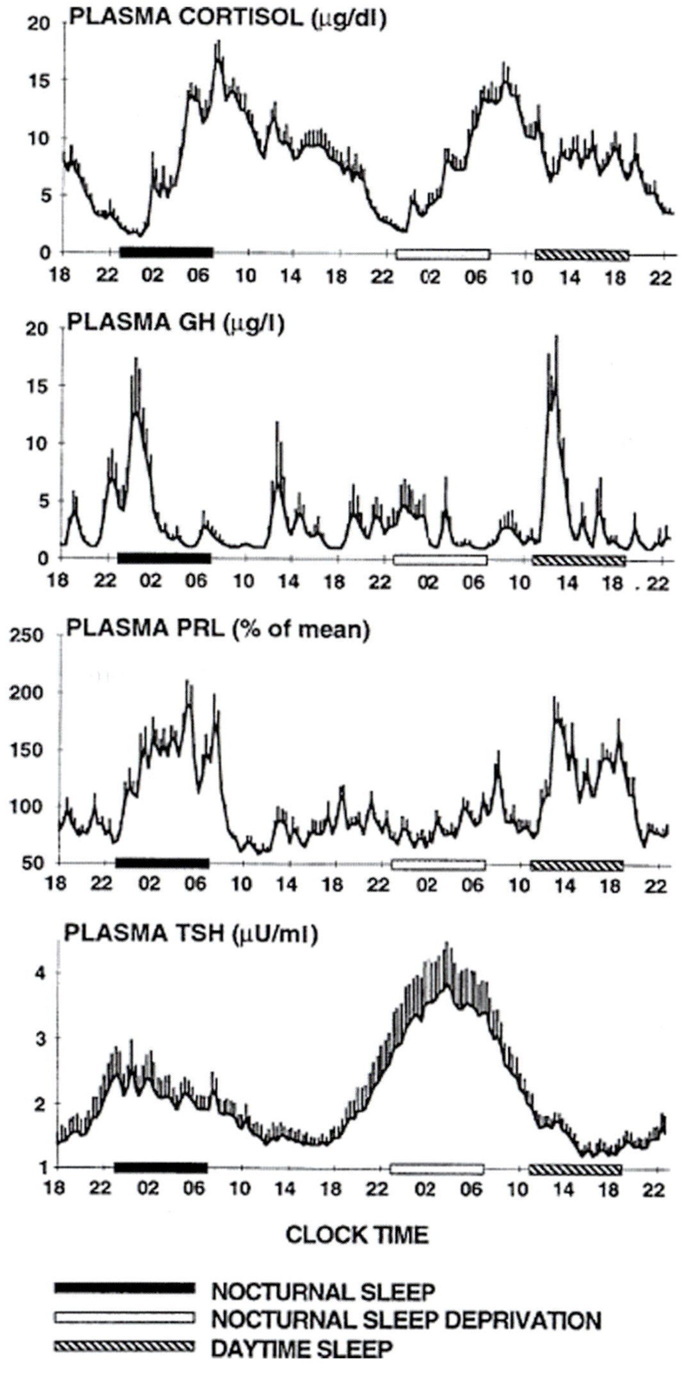

FIGURE 12.2 Mean (+ standard error of the mean (SEM)) 24-hour profiles of plasma cortisol, growth hormone (GH), prolactin (PRL) and thyrotropin (TSH) in a group of eight normal young men (20–27 years old) studied during a 53-hour period that include 8 hours of nocturnal sleep, 28 hours of sleep deprivation and 8 hours of daytime sleep. Black bars, sleep periods; open bars, period of nocturnal sleep deprivation; dashed bars, period of daytime sleep. Data were sampled at 20-minute intervals. *Reproduced from Van Cauter E, Spiegel K. Circadian and sleep control of endocrine secretions. In Turek FW, Zee PC, eds.* Neurobiology of Sleep and Circadian Rhythms, *Vol. 133. New York, NY: Marcel Dekker;1999:397–426, with permission.*

schedule was replaced by intravenous glucose infusion at a constant rate. As shown in Fig. 12.2, this drastic manipulation of sleep had only modest effects on the wave shape of the cortisol profile, in sharp contrast with the immediate shift of the GH and prolactin rhythms that followed the shift of the sleep–wake cycle. As will be reviewed in subsequent sections, numerous studies have indicated that the control of diurnal rhythms of corticotropic activity is primarily dependent on circadian timing, whereas sleep–wake homeostasis appears to be an important factor in the control of the 24-hour profiles of GH and prolactin.[69] Nevertheless, small modulatory effects of sleep–wake homeostasis on cortisol secretion and, conversely, influences of circadian timing on somatotropic function have been clearly demonstrated.[70] The diurnal variation of TSH levels includes an evening elevation thought to be under circadian control and nocturnal inhibition by sleep-dependent processes that is clearly demonstrated during sleep deprivation, when a large increase in nocturnal TSH levels is apparent, as shown in the lower panel of Fig. 12.2.[69]

Hormonal profiles are thus easily measurable reflections of central mechanisms of biological time-keeping. In clinical investigations of conditions of abnormal circadian rhythmicity, such as jet lag, and in human studies of the effects of exposure to natural or artificial zeitgebers, they are commonly used as markers of the status of the circadian clock and of its interactions with sleep.

Feeding–Fasting Rhythms

The timing of feeding, when restricted to a narrow temporal window, can affect the entrainment pattern of behavioral rhythms through what has been referred to as a "food-entrainable oscillator" that is independent of the SCN. Recent findings in rodents have led to a renewed interest in the role of feeding in the overall circadian organization in mammals, including that: (a) alterations in circadian clock genes can lead to obesity and other metabolic abnormalities;[2,18] (b) metabolic transcription factor and nuclear receptors involved in metabolism can alter the expression of circadian clock genes;[71,72] (c) the timing of feeding can alter rhythmicity in a number of peripheral circadian oscillators[18,73] and can lead to obesity;[74] and (d) a high-fat diet can alter behavioral as well as molecular circadian rhythms in central and peripheral tissues involved in the regulation of energy balance.[75] While controversy exists as to the location of a SCN-independent food entrainable oscillator,[76,77] a network of brain regions within the hypothalamus, instead of a single region, has recently been suggested to be acting as the FEO.[78,79] Nevertheless, the overwhelming evidence that the circadian and metabolic systems are linked together at molecular, cellular and behavioral levels has fueled great interest in the possible role of circadian disorganization in obesity, diabetes and other cardiometabolic disorders.[18,80,81]

GLUCOSE TOLERANCE AND INSULIN SECRETION

In normal man, glucose tolerance varies with the time of day. Fig. 12.3 shows circadian variations in glucose tolerance to oral glucose, identical meals, constant glucose infusion and continuous enteral nutrition. In all four conditions, plasma glucose levels are markedly higher in the evening than in the morning.[63] Studies of fasting during nocturnal sleep have consistently observed that, despite the prolonged fasting condition, glucose levels remain stable or decrease only minimally during the night, contrasting with a clear decrease during daytime fasting. Thus, a number of mechanisms operative during nocturnal sleep are likely to maintain stable glucose levels during the overnight fast. Experimental protocols involving intravenous glucose infusion or enteral nutrition while allowing for normal nocturnal sleep have shown that glucose tolerance deteriorates further as the evening progresses, reaches a minimum around mid-sleep, and then improves to return to morning levels.[82,83] During the first half of sleep, SWS is the dominant sleep stage. During SWS, cerebral glucose utilization is lower than during either wake or REM sleep.[84,85] A strong correlation between SWA and regional blood flow in the prefrontal regions was demonstrated by PET scans of subjects who were under continuous polygraphic recordings.[84,86] Brain glucose metabolism represents 30–50% of total body glucose utilization,[87,88] and therefore, a robust link between SWS and glucose tolerance should be expected. There also is evidence that the diurnal variation in glucose tolerance is partly driven by the wide and highly reproducible diurnal rhythm of plasma cortisol, an important counter-regulatory hormone.[62,89,90] Indeed, the diurnal variation in insulin secretion was found to be inversely related to the cortisol rhythm, with a significant correlation of the magnitudes of their morning to evening excursions. Rises in plasma levels of glucose and insulin following short-term elevations of plasma cortisol are more pronounced in the evening than in the morning.[90] Diminished insulin sensitivity and decreased β-cell responsiveness are also involved in reduced glucose tolerance later in the day. Under conditions of constant glucose infusion, sleep-associated rises in glucose were found to correlate with the amount of concomitant GH secreted. Thus, during the first part of the night decreased glucose tolerance is due to decreased glucose utilization both by peripheral tissues – resulting from muscle relaxation and rapid insulin-like effects of sleep-onset GH secretion – and by the brain.[85,91] During the second part of the night, these effects subside as sleep becomes shallow and more fragmented and GH is no longer secreted. Thus, complex interactions of

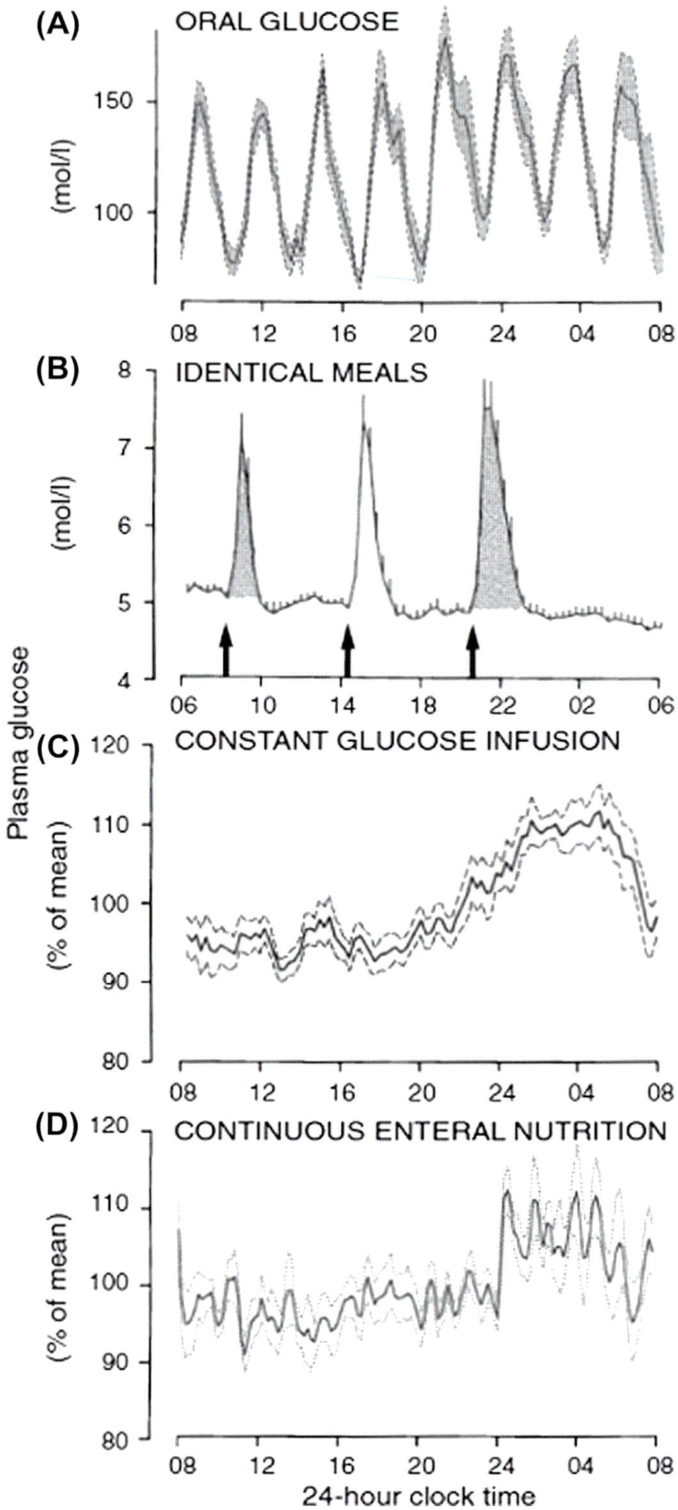

FIGURE 12.3 Mean (± SEM) 24-hour pattern of plasma glucose changes in response to oral glucose 50 g every 3 hours, identical meals, constant glucose infusion, and continuous enteral nutrition in normal young adults. *Reproduced from Van Cauter E, Polonsky KS, Scheen AJ. Roles of circadian rhythmicity and sleep in human glucose regulation.* Endocr Rev. *1997;18:716–738, with permission.*

circadian and sleep effects, possibly partly mediated by cortisol and GH, result in a consistent pattern of changes of set-point of glucose regulation over the 24-hour period. A recent study has demonstrated a continuous decline of plasma glucagon levels during night-time sleep in healthy non-diabetic subjects.[92] Because this nocturnal decline is preserved in type 1 diabetic patients, it has been suggested that the nocturnal regulation of spontaneous glucagon release is independent of circulating glucose and insulin levels.[92]

Consistent with the important modulatory effects of sleep on glucose regulation, recurrent sleep loss is associated with marked alterations of parameters of glucose tolerance. In a study of sleep curtailment (4 hours in bed per night for 6 nights) performed in young healthy subjects,[93,94] nocturnal glucose levels were similar to those recorded during an 8-hour bedtime sleep, but on the next morning the overall glucose response to breakfast was increased and the insulin sensitivity – as assessed by the homeostatic model assessment (HOMA) index, an index directly proportional to the product of insulin by glucose concentrations – was decreased (Fig. 12.4). Moreover, following an intravenous glucose tolerance test, insulin release was reduced by 30%, and glucose tolerance was found to fall in the range observed in older adults with impaired glucose tolerance. The deleterious impact of sleep restriction on glucose metabolism was subsequently confirmed in a follow-up study using a randomized cross-over design.[95] Twelve healthy men were investigated after 2 consecutive nights of 10 hours in bed (2200 to 0800 h), and after 2 consecutive nights of 4 hours in bed (0100 to 0500 h). After the second night of each bedtime condition, caloric intake was replaced by an intravenous glucose infusion at a constant rate. Even though sleep duration was manipulated for only 2 nights, the glucose and insulin profiles obtained during continuous glucose infusion were consistent with the results obtained in the previous study. Indeed, in the early part of the day, glucose levels were higher and insulin levels were lower following 2 days of short bedtimes than following 2 days of long bedtimes.[95] Recently, a laboratory study in young healthy adults demonstrated that reduced sleep quality, without change in sleep duration, also has a clear negative impact on glucose tolerance. Indeed, this study indicated that all-night selective suppression of SWS, without any change in total sleep time, resulted in marked decreases in insulin

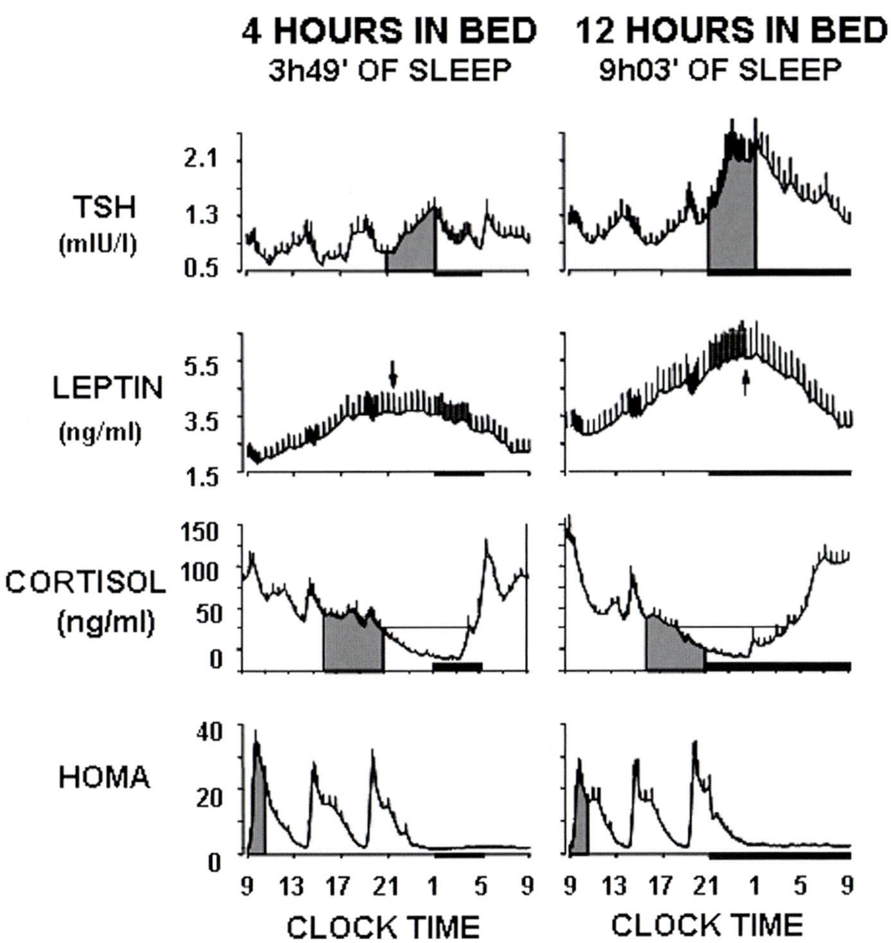

FIGURE 12.4 Impact of recurrent sleep curtailment (4 hours in bed for 6 nights) and of sleep recovery (12 hours in bed for 6 nights) on 24-hour plasma profiles of thyrotropin (TSH), leptin, and cortisol, and on homeostatic model assessment (HOMA) in healthy young men. HOMA was calculated as glucose concentration (mg/dl) × insulin concentration (μU/ml). *Adapted from Spiegel* et al. *(1999);*[93] *and Spiegel* et al. *(2004).*[94]

sensitivity without adequate compensatory increase in insulin release, leading to reduced glucose tolerance and increased diabetes risk.[96] To date, there have been seven prospective epidemiologic studies that found an association between short and/or poor sleep and diabetes risk, even after controlling for many covariates, such as BMI, shiftwork, hypertension, exercise and depression (reviewed in ref. 97).

Human insulin secretion is a complex oscillatory process involving rapid pulses of small amplitude recurring every 10–15 minutes superimposed on slower ultradian oscillations with periods in the 90- to 120-minute range.[98,99] The ultradian oscillations are tightly coupled to glucose, with a tendency for glucose pulses to lead insulin pulses by 10 minutes, and have been shown to promote more efficient glucose utilization.[100] They are best seen in conditions where insulin secretion is stimulated, including ingestion of meal, continuous enteral nutrition or constant intravenous glucose infusion.[83,98] Under these conditions, their relative amplitude is about 50–70% for insulin secretory pulses and 20% for plasma glucose. Their amplitude is maximal immediately after a meal, then decreases progressively. Moreover, the periodicity of the insulin secretory oscillations can be entrained to the period of an oscillatory glucose infusion,[101] supporting the concept that these ultradian oscillations are generated by the glucose–insulin feedback mechanism.[102] However, ultradian oscillations, but less regular and of smaller amplitude, are still present in fasting conditions. Stimulatory effects of sleep on insulin secretion are mediated by an increase in the amplitude of the oscillation.[83] During constant glucose infusion, REM sleep and wake episodes coincide significantly with decreasing levels of glucose and insulin, while increasing glucose levels occur during the deeper stages of non-REM sleep.[91]

The rapid 10- to 15-minute pulsations seem to have a different origin than the ultradian oscillations. Indeed, they may appear independently of glucose, since they were observed in the isolated perfused pancreas and in perifused islets.[98] Rapid insulin pulsations were also observed in perfused human islets.[103] Insulin administration by pulsatile infusion improves insulin-mediated glucose uptake.[104] The frequency, amplitude and regularity of rapid insulin pulses are decreased in aging.[105] Omental lipolysis is also pulsatile, with a rapid frequency, similar to that of insulin.[106] However, the oscillation of free fatty acids appears to be driven by the central nervous system, rather than by insulin.[106]

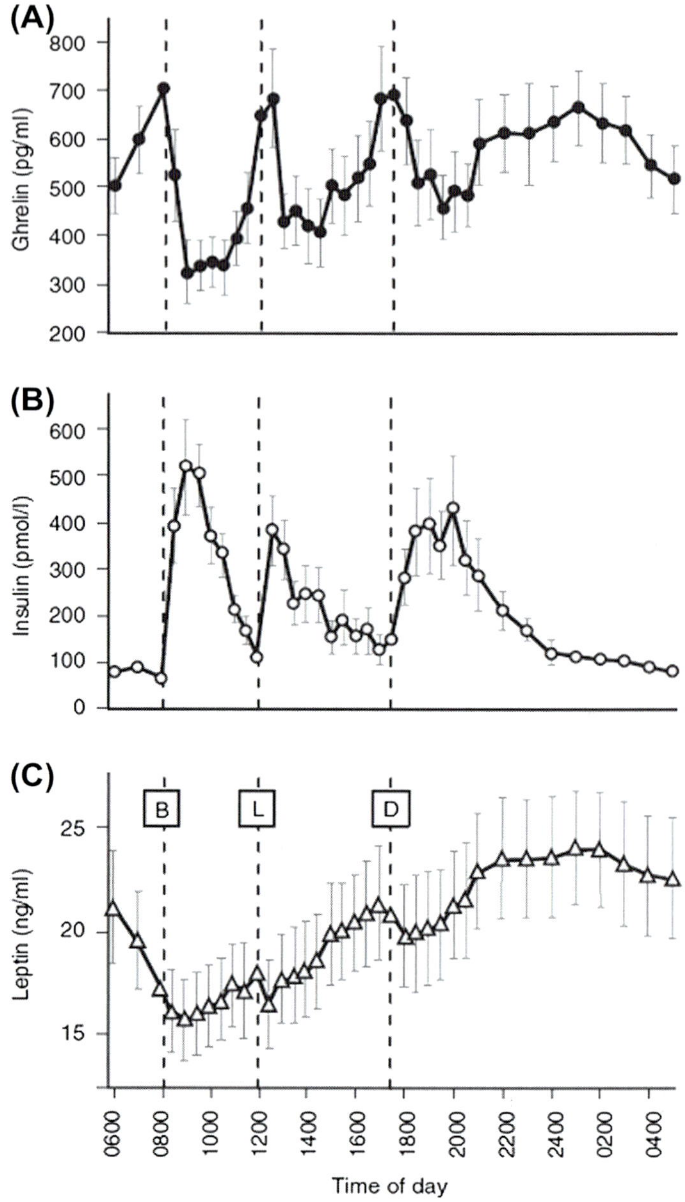

FIGURE 12.5 Mean (± SEM) 24-hour profiles of plasma ghrelin, insulin and leptin levels in 10 healthy subjects 29–64 years of age (body mass index (BMI) 22–30 kg/m^2) receiving breakfast, lunch and dinner at 0800, 1200 and 1730 h, respectively. *Adapted from, Cummings et al. (2001).*[145]

HORMONES INVOLVED IN APPETITE REGULATION

LEPTIN Leptin is an anorexigenic hormone, mainly released by the adipocytes, that provides information about energy status to hypothalamic regulatory centers.[107,108] As illustrated in Fig. 12.5, plasma leptin levels in normal lean men and women show a robust diurnal rhythm, with minimum values during the daytime, and a nocturnal rise with maximum values during early to mid-sleep.[109] The amplitude of the diurnal variation averages 25–30% of the mean level.[110]

The timing of the daily maximum of plasma leptin levels is markedly dependent on the timing of meals, as shifts in meal timings induce immediate shifts in leptin profiles: fasting and eating are associated with a decrease and an increase in leptin levels, respectively.[111] However, the diurnal rhythm was found to persist, albeit with a smaller amplitude, in subjects who received continuous enteral nutrition,[83] and in subjects receiving identical snacks at 2-hour intervals.[112] The rhythm also persisted in subjects submitted to 38 or to 88 consecutive hours of wakefulness.[112,113] Following an abrupt shift of the sleep period, nocturnal leptin levels rose despite the absence of sleep, and a second rise was observed following the onset of daytime recovery sleep.[83] Thus, leptin diurnal variations reflect the combined effects of the circadian pacemaker, the sleep—wake cycle and the food intake.

Several studies have reported that human leptin levels are pulsatile,[83,110,114–116] including in subjects receiving continuous enteral nutrition rather than separate meals.[83]

Leptin levels reflect cumulative energy balance, with a decline or increase in response to under- or overfeeding, respectively.[117,118] These changes have been found to be associated with reciprocal changes in hunger.[118] Circulating leptin concentrations are higher and the relative amplitude of their diurnal variation is lower in obese subjects than in normal-weight controls.[110] There are marked sex differences in 24-hour mean leptin levels, which are two- to ten-fold higher in women than in men, regardless of fat mass.[110,116] In anorexia nervosa, as in amenorrheic female athletes, leptin levels are low and diurnal variations are abolished.[119,120] Aging is associated with a dampening of the amplitude of the 24-hour rhythm of plasma leptin and an advance of the nocturnal acrophase.[121]

Both 2 nights and 6 nights of sleep restriction (4 hours bedtime per night), performed under controlled conditions of caloric intake and physical activity, were found associated with a 20—30% reduction in mean leptin levels, acrophase and amplitude of the diurnal variation.[94,122] The magnitude of this impact of sleep restriction on leptin levels is comparable to that observed in young adults under normal sleep conditions after 3 days of dietary restriction by approximately 900 kcal per day.[118] Consistent findings have been obtained in two epidemiologic studies that found association between short sleep and lower morning leptin levels after controlling for BMI[123] or the degree of adiposity.[124] The diurnal variations of leptin and cortisol levels are approximate mirror images,[115,116,125,126] and in fully rested subjects maximum leptin levels coincide with minimum cortisol levels,[94] consistent with the well-documented action of leptin on HPA activity.[108] Recurrent partial sleep deprivation results in an advance of the leptin acrophase such that high levels of leptin occur when cortisol concentrations are still elevated relative to the nocturnal nadir, possibly acting in concert to increase appetite at the end of the day.[94]

Normal 24-hour leptin levels and diurnal leptin variations were found in patients with primary adrenal failure.[126] Increased leptin levels,[127,128] caused by an equal amplification of basal and pulsatile secretion,[129] with a preservation of the diurnal pattern[127] have been reported in patients with Cushing's syndrome. A normal diurnal leptin pattern has also been reported in patients with GH deficiency[130] or with perinatal stalk-transection syndrome.[131]

Another secretory product of differentiated adipocytes is adiponectin, a hormone that enhances insulin sensitivity.[132,133] In normal-weight men, adiponectin levels exhibit ultradian pulsatility as well as diurnal variation, with a significant nocturnal decline, reaching minimum values in the early morning.[134] So far, possible relations between the adiponectin rhythm and sleep have not been reported. Adiponectin levels are decreased, adiponectin pulsatility is blunted and diurnal variations are abolished in obese subjects.[135–137] In severely obese subjects, massive weight loss coupled with the reversibility of insulin resistance is associated with a restoration of adiponectin pulsatility.[137]

GHRELIN Ghrelin is an orexigenic hormone secreted primarily by the stomach and the duodenum.[138–140] Ghrelin also stimulates GH secretion and displays ACTH- and prolactin-releasing activities.[140] One report claimed that ghrelin was found to promote SW sleep in man,[141] but others found suppressive effects on sleep in rats.[142] Ghrelin secretion is pulsatile.[136,143] Daytime profiles are primarily regulated by the schedule of food intake: levels rise sharply before each designated meal time and fall to trough levels within 1 hour after eating. A study examining spontaneous meal initiation in the absence of time- and food-related cues provided good evidence for a role for ghrelin in meal initiation.[144] This pattern seems to be exaggerated after the dinner meal, as ghrelin levels peak around 0100 h and remain elevated until the later part of the night, when they tend to spontaneously decrease[145–147] (Fig. 12.5). Ghrelin levels are increased in anorexia nervosa and decreased in young obese subjects.[146,148] In obese subjects, the diurnal pattern was found either to remain largely unaltered[146] or to be markedly blunted.[135] A diet-induced weight loss was associated with increased 24-hour ghrelin levels.[146]

Diurnal variations in ghrelin levels persist after 3 days of total fasting. However, the timings of acrophase and nadir are not consistent across studies.[149] Studies concord in indicating that ghrelin levels are not increased by prolonged fasting, and that women have higher levels than men.[149] The diurnal variations in ghrelin likely reflect the combined effects of the circadian pacemaker, of the schedule of food intake, and possibly of the sleep—wake cycle.

Acute total sleep deprivation was found to be associated with a blunted but prolonged nocturnal ghrelin

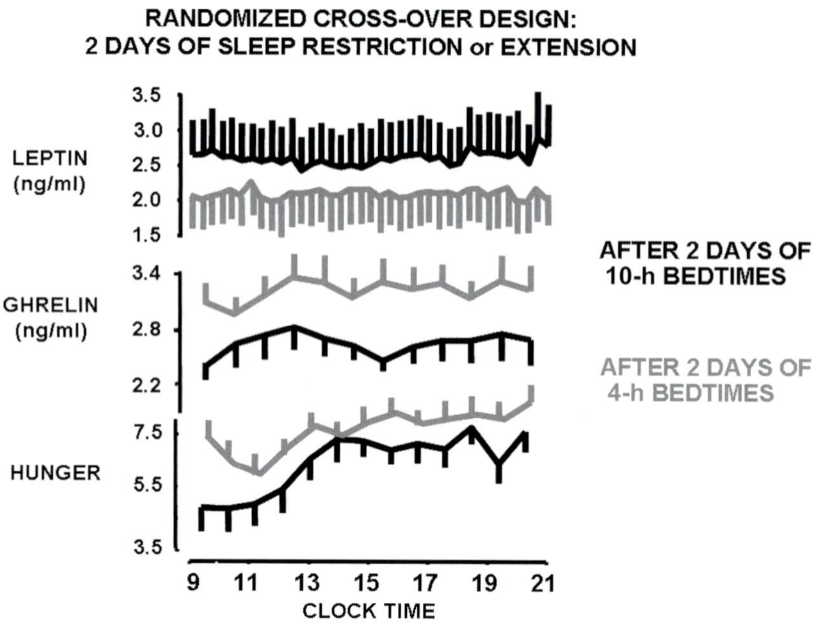

FIGURE 12.6 Mean (±SEM) daytime profiles of plasma leptin and ghrelin and of hunger ratings in healthy young men after 2 days of 10-hour bedtimes and 2 days of 4-hour bedtimes. *Data from Spiegel et al. (2004).*[122]

elevation.[150] A 2-day sleep restriction (4 hours bedtime per night) under controlled conditions of caloric intake and physical activity was reported to be associated with a nearly 30% elevation in daytime levels[122] (Fig. 12.6). Consistent with these laboratory findings is the recent observation of a 22% increase in plasma ghrelin levels after a single night of total sleep deprivation.[151] In a large epidemiologic study, short sleep was associated with higher ghrelin levels after controlling for sex, age and BMI.[12]

Circadian Clock Genes: The Molecular Genetic Basis for Mammalian Circadian Rhythms

The foundation of all circadian rhythms begins with the transcriptional–translational feedback of core circadian clock genes. Indeed, the molecular clock machinery is found in all organs and tissues of the body, including endocrine tissues (e.g., pituitary, pancreas). To fully understand the extent to which circadian rhythms influence neuroendocrine expression, this section focuses on the genetic basis of rhythms, including how the core circadian genes were identified.

The synthesis of proteins by each oscillatory cell is central to the mechanism for the generation of 24-hour circadian rhythms. Gene expression profile studies, in which expression levels are sampled at regular timepoints in constant darkness (free-running conditions), reveal that approximately one-third of the transcriptome is rhythmically expressed, even in peripheral tissues.[10–13] With so many genes exhibiting circadian expression, and competent oscillators present in such a variety of tissues, identifying a valid "clock gene" is a challenge. Nevertheless, multiple genes have now been identified as part of the core circadian clock mechanism.

The Positive Elements

CLOCK

In the early 1990s, the *Clock* gene was discovered while screening for mutations in over 300 progeny of mutagen-treated mice. One animal had a free-running period of about 24.8 hours, more than six standard deviations longer than the mean.[152] In the homozygous condition this mutation results in a dramatic lengthening of the period to about 28 hours, which is usually followed by the eventual loss of circadian rhythmicity (i.e., arrhythmicity) after about 1–3 weeks in DD. The affected gene was mapped to mouse chromosome 5, and named *Clock*.[152,153] The *Clock* gene encodes a transcriptional regulatory protein having a basic helix–loop–helix DNA-binding domain, a PAS dimerization domain and a Q-rich transactivation domain.

BMAL1

The presence of the PAS dimerization domain in CLOCK protein suggested that it may form a heterodimer similar to that of PER and the protein product of another *Drosophila* clock gene, TIM.[154] A screen for potential partners for the CLOCK protein, using the yeast two-hybrid system, revealed that a protein of unknown function, BMAL1 (Brain and Muscle ARNT-Like 1), was able to dimerize with the CLOCK protein.[155] Creation of mice harboring a null allele of *Bmal1* (also referred to as *MOP3*) demonstrated the critical role of

BOX 12.1

HOW DO WE KNOW ABOUT THE *CLOCK* GENE?

By the early 1990s, a circadian rhythm of locomotor activity had been identified in a variety of organisms, including humans, rodents, and the fruit fly *Drosophila*. However, few genes had been identified to regulate the circadian clock, and no molecular mechanisms could explain the rhythmic patterns of activity. A key advance occurred with the discovery of *Clock*, a gene found to lengthen circadian period when mutated.[1] The *Clock* mutation was discovered by using a combination of chemical mutagenesis, genetic mapping, and positional cloning. Importantly, a similar approach could be used today to elucidate genes associated with other complex behaviors.

To identify *Clock*, mutagenesis began with a single injection of *N*-ethyl-*N*-nitrosourea (ENU) into male C57BL/6J (B6) mice. The B6 mouse was selected due to its well-characterized genome and circadian period. The ENU drug leads to random point mutations, commonly nucleobase transversions, in approximately 1 out of every 700 gametes. Mutations occurring within the spermatogonial stem cells are heritable and can be passed on to progeny. The treated males are then mated with healthy, untreated B6 females to produce the first generation (G_1) of progeny which are heterozygous for possible mutations (see figure below).

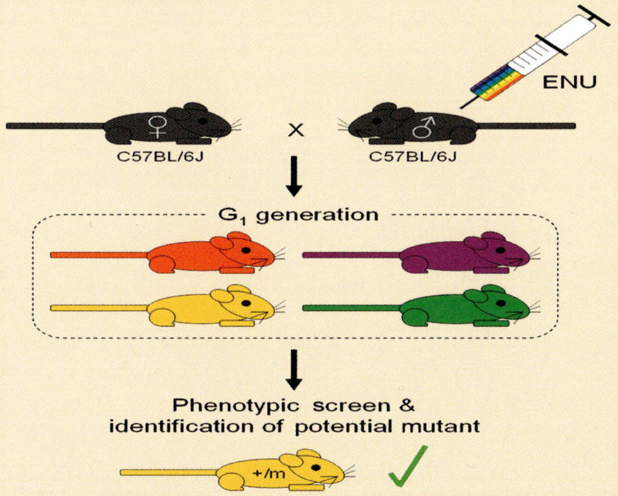

ENU mutagenesis and phenotypic screening.

At this point, the G_1 mice must be phenotypically screened. This portion of the mutagenesis screen can be custom designed for any behavior of interest. We chose to use the circadian free-running period because it is a fundamental clock property, relatively easy to measure (e.g., minimally invasive), and known to have a low variance in the wild-type B6 mouse, thus making mutants more likely to stand out. The average circadian period of the B6 mouse is between 23.3 and 23.8 hours, and follows a normal distribution. Any circadian period significantly different from the average could potentially indicate a mutation in a circadian gene. After screening 304 G_1 progeny, one male mouse (G_1-25) was found to have a long running period (24.8 hours) that was more than six standard deviations from the mean.

Following identification of a potential mutant (G_1-25), we next determined if the mutation was heritable. G_1-25 was mated with healthy B6 females to produce a backcross (N_2) generation. These N_2 mice were screened for circadian period and found to have a phenotypic ratio of 1 : 1 (long : wild-type period), suggesting that the mutation functioned as a single-locus, semidominant autosomal mutation. Heterozygous long period mice were also mated together (e.g. *Clock*/+ × *Clock*/+) to produce the F_2 generation. From the heterozygous cross, 1 : 4 of the F_2 progeny were found to have extremely long periods (26–29 hours), 2 : 4 were found to have long periods, and 1 : 4 were found to have wild-type periods. This was further indication of *Clock* segregating as a single gene.

To determine the location of the *Clock* gene, linkage analysis was used on an N_2 generation of (BALB × B6)F_1 crossed with wild-type B6 (see figure below). Since the *Clock* mutation derives from a B6 mouse, N_2 mice exhibiting a long period would be expected to have B6 alleles flanking the location of the *Clock* mutation, whereas other regions would be from the opposing BALB/cJ (BALB) inbred strain. This combination of B6 and BALB alleles allows the testing of multiple haplotypes using simple sequence length polymorphisms (SSLPs) as DNA markers to link to the long period phenotype. Positional mapping found a high association with SSLPs on chromosome 5, eventually determining that the *Clock* gene must lie on the mid-portion of chromosome 5, 6.0 ± 2.4 cm distal to SSLP marker *D5Mit83* and 7.0 ± 2.6 cm proximal to marker *D5Mit24*.

ENU mutagenesis and positional cloning can provide an effective approach to studying the genetic basis of complex behavior. While ENU mutagenesis assumes there are genes that, when mutated, will alter the phenotype, it does not make any assumptions about the identity of the genes or where they are located. For this reason, using the ENU mutagenesis approach can be particularly beneficial when little is known about the processes which lead to the behavior of interest.

BOX 12.1 cont'd

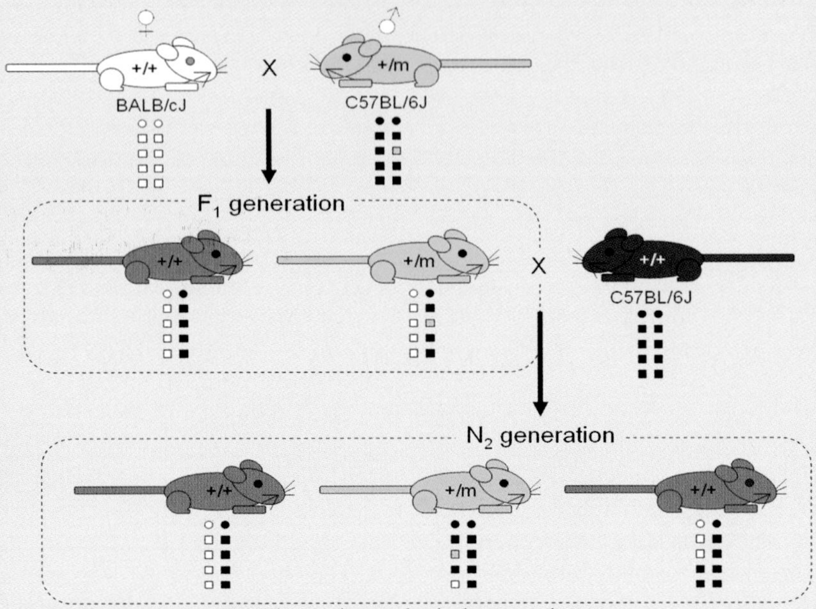

Breeding scheme for linkage analysis.

Reference

1. Vitaterna MH, King DP, Chang AM, et al. Mutagenesis and mapping of a mouse gene, CLOCK, essential for circadian behavior. *Science*. 1994;264(5159):719–725.

this gene in circadian rhythm generation. These mutant mice, while displaying light–dark responsive differences in activity level, become arrhythmic immediately upon release in constant darkness.

Recently, additional actions of the CLOCK–BMAL1 heterodimer have become clear. While *Clock* mRNA does not oscillate, its protein's nuclear vs cytoplasmic localization does.[156] By studying the intracellular localization of CLOCK and BMAL1 in fibroblasts of mouse embryos with mutations in different clock genes, and ectopically expressing the proteins, it was found that nuclear accumulation of CLOCK was dependent on formation of the CLOCK:BMAL1 dimer, as was phosphorylation of the complex and its degradation.[156] Other PAS domain containing proteins failed to affect the localization of CLOCK, indicating that these post-translational events are specific to the CLOCK–BMAL dimer.

Negative Elements

THE *PERIOD* GENES

The first identified gene that encodes a clock component, *period*, denoted with the symbol *per*, was discovered in 1971 in *Drosophila*.[157] The *per* gene of *Drosophila* contains an upstream regulatory element, the "clock control region," within which is contained a sequence needed for positive regulation of transcription, the E-box element (CACGTG).[158] Confirmation of the importance of the *per* gene as a central circadian clock component was the rescue of the mutant phenotype after introduction of the wild-type allele of the *per* gene into mutant flies.[159,160]

Following the identification of CLOCK:BMAL1 dimerization, the ability of this heterodimer to regulate transcription was tested using a reporter construct based on the upstream regulatory elements of the *per* gene. CLOCK–BMAL1 heterodimers were found to activate transcription of the *mPer* gene in a process that requires binding to the E-box element.[155] However, CLOCK Δ19 mutant protein was not able to activate transcription, consistent with the finding that exon 19, which is skipped in *Clock* mutant animals,[161] is necessary for transactivation. Thus, CLOCK protein interacts with the regulatory regions of the *per* gene to allow transcription of the *per* mRNA and eventual translation of PER protein. A similar activation of transcription of the *tim* gene by the CLOCK–BMAL1 heterodimer also occurs in flies.[162] However, this positive regulation alone will not produce an oscillation in *per* mRNA levels, which is known to be responsible for the oscillation in PER protein levels.[163] Findings that the *Clock* mutation dramatically decreases *per* genes' expression also confirms the positive regulation of CLOCK–BMAL1 on *per* transcription *in situ.* [47,164] Mice with null mutations of *mPer1*, *mPer2* or *mPer3* alone display altered circadian periods,[165,166] while mice with both *mPer1* and *mPer2* null mutations lose rhythmicity. *mPer3* null mutant mice exhibit only a subtle alteration in rhythmicity, and *mPer1/mPer3* or *mPer2/mPer3* double mutants are not substantially distinct from the *mPer1* or *mPer3* single mutants. These findings suggest there may be some compensation of function among the different mammalian *per* genes, and raise the question of the significance of *mPer3* for the generation of mammalian circadian rhythms.

CRYPTOCHROMES

Cryptochromes are blue-light responsive flavoprotein photopigments related to photolyases, so named because their function was cryptic when first identified. In mammals, two cryptochrome genes, *Cry1* and *Cry2*, have been identified, and were found to be highly expressed in the ganglion cells and inner nuclear layer of the retina as well as the SCN,[167] and their mRNA expression levels oscillate in these tissues. Targeted mutant mice lacking *Cry2* exhibit a lengthened circadian period, while mice lacking *Cry1* have a shortened circadian period; mice with both mutations have immediate loss of rhythmicity upon transfer to constant darkness.[168–170] Thus, like the mammalian *period* genes, the *cryptochrome* genes appear to have both distinct (given their opposite effects on circadian period) and compensatory (given that either gene can sustain rhythmicity in the absence of the other) functions.

Further evidence for a central clock function is the finding that the *cryptochromes* appear to share a number of regulatory features with the *period* genes. In *Clock* mutant mice, the mRNA levels of *Cry1* and *Cry2* are reduced in the SCN and in skeletal muscle,[171] suggesting that the *cryptochromes* also are induced by CLOCK:BMAL1 transactivation. Using mammalian (NIH 3T3 or COS7) cell lines, CRY1 and CRY2 were found by coimmunoprecipitation to interact with mPER1, mPER2 and mPER3, leading to nuclear localization of the CRY:PER dimer as indicated by co-transfection assays with epitope-tagged proteins.[171] Luciferase assays indicate that CRY:CRY or CRY:PER complexes were capable of inhibiting CLOCK:BMAL1 transactivation of *mPer1* or vasopressin transcription.[171] Thus, the CRYs as well as the PERs are capable of a negative feedback function, inhibiting CLOCK:BMAL1-induced transcription.

Since the identification of core circadian genes, a number of modulating genes have been found which together create the 24 circadian oscillations in mammalian cells. Indeed, our current knowledge of the molecular genetic interactions of circadian clock gene provides for a basic mechanism involving multiple feedback loops of clock proteins on transcription of clock gene. These core interactions are summarized in Fig. 12.7. Undoubtedly, additional clock genes will be part of the mechanism, and additional transcriptional, translational, and post-translational interactions among these genes and their proteins will be identified.

Circadian Clock Gene Summary

Circadian rhythms result from coordinated daily oscillations in the transcription and translation of several clock component genes. In mammals, central to the generation of these cycles are the levels of the proteins PER and CRY, which feed back to inhibit transcription of their own genes. This inhibition is exerted on the enhancement of transcription that results from binding of the CLOCK and BMAL1 proteins to E-box elements of the promoter regions of the *Per* and *Cry* genes. These four genes are thought to form the core feedback loop of the time-keeping mechanism.

Additional interactions between the protein products of these genes, as well as other proteins, appear to add to the complexity of the circadian system. The phosphorylation of PER by CKIε may lead to its degradation, and the association with BMAL1 appears needed for CLOCK to be present in the nucleus. Rhythmic transcription of *Bmal1* appears to result from regulation via the protein REV-ERBα, its transcription regulated by CLOCK: BMAL1 binding to E-box elements. Furthermore, a number of cyclic clock genes have been identified in key endocrine tissues and brain regions, and are directly involved in endocrine rhythms such as the LH induction of progesterone synthesis [172] and estrogen cyclicity.[16]

NEURENDOCRINE RHYTHMS

Diurnal and/or ultradian oscillations have been observed in essentially all endocrine systems. An exhaustive review of all such observations is not possible. The following summary of the findings will therefore be limited to the neuroendocrine rhythms of the hypothalamo-pituitary axes.

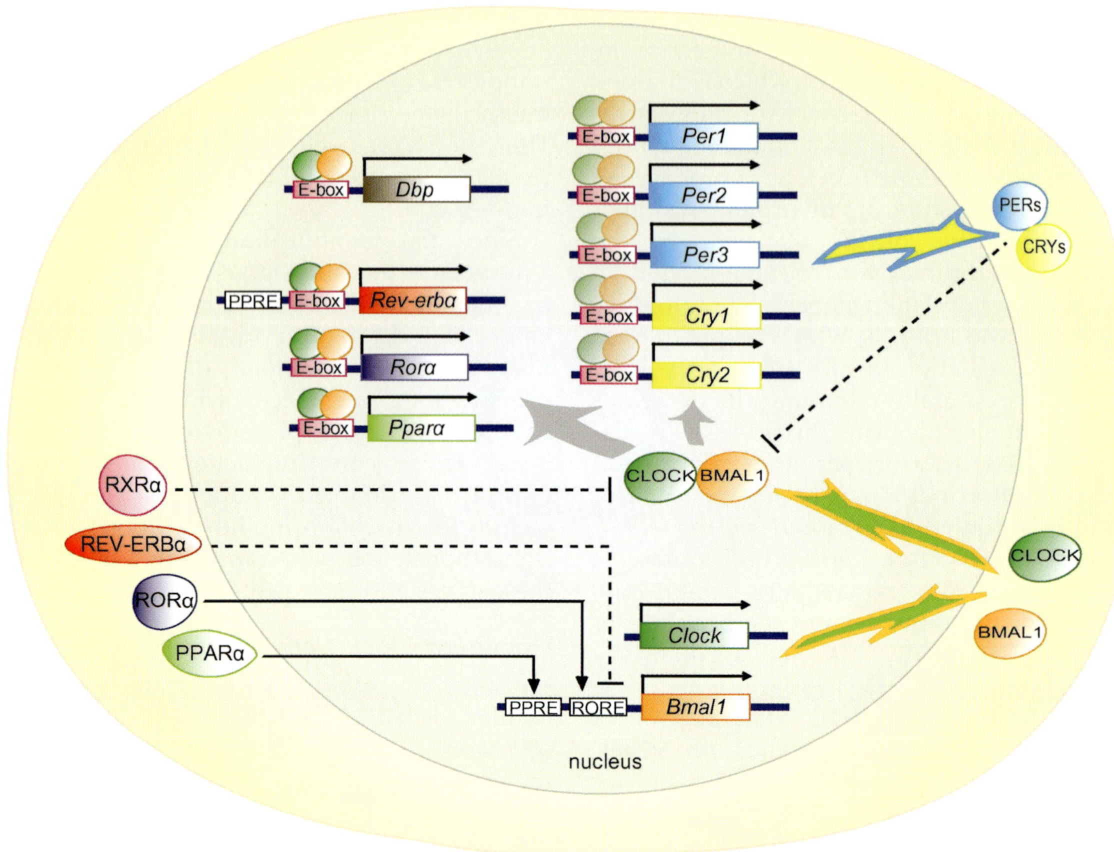

FIGURE 12.7 Transcriptional–translational feedback loop of core circadian clock genes. This diagram depicts the positive (CLOCK:BMAL1) and negative (PER:CRY) limbs of the circadian clock gene feedback loop that is present in virtually all cells throughout the body. REV-ERBα and RORα represent components of a secondary feedback loop that either activate (solid line with arrow) or inhibit (dashed line) *Bmal1* activity. The CLOCK:BMAL1 transcription complex has the ability to directly mediate the activity of genes (clock controlled genes) critically involved in energy metabolism, including Dbp and Pparα. In turn, PPARα can directly bind to *Bmal1* and influence activation of the positive limb of the feedback loop. BMAL1, brain muscle arnt like factor; CLOCK, circadian locomotor output cycles kaput; CRY, cryptochrome; DBP, albumin D-element binding protein; PER, period; PPARα, peroxisome proliferator-activated factor alpha; REV-ERBα, reverse erythroblastosis virus alpha; RORα, retinoic acid receptor-related orphan receptor alpha; RXRα, retinoid X receptor alpha. *Adapted from Kohsaka* et al. *(2007).*[75]

The Corticotropic Axis

Outputs from the SCN activate rhythmic release of corticotropin-releasing hormone (CRH) that stimulates circadian ACTH release. The 24-hour rhythm of adrenal secretion is primarily dependent on the diurnal pattern of ACTH release. In addition, neuronal signals generated by the SCN are transmitted by a multisynaptic neural pathway to the adrenal cortex.[173] The presence in the adrenal cortex of an intrinsic circadian oscillator consisting of interacting positive and negative feedback loops in circadian gene expression has been demonstrated in various animals, including monkeys,[174–176] and it has been shown that this adrenal circadian pacemaker gates the physiological adrenal response to ACTH – i.e., defines a time window during which the adrenal most effectively responds to ACTH.[177]

The 24-hour profiles of ACTH and cortisol show an early morning maximum, declining levels throughout daytime, a quiescent period of minimal secretory activity centered around midnight and an abrupt elevation during late sleep resulting in an early-morning maximum. Mathematical derivations of secretory rates from plasma concentrations have suggested that the 24-hour profile of plasma cortisol reflects a succession of secretory pulses of magnitude modulated by a circadian rhythm with no evidence of tonic secretion.[178,179] In normal conditions, the acrophase of the pituitary–adrenal periodicity occurs between 0600 and 1000 h. With a 15-min sampling interval, 12 to 18 significant pulses of plasma ACTH and cortisol per 24-hour span can be detected.[180] Circadian and pulsatile variations parallel to that of cortisol have been demonstrated for the plasma levels of several other adrenal steroids, in particular dehydroepiandrosterone (DHEA).[181] The temporal concomitance of 24-hour profiles of ACTH, cortisol and DHEA is illustrated in Fig. 12.8.

The profile shown in the upper panel of Fig. 12.2 illustrates the remarkable persistence of the cortisol and, by

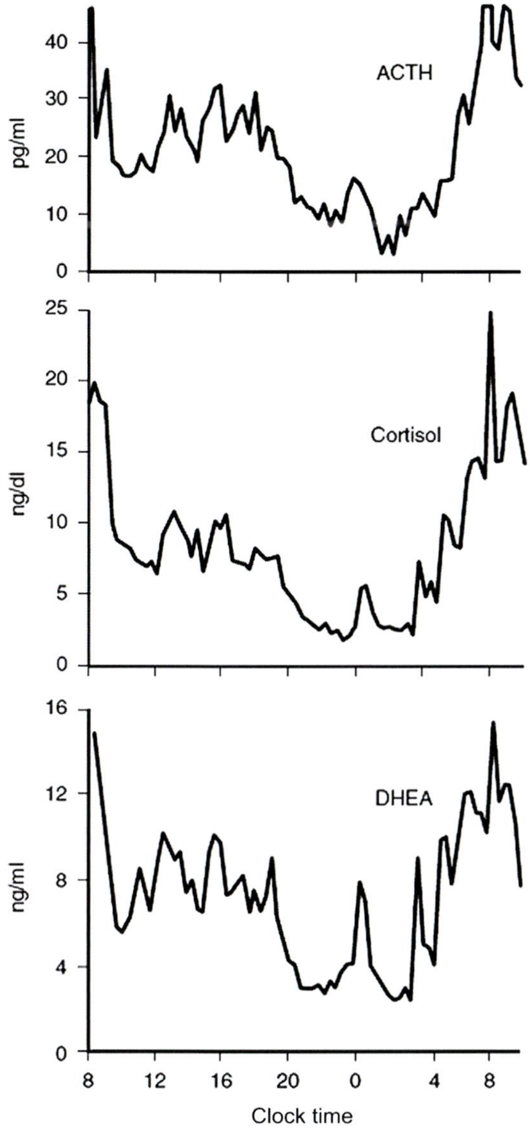

FIGURE 12.8 24-hour profiles of plasma adrenocorticotropic hormone (ACTH), cortisol and dehydroepiandrosterone (DHEA) levels sampled at 20-minute intervals in a healthy young man. Note the temporal concomitance of circadian and pulsatile variations of the three hormones. *Unpublished data kindly provided by Dr K. Spiegel.*

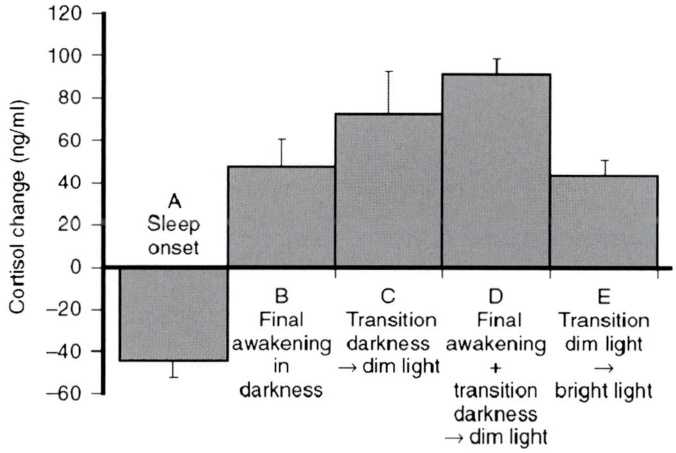

FIGURE 12.9 Mean (± SEM) changes in plasma cortisol levels: (**A**) Within 120 minutes following sleep onset at 1500 h ($n = 32$). (**B**) Within 20 minutes following final spontaneous awakening in darkness (scheduled sleep period 2300–0700 h or 1500–2300 h; $n = 10$). (**C**) Within 20 minutes following transition from darkness to dim light at 0700 or 2300 h in subjects awake at bed rest ($n = 10$). (**D**) Within 20 minutes following final awakening concomitant with transition from darkness to dim light at 0700 or 2300 h ($n = 38$). (**E**) Within 15 minutes following transition from dim to bright light at 0500 h in subjects awake at bed rest ($n = 8$). Panels **A–D**, data from Caufriez A, Moreno-Reyes R, Leproult R, et al. Immediate effects of an 8-h advance shift of the rest–activity cycle on 24-h profiles of cortisol. *Am J Physiol.* 2002;282:E1147–E1153; panel (**E**), data from Leproult R, Colecchia EF, L'Hermite-Balériaux M, et al. Transition from dim to bright light in the morning induces an immediate elevation of cortisol levels. *J Clin Endocrinol Metab.* 2001;86:151–157.

inference, ACTH secretory rhythm when sleep is manipulated. Indeed, the overall wave shape of the profile was not markedly affected by the absence of sleep or the presence of sleep at an abnormal time of day. Thus, this rhythm is primarily controlled by the circadian pacemaker.

Modulatory effects of sleep–wake homeostasis have, however, been clearly demonstrated. As illustrated in Fig. 12.9, sleep onset is consistently associated with a short-term inhibition of cortisol secretion (which may not be detectable when sleep is initiated in the morning – i.e., at the peak of corticotropic activity).[82,182–184] This inhibitory effect of sleep appears to be related to SW sleep stages.[185,186] Conversely, as illustrated in Fig. 12.9, final awakenings from sleep, as well as transient awakenings interrupting the sleep period, consistently trigger pulses of cortisol secretion,[82,178,185,187–189] and the number of nocturnal microarousals predicts morning plasma and saliva cortisol levels.[190] In an analysis of cortisol profiles during nocturnal sleep, it was observed that all transient awakenings interrupting sleep and lasting at least 10 minutes were followed within the next 20 minutes by significant bursts of cortisol secretion.[189] In addition, a temporal coupling between pulses of cortisol secretion and ultradian variations in an EEG marker of alertness has been reported.[191] Total nocturnal wake time is associated with increased 24-hour plasma cortisol concentrations,[192] chronic insomnia with reduced total sleep time is associated with higher cortisol levels across the night,[193] and daytime electroencephalographic alertness is associated with increased cortisol secretion.[191]

Modulatory effects of dark–light transitions have also been evidenced. Cortisol secretory pulses associated with morning awakening are enhanced by increasing light intensity.[194] Moreover, the transition from to darkness to dim light and from dim to bright light may also stimulate cortisol secretion in subjects who are awake at bed rest.[189,195] The stimulatory effects of bright light exposure on nocturnal cortisol levels appear to be dependent on the timing of exposure.[196] When dark–light and sleep–wake transitions occur concomitantly, associated cortisol elevations are nearly two times as high as when the final awakening occur in continuous darkness[189] (Fig. 12.9). Thus, in usual bedtime schedules both the sleep–wake and the dark–light transitions amplify the effects of circadian rhythmicity.

Studies of the 24-hour cortisol profile in the course of adaptation to shifts of the sleep–wake cycle have demonstrated that the end of the quiescent period, which coincides with the onset of the early morning rise, takes longer to adjust and appears to be a robust marker of circadian timing. Twin studies have demonstrated that the timing of the nadir is influenced by genetic factors,[197] providing evidence for a genetic control of human circadian phase. In contrast, the timing of the morning acrophase is more labile, and may be influenced by the timing of sleep offset,[188] the transition from dark to bright light,[194] and breakfast intake.[198] Finally, anticipation of the expected time of waking has been reported to be associated with a rise in ACTH, but not cortisol, levels during the end of the sleep period.[199]

In addition to the immediate modulatory effects of sleep–wake transitions on ACTH and cortisol levels, acute total or partial (4 hours in bed) nocturnal sleep deprivation results in elevated cortisol concentrations in late afternoon and evening on the following day.[200] The nadir of cortisol levels is higher (because of the absence of the inhibitory effects of sleep) and the acrophase is lower (because of the absence of the stimulatory action of morning awakening), so that the amplitude of the cortisol circadian variations is reduced by approximately 15%. Similarly, as illustrated in Fig. 12.4, recurrent partial sleep deprivation (4 hours in bed per night for 6 nights) also results in an elevation of cortisol levels in the late afternoon and the evening.[93] These disturbances, which were observed in young healthy subjects, are strikingly similar to those found in older healthy subjects with normal sleep schedules.[201–203] In any case, sleep loss appears to delay the return to quiescence of the hypothalamo-pituitary–adrenal axis (HPA) normally occurring in the evening. This suggests that sleep loss, similar to aging, may slow down the rate of recovery of the HPA axis response following a challenge, and could therefore facilitate the development of central and peripheral disturbances associated with glucocorticoid excess – in particular when cortisol concentrations are elevated at the time of the normal daily nadir – such as memory deficits, insulin resistance and osteoporosis.[90,204–207] Conversely, decreased HPA resiliency results in HPA hyperactivity that will inhibit SW sleep and promote nocturnal awakenings, initiating a feed-forward cascade of negative events generated by both HPA and sleep disruptions.

The circadian rhythm of cortisol persists throughout adulthood and has been observed through the ninth decade.[202,203] In young adults, 24-hour cortisol levels are slightly lower in women than in men, primarily because of lower morning maxima. With aging, evening cortisol levels increase progressively, both in men and women, so that the cortisol nadir is markedly higher in healthy subjects over 70 years of age than in young adults (Fig. 12.10). Interestingly, this elevation of evening cortisol levels occurs with a chronology similar to that observed for a progressive decrease in the duration of REM sleep[208] (Fig. 12.11). As a result, older subjects have elevated 24-hour mean cortisol levels and reduced amplitude of cortisol variations. In addition, the timing of the nadir is advanced by 1–2 hours, indicating that aging is associated with an advance of circadian phase.[202,203]

In pregnancy, total and (to a much lesser extent) free cortisol levels are elevated but the circadian pattern of secretion persists, albeit set at a higher level.[209] Interestingly, placental CRH is secreted into the maternal circulation in a pulsatile but not in a circadian fashion, and there is no correlation between maternal levels of CRH and ACTH. Remarkably, ACTH and cortisol concentrations remain strongly correlated with each other over time, suggesting that diurnal variation of maternal ACTH is probably driven by another ACTH secretagog, most likely arginine vasopressin.[210]

The Somatotropic Axis

Pituitary secretion of GH is stimulated by hypothalamic GH-releasing hormone (GHRH) and inhibited by somatostatin. In addition, the acylated form of ghrelin (acyl-ghrelin), a peptide produced predominantly by the stomach, binds to the GH-secretagog receptor and is therefore another potent endogenous stimulus of GH secretion.[140,211] In normal adult subjects, the 24-hour profile of plasma GH levels consists of stable low levels abruptly interrupted by bursts of secretion. The most reproducible pulse occurs shortly after sleep onset, in association with the first phase of SW sleep.[212] Other secretory pulses may occur in later sleep and during wakefulness, in the absence of any identifiable stimulus. Studies in young male twins have evidenced a major genetic effect on GH secretion during waking, but not during sleep.[213] In adult men, the sleep-onset GH pulse is generally the largest pulse observed over the 24-hour span. In normally cycling women, the 24-hour GH levels are higher than in age-matched men, daytime pulses are more frequent, and the sleep-associated pulse, while still present in most cases, does not generally account for the

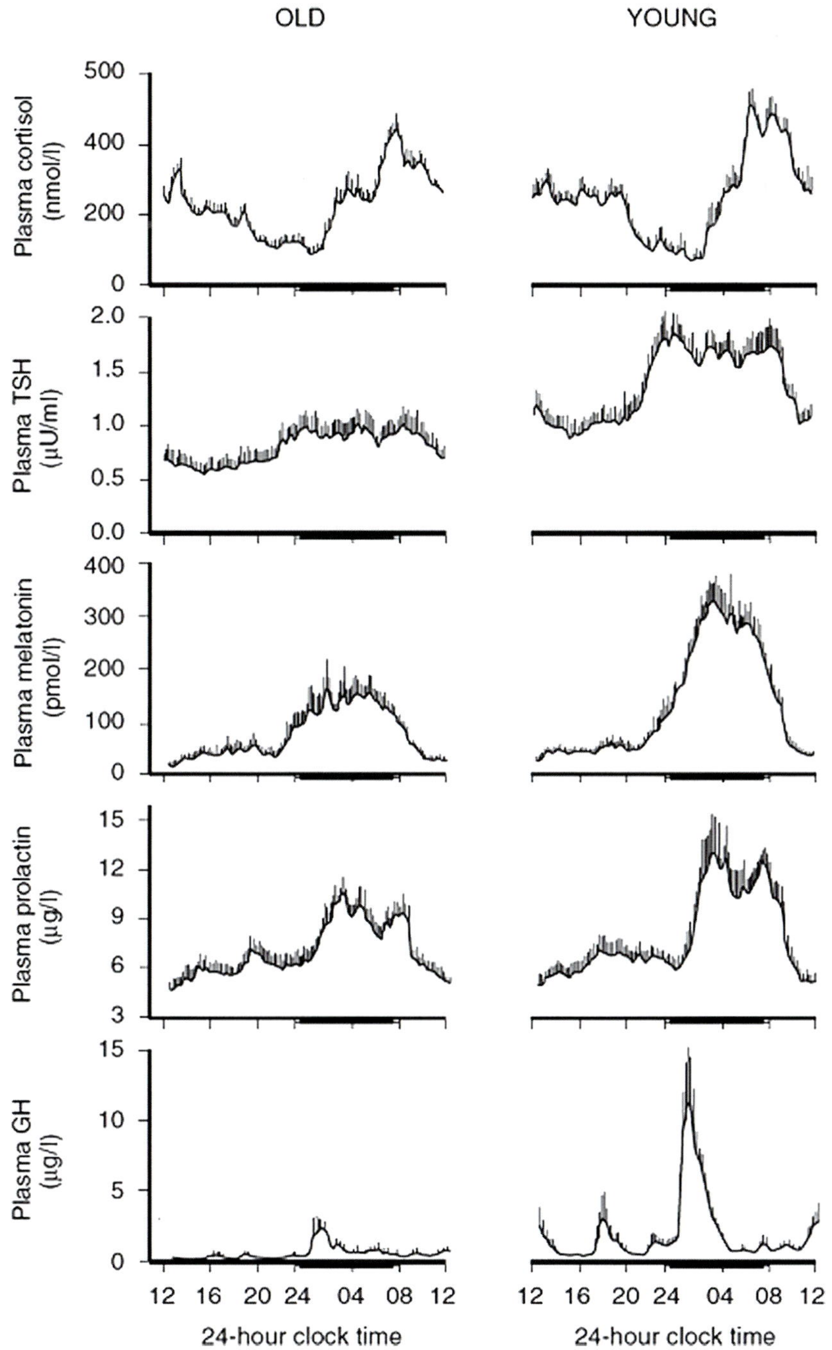

FIGURE 12.10 Mean (±SEM) 24-hour profiles of plasma cortisol, thyrotropin (TSH), melatonin, prolactin (PRL) and growth hormone (GH) levels in eight old (67–84 years) and eight young (20–27 years) subjects. Data were sampled at 15-minute intervals. Black bars, mean sleep period. *Reproduced from van Coevorden A, Mockel J, Laurent E,* et al. *Neuroendocrine rhythms and sleep in aging men.* Am J Physiol. *1991;260:E651–E661, with permission.*

majority of the 24-hour GH release.[214] Typical profiles of young men and women are shown in Fig. 12.10. Well-documented studies have demonstrated that in women the amplitude of GH secretory pulses is correlated with the circulating level of estradiol.[214,215] In normally cycling young women, it was also observed that daytime GH secretion was increased during the luteal phase as compared to the follicular phase, and that this elevation correlated positively with plasma levels of progesterone but not estradiol.[216]

Sleep onset will elicit a GH secretory pulse whether sleep is advanced, delayed, interrupted or fragmented.[212] Thus, as illustrated in Fig. 12.2, shifts of the sleep–wake cycle are immediately followed by parallel shifts of the GH rhythm.[212] In night workers, the main GH secretory episode occurs during the first half of the shifted sleep period.[217] The release of GH in early sleep is temporally and quantitatively associated with the amount of SW sleep.[218,219] Both SW sleep and GH levels are increased in the recovery night following

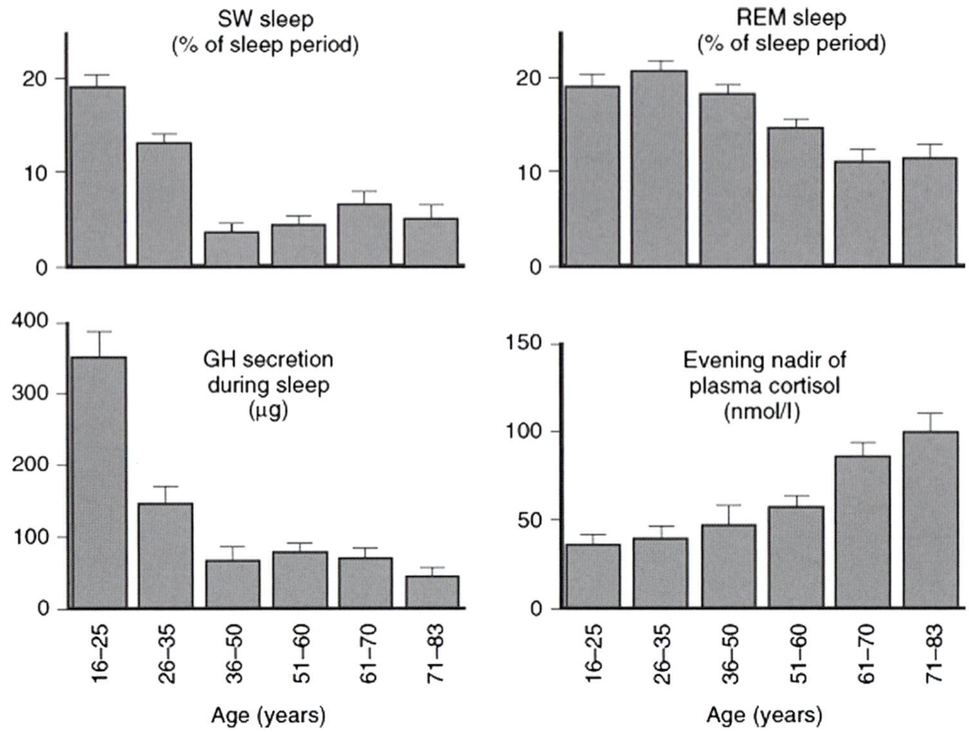

FIGURE 12.11 (Left) Slow-wave (SW) sleep and growth hormone (GH) secretion during sleep as a function of age. Note the temporal concomitance between the decrease in SW sleep and in GH secretion. (Right) Rapid eye movement (REM) sleep and level of evening nadir of plasma cortisol as a function of age. Note the temporal concomitance between the decrease in REM sleep and the increase in evening cortisol levels. Values shown are means (± SEM) for each age group. Data were obtained in 149 healthy men, ages 16–83 years. *Data from Van Cauter* et al. *(2000).*[208]

nocturnal total sleep deprivation as compared with the baseline pre-deprivation night, especially during the first 4 hours of sleep.[220] There is good evidence that the mechanisms underlying the relationship between SW sleep and GH release involve synchronous activity of at least two different populations of hypothalamic GHRH neurons.[221] Indeed, inhibition of endogenous GHRH, either by administration of a specific antagonist or by immunoneutralization, inhibits sleep as well as GH secretion.[222] Additional evidence for the existence of a robust relationship between SW activity and GH release is provided by studies using pharmacological stimulation of SW sleep. Indeed, enhancement of SW sleep by oral administration of low doses of gamma-hydroxybutyrate (GHB), a natural metabolite of GABA used in the treatment of narcolepsy, or of ritanserin, a selective 5HT2 antagonist, results in simultaneous and highly correlated increases in nocturnal GH secretion.[223,224] Conversely, transient awakenings during sleep inhibit GH secretion.[225] Thus, sleep fragmentation will generally decrease nocturnal GH release.

However, while sleep is clearly the major determinant of GH secretion in man, there is also evidence for the existence of a circadian modulation of the occurrence and amplitude of GH pulses. This may reflect decreased somatostatin inhibitory activity in the evening and during the night,[226] or could result from the nocturnal rise of ghrelin, which occurs even in the absence of sleep.[150] Thus, the major sleep-onset associated GH pulse is caused by a surge of hypothalamic GHRH coincident with a circadian period of relative somatostatin disinhibition.[212,222] In normal-weight subjects, fasting, even for only 1 day, enhances GH secretion via an increase in pulse amplitude.[227] Pre-sleep GH pulses, reported by some investigators in normal men,[228] may reflect the presence of a sleep debt, unmasking the circadian component of GH secretion.[229] A recent study provides evidence for a modulation of GH secretion by endogenous acyl-ghrelin leading to higher GH peaks before times of food intake in subjects given regular meals.[230]

Following a night of total sleep deprivation, a compensatory increase in GH release is observed during daytime, so that the overall 24-hour secretion is not significantly altered.[230] The mechanisms underlying this compensatory increase might involve decreased somatostatinergic tone and/or elevated ghrelin levels. Recurrent partial sleep restriction is consistently associated with the appearance of a pre-sleep GH pulse.[229]

Aging is associated with dramatic decreases in circulating levels of GH (see Figs 12.10, 12.11).[203,214] This reduction is achieved by a decrease in amplitude, rather than in frequency, of GH pulses.[203,231,232] It has also been reported that the orderliness of GH secretion is decreased in the elderly.[233] As illustrated in Fig. 12.11, this age-related GH decrease occurs in an exponential fashion between young adulthood and midlife, and follows the same chronology as the decrease in SW sleep. Despite the persistence of high levels of sex steroids, plasma concentrations and pulsatile secretion rates of GH fall in midlife to less than half the values achieved in young adulthood. Thereafter, smaller and more progressive decrements occur from midlife to old age.[208] In the elderly, GH secretory profiles are similar in men and in women.[214] The age-related reduction of GH secretion appears to result from increased somatostatin secretion and diminished GHRH responsiveness.[234]

Interestingly, during pregnancy, a placental GH variant, which substitutes for pituitary GH to regulate maternal IGF-I levels,[235] is released in a tonic rather than pulsatile fashion.[236]

The Lactotropic Axis

Under normal conditions, the 24-hour profile of prolactin levels exhibits minimal levels around noon and a modest increase in the afternoon, followed by a major nocturnal elevation starting shortly after sleep onset and culminating around mid-sleep[237,238] at levels corresponding to an average increase of more than 200% above minimum levels (see Fig. 12.2).[239,240] Episodic pulses occur throughout the 24-hour span, but their amplitude and frequency are higher during the night than during the day. Decreased dopaminergic inhibition of prolactin secretion during sleep is likely to be the primary mechanism underlying this nocturnal elevation. Mean prolactin levels, pulse amplitude and pulse frequency are higher in normally cycling women than in either postmenopausal women or normal young men.[241] In normally cycling young women, it was also observed that daytime prolactin pulsatility was enhanced during the luteal phase as compared with the follicular phase, leading to increased late afternoon and evening levels.[216] In the luteal phase, the magnitude of the evening prolactin rise correlated positively with both estradiol and progesterone levels.[216] These data indicate that endogenous estrogens and progesterone play a critical role in the differential regulation of prolactin secretion associated with sex and age. Deconvolution analysis has shown that the prolactin profile reflects both tonic and intermittent release.[242] Twin studies have revealed that genetic factors determine partially the temporal organization of prolactin secretion.[243]

Diurnal prolactin variations are primarily regulated by sleep–wake homeostasis. Sleep onset is invariably associated with an increase in prolactin secretion, irrespective of the time of the day. Thus, as illustrated in Fig. 12.2, shifts of the sleep–wake cycle are immediately followed by parallel shifts of the prolactin rhythm,[212] but the amplitude of the prolactin rise may be dampened when associated with daytime sleep as compared to nocturnal sleep.[244] Conversely, modest elevations of prolactin levels may occur during waking around the time of the usual sleep onset, particularly in women.[240] Thus, prolactin secretion appears to be modulated by circadian rhythmicity, and maximal secretion occurs when sleep and circadian effects are superimposed – i.e., at the usual bedtime.[239,240,245] Benzodiazepine (e.g., triazolam) and imidazopyridine (e.g., zolpidem) hypnotics taken at bedtime generally enhance the nocturnal prolactin elevation.[246,247]

A close temporal relationship has been evidenced between increased prolactin secretion and SW activity when sleep structure was characterized by power spectral analysis of the EEG.[248] Conversely, prolonged awakenings interrupting sleep are consistently associated with decreasing prolactin concentrations.[248] Thus, SW sleep is associated with elevated prolactin secretion, while shallow and fragmented sleep will generally be associated with a dampening of the nocturnal prolactin rise. This is indeed observed in elderly subjects, who have a nearly 50% dampening of the nocturnal prolactin elevation (see Fig. 12.10).[203,249]

During pregnancy, serum prolactin levels rise but the 24-hour pattern of secretion is maintained, albeit at a higher level. During the post-partum period prolactin secretory pulses follow suckling episodes, and the nocturnal rise, independent of suckling, is only evident once breast-feeding has ceased.[250]

The Gonadotropic Axis

Rhythms in the gonadotropic axis cover a wide range of frequencies, from episodic release in the ultradian range to diurnal rhythmicity and menstrual cycles. These various rhythms interact to provide a coordinated temporal program governing the development of the reproductive axis and its operation at every stage of maturation. The following description of the current state of knowledge in this area will be limited to 24-hour rhythms and their interaction with pulsatile release during adulthood.

Patterns of LH release in adult men exhibit episodic pulses with large inter-individual variability.[251] LH secretory episodes mainly reflect gonadotropin-releasing hormone (GnRH) pulsatility. A recent study indicates that LH pulsatility is inhibited by acyl-ghrelin, suggesting that the ghrelin system may play a centrally mediated inhibitory role on the gonadal axis.[252] The diurnal variation is dampened or even undetectable. During the sleep period, LH pulses appear to be temporally related to the

REM—non-REM cycle.[253] FSH profiles may show some occasional pulses, without any diurnal variation.

In contrast, a marked diurnal rhythm in circulating testosterone levels is present in young normal men, with minimal levels in the late evening and maximal levels in the early morning.[181,254] In young adult men, the amplitude of the testosterone rhythm averages 25%.[181] With a 15-minute sampling interval, 17 to 18 testosterone pulses per 24-hour span can be detected.[181] There is growing evidence that diurnal testosterone variations are primarily regulated by sleep—wake homeostasis. Experimental sleep fragmentation (a schedule allowing 7 minutes of sleep every 20 minutes) results in a dampening of the nocturnal testosterone rise, particularly in subjects who do not achieve REM sleep.[255] Daytime sleep, as well as nocturnal sleep, is associated with a robust rise in testosterone levels.[256] However a progressive elevation of testosterone levels persists during nocturnal wakefulness, albeit blunted as compared to nocturnal sleep,[256] indicating the existence of a circadian component that could reflect adrenal androgen secretion. Diurnal profiles of testosterone are paralleled by inhibin B variations, with peak values in the early morning and nadirs in the late afternoon, and significant cross-correlations between inhibin B and testosterone or estradiol were detected.[257]

A progressive decline in testosterone levels, together with an increase in sex hormone-binding globulin (SHBG) levels, is observed from 30 years of age onwards, so that the decrease in bioavailable testosterone is more important than the decline in total testosterone.[258] In elderly men, the diurnal variation of testosterone is still present, but may be markedly dampened.[254,259] A strong positive correlation has been evidenced between total sleep time and morning testosterone levels.[260] Pulsatile testosterone secretion is attenuated, suggesting a possible partial desensitization of Leydig cells to LH.[261] Mean LH levels are increased, but the amplitude of LH pulses is decreased[262,263] while their frequency is increased,[261] and no significant diurnal pattern can be detected.[260] In contrast, pulsatile FSH secretion is increased in older men.[264] In addition, older males secrete LH and testosterone more irregularly, and jointly more asynchronously, than younger men.[265]

In adult women, diurnal profiles of LH, FSH, estradiol and progesterone exhibit episodic pulses throughout the 24-hour span in follicular and in luteal phase.[266–271] The 24-hour variations in plasma LH are markedly modulated by the menstrual cycle.[266,272] In the early follicular phase LH pulses are large and infrequent, and a slowing of the frequency of secretory pulses occurs during the night. In the mid-follicular phase, pulse amplitude is decreased, pulse frequency is increased, and the nocturnal slowing of pulsatile LH secretion is less apparent. Pulse amplitude increases again by the late follicular phase. In the early luteal phase, the pulse amplitude is markedly increased, the pulse frequency is decreased, and nocturnal slowing of pulsatility is again evident. In the mid-luteal phase, pulse amplitude and frequency are decreased and there is no nocturnal slowing. Both pulse amplitude and frequency further decrease in the late luteal phase. During the luteal—follicular transition, there is a four- to fivefold increase in LH pulse frequency, which accompanies the selective FSH rise necessary for normal folliculogenesis.[273]

The nocturnal slowing of pulsatile LH secretion during the early follicular phase is specifically related to sleep rather than time of day, since it is also observed during daytime sleep, but not during night-time wake.[274] During periods of sleep, LH pulses were found to occur preferentially in association with brief awakenings, suggesting an inhibitory effect of sleep on pulsatile LH secretion.[274] Since night- and shift-work is consistently associated with shorter and more fragmented sleep, these results indicate that altered menstrual function frequently observed in night- and shift-workers could directly result from altered sleep patterns. An evening elevation of LH levels and LH pulse amplitude is observed in the absence of sleep, suggesting the existence of a circadian modulation of LH secretion.[274] Circulating levels of LH and FSH and LH pulse frequency increase with aging, and are higher in normal women older than 40 years of age with regular menstrual cycles than in women younger than age 35.[275] Gonadotropin levels remain elevated after menopause.

The Thyrotropic Axis

In normal adult men and women, TSH levels are low and relatively stable throughout the daytime and begin to increase in the late afternoon or early evening. Maximal levels occur around the beginning of the sleep period.[276] TSH levels progressively decline during the later part of sleep, and daytime values resume shortly after morning awakening. Because the onset of the nocturnal rise of TSH occurs well before sleep onset, it is believed to reflect a circadian effect. This 24-hour pattern of TSH levels appears to be generated by frequency as well as amplitude modulation of thyrotropin-releasing hormone (TRH)-driven secretory pulses.[277] Studies involving sleep deprivation and shifts of the sleep—wake cycle have consistently indicated that sleep exerts an inhibitory influence on TSH secretion, and sleep deprivation relieves this inhibition.[276,278] Interestingly, when sleep occurs during daytime, TSH secretion is not suppressed significantly below normal daytime levels.[279] Profiles of plasma TSH during normal nocturnal sleep, nocturnal sleep deprivation and daytime sleep, are illustrated in the lower panel of

Fig. 12.2. When the depth of sleep at the habitual time is enhanced by prior sleep deprivation, the inhibition of the nocturnal TSH rise is more pronounced than in basal conditions. Descending slopes of TSH concentrations during sleep are consistently associated with SW stages, and negative cross-correlations have been found between TSH fluctuations and SW activity,[280,281] suggesting that SW sleep is probably the primary determinant of the sleep-associated TSH decrease. Conversely, awakenings are frequently associated with TSH increments.[279] The timing of the TSH evening rise seems to be controlled by circadian rhythmicity, and shifts in concordance with the melatonin rhythm following exposure to light or nocturnal exercise.[282] Free triiodothyronine shows a diurnal rhythm that parallels TSH variations.[283]

Under conditions of sleep deprivation, the increased amplitude of the TSH rhythm may result in a detectable increase in plasma triiodothyronine levels, paralleling the nocturnal TSH rise,[279] although negative findings have also been reported.[284] If sleep deprivation is prolonged for a second night, the nocturnal rise of TSH is markedly diminished as compared to that occurring during the first night.[284] It is likely that, following the first night of sleep deprivation, the elevated thyroid hormone levels, which persist during the daytime period because of the prolonged half-life of these hormones, limit the subsequent TSH rise. A study involving 64 hours of sleep deprivation demonstrated, during the second night of sleep deprivation, a nocturnal increase in both triiodothyronine and thyroxine levels, contrasting with the decreases seen during normal sleep.[285] These data suggest that prolonged sleep loss may be associated with an upregulation of the thyroid axis. Consistent findings have been reported in a study of 6 days of partial sleep loss (4 hours in bed per night) where the nocturnal TSH rise was strikingly decreased and overall mean TSH levels were reduced by more than 30%, probably secondary to increased levels of thyroid hormones resulting from an initial TSH elevation at the beginning of sleep curtailment (Fig. 12.4).[93]

Because inhibitory effects of sleep on TSH secretion are time-dependent, elevations of plasma TSH levels may occur in conditions of misalignment of sleep and circadian timing. This is illustrated in Fig. 12.12, which shows the mean profiles of plasma TSH observed in a group of normal young men in the course of adaptation to simulated "jet lag" involving an abrupt 8-hour advance of the sleep–wake cycle and the dark period, following a 24-hour baseline period.[279] In the course of adaptation, TSH levels increased progressively because night-time wakefulness was associated with large circadian-dependent TSH elevations while daytime sleep failed to inhibit TSH. As a result, mean TSH levels were more than twofold higher following awakening from the second shifted period than during the same

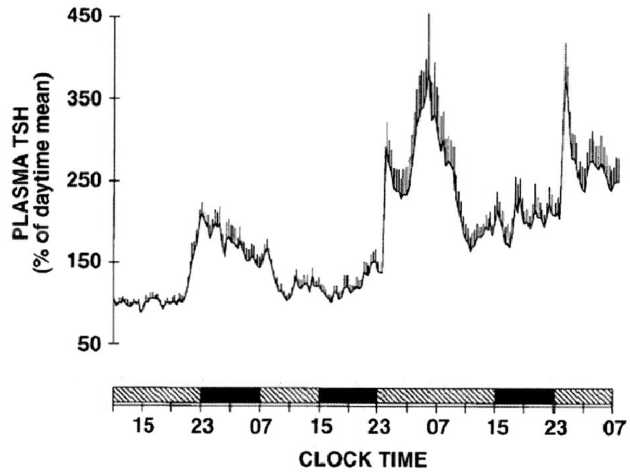

FIGURE 12.12 Mean (± SEM) profile of plasma thyrotropin (TSH) from eight normal young men submitted to an 8-hour advance of the sleep–wake and light–dark cycles. Black bars, bedtime periods. *Data from Hirschfeld* et al. *(1996).*[279]

time interval following normal nocturnal sleep. This study indicates that the subjective discomfort and fatigue associated with jet lag may involve a prolonged elevation of a hormonal concentration in the peripheral circulation.

Aging is associated with a progressive decrease in overall TSH secretion (which is achieved by a decrease in amplitude, rather than in frequency, of secretory pulses) and in circulating TSH levels, and with a dampening of the amplitude of the circadian variation.[203] In subjects in the seventh and eighth decades, TSH levels are lower than in young adults throughout the 24-hour span, although the difference is more marked during sleep than during the daytime period (see Fig. 12.10). In middle-age subjects, age-related decreases in TSH levels may be evidenced only in response to nocturnal sleep deprivation. Thus, it appears that the TSH secretory capacity declines progressively with aging.

CONDITIONS OF ALTERED SLEEP AND CIRCADIAN RHYTHMICITY

Circadian rhythms provide synchronization with the pronounced periodic fluctuations in the external environment and organize the internal milieu so that there is coordination and synchronization of internal processes. "External synchronization" is of obvious importance for the survival of the species, and ensures that the organism does the "right thing" at the right time of the day. Of equal, but perhaps less appreciated, importance is the fact that the circadian clock system provides internal temporal organization between the myriad of biochemical and physiological systems in the body. Lack of synchrony within the internal environment may lead to chronic difficulties, with serious consequences for the health and well-being of the organism. The physical and mental malaise occurring following rapid travel across time zones (i.e., jet-lag syndrome) and the pathologies associated with long-term shift work, are assumed to be due in part to an alteration in the normal phase relationships between various internal rhythms. In addition, it has been speculated that alterations of internal phase relationships between rhythms underlie certain forms of affective illness.

"Jet Lag"

Subjects who travel rapidly across time zones are confronted with a desynchronization between their internal circadian rhythms and the periodicity of the new external environment. Upon arrival, the timings of the light–dark cycle, social schedule and meals are abnormally matched to the phase of the physiological rhythms of the traveler. Associated with this lack of synchronization are symptoms of fatigue, subjective discomfort, sleep disturbances, reduced mental and psychomotor performance, and gastrointestinal disorders.

The rate of adaptation is generally slower for overt rhythms that are strongly dependent on the circadian system, such as those of cortisol and melatonin secretions, than for those that are markedly modulated by sleep–wake homeostasis, such as prolactin and GH secretion. As a result, during the period of adaptation, abnormal phase relationships between overt rhythms occur. Thus, jet-lag syndrome involves not only desynchronization between internal and external rhythms, but also a perturbation of internal temporal organization of physiologic functions.

Shift Work

Shift work, which is – voluntarily or not – performed by millions of workers, is a major health hazard, involving an increased risk of cardiovascular illness, gastrointestinal disorders, infertility and insomnia.[286–288] The medical consequences of shift work are associated with chronic misalignment of physiologic circadian rhythms and the activity–rest cycle. In addition, shift work almost invariably results in substantial sleep loss because daytime sleep is generally shorter and more fragmented than nocturnal sleep. Shift work usually creates conditions in which some zeitgebers (e.g., an artificial light–dark cycle) and additional phase-setting factors such as the rest–activity cycle are shifted, while others, such as the natural light–dark cycle and the routines of family life, remain unaltered. Shift workers thus live in a situation of conflicting zeitgebers that almost never allow a complete shift of the circadian system. Indeed, several studies have shown that workers on permanent or rotating night shifts do not adapt to these schedules, even after several years.[289–291] The conflict between the solar day and the abnormally phased activity–rest cycle (which includes perturbations to both the sleep–wake and feeding–fasting cycles), as experienced by the shift worker, can lead to abnormal neuroendocrine and endocrine rhythms, which may directly impact health.

Altered Feeding Rhythms

While humans typically have freedom to eat whenever they choose, often cultural or work-related schedules promote eating at specific times. For example, the night-shift worker will eat meals at night during scheduled breaks. During jet lag, social events based around meals encourage eating at specific times, even if those meal times do not align with habitual feeding times and appetite rhythms. Since the feeding–fasting cycle plays a substantial role in many neuroendocrine systems (e.g., insulin, leptin), it is possible that meal timing alone, independent of activity–rest or sleep–wake cycles, may directly affect rhythms.

Indeed, work with both human and animal models indicates that the feeding–fasting cycle can be a potent cue for circadian gene expression within brain and peripheral tissues, as well as hormonal rhythms such as glucose, insulin, leptin, and ghrelin. Recent research focusing on metabolic outcomes from "wrong" timed meals suggests that the feeding–fasting cycle must be aligned to the solar day for optimal health. Similarly, misaligned neuroendocrine rhythms caused from abnormally timed meals could be contributing to health impairments seen acutely in the jet-lag population, or more chronically in the shift worker.

Sleep Curtailment

Whether voluntary or not, sleep restriction is a hallmark of modern society. "Normal" sleep duration has decreased from approximately 8.5 hours in 1960 to an average of less than 7 hours today. Many individuals voluntarily choose to curtail their sleep to the shortest amount tolerable to maximize the time available for work and leisure activities, and more than 30% of American adult men and women between the ages of 30 and 64 years report sleeping less than 6 hours per night. To meet the demands of round-the-clock operations, millions of shift workers sleep on average less than 6 hours per day.

Despite the fact that sleep is a major modulator of metabolic and endocrine regulation, the consensus which prevailed until recently was that sleep loss results in increased sleepiness and decreased cognitive performance, but has little or no effect on peripheral function. However, as reported in previous sections, partial sleep curtailment induces potentially harmful alterations in hormonal profiles. Sleep restriction over 2 to 6 nights (4 hours per night) under controlled conditions of caloric intake and physical activity in healthy young men was found to result in the following:

- Elevated evening total and free cortisol concentrations,[94] strikingly similar to those observed in the elderly.[201–203] This disturbance may reflect decreased efficacy of the negative feedback regulation of the hypothalamo-pituitary–adrenal axis, and could promote the development of insulin resistance and memory impairments.[90,292]
- A clinically significant impairment of carbohydrate tolerance,[93–95] consistent with a state of impaired glucose tolerance as observed in older adults.
- A decrease in circulating levels of leptin,[94,122] an anorexigenic hormone, and a concomitant increase in circulating levels of ghrelin,[122] an orexigenic hormone (Fig. 12.6). Moreover, sleep curtailment was associated with an increase in hunger, and this increase in hunger was strongly correlated with the increase in ghrelin to leptin ratio.[122]

These data suggest that recurrent partial sleep curtailment may increase the risk of obesity and diabetes, and accelerate the senescence of endocrine and metabolic function. An obvious limitation of these studies is that the investigation period was not extended beyond 6 days. Thus, a progressive adaptation to chronic partial sleep deprivation cannot be excluded. However, the findings from these laboratory studies are consistent with the conclusions of epidemiologic studies. Several prospective cross-sectional epidemiologic studies, which varied considerably in geographical locations and subject populations, were remarkably consistent in indicating that short sleep may increase the risk of developing type 2 diabetes and/or obesity (reviewed in Knutson and Van Cauter[97]; Van Cauter and Knutsen[293]). One major limitation to all of these epidemiological studies is that they only used subjective reporting of sleep. Laboratory and epidemiologic studies now need to be complemented by large field studies incorporating objective measures of sleep duration and interventional methods to better understand the mechanisms linking sleep loss to endocrine and metabolic alterations. Considering the morbidity and mortality associated with obesity and diabetes, the identification of novel risk factors that are potentially modifiable, such as sleep curtailment, is particularly important.

Sleep Disorders

A recent laboratory study has shown that decreased sleep quality without change in sleep duration results in a marked decrease in insulin sensitivity, without appropriate compensation of insulin release.[96] Thus, diabetes risk was markedly increased. This study suppressed SWS, thus replacing deep non-REM sleep by shallow non-REM sleep, as occurs in the course of normal aging and in a variety of sleep disorders, including obstructive sleep apnea (OSA). Importantly, this study demonstrated that sleep disruption can cause hormonal and metabolic alterations.

Obstructive Sleep Apnea (OSA)

OSA is the most common sleep disorder, and its incidence is rapidly rising in parallel with the current epidemic of obesity. A few studies have examined pituitary hormonal release in patients with obstructive apnea before and after treatment.[294–296] The nocturnal release of the two pituitary hormones that are markedly dependent on sleep (i.e., GH and prolactin) is decreased in untreated apneic subjects. However, treatment with continuous positive airway pressure (CPAP) results in a clear increase in the amount of GH secreted during the first few hours of sleep.[294,295] The total amount of prolactin secreted during the sleep period is not modified by the CPAP treatment, but the frequency of prolactin pulses is restored to values similar to those observed in normal subjects.[296] Nocturnal LH and testosterone secretions are decreased in men with untreated OSA. These alterations are partially corrected during chronic CPAP treatment.[297]

Morning leptin levels are elevated in patients with obstructive apnea when compared to weight-matched non-apneic subjects, and CPAP treatment decreases morning leptin levels.[298–305]

While obesity is a major risk for OSA, OSA is now recognized as a risk for insulin resistance, independently of BMI, as supported by a large set of nine cross-sectional studies which have assessed OSA by polysomnography (reviewed in Tasali *et al.*[306]). All but the earliest study (which also involved the smallest sample size) found an association between increased severity of OSA and alterations in glucose metabolism consistent with an increased risk of diabetes. The only prospective study that used polysomnography to assess OSA did not find an independent relationship between severity of OSA at baseline and incident diabetes, but the duration of follow-up was only 4 years. The findings from clinic-based studies are largely consistent with those of epidemiologic studies. Indeed, despite differences in sample size, study design, measurement techniques, cut points, and control for possible confounders, the majority of clinic-based studies (10 out of 13) were consistent in finding an independent association between OSA and abnormal glucose metabolism.[306]

Moreover, there is accumulating evidence that suggests that metabolic abnormalities can be partially corrected by CPAP treatment, which supports the concept of a causal link between OSA and altered glucose control.[307,308] Using hyperinsulinemic euglycemic clamp evaluations in non-diabetic patients with OSA, it was found that CPAP significantly improves insulin sensitivity after 2 days of treatment, and the improvement still persists after nearly 3 years of treatment. In obese patients with type 2 diabetes, insulin sensitivity was improved after 3 months, but not after 2 days, of CPAP treatment. This finding suggests that the time-course of improvement may be longer in obese patients who are diabetic. Two studies found that postprandial interstitial glucose levels and elevated hemoglobin A_1C levels were reduced by CPAP use in type 2 diabetes. A population-based study showed reductions in fasting insulin levels and insulin resistance (estimated by HOMA) after 3 weeks of CPAP treatment in men with OSA compared with non-apneic controls followed over the same time-period without CPAP. However, several studies did not show a beneficial effect of CPAP treatment on glucose metabolism.[306] Conflicting results could be due to differences in sample sizes and populations, variable durations of therapy, variable compliance, and changes in body composition during the study period.

The pathophysiologic mechanisms leading to alterations in glucose metabolism in OSA are likely to be multiple. High sympathetic nervous system activity, intermittent hypoxia, low levels of SWS, sleep fragmentation and sleep loss, dysregulation of the hypothalamo-pituitary axis, endothelial dysfunction, and alterations in cytokine and adipokine release have all been proposed as potential mechanisms for abnormal glucose metabolism in OSA. In a recent report,[309] the overall prevalence of OSA in diabetic men was estimated at 23% compared with 6% in a community-based sample. However, preliminary analysis of cross-sectional data from a multicenter study revealed an exceptionally high prevalence of undiagnosed OSA in obese type 2 diabetics, with over 75% of patients having moderate to severe OSA diagnosed by polysomnography.[310] These remarkable associations raise the possibility that OSA may be a novel risk factor for type 2 diabetes and/or, conversely, that chronic hyperglycemia may promote OSA. Whether the treatment of OSA may delay the development or reduce the severity of type 2 diabetes is another important question.

Other Sleep Disorders

There is a paucity of data regarding endocrine and metabolic abnormalities in sleep disorders other than OSA.

Narcolepsy is a sleep disorder characterized by excessive daytime sleepiness, reduced quality of nocturnal sleep with sleep-onset REM episodes, and cataleptic attacks. Narcolepsy is caused by impaired orexin (hypocretin) neurotransmission. Consistent with the role of orexin in the control of energy balance, narcoleptic patients have increased BMI and lower basal metabolism. The 24-hour rhythm of ACTH and cortisol persists in narcolepsy, suggesting that the circadian clock is not affected.[311,312] In contrast, the 24-hour profiles of hormones known to be dependent on sleep–wake homeostasis, such as GH and prolactin, are markedly disrupted, with dampened or absent nocturnal GH and prolactin release.[311,313,314] Leptin levels are decreased and the nocturnal rise is abolished.[312]

Despite the high prevalence of insomnia in modern society, very little is known regarding the neuroendocrine and metabolic consequences of poor or insufficient sleep in this condition. The 24-hour profiles of ACTH and cortisol have been assessed in insomniacs who were monitored in a sleep laboratory for 4 consecutive nights. An increase in ACTH and cortisol secretion was observed in the evening and early part of the night in the patients who had objectively documented short total sleep time and poor sleep efficiency. However, insomniacs who had a normal sleep time did not show alterations in ACTH and cortisol profiles.[193] Certain, much less prevalent, forms of sleep disorders seem to originate from a disturbance in the circadian system. Delayed sleep-phase insomnia is characterized by a chronic inability to fall asleep at a normal bedtime and to awake in the morning. Non-pharmacological chronotherapy involving repeated scheduled exposure to bright light is the treatment of choice for this disorder.[315] In contrast, in the advanced sleep phase syndrome, the timing of the major sleep episode is advanced in relation to normal bedtime, resulting in symptoms of extreme evening sleepiness and early morning awakening. Familial forms of this syndrome could reflect an autosomal dominant mutation.[316]

CONCLUSION

Circadian rhythms may have developed "in the beginning" as a way for an organism to optimally time and predict changes in its environment. Well-timed rhythms may have resulted in increased fitness, and allowed endocrine processes increased complexity and signaling capabilities. Current research indicates that neuroendocrine rhythms are primarily influenced by both the endogenous circadian pacemaker (e.g., the SCN in mammals), and cyclic behaviors such as the sleep–wake and feeding–fasting cycles. While the relative influence of the endogenous versus the behavioral rhythm impacts each neuroendocrine rhythm differently, the core molecular circadian mechanism, the transcriptional–translation feedback loop, underlies all neuroendocrine rhythms. An increased understanding of how neuroendocrine rhythms integrate signals from the endogenous circadian pacemaker and cyclic behaviors is expected to lead to new insights and therapeutic treatments in endocrinology.

Acknowledgments

Portions of this chapter have been reproduced, with permission, from: Copinschi G, Turek FW, van Cauter E. Endocrine rhythms, the sleep−wake cycle, and biological clocks. In: Jameson JL, DeGroot LG, eds. *Endocrinology* (2-volume set), 6th ed. Philadelphia, Elsevier Saunders; 2010:Ch. 11.
Vitaterna MH, Turek FW. Circadian clock genes. In: Kryger MH, Roth T, Dement WC, eds. *Principles and Practice of Sleep Medicine*, 5th ed. Philadelphia, PA: Elsevier Saunders; 2010; Ch. 12.

References

1. Rosenwasser AM, Turek FW. Physiology of the mammalian circadian system. Section 4 − Chronobiology. In: Kryger MH, Roth T, Dement WC, eds. *Principles and Practices of Sleep Medicine*. New York: WB Saunders; 2005: 363−74.
2. Turek FW, Dugovic C, Laposky A. Master circadian clock, master circadian rhythm. Section 4-Chronobiology. In: Kryger MH, Roth T, Dement WC, eds. *Principles and Practices of Sleep Medicine*. New York: W B Saunders; 2005: 318−20.
3. Turek FW. Circadian rhythms. *Horm Res*. 1998;49:103−113.
4. Saper CB, Scammell TE, Lu J. Hypothalamic regulation of sleep and circadian rhythms. *Nature*. 2005;437(7063):1257−1263.
5. Lehman MN, Silver R, Gladstone WR, Kahn RM, Gibson M, Bittman EL. Circadian rhythmicity restored by neural transplant. Immunocytochemical characterization of the graft and its integration with the host brain. *J Neurosci*. 1987;7(6):1626−1638.
6. Ralph MR, Foster RG, Davis FC, Menaker M. Transplanted suprachiasmatic nucleus determines circadian period. *Science*. 1990;247(4945):975−978.
7. Abe M, Herzog ED, Yamazaki S, Straume M, Tei H, Sakaki Y, et al. Circadian rhythms in isolated brain regions. *J Neurosci*. 2002;22(1):350−356.
8. Earnest DJ, Sladek CD. Circadian vasopressin release from perifused rat suprachiasmatic explants *in vitro*: effects of acute stimulation. *Brain Res*. 1987;422(2):398−402.
9. Prosser RA, Gillette MU. Cyclic changes in cAMP concentration and phosphodiesterase activity in a mammalian circadian clock studied *in vitro*. *Brain Res*. 1991;568(1-2):185−192.
10. Panda S, Antoch MP, Miller BH, Su AI, Schook AB, Straume M, et al. Coordinated transcription of key pathways in the mouse by the circadian clock. *Cell*. 2002;109(3):307−320.
11. Welsh DK, Logothetis DE, Meister M, Reppert SM. Individual neurons dissociated from rat suprachiasmatic nucleus express independently phased circadian firing rhythms. *Neuron*. 1995;14(4):697−706.
12. Lee HS, Billings HJ, Lehman MN. The suprachiasmatic nucleus: a clock of multiple components. *J Biol Rhythms*. 2003;18(6):435−449.
13. Laposky A, Easton A, Dugovic C, Walisser J, Bradfield C, Turek F. Deletion of the mammalian circadian clock gene BMAL1/Mop3 alters baseline sleep architecture and the response to sleep deprivation. *Sleep*. 2005;28(4):395−409.
14. Lowrey PL, Takahashi JS. Mammalian circadian biology: elucidating genome-wide levels of temporal organization. *Annu Rev Genomics Hum Genet*. 2004;5:407−441.
15. Vitaterna MH, Pinto LH, Turek FW. Molecular genetic basis for mammalian circadian rhythms. Section 4 − Chronobiology. In: Kryger MH, Roth T, Dement WC, eds. *Principles and Practices of Sleep Medicine*. New York: WB Saunders; 2005: 363−74.
16. Miller BH, Olson SL, Turek FW, Levine JE, Horton TH, Takahashi JS. Circadian clock mutation disrupts estrous cyclicity and maintenance of pregnancy. *Curr Biol*. 2004;14(15):1367−1373.
17. Turek FW, Joshu C, Kohsaka A, Lin E, Ivanova G, McDearmon E, et al. Obesity and metabolic syndrome in circadian Clock mutant mice. *Science*. 2005;308(5724):1043−1045.
18. Laposky AD, Bass J, Kohsaka A, Turek FW. Sleep and circadian rhythms: key components in the regulation of energy metabolism. *FEBS Lett*. 2008;582(1):142−151.
19. Yoo SH, Yamazaki S, Lowrey PL, Shimomura K, Ko CH, Buhr ED, et al. PERIOD2:LUCIFERASE real-time reporting of circadian dynamics reveals persistent circadian oscillations in mouse peripheral tissues. *Proc Natl Acad Sci USA*. 2004;101(15): 5339−5346.
20. Czeisler CA, Duffy JF, Shanahan TL, Brown EN, Mitchell JF, Rimmer DW, et al. Stability, precision, and near-24-hour period of the human circadian pacemaker. *Science*. 1999;284:2177−2181.
21. Dijk DJ, Czeisler CA. Contribution of the circadian pacemaker and the sleep homeostat to sleep propensity, sleep structure, electroencephalographic slow waves, and sleep spindle activity in humans. *J Neurosc*. 1995;15:3526−3538.
22. Wright Jr KP, Hughes RJ, Kronauer RE, Dijk DJ, Czeisler CA. Intrinsic near-24-h pacemaker period determines limits of circadian entrainment to a weak synchronizer in humans. *Proc Natl Acad Sci USA*. 2001;98(24):14027−14032.
23. Card JP, Whealy ME, Robbins AK, Moore RY, Enquist LW. Two alpha-herpesvirus strains are transported differentially in the rodent visual system. *Neuron*. 1991;6(6):957−969.
24. Underwood H, Steele CT, Zivkovic B. Circadian organization and the role of the pineal in birds. *Microsc Res Tech*. 2001;53(1):48−62.
25. Foster RG, Provencio I, Hudson D, Fiske S, De Grip W, Menaker M. Circadian photoreception in the retinally degenerate mouse (rd/rd). *J Comp Physiol [A]*. 1991;169(1):39−50.
26. Panda S, Sato TK, Castrucci AM, Rollag MD, DeGrip WJ, Hogenesch JB, et al. Melanopsin (Opn4) requirement for normal light-induced circadian phase shifting. *Science*. 2002;298(5601): 2213−2216.
27. Panda S, Provencio I, Tu DC, Pires SS, Rollag MD, Castrucci AM, et al. Melanopsin is required for non-image-forming photic responses in blind mice. *Science*. 2003;301(5632):525−527.
28. Lupi D, Oster H, Thompson S, Foster RG. The acute light-induction of sleep is mediated by OPN4-based photoreception. *Nat Neurosci*. 2008;11(9):1068−1073.
29. Guler AD, Ecker JL, Lall GS, Haq S, Altimus CM, Liao HW, et al. Melanopsin cells are the principal conduits for rod-cone input to non-image-forming vision. *Nature*. 2008;453(7191):102−105.
30. Bernard DJ, Abuav-Nussbaum R, Horton TH, Turek FW. Photoperiodic effects on gonadotropin-releasing hormone (GnRH) content and the GnRH-immunoreactive neuronal system of male Siberian hamsters. *Biol Reprod*. 1999;60(2):272−276.
31. Foster RG. Seeing the light. in a new way. *J Neuroendocrinol*. 2004;16(2):179−180.
32. Klein T, Martens H, Dijk DJ, Kronauer RE, Seely EW, Czeisler CA. Circadian sleep regulation in the absence of light perception: chronic non-24-hour circadian rhythm sleep disorder in a blind man with a regular 24-hour sleep−wake schedule. *Sleep*. 1993;16(4):333−343.
33. Copinschi G, Turek FW, Van Cauter E. Endocrine rhythms, the sleep−wake cycle, and biological clocks. In: Jameson JL, De Groot LJ, eds. *Endocrinology*. 6th ed. Philadelphia: Elsevier Saunders; 2010. Ch. 11.
34. Rechtschaffen A, Kales A. *A manual of standardized terminology, techniques and scoring system for sleep stages of human subjects*. Los

Angels: UCLA Brain Information Service/Brain Research Institute; 1968.
35. Jones BE. Modulation of cortical activation and behavioral arousal by cholinergic and orexinergic systems. *Ann NY Acad Sci*. 2008;1129:26–34.
36. Szymusiak R, McGinty D. Hypothalamic regulation of sleep and arousal. *Ann NY Acad Sci*. 2008;1129:275–286.
37. Sherin JE, Shiromani PJ, McCarley RW, Saper CB. Activation of ventrolateral preoptic neurons during sleep. *Science*. 1996; 271(5246):216–219.
38. Gallopin T, Fort P, Eggermann E, Cauli B, Luppi PH, Rossier J, et al. Identification of sleep-promoting neurons *in vitro*. *Nature*. 2000;404(6781):992–995.
39. Gaus SE, Strecker RE, Tate BA, Parker RA, Saper CB. Ventrolateral preoptic nucleus contains sleep-active, galaninergic neurons in multiple mammalian species. *Neuroscience*. 2002;115(1):285–294.
40. Saper CB, Chou TC, Scammell TE. The sleep switch: hypothalamic control of sleep and wakefulness. *Trends Neurosci*. 2001;24(12):726–731.
41. Sakurai T. Roles of orexin/hypocretin in regulation of sleep/wakefulness and energy homeostasis. *Sleep Med Rev*. 2005; 9(4):231–241.
42. Aschoff J. Circadian rhythms: general features and endocrinological aspects. In: Krieger DT, ed. *Endocrine Rhythms*. New York: Raven Press; 1979: 1–61.
43. Aschoff J. On the perception of time during prolonged temporal isolation. *Hum Neurobiol*. 1985;4(1):41–52.
44. Aschoff J, von Goetz C, Wildgruber C, Wever RA. Meal timing in humans during isolation without time cues. *J Biol Rhythms*. 1986;1:151–162.
45. Aschoff J. On the dilatability of subjective time. *Persp Biol Med*. 1992;35:276–280.
46. Czeisler CA, Weitzman ED, Moore-Ede MC, Zimmerman JC, Knauer RS. Human sleep: its duration and organization depends on its circadian phase. *Science*. 1980;210:1264–1267.
47. Borbely AA. Processes underlying sleep regulation. *Horm Res*. 1998;49:114–117.
48. Chou TC, Bjorkum AA, Gaus SE, Lu J, Scammell TE, Saper CB. Afferents to the ventrolateral preoptic nucleus. *J Neurosci*. 2002;22(3):977–990.
49. Edgar DM, Dement WC, Fuller CA. Effect of SCN lesions on sleep in squirrel monkeys: evidence for opponent processes in sleep-wake regulation. *J Neurosci*. 1993;13(3):1065–1079.
50. Mendelson WB, Bergmann BM, Tung A. Baseline and post-deprivation recovery sleep in SCN-lesioned rats. *Brain Res*. 2003;980(2):185–190.
51. Aston-Jones G, Chen S, Zhu Y, Oshinsky ML. A neural circuit for circadian regulation of arousal. *Nat Neurosci*. 2001;4(7):732–738.
52. Naylor E, Bergmann BM, Krauski K, Zee PC, Takahashi JS, Vitaterna MH, et al. The circadian clock mutation alters sleep homeostasis in the mouse. *J Neurosci*. 2000;20(21):8138–8143.
53. Franken P, Thomason R, Heller HC, O'Hara BF. A non-circadian role for clock-genes in sleep homeostasis: a strain comparison. *BMC Neurosci*. 2007;8:87.
54. Viola AU, Archer SN, James LM, Groeger JA, Lo JC, Skene DJ, et al. PER3 polymorphism predicts sleep structure and waking performance. *Curr Biol*. 2007;17(7):613–618.
55. Landolt HP. Sleep homeostasis: a role for adenosine in humans? *Biochem Pharmacol*. 2008;75(11):2070–2079.
56. Monk TH, Buysse DJ, Reynolds III CF, Jarrett DB, Kupfer DJ. Rhythmic versus homeostatic influences on mood, activation and performance in the elderly. *J Gerontol*. 1991;47:221–227.
57. Monk TH, Buysse DJ, Reynolds CF, Berga SL, Jarrett DB, Begley AE, et al. Circadian rhythms in human performance and mood under constant conditions. *J Sleep Res*. 1997;6:9–18.
58. Dijk DJ, Czeisler CA. Paradoxical timing of the circadian rhythm of sleep propensity serves to consolidate sleep and wakefulness in humans. *Neurosc Lett*. 1994;166:63–68.
59. Folkard S, Hume KI, Minors DS, Waterhouse JM, Watson FL. Independence of the circadian rhythm in alertness from the sleep–wake cycle. *Nature*. 1985;313:678–679.
60. Leproult R, Van Reeth O, Byrne MM, Sturis J, Van Cauter E. Sleepiness, performance and neuroendocrine function during sleep deprivation: Effects of exposure to bright light or exercise. *J Biol Rhythms*. 1997;12:245–258.
61. Dijk DJ, Duffy JF. Circadian regulation of human sleep and age-related changes in its timing, consolidation and EEG characteristics. *Ann Med*. 1999;31(2):130–140.
62. Brandenberger G, Follenius M, Goichot B, Saini J, Ehrhart J, Simon C. Twenty-four hour profiles of plasma renin activity in relation to the sleep–wake cycle. *J Hypertension*. 1994;12: 277–283.
63. Van Cauter E, Polonsky KS, Scheen AJ. Roles of circadian rhythmicity and sleep in human glucose regulation. *Endocr Rev*. 1997;18:716–738.
64. Follenius M, Brandenberger G, Simon C, Schlienger JL. REM sleep in humans begins during decreased secretory activity of the anterior pituitary. *Sleep*. 1988;11:546–555.
65. Gronfier C, Luthringer R, Follenius M, Schaltenbrand N, Macher JP, Muzet A, et al. A quantitative evaluation of the relationships between growth hormone secretion and delta wave electroencephalographic activity during normal sleep and after enrichment in delta waves. *Sleep*. 1996;19(10):817–824.
66. Gronfier C, Luthringer R, Follenius M, Schaltenbrand N, Macher JP, Muzet A, et al. Temporal link between plasma thyrotropin levels and electroencephalographic activity in man. *Neurosci Lett*. 1995;200(2):97–100.
67. Gronfier C, Luthringer R, Follenius M, Schaltenbrand N, Macher JP, Muzet A, et al. Temporal relationships between pulsatile cortisol secretion and electroencephalographic activity during sleep in man. *Electroencephalogr Clin Neurophysiol*. 1997;103(3):405–408.
68. Spiegel K, Luthringer R, Follenius M, Schaltenbrand N, Macher JP, Muzet A, et al. Temporal relationship between prolactin secretion and slow-wave electroencephalic activity during sleep. *Sleep*. 1995;18(7):543–548.
69. Van Cauter E, Spiegel K. In: Turek FW, Zee PC, eds. *Neurobiology of Sleep and Circadian Rhythms. Circadian and sleep control of endocrine secretions*, Vol. 133. New York: Marcel Dekker; 1999: 397–426.
70. Van Cauter E, Copinschi G. Interactions between growth hormone secretion and sleep. In: Smith RG, Thorner MO, eds. *Human growth hormone secretion: Basic and clinical research*. Totowa, NJ: Humana Press, Inc; 1999.
71. Yin L, Wu N, Curtin JC, Qatanani M, Szwergold NR, Reid RA, et al. Rev-erbalpha, a heme sensor that coordinates metabolic and circadian pathways. *Science*. 2007;318(5857):1786–1789.
72. Duez H, Staels B. The nuclear receptors Rev-erbs and RORs integrate circadian rhythms and metabolism. *Diab Vasc Dis Res*. 2008;5(2):82–88.
73. Stokkan KA, Yamazaki S, Tei H, Sakaki Y, Menaker M. Entrainment of the circadian clock in the liver by feeding. *Science*. 2001;291(5503):490–493.
74. Arble DM, Bass J, Laposky AD, Vitaterna MH, Turek FW. Circadian timing of food intake contributes to weight gain. *Obesity (Silver Spring)*. 2009;17(11):2100–2102.
75. Kohsaka A, Laposky AD, Ramsey KM, Estrada C, Joshu C, Kobayashi Y, et al. High-fat diet disrupts behavioral and molecular circadian rhythms in mice. *Cell Metab*. 2007;6(5): 414–421.

76. Fuller PM, Lu J, Saper CB. Differential rescue of light-and food-entrainable circadian rhythms. *Science*. 2008;320(5879): 1074–1077.
77. Landry GJ, Yamakawa GR, Webb IC, Mear RJ, Mistlberger RE. The dorsomedial hypothalamic nucleus is not necessary for the expression of circadian food-anticipatory activity in rats. *J Biol Rhythms*. 2007;22(6):467–478.
78. Angeles-Castellanos M, Salgado-Delgado R, Rodriguez K, Buijs RM, Escobar C. The suprachiasmatic nucleus participates in food entrainment: a lesion study. *Neuroscience*. 2010;165(4):1115–1126.
79. Moriya T, Aida R, Kudo T, Akiyama M, Doi M, Hayasaka N, et al. The dorsomedial hypothalamic nucleus is not necessary for food-anticipatory circadian rhythms of behavior, temperature or clock gene expression in mice. *Eur J Neurosci*. 2009;29(7):1447–1460.
80. Martino TA, Oudit GY, Herzenberg AM, Tata N, Koletar MM, Kabir GM, et al. Circadian rhythm disorganization produces profound cardiovascular and renal disease in hamsters. *Am J Physiol Regul Integr Comp Physiol*. 2008;294(5):R1675–R1683.
81. Turek FW. Staying off the dance floor: when no rhythm is better than bad rhythm. *Am J Physiol Regul Integr Comp Physiol*. 2008;294(5):R1672–R1674.
82. Van Cauter E, Blackman JD, Roland D, Spire JP, Refetoff S, Polonsky KS. Modulation of glucose regulation and insulin secretion by circadian rhythmicity and sleep. *J Clin Invest*. 1991;88:934–942.
83. Simon C, Gronfier C, Schlienger JL, Brandenberger G. Circadian and ultradian variations of leptin in normal man under continuous enteral nutrition: relationship to sleep and body temperature. *J Clin Endocrinol Metab*. 1998;83(6):1893–1899.
84. Maquet P. Positron emission tomography studies of sleep and sleep disorders. *J Neurol*. 1997;244(4 Suppl 1):S23–S28.
85. Maquet P. Functional neuroimaging of normal human sleep by positron emission tomography. *J Sleep Res*. 2000;9(3):207–231.
86. Dang-Vu TT, Desseilles M, Laureys S, Degueldre C, Perrin F, Phillips C, et al. Cerebral correlates of delta waves during non-REM sleep revisited. *Neuroimage*. 2005;28(1):14–21.
87. DeFronzo RA. Pathogenesis of type 2 diabetes mellitus. *Med Clin North Am*. 2004;88(4):787–835. ix.
88. Magistretti PJ. Neuron-glia metabolic coupling and plasticity. *J Exp Biol*. 2006;209(Pt 12):2304–2311.
89. Plat L, Byrne MM, Sturis J, Polonsky KS, Mockel J, Féry F, et al. Effects of morning cortisol elevation on insulin secretion and glucose regulation in humans. *Am J Physiol*. 1996;270:E36–E42.
90. Plat L, Féry F, L'Hermite-Balériaux M, Mockel J, Van Cauter E. Metabolic effects of short-term physiological elevations of plasma cortisol are more pronounced in the evening than in the morning. *J Clin Endocrinol Metab*. 1999.
91. Scheen AJ, Byrne MM, Plat L, Van Cauter E. Relationships between sleep quality and glucose regulation in normal humans. *Am J Physiol*. 1996;271:E261–E270.
92. Jauch-Chara K, Hallschmid M, Schmid SM, Oltmanns KM, Peters A, Born J, et al. Plasma glucagon decreases during night-time sleep in Type 1 diabetic patients and healthy control subjects. *Diabet Med*. 2007;24(6):684–687.
93. Spiegel K, Leproult R, Van Cauter E. Impact of sleep debt on metabolic and endocrine function. *Lancet*. 1999;354:1435–1439.
94. Spiegel K, Leproult R, L'Hermite-Baleriaux M, Copinschi G, Penev PD, Van Cauter E. Leptin levels are dependent on sleep duration: relationships with sympathovagal balance, carbohydrate regulation, cortisol, and thyrotropin. *J Clin Endocrinol Metab*. 2004;89(11):5762–5771.
95. Spiegel K, Knutson K, Leproult R, Tasali E, Van Cauter E. Sleep loss: a novel risk factor for insulin resistance and Type 2 diabetes. *J Appl Physiol*. 2005;99(5):2008–2019.
96. Tasali E, Leproult R, Ehrmann DA, Van Cauter E. Slow-wave sleep and the risk of type 2 diabetes in humans. *Proc Natl Acad Sci USA*. 2008;105(3):1044–1049.
97. Knutson KL, Van Cauter E. Associations between sleep loss and increased risk of obesity and diabetes. *Ann NY Acad Sci*. 2008;1129:287–304.
98. Polonsky KS, Sturis J, Van Cauter E. Temporal profiles and clinical significance of pulsatile insulin secretion. *Horm Res*. 1998;49:178–184.
99. Simon C, Brandenberger G. Ultradian oscillations of insulin secretion in humans. *Diabetes*. 2002;(51 Suppl 1):S258–S261.
100. Sturis J, Scheen AJ, Leproult R, Polonsky KS, Van Cauter E. 24-hour glucose profiles during continuous or oscillatory insulin infusion. *J Clin Invest*. 1995;95:1464–1471.
101. Sturis J, Van Cauter E, Blackman JD, Polonsky KS. Entrainment of pulsatile insulin secretion by oscillatory glucose infusion. *J Clin Invest*. 1991;87:439–445.
102. Sturis J. *Possible mechanisms underlying slow oscillations of human insulin secretion*. The Technical University of Denmark; 1991.
103. Song SH, Kjems L, Ritzel R, McIntyre SM, Johnson ML, Veldhuis JD, et al. Pulsatile insulin secretion by human pancreatic islets. *J Clin Endocrinol Metab*. 2002;87(1):213–221.
104. Porksen N, Hollingdal M, Juhl C, Butler P, Veldhuis JD, Schmitz O. Pulsatile insulin secretion: detection, regulation, and role in diabetes. *Diabetes*. 2002;51(Suppl 1):S245–S254.
105. Meneilly GS, Veldhuis JD, Elahi D. Disruption of the pulsatile and entropic modes of insulin release during an unvarying glucose stimulus in elderly individuals. *J Clin Endocrinol Metab*. 1999;84(6):1938–1943.
106. Getty L, Panteleon AE, Mittelman SD, Dea MK, Bergman RN. Rapid oscillations in omental lipolysis are independent of changing insulin levels *in vivo*. *J Clin Invest*. 2000;106(3): 421–430.
107. Kershaw EE, Flier JS. Adipose tissue as an endocrine organ. *J Clin Endocrinol Metab*. 2004;89(6):2548–2556.
108. Flier JS. Obesity wars: molecular progress confronts an expanding epidemic. *Cell*. 2004;116(2):337–350.
109. Sinha MK, Ohannesian JP, Heiman ML, Kriauciunas A, Stephens TW, Magosin S, et al. Nocturnal rise of leptin in lean, obese and non-insulin-dependent diabetes mellitus subjects. *J Clin Invest*. 1996;97:1344–1347.
110. Saad MF, Riad-Gabriel MG, Khan A, Sharma A, Michael R, Jinagouda SD, et al. Diurnal and ultradian rhythmicity of plasma leptin: effects of gender and adiposity. *J Clin Endocrinol Metab*. 1998;83(2):453–459.
111. Schoeller DA, Cella LK, Sinha MK, Caro JF. Entrainment of the diurnal rhythm of plasma leptin to meal timing. *J Clin Invest*. 1997;100(7):1882–1887.
112. Shea SA, Hilton MF, Orlova C, Ayers RT, Mantzoros CS. Independent circadian and sleep/wake regulation of adipokines and glucose in humans. *J Clin Endocrinol Metab*. 2005;90(5): 2537–2544.
113. Mullington JM, Chan JL, Van Dongen HP, Szuba MP, Samaras J, Price NJ, et al. Sleep loss reduces diurnal rhythm amplitude of leptin in healthy men. *J Neuroendocrinol*. 2003;15(9):851–854.
114. Sinha MK, Sturis J, Ohannesian J, Magosin S, Stephens T, Heiman ML, et al. Ultradian oscillations of leptin secretion in humans. *Biochem Biophys Res Commun*. 1996;228(3):733–738.
115. Licinio J, Mantzoros C, Negrao AB, Cizza G, Wong ML, Bongiorno PB, et al. Human leptin levels are pulsatile and inversely related to pituitary–adrenal function. *Nat Med*. 1997;3(5):575–579.
116. Licinio J, Negrao AB, Mantzoros C, Kaklamani V, Wong ML, Bongiorno PB, et al. Sex differences in circulating human leptin

pulse amplitude: clinical implications. *J Clin Endocrinol Metab.* 1998;83(11):4140–4147.

117. Kolaczynski JW, Considine RV, Ohannesian J, Marco C, Opentanova I, Nyce MR, et al. Responses of leptin to short-term fasting and refeeding in humans: a link with ketogenesis but not ketones themselves. *Diabetes.* 1996;45(11):1511–1515.
118. Chin-Chance C, Polonsky KS, Schoeller DA. Twenty-four-hour leptin levels respond to cumulative short-term energy imbalance and predict subsequent intake. *J Clin Endocrinol Metab.* 2000;85(8):2685–2691.
119. Laughlin GA, Yen SS. Hypoleptinemia in women athletes: absence of a diurnal rhythm with amenorrhea. *J Clin Endocrinol Metab.* 1997;82(1):318–321.
120. Balligand JL, Brichard SM, Brichard V, Desager JP, Lambert M. Hypoleptinemia in patients with anorexia nervosa: loss of circadian rhythm and unresponsiveness to short-term refeeding. *Eur J Endocrinol.* 1998;138(4):415–420.
121. Franceschini R, Corsini G, Cataldi A, Fiorucci A, Tenerelli P, Rolandi E, et al. Twenty-four-hour variation in serum leptin in the elderly. *Metabolism.* 1999;48(8):1011–1014.
122. Spiegel K, Tasali E, Penev P, Van Cauter E. Brief communication: Sleep curtailment in healthy young men is associated with decreased leptin levels, elevated ghrelin levels, and increased hunger and appetite. *Ann Intern Med.* 2004;141(11):846–850.
123. Taheri S, Lin L, Austin D, Young T, Mignot E. Short sleep duration is associated with reduced leptin, elevated ghrelin, and increased body mass index. *PLoS Med.* 2004;1(3):e62.
124. Chaput JP, Despres JP, Bouchard C, Tremblay A. Short sleep duration is associated with reduced leptin levels and increased adiposity: Results from the Quebec family study. *Obesity (Silver Spring).* 2007;15(1):253–261.
125. Elimam A, Knutsson U, Bronnegard M, Stierna P, Albertsson-Wikland K, Marcus C. Variations in glucocorticoid levels within the physiological range affect plasma leptin levels. *Eur J Endocrinol.* 1998;139(6):615–620.
126. Purnell JQ, Samuels MH. Levels of leptin during hydrocortisone infusions that mimic normal and reversed diurnal cortisol levels in subjects with adrenal insufficiency. *J Clin Endocrinol Metab.* 1999;84(9):3125–3128.
127. Leal-Cerro A, Considine RV, Peino R, Venegas E, Astorga R, Casanueva FF, et al. Serum immunoreactive-leptin levels are increased in patients with Cushing's syndrome. *Horm Metab Res.* 1996;28(12):711–713.
128. Masuzaki H, Ogawa Y, Hosoda K, Miyawaki T, Hanaoka I, Hiraoka J, et al. Glucocorticoid regulation of leptin synthesis and secretion in humans: elevated plasma leptin levels in Cushing's syndrome. *J Clin Endocrinol Metab.* 1997;82(8): 2542–2547.
129. Veldman RG, Frolich M, Pincus SM, Veldhuis JD, Roelfsema F. Hyperleptinemia in women with Cushing's disease is driven by high-amplitude pulsatile, but orderly and eurhythmic, leptin secretion. *Eur J Endocrinol.* 2001;144(1):21–27.
130. Kousta E, Chrisoulidou A, Lawrence NJ, al-Shoumer KA, Parker KH, McCarthy MI, et al. The circadian rhythm of leptin is preserved in growth hormone deficient hypopituitary adults. *Clin Endocrinol (Oxf).* 1998;48(6):685–690.
131. Pombo M, Herrera-Justiniano E, Considine RV, Hermida RC, Galvez MJ, Martin T, et al. Nocturnal rise of leptin in normal prepubertal and pubertal children and in patients with perinatal stalk-transection syndrome. *J Clin Endocrinol Metab.* 1997;82(8): 2751–2754.
132. Yamauchi T, Kamon J, Waki H, Terauchi Y, Kubota N, Hara K, et al. The fat-derived hormone adiponectin reverses insulin resistance associated with both lipoatrophy and obesity. *Nat Med.* 2001;7(8):941–946.
133. Kubota N, Terauchi Y, Yamauchi T, Kubota T, Moroi M, Matsui J, et al. Disruption of adiponectin causes insulin resistance and neointimal formation. *J Biol Chem.* 2002;277(29): 25863–25866.
134. Gavrila A, Peng CK, Chan JL, Mietus JE, Goldberger AL, Mantzoros CS. Diurnal and ultradian dynamics of serum adiponectin in healthy men: comparison with leptin, circulating soluble leptin receptor, and cortisol patterns. *J Clin Endocrinol Metab.* 2003;88(6):2838–2843.
135. Weyer C, Funahashi T, Tanaka S, Hotta K, Matsuzawa Y, Pratley RE, et al. Hypoadiponectinemia in obesity and type 2 diabetes: close association with insulin resistance and hyperinsulinemia. *J Clin Endocrinol Metab.* 2001;86(5):1930–1935.
136. Yildiz BO, Suchard MA, Wong ML, McCann SM, Licinio J. Alterations in the dynamics of circulating ghrelin, adiponectin, and leptin in human obesity. *Proc Natl Acad Sci USA.* 2004;101(28):10434–10439.
137. Calvani M, Scarfone A, Granato L, Mora EV, Nanni G, Castagneto M, et al. Restoration of adiponectin pulsatility in severely obese subjects after weight loss. *Diabetes.* 2004;53(4): 939–947.
138. Date Y, Kojima M, Hosoda H, Sawaguchi A, Mondal MS, Suganuma T, et al. Ghrelin, a novel growth hormone-releasing acylated peptide, is synthesized in a distinct endocrine cell type in the gastrointestinal tracts of rats and humans. *Endocrinology.* 2000;141(11):4255–4261.
139. Wren AM, Seal LJ, Cohen MA, Brynes AE, Frost GS, Murphy KG, et al. Ghrelin enhances appetite and increases food intake in humans. *J Clin Endocrinol Metab.* 2001;86(12):5992.
140. van der Lely AJ, Tschop M, Heiman ML, Ghigo E. Biological, physiological, pathophysiological, and pharmacological aspects of ghrelin. *Endocr Rev.* 2004;25(3):426–457.
141. Weikel JC, Wichniak A, Ising M, Brunner H, Friess E, Held K, et al. Ghrelin promotes slow-wave sleep in humans. *Am J Physiol Endocrinol Metab.* 2003;284(2):E407–E415.
142. Szentirmai E, Hajdu I, Obal Jr F, Krueger JM. Ghrelin-induced sleep responses in *ad libitum* fed and food-restricted rats. *Brain Res.* 2006;1088(1):131–140.
143. Koutkia P, Canavan B, Breu J, Johnson ML, Grinspoon SK. Nocturnal ghrelin pulsatility and response to growth hormone secretagogues in healthy men. *Am J Physiol Endocrinol Metab.* 2004;287(3):E506–E512.
144. Cummings DE, Frayo RS, Marmonier C, Aubert R, Chapelot D. Plasma ghrelin levels and hunger scores in humans initiating meals voluntarily without time- and food-related cues. *Am J Physiol Endocrinol Metab.* 2004;287(2):E297–E304.
145. Cummings DE, Purnell JQ, Frayo RS, Schmidova K, Wisse BE, Weigle DS. A preprandial rise in plasma ghrelin levels suggests a role in meal initiation in humans. *Diabetes.* 2001;50(8): 1714–1719.
146. Cummings DE, Weigle DS, Frayo RS, Breen PA, Ma MK, Dellinger EP, et al. Plasma ghrelin levels after diet-induced weight loss or gastric bypass surgery. *N Engl J Med.* 2002;346(21):1623–1630.
147. Teff KL, Elliott SS, Tschop M, Kieffer TJ, Rader D, Heiman M, et al. Dietary fructose reduces circulating insulin and leptin, attenuates postprandial suppression of ghrelin, and increases triglycerides in women. *J Clin Endocrinol Metab.* 2004;89(6): 2963–2972.
148. Misra M, Miller KK, Kuo K, Griffin K, Stewart V, Hunter E, et al. Secretory dynamics of ghrelin in adolescent girls with anorexia nervosa and healthy adolescents. *Am J Physiol Endocrinol Metab.* 2005;289(2):E347–E356.
149. Espelund U, Hansen TK, Hojlund K, Beck-Nielsen H, Clausen JT, Hansen BS, et al. Fasting unmasks a strong inverse

association between ghrelin and cortisol in serum: studies in obese and normal-weight subjects. *J Clin Endocrinol Metab.* 2005;90(2):741–746.

150. Dzaja A, Dalal MA, Himmerich H, Uhr M, Pollmacher T, Schuld A. Sleep enhances nocturnal plasma ghrelin levels in healthy subjects. *Am J Physiol Endocrinol Metab.* 2004;286(6): E963–E967.
151. Schmid SM, Hallschmid M, Jauch-Chara K, Born J, Schultes B. A single night of sleep deprivation increases ghrelin levels and feelings of hunger in normal-weight healthy men. *J Sleep Res.* 2008;17(3):331–334.
152. Vitaterna MH, King DP, Chang AM, Kornhauser JM, Lowrey PL, McDonald JD, et al. Mutagenesis and mapping of a mouse gene, Clock, essential for circadian behavior. *Science.* 1994;264(5159):719–725.
153. King DP, Vitaterna MH, Chang AM, Dove WF, Pinto LH, Turek FW, et al. The mouse Clock mutation behaves as an antimorph and maps within the W19H deletion, distal of Kit. *Genetics.* 1997;146(3):1049–1060.
154. Huang ZJ, Edery I, Rosbash M. PAS is a dimerization domain common to Drosophila period and several transcription factors. *Nature.* 1993;364(6434):259–262.
155. Gekakis N, Staknis D, Nguyen HB, Davis FC, Wilsbacher LD, King DP, et al. Role of the CLOCK protein in the mammalian circadian mechanism. *Science.* 1998;280(5369):1564–1569.
156. Kondratov RV, Chernov MV, Kondratova AA, Gorbacheva VY, Gudkov AV, Antoch MP. BMAL1-dependent circadian oscillation of nuclear CLOCK: posttranslational events induced by dimerization of transcriptional activators of the mammalian clock system. *Genes Dev.* 2003;17(15):1921–1932.
157. Konopka RJ, Benzer S. Clock mutants of Drosophila melanogaster. *Proceedings of the Natl Acad Sci USA.* 1971;68(9): 2112–2116.
158. Hao H, Allen DL, Hardin PE. A circadian enhancer mediates PER-dependent mRNA cycling in Drosophila melanogaster. *Mol Cell Biol.* 1997;17(7):3687–3693.
159. Bargiello TA, Jackson FR, Young MW. Restoration of circadian behavioural rhythms by gene transfer in *Drosophila. Nature.* 1984;312(5996):752–754.
160. Zehring WA, Wheeler DA, Reddy P, Konopka RJ, Kyriacou CP, Rosbash M, et al. P-element transformation with period locus DNA restores rhythmicity to mutant, arrhythmic Drosophila melanogaster. *Cell.* 1984;39(2 Pt 1):369–376.
161. King DP, Zhao Y, Sangoram AM, Wilsbacher LD, Tanaka M, Antoch MP, et al. Positional cloning of the mouse circadian clock gene. *Cell.* 1997;89(4):641–653.
162. Darlington TK, Wager-Smith K, Ceriani MF, Staknis D, Gekakis N, Steeves TDL, et al. Closing the circadian loop: CLOCK-induced transcription of its own inhibitors *per* and *tim*. *Science.* 1998;280(5369):1599–1603.
163. Hardin PE, Hall JC, Rosbash M. Circadian oscillations in period gene mRNA levels are transcriptionally regulated. *Proc Natl Acad Sci USA.* 1992;89(24):11711–11715.
164. Bunger MK, Wilsbacher LD, Moran SM, Clendenin C, Radcliffe LA, Hogenesch JB, et al. Mop3 is an essential component of the master circadian pacemaker in mammals. *Cell.* 2000;103(7):1009–1017.
165. Bae K, Jin X, Maywood ES, Hastings MH, Reppert SM, Weaver DR. Differential functions of mPer1, mPer2, and mPer3 in the SCN circadian clock. *Neuron.* 2001;30(2):525–536.
166. Zheng B, Albrecht U, Kaasik K, Sage M, Lu W, Vaishnav S, et al. Nonredundant roles of the mPer1 and mPer2 genes in the mammalian circadian clock. *Cell.* 2001;105(5):683–694.
167. Miyamoto Y, Sancar A. Vitamin B2-based blue-light photoreceptors in the retinohypothalamic tract as the photoactive pigments for setting the circadian clock in mammals. *Proc Natl Acad Sci USA.* 1998;95(11):6097–6102.
168. Thresher RJ, Vitaterna MH, Miyamoto Y, Kazantsev A, Hsu DS, Petit C, et al. Role of mouse cryptochrome blue-light photoreceptor in circadian photoresponses. *Science.* 1998;282(5393):1490–1494.
169. Vitaterna MH, Selby CP, Todo T, Niwa H, Thompson C, Fruechte EM, et al. Differential regulation of mammalian period genes and circadian rhythmicity by cryptochromes 1 and 2. *Proc Natl Acad Sci USA.* 1999;96(21):12114–12119.
170. van der Horst GT, Muijtjens M, Kobayashi K, Takano R, Kanno S, Takao M, et al. Mammalian Cry1 and Cry2 are essential for maintenance of circadian rhythms. *Nature.* 1999;398(6728):627–630.
171. Kume K, Zylka MJ, Sriram S, Shearman LP, Weaver DR, Jin X, et al. mCRY1 and mCRY2 are essential components of the negative limb of the circadian clock feedback loop. *Cell.* 1999;98(2):193–205.
172. Nakao N, Yasuo S, Nishimura A, Yamamura T, Watanabe T, Anraku T, et al. Circadian clock gene regulation of steroidogenic acute regulatory protein gene expression in preovulatory ovarian follicles. *Endocrinology.* 2007;148(7):3031–3038.
173. Buijs RM, Wortel J, Van Heerikhuize JJ, Feenstra MG, Ter Horst GJ, Romijn HJ, et al. Anatomical and functional demonstration of a multisynaptic suprachiasmatic nucleus adrenal (cortex) pathway. *Eur J Neurosci.* 1999;11(5):1535–1544.
174. Lemos DR, Downs JL, Urbanski HF. Twenty-four-hour rhythmic gene expression in the rhesus macaque adrenal gland. *Mol Endocrinol.* 2006;20(5):1164–1176.
175. Torres-Farfan C, Rocco V, Monso C, Valenzuela FJ, Campino C, Germain A, et al. Maternal melatonin effects on clock gene expression in a nonhuman primate fetus. *Endocrinology.* 2006;147(10):4618–4626.
176. Valenzuela FJ, Torres-Farfan C, Richter HG, Mendez N, Campino C, Torrealba F, et al. Clock gene expression in adult primate suprachiasmatic nuclei and adrenal: is the adrenal a peripheral clock responsive to melatonin? *Endocrinology.* 2008;149(4):1454–1461.
177. Oster H, Damerow S, Kiessling S, Jakubcakova V, Abraham D, Tian J, et al. The circadian rhythm of glucocorticoids is regulated by a gating mechanism residing in the adrenal cortical clock. *Cell Metab.* 2006;4(2):163–173.
178. Van Cauter E, van Coevorden A, Blackman JD. Modulation of neuroendocrine release by sleep and circadian rhythmicity. In: Yen S, Vale W, eds. *Advances in neuroendocrine regulation of reproduction.* Norwell: Serono Symposia USA; 1990: 113–22.
179. Veldhuis JD, Iranmanesh A, Johnson ML, Lizarralde G. Amplitude, but not frequency, modulation of adrenocorticotropin secretory bursts gives rise to the nyctohemeral rhythm of the corticotropic axis in man. *J Clin Endocrinol Metab.* 1989;71:452–463.
180. Van Cauter E, Honinckx E. Pulsatility of pituitary hormones. *Exp Brain Res.* 1985;(Suppl 12):41–60.
181. Lejeune-Lenain C, Van Cauter E, Desir D, Beyloos M, Franckson JRM. Control of circadian and episodic variations of adrenal androgens secretion in man. *J Endocrinol Invest.* 1987;10:267–276.
182. Weitzman ED, Zimmerman JC, Czeisler CA, Ronda JM. Cortisol secretion is inhibited during sleep in normal man. *J Clin Endocrinol Metab.* 1983;56:352–358.
183. Born J, Muth S, Fehm HL. The significance of sleep onset and slow wave sleep for nocturnal release of growth hormone (GH) and cortisol. *Psychoneuroendocrinology.* 1988;13:233–243.
184. Weibel L, Follenius M, Spiegel K, Ehrhart J, Brandenberger G. Comparative effect of night and daytime sleep on the 24-hour cortisol secretory profile. *Sleep.* 1995;18:549–556.

185. Follenius M, Brandenberger G, Bardasept J, Libert J, Ehrhart J. Nocturnal cortisol release in relation to sleep structure. *Sleep*. 1992;15:21–27.
186. Gronfier C, Luthringer R, Follenius M, Schaltenbrand N, Macher JP, Muzet A, et al. Temporal relationships between pulsatile cortisol secretion and electroencephalographic activity during sleep in man. *Electroencephalogr Clin Neurophysiol*. 1997;103:405–408.
187. Spath-Schwalbe E, Gofferje M, Kern W, Born J, Fehm HL. Sleep disruption alters nocturnal ACTH and cortisol secretory patterns. *Biol Psychiatry*. 1991;29:575–584.
188. Pruessner JC, Wolf OT, Hellhammer DH, Buske-Kirschbaum A, von Auer K, Jobst S, et al. Free cortisol levels after awakening: a reliable biological marker for the assessment of adrenocortical activity. *Life Sci*. 1997;61:2539–2549.
189. Caufriez A, Moreno-Reyes R, Leproult R, Vertongen F, Van Cauter E, Copinschi G. Immediate effects of an 8-h advance shift of the rest-activity cycle on 24-h profiles of cortisol. *Am J Physiol Endocrinol Metab*. 2002;282(5):E1147–E1153.
190. Ekstedt M, Akerstedt T, Soderstrom M. Microarousals during sleep are associated with increased levels of lipids, cortisol, and blood pressure. *Psychosom Med*. 2004;66(6):925–931.
191. Chapotot F, Gronfier C, Jouny C, Muzet A, Brandenberger G. Cortisol secretion is related to electroencephalographic alertness in human subjects duirng daytime wakefulness. *J Clin Endocrinol Metab*. 1998;83:4263–4268.
192. Vgontzas AN, Zoumakis M, Bixler EO, Lin HM, Prolo P, Vela-Bueno A, et al. Impaired nighttime sleep in healthy old versus young adults is associated with elevated plasma interleukin-6 and cortisol levels: physiologic and therapeutic implications. *J Clin Endocrinol Metab*. 2003;88(5):2087–2095.
193. Vgontzas AN, Bixler EO, Lin HM, Prolo P, Mastorakos G, Vela-Bueno A, et al. Chronic insomnia is associated with nyctohemeral activation of the hypothalamic-pituitary-adrenal axis: clinical implications. *J Clin Endocrinol Metab*. 2001;86(8):3787–3794.
194. Scheer FA, Buijs RM. Light affects morning salivary cortisol in humans. *J Clin Endocrinol Metab*. 1999;84:3395–3398.
195. Leproult R, Colecchia EF, L'Hermite-Baleriaux M, Van Cauter E. Transition from dim to bright light in the morning induces an immediate elevation of cortisol levels. *J Clin Endocrinol Metab*. 2001;86:151–157.
196. Ruger M, Gordijn MC, Beersma DG, de Vries B, Daan S. Time-of-day-dependent effects of bright light exposure on human psychophysiology: comparison of daytime and nighttime exposure. *Am J Physiol Regul Integr Comp Physiol*. 2006;290(5):R1413–R1420.
197. Linkowski P, Van Onderbergen A, Kerkhofs M, Bosson D, Mendlewicz J, Van Cauter E. Twin study of the 24-h cortisol profile: evidence for genetic control of the human circadian clock. *Am J Physiol*. 1993;264:E173–E181.
198. Van Cauter E, Shapiro ET, Tillil H, Polonsky KS. Circadian modulation of glucose and insulin responses to meals: relationship to cortisol rhythm. *Am J Physiol*. 1992;262:E467–E475.
199. Born J, Hansen K, Marshall L, Molle M, Fehm HL. Timing the end of nocturnal sleep. *Nature*. 1999;397(6714):29–30.
200. Leproult R, Copinschi G, Buxton O, Van Cauter E. Sleep loss results in an elevation of cortisol levels the next evening. *Sleep*. 1997;20:865–870.
201. Sherman B, Wysham C, Pfohl B. Age-related changes in the circadian rhythm of plasma cortisol in man. *J Clin Endocrinol Metab*. 1985;61:439–443.
202. Van Cauter E, Leproult R, Kupfer DJ. Effects of gender and age on the levels and circadian rhythmicity of plasma cortisol. *J Clin Endocrinol Metab*. 1996;81:2468–2473.
203. van Coevorden A, Mockel J, Laurent E, Kerkhofs M, L'Hermite-Balériaux M, Decoster C, et al. Neuroendocrine rhythms and sleep in aging men. *Am J Physiol*. 1991;260:E651–E661.
204. McEwen BS, Stellar E. Stress and the individual. *Arch Intern Med*. 1993;153:2093–2101.
205. McEwen B. Protective and damaging effects of stress mediators. *N Engl J Med*. 1998;338:171–179.
206. Dallman MF, Strack AL, Akana SF, Bradbury MJ, Hanson ES, Scribner KA, et al. Feast and famine: Critical role of glucocorticoids with insulin in daily energy flow. *Frontiers in Neuroendocrinology*. 1993;14:303–347.
207. Dennison E, Hindmarsh P, Fall C, Kellingray S, Barker D, Phillips D, et al. Profiles of endogenous circulating cortisol and bone mineral density in healthy elderly men. *J Clin Endocrinol Metab*. 1999;84:3058–3063.
208. Van Cauter E, Leproult R, Plat L. Age-related changes in slow-wave sleep and REM sleep and relationship with growth hormone and cortisol levels in healthy men. *J Am Med Assoc*. 2000;284:861–868.
209. Nolten WE, Lindheimer MD, Rueckert PA, Oparil S, Ehrlich EN. Diurnal patterns and regulation of cortisol secretion in pregnancy. *J Clin Endocrinol Metab*. 1980;51:466–472.
210. Magiakou MA, Mastorakos G, Rabin D, Margioris AN, Dubbert B, Calogero AE, et al. The maternal hypothalamic-pituitary-adrenal axis in the third trimester of human pregnancy. *Clin Endocrinol (Oxf)*. 1996;44:419–428.
211. Muccioli G, Tschop M, Papotti M, Deghenghi R, Heiman M, Ghigo E. Neuroendocrine and peripheral activities of ghrelin: implications in metabolism and obesity. *Eur J Pharmacol*. 2002;440(2-3):235–254.
212. Van Cauter E, Plat L, Copinschi G. Interrelations between sleep and the somatotropic axis. *Sleep*. 1998;21:553–566.
213. Mendlewicz J, Linkowski P, Kerkhofs M, Leproult R, Copinschi G, Van Cauter E. Genetic control of 24-hour growth hormone secretion in man: A twin study. *J Clin Endocrinol Metab*. 1999;84(3):856–862.
214. Ho KY, Evans WS, Blizzard RM, Veldhuis JD, Merriam GR, Samojlik E, et al. Effects of sex and age on the 24-hour profile of growth hormone secretion in man: importance of endogenous estradiol concentrations. *J Clin Endocrinol Metab*. 1987;64:51–58.
215. Shah N, Evans WS, Veldhuis JD. Actions of estrogen on pulsatile, nyctohemeral, and entropic modes of growth hormone secretion. *Am J Physiol*. 1999;1999:R1351–R1358.
216. Caufriez A, Leproult R, L'Hermite-Baleriaux M, Moreno-Reyes R, Copinschi G. A potential role of endogenous progesterone in modulation of GH, prolactin and thyrotropin secretion during normal menstrual cycle. *Clin Endocrinol (Oxf)*. 2009;71:535–542.
217. Weibel L, Spiegel K, Gronfier C, Follenius M, Brandenberger G. Twenty-four-hour melatonin and core body temperature rhythms: their adaptation in night workers. *Am J Physiol*. 1997;272:R948–R954.
218. Van Cauter E, Kerkhofs M, Caufriez A, Van Onderbergen A, Thorner MO, Copinschi G. A quantitative estimation of GH secretion in normal man: reproducibility and relation to sleep and time of day. *J Clin Endocrinol Metab*. 1992;74:1441–1450.
219. Holl RW, Hartmann ML, Veldhuis JD, Taylor WM, Thorner MO. Thirty-second sampling of plasma growth hormone in man: correlation with sleep stages. *J Clin Endocrinol Metab*. 1991;72:854–861.
220. Vgontzas AN, Mastorakos G, Bixler EO, Kales A, Gold PW, Chrousos GP. Sleep deprivation effects on the activity of the hypothalamic–pituitary–adrenal and growth axes: potential clinical implications. *Clin Endocrinol (Oxf)*. 1999;51(2):205–215.

221. Obal Jr F, Krueger JM. Biochemical regulation of non-rapid-eye-movement sleep. *Front Biosci*. 2003;8:d520–d550.
222. Ocampo-Lim B, Guo W, DeMott Friberg R, Barkan AL, Jaffe CA. Nocturnal growth hormone (GH) secretion is eliminated by infusion of GH- releasing hormone antagonist. *J Clin Endocrinol Metab*. 1996;81:4396–4399.
223. Van Cauter E, Plat L, Scharf M, Leproult R, Cespedes S, L'Hermite-Balériaux M, et al. Simultaneous stimulation of slow-wave sleep and growth hormone secretion by gamma-hydroxybutyrate in normal young men. *J Clin Invest*. 1997;100:745–753.
224. Gronfier C, Luthringer R, Follenius M, Schaltenbrand N, Macher JP, Muzet A, et al. A quantitative evaluation of the relationships between growth hormone secretion and delta wave electroencephalographic activity during normal sleep and after enrichment in delta waves. *Sleep*. 1996;19:817–824.
225. Van Cauter E, Caufriez A, Kerkhofs M, Van Onderbergen A, Thorner MO, Copinschi G. Sleep, awakenings and insulin-like growth factor I modulate the growth hormone secretory response to growth hormone-releasing hormone. *J Clin Endocrinol Metab*. 1992;74:1451–1459.
226. Jaffe C, Turgeon D, DeMott Friberg R, Watkins P, Barkan A. Nocturnal augmentation of growth hormone (GH) secretion is preserved during repetitive bolus administration of GH-releasing hormone: potential involvement of endogenous somatostatin – A clinical research center study. *J Clin Endocrinol Metab*. 1995;80:3321–3326.
227. Nass R, Farhy LS, Liu J, Prudom CE, Johnson ML, Veldhuis P, et al. Evidence for acyl-ghrelin modulation of growth hormone release in the fed state. *J Clin Endocrinol Metab*. 2008;93(5):1988–1994.
228. Steiger A, Herth T, Holsboer F. Sleep-electroencephalography and the secretion of cortisol and growth hormone in normal controls. *Acta Endocrinol*. 1987;116:36–42.
229. Spiegel K, Leproult R, Colecchia EF, L'Hermite-Baleriaux M, Nie Z, Copinschi G, et al. Adaptation of the 24-h growth hormone profile to a state of sleep debt. *Am J Physiol Regul Integr Comp Physiol*. 2000;279(3):R874–R883.
230. Brandenberger G, Gronfier C, Chapotot F, Simon C, Piquard F. Effect of sleep deprivation on overall 24 h growth-hormone secretion. *Lancet*. 2000;356(9239):1408.
231. Vermeulen A. Nyctohemeral growth hormone profiles in young and aged men: correlation with somatomedin-C levels. *J Clin Endocrinol Metab*. 1987;64:884–888.
232. Veldhuis J, Liem A, South S, Weltman A, Weltman J, Clemmons D, et al. Differential impact of age, sex steroid hormones, and obesity on basal versus pulsatile growth hormone secretion in men as assessed in an ultrasensitive chemiluminescence assay. *J Clin Endocrinol Metab*. 1995;80:3209–3222.
233. Veldhuis JD, Iranmanesh A, Weltman A. Elements in the pathophysiology of diminished growth hormone (GH) secretion in aging humans. *Endocrine*. 1997;7:41–48.
234. Martin FC, Yeo AL, Sonksen PH. Growth hormone secretion in the elderly: ageing and the somatopause. *Baillieres Clin Endocrinol Metab*. 1997;11:223–250.
235. Caufriez A, Frankenne F, Hennen G, Copinschi G. Regulation of maternal IGF-I by placental GH in normal and abnormal human pregnancies. *Am J Physiol*. 1993;265:E572–E577.
236. Eriksson L, Frankenne F, Eden S, Hennen G, Von Schoultz B. Growth hormone 24-h serum profiles during pregnancy – lack of pulsatility for the secretion of the placental variant. *Br J Obstet Gynaecol*. 1989;1989:949–953.
237. Sassin J, Frantz A, Weitzman E, Kapen S. Human prolactin: 24-hour pattern with increased release during sleep. *Science*. 1972;177:1205–1207.
238. Van Cauter E, L'Hermite M, Copinschi G, Refetoff S, Desir D, Robyn C. Quantitative analysis of spontaneous variations of plasma prolactin in normal man. *Am J Physiol*. 1981;241: E355–E363.
239. Spiegel K, Follenius M, Simon C, Saini J, Ehrhart J, Brandenberger G. Prolactin secretion and sleep. *Sleep*. 1994;17:20–27.
240. Waldstreicher J, Duffy JF, Brown EN, Rogacz S, Allan JS, Czeisler CA. Gender differences in the temporal organization of prolactin (PRL) secretion: Evidence for a sleep-independent circadian rhythm of circulating PRL levels – A Clinical Research Center study. *J Clin Endocrinol Metab*. 1996;81: 1483–1487.
241. Katznelson L, Riskind P, N., Saxe VC, Klibanski A. Prolactin pulsatile characteristics in postmenopausal women. *J Clin Endocrinol Metab*. 1998;83:761–764.
242. Veldhuis JD, Johnson ML, Lizarralde G, Iranmanesh A. Rhythmic and nonrhythmic modes of anterior pituitary gland secretion. *Chronobiology International*. 1992;9:371–379.
243. Linkowski P, Spiegel K, Kerkhofs M, L'Hermite-Balériaux M, Van Onderbergen A, Leproult R, et al. Genetic and environmental influences on prolactin secretion during wake and during sleep. *Am J Physiol*. 1998;274:E909–E919.
244. Van Cauter E, Refetoff S. Multifactorial control of the 24-hour secretory profiles of pituitary hormones. *J Endocrinol Invest*. 1985;8:381–391.
245. Désir D, Van Cauter E, L'Hermite M, Refetoff S, Jadot C, Caufriez A, et al. Effects of "jet lag" on hormonal patterns. III. Demonstration of an intrinsic circadian rhythmicity in plasma prolactin. *J Clin Endocrinol Metab*. 1982;55:849–857.
246. Copinschi G, Van Onderbergen A, L'Hermite-Balériaux M, Szyper M, Caufriez A, Bosson D, et al. Effects of the short-acting benzodiazepine triazolam, taken at bedtime, on circadian and sleep-related hormonal profiles in normal men. *Sleep*. 1990;13: 232–244.
247. Copinschi G, Akseki E, Moreno-Reyes R, Leproult R, L'Hermite-Balériaux M, Caufriez A, et al. Effects of bedtime administration of zolpidem on circadian and sleep-related hormonal profiles in normal women. *Sleep*. 1995;18:417–424.
248. Spiegel K, Luthringer R, Follenius M, Schaltenbrandt N, Macher JP, Muzet A, et al. Temporal relationship between prolactin secretion and slow-wave electroencephalographic activity during sleep. *Sleep*. 1995;18:543–548.
249. Greenspan SL, Klibanski A, Rowe JW, Elahi D. Age alters pulsatile prolactin release: influence of dopaminergic inhibition. *Am J Physiol*. 1990;258:E799–E804.
250. Tay CC, Glasier AF, McNeilly AS. Twenty-four hour patterns of prolactin secretion during lactation and the relationship to suckling and the resumption of fertility in breast-feeding women. *Hum Reprod*. 1996;11:950–955.
251. Spratt DI, O'Dea LL, Schoenfeld D, Butler JP, Rao TN, Crowley WFJ. Neuroendocrine-gonadal axis in men: frequent sampling of LH, FSH and testosterone. *Am J Physiol*. 1988;254:E658–E666.
252. Lanfranco F, Bonelli L, Baldi M, Me E, Broglio F, Ghigo E. Acylated ghrelin inhibits spontaneous luteinizing hormone pulsatility and responsiveness to naloxone but not that to gonadotropin-releasing hormone in young men: evidence for a central inhibitory action of ghrelin on the gonadal axis. *J Clin Endocrinol Metab*. 2008;93(9):3633–3639.
253. Fehm HL, Clausing J, Kern W, Pietrowsky R, Born J. Sleep-associated augmentation and synchronization of luteinizing hormone pulses in adult men. *Neuroendocrinology*. 1991;54:192–195.
254. Bremner WJ, Vitiello MV, Prinz PN. Loss of circadian rhythmicity in blood testosterone levels with aging in normal men. *J Clin Endocrinol Metab*. 1983;56:1278–1280.
255. Luboshitzky R, Zabari Z, Shen-Orr Z, Herer P, Lavie P. Disruption of the nocturnal testosterone rhythm by sleep

fragmentation in normal men. *J Clin Endocrinol Metab.* 2001;86(3):1134–1139.
256. Axelsson J, Ingre M, Akerstedt T, Holmback U. Effects of acutely displaced sleep on testosterone. *J Clin Endocrinol Metab.* 2005;90(8):4530–4535.
257. Carlsen E, Olsson C, Petersen JH, Andersson AM, Skakkebaek NE. Diurnal rhythm in serum levels of inhibin B in normal men: relation to testicular steroids and gonadotropins. *J Clin Endocrinol Metab.* 1999;84(5):1664–1669.
258. Lejeune H, Dechaud H, Pugeat M. Contribution of bioavailable testosterone assay for the diagnosis of androgen deficiency in elderly men. *Ann Endocrinol (Paris).* 2003;64(2):117–125.
259. Tenover JS, Matsumoto AM, Clifton DK, Bremner WJ. Age-related alterations in the circadian rhythms of pulsatile luteinizing hormone and testosterone secretion in healthy men. *J Gerontol.* 1988;43:M163–M169.
260. Penev PD. Association between sleep and morning testosterone levels in older men. *Sleep.* 2007;30(4):427–432.
261. Mulligan T, Iranmanesh A, Gheorghiu S, Godschalk M, Veldhuis JD. Amplified nocturnal luteinizing hormone (LH) secretory burst frequency with selective attenuation of pulsatile (but not basal) testosterone secretion in healthy aged men: possible Leydig cell desensitization to endogenous LH signaling—a clinical research center study. *J Clin Endocrinol Metab.* 1995;80(10):3025–3031.
262. Vermeulen A, Deslypere JP, Kaukman JM. Influence of antiopioids on luteinizing hormone pulsatility in aging men. *J Clin Endocrinol Metab.* 1989;68:68–72.
263. Veldhuis JD, Urban RJ, Lizarralde G, Johnson ML, Iranmanesh A. Attenuation of luteininzing hormone secretory burst amplitude as a proximate basis for the hypoandrogenism of healthy aging in men. *J Clin Endocrinol Metab.* 1992;75:52–58.
264. Veldhuis JD, Iranmanesh A, Mulligan T, Pincus SM. Disruption of the young-adult synchrony between luteinizing hormone release and oscillations in follicle-stimulating hormone, prolactin, and nocturnal penile tumescence (NPT) in healthy older men. *J Clin Endocrinol Metab.* 1999;84(10):3498–3505.
265. Pincus SM, Mulligan T, Iranmanesh A, Gheorghiu S, Godschalk M, Veldhuis JD. Older males secrete luteinizing hormone and testosterone more irregularly, and jointly more asynchronously, than younger males. *Proc Natl Acad Sci USA.* 1996;93(24):14100–14105.
266. Filicori M, Santoro N, Merriam GR, Crowley WFJ. Characterization of the physiological pattern of episodic gonadotropin secretion throughout the menstrual cycle. *J Clin Endocrinol Metb.* 1986;62:1136–1144.
267. Samuels MH, Lillehei K, Kleinschmidt-Demasters BK, Stears J, Ridgway EC. Patterns of pulsatile pituitary glycoprotein secretion in central hypothyroidism and hypogonadism. *J Clin Endocrinol Metab.* 1990;70(2):391–395.
268. Licinio J, Negrao AB, Mantzoros C, Kaklamani V, Wong ML, Bongiorno PB, et al. Synchronicity of frequently sampled, 24-h concentrations of circulating leptin, luteinizing hormone, and estradiol in healthy women. *Proc Natl Acad Sci USA.* 1998; 95(5):2541–2546.
269. Judd S, Terry A, Petrucco M, White G. The source of pulsatile secretion of progesterone during the human follicular phase. *J Clin Endocrinol Metab.* 1992;74(2):299–305.
270. Veldhuis JD, Christiansen E, Evans WS, Kolp LA, Rogol AD, Johnson ML. Physiological profiles of episodic progesterone release during the midluteal phase of the human menstrual cycle: analysis of circadian and ultradian rhythms, discrete pulse properties, and correlations with simultaneous luteinizing hormone release. *J Clin Endocrinol Metab.* 1988;66(2):414–421.
271. Rossmanith WG, Laughlin GA, Mortola JF, Yen SS. Secretory dynamics of oestradiol (E2) and progesterone (P4) during periods of relative pituitary LH quiescence in the midluteal phase of the menstrual cycle. *Clin Endocrinol (Oxf).* 1990;32(1):13–23.
272. Reame N, Sauder SE, Kelch RP, Marshall JC. Pulsatile gonadotropin secretion during the human menstrual cycle: evidence for altered frequency of gonadotropin-releasing hormone secretion. *J Clin Endocrinol Metab.* 1984;59:328–337.
273. Hall JE, Schoenfeld DA, Martin KA, Crowley WFJ. Hypothalamic gonadotropin-releasing hormone secretion and follicle-stimulating hormone dynamics during the luteal-follicular transition. *J Clin Endocrinol Metab.* 1992;74:600–607.
274. Hall JE, Sullivan JP, Richardson GS. Brief wake episodes modulate sleep-inhibited luteinizing hormone secretion in the early follicular phase. *J Clin Endocrinol Metab.* 2005;90(4): 2050–2055.
275. Reame NE, Kelche RP, Beitins IZ, Yu MY, Zawacki CM, Padmanabhan V. Age effects of follicle-stimulating hormone and pulsatile luteinizing hormone secretion across the menstrual cycle of premenopausal women. *J Clin Endocrinol Metab.* 1996;81(4):1512–1518.
276. Brabant G, Prank K, Ranft U, Schuermeyer T, Wagner TOF, Hauser H, et al. Physiological regulation of circadian and pulsatile thyrotropin secretion in normal man and woman. *J Clin Endocrinol Metab.* 1990;70:403–409.
277. Veldhuis JD, Iranmanesh A, Johnson ML, Lizarralde G. Twenty-four-hour rhythms in plasma concentrations of adenohypophyseal hormones are generated by distinct amplitude and/or frequency modulation of underlying pituitary secretory bursts. *J Clin Endocrinol Metab.* 1990;71:1616–1623.
278. Parker DC, Rossman LG, Pekary AE, Hershman JM. Effect of 64-hour sleep deprivation on the circadian waveform of thyrotropin (TSH): further evidence of sleep-related inhibition of TSH release. *J Clin Endocrinol Metab.* 1987;64:157–161.
279. Hirschfeld U, Moreno-Reyes R, Akseki E, L'Hermite-Balériaux M, Leproult R, Copinschi G, et al. Progressive elevation of plasma thyrotropin during adaptation to simulated jet lag: effects of treatment with bright light or zolpidem. *J Clin Endocrinol Metab.* 1996;81:3270–3277.
280. Goichot B, Brandenberger G, Saini J, Wittersheim G, Follenius M. Nocturnal plasma thyrotropin variations are related to slow-wave sleep. *J Sleep Res.* 1992;1:186–190.
281. Gronfier C, Luthringer R, Follenius M, Schaltenbrand N, Macher JP, Muzet A, et al. Temporal link between plasma thyrotropin levels and electroencephalographic activity in man. *Neurosci Lett.* 1995;200:97–100.
282. Buxton OM, Frank SA, L'Hermite-Balériaux M, Leproult R, Turek FW, Van Cauter E. Roles of intensity and duration of nocturnal exercise in causing phase-shifts of human circadian rhythms. *Am J Physiol (Endocrinol Metab).* 1997;273:E536–E542.
283. Russell W, Harrison RF, Smith N, Darzy K, Shalet S, Weetman AP, et al. Free triiodothyronine has a distinct circadian rhythm that is delayed but parallels thyrotropin levels. *J Clin Endocrinol Metab.* 2008;93(6):2300–2306.
284. Allan JS, Czeisler CA. Persistence of the circadian thyrotropin rhythm under constant conditions and after light-induced shifts of circadian phase. *J Clin Endocrinol Metab.* 1994;79:508–512.
285. Gary KA, Winokur A, Douglas SD, Kapoor S, Zaugg L, Dinges DF. Total sleep deprivation and the thyroid axis: Effects of sleep and waking activity. *Aviat Space Environ Med.* 1996; 67:513–519.
286. Czeisler CA, Johnson MP, Duffy JF, Brown EN, Ronda JM, Kronauer RE. Exposure to bright light and darkness to treat

physiologic maladaptation to night work. *N Engl J Med.* 1990;322(18):1253–1259.
287. Knutsson A, Akerstedt T, Jonsson BG, Orth-Gomer K. Increased risk of ischaemic heart disease in shift workers. *Lancet.* 1986;2(8498):89–92.
288. Rosa RR. Extended workshifts and excessive fatigue. *J Sleep Res.* 1995;4(S2):51–56.
289. Roden M, Koller M, Pirich K, Vierhapper H, Waldhauser F. The circadian melatonin and cortisol secretion pattern in permanent night shift workers. *Am J Physiol.* 1993;265(1 Pt 2):R261–R267.
290. Weibel L, Spiegel K, Follenius M, Ehrhart J, Brandenberger G. Internal dissociation of the circadian markers of the cortisol rhythm in night workers. *Am J Physiol.* 1996;270(4 Pt 1): E608–E613.
291. Weibel L, Spiegel K, Gronfier C, Follenius M, Brandenberger G. Twenty-four-hour melatonin and core body temperature rhythms: their adaptation in night workers. *Am J Physiol.* 1997;272(3 Pt 2):R948–R954.
292. McEwen BS. Stress, adaptation, and disease. Allostasis and allostatic load. *Ann NY Acad Sci.* 1998;840:33–44.
293. Van Cauter E, Knutson K. Sleep and the epidemic of obesity in children and adults. *Eur J Endocrinol.* 2008;159(Suppl. 1): S59–S66.
294. Cooper BG, White JE, Ashworth LA, Alberti KG, Gibson GJ. Hormonal and metabolic profiles in subjects with obstructive sleep apnea syndrome and the acute effects of nasal continuous positive airway pressure (CPAP) treatment. *Sleep.* 1995;18(3):172–179.
295. Saini J, Krieger J, Brandenberger G, Wittersheim G, Simon C, Follenius M. Continuous positive airway pressure treatment. Effects on growth hormone, insulin and glucose profiles in obstructive sleep apnea patients. *Horm Metab Res.* 1993;25(7): 375–381.
296. Spiegel K, Follenius M, Krieger J, Sforza E, Brandenberger G. Prolactin secretion during sleep in obstructive sleep apnoea patients. *J Sleep Res.* 1995;4(1):56–62.
297. Luboshitzky R, Lavie L, Shen-Orr Z, Lavie P. Pituitary–gonadal function in men with obstructive sleep apnea. The effect of continuous positive airways pressure treatment. *Neuroendocrinol Lett.* 2003;24(6):463–467.
298. Ozturk L, Unal M, Tamer L, Celikoglu F. The association of the severity of obstructive sleep apnea with plasma leptin levels. *Arch Otolaryngol Head Neck Surg.* 2003;129(5):538–540.
299. Patel SR, Palmer LJ, Larkin EK, Jenny NS, White DP, Redline S. Relationship between obstructive sleep apnea and diurnal leptin rhythms. *Sleep.* 2004;27(2):235–239.
300. Phillips BG, Kato M, Narkiewicz K, Choe I, Somers VK. Increases in leptin levels, sympathetic drive, and weight gain in obstructive sleep apnea. *Am J Physiol Heart Circ Physiol.* 2000;279(1):H234–H237.
301. Ip MS, Lam KS, Ho C, Tsang KW, Lam W. Serum leptin and vascular risk factors in obstructive sleep apnea. *Chest.* 2000;118(3):580–586.
302. Sanner BM, Kollhosser P, Buechner N, Zidek W, Tepel M. Influence of treatment on leptin levels in patients with obstructive sleep apnoea. *Eur Respir J.* 2004;23(4):601–604.
303. Harsch IA, Konturek PC, Koebnick C, Kuehnlein PP, Fuchs FS, Pour Schahin S, et al. Leptin and ghrelin levels in patients with obstructive sleep apnoea: effect of CPAP treatment. *Eur Respir J.* 2003;22(2):251–257.
304. Shimizu K, Chin K, Nakamura T, Masuzaki H, Ogawa Y, Hosokawa R, et al. Plasma leptin levels and cardiac sympathetic function in patients with obstructive sleep apnoea-hypopnoea syndrome. *Thorax.* 2002;57(5):429–434.
305. Chin K, Shimizu K, Nakamura T, Narai N, Masuzaki H, Ogawa Y, et al. Changes in intra-abdominal visceral fat and serum leptin levels in patients with obstructive sleep apnea syndrome following nasal continuous positive airway pressure therapy. *Circulation.* 1999;100(7):706–712.
306. Tasali E, Mokhlesi B, Van Cauter E. Obstructive sleep apnea and type 2 diabetes: interacting epidemics. *Chest.* 2008;133(2): 496–506.
307. Schahin SP, Nechanitzky T, Dittel C, Fuchs FS, Hahn EG, Konturek PC, et al. Long-term improvement of insulin sensitivity during CPAP therapy in the obstructive sleep apnoea syndrome. *Med Sci Monit.* 2008;14(3):CR117–CR121.
308. Tasali E, Van Cauter E, Hoffman L, Ehrmann DA. Impact of obstructive sleep apnea on insulin resistance and glucose tolerance in women with polycystic ovary syndrome. *J Clin Endocrinol Metab.* 2008;93(10):3878–3884.
309. West SD, Nicoll DJ, Stradling JR. Prevalence of obstructive sleep apnoea in men with type 2 diabetes. *Thorax.* 2006;61(11): 945–950.
310. Foster GE, Hanly PJ, Ostrowski M, Poulin MJ. Ventilatory and blood pressure responses to isocapnic hypoxia in OSA patients. *Adv Exp Med Biol.* 2008;605:463–468.
311. Higuchi T, Takahashi Y, Takahashi K, Niimi Y, Miyasita A. Twenty-four-hour secretory patterns of growth hormone, prolactin, and cortisol in narcolepsy. *J Clin Endocrinol Metab.* 1979;49:197–204.
312. Kok SW, Meinders AE, Overeem S, Lammers GJ, Roelfsema F, Frolich M, et al. Reduction of plasma leptin levels and loss of its circadian rhythmicity in hypocretin (orexin)-deficient narcoleptic humans. *J Clin Endocrinol Metab.* 2002;87(2):805–809.
313. Higuchi T, Takahashi Y, Takahashi K, Niimi Y, Miyasita A. Twenty-four-hour secretory patterns of growth hormone, prolactin, and cortisol in narcolepsy. *J Clin Endocrinol Metab.* 1979;49(2):197–204.
314. Overeem S, Kok SW, Lammers GJ, Vein AA, Frolich M, Meinders AE, et al. Somatotropic axis in hypocretin-deficient narcoleptic humans: altered circadian distribution of GH-secretory events. *Am J Physiol Endocrinol Metab.* 2003;284(3): E641–E647.
315. Rosenthal NE, Joseph-Vanderpool JR, Levendosky AA, Johnston SH, Allen R, Kelly KA, et al. Phase-shifting effects of bright morning light as treatment for delayed sleep phase syndrome. *Sleep.* 1990;13(4):354–361.
316. Jones CR, Campbell SS, Zone SE, Cooper F, DeSano A, Murphy PJ, et al. Familial advanced sleep-phase syndrome: A short-period circadian rhythm variant in humans. *Nat Med.* 1999;5(9):1062–1065.

CHAPTER

13

Cardiovascular Neuroendocrinology

Gina L.C. Yosten, Willis K. Samson

Department of Pharmacological and Physiological Science, Saint Louis University School of Medicine, St Louis, MO, USA

OUTLINE

Summary

The hypothalamus serves as an interface between circulating messengers of cardiovascular function and central nervous system pathways controlling energy homeostasis, neuroendocrine and endocrine function, and autonomic regulation. In this manner, the hypothalamus, together with its partnered structures in brainstem cardiovascular centers, organizes appropriate responses to changes in fuel availability, volume and electrolyte status, and cardiac function, assuring the maintenance of circulatory homeostasis. This chapter discusses the confluent mechanisms by which circulating factors and afferent neuronal information interact, how that information gains access to the hypothalamus, and the consequences of the subsequent hypothalamic responses.

INTRODUCTION

The hypothalamus can be considered the computer chip of the brain, responsible for integrating enteroceptive

Handbook of Neuroendocrinology, DOI: 10.1016/B978-0-12-375097-6.10013-7

and exteroceptive information and generating the appropriate neuronal and hormonal responses to any challenges to homeostasis. Weighing only approximately 4.0 grams in the adult human, it is protected by the soft mass of the overlying thalamus and cerebral cortices, and is positioned to receive abundant arterial blood supply from the internal carotids and branches of the Circle of Willis. Afferent information reaches the hypothalamus from optic and olfactory relays, blood-borne chemicals and temperature, and afferent neuronal connections from not only cortical structures (particularly limbic areas of brain), but also brainstem medullary centers. The hypothalamus then responds with efferent output to limbic nuclei (thalamic and temporal lobe structures), medulla and spinal cord (predominantly autonomic centers), as well as the median eminence where those efferents exert primary control over hormone secretion from the adenohypophysis (anterior pituitary gland) and the neurohypophysis (posterior pituitary), the site of vasopressin and oxytocin release into the peripheral circulation. This chapter details the interactive nature of neuroendocrine and autonomic regulation exerted primarily by neurons in the paraventricular nucleus of the hypothalamus. It focuses on how signals reach neurons in this area, the integration of that information, and the consequences of that information processing.

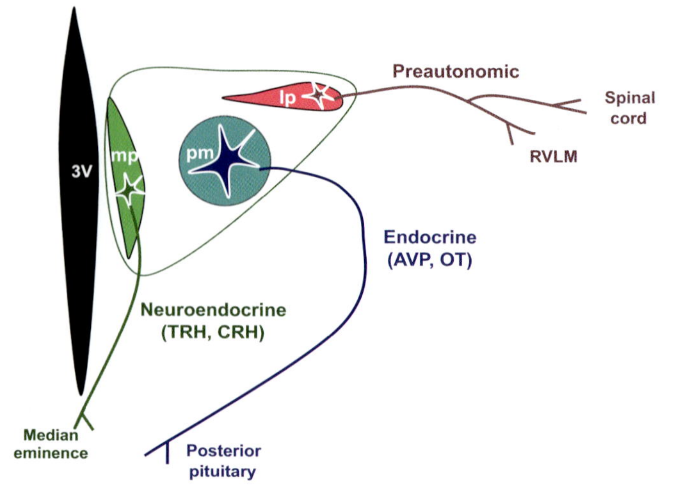

FIGURE 13.1 **Three populations of neurons in the paraventricular nucleus of the hypothalamus.** 3 V, third cerebral ventricle; lp, lateral parvocellular neurons; mp, medial parvocellular neurons; pm, magnocellular neurons; RVLM, rostral ventrolateral medulla; AVP, arginine vasopressin; OT, oxytocin; TRH, thyrotropin-releasing hormone; CRH, corticotropin-releasing hormone.

THE PARAVENTRICULAR NUCLEUS OF THE HYPOTHALAMUS

Three major classes of neurons project from the paraventricular nucleus (PVN) of the hypothalamus. These neurons also project to intra- and extrahypothalamic sites important in the control of social behaviors, energy homeostasis and cortical motor neurons. Intrahypothalamic axonal projections from these neurons are responsible for the coordinated control of neuroendocrine (in particular stress hormone secretion) and endocrine (e.g., vasopressin) function. Preautonomic fibers originate from parvocellular neurons in the dorsal and ventral parvocellular divisions of the nucleus and project to the nucleus tractus solitarius (NTS) in the medulla, and preganglionic neurons in the spinal cord. These neurons display characteristic electrophysiologic behavior and utilize a variety of neurotransmitters (peptides, amines) as their signals. When depolarized in hypothalamic slice preparations, these preautonomic neurons display a low threshold calcium spike resulting from activation of a T-type calcium conductance.[1,2] They can be further identified by anatomic techniques to project to brainstem cardiovascular areas and spinal cord by identification of tracking agents injected into those sites[3–8] (Fig. 13.1).

Magnocellular neurons are much larger in size, display a characteristic electrophysiologic footprint and, by releasing vasopressin into the general circulation via the neurohypophysis, serve in some cases to augment cardiovascular control exerted by the neighboring preautonomic neurons in the PVN. These neurons, located primarily in the ventrolateral aspects of the PVN, are designated type I neurons[1] and are characterized by a delayed return to baseline following depolarization, due to the presence of a dominant transient potassium conductance designated I_A.[1] They can also be identified by immunohistochemistry or *in situ* hybridization histochemistry (positive staining for vasopressin and/or oxytocin peptide or presence of mRNA encoding the peptides) and projection to the neurohypophysis, by antidromic activation[9,10] or retrograde track tracing.[4,7,11]

Neuroendocrine neurons, also parvocellular in nature, originate predominately from the medial parvocellular aspects of the nucleus and project to the median eminence,[11] where they release their signaling molecules, mainly peptides, into the interstitial spaces adjacent to the fenestrated capillary endothelium of the portal vessels. These neurons are short projection neurons characterized electrophysiologically by the absence of distinct electrical profiles common to the other two classes of PVN neurons. Because of their importance in the control of anterior pituitary hormone secretion, these neurons can be identified by the induction of early response genes following activation or inhibition of the secretion of the trophic hormones whose release they control (e.g., corticotropin releasing factor neurons in parvocellular PVN projecting to the median eminence that display c-*fos* activation to stress).[12]

The Blood–Brain Barrier and Paraventricular Nucleus Function

It is generally accepted that the PVN lies wholly behind the blood–brain barrier, and thus many circulating hormone or metabolites do not gain direct access, or at least that access is limited by a diffusion barrier that results in a gradient that decreases from plasma levels to those present in the cerebrospinal fluid. Some metabolites (e.g., glucose, free fatty acids, ketones)[13–16] as well as some hormones (e.g., leptin, insulin)[17] may cross the barrier; however, most circulating factors affect PVN function by indirect routes (e.g., adiponectin).[18]

Hormones, particularly those produced in the gut, can activate vagal afferents (e.g., cholecystokinin, ghrelin) to brain, which relay in the medulla to ascending fiber pathways that transmit their signals to the PVN.[19–23] Additionally, several circulating peptide hormones (e.g., angiotensin II, leptin, amylin) affect neuronal activity in the PVN after interacting with their receptors on brain sites devoid of the blood–brain barrier, known as the circumventricular organs (CVOs).[24,25] Some of those same circulating peptide hormones, originating primarily from non-CNS sites (e.g., heart, blood vessels, pancreas, adrenal gland), are also produced in brain neurons, particularly hypothalamic neurons.[26–28] They have been demonstrated to exert profound pharmacologic, and in some cases physiologic, actions on cardiovascular function, by interacting either directly with preautonomic neurons in parvocellular PVN, or with neurons located in CVOs that then project to the PVN (Fig. 13.2).

Signals Accessing the PVN via Blood–Brain Barrier-free Sites

Seven CVOs have been identified as areas of the CNS where the blood–brain barrier is breached, due to the presence of fenestrated capillary endothelial cells without glial cells arrayed in tight-junction barriers, allowing the passage of some (but not all) blood-borne substances to cross into a limited volume of the neurophil.[24,29] The CVO status of one of those, the subcommissural organ, adjacent to the pineal gland, has been questioned largely due to the absence of confirmatory histologic evidence of fenestrations.[30] Of the remainder, for the purposes of this chapter the most important are the organum vasculosum lamina terminalis (OVLT) and the subfornical organ (SFO), positioned at the ventral and dorsal aspects, respectively, of the lamina terminalis; and the area postrema (AP), positioned adjacent to the Obex, at the caudal floor of the fourth cerebroventricle.[31,32] Due to the presence of neurons in these CVOs that can respond to circulating factors and communicate those responses to neurons behind the blood–brain barrier, these barrier-free ports have been termed "sensory CVOs".[32]

The OVLT and SFO are important free-ports for the access of circulating peptides, inflammatory cytokines and plasma electrolytes to neurons that communicate peripheral signals to the PVN. While the AP is classically considered a site of chemoreception, it has recently been demonstrated to be the site where numerous gut- and adipocyte-derived hormones exert their CNS actions.[33] The OVLT and SFO, on the other hand, are major sites of detection of plasma electrolytes/osmolality and circulating factors affecting autonomic function.[24–36] Here is where circulating levels of angiotensin II (AII) are thought to elicit behavioral, neuroendocrine, endocrine and autonomic effects. Both the OVLT and the SFO, as well as the adjacent median preoptic nucleus (MnPO), are important integrators of plasma sodium and osmotic cues with appropriate hypothalamic and limbic centers.[37–39] These forebrain CVOs have been recognized as important relays of the information carried by cardiovascular hormones regarding plasma volume and electrolyte content.[32]

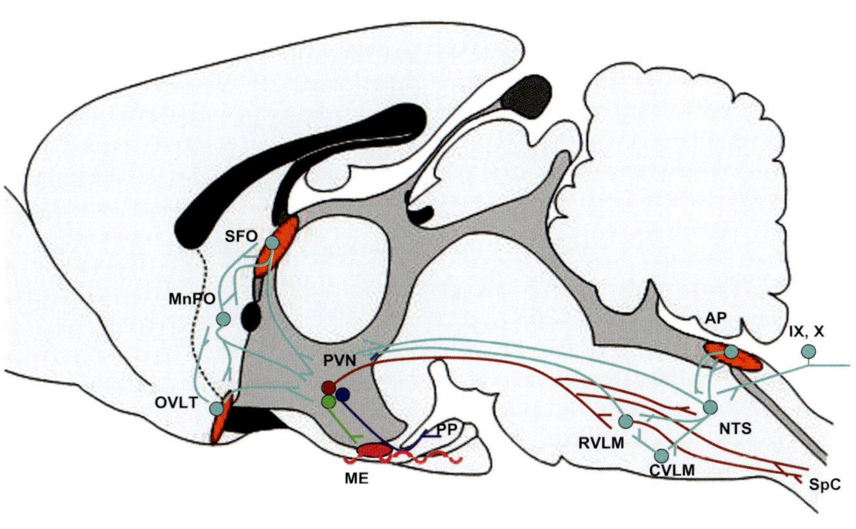

FIGURE 13.2 **Afferent and efferent connections of the hypothalamic paraventricular nucleus (PVN).** SFO, subfornical organ; MnPO, median preoptic nucleus; OVLT, organum vasculosum lamina terminalis; ME, median eminence; PP, posterior pituitary; RVLM, rostral ventrolateral medulla; CVLM, caudal ventrolateral medulla; NTS, nucleus tractus solitarius, AP, area postrema; SpC, spinal cord; IX, glossopharyngeal nerve; X, vagus nerve. Endocrine neurons in dark blue project to the posterior pituitary. Neuroendocrine neurons in green project to the median eminence. Preautonomic neurons in brown project to RVLM, NTS and spinal cord. Sites lacking a blood–brain barrier are indicated in red.

Indeed, the OVLT and SFO may prove to be critically important sites of the initial integration of volume, pressure and electrolyte status, leading to appropriate endocrine and autonomic responses maintaining cardiovascular homeostasis.

Afferent and Efferent Connections between the CVOs and the PVN

In addition to sensing circulating signals, the sensory CVOs receive abundant afferent input from a variety of brain structures known to be important in fluid and electrolyte homeostasis and cardiovascular regulation.[32] Importantly, axons originating in the PVN innervate the OVLT and SFO, and cardiovascular centers in brainstem send projections there as well (e.g., nucleus tractus solitarius, lateral parabrachial nucleus and locus coeruleus).[7] Via the nucleus tractus solitarius connections, the OVLT and SFO are informed of cardiovascular function because afferents to the NTS in cranial nerves IX and X carry baroreceptor information.[40–42] Reciprocal interconnections exist between the OVLT and SFO, and the adjacent median preoptic nucleus,[38] providing a local network for the integration of osmotic cues and hormonal signals (e.g., angiotensin II, leptin). Indeed, these three areas reside in the lamina terminalis, which is the most rostral, midline extension of the developing neural tube, and during development are connected by primitive mesenchymal elements that eventually give rise to the vascular portions of the OVLT and SFO, and the median eminence, another CVO.[43] Thus, there is also developmental evidence for the interactions of these tissues.

Efferents from the SFO and OVLT project within the region of the anteroventral third ventricle (AV3 V), particularly to the MnPO and prominently to the nearby PVN,[44,45] whose neurons then relay information to cardiovascular regulatory areas of brainstem (rostral ventrolateral medulla and nucleus of the solitary tract) and the intermediolateral cell column in the spinal cord.[37,38,46,47] This is the anatomic basis of the action of hyperosmolar stimuli to increase vasopressin release (via AV3 V relays to magnocellular PVN) and, via relays in parvocellular PVN projecting to RVLM (rostral ventrolateral medulla) and spinal cord, elevate sympathetic nerve activity (SNA).[36] Sensory elements in OVLT and SFO must have some intrinsic activity[48] because elevations and reductions in plasma osmolarity result in increased and decreased SNA, respectively,[49–51] and these responses are fully dependent on the integrity of those CVOs.[35,52]

How does the Paraventricular Nucleus Regulate Cardiovascular Function?

Through at least four distinct neuronal pathways, neurons in the PVN play a major role in fluid and electrolyte balance, vascular reactivity and glucose homeostasis, thus contributing to the maintenance of adequate tissue perfusion pressure and cardiac output. Via connections with cortical and limbic structures, the PVN organizes thirst and sodium appetite.[53,54] By releasing vasopressin (also known as antidiuretic hormone, ADH) into the peripheral circulation, the PVN plays an important role in protecting tissue perfusion pressure (plasma volume) and plasma osmolarity (see Chapter 16). Vasopressin released in response to hemorrhage or decreased central venous pressure contributes to the increases in total peripheral resistance that help to maintain venous return and perfusion of vital organs.[55,56] Through neuroendocrine mechanisms, parvocellular neurons (particularly those in the medial parvocellular PVN) control the release of counter-regulatory hormones such as ACTH and GH, which by their actions in the periphery help maintain plasma glucose levels during bouts of hypoglycemia, and via actions of ACTH on corticosteroid release help recover sodium from the tubular urine during hyponatremia and hypovolemia (see Chapters 8 and 14).

The PVN and Autonomic Nervous System Function

Using retrograde tracing techniques, Sawchenko and Swanson[57] demonstrated that neurons in the locus coeruleus, the dorsal vagal complex (A2 catecholamine neurons) and the ventrolateral medulla (A1 catecholamine neurons) project directly to the PVN and the locus coeruleus. They further identified distinct stereotypic innervation of the PVN by these brainstem nuclei, using an autoradiographic tracing approach. The locus coeruleus appeared to project primarily to the parvocellular cells adjacent to the third ventricle (neuroendocrine neurons). Cells in the A2 region of the dorsal vagal complex (including the NTS) projected mainly to parvocellular elements of the PVN adjacent to the third ventricle and in the dorsomedial PVN, but not the magnocellular neurons. On the other hand, the A1 catecholamine neurons in ventrolateral medulla appeared to innervate not only parvocellular elements, similar to the A2 projections, but also, uniquely, the magnocellular elements of the PVN. These classic studies further identified reciprocal innervation of the A1 and A2 cell groups, which is significant if only because the A2 cells in NTS are those receiving information for the baroreceptors via cranial nerves IX and X. Thus the anatomic framework was established for a circuitry which linked incoming cardiovascular information with the PVN and its return of commands to the autonomic centers in medulla and spinal cord.

Efferent projections from parvocellular preautonomic cells of the PVN form the basis of the hypothalamic control of cardiovascular function.[6,57–59] Those efferents innervate not only the intermediolateral cell column (preganglionic sympathetic neurons) but also the NTS and the dorsal motor nucleus of the vagus, thus influencing the activity of both the parasympathetic and sympathetic branches of the autonomic nervous system.[58] Some of those preautonomic projection neurons surprisingly contain oxytocin and vasopressin (in addition to other neuropeptides such as somatostatin, and the enkephalins), establishing their presence in not only magnocellular but also parvocellular neurons of the PVN.[58] More recently, retrograde tracing techniques utilizing tagged microspheres have revealed that some of the descending fibers from parvocellular PVN innervate both the pressor region of the rostral ventrolateral medulla (RVLM) and the intermediolateral cell column of the spinal cord.[6] In this way, command neurons in the parvocellular PVN can affect cardiovascular function through a balance of sympathetic (contractile actions in the vasculature and positive chronotropic/inotropic effects in heart) and parasympathetic (negative chronotropic actions in heart) actions, and via renal sympathetic afferents controlling sodium excretion.[60] The importance of the connectivity of neurons in the paraventricular nucleus of the hypothalamus with autonomic centers in medulla and spinal cord to the integrated control of blood pressure has recently been extensively reviewed elsewhere.[61]

The PVN responds not only to osmotic stimuli via the CVOs and afferent input from high (aortic and carotid) and low (atrial) pressure baroreceptors via the IXth and Xth cranial nerves via relays in the NTS and RVLM,[62] but also to vasoactive hormones present in the circulation and produced locally within the hypothalamus. Whether the production and/or release of these hormones centrally is coordinated to circulating levels is unknown; however, peptides from both sources exert potent actions on the hypothalamic control of

BOX 13.1

HOW DO WE KNOW THAT THE HYPOTHALAMIC PARAVENTRICULAR NUCLEUS LINKS TO BRAINSTEM AND SPINAL CORD NEURONS CONTROLLING CARDIOVASCULAR FUNCTION?

The importance of preautonomic, parvocellular neurons of the hypothalamic paraventricular nucleus in the control of cardiovascular homeostasis is based upon numerous evidences. However, a critical step in establishing that importance was the identification of a direct innervation by those neurons of medullary cardiovascular control centers and the intermediolateral cell column of the spinal cord. Employing anterograde (*Phaseolus vulgaris* leucoagglutinin, PHA-L) and retrograde (cholera toxin subunit b, CTb) labeling methods, Hosoya and Matsushita[1] established direct innervation of spinal preganglionic neurons in the upper thoracic spinal cord. Injection of PHA-L into the paraventricular nucleus resulted in transport of the marker to terminal fields in the intermediolateralis pars principalis and pars funicularis of the of the upper thoracic spinal cord, in close apposition to the diffuse dendritic fields of the resident, sympathetic preganglionic neurons. Those sympathetic preganglionic sympathetic neurons were identified by the presence within the cells of the retrograde marker (CTb) that had been injected into the superior cervical ganglion.

These preautonomic, parvocellular paraventricular neurons were later demonstrated to project to both sympathetic pregangionic neurons in the spinal cord, and, via axon collaterals, to neurons located in the rostral ventrolateral medulla (RVLM), an important brain stem cardiovascular center. This was accomplished again with the use of neuronal track tracing techniques. Shafton and colleagues[2] injected fluorescein-labeled microspheres into the intermediolateral cell column of the rat spinal cord and, in the same animals, rhodamine-labeled microspheres into the RVLM. The microspheres were taken up by the axon terminals and transported in a retrograde fashion to the cell bodies of the projection neurons in PVN. The green- and red-labeled microspheres were detected histologically in parvocellular PVN neurons 7 days later, indicating those neurons projecting to one or both injection sites. A significant population of the green, fluorescein-labeled cells that projected to spinal cord also contained red, rhodamine-labeled microspheres. Thus, preautonomic neurons of the hypothalamic PVN are capable of influencing cardiovascular function via both medullary and spinal cord autonomic centers.

References

1. Hosoya Y, Matsushita M. Identification and distribution of the spinal and hypophysial projection neurons in the paraventricular nucleus of the rat. A light and electron microscope study with the horseradish peroxidase method. *Exp Brain Res.* 1979;35:315–331.
2. Shafton AD, Ryan A, Badoer E. Neurons in the hypothalamic paraventricular nucleus send collaterals to the spinal cord and to the rostral ventrolateral medulla in the rat. *Brain Res.* 1998;801:239–243.

thirst, salt appetite, autonomic function and vasopressin release.[63–67]

How Does the Cardiovascular System Communicate with the PVN?

In addition to chemoreceptor and baroreceptor input to neurons in the dorsal vagal complex and osmotic information delivered to the CVOs, circulating hormones provide important information to the hypothalamus about cardiac function, vascular tone, and solute and volume delivery to the kidney. Because these hormones penetrate the blood–brain barrier poorly, it is thought that most of their actions are initiated at one or more of the sensory CVOs.[24] However, as mentioned above, many of the vasoactive peptides present in circulation as a result of production in the periphery are also produced within the CNS – indeed, in neurons of the hypothalamus in some cases. Thus, studies designed to establish the physiologic or pathophysiologic relevance of the pharmacologic actions of any one of these peptides must take into consideration not only the site of action but also the original source of the peptide. Here we review representative examples of three classes of vasoactive hormones: those produced from circulating prohormones or local tissue processing (the angiotensins), those produced in cardiomyocytes (the natriuretic peptides), and those produced in vascular smooth muscle cells (the adrenomedullins).

CIRCULATING AND LOCAL PRODUCTION OF THE ANGIOTENSINS

The renin–angiotensin system (RAS) is a multi-peptide, multi-enzyme, multi-receptor system that is important for the maintenance of arterial blood pressure and blood volume. In this system, a single precursor, angiotensinogen, is cleaved by renin or other enzymes to yield angiotensin I (AI). AI is subsequently converted by angiotensin converting enzyme (ACE) to angiotensin II (AII), which acts peripherally to increase aldosterone release and as a potent vasoconstrictor agent, and centrally to increase blood pressure, fluid intake and vasopressin secretion via the G-protein coupled receptor AT1.[68,69] An additional arm of the RAS has also been identified, in which another angiotensin converting enzyme, ACE2,[70,71] mediates the conversion of AI or AII to angiotensin1-7 (Ang(1-7)). In striking contrast to AII, Ang(1-7) acts through the Mas receptor to release nitric oxide (NO) and prostaglandins, leading to vasodilation and a reduction in blood pressure.[72] Although the first component of the RAS to be discovered, renin, was identified over 100 years ago, the RAS is still widely studied and pharmacologically manipulated for therapeutic purposes.[73] Classically, the major effector molecule of the RAS, AII, is formed in the plasma and circulates in the blood to cause vasoconstriction and increase arterial pressure. It is now known that there are tissue-specific renin–angiotensin systems, wherein all or most of the components of the RAS are produced and act locally within a specific tissue, such as adipose tissue and brain.[74,75] These tissue-specific renin–angiotensin systems, especially that in the brain, may have important implications for the development of cardiovascular disease.

Components of the Renin–Angiotensin–Aldosterone System (RAAS)

Angiotensinogen is an α2-globulin precursor from which all other angiotensin peptides are derived.[76] Angiotensinogen is produced primarily by the liver, but angiotensinogen mRNA has also been detected in adipocytes, the kidney, regions of the brain, the adrenal gland, the heart and blood vessels.[77] Renin, an enzyme synthesized by juxtaglomerular cells of the renal cortex, cleaves angiotensinogen to the decapeptide, angiotensin I. The conversion of angiotensinogen to angiotensin I is the rate-limiting step in the production of biologically active angiotensin peptides. Because of the abundance of angiotensinogen in the plasma, renin is usually the limiting factor, although changes in the plasma concentration of angiotensinogen can influence the synthesis of AII. The secretion of renin is tightly controlled, and is regulated primarily by renal arteriolar blood pressure, by the concentration of sodium in the renal tubular fluid as sensed by the macula densa, and by renal sympathetic nerve activity.[78] Renin release can also be inhibited via a direct short-loop negative feedback mechanism, wherein AII binds to AT1 receptors on renal juxtaglomerular cells, causing a decrease in renin secretion.

Receptors and Biological Actions of RAAS Peptides

AT1 Receptors

The activities of biologically active angiotensin peptides are mediated by at least two G-protein coupled receptors (GPCRs), AT1 and AT2.[79] Angiotensin II exerts its major actions, including vasoconstriction, sympathoactivation, stimulation of water intake and aldosterone secretion, via the primary angiotensin receptor AT1.[80–82] In humans, AT1 receptors have been detected in various regions of the brain and anterior pituitary, in blood vessels, heart, adrenal glands and kidneys.[83] In the human brain, AT1 receptors are expressed predominantly in areas that are devoid of a blood–brain barrier (circumventricular organs) and thus have access to blood-borne angiotensin II, including the subfornical organ, the organum vasculosum of the lamina terminalis, and the area postrema in the hindbrain. AT1 receptors are also localized in areas of the brain that lie behind the blood–brain barrier, including several nuclei of the hypothalamus, the nucleus tractus solitarius, and the ventrolateral medulla.[84] In the adrenal gland, AT1 receptors are localized in the chromaffin cells of the adrenal medulla and the zona glomerulosa of the cortex (the site of aldosterone production).[85] AT1 receptors have also been identified in the conducting system of the heart[86,87] and on smooth muscle cells of the vasculature, where angiotensin II binds to cause direct vasoconstriction.[83] AT1 receptors signal through at least five intracellular transduction mechanisms, including activation of L-type calcium channels, phospholipase A2, phospholipase C and phospholipase D, and inhibition of adenylate cyclase. Activation of phospholipases A2, C and D occurs via a $G_{q/11}$-dependent mechanism, and, of the signal transduction mechanisms initiated by activation of AT1 receptors, the phospholipase C signaling pathway is the most well described.[83] Activation of AT1 receptors has been linked to the opening of calcium channels, allowing extracellular calcium entry into the cytoplasm. Increases in cytoplasmic calcium have been linked not only to vasoconstriction of blood vessels, but also to the synthesis and secretion of aldosterone from the adrenal cortex.[88] In mice and rats there are two distinct isoforms of the AT1 receptor, AT1a and AT1b, which share 94% sequence homology.[89,90] AT1a receptors are expressed in the vasculature, kidney, lung and liver, and are thought to modulate the vasopressor effects of angiotensin II. AT1b receptors are predominantly expressed in brain and adrenals, and may play a role in the dipsogenic effect of angiotensin II.[83]

AT2 Receptors

While the AT2 receptor is abundantly expressed during fetal development, expression in the adult is markedly reduced.[91,92] However, AT2 receptors have been detected in the adult brain, kidney, heart and adrenal medulla, and in reproductive tissues.[83] Because of the high expression of the AT2 receptor in fetal tissues, it is thought to be important in fetal development. However, mice lacking AT2 receptors develop normally but exhibit a deficit in the drinking response induced by water deprivation, and a reduction in locomotor activity.[93,94] In the adult kidney, activation of AT2 receptors results in the production of cGMP, prostaglandin F2α and nitric oxide, suggesting a vasodilatory role for AT2 receptors in that tissue, thus opposing the effects of AT1 receptor activation.[95–97] Furthermore, AT2 knockout mice exhibit an elevation in blood pressure and an increased sensitivity to the pressor effect of AII, arguing in favor of a physiologically relevant protective role for the AT2 receptor in the vasculature.[93,94] In the systemic vasculature, AT2 receptors may be important for vascular repair, since balloon injury leads to an upregulation of AT2 receptors in the carotid artery,[98] and treatment with an AT2-specific ligand decreases neointimal formation.[99] Like the AT1 receptor, the AT2 receptor is a seven-transmembrane domain protein and has been linked to several downstream signaling pathways. Activation of AT2 receptors has been linked to the inhibition of T-type calcium channels and the stimulation of membrane potassium channels as a result of the induction of several phosphatases, including MAP kinase phosphatase 1, protein tyrosine phosphatase, SH2-domain-containing phosphatase, and serine/threonine phosphatase 2A. Additionally, AT2 receptors can be coupled to the NO-cGMP system and phospholipase A2. It is unclear if AT2 receptors bind G proteins, although there is evidence suggesting that heterotrimeric G_i proteins are involved in AT2 signaling mechanisms.[83]

Mas Receptor

The receptor for Ang(1-7), Mas, is a seven-transmembrane domain receptor that bears striking similarity to G-protein coupled receptors.[72] Ang(1-7) is produced in brain,[100] where it exerts effects opposing those of AII.[101–104] Mas was originally thought to be a protooncogene because of its tumorigenicity in nude mice. Mas expression has been identified in mouse brain, heart and kidney, as well as in other peripheral tissues. In the brain, Mas immunoreactivity has been observed in neurons of known cardiovascular centers, such as the NTS and RVLM, as well as various nuclei of the hypothalamus, including the PVN and SON. Activation of the Mas receptors on endothelial cells leads to the release of NO and prostaglandins, both of which lead to vasodilation.[105]

Tissue-Specific Renin–Angiotensin Systems

Classically, the renin–angiotensin system was thought of as a circulating endocrine system, in which angiotensinogen, produced by the liver, is cleaved by renally-derived renin in the plasma to angiotensin I. Angiotensin I is then further processed by ACE to angiotensin II, which activates AT1 and AT2 receptors in target tissues. In this traditional view of the RAS, AII exerts its major actions (hypertension, dipsogenesis and AVP secretion) in brain by binding to AT1 receptors in areas of the brain lacking a blood–brain barrier (circumventricular organs), particularly the subfornical organ. It is now clear that in addition to a circulating endocrine RAS, there exist local paracrine renin–angiotensin systems in many tissues, including adipose tissue,[106,107] blood vessels, kidney, heart, adrenal gland, reproductive tissues and brain.[84,86] A brain-specific RAS was first suggested by Ganten and colleagues, who demonstrated renin-like activity in brain.[108] It is now known that the brain produces all components of RAS, including angiotensinogen, renin, ACE and ACE2,[105] and that AII likely functions as a neurotransmitter, since AII-like immunoreactivity has been visualized in nerve terminals.[109,110] Angiotensinogen is produced mainly by astrocytes,[111] which constitutively secrete angiotensinogen into the extracellular fluid of the brain.[112] In contrast, renin is synthesized by neurons and exists as both a secreted form or as a non-secreted, intracellular enzyme.[75,113–117] ACE and ACE2 are produced by neurons and possibly endothelial cells, both in circumventricular organs and by areas of the brain protected by a blood–brain barrier, including the hypothalamus.[118]

The RAS in Brain: Lessons from Genetic Studies

Brain-specific overexpression of human angiotensinogen and human renin results in hypertension, polydipsia, and increased preference for salt in transgenic mice.[119] Furthermore, rats with a deficit in brain angiotensinogen exhibit a reduction in blood pressure and vasopressin secretion.[120] Central injection of either ACE inhibitors or AT1 receptor antagonists reduces blood pressure in several models of hypertension,[121] uncovering the potential importance of the counter-regulatory Ang(1-7)–Mas system.

Neuron-selective expression of the AT1A receptor in brain failed to alter baseline cardiovascular parameters; however, these NSE-AT1A mice exhibited an exaggerated increase in blood pressure when a pressor dose of AII was injected centrally. In addition, these mice had exaggerated drinking responses to AII and increased sodium appetite when sodium depleted.[82] These studies further emphasized the importance of the AT1 receptors in the NS actions of both centrally and peripherally produced AII.

One of the hallmark actions of AII in brain is a decrease in baroreflex sensitivity,[122] and yet Ang(1-7) enhances the reflex.[123] The importance of the counter-regulatory ACE2–Ang(1-7)–Mas pathway is further demonstrated by studies in which blood pressure was lowered in the spontaneously hypertensive rat (SHR) by overexpression of ACE2 in RVLM,[124] and another study in which brain-selective overexpression of ACE2 attenuated the development of hypertension during

BOX 13.2

HOW DO WE KNOW THAT BRAIN-DERIVED ANGIOTENSIN PEPTIDES CONTROL CARDIOVASCULAR FUNCTION?

The importance of angiotensin II in the development of hypertension is well established, regarding both its peripheral and central actions. Angiotensin receptor blockers (AT1 receptors) and angiotensin converting enzyme (ACE1) inhibitors are in extensive clinical use. However, it is now apparent that the second converting enzyme, ACE2, functions to produce a counter-regulatory form of angiotensin, Ang(1-7), which exerts counter-regulatory actions to those of angiotensin II. We now know that the activity of ACE2 in rostralventrolateral medulla (RVLM) plays an important role in blood pressure regulation. Raizada and colleagues[1] demonstrated a 40% decrease in ACE2 protein in the RVLM of spontaneously hypertensive rats. Using a lentiviral approach, they overexpressed ACE2 in the RVLM of these rats, resulting in long-term expression of the transgene in cells of the RVLM. Basal blood pressures were significantly attenuated compared to hypertensive controls, suggesting not only a role for endogenous ACE2 in the buffering of the hypertensive actions of angiotensin II, but also the possibility that the ACE2 enzyme might be a therapeutic target for control of hypertension.

This was further established by innovative approaches in the Lazartigues laboratory.[2] These investigators created transgenic mice with neuron-selective ACE2 overexpression, and demonstrated that the development of neurogenic hypertension (chronic, peripheral angiotensin II infusion) was significantly attenuated. Infusion of the Ang(1-7) receptor blocker reversed the salutary effect of ACE2 overexpression on spontaneous baroreflex sensitivity and parasympathetic tone, indicating the importance of the Ang(1-7) receptor (*Mas*) in the central integration of autonomic function.

References

1. Yamazato M, Yamazato Y, Sun C, Diez-Freire, Raizada MK. Overexpression of angiotensin-converting enzyme 2 in the rostral ventrolateral medulla causes long-term decrease in blood pressure in the spontaneously hypertensive rats. *Hypertension*. 2007;49:926–931.
2. Feng Y, Xia H, Cai Y, et al. Brain-selective overexpression of human angiotensin-converting enzyme type 2 attenuates neurogenic hypertension. *Circ Res*. 2010;106:373–382.

AII infusions and prevented the dampening of spontaneous baroreflex sensitivity observed in wild-type littermates.[125]

Mice bearing expression of the human renin and angiotensinogen genes are hypertensive[126] and display impaired baroreflex sensitivity.[127] In addition, ACE2 activity is decreased in those animals. When the AT1 receptor was blocked by losartan, blood pressure normalized, baroreflex sensitivity returned and ACE2 activity increased. Thus, AII acting via the AT1 receptor can actually downregulate the counter-regulatory Ang(1-7)—Mas system. Proof for this came in triple transgenic mice expressing the human renin and angiotensinogen genes (R + A + mice), along with the human ACE2 gene targeted to neurons, that displayed normalized water intakes, blood pressure and baroreflex sensitivity when compared to the R + A + mice without the third transgene.[127,128]

In summary, the renin—angiotensin systems, both in the periphery and within the central nervous system, are primary regulators of cardiovascular function through endocrine, neuroendocrine, behavioral and autonomic actions. The emergence of the ACE2—Ang(1-7)—Mas wing of the RAS has provided great hope for novel therapeutic strategies for the treatment of cardiovascular disease. Several other vasoactive hormones interact with the RAS, and thus play important roles in the cardiovascular neuroendocrinology.

CARDIAC HORMONES

Cardiac Hormones Act in Multiple Tissue Sites to Regulate Cardiovascular Function

It has long been recognized that cardiac myocytes, when examined under the electron microscope, contain what appeared to be secretory granules.[129] Thus, in addition to their contractile properties, they demonstrate an endocrine appearance.[130] It was also well known that atrial distension resulted in profound diuresis.[131] Indeed, extracts of cardiac tissue were observed to exert natriuretic actions in kidney,[132] and once the 28-amino acid peptide responsible for that activity was isolated from atrial extracts and characterized it was named atrial natriuretic peptide (ANP).[133] ANP is released primarily in response to atrial stretch (volume overload), but also in response to rapid atrial pacing (tension) and a variety of circulating factors.[134] Soon after the initial discovery of ANP, a second member of the family of natriuretic peptides was identified and given the unfortunate name "brain natriuretic peptide" (BNP). It emerged that very little, if any, of the 32-amino acid BNP is actually made in the brain, so the nomenclature was changed to B-type natriuretic peptide.[135,136] BNP shares virtually all of the biologic activities of ANP, and is produced predominantly in ventricular myocytes, particularly under volume overload states.[137] Indeed, BNP has become an excellent clinical marker for the presence and monitoring of heart failure in humans because of the remarkable increases observed in that clinical condition.[138] A third member of the family, C-type natriuretic peptide (CNP), was subsequently discovered to be the predominant form of natriuretic peptide produced in the brain.[139] CNP, a 22-amino acid peptide, shares some, but not all, the biologic activity of ANP and BNP.[140–142] The diversity of action of the individual members can be explained by the fact that three unique receptors[143–145] have been identified.

In addition to their potent pharmacologic actions in the periphery, the natriuretic peptides act both at the CVOs and behind the blood—brain barrier to influence cardiovascular function. All three receptors are expressed in brain, and it took a unique approach to understanding the importance of the NPR-B to understand that some, but not all, of the actions in the brain are a reflection of the effects of endogenous CNP.[146–148] The earliest indication of an action of the natriuretic peptides in the brain was the observation of the inhibitory effect of ANP on vasopressin secretion and thirst[149–151] – something that seemed well coordinated with its diuretic action in the kidney.[152–154] Soon thereafter, it became clear that ANP exerted additional brain actions that mirrored in a physiologic sense its diuretic and natriuretic actions. ANP inhibits thirst[151] and sodium appetite[155] at least in part by antagonizing the actions of angiotensin II. These behavioral actions of ANP are in all likelihood due to activation of NPR-A receptor expressing neurons, since CNP exerts an opposing action on thirst.[156] The natriuretic peptides also act in hypothalamus to alter autonomic function.[157–159] The sites of action of the natriuretic peptides to decrease autonomic tone probably reflect the distribution of the NPR-A receptor in circumventricular organs such as the OVLT and SFO,[160,161] or in brainstem cardiovascular centers.[159,162,163] Electrophysiologic studies have demonstrated direct neuronal actions of the natriuretic peptides in hypothalamus,[164–167] the subfornical organ[168,169] and the nucleus tractus solitarius.[170] Here their actions appear complementary to their vasodilatory actions in the periphery,[171,172] and to be physiologically important, because ANP knockout mice display increased blood pressure as adults when challenged with a high sodium diet.[173,174] The CNS actions of ANP on salt appetite are physiologically relevant, as demonstrated by selective cytotoxin targeting of the natriuretic receptors, which resulted in exaggerated sodium intakes.[155] In fact those same selective cell targeting

TABLE 13.1 The Opposing Actions of Angiotensin II and the Natriuretic Peptides

Tissue site of action	Angiotensin II effects	Natriuretic peptide effects
Kidney	Increase Na^+ reabsorption (antinatriuresis) Renal vasoconstriction	Increase Na^+ excretion (natriuresis) Increase water excretion (diuresis) Renal afferent arteriolar dilation and afferent arteriolar constriction
Adrenal	Increase aldosterone secretion	Decrease aldosterone secretion (basal and A II-stimulated)
Heart	Increase cardiac fibroblast proliferation leading to cardiac dysfunction	Increase coronary artery blood flow improving cardiac function
Vasculature	Increase smooth muscle proliferation Vasoconstriction Increase noradrenalin release	Decrease smooth muscle proliferation Vasodilation Decrease noradrenalin release
Pituitary	Increase vasopressin release Increase ACTH release (basal and CRH-stimulated)	Decrease vasopressin release Decrease ACTH release (basal and CRH-stimulated)
Central nervous system	Increase thirst Increase sodium appetite Increase sympathetic tone	Decrease thirst Decrease sodium appetite Decrease sympathetic tone

approaches revealed the importance of ANP in not only osmotically driven thirst in general, but also selective hypernatremia induced water drinking and salt appetite. These were some of the first evidences for the existence of a unique sodium receptor, independent of the osmoreceptors in the CVOs[155] (Table 13.1).

Opposing Actions of Angiotensin II and the Natriuretic Peptides

The opposing actions of AII and the natriuretic peptides are most readily observed when the CNS actions of the peptides are compared. Angiotensin II stimulates vasopressin secretion,[175,176] increases mean arterial pressure,[177,178] and induces water and salt intakes[179,180] by central actions either in the CVOs or within the brain itself, while ANP inhibits vasopressin secretion[149] and water and salt intakes,[151] and lowers mean arterial pressure by actions in similar brain regulatory centers.[157–159] In the periphery angiotensin II is a vasoconstrictor,[74] while the natriuretic peptides exert vasorelaxant effects.[181,182] Furthermore antimitogenic actions have been ascribed to the natriuretic peptides, particularly CNP,[183,184] while angiotensin II is promitogenic.[185] In the adrenal gland, angiotensin II stimulates[186] while ANP inhibits[187–189] aldosterone secretion. ANP exerts natriuretic and diuretic actions in the kidney through a combination of direct tubular and vascular actions to increase glomerular filtration[190] and, potentially, decrease renin secretion. Angiotensin II exerts opposing actions.[191] Thus, these two peptide families can be understood by their opposing actions; indeed, these unique biologic properties have led to several novel therapeutic strategies for the treatment of cardiovascular diseases.

VASCULAR HORMONES

Central and Peripheral Cardiovascular Actions of Peptides Produced in the Adrenal Gland and Vasculature

In 1993 a second family of vasorelaxant peptides was discovered, initially in extracts of a pheochromocytoma,[192] and was later found to be produced in many tissues, but predominantly in endothelial and smooth muscle cells of the vasculature.[65] Two vasorelaxant peptides are encoded in the same preprohomone: the 52-amino acid adrenomedullin (AM), and the smaller, 20-amino acid peptide proadrenomedullin N-terminal 20 (PAMP). Adrenomedullin exerts its vasorelaxant effect primarily via a paracrine action in the vasculature. Its release from endothelial and vascular smooth muscle cells is stimulated by tumor necrosis factor-α, interleukin-1β, cortisol, aldosterone and lipopolysaccharide (LPS). The ability of LPS to stimulate AM production and release led to the discovery that endogenous AM plays a critically important survival role in septic shock by its renoprotective actions.[193–195] The vasorelaxant effects of AM are due to multiple second messenger systems, dependent upon the vascular bed studied. To be sure, an important mediator of the vasorelaxant effect of AM is nitric oxide released from the endothelial cells.[196,197] PAMP, on the other hand, exerts its less potent vasorelaxant effect via a non-vasculature action, instead acting presynaptically on sympathetic nerve terminals innervating the vasculature, resulting in decreased norepinepherine release.[198,199]

Vasorelaxant effects (i.e., increased conductance) of AM have been observed in mesenteric, renal and hindlimb circulations.[200] In the kidney, AM increases blood flow via release of nitric oxide, resulting in dilation of both the afferent and efferent arterioles. Even in the face of profound hypotensive actions, AM infusion increases renal blood flow[201] leading to diuresis and natriuresis that is not entirely due to increased glomerular filtration, because tubular actions have been described as well.[154] Thus, in the kidney the actions of AM, like ANP, oppose those of angiotensin II. This also holds true in the adrenal gland, where AM inhibits angiotensin II and potassium-stimulated aldosterone secretion,[202] and in the anterior pituitary gland, where AM inhibits corticotropin-releasing hormone-stimulated adrenocorticotropin (ACTH) release.[203]

While in the periphery the actions of AM appear to oppose those of angiotensin II, in hypothalamus and associated brain areas some of the actions of AM and angiotensin II overlap. The diuretic and natriuretic actions of AM in the kidney are mirrored by inhibitory effects in the brain on angiotensin II- and fluid deprivation-induced water drinking[204,205] and on sodium appetite simulated by hypovolemic challenge.[206] These are in all likelihood reflective of the actions of endogenously produced AM, since compromise of its production selectively in the brain results in exaggerated fluid intakes.[205]

AM is produced in multiple brain sites known to be important in cardiovascular control, including thalamus, amygdala, medulla and hypothalamus,[207,208] where the peptide co-localizes with vasopressin and oxytocin in the PVN and SON.[209] In the hypothalamus, the actions of AM mirror those of angiotensin II. Central administration of AM leads to significant increases in plasma vasopressin levels.[210] Conversely, compromise of adrenomedullin production in hypothalamus (central administration of a ribozyme that catalyzes cleavage of the AM mRNA) resulted in decreased basal vasopressin secretion and vasopressin secretion stimulated by dehydration or hypovolemia.[210] The physiologic consequence of this compromise of AM production and decreased vasopressin release was reflected by increased loss of plasma volume in the ribozyme-treated animals during hypovolemic challenge. These data strongly suggest that endogenous brain-derived AM is a physiologically relevant regulator of vasopressin secretion and fluid homeostasis. Another action of AM in the brain that mirrors the actions of angiotensin II is its ability to raise mean arterial pressure in anesthetized[211] and conscious[212,213] animals when administered into the cerebroventricles. This pressor effect appears to be site-specific, and expressed by an action in the area postrema[214,215] or the NTS.[213] The co-produced peptide, PAMP, also raises mean arterial pressure when administered centrally, by increasing sympathetic tone.[213] Neuronal actions of AM in PVN have been demonstrated.[216] In fact, in hypothalamic slice preparations AM hyperpolarized type I magnocellular neurons, although the effect was absent in synaptic isolation (in the presence of tetrodotoxin), suggesting an action on inhibitory interneurons. Similarly, the stimulatory effect of AM on vasopressin release observed *in vivo*[210] was not observed in hypothalamo-neurohypophysial explants under basal conditions, and thus the major action of AM on vasopressin secretion is likely exerted outside of the PVN – perhaps in the SFO or adjacent tissues. In PVN slice preparations, Type II, parvocellular neurons directly depolarized to bath application of AM.[216] This may explain the ability of centrally administered AM to stimulate stress hormone secretion *in vivo*,[217] via increased corticotropin releasing hormone release into the median eminence for delivery to the adenohypophysis (Table 13.2).

TABLE 13.2 The Shared and Opposing Actions of Angiotensin II and the Adrenomedullin Peptides

Tissue site of action	Angiotensin II effects	Adrenomedullin effects
Kidney	Increase Na^+ reabsorption (antinatriuresis) Renal vasoconstriction (decrease renal blood flow)	Increase Na^+ excretion (natriuresis) Increase water excretion (diuresis) Renal arteriolar dilation (increase renal blood flow)
Adrenal	Increase aldosterone secretion	Decrease aldosterone secretion
Heart	Increase cardiac fibroblast proliferation leading to cardiac dysfunction	Increase cardiac contractile function Increase ANP secretion
Vasculature	Increase smooth muscle proliferation Vasoconstriction Increase noradrenalin release	Decrease smooth muscle proliferation Vasodilation Decrease noradrenalin release
Pituitary	Increase vasopressin release Increase ACTH release (basal and CRH-stimulated)	Increase vasopressin release Decrease ACTH release (CRH-stimulated)
Central nervous system	Increase thirst Increase sodium appetite Increase sympathetic tone	Decrease thirst Decrease sodium appetite Increase sympathetic tone

Are the Central Autonomic Effects of AM and PAMP Physiologically Relevant?

The physiologic relevance of endogenous AM in the hypothalamic control of vasopressin secretion and fluid/electrolyte homeostasis has been evidenced in animal studies,[205,210] and there is some indication that the peptide's actions on autonomic function are physiologically relevant as well.[218,219] However, convincing proof of the absolute importance of AM in central cardiovascular control is hard to obtain for several reasons. First, there are very few pharmacologic agents that selectively antagonize the action of AM or PAMP.[220] Second, AM belongs to a family of homologous peptides that includes calcitonin gene related peptide (CGRP), intermedin (IMD), calcitonin and amylin. What usually distinguishes homologous peptides are their individual cognate receptors; however, these peptides share common receptor activity modifying proteins (RAMPs) that make up part of their cell surface receptor complexes. There are at least three RAMPs. When RAMPs are associated with the calcitonin receptor, the complex functions as a receptor for amylin.[221] When RAMP1 is associated with the calcitonin receptor-like receptor (CLR), it functions as a CGRP receptor. When RAMP 2 or RAMP3 associate with the CLR, they function as receptors for AM.[78] Finally, intermedin (also known as adrenomedullin 2) binds to the CLR when the receptor is associated with any of the three RAMPs.[222] These overlaps explain why many of the actions of CGRP[223] and intermedin[63,224–226] mirror those of AM, and, probably, why it has proven difficult to develop selective AM antagonists. It also makes it difficult to use genetic methodologies to isolate the action of one or another family member by knocking out the gene for the CLR.[227]

Since CGRP, intermedin and AM are products of unique genes, it was thought that embryonic gene compromise could be used to identify the physiologic function of AM. However, $AM^{-/-}$ mice die *in utero* around day 14.5 of gestation.[228] The observed phenotype of these embryos is characterized by hydrops

BOX 13.3

HOW DO WE KNOW HOW TO PROVE THE PHYSIOLOGIC RELEVANCE OF A NEUROPEPTIDE WHEN GENE KNOCKOUT IS LETHAL?

Adrenomedullin (AM) producing neurons and their axon terminals have been localized to autonomic centers in brain, including the paraventricular nucleus (PVN) and rostralventrolateral medulla (RVLM). Injection of exogenous AM either into the cerebroventricles or directly into these autonomic centers resulted in not only increased mean arterial pressures, via activation of the sympathetic nervous system, but also vasopressin (AVP) release. Furthermore, the electrical activities of preautonomic neurons in PVN and RVLM are altered by direct application of AM in both *in vivo* and electrophysiologic studies. Those anatomic, electrophysiologic and pharmacologic evidences suggest a physiologically relevant role for endogenous AM in the central control of cardiovascular function. How do we know if such a role exists?

Adrenomedullin "knockout" animals do not survive gestation,[1] and no selective AM receptor antagonists are available for testing. Thus acute, temporal compromise of endogenous AM production was attempted to provide evidence for the role of brain-derived AM in central endocrine and autonomic control of cardiovascular function. Reversible compromise of AM production was accomplished by means of a central injection of a ribozyme, a self catalytic RNA molecule, that targeted for destruction the mRNA encoding preproadrenomedullin. Compromise of both mRNA abundance and AM protein was verified by Northern blotting and radioimmunoassay.[2] These animals displayed a decreased AVP response to dehydration or hypovolemia compared to controls, and a significant loss of plasma volume during the hypovolemic challenge.[2] In additional experiments using passive immunoneutralization with an antibody selective for AM, these same investigators demonstrated a compromise in baroreflex activity when the function of endogenous AM was acutely compromised.[3] These baroreflex activity results require confirmation with more molecular approaches; however, together with the ribozyme studies, evidence is developing for the importance of brain-derived AM in the integration of both the endocrine and autonomic control of cardiovascular function.

References

1. Caron KM, Smithies O. Extreme hydrops fetalis and cardiovascular abnormalities in mice lacking a functional adrenomedullin gene. *Proc Natl Acad Sci.* 2001;98:615–619.
2. Taylor MM, Baker JR, Samson WK. Brain-derived adrenomedullin controls blood volume through the regulation of arginine vasopressin production and release. *Am J Physiol.* 2005;288:R1203–R1210.
3. Taylor MM, Keown CA, Samson WK. Involvement of the central adrenomedullin peptides in the baroreflex. *Regul Pept.* 2003;112(1–3):87–93.

fetalis, and defective cardiac and vascular development. Defects in placentation have also been observed, even in hetereozygote mice ($AM^{+/-}$), thus pointing to a critically important role for endogenous AM in cardiovascular development.[229] Absent selective receptor antagonists, and lacking the ability to employ embryonic gene compromise, the only available tools for the study of AM's physiologic relevance are acute compromise of peptide action[219] or mRNA transcription or translation.[205,210] Because of AM's multiple sites of production and action, these techniques are difficult to exercise without selective tissue targeting.

GASTROINTESTINAL HORMONES

Several gut hormones are known to interact, via direct or indirect routes, with hypothalamic and brainstem networks controlling autonomic outflow and, therefore, cardiovascular function. Some of these hormones access the CNS via primary actions on vagal afferents,[22] while others act on neurons in the CVOs.[18,24,25,39] Indeed, some are able to cross, in limited amounts, the blood–brain barrier.[13–16,230,231] While studied mostly for their actions on food intake and energy homeostasis, several of these peptides also play a role in the integrated control of cardiovascular function. What they share in common is activation or inhibition of networks originating in or afferent to the PVN and neighboring arcuate nucleus. A pivotal role for the melanocortin system in the arcuate nucleus has been identified.[232]

The Central Melanocortin System and Cardiovascular Neuroendocrinology

Anatomy and Biological Activity

The central melanocortin system (CMS) is defined anatomically as neurons originating in the arcuate nucleus of the hypothalamus that produce either α-melanocyte stimulating hormone (α-MSH), a product of the proopiomelanocortin (POMC) prohormone, or neuropeptide Y (NPY) and agouti-related peptide (AgRP).[232] These neurons project to target neurons expressing melanocortin 3/4 receptors (MC3/4) located throughout the brain, including neurons within the arcuate nucleus as well as other hypothalamic nuclei, such as the paraventricular nucleus (PVN), periventricular nucleus, lateral hypothalamic area (LHA) and perifornical region. Arcuate POMC neurons also send projections to brainstem autonomic areas, including the nucleus of the solitary tract (NTS) and the rostroventrolateral medulla (RVLM), which account for approximately half of all α-MSH immunoreactivity in the brainstem.[233,234] The remaining half is indicative of a small population of POMC neurons that originate in the commissural NTS. These brainstem POMC neurons, which project to the rostral NTS, as well as to the lateral parabrachial nucleus, the locus coeruleus and the dorsal motor nucleus of the vagus, are strategically positioned to receive visceral and cardiovascular information via vagal afferents.[235,236]

Melanocortin receptors in brain (MC3R and MC4R) are G-protein coupled receptors coupled to G_s heterotrimeric proteins. Agonist binding to melanocortin receptors results in stimulation of adenylate cyclase and increased production of cAMP.[237] Activation of MC3/4 receptors by α-MSH results in a potent inhibition of food intake, and an increase in sympathetic nervous system activity and blood pressure. Antagonism of MC3/4 receptors by AgRP, a natural antagonist of melanocortin receptors, results in hypophagia and a decrease in blood pressure.[232]

The role of the CMS in modulating food intake and energy expenditure has been very well established. Mutations in the components of the CMS, particularly the MC4 receptor, result in extreme hyperphagia and profound obesity in experimental animals,[238,239] and may account for as many as 1 in 16 cases of human obesity.[240] The CMS also plays an important role in regulating blood pressure by modulating sympathetic activity. Central injection of melanocortin agonists increases sympathetic activity and mean arterial pressure in rabbits and rodents.[241,242] Furthermore, mice deficient in the MC4 receptor exhibit altered cardiovascular function.[243]

Input to the Central Melanocortin System

The CMS is thought by many to be the "final common pathway" through which many hormones must act to exert their effects on food intake and cardiovascular function. The most well-known hormone that is dependent upon the CMS to exert its activities is the adipocyte-derived hormone, leptin. Both the anorexigenic and sympathostimulatory activities of leptin are attenuated by pretreatment with MC3/4 receptor antagonists.[244,245] Many other peripherally and centrally-derived hormones also rely on melanocortin signaling. Intra-PVN injection of the melanocortin antagonist SHU9119 reverses the increase in sympathetic activity induced by insulin,[246] and the orexigenic effect of ghrelin is abolished by pretreatment with NPY receptor antagonists.[247] Furthermore, pretreatment with SHU9119 reverses the anorexigenic effect of two novel neuropeptides, nesfatin-1[248,249] and neuronostatin.[250]

Downstream Circuits Mediating CMS Action

There is now strong evidence to support central oxytocin signaling as an important downstream mediator of melanocortin action. Like melanocortin agonists, centrally-injected oxytocin inhibits food intake,[251,252] and oxytocin-deficient mice exhibit altered cardiovascular function.[253] Central injection of α-MSH leads to an increase in cFos accumulation in oxytocin immunoreactive neurons,[254] and induces the central release of oxytocin while inhibiting the peripheral release of the peptide.[255] Additionally, several peptides that are dependent upon the central melanocortin system to exert their effects are likewise dependent upon functional central oxytocin receptors, such as leptin[256] and nesfatin-1.[257]

The CMS likely utilizes other downstream circuits as well, since the neuropeptide neuronostatin, which is dependent upon melanocortin action, does not require oxytocin signaling.[250] Other possible circuits mediating melanocortin activities include central corticotropin-releasing hormone (CRH) and brain-derived neurotropic factor (BDNF). Melanocortin agonists stimulate the release of BDNF in the hypothalamus,[258] and the activity of both forebrain[105] and brainstem[259] POMC neurons requires the downstream activation of BDNF receptors.

CONCLUSIONS

Cardiovascular neuroendocrinology is an emerging discipline that seeks to understand the hypothalamic mechanisms by which endocrine, neuroendocrine, cognitive, behavioral and autonomic responses to changes in our external and internal environments are received and integrated, and how that integration generates appropriate adaptive responses. Three main sources of afferent information enter via: (a) the circumventricular organs (OVLT, SFO); (b) diffusion of metabolic substrates across the blood–brain barrier (e.g., glucose, free fatty acids, leptin, insulin); and (c) afferent neural input from exteroceptors (olfactory and optic) and enteroreceptors (baroreceptors, chemoreceptors). This information is then organized for appropriate efferent output to the median eminence (neuroendocrine control), neurohypophysis (endocrine responses), and other central nervous system structures, including cortical and limbic centers (cognition and behavior), and autonomic centers in the brainstem and spinal cord. In this manner, the hypothalamus serves as the central command center monitoring peripheral and central messengers of homeostasis.

FUTURE DIRECTIONS

While animal models are informative for the understanding of basic physiologic mechanisms underlying the hypothalamic control of both neuroendocrine and autonomic functions (i.e., cardiovascular neuroendocrinology), they only produce valid insights for known or recently discovered peptides, hormones and neurotransmitters. Just as our knowledge of the integrative mechanisms by which the hypothalamus and brainstem-medullary centers coordinate autonomic output is still incomplete, so too it appears is our knowledge of the chemical mediators of those neuronal communications. Modern molecular tools, when coupled with recently developed bioinformatic approaches, have revealed the existence of previously unrecognized neuropeptides that exert profound effects of behavior, endocrine, neuroendocrine and autonomic function via actions in hypothalamus and brainstem.

Reverse pharmacology was employed[248] to discover a potent anorexigenic peptide, named nesfatin, that also exerts significant actions on stress hormone secretion[260] and sympathetic outflow.[249] The physiologic significance of those actions is now recognized,[248,257] and thus this new peptide has to be incorporated into the existing models of central cardiovascular control. Will it turn out to be a critically important regulator of autonomic function? Only time will tell; however, its pharmacologic profile is promising. Will nesfatin and its yet to be discovered receptor become drug targets for the treatment of cardiovascular disease? Again, only time will tell.

A second approach to the discovery of previously unrecognized peptide hormones is the use of bioinformatic screening of published genomes and computer-based screening for evolutionary conservation across species.[261] Using this approach, a novel peptide encoded in the somatostatin precursor, named neuronostatin, was discovered in 2008.[262] In just 2 short years, neuronostatin has been shown to act within the brain to control mean arterial pressure by a combination of autonomic and neuroendocrine effects.[250,263] These appear to reflect a physiologically important role for the peptide in central cardiovascular control, and recent studies have begun to identify its interactions with the more established neuronal systems that are prime candidates for drug targeting. Indeed, both neuronostatin[250] and nesfatin[248] exert at least some of their initial actions by activating POMC neurons in the arcuate nucleus that project to hypothalamic and extrahypothalmic centers known to be the critical sites of neuroendocrine and autonomic integration.

In the end, the development of novel methodologies for reversible and selective cell and gene targeting coupled with the continued search for novel peptides and hormones will provide much needed insight into the hierarchy of neuronal networks responsible for the important integrative nature of the hypothalamus and its associated brain structures.

Acknowledgments

Drs Samson and Yosten are supported by grants from the NIH and the American Heart Association.

References

1. Tasker JG, Dudek FE. Electrophysiological properties of neurones in the region of the paraventricular nucleus in slices of rat hypothalamus. *J Physiol.* 1991;434:271–293.
2. Latchford KJ, Ferguson AV. Angiotensin depolarizes parvocellular neurons in the paraventricular nucleus through modulation of putative nonselective cationic and potassium conductances. *Am J Physiol.* 2005;289:R52–R58.
3. Stern JE. Electrophysiological and morphological properties of pre-autonomic neurones in the rat hypothalamic paraventricular nucleus. *J Physiol.* 2001;537(1):161–177.
4. Hosoya Y, Matsushita M. Identification and distribution of the spinal and hypophysial projection neurons in the paraventricular nucleus of the rat. A light and electron microscope study with the horseradish peroxidase method. *Exp Brain Res.* 1979;35:315–331.
5. Pyner S, Coote JH. Identification of branching paraventricular neurons of the hypothalamus that project to the rostroventrolateral medulla and spinal cord. *Neurosci.* 2000;100(3):549–556.
6. Shafton AD, Ryan A, Badoer E. Neurons in the hypothalamic paraventricular nucleus send collaterals to the spinal cord and to the rostral ventrolateral medulla in the rat. *Brain Res.* 1998;801:239–243.
7. Swanson LW, Kuypers HGJM. The paraventricular nucleus of the hypothalamus: cytoarchitectonic subdivisions and organization of projections to the pituitary, dorsal vagal complex, and spinal cord as demonstrated by retrograde fluorescence double-labeling methods. *J Comp Neurol.* 1980;194:555–570.
8. Swanson LW, Sawchenko PE, Wiegand SJ, Price JL. Separate neurons in the paraventricular nucleus project to the median eminence and to the medulla or spinal cord. *Brain Res.* 1980;197:207–212.
9. Moss RL, Dyball RE, Cross BA. Excitation of antidromically identified neurosecretory cells of the paraventricular nucleus by oxytocin applied iontophoretically. *Exp Neurol.* 1972;34(1): 95–102.
10. Clarke G, Merrick LP. Electrophysiological studies of the magnocellular neurons. In: Ganten D, Pfaff D, eds. *Neurobiology of Vasopressin*. Berlin: Springer-Verlag; 1985:17–60. Current Topics in Neuroendocrinology, vol. 4.
11. Luther JA, Daftary SS, Boudaba C, et al. Neurosecretory and non-neurosecretory parvocellular neurones of the hypothalamic paraventricular nucleus express distinct electrophysiological properties. *J Neuroendocrinol.* 2002;14:929–932.
12. Covenas R, de Leon M, Cintra A, Bjelke B, Gustafsson JA, Fuxe K. Coexistence of c-Fos and glucocorticoid receptor immunoreactivities in the CRF immunoreactive neurons of the paraventricular hypothalamic nucleus of the rat after acute immobilization stress. *Neurosci Lett.* 1993;149(2):149–152.
13. Le Foll C, Irani BG, Magnan C, Dunn-Meynell AA, Levin BE. Characteristics and mechanisms of hypothalamic neuronal fatty acid sensing. *Am J Physiol.* 2009;297:R655–R664.
14. Kang L, Sanders NM, Dunn-Meynell AA, et al. Prior hypoglycemia enhances glucose responsiveness in some ventromedial hypothalamic glucosensing neurons. *Am J Physiol.* 2008;294:R784–R792.
15. Dunn-Meynell AA, Sanders NM, Compton D, et al. Relationship among brain and blood glucose levels and spontaneous and glucoprivic feeding. *J Neurosci.* 2009;29:7015–7022.
16. Langhans W, Leitner C, Arnold M. Dietary fat sensing via fatty acid oxidation in enterocytes: possible role in the control of eating. *Am J Physiol.* 2011;300:R554–R565.
17. Urayama A, Banks WA. Starvation and triglycerides reverse the obesity-induced impairment of insulin transport at the blood-brain barrier. *Endocrinology.* 2008;149:3592–3597.
18. Spranger J, Verma S, Goehring I, et al. Adiponectin does not cross the blood–brain barrier but modifies cytokine expression of brain endothelial cells. *Diabetes.* 2006;55:141–147.
19. Brooks VL, Dampney RAL, Heesch CM. Pregnancy and the endocrine regulation of the baroreceptor reflex. *Am J Physiol.* 2010;299(1-3):R439–R451.
20. Williams DL. Minireview: Finding the sweet spot: Peripheral *versus* central glucagon-like peptide 1 action in feeding and glucose homeostasis. *Endocrinology.* 2009;150:2997–3001.
21. Dockray GJ. The versatility of the vagus. *Physiol Behav.* 2009;97: 531–536.
22. Berthoud HR. The vagus nerve, food intake and obesity. *Regul Pep.* 2008;149:15–25.
23. Raybould HE. Gut chemosensing: interactions between the gut endocrine cells and visceral afferents. *Auton Neurosci.* 2010;153(1-2):41–46.
24. Fry M, Hoyda TD, Ferguson AV. Making sense of it: roles of the sensory circumventricular organs in feeding and regulation of energy homeostasis. *Exp Biol Med.* 2007;232:14–26.
25. Smith PM, Ferguson AV. Circulating signals as critical regulators of autonomic state—central roles for the subfornical organ. *Am J Physiol.* 2010;299:R405–R415.
26. White MM, Samson WK. A possible relationship between brain-derived adrenomedullin and oxytocin in the regulation of sodium balance. *J Endocrinol.* 2009;203:253–262.
27. Gardner DG, Chen S, Glenn DJ, Grigsby CL. Molecular biology of the natriuretic peptide system: implications for physiology and hypertension. *Hypertension.* 2007;49:419–426.
28. Kohan DE, Rossi NF, Inscho EW, Pollock DM. Regulation of blood pressure and salt homeostasis by endothelin. *Physiol Rev.* 2011;91:1–77.
29. Wislocki GB, King LS. The permeability of the hypophysis and the hypothalamus to vital dyes, with a study of the hypophyseal vascular supply. *Am J Anat.* 1936;58:421–476.
30. Rodrigues EM, Herrera H, Peruzzo B, et al. Light- and electron-microscopic immunocytochemistry and lectin histochemistry of the subcommissural organ: evidence for processing of the secretory material. *Cell Tissue Res.* 1986;243: 545–559.
31. Nieuwenhuys R, Voogd J, van Huijzen C. *The Human Central Nervous System*. 3rd ed. Berlin: Springer Verlag; 1988.
32. Johnson AK, Gross PM. Sensory circumventricular organs and brain homeostatic pathways. *FASEB J.* 1993;7:678–686.
33. Lutz TA. The role of amylin in the central control of energy homeostasis. *Am J Physiol.* 2010;298:R1475–R1484.
34. Ferguson AV, Latchford KJ, Samson WK. The paraventricular nucleus of the hypothalamus – a potential target for the integrative treatment of autonomic dysfunction. *Expert Opin Ther Targets.* 2008;12(6):1–11.
35. McKinley MJ, Bicknell RJ, Hards D, et al. Efferent neural pathways of the lamina terminalis subserving osmoregulation. In: Ermisch A, Landgraf R, Ruhle H-J, eds. *Prog Brain Res.* 1992;91:395–402.
36. Toney GM, Stocker SD. Hyperosmotic activation of CNS sympathetic drive: implications for cardiovascular disease. *J Physiol.* 2010;588(18):3375–3384.
37. Lind RW. Neural connections of the subfornical organ. In: Gross PM, ed. *Circumventricular Organs and Body Fluids*. Boca Raton: CRC Press; 1987:27–42.

38. Phillips MI, Camacho A. Neural connections of the organum vasculosum lamina terminalis. In: Gross PM, ed. *Circumventricular Organs and Body Fluids*. Boca Raton: CRC Press; 1987:158–168.
39. Ferguson AV. The area postrema: a cardiovascular centre at the blood-brain interface? *Can J Physiol Pharmacol.* 1991;69: 1026–1034.
40. Norgren R. The central organization of the gustatory and visceral afferent systems in the nucleus of the solitary tract. In: Katsuki Y, Norgren R, Sato M, eds. *Brain Mechanisms of Sensation*. New York: Wiley & Sons; 1981:143–160.
41. Saper CB, Reis DJ, Joh T. Medullary catecholamine inputs to the anteroventral third ventricular cardiovascular regulatory region in the rat. *Neurosi Lett.* 1983;42:285–291.
42. Saper CB, Levisohn D. Afferent connections of the median preoptic nucleus in the rat: anatomical evidence for a cardiovascular integrative mechanism in the anteroventral third ventricular (AV3V) region. *Brain Res.* 1983;288:21–31.
43. Kuhlenbeck H. *The Central Nervous System of Vertebrates. Derivatives of the Prosencephalon, Diencephalon and Telencephalon*. Basel: Karger; 1977. Vol. 5(1).
44. Sawchenko PE, Swanson LW. The organization and biochemical specificity of afferent projections to the paraventricular and supraoptic nuclei. In: Cross BA, Leng G, eds. *The Neurohypophysis: Structure, Function and Control, Progress in Brain Research*. Amsterdam: Elsevier; 1983;60:19–29.
45. Ferguson AV. Angiotensinergic regulation of autonomic and neuroendocrine outputs: critical roles for the subfornical organ and the paraventricular nucleus. *Neurendocrinology.* 2009; 89(4):370–376.
46. Bains JS, Ferguson AV. Paraventricular nucleus neurons projecting to spinal cord receive excitatory input from the subfornical organ. *Am J Physiol.* 1995;268:R625–R633.
47. Vieira AA, Nahey D, Collister JP. Role of the organum vasculosum of the lamina terminalis for the chronic cardiovascular effects produced by endogenous and exogenous ANG II in conscious rats. *Am J Physiol.* 2010;299:R1564–R1571.
48. Bourque CW. Central mechanisms of osmosensation and systemic osmoregulation. *Nat Rev Neurosci.* 2008;9(7):519–531.
49. Weiss ML, Classen DE, Hirai T, Kenney MJ. Nonuniform sympathetic nerve responses to intravenous hypertonic saline infusion. *J Auton Nerv Syst.* 1996;57:109–115.
50. Shi P, Martinez MA, Calderon AS, et al. Intra-carotid hyperosmotic stimulation increases Fos staining in forebrain organum vasculosum laminae terminalis neurones that project to the hypothalamic paravenricular nucleus. *J Physiol.* 2008;586: 5231–5245.
51. Brooks VL, Qi Y, O'Donaughy. Increased osmolality of conscious water-deprived rats supports arterial pressure and sympathetic activity via a brain action. *Am J Physiol.* 2005; 288:R1248–R1255.
52. Adams JM, Bardgett ME, Stocker SD. Ventral lamina terminalis mediates enhanced cardiovascular responses of rostral ventrolateral medulla neurons during increased dietary salt. *Hypertension.* 2009;54:308–314.
53. Gottlieb HB, Ji LL, Jones H, et al. Differential effects of water and saline intake on water-deprivation-induced c-Fos staining in rat. *Am J Physiol.* 2006;290:R1251–R1261.
54. Johnson AK, Thunhorst RL. The neuroendocrinology of thirst and salt appetite: visceral sensory signals and mechanisms of central integration. *Front Neuroendocrinol.* 1997;18(3): 292–353.
55. Altura BM. DPAVP: a vasopressin analog with selective microvascular and RES actions for the treatment of circulatory shock in rats. *Eur J Pharmacol.* 1976;37(1):155–167.
56. Giusta-Paiva A, Martinez MR, Bispo-Da-Silva LB, et al. Vasopressin mediates the pressor effect of hypertonic saline solution in endotoxic shock. *Shock.* 2007;27:416–421.
57. Sawchenko PE, Swanson LW. Central noradrenergic pathways for the integration of hypothalamic neuroendocrine and autonomic responses. *Science.* 1981;214(6):685–687.
58. Sawchenko PE, Swanson LW. Immunohistochemical identification of neurons in the paraventricular nucleus of the hypothalamus that project to the medulla or to the spinal cord in the rat. *J Comp Neurol.* 1982;205:260–272.
59. Sawchenko PE, Swanson LW. The organization of forebrain afferents to the paraventricular and supraoptic nuclei of the rat. *J Comp Neurol.* 1983;218:121–144.
60. Coote JH. A role for the paraventricular nucleus of the hypothalamus in the autonomic control of heart and kidney. *Exp Physiol.* 2005;90:169–173.
61. Guyenet PG. The sympathetic control of blood pressure. *Nat Rev Neurosci.* 2006;7:335–346.
62. Rinaman L. Hindbrain noradrenergic A2 neurons: diverse roles in autonomic, endocrine, cognitive, and behavioral functions. *Am J Physiol.* 2011;300:R222–R235.
63. Taylor MM, Samson WK. Stress hormone secretion is altered by central administration of intermedin/adrenomedullin-2. *Brain Res.* 2005;1045(1-2):199–205.
64. Samson WK. Natriuretic peptides: a family of hormones. *Trends Endocr Metab.* 1992;3(3):86–90.
65. Samson WK. Adrenomedullin and the control of fluid and electrolyte homeostasis. *Ann Rev Physiol.* 1999;61:363–389.
66. Samson WK, Huang FI, Fulton RJ. C-type natriuretic peptide mediates the hypothalamic actions of the natriuretic peptides to inhibit luteinizing hormone secretion. *Endocrinology.* 1993; 132:504–509.
67. Levin ER, Gardner DG, Samson WK. Natriuretic peptides. *N Eng J Med.* 1998;339(5):321–328.
68. Phillips MI. Functions of angiotensin in the central nervous system. *Ann Rev Physiol.* 1987;49:413–435.
69. Higuchi S, Ohtsu H, Suzuki H, et al. Angiotensin II signal transduction through the AT1 receptor: novel insights into mechanisms and pathophysiology. *Clin Sci.* 2007;112(8): 417–428.
70. Donoghue M, Hsieh F, Baronas E, et al. A novel angiotensin-converting enzyme–related carboxypeptidase (ACE2) converts angiotensin I to angiotensin 1-9. *Circ Res.* 2000; 87:E1–E9.
71. Tipnis SR, Hooper NM, Hyde E, et al. A human homolog of angiotensin-converting enzyme. Cloning and functional expression as a captopril-insensitive carboxypeptidase. *J Biol Chem.* 2000;275:33238–33243.
72. Santos RAS, Simoes e Silva AC, Maric C, et al. Angiotensin-(1–7) is an endogenous ligand for the G protein-coupled receptor Mas. *Proc Natl Acad Sci.* 2003;100:8258–8263.
73. Ferreira AJ, Santos RAS, Bradford CN, et al. Therapeutic implications of the vasoprotective axis of the renin-angiotensin system in cardiovascular diseases. *Hypertension.* 2010;55: 207–213.
74. Lavoie JL, Sigmund CD. Minireview: Overview of the renin-angiotensin system—an endocrine and paracrine system. *Endocrinology.* 2003;194:2179–2183.
75. Lavoie JL, Liu X, Blanco RA, et al. Evidence supporting a functional role for intracellular renin in the brain. *Hypertension.* 2006;47:461–466.
76. Oparil S, Haber E. The renin-angiotensin system. *New Engl J Med.* 1974;291(8):389–401.
77. Dickson ME, Sigmund CD. Genetic basis of hypertension: revisiting angiotensinogen. *Hypertension.* 2006;48(1):14–20.

78. Gibbons C, Dackor R, Dunworth W, Fritz-Six K, Caron KM. Receptor activity modifying proteins: RAMPing up adrenomedullin signaling. *Mol Endocrinol.* 2007;21:783–796.
79. deGasparo M, Catt KJ, Inagami T, Wright JW, Unger T. *International Union of Pharmacology. XXIII. The angiotensin II receptors.* 2000;52:415.
80. Carey RM, Howell NL, Jin X-H, Siragy HM. Angiotensin type 2 receptor-mediated hypotension in angiotensin type-1 receptor-blocked rats. *Hypertension.* 2001;38:1272–1277.
81. Mehta PK, Griendling KK. Angiotensin II cell signaling: physiological and pathological effects in the cardiovascular system. *Am J Physiol.* 2007;292:C82–C97.
82. Lazartigues E, Sinnayah P, Augoyard G, et al. Enhanced water and salt intake in transgenic mice with brain-restricted overexpression of angiotensin (AT_1) receptors. *Am J Physiol.* 208;295:R1539-R1545.
83. Dinh DT, Frauman AG, Johnston CI, Fabiani ME. Angiotensin receptors: distribution, signalling and function. *Clin Sci.* 2001;100:481–492.
84. Allen AM, MacGregor DP, McKinley MJ, Mendelsohn FAO. Angiotensin II receptors in the human brain. *Regul Peptides.* 1999;79(1):1–7.
85. Montiel M, Barker S, Vinson GP, Jimenez E. Angiotensin II receptor isoforms in the rat adrenal gland: studies with the selective subtype antagonist DuP753 and CGP42112A. *J Mol Endocrinol.* 1993;11:69–75.
86. Allen AM, Yamada H, Mendelsohn FA. *In vitro* autoradiographic localization of binding to angiotensin receptors in the rat heart. *Intl J Cardiol.* 1990;28:25–33.
87. Saavedra JM, Viswanathan M, Shigematsu K. Localization of angiotensin AT1 receptors in the rat heart conduction system. *Eur J Pharmacol.* 1993;235:301–303.
88. Apfeldorf WJ, Rasmussen H. Simultaneous determination of intracellular free calcium and aldosterone production in bovine adrenal zona glomerulosa. *Cell Calcium.* 1988;9:71–80.
89. Sasamura H, Hein L, Krieger JE, Pratt RE, Kobilka BK, Dzau VJ. Cloning, characterization, and expression of two angiotensin receptor (AT-1) isoforms from the mouse genome. *Biochem Biophys Res Commun.* 1992;185:253–259.
90. Iwai N, Inagami T. Identification of two subtypes in the rat type I angiotensin II receptor. *FEBS Lett.* 1992;298:257–260.
91. Lazard D, Briend-Sutren MM, Villageois P, et al. Molecular characterization and chromosome localization of a human angiotensin II AT2 receptor gene highly expressed in fetal tissues. *Receptors Channels.* 1994;2:271–280.
92. Shanmugam S, Sandberg K. Ontogeny of angiotensin II receptors. *Cell Biol Intl.* 1996;20:169–176.
93. Hein L, Barsh GS, Pratt RE, Dzau VJ, Kobilka BK. Behavioural and cardiovascular effects of disrupting the angiotensin II type-2 receptor in mice. *Nature.* 1995;377(6551):744–747.
94. Ichiki T, Labosky PA, Shiota C, et al. Effects on blood pressure and exploratory behaviour of mice lacking angiotensin II type-2 receptor. *Nature.* 1995;377(6551):748–750.
95. Siragy HM, Carey RM. The subtype-2 (AT2) angiotensin receptor regulates renal cyclic guanosine 3′,5′-monophosphate and AT1 receptor-mediated prostaglandin E2 production in conscious rats. *J Clin Invest.* 1996;97:1978–1982.
96. Siragy HM, Carey RM. The subtype 2 (AT2) angiotensin receptor mediates renal production of nitric oxide in conscious rats. *J Clin Invest.* 1997;100:264–269.
97. Siragy HM, Carey RM. The subtype 2 (AT2) angiotensin receptor regulates renal prostaglandin F2alpha formation in conscious rats. *Am J Physiol.* 1997;273:R1103–R1107.
98. Viswanathan M, Saavedra JM. Expression of angiotensin II AT2 receptors in the rat skin during experimental wound healing. *Peptides.* 1992;13:783–786.
99. Janiak P, Pillon A, Prost JF, Vilaine JP. Role of angiotensin subtype 2 receptor in neointima formation after vascular injury. *Hypertension.* 1992;20:737–745.
100. Chappell MC, Brosnihan KB, Diz DI, Ferrario CM. Identification of angiotensin-(1-7) in rat brain. Evidence for differential processing of angiotensin peptides. *J Biol Chem.* 1989;264: 16518–16523.
101. Silva LC, Fontes MA, Campagnole-Santos MJ, et al. Cardiovascular effects produced by micro-injection of angiotensin-(1-7) on vasopressor and vasodepressor sites of the ventrolateral medulla. *Brain Res.* 1993;613(2):321–325.
102. Moriguchi A, Tallant EA, Matsumura K, et al. Opposing actions of angiotensin-(1-7) and angiotensin II in the brain of transgenic hypertensive rats. *Hypertension.* 1995;25:1260–1265.
103. Becker LK, Santos RA, Campagnole-Santos MJ. Cardiovascular effects of angiotensin II and angiotensin-(1-7) at the RVLM of trained normotensive rats. *Brain Res.* 2005;1040(1-2):121–128.
104. Sakima A, Averill DB, Gallagher PE, et al. Impaired heart rate baroreflex in older rats: role of endogenous angiotensin-(1-7) at the nucleus tractus solitarii. *Hypertension.* 2005;46:333–340.
105. Xu B, Goulding EH, Zang K, et al. Brain-derived neurotrophic factor regulates energy balance downstream of melanocortin-4 receptor. *Nature Neurosci.* 2003;6(7):736–742.
106. Giachetti G, Faloia E, Mariniello B, et al. Overexpression of the renin–angiotensin system in human visceral adipose tissue in normal and overweight subjects. *Am J Hypertens.* 2002;15(5): 381–388.
107. Engeli S, Negrel R, Sharma AM. Physiology and pathophysiology of the adipose tissue renin-angiotensin system. *Hypertension.* 2000;35:1270–1277.
108. Ganten D, Marquez-Julio A, Granger P, et al. Renin in dog brain. *Am J Physiol.* 1971;221(6):1733–1737.
109. Lind RW, Swanson LW, Bruhn TO, Ganten D. The distribution of angiotensin II-immunoreactive cells and fibers in the paraventriculo-hypophysial system of the rat. *Brain Res.* 1985; 338(1):81–89.
110. Lind RW, Swanson LW, Ganten D. Organization of angiotensin II immunoreactive cells and fibers in the rat central nervous system. An immunohistochemical study. *Neuroendocrinology.* 1985;40(1):2–24.
111. Milsted A, Barna BP, Ransohoff RM, Brosnihan KB, Ferrario CM. Astrocyte cultures derived from human brain tissue express angiotensinogen mRNA. *Proc Natl Acad Sci.* 1990;87:5720–5723.
112. Sernia C. Location and secretion of brain angiotensinogen. *Regul Peptides.* 1995;57(1):1–18.
113. Dzau VJ, Brenner A, Emmett NL. Evidence for renin in rat brain: differentiation from other renin like enzymes. *Am J Physiol.* 1982;242:E292–E29.
114. Morris BJ. Molecular biology of renin. I: Gene and protein structure, synthesis and processing. *J Hypertens.* 1992;10: 209–214.
115. Lee-Kirsch MA, Gaudet F, Cardoso MC, Lindpaintner K. Distinct renin isoforms generated by tissue-specific transcription initiation and alternative splicing. *Circ Res.* 1999;84:240–246.
116. Sinn PL, Sigmund CD. Identification of three human renin mRNA isoforms from alternative tissue-specific transcriptional initiation. *Physiol Genomics.* 2000;3:25–31.
117. Lavoie JL, Cassell MD, Gross KW, Sigmund CD. Adjacent expression of renin and angiotensinogen in the rostral ventrolateral medulla using a dual-reporter transgenic model. *Hypertension.* 2004;43:1116–1119.
118. Francis J, Wei SG, Weiss RM, Felder RB. Brain angiotensin-converting enzyme activity and autonomic regulation in heart failure. *Am J Physiol.* 2004;287:H2138–H2146.

119. Morimoto S, Cassell MD, Sigmund CD. Neuron-specific expression of human angiotensinogen in brain causes increased salt intake. *Physiol Genomics*. 2002;9:113–120.
120. Schinke M, Baltatu O, Boehm M, et al. Blood pressure reduction and diabetes insipidus in transgenic rats deficient in brain angiotensinogen. *Proc Natl Acad Sci*. 1999;96:3975–3980.
121. Veerasingham SJ, Raizada MK. Brain renin-angiotensin system dysfunction in hypertension: recent advances and perspectives. *Br J Pharmacol*. 2003;139(2):191–202.
122. Gao L, Wang W, Li Y-H, et al. Sympathoexcitation by central ANG II: Roles for AT1 receptor upregulation and NAD(P)H oxidase in RVLM. *Am J Physiol*. 2005;288:H2271–H2279.
123. Campagnole-Santos MJ, Heringer SB, Batista EN, Kholsa MC, Santos RA. Differential baroreceptor reflex modulation by centrally infused angiotensin peptides. *Am J Physiol*. 1992;263: R89–R94.
124. Yamazato M, Yamazato Y, Sun C, Diez-Freire, Raizada MK. Overexpression of angiotensin-converting enzyme 2 in the rostral ventrolateral medulla causes long-term decrease in blood pressure in the spontaneously hypertensive rats. *Hypertension*. 2007;49:926–931.
125. Feng Y, Xia H, Cai Y, et al. Brain-selective overexpression of human angiotensin-converting enzyme type 2 attenuates neurogenic hypertension. *Circ Res*. 2010;106:373–382.
126. Davisson RL. Physiological genomic analysis of the brain renin-angiotensin system. *Am J Physiol*. 2003;285:R498–R511.
127. Xia H, Feng Y, Obr TD, Hickman PJ, Lazartigues E. Angiotensin II type 1 receptor-mediated reduction of angiotensin-converting enzyme 2 activity in the brain impairs baroreflex function in hypertensive mice. *Hypertension*. 2009;53:210–216.
128. Xia H, Lazartigues E. Angiotensin-converting enzyme 2 in brain: properties and future directions. *J Neurochem*. 2008;107:1482–1494.
129. Kisch B. The sarcomeres of the heart. *J Cell Biol*. 1956;2:361–362.
130. Jamieson JD, Palade GD. Specific granules in atrial muscle cells. *J Cell Biol*. 1964;23:151.
131. Henry JP, Gauer OH, Reeves JL. Evidence of the atrial location of receptors influencing urine flow. *Circ Res*. 1956;4:85–90.
132. Sonnenberg H, Cupples WA, de Bold, Veress AT. Intrarenal localization of the natriuretic effect of cardiac atrial extract. *Can J Physiol Pharmacol*. 1982;60(9):1149–1152.
133. Flynn TG, de Bold ML, de Bold AJ. The amino acid sequence of an atrial peptide with potent diuretic and natriuretic properties. *Biochem Biophys Res Commun*. 1983;117(3):859–865.
134. Ruskoaho H. Atrial natriuretic peptide: synthesis, release, and metabolism. *Pharmacol Rev*. 1992;44:479–602.
135. Sudoh T, Kangawa K, Minamino N, Matsuo H. A new natriuretic peptide. *Nature*. 1988;332(6159):78–81.
136. Mukoyama M, Nakao K, Hosoda K, et al. Brain natriuretic peptide as a novel cardiac hormone in humans. Evidence for an exquisite dual natriuretic peptide system, atrial natriuretic peptide and brain natriuretic peptide. *J Clin Invest*. 1991;87(4):1402–1412.
137. Grantham JA, Borgeson DD, Burnett JC. BNP: pathophysiological and potential therapeutic roles in acute congestive heart failure. *Am J Physiol*. 1997;272:R1077–R1083.
138. Wang TJ, Larson MG, Levy D, et al. Plasma natriuretic peptide levels and the risk of cardiovascular events and death. *N Engl J Med*. 2004;350(7):655–663.
139. Sudoh T, Minamino N, Kangawa K. Matsuo. C-type natriuretic peptide (CNP): a new member of natriuretic peptide family identified in porcine brain. *Biochem Biophys Res Commun*. 1990;168(2):863–870.
140. Suga S, Nakao K, Itoh H, et al. Endothelial production of C-type natriuretic peptide and its marked augmentation by transforming growth factor-beta. Possible existence of "vascular natriuretic peptide system". *J Clin Invest*. 1992;90(3):1145–1149.
141. Barr CS, Rhodes P, Struthers AD. C-type natriuretic peptides. *Peptides*. 1996;17(7):1243–1251.
142. Scotland RS, Ahluwalia A, Hobbs AJ. C-type natriuretic peptide in vascular physiology and disease. *Pharmacol Ther*. 2005;105(2):85–93.
143. Koller KJ, Goeddel DV. Molecular biology of the natriuretic peptides and their receptors. *Circulation*. 1992;86:1081–1088.
144. Samson WK. Natriuretic peptides: a family of hormones. *Trends Endo Metabol*. 1992;3(3):86–90.
145. Pandey KN. Biology of the natriuretic peptides and their receptors. *Peptides*. 2005;26(6):901–932.
146. Samson WK, Alexander BD, Skala KD, Huang FL, Fulton RJ. Central peptidergic mechanisms controlling reproductive hormone secretion: novel methodology reveals a role for the natriuretic peptides. *Can J Physiol Pharmacol*. 1992;70(5): 773–778.
147. Samson WK, Alexander BD, Skala KD, Huang FL, Fulton RJ. Ricin-cytotoxin conjugate administration reveals a physiologically relevant role for oxytocin in the control of gonadotropin secretion. *Ann NY Acad Sci*. 1992;652:411–422.
148. Samson WK, Huang Fl, Fulton RJ. C-type natriuretic peptide mediates the hypothalamic actions of the natriuretic peptides to inhibit luteinizing hormone secretion. *Endocrinology*. 1993;132: 504–509.
149. Samson WK. Atrial natriuretic factor inhibits dehydration and hemorrhage-induced vasopressin release. *Neuroendocinology*. 1985;40(3):277–279.
150. Samson WK, Aguila MC, Matinovic J, Antunes-Rodrigues J, Norris M. Hypothalamic action of atrial natriuretic factor to inhibit vasopressin secretion. *Peptides*. 1987;8(3):449–454.
151. Antunes-Rodrigues J, McCann SM, Rogers LC, Samson WK. Atrial natriuretic factor inhibits dehydration- and angiotensin II-induced water intake in the conscious, unrestrained rat. *Proc Natl Acad Sci*. 1985;82:8720–8723.
152. Sonnenberg H, Honrath U, Chong CK, Wilson DR. Atrial natriuretic factor inhibits sodium transport in medullary collecting duct. *Am J Physiol*. 1986;250:F963–F966.
153. Zeidel ML, Seifter JL, Lear S, Brenner BM, Silva P. Atrial peptides inhibit oxygen consumption in kidney medullary collecting duct cells. *Am J Physiol*. 1986;251:F379–F383.
154. Jougasaki M, Wei CM, Aarhus LL, et al. Renal localization and actions of adrenomedullin: a natriuretic peptide. *Am J Physiol*. 1995;268:F657–F663.
155. Blackburn RE, Samson WK, Fulton RJ, Stricker EM, Verbalis JG. Central oxytocin and ANP receptors mediate osmotic inhibition of salt appetite in rats. *Am J Physiol*. 1995;269:R245–R251.
156. Samson WK, Skala KD, Huang FL. CNP-22 stimulates, rather than inhibits, water drinking in the rat: evidence for a unique biological action of the C-type natriuretic peptides. *Brain Res*. 1991;568(1-2):285–288.
157. Shimizu T, Katsuura G, Nakamura M, et al. Effect of intracerebroventricular atrial natriuretic polypeptide on blood pressure and urine production in rats. *Life Sci*. 1986;39: 1263–1270.
158. Shirakami G, Nakao K, Yamada T, et al. Inhibitory effect of brain natriuretic peptide on central angiotensin II-stimulated pressor response in conscious rats. *Neurosci Lett*. 1988;91:77–83.
159. Levin ER, Weber MA, Mills S. Atrial natriuretic factor-induced vasodepression occurs through central nervous system. *Am J Physiol*. 1988;255:H616–H622.
160. Quirion R, Dalpe M, Dam TV. Characterization and distribution of receptors for the atrial natriuretic peptides in mammalian brain. *Proc Natl Acad Sci*. 1986;83:174–178.

161. Konrad EM, Thibault G, Schriffrin EL, Cantin M. Atrial natriuretic factor receptor subtypes in the rat central nervous system. *Hypertension.* 1991;17:1144–1151.
162. Langub MC, Watson RE, Herman JP. Distribution of natriuretic peptide precursor mRNAs in the rat brain. *J Comp Neurol.* 1995;356(2):183–199.
163. Yang R, Jin H, Wyss JM, Chen YF, Oparil S. Salt supplementation does not alter the pressor effect of blocking atrial natriuretic peptide in nucleus tractus solitarii. *Hypertension.* 1992; 20:242–246.
164. Wong M, Samson WK, Dudley CA, Moss RL. Direct neuronal; action of atrial natiuretic factor in the rat brain. *Neuroendocrinology.* 1986;44(1):49–53.
165. Haskins JT, Zingaro GJ, Lappe RW. Rat atriopeptin III alters hypothalamic neuronal activiy. *Neurosci Lett.* 1986;67(3): 279–284.
166. Okuya S, Yamashita H. Effects of atrial natriuretic polypeptide on rat hypothalamic neurones *in vitro. J Physiol.* 1987;389: 717–728.
167. Standaert DG, Cechetto DF, Needleman P, Saper CB. Inhibition of the firing of vasopressin neurons by atriopeptin. *Nature.* 1987;329(6135):151–153.
168. Buranarugsa P, Hubbard JI. Excitatory effects of atrial natriuretic peptide on rat subfornical organ neurons *in vitro. Brain Res Bull.* 1988;20(5):627–631.
169. Hattori Y, Kasai M, Uesugi S, Kawata M, Yamashita H. Atrial natriuretic polypeptide depresses angiotensin II induced excitation of neurons in the rat subfornical organ *in vitro. Brain Res.* 1988;443(1-2):355–359.
170. Ermirio R, Ruggeri P, Cogo CE, Molinari C, Calaresu FR. Neuronal and cardiovascular responses to ANF microinjected into the solitary nucleus. *Am J Physiol.* 1989;256: R577–R582.
171. Currie MG, Geller DM, Cole BR, et al. Bioactive cardiac substances: potent vasorelaxant activity in mammalian atria. *Science.* 1983;221:71–73.
172. Grammer RT, Fukumi H, Inagami T, Misono KS. Rat atrial natriuretic factor. Purification and vasorelaxant activity. *Biochem Biophys Res Commun.* 1983;116(2):696–703.
173. John SW, Krege JH, Oliver PM, et al. Genetic decreases in atrial natriuretic peptide and salt-sensitive hypertension. *Science.* 1995;267:679–681.
174. John SW, Veress AT, Honrath U, et al. Blood pressure and fluid-electrolyte balance in mice with reduced or absent ANP. *Am J Physiol.* 1996;271:R109–R114.
175. Phillips MI. Angiotensin in brain. *Neuroendocrinology.* 1978;25(6): 354–377.
176. Suzuki H, Kondo K, Handa M, Lawabe H, Saruta T. Mechanism of pressor effects of intraventricular injection of angiotensin II in the rat: role of vasopressin and renal nerves. *Clin Sci.* 1982;63(3):275–279.
177. Falcon JC, Phillips MI, Hoffman WE, Brody MJ. Effects of intraventricular angiotensin II mediated by the sympathetic nervous system. *Am J Physiol.* 1978;235:H392–H399.
178. Weekley LB. Angiotensin-II acts centrally to alter renal sympathetic nerve activity and the intrarenal renin-angiotensin system. *Cardiovasc Res.* 1991;25:353–363.
179. Fitzsimons JT, Stricker EM. Sodium appetite and the renin–angiotensin system. *Nat New Biol.* 1977;231(19):58–60.
180. Antunes-Rodrigues J, de Castro M, Elias LLK, Valenca MW, McCann SM. Neuroendocrine control of body fluid metabolism. *Physiol Rev.* 2004;84:169–208.
181. Garcia R, Thibault G, Cantin M, Genest J. Effect of a purified atrial natriuretic factor on rat and rabbit vascular strips and vascular beds. *Am J physiol.* 1984;247:R34–R39.
182. Bolli P, Muller FB, Linder L, Raine AE, et al. The vasodilator potency of atrial natriuretic peptide in man. *Circulation.* 1987;75:221–228.
183. Levin ER, Weber MA, Mills S. Atrial natriuretic factor-induced vasodepression occurs through central nervous system. *Am J Physiol.* 1988;255:H616–H622.
184. Appel RG. Growth-regulatory properties of atrial natriuretic factor. *Am J Physiol.* 1992;262:F911–F918.
185. Booz GW, Baker KN. Molecular signaling mechanisms controlling growth in cardiac fibroblasts. *Cardiovasc Res.* 1995;30:537–543.
186. Aguilera G, Catt KJ. Regulation of aldosterone secretion by the renin-angiotensin system during sodium restriction in rats. *Proc Natl Acad Sci.* 1978;75:4057–4061.
187. Richards AM, McDonald D, Fitzpatrick MA, et al. Atrial natriuretic hormone has biological effects in man at physiological plasma concentrations. *J Clin Endocrinol Metab.* 1988;67: 1134–1139.
188. Inagami T, Takayangi R, Snajdas RM. Co-purification of atrial natriuretic factor receptor and guanylyl cyclase from adrenal cortex. *Methods Enzymol.* 1991;195:404–413.
189. Vesely DL, Chiou S, Douglass MA, et al. Kaliuretic peptide and long acting natriuretic peptide as well as atrial natriuretic factor inhibit aldosterone secretion. *J Endocrinol.* 1995;146:373–380.
190. Huang CL, Lewicki J, Johnson LK, Cogan MG. Renal mechanism of action of atrial natriuretic factor. *J Clin Invest.* 1985;75(2):69–773.
191. Schuster VL, Kokko JP, Jacobson HR. Angiotensin II directly stimulates sodium transport in rabbit proximal convoluted tubules. *J Clin Invest.* 1984;73(2):507–515.
192. Kitamura K, Kangawa K, Kawamoto M, et al. Adrenomedullin: a novel hypotensive peptide isolated from human pheochromocytoma. *Biochem Biophys Res Commun.* 1993;192(2):553–560.
193. Hirata Y, Mitaka C, Sato K, et al. Increased circulating adrenomedullin, a novel vasodilatory peptide, in sepsis. *J Clin Endocrinol Metab.* 1996;81:1449–1453.
194. Nishio K, Akai Y, Murao Y, et al. Increased plasma concentrations of adrenomedullin correlate with relaxation of vascular tone in patients with septic shock. *Crit Care Med.* 1997;25(6): 953–957.
195. Shimosawa T, Ito Y, Ando K, et al. Proadrenomedullin NH(2)-terminal 20 peptide, a new product of the adrenomedullin gene, inhibits norepinephrine overflow from nerve endings. *J Clin Invest.* 1995;96(3):1672–1676.
196. Nakamura K, Toda H, Terasako K, et al. Vasodilative effect of adrenomedullin in isolated arteries of the dog. *Jpn J Pharmacol.* 1995;67(3):259–262.
197. Miura K, Ebara T, Okumura M, et al. Attenuation of adrenomedullin-induced renal vasodilatation by NG-nitro L-arginine but not glibenclamide. *Br J Pharmacol.* 1995;115(6):917–924.
198. Kitamura K, Kangawa K, Ishiyama Y, et al. Identification and hypotensive activity of proadrenomedullin N-terminal 20 peptide (PAMP). *FEBS Lett.* 1994;351(1):35–37.
199. Shimosawa T, Ito Y, Ando K, et al. Proadrenomedullin NH(2)-terminal 20 peptide, a new product of the adrenomedullin gene, inhibits norepinephrine overflow from nerve endings. *J Clin Invest.* 1995;96(3):1672–1676.
200. Gardiner SM, Kemp PA, March JE, Bennett T. Regional hemodynamic effects of human and rats adrenomedullin in conscious rats. *Br J Pharmacol.* 1995;114(3):584–591.
201. Vari RC, Adkins SD, Samson WK. Renal effects of adrenomedullin in the rat. *Exp Biol Med.* 1996;211(2):178–183.
202. Yamaguchi T, Baba K, Doi Y, Yano K. Effect of adrenomedullin on aldosterone secretion by dispersed rat adrenal zona glomerulosa cells. *Life Sci.* 1995;56(6):379–387.

203. Samson WK, Murphy T, Schell DA. A novel vasoactive peptide, adrenomedullin, inhibits pituitary adrenocorticotropin release. *Endocrinology.* 1995;136:2349–2352.
204. Murphy TC, Samson WK. The novel vasoactive hormone, adrenomedullin, inhibits water drinking in the rat. *Endocrinology.* 1995;136:2459–2463.
205. Taylor MM, Samson WK. Ribozyme compromise of adrenomedullin mRNA reveals a physiological role in the regulation of water intake. *Am J Physiol.* 2002;282:R1739–R1745.
206. Samson WK, Bode AM, Murphy TC, Resch ZT. Antisense oligonucleotide treatment reveals a physiologically relevant role for adrenomedullin gene products in sodium intake. *Brain Res.* 1999;818(1):164–167.
207. Serrano J, Uttenhal LO, Martinez A, et al. Distribution of adrenomedullin-like immunoreactivity in the rat central nervous system by light and electron microscopy. *Brain Res.* 2000; 853(2):245–268.
208. Shan J, Krukoff TL. Distribution of preproadrenomedullin mRNA in the rat central nervous system and its modulation by physiological stressors. *J Comp Neurol.* 2001;432(1):88–100.
209. Ueta Y, Kitamura K, Isse T, et al. Adrenomedullin-immunoreactive neurons in the paraventricular and supraoptic nuclei of the rat. *Neurosci Lett.* 1995;202(1-2):37–40.
210. Taylor MM, Baker JR, Samson WK. Brain-derived adrenomedullin controls blood volume through the regulation of arginine vasopressin production and release. *Am J Physiol.* 2005;288: R1203–R1210.
211. Takahashi H, Watanabe TX, Nishimura M, et al. Centrally induced vasopressor and sympathetic responses to a novel endogenous peptide, adrenomedullin, in anesthetized rats. *Am J Hypertension.* 1994;7(5):478–482.
212. Saita M, Shimokawa A, Kunitake T, et al. Central actions of adrenomedullin on cardiovascular parameters and sympathetic outflow in conscious rats. *Am J Physiol.* 1998;274:R979–R984.
213. Samson WK, Murphy TC, Resch ZT. Central mechanisms for the hypertensive effects of preproadrenomedullin-derived peptides in conscious rats. *Am J Physiol.* 1998;274:R1505–R1509.
214. Allen MA, Ferguson AV. *In vitro* recordings from area postrema neurons demonstrate responsiveness to adrenomedullin. *Am J Physiol.* 1996;270:R920–R925.
215. Allen MA, Smith PA, Ferguson AV. Adrenomedullin microinjection into the area postrema increases blood pressure. *Am J Physiol.* 1997;272:R1698–R1703.
216. Follwell MJ, Ferguson AV. Adrenomedullin influences magnocellular and parvocellular neurons of paraventricular nucleus via separate mechanisms. *Am J Physiol.* 2002;283:R1293–R1302.
217. Taylor MM, Samson WK. A possible mechanism for the action of adrenomedullin in brain to stimulate stress hormone secretion. *Endocrinology.* 2004;145(11):4890–4896.
218. Matsumura K, Abe I, Tsuchihashi T, Fujishima M. Central adrenomedullin augments the baroreceptor reflex in conscious rabbits. *Hypertension.* 1999;33:992–997.
219. Taylor MM, Keown CA, Samson WK. Involvement of the central adrenomedullin peptides in the baroreflex. *Regul Pept.* 2003;112(1-3):87–93.
220. Dunworth WP, Caron KM. G protein-coupled receptors as potential drug targets for lymphangiogenesis and lymphatic vascular development. *Arterioscler Thromb Vasc Biol.* 2009; 29:650–656.
221. Young A. Receptor pharmacology. *Adv Pharmacol.* 2005;52: 47–65.
222. Hay DL, Christopoulos G, Christopoulos A, Poyner DR, Sexton PM. Pharmacological discrimination of calcitonin receptor: receptor activity-modifying protein complexes. *Mol Pharmacol.* 2005;67:1655–1665.
223. Brain SD, Grant AD. Vascular actions of calcitonin gene-related peptide and adrenomedullin. *Physiol Rev.* 2004;84:903–934.
224. Takei Y, Hyodo S, Katafuchi T, Minamino N. Novel fish-derived adrenomedullin in mammals: structure and possible function. *Peptides.* 2004;25(10):1643–1656.
225. Taylor MM, Bagley SL, Samson WK. Intermedin/adrenomedullin-2 inhibits growth hormone release from cultured, primary anterior pituitary cells. *Endocrinology.* 2006; 147:859–864.
226. Taylor MM, Bagley SL, Samson WK. Intermedin/adrenomedullin-2 acts within central nervous system to elevate blood pressure and inhibit food and water intake. *Am J Physiol.* 2005;288:R919–R927.
227. Dackor RT, Frit-Six K, Dunworth WP, et al. Hydrops fetalis, cardiovascular defects, and embryonic lethality in mice lacking the *calcitonin receptor-like receptor* gene. *Mol Cell Biol.* 2006;26: 2511–2518.
228. Caron KM, Smithies O. Extreme hydrops fetalis and cardiovascular abnormalities in mice lacking a functional adrenomedullin gene. *Proc Natl Acad Sci.* 2001;98:615–619.
229. Li M, Yee D, Magnuson TR, Smithies O, Caron KM. Reduced maternal expression of adrenomedullin disrupts fertility, placentation, and fetal growth in mice. *J Clin Invest.* 2006;116(10):2653–2662.
230. Banks WA, Kastin AJ, Huang W, Jaspan JB, Maness LM. Leptin enters the brain by a saturable system independent of insulin. *Peptides.* 1996;17(2):305–311.
231. Banks WA, Kastin AJ, Pan W. Uptake and degradation of blood-borne insulin by the olfactory bulb. *Peptides.* 1999;20(3):373–378.
232. Cone RD. Anatomy and regulation of the central melanocortin system. *Nature Neuroscience.* 2005;8(5):571–578.
233. Jacobowitz DM, O'Donohue TL. Alpha-melanocyte stimulating hormone: Immunohistochemical identification and mapping in neurons of rat brain. *Proc Natl Acad Sci.* 1978;75(12):6300–6304.
234. Watson SJ, Akil H, Richard CW, Barchas JD. Evidence for two separate opiate peptide neuronal systems. *Nature.* 1978; 275(5677):226–228.
235. Joseph SA, Michael GJ. Efferent ACTH-IR opiocortin projections from nucleus tractus solitarius: a hypothalamic deafferentation study. *Peptides.* 1988;9:193–201.
236. Pilcher WH, Joseph SA. Differential sensitivity of hypothalamic and medullary opiocortin and tyrosine hydroxylase neurons to the neurotoxic effects of monosodium glutamate (MSG). *Peptides.* 1986;7(5):783–789.
237. Konda Y, Grantz I, DelValle J, et al. Interaction of dual intracellular signalling pathways activated by the melanocortin-3 receptor. *J Biol Chem.* 1994;269:13162–13166.
238. Huszar D, Lynch CA, Fairchild-Huntress V, et al. Targeted disruption of the melanocortin-4 receptor results in obesity in mice. *Cell.* 1997;88(1):131–141.
239. Fan W, Boston BA, Kesterson RA, Hruby VJ, Cone RD. Role of melanocortinergic neurons in feeding and the agouti obesity syndrome. *Nature.* 1997;385(6612):165–168.
240. Loos RJF, Lindgren CM, Li S, et al. Common variants near MC4R are associated with fat mass weight and risk of obesity. *Nat Genetics.* 2008;40(6):768–775.
241. Kuo JJ, da Silva AA, Hall JE. Hypothalamic melanocortin receptors and chronic regulation of arterial pressure and renal function. *Hypertension.* 2003;41(3):768–774.
242. Matsumura K, Tsuchihashi T, Abe I, Mitsuo I. Central alpha-melanocoyte-stimulating hormone acts at melanocortin-4 receptor to activate sympathetic nervous system in conscious rabbits. *Brain Res.* 2002;948(1-2):145–148.
243. Tallam LS, Stec DE, Willis MA, da Silva AA, Hall JE. Melanocortin-4 receptor-deficient mice are not hypertensive or

salt-sensitive despite obesity, hyperinsulinermia, and hyperleptinemia. *Hypertension.* 2005;46(2):326–332.
244. Dunbar JC, Lu H. Leptin-induced increase in sympathetic nervous and cardiovascular tone is mediated by proopiomelanocortin (POMC) products. *Brain Res Bull.* 1999;50(3): 215–221.
245. da Silva AA. Kuo JJ, Hall JE. Role of hypothalamic melanocortin 3/4 receptors in mediating chronic cardiovascular, renal, and metabolic actions of leptin. *Hypertension.* 2004; 43(6):1312–1317.
246. Ward KR, Bardgett JF, Wolfgang L, Stocker SD. Sympathetic response to insulin is mediated by melanocortin 3/4 receptors in the hypothalamic paraventricular nucleus. *Hypertension.* 2011;57(3):435–441.
247. Shintani M, Ogawa Y, Ebihara K. Ghrelin, an endogenous growth hormone secretagogue, is a novel orexigenic peptide that antagonizes leptin action through the activation of hypothalamic neuropeptide Y/Y1 receptor pathway. *Diabetes.* 2001;50(2):227–232.
248. Oh-I S, Shimizu H, Satoh T, et al. Identification of nesfatin-1 as a satiety molecule in the hypothalamus. *Nature.* 2006;443(7112): 709–712.
249. Yosten GLC, Samson WK. Nesfatin-1 exerts cardiovascular actions in brain: possible interaction with the central melanocortin system. *Am J Physiol.* 2009;297:R330–R336.
250. Yosten GLC, Samson WK. The melanocortins, not oxytocin, mediate the anorexigenic and antidipsogenic effects of neuronostatin. *Peptides.* 2010;31(9):1711–1714.
251. Olson BR, Drutarosky MD, Chow MS, Hruby VJ, Stricker EM, Verbalis JG. Oxytocin and an oxytocin agonist administered centrally decrease food intake in rats. *Peptides.* 1991;12(1): 113–118.
252. Arletti R, Benelli A, Bertolini A. Influence of oxytocin on feeding behavior in the rat. *Peptides.* 1989;10(1):89–93.
253. Michelini LC, Marcelo MC, Amico J, Morris M. Oxytocinergic regulation of cardiovascular function: studies in oxytocin-deficient mice. *Am J Physiol.* 2003;284(6):H2269–H2276.
254. Caquineau C, Leng G, Guan XMM, et al. Effects of alpha-melanocyte-stimulating hormone on magnocellular oxytocin neurons and their activation at intromission in male rats. *J Neuroendocrinol.* 2006;18(9):685–691.
255. Sabatier N, Caquineau C, Dayanithi G, et al. Alpha-melanocyte-stimulating hormone stimulates oxytocin release from the dendrites of hypothalamic neurons while inhibiting oxytocin release from their terminals in the neurohypophysis. *J Neurosci.* 2003;23(32):10351–10358.
256. Blevins JE, Schwartz MW, Baskin DG. Evidence that paraventricular nucleus oxytocin neurons link hypothalamic leptin action to caudal brain stem nuclei controlling meal size. *Am J Physiol.* 2004;287(1):R87–R96.
257. Yosten GLC, Samson WK. The anorexigenic and hypertensive effects of nesfatin-1 are reversed by pretreatment with an oxytocin receptor antagonist. *Am J Physiol.* 2010;298: R1642–R1647.
258. Nicholson JR, Peter JC, Lecourt AC, Barde YA, Hofbauer KG. Melanocortin-4 receptor activation stimulates hypothalamic brain-derived neurotrophic factor to regulate food intake, body temperature and cardiovascular function. *J Neuroendocrinol.* 2007;19(12):974–982.
259. Bariohay B, Roux J, Tardivel C, et al. Brain-derived neurotrophic factor/tropomyosin-related kinase receptor type B signaling is a downstream effector of the brainstem melanocortin system in food intake control. *Endocrinology.* 2009;150(6):2646–2653.
260. Goebel M, Stengl A, Wang L, Tache Y. Restraint stress activates nesfatin-1-immunoreactive brain nuclei in rats. *Brain Res.* 2009;1300:114–124.
261. Ben-Shlomo I, Rauch R, Avsian-Kretchmer O, Hsueh AJW. Matching receptome genes with their ligands for surveying paracrine/autocrine signaling systems. *Mol Endocrinol.* 2007;21:2009–2014.
262. Samson WK, Zhang JV, Avsian-Kretchmer O, et al. Neuronostatin encoded by the somatostatin gene regulates neuronal, cardiovascular, and metabolic functions. *J Biol Chem.* 2008;283: 31949–31959.
263. Yosten GLC, Pate AT, Samson WK. Neuronostatin acts in brain to biphasically increase mean arterial pressure through sympatho-activation followed by vasopressin secretion: role of melanocortins. *Am J Physiol.* 2011. 10.1152/ajpregu.00849.2010.

PART III

HORMONES, BRAIN FUNCTION AND BEHAVIOR

CHAPTER

14

Neuroendocrine Regulation of Food Intake

Alan G. Watts

Department of Biological Sciences, USC College of Letters, Arts, and Sciences, University of Southern California, Los Angeles, CA, USA

OUTLINE

Summary

Maintaining adequate energy stores is essential for survival, and animals have evolved elaborate, powerful, and adaptable mechanisms for this purpose. Appropriate feeding behavior is the only way to obtain new energy supplies, and the brain organizes and controls the actions that constitute feeding. To do this it integrates a wealth of information, including sensory signals from inside and outside the body, arousal state and circadian rhythm status, and hormones. The brain is a target for hormones that reflect adiposity stores (primarily leptin and insulin), along with circulating peptides that relay gastrointestinal status (particularly ghrelin, CCK, and PYY). In addition to controlling behavior, the brain coordinates pre- and post-ingestive processes with ongoing behavior using endocrine and autonomic motor actions. These processes include the release of hormones from the pars distalis of the pituitary

Handbook of Neuroendocrinology, DOI: 10.1016/B978-0-12-375097-6.10014-9

gland that contribute to the endocrine response to compromised energy homeostasis. This chapter describes the neuroendocrinology of these processes. It emphasizes the importance of understanding the detailed structure of feeding behavior, and describes the contributions that sophisticated molecular genetics are making to our understanding of these essential functions.

INTRODUCTION

Ever since the brain was recognized as an important regulator of endocrine function, neuroendocrinology has traditionally described bidirectional interactions[1] – that is, how the brain regulates the pituitary gland and thereby the release of hormones from a wide range of endocrine organs; and how hormones interact with the brain to regulate cognitive, autonomic, and behavioral function. To do this, neurooendocrine motor neurons in the hypothalamus control hormone release from the pituitary gland, whereas many hormones interact with neurons in the brain in a way that generates adaptive responses.

This chapter will explore the regulation of feeding behavior and energy homeostasis from this classic neuroendocrine perspective. We will begin with a description of the temporal organization of all motivated behaviors, the importance of understanding meal structures, and the different types of feeding behavior. This leads into a discussion of the neural components responsible for integrating brain function so that it can appropriately organize the behavioral, autonomic and neuroendocrine motor aspects of energy metabolism. The resulting schema provides the framework for considering the principal components responsible for the neuroendocrine regulation of feeding behavior and metabolism in more detail.

We will then examine how hormones that are critical for regulating whole-body energy metabolism – particularly leptin and insulin, together with a set of gastrointestinal hormones – interact with the brain to control feeding behavior and metabolism. Next, we will consider how various metabolic "sensory" signals that arise from within the body interact with the neural mechanisms which control those anterior pituitary hormones that help regulate metabolism – primarily adrenocorticotropin (ACTH), growth hormone and thyrotropin (TSH).

Finally, by delineating the neural networks crucial for the neuroendocrine regulation of energy metabolism, we will see that they offer a way to examine their wider interactions with regulatory systems responsible for other behaviors. In particular, there are major interactions between metabolism and reproductive function; the two systems that provide for the survival of the individual and the species.[2,3] We will conclude by using the network models developed below to consider briefly how an integration between the neuroendocrine aspects of metabolism and reproductive function might be enabled.

WHAT IS FEEDING BEHAVIOR AND HOW IS IT RELATED TO ENERGY METABOLISM?

To understand how neuroendocrine processes regulate feeding behavior and the associated endocrine and autonomic events related to energy metabolism, we first need to clearly define the behavioral actions that constitute feeding. While a great deal of work has relied upon the amount of food eaten during a defined period (i.e., intake) as the experimental endpoint, this measure is not behavior and provides no insight into the complex mechanics or various motor sequences that constitute the actual behavior, or how experimental manipulations might affect these different aspects of feeding. To do this, more in-depth analyses are required to reveal the various complex components that make up feeding episodes. These analyses provide a much more detailed framework for exploring how hormones impact the underlying neural mechanisms of these various components.

A Common Behavioral Sequence for All Motivated Behaviors

A simple and useful scheme for the temporal organization of all motivated behaviors was first outlined almost 100 years ago.[4] It has subsequently been adopted and elaborated for investigating the neural bases of motivated behaviors, including feeding (see, for example, Grill and Kaplan[5]; Swanson and Mogenson[6]). This scheme has four phases that must be expressed in a specific temporal sequence for the successful execution of the complete behavior.

In brief, feeding behavior is initiated by the appropriate integration within the brain of the four input categories (Fig. 14.1). The outcome of this integration leads to a procurement (or appetitive) phase where food is actively sought. The procurement phase primarily involves foraging behaviors that are individualized for particular species, can be quite complex, and are controlled by the forebrain – i.e., they are absent in chronic decerebrate animals where all connections between the forebrain and hindbrain are severed.[7] When food is located, the procurement phase is followed by a consummatory phase where direct interaction with food occurs (licking, biting, chewing, swallowing, etc.). As the consummatory

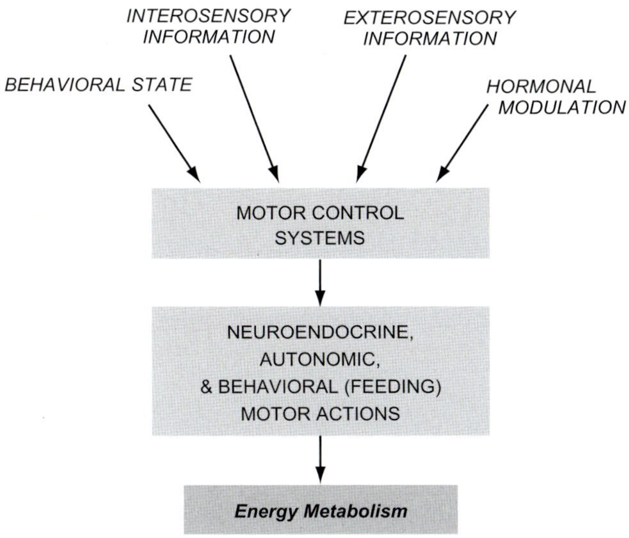

FIGURE 14.1 The four classes of inputs that are integrated by the motor control systems in the brain to control neuroendocrine, autonomic and behavioral actions for energy metabolism.

phase progresses and food is ingested, the nature of information that is directed to the brain from both the internal and external environments changes the performance of the ongoing behavior in a way that leads to its termination, often by satiety signals but possibly also by stimuli from the environment (e.g., the presence of a predator).

The importance of parsing feeding behavior in this manner means that there is a way to reveal the neural substrates underlying various experimental manipulations and determining how hormones or neuropeptide signals increase or decrease food intake (see "Metabolic Hormones and their Impact on the Brain," page 339). To provide a foundation for understanding how the brain shapes the complete feeding behavioral repertoire of an animal, and the influence of hormones, we first need to consider the different situations in which feeding behavior can be expressed.

BOX 14.1

HOW DO WE KNOW HOW ANIMALS EAT? THE STRUCTURED MEAL AS THE UNIT OF FEEDING BEHAVIOR

Although meals are the principle units of feeding behavior, the ability to look in detail at how they are constructed and how they relate to the signals known to influence food intake has proved difficult.[1–3] Two technical challenges were at the heart of the problem. The first was clarifying exactly which behavioral actions and movements are involved. This arose because of the difficulty of visualizing and analyzing the rapid and precise snout, facial, oral, and lingual movements required for feeding. The second was determining how these movements are temporally sequenced into the ingestive episodes we call meals. The main problem here is the large amounts of data that are required to obtain a clear picture of ingestive behavior across meaningful timeframes. The advent of video recorders (see, for example, Zeigler et al.[4]) and desktop computers in the early 1980s greatly facilitated collection and analysis of these data. Since then, our understanding of the movements that make up feeding behavior and how they are expressed has improved dramatically.[1]

Detailed analysis of meal structure in rats reveals that each meal is comprised of a number of small bouts that occur in close temporal proximity – generally within tens of seconds. Each meal is separated by a longer inter-meal interval, which is usually at least 5 minutes, but can be considerably longer. Over the years, a number of approaches have been used to gather the data needed to address this problem. Perhaps the most important has been the lickometers which allows investigators to measure and analyze the number and timing of the licks that make up a liquid meal.[1]

More recently, the development of instrumentation that facililates the detailed analysis of solid food meals together with the associated drinking (prandial drinking[5]) has added an extra dimension to our understanding of what comprises the behavioral mechanics of feeding.[6] With this type of approach it is possible to obtain detailed meal structures in terms of food *and* water intake. An example of such an analysis is shown in the figure. This approach generates measures that can then be used to compare the effects of an experimental intervention on different meal parameters. For example, does a treatment change meal size, meal frequency, or meal duration? Because each of these parameters is regulated by different mechanisms,[3] identifying which particular aspect of feeding behavior is modified by a hormone – or indeed by any factor – provides important clues about underlying mechanisms of action. For example, leptin decreases feeding primarily through direct actions on the brain (see Box 14.2). But simply measuring food intake gives no indication of *how* leptin suppresses intake. It could be achieved by eating the same number of smaller meals or fewer meals of the same size. (It turns out that leptin reduces meal size but not frequency or duration.[7])

(Continued)

BOX 14.1 *(cont'd)*

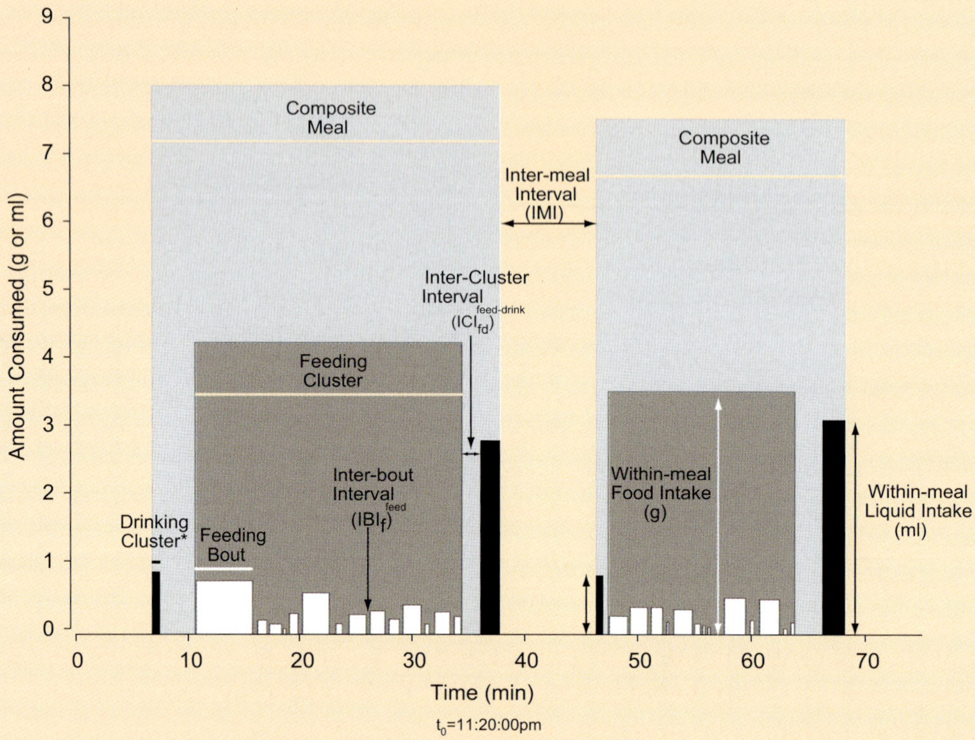

Detailed meal pattern analysis reveals how eating and drinking events are structured into meals. The figure shows the timing and amount of water and food consumed by a single rat during a 70-minute period in the middle of the dark phase. Black vertical bars show drinking clusters; white, feeding bouts; dark gray, feeding clusters; and light gray, composite meals. Individual bar widths and horizontal lines (with and without arrows) represent duration. Individual bar heights and vertical lines represent within-meal food (white) or liquid (black) intake in g or ml. Eating and drinking bouts are segmented into composite meals using the drinking-explicit method of Zorilla and colleagues.[6] *Figure adapted from Boyle* et al.[8]

References

1. Smith JC. Microstructure of the rat's intake of food, sucrose and saccharin in 24-hour tests. *Neurosci Biobehav Rev.* 2000;24:199–212.
2. Geary N. A new way of looking at eating. *Am J Physiol Regul Integr Comp Physiol.* 2005;288:R1444–R1446.
3. Woods SC. The control of food intake: behavioral versus molecular perspectives. *Cell Metab.* 2009;9:489–498.
4. Zeigler HP, Jacquin MF, Miller MG. Trigeminal sensorimotor mechanisms and ingestive behavior. *Neurosci Biobehav Rev.* 1984;8:415–423.
5. Kissileff HR. Food-associated drinking in the rat. *J Comp Physiol Psychol.* 1969;67:284–300.
6. Zorrilla EP, Inoue K, Fekete EM, et al. Measuring meals: structure of prandial food and water intake of rats. *Am J Physiol Regul Integr Comp Physiol.* 2005; 288:R1450–R1467.
7. Flynn MC, Scott TR, Pritchard TC, et al. Mode of action of OB protein (leptin) on feeding. *Am J Physiol.* 1998;275: R174–R179.
8. Boyle CN, Lorenzen SM, Compton D, et al. Dehydration-anorexia is a consequence of a reduction in meal size in rats. Physiol Behav. (under review).

Types of Feeding Behavior

Although feeding behavior is made up of a series of structured meals, it is not a unitary construct that is always expressed with the same organization. Instead, feeding is a highly complex set of behaviors that occur in many different situations. To help understand how the brain interacts with metabolic signals and hormones to control feeding behavior, it is useful to divide the different forms of feeding into three basic categories.

Spontaneous feeding usually occurs at consistent times of the day if food is readily available.[8] The timing and frequency of meals vary across species.

Spontaneous feeding is a proactive (anticipatory) event, in that it usually occurs in the absence of a significant energy deficit and the associated signals (particularly hypoglycemia and hypoleptinemia). Spontaneous feeding anticipates and prevents the deficit. Circadian timing and arousal state play significant roles in initiating meals in most mammals, while a host of complex social and sensory factors can modify meal size in humans.[9]

Deficit-induced (or homeostatic) feeding is reactive, and occurs in response to negative energy balance. The timing and frequency of meals depends on the availability of food. Interosensory signals play the primary role in initiating meals by activating central networks that generate the feeling of hunger and stimulate feeding.

Opportunistic feeding occurs in response to the immediate availability of, or the memory of a previous encounter with, highly favored food items. This type of feeding relies on processing information about previous feeding encounters that are encoded within the neural networks responsible for learning, memory, and reward assignment. This information leads to the activation of motor control systems that initiate either foraging or immediate ingestion. The timing and frequency of meals depends on the availability of food.

The great variety of circumstances and signals that can stimulate and regulate feeding behavior, as well as coordinate it with appropriate changes in metabolic processes are consistent with the existence of many and complex neural systems. These are discussed in the following section.

NETWORK MODELS FOR THE INTEGRATION OF THE NEUROENDOCRINE, AUTONOMIC, AND BEHAVIORAL MOTOR COMPONENTS

Network Inputs

The basic information flow through the networks that control feeding and energy metabolism is shown in Fig. 14.1. A wide variety of inputs bring four categories of information to the control systems in the brain responsible for the neuroendocrine, autonomic and behavioral motor actions that impact energy metabolism. Two sets of inputs bring sensory information from the animal's external or internal environment. This includes exterosensory modalities (taste, smell, etc.), as well as interosensory information about gastrointestinal state, and circulating glucose and other metabolic fuels, etc.

Fig. 14.2 describes these sensory inputs in more detail, along with their interactions with the specific categories of motor control systems in the brain. Exterosensory sensory modalities use "labeled line" inputs to bring information from sensory receptors to the brain. At this point information is processed by primary, secondary and polymodal association areas, and is used to regulate all behavior, including feeding. Interosensory information can be usefully categorized as neural influences, hormonal modulation, and the more or less direct inputs from circulating metabolic fuels such as glucose[10] and free fatty acids.[11] Cellular levels of ATP, ADP and AMP can also regulate feeding and energy metabolism through the changing activities of "fuel gauge" enzymes such as AMP kinase. These are found in specific neurons in the hypothalamus and hindbrain that link into motor control networks.[12–15]

Another category of inputs provides information about the animal's behavioral state: arousal, vigilance and attention, and circadian timing (Fig. 14.1). Hormonal modulation provides the fourth set of inputs (Figs 14.1, 14.2). This category includes hormones whose actions on the brain have often been thought of in terms of simple negative feedback signals (e.g., glucocorticoids), but others inform the brain about peripheral energy stores (insulin and leptin) and gastrointestinal status – for example, ghrelin and cholecystokinin (CCK). More recently, a great deal of attention has been focused on the effects of metabolism-related hormones that are not simply reporter signals which inform the brain about the status of peripheral energetics. For example, the notion that gonadal steroids play key organizational roles during neural development has been recognized for over half a century (see Gorski[16] and Harris[17]). But the fact that metabolic hormones can also influence the development of key neural networks is much more recent.[18]

Motor Control

Fig. 14.3 expands the schemas of Figs 14.1 and 14.2 to reveal more detailed relationships between the various brain components that enable the integrated neuroendocrine, autonomic and behavioral motor actions that control energy metabolism. However, before we discuss which networks in the brain control motor function, it is worth remembering that a motor neuron was originally defined as one that controls the activity of an effector located outside of the brain – i.e., striate (or somatic) muscle, cardiac muscle, or smooth muscle; a gland; or some other effector. Therefore, if we consider that the brain controls three sets of "motor" or effector systems – the somatic or striate musculature (for behavior), and the autonomic and neuroendocrine systems – then each of these three systems must have accompanying sets of motor neurons: somatic, autonomic and neuroendocrine motor neurons. Motor

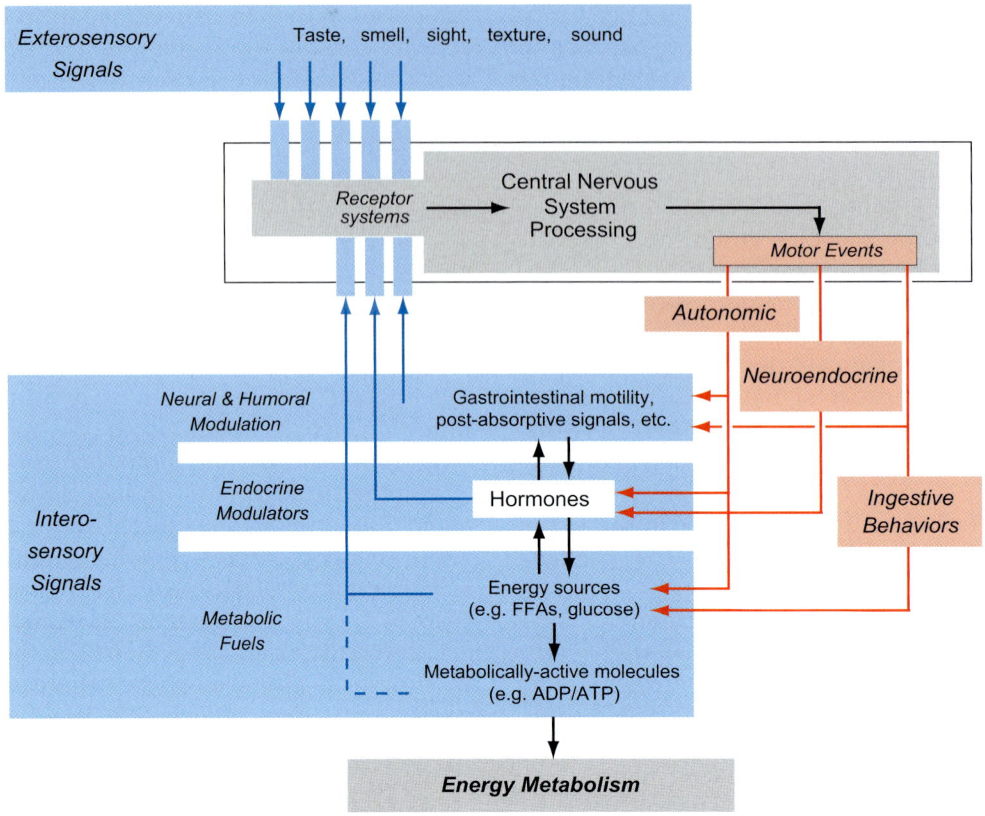

FIGURE 14.2 A schematic representation of the sensory signals (blue) that are integrated by the brain (gray) regulate the motor events (red) that control energy balance. Exterosensory inputs are the classic sensory modalities that each have their own "labeled line" access into the brain. Interosensory signals are categorized as non-hormonal feedback signals, hormones (metabolic modulators) and metabolic fuels. Autonomic motor actions regulate gastrointestinal, adrenal medullary and pancreatic functions; neuroendocrine motor output controls glucocorticoid, thyroid and growth hormones; while behavioral motor actions mediate eating, drinking and specialized appetites. ADP, adenosine diphosphate; ATP, adenosine triphosphate; FFA, free fatty acid. *Adapted from Watts and Swanson (2002).*[2]

control systems are organized so that each of these motor neuron systems is engaged in an appropriate and temporally coordinated manner for a particular behavior. In this way, feeding behavior not only requires the control of somatic motor neurons for the overt behavior; it also requires coordinated regulation of autonomic and neuroendocrine motor neurons.

The Organization of Motor Control Networks

At a first level of approximation, the motor control systems responsible for all motivated behaviors operate at three separate levels (Fig. 14.3): hypothalamic controllers, action selection and action execution.

HYPOTHALAMIC CONTROLLERS

The first level is a series of motor controllers in the hypothalamus, whose core is the behavior control column (28). This consists of a series of nuclei in the medial zone of the hypothalamus, each of which contains sets of drive networks whose output is integrated to set up and coordinate the specific motor events of a particular behavior. The role of the hypothalamus at this level is described in more detail in "The Hypothalamus and the Motor Control of Feeding Behavior and Metabolism" section, below.

ACTION SELECTION

A variety of networks receive direct inputs from the hypothalamic controllers to organize the planning, selection and maintenance of motor actions sequences appropriate for particular behaviors. They control diverse motor actions that are expressed by many types of behaviors to allow the animal to interact with different goal objects. These movements range from precise manipulation of goal objects by the forelimbs, to types of sequentially organized movements such as grooming (see Watts and Swanson[2]). Regions important for action selection are located in the prefrontal, motor and other parts of the cortex (including the hippocampus), the basal ganglia, parts of the lateral hypothalamic area (LHA) and subthalamic locomotor region, the "mesencephalic locomotor region," cerebellum, and hindbrain. Although these networks are topographically organized with regard

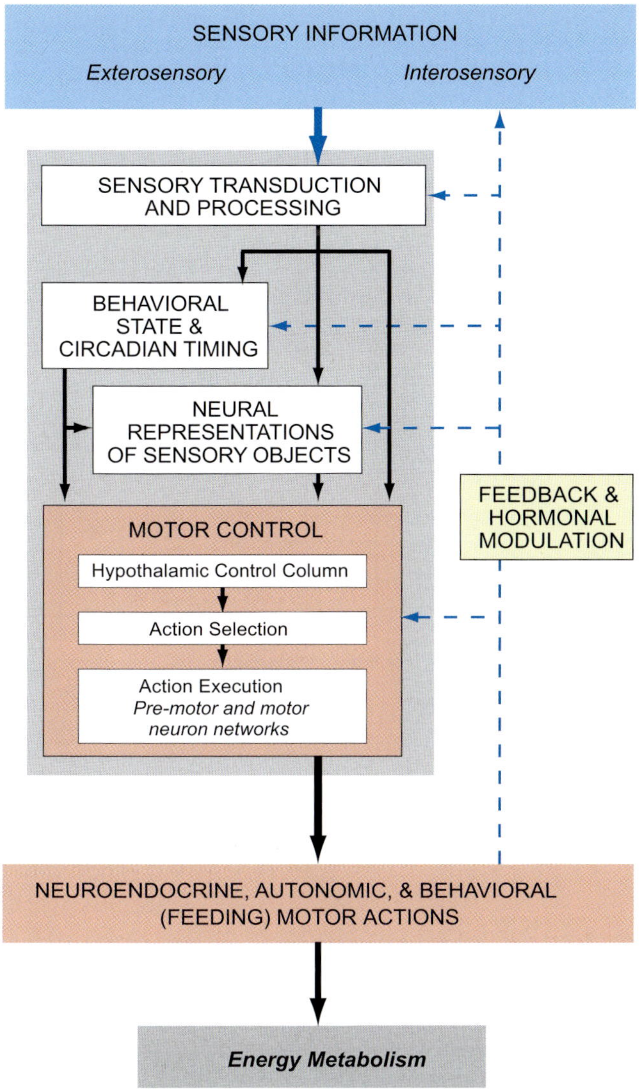

FIGURE 14.3 A hypothetical schema showing the interactions of the three sets of neural networks (white boxes) responsible for controlling all the motor actions (red boxes) – behavioral, neuroendocrine and autonomic – required for feeding and energy metabolism. These networks deal with the transduction and processing of sensory information; behavioral state, which involves the control of arousal state, attention and vigilance, and circadian timing; the neural representation of sensory objects, which comprises networks responsible for reward/punishment, learning, memory consolidation and retrieval; and the egocentric and allocentric spatial representation that are important for spatial awareness and navigation. Neurons in each of these networks project to the regions that control motor actions. Metabolic hormones and neural feedback signals from the gastrointestinal tract and other organs (yellow box, blue dashed lines) act on all three of these networks to modulate their function. *Adapted from Watts and Swanson (2002).*[2]

to the mapping of motor pattern generation and the contractions of specific muscles, they are not organized to control specific behaviors as are the hypothalamic controllers.

ACTION EXECUTION

Direct control of motor effector systems (striate and smooth muscle, the pituitary gland, adrenal medulla, etc.) is accomplished by pre-motor and motor neuron networks. α-Motor neurons in the ventral horn of the spinal cord and the corresponding cranial nerve nuclei control the entire striate musculature, and hence the expression of all behavior. Neuroendocrine motor neurons are primarily located in the periventricular hypothalamus, while the autonomic postganglionic neurons that directly control smooth muscle, adrenal medulla and pancreas are found in sets of autonomic ganglia. Motor neurons are innervated by sets of pre-motor neurons that enable oscillatory and more complex rhythmic firing patterns. Simple rhythmic motor patterns develop from an interaction between motor neurons, oscillatory rhythm generators, and pre-motor central pattern generators located somewhat more distally in the spinal cord, hindbrain, and hypothalamus (see Watts and Swanson[2]).

Information Flow into the Motor Control Networks

Three major systems comprise a brain network that informs the motor control networks (Fig. 14.3). We have already discussed sensory processing and behavioral state control, but there is a third system, which we have previously designated the "neural representation of sensory objects",[2,3] that is essential for the full expression of feeding behaviors (Fig. 14.3). This highly complex system interprets, prioritizes, stores and retrieves information collected by the sensory systems. It includes learning and memory mechanisms in the telencephalon and cerebellum; reward–aversion systems in the midbrain ventral tegmentum, parts of the basal forebrain (particularly the nucleus accumbens), the amygdala, and parts of the cortex, particularly prefrontal regions; and systems in the hippocampus, parts of the parietal cortex, and other brain regions containing the place- and head-direction neurons responsible for allocentric and egocentric spatial representation important for spatial navigation.[2] Fig. 14.3 has four other important features.

1. Motor control networks receive sensory information either directly with little intervening processing (i.e., in what might be called a reflex manner), or more indirectly after substantial integration with other processes.
2. Behavioral state and circadian timing can have major influences on the expression of the key. neuroendocrine, autonomic and behavioral events for energy metabolism.[19]
3. The important interactions between these three systems are mediated by complex and wide-ranging neural connections.[2,10,20–24]

4. Hormones can influence neural function at all levels of the control network, including the way that the brain stores and processes representations of objects in an animal's sensory world (e.g., glucocorticoid actions in the hippocampus, leptin's actions on those ventral tegmental dopaminergic neurons implicated in reward assignment). We will develop these themes in more detail later in the chapter.

The Hypothalamus and the Motor Control of Feeding Behavior and Metabolism

The importance of the hypothalamus for all motivated behaviors is encapsulated in a model proposed by Swanson[20] that derives from our increasingly detailed neuroanatomical knowledge of the connections exhibited by the various nuclei in the medial zone of the hypothalamus. The model posits the existence of a behavior control column in the medial hypothalamus, within which specific hypothalamic nuclei are associated with particular motivated behaviors. Its rostral segment contains sets of neurons that organize the full spectrum of motor events – autonomic, neuroendocrine and behavioral – associated with social and ingestive behaviors, whereas neurons in the caudal behavior control column are important for organizing exploratory and foraging behavior. The functional aspects of this model[2] derive from principles originally proposed by Elliot Stellar in the 1950s,[25] and posits the existence of distinct networks that either stimulate or inhibit feeding behaviors.[26]

Many studies have identified the paraventricular nucleus of the hypothalamus (PVH), and possibly parts of the LHA, as being key controllers for ingestive behaviors.[20] For this to occur, those PVH and LHA neurons that control ingestive behaviors must receive a variety of inputs to regulate their actions. These include exterosensory, viscerosensory, humerosensory, circadian timing information, and cognitive influences (Figs 14.1–14.4), all of which are brought to the PVH and LHA by the many and varied sets of afferent projections from the hindbrain, telencephalon and other parts of the hypothalamus.

The output of the stimulatory and inhibitory control networks in the hypothalamus must at some point be integrated in a way that generates the most appropriate feeding behavior for the ongoing situation. Mechanistic details of the integrative process are currently unclear, but, given the complex ascending and descending outputs from PVH and LHA neurons, it seems reasonable to assume that integration occurs both inside and outside of the hypothalamus, including locations in the telencephalon, midbrain and hindbrain. The outcome of this integration then influences the motor action selection circuits, so that the appropriate pre-motor and motor neurons are engaged to enable particular neuroendocrine, autonomic and behavioral actions (Fig. 14.3).

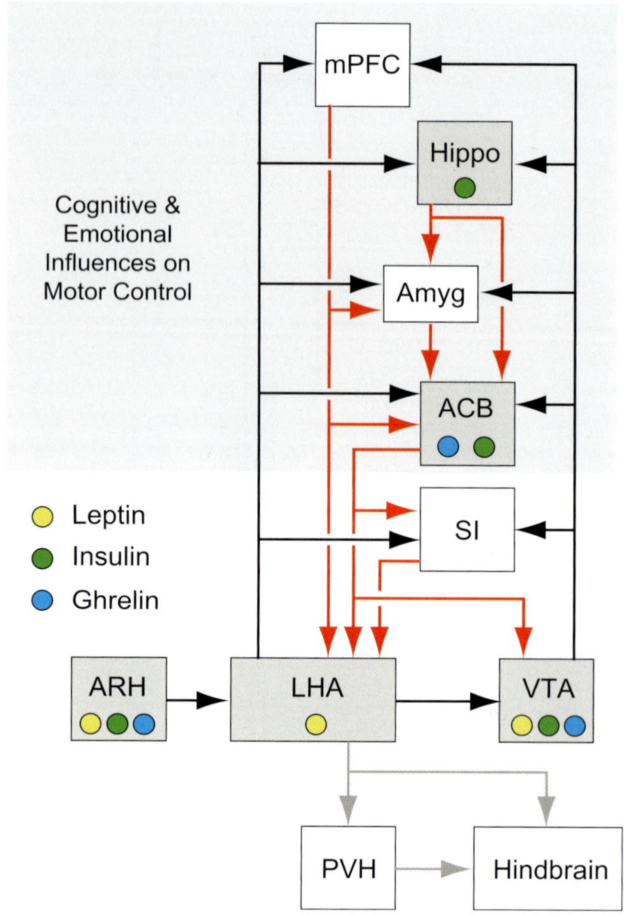

FIGURE 14.4 Leptin (yellow), insulin (green) and ghrelin (blue) all target multiple neurons in the forebrain (shown in gray). In particular, neurons in the arcuate nucleus (ARH), lateral hypothalamic area (LHA) and ventral tegmental area (VTA) contain receptors for these hormones that can alter their physiology. The LHA and VTA provide ascending projections (shown in black) to regions in the telencephalon that are important for cognitive and emotional influences on ingestive behaviors, including reward assignment. These regions include the medial prefrontal cortex (mPFC), hippocampus (Hippo), amygdala (Amyg), nucleus accumbens (ACB), and substantia innominata (SI) in the ventral pallidum. In turn, these telencephalic regions project (shown in red) back down to the areas that are important for motor control, which is enabled by further connections (gray arrows), including those to the paraventricular nucleus of the hypothalamus (PVH) and hindbrain. Note that not all connections of these regions are illustrated.

As we saw in Figs 14.2 and 14.3, circulating hormones are important signals that can interact with neurons throughout the brain to control feeding behavior and energy metabolism. The next section will describe these interactions in more detail.

METABOLIC HORMONES AND THEIR IMPACT ON THE BRAIN

Leptin

Leptin is the most important hormone secreted from adipocytes.[27] It was first characterized in 1994 by Jeffrey Friedman and his colleagues[28] – a discovery that in many respects galvanized the neuroendocrinology of feeding and metabolism by providing a new perspective for examining how metabolic hormones interact with the brain.

Leptin is a 167-amino acid peptide that has actions on the liver and skeletal muscle, where it helps regulate glucose and lipid metabolism by way of AMP kinase. It may also target pancreatic β-cells to suppress insulin secretion.[29] However, leptin's actions in the brain to suppress food intake and sympathetic activity have received the most attention. Because the leptin found in the general circulation derives almost exclusively from adipose tissue, circulating leptin provides an index of body adiposity stores for the brain. Circulating leptin concentrations are transduced by neurons with leptin receptors into signals that alter the activity of neural networks that control energy metabolism.[30] Many studies have focused on the consequences of elevated plasma leptin concentrations that accompany obesity, particularly leptin's ability to suppress food intake, and the development of central leptin resistance.[31] However, the more physiologically relevant signal is

BOX 14.2

HOW DO WE KNOW THAT LEPTIN ACTS IN THE BRAIN TO CONTROL METABOLISM?

The identification of leptin in 1994[1] came at a time when the powerful techniques that allow the controlled expression of particular genes were starting to be developed. This meant that for the first time the mode of action of a newly discovered hormone could be explored relatively quickly, and with a spatial and temporal precision not possible using the traditional lesion or targeted injection approach. In particular, these new techniques provided the opportunity to turn the expression of the genes encoding hormone receptors off or on in specific tissues.

The first studies to explore leptin's actions on the brain used classic targeted injections. Thus, central leptin injections suppressed food intake in *lep/lep* mice that lack the leptin gene[2]; more restricted leptin injections into the ventromedial hypothalamus also suppressed food intake more effectively than ventricular or raphe injections[3]; and peripheral leptin injections altered the expression of key neuropeptide genes in the hypothalamus associated with feeding and energy balance.[4] These studies quickly showed that the brain was a key site where leptin acted to regulated food intake and energy metabolism. However, they did not have the precision to implicate specific populations of neurons in leptin's actions.

The cloning of the leptin receptor gene (*Lepr*) in 1995[5] enabled a series of elegant studies that examined the contribution that neuronal leptin signaling – and particularly specific populations of neurons in the hypothalamic arcuate nucleus (ARH) – make to energy homeostasis. The first of these showed that *Lepr* knockout in neurons rather than hepatocytes accounted for the majority of the obese phenotype of *lepr/lepr* mice.[6] A second group of studies then found that replacing *Lepr* in neurons of *lepr/lepr* mice completely normalized their phenotype.[7,8] These studies added a molecular genetics twist to the classic endocrinological "extirpation and replacement" strategy to show the necessity of neuronal leptin signaling for the regulation of feeding and energy balance.

A more selective approach then used the previous characterization of the proopiomelanocortin (POMC) gene,[9] combined with techniques that targeted genes in particular cell types using cell-specific promoters and CRE-LOX methodology to investigate the role of POMC neurons – including those in the ARH – in leptin signaling. Removing *Lepr* from POMC neurons did indeed generate an obese mouse, but it did not fully recapitulate the phenotype of *lepr/lepr* (db/db) mice, meaning that other neurons must be involved.[10] This same group then went on to manipulate the expression of *Lepr* in ARH neurons using an inactive genetic construct of *Lepr* that could be reactivated by exogenous local delivery of FLPe-recombinase in an adeno-associated virus.[11] This powerful technique showed again that leptin signaling in ARH neurons was not sufficient for body-weight regulation, but that they were key for normal glucose homeostasis.

References

1. Zhang Y, Proenca R, Maffei M, Barone M, Leopold L, Friedman JM. Positional cloning of the mouse obese gene and its human homologue. *Nature.* 1994;372:425–432.
2. Campfield LA, Smith FJ, Guisez Y, et al. Recombinant mouse OB protein: evidence for a peripheral signal linking adiposity and central neural networks. *Science.* 1995;269:546–549.

(*Continued*)

BOX 14.2 *(cont'd)*

3. Jacob RJ, Dziura J, Medwick MB, et al. The effect of leptin is enhanced by microinjection into the ventromedial hypothalamus. *Diabetes.* 1997;46:150–152.
4. Schwartz MW, Seeley RJ, Campfield LA, et al. Identification of targets of leptin action in rat hypothalamus. *J Clin Invest.* 1996b;98:1101–1106.
5. Tartaglia LA, Dembski M, Weng X, et al. Identification and expression cloning of a leptin receptor, OB-R. *Cell.* 1995;83:1263–1271.
6. Cohen P, Zhao C, Cai X, et al. Selective deletion of leptin receptor in neurons leads to obesity. *J Clin Invest.* 2001;108:1113–1121.
7. Kowalski TJ, Liu SM, Leibel RL, et al. Transgenic complementation of leptin-receptor deficiency. I. Rescue of the obesity/diabetes phenotype of LEPR-null mice expressing a LEPR-B transgene. *Diabetes.* 2001;50: 425–435.
8. de Luca C, Kowalski TJ, Zhang Y, et al. Complete rescue of obesity, diabetes, and infertility in db/db mice by neuron-specific LEPR-B transgenes. *J Clin Invest.* 2005;115:3484–3493.
9. Nakanishi S, Inoue A, Kita T, et al. Nucleotide sequence of cloned cDNA for bovine corticotropin-beta-lipotropin precursor. *Nature.* 1979;278:423–427.
10. Balthasar N, Coppari R, McMinn J, et al. Leptin receptor signaling in POMC neurons is required for normal body weight homeostasis. *Neuron.* 2004;42:983–991.
11. Coppari R, Ichinose M, Lee CE, et al. The hypothalamic arcuate nucleus: a key site for mediating leptin's effects on glucose homeostasis and locomotor activity. *Cell Metab.* 2005;1:63–72.

the fall in circulating leptin that indicates reduced adiposity and the need to replenish energy stores.[32]

Leptin Receptors in the Brain

The discovery of leptin in 1994 was followed about a year later by the identification of its receptor (LepR) using a leptin-alkaline phosphatase fusion protein to screen a cDNA expression library derived from mouse choroid plexus.[33,34] Further evaluation resolved five splice variants of the leptin receptor gene.[32] Of these, the long receptor isoform (LepRb) is a member of the IL-6 family of class 1 cytokine receptors and mediates leptin signaling in neurons, while the short isoform (LepRa) is thought to contribute to the transport of leptin across the blood–brain barrier.[35,36]

Different techniques have been used to localize *Lepr* in the brain. The first studies used *in situ* hybridization to detect the LepR mRNA in rat brain, and found significant expression in parts of the hypothalamus, thalamus, hindbrain and cerebellum.[37] Further studies using *in situ* hybridization have substantiated the original findings, including reports in human and mouse brain.[38–40] In particular, *LepR* expression was prominent in the retrochiasmatic area, arcuate (ARH), dorsomedial (DMH), ventral premammillary (PMv) and ventromedial (VMH) nuclei of the hypothalamus. These regions have remained the focus of attention for the past several years, and a great deal of evidence now highlights the importance of these hypothalamic regions in mediating the actions of leptin on feeding and metabolism.

Another approach used various activation markers to identify neurons that are either directly targeted by leptin, or part of the downstream network from leptin-activated neurons. The first of these studies examined the location of Fos-containing neurons after peripheral leptin injections, and showed expression patterns that coincided with the locations of *Lepr*-expressing neurons in the hypothalamus and hindbrain.[41,42] However, Fos expression in regions such as the PVH, which only expresses very low levels of *Lepr*,[37,40] is the result of secondary activation from leptin-dependent processes, most likely in the ARH and DMH.[30,41,43]

More recently, novel and powerful transgenic tools have been developed by Myers and colleagues[44] that can identify the targets of neurons that express *Lepr* with great resolution and specificity. In particular, the generation of mice where the expression of farnesylated enhanced green fluorescent protein (EGFPfd) is restricted to neurons expressing *Lepr*[45] will likely prove a very useful tool for exploring the organization of leptin-sensitive networks. EGFPf is targeted to cell membranes, meaning that the projection network of *Lepr*-expressing neurons is revealed. This technique has recently been used to characterize two projections of *Lepr*-expressing neurons: the first is from the ventral premammillary nucleus (PMv) to the preoptic area and the gonadotropin-releasing hormone (GnRH) network that controls reproductive function[45]; the other is from *Lepr*-expressing neurons in the ventral tegmental area (VTA) to the amygdala and parts of the bed nucleus of the stria terminalis.[46]

Another technique has recently been used to trace the projections of leptin-sensitive neurons. This uses CRE recombinase-driven activation of an adeno-associated virus that encodes *Renilla* green fluorescent protein. Targeted injection of the virus into the brain of a mouse that only expresses *Cre* in *Lepr*-expressing neurons reveals their efferent projections. Gautron and colleagues have recently used this technique to trace the projections of *Lepr*-expressing neurons in the DMH.[43]

Together, this large body of experiments over the past 10 years has generated convincing and convergent data that the ARH, DMH, PMv and VMH of the hypothalamus, together with groups in neurons in the VTA and parts of the hindbrain (including the nucleus of solitary tract and dorsal raphe nuclei), are important leptin targets in the brain.

Leptin Receptor Signaling Pathways

The LepR recruits three intracellular signaling pathways to regulate neuronal function:[46,47] SHP2, which activates extracellular signal related kinase (ERK); STAT3, which leads to increased expression of suppressor of cytokine signaling 3 (SOCS3), a repressor of SHP2 signaling;[48] and STAT5. Each of these pathways is activated by increased Jak2 kinase activity that follows ligand binding, and each has distinct downstream functions that have been elucidated by Myers and colleagues using a series of mutant mice with defective recruitment in each pathway.[44] Thus, deleting the ability of LepR to engage SHP2 appears to amplify rather than impede leptin's actions on metabolism,[49] meaning that these functions are mediated by STAT3 and STAT5 pathways. Further work from Myers et al. has shown that STAT5 signaling appears to make only modest contributions to body-weight regulation,[50] while STAT3 signaling mediates the majority of leptin's actions on feeding, body weight, and energy metabolism.[51,53]

Leptin and the Regulation of Food Intake

The suppression of food intake by peripheral and central leptin administration was one of the first of its properties to be investigated,[32,53] and has remained a primary target for investigation ever since (see Box 14.2). Although much of the initial focus was on hypothalamic targets, subsequent work has embraced a wider appreciation of the neural systems engaged by leptin.[30,54–56] In particular, two observations have helped with this view. The first is that leptin reduces food intake by decreasing meal size rather than meal number,[15,57,58] which led to studies that use more behaviorally focused measures rather than simple intake as their endpoints (see Box 14.1). The second finding is that leptin can directly influence motivational processes. Both of these discoveries have provided new insights about the neural networks and systems that leptin uses to regulate feeding and metabolism. We will expand on these two topics later, but first we will discuss the hypothalamus as a target of leptin actions.

SITES OF ACTION: HYPOTHALAMUS

Two findings provided the foundation for understanding how leptin – as well as other hormones – interact with hypothalamic neurons to regulate feeding behavior and metabolism. The first was the discovery, about 30 years ago, that neuropeptide Y (NPY) injections into the hypothalamus increase food intake.[59–61] Further work has shown that NPY increases feeding by directing appetitive behaviors towards food sources (i.e., increases foraging) and by increasing meal size rather than activating consummatory actions.[62–64] The second was that melancortin receptor systems in the brain are important components of the neural networks that control feeding behavior.[65–69] The fact that the ARH in the ventral periventricular hypothalamus contains neurons that express NPY (which is co-localized with the melanocortin receptor inverse agonist, agouti-related peptide, AgRP) and POMC-derived peptides highlighted the ARH as a key hypothalamic contributor to these networks, and to how ARH neurons might be engaged by leptin.

One of the first indications that leptin might regulate feeding behavior by way of hypothalamic mechanisms was that it suppressed *Npy* and increased *Pomc* expression in the ARH of leptin-deficient *lep/lep* (ob/ob) mice.[70–72] The ARH was the only cell group where leptin could regulate *Npy* expression. These findings were followed by reports that leptin acts directly in the ARH, and specifically on POMC neurons, to change their firing rate.[73,74] Furthermore, the fact that knocking out *Npy* in *lep/lep* mice attenuated their hyperphagia and obesity emphasized the importance of ARH neuropeptides in mediating the actions of leptin.[75]

In addition to being obese and hyperphagic, *lep/lep* mice are diabetic – a phenotype they share with mice lacking *Lepr*. Thus, in addition to actions on feeding behavior, leptin acts on hypothalamic targets to regulate glucose and lipid metabolism. Restoring leptin in the ARH of *lepr/lepr* (db/db) mice or Koletsky rats, which also lack *Lepr*, dramatically improves glucose homeostasis[76,77] via effects that are mediated by the vagus nerve.[78] Furthermore, leptin may help improve insulin sensitivity by stimulating fatty acid oxidation and lipogenesis in adipose tissue through actions on mediobasal hypothalamic neurons.[79,80]

Since the original studies identifying ARH neurons as a key leptin target, others have reinforced the idea that these neurons are necessary but certainly not sufficient to account for all of leptin's actions on feeding behavior and metabolism.[30,31] Clearly other cell groups both in the hypothalamus and elsewhere in the brain are involved. In the hypothalamus; these include the VMH, which is heavily implicated in glucosensation[10] as well as female reproductive behavior (see Chapter 20). Leptin acts directly to change VMH neuronal firing rates.[81] Another hypothalamic target for leptin is the PMv,[44] which is involved in the regulation of reproductive function (see "Neuroendocrine Signals and the Integration of Energy Metabolism with Reproductive Function," page 345). However, leptin also targets the hindbrain, and the midbrain VTA to influence motivational processes associated with reward assignment.

SITES OF ACTION: HINDBRAIN

The fact that feeding behavior is comprised of a set of complex and coordinated muscle movements that must be organized in the correct temporal sequence is consistent with a function controlled by a broadly distributed neural network (see above). The fact that LepRbs are expressed throughout the brain means that leptin can target extra-hypothalamic regions to regulate feeding behavior. Of these regions, the hindbrain is been the focus of much attention, particularly since it has important roles in controlling the consummatory phase of feeding behavior. It is also the site of integration between descending projections from the forebrain and peripheral signals that reach the brain through the spinal cord or vagus nerve that together control meal size.[7,55] LepRb is expressed by neurons in the medial part of the nucleus of the solitary tract (NTS), and phospho-STAT3 (a marker of leptin activation) is found in this same region of the NTS following peripheral leptin administration.[82] Leptin administration into the NTS and local knockdown of LepRb both reduce food intake and body weight.[83,84] The integrative role of LepR-expressing hindbrain neurons is shown by the fact that the same NTS neurons are activated by leptin and gastric distension.[82]

Leptin and Reward

Although the mediobasal hypothalamic neurons that express *Lepr* are key contributors to leptin's actions on energy metabolism, these neurons do not account for the complete phenotype of *lep/lep* mice that completely lack leptin.[30,85,86] Other leptin-sensitive processes are required, and the fact that LepRb expression is much more widespread outside the mediobasal hypothalamus[37,40] shows that leptin can potentially affect many different neural systems. So one of the most exciting recent findings about leptin's role in regulating metabolism is its ability to interact with the neural systems that assign reward (or hedonic) value to food.[54,56,87] These processes provide an important channel for leptin's ability to regulate feeding (Fig. 14.4).

An important pointer to the fact that leptin can modify reward processes was a report that intracranial injections of leptin counteracted the ability of food restriction to enhance the sensitivity of particular LHA sites to electrical self-stimulation.[88] Leptin had no effect on LHA sites where rates of self-stimulation were unaltered by food restriction. These experiments were followed by studies showing that dopamine is a key neurotransmitter for maintaining normal food intake and metabolism, and for the hyperphagia shown by *lep/lep* mice.[89,90] Indeed, LepRb receptors are expressed by some of the VTA neurons that provide important dopaminergic inputs to the nucleus accumbens.[91,92]

Recent experiments from the DiLeone group[93] have shown that locally applying leptin into the VTA reduces both the firing rate of dopaminergic neurons and food intake by way of a JAK-STAT pathway. Specifically knocking out *Lepr* in VTA neurons increased food intake independent of body weight. These same knockout animals also increased their short-term consumption of 0.2% sucrose and ate more of a high-fat diet compared to controls.

A paper published simultaneously with the Hommel study[92] found that leptin-deficient *lep/lep* mice showed reduced locomotor responses to amphetamine together with an absence of the sensitizing effects of repeated amphetamine injections, which are both dopamine-dependent processes. These effects were reversed by leptin. Brain slices from *lep/lep* mice were also deficient in their ability to release dopamine following electrical stimulation, and had lower dopamine content and tyrosine hydroxylase expression than controls.

The complex way that leptin can influence the networks involved with feeding hedonics has recently been demonstrated by Leinninger and colleagues.[94] Using a series of novel mouse *Lepr* constructs, these workers showed that leptin acts directly on LHA neurons that project to the VTA to influence dopaminergic function and food intake. These leptin-sensitive LHA neurons do not contain orexin or MCH,[30] thus implicating other populations of peptidergic LHA neurons.

Leptin: Influences on Neural Development and Plasticity

The idea that hormones act as tropic or "organizing" factors for neural tissue was established over 50 years ago (see Gorski[16] and Harris[17] for reviews). However, the notion that metabolic hormones serve as signals for organizing neural development is much more recent. The key finding was that the development of the NPY/AgRP-containing pathway from the ARH to the PVH, DMH and LHA in rodents was leptin-sensitive.[18] This pathway is poorly developed at birth and does not mature to its adult form until into the second postnatal week, which contrasts with projections from the dorsomedial nucleus to the PVH and LHA, which are complete by the end of the first postnatal week.[95]

Using a combination of *in vivo* and *in vitro* techniques, Bouret and colleagues showed that ARH projections to the PVH required leptin for complete development.[18]

Thus, *lep/lep* mice had a projection density that was significantly compromised compared to wild-type mice, and leptin was able to restore this to control values. Furthermore, axonal outgrowth from *in vitro* organotypic cultures was significantly stimulated by leptin, consistent with a direct action of leptin on neurite outgrowth. The importance of these findings for understanding how metabolic dysfunction during the pre- and perinatal period can influence neural development is underscored by the fact that rats that are genetically predisposed to diet-induced obesity (DIO rats) have a defective ARH–PVH projection pattern.[96] Leptin is also much less effective at promoting neurite outgrowth from the ARH in the offspring of DIO rats compared to controls.

The effects of leptin on neuronal and synaptic architecture are not limited to the perinatal period. Electron-microscopic and electrophysiological data show that leptin can alter synaptic function in adults, including actions in the hypothalamus[97] and hippocampus.[98,99]

Insulin

Insulin is secreted by β-cells in the islets of Langerhans in the endocrine pancreas. It is the only hormone that can actively reduce blood glucose concentrations, which it does primarily by stimulating glucose uptake into muscle, liver and adipose tissue. Following almost 90 years of intensive study, insulin's peripheral mechanisms of action are now quite well understood.

The fact that changes in glycemia regulate the firing rate of hypothalamic neurons has been known for almost 50 years (see Watts and Donovan[10] for review). Because some of these early studies used insulin-induced hypoglycemia, the possibility that insulin itself could be a direct regulator was immediately recognized.[100] However, at that time it was difficult to separate insulin's direct actions on neurons from those mediated by its hypoglycemic effects. This problem began to be resolved about 30 years ago, when specific binding sites for insulin were found in the brain[101] and mechanisms for insulin transport across the blood–brain barrier were identified.[102] At about the same time, Stephen Woods and his colleagues showed that insulin could regulate food intake and body weight in baboons by direct actions on the brain rather than by way of hypoglycemia.[103] Importantly, these experiments also showed that central administration of glucagon had no effect on food intake and body weight.

A wealth of studies have built on these earlier findings to consolidate insulin's position as a major hormonal modulator of brain function. They include the identification and characterization of brain insulin receptors (IR) and their signaling pathways in neurons, and the downstream mechanisms underlying insulin's actions on the brain to influence feeding and metabolism.

Brain Insulin Receptors

Although the brain as a whole has historically been regarded as being insulin-insensitive with regard to glucose uptake, convergent sets of studies confirm that neurons expressing IRs are found at various locations, including in the olfactory bulb, cerebellum, parts of the cortex, hippocampus, choroid plexus, VTA and ARH.[91,104,105] The importance of these IRs for metabolic function is highlighted by the phenotype of neuronal insulin receptor knockout (NIRKO) mice. These animals lack IRs in neurons and related cell types, but have normal IR expression in non-neuronal tissues. They develop diet-sensitive obesity,[106] and have deficits in hypothalamic glucosensing[107] and sympathoadrenal counterregulatory responses.[108] As with leptin, ARH neurons have been a prime focus in the investigations of insulin's actions in the brain.[109] NPY and POMC neurons in the ARH are both thought to express IR.[110,111]

Insulin: Feeding Behavior and Metabolism

Insulin administered directly into the brain decreases food intake[103] and has effects on glucose homeostasis.[112] Three non-exclusive mechanisms have been proposed to help explain these findings, and each shares some overlap with the way that leptin regulates food intake. The first involves direct actions of insulin on ARH neurons, the second derives from insulin's ability to influence the neural systems that regulate learning, memory and reward, and the third involves interactions with CCK-dependent satiety mechanisms that rely on processes located in the hindbrain.

In support of the first mechanism, centrally injected insulin suppresses *Npy* expression and increases *Pomc* expression in the ARH.[110,113] As we have already discussed, these neurons have important functions in regulating many aspects of feeding and energy metabolism. Insulin suppression of food intake is blocked by a melanocortin receptor antagonist[110]; melanocortin signaling is important for insulin's central actions on energy homeostasis[114]; and insulin signaling in POMC neurons is required for insulin's central actions on glucose homeostasis.[115] Knockdown of hypothalamic IRs leads to hyperphagia.[116] Further studies have shown that phosphatidylinositol 3-kinase (PI3K) is a key downstream signaling enzyme that is required for insulin's actions in ARH neurons.[117]

The second mechanism is the ability of insulin to modify reward and motivational processes. In this regard, Figlewicz and Benoit have shown that insulin modulates a range of behavioral reporters for these processes.[118] The fact that the hippocampus contains significant numbers of IRs has focused attention on how insulin modifies hippocampal function, which is important for short-term memory and allocentric spatial navigation.[2] Thus, intra-hippocampal insulin administration alters glutamatergic neurotransmission[119,120] and enhances performance in spatial learning tasks in a PI3K-dependent manner.[121]

Third, the fact that insulin acts in the brain to reduce meal size suggests it interacts with satiety mechanisms. This idea is supported by the finding that centrally administered insulin potentiates the effects of CCK,[122] which is a circulating gastrointestinal-derived peptide that has powerful suppressive effects on meal size.

Glucocorticoids

Glucocorticoids are a group of steroids synthesized by cells mainly in the zona fasciculata of the adrenal cortex. Cortisol is the principal glucocorticoid in humans and other primates, while corticosterone is secreted in rats and mice. Glucocorticoids have a broad set of effects throughout the body, including actions on the immune system, development, and cognition (see Chapters 8 and 18).[123–125]

The metabolic actions of glucocorticoids are diverse and complex. As their name suggests, these steroids increase blood glucose levels, which they do by increasing the synthesis of liver enzymes required for gluconeogenesis. They also increase the mobilization of substrates for gluconeogenesis from other tissues, principally amino acids from muscle and fatty acids from lipolysis in adipose tissue. Glucocorticoids also function as permissive agents in the mechanisms of action of glucagon and catecholamines, and have important interactions with leptin and insulin.[126,127] The direct actions of glucocorticoid on liver, adipose tissue and muscle are relatively slow compared to those of insulin, epinephrine and glucagon, and involve changes in the expression of the genes encoding metabolic enzymes, rather direct control of enzyme activities.[128]

Glucocorticoids also act directly on the brain. They use the same receptors found elsewhere in the body: the high-affinity, low capacity mineralocorticoid receptor (MR), and the lower-affinity, higher capacity glucocorticoid receptor (GR).[125] Both receptors are expressed throughout the brain, and their topographical localization is known in some detail (see Chapters 8 and 18). Significant levels of GRs are found in the ARH, VMH, PVH, and catecholaminergic neurons that project to the hypothalamus.[129,130]

Whether glucocorticoids acting alone have direct actions on the brain to affect metabolism and feeding behavior is not clear. Adrenalectomized animals certainly show significant alterations in feeding and energy metabolism that can be reversed by glucocorticoid replacement.[126] However, because these animals also have compromised leptin and insulin secretion, it is difficult to determine which effects are mediated solely by glucocorticoids and which are the result of interactions with other metabolic hormones.

Gastrointestinal (GI) Hormones

Endocrine cells in the stomach, and proximal and distal intestine, are the principal source of peptide hormones that influence food intake and energy metabolism by directly targeting specific neurons in the brain and/or by modulating vagal nerve function.[131] In this way, they mostly act as short-term signals that influence meal size. They can also interact with insulin and leptin, which, as we have seen, are signals that act over longer timeframes to regulate metabolism. All but one of these peptides (ghrelin) are satiety signals and suppress food intake.

Cholecystokinin (CCK)

CCK is synthesized by proximal intestinal I cells, and is a powerful satiety signal that decreases meal size by acting on both vagal afferents and the hindbrain.[132,133] Leptin will augment CCK's actions on vagal afferents,[134] but can also interact with CCK through direct actions in the hindbrain.[55] At least part of leptin's interaction with CCK may be mediated by way of the hypothalamus and descending connections to the hindbrain.[135]

Ghrelin

Ghrelin is synthesized by the enteroendocrine cells in the stomach and proximal intestine, and is acylated by ghrelin O-acetyltransferase to generate the biologically active peptide.[136] Ghrelin is the only GI hormone that stimulates food intake.[137,138] The ghrelin receptor (GHSR1a) is G-protein coupled receptor that engages signal transduction pathways that regulate intracellular calcium.[136] GHSR1a is expressed in the pituitary, on vagal afferents, and in the brain, where it is found in many of the same regions as IRs and LeprRbs[139] (Fig. 14.4). Like insulin and leptin, ghrelin influences a broader range of functions than food intake, including glucose homeostasis and reward.[136,140] Ghrelin feeding requires ARH neurons that express NPY,[141] but not catecholaminergic inputs from the hindbrain.[142] Ghrelin, together with leptin and insulin, is being increasingly recognized as an important hormone that targets multiple brain sites and mechanisms to regulate feeding behavior and energy metabolism (Fig. 14.4).

Pancreatic Peptide YY_{3-36} (PYY), Glucagon-Like Peptide 1 (GLP-1) and Amylin

PYY and GLP-1 are both synthesized by L cells in the lower (distal) intestine. PYY belongs to the NPY family, while GLP-1 is product of the prepro-glucagon gene, and is released in response to nutrients in the gut.[143] Circulating PYY concentrations are proportional to meal energy content,[144] and will inhibit food intake when injected centrally, where it binds to the Y2 receptor.[145] The ARH is considered a major site of action for PYY.[146]

GLP-1 is an incretin that will augment insulin secretion from β-cells,[147] but it also reduces food intake.[148] GLP-1 can, like CCK, act on vagal afferents, and vagotomy abolishes the ability of peripheral GLP-1 to reduced food intake.[149] Although GLP-1 receptors are found in the brain, its very short half-life in the circulation questions whether circulating GLP-1 can directly affect brain function.

Amylin is synthesized by pancreatic β-cells and reduces food intake through actions on the area postrema in the hindbrain.[150] It also interacts with both short-term signals (CCK, GLP-1, PYY) and long-term (leptin) signals to regulate energy metabolism.[151]

METABOLIC SIGNALS AND THE HORMONES OF THE PARS DISTALIS OF THE PITUITARY GLAND

Changes in energy metabolism have major influences on the release of hormones from the pars distalis. ACTH, TSH and growth hormone all show modified secretory patterns as part of the adaptive responses to changing energy balance. By way of their regulatory actions on target endocrine organs, these three hormones have important influences on metabolism, including actions on the liver, muscle, and adipose tissue function. However, changes in energy metabolism also alter the release of LH and FSH in a way that suppresses or facilitates reproductive function depending on the ongoing status of energy stores, particularly adipose tissue.

This section will illustrate how the signals that derive from two metabolic challenges – starvation and hypoglycemia – modify ACTH, TSH and growth hormone secretion. The final section will then describe the broader issue of how metabolic hormones and gonadal steroids interact with overlapping neural networks in a way that enables the functional interactions between energy metabolism and reproduction.

Starvation and Negative Energy Balance

Starvation and the consequent drop in energy stores is one of the most powerful challenges an animal has to face. We have already seen that reductions in circulating leptin and insulin are important hormonal signals to the brain. These increase hunger and alter energy metabolism in a way that maintains blood glucose levels and conserves remaining energy stores. However, starvation also alters the release of hormones from the pars distalis to help maintain energy homeostasis as energy stores become depleted.

Rodent neuroendocrine responses are evident within 3 hours of food being absent. They include a failure to increase plasma ACTH and corticosterone concentrations that are usually seen at the onset of the dark period.[152] The signals that mediate these short-term changes probably involve the absence or a reversal of the increases in plasma insulin, glucose, and leptin concentrations that normally precede or occur just after food is eaten.[152]

Longer periods of starvation or food restriction reduce thyrotropin-releasing homone gene (*Trh*) expression in the PVH (see Chapter 11) and, subsequently, TSH secretion.[153–157] Decreased *Trh* expession is driven for the most part by falling leptin levels,[158,159] which signal both directly to TRH neurons in the PVH, and indirectly by way of leptin-sensitive projections from the ARH.[157,160,161]

The fall in thyroid hormone secretion has significant effects on basal metabolism, including a reduction in overall metabolic rate and alterations in the activity of liver enzymes associated with metabolism. Reduced thyroid hormone secretion also reduces the activity of the liver enzymes that catabolize circulating steroids, and thereby have indirect effects on circulating glucocorticoid concentrations. In this way, the half-life of plasma corticosterone is increased during negative energy balance,[162] resulting in increased circulating corticosterone without a concomitant increase in ACTH secretion.[163–166]

Hypoglycemia

Hypoglycemia is a major challenge to survival because it reduces the supply of glucose to the brain. A well-defined set of physical symptoms occurs as blood glucose concentrations fall out of the normal range, including sweating, hunger, disorientation, cognitive decline and seizures.[167,168] Neurons eventually die if low blood glucose levels (< 2.0 mM) are maintained for more than a few minutes.

Mammals use a set of well-described homeostatic mechanisms – counterregulation – to maintain blood glucose levels between approximately 4.0 and 6.5 mM (72–115 mg/dl), with a mean value of about 5.5 mM (100 mg/dl). Significant deviations outside this range initiate counterregulatory responses: increased insulin secretion with hyperglycemia; reduced insulin secretion;

and increased epinephrine, glucagon, ACTH and growth hormone (see Chapters 10 and 32) in response to hypoglycemia. There is no evidence that hypoglycemia alters TSH release in normal subjects.

Because endocrine counterregulatory responses are extremely effective at maintaining blood glucose concentrations within the control range, sustained hypoglycemia is not usually encountered for any extended period. However, this is not the case for patients with type 1 diabetes. Because insulin therapy is the only effective treatment for this condition, sustained hypoglycemia and its consequences can be a serious complication for these patients.[169]

As part of the rapid counterregulatory response, insulin- (β-cells) and glucagon- (α-cells) secreting endocrine cells in the islets of Langerhans of the pancreas are major targets for changing glycemia; however, the brain is the key initiator for epinephrine, glucocorticoid and growth hormone counterregulatory responses. As we have discussed earlier, circulating metabolic fuels, and particularly glucose, operate as interosensory signals that shape neuroendocrine responses (Fig. 14.2). To do this, the body contains distributed sets of glucosensors that inform the brain about changing blood-glucose dynamics. Although all cells respond to altered extracellular glucose concentrations, some neurons have specialized mechanisms that can transduce ambient glucose dynamics into altered firing rates. The cellular and molecular mechanisms responsible for this property are now quite well understood, and involve the enzyme glucokinase and an ATP-sensitive potassium channel.[170,171]

The most important sets of glucosensors for controlling counterregulatory responses are located in the hypothalamus, hindbrain, and mesenteric and hepatic portal veins.[10] Much work has focused on how this glucosensing network controls epinephrine and norepinephrine release from chromaffin cells in the adrenal medulla. However, we are now beginning to understand how hypoglycemia activates neuroendocrine systems in the hypothalamus, and particularly the hypothalamic–pituitary–adrenal (HPA) axis.

Compared with our understanding of the sympathoadrenal, islet cell and HPA responses to hypoglycemia, we know little about the mechanisms responsible for the altered growth hormone secretion that accompanies hypoglycemia. This may be partly related to the fact that the direction of the response is species-specific, with elevations occurring in humans, but either no effect or a suppression in rodents.[172,173] However, there is evidence that direct hypothalamic action of glucose can alter the release of growth hormone-releasing hormone (GHRH) and somatostatin,[172] and that catecholaminergic processes may be involved.[174]

Hypoglycemia and ACTH Release

Although it has been known for many years that insulin-induced hypoglycemia is a potent stimulus for ACTH and glucocorticoid release, recent studies are now beginning to reveal the mechanisms engaged by hypoglycemia to activate corticotropin-releasing hormone (CRH) neuroendocrine neurons in the PVH. This work has helped clarify the functional organization of the neural networks used by metabolic stimuli to drive neuroendocrine responses, as well as the intracellular mechanisms in CRH neurons that are used by afferent signals to control the synthesis and release of ACTH secretogogs.[175]

Three aspects of the ACTH response to hypoglycemia are worth noting. First, like all counterregulatory responses, ACTH release is only activated once a distinct plasma glucose threshold is reached.[167,168] In rats, this is between 3.00 and 3.15 mM[176] (Fig. 14.5). This threshold is lower than that for the suppression of insulin release (which is the first line of defense against hypoglycemia), but is higher than the threshold for serious cognitive deficits and pathological effects. This fact is consistent with the notion that the glucocorticoid contributions to counterregulation are considered to have slower onset but longer-lasting effects than those following altered insulin, epinephrine and glucagon secretion.

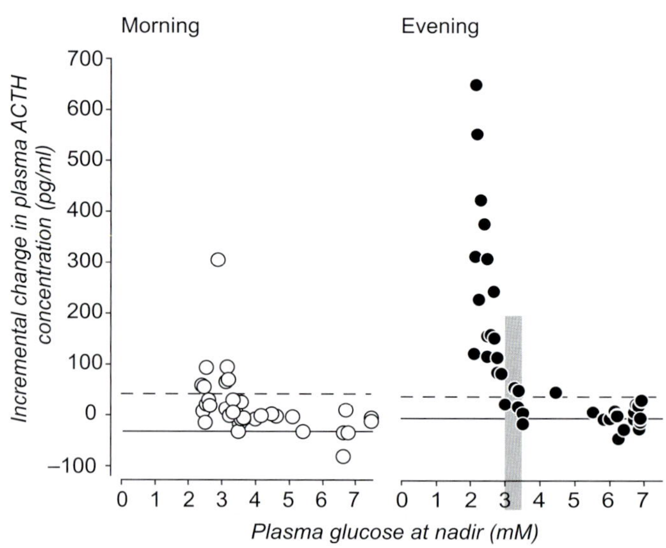

FIGURE 14.5 The magnitude of the ACTH response to hypoglycemia varies across the day. The two panels show the incremental change in rat plasma ACTH concentrations 20 min after an intravenous injection of insulin or vehicle given 2 h after lights on (Morning, open circles) or 1 h before lights off (Evening, closed circles). Each symbol represents the response of a single animal. Dashed horizontal lines represent 2 standard deviations above the mean increment of the saline-injected controls at that time (solid horizontal lines). The vertical shaded bar in the Evening denotes the range of glucose concentrations that contain the glycemic threshold. *Adapted from Gorton* et al. *(2007).*[176]

Second, the increased CRH neuronal activation that follows hypoglycemia relies on glucosensors in the hindbrain rather than hypothalamus. Thus, local hypothalamic glucopenia produced by 2-deoxyglucose (2-DG) is ineffective at increasing glucocorticoid secretion, whereas it is robustly increased by hindbrain injections.[177] Once the information about hypoglycemia has been processed by hindbrain glucosensing networks, it is relayed to CRH neurons almost exclusively by catacholaminergic inputs to the PVH.[178,179] This property makes hypoglycemia different to the majority of stressors in that a single afferent system is sufficient to drive CRH neurons. Catecholaminergic inputs are then coupled to CRH neuronal responses by MAP kinase pathways.[175,180]

Third, there is a clear interaction between the circadian timing system and hypoglycemia-activated ACTH release. This is evident from the highly significant difference in the ACTH and corticosterone responses to hypoglycemia across the day[176,181] (Fig. 14.5). Although there are a number of possible reasons for this marked difference in responsivity, it does not derive from an altered sensitivity of the glucosensing network to glucose. However, it does appear to depend on processes in the forebrain rather than the hindbrain[176] that may involve interactions with suprachiasmatic nucleus-dependent mechanisms.

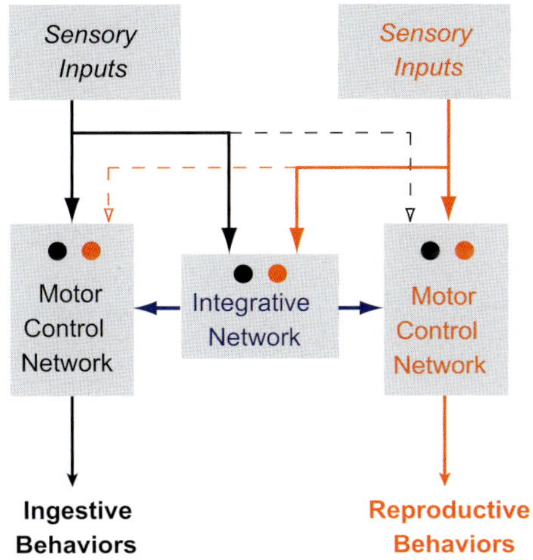

FIGURE 14.6 As shown in Figure 14.3, the motor control networks receive information from a variety of sensory inputs. Both the structure of the control networks and the information they receive (black and red arrows) are different for ingestive (black) and reproductive behaviors (red).[3,19] Interactions between the control networks may be mediated in two ways. First, the diversity of hormone receptor expression allows hormonal information that is primarily directed at one control network to influence the other. The expression of gonadal steroid receptors is illustrated as red dots, while leptin, insulin, ghrelin, and glucocorticoid receptors are shown as black dots. Second, a set of integrative nuclei (blue) receives diverse sensory information and expresses the receptors for both gonadal steroids and the hormones concerned with energy balance. The neural projections of the integrative network then target both the reproductive and ingestive behavior control networks (blue arrows). *Adapted from Watts and Swanson (2002).*[2]

NEUROENDOCRINE SIGNALS AND THE INTEGRATION OF ENERGY METABOLISM WITH REPRODUCTIVE FUNCTION

A key aspect to the organization of the neural networks that are responsible for the control of feeding behavior and energy metabolism is that they provide access to the networks that control other motivated behaviors. Perhaps the most important of these interactions is with the systems that control reproductive function. Pregnancy and lactation are the most energy-demanding of all physiological processes, and so information flow between the control networks responsible for specific ingestive and reproductive motor events is a fundamental aspect of individual and species survival. Figure 14.6 shows that control network interactions may be mediated by the actions of hormones and sensory inputs that, in addition to regulating the function of one network, can directly influence the other. A second mechanism uses sets of projections from intermediate (integrative) cell groups that contain mixed populations of receptors and have divergent sets of connections (see following sections).

How Hormones Act to Mediate Crosstalk between Metabolic and Reproductive Systems

Metabolic Hormones and the GnRH Control Network

INSULIN

In addition to its profound effects on energy metabolism, insulin influences reproductive function. Insulin appears to have a bimodal effect on LH pulsatility. Hypoglycemic doses of insulin suppress LH secretion.[182] Other studies, however, suggest that the frequency of LH pulses in a number of species positively correlates with plasma insulin concentrations as levels fall.[183] Chronic changes in insulin secretion may therefore contribute significantly to reduced reproductive competence during diabetes, or when insulin levels are low or absent during starvation. The mode and site of action for these effects have been controversial and difficult to determine because insulin may act directly on the neurons that control GnRH pulsatility, or may affect these neurons indirectly through changes in glucose metabolism.[183] Two studies suggest that insulin acts directly within

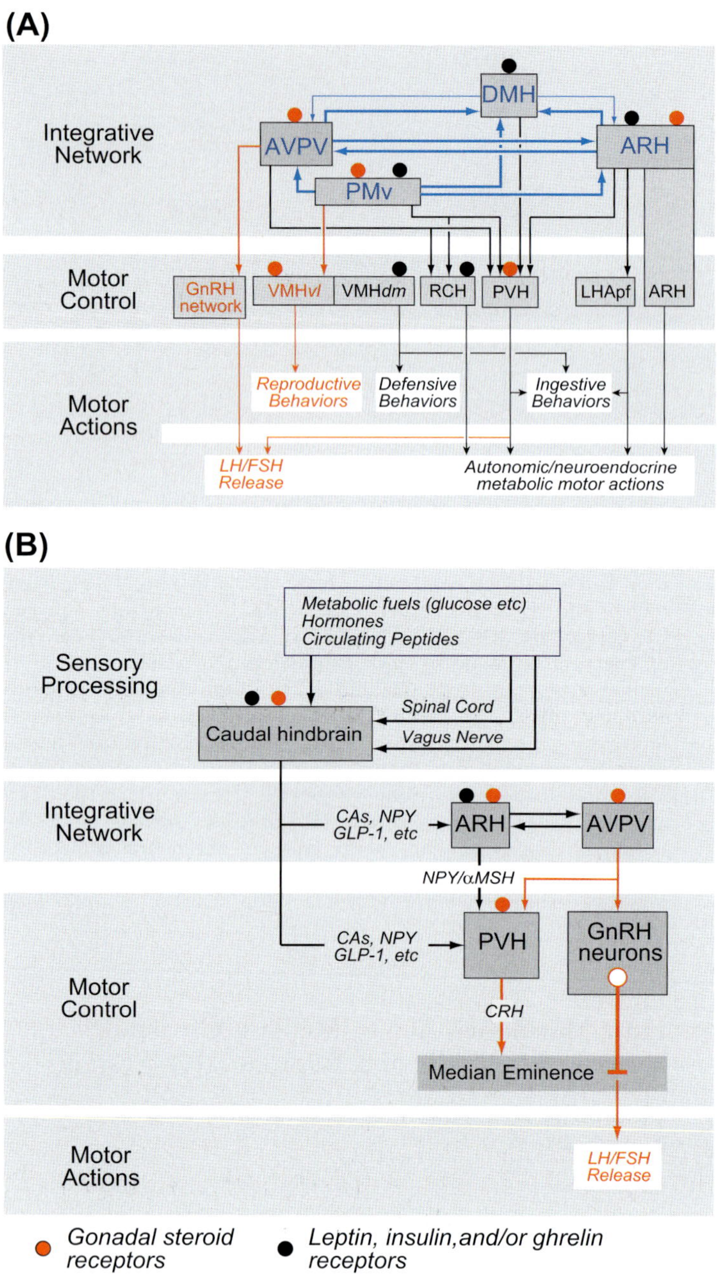

FIGURE 14.7 (**A**) Schematic of the integrative and motor control networks for the motor events of reproductive (LH/FSH release) and ingestive behaviors (the autonomic and neuroendocrine aspects of energy expenditure and metabolism). Sets of neurons in the gonadotropin-releasing hormone (GnRH) network, ventromedial nucleus (VMH), retrochiasmatic area (RCH), paraventricular nucleus (PVH), perifornical part of the lateral hypothalamic area (LHApf) and arcuate nucleus (ARH) control reproductive (red) or energy balance (black) motor actions directly. In turn, the ventrolateral part of the VMH (VMHvl) controls female reproductive behaviors, and the dorsomedial part of the VMH (VMHdm) controls defensive and some aspects of ingestive behavior. Collectively, all these motor control cell groups are then regulated by inputs from an integrative network (shown in blue; see also Fig. 14.6). The receptor expression patterns and the connections from neurons in the ARH, anteroventral periventricular nucleus (AVPV), dorsomedial nucleus (DMH) and ventral premammillary nucleus (PMv) provide evidence that they make up an integrating network that can influence neurons in the reproductive and energy balance motor control networks in a divergent manner. (**B**) Schematic showing sensory processing, integrative, and motor control networks for LH/FSH release. Processing of information about metabolic fuel availability from the periphery occurs in the hindbrain and is projected to forebrain areas, including the PVH and ARH, using catecholaminergic (CA) and other neurons. Estrogen receptors in the PVH could influence CRH neuronal activity in a way that modifies GnRH output in the median eminence. Expression of estrogen receptors (red dots), or leptin, insulin and ghrelin receptors (black dots), occurs in many of these areas. αMSH, α-melanocyte-stimulating hormone; CRH, corticotropin-releasing hormone; GLP-1, glucagon-like peptide-1;NPY, neuropeptide Y. *Adapted from Watts and Swanson (2002).*[2]

the brain to facilitate LH pulsatility. First, streptozotocin-induced diabetes reduces the frequency of LH pulses in male sheep,[184] which can be reinstated by intracerebroventricular insulin at doses that affect neither peripheral insulin nor glucose concentrations.[185] Second, female NIRKO mice show a 90% reduction in circulating LH levels, while males show a 60% reduction.[106]

LEPTIN

Fluctuating levels of circulating leptin have a major impact on reproductive competence.[3] LepRs are strongly expressed in the PMv and the ARH, both of which have major interactions with the GnRH network[30,44] (Fig. 14.7). Ablation of IRs and LepRs from POMC neurons disrupts reproduction function in female mice.[115] Furthermore, both nuclei project to the anteroventral periventricular nucleus (AVPV) (Fig. 14.7), which is essential for the pre-ovulatory GnRH surge and may be able to help mediate leptin-sensitive effects on reproduction. Indeed, some leptin receptor-containing neurons in the PMv contribute to these projections.[44] It is also worth noting that kisspeptin neurons in the ARH and AVPV are leptin-sensitive, and so provide another target for metabolic signals to regulate the GnRH network.[186–188]

Another important hypothalamic locus where leptin can influence reproductive function is the VMH. Although widely recognized as a site of glucosensation and the regulation of glucose homeostasis,[10] the VMH has long been known as a key region for female reproductive behavior (see Chapter 20). The VMH consists of three distinct subdivisions, each of which has somewhat different projection patterns[189] and different hormone receptor expression. Neurons in the dorsomedial VMH preferentially express LepRb, while those in the ventrolateral VMH express gonadal steroid receptors[190] (Fig. 14.7A).

Although not yet explored in any detail, the possibility remains that leptin's interactions with the reward-assignment systems may influence motivated behaviors other than feeding, including reproduction.

Effects of Gonadal Steroids on Circuits Concerned with Energy Balance

Much evidence shows that both gonadal steroids, but particularly estrogens, have significant effects on the neural networks regulating energy balance. The fact that some aspects of eating patterns, pathology and metabolism are sexually dimorphic[63,191–193] emphasizes the powerful effects gonadal steroids can have on energy balance. Furthermore, sex differences in the regulation of the HPA axis have been known for over 50 years, and these have implications for sex differences in energy metabolism. These range from variations in resting circulating glucocorticoid levels to differential responses to stress.[194]

Figure 14.7B provides a model for considering how energy balance can affect the GnRH network, and for evaluating the effects of gonadal steroids on energy balance. The direct actions of gonadal steroids on energy balance is suggested by the expression of ERs in the PVH (ERβ) and ARH (ERα)[195,196]), a small proportion of which express NPY,[197,198] and the fact that targeted silencing of the ERα gene in the VMH leads to obesity, hyperphagia, impaired tolerance to glucose, and reduced energy expenditure.[199] In most instances, however, there are no specific data showing which cell types are steroid-sensitive and where they project.

Gonadal steroids may also act indirectly to alter the activity of neurons that provide inputs to the drive networks for energy balance. These may involve AVPV, PMv, or catecholaminergic and other hindbrain projections to those neurons in the ARH and PVH that regulate energy balance[200] (Fig. 14.7B). Some of these effects may be mediated by steroid-sensitive inputs to CRH neuroendocrine motor neurons,[201,202] and by estrogen-mediated effects on the neural elements that contribute to HPA regulation.[203]

Interactions between Control Networks

The properties of certain hypothalamic nuclei make them potential components in an integrative network that can facilitate information flow between reproductive and energy metabolism motor control systems (Fig. 14.6). Some of these cell groups express mixed populations of hormone receptors, and all have robust connections to each of the neural networks responsible for controlling reproductive and ingestive behavioral motor actions. Four hypothalamic cell groups are particularly well placed to function in this manner (Fig. 14.7A): the DMH, AVPV, ARH and PMv. Of these, the ARH and PMv express gonadal steroid receptors together with LepRs and/or IRs.[30,42,104,190,195] On the other hand, neurons in the AVPV (gonadal steroid) and DMH (LepR) contain significant numbers of only one class of receptor, but have projection patterns that can mediate strong neural interactions with the more receptor-diverse ARH and PMv (Fig. 14.7A). Critically, each of these nuclei also has strong connections with cell groups such as the PVH or the GnRH network that are responsible for controlling specific reproductive or metabolic motor actions.[45,204–206]

CONCLUSION

The fact that the brain and endocrine systems interact to control feeding behavior and energy metabolism has been known for over 80 years; however, our current understanding of how this occurs has made enormous strides in the past 20 years or so. Much of this progress derives from the fact that we now understand the detailed structure of ingestive behaviors, meaning that we can relate the individual components of feeding to underlying neural processes. Together with a detailed structural description of neural connections, this has led to an appreciation that the neuroendocrine control of feeding, metabolism and energy balance depends on sets of brain-wide neural networks, rather than simply and exclusively on hypothalamic or hindbrain "centers". In terms of delineating the functional organization of these neural systems and their components, there is no doubt that the precision afforded by molecular genetics to target cellular mechanisms has opened up new vistas for how hormones regulate brain function. As we have seen, this is not just in terms of how hormones control autonomic and neuroendocrine function, but also about how they influence a range of complex cognitive and emotional processes that impact behavior.

References

1. Watts AG. Structure and function in the conceptual development of mammalian neuroendocrinology between 1920 and 1965. *Brain Res Rev.* 2011;66:174–204.
2. Watts AG, Swanson LW. Anatomy of motivational systems. In: Hal Pashler, Randy Gallistell, eds. *'Stevens' Handbook of Experimental Psychology, Volume 3, Learning, Motivation, and Emotion*. 3rd ed. New York: John Wiley & Sons; 2002:563–632.
3. Schneider JE, Watts AG. Energy partitioning, ingestive behavior and reproductive success. In: Pfaff DW, Arnold AP, Etgen AM, Fahrbach SE, Rubin RT, eds. *Hormones, Brain, and Behavior.* 2nd ed. Vol 1. New York: Academic Press; 2009.
4. Craig W. Appetites and aversions as constituents of instincts. *Biol Bull.* 1918;34:91–107.
5. Grill HJ, Kaplan JM. Caudal brainstem participates in the distributed neural control of feeding. In: Stricker Edward M, ed. *The Handbook of Behavioral Neurobiology. Food and Fluid Intake.* 1st ed. New York: Plenum Press; 1990:125–149.
6. Swanson LW, Mogenson GJ. Neural mechanisms for the functional coupling of autonomic, endocrine and somatomotor responses in adaptive behavior. *Brain Res.* 1981;228:1–34.
7. Grill HJ, Kaplan JM. The neuroanatomical axis for control of energy balance. *Front Neuroendocrinol.* 2002;23:2–40.
8. Strubbe JH, Woods SC. The timing of meals. *Psychol Rev.* 2004;111:128–141.
9. Wansink B. From mindless eating to mindlessly eating better. *Physiol Behav.* 2010;100:454–463.
10. Watts AG, Donovan CD. Sweet talk in the brain:glucosensing, neural networks, and hypoglycemic counterregulation. *Front Neuroendocrinol.* 2010;31:32–43.
11. Lam TKT, Schwartz GJ, Rossetti L. Hypothalamic sensing of fatty acids. *Nat Neurosci.* 2005;8:579–584.
12. Minokoshi Y, Alquier T, Furukawa N, et al. AMP-kinase regulates food intake by responding to hormonal and nutrient signals in the hypothalamus. *Nature.* 2004;428: 569–574.
13. Kahn BB, Alquier T, Carling D, et al. AMP-activated protein kinase: ancient energy gauge provides clues to modern understanding of metabolism. *Cell Metab.* 2005;1:15–25.
14. Xue B, Kahn BB. AMPK integrates nutrient and hormonal signals to regulate food intake and energy balance through effects in the hypothalamus and peripheral tissues. *J Physiol.* 2006;574:73–83.
15. Grill HJ, Hayes MR. The nucleus tractus solitarius: a portal for visceral afferent signal processing, energy status assessment and integration of their combined effects on food intake. *Intl J Obes (Lond).* 2009;33(Suppl 1):S11–S15.
16. Gorski RA. Gonadal hormones and the perinatal development of neuroendocrine function. In: Martini L, Ganong WF, eds. *Frontiers in Neuroendocrinology.* New York: Oxford University Press; 1971:237–290.
17. Harris GW. Hormonal differentiation of the developing central nervous system with respect to patterns of endocrine function. *Phil Trans R Soc Lond B.* 1970;259:165–177.
18. Bouret SG, Draper SJ, Simerly RB. Trophic action of leptin on hypothalamic neurons that regulate feeding. *Science.* 2004;304: 108–110.
19. Swanson LW. Cerebral hemisphere regulation of motivated behavior. *Brain Res.* 2000;886:113–164.
20. Saper CB. Staying awake for dinner: hypothalamic integration of sleep, feeding, and circadian rhythms. *Prog Brain Res.* 2006;153:243–252.
21. Kelley AE, Baldo BA, Pratt WE. A proposed hypothalamic–thalamic–striatal axis for the integration of energy balance, arousal, and food reward. *J Comp Neurol.* 2005;493: 72–85.
22. Berthoud HR. Interactions between the "cognitive" and "metabolic" brain in the control of food intake. *Physiol Behav.* 2007;91:486–498.
23. Saper CB, Chou TC, Elmquist JK. The need to feed: homeostatic and hedonic control of eating. *Neuron.* 2002;36:199–211.
24. Kampe J, Tschöp MH, Hollis JH, et al. An anatomic basis for the communication of hypothalamic, cortical and mesolimbic circuitry in the regulation of energy balance. *Eur J Neurosci.* 2009;30:415–430.
25. Stellar E. The physiology of motivation. *Psychol Rev.* 1954;61:5–22.
26. Watts AG, Boyle CN. The functional architecture of dehydration-anorexia. *Physiol Behav.* 2010;100:472–477.
27. Ahima RS, Flier JS. Leptin. *Annu Rev Physiol.* 2000;62:413–437.
28. Zhang Y, Proenca R, Maffei M, Barone M, Leopold L, Friedman JM. Positional cloning of the mouse obese gene and its human homologue. *Nature.* 1994;372:425–432.
29. Bjørbaek C, Kahn BB. Leptin signaling in the central nervous system and the periphery. *Recent Prog Horm Res.* 2004;59:305–331.
30. Myers Jr MG, Münzberg H, Leinninger GM, et al. The geometry of leptin action in the brain: more complicated than a simple ARC. *Cell Metab.* 2009;9:117–123.
31. Myers MG, Cowley MA, Münzberg H. Mechanisms of leptin action and leptin resistance. *Annu Rev Physiol.* 2008;70:537–556.
32. Ahima RS, Lazar MA. Adipokines and the peripheral and neural control of energy balance. *Mol Endocrinol.* 2008;22: 1023–1031.
33. Tartaglia LA, Dembski M, Weng X, et al. Identification and expression cloning of a leptin receptor, OB-R. *Cell.* 1995;83: 1263–1271.
34. Tartaglia LA. The leptin receptor. *J Biol Chem.* 1997;272:6093–6096.
35. Kastin AJ, Pan W, Maness LM, et al. Decreased transport of leptin across the blood–brain barrier in rats lacking the short form of the leptin receptor. *Peptides.* 1999;20:1449–1453.
36. Hileman SM, Pierroz DD, Masuzaki H, et al. Characterization of short isoforms of the leptin receptor in rat cerebral microvessels and of brain uptake of leptin in mouse models of obesity. *Endocrinology.* 2002;143:775–783.
37. Elmquist JK, Bjørbaek C, Ahima RS, et al. Distributions of leptin receptor mRNA isoforms in the rat brain. *J Comp Neurol.* 1998;395:535–547.
38. Burguera B, Couce ME, Long J, et al. The long form of the leptin receptor (OB-Rb) is widely expressed in the human brain. *Neuroendocrinology.* 2000;71:187–195.
39. Scott MM, Lachey JL, Sternson SM, et al. Leptin targets in the mouse brain. *J Comp Neurol.* 2009;514:518–532.
40. Caron E, Sachot C, Prevot V, et al. Distribution of leptin-sensitive cells in the postnatal and adult mouse brain. *J Comp Neurol.* 2010;518:459–476.
41. Elmquist JK, Ahima RS, Elias CF, et al. Leptin activates distinct projections from the dorsomedial and ventromedial hypothalamic nuclei. *Proc Natl Acad Sci USA.* 1998;95:741–746.
42. Elias CF, Kelly JF, Lee CE, et al. Chemical characterization of leptin-activated neurons in the rat brain. *J Comp Neurol.* 2000;423:261–281.
43. Gautron L, Lazarus M, Scott MM, et al. Identifying the efferent projections of leptin-responsive neurons in the dorsomedial hypothalamus using a novel conditional tracing approach. *J Comp Neurol.* 2010;518:2090–2108.
44. Leshan RL, Louis GW, Jo YH, et al. Direct innervation of GnRH neurons by metabolic- and sexual odorant-sensing leptin receptor neurons in the hypothalamic ventral premammillary nucleus. *J Neurosci.* 2009;29:3138–3147.

45. Leshan RL, Opland DM, Louis GW, et al. Ventral tegmental area leptin receptor neurons specifically project to and regulate cocaine- and amphetamine-regulated transcript neurons of the extended central amygdala. *J Neurosci.* 2010;30: 5713–5723.
46. Villanueva EC, Myers Jr MG. Leptin receptor signaling and the regulation of mammalian physiology. *Intl J Obes (Lond).* 2008;7(Suppl):S8–S12.
47. Myers Jr MG. Outstanding Scientific Achievement Award Lecture 2010: Deconstructing leptin: from signals to circuits. *Diabetes.* 2010;59:2708–2714.
48. Bjørbaek C, Elmquist JK, Frantz JD, et al. Identification of SOCS-3 as a potential mediator of central leptin resistance. *Mol Cell.* 1998;1:619–625.
49. Björnholm M, Münzberg H, Leshan RL, et al. Mice lacking inhibitory leptin receptor signals are lean with normal endocrine function. *J Clin Invest.* 2007;117:1354–1360.
50. Lee JY, Muenzberg H, Gavrilova O, et al. Loss of Cytokine-STAT5 signaling in the cns and pituitary gland alters energy balance and leads to obesity. *PLoS ONE.* 2008;3:e1639.
51. Bates SH, Kulkarni RN, Seifert M, et al. Roles for leptin receptor/STAT3-dependent and -independent signals in the regulation of glucose homeostasis. *Cell Metab.* 2005;1:169–178.
52. Bates SH, Stearns WH, Dundon TA, et al. STAT3 signaling is required for leptin regulation of energy balance but not reproduction. *Nature.* 2003;421:856–859.
53. Campfield LA, Smith FJ, Guisez Y, et al. Recombinant mouse OB protein: evidence for a peripheral signal linking adiposity and central neural networks. *Science.* 1995;269:546–549.
54. Fulton S. Appetite and reward. *Front Neuroendocrinol.* 2010;31:85–103.
55. Grill HJ. Leptin and the systems neuroscience of meal size control. *Front Neuroendocrinol.* 2010;31:61–78.
56. Narayanan NS, Guarnieri DJ, DiLeone RJ. Metabolic hormones, dopamine circuits, and feeding. *Front Neuroendocrinol.* 2010;31: 104–112.
57. Hulsey MG, Lu H, Wang T, et al. Intracerebroventricular (i.c.v.) administration of mouse leptin in rats: behavioral specificity and effects on meal patterns. *Physiol Behav.* 1998;65:445–455.
58. Kahler A, Geary N, Eckel LA, et al. Chronic administration of OB protein decreases food intake by selectively reducing meal size in male rats. *Am J Physiol.* 1998;275:R180–R185.
59. Clark JT, Kalra PS, Crowley WR, et al. Neuropeptide Y and human pancreatic polypeptide stimulate feeding behavior in rats. *Endocrinology.* 1984;115:427–429.
60. Levine AS, Morley JE. Neuropeptide Y: a potent inducer of consummatory behavior in rats. *Peptides.* 1984;5:1025–1029.
61. Stanley BG, Leibowitz SF. Neuropeptide Y: stimulation of feeding and drinking by injection into the paraventricular nucleus. *Life Sci.* 1984;35:2635–2642.
62. Baird JP, Rios C, Gray NE, et al. Effects of melanin-concentrating hormone on licking microstructure and brief-access taste responses. *Am J Physiol Regul Integr Comp Physiol.* 2006;291: R1265–R1274.
63. Södersten P, Nergårdh R, Bergh C, et al. Behavioral neuroendocrinology and treatment of anorexia nervosa. *Front Neuroendocrinol.* 2008;29:445–462.
64. Woods SC, Figlewicz DP, Madden L, et al. NPY and food intake: discrepancies in the model. *Regul Pept.* 1998;75–76:403–408.
65. Boston BA, Blaydon KM, Varnerin J, et al. Independent and additive effects of central POMC and leptin pathways on murine obesity. *Science.* 1997;278:1641–1644.
66. Fan W, Boston BA, Kesterson RA, et al. Role of melanocortinergic neurons in feeding and the agouti obesity syndrome. *Nature.* 1997;385:165–168.
67. Huszar D, Lynch CA, Fairchild-Huntress V, et al. Targeted disruption of the melanocortin-4 receptor results in obesity in mice. *Cell.* 1997;88:131–141.
68. Rossi M, Kim MS, Morgan DG, et al. A C-terminal fragment of Agouti-related protein increases feeding and antagonizes the effect of alpha-melanocyte stimulating hormone *in vivo*. *Endocrinology.* 1998;139:4428–4431.
69. Marsh DJ, Hollopeter G, Huszar D, et al. Response of melanocortin-4 receptor-deficient mice to anorectic and orexigenic peptides. *Nat Genet.* 1999;21:119–122.
70. Schwartz MW, Seeley RJ, Campfield LA, et al. Identification of targets of leptin action in rat hypothalamus. *J Clin Invest.* 1996;98:1101–1106.
71. Schwartz MW, Baskin DG, Bukowski TR, et al. Specificity of leptin action on elevated blood glucose levels and hypothalamic neuropeptide Y gene expression in ob/ob mice. *Diabetes.* 1996;45:531–535.
72. Thornton JE, Cheung CC, Clifton DK, et al. Regulation of hypothalamic proopiomelanocortin mRNA by leptin in ob/ob mice. *Endocrinology.* 1997;138:5063–5066.
73. Glaum SR, Hara M, Bindokas VP, et al. Leptin, the obese gene product, rapidly modulates synaptic transmission in the hypothalamus. *Mol Pharmacol.* 1996;50:230–235.
74. Cowley MA, Smart JL, Rubinstein M, et al. Leptin activates anorexigenic POMC neurons through a neural network in the arcuate nucleus. *Nature.* 2001;411:480–484.
75. Erickson J, Hollopeter G, Palmiter JD. Attenuation of the obesity syndrome of ob/ob mice by the loss of neuropeptide Y. *Science.* 1996;274:1704–1707.
76. Coppari R, Ichinose M, Lee CE, et al. The hypothalamic arcuate nucleus: a key site for mediating leptin's effects on glucose homeostasis and locomotor activity. *Cell Metab.* 2005;1:63–72.
77. Morton GJ, Gelling RW, Niswender KD, et al. Leptin regulates insulin sensitivity via phosphatidylinositol-3-OH kinase signaling in mediobasal hypothalamic neurons. *Cell Metab.* 2005;2:411–420.
78. German J, Kim F, Schwartz GJ, et al. Hypothalamic leptin signaling regulates hepatic insulin sensitivity via a neurocircuit involving the vagus nerve. *Endocrinology.* 2009;150:4502–4511.
79. Buettner C, Muse ED, Cheng A, et al. Leptin controls adipose tissue lipogenesis via central, STAT3-independent mechanisms. *Nat Med.* 2008;14:667–675.
80. Plum L, Rother E, Münzberg H, et al. Enhanced leptin-stimulated Pi3k activation in the CNS promotes white adipose tissue trans-differentiation. *Cell Metab.* 2007;6:431–445.
81. Dhillon H, Zigman JM, Ye C, et al. Leptin directly activates SF1 neurons in the VMH, and this action by leptin is required for normal body-weight homeostasis. *Neuron.* 2006;49:191–203.
82. Huo L, Maeng L, Bjørbaek C, et al. Leptin and the control of food intake: neurons in the nucleus of the solitary tract are activated by both gastric distension and leptin. *Endocrinology.* 2007;148:2189–2197.
83. Grill HJ, Schwartz MW, Kaplan JM, et al. Evidence that the caudal brainstem is a target for the inhibitory effect of leptin on food intake. *Endocrinology.* 2002;143:239–246.
84. Hayes MR, Skibicka KP, Leichner TM, et al. Endogenous leptin signaling in the caudal nucleus tractus solitarius and area postrema is required for energy balance regulation. *Cell Metab.* 2010;11:77–83.
85. Balthasar N, Coppari R, McMinn J, et al. Leptin receptor signaling in POMC neurons is required for normal body weight homeostasis. *Neuron.* 2004;42:983–991.
86. van de Wall E, Leshan R, Xu AW, et al. Collective and individual functions of leptin receptor modulated neurons

controlling metabolism and ingestion. *Endocrinology.* 2008;149: 1773–1785.
87. Figlewicz DP, Bennett JL, Naleid AM, et al. Intraventricular insulin and leptin decrease sucrose self-administration in rats. *Physiol Behav.* 2006;89:611–616.
88. Fulton S, Woodside B, Shizgal P. Modulation of brain reward circuitry by leptin. *Science.* 2000;287:125–128.
89. Szczypka MS, Rainey MA, Kim DS, et al. Feeding behavior in dopamine-deficient mice. *Proc Natl Acad Sci USA.* 1999;96: 12138–12143.
90. Szczypka MS, Rainey MA, Palmiter RD. Dopamine is required for hyperphagia in Lep(ob/ob) mice. *Nat Genet.* 2000;25: 102–104.
91. Figlewicz DP, Evans SB, Murphy J, et al. Expression of receptors for insulin and leptin in the ventral tegmental area/substantia nigra (VTA/SN) of the rat. *Brain Res.* 2003;964:107–115.
92. Fulton S, Pissios P, Manchon RP, et al. Leptin regulation of the mesoaccumbens dopamine pathway. *Neuron.* 2006;51:811–822.
93. Hommel JD, Trinko R, Sears RM, et al. Leptin receptor signaling in midbrain dopamine neurons regulates feeding. *Neuron.* 2006;51:801–810.
94. Leinninger GM, Jo YH, Leshan RL, et al. Leptin acts via leptin receptor-expressing lateral hypothalamic neurons to modulate the mesolimbic dopamine system and suppress feeding. *Cell Metab.* 2009;10:89–98.
95. Bouret SG, Draper SJ, Simerly RB. Formation of projection pathways from the arcuate nucleus of the hypothalamus to hypothalamic regions implicated in the neural control of feeding behavior in mice. *J Neurosci.* 2004;24:2797–2805.
96. Bouret SG, Gorski JN, Patterson CM, et al. Hypothalamic neural projections are permanently disrupted in diet-induced obese rats. *Cell Metab.* 2008;7:179–185.
97. Pinto S, Roseberry AG, Liu H, et al. Rapid rewiring of arcuate nucleus feeding circuits by leptin. *Science.* 2004;304:110–115.
98. Moult PR, Milojkovic B, Harvey J. Leptin reverses long-term potentiation at hippocampal CA1 synapses. *J Neurochem.* 2009;108:685–696.
99. O'Malley D, MacDonald N, Mizielinska S, et al. Leptin promotes rapid dynamic changes in hippocampal dendritic morphology. *Mol Cell Neurosci.* 2007;35:559–572.
100. Anand BK, Chhina GS, Sharma KN, et al. Activity of single neurons in the hypothalamic feeding centers: effect of glucose. *Am J Physiol.* 1964;207:1146–1154.
101. Havrankova J, Roth J, Brownstein M. Insulin receptors are widely distributed in the central nervous system of the rat. *Nature.* 1978;272:827–829.
102. Banks WA. The source of cerebral insulin. *Eur J Pharmacol.* 2004;490:5–12.
103. Woods SC, Lotter EC, McKay LD, et al. Chronic intracerebroventricular infusion of insulin reduces food intake and body weight of baboons. *Nature.* 1979;282:503–505.
104. Marks JL, Porte Jr D, Stahl WL, et al. Localization of insulin receptor mRNA in rat brain by *in situ* hybridization. *Endocrinology.* 1990;127:3234–3236.
105. Werther GA, Hogg A, Oldfield BJ, et al. Localization and characterization of insulin receptors in rat brain and pituitary gland using *in vitro* autoradiography and computerized densitometry. *Endocrinology.* 1987;121:1562–1570.
106. Brüning JC, Gautam D, Burks DJ, et al. Role of brain insulin receptor in control of body weight and reproduction. *Science.* 2000;289:2122–2125.
107. Diggs-Andrews KA, Zhang X, Song Z, et al. Brain insulin action regulates hypothalamic glucose sensing and the counter-regulatory response to hypoglycemia. *Diabetes.* 2010;59: 2271–2280.
108. Fisher SJ, Brüning JC, Lannon S, et al. Insulin signaling in the central nervous system is critical for the normal sympathoadrenal response to hypoglycemia. *Diabetes.* 2005;54:1447–1451.
109. Williams KW, Margatho LO, Lee CE, et al. Segregation of acute leptin and insulin effects in distinct populations of arcuate proopiomelanocortin neurons. *J Neurosci.* 2010;30:2472–2479.
110. Benoit SC, Air EL, Coolen LM, et al. The catabolic action of insulin in the brain is mediated by melanocortins. *J Neurosci.* 2002;22:9048–9052.
111. Plum L, Schubert M, Brüning JC. The role of insulin receptor signaling in the brain. *Trends Endocrinol Metab.* 2005;16:59–65.
112. Obici S, Zhang BB, Karkanias G, et al. Hypothalamic insulin signaling is required for inhibition of glucose production. *Nat Med.* 2002;8:1376–1382.
113. Schwartz MW, Sipols AJ, Marks JL, et al. Inhibition of hypothalamic neuropeptide Y gene expression by insulin. *Endocrinol.* 1992;130:3608–3616.
114. Obici S, Feng Z, Tan J, et al. Central melanocortin receptors regulate insulin action. *J Clin Invest.* 2001;108:1079–1085.
115. Hill JW, Elias CF, Fukuda M, et al. Direct insulin and leptin action on pro-opiomelanocortin neurons is required for normal glucose homeostasis and fertility. *Cell Metab.* 2010;11:286–297.
116. Obici S, Feng Z, Karkanias G, et al. Decreasing hypothalamic insulin receptors causes hyperphagia and insulin resistance in rats. *Nat Neurosci.* 2002;5:566–572.
117. Niswender KD, Morrison CD, Clegg DJ, et al. Insulin activation of phosphatidylinositol 3-kinase in the hypothalamic arcuate nucleus: a key mediator of insulin-induced anorexia. *Diabetes.* 2003;52:227–231.
118. Figlewicz DP, Benoit SC. Insulin, leptin, and food reward: update 2008. *Am J Physiol Regul Integr Comp Physiol.* 2008;296:R9–R19.
119. Huang CC, Lee CC, Hsu KS. An investigation into signal transduction mechanisms involved in insulin-induced long-term depression in the CA1 region of the hippocampus. *J Neurochem.* 2004;89:217–231.
120. van der Heide LP, Kamal A, Artola A, et al. Insulin modulates hippocampal activity-dependent synaptic plasticity in a N-methyl-d-aspartate receptor and phosphatidyl-inositol-3-kinase-dependent manner. *J Neurochem.* 2005;94:1158–1166.
121. McNay EC, Ong CT, McCrimmon RJ, et al. Hippocampal memory processes are modulated by insulin and high-fat-induced insulin resistance. *Neurobiol Learn Mem.* 2010;93: 546–553.
122. Figlewicz DP, Sipols AJ, Seeley RJ, et al. Intraventricular insulin enhances the meal-suppressive efficacy of intraventricular cholecystokinin octapeptide in the baboon. *Behav Neurosci.* 1995;109:567–569.
123. Sapolsky RM, Romero LM, Munck AU. How do glucocorticoids influence stress responses? Integrating permissive, suppressive, stimulatory, and preparative actions. *Endocr Rev.* 2000;21:55–89.
124. Roozendaal B, Okuda S, de Quervain DJ, et al. Glucocorticoids interact with emotion-induced noradrenergic activation in influencing different memory functions. *Neuroscience.* 2006;138: 901–910.
125. de Kloet ER, Joëls M, Holsboer F. Stress and the brain: from adaptation to disease. *Nat Rev Neurosci.* 2005;6:463–475.
126. Dallman MF, Akana SF, Bhatnagar S, et al. Starvation: early signals, sensors, and sequelae. *Endocrinology.* 1999;140:4015–4023.
127. Margetic S, Gazzola C, Pegg GG, et al. Leptin: a review of its peripheral actions and interactions. *Intl J Obes Relat Metab Disord.* 2002;26:1407–1433.
128. Barthel A, Schmoll D. Novel concepts in insulin regulation of hepatic gluconeogenesis. *Am J Physiol Endocrinol Metab.* 2003; 285:E685–E692.

129. Aronsson M, Fuxe K, Dong Y, et al. Localization of glucocorticoid receptor mRNA in the male rat brain by *in situ* hybridization. *Proc Natl Acad Sci USA.* 1988;85:9331–9335.
130. Sawchenko PE, Bohn MC. Glucocorticoid receptor-immunoreactivity in C1, C2, and C3 adrenergic neurons that project to the hypothalamus or to the spinal cord in the rat. *J Comp Neurol.* 1989;285:107–116.
131. Cummings DE, Overduin J. Gastrointestinal regulation of food intake. *J Clin Invest.* 2007;117:13–23.
132. Bray GA. Afferent signals regulating food intake. *Proc Nutr Soc.* 2000;59:373–384.
133. Moran TH, Kinzig KP. Gastrointestinal satiety signals II. Cholecystokinin. *Am J Physiol Gastrointest Liver Physiol.* 2004; 286:G183–G188.
134. Peters JH, Simasko SM, Ritter RC. Modulation of vagal afferent excitation and reduction of food intake by leptin and cholecystokinin. *Physiol Behav.* 2006;89:477–485.
135. Blevins JE, Schwartz MW, Baskin DG. Evidence that paraventricular nucleus oxytocin neurons link hypothalamic leptin action to caudal brain stem nuclei controlling meal size. *Am J Physiol Regul Integr Comp Physiol.* 2004;287:R87–R96.
136. Castañeda TR, Tong J, Datta R, et al. Ghrelin in the regulation of body weight and metabolism. *Front Neuroendocrinol.* 2010;31: 44–60.
137. Tschöp M, Smiley DL, Heiman ML. Ghrelin induces adiposity in rodents. *Nature.* 2000;407:908–913.
138. Wren AM, Seal LJ, Cohen MA, et al. Ghrelin enhances appetite and increases food intake in humans. *J Clin Endocrinol Metab.* 2001;86:5992.
139. Zigman JM, Jones JE, Lee CE, et al. Expression of ghrelin receptor mRNA in the rat and the mouse brain. *J Comp Neurol.* 2006;494:528–548.
140. Olszewski PK, Schiöth HB, Levine AS. Ghrelin in the CNS: from hunger to a rewarding and memorable meal? *Brain Res Rev.* 2008;58:160–170.
141. Luquet S, Phillips CT, Palmiter RD. NPY/AgRP neurons are not essential for feeding responses to glucoprivation. *Peptides.* 2007;28:214–225.
142. Emanuel AJ, Ritter S. Hindbrain catecholamine neurons modulate the growth hormone but not the feeding response to ghrelin. *Endocrinology.* 2010;151:3237–3246.
143. Brubaker PL, Anini Y. Direct and indirect mechanisms regulating secretion of glucagon-like peptide-1 and glucagon-like peptide-2. *Can J Physiol Pharmacol.* 2003;81:1005–1012.
144. Adrian TE, Ferri GL, Bacarese-Hamilton AJ, et al. Human distribution and release of a putative new gut hormone, peptide YY. *Gastroenterology.* 1985;89:1070–1077.
145. Ueno H, Yamaguchi H, Mizuta M, et al. The role of PYY in feeding regulation. *Regul Pept.* 2008;145:12–16.
146. Acuna-Goycolea C, van den Pol AN. Peptide YY(3-36) inhibits both anorexigenic proopiomelanocortin and orexigenic neuropeptide Y neurons: implications for hypothalamic regulation of energy homeostasis. *J Neurosci.* 2005;25: 10510–10519.
147. Kim W, Egan JM. The role of incretins in glucose homeostasis and diabetes treatment. *Pharmacol Rev.* 2008;60:470–512.
148. Turton MD, O'Shea D, Gunn I, et al. A role for glucagon-like peptide-1 in the central regulation of feeding. *Nature.* 1996;379:69–72.
149. Talsania T, Anini Y, Siu S, et al. Peripheral exendin-4 and peptide YY(3-36) synergistically reduce food intake through different mechanisms in mice. *Endocrinology.* 2005;146: 3748–3756.
150. Potes CS, Lutz TA. Brainstem mechanisms of amylin-induced anorexia. *Physiol Behav.* 2010;100:511–518.
151. Lutz TA. The role of amylin in the control of energy homeostasis. *Am J Physiol Regul Integr Comp Physiol.* 2010;298: R1475–R1484.
152. Dallman MF, Strack AM, Akana SF, et al. Feast and famine: critical role of glucocorticoids with insulin in daily energy flow. *Front Neuroendocrinol.* 1993;14:303–347.
153. Harris RC, Fang S-L, Azizi F, et al. Effect of starvation on hypothalamic–pituitary–thyroid function in the rat. *Metabolism.* 1978;27:1074–1083.
154. Kinlaw WB, Schwartz HL, Oppenheimer JL. Decreased serum triiodothyronine in starving rats is due primarily to diminished thyroidal secretion of thyroxine. *J Clin Invest.* 1985;75:1238–1241.
155. Blake NG, Eckland DJ, Foster OJ, et al. Inhibition of hypothalamic thyrotropin-releasing hormone messenger ribonucleic acid during food deprivation. *Endocrinol.* 1991;129:2714–2718.
156. van Haasteren GA, Linkels E, van Toor H, et al. Effects of long-term food reduction on the hypothalamus–pituitary–thyroid axis in male and female rats. *J Endocrinol.* 1996;150:169–178.
157. Harris M, Aschkenasi C, Elias CF, et al. Transcriptional regulation of the thyrotropin-releasing hormone gene by leptin and melanocortin signaling. *J Clin Invest.* 2001;107:111–112.
158. Ahima RS, Prabakaran D, Mantzoros C, et al. Role of leptin in the neuroendocrine response to fasting. *Nature.* 1996;382: 250–252.
159. Légrádi G, Emerson CH, Ahima RS, et al. Leptin prevents fasting-induced suppression of prothyrotropin-releasing hormone messenger ribonucleic acid in neurons of the hypothalamic paraventricular nucleus. *Endocrinology.* 1997;138: 2569–2576.
160. Ghamari-Langroudi M, Vella KR, Srisai D, et al. Regulation of thyrotropin-releasing hormone-expressing neurons in paraventricular nucleus of the hypothalamus by signals of adiposity. *Mol Endocrinol.* 2010;24:2366–2381.
161. Nillni EA, Vaslet C, Harris M, et al. Leptin regulates prothyrotropin-releasing hormone biosynthesis. Evidence for direct and indirect pathways. *J Biol Chem.* 2000;275:36124–36133.
162. Woodward CJ, Hervey GR, Oakey RE, et al. The effects of fasting on plasma corticosterone kinetics in rats. *Br J Nutr.* 1991;66:117–127.
163. Suemaru S, Hashimoto K, Hattori T, et al. Starvation-induced changes in rat brain corticotropin-releasing factor (CRF) and pituitary–adrenocortical response. *Life Sci.* 1986;39:1161–1166.
164. Watts AG. Disturbance of fluid homeostasis leads to temporally and anatomically distinct responses in the levels of neuropeptide and tyrosine hydroxylase mRNAs within the hypothalamic paraventricular and supraoptic nuclei of the rat. *Neuroscience.* 1992;46:859–879.
165. Watts AG. Understanding the neural control of ingestive behaviors: helping to separate cause from effect with dehydration-associated anorexia. *Horm Behav.* 2000;37:261–283.
166. Brady LS, Smith MA, Gold PW, et al. Altered expression of hypothalamic neuropeptide mRNAs in food-restricted and food-deprived rats. *Neuroendocrinology.* 1990;52:441–447.
167. Mitrakou A, Ryan C, Veneman T, et al. Hierarchy of glycemic thresholds for counterregulatory hormone secretion, symptoms, and cerebral dysfunction. *Am J Physiol.* 1991;260:E67–E74.
168. Tesfaye N, Seaquist ER. Neuroendocrine responses to hypoglycemia. *Ann NY Acad Sci.* 2010;1212:12–28.
169. McCrimmon RJ, Sherwin RS. Hypoglycemia in type 1 diabetes. *Diabetes.* 2010;59:2333–2339.
170. Levin BE, Routh VH, Kang L, et al. Neuronal glucosensing: what do we know after 50 years? *Diabetes.* 2004;53:2521–2528.
171. Parton LE, Ye CP, Coppari R, et al. Glucose sensing by POMC neurons regulates glucose homeostasis and is impaired in obesity. *Nature.* 2007;449:228–232.

172. Baes M, Vale WW. Characterization of the glucose-dependent release of growth hormone-releasing factor and somatostatin from superfused rat hypothalami. *Neuroendocrinology.* 1990;51: 202–207.
173. Tamaki M, Sato M, Niimi M, et al. Resistance of growth hormone secretion to hypoglycemia in the mouse. *J Neuroendocrinol.* 1995;7:371–376.
174. Chapman IM, Kapoor R, Willoughby JO. Endogenous catecholamines modulate growth hormone release in the conscious rat during hypoglycaemia but not in the basal state. *J Neuroendocrinol.* 1993;5:145–150.
175. Khan AM, Ponzio TA, Sanchez-Watts G, et al. Catecholaminergic control of MAP kinase signaling in paraventricular neuroendocrine neurons *in vivo* and *in vitro*: A proposed role during glycemic challenges. *J Neurosci.* 2007;27:7344–7360.
176. Gorton LM, Khan AM, Bohland MA, et al. A role for the forebrain in mediating time-of-day differences in glucocorticoid counterregulatory responses to hypoglycemia. *Endocrinology.* 2007;148:6026–6039.
177. Andrew SF, Dinh TT, Ritter S. Localized glucoprivation of hindbrain sites elicits corticosterone and glucagon secretion. *Am J Physiol Regul Integr Comp Physiol.* 2007;292:R1792–R1798.
178. Ritter S, Bugarith K, Dinh TT. Immunotoxic destruction of distinct catecholamine subgroups produces selective impairment of glucoregulatory responses and neuronal activation. *J Comp Neurol.* 2001;432:197–216.
179. Ritter S, Watts AG, Dinh TT, et al. Immunotoxin lesions of hypothalamically-projecting norepinephrine and epinephrine neurons differentially affect circadian and stressor-stimulated corticosterone secretion. *Endocrinology.* 2003;144:1357–1367.
180. Khan AM, Watts AG. Intravenous 2-deoxy-D-glucose injection rapidly elevates levels of the phosphorylated forms of p44/42 mitogen activated protein kinases (ERK1/2) in rat hypothalamic parvocellular paraventricular neurons. *Endocrinology.* 2004;145: 351–359.
181. Kalsbeek A, Ruiter M, La Fleur SE, et al. The diurnal modulation of hormonal responses in the rat varies with different stimuli. *J Neuroendocrinol.* 2003;15:1144–1155.
182. Medina CL, Nagatani S, Darling TA, et al. Glucose availability modulates the timing of the luteinizing hormone surge in the ewe. *J Neuroendocrinol.* 1998;10:785–792.
183. Dong Q, Lazarus RM, Wong LS, et al. Pulsatile LH secretion in streptozotocin-induced diabetes in the rat. *J Endocrinol.* 1991;131:49–55.
184. Bucholtz DC, Chiesa A, Pappano WN, et al. Regulation of pulsatile luteinizing hormone secretion by insulin in the diabetic male lamb. *Biol Reprod.* 2000;62:1248–1255.
185. Tanaka T, Nagatani S, Bucholtz DC, et al. Central action of insulin regulates pulsatile luteinizing hormone secretion in the diabetic sheep model. *Biol Reprod.* 2000;62:1256–1261.
186. Cravo RM, Margatho LO, Osborne-Lawrence S, Donato Jr J, Atkin S, Bookout AL, Rovinsky S, Frazão R, Lee CE, Gautron L, Zigman JM, Elias CF. Characterization of Kiss1 neurons using transgenic mouse models. *Neuroscience.* 2011;173:37–56.
187. Smith JT, Acohido BV, Clifton DK, Steiner RA. KiSS-1 neurones are direct targets for leptin in the ob/ob mouse. *J Neuroendocrinol.* 2006;18:298–303.
188. Quennell JH, Howell CS, Roa J, Augustine RA, Grattan DR, Anderson GM. Leptin deficiency and diet-induced obesity reduce hypothalamic kisspeptin expression in mice. *Endocrinology.* 2011 Feb 15 [Epub ahead of print].
189. Canteras NS, Simerly RB, Swanson LW. Organization of projections from the ventromedial nucleus of the hypothalamus: a *Phaseolus vulgaris*—leucoagglutinin study in the rat. *J Comp Neurol.* 1994;348:41–79.
190. Simerly RB, Chang C, Muramatsu M, et al. Distribution of androgen and estrogen receptor mRNA-containing cells in the rat brain: an *in situ* hybridization study. *J Comp Neurol.* 1990;294:76–95.
191. Asarian L, Geary N. Modulation of appetite by gonadal steroid hormones. *Philos Trans R Soc Lond B Biol Sci.* 2006;361: 1251–1263.
192. Shi H, Seeley RJ, Clegg DJ. Sexual differences in the control of energy homeostasis. *Front Neuroendocrinol.* 2009;30:396–404.
193. Brown LM, Clegg DJ. Central effects of estradiol in the regulation of food intake, body weight, and adiposity. *J Steroid Biochem Mol Biol.* 2010;122:65–73.
194. Viau V. Functional cross-talk between the hypothalamic—pituitary—gonadal and —adrenal axes. *J Neuroendocrinol.* 2002;14:506–513.
195. Laflamme N, Nappi RE, Drolet G, et al. Expression and neuropeptidergic characterization of estrogen receptors (ERalpha and ERbeta) throughout the rat brain: anatomical evidence of distinct roles of each subtype. *J Neurobiol.* 1998;36:357–378.
196. Shughrue PJ, Lane MV, Merchenthaler I. Comparative distribution of estrogen receptor-alpha and -beta mRNA in the rat central nervous. *J Comp Neurol.* 1997;388:507–525.
197. Sar M, Sahu A, Crowley WR, et al. Localization of neuropeptide-Y immunoreactivity in estradiol-concentrating cells in the hypothalamus. *Endocrinology.* 1990;127:2752–2756.
198. Simonian SX, Spratt DP, Herbison AE. Identification and characterization of estrogen receptor alpha-containing neurons projecting to the vicinity of the gonadotropin-releasing hormone perikarya in the rostral preoptic area of the rat. *J Comp Neurol.* 1999;411:346–358.
199. Musatov S, Chen W, Pfaff DW, Mobbs CV, Yang XJ, Clegg DJ, Kaplitt MG, Ogawa S. Silencing of estrogen receptor alpha in the ventromedial nucleus of hypothalamus leads to metabolic syndrome. *Proc Natl Acad Sci USA.* 2007;104(7):2501–2506.
200. Thammacharoen S, Lutz TA, Geary N, et al. Hindbrain administration of estradiol inhibits feeding and activates estrogen receptor-alpha-expressing cells in the nucleus tractus solitarius of ovariectomized rats. *Endocrinology.* 2008;149: 1609–1617.
201. Viau V, Soriano L, Dallman MF. Androgens alter corticotropin releasing hormone and arginine vasopressin mrna within forebrain sites known to regulate activity in the hypothalamic—pituitary—adrenal axis. *J Neuroendocrinol.* 2001;13: 442–452.
202. Williamson M, Viau V. Androgen receptor expressing neurons that project to the paraventricular nucleus of the hypothalamus in the male rat. *J Comp Neurol.* 2007;503:717–740.
203. Patchev VK, Almeida OF. Gonadal steroids exert facilitating and "buffering" effects on glucocorticoid-mediated transcriptional regulation of corticotropin-releasing hormone and corticosteroid receptor genes in rat brain. *J Neurosci.* 1996;16:7077–7084.
204. Canteras NS, Simerly RB, Swanson LW. The projections of the ventral premammillary nucleus. *J Comp Neurol.* 1992;324: 195–212.
205. Gu GB, Simerly RB. Projections of the sexually dimorphic anteroventral periventricular nucleus in the female rat. *J Comp Neurol.* 1997;384:142–164.
206. Thompson RH, Swanson LW. Structural characterization of a hypothalamic visceromotor pattern generator network. *Brain Res Rev.* 2003;41:153–202.

CHAPTER

15

Neuropeptide Regulation of Stress-Induced Behavior: Insights from the CRF/Urocortin Family

Yehezkel Sztainberg, Alon Chen

Department of Neurobiology, Weizmann Institute of Science, Rehovot, Israel

OUTLINE

Summary

The neuropeptide corticotropin-releasing factor (CRF) and the more recently identified family members, the urocortins (Ucns), are proposed to integrate the neuroendocrine, autonomic, metabolic and behavioral responses to stressors. Chronic hyperactivation of the CRF system has been linked to stress-related emotional disorders such as anxiety and depression. The intracerebral administration of CRF or Ucn1 results in behavioral responses that are similar to those observed when animals are exposed to a stressor. The behavioral effects include increased anxiety-like behavior, decreased food consumption, increased arousal, altered locomotor activity, diminished sexual behavior, and sleep disruption. The role of the CRF/Urocortin system in modulating the behavioral responses to stressors was further demonstrated based on the behavioral phenotypes of transgenic mice that overexpress, or are deficient in, the different CRF family members. The CRF–CRFR1 system is suggested as critical for initiating stress responses, while the urocortins–CRFR2 system is suggested to terminate stress responses and restore homeostasis.

INTRODUCTION: THE CRF FAMILY OF PEPTIDES AND RECEPTORS

Perception of physical or psychological stress by an organism is followed by a series of events that result in changes in emotional and cognitive functions, modulation of autonomic activities and the secretion of glucocorticoids from the adrenal cortex. Both the activation

Handbook of Neuroendocrinology, DOI: 10.1016/B978-0-12-375097-6.10015-0

and the termination of the behavioral, autonomic and adrenocortical stress responses are crucial for adaptation and survival. The neuropeptide corticotropin-releasing factor (CRF), expressed and secreted from the paraventricular nucleus (PVN) of the hypothalamus, represents the final common pathway for the integration of the neuroendocrine stress response in the brain.[1] CRF and, to a lesser extent, arginine vasopressin (AVP) play an important and well-established role in the regulation of the hypothalamus–pituitary–adrenal (HPA) axis under basal and stress conditions.[2,3] In addition to the PVN, CRF mRNA and peptide have been found in several extrahypothalamic brain nuclei, including the cerebral cortex, amygdala, bed nucleus of stria terminalis (BNST), and hippocampus, suggesting that, in addition to its hypophysiotropic action, CRF acts to integrate the neuroendocrine, behavioral, autonomic and metabolic responses to stressors.[4–6] Dysregulation of the stress response can have severe psychological and physiological consequences,[7] and chronic hyperactivation of the CRF system has been linked to stress-related emotional disorders such as anxiety, anorexia nervosa and melancholic depression.[8–10] In addition to CRF, the mammalian CRF-peptide family contains urocortin (Ucn) 1,[11] Ucn2[12,13] and Ucn3.[13,14] CRF and its related peptides signal through the activation of two high-affinity G-protein coupled receptors, CRF receptor type 1 (CRFR1)[15–17] and CRFR2.[18–22] Both CRFR1 and CRFR2 belong to the class B1 subfamily of seven-transmembrane domain receptors that signal by coupling to G proteins. CRFR1 has one known functional splice variant (α) expressed in the central nervous system and the periphery.[23,24] The CRFR1 mRNA is widely expressed in mammalian brain and pituitary, with high levels in the anterior pituitary, cerebral cortex, cerebellum, globus pallidus, amygdala, hippocampus and olfactory bulb.[25] In the periphery, CRFR1 is expressed in testes, ovary, skin and spleen. CRFR2 has three functional membrane splice variants in the human (α, β and γ) and two in the rodent (α and β).[18–22,24,26] CRFR type 2α (CRFR2α) is the major splice variant in the rodent brain.[27] It is expressed in a more discrete pattern than CRFR1, with highest densities in the lateral septum, BNST, ventromedial hypothalamic nucleus (VMH), olfactory bulb and mesencephalic dorsal raphe nucleus (DRN),[25] whereas CRFR type 2β (CRFR2β) is expressed primarily in peripheral tissues and the choroid plexus of the brain.[18] CRFR1 and CRFR2 differ pharmacologically: CRF has relatively lower affinity for CRFR2 compared to its affinity for CRFR1, Ucn1 has equal affinities for both receptors, and Ucn2 and Ucn3 appear to be selective for CRFR2.[11,12,14] In addition to both CRF receptors, CRF and Ucn1 can bind to the CRF-binding protein (CRF-BP) suggested to function as an endogenous "buffer," thus adding a further level of complexity to the control of the CRF-related ligands' actions.[28–30]

CENTRAL ADMINISTRATION OF CRF/Ucn1

The intracerebroventricular administration of CRF or Ucn1 results in behavioral responses that are similar to those observed when animals are exposed to a stressor (Fig. 15.1). The behavioral effects include increased anxiety-like behavior, decreased food consumption, increased arousal, altered locomotor activity, diminished sexual behavior, and sleep disruption (for detailed reviews, see references 5,31–35).

One of the earliest and best described effects of central CRF administration is increased arousal and altered locomotor activity which is dependent on the testing conditions. In non-stressed rats tested in a familiar environment, i.c.v. injection of CRF elicits a dose-dependent increase in general behavioral activation that includes elevated walking, excessive grooming and rearing.[4,36–38] This effect is apparently mediated by central pathways and independent of pituitary–adrenal activation, since CRF increases locomotor activity in both naïve and hypophysectomized animals,[39] and the effect is not blocked by dexamethasone administration.[40] The potent stimulatory effect of CRF in behavioral activation is evolutionary conserved, as has also been described in several vertebrate species.[35] CRF increases swimming and walking in amphibians[41–44] and fish,[45,46] stepping in birds,[47,48] and general motor activity in pigs,[49–52] mice[36,53,54] and primates.[55,56]

In contrast to the behavioral activating effect of CRF in animals under low arousal conditions, when animals are exposed to a novel stressful environment such as an open field or an elevated plus maze, CRF administration effects changes to behavioral inhibition. Following i.c.v. injection of CRF, rodents show decreased locomotion, rearing and inner square crossings in an open field test,[4,57] a reduced percentage of time spent on the open arms of the elevated plus maze,[58,59] and decreased investigatory behavior in a multi-compartment chamber test.[53,60,61] In addition to the behavioral suppression observed in these exploration tests, the anxiogenic-like profile of central CRF administration includes increased stress-induced freezing behavior,[62,63] reduced social interaction,[64] potentiated acoustic startle response[65,66] and increased defensive burying.[67] Other stress-related behaviors induced by i.c.v. CRF administration are decreased food consumption,[63,68–71] inhibition of maternal behavior,[72] decreased sexual behavior,[73] and the induction of place and taste aversion.[74,75] The i.c.v.

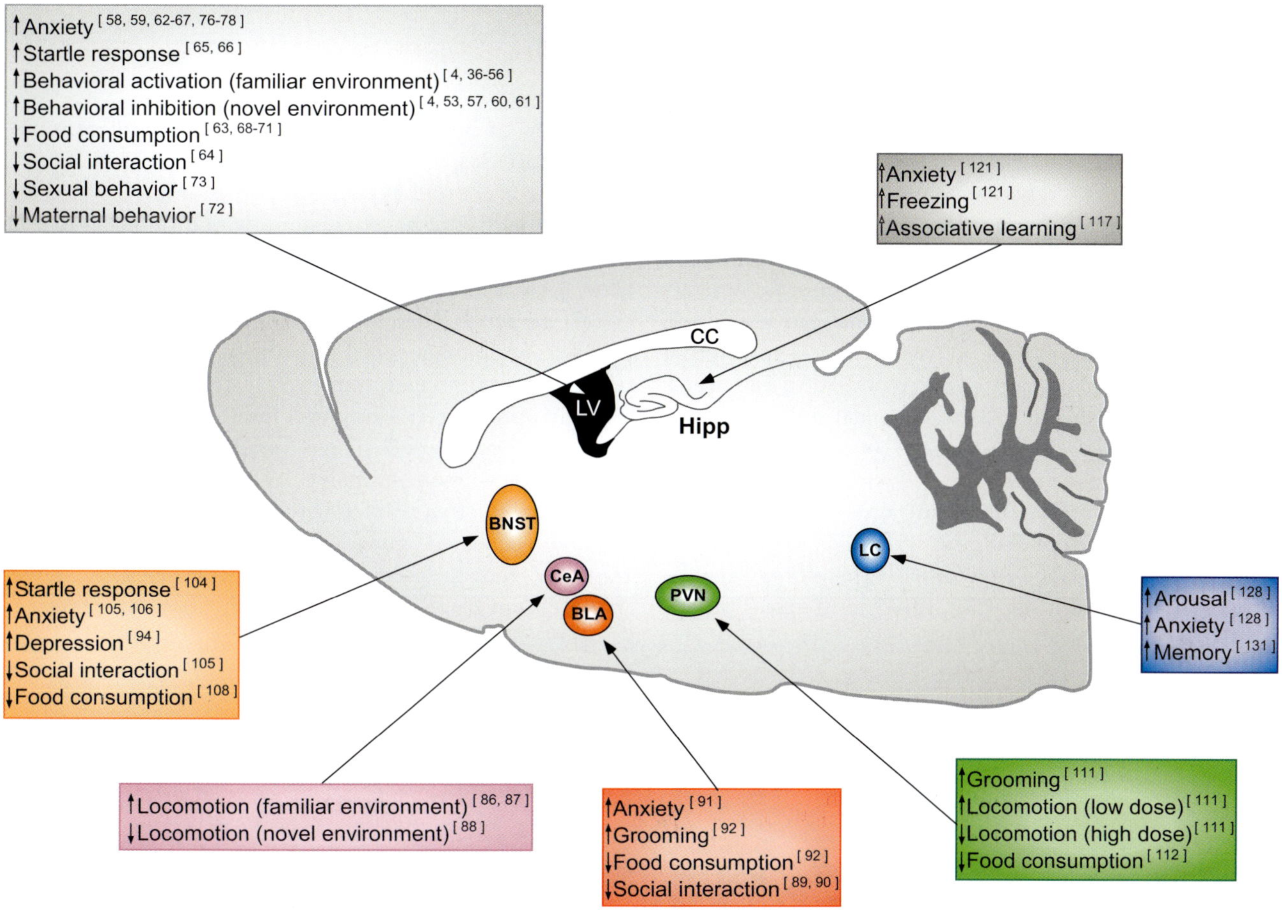

FIGURE 15.1 A schematic representation of a sagittal view of the rodent brain, showing the behavioral consequences of site-specific CRF/Urocortin administration. LV, lateral ventricle; CC, corpus callosum; Hipp, hippocampus; BNST, bed nucleus of stria terminalis; CeA, central amygdala; BLA, basolateral amygdala; PVN, paraventricular nucleus of the hypothalamus; LC, locus coeruleus.

BOX 15.1

HOW DO WE KNOW THE GENETIC APPROACH FOR INTRACEREBROVENTRICULAR DELIVERY?

Pharmacological administration of synthetic peptides or secreted recombinant proteins into the brain ventricles is a common method used by neuroscientists for exploring physiological and behavioral functions of novel or known gene products. Cerebroventricular (rather than systemic), administration of these proteins is required in order to bypass the blood–brain barrier (BBB) and allow the non-selective transport of peptides or proteins from the periphery into the central nervous system (CNS).

Current approaches for delivery of peptides or secreted proteins to the CNS include, for short-term acute administration, a stereotaxic injection into the ventricular space, known as intracerebroventricular (i.c.v. administration); and for chronic delivery, an i.c.v. microinjection pump. These methods rely heavily on the solubility and half-life of the injected substance, require chemical or *in vitro* synthesis of the administered ligand, and may involve different purification procedures. The i.c.v. administration is difficult to use in studies requiring repeated or prolonged administration of the ligand, since the microinjection pump procedure depends on the ligand stability and capacity of the reservoir, and complex surgical procedures are required for installation and manipulation of the pump. Furthermore, the current procedures require

(Continued)

BOX 15.1 *(cont'd)*

extensive handling of the experimental animals, which may create behavioral or physiological disturbances.

We have recently described a genetic approach for the delivery of secreted peptides or proteins into the cerebrospinal fluid (CSF) using a choroid plexus-specific and lentiviral-based genetic system.[1] The choroid plexus plays a critical role in the barrier mechanism regulating the exchange of molecules between the brain's internal milieu and the periphery. This blood–CSF barrier is composed of epithelial cells with apical tight junctions that restrict intercellular passage of molecules from fenestrated blood vessels. The CSF circulatory system's function is to provide micronutrients, neurotrophins, hormones, neuropeptides and growth factors extensively to neuronal networks. Therefore, neuromodulators directed to CSF can modify and adapt a variety of behavioral, neuroendocrine and immunologic processes.

Using a choroid plexus-specific promoter, we established a lentiviral-based system, which offers inducible and reversible delivery of a gene product into the CSF. The system is composed of two complementary lentiviral vectors. The "Effector" construct consists of a choroid plexus-specific promoter that drives the expression of reverse tetracycline transactivator (rtTA) protein and the reporter green fluorescent protein (GFP) (see A, upper panel). The "Target" construct includes the tetracycline-responsive element (TRE) DNA sequence, upstream to the nucleotide coding sequence of the requested gene of interest, followed by the reporter red florescent protein (RFP) (see A, lower panel). Transcription initiation of the gene of interest

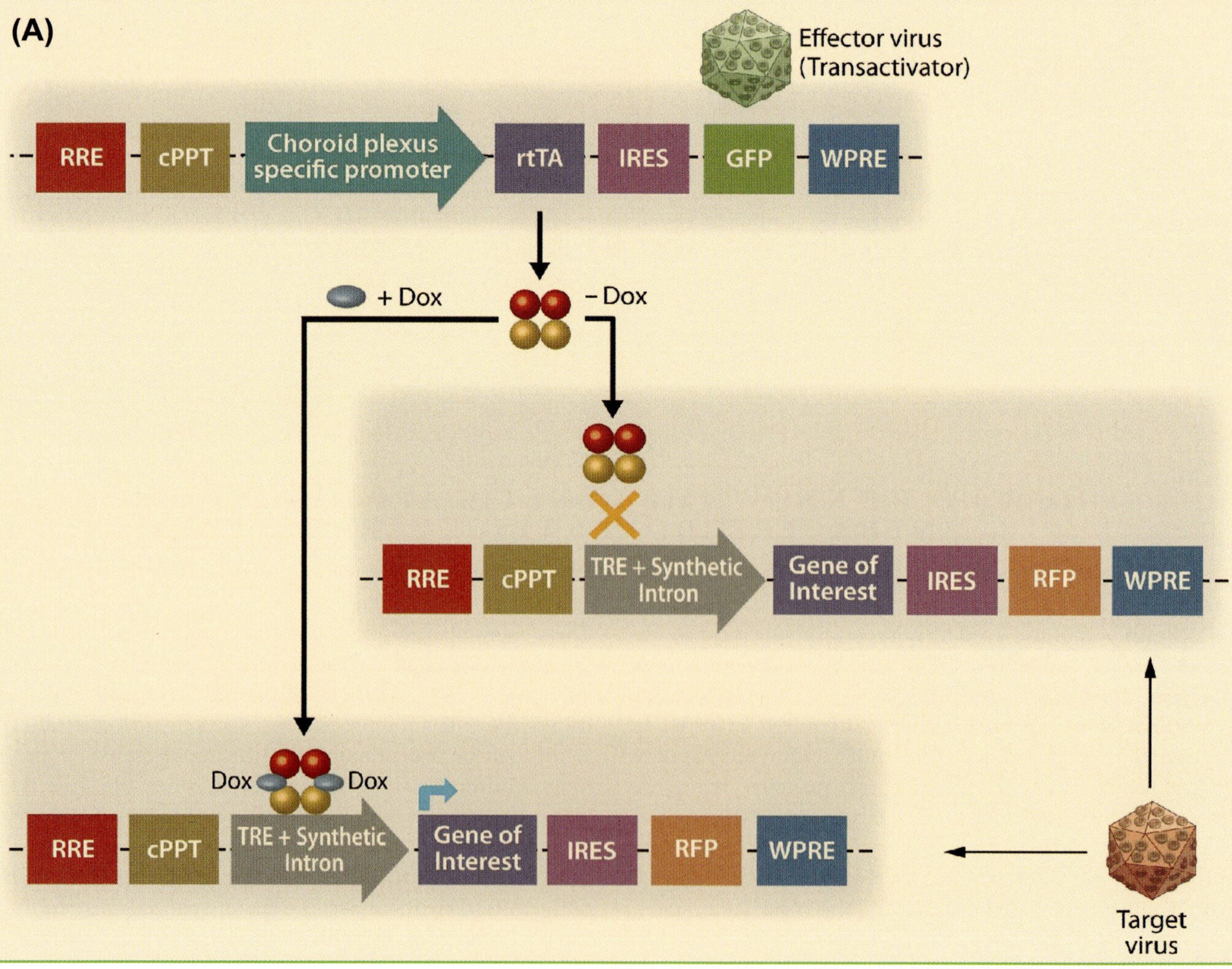

BOX 15.1 *(cont'd)*

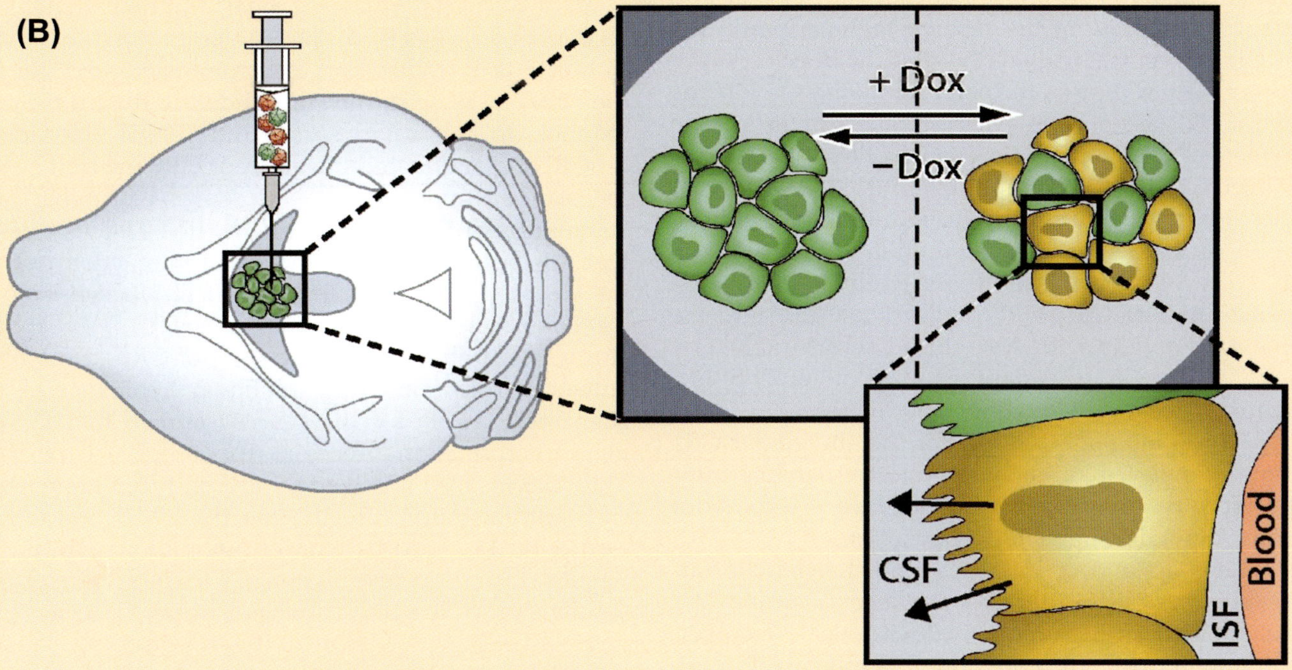

and the RFP is mediated only in the presence of the inducer, doxycycline (Dox). A mixture of the two lentiviruses is injected intracerebroventricularly, and the delivered genes are incorporated into the DNA of the choroid plexus cells. Initiation of transcription, limited to the choroid plexus cells by the choroid plexus-specific promoter, is induced by administrating Dox-containing drinking water, and results in secretion of the final processed gene product into the CSF (see B). Dox is the inducer of choice for our purposes, as it has been demonstrated to cross the blood–brain barrier. The functionality of this system was demonstrated using the overexpression of the two established neuropeptides, corticotropin-releasing factor and gonadotropin-releasing hormone, modulating anxiety-like behavior and estrous cycle, respectively.[1]

Reference

1. Regev L, Ezrielev E, Gershon E, et al. Genetic approach for intracerebroventricular delivery. *Proc Natl Acad Sci. USA.* 2010;107:4424–4429.

administration of Ucn1 results in a similar behavioral profile to that of CRF, including increased grooming and motor activity in a familiar environment,[38] and increased anxiety-like behavior in different paradigms, including the open field test, the elevated plus maze test, the light–dark transfer test, the defensive withdrawal test and the conflict test.[76,77]

Recently, a new genetic approach has been applied for the inducible and reversible secretion of CRF into the cerebrospinal fluid (CSF).[78] The choroid plexus, a group of modified ependymal cells in the ventricles of the brain that produce the CSF, was genetically targeted by i.c.v. injection of a lentiviral vector expressing CRF. The initiation of CRF transcription, limited to the choroid plexus cells by a choroid plexus-specific promoter, was induced by administrating the animals with doxycycline in the drinking water, which results in the secretion of the CRF peptide into the CSF. Mice conditionally overexpressing CRF at the choroid plexus showed an increase in anxiety-like behavior in the light–dark transfer, open field and elevated plus maze tests relative to the control group.[78] These results further support the principal role of CRF in the regulation of anxiety, previously demonstrated by pharmacological studies.

CNS SITES OF ACTION OF CRF/UCN1

Amygdala

The amygdala is an important mediator of fear and anxiety.[79–81] In view of the relative high concentration of CRF in the central nucleus[82,83] and the dense concentration of CRF receptors in the basal nucleus,[25,84,85] the amygdala has been largely investigated for its role in the effects of CRF on emotion-related behavior. In rats, CRF injected into the central amygdala (CeA) induces an increase in locomotor activity in a familiar environment,[86,87] but a decrease in locomotion in a novel open field[88] – results that resemble those obtained by i.c.v. administration of CRF. CRF or Ucn1 injected into the basolateral amygdala (BLA) reduced social interaction time,[89,90] led to a persistent increase in anxiety-like behavior,[91] decreased feeding, and increased grooming,[92] suggesting a specific role for the amygdalar CRF system in anxiety-like behavior. Two research groups have recently used a lentiviral-based system for chronic amygdala-specific overexpression of CRF. Keen-Rhinehart and colleagues found that female rats overexpressing CRF in the amygdala for a period of 2 weeks showed an increase in acoustic startle response and attenuated sexual motivation,[93] both results indicating an increase in anxiety levels. Regev and colleagues found that male mice overexpressing CRF in the amygdala for a period of 4 months showed reduced levels of anxiety-like behavior in response to acute stressful stimulation in the open field and light–dark transfer tests.[94] The discrepancy between the studies may be attributed to the difference in CRF overexpression duration, gender differences, species differences and behavioral tests.

Injections of α-hCRF$_{9-41}$, a synthetic competitive antagonist of CRF receptor, into the CeA reverses the social stress-induced decrease in time spent in the open arms of the EPM[95] and attenuates stress-induced freezing behavior.[96] Intra-CeA infusions of antisense oligodeoxynucleotides against CRFR1 reduce anxiety-like behavior in socially defeated rats.[97] Intra-BLA administration of astressin B, a CRFR1/2 antagonist, reverses the anxiogenic effects of Ucn1,[98] whereas intra-BLA administration of the CRFR1-specific antagonist antalarmin prevents defeat-induced defensive behavior in mice.[99] Finally, a recent study has demonstrated that the decrease in anxiety-like behavior in mice reared in an enriched environment is associated with very low levels of CRFR1 in the BLA. Knockdown of CRFR1 in the BLA mimicked the anxiolytic effect of environmental enrichment.[100]

BNST

Heavily innervated by the amygdala, and projecting to several regions involved in fear and anxiety responses, the BNST is considered an additional brain site that potentially mediates the behavioral effects of CRF/Ucn administration.[101] Expressing CRF[83] and both CRF1 and CRF2 receptors,[25,85,102] the BNST has been suggested to play a special role in longer-duration anxiety-like responses, in contrast to the CeA implicated in shorter-duration fear responses.[103] Microinfusions of CRF into the BNST increase startle amplitude in a dose-dependent manner. In the same study, intra-BNST injection of α-hCRF$_{9-41}$ attenuated the effect of CRF injected i.c.v. in the startle response test.[104] Moreover, "priming" with subanxiogenic doses of Ucn1 in the BNST elicits a persistent anxiogenic profile in the social interaction test and the elevated plus maze test, an effect blocked by prior local injection of CRFR1 antagonists.[105,106] In another study, post-training administration of CRF in the BNST enhanced retention in an inhibitory avoidance task,[107] suggesting a role for the CRF system in the BNST in memory formation processes for affective experience. The BNST is also involved in CRF-induced anorexia, as CRF administrated directly into the BNST, but not the CeA or the LC, induces a marked inhibition of feeding.[108] Finally, chronic overexpression of CRF in the dorsolateral subdivision of the BNST, but not in the CeA, increases depressive-like behavior without affecting anxiety levels.[94]

PVN

The paraventricular nucleus of the hypothalamus (PVN) contains a large number of CRF-immunopositive cell bodies and fibers,[83,109] as well as moderate concentrations of CRF receptors.[25,85,102] In addition to its role in activating the HPA axis by its projections to the median eminence, the CRF neurons in the PVN may be involved in CRF-induced behavioral inhibition through its projections to brainstem nuclei.[110] CRF administration directly into the PVN induces a dose-dependent increase in grooming, and an increase in locomotion at lower doses but a suppression of locomotion at higher doses of CRF.[111] In addition, intra-PVN injections of CRF decrease food intake,[112] whereas α-hCRF$_{9-41}$ injections in the PVN potentiate feeding induced by NPY injected in the same locus.[113] A recent study has demonstrated that lentiviral-mediated knockdown of CRF in the PVN attenuates social avoidance in chronic social-defeated mice.[114]

Hippocampus

The hippocampus contains scattered CRF-stained interneurons,[83] and moderate concentrations of CRF receptors.[25,85,102] The dorsal hippocampus has been suggested to be involved in context-dependent[115] and tone-dependent fear conditioning.[116] Injections of CRF in the dorsal hippocampus before or after training enhance both context-dependent and tone-dependent fear acquisition. This effect seems to be CRFR1-mediated, as it is prevented by local injections of astressin B, a non-selective CRF receptor antagonist, but not by antisauvagine-30, a specific CRFR2 antagonist.[117] Furthermore, recovery from stress-induced increased anxiety and impairment of context-dependent fear conditioning requires hippocampal CRFR1, as the recovery is prevented by intrahippocampal injections of DMP696, a highly selective CRFR1 antagonist, but not by antisauvagine-30.[118] Finally, the ventral hippocampus has been implicated in the modulation of anxiety-related behaviors.[119,120] Microinfusions of CRF into the ventral hippocampus increase anxiety-like behavior in the EPM test, as well as stress-induced freezing.[121]

Locus Coeruleus

Both physiologic and psychogenic stressors activate the noradrenergic cells in the locus coeruleus (LC).[122–124] These cells project to the forebrain and the spinal cord,[125–127] and induce a state of arousal required for the correct behavioral and physiological response to stress. Several studies suggest that CRF plays an important role in this effect. CRF infused into the LC increases non-ambulatory motor activity (shifting in body position), whereas locomotor activity is not altered. In addition, CRF infused into the LC produces a dose-dependent decrease in floating behavior in a modified Porsolt swim test, whereas struggling behavior is not affected. Both results suggest an increased arousal and agitation effect of CRF in the LC.[128] In the same study, CRF was injected into the LC decreased inner and outer crossings, free and wall rearing, as well as exploration outside of a darkened compartment in an open field test, indicating an anxiogenic effect.[128] Furthermore, injection of α-hCRF$_{9-41}$ into the LC reduces the levels of shock-induced freezing behavior[129] and immobilization stress-induced defensive withdrawal.[130] Finally, CRF injected directly into the LC improves memory retention in a passive-avoidance task,[131] suggesting involvement of the LC in the learning and memory-facilitating effects of CRF.

THE CRFR2/UROCORTIN CENTRAL SYSTEM

The CRF-peptide family includes, in addition to CRF, the three urocortin (Ucn) peptides (Ucn1, Ucn2 and Ucn3) that bind and activate the CRF receptor type 2 (CRFR2) with high affinity.[11–14,132–134] CRF has a relatively lower affinity for CRFR2 than for CRFR1, Ucn1 has equal affinities for both, and Ucns 2 and 3 appear to be selective for CRFR2.[11–14] These receptors are distributed differently throughout the brain, while CRFR1 is widely expressed, CRFR2 expression is more localized to selected stress-related brain nuclei, such as the medial amygdala, the bed nucleus of the stria terminalis (BNST), the lateral septum (LS) and the dorsal raphe nucleus (DRN).[25,102]

While the role of CRF and CRFR1 in the activation of the HPA axis and the regulation of emotional and cognitive functions following exposure to stressful challenge are well established,[9,135–138] the role of CRFR2 is still controversial. Studies exploring the role of CRFR2 using knockout mice models, antisense oligonucleotide, or antagonist administration are less clear, and show conflicting behavioral results. Central administration of Ucn2 or Ucn3,[139–142] CRFR2 antisense oligonucleotide,[143,144] and antagonist studies[117,145,146] have shown contrasting, dose-dependent and localization-dependent results. These results suggest a central role for CRFR2 that may vary between different brain nuclei and under different stress conditions. Several pharmacological studies suggest anxiolytic-like effects of CRFR2 activation. Intracerebroventricular administration of Ucn2[139,147,148] or Ucn3,[140,149] the endogenous ligands for CRFR2, exerts an anxiolytic-like phenotype in the tested rodents. In contrast, others have shown a possible anxiogenic role for CRFR2 activation, specifically in the LS,[117,141,150] the DRN,[151] and when using CRFR2-specific antisense oligonucleotides or CRFR2 agonists.[146,152–155] Furthermore, CRFR2 activity in the septum was shown to vary under different stress conditions, being less involved under low stress conditions, and significantly increasing anxiety-like behavior under high stress conditions.[140]

A growing body of evidence suggests that Ucn1 and Ucn2 may influence stress-related physiology and behavior by modulating the DRN serotonergic system.[156] Ucn2 i.c.v. administration caused an increase in c-Fos immunostaining in topographically organized subpopulations of serotonergic neurons in the DRN, specifically within the dorsal part of the mid-rostrocaudal DRN (DRD) and the caudal part of the DRN (DRC).[157,158] Moreover, while injection of Ucn2 to the DRC leads to increased 5-HT release in the BLA[159] and potentiation of conditioned fear, as well as escape deficits in a model of learned helplessness,[151] injection of the CRFR2 antagonist antisauvagine-30, and not of CRFR1 antagonists, showed anxiolytic effects, including reversal of the potentiation of conditioned fear and the escape deficits following exposure to inescapable stress.[146,151,154] Thus, these behaviors indicating heightened anxiety in response to uncontrollable stress seem to be mediated by CRFR2 receptors on serotonergic neurons in the DRC.[160]

GENETICALLY ALTERED MICE

A major approach for the study of the involvement of the CRF/urocortin system in the regulation of stress-induced behavior has been the use of transgenic mice models, overexpressing or knocking out the different family members. To date, all genes of the CRF/urocortin family have been successfully targeted, generating deficient animal models that have been crucial for the understanding of the role of the CRF/Urocortin family of peptides and receptors in mediating stress-related behaviors.

In the case of developmental transgenic mice model, it should be remarked that conclusions regarding their physiology and behavior must be drawn cautiously, due to compensatory changes in other related or non-related genes. In addition, in the case of gene overexpression the expression of the gene of interest is often driven under the control of a general promoter, resulting in overexpression of the gene of interest in non-endogenous brain regions and peripheral organs (for detailed reviews on CRF family mutant mice, see references 161–163) (Table 15.1).

CRF-OE

To study the effects of chronic exposure to CRF, transgenic mice overexpressing CRF under the control of

TABLE 15.1 Summary of Behavioral and Endocrinological Consequences of CRF System Gene Targeting

Genotype	Phenotype	Reference(s)
CRF OE (Stenzel-Poore *et al.*)	High plasma levels of ACTH and corticosterone. Increased anxiety and depression-like behavior, decrease in sexual behavior (only in females) and deficits in learning and attention.	164–69
CRF OE (Dirks *et al.*)	Reduced acoustic startle reactivity, impaired pre-pulse inhibition, and abnormal behavioral adaptation to a novel environment.	170–172
CRF OE (CNS restricted)	Stress-induced hypersecretion of ACTH and corticosterone (only in males). Reduced immobility in the forced swim test and tail suspension test.	173
CRF KO	Normal basal and stress-induced anxiety-like behavior, locomotor activity, exploration, startle response and learning.	175–176
CRFR1 KO	Disrupted HPA axis activation in response to stress. Reduced anxiety-like behavior, impaired spatial recognition memory, and deficiencies in nurturing behavior (females).	177–182
CRFR2 KO (Bale *et al.*)	Increased ACTH and corticosterone response to stress and early termination of ACTH release. Increased anxiety and depression-like behavior.	183, 186
CRFR2 KO (Coste *et al.*)	Increased ACTH and corticosterone response to stress and early termination of ACTH release. No differences in anxiety-like behavior.	184
CRFR2 KO (Kishimoto *et al.*)	Increased anxiety and depression-like behavior.	185, 187
CRF-BP OE	Increased locomotion and rearing in the open field test, and increased total arm entries in the EPM test.	188
CRF-BP KO	Increased anxiety-like behavior. Deficits in maternal aggression.	190,191
CRFR1/CRFR2 dKO	Impaired stress-induced HPA system activation. Reduced anxiety-like behavior in the EPM test (only in females).	192, 193
CRFR1$^{loxp/loxp}$CamKIIαCRE	Reduced anxiety-like behavior. Normal basal plasma ACTH and corticosterone levels.	196
Ucn1 KO (Vetter *et al.*)	Increased anxiety-like behavior.	197
Ucn1 KO (Wang *et al.*)	No effect in anxiety-like behavior.	198
Ucn2 KO	Increased nocturnal ACTH and corticosterone levels, and reduced depressive-like behavior (only in females). No effect in anxiety-like behavior.	199
Ucn3 KO	Normal HPA axis regulation, anxiety-like behavior and depressive-like behavior. Increased social recognition memory.	201
Ucn1/Ucn2 dKO	Increased stress-induced corticosterone response (only in males). Decreased anxiety-like behavior.	156
Ucn1/Ucn2/Ucn3 tKO	Increased anxiety-like behaviors 24 h following stress exposure. Increased stress-induced acoustic startle-response.	202

metallothionein (mMT1) promoter have been developed.[164] CRF-overexpressing mice (CRF-OE) exhibit high plasma levels of ACTH and corticosterone, and develop physical changes typical of Cushing's syndrome. At the behavioral level these mutants display a phenotype characteristic of increased anxiety and depression levels, such as decreased baseline and stress-induced locomotor activity in a novel environment, decreased time spent in the open arms of an elevated plus maze (EPM),[165] decreased time spent in the light area of a light–dark transfer box, reduced time spent rearing in home cages, and increased immobility in a forced swim test.[166] In terms of sexual behavior, CRF-OE female mice display a profound decrease in sexual receptivity, whereas the sexual performance of CRF-OE males remains intact.[167] CRF-OE mice also show significant deficits in learning,[168] and reduced attention.[169]

An additional CRF-OE transgenic mouse was generated using the murine Thy-1.2 promoter.[170] In contrast to the mMT1 CRF-OE mice, the overexpression in Thy1 CRF-OE mice starts 4–8 days after birth and is restricted to the central nervous system and the spinal cord.[170] These mice display reduced acoustic startle reactivity, impaired pre-pulse inhibition,[171] and abnormal behavioral adaptation to a novel environment.[172]

Finally, three different mouse models of conditional CRF overexpression have been generated using the nestin (Nes) promoter (CNS-restricted overexpression), CamKIIα promoter (forebrain-restricted overexpression) and Dlx5/6 promoter (forebrain GABAergic neurons-restricted overexpression).[173] CRF overexpression in the entire central nervous system, but not in specific forebrain regions, resulted in stress-induced hypersecretion of ACTH and corticosterone in males but not in females, and reduced immobility in the forced swim test and tail suspension test in both males and females.[173]

CRF-KO

The development of CRF knockout (CRF-KO) mice has been important in addressing the physiologic and pathologic roles of CRF.[3,174] However, in terms of behavioral effects, CRF-KO mice exhibit normal basal and stress-induced anxiety-like behavior. In addition, baseline locomotor activity, exploration, startle response and learning appear to be unaffected.[175,176] CRF receptor antagonists have an anxiolytic effect in CRF-KO mice, suggesting that whereas CRFR1 activation is crucial to induce anxiety, CRF itself is not.[176] Therefore, it has been proposed that another CRF-related peptide, such as Ucn1, which has high affinity to CRFR1, may contribute to the observed stress-induced behavior in these animals.[175,176]

CRFR1-KO

In order to elucidate the relative contribution of the CRF receptor type 1 to stress-induced physiology and behavior, two independent groups generated a mouse line null for the CRFR1 gene.[177,178] Mice deficient for CRFR1 (CRFR1-KO) display a disrupted HPA axis activation in response to stress, causing glucocorticoid deficiency.[177,178] However, basal plasma ACTH was found to be similar to that in wild-type controls, probably due to the action of AVP.[179] Behaviorally, CRFR1-KO mice display reduced anxiety-like behavior in the light–dark transfer test and the EPM test.[177,178,180] In addition, it has been reported that CRFR1-KO mice are impaired in spatial recognition memory in a Y-maze paradigm.[180] In terms of social behavior, male CRFR1-KO mice show intact isolation-induced inter-male aggression,[181] whereas female CRFR1-KO mice exhibit deficiencies in nurturing behavior, such as decreased time spent nursing, licking and grooming the pups.[182] It should be remarked that, in both studies,[181,182] CRFR1-KO mice with C57BL/6 X 129 background, have been crossed into an outbred hsd:ICR strain selectively bred to exhibit high levels of maternal aggression, to improve aggressive performance.

CRFR2-KO

In contrast to the clear and consistent phenotype of the two CRFR1-KO mouse lines, the three independently generated CRFR2-KO mouse lines[183–185] present several inconsistencies in their neuroendocrine and behavioral profiles. Coste and colleagues[184] found an increased ACTH and corticosterone response to stress and an early termination of ACTH release in CRFR2-KO mice. However, no differences in anxiety-like behavior[184] were found between CRFR2-KO and their wild-type littermates. Bale *et al.*[183] also showed increased ACTH and corticosterone response to stress and an early termination of ACTH release in their CRFR2-KO mouse line. However in contrast to the Coste *et al.*[184] findings, Bale and colleagues[183] found that CRFR2-KO mice display a significant increase in anxiety-like behavior in the EPM and the open field tests, but not in the light–dark transfer test.[183] Finally, Kishimoto *et al.*[185] showed that only male CRFR2-KO mice are more anxious in the EPM test and the light–dark transfer test; however, in the open field test, CRFR2-KO mice spent more time in the center of the arena.[185] These differences could be a result of the different genetic backgrounds of the mouse lines. Both the Bale *et al.*[183]

and the Kishimoto *et al.*[185] CRFR2-KO mouse lines exhibited a depression-like phenotype in the forced swim test and the tail suspension test.[186,187]

CRF-BP-OE

Two different mouse models of CRF binding protein overexpression (CRF-BP-OE) have been independently generated.[188,189] The CRF-BP-OE line created by Lovejoy and colleagues expresses the transgene not only in the brain and pituitary, but also ectopically in peripheral tissues.[189] However, this mouse line has not yet been behaviorally phenotyped. The CRF-BP-OE line created by Burrows *et al.* expresses the transgene under the control of the pituitary glycoprotein hormone α-subunit (α-GSU) promoter, resulting in specific overexpression of CRF-BP in the anterior pituitary gland.[188] These mice exhibit increased locomotion and rearing in the open field test, increased total arm entries in the EPM, and a tendency to spend more time in the open arms of the EPM.[188]

CRF-BP-KO

Mice deficient for CRF-BP (CRF-BP-KO) display increased anxiogenic-like behavior when tested in the EPM and the light–dark transfer test, and a trend toward increased anxiety-like behavior in the open field test.[190] These results suggest an increase in "free" CRF or Ucn1 levels, leading to an anxiogenic phenotype; however, this hypothesis has not yet been examined in these mutants. The same line of CRF-BP-KO mice has been crossed into an outbred hsd:ICR strain selectively bred to exhibit high levels of maternal aggression, to investigate aggressive behavior.[191] In this study, CRF-BP-KO mice exhibited significant deficits in maternal aggression (offspring protection) relative to wild-type mice. Female CRF-BP-KO mice were also tested in the forced swim test, where no differences were found between the groups. In the light–dark transfer test, female CRF-BP-KO mice exhibited higher levels of anxiety-like behavior. For males, no significant differences in light–dark transfer, swim test and isolation-induced resident–intruder male aggression were found between the groups. However, increased anxiety-like behavior in mutant males was detected in the approach to a novel object placed in the center of an open field arena.[191]

CRFR1/CRFR2 Double KO (dKO)

To further elucidate the role of both CRF receptors in the activation and modulation of the neuroendocrine stress response and anxiety-like behavior, two mouse models deficient of both CRFR1 and CRFR2 (CRFR1/CRFR2 double knockout; dKO) have been independently generated.[192,193] Both mouse lines display impairment in stress-induced HPA system activation. In terms of anxiety-like behavior, double-mutant mice are sexually dichotomous. Whereas no difference was found in anxiety-like behavior between male double-mutants and wild-type mice, female CRFR1/CRFR2 dKO mice display reduced anxiety-like behavior in the EPM test. However, no difference was found in the number of center visits in the open field test.[192] The behavioral phenotype of the Preil *et al.* mouse model[193] has not yet been reported.

CRFR1$^{loxp/loxp}$CamKIIαCRE

As glucocorticoids are known to be involved in the modulation of fear and anxiety-like behavior,[194,195] the anxiolytic-like profile observed in the developmental CRFR1 KO mice may result either from central CRFR1 deficiency, or from the low circulating levels of glucocorticoids. In order to differentiate the CRF/CRFR1 pathways that control stress-induced behavior from those regulating the HPA axis, a conditional CRFR1 knockout mouse line was generated using the *Cre/loxP* system. In these mutants the CRFR1 expression, driven by the calcium calmodulin-kinase IIα (CamKIIα) promoter, is postnatally ablated in the anterior forebrain and limbic system including the hippocampus, amygdala and neocortex, but remains intact in the pituitary gland.[196] The CRFR1 conditional mutants (CRFR1$^{loxp/loxp}$CamKIIαCRE) display reduced anxiety-like behavior levels in the EPM and the light–dark transfer tests. In contrast, the basal plasma ACTH and corticosterone levels are similar to those in wild-type mice. Moreover, the hormone levels in conditional mutants remain significantly elevated 30 and 90 min following restraint stress, indicating that CRFR1 conditional mutants are hypersensitive to stress, and that limbic CRFR1 is required for central control of HPA system feedback and hormonal adaptation to stress.[196]

Ucn1-KO

Two mouse lines deficient for the Ucn1 gene (Ucn1-KO) were independently generated.[197,198] Ucn1-KO mice show normal endocrine stress responses in both mouse lines,[197,198] supporting the view that Ucn1 does not play a role (or has only a minor one) in stress-induced HPA axis regulation. Similar to the case of CRFR2-KO, the behavioral phenotype of Ucn1-KO mice is controversial. Whereas Wang and colleagues showed no differences in anxiety-like behavior,[198] Vetter *et al.* demonstrated that Ucn1-KO mice have increased anxiety-like behavior in the EPM and the open field tests; however, no differences were found in the light–dark transfer test.[197]

Ucn2-KO

Mice deficient for Ucn2 (Ucn2-KO) exhibit gender-specific alterations in the basal circadian rhythms of ACTH and corticosterone secretion, and in depressive-like behavior.[199] The nocturnal ACTH and corticosterone levels were found to be higher in female Ucn2-KO mice compared to their wild-type littermates, but not in male. No differences were found in stress-induced hormone response in both genders.[199] In addition, female mutant mice display reduced depressive-like behavior assessed by both the forced swim test and the tail suspension test. The altered performance of female Ucn2 null mice in tests of antidepressant activity was behaviorally specific, because Ucn2 null mutant mice did not differ in their anxiety-like behavior in the EPM or light–dark transfer tests.[199] In addition, the differences observed in the forced swim test could not be attributed to deficits in learning abilities, as no differences were found between the genotypes in a cued contextual fear conditioning test.[199]

Ucn3-KO

Mice deficient for the Ucn3 gene (Ucn3-KO) were independently generated by two groups.[200,201] Li *et al.*[200] revealed an important role for Ucn3 in the regulation of glucose-induced insulin secretion in the pancreas,[200] while not reporting any behavioral phenotype. Deussing *et al.*[201] recently assessed the second Ucn3 KO mouse model in several behavioral tests. No differences were found in HPA axis regulation, anxiety-like behavior or depressive-like behavior between the mutants and their wild-type littermates. However, in a social discrimination task, Ucn3-KO mice were able to recognize previously encountered conspecifics for a longer time than wild-type littermates – an observation that was also confirmed in CRFR2-KO mice in the same study.[201] Taken together, these findings suggest a specific role of the Ucn3/CRFR2 system in social recognition.

Ucn1/Ucn2 Double KO (dKO)

To further explore the physiological role of Ucn1 and Ucn2 in mediating the central stress response, a double Ucn1 and Ucn2-deficient mouse line (Ucn1/Ucn2 dKO) was recently generated.[156] Ucn1/Ucn2 dKO male mice show a higher stress-induced corticosterone response and an anxiolytic profile in the EPM, the open field and the light–dark transfer tests, relative to wild-type mice. Female Ucn1/Ucn2 dKO mice show no significant changes in basal or stress-induced corticosterone levels. However, similar to male mutants, an anxiolytic-like phenotype was observed in the EPM and the open field tests.[156] Whereas both female and male wild-type mice exhibited an expected increase in anxiety-like behavior when subjected to restraint stress before the tests, Ucn1/Ucn2 dKO mice showed a significantly smaller stress-induced change in behavior.[156]

Ucn1/Ucn2/Ucn3 Triple KO (tKO)

To better understand the role of the endogenous CRFR2 ligands Ucn1, Ucn2 and Ucn3 in regulating the central stress response, a triple knockout (tKO) mouse model lacking all three urocortin genes has been generated.[202] Intriguingly, these urocortin tKO mice exhibit increased anxiety-like behaviors in the open field and light–dark transfer tests, 24 h following stress exposure, but not under unstressed conditions or immediately following exposure to acute stress. In addition, urocortin tKO mice exhibit a significantly higher stress-induced increase in acoustic startle-response. Together, these results suggest that lacking all urocortins has a limited effect on anxiety under non-challenged conditions but renders the mice susceptible to the effects of stress, possibly by impairing recovery mechanisms.

DYSREGULATION OF THE CRF SYSTEM: MOOD DISORDERS

Depression

Several human studies have suggested that abnormal CRF neurotransmission and CRFR1 receptor signal transduction play a central role in the pathophysiology of depression (reviewed in Holsboer and Ising[203]; Binder and Nemeroff.[204]). For instance, the CSF concentrations of CRF are elevated in depressive patients[205–207] as well as in suicide victims[208] when compared to normal controls. In contrast, electroconvulsive therapy[209] or antidepressant administration results in reduced levels of CRF in the CSF.[210–213] Following CRF administration, depressive patients display blunted ACTH release, suggesting a desensitization of CRF receptors caused by CRF hypersecretion.[214–216] Consistent with this hypothesis, it was found that CRF binding sites are reduced in the forebrain of depressive suicide victims,[217,218] whereas CRF concentrations are elevated in the PVN, pontine nucleus, locus coeruleus and cortex.[219–223] Moreover, in depressive patients Ucn1 is upregulated in the Edinger-Westphal nucleus[224] and Ucn3 in the prefrontal cortex,[225] suggesting a possible role for CRFR2 in the development of depression.

Several genetic association studies have linked polymorphisms in the CRF family genes with depression and suicidality.[226,227] A CRFR1 haplotype with alleles G-A-G for the single nucleotide polymorphisms (SNP) rs1876828, rs242939 and rs242941, respectively, was found to be significantly associated with a higher response to antidepressants in major depressive patients,[228,229] whereas the haplotype defined by the alleles G-G-T for the same SNPs was found to be overrepresented in major depressive patients.[230] Furthermore, the CRFR1 intronic SNP rs4792887 has been linked to suicidal attempts in major depressive men,[231,232] whereas the A-allele of rs110402 has been associated with onset and seasonal pattern of major depression among females[233] and depression in male suicide attempters.[232] Interestingly, multiple individual SNPs, as well as a common haplotype spanning the CRFR1 intron 1 locus, appear to attenuate the effect of childhood abuse on the risk for adult depressive symptomatology.[234] Regarding the CRFR2 gene, several positive associations have been reported. A CRFR2 haplotype consisting of three microsatellites has been associated with high severity suicidal behavior in bipolar disorder.[235] In addition, allele G carriers of the rs2270007 SNP in the CRFR2 gene showed a worse overall response to antidepressant treatment.[233] Finally, one study found that the rs10473984 SNP within the CRFBP locus showed a significant association with both remission and reduction in depressive symptoms following antidepressant treatment.[226]

A large number of selective CRFR1 antagonists from different pharmaceutical companies have been developed in the past two decades as novel therapeutics for anxiety and depression. However, few of them have entered clinical development and, despite major efforts, to date no CRFR1 antagonist has successfully completed a Phase III trial.[34,203] An open-label Phase IIa clinical trial assessed the effects of NBI-30775/R121919, a non-peptide high-affinity CRFR1 antagonist, in depressed patients.[236–238] The group receiving NBI-30775/R121919 exhibited a safety and tolerability profile,[237,238] increased slow-wave sleep and decreased REM density,[236] and reduced depression and anxiety scores.[238] Several other CRFR1 antagonists are currently in clinical development for the treatment of anxiety and depression.

Anxiety

The anxiogenic effects of CRF administration in laboratory animals have led to the suggestion that the CRF system may also be involved in the pathophysiology of anxiety disorders.[138,161,239] Altered HPA axis function has been reported in PTSD[240,241] and panic disorder patients.[242,243] In addition, several clinical studies have found elevated CRF concentrations in the CSF of post-traumatic stress disorder (PTSD) patients.[244–246] In contrast, no alterations in CSF CRF concentrations have been found in patients with generalized anxiety disorder (GAD),[247–249] panic disorder[247–249] or obsessive-compulsive disorder (OCD).[248,250]

A few association studies have linked polymorphisms in the CRF family with anxiety disorders.[204] A CRF-linked microsatellite marker situated 23 kb 3′ of the gene, and three SNPs, rs6999100, rs6159 and rs1870393, have been associated with behavioral inhibition, a heritable temperamental phenotype considered a risk factor for anxiety disorders.[251,252] This association was particularly evident in the families in which at least one of the parents was diagnosed with panic disorder.[251] Significant associations between SNPs in the CRFR1 gene and panic disorder have been reported in two independent samples.[253] Intriguingly, the results were even stronger when the effects of the CRFR1 rs878886 SNP were combined with the AVPR1B (vasopressin receptor 1B) rs28632197 SNP, suggesting a potential genetic interaction between CRFR1 and AVPR1B polymorphisms in the development of panic disorder.[253]

CONCLUDING REMARKS

Stress-related psychopathologies, such as depression and anxiety disorders, represent some of the most common and proliferating health problems worldwide.[254,255] Better understanding of the etiology of these disorders may fill the urgent requirement for development of novel and improved therapeutics. The current view on the etiology of mental disorders is of a complex interaction between genetic and environmental factors. Certain genetic backgrounds might constitute a predisposition for the development of mental disorder, which will onset with the occurrence of additional environmental factors.[256,257]

One of the key environmental risk factors that contribute to the onset of a variety of psychopathologies is stress.[7,258–260] When a situation is perceived as stressful, the brain activates many neuronal circuits, linking centers involved in sensory, motor, autonomic, neuroendocrine, cognitive and emotional functions in order to adapt to the demand. However, the details of the pathways by which the brain translates stressful stimuli into the final, integrated biological response are not completely understood. Nevertheless, it is clear that dysregulation of these physiological responses to stress can have severe psychological and physiological consequences, and there is growing evidence to suggest that inappropriate regulation, disproportional intensity, or chronic and/or irreversible activation of the stress response is linked to the etiology and pathophysiology of anxiety disorders and depression.[204,261,262]

Previous studies suggested that the CRF/urocortin systems in the brain have a unique role in mediating behavioral and physiological responses to diverse stressors.[1,5,8–10,137,162,204,263,264] However, while the involvement of the CRF/CRFR1 system in regulating the activation of the HPA axis and the regulation of stress-linked behaviors is well established, the role of the central urocortins/CRFR2 system is less understood. The urocortins/CRFR2 system may be particularly important in situations where an organism must mobilize not only the HPA system but also the central nervous system in response to environmental challenge. Clearly, dysfunction in such a fundamental brain-activating system may be the key to a variety of pathophysiological conditions involving abnormal responses to stressors, such as anxiety disorders, affective disorders and anorexia nervosa.

Evidence from studies employing competitive peptides, or small-molecule CRF/urocortin receptor antagonists, suggested that the brain CRF/urocortin systems play diverse roles in mediating behavioral responses to stress. Based on the complementary behavioral phenotypes of CRFR1- and CRFR2-deficient (KO) mice, opposing roles were suggested for the two CRF receptor systems in modulating anxiety-like behaviors. CRFR1-KO mice display decreased anxiety-like behaviors coupled with an impaired HPA axis stress response, while CRFR2-KO mice show increased anxiety-like behaviors and an accelerated HPA axis response to stress. Thus, the CRF–CRFR1 system was suggested as critical for initiating stress responses, while the urocortins–CRFR2 system was suggested to terminate it or restore allostasis. Nevertheless, the anxiety-related effects of CRFR2 agonist and antagonist administration into the cerebral ventricles or into specific brain regions were less consistent, with some evidence for brain-site or ligand specificity.

Mice lacking all three *urocortin* genes (tKO) exhibit increased anxiety-like behaviors 24 hours following stress exposure, but not under unstressed conditions or immediately following exposure to acute stress. This suggests that the urocortins might play an essential role in the stress-recovery process. Mouse models of Ucn1, Ucn2 and Ucn3 individual KOs have not indicated clear and robust changes in stress-related behaviors; this may reflect differences in the time point following the stress exposure at which these mice were tested. Therefore, further "dissection" of the contribution of each urocortin to the tKO phenotype using longitudinal comparative studies in both sexes, under different time points following stress exposure, may promote further understanding of the role of each *urocortin* gene product in the regulation of the central stress response.

Conventional gene knockout models, generated for the different urocortin genes, have provided important information toward elucidating the function of these genes. However, these mice showed significant changes in the expression levels of the other CRF family members in the CNS, likely due to developmental compensatory mechanisms, which may have contributed to the observed stress-related phenotypes. In order both to avoid the developmental compensatory changes and to genetically target the gene of interest within specific brain nuclei, future studies will need to use more specific transgenic mice models and viral tools that will allow the manipulation of both the levels and site of urocortin gene expression in adult mice.

References

1. Vale W, Spiess J, Rivier C, Rivier J. Characterization of a 41-residue ovine hypothalamic peptide that stimulates secretion of corticotropin and beta-endorphin. *Science.* 1981;213:1394–1397.
2. Rivier C, Vale W. Modulation of stress-induced ACTH release by corticotropin-releasing factor, catecholamines and vasopressin. *Nature.* 1983;305:325–327.
3. Muglia L, Jacobson L, Dikkes P, Majzoub JA. Corticotropin-releasing hormone deficiency reveals major fetal but not adult glucocorticoid need. *Nature.* 1995;373:427–432.
4. Sutton RE, Koob GF, Le Moal M, Rivier J, Vale W. Corticotropin releasing factor produces behavioural activation in rats. *Nature.* 1982;297:331–333.
5. Koob GF, Heinrichs SC. A role for corticotropin releasing factor and urocortin in behavioral responses to stressors. *Brain Res.* 1999;848:141–152.
6. Brown MR, Fisher LA, Spiess J, Rivier C, Rivier J, Vale W. Corticotropin-releasing factor: actions on the sympathetic nervous system and metabolism. *Endocrinology.* 1982;111: 928–931.
7. de Kloet ER, Joels M, Holsboer F. Stress and the brain: from adaptation to disease. *Nat Rev Neurosci.* 2005;6:463–475.
8. Chrousos GP, Gold PW. The concepts of stress and stress system disorders. Overview of physical and behavioral homeostasis. *J Am Med Assoc.* 1992;267:1244–1252.
9. Holsboer F. The rationale for corticotropin-releasing hormone receptor (CRH-R) antagonists to treat depression and anxiety. *J Psychiatr Res.* 1999;33:181–214.
10. Holmes A, Heilig M, Rupniak NM, Steckler T, Griebel G. Neuropeptide systems as novel therapeutic targets for depression and anxiety disorders. *Trends Pharmacol Sci.* 2003;24: 580–588.
11. Vaughan J, Donaldson C, Bittencourt J, Perrin MH, Lewis K, Sutton S, Chan R, Turnbull AV, Lovejoy D, Rivier C, et al. Urocortin, a mammalian neuropeptide related to fish urotensin I and to corticotropin-releasing factor. *Nature.* 1995;378:287–292.
12. Reyes TM, Lewis K, Perrin MH, Kunitake KS, Vaughan J, Arias CA, Hogenesch JB, Gulyas J, Rivier J, Vale WW, Sawchenko PE. Urocortin II: a member of the corticotropin-releasing factor (CRF) neuropeptide family that is selectively bound by type 2 CRF receptors. *Proc Natl Acad Sci USA.* 2001;98:2843–2848.
13. Hsu SY, Hsueh AJ. Human stresscopin and stresscopin-related peptide are selective ligands for the type 2 corticotropin-releasing hormone receptor. *Nat Med.* 2001;7:605–611.
14. Lewis K, Li C, Perrin MH, Blount A, Kunitake K, Donaldson C, Vaughan J, Reyes TM, Gulyas J, Fischer W, Bilezikjian L, Rivier J, Sawchenko PE, Vale WW. Identification of urocortin III, an additional member of the corticotropin-releasing factor (CRF) family with high affinity for the CRF2 receptor. *Proc Natl Acad Sci USA.* 2001;98:7570–7575.

15. Chen R, Lewis KA, Perrin MH, Vale WW. Expression cloning of a human corticotropin-releasing-factor receptor. *Proc Natl Acad Sci USA.* 1993;90:8967–8971.
16. Vita N, Laurent P, Lefort S, Chalon P, Lelias JM, Kaghad M, Le Fur G, Caput D, Ferrara P. Primary structure and functional expression of mouse pituitary and human brain corticotrophin releasing factor receptors. *FEBS Lett.* 1993;335:1–5.
17. Chang CP, Pearse II RV, O'Connell S, Rosenfeld MG. Identification of a seven transmembrane helix receptor for corticotropin-releasing factor and sauvagine in mammalian brain. *Neuron.* 1993;11:1187–1195.
18. Perrin M, Donaldson C, Chen R, Blount A, Berggren T, Bilezikjian L, Sawchenko P, Vale W. Identification of a second corticotropin-releasing factor receptor gene and characterization of a cDNA expressed in heart. *Proc Natl Acad Sci USA.* 1995;92:2969–2973.
19. Stenzel P, Kesterson R, Yeung W, Cone RD, Rittenberg MB, Stenzel-Poore MP. Identification of a novel murine receptor for corticotropin-releasing hormone expressed in the heart. *Mol Endocrinol.* 1995;9:637–645.
20. Kishimoto T, Pearse II RV, Lin CR, Rosenfeld MG. A sauvagine/corticotropin-releasing factor receptor expressed in heart and skeletal muscle. *Proc Natl Acad Sci USA.* 1995;92:1108–1112.
21. Lovenberg TW, Liaw CW, Grigoriadis DE, Clevenger W, Chalmers DT, De Souza EB, Oltersdorf T. Cloning and characterization of a functionally distinct corticotropin-releasing factor receptor subtype from rat brain. *Proc Natl Acad Sci USA.* 1995;92:836–840.
22. Chen A, Perrin M, Brar B, Li C, Jamieson P, Digruccio M, Lewis K, Vale W. Mouse corticotropin-releasing factor receptor type 2alpha gene: isolation, distribution, pharmacological characterization and regulation by stress and glucocorticoids. *Mol Endocrinol.* 2005;19:441–458.
23. Pisarchik A, Slominski A. Molecular and functional characterization of novel CRFR1 isoforms from the skin. *Eur J Biochem.* 2004;271:2821–2830.
24. Grammatopoulos DK, Dai Y, Randeva HS, Levine MA, Karteris E, Easton AJ, Hillhouse EW. A novel spliced variant of the type 1 corticotropin-releasing hormone receptor with a deletion in the seventh transmembrane domain present in the human pregnant term myometrium and fetal membranes. *Mol Endocrinol.* 1999;13:2189–2202.
25. Van Pett K, Viau V, Bittencourt JC, Chan RK, Li HY, Arias C, Prins GS, Perrin M, Vale W, Sawchenko PE. Distribution of mRNAs encoding CRF receptors in brain and pituitary of rat and mouse. *J Comp Neurol.* 2000;428:191–212.
26. Kostich WA, Chen A, Sperle K, Largent BL. Molecular identification and analysis of a novel human corticotropin-releasing factor (CRF) receptor: the CRF2gamma receptor. *Mol Endocrinol.* 1998;12:1077–1085.
27. Lovenberg TW, Chalmers DT, Liu C, De Souza EB. CRF2 alpha and CRF2 beta receptor mRNAs are differentially distributed between the rat central nervous system and peripheral tissues. *Endocrinology.* 1995;136:4139–4142.
28. Potter E, Behan DP, Fischer WH, Linton EA, Lowry PJ, Vale WW. Cloning and characterization of the cDNAs for human and rat corticotropin releasing factor-binding proteins. *Nature.* 1991;349:423–426.
29. Behan DP, De Souza EB, Lowry PJ, Potter E, Sawchenko P, Vale WW. Corticotropin releasing factor (CRF) binding protein: a novel regulator of CRF and related peptides. *Front Neuroendocrinol.* 1995;16:362–382.
30. Seasholtz AF, Valverde RA, Denver RJ. Corticotropin-releasing hormone-binding protein: biochemistry and function from fishes to mammals. *J Endocrinol.* 2002;175:89–97.
31. Dunn AJ, Berridge CW. Physiological and behavioral responses to corticotropin-releasing factor administration: is CRF a mediator of anxiety or stress responses? *Brain Res Brain Res Rev.* 1990;15:71–100.
32. Steckler T, Holsboer F. Corticotropin-releasing hormone receptor subtypes and emotion. *Biol Psychiatry.* 1999;46:1480–1508.
33. Heinrichs SC, Koob GF. Corticotropin-releasing factor in brain: a role in activation, arousal, and affect regulation. *J Pharmacol Exp Ther.* 2004;311:427–440.
34. Zorrilla EP, Koob GF. Progress in corticotropin-releasing factor-1 antagonist development. *Drug Discov Today* 15, 371-83.
35. Lowry CA, Moore FL. Regulation of behavioral responses by corticotropin-releasing factor. *Gen Comp Endocrinol.* 2006; 146:19–27.
36. Dunn AJ, Berridge CW, Lai YI, Yachabach TL. CRF-induced excessive grooming behavior in rats and mice. *Peptides.* 1987;8:841–844.
37. Sherman JE, Kalin NH. The effects of ICV-CRH on novelty-induced behavior. *Pharmacol Biochem Behav.* 1987;26:699–703.
38. Jones DN, Kortekaas R, Slade PD, Middlemiss DN, Hagan JJ. The behavioural effects of corticotropin-releasing factor-related peptides in rats. *Psychopharmacology (Berl).* 1998;138:124–132.
39. Eaves M, Thatcher-Britton K, Rivier J, Vale W, Koob GF. Effects of corticotropin releasing factor on locomotor activity in hypophysectomized rats. *Peptides.* 1985;6:923–926.
40. Britton DR, Varela M, Garcia A, Rosenthal M. Dexamethasone suppresses pituitary-adrenal but not behavioral effects of centrally administered CRF. *Life Sci.* 1986;38:211–216.
41. Moore FL, Miller LJ. Stress-induced inhibition of sexual behavior: corticosterone inhibits courtship behaviors of a male amphibian (*Taricha granulosa*). *Horm Behav.* 1984;18:400–410.
42. Lowry CA, Deviche P, Moore FL. Effects of corticotropin-releasing factor (CRF) and opiates on amphibian locomotion. *Brain Res.* 1990;513:94–100.
43. Lowry CA, Burke KA, Renner KJ, Moore FL, Orchinik M. Rapid changes in monoamine levels following administration of corticotropin-releasing factor or corticosterone are localized in the dorsomedial hypothalamus. *Horm Behav.* 2001; 39:195–205.
44. Lowry CA, Moore FL. Corticotropin-releasing factor (CRF) antagonist suppresses stress-induced locomotor activity in an amphibian. *Horm Behav.* 1991;25:84–96.
45. Clements S, Schreck CB, Larsen DA, Dickhoff WW. Central administration of corticotropin-releasing hormone stimulates locomotor activity in juvenile chinook salmon (*Oncorhynchus tshawytscha*). *Gen Comp Endocrinol.* 2002;125:319–327.
46. Clements S, Schreck CB. Central administration of corticotropin-releasing hormone alters downstream movement in an artificial stream in juvenile chinook salmon (*Oncorhynchus tshawytscha*). *Gen Comp Endocrinol.* 2004;137:1–8.
47. Ohgushi A, Bungo T, Shimojo M, Masuda Y, Denbow DM, Furuse M. Relationships between feeding and locomotion behaviors after central administration of CRF in chicks. *Physiol Behav.* 2001;72:287–289.
48. Zhang R, Nakanishi T, Ohgushi A, Ando R, Yoshimatsu T, Denbow DM, Furuse M. Interaction of corticotropin-releasing factor and glucagon-like peptide-1 on behaviors in chicks. *Eur J Pharmacol.* 2001;430:73–78.
49. Johnson RW, von Borell EH, Anderson LL, Kojic LD, Cunnick JE. Intracerebroventricular injection of corticotropin-releasing hormone in the pig: acute effects on behavior, adrenocorticotropin secretion, and immune suppression. *Endocrinology.* 1994;135:642–648.
50. Salak-Johnson JL, McGlone JJ, Whisnant CS, Norman RL, Kraeling RR. Intracerebroventricular porcine corticotropin-

releasing hormone and cortisol effects on pig immune measures and behavior. *Physiol Behav.* 1997;61:15–23.

51. Parrott RF, Vellucci SV, Goode JA. Behavioral and hormonal effects of centrally injected "anxiogenic" neuropeptides in growing pigs. *Pharmacol Biochem Behav.* 2000;65:123–129.
52. Parrott RF, Vellucci SV. Behaviour of pigs given corticotrophin-releasing hormone in combination with flumazenil or diazepam. *Pharmacol Biochem Behav.* 2000;67:465–471.
53. Berridge CW, Dunn AJ. Corticotropin-releasing factor elicits naloxone sensitive stress-like alterations in exploratory behavior in mice. *Regul Pept.* 1986;16:83–93.
54. Dunn AJ, Berridge CW. Corticotropin-releasing factor administration elicits a stress-like activation of cerebral catecholaminergic systems. *Pharmacol Biochem Behav.* 1987;27:685–691.
55. Kalin NH, Shelton SE, Kraemer GW, McKinney WT. Corticotropin-releasing factor administered intraventricularly to rhesus monkeys. *Peptides.* 1983;4:217–220.
56. Winslow JT, Newman JD, Insel TR. CRH and alpha-helical-CRH modulate behavioral measures of arousal in monkeys. *Pharmacol Biochem Behav.* 1989;32:919–926.
57. Takahashi LK, Kalin NH, Vanden Burgt JA, Sherman JE. Corticotropin-releasing factor modulates defensive-withdrawal and exploratory behavior in rats. *Behav Neurosci.* 1989;103: 648–654.
58. Baldwin HA, Rassnick S, Rivier J, Koob GF, Britton KT. CRF antagonist reverses the "anxiogenic" response to ethanol withdrawal in the rat. *Psychopharmacology (Berl).* 1991;103:227–232.
59. Momose K, Inui A, Asakawa A, Ueno N, Nakajima M, Fujimiya M, Kasuga M. Intracerebroventricularly administered corticotropin-releasing factor inhibits food intake and produces anxiety-like behaviour at very low doses in mice. *Diabetes Obes Metab.* 1999;1:281–284.
60. Berridge CW, Dunn AJ. CRF and restraint-stress decrease exploratory behavior in hypophysectomized mice. *Pharmacol Biochem Behav.* 1989;34:517–519.
61. Spadaro F, Berridge CW, Baldwin HA, Dunn AJ. Corticotropin-releasing factor acts via a third ventricle site to reduce exploratory behavior in rats. *Pharmacol Biochem Behav.* 1990;36:305–309.
62. Sherman JE, Kalin NH. ICV-CRH alters stress-induced freezing behavior without affecting pain sensitivity. *Pharmacol Biochem Behav.* 1988;30:801–807.
63. Pelleymounter MA, Joppa M, Carmouche M, Cullen MJ, Brown B, Murphy B, Grigoriadis DE, Ling N, Foster AC. Role of corticotropin-releasing factor (CRF) receptors in the anorexic syndrome induced by CRF. *J Pharmacol Exp Ther.* 2000;293:799–806.
64. Dunn AJ, File SE. Corticotropin-releasing factor has an anxiogenic action in the social interaction test. *Horm Behav.* 1987; 21:193–202.
65. Swerdlow NR, Geyer MA, Vale WW, Koob GF. Corticotropin-releasing factor potentiates acoustic startle in rats: blockade by chlordiazepoxide. *Psychopharmacology (Berl).* 1986;88:147–152.
66. Liang KC, Melia KR, Miserendino MJ, Falls WA, Campeau S, Davis M. Corticotropin-releasing factor: long-lasting facilitation of the acoustic startle reflex. *J Neurosci.* 1992;12:2303–2312.
67. Diamant M, Croiset G, de Wied D. The effect of corticotropin-releasing factor (CRF) on autonomic and behavioral responses during shock-prod burying test in rats. *Peptides.* 1992; 13:1149–1158.
68. Arase K, York DA, Shimizu H, Shargill N, Bray GA. Effects of corticotropin-releasing factor on food intake and brown adipose tissue thermogenesis in rats. *Am J Physiol.* 1988;255:E255–E259.
69. Krahn DD, Gosnell BA, Grace M, Levine AS. CRF antagonist partially reverses CRF- and stress-induced effects on feeding. *Brain Res Bull.* 1986;17:285–289.
70. Gosnell BA, Morley JE, Levine AS. A comparison of the effects of corticotropin releasing factor and sauvagine on food intake. *Pharmacol Biochem Behav.* 1983;19:771–775.
71. Benoit SC, Thiele TE, Heinrichs SC, Rushing PA, Blake KA, Steeley RJ. Comparison of central administration of corticotropin-releasing hormone and urocortin on food intake, conditioned taste aversion, and c-Fos expression. *Peptides.* 2000;21:345–351.
72. Pedersen CA, Caldwell JD, McGuire M, Evans DL. Corticotropin-releasing hormone inhibits maternal behavior and induces pup-killing. *Life Sci.* 1991;48:1537–1546.
73. Sirinathsinghji DJ. Inhibitory influence of corticotropin releasing factor on components of sexual behaviour in the male rat. *Brain Res.* 1987;407:185–190.
74. Cador M, Ahmed SH, Koob GF, Le Moal M, Stinus L. Corticotropin-releasing factor induces a place aversion independent of its neuroendocrine role. *Brain Res.* 1992;597:304–309.
75. Heinrichs SC, Britton KT, Koob GF. Both conditioned taste preference and aversion induced by corticotropin-releasing factor. *Pharmacol Biochem Behav.* 1991;40:717–721.
76. Moreau JL, Kilpatrick G, Jenck F. Urocortin, a novel neuropeptide with anxiogenic-like properties. *NeuroReport.* 1997;8: 1697–1701.
77. Spina MG, Merlo-Pich E, Akwa Y, Balducci C, Basso AM, Zorrilla EP, Britton KT, Rivier J, Vale WW, Koob GF. Time-dependent induction of anxiogenic-like effects after central infusion of urocortin or corticotropin-releasing factor in the rat. *Psychopharmacology (Berl).* 2002;160:113–121.
78. Regev L, Ezrielev E, Gershon E, Gil S, Chen A. Genetic approach for intracerebroventricular delivery. *Proc Natl Acad Sci USA.* 2010;107:4424–4429.
79. LeDoux J. The amygdala. *Curr Biol.* 2007;17:R868–R874.
80. Ehrlich I, Humeau Y, Grenier F, Ciocchi S, Herry C, Luthi A. Amygdala inhibitory circuits and the control of fear memory. *Neuron.* 2009;62:757–771.
81. Roozendaal B, McEwen BS, Chattarji S. Stress, memory and the amygdala. *Nat Rev Neurosci.* 2009;10:423–433.
82. Palkovits M, Brownstein MJ, Vale W. Distribution of corticotropin-releasing factor in rat brain. *Fed Proc.* 1985;44:215–219.
83. Swanson LW, Sawchenko PE, Rivier J, Vale WW. Organization of ovine corticotropin-releasing factor immunoreactive cells and fibers in the rat brain: an immunohistochemical study. *Neuroendocrinology.* 1983;36:165–186.
84. De Souza EB, Insel TR, Perrin MH, Rivier J, Vale WW, Kuhar MJ. Corticotropin-releasing factor receptors are widely distributed within the rat central nervous system: an autoradiographic study. *J Neurosci.* 1985;5:3189–3203.
85. Justice NJ, Yuan ZF, Sawchenko PE, Vale W. Type 1 corticotropin-releasing factor receptor expression reported in BAC transgenic mice: implications for reconciling ligand-receptor mismatch in the central corticotropin-releasing factor system. *J Comp Neurol.* 2008;511:479–496.
86. Tazi A, Swerdlow NR, LeMoal M, Rivier J, Vale W, Koob GF. Behavioral activation by CRF: evidence for the involvement of the ventral forebrain. *Life Sci.* 1987;41:41–49.
87. Wiersma A, Bohus B, Koolhaas JM. Corticotropin-releasing hormone microinfusion in the central amygdala diminishes a cardiac parasympathetic outflow under stress-free conditions. *Brain Res.* 1993;625:219–227.
88. Liang KC, Lee EH. Intra-amygdala injections of corticotropin releasing factor facilitate inhibitory avoidance learning and reduce exploratory behavior in rats. *Psychopharmacology (Berl).* 1988;96:232–236.
89. Sajdyk TJ, Schober DA, Gehlert DR, Shekhar A. Role of corticotropin-releasing factor and urocortin within the basolateral

amygdala of rats in anxiety and panic responses. *Behav Brain Res.* 1999;100:207–215.

90. Spiga F, Lightman SL, Shekhar A, Lowry CA. Injections of urocortin 1 into the basolateral amygdala induce anxiety-like behavior and c-Fos expression in brainstem serotonergic neurons. *Neuroscience.* 2006;138:1265–1276.
91. Rainnie DG, Bergeron R, Sajdyk TJ, Patil M, Gehlert DR, Shekhar A. Corticotrophin releasing factor-induced synaptic plasticity in the amygdala translates stress into emotional disorders. *J Neurosci.* 2004;24:3471–3479.
92. Jochman KA, Newman SM, Kalin NH, Bakshi VP. Corticotropin-releasing factor-1 receptors in the basolateral amygdala mediate stress-induced anorexia. *Behav Neurosci.* 2005;119: 1448–1458.
93. Keen-Rhinehart E, Michopoulos V, Toufexis DJ, Martin EI, Nair H, Ressler KJ, Davis M, Owens MJ, Nemeroff CB, Wilson ME. Continuous expression of corticotropin-releasing factor in the central nucleus of the amygdala emulates the dysregulation of the stress and reproductive axes. *Mol Psychiatry.* 2009;14:37–50.
94. Regev L, Neufeld-Cohen A, Tsoory M, Kuperman Y, Getselter D, Gil S, Chen A. Prolonged and site-specific over-expression of corticotropin-releasing factor reveals differential roles for extended amygdala nuclei in emotional regulation. *Mol Psychiatry*; 2011, in press.
95. Heinrichs SC, Pich EM, Miczek KA, Britton KT, Koob GF. Corticotropin-releasing factor antagonist reduces emotionality in socially defeated rats via direct neurotropic action. *Brain Res.* 1992;581:190–197.
96. Swiergiel AH, Takahashi LK, Kalin NH. Attenuation of stress-induced behavior by antagonism of corticotropin-releasing factor receptors in the central amygdala in the rat. *Brain Res.* 1993;623:229–234.
97. Liebsch G, Landgraf R, Gerstberger R, Probst JC, Wotjak CT, Engelmann M, Holsboer F, Montkowski A. Chronic infusion of a CRH1 receptor antisense oligodeoxynucleotide into the central nucleus of the amygdala reduced anxiety-related behavior in socially defeated rats. *Regul Pept.* 1995;59:229–239.
98. Sajdyk TJ, Gehlert DR. Astressin, a corticotropin releasing factor antagonist, reverses the anxiogenic effects of urocortin when administered into the basolateral amygdala. *Brain Res.* 2000;877:226–234.
99. Robison CL, Meyerhoff JL, Saviolakis GA, Chen WK, Rice KC, Lumley LA. A CRH1 antagonist into the amygdala of mice prevents defeat-induced defensive behavior. *Ann NY Acad Sci.* 2004;1032:324–327.
100. Sztainberg Y, Kuperman Y, Tsoory M, Lebow M, Chen A. The anxiolytic effect of environmental enrichment is mediated via amygdalar CRF receptor type 1. *Mol Psychiatry.* 2010;15:905–917.
101. Walker DL, Toufexis DJ, Davis M. Role of the bed nucleus of the stria terminalis versus the amygdala in fear, stress, and anxiety. *Eur J Pharmacol.* 2003;463:199–216.
102. Chalmers DT, Lovenberg TW, De Souza EB. Localization of novel corticotropin-releasing factor receptor (CRF2) mRNA expression to specific subcortical nuclei in rat brain: comparison with CRF1 receptor mRNA expression. *J Neurosci.* 1995;15: 6340–6350.
103. Walker DL, Miles LA, Davis M. Selective participation of the bed nucleus of the stria terminalis and CRF in sustained anxiety-like versus phasic fear-like responses. *Prog Neuro-psychopharmacol Biol Psychiatry.* 2009;33:1291–1308.
104. Lee Y, Davis M. Role of the hippocampus, the bed nucleus of the stria terminalis, and the amygdala in the excitatory effect of corticotropin-releasing hormone on the acoustic startle reflex. *J Neurosci.* 1997;17:6434–6446.
105. Lee Y, Fitz S, Johnson PL, Shekhar A. Repeated stimulation of CRF receptors in the BNST of rats selectively induces social but not panic-like anxiety. *Neuropsychopharmacology.* 2008;33: 2586–2594.
106. Sahuque LL, Kullberg EF, McGeehan AJ, Kinder JR, Hicks MP, Blanton MG, Janak PH, Olive MF. Anxiogenic and aversive effects of corticotropin-releasing factor (CRF) in the bed nucleus of the stria terminalis in the rat: role of CRF receptor subtypes. *Psychopharmacology (Berl).* 2006;186:122–132.
107. Liang KC, Chen HC, Chen DY. Posttraining infusion of norepinephrine and corticotropin releasing factor into the bed nucleus of the stria terminalis enhanced retention in an inhibitory avoidance task. *Chin J Physiol.* 2001;44:33–43.
108. Ciccocioppo R, Fedeli A, Economidou D, Policani F, Weiss F, Massi M. The bed nucleus is a neuroanatomical substrate for the anorectic effect of corticotropin-releasing factor and for its reversal by nociceptin/orphanin FQ. *J Neurosci.* 2003;23: 9445–9451.
109. Silverman AJ, Hou-Yu A, Chen WP. Corticotropin-releasing factor synapses within the paraventricular nucleus of the hypothalamus. *Neuroendocrinology.* 1989;49:291–299.
110. Gray TS, Magnuson DJ. Peptide immunoreactive neurons in the amygdala and the bed nucleus of the stria terminalis project to the midbrain central gray in the rat. *Peptides.* 1992;13:451–460.
111. Monnikes H, Heymann-Monnikes I, Tache Y. CRF in the paraventricular nucleus of the hypothalamus induces dose-related behavioral profile in rats. *Brain Res.* 1992;574:70–76.
112. Krahn DD, Gosnell BA, Levine AS, Morley JE. Behavioral effects of corticotropin-releasing factor: localization and characterization of central effects. *Brain Res.* 1988;443:63–69.
113. Heinrichs SC, Menzaghi F, Pich EM, Hauger RL, Koob GF. Corticotropin-releasing factor in the paraventricular nucleus modulates feeding induced by neuropeptide Y. *Brain Res.* 1993;611:18–24.
114. Elliott E, Ezra-Nevo G, Regev L, Neufeld-Cohen A, Chen A. Resilience to social stress coincides with functional DNA methylation of the Crf gene in adult mice. *Nat Neurosci.* 2010;13:1351–1353.
115. Phillips RG, LeDoux JE. Differential contribution of amygdala and hippocampus to cued and contextual fear conditioning. *Behav Neurosci.* 1992;106:274–285.
116. Maren S, Aharonov G, Fanselow MS. Neurotoxic lesions of the dorsal hippocampus and Pavlovian fear conditioning in rats. *Behav Brain Res.* 1997;88:261–274.
117. Radulovic J, Ruhmann A, Liepold T, Spiess J. Modulation of learning and anxiety by corticotropin-releasing factor (CRF) and stress: differential roles of CRF receptors 1 and 2. *J Neurosci.* 1999;19:5016–5025.
118. Todorovic C, Radulovic J, Jahn O, Radulovic M, Sherrin T, Hippel C, Spiess J. Differential activation of CRF receptor subtypes removes stress-induced memory deficit and anxiety. *Eur J Neurosci.* 2007;25:3385–3397.
119. Bertoglio LJ, Joca SR, Guimaraes FS. Further evidence that anxiety and memory are regionally dissociated within the hippocampus. *Behav Brain Res.* 2006;175:183–188.
120. Pentkowski NS, Blanchard DC, Lever C, Litvin Y, Blanchard RJ. Effects of lesions to the dorsal and ventral hippocampus on defensive behaviors in rats. *Eur J Neurosci.* 2006;23: 2185–2196.
121. Pentkowski NS, Litvin Y, Blanchard DC, Vasconcellos A, King LB, Blanchard RJ. Effects of acidic-astressin and ovine-CRF microinfusions into the ventral hippocampus on defensive behaviors in rats. *Horm Behav.* 2009;56:35–43.
122. Cassens G, Roffman M, Kuruc A, Orsulak PJ, Schildkraut JJ. Alterations in brain norepinephrine metabolism induced by

environmental stimuli previously paired with inescapable shock. *Science.* 1980;209:1138–1140.
123. Korf J, Aghajanian GK, Roth RH. Increased turnover of norepinephrine in the rat cerebral cortex during stress: role of the locus coeruleus. *Neuropharmacology.* 1973;12:933–938.
124. Cedarbaum JM, Aghajanian GK. Activation of locus coeruleus neurons by peripheral stimuli: modulation by a collateral inhibitory mechanism. *Life Sci.* 1978;23:1383–1392.
125. Grzanna R, Molliver ME. The locus coeruleus in the rat: an immunohistochemical delineation. *Neuroscience.* 1980;5:21–40.
126. Swanson LW. The locus coeruleus: a cytoarchitectonic, Golgi and immunohistochemical study in the albino rat. *Brain Res.* 1976;110:39–56.
127. Swanson LW, Hartman BK. The central adrenergic system. An immunofluorescence study of the location of cell bodies and their efferent connections in the rat utilizing dopamine-beta-hydroxylase as a marker. *J Comp Neurol.* 1975;163: 467–505.
128. Butler PD, Weiss JM, Stout JC, Nemeroff CB. Corticotropin-releasing factor produces fear-enhancing and behavioral activating effects following infusion into the locus coeruleus. *J Neurosci.* 1990;10:176–183.
129. Swiergiel AH, Takahashi LK, Rubin WW, Kalin NH. Antagonism of corticotropin-releasing factor receptors in the locus coeruleus attenuates shock-induced freezing in rats. *Brain Res.* 1992;587:263–268.
130. Smagin GN, Harris RB, Ryan DH. Corticotropin-releasing factor receptor antagonist infused into the locus coeruleus attenuates immobilization stress-induced defensive withdrawal in rats. *Neurosci Lett.* 1996;220:167–170.
131. Chen MF, Chiu TH, Lee EH. Noradrenergic mediation of the memory-enhancing effect of corticotropin-releasing factor in the locus coeruleus of rats. *Psychoneuroendocrinology.* 1992;17: 113–124.
132. Fekete EM, Zorrilla EP. Physiology, pharmacology, and therapeutic relevance of urocortins in mammals: ancient CRF paralogs. *Front Neuroendocrinol.* 2007;28:1–27.
133. Kuperman Y, Chen A. Urocortins: emerging metabolic and energy homeostasis perspectives. *Trends Endocrinol Metab.* 2008;19:122–129.
134. Joels M, Baram TZ. The neuro-symphony of stress. *Nat Rev Neurosci.* 2009;10:459–466.
135. Nemeroff CB. The role of corticotropin-releasing factor in the pathogenesis of major depression. *Pharmacopsychiatry.* 1988;21: 76–82.
136. Nemeroff CB. New vistas in neuropeptide research in neuropsychiatry: focus on corticotropin-releasing factor. *Neuropsychopharmacology.* 1992;6:69–75.
137. Arborelius L, Owens MJ, Plotsky PM, Nemeroff CB. The role of corticotropin-releasing factor in depression and anxiety disorders. *J Endocrinol.* 1999;160:1–12.
138. Reul JM, Holsboer F. Corticotropin-releasing factor receptors 1 and 2 in anxiety and depression. *Curr Opin Pharmacol.* 2002;2:23–33.
139. Valdez GR, Inoue K, Koob GF, Rivier J, Vale W, Zorrilla EP. Human urocortin II: mild locomotor suppressive and delayed anxiolytic-like effects of a novel corticotropin-releasing factor related peptide. *Brain Res.* 2002;943:142–150.
140. Valdez GR, Zorrilla EP, Rivier J, Vale WW, Koob GF. Locomotor suppressive and anxiolytic-like effects of urocortin 3, a highly selective type 2 corticotropin-releasing factor agonist. *Brain Res.* 2003;980:206–212.
141. Henry B, Vale W, Markou A. The effect of lateral septum corticotropin-releasing factor receptor 2 activation on anxiety is modulated by stress. *J Neurosci.* 2006;26:9142–9152.
142. Zhao Y, Valdez GR, Fekete EM, Rivier JE, Vale WW, Rice KC, Weiss F, Zorrilla EP. Subtype-selective corticotropin-releasing factor receptor agonists exert contrasting, but not opposite, effects on anxiety-related behavior in rats. *J Pharmacol Exp Ther.* 2007;323:846–854.
143. Heinrichs SC, Lapsansky J, Lovenberg TW, De Souza EB, Chalmers DT. Corticotropin-releasing factor CRF1, but not CRF2, receptors mediate anxiogenic-like behavior. *Regul Pept.* 1997;71:15–21.
144. Liebsch G, Landgraf R, Engelmann M, Lorscher P, Holsboer F. Differential behavioural effects of chronic infusion of CRH 1 and CRH 2 receptor antisense oligonucleotides into the rat brain. *J Psychiatr Res.* 1999;33:153–163.
145. Ruhmann A, Bonk I, Lin CR, Rosenfeld MG, Spiess J. Structural requirements for peptidic antagonists of the corticotropin-releasing factor receptor (CRFR): development of CRFR2beta-selective antisauvagine-30. *Proc Natl Acad Sci USA.* 1998;95:15264–15269.
146. Takahashi LK, Ho SP, Livanov V, Graciani N, Arneric SP. Antagonism of CRF(2) receptors produces anxiolytic behavior in animal models of anxiety. *Brain Res.* 2001;902: 135–142.
147. Ohata H, Shibasaki T. Effects of urocortin 2 and 3 on motor activity and food intake in rats. *Peptides.* 2004;25:1703–1709.
148. Zorrilla EP, Reinhardt LE, Valdez GR, Inoue K, Rivier JE, Vale WW, Koob GF. Human urocortin 2, a corticotropin-releasing factor (CRF)2 agonist, and ovine CRF, a CRF1 agonist, differentially alter feeding and motor activity. *J Pharmacol Exp Ther.* 2004;310:1027–1034.
149. Venihaki M, Sakihara S, Subramanian S, Dikkes P, Weninger SC, Liapakis G, Graf T, Majzoub JA. Urocortin III, a brain neuropeptide of the corticotropin-releasing hormone family: modulation by stress and attenuation of some anxiety-like behaviours. *J Neuroendocrinol.* 2004;16:411–422.
150. Bakshi VP, Smith-Roe S, Newman SM, Grigoriadis DE, Kalin NH. Reduction of stress-induced behavior by antagonism of corticotropin-releasing hormone 2 (CRH2) receptors in lateral septum or CRH1 receptors in amygdala. *J Neurosci.* 2002; 22:2926–2935.
151. Hammack SE, Schmid MJ, LoPresti ML, Der-Avakian A, Pellymounter MA, Foster AC, Watkins LR, Maier SF. Corticotropin releasing hormone type 2 receptors in the dorsal raphe nucleus mediate the behavioral consequences of uncontrollable stress. *J Neurosci.* 2003;23:1019–1025.
152. Ho SP, Takahashi LK, Livanov V, Spencer K, Lesher T, Maciag C, Smith MA, Rohrbach KW, Hartig PR, Arneric SP. Attenuation of fear conditioning by antisense inhibition of brain corticotropin releasing factor-2 receptor. *Brain Res Mol Brain Res.* 2001; 89:29–40.
153. Risbrough VB, Hauger RL, Pelleymounter MA, Geyer MA. Role of corticotropin releasing factor (CRF) receptors 1 and 2 in CRF-potentiated acoustic startle in mice. *Psychopharmacology (Berl).* 2003;170:178–187.
154. Pelleymounter MA, Joppa M, Ling N, Foster AC. Pharmacological evidence supporting a role for central corticotropin-releasing factor(2) receptors in behavioral, but not endocrine, response to environmental stress. *J Pharmacol Exp Ther.* 2002;302:145–152.
155. Pelleymounter MA, Joppa M, Ling N, Foster AC. Behavioral and neuroendocrine effects of the selective CRF2 receptor agonists urocortin II and urocortin III. *Peptides.* 2004;25: 659–666.
156. Neufeld-Cohen A, Evans AK, Getselter D, Spyroglou A, Hill A, Gil S, Tsoory M, Beuschlein F, Lowry CA, Vale W, Chen A. Urocortin-1 and -2 double-deficient mice show robust anxiolytic

phenotype and modified serotonergic activity in anxiety circuits. *Mol Psychiatry.* 15, 426-41, 339.
157. Staub DR, Evans AK, Lowry CA. Evidence supporting a role for corticotropin-releasing factor type 2 (CRF2) receptors in the regulation of subpopulations of serotonergic neurons. *Brain Res.* 2006;1070:77–89.
158. Staub DR, Spiga F, Lowry CA. Urocortin 2 increases c-Fos expression in topographically organized subpopulations of serotonergic neurons in the rat dorsal raphe nucleus. *Brain Res.* 2005;1044:176–189.
159. Amat J, Tamblyn JP, Paul ED, Bland ST, Amat P, Foster AC, Watkins LR, Maier SF. Microinjection of urocortin 2 into the dorsal raphe nucleus activates serotonergic neurons and increases extracellular serotonin in the basolateral amygdala. *Neuroscience.* 2004;129:509–519.
160. Maier SF, Watkins LR. Stressor controllability and learned helplessness: the roles of the dorsal raphe nucleus, serotonin, and corticotropin-releasing factor. *Neurosci Biobehav Rev.* 2005;29:829–841.
161. Bakshi VP, Kalin NH. Corticotropin-releasing hormone and animal models of anxiety: gene-environment interactions. *Biol Psychiatry.* 2000;48:1175–1198.
162. Bale TL, Vale WW. CRF and CRF receptors: role in stress responsivity and other behaviors. *Annu Rev Pharmacol Toxicol.* 2004;44:525–557.
163. Keck ME, Ohl F, Holsboer F, Muller MB. Listening to mutant mice: a spotlight on the role of CRF/CRF receptor systems in affective disorders. *Neurosci Biobehav Rev.* 2005;29: 867–889.
164. Stenzel-Poore MP, Cameron VA, Vaughan J, Sawchenko PE, Vale W. Development of Cushing's syndrome in corticotropin-releasing factor transgenic mice. *Endocrinology.* 1992;130: 3378–3386.
165. Stenzel-Poore MP, Heinrichs SC, Rivest S, Koob GF, Vale WW. Overproduction of corticotropin-releasing factor in transgenic mice: a genetic model of anxiogenic behavior. *J Neurosci.* 1994;14:2579–2584.
166. van Gaalen MM, Stenzel-Poore MP, Holsboer F, Steckler T. Effects of transgenic overproduction of CRH on anxiety-like behaviour. *Eur J Neurosci.* 2002;15:2007–2015.
167. Heinrichs SC, Min H, Tamraz S, Carmouche M, Boehme SA, Vale WW. Anti-sexual and anxiogenic behavioral consequences of corticotropin-releasing factor overexpression are centrally mediated. *Psychoneuroendocrinology.* 1997;22:215–224.
168. Heinrichs SC, Stenzel-Poore MP, Gold LH, Battenberg E, Bloom FE, Koob GF, Vale WW, Pich EM. Learning impairment in transgenic mice with central overexpression of corticotropin-releasing factor. *Neuroscience.* 1996;74:303–311.
169. van Gaalen MM, Stenzel-Poore M, Holsboer F, Steckler T. Reduced attention in mice overproducing corticotropin-releasing hormone. *Behav Brain Res.* 2003;142:69–79.
170. Dirks A, Groenink L, Bouwknecht JA, Hijzen TH, Van Der Gugten J, Ronken E, Verbeek JS, Veening JG, Dederen PJ, Korosi A, Schoolderman LF, Roubos EW, Olivier B. Overexpression of corticotropin-releasing hormone in transgenic mice and chronic stress-like autonomic and physiological alterations. *Eur J Neurosci.* 2002;16:1751–1760.
171. Dirks A, Groenink L, Schipholt MI, van der Gugten J, Hijzen TH, Geyer MA, Olivier B. Reduced startle reactivity and plasticity in transgenic mice overexpressing corticotropin-releasing hormone. *Biol Psychiatry.* 2002;51:583–590.
172. Kasahara M, Groenink L, Breuer M, Olivier B, Sarnyai Z. Altered behavioural adaptation in mice with neural corticotrophin-releasing factor overexpression. *Genes Brain Behav.* 2007;6:598–607.
173. Lu A, Steiner MA, Whittle N, Vogl AM, Walser SM, Ableitner M, Refojo D, Ekker M, Rubenstein JL, Stalla GK, Singewald N, Holsboer F, Wotjak CT, Wurst W, Deussing JM. Conditional mouse mutants highlight mechanisms of corticotropin-releasing hormone effects on stress-coping behavior. *Mol Psychiatry.* 2008;13:1028–1042.
174. Venihaki M, Majzoub JA. Animal models of CRH deficiency. *Front Neuroendocrinol.* 1999;20:122–145.
175. Dunn AJ, Swiergiel AH. Behavioral responses to stress are intact in CRF-deficient mice. *Brain Res.* 1999;845:14–20.
176. Weninger SC, Dunn AJ, Muglia LJ, Dikkes P, Miczek KA, Swiergiel AH, Berridge CW, Majzoub JA. Stress-induced behaviors require the corticotropin-releasing hormone (CRH) receptor, but not CRH. *Proc Natl Acad Sci USA.* 1999; 96:8283–8288.
177. Smith GW, Aubry JM, Dellu F, Contarino A, Bilezikjian LM, Gold LH, Chen R, Marchuk Y, Hauser C, Bentley CA, Sawchenko PE, Koob GF, Vale W, Lee KF. Corticotropin releasing factor receptor 1-deficient mice display decreased anxiety, impaired stress response, and aberrant neuroendocrine development. *Neuron.* 1998;20:1093–1102.
178. Timpl P, Spanagel R, Sillaber I, Kresse A, Reul JM, Stalla GK, Blanquet V, Steckler T, Holsboer F, Wurst W. Impaired stress response and reduced anxiety in mice lacking a functional corticotropin-releasing hormone receptor 1. *Nat Genet.* 1998; 19:162–166.
179. Turnbull AV, Smith GW, Lee S, Vale WW, Lee KF, Rivier C. CRF type I receptor-deficient mice exhibit a pronounced pituitary-adrenal response to local inflammation. *Endocrinology.* 1999;140:1013–1017.
180. Contarino A, Dellu F, Koob GF, Smith GW, Lee KF, Vale W, Gold LH. Reduced anxiety-like and cognitive performance in mice lacking the corticotropin-releasing factor receptor 1. *Brain Res.* 1999;835:1–9.
181. Gammie SC, Stevenson SA. Intermale aggression in corticotropin-releasing factor receptor 1 deficient mice. *Behav Brain Res.* 2006;171:63–69.
182. Gammie SC, Bethea ED, Stevenson SA. Altered maternal profiles in corticotropin-releasing factor receptor 1 deficient mice. *BMC Neurosci.* 2007;8:17.
183. Bale TL, Contarino A, Smith GW, Chan R, Gold LH, Sawchenko PE, Koob GF, Vale WW, Lee KF. Mice deficient for corticotropin-releasing hormone receptor-2 display anxiety-like behaviour and are hypersensitive to stress. *Nat Genet.* 2000;24:410–414.
184. Coste SC, Kesterson RA, Heldwein KA, Stevens SL, Heard AD, Hollis JH, Murray SE, Hill JK, Pantely GA, Hohimer AR, Hatton DC, Phillips TJ, Finn DA, Low MJ, Rittenberg MB, Stenzel P, Stenzel-Poore MP. Abnormal adaptations to stress and impaired cardiovascular function in mice lacking corticotropin-releasing hormone receptor-2. *Nat Genet.* 2000;24: 403–409.
185. Kishimoto T, Radulovic J, Radulovic M, Lin CR, Schrick C, Hooshmand F, Hermanson O, Rosenfeld MG, Spiess J. Deletion of crhr2 reveals an anxiolytic role for corticotropin-releasing hormone receptor-2. *Nat Genet.* 2000;24:415–419.
186. Bale TL, Vale WW. Increased depression-like behaviors in corticotropin-releasing factor receptor-2-deficient mice: sexually dichotomous responses. *J Neurosci.* 2003;23:5295–5301.
187. Todorovic C, Sherrin T, Pitts M, Hippel C, Rayner M, Spiess J. Suppression of the MEK/ERK signaling pathway reverses depression-like behaviors of CRF2-deficient mice. *Neuropsychopharmacology.* 2009;34:1416–1426.
188. Burrows HL, Nakajima M, Lesh JS, Goosens KA, Samuelson LC, Inui A, Camper SA, Seasholtz AF. Excess corticotropin releasing

hormone-binding protein in the hypothalamic-pituitary-adrenal axis in transgenic mice. *J Clin Invest.* 1998;101:1439–1447.

189. Lovejoy DA, Aubry JM, Turnbull A, Sutton S, Potter E, Yehling J, Rivier C, Vale WW. Ectopic expression of the CRF-binding protein: minor impact on HPA axis regulation but induction of sexually dimorphic weight gain. *J Neuroendocrinol.* 1998;10:483–491.
190. Karolyi IJ, Burrows HL, Ramesh TM, Nakajima M, Lesh JS, Seong E, Camper SA, Seasholtz AF. Altered anxiety and weight gain in corticotropin-releasing hormone-binding protein-deficient mice. *Proc Natl Acad Sci USA.* 1999;96: 11595–11600.
191. Gammie SC, Seasholtz AF, Stevenson SA. Deletion of corticotropin-releasing factor binding protein selectively impairs maternal, but not intermale aggression. *Neuroscience.* 2008; 157:502–512.
192. Bale TL, Picetti R, Contarino A, Koob GF, Vale WW, Lee KF. Mice deficient for both corticotropin-releasing factor receptor 1 (CRFR1) and CRFR2 have an impaired stress response and display sexually dichotomous anxiety-like behavior. *J Neurosci.* 2002;22:193–199.
193. Preil J, Muller MB, Gesing A, Reul JM, Sillaber I, van Gaalen MM, Landgrebe J, Holsboer F, Stenzel-Poore M, Wurst W. Regulation of the hypothalamic-pituitary-adrenocortical system in mice deficient for CRH receptors 1 and 2. *Endocrinology.* 2001;142:4946–4955.
194. Korte SM, Korte-Bouws GA, Koob GF, De Kloet ER, Bohus B. Mineralocorticoid and glucocorticoid receptor antagonists in animal models of anxiety. *Pharmacol Biochem Behav.* 1996;54:261–267.
195. Tronche F, Kellendonk C, Kretz O, Gass P, Anlag K, Orban PC, Bock R, Klein R, Schutz G. Disruption of the glucocorticoid receptor gene in the nervous system results in reduced anxiety. *Nat Genet.* 1999;23:99–103.
196. Muller MB, Zimmermann S, Sillaber I, Hagemeyer TP, Deussing JM, Timpl P, Kormann MS, Droste SK, Kuhn R, Reul JM, Holsboer F, Wurst W. Limbic corticotropin-releasing hormone receptor 1 mediates anxiety-related behavior and hormonal adaptation to stress. *Nat Neurosci.* 2003;6: 1100–1107.
197. Vetter DE, Li C, Zhao L, Contarino A, Liberman MC, Smith GW, Marchuk Y, Koob GF, Heinemann SF, Vale W, Lee KF. Urocortin-deficient mice show hearing impairment and increased anxiety-like behavior. *Nat Genet.* 2002;31:363–369.
198. Wang X, Su H, Copenhagen LD, Vaishnav S, Pieri F, Shope CD, Brownell WE, De Biasi M, Paylor R, Bradley A. Urocortin-deficient mice display normal stress-induced anxiety behavior and autonomic control but an impaired acoustic startle response. *Mol Cell Biol.* 2002;22:6605–6610.
199. Chen A, Zorrilla E, Smith S, Rousso D, Levy C, Vaughan J, Donaldson C, Roberts A, Lee KF, Vale W. Urocortin 2-deficient mice exhibit gender-specific alterations in circadian hypothalamus-pituitary-adrenal axis and depressive-like behavior. *J Neurosci.* 2006;26:5500–5510.
200. Li C, Chen P, Vaughan J, Lee KF, Vale W. Urocortin 3 regulates glucose-stimulated insulin secretion and energy homeostasis. *Proc Natl Acad Sci USA.* 2007;104:4206–4211.
201. Deussing JM, Breu J, Kuhne C, Kallnik M, Bunck M, Glasl L, Yen YC, Schmidt MV, Zurmuhlen R, Vogl AM, Gailus-Durner V, Fuchs H, Holter SM, Wotjak CT, Landgraf R, de Angelis MH, Holsboer F, Wurst W. Urocortin 3 modulates social discrimination abilities via corticotropin-releasing hormone receptor type 2. *J Neurosci.* 30, 9103-9116.
202. Neufeld-Cohen A, Tsoory MM, Evans AK, Getselter D, Gil S, Lowry CA, Vale WW, Chen A. A triple urocortin knockout mouse model reveals an essential role for urocortins in stress recovery. *Proc Natl Acad Sci USA.* 1902;107:0–5.
203. Holsboer F, Ising M. Central CRH system in depression and anxiety—evidence from clinical studies with CRH1 receptor antagonists. *Eur J Pharmacol.* 2008;583:350–357.
204. Binder EB, Nemeroff CB. The CRF system, stress, depression and anxiety-insights from human genetic studies. *Mol Psychiatry.* 15, 574-88.
205. Nemeroff CB, Widerlov E, Bissette G, Walleus H, Karlsson I, Eklund K, Kilts CD, Loosen PT, Vale W. Elevated concentrations of CSF corticotropin-releasing factor-like immunoreactivity in depressed patients. *Science.* 1984;226:1342–1344.
206. Banki CM, Bissette G, Arato M, O'Connor L, Nemeroff CB. CSF corticotropin-releasing factor-like immunoreactivity in depression and schizophrenia. *Am J Psychiatry.* 1987;144:873–877.
207. Hartline KM, Owens MJ, Nemeroff CB. Postmortem and cerebrospinal fluid studies of corticotropin-releasing factor in humans. *Ann NY Acad Sci.* 1996;780:96–105.
208. Arato M, Banki CM, Bissette G, Nemeroff CB. Elevated CSF CRF in suicide victims. *Biol Psychiatry.* 1989;25:355–359.
209. Nemeroff CB, Bissette G, Akil H, Fink M. Neuropeptide concentrations in the cerebrospinal fluid of depressed patients treated with electroconvulsive therapy. Corticotrophin-releasing factor, beta-endorphin and somatostatin. *Br J Psychiatry.* 1991;158: 59–63.
210. De Bellis MD, Gold PW, Geracioti Jr TD, Listwak SJ, Kling MA. Association of fluoxetine treatment with reductions in CSF concentrations of corticotropin-releasing hormone and arginine vasopressin in patients with major depression. *Am J Psychiatry.* 1993;150:656–657.
211. Veith RC, Lewis N, Langohr JI, Murburg MM, Ashleigh EA, Castillo S, Peskind ER, Pascualy M, Bissette G, Nemeroff CB, et al. Effect of desipramine on cerebrospinal fluid concentrations of corticotropin-releasing factor in human subjects. *Psychiatry Res.* 1993;46:1–8.
212. Heuser I, Bissette G, Dettling M, Schweiger U, Gotthardt U, Schmider J, Lammers CH, Nemeroff CB, Holsboer F. Cerebrospinal fluid concentrations of corticotropin-releasing hormone, vasopressin, and somatostatin in depressed patients and healthy controls: response to amitriptyline treatment. *Depress Anxiety.* 1998;8:71–79.
213. Banki CM, Karmacsi L, Bissette G, Nemeroff CB. CSF corticotropin-releasing hormone and somatostatin in major depression: response to antidepressant treatment and relapse. *Eur Neuropsychopharmacol.* 1992;2:107–113.
214. Gold PW, Chrousos GP. Clinical studies with corticotropin releasing factor: implications for the diagnosis and pathophysiology of depression, Cushing's disease, and adrenal insufficiency. *Psychoneuroendocrinology.* 1985;10:401–419.
215. Holsboer F, Gerken A, Stalla GK, Muller OA. Blunted aldosterone and ACTH release after human CRH administration in depressed patients. *Am J Psychiatry.* 1987;144:229–231.
216. Amsterdam JD, Maislin G, Winokur A, Kling M, Gold P. Pituitary and adrenocortical responses to the ovine corticotropin releasing hormone in depressed patients and healthy volunteers. *Arch Gen Psychiatry.* 1987;44:775–781.
217. Merali Z, Du L, Hrdina P, Palkovits M, Faludi G, Poulter MO, Anisman H. Dysregulation in the suicide brain: mRNA expression of corticotropin-releasing hormone receptors and GABA(A) receptor subunits in frontal cortical brain region. *J Neurosci.* 2004;24:1478–1485.
218. Nemeroff CB, Owens MJ, Bissette G, Andorn AC, Stanley M. Reduced corticotropin releasing factor binding sites in the frontal cortex of suicide victims. *Arch Gen Psychiatry.* 1988; 45:577–579.

219. Austin MC, Janosky JE, Murphy HA. Increased corticotropin-releasing hormone immunoreactivity in monoamine-containing pontine nuclei of depressed suicide men. *Mol Psychiatry.* 2003;8:324–332.
220. Merali Z, Kent P, Du L, Hrdina P, Palkovits M, Faludi G, Poulter MO, Bedard T, Anisman H. Corticotropin-releasing hormone, arginine vasopressin, gastrin-releasing peptide, and neuromedin B alterations in stress-relevant brain regions of suicides and control subjects. *Biol Psychiatry.* 2006;59:594–602.
221. Raadsheer FC, Hoogendijk WJ, Stam FC, Tilders FJ, Swaab DF. Increased numbers of corticotropin-releasing hormone expressing neurons in the hypothalamic paraventricular nucleus of depressed patients. *Neuroendocrinology.* 1994;60:436–444.
222. Bissette G, Klimek V, Pan J, Stockmeier C, Ordway G. Elevated concentrations of CRF in the locus coeruleus of depressed subjects. *Neuropsychopharmacology.* 2003;28:1328–1335.
223. Wang SS, Kamphuis W, Huitinga I, Zhou JN, Swaab DF. Gene expression analysis in the human hypothalamus in depression by laser microdissection and real-time PCR: the presence of multiple receptor imbalances. *Mol Psychiatry.* 2008;13:786–799. 741.
224. Kozicz T, Tilburg-Ouwens D, Faludi G, Palkovits M, Roubos E. Gender-related urocortin 1 and brain-derived neurotrophic factor expression in the adult human midbrain of suicide victims with major depression. *Neuroscience.* 2008;152:1015–1023.
225. Kang HJ, Adams DH, Simen A, Simen BB, Rajkowska G, Stockmeier CA, Overholser JC, Meltzer HY, Jurjus GJ, Konick LC, Newton SS, Duman RS. Gene expression profiling in postmortem prefrontal cortex of major depressive disorder. *J Neurosci.* 2007;27:13329–13340.
226. Binder EB, Owens MJ, Liu W, Deveau TC, Rush AJ, Trivedi MH, Fava M, Bradley B, Ressler KJ, Nemeroff CB. Association of polymorphisms in genes regulating the corticotropin-releasing factor system with antidepressant treatment response. *Arch Gen Psychiatry.* 67, 369-379.
227. Wasserman D, Wasserman J, Sokolowski M. Genetics of HPA-axis, depression and suicidality. *Eur Psychiatry.* 25, 278-280.
228. Licinio J, O'Kirwan F, Irizarry K, Merriman B, Thakur S, Jepson R, Lake S, Tantisira KG, Weiss ST, Wong ML. Association of a corticotropin-releasing hormone receptor 1 haplotype and antidepressant treatment response in Mexican-Americans. *Mol Psychiatry.* 2004;9:1075–1082.
229. Liu Z, Zhu F, Wang G, Xiao Z, Tang J, Liu W, Wang H, Liu H, Wang X, Wu Y, Cao Z, Li W. Association study of corticotropin-releasing hormone receptor1 gene polymorphisms and antidepressant response in major depressive disorders. *Neurosci Lett.* 2007;414:155–158.
230. Liu Z, Zhu F, Wang G, Xiao Z, Wang H, Tang J, Wang X, Qiu D, Liu W, Cao Z, Li W. Association of corticotropin-releasing hormone receptor1 gene SNP and haplotype with major depression. *Neurosci Lett.* 2006;404:358–362.
231. Wasserman D, Sokolowski M, Rozanov V, Wasserman J. The CRHR1 gene: a marker for suicidality in depressed males exposed to low stress. *Genes Brain Behav.* 2008;7:14–19.
232. Wasserman D, Wasserman J, Rozanov V, Sokolowski M. Depression in suicidal males: genetic risk variants in the CRHR1 gene. *Genes Brain Behav.* 2009;8:72–79.
233. Papiol S, Arias B, Gasto C, Gutierrez B, Catalan R, Fananas L. Genetic variability at HPA axis in major depression and clinical response to antidepressant treatment. *J Affect Disord.* 2007;104: 83–90.
234. Bradley RG, Binder EB, Epstein MP, Tang Y, Nair HP, Liu W, Gillespie CF, Berg T, Evces M, Newport DJ, Stowe ZN, Heim CM, Nemeroff CB, Schwartz A, Cubells JF, Ressler KJ. Influence of child abuse on adult depression: moderation by the corticotropin-releasing hormone receptor gene. *Arch Gen Psychiatry.* 2008;65:190–200.
235. De Luca V, Tharmalingam S, Kennedy JL. Association study between the corticotropin-releasing hormone receptor 2 gene and suicidality in bipolar disorder. *Eur Psychiatry.* 2007; 22:282–287.
236. Held K, Kunzel H, Ising M, Schmid DA, Zobel A, Murck H, Holsboer F, Steiger A. Treatment with the CRH1-receptor-antagonist R121919 improves sleep-EEG in patients with depression. *J Psychiatr Res.* 2004;38:129–136.
237. Kunzel HE, Zobel AW, Nickel T, Ackl N, Uhr M, Sonntag A, Ising M, Holsboer F. Treatment of depression with the CRH-1-receptor antagonist R121919: endocrine changes and side effects. *J Psychiatr Res.* 2003;37:525–533.
238. Zobel AW, Nickel T, Kunzel HE, Ackl N, Sonntag A, Ising M, Holsboer F. Effects of the high-affinity corticotropin-releasing hormone receptor 1 antagonist R121919 in major depression: the first 20 patients treated. *J Psychiatr Res.* 2000;34:171–181.
239. Mathew SJ, Price RB, Charney DS. Recent advances in the neurobiology of anxiety disorders: implications for novel therapeutics. *Am J Med Genet C Semin Med Genet.* 2008;148C:89–98.
240. Yehuda R, Giller EL, Southwick SM, Lowy MT, Mason JW. Hypothalamic–pituitary–adrenal dysfunction in posttraumatic stress disorder. *Biol Psychiatry.* 1991;30:1031–1048.
241. de Kloet CS, Vermetten E, Geuze E, Kavelaars A, Heijnen CJ, Westenberg HG. Assessment of HPA-axis function in posttraumatic stress disorder: pharmacological and non-pharmacological challenge tests, a review. *J Psychiatr Res.* 2006;40:550–567.
242. Erhardt A, Ising M, Unschuld PG, Kern N, Lucae S, Putz B, Uhr M, Binder EB, Holsboer F, Keck ME. Regulation of the hypothalamic-pituitary-adrenocortical system in patients with panic disorder. *Neuropsychopharmacology.* 2006;31:2515–2522.
243. Schreiber W, Lauer CJ, Krumrey K, Holsboer F, Krieg JC. Dysregulation of the hypothalamic-pituitary-adrenocortical system in panic disorder. *Neuropsychopharmacology.* 1996;15:7–15.
244. Bremner JD, Licinio J, Darnell A, Krystal JH, Owens MJ, Southwick SM, Nemeroff CB, Charney DS. Elevated CSF corticotropin-releasing factor concentrations in posttraumatic stress disorder. *Am J Psychiatry.* 1997;154:624–629.
245. Baker DG, West SA, Nicholson WE, Ekhator NN, Kasckow JW, Hill KK, Bruce AB, Orth DN, Geracioti Jr TD. Serial CSF corticotropin-releasing hormone levels and adrenocortical activity in combat veterans with posttraumatic stress disorder. *Am J Psychiatry.* 1999;156:585–588.
246. Sautter FJ, Bissette G, Wiley J, Manguno-Mire G, Schoenbachler B, Myers L, Johnson JE, Cerbone A, Malaspina D. Corticotropin-releasing factor in posttraumatic stress disorder (PTSD) with secondary psychotic symptoms, nonpsychotic PTSD, and healthy control subjects. *Biol Psychiatry.* 2003; 54:1382–1388.
247. Banki CM, Karmacsi L, Bissette G, Nemeroff CB. Cerebrospinal fluid neuropeptides in mood disorder and dementia. *J Affect Disord.* 1992;25:39–45.
248. Fossey MD, Lydiard RB, Ballenger JC, Laraia MT, Bissette G, Nemeroff CB. Cerebrospinal fluid corticotropin-releasing factor concentrations in patients with anxiety disorders and normal comparison subjects. *Biol Psychiatry.* 1996;39:703–707.
249. Jolkkonen J, Lepola U, Bissette G, Nemeroff C, Riekkinen P. CSF corticotropin-releasing factor is not affected in panic disorder. *Biol Psychiatry.* 1993;33:136–138.
250. Chappell P, Leckman J, Goodman W, Bissette G, Pauls D, Anderson G, Riddle M, Scahill L, McDougle C, Cohen D. Elevated cerebrospinal fluid corticotropin-releasing factor in Tourette's syndrome: comparison to obsessive compulsive disorder and normal controls. *Biol Psychiatry.* 1996;39:776–783.

251. Smoller JW, Rosenbaum JF, Biederman J, Kennedy J, Dai D, Racette SR, Laird NM, Kagan J, Snidman N, Hirshfeld-Becker D, Tsuang MT, Sklar PB, Slaugenhaupt SA. Association of a genetic marker at the corticotropin-releasing hormone locus with behavioral inhibition. *Biol Psychiatry.* 2003;54:1376–1381.
252. Smoller JW, Yamaki LH, Fagerness JA, Biederman J, Racette S, Laird NM, Kagan J, Snidman N, Faraone SV, Hirshfeld-Becker D, Tsuang MT, Slaugenhaupt SA, Rosenbaum JF, Sklar PB. The corticotropin-releasing hormone gene and behavioral inhibition in children at risk for panic disorder. *Biol Psychiatry.* 2005;57:1485–1492.
253. Keck ME, Kern N, Erhardt A, Unschuld PG, Ising M, Salyakina D, Muller MB, Knorr CC, Lieb R, Hohoff C, Krakowitzky P, Maier W, Bandelow B, Fritze J, Deckert J, Holsboer F, Muller-Myhsok B, Binder EB. Combined effects of exonic polymorphisms in CRHR1 and AVPR1B genes in a case/control study for panic disorder. *Am J Med Genet B Neuropsychiatr Genet.* 2008;147B:1196–1204.
254. Cryan JF, Holmes A. The ascent of mouse: advances in modelling human depression and anxiety. *Nat Rev Drug Discov.* 2005;4:775–790.
255. Wong ML, Licinio J. From monoamines to genomic targets: a paradigm shift for drug discovery in depression. *Nat Rev Drug. Discov.* 2004;3:136–151.
256. Krishnan V, Nestler EJ. The molecular neurobiology of depression. *Nature.* 2008;455:894–902.
257. Francis DD, Szegda K, Campbell G, Martin WD, Insel TR. Epigenetic sources of behavioral differences in mice. *Nat Neurosci.* 2003;6:445–446.
258. McEwen BS. Physiology and neurobiology of stress and adaptation: central role of the brain. *Physiol Rev.* 2007;87:873–904.
259. Ulrich-Lai YM, Herman JP. Neural regulation of endocrine and autonomic stress responses. *Nat Rev Neurosci.* 2009;10:397–409.
260. Lupien SJ, McEwen BS, Gunnar MR, Heim C. Effects of stress throughout the lifespan on the brain, behaviour and cognition. *Nat Rev Neurosci.* 2009;10:434–445.
261. McEwen BS. Stress, sex, and neural adaptation to a changing environment: mechanisms of neuronal remodeling. *Ann NY Acad Sci.* 1204 Suppl, E38-59.
262. Holsboer F, Ising M. Stress hormone regulation: biological role and translation into therapy. *Annu Rev Psychol.* 61, 81-109, C1-11.
263. Bale TL. Sensitivity to stress: dysregulation of CRF pathways and disease development. *Horm Behav.* 2005;48:1–10.
264. Zorrilla EP, Koob GF. The therapeutic potential of CRF1 antagonists for anxiety. *Expert Opin Investig Drugs.* 2004;13:799–828.

CHAPTER

16

Neuroendocrine Regulation of Body Water and Electrolytes: Thirst and Salt Appetite

Edward M. Stricker[1], *Joseph G. Verbalis*[2]

[1] Department of Neuroscience, University of Pittsburgh, Pittsburgh, PA, USA, [2] Department of Medicine, Georgetown University Medical Center, Washington, DC, USA

OUTLINE

Summary

For the past 30 years it has been recognized that thirst is stimulated by three separate but related signals: increased plasma osmolality, decreased plasma volume, and increased circulating levels of angiotensin II (as might result from arterial hypotension). It is also well known that decreased plasma osmolality and acute hypertension provide potent inhibitory signals of thirst. The present discussion updated those familiar elements in the control of thirst and added other elements for consideration that derived from recent studies of fluid processing in the alimentary canal, of salt appetite, and of multiple endocrine responses to alterations in body fluids. Three principles emerge that deserve emphasis. First, each of the three systemic signals for thirst additionally stimulates salt appetite as well, although NaCl intake may not become manifest because of overriding inhibitory factors. Second, central angiotensin II helps to mediate thirst and salt appetite in rats, while central atrial natriuretic peptide opposes those effects and central oxytocin helps to mediate inhibition of salt appetite. Third, many treatments that affect thirst or salt appetite actually provide mixed signals containing both excitatory and inhibitory components.

INTRODUCTION

Body fluids are the watery matrix in which the biochemical reactions of cellular metabolism occur. The concentration of substrates in cellular fluid is a key factor in determining the rate at which those reactions take place. All tissues depend on circulating blood to deliver the nutrients needed to support cellular metabolism, and to carry away unwanted metabolites for excretion. Thus, the maintenance of solute concentrations or osmolalities and the regulation of plasma volume are essential functions in the physiology of animals.

After decades of intensive and fruitful research, still ongoing, much of the basis of body fluid homeostasis is known. Because more than a single chapter would be required to provide a comprehensive presentation of this material, our goal is to provide an overview of the general principles and organization, and to indicate the key evidence that supports these formulations. We will focus on the hormonal and neuroendocrine contributions to body fluid homeostasis, with a special emphasis on the brain's control of thirst and salt appetite, and will close by mentioning several residual unsettled issues of significance.

THIRST

Osmoregulation

Body fluid osmolality is an expression of concentration, representing the ratio of solute to water in the fluid.

Handbook of Neuroendocrinology, DOI: 10.1016/B978-0-12-375097-6.10016-2

Osmolality increases when there is either a decrease in the denominator or an increase in the numerator of this ratio. Under each circumstance, water leaves body cells by osmosis, which has the adaptive effect of blunting the increased osmolality of extracellular fluid. The reverse effect occurs during overhydration; water enters cells by osmosis, thus blunting the decreased extracellular fluid osmolality. Hence, osmosis provides a rapid, intrinsic, first line of defense against large changes in body fluid osmolality.

When body fluid osmolality is altered, behavioral and physiological responses are stimulated to restore the basal state. Thirst is stimulated by a signal associated with cellular dehydration that results when the osmolality of extracellular fluid is increased – as, for example, following the administration of hypertonic NaCl solution. In contrast, much less drinking results from the delivery of equiosmolar solutions containing urea or glucose, which diffuse freely across cell membranes and therefore do not cause osmotic gradients that induce water redistribution.[1,2] An increase in extracellular Na^+ concentration is not the basis of the excitatory signal, because Na^+-free hyperosmotic solutions containing non-permeant solutes such as mannitol, sorbitol or sucrose are just as effective as equiosmotic NaCl solutions in stimulating thirst.[3,4] This stimulus increases linearly in proportion to increases in plasma osmolality (pOsm) above a threshold of 1–2%,[5,6] with intakes proportional to dehydration and sufficient to restore water balance.[1,3] In addition, the motivational properties of the stimulus in laboratory animals have been established by experiments in which dehydration elicits learned operant behaviors leading to water reinforcements.

Complementing water ingestion is neurohypophysial secretion of the antidiuretic hormone arginine vasopressin (AVP), which also occurs in response to increases in the effective osmolality of blood plasma.[7] AVP is synthesized in the magnocellular neurons of the supraoptic nucleus (SON) and paraventricular nucleus (PVN) in the hypothalamus, transported along axons projecting to the nearby posterior lobe of the pituitary gland, and secreted from there into the systemic circulation (Fig. 16.1). The hormone's main action is on the V_2 subtype of AVP receptors in the kidneys to promote the reabsorption of water from the renal tubules, thereby making urine more concentrated and reducing further water loss. Tiny amounts of the peptide are needed for this effect, because the hormone is very

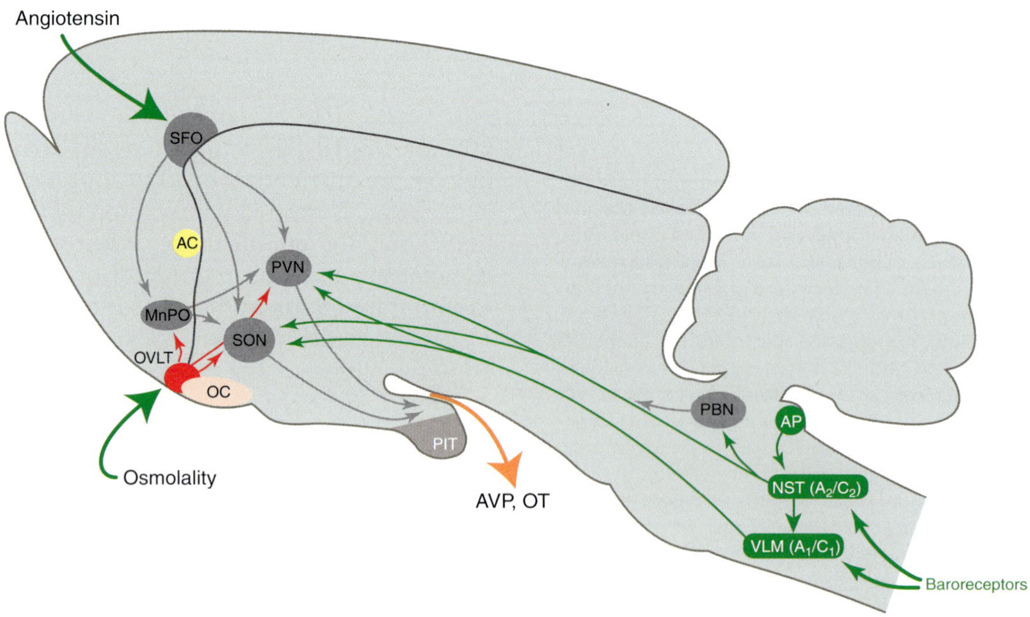

FIGURE 16.1 Summary of the main pathways that mediate secretion of arginine vasopressin (AVP) and oxytocin (OT) in rats. The vascular organ of the lamina terminalis (OVLT) is especially sensitive to the effective osmolality of plasma. Circulating angiotensin II (AngII) activates neurons of the subfornical organ (SFO), an essential site of AngII action, as well as cells throughout the lamina terminalis. In response to hyperosmolality or AngII, projections from the SFO and OVLT to the median preoptic nucleus (MnPO) activate excitatory and inhibitory interneurons that project to the supraoptic nucleus (SON) and paraventricular nucleus (PVN) to modulate direct inputs to these areas from the circumventricular organs. Baroreceptor-mediated stimuli, such as hypovolemia and hypotension, act on afferents that terminate in the nucleus of the solitary tract (NTS) and area postrema (AP). The major projection to magnocellular VP neurons appears to arise from A1 cells of the ventrolateral medulla (VLM) that are activated by excitatory interneurons from the NTS. Other areas, such as the parabrachial nucleus (PBN), may contribute multisynaptic projections. Circulating atrial natriuretic peptide (ANP) appears to activate neurons in the SFO and AP and thereby inhibit the effect of AngII. Not shown are the sodium-sensitive receptors in the three circumventricular organs, or the likely inhibitory effect of ANP on OVLT neurons. Similar (but not identical) signals and brain sites participate in the central control of thirst and salt appetite. AC, anterior commissure; OC, optic chiasm; PIT, anterior pituitary.

potent; for example, maximal antidiuresis is achieved in humans when blood levels of AVP reach 5–6 pg/ml (5–6 nmol/l), which is only a few trillionths of a gram per milliliter above basal levels of 1–2 pg/ml (1–2 nmol/l). Conversely, secretion of AVP is inhibited by osmotic dilution of body fluids, which allows excess water to be excreted in dilute urine. When AVP function is impaired, as in diabetes insipidus (a disorder that typically results from hypothalamic disease), human patients may excrete up to 25 liters of dilute urine per day and adaptively drink comparable volumes of water in order to maintain water balance. The same disorder in rats results in daily consumption of water equal in magnitude to their body weight. AVP plainly allows animals the freedom to pursue their normal activities without frequent interruption to drink or urinate.

The complementary contributions of thirst and neurohypophysial hormone secretion to osmoregulation have been noted in many different types of experiments. For example, when dogs were given a salt load orally by stomach tube, some animals increased water intake while conserving little urinary water, whereas other animals drank less water while excreting very concentrated urine.[4,8] In other studies, a NaCl load elicited a more substantial increase in water intake in rats after renal function was eliminated by bilateral nephrectomy,[3] whereas ingestion of a salt load by dogs stimulated larger increases in AVP secretion when daily water intake was limited.[9]

The pioneering investigations of Verney[10] and Andersson[11] suggested that osmoreceptor cells, located in the anterior hypothalamus, detected the increase in pOsm and initiated the regulatory responses. Subsequently, osmoreceptors were localized in the vascular organ of the lamina terminalis (OVLT), a midline structure adjacent to the third cerebral ventricle and ventral to the anterior commissure (Fig. 16.1). The OVLT is highly vascular and lacks a blood–brain barrier, which allows it to respond readily to changes in systemic pOsm. The OVLT sends neural afferents to the magnocellular hypothalamic neurons that secrete AVP, directly,[12,13] and indirectly via the median preoptic (MnPO) area.[14,15] The MnPO (which also is known as the ventral or subcommissural portion of the nucleus medianus), a midline structure located in the lamina terminalis along the rostral border of the third cerebral ventricle (Fig. 16.1), additionally is critical in the neural circuitry that mediates thirst.

Considerable evidence supports the crucial role of the anterior hypothalamus in osmoregulation. Thus, experimental damage to the anterior hypothalamus in rats significantly reduced water intake and secretion of AVP in response to the administration of hypertonic saline.[16,17] Similar observations have been made in other animals, including in dogs with more discrete surgical lesions of the OVLT[18,19] and in human patients with brain tumors that destroyed the OVLT and surrounding region.[20] Complementing these findings are the results of studies showing that electrolytic lesions of the MnPO in rats block drinking and attenuate AVP secretion in response to injections of hypertonic saline.[21–23] Finally, an increased expression of Fos (the protein product of the immediate-early gene, *c-fos*, which is a useful marker of cell activation)[24] occurred in OVLT and MnPO neurons after systemic administration of hyperosmotic solutions.[25–27]

Although ingested water does not rapidly dilute systemic pOsm and remove the excitatory signal for thirst, dehydrated animals do not drink amounts exceeding need. Thus, drinking must be inhibited by some early stimulus in anticipation of rehydration. In seminal studies that addressed this issue, dehydrated dogs were found to replace their water deficits precisely even though water ingestion ended within a few minutes, before a decrease in systemic pOsm became apparent.[28] That early signal, which simultaneously inhibited AVP secretion, also occurred when osmotic dilution was precluded by having dogs drink isotonic saline or by having the ingested water drain through an open gastric fistula, but it did not occur when a water load was delivered directly to the stomach. Indeed, it occurred even when dogs drank concentrated saline solution, although, as might be expected, the dogs became even thirstier and secreted more AVP once the saline was absorbed and pOsm was elevated.[29] The inhibitory effects appear to be mediated by an oropharyngeal signal associated with the rhythmic pattern of swallowing that occurs during drinking.[30] This signal is temporary, and is supplemented and then replaced by a later postgastric satiety signal associated with rehydration. The rapid inhibition of thirst and AVP secretion by water consumption, mediated by pre-systemic signals, also has been observed to occur during rehydration in other animals as well as in human subjects.[31–34]

In contrast, although an early inhibitory signal also was evident when thirsty rats drank water, no effect on plasma AVP was seen when rats consumed comparable quantities of isotonic saline solution or when water ingested by dehydrated rats drained through an open gastric fistula.[35,36] Because ingested water empties from the stomach very rapidly and begins to dilute systemic pOsm while the rat is still drinking, it seems likely that visceral osmoreceptors inhibit AVP secretion while diluting fluids pass through the small intestine. That hypothesis was first proposed by Baertschi and Pence,[37] who reported that gastric water loads rapidly decreased plasma AVP in dehydrated rats even when dilution of systemic pOsm was prevented by concurrent intravenous infusion of hypertonic saline. Subsequent parallel studies suggest that an early signal of hydration also can inhibit thirst in rats; thirsty rats ingested excessive amounts of water when peripheral sensory fibers were destroyed non-selectively by systemic injection of the neurotoxin capsaicin,[38] and when visceral signals were disrupted by lesions of the area postrema (AP),[39,40] a neural structure that is located at the dorsal surface of the medulla just ventral to the fourth cerebral ventricle (Fig. 16.1). Conversely, a gastric NaCl load was found to stimulate AVP secretion and drinking by rats before systemic pOsm was observed to increase.[41–43] Thus, just as cerebral osmoreceptors respond both to increases and decreases in systemic pOsm, splanchnic osmoreceptors respond to concentrations of intestinal fluid that are above or below isotonic. These putative osmoreceptors might be located in or near the duodenum and work together with stretch receptors on the stomach and small intestine, which collectively, via neural afferents to the caudal brainstem,[44] would be well positioned to influence behavioral and physiological responses before significant absorption of the ingested load into the general circulation allowed detection by cerebral osmoreceptors.

An additional factor contributes importantly to osmoregulation in rats, if not in other laboratory animals. Like AVP, neurohypophysial oxytocin (OT) is secreted when pOsm is elevated in rats.[45,46] OT is a potent stimulus of the secretion of atrial natriuretic peptide (ANP), and both hormones act in the kidneys to promote urinary Na^+ excretion in rats and thereby blunt increases in pOsm.[47–50]

Blood Volume and Pressure Regulation

Until the early 1960s, the only established stimulus of thirst was cellular dehydration associated with an increase in the effective osmolality of extracellular fluid. However, it certainly was recognized that elaborate neural and endocrine mechanisms had evolved to maintain blood volume and pressure, among which was the secretion of AVP in response to hypovolemia. It therefore was not surprising to discover that an experimental loss of blood volume also stimulated thirst, analogous with the signal for AVP secretion and thirst that is generated when pOsm is elevated).[51,52]

Among numerous and varied models of hypovolemia, extravascular administration of a hyperoncotic colloid solution in rats has been the most thoroughly investigated. The colloidal solution reliably induces plasma volume deficits and thirst by opposing the equilibrium of fluid movement across the capillary walls near the injection site, so that plasma leaving the blood vessels due to hydrostatic pressure (caused by the contraction of the heart) cannot return readily and isosmotic extracellular fluid gradually accumulates in the local interstitial fluid. The effect of colloid treatment on plasma volume is orderly; the larger the volume and more concentrated the solution that is injected, the larger the local edema and the greater the plasma volume deficit. This gradual sequestration of fluid prevents the normal restoration of plasma volume from endogenous extravascular reservoirs. Indeed, even ingested fluids are not at first retained in the circulation but are leached into the injection area, and so fluid volumes considerably larger than the net deficits must be consumed for plasma volume to be restored. Non-hypotensive hypovolemia produced in this way lasts for many hours, during which water intake is increased in proportion to the extravascular fluid accumulation[51] and the induced plasma volume deficit.[53] Beyond a threshold value of ~5%, water consumption is linear up to a plasma volume loss of at least 30%. Comparable intakes are obtained when colloid-treated rats are required to press a lever for water, demonstrating that hypovolemia elicits a motivational state of thirst rather than mere reflexive drinking.

Complementing thirst during hypovolemia is antidiuresis, resulting both from an exponential increase in AVP secretion[7,46] and from a progressive decrease in

BOX 16.1

HOW DO WE KNOW ABOUT WATER INTAKE AND VP SECRETION IN RESPONSE TO HYPOVOLEMIA IN RATS?

Subcutaneous injection of a colloidal solution in rats disrupts the Starling equilibrium and pulls protein-free plasma fluid into the local interstitium, thereby raising plasma protein (pProt) concentration. This treatment also increases water intake.[1] Consumption of water alone does not alter pProt concentration because ingested water, once it has been emptied from the stomach and absorbed from the small intestine, is distributed throughout the body so as to maintain osmotic equilibrium across all tissues. Thus, since approximately two-thirds of body water is found in cells, approximately two-thirds of ingested water will quickly move into cells by osmosis. The remaining third of the ingested fluid will remain extracellular; however, since so much extracellular fluid is sequestered in a subcutaneous edema at the injection site, rather little of the ingested fluid will remain in the intravascular space whose deficit prompted thirst and water intake in the first place.

BOX 16.1 *(cont'd)*

By simply measuring pProt concentration and water intake in individual rats, and expressing those values as M ± SE, it is evident that thirsty rats are hypovolemic. Much more information can be obtained from collected data, though, by plotting values of pProt concentration and water intake as a function of one another. It can be seen readily in the figure below that the induced water intake is proportional to the plasma volume deficit (estimated from the change in pProt concentration). In other words, the bigger the plasma volume deficits, the thirstier the rats were and the more water they drank.

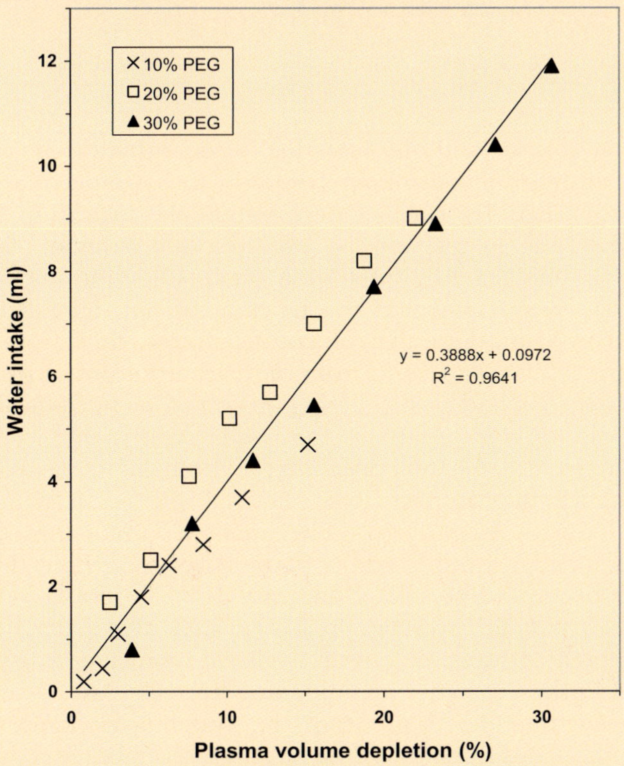

Estimated plasma volume deficits and cumulative water intakes of rats injected subcutaneously with 5 ml of 10, 20 or 30% solutions of polyethylene glycol (PEG). Mean values are shown from data collected at hourly intervals. Note that the 20% and 30% solutions produced volume depletions that were larger than those produced by the 10% solution. *From Stricker.*[1]

We have used scatter plots to show how variables relate to one another in many other circumstances. For example, one of the consequences of plasma volume depletion is the secretion of vasopressin (VP) from the posterior lobe of the pituitary gland. The secreted VP has a relatively short half-life of 4–6 minutes, which means that for its levels in plasma to remain elevated the animal must be continuously secreting the hormone. The second figure shows the plasma VP (pVP) levels in PEG-treated rats, plotted as a function of pProt concentrations in individual animals.[2] It is clear that the relation between these two variables is exponential, with rather small increases in plasma levels of the peptide observed when plasma volume deficits were 10–15% but increasingly large amounts of the peptide seen when hypovolemia was more severe. The signal for VP secretion is believed to result from a combination of neural signals (originating from cardiac baroreceptors) and endocrine signals (originating from AngII receptors in the SFO) to the PVN and SON of the hypothalamus. An understanding of the binding properties of the VP receptors allows additional meaning to be derived from this figure. The relatively small increase in pVP that occurred when plasma volume deficits were only 10–20% was still sufficient to stimulate the V2 receptors in the kidneys and thereby allow conservation of water in urine (i.e., antidiuresis). In contrast, the much higher levels of pVP associated with plasma volume deficits above 25% allowed VP to bind to the V1 receptors on blood vessels to cause vasoconstriction. These details would have been lost if pProt concentrations had been presented as M ± SE values.

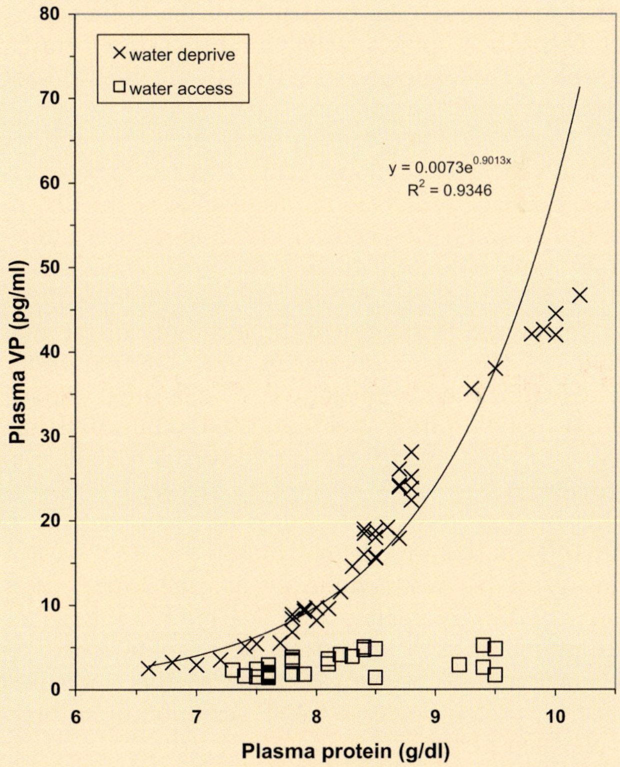

Plasma levels of vasopressin (VP) as a function of plasma protein concentration in individual rats after s.c. PEG treatment. Some rats were denied access to drinking water (×), some were allowed to drink (□). The exponential increase in pVP levels was suppressed by osmotic dilution. *From Stricker and Verbalis.*[2]

(Continued)

BOX 16.1 *(cont'd)*

The second figure also shows the apparent suppression of VP secretion when rats drink water.[2] Plasma VP levels in PEG-treated rats drinking water resembled the basal levels of untreated rats despite relatively large plasma volume deficits, which would have been expected to provoke a very large VP secretion. These data suggest that osmotic dilution provides an intense inhibitory stimulus that completely suppresses VP secretion even when stimulated by severe hypovolemia. Similar effects of hypovolemia and of osmotic dilution on plasma levels of oxytocin (OT) in PEG-treated rats are seen when pOT is plotted as a function of pProt concentration in individual PEG-treated rats.[2] In other words, osmotic dilution is a more potent inhibitory stimulus of neurohypophysial hormone secretion than hypovolemia is an excitatory stimulus in these circumstances.

References

1. Stricker EM. Some physiological and motivational properties of the hypovolemic stimulus for thirst. *Physiol Behav.* 1968;3:379–385.
2. Stricker EM, Verbalis JG. Interaction of osmotic and volume stimuli in regulation of neurohypophyseal secretion in rats. *Am J Physiol.* 1986;250:R267–R275.

renal blood flow and glomerular filtration. A parallel increase in OT secretion is also stimulated by hypovolemia,[46] although the expected natriuresis does not occur in part because of the dominant Na^+-conserving effects of aldosterone,[54] whose secretion is stimulated as well.[55] Instead, elevated plasma OT is one of many stimuli in rats that increase renin secretion from the kidneys,[56,57] which helps to support arterial blood pressure through the formation of angiotensin II (AngII), the most potent endogenous pressor agent.

The identification of hypovolemia as a stimulus of thirst allowed investigations of the interaction between osmoregulatory and volume regulatory controls of water intake and AVP secretion. When thirst was stimulated by concurrent treatments that increased pOsm while also decreasing plasma volume, the resultant water consumption suggested an additivity of the two stimuli.[58–60] The same was true of AVP secretion under these experimental circumstances.[7,46,61] In other experiments, osmotic dilution was found to inhibit thirst and secretion of AVP and OT despite severe hypovolemia. The magnitude of this effect was striking: a 5–7% osmotic dilution completely inhibited these responses despite plasma volume deficits of 30%.[46,60] In contrast, expansion of plasma volume by a large intravenous infusion of isotonic saline did not inhibit the water consumption of rats pretreated with an injection of hypertonic saline,[3,62] whereas much smaller increases in plasma volume decreased AVP secretion in animals with elevated pOsm.[61,63]

Baroreceptors located in the relatively distensible great veins and right atrium sense a reduction in blood volume and appear to mediate AVP secretion and thirst during hypovolemia. For example, destruction of the atrial baroreceptors in sheep eliminates the drinking response to hypovolemia produced by ultrafiltration of the blood without affecting osmoregulatory thirst.[64] Similarly, the drinking response to hypovolemia was abolished by cardiac denervation in dogs.[65] In rats, drinking in response to colloid-induced hypovolemia was abolished by inflating a small balloon in the superior vena cava near the junction of the right atrium, thereby masking the volume deficit from the baroreceptors.[66] Vagal afferents from cardiac baroreceptors project to the AP and to the lateral portions of the nucleus of the solitary tract (NTS) in the caudal brainstem,[67] which project to the magnocellular hypothalamic neurons[68] and to the MnPO.[69]

Forty years ago, Fitzsimons[70] discovered that systemic administration of renin or AngII produced a prompt, dose-related increase in water intake in rats. Subsequent investigations revealed that AngII has potent dipsogenic effects in birds, reptiles and fish, in addition to other mammals. Although AngII cannot penetrate the brain from the systemic circulation, it stimulates thirst by acting on AngII receptors in the subfornical organ (SFO),[71,72] which lacks a blood–brain barrier. The SFO is a highly vascularized midline structure that protrudes into the anterior dorsal portion of the third cerebral ventricle near the interventricular foramen, and sends neural projections to the MnPO and to the AVP- and OT-containing neurons in the hypothalamus (Fig. 16.1).[69,73] Drinking in rats can be stimulated by injection of AngII into the SFO in femtomolar amounts,[74] demonstrating the remarkable potency of the peptide as a dipsogen. In addition, drinking elicited by systemic injections of AngII was abolished or severely attenuated by SFO lesions[75,76] or by administration of AngII receptor blockers into the SFO.[74] Furthermore, expression of cFos is elevated in the SFO when AngII is injected into the circulation.[26,77] Curiously, AngII is also a neurotransmitter in efferent neural pathways from the SFO;[78] evidently, renin, angiotensinogen and angiotensin converting enzyme all are present in the brain.[79] This arrangement has several implications with regard to thirst. First, in addition to its action as a blood-borne hormone, AngII likely is dipsogenic as a neurotransmitter released in the brain when AngII of renal origin acts in the SFO. Second, if angiotensinergic efferent neurons projecting from the SFO do play a role in mediating thirst, then AngII receptors in the MnPO likely are activated when intracerebroventricular AngII stimulates water intake.[26] Third, the presence of angiotensin converting enzyme in the brain provides an explanation for paradoxical observations that systemic administration of captopril in rats may increase water intake during hypovolemia or arterial hypotension.[80–83] That is, the drug given in low doses blocks the conversion of AngI to AngII in the periphery but not in the SFO (or elsewhere in the brain), where production of AngII increases water intake.[84]

Although renin secretion increases exponentially in proportion to colloid-induced volume deficits in rats,[85] the contribution of the renin–angiotensin system to thirst during hypovolemia is uncertain. Observations that water intake in response to a range of plasma volume deficits is not affected either by NTS lesions[86] or by bilateral nephrectomy[51,55] indicate that neither the baroreceptor signal nor AngII is necessary for mediating hypovolemic thirst; presumably, when one signal is absent the other adaptively compensates by increasing its intensity. Findings that water intake was impaired after moderate hypovolemia in rats with SFO lesions[74] are difficult to interpret because a similar dysfunction occurred when hypertonic saline was infused.[87] Perhaps the most compelling approach involved electrolytic lesions of the septal area, which increase sensitivity to the dipsogenic effects of AngII.[88] If AngII provides an important signal for thirst during hypovolemia, then rats with septal area lesions should drink much more water than control rats in response to PEG treatment. In fact they do, but it is interesting to note when this drinking occurs; no increase in water intake was seen during the first 9 hours after subcutaneous injection of colloid solution, by which time plasma volume deficits were 25% and plasma renin activities increased to values 20–40 times greater than normal, but by 12 hours those animals drank (and retained) unusually large amounts of water.[89] These results suggest that AngII normally contributes little to thirst during progressive hypovolemia until plasma volume deficits become substantial. On the other hand, AngII provides a unique stimulus of thirst in response to drugs that reduce peripheral vascular resistance and thereby lower arterial blood pressure (ABP) without reducing blood volume.[90,91] Thus, neural input from arterial baroreceptors evidently does not directly elicit thirst during arterial hypotension.

It is important to note that the administration of AngII increases ABP and thereby provides a signal that inhibits thirst, thus attenuating its own dipsogenic effect.[92,93] This signal also is effective in inhibiting thirst induced by increased pOsm or by hypovolemia.[94] The inhibition of drinking was proportional to ABP up to 160 mmHg, beyond which further decreases in water intake did not occur (thus resembling the limits of acute hypertension that produce a reflexive reduction in heart rate). Consistent with the conclusion that exogenous AngII provides a mixed stimulus for thirst, with both excitatory and inhibitory components, water intake elicited by AngII was enhanced after surgical destruction of the sinoaortic baroreceptors,[95,96] which prevents detection of the induced increase in ABP. In contrast, the AngII-induced increase in AVP secretion, which is mediated by the SFO,[97] was not enhanced by chronic cardiac and arterial baroreceptor denervation[98] or by NTS lesions.[99] Thus, the regulatory controls of AVP secretion and thirst, while similar, clearly are not identical.

In summary, decreases in blood volume stimulate thirst and elicit water intake, which has an early inhibitory effect on thirst (and AVP secretion) as well as a sustained inhibitory effect associated with systemic osmotic dilution. However, in contrast to osmoregulatory thirst, water intake does not eliminate the excitatory stimulus of hypovolemic thirst because water is not very effective in repairing plasma volume deficits; ingested water is distributed as body fluids are distributed, which means that about two-thirds of the water moves into cells by osmosis while most of the remainder stays in the interstitial space. Instead of water, hypovolemic animals need isotonic saline, which remains extracellular and therefore can repair the fluid deficits effectively. Appropriate to their need, PEG-treated rats consume isotonic saline in large amounts sufficient to restore plasma volume without the inhibition of thirst that results when they drink and retain water.[100] Furthermore, and most remarkably, when given water and concentrated saline solution to drink in a two-bottle test, hypovolemic rats alternately ingest the two fluids in the amounts required to concoct an isotonic saline solution.[101,102] In other words, hypovolemia stimulates both thirst and salt appetite in rats, and the two allied motivations together direct the animal to consume the fluid mixture suitable for repairing the plasma volume deficit.

SALT APPETITE

A key point to make at the outset is that the stimulation of thirst and salt appetite by subcutaneous colloid treatment is strongly influenced by the diet on which the rats had been maintained. Thus, when rats are fed standard laboratory chow, salt appetite usually develops many hours after colloid treatment, well after thirst appears.[100] In contrast, when rats are fed NaCl-deficient diet, salt appetite becomes evident soon after the colloid treatment and before thirst appears.[101] The difference in outcomes draws attention to the variables that seem most relevant to the control of salt appetite in rats: increases in excitatory signals that appear to derive from blood-borne hormones, and decreases in a central inhibitory signal. Both developments usually are required before salt appetite becomes manifest.

Numerous observations support the hypothesis that salt appetite in rats is stimulated by AngII, whether acting as a blood-borne hormone or as a neurotransmitter in the brain. For example, the pronounced salt appetite in rats after bilateral adrenalectomy,[103] which is secondary to the uncontrolled loss of Na^+ in urine, is associated with a substantial increase in circulating AngII levels and was blunted by intracerebroventricular (i.c.v.) administration of an AngII receptor antagonist.[104] Furthermore, salt appetite was enhanced when AngII was administered into the brain of PEG-treated rats[105] and when PEG-treated rats had been maintained on NaCl-deficient diet (which augments the induced renin secretion), especially after pretreatment with a low dose of captopril.[55,106] Conversely, salt appetite was abolished in Na^+-depleted rats by large doses of captopril[107,108] and it was restored by intravenous AngII.[109,110] AngII seems to act in the SFO to stimulate salt appetite,[111,112] but a role for the MnPO also is possible.[113,114]

Even a day or two of NaCl deprivation is known to increase the steroidogenic effects of AngII in adrenal glomerulosa cells, secondary to an increase in the number

BOX 16.2

HOW DO WE KNOW ABOUT SALT APPETITE IN RESPONSE TO HYPOVOLEMIA IN RATS?

Numerous studies have shown that PEG treatment stimulates consumption of concentrated NaCl solution as well as water intake in rats. By measuring intake every hour after PEG treatment, and plotting the data as a function of time, it can be seen that these two effects occur in a different timescale; the increase in water intake occurs within 1–2 hours after the injection whereas the increase in saline intake occurs 5–10 hours later.[1] This separation in the onset of the two drinking responses suggests that the excitatory stimuli for thirst and salt appetite are not identical. Indeed, subsequent work suggests that the stimulus of thirst is largely neural (i.e., input from cardiac baroreceptors) whereas the stimulus of salt appetite is largely endocrine (i.e., AngII acting in the SFO, aldosterone acting in the NTS). A further insight was obtained by replotting the data (see figure below), showing the intakes of water and saline as a function of one another and independent of time.[2] More specifically, during the first 5 hours of the drinking test, when water was consumed almost exclusively, the plotted values fell on a steep line close to the *y*-axis, whereas once saline consumption began the two fluids were consumed in the same timeframe, resulting in the diagonal portion of the plot. Of particular significance is the slope of that diagonal line, representing the mixture of ingested water and 0.50-M NaCl; that fluid mixture is equivalent to a solution whose concentration is approximately 0.15-M NaCl. In other words, the hypovolemic rats are consuming an ideal fluid for restoring plasma volume deficits

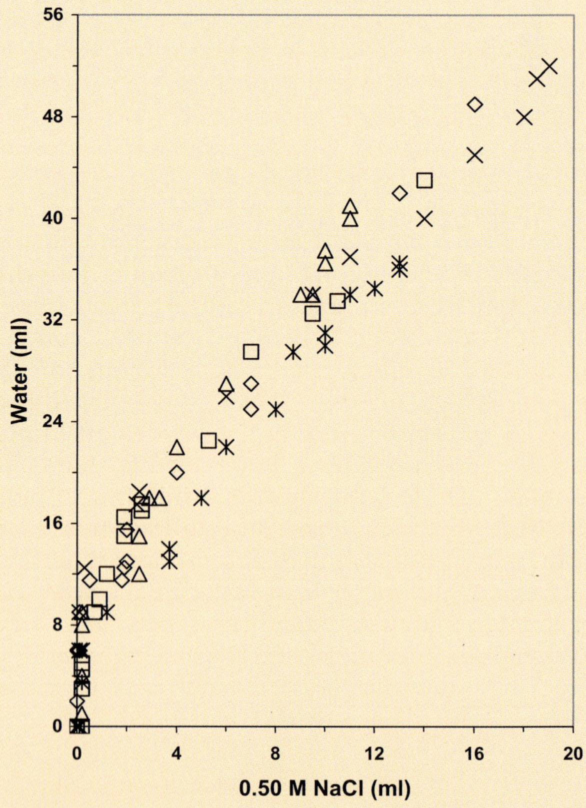

Intakes of water and 0.50-M NaCl, measured hourly, by individual rats after s.c. PEG treatment. Different symbols represent the hourly intakes of different animals. Rats drank water almost exclusively in the first 5 hours of the drinking test, but the fluid mixture consumed by each animal subsequently approximated 0.15-M NaCl. *From Stricker and Jalowiec.*[1]

BOX 16.2 *(cont'd)*

Other experiments revealed two additional features of this relation.[3] First, when the concentration of the ingested NaCl solution was 0.33 M instead of 0.50 M, the slope of the diagonal line in the plotted data from the two-bottle drinking test again approximately equaled 0.15-M NaCl. Second, when these tests were repeated in rats that had been maintained on NaCl-deficient diet instead of standard NaCl-rich laboratory chow, two differences were evident: (a) at first rats consumed 0.50-M NaCl alone, reflecting a predominant salt appetite initially, and (b) subsequent water consumption, diluted to isotonicity, meant that the hypertonic saline that ingested initially allowed the fluid mixture to empty rapidly from the stomach.[4] In these and other experiments, it is plain that hypovolemia stimulates both thirst and salt appetite in rats. However, the two behavioral responses are not independent of one another; the induced intakes subserve volume regulation, but the pattern of the two intakes reflects concurrent osmoregulation.

We have proposed the following arrangement of the interacting stimuli that influence the intakes of water and NaCl solution by PEG-treated rats, summarized schematically in the figure below.[5] In rats that had been maintained on standard lab chow, the plasma volume deficits induced by s.c. PEG treatment initially stimulate thirst and water intake. Hypovolemic rats drink water until osmotic dilution of body fluids is induced, which inhibits further water intake. That osmotic dilution inhibits secretion of OT in central neurons, which had been stimulated by hypovolemia and was inhibiting salt appetite. Consumption of concentrated NaCl solution raises plasma osmolality and thereby disinhibits thirst, allowing renewed water intake until osmotic dilution again occurs. But subsequent consumption of saline again removes osmotic dilution and thirst recurs. By thus alternating water and saline intakes, the hypovolemic rats can osmoregulate while restoring their plasma volume deficits.

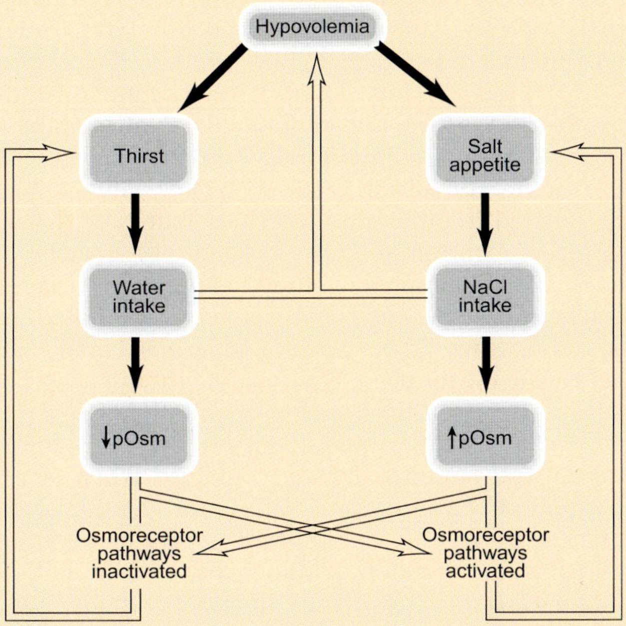

Schematic diagram of the mechanisms controlling thirst and salt appetite in hypovolemic rats. Solid arrows indicate stimulation, and unfilled arrows indicate inhibition. The combination of the effects of blood-borne AII on the brain and neural baroreceptor signals to the brainstem stimulates hypovolemic animals to drink water and concentrated saline solution. The rats alternately drink the two fluids in amounts that ultimately add up to a volume of isotonic saline sufficient to repair the volume deficit. When the animals have access to only one of the fluids, the intake of water and concentrated saline are limited by activation of the appropriate inhibitory osmoregulatory pathways. When rats drink isotonic saline instead of water and concentrated saline, neither inhibitory pathway is activated and consequently fluid intake is continuous. *From Stricker and Verbalis.*[5]

References

1. Stricker EM, Jalowiec JE. Restoration of intravascular fluid volume following acute hypovolemia in rats. *Am J Physiol.* 1970;218:191–196.
2. Stricker EM, Gannon KS, Smith JC. Thirst and salt appetite induced by hypovolemia in rats: analysis of drinking behavior. *Physiol Behav.* 1992;51:27–37.
3. Stricker EM. Thirst and sodium appetite after colloid treatment in rats. *J Comp Physiol Psychol.* 1981;95:1–25.
4. Stricker EM, Bykowski MR, Smith Hossler CA, et al. Gastric emptying and intestinal absorption of ingested water and saline by hypovolemic rats. *Physiol Behav.* 2009;98:570–578.
5. Stricker EM, Verbalis JG. Hormones and behavior: the biological bases of thirst and sodium appetite. *Am Scientist.* 1988;76:261–267.

and affinity of AngII receptors in these aldosterone-producing cells. Aldosterone appears to complement its potent Na^+-conserving properties by creating a synergy with AngII in stimulating salt appetite. Consistent with this proposal, i.c.v. administration of AngII increased NaCl intake in rats pretreated systemically with the mineralocorticoid desoxycorticosterone (DOC)[115] or maintained previously on NaCl-deficient diet (which enhances basal secretion of aldosterone).[116] In contrast, salt appetite was not enhanced by dietary NaCl deprivation in PEG-treated, hypophysectomized rats, which could augment secretion of renin but not aldosterone.[106] Similarly, salt appetite in Na^+-depleted rats was blunted by the administration of drugs that blocked either central mineralocorticoid receptors or AngII synthesis, and combined treatment eliminated the induced NaCl intake.[117] Because DOC treatment sufficient to induce salt appetite in rats increased AngII binding in the SFO,[118,119] it seems plausible that the lipid-soluble mineralocorticoids, which enter the brain readily, act in part by modifying the central actions of AngII.

Pharmacological doses of aldosterone and DOC elicit a dose-related increase in salt appetite[120,121] even though plasma AngII levels are low.[122] These steroids appear to act directly on neurons in the NTS with mineralocorticoid receptors,[123,124] which have the important additional property of containing the enzyme 11β-hydroxysteroid dehydrogenase type 2 (HSD2), which inactivates glucocorticoids that also bind to those receptors, and thereby allows the receptors specificity to mineralocorticoids. These HSD2-containing cells are activated by all treatments that elicit salt appetite in rats but are quickly inactivated once saline has been consumed.[123–125] In addition, it seems relevant that rats with lesions of the medial parabrachial nuclei (PBN), which may receive neural input from those cells in the NTS,[126] do not develop a salt appetite in response to diuretic-induced Na^+ deficiency[127,128] or DOCA treatment. Conversely, rats with discrete lesions of the AP, which sends inhibitory neural input to those cells in the NTS,[129] consumed prodigious amounts of concentrated NaCl solution, whereas rats with lesions that invaded the NTS had much smaller saline intakes.[130]

Numerous observations are consistent with the additional hypothesis that cerebral OT mediates an inhibitory signal in the control of salt appetite in rats. Since subcutaneous PEG treatment and i.c.v. AngII each stimulate OT secretion in addition to NaCl ingestion ,[46,131] it is important to note that salt appetite is proposed to occur only when the inhibitory component of these treatments is suppressed relative to the excitatory component. For example, administration of OT directly into the cerebral ventricles of rats eliminated the increased NaCl intake elicited by PEG treatment,[132] while pretreatment i.c.v. with an OT receptor antagonist enhanced NaCl intake in AngII-treated rats.[133] Moreover, salt appetite in PEG-treated rats was abolished by systemic injection of naloxone, which blocks the receptors of endogenous opioids and thereby removes their basal inhibition of OT secretion, and the inhibitory effect of naloxone on NaCl intake was prevented by pretreatment i.c.v. with an OT receptor antagonist.[132] Note that the central release of OT from a subset of parvocellular hypothalamic neurons, acting in the brain as a neurotransmitter, is thought to occur in parallel with activation of magnocellular OT-containing neurons,[134] and likely is correlated with the hormone's central effects.[135] Thus, saline ingestion appears to be most prominent when pituitary OT secretion is low, as it is in association with osmotic dilution,[46] after pretreatment maintenance of rats on NaCl-deficient diet,[136] after systemic treatment with alcohol[137] or with DOC,[135,138] or after bilateral adrenalectomy.[135]

It is interesting to note that mutant mice in which the gene for OT had been deleted did not have a spontaneous salt appetite, although they did show elevated intakes of NaCl solution after overnight water deprivation.[139,140] Similarly, basal NaCl intake did not increase in rats that had been injected i.c.v. with OT conjugated to the toxic A chain of the plant cytotoxin ricin (rA-OT), which was intended to destroy brain neurons that contained OT receptors.[141] Thus, it appears that central oxytocinergic pathways provide inhibition of NaCl intake only when excitatory stimuli for salt appetite are present.

Another dimension in the control of salt appetite was discovered in other studies of rats with rA-OT lesions. Whereas control animals injected systemically with hypertonic mannitol solution adaptively increased consumption of water but not of concentrated saline solution, as would be expected since NaCl intake would aggravate the induced hyperosmolality, rats with rA-OT lesions increased their intakes of both fluids. In contrast, they drank only water after injection of hypertonic saline. The same results were obtained when rats were pretreated i.c.v. with an OT receptor antagonist rather than with rA-OT.[141] Similarly, systemic injection of hypertonic mannitol solution did not inhibit NaCl intake during hypovolemia in these animals, whereas injection of equiosmotic saline solution did so.[142] Collectively, these findings indicate that there is differential sensing of plasma osmolality and Na^+ concentration in rats, and they suggest that hyperosmolality actually provides a mixed stimulus for salt appetite in addition to stimulating thirst. The latent excitatory component of the stimulus for salt appetite was revealed only under the unusual circumstance in which hypertonic mannitol solution was injected and the inhibitory effect of cerebral OT was eliminated.

These observations also suggest that central OT mediates the inhibition of salt appetite that is stimulated by cerebral osmoreceptors but not by Na^+-receptors. The inhibitory stimulus elicited by injection of hypertonic saline solution may be mediated by atrial natriuretic peptide (ANP), which is secreted from storage granules in atrial myocytes in response to blood volume expansion.[143] ANP has multiple regulatory effects during hypervolemia and arterial hypertension; specifically, the hormone is known to inhibit secretion of renin and aldosterone,[144] stimulate natriuresis[50] and oppose the physiological effects of AngII.[145,146] In addition, the SFO contains high numbers of ANP receptors[147,148] but sparse ANP-positive nerve terminals,[149] which suggests that it may respond to circulating ANP.

ANP is released from hypothalamic neurons in response to neural input from arterial baroreceptors and renal afferents during blood volume expansion in rats,[150] which is known to inhibit salt appetite.[151] One of its functions is to increase secretion of ANP from the cardiac atria,[152–154] and another may be to oppose the actions of AngII in the brain. For example, i.c.v. injection of ANP attenuated the water intake of rats in response to i.c.v. AngII[155] and reduced NaCl ingestion in Na^+-depleted rats,[156,157] while i.c.v. administration of ANP antiserum enhanced the drinking response to i.c.v. AngII.[158] Furthermore, the dipsogenic effect of AngII administered in the SFO was blunted by pretreatment with ANP in the SFO.[159] In addition, salt appetite was not inhibited by hypertonic saline in hypovolemic rats that had been injected i.c.v. with ANP conjugated to ricin, which was intended to destroy brain neurons that contained ANP receptors.[142] Because NaCl ingestion also was not inhibited after systemic injection of hypertonic mannitol solution in these animals, central ANP may be necessary both for Na^+-mediated and for osmolality-mediated inhibition of salt appetite.

Analogous to the control of thirst, presystemic signals appear to inhibit salt appetite in association with gastrointestinal distension and increased pOsm.[160–162] For example, DOC-treated rats drank excessive amounts of saline after peripheral sensory fibers were destroyed non-selectively by systemic injection of capsaicin.[38] Conversely, rats consumed saline in numerous but very short bouts whether PEG-treated, adrenalectomized or DOC-treated.[102,163] In each case, although basal OT secretion is low, thereby releasing NaCl intake from tonic inhibition, a phasic inhibition associated with stimulation of OT secretion presumably returns each time the animals ingest concentrated saline solution.[164] The visceral signals appear to project through the vagus to the NTS/AP and, in this regard, neurons from the AP influence the magnocellular cells in the hypothalamus indirectly, via the NTS[15,165] and the PBN.[166] In addition, pharmacological blockade in the PBN of the viscerosensory serotoninergic input from the AP[167] similarly led to remarkably large intakes of NaCl solution by rats.[168,169]

SUMMARY AND REMAINING QUESTIONS

The control of body fluid homeostasis is complex but not complicated. The excitatory stimuli for thirst are predominantly neural, although AngII makes an increasing contribution as hypovolemia becomes severe and provides a unique stimulus during arterial hypotension. In contrast, the excitatory stimuli for salt appetite appear to be predominantly endocrine, specifically AngII and aldosterone. However, overlapping neural inhibitory signals are also importantly involved in the control of both water intake and NaCl intake, with another hormone, ANP, also contributing to the latter.

Although this general formulation is straightforward, nonetheless there is a clear need for further research to understand remaining questions about the neuroendocrine regulation of thirst and salt appetite, including the following.

1. *Delineation of the neural circuitry that mediates thirst and salt appetite.* Aspiration lesions of the AP in rats impair AVP secretion and renal water conservation in response to systemic injection of hypertonic saline.[130,170] In contrast, AP lesions in rats did not affect AVP secretion in response to systemic injection of equiosmolar hypertonic mannitol solution.[171] Those disparate findings indicate that the neural circuits for control of AVP secretion must be more complex than the usual schema, which have focused exclusively on the hypothalamic osmoreceptors. New speculative formulations of the osmoregulatory control of AVP secretion will likely need to include the hypothesis that the rat forebrain contains both osmoreceptors and Na^+-receptors. In this regard, voltage-gated sodium channels that respond to alterations in extracellular Na^+ concentration but not to changes in osmolality have been located in the OVLT and SFO.[172,173] Consistent with their possible role as central Na^+-receptors that stimulate the inhibition of salt appetite are findings that mutant mice deficient in the gene for voltage-gated Na^+-channels in the SFO consumed concentrated NaCl solution in significantly greater amounts when Na^+-depleted than did control animals.[174]
2. *Cellular mechanisms underlying osmosensing and osmoregulation.* "Effective" solutes are those that penetrate cells slowly or not at all, thereby creating an osmotic gradient that causes an efflux of water from osmoreceptive cells. The resultant shrinkage of osmosensitive cells has been found to activate membrane non-selective cationic conductances that cause generation of an inward current; if of sufficient magnitude, the resulting depolarization of an osmoreceptor neuron can generate an action potential.[175] Electrophysiological studies of neurons in the OVLT show that they display changes in action potential firing rate that vary in proportion to the tonicity of extracellular fluid, supporting the likelihood that these cells represent osmosensory

neurons.[176] Osmotically-evoked changes in the firing rate of the OVLT neurons in turn synaptically regulate the electrical activity of downstream effector neurons, including the magnocellular AVP neurons in the SON and PVN and likely neurons that regulate thirst, through graded changes in release of excitatory neurotransmitters such as glutamate.

The cellular osmosensing mechanism utilized by OVLT osmosensitive cells appears to be an intrinsic depolarizing receptor potential, which these cells generate via a molecular transduction complex. Recent results suggest that this complex likely include members of the transient receptor potential vanilloid (TRPV) family of cation channel proteins.[177] These channels are generally activated by cell membrane stretch to cause a non-selective conductance of cations, with a preference for Ca^{2+}. Multiple studies have characterized various members of the TRPV family, particularly TRPV1, TRPV2 and TRPV4, as cellular mechanoreceptors in different tissues.[178,179] This mechanism of action is consistent with the concept that osmoreceptors detect the degree of membrane stretch at the cellular level, similar to the function of baroreceptors at the vascular level. Although many questions remain to be answered about the exact cellular mechanisms involved,[180] a strong case can already be made for the involvement of the TRPV family of cation channel proteins in the transduction of osmotic stimuli in the neural cells in the OVLT and surrounding hypothalamus that regulate osmotic homeostasis, which appears to have been highly conserved throughout evolution.[177]

3. *Species differences in thirst and salt appetite.* Studies of the human brain using positron emission tomography have allowed inquiries into the brain areas involved in the central circuitry mediating the sensation of thirst. This important work has suggested that the cingulate cortex and cerebellum may play a role in the genesis of thirst.[181–184] Because those brain sites have received little attention to date in studies of water ingestion by experimental animals, further investigations are needed to determine their significance.

Similarly, further research is needed to evaluate the central mechanisms mediating salt appetite in animals and humans.[185] Although a craving for salty foods and fluids can result from salt wasting associated with adrenal cortical dysfunction, as was noted 70 years ago,[186] salt appetite is surprisingly not universal in patients with Addison's disease, and nor is salt appetite commonly stimulated in people who lose excessive amounts of Na^+ in perspiration. In contrast, it is curious that in baboons an adaptive salt appetite appears readily,[187] and a synergy between central AngII and peripheral aldosterone in stimulating salt appetite has been demonstrated.[188] Although much has been learned about the central control of body fluid homeostasis from studies of salt appetite in laboratory rats, much more remains to be learned by extension of those studies to human subjects.

References

1. Adolph EF, Barker JP, Hoy PA. Multiple factors in thirst. *Am J Physiol.* 1954;178:538–562.
2. Gilman A. The relation between blood osmotic pressure, fluid distribution and voluntary water intake. *Am J Physiol.* 1937; 120:323–328.
3. Fitzsimons JT. Drinking by nephrectomized rats injected with various substances. *J Physiol.* 1961;155:563–579.
4. Holmes JH, Gregersen MI. Observations on drinking induced by hypertonic solutions. *Am J Physiol.* 1950;162:326–337.
5. Wolf AV. Osmometric analysis of thirst in man and dog. *Am J Physiol.* 1950;161:75–86.
6. Fitzsimons JT. The effects of slow infusions of hypertonic solutions on drinking and drinking thresholds in rats. *J Physiol.* 1963;167:344–354.
7. Dunn FL, Brennan TJ, Nelson AE, et al. The role of blood osmolality and volume in regulating vasopressin secretion in the rat. *J Clin Invest.* 1973;52:3212–3219.
8. Kanter GS. Excretion and drinking after salt loading in dogs. *Am J Physiol.* 1953;174:87–94.
9. Cowley Jr AW, Skelton MM, Merrill DC. Osmoregulation during high salt intake: relative importance of drinking and vasopressin secretion. *Am J Physiol.* 1986;251:R878–R886.
10. Verney EB. The antidiuretic hormone and the factors which determine its release. *Proc R Soc, London.* 1947;B135:25–106.
11. Andersson B. The effect of injections of hypertonic NaCl-solutions into different parts of the hypothalamus of goats. *Acta Physiol Scand.* 1953;28:188–201. 154.
12. Camacho A, Phillips MI. Horseradish peroxidase study in rat of the neural connections of the organum vasculosum of the lamina terminalis. *Neurosci Lett.* 1981;25:201–204.
13. Wilkin LD, Mitchell LD, Ganten D, et al. The supraoptic nucleus: afferents from areas involved in control of body fluid homeostasis. *Neuroscience.* 1989;28:573–584.
14. Oldfield BJ, Miselis RR, McKinley MJ. Median preoptic nucleus projections to vasopressin-containing neurons of the supraoptic nucleus in sheep. A light and electron microscopic study. *Brain Res.* 1991;542:193–200.
15. Tribollet E, Armstrong WE, Dubois-Dauphin M, et al. Extra-hypothalamic afferents input to the supraoptic nucleus of the rat as determined by retrograde and anterograde tracing techniques. *Neuroscience.* 1985;15:135–148.
16. Buggy J, Johnson AK. Preoptic-hypothalamic periventricular lesions: thirst deficits and hypernatremia. *Am J Physiol.* 1977; 233:R44–R52.
17. Johnson AK, Buggy J. Periventricular preoptic-hypothalamus is vital for thirst and normal water economy. *Am J Physiol.* 1978; 234:R122–R129.
18. Thrasher TN, Keil LC. Regulation of drinking and vasopressin secretion: role of organum vasculosum laminae terminalis. *Am J Physiol.* 1987;253:R108–R120.
19. Thrasher TN, Keil LC, Ramsay DJ. Lesions of the organum vasculosum of the lamina terminalis (OVLT) attenuate osmotically-induced drinking and vasopressin secretion in the dog. *Endocrinology.* 1982;110:1837–1839.

20. Robertson GL, Aycinena P, Zerbe RL. Neurogenic disorders of osmoregulation. *Am J Med*. 1982;72:339–353.
21. Gardiner TW, Stricker EM. Impaired drinking responses of rats with lesions of nucleus medianus: circadian dependence. *Am J Physiol*. 1985;248:R224–R230.
22. Gardiner TW, Verbalis JG, Stricker EM. Impaired secretion of vasopressin and oxytocin in rats after lesions of nucleus medianus. *Am J Physiol*. 1985;249:R681–R688.
23. Mangiapane ML, Thrasher TN, Keil LC, et al. Deficits in drinking and vasopressin secretion after lesions of the nucleus medianus. *Neuroendocrinology*. 1983;37:73–77.
24. Hoffman GE, Smith MS, Verbalis JG. c-Fos and related intermediate early gene products as markers of activity in neuroendocrine systems. *Front Neuroendocrinol*. 1993;14:173–213.
25. McAllen RM, Pennington GL, McKinley MJ. Osmoresponsive units in sheep median preoptic nucleus. *Am J Physiol*. 1990; 259:R593–R600.
26. Oldfield BJ, Badoer E, Hards DK, et al. Fos production in retrogradely labelled neurons of the lamina terminalis following intravenous infusion of either hypertonic saline or angiotensin II. *Neuroscience*. 1994;60:255–262.
27. Oldfield BJ, Bicknell RJ, McAllen RM, et al. Intravenous hypertonic saline induces Fos immunoreactivity in neurons throughout the lamina terminalis. *Brain Res*. 1991;561: 151–156.
28. Thrasher TN, Nistal-Herrera JF, Keil LC, et al. Satiety and inhibition of vasopressin secretion after drinking in dehydrated dogs. *Am J Physiol*. 1981;240:E394–E401.
29. Appelgren BH, Thrasher TN, Keil LC, et al. Mechanism of drinking-induced inhibition of vasopressin secretion in dehydrated dogs. *Am J Physiol*. 1991;261:R1226–R1233.
30. Thrasher TN, Keil LC, Ramsay DJ. Drinking, oropharyngeal signals, and inhibition of vasopressin secretion in dogs. *Am J Physiol*. 1987;253:R509–R515.
31. Bott E, Denton DA, Weller S. Water drinking in sheep with oesophageal fistulae. *J Physiol*. 1965;176:323–336.
32. Geelen G, Keil LC, Kravik SE, et al. Inhibition of plasma vasopressin after drinking in dehydrated humans. *Am J Physiol*. 1984;247:968–R971.
33. Rolls BJ, Wood RJ, Rolls ET, et al. Thirst following water deprivation in humans. *Am J Physiol*. 1980;239:R476–R482.
34. Wood RJ, Maddison S, Rolls ET, et al. Drinking in rhesus monkeys: Roles of presystemic and systemic factors in control of drinking. *J Comp Physiol Psychol*. 1980;94:1135–1148.
35. Huang W, Sved AF, Stricker EM. Water ingestion provides an early signal inhibiting osmotically stimulated vasopressin secretion in rats. *Am J Physiol*. 2000;279:R756–R760.
36. Stricker EM, Hoffmann ML. Inhibition of vasopressin secretion when dehydrated rats drink water. *Am J Physiol*. 2005; 289:R1238–R1243.
37. Baertschi AJ, Pence RA. Gut-brain signaling of water absorption inhibits vasopressin in rats. *Am J Physiol*. 1995;268: R236–R247.
38. Curtis KS, Stricker EM. Enhanced fluid intake by rats after capsaicin treatment. *Am J Physiol*. 1997;272:R704–R709.
39. Curtis KS, Verbalis JG, Stricker EM. Area postrema lesions in rats appear to disrupt rapid feedback inhibition of fluid intake. *Brain Res*. 1996;726:31–38.
40. Stricker EM, Curtis KS, Peacock KA, et al. Rats with area postrema lesions have lengthy eating and drinking bouts when fed *ad libitum*: Implications for feedback inhibition of ingestive behavior. *Behav Neurosci*. 1997;111:624–633.
41. Choi-Kwon S, Baertschi AJ. Splanchnic osmosensation and vasopressin: mechanisms and neural pathways. *Am J Physiol*. 1991;261:E18–E25.
42. Carlson SH, Beitz A, Osborn JW. Intragastric hypertonic saline increases vasopressin and central Fos immunoreactivity in conscious rats. *Am J Physiol*. 1997;272:R750–R758.
43. Manesh R, Hoffmann ML, Stricker EM. Water ingestion by rats fed high-salt diet may be mediated, in part, by visceral osmoreceptors. *Am J Physiol*. 2006;290:R1742–R1749.
44. Norgren R, Smith GP. Central distribution of the subdiaphragmatic vagal branches in the rat. *J Comp Neurol*. 1988; 273:207–223.
45. Brimble MJ, Dyball REJ, Forsling ML. Oxytocin release following osmotic activation of oxytocin neurones in the paraventricular and supraoptic nuclei. *J Physiol*. 1978;278:69–78.
46. Stricker EM, Verbalis JG. Interaction of osmotic and volume stimuli in regulation of neurohypophyseal secretion in rats. *Am J Physiol*. 1986;250:R267–R275.
47. Balment RJ, Brimble MJ, Forsling ML. Release of oxytocin induced by salt loading and its influence on renal excretion in the male rat. *J Physiol*. 1980;308:439–449.
48. Haanwinckel MA, Elias LK, Favaretto ALV, et al. Oxytocin mediates atrial natriuretic peptide release and natriuresis after volume expansion in the rat. *Proc Natl Acad Sci USA*. 1995; 92:7902–7906.
49. Huang W, Lee SL, Sjöquist M. Natriuretic role of endogenous oxytocin in male rats infused with hypertonic NaCl. *Am J Physiol*. 1995;268:R634–R640.
50. Soares TJ, Coimbra TM, Martins AR, et al. Atrial natriuretic peptide and oxytocin induce natriuresis by release of cGMP. *Proc Natl Acad Sci USA*. 1999;96:278–283.
51. Fitzsimons JT. Drinking by rats depleted of body fluid without increase in osmotic pressure. *J Physiol*. 1961;159:297–309.
52. Stricker EM. Extracellular fluid volume and thirst. *Am J Physiol*. 1966;211:232–238.
53. Stricker EM. Some physiological and motivational properties of the hypovolemic stimulus for thirst. *Physiol Behav*. 1968; 3:379–385.
54. Verbalis JG, Mangione MP, Stricker EM. Oxytocin produces natriuresis in rats at physiological plasma concentrations. *Endocrinology*. 1991;128:1317–1322.
55. Stricker EM, Vagnucci AH, McDonald RH, et al. Renin and aldosterone secretions during hypovolemia in rats: Relation to NaCl intake. *Am J Physiol*. 1979;237:R45–R51.
56. Huang W, Sjöquist M, Skott O, et al. Oxytocin-induced renin secretion in conscious rats. *Am J Physiol*. 2000;278:R226–R230.
57. Huang W, Sjöquist M, Skott O, et al. Oxytocin antagonist disrupts hypotension-evoked renin secretion and other responses in conscious rats. *Am J Physiol*. 2001;280: R760–R765.
58. Blass EM, Fitzsimons JT. Additivity of effect and interaction of a cellular and an extracellular stimulus of drinking. *J Comp Physiol Psychol*. 1970;70:200–205.
59. Corbit JD. Cellular dehydration and hypovolaemia are additive in producing thirst. *Nature*. 1968;218:886–887.
60. Stricker EM. Osmoregulation and volume regulation in rats: Inhibition of hypovolemic thirst by water. *Am J Physiol*. 1969; 217:98–105.
61. Quillen Jr EW, Cowley Jr AW. Influence of volume changes on osmolality–vasopressin relationships in conscious dogs. *Am J Physiol*. 1983;244:H73–H79.
62. Corbit JD. Effect of intravenous sodium chloride on drinking in the rat. *J Comp Physiol Psychol*. 1965;60:397–406.
63. Robertson GL, Shelton RL, Athar S. The osmoregulation of vasopressin. *Kidney Intl*. 1976;10:25–37.
64. Zimmerman MB, Blaine EH, Stricker EM. Water intake in hypovolemic sheep: Effects of crushing the left atrial appendage. *Science*. 1981;211:489–491.

65. Quillen Jr EW, Keil LC, Reid IA. Effects of baroreceptor denervation on endocrine and drinking responses to caval constriction in dogs. *Am J Physiol*. 1990;259:R618–R626.
66. Kaufman S. Role of right atrial receptors in the control of drinking in the rat. *J Physiol*. 1984;349:389–396.
67. Kalia M, Mesulam MM. Brain stem projections of sensory and motor components of the vagus complex in the cat: II. Laryngeal, tracheobronchial, pulmonary, cardiac, and gastrointestinal branches. *J Comp Neurol*. 1980;193:467–508. 8.
68. Ter Horst GJ, deBoer P, Luiten PGM, et al. Ascending projections from the solitary tract nucleus to the hypothalamus. A *Phaseolus vulgaris* lectin tracing study in the rat. *Neuroscience*. 1989;31:785–797.
69. Saper CB, Levisohn D. Afferent connections of the median preoptic nucleus in the rat: anatomical evidence for a cardiovascular integrative mechanism in the anteroventral third ventricular (AV3V) region. *Brain Res*. 1983;288:21–31.
70. Fitzsimons JT. The role of a renal thirst factor in drinking induced by extracellular stimuli. *J Physiol*. 1969;201: 349–368.
71. Mendelsohn FAO, Quirion R, Saavedra JM, et al. Autoradiographic localization of angiotensin II receptors in rat brain. *Proc Natl Acad Sci USA*. 1984;81:1575–1579.
72. van Houten M, Schiffrin EL, Mann JFE, et al. Radioautographic localization of specific binding sites for blood-borne angiotensin II in the rat brain. *Brain Res*. 1980;186:480–485.
73. Miselis RR. The efferent projections of the subfornical organ of the rat: a circumventricular organ with a neural network subserving water balance. *Brain Res*. 1981;230:1–23.
74. Simpson JB, Epstein AN, Camardo Jr JS. Localization of receptors for the dipsogenic action of angiotensin II in the subfornical organ of rat. *J Comp Physiol Psychol*. 1978;92: 581–608.
75. Abdelaal AE, Assaf SY, Kucharczyk J, et al. Effect of ablation of the subfornical organ on water intake elicited by systemically administered angiotensin-II. *Can J Physiol Pharmacol*. 1974; 52:1217–1220.
76. Simpson JB, Routtenberg A. Subfornical organ lesions reduce intravenous angiotensin-induced drinking. *Brain Res*. 1975;88:154–161.
77. Rowland NE, Li BH, Rozelle AK, et al. Localization of changes in immediate early genes in brain in relation to hydromineral balance: intravenous angiotensin II. *Brain Res Bull*. 1994; 33:427–436.
78. Lind RW, Swanson LW, Ganten D. Organization of angiotensin II immunoreactive cells and fibers in the rat central nervous system: An immunohistochemical study. *Neuroendocrinology*. 1985;40:2–24.
79. Wright JW, Harding JW. Regulatory role of brain angiotensins in the control of physiological and behavioral processes. *Brain Res Rev*. 1997;17:227–262.
80. Evered MD, Robinson MM. Increased or decreased thirst caused by inhibition of angiotensin-converting enzyme in the rat. *J Physiol*. 1984;348:573–588.
81. Fitzsimons JT, Elfont RM. Angiotensin does contribute to drinking induced by caval ligation in the rat. *Am J Physiol*. 1982;243:R558–R562.
82. Lehr D, Goldman HW, Casner P. Renin–angiotensin role in thirst: paradoxical enhancement of drinking by angiotensin converting enzyme inhibitor. *Science*. 1973;182:1031–1034.
83. Stricker EM, Bradshaw WG, McDonald Jr RH. The renin–angiotensin system and thirst: A reevaluation. *Science*. 1976;194:1169–1171.
84. Schiffrin EL, Genest J. Mechanism of captopril-induced drinking. *Am J Physiol*. 1982;242:R136–R140.
85. Leenen FH, Stricker EM. Plasma renin activity and thirst following hypovolemia or caval ligation in rats. *Am J Physiol*. 1974;226:1238–1242.
86. Schreihofer AM, Anderson BK, Schiltz JC, et al. Thirst and salt appetite elicited by hypovolemia in rats with chronic lesions of the nucleus of the solitary tract. *Am J Physiol*. 1999;276: R251–R258.
87. Hosutt JA, Rowland N, Stricker EM. Impaired drinking responses of rats with lesions of the subfornical organ. *J Comp Physiol Psychol*. 1981;95:104–113.
88. Blass EM, Nussbaum AI, Hanson DG. Septal hyperdipsia: Specific enhancement of drinking to angiotensin in rats. *J Comp Physiol Psychol*. 1974;87:422–439.
89. Stricker EM. Excessive drinking by rats with septal lesions during hypovolemia induced by subcutaneous colloid treatment. *Physiol Behav*. 1978;21:905–907.
90. Evered MD. Relationship between thirst and diazoxide-induced hypotension in rats. *Am J Physiol*. 1990;259:R362–R370.
91. Fitts DA. Angiotensin II receptors in SFO but not in OVLT mediate isoproterenol-induced thirst. *Am J Physiol*. 1994;267: R7–R15.
92. Evered MD, Robinson MM, Rose PA. Effect of arterial pressure on drinking and urinary responses to angiotensin II. *Am J Physiol*. 1988;254:R69–R74.
93. Robinson MM, Evered MD. Pressor action of intravenous angiotensin II reduces drinking response in rats. *Am J Physiol*. 1987;252:R754–R759.
94. Stocker SD, Stricker EM, Sved AF. Acute hypertension inhibits thirst stimulated by ANG II, hyperosmolality, or hypovolemia in rats. *Am J Physiol*. 2001;280:R214–R224.
95. Klingbeil CK, Brooks VL, Quillen Jr EW, et al. Effect of baroreceptor denervation on stimulation of drinking by angiotensin II in conscious dogs. *Am J Physiol*. 1991;260:E333–E337.
96. Stocker SD, Stricker EM, Sved AF. Arterial baroreceptors mediate the inhibitory effect of acute increases in arterial blood pressure on thirst. *Am J Physiol*. 2002;282:R1718–R1729.
97. Ferguson AV, Renaud LP. Systemic angiotensin acts at subfornical organ to facilitate activity of neurohypophysial neurons. *Am J Physiol*. 1986;251:R712–R717.
98. Brooks VL, Klingbeil CK, Quillen EW, et al. Effect of baroreceptor denervation on vasopressin and cortisol responses to angiotensin II infusion in conscious dogs. *Am J Physiol*. 1989;257:R1175–R1181.
99. Schreihofer AM, Stricker EM, Sved AF. Nucleus of the solitary tract lesions enhance drinking, but not vasopressin release, induced by angiotensin. *Am J Physiol*. 2000;279:R239–R247.
100. Stricker EM, Jalowiec JE. Restoration of intravascular fluid volume following acute hypovolemia in rats. *Am J Physiol*. 1970;218:191–196.
101. Stricker EM. Thirst and sodium appetite after colloid treatment in rats. *J Comp Physiol Psychol*. 1981;95:1–25.
102. Stricker EM, Gannon KS, Smith JC. Thirst and salt appetite induced by hypovolemia in rats: Analysis of drinking behavior. *Physiol Behav*. 1992;51:27–37.
103. Richter CP. Increased salt appetite in adrenalectomized rats. *Am J Physiol*. 1936;115:155–161.
104. Sakai RR, Epstein AN. Dependence of adrenalectomy-induced sodium appetite on the action of angiotensin II in the brain of the rat. *Behav Neurosci*. 1990;104:167–176.
105. Fitts DA, Thunhorst RL, Simpson JB. Modulation of salt appetite by lateral ventricular infusions of angiotensin II and carbachol during sodium depletion. *Brain Res*. 1985;346:273–280.
106. Stricker EM. Thirst and sodium appetite after colloid treatment in rats: Role of the renin–angiotensin–aldosterone system. *Behav Neurosci*. 1983;97:725–737.

107. Elfont RM, Fitzsimons JT. The effect of captopril on sodium appetite in adrenalectomized and deoxycorticosterone-treated rats. *J Physiol*. 1985;365:1–12.
108. Moe KE, Weiss ML, Epstein AN. Sodium appetite during captopril blockade of endogenous angiotensin II formation. *Am J Physiol*. 1984;247:R356–R365.
109. Fitts DA, Thunhorst RL. Rapid elicitation of salt appetite by an intravenous infusion of angiotensin II in rats. *Am J Physiol*. 1996;270:R1092–1098.
110. Schoorlemmer GHM, Johnson AK, Thunhorst RL. Circulating angiotensin II mediates sodium appetite in adrenalectomized rats. *Am J Physiol*. 2001;281:R723–R729.
111. Thunhorst RL, Beltz TG, Johnson AK. Effects of subfornical organ lesions on acutely induced thirst and salt appetite. *Am J Physiol*. 1999;277:R56–R65.
112. Weisinger RS, Denton DA, Di Nicolantonio R, et al. Subfornical organ lesion decreases sodium appetite in the sodium-depleted rat. *Brain Res*. 1990;526:23–30.
113. Avrith DB, Fitzsimons JT. Renin-induced sodium appetite: effects on sodium balance and mediation by angiotensin in the rat. *J Physiol*. 1983;337:479–496.
114. Fitts DA, Masson DB. Preoptic angiotensin and salt appetite. *Behav Neurosci*. 1990;104:643–650.
115. Fluharty SJ, Epstein AN. Sodium appetite elicited by intracerebroventricular infusion of angiotensin II in the rat. II. Synergistic interaction with systemic mineralocorticoids. *Behav Neurosci*. 1983;97:746–758.
116. Buggy J, Fisher AE. Evidence for a dual central role for angiotensin in water and sodium intake. *Nature*. 1974;250:733–735.
117. Sakai RR, Nicolaidis S, Epstein AN. Salt appetite is suppressed by interference with angiotensin II and aldosterone. *Am J Physiol*. 1986;251:R762–R768.
118. De Nicola AF, Seltzer A, Tsutsumi K, et al. Effects of deoxycorticosterone acetate (DOCA) and aldosterone on Sar1-angiotensin II binding and angiotensin-converting enzyme binding sites in brain. *Cell Mol Neurobiol*. 1993;13:529–539.
119. Shelat SG, King JL, Flanagan-Cato LM, et al. Mineralocorticoids and glucocorticoids cooperatively increase salt intake and angiotensin II receptor binding in rat brain. *Neuroendocrinology*. 1999;69:339–351.
120. Rice KK, Richter CP. Increased sodium chloride and water intake of normal rats treated with desoxycorticosterone acetate. *Endocrinology*. 1943;33:106–115.
121. Wolf G, Handal PJ. Aldosterone-induced sodium appetite: dose-response and specificity. *Endocrinology*. 1966;79:1120–1124.
122. Pettinger WA, Marchelle M, Augusto L. Renin suppression by DOC and NaCl in the rat. *Am J Physiol*. 1971;221:1071–1074.
123. Geerling JC, Engeland WC, Kawata M, et al. Aldosterone target neurons in the nucleus tractus solitarius drive sodium appetite. *J Neurosci*. 2006;26:411–417.
124. Geerling JC, Loewy AD. Sodium depletion activates the aldosterone-sensitive neurons in the NTS independently of thirst. *Am J Physiol*. 2007;292:R1338–R1348.
125. Geerling JC, Loewy AD. Aldosterone-sensitive NTS neurons are inhibited by saline ingestion during chronic mineralocorticoid treatment. *Brain Res*. 2006;1115:54–64.
126. Geerling JC, Loewy AD. Sodium deprivation and salt intake activate separate neuronal subpopulations in the nucleus of the solitary tract and the parabrachial complex. *J Comp Neurol*. 2007;504:379–403.
127. Flynn FW, Grill HJ, Schulkin J, Norgren R. Central gustatory lesions. II. Effects on sodium appetite, taste aversion learning, and feeding behaviors. *Behav Neurosci*. 1991;105:944–954.
128. Scalera G, Spector AC, Norgren R. Excitotoxic lesions of the parabrachial nuclei prevent conditioned taste aversions and sodium appetite in rats. *Behav Neurosci*. 1995;109:997–1008.
129. Sequeira SM, Geerling JC, Loewy AD. Local inputs to aldosterone-sensitive neurons of the nucleus tractus solitarius. *Neuroscience*. 2006;141:1995–2005.
130. Curtis KS, Huang W, Sved AF, et al. Impaired osmoregulatory responses in rats with area postrema lesions. *Am J Physiol*. 1999;277:R209–R219.
131. Lang RE, Rascher W, Heil J, et al. Angiotensin stimulates oxytocin release. *Life Sci*. 1981;29:1425–1428.
132. Blackburn RE, Stricker EM, Verbalis JG. Central oxytocin mediates inhibition of sodium appetite by naloxone in hypovolemic rats. *Neuroendocrinology*. 1992;56:255–263.
133. Blackburn RE, Demko AD, Hoffman GE, et al. Central oxytocin inhibition of angiotensin-induced salt appetite in rats. *Am J Physiol*. 1992;263:R1347–R1353.
134. Stricker EM, Verbalis JG. Sodium appetite. In: Stricker EM, ed. *Handbook of Behavioral Neurobiology, Vol. 10, Neurobiology of Food and Fluid Intake*. New York: Plenum; 1990:38–419.
135. Stricker EM, Verbalis JG. Central inhibitory control of sodium appetite in rats: Correlation with pituitary oxytocin secretion. *Behav Neurosci*. 1987;101:560–567.
136. Stricker EM, Schreihofer AM, Verbalis JG. Sodium deprivation blunts hypovolemia-induced pituitary secretion of vasopressin and oxytocin in rats. *Am J Physiol*. 1994;267:R1336–R1341.
137. Blackburn RE, Stricker EM, Verbalis JG. Acute effects of ethanol on ingestive behavior in rats. *Alcoholism Clin Exp Res*. 1994;18:924–930.
138. Roesch DM, Blackburn-Monro RE, Verbalis JG. Mineralocorticoid treatment attenuates activation of oxytocinergic and vasopressinergic neurons by icv ANG II. *Am J Physiol*. 2001;280:R1853–R1864.
139. Amico JA, Mantella RC, Vollmer RR. Consumption of solutions containing sodium chloride is enhanced in female oxytocin-deficient mice. *Behav Neurosci*. 2003;117:32–37.
140. Amico JA, Morris M, Vollmer RR. Mice deficient in oxytocin manifest increased saline consumption following overnight fluid deprivation. *Am J Physiol*. 2001;281:R1368–R1373.
141. Blackburn RE, Samson WK, Fulton RJ, et al. Central oxytocin inhibition of salt appetite in rats: Evidence for differential sensing of plasma sodium and osmolality. *Proc Natl Acad Sci USA*. 1993;90:10380–10384.
142. Blackburn RE, Samson WK, Fulton RJ, et al. Central oxytocin and atrial natriuretic peptide receptors mediate osmotic inhibition of salt appetite in rats. *Am J Physiol*. 1995;269:R245–R251.
143. Chiu PJS, Vemulapalli S, Policelli M, et al. The responses of atrial natriuretic factor concentrations to acute volume changes in conscious rats. *Life Sci*. 1987;41:2339–2347.
144. Atlas SA, Maack T. Effects of atrial natriuretic factor on the kidney and the renin-angiotensin-aldosterone system. *Endocrinol Metab Clinics N Am*. 1987;16:107–143.
145. McCann SM, Gutkowska J, Antunes-Rodrigues J. Neuroendocrine control of body fluid homeostasis. *Brazilian J Med Biol Res*. 2003;36:165–181.
146. Saavedra JM. Interactions between the circulating hormones angiotensin and atrial natriuretic peptide and their receptors in brain. In: Porter JC, Ježová D, eds. *Circulating Regulatory Factors and Neuroendocrine Function*. New York: Plenum; 1990:191–210.
147. Quirion R, Dalpé M, Dam T-V. Characterization and distribution of receptors for the atrial natriuretic peptides in mammalian brain. *Proc Natl Acad Sci USA*. 1986;83:174–178.
148. Saavedra JM. Regulation of atrial natriuretic peptide receptors in the rat brain. *Cell Mol Neurobiol*. 1987;7:151–173.

149. Skofitsch G, Jacobowitz DM, Eskay RL, et al. Distribution of atrial natriuretic factor-like immunoreactive neurons in the rat brain. *Neuroscience*. 1985;16:917–948.
150. Antunes-Rodrigues J, Machado BH, Andrade HA, et al. Carotid-aortic and renal baroreceptors mediate the atrial natriuretic peptide release induced by blood volume expansion. *Proc Natl Acad Sci USA*. 1992;89:6828–6831.
151. Toth E, Stelfox J, Kaufman S. Cardiac control of salt appetite. *Am J Physiol*. 1987;252:R925–R929.
152. Antunes-Rodrigues J, Picanço-Diniz DWL, Favaretto ALV, et al. Brain atrial natriuretic peptide neurons play an essential role in volume expansion-induced release of atrial natriuretic peptide and natriuresis. *Neuroendocrinology*. 1993;58:696–700.
153. Antunes-Rodrigues J, Ramalho MJ, Reis LC, et al. Lesions of the hypothalamus and pituitary inhibit volume-expansion-induced release of atrial natriuretic peptide. *Proc Natl Acad Sci USA*. 1991;88:2956–2960.
154. Rauch AL, Callahan MF, Buckalew Jr VM, et al. Regulation of plasma atrial natriuretic peptide by the central nervous system. *Am J Physiol*. 1990;258:R531–R535.
155. Antunes-Rodrigues J, McCann SM, Rogers LC, et al. Atrial natriuretic factor inhibits dehydration- and angiotensin II-induced water intake in the conscious, unrestrained rat. *Proc Natl Acad Sci USA*. 1985;82:8720–8723.
156. Antunes-Rodrigues J, McCann SM, Samson WK. Central administration of atrial natriuretic factor inhibits saline preference in the rat. *Endocrinology*. 1986;118:1726–1728.
157. Fitts DA, Thunhorst RL, Simpson JB. Diuresis and reduction of salt appetite by lateral ventricular infusions of atriopeptin II. *Brain Res*. 1985;348:118–124.
158. Katsuura G, Nakamura M, Inouye K, et al. Regulatory role of atrial natriuretic polypeptide in water drinking in rats. *Eur J Pharmacol*. 1986;121:285–287.
159. Ehrlich KJ, Fitts DA. Atrial natriuretic peptide in the subfornical organ reduces drinking induced by angiotensin or in response to water deprivation. *Behav Neurosci*. 1990;104:365–372.
160. Bykowski MR, Smith JC, Stricker EM. Regulation of NaCl solution intake and gastric emptying in adrenalectomized rats. *Physiol Behav*. 2007;92:781–789.
161. Smith CA, Curtis KS, Smith JC, et al. Presystemic influences on thirst, salt appetite, and vasopressin secretion in the hypovolemic rat. *Am J Physiol*. 2007;292:R2089–R2099.
162. Stricker EM, Bushey MA, Hoffmann ML, et al. Inhibition of NaCl appetite when DOCA-treated rats drink saline. *Am J Physiol*. 2007;292:R652–R662.
163. Stricker EM, Gannon KS, Smith JC. Salt appetite induced by DOCA treatment or adrenalectomy in rats: Analysis of ingestive behavior. *Physiol Behav*. 1992;52:793–802.
164. Stricker EM, Callahan JB, Huang W, et al. Early osmoregulatory stimulation of thirst and neurohypophyseal hormone secretion after gastric NaCl loads in rats. *Am J Physiol*. 2002;282:R1710–R1717.
165. Shapiro RE, Miselis RR. The central neural connections of the area postrema of the rat. *J Comp Neurol*. 1985;234:344–364.
166. Saper CB, Loewy AD. Efferent connections of the parabrachial nucleus in the rat. *Brain Res*. 1980;197:291–317.
167. Lança AJ, van der Kooy D. A serotonin-containing pathway from the area postrema to the parabrachial nucleus in the rat. *Neuroscience*. 1985;14:1117–1126.
168. Menani JV, Thunhorst RL, Johnson AK. Lateral parabrachial nucleus and serotonergic mechanisms in the control of salt appetite in rats. *Am J Physiol*. 1996;270:R162–R168.
169. Menani JV, De Luca Jr LA, Johnson AK. Lateral parabrachial nucleus serotonergic mechanisms and salt appetite induced by sodium depletion. *Am J Physiol*. 1998;274:R555–R560.
170. Arima H, Kondo K, Murase T, et al. Regulation of vasopressin synthesis and release by area postrema in rats. *Endocrinology*. 1998;139:1481–1486.
171. Huang W, Sved AF, Stricker EM. Vasopressin and oxytocin release evoked by NaCl loads are selectively blunted by area postrema lesions in rats. *Am J Physiol*. 2000;278:R732–R740.
172. Hiyama TY, Watanabe E, Ono K, Inenaga K, et al. Na_x channel involved in CNS sodium-level sensing. *Nat Neurosci*. 2002;5:511–512.
173. Watanabe E, Hiyama TY, Shimizu H, et al. Sodium-level-sensitive sodium channel Na(x) is expressed in glial laminate processes in the sensory circumventricular organs. *Am J Physiol*. 2006;290:R568–R576.
174. Watanabe E, Fujikawa A, Matsunaga H, et al. Nav2/NaG channel is involved in control of salt intake behavior in the CNS. *J Neurosci*. 2000;20:7743–7751.
175. Oliet SH, Bourque CW. Mechanosensitive channels transduce osmosensitivity in supraoptic neurons. *Nature*. 1993;364:341–343.
176. Bourque CW, Ciura S, Trudel E, Stachniak TJ, Sharif-Naeini R. Neurophysiological characterization of mammalian osmosensitive neurones. *Exp Physiol*. 2007;92:499–505.
177. Liedtke W. Role of TRPV ion channels in sensory transduction of osmotic stimuli in mammals. *Exp Physiol*. 2007;92:507–512.
178. Liedtke W, Friedman JM. Abnormal osmotic regulation in trpv4−/− mice. *Proc Natl Acad Sci USA*. 2003;100:13698–13703.
179. Becker D, Blase C, Bereiter-Hahn J, et al. TRPV4 exhibits a functional role in cell-volume regulation. *J Cell Sci*. 2005;118:2435–2440.
180. Verbalis JG. How does the brain sense osmolality? *J Am Soc Nephrol*. 2007;18:3056–3059.
181. Denton D, Shade R, Zamarippa F, Egan G, et al. Correlation of regional cerebral blood flow and change of plasma sodium concentration during genesis and satiation of thirst. *Proc Natl Acad Sci USA*. 1999;96:2532–2537.
182. Denton D, Shade R, Zamarippa F, Egan G, et al. Neuroimaging of genesis and satiation of thirst and an interoceptor-driven theory of origins of primary consciousness. *Proc Natl Acad Sci USA*. 1999;96:5304–5309.
183. Farrell MJ, Zamarripa F, Shade R, Phillips PA, et al. Effect of aging on regional cerebral blood flow responses associated with osmotic thirst and its satiation by water drinking: a PET study. *Proc Natl Acad Sci USA*. 2008;105:382–387.
184. Parsons LM, Denton D, Egan G, McKinley M, et al. Neuroimaging evidence implicating cerebellum in support of sensory/cognitive processes associated with thirst. *Proc Natl Acad Sci USA*. 2000;97:2332–2336.
185. Geerling JC, Loewy AD. Central regulation of sodium appetite. *Exp Physiol*. 2008;93:177–209.
186. Wilkins L, Richter CP. A great craving for salt by a child with cortico-adrenal insufficiency. *J Am Med Assoc*. 1940;114:866–868.
187. Blair-West JR, Carey KD, Denton DA, et al. Evidence that brain angiotensin II is involved in both thirst and sodium appetite in baboons. *Am J Physiol*. 1998;275:R1639–R1646.
188. Shade RE, Blair-West JR, Carey KD, et al. Synergy between angiotensin and aldosterone in evoking sodium appetite in baboons. *Am J Physiol*. 2002;283:R1070–R1078.

CHAPTER

17

Sexual Differentiation of Brain and Behavior

Margaret M. McCarthy

Departments of Physiology and Psychiatry, University of Maryland School of Medicine, Baltimore, MD, USA

OUTLINE

Summary

Sex differences in the brain are established early in development and subserve the need of the divergent reproductive strategies of adult males and females. Gonadal hormones mediate organizational changes to the developing brain to assure that activation of the same neural structure by hormones in adulthood is consistent with behavioral and physiological needs of that gender. Specific subregions of the brain regulate distinct reproductive and non-reproductive endpoints, and the cellular mechanisms establishing sex differences in each subregion are largely unique, which ultimately increases the overall within-gender variability of males and females and maximizes the exploitation of available resources.

INTRODUCTION

Boy or girl, man or woman, male or female. Sex or gender are the first things we register about another individual in our continuous effort to codify the world

Handbook of Neuroendocrinology, DOI: 10.1016/B978-0-12-375097-6.10017-4

around us, and it is pretty much the first thing we become aware of about ourselves. When considering humans we speak of gender, as only humans have sufficient self-awareness of their sex that it can become a source of internal or externally imposed conflict. For animals, we restrict our definition to sex, and even though sex might change, as in some species of tropical fish, we still delineate all animals as male or female. In humans, the study of sex differences remains a nascent field because we long denied the potential for important health consequences as a function of sex. However, a National Research Council study commissioned by the Institute of Medicine, an arm of the National Academy of Science in the United States, put the debate to rest with a definitive report in 2001 that clearly elucidated not only how important sex or gender is to overall human health, but also how little we know about it. This is equally true in the area of neuroscience and sex differences in the brain. Ironically, the study of sex differences in the brain actually began over 50 years ago with elucidation of the *Organizational/Activational Hypothesis* of sexual differentiation of behavior,[1] making it one of the oldest subdisciplines in neuroscience, but at that time it was not explicitly recognized as a study of sex differences in the brain. It is the genitalia and secondary sex characteristics we use to codify an individual's sex or gender, not their brain, and so it was not intuitively obvious that it was the brain that directed sex differences in behavior. The source of sex differences in behavior is no longer a matter of debate, but what is still a contentious and active area of research is the origins, magnitude, meaning and immutability of sex differences in brain and behavior. The goal of this chapter is to review the current state-of-the-art in the study of sex differences in animal models, and to highlight those areas where we need more information and/or a rethinking of past assumptions.

SEX DETERMINATION PRECEDES SEX DIFFERENTIATION

The determination of sex begins with the chromosomes. In mammals, males and females share 22 common autosomes but differ in one set called the sex chromosomes, X and Y. The presence of a Y chromosome – more specifically, a single gene on the Y chromosome called Sry (for sex-determining region of the Y chromosome) – will direct the formation of testes from the bipotential gonad and all else will stem from there. If Sry is mutated or lost an XY individual will become a female, and if Sry is translocated to the X chromosome, or an autosome, an XX individual will become a male. While the Sry gene is an essential trigger for male determination, it does not act in complete isolation, as it codes for a protein – sometimes referred to as tdf, for testis determining factor – that is a transcription factor which initiates a cascade of gene expression with precise temporal and cellular constraints. In response to various directives, the developing gonadal tissue will involute to form tubules encasing the primordial germ cells which will become the source of continuous sperm production and coordinate with a ductal system that expedites the removal of sperm from the body. This is the polar opposite of the developing ovary, in which primordial germ cells are engulfed in follicles on the surface of the organ so that they can be released into the body cavity and captured by the fimbria of the oviduct (Fallopian tubes) to pass into the uterus. Both the testis and the ovary require precise matching with the appropriate ductal system to be functional, and so ductal development is driven by gonadal development. Thus, once the testis is formed, the onset of significant androgen synthesis and production of anti-Müllerian hormone is fairly rapid. Combined, these hormones assure the survival of the Wollfian duct system and the demise of the Müllerian ducts, resulting in formation of the epididymus and vas deferens to match the testis. If the determinant path is towards the ovary, the lack of androgen and anti-Müllerian hormone synthesis allows for survival of the Müllerian ducts and regression of the Wollfian duct system, resulting in an oviduct, a uterus and the upper portion of the vagina. Subsequent hormonal production will finish the job in both cases, filling in the development of a penis and scrotum in males, and clitoris and labia in females. Additional gonad-specific hormone production peripubertally leads to the final masculine or feminine phenotype.

In the absence of Sry, the default developmental program is toward the female, with the bipotential gonad differentiating into an ovary and the Müllerian duct system persisting while the Wollfian duct system regresses due to the lack of androgen. There are both similarities and differences between the sexual differentiation of gonads and reproductive track and the sexual differentiation of the brain. The default phenotype of the brain is also female, and will develop in the absence of any perturbation. The male brain phenotype is actively induced by steroid hormones synthesized by the testis, making the brain similar to the undifferentiated gonad in this regard. Moreover, in a manner further analogous to the reproductive tract, there is also an active regression of components of the female brain in a process called *defeminization*. What does not appear to be analogous is the presence of two separate circuits that would be directly comparable to the Müllerian and Wollfian duct systems; instead there appears to be a neuronal network that controls sex behavior, and it is relatively weighted towards male behavior versus female behavior. Understanding the distinction requires that we first review components of the sexual differentiation process.

SEXUAL DIFFERENTIATION OF THE BRAIN OCCURS DURING A SENSITIVE PERIOD

The very early period of gonadal steroidogenesis orchestrates the coordinated development of the internal reproductive tract and external genitalia, and the process is essentially complete by the first half of gestation. A second period of steroidogenesis commences during the latter half of pregnancy, and it is this period of hormone production that differentiates the brain. In rodents, the production of androgens by the testis commences during the last few days of gestation and proceeds briefly into postnatal life – as briefly as a few hours, by some reports.[2,3] The onset of steroidogenesis in males defines the beginning of the sensitive period for sexual differentiation of the brain; in essence, the beginning of masculinization. Conversely, the termination of the sensitive period is operationally defined as the time-point at which females are no longer masculinized by administration of exogenous steroids (Fig. 17.1). The sensitivity of females varies by endpoint, in some cases exhibiting a gradual waning of sensitivity while in others exhibiting an abrupt shutting of the window of opportunity. Sensitive periods are a frequent topic for challenge and refinement, and are inherently difficult to define precisely, because a total lack of sensitivity is a rare thing. Recent efforts have defined additional sensitive periods, particularly associated with the onset of hormone production at puberty,[4] but it is still true that there is a perinatal period during which the brain is particularly sensitive to hormonally induced differentiation, and that much of that differentiation is enduring, if not permanent. Subsequent hormone production in adulthood serves to activate the previously organized neural substrate, with the activational effects being constrained by earlier organizational events – hence the Organizational/Activational Hypothesis.

THERE ARE SPECIFIC ROLES FOR DIFFERENT HORMONES IN SEXUAL DIFFERENTIATION OF THE BRAIN

Steroids form a unique and powerful class of signaling molecules. They are synthesized on demand in a variety of tissues as a function of the enzymatic profile and availability of precursor. The gonads and adrenals are the primary steroidogenic organs, and the quintessential male hormone is testosterone, while estrogens, chiefly estradiol, are irrevocably associated with femaleness. Steroidogenesis is a synthetic process that begins with cholesterol as a substrate and, through a series of enzymatic reactions, produces a wide array of bioactive interrelated signaling molecules. The specificity of steroid action is determined by a combination

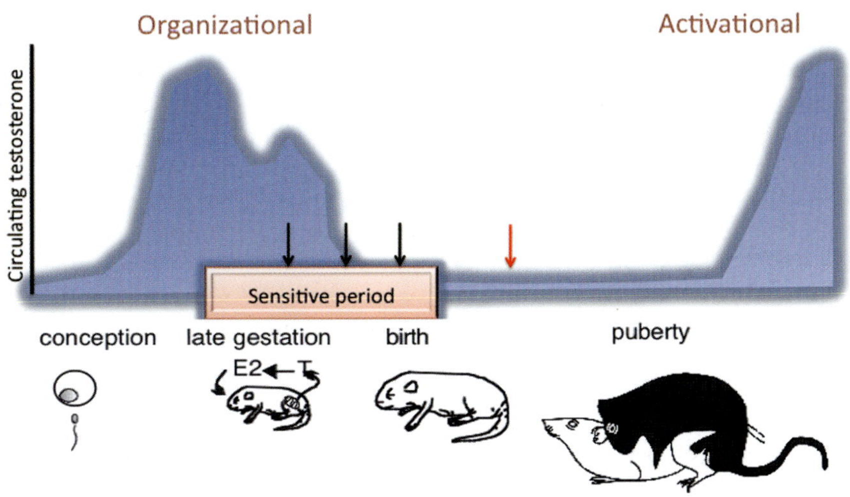

FIGURE 17.1 **Sexual differentiation of the brain occurs during a developmental sensitive period.** In the rodent the sensitive period begins prenatally with the onset of testicular testosterone production in males, which lasts briefly into postnatal life before declining to undetectable levels until early adulthood. During development the gonadally derived androgen gains access to the brain via the circulatory system, and is locally aromatized to estradiol within neurons. The actions of estradiol and androgen combined organize the neural substrate into a masculine phenotype. In the absence of high levels of gonadal hormones, the default brain phenotype is female. Treatment of females with either aromatizable androgens or estrogens during the sensitive period (black arrows) will induce masculinization of the brain, whereas the same treatment outside the sensitive period (red arrow) has no effect and defines the end of this developmental window. Gonadal steroids produced at puberty then act on the organized neural substrate to activate either male or female typical reproductive behavior and physiology.

of the ligand, the receptor, and a host of associated proteins called co-activators or co-repressors. Steroid receptors are members of a nuclear receptor superfamily, and function largely as transcription factors. Promoters of steroid-responsive genes either contain palindromic nucleotide sequences known as hormone-response elements (HRE) that bind specific steroid receptors directly, or response elements for other transcription factors that in turn associate with steroid receptors. Both mechanisms are known to mediate transcriptional regulation at steroid target gene promoters. In recent years it has become apparent that steroid receptor activity is not limited to the nucleus and regulation of gene expression, but that steroid receptors also associate with cell membranes and either interact directly with other membrane associated proteins, including neurotransmitter receptors and membrane anchored kinases, or indirectly activate signal transduction pathways through mechanisms not yet understood. Thus, the array of cellular responses regulated by steroids keeps expanding, and at times appears limitless.

Important to the study of sexual differentiation of the brain is an understanding of the direction of flow of steroid synthesis (Fig. 17.2). Androgens are an immediate precursor to estrogens, with the rate-limiting step being the aromatic conversion of the A ring by the enzyme aromatase, also known as estradiol synthetase. Estradiol is the most potent of the class of steroids known as estrogens, the others being estriol and estrone. Estrogens are the last metabolic step in steroid synthesis, with further conversions either rendering it biologically inactive or adding moieties that will increase water solubility and promote excretion by the kidneys. Testosterone is one member of the class of steroids called androgens, with dihydrotestosterone (DHT) being the most potent, and androstenedione being among the weakest. Testosterone is a critical nexus point in sex-steroid synthesis, as it is the immediate precursor to both estradiol, via aromatization, and DHT, via 5α reduction. Estradiol and DHT bind with high affinity only to their cognate receptors, estrogen receptor (ER) and androgen receptor (AR), thus diverging the two classes of hormones at the functional level.

Initial studies on the hormonal control of brain sexual differentiation focused on androgens, for obvious reasons; in particular, testosterone. Treatment of female rat pups with large doses of testosterone in the first few days of life was an effective means of masculinizing brain and behavior. However, treatment with DHT, an arguably more potent androgen, did not have the same effect as testosterone, and in fact had no effect on many endpoints. Moreover, treatment of females with estradiol proved to be even more effective at masculinizing sexual behavior than testosterone. This observation remained a mystery until the combined observations that: (a) there is a high density of ER found in brain regions relevant to control of reproductive behavior; and (b) aromatase is also expressed at high levels in the brain, particularly in those brain regions notable for high levels of ER, leading to the conclusion that local conversion of testosterone to estradiol within select neuronal populations in the brain provided the hormonal basis of masculinization. The missing piece of the puzzle was determining how both males and females avoided being masculinized by the extraordinarily high levels of estradiol found in the circulation of the pregnant dam. The solution was provided by α-fetoprotein, a steroid binding globulin that, in rodents, has a high affinity for estradiol but not for testosterone. Levels of α-fetoprotein are elevated in the circulatory system of fetuses and neonates at the time that the male testis is producing large quantities of testosterone. As a result, maternal or placental-derived estrogens are sequestered in the bloodstream of the developing pups of both sexes, while in males testosterone gains access to neurons and is aromatized to estradiol, which induces masculinization. Together, these processes constitute what has become known as the *Aromatization Hypothesis* of sexual differentiation of the brain.[5] In this scenario, α-fetoprotein acts as a selective sponge that keeps estradiol from inadvertently inducing indiscriminant masculinization, but there is persistent evidence that the steroid binding globulin may also serve as a selective delivery mechanism to certain populations of neurons and mediate processes of brain development not yet understood.[6]

The Aromatization Hypothesis is both elegant and appealing, but it is also apparently limited to rodents. Both direct and indirect evidence implicate androgens as the primary mediators of masculinization in the primate brain, including humans.[7] What role, if any, estrogens play in the development of sex differences in the primate brain remains unknown. The difference in estrogen versus androgen mediation of sexual differentiation leads some to argue that the rodent is not a good model for the human, and that no conclusions can be drawn regarding results in one species versus another. However, this argument breaks down when considering what the mechanism of action of the steroid hormone is. Steroids are signaling molecules; they act by initiating other cellular processes which then mediate the enduring consequences of sexual differentiation. Therefore, it is critical to identify the cellular processes mediated by steroids and determine if they are the same or different for estrogens versus androgens, rodents versus primates. This begins with identifying those cellular processes in rodents, as we cannot conduct experiments on humans and have only limited capacity to do so in non-human primates. Recent advances in

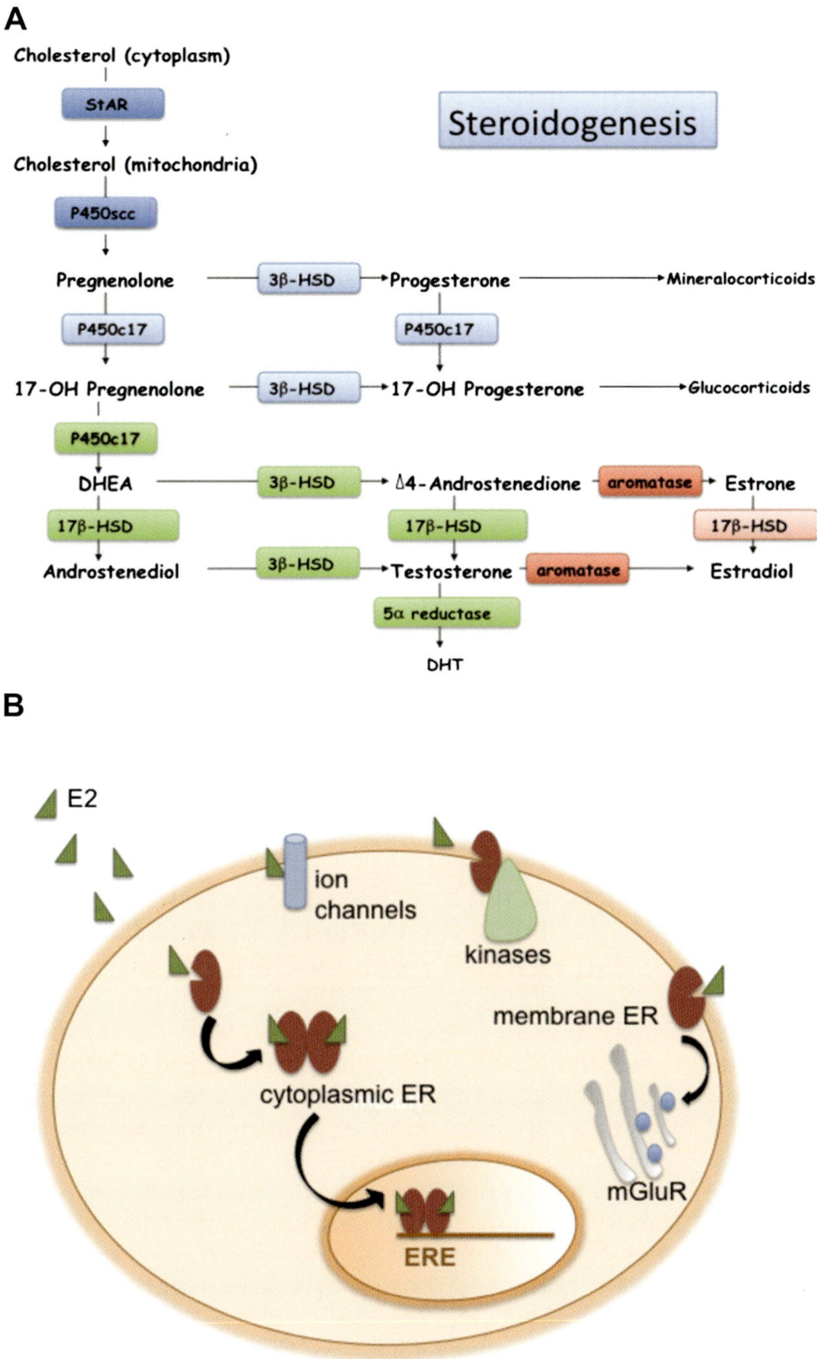

FIGURE 17.2 **Steroidogenesis and steroid action.** (**A**) The production of androgens and estrogens, so-called sex steroids or gonadal steroids, begins with the active transport of cholesterol into cells by StAR (steroidogenic acute regulatory protein) and subsequent enzymatic conversions which generally reduce the size of the previous precursor molecule by side-chain cleavage (scc). The androgen, testosterone, can be either 5α-reduced to dihydrotestosterone, which is the final active metabolite in that pathway, or aromatized to estradiol, which is the final active metabolite in that pathway. (**B**) Estradiol binds to an estrogen receptor (ER) which either presides in the membrane or in the cytoplasm. Once bound, the receptors in the cytoplasm homodimerize before moving into the nucleus, where they associate with the transcriptional complex at an ERE (estrogen response element) on the DNA and modulate gene transcription. ER in the membrane may associate directly with intracellular kinases, such as MAPKinase or PI3 kinase, or they may associate with mGluR located on the endoplasmic reticulum. Finally, estradiol itself may interact directly with ion channels, including voltage gated calcium channels, ionotropic glutamate receptors or GABA-A receptors.

elucidating the mechanisms of sexual differentiation in rodents will be discussed below.

Even in the rodent brain, not all sexually differentiated endpoints are determined by estrogens. Androgens are directly responsible for differential cell survival in the spinobulbocavernosus nucleus of the spinal cord,[8] a collection of motor neurons that controls the penis. Androgens mediate aspects of a sex difference in cell proliferation in the developing hippocampus,[9] as well as the size of the adult amgydala[10] and the number of vasopressin neurons of the bed nucleus of the stria terminalis.[11] Studies of rats and mice with naturally occurring mutations in the androgen receptor, referred to as tfm for testicularly feminized males, suggest the role of androgens in brain development is subtle, complex and poorly understood.[12] In humans, spontaneous mutations of the gene for the androgen receptor can render it completely non-functional, a condition referred

TABLE 17.1 Relationship Between Genetic, Gonadal, Body and Brain Sex

Genetic sex	Gonads	Body	Brain
XX	Ovaries	Female	Feminized
XY	Testes	Male	Masculinized
XY AIS (tfm)	Testes	Female	Feminized
XY with ER null mutation	Testes	Male	Masculinized
XY with aromatase null mutation	Testes	Male	Masculinized
XY with Sry null mutation	Ovaries	Female	Feminized
XX with CAH	Ovaries	Female with partially masculinized genitalia	Feminized with some masculinized aspects

In general, individuals with an XX genotype develop ovaries, a female body type and a feminized brain. The opposite is true for XY individuals, who develop testes, a male body type and a masculinized brain. However, if an XY individual has Androgen Insensitivity Syndrome (also referred to as tfm for testicular feminization), they will still develop testes but have a female body type and a feminized brain. Individuals with an XY genetic makeup but an inactivating mutation of the ER or aromatase enzyme will develop as normal males, but if the SrY gene is mutated, they will develop ovaries, a female body type and feminized brain. Females with Congenital Adrenal Hyperplasia (CAH) are exposed to high levels of androgens *in utero*, and exhibit both somatic and cognitive aspects of partial masculinization.

to as complete Androgen Insensitivity Syndrome (AIS). These XY individuals are externally phenotypically female, although internally they have undescended testis and no ductal system. These individuals are identified as female at birth and uniformly self-identify as female, with little to no gender identity conflict. Individuals with genetic mutations that only impair the androgen receptor are referred to as having partial AIS and, while much more phenotypically variable, generally self-identify as male (for reviews, see Breedlove[13] and Hines[14]). These observations confirm the importance of androgens to psychosexual differentiation in humans. Analogous individuals with null mutations of the estrogen receptor gene or the gene coding for the aromatase enzyme are much more rare, but again consistently self-identify as male, which is consistent with both their genetic and phenotypic sex. Table 17.1 presents a summary of the relationship between chromosomal, gonadal, hormonal and brain sex.

The majority of research emphasis in sexual differentiation of the brain is appropriately focused on estrogens and androgens, while progestins, as the name reflects, are relegated to adult female reproduction – in particular, pregnancy. Yet one of the principle gene products induced by estrogen receptor activation is progesterone receptor (PR), and this is true in the developing brain as well as the adult brain. Indeed, while the levels of ER and AR do not differ to any considerable extent in developing male and female brains, PR is markedly dimorphic, with males having high levels of PR in the developing POA and ventromedial hypothalamus, and females having little to none.[15] What remains unclear is whether the PR in the developing male brain is activated, either by progesterone or other ligands, such as dopamine, and whether the activation of PR plays any functional role in the process of sexual differentiation.[16,17] Answering this question has proven to be far more complicated than originally imagined, and remains an area of active interest.

REPRODUCTIVE BEHAVIORS ARE DETERMINED BY SEXUAL DIFFERENTIATION OF THE BRAIN

Reproductive behavior is a broad term encompassing solicitation and seeking behaviors involved in attracting and finding a mate, copulation for the exchange of gametes, and the care and nurturing of offspring after they are born. In studies of birds, rodents or non-human primates, sexual behavior is the most robustly dimorphic and readily quantifiable endpoint that is determined (or organized) by developmental steroid hormone exposure and subsequently activated in adulthood by gonadally derived steroids. Sex behavior of guinea pigs was the dependent measure used by Young and his students to generate the now iconic Organizational/Activational Hypothesis of sexual differentiation of the brain. That this is the case is intuitively obvious, given that evolutionary forces ensure that reproductive success is maximized by matching physiology and behavior. In other words, no matter how fertile an individual is, it is all for naught if an exchange of gametes with the opposite sex doesn't occur. In rodent and bird animal models we can identify three distinct developmental processes by which the adult profile of sex behavior is established; *masculinization*, *feminization* and *defeminization* (Fig. 17.3).

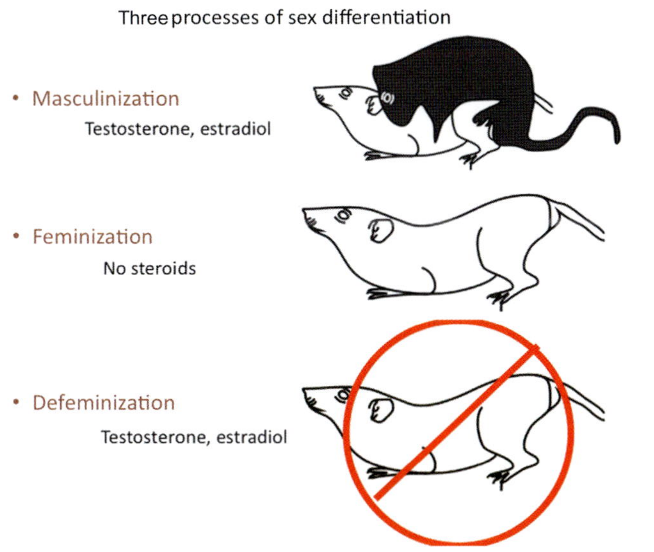

FIGURE 17.3 **Feminization, masculinization and defeminization.** Sexual differentiation of behavior involves three distinct processes. Masculinization is an organized process mediated by a combination of androgens and estrogens that results in an adult brain which is activated by steroids to reduce the threshold for expression of male sexual behavior. Feminization occurs in the absence of gonadal steroid secretion during the sensitive period, and results in an adult brain which is activated by steroids to reduce the threshold for expression of female sexual behavior. Defeminization is also a steroid-mediated organizational event which removes or suppresses the neural substrate of feminization so that the threshold for activated expression of female sexual behavior is substantially increased in adulthood.

As with the development of the bipotential gonad, feminization of the brain in regard to sex behavior is the default process that will occur in the absence of elevated androgen and estrogen exposure during the perinatal sensitive period. In rodents, sexual behavior is a combination of proceptivity, solicitations to attract males, and receptivity – the adopting of the lordosis posture to allow the male to mount and intromit for copulation. This behavior is activated by sequential exposure to estradiol over a period of 2 days, followed by rising progesterone around the time of ovulation, and the presence of a sexually active male. Lordosis is an evoked response following stimulation of the female's flanks by the forepaws of the male, thus it does not occur spontaneously, nor does it occur in the absence of the appropriate duration and quantity of hormonal exposure. The tight hormonal control of sexual receptivity in females assures that mating occurs only in the context of ovulation. The complexity of the female sexual behavior repertoire belies any notions that feminization is not an active developmental process, but determining what cellular processes mediate feminization of brain and behavior is challenging in the face of no clear genetic or hormonal trigger. Thus, we know relatively little about how the process of behavioral feminization occurs, other than to say it's not masculinization.

By contrast, masculinization is an active process that occurs normally in males in response to testicular androgen production, but which can be initiated in females with a single exposure to either testosterone or estradiol during the sensitive period. In rodents this extends into the first few days of life, providing easy accessibility for treatment and study. Masculinization is the establishment of a neural substrate which in adulthood promotes male sex behavior, consisting of the approach and solicitation of sexually receptive females and attempts to copulate with her. This behavioral set requires activation by androgens. Thus, prepubertal males show no interest in adult females, nor do adult males if they have had no prior experience and have been deprived of testosterone as adults. Steroidogenesis in adult males is a relatively continuous process, and so adult males are continuously sexually receptive (unless they have recently copulated to satiety). This, of course, matches well with the continuous production of gametes by males, as opposed to cyclical periodic ovulation in females. However, there is an interesting difference between the outcomes of masculinization and feminization. Adult males that have had sexual experience will continue to respond normally to receptive females for an extended period of time after castration, with only a gradual waning of interest that takes weeks to extinguish. By contrast, females exhibit robust sexual receptivity during a brief temporal window determined by the pattern of hormone exposure followed by a refractory period, and if hormones are removed then sexual receptivity is removed as well. The neurological basis for this difference is unknown, and the mystery is only heightened by our inability to identify distinct neural circuits controlling male versus female sexual behavior – a topic discussed in greater detail below.

The third process in behavioral sexual differentiation, defeminization, is perhaps the most mysterious of all. Defeminization is the active removal of the female phenotype from the male brain. Thus, a normal adult male treated with the hormonal milieu appropriate to females will not exhibit lordosis, but if defeminization is blocked during the perinatal sensitive period, a fully masculinized adult male will also behave as a female under the appropriate experimental circumstances (i.e., female hormones, presence of a sexually active male). The mystery is: why? Why has the system evolved to remove the female phenotype from the male, when a normal adult male will never experience sequential estradiol followed by progesterone? Is there a cost to retaining the female phenotype? Or is there another function of defeminization that we have not yet identified, and the loss of female sexual behavior is a mere side effect? These unanswered questions highlight how little we still understand about the neuroscience of sexual differentiation of the brain.

Following sexual behavior, there is pregnancy, delivery and parenting. In mammals, parental care is by definition sexually dimorphic, since only females can directly feed the offspring. Because males don't lactate, we tend to talk about maternal behavior and give relatively little attention to paternal behavior. The onset of maternal care is triggered by the hormonal milieu of pregnancy, and evidence suggests this is largely an activational event, but the potential for organizational effects on parental behavior is beginning to be explored.[18] The study of sex differences in parenting has provided us a different angle on the role of sexual differentiation. Quite naturally, our attention is drawn to how the process of sexual differentiation makes males and females different, but this has constrained our thinking in unexpected ways. When considering a biparental species of vole, Geert De Vries and colleagues noted that the vasopressin system is highly sexual dimorphic, with males having a dense vasopressinergic innervation of the septum, whereas females have relatively little. Given that males do not experience pregnancy or parturition (birth), De Vries considered whether the function of the enhanced vasopressinergic innervation in males might be for the purpose of promoting parental behavior – in other words, whether the role of sexual differentiation in this instance was to make the sexes more alike, not different.[19] This same concept has been applied to sex differences in cognition that are established by hormonal effects on the developing hippocampus,[20] although considerably more work needs to be done before we thoroughly understand the nature and origins of subtle sex differences in cognition.

REPRODUCTIVE PHYSIOLOGY IS DETERMINED BY SEXUAL DIFFERENTIATION OF THE BRAIN

The brain is an integral and critical component of reproduction as part of the hypothalamic–pituitary–gonadal axis. The GnRH neurons project to the median eminence

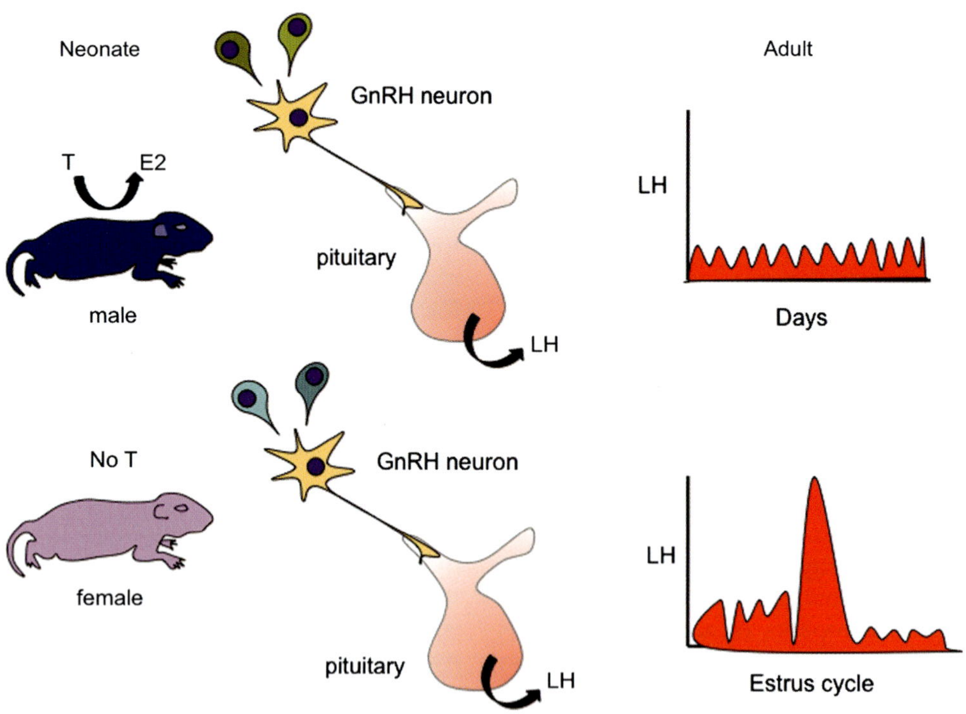

FIGURE 17.4 **Sexual differentiation of gonadotropin secretion.** The hypothalamic–pituitary–gonadal axis is fundamentally different in males and females in order to subserve the unique needs for continual gamete production in males and periodic ovulation in females. During the sensitive period of sexual differentiation the neural control of LH release by the anterior pituitary is organized for this purpose. In males, the steroid hormone control of the GnRH neuron is altered so that only negative feedback effects of gonadal steroids are realized and therefore LH release is pulsatile but continuous, resulting in elevated gonadal steroids and steady gamete production. In females, the control of LH release is under both positive and negative feedback control by estradiol, resulting in a changing periodicity of LH release and an ultimate surge of LH that is essential for ovulation, followed by a cyclical return to the previous pattern. As with the neural control of sexual behavior, the female pattern of LH release is the default and occurs in the absence of exposure to gonadal steroids during a sensitive developmental window. The male pattern is organized in response to steroid hormone exposure during the same critical window. The concordance in the hormonal profile mediating sexual differentiation of physiology and behavior assures that brain phenotype matches gonadal phenotype.

and release into the portal vasculature to gain quick access to the anterior pituitary and induce release of the gonadotropins LH and FSH (Fig. 17.4). These tropic factors travel via the circulatory system to the gonads, stimulating steroidogenesis and gamete development. The critical difference between males and females is the pattern of release of LH, which is pulsatile and generally constant in males, versus cyclical with a major surge in females. The LH surge is required for the induction of ovulation, and if prevented, the follicle will fully develop but never rupture. The surge in LH release is the result of positive feedback effects of estradiol over the preceding few days, followed by negative feedback to re-set homeostasis. Males are refractory to the positive feedback effects of estradiol, and instead exhibit only negative feedback and relatively stable levels of LH. Treatment of neonatal females with either testosterone or estradiol will masculinize the pattern of LH release by destroying the capacity for positive feedback in response to rising estradiol levels in adulthood, generating a sterile female which develops fully mature Graafian follicles that never ovulate.[21,22] While the full details of how estradiol exerts both positive and negative feedback control over LH synthesis and release have not been completely worked out, we do have a good understanding of how this process is sexually differentiated, and it involves estradiol action at several nodes in the neural circuit controlling ovulation, but, interestingly, not the GnRH neurons themselves. This leads us to the next section, on the neuronal underpinnings of sex differences in physiology and behavior.

SEX DIFFERENCES IN THE BRAIN UNDERLIE SEX DIFFERENCES IN REPRODUCTIVE BEHAVIOR AND PHYSIOLOGY

Up to this point, the discussion has focused on two major aspects of sexual differentiation of the brain: its functional consequences, and the hormones that mediate it. What has not been discussed in any detail is the neurological substrate that is both the target for hormonal modification and, presumably, the source of the functional consequences. The reason for saying "presumably" is that connecting morphological changes in the brain to changes in behavior is far less straightforward than presumed.[23] We can demonstrate that a particular brain region is larger or neurons are a different shape in one sex or the other, and that the function generally associated with that structure is different in males and females, but actually proving that the size of the structure or the shape of the neurons is the *cause* of the difference in behavior has not been achieved. At best, we only have correlations between form and function, not causation. This is not a shortcoming that particularly plagues neuroendocrinology, as the same can be said for other behavioral fields, such as learning and memory, stress and anxiety, or feeding and ingestive behaviors. Nonetheless, it is worth bearing in mind when considering the impact of steroids on the developing brain, if for no other reason than avoiding the mistake of bringing preconceived notions to the interpretation of new data.

There are multiple ways in which a brain region can differ between the sexes. It can be larger in one sex by either having more neurons and glia, or greater spacing between cells. Or the individual cells within a region can vary in the number and type of synapses they receive, which may be coordinated with the size of an afferent projection to the region. Or the neurochemical identity of the neurons making up a particular population of neurons might be systematically different in males versus females. Examples of each of these can be found in multiple brain regions, but some general principles have emerged regarding how specific types of neuroanatomical sex differences are created. These principles are best established for sex differences in brain regions directly relevant to reproduction, and what little information we have regarding other areas, such as the hippocampus, cortex, cerebellum and amygdala, suggests unique principles may be guiding the establishment of sex differences in these regions. We'll begin with those best established.

STEROID-MEDIATED CELL DEATH DURING A SENSITIVE PERIOD ESTABLISHES SEX DIFFERENCES IN THE SIZE OF PARTICULAR BRAIN REGIONS

In the mammalian brain, there are at least four distinct nuclei or subnuclei that are bigger in one sex versus the other. These are, in approximate order of magnitude of the size of the sex difference, the sexually dimorphic nucleus of the preoptic area (SDN-POA), the bulbocavernosus nucleus of the spinal cord (SNB), the anteroventral periventricular nucleus of the POA (AVPV) and the principle subdivision of the bed nucleus of the stria terminalis (pBNST). In each of these nuclei, males and females begin with the same approximate number of neurons, but during a perinatal sensitive window they die in one sex versus the other. With the exception of the AVPV, the volume of the nucleus is significantly larger in males than females, and with the exception of the SNB, it is estradiol that regulates cell death (Table 17.2). Thus, in the SDN-POA and pBNST estradiol protects neurons from cell death, leading to a larger volume nucleus in males, whereas in the AVPV estradiol actively kills neurons, resulting in a smaller volume nucleus in males. The SNB is

TABLE 17.2 Sex Differences in the Size of Subnuclei in the Mammalian Brain

Brain area	Sex difference	Mechanism	Function
SDN	M > F	Increased cell death in neonatal females	Defeminization, partner preference
AVPV	F > M	Increased cell death in neonatal males	Control of GnRH neurons, LH surge
SNB	M > F	Increased cell death in neonatal females	Motor neurons innervating the penis
BNST	M > F	Increased cell death in neonatal females	Projects to AVPV, neural circuit controlling LH secretion

Many subnuclei in the mammalian brain are larger in males than females, and this is frequently due to greater rates of cell death in females during a perinatal sensitive period. Most of these regions are strongly associated with reproduction. SDN, sexually dimorphic nucleus of the preoptic area; AVPV, anteroventral periventricular nucleus; SNB, spinal nucleus of the bulbocavernosus; BNST, bed nucleus of the stria terminalis.

a collection of motoneurons that innervate the penis, and androgen rescues these cells from death. Not surprisingly, females have no real use for the neurons, and they die due to lack of androgen during the perinatal period.

The cellular mechanisms by which steroids either promote or repress cell death are beginning to be elucidated, and appear to be unique in each brain region examined, although there is a unifying role for the signal transduction cascade associated with apoptosis.[8] The survival of the motor neurons of the SNB requires a specific growth factor and a complex cross-talk between the muscles being innervated and the neurons,[24] while activation of glutamate receptors and a subsequent gene expression cascade is important in the SDN.[25] The AVPV is particularly complex, with two separate but coordinated processes regulating cell death in distinct sets of neurons leading to the final composition. The dopamine neurons of the AVPV die in males due to activation of caspase-induced cell death,[26] whereas GABAergic neurons die because of downregulation of a pro-survival cytokine, TNFα, which signals through the NFκB receptor. The downregulation of TNFα is secondary to the upregulation of a negative modulator called TRIP, for TNF-receptor-associated-factor 2 inhibiting protein.[27] This pathway is not present in SDN neurons, and in this way estradiol can induce diametrically opposed effects – cell death versus cell survival – in two subnuclei located with the same brain region, the POA.

STEROIDS EITHER PROMOTE OR INHIBIT SYNAPTOGENESIS IN PARTICULAR BRAIN REGIONS DURING THE SENSITIVE PERIOD

The amount and type of synapses that a particular neuron receives determines how that neuron responds to stimuli transduced via afferent input. Generally, synapses are either excitatory or inhibitory. At excitatory synapses, the signal will either be propagated or even amplified before being passed on down the line. Inhibitory synapses do the opposite; the signal is attenuated or stopped. Most neurotransmitters promote excitation, although the location and types of receptors at a synapse can change the nature of the response to inhibitory. One class of neurotransmitters, the amino acid transmitters, is particularly important, because they are ubiquitous throughout the brain and constitute the main source of excitation (via glutamate) and inhibition (via GABA – gamma aminobutyric acid). The optimal location for an excitatory versus inhibitory synapse is different, so that most excitatory synapses are found on small protuberances on neurons called dendritic spines, whereas the majority of inhibitory synapses are located near the cell body, or soma. The location of excitatory synapses at the distal end of spines presumably provides a compartmentalization that is advantageous to signal integration by the neuron, but it also provides the fortuitous advantage of making this type of synapse relatively easy to visualize and count.

The identification of dendritic spines dates back to Ramon-y-Cajal and the use of Golgi impregnation – a process whereby pieces of brain are immersed in a soup of heavy metals for a period of time (days to weeks). For reasons that no one understands, some but not all of the neurons and glia will soak up the metal, which, when chemically precipitated, will make the entire neuron visible, including the small protuberances known as dendritic spines. This technique is still used today, but complemented by many additional strategies, including EM (electron microscopy), injecting individual neurons with dye, or visualization of neurons from transgenic animals engineered to express fluorescent markers. All of these approaches are effective, but they can be time-consuming, biased, and limited to only a small number of neurons. An additional valuable approach is to quantify proteins known to be specifically associated with synapses – for instance spinophilin, which is preferentially localized to dendritic spines and positively correlated with the total number of spines in a particular sample.[28,29] Synaptophysin or PSD-95 can also give an indication of total synapse number, but does not distinguish between excitatory versus inhibitory synapses.

We are fortunate to have a fairly comprehensive knowledge base of sex differences in synaptic patterns in the majority of brain regions subject to robust sexual differentiation. These include the preoptic area, arcuate nucleus, ventromedial nucleus and subdivisions of the amygdala. In a series of elegant and exhaustive studies conducted in the 1980s and 1990s under the direction of Yasumasa Arai in Japan, the relative number of synapses on dendritic spines, the dendritic shaft and the soma was quantified by EM in adult males and females, males neonatally gonadectomized, and females treated with testosterone as neonates (see Matsumoto[30] for review). Remarkable sex differences in the number of dendritic spine and soma synapses was observed in each region, but, surprisingly, the total number of synapses was generally the same. Thus, the ratio of excitatory to inhibitory input in a particular region is profoundly different in males versus females, and this difference is permanently established by gonadal hormones early in life. The next question to ask is: how do steroids permanently organize synaptic patterns?

When thinking about synapses, the obvious candidate to consider is, of course, neurotransmitters. Many investigators, including this one, hypothesized that hormonal modulation of either the synthesis of neurotransmitters or the receptors they bind to would be a fundamental basis of how sexually dimorphic synaptic patterns are established. Studies on serotonin, dopamine, acetycholine, etc., all found some effects, but none could fulfill the role of replacing the steroid – in other words, being THE mediator of the steroid. Keep in mind that steroids bind to receptors that are transcription factors, and the presumed first step in the process is the activation of particular gene(s). For a time it seemed that perhaps sexual differentiation of brain and behavior was a process mediated by steroid-induced changes in multiple different genes simultaneously. Each gene product would then be capable of activating additional genes or signal transduction pathways, resulting in an amplifying and redundant cascade. Disrupting any one step in the process would modestly impact on brain and behavior, and no one agent could substitute for the initiating steroid (Fig. 17.5). In the absence of any evidence to the contrary (i.e., identification of one critical component that was both necessary and sufficient for sexual differentiation beyond the initiating steroid), the scenario of multiple effectors and redundancy held sway.

Alternatively, there could be one critical effector that mediates all steroid-induced sexual differentiation, but it simply hadn't been identified yet (Fig. 17.5). The truth, as is so often the case, lies somewhere in the middle. There are multiple critical effectors, but each one is tied to a particular behavioral and/or physiological endpoint, with relatively little overlap between the distinct systems. Unraveling these various pathways began with the discovery of a novel critical effector which is not a neurotransmitter and had therefore not been previously considered as a potential mediator of sexual differentiation: a prostaglandin.

Male-specific Synaptic Patterning in the Preoptic Area is Determined by Prostaglandins

Prostaglandins were named after the tissue in which they were first identified, the prostate gland. They are a class of lipid derived signaling molecules generated by arachadonic acid derived from the cell membrane, which is then converted to prostaglandins by an initial cyclization reaction mediated by the cyclooxygenases, COX-1 and COX-2. Short-lived precursors are then converted into biologically active reagents via specific synthetic enzymes.[31] For instance, the prostaglandin PGE2 is synthesized by PGE synthase. Both PGE synthase and COX 1 &2 are broadly distributed throughout the brain and are predominantly, although not exclusively, found in neurons (Fig. 17.6). Steroids have been found to modulate the amount of prostaglandins in peripheral tissues, particularly those related to reproduction such as the ovary and uterus, but the potential for a similar relationship in the developing brain had not been explored. PGE2 is implicated in the neural control of puberty because of its positive effect on GnRH release,[32] and this led to an exploration of whether PGE2 might also play a role in the sexual differentiation of the preoptic area.

The density of dendritic spine synapses is two to three times greater on developing POA neurons in male brains compared to female brains within days of birth, and this sex difference is the result of elevated estradiol in the developing male POA compared to the female. That this is true is demonstrated by treating females with estradiol and observing an increase in dendritic spine density to the level seen in unmanipulated males. However, treating females with PGE2 also increased the number of dendritic spine synapses to that seen in males, and treating males with an inhibitor of the COX enzymes reduced dendritic spine density to that of females. Combining estradiol and PGE2 did not have an additive effect, and, more importantly, treating with estradiol plus a COX inhibitor blocked the effect of estradiol. This combination of results suggested that estradiol was stimulating either the production or the activity of the COX enzymes or PGE synthase, and this was confirmed by observation

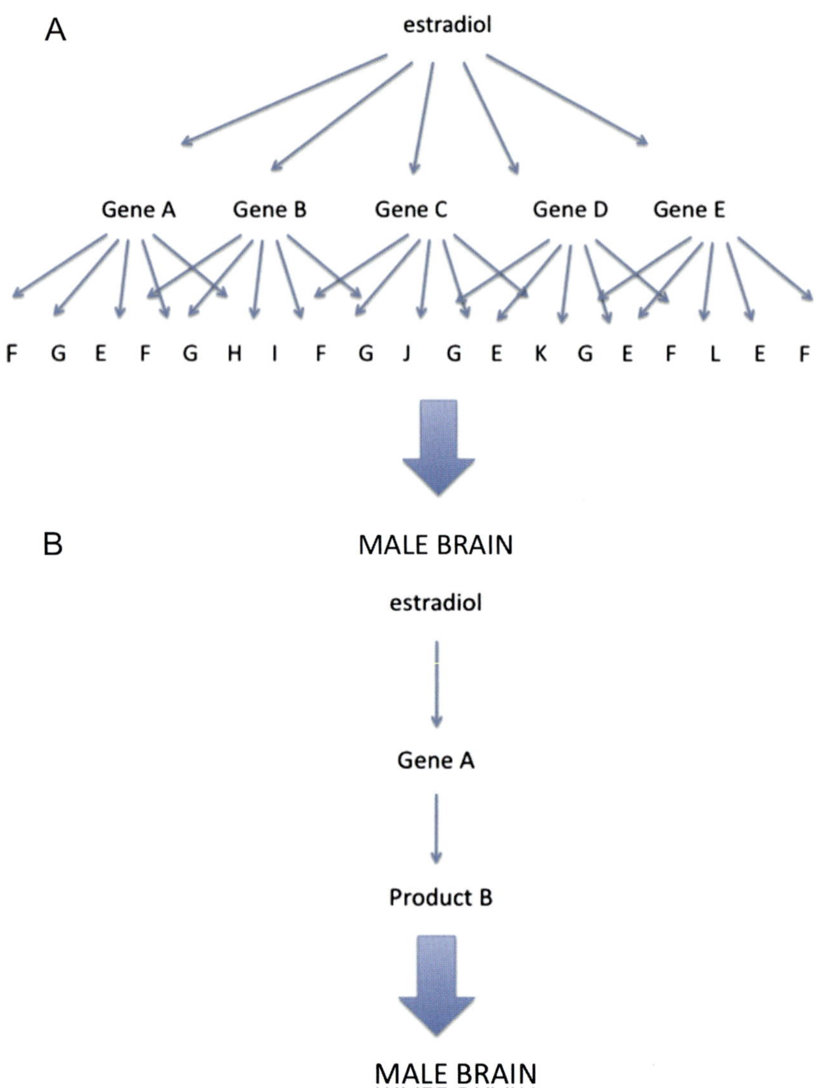

FIGURE 17.5 **Steroids induce gene expression to mediate masculinization of the brain.** The masculinizing action of a particular steroid, such as estradiol, could be due to (**A**) the simultaneous induction of expression of multiple genes which in turn alter the expression of still more genes and proteins, resulting in a cascade of changes. In this way, numerous small changes in gene expression could collectively mediate the process of masculinization and no one signaling molecule would be both necessary and sufficient. Alternatively, (**B**) there could be one critical gene which is induced by estradiol and that encodes for a specific signaling molecule which initiates the complete process of masculinization. Evidence that such a molecule exists would be that selectively blocking it prevents estradiol-induced masculinization and, more importantly, providing this molecule exogenously would substitute for estradiol and induce masculinization independent of steroid – i.e., this molecule would be both necessary and sufficient for steroid-induced masculinization. Current evidence indicates that the truth lies somewhere in the middle. At least one gene is critical for masculinization: COX2, that mediates the production of PGE2 which can induce behavioral and neuroanatomical masculinization of females. However, there are also multiple other gene products that are induced by steroids in the developing brain and play critical roles in sexual differentiation. Many of these genes are activated in a highly region-specific manner, leading to greater heterogeneity within the brains of males and females.

of an estradiol-induced two- to three-fold increase in the mRNA and protein levels of COX 1 and 2, and a seven-fold increase in the amount of PGE2. When females treated with PGE2 as neonates were raised to adulthood and assayed for male sexual behavior, they displayed a behavioral profile that was indistinguishable from that of normal males; conversely, males treated neonatally with a COX inhibitor displayed behavioral profiles indistinguishable from that of normal females (only in regard to male sex behavior, not female). This was the first time a signaling molecule was found to substitute for a steroid in the induction of a sexually differentiated endpoint, and led to the conclusion that PGE2 is both necessary and sufficient for masculinization of POA neuronal morphology and behavior.[28,33]

But how does a prostaglandin stimulate the formation of dendritic spine synapses? The answer to this question required first identifying which of the four receptors for PGE2, EP1–4, was responsible for transducing the PGE2 signal into a cellular pathway. Accomplishing this was much easier said than done, as this system is notoriously interconnected, redundant, overlapping, reciprocally inhibitory, and with every other possible complication one can imagine. After many years and convergent approaches, it became clear that two receptors, EP2 and EP4, both of which couple to adenyl cyclase, production of cAMP and activation of protein kinase A (PKA), were the critical transducing receptors.[34–36] Once PKA was identified as a key participant, the next question was to ask how this ubiquitous kinase specifically impacts on synapses. PKA is not randomly distributed throughout the cell, and different isoforms play different roles depending upon where they are located. One form of PKA is anchored to the actin

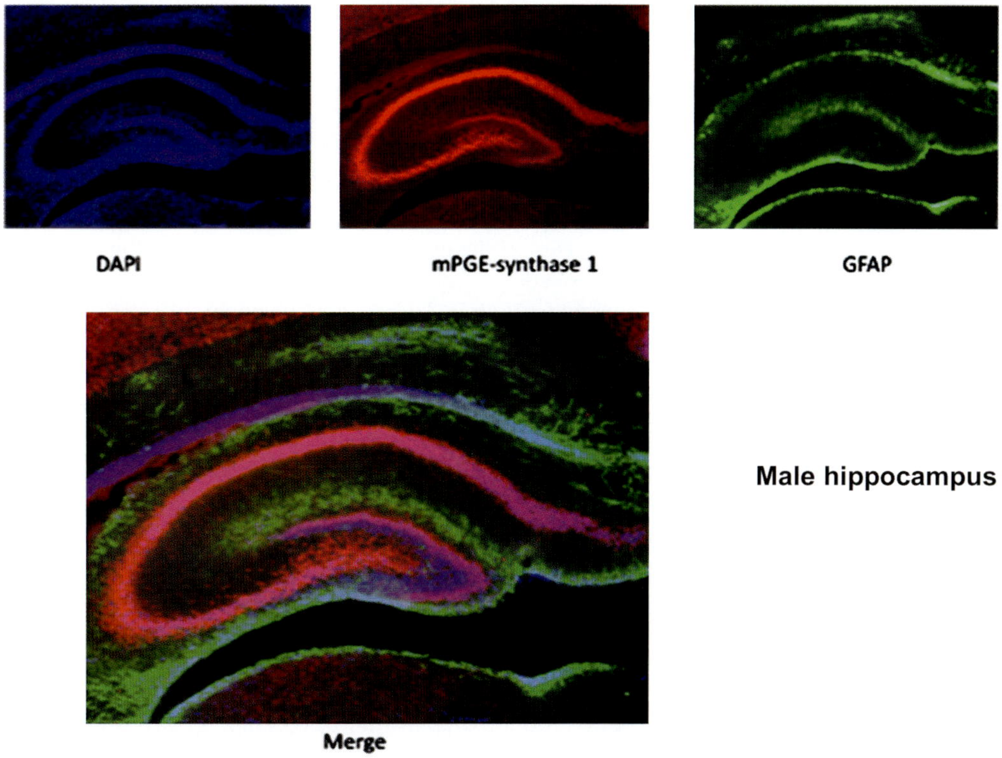

FIGURE 17.6 **PGE synthase is expressed in neurons.** Understanding which cells express specific genes relevant to sexual differentiation is critical to elucidating the sequence of cellular events underlying permanently organized changes in the neuronal architecture. The prostaglandin system plays a key role in masculinization of sexual behavior, and there are associated changes in the morphology of both neurons and neighboring astrocytes, a specialized type of glia. The use of double-label fluorescent immunohistochemistry allows for the ability to identify particular cell types co-expressing specific markers. In this instance, DAPI is used to stain the nuclei of all neurons, while an antibody for PGE synthase, the enzyme that produces the prostaglandin PGE2, is linked to a marker that produces light in the red spectrum. Glial fibrillary acidic protein (GFAP) is a marker specific for astrocytes, and is here visualized by an antibody linked to a marker that produces light in the green wavelength. When the images are combined, it is apparent that PGE synthase is expressed in neurons and not astrocytes. Thus, the increased production of prostaglandins induced by estradiol appears to be occurring selectively in the neuronal population. *Photograph courtesy of Katherine Lenz.*

matrix that is formed as a component of dendritic spine synapses, and there it serves the critical function of phosphorylating specific subunits of the glutamate AMPA receptors. If the GluR1 subunit is not phosphorylated, it will traffic out of the postsynaptic density and the synapses become silent, and, ultimately, the spine will collapse. Conversely, continued presence of functional AMPA receptors at the spine head is essential for the formation and maintenance of the dendritic spine (Fig. 17.7). Thus, the estradiol-induced increase in dendritic spine synapses of the POA begins with increased transcription of the COX enzymes, increased production of PGE2, activation of PKA, phosphorylation of GluR1, trafficking of AMPA receptors to the postsynaptic density, and the formation and stabilization of dendritic spine synapses. There are many more questions that remain to be answered, but this represents one of the most complete signal transduction pathways in the establishment of a sex difference in synaptic patterning in the developing brain.

Male-specific Synaptic Patterning in the Arcuate Nucleus is Determined by GABA

The arcuate nucleus also displays a marked sex difference in the balance of excitatory versus inhibitory synapses, only in this case males have far fewer excitatory dendritic spine synapses than females.[37] A specialized form of glia, called astrocytes, are found in abundance in the arcuate nucleus, and are highly responsive to external signals such as steroids and osmolality. Rapid changes in astrocyte morphology modulate the function of local synapses by either stripping them off the cell (in essence, physically covering them up) or unmasking previously blocked synapses. The morphology of astrocytes in the arcuate nucleus is also systematically different in males and females, with this difference appearing within days of birth. Males have highly complex astrocytes with many primary processes that branch frequently, whereas females have much more simple, almost bipolar, astrocytes, with fewer primary processes and infrequent

BOX 17.1

HOW DO WE KNOW ABOUT PROSTAGLANDINS AND THE DEVELOPING PREOPTIC AREA?

Prostaglandins are best known as inflammatory mediators that initiate fever in the short-term and can contribute to neurodegeneration in the long-term. They are membrane-derived signaling molecules synthesized from arachidonic acid by the rate-limiting enzymes cyclooxygenases 1 and 2. The cyclooxygenases, known as COX enzymes, are the targets of most anti-fever and analgesic medications, such as aspirin and NSAIDs (non-steroidal anti-inflammatory drugs). Those that don't act directly on COX act on additional synthetic enzymes just downstream. One particular prostaglandin, prostaglandin E2 (PGE2), is generated by the actions of the enzyme PGE synthase, the activity of which is closely yoked to that of the COX enzymes.[1] Determining that PGE2 mediated the masculinizing effect of estradiol involved a multistep process that included first treating unmasculinized newborn female rats with PGE2 injected directly into the preoptic area, and then measuring either the effect on dendritic spine density of POA neurons a few days later, or the ability to express male sexual behavior as an adult several months later. The next step was to do the opposite and disrupt endogenous prostaglandin synthesis with NSAIDs. Treatment of males or females masculinized with estradiol with the COX inhibitor indomethacin prevented both the formation of new dendritic spines and the expression of normal male sexual behavior in adulthood.[2,3] Further steps in the pathway were filled in by determining that estradiol increases the transcription and translation of COX-1 and COX-2 in the preoptic area, but not in the hippocampus, and that this leads to a several-fold increase in PGE2 levels within 48 hours. One of the puzzling aspects of these findings is that PGE2 is also the prostaglandin known to mediate fever, and the principle brain region controlling fever is the preoptic area.[1,4–6] How can it be that the same cellular signaling molecule that mediates the fever response also mediates masculinization of sexual behavior, and that these happen in the same brain region? The answer remains unknown, but one interesting distinction in the regulation of these two physiological responses is the receptor that transduces the prostaglandin signal. PGE2 binds to four different G-protein coupled receptors, referred to as EP1–4. A detailed chronological analyses revealed that all four receptors are present in the neonatal rodent brain, and that there is no sex difference or hormonal modulation of the amount of receptors.[7] However, there is a divergence in which receptor mediates which response. The control of fever is mediated primarily by the EP3 receptor, which is coupled to a G_i protein and a decrease in cAMP production.[8] Conversely, sexual differentiation of behavior is mediated by the EP2 and EP4 receptors, which are G_s coupled and therefore increase cAMP production, and indeed activation of PKA is a critical component of the differentiation cascade.[3] The relative affinities of the receptors do not differ remarkably; thus, one possibility is that, by sheer numbers, the activation of the EP2 and EP4 receptors induces an increase in cAMP that overwhelms the functionality of the EP3 receptor and prevents a fever from occurring at this developmental time-point. Alternatively, a disconnect in the downstream signaling from the EP receptors could mediate a temporally distinct process such that the EP2 and EP4 receptors are functional early in life, to be later replaced by the fever-mediating EP3 receptor. A distinction between these possibilities remains to be determined.

References

1. Hoffmann C. COX-2 in brain and spinal cord implications for therapeutic use. *Curr Med Chem.* 2000;7(11):1113–1120.
2. Amateau SK, McCarthy MM. Induction of PGE(2) by estradiol mediates developmental masculinization of sex behavior. *Nat Neurosci.* 2004;7(6):643–650.
3. Wright CL, McCarthy MM. Prostaglandin E2-induced masculinization of brain and behavior requires protein kinase A, AMPA/kainate, and metabotropic glutamate receptor signaling. *J Neurosci.* 2009;29(42):13274–13282.
4. Blatteis C, Li S, Li Z, Feleder C, Perlik V. Cytokines, PGE2 and endotoxic fever: a re-assessment. *Prostagland Other Lipid Mediat.* 2005;76:1–4.
5. Inoue W, Matsumura K, Yamagata K, Takemiya T, Shiraki T, Kobayashi S. Brain-specific endothelial induction of prostaglandin E_2 synthesis enzymes and its temporal relation to fever. *Neurosci Res.* 2002;44:51–61.
6. Ushikubi F, Sugimoto Y, Ichikawa A, Narumiya S. Roles of prostanoids revealed from studies using mice lacking specific prostanoid receptors. *Jpn J Pharmacol.* 2000;83(4):279–285.
7. Wright CL, Burks SR, McCarthy MM. Identification of prostaglandin E2 receptors mediating perinatal masculinization of adult sex behavior and neuroanatomical correlates. *Dev Neurobiol.* 2008:68.
8. Sugimoto Y, Narumiya S. Prostaglandin E receptors. *J Biol Chem.* 2007;282(16):11613–11617.

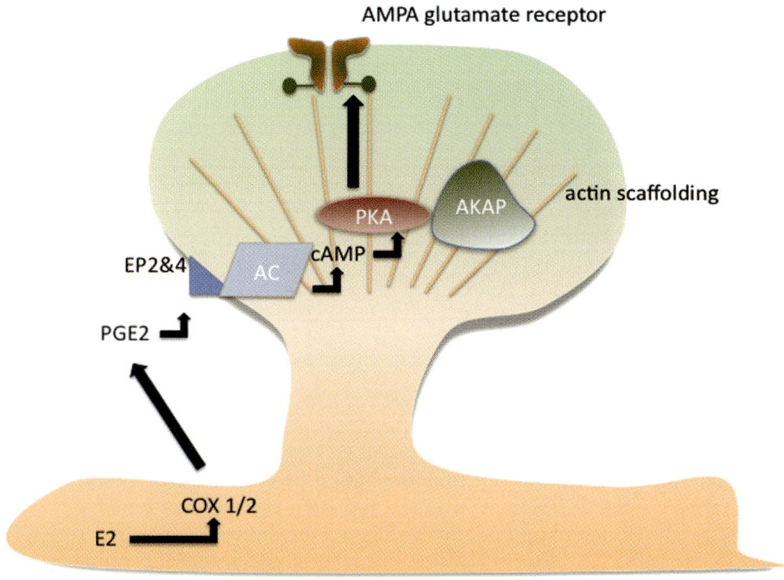

FIGURE 17.7 **PGE2 induces dendritic spine formation by activating PKA.** Estradiol increases the density of dendritic spine excitatory synapses in the developing brain, and the signal transduction pathway for this process has been extensively characterized. Estradiol (E2) increases the expression of the critical cyclooxygenases (COX) enzymes, which leads to elevated PGE2. The EP2 and 4 receptors for PGE2 are linked to adenyl cyclase (AC), and when activated induce cAMP production, which in turn activates Protein Kinase A (PKA). There are multiple subpopulations of PKA, one of which is anchored to the actin scaffolding in the dendritic spine head by AKAP. Activation of this population of PKA induces phosphorylation of specific subunits on the AMPA glutamate receptor, which then traffics those receptors to the postsynaptic density, making them available for synaptic transmission. Cellular processes subsequent to AMPA receptor activation are then key to maintaining and stabilizing these dendritic spine synapses.

branching. Treatment of females results in male-like astrocytes, and blocking estrogen action in males results in female-like astrocytes.[38] Such a dramatic effect of estradiol on astrocyte morphology predicts that the steroid is acting directly on the cell, but attempts to identify ER within arcuate astrocytes were not successful, suggesting that the initiating signal begins in the neuron and is transduced to the astrocytes secondarily. This turned out to be true, with the initiating signal being the amino acid transmitter GABA, which acts via GABA-A receptors that are found in abundance on the surface of astrocytes. The sequence of events in the arcuate goes like this: estradiol induces synthesis of the enzyme GAD (glutamate decarboxylase), which makes more GABA, which leaves the neuron and binds to GABA-A receptors on the neighboring astrocytes and stimulates the astrocytes to grow more primary processes with more branches. Next, via a process that is not well understood, these highly differentiated astrocytes apparently prevent the formation of dendritic spine synapses, and as a result males have far fewer of this type of synapse than females.[39] What this means functionally to the animal has not yet been determined, but this brain area illustrates two important principles of sexual differentiation: (a) within a specific brain region, steroids induce one gene product to initiate a cascade of cellular events, not multiple gene products at once; and (b) often the cells that are undergoing sexual differentiation are not the ones that express steroid receptors – in other words, steroid-mediated sexual differentiation is not cell autonomous. This same principle is even better illustrated in the next example.

Male-specific Synaptic Patterning in the Ventromedial Nucleus is Determined via Non-genomic Effects of Estradiol on Glutamate Release

The ventromedial nucleus is located in the basomedial hypothalamus, and is a critical node of the neural circuitry of female sexual behavior – so, not surprisingly, there is yet another pattern of synaptic profile here, and it is regulated by a completely different set of cellular events. What is surprising about this brain region is that the estradiol-initiated event does not involve gene transcription, but is instead a non-genomic activation of a specific kinase, PI3 kinase, which causes increased release of glutamate from the presynaptic terminal. We know that the initiating event is non-genomic because it still occurs in the presence of a protein synthesis inhibitor. Ultimately protein synthesis is required, but only in the postsynaptic neuron where the new dendritic spines are formed. Glutamate released from the presynaptic terminal binds to and activates both NMDA and AMPA receptors, leading to the activation of another protein kinase, MAP Kinase, which leads to the production of proteins required for the construction and maintenance of dendritic spine synapses, such as spinophilin.[29] One of the remarkable things about this sequence of cellular events is the clear demonstration that estrogen receptor activation is critical to the presynaptic cell but irrelevant to the postsynaptic cell, which is the one undergoing morphological change. Thus, the principle of cell-to-cell communication and that sexual differentiation is not cell autonomous is upheld (Fig. 17.8).

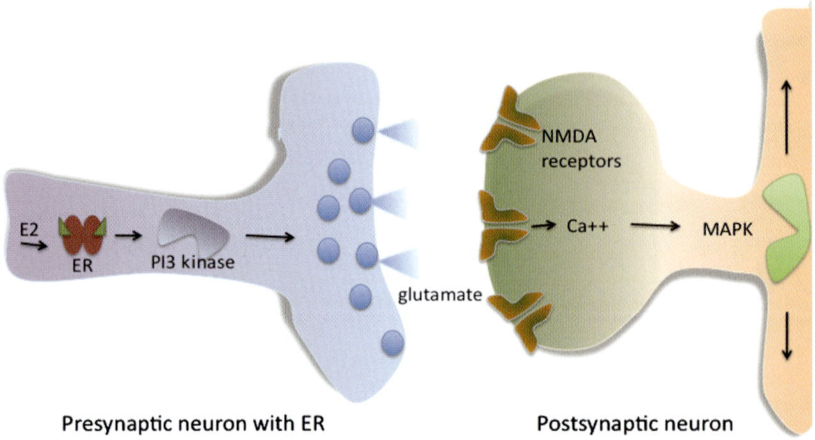

FIGURE 17.8 **Sex differences in neuronal morphology can be determined by cell-to-cell communication.** Steroids act by binding to their cognate receptors, making the presence of the receptor a limiting factor in steroid hormone action. It has been generally assumed that only a subset of neurons within a sexually dimorphic brain region express specific steroid receptors, such as ER, and therefore only those neurons will exhibit a morphological change in response to estradiol. However, recent evidence has revealed the importance of cell-to-cell communication in changing the neuronal morphology of an entire population of neurons within a specific brain region. In some regions this involves neuronal–glial communication, but in the example illustrated here a presynaptic neuron contains ER, which, when bound to E2, activates PI3 kinase and promotes glutamate release. Glutamate activates postsynaptic NMDA receptors and the subsequent calcium influx activates an additional kinase, MAPKinase, which ultimately leads to the formation and stabilization of the excitatory dendritic spine synapse. In this way the hormonal signal is transduced by one neuron to another neuron, resulting in an organizational change in synaptic patterning.

How do Changes in Synaptic Profile of the VMN Relate to Sex Differences in Behavior?

As discussed above, it is not possible to directly link changes in synaptic profiles to behavioral outcomes in the mammalian brain, but we can identify correlations. The POA has already been established as the key brain region for male sexual behavior, and the VMN is known to be critical to female sexual behavior. A logical extension is that the VMN is also the site of defeminization, as it naturally occurs in developing male brains. If the ability of estradiol to induce glutamate release is a critical component of defeminization, then mimicking the increased release of glutamate should also induce defeminization, and this indeed seems to be the case. Treatment of newborn females with glutamate receptor agonists reduces their expression of female sexual behavior as adults, whereas treating newborn females with PGE2, the agent of masculinization, has no effect on adult female sexual behavior. Thus, the cellular mechanisms and brain regions mediating masculinization and defeminization appear to be entirely separate. But does this mean there are separate neural circuits for male and female sexual behavior?

Are there Separate Neural Circuits for Male and Female Sexual Behavior?

We often discuss the neural circuits controlling male reproductive behavior or female reproductive behavior, and this has quite naturally led to the pervasive belief that there are two distinct circuits, one embedded in the male brain and the other embedded in the female brain. But what evidence do we have that this is really true? We know there is a clear, hormonally-mediated sexual differentiation of the brain, and this is at times compared to the reproductive track, where both sexes begin life with two sets of plumbing: the Müllerian duct system leading to the uterus and oviduct of females, and the Wollfian duct system leading to the vas deferens and associated structures in males. The female system is eliminated in males by the active production of anti-Müllerian hormone, while the Wollfian duct is retained in response to androgen production by the testis. In females, the Wollfian duct regresses due to the lack of androgen and the Müllerian duct is retained due to the lack of anti-Müllerian hormone. Given the elegance of this system it is tempting to assume something similar is happening in the mammalian brain, and the homology of the female brain being a default and active defeminization of the male brain makes the temptation even stronger. However, there is no evidence for separate male and female neural circuits.

Instead, what seems more likely is that there is a single neural network which contains nodes that are more heavily weighted in one sex versus the other to respond to stimuli in a manner that will evoke a male-specific or female-specific behavioral response. Variables that might contribute to the weight of a particular node would be the type and density of synapses, the frequency of dendritic branching, or the length of specific dendrites. The weighting of the nodes is a combination of organizational effects to set the synaptic pattern, and adult activational hormonal effects to set the excitatory tone. The stimuli are then a combination of olfactory, visual and, probably, auditory messages received by the animal and interpreted according to the characteristics of the neural net. Thus, in a normal male rat the odor of a sexually receptive female combined with her proceptive hopping and darting is interpreted by his neural net as an irresistible urge to approach, mount and mate. A normal female would be unresponsive to this particular set of stimuli, but the odor and site of a sexually receptive male would be interpreted by her brain as something she wants to be around, and even solicit his attention. But the neural network is the same in both cases, and it is possible that within one animal both responses could occur if the stimuli are sufficiently strong and the neural net is weighted in such a way that the information passes on down the line in each case. Animals that exhibit both male and female sexual behavior have been identified many times, and support this view, but they are the exception that proves the rule – which is that usually animals display only one sex-specific behavioral response.

STEROIDS MODULATE BRAIN REGIONS NOT DIRECTLY TIED TO SEXUAL DIFFERENTIATION, BUT THE MECHANISMS ARE UNIQUE

The hippocampus and amygdala are brain regions that are intricately involved in behavioral and physiological functions and that show small but consistent differences in males and females – i.e., in emotionality, aspects of learning and memory, stress responsiveness and affiliation. Both brain areas are responsive to steroids in the adult, and while there are some morphological sex differences in the hippocampus and amygdala, they are relatively small in comparison to the robust sex differences seen in the hypothalamus and preoptic area. Moreover, sex differences in the behaviors associated with these brain regions are equally unimpressive and are strongly subject to modification by extenuating factors such as experience, environment and current context (see McCarthy and Konkle[20] for review). Thus, it's not surprising that attempts to apply the rules governing the Organizational/Activational Hypothesis of sexual differentiation of sexual behavior have largely fallen short.

Nonetheless, progress is being made. There are effects of exogenous hormone treatment of neonates on adult hippocampal morphology and learning and memory[40,41] that follow along the standard lines of organizational/activational effects. Interestingly, the level of endogenous steroid in the developing hippocampus is exceedingly low compared to that of the POA and hypothalamus, and there is no discernible difference between males and females.[42] Yet the sensitivity of males and females appears to be markedly different, based on at least one cellular endpoint: cell genesis.

Unlike the nuclei listed in Table 17.2 in which cell death is the critical determinant of sex differences, rates of cell birth are markedly different in the male and female hippocampus and amygdala (Fig. 17.9). Neonatal males have about twice as many new hippocampal cells being born at any given time compared to females,[9] whereas females have higher rates of proliferation in the amygdala.[43] Both the ultimate fate of the new cells and what is regulating proliferation differs between the two brain regions. In the hippocampus most of the new cells will become neurons, and endogenous estradiol synthesis in males, but not females, is critical to the sex difference.[44] In the amygdala, the role of hormones is not yet known, but more of the newly born cells in females will become glia, as opposed to those in males. A novel new regulator in the arena of sex differences mediates the sex difference in cell proliferation in the amygdala: the endocannabinoids. Females have lower amounts of endocannabinoid than males, and treating females with exogenous ligands for the endocannabinod receptors reduces cell proliferation in the amygdala to that of males. This represents yet another unique cellular mechanism by which sex differences in the brain are determined.

What is the Value of Understanding the Mechanism?

At this point, the reader might be asking: Why does it matter that there are so many different mechanisms, and that sexual differentiation is cell autonomous? What difference does it make? Sometimes, addressing the proximate mechanisms of an issue can illuminate aspects of ultimate causation. The fact that steroid hormones simultaneously activate so many different signal transduction pathways in different brain regions and for different endpoints means that there is no one unifying mechanism across the brain, but instead there are multiple nodal points at which variability in the process can be introduced. Consider the preoptic area as an example. Allelic variations in the COX enzymes are known to exist, and result in variable production of PGE2 across the lifespan. There is also likely to be allelic variation in the EP receptors, and experience and environment exert profound

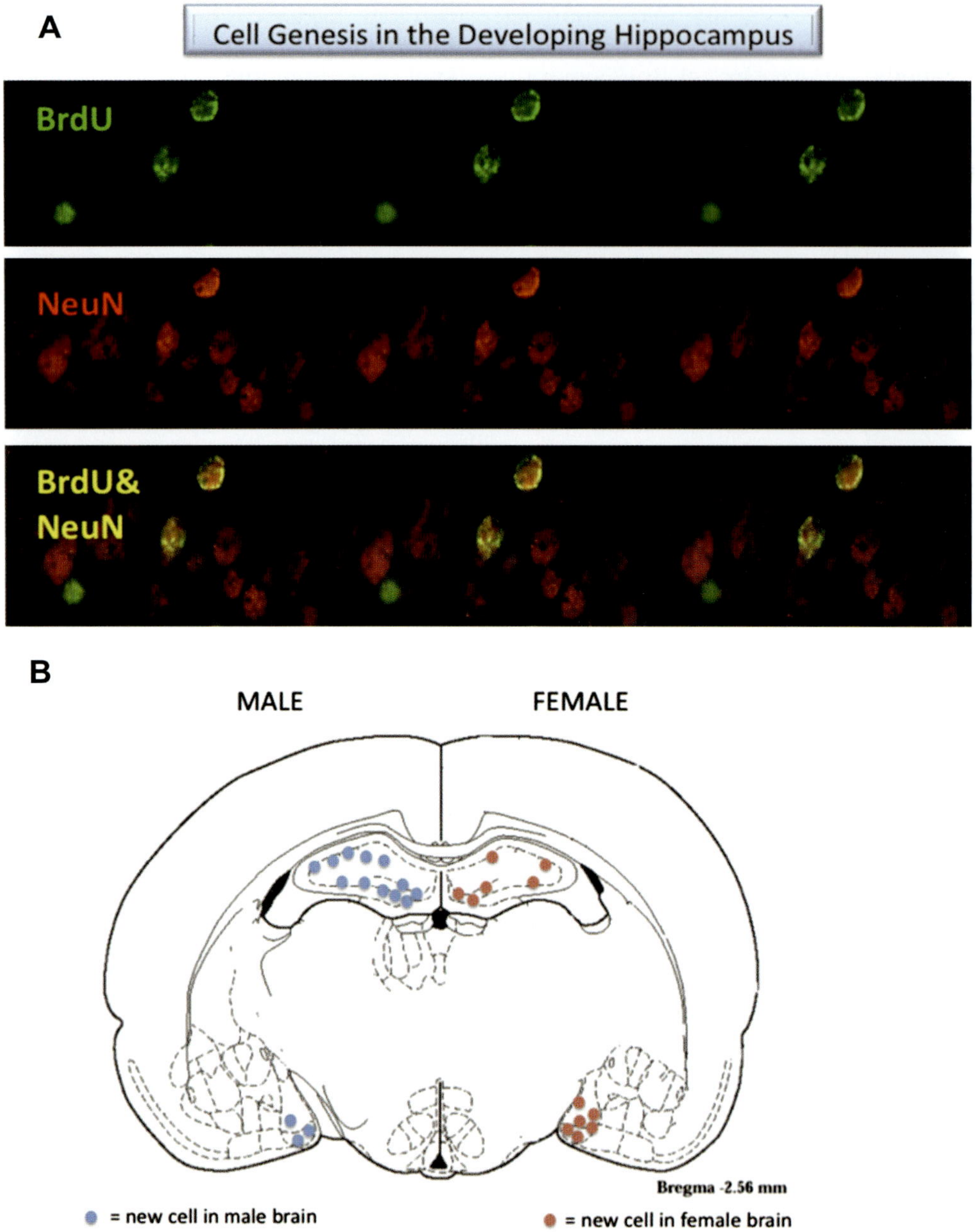

FIGURE 17.9 **Sex differences in cell birth in the developing hippocampus and amygdala.** New cells can be detected by immunocytochemical visualization of the thymidine analog bromo-deoxy-urocil (BrdU), which is incorporated into the DNA of a dividing cell during the S phase. The fate of the new-born cell can be determined several weeks later by simultaneous detection of specific markers, such as NeuN for neurons or GFAP for glia. (**A**) In this image from the hippocampus, cells that were born during the first few days of postnatal life are detected almost 3 weeks later by green fluorescence. Neurons are discerned by the presence of NeuN, which appears red, and the combination of BrdU and NeuN is yellow, indicating that some of the cells born 3 weeks earlier have now become mature neurons. (**B**) When males and females were compared for the number of new neurons being born during the early postnatal period, males were found to have almost twice as many new cells as females. Conversely, in the developing medial amygdala the opposite was true, with females having more new cells than males. In both cases the functional consequences and cellular mechanisms mediating these sex differences remain to be determined, but this is notably different from many other brain regions, where there are sex differences in cell death but no evidence for differences in cell birth.

influences on the phosphorylation state of AMPA receptors. Thus, the effect of estradiol on this cellular cascade in one individual is sure to be slightly different in another individual, resulting in a masculinized POA in both instances, but not exactly the same masculinized POA. Compound this across multiple brain regions, and with the tremendous genetic heterogeneity of humans combined with the unique complement of experience and environment of each person, and it is not hard to imagine that the final end product of sexual differentiation is never exactly the same for any two people. Moreover, it means that some aspects of one male's brain might be heavily weighted towards the masculine phenotype, while another region might be much less so — or even weighted toward the feminine phenotype. In other words, each individual brain is a mosaic of maleness and femaleness. The increased variability assures that all aspects of an environmental niche will be exploited to maximum advantage.

SUMMARY

In this chapter we have reviewed the phenomenon of steroid-mediated sexual differentiation of the brain. The developing brain is bipotential, and capable of taking on a male or female phenotype as a function of perinatal steroid exposure during a sensitive developmental window. In rodents, the dominant steroid is estradiol, derived by aromatization from testicular androgens, but androgens also have direct effects in some brain regions. The Organizational/Activational Hypothesis codifies the notion that early hormone exposure exerts enduring effects on the developing brain, thereby directing and restricting the responsiveness of the adult brain to circulating steroids following puberty. This now iconic concept continues to hold true for brain regions directly associated with reproduction, such as the preoptic area and hypothalamus, but may not be appropriate for non-reproductive related areas such as the hippocampus and amygdala. The cellular mechanisms by which steroids exert enduring effects appear to be unique for each brain region and/or endpoint, resulting in a far more diverse phenotype than would be obtained by a single unifying mechanism across a brain region. Moreover, cell-to-cell communication following initiation of hormonal signaling cascades assures that all cells within a particular brain region are altered by the steroid milieu, as opposed to just those cells that express steroid receptor.

OUTSTANDING QUESTIONS FOR THE FUTURE

The study of sexual differentiation of the brain began over 50 years ago, and yet remains as vibrant, surprising and relevant as it was from the start. Advances in mechanisms and reinterpretations of old discoveries in the light of new data are regular occurrences, making any review of the topic at least partially out of date before the ink is dry. Regardless, many questions persist.

How do Separate Sexually Differentiated Brain Regions Interact?

One of the points of emphasis of this discussion has been the unique mechanisms evoked by steroids for each sexually differentiated endpoint and/or brain region. Thus, it is possible for specific brain regions to respond to steroid hormone exposure completely independently from neighboring brain regions. This begs the question, how do we get to a coordinated, integrated and coherent whole? Ultimately there is only one brain residing in one body, which is either male or female, so how do these various brain regions interact and influence each others' functions? Advanced imaging techniques and mathematical modeling will be necessary approaches to gaining a full understanding of all the working components of a neural network, and the responses of various nodes to intrinsic and extrinsic stimuli.

What about Genetics and Sex Differences in the Brain?

This discussion has ignored an additional and important contributor to sex differences in the brain, and potentially the process of sexual differentiation: genetics. Every cell has a sex, which is determined by its chromosomal complement, so that in the mammalian male brain every neuron and glial cell is an XY while in the female every cell is XX. One of those X chromosomes is inactivated to avoid the consequences of twice as much gene expression in females as males, but this silencing is imperfect, and some genes are known to "escape" X inactivation. Moreover, there are Y-specific genes that females do not possess. How much do these genes alter brain functioning? This question has been on the minds of a small cadre of researchers that have provided valuable insight via the use of genetic models that allow for separation of genotype from gonadal phenotype.[45,46] While still in its earliest stages of discovery, one tentative conclusion is that genetics contributes more strongly to sex differences in brain regions that are not directly relevant to reproduction, such as the cortex, hippocampus and amygdala. Conversely, in those regions directly controlling reproductive physiology and behavior, such as the preoptic area and hypothalamus, hormones rule the day.

How do Organizational Effects of Hormones Endure Across the Lifespan?

Hormonal effects on the developing preoptic area and hypothalamus are so potent that they endure for a lifetime, predetermining adult physiological and behavioral responses. This is the "organizational" component of the Organizational/Activational Hypothesis. But how is it that these effects are permanent? We have generally assumed that it is because the cellular endpoints include cell death, cell differentiation and synaptic patterning, and that these things must be in essence permanent. When considered more carefully, however, we see that this assumption is false. Cell birth and cell death continue at least into puberty in those brain regions known to be permanently differentiated, such as the SDN, AVPV and BNST,[47] and synapses retract and reform on a continual basis, yet the number, type and placement of the synapses in sexually differentiated brain regions remains constant.[28] It is as if there is a cellular memory, and perhaps there is, in the form of epigenetic changes to the DNA. Epigenetics means literally above or beyond the genome, and is a means by which the expression of specific genes is regulated by alterations to the surrounding chromatin or directly on the DNA base-pairs. There are tantalizing hints that steroid-induced epigenetic changes may underlie the enduring consequences of early hormone exposure,[48] but definitive evidence has not been established.

Is the Human Brain Sexually Differentiated?

Finally, one of the most critical of outstanding questions is: what does this all mean to humans? The short answer to the question of whether the human brain is sexually differentiated is "yes" — we know this, based on individuals either blind to their own steroids due to genetic mutations, or individuals inappropriately exposed to steroids during a critical developmental window (see Table 17.1). However, when considering humans the impact of experience and environment is both so strong and so individual that it becomes exceedingly difficult to tease out the biological from the cultural, the nature from the nurture. The debate rages on as to whether sexual orientation is immutable, being the result of genetic and hormonal influences, or malleable, opportunistic and subject to influence. This topic alone succinctly illustrates how little we still understand about the nature of human sexuality. Because we cannot and should not conduct experiments on humans, we may never have all the answers. By the study of animals, we can eliminate the cultural influences and control the environmental and experiential variables so that a clear understanding of the biological contributions toward sexuality and sex differences in the brain can be gained, but we have a long way yet to go before achieving that goal.

References

1. Phoenix CH, Goy RW, Gerall AA, Young WC. Organizing action of prenatally administered testosterone proprionate on the tissues mediating mating behavior in the female guinea pig. *Endocrinology.* 1959;65:369–382.
2. Baum MJ, Brand T, Ooms MP, Vreeburg JT, Slob AK. Immediate postnatal rise in whole body androgen content in male rats: correlation with increased testicular content and reduced body clearance of testosterone. *Bio. Reprod.* 1988;38:980–986.
3. Weisz J, Ward IL. Plasma testosterone and progesterone titers of pregnant rats, their male and female fetuses and neonatal offspring. *Endocrinology.* 1980;106:306–313.
4. Sisk CL, Foster DL. The neural basis of puberty and adolescence. *Nat Neurosci.* 2004;7(10):1040–1047.
5. McEwen BS, Lieberburg I, Chaptal C, Krey LC. Aromatization: Important for sexual differentiation of the neonatal rat brain. *Hormones and Behavior.* 1977;9:249–263.
6. Toran-Allerand CD. Coexistence of alpha-fetoprotein, albumin and transferrin immunoreactivity in neurones of the developing mouse brain. *Nature.* 1980;286(5774):733–735.
7. Wallen K. Hormonal influences on sexually differentiated behavior in nonhuman primates. *Front Neuroendocrinol.* 2005; 26(1):7–26.
8. Forger NG. Cell death and sexual differentiation of the nervous system. *Neuroscience.* 2006;138(3):929–938.
9. Zhang J-M, Konkle ATM, Zup SL, McCarthy MM. Impact of sex and hormones on new cells in the developing rat hippocampus: A novel source of sex dimorphism? *Eur. J. Neurosci.* 2008;27: 791–800.
10. Cooke BM, Breedlove SM, Jordan CL. A brain sexual dimorphism controlled by adult circulating androgens. *Proc Natl Acad Sci.* 1999;96:7538–7540.
11. De Vries GJ, Panzica GC. Sexual differentiation of central vasopressin and vasotocin systems in vertebrates: different mechanisms, similar endpoints. *Neuroscience.* 2006;138(3):947–955.
12. Zuloaga DG, Puts DA, Jordan CL, Breedlove SM. The role of androgen receptors in the masculinization of brain and behavior: what we've learned from the testicular feminization mutation. *Horm Behav.* 2008;53(5):613–626.
13. Breedlove SM. Sexual differentiation of the human nervous system. *Annu Rev Psychol.* 1994;45:389–418.
14. Hines M. Sexual differentiation of human brain and behavior. In: Pfaff D, ed. *Hormones, Brain and Behavior.* Vol. 5. London, UK: Academic Press; 2002:425–462.
15. Wagner CK, Pfau JL, De Vries GJ, Merchenthaler IJ. Sex differences in progesterone receptor immunoreactivity in neonatal mouse brain depend on estrogen receptor alpha expression. *J Neurobiol.* 2001;47(3):176–182.
16. Quadros PS, Lopez V, De Vries GJ, Chung WC, Wagner CK. Progesterone receptors and the sexual differentiation of the medial preoptic nucleus. *J Neurobiol.* 2002;51(1):24–32.
17. Wagner CK. Potential role of maternal progesterone in the sexual differentiation of the brain. *Endocrinology.* 1998;139: 3658–3661.
18. Lonstein JS, De Vries GJ. Sex differences in the parental behavior of rodents. *Neurosci Biobehav Rev.* 2000;24(6):669–686.
19. De Vries GJ. Minireview: Sex differences in adult and developing brains: compensation, compensation, compensation. *Endocrinology.* 2004;145(3):1063–1068.
20. McCarthy MM, Konkle AT. When is a sex difference not a sex difference? *Front Neuroendocrinol.* 2005;26(2):85–102.
21. Barraclough CA. Production of anovulatory, sterile rats by single injections of testosterone propionate. *Endocrinology.* 1961;68: 62–67.

22. Barraclough CA, Gorski RA. Evidence that the hypothalamus is responsible for androgen-induced sterility in the female rat. *Endocrinology.* 1961;68:68–79.
23. De Vries GJ, Sodersten P. Sex differences in the brain: the relation between structure and function. *Horm Behav.* 2009;55(5):589–596.
24. Forger NG, Wong V, Breedlove SM. Ciliary neurotrophic factor arrests muscle and motoneuron degeneration in androgen-insensitive rats. *J Neurobiol.* 1995;28(3):354–362.
25. Hsu HK, Shao PL, Tsai KL, Shih HC, Lee TY, Hsu C. Gene regulation by NMDA receptor activation in the SDN-POA neurons of male rats during sexual development. *J Mol Endocrinol.* 2005;34(2):433–445.
26. Waters EM, Simerly RB. Estrogen induces caspase-dependent cell death during hypothalamic development. *J Neurosci.* 2009; 29(31):9714–9718.
27. Krishnan S, Intlekofer KA, Aggison LK, Petersen SL. Central role of TRAF-interacting protein in a new model of brain sexual differentiation. *Proc Natl Acad Sci USA.* 2009;106(39):16692–16697.
28. Amateau SK, McCarthy MM. Induction of PGE(2) by estradiol mediates developmental masculinization of sex behavior. *Nat Neurosci.* 2004;7(6):643–650.
29. Schwarz JM, Liang S-L, Thompson SM, McCarthy MM. Estradiol induces hypothalamic dendritic spines by enhancing glutamate release: A mechanism for organizational sex differences. *Neuron.* 2008;58:584–598.
30. Matsumoto A. *Sexual Differentiation of the Brain.* Boca Raton, FL: CRC Press; 2000.
31. Chell S, Kadi A, Williams AC, Paraskeva C. Mediators of PGE2 synthesis and signalling downstream of COX-2 represent potential targets for the prevention/treatment of colorectal cancer. *Biochim Biophys Acta.* 2006;1766(1):104–119.
32. Rage F, Lee BJ, Ma YJ, Ojeda SR. Estradiol enhances prostaglandin E2 receptor gene expression in luteinizing hormone-releasing hormone (LHRH) neurons and facilitates the LHRH response to PGE2 by activating a glia-to-neuron signaling pathway. *J Neurosci.* 1997;17(23):9145–9156.
33. Amateau SK, McCarthy MM. A novel mechanism of dendritic spine plasticity involving estradiol induction of prostglandin-E2. *J Neurosci.* 2002;22:8586–8596.
34. Burks SR, Wright CL, McCarthy MM. Exploration of prostanoid receptor subtype regulating estradiol and prostaglandin E2 induction of spinophilin in developing preoptic area neurons. *Neuroscience.* 2007;146(3):1117–1127.
35. Wright CL, Burks SR, McCarthy MM. Identification of prostaglandin E2 receptors mediating perinatal masculinization of adult sex behavior and neuroanatomical correlates. *Dev Neurobiol.* 2008:68.
36. Wright CL, McCarthy MM. Prostaglandin E2-induced masculinization of brain and behavior requires protein kinase A, AMPA/kainate, and metabotropic glutamate receptor signaling. *J Neurosci.* 2009;29(42):13274–13282.
37. Matsumoto A, Arai Y. Sexual dimorphism in 'wiring pattern' in the hypothalamic arcuate nucleus and its modification by neonatal hormonal environment. *Brain Res.* 1980;19(1):238–242.
38. Mong JA, Glaser E, McCarthy MM. Gonadal steroids promote glial differentiation and alter neuronal morphology in the developing hypothalamus in a regionally specific manner. *J Neurosci.* 1999;19(4):1464–1472.
39. Mong JA, Nunez JL, McCarthy MM. GABA mediates steroid-induced astrocyte differentiation in the neonatal rat hypothalamus. *J Neuroendocrinol.* 2002;14:1–16.
40. Williams CL, Meck WH. The organizational effects of gonadal steroids on sexually dimorphic spatial ability. *Psychoneuroendocrinology.* 1991;16:155–176.
41. Isgor C, Sengelaub DR. Prenatal gonadal steroids affect adult spatial behavior, CA1 and CA3 pyramidal cell morphology in rats. *Horm Behav.* 1998;34:183–198.
42. Amateau SK, Alt JJ, Stamps CL, McCarthy MM. Brain estradiol content in newborn rats: sex differences, regional heterogeneity, and possible de novo synthesis by the female telencephalon. *Endocrinology.* 2004;145(6):2906–2917.
43. Krebs-Kraft DL, Hill MN, Hillard CJ, McCarthy MM. Sex difference in cell proliferation in developing rat amygdala mediated by endocannabinoids has implications for social behavior. *Proc Natl Acad Sci USA.* 2010;107(47):20535–20540.
44. Bowers J, Waddell J, McCarthy M. A developmental sex difference in hippocampal neurogenesis is mediated by endogenous estradiol. *Bio Sex Diff.* 2010;1(1):8.
45. Arnold AP, Xu J, Grisham W, Chen X, Kim YH, Itoh Y. Minireview: Sex chromosomes and brain sexual differentiation. *Endocrinology.* 2004;145(3):1057–1062.
46. De Vries GJ, Rissman EF, Simerly RB, et al. A model system for study of sex chromosome effects on sexually dimorphic neural and behavioral traits. *J Neurosci.* 2002;22(20):9005–9014.
47. Ahmed EI, Zehr JL, Schulz KM, Lorenz BH, Doncarlos LL, Sisk CL. Pubertal hormones modulate the addition of new cells to sexually dimorphic brain regions. *Nat Neurosci.* 2008;11:995–997.
48. McCarthy MM, Auger AP, Bale TL, et al. The epigenetics of sex differences in the brain. *J Neurosci.* 2009;29(41):12815–12823.

CHAPTER

18

Corticosteroid Actions on Neurotransmission

Marian Joëls [1], *E. Ronald de Kloet* [2], *Henk Karst* [1]

[1] Rudolf Magnus Institute, UMC Utrecht, Utrecht, The Netherlands, [2] LACDR, Leiden University, Leiden, The Netherlands

OUTLINE

Summary

Shortly after stress the brain is exposed to a plethora of stress hormones, of which corticosteroids (corticosterone in most rodents and cortisol in humans) are prominent examples. Corticosteroid hormones are active in those cells that express one of two hormone receptor types – i.e., the high-affinity mineralocorticoid receptor and the lower-affinity glucocorticoid receptor. Principal cells in most hippocampal subregions and in the basolateral amygdala display high receptor levels. These receptors for corticosteroids mediate not only slow and long-lasting actions involving changes in gene transcription, but also rapid non-genomic effects. In both ways, the hormones can change responses to neurotransmitters as well as intrinsic neuronal properties which indirectly affect neurotransmission. In hippocampal cells, stress hormones (including corticosteroids) generally evoke a wave of enhanced activity, followed after some hours by suppression of neuronal activity. However, corticosteroid actions on neurotransmission are liable to regional differentiation. Overall, corticosteroid effects on brain function are a composite of the regional hormonal actions and the way in which these regions are interconnected.

INTRODUCTION

Neural communication mediated by corticosteroid hormones is essentially different from signal transduction through classical neurotransmitters. While neurotransmitters are generally involved in the precise transfer of signals from one location – usually the presynaptic terminal – to the site where membrane-located receptors are enriched, and become active within milliseconds to minutes at most, corticosteroids act as true hormones. They are synthesized in a site far removed from the brain – i.e., the adrenal cortex – and released in a pulsatile fashion.[1] The pulse amplitudes follow a circadian variation, with low pulse amplitudes early in the inactive period and high amplitude pulses just before the onset of the active period of the organism. These daily fluctuations play an important role in the coordination of many body functions. On top of this circadian rhythm, corticosteroids can be secreted in substantial amounts in response to any situation that is perceived as a potential threat to homeostasis, that is, stress. Corticosteroid hormones (corticosterone in most rodents and primarily cortisol in humans) are then transported through the circulation, bound to corticosteroid-binding globulin. The steroids pass the blood–brain barrier quite well, though cortisol is more restricted in its passage than corticosterone.[2,3] Within the brain they reach, in principle, every cell – neurons and glial cells alike. In contrast to

Handbook of Neuroendocrinology, DOI: 10.1016/B978-0-12-375097-6.10018-6

this non-discriminatory transport of the hormone to all parts of the brain, corticosteroid receptors display a distinct distribution, restricting the potential of corticosteroids to affect brain function. Local enzymatic conversion,[4] passage over the cell membrane[3] and cell-specific intracellular signaling pathways further fine-tune this picture. All in all, many cells in different parts of the brain are exposed to a wave of steroids shortly after stress. Adrenal secretion of these hormones can therefore exert a synchronized and coordinated change in the activity of all of the cells carrying receptors.

Unlike neurotransmitters, corticosteroids usually affect neural function in a very slow but persistent manner, since their effects involve gene transcription: corticosteroid receptors bind to specific recognition sites in the DNA, and thus alter transcriptional activity of responsive genes.[5] Two main receptor types – each with several splice variants and various isoforms involving different translation initiation sites – have been described. The first is the mineralocorticoid receptor (MR), which displays a high affinity for the endogenous ligands cortisol, corticosterone, and the less prevalent mineralocorticoid aldosterone[6]; this receptor has a restricted distribution in the brain, with high expression levels in all hippocampal subfields, the lateral septum, some amygdalar nuclei and motor nuclei in the brainstem. The second is the glucocorticoid receptor (GR), with an approximately 10-fold lower affinity for corticosterone and cortisol. This receptor is quite ubiquitous in its distribution, although particularly high expression levels are seen in the hippocampal CA1 area and dentate gyrus, and in the main feedback site, the paraventricular nucleus (PVN) of the hypothalamus.

Overall, neurotransmitters (including those involved in the stress response) quickly change brain function – changes that are subsequently reversed and normalized by the slower-acting corticosteroid hormones. It should be realized, though, that this is a simplification of the reality[7] (Fig. 18.1). For instance, it is now recognized that corticosteroids also affect brain function in a rapid, non-genomic fashion (see below), which shares elements of signaling pathways observed in classical neurotransmission. Conversely, it is widely accepted that neurotransmitters do not exclusively affect signal transduction over the course of milliseconds to minutes; their actions can evolve into more lasting responses, via secondary gene-mediated mechanisms. Although neurotransmitters and corticosteroid hormones have their preferential time-domains of action, there is a certain degree of overlap.

In this chapter we will explain the current insights in corticosteroid actions on neural activity, particularly on neurotransmission. We start by briefly summarizing the approaches that have been used to investigate these effects of corticosteroids in the brain. Next, we will highlight the rapid non-genomic pathway through which corticosteroids affect brain cells shortly after stress. The chapter proceeds by providing an overview of the slow gene-mediated actions exerted by corticosteroids;

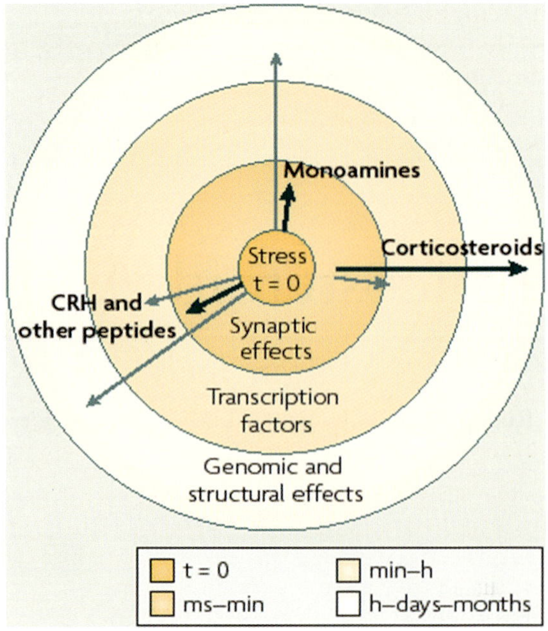

FIGURE 18.1 **The stress response is classically characterized by two major temporal domains.** The first includes rapid actions of noradrenaline, serotonin, dopamine, CRH (predominantly through $CRHR_1$), corticosteroids (via membrane-located MRs) and neurosteroids. These rapid actions of stress mediators promote vigilance, alertness, appraisal of the situation and choice of an optimal strategy to face the challenge. Because local increases of stress-mediator levels are short-lived, and their actions typically quickly subside, this first wave of events is not optimal for provoking the sustained, adaptive components of a stress response, such as the consolidation of information associated with (and thus memory of) the stressor. This latter function is accomplished through alterations of gene expression and cell function which have classically been attributed to corticosteroids, acting via GR. Recent work has established that, in addition to rapid synaptic effects, receptor activation by monoamines and peptides regulate transcription factors (such as CREB and AP1) within the second-to-minute timeframe. This rapid activation (or deactivation) of transcription factors may lead to genomic changes which can be sustained, helping the organism to respond to stresses of different magnitudes or durations, as well as to recurrent stress. Conversely, corticosterone via non-genomic pathways involving membrane-located MRs can act within minutes. Taken together, these findings blur the classical distinction between rapid and delayed responses to stress. In view of the above, the temporal profiles of action of stress mediators might best be considered as a series of vectors commencing at the onset of stress. Each hormone has a preferred temporal vector that is associated with a specific mode of cellular function (e.g., synaptic, transcriptional). In addition, most modulators exert stress-related actions via alternative molecular and cellular pathways, which take place in different temporal niches. Together, these contribute to the compound temporal profile of a given stress mediator, as illustrated by the lengths and widths of the vectors in the figure. *Reproduced from Joëls M, Baram TZ. The neuro-symphony of stress.* Nat Rev Neurosci. *2009;10:459–466, with permission.*

in particular, how they affect the actions of neurotransmitters. In the final section, we will briefly discuss how corticosteroid effects on neurotransmission are regulated by life events such as chronic unpredictable stress, or stressful experiences early in life.

MEASURING CORTICOSTEROID ACTIONS ON NEUROTRANSMISSION

To determine the effect of corticosteroid hormones on neurotransmission, many approaches are available. Neurochemical methods could be used to determine neurotransmitter synthesis, release and reuptake, both *in vitro* and *in vivo*. Effects of corticosteroids on neurotransmitter release are usually too fast to be established through a gene-mediated pathway (see the following section).

In addition to these changes in neurotransmitter availability, more delayed effects of stress on neurotransmitter receptor expression have been examined, using *in situ* hybridization or quantitative PCR, but also examination at the protein level with (semi-quantitative) immunohistochemistry, Western blotting or PET scans in humans. These studies will not be discussed here. Ultimately, the importance of corticosteroid modulation of neurotransmitter receptors is likely to be revealed by altered functionality of the transmitters, unless the number of receptors is the rate-limiting factor.

Functionality can be best established with electrophysiological techniques. Ideally, one would like to determine changes in the firing pattern while individuals are exposed to "natural" stress conditions. While this is possible (for example, by applying multi-electrode technology), data from using this technique are still scarce. An example of this approach is a recent study showing that the stability of firing rates of place cells recorded from rat dorsal hippocampal CA1 neurons is impaired in a 15-minute period following free foraging on a stressful novel open-field platform located in a familiar surrounding room.[8] Interestingly, this is compatible with a very early study using multi-unit recording techniques, showing a reduction in the hippocampal firing rate approximately 20 minutes after a peripheral injection with corticosteroids.[9] While the unique distinction of single neurons is not possible with the latter approach as opposed to tetrode recording technology, both studies point to impaired hippocampal information transfer after stress. The time of recording relative to elevation in corticosteroid levels was in both cases somewhat ambiguous, allowing development of both non-genomic and gene-mediated processes. Yet these *in vivo* recording techniques are extremely important, and the possibility of recording from ensembles of neurons during stressful conditions is, in particular, very promising, because this technology allows detection of temporal patterns of activity changes in brain circuits, particularly when tetrodes are placed in several interconnected brain regions. These experiments are, however, technically very challenging. Therefore, researchers have used easier but less precise approaches, recording responses of fields of neurons, usually in response to stimulation of afferent fibers. This is most informative in areas with a laminar distribution of cells, such as the hippocampus, leading to simple compound electrical signals.

Functional read-outs of all of these *in vivo* experiments are the firing rate of single neurons, or the amplitude and slope of field potentials. In many cases these parameters are linked to the functionality of glutamatergic synapses, but the exact mechanism of action is hard to determine from these indirect measures. For one thing, these techniques do not allow identification of the neurons from which the recording is made. Basal firing rates can give a rough indication of their nature (e.g., interneurons versus principal cells), and *post hoc* histological verification could specify the location (CA1 versus dentate gyrus). However, the latter is not easy in areas without a laminar organization and multiple types of neurons, such as the basolateral amygdala. Moreover, an increase in firing rate may result from enhanced glutamate release, but also from, for example, a slightly more depolarized resting membrane potential of the neurons, bringing them closer to the firing threshold. Other explanations are equally possible, such as a reduction in GABAergic inhibitory input or a change in modulatory inputs via monoamines. Although pharmacological manipulation and particular stimulation protocols *in vivo* allow some distinction between neurotransmitter influences and pre- versus postsynaptic mechanisms respectively, at present much more insight can be obtained when cells are studied *in vitro*, in brain slices.

RAPID EFFECTS OF CORTICOSTEROIDS ON NEUROTRANSMISSION

Rapid effects of corticosteroids have been described occasionally over the past decades. For instance, suppression of field potentials and long-term potentiation (LTP) by corticosterone in hippocampal slices was observed within 10–15 minutes[10,11] – a timeframe that is a bit slow for non-genomic signaling but rather rapid for gene-mediated events. Another example concerns a study showing that iontophoretically administered glucocorticoids quickly inhibit cell firing of the majority of cells tested in the PVN *in vivo*.[12] These observations remained anecdotal for many years, and did not provide much insight in the underlying mechanism.

In 2003, however, Di *et al.*[13] described that in parvocellular PVN neurons corticosterone quickly reduces the frequency of miniature excitatory postsynaptic currents (mEPSCs), each of which represents the spontaneous release of a glutamate-containing vesicle under conditions where sodium channels are blocked (i.e., in the absence of any presynaptic activation), by a process requiring G proteins. The effects were mimicked by a membrane-impermeable conjugate of corticosterone with bovine serum albumin. In view of the effectiveness of dexamethasone it was proposed that these rapid effects involve a membrane-located GR, although the GR-antagonist RU 38486 was ineffective. Interestingly, blockade of endocannabinoid signaling prevented the corticosteroid effects. Taken together, these data suggest that corticosterone, by binding to postsynaptically located G-protein coupled GRs inserted into the membrane, causes enhanced release of retrogradely transported endocannabinoids, which in turn suppress the probability that glutamate-containing vesicles fuse with the presynaptic membrane. This may contribute to the phenomenon of rapid feedback in the hypothalamo-pituitary–adrenal axis, which had been described more than a decade earlier.[14]

In agreement with a retrograde-messenger dependent process, the same group later reported that dexamethasone treatment of hypothalamic slices gives rise to a glucocorticoid-induced rapid increase in the levels of the endocannabinoids anandamide and 2-arachidonoylglycerol.[15] The reduced release probability of glutamate was found not only in parvocellular neurons but also in magnocellular neurons in the hypothalamus. Concomitantly, corticosterone rapidly enhanced the frequency of GABAergic miniature inhibitory postsynaptic currents (mIPSCs) in these cells, though via a different pathway, involving activation of neuronal nitric oxide synthase.[16] The frequency of mIPSCs in parvocellular neurons was not rapidly affected by corticosterone.[17]

Experiments in the hippocampus reported a few years later indicated that rapid non-genomic actions of corticosterone may be subject to regional differentiation. Thus, application of corticosterone (and the conjugate with bovine serum albumin) quickly and reversibly *enhanced* the frequency of mEPSCs and the release probability of glutamate-containing vesicles in CA1 pyramidal neurons[18] (Fig. 18.2). The underlying signaling pathway of the rapid corticosteroid actions in the hippocampus differs from that described for PVN neurons. Although the effective dose appeared to be within the affinity range of GRs, pharmacological and genetic approaches favored the involvement of membrane-located MRs in rapid corticosteroid effects on hippocampal neurons. Blockers of retrograde signals, such as endocannabinoids or nitric oxide, were ineffective.[19] It was concluded that the rapid corticosteroid effects in the hippocampus involve MRs localized on the presynaptic membrane, where they activate the ERK1/2 signaling pathway, causing more vesicles containing glutamate to be spontaneously released.

Effects of acute corticosterone administration on GABAergic transmission appear to be more complex. Thus, corticosterone does not rapidly affect the frequency of mIPSCs in CA1 neurons of the dorsal hippocampus,[19] but the synthetic glucocorticoid dexamethasone did increase the frequency of spontaneous IPSCs via a nitric-oxide dependent retrograde pathway implicating interneurons in one study,[20] though not another.[21] In the ventral-most part of the hippocampus, CA1 neurons show a decrease in spontaneous IPSC frequency, via an

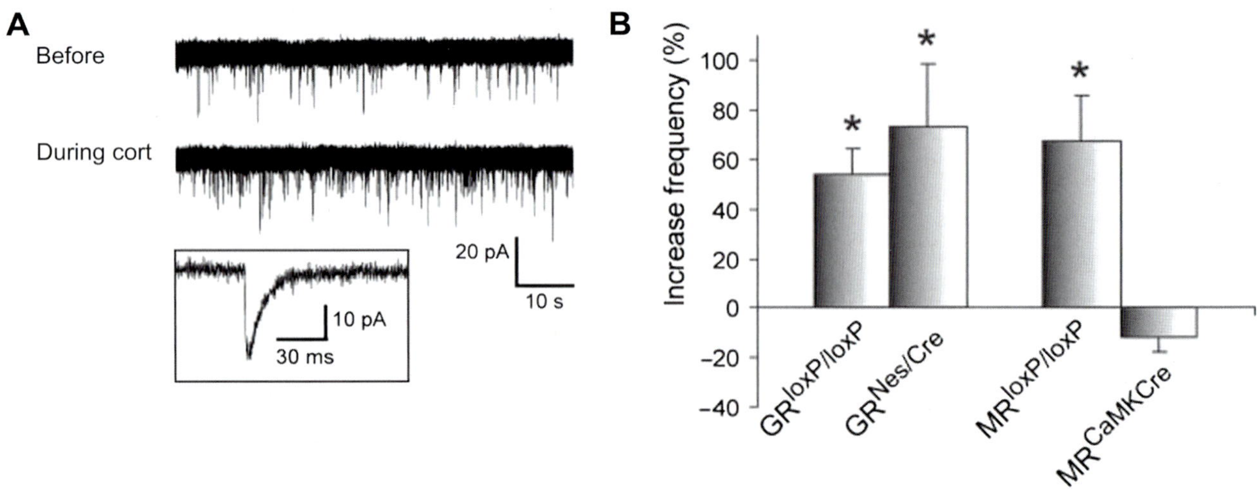

FIGURE 18.2 (A) Corticosterone increases the frequency of the miniature excitatory postsynaptic currents in CA1 hippocampal neurons (see inset for example). (B) The corticosterone-induced increase in frequency was observed in forebrain-specific GR knockout animals as well as (floxed) controls, but not in MR knockouts. *Based on Karst* et al. *(2005).*[18]

MR-dependent pathway.[21] Spontaneous IPSCs reflect the release of GABA-containing vesicles in response to spontaneous (action-potential induced) activation of presynaptic terminals. Since the latter situation is closer to *in vivo* conditions, this would predict that stress quickly changes extrasynaptic levels of, in particular, GABA. However, the results from microdialysis studies turn out to be equivocal. For instance, injecting rats intraperitoneally with corticosterone enhanced glutamate but not GABA release in the CA1 area via a quick and reversible non-genomic pathway.[22] Ether stress also resulted in a peak of glutamate in the hippocampus 30 minutes after stress, which was paralleled by a peak in the stress-induced rise of plasma corticosterone levels.[23] GABA release in the hippocampus can also be enhanced after stress, but only when the stressor is psychological in nature – for example, exposure to novelty or swim stress at 35°C[24]; a more physical stressor, such as swim stress at 25°C, decreased GABA release in the hippocampus. Most likely, the limited spatial resolution and access to synaptic processes of this technique and the interpretation of local phenomena in a system that is liable to all kinds of inputs from different parts of the brain contributes to these equivocal results. Nevertheless, this approach provides a behavioral relevance that detailed investigations in reduced *in vitro* preparations lack.

In addition to these rapid effects of corticosterone on release probability of neurotransmitters, the hormones also exert rapid effects postsynaptically in the hippocampus. Electrophysiological studies provided evidence that MRs are also located in the postsynaptic membrane. These MRs are coupled to G proteins, and (probably via mediation of ERK1/2) inhibit the function of a transient, voltage-dependent K-conductance, the I_A.[19] Corticosterone also quickly and non-genomically promotes lateral diffusion of the glutamate receptor subunit 2 (GluR2) to the postsynaptic density of cultured hippocampal cells, via MRs.[25] All of these MR-dependent processes – enhanced release probability of glutamate, reduction of I_A and enrichment of postsynaptic densities with GluR2 – will stimulate excitatory transmission in hippocampal cells shortly after corticosteroid levels rise, regardless of possible effects on GABAergic transmission. Thus, in the CA1 hippocampal area as opposed to the PVN, corticosterone seems to stimulate excitatory transmission within only a few minutes, in a transient manner. This fits with the observation that CA1 LTP is facilitated if corticosterone is present at the moment of induction.[26] Similarly, in the dentate gyrus corticosterone promotes the maintenance of LTP via an MR-dependent mechanism.[27]

Very recent studies in another limbic area, the basolateral amygdala (BLA), may explain these paradoxical findings. In slices prepared from animals under rest (with low circulating corticosteroid levels), *in vitro* application of corticosterone quickly enhanced mEPSC frequency, as in the hippocampal CA1 area.[28] This non-genomic process involved activation of MRs. However, contrary to the CA1, this rise in mEPSC frequency was not quickly reversible but lasted for hours after corticosterone application or stress exposure of the animals prior to preparation of the slices. This raised the question whether renewed exposure to high levels of corticosterone would evoke a comparable

BOX 18.1

HOW DO WE KNOW THAT CORTICOSTEROID HORMONES RAPIDLY CHANGE NEUROTRANSMITTER RELEASE?

Biochemical studies using microdialysis revealed that shortly after stress, circulating levels of many neurotransmitters (glutamate, GABA, monoamines) temporarily change. These levels are probably a reflection of extrasynaptic circulation, and may point to altered release, but also to differences in reuptake. The kinetic properties of these changes favor a non-genomic mechanism of action of stress hormones, because the relatively rapid onset (< 15 minutes) is hard to reconcile with transcriptional regulation. In many cases, these neurochemical changes were investigated in conjunction with stress exposure of the animal – e.g., the swim stress. Particularly in view of the rapid onset, it is necessary to perform follow-up experiments with corticosteroid receptor (ant)agonists, proving that corticosteroids rather than other stress hormones (e.g., monoamines or peptides like CRH) are responsible for the changes in circulating neurotransmitters.

Another approach is provided by patch clamp electrophysiological recording. With this method, the response of the postsynaptic cell to the spontaneous release of one vesicle containing a particular neurotransmitter can be recorded. If the frequency (but not amplitude) of these responses is changed, this can be an indication for an altered release probability of the presynaptic terminal. With this method, it was found that corticosteroids rapidly change glutamate release in the PVN, hippocampus and amygdala.

enhancement of the EPSC frequency or would induce a condition where BLA cells become resistant to rapid effects of corticosterone. Surprisingly, neither was the case. Instead, BLA neurons responded to renewed corticosterone application with a quick *reduction* in mEPSC frequency, much like the effect earlier described in the PVN. Moreover, the rapid non-genomic reduction in mEPSC frequency upon a second pulse of corticosterone involved GRs and was blocked by an endocannabinoid receptor-1 antagonist. This demonstrates that the rapid response of neurons to a pulse of corticosterone may strongly depend on the recent stress history of the organism – a phenomenon that we called the "metaplasticity of the corticosteroid response." It is not a general phenomenon, since CA1 pyramidal neurons respond similarly to two pulses of corticosterone, at least *in vitro*; *in vivo*, their response may be influenced by amygdalar inputs and hence also depend on the stress history of the animal. It is very possible that the principle of metaplasticity is also applicable to PVN neurons.

In conclusion, studies over the past decade have substantiated earlier reports that corticosteroids affect neurotransmission not only in a slow gene-dependent manner, but also quickly, via membrane-located receptors and G-protein coupled signaling pathways. These receptors are probably not different from the classical nuclear receptors. In agreement, preliminary reports show that both MR and GR can be discerned in the synaptic membrane.[29,30] Why part of the intracellular corticosteroid receptor pool translocates to the plasma membrane instead of the nucleus is at this time unresolved. It is, however, not unprecedented, since comparable observations have been made for estrogen and androgen receptors.[31–33]

SLOW GENE-MEDIATED EFFECTS OF CORTICOSTEROIDS ON NEUROTRANSMISSION

The notion that corticosteroid hormones affect neurotransmission in a slow gene-mediated manner was already firmly established two decades ago. Biochemical studies showed that glucocorticoids markedly alter the turnover of monoamines, most notably of serotonin. The 5-HT turnover in hippocampus is enhanced after acute exposure to corticosterone by increasing tryptophan availability and stimulation of tryptophan hydroxylase activity,[34,35] but reduced after sustained corticosterone activation.[36] The interaction of 5-HT and stress-induced HPA axis activity occurs at various levels. Acute stress enhances 5-HT turnover in hippocampus and 5-HT terminals innervating the CRH neurons are excitatory, activating the HPA axis.[37] 5-HT released in the raphe nuclei attenuates stress-induced activation by an auto-inhibitory feedback loop.[38]

Likewise, acute increases in corticosterone levels during stress enhance tyrosine hydroxylase and hence dopamine turnover/release, particularly in the mesolimbic system, but dopamine turnover is reduced after prolonged stress.[39] These effects of stress on the dopamine system are regionally different, because the nigrostriatal area is not affected while dopamine turnover in the hypothalamus is suppressed after acute stress, leading to enhanced prolactin release. In subsequent studies using either GR antagonists or site-specific GR knockout mice, Piazza, Tronche and co-workers demonstrated that glucocorticoids increase dopamine release in the nucleus accumbens mesolimbic terminal fields, enhancing locomotion and the propensity to self-administer cocaine.[40,41] Sustained (uncontrollable) stress ultimately attenuates dopaminergic activity. Finally, sustained stress and hypercortisolemia induced changes in the noradrenergic system that are comparable with those observed during depression. In particular, the reverberating positive feedback loop between the locus coeruleus and amygdala has received much attention – a feedback loop that is activated in synergy by CRH and glucocorticoids.[42]

Later it was shown that expression of neurotransmitter receptors is also largely influenced by exposure to corticosteroids and/or stress. The 5-HT1A receptors in hippocampus are downregulated by acute and chronically elevated levels of corticosterone and stress, particularly in the dentate gyrus.[43,44] The 5-HT2 receptors are desensitized by acute stress, but under conditions of chronic stress 5-HT2 receptor binding is increased[45] – particularly of the 5HT2A and 5HT2C receptors[46] – potentially contributing to heightened anxiety. Activation of 5-HT1A receptors in the median raphe facilitates the disconnection between aversive stimuli and anxiety reactions. This would allow normal behavior to reemerge despite continuation of the aversive conditions, a mechanism promoting resilience through dissociation.[47] Downregulation of this system would also lead to enhanced anxiety and impaired adaptation to stressors. Again, the interaction is bidirectional. Enhancing the serotonergic input to hippocampal cells using 5-HT reuptake blockers enhances the synthesis of MR and GR, which are known to mediate glucocorticoid facilitation of stress adaptation.[48,49] However, activation of the 5HT1A receptors by flesinoxan led to downregulation of hippocampal GR *in vivo*.[50]

In conclusion, these neurochemical studies on stress, glucocorticoids and mono-amines seem to suggest that chronic stress reduces 5HT metabolism, downregulates 5HT1A receptors and upregulates 5HT2 receptors. The effects on the noradrenergic system are variable, while reduction of mesolimbic dopaminergic activity is observed under conditions reminiscent of some forms of depression characterized by elevated glucocorticoids.

As with the rapid non-genomic effects, the influence of corticosteroid hormones on neurotransmission also needs to be tested at the functional level. Indirect evidence for corticosteroid-induced changes in the functionality of neurotransmitter systems was derived from behavioral studies, using peripheral or (preferably) local application of neurotransmitter analogs to determine which transmitter systems are essential in the accomplishment of behavioral responses to stress. The research line of James McGaugh and colleagues serves as an example. With a combination of lesions, local application of transmitter receptor (ant)agonists and relatively straightforward behavioral paradigms, this research indicated how corticosteroids might interact with neurotransmitters, resulting in behavioral adaptation.[51] McGaugh and colleagues proposed that monoamines, particularly noradrenaline, facilitate memory of an aversive situation, via β-adrenoceptors in the BLA and hippocampus. Corticosteroids play a permissive role in this process. Another example concerns the work by Steve Maier and colleagues,[52] who demonstrated that the medial prefrontal cortex exerts an inhibitory control over the raphe nuclei when a stressful situation is experienced as controllable – a system that fails when the stress is perceived as uncontrollable. This suggests that the uncontrollable aspect of the situation indirectly affects serotonergic transmission in the brain. These studies are extremely valuable, since they prove that stress hormones do (in)directly alter the functionality of neurotransmitter systems. To elucidate the underlying mechanism, however, approaches allowing investigation at the cellular level (e.g., by electrophysiology) are indispensable. We will provide here an overview of these detailed studies, discussing several neurotransmitter systems, with an emphasis on observations made in the hippocampus.

Intrinsic Properties

Although actions of corticosteroids on intrinsic membrane properties, such as input resistance, ion transporters and also specific ion channels, are not directly aimed at specific neurotransmitter systems, they nevertheless have a significant impact on the effect of transmitters. These intrinsic properties may alter the excitability of neurons, so that subsequent activation of a neurotransmitter receptor finds the cell less or more responsive.

One of the most conspicuous effects of corticosterone in hippocampal CA1 cells is a change in the amplitude of voltage-dependent Ca currents[53–55] (Fig. 18.3). In the absence of corticosterone, the Ca-current amplitude is large. Activation of MRs causes a slow reduction in the current amplitude. When GRs are also activated, Ca-current amplitude is again slowly increased, so that overall a U-shaped dose-dependency exists for the effect of corticosterone on this parameter[56] (Fig. 18.4). The slow enhancement seen after high doses of corticosterone depends on protein synthesis,[53] and involves DNA binding of GR homodimers.[55] *In vitro* it develops over the course of 1–2 hours, while subjecting an animal to stress changes the Ca-current amplitude subsequently measured *in vitro* with a delay of approximately 3 hours.[57] Sustained L-type Ca currents are particularly affected by corticosterone – much more so than, for example, N-type currents.[58] Exactly how

BOX 18.2

HOW DO WE KNOW THAT CORTICOSTEROID HORMONES SLOWLY CHANGE THE RESPONSE OF IDENTIFIED NEURONS TO A PARTICULAR NEUROTRANSMITTER?

In brain slices, neurons can be easily identified using optical methods (e.g., infra-red optics). Also, the cells can be clamped at every desired potential, allowing investigation of voltage-dependent currents. Finally, pharmacological stimulation or blockade of corticosteroid as well as specific neurotransmitter receptors can reveal much of the mechanism of action by which corticosteroid hormones might affect specific transmitter systems. An advantage of studying the stress system is that corticosteroid effects induced *in vivo* (e.g., by stress) probably persist after preparation of brain slices, due to the long-lasting genomic effects of the hormones. Most of the studies mentioned in this chapter used these *in vitro* recording techniques. These studies illustrate nicely how corticosteroid hormones could affect neurotransmitter responses in the brain, but it should always be remembered that what happens in artificial conditions in reduced preparations may in fact not happen *in vivo*. Obviously, a complete understanding of corticosteroid–neurotransmitter interactions will only be obtained with a combination of *in vitro* and *in vivo* recording techniques.

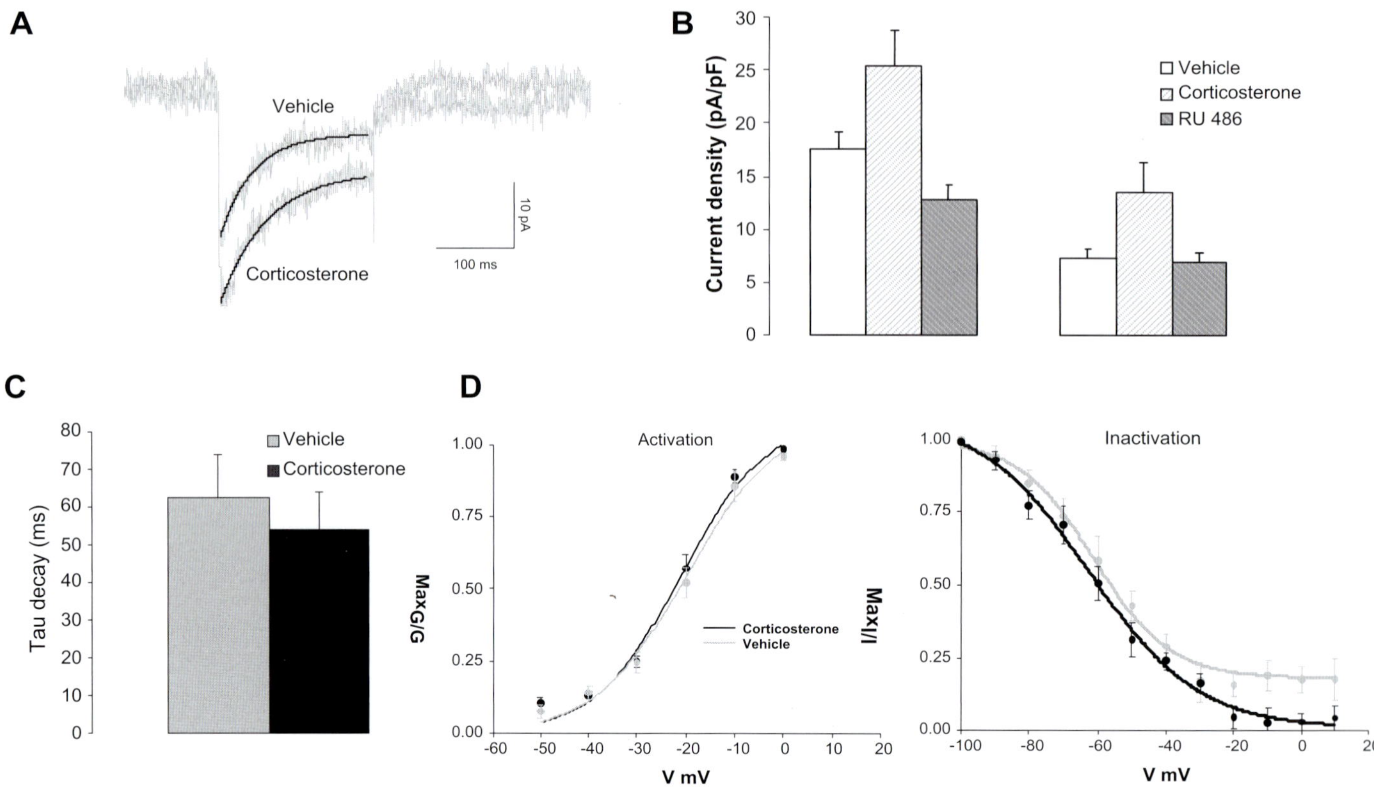

FIGURE 18.3 **Effect of corticosteroid on voltage-dependent calcium currents.** (**A**) Examples of calcium currents evoked by 200-ms depolarization to 0 mV from a holding potential of −85 mV in control conditions (vehicle) and after corticosterone treatment. Solid line depicts the fit with a single exponential. (**B**) Average currents densities evoked by voltage pulse to 0 mV in control conditions, after corticosterone treatment and when corticosterone was applied in the presence of the GR antagonist (RU 486). The means + SEM are given for the peak and the sustained currents. (**C**) Time constant of inactivation in control conditions and after corticosterone treatment was comparable (mean + SEM). The decay of the calcium current was fitted with a single exponential. (**D**) Voltage-dependent activation and inactivation of total calcium current in control conditions and after corticosterone treatment. The current was evoked by a 200-ms depolarizing pulse from a holding potential of −85 mV to different voltage potentials (from −50 mV to 0 mV). Conductance was calculated according to the equation $G(V) = I/(V_m - V_{rev})$, where I is the current, V_m the test pulse potential, V_{rev} the measured reversal potential. The averaged data were fitted with the Boltzmann equation: $G = G_{max}/(1 + \exp(V - V_h)/k)$. *Based on Chameau* et al. *(2007).*[58]

this effect is accomplished is only partly understood. Of all the Ca-channel subunits tested, none of the pore-forming α subunits of the L-type channels was consistently transcriptionally modulated by hormonal treatment. The steroid rather seems to target one of the supportive β subunits, the $Ca_v\beta4$,[58] which, among other things, is involved in surface expression of functional channels. With several techniques, and in rats as well as mice, it was observed that corticosteroids increase the expression level of this subunit. A comparable enhancement in sustained Ca-current amplitude was also reported for CA3 neurons[59] (note that GR involvement in this case is not evident, since expression levels of GRs are very low in the CA3 region) and pyramidal-like principal cells in the BLA.[60] Surprisingly, dentate granule cells do not respond to corticosterone application with a delayed enhancement in sustained Ca-current amplitude, despite the fact that these cells highly express GRs.[61] Up until the transcriptional level, responses to corticosterone are identical for dentate and CA1 neurons, yet at the protein level they start to diverge. The presence of receptors in a particular cell is therefore no guarantee that corticosterone really exerts an effect; this also depends on other intracellular messengers, such as proteins involved in the stability of gene products or factors involved in translational efficiency (e.g., micro-RNAs).

The enhanced influx of calcium is probably coupled to a decreased extrusion. A high dose of glucocorticoids was found to significantly reduce expression of the plasma membrane calcium pump isoform 1 (PMCA1), a plasma membrane calcium ATPase.[62] The functional relevance of this finding was underlined by the fact that glucocorticoid treatment of cultured hippocampal cells prolonged an agonist-evoked elevation of the intracellular calcium concentration, which in turn was

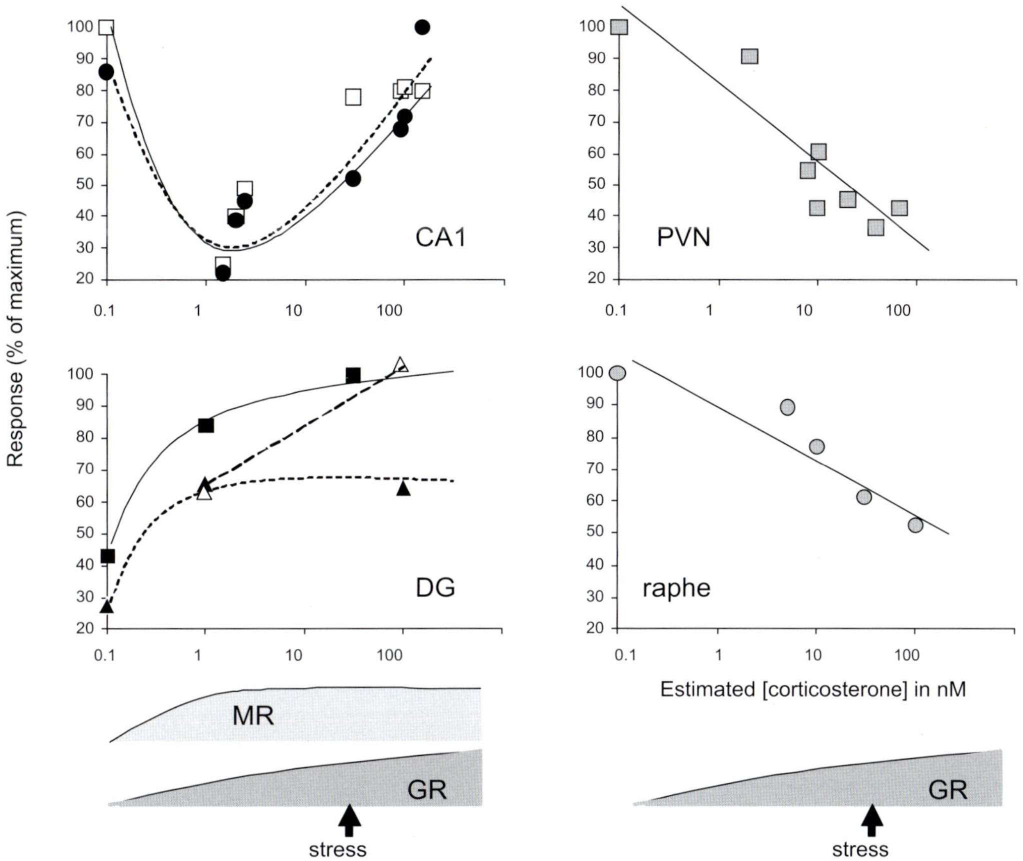

FIGURE 18.4 Dose–response relationships for cellular effects of corticosterone in the CA1 hippocampal area, the dentate gyrus (DG), the paraventricular nucleus of the hypothalamus (PVN) and the dorsal raphe nucleus. The diagrams show hormone responses expressed as a percentage of the maximal response, in various brain regions. The concentration of corticosterone is a rough estimate of the local concentration, based on the solutions perfused on *in vitro* preparations or derived from the plasma concentration when fluctuations in hormone levels were accomplished *in vivo*. In the CA1 area, both the amplitude of depolarization-induced calcium currents (open squares) and the hyperpolarization caused by serotonin-1A receptor activation (filled circles) display a clear U-shaped dose-dependency. The descending limb is linked to activation of MRs (see below), while the ascending limb is associated with gradual GR activation on top of already activated MRs, as occurs after stress. DG granule neurons show a clear effect on the field potential (filled squares) and single cell response (filled triangle) caused by activation of glutamatergic AMPA receptors; this effect is linked to MR activation. Although these cells also abundantly express GRs, high doses of corticosterone do not give additional changes in the signal, except when tested in chronically stressed rats (open triangles). Neurons in the PVN and raphe nucleus primarily express GRs. In these cells a clear linear dose-dependency can be seen for the frequency of spontaneous $GABA_A$-receptor mediated synaptic events (grey squares) and the inhibition caused by serotonin-1A receptor activation (grey circles) respectively. *Reproduced from Joëls M. Corticosteroid effects in the brain: U-shape it.* Trends Pharmacol Sci. *2006;27: 244–250, with permission.*

prevented by the expression of exogenous PMCA1.[63] This implies that some hours after exposure of hippocampal pyramidal cells to corticosterone, cells are exposed to much higher intracellular concentrations of calcium, particularly during depolarizations that result in activation of L-type channels.

Enhanced intracellular calcium concentrations may have multiple downstream consequences (see also IV.2). For instance, activation of channels that are calcium sensitive will be promoted. This is in line with early observations that some hours after activation of GRs, the amplitude of the slow Ca-dependent K-conductance is markedly enhanced.[64,65] This current causes hyperpolarization of cells when they are exposed to a steady excitatory input, so that firing frequency is attenuated; due to the slow deactivation of the channels, a lingering (after)hyperpolarization is observed when the steady excitatory input stops. Both phenomena potentially suppress the transfer of excitatory input through the CA1 area. This fits with the supposed normalizing effect of glucocorticoids after stress: with a delay of several hours (due to the slow gene-mediated signaling pathway) they reverse the excitatory information earlier facilitated by rapidly acting stress hormones, including corticosterone itself.

Although other ion conductances have not been investigated to the same degree as calcium currents, it is clear that they are less influenced by corticosterone and/or stress than L-type Ca currents. Subtle effects on specific Na- and K-conductances have been reported, but the functional relevance of these findings has mostly not been investigated (for review, see Joëls *et al.*[66]).

Amino Acids

For many years it was unclear if corticosteroids affected amino acid-mediated neurotransmission in a gene-dependent manner. Some studies reported changes after corticosterone administration in the amplitude of field potentials caused by glutamatergic transmission,[10,67] or in glutamatergic and GABAergic synaptic potentials.[68] However, the doses required to induce these effects were very high, or the effects seemed to develop only with (low-frequency) repeated stimulation of the afferent fibers, pointing to a metabolic problem rather than a specific effect on amino acid transmission.

Recently, though, some studies have appeared that show clear GR-dependent slow effects of corticosterone on amino acid transmission. One series was carried out in cultured hippocampal neurons.[25,69] These experiments consistently revealed enhanced surface expression of GluR2 subunits >3 hours after administration of corticosterone; changes in surface expression of GluR1 subunits were only observed with very high corticosterone concentrations. The effects on GluR2 were blocked by a protein synthesis inhibitor and by the GR-antagonist RU 38486. Under the same experimental conditions, the amplitude (but not frequency) of mEPSCs in cultured hippocampal cells was enhanced, although to date it has not been proven that these two observations are causally linked. A similarly enhanced mEPSC amplitude was also found in CA1 pyramidal neurons recorded in slices from adult mice, 1–4 hours after a brief application of the GR agonist RU 28362.[70]

AMPA receptor subunit trafficking is very important in the development of LTP.[71] Shortly after high-frequency stimulation of hippocampal CA1 afferents, the membrane surface is enriched with GluR1/GluR2 heterodimers. After approximately 20 minutes, these dimers are internalized and replaced by GluR2/GluR3 heterodimers. The delayed increased surface expression of GluR2 after GR activation may thus promote the strengthening of synaptic contacts involved in information processing of stressful events. This is corroborated by recent *in vivo* experiments showing that improved spatial learning under stressful conditions is associated with increased levels of GluR2 and GluR3 in hippocampal synaptoneurosomes.[72]

If GR signaling and LTP do share part of their signaling pathways and compete for the same proteins, it is expected that activation of the one pathway hampers the effects of the other. In agreement, chemical LTP in hippocampal cultures by application of picrotoxin and glycine increased surface expression of GluR2 subunits.[25] This was prevented by pretreatment with corticosterone. This is only one explanation for the well-documented finding that several hours after stress or corticosterone application *in vitro*, it is extremely hard to induce LTP in the hippocampal CA1 area.[73] A second explanation is provided by the fact that GR activation causes an increase in the firing frequency accommodation and the amplitude of the afterhyperpolarization. This has been described to impair the induction of LTP.[74] Finally, corticosterone causes a delayed increase in the intracellular Ca concentration, which may suppress the functionality of NMDA receptors. Both of these latter effects may not occur in the dentate gyrus, where corticosterone does not affect Ca influx. This could explain why effects of stress and/or corticosterone on LTP in the dentate are more ambiguous than those in CA1, and depend on the type and timing of the stressor (see, for example, Straube *et al.*[75]). Moreover, LTP in the dentate is strongly influenced by amygdalar input.[76] This may explain why corticosteroid effects in the BLA indirectly affect the possibility to induce LTP in the dentate.[77,78]

Collectively, the data indicate that some hours after stress, synapses in the CA1 hippocampal area (and possibly in the dentate) involved in the processing of stressful information are strengthened, while strengthening of other synapses at that time is strongly impaired. This would enhance the signal-to-noise ratio of stress-related versus background information. Interestingly, the ventral-most (20%) part of the hippocampus responds differently to corticosterone; there, LTP induced in the CA1 area is *facilitated* after stress, via an MR-dependent pathway.[79]

Reports about slow corticosteroid effects on GABAergic transmission in the hippocampus are rather sparse. Recently, it was shown that corticosterone slowly enhances the amplitude of spontaneous and miniature inhibitory postsynaptic currents in hippocampal CA1 neurons, via a GR-dependent mechanism.[21] In the dentate, acute stress was reported to induce an upregulation of $GABA_A$ receptor δ-subunit expression within 30 minutes, and to increase tonic inhibition.[80] These effects probably involve neurosteroids like 3α,5α-tetrahydrodeoxycorticosterone rather than corticosterone itself.

In the BLA, slow corticosteroid effects on GABAergic transmission are more documented. $GABA_A$-receptor mediated inhibitory postsynaptic potentials evoked by stimulation of the external capsule in the BLA were significantly reduced by CORT application.[81] This was not explained by a reduction in GABA release, but rather by a positive shift of the reversal potential for $GABA_A$-receptor mediated potentials. This shift may bring BLA neurons closer to the firing threshold, enhancing their excitability. This is in line with another observation in the BLA – that 1–4 hours after application of corticosterone, the firing frequency accommodation and the amplitude of the afterhyperpolarization are not increased[82] but, if anything, attenuated,[81] despite a larger Ca influx.[60] These findings illustrate that the same hormone (corticosterone) via the same receptor (GR) can exert very different effects in the hippocampus versus BLA, and probably other areas as well. The overall response to the wave of corticosteroids after stress is therefore a composite of local responses, each contributing in a different way to the activity of the networks to which they belong.

Catecholamines

Shortly after stress the firing patterns of noradrenergic neurons in the locus coeruleus are changed,[83] which will result in altered release of this transmitter in many brain areas, including, for example, the basolateral amygdala. It was shown that the amount of noradrenaline released is important for the formation of memory in a simple aversive learning task.[84] This involves activation of the β-adrenoceptor. Slightly later corticosteroids reach the very same areas exposed to enhanced noradrenaline concentrations. Interactions may therefore occur at the circuit and single-cell levels. This has indeed been demonstrated, not only with biochemical methods but also with single-cell electrophysiological methods. The timing of the hormone's presence was found to be very important.

In one set of experiments, β-adrenergic effects were examined on synaptic plasticity (i.e., LTP), and corticosterone was either given concomitantly or several hours earlier. The β-agonist isoproterenol can facilitate weak forms of LTP, for example in the dentate gyrus and BLA.[85] If corticosterone and isoproterenol were applied at the same time, corticosterone enhanced the effectiveness of isoproterenol to facilitate LTP in the dentate gyrus. In the BLA, corticosterone did not exert a rapid effect but instead gradually suppressed the ability of isoproterenol to facilitate LTP.[86] Interestingly, exposure of dentate or BLA cells to a pulse of corticosterone by itself did not affect the potency to induce LTP several hours later, but strongly suppressed the ability of isoproterenol to facilitate LTP at that time. This suggests that exposure of these limbic cells to noradrenaline as well as corticosterone in the short-term might not only lead to synergistic effects (at least in the dentate), but also make these regions gradually refractory to the effect of subsequent pulses of noradrenaline. The latter was also observed in CA1 neurons.[64] In these cells, noradrenaline suppresses the Ca-dependent K conductance, and therefore leads to more action potentials during periods of depolarization. The effectiveness of noradrenaline was suppressed by a pulse of the GR-agonist RU 28386 applied some hours before the noradrenaline. Neuronal activity raised by noradrenaline after stressful events is therefore reversed by slow and presumably gene-mediated corticosteroid actions; at the same time, the effect of renewed exposure to noradrenaline is suppressed, enhancing the signal-to-noise ratio of earlier encoded information over information acquired some hours after the initial stressful event.

Directly after stress, dopamine release is also thought to change, particularly in prefrontal and striatal regions. The stress hormone CRH has been reported to affect dopaminergic responses in these areas.[87] However, so far detailed electrophysiological studies on interactions between corticosteroids and dopamine are not available.

Serotonin

As outlined before, biochemical studies demonstrated that both the synthesis and turnover of serotonin (5-hydroxytryptamine, 5-HT) are affected by corticosteroids, involving the GR, which is expected to change the bioavailability of 5-HT. Moreover, the expression and responsiveness of 5-HT receptors was found to be modulated by corticosteroid hormones. Yet receptor expression and responsiveness are not always coupled: for instance, exposure to a high dose of corticosterone results in a strong 5-HT response of CA1 neurons,[55] but does not increase expression of the 5-HT receptors mediating this response (reviewed by Meijer and de Kloet[44]). We will focus here on the effect of corticosteroid hormones on 5-HT responses, as measured with electrophysiological techniques.

In CA1 hippocampal neurons, 5-HT mostly activates 5-HT_{1A} receptors, which, through a G protein, are coupled to an inwardly rectifying K channel.[88] Binding of 5-HT to this receptor therefore causes hyperpolarization of the membrane. The effects of corticosterone on 5-HT_{1A}-receptor mediated hyperpolarizations display a U-shaped dose-dependency in CA1 pyramidal cells, with large responses both in the absence of corticosterone and several hours after application of a high dose of the hormone, and small responses under conditions of predominant MR activation.[56] The enhancement of 5-HT_{1A}-receptor mediated responses observed several hours after brief administration of a high dose of corticosterone was found to depend on DNA binding of GR homodimers.[55] In dentate granule cells, effects of corticosterone on 5-HT_{1A}-receptor mediated responses were less pronounced.[89] The U-shaped dose-dependency, the involvement of GR homodimers and the regional specificity are highly reminiscent of the hormonal effects on Ca-current amplitude (see above, and Fig. 18.4), suggesting common elements in the intracellular signaling pathway. This is supported by the fact that transcriptional regulation of target genes directly involved in the signaling pathway of 5-HT (e.g., G proteins, the inwardly rectifying K channel), or modulators of this pathway (e.g., RGS4, SGK) could not explain the electrophysiological phenomena.[90–92]

In the raphe nucleus, corticosterone (see also[93,94]) and various stressors[95] resulted in a delayed GR-mediated attenuation of 5-HT1A-receptor dependent autoinhibition, causing disinhibition of 5-HT cell activity. In these cells a linear dose-dependency was found, with the strongest effect seen at the highest dose of the hormone. Although the mechanism appears to be different, the functional consequence is comparable to that seen in the hippocampus: upregulation of 5-HT activity some hours after stress. Interestingly, the response of raphe neurons to acute stress in fact depends on the expression level of the 5-HT1A autoreceptor; mice with high levels of this receptor showed a blunted physiological response to stress which was associated with increased behavioral despair and no behavioral response to antidepressants.[96]

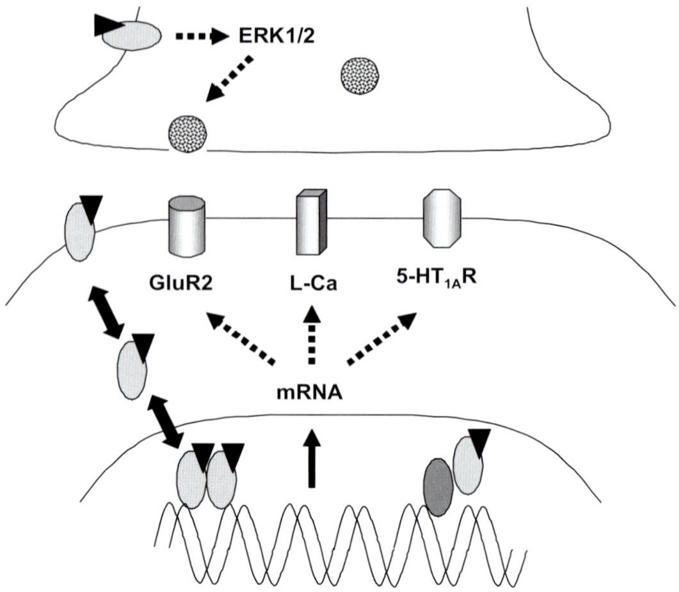

FIGURE 18.5 **Schematic overview of the signaling pathways of corticosteroid hormones.** Corticosterone (triangle) can change neuronal function of CA1 neurons in a rapid non-genomic manner via receptors presumably located in the (pre- and postsynaptic) membrane. Presynaptically, the receptor is linked to the ERK1/2 signaling pathway, causing release of glutamate-containing vesicles. In addition, corticosteroids affect neuronal excitability in a slow manner. This involves either binding of GR homodimers to the DNA or interaction of monomers with other transcription factors. Gene products may encompass many molecules important for neurotransmission, such as ionotropic receptors (e.g., the glutamate receptor subunit 2), ion channels (most conspiciously the L-type calcium channel) and metabotropic receptors (such as the 5-HT$_{1A}$ receptor). Light gray ellipsoid, corticosteroid receptor; dark gray ellipsoid, other transcription factor.

CONCLUDING REMARKS

We have described here how corticosteroid hormones change neurotransmitter responses in the rodent brain (see Fig. 18.5). The data were collected under highly controlled conditions – i.e., in young-adult male rodents raised under standard laboratory conditions, with timed administration of corticosterone at a known dosage, to reduced brain preparations. However, the reality, even in laboratory animals, is much more complex.

As argued above, corticosteroid hormones do not affect all brain cells in the same way. Some cells seem indifferent, while other cells even give opposite effects to those originally described for CA1 pyramidal neurons. This is only partly explained by the regional distribution of the corticosteroid receptors. The inter- and intracellular environment seems equally important. Overall, exposure of the brain to a surge of corticosteroids will lead to a mosaic of responses in the various brain areas. An important aspect that is not addressed in reduced slice preparation is that many of these different brain regions are interconnected. Effects in one area indirectly change the function of another; this was clearly demonstrated for corticosteroid actions in the BLA that indirectly change the functionality of the dentate gyrus.[77,78] It is therefore imperative to determine the functional implications of corticosteroid actions as seen in reduced preparations for the overall function of brain circuits and the behavioral processes in which these circuits are critically involved.

A second aspect that is not captured by the reduced approaches is the fact that stress does not exclusively lead to enhanced secretion of corticosteroids, but also to enhanced levels of other stress hormones, such as catecholamines and peptides like CRH and vasopressin. Many of these hormones converge on the same places, roughly at the same time. More likely than not, they affect each other's efficacy.[7] This has already been demonstrated for the interaction between corticosteroids and noradrenaline, but there is no reason to assume that this is not also true for other combinations of stress hormones. In the short-term (directly after stress) there might be spatial segregation of responses, due to the fact that receptors for the various stress hormones are probably localized at different compartments of the cell,[7] although interactions at the level of signaling pathways downstream of the receptor are likely to occur. In the longer term, corticosteroid hormones can certainly change the responsiveness to other stress hormones – for example, by altering the expression level of receptors for the latter via gene-mediated actions.

One fact that has hardly been addressed is the influence of sex and sex hormones. Nearly all experiments have been performed in male rodents. However, there are several examples in the literature illustrating that male and female brain cells can be affected by corticosterone in different (even opposite) directions. For instance, exposure to intermittent tailshocks results, 24 hours later, in enhanced spine density in the male hippocampus, but reduces spine density in the female hippocampus.[97]

The experiments have so far also largely neglected the existence of inter-individual differences. Variation between subjects in their response to stressful situations is known from human studies, but even in inbred rodent strains, where genetic variation is limited, considerable variation between organisms can be observed. Life history is a prominent factor determining these variations. The early life environment in particular was shown to be very important for the functionality of the stress system, and the response to stress hormones in adulthood.[98–101] This has been extensively studied in models manipulating the mother–pup interaction, one of the most relevant factors in the early life environment of rodents. For instance, low as opposed to high maternal care is linked to less efficient normalization of the stress response in adult offspring.[98] The low-care offspring also exhibited poor LTP in the CA1 area and dentate gyrus, as compared to the high-care offspring, when tissue was studied under conditions of low circulating levels of corticosterone.[102,103] The most interesting finding, however, was that brief exposure to corticosterone facilitates the induction of LTP in adult offspring of low-caring mothers – i.e., the opposite effect of what is found in naïve rodents and in animals with a high-caring background. Apparently the functional consequence of corticosterone exposure on this glutamate-dependent process can completely reverse, depending on the early life environment (see Fig. 18.6).

Similar observations were made with respect to repetitive stress in adulthood.[104] For instance, Ca-current amplitude is low under basal corticosterone conditions in CA1 neurons of naïve rodents, and increases slowly after GR activation.[55] In animals with a recent history of repetitive exposure to unpredictable stress, this phenotype is reversed: CA1 neurons from these animals show large Ca-current amplitude under basal corticosterone conditions, whereas the amplitude is reduced several hours after a brief pulse of corticosterone.[105] Many studies over the past years have supplied such examples, where not only structure and function of brain cells are altered after chronic stress, but also the way in which these cells respond to renewed exposure to corticosterone. The effect of a pulse of corticosterone on neurotransmission therefore depends not only on the area of investigation but also on the overall stress history of the individual.

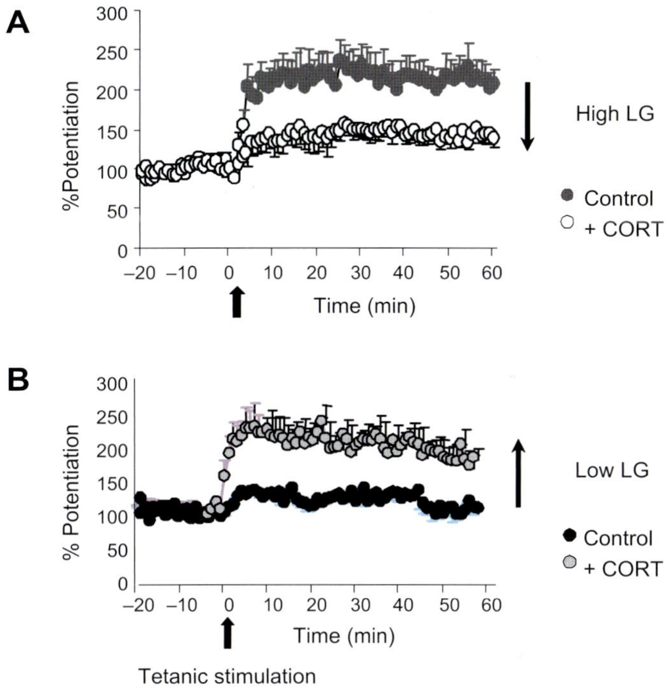

FIGURE 18.6 Long-term potentiation – strong under rest – is decreased several hours after a pulse of corticosterone in the CA1 area of adult rats which received high amounts of maternal care (licking-grooming, LG) early in life. In offspring from low LG mothers, the phenotype is reversed. Thus, LTP induction under control conditions is extremely poor, yet large several hours after exposure to high levels of corticosterone. *Based on Champagne* et al. *(2008)*.[102]

Obviously, all the experimental data presented in this chapter were derived from studies in animals – usually rats or mice. Whether the same principles hold for corticosteroid actions on neurotransmission in the human brain remains to be proven. Methods to address this issue in the human brain are, of course, limited. Investigation at the single-cell level or even in small groups of cells is still beyond the resolution of both optical and electrophysiological methods that can be applied in humans. Currently, one of the best possible approaches is to manipulate the availability of particular stress hormones by pharmacological means – for example, by giving extra cortisol or drugs that supposedly lead to enhanced release of noradrenaline in the brain, or alternatively stressing individuals pretreated with steroid synthesis inhibitors or blockers of corticosteroid/adrenergic receptors. Interestingly, nearly all studies that combined stress or intake of cortisol with functional neuroimaging techniques observed a decreased BOLD (blood oxygen level dependence) signal in limbic areas,[106–108] linked to improved memory performance. If we assume that the decreased BOLD signal represents reduced

BOX 18.3

HOW DO WE KNOW THE EFFECTS OF EARLY LIFE STRESS ON BRAIN FUNCTION?

Perinatal stress hits the brain at a sensitive period. First, this is a period during which brain circuits are being formed. Interventions at that time could have lasting consequences for how brain areas are wired. Secondly, many species, including rodents, undergo a period (in rodents, during the first two postnatal weeks) during which the adrenal is less responsive to stress. Mother–pup interactions are very important for this supposedly protective hypo-responsiveness. For instance, when pups are removed from the dam, they do show a clear stress response. While they adapt to this situation, their sensitivity to new stressors is nevertheless increased. Many models have been applied to study the influence of perinatal stress on stress responsiveness and brain structure or function later in life. These include: exposure of pregnant rodents to stress or dexamethasone, postnatal daily separation of pups from their mother (usually for 3 hours during 2 weeks) or the more severe deprivation protocol in which pups and mothers are separated for 24 hours. A more naturalistic model was developed by Michael Meaney and colleagues, who observed undisturbed maternal care of litters by the dam, according to a normally distributed variation.[1] In this model, epigenetic modification of critical genes during the first postnatal week was shown to have lasting consequences for stress responsivity and hippocampal function.[2]

References

1. Champagne FA, Francis DD, Mar A, Meaney MJ. Variations in maternal care in the rat as a mediating influence for the effects of environment on development. *Physiol Behav.* 2003;79:359–371.
2. Meaney MJ, Szyf M, Seckl JR. Epigenetic mechanisms of perinatal programming of hypothalamic–pituitary–adrenal function and health. *Trends Mol Med.* 2007;13:269–277.

neuronal activity (which is still under debate; see Logothetis & Wandell[109]), this partly fits with the slow suppressive mode of action seen for corticosterone in the hippocampal CA1 area. Distinction between the CA1 area and the dentate gyrus, though, has not been possible with the 3T imaging facilities used so far. This awaits further investigation with higher-field magnetic technology. However, given the large degree of homology in neurotransmitters, receptors and the overall stress system between rodents and humans, it seems likely that in the human brain, too, waves of stress hormones have a large impact on the function of specific neurotransmitters.

References

1. Lightman SL, Wiles CC, Atkinson HC, Henley DE, Russell GM, Leendertz JA, McKenna MA, Spiga F, Wood SA, Conway-Campbell BL. The significance of glucocorticoid pulsatility. *Eur J Pharmacol.* 2008;583:255–262.
2. Karssen AM, Meijer OC, van der Sandt IC, Lucassen PJ, de Lange EC, de Boer AG, de Kloet ER. Multidrug resistance P-glycoprotein hampers the access of cortisol but not of corticosterone to mouse and human brain. *Endocrinology.* 2001;142:2686–2694.
3. Pariante CM. The role of multi-drug resistance p-glycoprotein in glucocorticoid function: studies in animals and relevance in humans. *Eur J Pharmacol.* 2008;583:263–271.
4. Holmes MC, Seckl JR. The role of 11beta-hydroxysteroid dehydrogenases in the brain. *Mol Cell Endocrinol.* 2006;248:9–14.
5. Gross KL, Cidlowski JA. Tissue-specific glucocorticoid action: a family affair. *Trends Endocrinol Metab.* 2008;19:331–339.
6. De Kloet ER, Joëls M, Holsboer F. Stress and the brain: from adaptation to disease. *Nat Rev Neurosci.* 2005;6:463–475.
7. Joëls M, Baram TZ. The neuro-symphony of stress. *Nat Rev Neurosci.* 2009;10:459–466.
8. Kim JJ, Lee HJ, Welday AC, Song E, Cho J, Sharp PE, Jung MW, Blair HT. Stress-induced alterations in hippocampal plasticity, place cells, and spatial memory. *Proc Natl Acad Sci USA.* 2007;104:18297–18302.
9. Pfaff DW, Silva MT, Weiss JM. Telemetered recording of hormone effects on hippocampal neurons. *Science.* 1971;172:394–395.
10. Vidal C, Jordan W, Zieglgänsberger W. Corticosterone reduces the excitability of hippocampal pyramidal cells *in vitro*. *Brain Res.* 1986;383:54–59.
11. Dubrovsky B, Gijsbers K, Filipini D, Birmingham MK. Effects of adrenocortical steroids on long-term potentiation in the limbic system: basic mechanisms and behavioral consequences. *Cell Mol Neurobiol.* 1993;13:399–414.
12. Chen YZ, Hua SY, Wang CA, Wu LG, Gu Q, Xing BR. An electrophysiological study on the membrane receptor-mediated action of glucocorticoids in mammalian neurons. *Neuroendocrinology.* 1991;53(Suppl 1):25–30.
13. Di S, Malcher-Lopes R, Halmos KC, Tasker JG. Nongenomic glucocorticoid inhibition via endocannabinoid release in the hypothalamus: a fast feedback mechanism. *J Neurosci.* 2003;23:4850–4857.

14. Dallman MF. Fast glucocorticoid actions on brain: back to the future. *Front Neuroendocrinol.* 2005;26(3-4):103–108.
15. Di S, Malcher-Lopes R, Marcheselli VL, Bazan NG, Tasker JG. Rapid glucocorticoid-mediated endocannabinoid release and opposing regulation of glutamate and gamma-aminobutyric acid inputs to hypothalamic magnocellular neurons. *Endocrinology.* 2005;146:4292–4301.
16. Di S, Maxson MM, Franco A, Tasker JG. Glucocorticoids regulate glutamate and GABA synapse-specific retrograde transmission via divergent nongenomic signaling pathways. *J Neurosci.* 2009;29:393–401.
17. Verkuyl JM, Karst H, Joëls M. GABAergic transmission in the rat paraventricular nucleus of the hypothalamus is suppressed by corticosterone and stress. *Eur J Neurosci.* 2005;21:113–121.
18. Karst H, Berger S, Turiault M, Tronche F, Schütz G, Joëls M. Mineralocorticoid receptors are indispensable for nongenomic modulation of hippocampal glutamate transmission by corticosterone. *Proc Natl Acad Sci USA.* 2005;102:19204–19207.
19. Olijslagers JE, de Kloet ER, Elgersma Y, van Woerden GM, Joëls M, Karst H. Rapid changes in hippocampal CA1 pyramidal cell function via pre- as well as postsynaptic membrane mineralocorticoid receptors. *Eur J Neurosci.* 2008;27: 2542–2550.
20. Hu W, Zhang M, Czéh B, Flügge G, Zhang W. Stress impairs GABAergic network function in the hippocampus by activating nongenomic glucocorticoid receptors and affecting the integrity of the parvalbumin-expressing neuronal network. *Neuropsychopharmacology.* 2010;35:1697–1701.
21. Maggio N, Segal M. Differential corticosteroid modulation of inhibitory synaptic currents in the dorsal and ventral hippocampus. *J Neurosci.* 2009;29:2857–2866.
22. Venero C, Borrell J. Rapid glucocorticoid effects on excitatory amino acid levels in the hippocampus: a microdialysis study in freely moving rats. *Eur J Neurosci.* 1999;11:2465–2473.
23. Abrahám I, Juhász G, Kékesi KA, Kovács KJ. Corticosterone peak is responsible for stress-induced elevation of glutamate in the hippocampus. *Stress.* 1998;2:171–181.
24. De Groote L, Linthorst AC. Exposure to novelty and forced swimming evoke stressor-dependent changes in extracellular GABA in the rat hippocampus. *Neuroscience.* 2007;148:794–805.
25. Groc L, Choquet D, Chaouloff F. The stress hormone corticosterone conditions AMPAR surface trafficking and synaptic potentiation. *Nat Neurosci.* 2008;11:868–870.
26. Wiegert O, Joëls M, Krugers H. Timing is essential for rapid effects of corticosterone on synaptic potentiation in the mouse hippocampus. *Learn Mem.* 2006;13:110–113.
27. Korz V, Frey JU. Stress-related modulation of hippocampal long-term potentiation in rats: Involvement of adrenal steroid receptors. *J Neurosci.* 2003;23:7281–7287.
28. Karst H, Berger S, Erdmann G, Schütz G, Joëls M. Metaplasticity of amygdalar reponses to the stress hormone corticosterone. *Proc Natl Acad Sci USA.* 2010;107:14449–14454.
29. Johnson LR, Farb C, Morrison JH, McEwen BS, LeDoux JE. Localization of glucocorticoid receptors at postsynaptic membranes in the lateral amygdala. *Neuroscience.* 2005;136: 289–299.
30. Prager EM, Brielmaier J, Johnson LR. Anatomical evidence for extra nuclear mineralocorticoid receptors. *Soc Neurosci Abstr.* 2009;866:13.
31. Pedram A, Razandi M, Sainson RC, Kim JK, Hughes CC, Levin ER. A conserved mechanism for steroid receptor translocation to the plasma membrane. *J Biol Chem.* 2007;282: 22278–22288.
32. Foradori CD, Weiser MJ, Handa RJ. Non-genomic actions of androgens. *Front Neuroendocrinol.* 2008;29:169–181.
33. Kelly MJ, Rønnekleiv OK. Control of CNS neuronal excitability by estrogens via membrane-initiated signaling. *Mol Cell Endocrinol.* 2009;308:17–25.
34. De Kloet ER, Versteeg DHG, Kovacs GL. Aldosterone blocks the response to corticosterone in the raphe-hippocampal serotonin system. *Brain Res.* 1983;264:323–327.
35. Chaouloff F. Physiopharmacological interactions between stress hormones and central serotonergic systems. *Brain Res Rev.* 1993;18:1–32.
36. De Souza EB, Van Loon GR. Brain serotonin and catecholamine responses to repeated stress in rats. *Brain Res.* 1986;367:77–86.
37. Calogero AE, Bagdy G, Moncada ML, D'Agata R. Effect of selective serotonin agonists on basal, corticotrophin-releasing hormone- and vasopressin-induced ACTH release *in vitro* from rat pituitary cells. *J Endocrinol.* 1993;136:381–387.
38. Price ML, Curtis AL, Kirby LG, Valentino RJ, Lucki I. Effects of corticotropin-releasing factor on brain serotonergic activity. *Neuropsychopharmacology.* 1998;18:492–502.
39. Puglisi-Allegra S, Imperato A, Angelucci L, Cabib S. Acute stress induces time-dependent responses in dopamine mesolimbic system. *Brain Res.* 1991;554:217–222.
40. Deroche-Gamonet V, Sillaber I, Aouizerate B, Izawa R, Jaber M, Ghozland S, Kellendonk C, Le Moal M, Spanagel R, Schütz G, Tronche F, Piazza PV. The glucocorticoid receptor as a potential target to reduce cocaine abuse. *J Neurosci.* 2003;23:4785–4790.
41. Ambroggi F, Turiault M, Milet A, Deroche-Gamonet V, Parnaudeau S, Balado E, Barik J, van der Veen R, Maroteaux G, Lemberger T, Schütz G, Lazar M, Marinelli M, Piazza PV, Tronche F. Stress and addiction: glucocorticoid receptor in dopaminoceptive neurons facilitates cocaine seeking. *Nat Neurosci.* 2009;12:247–249.
42. Van Praag HM, de Kloet ER, Van Os J. *Stress, the Brain and Depression.* Cambridge: Cambridge University Press; 2004.
43. López JF, Chalmers DT, Little KY, Watson SJ. A.E. Bennett Research Award Regulation of serotonin1A, glucocorticoid, and mineralocorticoid receptor in rat and human hippocampus: implications for the neurobiology of depression. *Biol Psychiatry.* 1998;43:547–573.
44. Meijer OC, De Kloet ER. Corticosterone and serotonergic neurotransmission in the hippocampus: functional implications of central corticosteroid receptor diversity. *Critical Rev Neurobiol.* 1998;12:1–20.
45. Fernandes C, McKittrick CR, File SE, McEwen BS. Decreased 5-HT1a and increased 5-HT2a receptor binding after chronic corticosterone associated with a behavioural indication of depression but not anxiety. *Psychoneuroendocrinology.* 1997;22: 477–491.
46. Rënyi L, Evenden JL, Fowler CJ, Jerning E, Kelder D, Lake-Bakaar D, Larsson L-G, Mohell N, Sällemark M, Ross SB. The pharmacological profile of (R)-3,4-Dihydro-N-isopropyl-3-(N-isopropyl-N-propylamino)-2H-1-benzopyran-5-carboxamide, a selective 5- hydroxytryptamine 1A receptor agonist. *J Pharmacol Exper Ther.* 2001;299:883–893.
47. Graeff FG, Guimarães FS, De Andrade TG, Deakin JF. Role of 5-HT in stress, anxiety, and depression. *Pharmacol Biochem Behav.* 1996;54:129–141.
48. Seckl JR, Fink G. Antidepressants increase glucocorticoid and mineralocorticoid receptor mRNA expression in rat hippocampus *in vivo*. *Neuroendocrinology.* 1992;55:621–629.
49. Reul JM, Stec I, Söder M, Holsboer F. Chronic treatment of rats with the antidepressant amitriptyline attenuates the activity of the hypothalamic–pituitary adrenocortical system. *Endocrinology.* 1993;133:312–321.

50. Sibug RM, Compaan JC, Meijer OC, Van der Gugten J, Olivier B, De Kloet ER. Effects of flesinoxan on corticosteroid receptor expression in the rat hippocampus. *Eur J Pharmacol.* 2000;404: 111–119.
51. McGaugh JL, Roozendaal B. Drug enhancement of memory consolidation: historical perspective and neurobiological implications. *Psychopharmacology.* 2009;202:3–14.
52. Maier SF, Watkins LR. Stressor controllability and learned helplessness: the roles of the dorsal raphe nucleus, serotonin, and corticotropin-releasing factor. *Neurosci Biobehav Rev.* 2005;29:829–841.
53. Kerr DS, Campbell LW, Thibault O, Landfield PW. Hippocampal glucocorticoid receptor activation enhances voltage-dependent Ca^{2+} conductances: relevance to brain aging. *Proc Natl Acad Sci USA.* 1992;89:8527–8531.
54. Karst H, Wadman WJ, Joëls M. Corticosteroid receptor-dependent modulation of calcium currents in rat hippocampal CA1 neurons. *Brain Res.* 1994;649:234–242.
55. Karst H, Karten YJ, Reichardt HM, de Kloet ER, Schütz G, Joëls M. Corticosteroid actions in hippocampus require DNA binding of glucocorticoid receptor homodimers. *Nat Neurosci.* 2000;3:977–978.
56. Joëls M. Corticosteroid effects in the brain: U-shape it. *Trends Pharmacol Sci.* 2006;27:244–250.
57. Joëls M, Velzing E, Nair S, Verkuyl JM, Karst H. Acute stress increases calcium current amplitude in rat hippocampus: temporal changes in physiology and gene expression. *Eur J Neurosci.* 2003;18:1315–1824.
58. Chameau P, Qin Y, Spijker S, Smit G, Joëls M. Glucocorticoids specifically enhance L-type calcium current amplitude and affect calcium channel subunit expression in the mouse hippocampus. *J Neurophysiol.* 2007;97:5–14.
59. Kole MH, Koolhaas JM, Luiten PG, Fuchs E. High-voltage-activated Ca^{2+} currents and the excitability of pyramidal neurons in the hippocampal CA3 subfield in rats depend on corticosterone and time of day. *Neurosci Lett.* 2001;307:53–56.
60. Karst H, Nair S, Velzing E, Rumpff-van Essen L, Slagter E, Shinnick-Gallagher P, Joëls M. Glucocorticoids alter calcium conductances and calcium channel subunit expression in basolateral amygdala neurons. *Eur J Neurosci.* 2002;16:1083–1089.
61. Van Gemert NG, Carvalho DM, Karst H, van der Laan S, Zhang M, Meijer OC, Hell JW, Joëls M. Dissociation between rat hippocampal CA1 and dentate gyrus cells in their response to corticosterone: effects on calcium channel protein and current. *Endocrinology.* 2009;150:4615–4624.
62. Bhargava A, Meijer OC, Dallman MF, Pearce D. Plasma membrane calcium pump isoform 1 gene expression is repressed by corticosterone and stress in rat hippocampus. *J Neurosci.* 2000;20:3129–3138.
63. Bhargava A, Mathias RS, McCormick JA, Dallman MF, Pearce D. Glucocorticoids prolong Ca(2+) transients in hippocampal-derived H19-7 neurons by repressing the plasma membrane Ca(2+)-ATPase-1. *Mol Endocrinol.* 2002;16:1629–1637.
64. Joëls M, de Kloet ER. Effects of glucocorticoids and norepinephrine on the excitability in the hippocampus. *Science.* 1989;245:1502–1505.
65. Kerr DS, Campbell LW, Hao SY, Landfield PW. Corticosteroid modulation of hippocampal potentials: increased effect with aging. *Science.* 1989;245:1505–1509.
66. Joëls M, de Kloet ER. Corticosteroid actions on amino acid-mediated transmission in rat CA1 hippocampal cells. *J Neurosci.* 1993;13:4082–4090.
67. Talmi M, Carlier E, Rey M, Soumireu-Mourat B. Modulation of the *in vitro* electrophysiological effect of corticosterone by extracellular calcium in the hippocampus. *Neuroendocrinology.* 1992;55:257–263.
68. Joëls M, de Kloet ER. Corticosteroid actions on amino acid-mediated transmission in rat CA1 hippocampal cells. *J Neurosci.* 1993 Sep;13(9):4082–4090.
69. Martin S, Henley JM, Holman D, Zhou M, Wiegert O, van Spronsen M, Joëls M, Hoogenraad CC, Krugers HJ. Corticosterone alters AMPAR mobility and facilitates bidirectional synaptic plasticity. *PLoS One.* 2009;4:e4714.
70. Karst H, Joëls M. Corticosterone slowly enhances miniature excitatory postsynaptic current amplitude in mice CA1 hippocampal cells. *J Neurophysiol.* 2005;94:3479–3486.
71. Isaac JT, Ashby M, McBain CJ. The role of the GluR2 subunit in AMPA receptor function and synaptic plasticity. *Neuron.* 2007;54:859–871.
72. Conboy L, Sandi C. Stress at learning facilitates memory formation by regulating AMPA receptor trafficking through a glucocorticoid action. *Neuropsychopharmacology.* 2010;35: 674–685.
73. Kim JJ, Diamond DM. The stress hippocampus, synaptic plasticity and lost memories. *Nat Rev Neurosci.* 2002;3:453–462.
74. Sah P, Bekkers JM. Apical dendritic location of slow afterhyperpolarization current in hippocampal pyramidal neurons: implications for the integration of long-term potentiation. *J Neurosci.* 1996;16:4537–4542.
75. Straube T, Korz V, Frey JU. Bidirectional modulation of long-term potentiation by novelty-exploration in rat dentate gyrus. *Neurosci Lett.* 2003;344:5–8.
76. Nakao K, Matsuyama K, Matsuki N, Ikegaya Y. Amygdala stimulation modulates hippocampal synaptic plasticity. *Proc Natl Acad Sci USA.* 2004;101:14270–14275.
77. Akirav I, Richter-Levin G. Biphasic modulation of hippocampal plasticity by behavioral stress and basolateral amygdala stimulation in the rat. *J Neurosci.* 1999;19:10530–10535.
78. Akirav I, Richter-Levin G. Mechanisms of amygdala modulation of hippocampal plasticity. *J Neurosci.* 2002;22:9912–9921.
79. Maggio N, Segal M. Striking variations in corticosteroid modulation of long-term potentiation along the septotemporal axis of the hippocampus. *J Neurosci.* 2007;27:5757–5765.
80. Maguire J, Mody I. Neurosteroid synthesis-mediated regulation of GABA(A) receptors: relevance to the ovarian cycle and stress. *J Neurosci.* 2007;27:2155–2162.
81. Duvarci S, Paré D. Glucocorticoids enhance the excitability of principal basolateral amygdala neurons. *J Neurosci.* 2007;27: 4482–4491.
82. Liebmann L, Karst H, Sidiropoulou K, van Gemert N, Meijer OC, Poirazi P, Joëls M. Differential effects of corticosterone on the slow afterhyperpolarization in the basolateral amygdala and CA1 region: possible role of calcium channel subunits. *J Neurophysiol.* 2008;99:958–968.
83. Valentino RJ, Van Bockstaele E. Convergent regulation of locus coeruleus activity as an adaptive response to stress. *Eur J Pharmacol.* 2008;583:194–203.
84. McIntyre CK, Hatfield T, McGaugh JL. Amygdala norepinephrine levels after training predict inhibitory avoidance retention performance in rats. *Eur J Neurosci.* 2002;16:1223–1226.
85. Pu Z, Krugers HJ, Joëls M. Corticosterone time-dependently modulates beta-adrenergic effects on long-term potentiation in the hippocampal dentate gyrus. *Learn Mem.* 2007;14:359–367.
86. Pu Z, Krugers HJ, Joëls M. Beta-adrenergic facilitation of synaptic plasticity in the rat basolateral amygdala *in vitro* is gradually reversed by corticosterone. *Learn Mem.* 2009;16: 155–160.
87. Orozco-Cabal L, Liu J, Pollandt S, Schmidt K, Shinnick-Gallagher P, Gallagher JP. Dopamine and corticotropin-releasing

factor synergistically alter basolateral amygdala-to-medial prefrontal cortex synaptic transmission: functional switch after chronic cocaine administration. *J Neurosci.* 2008;28:529–542.

88. Andrade R, Nicoll RA. Pharmacologically distinct actions of serotonin on single pyramidal neurones of the rat hippocampus recorded *in vitro*. *J Physiol.* 1987;394:99–124.
89. Karten YJ, Stienstra CM, Joëls M. Corticosteroid effects on serotonin responses in granule cells of the rat dentate gyrus. *J Neuroendocrinol.* 2001;13:233–238.
90. Okuhara DY, Beck SG, Muma NA. Corticosterone alters G protein alpha-subunit levels in the rat hippocampus. *Brain Res.* 1997;745:144–151.
91. Muma NA, Beck SG. Corticosteroids alter G protein inwardly rectifying potassium channels protein levels in hippocampal subfields. *Brain Res.* 1999;839:331–335.
92. Van Gemert NG, Meijer OC, Morsink MC, Joëls M. Effect of brief corticosterone administration on SGK1 and RGS4 mRNA expression in rat hippocampus. *Stress.* 2006;9:165–170.
93. Laaris N, Haj-Dahmane S, Hamon M, Lanfumey L. Glucocorticoid receptor-mediated inhibition by corticosterone of 5-HT1A autoreceptor functioning in the rat dorsal raphe nucleus. *Neuropharmacology.* 1995;34:1201–1210.
94. Fairchild G, Leitch MM, Ingram CD. Acute and chronic effects of corticosterone on 5-HT1A receptor-mediated autoinhibition in the rat dorsal raphe nucleus. *Neuropharmacology.* 2003;45: 925–934.
95. Laaris N, Le Poul E, Laporte AM, Hamon M, Lanfumey L. Differential effects of stress on presynaptic and postsynaptic 5-hydroxytryptamine-1A receptors in the rat brain: an *in vitro* electrophysiological study. *Neuroscience.* 1999;91:947–958.
96. Richardson-Jones JW, Craige CP, Guiard BP, Stephen A, Metzger KL, Kung HF, Gardier AM, Dranovsky A, David DJ, Beck SG, Hen R, Leonardo ED. 5-HT1A autoreceptor levels determine vulnerability to stress and response to antidepressants. *Neuron.* 2010;65:40–52.
97. Shors TJ, Chua C, Falduto J. Sex differences and opposite effects of stress on dendritic spine density in the male versus female hippocampus. *J Neurosci.* 2001;21:6292–6297.
98. Liu D, Diorio J, Day JC, Francis DD, Meaney MJ. Maternal care, hippocampal synaptogenesis and cognitive development in rats. *Nat Neurosci.* 2000;3:799–806.
99. Liu D, Diorio J, Tannenbaum B, Caldji C, Francis D, Freedman A, Sharma S, Pearson D, Plotsky PM, Meaney MJ. Maternal care, hippocampal glucocorticoid receptors, and hypothalamic–pituitary–adrenal responses to stress. *Science.* 1997;277:1659–1662.
100. Meaney MJ, Szyf M, Seckl JR. Epigenetic mechanisms of perinatal programming of hypothalamic–pituitary–adrenal function and health. *Trends Mol Med.* 2007;13:269–277.
101. Levine S. Developmental determinants of sensitivity and resistance to stress. *Psychoneuroendocrinology.* 2005;30:939–946.
102. Champagne DL, Bagot RC, van Hasselt F, Ramakers G, Meaney MJ, de Kloet ER, Joëls M, Krugers H. Maternal care and hippocampal plasticity: evidence for experience-dependent structural plasticity, altered synaptic functioning, and differential responsiveness to glucocorticoids and stress. *J Neurosci.* 2008;28:6037–6045.
103. Bagot RC, van Hasselt FN, Champagne DL, Meaney MJ, Krugers HJ, Joëls M. Maternal care determines rapid effects of stress mediators on synaptic plasticity in adult rat hippocampal dentate gyrus. *Neurobiol Learn Mem.* 2009;92:292–300.
104. Joëls M, Karst H, Krugers HJ, Lucassen PJ. Chronic stress: implications for neuronal morphology, function and neurogenesis. *Front Neuroendocrinol.* 2007;28:72–96.
105. Karst H, Joëls M. Brief RU 38486 treatment normalizes the effects of chronic stress on calcium currents in rat hippocampal CA1 neurons. *Neuropsychopharmacology.* 2007;32:1830–1839.
106. Pruessner JC, Dedovic K, Khalili-Mahani N, Engert V, Pruessner M, Buss C, Renwick R, Dagher A, Meaney MJ, Lupien S. Deactivation of the limbic system during acute psychosocial stress: evidence from positron emission tomography and functional magnetic resonance imaging studies. *Biol Psychiatry.* 2008;63:234–240.
107. Henckens MJ, Hermans EJ, Pu Z, Joëls M, Fernández G. Stressed memories: how acute stress affects memory formation in humans. *J Neurosci.* 2009;29:10111–10119.
108. Van Stegeren AH, Roozendaal B, Kindt M, Wolf OT, Joëls M. Interacting noradrenergic and corticosteroid systems shift human brain activation patterns during encoding. *Neurobiol Learn Mem.* 2010;93:56–65.
109. Logothetis NK, Wandell BA. Interpreting the BOLD signal. *Annu Rev Physiol.* 2004;66:735–769.

CHAPTER

19

Neuroendocrine Mechanism of Puberty

Ei Terasawa [1,2], *Joseph R. Kurian* [1]

[1] Wisconsin National Primate Research Center, University of Wisconsin, Madison, WI, USA,

[2] Department of Pediatrics, University of Wisconsin, Madison, WI, USA

OUTLINE

Summary

Puberty is the transitional period between childhood and adulthood when major hormonal, physical and behavioral changes take place to attain full reproductive capability. The timing of puberty is influenced by several factors, such as genetic background, nutritional status, socio-economic background, climate (ambient temperatures and photo period), and environmental chemicals, such as bisphenol A. However, the underlying mechanism triggering puberty is still unknown. This is, in part, due to the complexities associated with maturation of the brain, pituitary gland and gonads, and interactions between the endocrine and nervous systems, as well as species differences in neuroendocrine mechanisms leading to puberty onset. In this chapter, we discuss the most

Handbook of Neuroendocrinology, DOI: 10.1016/B978-0-12-375097-6.10019-8

likely and compelling factors that could lead to the initiation of puberty. As it has become clear over the past 30 years that an increased release of gonadotropin-releasing hormone (GnRH) from the hypothalamus triggers the onset of puberty, we emphasize factors that could trigger the pubertal increase in GnRH release, with a focus on recent discoveries in the field.

INTRODUCTION

Puberty is the transitional period between childhood and adulthood. During puberty, major hormonal, physical and behavioral changes take place to attain full reproductive capability. In humans, changes during puberty are accompanied by psychological modifications, which significantly impact the adult life and role in society. Mammalian puberty requires full maturation of the hypothalamus, pituitary gland and gonads, and establishment of the feedback circuit between them. The age of puberty onset is characteristic for each species, as each species has its own time-course for hypothalamo-pituitary–gonadal development and distinct interactions between hormones from the hypothalamus, pituitary and gonads. The timing of puberty is influenced by several factors, such as genetic background, nutritional status, socioeconomic background, climate (ambient temperatures and photo period) and environmental chemicals (such as bisphenol A). However, the underlying mechanism triggering puberty is still unknown.

During the past 30 years it has become clear that an increased release of the gonadotropin-releasing hormone (GnRH, a synonymous term for LHRH, luteinizing hormone-releasing hormone) from the hypothalamus triggers the onset of puberty. Therefore, the most important question here is: What triggers the pubertal increase in GnRH release? With recent advancement of human genetic analyses, several genes associated with delaying or advancing the timing of puberty have been identified. Mutations of the G-protein coupled receptor 54 (GPR54, also called KiSS1r) and its ligand kisspeptin (*KiSS1*),[1,2] or the neurokinin B receptor (TACR3) and its ligand neurokinin B (TAC3),[3] all appear to cause an abnormal timing of puberty. Importantly, products of these genes (such as kisspeptin and neurokinin B) in the hypothalamus are intimately involved in the control of GnRH release. Nonetheless, the mechanism of puberty onset is complex, and kisspeptin and neurokinin B are not likely the initial signals responsible for ultimately triggering the pubertal increase in GnRH release. In this chapter, we will describe the mechanism of puberty onset, focusing on recent discoveries in the field.

DESCRIPTION OF PUBERTY

Puberty in Humans

Physical and hormonal changes associated with puberty are well-defined signs of sexual maturation, and indicative of hypothalamo-pituitary–gonadal development (gonadarche[4,5]) versus mature adrenal function (adrenarche). For example, the first appearance of pubic hair and enlargement of the breasts occurs at 8–10 years of age).[5–7] While pubic and axillary hair growth are primarily due to a pubertal and pituitary–gonadal independent increase in adrenal androgen, breast development is due to an increase in ovarian estrogens.[6] Based on development of the breasts and pubic hair, Tanner[8] has developed standards for puberty in girls. Because an increase in estrogen during the early stage of puberty stimulates epiphyseal growth, in individual girls accelerated growth (the pubertal increase in height velocity) is the first sign of puberty.[6] Menarche, due to a further increase in ovarian estrogens, occurs at an average age of 11–13 years[5–7] (Fig. 19.1).

In boys, an increase in testicular size occurs at 9.5–13.5 years (average 12 years) of age,[9] which is followed by the growth of pubic hair and of the penis.[8] Testicular growth is due to a thickening of the seminiferous tubular lining, the formation of a lumen, and the differentiation and growth of Sertoli and Leydig cells, resulting in a volume increase in the Leydig cells. Tanner[8] has also established standards for pubertal development in boys.

Puberty in Non-human Primates

In the female rhesus monkey, the first observable physical signs of puberty are a slight increase in growth velocity and nipple size, and development of perineal sex-skin color, which occurs at ~24–26 months of age.[10] The increase in nipple size and perineal sex skin development further progresses, and menarche occurs at 27–30 months of age. Several episodes of menstrual cycles occur after menarche[10] without ovulation, and first ovulation occurs at 36–48 months of age.[10–13] After first ovulation, the ovulatory cycles in the rhesus monkey often exhibit a short luteal phase. Previously, we designated the period prior to any signs of puberty as the prepubertal stage, the period between first signs of puberty and menarche as the early pubertal stage, the period between menarche and first ovulation as the midpubertal stage, and the period between first ovulation and full maturity as the late pubertal stage[10] (Fig. 19.1).

In male rhesus monkeys, testicular descent into the scrotum occurs slightly after 30 months of age. Subsequently, testicular volume starts to increase. This

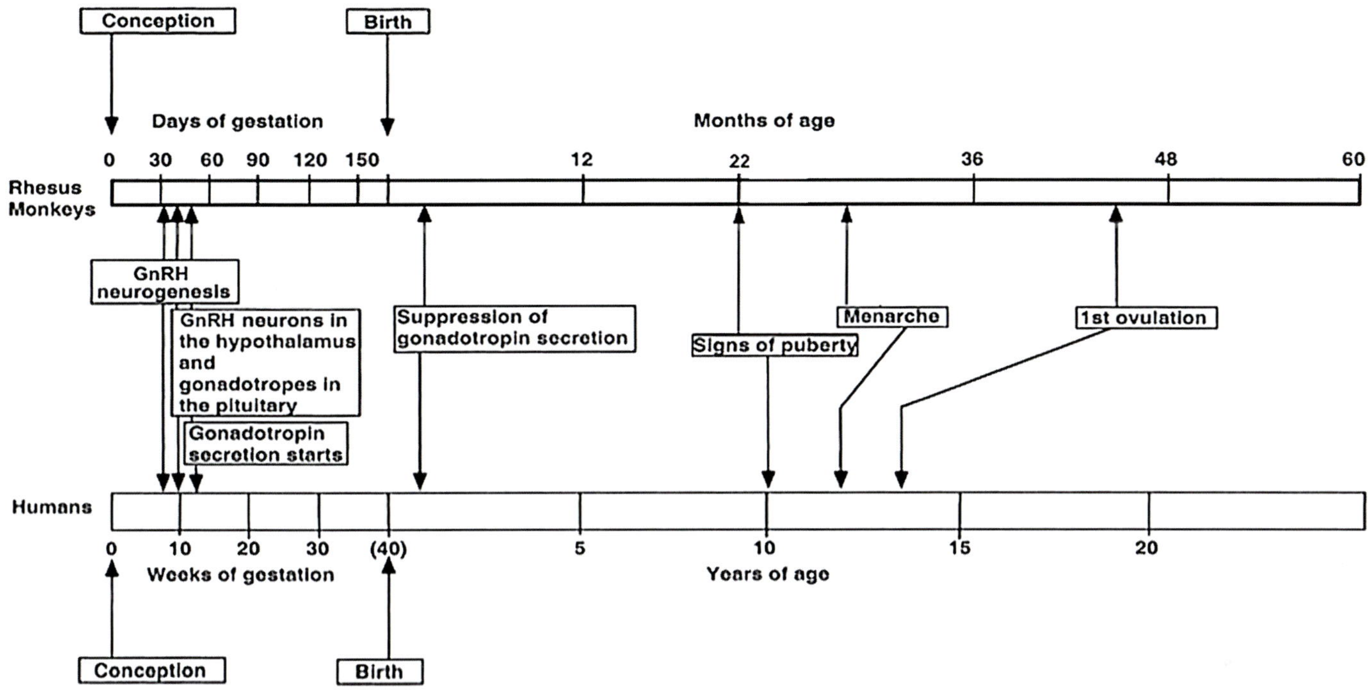

FIGURE 19.1 **Ontogeny and function of the gonadotropin-releasing hormone (GnRH)–pituitary–gonadal system from prior to birth until after the onset of puberty in female rhesus monkeys (top) and in female humans (bottom).** Days (gestational period) and months (after birth) are used as the scale in monkeys, whereas weeks (gestational period) and years (after birth) are used as the scale in humans. *Modified from Terasawa and Fernandez (2001),[616] with permission.*

increase continues until approximately 48 months of age, when full spermatogenesis is observed.[14]

Puberty in Rodents

In female rats, vaginal opening, which is usually accompanied by the first ovulation, occurs around postnatal days 35–37 (P35–P37), and is a marker of puberty. However, because vaginal opening often occurs without first ovulation under experimental conditions, direct observations of ova in oviducts or corpora lutea formation on the ovary ensure the documentation of true puberty. The time of puberty in male rats is less clear, although it is considered as the stage when fertility is apparent, which takes place between P42–P45. Externally, this stage is associated with preputial separation, or separation of the prepuce from the glans penis.[15]

Puberty in Sheep

The initiation of puberty in sheep (ewes and rams) occurs at approximately 6 months, with first estrus occurring 2–3 weeks prior to first ovulation in ewes. In ewes, observation of increased plasma luteinizing hormone pulse frequency and amplitude, as well as the first sign of sexual receptivity, verifies the pubertal transition. While sheep are long-lived mammals, their period from birth to sexual maturity is quite protracted, and associated with rapid body weight gain during the first 30 weeks of postnatal life.[16] This rapid postnatal development is similar to that of the rat, and distinct from humans and non-human primates. Sheep are also unique in their timing of puberty initiation, with the age of puberty onset in ewes heavily related to the month of birth. As seasonal breeders with reproductive capability largely influenced by photoperiod (long versus short days), sheep tend to reach puberty during a breeding season, regardless of the month of birth.[17,18] Consequently, depending on the month of birth in relation to breeding season, ewes can reach puberty at an early (before 6 months) or significantly later age.

DEVELOPMENT OF THE HYPOTHALAMO-PITUITARY–GONADAL AXIS

Development of GnRH Neurons, the Pituitary Gland, and Gonads

GnRH Neurons

The ontogeny of mammalian GnRH neurons is unique among other neurons in the brain. They originate from the nasal epithelium during the gestational period, and migrate into the preoptic area (POA) and medial basal hypothalamus (MBH).[19–23] In the rhesus monkey, GnRH neurons are found in the nasal placode as early as embryonic (E) day 32 (E32), and commonly at E34–E36.[21,22,24] GnRH neurons migrate along the nasal septum and then terminal nerve, enter the forebrain through the cribriform sieve at E38,[22] and subsequently migrate into the basal hypothalamus of the rhesus monkey by E47.[21] The basic distribution pattern of GnRH neurons in the brain is already established at E55,[21,22,25] although GnRH neurons continue migrating into the POA and MBH until the last trimester. Gonadotropes are found in the pituitary at E50,[21] and sex-specific gonadal steroids are detectable in the umbilical cord at E70,[26] suggesting that GnRH neurons are functioning at E50–E70. Recent *in vitro* studies in our laboratory further suggest that mRNA expression in GnRH neurons derived from E36 rhesus monkey embryos dramatically increases after 3 weeks in culture, which is equivalent to ~E56–E57 *in vivo*.[27] Activity of GnRH neurons appears to increase close to term; proGnRH mRNA is detectable in GnRH cells at E38–E50, but does not increase significantly until E135.[21] Additional systematic study is needed regarding this issue. Gonadectomy at E98–E104 results in the elevation of LH and FSH in male monkeys but not in female monkeys,[28] indicating that the negative feedback system is operative in the male hypothalamo-pituitary–gonadal axis during the second trimester.

A similar ontogenic picture of the reproductive neuroendocrine system has been described in humans.[29] In human fetuses, GnRH neurons are found in the nasal placode as early as E5.5, although a majority of them originate in the olfactory pit at E week 6.0–6.5. GnRH neurons enter the forebrain through the terminal nerve by E week 6.5, and they migrate into the hypothalamus by E week 9.0 (Refs 23,30; Quanbeck and Terasawa, unpublished observation) (Fig. 19.1).

Although the distribution pattern of GnRH neurons is already established before birth in most species in which it has been studied, the function, morphology and biosynthesis, as well as synaptic connectivity of GnRH neurons, may not be mature until the time of puberty. In rats, it has been shown that the number of GnRH neurons with a wrinkled contour increases, while the number of GnRH neurons with a smooth contour decreases, at the age of puberty.[31] A recent observation with a biocytin filling method shows that early postnatal mouse GnRH neurons have more dendritic branches and less spine density than adult GnRH neurons.[32] Interestingly, mouse GnRH neurons receive more glutamatergic input, but no changes in GABAergic input, across the time of puberty.[32] In rats and mice, GnRH mRNA levels increase gradually with age, and significant increases occur at P15–P30 depending on sex and experimental conditions.[33–37] Importantly, NMDA treatments accelerate the timing of the prepubertal increase in GnRH mRNA expression, resulting in precocious puberty.[35] Therefore, after birth GnRH neurons in rodents appear to undergo dendritic tree remodeling and receive more selective innervation from other neurons, such as glutamatergic neurons, leading to an increase in *GnRH* gene expression.

In contrast, data on primate GnRH neurons are somewhat controversial. For example, earlier data indicate no differences in the number[38] or the shape of GnRH neurons,[39] or in GnRH mRNA levels,[40] between the juvenile period and adulthood, whereas more recent data suggest that GnRH mRNA increases between the juvenile period and adulthood.[41,42] Nonetheless, GnRH neurons appear to be mature well before the onset of puberty, as elevated levels of LH, presumably due to elevated GnRH release, are seen at the neonatal period in gonadectomized monkeys[43,44] and agonadal humans.[45–48] Moreover, we have shown that electrical stimulation of the MBH in prepubertal monkeys, in which basal GnRH release is very low, induces GnRH release similar to that observed in pubertal monkeys.[49] It is possible that GnRH neurons in primates undergo subtle developmental changes during the juvenile period, as shown by two recent reports: (1) Perikarya of GnRH neurons located in the MBH of early pubertal female monkeys are more invested with glial sheath than in adult cyclic female monkeys,[50] and (2) a decrease in the area occupied by synapses onto GnRH neurons occurs in adult castrated male monkeys as compared to prepubertal castrated male monkeys.[51] A systematic study of ontogenic changes in morphology and gene expression of GnRH neurons in primates is urgently needed.

Pituitary (Gonadotropes)

FSH and LH are detectable in the human pituitary by E week 10, and their content increases until E week 25–29. The pituitary starts to release gonadotropins into the general circulation by E week 11–12. Circulating gonadotropins reach peak levels at midgestation, and both LH and FSH levels subsequently decline during late gestation.[52–55] The gonadotropes in human fetuses respond to GnRH by releasing LH and FSH both *in vivo* and *in vitro*.[56–59] A sex difference in gonadotropin levels is seen during midgestation: pituitary content and circulating concentrations of LH and FSH in female fetuses are higher than those in male fetuses.[52,60–63] Since circulating testosterone levels are quite high in male fetuses as compared to circulating estrogen levels in female fetuses during midgestation, both the sex difference in gonadotropin levels and the decrease in gonadotropin levels towards late gestation in fetuses are attributed to the development of the negative feedback mechanism by gonadal steroid hormones from the fetal gonads as well as from the placenta.[6] A report that castration in male rhesus monkeys at E98–E104 increases circulating gonadotropins to levels as high as those in ovarian-intact female monkeys at similar ages[26,28] supports this notion. In females, negative feedback by ovarian steroids is operative in the human fetus during late gestation, when estrogen secretion is elevated.[29,52] In sheep fetuses during late gestation, LH release is pulsatile, and orchidectomy in males, but not ovariectomy (OVX) in females, results in an increase in the pulse amplitude of LH.[62]

Gonads

THE TESTIS

Testicular growth is the earliest clinical sign of puberty in human males. The dramatic acceleration of testicular growth occurs in humans and non-human primates when serum gonadotropin levels rise, which takes place in humans between 9 and 13 years of age.[6,64]

Testicular steroid secreting cells (Leydig cells) are largely quiescent throughout prepubertal development,[65] with transient periods of activation. For example, Leydig cells are active from birth through 6 months of postnatal life, when serum gonadotropin levels are elevated. After the first 6 months, activity subsides and remains low until puberty; accordingly, plasma testosterone levels are elevated during the first 7 months of life, then remain low until puberty.[66]

The human testis secretes several steroids, including testosterone, dihydrotestosterone, estradiol and androstenedione, which is the predominate androgen produced before puberty. Between 10 and 17 years of age in humans, plasma testosterone levels rise dramatically – an increase associated with the pubertal increase in nocturnal LH.[67] In rats, the pubertal increase in testosterone secretion is also triggered by rising plasma LH. By 17 years of age in humans, adult testis size, plasma testosterone, FSH, LH and secondary sexual characteristics are well established.[66] Testosterone acts synergistically with FSH to promote the growth and maturation of seminiferous tubules, and plays an important role in promoting the earliest physical changes in male puberty, which include accelerated development of the penis, and axillary and pubic hair growth.[6,66] In rats, FSH stimulates spermatogenesis and promotes LH receptor expression in Leydig cells. Rising plasma FSH levels also stimulate human testicular production of inhibin, a peptidergic hormone secreted by Sertoli cells, which increases significantly during the initiation of puberty[6] and participates in a feedback loop to selectively inhibit FSH secretion from the anterior pituitary.

The testis is also innervated by sympathetic and sensory neurons, and this innervation density was found to increase at the time of puberty in rhesus monkeys.[68] As norepinephrine can stimulate testicular steroidogenesis, it is possible that, along with gonadotropin stimulation, increased noradrenergic tone may contribute to the pubertal activation of testicular function.[69]

THE OVARY

Prepubertal ovaries also actively release steroids, but do not appear to exhibit the same levels or periodicity of increased steroid secretion when compared to the testis. In humans, the ovary produces estrone, androstenedione, testosterone and other steroids, but mainly secretes estradiol,[70] which is already detectable at very high levels in the umbilical cord artery during the last trimester of gestation.[6,70] Unlike the extended (~7 month) elevation of plasma testosterone in males, estrogen levels in female plasma decline during the first postnatal week of life and subsequently remain below 10 pg/ml until puberty onset. Plasma estradiol begins to increase between 8 and 10 years of age, and levels off near 14;[71] this increase stimulates breast and reproductive organ development and triggers a redistribution of fat and bone maturation. Circulating levels of inhibin A and B produced by ovarian follicles significantly increase after serum FSH levels rise at puberty.[72]

The basic mechanisms regulating ovarian growth, secretory activity and gonadotropin control are similar between primates and rodents.[73,74] In immature rats, several pituitary hormones appear to be responsible for prepubertal ovarian development and increased responsiveness to gonadotropins; these include LH, FSH, prolactin and growth hormone.[75]

Mammalian ovaries, like testes, are also influenced by direct neuronal innervation, and the density of ovarian sympathetic innervation increases during primate prepubertal development. Neuronal norepinephrine and vasoactive intestinal peptide have a direct stimulatory effect on ovarian steroidogenesis, suggesting that increasing direct facilitatory adrenergic tone might influence ovarian development.[76]

Developmental Changes in Circulating Hormones

Hormonal Changes Before Puberty

Circulating LH levels in human male neonates abruptly increase within the first few minutes after birth, followed by an increase in serum concentrations of testosterone during the first 3–21 hours.[77] High levels of LH in the human male infant decline within 6 months, and remain low until the time of puberty. FSH levels in human males are slightly elevated for the first 3 months postnatally, after which they become low.[46,78] Circulating levels of testosterone are also postnatally elevated for 2–4 months.[79] In contrast, in female neonates LH levels are only slightly elevated during the first few months of postnatal life, but FSH levels are high for the first 5 months.[46,78] After the first 6 months of life, circulating levels of FSH, LH and gonadal steroids are all low, and the hypothalamo-pituitary–gonadal system enters a quiescent stage until the time of puberty.

Plasma concentrations of LH and FSH in infants with gonadal dysgenesis are strikingly elevated, due to the absence of the negative feedback loop.[46–48,78] However, the elevated level of gonadotropins in patients with gonadal dysgenesis declines during the juvenile period, as seen in eugonadal children.

In infantile male monkeys, circulating LH and testosterone are elevated during the first 2–3 postnatal months,[80,81] and diurnal variation of testosterone, quantitatively similar to that seen in sexually mature adults, is observed.[82] Bilateral orchidectomy at 1 week of age results in an increase in LH and FSH secretion, with a pulse pattern very similar to that in castrated adult males.[82] Consequently, it appears that the hypothalamo-pituitary axis governing testicular function in males is fully mature by the neonatal stage of development.[83] Both LH and FSH levels in male monkeys decrease after 3–4 months of age.[43,84] By contrast, in female monkeys circulating LH is only slightly elevated or not elevated during the first 3 months,[81] and the negative feedback mechanism appears to be only partially operative in the late gestational period through the neonatal period. A moderate elevation of estrogen levels is observed during late gestation through the neonatal period in females.[81] Ovariectomy in females at 1 week of age induces a truncated and abbreviated elevation of LH release with slower pulse frequency, when compared to those in orchidectomized male infants[44] (Fig. 19.2). It is possible that the GnRH neurosecretory system in female monkeys during the neonatal period is less mature than in males.[44] Alternatively, in females, central inhibition to the GnRH neurosecretory system (see below) may start earlier than in males, and may not be as complete as in males. Nevertheless, at 4–6 months of age the GnRH neurosecretory system in both sexes enters a quiescent state, which persists until the time of puberty.[10,43,83,85] This state of quiescence is unique to primates (Fig. 19.2) and represents a period of non-gonadal inhibition upon GnRH release, as discussed later in the chapter.

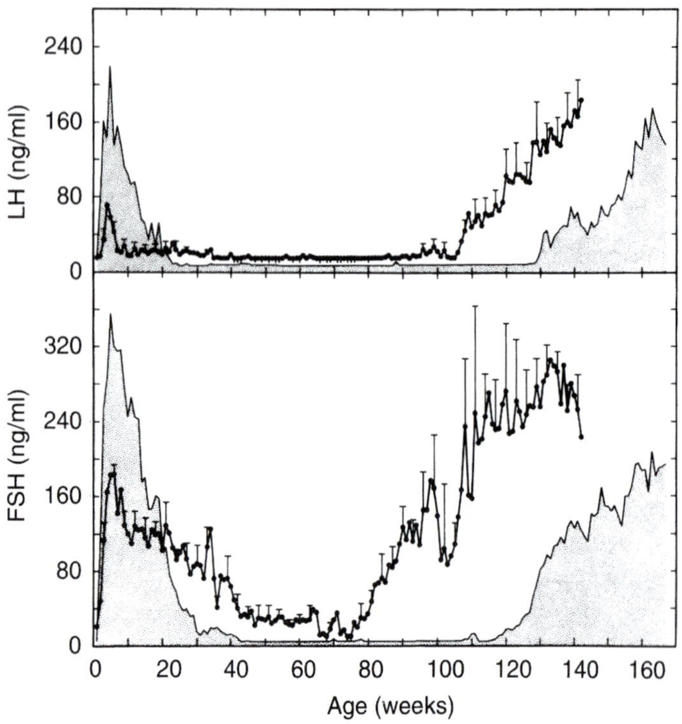

FIGURE 19.2 **Evidence for a central inhibition of the GnRH neurosecretory system in primates.** Developmental changes in circulating LH (top) and FSH (bottom) from orchidectomized males (fine lines with shadows) and ovariectomized females (heavy lines with closed circles) are shown. During the neonatal period, LH and FSH levels are elevated because of the absence of gonadal steroid feedback in both sexes, particularly in males. The elevated LH and FSH levels are subsequently suppressed during the juvenile period, despite the absence of gonadal steroids. At the onset of puberty, both LH and FSH start to increase. The age of pubertal onset in females is younger than that in males, and the pubertal increase in FSH precedes LH in both sexes. *Reproduced from Plant TM, Witchel SF. Puberty in nonhuman primates and primates. In J. Neill (ed.)* The Physiology of Reproduction, *Volume 2, 3rd edn. San Diego, CA: Academic Press; 2006:2177–2230, with permission.*

Hormonal Changes at the Onset of and During Puberty

Prior to the onset of puberty LH and FSH levels are low, but a highly sensitive assay indicates that circulating LH and FSH levels in prepubertal children are pulsatile, with slightly higher values at night than in the morning.[86–92] In both boys and girls, preceding the physical signs of puberty, LH and FSH levels become elevated, pulsatility of these hormones becomes more pronounced, and the nocturnal increase in gonadotropins is enhanced.[88,89,93–95] In addition, both pulse frequency and amplitude of LH release increase at this stage.[93]

Similarly, the first hormonal sign of puberty in female rhesus monkeys appears several months earlier than menarche.[10,85] During the early pubertal period mean FSH levels increase, followed by an increase in mean LH levels. Basal levels of LH and the amplitude and frequency of pulsatile LH release increase, and a circadian fluctuation (nocturnal increase) of LH emerges. The pubertal increase in gonadotropin release stimulates estrogen secretion from the ovary,

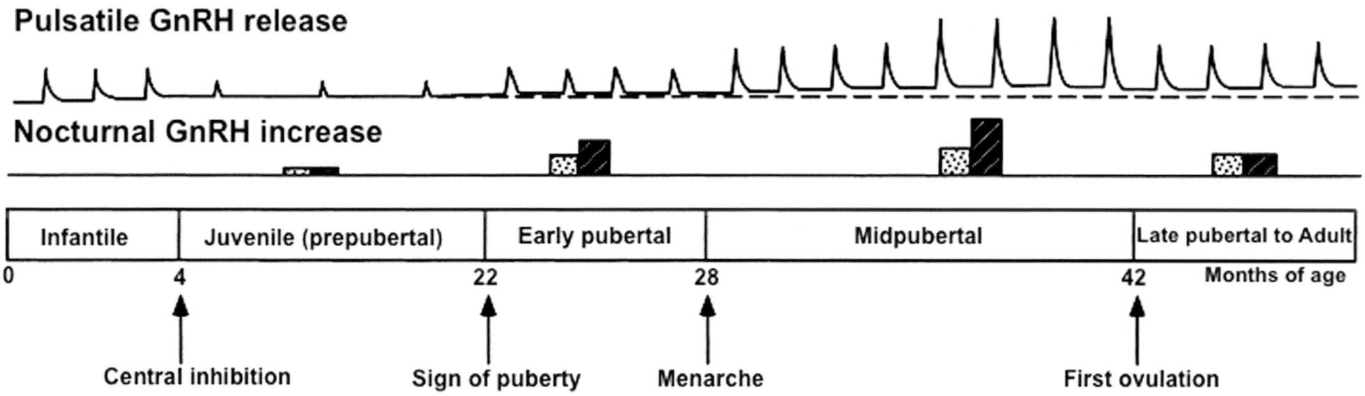

FIGURE 19.3 **Schematic illustration of postnatal changes in GnRH release in association with puberty in female non-human primates.** The GnRH neurosecretory system appears to be active during the infantile period, but it is suppressed by a central inhibition during the juvenile period, which can be seen as a low frequency and low amplitude of GnRH release. At the time of puberty, the pulse amplitude, pulse frequency and mean release of GnRH start to increase, and those changes are further augmented through puberty. A higher nocturnal level of GnRH release, shown by hatched bars (stippled bars indicate morning levels) becomes prominent at the time of puberty onset and the nocturnal increase in GnRH release continues until first ovulation, after which GnRH release is reduced to the adult level. *Modified from Terasawa and Fernandez (2001),[616] with permission.*

resulting in nipple growth, sex-skin development and, subsequently, menarche.[10,85] Due to an increase in circulating estrogen, FSH levels are suppressed. During the midpubertal period, the nocturnal increase in LH becomes more pronounced, and basal LH levels and the amplitude of LH pulses further increase. A periodic increase in estrogen causes menstrual cycles, although data with laparoscopy indicate that periodic vaginal bleeding is not due to cyclic ovulation.[85] Circulating estrogen during this developmental period sometimes reaches the preovulatory level, but LH surges do not occur until the first ovulation.[10] This phenomenon is referred to as pubescent infertility. After first ovulation, both basal LH levels and LH amplitude decline, and the nocturnal increase in LH release disappears.[10]

The pubertal changes in LH release patterns are the result of developmental changes in pulsatile GnRH release[96] (Fig. 19.3). The pulse frequency of GnRH release (measured in conscious rhesus monkeys by the push–pull perfusion technique) increases at the onset of puberty and then remains stable throughout the pubertal period,[96,97] while the pulse amplitude of GnRH release starts to increase at the pubertal transition, and continues to increase throughout puberty until first ovulation occurs[96,97] (Fig. 19.3). Moreover, the nocturnal increase in GnRH release is readily observed during the early pubertal period and becomes increasingly prominent during the midpubertal period, whereas nocturnal increases in GnRH release are seldom seen during the prepubertal period[96,97] (Fig. 19.3). Therefore, it is clear that: (1) hormonal changes occur gradually throughout puberty, and (2) development of the GnRH neurosecretory system or its regulatory systems are essential for the onset of puberty.

GENES CONTROLLING THE TIMING OF PUBERTY

Many genes in the brain are turned on or turned off to establish a complex series of physiological events leading to puberty. In human populations the timing of puberty varies, as it is influenced by environmental factors such as nutritional status, socioeconomic background, climate (ambient temperatures and photo period) and environmental chemicals (e.g., bisphenol A). However, the age of menarche is highly correlated within racial/ethnic groups,[98,99] and studies in human twins[100–102] suggest that the age of menarche is a highly heritable trait. In addition, genetic linkage studies within large human populations predict the presence of genes determining the timing of puberty. In this section we will review genes associated with abnormal reproductive function and timing of puberty in humans, and then discuss potential "master genes" which trigger a series of pubertal events in rodents (Fig. 19.4).

Genes Associated with GnRH Cell Migration (Kallmann Syndrome)

Kallmann syndrome is hypogondotropic hypogonadism with abnormal olfactory function (anosmia or hyposmia) in humans, which is caused by failed migration of GnRH neurons from the nasal placode into the brain. Mutations in several genes are shown to be associated with Kallmann syndrome. These include

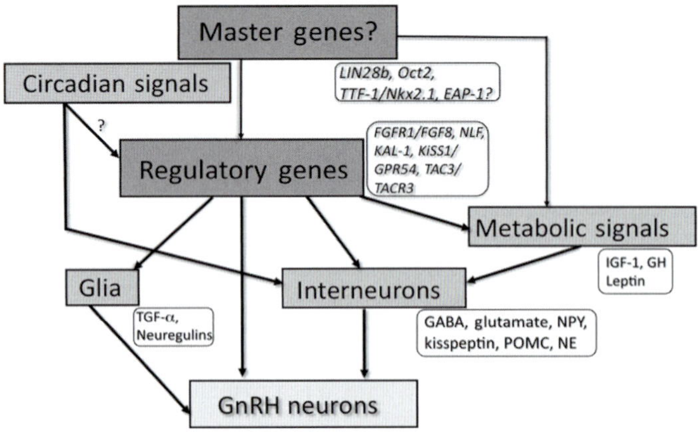

FIGURE 19.4 **Are there master genes for the onset of puberty?** Schematic illustration of a potential gene hierarchy that might be involved in puberty onset (see text for details).

Kallmann Syndrome 1 (*KAL-1*),[103–105] fibroblast growth factor receptor 1 (*FGFR1*) and the gene for its ligand fibroblast growth factor 8 (*FGF8*)[106–108] and the nasal embryonic luteinizing hormone releasing hormone factor (NELF).[109,110] Moreover, mutations in the prokineticin receptor-2 gene (*PKR2*), a G-protein coupled receptor, and the gene for its ligand prokineticin-2 (*PK2*), were identified in a subset of Kallmann syndrome patients.[111,112] Therefore, the *KAL-1*, *FGFR1/FGF8*, *NELF* and *PKR2/PK2* genes are important for the development of normal reproductive function, due to their control of GnRH cell migration from the nasal placode. Nonetheless, they are unlikely to play a role in the timing of puberty.

Genes Associated with Hypogonadotropic Hypogonadism

Hypogonadotropic hypogonadism also presents in patients with normal olfaction and GnRH neurons that have migrated into the brain. In some of these patients GnRH may be released into the portal circulation, but it is not functional because of mutations in the gonadotopin-releasing hormone receptor gene (*GnRHR*). For example, mutation in GnRHR causes a severe hypogonadotropic hypogonadism.[113–115] Apparently, however, genetic variations in *GnRH* or *GnRHR* genes are not common causes of delayed puberty, because of lack of reproductive function. Alternatively, in some patients with hypogonadotropic hypogonadism GnRH may not be sufficiently released because of mutations in genes that are involved in control of GnRH release. It has been reported that mutations in the genes encoding the G-protein coupled receptor 54 (GPR54) and its ligand kisspeptins (*KiSS1*),[1,2,116] or the neurokinin B receptor (TACR3) and its ligand neurokinin B (TAC3),[3] are often identified in hypogonadotropic hypogonadism patients and also cause abnormal timing of puberty. The GPR54/kisspeptin system is involved in the positive feedback mechanism of estrogens on GnRH release,[117] whereas the TACR3/TAC3 system is involved in pulsatile GnRH release.[118] We will further discuss the possible role of GPR54/kisspeptin in puberty in a later section.

Genes Associated with Body Size

Frisch and Revelle[119] were the first scientists to propose the critical body weight hypothesis for determining the timing of puberty, based on their observations that the age of menarche in girls declined over a 20-year period (1929–1950) yet the body weight at menarche was unchanged. This hypothesis has been repeatedly confirmed as a secular trend towards a younger age of puberty onset, and coincides with a trend towards obesity in human society.[120–123] Nonetheless, the mechanism and genes that might be responsible for transmitting body size information to the hypothalamus is unclear. Leptin may play a permissive role in puberty (see next section). Recent discoveries, however, suggest that *LIN28b*, a gene that appears to control body height,[124–127] could be more important for the timing of puberty.

Leptin

It has been proposed that leptin, a peptide synthesized in adipose tissue and secreted into the general circulation, is a key metabolic signal to the reproductive neuroendocrine axis. At puberty, circulating leptin concentrations increase and leptin binding activity decreases.[128–132] Leptin also induces precocious puberty in mice,[133,134] and prevents a delay in puberty onset in underfed female rats.[135,136] Leptin in food-deprived male rhesus monkeys stimulates an increase in LH concentrations as well as LH pulse frequency and amplitude,[137] indicating that leptin may provide information to the brain about nutritional status. The effect of leptin on GnRH release is likely indirect, as leptin receptor mRNA is not found in GnRH neurons, but it is found in NPY and proopiomelanocortin neurons, which innervate GnRH neurons.[137] Nonetheless, there is a large number of reports indicating that leptin concentrations do not change with puberty in monkeys or humans,[138–141] and the timing of puberty is not altered in patients with leptin deficiency who received recombinant leptin.[142] In addition, there is no correlation between leptin/leptin receptor gene mutations and timing of puberty.[143,144] Therefore, leptin may not be critical for the timing of puberty.

LIN28b

Recent studies with meta-analysis ($n = 17{,}000$) and genome-wide analysis ($n = 25{,}000$) revealed a strong interaction between the timing of puberty and genes related to physical size.[145] Specifically, two loci (6q21 and 9q31.2) were significantly correlated with the age of menarche and other signs of puberty, such as breast development in girls or pubic hair development in boys.[123–125,146] Interestingly, *LIN28b*, within chromosomal region 6q21 loci,[124] is associated with adult height,[147] and some variations (single nucleotide polymorphism) in *LIN28b* are associated with delayed pubertal growth and later puberty onset in girls.[125,146,148] Similarly, a two-nucleotide deletion immediately 5′ from the miRNA recognition element of *LIN28b* was associated with constitutional delay of growth and delayed puberty.[149] Finally, transgenic mice expressing LIN28a, a homolog of LIN28b, were recently produced[127] and found to exhibit an increased body size and delayed puberty. (Note that pubertal increase in estrogens causes epiphyseal closure, such that later puberty gives a longer period of long bone growth.) Thus, the *LIN28b* gene may be important for determining the timing of puberty.

Transcription Factors, Extrapolated from Experiments in Rodents

Ojeda and colleagues[150] suggest that three genes, *Oct-2*, TTF1 and *EAP*, may be involved in the timing of puberty through transactivation or repression of other downstream genes. Because Oct-2, a POU-II class protein,[151] is uniquely expressed in the postnatal hypothalamus,[151] these authors[150] examined the role of *Oct-2* POU domain genes in puberty. To summarize, they found that: (1) three alternatively spliced forms of the *Oct-2* gene, *Oct-2a*, *Oct-2b*, and *Oct-2c*, are found in the rat hypothalamus after lesions of the anterior hypothalamus, which are known to induce precocious puberty; (2) the time-course of changes in mRNA levels of each variant after the anterior hypothalamic lesion differs, expression of Oct-2a mRNA levels were maximal at 8 h after the lesion, Oct-2c transcripts transiently increased between 48 and 72 h, and Oct-2b mRNA levels only changed 4–5 days after the lesion; (3) anterior hypothalamic lesion-induced astroglia coexpress both TGFα and Oct-2 proteins; (4) an increase in Oct-2a mRNA levels occurs during the early phase of puberty onset, whereas an increase in Oct-2c mRNA levels coincides with the first LH surge; and (5) infusion of an antisense oligodeoxynucleotide for Oct-2 delays the timing of puberty in female rats. Because the TGFα promoter contains an octamer-like motif and a TAATGARAT (OCTA-) motif, the authors speculate that Oct-2a and Oct-2c may transactivate TGFα transcription, and in the case of Oct-2a lead to puberty. It is possible that Oct-1 (a Pou-II class protein) and Tst-1/SCIP, a POU-III class protein,[151] may be involved in the mechanism of the onset of puberty, since Oct-1 activates the neuron-specific enhancer of the *GnRH* gene,[152] Tst-1/SCIP represses transcriptional activity of the *GnRH* gene,[153] and both mRNAs are present in the adult brain.[150]

Thyroid transcription factor 1 (*TTF1*), another homeobox gene, may play a role in initiating puberty. *TTF1* is required for diencephalic morphogenesis,[154] and after birth it remains expressed in selected neuronal and glial populations of the hypothalamus.[155] Acting within diverse cellular subsets, TTF1 enhances the transcriptional activity of genes required for the stimulatory control of gonadotropin secretion, such as GnRH, erbB2 and KiSS1, while repressing the transcription of a gene involved in the inhibition of GnRH neuronal function, such as preproenkephalin.[156] Conditional deletion of the *Ttf1* gene in differentiated neurons resulted in delayed female puberty, irregular estrous cyclicity, reduced reproductive capacity and a shortened reproductive span, demonstrating the importance of this regulatory pathway for the normalcy of female reproductive function.[156]

Ojeda and colleagues[157] also recently examined the role of the *enhanced at puberty* gene (*EAP1*). This gene, C14ORF4, located in chromosome 14 (14q23–14q32), has been reported in association with precocious puberty in humans,[158,159] and is now renamed *EAP1*. Interestingly: (1) EAP1 containing cells were found in the monkey and rat arcuate nucleus (ARC) and ventromedial hypothalamic nucleus and the rat POA, (2) EAP1 expression increased at puberty in both monkeys and rats, and (3) inhibition of rat POA EAP1 expression via lentivirus-mediated delivery of EAP1 siRNAs delayed puberty, disrupted estrous cyclicity and resulted in other ovarian abnormalities. These authors speculate that because EAP1 contains a RING finger domain of the C3HC4 subclass, it may play a role in transactivation of the *GnRH* gene, while repressing the preproenkephalin gene, which is inhibitory to GnRH release.

While the results of these studies are exciting, and have provided novel concepts for the mechanism of pubertal development, several caveats should be kept in mind. The studies were mostly conducted in rats, in which the mechanism leading to the timing of puberty differs from that in primates. The results are inconclusive, and these genes may not necessarily be master genes, as further upstream gene(s) may be present. Nonetheless, the search for a master gene(s) that controls the timing of puberty will continue to be a major task.

NEUROENDOCRINE MECHANISM OF PUBERTY ONSET

Species Difference in the Mechanism of Puberty Onset

An increase in pulsatile GnRH release is essential for the pubertal increase in gonadotropin secretion and subsequent gonadal stimulation. Pulsatile infusion of GnRH into sexually immature monkeys[160] (Fig. 19.5) and guinea pigs[161] results in precocious puberty, and an increase in endogenous episodic GnRH release has been shown to occur at the onset of puberty in monkeys,[96] sheep[162] and rats.[163–165] Circadian rhythms of LH release and GnRH release become prominent in primates and rodents at puberty.[10,86,96,97,166,167]

Despite these similarities, mechanisms initiating puberty are distinctly different between primates and other species, such as rodents and sheep.[10,86,96,97,166,167] First, the effects of neonatal gonadectomy on the developmental pattern of gonadotropin release differ (Figs 19.2, 19.6). For example, LH and FSH are elevated in neonatally gonadectomized monkeys as well as in monkeys gonadectomized after the onset of puberty, but gonadectomy in monkeys during the juvenile period (before the onset of puberty) does not increase gonadotropin levels.[168] In contrast, in rats and sheep, gonadectomy-induced elevations of gonadotropin release occurs from the neonatal period throughout life.[75,162,169–172] Second, responses of the GnRH neurosecretory system to steroid hormones differ. For example, sensitivity of the rodent and sheep GnRH neurosecretory system to estradiol negative feedback is very high during juvenile development,[75,173–175] and decreases at the time of puberty.[172,176] In fact, Foster and colleagues hypothesize that the decreased negative feedback sensitivity of the GnRH neurosecretory system initiates puberty in sheep.[175,177,178] In contrast, gonadectomy in juvenile/prepubertal monkeys results in an increase of neither gonadotropin secretion nor GnRH release[24,43,97] (Fig. 19.7). Moreover, the prepubertal monkey GnRH neuronal system is insensitive to estradiol, but subsequently becomes sensitive to estradiol during the early stages of puberty, resulting in suppression of GnRH release.[179] Third, the consequence of precocious puberty induced by glutamatergic input differs. Pulsatile administration of either GnRH[160] or N-methyl-D-aspartic acid (NMDA)[180] to activate GnRH release in prepubertal monkeys precipitates the initiation of puberty, but neither manipulation results in establishment of adult reproductive function after discontinuation of the treatment. In contrast, NMDA-induced precocious puberty in rats leads to stable cyclic ovulation.[181,182] Fourth, restriction of the prepubertal GnRH neurosecretory system by a tonic central inhibition is unique to primates. The juvenile primate neurosecretory system is governed by a strong, gonadal-steroid feedback-independent, tonic inhibition, such as input from GABA neurons. It appears that reduction in this inhibition, as a consequence of maturational changes in neural substrates responsible for central inhibition, is essential for an increase in excitatory input to initiate the pubertal process.[183] Such an inhibition is not apparent in sheep, and thus it may not play a primary role in determining the timing of the pubertal activation of GnRH neurons. Although two recent preliminary reports in mice suggest that an estrogen negative feedback mechanism through kisspeptin neurons may play a role in prepubertal inhibition of GnRH release,[184,185] this mechanism may not be comparable to that in primates, as it depends on steroid hormones. Rather, a developmental switch of steroid-hormone sensitive neuronal mechanisms appears to be a critical factor for the initiation of puberty in these species.[75,186,187]

The species difference in the mechanism driving puberty onset may be attributable to species differences in neuroanatomical substrates responsible for preovulatory GnRH release. Several decades ago, neuroendocrinologists studying mature animals found that: (1) complete isolation of the MBH does not interfere with cyclic ovulation in rhesus monkeys,[188] whereas elimination of anterior afferents to the MBH causes persistent estrus in rats;[189] and (2) lesions of the SCN do not interfere with cyclic ovulation in sheep[190] or monkeys,[188] but result in persistent estrus in rats.[189]

These observations likely correspond to the neuroanatomical distribution pattern of GnRH neurons and kisspeptin neurons, which are major regulators of the preovulatory GnRH surge. In rats and mice GnRH neurons are primarily located in the MPOA and septal region, whereas in humans, monkeys and sheep they are found in the MBH as well as in the MPOA.[191,192] Moreover, in monkeys and sheep the kisspeptin perikarya are exclusively found in the ARC, whereas in rodents they are found in the ARC and the AVPV – a structure known to be critical for cyclic ovulation in rodents.[193] Therefore, it is speculated that species difference in the mechanism of puberty can be, in part, attributable to the species difference in neural and glial substrates involved in regulation of GnRH neurosecretion.

Neurocircuitry Signaling to GnRH Neurons at Puberty

GnRH neurons are capable of releasing the decapeptide well before puberty (Fig. 19.8). However, the pubertal increase in GnRH secretion does not occur until the time of puberty due to immaturity of the

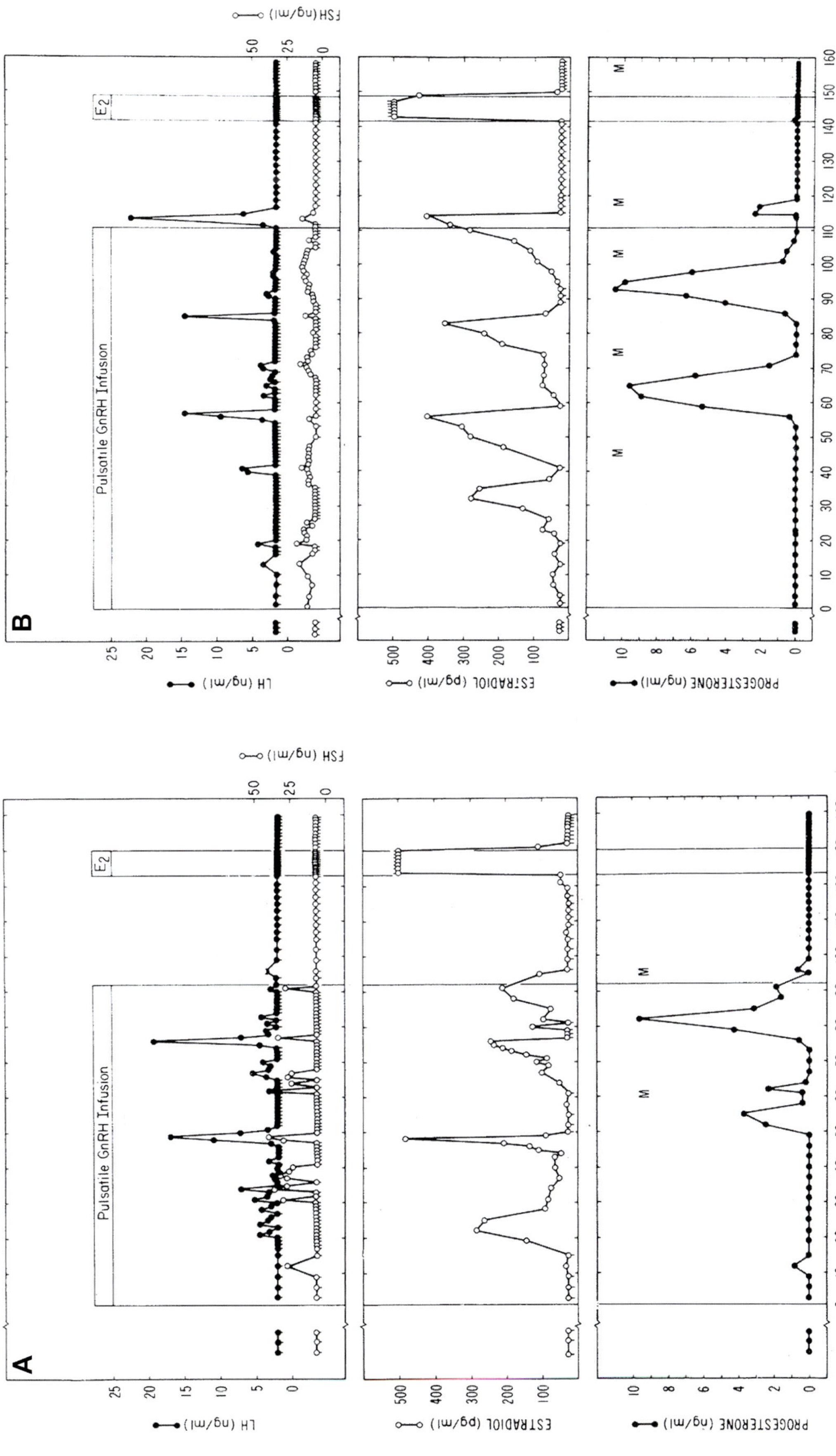

FIGURE 19.5 **Induction of puberty by chronic pulsatile infusion of GnRH into two prepubertal female monkeys (~14 months of age).** Changes in gonadotropin (LH and FSH, top), estradiol (middle) and progesterone (bottom) are shown. The GnRH infusion (shown in the top panel) was started on day 0 and continued until day 110. Ovulatory cycles, determined by a periodic estradiol increase followed by a progesterone increase, are seen approximately 2 months after the initiation of GnRH treatments. Note that after the termination of GnRH infusion, both females failed to respond to the estradiol challenge. *Reproduced from Wildt L, Marshall G, Knobil E. Experimental induction of puberty in the infantile female rhesus monkey.* Science. *1980;207:1373–1375, with permission.*

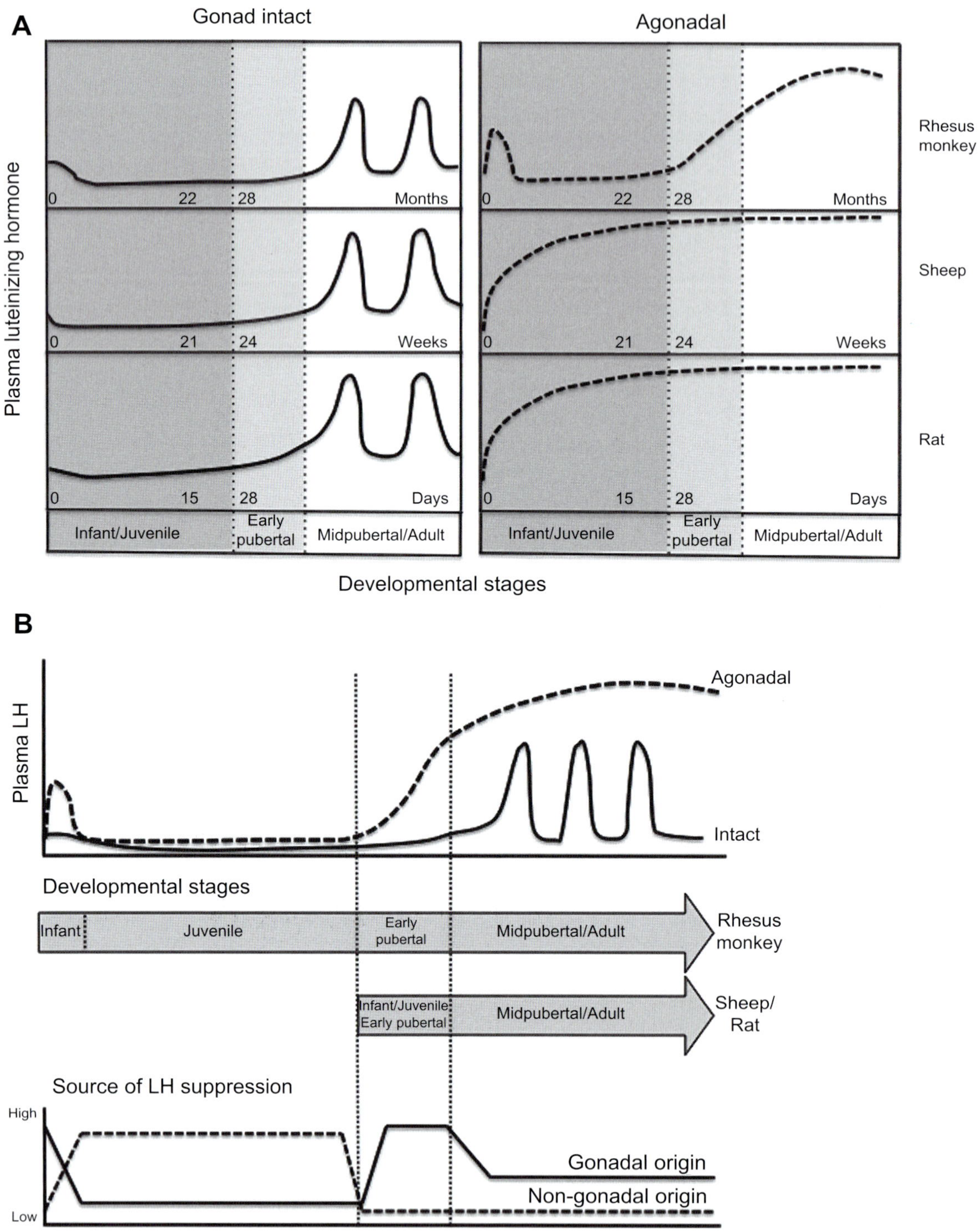

FIGURE 19.6 **A.** Schematic illustration of the ovarian impact on luteinizing hormone (LH) release, and presumably GnRH release, during postnatal development in rhesus monkeys (top), sheep (middle) and rats (bottom). Comparison of the LH developmental pattern between gonadal intact (left) and agonadal (right) females reveals distinct species differences. **B.** In primates, the juvenile stage is marked by a period of LH/GnRH release, which is controlled by a non-gonadal restraint mechanism. In contrast, in sheep and rats, prepubertal LH/GnRH release is controlled by gonadal steroids.

neurocircuitry controlling the secretory activity of GnRH neurons.[194,195] One of the most important mechanisms controlling GnRH neurons is synaptic input from other neurons. Whereas input from one type of neuron can be excitatory, other types of neural input can be inhibitory. Moreover, the same neural phenotype can be either excitatory or inhibitory, depending on the physiological condition. In this section we will review

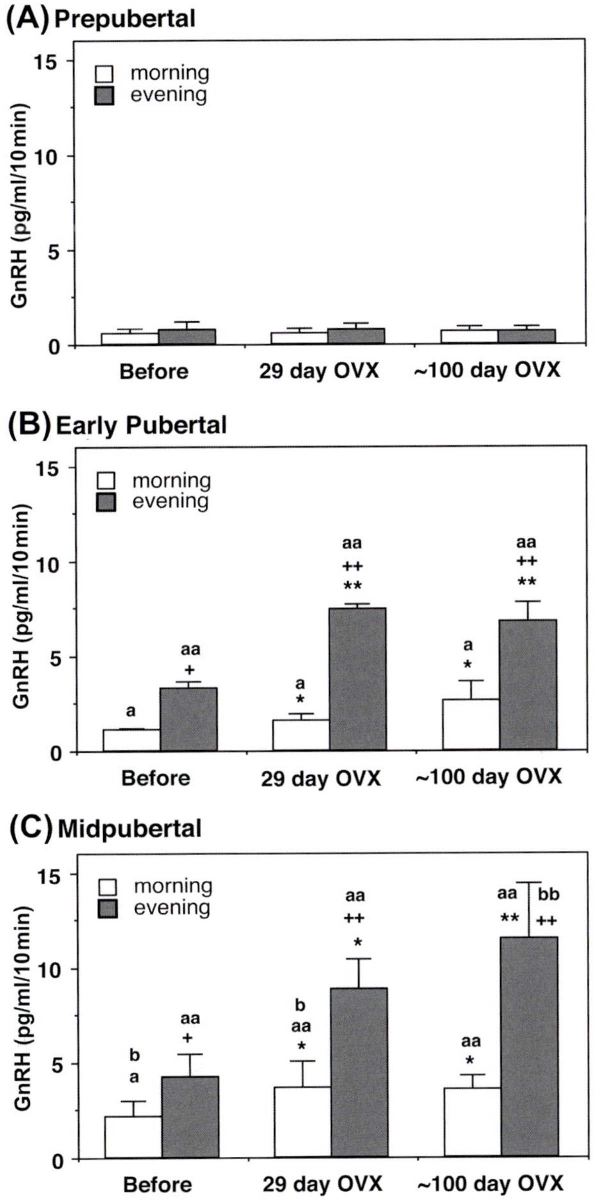

FIGURE 19.7 **Developmental changes in the effect of ovariectomy on GnRH release in female rhesus monkeys.** GnRH release from the stalk median eminence was measured by the push–pull perfusion method in prepubertal (**A**) early pubertal (**B**) and midpubertal (**C**) female monkeys. Group means (± SEM) over 6 hours in the morning (0600–1200 h, open bars) and 6 hours in the evening (1800–2400 h, filled bars) were assessed before ovariectomy, 29 days after ovariectomy and ~100 days after overiectomy. Note that the ovariectomy-induced increase in GnRH release is not seen in the prepubertal monkeys, whereas it is seen in early pubertal and midpubertal monkeys. *Reproduced from Chongthammakun S, Claypool LE, Terasawa E. Ovariectomy increases* in vivo *luteinizing hormone-releasing hormone release in pubertal, but not prepubertal, female rhesus monkeys.* J Neuroendocrinol. *1993; 5:41–50, with permission.*

several neuronal systems that contribute to the pubertal increase in GnRH release, and hence the onset of puberty.

Kisspeptin Neurons

In 2003, two groups reported a delay in or absence of puberty in patients with a single amino acid mutation of GPR54.[1,2,196] GPR54 is a member of the G-protein coupled seven-transmembrane receptor family, and the metastasis suppressor gene *KiSS-1* derived peptide, kisspeptin-54, is its ligand.[197–199] The C-terminal region of kisspeptin-54 is responsible for receptor binding; short forms of the peptide derived from this c-terminus (e.g., kisspeptin-10, -13, -14, and -15) are all biologically active.[197–199] Since the first reports on the possible role of kisspeptin signaling in reproductive function, and particularly in puberty, a significant number of papers and review articles have been published.[200–205]

Several important findings highlight the role of kisspeptin signaling in reproductive function and puberty. First, as seen in human patients, mice lacking the *GPR54* gene fail to undergo puberty and have immature gonads;[2,206] *KiSS-1* null mice also exhibited a lack of reproductive development, but the phenotype in these animals is not as severe as that of GPR54 knockout mice, suggesting the existence of additional ligands[207,208] or the presence of a kisspeptin-GPR54 independent GnRH regulatory mechanism.[113, 208] This notion is also supported by reports that low-amplitude LH pulses are observed in human patients with severe forms of GPR54 mutations.[2,209] Second, kisspeptin-expressing neurons are present in the ARC nucleus in monkeys, sheep, rats, mice and hamsters, and in the AVPV nucleus in mice and rats,[210–216] the brain structures known to be critical for control of the negative and positive feedback action of gonadal steroid hormones on GnRH secretion in rodents. Third, GnRH neurons express GPR54.[116,210,217,218] Kisspeptins stimulate the release of LH, FSH and GnRH,[210,219–221] and central administration of kisspeptin results in precocious LH elevation in prepubertal male monkeys[210,222] as well as precocious puberty in female rats,[223,224] whereas central administration of the kisspeptin antagonist, peptide 234, delays the timing of vaginal opening in female rats.[225] Fourth, hypothalamic expression of both GPR54 and KiSS-1 mRNA are significantly elevated around the time of puberty in male and female rats[219,226] and in ovarian intact pubertal female monkeys.[210] Similarly, the number of kisspeptin positive neurons in the AVPV of mice also increases at the age of puberty.[227,228] Interestingly, however, the developmental increase in the number of kisspeptin neurons in the rodent AVPV is dependent on the presence of circulating estradiol, as ovariectomy of prepubertal mice eliminates this developmental change.[229] Fifth, the number of GnRH neurons depolarized by kisspeptin-10 increases during postnatal development

BOX 19.1

HOW DO WE KNOW THAT, IN HUMANS AND NON-HUMAN PRIMATES, A JUVENILE HIATUS IN LH SECRETION IS CONTROLLED BY A MECHANISM THAT IS INDEPENDENT OF GONADAL INFLUENCE?

The concept that a non-gonadal mechanism inhibits the GnRH neurosecretory system prior to puberty onset in primates is based on findings in human and non-human primates. In humans, patients with gonadal dysgenesis or other forms of functional agonadism due to chromosomal abnormalities (e.g., Turner syndrome) exhibit elevated LH levels during the neonatal period, which decrease during childhood and consequently rise at the time of puberty.[1,2] Similarly, in a classic study by Plant,[3] male rhesus monkeys were either bilaterally orchidectomized (castrated) or kept intact (controls) at 1 week after birth and blood samples were taken on a weekly basis through to 300 days of age. After castration, plasma luteinizing hormone (LH) levels increased an order of magnitude in orchidectomized males compared to controls. This elevated secretion of LH persisted for nearly 4 months, at which point plasma LH levels dropped and stabilized at low levels, which were indistinguishable from gonadally intact controls. This period of low plasma LH levels is referred to as the "juvenile hiatus" in LH secretion.

Terasawa and colleagues, studying prepubertal female rhesus monkeys, also found that low circulating levels of LH and FSH [4] and GnRH release in the hypothalamus are independent of the presence or absence of ovarian steroid hormones, but then become sensitive to estradiol after the initiation of puberty.[5,6]

Later studies by Plant and colleagues characterized the developmental changes in circulating LH and FSH from orchidectomized males and ovariectomized females.[7] Again, they found that during the neonatal period LH and FSH levels were elevated when the gonads were absent. This elevation was more pronounced in males, but was still the case for both sexes. As seen in the earlier studies, a juvenile hiatus in gonadotropin secretion occurred in both sexes, but this time, samples were taken until the typical age of puberty onset. Similar to that observed in humans at the onset of puberty, both LH and FSH start to increase in gonadectomized animals of both sexes; however, the age of pubertal onset in females was younger than that in males, and the pubertal increase in FSH precedes LH in both sexes.

While LH secretion is also low during prepubertal–juvenile development of other mammals, including sheep and rats, it appears the female juvenile hiatus in LH secretion is due to elevated sensitivity to gonadal steroids (i.e., estradiol), which continues throughout prepubertal development. For example, Foster and colleagues [8] found that this juvenile hiatus is strikingly sensitive to estradiol, in female sheep. This also appears to be the case for rats. While similarly controlled studies are yet to be reported in rats, the relative absence of a juvenile non-gonadal brake on LH secretion in this model can be inferred from the work of Ojeda and Ramirez.[9]

References

1. Ross JL, Loriaux DL, Cutler Jr GB. Developmental changes in neuroendocrine regulation of gonadotropin secretion in gonadal dysgenesis. *J Clin Endocrinol Metab.* 1983;57:288–293.
2. Conte FA, Grumbach MM, Kaplan SL, Reiter EO. Correlation of luteinizing hormone-releasing factor-induced luteinizing hormone and follicle-stimulating hormone release from infancy to 19 years with the changing pattern of gonadotropin secretion in agonadal patients: relation to the restraint of puberty. *J Clin Endocrinol Metab.* 1980;50:163–168.
3. Plant TM. The effects of neonatal orchidectomy on the developmental pattern of gonadotropin secretion in the male rhesus monkey (Macaca mulatta). *Endocrinology.* 1980;106: 1451–1454.
4. Terasawa E, Nass TE, Yeoman RR, Loose MD, Schultz NJ. Hypothalamic Control of Puberty in the Female Rhesus Macaque. In: Norman RL, ed. *Neuroendocrine Aspects of Reproduction.* New York, NY: Academic Press; 1983:149–182.
5. Chongthammakun S, Terasawa E. Negative feedback effects of estrogen on luteinizing hormone-releasing hormonerelease occur in pubertal, but not prepubertal, ovariectomized female rhesus monkeys. *Endocrinology.* 1993;132:735–743.
6. Chongthammakun S, Claypool LE, Terasawa E. Ovariectomy increases *in vivo* luteinizing hormone-releasing hormone release in pubertal, but not prepubertal, female rhesus monkeys. *J Neuroendocrinol.* 1993;5:41–50.
7. Plant TM, Witchel SF. Puberty in nonhuman primates and primates. In: Neill J, ed. *The Physiology of Reproduction.* 3rd ed., Volume 2. San Diego, CA: Academic Press; 2006:2177–2230.
8. Foster DL, Ebling FJ, Vennerson LA, Wood RI, Fenner DE. Regulation of Puberty in the Lamb: Internal and External Cues. In: Imura H, Shizume K, Yoshida S, eds. *Progress in Endocrinology,* Vol. 2. Amsterdam: Elsevier; 1988:861–866.
9. Ojeda SR, Ramirez VD. Short-term steroid treatment on plasma LH and FSH in castrated rats from birth to puberty. *Neuroendocrinology.* 1973-1974;13:100–114.

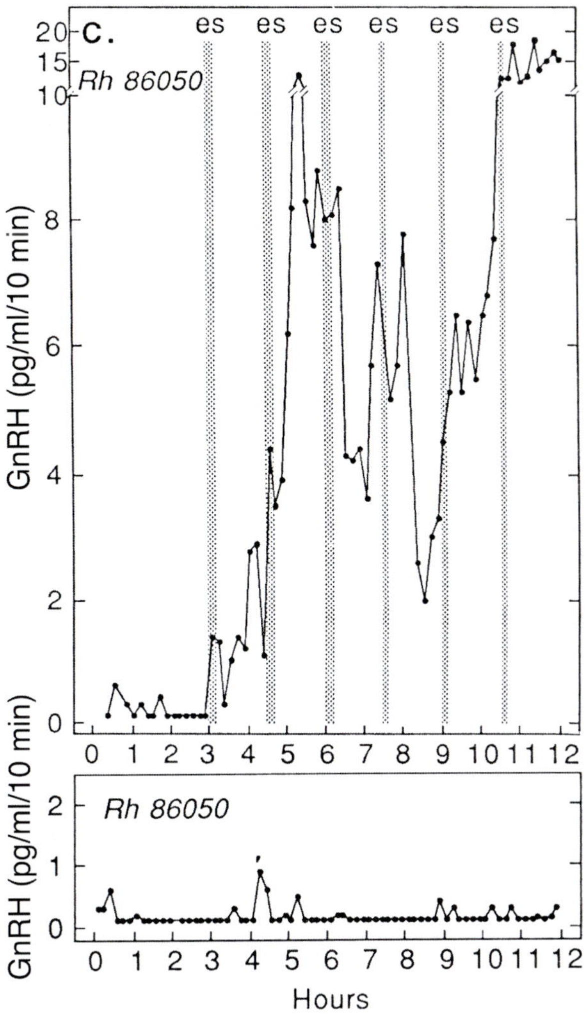

FIGURE 19.8 **Electrical stimulation of the medial basal hypothalamus induces an increase in GnRH release in prepubertal monkeys.** Hatched vertical lines in the top panel indicate the times at which electrical stimulation (es) was applied. *Reproduced from Claypool LE, Watanabe G, Terasawa, E. Effects of electrical stimulation of the medial basal hypothalamus on the* in vivo *release of luteinizing hormone-releasing hormone in the prepubertal and peripubertal female monkey.* Endocrinology. *1990;127:3014–3022, with permission.*

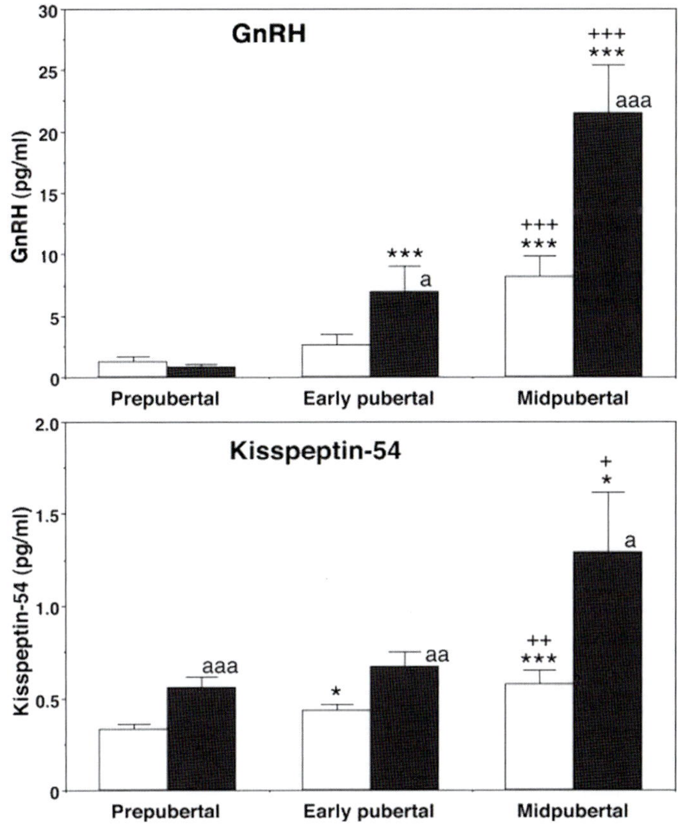

FIGURE 19.9 **Developmental changes in the release of GnRH (top) and kisspeptin-54 (bottom) in push–pull perfusates.** Both peptides were measured in the same samples. Kisspeptin-54 levels gradually increased along with the pubertal increase in GnRH release. The nocturnal increase in kisspeptin-54 release was already observed in prepubertal monkeys, and continued through the pubertal period. In contrast, the nocturnal increase in GnRH release occurred in early and midpubertal monkeys, but not in prepubertal monkeys. The numbers of animals at the prepubertal, early pubertal, and midpubertal stages were 6, 6, and 5, respectively. Morning values,open bar; evening values, closed bar. * $P < 0.05$ vs prepubertal; *** $P < 0.001$ vs prepubertal; + $P < 0.05$ vs early pubertal; ++ $P < 0.01$ vs early pubertal; +++ $P < 0.001$ vs early pubertal; a $P < 0.05$ vs AM; aa $P < 0.01$ vs AM; aaa $P < 0.001$ vs AM. *Reproduced from Keen KL, Wegner FH, Bloom SR, Ghatei MA, Terasawa E. An increase in kisspeptin-54 release occurs with the pubertal increase in luteinizing hormone-releasing hormone-1 release in the stalk median eminence of female rhesus monkeys* in vivo. Endocrinology. *2008;149:4151–4157, with permission.*

in mice and the LH response to central administration of kisspeptin in juvenile mice is higher than in adult mice,[227] which is parallel to a developmental increase in the number of GnRH neurons with kisspeptin fiber appositions,[228] indicating that sensitivity of GPR54 to kisspeptins may increase prior to puberty in rodents. In non-human primates, however, sensitivity of GnRH neurons to kisspeptin signals does not change across puberty onset.[230] Finally, a recent study in female monkeys suggests that kisspeptin-54 release increases along with the pubertal increase in GnRH release, and that the nocturnal increase in kisspeptin-54 release precedes the nocturnal increase in GnRH release[231] (Fig. 19.9).

Kisspeptin neurons in the ARC are indispensable for the negative feedback action of estrogen, whereas kisspeptin neurons in the AVPV appear to mediate at least part of the positive feedback effect of estrogen in rodents.[232] In fact, kisspeptin neurons in the AVPV, but not in the ARC, are sexually dimorphic in mice, with the number of kisspeptin neurons in females being much larger than that in males.[204,205,228] Because lesions in the AVPV result in the loss of ovulatory cycles in female rats,[193,233,234] kisspeptin neurons in the AVPV appear to be of critical importance for the occurrence of the preovulatory surge of GnRH/

gonadotropins.[204,205,229] Gonadal steroids modulate GPR54 and KiSS-1 mRNA expression in the brain: gonadectomy increases and steroid replacements decrease GPR54 and KiSS-1 mRNA abundance in the hypothalamus[235] and KiSS-1 mRNA in the ARC[212,213] of adult rats. These treatments cause the opposite change in KiSS-1 mRNA in the AVPV of adult rats.[212,213] Virtually all kisspeptin neurons in the rat and sheep ARC nucleus and AVPV express ERα.[236,237] In fact, estradiol fails to alter KiSS-1 mRNA expression in these neurons in ERα knockout mice.[236] Moreover, kisspeptin neurons in the ARC nucleus and the AVPV mediate negative and positive feedback effects, respectively, of estrogen on GnRH neurons in rodents.[212,213,238] Similarly, both KiSS-1 and GPR54 mRNA in the MBH of ovarian-intact female mice increase with sexual development, and in agonadal male monkeys KiSS-1, but not GPR54, mRNA expression in the MBH at the pubertal age is higher than that at the juvenile age.[210] Although the observation in agonadal male monkeys[210] indicates that the developmental increase in GPR54 signaling activity is independent from the pubertal increase in androgen, it is unclear whether developmental increases in both KiSS-1 and GPR54 mRNA in the MBH of ovarian-intact females are due to the pubertal increase in estrogen. Because a majority of kisspeptin neurons express ERα[236,237] and a significant sex difference in the negative feedback of GnRH release by gonadal steroids is present during the juvenile period,[239] an understanding of the role that kisspeptin plays in puberty requires further refinement.

Glutamate Neurons

Glutamate is the major excitatory amino acid neurotransmitter in the hypothalamus.[240,241] It exerts its biological actions via either ionotropic or metabotropic receptors. Ionotropic receptors are ligand-gated cation channels that mediate fast excitatory postsynaptic potentials; they have been divided into three distinct subtypes according to their preferred ligand: NMDA, kainate and α-amino-3-hydroxy-5-methyl-4-isoxazole propionic acid (AMPA) receptors.[242] Metabotropic receptors are large monomeric transmembrane proteins containing seven-transmembrane domains that initiate intracellular signaling via coupling to G proteins. All of these receptors are present in the hypothalamus.[243–246]

GnRH neurons receive direct glutamatergic innervation.[247,248] Activation of AMPA and NMDA receptors depolarizes the membrane and increases the firing rate in GFP-labeled mouse GnRH neurons, also suggesting that GnRH neurons receive a glutamatergic innervation, perhaps from the lateral ventromedial and premammillary nucleus of the hypothalamus.[249] An observation by Campbell and colleagues[250] confirms that GnRH neurons have long dendrites covered with a large number of spines, highly suggestive of an abundance of excitatory synaptic contacts.[251]

Activation of each of the three ionotropic receptor subtypes stimulates GnRH/LH release in adult animals: NMDA stimulates release of LH and GnRH in adult rats and monkeys *in vivo*,[252–258] while the NMDA receptor blocker MK-801 suppresses GnRH pulses as well as the preovulatory LH surge.[245,257,259,260] Glutamate stimulates GnRH release from hypothalamic fragments *in vitro* as well as the GT-1 GnRH-secreting cell line.[253,261–264]

Glutamate, NMDA, and kainate stimulate GnRH/LH release in sexually immature animals, including monkeys[265–267] (Fig. 19.10), rats[181,256,268,269] and sheep.[270,271] Importantly, stimulation of NMDA receptors results in precocious puberty in rats[181,182] and monkeys,[180] whereas administration of the NMDA receptor blockers MK-801 or 2-amino-5-phosphonovaleric acid (AP-5) delays the puberty onset in rats.[260,272–275] In contrast, the non-NMDA glutamate receptor antagonist 6,7-dinitro-quinoxaline-2,3-dione (DNQX) fails to alter the timing of puberty.[269]

Glutamine, a major glutamate precursor, elicits a dose-dependent release of GnRH from the developing rat hypothalamus *in vitro*.[186] This stimulatory effect is due to newly formed glutamate, because pharmacological blockade of the rate-limiting enzyme glutaminase, which catalyzes the conversion of glutamine into glutamate, prevents the stimulatory effect of glutamine, but not that of glutamate, on GnRH release.[186,187] Moreover, inhibition of glutamine-induced GnRH release requires a higher concentration of inhibitor in peripubertal than infantile hypothalami, indicating that glutamate biosynthesis from glutamine increases at the time of rat puberty.[186,187] Nonetheless, changes in glutaminase mRNA levels do not appear to change during pubertal development in the female rat hypothalamus.[276,277] The mechanism of the puberty-related increase in hypothalamic glutamate synthesis without an increase in glutaminase gene expression is unclear.

Consistent with a postnatal augmentation in hypothalamic glutamate synthesis, glutamate concentrations in the rat hypothalamus increase during prepubertal development,[278,279] reaching a maximum after the onset of puberty.[279] More detailed studies have shown that both glutamate concentrations in the POA/MBH and release of the amino acid from POA/MBH explants increase during juvenile development (P16 to P30) in female rats.[280,281] Importantly, the *in vivo* release of glutamate from the POA of female rats, measured in push–pull perfusates, also increases significantly during the juvenile–peripubertal periods,[282] and is markedly elevated prior to and during the preovulatory LH surge induced by either estrogen alone,[283] or the

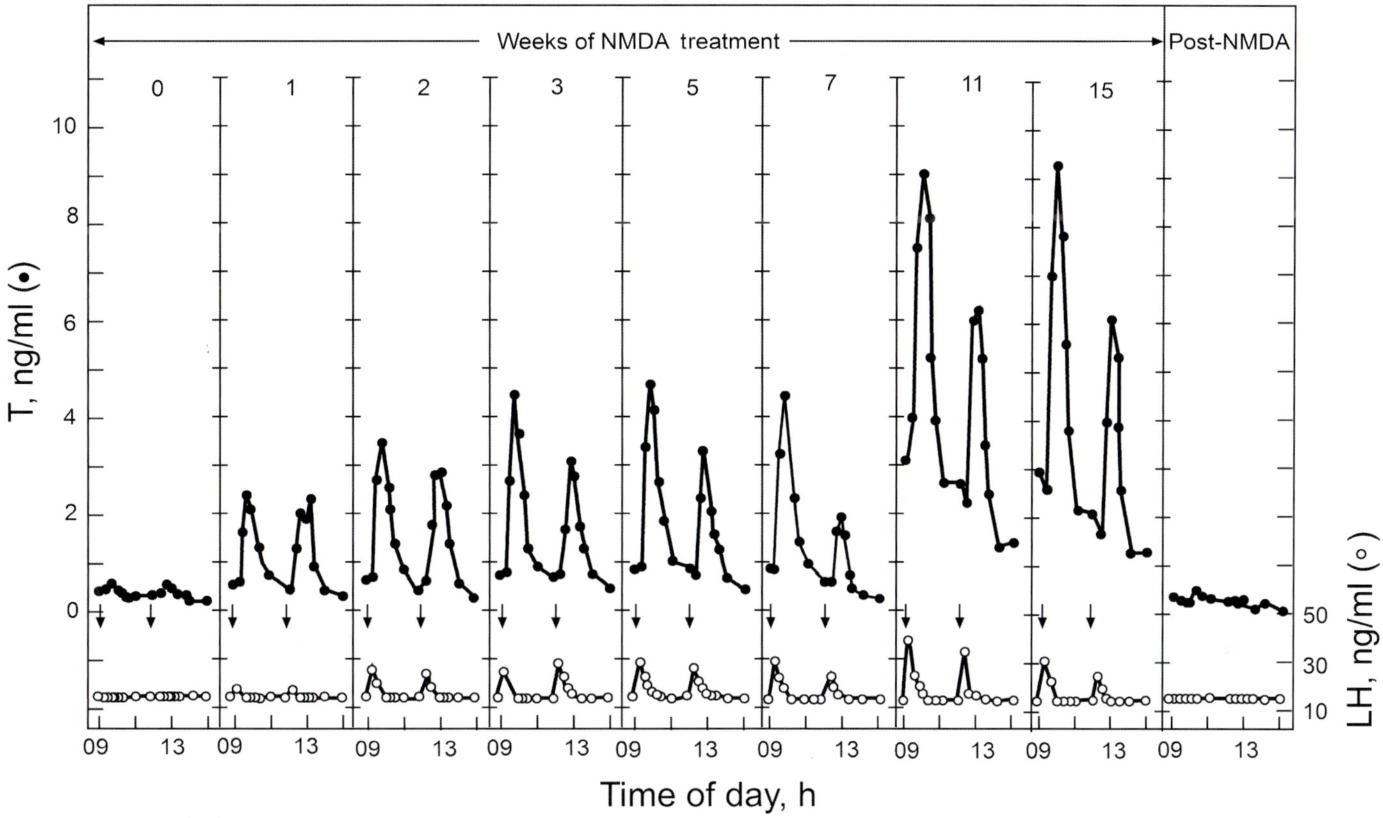

FIGURE 19.10 **Induction of premature hypothalamo-pituitary–testicular activity in juvenile male rhesus monkeys.** The excitatory amino acid receptor agonist, NMDA, was administered once every three hours starting at 15–16 months of age (1.5–2 years before the normal age of puberty in this species) and continued for 15 weeks. The arrows indicate the time of injection. Changes in testosterone (closed circles) and LH (open circles) in response to NMDA between 0900 and 1500 h are shown. *Reproduced from Plant TM, Gay VL, Marshall GR, Arslan M. Puberty in monkeys is triggered by chemical stimulation of the hypothalamus.* Proc Natl Acad Sci USA. *1989;86: 2506–2510, with permission.*

sequential treatment of estrogen plus progesterone.[284] In female rhesus monkeys, glutamate release measured by push–pull perfusion in the SME is very low during the prepubertal period, strikingly increases during early puberty, and remains elevated during midpuberty, though at levels slightly lower than at early puberty[25] (Fig. 19.11).

In vitro perfusion experiments with hypothalami from developing male rats have shown that the release of GnRH in response to glutamate stimulation increases transiently before puberty.[285] Similarly, GnRH released from POA/MBH explants in response to either NMDA or kainate demonstrated that both are more effective in peripubertal (P30) than prepubertal (P16) female rats.[280,281,286] Further, inhibition of GnRH secretion in response to the non-competitive NMDA receptor blocker MK-801 is more prominent in peripubertal animals,[285] suggesting that activation of hypothalamic NMDA receptors precedes the initial increase in GnRH release. Based on results from these *in vitro* experiments, it has been proposed[287] that a component of the NMDA receptor-mediated glutamatergic input to GnRH neurons is inhibitory to GnRH release in juvenile rats, gradually becoming stimulatory to GnRH release as the animals approach puberty. Importantly, the GnRH response to NMDA *in vitro* was significantly increased by *in vivo* pretreatment with a sequential regime of estrogen and progesterone of rats,[280,288] suggesting that a steroid-dependent potentiation of NMDA receptor-mediated excitatory signals is a part of the central mechanism underlying the first preovulatory surge of GnRH/gonadotropins.

Sensitivity to glutamatergic stimulation also increases at puberty in rhesus monkeys: infusion of NMDA into the SME stimulates GnRH release in pubertal monkeys at a dose as low as 10 nM, but a concentration of 100 nM is necessary to induce GnRH release in prepubertal monkeys.[289] In addition, an intravenous injection of NMDA results in more GnRH release and for a longer time in pubertal than in prepubertal monkeys.[289] It is, however, unclear whether the heightening in NMDA receptor sensitivity during puberty is due to the pubertal increase in gonadal steroids. Nonetheless, the developmental changes in NMDA and AMPA receptors are integral parts of changes leading to puberty.

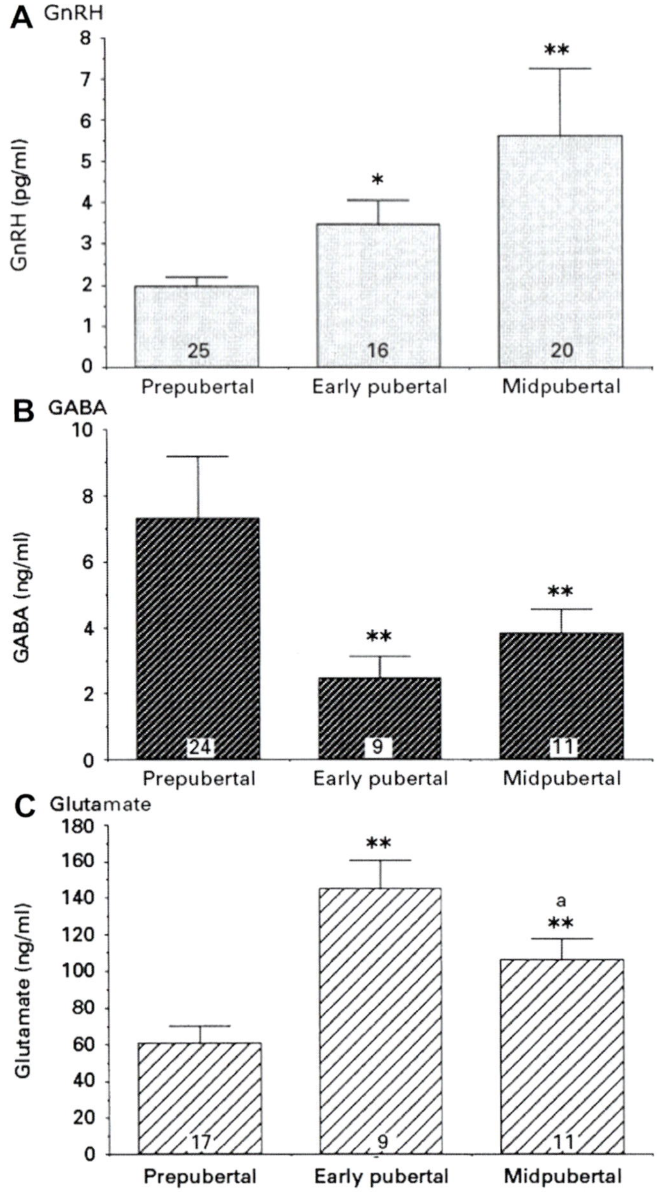

FIGURE 19.11 **Decrease in GABA release (middle) and increase in glutamate release (bottom) from the rhesus monkey hypothalamus coincides with the pubertal increase in GnRH release (top).** GnRH, GABA and glutamate levels in prepubertal, early pubertal and midpubertal female rhesus monkeys in the stalk median eminence were measured by the push–pull perfusion method. * $P < 0.05$ vs prepubertal and $P < 0.01$ vs prepubertal; a $P < 0.05$ vs early pubertal. *Reproduced from Terasawa E, Luchansky LL, Kasuya E, Nyberg CL. An increase in glutamate release follows a decrease in gamma aminobutyric acid and the pubertal increase in luteinizing hormone releasing hormone release in the female rhesus monkeys.* J Neuroendocrinol, *1999;11:275–282, with permission.*

GABA Neurons

GABA is the dominant inhibitory neurotransmitter in the hypothalamus.[290] GABA is primarily synthesized from glutamate via a decarboxylation reaction catalyzed by glutamic acid decarboxylase (GAD). Once synthesized, GABA is stored in vesicles and released by exocytosis upon depolarization, in the presence of extracellular Ca^{2+}.[291] Two isoforms of GADs, GAD-67 and GAD-65, encoded by different genes, have been reported.[292,293] However, changes in GAD enzyme mRNAs would not necessarily reflect alterations in synaptic GABA release, as GABA synthesis from glutamate is controlled by complex mechanisms, including the presence of both the active holo form and apo forms of GAD-67 and GAD-65 enzymes.[293–295]

Amino acid neurotransmitters exocytosed into the synaptic cleft are actively removed by specific transporters located on presynaptic terminals, and by glial cells surrounding the synapses, and recycled for further use. GABA transporters (GAT) are members of the Na^+ and Cl^--coupled transporter family. Four GABA transporters (GAT-1, GAT-2, GAT-3 and BGT-1) have been described,[296] and GAT-1 is the predominant GABA transporter in the mammalian brain.[296] Therefore, it is possible that the changes in GABA concentration in the SME at the onset of puberty [25] are due to a developmentally regulated alteration in GAD activity, as well as GABA transporter activity.

Three GABA receptors, $GABA_A$ $GABA_B$ and $GABA_C$, have been identified in the brain. $GABA_A$ receptors are pentameric membrane-spanning ligand-gated anion channels that permit chloride entry into the cell upon binding of GABA.[297–299] $GABA_B$ receptors are seven-transmembrane domain receptors negatively coupled to adenylate cyclase, and to calcium and potassium channels, via GTP-binding proteins.[300–302] $GABA_C$ receptors identified as chloride pores in the retina are similar to $GABA_A$, but contain $\rho1-\rho3$ subunits.[303]

$GABA_A$ receptors are modulated by allosteric agonists, benzodiazepines, barbiturates, neurosteroids (primarily metabolites of pregnane steroids) and ethanol, as well as polyvalent cations.[304,305] $GABA_A$ receptors are composed of at least 18 genetically distinct subunits ($\alpha1-\alpha6$, $\beta1-\beta4$, $\gamma1-\gamma4$, δ, ε, π and θ). An α subunit, β subunit and γ subunit are all required to form a fully functional $GABA_A$ receptor,[306,307] and the $\alpha1\beta2\gamma2$ arrangement is the most common in the brain. The subunit make-up of the $GABA_A$ receptor is important, because the functional and pharmacological characteristics of the receptor are determined by its subunit composition.[297,302,308] Subunit composition could differ between brain areas, but also from neuron to neuron.[309–312] Moreover, the pattern of subunit composition not only changes between prenatal and postnatal development, but also alters under specific physiological conditions.[310,313–317]

It is possible that developmental changes in $GABA_A$ receptor subunit composition alter the regulatory activity of GABA input to GnRH neurons. Studies performed in the rat brain indicate that, in general, α_2

subunit expression is very high at birth, declining thereafter towards adult levels. In contrast, α_1 expression is minimal in the fetal brain, and then gradually increases after birth towards adulthood.[314,316,318] Although a detailed observation in the hypothalamus has yet to be conducted, comparable developmental changes in the $GABA_A$ subunit composition of GnRH neurons may occur prior to the onset of puberty. In fact, it has been shown that a pattern of GnRH neurons expressing GABA subunits in the POA and medial septum of sexually immature mice is more heterogeneous than that in adults[319] – i.e., GnRH neurons in sexually immature mice (neonatal at P5 and juvenile at 15–20) express α_1–α_5, β_1–β_3, γ_2 and γ_3, whereas GnRH neurons obtained from adult mice have significantly reduced α_1, α_5, β_1 and γ_2 expression patterns. Interestingly, the same group reports that GnRH neurons of juvenile mice, but not those of adult animals, are depolarized by GABA,[319] and a switch from depolarization to hyperpolarization occurs around the time of puberty,[320] presumably due to a differential composition of $GABA_A$ receptor subunits between adult and immature animals. Other laboratories report that adult GnRH neurons express the β_3 but not the α_1 or β_2 $GABA_A$ subunits,[321] whereas in peripubertal female rats GnRH neurons express α_1, α_2, β_3 and γ_2 but not γ_1 subunits.[322] Some of the difference in the $GABA_A$ subunit composition in GnRH neurons between laboratories, however, might be attributed to the methods of detection employed (*in situ* hybridization vs single cell PCR) rather than developmental changes. In fact, a recent study shows that the GABAergic gene expression patterns assessed by microarray analysis and Q-PCR are not identical.[323] Nonetheless, using the gene array analysis, in ovariectomized adult female monkeys estradiol plus progesterone treatments increased α_1- and α_4-subunit mRNA levels in the MBH, whereas the treatment decreased levels of α_1- and α_4-subunit mRNA levels in the amygdala and hippocampus.[323] Interestingly, ε-subunit mRNA levels in the MBH were much higher than those in the amygdala and hippocampus, and ε-subunit mRNA expression levels were significantly increased by the estradiol alone or estradiol and progesterone treatments using Q-PCR analysis.[323] The developmental profile of GABA subunit composition in primate has not been reported. GAD-65 and GAD-67 mRNA content and GABA concentrations in the rat POA increase first between neonatal and infantile development,[278,324,325] and then again during the juvenile period.[276,279] With regard to developmental changes in GAD mRNA and/or GABA content within the MBH, the reports are more conflicting. Some[324] show an increase in GAD mRNA content in the ARC; others indicate that GABA concentrations in the rat MBH do not change during the juvenile period.[279] In rhesus monkeys the distribution pattern and concentration of GAD-67 and GAD-65 mRNA in hypothalamic nuclei, assessed by *in situ* hybridization in gonadally intact juvenile (~0.6 years of age) male rhesus monkeys, were not different from those in adult (~10 years of age) animals,[326] and GAD mRNA levels in the basal hypothalamus of juvenile castrated male rhesus monkeys did not differ from those of adult castrated male monkeys.[41] Consistent with the developmental changes, estrogen or estrogen plus progesterone treatment in ovariectomized monkeys did not alter GAD-67 and GAD-65 mRNA levels in the MBH, unlike clear suppression by the steroid treatments in the amygdala and hippocampus of ovariectomized monkeys.[323]

A developmental switch from excitation to inhibition by $GABA_A$ receptor activation has been described in hippocampal and spinal neurons, as a consequence of maturational changes in the regulation of intracellular chloride concentrations. Immature neurons, such as embryonic or neonatal neurons, lack the potassium-coupled chloride transporter KCC-2 and hyperpolarization-activated chloride channel CLC-2, but have the sodium- potassium-2-chloride cotransporter NKCC-1, which brings two chlorides along with sodium and potassium into the neuron. Consequently, they exhibit higher intracellular chloride concentrations (above the ion's electrochemical equilibrium), such that activation of $GABA_A$ receptors results in chloride efflux, leading to postsynaptic depolarization.[327–329] As neurons mature, the mechanisms for chloride extrusion are established and activation of $GABA_A$ receptors results in chloride influx, leading to postsynaptic hyperpolarization.[330–332] Therefore, the observation that adult GnRH neurons respond to $GABA_A$ receptor activation with excitation[333,334] can be interpreted to mean that GnRH neurons belong to the class of neurons in which chloride transporters do not undergo developmental changes. Rather, direct $GABA_A$ receptor-mediated inputs may induce postsynaptic activation even after maturation,[328] because GnRH neurons express neither KCC-2 nor CLC-2, but NKCC-1, in neonatal, juvenile and adult mice, and $GABA_A$ receptor activation is excitatory regardless of the age of mice.[335] Although it is possible that the developmentally controlled switch in GnRH response to $GABA_A$ receptor mediated inhibition is an event that takes place in interneurons connected to GnRH neurons, which contain more KCC-2 and CLC-2 and less NKCC-1, it remains a mystery how $GABA_A$ receptor activation is inhibitory in female rhesus monkeys, particularly during the prepubertal period.[336]

GABA has been shown to both stimulate and inhibit GnRH/LH secretion. Earlier studies indicate that systemic administration or intracerebroventricular infusion of GABA, and application of GABA to hypothalamic explants, stimulate GnRH and LH release,[337–339] whereas others show inhibitory or biphasic/triphasic effects.[261,338,340–342] More recent *in vivo* studies have yielded more consistent results with an inhibitory effect of GABA on pulsatile GnRH/LH release. For example, in adult ovariectomized rats GABA release in the MPOA is pulsatile, and a reduction of GABA release precedes each episode of LH release.[343] Direct infusion of GABA or the $GABA_A$ receptor agonist muscimol into the POA suppresses LH release in rats and sheep,[341,344–347] whereas GABA and muscimol inhibit LH pulses when injected into the third ventricle or the MPOA of female rats.[341,347] Interestingly, experiments in ovariectomized rats showed that the GABA agonist muscimol suppresses, whereas the $GABA_A$ receptor antagonist bicuculline enhances, hypothalamic multiunit activity.[348]

It is possible that inconsistencies of the *in vivo* results regarding the role of input through $GABA_A$ receptors is due to indirect GABA actions through interneurons on GnRH neurons, as results from single cell studies are generally excitatory. Recordings from embryonic mouse GnRH neurons[349] or the GnRH neuronal cell line GT-1[350,351] indicate that the direct action of GABA, mediated by $GABA_A$ receptors, on GnRH neurons is stimulatory. While it can be argued that this response is due to immaturity of GnRH neurons in these experimental preparations, because immature neurons always respond to $GABA_A$ receptor stimulation with excitation,[330] GABA in mature GnRH neurons is also excitatory. For example, electrophysiological or calcium imaging studies with GFP-labeled adult mouse GnRH neurons show that effects of GABA on adult GnRH neurons through $GABA_A$ receptors are exclusively excitatory.[333,335,352,353] Contrary to these observations, there are some reports, with a electrophysiological approach, indicating that GABA inhibits GFP-labeled adult mouse GnRH neurons through $GABA_A$ receptors, and that the $GABA_A$ antagonist bicuculline depolarizes the membrane and increases firing frequency.[354,355] It is plausible that GABAergic control of GnRH activity undergoes developmental changes in chloride reversal potential, which causes GABA to switch from depolarizing to hyperpolarizing effects on GnRH neurons, prior to puberty.[356] Regardless of these controversies, the question of how an excitatory or inhibitory $GABA_A$ postsynaptic signal modifies GnRH neurosecretion remains unknown. Nonetheless, the possibility that GABA may cause extra synaptic tonic inhibition[357] to GnRH neurons through $GABA_A$ receptors, as GnRH neurons express δ- and ε-subunits,[355,358] has not been examined in the context of puberty.

In contrast to $GABA_A$, there is good agreement that $GABA_B$ receptor activation is inhibitory to GnRH/LH release in rodents.[261,340,347,359] In fact, baclofen (a $GABA_B$ agonist) iontophoretically applied to adult GnRH neurons of guinea pigs causes immediate hyperpolarization via opening of potassium channels,[360] indicating that the direct $GABA_B$ receptor-mediated actions of GABA in these cells are inhibitory. In sheep, however, $GABA_B$ receptors appear to have a stimulatory role, as infusion of baclofen into the arcuate-ventromedial nucleus stimulated LH pulses in castrated rams during the non-breeding season,[361] increased LH release in ovariectomized and estrogen-treated ewes,[346] and reversed the inhibitory effect of testosterone on LH release in male sheep.[362]

ROLE OF GABA IN PUBERTY

There is a persistent skepticism regarding the inhibitory role of GABA prior to puberty, because of the absence of abnormal timing of puberty with mutation in GABA-related genes in humans[363] and mice.[364] It is difficult to examine developmental changes in $GABA_A$ receptor subunits using gene targeting approaches, because loss of some $GABA_A$ subunits is either lethal or results in loss or compensation by other subunits.[365] However, several findings in model species suggest an important role for GABA in the regulation of puberty onset.

In non-human primates, the GnRH neuronal system is tonically inhibited by GABA neurons during juvenile development. First, GABA levels in SME perfusate from prepubertal rhesus monkeys are much higher than in midpubertal monkeys[25,336] (Fig. 19.11). Second, bicuculline, a $GABA_A$ receptor antagonist, stimulates GnRH release in prepubertal monkeys, whereas exogenous GABA is not effective in suppressing GnRH release until after the onset of puberty, when endogenous SME levels are already reduced.[336] Third, infusion of antisense oligodeoxynucleotides to GAD-67 and GAD-65 mRNAs into the SME of prepubertal monkeys results in a dramatic increase in GnRH release,[336,366] likely as a consequence of a reduction in GABA synthesis[25,367] (Fig. 19.12). Finally, pulsatile infusion of bicuculline into the third ventricle of prepubertal monkeys results in precocious menarche, which occurs 6-8 weeks after the initiation of bicuculline infusion, and in precocious first ovulation, which occurs by 30 months, the age of menarche in control females[368] (Fig. 19.13). However, since the interval between menarche and first ovulation was not shortened by bicuculline infusion, additional mechanisms, such as the establishment of the excitatory neuronal input for pulsatile GnRH release, may be necessary for the progression of the pubertal process in female primates. Nonetheless, a case report that treatments

BOX 19.2

HOW DO WE KNOW THAT A SHIFT IN BALANCE OF EXCITATORY VERSUS INHIBITORY NEUROTRANSMITTERS OCCURS AT PUBERTY AND MAY BE RESPONSIBLE FOR THE PUBERTAL ACCELERATION OF HYPOTHALAMIC GNRH RELEASE

A series of experiments from the labs of Terasawa and Plant independently underscored the importance of a shift in sensitivity or exposure to excitatory versus inhibitory neurotransmitters as the pubertal process is initiated. Plant and colleagues found that administration of the excitatory amino acid receptor agonist NMDA once every 3 hours starting at 15–16 months of age (1.5–2 years before the normal age of puberty in this species) and continued for 15 weeks successfully induced premature hypothalamo-pituitary–testicular activity in juvenile male rhesus monkeys.[1] However, this does not explain the neurobiological mechanism of suppressed gonadotropin release during the juvenile period. Subsequently, Terasawa and colleagues found that GABA is an inhibitory neurotransmitter suppressing GnRH release before puberty onset.[2] In 1999, this group further reported that the pubertal increase in GnRH release from the female rhesus monkey hypothalamus (measured by push–pull perfusion) coincided with decreased GABA and increased glutamate release.[3]

Studies by this group later revealed a possible mechanism operating at the onset of puberty to control the balance between GABA and glutamate. They found that experimental suppression of GAD67 (an enzyme responsible for conversion of glutamate to GABA), by infusion of an antisense oligodeoxynucleotide for GAD67 mRNA into the stalk median eminence, suppressed GABA release, with a subsequent increase in glutamate. This shift in balance eventually led to an increase in GnRH release.[3]

Perhaps the most poignant studies implicating the balance between excitatory and inhibitory neurotransmission, and in particular a role for inhibitory control of GnRH neurosecretion by GABA prior to puberty, were reported by the Terasawa group in 1999. This group found that precocious puberty was induced by pulsatile infusion of bicuculline, a $GABA_A$ receptor antagonist, in immature female rhesus monkeys. Bicuculline or saline (control) was infused into the base of the third ventricle every 2 hours starting at 16 months of age until the first and second ovulations were confirmed by the detection of elevated serum levels of progesterone. Bicuculline increased circulating LH and estrogen levels shortly after the initiation of infusion, resulting in menarche at 18 months of age. In contrast, circulating LH and estrogen levels in saline-infused animals did not increase until much later, resulting in menarche at a typical age of about 31 months.[4]

While an attractive hypothesis may relate this shifting balance to the disappearance of non-gonadal juvenile restraint on LH release, there are currently no reported studies that confirm such a relationship.

References

1. Plant TM, Gay VL, Marshall GR, Arslan M. Puberty in monkeys is triggered by chemical stimulation of the hypothalamus. *Proceedings of the National Academy of Sciences of the United States of America.* 1989;86:2506–2510.
2. Mitsushima D, Hei DL, Terasawa E. Gamma-aminobutyric acid is an inhibitory neurotransmitter restricting the release of luteinizing hormone-releasing hormone before the onset of puberty. *Proc Natl Acad USA.* 1994;91:395–399.
3. Terasawa E, Luchansky LL, Kasuya E, Nyberg CL. An increase in glutamate release follows a decrease in gamma aminobutyric acid and the pubertal increase in luteinizing hormone releasing hormone release in the female rhesus monkeys. *J Neuroendocrinol.* 1999;11:275–282.
4. Keen KL, Burich AJ, Mitsushima D, Kasuya E, Terasawa E. Effects of pulsatile infusion of the GABA(A) receptor blocker bicuculline on the onset of puberty in female rhesus monkeys. *Endocrinology.* 1999;140:5257–5266.

with GABA agonists in an 11-month-old child showing precocious puberty with severe epileptic seizures regressed not only signs of precocious puberty but also seizure attacks,[369] indicates the involvement of GABA disinhibition in the mechanism of puberty in humans. Similarly, GABA in rats is inhibitory around the age of puberty,[262,264,370–372] although GABA is either stimulatory or without effect on LH release in mature rats, as discussed above. In fact, it has been shown that GABA release from the POA decreases during the prepubertal stage (P26–P35) in female rats.[279]

Catecholamine Neurons

NOREPINEPHRINE

Among many neurotransmitters and neuropeptides involved in GnRH release, adrenergic neurons have

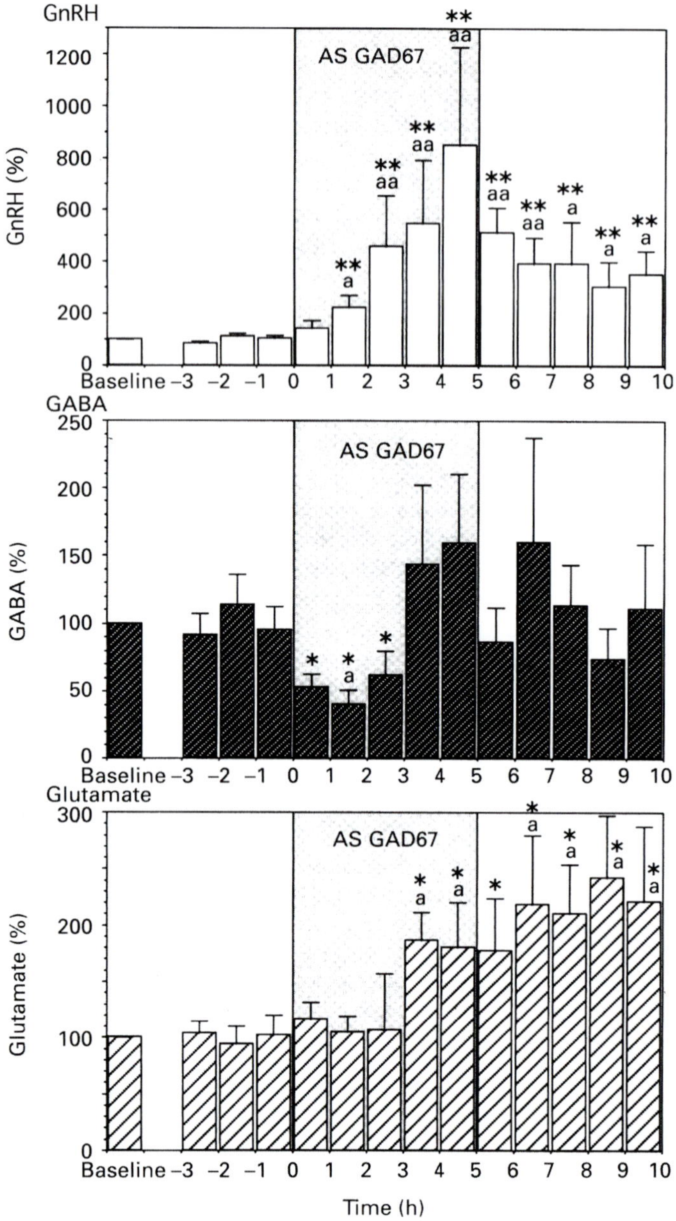

FIGURE 19.12 **Experimental suppression of GAD67 synthesis revealed a possible mechanism operating at the onset of puberty.** Infusion of an antisense oligodeoxynucleotide for GAD67 mRNA into the stalk median eminence suppressed GABA release (middle), with a subsequent increase in glutamate (bottom), prior to an increase in GnRH release (top). *Reproduced from Terasawa E, Luchansky LL, Kasuya E, Nyberg CL. An increase in glutamate release follows a decrease in gamma aminobutyric acid and the pubertal increase in luteinizing hormone releasing hormone release in the female rhesus monkeys.* J Neuroendocrinol. *1999; 11:275–282, with permission.*

been most extensively implicated in the control of pulsatile GnRH and LH release, and as mediators of the ovarian steroid-stimulated preovulatory gonadotropin surge in rodents. The interested reader is referred to several excellent reviews for a more detailed discussion of the subject in rodent models.[373–375]

There are two classes of adrenergic receptors, alpha and beta, and each class contains at least two receptor subtypes (α_1, α_2, β_1 and β_2). Each receptor subtype is affected by specific agonists and antagonists, and α receptors often have opposing effects to β receptors. In primates, activation of α_1-adrenergic receptors, but not α_2- and β-receptors, stimulates pulsatile release of LH and GnRH.[376–381] Conversely, infusion of the α_1-adrenergic blocker, prazosin, into the ME suppresses GnRH release.[377,382] Moreover, endogenous NE release from the SME is pulsatile and synchronous with GnRH secretion.[381] In female monkeys, an increase in NE release accompanies the preovulatory GnRH surge.[383] It is likely that NE modulation of GnRH neurons is through direct innervation of NE/epinephrine neurons, as Campbell and Herbison[384] have shown – using a conditional viral retrograde tract tracing method – that GnRH neurons receive a direct synaptic input from TH positive neurons located in the locus coeruleus and solitary nucleus.

A stimulatory role for catecholamines in puberty was first postulated by Weiner and Ganong,[385] based on the observation that treatment of female rats with reserpine, a monoamine synthesis blocker, dramatically reduced NE content in the hypothalamus and delayed vaginal opening. Supporting the involvement of NE in the developmental control of gonadotropin secretion, other investigators showed that treatment of prepubertal female rats with 6-hydroxydopamine, which selectively destroys catecholaminergic neurons, delays vaginal opening.[386,387] Inhibition of NE synthesis with FLA-63, a blocker of dopamine-β-hydroxylase (the enzyme that converts dopamine (DA) to NE), suppressed the postcastration-induced LH release in rats at all prepubertal ages examined between infantile (P7) and juvenile (P28) development.[388] Moreover, selective inhibition of NE synthesis with diethyldithiocarbamic acid, a dopamine-β-hydroxylase inhibitor, prevented the progesterone-induced LH surge as well as decreasing GnRH mRNA content in ovariectomized, prepubertal estrogen-primed rats.[389] The *in vitro* GnRH response to NE, examined in female rats, increases as the animal approaches puberty,[390,391] and this response is enhanced by estradiol,[391] suggesting an involvement of noradrenergic inputs in the progression of the pubertal process. Altogether, these results indicate that NE is an important neurotransmitter that contributes to facilitating the pubertal increase in LH/GnRH release. Also, NE may be inhibitory to GnRH release during embryonic development, instead of stimulatory. Both phenylephrine and clonidine (α_1- and α_2-adrenoreceptor agonists, respectively) suppressed GnRH release from primary embryonic hypothalamic cultures,[392,393] and blocking α_1- or α_2-adrenoreceptors or catecholamine synthesis – but not blocking DA receptors – resulted in an increase in GnRH release.[393]

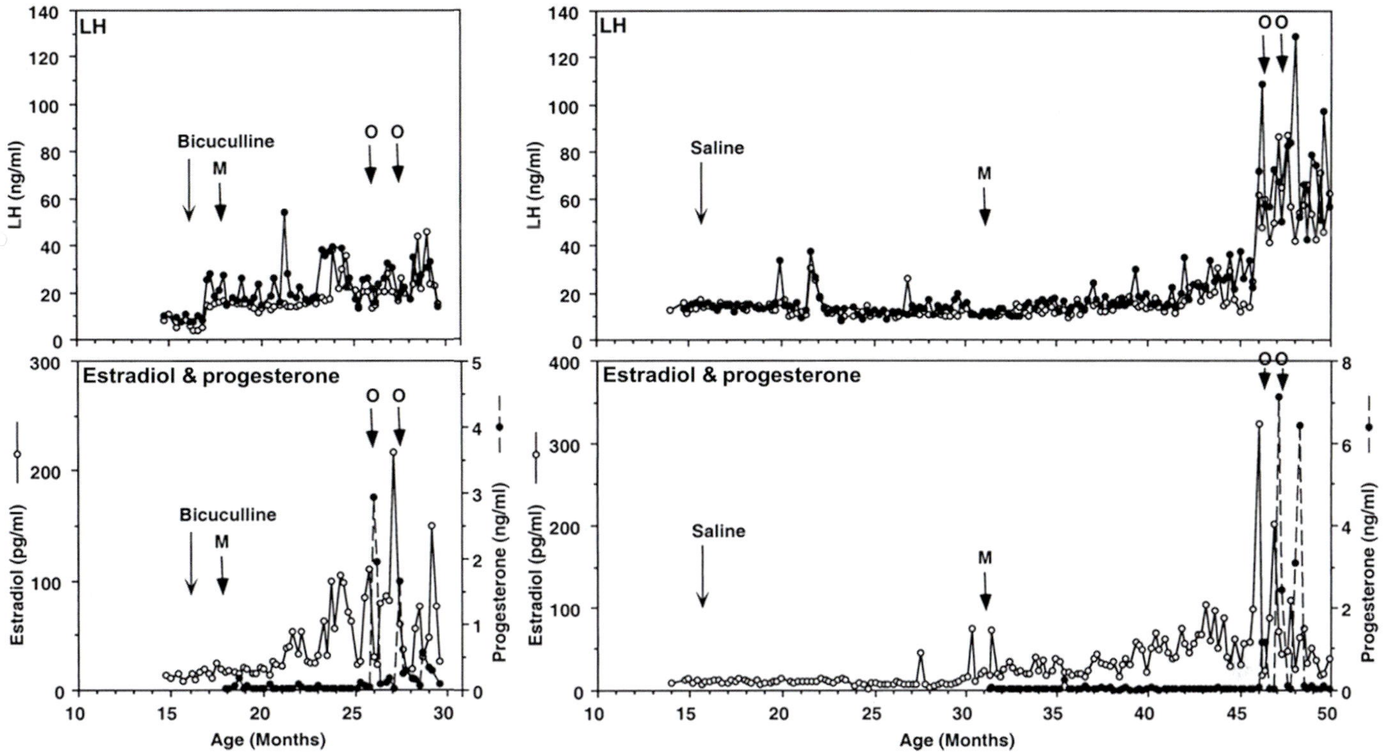

FIGURE 19.13 **Precocious puberty induced by pulsatile infusion of bicuculline, a $GABA_A$ receptor antagonist, in immature female rhesus monkeys.** Bicuculline (left), or saline (right) for control, was infused into the base of the third ventricle every 2 hours starting at 16 months of age until the first (O) and second (O) ovulations were confirmed by the detection of elevated serum levels of progesterone. In the top graphs, LH levels in the morning (open) and evening (closed circles) are shown; in the bottom graphs, estradiol (open circles) and progesterone (closed circles) levels are shown. Note that bicuculline increased circulating LH and estrogen levels shortly after the initiation of infusion, resulting in menarche (M) at 18 months of age. In contrast, circulating LH and estrogen levels in saline-infused animals did not increase until much later, resulting in menarche (M) at 31.3 months. *Modified from Keen* et al. *(1999).*[368]

In female rhesus monkeys, the α_1-noradrenergic neuronal system is involved in the stimulatory regulation of puberty.[394] When directly measured by push–pull perfusion of the SME, prepubertal monkeys exhibit the greatest GnRH increase in response to methoxamine, followed by a moderate increase in early puberty, and a low response in midpubertal monkeys.[394] However, endogenous NE levels in the SME increase only by midpuberty, suggesting that NE contributes to the progression of puberty, rather than to its initiation. In support of this view, an additional experiment has shown that transplantation of adrenal chromaffin cells, which contain high amounts of catecholamines and neuropeptide Y (NPY), into the third ventricle of juvenile 12- to 13-month-old monkeys did not alter the timing of menarche, but significantly advanced the age of first ovulation.[395] Nonetheless, the fact that the first ovulation – though earlier than normal – occurred more than a year after grafting the cells suggests that providing an exogenous source of NE/NPY to the hypothalamus is not sufficient, *per se*, to induce the pubertal activation of GnRH secretion.

Because the turnover rate may more faithfully reflect changes in neurotransmitter activity than alterations in catecholamine content, several researchers have measured NE turnover rates in the developing rat hypothalamus. The NE turnover rate in the medial POA has been shown to increase during sexual development of the female rat, reaching maximal values during the days preceding the first preovulatory surge of gonadotropins.[396–399]

Prostaglandin E_2 (PGE_2) has been shown to mediate NE-induced GnRH release[400] by an intra- and extracellular Ca^{2+}-dependent,[401,402] but not a calmodulin- and/or cAMP-dependent mechanism.[403] Exposure of ME fragments from late juvenile female rats to NE resulted in a dose-related increase in PGE_2 and GnRH release,[391] and *in vivo* treatment with estradiol increased the subsequent *in vitro* response of both PGE_2 and GnRH to NE. Since GnRH release from the ME of female rats increases from early juvenile (P22) to early pubertal development (P34) in response to PGE_2, and since the GnRH response to PGE_2 stimulation was blocked by ovariectomy and restored by estradiol, it was concluded that the stimulatory action of NE on GnRH release is mediated by an estradiol-sensitive PGE_2-dependent mechanism, which is further enhanced as puberty progresses.[391] Taken altogether, these observations suggest that in both rodents and primates, NE is involved in the developmental control of GnRH release but does not play a major role in initiating the pubertal process.

DOPAMINE (DA)

Despite the vast literature describing the importance of DA in the control of GnRH release (for reviews, see Kordon *et al.*[404]), the precise role that DA plays in the onset of puberty has not been established. GnRH neurons are innervated by TH-containing nerve fibers derived from diencephalic DA neurons, and not from noradrenergic neurons of the brainstem,[405] although another study[384] does not support this observation. What role DA plays in puberty is a matter of controversy. Chronic treatment with pergolide, a dopamine D_1/D_2 receptor agonist, advances the age of vaginal opening in rats[406], but implantation of pimozide, a DA receptor antagonist, in the MPOA of female rats also results in precocious puberty.[407,408] The results of a study examining the effects of selectively lesioning DA and NE neurons with neurotoxic agents at the onset of puberty further emphasized the complexity of the contribution that the dopaminergic system may have to the developmental control of gonadotropin secretion.[387] Depletion of DA in late juvenile (P31) female rats by intracisternal injection of 6-hydroxyDA and desipramine (a treatment that spares NE neurons) delayed vaginal opening in comparison to intact controls and rats injected with 6-hydroxyDA alone. The treatment also decreased the LH release induced by PMSG injection, suggesting a stimulatory role of extrahypothalamic DA in the mechanism of puberty.[387] However, i.v. injection of 6-hydroxyDA plus desipramine into PMSG-treated rats of the same age increased both LH and GnRH release, instead of decreasing it. Because 6-hydroxyDA given intravenously may preferentially affect the ME and other circumventricular organs, which are outside the blood–brain barrier, these results suggest that depletion of DA in the ME disinhibits, rather than diminishes, GnRH release.[387]

Hypothalamic DA concentrations and turnover rates have been shown to increase prior to and around the age of puberty.[397,398,404,409] However, unlike NE turnover rates, DA turnover rates in the MPOA did not change during an estrogen-induced LH surge in immature female rats, indicating that an increase in DA neurotransmission may not be required for the first GnRH surge to occur.[410] It is probable that the developmental increase in hypothalamic DA turnover rate observed in prepubertal rats is more related to the rise in circulating prolactin levels that occurs during prepubertal development in this species[410,411] than to a primary central event directly linked to the initiation of puberty.

Hypothalamic TH, the DA-synthesizing enzyme, activity increases in the hypothalamus of prepubertal rats of both sexes, reaching adult levels shortly after puberty.[412,413] Similar developmental changes in TH mRNA content and immunoreactive TH levels in selected hypothalamic nuclei, such as the ARC, occur in both males and females during prepubertal development, with TH mRNA levels increasing more prominently around the time of puberty in female rats.[411]

DA receptors appear to become functional in the rat brain by the end of gestation.[414] However, the receptors may not be coupled to GnRH release until later in postnatal life, as blockade of DA receptors had no effect on GnRH release from E16 hypothalamic cells cultured for 12 days *in vitro*.[393] This coupling may, however, occur after the first postnatal week of life in either an excitatory or an inhibitory fashion, as treatment with pergolide (a D_1 and D_2 agonist) between P11 and P20 induced precocious puberty in female rats,[406] whereas DA receptor antagonists induced GnRH and gonadotropin release in 12-day-old rats of both sexes.[406]

These observations indicate that while dopaminergic neurons may affect the initiation of puberty in both stimulatory and inhibitory manners, DA does not play a crucial role in setting the pubertal process in motion. To date, no studies have appeared describing the role of DA in the pubertal activation of GnRH or gonadotropin release during primate sexual development.

Neuropeptide Y (NPY) Neurons

NPY, a 36-amino acid peptide, is one of the most abundant peptides found in the mammalian central nervous system, and one of the most well-studied peptides in relation to neuroendocrine function and feeding behavior.[415,416] The physiological actions of NPY are mediated by a family of G-protein coupled receptors, consisting of at least six receptor subtypes: Y_1, Y_2, Y_3, Y_4, Y_5, and Y_6. These receptors also mediate the effects of the structurally related peptide YY (PYY). Different receptor subtypes are involved in different functions of NPY.[417]

The ontogeny of NPY neurons is species specific. For example, while in the chick a subset of NPY-positive neurons migrate into the brain along with GnRH neurons,[418] in rats the majority of NPY neurons originate within the brain, and are already present in the diencephalon and brainstem by E13–E14[419] – i.e., at the time when GnRH neurons are first found in the basal forebrain.[181] The hypothalamic content of NPY increases during postnatal development of the rat.[420,421] Since an age-related increase in the density of NPY fibers is observed in the hypothalamus of rats[422,423] and monkeys,[424] it would appear that as puberty approaches there is either an increased arborization of NPY terminal fields, or an increased transport of the peptide into neuronal processes. The observation that immunostaining of NPY perikarya in the adult hypothalamus often requires pretreatment with colchicine, an axonal transport blocker,[425,426] supports this speculation. In rats[422,423] and pigs,[427] the number of immunopositive NPY neurons in hypothalamic nuclei, such as the arcuate and paraventricular nucleus, declines during postnatal development. In monkeys, on the other hand, no changes in the number of hypothalamic immunopositive NPY neurons, including those that send their axons to the ME, appear to occur after birth.[428]

There is a significant increase in release of NPY from the ME into the portal blood at the time of the first preovulatory LH surge in rats,[420] and NPY release from the SME increases along with the pubertal increase in GnRH secretion in female rhesus monkeys,[429] suggesting an involvement of the neuropeptide in the control of the pubertal surge of GnRH secretion.

There are few reports describing developmental changes in brain NPY receptors, likely because specific antibodies for the different receptor subtypes are not available. Y_1 receptor mRNA is expressed in the rat brain as early as E12, while the Y_1/Y_4 preferred ligand [^{125}I][Leu31,Pro34]PYY shows specific binding by E14. Thereafter, both mRNA and ligand binding increase steadily, with Y_1 receptor mRNA increasing faster than the Y_1 protein during fetal life.[430] Y_1 mRNA levels reach a plateau around P0, remain high through P14, and display the adult distribution pattern by P21. Similarly, specific [^{125}I][Leu31,Pro34]PYY binding reaches a plateau by P21. Throughout ontogenesis, the relative abundance of Y_1 receptor mRNA and the binding profile of its encoded protein correlate well with the exception of some hypothalamic areas where relatively higher mRNA signals are observed.[430] In mice, RNase protection assays demonstrated that Y_1, Y_2 and Y_5 mRNAs are expressed very early in embryonic development in different regions of the brain.[431]

The role of NPY in the control of GnRH release in adult animals has been extensively discussed.[415,432–434] Numerous studies have shown that NPY is inhibitory to GnRH/LH secretion in ovariectomized rats, rabbits, sheep and monkeys,[435–439] whereas NPY stimulates GnRH/LH release in ovary-intact or estrogen-treated adult animals (for references see Levine *et al.*[433]). Studies in female rhesus monkeys have, however, shown that NPY is always stimulatory to GnRH release regardless of the gonadal state of the animal. For instance, endogenous NPY pulses in the SME precede the GnRH pulses by 5 minutes,[440] and NPY infusion into the SME increases GnRH release in a dose-related manner.[441] Conversely, infusion of an NPY antiserum into the SME suppressed GnRH pulsatile release.[440] That NPY may be an important factor for the progression of puberty is suggested by the ability of estrogen to facilitate the stimulatory effect of NPY on GnRH release.[441] Perhaps the difference between the results from Terasawa and colleagues and reports from other laboratories in gonadectomized animals[435,438,439] resides in the different route of NPY administration used by the latter investigators. When NPY is administered intraventricularly, instead of directly into the SME, it may stimulate neuronal systems inhibitory to GnRH release, such as β-endorphin neurons, by diffusion through the ventricular system. This interpretation is in keeping with results of studies showing that NPY is stimulatory when it is applied directly to the ME of either rats[442,443] or monkeys.[444] The obligatory role of NPY in full expression of the estrogen-induced LH surge has been suggested, based on the observation that the magnitude of the surge is significantly less in NPY knockout mice than in wild-type animals.[445]

Several studies indicate that NPY is involved in the mechanism underlying the onset of puberty in rats. For instance, the hypothalamic content of NPY increases during postnatal development of the rat, reaching maximal levels around the age of puberty.[420,421] An increase in NPY mRNA abundance and NPY peptide levels occurs in the MBH and MPOA of late (P29) juvenile rats treated with estrogen and progesterone to induce a preovulatory LH surge.[446,447] Intraventricular infusion of an NPY antiserum delays the age at puberty in female rats,[448] whereas a single intraventricular injection of NPY advances vaginal opening and first ovulation in these animals.[448] Moreover, NPY injected during a 5-day period into the lateral ventricle of chicks induces precocious puberty and an increase in food intake.[449] Despite these reports, it appears that NPY is modulatory and not critical for puberty onset, as no major reproductive deficiencies are found in NPY knockout as well as Y_1 and Y_5 receptor knockout mice.[450–452]

In rhesus monkeys, NPY appears to play a modulatory role in the mechanism of puberty. NPY infusion into the SME stimulates GnRH release in pubertal but not in prepubertal female monkeys.[429] NPY antiserum infusion into the SME suppresses GnRH release in pubertal but not in prepubertal female monkeys.[429] Like GnRH release, NPY release from the ME of prepubertal female monkeys is low, and increases concomitantly with the pubertal increase in GnRH secretion.[429] Grafting of adrenomedullary cells, which contain both NPY and NE-producing neurons, into the third ventricle of prepubertal female monkeys adjacent to GnRH neurons and their neuroterminals advanced the age of first ovulation, but not menarche,[395] indicating that NPY, like NE, contributes to facilitating the progression of puberty, instead of initiating the process.[194] This interpretation is supported by the failure of either NPY itself

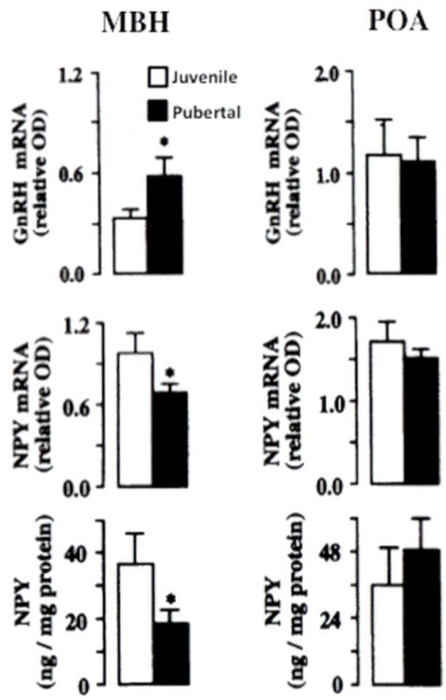

FIGURE 19.14 **Changes in GnRH mRNA (top), NPY mRNA (middle) and NPY protein (bottom) levels in the MBH and POA across puberty onset in male rhesus monkeys.** Closed bars represent pubertal monkeys, whereas open bars represent prepubertal monkeys. Note that in the MBH, not in the POA, both NPY mRNA and protein levels decrease while GnRH mRNA increases. *Modified from El Majdoubi* et al. *(2000),*[41] *with permission.*

or its antiserum to affect GnRH release in prepubertal monkeys.[394]

In contrast to these findings in female monkeys, NPY through Y_1 receptors appear to be inhibitory in prepubertal males.[41] NPY may directly modify activity of GnRH neurons, as GnRH neurons express Y_5 receptors and Y_1 receptors are found in close proximity to GnRH cell bodies in the preoptic area and fibers in the neuroterminals in rats.[453–455]

A negative modulatory role for NPY in the onset of puberty has been reported in rats[456–458] and, more recently, in rhesus monkeys.[41] These later authors also reported that although GnRH mRNA content in neonates was not different from that in juvenile animals, hypothalamic NPY mRNA abundance decreased in juvenile males from the higher levels detected in neonates when in gonadotropin output is lower (Fig. 19.14). Furthermore, blockade of Y_1 receptors by intraventricular injection of an NPY antagonist greatly stimulated LH release in prepubertal male monkeys.[41] While contradictory to previous observations indicating that NPY facilitates the pubertal process in female rhesus monkeys,[429,440,441] these findings are in keeping with results in rats showing that infusion of NPY into the lateral ventricle initiated during late juvenile development (P30) delays sexual maturation, concomitant with a decrease in food intake,[456] via activation of Y_5 receptors.[459] Chronic intracerebral administration of NPY also prevented the activation of puberty triggered by restoration of *ad libitum* feeding to immature female rats in which sexual maturation was delayed by food restriction.[458] Since NPY is found in some GABA neurons in the brain, such as neurons in the suprachiasmatic and arcuate nuclei,[460–462] it is plausible that NPY is co-released with GABA, exerting postsynaptic inhibitory effects. In fact, NPY has been shown to be inhibitory to hippocampal and suprachiasmatic neurons via presynaptic or postsynaptic receptors.[463–466] In harmony with these observations, studies in NPY knockout mice have shown that NPY is an endogenous antiepileptic agent that exerts its inhibitory effects via presynaptic inhibition of glutamate neurons.[467]

It is thus clear that NPY can exert both inhibitory and excitatory effects on hypothalamic neuronal activity. Whether the most important contribution of NPY to the pubertal increase in GnRH release is to act as a brake that restrains the timing of puberty,[41] or to act as a stimulatory signal that facilitates the progression of the pubertal process,[194] remains to be elucidated. It is possible that the divergent effects of NPY found between female and male monkeys are due to sex differences. Resolution of this controversy may require concomitant measurement of NPY and GnRH release from the male monkey hypothalamus at the time of puberty, and examination of the effect of blocking hypothalamic Y_1 receptors in the SME in both males and females. This route of administration would avoid activation of extrahypothalamic NPYergic systems that may be inhibitory to GnRH release. Also in need of investigation are several issues, such as: (1) Are NPY actions exerted directly on GnRH neurons or via interneurons? (2) Is the main site of NPY action on the GnRH perikarya or GnRH nerve terminals in the median eminence? (3) Are there two subsets of NPY neurons (one stimulatory and one inhibitory to GnRH release)?[41]

Opioid Peptides (Endorphins, Dynorphins and Enkephlins)

Because opioid peptides suppress pulsatile LH release,[468–470] and because the opioid receptor antagonist naloxone counteracts the negative feedback effect of estrogen on gonadotropins in adult animals,[471–477] a role for endogenous opioids in puberty has been proposed. In both immature rats and sheep, naloxone stimulates LH release[478–481] and administration of naloxone during early postnatal life has been shown to elicit precocious puberty in female rats.[482] However, there are several studies suggesting that a loss of opioid-mediated inhibition of GnRH release does not play a critical role in triggering puberty. For instance, blockade of opioid receptors causes opposite effects on

LH release during prepubertal development of male and females rats, inhibiting LH output in males but stimulating it in females.[75] In sheep, the decrease in sensitivity to the inhibitory feedback effect of estrogen that occurs at the time of the pubertal increase in LH release is unrelated to a loss of opioid inhibition.[483] Moreover, GnRH and β-endorphin (β-END) levels in the ME of lambs with delayed puberty induced by food restriction are lower than in normal lambs.[484] Similarly, in primates, opioid peptides are unlikely to play a major role in the onset of puberty, keeping GnRH release low before the onset of puberty. For instance, naloxone does not increase LH release until after the onset of puberty in chimpanzees,[485] and becomes more effective during the course of puberty.[486] Moreover, the release of β-END from the SME increases during puberty coincides with GnRH release.[487] Finally, pro-opiomelanocortin mRNA abundance in the ARC increases only after puberty has been initiated.[40] It is conceivable that an inhibitory effect of the opioid system on GnRH secretion is not fully operational before the pubertal increase in gonadal steroids, because opioid neurotransmission – such as that operating via μ-opioid receptors – is activated by estrogen.[488]

Therefore, opioid peptides as a group do not appear to play a role in restraining the initiation of puberty. Nonetheless, opioid peptides may complement an inhibitory role of GABA before puberty in providing homeostatic counterbalance to the cascade of excitatory events leading to the pubertal increase in GnRH output. In particular, the possibility exists that an inhibitory tone may be exerted by neuronal subsets of opioidergic neurons selectively using β-endorphin, dynorphin or met/leu-enkephalin as a neurotransmitter. Strategies to specifically disrupt the functions of each of these neuronal subsets are needed to resolve this issue.

Melatonin

Melatonin has been suggested as an inhibitory neurotransmitter for puberty onset. This is based on observations that: (1) melatonin delays the timing of puberty in hamsters and sheep;[489,490] (2) circulating melatonin concentration is elevated in early childhood, declines during late childhood, and remains stable from the early pubertal period to adulthood in humans and monkeys;[491–495] (3) some human patients with pineal tumors exhibit precocious puberty;[496] and (4) circadian melatonin rhythms in blind men differ from those in normal men,[497] and the timing of puberty in blind boys is delayed[498] whereas blind girls with no light perception have earlier onset of menarche than normal girls.[499,500]

However, pineal melatonin is not likely involved in the mechanism of gonadotropin suppression during prepuberty in primates.[501] Pineal tumors are often accompanied by lesions in the brain which may be responsible for precocious puberty,[502–505] and the tumor itself may release gonadotropic hormones or substances stimulatory to gonadotropin secretion.[506] Further, studies of male patients with GnRH deficiency compared with normal males suggest that a decrease in circulating melatonin prior to puberty appears to be due to a slight increase in circulating LH, which stimulates gonadal steroids.[507,508] Similarly, in female rhesus monkeys a pubertal increase in LH precedes a decrease in nocturnal melatonin levels by 2 months, and treatment with melatonin for 30 days does not delay pubertal increases in estradiol, IGF-1, and perineal coloration.[509] Finally and most importantly, pinealectomy during the neonatal period in castrated male rhesus monkeys[510] and at 20–23 months of age in OVX female rhesus monkeys[511] does not alter the pattern of developmental changes in gonadotropin secretion.

Metabolic Signals

Growth Hormone (GH)

GH is secreted episodically throughout life, and plays a major role in somatic growth. GH release is stimulated by GH-releasing hormone and inhibited by somatostatin. Circulating GH levels, especially the amplitude of GH secretory episodes, increase during puberty in both males and females.[61,75,512] This pubertal increase in GH secretion facilitates the progression of puberty, as individuals with isolated GH deficiency show delayed puberty, which is often prevented by treatment with GH.[65] Studies in female rats in which GH secretion was inhibited by pharmacological treatments[513,514] and mutant mice lacking the GH receptor[515] have shown that GH deficiency results in delayed puberty. However, the pubertal increase in GH is not responsible for triggering puberty; rather, it is caused by the rising levels of gonadal steroids after puberty onset. In fact, treatment with estradiol or androgen stimulates GH release.[516–518] The steroid responsible for the stimulation of GH secretion in both sexes is estradiol, as androgens have been shown to stimulate GH secretion after conversion to estrogen via peripheral aromatization.[65,75] The pubertal increase in GH secretion occurs chronologically earlier in girls than in boys. In girls, it coincides with the time of breast development; in boys it occurs during a more advanced phase of genital development (Tanner Stage IV). Nonetheless, Wilson and his colleagues[519] have shown that administration of a somatostain analog into prepubertal female monkeys, which suppresses circulating GH levels, results not only in reduced plasma IGF-1 levels and delayed skeletal maturation, but also in a delay of menarche and first ovulation.

Thus, Wilson proposes the hypothesis that metabolic hormones, such as GH, IGF-1 and leptin, play a permissive role in puberty, as levels of these metabolic hormones increase significantly during the evenings prior to the pubertal increase in LH in gonadectomized male rhesus monkeys.[520]

Insulin-like Growth Factor I (IGF-1)

Circulating IGF-I levels also increase at puberty in both sexes. In chimpanzees[521] and baboons,[522] the circulating levels of IGF-1 increase along with pubertal elevation of androgen in both males and females. In humans, serum IGF-I levels start to increase during the transition between Tanner Stages I and II[75] and reach a maximum around the peak of the pubertal growth spurt, suggesting the primary involvement of IGF-I in adolescent skeletal growth.[523] Because IGF-I is primarily synthesized in the liver under the control of GH, it has been speculated that the pubertal increase in circulating IGF-I is due to increased GH secretion, which is stimulated by gonadal steroids. However, studies in rats indicate that the pubertal increase in serum IGF-1 levels is due, at least in part, to a gonadal-steroid independent mechanism through the hypothalamus, as gonadectomy increases, rather than decreases, peripubertal serum IGF-1.[524] However, IGF-1 administration to prepubertal female monkeys advanced only first ovulation and not menarche,[525] and somatostatin administration to premenarcheal female monkeys, which suppressed circulating levels of both GH and IGF-1, delayed only first ovulation and not menarche,[519] indicating that increased IGF-1 is not responsible for the pubertal increase in GnRH release.

In addition to the liver, IGF-1 is also synthesized in the brain. Studies in rodents suggest that IGF-1 mRNA expression increases at puberty,[526,527] and that IGF-I stimulates GnRH release from the ME *in vitro*[528] and accelerates the timing of puberty when IGF-I is intracerebrally administered into sexually immature female rats.[529] In contrast, the intracerebral administration of specific antibodies to IGF-1 delayed sexual maturation in male rats.[530] Furthermore, the delayed puberty shown by mutant mice lacking the GH receptor gene is corrected by administration of IGF-1.[515] Altogether, these observations have led to the hypothesis that IGF-I is a metabolic signal that facilitates the onset of puberty in mammals.[528–531] In primates, IGF-1 treatment in prepubertal monkeys advances the timing of first ovulation (judged by the early age of the LH surge in response to estradiol challenge), but does not change the timing of menarche[531] (Fig. 19.15), further indicating that IGF-1 appears to be important for the pubertal progress during the early pubertal period.

Role of Glia and Growth Factors

Whereas glia were previously thought to play only supportive roles in central nervous system function, recent evidence suggests these cells can generate, process and convey information, much like neurons.[532,533] In fact, because glial cells possess a large array of neurotransmitter receptors, and many of the receptors are coupled to second-messenger systems that cause the release of Ca^{2+} from IP_3-sensitive stores, these cells can mediate the signal of one neuron to other neurons.[534,535] Glial cells may participate in puberty onset by modifying pulsatile GnRH release, which is likely achieved by (1) producing growth factors, (2) providing a signaling pathway from other neurons (e.g., the neuroligand, bradykinin, stimulates glutamate release from astrocytes, which in turn appears to signal back to bradykinin neurons[536]), and (3) modifying the local environment of GnRH release in the SME, which consists of an abundance of neuroterminals and glial cells, including tanycytes, but not GnRH cell bodies.[537–540]

Astroglia synthesize and release growth factors, such as transforming growth factor-α and -β (TGFα and TGFβ, respectively). The presence of basic fibroblast growth factor (bFGF), epidermal growth factor (EGF), insulin-like growth factor-1 (IGF-1), neuronal cell adhesion molecule (NCAM), cytokines (such as IL1 and IL6) and diffusible substances such as nitric oxide (NO) have been reported in astrocytes (see Ridet *et al.*[541]), and these substances appear to alter GnRH release. Because several excellent reviews are available on this topic,[542–547] in this chapter we only summarize the most recent relevant findings.

Transforming growth factor and EGF peptides are secreted by astrocytes, have potent effects on neuronal GnRH release, and may be associated with the mechanism of puberty onset. For example, TGFα mRNA in astrocytes increases with normal puberty as well as precocious puberty induced by hypothalamic lesions in female rats.[548,549] TGFα appears to stimulate GnRH release through increased glial secretion of PGE_2,[550] which then mediates the stimulatory effect of norepinephrine on GnRH release. Cultured hypothalamic astrocytes treated with estradiol yield a conditioned medium that stimulates GT1 cells, PGE_2 receptor production and GnRH release.[551] These effects appear to be mediated by as yet unidentified factors from astrocytes, as the astrocyte conditioned medium is more effective in inducing GnRH release than simply the addition of PGE_2 to the medium.[552]

EGF receptors have also been shown to mediate the effects of TGFα, and EGF receptor mRNA and EGF protein expression appear to decrease prior to the first

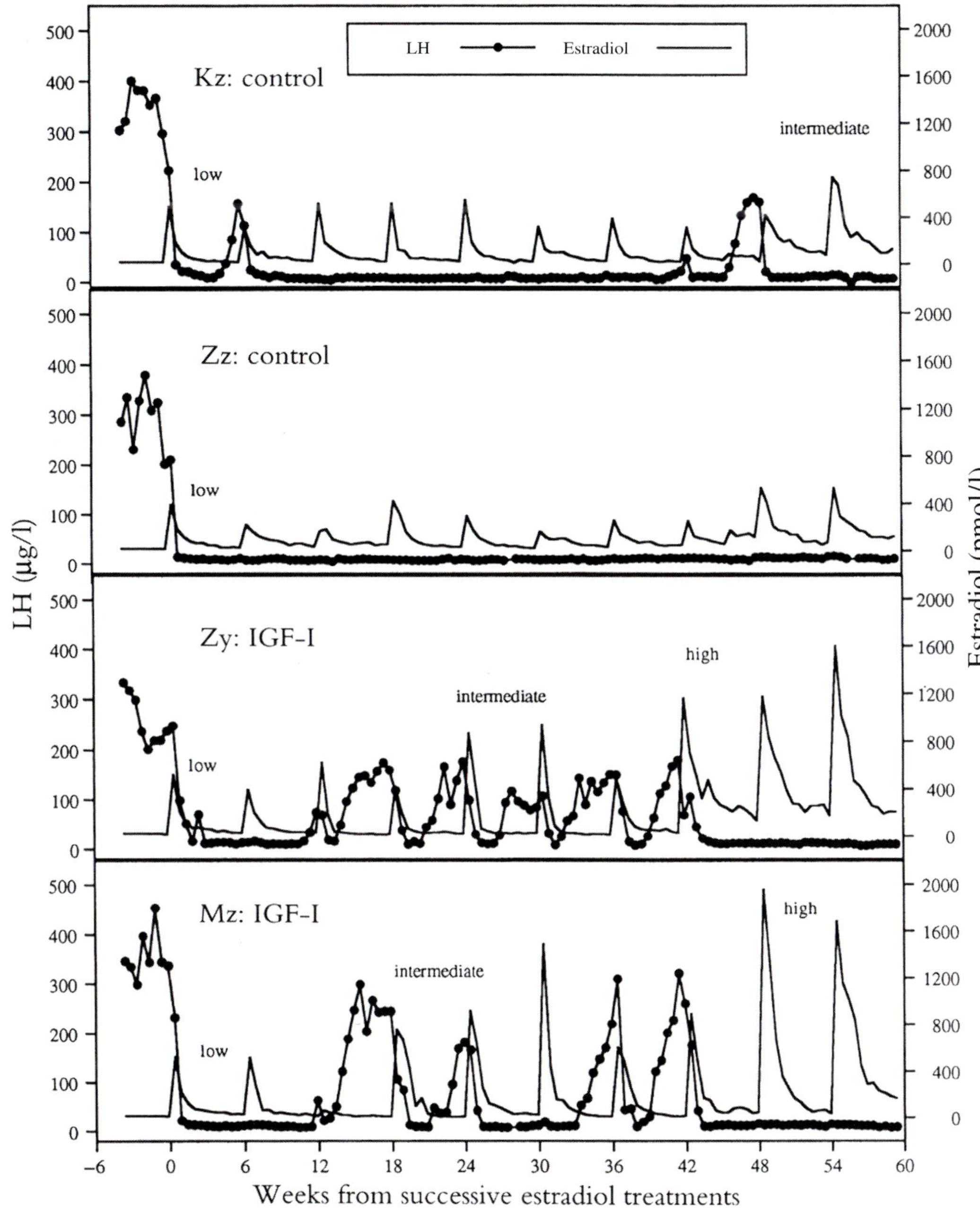

FIGURE 19.15 **IGF administration advances the timing of first ovulation, but not menarche, in female rhesus monkeys.** Changes in LH (closed circles with heavy line) and estradiol (thin line) in response to repeated estradiol challenges (every 6 weeks) in two IGF-1 (an intermediate dose) -treated females (bottom two panels) and two control females (top two panels) are shown. Twice-weekly treatment with three doses of IGF-1 is initiated at 13 months of age; both middle- and high-dose treatments are effective. *Reproduced from Wilson ME. IGF-I administration advances the decrease in hypersensitivity to oestradiol negative feedback inhibition of serum LH in adolescent female rhesus monkeys.* J Endocrinol. *1995;145:121–130, with permission.*

preovulatory GnRH surge, indicating a possible role for EGF in the onset of puberty.[553] Similarly, TGFβ derived from astroglia stimulates GnRH mRNA levels and release from GT1 cells.[554,555]

Basic fibroblast growth factor is a potent mitogen and neurotropic factor for hypothalamic neurons,[548,556,557] and also stimulates GnRH release from GT1 cells[558] *in vitro*. bFGF alters the secretion of GnRH from GT1 cells by inducing the secretion of intermediate products from the post-translational processing of GnRH rather than the completely processed decapeptide.[559] bFGF mRNA levels in the MBH, but not in the cerebral cortex, significantly decrease with puberty (first ovulation) induced by NMDA or PMSG in female rats, whereas bFGF receptor mRNA levels in the MBH and cortex show no change.[560] Further, the number of bFGF positive cells in the ARC is significantly higher in NMDA-treated rats before puberty (P29) as compared to saline-treated controls at the same age, and the number of bFGF positive cells is significantly lower in NMDA-treated rats after first ovulation (P33) as compared to saline-treated controls (P33).[560]

It is possible that glia also participate in the onset of puberty in rhesus monkeys. TGFα mRNA increases in the monkey hypothalamus with puberty.[561] In fact, hypothalamic TGFα expression changes in a biphasic manner during sexual maturation.[561] Specifically, expression is high during the infantile period when gonadotropin levels are high (prior to 6 months in rhesus monkeys), then decreases during the juvenile period and increases again at the time of puberty. A larger area of glial ensheathment is present on the perikaryal membrane of GnRH neurons in early pubertal monkeys than in adult monkeys.[50] However, additional information is needed to establish the critical involvement of glia in the mechanism of puberty, because it is uncertain if gliosis occurs in the primate hypothalamus before the onset of spontaneous puberty, and because the anatomical difference seen in prepubertal and adult monkeys could be due to the difference in circulating gonadal steroids, as a similar anatomical difference was observed in gonadectomized monkeys vs gonadal intact monkeys.[539] In addition, the importance of region-specific neuronal networks is not accounted for in the glia hypothesis of puberty. Lesions in two different regions of the hypothalamus induce similar massive gliosis, yet the effects on the timing of puberty are completely different. For instance, lesions in the posterior hypothalamus–premammillary area result in precocious puberty,[85,562] whereas lesions of an area that includes the posterior ME and ARC delays puberty, even though the GnRH neurosecretory system is spared.[563]

Radovick and colleagues have reported that the GnRH neurosecreting NLT (GN) cell line expresses IGF-1 receptors,[564] and that IGF-1 is capable of regulating GnRH mRNA expression directly through the AP-1 transcription factor.[564] Thus, it is plausible that the pubertal increase in IGF-1 may directly alter the activity of GnRH neurons, and this mechanism may contribute to the pubertal increase in GnRH release, although it is doubtful that astroglial-derived IGF-1 plays a role in triggering puberty onset. Specifically, while astrocytes are the primary source of IGF-1 from the hypothalamus,[565] it appears that peripheral sources of IGF-1 are more likely to contribute to GnRH neuronal network activation at puberty.[529]

Influence of Seasons and Circadian Rhythm

In both rodents[166] and primates,[11,86,92,96,566] the earliest hormonal change associated with the initiation of puberty is an amplification of a diurnal pattern of GnRH/LH release. Although it is clear that this diurnal rhythm is controlled by a neural circadian clock located in the suprachiasmatic nucleus (SCN) of the hypothalamus,[567] both the synaptic pathways conveying information from the circadian clock to the GnRH neuronal network and the identity of the transynaptic/cell–cell signaling molecules involved in the process remain incompletely understood. In contrast to this relative paucity of information, substantial progress has been made in recent years towards the elucidation of the molecular mechanisms underlying the oscillations of the master circadian clock entrained by the daily light–dark cycle (reviewed in Reppert[568] and Hall[569]). A central but still only partially answered question[570,571] concerns the identity of the structural components used by the SCN to entrain the diurnal fluctuations in GnRH output. Equally incomplete is our knowledge of the cell–cell signaling molecules and the transynaptically active molecules that, generated by SCN cells, may contribute to initiating, maintaining and enhancing the diurnal activity of GnRH neurons. Whereas immunocytochemical and retrograde tracing studies have shown that GnRH neurons receive a substantial input from vasointestinal peptide (VIP) neurons in the SCN,[66,571] co-culture of SCN and POA tissues shows that diurnal fluctuations in GnRH release coincide with arginine-vasopressin (AVP) release, rather than VIP.[572] This suggests that AVP, instead of VIP, may be one of the signals that link the SCN circadian oscillator to GnRH neurons. However, recent evidence based on the application of fluorescence and bioluminescence videomicroscopy to organotypic cultures from transgenic mice expressing GFP or luciferase under the control of the *mPer1* promoter, have demonstrated that VIP receptor mediated signaling is essential to maintain the function of clock cells in the SCN.[573] In addition to AVP and VIP, both excitatory and inhibitory amino acid neurotransmitters have been implicated in the generation of clock-dependent transynaptic output from the SCN.[574,575] Besides transynaptic modulation of GnRH neurons by SCN neurons, glial cells may be involved in this process. For instance, astroglial cells in the SCN are the cell type showing the most consistent oscillatory response to glutamate,[574] which is the predominant neurotransmitter mediating visual input to the SCN.[576]

While the search for conventional candidates is under way, new high-throughput and genetic approaches may lead to the discovery of other molecules involved in the control of circadian clocks. Two of these molecules, required for the maintenance of circadian rhythmicity, appear to be potential candidates for a role in SCN–neuroendocrine communication. The photolyase-like proteins Cry1 and Cry2 are mammalian homologs of a family of plant blue-light receptors known as cryptochromes found to be expressed not only in the SCN, but also in a variety of other tissues. These Cry proteins are thought to act as circadian photoreceptors able to detect light directly, and thus photoentrain the biological clock of different tissues.[577] Because in their absence there is a complete loss of free-running locomotor rhythmicity, they may also be integral components of the SCN circadian clock.[577]

While these considerations make evident the need for a more complete understanding of the mechanisms by which the SCN circadian clock entrains the diurnal changes in activity of the GnRH pulse-generating mechanism at the time of puberty, the presence of an SCN-independent endogenous clock in GnRH neurons has also been reported: (1) GT1 cells express several mammalian clock genes, including *Bmal*1, *Clock*, *mCry*1, *mCry*2, *mPer*1, *mPer*2, *mPer*3 and *CKLe*; (2) *Bmal* 1 and *mPer*1 protein levels exhibit circadian rhythms; (3) transfection with the dominant negative *Clock* gene, clockΔ19, severely dampened pulsatility of GnRH release, whereas overexpression of *mCry*1 increases the amplitude of GnRH pulses; and (4) somatic mutation in the *Clock* gene resulted in disruption of normal cyclic ovulation.[578] Therefore, not only the identification of the transynaptic and cell–cell communication pathways, which include functional characterization of the key molecules affecting this communication, but also maturational changes in internal clock mechanism within GnRH neurons are essential to understand the pubertal amplification of pulsatile GnRH release.

ADOLESCENT BRAIN

Many teenagers go through difficult times both emotionally and behaviorally during the transitional period between childhood and adulthood. Adolescence (behavioral maturation) occurs concurrently or shortly following the onset of reproductive maturation (puberty). During adolescence, matured cognition, correct decision-making strategies and controlled social behaviors need to be established for adult life. Puberty and adolescence are driven by different parts of the brain.[579] While reproductive puberty is due to hypothalamic maturation, adolescence is the result of maturation of neocortical and limbic circuits. Hormonal changes during puberty can greatly influence maturation of neocortical and limbic circuit formation, thereby modifying behavioral maturation, although these processes can also occur independently.

Maturation of higher cortical function continues after birth until early adulthood. Maturational changes specific to the adolescent brain can be monitored by MRI in humans. White matter volume increases from childhood through adolescence with a peak at 16 years of age, presumably due to increased myelination of cortical and subcortial fiber tracts.[580,581] A gradual volume increase in frontal and parietal gray matter peaks at the preadolescent age (10.5 and 12 years old in girls and boys, respectively), whereas an increase in temporal gray matter volume peaks during adolescence (16.5 years old) in both sexes.[582] Subsequently, these increases in gray matter volume decline to the adult level. A longitudinal study[583] further indicates that region-specific changes in gray matter volume occur during the adolescent period. Maturation of sensory-motor cortices precedes maturation of the prefrontal cortex, where cognitive functions are controlled, and the maturation of the prefrontal cortex is not completed until the early twenties (Fig. 19.16). Although the cellular basis of these adolescent changes has not been well studied, it is attributed to the second wave of synaptogenesis and dendritic arborization occurring just prior to or at the very early stage of puberty,[583] which is followed by synaptic pruning after neurocircuitry formation in early adulthood. This concept is, in part, supported by observations in hamsters.[584]

The first wave of synaptogenesis and dendritic arborization occurs during the neonatal period to early childhood, and overproduction of synaptogenesis and dendritic arborization is pruned during late childhood after the establishment of proper neurocircuitry formation. Synaptogenesis in the cerebral cortex of rhesus monkeys drastically increases after birth, reaching a plateau at 2–4 months of age, declining slightly through the juvenile period, then substantially decreasing through puberty to adulthood.[585] The prepubertal pruning of axospinous synapses and dendritic spines in the monkey frontal cortex has also been reported.[586] The tempo of changes in synaptic density in human cortices during early postnatal life through the adolescent period is not uniform among different cortical regions or different layers of the cerebral cortex,[587] although the general pattern is similar to that described in monkeys.[585,588] Therefore, presently, detailed neuroanatomical changes during the second wave of synaptogenesis and dendritic arborization observed by MRI remain to be determined.

It is conceivable that neuroanatomical maturation of the neocortex and limbic structures, such as described in humans by MRI,[583] is a consequence of the pubertal increase in gonadal steroids. It has been shown that exposure to gonadal steroids could induce neurogenesis, axonal sprouting, myelination, synaptogenesis, dendritic arborization and apoptosis.[589–591] For example, the seizure

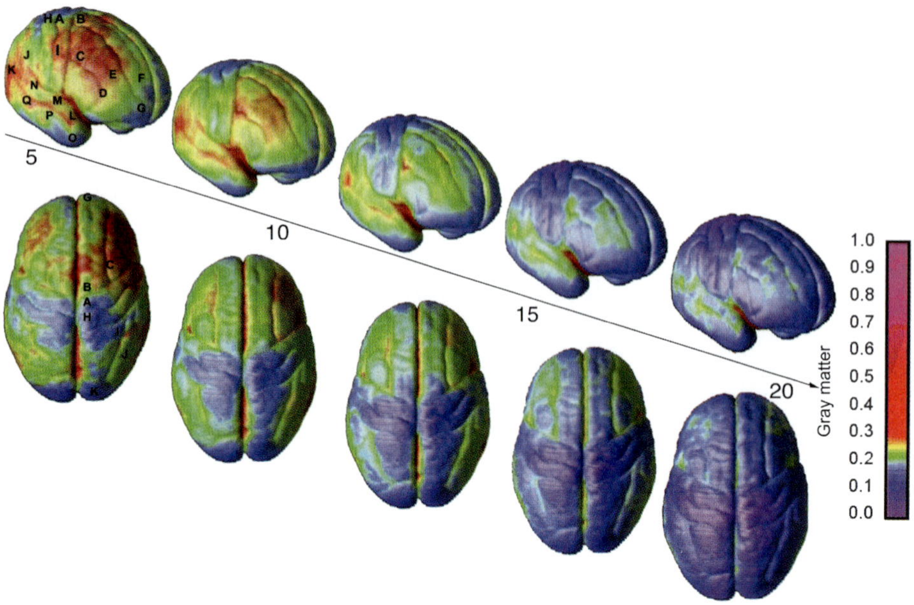

FIGURE 19.16 Representation of developmental changes in gray matters of the neocortex longitudinally assessed by brain imaging in 13 healthy boys at 4 to 21 years of age. Right lateral and top views of the cortical surface are shown. The side bar shows a color representation in units of gray matter volume. Note that development of the paraprefrontal cortex is not complete until after 20 years of age. *Modified from Gogtay* et al. *(2004).*[583]

threshold induced by hippocampal or amygdaloid stimulation becomes low at the time of puberty in female rats.[592] Rat cerebral white matter increases during puberty due to increases in the number of axons that become myelinated, even though the total number of axons decreases, and the cortical volume and number of neurons decreases in some cortical regions during the time of puberty in female rats.[593] These functional and anatomical changes during puberty are dependent on the presence of estrogen, as the age of the decreased seizure threshold can be accelerated by induction of precocious puberty[592] and the pubertal increase in white matter does not occur in ovariectomized animals prior to puberty.[593]

Exposure to gonadal steroids during puberty results in permanent changes in behavior. For example, adult hamsters in which the testes were removed prior to puberty do not exhibit full copulatory behavior even with testosterone replacement therapy, whereas this is not the case if castrated males are treated with testosterone before puberty.[594,595] Recent reports further indicate that in mice exposed to shipping stress at pubertal (4–6 weeks), but not postpubertal (8–12 weeks), age exhibit reduced sexual behavior after estradiol and progesterone treatments, tested in adulthood.[596,597] In humans, exposure to a pubertal elevation of gonadal steroids influences the neurocircuitry necessary for cognitive function: it has been reported that spatial cognition in adult male patients with congenital hypogonadotropic hypogonadism (no exposure to androgen during puberty) was significantly impaired as compared to spatial cognition in adults who acquired hypogonadotropic hypogonadism and in healthy control males.[598] These observations suggest that the absence of gonadal steroids during adolescent brain development results in a long-lasting impairment of testosterone-dependent and testosterone-independent behaviors. Conversely, the presence of androgen during male adolescent brain development enhances masculinization of behavioral responses. In a review, Sisk and Zehr[599] proposed the hypothesis that "organizational effects of steroid hormones" occur not only during the prenatal and early postnatal periods, during which sexual differentiation of the brain is induced, but also during the adolescent period.

It is, however, also possible that pubertal remodeling of neurocircuitry during the adolescent period can occur independent of gonadal steroids. For example, stress induces a prolonged elevation of corticosterone in prepubertal male and female rats as compared to that in adults, but gonadectomy does not eliminate the stress-induced differential effect of corticosterone elevations in prepubertal and adult rats of either sex. Thus, stress reactivity undergoes a maturational change during puberty, and this maturational change appears to be independent from the pubertal increase in gonadal steroids.[600,601] These authors further report that maturational changes in the corticosterone feedback system in the brain are not a consequence of developmental changes in the number of corticosterone-releasing hormone neurons in the paraventricular nucleus, or the number of glucocorticoid receptors in the hippocampus.[602,603]

In summary, significant neurocircuitry formation appears to occur during the adolescent period both with and without the pubertal elevation of gonadal steroids. However, much information is still missing, and many issues remain to be resolved. For example, to what extent are the pubertal changes in structures of

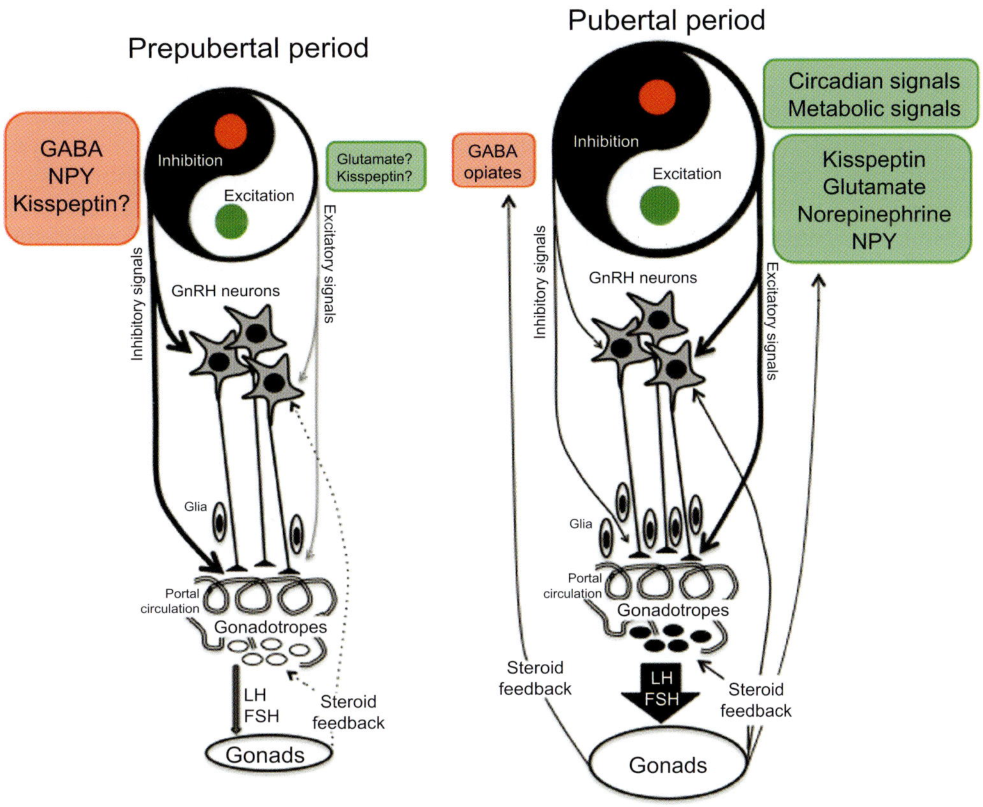

FIGURE 19.17 **Schematic illustration of possible mechanisms triggering puberty.** The yin–yang relationship among neurotransmitters and neuropeptides controlling GnRH release appears to be a key for the mechanism of puberty. While a gonadal-steroid insensitive mechanism is operative at the prepubertal and pubertal transition in primates, gonadal-steroid sensitive mechanisms are involved in rodents and sheep. In primates, predominant GABAergic inhibition on GnRH neurons through $GABA_A$ receptors appears to keep GnRH release low during the prepubertal period. During this period, GABA release is elevated in the stalk median eminence (SME), and differences in $GABA_A$ receptor subunit composition may contribute to tonic inhibition of GnRH release. Tonic GABA inhibition on GnRH neurons may also be direct or indirect through kisspetinergic and glutamatergic neurons. At the onset of puberty, a decrease in GABA tone in the SME allows excitatory kisspeptinergic and glutamatergic input to GnRH neurons to become dominant. This leads to the pubertal increase in GnRH release, which triggers a cascade of events at puberty. Although the mechanisms are completely unknown, circadian input and perhaps metabolic signals play a significant role in puberty. As puberty progresses, the stimulatory (kisspeptin, NPY, norepinephrine and glutamate) and the inhibitory (opioid and GABA) signals either strengthen or begin to participate in the control of GnRH release. During this period, gonadal steroids are also involved in the establishment of the adult pattern of GnRH release. In rodents, it is likely that a similar mechanism is operative at the pubertal transition, but currently it is unclear to what extent gonadal steroid hormones are involved in this mechanism. A possible role of glia in puberty is also illustrated.

the brain due to the pubertal elevation of gonadal steroids? Is neurocircuitry formation affected by environmental factors during the adolescent period, and, if so, is the adolescent period more sensitive to environmental factors? Finally, understanding of the adolescent brain may lead to more effective treatments for diseases and disorders which originate or worsen during this period, such as schizophrenia (known to occur at late adolescence) or epilepsy, autism and depression (known to worsen at the time of puberty).[604]

CONCLUSION

In this chapter, we have attempted to describe possible mechanisms leading to the initiation of puberty. Because pulsatile infusion of GnRH triggers puberty[160] and an increase in pulsatile GnRH release occurs at puberty,[96] the key question now is: what triggers the pubertal increase in GnRH release? Shortly prior to or at the time of puberty onset, GnRH release is profoundly influenced by circadian signals, such that the amount of GnRH release during the night is much larger than that during the day, although mechanisms for this circadian fluctuation are completely unknown. As discussed earlier, several neurotransmitters and neuromodulators have been shown to control pulsatile GnRH release (Fig. 19.17). Among them, we now know that kisspeptin plays a significant role in puberty. In fact, human patients with mutations of the kisspeptin receptor (GPR54 or kiSS1r) exhibit delayed puberty, no puberty at all or, in some cases, precocious puberty.[1,2,605] The variation of pubertal timing (advanced or delayed) by disrupted kisspeptin/GPR54 signaling, however, suggests the presence of additional upstream control mechanisms for the kisspeptin/GPR54–GnRH neuronal system. In support of this speculation, targeted deletion of ERα in kisspeptin neurons advanced, rather than delayed, puberty onset in mice (U. Boehm and J. Levine, personal communication).

As discussed earlier, the excitatory neurotransmitter glutamate and the inhibitory neurotransmitter GABA are prominently involved in the mechanism of puberty onset. In prepubertal monkeys, low levels of GnRH release during the prepubertal period are due to a tonic GABA inhibition,[336] and reduction of this tonic inhibition allows an increase in GnRH release.[368] It appears that a glutamate increase follows an initial decrease in GABA release, because when GABA synthesis is experimentally suppressed in the prepubertal hypothalamus, glutamate increases are observed only after the reduction in GABA levels became apparent.[25] While this finding suggests that, at puberty, an increase in glutamate release may be secondary to an initial decrease in GABA release in the SME of monkeys, it is possible in rats that the elevation in glutamate release may be responsible for the initiation of puberty.[606] This highlights an important point discussed earlier: that there are significant species differences in the mechanisms leading to puberty. Nonetheless, interactions between kisspeptin and glutamate/GABA neuronal systems appear to lead to the pubertal increase in GnRH release. Several questions arise: (1) Does the decrease in the prepubertal GABA inhibition precede the pubertal increase in kisspeptin release? (2) Does the pubertal increase in kisspeptin release suppress GABA inhibition prior to the onset of puberty? (3) Does the pubertal increase in glutamate release antedate the increase in kisspeptin release? (4) Are kisspeptin neurons subjected to glutamatergic and/or GABAergic transynaptic control?

Finally, we need to pay attention to a secular trend toward a younger age at menarche in humans.[607,608] Among possible causes, environmental contamination by industrial chemicals (such as bisphenol A) and increased body weight are worth consideration. Presently, relationships between the exposure to the estrogenic substance bisphenol A and precocious puberty in humans are limited to population studies,[609,610] and experimental data are primarily obtained from rodents.[611] In terms of body weight, a secular trend toward increased BMI and obesity in children, due to excessive calorie intake and lack of physical activity, has been extensively reported.[120,121,123,608] Preliminary data from our colony monkeys further indicate an increase in food intake leads to accelerated body weight gain and early menarche (Kurian, Shiel, Keen, and Terasawa, unpublished). Beyond the necessity for studying the mechanism of body-weight influence on puberty, the current trend towards obesity and early puberty onset are significant concerns in our society. Obesity onset before normal menarche age increases the risk of being obese in young adulthood,[121,612] which predisposes to many diseases, such as hypertension, hypercholesterolemia and the metabolic syndrome.[613,614]

The 125th Anniversary Issue of *Science* magazine highlighted 125 of the most compelling puzzles facing scientists, including "What triggers the onset of puberty?"[615] This remains a mystery, in part due to the complexities associated with maturation of the brain, pituitary gland and gonads, and interactions between the endocrine and nervous systems. During maturation, many genes are activated and repressed to establish the two systems, which constantly interact with each other through hormones and neuronal circuits. As a developmental event, there must be a master gene or complex of genes that triggers puberty. However, characterization of these genes is far from complete. In addition to these intrinsic factors, puberty onset is also profoundly influenced by environmental and metabolic factors. Consequently, while we have made significant progress in understanding the neuroendocrine regulation of puberty, plenty of challenges remain before an answer to the question of what triggers puberty is found.

Acknowledgments

This work is supported by NIH grants HD15433 and HD11355 and was possible through NIH support (P51RR000167, RR15459, and RR020141) to the Wisconsin National Primate Research Center.

References

1. de Roux N, Genin E, Carel J-C, Matsuda F, Chaussain J-L, Milgrom E. Hypogonadotropic hypogonadism due to loss of function of the KiSS1-derived peptide receptor GPR54. *Proc Natl Acad USA.* 2003;100:10972–10976.
2. Seminara SB, Messager S, Chatzidaki EE, Thresher RR, Acierno Jr JS, Shagoury JK, Bo-Abbas Y, Kuohung W, Schwinof KM, Hendrick AG, Zahn D, Dixon J, Kaiser UB, Slaugenhaupt SA, Gusella JF, O'Rahilly S, Carlton MB, Crowley Jr WF, Aparicio SA, Colledge WH. The GPR54 gene as a regulator of puberty. *N Engl J Med.* 2003;349:1614–1627.
3. Topaloglu AK, Reimann F, Guclu M, Yalin AS, Kotan LD, Porter KM, Serin A, Mungan NO, Cook JR, Ozbek MN, Imamoglu S, Akalin NS, Yuksel B, O'Rahilly S, Semple RK. TAC3 and TACR3 mutations in familial hypogonadotropic hypogonadism reveal a key role for Neurokinin B in the central control of reproduction. *Nat Genet.* 2009;41:354–358.
4. Cutler Jr GB, Schiebinger RJ, Albertson BD, Cassorla FG, Chrousos GP, Comite F, Booth JD, Levine J, Hobson WC, Loriaux DL. The Adrenarche (Human and Animal). In: Grumbach MM, Sizonenko PC, Aubert ML, eds. *Control of the Onset of Puberty.* Baltimore, MA: Williams & Wilkins; 1990: 506–533.
5. Tanner JM, Davies PS. Clinical longitudinal standards for height and height velocity for North American children. *J Pediatrics.* 1985;107:317–329.

6. Styne DM, Grumbach MM. Puberty, Ontogeny, Neuroendocrinology, Physiology, and Disorders. In: Kroenenberg H, Melmed S, Polonsky KS, Larsen PR, eds. *Williams Textbook of Endocrinology.* 11th ed. Philadelphia: Saunders Elsevier; 2008:970–1166.
7. Herman-Giddens ME, Slora EJ, Wasserman RC, Bourdony CJ, Bhapker MV, Koch GG, Hasemeier CM. Secondary sexual characteristics and menses in young girls seen in office practice: a study from the pediatrics research in office setting network. *Pediatrics.* 1997;99:505–512.
8. Tanner JM. *Growth at Adolescence.* Oxford: Blackwell Scientific Publications; 1962.
9. Marshall WA. Interrelationships of skeletal maturation, sexual development and somatic growth in man. *Ann Human Biol.* 1974;1:29–40.
10. Terasawa E, Nass TE, Yeoman RR, Loose MD, Schultz NJ. Hypothalamic control of puberty in the female rhesus macaque. In: Norman RL, ed. *Neuroendocrine Aspects of Reproduction.* New York, NY: Academic Press; 1983:149–182.
11. Rowell TE. Variation in age at puberty in monkeys. *Folia Primatol.* 1977;27:284–296.
12. Foster DL, Rapisarda JJ, Bergman KS, Lemons JA, Jaffee RB, Steiner RA, Wolf RC. Mechanisms limiting initiation of ovulation in the postmenarchial rhesus macaque. In: Norman RL, ed. *Neuroendocrine Aspects of Reproduction.* New York: Academic Press; 1983:103–132.
13. Wilson ME, Gordon TP, Collins DC. Ontogeny of luteinizing hormone secretion and first ovulation in seasonal breeding rhesus monkeys. *Endocrinology.* 1986;118:293–301.
14. Plant TM. Hypothalamic control of the pituitary–gonadal axis in higher primates: key advances over the last two decades. *J Neuroendocrinol.* 2008;20:719–726.
15. Korenbrot CC, Huhtaniemi IT, Weiner RI. Preputial separation as an external sign of pubertal development in the male rat. *Biol Reprod.* 1977;17:298–303.
16. Foster DL. Photoperiod and sexual maturation of the female lamb: early exposure to short days perturbs estradiol feedback inhibition of luteinizing hormone secretion and produces abnormal ovarian cycles. *Endocrinology.* 1983;112:11–17.
17. Boyd JD, Hamilton WJ, Hammond J. Transuterine migration of the ovum in sheep and other mammals. *J Anat.* 1944; 78:5–14.
18. Foster DL, Ebling FJ, Vennerson LA, Wood RI, Fenner DE. Regulation of puberty in the lamb: internal and external cues. *Prog Endocrinol.* 1988;2:861–866.
19. Schwanzel-Fukuda M, Pfaff DW. Origin of luteinizing hormone-releasing hormone neurons. *Nature.* 1989;338:161–164.
20. Wray S, Grant P, Gainer H. Evidence that cells expressing luteinizing hormone-releasing hormone mRNA in the mouse are derived from progenitor cells in the olfactory placode. *Proc Natl Acad USA.* 1989;86:8132–8136.
21. Ronnekleiv OK, Resko JA. Ontogeny of gonadotropin-releasing hormone-containing neurons in early fetal development of rhesus macaques. *Endocrinology.* 1990;126:498–511.
22. Quanbeck C, Sherwood NM, Millar RP, Terasawa E. Two populations of luteinizing hormone-releasing hormone neurons in the forebrain of the rhesus macaque during embryonic development. *J Comp Neurol.* 1997;380:293–309.
23. Schwanzel-Fukuda M, Crossin KL, Pfaff DW, Bouloux PM, Hardelin JP, Petit C. Migration of luteinizing hormone-releasing hormone (LHRH) neurons in early human embryos. *J Comp Neurol.* 1996;366:547–557.
24. Terasawa E, Quanbeck CD, Schultz CA, Burich AJ, Luchansky LL, Claude P. A primary cell culture system of luteinizing hormone releasing hormone neurons derived from embryonic olfactory placode in the rhesus monkey. *Endocrinology.* 1993;133:2379–2390.
25. Terasawa E, Luchansky LL, Kasuya E, Nyberg CL. An increase in glutamate release follows a decrease in gamma aminobutyric acid and the pubertal increase in luteinizing hormone releasing hormone release in the female rhesus monkeys. *J Neuroendocrinol.* 1999;11:275–282.
26. Resko JA, Ellinwood WE, Pasztor LM, Huhl AE. Sex steroids in the umbilical circulation of fetal rhesus monkeys from the time of gonadal differentiation. *J Clin Endocrinol Metab.* 1980;50: 900–905.
27. Kurian JR, Keen KL, Terasawa E. Epigenetic changes coincide with *in vitro* primate GnRH neuronal maturation. *Endocrinology.* 2010;151:5359–5368.
28. Resko JA, Ellinwood EE. Negative feedback regulation of gonadotropin secretion by androgens in fetal rhesus macaques. *Biol Reprod.* 1985;3:346–352.
29. Grumbach MM, Kaplan SL. The neuroendocrinology of human puberty, an ontogenetic perspective. In: Grumbach MM, Sizonenko PC, Aubert ML, eds. *Control of the Onset of Puberty.* Baltimore, MD: Williams & Wilkins; 1990:1–62.
30. Verney C, el Amraoui A, Zecevic N. Comigration of tyrosine hydroxylase- and gonadotropin-releasing hormone-immunoreactive neurons in the nasal area of human embryos. *Brain Res Dev Brain Res.* 1996;97:251–259.
31. Wray S, Hoffman G. Postnatal morphological changes in rat LHRH neurons correlated with sexual maturation. *Neuroendocrinology.* 1986;43:93–97.
32. Cottrell EC, Campbell RE, Han SK, Herbison AE. Postnatal remodeling of dendritic structure and spine density in gonadotropin-releasing hormone neurons. *Endocrinology.* 2006;147:3652–3661.
33. Dutlow CM, Rachman J, Jacobs TW, Millar RP. Prepubertal increases in gonadotropin-releasing hormone mRNA, gonadotropin-releasing hormone precursor, and subsequent maturation of precursor processing in male rats. *J Clin Invest.* 1992;90:2496–2501.
34. Jakubowski M, Blum M, Roberts JL. Postnatal development of gonadotropin-releasing hormone and cyclophilin gene expression in the female and male rat brain. *Endocrinology.* 1991;128:2702–2708.
35. Gore AC, Wu TJ, Rosenberg JJ, Roberts JL. Gonadotropin-releasing hormone and NMDA receptor gene expression and colocalization change during puberty in female rats. *J Neurosci.* 1996;16:5281–5289.
36. Gore AC, Roberts JL, Gibson MJ. Mechanisms for the regulation of gonadotropin-releasing hormone gene expression in the developing mouse. *Endocrinology.* 1999;140:2280–2287.
37. Gore AC. Diurnal rhythmicity of gonadotropin-releasing hormone gene expression in the rat. *Neuroendocrinology.* 1998;68: 257–263.
38. Goldsmith PC, Lambert R, Berizina LR. Gonadotropin-releasing hormone neurons and pathways in the primate hypothalamus and forebrain. In: Norman RL, ed. *Neuroendocrine Aspects of Reproduction.* New York, NY: Academic Press; 1983:7–45.
39. Cameron JL, McNeill TJ, Fraser HM, Bremmer WJ, Clifton DK, Steiner RA. The role of endogenous gonadotropin-releasing hormone in the control of luteinizing hormone and testosterone secretion in the juvenile male monkey, *Macaca fascicularis. Biol Reprod.* 1985;33:147–156.
40. Vician L, Adams LA, Clifton DK, Steiner RA. Pubertal changes in pro-opiomelanocortin and gonadotropin-releasing hormone gene expression in the brain of the male monkey. *Mol Cell Neurosci.* 1991;2:31–38.

41. El Majdoubi M, Sahu A, Ramaswamy S, Plant TM. Neuropeptide Y: A hypothalamic brake restraining the onset of puberty in primates. *Proc Natl Acad USA.* 2000;97:6179–6184.
42. Kurian JR, Auger AP, Terasawa E. Possible epigenomic relationship to a sex difference in rhesus monkey (*Macaca mulatta*) hypothalamic LHRH mRNA expression. Abstracts of the 39th Annual Meeting of the Society for Neuroscience, held at Chicago, IL; 2009. No. 703.8.
43. Plant TM. A study of the role of the postnatal testes in determining the ontogeny of gonadotropin secretion in the male rhesus monkey (*Macaca mulatta*). *Endocrinology.* 1985;116: 1341–1350.
44. Plant TM. A striking sex difference in the gonadotropin response to gonadectomy during infantile development in the rhesus monkey (*Macaca mulatta*). *Endocrinology.* 1986;119: 539–545.
45. Winter JSD, Faiman C. Serum gonadotropin in concentrations in agonadal children and adults. *J Clin Endocrinol Metab.* 1972; 35:561–564.
46. Winter JS, Faiman C, Hobson WC, Prasad AV, Reyes FI. Pituitary–gonadal relations in infancy. 1. Patterns of serum gonadotropin concentrations from birth to four years of age in man and chimpanzee. *J Clin Endocrinol Metab.* 1975;40:545–551.
47. Conte FA, Grumbach MM, Kaplan SL. A diphasic pattern of gonadotropin secretion in patients with the syndrome of gonadal dysgenesis. *J Clin Endocrinol Metab.* 1975;40:670–674.
48. Ross JL, Loriaux DL, Cutler Jr GB. Developmental changes in neuroendocrine regulation of gonadotropin secretion in gonadal dysgenesis. *J Clin Endocrinol Metab.* 1983;57:288–293.
49. Claypool LE, Watanabe G, Terasawa E. Effects of electrical stimulation of the medial basal hypothalamus on the *in vivo* release of luteinizing hormone-releasing hormone in the prepubertal and peripubertal female monkey. *Endocrinology.* 1990;127:3014–3022.
50. Witkin JW, O'Sullivan H, Miller R, Ferin M. GnRH perikarya in medial basal hypothalamus of pubertal female rhesus macaque are ensheathed with glia. *J Neuroendocrinol.* 1997;9:881–885.
51. Perera AD, Plant TM. Ultrastructural studies of neuronal correlates of the pubertal reaugmentation of hypothalamic gonadotropin-releasing hormone (GnRH) release in the rhesus monkey (*Macaca mulatta*). *J Comp Neurol.* 1997;385:71–82.
52. Grumbach MM, Roth JC, Kaplan SL, Kelch RP. Hypothalamic–pituitary regulation of puberty in man, evidence and concepts derived from clinical research. In: Grumbach MM, Grave GD, Mayer FE, eds. *Control of the Onset of Puberty,* Vol. 1. New York, NY: John Wiley & Sons; 1974:115–181.
53. Kaplan SL, Grumbach MM, Aubert ML. The ontogenesis of pituitary hormones and hypothalamic factors in the human fetus: maturation of central nervous system regulation of anterior pituitary function. *Recent Prog Horm Res..* 1976;32:161–243.
54. Clements JA, Reyes FI, Winter JS, Faiman C. Studies on human sexual development. III. Fetal pituitary and serum, and amniotic fluid concentrations of LH, CG, and FSH. *J Clin Endocrinol Metab.* 1976;42:9–19.
55. Kaplan SL, Grumbach MM. Pituitary and placental gonadotrophins and sex steroids in the human and sub-human primate fetus. *J Clin Endocrinol Metab.* 1978;7:487–511.
56. Grumbach MM. The central nervous system and the onset of puberty. In: Falkner F, Tanner JM, eds. *Human Growth.* New York: Plenum; 1978:215–238.
57. Mueller PL, Sklar CA, Gluckman PD, Kaplan SL, Grumbach MM. Hormone ontogeny in the ovine fetus. IX. Luteinizing hormone and follicle-stimulating hormone response to luteinizing hormone-releasing factor in mid- and late gestation and in the neonate. *Endocrinology.* 1981;108:881–886.
58. Matwijiw I, Faiman C. Control of gonadotropin secretion in the ovine fetus: the effects of a specific gonadotropin-releasing hormone antagonist on pulsatile luteinizing hormone secretion. *Endocrinology.* 1987;121:347–351.
59. Dumesic DA, Castillo RH, Bridson WE. Increase in follicle stimulating hormone content occurs in cultured human fetal pituitary cells exposed to gonadotropin-releasing hormone. *Life Sci.* 1991;48:1115–1122.
60. Castillo RH, Matteri RL, Dumesic DA. Luteinizing hormone synthesis in cultured fetal human pituitary cells exposed to gonadotropin-releasing hormone. *J Clin Endocrinol Metab.* 1992;75:318–322.
61. Matwijiw I, Faiman C. Control of gonadotropin secretion in the ovine fetus. II. A sex difference in pulsatile luteinizing hormone secretion after castration. *Endocrinology.* 1989;124:1352–1358.
62. Mesiano S, Hart CS, Heyer BW, Kaplan SL, Grumbach MM. Hormone ontogeny in the ovine fetus. XXVI. A sex difference in the effect of castration of the hypothalamic–pituitary gonadotropin unit in the ovine fetus. *Endocrinology.* 1991;129:3073–3079.
63. Winter JSD, Faiman C, Reyes FI. Sex steroid production by the human fetus: its role in morphogenesis and control by gonadotropins. *Birth Defects Original Article Series.* 1977;13:41–58.
64. de Kretser DM, Kerr JB. The cytology of the testis. In: Knobil E, Neill JD, eds. *Physiology of Reproduction.* 2nd ed. Vol 1. New York: Raven Press; 1994:1177–1290.
65. Grumbach MM, Styne DM. Puberty, Ontogeny, Neuroendocrinology, Physiology, and Disorders. In: Williams RH, Foster DW, Kroenenberg H, Larsen PR, Zorab R, eds. *Williams Textbook of Endocrinology.* 9th ed. Philadelphia: W.B. Saunders; 1998: 1509–1625.
66. Reppert SM, Uhl GR. Vasopressin messenger ribonucleic acid in supraoptic and suprachiasmatic nuclei: appearance and circadian regulation during development. *Endocrinology.* 1987;120: 2483–2487.
67. Boyar RM, Rosenfeld RS, Kapen S, Finkelstein JW, Roffwarg HP, Weitzman ED, Hellman L. Human puberty. Simultaneous augmented secretion of luteinizing hormone and testosterone during sleep. *J Clin Invest.* 1974;54:609–618.
68. Frungieri MB, Urbanski HF, Höhne-Zell B, Mayerhofer A. Neuronal elements in the testis of the rhesus monkey: ontogeny, characterization and relationship to testicular cells. *Neuroendocrinology.* 2000;71:43–50.
69. Ojeda SR, Skinner MK. Puberty in the rat. In: Neill JD, ed. *The Physiology of Reproduction.* 3rd ed. San Diego, CA: Academic Press/Elsevier; 2006:2061–2126.
70. Rosenfield RL. The ovary and female sexual maturation. In: Sperling MA, ed. *Pediatric Endocrinology.* Philadelphia: W.B. Saunders; 1996:329–385.
71. Jenner MR, Kelch RP, Kaplan SL, Grümbach MM. Hormonal changes in puberty. IV. Plasma estradiol, LH, and FSH in prepubertal children, pubertal females, and in precocious puberty, premature thelarche, hypogonadism, and in a child with a feminizing ovarian tumor. *J Clin Endocrinol Metab.* 1972; 34:521–530.
72. Raivio T, Dunkel L. Inhibins in childhood and puberty. *Best Pract Res Clin Endocrinol Metab.* 2002;16:43–52.
73. Greenwald GS, Roy SK. Follicular development and its control. In: Knobil E, Neill JD, eds. *The Physiology of Reproduction.* 2nd ed. New York, NY: Raven Press; 1994:629–724.
74. Peters H, Byskov AG, Himelstein-Braw R, Faber M. Follicular growth: the basic event in the mouse and human ovary. *J Reprod Fertil.* 1975;45:559–566.
75. Ojeda SR, Urbanski HF. Puberty in the rat. In: Knobil E, Neill JD, eds. *The Physiology of Reproduction.* 2nd ed. New York, NY: Raven Press; 1994:363–409.

76. Dissen GA, Paredes A, Romero C, Dees WL, Ojeda SR. Neural and neurotrophic control of ovarian development. In: Leung P, Adashi E, eds. *The Ovary.* 2nd ed. San Diego, CA: Academic Press; 2004:3–23.
77. Corbier P, Dehennin M, Castanier A, Mebazaa A, Edwards DA, Roffi J. Sex differences in serum luteinizing hormone and testosterone in the human neonate during the first few hours after birth. *J Clin Endocrinol Metab.* 1990;71:1344–1348.
78. Faiman C, Winter J.S.D. Gonadotropins and sex hormone patterns in puberty, clinical data. In: Grumbach MM, Grave GD, Mayer FE, eds. *Control of the Onset of Puberty.*, (pp 32–55) Vol. 1 New York, NY: John Wiley & Sons.
79. Andersson AM, Toppari J, Haavisto AM, Petersen JH, Simell T, Simell O, Skakkebæk NE. Longitudinal reproductive hormone profiles in infants: peak of inhibin B levels in infant boys exceeds levels in adult men. *J Clin Endocrinol Metab.* 1998;83: 675–681.
80. Frawley LS, Neill JD. Age related changes in serum levels of gonadotropins and testosterone in infantile male rhesus monkeys. *Biol Reprod.* 1979;20:1147–1151.
81. Robinson JA, Bridson WE. Neonatal hormone patterns in the macaque. I. Steroids. *Biol Reprod.* 1978;19:773–778.
82. Plant TM. Pulsatile luteinizing hormone secretion in the neonatal male rhesus monkey (*Macaca mulatta*). *J Endocrinol.* 1982;93:71–74.
83. Plant TM. Neuroendocrine basis of puberty in the rhesus monkey (*Macaca mulatta*). In: Martini L, Ganong WF, eds. *Frontiers in Neuroendocrinology.* New York: Raven Press Limited; 1988:215–238.
84. Steiner RA, Bremner WJ. Endocrine correlates of sexual development in the male monkey, *Macaca fascicularis. Endocrinology.* 1981;109:914–919.
85. Terasawa E, Bridson WE, Nass TE, Noonan JJ, Dierschke DJ. Developmental changes in the luteinizing hormone secretory pattern in peripubertal female rhesus monkeys: comparisons between gonadally intact and ovariectomized animals. *Endocrinology.* 1984;115:2233–2240.
86. Boyar R, Finkelstein J, Roffwarg H, Kapen S, Weitzman E, Hellman L. Synchronization of augmented luteinizing hormone secretion with sleep during puberty. *N Engl J Med.* 1972;287: 582–586.
87. Boyar RM, Finkelstein JW, Roffwarg H, Kapen S, Weitzman D, Hellman L. Twenty-four-hour luteinizing hormone and follicle-stimulating hormone secretory patterns in gonadal dysgenesis. *J Clin Endocrinol Metab.* 1973;37:521–525.
88. Judd HL, Parker DC, Siler TM, Yen SS. The nocturnal rise of plasma testosterone in pubertal boys. *J Clin Endocrinol Metab.* 1974;38:710–713.
89. Jakacki RI, Kelch RP, Sauder SE, Lloyd JS, Hopwood NJ, Marshall JC. Pulsatile secretion of luteinizing hormone in children. *J Clin Endocrinol Metab.* 1982;55:453–458.
90. Wu FCW, Butler GE, Kelnar CJH, Stirling HF, Huhtaniemi I. Patterns of pulsatile luteinizing hormone and follicle-stimulating hormone secretion in prepubertal (midchildhood) boys and girls and patients with idiopathic hypogonadotropic hypogonadism (Kallmann's syndrome): a study using an ultrasensitive time-resolved immunofluorometric assay. *J Clin Endocrinol Metab.* 1991;72:1229–1237.
91. Manasco PK, Umbach DM, Muly SM, Godwin DC, Negro-Vilar A, Culler MD, Underwood LE. Ontogeny of gonadotrophin and inhibin secretion in normal girls through puberty based on overnight serial sampling and a comparison with normal boys. *Human Reprod.* 1997;12:2108–2114.
92. Mitamura R, Yano K, Suzuki N, Ito Y, Makita Y, Okuno A. Diurnal rhythms of luteinizing hormone, follicle-stimulating hormone, and testosterone secretion before the onset of male puberty. *J Clin Endocrinol Metab.* 1999;84:29–37.
93. Wu FCW. GnRH Pulse generator activity during human puberty. In: Plant TM, Lee PA, eds. *The Neurobiology of Puberty.* Bristol: Journal of Endocrinology Ltd; 1995:185–197.
94. Wennink JMB, Delemarre Van-Der Waal HA, Schoemaker R, Schoemaker H, Schoemaker J. Luteinizing hormone and follicle stimulating hormone secretion patterns in boys throughout puberty measured using highly sensitive immunoradiometric assays. *Clin Endocrinol.* 1989;31:551–564.
95. Wennink JMB, Delemarre Van-Der Waal HA, Schoemaker R, Schoemaker H, Schoemaker J. Luteinizing hormone and follicle stimulating hormone secretion patterns in girls throughout puberty measured using highly sensitive immunoradiometric assays. *Clin Endocrinol.* 1990;33:333–344.
96. Watanabe G, Terasawa E. *In vivo* release of luteinizing hormone releasing hormone increases with puberty in the female rhesus monkey. *Endocrinology.* 1989;125:92–99.
97. Chongthammakun S, Claypool LE, Terasawa E. Ovariectomy increases *in vivo* luteinizing hormone-releasing hormone release in pubertal, but not prepubertal, female rhesus monkeys. *J Neuroendocrinol.* 1993;5:41–50.
98. Palmert MR, Hirschhorn JN. Genetic approaches to stature, pubertal timing, and other complex traits. *Mol Genet Metab.* 2003;80:1–10.
99. Gajdos ZK, Butler JL, Henderson KD, He C, Supelak PJ, Egyud M, Price A, Reich D, Clayton PE, Le Marchand L, Hunter DJ, Henderson BE, Palmert MR, Hirschhorn JN. Association studies of common variants in 10 hypogonadotropic hypogonadism genes with age at menarche. *J Clin Endocrinol Metab.* 2008;93:4290–4298.
100. Treloar SA, Martin NG. Age at menarche as a fitness trait: nonadditive genetic variance detected in a large twin sample. *Am J Human Genet.* 1990;47:137–148.
101. Towne B, Czerwinski SA, Demerath EW, Blangero J, Roche AF, Siervogel RM. Heritability of age at menarche in girls from the Fels Longitudinal Study. *Am J Physical Anthropol.* 2005;128: 210–219.
102. Eaves L, Silberg J, Foley D, Bulik C, Maes H, Erkanli A, Angold A, Costello EJ, Worthman C. Genetic and environmental influences on the relative timing of pubertal change. *Twin Res.* 2004;7:471–481.
103. Franco B, Guioli S, Pragliola A, Incerti B, Bardoni B, Tonlorenzi R, Carrozzo R, Maestrini E, Pieretti M, Taillon-Miller P, Brown CJ, Willard HF, Lawrence C, Graziella Persico M, Camerino G, Ballabio A. A gene deleted in Kallmann's syndrome shares homology with neural cell adhesion and axonal path-finding molecules. *Nature.* 1991;353: 529–536.
104. Legouis R, Hardelin JP, Levilliers J, Claverie JM, Compain S, Wunderle V, Millasseau P, Le Paslier D, Cohen D, Caterina D, et al. The candidate gene for the X-linked Kallmann syndrome encodes a protein related to adhesion molecules. *Cell.* 1991;67:423–435.
105. Hardelin JP, Levilliers J, Young J, Pholsena M, Legouis R, Kirk J, Bouloux P, Petit C, Schaison G. Xp22.3 deletions in isolated familial Kallmann's syndrome. *J Clin Endocrinol Metab.* 1993;76:827–831.
106. Dodé C, Levilliers J, Dupont JM, De Paepe A, Le Dû N, Soussi-Yanicostas N, Coimbra RS, Delmaghani S, Compain-Nouaille S, Baverel F, Pêcheux C, Le Tessier D, Cruaud C, Delpech M, Speleman F, Vermeulen S, Amalfitano A, Bachelot Y, Bouchard P, Cabrol S, Carel JC, Delemarre-van de Waal H, Goulet-Salmon B, Kottler ML, Richard O, Sanchez-Franco F, Saura R, Young J, Petit C, Hardelin JP. Loss-of-function

mutations in FGFR1 cause autosomal dominant Kallmann syndrome. *Nat Genet.* 2003;33:463–465.
107. Pitteloud N, Acierno Jr JS, Meysing AU, Dwyer AA, Hayes FJ, Crowley Jr WF. Reversible Kallmann syndrome, delayed puberty, and isolated anosmia occurring in a single family with a mutation in the fibroblast growth factor receptor 1 gene. *J Clin Endocrinol Metab.* 2005;90:1317–1322.
108. Pitteloud N, Meysing A, Quinton R, Acierno Jr JS, Dwyer AA, Plummer L, Fliers E, Boepple P, Hayes F, Seminara S, Hughes VA, Ma J, Bouloux P, Mohammadi M, Crowley Jr WF. Mutations in fibroblast growth factor receptor 1 cause Kallmann syndrome with a wide spectrum of reproductive phenotypes. *Mol Cell Endocrinol.* 2006;254-255:60–69.
109. Kramer PR, Wray S. Novel gene expressed in nasal region influences outgrowth of olfactory axons and migration of luteinizing hormone-releasing hormone (LHRH) neurons. *Genes Dev.* 2000;14:1824–1834.
110. Miura K, Acierno Jr JS, Seminara SB. Characterization of the human nasal embryonic LHRH factor gene, NELF, and a mutation screening among 65 patients with idiopathic hypogonadotropic hypogonadism (IHH). *J Human Genet.* 2004;49:265–268.
111. Dodé C, Teixeira L, Levilliers J, Fouveaut C, Bouchard P, Kottler ML, Lespinasse J, Lienhardt-Roussie A, Mathieu M, Moerman A, Morgan G, Murat A, Toublanc JE, Wolczynski S, Delpech M, Petit C, Young J, Hardelin JP. Kallmann syndrome: mutations in the genes encoding prokineticin-2 and prokineticin receptor-2. *PLoS Genetics.* 2006;2:e175.
112. Pitteloud N, Zhang C, Pignatelli D, Li JD, Raivio T, Cole LW, Plummer L, Jacobson-Dickman EE, Mellon PL, Zhou QY, Crowley Jr WF. Loss-of-function mutation in the prokineticin 2 gene causes Kallmann syndrome and normosmic idiopathic hypogonadotropic hypogonadism. *Proc Natl Acad USA.* 2007;104:17447–17452.
113. Chan YM, de Guillebon A, Lang-Muritano M, Plummer L, Cerrato F, Tsiaras S, Gaspert A, Lavoie HB, Wu CH, Crowley Jr WF, Amory JK, Pitteloud N, Seminara SB. GNRH1 mutations in patients with idiopathic hypogonadotropic hypogonadism. *Proc Natl Acad Science USA.* 2009;106:11703–11708.
114. Bouligand J, Ghervan C, Tello JA, Brailly-Tabard S, Salenave S, Chanson P, Lombès M, Millar RP, Guiochon-Mantel A, Young J. Isolated familial hypogonadotropic hypogonadism and a GNRH1 mutation. *N Engl J Med.* 2009;360:2742–2748.
115. Lin L, Conway GS, Hill NR, Dattani MT, Hindmarsh PC, Achermann JC. A homozygous R262Q mutation in the gonadotropin-releasing hormone receptor presenting as constitutional delay of growth and puberty with subsequent borderline oligospermia. *J Clin Endocrinol Metab.* 2006;91:5117–5121.
116. Messager S, Chatzidaki EE, Ma D, Hendrick AG, Zahn D, Dixon J, Thresher RR, Malinge I, Lomet D, Carlton MB, Colledge WH, Caraty A, Aparicio SA. Kisspeptin directly stimulates gonadotropin-releasing hormone release via G protein-coupled receptor 54. *Proc Natl Acad USA.* 2005;102:1761–1766.
117. Oakley AE, Clifton DK, Steiner RA. Kisspeptin signaling in the brain. *Endocrine Rev.* 2009;30:713–743.
118. Rance NE, Krajewski SJ, Smith MA, Cholanian M, Dacks PA. Neurokinin B and the hypothalamic regulation of reproduction. *Brain Res* 2010;1364:116–128.
119. Frisch RE, Revelle R. Height and weight at menarche and a hypothesis of menarche. *Arch Dis Childhood.* 1971;46:695–701.
120. Garn SM, LaVelle M, Rosenberg KR, Hawthorne VM. Maturational timing as a factor in female fatness and obesity. *Am J Clin Nutr.* 1986;43:879–883.
121. van Lenthe FJ, Kemper CG, van Mechelen W. Rapid maturation in adolescence results in greater obesity in adulthood: the Amsterdam Growth and Health Study. *Am J Clin Nutr.* 1996;64:18–24.
122. Demerath EW, Li J, Sun SS, Chumlea WC, Remsberg KE, Czerwinski SA, Towne B, Siervogel RM. Fifty-year trends in serial body mass index during adolescence in girls: the Fels Longitudinal Study. *Am J Clin Nutr.* 2004;80:441–446.
123. Bau AM, Ernert A, Schenk L, Wiegand S, Martus P, Grüters A, Krude H. Is there a further acceleration in the age at onset of menarche? A cross-sectional study in 1840 school children focusing on age and bodyweight at the onset of menarche. *Eur J Endocrinol.* 2009;160:107–113.
124. Perry JR, Stolk L, Franceschini N, Lunetta KL, Zhai G, McArdle PF, Smith AV, Aspelund T, Bandinelli S, Boerwinkle E, Cherkas L, Eiriksdottir G, Estrada K, Ferrucci L, Folsom AR, Garcia M, Gudnason V, Hofman A, Karasik D, Kiel DP, Launer LJ, van Meurs J, Nalls MA, Rivadeneira F, Shuldiner AR, Singleton A, Soranzo N, Tanaka T, Visser JA, Weedon MN, Wilson SG, Zhuang V, Streeten EA, Harris TB, Murray A, Spector TD, Demerath EW, Uitterlinden AG, Murabito JM. Meta-analysis of genome-wide association data identifies two loci influencing age at menarche. *Nat Genet.* 2009;41:648–650.
125. He C, Kraft P, Chen C, Buring JE, Paré G, Hankinson SE, Chanock SJ, Ridker PM, Hunter DJ, Chasman DI. Genome-wide association studies identify loci associated with age at menarche and age at natural menopause. *Nat Genet.* 2009;41:724–728.
126. Ong KK, Elks CE, Li S, Zhao JH, Luan J, Andersen LB, Bingham SA, Brage S, Smith GD, Ekelund U, Gillson CJ, Glaser B, Golding J, Hardy R, Khaw KT, Kuh D, Luben R, Marcus M, McGeehin MA, Ness AR, Northstone K, Ring SM, Rubin C, Sims MA, Song K, Strachan DP, Vollenweider P, Waeber G, Waterworth DM, Wong A, Deloukas P, Barroso I, Mooser V, Loos RJ, Wareham NJ. Genetic variation in LIN28B is associated with the timing of puberty. *Nat Genet.* 2009; 41:729–733.
127. Zhu H, Shah S, Shyh-Chang N, Shinoda G, Einhorn WS, Viswanathan SR, Takeuchi A, Grasemann C, Rinn JL, Lopez MF, Hirschhorn JN, Palmert MR, Daley GQ. Lin28a transgenic mice manifest size and puberty phenotypes identified in human genetic association studies. *Nat Genet.* 2010;42:626–630.
128. Ahmed ML, Ong KK, Morrell DJ, Cox L, Perry L, Preece MA, Dunger DB. Longitudinal study of leptin concentrations during puberty: sex differences and relationship to changes in body composition. *J Clin Endocrinol Metab.* 1999;84:899–905.
129. Garcia-Mayor RV, Andrade MA, Rios M, Lage M, Dieguez C, Casanueva FF. Serum leptin levels in normal children: relationship to age, gender, body mass index, pituitary-gonadal hormones, and pubertal stage. *J Clin Endocrinol Metab.* 1997;82:2849–2855.
130. Palmert MR, Radovick S, Boepple PA. Leptin levels in children with central precocious puberty. *J Clin Endocrinol Metab.* 1998;83:2260–2265.
131. Palmert MR, Radovick S, Boepple PA. The impact of reversible gonadal sex steroid suppression on serum leptin concentrations in children with central precocious puberty. *J Clin Endocrinol Metab.* 1998;83:1091–1096.
132. Quinton ND, Smith RF, Clayton PE, Gill MS, Shalet S, Justice SK, Simon SA, Walters S, Postel-Vinay MC, Blakemore AI, Ross RJ. Leptin binding activity changes with age: the link between leptin and puberty. *J Clin Endocrinol Metab.* 1999;84:2336–2341.
133. Ahima RS, Dushay J, Flier SN, Prabakaran D, Flier JS. Leptin accelerates the onset of puberty in normal female mice. *J Clin Endocrinol Metab.* 1997;99:391–395.
134. Chehab FF, Mounzih K, Lu R, Lim ME. Early onset of reproductive function in normal female mice treated with leptin. *Science.* 1997;275:88–90.

135. Cheung CC, Thornton JE, Kuijper JL, Weigle DS, Clifton DK, Steiner RA. Leptin is a metabolic gate for the onset of puberty in the female rat. *Endocrinology*. 1997;138:855–858.
136. Gruaz NM, Lalaoui M, Pierroz DD, Englaro P, Sizonenko PC, Blum WF, Aubert. ML. Chronic administration of leptin into the lateral ventricle induces sexual maturation in severely food-restricted female rats. *J Neuroendocrinol*. 1998;10:627–633.
137. Finn PD, Cunningham MJ, Pau K-YF, Spies HG, Clifton DK, Steiner RA. The stimulatory effect of leptin on the neuroendocrine reproductive axis of the monkey. *Endocrinology*. 1998; 139:4652–4662.
138. Plant TM, Durrant AR. Circulating leptin does not appear to provide a signal for triggering the initiation of puberty in the male rhesus monkey (*Macaca mulatta*). *Endocrinology*. 1997; 138:4505–4508.
139. Urbanski HF, Pau K-YF. A biphasic developmental pattern of circulating leptin in the male rhesus macaque (*Macaca mulatta*). *Endocrinology*. 1998;139:2284–2286.
140. Mann DR, Akinbami MA, Gould KG, Castracane VD. A longitudinal study of leptin during development in the male rhesus monkey: the effect of body composition and season on circulating leptin levels. *Biol Reprod*. 2000;62:285–291.
141. Witchel SF, Arslanian S, Lee PA. Leptin concentrations in precocious puberty or untimely puberty with and without GnRH analogue therapy. *J Ped Endocrinol Metab*. 1999;12:839–845.
142. Farooqi IS, Jebb SA, Langmack G, Lawrence E, Cheetham CH, Prentice AM, Hughes IA, McCamish MA, O'Rahilly S. Effects of recombinant leptin therapy in a child with congenital leptin deficiency. *N Engl J Med*. 1999;341:879–884.
143. Jackson RS, Creemers JW, Ohagi S, Raffin-Sanson ML, Sanders L, Montague CT, Hutton JC, O'Rahilly S. Obesity and impaired prohormone processing associated with mutations in the human prohormone convertase 1 gene. *Nat Genet*. 1997;16:303–306.
144. Silveira LF, MacColl GS, Bouloux PM. Hypogonadotropic hypogonadism. *Sem Reprod Med*. 2002;20:327–338.
145. Gajdos ZK, Henderson KD, Hirschhorn JN, Palmert MR. Genetic determinants of pubertal timing in the general population. *Mol Cell Endocrinol*. 2010;324:21–29.
146. Sulem P, Gudbjartsson DF, Rafnar T, Holm H, Olafsdottir EJ, Olafsdottir GH, Jonsson T, Alexandersen P, Feenstra B, Boyd HA, Aben KK, Verbeek AL, Roeleveld N, Jonasdottir A, Styrkarsdottir U, Steinthorsdottir V, Karason A, Stacey SN, Gudmundsson J, Jakobsdottir M, Thorleifsson G, Hardarson G, Gulcher J, Kong A, Kiemeney LA, Melbye M, Christiansen C, Tryggvadottir L, Thorsteinsdottir U, Stefansson K. Genome-wide association study identifies sequence variants on 6q21 associated with age at menarche. *Nat Genet*. 2009;41:734–738.
147. Lettre G, Jackson AU, Gieger C, Schumacher FR, Berndt SI, Sanna S, Eyheramendy S, Voight BF, Butler JL, Guiducci C, Illig T, Hackett R, Heid IM, Jacobs KB, Lyssenko V, Uda M, Diabetes Genetics Initiative; Fusion; KORA; Prostate, Lung Colorectal and Ovarian Cancer Screening Trial; Nurses' Health Study; SardiNIA, Boehnke M, Chanock SJ, Groop LC, Hu FB, Isomaa B, Kraft P, Peltonen L, Salomaa V, Schlessinger D, Hunter DJ, Hayes RB, Abecasis GR, Wichmann HE, Mohlke KL, Hirschhorn JN. Identification of ten loci associated with height highlights new biological pathways in human growth. *Nat Genet*. 2008;40:584–591.
148. Widén E, Ripatti S, Cousminer DL, Surakka I, Lappalainen T, Järvelin MR, Eriksson JG, Raitakari O, Salomaa V, Sovio U, Hartikainen AL, Pouta A, McCarthy MI, Osmond C, Kajantie E, Lehtimäki T, Viikari J, Kähönen M, Tyler-Smith C, Freimer N, Hirschhorn JN, Peltonen L, Palotie A. Distinct variants at LIN28B influence growth in height from birth to adulthood. *Am J Human Genet*. 2010;86:773–782.
149. Tommiska J, Wehkalampi K, Vaaralahti K, Laitinen EM, Raivio T, Dunkel L. LIN28B in constitutional delay of growth and puberty. *J Clin Endocrinol Metab*. 2010;95:3063–3066.
150. Ojeda SR, Hill J, Hill DF, Costa ME, Tapia V, Cornea A, Ma YJ. The Oct-2 POU domain gene in the neuroendocrine brain: a transcriptional regulator of mammalian puberty. *Endocrinology*. 1999;140:3774–3789.
151. He X, Treacy MN, Simmons DM, Ingraham HA, Swanson LW, Rosenfeld MG. Expression of a large family of POU-domain regulatory genes in mammalian brain development. *Nature*. 1989;340:35–42.
152. Clark ME, Mellon PL. The POU homeodomain transcription factor Oct-1 is essential for activity of the gonadotropin-releasing hormone neuron-specific enhancer. *Mol Cell Biol*. 1995;15:6169–6177.
153. Wierman ME, Xiong X, Kepa JK, Spaulding AJ, Jacobson BM, Fang Z, Nilaver G, Ojeda SR. Repression of gonadotropin-releasing hormone (GnRH) promoter activity by the POU homeodomain transcription factor, SCIP/Oct-6/Tst-1: a regulatory mechanism of phenotype expression? *Mol Cell Biol*. 1997;17:1652–1665.
154. Kimura S, Hara Y, Pineau T, Fernandez-Salguero P, Fox CH, Ward JM, Gonzalez FJ. The T/ebp null mouse: thyroid-specific enhancer-binding protein is essential for the organogenesis of the thyroid, lung, ventral forebrain, and pituitary. *Genes Dev*. 1996;10:60–69.
155. Lee BJ, Cho GJ, Norgren R, Junier M-P, Hill DF, Tapia V, Costa ME, Ojeda SR. TTF-1, a homeodomain gene required for diencephalic morphogenesis, is postnatally expressed in the neuroendocrine brain in a developmentally regulated and cell-specific fashion. *Mol Cell Neurosci*. 2001;17:107–126.
156. Mastronardi C, Smiley GG, Raber J, Kusakabe T, Kawaguchi A, Matagne V, Dietzel A, Heger S, Mungenast AE, Cabrera R, Kimura S, Ojeda SR. Deletion of the Ttf1 gene in differentiated neurons disrupts female reproduction without impairing basal ganglia function. *J Neurosci*. 2006;26:13167–13179.
157. Heger S, Mastronardi C, Dissen GA, Lomniczi A, Cabrera R, Roth CL, Jung H, Galimi F, Sippell W, Ojeda SR. Enhanced at puberty 1 (EAP1) is a new transcriptional regulator of the female neuroendocrine reproductive axis. *J Clin Invest*. 2007;117:2145–2154.
158. Sutton VR, Shaffer LG. Search for imprinted regions on chromosome 14: comparison of maternal and paternal UPD cases with cases of chromosome 14 deletion. *Am J Med Genet*. 2000;93:381–387.
159. Martin RA, Sabol DW, Rogan PK. Maternal uniparental disomy of chromosome 14 confined to an interstitial segment (14q23-14q24.2). *J Med Genet*. 1999;36:633–636.
160. Wildt L, Marshall G, Knobil E. Experimental induction of puberty in the infantile female rhesus monkey. *Science*. 1980;207:1373–1375.
161. Loose MD, Terasawa E. Pulsatile infusion of luteinizing hormone-releasing hormone induces precocious puberty (vaginal opening and first ovulation) in the immature female guinea pig. *Biol Reprod*. 1985;33:1084–1093.
162. Foster DL, Jackson LM. Puberty in the sheep. In: Neill JD, ed. *Physiology of Reproduction*. 3rd ed. San Diego, CA: Academic Press/Elsevier; 2006:1415–1482.
163. Sarkar DK, Fink G. Mechanism of the first spontaneous gonadotrophin surge and that induced by pregnant mare serum and effects of neonatal androgen in rats. *J Endocrinol*. 1979;83:339–354.
164. Sisk CL, Richardson HN, Chappell PE, Levine JE. *In vivo* gonadotropin-releasing hormone secretion in female rats during peripubertal development and on proestrus. *Endocrinology*. 2001;142:2929–2936.

165. Harris GC, Levine JE. Pubertal acceleration of pulsatile gonadotropin-releasing hormone release in male rats as revealed by microdialysis. *Endocrinology.* 2003;144:163–171.
166. Urbanski HF, Ojeda SR. The juvenile–peripubertal transition period in the female rat: establishment of a diurnal pattern of pulsatile luteinizing hormone secretion. *Endocrinology.* 1985; 117:644–649.
167. Bourguignon J-P, Lebrethon MC, Gérard A, Purnell G, Vandersmissen E, Parent AS, Yamanaka C. Amino acid neurotransmission and early ontogeny of pulsatile GnRH secretion from the rat hypothalamus. In: Bourguignon J-P, Plant TM, eds. *The Onset of Puberty in Perspective.* Amsterdam: Elsevier Science B.V; 2000:119–129.
168. Plant TM. The male monkey as a model for the study of the neurobiology of puberty onset in man. *Mol Cell Endocrinol.* 2006;254-255:97–102.
169. Ojeda SR, Ramirez VD. Short-term steroid treatment on plasma LH and FSH in castrated rats from birth to puberty. *Neuroendocrinology.* 1973;13:100–114.
170. Döhler KD, Wuttke W. Serum LH, FSH, prolactin and progesterone from birth to puberty in female and male rats. *Endocrinology.* 1974;94:1003–1008.
171. Döhler KD, Wuttke W. Changes with age in levels of serum gonadotropins, prolactin and gonadal steroids in prepubertal male and female rats. *Endocrinology.* 1975;97:898–907.
172. Jackson LM, Timmer KM, Foster DL. Organizational actions of postnatal estradiol in female sheep treated prenatally with testosterone: programming of prepubertal neuroendocrine function and the onset of puberty. *Endocrinology.* 2009;150: 2317–2324.
173. Steele RE, Weisz J. Changes in sensitivity of the estradiol-LH feedback system with puberty in the female rat. *Endocrinology.* 1974;95:513–520.
174. Bronson FH. The regulation of luteinizing hormone secretion by estrogen: relationships among negative feedback, surge potential, and male stimulation in juvenile, peripubertal, and adult female mice. *Endocrinology.* 1981;108:506–516.
175. Foster DL, Ryan KD. Endocrine mechanisms governing transition into adulthood: a marked decrease in inhibitory feedback action of estradiol on tonic secretion of luteinizing hormone in the lamb during puberty. *Endocrinology.* 1979;105:896–904.
176. Andrews WW, Advis JP, Ojeda SR. The maturation of estradiol-negative feedback in female rats: evidence that the resetting of the hypothalamic "gonadostat" does not precede the first preovulatory surge of gonadotropins. *Endocrinology.* 1981;109: 2022–2031.
177. Ebling FJ, Claypool LE, Foster DL. Neuroendocrine responsiveness to light during the neonatal period in the sheep. *J Endocrinol.* 1988;119:211–218.
178. Foster DL, Jackson LM, Padmanabhan V. Programming of GnRH feedback controls timing puberty and adult reproductive activity. *Mol Cell Endocrinol.* 2006;254–255:109–119.
179. Chongthammakun S, Terasawa E. Negative feedback effects of estrogen on luteinizing hormone-releasing hormone release occur in pubertal, but not prepubertal, ovariectomized female rhesus monkeys. *Endocrinology.* 1993;132:735–743.
180. Plant TM, Gay VL, Marshall GR, Arslan M. Puberty in monkeys is triggered by chemical stimulation of the hypothalamus. *Proc Natl Acad USA.* 1989;86:2506–2510.
181. Urbanski HF, Ojeda SR. Activation of luteinizing hormone-releasing hormone release advances the onset of female puberty. *Neuroendocrinology.* 1987;46:273–276.
182. Smyth C, Wilkinson M. A critical period for glutamate receptor-mediated induction of precocious puberty in female rats. *J Neuroendocrinol.* 1994;6:275–284.
183. Terasawa E. Mechanisms Controlling the onset of puberty in primates, the role of GABAergic neurons. In: Plant TM, Lee PA, eds. *The Neurobiology of Puberty.* Bristol: Journal of Endocrinology Ltd; 1995:139–151.
184. Acosta M, Mayer C, Dubois S, Wolfe A, Radovick S, Levine JE, & Boehm U. Kisspeptin-cell-specific deletion of estrogen receptor alpha (ERα) results in advanced pubertal onset and incomplete pubertal maturation in Female Mice. Abstracts of the 92nd Annual Meeting of the Endocrine Society; 2010 No. OR10–4.
185. Gill JC, Carroll RS, & Kaiser UB. Neurokinin B (NKB) Expression in the arcuate nucleus is an early marker for pubertal onset. Abstracts of the 92nd Annual Meeting of the Endocrine Society; 2010 No. OR10–5.
186. Bourguignon J-P, Gerard A, Alvarez-Gonzalez M-L, Purnelle G, Franchimont P. Endogenous glutamate involvement in pulsatile secretion of gonadotropin-releasing hormone: evidence from effect of glutamine and developmental changes. *Endocrinology.* 1995;136:911–916.
187. Bourguignon J-P, Gérard A, Gonzalez A, Purnelle G, Franchimont P. The role of excitatory amino acids in triggering the onset of puberty. In: Plant TM, Lee PA, eds. *The Neurobiology of Puberty.* Bristol, UK: Journal of Endocrinology Ltd; 1995: 129–138.
188. Krey LC, Butler WR, Knobil E. Surgical disconnection of the medial basal hypothalamus and pituitary function in the rhesus monkey. I. Gonadotropin secretion. *Endocrinology.* 1975;96: 1073–1087.
189. Halász B, Pupp L. Hormone secretion of the anterior pituitary gland after physical interruption of all nervous pathways to the hypophysiotrophic area. *Endocrinology.* 1965;77:553–562.
190. Przekop F, Domański E. Abnormalities in the seasonal course of oestrous cycles in ewes after lesions of the suprachiasmatic area of the hypothalamus. *J Endocrinol.* 1980;85:481–486.
191. Silverman AJ. The gonadotropin-releasing hormone (GnRH) neuronal systems: Immunocytochemistry. In: Knobil E, Neill JD, eds. *The Physiology of Reproduction.* New York: Raven Press; 1988:1283–1304.
192. Lehman MN, Goodman RL, Karsch FJ, Jackson GL, Berriman SJ, Jansen HT. The GnRH system of seasonal breeders: anatomy and plasticity. *Brain Res Bull.* 1997;44:445–457.
193. Wiegand SJ, Terasawa E. Discrete lesions reveal functional heterogeneity of suprachiasmatic structures in regulation of gonadotropin secretion in the female rat. *Neuroendocrinology.* 1982;34:395–404.
194. Terasawa E. Hypothalamic control of the onset of puberty. *Curr Opin Endocrinol Diabetes.* 1999;6:44–49.
195. Ojeda SR, Bilger M. Neuroendocrine regulation of puberty. In: Conn PM, Freeman M, eds. *Neuroendocrinology in Physiology and Medicine.* Totowa, NJ: Humana Press; 1999:197–224.
196. Semple RK, Achermann JC, Ellery J, Farooqi IS, Karet FE, Stanhope RG, O'Rahilly S, Aparicio SA. Two novel missense mutations in g protein-coupled receptor 54 in a patient with hypogonadotropic hypogonadism. *J Clin Endocrinol Metab.* 2005; 90:1849–1855.
197. Ohtaki T, Shintani Y, Honda S, Matsumoto H, Hori A, Kanehashi K, Terao Y, Kumano S, Takatsu Y, Masuda Y, Ishibashi Y, Watanabe T, Asada M, Yamada T, Suenaga M, Kitada C, Usuki S, Kurokawa T, Onda H, Nishimura O, Fujino M. Metastasis suppressor gene KiSS-1 encodes peptide ligand of a G-protein-coupled receptor. *Nature.* 2001;411:613–617.
198. Muir AI, Chamberlain L, Elshourbagy NA, Michalovich D, Moore DJ, Calamari A, Szekeres PG, Sarau HM, Chambers JK, Murdock P, Steplewski K, Shabon U, Miller JE, Middleton SE, Darker JG, Larminie CG, Wilson S, Bergsma DJ, Emson P,

Faull R, Philpott KL, Harrison DC. AXOR12, a novel human G protein-coupled receptor, activated by the peptide KiSS-1. *J Biol Chem.* 2001;276:28969–28975.

199. Kotani M, Detheux M, Vandenbogaerde A, Communi D, Vanderwinden JM, Le Poul E, Brezillon S, Tyldesley R, Suarez-Huerta N, Vandeput F, Blanpain C, Schiffmann SN, Vassart G, Parmentier M. The metastasis suppressor gene KiSS-1 encodes kisspeptins, the natural ligands of the orphan G protein-coupled receptor GPR54. *J Biol Chem.* 2001;276:34631–34636.
200. Colledge WH. GPR54 and puberty. *Trends Endocrinol Metabol.* 2004;15:448–453.
201. Kaiser UB, Kuohung W. KiSS-1 and GPR54 as new players in gonadotropin regulation and puberty. *Endocrine.* 2005;26:277–284.
202. Seminara SB. Metastin and its G protein-coupled receptor, GPR54: critical pathway modulating GnRH secretion. *Front Neuroendocrinol.* 2005;26:131–138.
203. Roa J, Aguilar E, Dieguez C, Pinilla L, Tena-Sempere M. New frontiers in kisspeptin/GPR54 physiology as fundamental gatekeepers of reproductive function. *Vit Horm.* 2007;71:1–25.
204. Kauffman AS, Clifton DK, Steiner RA. Emerging ideas about kisspeptin- GPR54 signaling in the neuroendocrine regulation of reproduction. *Trends Neurosci.* 2007;30:504–511.
205. Kauffman AS, Gottsch ML, Roa J, Byquist AC, Crown A, Clifton DK, Hoffman GE, Steiner RA, Tena-Sempere M. Sexual differentiation of Kiss1 gene expression in the brain of the rat. *Endocrinology.* 2007;148:1774–1783.
206. Funes S, Hedrick JA, Vassileva G, Markowitz L, Abbondanzo S, Golovko A, Yang S, Monsma FJ, Gustafson EL. The KiSS-1 receptor GPR54 is essential for the development of the murine reproductive system. *Biol Chem Res Comm.* 2003;312:1357–1363.
207. d'Anglemont de Tassigny X, Fagg LA, Dixon JP, Day K, Leitch HG, Hendrick AG, Zahn D, Franceschini I, Caraty A, Carlton MB, Aparicio SA, Colledge WH. Hypogonadotropic hypogonadism in mice lacking a functional Kiss1 gene. *Proc Natl Acad USA.* 2007;104:10714–10719.
208. Lapatto R, Pallais JC, Zhang D, Chan YM, Mahan A, Cerrato F, Wei Le W, Hoffman GE, Seminara SB. Postnatal development of kisspeptin neurons in mouse hypothalamus; sexual dimorphism and projections to gonadotropin-releasing hormone neurons. *Endocrinology.* 2007;147:5817–5825.
209. Tenenbaum-Rakover Y, Commenges-Ducos M, Iovane A, Aumas C, Admoni O, de Roux N. Neuroendocrine phenotype analysis in five patients with isolated hypogonadotropic hypogonadism due to a L102P inactivating mutation of GPR54. *J Clin Endocrinol Metab.* 2007;92:1137–1144.
210. Shahab M, Mastronardi C, Seminara SB, Crowley WF, Ojeda SR, Plant TM. Increased hypothalamic GPR54 signaling: a potential mechanism for initiation of puberty in primates. *Proc Natl Acad USA.* 2005;102:2129–2134.
211. Gottsch ML, Cunningham MJ, Smith JT, Popa SM, Acohido BV, Crowley WF, Seminara S, Clifton DK, Steiner RA. A role for kisspeptins in the regulation of gonadotropin secretion in the mouse. *Endocrinology.* 2004;145:4073–4077.
212. Smith JT, Cunningham MJ, Rissman EF, Clifton DK, Steiner RA. Regulation of Kiss1 gene expression in the brain of the female mouse. *Endocrinology.* 2005;146:3686–3692.
213. Smith JT, Dungan HM, Stoll EA, Gottsch ML, Braun RE, Eacker SM, Clifton DK, Steiner RA. Differential regulation of KiSS-1 mRNA expression by sex steroids in the brain of the male mouse. *Endocrinology.* 2005;146:2976–2984.
214. Pompolo S, Pereira A, Estrada KM, Clarke IJ. Colocalization of kisspeptin and gonadotropin-releasing hormone in the ovine brain. *Endocrinology.* 2006;147:804–810.
215. Tena-Sempere M. GPR54 and kisspeptin in reproduction. *Human Reprod Update.* 2006;12:631–639.
216. Maeda K, Adachi S, Inoue K, Ohkura S, Tsukamura H. Metastin/kisspeptin and control of estrous cycle in rats. *Rev Endocrine Metab Disord.* 2007;8:21–29.
217. Parhar IS, Ogawa S, Sakuma Y. Laser-captured single digoxigenin-labeled neurons of gonadotropin-releasing hormone types reveal a novel G protein-coupled receptor (Gpr54) during maturation in cichlid fish. *Endocrinology.* 2004;145:3613–3618.
218. Irwig MS, Fraley GS, Smith JT, Acohido BV, Popa SM, Cunningham MJ, Gottsch ML, Clifton DK, Steiner RA. Kisspeptin activation of gonadotropin releasing hormone neurons and regulation of KiSS-1 mRNA in the male rat. *Neuroendocrinology.* 2005;80:264–272.
219. Navarro VM, Castellano JM, Fernandez-Fernandez R, Tovar S, Roa J, Mayen A, Barreiro ML, Casanueva FF, Aguilar E, Dieguez C, Pinilla L, Tena-Sempere M. Effects of KiSS-1 peptide, the natural ligand of GPR54, on follicle-stimulating hormone secretion in the rat. *Endocrinology.* 2005;146:1689–1697.
220. Navarro VM, Castellano JM, Fernández-Fernández R, Tovar S, Roa J, Mayen A, Nogueiras R, Vazquez MJ, Barreiro ML, Magni P, Aguilar E, Dieguez C, Pinilla L, Tena-Sempere M. Characterization of the potent luteinizing hormone-releasing activity of KiSS-1 peptide, the natural ligand of GPR54. *Endocrinology.* 2005;146:156–163.
221. Dhillo WS, Chaudhri OB, Patterson M, Thompson EL, Murphy KG, Badman MK, McGowan BM, Amber V, Patel S, Ghatei MA, Bloom SR. Kisspeptin-54 stimulates the hypothalamic-pituitary gonadal axis in human males. *J Clin Endocrinol Metab.* 2005;90:6609–6615.
222. Plant TM, Ramaswamy S, Dipietro MJ. Repetitive activation of hypothalamic G protein-coupled receptor 54 with intravenous pulses of kisspeptin in the juvenile monkey (*Macaca mulatta*) elicits a sustained train of gonadotropin-releasing hormone discharges. *Endocrinology.* 2006;147:1007–1013.
223. Navarro VM, Fernandez-Fernandez R, Castellano JM, Roa. J, Mayen A, Barreiro ML, Gaytan F, Aguilar E, Pinilla L, Dieguez C, Tena-Sempere M. Advanced vaginal opening and precocious activation of the reproductive axis by KiSS-1 peptide, the endogenous ligand of GPR54. *J Physiol.* 2004; 561:379–386.
224. Castellano JM, Navarro VM, Fernandez-Fernandez R, Nogueiras R, Tovar S, Roa J, Vazquez MJ, Vigo E, Casanueva FF, Aguilar E, Pinilla L, Dieguez C, Tena-Sempere M. Changes in hypothalamic KiSS-1 system and restoration of pubertal activation of the reproductive axis by kisspeptin in undernutrition. *Endocrinology.* 2005;146:3917–3925.
225. Pineda R, Garcia-Galiano D, Roseweir A, Romero M, Sanchez-Garrido MA, Ruiz-Pino F, Morgan K, Pinilla L, Millar RP, Tena-Sempere M. Critical roles of kisspeptins in female puberty and preovulatory gonadotropin surges as revealed by a novel antagonist. *Endocrinology.* 2010;151:722–730.
226. Smith JT, Clarke IJ. Kisspeptin expression in the brain: catalyst for the initiation of puberty. *Rev Endocr Metab Disord.* 2007;8:1–9.
227. Han SK, Gottsch ML, Lee KJ, Popa SM, Smith JT, Jakawich SK, Clifton DK, Steiner RA, Herbison AE. Activation of gonadotropin-releasing hormone neurons by kisspeptin as a neuroendocrine switch for the onset of puberty. *J Neurosci.* 2005; 25:11349–11356.
228. Clarkson J, Herbison AE. Postnatal development of kisspeptin neurons in mouse hypothalamus; sexual dimorphism and projections to gonadotropin-releasing hormone neurons. *Endocrinology.* 2006;147:5817–5825.
229. Clarkson J, Herbison AE. Oestrogen, kisspeptin, GPR54 and the pre-ovulatory luteinising hormone surge. *J Neuroendocrinol.* 2009;21:305–311.

230. Guerriero KA, Keen KL, Roseweir AK, Millar RP, & Terasawa E. Effects of a kisspeptin agonist and antagonist on the release of luteinizing hormone-releasing hormone-1 (LHRH-1) in female rhesus monkeys *in vivo*. Abstracts of the 38 Annual Meeting of the Society for Neuroscience, held November 15-19, at Washington DC; 2008. No 618.1.
231. Keen KL, Wegner FH, Bloom SR, Ghatei MA, Terasawa E. An increase in kisspeptin-54 release occurs with the pubertal increase in luteinizing hormone-releasing hormone-1 release in the stalk median eminence of female rhesus monkeys *in vivo*. *Endocrinology*. 2008;149:4151–4157.
232. Dungan HM, Clifton DK, Steiner RA. Minireview: Kisspeptin neurons as central processors in the regulation of gonadotropin-releasing hormone secretion. *Endocrinology*. 2006;147:1154–1158.
233. Wiegand SJ, Terasawa E, Bridson WE. Persistent estrus and blockade of progesterone-induced LH release follows lesions which do not damage the suprachiasmatic nucleus. *Endocrinology*. 1978;102:1645–1648.
234. Terasawa E, Wiegand SJ, Bridson WE. A role for medial preoptic nucleus on afternoon of proestrus in female rats. *Am J Physiol*. 1980;238:E533–E539.
235. Navarro VM, Castellano JM, Fernández-Fernández R, Barreiro ML, Roa J, Sanchez-Criado JE, Aguilar E, Dieguez C, Pinilla L, Tena-Sempere M. Developmental and hormonally regulated messenger ribonucleic acid expression of KiSS-1 and its putative receptor, GPR54, in rat hypothalamus and potent luteinizing hormone-releasing activity of KiSS-1 peptide. *Endocrinology*. 2004;145:4565–4574.
236. Smith JT. Kisspeptin signalling in the brain: steroid regulation in the rodent and ewe. *Brain Res Dev*. 2008;57:288–298.
237. Franceschini I, Lomet D, Cateau M, Delsol G, Tillet Y, Caraty A. Kisspeptin immunoreactive cells of the ovine preoptic area and arcuate nucleus co-express estrogen receptor alpha. *Neurosci Lett*. 2006;401:225–230.
238. Clarkson J, Boon WC, Simpson ER, Herbison AE. Postnatal development of an estradiol-kisspeptin positive feedback mechanism implicated in puberty onset. *Endocrinology*. 2009; 150:3214–3220.
239. Plant TM, Witchel SF. Puberty in nonhuman primates and primates. In: Neill J, ed. *The Physiology of Reproduction*. 3rd ed. Vol. 2. San Diego, CA: Academic Press; 2006:2177–2230.
240. van den Pol AN, Wuarin J-P, Dudek FE. Glutamate, the dominant excitatory transmitter in neuroendocrine regulation. *Science*. 1990;250:1276–1278.
241. van den Pol AN, Trombley PQ. Glutamate neurons in hypothalamus regulate excitatory transmission. *J Neurosci*. 1993;13: 2829–2836.
242. Seeburg PH. The TINS/TiPS Lecture. The molecular biology of mammalian glutamate receptor channels. *Trends Neurosci*. 1993;16:359–365.
243. Meeker RB, Greenwood RS, Kayward JN. Glutamate receptors in the rat hypothalamus and pituitary. *Endocrinology*. 1994;134: 621–629.
244. van den Pol AN, Kogelman L, Ghost P, Liljelund P, Blackstone C. Developmental regulation of the hypothalamic metabotropic glutamate receptor mGluR1. *J Neurosci*. 1994;14: 3816–3834.
245. Brann DW, Mahesh VB. Endogenous excitatory amino acid involvement in the preovulatory and steroid-induced surge of gonadotropins in the female rat. *Endocrinology*. 1991;128: 1541–1547.
246. Brann DW. Glutamate: a major excitatory transmitter in neuroendocrine regulation. *Neuroendocrinology*. 1995;61:213–225.
247. Goldsmith PC, Thind KK, Perera AD, Plant TM. Glutamate-immunoreactive neurons and their gonadotropin-releasing hormone-neuronal interactions in the monkey hypothalamus. *Endocrinology*. 1994;134:858–868.
248. Leranth C, Naftolin F, Shanabrough M, Horvath TL. Neuronal circuits regulating gonadotropin release in the rat. In: Plant TM, Lee PA, eds. *The Neurobiology of Puberty*. Bristol, UK: Journal of Endocrinology Ltd; 1995:55–72.
249. Boehm U, Zou Z, Buck LB. Feedback loops link odor and pheromone signaling with reproduction. *Cell*. 2005;123:683–695.
250. Campbell RE, Han SK, Herbison AE. Biocytin filling of adult gonadotropin-releasing hormone neurons *in situ* reveals extensive, spiny, dendritic processes. *Endocrinology*. 2005;146: 1163–1169.
251. Herbison AE. Physiology of the gonadotropin-releasing hormone neuronal network. In: Neill JD, ed. *Physiology of Reproduction*. 3rd ed. San Diego, CA: Academic Press/Elsevier; 2006:1415–1482.
252. Price MT, Olney JW, Cicero TJ. Acute elevations of serum luteinizing hormone induced by kainic acid, N-methyl aspartic acid or homocysteic acid. *Neuroendocrinology*. 1978;26:352–358.
253. Donoso AO, López FJ, Negro-Vilar A. Glutamate receptors of the non-N-methyl-D-aspartic acid type mediate the increase in luteinizing hormone-releasing hormone release by excitatory amino acids *in vitro*. *Endocrinology*. 1990;126:414–420.
254. Ondo JG, Wheller DD, Dom RM. Hypothalamic site of action for N-methyl-D-aspartate (NMDA) on LH secretion. *Life Sci*. 1988;43:2283–2286.
255. Wilson RC, Knobil E. Acute effects of N-methyl-DL-aspartate on the release of pituitary gonadotropins and prolactin in the adult female rhesus monkey. *Brain Res*. 1982;248:177–179.
256. Olney JW, Cicero TJ, Meyer ER, De Gubareff T. Acute glutamate-induced elevations in serum testosterone and luteinizing hormone. *Brain Res*. 1976;112:420–424.
257. Bourguignon J-P, Gérard A, Frachimont P. Direct activation of gonadotropin-releasing hormone secretion through different receptors to neuroexcitatory amino acids. *Neuroendocrinology*. 1989;49:402–408.
258. Eyigor O, Jennes L. Kainate receptor subunit-positive gonadotropin-releasing hormone neurons express c-Fos during the steroid-induced luteinizing hormone surge in the female rat. *Endocrinology*. 2000;141:779–786.
259. Bourguignon J-P, Gérard A, Alvarez-Gonzalez M-L, Franchimont P. Neuroendocrine mechanism of onset of puberty. Sequential reduction in activity of inhibitory and facilitatory N-methyl-D-aspartate receptors. *J Clin Invest*. 1992;90:1736–1744.
260. Meijs-Roelofs HMA, Kramer P, van Leeuwen ECM. The N-methyl-D-aspartate receptor antagonist MK-801 delays the onset of puberty and may acutely block the first spontaneous LH surge and ovulation in the rat. *J Endocrinol*. 1991;131: 435–441.
261. Donoso AO, López FJ, Negro-Vilar A. Cross-talk between excitatory and inhibitory amino acids in the regulation of luteinizing hormone-releasing hormone secretion. *Endocrinology*. 1992;131:1559–1561.
262. Bourguignon J-P, Gérard A, Purnelle G, Czajkowski V, Yamanaka C, Lemaître M, Rigo J-M, Moonen G, Frachimont P. Duality of glutamatergic and GABAergic control of pulsatile GnRH secretion by rat hypothalamic explants: I. Effects of antisense oligodeoxynucleotides using explants including or excluding the preoptic area. *J Neuroendocrinol*. 1997;9:183–191.
263. Bourguignon J-P, Gérard A, Purnelle G, Czajkowski V, Yamanaka C, Lemaître M, Rigo J-M, Moonen G, Franchimont P. Duality of glutamatergic and GABAergic control of pulsatile GnRH secretion by rat hypothalamic explants: II. Reduced NR2C- and $GABA_A$-receptor-mediated inhibition at initiation of sexual maturation. *J Neuroendocrinol*. 1997;9:193–199.

264. Mahachoklertwattana P, Sanchez J, Kaplan SL, Grumbach MM. N-methyl-D-aspartate (NMDA) receptors mediate the release of gonadotropin-releasing hormone (GnRH) by NMDA in a hypothalamic GnRH neuronal cell line (GT1-1). *Endocrinology.* 1994;134:1023–1030.
265. Gay VL, Plant TM. N-methyl-D, L-aspartate elicits hypothalamic gonadotropin-releasing hormone release in prepubertal male rhesus monkeys (*Macaca mulatta*). *Endocrinology.* 1987;120: 2289–2296.
266. Medhamurthy R, Dichek HL, Plant TM, Bernardini I, Cutler Jr GB. Stimulation of gonadotropin secretion in prepubertal monkeys after hypothalamic excitation with aspartate and glutamate. *J Clin Endocrinol Metab.* 1990;71:1390–1392.
267. Medhamurthy R, Gay VL, Plant TM. Repetitive injections of L-glutamic acid, in contrast to those of N-methyl-D, L-aspartic acid, fail to elicit sustained hypothalamic GnRH release in the prepubertal male rhesus monkey (*Macaca mulatta*). *Neuroendocrinology.* 1992;55:660–666.
268. Cicero TJ, Meyer ER, Bell RD. Characterization and possible opioid modulation of N-methyl-D-aspartic acid induced increases in serum luteinizing hormone levels in the developing male rat. *Life Sci.* 1988;42:1725–1732.
269. Brann DW, Zamorano PL, Ping L, Mahesh VB. Role of excitatory amino acid neurotransmission during puberty in the female rat. *Mol Cell Neurosci.* 1993;4:107–112.
270. I'Anson H, Herbosa CG, Ebling FJ, Wood RI, Bucholtz DC, Mieher CD, Foster DL, Padmanabhan V. Hypothalamic versus pituitary stimulation of luteinizing hormone secretion in the prepubertal female lamb. *Neuroendocrinology.* 1993;57:467–475.
271. Bettendorf M, de Zegher F, Albers N, Hart CS, Kaplan SL, Grumbach MM. Acute N-methyl-D, L-aspartate administration stimulates the luteinizing hormone releasing hormone pulse generator in the ovine fetus. *Horm Res.* 1999;51:25–30.
272. Urbanski HF, Ojeda SR. A role for N-methyl-D-aspartate (NMDA) receptors in the control of LH secretion and initiation of female puberty. *Endocrinology.* 1990;126:1774–1776.
273. MacDonald MC, Wilkinson M. Peripubertal treatment with N-methyl-D-aspartic acid or neonatally with monosodium glutamate accelerates sexual maturation in female rats, an effect reversed by MK-801. *Neuroendocrinology.* 1990;52:143–149.
274. MacDonald MC, Robertson HA, Wilkinson M. Age- and dose-related NMDA induction of Fos-like immunoreactivity and c-fos mRNA in the arcuate nucleus of immature female rats. *Brain Res Dev Brain Res.* 1993;73:193–198.
275. Wu FCW, Howe DC, Naylor AM. N-Methyl-DL-aspartate receptor antagonism by D-2-amino-5-phosphonovaleric acid delays onset of puberty in the female rat. *J Neuroendocrinol.* 1990;2:627–631.
276. Roth C, Leonhardt S, Theiling K, Lakomek M, Jarry H, Wuttke W. Ontogeny of the GNRH-, glutaminase- and glutamate decarboxylase-gene expression in the hypothalamus of female rats. *Brain Res Dev Brain Res.* 1998;110:105–114.
277. Roth CL, McCormack AL, Lomniczi A, Mungenast AE, Ojeda SR. Quantitative proteomics identifies a change in glial glutamate metabolism at the time of female puberty. *Mol Cell Endocrinol.* 2006;254-255:51–59.
278. Flügge G, Wuttke W, Fuchs E. Postnatal development of transmitter systems: sexual differentiation of the GABAergic system and effects of muscimol. *Intl J Dev Neurosci.* 1986;4:319–326.
279. Goroll D, Arias P, Wuttke W. Ontogenic changes in the hypothalamic levels of amino acid neurotransmitters in the female rat. *Brain Res Dev Brain Res.* 1994;77:183–188.
280. Carbone S, Szwarcfarb B, Otera Losada M, Moguilevsky JA. Effects of ovarian steroids on the gonadotropin response to N-methyl-D-aspartate and on hypothalamic excitatory amino acid levels during sexual maturation in female rats. *Endocrinology.* 1992;130:1365–1370.
281. Carbone S, Szwarcfarb B, Losada M, Moguilevsky JA. Effect of ovarian hormones on the hypothalamic excitatory amino acids system during sexual maturation in female rats. *Neuroendocrinology.* 1995;61:235–242.
282. Goroll D, Arias P, Wuttke W. Preoptic release of amino acid neurotransmitters evaluated in peripubertal and young adult female rats by push–pull perfusion. *Neuroendocrinology.* 1993; 58:11–15.
283. Jarry H, Hirsch B, Leonhardt S, Wuttke W. Amino acid neurotransmitter release in the preoptic area of rats during the positive feedback actions of estradiol on LH release. *Neuroendocrinology.* 1992;56:133–140.
284. Ping L, Mahesh VB, Wiedmeier VT, Brann DW. Release of glutamate and aspartate from the preoptic area during the progesterone-induced LH surge: *in vivo* microdialysis studies. *Neuroendocrinology.* 1994;59:318–324.
285. Bourguignon J-P, Gerard A, Franchimont P. Maturation of the hypothalamic control of pulsatile gonadotropin-releasing hormone secretion at onset of puberty: II. Reduced potency of an inhibitory autofeedback. *Endocrinology.* 1990;127:2884–2890.
286. Moguilevsky JA, Carbone S, Szwarcfarb B, Rondina D, Scacchi P. Hypothalamic excitatory amino acid system during sexual maturation in female rats. *J Steroid Biochem Mol Biol.* 1995;53:337–341.
287. Bourguignon J-P, Gérard A, Alvarez-Gonzalez M-L, Fawe L, Franchimont P. Gonadal-independent developmental changes in activation of N-methyl-D-aspartate receptors involved in gonadotropin-releasing hormone secretion. *Neuroendocrinology.* 1992;55:634–641.
288. Carbone S, Szwarcfarb B, Rodina D, Feleder C, Moguilevsky JA. Differential effects of the N-methyl-D-aspartate and non-N-methyl-D-aspartate receptors of the excitatory amino acids system on LH and FSH secretion. Its effects on the hypothalamic luteinizing hormone releasing hormone during maturation in male rats. *Brain Res.* 1996;707:139–145.
289. Claypool LE, Kasuya E, Saitoh Y, Marzban F, Terasawa E. N-methyl D, L-aspartate induces the release of luteinizing hormone-releasing hormone in the prepubertal and pubertal female rhesus monkey as measured by *in vivo* push–pull perfusion in the stalk-median eminence. *Endocrinology.* 2000;141:219–228.
290. Decavel C, van den Pol AN. GABA: a dominant neurotransmitter in the hypothalamus. *J Comp Neurol.* 1990;302:1019–1037.
291. Rando RR, Bangerter FW, Farb DH. The inactivation of gamma-aminobutyric acid transaminase in dissociated neuronal cultures from spinal cord. *J Neurochem.* 1981;36:985–990.
292. Erlander MG, Tillakaratne NJK, Feldblum S, Patle N, Tobin AJ. Two genes encode distinct glutamate decarboxylases. *Neuron.* 1991;7:91–100.
293. Martin DL, Rimvall K. Regulation of gamma-aminobutyric acid synthesis in the brain. *J Neurochem.* 1993;60:395–407.
294. Erlander MG, Tobin A. The structural and functional heterogeneity of glutamic acid decarboxylase: a review. *Neurochem Res.* 1991;16:215–226.
295. Rimvall K, Martin DL. The level of GAD67 protein is highly sensitive to small increases in intraneuronal gamma-aminobutyric acid levels. *J Neurochem.* 1994;62:1375–1381.
296. Borden LA. GABA transporter heterogeneity: pharmacology and cellular localization. *Neurochem Intl.* 1996;29:335–356.
297. Wisden W, Seeburg PH. GABAA receptor channels: from subunits to functional entities. *Curr Opin Neurobiol.* 1992;2:263–269.
298. Sieghart W. Structure and pharmacology of gamma-aminobutyric acidA receptor subtypes. *Pharmacol Rev.* 1995;47: 181–234.

299. Rabow LE, Russek SJ, Farb DH. From ion currents to genomic analysis: recent advances in GABAA receptor research. *Synapse.* 1995;21:189–274.
300. Kaupmann K, Huggel K, Heid J, Flor PJ, Bischoff S, Mickel SJ, McMaster G, Angst C, Bittiger H, Froestl W, Bettler B. Expression cloning of GABA(B) receptors uncovers similarity to metabotropic glutamate receptors. *Nature.* 1997;386:239–246.
301. Bowery N. GABAB receptors and their significance in mammalian pharmacology. *Trends Pharmacol Sci.* 1989;10:401–407.
302. Bormann J. Electrophysiology of $GABA_A$ and $GABA_B$ receptor subtypes. *Trends Neurosci.* 1988;11:112–116.
303. Enz R, Cutting GR. Identification of 70 amino acids important for GABA(C) receptor rho1 subunit assembly. *Brain Res.* 1999; 846:177–185.
304. Gee KW, Bolger MB, Brinton RE, Coirini A, McEwen BS. Steroid modulation of the chloride ionophore in rat brain: structure-activity requirements, regional dependence and mechanism of action. *J Pharmacol Exp Ther.* 1988;246:803–812.
305. Puia G, Vicini S, Seeburg PH, Costa E. Influence of recombinant gamma-aminobutyric acid-A receptor subunit composition on the action of allosteric modulators of gamma-aminobutyric acid-gated Cl^- currents. *Mol Pharmacol.* 1991;39:691–696.
306. Pritchett DB, Luddens H, Seeburg PH. Type I and type II GABAA-benzodiazepine receptors produced in transfected cells. *Science.* 1989;245:1389–1392.
307. Benke D, Mertens S, Trzeciak A, Gillessen D, Mohler H. $GABA_A$ receptors display association of gamma 2-subunit with alpha 1- and beta 2/3-subunits. *J Biol Chem.* 1991;266:4478–4483.
308. Olsen RW, Tobin AJ. Molecular biology of $GABA_A$ receptors. *FASEB J.* 1990;4:1469–1480.
309. Wisden W, Morris BJ, Darlison MG, Hung SP, Barnard EA. Distinct GABAA receptor alpha subunit mRNAs show differential patterns of expression in bovine brain. *Neuron.* 1988;1:937–947.
310. Wisden W, Laurie DJ, Monyer H, Seeburg PH. The distribution of 13 GABAA receptor subunit mRNAs in the rat brain. I. Telencephalon, diencephalon, mesencephalon. *J Neurosci.* 1992; 12:1040–1062.
311. Fritschy J-M, Mohler H. $GABA_A$-receptor heterogeneity in the adult rat brain: differential regional and cellular distribution of seven major subunits. *J Comp Neurol.* 1995;359:154–194.
312. McKernan RM, Whiting PJ. Which $GABA_A$-receptor subtypes really occur in the brain? *Trends Neurosci.* 1996;19:139–143.
313. Tobin AJ, Brecha N, Chiang M-Y, Fundo S, Erlander MG, Feldblum S, Houser CR, Kaufman DL, Khrestchatistky M, MacLennan AJ, Olsen RW, Sternini C, Tillakaratne NJ. Alternative Forms of GAD and $GABA_A$ receptors. In: Biggio G, Concas A, Costa E, eds. *GABAergic Synaptic Transmission.* New York, NY: Raven Press; 1992:55–66.
314. Brooks-Kayal AR, Pritchett DB. Developmental changes in human gamma-aminobutyric acid A receptor subunit composition. *Ann Neurol.* 1993;34:687–693.
315. Hendrickson A, March D, Richards G, Erickson A, Shaw C. Coincidental appearance of the alpha 1 subunit of the GABA-A receptor and the type I benzodiazepine receptor near birth in macaque monkey visual cortex. *Intl J Dev Neurosci.* 1994;12: 299–314.
316. Fritschy J-M, Paysan J, Enna A, Mohler H. Switch in the expression of rat $GABA_A$-receptor subtypes during postnatal development: an immunohistochemical study. *J Neurosci.* 1994;14:5302–5324.
317. Brussaard AB, Kits KS, Baker RE, Willems WP, Leyting-Vermeulen JW, Voorn P, Smit AB, Bicknell RJ, Herbison AE. Plasticity in fast synaptic inhibition of adult oxytocin neurons caused by switch in GABA(A) receptor subunit expression. *Neuron.* 1997;19:1103–1114.
318. Laurie DJ, Wisden W, Seeburg PH. The distribution of thirteen $GABA_A$ receptor subunit mRNAs in the rat brain. III. Embryonic and postnatal development. *J Neurosci.* 1992;12:4151–4172.
319. Sim JA, Skynner MJ, Pape JR, Herbison AE. Late postnatal reorganization of GABA(A) receptor signalling in native GnRH neurons. *Eur J Neurosci.* 2000;12:3497–3504.
320. Han SK, Abraham IM, Herbison AE. Effect of GABA on GnRH neurons switches from depolarization to hyperpolarization at puberty in the female mouse. *Endocrinology.* 2002;143:1459–1466.
321. Peterson SL, McCrone S, Coy D, Adelman MP, Mahan LC. $GABA_A$ receptors subunit mRNAs in cells of the preoptic area: Colocalization with LHRH mRNA using dual-label *in situ* hybridization. *Endocrine.* 1993;J1:29–34.
322. Jung H, Danilchikm M, Fritschy J-M, Ojeda SR. Heterogeneity of $GABA_A$ receptor subunit expression in LHRH neurons of juvenile female rats. *Horm Res.* 1996;2(46 Suppl):110.
323. Noriega NC, Eghlidi DH, Garyfallou VT, Kohama SG, Kryger SG, Urbanski HF. Influence of 17beta-estradiol and progesterone on GABAergic gene expression in the arcuate nucleus, amygdala and hippocampus of the rhesus macaque. *Brain Res.* 2010;1307:28–42.
324. Davis AM, Grattan DR, Selmanoff M, McCarthy MM. Sex differences in glutamic acid decarboxylase mRNA in neonatal rat brain: implications for sexual differentiation. *Horm Behav.* 1996;30:538–552.
325. Davis AM, Ward SC, Selmanoff M, Herbison AE, McCarthy MM. Developmental sex differences in amino acid neurotransmitter levels in hypothalamic and limbic areas of rat brain. *Neuroscience.* 1999;90:1471–1482.
326. Urbanski HF, Rodrigues SM, Garyfallou VT, Kohama SG. Regional distribution of glutamic acid decarboxylase (GAD65 and GAD67) mRNA in the hypothalamus of male rhesus macaques before and after puberty. *Mol Brain Res.* 1998; 57:86–91.
327. Mladinic M, Becchetti A, Didelon F, Bradbury A, Cherubini E. Low expression of the ClC-2 chloride channel during postnatal development: a mechanism for the paradoxical depolarizing action of GABA and glycine in the hippocampus. *Proc R Soc B Biol Sci.* 1999;266:1207–1213.
328. Staley K, Smith R, Schaack J, Wilcox C, Jentsch TJ. Alteration of $GABA_A$ receptor function following gene transfer of the CLC-2 chloride channel. *Neuron.* 1996;17:543–551.
329. Smith RL, Clayton GH, Wilcox CL, Escudero KW, Staley KJ. Differential expression of an inwardly rectifying chloride conductance in rat brain neurons: a potential mechanism for cell-specific modulation of postsynaptic inhibition. *J Neurosci.* 1995;15:4057–4067.
330. Cherubini E, Gaiarsa JL, Ben-Ari Y. GABA: an excitatory transmitter in early postnatal life. *Trends Neurosci.* 1991;14:515–519.
331. Ben-Ari Y, Khazipov R, Leinekugel X, Caillard O, Gaiarsa J-L. GABAA, NMDA and AMPA receptors: a developmentally regulated "ménage à trois". *Trends Neurosciences.* 1997;20:523–529.
332. Rivera C, Viopio J, Payne JA, Ruusuvuori E, Lahtinen H, Lamsa K, Pirvola U, Saarma M, Kaila K. The K^+/Cl^- co-transporter KCC-2 renders GABA hyperpolarizing during neuronal maturation. *Nature.* 1999;397:251–255.
333. Moenter SM, DeFazio RA. Endogenous gamma-aminobutyric acid can excite gonadotropin-releasing hormone neurons. *Endocrinology.* 2005;146:5374–5379.
334. Moenter SM, DeFazio AR, Pitts GR, Nunemaker CS. Mechanisms underlying episodic gonadotropin-releasing hormone secretion. *Front Neuroendocrinol.* 2003;24:79–93.
335. DeFazio RA, Heger S, Ojeda SR, Moenter SM. Activation of A-type gamma-aminobutyric acid receptors excites gonadotropin-releasing hormone neurons. *Mol Endocrinol.* 2002;16:2872–2891.

336. Mitsushima D, Hei DL, Terasawa E. Gamma-aminobutyric acid is an inhibitory neurotransmitter restricting the release of luteinizing hormone-releasing hormone before the onset of puberty. *Proc Natl Acad USA.* 1994;91:395–399.
337. Ondo JG. Gamma-aminobutyric acid effects on pituitary gonadotropin secretion. *Science.* 1974;186:738–739.
338. Nikolarakis KE, Loeffler J-PH, Almeida OFX, Herz A. Pre- and postsynaptic actions of GABA on the release of hypothalamic gonadotropin-releasing hormone (GnRH). *Brain Res Bull.* 1988;21:677–683.
339. Vijayan E, McCann SM. The effects of intraventricular injection of gamma-aminobutyric acid (GABA) on prolactin and gonadotropin release in conscious female rats. *Brain Res.* 1978;155: 35–43.
340. Masotto C, Negro-Vilar A. Activation of gamma-aminobutyric acid B-receptors abolishes naloxone-stimulated luteinizing hormone release. *Endocrinology.* 1987;121:2251–2255.
341. Lamberts R, Vijayan E, Graf M, Mansky T, Wuttke W. Involvement of preoptic–anterior hypothalamic GABA neurons in the regulation of pituitary LH and prolactin release. *Exp Brain Res.* 1983;52:356–362.
342. Akema T, Kimura F. Modulation of pulsatile LH secretion by baclofen, a selective $GABA_B$ receptor agonist, in ovariectomized rats. *Neuroendocrinology.* 1992;56:141–147.
343. Jarry H, Perschl A, Wuttke W. Further evidence that preoptic anterior hypothalamic GABAergic neurons are part of the GnRH pulse and surge generator. *Acta Endocrinol.* 1988;118:573–579.
344. Herbison AE, Dyer RG. Effect on luteinizing hormone secretion of GABA receptor modulation in the medial preoptic area at the time of proestrous luteinizing hormone surge. *Neuroendocrinology.* 1991;53:317–320.
345. Jarry H, Leonhardt S, Wuttke W. Gamma-aminobutyric acid neurons in the preoptic/anterior hypothalamic area synchronize the phasic activity of the gonadotropin-releasing hormone pulse generator in ovariectomized rats. *Neuroendocrinology.* 1991;53:261–267.
346. Scott CJ, Clarke IJ. Evidence that changes in the function of the subtypes of the receptors for gamma-amino butyric acid may be involved in the seasonal changes in the negative-feedback effects of estrogen on gonadotropin-releasing hormone secretion and plasma luteinizing hormone levels in the ewe. *Endocrinology.* 1993;133:2904–2912.
347. Akema T, Chiba A, Kimura F. On the relationship between noradrenergic stimulatory and GABAergic inhibitory systems in the control of luteinizing hormone secretion in female rats. *Neuroendocrinology.* 1990;52:566–572.
348. Kimura F, Sano A, Hiruma H, Funabashi T. Effects of gamma-aminobutyric acid-A receptor antagonist, bicuculline, on the electrical activity of luteinizing hormone-releasing hormone pulse generator in the ovariectomized rat. *Neuroendocrinology.* 1993;57:605–614.
349. Kusano K, Fueshko S, Gainer H, Wray S. Electrical and synaptic properties of embryonic luteinizing hormone-releasing hormone neurons in explant cultures. *Proc Natl Acad USA.* 1995;92:3918–3922.
350. Hales TG, Sanderson MJ, Charles AC. GABA has excitatory actions on GnRH-secreting immortalized hypothalamic (GT1-7) neurons. *Neuroendocrinology.* 1994;59:297–308.
351. Martínez de la Escalera G, Choi ALH, Weiner RI. Biphasic gabaergic regulation of GnRH secretion in GT1 cell lines. *Neuroendocrinology.* 1994;59:420–425.
352. Spergel DJ, Krüth U, Hanley DF, Sprengel R, Seeburg PH. GABA- and glutamate-activated channels in green fluorescent protein-tagged gonadotropin-releasing hormone neurons in transgenic mice. *J Neurosci.* 1999;19:2037–2050.
353. Romanò N, Lee K, Abrahám IM, Jasoni CL, Herbison AE. Nonclassical estrogen modulation of presynaptic GABA terminals modulates calcium dynamics in gonadotropin-releasing hormone neurons. *Endocrinology.* 2008;149:5335–5344.
354. Sim JA, Skynner MJ, Herbison AE. Heterogeneity in the basic membrane properties of postnatal gonadotropin-releasing hormone neurons in the mouse. *J Neurosci.* 2001;21:1067–1075.
355. Penatti CA, Davis MC, Porter DM, Henderson LP. Altered $GABA_A$ receptor-mediated synaptic transmission disrupts the firing of gonadotropin-releasing hormone neurons in male mice under conditions that mimic steroid abuse. *J Neurosci.* 2010;30: 6497–6506.
356. Han SK, Todman MG, Herbison AE. Endogenous GABA release inhibits the firing of adult gonadotropin-releasing hormone neurons. *Endocrinology.* 2004;145:495–499.
357. Semyanov A, Walker MC, Kullmann DM, Silver RA. Tonically active $GABA_A$ receptors: modulating gain and maintaining the tone. *Trends Neurosci.* 2004;27:262–269.
358. Moragues N, Ciofi P, Lafon P, Tramu G, Garret M. $GABA_A$ receptor epsilon subunit expression in identified peptidergic neurons of the rat hypothalamus. *Brain Res.* 2003;967:285–289.
359. Adler BA, Crowley WR. Evidence for gamma-aminobutyric acid modulation of ovarian hormonal effects on luteinizing hormone secretion and hypothalamic catecholamine activity in the female rat. *Endocrinology.* 1986;118:91–97.
360. Lagrange AH, Ronnekleiv OK, Kelly MJ. Estradiol-17 beta and mu-opioid peptides rapidly hyperpolarize GnRH neurons: a cellular mechanism of negative feedback? *Endocrinology.* 1995;136:2341–2344.
361. Ferreira SA, Scott CJ, Kuehl DE, Jackson GL. Differential regulation of luteinizing hormone release by gamma-aminobutyric acid receptor subtypes in the arcuate–ventromedial region of the castrated ram. *Endocrinology.* 1996;137:3453–3460.
362. Jackson GL, Wood SG, Kuehl DE. A gamma-aminobutyric acidB agonist reverses the negative feedback effect of testosterone on gonadotropin-releasing hormone and luteinizing hormone secretion in the male sheep. *Endocrinology.* 2000;141: 3940–3945.
363. Brito VN, Mendonca BB, Guilhoto LM, Freitas KC, Arnhold IJ, Latronico AC. Allelic variants of the gamma-aminobutyric acid-A receptor alpha1-subunit gene (GABRA1) are not associated with idiopathic gonadotropin-dependent precocious puberty in girls with and without electroencephalographic abnormalities. *J Clin Endocrinol Metab.* 2006;91:2432–2436.
364. Lee K, Porteous R, Campbell RE, Lüscher B, Herbison AE. Knockdown of GABA(A) receptor signaling in GnRH neurons has minimal effects upon fertility. *Endocrinology.* 2010;151: 4428–4436.
365. Rudolph U, Möhler H. Analysis of $GABA_A$ receptor function and dissection of the pharmacology of benzodiazepines and general anesthetics through mouse genetics. *Annu Rev Pharmacol Toxicol.* 2004;44:475–498.
366. Kasuya E, Nyberg CL, Mogi K, Terasawa E. A role of gamma-amino butyric acid (GABA) and glutamate in control of puberty in female rhesus monkeys: effect of an antisense oligodeoxynucleotide for GAD67 messenger ribonucleic acid and MK801 on luteinizing hormone-releasing hormone release. *Endocrinology.* 1999;140:705–712.
367. Mitsushima D, Marzban F, Luchansky LL, Burich AJ, Keen KL, Durning M, Golos TG, Terasawa E. Role of glutamic acid decarboxylase in the prepubertal inhibition of the luteinizing hormone releasing hormone release in female rhesus monkeys. *J Neurosci.* 1996;16:2563–2573.
368. Keen KL, Burich AJ, Mitsushima D, Kasuya E, Terasawa E. Effects of pulsatile infusion of the GABA(A) receptor blocker

bicuculline on the onset of puberty in female rhesus monkeys. *Endocrinology.* 1999;140:5257–5266.
369. Bourguignon J-P, Jaeken J, Gerard A, de Zegher F. Amino acid neurotransmission and initiation of puberty: evidence from nonketotic hyperglycinemia in a female infant and gonadotropin-releasing hormone secretion by rat hypothalamic explants. *J Clin Endocrinol Metab.* 1997;82:1899–1903.
370. Moguilevsky JA, Carbone S, Szwarcfarb B, Rondina D. Sexual maturation modifies the GABAergic control of gonadotrophin secretion in female rats. *Brain Res.* 1991;563:12–16.
371. Mitsushima D, Kimura F. The maturation of GABA(A) receptor-mediated control of luteinizing hormone secretion in immature male rats. *Brain Res.* 1997;748:258–262.
372. Feleder C, Jarry H, Leonhardt S, Wuttke W, Moguilevsky JA. The GABAergic control of gonadotropin-releasing hormone secretion in male rats during sexual maturation involves effects on hypothalamic excitatory and inhibitory amino acid systems. *Neuroendocrinology.* 1996;64:305–312.
373. Kalra SP. Neural circuitry involved in the control of LHRH secretion, a model for preovulatory LH release. *Front Neuroendocrinol.* 1986;9:31–75.
374. Barraclough CA, Wise PM. The role of catecholamines in the regulation of pituitary luteinizing hormone and follicle-stimulating hormone secretion. *Endocrine Rev.* 1982;3:91–119.
375. Ramirez VD, Feder HH, Sawyer CH. The role of brain catecholamines in the regulation of LH secretion, A critical inquiry. *Front Neuroendocrinol.* 1984;8:27–84.
376. Gearing M, Terasawa E. Prostaglandin E2 mediates the stimulatory effect of methoxamine on *in vivo* luteinizing hormone-releasing hormone (LH-RH) release in the ovariectomized female rhesus monkey. *Brain Res.* 1991;560:276–281.
377. Gearing M, Terasawa E. The alpha-1-adrenergic neuronal system is involved in the pulsatile release of luteinizing hormone-releasing hormone in the ovariectomized female rhesus monkey. *Neuroendocrinology.* 1991;53:373–381.
378. Bhattacharya AN, Dierschke DJ, Yamaji T, Knobil E. The pharmacologic blockade of the circhoral mode of LH secretion in the ovariectomized rhesus monkey. *Endocrinology.* 1972;90: 778–786.
379. Plant TM, Nakai Y, Belchetz P, Keogh E, Knobil E. The sites of action of estradiol and phentolamine in the inhibition of the pulsatile, circhoral discharges of LH in the rhesus monkey (*Macaca mulatta*). *Endocrinology.* 1978;102:1015–1018.
380. Kaufman J-M, Kesner JS, Wilson RC, Knobil E. Electrophysiological manifestation of luteinizing hormone-releasing hormone pulse generator activity in the rhesus monkey: influence of alpha-adrenergic and dopaminergic blocking agents. *Endocrinology.* 1985;116:1327–1333.
381. Terasawa E, Krook C, Hei D, Gearing M, Schultz NJ, Davis GA. Norepinephrine is a possible neurotransmitter stimulating pulsatile release of luteinizing hormone-releasing hormone in the rhesus monkey. *Endocrinology.* 1988;123:1808–1816.
382. Gearing M, Terasawa E. Suppression of luteinizing hormone (LH) release by the α1-adrenergic receptor antagonist prazosin in the ovariectomized female rhesus monkey. *Am J Primatol.* 1991;25:23–33.
383. Pau K-YF, Hess DL, Kohama S, Bao J, Pau CY, Spies H. Oestrogen upregulates noradrenaline release in the mediobasal hypothalamus and tyrosine hydroxylase gene expression in the brainstem of ovariectomized rhesus macaques. *J Neuroendocrinol.* 2000;12:899–909.
384. Campbell RE, Herbison AE. Definition of brainstem afferents to gonadotropin-releasing hormone neurons in the mouse using conditional viral tract tracing. *Endocrinology.* 2007;148: 5884–5890.
385. Weiner RI, Ganong WF. Effect of the depletion of brain catecholamines on puberty and the estrous cycle in the rat. *Neuroendocrinology.* 1971;8:125–135.
386. Ruf KB, Holmes MJ. Delayed vaginal opening in rats after an intraventricular injection of 6-hydroxydopamine. *J Endocrinol.* 1974;60:383–384.
387. Sarkar DK, Smith GC, Fink G. Effect of manipulating central catecholamines on puberty and the surge of luteinizing hormone and gonadotropin releasing hormone induced by pregnant mare serum gonadotropin in female rats. *Brain Res.* 1981;213:335–349.
388. Cocchi D, Fraschini F, Jalanbo H, Müller EE. Role of brain catecholamines in the postcastration rise in plasma LH of prepuberal rats. *Endocrinology.* 1974;95:1649–1657.
389. Kim K, Lee BJ, Cho RN, Kang SS, Choi WS, Park SD, Lee CC, Cho WK, Wuttke W. Blockade of noradrenergic neurotransmission with diethyldithiocarbamic acid decreases the mRNA level of gonadotropin-releasing hormone in the hypothalamus of ovariectomized, steroid-treated prepubertal rats. *Neuroendocrinology.* 1994;59:539–544.
390. Clough RW, Hoffman GE, Sladek CD. Peripubertal development of noradrenergic stimulation of luteinizing hormone-releasing hormone neurosecretion *in vitro*. *Brain Res.* 1988;446: 121–132.
391. Ojeda SR, Urbanski HF, Katz KH, Costa ME. Activation of estradiol-positive feedback at puberty: estradiol sensitizes the LHRH-releasing system at two different biochemical steps. *Neuroendocrinology.* 1986;43:259–265.
392. Weiss B, Chen JF, Zhang S, Zhou L-W. Developmental and age-related changes in the D2 dopamine receptor mRNA subtypes in rat brain. *Neurochem Intl.* 1992;20(Suppl):49S–58S.
393. Hery M, Francois-Bellan AM, Deprez P, Faudon M, Hery F. Evidence for the presence of noradrenergic neurons and their inhibitory action on luteinizing hormone-releasing hormone release in cultured fetal rat hypothalamic cells. *Life Sci.* 1993;52: 2017–2025.
394. Gore AC, Terasawa E. A role for norepinephrine in the control of puberty in the female rhesus monkey, *Macaca mulatta*. *Endocrinology.* 1991;129:3009–3017.
395. Gore AC, Saitoh Y, Terasawa E. Effects of adrenal medulla transplantation into the third ventricle on the onset of puberty in female rhesus monkeys. *Exp. Neurol.* 1996;140:172–183.
396. Advis JP, Simpkins JW, Chen HT, Meites J. Relation of biogenic amines to onset of puberty in the female rat. *Endocrinology.* 1978;103:11–16.
397. Höhn K-G, Wuttke W. Ontogeny of catecholamine turnover rates in limbic and hypothalamic structures in relation to serum prolactin and gonadotropin levels. *Brain Res.* 1979;179:281–293.
398. Donoso AO, Banzan AM. Brain and hypothalamic catecholamine turnover in relation with puberty in the female rat. *Acta Physiol Latino Americana.* 1975;25:37–44.
399. Raum WJ, Glass AR, Swerdloff RS. Changes in hypothalamic catecholamine neurotransmitters and pituitary gonadotropins in the immature female rat: relationships to the gonadostat theory of puberty onset. *Endocrinology.* 1980;106:1253–1258.
400. Ojeda SR, Negro-Vilar A, McCann SM. Release of prostaglandin Es by hypothalamic tissue: evidence for their involvement in catecholamine-induced luteinizing hormone-releasing hormone release. *Endocrinology.* 1979;104:617–624.
401. Ojeda SR, Negro-Vilar A. Release of prostaglandin E2 from the hypothalamus depends on extracellular Ca^{2+} availability: relation to LHRH release. *Neuroendocrinology.* 1984;39:442–447.
402. Ojeda SR, Negro-Vilar A. Prostaglandin E2-induced luteinizing hormone-releasing hormone release involves mobilization of intracellular Ca^{2+}. *Endocrinology.* 1985;116:1763–1770.

403. Ojeda SR, Urbanski HF. Intracellular regulatory mechanisms of LHRH secretion and the onset of female puberty. In: Imura H, ed. *Neuroendocrine Control of the Hypothalamic–Pituitary System*. Tokyo, Japan: Japan Scientific Press; 1988:49–64.
404. Kordon C, Drouva SV, Martínez de la Escalera G, Weiner RI. Role of classic and peptide neuromediators in the neuroendocrine regulation of luteinizing hormone and prolactin. In: Knobil E, Neill JD, eds. *The Physiology of Reproduction*. 2nd ed. Vol. 1. New York, NY: Raven Press; 1994:1621–1681.
405. Leranth C, MacLusky NJ, Shanabrough M, Naftolin F. Catecholaminergic innervation of luteinizing hormone-releasing hormone and glutamic acid decarboxylase immunopositive neurons in the rat medial preoptic area. An electron-microscopic double immunostaining and degeneration study. *Neuroendocrinology*. 1988;48:591–602.
406. Lacau de Mengido IM, Becú-Villalobos D, Diaz G, Libertun C. Chronic activation of dopamine receptors in the female infantile rat: effect on hypophyseal hormones and on the onset of puberty. *Endocrinology*. 1989;124:746–753.
407. Gerber P, Döcke F, Rohde W, Dörner G. Evidence that inhibition of medial preoptic dopaminergic activity may be involved in the prepubertal desensitization to the negative oestrogen feedback in female rats. *Exp Clin Endocrinol*. 1984;84:7–12.
408. Döcke F, Rohde W, Oelssner W, Schleussner E, Gutenschwager I, Dörner G. Influence of the medial preoptic dopaminergic activity on the efficiency of the negative estrogen feedback in prepubertal and cyclic female rats. *Neuroendocrinology*. 1987;46:445–452.
409. Ohkura T, Hagino N, Yaoi Y. Changes in monoamine concentrations in developing female rat brains under continuous light. Nippon Sanka Fujinka Gakkai Zasshi. *Acta Obstet Gynecol Jpn*. 1995;47:225–230.
410. Wuttke W, Honma K, Lamberts R, Hohn KG. The role of monoamines in female puberty. *Fed Proc. FASEB J*. 1980;39: 2378–2383.
411. Arbogast LA, Voogt JL. Ontogeny of tyrosine hydroxylase mRNA signal levels in central dopaminergic neurons: development of a gender difference in the arcuate nuclei. *Brain Res Dev Brain Res*. 1991;63:151–161.
412. Krieger A, Wuttke W. Ontogeny of tyrosine hydroxylase and dopamine-beta-hydroxylase activity in discrete limbic and hypothalamic structures of female rats. *Brain Res*. 1980;193: 181–188.
413. Arbogast LA, Voogt JL. Hyperprolactinemia increases and hypoprolactinemia decreases tyrosine hydroxylase messenger ribonucleic acid levels in the arcuate nuclei, but not the substantia nigra or zona incerta. *Endocrinology*. 1991;128:997–1005.
414. Sales N, Martres MP, Bouthenet ML, Schwartz JC. Ontogeny of dopaminergic D-2 receptors in the rat nervous system: characterization and detailed autoradiographic mapping with [125I] iodosulpride. *Neuroscience*. 1989;28:673–700.
415. Kalra SP, Crowley WR. Neuropeptide Y, A novel neuroendocrine peptide in the control of pituitary hormone secretion, and its relation to luteinizing hormone. In: Ganong WF, Martini L, eds. *Frontiers in Neuroendocrinology*., Vol. 13. New York, NY: Raven Press; 1992:1–46.
416. Kalra SP, Pu S, Horvath TL, Kalra PS. Leptin and NPY Regulation of GnRH secretion and Energy Homeostasis. In: Bourguignon J-P, Plant TM, eds. *The Onset of Puberty in Perspective*. Amsterdam: Elsevier Science, B.V; 2000:317–327.
417. Blomqvist AG, Herzog H. Y-receptor subtypes—how many more? *Trends Neurosci*. 1997;20:294–298.
418. Hilal EM, Chen JH, Silverman A-J. Joint migration of gonado tropin-releasing hormone (GnRH) and neuropeptide Y (NPY) neurons from olfactory placode to central nervous system. *J Neurobiol*. 1996;31:487–502.
419. Allen JM, McGregor GP, Woodhams PL, Polak JM, Bloom SR. Ontogeny of a novel peptide, neuropeptide Y (NPY) in rat brain. *Brain Res*. 1984;303:197–200.
420. Sutton SW, Mitsugi N, Plotsky PM, Sarkar DK. Neuropeptide Y (NPY): a possible role in the initiation of puberty. *Endocrinology*. 1988;123:2152–2154.
421. Corder R, Walker CD, Gaillard RC, Aubert ML. Inhibition of sexual maturation in male rats by melatonin: Evidence linking the mechanism of action to changes in the regulation of hypothalamic neuropeptide Y. *J Neuroendocrinol*. 1992;4:1–7.
422. Woodhams PL, Allen YS, McGovern J, Allen JM, Bloom SR, Balázs R, Polak JM. Immunohistochemical analysis of the early ontogeny of the neuropeptide Y system in rat brain. *Neuroscience*. 1985;15:173–202.
423. Kagotani Y, Hashimoto T, Tsuruo Y, Kawano H, Daikoku S, Chihara K. Development of the neuronal system containing neuropeptide Y in the rat hypothalamus. *Intl J Dev Neurosci*. 1989;7:359–374.
424. McDonald JK, Tigges J, Tigges M, Reich C. Developmental study of neuropeptide Y-like immunoreactivity in the neurohypophysis and intermediate lobe of the rhesus monkey (*Macaca mulatta*). *Cell Tissue Res*. 1988;254:499–509.
425. de Quidt ME, Emson PC. Distribution of neuropeptide Y-like immunoreactivity in the rat central nervous system – II. Immunohistochemical analysis. *Neuroscience*. 1986;18:545–618.
426. Reuss S, Hurlbut EC, Speth JC, Moore RY. Neuropeptide Y localization in telencephalic and diencephalic structures of the ground squirrel brain. *Am J Anat*. 1990;188:163–174.
427. Pearson PL, Anderson LL, Jacobson CD. The prepubertal ontogeny of neuropeptide Y-like immunoreactivity in the male Meishan pig brain. *Brain Res Dev Brain Res*. 1996;91:41–69.
428. Thind KK, Boggan JE, Goldsmith PC. Neuropeptide Y system of the female monkey hypothalamus: retrograde tracing and immunostaining. *Neuroendocrinology*. 1993;57:289–298.
429. Gore AC, Mitsushima D, Terasawa E. A possible role of neuropeptide Y in the control of the onset of puberty in female rhesus monkeys. *Neuroendocrinology*. 1993;58:23–34.
430. Tong Y, Dumont Y, Shen SH, Quirion R. Comparative developmental profile of the neuropeptide Y Y1 receptor gene and protein in the rat brain. *Brain Res Mol Brain Res*. 1997;48: 323–332.
431. Naveilhan P, Neveu I, Arenas E, Ernfors P. Complementary and overlapping expression of Y1, Y2 and Y5 receptors in the developing and adult mouse nervous system. *Neuroscience*. 1998;87:289–302.
432. Kalra SP. Mandatory neuropeptide-steroid signaling for the preovulatory luteinizing hormone-releasing hormone discharge. *Endocrine Rev*. 1993;14:507–538.
433. Levine JE, Bauer-Dantoin AC, Besecke LM, Conaghan LA, Legan SJ, Meredith JM, Strobl FJ, Urban JH, Vogelsong KM, Wolfe AM. Neuroendocrine regulation of the luteinizing hormone-releasing hormone pulse generator in the rat. *Recent Prog Horm Res*. 1991;47:97–153.
434. Levine JE, Chappell P, Besecke LM, Bauer-Dantoin AC, Wolfe AM, Porkka-Heiskanen T, Urban JH. Amplitude and frequency modulation of pulsatile luteinizing hormone-releasing hormone release. *Cell Mol Neurobiol*. 1995;15:117–139.
435. Kaynard AH, Pau K-YF, Hess DL, Spies HG. Third-ventricular infusion of neuropeptide Y suppresses luteinizing hormone secretion in ovariectomized rhesus macaques. *Endocrinology*. 1990;127:2437–2444.
436. Kalra SP, Crowley WR. Norepinephrine-like effects of neuropeptide Y on LH release in the rat. *Life Sci*. 1984;35:1173–1176.

437. McDonald JK, Lumpkin MD, Samson WK, McCann SM. Neuropeptide Y affects secretion of luteinizing hormone and growth hormone in ovariectomized rats. *Proc Natl Acad USA.* 1985;82:561–564.
438. Khorram O, Pau K-YF, Spies HG. Bimodal effects of neuropeptide Y on hypothalamic release of gonadotropin-releasing hormone in conscious rabbits. *Neuroendocrinology.* 1987;45:290–297.
439. McShane TM, May T, Miner JL, Keisler. DH. Central actions of neuropeptide-Y may provide a neuromodulatory link between nutrition and reproduction. *Biol Reprod.* 1992;46: 1151–1157.
440. Woller MJ, McDonald JK, Reboussin DM, Terasawa E. Neuropeptide Y is a neuromodulator of pulsatile luteinizing hormone-releasing hormone release in the gonadectomized rhesus monkey. *Endocrinology.* 1992;130:2333–2342.
441. Woller MJ, Terasawa E. Estradiol enhances the action of neuropeptide Y on *in vivo* luteinizing hormone-releasing hormone release in the ovariectomized rhesus monkey. *Neuroendocrinology.* 1992;56:921–925.
442. Besecke LM, Levine JE. Acute increase in responsiveness of luteinizing hormone (LH)-releasing hormone nerve terminals to neuropeptide-Y stimulation before the preovulatory LH surge. *Endocrinology.* 1994;135:63–66.
443. Xu B, Pu S, Kalra PS, Hyd JF, Crowley WR, Kalra SP. An interactive physiological role of neuropeptide Y and galanin in pulsatile pituitary luteinizing hormone secretion. *Endocrinology.* 1996;137:5297–5302.
444. Pau K-YF, Kaynard AH, Hess DL, Spies HG. Effects of neuropeptide Y on the *in vitro* release of gonadotropin-releasing hormone, luteinizing hormone, and beta-endorphin and pituitary responsiveness to gonadotropin-releasing hormone in female macaques. *Neuroendocrinology.* 1991;53:396–403.
445. Xu M, Hill JW, Levine JE. Attenuation of luteinizing hormone surges in neuropeptide Y knockout mice. *Neuroendocrinology.* 2000;72:263–271.
446. O'Connor JL, Wade MF, Brann DW, Mahesh VB. Evidence that progesterone modulates anterior pituitary neuropeptide Y levels during the progesterone-induced gonadotropin surge in the estrogen-primed intact immature female rat. *J Steroid Biochem Mol Biol.* 1995;52:497–504.
447. Brann DW, McDonald JK, Putnam CD, Mahesh VB. Regulation of hypothalamic gonadotropin-releasing hormone and neuropeptide Y concentrations by progesterone and corticosteroids in immature rats: correlation with luteinizing hormone and follicle-stimulating hormone release. *Neuroendocrinology.* 1991;54: 425–432.
448. Minami S, Frautschy SA, Plotsky PM, Sutton SW, Sarkar DK. Facilitatory role of neuropeptide Y on the onset of puberty: effect of immunoneutralization of neuropeptide Y on the release of luteinizing hormone and luteinizing-hormone-releasing hormone. *Neuroendocrinology.* 1990;52:112–115.
449. Fraley GS, Kuenzel WJY. Precocious puberty in chicks (Gallus domesticus) induced by central injections of neuropeptide Y. *Life Sci.* 1993;52:1649–1656.
450. Palmiter RD, Erickson JC, Hollopeter G, Baraban SC, Schwartz MW. Life without neuropeptide Y. *Recent Prog Horm Res.* 1998;53:163–199.
451. Marsh DJ, Baraban SC, Hollopeter G, Palmiter RD. Role of the Y5 neuropeptide Y receptor in limbic seizures. *Proc Natl Acad USA.* 1999;96:13518–13523.
452. Kushi A, Sasai H, Koizumi H, Takeda N, Yokoyama M, Nakamura M. Obesity and mild hyperinsulinemia found in neuropeptide Y-Y1 receptor-deficient mice. *Proc Natl Acad USA.* 1998;95:15659–15664.
453. Li C, Chen P, Smith MS. Morphological evidence for direct interaction between arcuate nucleus neuropeptide Y (NPY) neurons and gonadotropin-releasing hormone neurons and the possible involvement of NPY Y1 receptors. *Endocrinology.* 1999;140:5382–5390.
454. Campbell RE, ffrench-Mullen JM, Cowley MA, Smith MS, Grove KL. Hypothalamic circuitry of neuropeptide Y regulation of neuroendocrine function and food intake via the Y5 receptor subtype. *Neuroendocrinology.* 2001;74:106–119.
455. Smith MS, Grove KL. Integration of the regulation of reproductive function and energy balance: lactation as a model. *Front Neuroendocrinol.* 2002;23:225–256.
456. Pierroz DD, Gruaz NM, d'Alieves V, Aubert ML. Chronic administration of neuropeptide Y into the lateral ventricle starting at 30 days of life delays sexual maturation in the female rat. *Neuroendocrinology.* 1995;61:293–300.
457. Aubert ML, Gruaz NM, d'Alleves V, Pierroz DD, Catzeflis C, Sizonenko PC. The control of the onset of puberty in the rat: potential role of neuropeptide Y. In: Plant TM, Lee PA, eds. *The Neurobiology of Puberty.* Bristol UK: Journal of Endocrinology Ltd; 1995:119–127.
458. Gruaz NM, Pierroz DD, Rohner-Jeanrenaud F, Sizonenko PC, Aubert ML. Evidence that neuropeptide Y could represent a neuroendocrine inhibitor of sexual maturation in unfavorable metabolic conditions in the rat. *Endocrinology.* 1993;133:1891–1894.
459. Raposinho PD, Broqua P, Pierroz DD, Hyward A, Dumont Y, Quirion R, Junien J-L, Aubert ML. Evidence that the inhibition of luteinizing hormone secretion exerted by central administration of neuropeptide Y (NPY) in the rat is predominantly mediated by the NPY-Y5 receptor subtype. *Endocrinology.* 1999;140:4046–4055.
460. Hendry SH, Jones EG, DeFelipe J, Schmechel D, Brandon C, Emson PC. Neuropeptide-containing neurons of the cerebral cortex are also GABAergic. *Proc Natl Acad USA.* 1984;81: 6526–6530.
461. Francois-Bellan AM, Kachidian P, Dusticier G, Tonon MC, Vaudry H, Bosler O. GABA neurons in the rat suprachiasmatic nucleus: involvement in chemospecific synaptic circuitry and evidence for GAD-peptide colocalization. *J Neurocytol.* 1990;19:937–947.
462. Horvath TL, Bechman I, Naftolin F, Kalra SP, Leranth C. Heterogeneity in the neuropeptide Y-containing neurons of the rat arcuate nucleus: GABAergic and non-GABAergic subpopulations. *Brain Res.* 1997;756:283–286.
463. Obrietan K, van den Pol AN. Neuropeptide Y depresses GABA-mediated calcium transients in developing suprachiasmatic nucleus neurons: a novel form of calcium long-term depression. *J Neurosci.* 1996;16:3521–3533.
464. van den Pol AN, Obrietan K, Chen G, Belousov AB. Neuropeptide Y-mediated long-term depression of excitatory activity in suprachiasmatic nucleus neurons. *J Neurosci.* 1996;16:5883–5895.
465. Chen G, van den Pol AN. Multiple NPY receptors coexist in pre- and postsynaptic sites: inhibition of GABA release in isolated self-innervating SCN neurons. *J Neurosci.* 1996;16:7711–7724.
466. Qian J, Colmers F, Saggau P. Inhibition of synaptic transmission by neuropeptide Y in rat hippocampal area CA1: modulation of presynaptic Ca^{2+} entry. *J Neurosci.* 1997;17:8169–8177.
467. Baraban SC, Hollopeter G, Erickson JC, Schwartzkroin PA, Palmiter RD. Knock-out mice reveal a critical antiepileptic role for neuropeptide Y. *J Neurosci.* 1997;17:8927–8936.
468. Ferin M, Wehrenberg WB, Lam NY, Alston EJ, Vande Wiele RL. Effects and site of action of morphine on gonadotropin secretion in the female rhesus monkey. *Endocrinology.* 1982;111:1652–1656.
469. Mallory DS, Bona-Gallo A, Gallo RV. Neurotransmitter involvement in naloxone-induced stimulation of pulsatile LH

release on day 8 of pregnancy in the rat. *Brain Res Bull.* 1989;22:1015–1021.
470. Kesner JS, Kaufman JM, Wilson RC, Kuroda G, Knobil E. The effect of morphine on the electrophysiological activity of the hypothalamic luteinizing hormone-releasing hormone pulse generator in the rhesus monkey. *Neuroendocrinology.* 1986;43: 686–688.
471. Van Vugt DA, Lam NY, Ferin M. Reduced frequency of pulsatile luteinizing hormone secretion in the luteal phase of the rhesus monkey. Involvement of endogenous opiates. *Endocrinology.* 1984;115:1095–1101.
472. Orstead KM, Hess DL, Spies HG. Opiatergic inhibition of pulsatile luteinizing hormone release during the menstrual cycle of rhesus macaques. *Proc Soc Exp Biol Med.* 1987;184:312–319.
473. Devorshak-Harvey E, Bona-Gallo A, Gallo RV. Endogenous opioid peptide regulation of pulsatile luteinizing hormone secretion during pregnancy in the rat. *Neuroendocrinology.* 1987;46:369–378.
474. Rossmanith WG, Mortola JF, Yen SS. Role of endogenous opioid peptides in the initiation of the midcycle luteinizing hormone surge in normal cycling women. *J Clin Endocrinol Metab.* 1988;67:695–700.
475. Ji W-Z, Kaynard AH, Pau K-YF, Hess DL, Baughman WL, Spies HG. Endogenous opiates regulate the nocturnal reduction in luteinizing hormone pulse frequency during the luteal phase of the macaque menstrual cycle. *Biol Reprod.* 1989;41:1024–1033.
476. Evans WS, Weltman ML, Johnson ML, Weltman A, Veldhuis JD, Rogol AD. Effects of opioid receptor blockade on luteinizing hormone (LH) pulses and interpulse LH concentrations in normal women during the early phase of the menstrual cycle. *J Endocrinol Invest.* 1992;15:525–531.
477. Grosser PM, O'Byrne KT, Williams CL, Thalabard JC, Hotchkiss J, Knobil E. Effects of naloxone on estrogen-induced changes in hypothalamic gonadotropin-releasing hormone pulse generator activity in the rhesus monkey. *Neuroendocrinology.* 1993;57:115–119.
478. Blank MS, Panerai AE, Friesen HG. Opioid peptides modulate luteinizing hormone secretion during sexual maturation. *Science.* 1979;203:1129–1131.
479. Wilkinson M, Bhanot R. A puberty-related attenuation of opiate peptide-induced inhibition of LH secretion. *Endocrinology.* 1982;110:1046–1048.
480. Ebling FJ, Schwartz ML, Foster DL. Endogenous opioid regulation of pulsatile luteinizing hormone secretion during sexual maturation in the female sheep. *Endocrinology.* 1989;125: 369–383.
481. Cutler L, Egli CA, Styne DM, Kaplan SL, Grumbach MM. Hormone ontogeny in the ovine fetus. XVIII. The effect of an opioid antagonist on luteinizing hormone secretion. *Endocrinology.* 1985;116:1997–2002.
482. Sirinathsinghji DJS, Motta M, Martini L. Induction of precocious puberty in the female rat after chronic naloxone administration during the neonatal period: the opiate 'brake' on prepubertal gonadotrophin secretion. *J Endocrinol.* 1985;104:299–307.
483. Wood RI, I'Anson H, Ebling FJ, Foster DL. Opioid inhibition of luteinizing hormone secretion compared in developing male and female sheep. *Neuroendocrinology.* 1992;56:822–830.
484. Prasad BM, Conover CD, Sarkar DK, Rabii J, Advis JP. Feed restriction in prepubertal lambs: effect on puberty onset and on *in vivo* release of luteinizing hormone-releasing hormone, neuropeptide Y and beta-endorphin from the posterior–lateral median eminence. *Neuroendocrinology.* 1993;57:1171–1181.
485. Blank MS, Murphy JR. Luteinizing hormone sensitivity to naloxone in maturing male chimpanzees. *Brain Res Bull.* 1991;27:241–245.
486. Petraglia F, Bernasconi S, Inghetti L, Loche S, Romanini F, Facchinetti F, Marcellini C, Genazzani AR. Naloxone-induced luteinizing hormone secretion in normal, precocious, and delayed puberty. *J Clin Endocrinol Metab.* 1986;63:1112–1116.
487. Terasawa E, Chongthammakun S. Annual Meeting of the Society for Neuroscience., Nov. 10–15, 1991, New Orleans, p. 906. (abstr.).
488. Eckersell CB, Popper P, Micevych PE. Estrogen-induced alteration of mu-opioid receptor immunoreactivity in the medial preoptic nucleus and medial amygdala. *J Neurosci.* 1998;18: 3967–3976.
489. Tamarkin L, Westrom WK, Hamill AI, Goldman BD. Effect of melatonin on the reproductive systems in syrian male and female hamsters: a diurnal rhythm in sensitivity to melatonin. *Endocrinology.* 1976;99:1534–1541.
490. Bittman EL, Karsch FJ. Nightly duration of pineal melatonin secretion determines the reproductive response to inhibitory day length in the ewe. *Biol Reprod.* 1984;30:585–593.
491. Waldhauser F, Boepple PA, Schemper M, Mansfeild MJ, Crowley Jr WF. Serum melatonin in central precocious puberty is lower than in age-matched prepubertal children. *J Clin Endocrinol Metab.* 1991;73:793–796.
492. Waldhauser F, Weiszenbacher G, Frisch H, Zeitlhuber U, Waldhauser M, Wurtman RJ. Fall in nocturnal serum melatonin during prepuberty and pubescence. *Lancet.* 1984;8373: 362–365.
493. Attanasio A, Borrelli P, Gupta D. Circadian rhythms in serum melatonin from infancy to adolescence. *J Clin Endocrinol Metab.* 1985;61:388–390.
494. Wilson ME, Gordon TP. Nocturnal changes in serum melatonin during female puberty in rhesus monkeys: a longitudinal study. *J Endocrinol.* 1989;121:553–562.
495. Cavallo A, Dolan LM. 6-hydroxymelatonin sulfate excretion in human puberty. *J Pineal Res.* 1996;21:225–230.
496. Kitay JI, Altshule MD. *The Pineal Gland. A review of the physiological literature.* Cambridge, MA: Harvard Press; 1954.
497. Sack RL, Lewy AJ, Blood ML, Keith LD, Nakagawa H. Circadian rhythm abnormalities in totally blind people: incidence and clinical significance. *J Clin Endocrinol Metab.* 1992;75: 127–134.
498. Bellastella A, Criscuolo T, Sinisi AA, Iorio S, Mazzuca A, Parlato F, Perrone L, Faggiano M. Influence of blindness on plasma luteinizing hormone, follicle-stimulating hormone, prolactin, and testosterone levels in prepubertal boys. *J Clin Endocrinol Metab.* 1987;64:862–864.
499. Zachrias L, Wurtman RJ. Blindness: Its relation to age of menarche. *Science.* 1964;144:1154–1155.
500. Magee K, Basinska J, Quarrington B, Stancer HC. Blindness and menarche. *Life Sci.* 1970;9:7–12.
501. Cavallo A. Melatonin and human puberty: current perspectives. *J Pineal Res.* 1993;15:115–121.
502. Packer RJ, Sutton LN, Rosenstock JG, Rorke LB, Bilaniuk LT, Zimmerman RA, Littman PA, Bruce DA, Schut L. Pineal region tumors of childhood. *Pediatrics.* 1984;74:97–102.
503. Cardinali DP, Vacas MI. Pineal gland, photoperiodic responses, and puberty. *J Endocrinol Invest.* 1984;7:157–165.
504. Edwards MS, Hudgins RJ, Wilson CB, Levin VA, Wara WM. Pineal region tumors in children. *J Neurosurg.* 1988;68:689–697.
505. Reiter RJ. Pineal control of reproduction. *Prog Clin Biol Res.* 1981;59B:349–355.
506. Siegel-Witchel S. CNS lesions, neurologic disorders, and puberty in man. In: Plant TM, Lee PA, eds. *The Neurobiology of Puberty.* Bristol: Journal of Endocrinology Ltd; 1995:229–239.
507. Luboshitzky R, Lavi S, Thuma I, Lavie P. Increased nocturnal melatonin secretion in male patients with hypogonadotropic

hypogonadism and delayed puberty. *J Clin Endocrinol Metab.* 1995;80:2144–2148.
508. Luboshitzky R, Lavi S, Thuma I, Herer P, Lavie P. Nocturnal secretor patterns of melatonin, luteinizing hormone, prolactin and cortisol in male patients with gonadotropin-releasing hormone deficiency. *J Pineal Res.* 1996;21:49–54.
509. Wilson ME, Lackey S, Chikazawa K, Gordon TP. The amplitude of nocturnal melatonin concentrations is not decreased by oestradiol and does not alter reproductive function in adolescent or adult female rhesus monkeys. *J Endocrinol.* 1993;137:229–309.
510. Plant TM, Zorub DS. Pinealectomy in agonadal infantile male rhesus monkeys (*Macaca mulatta*) does not interrupt initiation of the prepubertal hiatus in gonadotropin secretion. *Endocrinology.* 1986;118:227–232.
511. Dierschke DJ, Weiss G, Knobil E. Sexual maturation in the rhesus monkey and the development of estrogen-induced gonadotropic hormone release. *Endocrinology.* 1974;94:198–206.
512. Rosenfeld RG. Disorders of growth hormone and insulin-like growth factor secretion and action. In: Sperling MA, ed. *Pediatric Endocrinology.* Philadelphia: W.B. Saunders; 1996: 117–169.
513. Ramaley JA, Phares CK. Delay of puberty onset in females due to suppression of growth hormone. *Endocrinology.* 1980;106: 1989–1993.
514. Advis JP, White S, Ojeda SR. Activation of growth hormone short loop negative feedback delays puberty in the female rat. *Endocrinology.* 1981;108:1343–1352.
515. Danilovich N, Wernsing D, Coschigano KT, Kopchick JJ, Bartke A. Deficits in female reproductive function in GH-R-KO mice; role of IGF-I. *Endocrinology.* 1999;140:2637–2640.
516. Müller EE, Locatelli V, Cocchi D. Neuroendocrine control of growth hormone secretion. *Physiol Rev.* 1999;79:511–607.
517. Finkelstein JW, Roffwarg HP, Boyar RM, Kream J, Hellman L. Age-related change in the twenty-four-hour spontaneous secretion of growth hormone. *J Clin Endocrinol Metab.* 1972;35: 665–670.
518. Miller JD, Tannenbaum GS, Colle E, Guyda HJ. Daytime pulsatile growth hormone secretion during childhood and adolescence. *J Clin Endocrinol Metab.* 1982;55:989–994.
519. Wilson ME, Tanner JM. Somatostatin analog treatment slows growth and the tempo of reproductive maturation in female rhesus monkeys. *J Clin Endocrinol Metab.* 1994;79:495–501.
520. Suter KJ, Pohl CR, Wilson ME. Circulating concentrations of nocturnal leptin, growth hormone, and insulin-like growth factor-I increase before the onset of puberty in agonadal male monkeys: potential signals for the initiation of puberty. *J Clin Endocrinol Metab.* 2000;85:808–814.
521. Copeland KC, Eichberg JW, Parker Jr CR, Bartke A. Puberty in the chimpanzee: somatomedin-C and its relationship to somatic growth and steroid hormone concentrations. *J Clin Endocrinol Metab.* 1985;60:1154–1160.
522. Copeland KC, Kuehl TJ, Castracane VD. Pubertal endocrinology of the baboon: elevated somatomedin-C/insulin-like growth factor I at puberty. *J Clin Endocrinol Metab.* 1982;55:1198–1201.
523. Luna AM, Wilson DM, Wibbelsman CJ, Brown RC, Nagashima RJ, Hintz RL, Rosenfeld RG. Somatomedins in adolescence: a cross-sectional study of the effect of puberty on plasma insulin-like growth factor I and II levels. *J Clin Endocrinol Metab.* 1983;57:268–271.
524. Handelsman DJ, Spaliviero JA, Scott CD, Baxter RC. Hormonal regulation of the peripubertal surge of insulin-like growth factor-I in the rat. *Endocrinology.* 1987;120:491–496.
525. Wilson ME. IGF-I administration advances the decrease in hypersensitivity to oestradiol negative feedback inhibition of serum LH in adolescent female rhesus monkeys. *J Endocrinol.* 1995;145:121–130.
526. Miller BH, Gore AC. Alterations in hypothalamic insulin-like growth factor-I and its associations with gonadotropin releasing hormone neurones during reproductive development and ageing. *J Neuroendocrinol.* 2001;13:728–736.
527. Daftary SS, Gore AC. The hypothalamic insulin-like growth factor-1 receptor and its relationship to gonadotropin-releasing hormones neurones during postnatal development. *J Neuroendocrinol.* 2004;16:160–169.
528. Hiney JK, Ojeda SR, Dees WL. Insulin-like growth factor I: a possible metabolic signal involved in the regulation of female puberty. *Neuroendocrinology.* 1991;54:420–423.
529. Hiney JK, Srivastava V, Nyberg CL, Ojeda SR, Dees WL. Insulin-like growth factor I of peripheral origin acts centrally to accelerate the initiation of female puberty. *Endocrinology.* 1996; 137:3717–3728.
530. Pazos F, Sánchez-Franco F, Balsa J, López-Fernandez J, Escalada J, Cacicedo L. Regulation of gonadal and somatotropic axis by chronic intraventricular infusion of insulin-like growth factor 1 antibody at the initiation of puberty in male rats. *Neuroendocrinology.* 1999;69:408–416.
531. Wilson ME. Premature elevation in serum insulin-like growth factor-I advances first ovulation in rhesus monkeys. *J Endocrinol.* 1998;158:247–257.
532. Ransom B, Behar T, Nedergaard M. New roles for astrocytes (stars at last). *Trends Neurosci.* 2003;26:520–522.
533. Haydon PG. GLIA: listening and talking to the synapse. *Nat Rev Neurosci.* 2001;2:185–193.
534. Araque A, Parpura V, Sanzgiri RP, Haydon PG. Tripartite synapses: glia, the unacknowledged partner. *Trends Neurosci.* 1999;22:208–215.
535. Ma YJ, Hill DF, Creswick KE, Costa ME, Cornea A, Lioubin MN, Plowman GD, Ojeda SR. Neuregulins signaling via a glial erbB-2-erbB-4 receptor complex contribute to the neuroendocrine control of mammalian sexual development. *J Neurosci.* 1999;19:9913–9927.
536. Parpura V, Basarsky TA, Liu F, Jeftinija K, Jeftinija S, Haydon PG. Glutamate-mediated astrocyte-neuron signaling. *Nature.* 1994;369:744–747.
537. Kobayashi H, Wada M, Uemura H. *The hypothalamic median eminence as a neuroendocrine organ.*, 23. Med J Osaka University; 1972:43-55.
538. Kozlowski GP, Coats PW. Ependymoneuronal specializations between LHRH fibers and cells of the cerebroventricular system. *Cell Tissue Res.* 1985;242:301–311.
539. Witkin JW, Ferin M, Popilskis SJ, Silverman AJ. Effects of gonadal steroids on the ultrastructure of GnRH neurons in the rhesus monkey: synaptic input and glial apposition. *Endocrinology.* 1991;129:1083–1092.
540. King JC, Rubin BS. Dynamic changes in LHRH neurovascular terminals with various endocrine conditions in adults. *Horm Behav.* 1994;28:349–356.
541. Ridet JL, Malhorta SK, Privat A, Gage FH. Reactive astrocytes: cellular and molecular cues to biological function. *Trends Neurosci.* 1997;20:570–577.
542. Ojeda SR, Ma YJ, Rage F. The transforming growth factor alpha gene family is involved in the neuroendocrine control of mammalian puberty. *Mol Psychiatry.* 1997;2:355–358.
543. Ojeda SR, Ma YJ. Epidermal growth factor tyrosine kinase receptors and the neuroendocrine control of mammalian puberty. *Mol Cell Endocrinol.* 1998;140:101–106.
544. Marchetti B. Cross-talk signals in the CNS: role of neurotrophic and hormonal factors, adhesion molecules and intracellular signaling agents in luteinizing hormone-releasing hormone (LHRH)-astroglial interactive network. *Front Biosci.* 1997;2:1–32.

545. Marchetti B, Gallo F, Farinella Z, Romeo C, Morale MC. Luteinizing hormone-releasing hormone (LHRH) receptors in the neuroendocrine-immune network. Biochemical bases and implications for reproductive physiopathology. *Ann NY Acad Sci.* 1996;784:209–236.
546. Garcia-Segura LM, Chowen JA, Duenas M, Parducz A, Naftolin F. Gonadal steroids and astroglial plasticity. *Cell Mol Neurobiol.* 1996;16:225–237.
547. Martini L, Motta M, Piva F, Zanisi M. LHRF, LHRH, GnRH – What controls the secretion of this hormone? *Mol Psychiatry.* 1996;2:373–376.
548. Ojeda SR, Urbanski HF, Costa M, Hill D, Moholt-Siebert M. Involvement of transforming growth factor alpha in the release of LHRH from the developing female hypothalamus. *Proc Natl Acad USA.* 1990;87:9698–9702.
549. Junier M-P, Ma YJ, Costa ME, Hoffman G, Hill DF, Ojeda SR. Transforming growth factor contributes to the mechanism by which hypothalamic injury induces precocious puberty. *Proc Natl Acad USA.* 1991;88:9743–9747.
550. Ojeda SR, Ma YJ. Glial–neuronal interactions in the neuroendocrine control of mammalian puberty: facilitatory effects of gonadal steroids. *J Neurobiol.* 1999;40:528–540.
551. Rage F, Hill DF, Senaesteves M, Breakefield XO, Coffey RJ, Costa ME, McCann SM, Ojeda SR. Targeting transforming growth factor alpha expression to discrete loci of the neuroendocrine brain induces female sexual precocity. *Proc Natl Acad USA.* 1997;94:2735–2740.
552. Ma YJ, Berg-von der Emde K, Rage F, Wetsel WC, Ojeda SR. Hypothalamic astrocytes respond to transforming growth factor-alpha with the secretion of neuroactive substances that stimulate the release of luteinizing hormone-releasing hormone. *Endocrinology.* 1997;138:19–25.
553. Ma YJ, Hill DF, Junier M, Costa ME, Felder SE, Ojeda SR. Expression of epidermal growth factor receptor changes in the hypothalamus during the onset of female puberty. *Mol Cell Neurosci.* 1994;5:246–262.
554. Galbiati M, Zanisi E, Messi E, Cavarretta I, Martini L, Melcangi RC. Transforming growth factor-b and astrocytic conditioned medium influence luteinizing hormone-releasing hormone gene expression in the hypothalamic cell line GT1. *Endocrinology.* 1996;137:5605–5609.
555. Melcangi RC, Galbiati M, Messi E, Magnaghi V, Cavarretta I, Riva MA, Zansi M. Astrocyte-neuron interactions *in vitro*: role of growth factors and steroids on LHRH dynamics. *Brain Res Bull.* 1997;44:465–469.
556. Torres-Aleman I, Naftolin F, Robbins RJ. Trophic effects of basic fibroblast growth factor on fetal rat hypothalamic cells: interactions with insulin-like growth factors I. *Brain Res Dev Brain Res.* 1990;52:253–257.
557. Ishikawa K, Ohe Y, Okutomi Y, Takeuchi T, Suzuki M. Neurotrophic effects of fibroblast growth factors on peptide-containing neurons in culture from postnatal rat hypothalamus. *Neuroendocrinology.* 1992;55:193–198.
558. Tsai P, Werner S, Weiner R. bFGF is a neurotropic factor in GT1 GnRH neuronal cell lines. *Endocrinology.* 1995;136:3831–3838.
559. Wetsel WC, Hill DF, Ojeda SR. Basic fibroblast growth factor regulates the conversion of pro-luteinizing hormone (Pro-LHRH) to LHRH in immortalized hypothalamic neurons. *Endocrinology.* 1996;137:2606–2616.
560. Smyth C, Knee R, Wilkinson M, Murphy PR. Decline in basic fibroblast growth factor (FGF-2) mRNA expression in female rat hypothalamus at puberty. *J Neuroendocrinol.* 1997;9:151–159.
561. Ma YJ, Costa ME, Ojeda SR. Developmental expression of the genes encoding transforming growth factor alpha and its receptor in the hypothalamus of female rhesus macaques. *Neuroendocrinology.* 1994;60:346–359.
562. Schultz NJ, Terasawa E. Posterior hypothalamic lesions advance the time of the pubertal changes in luteinizing hormone release in ovariectomized female rhesus monkeys. *Endocrinology.* 1988;123:445–455.
563. Terasawa E, Claypool LE, Gore AC, Watanabe G. The timing of the onset of puberty in the female rhesus monkey. In: Delemarre-van de Waal HA, Plant TM, van Rees GP, Schoemaker J, eds. *Control of the Onset of Puberty III*. New York, NY: Elsevier Science Publishers B.V; 1989:123–136.
564. Zhen S, Zakaria M, Wolfe A, Radovick S. Regulation of gonadotropin releasing hormone (GnRH) gene expression by insulin-like growth factor I in a cultured GnRH-expressing neuronal cell line. *Mol Endocrinol.* 1997;11:1145–1155.
565. Balloti R, Nielsen FC, Pringle N, Kowalski A, Richardson WD, Van Obberghen E, Gammeltoft S. Insulin-like growth factor I in cultured rat astrocytes: expression of the gene, and receptor tyrosine kinase. *EMBO J.* 1987;6:3633–3639.
566. Mitamura R, Yano K, Suzuki N, Ito Y, Makita Y, Okuno A. Diurnal rhythms of luteinizing hormone, follicle-stimulating hormone, testosterone, and estradiol secretion before the onset of female puberty in short children. *J Clin Endocrinol Metab.* 2000;85:1074–1080.
567. Turek FW, Van Cauter E. Rhythms in reproduction. In: Knobil E, Neill JD, eds. *The Physiology of Reproduction*. 2nd ed. New York, NY: Raven Press; 1994:487–540.
568. Reppert SM. A clockwork explosion! *Neuron.* 1998;21:1–4.
569. Hall JC. Tripping along the trail to the molecular mechanisms of biological clocks. *Trends Neurosci.* 1995;18:230–240.
570. Watts AG. The Efferent projections of the suprachiasmatic nucleus, anatomical insights into the control of circadian rhythms. In: Klein DC, Moore RY, Reppert SM, eds. *Suprachiasmatic Nucleus, The Mind's Clock*. New York, NY: Oxford University Press; 1991:77–106.
571. Van der Beek EM. Circadian control of reproduction in the female rat. *Prog Brain Res.* 1996;111:295–320.
572. Funabashi T, Shinohara K, Mitsushima D, Kimura F. Gonadotropin-releasing hormone exhibits circadian rhythm in phase with arginine-vasopressin in co-cultures of the female rat preoptic area and suprachiasmatic nucleus. *J Neuroendocrinol.* 2000;12:521–528.
573. Maywood ES, O'Neill JS, Chesham JE, Hastings MH. Minireview: The circadian clockwork of the suprachiasmatic nuclei–analysis of a cellular oscillator that drives endocrine rhythms. *Endocrinology.* 2007;148:5624–5634.
574. van den Pol AN, Finkbeiner SM, Cornell-Bell AH. Calcium excitability and oscillations in suprachiasmatic nucleus neurons and glia *in vitro*. *J Neurosci.* 1992;12:2648–2664.
575. Cagampang FRA, Rattray M, Powell JF, Campbell IC, Coen CW. Circadian changes of glutamate decarboxylase 65 and 67 mRNA in the rat suprachiasmatic nuclei. *NeuroReport.* 1996;7:1925–1928.
576. Shibata S, Liou SY, Ueki S. Influence of excitatory amino acid receptor antagonists and of baclofen on synaptic transmission in the optic nerve to the suprachiasmatic nucleus in slices of rat hypothalamus. *Neuropharmacology.* 1986;25:403–409.
577. van der Horst GTJ, Muijtjens M, Kobayashi K, Takano R, Kanno S-I, Takao M, de Wit J, Verkerk A, Eker APM, van Leenen D, Buijs R, Bootsma D, Hoeijmakers JHJ, Yasui A. Mammalian Cry1 and Cry2 are essential for maintenance of circadian rhythms. *Nature.* 1999;398:627–630.
578. Chappell PE, White RS, Mellon PL. Circadian gene expression regulates pulsatile gonadotropin-releasing hormone (GnRH) secretory patterns in the hypothalamic GnRH-secreting GT1-7 cell line. *J Neurosci.* 2003;23:11202–11213.

579. Sisk CL, Foster DL. The neural basis of puberty and adolescence. *Nature Neurosci.* 2004;7:1040–1047.
580. Paus T, Zijdenbos A, Worsley K, Collins DL, Blumenthal J, Giedd JN, Rapoport JL, Evans AC. Structural maturation of neural pathways in children and adolescents: *in vivo* study. *Science.* 1999;283:1908–1911.
581. Paus T, Collins DL, Evans AC, Leonard G, Pike B, Zijdenbos A. Maturation of white matter in the human brain: a review of magnetic resonance studies. *Brain Res Bull.* 2001;54:255–266.
582. Giedd JN, Blumenthal J, Jeffries NO, Castellanos FX, Liu H, Zijdenbos A, Paus T, Evans AC, Rapoport JL. Brain development during childhood and adolescence: a longitudinal MRI study. *Nat Neurosci.* 1999;2:861–863.
583. Gogtay N, Giedd JN, Lusk L, Hayashi KM, Greenstein D, Vaituzis AC, Nugent III TF, Herman DH, Clasen LS, Toga AW, Rapoport JL, Thompson PM. Dynamic mapping of human cortical development during childhood through early adulthood. *Proc Natl Acad USA.* 2004;101:8174–8179.
584. Zehr JL, Todd BJ, Schulz KM, McCarthy MM, Sisk CL. Dendritic pruning of the medial amygdala during pubertal development of the male Syrian hamster. *J Neurobiol.* 2006;66:578–590.
585. Lidow MS, Goldman-Rakic PS, Rakic P. Synchronized overproduction of neurotransmitter receptors in diverse regions of the primate cerebral cortex. *Proc Natl Acad USA.* 1991;88: 10218–10221.
586. Woo TU, Pucak ML, Kye CH, Matus CV, Lewis DA. Peripubertal refinement of the intrinsic and associational circuitry in monkey prefrontal cortex. *Neuroscience.* 1997;80:1149–1158.
587. Huttenlocher PR, Dabholkar AS. Regional differences in synaptogenesis in human cerebral cortex. *J Comp Neurol.* 1997; 387:167–178.
588. Bourgeois JP, Goldman-Rakic PS, Rakic P. Synaptogenesis in the prefrontal cortex of rhesus monkeys. *Cerebral Cortex.* 1994; 4:78–96.
589. McEwen B. Estrogen actions throughout the brain. *Recent Prog Horm Res.* 2002;57:357–384.
590. Olmos G, Aguilera P, Tranque P, Naftolin F, Garcia-Segura LM. Estrogen-induced synaptic remodelling in adult rat brain is accompanied by the reorganization of neuronal membranes. *Brain Res.* 1987;425:57–64.
591. Naftolin F, Leranth C, Garcia-Segura LM. Ultrastructural changes in hypothalamic cells during estrogen-induced gonadotrophin feedback. Neuroprotocols. *Companion Methods Neurosci.* 1992;1:16–26.
592. Terasawa E, Timiras PS. Electrophysiological study of the limbic system in the rat at onset of puberty. *Am J Physiol.* 1968; 215:1462–1467.
593. Juraska JM, Markham JA. The cellular basis for volume changes in the rat cortex during puberty: white and gray matter. *Ann NY Acad Sci.* 2004;1021:431–435.
594. Schulz KM, Richardson HN, Zehr JL, Osetek AJ, Menard TA, Sisk CL. Gonadal hormones masculinize and defeminize reproductive behaviors during puberty in the male Syrian hamster. *Horm Behav.* 2004;45:242–249.
595. Schulz KM, Sisk CL. Pubertal hormones, the adolescent brain, and the maturation of social behaviors: Lessons from the Syrian hamster. *Mol Cell Endocrinol.* 2006;254-255:120–126.
596. Laroche J, Gasbarro L, Herman JP, Blaustein JD. Reduced behavioral response to gonadal hormones in mice shipped during the peripubertal/adolescent period. *Endocrinology.* 2009;150:2351–2358.
597. Laroche J, Gasbarro L, Herman JP, Blaustein JD. Enduring influences of peripubertal/adolescent stressors on behavioral response to estradiol and progesterone in adult female mice. *Endocrinology.* 2009;150:3717–3725.
598. Hier DB, Crowley Jr WF. Spatial ability in androgen-deficient men. *N Engl J Med.* 1982;306:1202–1205.
599. Sisk CL, Zehr JL. Pubertal hormones organize the adolescent brain and behavior. *Front Neuroendocrinol.* 2005;26:163–174.
600. Romeo RD, Lee SJ, McEwen BS. Differential stress reactivity in intact and ovariectomized prepubertal and adult female rats. *Neuroendocrinology.* 2004;80:3873–3893.
601. Romeo RD, Karatsoreos IN, McEwen BS. Pubertal maturation and time of day differentially affect behavioral and neuroendocrine responses following an acute stressor. *Horm Behav.* 2006;50:463–468.
602. Romeo RD, Ali FS, Karatsoreos IN, Bellani R, Chhua N, Vernov M, McEwen BS. Glucocorticoid receptor mRNA expression in the hippocampal formation of male rats before and after pubertal development in response to acute or repeated stress. *Neuroendocrinology.* 2007;87:160–167.
603. Romeo RD, Karatsoreos IN, Jasnow AM, McEwen BS. Age- and stress-induced changes in corticotropin-releasing hormone mRNA expression in the paraventricular nucleus of the hypothalamus. *Neuroendocrinology.* 2007;85:199–206.
604. Terasawa E. Role of GABA in the mechanism of the onset of puberty in non-human primates. *Intl Rev Neurobiol.* 2005;71: 113–129.
605. Teles MG, Bianco SD, Brito VN, Trarbach EB, Kuohung W, Xu S, Seminara SB, Mendonca BB, Kaiser UB, Latronico AC. A GPR54-activating mutation in a patient with central precocious puberty. *N Engl J Med.* 2008;358:709–715.
606. Ojeda SR, Dubay C, Lomniczi A, Kaidar G, Matagne V, Sandau US, Dissen GA. Gene networks and the neuroendocrine regulation of puberty. *Mol Cell Endocrinol.* 2010;324:3–11.
607. Parent AS, Teilmann G, Juul A, Skakkebaek NE, Toppari J, Bourguignon JP. The timing of normal puberty and the age limits of sexual precocity: variations around the world, secular trends, and changes after migration. *Endocrine Rev.* 2003;24: 668–693.
608. Demerath EW, Towne B, Chumlea WC, Sun SS, Czerwinski SA, Remsberg KE, Siervogel RM. Recent decline in age at menarche: the Fels Longitudinal Study. *Am J Human Biol.* 2004;16:453–457.
609. Rasier G, Toppari J, Parent AS, Bourguignon JP. Female sexual maturation and reproduction after prepubertal exposure to estrogens and endocrine disrupting chemicals: a review of rodent and human data. *Mol Cell Endocrinol.* 2006;254–255: 187–201.
610. Roy JR, Chakraborty S, Chakraborty TR. Estrogen-like endocrine disrupting chemicals affecting puberty in humans – a review. *Medical Sci Mon.* 2009;15:RA137–RA145.
611. Dickerson SM, Gore AC. Estrogenic environmental endocrine-disrupting chemical effects on reproductive neuroendocrine function and dysfunction across the life cycle. *Rev Endocrine Metab Disord.* 2007;8:143–159.
612. Thompson DR, Obarzanek E, Franko DL, Barton BA, Morrison J, Biro FM, Daniels SR, Striegel-Moore RH. Childhood overweight and cardiovascular disease risk factors: the National Heart, Lung, and Blood Institute Growth and Health Study. *J Pediatrics.* 2007;150:18–25.
613. Wardle J, Brodersen NH, Cole TJ, Jarvis MJ, Boniface DR. Development of adiposity in adolescence: five year longitudinal study of an ethnically and socio-economically diverse sample of young people in Britain. *Br Med J.* 2006;332:1130–1135.
614. Jasik CB, Lustig RH. Adolescent obesity and puberty: the "perfect storm". *Ann NY Acad Sci.* 2008;1135:265–279.
615. Kennedy D, Norman C. What don't we know? *Science.* 2005;309:90.
616. Terasawa E, Fernandez DL. Neurobiological mechanisms of the onset of puberty in primates. *Endocrine Rev.* 2001;22:111–151.

CHAPTER

20

Sexual Behaviors

Marilyn Y. McGinnis [1], *Donald W. Pfaff* [2]

[1] Department of Pharmacology and Center for Biomedical Neuroscience, University of Texas Health Sciences Center at San Antonio, San Antonio, TX, USA, [2] Neurobiology and Behavior, The Rockefeller University, New York, NY, USA

OUTLINE

Summary

This chapter provides a brief overview of several key issues related to the display of male and female sexual behaviors, with the focus on evidence primarily derived from studies in rodents and humans. The role of gonadal steroid hormones in mediating sexual behaviors is discussed within the context of both the motivation and performance aspects of sexual behavior. The neural sites of action for androgens in males, and estrogens in females, are described, with particular emphasis on the neural circuitry that has been established. Identifying mechanisms of gonadal steroid hormone action in brain is crucial to understanding how gonadal hormones affect sexual behavior. Both genomic and non-genomic mechanisms as well as genes that are activated by hormone action in brain are discussed. Finally, the important question of clinical relevance is addressed, not only with respect to to sexual behavior, but also in relation to diseases and syndromes that are affected by gonadal hormone action.

INTRODUCTION

For many humans, the goal of sexual behavior is to obtain pleasure. However, the real goal of sexual behavior is to pass on one's DNA. The pleasurable aspect exists in order to facilitate copulation. Species that use sex as a means of reproduction depend on the presence of steroid hormones secreted from the gonads. These hormones are of paramount importance for promoting sexual behavior in both males and females. Questions to be addressed in this chapter are:

1. Which hormones are involved in mediating male and female sexual behavior?

Handbook of Neuroendocrinology, DOI: 10.1016/B978-0-12-375097-6.10020-4

2. Where in the brain do hormones act to facilitate sexual behavior in males and females?
3. What is known about the neural circuitry underlying male and female sexual behavior?
4. How do hormones turn on genes involved in facilitating male and female sexual behavior?
5. What is the evidence for extranuclear (non-genomic) actions of gonadal steroid hormones in brain?
6. What is the clinical relevance of understanding how gonadal hormones act in brain to modulate sexual behavior?

The focus will be on mammals, relying on evidence derived primarily from rodent models.

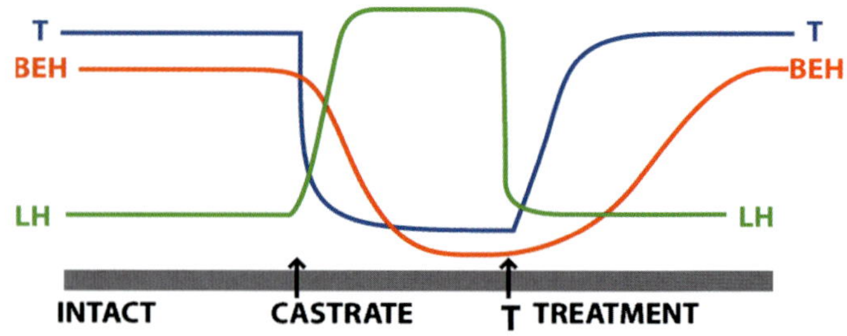

FIGURE 20.1 **Tonic pattern of hormone secretion and sexual behavior in males.** The schematic diagram shows the tonic pattern of serum testosterone (T) and luteinizing hormone (LH) release in males. Gonadally intact males will display sexual behavior whenever a female is available (BEH). Following castration, T levels fall and LH levels rise within hours. Sexual behavior drops off over a longer period (weeks to months). Testosterone replacement restores T levels and suppresses LH levels in serum. Sexual behavior is gradually restored.

WHICH HORMONES ARE INVOLVED IN MEDIATING MALE AND FEMALE SEXUAL BEHAVIOR?

Sexual behavior in males and females has three components: attraction, motivation and performance. In most species, either the male or the female has evolved elaborate displays to attract the opposite sex. Without motivation, the desire to seek out sex is low. The copulatory act itself is the process of DNA exchange. All of these aspects of sexual behavior are under hormonal control.

Males

A classic study by Berthold showed, in 1849, that a hormone secreted from the testes was responsible for comb growth in roosters.[1] It was not until 1927 that castration was shown to eliminate sexual behavior, suggesting that a testicular substance was necessary for male sexual behavior. The isolation of a testicular hormone, named testosterone, in 1935 opened the door for studies on the hormonal basis for reproductive behavior in males. Once testosterone was synthesized, it was used to study male sexual behavior. Studies in the late 1930s and early 1940s were instrumental in showing that testosterone restores male sexual behavior. Another study, by Beach and Holtz-Tucker in 1949,[2] demonstrated that castration followed by exogenous testosterone treatment maintained sexual behavior. These studies confirmed a critical role of testosterone in mediating male sexual behavior, and illustrated the point that the dose of testosterone needed to maintain existing sexual behavior was considerably less than the dose needed to restore it following castration.

The pattern of hormone secretion in males is tonic (Fig. 20.1), in the sense that serum testosterone, though pulsatile in nature, is maintained at relatively constant levels. Males are sexually receptive as long as testosterone levels are high. Castration results in a gradual decline in sexual behavior that, in rats, takes approximately 10–14 days. Restoration of male sexual behavior by testosterone is also gradual and takes about 10–14 days. The reason for this lengthy time for hormone action is not well understood, especially since testosterone levels fall to undetectable levels within 24 hours of castration.[3] However, one important component of the sexual response is experience. The more sexual experience the animal has, whether rodent or human, the longer sexual behavior is exhibited in the absence of testosterone. In animals that are seasonal breeders, testosterone levels, while basically tonic in nature, rise and fall with the seasons. Sexual interest rises and falls in concert with hormone levels.

The importance of testosterone as the male sex hormone was challenged in the 1960s and 1970s by studies showing that estradiol can also increase male sexual behavior in castrated males.[4] Testosterone is a steroid hormone that has two major metabolites: one is estradiol, which is formed within the brain via the P450 aromatase enzyme; the other is dihydrotestosterone (DHT), which is formed by the enzyme 5α reductase and is not aromatizable to an estrogen. These earlier studies showed that DHT was minimally effective in restoring male sexual behavior. However, estradiol restored some aspects of male sexual behavior. The potential importance of estrogen was bolstered by studies showing that administration of estrogen receptor blockers inhibited male sexual behaviors. Finally, the aromatase inhibitor ATD (1,4,6-androstatriene-3,17-dione) reportedly prevented restoration of sexual behavior in males.[4] ATD acts as potent inhibitor of estrogen synthesis by permanently binding to the enzyme aromatase, thereby blocking the metabolic conversion of testosterone to estradiol. The results of these experiments pointed to an important role for

BOX 20.1

HOW DO WE KNOW ABOUT MALE SEXUAL BEHAVIOR?

One other factor that is important for understanding male sexual behavior is the distinction between motivation and performance, both of which are under hormonal control. Motivation, often referred to as "libido," "desire" or "sex drive" in humans, energizes and directs behavior.[1] Performance, on the other hand, is the copulatory act itself. Performance can be measured simply by observing the animals' behavior. Motivation is not directly observable, but is inferred. However, several methods have been devised to measure sexual motivation. One example is the place preference test. This test employs a three-chambered apparatus with a sexually receptive female confined to one chamber and an ovariectomized (non-receptive) female confined to the opposite chamber. Gonadally intact and testosterone-treated males will spend more time with the receptive female, whereas castrated males show no preference.[2] Studies such as these are valuable for our understanding of human problems such as impotence, because in many instances desire is present but the ability to copulate is compromised. In some cases, both motivation and performance are absent.

References

1. Pfaff DW. *Drive. Neurobiological and Molecular Mechanisms of Sexual Motivation*. Cambridge, MA: The MIT Press; 1999.
2. Hughes AM, Everitt BJ, Herbert J. Comparative effects of preoptic area infusions of opioid peptides, lesions and castration on sexual behaviour in male rats: Studies of instrumental behaviour, conditioned place preference and partner preference. *Psychopharmacology*. 1990;102:243–256.

estrogen in the neuromodulation of sexual behavior in males. It was thought that testosterone was a prohormone and that estradiol was the active agent in stimulating male sexual behavior. This view came to be known as the "aromatization hypothesis."

Later studies have provided evidence supporting the view that testosterone is, in fact, necessary for the display of male sexual behavior; however, at least a small amount of estrogen must also be present. Most of the earlier studies employed high doses of estradiol. Later studies reported that silastic capsules filled with estradiol, which provided physiological levels of hormone, were ineffective in restoring male sexual behavior.[4] In contrast, capsules providing physiological levels of testosterone restored male sexual behavior. The early studies using ATD were brought into question by the finding that ATD is also an effective androgen receptor blocker.[5] Therefore, ATD's effect in blocking male sexual behavior can be accounted for without invoking a blockage of estrogenic effects. With the development of specific aromatase inhibitors that do not bind to androgen receptors, it was possible to re-examine this issue. Using the specific aromatase inhibitor fadrozole, it was found that blocking the formation of estrogen does indeed prevent the restoration of male sexual behavior by testosterone. However, giving a very small amount of estrogen in a silastic capsule (1% estradiol) was sufficient to restore male sexual behavior in males treated with testosterone plus fadrozole.[4] These data suggested that neither testosterone nor estrogen alone are sufficient to restore copulation, and that both androgens and estrogens are needed for the full restoration of male sexual behavior.

Females

There are four epochs of time in the lives of mammalian females during which fluctuations in ovarian hormones are behaviorally important: puberty, the estrous (or menstrual) cycle, pregnancy and menopause. For inducing courtship and mating behaviors in laboratory mice and rats, the optimal pattern and timing of ovarian hormone administration mimics the natural estrous cycle. It is the cycle itself that distinguishes female-type neuroendocrinology from male-type (Fig. 20.2). These cycling hormones are absolutely essential for female laboratory animals to perform female-typical sexual behaviors. A period of more-or-less constant exposure to an estrogen such as estradiol, preferably 48 hours or longer, is followed by an injection of progesterone, followed 2–5 hours later by behavioral tests. Progesterone used this way amplifies the positive effect of estrogens on female reproductive behavior.

We note that, for some social and sexual behaviors, metabolites of progesterone also play a role. In particular, metabolites of progesterone that reduce anxiety make it easier for female laboratory animals to initiate courtship responses.

Thus, the data for female sexual behaviors illustrate three principles of hormone action on the brain: (a) that combinations of hormones, rather than just an individual hormone, may be required; (b) that temporal

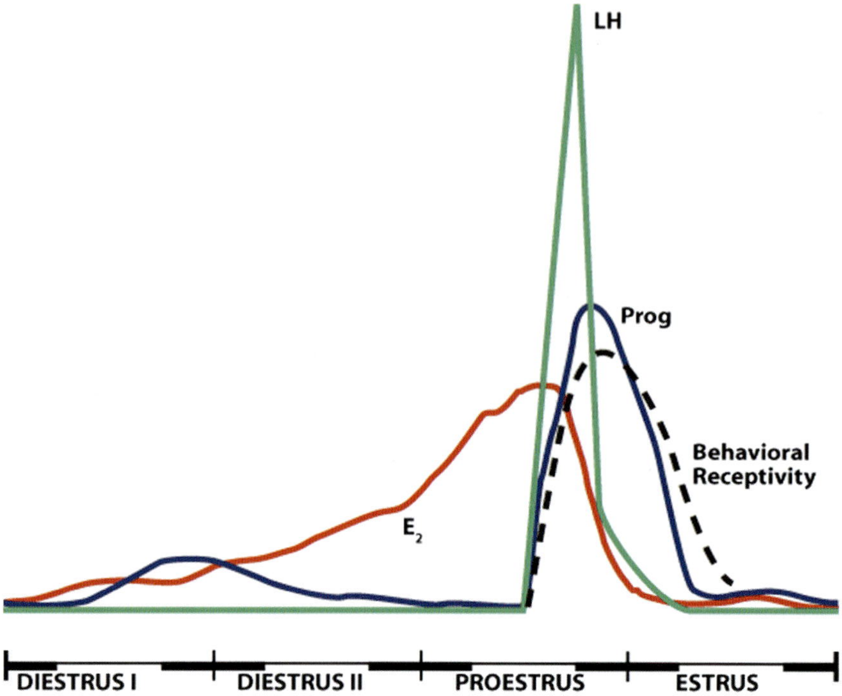

FIGURE 20.2 Cyclic pattern of hormone secretion and sexual behavior in females. The schematic diagram shows the changes in levels in the blood of ovarian hormones, estrogens and progestins, plotted across the 4-day estrus cycle of a laboratory animal such as a rat or a mouse. Of course, a human menstrual cycle would be a lot longer, at about 28 days. In many lower animals, mating behaviors and ovulation are linked tightly to levels of sex hormones in the blood.

BOX 20.2

HOW DO WE KNOW HOW HORMONES ACTIVATE FEMALE SEXUAL BEHAVIOR?

In the initial studies to determine which hormones were essential to reinstate sexual behavior in ovariectomized female rodents, it was found that both estrogen and progesterone were needed. It was also found that estrogen must be present first, given that progesterone followed by estrogen had no effect. We now know that one important reason for this sequence is that estrogen induces progestin receptors. Without their receptors, progestins have no means of activating cells in their target tissues, including brain. This relationship between estrogen and progesterone acting in sequence holds true for many species, including humans.[1]

References

1. Blaustein JD, Erskine MS. In: Pfaff DW, Arnold AP, Etgen AM, Fahrbach SE, Rubin RT, eds. *Feminine sexual behavior: Cellular integration of hormonal and afferent information in the rodent forebrain*, vol. 1. San Diego: Academic Press; 2002:139–214.

patterns of hormone administration may be important; and (3) that, sometimes, metabolites of the hormone administered may play a role.

WHERE IN THE BRAIN DO HORMONES ACT TO FACILITATE SEXUAL BEHAVIOR IN MALES AND FEMALES?

Many studies, employing a variety of techniques, have been aimed at determining the neural sites involved in mediating male and female sexual behavior. Two areas that both males and females have in common are the hypothalamus and the preoptic area. However, the extent of overlap is not striking. In addition, studies have revealed that several other brain areas are involved in mediating sexual behavior, and that differences between males and females are present with regard to the specific brain sites involved.

Males

In an attempt to demonstrate that the hormonal influence on male sexual behavior was mediated by neural mechanisms, early studies used electrolytic lesions[4] of large areas of the brain, which did, in fact, disrupt male sexual behavior. This was rather a sledge-hammer approach, but it served the purpose of stimulating research in this area. During the 1960s, it was shown that lesions of the "anterior hypothalamic–medial preoptic continuum" disrupted sexual performance. Later studies in numerous species identified the medial preoptic area (MPOA) as the primary site of action for testosterone in mediating male sexual behavior. The evidence is compelling, as lesions here eliminate sexual behavior and brain stimulation reinstates it in virtually all species studied. Other brain areas in which lesions have been effective in disrupting performance of male sexual behavior have been reported. Two areas where consistent deficits have been found are the olfactory bulbs, which have projections to the amygdala, and the corticomedial amygdala. Copulation can be stimulated by electric shock to olfactory-bulbectomized males. This led to the postulation that the copulatory deficits are due to loss of arousal. Since the amygdala is involved in arousal, this is a possibility. However, olfactory bulbectomy also decreases androgen receptors in the amygdala, and this could account for the loss of behavior as well. Lesions of the bed nucleus of the stria terminalis (BNST) also decrease copulation, but since the BNST is the main connection between the amygdala and the MPOA, it is thought that the effects on copulatory behavior are due to destruction of this pathway, rather than to specific neurons involved in mediating male sexual behavior. Lesions and stimulation of the mesocorticolimbic tract and the nigrostriatal tract, both of which are involved in dopaminergic mechanisms have also been found to affect male sexual behavior.[4]

Of course, copulatory performance is just one component of sexual behavior, yet far fewer studies have examined the effects of lesions on sexual motivation. Most studies have focused on the MPOA, and the results suggest a role for the MPOA in sexual motivation.[6] Though there is some dispute regarding the exact MPOA site for sexual motivation, in most cases the effective site for sexual motivation is different from that involved in sexual performance. The data suggest that these two components of male sexual behavior are initiated in separate, but related, brain areas. In agreement with this is the finding that removal of the basolateral nuclei of the amygdala affects sexual motivation, but not sexual performance.[7] Lesions of the olfactory bulbs, which decrease sexual performance, do not decrease sexual motivation.

With the advent of autoradiographic techniques in the early 1970s it became possible to identify brain regions where androgens are taken up, which signified putative androgen receptors (ARs). Later, biochemical techniques for measuring androgen receptors in brain were developed. These studies clearly showed that the areas identified using autoradiography contained high concentrations of ARs. Virtually all of the brain regions that have been implicated in the neural control of male sexual behavior contain androgen receptors. The highest AR concentrations are in the MPOA, the ventromedial hypothalamus (VMN) and the amygdala.[8] As a result, these have been the areas most commonly used in testosterone implant studies. Testosterone implants into the MPOA reinstate male sexual behavior in castrates of several rodent species.[4] Few studies have examined the role of the VMN in male sexual behavior, despite its high concentration of androgen receptors. In one study, testosterone implanted into the VMN did not restore copulation in castrated male rats.[9] Notably, however, these implants were effective in restoring sexual motivation.

Many early studies used anti-androgens such as cyproterone acetate (CA), and the results were equivocal. It turned out that CA also had androgenic properties. The newer anti-androgens, such as flutamide and hydroxyflutamide, are pure anti-androgens, and have been shown to be very effective in identifying AR effects in brain. The technique of implanting anti-androgens directly into the brain has the advantage that the entire brain is exposed to testosterone except for the region of the implant. Thus, any effect can be attributed to the blockade of AR action at that site. Congruent with lesion studies, blocking ARs in the MPOA has been shown to affect both copulation and motivation.[4] However, there does appear to be some site-specificity, in that copulation was decreased when implants were located in the anterior MPOA, whereas implants located in the posterior MPOA decreased sexual motivation. Implants of AR blocker into the VMN decreased sexual behavior, with anterior VMN implants being more effective than posterior VMN implants. Blocking ARs in the VMN also significantly decreased sexual motivation, suggesting a role for androgens acting in this region in mediating both sexual motivation and sexual performance. While lesions of the amygdala decrease male sexual behavior, AR blockade in the medial amygdala is only partially effective in suppressing copulation. One factor that is illustrated by implant studies is that location is of critical importance. Differences between results may signify subtle differences in the implant/lesion site.

Although estrogens appear to be important for the display of male sexual behavior, it is not clear that estrogen receptor (ER) activation is needed. One problem has been the lack of availability of a good ER antagonist. Many ER anatagonists have estrogenic effects and lack specificity. One study used a relatively specific ER antagonist, RU58668, in conjunction with hydroxyflutamide to block male sexual behavior.[4] It was found that although blocking ARs decreased male sexual behavior and sexual motivation, blocking ERs did not. The assumption has been that estrogen's action in modulating male sexual behavior is via ERs, but this may not be the case. It is possible that estrogen acts via non-genomic mechanisms to stimulate male sexual behavior.

Females

Four lines of evidence prove that estrogens act in the brain to promote female sexual behaviors. First, estrogens circulating in the blood are bound and retained in the cell nuclei of the same hypothalamic nerve cells whose destruction abolishes lordosis behavior, the primary female rodent sex behavior. Secondly, in surgically ovariectomized female rats that do not have circulating estrogens, implanting tiny amounts of estrogen amongst their hypothalamic neurons permits lordosis behavior to occur. Thirdly, the estrogenic actions on the brain facilitate sex behavior directly, and not through the pituitary gland, because estrogens are behaviorally effective in females in which the pituitary gland had been surgically removed. Finally, in females with estrogens circulating in the blood, implanting an anti-estrogen in the hypothalamus significantly reduced female reproductive behavior.

The most straightforward differential comparison of cells involved in regulating female and male sex behaviors comes from the study of the diencephalon. The neurons there required for activating lordosis behavior are in the ventromedial nucleus of the hypothalamus, while those most obviously essential for permitting male sex behavior are in the preoptic area. Conversely, destroying preoptic neurons in the way that would abolish male sex behavior actually increases lordosis behavior.

WHAT IS KNOWN ABOUT THE NEURAL CIRCUITRY UNDERLYING MALE AND FEMALE SEXUAL BEHAVIOR?

The main idea emerging from these earlier studies was that sexual behavior does not involve a single brain area. Indeed, the complexity of the sexual response of both males and females requires integration of several brain regions as well as spinal mechanisms. This led to the concept of a neural circuit consisting of several brain regions that act in concert to modulate sexual behavior.

Males

A substantial body of research, employing a variety of techniques, has been aimed at identifying the androgen-concentrating brain regions and their neural connections that are important in the regulation of male sexual behavior.[4] In forebrain, the olfactory bulbs that are a major sensory modality, the MPOA, the medial amygdala and the VMN are known to be critical sites. It is significant that these areas are all interconnected. The olfactory bulbs have connections with the amygdala via both the main olfactory bulbs and the accessory olfactory bulbs. In turn, the amygdala has a large projection to the MPOA via the BNST. The amygdala also has reciprocal connections with the VMN, and the VMN has projections to the MPOA.

Efforts to link the forebrain areas to the brainstem and spinal cord are necessary to establish a neural circuit that includes the motor components of sexual performance, and to potentially couple this with sexual motivation. While a number of brain and spinal cord areas have been investigated, a circuit for male sexual behavior has not been conclusively established. Brain areas that have been implicated thus far include the midbrain tegmentum, nucleus paragigantocellularis and periaqueductal gray.[4] Despite the large body of data available, confirmation and details of a neural circuit for male sexual behavior are not well integrated. One thing that is clear is that virtually all the areas examined are connected to the MPOA either directly or indirectly, and most contain androgen receptors. This suggests that the MPOA plays an influential role in integrating neural signaling to and from the genitalia.

Females

The basic neural circuitry for producing lordosis behavior in a laboratory animal is illustrated in Fig. 20.3. Two features of the lordosis circuit, the first reported for any vertebrate animal, are as follows: (a) The circuit is *modular.* Neuroanatomical and electrophysiological evidence tells us that the spinal module handles local reflexes, the lower brainstem module coordinates activity across spinal segments, the hypothalamic module adds hormonal dependency to the circuit, and the midbrain module translates hormonal signaling into sensory/motor signaling. (b) The circuit is *hierarchical.* Without estrogenic stimulation of the appropriate hypothalamic cells and subsequent signaling to the midbrain, the midbrain cannot activate reticulospinal neurons that, in turn, will control spinal mechanisms for lordosis behavior.

On the other hand, some neural circuits that are likely to be important for both female and male sex behavior are those having to do with generalized arousal of the entire CNS.[10] During the initiation of courtship behaviors, both male and female laboratory animals show evidence of behavioral arousal. Males stir themselves from sedentary positions to chase females and investigate their sensory properties in a manner specific to each species. Females leave their homes, defined in a species-specific manner, locomote and otherwise communicate in a manner that signals their sexual readiness. Thus, both males and females are aroused from conditions of relative inactivity to initiate a series of active behavioral responses. Pharmaceutical treatments

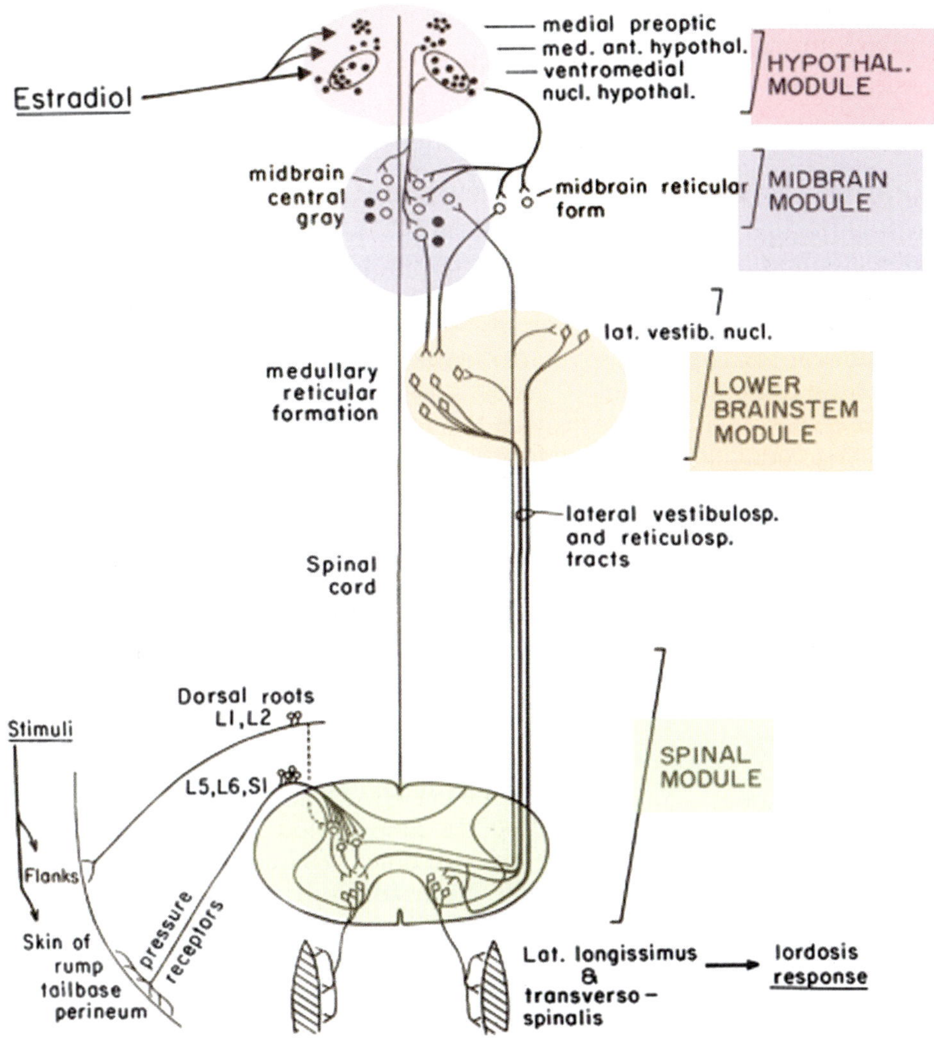

FIGURE 20.3 **Neural circuit for female sexual behavior.** This is the basic circuit of nerve cells responsible for producing the most elementary sex behavior in female laboratory animals. The behavior, lordosis, comprises an arching of the back that is necessary to permit fertilization by the male. The circuit is governed by estradiol-binding nerve cells in the hypothalamus, which then regulates the rest of the circuit, which is modular. Note that the behavior is triggered by somatosensory input due to the male's mounting, and these inputs impact the spinal cord (lower left). Despite the fact that the spinal circuitry involved is robust and complex, spinally-transected females can not do lordosis behaviour. A spinal-brainstem-spinal loop is required. Activity in this loop is regulated by the hypothalamic output (top of figure). Then, the final result of this loop's activity is to heighten excitation in the motoneurons whose axons projecting to deep back muscles cause lordosis behavior to occur (lower right). (Adapted, with permission, from Pfaff et al, 2002.)[19]

that increase arousal enhance females' sex behavior.[11] Further, breeding male mice for high generalized arousal potentiates their male sex behavior. The chemistry and pharmacology of such arousal regulation depends on the sex, and on the chemistry of the neurons involved in each relevant brain region.

HOW DO HORMONES TURN ON GENES INVOLVED IN FACILITATING MALE AND FEMALE SEXUAL BEHAVIOR?

The classical nuclear action of steroid hormones in brain is via genomic mechanisms.[12] Since androgens and estrogens are lipid-based molecules, they can enter cells by passive diffusion to bind to intracellular receptors. The binding of gonadal steroids to their respective receptors induces a conformational change in the receptor that involves dissociation of chaperones such as heat shock proteins. This process results in activation of the receptor to a form that can bind to specific DNA sites, referred to as hormone response elements (HRE). The interaction of the receptor with the HRE, in conjunction with nuclear proteins such as co-factors, directs the production of proteins that alter cellular function. This ultimately effectuates changes in behavior. This is commonly referred to as a genomic effect of hormones because of the direct action at the genome. Receptors for gonadal steroid hormones are evolutionarily conserved throughout the brains of vertebrates.

Males

Genomic effects of steroid hormones usually require minutes to hours, as it is thought that time is needed for transcription and translation of new proteins. The lengthy temporal requirements for restoration of male sexual behavior strongly suggest a genomic influence in the neuromodulation of male sexual behavior. Protein synthesis inhibitors such as anisomycin inhibit male sexual behavior, confirming that a genomic mechanism is responsible for facilitating copulation.[4] The need for prolonged exposure to testosterone (at least 21 hours per day) to restore male sexual behavior also suggests a genomic effect.[13] It is remarkable, however, that while gene products of androgen action have been identified in accessory tissue such as prostate, thus far virtually no androgen-regulated gene products have been confirmed in brain.

Females

In the brains of female laboratory mammals, estrogens facilitate behaviors by processes that require new messenger RNA and protein synthesis. Figure 20.4 shows a list of genes that have two properties: (a) estrogens elevate their messenger RNA levels, and (b) their gene products foster female reproductive behaviors. The exception is the "disinhibition" routes through prostaglandin D synthase. Mong and Pfaff[14] have theorized that the individual gene products activate functional modules that contribute in distinguishably different ways to the performance of female sex behaviors.

The transcriptional activation initiated by estrogens required for normal lordosis behavior is highly likely to involve modification of the chromatin covering the promoters of sex behavior-relevant genes. These chemical modifications are on the tails of specific histones, relatively small proteins that control access to DNA. Estrogenic treatment has recently been shown to cause chemical modifications of histones in the ventromedial hypothalamic neurons that regulate lordosis behavior.

WHAT IS THE EVIDENCE FOR EXTRANUCLEAR (NON-GENOMIC) ACTIONS OF GONADAL STEROID HORMONES IN BRAIN?

Several lines of evidence suggest that some actions of steroid hormone are non-genomic. One criterion is that the effect should occur within seconds to minutes – a time period that is too short for genomic effects. Another characteristic of non-genomic actions of steroids is that they occur in the presence of inhibitors of transcription and translation. Finally, the steroid effects may occur in brain areas lacking in classical (genomic) steroid receptors.

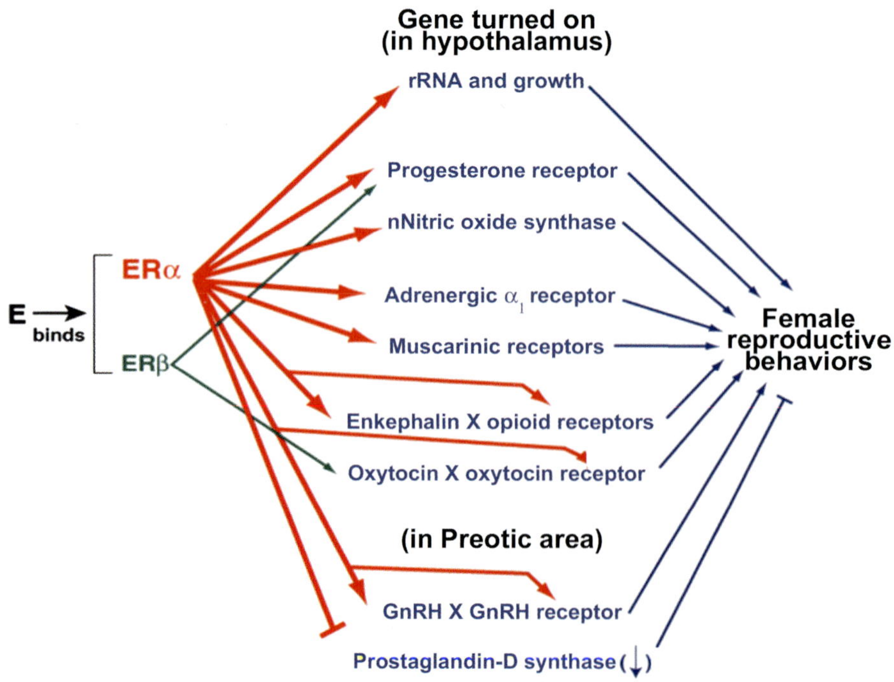

FIGURE 20.4 List of genes activated by estrogens. The genes have two properties. First, estrogens (E), having bound to estrogen receptors (ER), elevate their gene expression level (except for the bottom example). Secondly, those gene products foster some aspects of female reproductive behaviors.

Males

While membrane actions of female sex hormones have been known for years, membrane-initiated actions of testosterone have received far less attention – particularly with regard to the neuroregulation of sexual behavior.[15] In one recent study,[16] male rats displayed a rapid (30- to 60-minute) reduction in mount latency in response to a pulse of testosterone, suggesting a possible membrane effect. In contrast to other systems, putative membrane receptors for androgens have not yet been identified. Neuroanatomical studies have provided intriguing evidence for membrane ARs. For example, the presence of AR immunoreactivity has been reported in extranuclear sites such as axons and dendrites of neurons of the hippocampus, as well as areas lacking in classical ARs.[17] Thus, while the evidence for a membrane effect on sexual behavior may not be particularly persuasive, the available data suggest that membrane ARs probably do exist.

Females

In female laboratory animals, too, rapid actions of sex steroids are known to occur and to be important for behavioral changes. The crucial behavioral evidence came from a "two-pulse" design of estrogenic administration, using small doses applied directly to the hypothalamus. The first pulse, composed of an estrogen that was chemically limited to rapid actions that are initiated at the nerve cell membrane, potentiated the ability of a second pulse of estradiol to cause lordosis behavior. In parallel studies using the techniques of molecular biology, estrogens chemically modified to be limited to the cell membrane potentiated the ability of estradiol to cause hormone-regulated gene transcription. This transcriptional potentiation may be at the basis of the behavioral effect or rapid estrogenic actions.

WHAT IS THE CLINICAL RELEVANCE OF UNDERSTANDING HOW GONADAL HORMONES ACT IN BRAIN TO MODULATE SEXUAL BEHAVIOR?

A great deal of research has been conducted using laboratory rats and mice – but what clinical relevance do these studies have? In some cases scientists are simply doing basic research, attempting to understanding how gonadal hormones act, and the results become clinically relevant. For example, flutamide and tamoxifen were initially employed as hormone receptor blockers to understand fundamental concepts of hormone action, and these drugs are now widely used to treat breast and prostate cancer.

Further, since we know that GnRH released from the brain in a pulsatile fashion is necessary for the pituitary to send gonadotropic hormones to the testes and thus cause the secretion of testosterone, putting in a long-duration-acting GnRH receptor agonist such as Lupron effectively shuts down the testes and reduces testosterone levels. This treatment has been useful for suppressing criminal sexual behaviors by pedophiles. Such studies illustrate the many benefits of animal research on hormone action in brain.

Males

Experiments using mice and rats with androgen receptor mutations have proven to be valuable in understanding sexual orientation and behavioral patterns of humans with androgen receptor mutations. Point mutations in the androgen receptor have been reported in several species, including humans.[18] In humans, the androgen insensitivity syndrome (AIS) can range from complete insensitivity with a feminine phenotype, to a relatively mild insensitivity with somewhat ambiguous genitalia. Depending on the severity of the AIS, XY male infants with ambiguous genitalia may be surgically altered and raised as females. Since it is the presence of androgens during neonatal development that determines the sex of the brain as well as the genitalia, it is critical that the sex of the gonads matches the sex of the brain. Studies using animal models have contributed to our knowledge and understanding of this complicated issue. The two most commonly studied animal models are the testicular feminized male (Tfm) rat and mouse. The Tfm mutation is on the X chromosome, and was first described in 1970.[18] Tfm mice have greatly decreased levels of androgen receptors in all androgen-sensitive tissues, including brain. These animals fail to copulate, displaying neither mounts, intromissions nor ejaculations, and serve as a useful model for studying complete AIS.

Females

One of the most prominent issues relating neuroendocrinology in females to current human affairs has to do with hormone replacement therapy (HRT) for women during and after the years of menopause. Estrogens are widely recognized to raise arousal and mood in such women, and thus to have true psychological value for them. On the other hand, these hormones must be used cautiously because of the possibility they might hasten the growth of estrogen-dependent cancers of the breast or the endometrium. The only major clinical trial that tried to estimate the balance between the positive and negative effects of HRT was criticized because it started with women who were too old and had been without

BOX 20.3

HOW DO WE KNOW ABOUT THE ROLE OF ANDROGEN RECEPTORS IN MODULATING SEXUAL BEHAVIOR?

The experiments of Olsen[1] used the Tfm rat to investigate the role of androgen receptors in modulating sexual behavior. Since the defect is present before birth, these animals are feminized. However, in contrast to the Tfm mouse, the mutation in rats results in incomplete (partial) androgen receptor insensitivity. Thus, androgen receptors in the brains of Tfm rats are not completely absent, but are decreased to 10–15% of normal. Adult Tfm male rats also show some behavioral responses to androgens. For example, they display mounting, though at lower than normal levels, and a few intromissions and ejaculations. These factors make the Tfm rat a valuable model for the study of partial androgen insensitivity, and establish the importance of functional androgen receptor in the neuromodulation of male sexual behavior.

References

1. Olsen KL. Androgen-insensitive rats are defeminized by their testes. *Nature.* 1979;279:238–239.

estrogens of any sort for too long, and the trial was terminated prematurely. Now, doctors tend to say that HRT is particularly useful for women suffering psychological and physical discomfort during the years of menopause and just after – especially if they do not have a family history of estrogen-dependent cancers. A powerful field of current pharmacologic research explores the possibility that artificial estrogens could be designed that benefit hormone-dependent activities in the brain without stimulating ERα molecules in the breast or the uterus.

SUMMARY

The evidence shows that gonadal steroid hormones play a powerful role in facilitating sexual behaviors in both males and females. Males require the presence of testosterone and estrogen, whereas females require estrogen plus progesterone. The medial preoptic area and the ventral hypothalamus are the primary brain regions involved in modulating sexual behavior. Androgens, estrogens and progestins act at the genome to alter the synthesis of new proteins in brain which are thought to be responsible for changes in sexual behaviors. However, non-genomic actions of gonadal hormones are becoming more apparent. Studies using animal models have been instrumental in advancing our understanding of human issues such as androgen insensitivity, hormone replacement therapy and treatment for hormone responsive tumors in cancer patients.

OUTSTANDING NEW QUESTIONS FOR THE FUTURE

As we move forward, we face challenges which reflect questions that have remained unanswered for decades, as well as questions that were not even thought of until recently. For example, we have known for a long time that estrogens can activate female sexual behavior within hours, but the activation of male sexual behavior by androgens requires 1–2 weeks of hormone exposure. Why males require prolonged androgen exposure, and its corollary, why it takes 1–2 weeks for behavior to fall off after castration, have been longstanding questions in behavioral neuroendocrinology. The possibility that the identification of androgen-regulated gene products in brain will solve this puzzle is appealing. Alternatively, a concept yet to be conceived may prove to be the underlying mechanism. For many years, androgens and estrogens were believed to act via genomic mechanisms. This model was challenged with mounting evidence that estrogens may act via non-genomic pathways. We are just beginning to realize that the actions of gonadal hormones in classical hormone receptor-containing neurons may be only part of the story. Future studies on the role of non-genomic hormone receptor action could dramatically transform the way we look at hormone effects on brain function in the facilitation of sexual behaviors.

Other questions have to do with the implications of basic neuroendocrine work on animal brain and behavior for human feelings, motivations and behaviors. Even though human mating behavior has, to some degree, been emancipated from a slavish dependence on sex hormone levels, it is nevertheless true that, for example, a certain amount of testosterone is necessary for male sexual motivation. That is why eunuchs were allowed to guard the harem. The more complex questions will address how powerfully hormones can influence emotions in men and women that are related to sociosexual behaviors but that have little to do with reproduction itself.

References

1. Beach FA. Historical origins of modern research on hormones and behavior. *Horm Behav.* 1981;15:325–376.
2. Beach FA, Holtz-Tucker AM. Effects of different concentrations of androgen upon sexual behavior in castrated male rats. *J Comp Physiol Psychol.* 1949;42:433–453.
3. Krey LC, McGinnis MY. Time-courses of the appearance/disappearance of nuclear androgen-receptor complexes in the brain and adenohypophysis following testosterone administration/withdrawal to castrated male rats: Relationships with gonadotropin secretion. *Jf Steroid Biochem.* 1990;35:403–408.
4. Hull EM, Meisel RL, Sachs BD. Male sexual behavior. In: Pfaff DW, Arnold AP, Etgen AM, Fahrbach SE, Rubin RT, eds. *Hormones, Brain, and Behavior*, vol. 1. San Diego: Academic Press; 2002:3–137.
5. Kaplan ME, McGinnis MY. Effects of ATD (1,4,6-androstatriene-3,17-dione) on male sexual behavior and androgen receptor binding: A re-examination of the aromatization hypothesis. *Horm Behav.* 1989;23:10–26.
6. Paredes RG, Highland L, Karam P. Socio-sexual behavior in male rats after lesions of the medial preoptic area: evidence for reduced sexual motivation. *Brain Res.* 1993;618:271–276.
7. Everitt BJ. Sexual motivation: a neural and behavioural analysis of the mechanisms underlying appetitive and copulatory responses of male rats. *Neurosci Biobehav Rev.* 1990;14:217–232.
8. McGinnis MY, Davis PG, Meaney MJ, Singer M, McEwen BS. *In vitro* measurement of cytosol and cell nuclear androgen receptors in male rat brain and pituitary. *Brain Res.* 1983;275:75–82.
9. Harding SM, McGinnis MY. Androgen receptor blockade in the MPOA or VMN: Effects on sociosexual behaviors. *Physiol Behav.* 2004;81:671–680.
10. Pfaff DW. *Brain Arousal and Information Theory: Neural and Genetic Mechanisms*. Cambridge: Harvard University Press; 2006.
11. Holder MK, Hadjimarkou MM, Zup SL, Blutstein T, Benham RS, McCarthy MM, Mong JA. Methamphetamine facilitates female sexual behavior and enhances neuronal activation in the medial amygdala and ventromedial nucleus of the hypothalamus. *Psychoneuroendocrinology.* 2010;35:197–208.
12. McEwen BS. Steroid hormone actions on the brain: when is the genome involved? *Horm Behav.* 1994;28:396–405.
13 McGinnis MY, Mirth MC, Zebrowski AF, Dreifuss RM. Critical exposure time for androgen activation of male sexual behavior in rats. *Physiol Behav.* 1989;46:159–165.
14. Mong JA, Pfaff DW. Hormonal symphony: steroid orchestration of gene modules for sociosexual behaviours. *Mol Psych.* 2004;9:550–556.
15. Foradori CD, Weiser MJ, Handa RJ. Non-genomic actions of androgens. *Front Neuroendocrinol.* 2008;29:169–181.
16. Nyby JG. Reflexive testosterone release: a model system for stufying the nongenomic effects of testosterone upon male behavior. *Front Neuroendocrinol.* 2008;29:199–210.
17. Tabori NE, Stewart LS, Znamensky V, Romeo RD, Alves SE, McEwen BS, Milner TA. Ultrastructural evidence that androgen receptors are located at extranuclear sites in the rat hippocampal formation. *Neuroscience.* 2005;130:151–163.
18. McGinnis MY, Marcelli M, Lamb DJ. Consequences of mutations in androgen receptor genes: molecular biology and behavior. In: Pfaff DW, Arnold AP, Etgen AM, Fahrbach SE, Rubin RT, eds. *Hormones, Brain, and Behavior*, vol. 5. San Diego: Academic Press; 2002:347–80.
19. Pfaff DW, et al. Female sexual behavior. In: Pfaff D, et al., eds. *Hormones, Brain, and Behavior.* San Diego: Academic Press (Elsevier); 2002.

CHAPTER

21

Stress and Brain Function

Ilia N. Karatsoreos, Bruce S. McEwen

Harold and Margaret Milliken Hatch, Laboratory of Neuroendocrinology, The Rockefeller University, New York, NY, USA

OUTLINE

Summary

Stress is a necessary part of life. It helps to ensure that an organism can deal with changes in the environment that may pose a threat to homeostasis. The brain regulates this stress response by evaluating threats in the environment, engaging neural and hormonal countermeasures to deal with the threats, and terminating the stress response when the threat has passed. Thus, the brain is both the generator of the stress response, and also a target. Numerous brain regions are impacted by stress and stress mediators, and can impact behavior. This chapter will highlight how the hippocampus, prefrontal cortex and amygdala are affected by stress, and how these stress-induced changes relate to changes in behavior. The chapter concludes with a discussion about stressors in the modern world, how they can impact brain structure and function, and what this means for mental and physical health.

INTRODUCTION

The brain is the most plastic and adaptable organ in the body. This capacity for plasticity is necessary for the brain to do its job – to learn about the external environment, adapt behavior and physiology to environmental "stressors," and ensure the survival of the organism. It is in this context that an important fact about brain function is revealed: the brain is not a passive observer of changes in the environment; it is active in adapting to a changing environment, and guiding an organism safely through life. But how does the brain detect, classify and respond to stressors in the environment? How do the physiological and behavioral responses the brain makes in response to stress feed back to alter its structure and function? What are the ramifications of these changes for behavior? These are questions that will be addressed in this chapter. However, before we delve further, we must explore and define what we mean by "stress," how the study of stress and the brain has developed, and the place that stress holds in nature.

DEFINING STRESS, AND THE HISTORY OF STRESS RESEARCH

The term "stress" carries with it many negative connotations – but what does this term really mean? How we define stress and stressors can influence how we study

Handbook of Neuroendocrinology, DOI: 10.1016/B978-0-12-375097-6.10021-6

the way in which the brain and body respond to stressors in the environment. In biology, the term "stress" was adapted from the engineering parlance by Hans Selye in the 1930s. In engineering, stress is defined as "a measure of the internal forces acting within a deformable body." In his translation to biology, Selye defined stress as the result of an organism's failed attempt to respond appropriately to a physical challenge.[1] In the intervening years this definition has been further elaborated to incorporate psychological threats, including anticipation of both physical and psychological stressors.[2] In this regard, the pioneering work of John Mason on the power of psychological stress was particularly seminal.[3]

As such, a useful definition of the stress response is that set of physiological and behavioral countermeasures deployed by the brain and body to protect organisms against actual or perceived threats to homeostasis. This set of responses can also be described as the "fight-or-flight" response – a set of changes in the body that prepare an animal to confront the stressor head-on, or to flee and thereby remove itself from the stressful situation. When these responses are appropriately engaged, regulated, and finally terminated once the stressor has been removed, they serve an important adaptive role, helping an animal to cope with an unexpected change in the environment.

These adaptive changes include the redirection of blood flow from the extremities to the core, increased respiratory rate, increased energy availability and a stimulated immune system. Thus, the organism would be in the optimal physiological state to expend large amounts of energy to escape a stressful encounter with a predator or a conspecific. In this case, the stress response is helpful and necessary for survival. However, if the stress response is inappropriately engaged, poorly regulated, or not properly terminated after the stressor has been removed, the effects of stress hormones and factors such as cortisol (corticosterone in rodents; CORT) begin to have negative effects on the brain and body. This produces a situation in which those same factors that were released in order to help survival end up having the opposite effect, and eventually increase wear and tear on the brain and body. The remainder of the chapter will be focused on understanding how the brain and body regulate the stress axis, how the mediators of the stress response (particularly CORT) can modify the structure and function of the brain, and what these effects on neural function mean for behavior.

THE STRESS AXIS – FROM BRAIN TO ADRENAL AND BACK

As discussed in earlier chapters, the primary neuroendocrine axis that regulates the physiological stress response is the hypothalamic–pituitary–adrenal (HPA) axis. This axis responds to stressful events, as well as to the daily cycle of rest–activity, by regulating the secretion of glucocorticoids from the adrenal cortex. Thus, the HPA does more than just respond to "stressors," even though it is frequently referred to as the "stress response system." The HPA axis is a classic example of a neuroendocrine negative feedback loop. Upon activation, corticotropin-releasing hormone (CRH) is released from the paraventricular nucleus (PVN) of the hypothalamus into the hypophysial portal system. This then causes adrenocorticotropic hormone (ACTH) to be released from the anterior lobe of the pituitary gland into general circulation. This signal travels through the bloodstream and then increases the synthesis and release of glucocorticoid hormones, such as CORT from the adrenal cortex, which enter the circulation.

While the specifics of the neuroendocrine control of the stress response have been discussed previously in this text, we will focus most of our discussion on how the brain responds, structurally and functionally, to stress and stress mediators.

EFFECTS OF STRESS HORMONES ON BRAIN FUNCTION

The primary purpose of the stress response is to protect the organism from a threat to homeostasis. The brain is a key player in this response, in that it not only regulates the neuroendocrine stress response but also, as a consequence of these stress mediators, can be altered both structurally and functionally. In the following section, we will explore the effects of stress on the structure and function of brain by highlighting the changes that occur in the hippocampus, amygdala and prefrontal cortex – brain regions critical for cognition and emotionality.

Hippocampus

Function, Structure and Characteristics of the Hippocampus

Located in the medial temporal lobe, the hippocampus is one of the most plastic and sensitive regions of the brain. It is crucially important in both cognitive function and emotionality. The hippocampus integrates patterns in the environment into a cohesive unit, which forms the initial basis for long-term memory. Thus, pattern combination and pattern separation are key functions of the hippocampus. The hippocampus is able to perform this tremendous feat of computation largely due to its anatomical "wiring." A primary input to the hippocampus is from the entorhinal cortex, which projects to the dentate gyrus. The dentate then makes connections with CA3 pyramidal neurons. This connection is remarkably amplified, with one dentate granule neuron innervating 12 CA3 neurons (on average), with excitatory axon collaterals of these CA3 neurons further innervating up to 50 other CA3 neurons. To further complicate this circuit, each CA3 cell can also innervate up to 25 inhibitory cells. This complex wiring results in a net 600-fold amplification of excitation while also providing a 300-fold amplification of inhibition, forming the basis of the accelerator and brake pedals in the hippocampus and thereby controlling the overall activity of this brain region.[4] While such wiring provides the hippocampus with an exquisite circuit to process and integrate stimuli from the environment, it is also its Achille's heel in that the exceptional degree of amplification and high levels of activity in this circuit make it particularly vulnerable to damage.

Stress Effects on the Hippocampus

In addition to affecting neurogenesis, chronic stressors can lead to changes in the morphology of existing hippocampal cells. Specifically, CA3 pyramidal cells undergo remodeling of their dendrites in conditions such as hibernation and chronic stress.[5–7] What is important in all of these cases is that this remodeling seems to be reversible – that is, given a period of recovery, these cells return to what appears to be their pre-stress state. The role of this plasticity may be to protect against permanent damage, and is not merely a result of chronic physiological stress. For instance, CA3 dendritic reorganization is found in both dominant and subordinate rats in the visible burrow system (which is a model of chronic social stress), but these changes are independent of adrenal size.[8] Moreover, these changes are in some cases quickly reversible. In species of mammals that hibernate, dendritic remodeling is a reversible process and occurs within hours of the onset of hibernation. It is also reversible within hours of the animals waking,[5–7,9] suggesting not only that cytoskeletal reorganization is taking place rapidly, but also that it can be reversed. This also indicates that changes in dendrite length and branching are not "damage," but a form of adaptive structural plasticity. Together, these data imply that it is not the presumed amount of physiological stress that determines dendritic remodeling, but instead it is a complex set of other factors that modulate neuronal structure, likely in an attempt to protect the hippocampus from permanent damage caused by excitotoxicity or increased oxidative stress. But how could chronic stress actually lead to cell damage?

The hippocampus was one of the first "higher" brain regions that was identified as a target of stress hormones. Because of this early start, we know much about how these mediators of the stress response can affect hippocampal function, and this understanding has helped scientists to begin to understand how stress affects the brain more generally. One of the findings that seems to apply to other brain areas, including the prefrontal cortex and amygdala (discussed later), is that stress seems to have a biphasic effect on the function and structure of the hippocampus, particularly in the CA1 region. That is, the effects of stress hormones can be visualized as an inverted U-shaped curve, where too small an amount of stress hormones can have deleterious effects on the function of the circuit, while too large an amount can also lead to a dysfunctional hippocampus.

One of the hypotheses that has been proposed to explain this biphasic effect of stress hormones is the pattern of expression of the cellular receptors of these hormones, the type I (or mineralocorticoid receptor, MR) and type II (or glucocorticoid receptor, GR). There are numerous mechanisms by which these receptors could play a role in the biphasic response of the CA1 region. One such mechanism could be that the co-expression of MR and GR in the same cells could lead to heterodimerization of the receptors following activation, which could then lead to a different pattern of genomic activation when compared to MR or GR homodimers. However, it is also plausible that other non-genomic mechanisms are at work here, since the MR is required for the glucocorticoid regulation of excitatory glutamatergic transmission.

A non-genomic role for glucocorticoids in regulating the function of the hippocampal circuit is further substantiated by findings that glucocorticoid receptors have been implicated in rapid actions at the synapse, since there are reports of GR localization in non-perinuclear regions of the cell, such as the postsynaptic density[10] and findings that have demonstrated a subset of adrenal steroid receptors being G-protein coupled.[11] Glucocorticoids also have effects on the endocannabinoid system, a potent modulatory element of synaptic communication.[11,12]

Stress Effects on the Hippocampus: Not Just Glucocorticoids

While glucocorticoids are a key part of stress-induced changes in hippocampal structure and function, particularly during periods of repeated or chronic stress, they can have effects that are independent of an actual stressor. This is evidenced by the effects of treating animals with exogenous adrenal steroids, such as via repeated injections, subcutaneous mini-pumps, or even in drinking water. In many cases these effects are very similar to those observed with actual stress, but in some cases exposure to adrenal steroids without concomitant stress can lead to changes that are opposite. Making matters more complicated, combination of stress with additional exogenous treatment with adrenal hormones can result in the two treatments cancelling each other out! This highlights the fact that the adrenal stress hormones can interact with many different neurotransmitter systems in the hippocampus, including excitatory amino acids such as glutamate, inhibitory neurotransmitters such as GABA, and monoamines such as serotonin.[4,13]

Taking into account our previous discussion on the potential for damage induced by excitatory amino acids (EAAs), we will focus much of this section on how stress and stress mediators interact with this system.

EAAs are released by the mossy fiber pathway, and can be regulated by adrenal steroids. For instance, one consequence of restraint stress (where an animal's mobility is severely limited by being placed inside a plastic tube) is significantly elevated extracellular glutamate. This elevation of glutamate is accompanied by increased glutamate transporters in glial cells, as well as induction of phosphorylated CREB, a nuclear transcription factor.[14] When restraint stress is applied chronically (21 days), many signs of increased glutamate release are observed, including increased expression of presynaptic proteins involved in vesicle release and a concentration of vesicles near active synaptic zones. Interestingly, there is also an increase in the number of mitochondria in the giant mossy fiber terminals of chronically stressed rats.[15]

Stress and Hippocampal Neurogenesis

As discussed above, the anatomy of the hippocampus allows for its tremendous ability to integrate stimuli in the environment, but also renders it exquisitely sensitive, since unchecked excitatory tone can lead to epileptiform activity, resulting in cell damage or even cell death. Perhaps in part to replace neurons that are damaged by these high levels of activity, or perhaps to add new nodes to aid in the complex processing of information, the dentate gyrus of the hippocampus is also one of the few regions of the adult brain where new neurons continue to be produced. Specifically, in the subgranular layer of the dentate gyrus, cells that have some properties of astrocytes will eventually develop into mature granule neurons.[16–18] These newly born cells cluster in the inner part of the granule cell layer, where a substantial number of them will go on to differentiate into granule neurons within as little as 1 week. In the adult rat, 9000 new neurons are born per day and survive with a half-life of 28 days.[19] Importantly, stress and mediators of the stress response can reduce both the rate of proliferation of these cells and their survival, suggesting that stress can alter the complement of new cells within the hippocampal circuit.

BOX 21.1

HOW DO WE KNOW THAT STRESS AFFECTS HIPPOCAMPAL NEUROGENESIS?

The hippocampus was one of the first regions that was heavily studied as a locus for the neural effects of stress, as it plays an important role in the regulation of the HPA axis, contains both mineralocorticoid and glucocorticoid receptors, is sensitive to other types of neural damage (such as hypoxia), and is central in the regulation of cognition and emotionality. The early work of Landfield showed that the hippocampus loses some hippocampal pyramidal neurons in aging, but that this age-related decline can be mitigated (though not completely blocked) by adrenalectomy.[1] This work suggested that corticosteroids may be related to cell loss in the hippocampus. How did we test this hypothesis? Sapolsky was able to mimic this age-dependent loss of pyramidal neurons by giving young rats exogenous corticosterone for a period of 12 weeks,[2] and further showed that such CORT treatment could alter the resilience of these neurons by making them more sensitive to damage by ischemia or excitotoxicity.[3] He also showed that cultured hippocampal cells showed increased cell death when exposed to high levels of CORT *in vitro*.[3] This wasn't the end of the road, though. While stress and CORT can lead to cell death in the hippocampus, new neurons are constantly being "born" in the subgranular zone of the dentate gyrus. The focus then moved to ask how these factors modulated neurogenesis in this brain region.

Work in the 1990s showed that new neurons are born in the dentate gyrus of the hippocampus. This constant turnover of neurons in the hippocampus is likely related to its functions in adapting to changing environmental demands and learning – providing the hippocampus with an enhanced capacity for plasticity. However, glucocorticoids and stress can alter the rate of birth and the rate of survival of these neurons. How do we know this? By injecting the synthetic nucleoside bromodeoxyuridine (5-bromo-2-deoxyuridine, BrdU) into animals as they are placed into a stress experiment, dividing cells have BrdU incorporated into the newly synthesized DNA, substituting the BrdU for thymidine. After chronic stress, antibodies raised against BrdU can be used to determine the extent of incorporation, which indicates which cells had actively replicating DNA. A series of studies by Elizabeth Gould and co-workers in the early 1990s showed that neurogenesis decreased following chronic stress, suggesting that there are fewer new neurons born during the stress paradigm. In a similar vein, injecting the animals with BrdU prior to stress, and comparing BrdU staining in the hippocampus of stress or control animals,

BOX 21.1 *(cont'd)*

numerous labs have shown that the survival rate of new neurons is similarly decreased due to stress.[4,5] This shows that chronic stress and adrenal steroids can both decrease the amount of neurogenesis as well as impact the survival of newly born cells in the hippocampus.

References

1. Landfield PW. Modulation of brain aging correlates by long-term alterations of adrenal steroids and neurally-active peptides. *Prog Brain Res.* 1987;72:279–300.
2. Sapolsky RM, Krey LC, McEwen BS. Prolonged glucocorticoid exposure reduces hippocampal neuron number: implications for aging. *J Neurosci.* 1985;5:1222–1227.
3. Sapolsky, R.M. *Stress, the Aging Brain, and the Mechanisms of Neuron Death*, MIT Press, Cambridge, MA, 1992.
4. Gould E, McEwen BS, Tanapat P, et al. Neurogenesis in the dentate gyrus of the adult tree shrew is regulated by psychosocial stress and NMDA receptor activation. *J Neurosci.* 1997;17:2492–2498.
5. Gould E, Woolley CS, Cameron HA, et al. Adrenal steroids regulate postnatal development of the rat dentate gyrus: II. Effects of glucocorticoids and mineralocorticoids on cell birth. *J Comp Neurol.* 1991;313:486–493.

The relationships between adrenal steroids and hippocampal neurogenesis are not simple, however, and sometimes they are paradoxical. For instance, while acute glucocorticoid injection suppresses cell proliferation, stress-induced elevation does not necessarily result in the same suppression; also, even though glucocorticoid levels are elevated in enriched environments or during exercise, both of these treatments result in increased levels of neurogenesis because of other as-yet undefined mediators.[20] Moreover, isolation housing impairs the neurogenesis enhancing effects of exercise in a manner independent of glucocorticoid levels.[21] Conversely, sexual experience, while elevating glucocorticoids, enhances dentate gyrus neurogenesis.[22] On the other hand, chronic restraint stress reduces proliferation in the dentate gyrus, but acute restraint does not seem to result in any observable effect.[23] Perhaps some of these effects can be explained by the effects of interactions between glutamate neurotransmission and adrenal steroids, as discussed above. It is clear that NMDA receptors play a very important role in the regulation of neurogenesis, and blockade of these receptors can prevent the effects of glucocorticoids on neurogenesis.[24]

Prefrontal Cortex and Amygdala

While the hippocampus has been the focus of much of the initial work on stress-induced changes in brain structure and function, work over the past decade has shown that other brain areas are also involved. In this section, we will discuss two such regions: the prefrontal cortex and the amygdala. While the hippocampus plays a central role in learning and memory, as well as in mood, the prefrontal cortex (PFC) is a key brain region that regulates executive function, inhibitory control and cognitive flexibility. The amygdala (AMY) is a key brain region that regulates emotionality, aggression and affect-based learning and memory, such as fear conditioning. Interestingly, these brain regions show somewhat divergent responses to stress and stress mediators, in that they don't always respond in the same way as the hippocampus – that is, they don't only show dendritic retraction. Importantly, the behavioral consequences of chronic stress exposure closely match the changes in underlying brain morphology and function.

As mentioned, the PFC is a key brain region that mediates many "higher cognitive" functions in humans and non-human animals. While the human PFC is large and extremely complex compared to other brain regions, in rodents such as rats and mice, it is much simpler. Yet even in these species the PFC exerts clear control over inhibition, attention and cognitive flexibility.[25] Following chronic stress exposure, neurons of the prelimbic medial PFC (PL) show dramatic shrinkage and loss of complexity . However, neurons of the orbitofrontal PFC (OF) show a clear increase in complexity following chronic stress,[26] which is interesting when one considers that this part of the PFC is centrally important in processing the affective aspect of cognitive behaviors, and given the changes in the AMY (discussed below) provides an anatomical substrate for the changes in emotionality observed following chronic stress. In both the PL and the OF, the changes in dendritic complexity are accompanied by concomitant changes in dendritic spines, which largely decrease in the PL and increase in the OF.

In this sense, not only are there fewer and less complex dendrites in the PL, but those dendrites also seem to have fewer dendritic spines, suggesting even less connectivity. In the OF, the effect is the opposite: hypertrophied neurons with an increase in spine density. Along the lines observed in the PFC, in response to chronic stress the AMY shows hypertrophy of neurons in the basolateral aspect of the nucleus,[27] as well as an increase in the number of dendritic spines in this region, thus greatly increasing the activity of the AMY and responses to emotionally charged stimuli in the environment.

It is important to highlight that these stress-induced changes in each of these brain regions is not occurring in a vacuum. That is, these brain regions are interconnected, and have interacting effects on behavioral output. For instance, the processing of memories with an emotional valence with contextual information in the environment requires intact amygdala–hippocampal connections, while, on the other hand, the PFC can inhibit AMY activity, and plays a key role in the extinction of fear learning. Thus, when taken as a circuit, one can see how decreased inhibitory outflow from the PFC and increased activity at the AMY can result in a circuit that is fundamentally altered and no longer responds appropriately to stimuli in the environment – particularly emotionally charged stimuli.

STRESS EFFECTS ON BEHAVIOR

Given the clear effects of stress and stress mediators on the structure of the hippocampus, PFC and AMY, it is not surprising that both acute and chronic stress can have large effects on behavior. In terms of acute stress, numerous findings have shown that, depending on when the stress is encountered, it can actually improve certain types of memory performance.[28] This makes sense considering that, evolutionarily speaking, remembering where and when you encountered a lion or tiger or bear that almost killed you allows you to adjust your behavior so that you don't place yourself in that risky situation again. In other cases, acute stress (if severe enough) can disrupt the consolidation of memories, and impair performance.[4,29] These effects are likely based in both rapid non-genomic effects as well as slower genomic effects of glucocorticoids, potentially even at the level of the synapse. Numerous studies have shown that acute stress changes excitatory signaling at the synapse, and can occur quite rapidly.[30] Thus, as in many things in stress research, the effects of a stressor are multifaceted and in many cases contextually modulated.

The previous section dealt with the anatomical and morphological changes that occur in neural circuits, particularly the hippocampus, PFC and AMY during and after chronic stress. These changes have significant and wide-ranging effects on behavior, especially given the interactive nature of the connections between these brain regions. For instance, chronic stress can lead to impairment in memory in the radial arm maze and Y-maze tasks, both of which are hippocampal-based behavioral tests. Remarkably, these behavioral effects can be minimized by treatment with drugs such as Dilantin™ or the antidepressant tianeptine, both of which actually prevent stress-induced remodeling of CA3 hippocampal dendrites.

A set of elegant studies has examined how chronic stress can modulate PFC-regulated cognitive behaviors. Using an attentional set-shifting task, which probes cognitive flexibility by having the animal learn an association and then shift this learning to another modality, Liston and colleagues demonstrated that chronic stress produced impairments in this task that were accompanied by reduction in complexity of pyramidal cells in the medial prefrontal cortex (mPFC).[31] In an extension of this study, they further demonstrated that psychosocial stress in humans (medical students studying for an important academic exam for 4 weeks) caused measurable changes in their perceived stress levels, decreased performance in an attentional control task similar to the set-shifting task used in rodents, and also impaired processing in the mPFC as measured by fMRI,[32] further providing evidence for the effects of chronic stress on both brain and cognitive behavior in humans and non-human animals.

In addition to the effects on cognition, the effects of stress on emotional behaviors are also well characterized. Several stress-related psychopathologies have been linked to a change in amygdalar functioning, including anxiety, post-traumatic stress disorder, social phobia, depression and impulsive aggression.[33] As in the hippocampus and PFC, the changes in emotional behavior are associated with the neuroanatomical changes observed in the amygdala, which include hypertrophy of dendrites in the basolateral amygdala, as well as an increase in spines. Combined with the decrease in inhibitory influence from the PFC, the increased AMY activity can tilt the balance to an amygdala-centric tone in which signals and stimuli in the environment that should not be considered threatening are interpreted as such.

ALLOSTASIS AND ALLOSTATIC OVERLOAD

To this point in the chapter, we have discussed stress in the context of its effects on brain circuits and the behaviors that these circuits underlie. We have noted both positive and negative effects of adrenal steroids and stress mediators on these circuits, demonstrating that all stress isn't "toxic," and that in some cases stress can have a facilitative role in both physiology and behavior. As presented in the introduction, stress has an important role to play in the survival of an organism, and should be considered an adaptive response when functioning optimally because it protects homeostasis. While homeostasis is defined as maintaining the right conditions for various physiological and biochemical systems to operate within optimal parameters, an additional term, *allostasis*, can be introduced. Allostasis is defined as the process of maintaining stability through active intervention. That is, allostasis encompasses the idea that changing the set point of a physiological system is a way to actively maintain homeostasis. In this context, the systems that vary according to demand, like the hypothalamic–pituitary–adrenal stress (HPA) axis or the autonomic nervous system (ANS), help to maintain other physiological systems that are truly "homeostatic," such as oxygen tension or pH. In that context, when necessary, activation of the HPA during times of stress can help maintain the integrity of other homeostatic systems, thus ensuring that an organism will survive. However, these allostatic responses are not free, in that there are certain costs incurred by the body when these systems are engaged. The term *allostatic load* can be used to describe the cumulative cost to the body from homeostatic processes and any unpredictable perturbations.[34,35]

Allostatic overload can be thought of as representing a state in which the sum of the costs of "regular" daily routines and the costs of responding to any perturbations in the environment exceed the available "fuel." In the short-term, allostatic overload could be a necessary response in order to ensure survival, but if it persists for too long and becomes chronic, serious negative effects on brain and peripheral physiology can occur. It is helpful to think of there being two different kinds of allostatic overload. Type 1 overload occurs when the animal faces an "emergency" energy shortfall due to unpredictable environmental events, such as a food shortage; the energy required exceeds the energy available. These events trigger an HPA response, and the organism responds to resist or avoid the effects of the stress. On the other hand, type 2 allostatic overload occurs when there is an environmental stressor that does not directly involve energy supply, such as social conflict. In this case, there is excess energy available, and the rise in stress mediators results in alterations in metabolism and energy usage, leading to such things as sustained increases in food consumption, glucocorticoid dysregulation, increased levels of insulin and the increased storage of energy as fat.

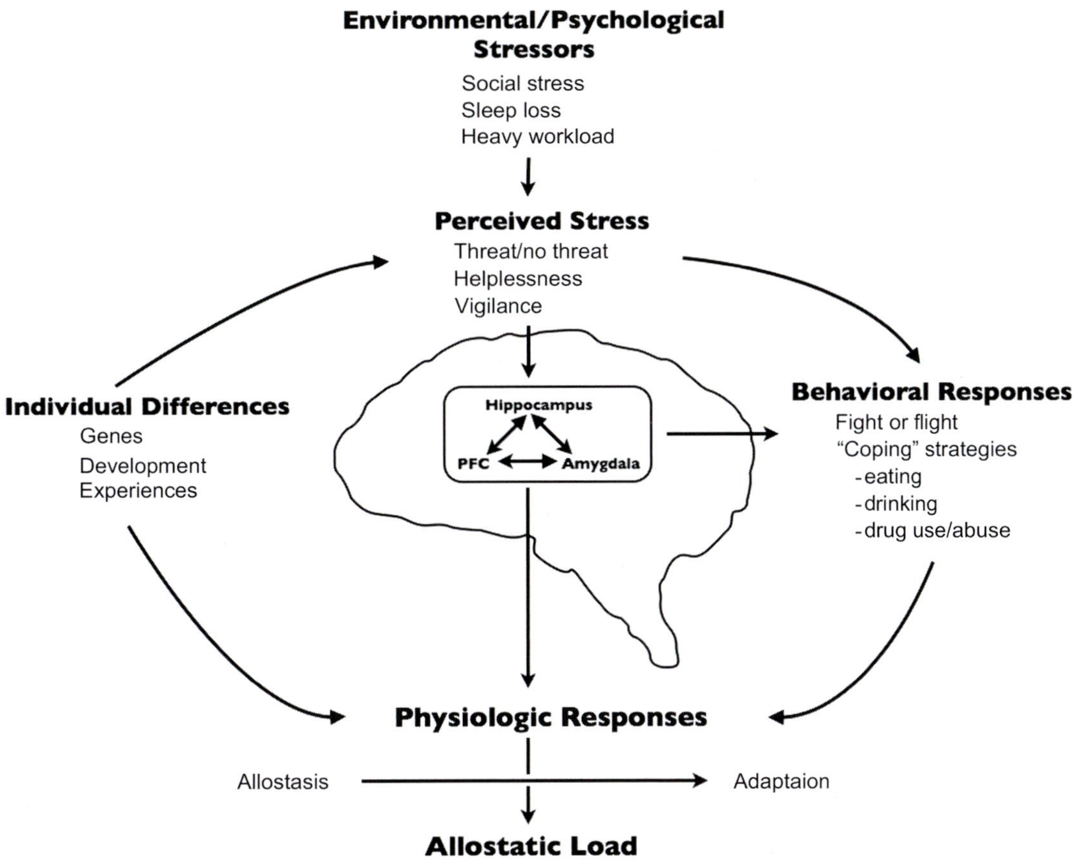

FIGURE 21.1 **Stress and the brain.** The brain is a key modulator of the stress response, and a target of its effects. An individual's genetics and developmental experiences modulate the way the brain interprets and perceives environmental stressors. Behavioral responses are mediated by an interaction between the hippocampus, amygdala and prefrontal cortex, as well as peripheral responses that mobilize the physiological stress response. Allostatic responses lead to adaptive changes that allow the organism to cope with these stressors, but if these responses are not appropriately terminated, or engaged repeatedly, they can lead to allostatic load and, eventually, overload.

Using an ecological example, consider a bear that is preparing to hibernate for the winter. The lack of food triggers changes in the animal that allow it to cope with the environmental stress it is facing, and is adaptive as these allostatic responses enable the animal to survive the winter – an example of type 1 allostatic overload. Consider a similar example: the same bear, but being raised in captivity, where energy availability (in the form of daily meals provided by the zookeeper, or snacks given by tourists) is never in short supply, but the social stresses of captivity are present and result in severe weight gain and behavioral changes that are the hallmarks of type 2 allostatic overload. While useful in terms of describing ecological situations in animals, we must also realize that similar conditions can face humans as well. For instance, intense military training (such as Special Forces' training) can result in short-term physical and psychological stress, extreme physical exertion and low food availability, and thus type 1 allostatic overload. Chronic stress in another context – say, due to low socioeconomic status (see more below) or social conflict, in the face of readily available high-calorie food without adequate energy expenditure through exercise – can lead to type 2 allostatic load. This should give us pause in considering how systems that evolved to protect our brains and bodies throughout our life history are responding in the face of rapid human adoption of technology and modern convenience, and can lead to mental and physical health problems (Fig. 21.1).

STRESSES OF MODERN LIFE

In this and in previous chapters, the stress response has been discussed in the context of its evolutionary relevance. An organism needs to be able to respond to an environmental challenge, and adjust its physiology and behavior to cope appropriately with the challenge in order to ensure survival. The classic example used is that of the zebra in the Serengeti, running from the lion that is chasing it in the hope of a tasty meal. Suffice it to say, this experience causes the zebra great alarm and distress, and engages body systems that are designed to help the zebra run away as fast as possible (e.g., increased blood flow to muscles, increased oxygen consumption), to prepare for possible injury (e.g., stimulation of the immune system), and to recall the place it encountered the lion in order to try to avoid experiencing the event again (e.g., increased learning). At the same time, functions of non-essential systems (i.e., body systems not immediately necessary for the task at hand) are reduced, such as the reproductive or digestive systems (there is no sense expending energy on digestion if you may not make it to the next meal!) are reduced. This is but one simplistic example of the adaptive significance of the stress response in the context of the "real world." However, while our ancestors, thousands of years ago, may have faced similar exceptional situations, modern civilization has largely reduced or eliminated such environmental stressors for most of the planet. In their place new environmental challenges have arisen; in many cases they are arguably more chronic than in times past and often less obvious, but they may engage the stress response even though the quality and quantity of these stressors has dramatically changed. Thus, while modern life protects many of us from these threats to existence, it has introduced a perhaps more insidious threat that can have significant physiological and behavioral ramifications over the long run.

One of these "modern day" threats that is beginning to receive more attention as a potential "stressor" is the 24/7 lifestyle that many have become accustomed to. While our bodies' biological clocks, which are evolutionarily ancient and share similar characteristics at the molecular level with those of simple organisms such as the fruit fly (*Drosophila melanogaster*), are set to keep us on a regular cycle of night and day that is in tune with our rest–activity cycles, modern society has damaged, or even broken, this relationship. For instance, all variety of shift workers, from assembly line workers and power plant operators to medical practitioners and airline pilots, live and work in a world where they are forced to operate outside of the "natural" day–night cycle, and have to adjust their clocks to try to stay "in sync." It is becoming clear that this type of chronic, repeated, and sometimes unpredictable setting and resetting can have health consequences that are similar to those observed in situations which people describe as "stressful". Anyone who has traveled across multiple time zones can attest to the general feelings of malaise and mental "slowness" that materialize in the hours and days after arriving at a destination, and that it takes several days for this to pass. Once the body clock has resynchronized to the local environment and gotten things back to normal, one gets back on the plane and heads in the opposite direction, repeating the cycle over again. A similar set of problems faces the shift worker, who is required to work nights one week and days the next. For the occasional traveler or shift worker this may seem more like a nuisance than a major problem; however, there is burgeoning evidence showing that a lifestyle of disrupted sleep–wake cycles can lead to physiological, behavioral and neural changes resembling those of chronic stress. Physiologically, a longitudinal study in a cohort of nurses in nightshift work found that exposure to night work can contribute to weight gain and obesity.[36] Moreover, alternating shift work was found to be an independent risk factor for the development of obesity in a large longitudinal study

BOX 21.2

HOW DO WE KNOW THAT SLEEP AND CIRCADIAN DISRUPTION MAY BE A POTENTIAL STRESSOR?

The modern world has challenges that our ancient ancestors would not even have dreamed of. Such challenges include technology that has allowed us to break the relationship between our sleep–wake cycles and the solar day, or being able to travel over many time zones in a few short hours. In the short-term, these challenges are likely no more than nuisances. However, in the long-term, disruption of the brain and body clocks could lead to changes in behavior and physiology that are similar to those of chronic stress. How do we know this? In a study conducted in 2001, Cho[1] was able to image the brains of flight crews that were classified as either "short recovery" or "long recovery." While long-recovery flight crews would work for many cycles within the same time zone (for instance, flying north and south within a time zone), short-recovery crews had only a few days recovery before continuing on further transmeridian flights. Remarkably, the right temporal lobes (the part of the brain that contains the hippocampus) were shrunken in the short-recovery crews, and this was accompanied by behavioral deficits that included slower reaction times on a cognitive task, as well as more errors. So, the short-recovery crews were slower to react, and made more mistakes when they did react. Interestingly, there was a significant correlation between the volume of the temporal lobe and salivary cortisol levels, but only in the short-recovery crew. In a different study, Karatsoreos and colleagues[2] asked what would happen to the brain and behavior of mice that rather than being flown across times zones, were chronically exposed to a light–dark cycle that was artificially shortened, to 20 hours rather than the normal 24 hours. In these mice, following long-term circadian disruption, neurons in the medial prefrontal cortex became less complex compared to controls, and this was associated with changes in behaviors mediated by the mPFC, like cognitive flexibility and impulse control. Together, the human and animal work helps to construct a picture which suggests that long-term disruption of the circadian clock can lead to changes in the brain and behavior.

References

1. Cho K. Chronic 'jet lag' produces temporal lobe atrophy and spatial cognitive deficits. *Nat Neurosci.* 2001;4:567–568.
2. Karatsoreos IN, Bhagat S, Bloss EB, et al. Disruption of circadian clocks has ramifications for metabolism, brain, and behavior. *Proc Natl Acad Sci USA.* 2011;108:1657–1662.

of male Japanese shift workers.[37] Neurobehaviorally, in a study of long-recovery vs short-recovery flight crews, it was found that short-recovery crews had impaired performance in a psychomotor task, reacting more slowly and with more errors when compared to a long-recovery crew.[38] In an animal model of circadian disruption, which may represent a chronic low-level allostatic load, remodeling of neurons in the prelimbic region of the medial prefrontal cortex of mice is observed, and these circadian-disrupted animals show impaired cognitive flexibility.[39] This suggests that while chronic circadian disruption may not impair simple cognitive tasks, more complex tasks requiring flexibility may be impaired. When one considers the other interacting stressors in the environment, such as immune and metabolic challenges, individuals who are chronically circadian-disrupted may be setting up the brain and body for significant damage.

Another important factor to consider as a stressor in the modern world that has impacts on brain function is socioeconomic status (SES). It is largely the case that individuals with lower SES are exposed to more environmental toxins in the environment because of where they live or the jobs they do, lower quality foods available in the neighborhoods, and high emotional stress and chaos at work and in the home. This is coupled with poorer access to good quality mental and physical healthcare. Numerous lines of evidence suggest that in humans, low SES is associated with changes in the brain, with greater amygdalar activation in response to threatening faces, and reduced gray matter in the anterior cingulate cortex. These findings are significant, as they suggest that, in both children and adults, the stressors of modern society as embodied by lower SES can lead to significant shifts in behavior and neural function.

SUMMARY AND CONCLUSIONS

Stress is a term that was borrowed by biologists from engineers to describe how the brain and body respond to threatening stimuli in the environment. While it is sometimes difficult to describe formally what a stressor is, a useful definition is 'a real or perceived threat to homeostasis.' The brain and body have devised systems to protect and adapt to stressors, and, through a process known as allostasis, they are able to protect the organism by allowing it to adapt to the changed environment. However, prolonged exposure to stress, or the mismanagement of the responses to stress, can lead to allostatic overload – a situation which contributes to wear and tear on brain and body systems, and can have significant effects on mental and physical health. In the brain, we have learned that glucocorticoids, and other stress mediators, can modify both the structure and the function of key neural centers that regulate affect and cognition, including the hippocampus, prefrontal cortex and amygdala. Interestingly, the effects of stress on these brain regions are not unidirectional, or explainable by simple toxicity of the stress mediators.

Specifically, while it seems that stress and stress mediators can result in the shrinkage of neurons in the prelimbic region of the prefrontal cortex, as well as the CA3 and CA1 regions of the hippocampus, they lead to a reliable increase in the exuberance of neurons in the orbitofrontal cortex and the amygdala. These neural changes are reflected in changes in behavior, with chronic stress resulting in impairments in memory, attention and executive function, while at the same time increasing fear and aggressive responses.

While the stress system seems to be an evolutionarily old system that developed to ensure survival of the organism in the face of environmental challenges, our modern society has alleviated many of our daily threats to existence. However, humans have been skillful in creating new and potentially insidious threats to our mental and physical health, and while our industrialized society has provided us protection from many of the threats to existence that our ancestors had to face on a daily basis, other threats – such as worry about jobs, chronic disruption of our sleep–wake cycles, exposure to foods that are easy to access and full of unneeded calories, and perceived social and economic stressors – means that these neural and physiological systems are still engaged by modern humans, perhaps too often and for too long. The costs to our mental and physical health are now being appreciated and actively studied. Further study is surely needed to understand how best to protect our brains from the negative aspects of modern stressors.

In many cases, the balance between an adaptive stress response at the level of brain and behavior, and maladaptive allostatic overload, can be described using an inverted U-shaped curve heuristic. That is, a failure to engage the stress system can have equally detrimental effects on the survival and health of an organism as an overreaction or delayed shut-down of the response can do. Restated plainly, an overly *hypo*active hypothalamic–pituitary–adrenal axis can be just as bad as a *hyperactive* HPA.

QUESTIONS ARISING

One of the key new questions that is being explored regards individual differences in susceptibility or resilience in response to stressors on brain and behavior. Numerous studies have been exploring whether there are gene and gene × environment interactions that can help explain why some individuals respond to stress with positive coping strategies while others seem to respond with negative strategies.[40] In investigating these differences, it is important to acknowledge that a person's genetic composition may predispose them to negative neural effects in response to stress, as exemplified by the recent studies of the BDNF Val66Met polymorphism in humans and in a transgenic mouse model.[41]

Moreover, the same genetic variations that predispose to negative outcomes in adverse, stressful environments may predispose the same genotype to positive outcomes in an enriched and nurturing environment during development. Such individuals are referred to as "orchid children," whereas those who are less susceptible to the negative influences are referred to as "dandelion children".[42,43]

Another key future question regards recovery following stress. As noted, numerous animal models indicate that the brain shows an adaptive capacity to respond to stress (e.g., by shrinking dendrites in the prelimbic cortex and hippocampus, and growing dendrites in the basolateral amygdala and orbitofrontal cortex), but that many of these changes can also recover to a more "normal" state if a suitable period of rest is granted.[44–47] However, what is also being uncovered is that while this may be true in younger animals, older animals show a greatly blunted recovery, suggesting that with age comes a reduced capacity to "bounce back" after chronic stress.[45] How this is mediated at the cellular level, and what this means for repeated bouts of stress, is as yet unknown, but it is clear that this will be an area of active research for years to come.

References

1. Selye H. A syndrome produced by diverse nocuous agents. *Nature.* 1936;138:32.
2. Schulkin J, McEwen BS, Gold PW. Allostasis, amygdala, and anticipatory angst. *Neurosci Biobehav Rev.* 1994;18:385–396.
3. Mason J. In: Pincus G, ed. *Recent Progress in Hormone Research.* New York: Academic Press; 1959:345–389.
4. McEwen BS. Stress and hippocampal plasticity. *Annu Rev Neurosci.* 1999;22:105–122.

5. Popov VI, Bocharova LS, Bragin AG. Repeated changes of dendritic morphology in the hippocampus of ground squirrels in the course of hibernation. *Neuroscience.* 1992;48:45–51.
6. Popov VI, Bocharova LS. Hibernation-induced structural changes in synaptic contacts between mossy fibres and hippocampal pyramidal neurons. *Neuroscience.* 1992;48:53–62.
7. Magarinos AM, McEwen BS, Saboureau M, et al. Rapid and reversible changes in intrahippocampal connectivity during the course of hibernation in European hamsters. *Proc Natl Acad Sci USA.* 2006;103:18775–18780.
8. McKittrick CR, Magarinos AM, Blanchard DC, et al. Chronic social stress reduces dendritic arbors in CA3 of hippocampus and decreases binding to serotonin transporter sites. *Synapse.* 2000;36:85–94.
9. Arendt T, Stieler J, Strijkstra AM, et al. Reversible paired helical filament-like phosphorylation of tau is an adaptive process associated with neuronal plasticity in hibernating animals. *J Neurosci.* 2003;23:6972–6981.
10. Johnson LR, Farb C, Morrison JH, et al. Localization of glucocorticoid receptors at postsynaptic membranes in the lateral amygdala. *Neuroscience.* 2005;136:289–299.
11. Tasker JG, Di S, Malcher-Lopes R. Minireview: Rapid glucocorticoid signaling via membrane-associated receptors. *Endocrinology.* 2006;147:5549–5556.
12. Hill MN, McEwen BS. Involvement of the endocannabinoid system in the neurobehavioural effects of stress and glucocorticoids. *Prog Neuropsychopharmacol Biol Psychiatry.* 2010;34: 791–797.
13. McEwen BS, Chattarji S. Molecular mechanisms of neuroplasticity and pharmacological implications: the example of tianeptine. *Eur Neuropsychopharmacol.* 2004;14(Suppl 5): S497–S502.
14. Wood GE, Young LT, Reagan LP, et al. Stress-induced structural remodeling in hippocampus: prevention by lithium treatment. *Proc Natl Acad Sci USA.* 2004;101:3973–3978.
15. Magarinos AM, Verdugo JM, McEwen BS. Chronic stress alters synaptic terminal structure in hippocampus. *Proc Natl Acad Sci USA.* 1997;94:14002–14008.
16. Cameron H, Woolley C, McEwen BS, et al. Differentiation of newly born neurons and glia in the dentate gyrus of the adult rat. *Neuroscience.* 1993;56:337–344.
17. Kempermann G, Gage FH. New nerve cells for the adult brain. *Sci Am.* 1999;280:48–53.
18. Seri B, Garcia-Verdugo JM, McEwen BS, et al. Astrocytes give rise to new neurons in the adult mammalian hippocampus. *J Neurosci.* 2001;21:7153–7160.
19. Cameron HA, McKay RD. Adult neurogenesis produces a large pool of new granule cells in the dentate gyrus. *J Comp Neurol.* 2001;435:406–417.
20. Mirescu C, Gould E. Stress and adult neurogenesis. *Hippocampus* 2006;16:233–238.
21. Stranahan AM, Khalil D, Gould E. Social isolation delays the positive effects of running on adult neurogenesis. *Nat Neurosci.* 2006;9:526–533.
22. Leuner B, Glasper ER, Gould E. Sexual experience promotes adult neurogenesis in the hippocampus despite an initial elevation in stress hormones. *PLoS One.* 2010;5:e11597.
23. Pham K, Nacher J, Hof PR, et al. Repeated restraint stress suppresses neurogenesis and induces biphasic PSA-NCAM expression in the adult rat dentate gyrus. *Eur J Neurosci.* 2003;17: 879–886.
24. Cameron HA, Hazel TG, McKay RD. Regulation of neurogenesis by growth factors and neurotransmitters. *J Neurobiol.* 1998;36: 287–306.
25. McEwen BS, Gianaros P. Plasticity of the brain in relationship to stress. *Annu Rev Med.* 2010.
26. Leuner B, Falduto J, Shors TJ. Associative memory formation increases the observation of dendritic spines in the hippocampus. *J Neurosci.* 2003;23:659–665.
27. Vyas A, Mitra R, Rao BSS, et al. Chronic stress induces contrasting patterns of dendritic remodeling in hippocampal and amygdaloid neurons. *J Neurosci.* 2002;22:6810–6818.
28. Roozendaal B, Okuda S, Van der Zee EA, et al. Glucocorticoid enhancement of memory requires arousal-induced noradrenergic activation in the basolateral amygdala. *Proc Natl Acad Sci USA.* 2006;103:6741–6746.
29. DeKloet ER, Vreugdenhil E, Oitzl MS, et al. Brain corticosteroid receptor balance in health and disease. *Endocrine Rev.* 1998;19:269–301.
30. Joels M. Corticosteroid effects in the brain: U-shape it. *Trends Pharmacol Sci.* 2006;27:244–250.
31. Liston C, Miller MM, Goldwater DS, et al. Stress-induced alterations in prefrontal cortical dendritic morphology predict selective impairments in perceptual attentional set-shifting. *J Neurosci.* 2006;26:7870–7874.
32. Liston C, McEwen BS, Casey BJ. Psychosocial stress reversibly disrupts prefrontal processing and attentional control. *Proc Natl Acad Sci USA.* 2009;106:912–917.
33. Holzel R. Dielectric and dielectrophoretic properties of DNA. *IET Nanobiotechnol.* 2009;3:28–45.
34. McEwen BS. Stress, adaptation, and disease. Allostasis and allostatic load. *Ann NY Acad Sci.* 1998;840:33–44.
35. McEwen BS, Wingfield JC. The concept of allostasis in biology and biomedicine. *Horm Behav.* 2003;43:2–15.
36. Niedhammer I, Lert F, Marne MJ. Prevalence of overweight and weight gain in relation to night work in a nurses' cohort. *Intl J Obes Relat Metab Disord.* 1996;20:625–633.
37. Suwazono Y, Dochi M, Sakata K, et al. A longitudinal study on the effect of shift work on weight gain in male Japanese workers. *Obesity (Silver Spring).* 2008;16:1887–1893.
38. Cho K. Chronic 'jet lag' produces temporal lobe atrophy and spatial cognitive deficits. *Nat Neurosci.* 2001;4:567–568.
39. Karatsoreos IN, Bhagat S, Bloss EB, et al. Disruption of circadian clocks has ramifications for metabolism, brain, and behavior. *Proc Natl Acad Sci USA.* 2011;108:1657–1662.
40. Kendler KS. Major depression and the environment: A psychiatric genetic perspective. *Pharmacopsychiatry.* 1998;31:5–9.
41. Soliman F, Glatt CE, Bath KG, et al. A genetic variant BDNF polymorphism alters extinction learning in both mouse and human. *Science.* 2010;327:863–866.
42. Boyce WT, Ellis BJ. Biological sensitivity to context: I. An evolutionary-developmental theory of the origins and functions of stress reactivity. *Dev Psychopathol.* 2005;17:271–301.
43. Dobbs D. Orchid children. *The Atlantic.* 2009:60–68.
44. Radley JJ, Rocher AB, Janssen WGM, et al. Reversibility of apical dendritic retraction in the rat medial prefrontal cortex following repeated stress. *Exp. Neurol.* 2005;196:199–203.
45. Bloss EB, Janssen WG, McEwen BS, et al. Interactive effects of stress and aging on structural plasticity in the prefrontal cortex. *J Neurosci.* 2010;30:6726–6731.
46. Goldwater DS, Pavlides C, Hunter RG, et al. Structural and functional alterations to rat medial prefrontal cortex following chronic restraint stress and recovery. *Neuroscience.* 2009;164: 798–808.
47. Conrad CD, Magarinos AM, LeDoux JE, et al. Repeated restraint stress facilitates fear conditioning independently of causing hippocampal CA3 dendritic atrophy. *Behav Neurosci.* 1999;113: 902–913.

CHAPTER

22

Neuroendocrinology of Aggression

Brian C. Trainor [1], *Randy J. Nelson* [2]

[1] Department of Psychology, University of California, Davis, CA, USA, [2] Departments of Neuroscience and Psychology, Ohio State University, Columbus, OH, USA

OUTLINE

Summary

Aggressive behaviors are often assumed to be regulated by testosterone. However, after a century of research it is becoming clear that there is a complex network of neuroendocrine pathways that can regulate aggressive behaviors. This network of steroid and peptide hormones allows for aggressive behaviors to be modified in response seasonal, reproductive and social information. We review the effects of steroids (gonadal, adrenal and neurosteroids) and neuropeptide hormones on aggressive behaviors. Many of these hormones only exert an effect on behavior within a defined environmental context, and we explore the complex interactions between hormones, the environment and behavior. Studies of the neuroendocrine bases of aggressive behavior have provided important insights into how complex behaviors are regulated by hormones. These discoveries may also eventually lead to practical applications, as dysregulated aggression is a component of several mental disorders.

INTRODUCTION

Neuroendocrinology is the interaction between the nervous and endocrine systems. Traditionally, this field has relied on behavioral changes as the "read-out" in response to manipulations of endocrine signals. Aggressive behaviors have been a focus of many neuroendocrine studies for a variety of reasons, including sex differences in aggression, increased aggression at puberty, and coincidence of aggression with the onset of the breeding season. Males are more aggressive than females in most vertebrate species (although exceptions have been reported), and aggression is associated with the breeding season (again with some notable exceptions, detailed below). Indeed, in the first documented neuroendocrine experiment, Professor Arnold Adolph Berthold, at the University of Göttingen, demonstrated experimentally that a product of the testes was necessary for young roosters to develop typical mating, vocalizing and aggressive behaviors in adulthood.[1]

Aggression is a social interaction to determine the outcome of a conflict over resources. Aggressive behaviors take many forms, despite evidence that the regulatory factors and brain mechanisms are conserved across vertebrate species.[2] Importantly, aggression is not a unified behavioral construct. There are many types of aggression,[3] and to understand the neuroendocrine contribution to the regulation of aggression it is critical

Handbook of Neuroendocrinology, DOI: 10.1016/B978-0-12-375097-6.10022-8

to discriminate the types of aggression under study. It is likely that very different hormones and brain circuits underlie predatory aggression, compared to parental aggression or territorial aggression. Much of what we know about the neuroendocrine regulation of aggression has been learned from studies of mice (*Mus musculus*). However, mice have been bred to be quite docile[4] so they must often be placed in isolation or electrically shocked to elicit aggressive behaviors, which may elicit so-called irritable or defensive aggression, even in the context of a resident–intruder lab test, which is considered a test of territorial aggression.[5]

Studies of aggression are important because it is critical to understand and potentially control aggression. In the 21st century, our only treatment for a dog that bites a person is to euthanize the dog. Human violence accounts for untold suffering for the victims, their families, and often the perpetrators and their families. Yet effective treatment for human violence or inappropriate aggression is primitive at best: castration, incarceration or death penalties. Because human violence is primarily a male-dominated activity, animal studies of aggression have focused appropriately on testosterone-dependent offensive inter-male aggression. We examine these types of studies in this chapter, as well as the influence of other steroid and protein hormones in the regulation of aggression.

NEURAL CIRCUITS OF AGGRESSION

Hormones modulate aggressive behaviors by regulating neural circuits in the brain. Based on data from lesion and neural activity studies, it has been hypothesized that aggressive behaviors are controlled by a social behavior circuit that regulates many kinds of social behaviors.[3,6,7] This circuit has been best characterized in rodents, where input from the olfactory bulbs[8] is sent to the medial amygdala (MEA) and then relayed to other hypothalamic and limbic brain regions, including the bed nucleus of the stria terminalis (BNST), lateral septum (LS), anterior hypothalamus (AHA), ventromedial hypothalamus (VMH) and periaqueductal gray (PAG)[9] (Fig. 22.1). This pathway is not linear, as there are many interconnections among these nuclei. Most of these brain regions express high levels of androgen receptor and estrogen receptors.[10,11] Recent data have suggested that the nucleus accumbens, which expresses androgen receptor, may contribute to reinforcement of aggressive behavior in certain contexts.[12] Although there is strong evidence for a causal link between testosterone and aggression in non-human animals, results from human and non-human primate studies are mixed, and appear to be context dependent. It is usually assumed that in primates the highly developed frontal cortex plays a more important role in inhibiting aggressive behaviors compared to rodents. Interestingly, small populations of neurons and glial cells in the rhesus monkey prefrontal cortex express androgen receptor.[13] This suggests that behavioral effects of androgens may not be limited to hypothalamic and limbic regions.

EFFECTS OF STEROIDS ON AGGRESSIVE BEHAVIOR

Androgens

Testosterone was one of the first substances identified to regulate aggressive behaviors. Berthold showed that

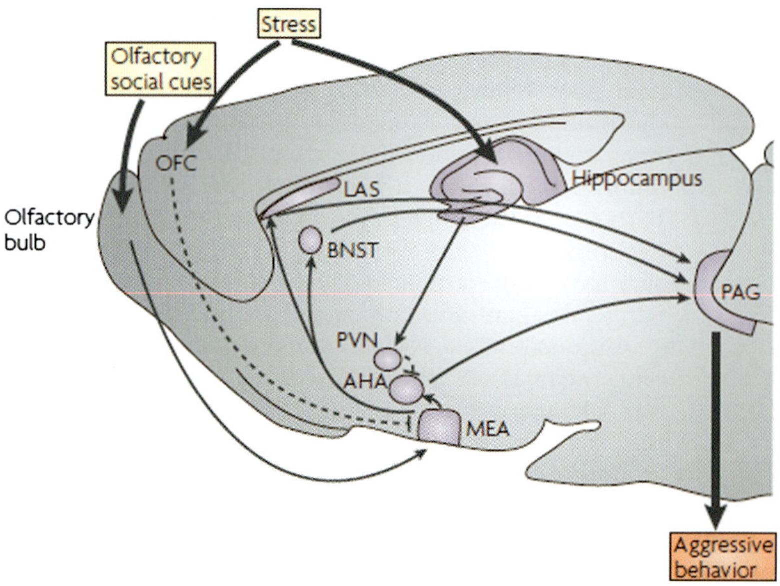

FIGURE 22.1 **Neuroanatomical pathways of aggression in rodents.** Information from the olfactory bulb is processed by the medial amygdala and sent to the lateral septum (LAS), bed nucleus of the stria terminalis (BNST) and anterior hypothalamus (AHA). These brain regions express both androgen and estrogen receptors. These brain areas are thought to prompt the periaquaductal gray into promoting species-specific aggressive behaviours. Stress can inhibit aggression via inhibitory inputs from orbital frontal cortex and the paraventricular nucleus (PVN). *Reproduced from Nelson RJ, Trainor BC. Neural mechanisms of aggression.* Nat Rev Neurosci. *2007;8:536–546, with permission from Macmillan Publishers Ltd ©2007.*

BOX 22.1

HOW DO WE KNOW ABOUT ESTABLISHING NEUROENDOCRINE–BEHAVIOR INTERACTIONS?

What sort of evidence would be sufficient to establish that a particular hormone affected a specific behavior, such as aggression, or that a specific behavior changed hormone concentrations? Experiments to test hypotheses about the effects of hormones on behavior must be carefully designed, and, generally, three conditions must be satisfied by the experimental results to establish a causal link between hormones and behavior[1]: (a) a hormonally dependent behavior should disappear when the source of the hormone is removed or the actions of the hormone are blocked; (b) after the behavior stops, restoration of the missing hormonal source or its hormone should reinstate the absent behavior, and (c) hormone concentrations and the behavior in question should be covariant – that is, the behavior should be observed only when hormone concentrations are relatively high and never or rarely observed when hormone concentrations are low.

The third class of evidence has proved difficult to obtain because hormones may have a long latency of action, and because many hormones are released in a pulsatile manner. Also, some pharmaceutical grades of steroids (e.g., esterfied steroids) have been altered to remain in circulation longer than endogenous steroids. Pulsatile secretion of hormones presents difficulties with making hormone–behavior inferences. For example, if a pulse of hormone is released into the blood, and then no more is released for an hour or so, a single blood sample will not provide an accurate picture of the endocrine status of the animal under study. Completely different conclusions about the effect of a hormone on behavior could be obtained if hormone concentrations were assessed at their peak rather than at their nadir. This problem can be overcome by obtaining measures in several animals, or by taking several sequential blood samples from the same animal and averaging across peaks and valleys. Another problem is that biologically effective amounts of hormones are vanishingly small, and difficult to measure accurately. Effective concentrations of hormones are usually measured in micrograms (μg, 10^{-6} g), nanograms (ng, 10^{-9} g), or picograms (pg, 10^{-12} g). The development of techniques, such as the radioimmunoassay, has increased the precision with which hormone concentrations can be measured, but, because of the multiple difficulties associated with obtaining reliable covariant hormone–behavior measures, obtaining the first two classes of evidence usually has been considered sufficient to establish a causal link in hormone–behavior relations.

Reference

1. Nelson RJ. *An Introduction to Behavioral Endocrinology.* Sunderland: Sinauer; 2005.

substances produced by the testes promoted aggression in male chickens,[1] and later testosterone was identified as a primary hormone secreted by the testes. Currently, a "gold standard" for determining whether testosterone promotes aggressive behavior is to use a combination of castration and hormone replacement.

Castrated animals are expected to show reduced aggression levels, whereas castrated animals with testosterone replaced via an implant are expected to display aggression levels similar to those of intact animals. Using this approach, a wide variety of species has been documented to display testosterone-dependent aggression, including tree lizards,[14] zebra finches,[15] cichlid fish,[16] house mice[17] and red deer.[18] Numerous species, however, have been identified in which castration does not reduce aggression.[19–22] In other species, castration affects aggression at one time of the year but not at others – for example, castration reduces song sparrow (*Melospiza melodia morphna*) aggression during the breeding season, but has no effect on aggression during the non-breeding season.[23] It has become apparent that testosterone does not simply increase aggression. One possible explanation for these varying results is that testosterone acts a biological signal for integrating reproductive and social context.

The "Challenge Hypothesis" predicted that the relationship between baseline testosterone and aggression would be stronger in species in which aggressive behavior was tightly linked with a relatively short breeding season, and weaker in species in which aggressive behaviors are observed across a longer time-period.[24] High testosterone concentrations are metabolically costly, and may have other negative physiological effects.[25,26] It is thought that when aggressive behaviors are expressed over a long time-period, aggressive behaviors are less dependent on high baseline testosterone concentrations due to these negative side effects. These original predictions were based primarily on data from free-living birds, but subsequent studies have provided support for this hypothesis in a variety of species.[27] For example, in ring-tailed lemurs, fecal testosterone concentrations were positively correlated with male aggressive behaviors during the mating season, but not during the pre-mating season.[28] An additional consideration is that social interactions themselves can have important effects on testosterone concentrations.[29,30] In particular, engaging in aggression, or even observing competitive interactions, can rapidly increase testosterone.[31,32] The function of these rapid increases in testosterone is poorly understood, although there are some obvious directions for further study. A challenge-induced increase in testosterone could serve to regulate aggression in future encounters, as it does in California mice.[33] An additional possibility is that an increase in testosterone could act rapidly to increase aggression. Although is often assumed that the effects of steroids on behavior require several hours or days, it is clear that testosterone and other steroids can affect behavior within minutes.[34,35] For example, testosterone acts rapidly to lower measures of anxiety-like behavior in male rats.[36] Very little is known about the rapid effects of testosterone on aggression, so this is clearly a topic with potential for new discoveries.

In humans, data linking testosterone and aggression are inconsistent.[37,38] However, it may be that previous studies have not taken into account the effects of reproductive or social experience on testosterone concentrations. In humans, testosterone levels fluctuate widely and are very sensitive to sexual[39] and agonistic experiences.[40] Even vicarious experiences, such as watching sports, can have large effects on testosterone concentrations.[31] Although it is possible that human aggressive behavior is insensitive to testosterone, an alternative possibility is that variation in behavior is more closely linked to peak testosterone levels in response to specific social experiences. It has been suggested that future research should concentrate on testing whether hypotheses describing the relationship between testosterone and aggression in non-human animals can explain variation in human behavior.[41]

Estrogens

Androgens such as testosterone can be converted to estrogens by the aromatase enzyme, and many brain regions known to regulate aggression express aromatase.[42–44] Thus, in some cases the effects of testosterone on aggression may actually be mediated by estrogenic metabolites. Male aromatase knock-out mice show greatly reduced aggression levels in resident–intruder tests, and estradiol replacement therapy during the first 2 weeks post-partum restores normal aggression levels.[45,46] The creation of classical estrogen receptor knockout mice focused attention on the effects of estrogen receptor α (ERα) and estrogen receptor β (ERβ). Male–male aggressive behavior is reduced in ERα knockout mice,[47,48] and the number of ERα-positive cells in the several brain regions, including the bed nucleus of the stria terminalis and anterior hypothalamus, is positively correlated with aggression.[49] In contrast, deletion of the ERβ gene is generally associated with increased aggression,[50,51] although this effect was stronger in younger mice (12 weeks old) and absent in older adults (18 weeks old).[52]

In addition to age, there is evidence that photoperiod (day length) can influence how estrogen receptors function. In old field mice (*Peromyscus polionotus*), estradiol decreases aggression under long days (16 light : 8 dark) and increases aggression in short days (8L : 16D). This does not appear to be due to differential nuclear estrogen receptor expression. The selective estrogen receptor agonists propyl pyrazole triol (an ERα-selective agonist) and diarylpropionitrile (an ERβ-selective agonist) both had the same effects as estradiol, decreasing aggression in long days and increasing aggression in short days.[53] Photoperiod appears to affect the molecular mechanisms activated by estrogens. In long days, estrogen-dependent gene expression is increased in the bed nucleus of the stria terminalis. Furthermore estradiol acts rapidly to increase aggression under short days, but not long days.[53] This suggests that estrogens may increase aggression under short days via non-genomic mechanisms. Interestingly, in song sparrows estrogens also increase aggression during the non-breeding season.[54] Aggression also appears to be more tightly regulated in the non-breeding season. During the non-breeding season male aggressive displays decline rapidly after an encounter, whereas during the breeding season aggressive displays continue for several hours after an encounter.[55] These data from *Peromyscus* and sparrows suggest that rapid actions of estradiol in the brain could play an important role in fine-tuning levels of aggressive behavior. This also raises the question of how and where estrogens are synthesized, as there is substantial evidence that behavioral active estrogens are synthesized in the brain.

Neurosteroids

As discussed above, the brain is capable of metabolizing steroid precursors (e.g., converting androgens to estrogens). However, the brain also possesses the capability of synthesizing neurosteroids, or steroids produced *de novo* in the brain from cholesterol (Fig. 22.2).[56] The ability of the brain to produce steroids independently of peripheral glands has important

BOX 22.2

HOW DO WE KNOW HOW TO IDENTIFY NEUROSTEROIDS IN BRAIN TISSUE?

The identification of neurosteroids in a given brain region can be accomplished at multiple levels, but there are special challenges to detecting steroids in brain tissue. Brain samples have high lipid content, which can interfere with the extraction of steroids. This obstacle can be overcome by using solid phase extractions followed by radioimmunoassay.[1] An alternative approach is to sample extracellular fluid via microdialysis, and measure steroids via radio- or enzyme immunoassays. Although technically challenging, this method has a unique advantage in that steroid production can be tracked during dynamic behavioral interactions.[2] To determine whether steroids in the brain are synthesized locally, the relevant steroid synthesizing enzymes must be examined. This can be accomplished using real-time PCR to determine the level of mRNA expression or Western blotting to determine the level of protein. However, the presence of enzymes alone does not constitute evidence of steroid production. Recent discoveries have illustrated how modifications such as phosphorylation can regulate the activity of enzymes such as aromatase.[3] Enzyme activity assays examine how brain homogenate samples convert a radiolabeled substrate, such as androstenedione, to other behaviorally active steroids such as androgens and estrogens.[4,5] To determine cause and effect relationships, neurosteroids need to be experimentally manipulated. In some cases it is possible to use peripheral manipulations to gain insights into the actions of neurosteroids. For example, estradiol levels in California mice are virtually undetectable,[6] so behavioral effects of peripheral injections of estradiol are most likely activating mechanisms regulated by centrally produced estradiol.[7] Ideally, manipulations of neurosteroids should be anatomically specific.[8] By combining brain hormone measurements, enzyme activity assays and functional manipulations, it is possible to obtain a comprehensive picture of how neurosteroids regulate behavior.

References

1. Newman AEM, Chin EH, Schmidt KL, Bond L, Wynne-Edwards KE, Soma KK. Analysis of steroids in songbird plasma and brain by coupling solid phase extraction to radioimmunoassay. *Gen Comp Endocrinol.* 2008;155:503–510.
2. Remage-Healey L, Maidment NT, Schlinger BA. Forebrain steroid levels fluctuate rapidly during social interactions. *Nat Neurosci.* 2008;11:1327–1334.
3. Balthazart J, Cornil CA, Taziaux M, Charlier TD, Baillien M, Ball GF. Rapid changes in production and behavioral action of estrogens. *Neuroscience.* 2006;138:783–791.
4. Roselli CE, Horton LE, Resko JA. Distribution and regulation of aromatase activity in the rat hypothalamus and limbic system. *Endocrinology.* 1985;117:2471–2477.
5. Schlinger BA, Arnold AP. Brain is the major site of estrogen synthesis in a male songbird. *Proc Natl Acad Sci USA.* 1991;88:4191–4194.
6. Trainor BC, Marler CA. Testosterone promotes paternal behaviour in a monogamous mammal via conversion to oestrogen. *Proc R Soc Lond B Biol Sci.* 2002;269:823–829.
7. Trainor BC, Finy MS, Nelson RJ. Rapid effects of estradiol on male aggression depend on photoperiod in reproductively non-responsive mice. *Horm Behav.* 2008;53:192–199.
8. Frye CA, Walf AA. Changes in progesterone metabolites in hippocampus can modulate open field and forced swim test behavior of proestrous rats. *Horm Behav.* 2002;41:306–315.

consequences for behavior, as it allows for steroid-sensitive pathways of behavior to become unlinked from other processes, such as reproduction.[57] Two of the most abundant steroids in the brain are dehydroepiandrosterone (DHEA) (see below) and allopregnenalone, and there is growing evidence that a significant portion of allopregnenalone is synthesized within the brain. Overall, it appears that neurosteroids are produced primarily in neurons. For example, in the hippocampus neurons express P450 side-chain cleavage (the enzyme needed to convert cholesterol to pregnenolone),[58] as well as the enzymes needed to convert pregnenolone to estradiol.[59] However, in certain cases the expression of steroid-synthesizing enzymes can be induced in glial cells.[60,61]

Although neurosteroids can bind directly to steroid receptors, others can interact directly with neurotransmitter receptors. One of the best described interactions is between allopregnanolone and the $GABA_A$ receptors,[62] with allopregnanolone inhibiting neuronal activity by selectively enhancing $GABA_A$ receptor activity. In rats, treatment with exogenous allopregnanolone increases aggression at moderate doses.[63] However, the effects of allopregnanolone on aggression can be substantially different in the presence of drugs of abuse, such as alcohol. For example, in rats that respond to alcohol with exaggerated aggression, the same doses of allopregnanolone that increase aggression in the absence of alcohol reduce aggression in the presence of alcohol.[63] The $GABA_A$ receptors also

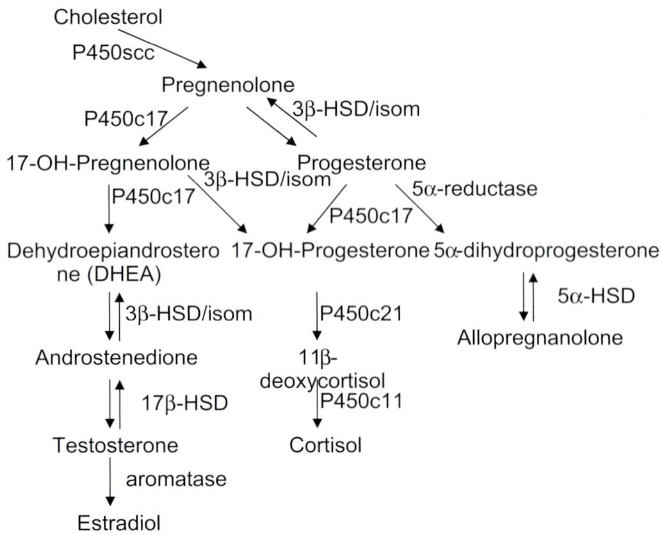

FIGURE 22.2 Synthesis pathways for neurosteroids.

appear to be targets of anabolic androgenic steroids (AAS), which are widely abused by professional athletes as well as adolescents. Studies in rats indicate that acute administration of AAS is associated with allosteric enhancement of $GABA_A$ function in the ventromedial hypothalamus.[64] In contrast, chronic treatment with high doses of AAS alters the composition of the $GABA_A$ receptor, specifically reducing expression of α subunits.[65] These patterns might help explain how abuse of AAS is reported initially to decrease anxiety and increase sense of well–being, whereas after chronic abuse there is an increase in anxiety and aggression.[66]

DHEA

Dehydroepiandrosterone (DHEA) is an endogenous steroid that circulates at relatively high concentrations among humans. DHEA is the most abundant androgen produced by adrenal glands, but it is also synthesized *de novo* in the brain, and thus can serve as a precursor for other steroids in the brain.[67] DHEA has been reported to have numerous effects on aggression. Treatment of male mice with DHEA reduced aggression towards a pregnant female.[68,69] DHEA has the potential to be converted to testosterone, so it was originally assumed that any behavioral effects of DHEA must be mediated by subsequent conversion to androgens or estrogens. However, a synthetic form of DHEA that cannot be converted to androgens was found to reduce male aggression directed at a pregnant female,[68] suggesting that DHEA can affect behavior independently of steroid receptors. Although a specific DHEA receptor has not been identified, DHEA can modulate the effects of $GABA_A$ and NMDA receptors.[70] This could potentially explain some of the androgen-independent actions of DHEA.

As noted above, song sparrows (*Melospiza melodia morphna*) are highly aggressive outside of their breeding season, even though their testes are regressed and their circulating concentrations of androgens and estrogens are basal.[71] However, DHEA circulates at detectable concentrations during the non-breeding season. Treatment with this hormone provoked increased aggressive interactions in wild non-breeding male sparrows. Similarly, in the spotted antbird (*Hylophylax n. naevioides*), a tropical bird that shows territorial aggression year-round in Panama, plasma testosterone is undetectable throughout the year. In males, blood concentrations of DHEA were correlated with aggressive encounters and vocalizations.[72] It was hypothesized that DHEA provides a steroidal precursor to *de novo* steroid production in the brain to regulate aggressive behaviors in birds that must defend resources outside of the breeding season.

In seasonally-breeding rodents such as Siberian hamsters (*Phodopus sungorus*), DHEA may regulate aggression in the non-breeding season when circulating androgens are basal as well. Treatment of Siberian hamsters with exogenous DHEA did not affect male aggression.[22] However, assessment of circulating DHEA and testosterone revealed that at night, but not during the day, aggressive interactions rapidly decreased serum DHEA and increased serum testosterone.[73] It remains possible that changes in DHEA influence local steroid concentrations in neurons that are part of the neural circuitry regulating aggression.

In humans, the relationships between DHEA and aggression has been studied in juvenile populations and, less so, in adults. In peri-pubertal children, severity of aggression in conduct disorder is positively correlated to circulating concentrations of DHEA, but not testosterone.[74] Another study reported that aggressive behaviors in girls diagnosed with conduct disorder are negatively correlated with the ratio of cortisol to DHEA.[75] One possible mechanism for this correlation is a shift in the adrenal gland from the production of cortisol to DHEA. This hypothesis is supported by data from studies on congenital adrenal hyperplasia (CAH), which is most commonly caused by a defect in the P450 21-hydroxylase enzyme and results in a loss of cortisol synthesis and an increase in androgens, including DHEA. Girls diagnosed with CAH show a positive association between aggressive behaviors and elevated DHEA.[76] A non-clinical sample of 5-year-old boys and girls detected no relationship between DHEA and aggressive behaviors,[77] which suggests that a link between DHEA and aggressive behaviors is limited to certain populations. Additional work is necessary to uncover the role of DHEA in aggressive behavior.

Glucocorticoids

Similar to androgens, the relationship between glucocorticoids and aggressive behaviors is complex.[78] Generally, chronic exposure to stressors, exogenous treatment with glucocorticoids, or pathological increase in circulating glucocorticoid concentrations inhibits aggressive behaviors,[79–81] whereas chronic deficiencies in glucocorticoid concentrations are associated with elevated aggression.[82] Sustained high glucocorticoid concentrations reduce testosterone secretion, which could contribute to reduced aggression. However, the acute effect of a transient increase in glucocorticoids on aggression differs diametrically from a chronic increase in baseline concentrations. For example, an acute injection of corticosterone was found to cause a rapid increase in aggression in rats.[83] Elevated aggression in response to acutely elevated corticosterone is particularly salient in challenge situations – for example, when confronted with unfamiliar opponents or other novel situations.[84] In roughskinned newts (*Taricha granulosa*), corticosterone acts rapidly to promote mating behavior[85] – an effect that has been linked to non-genomic hormone action.[86] This raises the possibility that the effects of chronic elevation of glucocorticoids on aggressive behavior are mediated by changes in gene expression (via activation of mineralocorticoid and glucocorticoid receptors), whereas the effects of a transient increase in glucocorticoids may be mediated by non-genomic responses of membrane receptors, as described above for estrogens and its receptors.

Environmental factors can also affect the development of aggressive phenotypes. For example, social interactions in the form of juvenile play appear to buffer against heightened aggressive responses to provocation in Syrian hamsters, and other species as well.[5] During puberty in male Syrian hamsters, the frequency of attacks diminishes and "play fighting" morphs into the adult version of aggression.[87] Prior to puberty (from 20–40 days of age) hamsters engage in agonistic behavior mainly through play behavior, which is characterized by attacking the face and cheeks of their opponents.[88] From about 40–50 days of age (mid-puberty), males undergo a transitional period of behavior during which the attacks are aimed at the flanks. Finally, as they emerge from the pubertal period, males perform adult-like offensive aggressive responses, which are characterized by attacks to the underbelly and rear end.[88] Exposure to social stress during puberty alters adult aggressive behavior in at least two ways. Repeated exposure of male peripubertal hamsters to aggressive adults hastened the onset of adult-like offensive, but not defensive, aggression. If the subjugation occurred early in puberty, then the hamsters were most likely to be aggressive, especially towards smaller opponents, post-puberty.[88] Furthermore, repeated social stress slowed the reduction in attack frequency normally seen during puberty. In contrast to males, female Syrian hamsters displayed a stable frequency of attacks through puberty and into adulthood, there was no obvious mid-puberty transitional period, and cortisol habituated to repeated exposures to aggressive adults.[89]

Male hamsters do not appear to habituate to repeated exposure to social subjugation.[87] After 2 weeks of repeated exposure to aggressive adults, male hamsters persistently displayed elevated plasma cortisol concentrations – an endocrine marker of the maturation of agonistic behavior from play fighting to adult-like attacks.[88] Although castration has well-known effects on aggression in adults, castration of peripubertal hamsters does not affect the transition from play fighting to adult-like fighting. The maturation of aggressive behavior is also correlated with maturation of the HPA axis, or adrenarche, during puberty, and activity of the HPA axis appears to control pre-adult agonistic behavior.[87] Hamsters exposed to adult aggression begin to show adult-like aggressive behavior earlier in life. These animals undergo normal gonadal development, but they are also very aggressive towards smaller and younger individuals.[88] In effect, they become bullies.

As mentioned, in most encounters of matched combatants the resident has a "home-field" advantage. After defeat in the home cage of an aggressive conspecific, male hamsters (*Mesocricetus auratus*) will subsequently fail to defend their home territory even if the intruder is a smaller, non-aggressive male.[90] This phenomenon has been called conditioned defeat, and appears to evoke a stress response via fear conditioning. The physiological effects of defeat include elevated HPA axis activity, such as increased plasma ACTH, β-endorphin, cortisol and corticosterone concentrations, as well as decreased plasma testosterone and prolactin concentrations.[91,92] This endocrine profile is observed among previously defeated hamsters upon re-exposure to another animal, even when the new opponent is blocked by a physical barrier. The physiological and behavioral consequences of conditioned social defeat persist for at least 33 days,[93] and perhaps throughout adulthood.[87] Few female hamsters exhibit conditioned social defeat, although ACTH concentrations were reduced in those females that displayed low levels of submissive/defensive behavior.[93] In contrast to males, the conditioned defeat response did not persist beyond the first test among female hamsters. These results suggest that in male hamsters conditioned defeat is a profound, persistent behavioral change characterized by a total absence of territorial aggression and by the frequent display of submissive and defensive behaviors.[93]

EFFECTS OF PROTEIN AND PEPTIDE HORMONES ON AGGRESSIVE BEHAVIOR

Vasopressin

The neuropeptides vasopressin and oxytocin have complex effects on aggressive behaviors.[94] Vasopressin neurons in the bed nucleus of the stria terminalis (BNST) and medial amygdala send projections to the lateral septum.[95] There are also dense vasopressin fibers in the anterior hypothalamus,[96] although the origins of these fibers are less clear. The effects of these vasopressin projections are site-specific, and vary across species. In the anterior hypothalamus, injections of vasopressin increase aggression in male hamsters,[97] and injections of a vasopressin V1A receptor antagonist inhibit aggression.[98,99] In the lateral septum the effects of vasopressin manipulations are more variable. Infusion of vasopressin into the lateral septum increased male aggression in hamsters,[99] whereas vasopressin antagonists and agonists infused into the lateral septum had no effect on aggression in male rats.[100] It is possible that differences in the social systems of these species may contribute to the contrasting results, as there is strong evidence that the effects of vasopressin on aggression depend on social systems. In colonial zebra finches, infusions of vasotocin (the avian homolog to vasopressin) into the lateral septum increased aggressive behaviors, but these same infusions decreased aggression in territorial field sparrows and violet-eared waxbills.[101–103] The V1B receptor has been more difficult to study, because V1B-selective antagonists have only become available recently. However, the selective deletion of the V1B gene in mice resulted in a substantial decrease in aggression both in resident–intruder tests and in neutral aggression tests.[104]

In humans, measurements of vasopressin in cerebrospinal fluid have been used to estimate central vasopressin activity. One study used the Life History of Aggression survey to estimate aggression, which is based on a semistructured interview with a clinician.[105] Vasopressin in CSF was positively correlated with aggression scores. The hypothesis that vasopressin facilitates aggressive behaviors was tested in a later study that administered exogenous vasopressin intranasally (which allows access to the brain). The effect of vasopressin on behavioral responses to pictures of neutral faces was sex-specific, increasing agonistic facial motor patterns in men and decreasing these same responses in women.[106] Although increased aggression *per se* was not observed, these data suggest that vasopressinergic activity could modulate behavior responses that could be associated with aggressive behaviors in a more naturalistic context.

Oxytocin

There is strong evidence for effects of oxytocin on aggression in the context of maternal defense. Oxytocin was measured, using microdialysis, in the paraventricular nucleus and central nucleus of the amygdala during maternal aggression tests. Oxytocin release in these brain regions increased during maternal aggression tests, and individual variation in aggression levels was positively correlated with oxytocin release in both regions.[107] A causal relationship between oxytocin and maternal aggression is established by data showing that oxytocin antagonists infused into the brain inhibit maternal aggression.[107,108] In males, the effects of oxytocin appear to be more subtle. Male oxytocin knockout mice fight as frequently as wild-type mice, although the duration of aggressive encounters is reduced.[109] It is possible that oxytocin could affect aggressive behavior indirectly by modulating processes such as social recognition or anxiety. In humans, CSF concentrations were oxytocin negatively correlated with scores on the Lifetime History of Aggression survey.[110] These data are consistent with findings from studies that reported that exogenous oxytocin increases the expression of pro-social behaviors,[111,112] which could reduce the probability that aggressive behaviors are expressed.

Prolactin

Prolactin has been reported to influence aggression in maternal aggression or so-called nest defense. For example, exogenous treatment of ovariectomized and adrenalectomized females with prolactin increased aggressive behaviors in *Peromyscus leucopus*.[113] Maternal aggression and maternal care are closely related in their timing of onset, and share a similar endocrine basis in rats. However, in rats[114] and mice[115] maternal care appears to require prolactin and reduced circulating progesterone values, whereas maternal aggression appears to be independent of pituitary hormones and progesterone concentrations. Neurons outside of the pituitary also produce prolactin, so the role of centrally-produced prolactin in maternal aggression of mice and rats cannot be completely excluded.[116] In women, high levels of circulating prolactin were significantly correlated with reduced anxiety and aggression variables.[117] Similarly, rat dams that engaged in aggressive behaviors displayed reduced circulating prolactin levels; prolactin receptor numbers in the MPOA did not change.[118]

CONCLUSIONS

This brief review should make it apparent that many neuroendocrine systems have likely evolved in vertebrates, including humans, to regulate species-specific aggressive behaviors. Thus, despite the differences in situational contexts and behavioral outputs between mice and humans, similar neuroendocrine and neuroanatomical pathways are likely activated. A number of difficult questions remain to be answered. For example, to what extent do different environmental factors, including social stressors, influence the development of these neuroendocrine and neuroanatomical pathways, and to what extent are they activated by observing aggression? Considerable debate has arisen in the media over the effects of violence on aggression, and many confounding factors make it difficult to study these putative effects. Recent studies have demonstrated that aggression is increased in animals that observe conflicts among other individuals.[119,120] Similar data show that vicarious experiences such as watching sports can have important biological effects in humans, such as increasing or decreasing testosterone.[18] Going forward, studies examining the relationship between hormones and aggression need to consider carefully the environmental and social contexts in which data are collected.

References

1. Soma KK. Testosterone and aggression: Berthold, birds, and beyond. *J Neuroendocrinol.* 2006;18:543–551.
2. Nelson RJ, Chiavegatto S. Molecular basis of aggression. *Trends Neurosci.* 2001;24:713–719.
3. Nelson RJ, Trainor BC. Neural mechanisms of aggression. *Nat Rev Neurosci.* 2007;8:536–546.
4. Rader K. *Making Mice: Standardizing Animals for American Biomedical Research, 1900-1955.* Princeton: Princeton University Press; 2004.
5. Trainor BC, Sisk CL, Nelson RJ. 2009. Hormones and the development and expression of aggressive behavior. In: Pfaff DW, et al., eds. *Hormones, Brain, and Behavior.* San Diego: Academic Press; 2009:167–203.
6. Newman S. The medial extended amygdala in male reproductive behavior. A node in the mammalian social behavior network. *Ann NY Acad Sci.* 1999;877:242–257.
7. Goodson JL. The vertebrate social behavior network: evolutionary themes and variations. *Horm Behav.* 2005;48:11–22.
8. DaVanzo JP, Sydow M, Garris DR. Influence of isolation and training on fighting in mice with olfactory bulb lesions. *Physiol Behav.* 1983;31:857–860.
9. Delville Y, De Vries GJ, Ferris CF. Neural connections of the anterior hypothalamus and agonistic behavior in golden hamsters. *Brain Behav Evol.* 2000;55:53–76.
10. Simerly R, Chang C, Muramatsu M, Swanson L. Distribution of androgen and estrogen receptor mRNA-containing cells in the rat brain: an *in situ* hybridization study. *J Comp Neurol.* 1990;294:76–95.
11. Mitra SW, Hoskin E, Yudkovitz J, Pear L, Wilkinson HA, Hayashi S, Pfaff DW, Ogawa S, Roher SP, Schaefer JM, McEwen BS, Alves SE. Immunolocalization of estrogen receptor beta in the mouse brain: Comparison with estrogen receptor alpha. *Endocrinology.* 2003;144:2055–2067.
12. Fuxjager MJ, Forbes-Lorman RM, Coss DJ, Auger CJ, Auger AP, Marler CA. Winning territorial disputes selectively enhances androgen sensitivity in neural pathways related to motivation and social aggression. *Proc Natl Acad Sci USA.* 2010;107:12393–12398.
13. Finley SK, Kritzer MF. Immunoreactivity for intracellular androgen receptors in identified subpopulations of neurons, astrocytes and oligodendrocytes in primate prefrontal cortex. *J Neurobiol.* 1999;40:446–457.
14. Weiss SL, Moore MC. Activation of aggressive behavior by progesterone and testosterone in male tree lizards, Urosaurus ornatus. *Gen Comp Endocrinol.* 2004;136:282–288.
15. Arthur PA. The effects of castration and androgen replacement on song, courtship, and aggression in zebra finches. *Poephila guttata J Exp Zool.* 1975;191:309–325.
16. Francis RC, Jacobson B, Wingfield JC, Fernald RD. Castration lowers aggression but not social-dominance in male Haplochromis burtoni (Cichlidae). *Ethology.* 1992;90:247.
17. Barkley MS, Goldman BD. The effects of castration and silastic implants of testosterone on intermal aggression in the mouse. *Horm Behav.* 1977;9:32–48.
18. Lincoln GA, Guinness F, Short RV. The way in which testosterone controls the social and sexual behavior of the Red Deer stag (*Cervus Elaphus*). *Horm Behav.* 1972;3:375–396.
19. Trainor BC, Marler CA. Testosterone, paternal behavior, and aggression in the monogamous California mouse (*Peromyscus californicus*). *Horm Behav.* 2001;40:32–42.
20. Demas GE, Moffatt CA, Drazen DL, Nelson RJ. Castration does not inhibit aggressive behavior in adult male prairie voles (*Microtus ochrogaster*). *Physiol Behav.* 1999;66:59–62.
21. Caldwell GS, Glickman SE, Smith ER. Seasonal aggression independent of seasonal testosterone in wood rats. *Proc Natl Acad Sci USA.* 1984;81:5255–5257.
22. Scotti MA, Belen J, Jackson JE, Demas GE. The role of androgens in the mediation of seasonal territorial aggression in male Siberian hamsters (*Phodopus sungourus*). *Physiol Behav.* 2008;95:633–640.
23. Wingfield JC. Regulation of territorial behavior in the sedentary song sparrow. *Melospiza melodia morphna Horm Behav.* 1994; 28:1–15.
24. Wingfield JC, Hegner RE, Dufty Jr AM, Ball GF. The 'challenge hypothesis': Theoretical implications for patterns of testosterone secretion, mating systems, and breeding strategies. *American Naturalist.* 1990;136:829–846.
25. Marler CA, Moore MC. Time and energy costs of aggression in testosterone-implanted free-living male mountain spiny lizards (*Sceloporus jarrovi*). *Physiol Zool.* 1989;62:1334–1350.
26. Ashley NT, Hays QR, Bentley GE, Wingfield JC. Testosterone treatment diminishes sickness behavior in male songbirds. *Horm Behav.* 2009;56:169–176.
27. Hirschenhauser K, Oliveira RF. Social modulation of androgens in male vertebrates: meta-analyses of the challenge hypothesis. *Anim Behav.* 2006;71:265–277.
28. Cavigelli SA, Pereira ME. Mating season aggression and fecal testosterone levels in male ring-tailed lemurs (*Lemur catta*). *Horm Behav.* 2000;37:246–255.
29. Gleason ED, Fuxjager MJ, Oyegbile TO, Marler CA. Testosterone release and social context: When it occurs and why. *Front Neuroendocrinol.* 2009;30:460–469.
30. Goymann W, Landys MM, Wingfield JC. Distinguishing seasonal androgen responses from male-male androgen responsiveness – revisiting the Challenge Hypothesis. *Horm Behav.* 2007;51:463–476.
31. Bernhardt PC, Dabbs JM, Fielden JA. Testosterone changes during vicarious experiences of winning and losing among fans at sporting events. *Physiol Behav.* 1998;65:59–62.
32. Oliveira RF, Lopos M, Carneiro LA, Canario AVM. Watching fights raises fish hormone levels. *Nature.* 2001;409:784.

33. Trainor BC, Bird IM, Marler CA. Opposing hormonal mechanisms of aggression revealed through short-lived testosterone manipulations and multiple winning experiences. *Horm Behav.* 2004;45:115–121.
34. Vasudevan N, Pfaff DW. Membrane initiated actions of estrogens in neuroendocrinology: emerging principles. *Endocrine Rev.* 2006;28:1–19.
35. Lord LD, Bond J, Thompson RR. Rapid steroid influences on visually guided sexual behavior in male goldfish. *Horm Behav.* 2009;56:519–526.
36. Edinger KL, Frye CA. Testosterone's anti-anxiety and analgesic effects may be due in part to actions of its 5α-reduced metabolites in the hippocampus. *Psychoneuroendocrinology.* 2005;30:418–430.
37. Archer J. The influence of testosterone on human aggression. *Br J Psychol.* 1991;82:1–28.
38. Albert DJ, Walsh ML, Jonik RH. Aggression in humans: what is its biological foundation? *Neurosci Biobehav Rev.* 1993;17:405–425.
39. Hirschenhauser K, Frigerio D, Grammer K, Magnusson MS. Monthly patterns of testosterone and behavior in prospective fathers. *Horm Behav.* 2002;42:172–181.
40. Pound N, Penton-Voak IS, Surridge AK. Testosterone responses to competition in men are related to facial masculinity. *Proc R Soc B.* 2009;276:153–159.
41. Archer J. Testosterone and human aggression: an evaluation of the challenge hypothesis. *Neurosci Biobehav Rev.* 2006;30:319–345.
42. Roselli CE, Horton LE, Resko JA. Distribution and regulation of aromatase activity in the rat hypothalamus and limbic system. *Endocrinology.* 1985;117:2471–2477.
43. Balthazart J, Foidart A, Surlemont C, Vockel A, Harada N. Distribution of aromatase in the brain of the Japanses quail, ring dove, and Zebra finch: An immunocytochemical study. *J Comp Neurol.* 1990;301:276–288.
44. Roselli CE, Resko JA. Cytochrome P450 aromatase (CYP19) in the non-human primate brain: distribution, regulation, and functional significance. *J Steroid Biochem Mol Biol.* 2001;79:247–253.
45. Toda K, Saibara T, Okada T, Onishi S, Shizuta Y. A loss of aggressive behaviour and its reinstatement by oestrogen in mice lacking the aromatase gene (Cyp19). *J Endocrinol.* 2001;168:217–220.
46. Matsumoto T, Honda S, Harada N. Alteration in sex-specific behaviors in male mice lacking the aromatase gene. *Neuroendocrinology.* 2003;77:416–424.
47. Ogawa S, Lubahn DB, Korach KS, Pfaff DW. Behavioral effects of estrogen receptor gene disruption in male mice. *Proc Natl Acad Sci USA.* 1997;94:1476–1481.
48. Scordalakes EM, Rissman EF. Aggression in male mice lacking functional estrogen receptor α. *Behav Neurosci.* 2003;117:38–45.
49. Trainor BC, Greiwe KM, Nelson RJ. Individual differences in estrogen receptor α in select brain nuclei are associated with individual differences in aggression. *Horm Behav.* 2006;50:338–345.
50. Nomura M, Andersson S, Korach KS, Gustafsson JÅ, Pfaff DW, Ogawa S. Estrogen receptor-b gene disruption potentiates estrogen-inducible aggression but not sexual behaviour in male mice. *Eur J Neurosci.* 2006;23:1860–1868.
51. Ogawa S, Chan J, Chester AE, Gustafsson J, Korach KS, Pfaff DW. Survival of reproductive behaviors in estrogen receptor beta gene-deficient (βERKO) male and female mice. *Proc Natl Acad Sci USA.* 1999;96:12887–12892.
52. Nomura M, Durbak I, Chan J, Gustafsson JA, Smithies O, Korach KS, Pfaff DW, Ogawa S. Genotype/age interactions on aggressive behavior in gonadally intact estrogen receptor beta knockout (βERKO) male mice. *Horm Behav.* 2002;41:288–296.
53. Trainor BC, Lin S, Finy MS, Rowland MR, Nelson RJ. Photoperiod reverses the effects of estrogens on male aggression via genomic and non-genomic pathways. *Proc Natl Acad Sci USA.* 2007;104:9840–9845.
54. Soma KK, Tramontin AD, Wingfield JC. Oestrogen regulates male aggression in the non-breeding season. *Proc R Soc Lond B Biol Sci.* 2000;267:1089–1096.
55. Soma KK, Wingfield JC. Spring and autumn territoriality in song sparrows: same behavior, different mechanisms? *Integ Comp Biol.* 2002;42:11–20.
56. Baulieu E-E, Robel P. Neurosteroids: a new brain function? *J Steroid Biochem Mol Biol.* 1990;37:395–403.
57. Schmidt KL, Pradhan DS, Shah AH, Charlier TD, Chin EH, Soma KK. Neurosteroids, immunosteroids, and the Balkanization of endocrinology. *Gen Comp Endocrinol.* 2008;157:266–274.
58. Kimoto T, Tsurugizawa T, Ohta Y, Makino Jy, Tamura H-o, Hojo Y, Takata N, Kawato S. Neurosteroid synthesis by cytochrome P450-containing systems localized in the rat brain hippocampal neurons: N-Methyl-D-aspartate and calcium-dependent synthesis. *Endocrinology.* 2001;142:3578–3589.
59. Hojo Y, Hattori T, Enami T, Furukawa A, Suzuki S, Ishii H, Mukai H, Morrison JH, Janssen WGM, Kominami S, Harada N, Kimoto T, Kawato S. Adult male rat hippocampus synthesizes estradiol from pregnenolone by cytochromes P45017a and P450 aromatase localized in neurons. *Proc Natl Acad Sci USA.* 2004;101:865–870.
60. Garcia-Segura LM, Wozniak A, Azcoitia I, Rodriguez JR, Hutchison RE, Hutchison JB. Aromatase expression by astrocytes after brain injury: implications for local estrogen formation in brain repair. *Neuroscience.* 1999;89:567–578.
61. Peterson RS, Saldanha CJ, Schlinger BA. Rapid upregulation of aromatase mRNA and protein following neural injury in the zebra finch (*Taeniopygia guttata*). *J Neuroendocrinol.* 2001;13:317–323.
62. Belelli D, Herd MB, Mitchell EA, Peden DR, Vardy AW, Gentet L, Lambert JL. Neuroactive steroids and inhibitory neurotransmission: mechanisms of action and physiological relevance. *Neuroscience.* 2006;138:821–829.
63. Fish EW, Faccidomo S, DeBold JF, Miczek KA. Alcohol, allopregnanolone and aggression in mice. *Psychopharmacology.* 2001;153:473–483.
64. Jorge-Rivera JC, McIntyre KL, Henderson LP. Anabolic steroids induce region and subunit-specific modulation of $GABA_A$ receptor mediated currents in the rat forebrain. *J Neurophysiol.* 2000;83:3299–3309.
65. McIntyre KL, Porter DM, Henderson LP. Anabolic androgenic steroids induce age-, sex-, and dose-dependent changes in $GABA_A$ receptor subunit mRNAs in the mouse forebrain. *Neuropharmacology.* 2002;43:634–645.
66. Clark AS, Henderson LP. Behavioral and physiological responses to anabolic-androgenic steroids. *Neurosci Biobehav Rev.* 2003;27:413–436.
67. Soma KK, Scotti M-AL, Newman AEM, Charlier TD, Demas GE. Novel mechanisms for neuroendocrine regulation of aggression. *Front Neuroendocrinol.* 2008;29:476–489.
68. Haug M, Ouss-Schlegel ML, Spetz JF, Brain PF, Simon V, Baulieu EE, Robel P. Suppressive effects of dehydroepiandrosterone and 3-beta-methylandrost-5-en-17-one on attack towards lactating female intruders by castrated male mice. *Physiol Behav.* 1989;46:955–959.
69. Haug M, Spetz JF, Schlegel ML, Robel P. Dehydroepiandrosterone inhibits aggressive behavior in castrated male mice.

Comptes Rendus Acad Sci Ser. III—Sci. Vie-Life Sci. 1983;296: 975–977.
70. Maninger N, Wolkowitz OM, Reus VI, Eperl ES, Mellon SH. Neurobiological and neuropsychiatric effects of dehydroepiandrosterone (DHEA) and DHEA sulfate (DHEAS). *Front Neuroendocrinol.* 2008;30:65–91.
71. Soma KK, Wissman AM, Brenowitz EA, Wingfield JC. Dehydroepiandrosterone (DHEA) increases territorial song and the size of an associated brain region in a male songbird. *Horm Behav.* 2002;41:203–212.
72. Hau M, Stoddard ST, Soma KK. Territorial aggression and hormones during the non-breeding season in a tropical bird. *Horm Behav.* 2004;45:40–49.
73. Scotti MA, Schmidt KL, Newman AE, Bonu T, Soma KK, Demas GE. Aggressive encounters differentially affect serum dehydroepiandrosterone and testosterone concentrations in male Siberian hamsters (*Phodopus sungorus*). *Horm Behav.* 2009;56:376–381.
74. Van Goozen SHM, Matthys W, Cohen-Ketterus PT, Thijssen JHH, van Engeland H. Adrenal androgens and aggression in conduct disorder prepubertal boys and normal controls. *Biol Psychiatry.* 1998;43:156–158.
75. Pajer K, Tabbah R, Gardner W, Rubin RT, Czambel RK, Wang Y. Adrenal androgen and gonadal hormone levels in adolescent girls with conduct disorder. *Psychoneuroendocrinology.* 2006;31:1245–1256.
76. Herzog AG, Edelheit PB, Jacobs AR. Low salivary cortisol levels and aggressive behavior. *Arch Gen Psychiatry.* 2001;58:513–514.
77. Azurmendi A, Braza F, Garcia A, Braza P, Munoz JM, Sanchez-Martin JR. Aggression, dominance, and affiliation: their relationships with androgen levels and intelligence in 5-year-old children. *Horm Behav.* 2006;50:132–140.
78. Demas GE, Cooper MA, Albers HE, Soma KK. 2007. Novel mechanisms underlying neuroendocrine regulation of aggression: a synthesis of rodent, avian, and primate studies. In: Blaustein J, ed. *Handbook of Neurochemistry and Molecular Neurobiology.* New York: Springer; 2007:338–372.
79. Maestripieri D, Badiani A, Puglisi-Allegra S. Prepartal chronic stress increases anxiety and decreases aggression in lactating female mice. *Behav Neurosci.* 1991;105:663–668.
80. Leshner AI, Korn SJ, Mixon JF, Rosenthal C, Besser AK. Effects of corticosterone on submissiveness in mice: some temporal and theoretical considerations. *Physiol Behav.* 1980;24:283–288.
81. Summers CH, Watt MJ, Ling TL, Forster GL, Carpenter RE, Korzan WJ, Lukkes JL, Overli O. Glucocorticoid interaction with aggression in non-mammalian vertebrates: reciprocal action. *Eur J Neurosci.* 2005;526:21–35.
82. Haller J, Halasz J, Mikics E, Kruk MR. Chronic glucocorticoid deficiency-induced abnormal aggression, autonomic hypoarousal, and social deficit in rats. *J Neuroendocrinol.* 2004;16:550–557.
83. Mikics E, Kruk MR, Haller J. Genomic and non-genomic effects of glucocorticoids on aggressive behavior in male rats. *Psychoneuroendocrinology.* 2004;29:618–635.
84. Mikics E, Barsy B, Haller J. The effect [sic] glucocorticoids on aggressiveness in established colonies of rats. *Psychoneuroendocrinology.* 2007;32:160–170.
85. Moore FL, Miller LJ. Stress-induced inhibition of sexual behavior: corticosterone inhibits courtship behaviors of a male amphibian (*Taricha granulosa*). *Horm Behav.* 1984;18:400–410.
86. Orchinik M, Murray TF, Moore FL. A corticosteroid receptor in neuronal membranes. *Science.* 1991;252:1848–1851.
87. Delville Y, Newman ML, Wommack JC, Taravosh-Lahn K, Cervantes MC. 2005. Developlment of aggression. In: Nelson RJ, ed. *Biology of Aggression.* New York: Oxford University Press; 2005:327–350.
88. Wommack JC, Taravosh-Lahn K, David JT, Delville Y. Repeated exposure to social stress alters the development of agonistic behavior in male golden hamsters. *Horm Behav.* 2003;43:229–236.
89. Taravosh-Lahn K, Delville Y. Aggressive behavior in female golden hamsters: development and the effect of repeated social stress. *Horm Behav.* 2004;46:428–435.
90. Huhman KL, Jasnow AM. Conditioned defeat. In: Nelson RJ, ed. *Biology of Aggression.* Oxford: Oxford University Press; 2005:295–326.
91. Huhman KL, Bunnell BN, Mougey EH, Meyerhoff JL. Effects of social conflict on POMC-derived peptides and glucocorticoids in male golden hamsters. *Physiol Behav.* 1990;47:949–956.
92. Huhman KL, Moore TO, Ferris CF, Mougey EH, Meyerhoff JL. Acute and repeated exposure to social conflict in male golden hamsters: increases in plasma POMC peptide and cortisol and decreases in plasma testosterone. *Horm Behav.* 1991;25:206–216.
93. Huhman KL, Solomon MB, Janicki M, Harmon AC, Lin SM, Israel JE, Jasnow AM. Conditioned defeat in male and female Syrian hamsters. *Horm Behav.* 2003;44:293–299.
94. Veenema AH, Neumann ID. Central vasopressin and oxytocin release: regulation of complex social behaviours. *Prog Brain Res.* 2008;170:261–276.
95. De Vries GJ, Panzica GC. Sexual differentiation of central vasopressin and vasotocin systems in vertebrates: different mechanisms and similar endpoints. *Neuroscience.* 2006;138:947–955.
96. Gobrogge KL, Liu Y, Jia X, Wang Z. Anterior hypothalamic neural activation and neurochemical associations iwth aggression in pair-bonded male prairie voles. *J Comp Neurol.* 2007;502:1109–1122.
97. Ferris CF, Albers HE, Wesolowski SM, Goldman BD, Luman SE. Vasopressin injected into the hypothalamus triggers a stereotypic behavior in godlen hamsters. *Science.* 1984;224:521–523.
98. Ferris CF, Potegal M. Vasopressin receptor blockade in the anterior hypothalamus suppresses aggression in hamsters. *Physiol Behav.* 1988;44:235–239.
99. Ferris CF, Delville Y. Vasopressin and serotonin interactions in the control of agonistic behavior. *Psychoneuroendocrinology.* 1994;19(5-7):593–601.
100. Beiderbeck DI, Neumann ID, Veenema AH. Differences in intermale aggression are accompanied by opposite vasopressin release patterns within the septum in rats bred for low and high anxiety. *Eur J Neurosci.* 2007;26:3597–3605.
101. Goodson JL. Vasotocin and vasoactive intestinal polypeptide modulate aggression in a territorial songbrid, the violet-eared waxbill (Estrildidae: *Uraeginthus granatina*). *Gen Comp Endocrinol.* 1998;111:233–244.
102. Goodson JL. Territorial aggression and dawn song are modulated by septal vasotocin and vasoactive intestinal polypeptide in male field sparrows (*Spizella pusilla*). *Horm Behav.* 1998;34:67–77.
103. Goodson JL, Adkins-Regan E. Effect of intraseptal vasotocin and vasoactive intestinal polypeptide infusions on courtship song and aggression in the male zebra finch (*Taenipygia guttata*). *J Neuroendocrinol.* 1999;11:19–25.
104. Wersinger SR, Ginns EI, O'Carroll AM, Lolait SJ, Young WSI. Vasopressin V1b receptor knockout reduces aggressive behavior in male mice. *Mol Psychiatry.* 2002;7:975–984.
105. Coccaro EF, Kavoussi RJ, Hauger RL, Cooper TB, Ferris CF. Cerebrospinal fluid vasopressin levels – Correlates with aggression and serotonin function in personality-disordered subjects. *Arch Gen Psychiatry.* 1998;55:708–714.
106. Thompson RR, George K, Walton JC, Orr SP, Benson J. Sex-specific influences of vasopressin on human social communication. *Proc Natl Acad Sci USA.* 2006;103:7889–7894.
107. Bosch OJ, Meddle SL, Beiderbeck DI, Douglas AJ, Neumann ID. Brain oxytocin correlates with maternal aggression: Link to anxiety. *J Neurosci.* 2005;25:6807–6815.

108. Ferris CF, Foote KB, Meltser HM, Plenby MG, Smith KL, Insel TR. Oxytocin in the amygdala facilitates maternal aggression. *Ann NY Acad Sci.* 1992;652:456–457.
109. DeVries AC, Young SW, Nelson RJ. Reduced aggressive behavior in mice with targeted disruption of the oxytocin gene. *J Neuroendocrinol.* 1997;9:363–368.
110. Lee R, Ferris C, Van de Kar LD, Coccaro EF. Cerebrospinal fluid oxytocin, life history of aggression, and personality disorder. *Psychoneuroendocrinology.* 2009;34:1567–1573.
111. Kosfeld M, Heinrichs M, Zak PJ, Fischbacher U, Fehr E. Oxytocin increases trust in humans. *Nature.* 2005;435:673–676.
112. Domes G, Heinrichs M, Michel A, Berger C, Herpetz SC. Oxytocin improves "mind-reading" in humans. *Biol Psychiatry.* 2007;61:731–733.
113. Gleason PE, Michael SD, Christian JJ. Prolactin-induced aggression in female *Peromyscus leucopus*. *Behav Neural Biol.* 1981;33:243–248.
114. Rosenblatt JS, Factor EM, Mayer AD. Relationship between maternal aggression and maternal care in the rat. *Aggressive Behav.* 2006;20:243–255.
115. Broida J, Michael SD, Svare B. Plasma prolactin levels are not related to the initiation, maintenance, and decline of postpartum aggression in mice. *Behav Neural Biol.* 1981;32:121–125.
116. Lonstein JS, Gammie SC. Sensory, hormonal, and neural control of maternal aggression in laboratory rodents. *Neurosci Biobehav Rev.* 2002;26:869–888.
117. Uvnäs-Mobcrg K, Widström A-M, Nissen E, Björvell H. Personality traits in women 4 days postpartum and their correlation with plasma levels of oxytocin and prolactin. *J Psychosom Obst Gynecol.* 1990;11:261–273.
118. Consiglio AR, Bridges RS. Circulating prolactin, MPOA prolactin receptor expression and maternal aggression in lactating rats. *Behav Brain Res.* 2008;197:97–102.
119. Peake TM, Terry AMR, McGregor PK, Dabelsteen T. Do great tits assess rivals by combining direct experience with information gathered by eavesdropping? *Proc R Soc Lond B.* 2002;269:1925–1929.
120. Earley RL, Dugatkin LA. Eavesdropping on visual cues in green swordtail (*Xiphophorus helleri*) fights: a case for networking. *Proc R Soc Lond B.* 2002;269:943–952.

CHAPTER

23

Chronic Social Stress: Effects on Neuroendocrine Function

Karen A. Scott[1], *Kellie L.K. Tamashiro*[2], *Randall R. Sakai*[1]

[1] University of Cincinnati College of Medicine, Department of Psychiatry and Behavioral Neurosciences, Cincinnati, OH, USA, [2] The Johns Hopkins University School of Medicine, Department of Psychiatry and Behavioral Sciences, Baltimore, MD, USA

OUTLINE

Summary

Stress is experienced on a daily basis. While most individuals are able to adapt to acute challenges to homeostasis, chronic activation can lead to dysregulated or maladaptive responses that leave them more susceptible to disease. Social stress is one of the most salient stressors to humans, and several animal models have been developed to determine how chronic stress can increase susceptibility to or exacerbate existing cardiometabolic, immune, reproductive and/or affective disorders. In this chapter, we discuss the importance of neuroethologically valid animal models of social stress, what information these models have produced, and how they may relate to the human condition. Individual variation in vulnerability and resilience to social stress is of great interest, and has gained more attention in recent years. This has led to a more multidisciplinary approach which takes into account genetics, epigenetics and environmental factors that affect how an individual copes with and adapts to stress.

INTRODUCTION

Stress is a part of daily life. For humans, commonly encountered stressors tend to be psychological and of social nature.[1] Interpersonal conflict, hostile work environments, perception of low socioeconomic status, or lack of social support can result in social stress exposure, which can negatively affect health. Exposure to chronic stress is associated with a higher risk for affective disorders, impaired immune and reproductive function, and cardiometabolic dysfunction.[1,2] It is important, therefore, to identify mechanisms by which stress can contribute to the development of these disorders, and this may ultimately lead to new interventions to prevent or limit the severity of stress-related disease.

Handbook of Neuroendocrinology, DOI: 10.1016/B978-0-12-375097-6.10023-X

In the 1930s, Walter Cannon introduced the term "homeostasis," building upon Claude Bernard's observations that there are mechanisms that allow the body to adapt and respond to changes within the environment.[3,4] Cannon proposed that the sympathoadrenal system is rapidly activated in order to re-establish homeostasis.[5] "Stress" is generally defined as a real or perceived threat (internal or external) to homeostasis.[5] Too often the stress response is thought of in negative terms, but the stress response is necessary for survival. The response to acute stress, rapid activation of the sympathetic nervous system followed by the hypothalamic–pituitary–adrenal (HPA) axis, acts to facilitate adaptation to the stressor and helps to maintain homeostasis. Cardiovascular tone is increased, energy stores are mobilized and functions that are not immediately necessary (including growth and reproduction) are suppressed. Under normal conditions the stress response is quickly terminated when the threat to homeostasis has passed (such as escape from a predator), and the functions that were suspended resume.

However, when the stress response is chronically activated, pathophysiological states can arise in susceptible individuals.[1,6] Chronic stress is associated with neuroendocrine dysfunction, which can induce vulnerability to, or exacerbate, multiple conditions, including affective, cardiovascular, metabolic and immune disorders.[7] The ideal organism for studying the effects of chronic stress on human health is, of course, the human. Indeed, many studies have investigated the effects of different stressful life experiences on neuroendocrine function in humans. Epidemiological studies can identify differences between groups that have or have not experienced the stressor(s) of interest, but there are obvious limitations to the interpretation of results. Clearly, it is impossible to control for individual life experiences, which makes the majority of these studies correlational; a true causal relationship is difficult to establish.

Human research can also be conducted in a controlled clinical or laboratory setting. In a clinical setting, one method is to assign different groups to an intervention or control condition, and then compare behavioral and physiological responses to treatment.[7] One commonly used laboratory paradigm is the Trier Social Stress Test (TSST), in which the participant is required to perform a psychologically stressful activity (such as giving a speech or performing difficult mathematic exercises) in front of an audience following a short preparatory period.[8] This paradigm reliably induces a robust stress response in participants, including increased heart rate, elevations in adrenocorticotropic hormone (ACTH) and cortisol, and transient increases in pro-inflammatory cytokines.[8,9] The TSST is useful in assessing responses to acute social stress, but is not designed for studying chronic social stress. There are obvious ethical and logistical difficulties in designing and conducting chronic social stress studies on humans within a laboratory setting. Because of these issues, animal studies are frequently used to model aspects of human disease.

The most commonly used species to study chronic stress are rodents and non-human primates, as they share many of the same neuroendocrine systems and anatomical structures (*construct validity*). They also share many similar behavioral and physiological responses to chronic stress, and often respond to pharmacological compounds in a similar fashion to humans (*predictive validity*). Because maladaptation to stress is associated with psychopathology and physiological dysfunction, many labs utilize animal models incorporating repeated exposure to acute stressors such as footshock, restraint and hypoxia. While these models undoubtedly elicit a stress response, and have helped to elucidate mechanisms involved in stress pathology, they lack *face validity* – that is, they do not reflect the types of stressors typically encountered in the animals' natural environment. Hans Selye introduced the concept of general adaptation syndrome (GAS) based on the observation that chronic exposure to different noxious stimuli elicited the same physiological response in laboratory animals, such as adrenal hypertrophy, thymic involution and gastric ulceration.[10–12] Over the years our understanding of stress has evolved, and we now recognize that while different stressors may elicit similar responses, not all stressors are alike. Physiological and psychological stressors can utilize different neural circuits.[13] In fact, endocrine and autonomic responses to different physiological stressors, such as hemorrhage, hypoglycemia and cold stress, also differ.[13,14] The response to stress varies not only with the type of stressor, but also with the duration or frequency of exposure, the age, sex and hormonal status of the organism, and the time period over which the stressor is applied. The response also varies widely among individuals; genetics and past and current environments can confer resilience or vulnerability to stress, depending on the nature of the exposures. For these reasons, it is preferable to utilize ethologically relevant animal models of social stress.

Like humans, many species used in laboratory research are social animals, and interactions with conspecifics are common and can be considered a source of both support and stress in their natural environments.[7] In humans, interpersonal conflict, hostile work environments, perception of low socioeconomic status, and lack of social support are commonly experienced chronic stressors, which are associated with a higher risk of affective disorders, impaired immune function and cardiometabolic dysfunction.[6,15–20] For these reasons, development of appropriate animal models of psychosocial stress may allow us to investigate the complex interactions of stress and neuroendocrine function.

As previously noted the majority of psychosocial stress paradigms utilize non-human primates and social rodents. Non-human primates, as the closest relatives of humans, have provided valuable insight into the effects of social stress on physiology. Most non-human primate species are social and form dominance hierarchies. They are most similar to humans in neuroanatomy and neuroendocrine function, and have a long life span, allowing longitudinal studies to be conducted. Additionally, humans and non-human primates share similar genetic polymorphisms, including those in genes that encode the serotonin transporter[21,22] and tryptophan hydroxylase,[23] which both have been found to increase vulnerability to social stress, and its negative consequences on health.

Although rodents are more distantly related to humans, they still share similar neuroanatomical and endocrine systems to those of humans, and there is a vast amount of literature detailing the physiology and behavior of the species. They require significantly less space and financial resources than non-human primates, and can afford longitudinal studies to be performed over a relatively short period, due to their shorter life spans. Through years of selective breeding, strains exist which can be used to determine how different traits related to temperament and coping styles affect social interactions and reactions to psychosocial stress.[6,24,25] In addition, the development of transgenic models with selective overexpression or knockdown of particular genes allows researchers to investigate specific genetic contributions to behavior and physiology.

MODELS OF SOCIAL STRESS

It is important to choose an appropriate species and model when testing hypotheses. Acknowledgement of the species' natural environment and behavioral repertoire has contributed greatly to our understanding of psychosocial stress. There are many models of chronic social stress, and the number has greatly increased in the past two decades, with greater acceptance of the need for models with face validity.[26] In this chapter we will focus on commonly used social stress paradigms, including social defeat, hierarchy formation, social instability and social isolation. This review is by no means exhaustive, but illustrates commonly used paradigms of social stress. Because of the great complexities in social behavior and number of endocrine systems affected by social stress, there are numerous unclear and sometimes conflicting findings. The purpose of this chapter is to outline the general findings, and how they may help us better understand how social stress affects the human condition.

Social Defeat

Rats and mice are social animals commonly used in social defeat or resident–intruder paradigms. This paradigm utilizes the inherent territorial nature of a species. The "resident" is a male that vigorously defends its territory against another, usually smaller, male "intruder." The intruder is introduced into the resident's home cage for a short period of time, during which the intruder is attacked and defeated by the resident. Housing the resident with a female often enhances this aggressiveness. After experience of defeat, proximity to the resident is often sufficient to elicit a stress response in the intruder. Some models house the dominant (resident) and subordinate (intruder) near one another – for instance, together in a cage with a partition, so that they are in auditory, olfactory and visual contact with one another, but physical attack is prevented. This type of housing is more of a chronic stressor, because of the continuous exposure to the dominant animal. Some paradigms involve repeated periods of social defeat, while others return the intruder to the dominant's cage but with a barrier, such as a smaller mesh cage, so that the dominant is not able physically to attack the subordinate. Social defeat is often used to study mechanisms that may contribute to affective and metabolic disorders. The interactions between the resident and intruder are probably most similar to bullying or experience of attack in humans, which may also make it useful for studying post-traumatic stress disorder (PTSD).[16,27,28]

Syrian hamsters and tree shrews are also used in social defeat studies, although they are less social than rats and mice. They tend to be more solitary, and most interactions with conspecifics occur during defense of territory, or during courtship and mating. While this lack of regular social interaction has some drawbacks regarding modeling stress in social organisms, there are also some particular strengths. Both male and female hamsters are territorial, so they allow the investigator to study sex differences. The tree shrew is phylogenetically closer to humans than rats are, and shares more similarities with human anatomy and physiology.[29–31] Female rats and mice do not display the clear dominance or subordination observed in males with resident–intruder paradigms. However, a lactating female will attack a conspecific that is introduced into her home cage,[35] and this is sometimes used as an acute or repeated stressor.

BOX 23.1

HOW DO WE KNOW THAT SUBORDINATE RATS ARE STRESSED?

A number of behavioral, physiological and neuroendocrine parameters are used to determine status in the visible burrow system (VBS). The VBS is designed with a series of tunnels and chambers that remain in constant darkness, simulating a burrow environment, and an open area that is on a 12 : 12 light : dark cycle. Infrared lighting allows for behavioral observations even under complete darkness. An ethogram is used to quantify different behaviors observed in dominant (DOM) and subordinate (SUB) rats. DOM males are more likely to engage in offensive behaviors such as chasing, standing over and biting. They are also more likely to spend time in the open surface area, and mount females more than subordinate males. SUB males express more defensive behaviors, such as flight and submissive supine postures. They also tend to incur more bites, which are primarily located on the hindquarters (reflecting flight from an attacking DOM). Although both DOM and SUB males lose body weight during the initial days of VBS housing, SUB continue to lose weight throughout the duration of housing while DOM either maintain or gain weight. SUB tend to have higher basal levels of corticosterone (CORT) in comparison with DOM and control (CON) males. Interestingly, there appears to be a subset of SUB rats that have a *blunted* response to restraint stress, and are therefore dubbed non-responsive subordinates (NRS). This may be a result or cause of different coping strategies, and is a current focus of research in our lab. The table summarizes some of the general characteristics and consequences of subordination stress in male rats and mice exposed to chronic social stress.

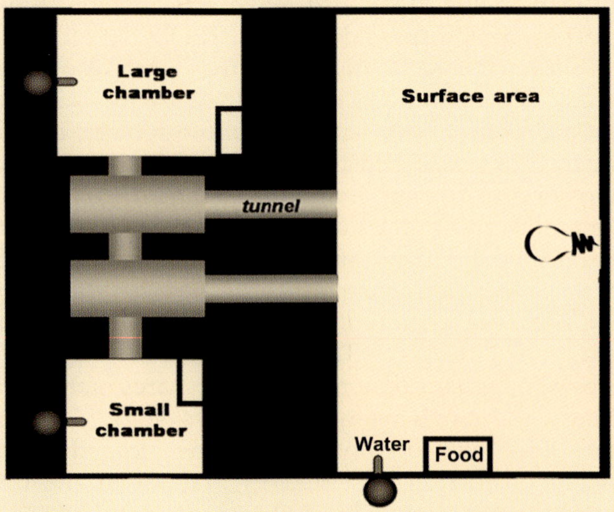

Schematic of the visible burrow system (VBS).

Characteristics and Consequences of Chronic Social Subordination of Male Rats and Mice

Parameter	Effect	Reference
BEHAVIORAL		
Offensive	↓SUB	1–3
Defensive	↑SUB	1–3
Reproductive	↓SUB	2
Bites received	↑SUB	2,4
Food intake	↓SUB	2,5,6
ORGAN WEIGHTS		
Adrenal weight	↑SUB	4,7
Thymus weight	↓SUB	7–9
METABOLISM		
Body weight	↓SUB	3,6,10,11
Adipose mass	↓SUB	6,10,11
Plasma glucose	↓SUB	12
Plasma insulin (non-fasted)	↓SUB	11
Plasma leptin	↓SUB	11
HPA AXIS		
Basal corticosterone (CORT)	↑SUB	2,7
Corticotropin-releasing hormone (CRH)	↑SUB (PVN and CeA)	13
Hippocampal GR/MR gene expression	↓SUB	7,14
Corticosteroid binding globulin (CBG)	↓SUB	2,15
11 beta-hydroxysteroid dehydrogenase (11β-HSD)	↓SUB (testes)	16
HPG AXIS		
Testosterone (T)	↓SUB	1,2,16
Luteinizing hormone (LH)	↓SUB	16

References

1. Nguyen MM, Tamashiro KL, Melhorn SJ, Ma LY, Gardner SR, Sakai RR. Androgenic influences on behavior, body weight, and body composition in a model of chronic social stress. *Endocrinology*. 2007;148(12):6145–6156.
2. Blanchard DC, Sakai RR, McEwen B, Weiss SM, Blanchard RJ. Subordination stress: behavioral, brain, and neuroendocrine correlates. *Behav Brain Res*. 1993;58 (1-2):113–121.

BOX 23.1 *(cont'd)*

3. Stefanski V, Knopf G, Schulz S. Long-term colony housing in Long Evans rats: immunological, hormonal, and behavioral consequences. *J Neuroimmunol.* 2001;114(1-2): 122–130.
4. Blanchard DC, Spencer RL, Weiss SM, Blanchard RJ, McEwen B, Sakai RR. Visible burrow system as a model of chronic social stress: behavioral and neuroendocrine correlates. *Psychoneuroendocrinology.* 1995;20(2):117–134.
5. Tamashiro KL, Nguyen MM, Ostrander MM, et al. Social stress and recovery: implications for body weight and body composition. *Am J Physiol Regul Integr Comp Physiol.* 2007;293(5):R1864–R1874.
6. Melhorn SJ, Krause EG, Scott KA, et al. Meal patterns and hypothalamic NPY expression during chronic social stress and recovery. *Am J Physiol Regul Integr Comp Physiol.* 2010;299(3):R813–R822.
7. Schmidt MV, Sterlemann V, Muller MB. Chronic stress and individual vulnerability. *Ann NY Acad Sci.* 2008;1148: 174–183.
8. Bartolomucci A. Social stress, immune functions and disease in rodents. *Front Neuroendocrinol.* 2007;28(1):28–49.
9. Engler H, Stefanski V. Social stress and T cell maturation in male rats: transient and persistent alterations in thymic function. *Psychoneuroendocrinology.* 2003;28(8):951–969.
10. Tamashiro KL, Hegeman MA, Nguyen MM, et al. Dynamic body weight and body composition changes in response to subordination stress. *Physiol Behav.* 2007;91(4):440–448.
11. Tamashiro KL, Nguyen MM, Fujikawa T, et al. Metabolic and endocrine consequences of social stress in a visible burrow system. *Physiol Behav.* 2004;80(5):683–693.
12. McKittrick CR. *Physiological, Endocrine and Neurochemical Consequences of Chronic Social Stress.* New York: The Rockefeller University; 1996;207.
13. Albeck DS, McKittrick CR, Blanchard DC, et al. Chronic social stress alters levels of corticotropin-releasing factor and arginine vasopressin mRNA in rat brain. *J Neurosci.* 1997;17(12):4895–4903.
14. Chao HM, Blanchard DC, Blanchard RJ, McEwen B, Sakai RR. The effect of social stress on hippocampal gene expression. *Molecular and Cellular Neuroscience.* 1993;4: 543–548.
15. Stefanski V. Social stress in laboratory rats: hormonal responses and immune cell distribution. *Psychoneuroendocrinology.* 2000;25(4):389–406.
16. Hardy MP, Sottas CM, Ge R, et al. Trends of reproductive hormones in male rats during psychosocial stress: role of glucocorticoid metabolism in behavioral dominance. *Biol Reprod.* 2002;67(6):1750–1755.

Social Hierarchy Models of Social Stress

In addition to resident–intruder paradigms or dominant–subordinate relationships between paired conspecifics, many species of animals live in more complex social environments, such as colonies. Dominance hierarchies in social species generally arise from competition for resources, including access to reproductive partners, food or water. Field research has provided valuable insight into the consequences of social status on health.[36–38] However, it is often challenging to study the species in the wild, and therefore colony housing in a laboratory environment is often used. Because dominant and subordinate animals are continuously housed together, colony housing employs a true chronic social stressor, in comparison with models of repeated, intermittent social defeat. An additional strength of this model is that the stress is dictated by the animals themselves, with little experimenter influence.

Non-human Primates

In the laboratory many non-human primate species, including macaques, vervets and squirrel monkeys, form dominance hierarchies, which are useful in understanding the physiological consequences of chronic social stress. In addition to males, female non-human primates also form social hierarchies, which may be either earned or inherited. This makes non-human primates particularly attractive in studying social factors involved in the development of depression and other disorders which afflict a disproportionate percentage of women.[19] Additionally, female non-human primates have reproductive cycles similar to that of humans, providing a better means for studying the interactions of reproductive hormones and social stress and how these influence fertility, affect and cardiometabolic function.[39–42]

Rodents

Many rodent species also form dominance hierarchies. The VBS, developed at the University of Hawaii by Drs. Bob and Caroline Blanchard, is one paradigm used to study dominance hierarchies in laboratory rodents. The VBS is a neuroethologically-based, semi-natural social environment consisting of a series of tunnels and chambers. When housed in mixed-sex groups, male rats form a dominance hierarchy, with one male becoming dominant and the others subordinate. Dominance in male rats is determined through analysis of agonistic interactions between males, weight loss, and bite patterns. These hierarchies are stable following formation, and subordinate males do not appear to habituate to the stress of social subordination, as stress-induced behavioral and physiological changes persist through the duration of social housing. For a more detailed description of the VBS paradigm, see references 43 to 48.

In mice, a recently introduced paradigm involves housing a larger dominant male with smaller mice that become subordinate. This is different from the VBS, in that there is a designated dominant animal that has been selected for size and aggressiveness prior to colony formation. In addition, the dominant animal is replaced with another dominant every week to decrease the likelihood of habituation.[49] This model borrows some aspects of social hierarchy formation, but also of social defeat and social instability (described below).

Guinea pigs are also social rodents, and both males and females form dominance hierarchies, although these are not as pronounced in females.[50] The majority of guinea-pig stress studies focus on either male social structure or a social instability model in females, although there are reports of female hierarchy differences according to housing density and presence of a male.[51]

Social Instability and Social Isolation

The VBS is very useful for studying social stress in male rodents, as a clear dominance hierarchy is formed, producing robust behavioral, physiological and neuroendocrine effects in both dominant and subordinate members according to their social status. However, a similar hierarchy situation has not yet been observed in female rats. Group-housed females are less aggressive and engage in fewer agonistic interactions with other females, and there is little physical wounding except in the case of naïve females introduced to a lactating female's housing.[52] This does not mean that females do not form hierarchies or are immune to social stress; it suggests that dominance may exist in a more subtle form in females that is not detectable using standard criteria originally developed from male colonies. In addition, they may be more susceptible to other stressors and express symptoms of stress in a different manner than males. Indeed, female rats are particularly sensitive to social instability arising from varying housing conditions, from social isolation and/or changing the composition of the group. Whereas group housing tends to elicit a strong stress response in males, single housing of females appears more stressful than paired housing.[53]

Social instability created by frequently changing the composition of groups is also used as a chronic stressor for rodents. This is similar to the VBS model of housing, but the make-up of the group is changed on a daily basis, so there is no point where a stable hierarchy is formed with clear dominant and subordinate roles.[54]

CONSEQUENCES OF SOCIAL STRESS

HPA Axis

Social stress is a potent activator of the HPA axis, and chronic social stress can lead to HPA dysregulation, which is thought to trigger the development of multiple pathologies. Whereas the HPA axis is typically kept in check through negative feedback loops, chronic social stress impairs these systems. Subordination stress is associated with elevations in basal circulating levels of the glucocorticoids corticosterone (the primary glucocorticoid in rats and mice) and cortisol (the primary glucocorticoid in primates), which will be referred to hereafter as "CORT." A hallmark of chronic HPA activation and elevations in glucocorticoids is enlargement of the adrenal glands, due to an increase in number and/or size of cells in the zona fasiculata which produce CORT.[55] Enlarged adrenal glands are observed in both non-human primate and rodent models of social stress.[44,56–58] Additionally, chronic CORT release induces thymic involution, indicative of suppressed immune function.[56,59]

In the VBS model of chronic social stress, subordinate males have greater expression of corticotropin-releasing hormone (CRH) mRNA expression in the paraventricular nucleus of the hypothalamus (PVN) and the central nucleus of the amygdala (CeA) consistent with enhanced HPA axis drive. Similarly, subordinate male and female non-human primates tend to have higher plasma cortisol levels than dominant conspecifics, and are less efficient in suppressing cortisol release following dexamethasone administration, suggesting impaired negative feedback.[60]

It is important to note that dominant animals are not "stress-free." In some situations, the dominant animal actually may have higher glucocorticoid levels than subordinate animals. In studies of non-human primates, Dr. Robert Sapolsky and colleagues have noted that this often depends on the stability of the hierarchy, aggression levels, and the composition of the group.[58] For instance, the dominant animal may feel significant amounts of stress due to the physical and psychological demands required in maintaining status. Subordinates, in turn, may develop coping mechanisms which allow them to adapt to stress associated with low social status. In the VBS model, dominant males often exhibit elevated plasma CORT and PVN CRH expression when compared with control animals. However, it appears that they possess mechanisms that allow more effective handling and termination of the stress response, such as higher levels of plasma corticosterone-binding globulin (CBG) than subordinates, which prevents CORT from binding to corticosteroid receptors, thus reducing the biological activity of circulating CORT. Similarly, male rats exposed to social defeat also have lower levels of circulating CBG in comparison with dominant males.[61] Therefore, although *total* levels of circulating CORT may be similar, less *free* CORT is available to activate receptors of the dominant animal. Differential expression of mineralocorticoid receptor

(MR) and glucocorticoid receptor (GR) is also observed according to social status. Hippocampal GR and MR are thought to participate in the termination of the HPA axis response, and subordinate males have lower levels of mRNA encoding these receptors, suggesting impaired negative feedback.[62] Dominant non-human primates occasionally have elevated glucocorticoid levels, but, as reported in rodents, the dominant typically does not exhibit impaired negative feedback in response to a dexamethasone suppression test (DST).[58,63] In non-human primates, the dominant male may have higher levels of circulating CORT when there are higher levels of aggression within the colony or there is instability, such as when the dominant animal's status is challenged.[63,64]

Reproduction

It has long been observed that chronic social stress can negatively affect reproductive function of both males and females via interactions between the HPA and hypothalamic–pituitary–gonadal (HPG) axes.[65] Stress hormones can act at the level of the hypothalamus to inhibit release of gonadotropin-releasing hormone (GnRH), at the pituitary level to suppress the production and/or pulsatile release of luteinizing hormone (LH) and follicle-stimulating hormone (FSH), as well as at the gonadal level by inhibiting sex steroidogenesis and release.[65,66] There are reciprocal interactions between the HPG and HPA axes. For instance, while CRH can suppress synthesis and release of LH, central injections of LHRH stimulate glucocorticoid release in rodents. Additionally, gonadectomy alters the circadian rhythm of glucocorticoid release.[65]

Males

Subordinate male rats and non-human primates tend to have lower testes weights and/or lower levels of testosterone (T).[37,42] In the VBS model of chronic social stress, male rats form a dominance hierarchy only in the presence of females. This is androgen-dependent; gonadectomy of males prevents hierarchy formation, whereas gonadectomy with replacement of T or 5α-dihydrotestosterone (DHT) results in hierarchy formation similar to that of intact males.[47] Hierarchies form in the androgen replacement burrows despite similar basal androgen concentrations, supporting previous findings that elevated androgen levels do not predict dominance status but, rather, are a consequence of social status.

In the VBS, T and LH levels are lower in subordinate male rats, whereas T levels are maintained or increased in dominant males during hierarchy formation. These reductions in T are most likely a result of actions at the gonadal level, as these changes occur prior to changes in LH. Additionally, these changes in T secretion occur even in the absence of elevated plasma CORT levels, suggesting that they may be due to tissue-specific differences in glucocorticoid activity. The Leydig cells that produce T also express 11β-hydroxysteroid dehydrogenase (11β-HSD), a bidirectional enzyme capable of activating or inactivating CORT. This is expressed at lower levels in subordinate males in comparison with control and dominant males, and may therefore expose result in greater inhibitory effects of CORT on T production.[48] Subordinate males also express lower levels of arginine vasopressin (AVP) in the medial amygdala (MeA).[67] AVP arising from the MeA mediates aggressive behavior, which may explain decreases in offensive behaviors observed in subordinate rats. This effect is most likely a secondary effect from reduced T levels, as T modulates AVP levels in this structure.[68] Similar effects of chronic social stress are observed in non-human primates. As in rats, T also does not predict, but rather reflects, dominance status of male non-human primates. The male that becomes dominant typically experiences a marked increase in T, while T drops in SUB males following hierarchy formation.[69] Multiple factors, such as the composition of the group, mating season, aggression levels and stability of the hierarchy, can have great effects on T.[42] Although T levels tend to be lower in subordinate males they are often within a normal physiological range, which would not severely affect fertility.[58]

Females

Females generally have higher HPA activity, which is directly related to estrogen levels. Clamping estrogen at low levels results in a less robust ACTH response to stress.[65] In female primates, subordination stress is also associated with reproductive dysfunction and reduced progesterone levels. In young females, puberty can be delayed. In cycling adults, some of the most severely stressed subordinate females cease to ovulate, while others continue to ovulate but have significantly lower progesterone levels.[19,41,60,70] This dysfunction is reversible; when female hierarchies are manipulated to facilitate a switch from subordinate to dominant status, reproductive function is improved.[60] Chronic stress and ovarian steroids can also affect the limbic–HPA axis response.

Cardiometabolic Function

In humans, low socioeconomic status (SES) and lack of emotional support is associated with obesity, insulin resistance, type 2 diabetes, cardiovascular disease, and other characteristics of the metabolic syndrome.[71] The Whitehall II study is a longitudinal set of experiments that has examined the relationship between SES and metabolic and cardiovascular outcomes in a large set of men and women in the British Civil Service.[72] Data from the Whitehall II study suggest that there are significant correlations between low SES and altered metabolic and cardiovascular health measures.[73,74] One of the strengths of the Whitehall II study is that, rather than a cross-sectional design, the subjects were followed prospectively, and thus the findings can be viewed as an indication of causal mechanisms rather than merely correlational. However, the actual mechanisms remain to be identified and manipulated in order to demonstrate causality. Rodent and non-human primate models have afforded the ability to investigate the interactions between psychosocial stress and cardiovascular and metabolic health.

The relationship between social stress and obesity is complicated and somewhat paradoxical, since stress has been found to be associated with both decreased and increased food intake and body weight in animals.[43,75–78] Differences observed in the literature are likely due to variations in the type, severity and duration of the stressor applied. Likewise, in humans, some report increases in food intake in response to stress, while others react by reducing food intake.[79–81] In general, rodents lose a significant amount of weight and reduce food intake during the stress period, and they do so by reducing the number of meals that they take as well as the size of those meals. Further analysis also suggests that meals are spread throughout the dark and into the light cycle – a pattern that suggests disruptions in sleep–wake cycles in socially stressed rodents as well.[77] Together, these data suggest that social stress produces altered feeding patterns that may contribute to changes in energy homeostasis and can have long-term adverse consequences for the animal.

Recovery of body weight and post-stress meal patterns have been measured upon removal from the VBS.[46,77] Socially subordinate rats are immediately hyperphagic, and rapidly regain weight. Analysis of meal patterns during this time period indicates that meal size is increased with social stress, with fewer meals taken.[77] Increased meal size has been associated with increased weight gain with the same caloric consumption, underlining the importance not only of total calories consumed, but also of the pattern of food consumption.[82] The implication of these data is that stress alters meal patterns of SUB rats to facilitate deposition of fat and development of obesity following chronic social stress exposure.

Chronic elevations in circulating glucocorticoids are believed to be responsible for some of the effects of psychosocial stress and obesity. Adipocytes express glucocorticoid receptors, and visceral adipose depots (those surrounding internal organs) express more GR than subcutaneous depots (adipose tissue that is located beneath the skin).[83,84] Additionally, stromal cells of visceral adipose tissue produce 11β-hydroxysteroid dehydrogenase 1 (11β–HSD1), which can act to increase the amount of biologically active CORT in the visceral depots by converting inactive glucocorticoid (cortisone) to the active form, cortisol or corticosterone.[85] Activation of the GR in adipose depots upregulates lipoprotein lipase, which in turn promotes fat accumulation.[85–87] Indeed, providing mice with corticosterone in their drinking water to elevate their levels of circulating glucocorticoids results in greater body weight, adiposity, hyperinsulinemia, hyperglycemia, hyperleptinemia and dyslipidemia characteristic of the metabolic syndrome.[88,89]

Furthermore, during stress and for a period of time following removal of the stressor the SUB animals are in prime endocrine status to rapidly gain weight as adipose tissue with low testosterone and high glucocorticoid levels. This weight gain is preferentially gained as adipose tissue,[46,90] which correlates with higher levels of leptin and insulin in SUB compared to DOM and control males.[91] This increase in insulin may promote glucose uptake and storage in adipose tissue, as recent *in vitro* studies have indicated that while muscle becomes insulin-resistant with chronic glucocorticoid exposure, human adipocytes become more sensitive.[92,93] Thus, intermittent chronic stress exposure creates a "perfect storm" period during which socially subordinate animals are in the perfect behavioral and endocrine state to rapidly gain weight and develop obesity.

Symptoms of the metabolic syndrome appear to be consistent outcomes among different models of stress, and in multiple animal species. The metabolic syndrome increases risk of coronary heart disease (CHD). Studies in cynomolgus monkeys indicate that social stress (subordination stress) results in hypercortisolemia and coronary artery atherogenesis (CAA) which is associated with visceral obesity in subordinate subjects to a greater degree compared to those of a dominant social status. While chronic social stress has been linked with increased CHD risk, what is not clear is whether this is a direct pathway, or whether greater visceral adiposity and metabolic syndrome first promote CAA to increase susceptibility to CHD.

Immune Function

Social stress can have profound and paradoxical effects on the immune system, as the HPA axis and immune system have reciprocal interactions. While chronic exposure to social stress and glucocorticoids has been shown to blunt inflammatory responses, slow healing and leave individuals susceptible to infection or disease, it can also increase the risk of or exacerbate the symptoms of inflammatory disease, including atherosclerosis, asthma, diabetes and other autoimmune disorders.[94–97]

As Selye noted, one commonly observed consequence of chronic stress exposure is thymic involution, which is believed to be due to apoptosis and reduced proliferation of thymocytes in response to elevated glucocorticoid levels.[61,98,99] In the VBS model of chronic social stress, thymic involution is routinely observed in subordinate male rats.[44,56] Similarly, rats exposed to repeated social defeat have smaller thymi containing fewer immature thymocytes, and this effect is blocked by adrenalectomy.[98] Glucocorticoids can also inhibit the expression of proinflammatory cytokines[100] or alter the trafficking of immune cells.[101,102]

The ability of chronic social stress to increase susceptibility to or exacerbate symptoms of autoimmune disease appears to be due in part to selective impairments in glucocorticoid signaling. It is hypothesized that in some situations immune cells become desensitized to the effects of stress hormones through downregulation of glucocorticoid receptors, similar to the effect that chronic stress has on negative feedback regions of the brain. This, in turn, would result in prolonged immune response and inflammation.[7] Lymphocytes of chronic socially stressed male rhesus monkeys become desensitized to the actions of stress hormones.[94] Recent studies have indicated that interactions between social stress and serotonin transporter (5HTT) polymorphisms in female rhesus monkeys may be responsible for some differences in immune response.[97]

SOCIAL STRESS AND AFFECTIVE DISORDERS

Animals exposed to chronic social stress develop many behavioral, neurochemical and neuroendocrine profiles that correlate with affective disorders, such as Major Depressive Disorder (MDD), anxiety, and post-traumatic stress disorder (PTSD), in humans. Although there are clearly some complex cognitive aspects of affective disorders that cannot be reproduced in animal models (such as suicidality, shame and guilt), other indices, such as anhedonia, sleep disruption, and neurochemical and nedurоendocrine changes, can be assessed in laboratory animals.[103,104] While it is impossible to recreate an entire constellation of symptoms associated with a particular mood disorder, certain aspects can be modeled in animals, and may allow us to identify mechanisms that are involved in the development of symptomatology associated with affective disorders.[104] It is also important to realize that these conditions are not black and white, and are not mutually exclusive; for instance, many with MDD also exhibit symptoms of anxiety disorders.

Animal models of social stress may allow us to investigate how altered HPA function associated with social stress affects neurotransmission in psychopathological conditions. The serotonergic and dopaminergic neurotransmitter systems are greatly affected by stress, and have been implicated in numerous human psychopathologies. These two neurotransmitter systems are the target of many psychotherapeutic drugs.

Serotonergic function is frequently altered in depressed patients as well as animals exposed to chronic social stress, and there is much cross-talk between the serotonergic system and HPA axis.[18,105] The main serotonergic neurons arise from the dorsal and median raphe nuclei. These cells have numerous projections throughout the brain, and can modulate limbic–HPA activity. In turn, cells of the raphe nuclei express CRH and GR, a means by which the HPA axis can control the serotonergic neurons. Chronic stress and prolonged CORT exposure also affect serotonin receptors, of which 5-HT1A and 5-HT2 are believed to have a significant role in the development of depression.[18,105] Imaging studies have revealed a reduction in 5-HT1A receptor presynaptic binding (to autoreceptors on the serotonergic neurons of the raphe nuclei) and postsynaptically, with reduced binding in cortical and hippocampal regions in depressed patients.[106,107] This has also been observed in post-mortem analyses of depressed patients. Other studies have found increased 5-HT2A receptor binding in cortex of depressed patients, as well as those with borderline personality disorder.[105,108,109] Similar changes in serotonin receptor binding have been observed in animal models of social stress. In the VBS, socially stressed rats have increased 5-HT2A receptor binding in the parietal cortex, as well as decreased 5-HT1A binding in the hippocampus.[110,111]

Dopamine (DA) is another neuro transmitter that is affected by social stress and is implicated in psychopathologies, including depression, schizophrenia and addiction. Stress has been shown to affect dopaminergic transmission in rodents and non-human primates, via alterations in dopamine release, dopamine receptor expression and/or dopamine transporter (DAT) binding in mesocortical and mesolimbic regions of the brain.[112,113] The effects of psychosocial stress on dopaminergic transmission are not well understood, and there are many conflicting findings as to which regions of the mesocorticolimbic are most affected. Both male and female subordinate cynomolgus monkeys have lower levels of homovanillic acid (HVA), a dopamine metabolite, in cerebrospinal fluid.[114] Dominant female cynomolgus monkeys have an increase in dopamine D2 receptors, and are less vulnerable to the reinforcing effects of self-administered cocaine than are subordinates.[115]

Major depressive disorder (MDD) is a debilitating condition that affects approximately 20% of humans at some point during their lifetime, and is one of the most common causes of disability.[116] Although there are many different symptoms of MDD, the most consistent physiological symptom is hypercortisolemia, which exists in up to 50% of depressed patients. It is believed that this is one of the causative factors involved in development of depression, as elevations in circulating glucocorticoids have been shown to precede the onset of depressive symptoms.[117–119] This HPA hyperactivity is also observed in the majority of socially subordinate animals, and appears to be a consequence of social stress, as CORT levels do not predict dominance or subordination. The majority of subordinate rodents and non-human primates have elevated CORT levels during and at the end of chronic social stress exposure, as do the majority of rodents and male tree shrews exposed to social defeat. This hyperactivity of the HPA axis observed in a subset of depressed humans and socially subordinate animals appears to be due to impaired glucocorticoid negative feedback, as assessed by DST. Consistent with this finding, post-mortem studies have found that some depressed patients have reduced GR binding in the brain, as was described earlier in subordinate male rats.[62] Chronically socially-stressed animals with hyperactivity of the HPA axis are also more likely to engage in depressive-like behaviors. Socially subordinate male rats will consume less of a dilute sucrose solution following removal from the VBS, and will not work as hard for a palatable food reward.[43,120] These changes in behavior are suggestive of anhedonia observed in human depression, in which normally pleasurable activities are less rewarding. Subordinate female cynomolgus monkeys with reduced suppression of cortisol following the DST are more likely to exhibit postures associated with depressive-like behaviors, whereas subordinate females that are more efficient at suppressing cortisol release in response to the DST are less likely to exhibit these postures.[60] Chronic social stress has also been correlated with anxiety disorders. Many of the symptoms of anxiety disorders overlap with those of depression, including disrupted sleep, fatigue and irritability.[121] While MDD is associated with a "learned helplessness" condition and depressed mood, the hallmark of anxiety is hypervigilance. PTSD is an anxiety disorder also associated with HPA axis dysfunction. It appears to arise following an experience of a particularly traumatic, terrifying experience, including combat, assault or accidents. Symptoms include hyperarousal, avoidance of stimuli that may be associated with the traumatic event, and recurrent dreams or memories of the event.[122]

INDIVIDUALITY AND COPING MECHANISMS

The perception of and response to stress is not homogeneous between or even within a species. It is important to remember that chronic social stress can, but does not always, precede the onset of pathology. Likewise, two individuals can be exposed to similar aversive experiences, but only one may develop a stress-related disorder while the other does not. Individual coping styles and existence or lack of social support can also affect the development, duration and severity of pathological conditions. Differences in sex, genetic make-up, environment and personal life experience must be accounted for. Indeed, different gene polymorphisms have been identified in humans and other animals, which may confer a susceptibility or resilience to stress and stress-induced pathology. For example, rhesus macaques expressing the short form of the serotonin transporter are more likely to exhibit signs of anxiety- and/or depression-like behavior. Humans with affective disorders are more likely to have either one or two copies of the short form of this allele.[123] Polymorphisms of corticosteroid and CRH receptors as well as transcription factors, chaperones and microRNAs which can alter expression or translation of genes are associated with hyper-reactivity to stress and susceptibility to psychopathology.[123–125] Therefore, it is important to take into account "the whole picture."

CONCLUSIONS AND FUTURE DIRECTIONS

This chapter serves as an overview of how the body responds to social stress; it is by no means complete, and it must also be noted that, due to the complex interactions which exist between social animals, variations in behavioral and physiological responses will occur. Animal models have helped us to gain a greater understanding of the role of social stress in behavioral and physiological function. It is of utmost importance to take into consideration the animal's natural environment and behavioral repertoire. If studying the effects of chronic exposure to social subordination, it makes sense to choose a species that lives in a social setting and forms dominance hierarchies. Conversely, social defeat models may be more relevant in studying the consequences of acute episodes of social stress, such as those occurring when an individual is bullied or physically attacked.

As noted earlier, an individual's response to social stress is not clear cut – some are more resilient to, while others are more susceptible to, certain stressors. The ways in which an individual is psychologically and physiologically able to respond, adapt to and cope with perceived stress will determine vulnerability to pathological conditions. Individual variation is an inherent characteristic of any population, and, in order to reduce the degree of variability among laboratory rodents, inbred strains have been created. In addition, use of artificial stressors which can be quantified and meted out on specific schedules are used to simplify experimental designs and reduce the number of animals required.[25] While that strategy has facilitated identification of basic behavioral, physiological and neurobiological mediators of stress consequences, it is clear that not all humans respond similarly to the same stressors. Naturally occurring variations in coping strategy and adaptability to stressful situations will aid discovery of underlying mechanisms that lead to pathological conditions in some humans and not others.[24,25] Genetic, epigenetic and environmental factors may underlie individual variability and represent exciting new directions for research in stress and neuropsychiatric disorders that ultimately may help in developing new approaches for the treatment of stress-related disorders.

Acknowledgments

This research has been supported by NIH grants MH088230 (KAS), HD055030 (KLKT) and DK068273 (RRS).

References

1. McEwen BS, Wingfield JC. The concept of allostasis in biology and biomedicine. *Horm Behav.* 2003;43(1):2–15.
2. McEwen BS. From molecules to mind. Stress, individual differences, and the social environment. *Ann NY Acad Sci.* 2001;935:42–49.
3. Cannon WB. *The Wisdom of the Body.* New York: Norton; 1939.
4. Kopin IJ. Definitions of stress and sympathetic neuronal responses. *Ann NY Acad Sci.* 1995;771:19–30.
5. Goldstein DS, Kopin IJ. Evolution of concepts of stress. *Stress.* 2007;10(2):109–120.
6. de Kloet ER, Joels M, Holsboer F. Stress and the brain: from adaptation to disease. *Nat Rev Neurosci.* 2005;6(6):463–475.
7. Miller G, Chen E, Cole SW. Health psychology: developing biologically plausible models linking the social world and physical health. *Annu Rev Psychol.* 2009;60:501–524.
8. Kirschbaum C, Pirke KM, Hellhammer DH. The Trier Social Stress Test – a tool for investigating psychobiological stress responses in a laboratory setting. *Neuropsychobiology.* 1993;28 (1-2):76–81.
9. Yamakawa K, Matsunaga M, Isowa T, et al. Transient responses of inflammatory cytokines in acute stress. *Biol Psychol.* 2009; 82(1):25–32.
10. Selye H. The general-adaptation-syndrome. *Annu Rev Med.* 1951;2:327–342.
11. Selye H. *The Stress of Life.* revised ed. New York: McGraw-Hill; 1956.
12. Selye H. A syndrome produced by diverse nocuous agents. *Nature.* 1936;138:32.
13. Pacak K, Palkovits M, Kvetnansky R, Yadid G, Kopin IJ, Goldstein DS. Effects of various stressors on *in vivo* norepinephrine release in the hypothalamic paraventricular nucleus and on the pituitary–adrenocortical axis. *Ann NY Acad Sci.* 1995;771:115–130.
14. Pacak K, Palkovits M. Stressor specificity of central neuroendocrine responses: implications for stress-related disorders. *Endocrine Rev.* 2001;22(4):502–548.
15. Wang JL. Perceived work stress, imbalance between work and family/personal lives, and mental disorders. *Soc Psychiatry Psychiatr Epidemiol.* 2006;41(7):541–548.
16. Huhman KL. Social conflict models: can they inform us about human psychopathology? *Horm Behav.* 2006;50(4):640–646.
17. Cohen S. Keynote Presentation at the Eight International Congress of Behavioral Medicine: the Pittsburgh common cold studies: psychosocial predictors of susceptibility to respiratory infectious illness. *Intl J Behav Med.* 2005;12(3):123–131.
18. Leonard BE. HPA and immune axes in stress: involvement of the serotonergic system. *Neuroimmunomodulation.* 2006;13 (5-6):268–276.
19. Shively CA, Musselman DL, Willard SL. Stress, depression, and coronary artery disease: modeling comorbidity in female primates. *Neurosci Biobehav Rev.* 2009;33(2):133–144.
20. McEwen BS, Gianaros PJ. Central role of the brain in stress and adaptation: links to socioeconomic status, health, and disease. *Ann NY Acad Sci.* 2010;1186:190–222.
21. Barr CS, Newman TK, Becker ML, et al. The utility of the non-human primate; model for studying gene by environment interactions in behavioral research. *Genes Brain Behav.* 2003; 2(6):336–340.
22. Jarrell H, Hoffman JB, Kaplan JR, Berga S, Kinkead B, Wilson ME. Polymorphisms in the serotonin reuptake transporter gene modify the consequences of social status on metabolic health in female rhesus monkeys. *Physiol Behav.* 2008;93(4-5):807–819.
23. Chen GL, Novak MA, Meyer JS, Kelly BJ, Vallender EJ, Miller GM. The effect of rearing experience and TPH2 genotype on HPA axis function and aggression in rhesus monkeys: a retrospective analysis. *Horm Behav.* 57(2):184–191.
24. Koolhaas JM, de Boer SF, Buwalda B. Stress and Adaptation. *Curr Dir Psychol Sci.* 2006;15(3):109–112.
25. Koolhaas JM, de Boer SF, Coppens CM, Buwalda B. Neuroendocrinology of coping styles: towards understanding the biology of individual variation. *Front Neuroendocrinol.* 2010; 31(3):307–321.
26. Willner P. The validity of animal models of depression. *Psychopharmacology (Berl).* 1984;83(1):1–16.
27. Tamashiro KL, Nguyen MM, Sakai RR. Social stress: from rodents to primates. *Front Neuroendocrinol.* 2005;26(1):27–40.
28. Gattermann R, Johnston RE, Yigit N, et al. Golden hamsters are nocturnal in captivity but diurnal in nature. *Biol Lett.* 2008; 4(3):253–255.
29. Fuchs E, Flugge G. Social stress in tree shrews: effects on physiology, brain function, and behavior of subordinate individuals. *Pharmacol Biochem Behav.* 2002;73(1):247–258.
30. van Kampen M, Kramer M, Hiemke C, Flugge G, Fuchs E. The chronic psychosocial stress paradigm in male tree shrews: evaluation of a novel animal model for depressive disorders. *Stress.* 2002;5(1):37–46.
31. Kawamichi T, Kawamichi M. Spatial organization and territory of tree shrews (*Tupaia glis*). *Anim Behav.* 1979;27:381–393.
32. Solomon MB, Karom MC, Huhman KL. Sex and estrous cycle differences in the display of conditioned defeat in Syrian hamsters. *Horm Behav.* 2007;52(2):211–219.
33. Faruzzi AN, Solomon MB, Demas GE, Huhman KL. Gonadal hormones modulate the display of submissive behavior in socially defeated female Syrian hamsters. *Horm Behav.* 2005;47(5):569–575.
34. Huhman KL, Solomon MB, Janicki M, et al. Conditioned defeat in male and female Syrian hamsters. *Horm Behav.* 2003;44(3):293–299.
35. Brunton PJ, Russell JA. Prenatal social stress in the rat programmes neuroendocrine and behavioural responses to stress in the adult offspring: sex-specific effects. *J Neuroendocrinol.* 2010; 22(4):258–271.

36. Sapolsky RM. The endocrine stress-response and social status in the wild baboon. *Horm Behav.* 1982;16(3):279–292.
37. Sapolsky RM. Stress-induced suppression of testicular function in the wild baboon: role of glucocorticoids. *Endocrinology.* 1985;116(6):2273–2278.
38. Sapolsky RM, Alberts SC, Altmann J. Hypercortisolism associated with social subordinance or social isolation among wild baboons. *Arch Gen Psychiatry.* 1997;54(12):1137–1143.
39. Shively CA, Grant KA, Ehrenkaufer RL, Mach RH, Nader MA. Social stress, depression, and brain dopamine in female cynomolgus monkeys. *Ann NY Acad Sci.* 1997;807:574–577.
40. Shively CA, Kaplan JR. Stability of social status rankings of female cynomolgus monkeys, of varying reproductive condition, in different social groups. *Am J Primatol.* 1991;23:239–245.
41. Shively CA, Register TC, Friedman DP, Morgan TM, Thompson J, Lanier T. Social stress-associated depression in adult female cynomolgus monkeys (*Macaca fascicularis*). *Biol Psychol.* 2005;69(1):67–84.
42. Cameron JL. Stress and behaviorally induced reproductive dysfunction in primates. *Semin Reprod Endocrinol.* 1997;15(1):37–45.
43. Tamashiro KL, Hegeman MA, Nguyen MM, et al. Dynamic body weight and body composition changes in response to subordination stress. *Physiol Behav.* 2007;91(4):440–448.
44. Blanchard DC, Spencer RL, Weiss SM, Blanchard RJ, McEwen B, Sakai RR. Visible burrow system as a model of chronic social stress: behavioral and neuroendocrine correlates. *Psychoneuroendocrinology.* 1995;20(2):117–134.
45. McKittrick CR, Magarinos AM, Blanchard DC, Blanchard RJ, McEwen BS, Sakai RR. Chronic social stress reduces dendritic arbors in CA3 of hippocampus and decreases binding to serotonin transporter sites. *Synapse.* 2000;36(2):85–94.
46. Tamashiro KL, Nguyen MM, Ostrander MM, et al. Social stress and recovery: implications for body weight and body composition. *Am J Physiol Regul Integr Comp Physiol.* 2007;293(5):R1864–R1874.
47. Nguyen MM, Tamashiro KL, Melhorn SJ, Ma LY, Gardner SR, Sakai RR. Androgenic influences on behavior, body weight, and body composition in a model of chronic social stress. *Endocrinology.* 2007;148(12):6145–6156.
48. Hardy MP, Sottas CM, Ge R, et al. Trends of reproductive hormones in male rats during psychosocial stress: role of glucocorticoid metabolism in behavioral dominance. *Biol Reprod.* 2002;67(6):1750–1755.
49. Singewald GM, Nguyen NK, Neumann ID, Singewald N, Reber SO. Effect of chronic psychosocial stress-induced by subordinate colony (CSC) housing on brain neuronal activity patterns in mice. *Stress.* 2009;12(1):58–69.
50. Sachser N, Durschlag M, Hirzel D. Social relationships and the management of stress. *Psychoneuroendocrinology.* 1998;23(8):891–904.
51. Yvonne T, Hubert H. Differences in behavior and social organization of female guinea pigs as a function of the presence of a Male. *Ethology.* 1990;85(1):25–34.
52. Mayer AD, Reisbick S, Siegel HI, Rosenblatt JS. Maternal aggression in rats: changes over pregnancy and lactation in a Sprague-Dawley Strain. *Aggressive Behav.* 1987;13(1):29.
53. Haller J, Fuchs E, Halasz J, Makara GB. Defeat is a major stressor in males while social instability is stressful mainly in females: towards the development of a social stress model in female rats. *Brain Res Bull.* 1999;50(1):33–39.
54. Schmidt MV, Sterlemann V, Muller MB. Chronic stress and individual vulnerability. *Ann NY Acad Sci.* 2008;1148:174–183.
55. Ulrich-Lai YM, Figueiredo HF, Ostrander MM, Choi DC, Engeland WC, Herman JP. Chronic stress induces adrenal hyperplasia and hypertrophy in a subregion-specific manner. *Am J Physiol Endocrinol Metab.* 2006;291(5):E965–E973.
56. Blanchard DC, Sakai RR, McEwen B, Weiss SM, Blanchard RJ. Subordination stress: behavioral, brain, and neuroendocrine correlates. *Behav Brain Res.* 1993;58(1-2):113–121.
57. Shively C, Kaplan J. Effects of social factors on adrenal weight and related physiology of *Macaca fascicularis*. *Physiol Behav.* 1984;33(5):777–782.
58. Sapolsky RM. The influence of social hierarchy on primate health. *Science.* 2005;308(5722):648–652.
59. Bartolomucci A. Social stress, immune functions and disease in rodents. *Front Neuroendocrinol.* 2007;28(1):28–49.
60. Shively CA, Laber-Laird K, Anton RF. Behavior and physiology of social stress and depression in female cynomolgus monkeys. *Biol Psychiatry.* 1997;41(8):871–882.
61. Stefanski V. Social stress in laboratory rats: hormonal responses and immune cell distribution. *Psychoneuroendocrinology.* 2000;25(4):389–406.
62. Chao HM, Blanchard DC, Blanchard RJ, McEwen B, Sakai RR. The effect of social stress on hippocampal gene expression. *Mol Cell Neurosci.* 1993;4:543–548.
63. Czoty PW, Gould RW, Nader MA. Relationship between social rank and cortisol and testosterone concentrations in male cynomolgus monkeys (*Macaca fascicularis*). *J Neuroendocrinol.* 2009;21(1):68–76.
64. Sapolsky RM. Testicular function, social rank and personality among wild baboons. *Psychoneuroendocrinology.* 1991;16(4):281–293.
65. Viau V. Functional cross-talk between the hypothalamic–pituitary–gonadal and –adrenal axes. *J Neuroendocrinol.* 2002;14(6):506–513.
66. Kalantaridou SN, Makrigiannakis A, Zoumakis E, Chrousos GP. Stress and the female reproductive system. *J Reprod Immunol.* 2004;62(1-2):61–68.
67. Albeck DS, McKittrick CR, Blanchard DC, et al. Chronic social stress alters levels of corticotropin-releasing factor and arginine vasopressin mRNA in rat brain. *J Neurosci.* 1997;17(12):4895–4903.
68. Koolhaas JM, van den Brink THC, Roozendaal B, Boorsma. Medial Amygdala F. Aggressive behavior: interaction between testosterone and vasopressin. *Aggressive Behav.* 1990;16(3/4):223.
69. Rose RM, Berstein IS, Gordon TP. Consequences of social conflict on plasma testosterone levels in rhesus monkeys. *Psychosom Med.* 1975;37(1):50–61.
70. Adams MR, Kaplan JR, Koritnik DR. Psychosocial influences on ovarian endocrine and ovulatory function in *Macaca fascicularis*. *Physiol Behav.* 1985;35(6):935–940.
71. Abraham NG, Brunner EJ, Eriksson JW, Robertson RP. Metabolic syndrome: psychosocial, neuroendocrine, and classical risk factors in type 2 diabetes. *Ann NY Acad Sci.* 2007;1113:256–275.
72. Marmot M, Brunner E. Cohort Profile: the Whitehall II study. *Intl J Epidemiol.* 2005;34(2):251–256.
73. Brunner EJ, Hemingway H, Walker BR, et al. Adrenocortical, autonomic, and inflammatory causes of the metabolic syndrome: nested case-control study. *Circulation.* 2002;106(21):2659–2665.
74. Steptoe A, Feldman PJ, Kunz S, Owen N, Willemsen G, Marmot M. Stress responsivity and socioeconomic status: a mechanism for increased cardiovascular disease risk? *Eur Heart J.* 2002;23(22):1757–1763.
75. Solomon MB, Foster MT, Bartness TJ, Huhman KL. Social defeat and footshock increase body mass and adiposity in male Syrian hamsters. *Am J Physiol Regul Integr Comp Physiol.* 2007;292(1):R283–R290.

76. Foster MT, Solomon MB, Huhman KL, Bartness TJ. Social defeat increases food intake, body mass, and adiposity in Syrian hamsters. *Am J Physiol Regul Integr Comp Physiol.* 2006; 290(5):R1284–R1293.
77. Melhorn SJ, Krause EG, Scott KA, et al. Meal patterns and hypothalamic NPY expression during chronic social stress and recovery. *Am J Physiol Regul Integr Comp Physiol.* 2010; 299(3):R813–R822.
78. Shively CA, Clarkson TB. Regional obesity and coronary artery atherosclerosis in females: a non-human primate model. *Acta Med Scand. (Suppl)* 1988;723:71–78.
79. Weinstein SE, Shide DJ, Rolls BJ. Changes in food intake in response to stress in men and women: psychological factors. *Appetite.* 1997;28(1):7–18.
80. Wallis DJ, Hetherington MM. Emotions and eating. Self-reported and experimentally induced changes in food intake under stress. *Appetite.* 2009;52(2):355–362.
81. Torres SJ, Nowson CA. Relationship between stress, eating behavior, and obesity. *Nutrition.* 2007;23(11-12):887–894.
82. Fabry P, Tepperman J. Meal frequency—a possible factor in human pathology. *Am J Clin Nutr.* 1970;23(8):1059–1068.
83. Bjorntorp P. Adipose tissue distribution and function. *Intl J Obes.* 1991;15(Suppl 2):67–81.
84. Bjorntorp P. Metabolic implications of body fat distribution. *Diabetes Care.* 1991;14(12):1132–1143.
85. Shively CA, Clarkson TB. The unique value of primate models in translational research. Nonhuman primate models of women's health: introduction and overview. *Am J Primatol.* 2009;71(9):715–721.
86. Rebuffe-Scrive M, Walsh UA, McEwen B, Rodin J. Effect of chronic stress and exogenous glucocorticoids on regional fat distribution and metabolism. *Physiol Behav.* 1992;52(3):583–590.
87. Fried SK, Russell CD, Grauso NL, Brolin RE. Lipoprotein lipase regulation by insulin and glucocorticoid in subcutaneous and omental adipose tissues of obese women and men. *J Clin Invest.* 1993;92(5):2191–2198.
88. Karatsoreos IN, Bhagat SM, Bowles NP, Weil ZM, Pfaff DW, McEwen BS. Endocrine and physiological changes in response to chronic corticosterone: a potential model of the metabolic syndrome in mouse. *Endocrinology.* 151(5):2117–2127.
89. Lee RS, Tamashiro KL, Yang X, et al. Chronic corticosterone exposure increases expression and decreases deoxyribonucleic acid methylation of Fkbp5 in mice. *Endocrinology.* 2010; 151(9):4332–4343.
90. Tamashiro KL, Nguyen MM, Fujikawa T, et al. Metabolic and endocrine consequences of social stress in a visible burrow system. *Physiol Behav.* 2004;80(5):683–693.
91. Tamashiro KL, Hegeman MA, Sakai RR. Chronic social stress in a changing dietary environment. *Physiol Behav.* 2006;89(4): 536–542.
92. Morgan SA, Sherlock M, Gathercole LL, et al. 11Beta-hydroxysteroid dehydrogenase type 1 regulates glucocorticoid-induced insulin resistance in skeletal muscle. *Diabetes.* 2009;58(11): 2506–2515.
93. Gathercole LL, Bujalska IJ, Stewart PM, Tomlinson JW. Glucocorticoid modulation of insulin signaling in human subcutaneous adipose tissue. *J Clin Endocrinol Metab.* 2007;92(11):4332–4339.
94. Cole SW, Mendoza SP, Capitanio JP. Social stress desensitizes lymphocytes to regulation by endogenous glucocorticoids: insights from *in vivo* cell trafficking dynamics in rhesus macaques. *Psychosom Med.* 2009;71(6):591–597.
95. Miller GE, Cohen S, Ritchey AK. Chronic psychological stress and the regulation of pro-inflammatory cytokines: a glucocorticoid-resistance model. *Health Psychol.* 2002;21(6):531–541.
96. Gust DA, Gordon TP, Wilson ME, Ahmed-Ansari A, Brodie AR, McClure HM. Formation of a new social group of unfamiliar female rhesus monkeys affects the immune and pituitary adrenocortical systems. *Brain Behav Immun.* 1991;5(3):296–307.
97. Paiardini M, Hoffman J, Cervasi B, et al. T-cell phenotypic and functional changes associated with social subordination and gene polymorphisms in the serotonin reuptake transporter in female rhesus monkeys. *Brain Behav Immun.* 2009;23(2):286–293.
98. Engler H, Stefanski V. Social stress and T cell maturation in male rats: transient and persistent alterations in thymic function. *Psychoneuroendocrinology.* 2003;28(8):951–969.
99. Stefanski V, Knopf G, Schulz S. Long-term colony housing in Long Evans rats: immunological, hormonal, and behavioral consequences. *J Neuroimmunol.* 2001;114(1-2):122–130.
100. Kemeny ME, Schedlowski M. Understanding the interaction between psychosocial stress and immune-related diseases: a stepwise progression. *Brain Behav Immun.* 2007;21(8):1009–1018.
101. Dhabhar FS, Miller AH, McEwen BS, Spencer RL. Effects of stress on immune cell distribution. Dynamics and hormonal mechanisms. *J Immunol.* 1995;154(10):5511–5527.
102. McEwen BS, Biron CA, Brunson KW, et al. The role of adrenocorticoids as modulators of immune function in health and disease: neural, endocrine and immune interactions. *Brain Res Brain Res Rev.* 1997;23(1-2):79–133.
103. Anisman H, Matheson K. Stress, depression, and anhedonia: caveats concerning animal models. *Neurosci Biobehav Rev.* 2005;29(4-5):525–546.
104. Lapiz-Bluhm MD, Bondi CO, Doyen J, Rodriguez GA, Bedard-Arana T, Morilak DA. Behavioural assays to model cognitive and affective dimensions of depression and anxiety in rats. *J Neuroendocrinol.* 2008;20(10):1115–1137.
105. van Praag HM. Can stress cause depression? *Prog Neuro-psychopharmacol Biol Psychiatry.* 2004;28(5):891–907.
106. Drevets WC, Thase ME, Moses-Kolko EL, et al. Serotonin-1A receptor imaging in recurrent depression: replication and literature review. *Nucl Med Biol.* 2007;34(7):865–877.
107. Savitz J, Lucki I, Drevets WC. 5-HT(1A) receptor function in major depressive disorder. *Prog Neurobiol.* 2009;88(1):17–31.
108. Soloff PH, Price JC, Meltzer CC, Fabio A, Frank GK, Kaye WH. 5HT2A receptor binding is increased in borderline personality disorder. *Biol Psychiatry.* 2007;62(6):580–587.
109. Bhagwagar Z, Hinz R, Taylor M, Fancy S, Cowen P, Grasby P. Increased 5-HT(2A) receptor binding in euthymic, medication-free patients recovered from depression: a positron emission study with [(11)C]MDL 100,907. *Am J Psychiatry.* 2006;163(9): 1580–1587.
110. McKittrick CR. *Physiological, Endocrine and Neurochemical Consequences of Chronic Social Stress.* New York: The Rockefeller University; 1996:207.
111. McKittrick CR, Blanchard DC, Blanchard RJ, McEwen BS, Sakai RR. Serotonin receptor binding in a colony model of chronic social stress. *Biol Psychiatry.* 1995;37(6):383–393.
112. Puglisi-Allegra S, Kempf E, Schleef C, Cabib S. Repeated stressful experiences differently affect brain dopamine receptor subtypes. *Life Sci.* 1991;48(13):1263–1268.
113. Isovich E, Mijnster MJ, Flugge G, Fuchs E. Chronic psychosocial stress reduces the density of dopamine transporters. *Eur J Neurosci.* 2000;12(3):1071–1078.
114. Kaplan JR, Manuck SB, Fontenot MB, Mann JJ. Central nervous system monoamine correlates of social dominance in cynomolgus monkeys (*Macaca fascicularis*). *Neuropsychopharmacology.* 2002; 26(4):431–443.
115. Morgan D, Grant KA, Gage HD, et al. Social dominance in monkeys: dopamine D2 receptors and cocaine self-administration. *Nat Neurosci.* 2002;5(2):169–174.

116. Nestler EJ, Barrot M, DiLeone RJ, Eisch AJ, Gold SJ, Monteggia LM. Neurobiology of depression. *Neuron.* 2002;34(1):13–25.
117. de Kloet ER, Derijk RH, Meijer OC. Therapy Insight: is there an imbalanced response of mineralocorticoid and glucocorticoid receptors in depression? *Nat Clin Pract Endocrinol Metab.* 2007;3(2):168–179.
118. Armario A, Valles A, Dal-Zotto S, Marquez C, Belda X. A single exposure to severe stressors causes long-term desensitisation of the physiological response to the homotypic stressor. *Stress.* 2004;7(3):157–172.
119. Hammen C. Stress and depression. *Annu Rev Clin Psychol.* 2005;1:293–319.
120. Davis JF, Krause EG, Melhorn SJ, Benoit SC, Sakai RR. Dominant rats are natural risk takers and display increased motivation for food reward. *Neuroscience.* 2009;162(1):23–30.
121. Ressler KJ, Nemeroff CB. Role of serotonergic and noradrenergic systems in the pathophysiology of depression and anxiety disorders. *Depress Anxiety.* 2000;12(Suppl 1):2–19.
122. Pervanidou P, Chrousos GP. Neuroendocrinology of post-traumatic stress disorder. *Prog Brain Res.* 182:149–160.
123. Haglund ME, Nestadt PS, Cooper NS, Southwick SM, Charney DS. Psychobiological mechanisms of resilience: relevance to prevention and treatment of stress-related psychopathology. *Dev Psychopathol.* 2007;19(3):889–920.
124. Derijk RH, de Kloet ER. Corticosteroid receptor polymorphisms: determinants of vulnerability and resilience. *Eur J Pharmacol.* 2008;583(2-3):303–311.
125. Gillespie CF, Phifer J, Bradley B, Ressler KJ. Risk and resilience: genetic and environmental influences on development of the stress response. *Depress Anxiety.* 2009;26(11):984–992.

CHAPTER

24

Estrogen and Cognitive Aging in Women: The Critical Period Hypothesis

Barbara B. Sherwin

Department of Psychology, McGill University, Montreal, Quebec, Canada

OUTLINE

Summary

Research findings on the influence of estrogen on cognition in women are inconsistent. In part this may be due to differences in the age and health characteristics of the participants, to the variety of estrogen preparations and routes of administration used in these studies, as well as to confounds due to the simultaneous treatment with a progestin in naturally menopausal women. Indeed, there is accumulating evidence to support the Critical Period Hypothesis, which holds that exogenous estrogen provides neuroprotection when treatment is initiated at the time of menopause while the initiation of estrogen therapy decades after the menopause to older women will be ineffective in attenuating cognitive aging and may even cause harm. Although considerable evidence from basic neuroscience and from animal behavioral studies is available in support of the Critical Period Hypothesis, more human research is required before it can be concluded that these findings generalize to women.

INTRODUCTION

One of the most significant advances in industrialized countries during the past century has been the considerable increase in life expectancy. This has occurred partly because of improvements in public health measures, the prevention of numerous infectious diseases through vaccination, new medical treatments and technologies, improvements in the standard of living and, for women, a drastic reduction in maternal mortality associated with childbirth. This increase in life expectancy means that women in industrialized countries now live to an average age of 80 years compared to an average life expectancy of 54 years in 1900. Since menopause occurs at an average age of 51

Handbook of Neuroendocrinology, DOI: 10.1016/B978-0-12-375097-6.10024-1

years, this mean that women now live one-third of their lives following the cessation of ovarian function. This increase in female longevity has led to a heightened interest in the discovery of preventative strategies against degenerative diseases that occur with aging, such as cardiovascular disease, osteoporosis, cerebrovascular disease and cognitive decline, in order to maintain the health and functionality of aging individuals during the latter part of their lifespan. This chapter reviews the evidence that estrogen may protect against cognitive decline in aging women, and describes what is currently known of the circumstances under which treatment with exogenous estrogen may either protect or potentially harm cognitive functioning in postmenopausal women, or be ineffective. Prior to a discussion of that evidence, some background on the neurobiology of estrogen, the components of cognition, cognitive aging, brain aging, and hormonal changes at menopause will be presented in order to facilitate understanding of the issues involved.

THE NEUROBIOLOGY OF ESTROGEN

There is considerable biological plausibility for the idea that estrogen protects aspects of cognitive functioning in women. Although a comprehensive account of the mechanisms of action of estrogen on the central nervous system (CNS) is beyond the scope of this chapter, several recent reviews of the area are available.[1,2] The mechanisms of estrogenic actions on the brain that are likely the most relevant for cognitive functioning are briefly described.

In 1980, the identification and mapping of estrogen receptors (ERs) in the brain led to the discovery that these proteins were concentrated in the hypothalamus and the pituitary,[3] and, later on, in the hippocampus, the cerebral cortex, midbrain and brainstem. Two types of intracellular ERs have now been identified, ERα and ERβ (see McEwen[1] for a review). Although the distribution of ERα in brain has been fairly well established by steroid autoradiography, immunocytochemistry and *in situ* hybridization,[3,4] there is less certainty concerning the localization of ERβ. Recently, the co-localization of ERβ mRNA with cell nuclear ERβ immunoreactivity was demonstrated in the rat cerebral cortex, in the paraventricular nucleus, and in the preoptic area of the hypothalamus.[5] Importantly, the use of I^{125} estrogen, which labels ERs with a higher specific radioactivity than ^{3}H estradiol (E_2), led to detection of label in pyramidal cells of CA1 and CA3 in the ventral hippocampus[6] – areas shown to be important for memory. In addition to its ability to effect both direct genomic actions, estrogen can also act in the CNS via non-nuclear receptors that involve interactions of ERs with second messenger systems.[7] Third, there is *in vitro* evidence that E_2 can modulate the functions of neural and vascular cells via non-genomic actions.[8] Although estrogen can have rapid and/or membrane-mediated effects in the hippocampus, intracellular ERs, particularly the β isoform of ERs, may be important targets for estrogen's functional effects for hippocampal processes.[9] Indeed, selective ERβ agonists increased key synaptic proteins *in vivo* and these effects were absent in ERβ knockout mice, demonstrating that activation of ERβ can regulate hippocampal synaptic plasticity, which could improve hippocampal-dependent cognitive functions such as memory.[10,11]

Due to the widespread presence of the ERs in their various forms throughout the brain, estrogen actions are also widespread and affect many neurotransmitter systems, including the catecholaminergic, serotonergic, cholinergic, and γ-aminobutyric acidergic systems.[1] Several well described effects of estrogen on brain structure and function offer possible explanations for the mechanisms of action by which this steroid hormone may influence cognitive functioning in women. For example, estrogen increases dendritic spine density on CA1 pyramidal neurons in the hippocampus within 24–72 h after acute administration.[12,13] When progesterone (P) is administered after E_2 priming, spine density increases during the first 6–8 h, followed by a rapid return to low baseline levels. Interestingly, changes in memory function in rats show a strong temporal relationship with these hormonal alterations in neuronal structure. When ovariectomized rats were primed with 10-μg injections of E_2 72 and 48 h before testing, memory retention was improved compared with retention in rats primed with oil.[14] An injection of P maintained the memory enhancement if testing occurred within 8 h of the P injection, but not when testing occurred more than 24 h after the P injection. Moreover, in OVX African green monkeys, spine density of CA1 pyramidal cells was significantly lower in the control compared to the estrogen-treated animals.[15]

Second, estrogen increases the concentration of choline acetyltransferase (ChAT), the synthetic enzyme for acetylcholine, a neurotransmitter critically implicated in memory functions[16] and whose levels are markedly reduced in Alzheimer's disease (AD).[17] Estrogen's numerous neurotrophic effects could also explain how it protects against declines in cognition with aging. Widespread co-localizations of ERs and nerve growth factor are found mainly in neurons of the cerebral cortex, hypothalamus, hippocampus and sensory ganglia, implying that estrogen facilitates neurotrophin responses.[18] Estrogen may also exert neuroprotective effects via its modulation of molecules involved in apoptosis[19] and via its action as an antioxidant.[20] Both 17β- and 17α-E_2 provide neuroprotection against oxidative stress via an antioxidant effect.[21] Moreover, E_2, *in vitro*, promotes the breakdown of the β-amyloid precursor protein to fragments that are less likely to accumulate as β-amyloid.[22] Estrogenic effects on neural systems that affect mood, fine motor control and pain have also been described.[7] The potential for these myriad mechanisms of action of estrogen to influence the structure and function of brain areas that subserve a variety of cognitive functions, described here only briefly, provides biological plausibility for the hypothesis that estrogen could help to maintain aspects of cognition in aging women.

COMPONENTS OF COGNITION

Several concepts concerning the components of cognition are important for an understanding of the results of the clinical studies on sex hormones and cognition, the focus of this review. Cognition is an umbrella term for the totality of human information processing; it is multidimensional, and includes such functions as attention, memory, learning, language processing, problem solving, abstract reasoning, higher-order intellectual functioning and psychomotor skills. Memory, a critical aspect of cognition, is composed of numerous component processes that localize to different anatomical sites.[23] Human memory may be regarded as a system that stores and retrieves information acquired through the senses. Visual and auditory systems have been the most thoroughly explored. Short-term visual memory implies that the visual memory trace decays after a fraction of a second to several seconds,[24] and long-term visual memory describes the ability to identify a visual stimulus for days, months or years after exposure. Similarly, short-term verbal memory describes the situation in which a verbal stimulus (for example, a word or phrase) is held in memory for seconds or minutes, whereas long-term verbal memory describes the ability to recognize or remember a verbal stimulus or association over an extended period of time. While the hippocampus is critical for acquiring a short-term memory, long-term memories are stored elsewhere in the brain.

Working memory refers to the ability to "hold in mind" and flexibly manipulate information over a short period of time to make a response.[24] During the past decade, considerable progress has been made toward specifying the neural mechanisms underlying working memory in humans.[25,26] In functional neuroimaging studies, prefrontal cortical (PFC) activation during the performance of working memory tasks occurred.[27,28] Moreover, there is evidence that working memory efficiency decreases with increasing age[29,30] in association with the decrease in the volume of the PFC with aging.[31]

Findings from studies on rats, monkeys, and humans underscore the importance of medial temporal-lobe structures in the acquisition of new information as well as in the retrieval of previously learned information.[32] Based on the information that estrogen induces cyclic changes in synapse formation and spine density of the hippocampus,[12,13] and in embryonic hippocampal neurons in cell culture,[33] and also increases the synthesis of acetylcholine in basal forebrain and the cholinergic neurons that project to the hippocampus and cortex,[16,33] it might be predicted that estrogen would have its most profound effect on hippocampal-dependent cognitive functions such as memory and learning, which involve the acquisition, encoding and consolidation of new information, as well as on cognitive abilities that depend on the prefrontal cortex, such as working memory.

In summary, numerous human cognitive functions are differentially mediated by specialized, anatomically distinct brain areas. Certain neurotransmitters and brain structures, along with neural pathways and projections, are also critical for cognitive function. The extreme complexity of this system makes it unlikely that any one neuroactive chemical could influence the totality of cognitive functions; rather, it is probable that any given neuroactive compound would exert a specific action on certain domains of cognitive functions. Estrogen is a neuroactive hormone; the largest concentrations of ERs are in the hypothalamus, the amygdala and the hippocampus, and the neurotransmitter that estrogen upregulates most profoundly is acetylcholine,[33] although it affects the serotonergic, noradrenergic and dopaminergic systems as well.[1] Moreover, the hippocampus itself has been shown to be critical for explicit or declarative memory. A hypothesis that derives from these findings, taken together, is that estrogen would have its most profound effects on verbal memory and on working memory, although it does not exclude the possibility that it might influence other cognitive functions as well.

There are established sex differences in cognitive function. Although there are no qualitative differences in cognitive skills between the sexes, quantitative differences have been consistently found. Whereas women tend to excel on tasks of verbal skills and memory, on perceptual speed and accuracy, and on fine motor skills (female-typical skills), men tend to outperform women on tests of visual memory and on mathematical and spatial ability (male-typical skills).[34] Although the effect sizes of these sexual dimorphisms in cognitive function are modest (0.5–1.0 SD), they have been found consistently in studies that have attempted to document them, and are thought to occur as a result of the exposure of the fetal brain to differential levels of the sex hormones during prenatal life. It is hypothesized that these so-called organizational effects of sex hormones permanently alter the structure and/or function of specific brain areas during fetal life, perhaps by directing the development of certain neural pathways. It is thought that, post-pubertally, the increased production of each of the sex hormones amplifies the neural "hard-wiring" laid down prenatally under its influence, usually referred to as the activational effect of that hormone. Therefore, this psychoendocrine theory proposes that, during prenatal life, the presence of significant quantities of a sex hormone organizes neural substrates for a certain behavior or function that becomes manifest after puberty under the influence of high circulating levels of that same hormone. A hypothesis that derives from the literature on the organizational and activational effects of sex hormones on cognition is that, in women, estrogen would have its most profound effect on cognitive tasks, such as verbal skills and memory, perceptual speed and accuracy, and fine motor skills, in which females are known to excel. If this is true, then the administration of estrogen to postmenopausal women should preferentially enhance female-typical cognitive skills.

Normal Human Cognitive Aging

There is now considerable evidence to suggest that changes in cognition occur with normal aging. While age-related cognitive decline generally does not occur before the age of 60 years,[35] in cross-sectional studies of cognitively healthy people linear declines in cognition occurred across the lifespan, specifically in processing speed, working memory and episodic memory.[36] From a clinical perspective, the fact that non-dementia memory decline can interfere with an older person's daily functioning renders it a meaningful clinical issue.[37]

It has also become clear that age-related cognitive changes are selective rather than diffuse. Across cognitive domains, memory performance, in particular, undergoes decline with increasing age.[36,37] Specifically, while there is little evidence that declines occur in tests of language, visuospatial ability and abstract reasoning, memory for the acquisition and early retrieval of new information[38] and working memory[39] become somewhat compromised with increasing age. Well-practiced tasks or tasks that involve knowledge show little or no decline in performance until very late in life, while short-term memory ability declines slightly until the age of 70 years and more sharply thereafter.[40] The accelerated declines in these cognitive functions during late life may be due to the increasing influence of disease processes in the elderly.[41]

Brain Aging

The integrity of some brain areas is more vulnerable than others to aging processes. Changes tend to occur most profoundly in the prefrontal cortex, in the hippocampus and in the parietal regions of the brain;[42] the same anatomical locations that subserve the specific cognitive functions that decline with normal aging, including verbal memory and working memory. The underlying pathologic basis of cognitive decline is inevitably related to the loss of synapses, neurotransmitters and neural networks.[43] For example, in medically and cognitively healthy older adults, longitudinal MRI scans showed that the annual rate of brain tissue loss was $5.4 \pm 0.3\ cm^3$ per year, with the frontal and parietal lobar regions showing the greatest decline.[10] However, neuronal loss may not be an inevitable feature of normal brain aging; indeed, there is evidence that neurogenesis continues throughout life, including old age.[44]

Microscopic studies found loss of neuronal bodies in the neocortex, the hippocampus and the cerebellum,[31] loss of dendritic spines, cumulative mitochondrial damage, reduction in DNA repair ability and reduction in synaptic density occur with increasing age.[43] Brain volume also decreases with age.[45,46] Neuron size also decreases with age, and a 46% reduction in dendritic spine numbers was found in humans over the age of 50 years.[47]

Subregions of the temporal lobes important for memory, such as the entorhinal cortex, the hippocampus and the parahippocampal gyrus, undergo differential rates of decline with aging. The rate of hippocampal atrophy increases with age,[48] and while the hippocampus showed substantial atrophy in healthy older adults, the entorhinal cortex did not.[49] There do not appear to be gender differences in volumetric decline across time except, possibly, in the caudate, which shows minimally greater decline in women compared to men,[49] and this volume loss may be attenuated in the healthiest women. In addition, multiple hyperintense inclusions in the white matter (white matter hyperintensities, WMH) are evident in asymptomatic older adults.[31,50] While age is the most robust predictor of WMH, risk factors for cerebrovascular disease, such as hypertension and history of transient ischemic attacks, are also associated with increased WMH burden. Magnetic resonance spectrometry (MRS) studies have provided limited evidence of a decline in metabolic markers of neuron viability in selected regions of the gray, but not the white, matter.[31] Although this suggests that the age-related shrinkage of some cortical regions might reflect a reduction in size and/or number of neurons, future studies are needed to replicate these findings.

CHANGES IN THE PRODUCTION OF SEX HORMONES AT MENOPAUSE

Although women in industrialized countries now live to an average age of 80 years, the mean age of menopause (51.8 years) has remained unchanged since recorded history. Factors that influence the induction of menopause, the complete cessation of menstrual cycles, occur at each level of the hypothalamic–pituitary–gonadal (HPG) axis. The loss of the limited pool of ovarian follicles causes a drastic decline in the production of E_2, which in turn causes most of the major symptoms associated with the menopause, such as hot flushes, atrophic vaginitis and a decrease in bone density, among others. In turn, the absence of negative feedback to the hypothalamus and pituitary as a result of very low levels of circulating E_2 results in increases in luteinizing hormone (LH) and follicle stimulating hormone (FSH), the classic endocrinological signs of the menopause.[50] However, there is also evidence that the circuitry driving GnRH hypothalamic neurons undergoes age-related changes even before disruptions in the menstrual cycle occur. It is thought that the temporal pattern of neural signals is altered during middle age, leading to cessation of reproductive cycles, and that the complex interplay of ovarian and hypothalamic/pituitary pacemakers becomes increasingly dysfunctional with aging, ultimately resulting in menopause. In female rats, activation of GnRH neurons during the preovulatory GnRH/LH surge declines before reproductive senescence has occurred.[51] Although GnRH cell numbers do not change appreciably with aging, inputs to GnRH neurons from other neurotransmitters and glia in the CNS influence age-associated changes in GnRH output in female rodents.[52] Indeed, the finding that the administration of a physiological dose of E_2 to a postmenopausal women decreases LH and FSH levels but does not restore them to the premenopausal range[53] provides support for the idea that all three levels of the HPG neuroendocrine axis are compromised with aging.

In summary, there is now considerable evidence to show that estrogen influences the structure and function of neurons and the metabolism of neurotransmitters in areas of the brain known to be important for aspects of memory that decline with increasing age. While the ovarian production of estrogen ceases at a mean age of 51.8 years in women in industrialized countries, female longevity has increased significantly over the past century, so that women now spend approximately 30 years in the postmenopausal state with very low levels of sex hormones. This raises the question as to whether increasing the duration of the exposure of the brain to estrogen beyond the natural or spontaneous menopause via the administration of exogenous estrogen following the menopause could provide neuroprotection for aging women – an issue that has both theoretical and clinical importance.

COGNITIVE CHANGES DURING THE MENOPAUSAL TRANSITION

Ovarian estrogen production begins to decline a few years before menopause, and levels of E_2 decrease to very low levels about 2 years following cessation of ovarian function. The period between the onset of the waning of ovarian function to almost total cessation of ovarian estrogen production is known as the perimenopause or the menopausal transition, and roughly encompasses the period 48–52 years of age. Because estrogen levels fall dramatically during the menopausal transition it might be expected that declines in memory may also occur, and indeed subjective memory complaints are frequent during the transition. Several US studies failed to find that serum E_2 levels were related to episodic memory performance during the transition,[54,55] while investigators from Australia, the UK and Taiwan concluded that any decline in memory that occurred during the menopausal transition was transitory and not clinically important.[56–58] Similarly, in a recent longitudinal study of over 2300 women who experienced a decrease in verbal memory during the perimenopause, memory scores rebounded to premenopausal levels by the time they reached postmenopause.[59] It would therefore seem that during the menopausal transition, declines in memory that sometimes occur coincident with declines in E_2 levels are transitory even though serum levels of E_2 remain at a stable nadir in the postmenopause. This conclusion gains support from the finding that when healthy 52-year-old women with subjective cognitive complaints were randomly treated with either conjugated equine estrogen (CEE) plus medroxyprogesterone acetate (MPA) or placebo for 4 months, there were no between-group differences in performance on memory and other cognitive tests at post-treatment.[60]

STUDIES ON ESTROGEN AND COGNITION IN POSTMENOPAUSAL WOMEN

There is now a voluminous literature on estrogen and cognitive functioning in postmenopausal women. Although the review is not exhaustive, the relevant randomized controlled trials, cross-sectional and longitudinal studies are reviewed below.

Randomized Controlled Trials (RCTS)

The earliest RCT that involved the administration of estrogen or placebo to 75-year-old women living in a nursing home found improvement on the verbal but not on the performance subscales of the Wechsler Bellevue Intelligence Scale.[61,62] Other early prospective RCTs undertaken during the 1970s to investigate whether estrogen influences aspects of cognitive functioning in women generally found beneficial effects of the hormone therapy compared with placebo, although they failed to use standardized tests of cognition[63] or had very small and heterogeneous sample sizes with regard to age.[64,65] Others, however, failed to find a beneficial effect of estrogen on a variety of cognitive tests in postmenopausal women.[66,67] While the findings of these early studies made an important contribution, they are difficult to interpret because each study used a different oral estrogen preparation in different doses, failed to measure serum levels of sex hormones, and used different psychometric measures to measure a restricted number of cognitive domains.

In the mid-1980s, premenopausal women who needed to have their uterus and ovaries surgically removed for benign disease were tested preoperatively.[68] Following their surgery they randomly received either estrogen-alone (E_2 valerate, 10 mg i.m. every 4 weeks), androgen-alone, a combined estrogen–androgen drug, or placebo. Women who received any of the active hormonal drugs maintained their performance on a test of verbal memory, whereas scores decreased significantly in those who had received a placebo. Moreover, the estrogenic protection of verbal memory in that study did not occur secondary to the alleviation of hot flashes. In a replication which used a more comprehensive battery of neuropsychological tests, surgically menopausal women who randomly received E_2 valerate (10 mg i.m. every 4 weeks) following the surgical removal of their ovaries had significantly higher scores on tests of short- and long-term verbal memory compared to women who had received placebo, when no differences in their performance had been evident prior to surgery.[69] In neither study did exogenous estrogen enhance performance on tests of visual memory or spatial abilities. When 65-year-old postmenopausal women who had never been treated with estrogen therapy (ET) randomly received either ET or placebo for 3 weeks, those who received 7.8 mg E_2 per week transdermally experienced a selective improvement in performance on tasks of visual learning and memory, while scores of placebo-treated women did not change compared with pre-treatment.[70] No tests of verbal memory were administered in that study.

In the Women's Estrogen for Stroke Trial (WEST), postmenopausal women were enrolled within 3 months of a non-disabling ischemic stroke or transient ischemic attack (TIA) and randomized to treatment with E_2 1 mg/day or placebo for approximately 3½ years.[71] The mean age of the 461 women was 70 years, and 39% met criteria for clinical depression. Although there were no differences in performance on any of the cognitive tests between the groups at post-treatment, women who had been randomly assigned to the estrogen group experienced a significantly smaller reduction and lower rate of decline over time on the Modified Mini Mental State Examination (3MSE) than those in the placebo group. Importantly, no benefit of ET on cognitive functioning occurred in estrogen-treated women who scored below normal on the 3MSE at baseline, suggesting that benefits of ET accrue only to cognitively-intact women. It must be considered, however, that these women had established cerebrovascular disease.

In perimenopausal and early postmenopausal women whose mean age was 51 years, those who randomly received 17β E_2 0.05 mg/day transdermally experienced a significant improvement on a frontally-mediated test of executive function compared to women who received placebo.[72] Fourteen women (mean age of 58 years) who received either 100 μg transdermal E_2 (which induced serum E_2 levels comparable to those at mid-cycle) for 3 days in a double-blind cross-over design to examine whether the induction of a single estrogen peak would improve their cognitive performance[73] had significantly increased scores on two tests of working memory (a PFC function) and on a test of short-term verbal memory (a hippocampal function) during the estrogen treatment phase, whereas no changes occurred in women given placebo.

Four RCTs failed to find any beneficial effects of estrogen on cognition in postmenopausal women. In one, 53-year-old asymptomatic women who had undergone a surgical menopause several years earlier and were randomly treated with either conjugated equine estrogen (CEE) 0.625 mg or CEE 1.25 mg, or with placebo, for 3 months experienced no change in their performance on two tests of attention, the only tests of cognitive function administered.[74] Neither were there any differences on a test of working memory when 69-year-old women were treated with either CEE 0.625 mg or placebo.[75] Finally, performance on tests of verbal and visual memory, attention, verbal fluency or psychomotor speed failed to differ in 81-year-old women treated with either CEE 0.625 mg or placebo for 9 months.[76] In a recent study of 52-year-old women who had been postmenopausal for approximately 3 years, scores on measures of verbal fluency and verbal memory did not differ following 8 weeks of randomized treatment with either E_2 2 mg/day or placebo.[77]

Using a different endocrine manipulation, the possible causal relationship between estrogen and cognitive function was tested in another RCT. Infertile 32-year-old women with a large uterine myoma but with normal ovarian function were tested at pretreatment baseline, and again after 12 weeks of treatment, with 3.75 mg of a GnRH, leuprolide acetate depot (LAD), i.m. every 4 weeks for 12 weeks.[78] Then, all women continued to receive LAD every 4 weeks, with the random addition of either add-back CEE 0.625 mg/day or add-back placebo daily for another 8 weeks. Scores on tests of verbal memory and learning but not on visual memory decreased following 12 weeks of ovarian suppression in these young women, and the deficit was reversed only in the group that randomly received add-back CEE plus LAD. These findings underscore the specificity of estrogen on verbal memory, and also demonstrate that estrogen may be important for the maintenance of aspects of cognition in young women.

Several factors distinguish the RCTs that found beneficial effects of estrogen on aspects of cognition in postmenopausal women from those that did not. First, the positive RCTs all found a specificity of the estrogenic effect on short- and long-term verbal memory and working memory, and, importantly, three of the four RCTs that failed to find a beneficial effect of estrogen on cognition also failed to administer tests of verbal and/or working memory.[70,74,75] Second, the women in two of the three RCTs that found a protective effect of estrogen on verbal memory were a mean age of 45 years at the time their ovaries were surgically removed and ET was administered immediately after their surgery,[70,74,75] whereas in the negative studies estrogen was administered *de novo* to women several years following their surgical menopause,[74] to women who were 3 years postmenopausal,[77] or to older women over the age of 65 years.[75,76] This suggests that ET may protect against cognitive aging only when it is initiated at the time of menopause. Third, in the positive RCTs, 17β E_2 was administered either intramuscularly[68–70] or transdermally,[70] whereas the RCTs that failed to find a beneficial effect of estrogen on cognition had all administered CEE orally. This raises two possible pharmacologic issues, one involving the possible superior efficacy of 17β E_2 over CEE (whose major metabolite is E_1 sulfate), and/or the possible superior efficacy of non-oral routes of administration of estrogen with regard to its effects on brain. Future studies are needed to test these possibilities.

Cross-sectional Studies

Cross-sectional studies involve comparing cognitive performance between a group of postmenopausal women who use estrogen and a group of estrogen non-users who were matched for relevant control variables with the estrogen users. The majority of these studies found that estrogen users performed significantly better than non-users on tests of verbal fluency,[79–81] verbal memory,[82–84] and verbal[71,85] and spatial[86] working memory. Although these findings are largely consistent with those of the RCTs, the quality of the evidence is less robust since methodological weaknesses are, of course, inherent in cross-sectional designs. It should be noted, though, that the majority of the cross-sectional studies discussed above attempted to control statistically for participant characteristics that are known to independently influence cognitive functioning in older people, such as age, level of education and socioeconomic status (see Sherwin and Henry[79] for a more comprehensive review).

Longitudinal Studies

Longitudinal designs involve the recruitment of a group of women who are estrogen users and a group who are non-users selected from the same community. They are tested at the time of recruitment and then one or more times several years apart to determine whether aspects of cognitive functioning decline at different rates in the estrogen users and non-users with increasing age. Typically, these studies contain several hundred to several thousand women, which increases their reliability. They have generally found that estrogen users over the age of 65 years have significantly higher scores on verbal memory[47] verbal fluency[87] and visual memory,[88] and higher scores on the 3MSE,[89,90] compared with age-matched non-users. In one study, past-users who had initiated ET earlier after their menopause showed less cognitive decline across time than current-users and never-users.[89] Although the authors attributed this finding to the characteristics of the women rather than to beneficial effects of ET, it is also possible that the past-users were somewhat protected by the fact that they had started ET at a younger age than the current-users. Another study found less cognitive decline over time in women who had higher serum E_2 levels.[91] Four longitudinal studies failed to find a difference in cognitive functioning between estrogen users and non-users over time.[92–95] However, in one study[92] the women were an average age of 56 years at follow-up – considerably younger than the age at which a decline in verbal memory usually becomes manifest. In a longitudinal study of 75-year-old women tested 2 years later, those with lower E_2 levels were more likely to experience decline in global cognitive functioning and in verbal memory compared to women with higher endogenous E_2 levels.[96]

Although the preponderance of findings from the longitudinal studies show that estrogen users performed

BOX 24.1

HOW DO WE KNOW THAT THE EFFECT OF ESTROGEN TREATMENT (ET) ON COGNITION IN WOMEN DEPENDS ON THE TIMING OF THE INITIATION OF THERAPY?

There is some evidence that the inconsistencies in the literature with regard to the neuroprotective effects of estrogen therapy (ET) on aspects of memory function in postmenopausal women may be due to the age of the woman at the time of the initiation of ET.

In the first randomized controlled trial (RCT) of 45-year-old premenopausal women who needed to undergo a hysterectomy and bilateral oophorectomy for benign disease, women were tested before surgery, and again 3 months postoperatively after randomly receiving treatment with either estrogen or placebo.[1] Whereas scores on a test of verbal memory were maintained pre- to post-operatively in women who had randomly received ET, scores decreased significantly on the same test of memory in women treated with placebo postoperatively. In this case, ET, given to 45-year-old premenopausal women immediately following the surgical removal of their ovaries, was neuroprotective. Findings of other RCTs supported the view that estrogen maintained aspects of memory in postmenopausal women.[2,3]

In contrast, when ET or placebo was randomly given to hysterectomized women participating in the Women's Health Initiative Memory Study (WHIMS) whose average age was 72 years (range of 65–79 years), those treated with ET had a slightly but significantly lower score on a test of global cognitive functioning after 5 years compared to the placebo-treated women ($P = 0.04$).[4] When women who had developed cognitive impairment or dementia during the trial were excluded from the analyses, the mean difference in cognitive scores was no longer statistically significant. It is noteworthy that the WHIMS used only a global test of cognitive functioning (the Mini-Mental State Examination) which does not provide a measure of memory performance. Furthermore, in the WHIMS, the participants were treated with conjugated equine estrogen, Premarin (comprised of at least 10 estrogens[5]), rather than estradiol, which had been administered in the positive RCTs discussed above.

More evidence from basic neuroscience and from animal and human studies is required to support the intriguing hypothesis that ET is neuroprotective for middle-aged women but may be harmful for aspects of cognitive functioning when initiated in elderly women. The rationale that underlies this hypothesis is likely based on the differential response of middle-aged and aged brains to exposure to estrogen.[6]

References

1. Sherwin BB. Estrogen and/or androgen replacement therapy and cognitive functioning in surgically menopausal women. *Psychoneuroendocrinology.* 1988;13:345–357.
2. Duka T, Tasker R, McGowen JF. The effects of a 3-week estrogen hormone replacement on cognition in healthy elderly females. *Psychopharmacology.* 2000;149:129–139.
3. Phillips SM, Sherwin BB. Effects of estrogen on memory function in surgically menopausal women. *Psychoneuroendocrinology.* 1992;17:485–495.
4. Espeland MA, Rapp SR, Shumaker SA, et al. Conjugated equine estrogens and global cognitive function in postmenopausal women. *J Am Med Assoc.* 2004;291: 2959–2968.
5. Bhavnani BR. Pharmacokinetics and pharmacodynamics of conjugated equine estrogens: chemistry and metabolism. *Proc Soc Exp Biol Med.* 1998;217:6–16.
6. Gibbs RB. Effects of aging and long-term hormone replacement on cholinergic neurons in the medial septum and the nucleus basalis magnocellularis of ovariectomized rats. *J Neuroendocrinol.* 2003;15:475–485.

better on cognitive tests and experienced less deterioration in aspects of cognition with increasing age compared to the non-users, it must be considered that these studies are replete with several sources of bias. Their participant samples reflect the frequent observation that women who take ET after the menopause are younger, better educated and generally healthier than non-users.[89,95,97] This bias is particularly serious in studies of cognitive function and aging, since younger age and higher educational levels are themselves independent predictors of cognitive aging. Although most of the studies controlled statistically for age and education, statistical adjustment cannot completely account for these sources of bias. A second and related problem is that, in five repeated-measures studies, the 25–33% of women who failed to return for cognitive testing were older, had less education, had lower 3MSE scores at baseline, and were more likely to be estrogen non-users.[89–91,98] The characteristics of these women who failed to return for follow-up testing would have led to an underestimation of the strength of the associations that were found between ET and the preservation of cognitive functioning. Additionally, women who participated in the cross-sectional and longitudinal studies were between 60 and 85 years of age, considerably older than those who participated in the RCTs, and few investigators were able to objectively determine when the women had started to take ET in relation to their menopause. Finally, most of these samples contained predominantly naturally menopausal women who were taking a synthetic progestin along with estrogen in order to protect against endometrial hyperplasia. In many cases, data from E-alone users and from E + P users were analyzed together as the "estrogen group" and their scores on cognitive tests were compared with the never-users, constituting a failure to acknowledge that progestins themselves have psychoactive properties which sometimes attenuate or even reverse the beneficial effects of estrogen on brain structure and function.

The Women's Health Initiative Memory Study (WHIMS)

The Women's Health Initiative Memory Study (WHIMS), an ancillary study to the Women's Health Initiative (WHI), was the largest RCT to examine the effects of the continuous administration of estrogen plus progestin (CEE 0.625 mg + MPA 2.5 mg; E + P) or placebo to naturally postmenopausal women, and either estrogen-alone (CEE 0.625 mg; E) or placebo to hysterectomized women, on probable dementia and, secondarily, on mild cognitive impairment (MCI) and global cognitive abilities. Over 4500 women participated in WHIMS. In the E + P study, there was a significantly higher risk of probable, all-cause dementia in the active group compared with placebo (HR = 2.05, 95% CI = 1.21−3.48), but no difference between the groups in the incidence of MCI, thought to be a precursor to AD.[99] In the CEE-alone study, there were no significant differences in the incidences of probable dementia or MCI between the hormone and placebo groups, although there was a non-significant increase in probable dementia in the CEE-alone group (37 vs 25 per 10,000 person years). Not surprisingly, women in the CEE arm who had lower 3MSE scores at baseline were at greater risk of decline compared to the placebo group.[100] Therefore, the findings from the largest RCT to have ever been undertaken failed to confirm previous evidence from smaller RCTs and from longitudinal and cross-sectional studies that estrogen protected against cognitive aging in postmenopausal women.

In 2006, findings from an ancillary study to the WHI and to WHIMS, the WHI Study of Cognitive Aging (WHISCA), appeared.[101] This longitudinal study, initiated 3 years after the women in the WHI had been randomized to active treatment or placebo, assessed whether treatment with CEE + MPA affects the rates of age-related changes in specific cognitive functions compared to placebo. The 1416 women, aged approximately 74 years at WHISCA baseline, were administered neuropsychological tests annually until the WHI was halted in 2002. In contradistinction to the WHIMS study, which found a decrease in global cognitive scores in the CEE + MPA group over time,[100] 3MSE scores of the CEE + P-treated women in the WHISCA study remained stable across the 3 years between the time of their randomization at the start of the WHI to the time of their recruitment into WHISCA 3 years later. In WHISCA, CEE + MPA had a negative impact on verbal memory and a positive impact on figural memory over time compared with placebo, but other cognitive domains were unaffected. Only 47% of the women on CEE + MPA were adherent by the 2-year follow-up.

The considerable inconsistencies between these studies on the putative protective effect of estrogen on cognitive aging in women are large and require explanation. At first glance the results of the WHIMS[99,100] were somewhat surprising, since it had been expected that the findings from this large, multi-center RCT would mirror earlier findings from the small RCTs discussed above. At second glance, however, the numerous discrepancies in methodology and in the populations studied between the earlier, smaller RCTs and the WHIMS became apparent. For example, whereas the earlier, positive studies had administered parenteral E_2, women in the WHIMS were administered oral CEE alone or in combination with MPA. Second, the 3MSE was the only measure of cognitive function used in WHIMS, so that any effect of the treatment on verbal memory and working memory (the cognitive domains found in previous studies to be among the most estrogen-sensitive) could not be assessed. Third, participants in the WHIMS were between 65 and 79 years of age (mean age of approximately 72 years) at the time of their recruitment into this study and had been postmenopausal for approximately 21 years following their presumed spontaneous menopause at an average age of 51 years − probably for a considerably longer period of time for women who had undergone a surgical menopause. As might be expected in such an elderly sample, approximately 55% were being treated for hypertension, 11% were diabetic, 12% were on statins, 23% were obese and a further 16% were morbidly obese at baseline. Therefore, the participants in WHIMS were older women at high risk for cardiovascular and cerebrovascular disease, which would, in turn, have a negative influence on cognitive functioning irrespective of hormonal status. However, it should also be noted that women assigned to CEE or placebo had comparable distributions of health risk factors.[102] Nonetheless, their age and health profile differed dramatically from the younger, healthy surgically menopausal women in the smaller RCTs.[78,79]

THE CRITICAL PERIOD HYPOTHESIS

Perhaps the most compelling explanation for the failure of WHIMS to find a beneficial effect of estrogen on cognition is the advanced age of the women at the time estrogen treatment was initiated. This raised the issue of whether estrogen had protective effects on some cognitive domains only when it was initiated closely in time to a natural or to a surgical menopause, but not when it was initiated years or decades after the menopause had occurred. This idea gave rise to the Critical Period Hypothesis, which holds that ET confers optimal cognitive protection only when it is initiated closely in time to the menopause, and will be ineffective or may even cause harm when initiated decades after the menopause.[103–106] If this theory is correct, it would suggest that there is a window of opportunity around the time of menopause when the initiation of treatment with estrogen would protect against aspects of cognitive aging in women. Indeed, there is now a considerable amount of evidence from different levels of analysis to support the Critical Period Hypothesis.

Evidence from Basic Neuroscience

The finding that estrogen enhanced synaptogenesis by increasing the density of spines on neurons in the CA1 area of the hippocampus[13] provided a rationale for the clinical studies that had documented a protective effect of estrogen on aspects of memory in women. However, whereas the administration of E_2 to young OVX rats increased axospinous synapse density, it was without effect on dendritic spine density in aged OVX animals.[107] This suggests that estrogen-induced morphological plasticity is attenuated in aged rats compared with young animals. However, in both young and aged monkeys the administration of E_2 caused a 35% increase in total spines in the CA1, demonstrating that, unlike the rat, aged female monkeys retain the capacity for an increase in spines in response to E_2.[108] On the other hand, the aged monkeys in that study were perimenopausal when they underwent OVX,and E_2 was administered soon after surgery,[43] similar to the experimental model used in the positive human RCTs.[68,69]

Structural changes to hippocampal neurons were investigated as a function of the timing of initiation of ET in rats. When young rats underwent OVX and ET was initiated 4 days following surgery, synaptic density in three CA1 strata of the hippocampus was significantly greater than that in rats from whom ET had been withheld for 12 days following their OVX.[109,110] Therefore, the delay between OVX and the initiation of ET impaired the ability of the hormone to modulate synaptic density. Finally, the ability of estrogen to enhance basal forebrain cholinergic function also declines with age.[109]

Evidence from Rat Behavioral Studies

Animal behavioral studies also provide supportive evidence that estrogenic effects on memory may be dependent on the age of the animal and on the timing of the initiation of ET following the loss of ovarian function. When E_2 or $E_2 + P$ was initiated either immediately or 3 months following OVX in aged rats, they made significantly fewer errors on a test of spatial memory than OVX control rats and rats in whom E_2 was initiated after a 10-month delay following surgery.[111] Similarly, when E_2 was administered to rats immediately following OVX when they were either 12 or 17 months of age, a significant improvement in performance of acquisition and delay trials of a working memory task occurred compared to the controls.[112] However, when E_2 was administered 5 months following OVX undertaken in 17-month-old animals, no enhancements in memory performance were evident. Likewise, ET enhanced performance on a spatial memory task in young and middle-aged OVX rats, but not in aged OVX rats.[113]

Evidence from Non-human Primate Studies

When rhesus monkeys underwent OVX when they were pre- or perimenopausal and were then treated with E_2 or vehicle 30 weeks later, ET reversed the marked age-related impairment exhibited by OVX vehicle-treated monkeys on a delayed-response test of spatial working memory.[114] However, aged animals who had undergone OVX 10 or more years prior to the initiation of treatment with E_2 showed much weaker and more circumscribed effects of the hormone treatment on cognitive performance,[115] suggesting that there may be a critical window for the initiation of ET for it to optimally enhance cognitive functioning.

Evidence from Human Studies

Although no studies have been undertaken to directly test the Critical Period Hypothesis in women, some indirect evidence is available. In a cross-sectional study, ET was initiated in 10 women within 3 months of their surgical menopause at a mean age of 42 years.[116] When neuropsychological testing was administered when the women were a mean age of 75 years, the E-treated women performed significantly better than an age- and education-matched control group on tests of verbal memory and constructional ability. Likewise, an Australian study[117] examined cognitive functioning subsequent to early initiation of hormone therapy (HT) (the commencement of treatment before the age of 56 years for at least 12 months duration in naturally menopausal women or 5 years or less after bilateral oophorectomy), and late initiation (the commencement of HT at age > 56 years, or > 5 years following a surgical menopause). Early initiators of HT achieved higher scores on a test of global cognitive function, and were significantly faster on a test of attention compared to the HT late-initiators. Older HT early-initiators aged 70–79 years also scored significantly higher than the late-initiators on a test of verbal fluency. Therefore, in this well-designed study, early initiation of HT seemed to have a beneficial effect on most cognitive tests in these elderly women, whereas late initiation had a detrimental effect.

In a population-based study, women who began HT during the menopausal transition had better memory scores than women who began HT after their final menstrual period.[118] Matthews and colleagues[89] reported significantly less cognitive decline in elderly women who initiated ET very close in time to their spontaneous menopause, but those who were older when they initiated ET were not protected. Likewise, in the Cache County longitudinal study of elderly women and men, prior users of ET who had initiated treatment in close proximity to the menopause had a significantly reduced risk of Alzheimer's disease (AD) whereas current use of estrogen was not protective in these older women unless such use had exceeded 10 years.[119] Because the initiation of HT was based on recall in all the human studies, these findings should be interpreted cautiously.

In summary, evidence from basic neuroscience and from rodent, non-human primate and human studies converges to support the idea that E_2 may prevent the deleterious effects of cognitive aging only when it is administered at the time of, or soon after, the cessation of ovarian function, and may actually cause harm when treatment is initiated decades after the menopause. However, it is also important to note that some findings are inconsistent with the Critical Period Hypothesis. For example, in several RCTs, ET initiated in older women enhanced their performance on tests of verbal memory.[70,91] Although the Critical Period Hypothesis would predict that ET would fail to protect memory in these older women because their spontaneous menopause had occurred at least 15 years prior to the initiation of ET, it is also potentially interesting and important to note that the positive effects on memory in both aforementioned studies occurred with the use of a transdermal, not an oral, E_2 preparation. This raises the possibility that the specific estrogenic compound used as well as its route of administration may play an important role in the attempt to understand the salient factors that likely interact to promote a protective effect of estrogen on aspects of cognitive functioning in aging women.

Although it is not certain why estrogen might protect neurons involved in memory processes when it is given soon after menopause but not when significant time has elapsed since the cessation of ovarian function, several explanations might be considered. It is possible that neurons become less sensitive to estrogen following a prolonged absence of exposure to the hormone and/or that older neurons have reduced responsivity to the hormone. Indeed, although young-adult rats responded to estrogen with increased expression of brain-derived neurotrophic factor (BDNF), which is important for the maintenance of plasticity in the aging brain, the administration of estrogen to senescent rats actually decreased BDNF expression in the olfactory bulb and basal forebrain, suggesting that there is a general decline in hormonal responsiveness of neurotrophin receptors and ligands in older, reproductively senescent animals compared with younger animals.[120] Also, prolonged absence of ovarian hormones was associated with a loss of neurogenic response to E_2 and a decrease in the number of new cells expressing a neuronal phenotype.[121] If this is also true for humans, it might explain why the 72-year-old hormone-treated women in the WHIMS study were not protected against cognitive aging or AD.

Second, the relatively recent concept of the "healthy cell bias" of estrogen action proposes that there is a beneficial effect of estrogen on the survival of neurons that are healthy at the time of their exposure to estrogen, but an exacerbation of neurological demise occurs if exposure to estrogen takes place when neurons are already compromised.[122] The findings that physiological doses of estrogen attenuate neuronal death resulting from an ischemic brain injury only when the hormone is administered to an animal that is healthy prior to the occurrence of the injury[123] would seem to support the healthy cell bias hypothesis. If this is also true for women, it could explain the differential response of older women to ET. For example, three RCTs have found that older women given ET who had scored at or above the average on tests of memory at pretreatment showed significantly less decline in verbal memory following several years of treatment compared to those given placebo, while no beneficial effects of ET on memory occurred in women who demonstrated memory deficits at pretreatment.[71,100,124] Because it is likely that older women who score below average on memory tests may have an undiagnosed neuropathological process underway, the healthy cell bias hypothesis would predict that these older, mildly cognitive-impaired women would respond poorly or not at all to ET. Although more human studies are needed to further test this hypothesis, for the time being it can be tentatively concluded that estrogen provides neuroprotection to young and middle-aged brains, but not to elderly brains that have already experienced neuronal loss, decreases in brain volume, and alterations in neurotransmitter systems, and/or that may be affected by neuropathology.

The possibility that ET initiated at the time of menopause and continued for a few years is neuroprotective raises the important clinical issue as to whether such treatment will have enduring protective effects on cognitive functioning 20 or 30 years later, when cognitive aging begins to be manifest. Indeed, there is some evidence to support that possibility. In a cross-sectional study, women who started ET within 3 months of their surgical menopause at a mean age of 42 years performed significantly better than an age- and education-matched control group on tests of verbal memory and constructional ability when the women were a mean age of 75 years.[116] In a longitudinal study, early initiators of ET were treated for an average duration of 4.5 years and tested 23 years later at a mean age of 74 years.[125] Scores of the ET ever-users were significantly higher than those of the ET never-users on tests of short- and long-term verbal memory, verbal abstract reasoning and concept formation, and verbal fluency; 30 months later, scores on the short- and long-term verbal memory test had improved in the ET ever-users and had declined in the ET never-users. In a second longitudinal study, nondemented 72-year-old women were administered cognitive tests twice, 4–6 years apart.[89] Past E users were significantly younger than the current E users when they initiated ET (48 years versus 52 years), and the current and past-users demonstrated better performance on the 3MSE than the never-users at baseline. After 4–6 years, when they were an average age of 76 and 78 years, respectively, the ET past-users exhibited less decline on the 3MSE compared to both the current users and the never-users.

In summary, both the cross-sectional and longitudinal studies that contained information on the timing of initiation of ET provide evidence that the early initiation of ET around the time of menopause is associated with less cognitive decline 20–25 years later, while the late initiation of ET is not protective. However, because there are so few studies, because the women's recall of the timing of their initiation of treatment may be unreliable, and because they had not been randomly assigned to ET, a definitive conclusion as to whether the early initiation of ET for a few years will have enduring protective effects on cognitive aging is premature at the present time.

Only one RCT tested women 15 years after they had been randomly assigned to estrogen or placebo at the time of their menopause. Here, 261 women had randomly received E + P or placebo for 2–3 years when they were a mean age of 54 years,[126] and the short Blessed test, a measure of global cognitive function, was administered 5–15 years later when the women were a mean age of 65 years. Women who had been randomized to E + P for 2–3 years and then discontinued treatment had a significantly lower risk of cognitive impairment 5–15 years later compared to women who had originally received placebo (never-users) (OR: 0.36, 95% CI, 0.15–0.90; $P < 0.03$). The risk of cognitive impairment for the long-term users (those who had been randomly assigned to E + P 5–15 years earlier and continued to take it) was also significantly lower than that of the never-users (OR: 0.34; 95% CI, 0.11–0.80; $P < 0.01$). These findings constitute the only evidence available from a RCT showing that the initiation of HT in the early postmenopause for 2–3 years had an enduring protective effect against cognitive decline 5–15 years later, when some cognitive decline would be expected due to brain aging.[127] In view of its potential clinical importance, much more work is needed in this area.

THE WHI AND WHIMS FINDINGS REVISITED

The preliminary reanalysis of the WHIMS data by age found that women who took ET before the age of 65 years (prior to their enrollment in the WHI) were 50% less likely to develop AD and all-cause dementia compared to non-users.[128] Indeed, young, but not aged, OVX rats experience an increase in dendritic spine density[129] and an increase in ChAT,[111] and perform better on tests of memory when they are treated soon after OVX,[130] and there is some indirect evidence that this is also true for women. Therefore, the currently available findings from the WHI and WHIMS reanalyses by age accord with the Critical Period Hypothesis by showing that ET has protective effects when taken by younger, healthier women, and may cause harm when given to older women.

In view of the fact that normal aging is associated with continuous, adverse changes in brain structure and function, it does not seem reasonable to compare, or to consider together, studies on estrogen and cognition that administered the hormone to women of very different ages, since the observations in elderly women clearly occur against a background of greater accrued negative consequences of brain aging. Furthermore, the development of neuropathology is also associated with increasing age and may be evident in the brain for some time prior to the manifestation of clinical symptoms,[131] suggesting that even though the older women in some studies may not have had cognitive deficits at the time of their recruitment, they could have had undocumented brain pathology.

POSSIBLE INFLUENCE OF UNCONTROLLED VARIABLES

It is important for the design of future studies to acknowledge that methodologic issues in the studies reviewed above complicate the effort to draw conclusions among them. For example, some studies administered E alone whereas others administered E + P to postmenopausal women, and others investigated women who were on different hormone regimens and analyzed the data together. MPA was the progestin most commonly used in combined therapy, including in the WHIMS. This is important because of the evidence that progestins, and MPA in particular, might have a negative effect on cognition and/or may attenuate the beneficial effect of the co-administered estrogen. For example, the decrease in apical dendritic spine density in the CA1 area of the hippocampus that occurs during the estrus phase of the rat cycle was prevented by the administration of RU-486, a progesterone (P) receptor antagonist, suggesting that P causes the decrease in spine density.[12] Second, route of administration of E may also be important. Oral, but not transdermal, administration of estrogen increases the hepatic production of sex hormone-binding globulin (SHBG) and results in lower concentrations of free E_2.[132] Also, CEE, the drug used in most studies, including the WHIMS, is predominantly comprised of estrone (E_1) and at least 10 other hormones.[133] Since E_1 is biologically weaker than E_2, CEE may not be bioequivalent to either endogenous or exogenous E_2, as the affinity of E_1 for the estrogen receptor (ER) is approximately half of that of E_2.[134] Finally, many studies used only a global measure of cognitive function, which does not allow for the evaluation of performance in specific cognitive domains. This is a considerable failing in view of the fact that numerous studies have found a specificity of the estrogenic effect on verbal memory and on working memory using specific neuropsychological tests.

CONCLUSION

Although there is now a considerable amount of literature on estrogen and cognitive functioning in women, taken together the findings are inconsistent. This is most likely due to the fact that the characteristics of the populations studied, the specific drugs and their routes of administration, the neuropsychological tests used and, likely most importantly, the age of the women at the time of the initiation of ET or HT differed considerably between studies. Moreover, it is also clear that unopposed E has different effects on the brain and on other organ systems than the administration of E + P. Future studies are needed to test whether initiating ET around the time of menopause in a healthy women for a few years will prevent or delay cognitive aging 25 years later, thereby safeguarding the quality of life for women during the latter third of their lifespan.

References

1. McEwen BS. Estrogen action throughout the brain. *Recent Prog Horm Res.* 2002;57:357–384.
2. McEwen BS, Alves SE. Estrogen actions in the central nervous system. *Endocr Rev.* 1999;20:279–307.
3. Pfaff DW. *Estrogen and Brain Function.* New York: Springer-Verlag; 1980.
4. Stumpf WE, Sar M. Steroid hormone target sites in the brain: the differential distribution of estrogen, progestin, androgen, and glucocorticosteroid. *J Steroid Biochem.* 1976;7:1163–1170.
5. Shughrue PJ, Merchenthaler I. Distribution of estrogen receptor β immunoreactivity in the rat central nervous system. *J Comp Neurol.* 2001;436:64–81.
6. Shughrue PJ, Scrimo PJ, Merchenthaler I. Estrogen binding and estrogen receptor characterization (ER-α and ER-β) in the cholingergic neurons of the rat basal forebrain. *Neurosci.* 2000;96:41–49.
7. Lee SJ, McEwen BS. Neurotrophic and neuroprotective actions of estrogens and their therapeutic implications. *Ann Rev Pharmacol Toxicol.* 2001;41:569–591.
8. Kelly MJ, Levin ER. Rapid actions of plasma membrane estrogen receptors. *Trends Endocrinol Metab.* 2001;12:152–156.
9. Walf AA, Frye CA. Rapid and estrogen receptor beta mediated actions in the hippocampus mediate some functional aspects of estrogen. *Steroids.* 2008;73:997–1007.
10. Liu F, Day M, Muniz LC, et al. Activation of estrogen receptor beta regulates hippocampal synaptic plasticity and improves memory. *Nat Neurosci.* 2008;11:334–342.
11. Brinton RD. Estrogen-induced plasticity from cells to circuits: predictions for cognitive function. *Trends Pharmacol Sci.* 2009;30:212–222.
12. Gould E, Woolley CS, Frankfurt M, McEwen BS. Gonadal steroids regulate dendritic spine density in hippocampal pyramidal cells in adulthood. *J Neurosci.* 1990;10:1286–1291.
13. Woolley CS, Gould E, Frankfurt M, McEwen BS. Naturally occurring fluctuation in dendritic spine density on adult hippocampal pyramidal neurons. *J Neurosci.* 1990;10:4035–4039.
14. Sandstrom NJ, Williams CL. Memory retention is modulated by acute estradiol and progesterone replacement. *Behav Neurosci.* 2001;115:384–393.
15. Leranth C, Shanabrough M, Redmond Jr DE. Gonadal hormones are responsible for maintaining the integrity of spine synapses in the CA1 hippocampal subfield of female nonhuman primates. *J Comp Neurol.* 2002;447:34–42.
16. Luine VN. Estradiol increases choline acetyltransferase activity in specific basal forebrain nuclei and projection areas of female rats. *Exp Neurol.* 1985;89:484–490.
17. Davies P, Maloney AJF. Selective loss of central cholinergic neurons in Alzheimer's disease. *Lancet.* 1976;2:1402–1406.
18. Toran-Allerand CD, Singh M, Setalo G. Novel mechanisms of estrogen action in the brain: new players in an old story. *Front Neuroendocrinol.* 1999;20:97–121.
19. Pike C. Estrogen modulates neuronal Bel-xl expression and β-amyloid-induced apoptosis: relevance to Alzheimer's disease. *J Neurochem.* 1999;72:1552–1563.
20. Greene RA. Estrogen and cerebral blood flow: a mechanism to explain the impact of estrogen on the incidence and treatment of Alzheimer's disease. *Intl J Fertil Womens Med.* 2000;45:253–257.
21. Green PS, Gridley KE, Simpkins JW. Nuclear estrogen receptor-independent neuroprotection by estratrienes: a novel interaction with glutathione. *Neuroscience.* 1998;84:7–10.
22. Jaffe AB, Toran-Allerand CD, Greengard P, et al. Estrogen regulates metabolism of Alzheimer amyloid beta precursor protein. *J Biol Chem.* 1994;269:13065–13068.

23. Baddeley AD, Warrington EK. Amnesia and the distinction between long- and short-term memory. *J Verb Learning Verb Behav.* 1970;9:176–189.
24. Baddeley AD. *Human Memory: Theory and Practice.* Boston: Allyn & Bacon; 1990.
25. Goldman-Rakic PS. Working memory and the mind. *Scientific Am.* 1992;267:110–117.
26. Smith EE, Jonides J, Marshuetz C, et al. Components of verbal working memory: evidence from neuroimaging. *Proc Nat Acad Sci USA.* 1998;95:876–882.
27. Cohen JD, Perlstein WM, Braver TS, et al. Temporal dynamics of brain activation during a working memory task. *Nature.* 1997;386:604–608.
28. Petrides M, Alivisatos B, Meyer E, et al. Functional activation of the human frontal cortex during the performance of verbal working memory tasks. *Proc Natl Acad Sci USA.* 1993;90:878–882.
29. Grady CL, Craik FIM. Changes in memory processing with age. *Curr Opin Neurobiol.* 2000;10:224–231.
30. Just MA, Carpenter PA, Keller TA. The capacity theory of comprehension: new frontiers of evidence and arguments. *Psychol Rev.* 1996;103:773–780.
31. Raz NK, Rodrigue KM. Differential aging of the brain: patterns, cognitive correlates and modifiers. *Neurosci Biobehav Rev.* 2006;30:730–748.
32. Squire LR, Zola-Morgan S, Cave CB, et al. Memory: organization of brain systems and cognition. In: Meyer DE, Kornblurn S, eds. *Attention and Performance.* Cambridge: MIT Press; 1993:393–423.
33. Gibbs RB, Wu D, Hersh LB, et al. Effects of estrogen replacement on the relative levels of choline acetyltransferase, trkA, and nerve growth factor messenger RNAs in the basal forebrain and hippocampal formation of adult rats. *Exp Neurol.* 1994;129:70–80.
34. Halpern DF. *Sex differences in cognitive abilities.* Hillsdale: Lawrence Erlbaum Associates; 1992.
35. Zelinski EM, Burnright KP. Sixteen-year longitudinal and time lag changes in memory and cognition in older adults. *Psychol Aging.* 1997;12:503–513.
36. Park DC, Welsh RC, Marshuetz C, et al. Working memory for complex scenes: age differences in frontal and hippocampal actvations. *J Cogn Neuroci.* 2003;15:1122–1134.
37. Small SA, Stern Y, Tang M, et al. Selective decline in memory function among healthy elderly. *Neurology.* 1999;52:1392–1396.
38. Zelinski EM, Gilewski MJ, Schaie KW. Individual differences in cross-sectional and 3-year longitudinal memory performance across the adult life span. *Psychol Aging.* 1993;8:176–186.
39. Salthouse TA. Mediation of adult age differences in cognition by reduction in working memory and speed of processing. *Psychol Sci.* 1991;2:179–183.
40. Gregoire J, Van der Linden M. Effects of age on forward and backward digit span. Aging. *Neuropsychol Cognition.* 1997;4: 140–149.
41. Hedden T, Gabrieli JDE. Insights into the ageing mind: a view from cognitive neuroscience. *Nat Rev Neurosci.* 2004;5:87–96.
42. Esiri MM. Ageing and the brain. *J Pathol.* 2007;211:181–187.
43. Morrison JH, Hoff PR. Life and death of neurons in the aging brain. *Science.* 1997;278:412–419.
44. Eriksson PS, Perfilieva E, Björk-Eriksson ET, et al. Neurogenesis in the adult human hippocampus. *Nat Med.* 1998;4:1313–1317.
45. Coffey CE, Wilkinson WE, Parashos LA, et al. Quantitative cerebral anatomy of the aging human brain. *Neurol.* 1992;42:527–536.
46. Pfefferbaum A, Mathalon DH, Sullivan EV, et al. A quantitative magnetic resonance imaging study of changes in brain morphology from infancy to late adulthood. *Arch Neurol.* 1994;51:874–887.
47. Jacobs B, Driscoll L, Schall M. Life span dendritic and spine changes in areas 10 and 18 of human cortex: a quantitative Golgi study. *J Comp Neurol.* 1997;386:661–680.
48. Raz N. The aging brain observed *in vivo*: Differential changes and their modifications. In: Cabeza Nyberg L, Park D, eds. *Cognitive Neuroscience of Aging.* New York: Oxford University Press; 2005:19–57.
49. Raz N, Lindenberger U, Rodrigue KM, et al. Regional brain changes in aging healthy adults: general trends, individual differences and modifiers. *Cereb Cortex.* 2005;15:1676.
50. Santoro N. The menopausal transition. *Am J Med.* 2005;118:8–13.
51. Gore AC, Oung T, Yung S, et al. Neuroendocrine mechanisms for reproductive senescence in the female rat: gonadotropin-releasing hormone neurons. *Endocrine.* 2000;3:315–323.
52. Gore AC. Gonadotropin-releasing hormone neurons: multiple inputs, multiple outputs. *Endocrinology.* 2004;145:4016–4017.
53. Bohler CS, Greenblatt RB. The pathophysiology of the hot flush. In: Greenblatt RB, Mahesh VB, McDonough PG, eds. *The Menopausal Syndrome.* New York: Medcom Press; 1974:29–38.
54. Henderson VW, Dudley EC, Gutherie JR, et al. Estrogen exposures and memory at midlife: a population-based study of women. *Neurology.* 2003;60:1369–1371.
55. Herlitz A, Thilers P, Habib R. Endogenous estrogen is not associated with cognitive performance before, during, or after menopause. *Menopause.* 2007;14:425–431.
56. Henderson VW. Cognitive changes after menopause: influence of estrogen. *Clin Obstet Gynecol.* 2008;51:618–626.
57. Kok HS, Kuh D, Cooper R, et al. Cognitive function throughout the life course and the menopausal transition in a British birth cohort. *Menopause.* 2006;13:19–27.
58. Fuh J-L, Wang S-J, Lee S-J, et al. A longitudinal study of cognition change during early menopausal transition in a rural community. *Maturitas.* 2006;53:447–453.
59. Greendale GA, Huang M-H, Wight RG, et al. Effects of the menopause transition and hormone use on cognitive performance in midlife women. *Neurol.* 2009;72:1850–1857.
60. Maki PM, Gast MJ, Vieweg AJ, et al. Hormone therapy in menopausal women with cognitive complaints: a randomized, double-blind trial. *Neurology.* 2007;69:1322–1330.
61. Caldwell BM, Watson RI. An evaluation of psychological effects of sex hormone administration in aged women, results of therapy after six months. *J Gerontol.* 1952;7:228–244.
62. Wechsler D. *WMS-III: Wechsler Memory Scale, 3rd edition.* San Antonio, USA: Harcourt Brace & Company; 1997.
63. Campbell S, Whitehead M. Oestrogen therapy and the menopausal syndrome. *Clin Obstet Gynecol.* 1977;4:31–47.
64. Fedor-Freybergh P. The influence of oestrogens on the wellbeing and mental performance in climacteric and postemenopausal women. *Acta Obstet Gynecol Scand.* 1997;64(Suppl):1–91.
65. Hackman BW, Galbraith D. Six-month study of oestrogen therapy with piperazine oestrone sulphate and its effects on memory. *Curr Med Res Opin.* 1977;4:21–27.
66. Rauramo L, Lagerspetz KP, Engblom P, et al. The effect of castration and peroral estrogen therapy on some psychological functions. *Front Horm Res.* 1975;3:94–104.
67. Van Hulle G, Demol R. A double-blind study into the influence of estriol on a number of psychological tests in post-menopausal women. In: van Keep PA, Greenblat RB, Albeaux-Fernet M, eds. *Consensus on Menopause Research: A Summary of International Opinion.* Baltimore: University Park Press; 1976:94–9.
68. Sherwin BB. Estrogen and/or androgen replacement therapy and cognitive functioning in surgically menopausal women. *Psychoneuroendocrinology.* 1988;13:345–357.

69. Phillips SM, Sherwin BB. Effects of estrogen on memory function in surgically menopausal women. *Psychoneuroendocrinology.* 1992;17:485–495.
70. Duka T, Tasker R, McGowan JF. The effects of 3-week estrogen hormone replacement on cognition in elderly healthy females. *Psychopharmacology.* 2000;149:129–139.
71. Viscoli CM, Brass LM, Kernan WN, et al. Estrogen therapy and the risk of cognitive decline: results from the Women's Estrogen for Stroke Trial (WEST). *Am J Obstet Gynecol.* 2005;192:387–393.
72. Joffe H, Hall JE, Gruber S, et al. Estrogen therapy selectively enhances prefrontal cognitive processes: a randomized, double-blind, placebo-controlled study with functional magnetic resonance imaging in perimenopausal women and recently postmenopausal women. *Menopause.* 2006;13:411–422.
73. Krug R, Born J, Rasch B. A 3-day estrogen treatment improves prefrontal cortex-dependent cognitive function in postmenopausal women. *Psychoneuroendocrinology.* 2006;31:965–975.
74. Ditkoff EC, Crary WG, Cristo M, et al. Estrogen improves psychological function in asymptomatic postmenopausal women. *Obstet Gynecol.* 1991;78:991–995.
75. Janowsky SJ, Chavez B, Orwoll F. Sex steroids modify working memory. *J Cogn Neurosci.* 2000;12:407–414.
76. Binder EF, SchectmanKB, Birge SJ, et al. Effects of hormone replacement therapy on cognitive performance in elderly women. *Maturitas.* 2001;38:137–146.
77. LeBlanc ES, Neiss MB, Carello PE, et al. Hot flashes and estrogen therapy do not influence cognition in early menopausal women. *Menopause.* 2007;1(4):191–202.
78. Sherwin BB, Tulandi T. "Add-back" estrogen reverses cognitive deficits induced by a gonadotropin-releasing hormone agonist in women with *leiomyomata uteri. J Clin Endocrinol Metab.* 1996;81:2545–2549.
79. Sherwin BB, Henry JF. Brain aging modulates the neuroprotective effects of estrogen on selective aspects of cognition in women: A critical review. *Front Neuroendocrinol.* 2008;29:88–113.
80. Barrett-Connor E, Kritz-Silverstein D. Estrogen replacement therapy and cognitive function in older women. *J Am Med Assoc.* 1993;269:2637–2641.
81. Hogervorst E, Boshuisen M, Riedel W, et al. The effect of hormone replacement therapy on cognitive function in elderly women. *Psychoneuroendocrinology.* 1999;24:43–68.
82. Miller KL, Conney JC, Rasgon NL, et al. Mood symptoms and cognitive performance in women estrogen users and nonusers and men. *J Am Geriatr Soc.* 2002;50:1826–1830.
83. Kampen DL, Sherwin BB. Estrogen use and verbal memory in healthy postmenopausal women. *Obstet Gynecol.* 1994;83:979–983.
84. Kimura D. Estrogen replacement therapy may protect against intellectual decline in postmenopausal women. *Horm Behav.* 1995;29:312–321.
85. Robinson D, Friedman L, Marcus R, et al. Estrogen replacement therapy and memory in older women. *J Am Geriatr Soc.* 1994;42:919–922.
86. Grigorova M, Sherwin BB, Tulandi T. Effects of treatment with leuprolide acetate depot on working memory and executive functions in young premenopausal women. *Psychoneuroendocrinology.* 2006;31:935–947.
87. Grodstein F, Chen J, Pollen DA, et al. Postmenopausal hormone therapy and cognitive function in healthy older women. *J Am Geriatr Soc.* 2000;48:746–752.
88. Resnick SM, Metter EJ, Zonderman AB. Estrogen replacement therapy and longitudinal decline in visual memory: a possible protective effect? *Neurology.* 1997;49:1491–1497.
89. Matthews K, Cauley J, Yaffe K, et al. Estrogen replacement therapy and cognitive decline in older community women. *J Am Geriatr Soc.* 1999;47:518–523.
90. Steffens DC, Norton MC, Plassman BL, et al. Enhanced cognitive performance with estrogen use in nondemented community dwelling older women. *J Am Geriatr Soc.* 1999;47:1171–1175.
91. Yaffe K, Lui L-Y, Grady D, et al. Cognitive decline in women in relation to non-protein-bound oestradiol concentrations. *Lancet.* 2000;356:708–712.
92. de Moraes SA, Szklo M, Knopman D, et al. Prospective assessment of estrogen replacement therapy and cognitive functioning: Atherosclerosis Risk in Communities Study. *Am J Epidemiol.* 2001;154:733–739.
93. File SE, Heard JE, Rymer J. Trough oestradiol levels associated with cognitive impairment in post-menopausal women after 10 years of oestradiol implants. *Psychopharmacologia.* 2002;161:107–112.
94. Fillenbaum GG, Hanlon JT, Landerman LR, et al. Impact of estrogen use on decline in cognitive function in a representative sample of older community-resident women. *Am J Epidemiol.* 2001;153:137–144.
95. O'Hara R, Schroder CM, Bloss C, et al. Hormone replacement therapy and longitudinal cognitive performance in postmenopausal women. *Am J Geriatr Psychiatr.* 2005;13:1107–1110.
96. Yaffe K, Barnes D, Lindquist K, et al. Health ABC Investigators, Endogenous sex hormone levels and risk of cognitive decline in an older biracial cohort. *Neurobiol Aging.* 2007;28:171–178.
97. Matthews KA, Kuller LH, Wing RR, et al. Prior to use of estrogen replacement therapy, are users healthier than nonusers? *Am J Epidemiol.* 11996;43:971–978.
98. Rice MM, Graves AB, McCurry LE, et al. Postmenopausal estrogen and estrogen-progestin use and 2-year rate of cognitive change in a cohort of older Japanese American women: the Kame project. *Arch Intern Med.* 12000;60:1641–1649.
99. Shumaker SA, Legault C, Kuller L, et al. for the Women's Health Initiative Memory Study, Conjugated equine estrogens and incidence of probable dementia and mild cognitive impairment in postmenopausal women. *J Am Med Assoc.* 2004;291:2947–2958.
100. Espeland MA, Rapp SR, Shumaker SA, et al. Conjugated equine estrogens and global cognitive function in postmenopausal women. *J Am Med Assoc.* 2004;291:2959–2968.
101. Resnick SM, Maki PM, Rapp SR, et al. for the Women's Health Initiative Study of Cognitive Aging Investigators, Effects of combination estrogen plus progestin hormone treatment on cognition and affect. *J Clin Endocrinol Metab.* 2006; 91:1802–1810.
102. Writing Group for the Women's Health Initiative Investigators. Risks and benefits of estrogen plus progestin in healthy postmenopausal women. *J Am Med Assoc.* 2002;288:321–333.
103. Gibbs RB, Gabor R. Estrogen and cognition: applying preclinical findings to clinical perspectives. *J Neurosci Res.* 2003;74:637–643.
104. Maki PM. Hormone therapy and cognitive function: is there a critical period for benefit? *Neuroscience.* 2006;138:1027–1030.
105. Resnick SM, Henderson VW. Hormone therapy and risk of Alzheimer disease: a critical time. *J Am Med Assoc.* 2002;288: 2170–2172.
106. Sherwin BB. Estrogen and memory in women: how can we reconcile the findings? *Horm Behav.* 2005;47:371–375.
107. Adams MM, Shah RA, Janssen WGM, et al. Different modes of hippocampal plasticity in response to estrogen in young and aged female rats. *Proc Natl Acad Sci USA.* 2001;98:8071–8076.
108. Hao J, Janssen WGM, Tang Y, et al. Estrogen increases the number of spinophillin-immunoreactive spines in the hippocampus of young and aged female rhesus monkeys. *J Comp Neurol.* 2003;465:540–550.
109. Gibbs RB. Effects of gonadal hormone replacement on measures of basal forebrain cholinergic function. *Neuroscience.* 2000;101:931–938.
110. Silva I, Mello LE, Freymuller E, et al. Onset of estrogen replacement has a critical effect on synaptic density of CA1

hippocampus in ovariectomized adult rats. *Menopause.* 2003;10:406–411.

111. Gibbs RB. Long-term treatment with estrogen and progesterone enhances acquisition of a spatial memory task by ovariectomized aged rats. *Neurobiol Aging.* 2000;21:107–116.
112. Daniel JM, Hulst JL, Berbling J. Estradiol replacement enhances working memory in middle-age rats when initiated immmediately after ovariectomy but not after a long-term period of ovarian hormone deprivation. *Endocrinology.* 2006;147:607–614.
113. Talboom JS, Williams BJ, Baxley ER, et al. Higher levels of estradiol replacement correlate with better spatial memory in surgically menopausal young and middle-aged rats. *Neurobiol Learn Mem.* 2008;90:155–163.
114. Rapp PR, Morrison JH, Roberts JA. Cyclic estrogen replacement improves cognitive function in aged ovariectomized rhesus monkeys. *J Neurosci.* 2003;23:5708–5714.
115. Lacreuse A, Wilson ME, Herndon JG. Estradiol, but not raloxifene, improves aspects of spatial working memory in aged ovariectomized rhesus monkeys. *Neurobiol Aging.* 2002;23:589–600.
116. Verghese J, Kuslansky G, Katz MJ, et al. Cognitive performance on surgically menopausal women on estrogen. *Neurology.* 2000;55:872–874.
117. MacLennan AH, Henderson VW, Paine BJ, et al. Hormone therapy, timing of initiation, and cognition in women aged older than 60 years: the REMEMBER pilot study. *Menopause.* 2006; 13:28–36.
118. Henderson VW, Guthrie JR, Dudley EC, et al. Estrogen exposures and memory at midlife. *Neurology.* 2003;60:1369–1371.
119. Zandi PP, Carlson ML, Plassman BL, et al. Hormone replacement therapy and incidence of Alzheimer's disease in older women: the Cache County Study. *J Am Med Assoc.* 2002;88:2123–2129.
120. Jezierski MK, Sohrabji F. Neurotrophin expression in the reproductively senescent forebrain is refractory to estrogen stimulation. *Neurobiol Aging.* 2001;22:311–321.
121. Tanapat P. Ovarian steroids influence cell proliferation in the dentate gyrus of the adult female rat in a dose- and time-dependent manner. *J Comp Neurol.* 2005;481:252–265.
122. Chen S, Nilsen J, Brinton RD. Dose and temporal pattern of estrogen exposure determines neuroprotective outcome in hippocampal neurons: therapeutic implications. *Endocrinology.* 2006;147:5303–5313.
123. Wise PM, Dubal DB, Rau SW, et al. Are estrogens protective or risk factors in brain injury and neurodegeneration? Reevaluation after the Women's Health Initiative. *Endocr Rev.* 2005;26: 308–312.
124. Tierney MC, Oh P, Moineddin R, Greenblatt EM, et al. A randomized double-blind trial of the effects of hormone therapy on delayed verbal recall in older women. *Psychoneuroendocrinology.* 2009;34:1065–1074.
125. Jacobs DM, Tang M-X, Stern Y, et al. Cognitive function in nondemented older women who took estrogen after menopause. *Neurol.* 1998;50:368–373.
126. Bagger YZ, Tanko LB, Alexandersen P, et al. PERF Study Group, Early postmenopausal hormone therapy may prevent cognitive impairment later in life. *Menopause.* 2005;12:12–17.
127. Sherwin BB. Estrogen therapy: is time of initiation critical for neuroprotection? *Nat Rev Endocrinol.* 2009;5:620–627.
128. Henderson VW, Espeland MA, Hogan PE, et al. Prior use of hormone therapy and incident Alzheimer's disease in the Women's Health Initiative Study. *Neurol.* 2007;68: A205.
129. Markowska AL, Savonenko AV. Effectiveness of estrogen replacement in restoration of cognitive function after long-term estrogen withdrawal in aging rats. *J Neurosci.* 2002;22:10985–10995.
130. Galea LAM, Wide JK, Paine TA, et al. High levels of estradiol disrupt conditioned place preference learning, stimulus response learning and reference memory but have limited effects of working memory. *Behav Brain Res.* 2001;126:115–126.
131. Koga H, Yuzuriha T, Yao H, et al. Quantitative MRI findings and cognitive impairment among community dwelling elderly subjects. *J Neurol Neurosurg Psychiatry.* 2002;72:737–741.
132. Nachtigall LE, Raju U, Banerjee S, et al. Serum estradiol binding profiles in postmenopausal women undergoing three common estrogen replacement therapies: associations with sex hormone-binding globulin, estradiol and estrogen levels. *Menopause.* 2000;7:243–250.
133. Bhavnani BR. Pharmacokinetics and pharmacodynamics of conjugated equine estrogens: chemistry and metabolism. *Proc Soc Exp Biol Med.* 1998;217:6–16.
134. Kuiper GG, Carlsson B, Grandien K, et al. Comparison of the ligand and binding specificity and transcript tissue distribution of estrogen receptors alpha and beta. *Endocrinology.* 1997;138: 863–870.

CHAPTER

25

Hormones, Mood and Affect

Sarah L. Berga[1], Yolanda R. Smith[2]

[1] Division of Reproductive Endocrinology and Infertility, Departments of Gynecology and Obstetrics and Psychiatry, Emory University School of Medicine, Atlanta, GA, USA

[2] Division of Reproductive Endocrinology and Infertility, Department of Obstetrics and Gynecology, University of Michigan School of Medicine, Ann Arbor, MI, USA

OUTLINE

Summary

This chapter focuses on the role of sex hormones, particularly estrogens, androgens and progestins, as mediators of mood and affect. We explore clinical controversies, such as whether commonly used steroid contraceptives alter mood, affect and libido, and whether postmenopausal hormone use ameliorates midlife affective disorders or provides neuroprotection. The subjective nature of mood and affect fuels a tendency to dismiss or mythologize the role of hormones as mediators and modulators of feelings and behaviors. However, many clinical presentations illustrate that the brain is a critical target of sex steroids. Concepts emerging from molecular, cellular and animal investigations provide a scientific grounding for understanding the neural and other mechanisms subserving the links between hormones, mood and behavior. New and emerging neuroimaging methodologies provide additional mechanistic insights in human models.

NEURAL CIRCUITRY MEDIATING MOOD AND AFFECT, AND ASSOCIATED BEHAVIORS

The limbic system, a group of brain regions highly preserved across species, is thought to be primarily responsible for the representation and processing of stimuli with emotional significance. These are highly interconnected regions forming neural circuits involved in motivated behavior. The primary brain regions involved in emotion processing include the orbital, medial and lateral prefrontal cortex (PFC), temporal cortex, anterior cingulate, insular cortex, thalamic nuclei, ventral striatum, caudate nucleus, ventral pallidum and

Handbook of Neuroendocrinology, DOI: 10.1016/B978-0-12-375097-6.10025-3

amygdala. Two overlapping circuits have been identified within these limbic structures: the medial prefrontal cortico-striato-pallido-thalamic and amygdalo-striato-pallido-thalamic. The medial network in the orbital and medial PFC, including the anterior cingulate cortex, is believed to be particularly important for the pathophysiology of the mood disorders. It includes projections to visceral control areas, as well as to the superior temporal gyrus, posterior cingulate cortex, and entorhinal and parahippocampal cortex. In addition, a reciprocal relationship exists between the medial PFC and amygdala, with fear-conditioned stimuli resulting in increased amygdala and decreased medial PFC neuronal activity, a pattern also seen in depressed individuals.[1,2]

Mood disorders have been associated with volumetric changes and reduced glial cell density in limbic regions, and alterations in synaptic transmission through the limbic circuits.[3] In rodent models repeated stress is associated with a similar picture, with dendritic atrophy (medial PFC, hippocampus, amygdala), reduced glial cell counts and altered behavioral responses to emotional stimuli.[4] This dendritic remodeling may be related to elevated excitatory glutaminergic transmission and cortisol secretion.[5]

The neuroendocrine and autonomic abnormalities (as well as the visceral and behavioral responses) identified in the mood disorders, such as CRF, ACTH and cortisol hypersecretion, may be related to projections from the medial and dorsal lateral PFC and amygdala to hypothalamus, periaqueductal gray, locus coeruleus, raphe and other autonomic areas.[1] Specifically, in one model of mood neurocircuitry and neuroendocrine interactions, elevated amygdala activity increases neurotransmission to the hypothalamus and brainstem, with subsequent hypothalamic CRF release and autonomic responses, while altered anterior cingulate cortex activity also contributes to HPA-axis dysfunction.[3] Indeed, Major Depression is associated with excessive CRF release, elevated CSF levels of CRF, impaired diurnal cortisol pattern, impairment in cortisol central negative feedback, and hypersecretion of cortisol in response to stress.[3,6,7] Circulating glucocorticoids alter the sensitivity of noradrenergic receptors, and regulate the β-adrenoceptor-coupled adenylate cyclase system in the brain.[8] The sustained elevated cortisol levels may additionally predispose to insulin resistance and obesity.[9,10]

HOW HORMONES IMPACT THE NEURAL CIRCUITRY MEDIATING MOOD AND AFFECT

A few key principles deserve emphasis. First, sex steroids *organize* the brain, sculpting and amplifying sexual dimorphism, and then *activate* the already imprinted and sexually dimorphic brain, leading to sex-specific hormone actions and behaviors. Since brain function is amazingly and intriguingly complex, the specifics of these general principles are still under investigation[11]; however, it appears that as late as adolescence, sex steroids impact brain organization while also activating already imprinted brain functions.[12]

Second, because the brain is sexually dimorphic, some of the actions of sex steroids and neurotransmitters are sex-specific. This means that testosterone or estrogen exposure could elicit one set of effects in men and a differing set of effects in women. Although the principle of sex-specific hormone and neurotransmitter action has been established, we are just beginning to understand the behavioral and health implications that derive from sex-specific hormone action and brain sexual dimorphism. One possibility is that testosterone, which is a major determinant of libido in men, may be less important for libido in women. Since estrogen exposure drives hippocampal synapse formation in women, while androgens and not estrogens do so in men,[13] neuroprevention strategies may differ for men and women. The GABA (gamma-aminobutyric acid)-A agonist muscimol was excitatory in male newborn rats but inhibitory in female newborn rats.[14] The excitotoxic effect in male rats has been linked to androgen exposure.[15] This observation has been postulated by playing a role in the sex difference in neurodevelopmental disorders. Epidemiological studies reveal that women are at excess risk compared to men for depression and Alzheimer's dementia, but not for neurovascular dementia; however, we do not understand the differential roles of hormones and sex as mediators of this sex difference. Understanding sex differences in brain function has been invigorated by the development of state-of-the-art neuroimaging modalities that can be combined with psychomotor and behavioral testing.

Third, not all sex-specific differences and behaviors are due to hormonal influences. Genes play a role even in the absence of expected hormonal exposures.[16,17] In humans, sociocultural expectations play a role in gating mood and affect. Thus, there may be sex-specific and gender-specific but hormone-independent moods, feelings and associated behaviors.

Fourth, hormone action is likely to be gated by age. Neuroplasticity within emotional circuitry and the role of hormones in conferring neuroplasticity is an important area of investigation, because hormones such as estrogen have been touted as agents for protecting against later-in-life mood disorders and neurodegeneration in women.

Finally, the bidirectional interaction between hormones and behaviors must be emphasized. Hormones affect mood and associated behaviors, but behaviors also influence hormones. A classic example is stress-induced gonadal compromise. Through central mechanisms culminating in reduced GnRH drive, stress impairs ovarian secretion of estradiol, progesterone and androgens in women, and testicular secretion of testosterone in men. Stress-related hormonal changes such as increased cortisol and reduced thyronine and thyroxine may also modify sex hormone action at the cellular and molecular levels. Common diseases and conditions linked to attenuation or amplification of central sex-steroid action include depression, anxiety, stress, premenstrual syndrome, stroke, dementia and other neurodegenerative disorders. Immune dysfunction and inflammation may also interfere with gonadal function and with central sex-steroid actions at cellular and molecular levels.[18]

The spatiotemporal patterning of the brain in action leads to thoughts, feelings and behaviors. Brain health implies not only the prevention of recognized diseases, but also the support of a vast array of brain functions and faculties. The brain functions impacted by sex steroids encompass critical domains, including motor skills, speed of processing, emotionality, social awareness and cognitive attributes, the sum of which is recognized as personality, mind or self, depending on the context. It is important to consider the role sex steroids play in gating behavior, as well as in gating diseases such as dementia, to be able to address common complaints voiced by menopausal women, such as "not feeling the same" before and after menopause, or loss of libido while taking oral contraceptive pills.

Some of the sexual differentiation of the central nervous system (CNS) is hormonally mediated, and begins *in utero* when testosterone from the fetal testes targets areas within the developing brain. The fetal testes begin to secrete testosterone in the late first trimester in response to placental human chorionic gonadotropin (hCG). In contrast, the fetal ovary does not develop functional follicles within the fetal ovary until the late second trimester. In late pregnancy, gonadotropin secretion is restrained by placental steroid production. When that restraint is lost at the time of birth, gonadotropin secretion rises dramatically in both sexes. In boys, the gonadotropin rise is followed by an elevation of testosterone concentrations to adult levels, with ongoing influences on brain development for a short period postnatally that does not last more than a year. In girls, estrogens from fetal or placental sources do not readily cross the blood–brain barrier because of binding proteins in the circulation, and thus circulating estrogens have limited significant access and less opportunity to imprint the developing CNS. However, because the pattern of postnatal gonadotropin suppression differs between boys and girls, ovarian and testicular activity and sex steroid secretion also differs in time course and pattern. Thus, by 2 years of age, the brains and bodies of girls and boys have been exposed to dramatically different patterns of sex-steroid secretion.

The organizational effects of testosterone on the developing CNS are thought to depend primarily on *in situ* aromatization (the conversion of androgens to estrogens by the enzyme aromatase) of testosterone to estradiol. Also, testosterone may bind directly (without conversion to estradiol) to androgen receptors in the CNS. Therefore, sexual dimorphism of brain structure may reflect not only androgen exposure but also central dimorphism in key brain areas in terms of both high aromatase activity (which, thus, can convert testosterone to estradiol) and those with androgen receptors.[19] In the non-human primate brain, androgen receptors are located in the hypothalamus, amygdala, and prefrontal visual and somatosensory cortices, whereas estrogen receptors are more highly concentrated in hippocampus in females relative to males. Interestingly, early postnatal blockade of either androgen receptors or aromatase heightened HPA reactivity to repeated stress in male rats while adult castration had no effect. This implies that androgen exposure during the organizational window blunts stress reactivity, and that both androgen receptor binding and blockade of the conversion of androgens to estrogens play a role in modulating stress reactivity.[20] The degree to which gender-related behavioral asymmetries are accounted for by differences in fetal/infant steroid hormone exposure is open to debate, but, clearly, multiple mechanisms for inducing differences exist.[18]

Morphometric studies of the developing and adult brain support the sex-steroid theory of CNS sexual dimorphism[12] (see Box 25.1). Sex differences in both cortical and subcortical regions appear to cluster in brain regions with developmentally rich sex-steroid receptor distribution (Fig. 25.1). Sexual dimorphism was prominent in the cortex, which may be a result of rapid postnatal decreases in sex-steroid receptor concentrations. By the age of 6 years, the brain volume is 95% that of an adult; however, boys and girls follow different trajectories in brain tissue changes between the age of 6 and adulthood. By adulthood, men have 10% larger brain volumes relative to women, larger corpus callosum genu, greater overall white matter volume, and increased ventricular size. By adulthood, women have larger cortical volume relative to men, higher cortical gray matter volume, and larger volumes in Broca's area, the superior temporal cortex, hippocampus, caudate, thalamus, anterior cingulate gyrus, dorsolateral prefrontal cortex, and interhemispheric white matter connections. Research is underway to explicate whether structural brain dimorphism is associated with behavioral dimorphism.

Finally, the coming years will reveal greater understanding of how sex chromosome genes, independent of gonadal hormones, play a role in sexual differentiation

BOX 25.1

HOW DO WE KNOW THAT THE HUMAN BRAIN IS SEXUALLY DIMORPHIC?

Convincing evidence of sexual dimorphism in the human brain in humans was provided by volumetric analysis of various brain regions using magnetic resonance imaging. The brain regions displaying the greatest volumetric dimorphism were regions that also showed greater levels of sex-steroid receptors during critical periods of brain development.[1]

Reference

1. Goldstein JM, Seidman LJ, Horton NJ, Makris N, Kennedy DN, Caviness Jr VS, Faraone SV, Tsuang MT. Normal sexual dimorphism of the adult human brain assessed by *in vivo* magnetic resonance imaging. *Cereb Cortex*. 2001;11:490–497.

of the brain.[16,21] The concept of sex-specific hormone action has received investigative attention (for review, see Gillies and McArthur[13]). In the clinic, human experiments of nature in which individuals are exposed to aberrant or absent gonadal function due to gonadal dysgenesis suggest the salience of this concept for understanding gender behaviors. Further insights have been gleaned through animals engineered to be discordant for genotypic and gonadal sex. For example, sex chromosomal status has been identified as a key determinant of the midbrain dopamine system. New models are under development to guide further exploration of genetic contributions to CNS sexual differentiation, including the role of prostaglandins and epigenetic factors. Emerging evidence suggests that the brain epigenome is sexually dimorphic, and that these differences contribute to sex differences in behavior and risk for neurodegenerative disorders.[22]

The ontogeny of female reproduction is marked by physiological events such as puberty, menarche, pregnancy, lactation and menopause. Reproductive events are accompanied by characteristic hormonal patterns

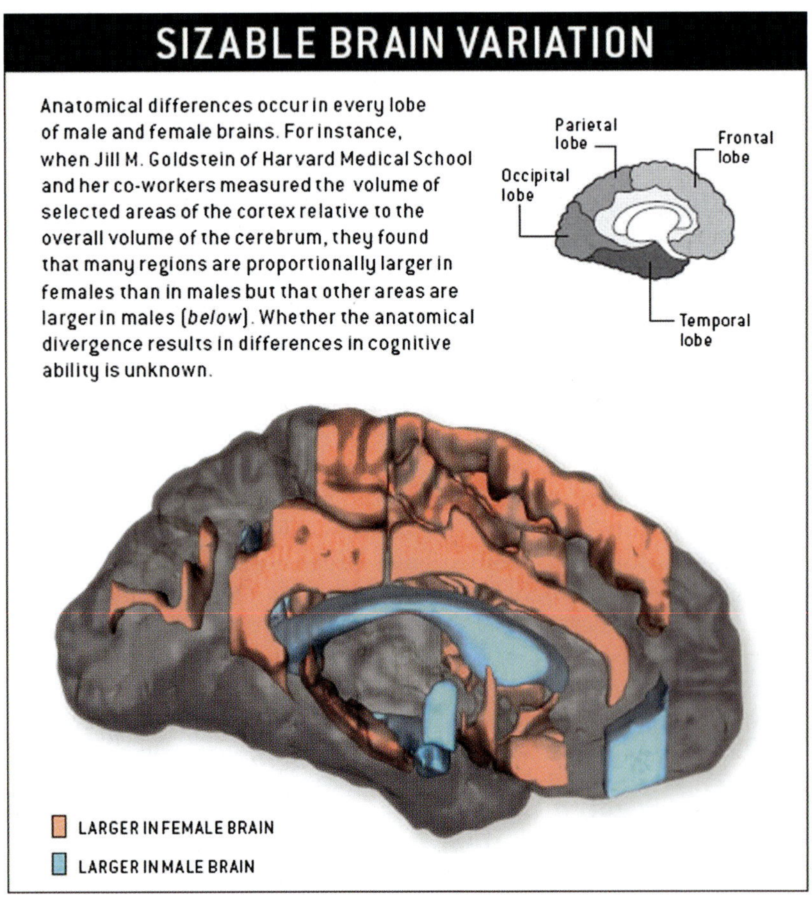

FIGURE 25.1 Anatomical differences of male and female brains. *Reproduced from Cahill L. His brain, her brain.* Scientific American *2005;292:40–47, based on Goldstein* et al. *(2001).*[12]

that then impact the brain in ways that foster reproductive functioning, including parenting. The relationship between hormone exposure and reproductive behaviors is well delineated in animal models but less stereotypic in humans, so seemingly subject to volition and acculturation. By convention, we refer to the genotype of an individual as his or her sex. The term "gender" refers to sex-linked or sex-specific behaviors and attributes. In humans, the postnatal gonadal hiatus entails suppression of the hypothalamic gonadotropin-releasing hormone (GnRH) pulse generator by undetermined central mechanisms that are largely gonad-independent. Puberty, the postnatal resumption of gonadal activity, also is a centrally initiated process that depends on synchronization of the pulsatile release of hypothalamic GnRH, which is likely triggered at least in part by expression of the KISS-1 gene. Although prenatal exposure to sex steroids has organizational effects on the developing brain, the reinitiation of gonadal activity during puberty is viewed as exerting primarily activational influences on the brain and behavior. Recent neuroimaging data suggest, however, that there are organizational effects that augment the pre-existing sexual dimorphism of the brain. Issues that remain to be resolved include whether the timing of puberty, both early and late, alters CNS substrate differentially and promotes behavioral maladjustment, and whether menstrual cycle disorders, such as functional hypothalamic amenorrhea and polycystic ovary syndrome, alter brain development and function and, if so, by what mechanisms and in what functional and behavioral domains.

Abundant evidence supports the notion that the brain is a primary target tissue for sex steroids.[18,22–24] The role of sex steroids in gating traditional reproductive processes such as implantation, lactation and reproductive behavior has been extensively chronicled and is widely appreciated. The commonplace symptoms that accompany menopause, namely, hot flashes, sleep disturbances and irritability, suggest that parts of the brain subserving functions other than solely reproductive ones are critically impacted by the presence or absence of sex steroids.[25,26] At a minimum, the following central nervous system outputs are modulated by sex-steroid exposure:

- Affective state and mood, including sleep,[27,28] libido and sexual attraction,[29–32] and complex social behavior;[33,34]
- Cognitive functioning, especially verbal memory, speed of processing, and spatial ability;[24,35–37]
- Metabolism[38] (see also Fig. 25.4, below);
- Motor coordination, including balance and psychomotor speed;[24,36]
- Stress sensitivity;[39]
- Neuronal excitability and vulnerability to epilepsy;[36]
- Pain sensitivity;[36,40] and
- Auditory threshold (which is lower in women than men).

Multiple mechanisms mediate the impact of sex steroids upon neuronal plasticity, function and integrity. Irwin *et al.*[41] found that both estradiol and progesterone independently enhanced oxidative metabolism of neuronal mitochondria. The increased respiratory activity was coupled with a decreased rate of reactive oxygen leak and reduced lipid peroxidation, representing a systematic enhancement of brain mitochondria efficiency. Sex steroids modulate the following neurotransmitter systems:

- Basal forebrain dopaminergic system, which gates attention;
- Forebrain cholinergic system, which regulates memory and learning;[42,44]
- Midbrain noradrenergic system, which is implicated in the modulation of arousal;
- Midbrain and cortical serotonergic system,[45] which subserves mood, affect, cognition, attention, and the integration of cognition and emotionality into complex social behaviors;
- Midbrain dopaminergic system, which mediates reward.

Sex steroids also regulate synaptic density and neuronal and glial viability by a variety of mechanisms (for reviews, see refs.[13,18,22,24,36,39,42,43] In short, sex steroids have important organizational (anatomical) and activational (functional) effects upon neural substrates relevant to mood and affect. It is these mechanisms that are responsible for the link between reproductive physiology and pathophysiology, and a host of behavioral sequelae.

INVESTIGATIVE MODELS

The complex nature of brain function makes it challenging to study, particularly in humans. Several investigative models exist to examine different levels of the relationship between sex steroids and affect regulation. Information that is pertinent and critical to understanding hormone actions on brain circuitry and human behavior is attained from animal studies, human studies, clinical trials and epidemiologic studies.[46] No one investigative approach is superior. Integrating the information learned from these various models provides the most comprehensive view of hormonal influences on mood function.

Cellular and Animal Studies

Animal models allow for the systematic and controlled study of hormone actions on specific neurotransmitter systems and brain circuitry, but not on behavioral complexity. Standardized models of animal stress exist to simulate depression through learned helplessness. One such well-known model is the forced swimming test, in which rodents are forced to swim for lengthy periods with no escape and eventually become immobile and float.[47,48] Animal models provide tissue for a detailed evaluation of relevant molecular mechanisms, which is rarely possible in human studies. However, the complex and sometimes erratic hormone environments that are present during human reproductive transitions most associated with mood dysregulation (such as puberty, perimenopause, etc.) are difficult to replicate in rodent models.

Non-human Primate Studies

The non-human primate model is a useful human surrogate for studying the influence of hormones on behavior, allowing for controlled experimentation and more invasive outcome measures than would be feasible in humans. The non-human primates have reproductive function and dysfunction that are similar to humans, and complex behaviors and emotional expression.[49–52] They rely on social structures and attachment bonds, and they exhibit depressive responses with loss of an attachment bond or with social subordination, providing experimental paradigms that mimic depression.[53]

Monkey studies are few, but those that have been done provide highly relevant information. Using a monkey model, Mook *et al.*[54] looked at behavior in female monkeys who were oophorectomized and then randomized to placebo, estradiol, tamoxifen, or a combination. In monkeys, tamoxifen was anxiogenic and antagonized the anxiolytic effects of estradiol, which may explain in part why women receiving tamoxifen are 42% more likely to develop depression than those not treated.[55] Further, tamoxifen inhibited sexual activity and prevented the increase in serotonergic activity conferred by estradiol.[54,56] Estradiol reduced depression in monkeys, and tamoxifen counteracted this effect. To discern the effects of progestins, Pazol *et al.*[57] studied oophorectomized monkeys randomized to placebo, estradiol alone, estradiol + progesterone, and estradiol + medroxyprogesterone acetate (MPA). Female initiation of sexual activity was greatest with estradiol alone, nil with placebo, intermediate with estradiol + progesterone, and nil with estradiol + MPA. Female-to-female antagonism in monkeys who received estradiol alone was least, intermediate in estradiol + progesterone and placebo, and maximal in monkeys receiving estradiol + MPA.

Human Studies

Interventional Trials

Interventional trials in which various hormonal preparations are administered and relevant outcome variables measured are difficult to conduct, because women with uteri need progestin exposure and may bleed, which interferes with blinding. Women who have had a hysterectomy can receive estrogen only, but it has been assumed that women who needed a hysterectomy are not the same in terms of study design as those who did not. A variety of estrogen preparations have been administered to assess efficacy in clinical conditions such as mild cognitive impairment, midlife depression, postpartum depression, and insomnia. Head-to-head comparisons of hormonal preparations have not been conducted, and nor has a target range been delineated. Since cognitive impairment is almost always accompanied by mood dysfunction and *vice versa*, a pure test of the impact of hormones upon mood and affect in humans is not theoretically possible. Thus, our understanding of the relationship in women between hormones, mood and affect is limited.

Longitudinal Cohorts

Epidemiological evidence from longitudinal cohort studies provides valuable information regarding changes over time, outcomes, and identification of risk factors associated with the development of mood disorders. Additionally, these studies have determined the relative risk of mood disorders during various reproductive transitions, and have quantified sex differences in mood disorders. Although, cellular, molecular and neuroimaging data consistently support the notion that sex steroids, particularly estradiol, modulate and preserve brain function, the epidemiological evidence on the neuroprotective effects of estrogen reveal a mixed picture. At least part of the confusion is due to the variety of types and routes of estrogens and progestins to which women have been exposed. The power of prospective observational trials is also hindered by the lag time between exposure and the detection of clinically significant outcomes.

Randomized Clinical Trials

Most clinical trials have been unable to clearly differentiate the effects of different estrogens or the impact of progestagens. Some information has been gleaned from the two arms of the WHI, but the two groups were not comparable prior to enrollment, so direct comparison between the CEE and CEE/MPA arms was felt not to be valid. Additional clues about the differential effects of various hormones have been provided by looking at women taking the selective estrogen receptor modulators (SERMS), such as tamoxifen, raloxifene, and aromatase inhibitors. Further studies are warranted to elucidate the neurobiological mechanisms subserving the observed behavioral differences; for understanding the impact of various sex steroids upon neural processes such as metabolism, synaptogenesis and neurotransmitter dynamics; and to further delineate the similarities and differences of various estrogens and progestins in the support of behavioral outputs such as mood and cognition, and risk of dementia. Indeed, the above studies confirm the validity and feasibility of monkey study in extending and complementing investigative paradigms with humans.

Imaging Studies

Non-invasive neuroimaging techniques provide a direct assessment of neuronal activity in women's brains.[45,58,59] Current neuroimaging techniques allow investigators to directly study the effects of sex steroids and mood disorders on various measures of neuronal function and neurochemistry, untangling the complexities of hormonal influences on neural circuitry and receptor changes. The techniques of positron emission tomography (PET) and single photon emission computed tomography (SPECT) have allowed neurochemical measures such as receptor concentrations, neurotransmitter release and enzymatic activity to be quantified. Similar to pituitary function testing, neurotransmitter release during a controlled stressful stimulus often provides more useful information about system function than baseline values. Furthermore, functional magnetic resonance imaging (fMRI) allows brain circuitry to be evaluated during the performance of affect-related tasks, while markers reflecting cellular metabolism and neuronal integrity can be examined with magnetic resonance spectroscopy.

PUTATIVE ROLES OF SEX HORMONES IN CLINICAL PRESENTATIONS

Overview

Affective and mood disorders differentially impact women and men throughout the life span and cross-nationally. Before puberty, boys are more vulnerable than girls are to depressive illness. Over the course of adolescence, however, a dramatic shift occurs, with girls displaying a precipitous rise in depression that far outstrips the negligible (if existent) increase displayed by adolescent boys. By age 15, girls are twice as likely as boys are to have experienced a lifetime episode of major depression. Angold and colleagues found that the onset of depression was linked to the onset of puberty rather than to chronological age. The elevated risk has been linked to menarche, advanced Tanner staging, and changes in estradiol and testosterone levels.[60–63] Girls are also at higher risk for the persistence of depressive symptoms.[63,64] This sex/gender difference persists for the next 35 to 40 years, essentially spanning female reproductive years. The predominance of depression in women and the adolescent onset of this sex/gender difference have now been replicated cross-nationally. Indeed, the gender difference in lifetime rates of major depressive disorder is considered to be one of the most robust findings in the field of psychiatric epidemiology.

Puberty and Adolescence

Sex differences in the rates of depression begin to appear in adolescence, a time of major change in the neuroendocrine reproductive axis. Thus, the fluctuation of ovarian steroids during specific phases of the menstrual cycle may bear some relationship to the particular vulnerability of women for mood disorders. The reproductive hormones could exert their effects on mood directly, or indirectly by their effect on neurotransmitter, neuroendocrine or circadian systems, all of which have been implicated in the pathogenesis of affective illness.

Much of adolescent brain development is in systems critical to regulation of behavior and emotion, as well as in areas important to perception and evaluation of risk.[65] While reproductive hormones secreted during puberty underlie this remodeling, it is difficult to determine which brain changes are related specifically to neuroendocrine changes. Cognitive maturation involves development of goal-directed thoughts and behaviors while suppressing inappropriate impulses. The prefrontal cortex, which mediates executive functions such as reasoning, planning and behavioral control, develops linearly as children mature into adults.[66,67] With development of the prefrontal cortex, impulsivity diminishes. On the contrary, risky behavior is increased in adolescents, and is associated with increased activation of the reward and affective processing areas (i.e., nucleus accumbens and amygdala). Therefore, the prefrontal cortical impulse control system (which develops slowly) must override inappropriate behaviors arising from increased activity in limbic areas (amygdala and nucleus accumbens) during adolescence.[68] The imbalance between increased novelty/reward seeking and immature self-control systems leads to emotional reactivity and risky behavior. This imbalance, exacerbated by puberty, may predispose to affective diseases.

Also important to affective regulation in adolescence is expanding connectivity between brain areas through continued myelination of nerve fibers and synaptic pruning.[69–73] Animal studies reveal a pattern of increased connectivity during adolescence, with elimination of unnecessary synapses and strengthening of important connections. In early adolescence there is an overproduction of axons and synapses, with dendritic pruning in later adolescence, specifically in the amygdala, nucleus accumbens and prefrontal cortex, with an increase in the density of fibers connecting the amygdala and prefrontal cortex.[74–76] Thus, in addition to development of cognitive and emotional brain regions in adolescence, there is improved connectivity between prefrontal cortex and the limbic systems. However, it is not known if this is specifically linked to the changes in reproductive hormones.[77,78]

Several mechanisms have been posited regarding the role of pubertal hormone changes in adolescent mood disorders. First, the serotonergic system may mature earlier than the noradrenergic system,[79] and serotonergic receptor changes, under the influence of gonadal hormones, may predispose to mood instability.[80] Second, gonadal hormones may alter adolescent biorhythms, such as sleep, increasing risk for mood disorders.[81] Third, changes in the stress axis during adolescence may also be important, as stressful life events may precipitate depression.[82] Reproductive transitions are associated with increased reactivity of the stress response system.[83] Gonadal steroids generally enhance stress responses in puberty, but this relationship is complex.[84,85] During adolescence the hypothalamic–pituitary–adrenal axis undergoes maturation, and stress response system dysregulation during this time may increase vulnerability to negative life events and chronic psychosocial stressors.[80]

Additional theories regarding the influence of pubertal hormone changes on adolescent mood vulnerability include specific hormone-driven behavior changes. Changes in the reproductive neuroendocrine axis in girls and boys at puberty may set the stage for alterations in the power balance of relationships, which may affect the person's sense of self, self-esteem, and mental health. Interestingly, in a study of hormone replacement given to adolescent boys and girls with delayed puberty, testosterone was associated with aggression and self-perceptions of athletic abilities in boys, whereas estrogen was associated with aggression and self-perceptions of romantic appeal and close friendships in girls. These data suggest that estrogen is unlikely to be the cause of the increased depression and loss of self-esteem observed in adolescent girls.

Another candidate hormone regulated by sex steroids and implicated in affective regulation is the hypothalamic neurohormone oxytocin. It may mediate or intensify affiliative behaviors such as maternal caregiving and adult pair-bond formation in mammalian females. Oxytocin stimulates female reproductive functions such as milk ejection during lactation and uterine contractions at parturition, and its release is regulated by estrogen and progesterone. The oxytocin-mediated affiliative system lies at the core of a uniquely female biobehavioral response to stress, the *tend-and-befriend response*. The fight-or-flight response has been regarded as the prototypic model of the human stress response in men. Because mammalian females typically bear a greater role in the care and protection of young offspring, quieting and tending to the needs of offspring when under stress and affiliating with other adults to promote the protection of oneself and one's offspring may represent a more adaptive female response to environmental threats. Behavioral evidence in humans supports gender-linked differences in affiliation under conditions of stress. The concept of a uniquely female biobehavioral response to stress would have broad implications for our understanding of stress responses and risk of psychopathology in women.[86]

Hormonal Contraception

The use of hormonal contraception is often accompanied by a variety of psychological symptoms, such as

BOX 25.2

HOW DO WE KNOW THAT THE SYNTHETIC HORMONES USED IN ORAL CONTRACEPTIVES MODULATE THE NEURAL CIRCUITRY OF EMOTION OF WOMEN?

Women using oral contraceptives often complain of reduced libido or increased irritability. Clinicians often find it hard to put patient reports into perspective. Given that oral contraceptives are highly effective in preventing pregnancy, it is often deemed that the reduction in wellbeing is an acceptable trade-off. Further, in the absence of a demonstration that the neural circuitry of human female brain was altered by exposure to the synthetic steroids used in oral contraceptives, clinicians were cautious about implicating oral contraceptives as the cause. However, using functional neuroimaging, Love *et al.*[1] convincingly showed that women taking a combination of ethinyl estradiol and norethindrone acetate had altered patterns of brain activation during negative and positive emotional presentations.

Reference

1. Love T, Smith YR, Persad CC, Tkaczyk A, Zubieta J- K. Short-term hormone treatment modulates emotion response circuitry in postmenopausal women. *Fertil Steril.* 2010;93: 1929–1937.

irritability, depression and loss of libido (see Box 25.2). Practitioners often dismiss patient reports of altered quality of life as minor, or view the alterations as unfortunate but necessary side effects associated with use of reliable contraception. Because it is not widely appreciated that the steroids in hormonal contraceptive formulations impact brain, mood, affect and behavior, women who experience psychological side effects may not be counseled about the potential link. Understanding that the side effects are "real" may help rather than hinder compliance. In support of the notion that exogenous hormonal compounds alter emotionality, Love *et al.*[28] found that women taking a hormonal preparation containing the estrogen ethinyl estradiol, which is the most common estrogen used in oral contraceptives, in combination with a 19-nortestosterone progestin, norethindrone acetate, displayed greater activation in brain regions mediating reactivity to negative emotion than positive emotion (Fig. 25.2). It is likely that women with pre-existing affective dysregulation are more likely to experience enhancement of negative affective side effects, and it is recommended that women who report increased affective negativity be counseled about the potential link to hormonal contraception so that they can decide if they want to continue. The progestin component may mediate emotional negativity, as estrogen has consistently been linked to increased sexual interest and arousal in animal and human reports. Although it is difficult to do in women with intact uteri, pharmacologic formulations that limit CNS exposure to progestins should be considered in women with altered quality of life associated with hormonal contraception.

Libido

It has been said that the most important sexual organ is the brain. This truism is meant to emphasize the essential role of desire and comfort in mediating sexual awareness and expression. Brain centers mediate libido in a general way. Other brain centers influence the selection of sexual partners and the circumstances under which sexual longings are expressed. Hormones made by the ovaries and testes, such as estrogens and androgens, bind in certain brain centers, and increase sexual interest. Sexual interest can be inhibited by any number of negative emotions, including anger, fear, worry and dislike, and by a marked decline in health or hormones. Aging may be associated with changes in sexuality. Sexual dysfunction can be due to psychological, hormonal, anatomical and medication-related causes. Hormonal changes can affect both libido and the reproductive tract. Loss of estrogen with menopause, or from surgical removal of the ovaries, can lead to loss of vaginal lubrication. Intercourse may be painful, or, more typically, there may be less sexual sensation and decreased arousal. Psychiatric conditions often impact sexuality and sexual behavior. Typically, depression and anxiety suppress libido and sexual performance.

Medications that may interfere with sexual responsivity or sexual interest include antidepressants, combined hormonal contraceptives, progestin-only contraceptives such as medroxyprogesterone (Depo-Provera), and GnRH agonists. Contraceptive agents and GnRH agonists suppress endogenous ovarian function, while progestin-only contraceptives also give a progestin-dominant exogenous hormonal exposure. In general,

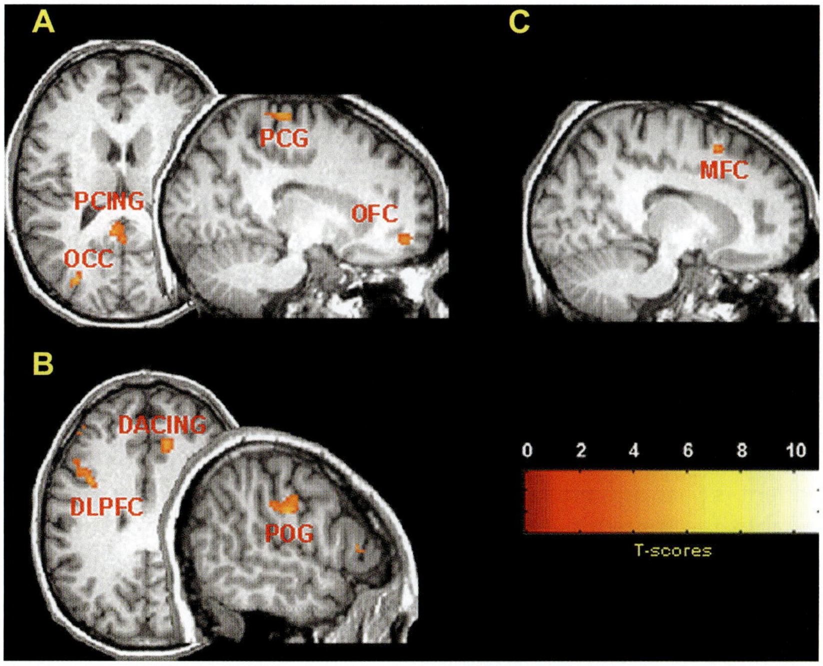

FIGURE 25.2 **Effects of HT on emotional image presentation.** Images display areas where (**A**) greater BOLD activity was detected in the HT condition relative to the placebo condition during negative image presentation (negative−neutral) and (**B**) lesser BOLD activity was detected in the HT condition relative to the placebo condition during negative image presentation (negative −neutral) or (**C**) during positive image presentation. DLPFC, dorsolateral prefrontal cortex; MFC, medial frontal cortex; PCG, precentral gyrus; POG, postcentral gyrus; DAC, dorsal anterior cingulate; PC, posterior cingulate; OCC, occipital cortex; OFC, orbitofrontal cortex. *Reproduced from Love T, Smith YR, Persad CC, Tkaczyk A, Zubieta J-K. Short-term hormone treatment modulates emotion response circuitry in postmenopausal women.* Fertil Steril. *2010;93:1929–1937.*

progestins diminish libido and sexual responsivity, while estrogens increase both. Antidepressants decrease sexual drive and responsivity by acting on important brain centers. Selective serotonin reuptake inhibitors (SSRIs) have been associated with decreases in orgasmic function, and may lessen sexual interest.

Sexual Orientation

The extent to which sexual orientation is conferred by hormonal imprinting or exposures remains a controversial topic. Rahmen[87] suggested that there are alterations in neural circuitry that reflect an interaction of genetic, hormonal, and behavioral factors. Human studies are difficult to conduct. However, Savic and colleagues have utilized neuroimaging paradigms to demonstrate that male and female pheromones elicit differential CNS activation in brain regions subserving affect and libido in men and women,[31] and that neurocircuitry differs among homosexual men, heterosexual men, homosexual women, and heterosexual women.[88] Also, homosexual men react to male pheromones while heterosexual men do not, and homosexual women display different activation patterns than heterosexual women.[89,90] On the other hand, women exposed *in utero* to androgens display activation patterns similar to those of heterosexual women without such exposures.[91] While the extent of neurobiological plasticity in this neural circuitry has received little investigative attention, it is clear that steroids, particularly those that are converted to pheromones, including testosterone and estradiol, exhibit activational influences that drive sexual orientation and libido.

Stress-Induced Anovulation

Functional hypothalamic anovulation or *amenorrhea* (FHA) is a consequence of a non-organic reduction in GnRH secretion that results in reduced pituitary secretion of gonadotropins and subsequent anovulation due to insufficient gonadotropin secretion to support folliculogenesis. Functional hypothalamic anovulation is a diagnosis of exclusion, and may be an accompaniment to excessive exercise, depression, and eating disorders. The term "functional" implies an absence of organic pathology. As such, the ovaries are capable of responding to appropriate gonadotropin input of either endogenous or exogenous origin; there are no structural abnormalities of the thyroid, adrenal, pituitary, or brain; the use of drugs, including antipsychotic medications, does not account for the suppression of GnRH; and the patient is not pregnant. Neuroimaging may be needed to establish that there are no significant anatomical lesions of the brain or pituitary. Two major questions concerning the pathogenesis of functional hypothalamic anovulation remain unresolved. First, the peripheral and central signals that disrupt GnRH pulsatility are poorly understood. Second, how do behavioral, cognitive and personality variables activate the neural systems to disrupt GnRH secretion? Physicians and patients alike wonder what lifestyle variables provoke or contribute to this type of ovulatory dysfunction, and what pharmacological and non-pharmacological treatment interventions should be considered.

The best biochemical evidence in support of the concept that stress impairs GnRH release in women with functional hypothalamic amenorrhea is the consistent demonstration that the activity of the HPA axis is increased only in women with FHA and not in women with other forms of anovulation, such as polycystic ovary syndrome (PCOS).[92] A prospective study found that young American women who developed transient amenorrhea while studying in Israel had higher urinary cortisol concentrations on arrival than did those whose menses remained regular.

There appears to be a dose–response relationship between the type, severity and number of stresses on one hand, and the proportion of women who develop anovulation on the other hand. Biological and psychological predispositions may confer resistance or sensitivity to various stressors. Exercise, low weight, weight loss, affective and eating disorders, various personality characteristics such as perfectionism and unrealistic expectations, drug use, and a variety of external and intrapersonal stresses have been linked to the development of anovulation.[93,94] Most women with functional hypothalamic anovulation – when carefully evaluated – display more than one of these traits or behaviors. Recent evidence suggests synergism between metabolic stressors that suppress the hypothalamic–pituitary–thyroidal axis, such as excessive exercise and nutritional restriction, and psychosocial challenges that activate the HPA axis.[95] Thus, multiple chronic stressors are more likely to induce functional hypothalamic anovulation than is a single larger-magnitude stressor.

Recovery is possible if response patterns to ongoing psychosocial demands are developed that are less likely to activate the central and metabolic processes that disrupt pulsatile GnRH release. The current standard of practice, other than observation, is to offer pharmacological interventions, such as oral contraceptives or hormonal replacement, if fertility is not desired, and pharmacological ovulation induction if fertility is sought. However, pharmacological intervention alone does not lead to spontaneous recovery, and cannot be expected to ameliorate stress-induced alterations in central neurotransmission and hypothalamic function, or to reverse ongoing metabolic derangements secondary to exercise or weight loss. For instance, bone accretion does not proceed apace in the face of metabolic compromise, even if exogenous hormone replacement is given in supraphysiological doses. Thus, it is not surprising to find that women with amenorrhea due to decreased GnRH drive, regardless of cause, have lower bone mineral density. Absent behavioral change, the use of exogenous hormones alone may mask recognition of psychological dysfunction and forestall the development of more effective response patterns. Also, ovulation induction may place low-weight women with functional hypothalamic amenorrhea who conceive at risk for premature labor and intrauterine growth retardation. Further, increased cortisol secretion suppresses thyroid hormone secretion and action. Fetal neuropsychological development may be compromised by the concomitant presence of hypothalamic hypothyroidism (euthyroid sick syndrome) or hypercortisolemia if ovulation is pharmacologically induced. If the parenting skills of women with functional hypothalamic are impaired by ongoing stress, their children may be at risk for poor psychosocial development. Clearly, treatment strategies need to consider that stress and mild psychological dysfunction can play an important role in the genesis of this form of anovulation, and cognitive behavior therapy aimed at ameliorating problematic attitudes has been shown to restore ovarian function.[96] If an eating disorder or depression is recognized, specialized psychiatric treatment is indicated. Psychobiological characteristics can predispose a woman to chronic activation of central neural processes in response to commonplace events. Global hypothalamic dysfunction, including inhibition of GnRH pulsatility, can result. Although identifying the neuromodulators that mediate the development and maintenance of the disruption of GnRH is of academic interest, clinicians must make sure that all other possible causes of anovulation and amenorrhea have been excluded before recommending psychological, psychiatric or behavioral interventions.

Postpartum Mood Disorders

The most common constellation of mood symptoms experienced by women in the immediate postpartum period is typically referred to as the *postpartum blues* or *baby blues*. A relatively common phenomenon (occurring in 50–80% of women), postpartum blues includes transient symptoms and rapid mood shifts, including tearfulness, irritability, anxiety, insomnia, lack of energy, loss of appetite, and the general experience of feeling overwhelmed. Onset typically occurs after the third postpartum day, after hospital discharge. Symptoms typically peak by day 5 and spontaneously resolve by day 10 postpartum. It has been estimated that 75 percent of women who experience symptoms of postpartum blues will display a time-limited course; however, 20–25% may progress to major postpartum depression. A number of studies also suggest a higher incidence of anxiety symptom co-morbidities with postpartum versus non-postpartum depression. Primary risk factors for the development of postpartum depression include a personal or family history of depression in general, as well as a personal history of previous depressions related to reproductive events (such as premenstrual dysphoric disorder or previous postpartum depressions). These and other findings suggest that a subgroup of women, particularly those with a history of recurrent depression, may display a heightened sensitivity to changes in reproductive hormone levels, such as those occurring premenstrually, as well as to changes that occur during postpartum and perimenopausal reproductive transitions. Although personal and family histories of psychiatric disorders are major risk factors, psychosocial variables also influence the risk for the onset of depression during the postpartum period. Research indicates that psychosocial stressors may sensitize the brain to subsequent metabolic or hormonal changes, and *vice versa*. Of note, lactation appears to diminish stress reactivity and enhance sense of well-being.[97]

Postpartum depression (PPD) coincides with a major life change – parenthood. PPD also occurs in the context of neuroendocrine events of unparalleled magnitude, which include large-scale fluctuations of steroid hormones and neuropeptides. Many have hypothesized a role for steroid/neuropeptide modulation of brain monoamine systems in the onset of postpartum mood disorders; however, these hypotheses have only begun to be tested in human studies. Neuroimaging studies of the serotonin 1A system and dopamine-2 receptor systems support the notion that brain systems are altered in the puerperium both in healthy women compared to non-postpartum women, and in postpartum depressed compared to postpartum control women. In randomized controlled trials, response rates to antidepressants vary from 43 to 87%, while remission rates are 27–48%.[98] In a randomized comparison of two antidepressants, sertraline and nortriptyline, no significant differences were noted in treatment response.[99] Additionally, recent findings suggest that low levels of n-3 (omega-3) polyunsaturated fatty acids, particularly docosahexaenoic acid (DHA), are associated with postpartum depression.[100] In animal models, decreased postpartum brain DHA is associated with neurobiological changes, including heightened stress responses. Based on large epidemiological studies, obstetric variables are not major risk factors.

Motherhood is a role transition that requires marked psychological adaptation. By necessity, a highly dependent human being becomes the focus of one's time and energy. Establishing an effective balance between the equally important roles of autonomy and dependence within each of the parents and between the spouses requires major shifts and reorganization. Risk of depression is particularly high for women during the perinatal period, with 10–15% of postpartum women experiencing a major depressive episode during this time. Postpartum psychiatric disorders may include transient depressive symptoms (i.e., the postpartum blues), postpartum depression and postpartum psychosis. Although psychosocial factors are implicated as risk factors for postpartum depression, epidemiological studies have not found an increased rate of depression in men who are new parents. The latter observation argues in favor of an interaction of biological and psychosocial factors.

The most severe of the perinatal mood disorders, postpartum psychosis, is rare, affecting approximately 0.1– 0.2% (1 in 1000 to 1 in 500) of postpartum women. Current evidence indicates that postpartum psychosis is most commonly associated with the diagnosis of bipolar disorder. This disorder tends to show onset within 2 weeks postpartum; however, it may present later in the course of the illness as a depression (i.e., 3–5 months postpartum). The delirium-like character of this illness, which is associated with severe impairment in judgment and nihilistic delusions, can result in the tragic outcomes of infanticide (in 5% of women with psychosis) and suicide.

To investigate the hormonal basis of postpartum mood disorders, one study simulated the decline in reproductive hormones after delivery in non-pregnant women with the use of leuprolide to induce a hypogonadal state. The women were then treated with supraphysiological doses of estradiol and progesterone, and then both steroids were withdrawn under double-blind conditions. Five of eight women with a history of postpartum depression, but none of eight women without previous depression, had mood changes. Thus, women with postpartum depression appear differentially sensitive to the effects of the withdrawal of gonadal steroids on mood.

Postpartum depression has been treated with estrogen and progesterone. Transdermal estradiol does not interfere significantly with lactation. In a study comparing estradiol (200 mg per day) with placebo, the estradiol-treated group had a significant reduction in depression scores during the first month.[101] Although half of the women were on antidepressant treatment, this did not alter outcome. Given the lack of randomized controlled trials, use of progesterone as a treatment or prophylaxis in the management of postpartum mood disorders is not advised. In fact, prophylactic administration of a progestogen after delivery increased the risk of postpartum depression as compared with placebo.

PMDD and PMS

Premenstrual mood changes have been described since the time of Hippocrates. In 1931, Frank delineated a constellation of physical, psychological and behavioral changes associated with menstrual cycles.[102] Because so many women report premenstrual and moliminal symptoms, the criteria for the diagnosis of premenstrual syndrome (PMS) and premenstrual dysphoric disorder (PMDD) remain controversial and vary among psychiatrists, psychologists, gynecologists, epidemiologists and sociologists. It is difficult to distinguish normal from pathological premenstrual mood disturbances, given the spectrum of severity and the variety of symptoms. PMDD is a psychiatric diagnosis, and includes criteria common to major depressive disorder, such as depressed mood, eating and sleeping disturbances, decreased interest in usual activities, lethargy, difficulty concentrating, and feelings of hopelessness or anxiety. Women with PMDD also may experience increased sensitivity to rejection, irritability, a sense of being out of control, and concomitant physical symptoms, such as headaches, breast tenderness and bloating. In aggregate, the symptoms suggest increased central and peripheral reactivity or sensitivity to the customary sex-steroid excursions that accompany an ovulatory menstrual cycle. Although neurotransmitter, neuroendocrine and chronobiological disturbances have been implicated in mood disorders and in PMDD, no one theory has been well-substantiated; neither has one treatment been shown to be consistently effective. Given the similarity of PMDD and MDD, it is not uncommon for psychiatrists and other physicians to employ psychotropic medications to treat women who meet DSM criteria for PMDD.

In addition to disturbances in mood, women with PMDD often report physical and cognitive deficits such as decreased concentration, motor inefficiency, forgetfulness and indecisiveness. Several factors have been linked to PMS/PMDD, including a personal or family history of mood disorders. Women with a personal or family history of major depressive disorders are more likely to have premenstrual depression, and, alternatively, women with premenstrual syndrome may later develop major mood disorders. Genetic factors appear to predispose to PMS/PMDD. Daughters of women who report premenstrual "tension" are more likely to develop premenstrual mood symptoms than daughters of unaffected mothers, and concordance rates for PMS symptoms are significantly higher in monozygotic twins (93%) compared with dizygotic twins or sibling controls. In addition, women with PMS are at greater risk for menopausal symptoms[103] and postpartum depression.[104] Premenstrual mood disturbances tend to become more severe with increasing age, likely triggered by altered hormonal exposures due to ovarian and brain aging. The initial presentation of ovarian aging entails preserved menstrual cyclicity with increased estradiol secretion and reduced progesterone secretion.[105] As menopause approaches, there are episodes of erratic ovarian secretion with high and low levels of estradiol and progesterone accompanied by irregular cycle intervals.

Regardless of the etiological model, hormonal changes associated with the menstrual cycle are likely to serve as contributory factors to mood disruption, even though ovarian function and hormone levels do not differ between women with and without PMS/PMDD.[106] Both expected and erratic hormonal excursions impact the central nervous systems, including classic neurotransmitter systems. Schmidt administered a sex hormone challenge to women with and without a diagnosis of PMDD.[107] Women with a diagnosis of PMDD experienced customary symptoms of PMDD, whereas those without PMDD were free of symptoms. In essence, the brains of women with PMS/PMDD displayed greater sensitivity to usual hormone exposures.

In keeping with the notion of increased central reactivity, Smith[108] observed an abnormal central response to progesterone during transcranial magnetic stimulation of the motor cortex in PMS as compared to control subjects. Women with PMS demonstrated relative facilitation in the luteal but not the follicular phase, indicating an abnormal brain reaction to progesterone, perhaps mediated by GABA. Dysregulation of the neuroactive progesterone metabolite, allogregnanolone, also has been reported in PMDD.[109] Allopregnanolone decreased when symptoms improved in women with severe PMS treated with a gonadotropin-releasing hormone (GnRH) agonist or placebo.[110] The depressogenic effects of progesterone administration may be mediated by allopregnanolone and γ-aminobututyric acid (GABA), systems also implicated in the pathogenesis of depression.[111] Using serial proton magnetic resonance spectroscope to measure occipital cortex GABA levels across the menstrual cycle, Epperson[112] found reduced GABA levels across the menstrual cycle in controls, and increased GABA levels from the follicular phase to the mid- and late-luteal phases and reduced follicular cortical GABA levels in women with PMDD. The benzodiazepine site antagonist flumazenil induced a greater panic response in women with PMDD than in control women.[113] Women with PMDD manifested chronobiological abnormalities of melatonin secretion[114] and responded to specific treatments that affect circadian physiology, such as sleep deprivation and bright light. Taken together, these findings suggest that progesterone interacts with serotonin, GABA and the central circadian clock, and that this interaction differs between women with and without PMS/PMDD.

In summary, there are few data to support the notion that PMDD is primarily a result of aberrant ovarian hormone secretion. However, women with PMS and PMDD appear to be more sensitive or more reactive to both expected and erratic hormonal exposures associated with reproductive physiology, reproductive events such as pregnancy, and reproductive aging. Thus, PMS and PMDD are envisioned as resulting from the impact of expected hormonal exposures upon vulnerable central processes mediating mood and behavior. Further evidence includes the observation that oophorectomy (both surgical and medical) interrupts cyclic symptomatology, that gonadal steroid doses that approximate ovarian secretion provoke affective symptomatology in predisposed individuals, that women with PMDD appear to be at high risk for the development of major mood disorders, and that psychotropics such as SSRIs ameliorate affective symptoms in women with PMS and PMDD.

Menopause and Hormone Therapy

Commonplace symptoms that accompany the menopause transition – namely, hot flashes, sleep disturbances and irritability – suggest that parts of the brain subserving functions other than solely reproductive ones are critically impacted by the presence or absence of sex steroids. Sex steroids modulate synaptic density, neuronal and glial viability, and neurotransmitter systems that underpin cognitive, emotional and motor functions. A wealth of molecular, cellular and physiological investigations, clinical observations and trials, and epidemiological studies support the notion that sex steroids modulate brain and behavior. Whether progestational agents counteract the putative benefits of estrogens remains to be more fully investigated, but available cellular and monkey studies suggest that progestins negatively impact neural substrates, and may heighten the risk for depression and anxiety.[57]

It is reasonable to posit that a minimal concentration of estradiol at the level of the brain and in the cerebrospinal fluid will promote neuronal survival, synaptic density, neurotransmitter function, and speed of processing, and confer overall neuroprotection. It is not commonly appreciated, but premenopausal women and men have roughly equivalent levels of estradiol in their cerebrospinal fluid. While men aromatize androgens *in situ* from a pool that only modestly declines with age, postmenopausal women have loss of both ovarian estrogens and androgens as they age. Thus, after menopause, women would be expected to have much lower CSF levels of estradiol and androgen than do men. Estrogens interact centrally with GABAergic and cholinergic systems to mediate synaptic integrity and plasticity in females. In females, but not in males, aromatase blocked the increase in hippocampal spine density induced by testosterone exposure.[13] Sex differences in connectivity and estrogen responsiveness likely explain at least in part the sex-specific vulnerabilities to psychiatric and neurodegenerative conditions, especially under conditions of stress. The most neuroprotective estrogen appears to be estradiol administered transdermally.[115] Route of administration matters in older women. As the cardiovascular tree ages and loses compliance, and develops plaque and stenosis, orally administered estrogens uniformly increase the risk of stroke.

New-onset depression affects 20–30% of women during the menopausal transition.[116] The perimenopausal transition is associated with a two- to four-fold increased risk of depressive symptoms, and a two-fold increased risk of a depressive disorder, in medically and reproductively healthy women *without* prior history of MDD.[117] These increased risks remained after correction for numerous factors, including hot flashes, poor sleep and adverse life events – factors that have challenged the interpretation of earlier studies. While there was no association between mean hormone levels and mood, higher variability of between-visit estradiol concentrations in women relative to their non-depressed, premenopausal baseline was the strongest risk factor for the onset of a depressive disorder. This strongly argues for a mechanistic role of estradiol fluctuations in the onset of perimenopausal depression. Hot flashes were 1.3- to 2.2-fold more likely in women with perimenopausal-onset depressive symptoms and depressive disorder, a factor that argues for a shared neurophysiological mechanism of perimenopausal depression and vasomotor symptoms. Other risk factors for perimenopausal depression include personal history of mood disorder, previous reproductive-related mood disturbance (such as postpartum and/or premenstrual mood disorder), and psychosocial stress.

Antidepressant drugs are increasingly prescribed for hot flashes based upon clinical trials which show benefit with low dose serotonin/serotonin-norepinephrine reuptake inhibitors (i.e., escitalopram, venlafaxine, duloxetine, but not sertraline). It is proposed that antidepressants may improve distress tolerance of hot flashes via symptomatic improvement of coexisting mild insomnia, anxiety, dysphoria and/or irritability. Conversely, estradiol has proven utility in the treatment of minor and major depression.

Another unsettled issue is whether hormones, endogenous or exogenous, buffer the brain from stress and protect from depression or neurodegenerative conditions. In a systematic review and meta-analysis, LeBlanc *et al.*[118] found that symptomatic women improved most when given hormone therapy. These same investigators were unable to confirm this in a subsequent small prospective study.[119] However, in the HERS trial, symptomatic postmenopausal women were found to benefit most in the short-term from HRT.[120]

Women in the HERS trial had active cardiovascular disease, and were randomized to placebo versus continuous 0.625 mg conjugated equine estrogens combined with 2.5 mg medroxyprogesterone acetate. Participants were followed for 3 years, and quality of life was monitored with three inventories: the Duke Activity Status, the RAND scale, and the Burnham screening scale. Women with hot flashes who were given HRT had improved mental health and fewer depressive symptoms as compared to women given placebo. Asymptomatic women given HRT had greater declines in physical functioning and energy, with no change in depressive symptoms, when compared to those given placebo. Of note, the asymptomatic women also had very low depression scores at the time of randomization. Thus, it was unlikely that their mood would further improve (a ceiling effect), and it is noteworthy that PERT did not worsen mood (for editorial, see Berga, ref 121). With regard to mood, a cross-sectional analysis of 6602 white women over age 71 years, which employed the short form of the Geriatric Depression Scale, found the odds ratio of depression in ERT users to be 0.7, with a confidence interval of 0.5–0.9.[122] For PERT users, the odds ratio was 0.8 (CI 0.5–1.4). These and other studies have led to the view that estrogen therapy may improve mood in those in whom it is compromised, and may reduce the risk, or aid in the treatment, of late-life depression.

There is only a handful of neuroimaging studies relevant to understanding the relationship between hormone use and brain health in postmenopausal women, but these studies contribute a great deal to understanding the impact of sex steroids (see Box 25.3). While some studies were longitudinal in design,[35,38,123] most were cross-sectional.[46,124] To the best of our knowledge, only one compared women before hormones, after estradiol alone, and then after estradiol + progesterone.[125] Moses *et al.* gave recently postmenopausal women ages 50–55 years transdermal estradiol to approximate a circulating estradiol of 80 pg/ml for 12 weeks followed by 14 days of 200 mg of micronized progesterone. Combining MRI for neuroanatomical localization and PET (positron emission tomography) with a serotonin 2A receptor agonist for neurochemical detection with a region-of-interest analysis revealed that estradiol enhanced serotonergic binding in cortical regions, particularly in the dorsolateral prefrontal cortex and anterior cingulate cortex.[60] A subsequent analysis of the same brain images using the technique of statistical parametric mapping, which is not confined to region of interest or neuroanatomical localization, showed that estradiol increased serotonergic binding throughout the brain, and that progesterone markedly enhanced serotonergic binding throughout the brain relative to estradiol (Fig. 25.3). There are few comparative studies to discern the extent to which the impact of hormones depends on dose or type of estrogen and progestin being administered. However, Silverman *et al.*[38] randomized 53 women at

BOX 25.3

HOW DO WE KNOW THAT ESTRADIOL AND PROGESTERONE HAVE INDEPENDENT EFFECTS UPON KEY NEUROTRANSMITTER SYSTEMS THAT MEDIATE AFFECT AND EMOTION

Using a specific serotonergic ligand and combined neuroimaging with magnetic resonance and positron emission tomography (PET), Moses *et al.*[1] and Moses-Kolko *et al.*[2]) showed that women taking a physiological dose of transdermal estradiol displayed increased serotonergic binding in key brain areas mediating affect and cognition, and that the addition of progesterone markedly increased serotonergic binding throughout the brain.

References

1. Moses EL, Drevets WC, Smith G, Mathis CA, Kalro BN, Butters MA, Leondires MP, Greer PJ, Lopresti B, Loucks TL, Berga SL. Effects of estradiol and progesterone administration on human serotonin 2A receptor binding: a PET study. *Biol Psychiatry.* 2000;48:854–860.
2. Moses-Kolko EL, Berga SL, Greer PJ, Smith G, Cidis Meltzer C, Drevets WC. Widespread increases of cortical serotonin type 2A receptor availability after hormone therapy in euthymic postmenopausal women. *Fertil Steril.* 2003;80:554–559.

risk for dementia and who were taking hormones postmenopausally to continuation versus discontinuation. Neuroimaging to discern regional brain metabolism with FDG-PET scans was done before and after. Hormone exposure correlated with brain metabolism in select cortical areas. In particular, estradiol users were significantly better in verbal performance than women taking conjugated equine estrogen (Premarin®) (Fig. 25.4), which was the estrogen used in the Women's Health Initiative. Further, those using an estrogen combined with a progestin had lower metabolism than those using only estrogen. Of interest, in the Olmstead County longitudinal cohort, women who experienced premature withdrawal of ovarian function due to surgical oophorectomy had a higher risk of depression and neurodegenerative disorders if they did not receive exogenous hormone replacement until the time of menopause.[126–128] The Cache County study[129] showed that exogenous hormone use begun at the time of menopause reduced the risk of dementia to that of men, but the investigators have not reported whether the risk of affective symptoms or disorders was reduced to the same extent. However, since hormones impact areas subserving both cognition and emotionality, and since cognitive impairment often manifests as depression and mood disorders and vice versa, it is reasonable to suggest that exogenous hormone exposure, when given in a manner that promotes metabolism[38] and does not

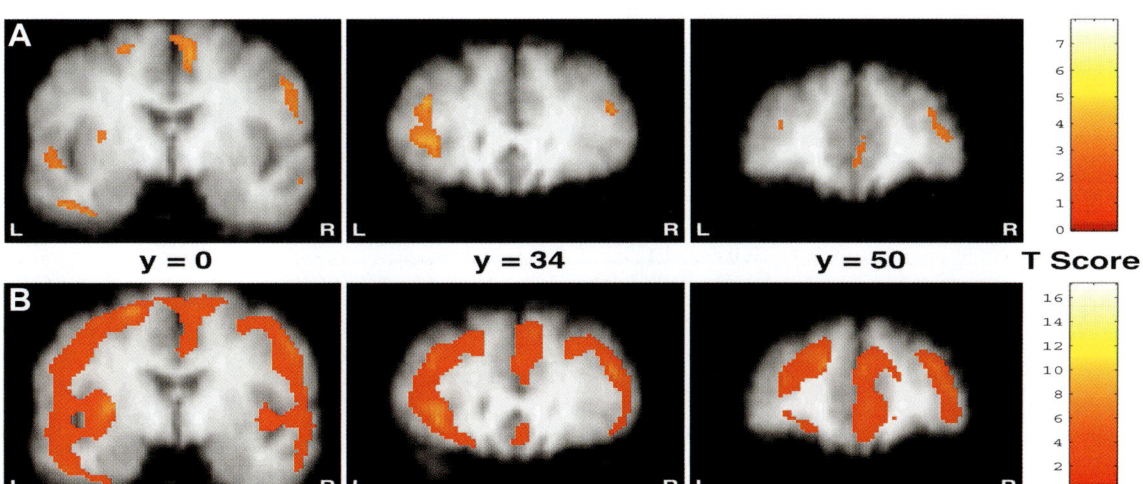

FIGURE 25.3 Regions where serotonergic (5HT2A) binding potential increased following 12 weeks of transdermal estradiol to approximate 80 pg/ml in recently postmenopausal women off hormones for at least 3 months prior to enrollment (**A**) and after 2 weeks of estradiol + oral micronized progesterone (**B**). *Reproduced from Moses-Kolko EL, Berga SL, Greer PJ, Smith G, Cidis Meltzer C, Drevets WC. Widespread increases of cortical serotonin type 2A receptor availability after hormone therapy in euthymic postmenopausal women.* Fertil Steril. *2003;80:554–559.*

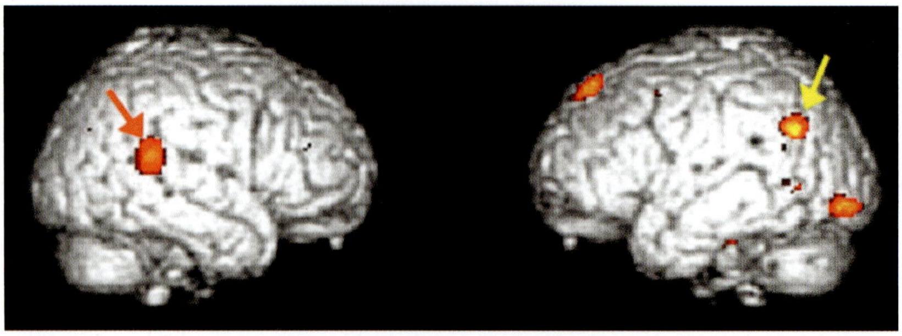

FIGURE 25.4 **Specific areas of greater metabolism in postemenopausal women taking estradiol compared to those taking conjugated equine estrogen (CEE, Premarin®).** Metabolism in the left Wernicke's area (yellow arrow) and right superior temporal gyrus (red arrow) was greater in estradiol users compared to CEE users ($P < 0.0005$). Voxels with significance < 0.0005 are shown above in color. *Reproduced from Silverman DHS, Geist CL, Kenna HA, Williams K, Wroolie T, Powers B, Brooks J, Rasgon NL. Differences in regional brain metabolism associated with specific formulations of hormone therapy in postmenopausal women at risk for AD.* Psychoneuroendocrinology. *2010;36:502–513.*

increase the risk of stroke,[115] will buffer the brain from stress, depression and neurodegeneration (see Box 25.4).

It is important to emphasize that hormonal interventions are sex-specific and are likely hormone-specific. The brain converts testosterone to estradiol, and men typically have at all times in their lives, except when women are pregnant, as much estradiol bathing their neurons and glia as women do. Because men do not experience gonadal cessation at midlife, as women do, their brains continue to be exposed to both estradiol and testosterone indefinitely. As noted earlier, androgens may be more important for synapse formation and maintenance in men than women, so it is premature to conclude that androgen therapy would benefit women. Because basic research substantiates that estrogens have both neurotropic and neuroprotective properties in women, it could be that aromatizable androgens will benefit women, but this has not been proved. It could also be that certain selective estrogen receptor modulators will benefit women. It seems unlikely that aromatase inhibitors will benefit women. These extrapolations are based on our understanding of hormone mechanisms in the brain. The impact of estrogens upon the brains of women is mediated by a variety of mechanisms, including via classical genomic functions of estrogen receptors (ER). In addition, cell-membrane ERs activate second messenger cascades and highly amplify gene transcription. Selective estrogen receptor modulators (SERMs) target the same receptors as estrogen, estrogen receptor (ER) subtypes ERα and ERβ, but differ in transcriptional ability based upon unique ligand-induced conformational change of the estrogen receptor (ER) to facilitate interaction of ER with co-activators, and consequent agonism or antagonism of ER-mediated DNA transcription. Antagonism of ERβ may pose risk for heightened depression and anxiety, given the demonstration of ERβ facilitory effects on serotonergic activity, social interaction and positive behaviors in rodent models of depression. Both ER subtypes are involved in neuroprotective functions. The majority of studies find no increased risk of depressive symptoms in SERM-treated women, and some find SERM-associated amelioration of depressive symptoms, but this topic remains a subject open for investigative studies that employ neuroimaging, or that include women with affective burden or risk.

BOX 25.4

HOW DO WE KNOW THAT THE POSTMENOPAUSAL USE OF EXOGENOUS HORMONES ALTERS BRAIN METABOLISM?

Silverman *et al.*[1] showed that postmenopausal women who were randomized to stop various hormone therapy formulations showed reduced brain metabolism. Also, women taking estradiol performed significantly better in verbal memory tasks than those taking conjugated equine estrogen, and verbal memory performance positively correlated with metabolism in key brain areas.

Reference

1. Silverman DHS, Geist CL, Kenna HA, Williams K, Wroolie T, Powers B, Brooks J, Rasgon NL. Differences in regional brain metabolism associated with specific formulations of hormone therapy in postmenopausal women at risk for AD. *Psychoneuroendocrinology.* 2010;36: 502–513.

Androgen therapy is more controversial. First, a tested, safe, physiological replacement product is not commercially available in most parts of the world. The adrenal hormone dehydroepiandrosterone (DHEA) is an androgen currently being promoted as a dietary supplement to restore a sense of well-being, but few data are available on its long-term safety and effects. The same holds for the androgen androstenedione, which is often promoted to enhance athletic performance or support libido. There are no commercially available testosterone products for women, but one hormone product combines estrogen and methyltestosterone. Its use is recommended to enhance libido. It remains to be demonstrated that methyltestosterone transgresses the blood–brain barrier and gains access to the key brain centers that drive libido. To the best of our knowledge, there are no studies showing that androgens enhance mood or protect from depression in women, but there is scant evidence to address these important knowledge gaps.

Neuroprotection

Neurodegenerative disorders may masquerade as mood disorders, and *vice versa*. The brain centers subserving cognition express both types of estrogen receptors, namely ERα and ERβ.[43] Using the powerful technique of selectively deleting each estrogen receptor subtype to create a knockout mouse model, Dubal *et al.*[130] showed that ERα occupation was required to limit neuronal death in the face of ischemic insult. Cholinergic neurons are critically involved in memory and learning, and are heavily invested with ERα. In a rodent model, cholinergic neurons mediated estrogen's preservation of memory after ovariectomy.[44] These and other molecular data have led to the notion that full neuroprotection requires estrogen occupation and activation of ERα.

Much of the confusion regarding the neuroprotective effects of estrogen comes from the trial in which conjugated equine estrogens (which may not readily transgress the blood–brain barrier because they are sulfated) were given to women with established mild dementia.[131] No benefit was observed. However, by the time cognitive impairment or dementia is clinically evident there has been a significant loss of synapses,[132] and, with advancing loss and age, the extent of neural plasticity may be so compromised that brain architecture cannot be restored. Thus, the more relevant question is whether ERT or PERT has a role in protecting from the development of dementia or synaptic loss. The studies by Rocca *et al.* and Zandi *et al.* suggest that timing of initiation matters, but it may be that any hiatus in estrogen exposure is somewhat deleterious.[126–129] However, these observational trials cannot be used to exclude the notion that brain plasticity is nil if estrogen use is not begun immediately after menopause or premenopausal oophorectomy. For instance, short-term transdermal estradiol therapy has been demonstrated to enhance memory after many years of hypoestrogenism. In a study of elderly postmenopausal women who had not used hormones for at least 7 years, Wolf and colleagues found that a 2-week exposure to transdermal estradiol increased memory, but only when circulating estradiol levels exceeded 29 pg/ml.[133] As noted earlier, Shaywitz *et al.*[37] studied 60 postmenopausal women randomized to placebo versus 1.25 mg CEE for 21 days. CEE improved oral reading and short-term verbal memory, but not spatial ability or vocabulary. Functional imaging studies showed that estrogen activated the inferior parietal lobe, a region sensitive to phonological demands. REMEMBER, a large representative population-based study investigating the effect of hormone use after menopause in Adelaide, Australia, found that women who initiated hormones before age 56 years performed better that later initiators, and were faster on psychometric testing that never users. Those who used hormones longer than 11 years did best, and those who used estrogen alone performed faster than those using a combination of estrogen and a progestin.[134]

In aggregate, data support the concept that postmenopausal hormone use, particularly transdermal estradiol, may confer modest neuroprotection, especially if begun promptly after menopause or oophorectomy. Hormones cannot be used for secondary prevention once dementia is clinically apparent. Healthy women with healthy habits may experience beneficial synergism, whereas unhealthy women with poor health habits may fail to benefit or may even be harmed by hormone use, especially if the hormone preparation is not a full estrogen and promotes stroke and venous thromboembolism.[115]

CONCLUSIONS

In summary, sex steroids organize the brain, rendering it sexually dimorphic, as well as activate and maintain the sexually differentiated brain. Sex steroids are implicated in the modulation of critical behaviors, including mood, libido, cognition and motor functions, and in the risk for brain diseases such as dementia and stroke. Conditions such as postpartum mood disorders and premenstrual syndrome reflect the interplay of sex steroids upon a brain vulnerable to or sensitized by psychosocial and biological factors. The exact contribution of sex-steroid exposure after menopause for neuroprotection, both cognitive and affective, remains uncertain. This uncertainty reflects the lack of appropriate investigative methodologies, the sheer complexity that arises when a multiplicity of determinants interact and synergize, the vast array of behavioral outputs conferred by the spatiotemporal patterning of the brain in action, and the late-life presentation of neurodegenerative disorders.

References

1. Drevets WC, Gadde K, Krishnan KRR. Neuroimaging of mood disorders. In: Charney DS, Nestler EJ, eds. *Neurobiology of Mental Illness*. 2nd ed. New York: Oxford University Press; 2004:461–480.
2. Garcia R, Vouimba RM, Baudry M, Thompson RF. The amygdala modulates prefrontal cortex activity relative to conditioned fear. *Nature*. 1999;402:294–296.
3. Price JL, Drevets WC. Neurocircuitry of mood disorders. *Neuropsychopharmacology*. 2010;35:192–216.
4. Izquierdo A, Wellman CL, Holmes A. Brief uncontrollable stress causes dendritic retraction in infralimbic cortex and resistance to fear extinction in mice. *J Neurosci*. 2006;26:5733–5738.
5. Drevets WC, Price JL, Bardgett MD, Reich T, Todd RD, Raichle ME. Glucose metabolism in the amygdala in depression: relationship to diagnostic subtype and plasma cortisol levels. *Pharmacol Biochem Behav*. 2002;71:4331–4447.
6. Nemeroff CB, Widerlov E, Bissette G, Walleus H, Karlsson I, Eklund K, Kilts CD, Loosen PT, Vale W. Elevated concentrations of CSF corticotrophin-releasing factor-like immunoreactivity in depressed patients. *Science*. 1984;226:1342–1344.
7. Keller J, Flores B, Gomez RG, Solvason HB, kenna H, Williams GH, Schatzberg AF. Cortisol circadian rhythm alterations in psychotic major depression. *Biol Psychiatry*. 2006;60:275–281.
8. McEwen BS. Glucocorticoids, depression, mood disorders: structural remodeling in the brain. *Metabolism*. 2005;54(Suppl):20–23.
9. McEwen BS. Physiology and neurobiology of stress and adaptation: central role of the brain. *Physiol Rev*. 2007;87:873–904.
10. Golden SH. A review of the evidence for a neuroendocrine link between stress, depression, and diabetes mellitus. *Curr Diabetes Rev*. 2007;3:252–259.
11. Tobet S, Knoll JG, Hartshorn C, Aurand E, Stratton M, Kumar P, Searcy B, McClellan K. Brain sex differences and hormone influences: a moving experience? *J Neuroendocrinol*. 2009;21:387–392.
12. Goldstein JM, Seidman LJ, Horton NJ, Makris N, Kennedy DN, Caviness Jr VS, Faraone SV, Tsuang MT. Normal sexual dimorphism of the adult human brain assessed by *in vivo* magnetic resonance imaging. *Cereb Cortex*. 2001;11:490–497.
13. Gillies GE, McArthur S. Estrogen actions in the brain and the basis for differential action in men and women: a case for sex-specific medicines. *Pharmacol Rev*. 2010;62:155–198.
14. Auger AP, Perrot-Sinai TS, McCarthy MM. Excitatory versus inhibitory GABA as a divergence point in steroid-mediated sexual differentiation of the brain. *Proc Natl Acad Sci USA*. 2001;98:8059–8064.
15. Nuñez JL, McCarthy MM. Androgens predispose males to $GABA_A$-mediated excitotoxicity in the developing hippocampus. *Exp Neurol*. 2008;10:699–708.
16. Arnold AP. Sex chromosomes and brain gender. *Nat Rev Neuroscience*. 2004;5:701–708.
17. Carruth LL, Reisert I, Arnold AP. Sex chromosome genes directly affect brain sexual differentiation. *Nat Neurosci*. 2002;10:933–934.
18. McCarthy MM, Wright CL, Schwarz JM. New tricks by an old dogma: mechanisms of the organizational/activational hypothesis of steroid-mediated sexual differentiation of the brain and behavior. *Horm Behav*. 2009;55:655–665.
19. Yague JG, Wang AC-J, Janssen WGM, Hof PR, Garcia-Segura LM, Azcoita I, Morrison JH. Aromatase distribution in the monkey temporal neocortex and hippocampus. *Brain Res*. 2008;1209:115–127. doi:10.1016/j.brainres.2008.02.061.
20. Bingham B, Gray M, Sun T, Viau V. Postnatal blockade of androgen receptors or aromatase impair the expression of stress hypothalamic–pituitary–adrenal axis habituation in adult male rats. *Psychoneuroendocrinology*; 2010. doi:10.1016/j.psyneuen.2010.07.015.
21. Arnold AP, Burgoyne PS. Are XX and XY brain cells intrinsically different? *Trends Endocrinol Metab*. 2004;15:6–11.
22. McCarthy MM, Auger AP, Bale TL, De Vries GJ, Dunn GA, Forger NG, Murray EK, Nugent BM, Schwarz JM, Wilson ME. The epigenetics of sex differences in the brain. *J Neuroscience*. 2009;29:12815–12823.
23. Morrison JH, Brinton RD, Schmidt PJ, Gore AC. Estrogen, menopause, and the aging brain: how basic neuroscience can inform hormone therapy in women. *J Neurosci*. 2006;26:10332–10348.
24. McEwen BS. Estrogen actions throughout the brain. *Recent Prog Horm Res*. 2002;57:357–384.
25. Green R, Santoro N. Menopausal symptoms and ethnicity: the Study of Women's Health Across the Nation. *Women's Health (Lond Engl)*. 2009;5:127–133.
26. Stadberg E, Mattson LA, Milsom I. The prevalence and severity of climacteric symptoms and the use of different treatment regimens in a Swedish population. *Acta Ob Gyn Scand*. 1997;76:442–448.
27. Hollander LE, Freeman EW, Sammel MD, Berlin JA, Grisso JA, Battistini M. Sleep quality, estradiol levels, and behavioral factors in late reproductive age women. *Obstet Gynecol*. 2001;98:391–397.
28. Love T, Smith YR, Persad CC, Tkaczyk A, Zubieta J- K. Short-term hormone treatment modulates emotion response circuitry in postmenopausal women. *Fertil Steril*. 2010;93:1929–1937.
29. Matteo S, Rissman EF. Increased sexual activity during the midcycle portion of the human menstrual cycle. *Horm Behav*. 1984;18:249–255.
30. Pfaff DW. Hormones, genes, and behavior. *Proc Natl Acad Sci USA*. 1997;94:14213–14216.
31. Savic I, Berglund H, Gulyas B, Roland P. Smelling of odorous sex hormone-like compounds causes sex-differentiated hypothalamic activations in humans. *Neuron*. 2001;31:661–668.
32. Warner P, Bancroft J. Mood, sexuality, oral contraceptives, and the menstrual cycle. *J Psychosom Res*. 1988;32:417–427.
33. Finkelstein JW, Susman EJ, Chinchilli VM, Kunselman SJ, D'Arcangelo MR, Schwab J, Demers LM, Liben LS, Lookingbill G, Kulin HE. Estrogen or testosterone increases self-reported aggressive behaviors in hypogonadal adolescents. *J Clin Endocrinol Metab*. 1997;82:2433–2438.
34. Finkelstein JW, Susman EJ, Chinchilli VM, D'Arcangelo MR, Kunselman SJ, Schwab J, Demers LM, Liben LS, Kulin HE. Effects of estrogen or testosterone on self-reported sexual responses and behaviors in hypogonadal adolescents. *J Clin Endocrinol Metab*. 1998;83:2281–2285.
35. Maki PM, Resnick SM. Longitudinal effects of estrogen replacement therapy on PET cerebral blood flow and cognition. *Neurobiol Aging*. 2000;21:373–383.
36. McEwen BS. The molecular and neuroanatomical basis for estrogen effects in the central nervous system. *J Clin Endocrinol Metab*. 1999;84:1790–1797.
37. Shaywitz SE, Naftolin F, Zelterman D, Marchione KE, Holahan JM, Palter SF, Shaywitz BA. Better oral reading and short-term memory in midlife, postmenopausal women taking estrogen. *Menopause*. 2003;10:420–426.

38. Silverman DHS, Geist CL, Kenna HA, Williams K, Wroolie T, Powers B, Brooks J, Rasgon NL. Differences in regional brain metabolism associated with specific formulations of hormone therapy in postmenopausal women at risk for AD. *Psychoneuroendocrinology.* 2010;36:502–513.
39. McEwen BS. Stress, sex, and neural adaptation to a changing environment: mechanisms of neuronal remodeling. *Ann NY Acad Sci.* 2010;1204:E38–E59.
40. Smith YR, Stohler CS, Nichols TE, Bueller JA, Koeppe RA, Zubieta JK. Pronociceptive and antinociceptive effects of estradiol through endogenous opioid neurotransmisson in women. *J Neurosci.* 2006;26(21):5777–5785.
41. Irwin RW, Yao J, Hamilton RT, Cadenas E, Brinton RD, Nilsen J. Progesterone and estrogen regulate oxidative metabolism in brain mitochondria. *Endocrinology.* 2008;149:3167–3175.
42. Smith YR, Zubieta JK. Neuroimaging of aging and estrogen effects on central nervous system physiology. *Fertil Steril.* 2001;76:651–659.
43. Wang L, Andersson S, Warner M, Gustafsson JA. Estrogen actions in the brain. *Sci STKE.* 2002;138:29.
44. Gibbs RB. Basal forebrain cholinergic neurons are necessary for estrogen to enhance acquisition of a delayed matching-to-position T-maze task. *Horm Behav.* 2002;42:245–257.
45. Smith YR, Minoshima S, Kuhl DE, Zubieta JK. Effects of long-term hormone therapy on cholinergic synaptic concentrations in healthy postmenopausal women. *J Clin Endocrinol Metab.* 2001; 86:679–684.
46. Deecher D, Andree TH, Sloan D, Schechter LE. From menarche to menopause: Exploring the underlying biology of depression in women experiencing hormonal changes. *Psychoneuroendocrinology.* 2008;33:3–17.
47. Porsolt RD, Bertin A, Jalfre M. Behavioral despair in mice: a primary screening test for antidepressants. *Arch Intl Pharmacodyn Ther.* 1977;229:327–336.
48. Petit-Demouliere B, Chenu F, Bourin M. Forced swimming test in mice: a review of antidepressant activity. *Psychopharmacology.* 2005;177:245–255.
49. Corner GW. Ovulation and menstruation in *Macacus rhesus*. *Contrib Embryol.* 1923;75:77–101.
50. Knobil E. The neuroendocrine control of ovulation. *Hum Reprod.* 1988;3:469–472.
51. Zuckerman S. The menstrual cycle of the primates. I. General nature and homology. *Proc Zool Soc Lond p.* 1930:691–754.
52. Williams RF, Hodgen GD. The reproductive cycle in female macaques. *Am J Primatol Suppl.* 1982;1:181–192.
53. Shively CA, Bethea CL. Cognition, mood disorders, and sex hormones. *ILAR J.* 2004;45(2):189–199.
54. Mook D, Felger J, Graves F, Wallen K, Wilson ME. Tamoxifen fails to affect central serotonergic tone but increases indices of anxiety in female rhesus macaques. *Psychoneuroendocrinology.* 2005;30:273–283.
55. Breuer B, Anderson R. The relationship of tamoxifen with dementia, depression, and dependence in activities of daily living in elderly nursing home residents. *Women Health.* 2000;31:71–85.
56. Wilson ME, Mook D, Graves FC, Gleger J, Bielsky IR, Wallen K. Tamoxifen is an estrogen antagonist on gonadotropin secretion and responsiveness of the hypothalamic–pituitary–adrenal axis in female monkeys. *Endocrine.* 2003;22:305–315.
57. Pazol K, Wilson ME, Wallen K. Medroxyprogesterone acetate antagonizes the effects of estrogen treatment on social and sexual behavior in female macaques. *J Clin Endocrinol Metab.* 2004;89:2998–3006.
58. Craig MC, Murphy DG. Estrogen: effects on normal brain function and neuropsychiatric disorders. *Climacteric.* 2007; 10(Suppl 2):97–104.
59. Moses EL, Drevets WC, Smith G, Mathis CA, Kalro BN, Butters MA, Leondires MP, Greer PJ, Lopresti B, Loucks TL, Berga SL. Effects of estradiol and progesterone administration on human serotonin 2A receptor binding: a PET study. *Biol Psychiatry.* 2000;48:854–860.
60. Angold A, Costello EJ, Worthman CM. Puberty and depression: the roles of age, pubertal status and pubertal timing. *Psychol Med.* 1998;28:51–61.
61. Angold A, Costello EJ, Erkanli A, Worthman CM. Pubertal changes in hormone levels and depression in girls. *Psychol Med.* 1999;29:1043–1053.
62. Patton GC, Hibbert ME, Carlin J, Shao Q, Rosier M, Caust J, Bowes G. Menarche and the onset of depression and anxiety in Victoria, Australia. *J Epidemiol Comm Health.* 1996;50:661–666.
63. Patton GC, Olsson C, Bond L, Toumbourou JW, Carlin JB, Hemphill SA, Catalano RF. Predicting female depression across puberty: a two-nation longitudinal study. *J Am Acad Child Adolesc Psychiatry.* 2008;47:1424–1432.
64. Essau CA, Lewinsohn PM, Seeley JR, Sasagawa S. Gender differences in the developmental course of depression. *J Affect Disord.* 2010;127:185–190.
65. Steinberg L. Psychological control: style or substance? *New Dir Child Adolesc Dev.* 2005;108:71–78.
66. Adleman NE, Menon V, Blasey CM, White CD, Warsofsky IS, Glover GH, Reiss AL. A developmental fMRI study of the Stroop color-word task. *Neuroimage.* 2002;16:61–75.
67. Killgore WD, Yurgelun-Todd DA. Sex-related developmental differences in the lateralized activation of the prefrontal cortex and amygdala during perception of facial affect. *Percept Mot Skills.* 2004 Oct;99:371–391.
68. Casey BJ, Jones RM, Hare TA. The adolescent brain. *Ann NY Acad Sci.* 2008;1124:111–126.
69. Giedd JN, Blumenthal J, Jeffries NO, Castellanos FX, Liu H, Zijdenbos A, Paus T, Evans AC, Rapoport JL. Brain development during childhood and adolescence: a longitudinal MRI study. *Nat Neurosci.* 1999;2:861–863.
70. Luna B, Thulborn KR, Munoz DP, Merriam EP, Garver KE, Minshew NJ, Keshavan MS, Genovese CR, Eddy WF, Sweeney JA. Maturation of widely distributed brain function subserves cognitive development. *Neuroimage.* 2001;13:786–793.
71. Paus T, Zijdenbos A, Worsley K, Collins DL, Blumenthal J, Giedd JN, Rapoport JL, Evans AC. Structural maturation of neural pathways in children and adolescents: *in vivo* study. *Science.* 1999;283:1908–1911.
72. Sowell ER, Thompson PM, Tessner KD, Toga AW. Mapping continued brain growth and gray matter density reduction in dorsal frontal cortex: Inverse relationships during post-adolescent brain maturation. *J Neurosci.* 2001 Nov 15;21: 8819–8829.
73. Sowell ER, Trauner DA, Gamst A, Jernigan TL. Development of cortical and subcortical brain structures in childhood and adolescence: a structural MRI study. *Dev Med Child Neurol.* 2002;44(1):4–16.
74. Cunningham MG, Bhattacharyya S, Benes FM. Amygdalo-cortical sprouting continues into early adulthood: implications for the development of normal and abnormal function during adolescence. *J Comp Neurol.* 2002;453:116–130.
75. Teicher MH, Anderson SL, Hostetter Jr JC. Evidence for dopamine receptor pruning between adolescence and adulthood in striatum but not nucleus accumbens. *Brain Res Dev Brain Res.* 1995;89:167–172.
76. Zehr JL, Todd BJ, Schultz KM, McCarthy MM, Sisk CL. Dendritic pruning of the medial amygdala during pubertal development of the male Syrian hamster. *J Neurobiol.* 2006; 66:578–590.

77. Spear P. The adolescent brain and age-related behavioral manifestations. *Neurosci Rev.* 2000;24:417–463.
78. Steinberg L. Cognitive and affective development in adolescence. *Trends Cogn Sci.* 2005;9:69–74.
79. Ryan ND, Varma D. Child and adolescent mood disorders: experience with serotonin-based therapies. *Biol Psychiatry.* 1998; 44:336–340.
80. Becker JB, Berkley KJ, Geary N, Hampson E, Herman JP, Young EA, eds. *Sex Differences in the Brain from Genes to Behavior. Chapter 21 Hormones and Mood by M Steiner and EA Young,* p. 405–426. New York, NY: Oxford University Press Inc; 2008.
81. Ehler CL, Fran KE, Kupfer DJ. Social zeitgebers and biological rhythms. A unified approach to understanding the etiology of depression. *Arch Gen Psychiatry.* 1988;45:948–952.
82. Cyranowski JM, Frank E, Young E, Shear MK. Adolescent onset of the gender difference in lifetime rates of major depression: a theoretical model. *Arch Gen Psychiatry.* 2000;57:21–27.
83. Dorn LD, Chrousos GP. The neurobiology of stress: Understanding regulation of affect during female biological transitions. *Semin Reprod Endocrinol.* 1997;15:19–35.
84. Young EA, Altemus M. Puberty, ovarian steroids, and stress. *Ann NY Acad Sci.* 2004;1021:124–133.
85. Young E, Korsun A. Sex, trauma, stress hormones, and depression. *Mol Psychiatry.* 2010;15:23–28.
86. Taylor SE, Klein LC, Lewis BP, Gruenewald TL, Gurung RA, Updegraff JA. Biobehavioral responses to stress in females: tend-and-befriend, not fight-or-flight. *Psychol Rev.* 2000;107: 411–429.
87. Rahmen Q. The neurodevelopment of human sexual orientation. *Neurosci Biobehav Rev.* 2005;29:1057–1066.
88. Savic I, Lindstrom P. PET and MRI show differences in cerebral asymmetry and functional connectivity between homo- and heterosexual subjects. *Proc Natl Acad Sci.* 2008;105:9403–9408.
89. Savic I, Berglund P, Lindstrom P. Brain response to putative pheromones in homosexual men. *Proc Natl Acad Sci.* 2005;102: 7356–7361.
90. Berglund H, Lindstrom P, Savic I. Brain response to putative pheromones in lesbian women. *Proc Natl Acad Sci.* 2006;103: 8269–8274.
91. Ciumas C, Hirschberg AL, Savic I. High fetal testosterone and sexually dimorphic cerebral networks in females. *Cereb Cortex.* 2009;19:1167–1174.
92. Berga SL, Daniels TL, Giles DE. Women with functional hypothalamic amenorrhea but not other forms of anovulation display amplified cortisol concentrations. *Fertil Steril.* 1997;67: 1024–1030.
93. Giles DE, Berga SL. Cognitive and psychiatric correlates of functional hypothalamic amenorrhea: A controlled comparison. *Fertil Steril.* 1993;60:486–492.
94. Marcus MD, Loucks TL, Berga SL. Psychological correlates of functional hypothalamic amenorrhea. *Fertil Steril.* 2001;76: 310–316.
95. Williams NI, Berga SL, Cameron JL. Synergistic effects of multiple sub-threshold stressors on menstrual cyclicity in cynomolgus monkeys. *Am J Physiol Endocrinol Metab.* 2007;293: E270–E276.
96. Berga SL, Marcus MD, Loucks TL, Hlastala S, Ringham R, Krohn MA. Recovery of ovarian activity in women with functional hypothalamic amenorrhea (FHA) treated with cognitive behavior therapy (CBT). *Fertil Steril.* 2003;80:976–981.
97. Altemus M, Deuster PA, Galliven E, Carter CS, Gold PW. Suppression of hypothalamic–pituitary–adrenal axis responses to stress in lactating women. *J Clin Endocrinol Metab.* 1995;80: 2954–2959.
98. Ng RC, Hirata CK, Yeung W, Haller E, Finley PR. Pharmacologic treatment for postpartum depression: a systematic review. *Pharmacotherapy.* 2010;30:928–941.
99. Wisner KL, Hanusa BH, Perel JM, et al. Postpartum depression: a randomized trial of sertraline versus nortriptyline. *J Clin Psychopharmacol.* 2006;26:353–360.
100. Levant B. N-3 (omega-3) Fatty acids in postpartum depression: implications for prevention and treatment. *Depress Res Treat* 2011:467349 [Epub ahead of print].
101. Gregoire AJ, Kumar R, Everitt B, et al. Transdermal oestrogen for treatment of severe postnatal depression. *Lancet.* 1996;347: 930–933.
102. Frank RT. The hormonal causes of premenstrual tension. *Arch Neurol Psychiatry.* 1931;26:1053–1057.
103. Freeman EW, Sammel MD, Rinaudo PJ, Sheng L. Premenstrual syndrome as a predictor of menopausal symptoms. *Obstet Gynecol.* 2004;103:960–966.
104. Chuong CJ, Burgos DM. Medical history in women with premenstrual syndrome. *J Psychosom Obstet Gynaecol.* 1995;16: 21–27.
105. Santoro N, Brown JR, Adel T, Skurnick JH. Characterization of reproductive hormonal dynamics in the perimenopause. *J Clin Endocrinol Metab.* 1996;81:1495–1501.
106. Rubinow DR, Hoban MC, Grover GN, Galloway DS, Roy-Byrne P, Andersen R, Merriam GR. Changes in plasma hormones across the menstrual cycle in patients with menstrually related mood disorder and in control subjects. *Am J Obstet Gynecol.* 1988;158:5–11.
107. Schmidt PJ, Nieman LK, Grover GN, Muller KL, Merriam GR, Rubinow DR. Lack of effect of induced menses on symptoms in women with premenstrual syndrome. *N Engl J Med.* 1991;324: 1174–1179.
108. Smith MJ, Adams LF, Schmidt PJ, Rubinow DR, Wassermann EM. Abnormal luteal phase excitability of the motor cortex in women with premenstrual syndrome. *Biol Psychiatry.* 2003;54:757–762.
109. Girdler SS, Straneva PA, Light KC, Pedersen CA, Morrow AL. Allopregnanolone levels and reactivity to mental stress in premenstrual dysphoric disorder. *Biol Psychiatry.* 2001;49: 788–797.
110. Oberlander TF, Warburton W, Misri S, Aghajanian J, Hertzman C. Neonatal outcomes after prenatal exposure to selective serotonin reuptake inhibitor antidepressants and maternal depression using population-based linked health data. *Arch Gen Psychiatry.* 2006;63:898–906.
111. Klatzkin RR, Morrow AL, Light KC, Pedersen CA, Girdler SS. Associations of histories of depression and PMDD diagnosis with allopregnanolone concentrations following the oral administration of micronized progesterone. *Psychoneuroendocrinology.* 2006;31:1208–1219.
112. Epperson CN, Haga K, Mason GF, Sellers E, Gueorguieva R, Zhang W, Weiss E, Rothman DL, Krystal JH. Cortical gamma-aminobutyric acid levels across the menstrual cycle in healthy women and those with premenstrual dysphoric disorder: a proton magnetic resonance spectroscopy study. *Arch Gen Psychiatry.* 2002;59:851–858.
113. Le Mellédo JM, Van Driel M, Coupland NJ, Lott P, Jhangri GS. Response to flumazenil in women with premenstrual dysphoric disorder. *Am J Psychiatry.* 2000;157:821–823.
114. Parry BL, Berga SL, Mostofi N, Sependa PA, Kripke DF, Gillin JC. Morning versus evening bright light treatment of late luteal phase dysphoric disorder. *Am J Psychiatry.* 1989;146:1215–1217.
115. Renoux C, Dell'aniello S, Garbe E, Suissa S. Transdermal and oral hormone replacement therapy and the risk of stroke: a nested case-control study. *Br Med J.* 2010;340:c2519.

116. Schmidt PJ, Haq N, Rubinow DR. A longitudinal evaluation of the relationship between reproductive status and mood in perimenopausal women. *Am J Psychiatry.* 2004;161:2238–2244.
117. Bromberger JT, Kravitz HM, Chang YF, Cyranowski JM, Brown C, Matthews KA. Major depression during and after the menopausal transition: Study of Women's Health Across the Nation (SWAN). *Psychol Med.* 2011;9:1–10.
118. LeBlanc ES, Janowsky J, Chan BKS, Nelson HD. Hormone replacement therapy and cognition. *J Am Med Assoc.* 2001;285:1489–1499.
119. LeBlanc ES, Neiss MB, Carello PE, Samuels MH, Janowsky JS. Hot flashes and estrogen therapy do not influence cognition in early menopausal women. *Menopause.* 2007;14:191–200.
120. Hlatky MA, Boothroyd D, Vittinghoff E, Sharp P, Whooley MA. Quality-of-life and depressive symptoms in postmenopausal women after receiving hormone therapy. *J Am Med Assoc.* 2002;287:591–597.
121. Berga SL. Does estrogen use support cognition in postmenopausal women? *Menopause.* 2007;14:163–165.
122. Whooley MA, Grady D, Cauley JA. Postmenopausal estrogen therapy and depressive symptoms in older women. *J Gen Intern Med.* 2000;15:535–541.
123. Kugaya A, Epperson CN, Zoghbi S, van Dyck CH, Hou Y, Fujita M, Staley JK, Garg PK, Seibyl JP, Innis RB. Increase in prefrontal cortex serotonin 2A receptors following estrogen treatment in postmenopausal women. *Am J Psychiatry.* 2003;160:1522–1524.
124. Rasgon NL, Silverman D, Siddarth P, Miller K, Ercoli LM, Elman S, Lavretsky H, Huang SC, Phelps ME, Small GW. Estrogen use and brain metabolic change in postmenopausal women. *Neurobiol Aging.* 2005;26:229–235.
125. Moses-Kolko EL, Berga SL, Greer PJ, Smith G, Cidis Meltzer C, Drevets WC. Widespread increases of cortical serotonin type 2A receptor availability after hormone therapy in euthymic postmenopausal women. *Fertil Steril.* 2003;80:554–559.
126. Rocca WA, Grossardt BR, de Andrade M, Malkasian GD, Melton III LJ. Survival patterns after oophorectomy in premenopausal women: a population-based cohort study. *Lancet Oncol.* 2006;10:821–828.
127. Rocca WA, Bower JH, Maraganore DM, Ahlskog JE, Grossardt BR, de Andrade M, Melton III LJ. Increased risk of cognitive impairment or dementia in women who underwent oophorectomy before menopause. *Neurology.* 2007;69:1074–1083.
128. Rocca WA, Bower JH, Maraganore DM, Ahiskog JE, Grosshardt BR, de Andrade M, Melton III LJ. Increased risk of parkinsonism in women who underwent oophorectomy before menopause. *Neurology.* 2008;70:200–209.
129. Zandi PP, Carlson MC, Plassman BL, Welsh-Bohmer KA, Mayer LS, Steffens DC, Breitner JC. Cache County Memory Study Investigators. Hormone replacement therapy and incidence of Alzheimer disease in older women: the Cache County Study. *J Am Med Assoc.* 2002;288:2123–2129.
130. Dubal DB, Zhu H, Yu J, Rau SW, Shughrue PJ, Merchenthaler I, Kindy MS, Wise PM. Estrogen receptor alpha, not beta, is a critical link in estradiol-mediated protection against brain injury. *Proc Natl Acad Sci.* 2001;98:1952–1957.
131. Mulnard RA, Cotman CW, Kawas C, van Dyck CH, Sano M, Doody R, Koss E, Pfeiffer E, Jin S, Gamst A, Grundman M, Thomas R, Thal LJ. Estrogen replacement therapy for treatment of mild to moderate Alzheimer disease: a randomized, clinical trial. Alzheimer's Disease Cooperative Study. *J Am Med Assoc.* 2000;283:1007–1015.
132. Querfurth H, LaFerla F. Alzheimer's disease. *N Engl J Med.* 2010;362:329–344.
133. Wolf OT, Kudielka BM, Hellhammer DH, Torber S, McEwen BS, Kirschbaum C. Two weeks of transdermal estradiol treatment in postmenopausal elderly women and its effect on memory and mood: verbal memory changes are associated with the treatment induced estradiol levels. *Psychoneuroendocrinology.* 1999;24:727–741.
134. MacLennan AH, Henderson VW, Paine BJ, Mathias J, Ramsay EN, Ryan P, Stocks NP, Taylor AW. Hormone therapy, timing of initiation, and cognition in women aged older than 60 years: the REMEMBER pilot study. *Menopause.* 2006;13:28–36.

CHAPTER

26

Hormones and the Aging Brain

Iñigo Azcoitia[1], *Luis M. Garcia-Segura*[2]

[1] Departamento de Biología Celular, Facultad de Biología, Universidad Complutense de Madrid, Madrid, Spain,
[2] Instituto Cajal, Consejo Superior de Investigaciones Científicas, Madrid, Spain

OUTLINE

Summary

The immune, endocrine and nervous systems undergo a remarkable adjustment with aging. Changes in immune function, endocrine secretions or endocrine signaling with aging affect the activity of different brain regions involved in cognition and emotions, such as the hippocampus, the amygdala and the frontal cortex. In turn, the reorganization of circuits controlling hormonal secretions in the hypothalamus and other brain regions with aging may further affect the function of the endocrine and the immune systems. Although the modifications in the endocrine, immune and nervous systems with aging may in part represent an allostatic adaptation to maintain stable body function, in some cases they may result in non-adaptive responses caused by the impossibility of reaching adequate homeostatic equilibrium. These non-adaptive modifications may finally cause pathological alterations that, for the brain, may be manifested as cognitive decline, depressive symptoms or psychological disturbances. Hormone therapies for elderly people may potentially help the brain to adapt to the endocrine changes associated with aging. However, in general there is limited information on the risks and benefits associated with these therapies in humans. There is also limited and fragmentary information on whether the responsiveness of the human brain to hormones is or is not affected by aging. Therefore, we do not yet have enough information to design rational protocols for hormonal therapies to protect the aging brain.

INTRODUCTION: THE AGING NERVOUS SYSTEM

General Considerations

Aging is a period of life in which the nervous system undergoes numerous functional modifications. Some of

Handbook of Neuroendocrinology, DOI: 10.1016/B978-0-12-375097-6.10026-5

these modifications probably reflect physiological adaptations to the new body conditions. However, aging is also associated with an increase in the incidence of neurodegenerative diseases. Therefore, one of the difficulties in interpreting changes in the nervous system with aging is distinguishing those aspects of neural remodeling that correspond to physiological adaptive modifications from those that reflect pathological alterations. This is also true when considering the modifications in the interactions of endocrine organs and the nervous system.

In the next two subsections we will summarize the modifications that the peripheral and central nervous system can sustain in elderly individuals. We will then consider changes in the internal milieu with aging, focusing on aging-associated changes in the immune and endocrine systems. After this, we will examine how the cross-talk of endocrine organs and the brain is affected by aging, before analyzing the influence of hormones on longevity and the implications for brain aging. Finally, we will address the pros and cons of the possible use of hormonal therapies to promote healthy brain aging, before concluding with questions for the future.

The Peripheral Nervous System

The aging process induces important biochemical and morphological changes in peripheral nerves, and the prevalence of peripheral neuropathy in humans rises from about 2.4% to 8% with aging. Aging is associated with a decrease in the synthesis of myelin proteins, such as P0 and PMP22. In parallel, large myelinated fibers undergo atrophy, while myelin sheaths increase in thickness and show various irregularities, such as myelin ballooning, splitting, infolding and reduplication (Fig. 26.1). A reduction in the number of myelinated fibers has been reported with aging in peripheral nerves of several animal species, and this effect is particularly evident in myelinated fibers of small caliber. Indeed, more than 60% of the myelinated fibers with a diameter of less than 5 μm are lost in aged animals. Moreover, alterations in the size and shape of myelinated fibers also occur with aging.[1–3]

The Central Nervous System

Aging is accompanied by remarkable structural remodeling and volume changes of different brain regions. Studies using non-invasive brain imaging of living subjects have revealed that the human brain decreases in volume with aging. In particular, very interesting changes have been detected in the white matter, which shows a decline in volume in the later years of life. Myelin may represent one of the most vulnerable structures to aging,[4,5] and myelin loss implies less functional neuronal connections; thus, it may have a strong impact on brain function and cognition. While the decrease in volume of the white matter with aging seems to be uniform, some regions of the gray matter are more affected by aging than others[6,7] – for instance, the frontal and parietal regions of the cerebral cortex seem to be more affected than the temporal and occipital cortical regions.

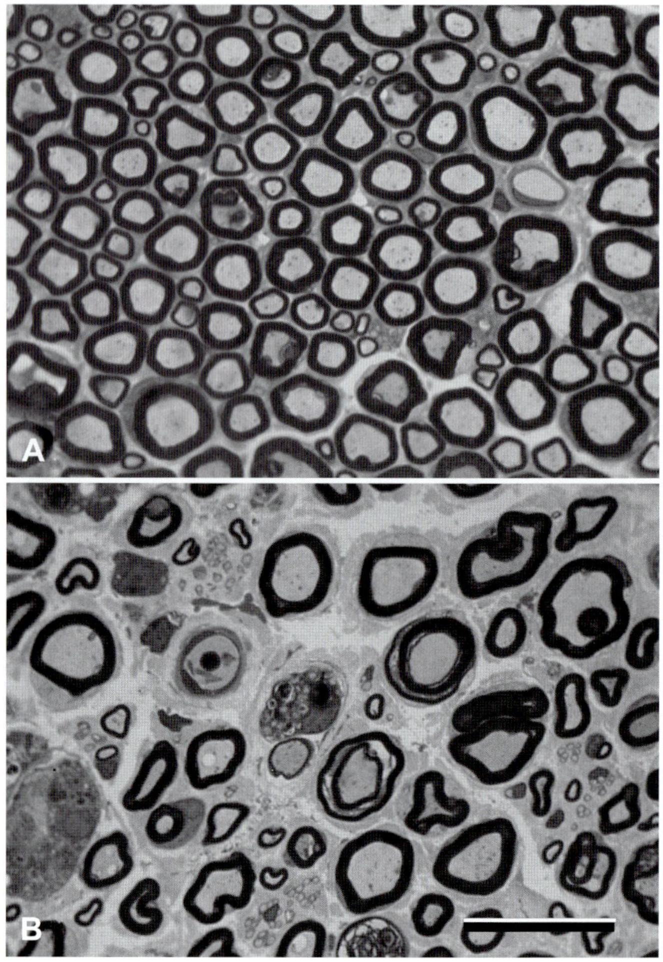

FIGURE 26.1 Photomicrographs from the sciatic nerves of 3-month-old (**A**) and 24-month-old (B) male Sprague-Dawley rats. The sciatic nerve from the older rat shows a decrease in the number of myelinated axons and several alterations in myelin structure, such as myelin invaginations in the axoplasm (infoldings) and myelin evaginations in the Schwann cell cytoplasm (outfoldings), as well as alterations in myelin compaction, such as abnormally wide incisures and abnormal separation of myelin lamellae. Scale bar, 25 μm.

Histological studies have also detected structural changes associated with aging in the brain of different species, such as loss of neurons and myelinated fibers, increased numbers of astrocytes and microglia, reduced neurogenesis in the subventricular zone and in the dentate gyrus of the hippocampus, reduced dendritic branching, reduced the number of dendritic spines and synapses, and reduced size of perforated postsynaptic densities.[6,8–18] In accordance with the abovementioned findings using non-invasive imaging techniques, histological studies in the brains of animals and humans have also detected that not all brain regions appear to be equally affected by aging, suggesting that some neuronal populations or brain regions are more vulnerable than others. This may in part be due to differences in vascularization or metabolic activity. However, there is a striking concordance between the brain regions that are more affected by aging and those in which hormones exert more profound regulatory actions on the plasticity of neural tissue.

The hippocampus, amygdala and prefrontal cortex are highly sensitive to hormones, and also show strong structural remodeling with aging. For instance, synaptic loss and decreased dendritic branching has been detected in the prefrontal cortex, medial temporal lobe structures and hippocampus of older humans and rodents,[8,9,15,19–25] and moderate neuronal loss has been detected in the prefrontal cortex of aging monkeys[26] and in the medial prefrontal cortex, visual cortex and hilus of the dentate gyrus of the hippocampus of aged rats[27,28] (Fig. 26.2), although generalized neuronal loss with aging is an uncommon finding.[29–37]

Most probably, non-pathological changes in brain function during aging are predominantly related to decreased myelination, selective and regional specific dendritic remodeling, regional modifications in the number of synaptic inputs, and regional modifications of synaptic function and plasticity, and are not the consequence of massive neuronal or synaptic loss. For instance, aging is associated with a decrease in the number of dendritic spines in the apical dendrites of layer V cortical pyramidal neurons of the rat visual cortex. However, in older individuals this decrease is not homogeneous along the whole length of the apical dendritic shaft, being more pronounced in the segments of the dendrite crossing layers II/III and IV than in dendritic segments crossing deep layers.[38] This probably reflects a selective reorganization of neuronal circuits with aging, rather than an unspecific alteration in all synaptic contacts on dendritic spines. The selective reorganization of neuronal connectivity is associated with selective functional modifications in neuronal excitability, calcium homeodynamics and functional synaptic plasticity.[25,39–42] For instance, synaptic long-term potentiation is impaired in the brain of aged rats, although with regional specificity.[25,41,43–45]

Not all types of cognitive functions are equally affected by aging. Cognitive processes that rely on the medial temporal lobe and prefrontal cortex, such as learning, memory and executive function, show considerable aging-related impairments.[44] Spatial memory is one of the modalities more affected in the elderly, and this may reflect an increased vulnerability of the hippocampus. Altered synaptic function and plasticity in the hippocampus with aging may impair the process of encoding new information,[46] while decreased neurogenesis may also contribute to impaired hippocampal function with aging.[47,48] Working memory (which is dependent on the prefrontal cortex) and declarative memory (which is dependent on the hippocampus and other medial temporal lobe regions) are affected in elderly people, who may also be affected by depression – especially those with chronic illnesses and cognitive impairment.[49] Sensation and motor control are affected in older age groups. All these functional losses may be associated with alterations in brain plasticity with aging.[50]

Although the probability of developing a neurodegenerative disorder, such as Alzheimer's or Parkinson's disease, increases in the elderly, age *per se* is not the cause of brain pathological alterations. If we compare different older people of the same age, we will find great variability in the degrees of cognitive, neurological or affective alterations. Some people will have a completely normal brain function, while in others we will find a neurodegenerative disorder. For instance, cognitive decline in elderly people ranges from normal cognitive aging, through mild cognitive impairment, to the dementias. Thus, there is an obvious individual variability in terms of brain deterioration with aging.[51,52]

Several factors impact on the variability on normal cognitive aging. According to the Scottish Mental Surveys of 1932 and 1947, the most important

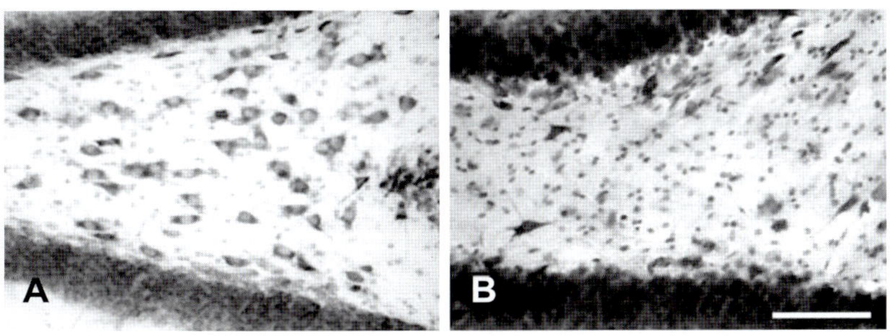

FIGURE 26.2 Nissl-stained histological sections from the septal region of the hilus of the dentate gyrus of 12-month-old (**A**) and 24-month-old (**B**) female Wistar rats. A reduction in neuronal content together with an increased gliosis (small cell nuclei) is observed in the 24-month-old rat. Scale bar, 140 μm.

determining factor for cognitive aging is childhood intelligence.[53] In addition, a variety of medical, psychological, social and lifestyle factors, interacting with genetic contributions, influence normal cognitive aging. Early events during fetal life[54] may impact on late cognitive decline. The development of neurodegenerative alterations also depends on different interactions between life events and environmental and genetic factors. Restricted caloric diet, exercise, and exposure to a rich environment are among the environmental factors that may promote healthy brain aging. Finally, some genes appear to be associated with healthy brain aging while other genes increase the risk of neurodegeneration.[55]

THE AGING BRAIN IN AN AGING BODY: ALTERED COMMUNICATION BETWEEN THE IMMUNE, ENDOCRINE AND NERVOUS SYSTEMS

Changes in Immune Function with Aging

One of the landmarks of the aging process is a decline in immune function.[56,57] The involution of the thymus with aging is the cause of alterations in cellular and humoral immunity,[58] which affects all tissues and organs, including the nervous system. The brain is also affected by changes in hormonal levels during aging. In turn, aging-associated brain remodeling may be involved in the alterations of the endocrine and immune systems. Probably the immune, endocrine and nervous systems closely interact with mutual influences during the aging process (Fig. 26.3).

The brain, via the autonomic nervous system and the hypothalamic regulation of pituitary hormones such as growth hormone, may affect thymic involution and T-cell mediated immune responses in aged animals.[59–62]

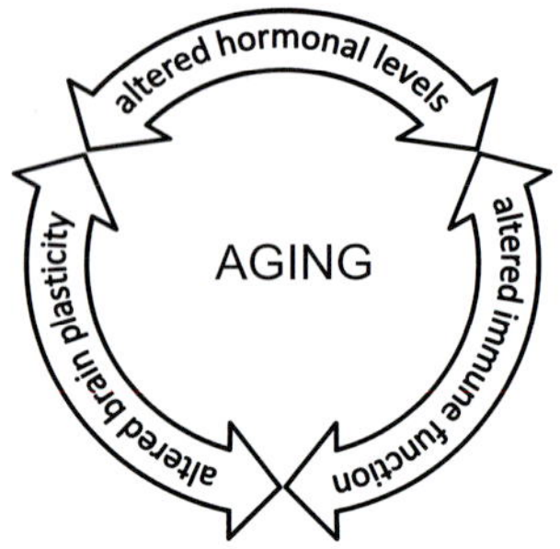

FIGURE 26.3 Aging is associated with alterations in brain plasticity, hormonal levels and immune function. Alterations in hormonal levels may impact on the immune system, and alterations in immune function impact on the activity of endocrine glands. Brain plasticity is affected by the modifications in hormonal levels and in endocrine function. In turn, aging-associated changes in brain plasticity affect the activity of endocrine glands and the immune system.

Immune responses and cytokines released by thymus, in turn, may affect the endocrine glands and the brain. Aging is associated with increased levels of various cytokines in plasma,[63–66] and there is substantial evidence that peripheral cytokines can cross the blood–brain barrier and affect brain function directly,[67–69] or indirectly via vagal afferents.[70,71]

Acting on the brain, cytokines may induce reactive gliosis and central inflammatory responses, alter synaptic function and synaptic plasticity, affect behavior, impair cognition, alter sleep and promote depression.[72] In addition, the endocrine thymus produces hormones, such as thymulin, homeostatic thymus hormone, thymus factor, thymosin fraction 5, thymosin α1 and thymosin β4, which may affect the activity of the hypothalamus–pituitary axis acting on β-endorphin, adrenocorticotropic hormone (ACTH), glucocorticoids, luteinizing-hormone releasing hormone and luteinizing hormone secretion.[73,74] Therefore, thymic involution with aging may also result in alterations in pituitary hormonal secretions.

Several hormones may affect immune responses with aging, including melatonin and thyroid hormones. In addition, it is well established that the hypothalamic–pituitary–adrenal axis, which shows activation with aging and is known to modulate brain plasticity, exerts a strong regulation of immune function.[75] Indeed, glucocorticoids have been used as immunosuppressive agents since the 1940s.[76,77] Emotional stress, which may also be enhanced in elderly people showing impaired social cognition, affects immune function[75,78] and regulates brain plasticity. Therefore, brain remodeling, alteration of hormonal levels and modifications of immune responses occur in parallel during aging, and are probably closely interrelated. Although our attention in this chapter will be focused on the relationship between hormonal changes and brain aging, it is important to keep in mind that the reshaping of neuroendocrine equilibrium with aging occurs in a context in which the immune system is also undergoing significant reorganization, which may be actively involved in endocrine and neural modifications.

Modifications in Hormonal Levels and Hormonal Signaling with Aging

The function of endocrine organs is also affected by aging. In consequence, the levels of several hormones change in plasma with aging: some decrease, others increase, and others remain the same. It should be noted that the change in the levels of hormones with aging and the nature of the changes is not identical in all species. For instance, menopause in women is associated with a dramatic decrease of estradiol levels in plasma; in contrast, reproductive aging in rats is often associated with high estradiol levels in plasma.

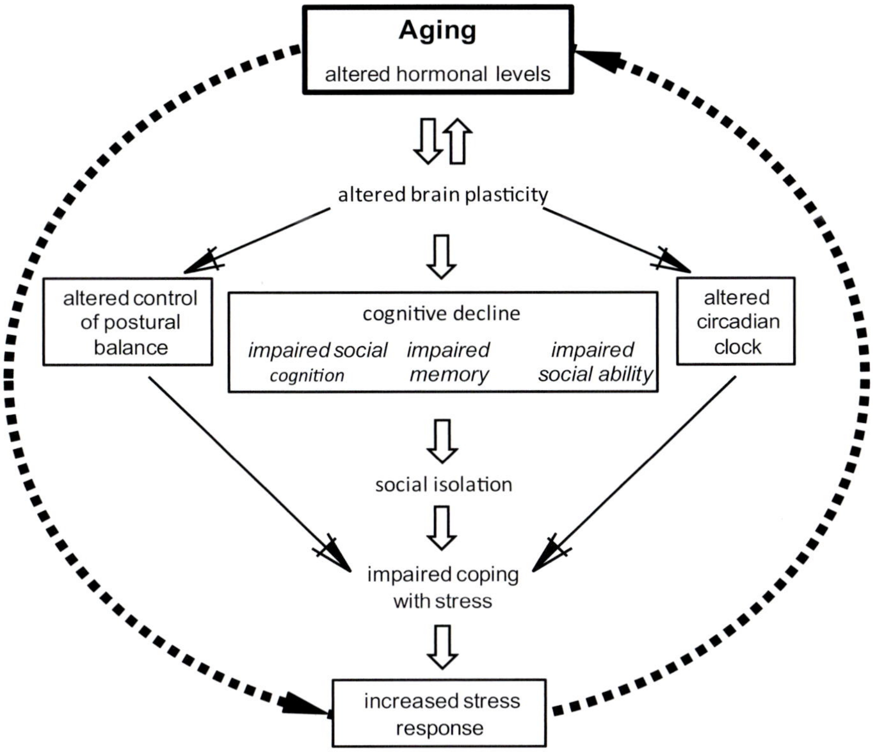

FIGURE 26.4 Modifications in brain plasticity in older people and in their ability to process sensory information may cause impairments in their cognitive skills, affecting social cognition, memory, and their ability to interact with other individuals. Alterations in physical strength and postural balance, together with alterations in the circadian clock, further contribute to social isolation, which causes an impaired ability to cope with stress and an increased stress response. This generates a vicious circle that results in further instability of the system: the increased stress response results in hormonal changes that will further affect brain plasticity, physical strength, motor control, circadian rhythms and cognition.

Dehydroepiandrosterone (DHEA) levels decrease with aging in humans; this change does not occur in rodents, where DHEA levels are undetectable in plasma. In humans, growth hormone, ghrelin, insulin-like growth factor I (IGF-I), DHEA, sex hormones and thyroid-stimulating hormone are among the hormones that decrease with aging. In contrast, other hormones, like parathyroid hormone and prolactin, follicle-stimulating hormone and luteinizing hormone, increase in older people.[79]

Aging-associated changes in hormonal levels may involve differences in secretion, metabolism, or both. For instance, both thyroid hormone secretion and metabolism decrease with aging, and serum thyroxine (T4) concentrations are not affected. However, as a result of the decreased metabolism of T4 into triiodothyronine (T3), the levels of T3 are decreased with aging.[80] On the other hand, hormonal changes with aging are not necessarily linear. Good examples of this are the hormonal changes during menopause in women.[81–83] Compared to younger women, estradiol and inhibin B levels are lower, follicle-stimulating hormone concentrations are higher and luteinizing hormone levels are unchanged during the follicular phase in women of advanced reproductive age with ovulatory cycles. After menopause, estrogen concentrations fall and follicle-stimulating hormone and luteinizing hormone levels rise above premenopausal concentrations. Follicle-stimulating hormone and luteinizing hormone levels also show a tendency to increase with aging in men.

In addition to changes in hormonal levels with aging, there are also alterations in circadian hormonal secretion rhythms. This is the case for testosterone in men, where the circadian rhythm of plasma levels (higher in the morning than in the evening) is generally lost with aging.[84] Amplitude of the nocturnal pulses of thyroid-stimulating hormone secretion is also lower in older subjects,[85] and there are also alterations in the circadian rhythms of plasma cortisol in the elderly.[86,87]

It is important to consider that hormonal signaling and the cross-talk between the brain and the endocrine glands in aging individuals may be altered without major changes in basal hormonal levels. One example is insulin resistance in the aged brain,[88,89] which may cause cerebral atrophy and result in deleterious consequences for cognitive function.[90] Another example is brain resistance to leptin,[91–93] which may also potentially impact on the function of aging brain. This is again the case with hormones regulated by the hypothalamo-pituitary–adrenal axis. Although in humans there are no apparent changes in the levels of adrenal corticosteroids with aging,[94] the response to stress may be altered, becoming accentuated in older individuals.[95,96] Alterations in their capacity to process incoming information and to adequately adapt their responses to the changing environmental and social conditions are among the causes and consequences of their difficulties in coping with stressors (Fig. 26.4). Thus, decreases in the volume of the hippocampus in elderly people may reduce their capacity to process information and to adapt to the social and environmental context.[97]

Changes in hormonal levels and/or in hormonal responsiveness with aging are associated, in time, with the progression of neurodegenerative disorders, increased depressive symptoms and other psychological disturbances. This suggests that alteration of the cross-talk between the endocrine glands and the brain may be a risk factor for the development of neurodegenerative diseases or affective disorders of different etiologies in the elderly. Thus, the modification in hormone levels or hormonal signaling with aging may have a negative impact on the regulation of brain function, and may in some way contribute to brain alterations with aging.

MUTUAL INTERRELATIONS BETWEEN BRAIN AGING AND ENDOCRINE AGING

Brain Aging Generates Changes in the Activity of Endocrine Organs

The interrelationship of hormones, brain and aging has two complementary faces: hormonal changes associated with the aging of endocrine glands may affect the brain, while, conversely, brain remodeling associated with the aging of the nervous tissue may alter hormonal production. Since the brain is the most important center for endocrine control, brain aging may have a strong impact on hormonal levels. Brain modifications with aging may result in reorganization of the circuits involved in sensing hormones and controlling hormone release. Indeed, there is evidence indicating that the hypothalamic circuits regulating hormonal production suffer marked alterations with aging.

During aging there is a decrease in the ordered rhythmic pattern of hormone release that probably reflects a modification of neuronal circuits regulating pituitary secretions.[98–101] Indeed, aging is associated with a dramatic reorganization of the serotonergic and noradrenergic hypothalamic systems, and of the tuberoinfundibular dopaminergic system.[102–105] There is also a decrease in the expression of receptors for hormones, neuropeptides and neurotransmitters,[106–108] reorganization on neural membranes,[109] and alterations in glial plasticity[110] in brain regions regulating hormonal secretions.

All these changes affect the secretion of several hormones, such as luteinizing hormone and growth hormone. The pulse amplitude of these hormones is affected due to the hypothalamic alterations of the secretion of gonadotropin-releasing hormone and growth-hormone releasing hormone (GHRH), respectively. In turn, aging-associated modifications in growth hormone result in reduced production of IGF-I in the liver, and alterations in gonadotropins affect the release of sex-steroid hormones. Thus, it is probable that the reorganization of specific neuronal circuits may initiate the hormonal changes associated with aging. If this were the case, the hormonal changes would be a consequence of brain aging. Reproductive aging in rodents is an example of how the aging of the brain may affect hormonal secretion. Reproductive aging in rodents is believed to be caused by hypothalamic failure to regulate gonadotropin secretion and, thus, ovarian function. In rodents, areas of the hypothalamus that regulate gonadotropin secretion show structural and functional changes in aged females;[105–109] the hypothalamus is then unable to support cyclic regulation of ovarian function, and the animal enters the period of reproductive aging. Consequently, ovarian hormones do not show the cyclic changes observed in young females. Therefore, in this case, brain remodeling is the cause of the hormonal changes. Hypothalamic reorganization may also be the cause of changes in the control of gonadotropins that are involved in the irregularity of menstrual cycles in the perimenopausal period in women.[99,100]

Aging of Endocrine Organs Generates Changes in Brain Function

According to the previous section, we may assume that some hormonal changes associated with aging are initiated by specific plastic alterations in the brain circuits that regulate endocrine secretions. The question then is: what triggers these brain modifications? Are we looking at an endogenous mechanism of brain aging, or an adaptation of neuronal networks to previous hormonal changes? We may imagine a scenario in which subtle hormonal changes initiate modification in brain plasticity, and the reorganized neuronal circuits will then result in a modification in the regulation of endocrine secretions. Subsequently, the distorted hormone levels will in turn produce new modifications in brain plasticity that will finally result in an amplification of the hormonal changes. This may reflect a situation where equilibrium is broken at a temporal point of maximal instability. The temporal point at which equilibrium is broken may be reached by the cumulative sum of small changes in hormonal levels and brain plastic modifications. For instance, Naftolin and Brawer proposed that the hypothalamic failure to maintain ovarian cycles in rodents is the result of an accumulative estrogen input to the hypothalamic arcuate nucleus, which is involved in the cycling regulation of gonadotropin secretion.[111,112] According to this hypothesis, accumulative regulatory actions of estradiol on synaptic and glial plasticity in the arcuate nucleus during each estrous cycle would gradually alter future synaptic and glial plastic responses to ovarian hormones in this brain region.[105,113,114] This would finally result in the impaired function of specific neuronal circuits that regulate gonadotropin secretion, such as those in which β-endorphin neurons are involved,[115] leading to altered regulation of ovarian secretions, which finally causes reproductive aging.

The complexity of the cross-talk between the hormones and brain plasticity in the aging process is also well illustrated by the effects of stress on brain aging. Substantial evidence supports the existence of a link between stress, hypothalamo-pituitary–adrenal axis dysfunction, memory disorders and aging.[116,117] Here, it is important to consider the different effects that stress hormones have on the brain, depending on whether the stress is acute, or chronic and unpredictable.

Two important concepts may help us to understand the different effects of acute and repetitive stress: allostasis and allostatic load.[117] *Allostasis* is the adaptation exerted by the neuroendocine system, autonomic nervous system and the immune system in response to challenges of daily life. These systems maintain stability through change, and their action is initiated when needed and shut off when no longer needed. These changes, such as the acute stress response, are adaptive in the short-term. However, the action of the regulatory systems may be damaging if it is not shut off at the appropriate time, and the regulatory signals may then cause alterations in the target tissues that can lead to pathological damage. This is the situation termed *allostatic load*. Thus, acute actions of glucocorticoids may represent an allostatic response and maintain proper function of the brain. However, if the action of glucocorticoids is not efficiently terminated – i.e., if we are unable to overcome stress – our brain and its capacity to program adaptive control of endocrine glands and behavior will be affected. The new endocrine adjustments will then further affect the brain, initiating a cascade of alterations that will finally lead to brain damage and permanent hormonal imbalance.

This is the basis of the so called "glucocorticoid cascade hypothesis" of aging – a theory which postulates that changes in specific brain areas, as a result of allostatic load, initiate the process of aging.[118] According to this hypothesis, one of the key brain regions involved in this process is the hippocampus, which is involved in episodic, declarative, spatial and contextual memory, and is also an important integration center for the regulation of autonomic, neuroendocrine and immune responses. Interestingly, the hippocampus is one of the brain regions more vulnerable to damage caused by repetitive stress. Neuronal death in the hippocampus after chronic glucocorticoid exposure may occur in some species and under certain conditions.[118,119] However, hippocampal cell loss after chronic glucocorticoid exposure is not a generalized finding in all species,[120–123] and does not occur in humans.[124,125] This may be due to the fact that the primate hippocampus expresses lower levels of glucocorticoid receptors than the hippocampus of rodents,[126] which shows neuronal death after chronic glucocorticoid exposure.[118,119]

Although neuronal loss is not involved in all species, structural plastic changes in the hippocampus are probably a generalized response to the allostatic load caused by glucocorticoids. Chronic stress produces remodeling of dendrites and dendritic spines in the hippocampal neurons, decreases hippocampal neurogenesis, induces gliosis and, finally, may promote the impairment of hippocampal function.[117,119,122,127] As a consequence, memory, cognition, autonomic control, endocrine secretions and the immune system will be altered. It is not only unpredictable repetitive stress in adulthood that may increase hippocampal damage; even repetitive stress during the prenatal period may result in alterations in hippocampal synaptic plasticity and neurogenesis at older ages. In rats, prenatal stress induces increased anxiety-like behavior and permanent dysfunction of the hypothalamo-pituitary–adrenal axis system in adult life. Aged rats that were stressed during prenatal life show spatial hippocampal-dependent learning impairments[48,128] compared to normal older animals.

Stress hormones affect other brain regions, such as the cerebral cortex and the amygdala. Repeated stress causes dendritic remodeling in the amygdala, increasing dendritic length and spine synapse formation. In contrast, dendritic length in neurons from the medial prefrontal cortex is decreased by chronic stress.[129,130] It is noteworthy that the brain regions that are more vulnerable to chronic stress, such as the hippocampus, amygdala and prefrontal cortex, are also those regions that are more affected by aging.

BRAIN AGING, HORMONES AND LONGEVITY

General Considerations

Animal studies have shown that several hormones may protect neurons from different neurodegenerative stimuli. These include growth hormone, IGF-I, DHEA and sex hormones. All these hormones have been shown to regulate plasticity and exert neuroprotective actions in the aging brain. Therefore, it is possible that the decrease in the levels of these hormones in older people may result in altered brain plasticity and decreased protection against the environmental and genetic factors that promote neurodegeneration. Some hormones, such as estradiol and IGF-I, antagonize the damaging effects of adrenal steroids. IGF-I attenuates spatial learning deficits in aged rats, and promotes neurogenesis in the hippocampus. Chronic IGF-I infusion into the brain restores the spatial learning abilities of aged rats that were stressed during prenatal life. IGF-I also upregulates neurogenesis in the hippocampus of these animals, and reduces their hypothalamo-pituitary–adrenal axis dysfunction.[48] Interestingly, IGF-I increases estradiol levels in the plasma of aged rats that were submitted to prenatal stress. Estradiol, in turn, is known to increase neurogenesis in the hippocampus of old rats,[131] and prevents hippocampal damage induced by excitotoxic injuries.[132] In addition, the signaling of estradiol and IGF-I interacts to promote neuroprotection[133–135] and to regulate hippocampal neurogenesis.[136] Therefore, different hormones affected by the aging process may act in cooperation or antagonistically. In consequence, decreased levels of protective hormones (such as IGF-I) with aging may increase the risk of neural damage induced by the allostatic load of stress hormones.

BOX 26.1

HOW DO WE KNOW THAT SOME HORMONES PROTECT THE NERVOUS SYSTEM FROM AGING-ASSOCIATED DETERIORATION?

Some studies have assessed the efficacy of hormonal therapy to protect the nervous system from aging-associated deterioration. These are a few examples on how protective effects of hormones have been identified in the peripheral and central nervous system of aged animals.

Treatment with progesterone or with its derivatives, dihydroprogesterone and tetrahydroprogesterone has been shown to reduce myelin abnormalities in the sciatic nerve of 22-month-old male rats[1] (see figure). Regarding the central nervous system, treatment of 22-month-old male and female rats with recombinant human growth hormone for 10 weeks prevents aging-associated neuronal loss in the hilus of the dentate gyrus.[2] Behavioral studies also indicate that some hormones protect brain function in aged animals. For instance, infusion of insulin-like growth factor type I (IGF-I) into brain lateral ventricle in aged rats that were prenatally stressed restores their spatial learning abilities.[3] Some hormones not only protect the brain from aging-associated deficits but also protect the aged brain from neurodegenerative damage. This is the case with estradiol, which has been shown to protect the brain tissue from stroke-related injury in middle-aged rats.[4]

Photomicrographs from the sciatic nerves of male Sprague-Dawley rats. (**A**) 24-month-old rat injected with vehicle. (**B**) 24-month-old rat injected dihydroprogesterone, a reduced progesterone metabolite. The sciatic nerve from the animal injected with dihydroprogesterone shows an increase in the number of myelinated fibers and a decrease in the proportion of nerve fibers with alterations in myelin structure. Scale bar, 25 μm.

References

1. Azcoitia I, Leonelli E, Magnaghi V, Veiga S, Garcia-Segura LM, Melcangi RC. Progesterone and its derivatives dihydroprogesterone and tetrahydroprogesterone reduce myelin fiber morphological abnormalities and myelin fiber loss in the sciatic nerve of aged rats. *Neurobiol Aging* 2003;24:853–860.
2. Azcoitia I, Perez-Martin M, Salazar V, Castillo C, Ariznavarreta C, Garcia-Segura LM, Tresguerres JA. Growth hormone prevents neuronal loss in the aged rat hippocampus. *Neurobiol Aging* 2005;26:697–703.
3. Darnaudery M, Perez-Martin M, Belizaire G, Maccari S, Garcia-Segura LM. Insulin-like growth factor 1 reduces age-related disorders induced by prenatal stress in female rats. *Neurobiol Aging* 2006;27:119–127.
4. Dubal DB, Wise PM. Neuroprotective effects of estradiol in middle-aged female rats. *Endocrinology* 2001;142:43–48.

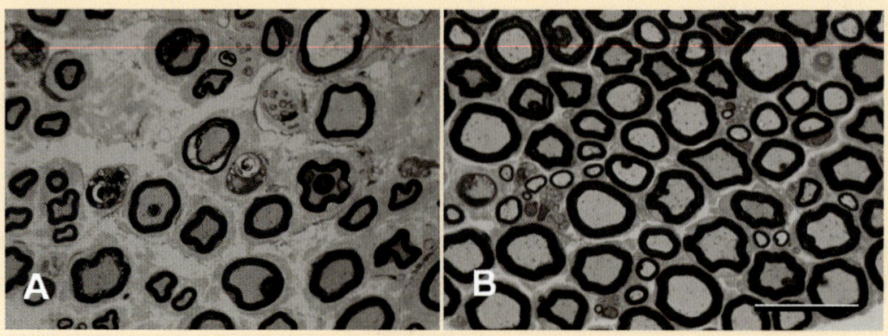

Thus, we may conclude that the decrease in plasma levels of some hormones in elderly individuals may have a negative impact on healthy brain aging. We might also be tempted to interpret the change in levels of some hormones with aging as a pathological alteration that may in turn result in secondary pathology as a consequence of the hormonal misbalance. However, the situation is not so simple. First, the decrease in some neuroprotective hormones with aging may hardly be defined as pathological – for instance, the levels of the neuroprotective hormones estradiol and progesterone drop with menopause, which is not a pathological situation. Furthermore, there is reason to suspect that the decrease in some hormones with aging may represent a positive adaptation to increase lifespan.

IGF/Insulin Signaling

Pioneering work of Kenyon and coworkers[137] showed that mutations in single genes related to the IGF/insulin signaling pathways could considerably enhance lifespan in the nematode *Caenorhabditis elegans*. We know today that, in *C. elegans*, insulin/IGF-I-like molecules may, through the activity of the DAF-2/insulin/IGF-I-like receptor and the DAF-16/FKHRL1/FOXO transcription factor, regulate lifespan. Targeted genetic manipulation of homologous genes in *Drosophila* has confirmed the involvement of this signaling pathway in the regulation of longevity.[138] Similar genes in vertebrates may have the same function[139–142] – for instance, growth hormone/IGF-I deficiency in dwarf mice is associated with increased longevity. Let us for a moment examine what dwarf mice may tell us about the relations of hormones and longevity.

In 1929, George Snell described a recessive autosomal mutation in mice that results in greatly decreased postnatal growth and adult body size. We know today that the cause for this dwarfism is a loss-of-function mutation of pituitary factor 1 (pit-1), a homeobox transcription factor involved in the differentiation of the somatotropes, lactotropes and thyrotropes.[143] Therefore, Snell dwarf mice have defective development of the anterior pituitary, and a deficit in growth hormone, thyroid-stimulating hormone and prolactin. As a consequence of the deficit in growth hormone, Snell dwarf mice have reduced postnatal growth and undetectable peripheral levels of IGF-I. These animals are hypothyroid due to the lack of thyroid-stimulating hormone, and females are sterile due to failure of luteal function as a consequence of the deficit in prolactin.[144,145]

Ames dwarf mice were discovered by Schaible and Gowen[146] at Iowa State University in Ames, Iowa. These animals have a mutation in a gene (Prop-1) that is expressed before Pit-1 and is necessary for the development of Pit-1 expressing cells that will differentiate into somatotropes, lactotropes and thyrotropes. Ames dwarf mice have similar deficits in growth hormone, prolactin and thyroid-stimulating hormone as do Snell mice.[145]

Another type of dwarfism in mice was obtained by targeted disruption of the growth hormone receptor/growth hormone binding protein (GHR/GHBP) gene.[147] These mice have the same deficit as people with Laron syndrome: they lack growth hormone receptor and thus are growth-hormone resistant. In consequence, IGF-I levels in plasma are dramatically decreased in these mice.[145]

Snell, Ames and Laron mice have something else in common, in addition to having hormonal deficits and being dwarf: they live longer than normal mice. Both average lifespan and maximal lifespan are increased in these mice.[145] As mentioned above, mutations of genes that codify for proteins analogous to those involved in mammalian IGF-I signaling increase longevity in *C. elegans* and the fruit fly, *Drosophila melanogaster*. In Snell, Ames and Laron mice IGF-I is undetectable in plasma, and therefore reduced IGF-I signaling may contribute to the increased longevity of these animals. Indeed, longevity is also increased in heterozygous IGF-I receptor knockout mice (homozygous IGF-I receptor knockout mice are not viable), even if these mice are not dwarf. Snell, Ames and Laron mice also have reduced plasma insulin and glucose levels, and this may also be a contributing factor for the increase in longevity. Reduced thyroid hormones may also be involved in the increased lifespan. Changes in metabolism, oxidative stress, and decreased incidence of cancer or other causes may account for the increased longevity of these mice. However, independently of whatever the causes of the increased longevity are, it seems that reduced levels of certain hormones may have a positive impact on this parameter. Thus, it is tempting to speculate on the possibility that the decrease in the levels of some hormones with aging may be a positive adaptation.

Longevity vs Neuroprotection

Increased lifespan does not necessarily imply decreased brain aging. Since IGF-I and IGF-I signaling are neuroprotective, and decreased IGF-I levels in plasma are associated with an increase in neurological alterations, it might be expected that Snell, Ames and Laron mice, with undetectable IGF-I levels in plasma, would show increased brain deterioration. We would then have mice that live longer but suffer more neurological alterations: increased brain deterioration would be the price for increased longevity. This prediction sounds reasonable, and yet is completely wrong.

What is surprising and unexpected is that Ames and Laron mice appear to have delayed brain aging.

Age-related decline in learning and memory is postponed in these animals. In addition, Ames dwarf mice exhibit an increase in hippocampal neurogenesis.[148] All this is puzzling, since we know that IGF-I administration to normal animals increases hippocampal neurogenesis and prevents aging-associated cognitive decline, and that IGF-I is highly protective for the brain. Furthermore, reduced IGF-I levels in plasma are a common marker for several neurodegenerative diseases.[149] Yet these dwarf mice have undetectable levels of IGF-I in plasma, but also have reduced brain aging. How can this be? Growth hormone is also neuroprotective, has positive effects on learning and memory and prevents brain aging – yet Ames and Laron mice have undetectable levels of growth hormone in plasma and, in spite of this profound hormonal deficit, these animals exhibit apparently normal cognitive functions and maintain them into advanced age. How can we reconcile the positive effects that IGF-I and growth hormone administration have for brain aging with the lack of a negative impact of IGF-I and growth hormone deficiency in the brain of aged dwarf mice?

Andrzej Bartke and colleagues have proposed a solution to this problem. They decided to assess growth hormone and IGF-I levels in the hippocampus of Ames dwarf mice. (It should be mentioned here that growth hormone and IGF-I are locally produced in the brain; thus, in addition to being hormones, they are locally produced factors.) Bartke *et al.* found that growth hormone and IGF-I protein levels were increased in the hippocampus of Ames dwarf mice compared with normal mice. They also assessed the signaling of growth hormone and IGF-I, and discovered that Ames dwarf mice had 70% higher levels of phosphorylated Akt in the hippocampus when compared to normal controls. A significant increase in the phosphorylation of cyclic AMP response element-binding protein (CREB) was also detected in the hippocampus of Ames dwarf mice.[148] Thus, these mice have an increased synthesis and signaling of growth hormone and IGF-I in the hippocampus. In addition, this increased local production of growth hormone and IGF-I in the hippocampus of Ames dwarf mice is associated with increased neurogenesis in the dentate gyrus and increased activation of anti-apoptotic signaling, which may contribute to maintaining hippocampal function.[150,151] Therefore, Bartke *et al.* have proposed that the increase in hippocampal growth hormone and IGF-I expression, and the subsequent activation of phosphoinositide 3-kinase (PI3K)/Akt-CREB signal transduction cascade, might contribute to the maintenance of cognitive function and neuronal structure in aged dwarf mice.

Thus, the brain may adapt its local synthesis of growth hormone and IGF-I, as a compensatory mechanism, to the deficit in the peripheral levels of these hormones. This may explain why Ames dwarf mice (and probably the other dwarf mice deficient in growth hormone and IGF-I) do not show brain deterioration in spite of decreased plasma levels of growth hormone and IGF-I. These mice may have increased longevity due to the decreased plasma levels of growth hormone and IGF-I, and the consequent changes in metabolism, oxidative stress and the growth of tumor cells. At the same time, the loss of growth hormone and IGF-I in plasma does not affect the brain of these animals, since this organ is able to increase its local production of these factors that are so important in maintaining brain function.

Of course, it is possible that Ames dwarf mice, which are permanently deprived of growth hormone and IGF-I during their life, may develop compensatory mechanisms for local production of these molecules in their brains that normal animals do not usually develop. Therefore, deprivation of growth hormone and IGF-I with aging in normal animals may have a stronger negative impact for brain function. However, the data obtained by Bartke and colleagues suggests that, at least during some specific circumstances, the brain may modulate the local synthesis of some hormones, such as growth hormone and IGF-I, in adaptation to plasma levels of these molecules.

HORMONE THERAPY AND HEALTHY BRAIN AGING

General Considerations

In previous sections of this chapter we have seen that hormonal alterations, such as those associated with repetitive stress, may promote alterations in brain plasticity and function during aging. However, the decrease with aging of some hormones may, at least in part, reflect a physiological adaptation to minimize aging-associated pathological alterations in metabolism, tumor growth and oxidative stress. Thus, changes in the levels of some hormones with aging may reflect an allostatic response aimed at compensating for other imbalances associated with aging, and therefore, at least in theory, administration of these hormones could be an inadequate therapeutic approach to taking care of aging-associated symptoms. However, once more, the situation is not that simple. In fact, the decline in the levels of certain hormones during aging may potentially have negative impacts for some tissues, like the bone or the brain. Actually, the same hormones that may potentially promote tumor growth, such growth hormone, IGF-I or estradiol, are also protective growth factors for normal cells. Thus, hormones that may promote tumor growth in aged animals may also promote neuronal survival and bone calcification. For instance, decreased levels of estradiol and IGF-I in plasma after menopause may be a protective physiological adaptation to minimize the risk of mammary tumors. However, at the same time, the decreased levels of these hormones in plasma are a risk factor for osteoporosis, and probably also for neurodegenerative diseases. Therefore, hormonal therapies may be justified to treat or prevent the collateral effects of the hormonal changes with aging. The problem, and it is not an easy one, is to determine the age for treatment, the hormonal composition, dose, pattern of administration and length of treatment that result in positive effects. Unfortunately, we know very little regarding the effects of hormonal treatments on brain function in older humans.

While the basic studies on animals strongly suggest that several hormones are neuroprotective, the situation is not so clear when data collected from human studies are analyzed. The effect of different hormonal therapies on brain function has been assessed in elderly men and women with uncertain results. Here, we briefly examine the available information for growth hormone, IGF-I, melatonin, DHEA, testosterone and ovarian hormones.

Growth Hormone and IGF-I

Growth hormone deficiency in humans is associated with sleep disturbances, memory loss, feeling of diminished well-being, and other cognitive impairments. Several studies have reported a positive effect of growth hormone therapy in the cognitive performances of growth-hormone deficient patients. In animal models, growth hormone has been shown to protect the brain and the spinal cord from different forms of neurodegenerative stimuli, and to promote neuronal survival after hypoxic-ischemic injury. Growth hormone deficiency in rats is associated with cognitive deficits, such as impaired spatial learning, and short-term synaptic plasticity and growth hormone therapy attenuates these deficits.[152] In addition, growth hormone treatment increases microvascular density and decreases neurodegenerative changes in the brain of aged animals.[27,153,154] A potential alternative to hormonal therapy with growth hormone is ghrelin, the endogenous ligand for the growth hormone secretagogue receptor. Ghrelin has potent growth-hormone releasing activity, and ghrelin receptor agonists may represent a potential treatment for alterations in the aging brain.[155,156]

Neuroprotective effects of growth hormone may be mediated by direct actions on growth hormone receptors in the brain, or by its action in the liver, increasing the production of IGF-I, a potent neuroprotective hormone. In addition, administration of growth hormone to elderly rats increases IGF-I expression in the brain.[154,157,158] Since the fall in IGF-I levels with aging correlates with cognitive decline, and since IGF-I levels are decreased in several neurodegenerative diseases, we may assume that IGF-I replacement therapy in elderly people may have a positive impact on brain function. Indeed, administration of IGF-I to aged rats exerts protective effects in the brain.[159,160] However, unfortunately we do not have enough data yet regarding IGF-I effects on the brains of elderly people.

Melatonin

Melatonin is another neuroprotective hormone that may affect brain function in aged individuals.[161] Some studies have reported a decrease production of melatonin with aging, which is also associated with a significant reduction in the quality and continuity of sleep. Diminished nocturnal melatonin secretion has been detected in patients with Alzheimer's disease. Some of these patients also show severe disturbances in their sleep–wake cycle. Clinical trials involving elderly insomniacs and Alzheimer's patients suffering from sleep disturbances suggest that treatment with melatonin may be effective in improving sleep. Interestingly, treatment with melatonin appears to be more effective in those individuals with low endogenous levels of melatonin, pointing to the possible influence of endogenous hormone levels in the outcome of hormonal therapies – an interesting theory that will be analyzed later on in this chapter.

Dehydroepiandrosterone

Plasma levels of DHEA show a strong decline in humans with aging.[162,163] DHEA has memory-enhancing effects in aging rodents, and counteracts the negative actions of glucocorticoids on memory.[164] DHEA therapy has considerable effects on mood, well-being and sexuality in patients with adrenal insufficiency, and some studies suggested that this steroid may improve cognition, promote a sense of well-being and reduce depressive symptoms in normal elderly men and women,[165] although other studies do not support this conclusion.[166–169] There are also conflicting results regarding the possible association between DHEA-sulfate levels in human plasma and cognition.[164]

Testosterone

Testosterone declines in plasma with aging in both men and women. Testosterone deprivation has been associated with poor memory in men, and some studies have suggested that testosterone loss may be a risk factor for cognitive decline and possibly for dementia. Thus, men with low levels of free circulating testosterone appear to be at higher risk of developing Alzheimer's disease than men with higher serum levels of this hormone.[170] Higher testosterone levels are associated with better cognitive performance in elderly men,[171] and some studies have shown that testosterone supplementation enhances working memory and spatial cognition in this group.[172] However, excessively high levels of testosterone may be unable to improve verbal and spatial memory.[173] Although testosterone therapy may have positive effects on cognition, it is unclear whether it is an effective treatment for depressive symptoms in elderly men.[174,175]

Ovarian Hormones

Progesterone in the Peripheral Nervous System

As mentioned at the beginning of this chapter, myelin loss may be one of the most important contributors to brain dysfunction with aging.[4,5] Therefore, it is important to search for potential treatments to promote remyelination in the aged brain. In this regard, it is highly relevant that progesterone may exert a modest but significant reversal of the age-associated decline in brain remyelination in middle-aged rats.[176] Peripheral myelin is also responsive to progesterone in aged rats, although aging affects the responsiveness of peripheral nerves to the hormone.

Treatment with progesterone or with its reduced metabolite dihydroprogesterone is able to increase the levels of the myelin protein P0 in the sciatic nerve of aged male rats, while tetrahydroprogesterone significantly increases the protein levels of PMP22.[2,3] Moreover, treatments with progesterone or its metabolites have clear effects on the number and shape of myelinated fibers, as well as on the frequency of myelin abnormalities. One of the most striking effects of the steroids is on the number of myelinated fibers of small caliber ($< 5\ \mu m$), which is significantly increased by progesterone and its metabolites in the nerves of elderly rats. This is accompanied by a decrease of similar magnitude in the number of unmyelinated axons and, in particular, of large ($> 3\ \mu m$) unmyelinated axons. Moreover, the morphometric analysis suggests that the increase in the number of myelinated fibers reflects an increased remyelination of small fibers in aged sciatic nerves. Another significant effect of the treatment of elderly rats with progesterone, dihydroprogesterone and tetrahydroprogesterone is a reduction in the frequency of axons with myelin abnormalities. These steroids significantly reduce the frequency of axons with myelin infoldings, one of the structural alterations that are increased in aged nerves, and reduce the proportion of fibers with irregular shapes. Therefore, progesterone therapy may contribute to preserving the functional and structural integrity of peripheral nerves with aging.[1–3]

Estrogens and Progestins in the Central Nervous System

Many studies have analyzed the cognitive effects of hormone therapies with female hormones. Therapies with estrogens and progestins (HT) or estrogen only (ET) have been used for many years for the treatment of symptoms of menopause. Many postmenopausal women have received HT or ET, and therefore we have much more information on the effects on aged human brain for female sex hormones than for other hormones. Both progesterone and estradiol have been shown to be neuroprotective in animal studies and to prevent cognitive and neuronal loss in several experimental animal models of neurodegeneration. ET in humans is associated with an increase in volume of the hippocampus,[177] and most studies suggest that ET increases memory and cognitive function in healthy women.[178,179] For example, in the Baltimore Longitudinal Study of Aging, nondemented postmenopausal women receiving HT performed better on tests of verbal and visual memory compared with never-treated women in samples in which both groups of women were comparable with respect to educational attainment, general medical health, and performance on a test of verbal knowledge.[180] In addition, short-term ET has been shown to improve prefrontal cortex-dependent cognitive functions in postmenopausal women.[181]

HT may also be an effective treatment of depression for perimenopausal women,[182] and in postmenopausal women may reduce negative symptoms in schizophrenia.[183] Some studies have suggested that estrogen may reduce the motor disability associated with Parkinson's disease.[184–187] There is also some evidence, from observational studies, suggesting that moderate exposure to exogenous estrogen may decrease the risk of stroke in postmenopausal women.[188,189] Furthermore, some studies suggest that ET may prevent or delay the onset of Alzheimer's disease,[190] and some randomized trials, using estradiol, also suggest that the hormone may improve cognition for women with Alzheimer's disease.[191] However, the evidence for a protective effect of estrogens in the human brain is not without controversy, and some studies indicate that ET or HT has no effect on Parkinson disease.[192] Other studies indicate that estradiol does not reduce mortality or the recurrence of stroke in postmenopausal women with cerebrovascular disease.[193] In addition, some randomized trials do not support the protective effects of estrogens for Alzheimer's disease.[179,194,195] Some studies even suggest that HT may have a negative impact on cognition in postmenopausal women with Alzheimer's disease.[196]

Furthermore, the Women's Health Initiative (WHI) randomized trial, where participants received a single daily tablet of 0.625 mg of conjugated equine estrogens plus 2.5 mg of medroxyprogesterone acetate, or a matching placebo, suggests an increased risk of dementia and stroke as a result of long-lasting hormonal treatment several years after the menopause.[197–199] Therefore, there is an apparent discrepancy between the potent neuroprotective effect of sex steroids in animal models and the high variability of results in human studies, and the question as to the source of these discrepancies naturally arises.

One potential source for the discrepancies in the literature is that there is a considerable variation in the exact hormonal composition and pattern of administration in HT in humans. Usually, a mixture of different natural or synthetic estrogens and natural progestins is administered. Differences in formulation, dose, route of administration, length of treatment and sample size have been proposed as explanations for the disparity of results between different studies. For instance, different progestins may have very different effects on the brain. Medroxyprogesterone acetate, used in the WHI study, may have several undesirable effects. In animals, the neuroprotective effects of progesterone are blocked by the inhibition of its metabolism to dihydroprogesterone and tetrahydroprogesterone, and both progesterone metabolites seem necessary to exert neuroprotection.[200] In contrast, medroxyprogesterone acetate cannot be metabolized to dihydroprogesterone and tetrahydroprogesterone, and it is not neuroprotective in animals[200] and may even block neuroprotection by estradiol.[201,202] It therefore cannot be excluded that natural progesterone, being able to be transformed into the neuroprotective metabolites dihydroprogesterone and tetrahydroprogesterone by neural tissue, may have better cognitive effects in the human brain than medroxyprogesterone acetate.

Another parameter that may be highly relevant to explaining the different results is the age of the women receiving the treatment, or the previous duration of hormonal deprivation. The focus of the WHI study on women already many years beyond the onset of the menopause is a serious limitation, since there are reasons to think that the perimenopause may be a critical period for the highest efficacy of hormonal therapy on the prevention of brain disorders. Indeed, most studies that have analyzed the outcome of HT or ET treatments during the perimenopausal period have found positive cognitive effects.[203–205] It is conceivable that during this critical period the brain may be adapting to the changing hormonal conditions.

Another important consideration is the influence of an extended period of gonadal hormone deprivation on the neuroprotective effects of estrogen. Wise *et al.*[206] have reported that estradiol loses neuroprotective and anti-inflammatory actions against brain ischemia in mice after a long period of hypoestrogenicity. Estradiol exerts neuroprotective and anti-inflammatory actions when administered immediately after ovariectomy, but not when administered 10 weeks later.[206] These results are important in explaining the discrepancy between the findings of basic research and those of the WHI study, in which the majority of women were treated with hormone therapy after an extended period of hormonal deprivation. Therefore, the length of hormonal deprivation may affect brain responsiveness to ovarian hormones. Age is another important factor to consider. It should be noted that the WHI data are relevant for long-term hormonal therapy started in women aged 65 or over, and it could be that the brain also loses responsiveness to hormones with age. We will analyze this question below.

AGING AFFECTS BRAIN RESPONSIVENESS TO HORMONAL THERAPY

Most studies in animal models have used young adult rodents submitted to different forms of brain injury to assess the neuroprotective effects of hormones. Very little is known about the effects of hormone therapies in the brain of older animals. As mentioned earlier in the chapter, aging is associated not only with modifications in hormonal levels but also with modifications in hormonal signaling. Indeed, the aged brain shows a reduced sensitivity to several hormones.[101] For instance, the brain of old rats is less sensitive to leptin. The infusion of this hormone is more effective in reducing food consumption in young rats than in old rats; in addition, leptin induces an increase in oxygen consumption in young rats but not in old rats. The different effects of leptin in young and aged animals may be related to the fact that the hormone reduces neuropeptide Y levels in the hypothalamus of young animals and not in old animals. The aging brain is also less sensitive to ghrelin. The administration of ghrelin increases growth hormone levels in plasma. However, this response to ghrelin is reduced in elderly people due to age-dependent changes in hypothalamic responsiveness to the hormone.[207,208]

The limited evidence available suggests that, at least in rodents, sex steroids may still exert some neuroprotection in the aged brain.[209] Sex steroids, and their precursors pregnenolone and DHEA, decrease age-related memory deficits in rodents.[164] Progesterone exerts neuroprotective effects against traumatic brain injury in aged rats,[210] and motoneurons of aged male rats retain synaptic plasticity in response to androgens.[211] Furthermore, it has been shown that estradiol decreases gliosis in the brain of aged mice and rats,[212–214] and protects the brain of middle-aged rats (9–12 months) from middle cerebral artery occlusion.[215] Estradiol has similar effects on long-term synaptic potentiation and long-term synaptic depression in the hippocampus of adult (3–5 months) and aged (18–24 months) male rats, although the effect on long-term synaptic potentiation is smaller in aged rats.[216] Furthermore, estradiol is still protective in aged rats against the effects of behavioral stress on synaptic plasticity.[216] However, other studies suggest that some forms of plasticity in response to estrogen are abolished in the brain of old rats.[217] Estradiol is also able to promote hippocampal neurogenesis in middle-aged mice[213] and old rats,[131] although it is not able to restore the rate of hippocampal neurogenesis to the level observed in young animals.[131]

The peripheral nerves of aged rats, although responsive to progesterone, are less sensitive to the neuroprotective effects of the hormone than are the nerves of young rats. Therefore, aging may affect the responsiveness of peripheral nervous system to progesterone, due to the reduced capacity of the aged nerves to metabolize progesterone in the neuroprotective metabolites dihydroprogesterone and tetrahydroprogesterone.[218] A similar situation may occur in the central nervous system. It is clear that we need more studies to determine to what extent sex-steroid metabolism is altered in the brain of old animals.

In addition, neuroprotective effects of sex steroids that depend on the activation of their nuclear receptors may be impaired in the aged nervous system, because aging may affect the expression of steroid receptors and steroid receptor co-activators.[219,220] Therefore, sex-steroid receptor signaling may be very different in young and older brains, and in consequence the effects of sex steroids in the brain of young animals may not be predictive of the effects of the same molecules in aged brains. Aging might also deplete other substances that may be necessary for the effects of sex steroids, including growth factors, neuromodulators, neurotransmitters and their receptors. For instance, a growth factor involved in the action of sex steroids is IGF-I. Evidence has accumulated to support the idea that the actions of estrogen and IGF-I in the brain are interdependent. The interdependence between estrogen and IGF-I, or between estrogen receptors and IGF-I receptors, has been documented for neuronal differentiation, synaptic plasticity, adult neurogenesis and neuroprotection.[221–224] Plasma levels of IGF-I decrease with aging, and treatment of old rats with IGF-I ameliorates several age-related deficits in the brain.[159,160] Since brain IGF-I and IGF-I receptor levels are affected by aging,[225] the effect of estrogen receptor activation may be very different in young and old brains because aging decreases the availability of this key synergist.

QUESTIONS FOR THE FUTURE

We are still a long way from being able to design rational protocols for hormonal therapies to protect the brain from aging. Further studies to understand the mechanisms of action of hormones in the aging brain are necessary. We need more information on the interaction of hormones with other neuronal survival factors that are affected by the aging process. Moreover, it is not clear yet whether hormones may be protective for some types of damage but deleterious in the case of other types of brain insult. Alternative strategies to hormonal therapies that may be clinically more effective can be experimentally tested. Gene therapy directed to increase the levels of peptide hormones in specific brain areas may be experimentally tested as a therapeutic approach to reduce brain deficits associated with aging. Experimental gene therapy in rats, by the stereotaxic administration of an adenoviral vector expressing IGF-I, has shown positive results in preventing loss of hypothalamic tuberoinfundibular dopaminergic neurons with aging.[226,227]

It may also be useful to exploit the endogenous capacity of the brain to synthesize, as local paracrine or autocrine factors, some of the hormones affected by aging, such as growth hormone, ghrelin, IGF-I and sex steroids. For instance, one approach to increasing growth hormone and IGF-I levels in aged people is the administration of GHRH or growth-hormone releasing peptide analogs, or the use of molecules that increase the production of GHRH in the hypothalamus. For instance, L-dopa increases GHRH release in elderly rats, resulting in a pulsatile growth hormone profile typical to that observed in young rats.[101] IGF-I local synthesis in the brain may be enhanced by the administration of growth hormone. The use of new synthetic agonists for the growth hormone secretagogue receptor may represent an option to increase growth hormone and IGF-I levels in elder people.

For steroids, possible targets for therapeutic approaches are the proteins that participate in the transport of cholesterol from the cytoplasm to the inner mitochondrial membrane. Cholesterol is converted to pregnenolone in the mitochondria, and pregnenolone is then converted to DHEA or progesterone in the endoplasmic reticulum. The transport of cholesterol to the inner mitochondrial membrane is the rate-limiting step for sex-steroid synthesis, and is highly regulated. Two proteins located in the mitochondrial membrane, the steroidogenic acute regulatory protein (StAR)[228,229] and the peripheral-type benzodiazepine receptor (recently renamed as translocator protein (18 kDa, TSPO)),[230,231] are involved in this transport and may represent good candidates for pharmacological treatments to increase steroid synthesis in the aging brain. In agreement with this hypothesis, it has been shown that the TSPO ligand exerts neuroprotective and anti-inflammatory effects in the brain of young animals[232,233] and a beneficial effect on morphological parameters of the sciatic nerve of aged male rats.[234] In addition to increasing the local synthesis of neuroprotective steroids, other functions of TSPO may also be important to maintain neural tissue integrity. These include the regulation of oxidative processes, since TSPO may modulate mitochondrial sensitivity to reactive oxygen species. Reactive oxygen species have been implicated in several neurodegenerative events, and these generally worsen during the aging process. For instance, it is well known that the rate and degree of recovery after peripheral nerve lesions decline with age, and it has been proposed that reactive oxygen species contribute to this delayed recovery. Consequently, TSPO ligands may in part protect peripheral nerves by reducing sensitivity to reactive oxygen species-induced damage. Given these promising results, TSPO ligands and other molecules that may enhance steroidogenesis in the nervous system should be systematically tested in aged animals. Compounds that regulate StAR activity may also be explored and developed.

Other proteins of potential therapeutic interest are the enzymes responsible for sex-steroid formation and metabolism, such as 5α-reductase and aromatase. We need more information on the levels of expression of these enzymes in the aged human brain, since a potential mechanism to locally increase sex-steroid levels in the brain is to change the expression or activity of these enzymes specifically in the nervous system. For instance, the aromatase gene is under the control of different tissue-specific promoters,[235,236] and it is conceivable that it will be possible to develop specific selective aromatase modulators that will enhance the expression of this enzyme in brain but not in other tissues.[237] In addition, gene therapy can be envisaged to increase the production of steroidogenic enzymes in specific brain regions.

Finally, other potential candidates for pharmacological targeting are the hormone receptors. Selective modulators for androgen and estrogen receptors are under development. Several studies have shown that some synthetic selective estrogen receptor modulators (SERMs), such as tamoxifen, raloxifene or bazedoxifene, and some natural SERMs, such as genistein, are neuroprotective *in vitro* and *in vivo*.[214,238–251] SERMs interact with estrogen receptors and have tissue-specific effects distinct from those of estradiol, acting as estrogen receptor agonists in some tissues and as antagonists in others. Therefore, some SERMs may have estrogenic actions in some brain regions, but an absence of estrogenic actions (or even anti-estrogenic effects) in other brain regions or peripheral tissues. However, before the therapeutic use of these drugs as neuroprotectants is considered, it is essential to learn much more about the expression and regulation of sex-steroid receptors and their co-factors in the aging brain, and also about the impact of aging on the convergence of sex-steroid receptor signaling with other signaling pathways. In addition, SERMs may also exert protection by antioxidant actions.[252] Antioxidant non-genomic effects of estradiol and other estrogenic compounds are the basis for the development of estrogen derivatives with neuroprotective potency.[253] Among these, a group of interesting molecules are the so-called non-feminizing estrogens, such as 17α-estradiol and estratriene derivatives, which share the neuroprotective effects of 17β-estradiol but do not activate estrogen receptors in the reproductive tract.[254–256]

In conclusion, potential alternatives to treatment with neuroprotective hormonal factors might be treatment with molecules able to induce their local synthesis in the brain, or with molecules able to interact with their receptors and signaling cascades, or a combination of both. However, before considering broad clinical applications, more basic research is necessary to clarify the mechanisms of action and potential risks of some of these treatments. More importantly, we need to distinguish those aspects of brain remodeling with aging that represent a positive adaptive response from those that reflect a pathological alteration, to better define the adequate goals of therapeutic interventions.

References

1. Azcoitia I, Leonelli E, Magnaghi V, Veiga S, Garcia-Segura LM, Melcangi RC. Progesterone and its derivatives dihydroprogesterone and tetrahydroprogesterone reduce myelin fiber morphological abnormalities and myelin fiber loss in the sciatic nerve of aged rats. *Neurobiol Aging.* 2003;24:853–860.
2. Melcangi RC, Azcoitia I, Ballabio M, Cavarretta I, Gonzalez LC, Leonelli E, Magnaghi V, Veiga S, Garcia-Segura LM. Neuroactive steroids influence peripheral myelination: a promising opportunity for preventing or treating age-dependent dysfunctions of peripheral nerves. *Prog Neurobiol.* 2003;71:57–66.
3. Melcangi RC, Cavarretta IT, Ballabio M, Leonelli E, Schenone A, Azcoitia I, Garcia-Segura LM, Magnaghi V. Peripheral nerves: a target for the action of neuroactive steroids. *Brain Res Brain Res Rev.* 2005;48:328–338.
4. Bartzokis G. Related age-related myelin breakdown: a developmental model of cognitive decline and Alzheimer's disease. *Neurobiol Aging.* 2004;25:5–18.
5. Salat DH, Tuch DS, Greve DN, van der Kouwe AJ, Hevelone ND, Zaleta AK, Rosen BR, Fischl B, Corkin S, Rosas HD, Dale AM. Age-related alterations in white matter microstructure measured by diffusion tensor imaging. *Neurobiol Aging.* 2005;26:1215–1227.

6. Esiri MM. Ageing and the brain. *J Pathol.* 2007;211:181–187.
7. Alexander GE, Chen K, Aschenbrenner M, Merkley TL, Santerre-Lemmon LE, Shamy JL, Skaggs WE, Buonocore MH, Rapp PR, Barnes CA. Age-related regional network of magnetic resonance imaging gray matter in the rhesus macaque. *J Neurosci.* 2008;28:2710–2718.
8. Scheibel ME, Lindsay RD, Tomiyasu U, Scheibel AB. Progressive dendritic changes in the aging human limbic system. *Exp Neurol.* 1976;53:420–430.
9. Scheibel AB. The hippocampus: organizational patterns in health and senescence. *Mech Ageing Dev.* 1979;9:89–102.
10. Seki T, Arai Y. Age-related production of new granule cells in the adult dentate gyrus. *NeuroReport.* 1995;6:2479–2482.
11. Kuhn HG, Dickinson-Anson H, Gage FH. Neurogenesis in the dentate gyrus of the adult rat: age-related decrease of neuronal progenitor proliferation. *J Neurosci.* 1996;16:2027–2033.
12. Kempermann G, Kuhn HG, Gage FH. Experience-induced neurogenesis in the senescent dentate gyrus. *J Neurosci.* 1998;18:3206–3212.
13. Geinisman Y, de Toledo-Morrell L, Morrell F, Persina IS, Rossi M. Age-related loss of axospinous synapses formed by two afferent systems in the rat dentate gyrus as revealed by the unbiased stereological dissector technique. *Hippocampus.* 1992;2:437–444.
14. Finch CE. Neurons, glia, and plasticity in normal brain aging. *Neurobiol Aging.* 2003;24(Suppl 1):S123–S127.
15. Hof PR, Morrison JH. The aging brain morphomolecular senescence of cortical circuits. *Trends Neurosci.* 2004;27:607–613.
16. Nicholson DA, Yoshida R, Berry RW, Gallagher M, Geinisman Y. Reduction in size of perforated postsynaptic densities in hippocampal axospinous synapses and age-related spatial learning impairments. *J Neurosci.* 2004;24:7648–7653.
17. Luo J, Daniels SB, Lennington JB, Notti RQ, Conover JC. The aging neurogenic subventricular zone. *Aging Cell.* 2006;5:139–152.
18. Masliah E, Crews L, Hansen L. Synaptic remodeling during aging and in Alzheimer's disease. *J Alzheimer Dis.* 2006;9(3 Suppl):91–99.
19. Bondareff W, Geinisman Y. Loss of synapses in the dentate gyrus of the senescent rat. *Am J Anat.* 1976;145:129–136.
20. GeinismanY, de Toledo-Morrell L, Morrell F. Loss of perforated synapses in the dentate gyrus: morphological substrate of memory deficit in aged rats. *Proc Natl Acad Sci USA.* 1986;83:3027–3031.
21. de Brabander JM, Kramers RJ, Uylings HB. Layer-specific dendritic regression of pyramidal cells with ageing in the human prefrontal cortex. *Eur J Neurosci.* 1998;10:1261–1269.
22. Grill JD, Riddle DR. Age-related and laminar-specific dendritic changes in the medial frontal cortex of the rat. *Brain Res.* 2002;937:8–21.
23. Markham JA, Juraska JM. Aging and sex influence the anatomy of the rat anterior cingulate cortex. *Neurobiol Aging.* 2002;23:579–588.
24. Uylings HB, de Brabander JM. Neuronal changes in normal human aging and Alzheimer's disease. *Brain Cogn.* 2002;49:268–276.
25. Rosenzweig ES, Barnes CA. Impact of aging on hippocampal function: plasticity, network dynamics, and cognition. *Prog Neurobiol.* 2003;69:143–179.
26. Smith DE, Rapp PR, McKay HM, Roberts JA, Tuszynski MH. Memory impairment in aged primates is associated with focal death of cortical neurons and atrophy of subcortical neurons. *J Neurosci.* 2004;24:4373–4381.
27. Azcoitia I, Perez-Martin M, Salazar V, Castillo C, Ariznavarreta C, Garcia-Segura LM, Tresguerres JA. Growth hormone prevents neuronal loss in the aged rat hippocampus. *Neurobiol Aging.* 2005;26:697–703.
28. Yates MA, Juraska JM. Pubertal ovarian hormone exposure reduces the number of myelinated axons in the splenium of the rat corpus callosum. *Exp Neurol.* 2008;209:284–287.
29. Peters A, Leahu D, Moss MB, McNally KJ. The effects of aging on area 46 of the frontal cortex of the rhesus monkey. *Cereb Cortex.* 1994;4:621–635.
30. West MJ, Coleman PD, Flood DG, Troncoso JC. Differences in the pattern of hippocampal neuronal loss in normal ageing and Alzheimer's disease. *Lancet.* 1994;344:769–772.
31. Rapp PR, Gallagher M. Preserved neuron number in the hippocampus of aged rats with spatial learning deficits. *Proc Natl Acad Sci USA.* 1996;93:9926–9930.
32. Rasmussen T, Schliemann T, Sorensen JC, Zimmer J, West MJ. Memory impaired aged rats: no loss of principal hippocampal and subicular neurons. *Neurobiol Aging.* 1996;17:143–147.
33. Gazzaley AH, Thakker MM, Hof PR, Morrison JH. Preserved number of entorhinal cortex layer II neurons in aged macaque monkeys. *Neurobiol Aging.* 1997;18:549–553.
34. Pakkenberg B, Gundersen HJ. Neocortical neuron number in humans: effect of sex and age. *J Comp Neurol.* 1997;384:312–320.
35. Merrill DA, Roberts JA, Tuszynski MH. Conservation of neuron number and size in entorhinal cortex layers II, III, and V/VI of aged primates. *J Comp Neurol.* 2000;422:396–401.
36. Merrill DA, Chiba AA, Tuszynski MH. Conservation of neuronal number and size in the entorhinal cortex of behaviorally characterized aged rats. *J Comp Neurol.* 2001;438:445–456.
37. Keuker JI, Luiten PG, Fuchs E. Preservation of hippocampal neuron numbers in aged rhesus monkeys. *Neurobiol Aging.* 2003;24:157–165.
38. Ruiz-Marcos A, Sanchez-Toscano F, Muñoz-Cueto JA. Aging reverts to juvenile conditions the synaptic connectivity of cerebral cortical pyramidal shafts. *Dev Brain Res.* 1992;69:41–49.
39. Landfield PW. Hippocampal neurobiological mechanisms of age-related memory dysfunction. *Neurobiol Aging.* 1988;9:571–579.
40. Foster TC, Norris CM. Age-associated changes in Ca^{2+}-dependent processes: relation to hippocampal synaptic plasticity. *Hippocampus.* 1997;7:602–612.
41. Barnes CA. Long-term potentiation and the ageing brain. *Philos Trans R Soc Lond B Biol Sci.* 2003;358:765–772.
42. Chapman PF. Cognitive aging: recapturing the excitation of youth? *Curr Biol.* 2005;15:R31–R33.
43. Toescu EC, Verkhratsky A, Landfield PW. Ca^{2+} regulation and gene expression in normal brain aging. *Trends Neurosci.* 2004;27:614–620.
44. Burke SN, Barnes CA. Neural plasticity in the ageing brain. *Nat Rev Neurosci.* 2006;7:30–40.
45. Lynch G, Rex CS, Gall CM. Synaptic plasticity in early aging. *Ageing Res Rev.* 2006;5:255–280.
46. Wilson IA, Gallagher M, Eichenbaum H, Tanila H. Neurocognitive aging: prior memories hinder new hippocampal encoding. *Trends Neurosci.* 2006;29:662–670.
47. Drapeau E, Mayo W, Aurousseau C, Le Moal M, Piazza PV, Abrous DN. Spatial memory performances of aged rats in the water maze predict levels of hippocampal neurogenesis. *Proc Natl Acad Sci USA.* 2003;100:14385–14390.
48. Darnaudery M, Perez-Martin M, Belizaire G, Maccari S, Garcia-Segura LM. Insulin-like growth factor 1 reduces age-related disorders induced by prenatal stress in female rats. *Neurobiol Aging.* 2006;27:119–127.
49. Alexopoulos GS. Depression in the elderly. *Lancet.* 2005;365:1961–1970.

50. Mahncke HW, Bronstone A, Merzenich MM. Brain plasticity and functional losses in the aged: scientific bases for a novel intervention. *Prog Brain Res.* 2006;157:81–109.
51. Wilson RS, Beckett LA, Barnes LL, Schneider JA, Bach J, Evans DA, Bennett DA. Individual differences in rates of change in cognitive abilities of older persons. *Psychol Aging.* 2002;17:179–193.
52. Perls T. Centenarians who avoid dementia. *Trends Neurosci.* 2004;27:633–636.
53. Deary IJ, Whiteman MC, Starr JM, Whalley LJ, Fox HC. The impact of childhood intelligence on later life: following up the Scottish Mental Surveys of 1932 and 1947. *J Pers Soc Psychol.* 2004;86:130–147.
54. Whalley LJ, Dick FD, McNeill G. A life-course approach to the aetiology of late-onset dementias. *Lancet Neurol.* 2006;5:87–96.
55. Mattson MP, Magnus T. Ageing and neuronal vulnerability. *Nat Rev Neurosci.* 2006;7:278–294.
56. Miller RA. The aging immune system: primer and prospectus. *Science.* 1996;273:70–74.
57. Grubeck-Loebenstein B, Wick G. The aging of the immune system. *Adv Immunol.* 2002;80:243–284.
58. Gruver AL, Hudson LL, Sempowski GD. Immunosenescence of ageing. *J Pathol.* 2007;211:144–156.
59. Kelley KW, Davila DR, Brief S, Simon J, Arkins S. A pituitary-thymus connection during aging. *Ann NY Acad Sci.* 1988;521:88–98.
60. Fabris N, Mocchegiani E, Provinciali M. Pituitary–thyroid axis and immune system: a reciprocal neuroendocrine–immune interaction. *Horm Res.* 1995;43:29–38.
61. Burgess W, Liu Q, Zhou J, Tang Q, Ozawa A, VanHoy R, Arkins S, Dantzer R, Kelley KW. The immune–endocrine loop during aging: role of growth hormone and insulin-like growth factor-I. *Neuroimmunomodulation.* 1999;6:56–68.
62. Savino W, Arzt E, Dardenne M. Immunoneuroendocrine connectivity: the paradigm of the thymus–hypothalamus/pituitary axis. *Neuroimmunomodulation.* 1999;6:126–136.
63. Wei J, Xu H, Davies JL, Hemmings GP. Increase of plasma IL-6 concentration with age in healthy subjects. *Life Sci.* 1992; 51:1253–1956.
64. Ershler WB. Interleukin-6: A cytokine for gerontologists. *J Am Geriatr Soc.* 1993;41:176–181.
65. Hager K, Machein U, Krieger S, Platt D, Seefried G, Bauer J. Interleukin-6 and selected plasma proteins in healthy persons of different ages. *Neurobiol Aging.* 1994;15:771–772.
66. Roubenoff R, Harris TB, Abad LW, Wilson PW, Dallal GE, Dinarello CA. Monocyte cytokine production in an elderly population: effect of age and inflammation. *J Gerontol A Biol Sci Med Sci.* 1998;53:M20–M26.
67. Dunn AJ. Endotoxin-induced activation of cerebral catecholamine and serotonin metabolism: Comparison with interleukin-1. *J Pharmacol Exp Ther.* 1992;261:964–969.
68. Gutierrez EG, Banks WA, Kastin AJ. Murine tumor necrosis factor alpha is transported from blood to brain in the mouse. *J Neuroimmunol.* 1993;47:169–176.
69. Gutierrez EG, Banks WA, Kastin AJ. Blood-borne interleukin-1 receptor antagonist crosses the blood brain barrier. *J Neuroimmunol.* 1994;55:153–160.
70. Dantzer R, Bluthe RM, Gheusi G, Cremona S, Laye S, Parnet P, Kelley KW. Molecular basis of sickness behavior. *Ann NY Acad Sci.* 1998;856:132–138.
71. Hansen MK, Taishi P, Chen Z, Krueger JM. Vagotomy blocks the induction of interleukin-1beta (IL-1beta) mRNA in the brain of rats in response to systemic IL-1beta. *J Neurosci.* 1998;18:2247–2253.
72. Craig J. Wilson, Caleb E. Finch, Harvey J. Cohen. Cytokines and cognition – the case for a head-to-toe inflammatory paradigm. *J Am Geriatr Soc.* 2002;50:2041–2056.
73. Goya RG, Bolognani F. Homeostasis, thymic hormones and aging. *Gerontology.* 1999;45:174–178.
74. Goya RG, Brown OA, Bolognani F. The thymus–pituitary axis and its changes during aging. *Neuroimmunomodulation.* 1999;6:137–142.
75. McEwen BS, Biron CA, Brunson KW, Bulloch K, Chambers WH, Dhabhar FS, Goldfarb RH, Kitson RP, Miller AH, Spencer RL, Weiss JM. The role of adrenocorticoids as modulators of immune function in health and disease: neural, endocrine and immune interactions. *Brain Res Brain Res Rev.* 1997;23:79–133.
76. Hench PS. The reversibility of certain rheumatic and non-rheumatic conditions by the use of cortisone or of the pituitary adrenocorticotropic hormone. *Ann Internal Med.* 1952;36:1–25.
77. Webster JI, Tonelli l, Sternberg EM. Neuroendocrine regulation of immunity. *Annu Rev Immunol.* 2002;20:125–163.
78. Butcher SK, Lord JM. Stress responses and innate immunity: aging as a contributory factor. *Aging Cell.* 2004;3:151–160.
79. Chahal HS, Drake WM. The endocrine system and ageing. *J Pathol.* 2007;211:173–180.
80. Mariotti S, Franceschi C, Cossarizza A, Pinchera A. The aging thyroid. *Endocr Rev.* 1995;16:686–715.
81. Sherman BM, West JH, Korenman SG. The menopausal transition: analysis of LH, FSH, estradiol, and progesterone concentrations during menstrual cycles of older women. *J Clin Endocrinol Metab.* 1976;42:629–636.
82. Johnson SR. Menopause and hormone replacement therapy. *Med Clin North Am.* 1998;82:297–320.
83. Djahanbakhch O, Ezzati M, Zosmer A. Reproductive ageing in women. *J Pathol.* 2007;211:219–231.
84. Bremner WJ, Vitiello MV, Prinz PN. Loss of circadian rhythmicity in blood testosterone levels with aging in normal men. *J Clin Endocrinol Metab.* 1983;56:1278–1281.
85. Greenspan SL, Klibanski A, Rowe JW, Elahi D. Age-related alterations in pulsatile secretion of TSH: role of dopaminergic regulation. *Am J Physiol.* 1991;260:E486–E491.
86. Sherman B, Wysham C, Pfohl B. Age-related changes in the circadian rhythm of plasma cortisol in man. *J Clin Endocrinol Metab.* 1985;61:439–443.
87. Van CE, Leproult R, Kupfer DJ. Effects of gender and age on the levels and circadian rhythmicity of plasma cortisol. *J Clin Endocrinol Metab.* 1996;81:2468–2473.
88. Cole GM, Frautschy SA. The role of insulin and neurotrophic factor signaling in brain aging and Alzheimer's Disease. *Exp Gerontol.* 2007;42:10–21.
89. García-San Frutos M, Fernández-Agulló T, De Solís AJ, Andrés A, Arribas C, Carrascosa JM, Ros M. Impaired central insulin response in aged Wistar rats: role of adiposity. *Endocrinology.* 2007;148. 5238-4247.
90. Reagan LP. Insulin signaling effects on memory and mood. *Curr Opin Pharmacol.* 2007;7:633–637.
91. Scarpace PJ, Matheny M, Tümer N. Hypothalamic leptin resistance is associated with impaired leptin signal transduction in aged obese rats. *Neuroscience.* 2001;104:1111–1117.
92. Fernandez-Galaz C, Fernández-Agulló T, Campoy F, Arribas C, Gallardo N, Andrés A, Ros M, Carrascosa JM. Decreased leptin uptake in hypothalamic nuclei with ageing in Wistar rats. *J Endocrinol.* 2001;171:23–32.
93. Muzumdar RH, Ma X, Yang X, Atzmon G, Barzilai N. Central resistance to the inhibitory effects of leptin on stimulated insulin secretion with aging. *Neurobiol Aging.* 2006;27:1308–1314.
94. Waltman C, Blackman MR, Chrousos GP, Riemann C, Harman SM. Spontaneous and glucocorticoid-inhibited

adrenocorticotropic hormone and cortisol secretion are similar in healthy young and old men. *J Clin Endocrinol Metab.* 1991;73:495–502.
95. Seeman TE, Robbins RJ. Aging and hypothalamic–pituitary–adrenal response to challenge in humans. *Endocr Rev.* 1994;15:233–260.
96. Bergendahl M, Iranmanesh A, Mulligan T, Veldhuis JD. Impact of age on cortisol secretory dynamics basally and as driven by nutrient-withdrawal stress. *J Clin Endocrinol Metab.* 2000; 85:2203–2214.
97. Lupien SJ, Maheu F, Tu M, Fiocco A, Schramek TE. The effects of stress and stress hormones on human cognition: Implications for the field of brain and cognition. *Brain Cogn.* 2007;65:209–237.
98. Wise PM. The role of the hypothalamus in aging of the female reproductive system. *J Steroid Biochem.* 1987;27:713–719.
99. Wise PM. Neuroendocrine modulation of the "menopause": insights into the aging brain. *Am J Physiol.* 1999;277:E965–E970.
100. Wise PM, Smith MJ, Dubal DB, Wilson ME, Krajnak KM, Rosewell KL. Neuroendocrine influences and repercussions of the menopause. *Endocr Rev.* 1999;20:243–248.
101. Smith RG, Betancourt L, Sun Y. Molecular endocrinology and physiology of the aging nervous system. *Endocr Rev.* 2005;26:203–250.
102. Weiland NG, Cohen IR, Wise PM. Age-associated alterations in catecholaminergic concentrations, neuronal activity, and alpha 1 receptor densities in female rats. *Neurobiol Aging.* 1989; 10:323–329.
103. Reymond MJ. Age-related loss of the responsiveness of the tuberoinfundibular dopaminergic neurons to prolactin in the female rat. *Neuroendocrinology.* 1990;52:490–496.
104. Rossi GL, Bestetti GE, Reymond MJ. Tuberoinfundibular dopaminergic neurons and lactotropes in young and old female rats. *Neurobiol Aging.* 1992;13:275–281.
105. Hung AJ, Stanbury MG, Shanabrough M, Horvath TL, Garcia-Segura LM, Naftolin F. Estrogen, synaptic plasticity and hypothalamic reproductive aging. *Exp Gerontol.* 2003;38:53–59.
106. Sadow TF, Rubin RT. Effects of hypothalamic peptides on the aging brain. *Psychoneuroendocrinology.* 1992;17:293–314.
107. Gore AC. Gonadotropin-releasing hormone neurons, NMDA receptors, and their regulation by steroid hormones across the reproductive life cycle. *Brain Res Brain Res Rev.* 2001;37:235–248.
108. Brann DW, Mahesh VB. The aging reproductive neuroendocrine axis. *Steroids.* 2005;70:273–283.
109. Garcia-Segura LM, Perez J, Jones E, Naftolin F. Loss of sexual dimorphism in rat arcuate nucleus neuronal membranes with reproductive aging. *Expl Neurol.* 1991;112:125–128.
110. Cashion AB, Smith MJ, Wise PM. The morphometry of astrocytes in the rostral preoptic area exhibits a diurnal rhythm on proestrus: relationship to the luteinizing hormone surge and effects of age. *Endocrinology.* 2003;144:274–280.
111. Naftolin F, Brawer JR. The effect of estrogens on hypothalamic structure and function. *Am J Obstet Gynecol.* 1978;132:758–765.
112. Brawer JR, Schipper H, Naftolin F. Ovary-dependent degeneration in the hypothalamic arcuate nucleus. *Endocrinology.* 1980;107:274–279.
113. Leedom L, Lewis C, Garcia-Segura LM, Naftolin F. Regulation of arcuate nucleus synaptology by estrogen. *Ann NY Acad Sci.* 1994;743:61–71.
114. Horvath TL, Garcia-Segura LM, Naftolin F. Control of gonadotropin feedback: the possible role of estrogen-induced hypothalamic synaptic plasticity. *Gynecol Endocrinol.* 1997;11:139–143.
115. Desjardins GC, Beaudet A, Meaney MJ, Brawer JR. Estrogen-induced hypothalamic beta-endorphin neuron loss: a possible model of hypothalamic aging. *Exp Gerontol.* 1995;30:253–267.
116. Sapolsky RM, Krey LC, McEwen BS. The neuroendocrinology of stress and aging: the glucocorticoid cascade hypothesis. *Endocr Rev.* 1986;7:284–301.
117. McEwen BS. Sex, stress and the hippocampus: allostasis, allostatic load and the aging process. *Neurobiol Aging.* 2002; 23:921–939.
118. Sapolsky RM. *Stress, the Aging Brain and the Mechanisms of Neuron Death.* Cambridge: MIT Press; 1992:1–429.
119. Sapolsky RM, Krey LC, McEwen BS. Prolonged glucocorticoid exposure reduces hippocampal neuron number: implications for aging. *J Neurosci.* 1985;5:1222–1227.
120. Vollmann-Honsdorf GK, Flügge G, Fuchs E. Chronic psychosocial stress does not affect the number of pyramidal neurons in tree shrew hippocampus. *Neurosci Lett.* 1997;233:121–124.
121. Leverenz JB, Wilkinson CW, Wamble M, Corbin S, Grabber JE, Raskind MA, Peskind ER. Effect of chronic high-dose exogenous cortisol on hippocampal neuronal number in aged nonhuman primates. *J Neurosci.* 1999;19:2356–2361.
122. Fuchs E, Flügge G, Ohl F, Lucassen P, Vollmann-Honsdorf GK, Michaelis T. Psychosocial stress, glucocorticoids, and structural alterations in the tree shrew hippocampus. *Physiol Behav.* 2001;73:285–291.
123. Conrad CD. What is the functional significance of chronic stress-induced CA3 dendritic retraction within the hippocampus? *Behav Cogn Neurosci Rev.* 2006;5:41–60.
124. Swaab DF, Bao AM, Lucassen PJ. The stress system in the human brain in depression and neurodegeneration. *Ageing Res Rev.* 2005;4:141–194.
125. Bao AM, Meynen G, Swaab DF. The stress system in depression and neurodegeneration: focus on the human hypothalamus. *Brain Res Rev.* 2008;57:531–553.
126. Sanchez MM, Young LJ, Plotsky PM, Insel TR. Distribution of corticosteroid receptors in the rhesus brain: relative absence of glucocorticoid receptors in the hippocampal formation. *J Neurosci.* 2000;20:4657–4668.
127. Mirescu C, Gould E. Stress and adult neurogenesis. *Hippocampus.* 2006;16:233–238.
128. Vallée M, MacCari S, Dellu F, Simon H, Le Moal M, Mayo W. Long-term effects of prenatal stress and postnatal handling on age-related glucocorticoid secretion and cognitive performance: a longitudinal study in the rat. *Eur J Neurosci.* 1999;11:2906–2916.
129. McEwen BS. Stress and hippocampal plasticity. *Annu Rev Neurosci.* 1999;22:105–122.
130. McEwen BS. Glucocorticoids, depression, and mood disorders: structural remodeling in the brain. *Metabolism.* 2005;54(Suppl 1):20–23.
131. Perez-Martin M, Salazar V, Castillo C, Ariznavarreta C, Azcoitia I, Garcia-Segura LM, Tresguerres JA. Estradiol and soy extract increase the production of new cells in the dentate gyrus of old rats. *Exp Gerontol.* 2005;40:450–453.
132. Azcoitia I, Sierra A, Garcia-Segura LM. Estradiol prevents kainic acid-induced neuronal loss in the rat dentate gyrus. *NeuroReport.* 1998;9:3075–3079.
133. Azcoitia I, Sierra A, Garcia-Segura LM. Neuroprotective effects of estradiol in the adult rat hippocampus: Interaction with insulin-like growth factor-I signalling. *J Neurosci Res.* 1999;58:815–822.
134. Quesada A, Micevych PE. Estrogen interacts with the IGF-1 system to protect nigrostriatal dopamine and maintain motoric behavior after 6-hydroxdopamine lesions. *J Neurosci Res.* 2004;75:107–116.
135. Garcia-Segura LM, Sanz A, Mendez P. Cross-talk between IGF-I and estradiol in the brain: focus on neuroprotection. *Neuroendocrinology.* 2006;84:275–279.

136. Perez-Martin M, Azcoitia I, Trejo JL, Sierra A, Garcia-Segura LM. An antagonist of estrogen receptors blocks the induction of adult neurogenesis by insulin-like growth factor-I in the dentate gyrus of adult female rat. *Eur J Neurosci.* 2003;18:923–930.
137. Kenyon C, Chang J, Gensch E, Rudner A, Tabtiang R. A *C. elegans* mutant that lives twice as long as wild type. *Nature.* 1993;366:461–464.
138. Tatar M, Kopelman A, Epstein D, Tu MP, Yin CM, Garofalo RS. A mutant *Drosophila* insulin receptor homolog that extends life-span and impairs neuroendocrine function. *Science.* 2001;292:107–110.
139. Bluher M, Kahn BB, Kahn CR. Extended longevity in mice lacking the insulin receptor in adipose tissue. *Science.* 2003;299:572–574.
140. Holzenberger M, Dupont J, Ducos B, Leneuve P, Geloen A, Even PC, Cervera P, Le Bouc Y. IGF-1 receptor regulates lifespan and resistance to oxidative stress in mice. *Nature.* 2003;421: 182–187.
141. Tatar M, Bartke A, Antebi A. The endocrine regulation of aging by insulin-like signals. *Science.* 2003;299:1346–1351.
142. Kenyon C. The plasticity of aging: insights from long-lived mutants. *Cell.* 2005;120:449–460.
143. Li S, Crenshaw BE, Rawson EJ, Simmons DM, Swanson LW, Rosenfeld MG. Dwarf locus mutants lacking three pituitary cell types result from mutations in the POU-domain gene pit-1. *Nature.* 1990;347:528–533.
144. Sinha YN, Salocks CB, Vanderlaan WP. Pituitary and serum concentrations of prolactin and GH in Snell dwarf mice. *Proc Soc Exp Biol Med.* 1975;150:207–210.
145. Bartke A, Brown-Borg H. Life extension in the dwarf mouse. *Curr Top Dev Biol.* 2004;63:189–225.
146. Schaible R, Gowen JW. A new dwarf mouse. *Genetics.* 1961;46:896.
147. Zhou Y, Xu BC, Maheshwari HG, He L, Reed M, Lozykowski M, Okada S, Wagner TE, Cataldo LA, Coschigano K, Baumann G, Kopchick JJ. A mammalian model for Laron syndrome produced by targeted disruption of the mouse growth hormone receptor/binding protein gene (the Laron mouse). *Proc Natl Acad Sci. USA.* 1997;94:13215–13220.
148. Sun LY, Al-Regaiey K, Masternak MM, Wang J, Bartke A. Local expression of GH and IGF-1 in the hippocampus of GH-deficient long-lived mice. *Neurobiol Aging.* 2005; 26:929–937.
149. Carro E, Trejo JL, Gomez-Isla T, LeRoith D, Torres-Aleman I. Serum insulin-like growth factor I regulates brain amyloid-beta levels. *Nat Med.* 2002;8:1390–1397.
150. Sun LY, Evans MS, Hsieh J, Panici J, Bartke A. Increased neurogenesis in dentate gyrus of long-lived Ames dwarf mice. *Endocrinology.* 2005b;146:1138–1144.
151. Sun LY, Bartke A. Adult neurogenesis in the hippocampus of long-lived mice during aging. *J Gerontol A Biol Sci Med Sci.* 2007;62:117–125.
152. Ramsey MM, Weiner JL, Moore TP, Carter CS, Sonntag WE. Growth hormone treatment attenuates age-related changes in hippocampal short-term plasticity and spatial learning. *Neuroscience.* 2004;129:119–127.
153. Sonntag WE, Lynch CD, Cooney PT, Hutchins PM. Decreases in cerebral microvasculature with age are associated with the decline in growth hormone and insulin-like growth factor 1. *Endocrinology.* 1997;138:3515–3520.
154. Sonntag WE, Lynch C, Thornton P, Khan A, Bennett S, Ingram R. The effects of growth hormone and IGF-1 deficiency on cerebrovascular and brain ageing. *J Anat.* 2000;197:575–585.
155. Delgado-Rubin de Celix A, Chowen JA, Argente J, Frago LM. Growth hormone releasing peptide-6 acts as a survival factor in glutamate-induced excitotoxicity. *J Neurochem.* 2006;99:839–849.
156. Smith RG, Sun Y, Jiang H, Albarran-Zeckler R, Timchenko N. Ghrelin receptor (GHS-R1A) agonists show potential as interventive agents during aging. *Ann NY Acad Sci.* 2007;1119: 147–164.
157. Lopez-Fernandez J, Sanchez-Franco F, Velasco B, Tolon RM, Pazos F, Cacicedo L. Growth hormone induces somatostatin and insulin-like growth factor I gene expression in the cerebral hemispheres of aging rats. *Endocrinology.* 1996;137:4384–4391.
158. Frago LM, Paneda C, Dickson SL, Hewson AK, Argente J, Chowen JA. Growth hormone (GH) and GH-releasing peptide-6 increase brain insulin-like growth factor-I expression and activate intracellular signaling pathways involved in neuroprotection. *Endocrinology.* 2002;143:4113–4122.
159. Lichtenwalner RJ, Forbes ME, Bennett SA, Lynch CD, Sonntag WE, Riddle DR. Intracerebroventricular infusion of insulin-like growth factor-I ameliorates the age-related decline in hippocampal neurogenesis. *Neuroscience.* 2001;107:603–613.
160. Lynch CD, Lyons D, Khan A, Bennett SA, Sonntag WE. Insulin-like growth factor-1 selectively increases glucose utilization in brains of aged animals. *Endocrinology.* 2001;142:506–509.
161. Srinivasan V, Pandi-Perumal SR, Maestroni GJ, Esquifino AI, Hardeland R, Cardinali DP. Role of melatonin in neurodegenerative diseases. *Neurotox Res.* 2005;7:293–318.
162. Lamberts SWJ, van den Beld AW, van der Lely AJ. The endocrinology of aging. *Science.* 1997;278:419–424.
163. Ferrari E, Casarotti D, Muzzoni B, Albertelli N, Cravello L, Fioravanti M, Solerte SB, Magri F. Age-related changes of the adrenal secretory pattern: possible role in pathological brain aging. *Brain Res Rev.* 2001;37:294–300.
164. Vallée M, Mayo W, Le Moal M. Role of pregnenolone, dehydroepiandrosterone and their sulfate esters on learning and memory in cognitive aging. *Brain Res Brain Res Rev.* 2001;37:301–312.
165. Morales AJ, Nolan JJ, Nelson JC, Yen SS. Effects of replacement dose of dehydroepiandrosterone in men and women of advancing age. *J Clin Endocrinol Metab.* 1994;78:1360–1367.
166. Baulieu EE, Thomas G, Legrain S, Lahlou N, Roger M, Debuire B, Faucounau V, Girard L, Hervy MP, Latour F, Leaud MC, Mokrane A, Pitti-Ferrandi H, Trivalle C, de Lacharriere O, Nouveau S, Rakoto-Arison B, Souberbielle JC, Raison J, Le Bouc Y, Raynaud A, Girerd X, Forette F. Dehydroepiandrosterone (DHEA), DHEA sulfate, and aging: contribution of the DHEAge Study to a sociobiomedical issue. *Proc Natl Acad Sci. USA.* 2000;97:4279–4284.
167. van Niekerk JK, Huppert FA, Herbert J. Salivary cortisol and DHEA: association with measures of cognition and well-being in normal older men, and effects of three months of DHEA supplementation. *Psychoneuroendocrinology.* 2001;26:591–612.
168. Huppert FA, Van Niekerk JK, Herbert J. Dehydroepiandrosterone (DHEA) supplementation for cognition and well-being. *Cochrane Database Syst Rev.* 2000;(2). CD000304.
169. Allolio B, Arlt W. DHEA treatment: myth or reality? *Trends Endocrinol Metab.* 2002;13:288–294.
170. Moffat SD, Zonderman AB, Metter EJ, Kawas C, Blackman MR, Harman SM, Resnick SM. Free testosterone and risk for Alzheimer disease in older men. *Neurology.* 2004;62:188–193.
171. Muller M, Aleman A, Grobbee DE, de Haan EH, van der Schouw YT. Endogenous sex hormone levels and cognitive function in aging men is there an optimal level? *Neurology.* 2005;64:866–871.
172. Janowsky JS. The role of androgens in cognition and brain aging in men. *Neuroscience.* 2006;138:1015–1020.

173. Cherrier MM, Matsumoto AM, Amory JK, Johnson M, Craft S, Peskind ER, Raskind MA. Characterization of verbal and spatial memory changes from moderate to supraphysiological increases in serum testosterone in healthy older men. *Psychoneuroendocrinology.* 2007;32:72–79.
174. Carnahan RM, Perry PJ. Depression in aging men: the role of testosterone. *Drugs Aging.* 2004;21:361–376.
175. Shamlian NT, Cole MG. Androgen treatment of depressive symptoms in older men: a systematic review of feasibility and effectiveness. *Can J Psychiatry.* 2006;51:295–299.
176. Ibanez C, Shields SA, El-Etr M, Baulieu EE, Schumacher M, Franklin RJ. Systemic progesterone administration results in a partial reversal of the age-associated decline in CNS remyelination following toxin-induced demyelination in male rats. *Neuropathol Appl Neurobiol.* 2004;30:80–89.
177. Lord C, Buss C, Lupien SJ, Pruessner JC. Hippocampal volumes are larger in postmenopausal women using estrogen therapy compared to past users, never users and men: A possible window of opportunity effect. *Neurobiol Aging.* 2008;29:95–101.
178. Yaffe K, Sawaya G, Lieberburg I, Grady D. Estrogen therapy in postmenopausal women: effects on cognitive function and dementia. *J Am Med Assoc.* 1998;279:688–695.
179. Henderson VW, Paganini-Hill A, Miller BL, Elble RJ, Reyes PF, Shoupe D, McCleary CA, Klein RA, Hake AM, Farlow MR. Estrogen for Alzheimer's disease in women: randomized, double-blind, placebo-controlled trial. *Neurology.* 2000;54: 295–301.
180. Resnick SM, Maki PM. Effects of hormone replacement therapy on cognitive and brain aging. *Ann NY Acad Sci.* 2001;949:203–214.
181. Krug R, Born J, Rasch B. A 3-day estrogen treatment improves prefrontal cortex-dependent cognitive function in postmenopausal women. *Psychoneuroendocrinology.* 2006;31:965–975.
182. Birkhauser M. Depression, menopause and estrogens: is there a correlation? *Maturitas.* 2002;41(Suppl 1):3–8.
183. Stevens JR. Schizophrenia: reproductive hormones and the brain. *Am J Psychiatry.* 2002;159:713–719.
184. Saunders-Pullman R, Gordon-Elliott J, Parides M, Fahn S, Saunders HR, Bressman S. The effect of estrogen replacement on early Parkinson's disease. *Neurology.* 1999;52:1417–1421.
185. Tsang KL, Ho SL, Lo SK. Estrogen improves motor disability in parkinsonian postmenopausal women with motor fluctuations. *Neurology.* 2000;54:2292–2298.
186. Cyr M, Calon F, Morissette M, Di Paolo T. Estrogenic modulation of brain activity: implications for schizophrenia and Parkinson's disease. *J Psychiatry Neurosci.* 2002;27:12–27.
187. Dluzen D, Horstink M. Estrogen as neuroprotectant of nigrostriatal dopaminergic system: laboratory and clinical studies. *Endocrine.* 2003;21:67–75.
188. Paganini-Hill A. Estrogen replacement therapy and stroke. *Prog Cardiovasc Dis.* 1995;38:223–242.
189. Paganini-Hill A. Hormone replacement therapy and stroke: risk, protection or no effect? *Maturitas.* 2001;38:243–261.
190. Paganini-Hill A, Henderson VW. Estrogen replacement therapy and risk of Alzheimer disease. *Arch Intern Med.* 1996;156:2213–2217.
191. Asthana S, Baker LD, Craft S, Stanczyk FZ, Veith RC, Raskind MA, Plymate SR. High-dose estradiol improves cognition for women with AD: results of a randomized study. *Neurology.* 2001;57:605–612.
192. Saunders-Pullman R. Estrogens and Parkinson disease: neuroprotective, symptomatic, neither, or both? *Endocrine.* 2003; 21:81–87.
193. Viscoli CM, Brass LM, Kernan WN, Sarrel PM, Suissa S, Horwitz RI. A clinical trial of estrogen-replacement therapy after ischemic stroke. *N Engl J Med.* 2001;345:1243–1249.
194. Mulnard RA, Cotman CW, Kawas C, van Dyck CH, Sano M, Doody R, Koss E, Pfeiffer E, Jin S, Gamst A, Grundman M, Thomas R, Thal LJ. Estrogen replacement therapy for treatment of mild to moderate Alzheimer disease: a randomized controlled trial. Alzheimer's Disease Cooperative Study. *J Am Med Assoc.* 2000;283:1007–1015.
195. Wang PN, Liao SQ, Liu RS, Liu CY, Chao HT, Lu SR, Yu HY, Wang SJ, Liu HC. Effects of estrogen on cognition, mood, and cerebral blood flow in AD: a controlled study. *Neurology.* 2000;54:2061–2066.
196. Shaywitz BA, Shaywitz SE. Estrogen and Alzheimer disease: plausible theory, negative clinical trial. *J Am Med Assoc.* 2000;283:1055–1056.
197. Yaffe K. Hormone therapy and the brain: *déja vu* all over again? *J Am Med Assoc.* 2003;289:2717–2719.
198. Shumaker SA, Legault C, Rapp SR, Thal L, Wallace RB, Ockene JK, Hendrix SL, Jones III BN, Assaf AR, Jackson RD, Kotchen JM, Wassertheil-Smoller S, Wactawski-Wende J. Estrogen plus progestin and the incidence of dementia and mild cognitive impairment in postmenopausal women: the Women's Health Initiative Memory Study: a randomized controlled trial. *J Am Med Assoc.* 2003;289:2651–2662.
199. Wassertheil-Smoller S, Hendrix SL, Limacher M, Heiss G, Kooperberg C, Baird A, Kotchen T, Curb JD, Black H, Rossouw JE, Aragaki A, Safford M, Stein E, Laowattana S, Mysiw WJ. Effect of estrogen plus progestin on stroke in postmenopausal women: the Women's Health Initiative: a randomized trial. *J Am Med Assoc.* 2003;289:2673–2684.
200. Ciriza I, Carrero P, Frye CA, Garcia-Segura LM. Reduced metabolites mediate neuroprotective effects of progesterone in the adult rat hippocampus. The synthetic progestin medroxyprogesterone acetate (Provera) is not neuroprotective. *J Neurobiol.* 2006;66:916–928.
201. Nilsen J, Brinton RD. Impact of progestins on estrogen-induced neuroprotection: synergy by progesterone and 19-norprogesterone and antagonism by medroxyprogesterone acetate. *Endocrinology.* 2002;143:205–212.
202. Nilsen J, Morales A, Brinton RD. Medroxyprogesterone acetate exacerbates glutamate excitotoxicity. *Gynecol Endocrinol.* 2006;22:355–361.
203. Henderson VW, Benke KS, Green RC, Cupples LA, Farrer LA, Study Group MIRAGE. Postmenopausal hormone therapy and Alzheimer's disease risk: interaction with age. *J Neurol Neurosurg Psychiatry.* 2005;76:103–105.
204. Maki PM. Hormone therapy and cognitive function: is there a critical period for benefit? *Neuroscience.* 2006;138:1027–1030.
205. Maki PM. Potential importance of early initiation of hormone therapy for cognitive benefit. *Menopause.* 2006;13:6–7.
206. Suzuki S, Brown CM, Dela Cruz CD, Yang E, Bridwell DA, Wise PM. Timing of estrogen therapy after ovariectomy dictates the efficacy of its neuroprotective and antiinflammatory actions. *Proc Natl Acad Sci USA.* 2007;104:6013–6018.
207. Frutos MG, Cacicedo L, Fernández C, Vicent D, Velasco B, Zapatero H, Sánchez-Franco F. Insights into a role of GH secretagogues in reversing the age-related decline in the GH/IGF-I axis. *Am J Physiol Endocrinol Metab.* 2007;293:E1140–E1152.
208. Kok P, Paulo RC, Cosma M, Mielke KL, Miles JM, Bowers CY, Veldhuis JD. Estrogen supplementation selectively enhances hypothalamo-pituitary sensitivity to ghrelin in postmenopausal women. *J Clin Endocrinol Metab.* 2008;93:4020–4026.
209. Wise PM. Estrogen therapy: does it help or hurt the adult and aging brain? Insights derived from animal models. *Neuroscience.* 2006;138:831–835.

210. Cutler SM, Cekic M, Miller DM, Wali B, Vanlandingham JW, Stein DG. Progesterone improves acute recovery after traumatic brain injury in the aged rat. *J Neurotrauma.* 2007;24:1475–1486.
211. Matsumoto A. Androgen stimulates neuronal plasticity in the perineal motoneurons of aged male rats. *J Comp Neurol.* 2001;430:389–395.
212. Lei DL, Long JM, Hengemihle J, O'Neill J, Manaye KF, Ingram DK, Mouton PR. Effects of estrogen and raloxifene on neuroglia number and morphology in the hippocampus of aged female mice. *Neuroscience.* 2003;121:659–666.
213. Saravia F, Beauquis J, Pietranera L, De Nicola AF. Neuroprotective effects of estradiol in hippocampal neurons and glia of middle age mice. *Psychoneuroendocrinology.* 2007;32:480–492.
214. Barreto G, Santos-Galindo M, Diz-Chaves Y, Pernía O, Carrero P, Azcoitia I, Garcia-Segura LM. Selective estrogen receptor modulators decrease reactive astrogliosis in the injured brain: effects of aging and prolonged depletion of ovarian hormones. *Endocrinology.* 2009;150:5010–5015.
215. Dubal DB, Wise PM. Neuroprotective effects of estradiol in middle-aged female rats. *Endocrinology.* 2001;142:43–48.
216. Foy MR, Baudry M, Foy JG, Thompson RF. 17beta-estradiol modifies stress-induced and age-related changes in hippocampal synaptic plasticity. *Behav Neurosci.* 2008;122:301–309.
217. Adams MM, Shah RA, Janssen WG, Morrison JH. Different modes of hippocampal plasticity in response to estrogen in young and aged female rats. *Proc Natl Acad Sci USA.* 2001;98:8071–8076.
218. Melcangi RC, Magnaghi V, Martini L. Aging in peripheral nerves: regulation of myelin protein genes by steroid hormones. *Prog Neurobiol.* 2000;60:291–308.
219. Matsumoto A, Prins GS. Androgenic regulation of expression of androgen receptor protein in the perineal motoneurons of aged male rats. *J Comp Neurol.* 2002;443:383–387.
220. Jezierski MK, Sohrabji F. Neurotrophin expression in the reproductively senescent forebrain is refractory to estrogen stimulation. *Neurobiol Aging.* 2001;22:309–319.
221. Cardona-Gomez GP, Mendez P, DonCarlos LL, Azcoitia I, Garcia-Segura LM. Interactions of estrogens and insulin-like growth factor-I in the brain: implications for neuroprotection. *Brain Res Rev.* 2001;37:320–334.
222. Mendez P, Azcoitia I, Garcia-Segura LM. Interdependence of oestrogen and insulin-like growth factor-I in the brain: potential for analysing neuroprotective mechanisms. *J Endocrinol.* 2005;185:11–17.
223. Mendez P, Wandosell F, Garcia-Segura LM. Cross-talk between estrogen receptors and insulin-like growth factor-I receptor in the brain: cellular and molecular mechanisms. *Front Neuroendocrinol.* 2006;27:391–403.
224. Garcia-Segura LM, Diz-Chaves Y, Perez-Martin M, Darnaudéry M. Estradiol, insulin-like growth factor-I and brain aging. *Psychoneuroendocrinology.* 2007;32(Suppl 1):S57–S61.
225. Sonntag WE, Lynch CD, Bennett SA, Khan AS, Thornton PL, Cooney PT, Ingram RL, McShane T, Brunso-Bechtold JK. Alterations in insulin-like growth factor-1 gene and protein expression and type 1 insulin-like growth factor receptors in the brains of ageing rats. *Neuroscience.* 1999;88:269–279.
226. Herenu CB, Brown OA, Sosa YE, Morel GR, Reggiani PC, Bellini MJ, Goya RG. The neuroendocrine system as a model to evaluate experimental gene therapy. *Curr Gene Ther.* 2006;6:125–129.
227. Herenu CB, Cristina C, Rimoldi OJ, Becu-Villalobos D, Cambiaggi V, Portiansky EL, Goya RG. Restorative effect of insulin-like growth factor-I gene therapy in the hypothalamus of senile rats with dopaminergic dysfunction. *Gene Ther.* 2007;14:237–245.
228. Stocco DM. StAR protein and the regulation of steroid hormone biosynthesis. *Annu Rev Physiol.* 2001;63:193–213.
229. Lavaque E, Sierra A, Azcoitia I, Garcia-Segura LM. Steroidogenic acute regulatory protein in the brain. *Neuroscience.* 2006;138:741–747.
230. Lacor P, Gandolfo P, Tonon MC, Brault E, Dalibert I, Schumacher M, Benavides J, Ferzaz B. Regulation of the expression of peripheral benzodiazepine receptors and their endogenous ligands during rat sciatic nerve degeneration and regeneration: a role for PBR in neurosteroidogenesis. *Brain Res.* 1999;815:70–80.
231. Papadopoulos V, Baraldi M, Guilarte TR, Knudsen TB, Lacapere JJ, Lindemann P, Norenberg MD, Nutt D, Weizman A, Zhang MR, Gavish M. Translocator protein (18kDa): new nomenclature for the peripheral-type benzodiazepine receptor based on its structure and molecular function. *Trends Pharmacol Sci.* 2006;27:402–409.
232. Veiga S, Azcoitia I, Garcia-Segura LM. Ro5-4864, a peripheral benzodiazepine receptor ligand, reduces reactive gliosis and protects hippocampal hilar neurons from kainic acid excitotoxicity. *J Neurosci Res.* 2005;80:129–137.
233. Veiga S, Carrero P, Pernia O, Azcoitia I, Garcia-Segura LM. Translocator protein (18 kDa) is involved in the regulation of reactive gliosis. *Glia.* 2007;55:1426–1436.
234. Leonelli E, Yague JG, Ballabio M, Azcoitia I, Magnaghi V, Schumacher M, Garcia-Segura LM, Melcangi RC. Ro5-4864, a synthetic ligand of peripheral benzodiazepine receptor, reduces aging-associated myelin degeneration in the sciatic nerve of male rats. *Mech Ageing Dev.* 2005;126:1159–1163.
235. Honda S, Harada N, Abe-Dohmae S, Takagi Y. Identification of cis-acting elements in the proximal promoter region for brain-specific exon 1 of the mouse aromatase gene. *Mol Brain Res.* 1999;66:122–132.
236. Yague JG, Munoz A, de Monasterio-Schrader P, Defelipe J, Garcia-Segura LM, Azcoitia I. Aromatase expression in the human temporal cortex. *Neuroscience.* 2006;138:389–401.
237. Yague JG, Garcia-Segura LM, Azcoitia I. Selective transcriptional regulation of aromatase gene by vitamin D, dexamethasone, and mifepristone in human glioma cells. *Endocrine.* 2009;35:252–261.
238. Grandbois M, Morisette M, Callier S, Di Paolo T. Ovarian steroids and raloxifene prevent MPTP-induced dopamine depletion in mice. *NeuroReport.* 2000;11:343–346.
239. Rossberg MI, Murphy SJ, Traystman RJ, Hurn PD. LY353381.HCl, a selective estrogen receptor modulator, and experimental stroke. *Stroke.* 2000;31:3041–3046.
240. Callier S, Morissette M, Grandbois M, Pelaprat D, Di Paolo T. Neuroprotective properties of 17beta-estradiol, progesterone, and raloxifene in MPTP C57Bl/6 mice. *Synapse.* 2001;41:131–138.
241. Dhandapani KM, Brann DW. Protective effects of estrogen and selective estrogen receptor modulators in the brain. *Biol Reprod.* 2002;67:1379–1385.
242. Kimelberg HK, Jin Y, Charniga C, Feustel PJ. Neuroprotective activity of tamoxifen in permanent focal ischemia. *J Neurosurg.* 2003;99:138–142.
243. Mehta SH, Dhandapani KM, De Sevilla LM, Webb RC, Mahesh VB, Brann DW. Tamoxifen, a selective estrogen receptor modulator, reduces ischemic damage caused by middle cerebral artery occlusion in the ovariectomized female rat. *Neuroendocrinology.* 2003;77:44–50.
244. Ciriza I, Carrero P, Azcoitia I, Lundeen SG, Garcia-Segura LM. Selective estrogen receptor modulators protect hippocampal

neurons from kainic acid excitotoxicity: differences with the effect of estradiol. *J Neurobiol.* 2004;61:209–221.
245. Mickley KR, Dluzen DE. Dose-response effects of estrogen and tamoxifen upon methamphetamine-induced behavioral responses and neurotoxicity of the nigrostriatal dopaminergic system in female mice. *Neuroendocrinology.* 2004;79:305–316.
246. Zhao L, O'Neill K, Brinton RD. Selective estrogen receptor modulators (SERMs) for the brain: current status and remaining challenges for developing NeuroSERMs. *Brain Res Brain Res Rev.* 2005;49:472–493.
247. Zhao L, O'Neill K, Brinton RD. Estrogenic agonist activity of ICI 182,780 (Faslodex) in hippocampal neurons: implications for basic science understanding of estrogen signaling and development of estrogen modulators with a dual therapeutic profile. *J Pharmacol Exp Ther.* 2006;319:1124–1132.
248. Azcoitia I, Moreno A, Carrero P, Palacios S, Garcia-Segura LM. Neuroprotective effects of soy phytoestrogens in the rat brain. *Gynecol Endocrinol.* 2006;22:63–69.
249. Kokiko ON, Murashov AK, Hoane MR. Administration of raloxifene reduces sensorimotor and working memory deficits following traumatic brain injury. *Behav Brain Res.* 2006; 170:233–240.
250. Tapia-Gonzalez S, Carrero P, Pernia O, Garcia-Segura LM, Diz-Chaves Y. Selective oestrogen receptor (ER) modulators reduce microglia reactivity in vivo after peripheral inflammation: potential role of microglial ERs. *J Endocrinol.* 2008;198:219–230.
251. Cerciat M, Unkila M, Garcia-Segura LM, Arevalo MA. Selective estrogen receptor modulators decrease the production of interleukin-6 and interferon-gamma-inducible protein-10 by astrocytes exposed to inflammatory challenge *in vitro*. *Glia.* 2010;58:93–102.
252. Zhang Y, Milatovic D, Aschner M, Feustel PJ, Kimelberg HK. Neuroprotection by tamoxifen in focal cerebral ischemia is not mediated by an agonist action at estrogen receptors but is associated with antioxidant activity. *Exp Neurol.* 2007; 204:819–827.
253. Prokai L, Simpkins JW. Structure-nongenomic neuroprotection relationship of estrogens and estrogen-derived compounds. *Pharmacol Ther.* 2007;114:1–12.
254. Dykens JA, Simpkins JW, Wang J, Gordon K. Polycyclic phenols, estrogens and neuroprotection: a proposed mitochondrial mechanism. *Exp Gerontol.* 2003;38:101–107.
255. Simpkins JW, Yang SH, Liu R, Perez E, Cai ZY, Covey DF, Green PS. Estrogen-like compounds for ischemic neuroprotection. *Stroke.* 2004;35(11 Suppl 1):2648–2651.
256. Simpkins JW, Wang J, Wang X, Perez E, Prokai L, Dykens JA. Mitochondria play a central role in estrogen-induced neuroprotection. *Curr Drug Targets CNS Neurol Disord.* 2005;4:69–83.

PART IV

NEUROENDOCRINE PATHOLOGY

CHAPTER

27

Kallmann Syndrome and Other Causes of Hypothalamic Hypogonadism and Related Development Disorders

Thang S. Han, Pierre M.G. Bouloux

Centre for Neuroendocrinology, University College London Medical School, Royal Free Hospital, Hampstead, London, UK

OUTLINE

Handbook of Neuroendocrinology, DOI: 10.1016/B978-0-12-375097-6.10027-7

Summary

Hypogonadotropic hypogonadism (HH) is characterized by gonadal failure and absent/delayed or arrested pubertal sexual maturation secondary to absent or severely reduced circulating levels of luteinizing and follicle-stimulating hormones. Congenital abnormalities leading to HH may result from diminished GnRH secretion, occurring in isolation either with euosmia (idiopathic, IHH) or in association with anosmia/hyposmia (Kallmann syndrome), or as part of multiple pituitary hormone deficiencies secondary to developmental failure of the adenohypophysis, and can also occur as part of other ill-characterized developmental disorders of the central nervous system. In this chapter, the neuroendocrine organization and regulation of the hypothalamic–pituitary–gonadal axis is first considered, followed by a review of current concepts in the molecular pathogenesis of hypogonadotropic hypogonadism.

The ontogeny of GnRH neurons is complex and incompletely understood, and current knowledge of the pathogenesis of these disorders draws from evidence from several animal models and human observations, which have elucidated the role of several genetic loci which, if disrupted, lead to abnormalities in GnRH neuronal migration or connections of signals essential for GnRH neuron survival, or perturbation of GnRH secretion or action, or of pituitary development. In the human, mutations in more than 10 separate gene loci, including *FGFR1*, *FGF8*, *PROK2* and *PROKR2*, are responsible for both Kallmann syndrome and IHH; it has become evident that there is considerable pathogenetic overlap between these two entities, and that within affected pedigrees the same genetic defect may cause either phenotype. Moreover, transmission does not always follow classical Mendelian inheritance patterns, and there is growing evidence for interplay between several genes (oligogenic inheritance) in producing the phenotype. Recently described genetic lesions in the kisspeptin/GPR54 system in patients with HH or severe pubertal delay, and of the critical role of neurokinin B and its receptor in normal sexual maturation, have extended our understanding of GnRH neuronal secretion, and may have future therapeutic implications in the human.

INTRODUCTION

Hypogonadotropic hypogonadism (HH) may result from either absent or inadequate hypothalamic gonadotropin-releasing hormone (GnRH) secretion. HH is characterized by deficiency of gonadal androgens in males and estrogens in females, and absent/delayed or arrested pubertal sexual maturation. Congenital abnormalities leading to HH are usually due to diminished GnRH secretion, occurring in isolation either with euosmia (idiopathic, IHH) or in association with anosmia/hyposmia (Kallmann syndrome, KS), or as part of multiple pituitary hormone deficiencies. Most presentations of isolated HH are idiopathic, with defects identified to date in more than 10 separate genetic loci (Table 27.1), accounting for monogenic and occasionally oligogenic presentations in about 30–40% of cases. Multiple modes of inheritance have been described in familial cases, including autosomal dominant, autosomal recessive or X-linked recessive. More recently, further evidence for a genetic basis of euosmic IHH has been accumulated through homozygosity mapping in consanguineous families, including the identification of a key role for kisspeptin in triggering GnRH release and of a critical role for neurokinin B (NKB) in normal sexual maturation.

Isolated HH is a heterogeneous condition with poor genotypic and phenotypic correlations. For the same monogenic gene mutation, highly variable reproductive and accessory phenotypes among members of the same family or across families are often observed. Indeed, the genes *FGF8*, *FGFR1*, *PROK2* and *PROKR2* have been shown to associate both with hyposmic/anosmic and with euosmic phenotypes. This variable expressivity suggests that isolated GnRH deficiency comprises a broad diagnostic spectrum; therefore, the commonly used classification of IHH based on olfactory status may have limitations in clinical practice.

This chapter focuses primarily on a review of the genetic basis of KS and other causes of IHH and related development disorders. To appreciate the pathogenic complexities of these conditions, the neuroendocrinology of the hypothalamic–pituitary–gonadal (HPG) system and developmental consequences of abnormal regulation are described. These concepts form the basic principles for specific investigations and therapy of individuals with IHH.

NEUROENDOCRINE REGULATION OF THE HYPOTHALAMIC–PITUITARY–GONADAL SYSTEM

Regulation and Action of Gonadotropin-releasing Hormone

Embryologically, olfactory and GnRH neurons are born in the nasal pit. Olfactory axons migrate and establish contact with the olfactory bulb anlage, a prerequisite

TABLE 27.1 Etiology of Congenital Hypogonadotropic Hypogonadism

Monogenic/oligogenic mutations associated with anosmia, hyposmia or euosmia	Monogenic/oligogenic mutations associated with euosmia
Anosmin 1 (*KAL1*)	Kisspeptin 1 (*KISS1*)
Fibroblast growth factor 8 (*FGF8*)	G-protein coupled receptor 54 (*GPR54 or KISS1R*)
Fibroblast growth factor receptor 1 (*FGFR1*)	Leptin (*LEP*)
Prokineticin 2 (*PROK2*)	Leptin receptor (*LEPR*)
Prokineticin receptor 2 (*PROKR2*)	Luteinizing hormone β subunit (*LHβ*)
Nasal embryonic LHRH factor (*NELF*)	Prohormone convertase 1 (*PC1*)
Heparan sulfate 6 "O" sulfotransferase 2 (HS6ST2)	Gonadotropin releasing hormone 1 (*GNRH1*)
	Gonadotropin releasing hormone receptor (*GNRHR*)
	Tachykinin 3 (TAC3)
	Tachykinin receptor 3 (TACR3)
	Dosage-sensitive sex reversal, adrenal hypoplasia congenita, critical region on the X chromosome, gene 1 (*DAX1*)
DEFECTIVE TRANSCRIPTION FACTOR GENES OF PITUITARY DIFFERENTIATION CAUSING COMBINED PITUITARY HORMONE DEFICIENCY	
Pituitary-specific positive transcription factor 1 (*POU1F1*)	
Homeobox protein prophet of Pit-1 (*PROP1*)	
Homeobox expressed in ES cells 1 (*HESX1*)	
LIM/homeobox protein Lhx3 (*LHX3*)	
Transcription factor SOX-3 (*SOX3*)	
CONGENITAL HYPOGONADOTROPIC HYPOGONADISM ASSOCIATED WITH OTHER CENTRAL NERVOUS DISORDERS	
CHARGE (coloboma, heart defect, choanal atresia, growth retardation, ear abnormalities) syndrome	
Gordon Holmes spinocerebellar ataxia syndrome	
Laurence-Moon-Bardet-Biedl syndrome	
Möbius syndrome	
Prader-Willi syndrome	
Rud syndrome	

for olfactory bulb morphogenesis.[1] There are two waves of GnRH neuronal migration into the brain: one precedes olfactory bulb formation, the other occurs in association with olfactory bulb formation[2] (Fig. 27.1). In this later wave of migration, GnRH-secreting neurons migrate into the hypothalamus along a nasal mesenchymal scaffold of olfactory, vomeronasal and terminal nerves, expressing neural cell adhesion molecules along their central processes, prior to dispersing in the mediobasal part of the anterior hypothalamus.[3] Their axons innervate the median eminence and secrete pulsatile GnRH into the pituitary portal system. GnRH is initially produced as a 92-amino acid (AA) pre-pro-GnRH, encoded by the *GNRH* gene, mapped to 8p21-p11.2. Post-translational cleavage by prohormone convertase-1 (PC1) of the leading 23 AAs (pre portion) and the supporting gonadotropin associated peptide (GAP) 56 AAs (pro portion) results in the decapeptide GnRH.

GnRH has a specific receptor (GnRHR1: a G-protein-bound receptor with seven transmembrane loops) on pituitary gonadotropes. Binding of GnRH to GnRHR1 initiates gene expression of α- and β-chains of luteinizing and follicle-stimulating (FSH) hormones, and promotes their secretion by inducing inositol-1,4,5,-triphosphate (IP_3) and diacylglycerol (DAG) generation, resulting in the mobilization of intracellular calcium and protein kinase C (PKC), respectively.

GnRH neurons discharge at an intrinsic frequency of 90–120 minutes, releasing GnRH in a pulsatile manner. The amplitude and frequency of discharge are modulated by a number of classical and non-classical neurotransmitters. Whilst α-adrenergic impulses have a stimulatory action, β-adrenergic, dopamine and endorphinergic impulses have an inhibitory action on GnRH secretion. Testosterone and progesterone, probably via β-endorphins, attenuate pulse frequency.

Reproduction requires high energy consumption, and functions optimally when nutrition and energy storage are readily available.[4–6] GnRH plays a critical regulatory role in this process, particularly in anticipation and sustenance of puberty, pregnancy, and lactation.[7] It is now known that the hypothalamus receives signals of energy status by a range of peripheral hormones, including insulin, leptin, and ghrelin, as well as by centrally released peptides such as neuropeptide Y (NPY), melanocortins, melanin-concentrating hormone and the recently discovered kisspeptin. These signals, singly or together, can directly or indirectly

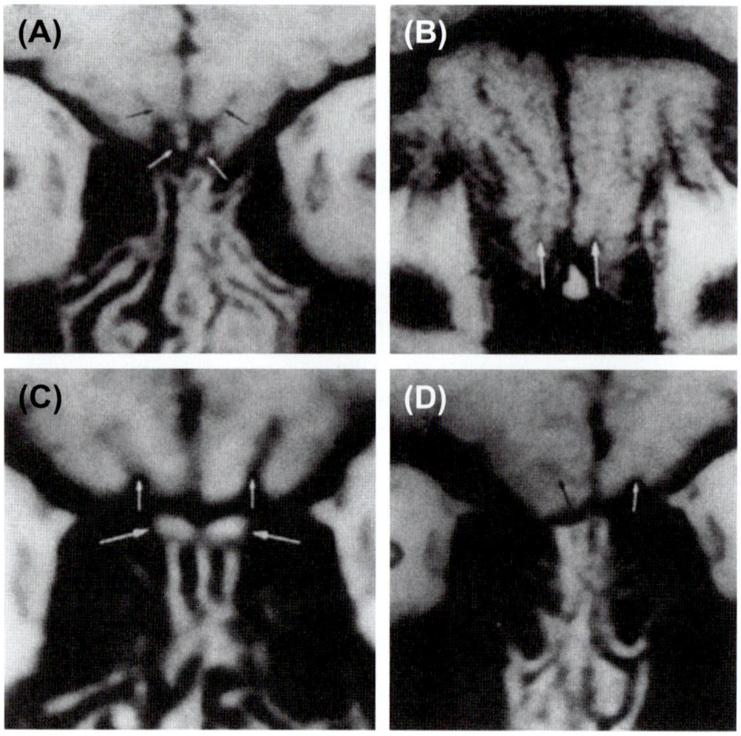

FIGURE 27.1 **MRI of the brain in patients with KS and IHH. (A)** Coronal T1-weighted image of a male with KS showing abnormal angulated olfactory sulci (black arrows) and normal olfactory bulbs (white arrows). (**B**) Axial T1-weighted image of the same patient showing normal olfactory sulci (white arrows). (**C**) Coronal T1-weighted image of a female with IHH showing normal olfactory bulbs (large arrows) and sulci (small arrows). (**D**) Coronal T1-weighted image of a female with KS showing absent olfactory bulbs with shallow olfactory sulci (arrows). *Images reproduced from Quinton R, Duke VM, de Zoysa PA, Platts AD, Valentine A, Kendall B, Pickman S, Kirk JM, Besser GM, Jacobs HS, Bouloux PM. The neuroradiology of Kallmann's syndrome: a genotypic and phenotypic analysis.* J Clin Endocrinol Metab. *1996 ;81:3010–3017, with permission from The Endocrine Society.*

integrate energy balance with reproduction at any level of the HPG axis.

Evidence suggests that the interactions between leptin and kisspeptin are vital for the initiation of puberty, reflecting the close links between metabolic and sexual hormones that are required for triggering the *KISS1/GPR54* signaling to GnRH neurons, thereby regulating reproductive activity. Leptin, a hormone produced by adipocytes, whose gene *LEP* has been cloned,[8] has specific receptors (LEPR) in the hypothalamus. It probably acts on the hypothalamus indirectly via NPY, by intervening with the feedback mechanism to signal the nutritional status (adipose tissue reserve). In anorexia nervosa, low leptin levels are associated with reduced GnRH levels and amenorrhoea – an effect that can be reversed by leptin administration.[9]

On the other hand, centrally, kisspeptin is a potent activator of GnRH neurons,[10–13] enhancing pituitary gonadotropin release, triggering a cascade that is necessary for entering puberty and normal sexual maturation[14,15] and maintaining ovulation and fertility,[16] while downregulation of *KISS1* has been suggested to mediate some of the HPG suppression seen in severe nutritional deprivation.[17] A study of primary culture of human fetal GnRH-secreting neuroblasts (FNC-B4) has shown that kisspeptin induces GnRH secretion, and kisspeptin *per se* is suppressed by estradiol and stimulated by androgens. Furthermore, leptin has also been shown to increase KISS1/GPR54 expression in FNC-B4.[18] Kisspeptin has been implicated in both negative and positive central feedback of sex steroids on GnRH production.[19,20]

Regulation and Action of Luteinizing Hormone

Like thyroid-stimulating hormone (TSH) and human chorionic gonadotropin (hCG), LH and FSH are large glycoproteins. They comprise two non-covalently binding peptide chains (α and β) forming a heterodimer. The α-chain is identical in all four glycoprotein hormones, the β-chain being unique to each hormone and mediating the biological effect. The α-chain is encoded by a gene on 6q12.21, FSH-β on 11p13, and LH-β and hCG-β on 19q13.32. The genes of LH receptors (*LHR*) and FSH receptors (*FSHR*) both locate to chromosome 2 (p21).

LH binds to specific G-protein coupled receptors on the surface of Leydig cells of the testis and the thecal cells of the ovary, activating adenyl cyclase and generating the intracellular second messenger cyclic adenosine monophosphate (cAMP), which, acting via protein kinase A (PKA), leads to gonadal steroid biosynthesis. LH influences the conversion of cholesterol to pregnenolone, the most critical step in testosterone biosynthesis, via two mechanisms: (a) cAMP-mediated stimulation of synthesis and activity of mitochondrial cytochrome P450-linked side-chain cleavage (P450scc) enzyme (or 20,22-desmolase), and (b) a PKC-mediated increase in the production of cholesterol by activation of hydrolase.

Regulation and Action of Follicle-stimulating Hormone

FSH binds to specific receptors on Sertoli cells in the testis and granulosa cells in the ovary to stimulate folliculogenesis. FSH stimulates the activity of aromatase, an enzyme catalyzing the conversion of androgens to estrogens. The hormones inhibin B and activin are formed in Sertoli and granulosa cells in an FSH-dependent manner. Inhibin B is a heterodimeric glycoprotein hormone comprising α and β subunits; the latter has two variants (β_A and β_B). Homodimers of the β-chains are called activin A ($\beta_A-\beta_A$) and activin B ($\beta_B-\beta_B$). On the one hand, inhibin B exerts a specific negative feedback inhibition on pituitary FSH secretion. Loss of inhibin B in conditions such as Sertoli-cell-only syndrome or following radiotherapy or chemotherapy augments FSH levels whilst LH remains unchanged. On the other hand, both activins A and B stimulate FSH secretion. A structurally unrelated protein termed follistatin has the capacity to suppress FSH, probably through binding and neutralizing the actions of activin. Testosterone and estradiol inhibit FSH either directly, or through GnRH suppression.

PATHOPHYSIOLOGY OF HYPOGONADOTROPIC HYPOGONADISM

Clinical Presentation of Hypogonadotropic Hypogonadism

In normal male infants, serum LH and FSH levels begin to rise from the second postnatal week, peaking at 4–10 weeks, until about 6 months of life. In normal female infants, the increase in LH and FSH (and inconstantly estradiol) is highly variable and may persist for 2–3 years. After these rises, sex hormones decline and enter a quiescent phase until the onset of puberty, when a significant intrinsic surge of gonadotropins occurs. The underlying mechanisms that retrigger this second surge remain elusive, but leptin and the kisspeptin/GPR54 system are thought to serve as key factors in this process. Male IHH caused by GnRH deficiency is associated with full sexual differentiation at birth, since early testosterone production (12–20 weeks of gestation) is stimulated by placental hCG. The lack of a GnRH (and therefore gonadotropin) surge at the late fetal/early neonatal period results in aberrant Leydig and Sertoli cell proliferation, testicular growth and scrotal descent in males. When GnRH release is absent or diminished at puberty, impaired secondary sexual development and failed reproductive maturity occur in both sexes. The clinical manifestations therefore depend on the stage of development at which the deficiency in the reproductive axis occurs.

Fetal Stage

Androgen deficiency during sexual differentiation (ninth to fourteenth weeks of gestation) leads to a disorder of sexual development with insufficient masculinization of male external genitalia, ranging from a completely female phenotype to external male genitalia with distal hypospadias. During fetal development, androgen deficiency also results in malpositioning of the testes and microphallus. Sex hormone deficiency does not appear to affect the fetal development of the female.

Before Puberty

Sex hormone deficiency occurring during this period leads to inadequate secondary sex maturation, including lack of virilization in males and breast development in females, with an associated syndrome of eunuchoidism with typical long limbs and increased arm span due to late epiphyseal fusion of the appendicular skeleton. The rate of long-bone growth is faster than that of the spine, resulting in a low ratio of upper body to lower body length. In males, maldevelopment of other androgen-dependent organs such the larynx (high pitched voice) occurs, there is reduced facial and pubic hair growth, and testes remain small.[21]

Adulthood

Although voice, body proportions and size of the penis are unaffected by androgen deficiency at this stage, body and facial hair growth dwindles, requiring infrequent shaving. Symptoms include decreased sexual potency, loss of libido and infertility.[22] Testicular volume may be reduced. Long-term health complications include osteoporosis due to a lack of stimulation of osteoblasts by androgens and estrogen, sarcopenia, and anemia as a result of decreased stimulation of erythropoiesis by testosterone. Quality of life is also impaired due to loss of strength from muscle atrophy and reduced vitality. Subfertility occurs due to loss of testosterone action on spermatogenesis.

Patterns of GnRH and Gonadotropin Secretion

The levels of GnRH in the peripheral circulation do not accurately reflect GnRH secretion, due to its short half-life of 2–4 minutes. Direct sampling of GnRH is not possible in the human because of its confinement within the hypophysial portal circulation. For these reasons, LH has been used as a surrogate marker of GnRH secretion. Pulsatile exposure of gonadotropes to GnRH is essential for gonadotropin secretion. Continuous exposure to GnRH attenuates gonadotropin secretion because of GnRH receptor downregulation. This is exploited in the treatment of steroid-dependent conditions such as endometriosis, prostate carcinoma and precocious puberty, where long-acting potent GnRH analogs are used. In contrast, correct pulsatile administration of GnRH induces LH and FSH release and thus normalizes reproductive function, forming the basis of one form of treatment for IHH. GnRH pulse frequency influences the pattern of LH and FSH secretion. Thus, slow pulse "oscillations" favor predominance of FSH production and release over LH, whereas at faster pulse frequency LH production and release is demonstrable.

It is well recognized that stress can have an adverse effect on reproductive function, particularly in females. Many factors are involved in this process, including corticotropin-releasing hormone (CRH), which inhibits GnRH through direct neuronal contact between the paraventricular nucleus and preoptic region. Elevated prolactin levels in response to stress further suppress GnRH pulsatility. Observations in a wide variety of species indicate that intermediary cytokines such as interleukin 1 (IL-1) or tumor necrosis factor-α (TNFα) may also have an inhibitory effect.[23,24]

The clinical and biochemical heterogeneity of IHH reflects the wide spectrum of abnormalities in GnRH secretion or action. A study of pulsatile gonadotropin secretion conducted in 50 men with isolated GnRH deficiency revealed several distinct subsets: the majority of patients (84%) demonstrated an apulsatile pattern, which occurred in those with the most severe form of GnRH deficiency (lacking evidence of puberty); a smaller subset of patients showed predominantly nocturnal LH pulsations, had some testicular growth and a history consistent with an arrest of puberty; and another smaller subset, despite having normal LH pulse frequency, had markedly diminished LH amplitude, suggesting either diminished hypothalamic release of GnRH or a state of GnRH resistance resulting from partial GnRHR defect.

KALLMANN SYNDROME

Historical Perspectives of Kallmann Syndrome

An adult male with a "Kallmann phenotype" (prepubertal testes and absent olfactory bulbs) was first described in the mid-nineteenth century.[25] Almost a hundred years later Kallmann identified a similar phenotype in two families with anosmia,[26] which led to the assumption of a genetic basis for this syndrome. Further associated neuropathological features were later described by de Morsier and Gauthier.[27] The discovery of GnRH in the early 1970s[28,29] revealed a unifying pathological origin of hypogonadism and anosmia in KS.[30]

Clinical Manifestations of Kallmann Syndrome

Anosmia or hyposmia is a characteristic manifestation of KS due to aplasia or hypoplasia of the olfactory bulbs and/or sulci in association with GnRH deficiency (Fig. 27.1).

Individuals may exhibit clues to the diagnosis of KS in early childhood, although some patients may not be aware of their olfactory deficiency. Subtle signs such as micropenis (stretched penile length <1.9 cm) and cryptorchidism associated with suppressed serum concentrations of gonadotropins and testosterone in the first months of life can be manifestations of an early impairment in the reproductive axis in the male infant.[31] Low levels of gonadotropins and estradiol are also seen in the female infant, but clinical signs are not apparent at this stage, only appearing when the expected sexual maturation fails to occur at pubertal age.

Adult males with KS typically present with prepubertal testicular volume (<4 cc), absence of secondary sexual characteristics, reduced muscle mass, diminished libido, erectile dysfunction and infertility. Adult females with KS usually present with poor breast development and primary amenorrhoea. In KS, the adrenals mature normally, producing low but sufficient levels of adrenal androgens to allow normal onset of pubic hair growth (adrenarche) in both sexes (Tanner Stages II–III). In the absence/diminution of gonadal sex hormones epiphyseal plates fail to fuse, resulting in a eunuchoidal habitus with arm span exceeding height by more than 5 cm. Although skeletal maturation is delayed, the rate of linear growth is usually normal (except for the absence of a distinct pubertal growth spurt).[32] Additional accessory non-reproductive manifestations in individuals with KS include bimanual synkinesia, sensorineural hearing loss (rare in the X-linked form), cleft lip and/or palate, imperfect facial fusion, seizure disorders, dentate agenesis, brachydactyly, syndactyly, pes cavus, cerebellar ataxia and nystagmus, ocular motor abnormalities, choanal atresia, and magnetic resonance imaging (MRI) evidence of agenesis of the corpus callosum, present in a minority. Upper limb mirror movements (bimanual synkinesia) occurs in 75% of patients with X-linked Kallmann syndrome, but has also been documented in autosomal dominant forms of the disorder. Renal agenesis occurs in 30% of patients with X-linked Kallmann syndrome.[33]

Incidence of Kallmann Syndrome

The estimated incidence of IHH varies widely; this is due partly to it being a rare condition, and partly to sample selection in previous surveys. The incidence of IHH was reported to be 1 in 10,000 men in a study of French conscripts.[34] Among Sardinian conscripts, testicular atrophy was found in 344 out of 600,000 men. Although not all of the affected men could be investigated to determine the cause, 7 of the 265 men examined were anosmic, leading to an extrapolated incidence of KS of 1 : 86,000.[35] A study of 24 individuals with anosmia identified one case of KS, indicating that the incidence of KS may be high among individuals with anosmia.[36] Based on a review of 250 consecutive cases seen at the Massachusetts General Hospital, IHH demonstrates a male predominance, with a male to female ratio of almost 4 : 1.[37]

Genetic Defects in Kallmann Syndrome Development

The first underlying genetic defect was identified five decades after Kallmann's initial report;[38] since then, several more genes causing KS have been identified (Table 27.1). X-linked recessive, autosomal dominant and autosomal recessive modes of inheritance are seen, but the majority of cases are sporadic (70%) or follow a non-Mendelian inheritance pattern, suggesting either that the frequency of spontaneous mutations in this disorder is high, or that the etiology of many cases may not be genetic.[39]

KAL1

The *KAL1* gene locates in the Xp22.31 region and encodes the extracellular glycoprotein anosmin 1,[40] which is believed to coordinate the early migration of GnRH neurons[39] (Fig. 27.2). Anosmin 1 shares homology with molecules involved in neural development, and contains four contiguous fibronectin type III repeats found in neural cell adhesion molecules.[41] Anosmin 1 associates with the cell membrane via heparan sulfate proteoglycans (HSPGs), and has multiple protein interaction domains suitable for a role in facilitating cell surface signaling. The causative role of the *KAL1* gene in KS first came to light from a study of a 19-week-old male KS fetus with a complete *KAL1* gene deletion from Xp22.31 to Xpter, resulting in the arrested migration of the GnRH and central projections of olfactory axons at the cribriform plate, forming a neuronal tangle.[42] This evidence suggests that mutations in *KAL1* cause premature termination of migration of both the olfactory and GnRH neurons to their final destinations, leading to anosmia and IHH. The role of anosmin 1 in olfactory bulb histogenesis is not clear, although anosmin-1 is expressed in the outer layers of the developing bulb, in association with cells expressing FGFR1, and anosmin-1 has been shown to modulate anosmin-1 signaling.

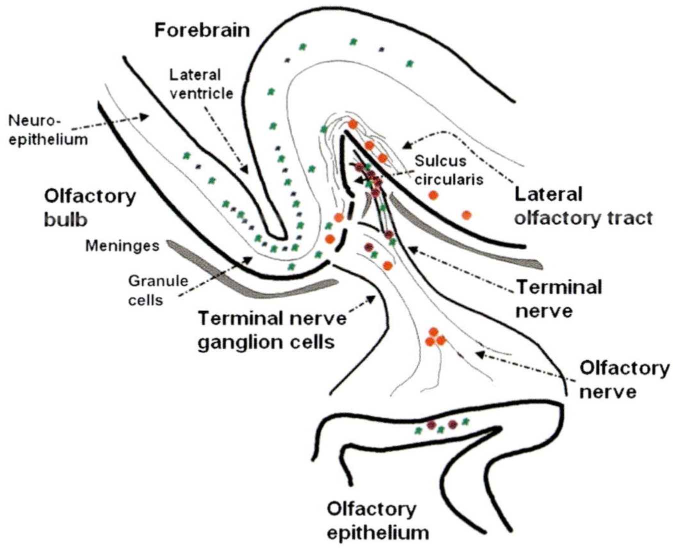

FIGURE 27.2 Schematic drawing depicting anosmin-1 (green star), FGFR1 (blue star), and GnRH (red circle) immunoreactivity in the olfactory system and rostral forebrain during human embryogenesis after 53–54 days (CS21). Red circles with blue stars at their centre represent GnRH cells that co-express FGFR1.

About 10–20% of males with KS carry *KAL1* mutations or intragenic microdeletions.[43–45] Full deletions of *KAL1* are most pathogenic, and lead to an absent protein, but are an extremely rare cause of KS. Mutations in the *KAL1* gene are distributed throughout the gene, although most point mutations cluster in the four fibronectin type III repeat domains.[46–48] The majority of X-linked KS mutations cause alteration of splicing, frameshift or stop codons leading to synthesis of truncated anosmin.[40] More rarely, missense mutations occur.[40,49]

Several lines of evidence suggest that, in familial cases, the X-linked form of KS is far less common than the autosomal form. A study of 36 familial cases with GnRH deficiency at Massachusetts General Hospital revealed only 21% to be attributable to X-linkage.[50] When the analysis was extended to include surrogate markers of IHH (isolated congenital anosmia and delayed puberty), the X-linked pedigrees comprised only 11%, whereas 25% were autosomal recessive and 64% autosomal dominant.

FGFR1 and *FGF8*

Evidence from a study of two patients with KS who were affected by different contiguous gene syndromes led to the discovery of *FGFR1*, a new autosomal dominant gene for KS.[51] The presence of overlapping deletions at 8p12-p11 prompted analysis of the FGFR1 in this critical interval. Subsequent determination of the nucleotide sequence of *FGFR1* in 129 unrelated individuals with KS led to the identification of heterozygous mutations in four familial and eight sporadic cases consistent with an autosomal dominant pattern of inheritance. In addition, one KS patient born to consanguineous parents who had a severe phenotype characterized by cleft palate, agenesis of the corpus callosum, unilateral hearing loss, and fusion of the fourth and fifth metacarpal bones was found to be homozygous for a missense mutation.

Around 10% of individuals with IHH have mutations in *FGFR1*. FGFR1 signaling is highly complex, involving various co-receptors and around 11 different FGF ligands. It is thought that FGFR1 signaling is facilitated by HSPG (a co-receptor) and anosmin 1.[52] In the past few years, several lines of evidence have implicated FGF8 as one of the relevant endogenous ligands.[53] Sequencing of *FGF8* revealed heterozygous loss-of-function mutations in 6 out of 461 individuals with HH. *FGFR1* mutations were found to occur in two familial KS, one in familial euosmic IHH, two in sporadic euosmic IHH, and one in the rare and milder adult-onset euosmic IHH.[54] There are no pathognomonic accessory features of *FGF8/FGFR1* loss-of-function mutations, but cleft palate occurs in about a third of patients.[51] On the other hand, the less frequently occurring cartilage abnormalities in ear or nose and some digits appear to be more specific to defects in *FGFR1*.[33] It is likely that FGFR1 plays a key role in the ontogeny of GnRH neurons. As such, alteration in factors/co-factors regulating FGFR1 signaling (either pre-receptor or post-receptor) may prove a fruitful line of investigation in unraveling other genetic causes of KS. Moreover, recent data suggest that there exists a functional interaction between KAL1 and KAL2 such that the anosmin 1 (KAL1) appears to be a modulator of FGF signaling at the FGFR1.[2]

PROKR2 and *PROK2*

The *PROK2* gene encodes prokinetecin 2, an 81-AA peptide that signals via the G-protein coupled product of the *PROKR2* gene termed prokineticin receptor 2. Initially, this ligand–receptor complex was implicated in diverse functions.[55–58] Subsequent observations from animal studies indicated the importance of this complex in normal olfactory bulb formation. The olfactory bulbs of mice with *Prok2* deficiency are markedly reduced in size, with abnormal architecture.[59] *Prokr2* knockout mice exhibit olfactory bulb hypoplasia with associated severe atrophy of the testis, ovary, uterus, vagina and mammary gland.[60] These *Prokr2* knockout mice have decreased plasma levels of testosterone and FSH, and immunohistochemical analysis revealed absent hypothalamic GnRH neurons. These observations suggest *PROKR2/PROK2* genes as likely candidates for KS. Recently, Dodé *et al.*[61] reported several DNA sequence changes in the *PROKR2* and *PROK2* genes in persons with KS. Using sequence analysis, about 5% of individuals with KS have mutations in *PROKR2* and fewer than 5% had mutations in *PROK2*, most of which were heterozygous; however, homozygous and compound heterozygous mutations were also described.

Several recent reports have shown patients with *PROKR2* or *PROK2* defects to have substantial variation in phenotypic expression, affecting 6% of KS and 3% of euosmic IHH patients.[62–64] Accessory features reported, including obesity, sleep disorder, fibrous dysplasia, synkinesia and epilepsy, are rare and not reliably pathognomonic.

NELF

The mouse nasal embryonic LHRH factor (*Nelf*) gene encodes a guidance molecule required for olfactory axonal outgrowth and GnRH neuronal migration in mice. The human ortholog, *NELF*, therefore seemed to be a likely candidate as a KS-associated locus. However, although rare sequence variants have been reported in KS,[62,65] no functional studies have been reported and there has been no definitive link to KS so far, other than as a potential contributor to a digenic cause of KS.

WDR11

Interstitial and terminal deletions of the chromosome 10q26 region have previously been associated with abnormal male genital development, including cryptorchidism, small testes, micropenis and hypogonadism,[66] and a trisomy involving 10q26 has been reported in a patient with an unbalanced chromosomal translocation.[67] Following investigation of a male KS patient with 46,XY,t(10;12)(q26.3;q13.1)dn, it has recently been reported that the gene *WDR11* (WD-repeat domain 11), located in the chromosome 10q25-26 region, is involved in the regulation of human puberty.[68] Moreover, its encoded WDR11 protein product was shown to interact with EMX1 (empty spiracles homeobox 1), a homeodomain transcription factor involved in the development of olfactory neurons. Missense mutations within the WD domains in patients with KS reduce or abolish this interaction.[68] Thus, *WDR11* may well represent a further gene.

Genotype–Phenotype Correlations in Kallmann Syndrome

The phenotype of X-linked KS produced by *KAL1* mutations appears to represent the most severe and least variable form of IHH. At presentation, males with *KAL1* mutations have complete absence of GnRH pulsations (as reflected by frequent sampling of serum LH) and low inhibin B levels, small and histologically immature testes, and higher rates of cryptorchidism and microphallus, than males with *FGFR1* mutations.[43,69,70] The penetrance for the IHH phenotype is almost complete in anosmic males with *KAL1* mutations, with only one case reported to date to be euosmic.[71] In contrast, penetrance is incomplete in individuals with non-X-linked forms of KS in that there is considerable variation in phenotypic expression within and across families.[72,73] This variability has been well established in patients harboring *FGFR1* loss-of-function mutations. Patients with non-X-linked KS may present with delayed puberty or as acquired adult-onset IHH,[37,73,74] and be associated with euosmia, hyposmia or anosmia. During familial investigations of their anosmic IHH sons, Salenave *et al.*[70] identified three men with *FGFR1* mutations but with normal pubertal development and gonadal status. This observation underlines the variable penetrance of this genetic form of KS, expressing in a broad spectrum of pubertal development, with individuals less severely affected than those with *KAL1* mutations.

There are a number of factors that may explain the disparity between genotype and phenotype commonly observed in KS, including environmental factors, some of which may be exerted by epigenetic effects on gene expression or the coexistence within families of defects in two or more different genes which interact functionally. In KS, this concept has received strong and growing support since the first report of digenicity in HH,[62] with the identification of multiple families with mutations in more than one of the KS genes. This means that many family histories of KS patients fail to conform to classical Mendelian inheritance patterns, a phenomenon now increasingly recognized in other classic "Mendelian" genetic diseases.[75]

Such genetic interactions commonly mirror functional interactions among encoded proteins, and although many questions remain to be answered about the precise biology of KS genes, well-reviewed evidence supports the contention that the gene products are involved in axonal migration and survival of GnRH neurons during their journey from the olfactory placode to the mediobasal hypothalamus, with many of them interacting physically at the cell membrane to form part of the axon guidance and survival machinery.[76]

Diagnosis of Kallmann Syndrome

The diagnosis of KS is based on clinical findings, low or normal serum concentrations of gonadotropins in the setting of inappropriately low circulating concentrations of sex steroids, normal pituitary and hypothalamus on MRI, and absence of other hypothalamic or pituitary abnormalities. Mutations within the genes *KAL1*, *FGFR1*, *FGF8*, *PROKR2*, *PROK2*, *NELF* and *WDR11* may account for about 20–25% of individuals with KS, and genes accounting for the remaining KS are unknown and unmapped. Deletion of *KAL1* can be detected by fluorescent *in situ* hybridization (FISH) or array comparative genomic hybridization (CGH).[77] Sequence analysis is commonly used to detect mutations in *FGFR1*, and several individuals with autosomal dominant KS have been reported.[51,71,78]

Mutational Analysis in Kallmann Syndrome Probands

Although most cases of KS are sporadic, it is worthwhile taking a careful family history, recognizing that there may be incomplete penetrance of the disorder within the same, and across, generations. If X-linked KS is suspected (strong penetrance, males affected only, hereditary bimanual synkinesis, unilateral renal agenesis) the 14 exons of the *KAL1* gene may be sequenced, and this will enable female carriers who are phenotypically normal to be identified, as well as male neonates in whom there is suspicion of HH (e.g., micropenis, undescended testes) or in males with delayed puberty. Carrier testing for at-risk relatives for KS requires identification of the disease-causing *KAL1* mutation in an affected family member. In childhood, formal smell-testing may be useful. Genetic counseling should be available, however. Where an autosomal dominant pattern of inheritance is suspected, or in families with both anosmic and euosmic IHH, testing of *FGFR1* is more likely to be of value. Coexisting midline lesions and dental agenesis should prompt the clinician to focus on the *FGFR1* locus. Carrier testing is relevant for an X-linked disorder, but not for an autosomal dominant disorder. Prenatal diagnosis for at-risk pregnancies requires prior identification of the disease-causing mutation in an affected family member.

Oligogenicity: A New Concept in the Genetic Basis of Isolated GnRH Deficiency

Although traditionally considered a monogenic group of disorders (X-linked, autosomal dominant and recessive) resulting in complete and permanent failure of sexual maturation due to silencing of the GnRH pulse generator,[37] recent reports have documented a number of patients who harbor pathogenic rare variants in more than one gene in the GnRH ontogenic cascade,[61,62] suggesting that the severity of the phenotype might result from the additive effects of subtle alterations in function of several loci in the GnRH neuronal developmental pathway. Thus, HH may not solely result from congenital arrest of GnRH neuronal migration, as previously demonstrated in the autopsy findings from a single fetus with a complete deletion of the X-linked KS locus,[3] but rather the phenotype may be caused by disruption at multiple points in the GnRH ontogenic cascade.

For example, Dodé *et al.*[61] have reported several DNA sequence changes in the *PROKR2* and *PROK2* genes in patients with KS. Using sequence analysis, about 5% of individuals with KS were found to have mutations in *PROKR2* and fewer than 5% had mutations in *PROK2*, most of which were heterozygous. Subsequent reports showed patients with *PROKR2* or *PROK2* defects to have substantial variation in phenotypic expression, with 6% of KS patients and also 3% of euosmic IHH patients affected in one study.[62–64] The variable expressivity and incomplete penetrance in several genes causing GnRH deficiency (including *FGFR1*,[79] *FGF8*,[53] *NELF*,[62] *GNRH1*[62] and *GNRHR*[79]) suggest that, far from being a monogenic disorder, the phenotype may result from gene–gene interactions, invoking the existence of digenic/oligogenic mechanisms.

Sykiotis *et al.*[80] examined genetic variants in the aforementioned genes in 397 well-phenotyped patients, family members, and 179 unaffected controls. A single affected allele (monoallelic) was found in 17% of patients and 10% in unaffected controls, whereas two affected alleles (biallelic) were found in 2.5% of patients but none in controls, and two different affected genes (digenic) were also found in 2.5% of patients compared to none among controls. These findings indicate that (a) oligogenicity in GnRH deficiency is more prevalent than previously thought; (b) in many cases more than one affected locus is required to cause the disorder, and (c) it is likely that other as unidentified loci contribute to disease pathogenesis in "monoallelic" patients. This new concept of oligogenic GnRH deficiency has wide implications for understanding the pathogenesis of a disorder that presents with a continuum of reproductive and non-reproductive phenotypes. It is also possible that rare genetic variants that underlie human GnRH deficiency may eventually explain the heritability of delayed puberty seen within the normal population.

OTHER DISORDERS OF ISOLATED HYPOTHALAMIC HYPOGONADISM

Monogenic/Oligogenic Mutations Associated with Euosmia

GNRHR/GNRH1

The product of the gene *GNRHR* is a G-protein coupled receptor with seven transmembrane segments termed GnRHR.[81,82] Activation of this receptor results in increased activity of phospholipase C and mobilization of intracellular calcium by means of the Gq/G11 group of G proteins. Defects in *GNRHR* emerged as the first autosomal cause of IHH.[83–88] Recently, genetic defects in the *GNRH1* gene *per se* have been reported for the first time.[89,90] Both *GNRH1* and *GNRHR* defects produce autosomal recessive euosmic IHH,[91,92] and in large series *GNRHR* mutations have been suggested to account for about 40–50% of familial autosomal recessive euosmic IHH and around 17% of sporadic euosmic IHH.[91] Recent evidence suggests that *GNRH1* mutations may also produce autosomal dominant IHH.[90]

Variant or Partial Forms of GnRH Deficiency

Although a *GNRHR* gene mutation would be expected to produce a complete functional defect, considerable variability in the clinical phenotype of *GNRHR* mutations within families has been described. Kindred with a partially inactivating *GNRHR* mutation were first described by de Roux.[83] The affected male had some testicular growth (8 cc), detectable gonadotropins, and a normal response to a single pharmacologic dose of GnRH. Since then several more mutations of this gene have been identified, including mutations in the transmembrane domains resulting in impaired GnRH binding and/or signaling.[85,86,93] These variable genotypes result in a wide phenotypic spectrum, ranging from the fertile eunuch syndrome to partial IHH[83,87,89,94,95] to the most complete form of GnRH deficiency.[87,88,93]

FERTILE EUNUCH SYNDROME

McCullagh *et al.*[96] first described a patient with eunuchoidal habitus and a lack of secondary sexual characteristics but whose testes were of normal size with preservation of spermatogenesis. This phenotype was subsequently recognized as a variant of IHH, also known as Pasqualini's syndrome. Although these individuals have reduced secretion of GnRH, there is sufficient to stimulate pituitary FSH and a minute amount of LH secretion, and thereby a rise in intratesticular testosterone, though not necessarily causing a significant rise in systemic testosterone concentrations. As a consequence, these patients are fertile but demonstrate eunuchoid body segments.

Fertile eunuch syndrome comprises a clinical picture as well as hormonal pattern similar to that of mid-pubertal normal boys. Analyses of the pulsatile patterns in two men with the fertile eunuch syndrome have shown a nocturnal rise of LH and testosterone secretion synchronous with sleep that is normally observed in mid-puberty boys.[97] The concept of partial GnRH deficiency as an underlying cause of this disorder is supported by the identification of partially inactivating mutation of the *GNRHR* gene in a patient with the fertile eunuch syndrome, who also displayed a detectable but apulsatile pattern of LH secretion.[94]

BOX 27.1

HOW DO WE KNOW ABOUT OLIGOGENICITY – A NEW CONCEPT IN THE GENETIC BASIS OF ISOLATED GNRH DEFICIENCY?

Isolated gonadotropin-releasing hormone (GnRH) deficiency is a rare genetic disorder comprising a spectrum of different reproductive and non-reproductive phenotypes ranging from classical anosmic Kallmann syndrome (KS) to congenital euosmic GnRH deficiency and adult-onset hypogonadotropic hypogonadism. The severity of the hypogonadotropic hypogonadism may also be variable. Although traditionally considered a monogenic group of disorders (X-linked, autosomal dominant and recessive), resulting in complete and permanent failure of sexual maturation due to silencing of the GnRH pulse generator,[1] recent reports have documented a number of patients who harbor pathogenic rare variants in more than one gene in the GnRH ontogenic cascade,[2–4] suggesting that the severity of the phenotype might result from the additive effects of subtle alterations in function of several loci in the GnRH neuronal developmental pathway. Thus HH may not solely result from congenital arrest of GnRH neuronal migration, as previously demonstrated in the autopsy findings from a single fetus with a complete deletion of the X-linked KS locus.[5]

For example, Dodé *et al.*[2] have reported several DNA sequence changes in the *PROKR2* and *PROK2* genes in patients with KS. Using sequence analysis, about 5% of individuals with KS were found to have mutations in *PROKR2* and fewer than 5% had mutations in *PROK2*, most of which were heterozygous. Subsequent reports showed patients with *PROKR2* or *PROK2* defects to have substantial variation in phenotypic expression,[4,6,7] affecting 6% of KS and 3% of euosmic IHH patients.[7] The variable expressivity and incomplete penetrance in several genes causing GnRH deficiency (including *FGFR1*[8] *FGF8*,[3] *NELF*,[4] *GNRH1*[8] *GNRHR*[9]) suggest that far from being a monogenic disorder, the phenotype may result from gene–gene interactions, and the existence of digenic/oligogenic mechanisms.

Sykiotis *et al.*[10] examined genetic variants in the aforementioned genes in 397 well-phenotyped patients, family members, and 179 unaffected controls. A single affected allele (monoallelic) was found in 17% of patients and 10% in unaffected controls, whereas two affected alleles (biallelic) were found in 2.5% of patients but none in controls and two different affected genes (digenic) were also found in 2.5% of patients compared to none among controls.

These findings indicate that: (a) oligogenicity in GnRH deficiency is more prevalent than previously thought; (b) in many cases more than one affected locus is required to cause the disorder, and (c) it is likely that other as unidentified loci contribute to disease pathogenesis in "monoallelic" patients. This new concept of oligogenic GnRH deficiency has wide implications for understanding the pathogenesis of a disorder that presents with a continuum of reproductive and non-reproductive phenotypes. It is also possible that rare genetic variants that underlie human GnRH deficiency may eventually explain the heritability of delayed puberty seen within the normal population.

References

1. Seminara SB, Hayes FJ, Crowley Jr WF. Gonadotropin-releasing hormone deficiency in the human (idiopathic hypogonadotropic hypogonadism and Kallmann's syndrome): pathophysiological and genetic considerations. *Endocr Rev.* 1998;19:521–539.
2. Dodé C, Teixeira L, Levilliers J, Fouveaut C, Bouchard P, Kottler ML, Lespinasse J, Lienhardt-Roussie A, Mathieu M, Moerman A, Morgan G, Murat A, Toublanc JE, Wolczynski S, Delpech M, Petit C, Young J, Hardelin J. Kallmann syndrome: mutations in the genes encoding prokineticin-2 and prokineticin receptor-2. *PLoS Genet.* 2006;20:2. e175.
3. Falardeau J, Chung WC, Beenken A, Raivio T, Plummer L, Sidis Y, Jacobson-Dickman EE, Eliseenkova AV, Ma J, Dwyer A, Quinton R, Na S, Hall JE, Huot C, Alois N, Pearce SH, Cole LW, Hughes V, Mohammadi M, Tsai P, Pitteloud N. Decreased FGF8 signaling causes deficiency of gonadotropin-releasing hormone in humans and mice. *J Clin Invest.* 2008;118:2822–2831.
4. Pitteloud N, Quinton R, Pearce S, Raivio T, Acierno J, Dwyer A, Plummer L, Hughes V, Seminara S, Cheng YZ, Li WP, Maccoll G, Eliseenkova AV, Olsen SK, Ibrahimi OA, Hayes FJ, Boepple P, Hall JE, Bouloux P, Mohammadi M, Crowley W. Digenic mutations account for variable phenotypes in idiopathic hypogonadotropic hypogonadism. *J Clin Invest.* 2007;117:457–763.
5. Schwanzel-Fukuda M, Bick D, Pfaff DW. Luteinizing hormonereleasing hormone (LHRH)-expressing cells do not migrate normally in an inherited hypogonadal (Kallmann) syndrome. *Brain Res Mol Brain Res.* 1989;6:311–326.
6. Abreu AP, Trarbach EB, de Castro M, Frade Costa EM, Versiani B, Matias Baptista MT, Garmes HM, Mendonca BB, Latronico AC. Loss-of-function mutations in the genes encoding prokineticin-2 or prokineticin receptor-2 cause autosomal recessive Kallmann syndrome. *J Clin Endocrinol Metab.* 2008;93:4113–4418.
7. Cole LW, Sidis Y, Zhang C, Quinton R, Plummer L, Pignatelli D, Hughes VA, Dwyer AA, Raivio T, Hayes FJ, Seminara SB, Huot C, Alos N, Speiser P, Takeshita A, Van Vliet G, Pearce S, Crowley Jr WF, Zhou QY, Pitteloud N. Mutations in prokineticin 2 and prokineticin receptor 2 genes in human gonadotrophin-releasing hormone deficiency: molecular genetics and clinical spectrum. *J Clin Endocrinol Metab.* 2008;93:3551–3559.

(*Continued*)

BOX 27.1 *(cont'd)*

8. Chan YM, de Guillebon A, Lang-Muritano M, Plummer L, Cerrato F, Tsiaras S, Gaspert A, Lavoie HB, Wu CH, Crowley Jr WF, Amory JK, Pitteloud N. Semina GNRH1 mutations in patients with idiopathic hypogonadotropic hypogonadism. *Proc Natl Acad Sci USA.* 2009;106: 11703–11708.
9. Pitteloud N, Acierno Jr JS, Meysing A, Eliseenkova AV, Ma J, Ibrahimi OA, Metzger DL, Hayes FJ, Dwyer AA, Hughes VA, Yialamas M, Hall JE, Grant E, Mohammadi M, Crowley Jr WF. Mutations in fibroblast growth factor receptor 1 cause both Kallmann syndrome and normosmic idiopathic hypogonadotropic hypogonadism. *Proc Natl Acad Sci USA.* 2006;103:6281–6286.
10. Sykiotis GP, Plummer L, Hughes VA, Au M, Durrani S, Nayak-Young S, Dwyer AA, Quinton R, Hall JE, Gusella JF, Seminara SB, Crowley Jr WF, Pitteloud N. Oligogenic basis of isolated gonadotropin-releasing hormone deficiency. *Proc Natl Acad Sci USA.* 2010;107:15140–15144.

ADULT-ONSET IHH

This is an acquired form of isolated GnRH deficiency[74] in which puberty occurs normally but then the patient develops progressive decrease in libido, sexual function and fertility. As with congenital GnRH deficiency, this variant has an apulsatile pattern of LH secretion and low serum testosterone levels. This disorder is likely to arise from a hypothalamic defect, since GnRH replacement restores the pituitary–gonadal axis in more than 90% of cases. Unlike patients with functional GnRH deficiency, where impaired GnRH secretion is transient and associated with stress, excessive exercise or weight loss, and is reversible when these factors are eliminated, adult onset IHH is not associated with these factors and GnRH secretion appears to be permanently impaired. The clinical presentation in adult-onset IHH and fertile eunuch syndrome is similar in that both conditions are characterized by GnRH deficiency in association with near-normal testicular size. However, fertile eunuchs are distinguished by the preservation of spermatogenesis sufficient for fertility to be achieved by testosterone or hCG therapy alone.[98,99]

DELAYED PUBERTY

The history of delayed but otherwise normal puberty is more often reported among the families of patients with IHH.[50] The incidence of delayed puberty is less than 1%[100] in the general population, compared to 12% of 106 patients with IHH whose relatives were reported to have a history of delayed puberty.[50] These data suggest that delay in initiation of puberty but subsequent normal progression through puberty may represent the mildest end of the phenotypic spectrum of IHH.

REVERSIBLE IHH

Reversible genetic hypogonadotropic hypogonadism is a recently described entity which has implications for the long-term management of patients with HH. A small subset of patients may have a reversible phenotype, and this may enable withdrawal of therapy in up to 10% of cases. As it is not possible to identify this phenotype at the time of diagnosis, male patients treated with sex-steroid replacement should be closely monitored for spontaneous testicular growth, which might indicate resumption of gonadotropin secretion. Although the mechanisms underlying reversal are unclear, prior treatment with androgens or hCG appears to be a common denominator and may be a critical requirement. Reversal has been observed in KS associated with mutations of *FGF8* and *FGFR1*.

LEPR/LEP

Homozygous gene mutations of either *LEP* or *LEPR* lead to autosomal recessive IHH and early-onset obesity secondary to a hypothalamic defect in GnRH secretion.[101] In contrast to patients with Kallmann syndrome, as well as the vast majority of IHH cases, individuals with *LEPR* mutations have central hypothyroidism as well as decreased growth hormone (GH) secretion, presumably on the basis of hypothalamic dysfunction.

GPR54 (KISS1R)/KISS1

The G-protein coupled receptor 54 (*GPR54*, also known as *KISS1R*) and its natural ligand kisspeptin (KISS1), a 54-AA neuropeptide, are encoded by *KISS1R* and *KISS1* genes, respectively.[102] KISS1 and GPR54 are expressed in the arcuate (Arc) and anteroventral periventricular (AVPV) nuclei of the forebrain and implicated in the neuroendocrine regulation of reproduction. The *KISS1* gene had been identified in a screen for putative metastasis suppressors, but it had not previously been implicated in HPG function until its discovery from two simultaneous studies in 2003.[14,15] Using linkage mapping of loss-of-function mutations in GPR54 of five siblings with familial euosmic IHH in a large consanguineous family, a novel locus for autosomal recessive IHH on 19p13 was identified.[15] Sequencing of several genes located within this region showed that all affected siblings had a homozygous 155-bp deletion in *GPR54*, while the deletion was absent or present on only 1 allele in unaffected family members.[15] The role of GPR54 as a regulator of puberty was confirmed by Seminara *et al.*[14] as a homozygous mutation (leu148ser) in the *GPR54* gene and identified in 6 of 19 offspring of a large inbred Saudi Arabian family. Subsequent analysis of 63 unrelated patients with euosmic IHH and 20 patients with KS revealed a compound heterozygous mutation in the *GPR54* gene in an African-American male.[14] The *GPR54* gene has since been redesignated as *KISS1R*.

It is unclear whether *KISS1/GPR54* represents the upregulated input to GnRH neurons which determines the timing of puberty. Studies have shown that, during the peripubertal period, the hypothalamic levels of *KISS1* and *GPR54* mRNA surge dramatically in the hypothalamus in both primates[103] and rodents,[17,104] and that exogenous kisspeptin can induce precocious sexual maturation in mice, whilst central kisspeptin can maintain normal adult HPG function after priming of prepubertal rhesus monkeys.[103] A study of healthy male volunteers has shown that kisspeptin administration results in increased levels of gonadotropins and testosterone.[105] Finally, a gain-of-function *KISS1R* mutation has been associated with precocious puberty in a single patient.[106]

The critical roles of the KISS1/GPR54 system in reproduction[90,107,108] have been demonstrated in several species of animals[103,109–111] and in humans.[112] *Gpr54* knockout mice, and compound heterozygote missense mutations in the *GPR54* gene in humans, have HH.[14,15] Furthermore multiple independent mouse knockouts of *Kiss1r* and *Kiss1* result in the euosmic IHH phenotype, confirming KISS1R/KISS1 to be robust proximal regulators of GnRH release.[90] Elevated expression of *KISS1* and increased connectivity between KISS1 and GnRH neurons has been suggested to be one of the effector mechanisms at puberty.[103,104]

The loss-of-function mutations in *GPR54* are an exceedingly rare cause of IHH and are associated with autosomal recessive euosmic IHH. Only 19 individuals from 6 unrelated families have been confirmed,[14,15,113–115] and in another large screening study, uncharacterized rare *GPR54* variants were found in $<3\%$ of patients with euosmic IHH.[116] Individuals with euosmic IHH due to *GPR54* mutations have been shown to have near-normal pulse frequency but severely reduced LH pulse amplitude.[14,115] Pregnancy has been achieved by exogenous induction of fertility in a patient with *GPR54* mutations, suggesting that *GPR54* function is not critical for placental function despite high levels of placental expression.[117] The KISS1/GPR54 system is unlikely to play a significant role in GnRH neuronal migration, since neither *Gpr54* nor *Kiss1* knockout mice have abnormal GnRH neuronal migration patterns, and appear to project normally to the median eminence.

TAC3/TACR3

The neurokinin 3 receptor (NK3R) is mainly expressed in the central nervous system and is the most selective of the tachykinin receptors, with highly preferential binding and activation by its endogenous ligand NKB (TAC3).[118,119] Missense loss-of-function mutations in *TAC3* and *TACR3*, the genes encoding NKB and NK3R, respectively, discovered recently in consanguineous kindreds of Turkish origin with nonsyndromic euosmic HH[120,121] have provided novel insights into HPG regulation. Defects in either *TACR3* or *TAC3* were found in 11 patients from 5 out of 10 families, but in none of 50 sporadic cases. Since micropenis and cryptorchidism have been found in male patients with *TACR3* mutations, intact function of NKB/NK3R complex is thought to be necessary both for normal HPG activation at puberty as well as for fetal gonadotropin secretion. The exact role of the NKB/NK3R system in the regulation of GnRH physiology and of reproduction remains to be elucidated, but may involve hypothalamic modulation of GnRH secretion, since NK3R is expressed by rodent GnRH-expressing neurons and NKB is strongly expressed by hypothalamic neurons that also express kisspeptin 1.[122,123] This suggestion is supported by a recent study by Young *et al.*[124] of three unrelated patients with the same homozygous substitution in the *TAC3* intron 3 acceptor splicing site (c.209-1G>C), and three siblings who bore a homozygous mutation in the *TACR3* intron 2 acceptor splicing site (c.738-1G>A), resulting in deleted NKB and truncated NK3R, respectively. These patients had very low LH with normal or near-normal FSH levels which responded excessively to the GnRH challenge test. Pulsatile GnRH administration to these patients normalized circulating sex steroids and LH release, and restored fertility in one subject. This study has demonstrated that the gonadotropin deficiency in these genetic forms of euosmic HH is of hypothalamic origin, and that NKB and NK3R both play a crucial role in hypothalamic GnRH release in humans.

DAX1

GnRH-producing neurons also appear to be affected by mutations and deletions of the *DAX1* gene (dosage-sensitive sex reversal, adrenal hypoplasia congenita critical region on the X-chromosome gene 1). The *DAX1* gene is located on Xp21, and is expressed in the hypothalamus, pituitary, adrenal glands, gonads and gonadal structures during fetal development. It encodes a 470-AA that belongs to the nuclear hormone receptor superfamily with a novel DNA-binding domain,[125] and regulates expression of hormones involved in sexual differentiation. Mutations in the *C*-terminal end of *DAX-1* lead to HH as well as X-linked adrenal hypoplasia congenita (AHC).[126] The AHC phenotype is also marked by clinical heterogeneity, ranging from IHH with normal adrenal function[127] to a complete syndrome of adrenal insufficiency in childhood with eventual HH at puberty.[128] The response of patients with AHC to pulsatile GnRH therapy is variable, suggesting that pathology in certain individuals may arise from either the hypothalamus or the pituitary.[85] In addition, the failure of hCG to induce normal spermatogenesis points to the presence of a concomitant testicular defect.[85,129,130]

DEFECTIVE TRANSCRIPTION FACTOR GENES OF PITUITARY DIFFERENTIATION CAUSING COMBINED PITUITARY HORMONE DEFICIENCY

PROP1 and *POU1F1*

Combined pituitary hormone deficiency has been linked with rare abnormalities in genes encoding transcription factors necessary for pituitary development. The pituitary-specific positive transcription factor-1 (*POU1F1*) and prophet of Pit-1 (*PROP1*) genes play an important role in the ontogenesis, differentiation and function of somatotropic, lactotropic, thyrotropic and gonadotropic cells. Mutations in *PROP1* appear to be the most common cause of both familial and sporadic congenital combined pituitary hormone deficiency, and follow a recessive pattern of inheritance. The hormonal phenotype involves deficiencies of gonadotropins, GH, prolactin and TSH, with consequent delayed puberty and hypogonadism.[131] Although hypogonadism is a characteristic feature of patients with *PROP1* mutations, there is variability in its clinical and hormonal expression. Most patients fail to enter puberty, and show consistently low LH and FSH responses to GnRH stimulation.[132] However, Flück and colleagues described two sibships with homozygosity for an Arg120Cys mutation in *PROP1* in which the affected children entered spontaneous puberty following GnRH initiation.[133] Similar observation was made in several children with complete loss-of-function mutations.[134]

HESX1

The *HESX1* gene, expressed in embryonic brain and pituitary, is a transcription factor and homeobox gene. *Hesx1* is a homeobox gene that is expressed in early forebrain and pituitary development. Targeted mutagenesis of *Hesx1* in mice causes variable anterior central nervous system defects and pituitary dysplasia. A homozygous R53C missense mutation within the DNA binding homodomain of *HESX1* was reported in two children with familial septo-optic dysplasia and panhypopituitarism.[135] Although these children were prepubertal, investigations suggested that gonadotropin release was impaired.

LHX3

Other transcription factors that regulate pituitary differentiation include LHX3, a homeodomain transcription factor. Targeted mutagenesis of *Lhx3*/*Lim3* in mice produces panhypopituitarism, suggesting that *LHX3* is a candidate gene for disorders of human pituitary development. This has been confirmed by Netchine *et al.*[136] who described *LHX3* mutations in four patients from two families with combined pituitary hormone deficiency, but preserved corticotrope function. In addition, limited head rotation due to cervical spine rigidity was reported. Evaluation of HPG function showed impaired gonadotropin release. One boy presented with cryptorchidism and micropenis at birth, and the three oldest patients failed to show any signs of pubertal development by 15 years of age.

CONGENITAL HYPOGONADISM ASSOCIATED WITH OTHER CENTRAL NERVOUS DISORDERS

CHARGE Syndrome (Colobomata, Heart Anomalies, Choanal Atresia, Retardation, Genital and Ear Anomalies)

CHARGE syndrome includes hyposmic IHH as part of a wider spectrum of abnormality. Although mostly sporadic, occasional autosomal dominant familial cases have been reported, but these tend to have the milder phenotype. The gene *CHD7* encodes a chromatin-remodeling factor (chromodomain helicase DNA binding protein 7), and its mutations that are responsible for 60% of cases of CHARGE syndrome have been identified.[137] The gene *CHD7* is expressed in many tissues during development, including the semicircular canals, forebrain, pituitary gland, olfactory bulbs and nerves, nasal epithelia, neural retina and optic nerve sheaths. Gonadotropin deficiency is responsible for genital hypoplasia in male patients, and for the lack of secondary sexual development in both sexes. In a study by Kim *et al.*[138] *CHD7* was screened in nearly 200 patients with KS or euosmic IHH, and heterozygous, likely loss-of-function, mutations were identified in 3 KS and 4 euosmic IHH patients. These patients may be identifiable clinically due to the occurrence of mild features of CHARGE syndrome, including particularly deafness and hypoplasia of the semicircular canals, in conjunction with HH.[139] One possibility is that CHD7 may control the transcription of one or more of the proteins involved in the regulation of the FGF—FGFR1 signaling pathway, but this remains an area of active investigation.

Prader-Willi Syndrome

Prader-Willi syndrome (PWS) results from a genetic imprinting defect caused by a seven exon gene deletion or inactivation of the paternally inherited chromosome 15 (15q11-13), whereby the maternally inherited copy is silent. PWS may also be caused by uniparental disomy from inheritance of two copies of the inactive maternal gene. The incidence is 1 in 15,000—20,000 live births. The syndrome is characterized by hypotonia at birth, hypogonadism, short stature, facial dysmorphism, learning difficulties, hyperphagia and obesity.[140] Hypogonadism in such patients is due to a disturbance in GnRH secretion resulting in undescended testes (90%) and microphallus in boys at birth, clitoral and labial hypoplasia in girls (75%), as well as failure of onset of, or progression through, puberty. A premature adrenarche is common, most likely due to increased adiposity. Boys progress into puberty spontaneously, but tend to arrest in mid-puberty with low or low-normal testosterones levels. Girls progress into puberty spontaneously, although this is frequently delayed. Breast development can be complete, although the breast is typically small or flat. Periods are absent or irregular. Central estrogen antagonists (e.g., clomiphene citrate) or aromatase inhibitors can stimulate gonadotropin secretion in affected individuals, leading to ovulation in females and spermatogenesis in males. Hyperphagia leads to childhood obesity, and up to 10% go on to develop diabetes.[141]

Laurence-Moon-Bardet-Biedl Syndrome

Laurence-Moon-Bardet-Biedl syndrome (LMBBS) is a rare autosomal-recessive heterogeneous genetic disorder (prevalence $<1:100{,}000$). At least 15 of the genes (*BBS1* to *BBS15*) responsible for LMBBS have been identified (http://www.ncbi.nlm.nih.gov/omim/209900). LMBBS is characterized by learning difficulties, retinitis pigmentosa, polydactyly, obesity, diabetes, renal abnormalities, and hypogonadism secondary to hypothalamic dysfunction. There has been longstanding uncertainty as to whether the Laurence-Moon syndrome and the Bardet-Biedl syndrome are two separate conditions. Klein and Ammann[142] suggested that the patients of Laurence and Moon represented a distinct disorder with paraplegia and without polydactyly and obesity, but a 22-year prospective cohort study of 46 patients from 26 Newfoundland families conducted by Moore *et al.*[143] found no correlation between clinical/dysmorphic features and genotype.

Gordon Holmes Spinocerebellar Ataxia Syndrome

In a report of three affected siblings with Gordon Holmes spinocerebellar ataxia syndrome, whose parents were first cousins (suggesting that the disease was inherited as an autosomal recessive trait), low serum levels of sex steroids and gonadotropins (consistent with hypogonadotropic hypogonadism), progressive ataxia, and dementia were documented. Extended treatment with physiological doses of pulsatile GnRH failed to stimulate gonadotropin response. Brain imaging revealed volume loss in the cerebellum, with extensive abnormalities in the cerebral white matter. Evidence from this unique family suggested that a common genetic mechanism is responsible for the syndrome of progressive hypogonadotropism and cerebellar ataxia, but with gonadotropes either unresponsive to GnRH or perhaps even absent.[144]

CONCLUSIONS

The dramatic increase in the discovery of genes associated with HH has contributed further insights into the pathogenesis of KS and other causes of hypothalamic hypogonadism. Recent reports of patients who harbor pathogenic mutations in more than one gene have suggested that the disorder is monogenic as well as oligogenic. Further research into how oligogenic inheritance and mechanisms of genetic interactions contribute to GnRH deficiency is needed to elucidate the pathogenesis of this group of disorders.

FUTURE STUDIES

Currently, gene defects have been identified in only 30—40% of HH cases, and more studies will be needed to uncover the remaining genetic loci whose disruption led to the KS/IHH phenotype. In this respect, given the central role of FGFR1 as a convergent pathway for several of the genes already identified, identification of genetic loci regulating *FGFR1* expression and signaling are likely to yield new candidate genes for KS/IHH.

Future investigation, including the use of both *in vivo* and *ex vivo* animal models, will be needed to target the identification of novel physiological regulators of GnRH and olfactory bulbs, neuronal differentiation, proliferation and migration, thereby generating new candidate genes for KS and IHH.

Since KS and IHH are rare conditions, genetic studies rely heavily on family size and consanguinity in order to provide substantial statistical power for linkage. Identification of chromosomal rearrangements as a basis for disease gene identification has already yielded success (e.g., *FGFR1, WDR11*), and, using such approaches, consortia with access to large cohorts of patients with KS/IHH may contribute to locating new disease-associated loci in future. Gene mutations in this disorder have been reported (more often case reports or small studies) in heterozygous, homozygous and compound heterozygous states, suggesting that they are likely to be involved both in monogenic recessive and digenic/oligogenic transmission modes, but the precise genetic architecture of oligogenic diseases remains largely obscure.

References

1. Graziadei PP, Monti Graziadei GA. Principles of organization of the vertebrate olfactory glomerulus: an hypothesis. *Neuroscience.* 1986;19:1025–1035.
2. González-Martínez D, Kim SH, Hu Y, Guimond S, Schofield J, Winyard P, Vannelli GB, Turnbull J, Bouloux P. Anosmin-1 modulates fibroblast growth factor receptor 1 signaling in human gonadotropin-releasing hormone olfactory neuroblasts through a heparan sulfate-dependent mechanism. *J Neurosci.* 2004;24:10384–10392.
3. Schwanzel-Fukuda M, Bick D, Pfaff DW. Luteinizing hormonereleasing hormone (LHRH)-expressing cells do not migrate normally in an inherited hypogonadal (Kallmann) syndrome. *Brain Res Mol Brain Res.* 1989;6:311–326.
4. Schneider JE. Energy balance and reproduction. *Physiol Behav.* 2004;81:289–317.
5. Crown A, Clifton DK, Steiner RA. Neuropeptide signaling in the integration of metabolism and reproduction. *Neuroendocrinology.* 2007;86:175–182.
6. Hill JW, Elmquist JK, Elias CF. Hypothalamic pathways linking energy balance and reproduction. *Am J Physiol Endocrinol Metab.* 2008;294:E827–E832.
7. Mircea CN, Lujan ME, Pierson RA. Metabolic fuel and clinical implications for female reproduction. *J Obstet Gynaecol Can.* 2007;29:887–902.
8. Zhang Y, Proenca R, Maffei M, Barone M, Leopold L, Friedman JM. Positional cloning of the mouse obese gene and its human homologue. *Nature.* 1994;372:425–432.
9. Baranowska B, Baranowska-Bik A, Bik W, Martynska L. The role of leptin and orexins in the dysfunction of hypothalamo-pituitary–gonadal regulation and in the mechanism of hyperactivity in patients with anorexia nervosa. *Neuro Endocrinol Lett.* 2008;29:37–40.
10. Han SK, Gottsch ML, Lee KJ, Popa SM, Smith JT, Jakawich SK, Clifton DK, Steiner RA, Herbison AE. Activation of gonadotropin-releasing hormone neurons by kisspeptin as a neuroendocrine switch for the onset of puberty. *J Neurosci.* 2005;25:11349–11356.
11. Pielecka-Fortuna J, Chu Z, Moenter SM. Kisspeptin acts directly and indirectly to increase GnRH neuron activity and its effects are modulated by estradiol. *Endocrinology.* 2008;149:1979–1986.
12. Zhang C, Roepke TA, Kelly MJ, Rønnekleiv OK. Kisspeptin depolarizes gonadotropin-releasing hormone neurons through activation of TRPC-like cationic channels. *J Neurosci.* 2008;28:4423–4434.
13. Dumalska I, Wu M, Morozova E, Liu R, van den Pol A, Alreja M. Excitatory effects of the puberty-initiating peptide kisspeptin and group I metabotropic glutamate receptor agonists differentiate two distinct subpopulations of gonadotropin-releasing hormone neurons. *J Neurosci.* 2008;28:8003–8013.
14. Seminara SB, Messager S, Chatzidaki EE, Thresher RR, Acierno Jr JS, Shagoury JK, Bo-Abbas Y, Kuohung W, Schwinof KM, Hendrick AG, Zahn D, Dixon J, Kaiser UB, Slaugenhaupt SA, Gusella JF, O'Rahilly S, Carlton MB, Crowley Jr WF, Aparicio SA, Colledge WH. The GPR54 gene as a regulator of puberty. *N Engl J Med.* 2003;349:1614–1627.
15. de Roux N, Genin E, Carel JC, Matsuda F, Chaussain JL, Milgrom E. Hypogonadotropic hypogonadism due to loss of function of the KiSS1-derived peptide receptor GPR54. *Proc Natl Acad Sci USA.* 2003;100:10972–10976.
16. Clarkson J, d'Anglemont de Tassigny X, Moreno AS, Colledge WH, Herbison AE. Kisspeptin-GPR54 signaling is essential for preovulatory gonadotropin-releasing hormone neuron activation and the luteinizing hormone surge. *J Neurosci.* 2008;28:8691–8697.
17. Castellano JM, Roa J, Luque RM, Dieguez C, Aguilar E, Pinilla L, Tena-Sempere M. KiSS-1/kisspeptins and the metabolic control of reproduction: physiologic roles and putative physiopathological implications. *Peptides.* 2009;30:139–145.
18. Morelli A, Marini M, Mancina R, Luconi M, Vignozzi L, Fibbi B, Filippi S, Pezzatini A, Forti G, Vannelli GB, Maggi M. Sex steroids and leptin regulate the "first Kiss" (KiSS 1/G-protein-coupled receptor 54 system) in human gonadotropin-releasing-hormone-secreting neuroblasts. *J Sex Med.* 2008;5:1097–1113.
19. Clarkson J, Herbison AE. Oestrogen, kisspeptin, GPR54 and the preovulatory luteinising hormone surge. *J Neuroendocrinol.* 2009;21:305–311.
20. Rance NE. Menopause and the human hypothalamus: evidence for the role of kisspeptin/neurokinin B neurons in the regulation of estrogen negative feedback. *Peptides.* 2009;30:111–122.
21. Han TS, Bouloux PM. What is the optimal therapy for young males with hypogonadotropic hypogonadism? *Clin Endocrinol (Oxf).* 2010;72:731–737.
22. Wu FC, Tajar A, Beynon JM, Pye SR, Silman AJ, Finn JD, O'Neill TW, Bartfai G, Casanueva FF, Forti G, Giwercman A, Han TS, Kula K, Lean ME, Pendleton N, Punab M, Boonen S, Vanderschueren D, Labrie F, Huhtaniemi IT. EMAS Group. Identification of late-onset hypogonadism in middle-aged and elderly men. *N Engl J Med.* 2010;363:123–135.
23. Kalra PS, Sahu A, Kalra SP. Interleukin-1 inhibits the ovarian steroid-induced luteinizing hormone surge and release of hypothalamic luteinizing hormone-releasing hormone in rats. *Endocrinology.* 1990;126:2145–2152.
24. Yoo MJ, Nishihara M, Takahashi M. Tumor necrosis factor-alpha mediates endotoxin induced suppression of gonadotropin-releasing hormone pulse generator activity in the rat. *Endocr J.* 1997;44:141–148.

25. Maestre de San Juan A. Falta total de los nervios olfactorios con anosmia en un individuo en quien existia una atrofia congenita de los testiculos y miembro viril. *Siglio Medico.* 1856;131:211.
26. Kallmann FJ, Schoenfeld WA, Barrera SE. The genetic aspects of primary eunuchoidism. *Am J Mental Defic.* 1944;48:203–236.
27. de Morsier G, Gauthier G. La dysplasie olfactogenitale. *Pathol Biol.* 1963;11:1267–1272.
28. Amoss M, Burgus R, Blackwell R, Vale W, Fellows R, Guillemin R. Purification, amino acid composition and N-terminus of the hypothalamic luteinizing hormone releasing factor (LRF) of ovine origin. *Biochem Biophys Res Commun.* 1971;44:205–210.
29. Schally AV, Arimura A, Baba Y, Nair RM, Matsuo H, Redding TW, Debeljuk L. Isolation and properties of the FSH and LH-releasing hormone. *Biochem Biophys Res Commun.* 1971;43:393–399.
30. Naftolin F, Harris GW, Bobrow M. Effect of purified luteinizing hormone releasing factor on normal and hypogonadotrophic anosmic men. *Nature.* 1971;232:496–497.
31. Grumbach MM. A window of opportunity: the diagnosis of gonadotropin deficiency in the male infant. *J Clin Endocrinol Metab.* 2005;90:3122–3127.
32. Van Dop C, Burstein S, Conte FA, Grumbach MM. Isolated gonadotropin deficiency in boys: clinical characteristics and growth. *J Pediatr.* 1987;111:684–692.
33. Tsai PS, Gill JC. Mechanisms of disease: Insights into X-linked and autosomal-dominant Kallmann syndrome. *Nat Clin Pract Endocrinol Metab.* 2006;2:160–171.
34. Fromantin M, Gineste J, Didier A, Rouvier J. Les impuberismes et les hypogonadismes a l'incorporation. Etude statistique. *Probl Actuels Endocrinol Nutr.* 1973;16:179–199.
35. Filippi G. Klinefelter's syndrome in Sardinia. Clinical report of 265 hypogonadic males detected at the time of military check-up. *Clin Genet.* 1986;30:276–284.
36. Pawlowitzki IH, Diekstall P, Schadel A, Miny P. Estimating frequency of Kallmann syndrome among hypogonadic and among anosmic patients. *Am J Med Genet.* 1987;26:473–479.
37. Seminara SB, Hayes FJ, Crowley Jr WF. Gonadotropin-releasing hormone deficiency in the human (idiopathic hypogonadotropic hypogonadism and Kallmann's syndrome): pathophysiological and genetic considerations. *Endocr Rev.* 1998;19:521–539.
38. Franco B, Guioli S, Pragliola A, Incerti B, Bardoni B, Tonlorenzi R, Carrozzo R, Maestrini E, Pieretti M, Taillon-Miller P, Brown CJ, Willard H, Lawrence C, Perisco MG, Camerino G, Ballabio A. A gene deleted in Kallmann's syndrome shares homology with neural cell adhesion and axonal path-finding molecules. *Nature.* 1991;353:529–536.
39. Cadman SM, Kim SH, Hu Y, González-Martínez D, Bouloux PM. Molecular pathogenesis of Kallmann's syndrome. *Horm Res.* 2007;67:231–242.
40. Hardelin JP, Levilliers J, Blanchard S, Carel JC, Leutenegger M, Pinard-Bertelletto JP, Bouloux P, Petit C. Heterogeneity in the mutations responsible for X chromosome linked Kallmann syndrome. *Hum Mol Genet.* 1993;2:373–377.
41. Edelman GM, Crossin KL. Cell adhesion molecules: implications for a molecular histology. *Annu Rev Biochem.* 1991;60:155–190.
42. Schwanzel-Fukuda M, Zheng LM, Bergen H, Weesner G, Pfaff DW. LHRH neurons: functions and development. *Prog Brain Res.* 1992;93:189–201.
43. Oliveira LM, Seminara SB, Beranova M, Hayes FJ, Valkenburgh SB, Schipani E, Costa EM, Latronico AC, Crowley Jr WF, Vallejo M. The importance of autosomal genes in Kallmann syndrome: genotype–phenotype correlations and neuroendocrine characteristics. *J Clin Endocrinol Metab.* 2001;86:1532–1538.
44. Albuisson J, Pêcheux C, Carel JC, Lacombe D, Leheup B, Lapuzina P, Bouchard P, Legius E, Matthijs G, Wasniewska M, Delpech M, Young J, Hardelin JP, Dodé C. Kallmann syndrome: 14 novel mutations in KAL1 and FGFR1 (KAL2). *Hum Mutat.* 2005;25:98–99.
45. Pedersen-White JR, Chorich LP, Bick DP, Sherins RJ, Layman LC. The prevalence of intragenic deletions in patients with idiopathic hypogonadotropic hypogonadism and Kallmann syndrome. *Mol Hum Reprod.* 2008;14:367–370.
46. Ballabio A, Bardoni B, Carrozzo R, Andria G, Bick D, Campbell L, Hamel B, Ferguson-Smith MA, Gimelli G, Fraccaro M. Contiguous gene syndromes due to deletions in the distal short arm of the human X chromosome. *Proc Natl Acad Sci USA.* 1989;86:10001–10005.
47. Clark SJ, Hauffa BP, Rodens KP, Styne DL, Kaplan SL, Grumbach MM. Hormone ontogeny in the ovine fetus: XIX: The effect of a potent luteinizing hormone-releasing factor agonist on gonadotropin and testosterone release in the fetus and neonate. *Pediatric Res.* 1989;25:347–352.
48. Cohen-Salmon M, Tronche F, del Castillo I, Petit C. Characterization of the promoter of the human KAL gene, responsible for the X-chromosome-linked Kallmann syndrome. *Gene.* 1995;164:235–242.
49. Georgopoulos NA, Pralong FP, Seidman CE, Seidman JG, Crowley Jr WF, Vallejo M. Genetic heterogeneity evidenced by low incidence of KAL-1 gene mutations in sporadic cases of gonadotropin-releasing hormone deficiency. *J Clin Endocrinol Metab.* 1997;82:213–217.
50. Waldstreicher J, Seminara SB, Jameson JL, Geyer A, Nachtigall LB, Boepple PA, Holmes LB, Crowley Jr WF. The genetic and clinical heterogeneity of gonadotropin-releasing hormone deficiency in the human. *J Clin Endocrinol Metab.* 1996;81:4388–4395.
51. Dodé C, Teixeira L, Levilliers J, Fouveaut C, Bouchard P, Kottler ML, Lespinasse J, Lienhardt-Roussie A, Mathieu M, Moerman A, Morgan G, Murat A, Toublanc JE, Wolczynski S, Delpech M, Petit C, Young J, Hardelin JP. Loss-of-function mutations in FGFR1 cause autosomal dominant Kallmann syndrome. *Nat Genet.* 2003;33:463–465.
52. Dodé C, Hardelin JP. Clinical genetics of Kallmann syndrome. *Ann Endocrinol (Paris).* 2010;71:149–157.
53. Falardeau J, Chung WC, Beenken A, Raivio T, Plummer L, Sidis Y, Jacobson-Dickman EE, Eliseenkova AV, Ma J, Dwyer A, Quinton R, Na S, Hall JE, Huot C, Alois N, Pearce SH, Cole LW, Hughes V, Mohammadi M, Tsai P, Pitteloud N. Decreased FGF8 signaling causes deficiency of gonadotropin-releasing hormone in humans and mice. *J Clin Invest.* 2008;118:2822–2831.
54. Xu N, Qin Y, Reindollar RH, Tho SP, McDonough PG, Layman LC. A mutation in the fibroblast growth factor receptor 1 gene causes fully penetrant normosmic isolated hypogonadotropic hypogonadism. *J Clin Endocrinol Metab.* 2007;92:1155–1158.
55. Schweitz H, Pacaud P, Diochot S, Moinier D, Lazdunski M. MIT(1), a black mamba toxin with a new and highly potent activity on intestinal contraction. *FEBS Lett.* 1999;461:183–188.
56. Cheng MY, Bullock CM, Li C, Lee AG, Bermak JC, Belluzzi J, Weaver DR, Leslie FM, Zhou QY. Prokineticin 2 transmits the behavioural circadian rhythm of the suprachiasmatic nucleus. *Nature.* 2002;417:405–410.
57. LeCouter J, Lin R, Ferrara N. Endocrine gland-derived VEGF and the emerging hypothesis of organ-specific regulation of angiogenesis. *Nat Med.* 2002;8:913–917.

58. Lecouter J, Lin R, Ferrara N. EG-VEGF: a novel mediator of endocrine-specific angiogenesis, endothelial phenotype, and function. *Ann NY Acad Sci.* 2004;1014:50–57.
59. Ng KL, Li JD, Cheng MY, Leslie FM, Lee AG, Zhou QY. Dependence of olfactory bulb neurogenesis on prokineticin 2 signaling. *Science.* 2005;308:1923–1927.
60. Matsumoto S, Yamazaki C, Masumoto KH, Nagano M, Naito M, Soga T, Hiyama H, Matsumoto M, Takasaki J, Kamohara M, Matsuo A, Ishii H, Kobori M, Katoh M, Matsushime H, Furuichi K, Shigeyoshi Y. Abnormal development of the olfactory bulb and reproductive system in mice lacking prokineticin receptor PKR2. *Proc Natl Acad Sci USA.* 2006;103:4140–4145.
61. Dodé C, Teixeira L, Levilliers J, Fouveaut C, Bouchard P, Kottler ML, Lespinasse J, Lienhardt-Roussie A, Mathieu M, Moerman A, Morgan G, Murat A, Toublanc JE, Wolczynski S, Delpech M, Petit C, Young J. Hardelin. Kallmann syndrome: mutations in the genes encoding prokineticin-2 and prokineticin receptor-2. *PLoS Genet.* 2006;2. e175.
62. Pitteloud N, Quinton R, Pearce S, Raivio T, Acierno J, Dwyer A, Plummer L, Hughes V, Seminara S, Cheng YZ, Li WP, Maccoll G, Eliseenkova AV, Olsen SK, Ibrahimi OA, Hayes FJ, Boepple P, Hall JE, Bouloux P, Mohammadi M, Crowley W. Digenic mutations account for variable phenotypes in idiopathic hypogonadotropic hypogonadism. *J Clin Invest.* 2007;117:457–463.
63. Abreu AP, Trarbach EB, de Castro M, Frade Costa EM, Versiani B, Matias Baptista MT, Garmes HM, Mendonca BB, Latronico AC. Loss-of-function mutations in the genes encoding prokineticin-2 or prokineticin receptor-2 cause autosomal recessive Kallmann syndrome. *J Clin Endocrinol Metab.* 2008;93:4113–4118.
64. Cole LW, Sidis Y, Zhang C, Quinton R, Plummer L, Pignatelli D, Hughes VA, Dwyer AA, Raivio T, Hayes FJ, Seminara SB, Huot C, Alos N, Speiser P, Takeshita A, Van Vliet G, Pearce S, Crowley Jr WF, Zhou QY, Pitteloud N. Mutations in prokineticin 2 and prokineticin receptor 2 genes in human gonadotrophin-releasing hormone deficiency: molecular genetics and clinical spectrum. *J Clin Endocrinol Metab.* 2008;93:3551–3559.
65. Miura K, Acierno Jr JS, Seminara SB. Characterization of the human nasal embryonic LHRH factor gene, NELF, and a mutation screening among 65 patients with idiopathic hypogonadotropic hypogonadism (IHH). *J Hum Genet.* 2004;49:265–268.
66. Suzuki Y, Sasagawa I, Nakada T, Onmura Y. Bilateral cryptorchidism associated with terminal deletion of 10q. *Urol Intl.* 1998;61(3):186–187.
67. Schinzel A, Lorda-Sanchez I, Binkert F, Carter NP, Bebb CE, Ferguson-Smith MA, Eiholzer U, Zachmann M, Robinson WP. Kallmann syndrome in a boy with a t(1;10) translocation detected by reverse chromosome painting. *J Med Genet.* 1995 Dec;32(12):957–961.
68. Kim HG, Ahn JW, Kurth I, Ullmann R, Kim HT, Kulharya A, Ha KS, Itokawa Y, Meliciani I, Wenzel W, Lee D, Rosenberger G, Ozata M, Bick DP, Sherins RJ, Nagase T, Tekin M, Kim SH, Kim CH, Ropers HH, Gusella JF, Kalscheuer V, Choi CY, Layman LC. WDR11, a WD protein that interacts with transcription factor EMX1, is mutated in idiopathic hypogonadotropic hypogonadism and Kallmann syndrome. *Am J Hum Genet.* 2010 Oct 8;87(4):465–479.
69. Pitteloud N, Hayes FJ, Boeppk PA, DeCruz S, Seminere SB, McLaughlin DT, Crowley Jr WF. The role of prior pubertal development, biochemical markers of testicular maturation, and genetics in deciding the phenotypeheterogeneity of idiopathic hypogonadotrpic hypogonadism. *J Clin Endocrinol Metab.* 2002;87:152–160.
70. Salenave S, Chanson P, Bry H, Pugeat M, Cabrol S, Carel JC, Murat A, Lecomte P, Brailly S, Hardelin JP, Dodé C, Young J. Kallmann's syndrome: a comparison of the reproductive phenotypes in men carrying KAL1 and FGFR1/KAL2 mutations. *J Clin Endocrinol Metab.* 2008;93:758–763.
71. Sato N, Katsumata N, Kagami M, Hasegawa T, Hori N, Kawakita S, Minowada S, Shimotsuka A, Shishiba Y, Yokozawa M, Yasuda T, Nagasaki K, Hasegawa D, Hasegawa Y, Tachibana K, Naiki Y, Horikawa R, Tanaka T, Ogata T. Clinical assessment and mutation analysis of Kallmann syndrome 1 (KAL1) and fibroblast growth factor receptor 1 (FGFR1, or KAL2) in five families and 18 sporadic patients. *J Clin Endocrinol Metab.* 2004;89:1079–1088.
72. Parenti G, Rizzolo MG, Ghezzi M, Di Maio S, Sperandeo MP, Incerti B, Franco B, Ballabio A, Andria G. Variable penetrance of hypogonadism in a sibship with Kallmann syndrome due to a deletion of the KAL gene. *Am J Med Genet.* 1995;57:476–478.
73. Quinton R, Duke VM, de Zoysa PA, Platts AD, Valentine A, Kendall B, Pickman S, Kirk JM, Besser GM, Jacobs HS, Bouloux PM. The neuroradiology of Kallmann's syndrome: a genotypic and phenotypic analysis. *J Clin Endocrinol Metab.* 1996;81:3010–3017.
74. Nachtigall LB, Boepple PA, Pralong FP, Crowley Jr WF. Adult onset idiopathic hypogonadotropic hypogonadism — a treatable form of male infertility. *N Engl J Med.* 1997;336:410–415.
75. Badano JL, Katsanis N. Beyond Mendel: an evolving view of human genetic disease transmission. *Nat Rev Genet.* 2002;3:779–789.
76. Hardelin JP, Dodé C. The complex genetics of Kallmann syndrome: KAL1, FGFR1, FGF8, PROKR2, PROK2, et al. *Sex Dev.* 2008;2:181–193.
77. Hou JW, Tsai WY, Wang TR. Detection of KAL-1 gene deletion with fluorescence *in situ* hybridization. *J Formos Med Assoc.* 1999;98:448–451.
78. Trarbach EB, Costa EM, Versiani B, de Castro M, Baptista MT, Garmes HM, de Mendonca BB, Latronico AC. Novel fibroblast growth factor receptor 1 mutations in patients with congenital hypogonadotropic hypogonadism with and without anosmia. *J Clin Endocrinol Metab.* 2006;91:4006–4012.
79. Pitteloud N, Acierno Jr JS, Meysing A, Eliseenkova AV, Ma J, Ibrahimi OA, Metzger DL, Hayes FJ, Dwyer AA, Hughes VA, Yialamas M, Hall JE, Grant E, Mohammadi M, Crowley Jr WF. Mutations in fibroblast growth factor receptor 1 cause both Kallmann syndrome and normosmic idiopathic hypogonadotropic hypogonadism. *Proc Natl Acad Sci USA.* 2006;103:6281–6286.
80. Sykiotis GP, Plummer L, Hughes VA, Au M, Durrani S, Nayak-Young S, Dwyer AA, Quinton R, Hall JE, Gusella JF, Seminara SB, Crowley Jr WF, Pitteloud N. Oligogenic basis of isolated gonadotropin-releasing hormone deficiency. *Proc Natl Acad Sci USA.* 2010;107:1514.
81. Stojilkovic SS, Reinhart J, Catt KJ. Gonadotropin-releasing hormone receptors: structure and signal transduction pathways. *Endocr Rev.* 1994;15:462–499.
82. Fan NC, Jeung E, Peng C, Olofsson JI, Krisinger J, Leung PCK. The human gonadotropin-releasing hormone (GnRH) receptor gene: cloning, genomic organization and chromosomal assignment. *Mol Cell Endocrinol.* 1994;103:R1–R6.
83. de Roux N, Young J, Misrahi M, Genet R, Chanson P, Schaison G, Milgrom E. A family with hypogonadotropic hypogonadism and mutations in the gonadotropin-releasing hormone receptor. *N Engl J Med.* 1997;337:1597–1602.
84. Layman LC, Peak DB, Xie J, Sohn SH, Reindollar RH, Gray MR. Mutation analysis of the gonadotropin releasing hormone receptor gene in idiopathic hypogonadotropic hypogonadism. *Fertil Steril.* 1997;68:1079–1085.

85. Caron P, Imbeaud S, Bennet A, Plantavid M, Camerino G, Rochiccioli P. Combined hypothalamic-pituitary-gonadal defect in a hypogonadic man with a novel mutation in the DAX-1 gene. *J Clin Endocrinol Metab.* 1999;84:3563–3569.
86. Pralong FP, Gomez F, Castillo E, Cotecchia S, Abuin L, Aubert ML, Portmann L, Gaillard RC. Complete hypogonadotropic hypogonadism associated with a novel inactivating mutation of the gonadotropin-releasing hormone receptor. *J Clin Endocrinol Metab.* 1999;84:3811–3816.
87. de Roux N, Young J, Brailly-Tabard S, Misrahi M, Milgrom E, Schaison G. The same molecular defects of the gonadotropin-releasing hormone receptor determine a variable degree of hypogonadism in affected kindred. *J Clin Endocrinol Metab.* 1999;84:567–572.
88. Seminara SB, Beranova M, Oliveira LM, Martin KA, Crowley Jr WF, Hall JE. Successful use of pulsatile gonadotropin-releasing hormone (GnRH) for ovulation induction and pregnancy in a patient with GnRH receptor mutations. *J Clin Endocrinol Metab.* 2000;85:556–562.
89. Bouligand J, Ghervan C, Tello JA, Brailly-Tabard S, Salenave S, Chanson P, Lombès M, Millar RP, Guiochon-Mantel A, Young J. Isolated familial hypogonadotropic hypogonadism and a GNRH1 mutation. *N Engl J Med.* 2009;360:2742–2748.
90. Chan YM, de Guillebon A, Lang-Muritano M, Plummer L, Cerrato F, Tsiaras S, Gaspert A, Lavoie HB, Wu CH, Crowley Jr WF, Amory JK, Pitteloud N. Semina GNRH1 mutations in patients with idiopathic hypogonadotropic hypogonadism. *Proc Natl Acad Sci USA.* 2009;106: 11703–11708.
91. Beranova M, Oliveira LM, Bédécarrats GY, Schipani E, Vallejo M, Ammini AC, Quintos JB, Hall JE, Martin KA, Hayes FJ, Pitteloud N, Kaiser UB, Crowley Jr WF, Seminara SB. Prevalence, phenotypic spectrum, and modes of inheritance of gonadotropin-releasing hormone receptor mutations in idiopathic hypogonadotropic hypogonadism. *J Clin Endocrinol Metab.* 2001;86:1580–1588.
92. de Roux N. GnRH receptor and GPR54 inactivation in isolated gonadotropic deficiency. *Best Pract Res Clin Endocrinol Metab.* 2006;20:515–528.
93. Layman LC, Cohen DP, Jin M, Xie J, Li Z, Reindollar RH, Bolbolan S, Bick DP, Sherins RR, Duck LW, Musgrove LC, Sellers JC. Neill JD. Mutations in gonadotropin-releasing hormone receptor gene cause hypogonadotropic hypogonadism. *Nat Genet.* 1998;18:14–15.
94. Pitteloud N, Boepple PA, DeCruz S, Valkenburgh SB, Crowley Jr WF, Hayes F. The fertile eunuch variant of idiopathic hypogonadotropic hypogonadism: spontaneous reversal associated with a homozygous mutation in the gonadotropin-releasing hormone receptor. *J Clin Endocrinol Metab.* 2001;86:2470–2475.
95. Lin L, Conway GS, Hill NR, Dattani MT, Hindmarsh PC, Achermann JC. A homozygous R262Q mutation in the gonadotropin-releasing hormone receptor presenting as constitutional delay of growth and puberty with subsequent borderline oligospermia. *J Clin Endocrinol Metab.* 2006;91:5117–5121.
96. McCullagh EP, Beck JC, Schaffenburg CA. A syndrome of eunuchoidism with spermatogenesis, normal urinary FSH and low or normal ICSH: ("Fertile Eunuchs"). *J Clin Endocrinol Metab.* 1953;13:489–509.
97. Boyar RM, Wu RH, Kapen S, Hellman L, Weitzman ED, Finkelstein JW. Clinical and laboratory heterogeneity in idiopathic hypogonadotropic hypogonadism. *J Clin Endocrinol Metab.* 1976;43:1268–1275.
98. Rogol AD, Mittal KK, White BJ, McGinniss MH, Lieblich JM, Rosen SW. HLA-compatible paternity in two"fertile eunuchs" with congenital hypogonadotropic hypogonadism and anosmia (the Kallmann syndrome). *J Clin Endocrinol Metab.* 1980;51:275–279.
99. Smals AG, Kloppenborg PW, van Haelst UJ, Lequin R, Benraad TJ. Fertile eunuch syndrome versus classic hypogonadotrophic hypogonadism. *Acta Endocrinol (Copenh).* 1978;87:389–399.
100. Styne DM. Puberty and its disorders in boys. *Endocrinol Metab Clin North Am.* 1991;20:43–69.
101. Clément K, Vaisse C, Lahlou N, Cabrol S, Pelloux V, Cassuto D, Gourmelen M, Dina C, Chambaz J, Lacorte JM, Basdevant A, Bougnères P, Lebouc Y, Froguel P, Guy-Grand B. A mutation in the human leptin receptor gene causes obesity and pituitary dysfunction. *Nature.* 1998;392:398–401.
102. Kotani M, Detheux M, Vandenbogaerde A, Communi D, Vanderwinden JM, Le Poul E, Brezillon S, Tyldesley R, Suarez-Huerta N, Vandeput F, Blanpain C, Schiffmann SN, Vassart G, Parmentier M. The metastasis suppressor gene KiSS-1 encodes kisspeptins, the natural ligands of the orphan G protein-coupled receptor GPR54. *J Biol Chem.* 2001;276: 34631–34636.
103. Shahab M, Mastronardi C, Seminara SB, Crowley WF, Ojeda SR, Plant TM. Increased hypothalamic GPR54 signaling: a potential mechanism for initiation of puberty in primates. *Proc Natl Acad Sci USA.* 2005;102:2129–2134.
104. Clarkson J, Herbison AE. Postnatal development of kisspeptin neurons in mouse hypothalamus; sexual dimorphism and projections to gonadotropin releasing hormone neurons. *Endocrinology.* 2006;147:5817–5825.
105. Dhillo WS, Murphy KG, Bloom SR. The neuroendocrine physiology of kisspeptin in the human. *Rev Endocr Metab Disord.* 2007;8:41–46.
106. Teles MG, Bianco SD, Brito VN, Trarbach EB, Kuohung W, Xu S, Seminara SB, Mendonca BB, Kaiser UB, Latronico AC. A GPR54-activating mutation in a patient with central precocious puberty. *N Engl J Med.* 2008;358:709–715.
107. Kauffman AS, Clifton DK, Steiner RA. Emerging ideas about kisspeptin-GPR54 signaling in the neuroendocrine regulation of reproduction. *Trends Neurosci.* 2007;30:504–511.
108. Gottsch ML, Clifton DK, Steiner RA. From KISS1 to kisspeptins: An historical perspective and suggested nomenclature. *Peptides.* 2009;30:4–9.
109. Gottsch ML, Cunningham MJ, Smith JT, Popa SM, Acohido BV, Crowley WF, Seminara S, Clifton DK, Steiner RA. A role for kisspeptins in the regulation of gonadotropin secretion in the mouse. *Endocrinology.* 2004;145:4073–4077.
110. Matsui H, Takatsu Y, Kumano S, Matsumoto H, Ohtaki T. Peripheral administration of metastin induces marked gonadotropin release and ovulation in the rat. *Biochem Biophys Res Commun.* 2004;320:383–388.
111. Messager S, Chatzidaki EE, Ma D, Hendrick AG, Zahn D, Dixon J, Thresher RR, Malinge I, Lomet D, Carlton MB, Colledge WH, Caraty A, Aparicio SA. Kisspeptin directly stimulates gonadotropin-releasing hormone release via G protein-coupled receptor 54. *Proc Natl Acad Sci USA.* 2005;102:1761–1766.
112. Dhillo WS, Chaudhri OB, Patterson M, Thompson EL, Murphy KG, Badman MK, McGowan BM, Amber V, Patel S, Ghatei MA, Bloom SR. Kisspeptin-54 stimulates the hypothalamic-pituitary gonadal axis in human males. *J Clin Endocrinol Metab.* 2005;90:6609–6615.
113. Lanfranco F, Gromoll J, von Eckardstein S, Herding EM, Nieschlag E, Simoni M. Role of sequence variations of the GnRH receptor and G protein-coupled receptor 54 gene in male idiopathic hypogonadotropic hypogonadism. *Eur J Endocrinol.* 2005;153:845–852.

114. Semple RK, Achermann JC, Ellery J, Farooqi IS, Karet FE, Stanhope RG, O'rahilly S, Aparicio SA. Two novel missense mutations in g protein-coupled receptor 54 in a patient with hypogonadotrophic hypogonadism. *J Clin Endocrinol Metab.* 2005;90:1849–1855.
115. Tenenbaum-Rakover Y, Commenges-Ducos M, Iovane A, Aumas C, Admoni O, de Roux N. Neuroendocrine phenotype analysis in five patients with isolated hypogonadotrophic hypogonadism due to a L102P inactivating mutation of GPR54. *J Clin Endocrinol Metab.* 2007;92:1137–1144.
116. Cerrato F, Shagoury J, Kralickova M, Dwyer A, Falardeau J, Ozata M, Van Vliet G, Bouloux P, Hall JE, Hayes FJ, Pitteloud N, Martin KA, Welt C, Seminara SB. Coding sequence analysis of GNRHR and GPR54 in patients with congenital and adult-onset forms of hypogonadotropic hypogonadism. *Eur J Endocrinol.* 2006;155(Suppl. 1):S3–S10.
117. Pallais JC, Bo-Abbas Y, Pitteloud N, Crowley Jr WF, Seminara SB. Neuroendocrine, gonadal, placental, and obstetric phenotypes in patients with IHH and mutations in the G-protein coupled receptor, GPR54. *Mol Cell Endocrinol.* 2006; 254-255:70–77.
118 Maggi CA, Schwartz TW. The dual nature of the tachykinin NK1 receptor. *Trends Pharmacol Sci.* 1997;18:351–355.
119. Sandoval-Guzmán T, Rance NE. Central injection of senktide, an NK3 receptor agonist, or neuropeptide Y inhibits LH secretion and induces different patterns of Fos expression in the rat hypothalamus. *Brain Res.* 2004;1026:307–312.
120. Topaloglu AK, Reimann F, Guclu M, Yalin AS, Kotan LD, Porter KM, Serin A, Mungan NO, Cook JR, Ozbek MN, Imamoglu S, Akalin NS, Yuksel B, O'Rahilly S, Semple RK. TAC3 and TACR3 mutations in familial hypogonadotropic hypogonadism reveal a key role for Neurokinin B in the central control of reproduction. *Nat Genet.* 2009;41:354–358.
121. Guran T, Tolhurst G, Bereket A, Rocha N, Porter K, Turan S, Gribble FM, Kotan LD, Akcay T, Atay Z, Canan H, Serin A, O'Rahilly S, Reimann F, Semple RK, Topaloglu AK. Hypogonadotropic hypogonadism due to a novel missense mutation in the first extracellular loop of the neurokinin B receptor. *J Clin Endocrinol Metab.* 2009;94:3633–3639.
122. Goodman RL, Lehman MN, Smith JT, Coolen LM, de Oliveira CV, Jafarzadehshirazi MR, Pereira A, Iqbal J, Caraty A, Ciofi P, Clarke IJ. Kisspeptin neurons in the arcuate nucleus of the ewe express both dynorphin A and neurokinin B. *Endocrinology.* 2007;148:5752–5760.
123. Navarro VM, Gottsch ML, Chavkin C, Okamura H, Clifton DK, Steiner RA. Regulation of gonadotropin-releasing hormone secretion by kisspeptin/dynorphin/neurokinin B neurons in the arcuate nucleus of the mouse. *J Neurosci.* 2009;29:11859–11866.
124. Young J, Bouligand J, Francou B, Raffin-Sanson ML, Gaillez S, Jeanpierre M, Grynberg M, Kamenicky P, Chanson P, Brailly-Tabard S. Guiochon-Mantel A. TAC3 and TACR3 defects cause hypothalamic congenital hypogonadotropic hypogonadism in humans. *J Clin Endocrinol Metab.* 2010;95:2287–2295.
125. Zanaria E, Muscatelli F, Bardoni B, Strom TM, Guioli S, Guo W, Lalli E, Moser C, Walker AP, McCabe ER, Meitinger T, Monaco AP, Sassone-Corsi P, Camerino G. An unusual member of the nuclear hormone receptor superfamily responsible for X-linked adrenal hypoplasia congenita. *Nature.* 1994;372:635–641.
126. Calvari V, Alpigiani MG, Poggi E, Podesta B, Camerino G, Lorini R. X-linked adrenal hypoplasia congenita and hypogonadotropic hypogonadism: report on new mutation of the DAX-1 gene in two siblings. *J Endocrinol Invest.* 2006;29:41–47.
127. Merke DP, Tajima T, Baron J, Cutler Jr GBGB. Hypogonadotropic hypogonadism in a female caused by an X-linked recessive mutation in the DAX1 gene. *N Engl J Med.* 1999;340:1248–1252.
128. Habiby RL, Boepple P, Nachtigall L, Sluss PM, Crowley Jr WF, Jameson JL. Adrenal hypoplasia congenita with hypogonadotropic hypogonadism: evidence that DAX-1 mutations lead to combined hypothalamic and pituitary defects in gonadotropin production. *J Clin Invest.* 1996;98:1055–1.
129. Seminara SB, Achermann JC, Genel M, Jameson JL, Crowley Jr WF. X-linked adrenal hypoplasia congenita: a mutation in DAX1 expands the phenotypic spectrum in males and females. *J Clin Endocrinol Metab.* 1999;84:4501–4509.
130. Tabarin A, Achermann JC, Recan D, Bex V, Bertagna X, Christin-Maitre S, Ito M, Jameson JL, Bouchard P. A novel mutation in DAX1 causes delayed-onset adrenal insufficiency and incomplete hypogonadotropic hypogonadism. *J Clin Invest.* 2000;105:321–328.
131. Pfäffle RW, Blankenstein O, Wüller S. Kentrup H. Combined pituitary hormone deficiency: role of Pit-1 and Prop-1. *Acta Paediatr Suppl.* 1999;88:33–41.
132. Parks JS, Brown MR, Hurley DL, Phelps CJ, Wajnrajch MP. Heritable disorders of pituitary development. *J Clin Endocrinol Metab.* 1999;84:4362–4370.
133. Flück C, Deladoey J, Rutishauser K, Eblé A, Marti U, Wu W, Mullis PE. Phenotypic variability in familial combined pituitary hormone deficiency caused by a PROP1 gene mutation resulting in the substitution of arg-cys at codon 120 (R120). *J Clin Endocrinol Metab.* 1998;83:3727–3734.
134. Deladoëy J, Flück C, Büyükgebiz A, Kuhlmann BV, Eblé A, Hindmarsh PC, Wu W, Mullis PE. "Hot spot" in the PROP1 gene responsible for combined pituitary hormone deficiency. *J Clin Endocr Metab.* 1999;84:1645–1650.
135. Dattani MT, Martinez-Barbera JP, Thomas PQ, Brickman JM, Gupta R, Martensson IL, Toresson H, Fox M, Wales JK, Hindmarsh PC, Krauss S, Beddington RS, Robinson IC. Mutations in the homeobox gene HESX1/Hesx1 associated with septo-optic dysplasia in human and mouse. *Nat Genet.* 1998;19:125–133.
136. Netchine I, Sobrier ML, Krude H, Schnabel D, Maghnie M, Marcos E, Duriez B, Cacheux V, Moers A, Goossens M, Gruters A, Amselem S. Mutations in LHX3 result in a new syndrome revealed by combined pituitary hormone deficiency. *Nat Genet.* 2000;25:182–186.
137. Vissers LE, van Ravenswaaij CM, Admiraal R, Hurst JA, de Vries BB, Janssen IM, van der Vliet WA, Huys EH, de Jong PJ, Hamel BC, Schoenmakers EF, Brunner HG, Veltman JA, van Kessel AG. Mutations in a new member of the chromodomain gene family cause CHARGE syndrome. *Nat Genet.* 2004;36:955–957.
138. Kim HG, Kurth I, Lan F, Meliciani I, Wenzel W, Eom SH, Kang GB, Rosenberger G, Tekin M, Ozata M, Bick DP, Sherins RJ, Walker SL, Shi Y, Gusella JF, Layman LC. Mutations in CHD7, encoding a chromatin-remodeling protein, cause idiopathic hypogonadotropic hypogonadism and Kallmann syndrome. *Am J Hum Genet.* 2008;83: 511–519.
139. Jongmans MC, van Ravenswaaij-Arts CM, Pitteloud N, Ogata T, Sato N. Claahsen-van der Grinten HL, van der Donk K, Seminara S, Bergman JE, Brunner HG, Crowley WF Jr, Hoefsloot LH. CHD7 mutations in patients initially diagnosed with Kallmann syndrome – the clinical overlap with CHARGE syndrome. *Clin Genet.* 2009;75:65–71.
140. Gunay-Aygun M, Schwartz S, Heeger S, O'Riordan MA, Cassidy SB. The changing purpose of Prader-Willi syndrome clinical diagnostic criteria and proposed revised criteria. *Pediatrics.* 2001;108:E92.

141. Nagai T, Mori M. Prader-Willi syndrome, diabetes mellitus and hypogonadism. *Biomed Pharmacother.* 1999;53:452–454.
142. Klein D, Ammann F. The syndrome of Laurence-Moon-Bardet-Biedl and allied diseases in Switzerland. Clinical, genetic and epidemiological studies. *J Neurol Sci.* 1969;9:479–513.
143. Moore SJ, Green JS, Fan Y, Bhogal AK, Dicks E, Fernandez BA, Stefanelli M, Murphy C, Cramer BC, Dean JC, Beales PL, Katsanis N, Bassett AS, Davidson WS, Parfrey. Clinical and genetic epidemiology of Bardet-Biedl syndrome in Newfoundland: a 22-year prospective, population-based, cohort study. *Am J Med Genet A.* 2005;132:352–360.
144. Seminara SB, Acierno Jr JS, Abdulwahid NA, Crowley Jr WF, Margolin DH. Hypogonadotropic hypogonadism and cerebellar ataxia: detailed phenotypic characterization of a large, extended kindred. *J Clin Endocrinol Metab.* 2002;87:1607–1612.

CHAPTER

28

Parental Behavior and the Perinatal Programming of Infant Development

Frances A. Champagne, James P. Curley

Department of Psychology, Columbia University, New York, NY, USA

OUTLINE

Summary

Parental care consists of the dynamic physiological and behavioral interactions between parents and offspring. These interactions can take on many different forms both within and across species, and vary dependent on the particular developmental needs of offspring. Studies of the neuroendocrine factors that shape the parental brain indicate activation within multiple hormonal, neuropeptide and neuroanatomical pathways. In the first part of this chapter, we will highlight specific hormone, hormone-receptor and neurobiological circuits involved in maternal and paternal behavior, with a particular emphasis on estrogen, oxytocin, prolactin and dopaminergic influences. We will also discuss evidence for genetic and environmental effects on the parental brain and behavior. In the second part of the chapter, we will illustrate the consequences of variation in prenatal and postnatal parent–offspring interactions in the context of perinatal programming hypotheses of development, and describe potential molecular mechanisms that may contribute to the long-term influence of parental care.

INTRODUCTION

Parental behaviors consist of the interactions between parents and offspring that promote the survival and development of those offspring. Each species has evolved a particular strategy to generate offspring and ensure that sufficient amounts of resources are available to allow the offspring to make a successful transition from infancy to adulthood (see Fig. 28.1). In some species this strategy involves both mothers and fathers participating in the care of infants, referred to as biparental care. This type of parental strategy is observed in many bird species and in some mammals (including humans). In some species of fish (e.g., seahorses), fathers are observed to provide the majority of care for offspring. However, in most species, and particularly in mammals, the mother is either the primary or the exclusive caregiver. Parental behaviors can be further divided into the periods in which they occur. For example, in egg-laying species, nutrients and hormones are placed in the yolk of the egg at the time of laying by the female to provide nutrients for growth, and, after laying, the eggs must usually be kept warm by the parents. In mammals, fetal development occurs within the mother, and the prenatal development of the offspring is dependent on the mother's provisioning of

Handbook of Neuroendocrinology, DOI: 10.1016/B978-0-12-375097-6.10028-9

FIGURE 28.1 **Parental care consists of a broad range of behaviors that vary from species to species to promote the survival and growth of offspring.** In birds (top and bottom left), a biparental strategy is used where both mothers and fathers provide warmth and food to chicks. In contrast, in laboratory rats and mice (top middle), males typically do not participate in the care of pups, and females nurse and lick pups to stimulate growth. In primates and humans (bottom middle and right) females are often the primary caregiver, though alloparenting by related kin or paternal care is also common. Prolonged parent–offspring contact following birth is typical in these species. In sheep, ewes are the exclusive caregivers for lambs, and form a highly selective bond with their offspring.

glucose and oxygen through her own circulatory system. After birth, many mammalian species require extended periods of contact with parents to survive – newborn infants need to be fed and kept warm.

The way in which males and females interact with infants can change very dramatically when they become parents. For example, if a newborn pup is placed in close proximity to a virgin female rat, the female is likely to avoid or behave aggressively toward the pup.[1,2] However, if that female becomes pregnant and gives birth, she will rapidly provide care for newborn pups (even if they are not her own). Parents will often invest care in offspring at a cost to their own health. Amongst penguins rearing their offspring in the Antarctic, incubating the egg and feeding the newly hatched chicks may require parents to fast for months, sacrifice a significant percentage of their own body weight, and engage in a treacherous journey to forage for food.[3]

Studies suggest that altered hormone levels may account for the behavioral and physiological changes observed in parents. Like most complex behaviors, parenting involves multiple hormones, neuropeptides, neurotransmitters and receptors acting in numerous brain regions and neural circuits to coordinate the interactions with offspring.[4,5] These neuroendocrine pathways can allow for transition of individuals from a non-parental to parental state, and also lead to variations in parental behavior. In most species, there are individual differences in the quality or quantity of parental investment in offspring. Studies of the neuroendocrine basis of these differences have contributed significantly to our understanding of the environmental, neurobiological and molecular mechanisms which influence parental behavior. In the first part of this chapter, we will discuss the biological basis of parental behavior and the diversity of mechanisms that shape the parental brain.

Parental influences on offspring are a critical feature of the process of development. In many cases, the complete withdrawal of parental care can result in offspring mortality. For example, in biparental species, removal of the father can reduce the percentage of surviving offspring by as much as 50%, particularly when the ecological demands of the environment are high.[6,7] If penguins abandon their eggs, infants will fail to survive in the harsh climate of the Antarctic.[8] In Harry Harlow's studies of rhesus macaques in the 1950–1960's, infants reared without a mother were found to survive if they were given adequate food and warmth, but displayed numerous behavioral and neuroendocrine abnormalities.[9,10] Variations in the quality of parental care can also lead to significant changes in offspring development. In birds, for example, if high levels of testosterone are placed in the yolk the chicks develop more rapidly, allowing them to compete more aggressively for nutritional resources.[11] In primates, the frequency of contact observed between mothers and infants will predict the social and exploratory behavior of those infants.[12] In humans, variation in the degree of parental bonding, measured using a questionnaire, can predict an individual's response to stress and risk of psychiatric dysfunction.[13,14] Studies of the mechanisms of parental influence suggest that there are long-term neuroendocrine consequences of variation in the quality of parent–offspring interactions. These effects can be observed on measures of metabolic activity, hypothalamic–pituitary–adrenal response to stress, and multiple indices of reproductive behavior. These findings suggest that parental influences may lead to differential susceptibility to physical and psychiatric disease, and also shape the parental behavior of offspring. In the second part of this chapter, we will discuss the consequences of parental behavior for offspring development and explore the mechanisms through which these parental effects are achieved.

PART 1: PARENTAL BEHAVIOR

Hormonal Influences on Parental Behavior

Parental behavior, like other aspects of reproduction, can be induced by changes in the level of circulating hormones. In mammals, females experience dramatic changes in estrogen, progesterone and prolactin during pregnancy that prime the maternal brain. Following fertilization and implantation of the developing embryo, the placenta develops and serves as an endocrine gland for the duration of pregnancy, producing progesterone, estrogen and placental lactogens to maintain the pregnancy and prepare mothers for the process of parturition, lactation and postnatal maternal care.[15–17] During gestation, the energy demands of the rapidly growing fetus are met through the influences of these hormones on energy metabolism, and maternal responsiveness to these hormonal cues may be important for offspring growth and survival. Fig. 28.2 illustrates the changes in progesterone, estrogen and prolactin during the 3-week gestational period of the rat (adapted from Siegel[18]). Note the sharp changes in the levels of these hormones at the time of parturition that mark the transition of females from one stage of reproduction (gestation) to the next (lactation). Overall, gestational hormones are thought to prime the maternal brain and prepare the

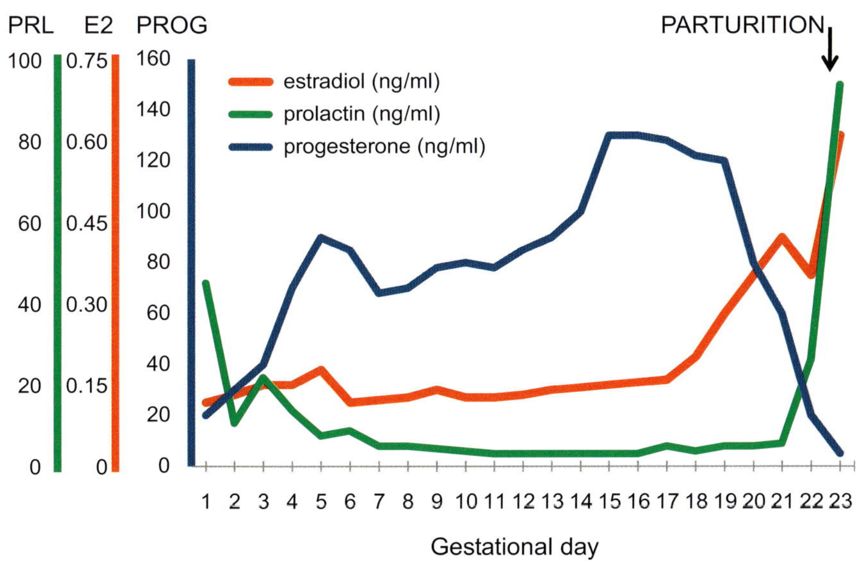

FIGURE 28.2 Schematic illustrating the changes in plasma estradiol (E2), prolactin (PRL) and progesterone (PROG) in female rats during gestation and at the onset of parturition. *Adapted from Siegel.*[18]

female for the behavioral changes needed to care for offspring.[19]

Experimental studies of the influence of hormones on maternal behavior have been conducted in a number of species. In laboratory rats, one approach that has formed the foundation for studies in maternal behavior involves determination of latency to maternal responsiveness in non-lactating females who are injected with estrogen and progesterone (see Box 28.1). These hormonal treatments can induce females to approach and care for pups rather than display the high level of timidity that is typical of inexperienced and non-lactating females.[20] Prolactin and oxytocin, hormones that are critical for the production and ejection of milk during lactation, can also increase maternal responsivity.[21] In primates and sheep, similar hormonal influences can be observed, and studies of these species suggest that stimulation of opioid receptors may also promote social interactions between mothers and infants.[22,23] Estrogen–progesterone influences on avian nest-building and incubation of eggs[24] and prolactin-induced hyperphagia and feeding of young[25,26] have also been observed in studies of female ring doves. Though these studies also highlight the influence of experience and environmental factors in the expression of maternal behavior, fluctuating hormones are a typical correlate to the initiation and, in some cases, the maintenance of mother–infant interactions.

In species where the male participates in the care of offspring, hormonal influences on the emergence of paternal care have also been observed. In most birds, increasing prolactin is associated with increased indices of paternal care, including incubation of eggs and post-hatching brooding.[25,27,28] The transition from copulatory to paternal behavior is also marked by a decrease in testosterone (see Fig. 28.3). In studies of dwarf hamsters, comparison of the change in prolactin levels in males during the postnatal period suggest that amongst biparental males prolactin levels increase, whereas in males that do not participate in the care of offspring this increase is not observed.[29] In meerkats, elevated levels of prolactin precede the onset of offspring care by males, and elevated cortisol is associated with the amount of observed pup feeding.[30] Cortisol levels were also

BOX 28.1

HOW DO WE KNOW THAT HORMONES INCREASE MATERNAL RESPONSIVITY?

The hormonal influence on maternal behavior has been studied using a variety of techniques. In the laboratory, an experimental protocol which has been used extensively to determine maternal responsivity in rodents involves monitoring the latency for females to retrieve and crouch over donor pups following various hormonal treatments. Non-treated virgin rats will initially avoid pups (3–4 days old) that are placed in their home cage. However, following constant exposure to pups, females can be induced to become maternal. The latency in days for this to occur is the dependent measure, and group differences are often expressed as the percentage of females displaying maternal behavior on each day of testing. Treatment with estradiol (E2) combined with progesterone (P) significantly reduces the latency to display maternal care, and combined E2-prolactin (PRL) or E2-oxytocin (OXY) can induce immediate maternal responsivity following exposure to pups. Tests of latency to engage in maternal behavior – whether it is over a period of minutes or days – have become the standard protocol for assessing genetic, neuroanatomical and pharmacological influences on maternal behavior in laboratory rats and mice

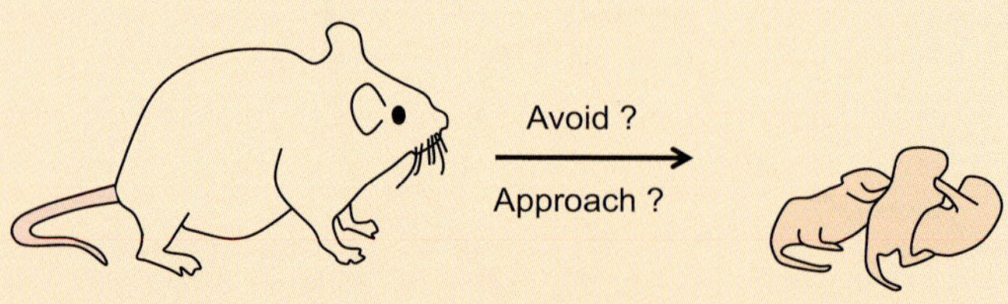

Group	Treatment	Latency (days)
HO	hysterectomized–ovariectomized	5–7
E2-P	HO + 100 μg/kg estradiol + 0.5 mg progesterone	3–4
E2-P-PRL	HO + E2-P + prolactin	<2
E2-P-OXY	HO + E2-P + oxytocin	<2

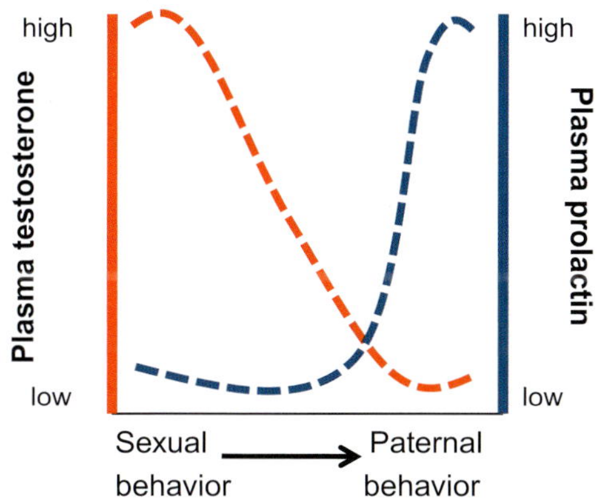

FIGURE 28.3 **Illustration of the combined effects of plasma testosterone and prolactin on paternal behavior.** During periods of heightened sexual behavior, testosterone levels are elevated and may stay elevated to promote territorial aggression and nest defence. However, the onset of nurturing behaviour toward offspring is associated with reduced testosterone and increased prolactin.

observed to increase in response to pup begging calls. Though studies of the role of prolactin in primate paternal behavior have been less supportive of the role of this hormone in the care of infants,[31,32] in human fathers a more positive response to infant cries is associated with elevated levels of prolactin.[33]

Interplay Between Hormones and Receptors in the Parental Brain

Correlations between circulating hormones and behavioral changes can be informative regarding the neuroendocrine pathways that are critical for the onset or maintenance of parental behavior. However, an important link between these endocrine factors and neurobiological function is the hormonal interactions with steroid and neuropeptide receptors in the brain. Elevated levels of circulating estrogen observed late in the gestation of mammals induce an upregulation of gene expression in a broad range of protein/receptor systems, with particular effects on oxytocin[34–36] and prolactin receptors.[37] Estrogen-induced upregulation of oxytocin receptors in the uterus and mammary gland is a critical feature of oxytocin-induced contraction of smooth muscles resulting in the uterine contractions associated with childbirth and milk let-down during lactation.[38] The oxytocin system exhibits an amazing degree of plasticity in response to hormones and experience. In addition to the changes in oxytocin receptor density that are observed in the periphery, postpartum females have elevated oxytocin receptors in several brain regions[39,40] which may regulate maternal behavior (discussed in more detail in the next section). Likewise, prolactin receptors increase within the mammary gland and hypothalamus during pregnancy and the postpartum period, associated with increased sensitivity to prolactin in these tissues.[41] Inhibiting the prolactin receptor can induce increased latencies to engage in maternal behavior by postpartum females (see Box 28.2).

One of the pathways through which estrogen can alter the level of oxytocin receptors (and other genes relevant for maternal behavior) involves interaction with nuclear estrogen receptors. Estrogen interactions with the hormone binding domain of the estrogen receptor result in an activated transcription factor that can bind to estrogen response elements within the promoter regions of genes and induce transcription.[42] There are two isoforms of the nuclear estrogen receptor: ERα and ERβ. Both of these receptor isoforms are capable of inducing genomic effects,[42] and studies of ERα and ERβ knockout mice indicate that both isoforms are involved in estrogen-mediated upregulation of genes involved in reproduction.[43,44] In the case of ERα, deletion of this gene results in significant impairments in maternal behavior[45] and an insensitivity to estrogen-mediated upregulation of oxytocin receptors in several hypothalamic brain regions.[44] Thus, the capacity of hormones to prime the parental brain is dependent on the density of steroid receptors in particular brain regions. Sex differences in the distribution of estrogen receptors are evident,[46–48] as are effects on this distribution of reproductive state and experience,[49,50] suggesting a possible mechanism for variability in hormonally-induced parental care.

Neural Basis of Parental Behaviors

The neurobiological basis of parental behavior is a topic that could easily fill the pages of an entire textbook. Early studies of the brain regions that either promote or inhibit maternal behavior in rodents suggested the importance of olfactory input. For rodents, olfaction is the primary sensory modality with which individuals interact with their environment. This is most certainly the case with parent–offspring interactions. Rodents use olfactory cues to identify pups and to stimulate neural circuits which will ultimately lead them to approach and care for offspring.[51,52] As noted previously, virgin female rats will not display immediate care of pups and will often avoid any contact with them. However, if the olfactory bulb (OB) is removed or input into the OB is blocked, and the females are rendered anosmic, females will display an earlier onset of maternal care and will be less inhibited by the pups.[53] In contrast, amongst females who are hormonally primed to engage in maternal behavior, removal of the olfactory bulb can reduce maternal responsivity.[54] Lesioning studies also implicate projections from the olfactory bulb to the amygdala in the

BOX 28.2

HOW DO WE KNOW THAT PROLACTIN RECEPTORS ARE INVOLVED IN MATERNAL BEHAVIOR?

Molecular techniques for altering neuropeptide receptor function allow for a high degree of target specificity, and can be used to determine the behavioral and neuroendocrine consequences of a particular gene or gene product. The role of the prolactin receptor (PRL) in maternal behavior has been studied through targeting of the mRNA for this receptor using an antisense (AS) oligonucleotide. When administered, the PRL-AS forms a duplex with the PRL mRNA due to the complementary nucleotide base pairs in these two molecules. This duplex prevents the mRNA from being translated into the receptor protein, thus reducing the number of prolactin receptors. The effects of the AS can be compared to a "mixed-base" (MB) oligonucleotide which consists of molecules containing sequences of base pairs that do not match the PRL. When AS and MB are injected into the lateral ventricle of female rats, there is a resulting decrease in maternal behavior in AS- but not MB-treated females. Lactating females treated with the PRL-AS display longer latencies to carry pups following a separation from their litter.[1]

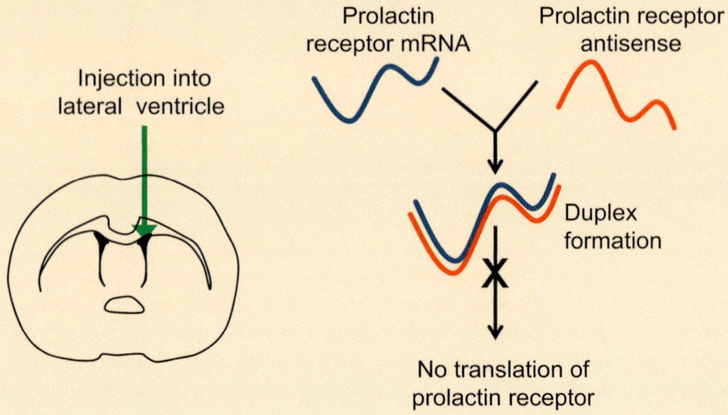

Reference

1. Torner L, Toschi N, Nava G, Clapp C, Neumann ID. Increased hypothalamic expression of prolactin in lactation: involvement in behavioural and neuroendocrine stress responses. *Eur J Neurosci* 2002;15:1381–1389.

inhibition of maternal care in virgin females.[55,56] However, regardless of the state of these inputs, damage to the medial preoptic area (MPOA) of the hypothalamus will inhibit maternal behavior.[57–59] The MPOA is a region of heightened hormonal sensitivity, due to the density of estrogen receptors, and elevated levels of oxytocin receptors can be observed in this region following hormonal priming.[60] Studies using the immediate early gene *c-fos* to determine activity in specific brain regions correlated with maternal behavior toward pups have consistently implicated the MPOA as a site of high activation[61,62] (see Fig. 28.4). These studies have also identified other sites that are activated during interactions with pups, including the nucleus accumbens (NAc), lateral septum (LS), bed nucleus of the stria terminalis (BNST) and somatosensory cortex.[63,64] Across-species comparisons provide further support for the influence of the MPOA on maternal behavior. In sheep, inactivation of the MPOA with the anesthetic lidocaine results in impaired maternal responsivity in ewes, particularly amongst primiparous females.[65] In ring doves, contact with young induces elevated c-fos in the preoptic area and lateral hypothalamus.[66] Although there are certainly many brain regions that become activated during parent–offspring interactions, the MPOA emerges as a critical site in the neurobiological basis of parental behavior.

Parental behavior involves pup-directed behaviors and responsiveness to the solicitation of care-giving by offspring cues. Thus, parents are motivated to interact with their offspring. Studies of the neurobiological basis

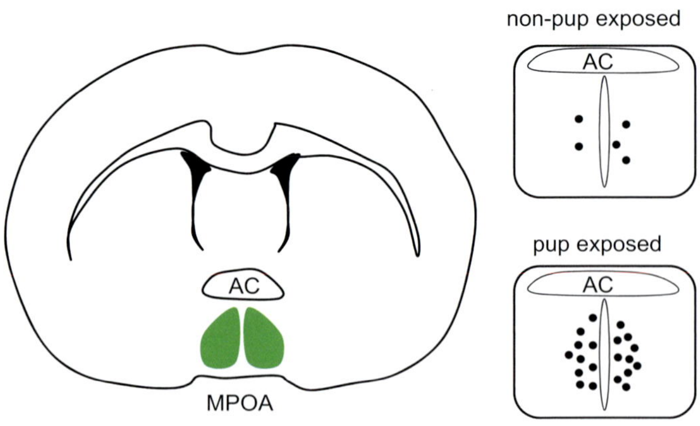

FIGURE 28.4 Illustration of a coronal section through the rat brain at the level of the medial preoptic area (MPOA), and c-fos activation of this region in response to mother–pup interactions. Lactating female rats have elevated c-fos (black dots in right panels) in the MPOA following exposure to pups. AC, anterior commissure.

of motivated behavior have revealed a significant role for activity within the mesolimbic dopaminergic pathways projecting from the ventral tegmental area (VTA) to the NAc.[67,68] Dopamine release by axon terminals projecting to the NAc is observed under conditions of arousal, in anticipation of food rewards, and in mothers when they are interacting with their offspring. Mother–pup interactions in the rat are associated with an increase in the number of c-fos-positive cells in the VTA and NAc[64,69,] and dopamine-depleting 6-hydroxydopamine (6-OHDA) or electrolytic lesions of the NAc or the dopamine cell bodies in the VTA significantly disrupt maternal behavior.[70–73] Infusion of dopamine D1/D2 receptor antagonists into the NAc also reduces pup retrieval and licking/grooming.[74,75] In microdialysis studies, presentation of pups to a lactating dam increases extracellular dopamine levels.[76] The mesolimbic dopamine system is a potential downstream target for oxytocinergic regulation, and oxytocin neurons within the MPOA appear to project directly to the VTA.[77] Severing the MPOA connections to the VTA can impair maternal behavior,[72] as can administration of an oxytocin receptor antagonist into the VTA.[78] Interestingly, postpartum females will lever-press to obtain access to pups, and this response is decreased significantly with lesions to the MPOA [79] (see Fig. 28.5). These studies clearly implicate the mesolimbic dopamine system in the expression of maternal behavior, and highlight the importance of the connectivity with hypothalamic regions in shaping mother–pup interactions. Moreover, these effects can be observed across species. Amongst biparental prairie voles, administration of a dopamine receptor antagonist leads to a reduction in licking/grooming behavior and an increased latency to come into contact with pups.[80] In humans, neuroimaging studies have revealed significant brain activation within regions associated with dopamine and reward, including the VTA, striatum and prefrontal cortex, in response to visual stimuli of their infant.[81,82] This activation is particularly significant when the infant is exhibiting a happy emotional state (see Box 28.3), highlighting the importance of considering the reciprocal interactions between mothers and infants when analyzing the contribution of particular neurobiological substrates of maternal behavior.

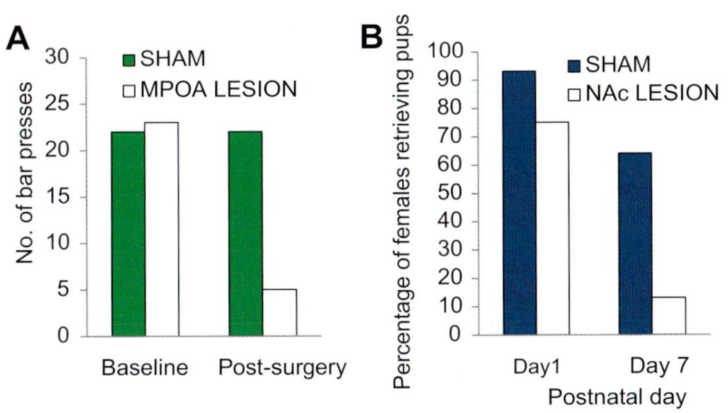

FIGURE 28.5 **Illustration of the effects of MPOA and NAc lesions on motivation to bar press to retrieve pups.** (A) MPOA lesions in multiparous cycling females reduced the number of bar presses to obtain pups observed during a 30-minute session. Each bar press was rewarded with the delivery of a 3-day old pup. (B) Lesions to the NAc were found to reduce the percentage of females observed to retrieve pups to the nest during the postpartum period.[79] SHAM surgeries in these studies involved lowering an electrode into the region but not inducing a lesion.

Studies of paternal behavior have likewise implicated the involvement of hypothalamic and dopaminergic activity. In biparental prairie voles, males who engage in paternal interactions with pups have elevated c-fos activation in several brain regions, including the LS, MPOA, BNST and amygdala.[83] Amongst biparental California mice (*P. californicus*), lesions to the MPOA, amygdala and NAc lead to reductions in paternal behavior[84,] and hormonal activation within the MPOA is associated with the onset of paternal behavior.[85] In ring doves, brooding males have higher levels of dopaminergic cells than non-brooding males.[86] In human neuroimaging studies, though sex-specific effects of brain activation in response to infant cues have been reported,[87] most fMRI studies report that parents, regardless of sex, exhibit infant-induced increases in activation of limbic and cortical brain regions.[81]

Though typical studies of the parental brain have compared non-parental to parental individuals, another approach to understanding the neurobiology of parenting is to examine the biological substrates of individual differences in parental behavior. Amongst most species, there are individual differences in the quality or quantity of observed parental behavior. In rats, variability in licking/grooming (LG) of pups during the first week postpartum is a stable and normally distributed trait.[88] Comparisons of dams who exhibit low vs high levels of LG have revealed that variations within hypothalamic regions such as the MPOA, and in mesolimbic dopamine circuits, may account for the behavioral differences between these females. During the postpartum period, lactating females who are high in LG have elevated levels of oxytocin receptors in the MPOA, LS, BNST and amygdala.[89] ERα is also elevated within the MPOA of high-LG dams, which may account for the elevated levels of oxytocin receptors within this region during the postpartum period.[90] Infusion of an oxytocin receptor antagonist into the lateral ventricle of high-LG dams leads to a significant decrease in LG.[89] Dopamine release in the NAc, as measured using *in vivo* voltammetry, is correlated with LG behavior, and females who exhibit high LG have higher peaks in dopamine release as well as prolonged levels of DA activity during bouts of maternal care[91] (see Fig. 28.6). Pharmacological manipulations that

BOX 28.3

HOW DO WE KNOW THAT INFANTS ACTIVATE REWARD PATHWAYS IN THE BRAIN OF PARENTS?

Neuroimaging studies of human parents have been used to determine the brain regions which become activated when parents are exposed to stimuli that are associated with their infant. Though many types of stimuli can be used, including infant smells and cries, a typical approach involves showing pictures or video of the infant to parents while they are in the scanner. Using functional Magnetic Resonance Imaging (fMRI), which determines blood flow to specific brain regions, this strategy has been used to assess brain activation in response to the subject's own infant *vs* an unknown infant, and also the activation in response to different emotional states of the infant. In a study of 28 first-time mothers, subjects were shown pictures of their own 5- to 10-month-old infant and a matched unknown infant displaying either a happy, neutral or sad facial expression. Pictures were shown for 2 seconds each while mothers were in the scanner. Results indicate that brain activation was higher in the VTA, striatum, and frontal lobe regions when women were shown pictures of their own infant vs an unknown infant. Interestingly, the emotion of the infant was a significant predictor of activation within dopaminergic (DA) pathways (including the substantia nigra and dorsal putamen), with happy but not neutral or sad images activating these regions.[1]

Reference

1. Strathearn L, Li J, Fonagy P, Montague PR. What's in a smile? Maternal brain responses to infant facial cues. *Pediatrics* 2008;122:40–51.

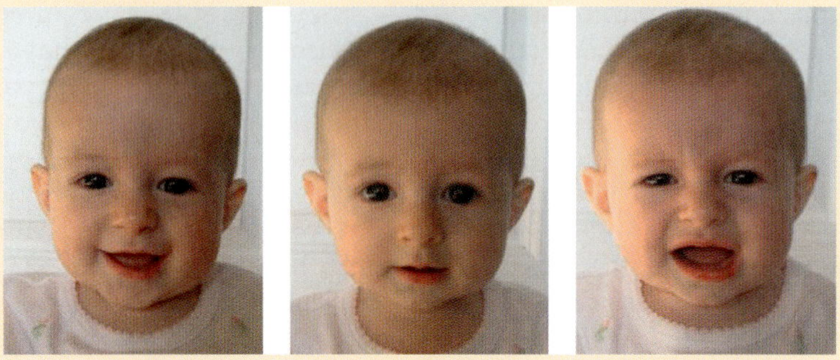

happy neutral sad

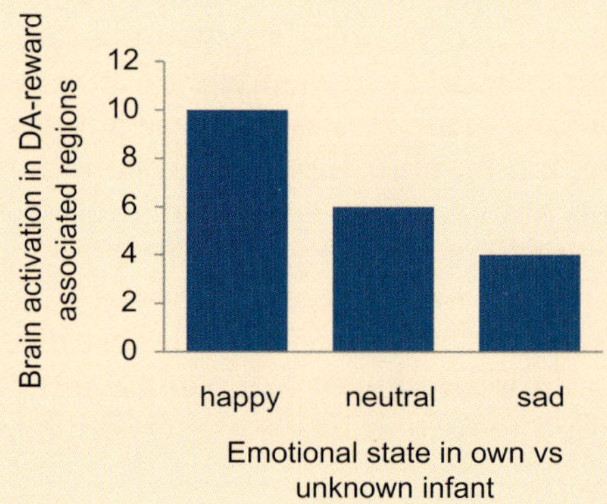

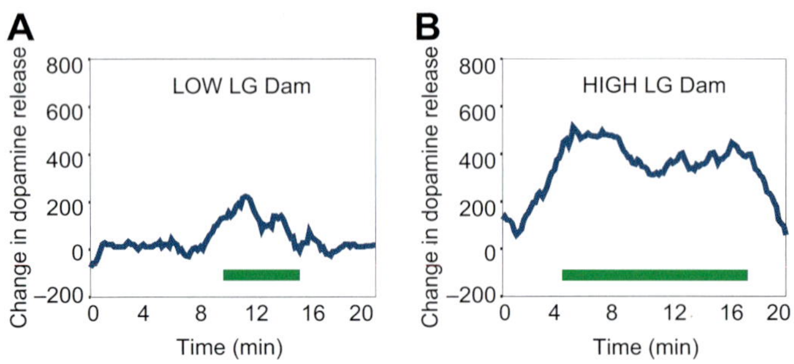

FIGURE 28.6 **Representative illustrations of the change in nM dopamine (DA) equivalent signal assessed using *in vivo* voltammetry in the NAc of (A) low and (B) high licking/grooming (LG) dams during bouts of LG.** The green bar indicates the duration of the LG bout (in minutes). Low LG is associated with reduced DA activity, and the duration of DA release is associated with the length of time the dam will spend licking pups.[91]

increase NAc dopamine activity lead to elevated LG, whereas inhibiting dopaminergic pathways leads to a decrease in LG.[91] Infusion of oxytocin into the MPOA leads to increased levels of dopamine release in the NAc, and amongst high-LG dams, infusion of an oxytocin receptor antagonist into the VTA leads to reduced NAc dopamine and reduced LG behavior.[92] These studies suggest that the mechanisms which are involved in the transition from a non-parental to a parental state may also account for individual differences amongst parents in the frequency of care-giving.

Genetic and Environmental Influences on Parental Behavior

Genetic studies of paternal behavior suggest that there are genes whose absence or variation can have a significant impact on the quality of parental care. Targeted gene knockout studies in laboratory mice have identified several genes that are critical for maternal behavior. For example, deletion of ERα,[45] fosB,[93] prolactin receptor,[94] and the paternally expressed genes Peg1[95] and Peg3[96] can alter various indices of postpartum behavior, including placentophagia, lactation/nursing, motivation to retrieve pups, nest-building, and pup LG. The functional loss of these genes is associated with changes in the developing hypothalamus which lead to decreased sensitivity to the priming effects on brain and behavior of circulating hormones during the gestational period. In male mice that do not typically exhibit paternal behavior, deletion of the progesterone receptor gene leads to reduced aggression towards pups and increased paternal care.[97] Amongst various mouse strains, variation in maternal behavior has also been noted[98,99] and is typically attributed to the genetic divergence between strains, though the exact nature of these genetic differences and how they contribute to maternal care have yet to be explored. Selective breeding studies and quantitative trait loci (QTL) analysis in rodents also suggest that genetic factors contribute to various aspects of parental behavior, including litter survival and growth, maternal aggression, and variations in postpartum care.[100] Amongst voles, comparative studies of biparental vs polygamous (non-biparental) species have indicated that genetic variations in the vasopressin 1a receptor (V1aR) gene promoter may account for the differences in social structure and parenting observed in these species.[101] In humans, polymorphisms in the oxytocin receptor (see Fig. 28.7) and serotonin transporter genes are associated with maternal sensitivity,[102] and amongst parents with genetic polymorphisms that alter dopaminergic efficiency the sensitivity of parenting was associated with the degree of daily hassles,[103] suggesting a gene–environment interaction.

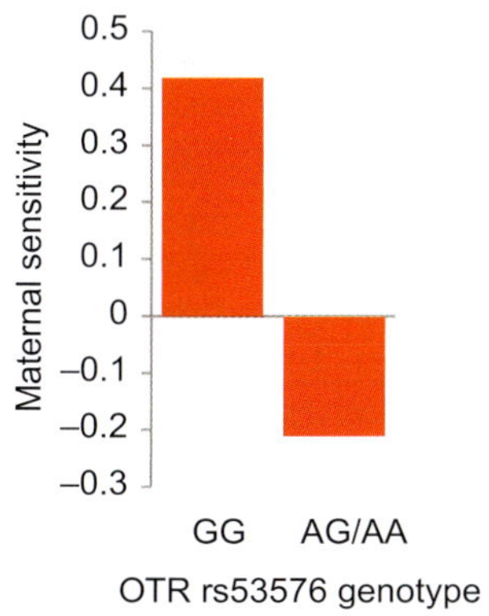

FIGURE 28.7 **Mean differences in maternal sensitivity as a function of oxytocin receptor (OTR) genotype.** A single nucleotide (A/G) polymorphism in the OTR gene (rs53576) predicts the degree of maternal sensitivity in human mothers. The AA/AG genotype is associated with decreased sensitivity compared to the GG genotype.[102]

Parental behavior is also subject to variation in response to the quality of the environment, as are the neurobiological and hormonal systems that regulate parenting. Stress can be induced in response to unpredictability in the environment, and can lead to alterations in parent–offspring interactions. In laboratory rodents, stress during gestation leads to reduced levels of postpartum mother–infant interactions,[104,105] even when females are rearing fostered non-stressed pups. Amongst females who have previously been characterized as exhibiting high LG toward their pups, the experience of chronic gestational stress suppresses the frequency of postpartum LG associated with stress-induced reductions in OTR in the MPOA and amygdala.[104] Reduction in LG persists when these females are re-mated, suggesting a long-term downregulation of the neuroendocrine circuits underlying maternal behavior. Unpredictability in the environment can also be induced experimentally by imposing a variable foraging demand (VFD). During the VFD manipulation, the duration of time that parents will require in order to find food is varied daily or weekly. In primates and rodents, VFD leads to reduced mother–infant interactions and decreased maternal sensitivity to infant cues.[106–109] Conversely, maternal behavior can be enhanced through changes in the social context of the postpartum environment. Though standard laboratory rearing of rodents consists of a single lactating dam and litter, a more frequently observed rearing strategy in the wild or under semi-naturalistic conditions, consists of multiple lactating females caring for a communal nest of pups.[110–112] Amongst lactating females who are rearing pups in a communal nest, there is an increased frequency of nursing and LG exhibited by each individual female,[113] suggesting that shifts in maternal behavior can be induced very rapidly through changes in the postpartum environment.

Environmental conditions, including photoperiod, timing of reproduction, and temperature can also lead to altered parental behavior. Amongst seasonal breeders, photoperiod is associated with fluctuations in gonadal hormones, with consequences for the transition to sexual receptivity and parenthood.[114] The temperature of the rearing environment (e.g. arctic vs temperate) will also change the duration and frequency of parent–offspring contact.[115] In birds, brooding behavior is influenced by the ambient and nest temperature, and studies of the effect of incubation temperature suggest that there is an optimal temperature for promoting growth and survival of offspring.[116] In rodents, the dynamic interactions between mothers and pups can be altered by changing the nest temperature. Dams will typically nurse pups and provide thermoregulation until the nest temperature rises beyond an optimal level, at which time females will leave the nest to allow the temperature to drop.[117] If the ambient room temperature or the temperature of the pups is increased, female nursing and crouching over the pups will be reduced.[118] These studies highlight the broader ecological variables that can influence parental behavior.

Perhaps one of the most significant influences on parental behavior and the underlying neuroendocrine circuits that influence parent–offspring interactions is the reproductive experience of mothers and fathers. Parental experience produces long-term changes in the maternal brain and behavior. Multiparous females are more sensitive to pup cues, exhibit shorter latencies to engage in maternal care, and may not be as dependent on hormones to induce maternal responsivity.[56,119] Experienced females have a heightened level of pup-induced dopamine release from the NAc,[120] and damage to this brain region amongst primiparous females can prevent the formation of this "maternal memory".[121] In sheep, damage to the MPOA amongst multiparous ewes does not impair maternal behavior as significantly as is observed in primiparous females,[65] suggesting that once this memory is formed amongst experienced females, maternal behavior becomes more liberated from neuroendocrine circuits. In rodents and sheep, parity leads to increases in ERα within the MPOA[50,122] and there are widespread changes in multiple neurotransmitter systems.[123,124] Reproductive experience can impact brain plasticity more broadly, and multiparous females are observed to have enhanced learning and memory ability.[125] Neurogenesis occurring within the postpartum brain of primiparous females may contribute to this long-term effect, and this neuronal proliferation is also observed in the paternal brain following experience with pups, associated with prolactin-induced neurogenesis.[126,127] Thus, parenthood influences the parental brain through multiple neuroanatomical and neuroendocrine pathways.

PART 2: CONSEQUENCES OF PARENTAL BEHAVIOR FOR OFFSPRING DEVELOPMENT

The goal of parental behavior is to promote the survival and development of offspring. However, there are a variety of strategies to accomplish this goal. Studies of within- and across-species variation in parental behavior suggest that, in addition to the neuroendocrine basis of this variability highlighted in the first part of this chapter, there are neuroendocrine consequences of disruptions or variation in parent–offspring interactions. These effects are typically induced early in development and persist into adulthood, and are collectively referred to as perinatal programming. In this part of the chapter we will highlight research findings that demonstrate the influence of parental effects on the perinatal origins of metabolism, stress sensitivity and reproduction, and the neuroendocrine pathways which are altered by these early life experiences. Here, we will focus primarily on studies of laboratory rodents and epidemiological evidence from humans. However, similar to the neuroendocrine regulation of parental behavior, the neuroendocrine consequences of parental care are likely to be generalizable to a broad range of species, involving common mechanisms. In the final section of this part of the chapter, we will discuss the molecular mechanisms that may be involved in the dynamic yet stable changes in offspring development that emerge in response to variation in the perinatal environment.

Parental Influence on Metabolism

In mammals, the energetic demands of the developing fetus are met through the transfer from mother to fetus of glucose and oxygen. When those energetic needs are not met, or when there are surplus levels of glucose, long-term consequences for metabolic rate and risk of cardiovascular disease, diabetes and obesity may emerge. Epidemiological studies have provided support for the fetal programming hypothesis of metabolic disorder. Longitudinal analyses of disease risk following exposure to famine during fetal development indicate elevated levels of obesity and impaired glucose tolerance amongst famine-exposed individuals, and suggest that the timing of exposure is a critical variable in predicting this effect.[128,129] The Hertfordshire studies of the long-term effects of birth weight have provided evidence that low birth weight (likely indicating prenatal growth restriction) is associated with glucose intolerance, diabetes, hypertension and cardiovascular disease.[130] These findings have contributed to the "Thrifty Phenotype" hypothesis of diabetes and other types of metabolic dysfunction (see Fig. 28.8). This hypothesis suggests than in response to maternal undernutrition, there are sequelae of changes which alter the function of insulin-secreting cells in the pancreas, leading to type 2 diabetes.[131] Since the proposal of this hypothesis it has become evident that maternal nutritional programming can alter multiple target systems, including kidney, muscle, adipose and liver.[132] These nutritionally-induced changes have been confirmed in experimental studies in animals, where maternal undernutrition during pregnancy is found to predict offspring hyperphagia, weight gain, and increased fat deposits.[133] In ewes, undernutrition of pregnant females is found to increased peroxisome proliferator-activated receptor (PPARα) mRNA in

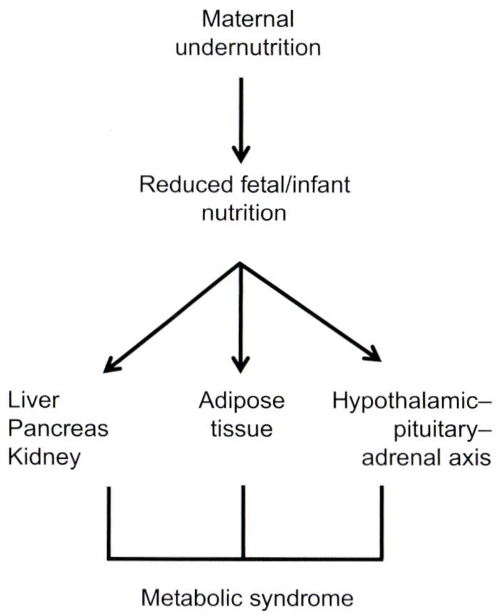

FIGURE 28.8 **Pathways linking maternal undernutrition to offspring metabolic syndrome as described within the "Thrifty Phenotype" hypothesis.** Maternal undernutrition during fetal and infant development may lead to a cascade of changes in liver, pancreas, kidney, adipose tissue and hypothalamus resulting in glucose intolerance and hypertension, leading to a metabolic syndrome associated with a heightened risk of obesity, diabetes and cardiovascular disease.[132]

offspring adipose tissue.[134] This gene regulates fatty acid oxidation and contributes to the formation of fat deposits in adults.[135] Undernutrition effects can also be observed on the hypothalamic response to stress through targeting of genes that regulate sensitivity to the effects of glucocorticoids[136] and thus contribute to the parental programming of stress responses (to be described in the next section). Overall, these studies illustrate the persistent effects of perinatal nutrition, and there is increasing support for the hypothesis that these developmental changes may be adaptive when the perinatal environment is predictive of the quality of the adult environment.[137]

Parental Programming of the Stress Response

Heightened sensitivity to stress is a risk factor common to many physical and psychiatric disorders. This sensitivity can be induced through changes in several neuroendocrine targets, including elevation in hypothalamic corticotropin-releasing hormone (CRH) and downregulation of hippocampal glucocorticoid receptors (GR). Prenatal stress and variation in the quality of postnatal mother–infant interactions have been observed to induce heightened stress responsivity through persistent changes in the hypothalamic–pituitary–adrenal (HPA) axis mediated by CRH and GR. Offspring of dams subjected to prenatal stress have elevated HPA activity, elevated anxiety-like behavior, increased CRH mRNA in the hypothalamus and amygdala, and reduced GR in the hippocampus.[138,139] These changes appear to be induced by fetal exposure to maternal glucocorticoids, as these phenotypic changes can be modified by adrenalectomizing the dam.[140] Direct administration of glucocorticoids to females during gestation can induce stress-programming effects in offspring,[141] and this may be an issue of concern when using synthetic glucocorticoids to prevent pre-term labor. The fetus is typically protected from the high levels of glucocorticoids in maternal circulation by the metabolizing enzyme 11-beta hydroxysteroid dehydrogenase-2 (11β-HSD-2), which is expressed within the placenta[142] (see Box 28.4). However, under conditions of increased prenatal stress there is increased maternal glucocorticoid production and reduced 11β-HSD-2 production, exposing the fetus to abnormally high levels of glucocorticoids.[143] Downregulation of 11β-HSD-2 is also observed under conditions of nutrient restriction,[144] and may account for the metabolic and HPA hyperactivity observed following prenatal exposure to famine. The effects of prenatal stress may also alter the level of parent–offspring interactions during the postnatal period, providing an alternative route for these programming effects. Prenatal stress can induce a decrease in pup LG by lactating dams,[104,105] and amongst cross-fostered non-stressed offspring who are reared by a stressed dam, a hyperactive phenotype can be observed.[105] The developmental consequences of prenatal stress can be reversed by adoption or neonatal handling of pups, which has been found to increase maternal behavior in both stressed and non-stressed dams.[145,146] Thus, when considering the long-term effects of prenatal adversity on the neuroendocrine and behavioral characteristics of offspring, it may be necessary to consider the interaction between prenatal and postnatal parent–infant interactions.

Epidemiological studies of the adverse consequences of childhood abuse and neglect indicate the heightened risk of disorder that emerges when there are disruptions to the postnatal environment.[147] Childhood neglect and abuse are associated with increased HPA activity and increased pituitary volume.[148–150] Retrospective reports of variations in parental care have also yielded significant negative associations with cerebrospinal levels of CRH.[151] In non-clinical subjects, low levels of maternal care are associated with increased trait anxiety and increased salivary cortisol in response to stress.[152] These findings have been supported by laboratory studies in primates and rodents. Heightened stress-induced cortisol and decreased rhythmicity in basal cortisol release are common features of rhesus monkeys reared in the absence of maternal stimulation, typically using a peer-rearing strategy.[153,154] Compared to mother-reared rhesus monkeys, peer-reared infants

BOX 28.4

HOW DO WE KNOW THAT 11-β HSD IN THE PLACENTA IS INVOLVED IN PRENATAL PROGRAMMING OF THE STRESS RESPONSE?

The 11β-hydroxysteroid dehydrogenase (11-β HSD) enzymes play a critical role in metabolizing glucocoticoids, either converting cortisone into active cortisol/corticosterone or, in the case of 11-β HSD-2, inactivating cortisol/corticosterone. 11-β HSD-2 is highly expressed in the placenta and functions to reduce the exposure of the developing fetus to the elevated plasma levels of glucocorticoids that are present in maternal circulation (particularly late in gestation). However, there are many factors which can lead to a decrease in the levels of this enzyme resulting in heightened fetal exposure to glucocorticoids and long-term increases in stress responses of offspring (through changes in hippocampal GR and hypothalamic CRH). Maternal exposure to stress or undernutrition can lead to reduced levels of 11-β HSD-2 in the placenta. This effect can also be achieved through pharmacological administration of carbenoxolone (CBX), a compound found in licorice. Epidemiological studies in Finland, where licorice consumption is relatively high, suggest that high licorice (≥500mg/day) compared to low licorice consumption by mothers during pregnancy is associated with increased salivary cortisol levels in offspring.[1] Thus, functioning of the placenta, and of the proteins and enzymes in this tissue that regulate the transfer of nutrients and hormones between the mother and fetus, is a critical feature of offspring prenatal development.

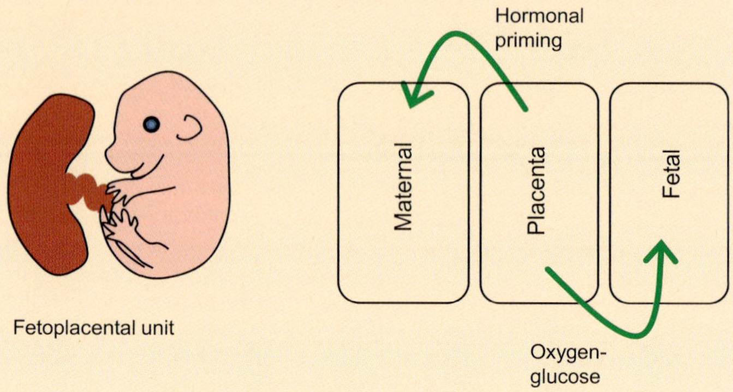

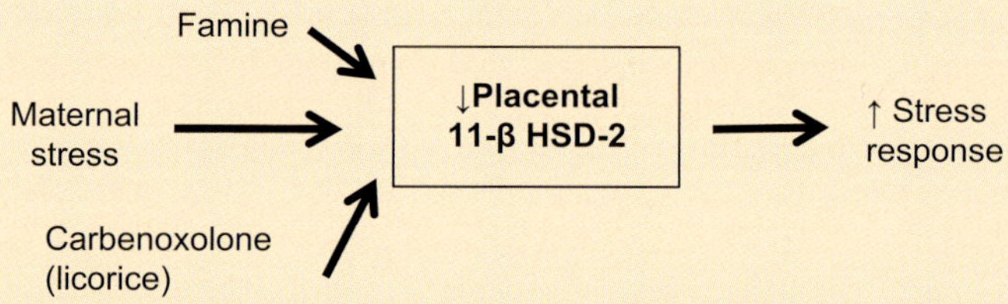

Reference

1. Raikkonen K, Seckl JR, Heinonen K, Pyhala R, Feldt K, Jones A, Pesonen AK, Phillips DI, Lahti J, Jarvenpaa AL, Eriksson JG, Matthews KA, Strandberg TE, Kajantie E. Maternal prenatal licorice consumption alters hypothalamic–pituitary–adrenocortical axis function in children. *Psychoneuroendocrinology*. 2010;35:1587–1593.

show increases in the volume of stress-sensitive brain regions, such as the dorsomedial prefrontal cortex and dorsal anterior cingulate cortex, in later life.[155] Disruptions of mother–infant interactions through use of a variable foraging demand has been demonstrated to increased levels of CRH.[156] In rodents, high levels of tactile stimulation of pups by the mother are associated with a long-term reduction in CRH-mediated HPA stress reactivity and emotionality. Adult rats reared by dams that provided high levels of LG have an attenuated stress response, reduced CRH mRNA, and higher expression of GRs in the hippocampus, compared to rats reared by dams that provide low levels of tactile stimulation during the postnatal period.[157–159] Likewise, mice and rats that undergo daily brief maternal separations during the first weeks of life have been shown to have an increased resilience to developing defensive behaviors, improved learning and memory, and an attenuated stress response.[160–162] These behavioral changes are associated with reduced expression of CRH mRNA in the hypothalamus and increased GR mRNA expression in the hippocampus,[163,164] and are induced in part through increased sensory stimulation derived from the mother on reunion with her pups.[158,165] Conversely, postnatal maternal separation induces increased stress-reactivity, anxiety, and depressive-like behaviors[161,166] associated with elevated CRH levels in the hypothalamus, and reduced GR in the hypothalamus and hippocampus.[161,167,168] Finally, another form of maternal separation, the early permanent removal of mouse pups from their mother on postnatal day 14 (early weaning), elevates HPA activity in adult males under basal and stressful conditions and induces a reduction in GR mRNA expression in the hippocampus.[169] Though plasticity in the HPA axis may continue in response to juvenile and adult environmental experiences, the quality of the perinatal environment clearly has a profound influence on the stress response that may have long-term consequences for physical and psychiatric health.

Parental Influence on the Neuroendocrine Basis of Reproductive Behavior

In addition and in relation to variations in metabolism and stress responsivity that are shaped by the quality of parent–offspring interactions, neuroendocrine targets that regulate maternal and sexual behavior have been found to be influenced by parental care. In the case of maternal behavior, this may account for the transgenerational continuity in parental styles that are observed across species.[170] Postnatal disruption to the quality of mother–infant interactions in rodents can lead to reduced maternal care amongst female offspring. Females reared under conditions of complete maternal deprivation are observed to provide reduced LG toward their own offspring,[171,172] and investigations of the mechanism of this effect suggest that artificially-reared females are less responsive to hormonally-induced increases in maternal behavior.[173] Reduced LG is also observed in females who experienced prolonged maternal separation during infancy[174] or who were reared by females engaging in low levels of LG.[88,157] The neuroendocrine basis of this maternal effect appears to involve sensitivity to hormone-induced upregulation of hypothalamic oxytocin receptors (OTR). Post-parturient offspring reared by high LG dams have elevated levels of oxytocin receptors in numerous brain regions, including the medial preoptic area (MPOA) of the hypothalamus.[89,175] Amongst ovariectomized females, the experience of high LG during postnatal development predicts increasing levels of OTR and c-fos in the MPOA in response to increasing levels of estradiol, whereas amongst offspring reared by low LG dams there is insensitivity to these estrogen-mediated effects.[89,90] Amongst adult female offspring reared by low-LG compared to high-LG dams, there are reductions in ERα gene expression within the MPOA, and these rearing effects emerge during the first week postpartum.[90,176] Cross-fostering studies have confirmed that this early-life programming of the maternal brain is associated with the quality of the postnatal rearing experience.

Interestingly, there appears to be a trade-off between maternal and sexual behavior in response to low vs high levels of parental investment. Though offspring of high-LG dams do provide high levels of maternal LG toward their own offspring, when compared to the offspring of low-LG dams, they are far less sexually receptive.[177–179] Despite exhibiting enhanced estrogen sensitivity within maternal circuits, female offspring of high-LG dams are less responsive to estrogen-induced luteinizing hormone (LH) levels, hypothalamic gonadotropin-releasing hormone (GnRH) and phosphorylated ERα within the anteroventral paraventricular nucleus of the hypothalamus (AVPVn).[177] In contrast to the MPOA, in which there are elevated levels of ERα mRNA, offspring of high-LG dams have reduced ERα mRNA in the AVPVn compared to offspring of low-LG dams. Reduced ovulation and reproductive success following paced mating have also been observed in the offspring of high-LG compared to low-LG dams.[178,179] These data suggest that the cost to enhanced maternal LG may be evident in the decreased sexual behavior of high-LG offspring, and there appears to be accelerated sexual maturation in the offspring who receive low LG. Puberty onset is earlier amongst offspring of low-LG dams.[178] Overall, these studies suggest a behavioral and neuroendocrine basis for the programming of reproductive strategies in response to parental behavior (see Fig. 28.9).

Molecular Mechanisms of Parental Influence

Throughout this chapter, there have been illustrations of the persistent effects of experience – whether it is the

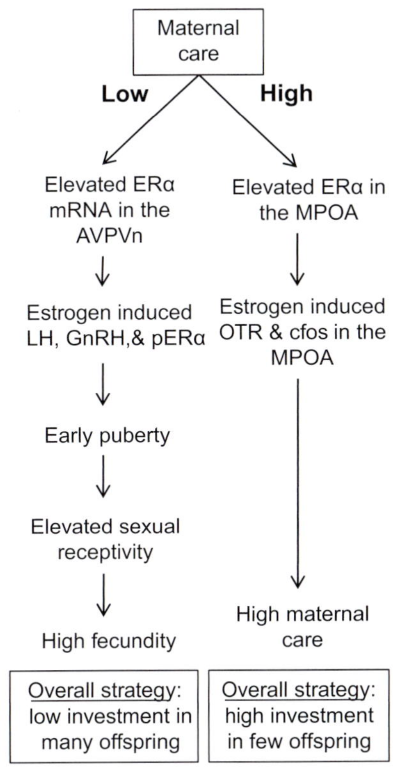

FIGURE 28.9 Illustration of the reproductive strategy that is associated with low vs high levels of maternal licking/grooming (LG). Low levels of maternal LG lead to neuroendocrine changes in female offspring that induce early puberty, high sexual receptivity, and enhanced reproductive success following mating. In contrast, in response to high levels of maternal LG, female offspring develop estrogen sensitivity within maternal circuits and inherit the high LG phenotype.

environmental- or parity-induced changes in the parental brain or the changes induced in offspring as a function of parental behavior. A question raised by these findings is: What are the mechanisms leading to persistent changes in the brain in response to experience? In the context of the evidence for early-life programming of offspring metabolism, stress response and reproduction, the answer to this question may be important in gaining an understanding of how changes in gene expression within particular neuroendocrine circuits is maintained. Recent advances in molecular biology have highlighted the effects of DNA methylation as a particularly stable modification which alters the transcriptional activity of genes. DNA methylation is a covalent modification to the DNA molecule whereby a methyl group is added to cytosine nucleotides within the gene sequence. DNA methyltransferases (DNMT1 & DNMT3a/b) mediate this chemical modification, which also attracts protein complexes, including methyl binding proteins, that add further "congestion" around the DNA and reduce gene transcription.[180–182] DNA methylation is considered an epigenetic "mark" as it is a potentially heritable chemical modification of the DNA. DNA methylation serves a critical role in stably maintaining cell-specific gene expression patterns during cellular differentiation, and these epigenetic marks are transmitted to daughter cells during the process of mitosis.[183,184] Without the capacity to induce this epigenetic modification, development would not proceed beyond the early fetal/postnatal stages.

Recent findings suggest that DNA methylation may be a critical molecular mechanism linking variations in the quality of the parental environment and long-term neuroendocrine and behavioral outcomes. Gestational stress, variation in maternal caloric/nutrient intake, and various toxicological exposures during fetal development have been associated with altered DNA methylation within the brain and placenta leading to changes in gene expression of numerous target genes, including CRH, GR, DNMT1, brain derived neurotropic factor (BDNF) and PPARα.[185–189] During postnatal development, the effects of prolonged maternal separation on vasopressin (AVP) mRNA in the paraventricular nucleus have been associated with separation-induced reductions in DNA methylation within the AVP gene that can be observed 6 weeks postnatally and persist into adulthood.[190] Daily exposure to abusive mother interactions leads to reduced expression of BDNF in the prefrontal cortex in adulthood associated with increased DNA methylation within the BDNF IV promoter region.[191] Natural variations in postnatal maternal LG in the rat are associated with changes in numerous receptor pathways, with effects on hippocampal GR being implicated in the high levels of HPA reactivity observed amongst offspring of low-LG dams.[158] Analysis of the GR promoter region suggests that low levels of LG are associated with increased methylation within the GR promoter, decreased GR expression and an increased HPA response to stress.[192] Time-course analysis has indicated that these maternally induced epigenetic profiles emerge during the postnatal period

BOX 28.5

HOW DO WE KNOW THAT MATERNAL CARE INFLUENCES DNA METHYLATION DURING POSTNATAL DEVELOPMENT?

Early life experiences can have a long-term impact on gene expression within the brain leading to altered neuroendocrine function and behavior. Studies of natural variation in maternal licking/grooming (LG) in the rat and the consequences of this maternal behavior for offspring development suggest that changes in DNA methylation may emerge in response to variations in maternal care. Using sodium bisulfate mapping of the 1_7 promoter region of the glucocorticoid receptor (GR) gene, offspring reared by a low-LG dam were found to have elevated levels of GR promoter methylation (when compared to offspring reared by high-LG dams). Cross-fostering studies indicate that this change in DNA methylation is associated with the quality of the rearing environment, and a time-course analysis indicates that group differences in DNA methylation are not evident prenatally or at the time of birth. These epigenetic changes appear during early postnatal period and then persist into adulthood. Dynamic yet stable changes in molecular "marks" such as DNA methylation may be a common feature of many early life experiences and provide a biological pathway linking parental behavior to offspring developmental outcomes.

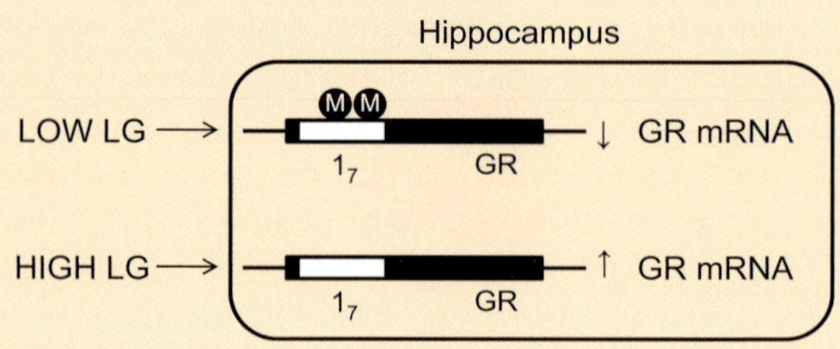

and are sustained into adulthood (see Box 28.5). Maternal LG also affects γ-aminobutyric acid (GABA) circuits and receptor subunit composition,[193,194] and in a recent study reduced hippocampal levels of glutamic acid decarboxylase (GAD1), the rate-limiting enzyme in GABA synthesis, were found in the male offspring of low-LG dams associated with increased DNA methylation within the GAD1 promoter.[195] Maternal influences on ERα in the MPOA of female offspring have likewise been associated with DNA methylation in the promoter region of this gene,[176] and the stable epigenetic regulation of ERα and perhaps other genes involved in parental behavior may account for the transmission of parental behavior across generations.[196] Though DNA methylation is likely to be only one of many molecular mechanisms involved in maintaining the persistent effects of early life experience, studies of this epigenetic mark continue to expand our understanding of interplay between genes and the environment in shaping neuroendocrine outcomes.

SUMMARY AND IMPLICATIONS

The parental brain is influenced by multiple genetic, hormonal and environmental factors that converge on specific neurendocrine circuits, leading to variations in parental care. This variation can have significant implications for offspring development and lead to multiple neurobiological and behavioral outcomes, including (but not limited to) effects on metabolism, stress and reproduction. Though the study of the neuroendocrine pathways involved in parent and offspring behavior has been primarily based on experiments in laboratory rodents, advances in neuroimaging techniques have allowed for the exploration of the human parental brain. However, across species there appears to be a high degree of homology in the mechanisms that influence the onset and maintenance of parental behavior. Animal studies of the molecular mechanisms of perinatal programming effects may likewise be relevant to human development. There is increasing evidence that DNA methylation of target genes is associated with exposure to prenatal famine or disruptions of parent–offspring interactions.[197,198] Thus, continued study of the neuroendocrine basis of parent–offspring interactions may advance our understanding of the mechanisms of disease risk and the environmental factors that contribute to susceptibility to physical and psychiatric dysfunction.

References

1. Fleming AS, Luebke C. Timidity prevents the virgin female rat from being a good mother: emotionality differences between nulliparous and parturient females. *Physiol Behav.* 1981;27:863–868.
2. Fleming AS, Rosenblatt JS. Maternal behavior in the virgin and lactating rat. *J Comp Physiol Psychol.* 1974;86:957–972.
3. Groscolas R, Robin JP. Long-term fasting and re-feeding in penguins. *Comp Biochem Physiol A Mol Integr Physiol.* 2001;128:645–655.
4. Bridges RS. Biochemical basis of parental behavior in the rat. *Adv Study Behav.* 1996;25:215–242.
5. Numan M. A neural circuitry analysis of maternal behavior in the rat. *Acta Paediatr Suppl.* 1994;397:19–28.
6. Wynne-Edwards KE, Lisk RD. Differential effects of paternal presence on pup survival in two species of dwarf hamster (*Phodopus sungorus* and *Phodopus campbelli*). *Physiol Behav.* 1989;45:465–469.
7. Wright SL, Brown RE. The importance of paternal care on pup survival and pup growth in *Peromyscus californicus* when required to work for food. *Behav Processes.* 2002;60:41–52.
8. Carlisle TR. Brood success in variable environments: Implications for parental care allocation. *Anim Behav.* 1982;30:824–836.
9. Harlow HF, Dodsworth RO, Harlow MK. Total social isolation in monkeys. *Proc Natl Acad Sci USA.* 1965;54:90–97.
10. Seay B, Harlow HF. Maternal separation in the rhesus monkey. *J Nerv Mental Dis.* 1965;140:434–441.
11. Schwabl H. Maternal testosterone in the avian egg enhances postnatal growth. *Comp Biochem Physiol A Physiol.* 1996;114:271–276.
12. Fairbanks LA, McGuire MT. Long-term effects of early mothering behavior on responsiveness to the environment in vervet monkeys. *Dev Psychobiol.* 1988;21:711–724.
13. Parker G. Parental 'affectionless control' as an antecedent to adult depression. A risk factor delineated. *Arch Gen Psychiatry.* 1983;40:956–960.
14. Parker G. The parental bonding instrument: a decade of research. *Soc Psychiatry Psychiatr Epidemiol.* 1990;25:281–282.
15. Petraglia F, Imperatore A, Challis JR. Neuroendocrine mechanisms in pregnancy and parturition. *Endocr Rev.* 2010;31:783–816.
16. Reis FM, Petraglia F. The placenta as a neuroendocrine organ. *Front Horm Res.* 2001;27:216–228.
17. Haig D. Genetic conflicts in human pregnancy. *Q Rev Biol.* 1993;68:495–532.
18. Siegel HI. Hormonal basis of maternal behavior in the rat. *Ann NY Acad Sci.* 1986;474:202–215.
19. Brunton PJ, Russell JA. The expectant brain: adapting for motherhood. *Nat Rev Neurosci.* 2008;9:11–25.
20. Siegel HI, Rosenblatt JS. Estrogen-induced maternal behavior in hysterectomized-overiectomized virgin rats. *Physiol Behav.* 1975;14:465–471.
21. Rosenblatt JS, Mayer AD, Giordano AL. Hormonal basis during pregnancy for the onset of maternal behavior in the rat. *Psychoneuroendocrinology.* 1988;13:29–46.
22. Saltzman W, Maestripieri D. The neuroendocrinology of primate maternal behavior. *Prog Neuropsychopharmacol Biol Psychiatry;* 2010. in press.
23. Keverne EB, Kendrick KM. Maternal behaviour in sheep and its neuroendocrine regulation. *Acta Paediatr Suppl.* 1994;397:47–56.
24. Lehrman DS. Effect of female sex hormones on incubation behavior in the ring dove *(Streptopelia risoria). J Comp Physiol Psychol.* 1958;51:142–145.
25. Buntin JD, Becker GM, Ruzycki E. Facilitation of parental behavior in ring doves by systemic or intracranial injections of prolactin. *Horm Behav.* 1991;25:424–444.
26. Buntin JD. Time course and response specificity of prolactin-induced hyperphagia in ring doves. *Physiol Behav.* 1989;45:903–909.

27. Van Roo BL, Ketterson ED, Sharp PJ. Testosterone and prolactin in two songbirds that differ in paternal care: the Blue-headed Vireo and the Red-eyed Vireo. *Horm Behav.* 2003;44:435–441.
28. Khan MZ, McNabb FM, Walters JR, Sharp PJ. Patterns of testosterone and prolactin concentrations and reproductive behavior of helpers and breeders in the cooperatively breeding red-cockaded woodpecker (*Picoides borealis*). *Horm Behav.* 2001;40:1–13.
29. Reburn CJ, Wynne-Edwards KE. Hormonal changes in males of a naturally biparental and a uniparental mammal. *Horm Behav.* 1999;35:163–176.
30. Carlson AA, Russell AF, Young AJ, Jordan NR, McNeilly AS, Parlow AF, Clutton-Brock T. Elevated prolactin levels immediately precede decisions to babysit by male meerkat helpers. *Horm Behav.* 2006;50:94–100.
31. Schradin C, Anzenberger G. Development of prolactin levels in marmoset males: from adult son to first-time father. *Horm Behav.* 2004;46:670–677.
32. Ziegler TE, Prudom SL, Zahed SR, Parlow AF, Wegner F. Prolactin's mediative role in male parenting in parentally experienced marmosets (*Callithrix jacchus*, *Horm Behav.* 2009;56: 436–443.
33. Fleming AS, Corter C, Stallings J, Steiner M. Testosterone and prolactin are associated with emotional responses to infant cries in new fathers. *Horm Behav.* 2002;42:399–413.
34. Breton C, Di Scala-Guenot D, Zingg HH. Oxytocin receptor gene expression in rat mammary gland: structural characterization and regulation. *J Mol Endocrinol.* 2001;27:175–189.
35. Rozen F, Russo C, Banville D, Zingg HH. Structure, characterization, and expression of the rat oxytocin receptor gene. *Proc Natl Acad Sci USA.* 1995;92:200–204.
36. Zingg HH, Lefebvre DL, Giaid A. Uterine oxytocin gene expression: a novel framework for oxytocin action. *Regul Pept.* 1993;45:43–46.
37. Jacobs LS. The role of prolactin in mammogenesis and lactogenesis. *Adv Exp Med Biol.* 1977;80:173–191.
38. Gimpl G, Fahrenholz F. The oxytocin receptor system: structure, function, and regulation. *Physiol Rev.* 2001;81:629–683.
39. Young LJ, Muns S, Wang Z, Insel TR. Changes in oxytocin receptor mRNA in rat brain during pregnancy and the effects of estrogen and interleukin-6. *J Neuroendocrinology.* 1997;9: 859–865.
40. Insel TR. Postpartum increases in brain oxytocin binding. *Neuroendocrinology.* 1986;44:515–518.
41. Grattan DR. The actions of prolactin in the brain during pregnancy and lactation. *Prog Brain Res.* 2001;133:153–171.
42. Nilsson S, Makela S, Treuter E, Tujague M, Thomsen J, Andersson G, Enmark E, Pettersson K, Warner M, Gustafsson JA. Mechanisms of estrogen action. *Physiol Rev.* 2001; 81:1535–1565.
43. Helena C, Gustafsson JA, Korach K, Pfaff D, Anselmo-Franci JA, Ogawa S. Effects of estrogen receptor alpha and beta gene deletion on estrogenic induction of progesterone receptors in the locus coeruleus in female mice. *Endocrine.* 2009;36: 169–177.
44. Young LJ, Wang Z, Donaldson R, Rissman EF. Estrogen receptor alpha is essential for induction of oxytocin receptor by estrogen. *Neuroreport.* 1998;9:933–936.
45. Ogawa S, Eng V, Taylor J, Lubahn DB, Korach KS, Pfaff DW. Roles of estrogen receptor-alpha gene expression in reproduction-related behaviors in female mice. *Endocrinology.* 1998; 139:5070–5081.
46. Scott CJ, Tilbrook AJ, Simmons DM, Rawson JA, Chu S, Fuller PJ, Ing NH, Clarke IJ. The distribution of cells containing estrogen receptor-alpha (ERalpha) and ERbeta messenger ribonucleic acid in the preoptic area and hypothalamus of the sheep: comparison of males and females. *Endocrinology.* 2000;141:2951–2962.
47. Yokosuka M, Okamura H, Hayashi S. Postnatal development and sex difference in neurons containing estrogen receptor-alpha immunoreactivity in the preoptic brain, the diencephalon, and the amygdala in the rat. *J Comp Neurol.* 1997;389:81–93.
48. Sheng Z, Kawano J, Yanai A, Fujinaga R, Tanaka M, Watanabe Y, Shinoda K. Expression of estrogen receptors (alpha, beta) and androgen receptor in serotonin neurons of the rat and mouse dorsal raphe nuclei; sex and species differences. *Neurosci Res.* 2004;49:185–196.
49. Ehret G, Buckenmaier J. Estrogen-receptor occurrence in the female mouse brain: effects of maternal experience, ovariectomy, estrogen and anosmia. *J Physiol Paris.* 1994;88:315–329.
50. Byrnes EM, Babb JA, Bridges RS. Differential expression of oestrogen receptor alpha following reproductive experience in young and middle-aged female rats. *J Neuroendocrinol.* 2009;21: 550–557.
51. Levy F, Keller M. Olfactory mediation of maternal behavior in selected mammalian species. *Behav Brain Res.* 2009;200:336–345.
52. Larsson K. The psychobiology of parenting in mammals. *Scand J Psychol.* 1994;35:97–143.
53. Fleming A, Vaccarino F, Tambosso L, Chee P. Vomeronasal and olfactory system modulation of maternal behavior in the rat. *Science.* 1979;203:372–374.
54. Benuck I, Rowe FA. Centrally and peripherally induced anosmia: influences on maternal behavior in lactating female rats. *Physiol Behav.* 1975;14:439–447.
55. Fleming AS, Vaccarino F, Luebke C. Amygdaloid inhibition of maternal behavior in the nulliparous female rat. *Physiol Behav.* 1980;25:731–743.
56. Fleming A. Psychobiology of rat maternal behavior: how and where hormones act to promote maternal behavior at parturition. *Ann NY Acad Sci.* 1986;474:234–251.
57. Cohn J, Gerall AA. Pre- and postpubertal medial preoptic area lesions and maternal behavior in the rat. *Physiol Behav.* 1989;46:333–336.
58. Jacobson CD, Terkel J, Gorski RA, Sawyer CH. Effects of small medial preoptic area lesions on maternal behavior: retrieving and nest building in the rat. *Brain Res.* 1980;194:471–478.
59. Oxley G, Fleming AS. The effects of medial preoptic area and amygdala lesions on maternal behavior in the juvenile rat. *Dev Psychobiol.* 2000;37:253–265.
60. Caldwell JD, Walker CH, Pedersen CA, Barakat AS, Mason GA. Estrogen increases affinity of oxytocin receptors in the medial preoptic area–anterior hypothalamus. *Peptides.* 1994;15: 1079–1084.
61. Fleming AS, Suh EJ, Korsmit M, Rusak B. Activation of Fos-like immunoreactivity in the medial preoptic area and limbic structures by maternal and social interactions in rats. *Behav Neurosci.* 1994;108:724–734.
62. Numan M, Numan MJ, Marzella SR, Palumbo A. Expression of c-fos, fos B, and egr-1 in the medial preoptic area and bed nucleus of the stria terminalis during maternal behavior in rats. *Brain Res.* 1998;792:348–352.
63. Fleming AS, Walsh C. Neuropsychology of maternal behavior in the rat: c-fos expression during mother-litter interactions. *Psychoneuroendocrinology.* 1994;19:429–443.
64. Lonstein JS, Simmons DA, Swann JM, Stern JM. Forebrain expression of c-fos due to active maternal behaviour in lactating rats. *Neuroscience.* 1998;82:267–281.
65. Perrin G, Meurisse M, Levy F. Inactivation of the medial preoptic area or the bed nucleus of the stria terminalis differentially disrupts maternal behavior in sheep. *Horm Behav.* 2007;52: 461–473.

66. Sharp PJ, Li Q, Georgiou G, Lea RW. Expression of fos-like immunoreactivity in the hypothalamus of the ring dove (*Streptopelia risoria*) at the onset of incubation. *J Neuroendocrinol.* 1996;8:291–298.
67. Robbins TW, Everitt BJ. Neurobehavioural mechanisms of reward and motivation. *Curr Opin Neurobiol.* 1996;6:228–236.
68. Ikemoto S, Panksepp J. The role of nucleus accumbens dopamine in motivated behavior: a unifying interpretation with special reference to reward-seeking. *Brain Res Rev.* 1999; 31:6–41.
69. Lin SH, Miyata S, Weng W, Matsunaga W, Ichikawa J, Furuya K, Nakashima T, Kiyohara T. Comparison of the expression of two immediate early gene proteins, FosB and Fos in the rat preoptic area, hypothalamus and brainstem during pregnancy, parturition and lactation. *Neurosci Res.* 1998;32:333–341.
70. Hansen S. Maternal behavior of female rats with 6-OHDA lesions in the ventral striatum: characterization of the pup retrieval deficit. *Physiol Behav.* 1994;55:615–620.
71. Hansen S, Harthon C, Wallin E, Lofberg L, Svensson K. The effects of 6-OHDA-induced dopamine depletions in the ventral or dorsal striatum on maternal and sexual behavior in the female rat. *Pharmacol Biochem Behav.* 1991;39:71–77.
72. Numan M, Smith HG. Maternal behavior in rats: evidence for the involvement of preoptic projections to the ventral tegmental area. *Behav Neurosci.* 1984;98:712–727.
73. Gaffori O, Le Moal M. Disruption of maternal behavior and appearance of cannibalism after ventral mesencephalic tegmentum lesions. *Physiol Behav.* 1979;23:317–323.
74. Giordano AL, Johnson AE, Rosenblatt JS. Haloperidol-induced disruption of retrieval behavior and reversal with apomorphine in lactating rats. *Physiol Behav.* 1990;48:211–214.
75. Keer SE, Stern JM. Dopamine receptor blockade in the nucleus accumbens inhibits maternal retrieval and licking, but enhances nursing behavior in lactating rats. *Physiol Behav.* 1999;67: 659–669.
76. Hansen S, Bergvall AH, Nyiredi S. Interaction with pups enhances dopamine release in the ventral striatum of maternal rats: a microdialysis study. *Pharmacol Biochem Behav.* 1993;45:673–676.
77. Numan M, Sheehan TP. Neuroanatomical circuitry for mammalian maternal behavior. *Ann NY Acad Sci.* 1997;807: 101–125.
78. Pedersen CA, Caldwell JD, Walker C, Ayers G, Mason GA. Oxytocin activates the postpartum onset of rat maternal behavior in the ventral tegmental and medial preoptic areas. *Behav Neurosci.* 1994;108:1163–1171.
79. Lee A, Clancy S, Fleming AS. Mother rats bar-press for pups: effects of lesions of the MPOA and limbic sites on maternal behavior and operant responding for pup-reinforcement. *Behav Brain Res.* 2000;108:215–231.
80. Lonstein JS. Effects of dopamine receptor antagonism with haloperidol on nurturing behavior in the biparental prairie vole. *Pharmacol Biochem Behav.* 2002;74:11–19.
81. Swain JE. The human parental brain: *In vivo* neuroimaging. *Prog Neuropsychopharmacol Biol Psychiatry*; 2010. in press.
82. Strathearn L, Li J, Fonagy P, Montague PR. What's in a smile? Maternal brain responses to infant facial cues. *Pediatrics.* 2008;122:40–51.
83. Kirkpatrick B, Kim JW, Insel TR. Limbic system fos expression associated with paternal behavior. *Brain Res.* 1994;658: 112–118.
84. Lee AW, Brown RE. Comparison of medial preoptic, amygdala, and nucleus accumbens lesions on parental behavior in California mice (*Peromyscus californicus*). *Physiol Behav.* 2007;92: 617–628.
85. Trainor BC, Bird IM, Alday NA, Schlinger BA, Marler CA. Variation in aromatase activity in the medial preoptic area and plasma progesterone is associated with the onset of paternal behavior. *Neuroendocrinology.* 2003;78:36–44.
86. Lea RW, Clark JA, Tsutsui K. Changes in central steroid receptor expression, steroid synthesis, and dopaminergic activity related to the reproductive cycle of the ring dove. *Microsc Res Tech.* 2001;55:12–26.
87. Seifritz E, Esposito F, Neuhoff JG, Luthi A, Mustovic H, Dammann G, von Bardeleben U, Radue EW, Cirillo S, Tedeschi G, Di Salle F. Differential sex-independent amygdala response to infant crying and laughing in parents versus non-parents. *Biol Psychiatry.* 2003;54:1367–1375.
88. Champagne FA, Francis DD, Mar A, Meaney MJ. Variations in maternal care in the rat as a mediating influence for the effects of environment on development. *Physiol Behav.* 2003;79: 359–371.
89. Champagne F, Diorio J, Sharma S, Meaney MJ. Naturally occurring variations in maternal behavior in the rat are associated with differences in estrogen-inducible central oxytocin receptors. *Proc Natl Acad Sci USA.* 2001;98:12736–12741.
90. Champagne FA, Weaver IC, Diorio J, Sharma S, Meaney MJ. Natural variations in maternal care are associated with estrogen receptor alpha expression and estrogen sensitivity in the medial preoptic area. *Endocrinology.* 2003;144:4720–4724.
91. Champagne FA, Chretien P, Stevenson CW, Zhang TY, Gratton A, Meaney MJ. Variations in nucleus accumbens dopamine associated with individual differences in maternal behavior in the rat. *J Neurosci.* 2004;24:4113–4123.
92. Shahrokh DK, Zhang TY, Diorio J, Gratton A, Meaney MJ. Oxytocin-dopamine interactions mediate variations in maternal behavior in the rat. *Endocrinology.* 2010;151:2276–2286.
93. Brown JR, Ye H, Bronson RT, Dikkes P, Greenberg ME. A defect in nurturing in mice lacking the immediate early gene fosB. *Cell.* 1996;86:297–309.
94. Lucas BK, Ormandy CJ, Binart N, Bridges RS, Kelly PA. Null mutation of the prolactin receptor gene produces a defect in maternal behavior. *Endocrinology.* 1998;139:4102–4107.
95. Lefebvre L, Viville S, Barton SC, Ishino F, Keverne EB, Surani MA. Abnormal maternal behaviour and growth retardation associated with loss of the imprinted gene Mest. *Nat Genet.* 1998;20:163–169.
96. Li L, Keverne EB, Aparicio SA, Ishino F, Barton SC, Surani MA. Regulation of maternal behavior and offspring growth by paternally expressed Peg3. *Science.* 1999;284:330–333.
97. Schneider JS, Stone MK, Wynne-Edwards KE, Horton TH, Lydon J, O'Malley B, Levine JE. Progesterone receptors mediate male aggression toward infants. *Proc Natl Acad Sci USA.* 2003;100:2951–2956.
98. Champagne FA, Curley JP, Keverne EB, Bateson PP. Natural variations in postpartum maternal care in inbred and outbred mice. *Physiol Behav.* 2007;91:325–334.
99. Wainwright PE. Maternal performance of inbred and hybrid laboratory mice (*Mus musculus*). *J Comp Physiol Psychol.* 1982;94:694–707.
100. Peripato AC, De Brito RA, Vaughn TT, Pletscher LS, Matioli SR, Cheverud JM. Quantitative trait loci for maternal performance for offspring survival in mice. *Genetics.* 2002;162:1341–1353.
101. Lim MM, Hammock EA, Young LJ. The role of vasopressin in the genetic and neural regulation of monogamy. *J Neuroendocrinol.* 2004;16:325–332.
102. Bakermans-Kranenburg MJ, van Ijzendoorn MH. Oxytocin receptor (OXTR) and serotonin transporter (5-HTT) genes associated with observed parenting. *Soc Cogn Affect Neurosci.* 2008;3:128–134.

103. van IMH, Bakermans-Kranenburg MJ, Mesman J. Dopamine system genes associated with parenting in the context of daily hassles. *Genes Brain Behav.* 2008;7:403–410.
104. Champagne FA, Meaney MJ. Stress during gestation alters postpartum maternal care and the development of the offspring in a rodent model. *Biol Psychiatry.* 2006;59:1227–1235.
105. Moore CL, Power KL. Prenatal stress affects mother–infant interaction in Norway rats. *Dev Psychobiol.* 1986;19:235–245.
106. Andrews MW, Rosenblum LA. Assessment of attachment in differentially reared infant monkeys (*Macaca radiata*): response to separation and a novel environment. *J Comp Psychol.* 1993;107:84–90.
107. Coplan JD, Smith EL, Altemus M, Mathew SJ, Perera T, Kral JG, Gorman JM, Owens MJ, Nemeroff CB, Rosenblum LA. Maternal–infant response to variable foraging demand in nonhuman primates: effects of timing of stressor on cerebrospinal fluid corticotropin-releasing factor and circulating glucocorticoid concentrations. *Ann NY Acad Sci.* 2006;1071: 525–533.
108. Bredy TW, Brown RE, Meaney MJ. Effect of resource availability on biparental care, and offspring neural and behavioral development in the California mouse (*Peromyscus californicus*). *Eur J Neurosci.* 2007;25:567–575.
109. Coutellier L, Friedrich AC, Failing K, Marashi V, Wurbel H. Effects of foraging demand on maternal behaviour and adult offspring anxiety and stress response in C57BL/6 mice. *Behav Brain Res.* 2009;196:192–199.
110. Mennella AM, Blumberg MS, McCiintock MK, Moltz H. Inter-litter competition and communal nursing among Norway rats: advantages of birth synchrony. *Behav Ecol Sociobiol.* 1990; 27:183–190.
111. Schultz LA, Lore RK. Communal reproductive success in rats (Rattus norvegicus): effects of group composition and prior social experience. *J Comp Psychol.* 1993;107:216–222.
112. Crowcroft P, Rowe FP. Social organization and territorial behavior in the wild house mice (*Mus musculus* L.). *Proc Zool Soc London.* 1963;140:517–531.
113. Curley JP, Davidson S, Bateson P, Champagne FA. Social enrichment during postnatal development induces transgenerational effects on emotional and reproductive behavior in mice. *Front Behav Neurosci.* 2009;3:25.
114. Hoffmann K. Photoperiod, pineal, melatonin and reproduction in hamsters. *Prog Brain Res.* 1979;52:397–415.
115. Schekkerman H, Tulp I, Piersma T, Visser GH. Mechanisms promoting higher growth rate in arctic than in temperate shorebirds. *Oecologia.* 2003;134:332–342.
116. Arieli Y, Peltonen L, Marder J. Reproduction of rock pigeon exposed to extreme ambient temperatures. *Comp Biochem Physiol A Comp Physiol.* 1988;90:497–500.
117. Croskerry PG, Smith GK, Leon M. Thermoregulation and the maternal behaviour of the rat. *Nature.* 1978;273:299–300.
118. Stern JM, Lonstein JS. Nursing behavior in rats is impaired in a small nestbox and with hyperthermic pups. *Dev Psychobiol.* 1996;29:101–122.
119. Cohen-Salmon C. Differences in patterns of pup care in *Mus musculus* domesticus. VIII. Effects of previous experience and parity in XLII inbred mice. *Physiol Behav.* 1987;40:177–180.
120. Afonso VM, Grella SL, Chatterjee D, Fleming AS. Previous maternal experience affects accumbal dopaminergic responses to pup-stimuli. *Brain Res.* 2008;1198:115–123.
121. Lee A, Li M, Watchus J, Fleming AS. Neuroanatomical basis of maternal memory in postpartum rats: selective role for the nucleus accumbens. *Behav Neurosci.* 1999;113:523–538.
122. Meurisse M, Gonzalez A, Delsol G, Caba M, Levy F, Poindron P. Estradiol receptor-alpha expression in hypothalamic and limbic regions of ewes is influenced by physiological state and maternal experience. *Horm Behav.* 2005;48:34–43.
123. Keverne EB, Levy F, Guevara-Guzman R, Kendrick KM. Influence of birth and maternal experience on olfactory bulb neurotransmitter release. *Neuroscience.* 1993;56:557–565.
124. Macbeth AH, Scharfman HE, Maclusky NJ, Gautreaux C, Luine VN. Effects of multiparity on recognition memory, monoaminergic neurotransmitters, and brain-derived neurotrophic factor (BDNF). *Horm Behav.* 2008;54:7–17.
125. Kinsley CH, Lambert KG. Reproduction-induced neuroplasticity: natural behavioural and neuronal alterations associated with the production and care of offspring. *J Neuroendocrinol.* 2008;20:515–525.
126. Shingo T, Gregg C, Enwere E, Fujikawa H, Hassam R, Geary C, Cross JC, Weiss S. Pregnancy-stimulated neurogenesis in the adult female forebrain mediated by prolactin. *Science.* 2003;299:117–120.
127. Mak GK, Weiss S. Paternal recognition of adult offspring mediated by newly generated CNS neurons. *Nat Neurosci.* 2010;13:753–758.
128. Ravelli AC, van Der Meulen JH, Osmond C, Barker DJ, Bleker OP. Obesity at the age of 50 y in men and women exposed to famine prenatally. *Am J Clin Nutr.* 1999;70:811–816.
129. Ravelli AC, van der Meulen JH, Michels RP, Osmond C, Barker DJ, Hales CN, Bleker OP. Glucose tolerance in adults after prenatal exposure to famine. *Lancet.* 1998;351:173–177.
130. Godfrey KM, Barker DJ. Fetal nutrition and adult disease. *Am J Clin Nutr.* 2000;71:1344S–1352S.
131. Hales CN, Barker DJ. Type 2 (non-insulin-dependent) diabetes mellitus: the thrifty phenotype hypothesis. *Diabetologia.* 1992;35:595–601.
132. Hales CN, Barker DJ. The thrifty phenotype hypothesis. *Br Med Bull.* 2001;60:5–20.
133. Jones AP, Friedman MI. Obesity and adipocyte abnormalities in offspring of rats undernourished during pregnancy. *Science.* 1982;215:1518–1519.
134. Bispham J, Gardner DS, Gnanalingham MG, Stephenson T, Symonds ME, Budge H. Maternal nutritional programming of fetal adipose tissue development: differential effects on messenger ribonucleic acid abundance for uncoupling proteins and peroxisome proliferator-activated and prolactin receptors. *Endocrinology.* 2005;146:3943–3949.
135. Reddy JK, Hashimoto T. Peroxisomal beta-oxidation and peroxisome proliferator-activated receptor alpha: an adaptive metabolic system. *Annu Rev Nutr.* 2001;21:193–230.
136. Bertram C, Trowern AR, Copin N, Jackson AA, Whorwood CB. The maternal diet during pregnancy programs altered expression of the glucocorticoid receptor and type 2 11beta-hydroxysteroid dehydrogenase: potential molecular mechanisms underlying the programming of hypertension *in utero.* *Endocrinology.* 2001;142:2841–2853.
137. Gluckman PD, Hanson MA, Spencer HG, Bateson P. Environmental influences during development and their later consequences for health and disease: implications for the interpretation of empirical studies. *Proc Biol Sci.* 2005; 272:671–677.
138. Brunton PJ, Russell JA. Prenatal social stress in the rat programmes neuroendocrine and behavioural responses to stress in the adult offspring: sex-specific effects. *J Neuroendocrinol.* 2010; 22:258–271.
139. Bosch OJ, Musch W, Bredewold R, Slattery DA, Neumann ID. Prenatal stress increases HPA axis activity and impairs maternal care in lactating female offspring: implications for postpartum mood disorder. *Psychoneuroendocrinology.* 2007;32: 267–278.

140. Barbazanges A, Piazza PV, Le Moal M, Maccari S. Maternal glucocorticoid secretion mediates long-term effects of prenatal stress. *J Neurosci.* 1996;16:3943–3949.
141. Dean F, Yu C, Lingas RI, Matthews SG. Prenatal glucocorticoid modifies hypothalamo-pituitary-adrenal regulation in prepubertal guinea pigs. *Neuroendocrinology.* 2001;73:194–202.
142. Benediktsson R, Calder AA, Edwards CR, Seckl JR. Placental 11 beta-hydroxysteroid dehydrogenase: a key regulator of fetal glucocorticoid exposure. *Clin Endocrinol (Oxf).* 1997;46:161–166.
143. Cottrell EC, Seckl JR. Prenatal stress, glucocorticoids and the programming of adult disease. *Front Behav Neurosci.* 2009;3:19.
144. Langley-Evans SC, Phillips GJ, Benediktsson R, Gardner DS, Edwards CR, Jackson AA, Seckl JR. Protein intake in pregnancy, placental glucocorticoid metabolism and the programming of hypertension in the rat. *Placenta.* 1996;17:169–172.
145. Maccari S, Piazza PV, Kabbaj M, Barbazanges A, Simon H, Le Moal M. Adoption reverses the long-term impairment in glucocorticoid feedback induced by prenatal stress. *J Neurosci.* 1995;15:110–116.
146. Wakshlak A, Weinstock M. Neonatal handling reverses behavioral abnormalities induced in rats by prenatal stress. *Physiol Behav.* 1990;48:289–292.
147. Trickett P, McBride-Chang C. The developmental impact of different forms of child abuse and neglect. *Dev Rev.* 1995;15: 11–37.
148. Fries AB, Shirtcliff EA, Pollak SD. Neuroendocrine dysregulation following early social deprivation in children. *Dev Psychobiol.* 2008;50:588–599.
149. Gerra G, Leonardi C, Cortese E, Zaimovic A, Dell'Agnello G, Manfredini M, Somaini L, Petracca F, Caretti V, Baroni C, Donnini C. Adrenocorticotropic hormone and cortisol plasma levels directly correlate with childhood neglect and depression measures in addicted patients. *Addict Biol.* 2008;13:95–104.
150. Neigh GN, Gillespie CF, Nemeroff CB. The neurobiological toll of child abuse and neglect. *Trauma Violence Abuse.* 2009;10: 389–410.
151. Lee RJ, Gollan J, Kasckow J, Geracioti T, Coccaro EF. CSF corticotropin-releasing factor in personality disorder: relationship with self-reported parental care. *Neuropsychopharmacology.* 2006;31:2289–2295.
152. Pruessner JC, Champagne F, Meaney MJ, Dagher A. Dopamine release in response to a psychological stress in humans and its relationship to early life maternal care: a positron emission tomography study using [11C]raclopride. *J Neurosci.* 2004;24: 2825–2831.
153. Barr CS, Newman TK, Shannon C, Parker C, Dvoskin RL, Becker ML, Schwandt M, Champoux M, Lesch KP, Goldman D, Suomi SJ, Higley JD. Rearing condition and rh5-HTTLPR interact to influence limbic–hypothalamic–pituitary–adrenal axis response to stress in infant macaques. *Biol Psychiatry.* 2004;55:733–738.
154. Suomi SJ. Early stress and adult emotional reactivity in rhesus monkeys. *CIBA Found Symp.* 1991;156:171–183. discussion 183–178.
155. Spinelli S, Chefer S, Suomi SJ, Higley JD, Barr CS, Stein E. Early-life stress induces long-term morphologic changes in primate brain. *Arch Gen Psychiatry.* 2009;66:658–665.
156. Coplan JD, Smith EL, Altemus M, Scharf BA, Owens MJ, Nemeroff CB, Gorman JM, Rosenblum LA. Variable foraging demand rearing: sustained elevations in cisternal cerebrospinal fluid corticotropin-releasing factor concentrations in adult primates. *Biol Psychiatry.* 2001;50:200–204.
157. Francis D, Diorio J, Liu D, Meaney MJ. Nongenomic transmission across generations of maternal behavior and stress responses in the rat. *Science.* 1999;286:1155–1158.
158. Liu D, Diorio J, Tannenbaum B, Caldji C, Francis D, Freedman A, Sharma S, Pearson D, Plotsky PM, Meaney MJ. Maternal care, hippocampal glucocorticoid receptors, and hypothalamic–pituitary–adrenal responses to stress. *Science.* 1997;277:1659–1662.
159. Meaney MJ. Maternal care, gene expression, and the transmission of individual differences in stress reactivity across generations. *Annu Rev Neurosci.* 2001;24:1161–1192.
160. Meaney MJ, Aitken DH, Bhatnagar S, Sapolsky RM. Postnatal handling attenuates certain neuroendocrine, anatomical, and cognitive dysfunctions associated with aging in female rats. *Neurobiol Aging.* 1991;12:31–38.
161. Plotsky PM, Meaney MJ. Early, postnatal experience alters hypothalamic corticotropin-releasing factor (CRF) mRNA, median eminence CRF content and stress-induced release in adult rats. *Brain Res Mol Brain Res.* 1993;18:195–200.
162. Viau V, Sharma S, Plotsky PM, Meaney MJ. Increased plasma ACTH responses to stress in nonhandled compared with handled rats require basal levels of corticosterone and are associated with increased levels of ACTH secretagogues in the median eminence. *J Neurosci.* 1993;13:1097–1105.
163. Avishai-Eliner S, Eghbal-Ahmadi M, Tabachnik E, Brunson KL, Baram TZ. Down-regulation of hypothalamic corticotropin-releasing hormone messenger ribonucleic acid (mRNA) precedes early-life experience-induced changes in hippocampal glucocorticoid receptor mRNA. *Endocrinology.* 2001;142: 89–97.
164. Fenoglio KA, Brunson KL, Avishai-Eliner S, Stone BA, Kapadia BJ, Baram TZ. Enduring, handling-evoked enhancement of hippocampal memory function and glucocorticoid receptor expression involves activation of the corticotropin-releasing factor type 1 receptor. *Endocrinology.* 2005;146: 4090–4096.
165. Fenoglio KA, Chen Y, Baram TZ. Neuroplasticity of the hypothalamic-pituitary-adrenal axis early in life requires recurrent recruitment of stress-regulating brain regions. *J Neurosci.* 2006;26:2434–2442.
166. Wigger A, Neumann ID. Periodic maternal deprivation induces gender-dependent alterations in behavioral and neuroendocrine responses to emotional stress in adult rats. *Physiol Behav.* 1999;66:293–302.
167. Plotsky PM, Thrivikraman KV, Nemeroff CB, Caldji C, Sharma S, Meaney MJ. Long-term consequences of neonatal rearing on central corticotropin-releasing factor systems in adult male rat offspring. *Neuropsychopharmacology.* 2005;30: 2192–2204.
168. Ladd CO, Thrivikraman KV, Huot RL, Plotsky PM. Differential neuroendocrine responses to chronic variable stress in adult Long Evans rats exposed to handling-maternal separation as neonates. *Psychoneuroendocrinology.* 2005;30:520–533.
169. Kikusui T, Nakamura K, Kakuma Y, Mori Y. Early weaning augments neuroendocrine stress responses in mice. *Behav Brain Res.* 2006;175:96–103.
170. Champagne F, Meaney MJ. Like mother, like daughter: evidence for non-genomic transmission of parental behavior and stress responsivity. *Prog Brain Res.* 2001;133:287–302.
171. Gonzalez A, Lovic V, Ward GR, Wainwright PE, Fleming AS. Intergenerational effects of complete maternal deprivation and replacement stimulation on maternal behavior and emotionality in female rats. *Dev Psychobiol.* 2001;38:11–32.
172. Melo AI, Lovic V, Gonzalez A, Madden M, Sinopoli K, Fleming AS. Maternal and littermate deprivation disrupts maternal behavior and social-learning of food preference in adulthood: tactile stimulation, nest odor, and social rearing prevent these effects. *Dev Psychobiol.* 2006;48:209–219.

173. Novakov M, Fleming AS. The effects of early rearing environment on the hormonal induction of maternal behavior in virgin rats. *Horm Behav.* 2005;48:528–536.
174. Boccia ML, Pedersen CA. Brief vs long maternal separations in infancy: contrasting relationships with adult maternal behavior and lactation levels of aggression and anxiety. *Psychoneuroendocrinology.* 2001;26:657–672.
175. Francis DD, Champagne FC, Meaney MJ. Variations in maternal behaviour are associated with differences in oxytocin receptor levels in the rat. *J Neuroendocrinol.* 2000;12:1145–1148.
176. Champagne FA, Weaver IC, Diorio J, Dymov S, Szyf M, Meaney MJ. Maternal care associated with methylation of the estrogen receptor-alpha1b promoter and estrogen receptor-alpha expression in the medial preoptic area of female offspring. *Endocrinology.* 2006;147:2909–2915.
177. Cameron N, Del Corpo A, Diorio J, McAllister K, Sharma S, Meaney MJ. Maternal programming of sexual behavior and hypothalamic–pituitary–gonadal function in the female rat. *PLoS One.* 2008;3:e2210.
178. Cameron NM, Fish EW, Meaney MJ. Maternal influences on the sexual behavior and reproductive success of the female rat. *Horm Behav.* 2008;54:178–184.
179. Uriarte N, Breigeiron MK, Benetti F, Rosa XF, Lucion AB. Effects of maternal care on the development, emotionality, and reproductive functions in male and female rats. *Dev Psychobiol.* 2007;49:451–462.
180. Fan G, Hutnick L. Methyl-CpG binding proteins in the nervous system. *Cell Res.* 2005;15:255–261.
181. Feng J, Fouse S, Fan G. Epigenetic regulation of neural gene expression and neuronal function. *Pediatr Res.* 2007;61:58R–63R.
182. Turner B. *Chromatin and Gene Regulation.* Oxford: Blackwell Science Ltd; 2001.
183. Fukuda S, Taga T. Cell fate determination regulated by a transcriptional signal network in the developing mouse brain. *Anat Sci Intl Jpn Assoc Anat.* 2005;80:12–18.
184. Jones PA, Taylor SM. Cellular differentiation, cytidine analogs and DNA methylation. *Cell.* 1980;20:85–93.
185. Kovacheva VP, Mellott TJ, Davison JM, Wagner N, Lopez-Coviella I, Schnitzler AC, Blusztajn JK. Gestational choline deficiency causes global- and Igf2 gene- DNA hypermethylation by upregulation of Dnmt1 expression. *J Biol Chem.* 2007;282:31777–31788.
186. Lillycrop KA, Phillips ES, Jackson AA, Hanson MA, Burdge GC. Dietary protein restriction of pregnant rats induces and folic acid supplementation prevents epigenetic modification of hepatic gene expression in the offspring. *J Nutr.* 2005;135:1382–1386.
187. Mueller BR, Bale TL. Sex-specific programming of offspring emotionality after stress early in pregnancy. *J Neurosci.* 2008;28:9055–9065.
188. Novikova SI, He F, Bai J, Cutrufello NJ, Lidow MS, Undieh AS. Maternal cocaine administration in mice alters DNA methylation and gene expression in hippocampal neurons of neonatal and prepubertal offspring. *PLoS One.* 2008;3:e1919.
189. Onishchenko N, Karpova N, Sabri F, Castren E, Ceccatelli S. Long-lasting depression-like behavior and epigenetic changes of BDNF gene expression induced by perinatal exposure to methylmercury. *J Neurochem.* 2008;106:1378–1387.
190. Murgatroyd C, Patchev AV, Wu Y, Micale V, Bockmuhl Y, Fischer D, Holsboer F, Wotjak CT, Almeida OF, Spengler D. Dynamic DNA methylation programs persistent adverse effects of early-life stress. *Nat Neurosci.* 2009;12:1559–1566.
191. Roth TL, Lubin FD, Funk AJ, Sweatt JD. Lasting epigenetic influence of early-life adversity on the BDNF gene. *Biol Psychiatry.* 2009;65:760–769.
192. Weaver IC, Cervoni N, Champagne FA, D'Alessio AC, Sharma S, Seckl JR, Dymov S, Szyf M, Meaney MJ. Epigenetic programming by maternal behavior. *Nat Neurosci.* 2004;7:847–854.
193. Caldji C, Diorio J, Meaney MJ. Variations in maternal care alter GABA(A) receptor subunit expression in brain regions associated with fear. *Neuropsychopharmacology.* 2003;28:1950–1959.
194. Caldji C, Francis D, Sharma S, Plotsky PM, Meaney MJ. The effects of early rearing environment on the development of $GABA_A$ and central benzodiazepine receptor levels and novelty-induced fearfulness in the rat. *Neuropsychopharmacology.* 2000;22:219–229.
195. Zhang TY, Hellstrom IC, Bagot RC, Wen X, Diorio J, Meaney MJ. Maternal care and DNA methylation of a glutamic acid decarboxylase 1 promoter in rat hippocampus. *J Neurosci.* 2010;30:13130–13137.
196. Champagne FA. Epigenetic mechanisms and the transgenerational effects of maternal care. *Front Neuroendocrinol.* 2008;29:386–397.
197. McGowan PO, Sasaki A, D'Alessio AC, Dymov S, Labonte B, Szyf M, Turecki G, Meaney MJ. Epigenetic regulation of the glucocorticoid receptor in human brain associates with childhood abuse. *Nat Neurosci.* 2009;12:342–348.
198. Heijmans BT, Tobi EW, Stein AD, Putter H, Blauw GJ, Susser ES, Slagboom PE, Lumey LH. Persistent epigenetic differences associated with prenatal exposure to famine in humans. *Proc Natl Acad Sci USA.* 2008;105:17046–17049.

CHAPTER

29

Disorders of the Hypothalamic–Pituitary–Adrenocortical System

Tomoshige Kino [1], *Evangelia Charmandari* [2,3], *George P. Chrousos* [2,3]

[1] Unit on Molecular Hormone Action, Program in Reproductive and Adult Endocrinology, *Eunice Kennedy Shriver* National Institute of Child Health and Human Development, National Institutes of Health, Bethesda, MD, USA, [2] First Department of Pediatrics, University of Athens Medical School, "Aghia Sophia" Children's Hospital, Athens, Greece, [3] Division of Endocrinology and Metabolism, Biomedical Research Foundation of the Academy of Athens, Athens, Greece

OUTLINE

Summary

The hypothalamic–pituitary–adrenocortical (HPA) system is essential for the maintenance of internal homeostasis and a major mediator component of the body's adaptive response to internal or external stressors. This system consists of three components, the hypothalamus, the pituitary gland and the adrenal cortex, which respectively secrete their signaling molecules: the corticotropin-releasing hormone (CRH) and arginine vasopressin (AVP), the adrenocorticotropic hormone (ACTH), and the glucocorticoid steroid cortisol. Importantly, various pathologic conditions affecting any of the three components, whether hereditary or acquired, result in dysfunction in this system. The clinical manifestations of impaired function of the HPA system are evident either as

Handbook of Neuroendocrinology, DOI: 10.1016/B978-0-12-375097-6.10029-0

glucocorticoid excess or as glucocorticoid insufficiency, and may have detrimental effects on several organs and systems. In this chapter, we describe the pathologic origins, the clinical manifestations and the management of the most common disorders of the HPA system.

INTRODUCTION

Human beings face continuous internal or external challenges, called "stressors." The stressors may be short- or long-term; they can be internal, in connection with internally physiologic and pathologic changes (e.g., aging, hurtful memories, splachnic injuries and neoplasias), or external (for instance, excessive exercise, or starvation). To adapt to these stressful stimuli, human beings have developed a highly conserved regulatory system, the stress system, which receives information through various sensory organs and adjusts the functions of the central nervous system (CNS) and peripheral organs.[1] The hypothalamic–pituitary–adrenocortical (HPA) system is one of the two major arms of the stress system that functions together with the locus caeruleus/norepinephrine–autonomic nervous system to adjust homeostasis. Glucocorticoids, the end-effectors of the HPA system, are secreted from the adrenocortical cells and exert their diverse actions on many biological functions, including those of the CNS, intermediary metabolism, immunity and reproduction.[1,2] Thus, dysfunction of the HPA system may result in the development of several pathologic conditions, which can strongly impact on health and quality of life. Below we describe the components of the HPA system, the hormonal effects of glucocorticoids and, finally, the most common/representative disorders associated with dysfunction of this system.

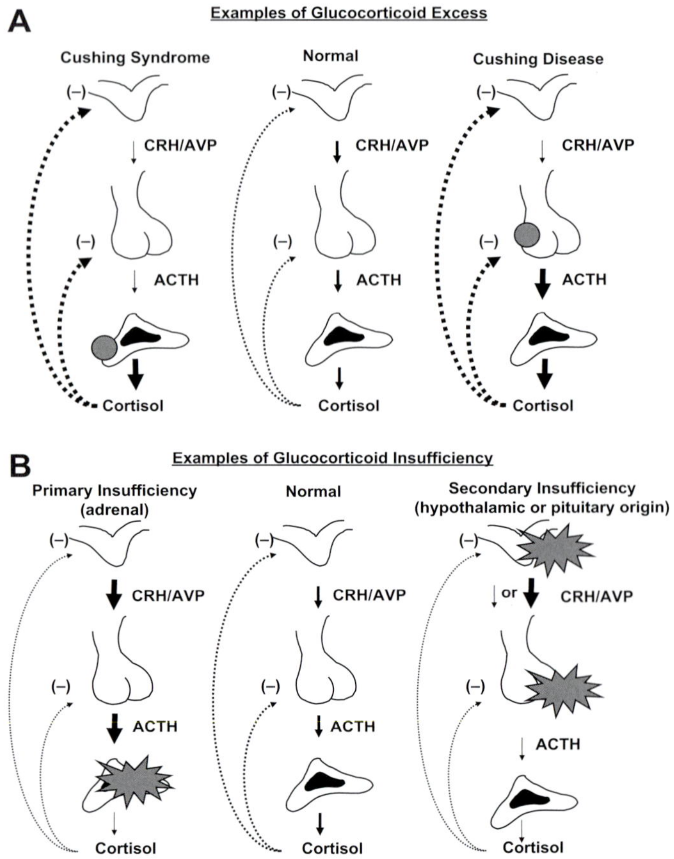

FIGURE 29.1 **Feedback regulation of the HPA system and its compensatory changes in various pathologic conditions.** The HPA system consists of three components, the hypothalamus, pituitary corticotropes and the adrenal cortex, which respectively secrete CRH/AVP, ACTH and cortisol, ultimately regulating downstream cortisol-responsive genes. Cortisol suppresses the upper regulatory centers of the HPA system, forming a closed regulatory feedback circle. A cortisol-producing adrenal adenoma and an ACTH-producing pituitary adenoma are shown as examples of the conditions associated with glucocorticoid excess (**A**), while destruction of the adrenal glands, the pituitary gland or the hypothalamic centers are shown as representatives for adrenal insufficiency (**B**). ACTH, adrenocorticotropic hormone; AVP, arginine vasopressin; CRH, corticotropin-releasing hormone.

The HPA System

The HPA system consists of three components: the paraventricular nucleus (PVN) located in the hypothalamus, the pituitary gland, and the adrenal cortices (Fig. 29.1). The CNS detects and integrates internal and external signals from various parts of the body, transduces this information to the PVN and other brain areas, such as the hypothalamic suprachiasmatic nucleus (SCN), the amygdala and the raphe nuclei of the brainstem, and regulates the activities of the PVN, a nodal center of the HPA system.[3] In particular, the SCNs also function as a central circadian rhythm center (CLOCK), under the control of light information from the eyes, and generate the circadian rhythmicity of the HPA system through neural connections with the PVN. This explains the diurnal fluctuation of circulating glucocorticoids.[4] Neurons located in the PVN produce a 41-amino acid peptide, the corticotropin-releasing hormone (CRH), which is released into the hypophysial portal system from the axon terminals projecting to the median eminence of the hypothalamus.[3,5] CRH then stimulates the corticotropes of the anterior pituitary gland to secrete adrenocorticotropic hormone (ACTH) in synergy with PVN arginine-vasopressin (AVP) and other molecules.[6–8] CRH is the strongest and most important stimulator of ACTH secretion from the pituitary gland.[6] ACTH is a 39-amino acid polypeptide synthesized as a part of the large precursor 214-amino acid molecule, called proopiomelanocortin (POMC), which is also the source of the β- or γ-lipotropins, β-endorphin and α-melanocyte-stimulating hormone (MSH).[9] The corticotropes secrete ACTH into the systemic circulation in a pulsatile fashion, and this hormone stimulates production of glucocorticoids by binding to its specific G-protein coupled seven-transmembrane receptor of the adrenocortical cells of the *zona fasciculata* of the adrenal cortices.[10] Cortisol is the major glucocorticoid secreted from the adrenal glands in humans, while corticosterone is the natural glucocorticoid in rodents.[11] These hormones are secreted into the systemic circulation in a pulsatile and circadian fashion, and exert their diverse actions in virtually all organs and tissues. Glucocorticoids exert negative feedback effects on both PVN CRH and AVP synthesis and secretion, and inhibit pituitary synthesis of POMC as well as ACTH secretion, and thus ultimately create a closed negative feedback regulatory circle.[3,12]

Glucocorticoid and Glucocorticoid Receptor (GR) Actions

Glucocorticoids regulate a variety of biologic processes and exert profound influences on many physiologic functions.[11] They are essential for the maintenance of basal and stress-related homeostasis, while at pharmacologic doses they exert catabolic effects, mainly on the skeletal muscles and the bones, and potent, mostly suppressive effects on the immune system. The anti-inflammatory actions of glucocorticoids have made them useful drugs for the management of many inflammatory, autoimmune and lymphoproliferative diseases.[13] In the CNS, glucocorticoids influence cognition, memory, mood and sleep.[14] They also affect anatomical structures of the CNS, causing a reduction of hippocampal volume, ventricular enlargement, and reversible cortical atrophy.[15–17] Glucocorticoids modulate diverse metabolic activities in humans, including carbohydrate, lipid and protein metabolism.[18] The direction of changes induced by stress levels or pharmacological doses of glucocorticoids is to increase the circulating levels of energy fuel, such as glucose, free fatty acids, amino acids and glycerol, by facilitating catabolism of hepatic glycogen and skeletal muscle protein, and by stimulating fat lipolysis. Glucocorticoids promote liver gluconeogenesis and glucose secretion.

In the bones, physiologic amounts of glucocorticoids appear to be necessary for the maintenance of normal bone function, turnover and growth,[19,20] while their excess and/or prolonged administration frequently causes osteopenia and osteoporosis, which are major side effects of long-term glucocorticoid therapy.[21] In the cardiovascular system, normal daily secretory rates of glucocorticoids are essential for maintaining normal vascular tone and heart contractility. Glucocorticoids are also necessary for renal tubular function, the latter being responsible for the reabsorption of water and electrolytes.[22] Excess amounts of glucocorticoids, on the other hand, increase peripheral vascular resistance and may result in the development of hypertension.[22,23]

In the immune system, glucocorticoids exert quite diverse actions on nearly all components of the immune and inflammatory reaction, including the innate, cellular and humoral immunity.[24,25] Glucocorticoids have complex final effects, but generally they cause: (a) suppression of inflammatory mediator secretion, including cytokines and eicosanoids; (b) inhibition of monocyte/granulocyte migration at inflamed tissues; (c) suppression of cellular immunity; and (d) stimulation of secondary humoral immunity (T helper (Th)-1 to Th-2 shift).[26,27]

At the cellular level, the actions of glucocorticoids are mediated by an intracellular receptor protein, the glucocorticoid receptor (GR).[28] The GR is also designated as nuclear receptor superfamily 3, group C, member 1 (NR3C1), and is a member of the steroid/sterol/thyroid/retinoid/orphan nuclear receptor superfamily, which consists of 49 members in mice and 48 in humans.[28,29] The GR functions as a hormone-activated transcription factor, which regulates the expression of glucocorticoid-responsive genes that represent 5–20% of the expressed human genome.[28,30] The GR is ubiquitously expressed in almost all human tissues and organs. The human (h) GR, a single polypeptide chain of 777 amino acid residues, has three major distinct functional domains: the N-terminal or immunogenic (NTD), the DNA-binding (DBD), and the ligand-binding (LBD) domains[31] (Fig. 29.2). The GR is located primarily in the cytoplasm in the absence of glucocorticoid ligand, as part of a hetero-oligomeric complex containing heat shock proteins (HSPs) 90, 70, 50 and 20, and possibly, other proteins.[29,31] Following binding to its agonist ligand, the GR undergoes conformational changes, dissociates from the heat shock proteins, homo- or heterodimerizes, and translocates as a dimer and/or monomer into the nucleus through the nuclear pore via an active ATP-dependent process mediated by its nuclear localization signal (NL)-1 and -2.[28,32] NL-1 is located in the junction of DBD and its C-terminally positioned hinge region, while NL-2 spans the entire LBD[32] (Fig. 29.2).

Inside the nucleus, the ligand-activated GR directly interacts as a homo- or heterodimer with specific DNA sequences, the glucocorticoid response elements (GREs), in the promoter regions of target genes.[28,29] The GR contains two transactivation domains, activation function (AF)-1 and -2, located at its NTD and LBD, respectively (Fig. 29.2), through which it interacts with many proteins and protein complexes, such as the nuclear receptor co-activators p160, p300/CREB-binding protein (CBP) and p300/CBP-associated factor (p/CAF) complexes, and SWI/SNF and vitamin D receptor-interacting protein/thyroid hormone receptor-associated protein (DRIP/TRAP) chromatin-remodeling complexes, and influences the activity of the RNA polymerase II and its ancillary factors, thereby modulating the transcription rates of glucocorticoid-responsive genes.[29,31]

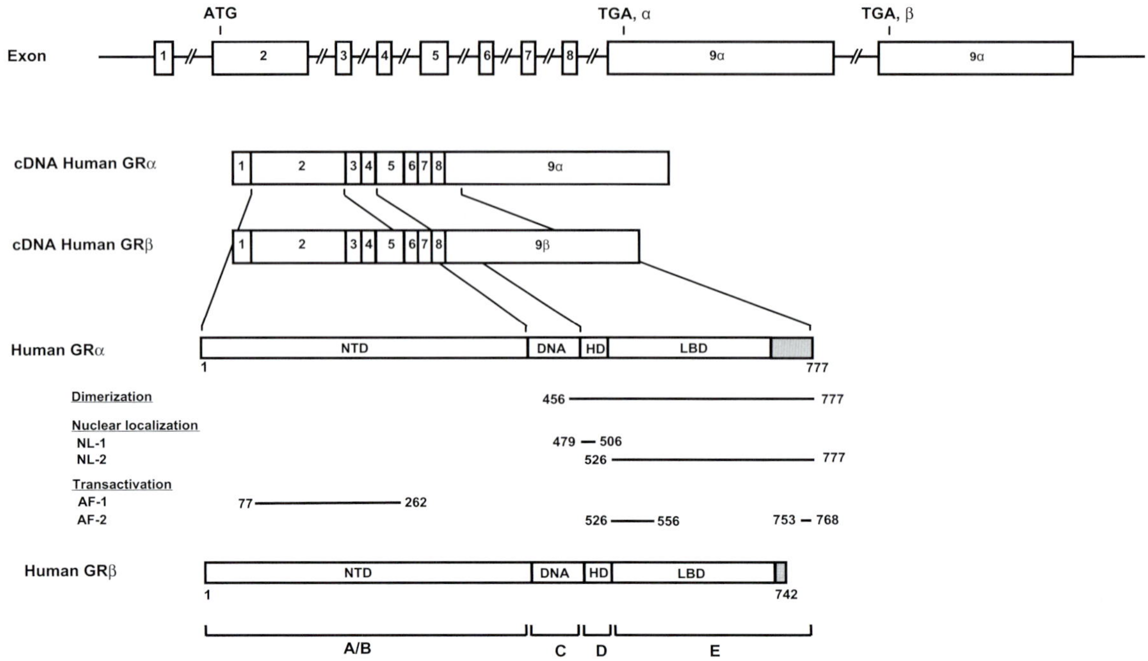

FIGURE 29.2 **Genomic and complementary DNA, and protein structures of the human GR, functional domains distribution of GRα, and the GR isoform produced through alternative splicing.** The human GR gene consists of nine exons: exon 1 is an untranslated region, exon 2 codes for the immunogenic domain (A/B), exons 3 and 4 code for the DNA-binding domain (C), and exons 5–9 code for the hinge region (D) and the ligand-binding domain (E). GR does not contain an F region, in contrast to the other steroid hormone receptors. The human GR gene contains two terminal exons 9 (exon 9α and 9β), alternatively spliced to produce the classic GRα and the non-ligand-binding GRβ. C-terminal gray-colored domains in GRα and GRβ show their specific portions. Locations of several functional domains are also indicated. In this chapter, we describe GRα as "GR" and do not discuss the potential physiologic functions and pathophysiological importance of isoform GRβ. AF-1 and -2, activation function 1 and 2; DBD, DNA-binding domain; HD, hinge region; LBD, ligand-binding domain; NTD, N-terminal region; NL-1 and-2, nuclear translocation signal 1 and 2.

DISORDERS ASSOCIATED WITH ALTERATIONS IN THE FUNCTION OF THE HPA SYSTEM

Overview

Given that glucocorticoids are the end-products and effector hormones of the HPA system, disorders caused by dysfunction of the HPA system result in either elevated or reduced concentrations of circulating glucocorticoids, and clinical manifestations are associated with, respectively, increased or decreased actions of these hormones. These disorders can also be categorized according to the component(s) of the HPA system primarily affected: (a) the hypothalamus, (b) the pituitary gland, (c) the adrenal glands, and (d) other tissues and organs, including alterations of glucocorticoid action or availability at local glucocorticoid-responsive tissues. Thus, they can be generally categorized into eight groups (Table 29.1).

Depending on the location of the primary cause, the circulating concentrations of the signaling hormones (CRH, ACTH and cortisol) of the HPA system demonstrate distinct secretory patterns (Fig. 29.1). Clinical manifestations owing to alterations in the HPA system reflect the alterations in glucocorticoid availability, as well as the local effects of their primary causal lesions. Symptoms and signs of excess or reduced glucocorticoid action are listed in Table 29.2. Generally, development and severity are strongly associated with the duration and magnitude of changes of serum glucocorticoid concentrations. Typically, chronic glucocorticoid excess results in central obesity, hypertension, purple striae of the skin, osteoporosis/multiple fractures/aseptic necrosis of femoral/humeral head, buffalo hump, glucose intolerance/overt diabetes mellitus, dyslipidemia and leukocytosis, all of which are referred to as the "Cushing phenotype."[13] Cardiovascular complications, such as ischemic heart/cerebral disease, are a major cause of morbidity in patients with excess exposure to glucocorticoids.[13] Increased susceptibility to infections secondary to any pathogens can also occur in these patients, since glucocorticoids have strong immunosuppressive effects. In contrast, glucocorticoid deficiency or adrenal insufficiency may result in fatigue, depression, nausea, vomiting, hypotension, hypoglycemia, hyponatremia and eosinophilia. Frequently, primary adrenal insufficiency is associated with both cortisol and aldosterone deficiency, with hyperkalemia and increased plasma renin activity complementing

TABLE 29.1 Disorders Associated with Changes in the HPA System

	Organs affected			
Serum glucocorticoid levels	**Hypothalamus**	**Pituitary gland**	**Adrenal glands**	**Others**
High	Paroxysmal ACTH discharge (Wolff syndrome) Melancholic major depression Post-traumatic stress disorder Anorexia nervosa Dysfunction of circadian rhythm (night shift work, trans-time-zone travel)	Cushing disease (Pituitary adenoma)	Cushing syndrome (Adrenal adenoma and adenocarcinoma)	Generalized glucocorticoid resistance syndrome (Chrousos syndrome) Ectopic ACTH-producing tumors Ectopic CRH-producing tumors Iatrogenic Cushing syndrome (Use of excess glucocorticoids)
Low	Craniopharyngioma Rathke's pouch cysts Basilar meningitis Encephalitis Head trauma Sarcoidosis Histiocytosis X Hypothalamic tumors Pinealoma Surgical destruction of the pituitary stalk Prader-Willi syndrome	Pituitary apoplexy/ Sheehan's syndrome Lymphocytic hypophysitis Destructive pituitary tumors Isolated ACTH deficiency syndrome Hereditary disorders caused by mutations in the factors necessary for the pituitary development/ function POMC deficiency syndrome	Addison's disease Autoimmune adrenalitis Adrenal hemorrhage (Waterhouse-Friederichen syndrome) Congenital adrenal hyperplasia Lipoid congenital adrenal hyperplasia Wolman's disease Adrenoleukodystrophy Familial glucocorticoid deficiency 1 and 2 Triple-A syndrome	Glucocorticoid hypersensitivity syndrome Drug-induced cortisol deficiency (Aminoglutethimide, trilostane, ketoconazol, carbamazepine, imipramine, mefipristone)

the clinical picture. The acute form of adrenal insufficiency is known as "adrenal crisis."[13] In the case of chronic primary adrenal insufficiency caused by defects in the adrenal glands, hyperpigmentation of the skin and buccal mucosa occurs due to increased circulating ACTH concentrations.

Hypothalamic Disorders Associated with Alterations in the HPA System

Many lesions of the hypothalamus, such as craniopharyngioma, Rathke's pouch cysts, head trauma, inflammatory or infiltrative processes (meningitis, encephalitis, sarcoidosis, histiocytosis X) and brain tumors (including metastatic tumors and pinealomas) result in hypofunction of the HPA system secondary to destruction of the CRH-producing PVN neurons or to dissection of the pituitary stalk that transfers CRH from the hypothalamic median eminence to the pituitary gland.[33–35] These pathologic conditions often result in impaired function of other hormonal axes, such as the hypothalamic–pituitary–gonadal or -thyroid axes, the growth hormone/insulin-like growth factor-I axis, and the water balance (antidiuretic hormone hyper- or hyposecretion).

Prader-Willi syndrome is a rare genetic disorder that is caused by deletion or non-expression of several genes (those for the small nuclear ribonucleoprotein polypeptide N, the nectidin, and a cluster of the small nucleolar (sno) RNAs) encoded on chromosome 15 (q11-13) (known as the PWS/AS region). It is characterized by a variety of symptoms associated with hypothalamic dysfunction, such as hypogonadism, hyperphagia, obesity, sleep disorders, learning disabilities/borderline intellectual functioning, and hypotonia.[36,37] These patients (~60%) develop mild adrenal insufficiency, while overt adrenal insufficiency is relatively rare.[38]

A syndrome of paroxysmal periodic discharge, also known as Wolff syndrome, results in activation of the HPA system and subsequent elevation of circulating ACTH and cortisol concentrations.[39] The following case was reported in 1964. The patient, a 14-year-old boy, demonstrated cyclic manifestations of nausea, vomiting, fever, emotional disturbances and marked weight loss.[39] During the episode, he also showed elevated plasma ACTH concentrations and increased urinary excretion of glucocorticoid metabolites associated with hypertension, reduced glucose tolerance, and elevated serum cholesterol and free fatty acid concentrations. Activation of the HPA system appears to be secondary

TABLE 29.2 Manifestations Associated with Glucocorticoid Excess and Deficiency

Manifestations associated with glucocorticoid excess	Manifestations associated with glucocorticoid insufficiency
Central obesity	Weakness, fatigue
Moon face	Anorexia
"Buffalo hump"	Nausea
Facial plethora	Vomiting
Muscle weakness, proximal myopathy	Constipation
Hypertension	Abdominal pain
Psychological changes	Diarrhea
Depression	Postural dizziness
Easy bruising	Muscle or joint pain
Oligomenorrhea	Weight loss
Amenorrhea	Hyperpigmentation
Impotence	Hypotension
Acne, oily skin	Vitiligo
Abdominal striae	Auricular calcification
Edema	Hyponatremia
Osteoporosis	Hypercalcemia
Backache	Azotemia
Bone fracture	Anemia
Polydipsia, polyuria	Eosinophilia
Headache	Hypocholesterolemia
Various infectious diseases	Hypoglycemia
Glucose intolerance/overt diabetes mellitus	
Hyperlipidemia	
Leukocytosis	
Hypokalemia	

to the yet unknown underlying mechanisms of hypothalamic dysfunction. Similar cases, however, have not been reported thereafter.

Patients with major depression, particularly the melancholic type, demonstrate sustained hyperactivity of the HPA system and subsequent elevation of cerebrospinal fluid CRH, plasma ACTH and serum cortisol concentrations, possibly caused by mental stress.[40,41] Activation of the HPA system in these patients occurs at the level of hypothalamus or higher brain centers. Patients with major depression have chronically elevated serum cortisol concentrations and therefore higher incidence of central obesity, hypertension, impaired glucose tolerance, decreased libido and osteoporosis.

Anorexia nervosa is also characterized by elevation of circulating cortisol concentrations through stimulation of the HPA system caused by strict energy restriction, as well as by mental stress.[42,43] These patients demonstrate elevations of CRH in their cerebrospinal fluid,[43] suggesting that the underlying activation of the HPA system is to some extent similar to that observed in major depression. In contrast to major depression, however, patients with anorexia nervosa do not show any clinical characteristics of glucocorticoid excess but rather present with those associated with reduced action of glucocorticoids, such as hypotension, hypoglycemia and decreased lipid concentrations. Thus, the biologic actions of circulating glucocorticoids are not manifested in the glucocorticoid target tissues of these patients possibly because of the lack of food substrate and changes of proteins that may influence the actions of glucocortcoids, such as the AMP-activated protein kinase and the forkhead transcription factors.[44,45]

The conditions associated with uncoupling of the circadian rhythm between the CNS and peripheral tissues, caused, for example, by frequent trans-time-zone travel and night-shift work, are likely to result in increased action of glucocorticoids in local glucocorticoid target tissues, which predisposes subjects to cardiovascular morbidity and mortality.[4] It was recently reported that the Clock transcription factor, an essential component for generating circadian rhythm in the CNS (central CLOCK) and peripheral tissues (peripheral CLOCK), regulates GR transcriptional activity, and this effect of CLOCK on GR may cause "functional" glucocorticoid hypersensitivity through uncoupling of circulating cortisol concentrations and peripheral target-tissue sensitivity to glucocorticoids.[4,46]

Pituitary Disorders Associated with Alterations in the HPA System

Cushing Disease (and Syndrome)

Cushing disease is part of Cushing syndrome, which is characterized by elevation of circulating cortisol concentrations and clinical manifestations of glucocorticoid excess. This disease was first reported by Harvey Cushing in a 23-year-old female with "obesity, hypertrichosis and amenorrhea."[18] Generally, "Cushing disease" indicates a subset of endogenous Cushing syndrome (~70%) caused by autonomous (neoplasia of corticotropes: major); (hyperplasia of corticotropes: rare) oversecretion of ACTH from the pituitary gland. In contrast, the term "Cushing syndrome" usually indicates the disorder of adrenal origin, caused by either adrenocortical adenoma or carcinoma (~15%). The term "Cushing syndrome" also includes the ectopic ACTH syndrome (~15%), which is caused by ectopic production of ACTH from various non-pituitary tissues, such as small cell carcinoma of the lung and intestinal carcinoids, and other several disorders that cause excessive production of glucocorticoids.[47] Cushing disease is observed more frequently in females than in males (ratio: 8 : 1). Absence of the diurnal rhythm in plasma ACTH and cortisol secretion may be the first manifestation in patients with Cushing disease; therefore, measurement of midnight serum cortisol concentration is helpful for the diagnosis. Once elevated circulating cortisol concentrations are documented following the overnight dexamethasone suppression test (1 mg dexamethasone, administration (p.o.) at 2200 h and measurement of serum cortisol the following morning at 0800 h), Cushing syndrome of adrenal origin can be distinguished from that caused by excess ACTH and cortisol, secretion by measuring plasma concentrations of ACTH, which are suppressed.[48] Cushing disease of pituitary origin can be further differentiated from ectopic ACTH syndrome by using the 8-mg (high-dose) dexamethasone suppression test either as a single dose at midnight and follow-up measurement of serum cortisol at 0800h the next morning, or as 2-mg dexamethasone administration every 6 hours for 2 days and measurement of urinary free cortisol, serum cortisol and plasma ACTH. ACTH secretion is partially responsive to dexamethasone in Cushing disease, while there is virtually no response in the ectopic ACTH syndrome.[48] The ectopic ACTH syndrome is also frequently characterized by hypokalemic alkalosis, while Cushing disease usually is not. The CRH stimulation test is helpful to distinguish the subtypes of Cushing syndrome.

Cushing disease and Cushing syndrome share many clinical characteristics – the "Cushingoid features" – such as central obesity, moon face, "buffalo hump," skin striae, hypertension, insulin resistance/overt diabetes mellitus, mood/psychiatric changes, menstrual irregularity and osteoporosis.[18] Florid cases of Cushing syndrome are quite obvious; however, some demonstrate subtle changes, and the diagnosis can be quite difficult. The diagnosis should be made after careful evaluation of the clinical manifestations, the results of endocrine tests and imaging studies. Cushing disease is mostly caused by pituitary adenomas. They usually have a diameter of less than 1 cm (microadenoma), and are sometimes very difficult to detect even in the MRI scan with contrast material.[47] Bilateral inferior petrosal sinus sampling with CRH stimulation is helpful in some cases to localize pituitary adenomas inside the pituitary gland. Pituitary carcinoma that results in Cushing disease is rare, and can only be identified by invasion to extrasellar tissues and distant metastasis. Hyperplasia of the corticotropes may also cause Cushing disease, based on the evidence that CRH-producing extra-pituitary tumors develop hyperplasia of corticotropes and Cushing phenotype, but its incidence, as well as its clinical course, are not well known.[49] Resection of the ACTH-secreting pituitary adenoma by transsphenoidal surgery is the first-line treatment for Cushing disease.[50] Initial remission by this surgical procedure is achieved in approximately 70–85% of the patients; however, 10–15% of patients with remission may recur in months or years. Bilateral adrenalectomy may be performed in cases in which tumors were not successfully resected.[18] Radiation may also be applied to the pituitary gland in combination with medical therapy, usually with mitotane or ketoconazol. Nelson syndrome (i.e., appearance of a pituitary macroadenoma after bilateral adrenalectomy) may occur, and is characterized by hyperpigmentation due to elevated circulating ACTH and symptoms from the space-occupying lesion. The gamma-knife, a conventional or proton accelerator with stereotactic multiple arc radiators, and a newer frameless sterotactic radiosurgery system, the cyber-nife, are also used for the treatment of Cushing disease as a secondary therapeutic approach in patients with surgical failure.[51] Some chemical compounds, such as ketoconazole, metyrapone, dopamine (D2) receptor agonists, and somatostatin analogs with the somatostatin receptor 5 (SSTR5) specificity, are also used with limited effectiveness.

Pituitary Disorders that Result in Hypofunction of the HPA System

Destructive lesions in the pituitary gland can cause hypofunction of the HPA system and secondary adrenal insufficiency. Sheehan syndrome, which is caused by pituitary apoplexy after labor and consequent pan-hypopituitarism, used to be a major cause of adrenal insufficiency of pituitary origin, but its incidence has been dramatically reduced due to the improvement of maternal care in Western countries.[52,53] Instead, lymphocyte infiltration and formation of a pituitary mass followed by partial or complete hypopituitarism has been observed among women in pregnancy or postpartum.[54–56] This disorder, called lymphocytic hypophysitis, was first reported in autopsy cases in 1962. The patients presented with adrenal insufficiency, other pituitary deficiencies, and symptoms associated with mass effect, such as headache and loss of visual field. The cause of this disease is not known but is apparently mediated by immune mechanisms, as some patients have anti-pituitary antibodies and other autoimmune endocrine disorders, such as thyroiditis and adrenalitis. Virus-mediated activation of an autoimmune reaction against the pituitary gland has also been suggested.

Isolated ACTH deficiency is a rare disorder characterized by decreased secretion of ACTH and subsequent development of adrenal insufficiency, but normal secretion of other pituitary hormones.[57] Some of these patients have associated autoimmune disorders, such as Graves disease, Hashimoto disease and lymphocytic hypophysitis; therefore, immune-mediated destruction of corticotropes may be one of the causative mechanisms for this disorder.[57] Patients with isolated ACTH deficiency tend to be relatively well during unstressed periods, but may develop an acute adrenal crisis after triggering events.[57] Similarly to what happens with hypothalamic lesions, sarcoidosis, histiocytosis and tuberculosis may also destroy the pituitary gland and can lead to secondaray adrenal insufficiency.

Hereditary Disorders Caused by Mutations in Molecules Necessary for Pituitary Development/ Function Resulting in Adrenal Insufficiency

The anterior pituitary gland originates from evagination of the stomodeal ectoderm (Rathke's pouch), and fuses to the neural component posterior lobe originated from the bottom of diencephalon to form the final pituitary gland.[58] Inside the anterior pituitary gland, the specific cell lineages secreting the pituitary hormones are also differentiated from common precursor cells.[58] This morphological and cellular differentiation of the anterior pituitary gland becomes possible under the influence of various extracellular factors, including members of the bone morphogenetic proteins and the fibroblast growth factors, with coordinated expression of specific transcription factors, including classes of factors linked to development of other organs, such as LIM homeodomain, paired-like homeodomain, bicoid-like homeodomain, and sine oculus-related factors.[58] In corticotropes, the production and secretion of ACTH and its precursor POMC are essential for their function. Thus, gene mutations of such transcription factors and the molecules essential for corticotrope functions cause developmental defects and/or dysfunction in these cells, and lead to a functional deficit in the HPA system. Since some of the transcription factors important for corticotrope differentiation also play a role in the development of other pituitary cell lineages, as well as that of brain and other organs, patients harboring inactivating mutations of these genes demonstrate manifestations of other pituitary hormone deficits and/or impaired development and function of the respective organs. So far, pathologic mutations that cause adrenal insufficiency are reported in the genes encoding the HESX homolog 1 (HESX1), orthodenticle homeobox 2 (OTX2), LIM homeobox 4 (LHX4), SRY (sex-determining region Y)-box 2 and 3 (SOX2 and SOX3), PROP paired-like homeobox 1 (PROP1) and T-box 19 (TBX19).[59–61] Mutations of these genes result in dysfunction of other pituitary hormones, such as growth hormone, thyrotropin and the gonadotropins, while some also result in short stature, metabolic disorders, craniofacial abnormalities and learning difficulties/mental retardation associated with abnormalities in the CNS.[62,63] The onset of adrenal insufficiency occurs generally early in life, whereas adrenal insufficiency owing to PROP1 mutations usually manifests in adolescence.[61–63]

Inactivating mutations in the POMC gene also result in adrenal insufficiency. Patients harboring these mutations develop a characteristic syndrome of hypocortisolism, early-onset morbid obesity, and alterations in skin and hair pigmentation.[64] Some patients also develop adolescent-onset deficiency of TSH, gonadotropins and growth hormone.[65–67] These findings indicate that POMC or the molecules produced from POMC, such as melanocortin-stimulating hormones, may play a role in the regulation of hypothalamic synthesis of the thyrotropin- and/or gonadotropin-releasing hormones.[64]

Adrenal Disorders Associated with Alterations in the HPA System

Cushing Syndrome

Cushing syndrome is caused by glucocorticoid-producing adenomas, micro-nodular dysplasia, ACTH-independent macro-nodular hyperplasia or adenocarcinoma of the adrenal cortex.[13,18] Patients with this disorder develop symptoms and signs associated with glucocorticoid excess; however, their manifestations depend on the degree and duration of hypercortisolism. As some tumors (especially adrenal carcinomas) may also produce androgens, manifestations associated with androgen excess, such as hirsutism and seborrhea, can be observed.[68,69] Some tumors may also produce aldosterone precursors, such as deoxycorticosterone (DOC), 18-hydroxy-DOC, corticosterone and 18-hydroxy corticosterone. Cushing syndrome due to adenocarcinoma is characterized by large tumors (often exceeding 5 cm in diameter), which have an acute and rapidly progressing clinical course, while adenomas are usually small (less than 3 cm) and have a generally insidious onset and chronic course.[68,69] First-choice treatment for adrenal Cushing syndrome is to remove the tumors surgically. Laparoscopic surgery is now the approach of choice for the treatment of Cushing syndrome caused by adenomas.[70] For cases with failure of resection of the entire tumors (usually adenocarcinomas), or those with metastatic lesions, mitotane (o,p′-DDD), an isomer of the insecticide para (p′) p′-DDD and a chemical congener of the insecticide dichlorodiphenyltrichloroethane, is used for suppressing both tumor growth and production of steroids.[68] Mitotane suppresses steroid synthesis by inhibition of the cholesterol side-chain cleavage (SCC) enzyme (human P450 CYP, cholesterol desmolase, or 20, 22 desmolase) and 11β-hydroxylase (P450 11β or CYP11B1)[47] (Fig. 29.3). Mitotane also has adrenotoxic effects through metabolic transformation of tumors and by inducing oxidative damage.[47,71] This chemical compound is also used as an adjunctive medicine for chemotherapy of adrenocortical carcinomas. Adrenal enzyme inhibitors, such as ketoconazole, metyrapone and etomidate, are also used in patients with surgical failure who continue to be hypercortisolemic.[47]

Adrenal Disorders that Result in Hypofunction of the HPA System

Hypofunction of the HPA system caused by disorders of the adrenal cortex is called primary adrenal insufficiency, or Addison disease. Any pathologic condition associated with destruction of the adrenal cortex can cause this disease. Addison disease is usually characterized by destruction of all three layers of the adrenal cortex; therefore, patients have decreased secretion of glucocorticoids, mineralocorticoids and adrenal androgens. Symptoms include those owing to glucocorticoid deficiency (fatigue, lethargy, nausea, vomiting, constipation, abdominal pain, diarrhea, anorexia and weight loss, hypoglycemia, pigmentation and eosinophilia), mineralocorticoid deficiency (hyponatremia, hypokalemia, hypotension), and adrenal androgen deficiency (only in women: loss of pubic and axillary hair, decreased libido).[18] The disorders causing Addison disease can be systemic or local: They include autoimmune adrenalitis (either isolated or as part of polyendocrine autoimmune syndrome: most frequent), and infectious diseases destroying the adrenal cortex, such as those caused by tuberculosis and cytomegalovirus. The former (tuberculosis) was the most common cause of Addison disease previously, but now represents 7–20% in Western countries.[61] Both are now frequently observed in patients with acquired immunodeficiency syndrome (AIDS) caused by infection with the human immunodeficiency type-1 virus. Bilateral adrenal hemorrhage caused by sepsis of meningococci, staphylococci and other bacteria (Waterhouse-Friederichen syndrome), histoplasmosis and metastasis of malignant tumors, such as lung and breast cancers, also destroy adrenal glands and may cause adrenal insufficiency.[61] The polygrandular autoimmune syndrome (PGA) consists of two subtypes, PGA types I and II. The former, also called autoimmune polyendocrinopathy–candidiasis–ectodermal dysplasia (APECED), is a rare autosomal recessive hereditary disorder caused by mutations of the autoimmune regulator (AIRE) gene, which produces a transcription factor that plays an essential role in the ectopic expression of peripheral tissue-restricted antigens in the thymic medullary cells for developing immune self-tolerance.[72–74] In addition to autoimmune adrenalitis, patients with this disease develop recurrent mucocutaneous candidiasis, hypoparathyroidism and ectodermal dysplasia, such as dental enamel hypoplasia, pitted dystrophy of the nails, keratopathy, and calcified plaques of the tympanic membranes. They also develop type 1 diabetes mellitus and autoimmune thyroiditis (Hashimoto disease), although these are much less frequent. PGA type II is characterized by primary adrenal insufficiency as the principal manifestation, but may also be accompanied by primary hypothyroidism, goiter, Graves disease,

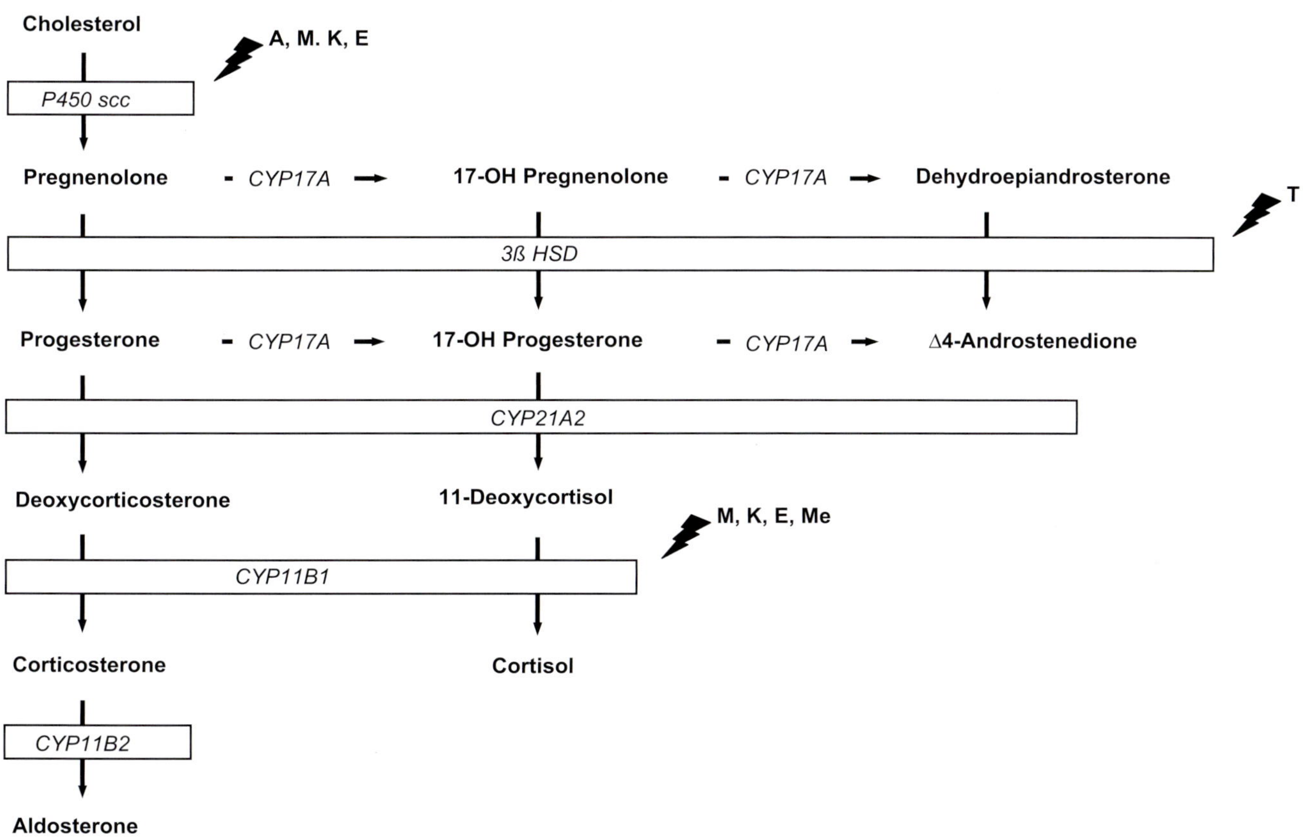

FIGURE 29.3 **Biosynthetic pathways of steroid hormones and their precursors and steroidogenic enzymes in the adrenal cortex.** Portion of the steroid biosynthesis where chemical compounds inhibit are shown. A, aminoglutethimide; CYP11B1, 11β-hydroxylase; CYP11B2, aldosterone synthase; CYP17A, 17α-hydroxylase/17, 20-lyase; CYP21A2, 21-hydroxylase; E, etomidate; 3βHSD, 3β-hydroxysteroid dehydrogenase; K, ketoconazole; M, mitotane; Me, metyrapone; P450 scc, P450 side chain-cleavage enzyme/desmolase; T, trilostane.

diabetes mellitus, hypoparathyroidism, pernicious anemia, Sjögren syndrome and celiac disease.[72] About half of the cases are familial but their modes of inheritance are heterogeneous, indicating that multiple genes could contribute to the development of this disease. An association of PGA type II with HLA DR3/4 region was reported.[72,75]

Hereditary Disorders Caused by Mutations in Molecules Necessary for Adrenal Development/Function, Resulting in Adrenal Insufficiency

Similar to hereditary diseases of the pituitary gland, inactivating mutations in the genes whose products play roles in the development and/or function of the adrenal cortex (such as those sensing signal of circulating ACTH or involved in steroid hormone production) cause adrenal dysfunction and lead to primary adrenal insufficiency.

Congenital adrenal hyperplasia (CAH) is one of the most well-known causes of hereditary adrenal disorders, with an overall incidence of 1 case in 15,000 live births.[61] The disease is due to mutations in the genes encoding several key enzymes for the steroidogensis, such as 3β-hydroxysteroid dehydrogenase (3βHSD), 21-hydroxylase (CYP21A2), steroid 11β-hydroxylase (CYP11B1), steroid 17α-hydroxylase (CYP17A) and P450 (cytochrome) oxidoreductase (POR)[76,77] (Fig. 29.3). Affected subjects have decreased serum cortisol and increased plasma ACTH concentrations. Additional manifestations depend on the specific enzymatic deficiency, and may include increased or decreased secretion of adrenal androgens and/or precursors with mineralocorticoid effects.[76,77]

Congenital adrenal hypoplasia is a hereditary disorder of the adrenal gland development, with an overall estimated frequency of 1 in 12,500 live births.[78,79] There are at least five genetic subtypes that have been characterized clinically. Among them, a cause of the X-linked cytomegalic form, which is also associated with hypogonadotropic hypogonadism, is ascribed to mutations in the DAX1 (dosage-sensitive sex reversal, adrenal hypoplasia critical region, on chromosome X, gene 1) gene, which is located in chromosome Xp21 and whose product is a member of the nuclear receptor family that plays an important role in the development and function of the adrenal glands, pituitary gland and hypothalamus.[78] Mutations in the steroidogenic factor-1 (SF1), another nuclear receptor family protein playing essential roles in the development and function of the adrenal gland and reproductive organs, also cause autosomally transmitted congenital adrenal hypoplasia with complete phenotypic sex-reversal, including persistent Müllerian structures.[80,81] Congenital lipoid adrenal hyperplasia (lipoid CAH) is an autosomal recessive disorder characterized by severe adrenal insufficiency and male sex reversal. This disorder is caused by mutations in either of the two genes encoding the steroidogenic acute regulatory protein (StAR) or the cytochrome P450 side-chain cleavage (scc) enzyme.[82] The former protein is essential for transfer of substrate cholesterol for steroidogenesis from the outer to the inner membrane of the mitochondria, while the latter catalyzes the conversion of cholesterol to pregnenolone (the first step of steroidogenesis); thus, they are essential for biosynthesis of all steroid hormones in the adrenal glands, testis and ovary.[82]

Wolman disease is a rare autosomal recessive lipid storage disease caused by mutations in the lipase A, lysosomal acid, cholesterol esterase (LIPA) gene, developing massive accumulation of cholesterol esters and triglycerides, and subsequent tissue damage.[83] Patients with this disease are characterized by adrenal insufficiency with calcification of the adrenal cortex, progressive mental retardation, hepatosplenomegaly, jaundice and steatorrhea.[83]

Adrenoleukodystrophy or adrenomyeloneuropathy is characterized by primary adrenal insufficiency and progressive neurologic dysfunction with increased plasma levels of very-long-chain saturated fatty acids and massive accumulation of their cholesterol esters and gangliosides in the adrenal cortex, CNS and other organs.[84] They are X-linked recessive hereditary disorders caused by mutations in the ATP-binding cassette, subfamily D, member 1 (ABCD1) or member 2 (ABCD2) genes, whose products catalyze transport of fatty acids in the peroxisome.[84–86] Mutations in the former gene lead to the severe phenotype, while those of the latter represent the milder form.

Mutations in the ACTH receptor gene cause primary adrenal insufficiency as a result of impaired responsiveness to circulating ACTH (familial glucocorticoid deficiency (FGD) type 1).[87,88] In addition to low or undetectable circulating cortisol and increased plasma ACTH concentrations, this syndrome is characterized by defects in adrenal androgens, while the secretion of mineralocorticoids is preserved, as this steroid production is also regulated by the renin/angiotensin system. Patients with this syndrome present with adrenal insufficiency, hyperpigmentation, hypertelorism, frontal bossing and muscle weakness. A case of activating mutations in the ACTH gene has been reported (ACTH hypersensitivity syndrome).[89] This patient had undetectable concentrations of plasma ACTH, but her serum cortisol concentrations were normal. Mutations in the ACTH receptor gene account only for approximately 25% of all familial glucocorticoid deficiency cases, while mutations in the melanocortin 2 receptor accessory protein (MRAP) gene also result in this hereditary disorder (FGD type 2) due to defective transfer of the synthesized ACTH receptor protein from the endoplasmic reticulum to the cell surface.[90,91] FGD type 2 patients tend to develop symptoms earlier than those with FGD type 1, and do not present with tall stature.[90]

Triple-A syndrome, also known as Achalasia–Adrenal insufficiency–Alacrima syndrome or Allgrove syndrome, is a rare autosomal recessive disorder characterized by ACTH-resistant adrenal insufficiency, alacrima (absence of tear secretion), achalasia (dilatation of thoracic esophagus), autonomic dysfunction and neurodegeneration.[58] This syndrome is caused by mutations in the AAAS gene, which encodes the nuclear pore protein ALADIN with as yet unknown functions.[58,92] Adrenal insufficiency develops gradually within the first 10 years of life, but sometimes becomes evident much later. It may present in some cases as an adrenal crisis with hypoglycemia and seizures.[92]

Other Disorders Associated with Changes in the HPA System

Generalized Glucocorticoid Resistance Syndrome (Chrousos Syndrome)

Mutations in the GR gene result in primary generalized glucocorticoid resistance, also known as Chrousos syndrome (see recent review by Charmandari and Kino[93]). The condition is characterized by hypercortisolism caused by inactivating mutations of the GR gene without Cushingoid features.[94,95] To compensate for the reduced target-tissue sensitivity to glucocorticoids, affected subjects have elevated circulating concentrations of cortisol and ACTH, and resistance of the HPA system to dexamethasone suppression, but no clinical manifestations of hypercortisolism. The excess ACTH secretion results in increased production of adrenal steroids with mineralocorticoid activity and/or androgenic activity; the former accounts for symptoms and signs of mineralocorticoid excess, while the latter accounts for manifestations of androgen excess. In adult cases, patients with this syndrome do not develop adrenal insufficiency due to compensatorily increased ACTH and cortisol secretion; however, symptoms suggestive of adrenal insufficiency have been reported in a child and a newborn baby with the condition.[96,97]

More than 10 kindreds and sporadic cases with abnormalities in the GR number, affinity for glucocorticoid, stability, and translocation into the nucleus have been reported.[93] The molecular defects have been elucidated in six kindreds and seven sporadic cases (Fig. 29.4). Most of the mutations are found in the GR LBD, which alters functions of this subdomain, such as ligand-binding, activation of transcription and nuclear translocation. So far, only one mutation has been reported in the DBD of GR, that replaces arginine at amino acid 477 by histidine (R477H).[98] This mutant receptor does not bind to GREs, but preserves normal affinity for dexamethasone.[98,99] Only one mutation has also been reported in the GR NTD, that replaces aspartic acid at amino acid 401 by histidine (D401H).[100] The patient harboring this heterozygous mutation

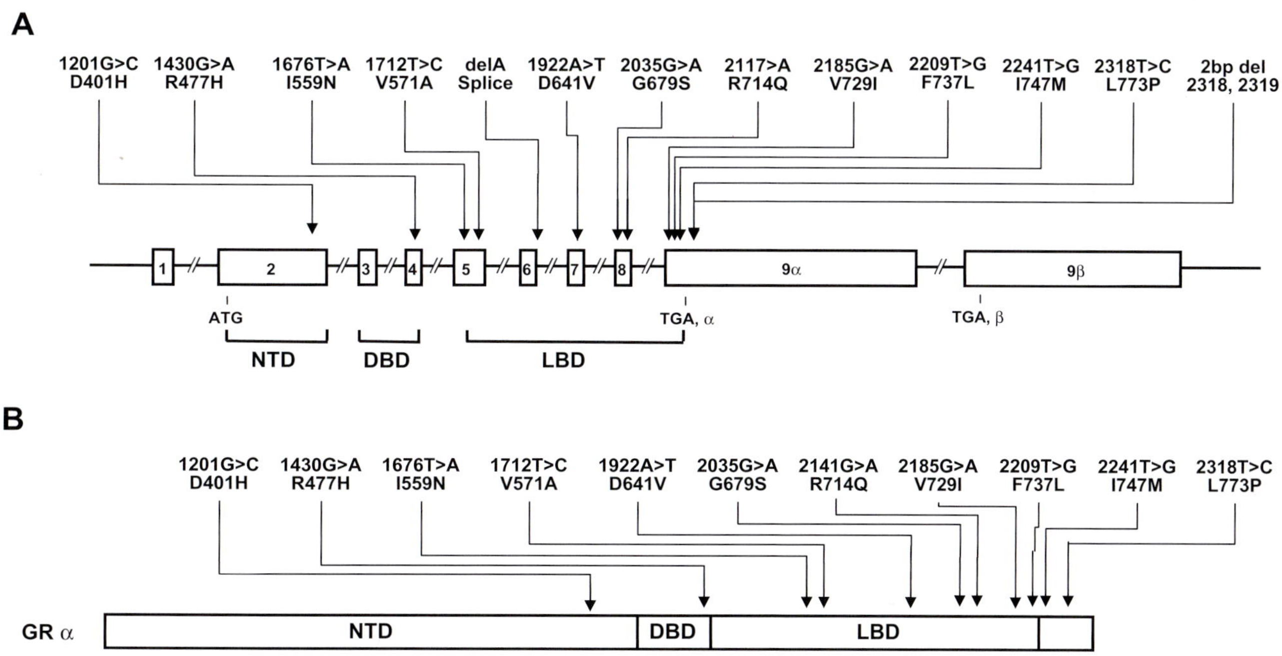

FIGURE 29.4 Location of the known mutations of the human GR in the genomic structure (A) and in its linearized protein molecule in patients with Chrousos syndrome or its mirror image, sporadic glucocorticoid hypersensitivity (B). Nucleoside numbers for mutated sites are determined by the definition of the adenine of the translation initiation site as number 1. *Modified from Charmandari and Kino (2010).*[93]

demonstrated manifestations consistent with glucocorticoid hypersensitivity, in accordance with the *in vitro* result that the mutant receptor demonstrated a 2.4-fold increase in its ability to transactivate glucocorticoid-responsive promoters. This condition represents the mirror image of the generalized glucocorticoid resistance syndrome. All reported pathologic mutations causing familial or sporadic generalized glucocorticoid resistance syndrome, also known as Chrousos syndrome, are listed in Table 29.3.[93]

GR Polymorphisms

Polymorphisms of the GR gene have also been reported. A heterozygous polymorphism replacing aspartic acid to serine at amino acid 363 that mildly increases the transcriptional activity of the affected receptor *in vitro* is associated with increased sensitivity to glucocorticoids, weakly correlating with the development of central obesity, and thus influencing the metabolic profile and the longevity of humans in a negative fashion.[101–103] This polymorphism found at amino

TABLE 29.3 Mutations in the Human Glucocorticoid Receptor Gene causing Generalized Glucocorticoid Resistance Syndrome (Chrousos Syndrome)

	Mutation position					
Authors	**cDNA***	**Amino acid**	**Molecular defects**	**Genotype**	**Phenotype**	**Reference(s)**
Chrousos *et al.* Hurley *et al.*	1922A>T	D641V	Transactivation↓ Affinity to ligand↓ (×3) Nuclear translocation: 22 min Abnormal interaction with GRIP1	Homozygous	Hypertension Hypokalemic alkalosis	95,121
Karl *et al.*	4-bp deletion in exon-intron 6		GRα number: 50% reduction Inactivation of affected allele	Heterozygous	Hirsutism Male-pattern hair-loss Menstrual irregularities	104
Malchoff *et al.*	2185G>A	V729I	Transactivation↓ Affinity to ligand↓ (×4) Nuclear translocation: 120 min Abnormal interaction with GRIP1	Homozygous	Precocious puberty Hyperandrogenism	122
Karl *et al.* Kino *et al.*	1676T>A	I559 N	Transactivation↓ Transdominance (+) Decrease in GR binding sites Nuclear translocation: 180 < min Abnormal interaction with GRIP1	Heterozygous	Hypertension Oligospermia Infertility	123,124
Ruiz *et al.* Charmandari *et al.*	1430G>A	R477H	Transactivation↓ No GREs binding Decrease in GR binding sites Nuclear translocation: 20 min	Heterozygous	Hirsutism Fatigue Hypertension	98,99
Ruiz *et al.* Charmandari *et al.*	2035G>A	G679S	Transactivation↓ Affinity to ligand↓ (×2) Nuclear translocation: 30 min Abnormal interaction with GRIP1	Heterozygous	Hirsutism Fatigue Hypertension	98,99
Mendonca *et al.*	1712T>C	V571A	Transactivation↓ Affinity to ligand↓ (×6) Nuclear translocation: 25 min Abnormal interaction with GRIP1	Homozygous	Ambiguous genitalia Hypertension Hypokalemia Oligo-amenorrhea	125
Vottero *et al.*	2241T>G	I747 M	Transactivation↓ Transdominance (+) Affinity to ligand↓ (×2) Nuclear translocation↓ Abnormal interaction with GRIP1	Heterozygous	Cystic acme Hirsutism Oligo-amenorrhea	126

TABLE 29.3 Mutations in the Human Glucocorticoid Receptor Gene causing Generalized Glucocorticoid Resistance Syndrome (Chrousos Syndrome)—cont'd

	Mutation position					
Authors	cDNA*	Amino acid	Molecular defects	Genotype	Phenotype	Reference(s)
Charmandari *et al.*	2318T>C	L773P	Transactivation↓ Transdominance (+) Affinity to ligand↓ (×2.6) Nuclear translocation: 30 min Abnormal interaction with GRIP1	Heterozygous	Fatigue Anxiety Acne Hirsutism Hypertension	127
Charmandari *et al.*	2209T>G	F737L	Transactivation↓ Transdominance (time-dependent) (+) Affinity to ligand↓ (×1.5) Nuclear translocation :180 min	Heterozygous	Hypertension Hypokalemia	128
Charmandari *et al.*	1201G>C	D401H	Transactivation↑ Transdominance (−) Affinity to ligand: no chnage Nuclear translocation: Normal	Heterozygous	Hypertenstion Diabetes mellitus Accumulation of visceral fat	100
McMahon *et al.*	2-bp (TG) deletion at 2318 and 2319	F774SfsX24	No transactivation activity No ligand-binding activity	Homozygous	Severe hypoglycemia developed 1 day after birth Hypertension Fatigues with feeding	97
Nader *et al.*	2141G>A	R714Q	Transactivation↓ Transdominance (+) Affinity to ligand↓ (×2.0) Nuclear translocation: 20 min Abnormal interaction with GRIP1	Heterozygous	Hypoglycemia developed at age 2 years and 10 months Hypertension Accelerated bone age Mild clitoromegaly	96

* *Nucleoside numbers for mutated sites are determined by the definition of the adenine of the translation initiation site as number 1.*
Reproduced from Charmandari E. Kino T. Chrousos syndrome: a seminal report, a phylogenetic enigma and the clinical implications of glucocorticoid signalling changes. Eur J Clin Invest. *2010;40, 932–942, with permission.*

acid 363 was first described by Karl *et al.*[104] The polymorphism in the GR gene that causes arginine to lysine replacement at amino acid 23 (ER22/23EK: GAG AGG to GAA AAG) is associated with relative glucocorticoid resistance by altering the expression levels of GR translational isoforms.[105] This polymorphism increases muscle mass in males and reduces waist to hip ratio in females, and is associated with greater insulin sensitivity, and lower total and low-density lipoprotein cholesterol levels, indicating that this polymorphism has a beneficial effect on longevity by reducing glucocorticoid action.[106,107]

Ectopic ACTH- or CRH-Producing Tumors

Ectopic ACTH-producing tumors account for ~15% of Cushing syndrome.[49,108,109] Most of the tumors developing ectopic ACTH syndrome are malignant, and are seen more frequently in men than in women. These tumors are usually of neuroendocrine cell origin, but adenocarcinoma and squamous cell carcinoma may also result in this syndrome.[108,110] Approximately half of the cases are caused by small cell carcinoma of the lung. Carcinoids originating from the thymus, bronchi and other organs, pancreatic islet tumors, pheochromocytomas and ovarian adenocarcinomas may also be associated with this syndrome.[110] Patients with this syndrome demonstrate elevated concentrations of serum cortisol and plasma ACTH, while serum concentrations of other POMC-related peptides and bioactive molecules, such as other pituitary hormones, calcitonin and somatostatin are sometimes elevated due to their ectopic production by the tumors.[110] Hypokalemia occurs in many cases with this syndrome.[18] The onset of manifestations is sudden, and their progression is usually rapid and sometimes not responsive to treatment. Neuroendocrine tumors, such as small cell carcinoma of the prostate and lung, medullary carcinoma of the thyroid gland and hypothalamic gangliocytoma, can lead to Cushing syndrome by ectopically secreting CRH, but

BOX 29.1

HOW DO WE KNOW ABOUT FAMILIAL OR SPORADIC GENERALIZED GLUCOCORTICOID RESISTANCE SYNDROME OR CHROUSOS SYNDROME?

Cushing syndrome and Addison disease are representative pathologic entities associated with glucocorticoid excess and deficiency, respectively. The former was first identified and elucidated by Harvey Cushing in 1912,[1] while the latter was discovered by Thomas Addison 60 years earlier, in 1849. Almost 70 years after the description of Cushing syndrome, George P. Chrousos first reported the generalized glucocorticoid resistance syndrome.[2] Two patients, a severely hypertensive and alkalotic father and his asymptomatic son, demonstrated, respectively, severely and mildly elevated levels of circulating cortisol and ACTH, similarly to patients with Cushing syndrome, but not the typical clinical manifestations and phenotype associated with this disorder. The patients preserved the circadian rhythmicity of their circulating concentrations of cortisol and ACTH, while their glucocorticoid receptors per white cell were normal in number but had a reduced affinity for radiolabeled dexamethasone – more reduced in the father than in his son (see figure below).

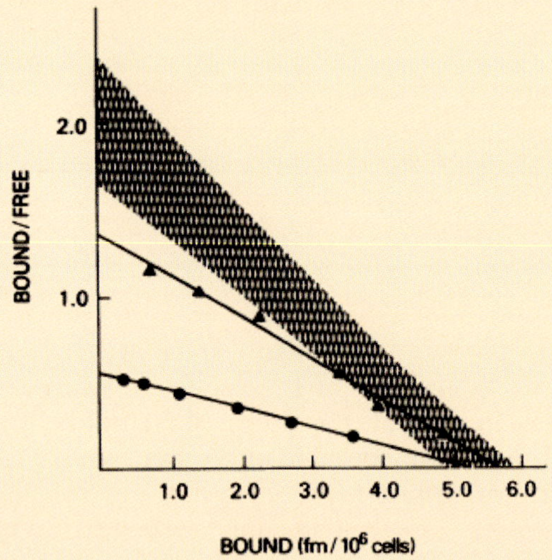

The results of the Scatchard analysis of dexamethasone binding to the glucocorticoid receptor of circulating mononuclear leukocytes in two patients with Chrousos syndrome, the propositi of the family, a severely symptomatic father and his asymptomatic son, and 10 controls. The shaded area represents the normal range. The concentration of receptors indicated by the x-axis intercept is within the normal range. The apparent affinity indicated by the slope of the line (slope = −1/Kd) is clearly different in the father (●) and just outside the normal range in his son (▲). *Reproduced from Chrousos GP, Vingerhoeds A, Brandon D, Eil C, Pugeat M, DeVroede M, Loriaux DL, Lipsett MB. Primary cortisol resistance in man. A glucocorticoid receptor-mediated disease.* J Clin Invest. *1982;69:1261–1269, with permission.*

These data suggested that the primary cause of the cortisol elevations and/or the clinical manifestations of the patients was the reduced ability of their glucocorticoid receptors to transduce the glucocorticoid signal in target tissues, including the hypothalamus and the pituitary. After the human GR was cloned by Ronald M. Evans and co-workers in 1985,[3] Chrousos' group identified the first mutation of the GR gene causing the generalized glucocorticoid resistance syndrome (see figure below).[4]

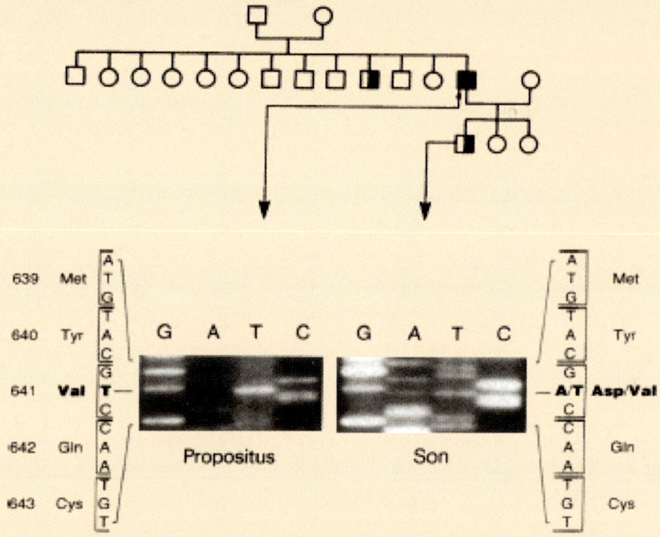

Nucleotide sequenceof hGR cDNA from the propositus with Chrousos syndrome and his son. The propositus has T in place of A at nucleotide 2054, changing the aspartate codon GAC, normally present at position 641, to the valine codon GTC, indicating homozygocity. Both A and T are present at this position in the son's sequence, indicating heterozygosity. *Reproduced from Hurley DM, Accili D, Stratakis CA, Karl M, Vamvakopoulos N, Rorer E, Constantine K, Taylor SI, Chrousos GP. Point mutation causing a single amino acid substitution in the hormone binding domain of the glucocorticoid receptor in familial glucocorticoid resistance.* J Clin Invest. *1991;87:680–686, with permission.*

The differential diagnosis of the generalized glucocorticoid resistance syndrome includes mild or early Cushing syndrome, as it shares with this syndrome the elevation of circulating cortisol and ACTH, the increased urinary free cortisol excretion, the resistance to dexamethasone suppression, and the elevated arterial blood pressure. In contrast to Cushing syndrome, however, patients with Chrousos syndrome preserve the diurnal fluctuation of serum ACTH and cortisol and the response of these hormones to stress, albeit at higher concentrations.

BOX 29.1 (cont'd)

References

1. Orth DN, Kovacs WJ, DeBold CR. The adrenal cortex. In: Wilson JD, Foster DW, Kronenberg HM, Larsen R, eds. *Textbook of Endocrinology.* 9th ed. Philadelphia: W.B. Saunders Co; 1998;517–664.
2. Chrousos GP, Vingerhoeds A, Brandon D, Eil C, Pugeat M, DeVroede M, Loriaux DL, Lipsett MB. Primary cortisol resistance in man. A glucocorticoid receptor-mediated disease. *J Clin Invest.* 1982;69:1261–1269.
3. Hollenberg SM, Weinberger C, Ong ES, Cerelli G, Oro A, Lebo R, Thompson EB, Rosenfeld MG, Evans RM. Primary structure and expression of a functional human glucocorticoid receptor cDNA. *Nature* 1985;318:635–641.
4. Hurley DM, Accili D, Stratakis CA, Karl M, Vamvakopoulos N, Rorer E, Constantine K, Taylor SI, Chrousos GP. Point mutation causing a single amino acid substitution in the hormone binding domain of the glucocorticoid receptor in familial glucocorticoid resistance. *J Clin Invest.* 1991;87:680–686.

this condition is very rare.[111,112] These tumors sometimes produce ACTH in addition to CRH, and clear-cut diagnosis of this syndrome may be difficult.

Alterations of the HPA System by Exogenous Glucocorticoids and Chemical Compounds

Iatrogenic Cushing syndrome due to administration of increased doses of glucocorticoids for the treatment of allergic, autoimmune, inflammatory and lymphoproliferative diseases is the most common among all conditions associated with glucocorticoid excess. Indeed, asthma affects over 30 million people in the United States, while autoimmune diseases, including rheumatoid arthritis, systemic lupus erythematosus, glomerulonephritis, psoriasis and autoimmune thyroiditis, affect almost 10 million Americans. Glucocorticoids are the most frequently prescribed medicines for all these conditions.[113] Pharmacologic amounts of glucocorticoids suppress the HPA system, and result in the development of manifestations associated with glucocorticoid excess or Cushingoid features. Among them, osteoporosis, muscle wasting, hypertension and insulin resistance/overt diabetes mellitus are the most frequently observed and problematic side effects.[113,114] To overcome these complications of excess glucocorticoids, several selective GR agonists of steroid or non-steroidal origin, such as RU24858, RU40066, RU24782, AL-438, ZK216348 and compound A, have been developed and tested with limited efficacy.[114,115]

Several compounds inhibiting key enzymes of steroidogenesis result in adrenal insufficiency by suppressing production of cortisol from the adrenal glands. They are usually used as test compounds for evaluating the HPA system for differential diagnosis of Cushing syndrome, for suppression of cortisol and other steroid production from cortisol-producing adrenal carcinoma, and/or for treatment of hormone-sensitive breast cancers. Among them, aminoglutethimide inhibits the P450scc enzyme and suppresses synthesis of all steroid hormones, while trilostane inhibits the 3β-hydroxysteroid dehydrogenase. Metyrapone inhibits 17β-hydroxylase and reduces production of cortisol and aldosterone, while etomidate suppresses enzymatic activity of 17β-hydroxylase and P450scc[47,116] (Fig. 29.3). Mitotane was discussed above, regarding Cushing disease and syndrome. Ketoconazole is an antifungal compound, which can suppress steroidogenesis by inhibiting P450scc and 17β-hydroxylase.[47,116] Carbamazepine is a frequently used therapeutic agent for mood disorders and seizures, and stimulates the HPA system and increases circulating cortisol levels by enhancing ACTH response to CRH at the pituitary level.[117] Imipramine is a tricyclic antidepressant and modulates GR transcriptional activity by acting as a partial GR agonist,[118,119] while mefipristone (RU 486), a potent progesterone antagonist used for medical termination of intrauterine pregnancies, also inhibits GR activity by acting as a potent GR antagonist.[120]

CONCLUSIONS

The HPA system plays a central role in the adaptive response to stress, and is essential for the maintenance of internal homeostasis of virtually all organs and tissues through its end-effectors, the glucocorticoids. Dysfunction in any of the components of this regulatory system can cause manifestations associated with alterations expressed either as glucocorticoid excess or deficiency. Causes of such dysfunction may be hereditary or acquired. The latter consist of numerous types of disorders, including trauma-related dysfunction, and autoimmune, metabolic, psychiatric and neoplastic diseases. The approach to the differential diagnosis of disorders of the HPA system includes hormonal and imaging evaluations. Accordingly, treatment is designed for correction of both the primary cause and the associated hormonal abnormality. During the past several decades, significant progress has been made in the clinical care of patients with disorders of the HPA system, but there are still many issues to be resolved.

Intensive efforts should be continuously made to further elucidate the pathophysiology, improve the differential diagnosis and refine the therapy of some of these disorders.

Acknowledgments

Literary work of this article was funded partly by the Intramural Research Program of the *Eunice Kennedy Shriver* National Institute of Child Health and Human Development, National Institutes of Health, Bethesda, MD, the Biomedical Research Foundation of the Academy of Athens, and the University of Athens, Athens, Greece.

References

1. Chrousos GP. Stress and disorders of the stress system. *Nat Rev Endocrinol.* 2009;5:374–381.
2. Chrousos GP, Kino T. Glucocorticoid action networks and complex psychiatric and/or somatic disorders. *Stress.* 2007;10: 213–219.
3. Chrousos GP. The hypothalamic–pituitary–adrenal axis and immune-mediated inflammation. *N Engl J Med.* 1995;332: 1351–1362.
4. Nader N, Chrousos GP, Kino T. Interactions of the circadian CLOCK system and the HPA axis. *Trends Endocrinol Metab.* 2010;21:277–286.
5. Chrousos GP, Calabrese JR, Avgerinos P, Kling MA, Rubinow D, Oldfield EH, Schuermeyer T, Kellner CH, Cutler Jr GB, Loriaux DL, et al. Corticotropin releasing factor: basic studies and clinical applications. *Prog Neuropsychopharmacol Biol Psychiatry.* 1985;9:349–359.
6. Sawchenko PE, Imaki T, Potter E, Kovacs K, Imaki J, Vale W. The functional neuroanatomy of corticotropin-releasing factor. *CIBA Found Symp.* 1993;172:5–21. discussion 21–29.
7. Calogero AE, Bernardini R, Gold PW, Chrousos GP. Regulation of rat hypothalamic corticotropin-releasing hormone secretion *in vitro*: potential clinical implications. *Adv Exp Med Biol.* 1988;245:167–181.
8. Smith MA, Kling MA, Whitfield HJ, Brandt HA, Demitrack MA, Geracioti TD, Chrousos GP, Gold PW. Corticotropin-releasing hormone: from endocrinology to psychobiology. *Horm Res.* 1989;31:66–71.
9. Stevens, A. amp; White, A. ACTH: cellular peptide hormone synthesis and secretory pathways. *Results Probl Cell Differ.* 50: 63-84.
10. Bornstein SR, Chrousos GP. Clinical review 104: Adrenocorticotropin (ACTH)- and non-ACTH-mediated regulation of the adrenal cortex: neural and immune inputs. *J Clin Endocrinol Metab.* 1999;84:1729–1736.
11. Kino T, Chrousos GP. Glucocorticoid effect on gene expression. In: Steckler T, Kalin NH, Reul JMHM, ed. *Handbook on Stress and the Brain.* Amsterdam: Elsevier BV; 2005;295–312.
12. Holsboer F, Barden N. Antidepressants and hypothalamic–pituitary–adrenocortical regulation. *Endocr Rev.* 1996;17: 187–205.
13. Chrousos GP. Glucocorticoid therapy. In: Felig P, Frohman LA, eds. *Endocrinology & Metabolism.* 4th edn. New York: McGraw-Hill; 2001;609–632.
14. Belanoff JK, Gross K, Yager A, Schatzberg AF. Corticosteroids and cognition. *J Psychiatr Res.* 2001;35:127–145.
15. Fuchs E, Flugge G, Ohl F, Lucassen P, Vollmann-Honsdorf GK, Michaelis T. Psychosocial stress, glucocorticoids, and structural alterations in the tree shrew hippocampus. *Physiol Behav.* 2001;73:285–291.
16. Sapolsky RM. Glucocorticoids and hippocampal atrophy in neuropsychiatric disorders. *Arch Gen Psychiatry.* 2000;57: 925–935.
17. McEwen BS, Conrad CD, Kuroda Y, Frankfurt M, Magarinos AM, McKittrick C. Prevention of stress-induced morphological and cognitive consequences. *Eur Neuropsychopharmacol.* 1997;7(Suppl 3):S323–S328.
18. Orth DN, Kovacs WJ, DeBold CR. The adrenal cortex. In: Wilson JD, Foster DW, Kronenberg HM, Larsen R, eds. *Textbook of Endocrinology.* 9th edn. Philadelphia: W.B. Saunders Co.; 1998;517–664.
19. Canalis E, Delany AM. Mechanisms of glucocorticoid action in bone. *Ann NY Acad Sci.* 2002;966:73–81.
20. Raisz LG, Bingham PJ. Effect of hormones on bone development. *Annu Rev Pharmacol.* 1972;12:337–352.
21. Clowes JA, Peel N, Eastell R. Glucocorticoid-induced osteoporosis. *Curr Opin Rheumatol.* 2001;13:326–332.
22. Brem AS. Insights into glucocorticoid-associated hypertension. *Am J Kidney Dis.* 2001;37:1–10.
23. Ullian ME. The role of corticosteriods in the regulation of vascular tone. *Cardiovasc Res.* 1999;41:55–64.
24. Fauci AS, Dale DC, Balow JE. Glucocorticosteroid therapy: mechanisms of action and clinical considerations. *Ann Intern Med.* 1976;84:304–315.
25. Boumpas DT, Chrousos GP, Wilder RL, Cupps TR, Balow JE. Glucocorticoid therapy for immune-mediated diseases: basic and clinical correlates. *Ann Intern Med.* 1993;119:1198–1208.
26. Elenkov IJ. Glucocorticoids and the Th1/Th2 balance. *Ann NY Acad Sci.* 2004;1024:138–146.
27. Elenkov IJ, Chrousos GP. Stress system–organization, physiology and immunoregulation. *Neuroimmunomodulation.* 2006;13: 257–267.
28. Kino T, De Martino MU, Charmandari E, Mirani M, Chrousos GP. Tissue glucocorticoid resistance/hypersensitivity syndromes. *J Steroid Biochem Mol Biol.* 2003;85:457–467.
29. Chrousos GP, Kino T. Intracellular glucocorticoid signaling: a formerly simple system turns stochastic. *Sci STKE* 2005;2005: pe48.
30. Galon J, Franchimont D, Hiroi N, Frey G, Boettner A, Ehrhart-Bornstein M, O'Shea JJ, Chrousos GP, Bornstein SR. Gene profiling reveals unknown enhancing and suppressive actions of glucocorticoids on immune cells. *FASEB J.* 2002;16:61–71.
31. Kino T, Chrousos GP. Glucocorticoid and mineralocorticoid receptors and associated diseases. *Essays Biochem.* 2004;40: 137–155.
32. Savory JG, Hsu B, Laquian IR, Giffin W, Reich T, Hache RJ, Lefebvre YA. Discrimination between NL1- and NL2-mediated nuclear localization of the glucocorticoid receptor. *Mol Cell Biol.* 1999;19:1025–1037.
33. Burke CW. Adrenocortical insufficiency. *Clin Endocrinol Metab.* 1985;14:947–976.
34. Grinspoon SK, Biller BM. Clinical review 62: Laboratory assessment of adrenal insufficiency. *J Clin Endocrinol Metab.* 1994;79:923–931.
35. Woolf PD. Hormonal responses to trauma. *Crit Care Med.* 1992;20:216–226.
36. Bruni O, Verrillo E, Novelli L, Ferri R. Prader-Willi syndrome: sorting out the relationships between obesity, hypersomnia, and sleep apnea. *Curr Opin Pulm Med.* 2010;16:568–573.
37. Kitsiou-Tzeli S, Tzetis M, Sofocleous C, Vrettou C, Xaidara A, Giannikou K, Pampanos A, Mavrou A, Kanavakis E. *De novo* interstitial duplication of the 15q11.2-q14 PWS/AS region of maternal origin: Clinical description, array CGH analysis, and review of the literature. *Am J Med Genet A.* 2010;152A: 1925–1932.

38. de Lind van Wijngaarden RF, Otten BJ, Festen DA, Joosten KF, de Jong FH, Sweep FC, Hokken-Koelega AC. High prevalence of central adrenal insufficiency in patients with Prader-Willi syndrome. *J Clin Endocrinol Metab.* 2008;93: 1649–1654.
39. Wolff SM, Adler RC, Buskirk ER, Thompson RH. A syndrome of perodic hypothalamic discharge. *Am J Med.* 1964;36:956–967.
40. Gold PW, Chrousos GP. The endocrinology of melancholic and atypical depression: relation to neurocircuitry and somatic consequences. *Proc Assoc Am Physicians.* 1999;111:22–34.
41. Gold PW, Goodwin FK, Chrousos GP. Clinical and biochemical manifestations of depression. Relation to the neurobiology of stress (2). *N Engl J Med.* 1988;319:413–420.
42. Attia E. Anorexia nervosa: current status and future directions. *Annu Rev Med.* 2010;61:425–435.
43. Kaye WH, Gwirtsman HE, George DT, Ebert MH, Jimerson DC, Tomai TP, Chrousos GP, Gold PW. Elevated cerebrospinal fluid levels of immunoreactive corticotropin-releasing hormone in anorexia nervosa: relation to state of nutrition, adrenal function, and intensity of depression. *J Clin Endocrinol Metab.* 1987;64: 203–208.
44. Nader N, Ng SS, Lambrou GI, Pervanidou P, Wang Y, Chrousos GP, Kino T. AMPK regulates metabolic actions of glucocorticoids by phosphorylating the glucocorticoid receptor through p38 MAPK. *Mol Endocrinol.* 2010;24:1748–1764.
45. Obendorf M, Meyer R, Henning K, Mitev YA, Schroder J, Patchev VK, Wolf SS. FoxG1, a member of the forkhead family, is a corepressor of the androgen receptor. *J Steroid Biochem Mol Biol.* 2007;104:195–207.
46. Nader N, Chrousos GP, Kino T. Circadian rhythm transcription factor CLOCK regulates the transcriptional activity of the glucocorticoid receptor by acetylating its hinge region lysine cluster: potential physiological implications. *FASEB J.* 2009; 23:1572–1583.
47. Veytsman I, Nieman L, Fojo T. Management of endocrine manifestations and the use of mitotane as a chemotherapeutic agent for adrenocortical carcinoma. *J Clin Oncol.* 2009;27: 4619–4629.
48. Tsigos C, Chrousos GP. Differential diagnosis and management of Cushing syndrome. *Annu Rev Med.* 1996;47:443–461.
49. Miyachi Y. Pathophysiology and diagnosis of Cushing syndrome. *Biomed Pharmacother.* 2000;54(Suppl 1):113s–117s.
50. Aghi MK. Management of recurrent and refractory Cushing disease. *Nat Clin Pract Endocrinol Metab.* 2008;4:560–568.
51. Patil CG, Hayden M, Katznelson L, Chang SD. Non-surgical management of hormone-secreting pituitary tumors. *J Clin Neurosci.* 2009;16:985–993.
52. Grimes HG, Brooks MH. Pregnancy in Sheehan's syndrome. Report of a case and review. *Obstet Gynecol Surv.* 1980;35:481–488.
53. Guerrero AF, Alfonso A. Traumatic brain injury-related hypopituitarism: a review and recommendations for screening combat veterans. *Mil Med.* 2010;175:574–580.
54. Molitch ME. Endocrine emergencies in pregnancy. *Baill Clin Endocrinol Metab.* 1992;6:167–191.
55. Nader S. Pituitary disorders and pregnancy. *Semin Perinatol.* 1990;14:24–33.
56. Pestell RG, Best JD, Alford FP. Lymphocytic hypophysitis. The clinical spectrum of the disorder and evidence for an autoimmune pathogenesis. *Clin Endocrinol (Oxf).* 1990;33:457–466.
57. Andrioli M, Pecori Giraldi F, Cavagnini F. Isolated corticotrophin deficiency. *Pituitary.* 2006;9:289–295.
58. Scully KM, Rosenfeld MG. Pituitary development: regulatory codes in mammalian organogenesis. *Science* 2002;295:2231–2235.
59. Kelberman D, Dattani MT. Hypopituitarism oddities: congenital causes. *Horm Res.* 2007;68(Suppl 5):138–144.
60. Colvin SC, Mullen RD, Pfaeffle RW, Rhodes SJ. LHX3 and LHX4 transcription factors in pituitary development and disease. *Pediatr Endocrinol Rev.* 2009;6(Suppl 2):283–290.
61. Bornstein SR. Predisposing factors for adrenal insufficiency. *N Engl J Med.* 2009;360:2328–2339.
62. Mehta A, Dattani MT. Developmental disorders of the hypothalamus and pituitary gland associated with congenital hypopituitarism. *Best Pract Res Clin Endocrinol Metab.* 2008;22:191–206.
63. Turki ZM, Hamaoui RB, Ben Slama C. Molecular basis of pituitary development defects. *Tunis Med.* 2007;85:999–1003.
64. Neary N, Nieman L. Adrenal insufficiency: etiology, diagnosis and treatment. *Curr Opin Endocrinol Diabetes Obes.* 2010;17: 217–223.
65. Krude H, Biebermann H, Luck W, Horn R, Brabant G, Gruters A. Severe early-onset obesity, adrenal insufficiency and red hair pigmentation caused by POMC mutations in humans. *Nat Genet.* 1998;19:155–157.
66. Krude H, Biebermann H, Schnabel D, Tansek MZ, Theunissen P, Mullis PE, Gruters A. Obesity due to proopiomelanocortin deficiency: three new cases and treatment trials with thyroid hormone and ACTH4-10. *J Clin Endocrinol Metab.* 2003;88: 4633–4640.
67. Clement K, Dubern B, Mencarelli M, Czernichow P, Ito S, Wakamatsu K, Barsh GS, Vaisse C, Leger J. Unexpected endocrine features and normal pigmentation in a young adult patient carrying a novel homozygous mutation in the POMC gene. *J Clin Endocrinol Metab.* 2008;93:4955–4962.
68. Patalano A, Brancato V, Mantero F. Adrenocortical cancer treatment. *Horm Res.* 2009;71(Suppl 1):99–104.
69. Wajchenberg BL, Albergaria Pereira MA, Medonca BB, Latronico AC, Campos Carneiro P, Alves VA, Zerbini MC, Liberman B, Carlos Gomes G, Kirschner MA. Adrenocortical carcinoma: clinical and laboratory observations. *Cancer.* 2000;88:711–736.
70. Porterfield JR, Thompson GB, Young Jr WF, Chow JT, Fryrear RS, van Heerden JA, Farley DR, Atkinson JL, Meyer FB, Abboud CF, Nippoldt TB, Natt N, Erickson D, Vella A, Carpenter PC, Richards M, Carney JA, Larson D, Schleck C, Churchward M, Grant CS. Surgery for Cushing syndrome: an historical review and recent ten-year experience. *World J Surg.* 2008;32:659–677.
71. Terzolo M, Angeli A, Fassnacht M, Daffara F, Tauchmanova L, Conton PA, Rossetto R, Buci L, Sperone P, Grossrubatscher E, Reimondo G, Bollito E, Papotti M, Saeger W, Hahner S, Koschker AC, Arvat E, Ambrosi B, Loli P, Lombardi G, Mannelli M, Bruzzi P, Mantero F, Allolio B, Dogliotti L, Berruti A. Adjuvant mitotane treatment for adrenocortical carcinoma. *N Engl J Med.* 2007;356:2372–2380.
72. Michels AW, Eisenbarth GS. Autoimmune polyendocrine syndrome type 1 (APS-1) as a model for understanding autoimmune polyendocrine syndrome type 2 (APS-2). *J Intern Med.* 2009;265:530–540.
73. Anderson MS, Venanzi ES, Chen Z, Berzins SP, Benoist C, Mathis D. The cellular mechanism of Aire control of T cell tolerance. *Immunity.* 2005;23:227–239.
74. Anderson MS, Venanzi ES, Klein L, Chen Z, Berzins SP, Turley SJ, von Boehmer H, Bronson R, Dierich A, Benoist C, Mathis D. Projection of an immunological self shadow within the thymus by the aire protein. *Science.* 2002;298:1395–1401.
75. Robles DT, Fain PR, Gottlieb PA, Eisenbarth GS. The genetics of autoimmune polyendocrine syndrome type II. *Endocrinol Metab Clin North Am.* 2002;31:353–368.
76. Carlson AD, Obeid JS, Kanellopoulou N, Wilson RC, New MI. Congenital adrenal hyperplasia: update on prenatal diagnosis and treatment. *J Steroid Biochem Mol Biol.* 1999;69:19–29.

77. Nimkarn S, New MI. Prenatal diagnosis and treatment of congenital adrenal hyperplasia owing to 21-hydroxylase deficiency. *Nat Clin Pract Endocrinol Metab.* 2007;3:405–413.
78. Phelan JK, McCabe ER. Mutations in NR0B1 (DAX1) and NR5A1 (SF1) responsible for adrenal hypoplasia congenita. *Hum Mutat.* 2001;18:472–487.
79. Niakan KK, McCabe ER. DAX1 origin, function, and novel role. *Mol Genet Metab.* 2005;86:70–83.
80. Achermann JC, Ito M, Hindmarsh PC, Jameson JL. A mutation in the gene encoding steroidogenic factor-1 causes XY sex reversal and adrenal failure in humans. *Nat Genet.* 1999;22:125–126.
81. Ozisik G, Achermann JC, Jameson JL. The role of SF1 in adrenal and reproductive function: insight from naturally occurring mutations in humans. *Mol Genet Metab.* 2002;76:85–91.
82. Bhangoo A, Anhalt H, Ten S, King SR. Phenotypic variations in lipoid congenital adrenal hyperplasia. *Pediatr Endocrinol Rev.* 2006;3:258–271.
83. Raafat F, Hashemian MP, Abrishami MA. Wolman's disease: report of two new cases, with a review of the literature. *Am J Clin Pathol.* 1973;59:490–497.
84. Gottesman MM, Ambudkar SV. Overview: ABC transporters and human disease. *J Bioenerg Biomembr.* 2001;33:453–458.
85. Hettema EH, Tabak HF. Transport of fatty acids and metabolites across the peroxisomal membrane. *Biochim Biophys Acta.* 2000;1486:18–27.
86. Kemp S, Pujol A, Waterham HR, van Geel BM, Boehm CD, Raymond GV, Cutting GR, Wanders RJ, Moser HW. ABCD1 mutations and the X-linked adrenoleukodystrophy mutation database: role in diagnosis and clinical correlations. *Hum Mutat.* 2001;18:499–515.
87. Clark AJ, McLoughlin L, Grossman A. Familial glucocorticoid deficiency associated with point mutation in the adrenocorticotropin receptor. *Lancet.* 1993;341:461–462.
88. Tsigos C, Arai K, Hung W, Chrousos GP. Hereditary isolated glucocorticoid deficiency is associated with abnormalities of the adrenocorticotropin receptor gene. *J Clin Invest.* 1993;92: 2458–2461.
89. Hiroi N, Yakushiji F, Shimojo M, Watanabe S, Sugano S, Yamaguchi N, Miyachi Y. Human ACTH hypersensitivity syndrome associated with abnormalities of the ACTH receptor gene. *Clin Endocrinol (Oxf).* 1998;48:129–134.
90. Chung TT, Chan LF, Metherell LA, Clark AJ. Phenotypic characteristics of familial glucocorticoid deficiency (FGD) type 1 and 2. *Clin Endocrinol (Oxf).* 2010;72:589–594.
91. Metherell LA, Chapple JP, Cooray S, David A, Becker C, Ruschendorf F, Naville D, Begeot M, Khoo B, Nurnberg P, Huebner A, Cheetham ME, Clark AJ. Mutations in MRAP, encoding a new interacting partner of the ACTH receptor, cause familial glucocorticoid deficiency type 2. *Nat Genet.* 2005; 37:166–170.
92. Cronshaw JM, Matunis MJ. The nuclear pore complex: disease associations and functional correlations. *Trends Endocrinol Metab.* 2004;15:34–39.
93. Charmandari E, Kino T. Chrousos syndrome: a seminal report, a phylogenetic enigma and the clinical implications of glucocorticoid signalling changes. *Eur J Clin Invest.* 2010;40:932–942.
94. Vingerhoeds ACM, Thijssen JHH, Schwarts F. Spontaneous hypercortisolism without Cushing syndrome. *J Clin Endocrinol Metab.* 1976;43:1128–1133.
95. Chrousos GP, Vingerhoeds A, Brandon D, Eil C, Pugeat M, DeVroede M, Loriaux DL, Lipsett MB. Primary cortisol resistance in man. A glucocorticoid receptor-mediated disease. *J Clin Invest.* 1982;69:1261–1269.
96. Nader N, Bachrach BE, Hurt DE, Gajula S, Pittman A, Lescher R, Kino T. A novel point mutation in helix 10 of the human glucocorticoid receptor causes generalized glucocorticoid resistance by disrupting the structure of the ligand-binding domain. *J Clin Endocrinol Metab.* 2010;95:2281–2285.
97. McMahon SK, Pretorius CJ, Ungerer JP, Salmon NJ, Conwell LS, Pearen MA, Batch JA. Neonatal complete generalized glucocorticoid resistance and growth hormone deficiency caused by a novel homozygous mutation in Helix 12 of the ligand binding domain of the glucocorticoid receptor gene (NR3C1). *J Clin Endocrinol Metab.* 2010;95:297–302.
98. Ruiz M, Lind U, Gafvels M, Eggertsen G, Carlstedt-Duke J, Nilsson L, Holtmann M, Stierna P, Wikstrom AC, Werner S. Characterization of two novel mutations in the glucocorticoid receptor gene in patients with primary cortisol resistance. *Clin Endocrinol (Oxf).* 2001;55:363–371.
99. Charmandari E, Kino T, Ichijo T, Zachman K, Alatsatianos A, Chrousos GP. Functional characterization of the natural human glucocorticoid receptor (hGR) mutants hGRαR477H and hGRαG679S associated with generalized glucocorticoid resistance. *J Clin Endocrinol Metab.* 2006;91:1535–1543.
100. Charmandari E, Ichijo T, Jubiz W, Baid S, Zachman K, Chrousos GP, Kino T. A novel point mutation in the amino terminal domain of the human glucocorticoid receptor (hGR) gene enhancing hGR-mediated gene expression. *J Clin Endocrinol Metab.* 2008;93:4963–4968.
101. Rosmond R, Bouchard C, Bjorntorp P. Tsp509I polymorphism in exon 2 of the glucocorticoid receptor gene in relation to obesity and cortisol secretion: cohort study. *BMJ.* 2001;322:652–653.
102. Huizenga NA, Koper JW, De Lange P, Pols HA, Stolk RP, Burger H, Grobbee DE, Brinkmann AO, De Jong FH, Lamberts SW. A polymorphism in the glucocorticoid receptor gene may be associated with and increased sensitivity to glucocorticoids *in vivo*. *J Clin Endocrinol Metab.* 1998;83: 144–151.
103. Dobson MG, Redfern CP, Unwin N, Weaver JU. The N363S polymorphism of the glucocorticoid receptor: potential contribution to central obesity in men and lack of association with other risk factors for coronary heart disease and diabetes mellitus. *J Clin Endocrinol Metab.* 2001;86:2270–2274.
104. Karl M, Lamberts SW, Detera-Wadleigh SD, Encio IJ, Stratakis CA, Hurley DM, Accili D, Chrousos GP. Familial glucocorticoid resistance caused by a splice site deletion in the human glucocorticoid receptor gene. *J Clin Endocrinol Metab.* 1993;76:683–689.
105. Russcher H, van Rossum EF, de Jong FH, Brinkmann AO, Lamberts SW, Koper JW. Increased expression of the glucocorticoid receptor-A translational isoform as a result of the ER22/23EK polymorphism. *Mol Endocrinol.* 2005;19: 1687–1696.
106. van Rossum EF, Koper JW, Huizenga NA, Uitterlinden AG, Janssen JA, Brinkmann AO, Grobbee DE, de Jong FH, van Duyn CM, Pols HA, Lamberts SW. A polymorphism in the glucocorticoid receptor gene, which decreases sensitivity to glucocorticoids *in vivo*, is associated with low insulin and cholesterol levels. *Diabetes.* 2002;51:3128–3134.
107. van Rossum EF, Voorhoeve PG, te Velde SJ, Koper JW, Delemarre-van de Waal HA, Kemper HC, Lamberts SW. The ER22/23EK polymorphism in the glucocorticoid receptor gene is associated with a beneficial body composition and muscle strength in young adults. *J Clin Endocrinol Metab.* 2004;89: 4004–4009.
108. Nieman LK, Ilias I. Evaluation and treatment of Cushing syndrome. *Am J Med.* 2005;118:1340–1346.
109. Wajchenberg BL, Mendonca B, Liberman B, Adelaide M, Pereira A, Kirschner MA. Ectopic ACTH syndrome. *J Steroid Biochem Mol Biol.* 1995;53:139–151.

110. Isidori AM, Kaltsas GA, Pozza C, Frajese V, Newell-Price J, Reznek RH, Jenkins PJ, Monson JP, Grossman AB, Besser GM. The ectopic adrenocorticotropin syndrome: clinical features, diagnosis, management, and long-term follow-up. *J Clin Endocrinol Metab.* 2006;91:371–377.
111. Bayraktar F, Kebapcilar L, Kocdor MA, Asa SL, Yesil S, Canda S, Demir T, Saklamaz A, Secil M, Akinci B, Yener S, Comlekci A. Cushing syndrome due to ectopic CRH secretion by adrenal pheochromocytoma accompanied by renal infarction. *Exp Clin Endocrinol Diabetes.* 2006;114:444–447.
112. Zangeneh F, Young Jr WF, Lloyd RV, Chiang M, Kurczynski E. Cushing syndrome due to ectopic production of corticotropin-releasing hormone in an infant with ganglioneuroblastoma. *Endocr Pract.* 2003;9:394–399.
113. Rhen T, Cidlowski JA. Antiinflammatory action of glucocorticoids – new mechanisms for old drugs. *N Engl J Med.* 2005;353: 1711–1723.
114. Rosen J, Miner JN. The search for safer glucocorticoid receptor ligands. *Endocr Rev.* 2005;26:452–464.
115. Schacke H, Schottelius A, Docke WD, Strehlke P, Jaroch S, Schmees N, Rehwinkel H, Hennekes H, Asadullah K. Dissociation of transactivation from transrepression by a selective glucocorticoid receptor agonist leads to separation of therapeutic effects from side effects. *Proc Natl Acad Sci USA.* 2004;101:227–232.
116. Fassnacht M, Hahner S, Beuschlein F, Klink A, Reincke M, Allolio B. New mechanisms of adrenostatic compounds in a human adrenocortical cancer cell line. *Eur J Clin Invest.* 2000;30(Suppl 3):76–82.
117. Perini GI, Devinsky O, Hauser P, Gallucci WT, Theodore WH, Chrousos GP, Gold PW, Kling MA. Effects of carbamazepine on pituitary-adrenal function in healthy volunteers. *J Clin Endocrinol Metab.* 1992;74:406–412.
118. Mukherjee K, Knisely A, Jacobson L. Partial glucocorticoid agonist-like effects of imipramine on hypothalamic–pituitary–adrenocortical activity, thymus weight, and hippocampal glucocorticoid receptors in male C57BL/6 mice. *Endocrinology.* 2004;145:4185–4191.
119. Budziszewska B, Jaworska-Feil L, Kajta M, Lason W. Antidepressant drugs inhibit glucocorticoid receptor-mediated gene transcription – a possible mechanism. *Br J Pharmacol.* 2000;130: 1385–1393.
120. Castinetti F, Conte-Devolx B, Brue T. Medical treatment of Cushing syndrome: glucocorticoid receptor antagonists and mifepristone. *Neuroendocrinology.* 2010;92(Suppl 1):125–130.
121. Hurley DM, Accili D, Stratakis CA, Karl M, Vamvakopoulos N, Rorer E, Constantine K, Taylor SI, Chrousos GP. Point mutation causing a single amino acid substitution in the hormone binding domain of the glucocorticoid receptor in familial glucocorticoid resistance. *J Clin Invest.* 1991;87:680–686.
122. Malchoff DM, Brufsky A, Reardon G, McDermott P, Javier EC, Bergh CH, Rowe D, Malchoff CD. A mutation of the glucocorticoid receptor in primary cortisol resistance. *J Clin Invest.* 1993;91:1918–1925.
123. Karl M, Lamberts SW, Koper JW, Katz DA, Huizenga NE, Kino T, Haddad BR, Hughes MR, Chrousos GP. Cushing disease preceded by generalized glucocorticoid resistance: clinical consequences of a novel, dominant-negative glucocorticoid receptor mutation. *Proc Assoc Am Physicians.* 1996;108: 296–307.
124. Kino T, Stauber RH, Resau JH, Pavlakis GN, Chrousos GP. Pathologic human GR mutant has a transdominant negative effect on the wild-type GR by inhibiting its translocation into the nucleus: Importance of the ligand-binding domain for intracellular GR trafficking. *J Clin Endocrinol Metab.* 2001; 86:5600–5608.
125. Mendonca BB, Leite MV, de Castro M, Kino T, Elias LL, Bachega TA, Arnhold IJ, Chrousos GP, Latronico AC. Female pseudohermaphroditism caused by a novel homozygous missense mutation of the GR gene. *J Clin Endocrinol Metab.* 2002;87:1805–1809.
126. Vottero A, Kino T, Combe H, Lecomte P, Chrousos GP. A novel, C-terminal dominant negative mutation of the GR causes familial glucocorticoid resistance through abnormal interactions with p160 steroid receptor coactivators. *J Clin Endocrinol Metab.* 2002;87:2658–2667.
127. Charmandari E, Raji A, Kino T, Ichijo T, Tiulpakov A, Zachman K, Chrousos GP. A novel point mutation in the ligand-binding domain (LBD) of the human glucocorticoid receptor (hGR) causing generalized glucocorticoid resistance: the importance of the C terminus of hGR LBD in conferring transactivational activity. *J Clin Endocrinol Metab.* 2005;90: 3696–3705.
128. Charmandari E, Kino T, Ichijo T, Jubiz W, Mejia L, Zachman K, Chrousos GP. A novel point mutation in helix 11 of the ligand-binding domain of the human glucocorticoid receptor gene causing generalized glucocorticoid resistance. *J Clin Endocrinol Metab.* 2007;92:3986–3990.

CHAPTER

30

Disorders of the Hypothalamic–Pituitary–Gonadal Axis

Lawrence C. Layman

Section of Reproductive Endocrinology, Infertility, & Genetics, Dept Obstetrics and Gynecology, Developmental Neurobiology Program, Institute of Molecular Medicine and Genetics, Neuroscience Program, Georgia Health Sciences University, Augusta, GA, USA

OUTLINE

Summary

Pubertal development and subsequent reproductive competence in mammals depend upon normal function of the hypothalamic–pituitary–gonadal (HPG) axis. Dysfunction of the HPG axis can result in a wide range of disorders in both males and females. These disorders are somewhat arbitrarily categorized into hypothalamic, pituitary, gonadal and outflow (genital) tract abnormalities. Here we review the molecular basis of reproductive disorders in the human, including those that result in hypogonadism (both hypergonadotropic and hypogonadotropic), as well as eugonadism. The diagnosis and clinical features of HPG axis disorders are discussed. Advances in our understanding of the chromosomal and genetic abnormalities that underlie HPG dysfunction will enhance our understanding of the pathophysiology of HPG disorders, and thereby assist in directing genetic counseling and, ultimately, treatment.

THE HYPOTHALAMIC–PITUITARY–GONADAL AXIS

The coordinated efforts of the hypothalamic–pituitary–gonadal (HPG) axis control sexual development and reproductive capacity in mammals. Although gonadotropin-releasing hormone (GnRH) neurons are dispersed throughout the hypothalamus (infundibulum, ventromedial nucleus, preoptic area, arcuate nucleus and lateral hypothalamus), most affecting reproduction reside in the arcuate nucleus of the hypothalamus. Here, GnRH is synthesized and secreted into the hypophysial portal vessels. From these capillaries GnRH is delivered to the anterior pituitary

Handbook of Neuroendocrinology, DOI: 10.1016/B978-0-12-375097-6.10030-7

TABLE 30.1 Diagnostic Categories of Reproductive Dysfunction in Males and Females

I. Hypogonadism
- A. Hypergonadotropic hypogonadism
  - 1. Abnormal chromosomes
    - a. Females: 46,X (with or without mosaicism); 46,XY
    - b. Males: 47,XXY; 46,XY
  - 2. Normal chromosomes
- B. Hypogonadotropic hypogonadism
  - 1. Reversible
  - 2. Irreversible

II. Eugonadism
- 1. Gamete dysfunction
  - a) Spermatogenic disorders (males)
  - b) Ovulation disorders (females)
- 2. Outflow obstruction
  - a) Congenital absence of the vas deferens (males)
  - b) Uterine and vaginal obstruction (females)

gland, where it binds to its G-protein coupled transmembrane receptor in gonadotrope cells, which then synthesize and secrete follicle-stimulating hormone (FSH) and luteinizing hormone (LH). These gonadotropins then stimulate the synthesis and secretion of sex steroids in the gonads: estradiol and progesterone in females, and testosterone in males. Negative feedback of sex steroids in both sexes causes inhibition at both the hypothalamus and the pituitary.[1] A number of growth factors, such as inhibins A and B, activins and insulin-like growth factors, also play a role in normal HPG function. Additionally, a number of neuropeptides, as well as alterations in other endocrine systems, such as the adrenal and thyroid glands, regulate GnRH pulsatility and can affect function. Recently, evidence for gonadotropin-inhibitory releasing hormone (GnIH) has been reported, and indicates that this peptide appears to inhibit gonadal function at multiple levels, including the hypothalamus, pituitary and gonads.[2]

Dysfunction of the HPG axis can result in a wide range of disorders in both males and females (see Table 30.1). They are somewhat arbitrarily categorized into hypothalamic, pituitary, gonadal and outflow (genital) tract abnormalities. First a systematic approach to the diagnosis of reproductive dysfunction will be outlined, as shown in Fig. 30.1, followed by a discussion of the major entities comprising each category for their effect upon normal reproduction function. Although the hypothalamic disorders idiopathic hypogonadotropic hypogonadism (IHH) and Kallmann syndrome (KS) will be briefly mentioned in the context of diagnosis, a separate chapter is devoted to these entities. Patients with reproductive disorders may be classified according to their gonadal status – as hypogonadal or eugonadal. As might be expected, hypogonadal patients usually have a more severe phenotype, manifesting as

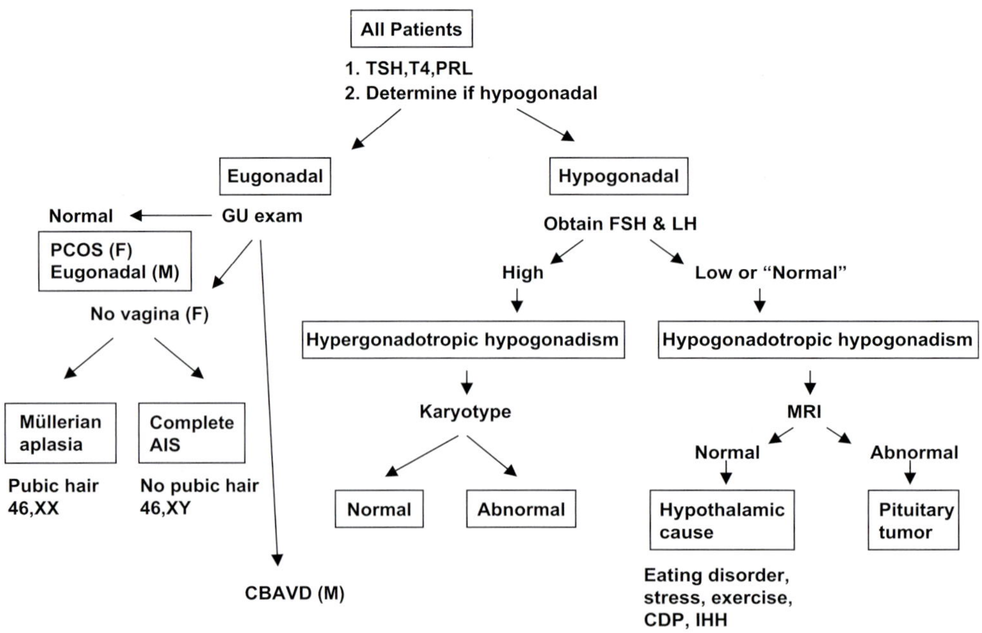

FIGURE 30.1 **An overview of the diagnostic steps in males and females with reproductive dysfunction.** The two most common causes of outflow obstruction causing primary amenorrhea are complete androgen insensitivity syndrome and Müllerian aplasia (Mayer-Rokitansky-Kuster-Hauser syndrome). PCOS, polycystic ovary syndrome; T4, thyroxin; TSH, thyroid stimulating hormone; PRL, prolactin.

delayed puberty and infertility, vs those who have attained normal pubertal milestones but are now infertile.

HYPOGONADISM

Diagnosis of Hypogonadism

The first step in elucidating the etiology of reproductive dysfunction is to determine if the patient is hypogonadal.[3] This is much more easily done in the male, with a serum total testosterone. Although testosterone assays in the very low range in males (and normal range for females of 30–75 ng/dl) have not been shown to be reliable,[4] normal male levels tend to have a broad range, from 275–300 ng/dl to 850–1100 ng/dl. Therefore, if a male has a testosterone level less than 100–125 ng/dl he clearly has hypogonadism, but the etiology needs to be determined (see below). For a female, serum estradiol levels are not very reproducible in the lower range, so it is difficult to distinguish hypoestrogenism (< 30 pg/ml) vs early follicular phase levels (30–50 pg/ml). The menstrual cycle fluctuations also clearly affect what a normal estradiol level will be. The periovulatory serum levels of estradiol approximate 200–300 pg/ml per mature follicle. Given these normal physiologic changes, other techniques can be employed. For adolescents, the presence of breast development can serve as a very valuable bioassay for indicating estrogen exposure. Conversely, the absence of breast development provides important clinical evidence of hypoestrogenism.

In females who have had breast development and never menstruated (primary amenorrhea), and in those with normal breast development who have had menarche but have stopped menstruating (secondary amenorrhea), the determination of estrogen status is quite important.[3] Just because they had normal estrogen levels at some time in the past does not mean that estrogen status is normal now. There are several different methods that could be considered. The first is a vaginal smear (also known as a vaginal maturation index); the second is a progestin challenge test. The vaginal smear can be simply performed at the first visit by inserting a Q-tip into the vagina and rolling it in the fornices of the vagina to obtain epithelial cells. If the patient has never had a pelvic exam, the use of a speculum may not be necessary. The epithelial cells are then gently rolled on to a slide, stained with a quick prep stain and visualized under a light microscope. Alternatively, the sample could be placed into a liquid fixative used for pap smears, which may allow better dispersion of cells for examination. The ratio of parabasal, intermediate and superficial (P/I/S) cells can be counted in ~100 cells (Fig. 30.2). Parabasal cells are small

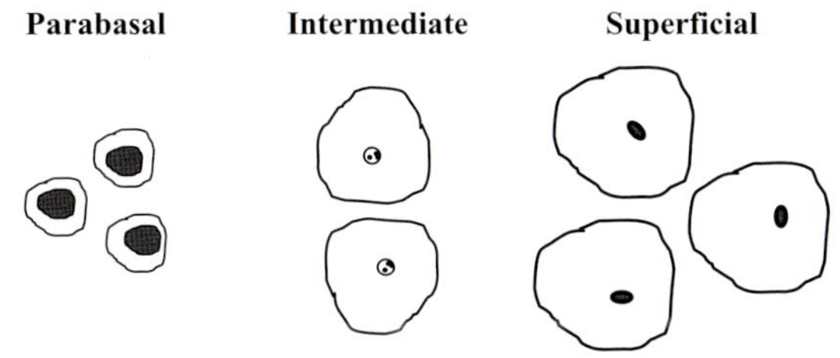

FIGURE 30.2 A vaginal maturation index can show parabasal, intermediate or superficial cells. Superficial cells are large cells with a small pyknotic nucleus; intermediate cells are similar, but have distinct chromatin seen within the nucleus; parabasal cells are small cells with a large nucleus.

cells with a large nucleus and a high nuclear/cytoplasmic ratio, while superficial cells display a large cytoplasm with a small pyknotic nucleus. If superficial cells are present (as in a 0 : 25 : 75 ratio), the patient is likely making estrogen and is probably eugonadal. In contrast, a predominance of parabasal cells (as with a 70 : 30 : 0 ratio) indicates hypoestrogenism, and therefore hypogonadism. Similar to the hypogonadal male, the demonstration of hypoestrogenism does not define the specific etiology of the hypogonadism – just that hypogonadism is present.[3]

Another method to determine the presence of hypogonadism is the progestin challenge test. If the patient has had recent estrogen exposure which should stimulate endometrial glandular development (and thicken the endometrium), then the administration of a natural or synthetic progestogen should result in bleeding within a week of stopping the last pill. In contrast, if the patient is hypogonadal, giving a progestin should not result in any bleeding or result only in minimal spotting. This test is usually performed by administering a synthetic progesterone such as medroxyprogesterone acetate 10 mg for 5–10 days following a negative pregnancy test. A normal menstrual bleed after the medication is completed indicates that sufficient endogenous estrogen is being produced to prime the endometrium (if only vaginal spotting occurs this is likely to be a negative test, as it can occur with hypogonadism). This test does not need to be performed in patients without breast development or those who have a hypoestrogenic-appearing vaginal smear. In a minority of patients, the results of either the vaginal smear or progestogen challenge test may be ambiguous. Nevertheless, they remain useful components for the clinical evaluation of hypogonadism.

Timing of Hypogonadism

If hypogonadism occurs prior to or during puberty, pubertal development can be delayed or arrested. In females, delayed puberty is defined as the absence of either breast development (thelarche) by age of 13 or the onset of menses (menarche) by age 15, both of which are 2.5 standard deviations above the mean for North American adolescents. For males, if no puberty occurs by age 14, it is usually considered delayed. However, hypogonadism can also occur after puberty in both sexes. In males or females who are truly hypogonadal, the most common differential diagnoses are shown in Table 30.1.

Patients presenting with delayed puberty, menstrual abnormalities after puberty, or suspicion for hypogonadism after puberty should have thyroid studies to exclude hypothyroidism, which is more common than hyperthyroidism.[3] Typically, a thyroid-stimulating hormone (TSH) and total thyroxine (T4) will exclude most thyroid disease. A total T4 is a more robust assay than free T4 unless it is done by dialysis and subsequent immunoassay. A serum prolactin will document the presence of hyperprolactinemia, but a detailed psychosocial history and history of medications should be ascertained, since many will raise prolactin. Hypothyroidism, hyperprolactinemia and hypothalamic disorders (such as eating disorders, excessive exercise and stress) may occur in patients who are either eugonadal or hypogonadal, depending upon how long the insult has occurred.

Categories of Hypogonadism

When hypogonadism is diagnosed, additional studies are needed to elucidate the precise cause of hypothalamic–pituitary–gonadal dysfunction. With the removal of gonadal steroid negative feedback in the hypogonadal patient, the performance of serum FSH and LH levels will help determine if the defect is hypothalamic/pituitary versus gonadal in origin[3] (Fig. 30.3). If the hypothalamus and pituitary are intact, then pulsatile GnRH should stimulate pituitary gonadotropins to appropriately elevated levels, which indicates a gonadal etiology. If gonadotropins remain elevated on at least two occasions several weeks apart, *hypergonadotropic hypogonadism* (or gonadal failure) is present. A karyotype should be considered in both males and females with elevated gonadotropins who fail to undergo normal puberty, to rule out a chromosomal abnormality. Karyotypic abnormalities will most commonly consist of a 45,X with or without mosaicism or a pure 46,XY cell line in a phenotypic female; or a 47,XXY or 46,XX in a phenotypic male. In females with absent puberty and gonadal failure (who therefore have primary amenorrhea), the prevalence of chromosomal abnormalities is much greater than in those with secondary amenorrhea (see below). About 10–15% of males with hypergonadotropic hypogonadism will have a chromosomal abnormality (see below).

If low serum gonadotropins are observed in the hypogonadal patient, this indicates a subnormal response of the hypothalamus or pituitary and so *hypogonadotropic hypogonadism* is present. It should also be emphasized that if serum FSH and LH are in the normal range in a hypogonadal individual, this is also inappropriate given low sex steroids, and indicates hypothalamic or pituitary dysfunction (and is also considered hypogonadotropic hypogonadism).[3,5] Although the prevalence of chromosomal abnormalities is much lower than that of gonadal failure, approximately 2–3% of patients with hypogonadotropic hypogonadism could have chromosomal rearrangements.[6] Since sex steroids are necessary for growth, it is reasonable to perform a hand film X-ray for a bone age in a hypogonadal patient with delayed puberty, as these patients typically have a delayed bone age compared to chronological age. Usually the

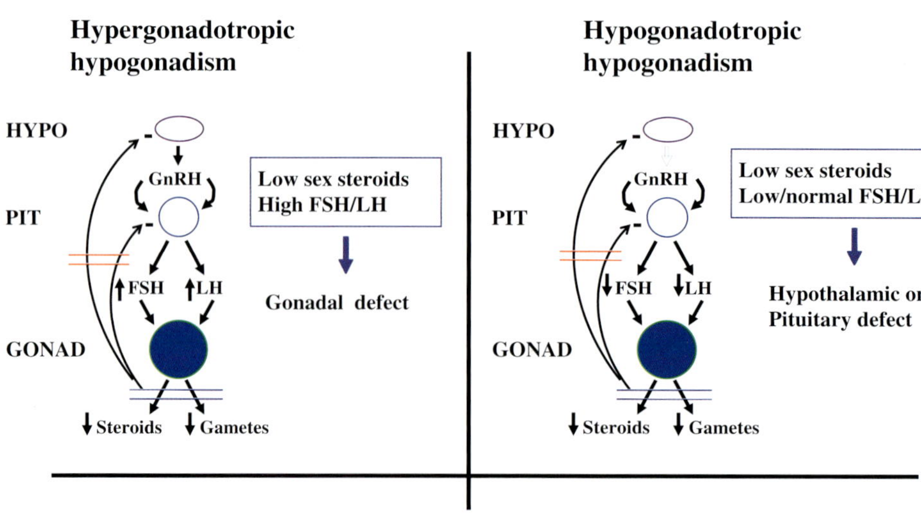

FIGURE 30.3 **Hypergonadotropic (left) and hypogonadotropic hypogonadism (right) are shown schematically.** Arrows indicate if serum gonadotropins are elevated or low. HYPO, hypothalamus; PIT, pituitary.

bone age is not less than 11–12 years for patients presenting with delayed puberty. However, if bone age is markedly delayed, growth hormone deficiency and/or hypothyroidism should be strongly considered. This could suggest the presence of panhypopituitarism or combined pituitary hormone deficiency (CPHD).

EUGONADAL INFERTILITY

If there is evidence of normal endogenous estrogen production in females, then anovulation, most commonly resulting from polycystic ovary syndrome (PCOS), will be present if thyroid disease or hyperprolactinemia are absent. In males that have normal endocrine status (testosterone, LH, and FSH levels), infertility with oligospermia fits into this eugonadal category. In fact, this is the most common diagnosis of male factor infertility.

Less commonly, obstruction of the outflow tract (uterus and/or vagina in females; vas deferens in males) is diagnosed. In women with primary amenorrhea, the two most common disorders include congenital absence of the uterus and vagina (CAUV), and complete androgen insensitivity syndrome (CAIS).[7] These disorders are ascertained by the finding of an absent vagina on pelvic exam. For women with secondary amenorrhea due to outflow obstruction, the reproductive tract was by definition patent initially and is now obstructed.[8] The most common cause of outflow obstruction in these women is related to endometrial adhesions after a vigorous surgical procedure to evacuate the pregnant uterus complicated by bleeding, such as a dilatation and curettage postpartum or after an abortion. Outflow obstruction in males is principally due to congenital bilateral absence of the vas deferens (CBAVD), a mild allelic variant of cystic fibrosis (Fig. 30.1).

SPECIFIC DISORDERS IN EACH DIAGNOSTIC CATEGORY

Now that the diagnosis of hypogonadism, whether hypergonadotropic (gonadal failure) or hypogonadotropic (hypothalamic–pituitary insufficiency), or eugonadism has been established, the pathophysiology of the most common neuroendocrine disorders affecting reproduction will be detailed. Hypogonadism will be discussed, first hypergonadotropic hypogonadism followed by hypogonadotropic hypogonadism. Lastly, disorders in eugonadal individuals and those resulting in outflow obstruction in both sexes will be discussed. Gender differences will be addressed where appropriate, particularly in disorders of gonadal function and outflow obstruction.

HYPOGONADISM

Hypergonadotropic Hypogonadism

Phenotypic Females

Women with gonadal failure may have a normal 46,XX karyotype; alternatively, they may have aneuploidy or a structurally abnormal X chromosome. Clearly, the risk for an abnormal karyotype is greater in women who have never menstruated (primary amenorrhea) compared with those who had menarche but became amenorrheic (secondary amenorrhea). Several very large series of patients with primary and secondary amenorrhea are shown in Table 30.2. More than two-thirds of patients with primary amenorrhea had a karyotypic abnormality, compared with only about 10% of patients with secondary amenorrhea.

X CHROMOSOME ABNORMALITIES

Women who completely lack one X chromosome include patients with a pure 45,X karyotype, but some patients may also possess another cell line, such as 46,XY, 46,XX, 47,XXX or 46,X iXq. About 90% of patients with a 45,X cell line with or without mosaicism present with primary amenorrhea, the complete lack of sexual development, and irreversible ovarian failure.[9] Only about 5–10% of 45,X women undergo puberty and menstruate, but menses are usually short-lived. For those 45,X patients who menstruate, fertility can occur, but reproductive wastage in the form of spontaneous abortions, stillbirths and chromosomal abnormalities (including Down syndrome and 45,X cell lines) is common.[9] Patients with a 45,X cell line are often referred to as having Turner syndrome, which includes

TABLE 30.2 Prevalence of Diagnostic Categories of both Primary Amenorrhea and Secondary Amenorrhea in Females

	Primary amenorrhea	Secondary amenorrhea
I. Hypogonadism		
A. Hypergonadotropic	43%	11%
1. Abnormal chromosomes	27%	0.5%
2. Normal chromosomes	16%	10%
B. Hypogonadotropic	31%	42%
1. Reversible	19%	39%
2. Irreversible	12%	3%
II. Eugonadism	26%	46%
1. Ovulation disorder	8%	37%
2. Outflow obstruction	18%	18%

phenotypic features such as short stature, webbed neck, shield chest, widely spaced nipples, multiple skin nevi and a short fourth metacarpal. The most constant feature is short stature, with heights that are nearly always less than 5 feet (152 cm). The most serious associated somatic anomalies include cardiac (in about 50%) and renal abnormalities. If a Y chromosome is present with a 45,X cell line, gonadoblastomas may occur within the streak gonads, which may give rise to even more serious germ cell tumors. Therefore, gonads should be removed in an individual with a Y cell line.

Patients with a 45,X/46,XY karyotype may clinically present with varying phenotypes. If they have bilateral abdominal streak gonads, they present similarly to other 45,X females, with short stature, absent breasts, but with an intact vagina and Müllerian system. If they have an intra-abdominal streak and a contralateral testis either in the abdomen or in the labioscrotum, sexual ambiguity will be present. Rarely, 45,X/46,XY patients may have bilateral scrotal testes, thereby rendering them an unambiguous male phenotype.[10]

Turner syndrome is likely due to haploinsufficiency of multiple genes on the X chromosome that affect embryologic development, ovarian function and stature. This phenotype suggests the presence of statural determinant genes on the X chromosome. In fact, the *SHOX* (short stature homeobox gene) located on Xp is a transcription factor expressed primarily in osteogenic cells. Rarely, patients with idiopathic short stature (without Turner syndrome) may harbor mutations in the *SHOX* gene. The phenotype of patients with *SHOX* mutations varies depending upon the type of mutation — either Leri-Weill dyschondrosteosis, a skeletal dysplasia with disproportionate short stature, mesomelic limbs, and the Madelung deformity (a radial bone anomaly also sometimes seen in Turner syndrome) caused by nonsense mutations; or Langer mesomelic dysplasia resulting from deletions.[11]

Partial X chromosome deletions may also impair ovarian function. In general, deletions affecting Xp11 result in ovarian failure in about half of women, and menstrual function in the other half[9] (Fig. 30.4). Even in those with normal menstruation, fertility is typically impaired. If the deletion is more distal, such as at X p21, the phenotype is usually less severe. Most women with Xp deletions are short, regardless of their ovarian function, further supporting that statural determinant genes likely lie within these regions. Several familial Xp deletions have also been reported.[9]

Deletions involving the long (q) arm of X also often result in ovarian failure. Similar to short-arm deletions, proximal (Xq13) deletions are usually more severe, and women have absent thelarche, primary amenorrhea and gonadal failure. At the more distal end of the long arm, menarche may occur, with or without ovarian failure (Fig. 30.4). Familial forms of Xq deletions manifesting ovarian failure have also been reported. The mechanism of ovarian failure in patients with X chromosome abnormalities could involve a dosage phenomenon, particularly if the gene/genes involved do not escape X-inactivation. Deleted/disrupted ovarian determinant genes probably increase follicular atresia, but perhaps

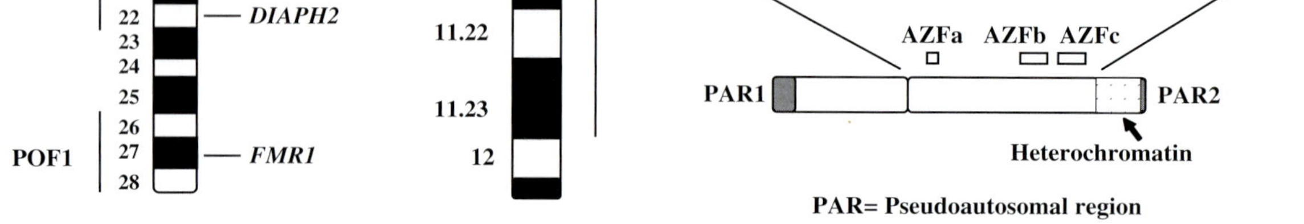

FIGURE 30.4 **Schematic diagrams of the X and Y chromosomes.** (**A**) For X chromosome deletions, the more severe phenotypes occur when the deletions are more proximal to the centromere on both the long (q) and short (p) arms. SRY is located on distal Yp, whereas spermatogenesis genes (AZFa–c) are on Yq. (**B**) (Left) The Y chromosome is shown schematically by its G-banded karyotype (left). The vertical line parallel and to the right of the Y chromosome contains the three putative spermatogenesis regions (AZFa–c), which are shown in more detail on the right. (Right) At the bottom, the Y chromosome is rotated 90° such that the tip of the short arm (which contains PAR1) is toward the left and the distal end of the long arm (which contains PAR2) is to the right. The three spermatogenesis regions AZFa–c, which were originally thought to be non-overlapping, are indicated above this figure in Yq11. At the top (as determined by the Genome Project), it can be seen that AZFb and AZFc actually overlap each other; and Yq11 consists of paired palindromes (indicated by P1–P8).

not to the extent of that seen in patients completly missing an entire X chromosome. It is also possible that the Xq deletion contains genes that might affect mitosis or meiosis, which could result in enhanced follicular atresia. Chromosomal rearrangements involving the X chromosome may also disrupt gene function.[9]

Y CHROMOSOME DISORDER – SWYER SYNDROME

Phenotypic females with ovarian failure can have a pure 46,XY karyotype without a coexistent 45,X cell line – Swyer syndrome. In this case, women will have bilateral streak gonads, normal stature, and a normal uterus and vagina without sexual ambiguity. These patients have the highest risk of developing tumors in their dysgenetic gonads (up to 25%); therefore, extirpating the gonads should be performed after the diagnosis.[12]

The gene for the sex-determining region of the Y chromosome (SRY) resides on the distal short arm of the Y chromosome (Fig. 30.4). When SRY is present the undifferentiated gonad will develop into a testis, which will first elaborate anti-Müllerian hormone (AMH) to remove the Müllerian system, and then testosterone for sexual differentiation. In Swyer syndrome, the gonad is dysfunctional (and replaced by fibrous streaks) so that AMH and testosterone are not produced. The lack of AMH leads to the retention of the uterus and vagina, while the reduction in normal Leydig cell function results in low testosterone and absent male sexual development. Therefore, the patient presents as a phenotypic female without breast development. Most cases of Swyer syndrome are sporadic and thought to be caused by *SRY* mutations, which is a single-exon gene on Yp encoding a transcription factor with a conserved HMG (high mobility group) domain. However, only about 15% of these patients actually have *SRY* mutations, suggesting that other genes must be involved.[12] Certainly, other genes may cause disorders of sexual development resulting in sexual ambiguity, but they will be covered in Chapter 27. Swyer syndrome is detailed here because it results in gonadal failure in phenotypic females (genetic males).

46,XX OVARIAN FAILURE

Women with hypergonadotropic hypogonadism are said to have premature ovarian failure (POF) when this occurs under age 40. Ovarian function may wax and wane – pregnancy may even occasionally occur. These patients may sometimes bleed when the progestin withdrawal test is done,[13] and this has prompted some investigators to suggest that it should be called hypergonadotropic amenorrhea or primary ovarian insufficiency.[14] Nevertheless, most patients persist with hypergonadotropic hypogonadism. For practical purposes, the molecular basis for most patients with amenorrhea, elevated gonadotropins on two occasions and a 46,XX karyotype is unknown, if known causes such as chemotherapy, pelvic radiation or surgical extirpation are excluded. Clinical studies have indicated that many of these women may have an autoimmune basis with other associated endocrinopathies including hypothyroidism, adrenal insufficiency and diabetes. However, a small percentage of patients have been found to have single gene disorders.

SINGLE GENE DISORDERS CAUSING OVARIAN FAILURE AND INFERTILITY

Although it has been previously hypothesized that Xq13-Xq26 could contain ovarian determinant genes, surprisingly few have been identified. Based upon these findings, the region Xq26-q28 has been designated POF1 while Xq13.3-q21.1 has been identified as POF2, as shown in Fig. 30.4. As outlined in Mendelian Inheritance in Man (MIM), there are seven forms of POF – POF1 (MIM 311360), POF2A and –POF2b (MIM 300511 and MIM 300604, respectively), POF3 (MIM 608996), POF4 (300510), POF5 (MIM 611548), POF6 (MIM 612310) and POF7 (MIM 612964). The *FMR1* gene causing Fragile X syndrome resides within POF1 region, whereas *DIAHP2* and *POF1B* genes reside within the POF2 region (see below). POF3–7 represent regions on autosomes (except POF4, which is on the X chromosome) with identified genes possessing mutations and backed by functional studies (Table 30.3). First the X chromosome genes will be reviewed, followed by the autosomal genes. If the gene is relevant to males, this will also be discussed; otherwise, male-specific disorders will be covered under testicular failure.

SINGLE GENE DISORDERS OF THE X CHROMOSOME ASSOCIATED WITH OVARIAN FAILURE It should be noted that nearly all of these are rare in patients with 46,XX ovarian failure except Fragile X syndrome, but they can provide new potential mechanisms for normal HPG function and disease.

The Diaphanous 2 (DIAPH2) Gene in the POF2 Region The first gene suggested to be an ovarian determinant gene was the *DIAPH2* gene. Although a woman with POF who had a balanced X;12 translocation, t(X;12)(q21;p1.3), that disrupted the diaphanous 2 (*DIAPH2*) gene was described, the importance of this gene remains speculative since no point mutations have yet been described.[15] The *DIAPH2* gene disruption was proposed to be causative since this gene has high homology to the *dia* gene of *Drosophila*, which is expressed in the testes and ovary, and results in sterility when mutations occur.[15] Interestingly, the other gene in this region (*POF1B*) has not conclusively been shown to possess mutations that are not seen in controls with normal ovarian function.

TABLE 30.3 Single Gene Disorders in Hypergonadotropic Hypogonadism

Gene	Localization	Phenotype	Inheritance
POF1	Xq26q28	Ovarian failure (F)	Sporadic or XLD
POF2	Xq13.3-q22	Ovarian failure (F)	Sporadic or XLD
DIAPH2	Xq22	Ovarian failure (F)	Disruption in X-autosome translocation
FMR1	Xq27.3	Fragile X syndrome (M); ovarian failure (F)	XLD
BMP15	Xp11.2	Ovarian failure (F)	XLR
SRY	Yp11.3	Swyer syndrome (genetic males)	Sporadic, Y linked
*FSHR**	2p21p16	Primary amenorrhea (F); Oligospermia (M)	AR
AIRE	21q22.3	APECED	AR
POF3 (*FOXL2*)	3q23	BPES (M & F); with ovarian failure (M)	AD
FIGLA	2p12	Ovarian failure	Sporadic, possible AD
NR5A1	9q33	Adrenal failure/sex reversal (M); POF (F)	AR
GALT	9p13	Galactosemia (with ovarian failure)	AR
EIFB2	14q24	Ovarioleukodystrophy	AR
EIFB4	2p23.3	Ovarioleukodystrophy	AR
EIFB5	3q27	Ovarioleukodystrophy	AR
CYP17A1	10q24.3	17-hydroxylase deficiency	AR
CYP19 A1	15q21.1	Aromatase deficiency	AR

* *LHR mutations can also cause gonadal failure in males with varying degrees of sexual ambiguity; in females it causes anovulation and is included in Table 30.6.*
F, females; M, males; XLD, X-linked dominant; XLR, X-linked recessive; AR, autosomal recessive; AD, autosomal dominant.

The FMR1 Gene in the POF1 Region The most frequently recognized single gene disorder with clinical implications that includes POF is Fragile X syndrome. Fragile X syndrome is an X-linked dominant disorder with incomplete penetrance characterized by large ears and jaws, macroorchidism, and varying degrees of mental deficiency in affected males. The mental deficiency may be extremely subtle or manifest solely as autism. The *FRMR1* gene resides at a fragile site of Xq27 (in the POF1 region), and contains a triplet repeat of CGG nucleotides that range from 6–50 copies in unaffected individuals. Affected males have a full expansion to ≥ 200 repeats, while carrier females have what is called a premutation allele of 50–200 repeats. These premutation alleles are unstable and may expand in meiosis in carrier females so that the full expansion results, giving rise to an affected male. Some females with premutation alleles may also be affected with mild degrees of mental deficiency or learning disability.

Males with Fragile X syndrome have macroorchidism and normal testicular histology, but approximately 15% of females with premutation alleles ascertained through Fragile X families may have POF.[16] If large cohorts of females with POF are studied (i.e., not Fragile X families), there has also been an increased risk of identifying the premutation allele. If there is no family history of POF, the patient has about a 3–4% risk of harboring an FMR1 premutation, but this rises to 12–15% if there are two or more females in a family affected with POF.[9,17] Interestingly, no women with full *FMR1* mutations had POF; it was only the carrier females. The Fragile X full mutation is a null allele with aberrant methylation and inactivation of the protein, whereas the premutation allele is associated with overexpression of the mutant mRNA, which has been found to be expressed in the ovary. Therefore, appropriate genetic counseling must be provided to women with POF and an *FMR1* premutation allele. If she becomes pregnant, half of her sons will have Fragile X syndrome and half of her daughters will be carriers. Even if she does not conceive, other family members could potentially be carriers and be at risk for Fragile X syndrome. Recently, it has also become appreciated that males with premutation alleles could later manifest an adult-onset tremor ataxia syndrome.[14]

Bone Morphogenic Protein-15 (BMP15) Gene in POF4 Mutations in the *BMP15* gene were first identified in a family with two young women with primary amenorrhea (ages 23 and 18), hypoplastic ovaries and elevated gonadotropin levels. This heterozygous missense *BMP15* mutation was identified in both of the affected females, transmitted from the father. This mutation was not seen in 210 controls, and there was *in vitro* analysis to support a functional effect upon protein action (abnormal processing, impaired granulosa cell growth, and antagonism of the wild-type *BMP15* stimulated granulosa cell growth).[18] Additional studies of large cohorts of POF suggest that the prevalence of *BMP15* mutations in POF is less than 1%.[19] Most mutations are missense mutations, and those that have been studied *in vitro* appear to impair secretion.

SINGLE GENE DISORDERS OF THE AUTOSOMES ASSOCIATED WITH OVARIAN FAILURE It is interesting that some autosomal genes possessing mutations in POF, which were identified long before POF 3, 5, 6 and 7, are not designated in such a manner in OMIM. Here, an attempt will be made to cover known human autosomal genes resulting in ovarian failure.

Follicle-stimulating Hormone Receptor (FSHR) Gene Mutations in the *FSHR* gene have been observed in some women with POF, and are transmitted in an autosomal recessive mode of inheritance.[20] The FSH ligand binds to this heptohelical G-protein coupled receptor located on the membrane surface of ovarian granulosa cells and stimulates estradiol production. Affected females usually present with primary amenorrhea, and about half have breast development while the others completely lack thelarche. Some women may have normal puberty and may have menstrual periods before becoming amenorrheic and hypogonadal. Ovarian follicles range from primordial to mature, suggesting that the defect is not quite as severe as mutations of the *FSHB* ligand (see pituitary disorders below). Males with *FSHR* mutations are fertile. Mutations may affect binding or signal transduction. Mutations in the LHR gene generally cause anovulation in females (and are discussed under eugonadism).

Autoimmune Regulatory (AIRE) Gene The autoimmune polyglandular syndrome type 1, also known as autoimmune polyendocrinopathy-candidiasis-ectodermal dystrophy (APECED), is an autosomal recessive disorder in which multisystem autoimmune disease is observed. APECED is more common in the Finnish population and in Iranian Jews. Moniliasis is a common initial presenting feature in more than half of patients, but can occur at anytime. Hypoparathyroidism (80%) and adrenal failure (70%) represent the most frequent endocrine manifestations, but ovarian failure (60%) and testicular failure (14%) may also occur with sufficient frequency to be an indication of APECED.[21,22]

Mutations in *AIRE* (autoimmune regulator), a transcriptional factor with two PHD-type zinc-finger motifs, have been identified in patients with APECED. The identification and characterization of *AIRE* mutations represented the first report of a single gene defect causing systemic autoimmune disease in humans. Approximately 80% of mutations in the Finnish patients consist of a single nonsense mutation (Arg257X), while a 13-bp deletion accounts for more than half of mutations in North American patients, and a single nonsense mutation (Arg139X) occurs in about 80% of Sardinian *AIRE* alleles.[21,22] Currently, it is unclear whether certain mutations increase the likelihood of ovarian failure vs other endocrinopathies.

Forkhead Transcription Factor (FOXL2) Gene in POF3 In an interesting syndrome comprised of blepharophimosis (narrowing of horizontal opening of the eyelids), ptosis and epicanthus inversus (BPES), with POF (BPES type I) or without (BPES type II), heterozygous mutations in the *FOXL2* gene have been identified. Both forms are inherited in an autosomal dominant fashion. It is interesting that mutations of this gene have eyelid effects as well as being important in normal ovarian function, while sparing testicular function (gonadal failure is sex-limited). Most patients with BPES (types I and II) have mutations in the *FOXL2* gene on chromosome 3q23.[23] Two mutational hotspots appear to be present in the gene: 30% lead to polyalanine expansions, and 13% are frameshift duplications. The prevalence of *FOXL2* mutations in isolated POF without BPES is not clearly known, but is likely to be rare. There appears to be at least one mutation in a POF patient with supporting functional analysis.[24]

Newborn Ovary Homeobox, Mouse, Homolog of (NOBOX) *NOBOX* is a homeobox gene localized to chromosome 7q35 which is expressed predominantly in the ovary and testis. Interestingly, male *Nobox* knockout mice were fertile, but females had atrophic ovaries. *Nobox* is preferentially expressed in the oocyte, and in the mouse has a role in folliculogenesis and regulation of oocyte-specific genes. One of 96 Caucasian POF women had a heterozygous missense mutation not present in 278 ethnically-matched controls. Its role in POF is currently unclear, but is probably minimal.[25]

Factor in Germline Alpha (FIGLA) Gene in POF6 Two of 100 Chinese POF women recently were found to have deletions (3 bp and 22 bp) in the *FIGLA* gene on chromosome 2p12 that were absent in 340 female controls with regular menses and no history of infertility.[26] *FIGLA* encodes for a germ cell-specific transcription factor that regulates expression of genes in the oocyte. Both women with the mutations had secondary amenorrhea and elevated gonadotropins. One of the deletions disrupted FIGLA binding to the TCF3 helix–loop–helix domain, and the other resulted in a frame shift.[26]

Nuclear Receptor Subfamily 5, Group A, Member 1 (NR5A1) Gene in POF7 Steroidogenic factor 1 (SF1), a transcription factor important in steroidogenesis in the gonad and adrenal glands, is encoded by *NR5A1*, previously known as *FTZF1*. Similar to targeted disruption of the homologous gene in the mouse, a mutation in the *NR5A1* gene caused undermasculinization of 46,XY males, and adrenal failure.[27] It has now become recognized that *NR5A1* mutations cause a wide range of disorders of sexual differentiation with and without adrenal failure.[28] NR5A1 mutations may be inherited as autosomal dominant or autosomal recessive traits.

The phenotype in genetic females would be expected to consist of adrenal failure, delayed puberty with absent breast development and primary amenorrhea, with elevated gonadotropins. In fact, there have now been demonstrated mutations that do result in 46,XX ovarian failure with or without adrenal failure. These individuals with POF can be identified in families with affected males. Two of 25 (8%) POF women with normal adrenal function were found to harbor heterozygous mutations consistent with autosomal dominant inheritance.[28]

Galactose-1-phosphate Uridyltransferase (GALT) Gene Galactosemia is an autosomal recessive disorder in which galactose cannot be properly metabolized to glucose. The phenotype typically consists of failure to thrive, nausea, vomiting, cataracts, hepatomegaly, mental retardation, speech abnormalities and hemolytic anemia. If treated by a galactose-free diet, the prognosis for liver function and mental capacity is improved, but not to the level expected. Although several enzymes are involved in the process, galactosemia caused by mutations in *GALT* has been shown to have a sex-specific effect upon reproduction in that females, but not males, develop gonadal failure.[9] The enzyme encoded by *GALT* is responsible for the conversion of galactose-1-phosphate and UDP-glucose to UDP-galactose and glucose-1-phosphate.

Up to two-thirds of women with galactosemia had premature ovarian failure, while none of eight men had testicular failure.[29] Females had normal pubertal development, but half presented with primary amenorrhea while the other half had secondary amenorrhea. However, gonadotropins were elevated in both groups. The exact cause of the ovarian failure is unknown, but could represent a detrimental metabolic defect during development caused by galactose-1-phosphate or by abnormal glycosylation of gonadotropin glycoproteins or their receptors. At least several hundred *GALT* mutations have been described, many of which are missense mutations. Heterozygotes do not develop gonadal failure as homozygotes do. Of interest, the *GALT* knockout mouse has a less severe phenotype, and no effect upon gonadal development and function.

Eukaryotic Translation Initiation Factors 2, 4 and 5 Leukoencephalopathy with vanishing white matter can be caused by any of the five subunits of the translation initiation factor eIF2B. It has also been reported that patients with leukoencephalopathy with vanishing white matter and ovarian failure, also known as ovarioleukodystrophy, have mutations in three of these subunits (2, 4 or 5).[30] These are all transmitted in an autosomal recessive fashion. Patients may present with primary or secondary amenorrhea, elevated gonadotropins and a variety of neurologic abnormalities, including gait and speech problems, spasticity, need for a walker, dysarthria, optic atrophy and reduced cognitive function. When 93 women with isolated POF (without neurologic abnormalities) were screened for mutations in the *EIF2B2* gene, none were identified, suggesting that they are not a common cause of POF.[30] These findings suggest that mutations either are not present or occur at a low prevalence.

GENES IN STEROID ENZYME PATHWAY THAT RESULT IN GONADAL FAILURE The biochemical pathway utilizes 27-carbon cholesterol as a precursor for the production of sex steroids, as well adrenal mineralocorticoids and glucocorticoids (Fig. 30.5). Some mutations cause both disorders of adrenal function and disorders of sexual development (congenital adrenal hyperplasia), and these will not be discussed here. However, selected genes of the steroid enzyme pathway may result in gonadal failure.

Cytochrome P450, Family 17, Subfamily A, Polypeptide 1 (CYP17A1) Gene A cytochrome P450 encoded by *CYP17A1* is an enzyme with at least two functions: a 17-hydroxylase activity (converting progesterone to 17-hydroxyprogesterone and pregnenolone to 17-hydroxypregnenolone), and a 17–20 desmolase activity (converting 17-hydroxypregnenolone to dehydroepiandrosterone (DHEA) and 17-hydroxyprogesterone to androstenedione).[31] *CYP17A1* mutations cause autosomal recessive deficiency in the production of androgens, progestins and estrogens. Cortisol is deficient, since it is derived from 17-hydroxyprogesterone; however, mineralocorticoids such as 11-deoxycorticosterone (DOC) and corticosterone may be elevated, with associated hypertension and hypokalemic alkalosis (Fig. 30.5).

The phenotype of CYP17 deficiency in 46,XX females includes delayed puberty with absent breast development, primary amenorrhea and elevated gonadotropins. The vagina, uterus and ovaries are present, but they are hypoestrogenic. 46,XY males have a similar phenotype, except that they will not have a uterus or upper vagina, since AMH is produced from their normal testes. Some males with partial deficiency may have sexual ambiguity. A variety of mutations in *CYP17A1* have been identified, consisting predominantly of deletions and insertions.[31]

Cytochrome P450, Family 19, Subfamily A, Polypeptide 1 (CYP19A1) Gene Encoding Aromatase There have been several reports that mutations in *CYP19A1*, which encodes for the cytochrome P450

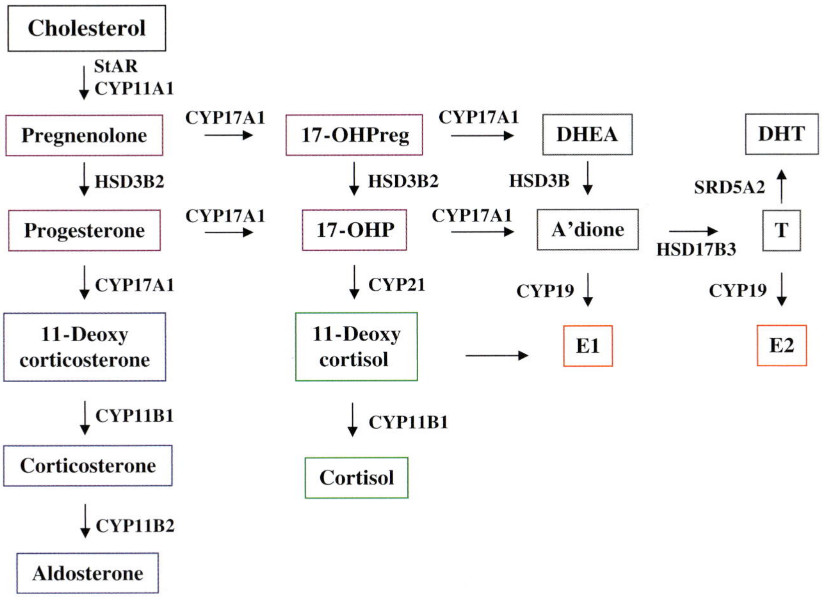

FIGURE 30.5 The biochemical pathway utilizes 27-carbon cholesterol as a precursor for the production of sex steroids, as well as adrenal mineralocorticoids and glucocorticoids.

enzyme aromatase, cause autosomal recessive aromatase deficiency. Females with aromatase deficiency cannot convert the androgens (testosterone and androstenedione) to the estrogens (estradiol and estrone, respectively).[32] At birth, females have sexual ambiguity with clitoromegaly, but since they have ovaries, they do not have labioscrotal gonads. When these children approach puberty, FSH rises and multicystic ovaries develop.[32] However, the increased FSH cannot elicit the production of sufficient ovarian estrogen so that no breast development or menses occur. Serum LH levels are also usually elevated in these patients.

An interesting feature of aromatase deficiency is that the heterozygous mother carrying an affected female fetus can also develop hyperandrogenism and hirsutism.[32] This is because the fetal placenta (derived from the fetus, having the same genotype as the fetus) cannot adequately convert androgens to estrogens. Normally, the placenta has quite robust aromatase activity, which protects the mother from hyperandrogenism.

Males with Hypergonadotropic Hypogonadism

Phenotypic males with hypergonadotropic hypogonadism usually have a 46,XY karyotype. However, approximately 10–15% may have a chromosomal abnormality, most commonly 47,XXY (Klinefelter syndrome), but some males will have a 46,XX karyotype (Table 30.1).

47,XXY (KLINEFELTER SYNDROME)

Males with Klinefelter syndrome typically manifest varying degrees of hypogonadism, but they do not uniformly have low serum testosterone and elevated gonadotropins. Perhaps for this reason, it is estimated that only about 25% are diagnosed.[33] Although puberty may be delayed, most will initiate puberty and then develop hypergonadotropic hypogonadism. This is not true for all patients, as many will have fluctuating low to low-normal levels of serum testosterone. Other features, such as mental retardation, gynecomastia, sparse hair and tall stature, are quite variable. Other associated congenital anomalies include clinodactyly, cleft palate and cryptorchidism. However, infertility with small atrophic testes and azoospermia is the most common presentation. It appears that testis size is smaller than eugonadal males even prior to puberty.

Some 47,XXY patients may have sperm that can be retrieved by testicular aspiration to be used for *in vitro* fertilization (by intracytoplasmic sperm injection into the oocyte), and normal pregnancies have been described.[33] However, there may still be offspring that will be 47,XXY, so this must be considered in counseling. Advanced maternal age in mothers of Klinefelter males predisposes to meiotic non-disjunction that can result in 47,XXY. Other associated medical problems, such as diabetes, osteoporosis, metabolic syndrome and cancer (breast cancer, mediastinal germ-cell tumors and non-Hodgkin's lymphomas), are increased, requiring medical surveillance.[33]

46,XX (SEX-REVERSED) MALES

Some males will have a 46,XX karyotype if there has been exchange between the X and Y chromosomes in paternal meiosis. The father, who now has *SRY* on the X chromosome, transmits this to his offspring, who also get an X from their mother. Since *SRY* is present, the undifferentiated gonad will become a testis and secrete AMH and testosterone. However, testis function is often deficient, and these males develop hypergonadotropic hypogonadism. They usually have bilaterally descended testes, although the testes and penis may be small. Infertility due to oligospermia is usually present. This is thought to occur because the spermatogenesis genes are on the long arm of the Y chromosome, which is not present in these patients.

Interestingly, some 46,XX males are *SRY* negative yet develop as phenotypic males. Recently, mutations in the gene for R-spondin-1 (*RSPO1*) have been identified in several of these males, who also have skin lesions on the palms and soles – palmoplantar hyperkeratosis (PPK) – that may develop into squamous cell cancer.[34] There is evidence that R-spondin modulates β-catenin stabilization, which may be a common thread. How mutations in this gene cause both seemingly different phenotypic features is currently unknown.

Hypogonadotropic Hypogonadism

In patients with hypogonadotropic hypogonadism, an MRI of the brain with and without gadolinium is necessary to exclude a CNS tumor, most commonly a prolactinoma or craniopharygioma.[3] It is also reasonable to visualize the olfactory tracts and bulb; if they are absent or hypoplastic, this suggests Kallmann syndrome. If a tumor is not present, the cause is usually considered to be hypothalamic by exclusion. Although it is possible to perform a triple test (insulin-induced hypoglycemia, GnRH, and TRH stimulation, and check hormone levels of TSH, prolactin, cortisol, LH, FSH and GH at baseline and every 15 minutes for 1–2 hours), this is expensive and the yield is very low except in patients who have extreme short stature, which could suggest pituitary failure. However, it is prudent to obtain an 8 am cortisol, since undiagnosed adrenal failure is much more serious and could indicate pituitary insufficiency with reduced ACTH secretion. If the patient has extremely short stature (with a height below the fifth percentile for age), particularly with a family history of pituitary failure, then combined pituitary hormone deficiency (CPHD) should be considered. Genetic counseling and testing for mutations in genes such as *PROP1*, *HESX1*, *SOX2*, *SOX3*, *LHX3* or *LHX4* could be considered in these patients (see below). In the absence of a tumor, strong consideration must be given to the history and physical examination, with particular attention to BMI, eating habits, exercise patterns and stress.

For patients with hypogonadotropic hypogonadism who have no pituitary tumor and are of normal weight, two diagnoses must be entertained. The first is constitutional delay of puberty (CDP), which can really only be documented in retrospect if girls subsequently initiate spontaneous puberty before age 17 or boys do so by age 18 (CDP is more common in boys). If delayed puberty persists, then the diagnosis is idiopathic or isolated hypogonadotropic hypogonadism (IHH). A history should be sought for anosmia/hyposmia, midline facial defects, associated neurologic deficits such as synkinesia (tested by raising both arms simultaneously when instructed to raise one), hearing loss or visual abnormalities. When IHH is combined with anosmia/hyposmia, the patient has Kallmann syndrome (KS). Although IHH/KS is typically irreversible, up to 10% of patients in a small series have been found to have spontaneous reversal. Mutations have been identified in approximately 30% of IHH/KS families, including *KAL1*, *GNRHR*, *NR0B1*, *FGFR1*, *KISSR1*, *PROK2*, *PROKR2*, *CHD7*, *FGF8*, *TAC3*, *TAC3R*, *NELF* and *GNRH1* (reviewed in Kim *et al.*,[1] and further discussed in Chapter 27). It is very important to ascertain history for eating disorders (anorexia or bulimia), extreme stress, or strenuous exercise, all of which may result in hypogonadotropic hypogonadism.

Hypothalamic Causes of Hypogonadotropic Hypogonadism

EATING DISORDERS

Eating disorders such as anorexia nervosa and bulimia may have a profound effect upon the reproductive axis.[35,36] A critical weight has previously been suggested as being important for the initiation of puberty; if weight is less, puberty can be delayed. Although this is not always true, weight is important, as indicated by eating disorders, which are about 10 times as common in females as in males. Anorexia nervosa consists of amenorrhea, weight loss and psychiatric disease, with an estimated prevalence of 0.3–1%. The diagnosis of anorexia nervosa, as defined by the DSM-IV (*Diagnostic and Statistical Manual of Mental Disorders*), consists of (a) the inability to maintain weight within 15% of ideal body weight; (b) abnormal fear of being obese; (c) an unrealistic body image regarding weight; and (d) amenorrhea for three consecutive months. Patients with anorexia may have either the restrictive type of reduced eating or the binge/purge type, in which periods of eating may be followed by vomiting to control weight.[35]

Bulimia nervosa is characterized by (a) recurring and uncontrollable binge eating; (b) recurrent behavior to reduce weight following binge eating, such as vomiting, or laxative or diuretic use; (c) regularity of binge/compensatory behavior (at least twice/week for 3 months); and (d) abnormal body image. The prevalence of bulimia, which can also be categorized into purging and non-purging subtypes, is about 1%. In addition, another 3–5% have been estimated to have eating disorders not otherwise specified (NOS).[35]

These patients may have a variety of physical findings in addition to the weight loss, such as cold intolerance, dry skin, hair loss, constipation, yellowish skin (carotinemia), bradycardia, hypothermia or hypotension.[35] They may also demonstrate some physical findings that could suggest eating disorders, such as parotid gland enlargement (from vomiting), enamel erosion of the teeth from acid exposure from vomiting, and skin lesions on the tips of fingers from inducing vomiting (Russell's sign). Laboratory studies may be normal, but might reveal electrolyte abnormalities (hyponatremia or hyperkalemia) or dehydration (abnormal renal function).[35]

The endocrinology of eating disorders is complex, and involves the HPG axis, HPA (hypothalamic–pituitary–adrenal) axis, thyroid, growth hormone (GH), and appetite hormones (Table 30.4).[36] GnRH pulse amplitude and/or frequency may be impaired so pituitary gonadotropins are decreased, as are subsequent gonadal steroids. Prolonged hypoestrogenism can lead to decreased bone density with the potential for fractures. Excessive adrenal stimulation results in hypersecretion of cortisol without its normal diurnal variation. CRH is also elevated in the cerebrospinal fluid, suggesting activation through the HPA axis.[36] Thyroid dysfunction manifested as the sick euthyroid syndrome may also occur in eating disorders. Thyroid-stimulating hormone (TSH) is often normal to low normal; instead of T4 being converted into T3, the less active reverse T3 (rT3) is preferred. A state of growth hormone resistance occurs in eating disorders, with increased bursts of GH and reduced IGF-1. Altered appetite-regulating hormone secretion of orexigenic and anorexigenic peptides also occurs. All of these endocrine effects are responses to starvation (Table 30.4), although some abnormalities may persist after successful treatment (weight gain), suggesting they could predate/predispose to eating disorders. Although leptin has been used for successful treatment in some patients, it is not uniformly beneficial.[36]

There is also recent evidence that kisspeptin, the ligand for GPR54, could play a role in hypothalamic dysfunction.[37] As mentioned above, mutations in the human gene for GPR54 (*KISS1R*) were identified by linkage analysis in patients with GnRH deficiency, and findings in the *Kiss1r* knockout mouse corroborated the phenotype. Kisspeptin stimulates gonadotropin secretion and gonadal sex-steroid production in animals, whether administered centrally or peripherally.[37] Since gonadotropin secretion can be prevented by co-administration of a GnRH antagonist, it suggests that kisspeptin acts by stimulating GnRH secretion. Giving kisspeptin increases circulating gonadotropins in both men and women (particularly in the

TABLE 30.4 Endocrine Abnormalities in Eating Disorders

Endocrine system	Abnormality	Effect	Proposed mechanisms
HPG Axis	Hypoestrogenism Androgen deficiency	Amenorrhea, bone loss	Decreased GnRH pulse frequency &/or amplitude
HPA Axis	Hypercortisolemia with loss of diurnal rhythm; high CRH in CSF fluid	Mood disorders; amenorrhea; cognitive defects; myopathy & bone loss	Increased cortisol pulse frequency; impaired feedback on hypothalamus and/or pituitary
Growth hormone	High GH; low IGF-1	Muscle atrophy; growth failure; bone loss	Increased basal secretory rates; increased frequency of GH secretory bursts; starvation reduces IGF-1 secretion in liver
Thyroid	Low T3; high rT3; high T4/T3 ratio; low/normal T4; low/normal TSH;	Sick euthyroid syndrome; protective against starvation	Preferential conversion of T4 to rT3 rather than more potent T3; other hypothalamic–pituitary abnormalities
Appetite regulation	Anorexigenic hormones: leptin decreased; PYY increased	Decreased food intake	Mixed signaling hypothesis favoring feeding inhibition
Appetite regulation	Orexigenic hormone: ghrelin increased, which activates GH and cortisol secretion	Decreased food intake	Increased secretory burst mass and amplitude; lack of response to eat is unknown

preovulatory period in women).[37] Kisspeptin antagonists do not reduce basal LH secretion, but have been shown in animal studies to reduce GnRH neuron firing, as well as inhibit LH secretion, pulse amplitude and pulse frequency. Therefore, in addition to its role in pubertal development, kisspeptin seems to have a role in GnRH neuron firing, GnRH pulsatile secretion, steroid negative feedback and, possibly, the LH surge.[37] Importantly, leptin receptors are present on kisspeptin neurons, and leptin stimulates *Kiss1* gene expression and increases gonadotropins in leptin-deficient mice, as well as in mice exposed to nutritional deprivation. Therefore, kisspeptin's role in hypothalamic disorders may not only involve a direct GnRH action, but also be mediated vs leptin.[37]

STRESS, EXERCISE AND FUNCTIONAL HYPOTHALAMIC AMENORRHEA

Excessive stress and/or exercise can also cause endocrine dysfunction similar to eating disorders, although usually less severe.[35] Women with functional hypothalamic amenorrhea (FHA) have persistent hypoestrogenic amenorrhea but without noticeable eating disorders, stress, or excessive exercise.[38] These women also have impaired GnRH secretion manifested in aberrant LH pulses: (a) decreased amplitude, (b) decreased frequency, (c) decreased amplitude and frequency, or (d) apulsatility of LH (some even have normal LH pulses).[39] However, when these patients are followed for longer periods of time, it becomes evident that at least half will have an eating disorder, have stress or take excessive exercise, which portend a more favorable outcome than those without any inciting event. Therefore, it is likely that functional hypothalamic amenorrhea might actually be caused in many cases by a known mechanism.[39] There is also evidence that stress and excessive exercise can affect males, although this is not as commonly recognized. An adult-onset form of IHH may be analogous to hypothalamic amenorrhea in women.[40]

Pituitary Causes of Hypogonadism

PITUITARY TUMORS

Pituitary tumors can impair reproductive function and cause other non-endocrine symptoms such as visual field abnormalities and headaches. Prolactin-producing tumors are the most common tumors, and are nearly always benign. Prolactin secretion can impair the pulsatile release of GnRH, which can impair menstrual function, with eventual amenorrhea and hypoestrogenism.[41] Craniopharygiomas, which can be malignant, and suprasellar tumors can also occur.[9] Whether malignant or benign, these tumors can impede the pathway of GnRH neurons in the tuberoinfundibular tract to the median eminence. Therefore, less GnRH might be released into the hypophysial portal system, resulting in hypothalamic–pituitary dysfunction.

PITUITARY HYPOFUNCTION

Mutations in some genes cause hypopituitarism in a minority of patients, but several of these disorders are clinically significant and should be considered.

GONADOTROPIN-RELEASING HORMONE RECEPTOR The pituitary-expressed GnRH receptor (*GNRHR*) gene represented the first gene found to be involved in normosmic, autosomal recessive hypogonadotropic hypogonadism.[5,42] The *GNRHR* belongs to the G-protein coupled receptor class of receptors, most of which contain an extracellular domain for ligand binding, a seven-transmembrane domain, three extracellular loops, three intracellular loops, and an intracellular carboxy-terminal tail (the *GNRHR* has no C-terminal tail). Since this will be covered elsewhere, suffice it to say that *GNRHR* mutations have been found in both males and females with either completely absent puberty and/or partial pubertal development. Because the tropic stimulus (i.e., gonadotropins) is deficient, infertility occurs if left untreated. Treatment is typically with exogenous subcutaneous gonadotropins (or GnRH). The typical prevalence is approximately 3–5% of normosmic IHH patients.[43]

LUTEINIZING HORMONE-BETA (LHB) Isolated deficiencies of LH and FSH are rare. The pituitary dimeric glycoprotein hormones consist of a common α subunit encoded by a single chorionic gonadotropin-α (*CGA*) gene and a specific β-subunit gene that confers specificity for the four proteins – LH, hCG, FSH and TSH. No human *CGA* mutations have been described but, based on the knockout mouse, the phenotype would be expected to include hypogonadotropic hypogonadism and hypothyroidism due to gonadotropin and TSH deficiency. Homozygous *Cga* knockout mice have normal-appearing neonatal gonads but prepubertal-appearing adult gonads, which suggests that gonadotropins are not necessary for prenatal sexual differentiation.[9]

The *LHB/CGB* gene complex in humans consists of six *CGB* genes and one *LHB* gene, which are highly polymorphic. However, there have only been several human *LHB* mutations, and these are inherited in an autosomal recessive fashion. At least five males have now been reported, and all have had small testes, oligospermia and deficient testosterone.[44] Interestingly, the first reported patient had elevated immunoreactive, but reduced biologically active, serum LH levels.[45] Others had low/undetectable LH levels and measurable to elevated FSH levels. One male who had no evidence of puberty had a normal semen analysis. Two reported females actually had normal pubertal development, presumably because of normal FSH secretion, but became amenorrheic with low LH and elevated FSH levels.[44]

FOLLICLE-STIMULATING HORMONE BETA (FSHB)

Mutations in *FSHB* have been described in several males and females, inherited as an autosomal recessive trait.[46,47] All mutations studied have been shown *in vitro* to result in low immunoreactive and/or low bioactive FSH levels.[46,48] Most females completely lack sexual development, although partial breast development has been described. All females have presented with primary amenorrhea with low serum estradiol levels, very low or unmeasurable serum FSH, and elevated LH levels. This endocrinologic profile indicates that in the face of hypogonadism GnRH must rise, and stimulates LH, resulting in elevated serum LH levels. Since there is a mutation in *FSHB*, dimeric FSH is reduced, resulting in isolated FSH deficiency. The several males with *FSHB* mutations have either had normal pubertal development or delayed puberty, but all have been azoospermic. Testosterone was either low or normal, with low FSH and normal LH levels.[48,49]

Although females with isolated FSH deficiency have elevated serum LH and exaggerated LH pulses, they do not exhibit hyperandrogenism or hirsutism.[50,51] This finding was unexpected, since women with polycystic ovary syndrome (PCOS), defined as chronic hyperandrogenic anovulation (see below), have elevated LH levels and hyperandrogenism. These findings question the two-gonadotropin two-cell hypothesis, which dictates that LH stimulates ovarian thecal cells to produce androgens, which serve as precursors for estrogens in granulosa cells, where FSH induces aromatization to estrogens.[50,51] The fact that isolated FSH deficiency is associated with low androgens suggests that FSH could play an important role in ovarian androgen production, perhaps by increasing LH receptors, or by inducing CYP17 enzyme activity (an androgen-dependent enzyme), inhibin or growth factors.[50,51] Ovarian follicles typically show primordial, primary and antral follicles, and normal fertility may be restored with FSH treatment.

BOX 30.1

HOW DO WE KNOW IF FSH IS NECESSARY FOR ANDROGEN PRODUCTION?

Several human mutations in the β-subunit gene of FSH (*FSHB*) have been described. One such patient provided important previously unconsidered clues into the function of FSH.[1] She was a 15 year 11 month female who presented with absent breast development, no menstrual periods and no evidence of hirsutism. Her serum FSH was undetectable, LH was moderately elevated, testosterone was in the low-normal range, and thyroid studies were normal. Upon GnRH stimulation, her FSH did not rise, but her LH was markedly elevated. These findings suggested the possibility of isolated FSH deficiency due to mutations in the *FSHB* gene. If there had been mutations in the α-subunit gene (*CGA*) this patient would have been expected to also possess hypothyroidism, which she did not have (and low FSH, LH and TSH, since they utilize the same α-subunit). This patient was found to harbor two different *FSHB* mutations, both of which resulted in absent immunoreactive and bioactive FSH levels *in vitro*.[1] Therefore, these regions of the FSH protein were deemed to be important in dimer formation and secretion.

What was very striking about this patient was that she was not hirsute, yet she had elevated LH levels, both basally and after stimulation. When she was brought into the CRC at the University of Chicago for overnight serial LH levels every 10 minutes, her pulsatile pattern appeared remarkably like a patient with hyperandrogenism/PCOS (see figure).[2,3] Could FSH play a role in ovarian androgen production? To test this hypothesis, she was given hCG (which is very similar in structure to LH but has a longer half-life) and steroid levels were repeated in 24 hours. A month later, she was given FSH followed by a similar hCG test.[2,3] After hCG, she had little change in estradiol (which was expected since it is FSH-induced), but her androgens did not change. However, when pretreated with FSH, her hCG stimulated endocrine profile revealed

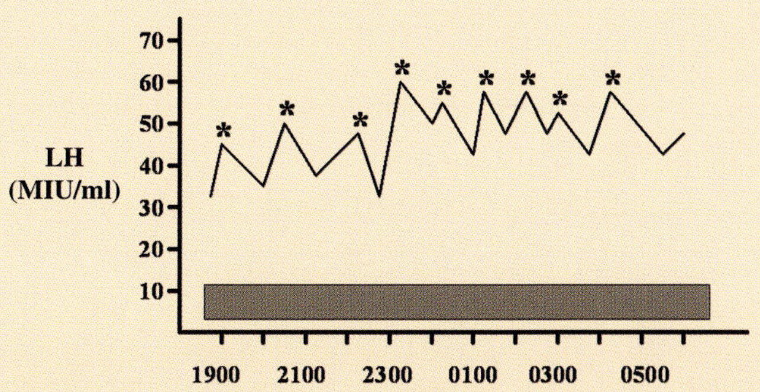

Overnight serum LH levels were measured every 10 minutes in a patient with isolated FSH deficiency due to compound heterozygous *FSHB* mutations. LH levels are measured in mIU/ml. The normal range (excluding the LH surge) is indicated as a shaded box. Each pulse is indicated by an asterisk. *Modified from Barnes* et al. *(2000)*.[2]

(*Continued*)

BOX 30.1 (*cont'd*)

that her serum testosterone tripled! These findings in the laboratory explained the clinical findings in the patient (i.e., no hirsutism); and suggested the possibility that FSH is important along with the androgen-dependent gonadotropin (LH) for adequate androgen production.[2,3]

References

1. Layman LC, Lee EJ, Peak DB, et al. Delayed puberty and hypogonadism caused by a mutation in the follicle stimulating hormone β-subunit gene. *N Engl J Med.* 1997;337:607–611.
2. Barnes RB, Namnoum A, Rosenfield RL, Layman LC. Effects of follicle-stimulating hormone on ovarian androgen production in a woman with isolated follicle-stimulating hormone deficiency. *N Engl J Med.* 2000;343:1197–1198.
3. Barnes RB, Namnoum A, Rosenfield RL, Layman LC. The role of LH and FSH in ovarian androgen secretion and ovarian follicular development: Clinical studies in a patient with isolated FSH deficiency and multicystic ovaries. *Hum Reprod.* 2002;17:88–91.

COMBINED PITUITARY HORMONE DEFICIENCY

Combined Pituitary Hormone Deficiency (CPHD) is usually defined as the deficiency of growth hormone (GH) and at least one additional pituitary hormone – ACTH, TSH, prolactin, FSH or LH. MIM lists CPHD1 (MIM 613038), CPHD2 (MIM 262600), CPHD3 (MIM 221750) and CPHD4 (MIM 262700). All except CPDH1 (*POU1F1*) affect gonadotropins. If all hormones are deficient, the patient may exhibit panhypopituitarism, which can be fatal unless treated, particularly because of resultant adrenal insufficiency. If this is congenital, the patient would be expected to have extremely short stature with deficiency of TSH and GH. Acquired causes of panhypopituitarism can include postpartum hemorrhage with impaired blood supply to the anterior pituitary (Sheehan's syndrome), pituitary tumor, infiltrative processes such inflammation or malignancy, and radiation exposure from cancer treatment. Genes known to be involved in CPHD that affect the HPG axis will now be reviewed (Table 30.5).

TABLE 30.5 Gene Mutations Affecting Pituitary Function

Gene	Localization	Phenotype	Inheritance
GNRHR	4q21.2	Normosmic HH	AR
LHB	19q13.3	Isolated LH deficiency	AR
FSHB	11p13	Isolated FSH deficiency	AR
PROP1	5q	CPHD2 – short stature, hypothyroid, HH	AR
HESX1	3p21.1-21.2	Septo-optic dysplasia	AR
LHX3	9q34.3	CPHD3	AR
LHX4	1q25	CPHD4	AD

AR, autosomal recessive; AD, autosomal dominant; CPHD, combined pituitary hormone deficiency.
Note that CPHD1 due to *POU1F1* mutations is not shown here since it does not affect gonadotropins.

PROP1 (CPHD2) Mutations in pituitary transcription factors may potentially cause hypogonadotropic hypogonadism either singly or in association with deficiencies of other pituitary hormones. *PROP1* appears to be important for early pituitary development, since mutations of *PROP1* have been shown to cause CPHD in humans. *PROP1*-deficient patients have an autosomal recessive form of GH, TSH, prolactin and gonadotropin deficiency.[52,53] They display extremely short stature secondary to deficient TSH and GH levels. Affected individuals from several different families demonstrated a failure of the pituitary to respond normally to GHRH, TRH and GnRH stimulation. A variety of mutations have been identified, including missense mutations and small deletions.[52,53] Mutations of the mouse ortholog, *Prop1*, cause a similar phenotype in the Ames dwarf mouse.

Less severe mutations of *PROP1* might be expected to cause gonadotropin deficiency, sparing GH, TSH and ACTH. However, in a large study of 175 patients with idiopathic hypogonadotropic hypogonadism (IHH), no *PROP1* mutations were identified by denaturing gradient gel electrophoresis of GC-clamped PCR products, suggesting that *PROP1* gene mutations rarely result in this phenotype.[54]

HESX1 Heterozygous and homozygous mutations in the *HESX1* gene, which encodes a transcription factor, have been identified in families with affected individuals with septo-optic dysplasia, a disorder characterized by panhypopituitarism, optic nerve atrophy and other midline CNS abnormalities, such as agenesis of the corpus callosum and septum pellucidum (Table 30.5).[55,56] The mouse ortholog *Hesx1* is expressed in the early forebrain. Later in development, *Hesx1* expression is restricted to Rathke's pouch, which ultimately becomes the anterior pituitary gland. Since panhypopituitarism occurs, IHH is a common feature of septo-optic dysplasia.[56]

***LHX3* (*CPHD3*) AND *LHX4* (*CPHD4*)[57]** *LHX3* and *LHX4* are two other genes that cause autosomal recessive combined pituitary hormone deficiency syndromes (CPHD). Mutations in *LHX3* cause autosomal recessive CPHD,[58] while mutations in the *LHX4* gene cause an autosomal dominant form. The phenotype of patients with *LHX3* gene mutations consists of severe growth retardation (due to GH and TSH deficiency) and hypogonadotropic hypogonadism. All pituitary hormones (TSH, prolactin, GH, FSH and LH) except ACTH are deficient.[58] Interestingly, there is severe restriction of rotation of the cervical spine, elevating and anteverting the shoulders similar to what is seen in the mouse ortholog. Although a family with GH, TSH and ACTH deficiency due to a heterozygous mutation in the *LHX4* gene has been described, the child was prepubertal, so gonadotropin levels were not investigated.[59] Theoretically, *LHX4* should also cause gonadotropin deficiency, and this was recently observed as affected children were followed into pubertal age.[60,61]

Treatment of Hypogonadism

Treatment of each specific entity involves understanding the inciting event if known. It is beyond the scope of this chapter to detail all types of treatment. For hypergonadotropic hypogonadism in phenotypic females, treatment depends upon the karyotype.[3] If a 45,X cell line is present, surveillance for cardiac disease is necessary; if a 46,XY cell line is present, gonads need to be removed to prevent malignancy. Women with 46,XX POF must be followed for autoimmune diseases such as hypothyroidism, adrenal failure and diabetes mellitus.[14] Although phenotypic males with 47,XXY are at risk for diabetes and some testicular tumors, they are generally healthy and can be followed like 46,XY males.[33] For the males with a 46,XX karyotype who have skin lesions, close observation to prevent squamous cell cancer of the hands and feet is warranted. For patients with hypogonadotropic hypogonadism, correction of the underlying eating disorder or stress event and modifying the exercise regimen should be beneficial. In patients with IHH/Kallmann syndrome, the presence of associated anomalies such as cleft lip/palate, unilateral renal agenesis, ataxia and neurologic disorders may require specific treatment.

Treatment for the reproductive system in patients with hypogonadism (whether hypergonadotropic or hypogonadotropic) involves the replacement of sex steroids.[57] For girls who do not have adult breast development, the administration of estrogen-only preparations is started at a low dose and increased gradually until normal breast development is reached (or vaginal bleeding occurs). An oral preparation containing 0.3 mg of conjugated equine estrogens that is increased every 3–6 months to 1.25 mg/day could be used. Alternatively, 0.5 mg of estradiol could be administered and increased by 0.5 mg increments until the desired result is observed. The treatment time for estrogen-only treatment is usually 1–1.5 years. Once adequate breast development is achieved a progestin is added for 7–10 days per month, or, for ease of administration, a combined estrogen–progestogen contraceptive preparation could be prescribed. The reason for adding the progestin is to prevent endometrial cancer from prolonged, unopposed estrogen stimulation. Boys can be treated with depot forms of testosterone, such as testosterone enanthate or propionate at doses of 50–100 mg i.m. every 2–3 weeks, titrated up to 200- to 300-mg doses. Alternatively, daily testosterone patches or gels can be used.

When fertility is desired, it will depend whether the patient has hypergonadotropic or hypogonadotropic hypogonadism, since only the latter is amenable to successful treatment. If there is uncertainty about potential fertility, sex-steroid treatments can be discontinued and serum levels of testosterone, FSH and LH can be obtained 1–2 months later in males. In females, menstrual cyclicity can be evaluated. In those women who remain amenorrheic, a progestin challenge test or vaginal maturation index can be performed as described above. For those that remain hypoestrogenic, serum FSH and LH levels can be obtained. Since up to 5–10% of patients with hypogonadotropic hypogonadism may have spontaneous reversal of their hypogonadism, a short follow-up period is advisable.[62] If they remain hypogonadotropic, treatment with gonadotropins for 10–15 days may induce ovulation in females, while longer-term treatment of 6–12 months of gonadotropins given three times weekly may be considered in males.[63] For men and women with hypergonadotropic hypogonadism, the use of donor gametes offers the best and most realistic chance for fertility.

EUGONADAL CAUSES OF INFERTILITY

Females

Only disorders that affect ovulation or development of the reproductive system will be discussed here (acquired disorders such as tubal disease or endometriosis will not be covered).

POLYCYSTIC OVARY SYNDROME (PCOS)

Polycystic ovary syndrome (PCOS) is the most common cause of anovulation in women, comprising more than two-thirds of all cases.[64] Although the

TABLE 30.6 Endocrine Derangements in Polycystic Ovary Syndrome

Hormone	Abnormality	Effect(s)
Estrogens	E2/E1 ratio decreased; tonic estrogen production	Weaker E1 is increased; inability to induce LH surge; stimulation of androgens; increased free estrogens (unopposed by progestins because of anovulation) raise risk for endometrial hyperplasia/cancer
Testosterone & androstenedione	Increased production & free levels	Hyperandrogenemia (hirsutism); increase in follicular atresia; decrease in SHBG
LH	Increased pulses & amplitude; increased LH/FSH ratio	Increased androgen production
FSH	Suppressed levels due to tonic estrogens	Small antral follicles that do not progress to maturity
SHBG	Decreased	Increased free androgens & estrogens
Insulin	Increased	Increased androgen production & hirsutism; impaired glucose tolerance & risk for diabetes
IGFBP-1	Decreased	Increase in free insulin

PCOS definitions: All exclude other known disorders.[64]
NIH criteria: Anovulation and hyperandrogenemia and/or hirsutism.[64]
Rotterdam criteria: two of three – anovulation, hyperandrogenemia/hirsutism, polycystic appearing ovaries.[65]
Androgen Excess/PCOS Society: hyperandrogenemia/hirsutism with either anovulation or polycystic appearing ovaries.[66]

definition may vary, generally it is defined as hyperandrogenic anovulation with or without polycystic-appearing ovaries on ultrasound (Table 30.6).[64–66] The phenotype of women with PCOS may include many of the following: menstrual abnormalities (amenorrhea, oligomenorrhea and/or menometrorrhagia), variable degrees of hirsutism or elevated androgens (testosterone and/or dehydroepiandrosterone sulfate), obesity, and hyperinsulinemia with an increased risk for developing type 2 diabetes mellitus. These patients may have hypercholesterolemia and hypertriglyceridemia. The combination of hypertension, increased waist/hip ratio, impaired glucose tolerance and hyperlipidemia comprise the metabolic syndrome, which has profound risks for future cardiovascular disease.[64]

In order to understand the derangements of PCOS, it is important to grasp the concepts of the normal human menstrual cycle, which is beyond the scope of this chapter. Suffice it to say that the menstrual cycle is divided into the follicular phase followed by the luteal phase. In the early follicular phase FSH rises, which then stimulates granulosa cell estrogen production in the dominant follicle destined to ovulate. This then causes negative feedback to the hypothalamus and pituitary to remove FSH stimulation to other follicles so that unifollicular ovulation occurs. Increasing estrogen stimulation by the dominant follicle results in positive feedback in the hypothalamus–pituitary for the LH surge, which then induces ovulation (the release of oocyte and production of progesterone). The ovulated follicle becomes the corpus luteum, which is maintained by LH and can only continue with pregnancy (and the production of human chorionic gonadotropin). Gonadal production of estrogen and progesterone in the luteal phase prepares the endometrium for implantation and also lowers gonadotropin secretion in the luteal phase and pregnancy by negative feedback.

The pathophysiology of PCOS is quite complex, and the salient features are summarized in Table 30.6.[64–66] The normal menstrual cycle is a dynamic process with positive and negative feedback. However, in PCOS there is tonic estrogen production, which does not elicit an LH surge (although it can happen occasionally). Instead of greater estradiol (E2) than estrone (E1) production, as in the normal menstrual cycle, PCOS is characterized by a reduced E2/E1 ratio. Sex hormone binding globulin (SHBG) is reduced in PCOS, and several factors may contribute to this – hyperandrogenemia, hyperinsulinemia and obesity (about 75% of PCOS women are obese). Decreased SHBG results in increased free estrogens and androgens. The increased free estrogens stimulate LH tonically, further producing hyperandrogenemia, which results in follicular atresia and the clinical signs and symptoms of hirsutism. Insulin is also known to increase the production of androgens, further exacerbating hirsutism and predisposing to diabetes. It is also tempting to speculate that kisspeptin could play a role in the LH surge since, in rodents, kisspeptin neurons in the anterioventricular paraventricular nucleus (AVPV) affect positive feedback, while those in the arcuate nucleus mediate steroid negative feedback.[37] Currently, it is not clear what function kisspeptin has in the anovulation of women with PCOS.

What incites this acyclicity, as well as the molecular basis of PCOS, is currently unknown. Evidence to date indicates that PCOS is a complex disease, which has largely been investigated through the use of association studies.[67] Genes such as insulin, the insulin receptor, steroid enzymes and follistatin have been implicated, but to date there is no evidence for a known, causative molecular defect. An association study is a case–control study in which alleles from cases (PCOS) are compared with controls. Individual candidate genes or the whole genome can be studied. If association is demonstrated, there are three principle possibilities – true association (and possible causation), population stratification (the allele happens to segregate with a particular ethnic group rather than being disease-causing), or chance alone. Therefore, any association study needs be confirmed by a follow-up study in an entirely new sample of cases and controls.[68] This has been a particular problem in PCOS, where some genes have demonstrated association, but have not been able to be independently confirmed. More recent evidence suggests the most significant association of PCOS is with fibrillin 3 (*FBN3*) on chromosome 19p13.2 and pro-opiomelanocortin (*POMC*) on chromosome on 2p23.3, and with lesser degrees to several other genes (Table 30.7), but causative mechanisms have not been determined.[67]

An animal model for PCOS would open new avenues of investigation that could improve our understanding of this common clinical condition. Interestingly, rhesus monkeys exposed to androgens prenatally give birth to offspring with a PCOS phenotype.[69] These monkeys demonstrate many of the features of PCOS women, including elevated LH levels and hyperandrogenism, and this state persists into early adulthood. They then develop abdominal adiposity and hyperlipidemia, and have an increased risk having type 2 diabetes.[69] The generation of this model should allow investigators to observe the pathophysiologic effects of this disorder from childhood through adulthood, and study interventions for treatment.

LUTEINIZING HORMONE RECEPTOR (LHR) GENE

A very rare cause of anovulation is mutation of the LHR, a G-protein coupled receptor expressed in the thecal cells of the ovary and the Leydig cells of the testis. The phenotype of patients with *LHR* gene mutations depends upon the karyotype.[70–72] Genetic females with *LHR* gene mutations have normal puberty with normal breast development and menarche. However, they become anovulatory, with usually elevated (or sometimes normal) LH and FSH levels.[70–72]

Genetic males with severe loss-of-function *LHR* mutations also usually present as phenotypic females, but have delayed puberty with absent breast development and normal-appearing external genitalia, but with a blind vaginal pouch.[70] Since the testes are normal and produce AMH, no uterus or upper two-thirds of the vagina is present. Therefore, the patient will present as a phenotypic female with no breast development. The external genitalia phenotype, however, depends upon the severity of the mutation, and ranges from normal-appearing female (severe mutations) to sexual ambiguity to bilaterally descended testes (mild mutations). Serum gonadotropins are both usually elevated, suggesting a gonadal etiology for the LH resistance. The testes are small and devoid of Leydig cells. Inactivating *LHR* mutations are inherited in an autosomal recessive fashion, in contrast to activating mutations of the *LHR*, which cause autosomal dominant familial male precocious puberty.

THE *HOXA13* GENE

Heterozygous mutations in the *HOXA13* homeobox gene have been identified in women with the hand-foot-uterus syndrome.[73] These patients do not necessarily have infertility, despite developmental abnormalities of the urogenital system. The phenotype of individuals with this autosomal dominant disorder typically includes small hands and feet along with a double uterus. Uterine anomalies consist of either bicornuate or didelphic abnormalities of midline Müllerian fusion. Associated urinary anomalies, such as displaced urethral or ureteral openings, may also be present in females, while males may have hypospadius.[73] Hand anomalies include delayed ossification or fusion of wrist bones, short first metacarpals, short middle phalanges of the fifth fingers, and small distal phalanges of the thumbs. The great toe may demonstrate a short first metatarsal and a small pointed distal phalanx. Only several *HOXA13* mutations have been described.[74]

Males

Male causes of infertility may be present in 40–50% of all couples, and most patients are eugonadal. A normal semen analysis is defined as ≥ 20 million/ml, $\geq 50\%$ motility and $\geq 30\%$ motility ($\geq 14\%$ by Kruger strict morphology). Azoospermia (the absence of sperm) or oligospermia (concentration ≤ 20 million/ml), asthenospermia (decreased motility) or teratospermia (decreased normal morphology) may be present alone or in combination. Although a number of genes have been involved in abnormal semen analyses (Table 30.7), the prevalence is unknown and probably rare. The cause of most disorders of abnormal sperm function is unknown.

TABLE 30.7 Gene Mutation Abnormalities in Eugonadism

Gene	Localization	Phenotype	Inheritance
Non-obstructive infertility			
Females			
*LHR**	2p21	Anovulation (F); undermasculinization (M)*	AR
HOXA13	7p15-p14.2	Hand–foot–uterus syndrome (also hypospadius in males)	AD
FBN3	19p13.2	PCOS	Association
POMC	2p23.3	PCOS	Association
ACVR2A	2q22.2-q23.3	PCOS	Association
FEM1B	5q23.1	PCOS	Association
SGTA	19p13	PCOS	Association
Males			
AZFa-c	Yq11.2	Non-obstructive spermatogenic failure	YL
9 genes & 4 loci#	Multiple loci	Primary ciliary dyskinesia (CILD1-13); Kartagener syndrome; immotile cilia syndrome; including situs inversus	AR
CATSPER1	11q13.1	Non-syndromic male infertility	AR
SPATA16	6q21	Globozoopsermia (acrosome malformation)	AR
STRC & CATSPER2	15q15.3	Sensorineural deafness and male infertility; Contiguous gene deletion—2 genes included	AR
SYCP3	12q23	Azoospermia with maturation arrest	AD
AURKC	19q13.43	Infertility associated with multi-tailed spermatozoa and excessive DNA	AR
Obstructive infertility			
Females			
AR	Xq11q12	Complete androgen insensitivity syndrome (M)	XLR
WNT4	1p35	Müllerian aplasia & hyperandrogenism	Sporadic, possible AD
Males			
CFTR	7q31.2	Congenital bilateral absence of vas deferens	AR

* *Causes sexual ambiguity in males (not discussed here).*
#*There are 13 known forms of primary ciliary dyskinesia, 9 genes are known and 4 loci for the remainder: CILD1 (MIM 24440)* – DNAI1; *CILD2 (MIM 612250) – 19q13.3-qter; CILD3 (MIM 603335)* – DNAH5; *CILD4 (MIM 608646) – 15q13; CILD5 (MIM 608647) – 16p12.2-p12.1; CILD6 (MIM 610852)* – TXNDC3; *CILD7 (MIM 611884)* – DNAH11; *CILD8 (MIM 612274) – 15q24-q25; CILD9 (MIM 612444)* – DNAI2; *CILD10 (MIM 612518)* – KTU; *CILD11 (MIM 612649)* – RSPH4A; *CILD12 (MIM 612650)* – RSPH9; *CILD13 (MIM 613190)* – RRC50.
XLR, X-linked recessive; AR, autosomal recessive; AD, autosomal dominant; YL, Y-linked.
DAZL, USP9Y, RBMY, DAZ are not included because the phenotype is not consistent and DNA sequence differences may be seen in both cases (infertility) and controls (fertile males).

AZF DELETIONS

Approximately 5–10% of males with azoospermia or severe oligospermia have deletions of the azoospermia factor (AZF) region on Yq11, which contains spermatogenesis genes.[75] Three different AZF regions (a–c) are localized within this region from centromere to telomere (Fig. 30.4). Deletions of AZF_a and AZF_b are the most severe, and often result in Sertoli only syndrome (absence of germ cells), while the more common deletions of AZF_c do not necessarily result in infertility. These regions were originally thought to be separate, but upon complete sequencing of the Y chromosome it became apparent that AZF_b and AZF_c overlap (Fig. 30.4).[76] Yq11 is comprised of inverted palindrome pairs which predispose to recombination and may result in deletions. Although a number of genes, such as multicopy *DAZ* and *RBMY*, have been implicated in male infertility, the identification of a gene that consistently causes infertility when mutated has been elusive. *USP9Y* within AZF_a was considered such a gene until a patient with an inactivating *USP9Y* mutation was found to exhibit normal spermatogenesis.[77] This complex, small chromosome directs sexual differentiation and appears to have spermatogenesis genes, but the exact etiologic genes remain unknown. The significance of this region lies in the presumed paradox of "inherited" infertility, in which sons of males with deletions (born via IVF/ICSI) will have the same microdeletion resulting in infertility.

CHROMOSOMAL ABNORMALITIES

Some males with infertility, particularly with severe oligospermia or azoospermia, may have a karyotypic abnormality even though they do not have complete gonadal failure. Endocrine function of the testes is normal, but gamete production/function is abnormal. Balanced chromosomal rearrangements such as translocations or inversions are found in 3–5% of such males.[78] Although these are unlikely to be causative, they are the subject of investigation to identify new genes involved in human disease. Immediate clinical consequences for carriers of balanced translocations are that they may predispose to unbalanced offspring with mental retardation and other congenital anomalies.

OTHER SINGLE GENE DISORDERS AFFECTING SPERM

As shown in Table 30.7, there are some other rare disorders, transmitted most commonly in an autosomal recessive mode of inheritance, that affect sperm. Primary ciliary dyskinesia is an autosomal recessive disorder affecting ciliary function – most commonly the dynein arms. Kartagener syndrome consists of immotile cilia syndrome and situs inversus (random patterning of visceral organs rather than right to left). There are at least 13 genes (*CILD1–13*) involved in this heterogeneous group of disorders. In particular sperm motility is decreased, but it is not absolute sterility, as fertility can occur in both sexes. Additional uncommon Mendelian forms of infertility affecting sperm concentration, motility and/or morphology are shown in Table 30.7, most of which are autosomal recessive and have only been reported once or a few times. These include: non-syndromic male infertility (MIM 612997) due to *CATSPER1* mutations; globoozospermia (MIM 102530) due to *SPATA16* mutations; sensorineural deafness and male infertility (MIM 611102) due to a deletion on 15q15.3 (including *STRC* and *CATSPER2* genes); azoospermia due to perturbations of meiosis (MIM 270960) due to *SYCP3* mutations; and male infertility with large-headed, multiflagellar, polyploid spermatozoa (MIM 243060) due to *AURKC* mutations.

Disorders of Outflow Obstruction

Females

COMPLETE ANDROGEN INSENSITIVITY

The androgen receptor (AR) belongs to the steroid superfamily of nuclear hormone receptors, and is encoded by an 8-exon gene, *AR*. The protein has an amino-terminus, a DNA binding domain and a carboxy-terminal androgen binding domain.[79] 46,XY males with complete androgen insensitivity syndrome (CAIS) due to mutations in the *AR* gene present as phenotypic females with normal breast development, minimal or no axillary and pubic hair, and a blind vaginal pouch. The phenotype results from the absence of normal androgen interaction with its receptor at the target level. Since testes are normal, anti-Müllerian hormone (AMH) is produced, which inhibits the formation of the uterus and upper vagina. The vagina appears as a blind vaginal pouch, and no cervix is identified. The testes, which may be intra-abdominal or inguinal, are capable of making testosterone, and normal adult male levels are produced (300–1100 ng/dl). Incomplete forms of AIS have also been described, but these will not be discussed here since they cause sexual ambiguity in males.

More than 300 different mutations, most commonly missense mutations, in *AR* have been reported to cause AIS.[80,81] Interestingly, nearly all of the exon-1 mutations cause CAIS or incomplete AIS, and most of these produce a premature stop codon. However, most of the mutations causing AIS occur in exons 2–8, despite the fact that exon 1 encodes for more than half of the protein.[81]

MÜLLERIAN APLASIA

Mayer-Rokitansky-Kuster-Hauser syndrome (MRKH) is a severe reproductive developmental disorder that results in congenital absence of the uterus and vagina (CAUV). It affects approximately 1 in 5,000 females, and is the second most common cause of primary amenorrhea in girls.[82] During embryologic development, both Müllerian ducts migrate into the proper anatomic position and give rise to the uterus, fallopian tubes and upper vagina. Renal development is intimately related to urogenital development. Therefore, disruption of either system can affect the development of the other. Patients with CAUV are 46,XX females that undergo normal puberty, since they have normal ovarian function. On physical exam, they have normal adult breast development and pubic hair, and either an absent vagina or a small blind vaginal pouch (a urogenital sinus derivative). Normally, the uterus is either absent or very hypoplastic, although fallopian tubes may be present, suggesting that development of the caudal portions of the Müllerian duct is more adversely affected. Rarely, the endometrium may be functional enough to cause pelvic pain due to genital tract obstruction. Pregnancy is only possible in these patients if oocytes are retrieved for *in vitro* fertilization with embryo transfer into a surrogate.

If only Müllerian aplasia is present, which seems to occur in about two-thirds of patients, it is classified as type I.[82] Approximately one-third of patients with CAUV may have unilateral renal agenesis, and another 10–15% may have skeletal abnormalities, including the Klippel-Feil sequence (fusion of cervical vertebrae). This has been classified either as type II or under the acronym MURCS (Müllerian duct aplasia, unilateral renal agenesis and cervical somite anomalies), which may also include deafness.[82] Cardiac anomalies have been reported in ~10–12% of all cases (type III). In fact, CAUV has been associated with other syndromes, including Holt-Oram syndrome (autosomal dominant due to *TBX5* mutations) and thrombocytopenia-absent radius (TAR) syndrome (autosomal recessive), both of which consist of associated cardiac and skeletal abnormalities. It has been suggested that perhaps CAUV, Holt-Oram syndrome, TAR syndrome and Klippel-Feil could have a common pathway of defective T-box transcription.[83]

The molecular basis of uterine and vaginal development is largely unknown, as it is difficult to identify families since pregnancy and vertical transmission will not occur (unless transmitted by a surrogate or female-limited autosomal dominant inheritance). Although a number of genes have been implicated and studied, including *CFTR*, *GALT*, *HOXA7*, *HOXA13*, *PBX1*, *HOXA10*, *AMH*, *AMHR*, *RARG* and *RXRA*, only mutations in *WNT4* appear to be causative in humans.[84] These patients have also been reported to have mild excess androgen production resulting in hair growth (hirsutism). Several chromosomal translocations and genomic imbalances have been reported, which could be useful in uncovering the molecular basis.

Males

CONGENITAL BILATERAL ABSENCE OF THE VAS DEFERENS (CBAVD)

Males with congenital bilateral absence of the vas deferens (CBAVD) have normal sexual development, but present with infertility due to obstruction of sperm egress.[85] Testes size and serum endocrine studies (testosterone, LH, and FSH) are normal. Most affected males have azoospermia, but some may have severe oligospermia. The absence of the vas deferens can be determined by individuals experienced in this examination. Performing a reflex fructose analysis of the semen when azoospermia is present should help identify the affected patients (fructose is made in the seminal vesicle, and so CBAVD men should be fructose-negative). CBAVD occurs in about 1–2% of infertile men, and is considered a mild allelic variant of cystic fibrosis (CF).[85] In fact, most affected men have two mutations in the cystic fibrosis transmembrane regulator (*CFTR*) gene that causes CF.

Genetic counseling of these couples when the male has CBAVD centers around the more concerning risk for having a child with classical CF, manifesting pulmonary and pancreatic insufficiency. Both members of the couple should be screened for *CFTR* mutations, and appropriate counseling performed. Some studies have indicated that >90% of CBAVD men have *CFTR* mutations, but even if only one is demonstrated, it is wise to consider that an unidentified second mutation could be present. It should also be stressed that universal screening of the most common CF mutations was designed to detect classical CF – not the milder allelic variant CBAVD. If CBAVD is suspected, the analysis of causative CFTR mutations is different. In addition, about 20% of men with CBAVD have renal anomalies but no *CFTR* mutations, suggesting this is a separate entity.[85]

TREATMENT OF EUGONADISM/OUTFLOW OBSTRUCTION

For males with eugonadism and infertility, treatment usually consists of intrauterine insemination if sufficient sperm are available. Otherwise, *in vitro* fertilization (IVF) or donor sperm is necessary. For women with ovulation disorders, treatment depends upon whether or not pregnancy is desired. Usually, progestin-containing agents are used to prevent endometrial hyperplasia/cancer from unopposed estrogen stimulation. Since estrogen–progestin preparations increase SHBG, they are useful for hirsutism/hyperandrogenemia, often in combination with an antiandrogen such as spironolactone. If pregnancy is desired, ovulation induction with clomiphene citrate is usually the first line of therapy. Metformin (an insulin sensitizer) added to clomiphene, letrozole (an aromatase inhibitor) or gonadotropins may be useful if clomiphene alone is unsuccessful. Women with PCOS who are overweight or obese should be encouraged to make lifestyle changes to lose weight and exercise. For women with hyperprolactinemia (with or without a pituitary tumor), treatment with a dopamine agonist such as bromocriptine or cabergoline may be used to normalize prolactin, while thyroid replacement may be utilized for hypothyroidism.

For males with outflow obstruction due to CBAVD, sperm may be retrieved from the epididymis or testis for intracytoplasmic sperm injection (ICSI) in *in vitro* fertilization (IVF). Careful genetic counseling and testing the female partner for carrier status are of paramount importance. For females with outflow obstruction who have primary amenorrhea due to complete androgen insensitivity syndrome (CAIS) or Müllerian aplasia, creation of a vagina using dilators (as first-line therapy) or surgery may be performed. For outflow obstruction due to uterine adhesions, a hysteroscopic surgery to remove them may become necessary to restore menses and permit pregnancy.

SUMMARY AND CONCLUSIONS

Disruption of the HPG axis may result in human disease at the level of the hypothalamus, pituitary, gonads or outflow tracts. The contribution from humans with reproductive disorders who possess chromosomal abnormalities or mutations in new genes has enhanced our understanding of the pathophysiology of the disorder and will continue to do so, assisting in directing genetic counseling and, ultimately, in treatment.

References

1. Kim HG, Bhagavath B, Layman LC. Clinical manifestations of impaired GnRH neuron development and function. *Neurosignals.* 2008;16:165–182.
2. Bentley GE, Ubuka T, McGuire NL, et al. Gonadotrophin-inhibitory hormone: a multifunctional neuropeptide. *J Neuroendocrinol.* 2009;21:276–281.
3. Layman LC, Reindollar RH. The diagnosis and treatment of pubertal disorders. Adolesc Med: *State Art Rev.* 1994;5:37–55.
4. Rosner W, Auchus RJ, Azziz R, Sluss PM, Raff H. Position statement: Utility, limitations, and pitfalls in measuring testosterone: an Endocrine Society position statement. *J Clin Endocrinol Metab.* 2007;92:405–413.
5. Layman LC, Cohen DP, Jin M, et al. Mutations in the gonadotropin-releasing hormone receptor gene cause hypogonadotropic hypogonadism. *Nat Genet.* 1998;18:14–15.
6. Bhagavath B, Podolsky RH, Ozata M, et al. Clinical and molecular characterization of a large sample of patients with hypogonadotropic hypogonadism. *Fertil Steril.* 2006;85:706–713.
7. Reindollar RH, Byrd JR, McDonough PG. Delayed sexual development: study of 252 patients. *Am J Obstet Gynecol.* 1981;140:371–380.
8. Reindollar RH, Novak M, Tho SPT, McDonough PG. Adult-onset amenorrhea: a study of 262 patients. *Am J Obstet Gynecol.* 1986;155:531–543.
9. Layman LC. Human gene mutations causing infertility. *J Med Genet.* 2002;39:153–161.
10. Layman LC, Tho SP, Clark AD, Kulharya A, McDonough PG. Phenotypic spectrum of 45,X/46,XY males with a ring Y chromosome and bilaterally descended testes. *Fertil Steril.* 2009;91:791–797.
11. Cormier-Daire V, Belin V, Cusin V, et al. SHOX gene mutations and deletions in dyschondrosteosis or Leri-Weill syndrome. *Acta Paediatr Suppl.* 1999;88:55–59.
12. Jager RJ, Anvret M, Hall K, Scherer G. A human XY female with a frame shift mutation in the candidate testis-determining gene SRY. *Nature.* 1990;348:452–454.
13. Rebar RW, Erickson GF, Yen SS. Idiopathic premature ovarian failure: clinical and endocrine characteristics. *Fertil Steril.* 1982;37:35–41.
14. Nelson LM. Clinical practice. Primary ovarian insufficiency. *N Engl J Med.* 2009;360:606–614.
15. Bione S, Sala C, Manzini C, et al. A human homologue of the *Drosophila melanogaster* diaphenous gene is disrupted in a patient with premature ovarian failure: evidence for conserved function in oogenesis and implications for human sterility. *Am J Hum Genet.* 1998;62:533–541.
16. Allingham-Hawkins DJ, Babul-Hirji R, Chitayat D, et al. Fragile X premutation is a significant risk factor for premature ovarian failure: the International Collaborative POF in Fragile X study – preliminary data. *Am J Med Genet.* 1999;83:322–325.
17. Conway GS, Payne NN, Webb J, Murray A, Jacobs PA. Fragile X premutation screening in women with premature ovarian failure. *Hum Reprod.* 1998;13:1184–1187.
18. Di Pasquale E, Beck-Peccoz P, Persani L. Hypergonadotropic ovarian failure associated with an inherited mutation of human bone morphogenetic protein-15 (BMP15) gene. *Am J Hum Genet.* 2004;75:106–111.
19. Rossetti R, Di Pasquale E, Marozzi A, et al. BMP15 mutations associated with primary ovarian insufficiency cause a defective production of bioactive protein. *Hum Mutat.* 2009;30:804–810.
20. Aittomaki K, Lucena JLD, Pakarinen P, et al. Mutation in the follicle-stimulating hormone receptor gene causes hereditary hypergonadotropic hypogonadism. *Cell.* 1995;82:959–968.
21. Consortium F-GA. An autoimmune disease, APECED, caused by mutations in a novel gene featuring two PHD-type zinc-finger domains. *Nat Genet.* 1997;17:399–403.
22. Nagamine K, Peterson P, Scott HS, et al. Positional cloning of the APECED gene. *Nat Genet.* 1997;17:393–398.
23. Crisponi L, Deiana M, Loi A, et al. The putative forkhead transcription factor FOXL2 is mutated in blepharophimosis/ptosis/epicanthus inversus syndrome. *Nat Genet.* 2001;27:159–166.
24. Laissue P, Lakhal B, Benayoun BA, et al. Functional evidence implicating FOXL2 in non-syndromic premature ovarian failure and in the regulation of the transcription factor OSR2. *J Med Genet.* 2009;46:455–457.
25. Qin Y, Choi Y, Zhao H, Simpson JL, Chen ZJ, Rajkovic A. NOBOX homeobox mutation causes premature ovarian failure. *Am J Hum Genet.* 2007;81:576–581.
26. Zhao H, Chen ZJ, Qin Y, et al. Transcription factor FIGLA is mutated in patients with premature ovarian failure. *Am J Hum Genet.* 2008;82:1342–1348.
27. Achermann JC, Ito M, Ito M, Hindmarsh PC, Jameson JL. A mutation in the gene encoding steroidogenic factor-1 causes XY sex reversal and adrenal failure in humans. *Nat Genet.* 1999;22:125–126.
28. Lourenco D, Brauner R, Lin L, et al. Mutations in NR5A1 associated with ovarian insufficiency. *N Engl J Med.* 2009;360:1200–1210.
29. Kaufman FR, Kogut MD, Donnell GN, Goebelsmann U, March C, Koch R. Hypergonadotropic hypogonadism in female patients with galactosemia. *N Engl J Med.* 1979;304:994–998.
30. Fogli A, Gauthier-Barichard F, Schiffmann R, et al. Screening for known mutations in EIF2B genes in a large panel of patients with premature ovarian failure. *BMC Women's Health.* 2004;4:8.
31. Yanase T, Simpson ER, Waterman MR. 17-Alpha-hydroxylase/17,20-lyase deficiency: from clinical investigation to molecular definition. *Endocr Rev.* 1991;12:91–108.
32. Ito Y, Fisher CR, Conte FA, Grumbach MM, Simpson ER. Molecular basis of aromatase deficiency in an adult female with

sexual infantilism and polycystic ovaries. *Proc Nat Acad Sci.* 1993;90:11673–11677.

33. Bojesen A, Gravholt CH. Klinefelter syndrome in clinical practice. *Nat Clin Pract Urol.* 2007;4:192–204.
34. Parma P, Radi O, Vidal V, et al. R-spondin1 is essential in sex determination, skin differentiation and malignancy. *Nat Genet.* 2006;38:1304–1309.
35. Mendelsohn FA, Warren MP. Anorexia, bulimia, and the female athlete triad: evaluation and management. *Endocrinol Metab Clin North Am.* 2010;39:155–167. x.
36. Lawson EA, Klibanski A. Endocrine abnormalities in anorexia nervosa. *Nat Clin Pract Endocrinol Metab.* 2008;4:407–414.
37. Millar RP, Roseweir AK, Tello JA, et al. Kisspeptin antagonists: unraveling the role of kisspeptin in reproductive physiology. *Brain Res.* 2010;1364:81–89.
38. Genazzani AD, Chierchia E, Santagni S, Rattighieri E, Farinetti A, Lanzoni C. Hypothalamic amenorrhea: from diagnosis to therapeutical approach. *Ann Endocrinol (Paris).* 2010;71:163–169.
39. Perkins RB, Hall JE, Martin KA. Neuroendocrine abnormalities in hypothalamic amenorrhea: spectrum, stability, and response to neurotransmitter modulation. *J Clin Endocrinol Metab.* 1999;84:1905–1911.
40. Nachtigall LB, Boepple PA, Pralong FP, Crowley WFJ. Adult-onset idiopathic hypogonadotropic hypogonadism–a treatable form of male infertility. *N Engl J Med.* 1997;336:410–415.
41. Klibanski A. Clinical practice. Prolactinomas. *N Engl J Med.* 2010;362:1219–1226.
42. de Roux N, Young J, Misrahi M, et al. A family with hypogonadotropic hypogonadism and mutations in the gonadotropin-releasing hormone receptor. *N Engl J Med.* 1997;337:1597–1602.
43. Bhagavath B, Ozata M, Ozdemir IC, et al. The prevalence of gonadotropin-releasing hormone receptor mutations in a large cohort of patients with hypogonadotropic hypogonadism. *Fertil Steril.* 2005;84:951–957.
44. Lofrano-Porto A, Barra GB, Giacomini LA, et al. Luteinizing hormone beta mutation and hypogonadism in men and women. *N Engl J Med.* 2007;357:897–904.
45. Weiss J, Axelrod L, Whitcomb RW, Harris PE, Crowley Jr WF, Jameson JL. Hypogonadism caused by a single amino acid substituion in the β subunit of luteinizing hormone. *N Engl J Med.* 1992;326:179–183.
46. Layman LC, Lee EJ, Peak DB, et al. Delayed puberty and hypogonadism caused by a mutation in the follicle stimulating hormone β-subunit gene. *N Engl J Med.* 1997;337:607–611.
47. Matthews C, Chatterjee VK. Isolated deficiency of follicle-stimulating hormone revisited. *N Engl J Med.* 1997;337:642.
48. Layman LC, Porto AL, Xie J, et al. FSH beta gene mutations in a female with partial breast development and a male sibling with normal puberty and azoospermia. *J Clin Endocrinol Metab.* 2002;87:3702–3707.
49. Porto AL, Xie J, da Motta LDC, da Motta LACR, Layman LC. A new follicle stimulating hormone-beta gene mutation in a Brazilian kindred results in normal puberty with azoospermia in a male and evidence of breast development with primary amenorrhea in a female. *Endocrine Soc.* 1999;P2-437:372.
50. Barnes RB, Namnoum A, Rosenfield RL, Layman LC. Effects of follicle-stimulating hormone on ovarian androgen production in a woman with isolated follicle-stimulating hormone deficiency. *N Engl J Med.* 2000;343:1197–1198.
51. Barnes RB, Namnoum A, Rosenfield RL, Layman LC. The role of LH and FSH in ovarian androgen secretion and ovarian follicular development: Clinical studies in a patient with isolated FSH deficiency and multicystic ovaries. *Hum Reprod.* 2002;17:88–91.
52. Cogan JD, Wu W, Phillips JAI, et al. The PROP1 2-base pair deletion is a common cause of combined pituitary hormone deficiency. *J Clin Endocrinol Metab.* 1998;83:3346–3349.
53. Wu W, Cogan JD, Pfaffle RW, et al. Mutations in PROP1 cause familial combined pituitary hormone deficiency. *Nat Genet.* 1998;18:147–149.
54. Park JK, Ozata M, Chorich LP, et al. Analysis of the PROP1 gene in a large cohort of patients with idiopathic hypogonadotropic hypogonadism. *Clin Endocrinol (Oxf).* 2004;60:147–149.
55. Carvalho LR, Woods KS, Mendonca BB, et al. A homozygous mutation in HESX1 is associated with evolving hypopituitarism due to impaired repressor-corepressor interaction. *J Clin Invest.* 2003;112:1192–1201.
56. Dattani MT, Martinez-Barbera J-P, Thomas PQ, et al. Mutations in the homeobox gene HESX1/Hesx1 associated with septo-optic dysplasia in human and mouse. *Nat Genet.* 1998;19:125–133.
57. Layman LC. The genetics of hypogonadotropic hypogonadism. *Endocrinologist.* 1999;9:366–370.
58. Netchine I, Sobrier ML, Krude H, et al. Mutations in LHX3 result in a new syndrome revealed by combined pituitary hormone deficiency. *Nat Genet.* 2000;25:182–186.
59. Machinis K, Pantel J, Netchine I, et al. Syndromic short stature in patients with a germline mutation in the LIM homeobox LHX4. *Am J Hum Genet.* 2001;69:961–968.
60. Dateki S, Fukami M, Uematsu A, et al. Mutation and gene copy number analyses of six pituitary transcription factor genes in 71 patients with combined pituitary hormone deficiency: Identification of a single patient with LHX4 deletion. *J Clin Endocrinol Metab.* 2010;95:4043–4047.
61. Tajima T, Yorifuji T, Ishizu K, Fujieda K. A novel mutation (V101A) of the LHX4 gene in a Japanese patient with combined pituitary hormone deficiency. *Exp Clin Endocrinol Diabetes.* 2010;118:405–409.
62. Pitteloud N, Acierno Jr JS, Meysing AU, Dwyer AA, Hayes FJ, Crowley Jr WF. Reversible Kallmann syndrome, delayed puberty, and isolated anosmia occurring in a single family with a mutation in the fibroblast growth factor receptor 1 gene. *J Clin Endocrinol Metab.* 2005;90:1317–1322.
63. Whitcomb RW, Crowley Jr WF. Clinical review 4: Diagnosis and treatment of isolated gonadotropin-releasing hormone deficiency in men. *J Clin Endocrinol Metab.* 1990;70:3–7.
64. ACOG Practice Bulletin No. 108: Polycystic ovary syndrome. *Obstet Gynecol.* 2009;114:936–949.
65. Revised 2003 consensus on diagnostic criteria and long-term health risks related to polycystic ovary syndrome (PCOS). *Hum Reprod.* 2004;19:41–47.
66. Azziz R, Carmina E, Dewailly D, et al. Positions statement: criteria for defining polycystic ovary syndrome as a predominantly hyperandrogenic syndrome: an Androgen Excess Society guideline. *J Clin Endocrinol Metab.* 2006;91:4237–4245.
67. Ewens KG, Stewart DR, Ankener W, et al. Family-based analysis of candidate genes for polycystic ovary syndrome. *J Clin Endocrinol Metab.* 2010;95:2306–2315.
68. Frazer KA, Murray SS, Schork NJ, Topol EJ. Human genetic variation and its contribution to complex traits. *Nat Rev Genet.* 2009;10:241–251.
69. Abbott DH, Tarantal AF, Dumesic DA. Fetal, infant, adolescent and adult phenotypes of polycystic ovary syndrome in prenatally androgenized female rhesus monkeys. *Am J Primatol.* 2009;71:776–784.
70. Kremer H, Kraaij R, Toledo SPA, et al. Male pseudohermaphroditism due to a homozygous missense mutation of the luteinizing hormone receptor gene. *Nat Genet.* 1995;9:160–164.
71. Latronico AC, Anasti J, Arnhold IJP, et al. Brief report: testicular and ovarian resistance to luteinizing hormone caused by

inactivating mutations of the luteinizing hormone-receptor gene. *N Engl J Med.* 1996;334:507–512.
72. Toledo SPA, Brunner HG, Kraaij R, et al. An inactivating mutation of the luteinizing hormone receptor causes amenorrhea in a 46,XX female. *J Clin Endocrinol Metab.* 1996;81:3850–3854.
73. Mortlock DP, Innis JW. Mutation of HOXA13 in hand-foot-genital syndrome. *Nat Genet.* 1997;15:179–181.
74. Jorgensen EM, Ruman JI, Doherty L, Taylor HS. A novel mutation of HOXA13 in a family with hand-foot-genital syndrome and the role of polyalanine expansions in the spectrum of Mullerian fusion anomalies. *Fertil Steril.* 2009;94:1235–1238.
75. Ferlin A, Arredi B, Speltra E, et al. Molecular and clinical characterization of Y chromosome microdeletions in infertile men: a 10-year experience in Italy. *J Clin Endocrinol Metab.* 2007;92:762–770.
76. Skaletsky H, Kuroda-Kawaguchi T, Minx PJ, et al. The male-specific region of the human Y chromosome is a mosaic of discrete sequence classes. *Nature.* 2003;423:825–837.
77. Luddi A, Margollicci M, Gambera L, et al. Spermatogenesis in a man with complete deletion of USP9Y. *N Engl J Med.* 2009;360:881–885.
78. Layman LC. Genetic causes of human infertility. *Endocrinol Metab Clin North Am.* 2003;32:549–572.
79. Brown TR, Lubahn DB, Wilson EM, Joseph DR, French FS, Migeon CJ. Deletion of the steroid-binding domain of the human androgen receptor gene in one family with complete androgen insensitivity syndrome: evidence for further genetic heterogeneity in this syndrome. *Proc Nat Acad Sci.* 1988;85:8151–8155.
80. Patterson MNH, I. A., Gottlieb B, Pinsky L. The androgen receptor gene mutations database. *Nucleic Acids Res.* 1994;22:3560–3562.
81. Gottlieb B, Pinsky L, Beitel LK, Trifiro M. Androgen insensitivity. *Am J Med Genet.* 1999;89:210–217.
82. Sultan C, Biason-Lauber A, Philibert P. Mayer-Rokitansky-Kuster-Hauser syndrome: recent clinical and genetic findings. *Gynecol Endocrinol.* 2009;25:8–11.
83. McDonough PG. Hot clues to the etiology of Mayer-Rokitansky-Kuster-Hauser syndrome? *Fertil Steril.* 2005;84:545.
84. Biason-Lauber A, Konrad D, Navratil F, Schoenle EJ. A WNT4 mutation associated with Müllerian-duct regression and virilization in a 46,XX woman. *N Engl J Med.* 2004;351:792–798.
85. Radpour R, Gourabi H, Dizaj AV, Holzgreve W, Zhong XY. Genetic investigations of CFTR mutations in congenital absence of vas deferens, uterus, and vagina as a cause of infertility. *J Androl.* 2008;29:506–513.

CHAPTER

31

Disorders of the Hypothalamic–Pituitary–Thyroid Axis

Aniket R. Sidhaye, Fredric E. Wondisford

Division of Metabolism, Departments of Pediatrics and Medicine, Johns Hopkins University School of Medicine, Baltimore, MD, USA

OUTLINE

Summary

Thyroid hormone production and secretion is finely regulated by a feedback system involving the hypothalamus, pituitary and thyroid to maintain circulating levels of thyroid hormone within a very narrow range. In this chapter we review those disorders that are not primarily related to thyroid gland pathology, as they create unique clinical scenarios.

INTRODUCTION

The appropriate synthesis and secretion of thyroid hormone depends on a normally functioning hypothalamic–pituitary–thyroid (HPT) axis. Thyroid hormone synthesis and secretion is stimulated by the action of thyrotropin (TSH), which is secreted into the general circulation from the anterior pituitary thyrotrope cells. Pituitary thyrotropes are, in turn, stimulated to secrete TSH upon the action of thyrotropin-releasing hormone (TRH), which is synthesized in the hypothalamus. This system is regulated chiefly based on the negative feedback of thyroid hormones thyroxine (T4) and tri-iodothyronine (T3) on both hypophysiotropic TRH neurons and pituitary thyrotropes. In this chapter, we will briefly review the important aspects of normal

Handbook of Neuroendocrinology, DOI: 10.1016/B978-0-12-375097-6.10031-9

HPT axis physiology (see Fig. 31.1), and those disorders of the HPT axis which cause either hyper- or hyposecretion of thyroid hormone. Although disorders such as Graves' disease, Hashimoto's thyroiditis or cretinism reflect thyroid dysfunction, they are not neuroendocrine disorders. For an account of these disorders, the reader is referred to standard textbooks of endocrinology.

REVIEW OF NORMAL HPT AXIS STRUCTURE AND FUNCTION

TRH

TRH is a tripeptide (pyro-Glu-His-Pro-NH2) synthesized by the proteolytic cleavage of a larger 242-amino acid precursor (pro-TRH) that contains six copies of TRH.[1–3] Mature TRH is synthesized in distinct cell populations of the paraventricular nuclei (PVN) of the hypothalamus, whose neuronal axons project to the median eminence where TRH is released into the hypothalamic–pituitary portal circulation.[4] TRH signals through a G-protein coupled receptor located on pituitary thyrotropes, and its principal action is to stimulate TSH production. It also regulates carbohydrate modifications of TSH that are important for the biological activity of TSH (discussed below). In addition, TRH receptors are present on lactotropes, accounting for the elevation in prolactin levels that sometimes accompanies primary hypothyroidism.[5]

It is important to note that TRH is synthesized in other regions of the hypothalamus, brain, pancreas and heart.[5] Further, TRH receptors are widely

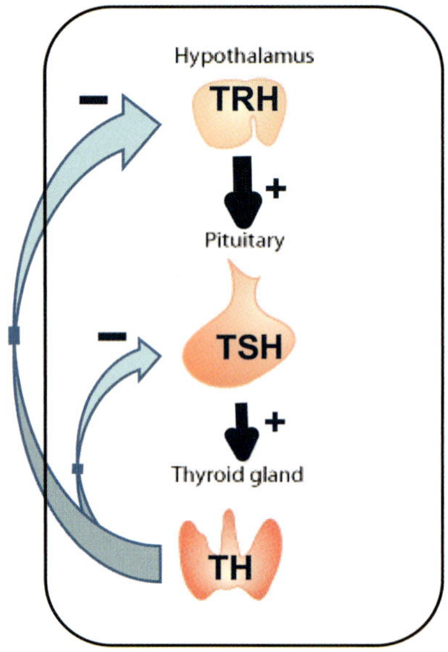

FIGURE 31.1 Diagram of the hypothalamic–pituitary–thyroid axis (TRH, thyrotropin releasing hormone; TSH, thyrotropin; TH, thyroid hormone).

distributed through the brain.[6] Although the precise role of neither TRH nor its receptor in these locations is known, it suggests that TRH may have other functions beyond regulation of TSH production.

TRH is Regulated by Thyroid Hormone, Nutritional Status and in Severe Illness

Thyroid hormones, in particular T3, regulate TRH mRNA production specifically in the PVN such that hypothyroidism causes increased TRH mRNA and hyperthyroidism causes decreased TRH mRNA.[7–9] It appears that the negative regulation of the HPT axis by T3 is dependent on thyroid hormone receptor isoform β2.[10] Further, the regulation of TRH by T3 also occurs by the T3-mediated regulation of the enzymes responsible for proteolytic cleavage of proTRH into mature TRH, prohormone convertase 1 (PC1) and prohormone convertase 2 (PC2).[5]

T3 is the most important regulator of TRH production from the PVN. However, other signaling mechanisms also play a role. Specifically, it is known that long-term fasting in humans is associated with decreased TSH and T3 levels.[11] Recently, evidence from rodents has shown that leptin, a hormone secreted from adipocytes in proportion to fat mass (and thereby a measure of nutritional status), can regulate TRH mRNA levels.[12,13] Indeed, TRH mRNA can be suppressed even in thyroid hormone receptor β2 knockout mice that have been fasted, suggesting that mechanisms which do not involve T3 and its receptor are involved in regulating T3.

Similar to human fasting, the sick-euthyroid syndrome is another situation that highlights how other factors can affect TRH. Critical illness in humans is associated with decreased T3, increased reverse T3 and, in more severely ill individuals, reductions in T4 and TSH.[14] Human studies have shown that patients in the ICU who succumb to severe illness have low T3 and low TRH mRNA in the PVN, but those patients who died acutely from cardiac illness, where T3 levels were not low, did not have low TRH mRNA levels.[15] It has been suggested that cytokines such as interleukin-1 (IL-1), IL6 and IL-1β may play a role in this TRH mRNA.[16] Finally, in rodents, lipopolysaccharide can suppress TRH mRNA.[17] In all of the above cases, definite mechanisms have not been proven in humans.

TSH

Thyrotropin (TSH) is a member of the glycoprotein hormone family, which includes luteinizing hormone (LH), follicle-stimulating hormone (FSH) and human chorionic gonadotropin (hCG). All these hormones are heterodimers consisting of non-covalently linked alpha- and beta-chains. The α-chain is common to all the hormones, while the β subunit is unique to each hormone.[18] The latter carries biological specificity, based on binding to a cognate receptor.[19] TSH is secreted by the anterior pituitary, and in the placenta during pregnancy.[20] TSH exerts its effect primarily by binding the TSH receptor (TSH-R) located on the basolateral surface of thyroid follicular cells. This binding results in the production of cAMP and phosphatidylinositols, which engage their respective transduction pathways, ultimately resulting in the synthesis and secretion of thyroid hormones and modulation of thyroid hormone responsive genes.[21]

TRH is the Main Positive Regulator of TSH Subunit Gene Expression[22]

The transcription factor Pit-1 is critical for the transcriptional upregulation of TSH-β subunit gene by TRH.[23] In contrast, for the α subunit gene there are no Pit-1 DNA binding sites, and TRH-dependent binding of the CBP to Lhx3 contributes to α subunit gene expression.[24] Despite its predominant role in positive regulation of TSH, TRH is not required for TSH subunit gene transcription; TRH knockout mice express slightly higher than normal levels of TSH, though this TSH has reduced biological activity. Furthermore, patients with mutations in TRH-R have central hypothyroidism. In contrast, TRH is absolutely required for the up regulation of TSH in the setting of primary hypothyroidism.[25–27]

Carbohydrate Modifications of TSH are Important for its Biological Activity

An important point about TSH is that for full biological activity, both subunits must have the correct pattern of carbohydrate modifications. This glycosylation occurs co-translationally in the endoplasmic reticulum, and is necessary for the proper folding, heterodimer formation and secretion of TSH. If proper oligosaccharide attachment is inhibited, TSH aggregates in the cell and is degraded.[28–30] Furthermore, the ratio of sialylation to sulfation determines metabolic clearance of TSH.[31–33] TRH-mediated regulation of TSH glycosylation is an important contributor to TSH biological activity.

T4/T3 and Thyroid Hormone Receptors

Thyroid Hormone is the Major Negative Regulator of TSH Subunit Gene Expression

The relationship between TSH and T4 levels has been described as log-linear (Fig. 31.2).[34] This occurs by direct regulation of TSH message levels, as well as indirectly by reducing TRH production.[22] Based on studies in mice, it appears that T3 binding to its receptors is required for this negative regulation; however, the precise mechanism by which this occurs is not known. The thyroid hormone receptors (TRs) are nuclear hormone receptors. There are four isoforms (TRα1, TRα2, TRβ1, TRβ2) of the thyroid hormone receptor produced from two separate gene loci, *Thra* and *Thrb*.[35] Of these receptors, TRβ2 is almost exclusively expressed in the hypothalamus, pituitary, auditory system and retina, and is the predominant mediator of negative feedback in the hypothalamus, based on mouse knock-out studies.[36–40] Studies in TRβ null mice involving the selective dominant negative inhibition of thyroid hormone action in the pituitary suggest TRα1 and TRβ1 also have a role in thyroid hormone regulation of TSH.[37] In addition, mouse models suggest that, for complete abrogation of TSH regulation, knockout of all TRs is required.[41] Finally, mouse studies with knockin mutations in the DNA-binding domain of the TR suggest that DNA binding is necessary for T3-mediated negative regulation of TSH, as these mice had elevated

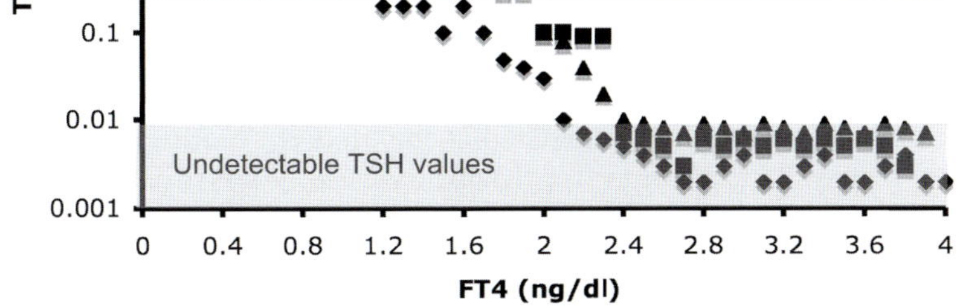

FIGURE 31.2 Log-linear relationship of thyrotropin (TSH) and free thyroid hormone (FT4). *Modified from Larsen* et al. *(2002).*[34]

levels of thyroid hormone in the serum in conjunction with non-suppressed TSH.[42,43]

On examining the HPT axis, disorders that occur primarily because of a disruption of the axis rather than because of altered thyroid gland function could be broadly caused by:

1. autonomous production of either TRH or TSH;
2. resistance to the effects of thyroid hormone;
3. diminished production of TRH or TSH.

The first two causes are classified as disorders with elevated thyroid hormone levels and inappropriately normal or elevated TSH levels, while the last category is characterized by low thyroid hormone levels and low, normal, or even mildly elevated TSH levels. The remainder of the chapter reviews these disorders.

DISORDERS WITH ELEVATED THYROID HORMONE LEVELS AND INAPPROPRIATELY NORMAL OR ELEVATED TSH LEVELS

Hyperthyroxinemia in the setting of a non-suppressed TSH level is secondary to two different clinical disorders: resistance to thyroid hormone (RTH), and TSH-secreting pituitary adenoma (TSHoma). Thus, in these disorders, normal negative feedback is disrupted. Both are extremely rare disorders, and are clinically distinguished by the fact that patients with TSHoma have signs and symptoms of hyperthyroidism, while patients with RTH are euthyroid in the majority of cases (generalized RTH or GRTH), with a minority showing signs of hyperthyroidism in some organs but not others (pituitary RTH or PRTH). Some authors classify these entities as central hyperthyroidism. The presence of ultrasensitive TSH assays has vastly improved the work-up of hyperthyroidism, and has made the recognition of this rare biochemical situation more facile.

Excluding Euthyroid Hyperthyroxinemia

In the clinical setting of elevated thyroid hormone levels with inappropriately normal or elevated TSH levels, it must first be confirmed that there is in fact "central hyperthyroidism." This is done by excluding those clinical scenarios that result in falsely elevated hormone values, or so-called euthyroid hyperthyroxinemia. The causes for euthyroid hyperthyroxinemia include the following.

1. Abnormalities of protein binding. T4 and T3 circulate bound to proteins in the blood, including albumin, thyroxine-binding globulin (TBG), and trans-thyretin (TTR). It is the free fraction of T4 and T3 that binds to receptors. Therefore, protein binding abnormalities can result in elevated total T4 or T3 levels, but since circulating free T4 or T3 levels are normal, TSH is normal in this clinical setting.
   a. TBG excess is the most common binding abnormality. From a clinical perspective, a common cause is estrogen, which increases the synthesis of TBG. This is seen in pregnancy, in women taking oral contraceptives or hormone replacement therapy after menopause, in patients with estradiol-secreting tumors, and in some taking the SERMs tamoxifen and raloxifene. Alternatively, TBG excess can be hereditary due to an X-linked dominant disorder causing increased synthesis of TBG. Hepatitis is associated with increased TBG. Several drugs increase TBG, including heroin, methadone, 5-fluorouracil, clofibrate and perphenazine. Finally, acute intermittent porphyria is associated with increased TBG through an unknown mechanism.
   b. Familial dysalbuminemic hyperthyroxinemia (FDH) is a genetic disorder where a mutant albumin is produced which has a high capacity but low affinity for T4, but not T3. It is seen most often in Hispanics, occurring in 0.2% of that population. FDH can be distinguished from TBG excess based on the T3 resin uptake, which is anticipated to be high in patients with TBG excess but is normal in patients with FDH. Interestingly, a single Thai family with a 40-fold increase in albumin capacity for T3 but only a 2-fold increase for T4 has been described, and this disorder is termed familial dysalbuminemic tri-iodothyroninemia.
   c. Abnormal TTR binding occurs in some families where the mutant TTR has abnormal binding to T4 but not T3.
2. Anti-T4 and -T3 immunoglobulins have been rarely described.
3. Drugs that alter T4 to T3 to conversion peripherally result in relatively increased T4 levels but normal TSH. These drugs include amiodarone, iodinated contrast agents and high-dose propranolol.
4. Symptomatic hyponatremia, high altitude and amphetamines have also shown increased total T4 levels with normal TSH.

Resistance to Thyroid Hormone

Resistance to thyroid hormone (RTH) is a syndrome characterized by reduced intracellular action of T3. Typically this is associated with elevated circulating levels of free T4 and often free T3 and normal or slightly elevated levels of TSH. This "inappropriate" TSH secretion represents a loss of negative feedback at the level of the hypothalamus and pituitary. Clinically, the typical signs and symptoms that may be expected with elevated free thyroid hormone levels are often absent in what is termed generalized resistance to thyroid hormone (GRTH). Less commonly, symptoms such as tachycardia, representing peripheral tissue sensitivity to thyroid hormone, may occur, termed central resistance to thyroid hormone (CRTH).

Pathogenesis

RTH was originally described in 1967, and is inherited as an autosomal dominant disorder.[44] Subsequent to the cloning of thyroid hormone receptors, disease-causing mutations in thyroid hormone receptor beta (TRβ) have been identified in 85% of patients studied. As a result, the etiology of RTH is taken to be due to TRβ mutations.[45–47] No mutations in TRα have been identified in RTH patients. It is conceivable that decreased intracellular effects of thyroid hormones may be due to defects in thyroid hormone transport or metabolism as well, and this is briefly discussed later.

Mutations in TRβ occur in the carboxy terminus, and interfere with binding of TR to its ligand or disrupt the normal interaction of co-regulator molecules with TR.[48–53] Mutations that disrupt DNA-binding have not been identified, and the currently characterized mutations are able to bind DNA and homodimerize with normal TR or heterodimerize with RXR. The mutant TRs act as dominant negatives, and this explains the dominant mode of inheritance. Interestingly, there is a family where the entire TRβ gene is deleted, and in this case the mode of inheritance is recessive.[54]

Epidemiology and Clinical Manifestations

RTH occurs with a prevalence of about 1/40,000, based on studies measuring T4 in newborns.[55] Females and males are equally affected. Clinically, most patients

BOX 31.1

HOW DO WE KNOW ABOUT THE PATHOGENESIS OF RTH AND THE FUNCTION OF THYROID HORMONE RECEPTORS?

The TRβ knockout mouse replicates the phenotype of humans with TRβ deletion. Therefore, heterozygotes have normal thyroid function, while homozygous mice have elevated thyroid hormone levels and inappropriate TSH levels. In addition, there is sensorineural deafness and monochromatic vision.[1–3] In contrast, mice with knockin mutations replicating the disease-causing mutations found in humans with RTH are true models of RTH. The mode of disease inheritance is autosomal dominant, with heterozygotes having a typical hormonal profile. Furthermore, the TRα1 knockout or common TRα knockout mice have minor alterations in the thyroid axis, as measured by thyroid hormone levels and TSH levels in the blood. If anything, thyroid hormone and TSH levels are slightly lower than in wild-type, suggesting a fundamentally different role for TRα compared to TRβ.[4,5] Interestingly, when human mutations of TRβ are introduced into homologous regions of the mouse TRα gene (TRα knockin), the phenotype is variable. Heterozygotes have growth retardation and increased body fat, and develop insulin resistance. There is also postnatal delay. RTH is not a feature. Homozygous TRα1 knockin is embryonic lethal.[6–8]

The combined TRα and TRβ knockout is interesting because it is compatible with life.[1] In contrast, athyreotic animals produced by the Pax8 KO mouse die prior to weaning. Further, if the TRα is knocked out in Pax8 KO mice, the mice are rescued. This manipulation also prevents cerebellar abnormalities that occur because of neonatal hypothyroidism.

Since some RTH-causing mutations of TRβ have been found to interfere with co-regulator association, it is interesting to consider mouse models of the same. Knockout of the co-activator SRC-1 produces RTH in mice, in addition to resistance to sex hormones.[9] In contrast, a mouse model where the interaction between the co-repressor complex NCoR and HDAC3 complex is disrupted (DADm) shows modestly increased basal circulating TSH in the presence of increased pituitary TSHα mRNA. Circulating thyroid hormone levels are similar to those in wild-type animals, thus producing a phentoype different from typical RTH.[10]

References

1. Gauthier K, Chassande O, Plateroti M, et al. Different functions for the thyroid hormone receptors TRalpha and TRbeta in the control of thyroid hormone production and post-natal development. *EMBO J.* 1999;18(3): 623–631.
2. Forrest D, Hanebuth E, Smeyne RJ, et al. Recessive resistance to thyroid hormone in mice lacking thyroid hormone receptor beta: evidence for tissue-specific modulation of receptor function. *EMBO J.* 1996;15(12): 3006–3015.
3. Forrest D, Erway LC, Ng L, Altschuler R, Curran T. Thyroid hormone receptor beta is essential for development of auditory function. *Nat Genet.* 1996;13(3):354–357.

(*Continued*)

BOX 31.1 *(cont'd)*

4. Macchia PE, Takeuchi Y, Kawai T, et al. Increased sensitivity to thyroid hormone in mice with complete deficiency of thyroid hormone receptor alpha. *Proc Natl Acad Sci USA.* 2001;98(1):349–354.
5. Wikstrom L, Johansson C, Salto C, et al. Abnormal heart rate and body temperature in mice lacking thyroid hormone receptor alpha 1. *EMBO J.* 1998;17(2):455–461.
6. Kaneshige M, Suzuki H, Kaneshige K, et al. A targeted dominant negative mutation of the thyroid hormone alpha 1 receptor causes increased mortality, infertility, and dwarfism in mice. *Proc Natl Acad Sci USA.* 2001;98(26): 15095–15100.
7. Liu YY, Schultz JJ, Brent GA. A thyroid hormone receptor alpha gene mutation (P398H) is associated with visceral adiposity and impaired catecholamine-stimulated lipolysis in mice. *J Biol Chem.* 2003;278(40):38913–38920.
8. Tinnikov A, Nordstrom K, Thoren P, et al. Retardation of post-natal development caused by a negatively acting thyroid hormone receptor alpha1. *EMBO J.* 2002;21(19): 5079–5087.
9. Weiss RE, Xu J, Ning G, Pohlenz J, O'Malley BW, Refetoff S. Mice deficient in the steroid receptor co-activator 1 (SRC-1) are resistant to thyroid hormone. *EMBO J.* 1999;18(7): 1900–1904.
10. You SH, Liao X, Weiss RE, Lazar MA. The interaction between nuclear receptor corepressor and histone deacetylase 3 regulates both positive and negative thyroid hormone action. *in vivo. Mol Endocrinol.* 2010;24(7):1359–1367.

are euthyroid.[54] Presumably this is because the resistance of thyroid hormone action is overcome by higher levels of T4 and T3, which occur as a result of increased TSH secretion. Indeed, typical symptoms of hyper- or hypothyroidism are absent, or are present variably and may even co-exist. Common clinical findings include goiter (65–95%), tachycardia (33–75%) and hyperkinetic behavior (33–68%).[56,57] The tachycardia is a result of the interaction between TRα with increased thyroid hormone levels, in heart tissue where TRβ is not the major isoform. The above clinical findings prompt evaluation for thyroid hormone abnormalities, and suggest hyperthyroidism; unfortunately, misunderstanding of the accompanying non-suppressed TSH leads to a mistaken diagnosis of primary hyperthyroidism. If treated in a manner to achieve "normal" thyroid hormone levels (either by surgery, anti-thyroid medications or radio-iodine), such a situation results in bradycardia, somnolence, weight gain, depression and fatigue. In children, this is extremely undesirable because of ensuing growth retardation. Such circumstances are now less common, with the generation of more sensitive TSH assays and increasing physician recognition of RTH.

Less common clinical findings include emotional disturbances (60%), attention deficit/hyperactive disorder (40–60%), learning disability (30%), mental retardation (4–16%), sensorineural hearing loss (10–22%) and recurrent viral and throat infections (55%).[57,58] Deafness, however, is rare, and is reported only in those families with deletion of TRβ. Patients must be monitored for osteoporosis, as bone contains the TRα receptor and increased bone turnover is commonly seen.[59] In children, failure to thrive and growth delay is not infrequent, but short stature is uncommon (18–25%). However, a delay in bone age and low BMI may occur more commonly (29–47%).[57]

Pregnancy creates a challenging clinical situation. The study of a large kindred in the Azores has made studies of fertility and pregnancy possible. Fertility does not appear to be adversely affected in either females or males harboring the TRβ mutation. However, the miscarriage rate in affected mothers is three to four times that of spouses of affected males or unaffected relatives. Furthermore, it appears that miscarriage occurs more in normal fetuses than affected ones.[60,61] It is believed that this is the consequence of fetal exposure to elevated thyroid hormones. There appears to be evidence of this in normal infants born to affected mothers, as low birth weights are common when compared to affected siblings, and blood TSH levels are suppressed at birth. In contrast to pregnant women with hyperthyroidism, affected women with RTH do not have an increased rate of other pregnancy complications, such as pre-term labor, pre-eclampsia, still births or perinatal loss.[62]

The clinical course is often benign, with no decrease in life span noted. With age it appears that levels of T4 and T3 may decline, suggesting either a relaxation of the degree of resistance or simply an exaggeration of what occurs in normal individuals.

Diagnosis

An elevated free T4 in association with non-suppressed TSH suggests RTH. Reverse T3 (rT3) and thyroglobulin levels are also increased; the latter reflects degree of TSH increase. Radioactive iodine uptake in the thyroid is increased, reflecting increased TSH or increased TSH bioactivity. Thyroid ultrasonography often reveals a diffusely enlarged gland or multinodular goiter.[63]

The differential diagnosis includes defects in transport of thyroid hormone, defects in T4 to T3 conversion, defects in binding of thyroid hormones to serum proteins, interference in measurement of thyroid hormone or TSH by circulating substances (such as antibodies), and thyrotropin-secreting tumors. The exclusion of other causes of euthyroid hyperthyroxinemia has been discussed above. The clinical scenario that is often difficult to differentiate from RTH is TSHoma (discussed in the following section). Therefore, it is helpful to carry out other confirmatory tests.

In patients with RTH, serum glycoprotein α subunit is often not elevated, suggesting that the increase in TSH secretion is balanced. Also, measures of peripheral thyroid hormone action such as SHBG are normal, reflecting the clinical euthyroid state. Dynamic tests include the TRH stimulation test and tri-iodothyronine (T3) suppression test.

The TRH stimulation test is based on the rationale that TSH will remain sensitive to the action of synthetic, exogenous TRH in patients with RTH. The test is performed by measuring TSH before and after the intravenous injection of TRH (200 μg/1.73 m^2). A measurement at baseline and at 30 minutes after injection is usually sufficient to verify the TSH response.[64] Currently TRH is not available in the United States, and therefore this test is not practicable there.

The T3 suppression test is based on the rationale that, while resistant to the effects of thyroid hormone, use of a high enough dose will result in suppression of TSH in RTH patients, since there is no autonomous production of TSH. In adults, twice-daily divided of doses of 50 μg/day × 3 days, 100 μg/day × 3 days and 200 μg/day × 3 days are administered. Blood samples to measure TSH and prolactin responses are taken, and measurement of thyroglobulin and T4 will verify the degree of thyroid gland suppression produced by T3 administration. Peripheral tissue responses are assessed by measuring ferritin, SHBG (both increased), serum cholesterol and osteocalcin (both decreased). A detailed protocol has been published, and the reader is referred to http://www.thyroidmanager.org/Chapter16/16d-frame.htm for an excellent diagram demonstrating the anticipated response to this test in patients who are normal, in those who have RTH with a defined TRβ mutation, and in patients with RTH in whom no TR mutation has been found. One report suggests that during the T3 suppression test it is possible to perform color flow Doppler sonography to help distinguish RTH from TSHoma. In patients with RTH it would be anticipated that peak velocity will normalize; however, this would not be case in patients with TSHoma.[65]

Given the autosomal dominant nature of this syndrome, it is advisable to verify the presence of the abnormal biochemical pattern in parents or siblings, keeping in mind that the occurrence of a *de novo* germline mutation in TRβ is possible. Identification of a family with RTH would be important to help identify affected members later on, and thus avoid delay in diagnosis or unnecessary procedures. Identification of specific mutations responsible for disease does not have a significant impact on subsequent treatment decisions. Although the degree of T3 binding defect seems to correlate with biochemical presentation, this correlation is not strong enough to be clinically useful. Thus, TR sequencing to identify a mutation can be undertaken if research protocols are available. Several research and commercial laboratories offer genetic testing (based primarily on sequence analysis), and can be found at http://www.ncbi.nlm.nih.gov/sites/GeneTests/lab/disease/Thyroid%20hormone%20resistance?db=genetests&search_param=suggestion.

Routine prenatal diagnosis of RTH, using samples of amniotic fluid, has been reported but is not routinely performed.[62] Given the possibility of early fetal loss, especially in normal fetuses of mothers with known RTH, this form of testing may be useful.

Treatment

The cause of RTH is a germline defect, and as such there is no cure for this condition. As many patients are clinically euthyroid, those compensated patients without symptoms do not require any treatment. Specifically, anti-thyroid treatments aimed at "normalizing" thyroid hormone levels should be avoided, as this can precipitate tissue hypothyroidism. There is no role for "suppression" of TSH with exogenous thyroxine to avoid the development of pituitary adenoma, as this has only been reported once in the setting of RTH.[66] Similarly, there is no known increased incidence of thyroid cancer in humans, or of large compressive goiters in humans.

A common symptom is sinus tachycardia, which suggests hyperthyroidism and represents the effect of elevated thyroid hormones on heart tissue where TRα is the predominant isoform governing heart rate. In such cases, beta-blockers are advisable and effective. Some authors advise atenolol as compared to propranolol, which has the effect of decreasing T4 to T3 conversion. This may contribute to resistant tissues becoming de-compensated. In rare cases other symptoms of hyperthyroidism, such as sweating, anxiety, heat intolerance and tremor, are not alleviated by atenolol treatment. In such cases, experience with other treatment is limited. One option is to reduce TSH secretion. Glucocorticoids, dopamine agonists and somatostatin are all options. Practically, glucocorticoids have unacceptable side effects, while dopamine agonists have decreased effectiveness over time. Somatostatin analogs seem to have a transient and modest effect in reducing TSH.[64] Another option is to decrease thyroid hormone action in sensitive tissues. For this, the thyroid hormone analog tri-iodothyroiacetic acid (TRIAC) has been investigated, as, *in vitro*, TRIAC can transactivate mutant TRβs to a greater extent than T3, thus potentially overcoming the dominant negative nature of the mutant TRβ.[67–69] In longer-term human studies, TRIAC reduced TSH as well as T4 and T3 levels, but measures of thyroid hormone action were unchanged.[70–72] Questions have arisen about the specificity of observed effects, and TRIAC is not available in the United States. Similarly, studies with dextrothyroxine (D-T4) have suggested a minimal clinical effect of unclear specificity given the small amounts of L-T4 in D-T4 preparations.[73,74]

As opposed to treatment for symptoms of apparent thyroid hormone excess described above, patients may have real or apparent thyroid hormone deficiency. That is, if treated with anti-thyroid agents or thyroid ablative therapy, patients may have objective signs of thyroid hormone deficiency for which intervention is recommended. Anti-thyroid agents should be stopped. For those patients with prior surgery or radio-iodine therapy, thyroid hormone replacement with supraphysiological doses may be required. In this situation, the dose must be incrementally increased to normalize serum TSH. Doses in the range of 500—1000 μg/day may be required.[64] Further, if tachycardia occurs but the TSH remains elevated, it should be managed with atenolol.

In general, goiters are not large. Surgical treatment has been found to be ineffective over the long-term because of recurrence. As a result, suppression of the goiter with thyroid hormone treatment is preferred. There is a report of regression of a large goiter in a patient with RTH who was given a dose of 250 μg of L-T3 every other day.[75] It appears that this was done safely and with good effect.

In children with apparent thyroid hormone deficiency manifest by failure to thrive, growth retardation or delayed bone age, treatment with L-T4 is desirable. To guide dose increases, clinical parameters such as growth, bone age and mental development may be used.[64] In addition, nitrogen balance, BMR and serum SHBG should be used to help guide therapy and ensure overtreatment does not occur.

Treatment of infants is more problematic, as long-term outcome studies are lacking. However, if the following are present, some authors suggest treating infants and children with L-T4: (1) family members with adverse outcomes such as mental retardation; (2) failure to thrive; (3) growth retardation; (4) developmental delay; (5) marked TSH elevation.

Finally, as noted earlier, there is an increased incidence of fetal loss, especially in unaffected fetuses of mothers with RTH. Although no formal studies are available to form the basis for a recommendation, one strategy that has been suggested is to use PTU conservatively to reduce the free T4 levels to not more than 20% above the upper limit of normal.[62]

TSH-secreting Pituitary Adenoma

TSH-secreting pituitary adenoma is a rare tumor of the anterior pituitary. Nearly three-quarters of all TSHomas secrete only TSH. One quarter are mixed adenomas. GH is the most-frequently co-secreted hormone (16%), followed by prolactin (10%). There are reports of gonadotropins being co-secreted with TSH, but not with ACTH.[76] These are benign tumors, and there is only one report of a TSH-secreting pituitary carcinoma.[77]

The molecular mechanism underlying the formation of pituitary adenomas is not known. Evidence points to the clonal expansion of a single, initially transformed cell.[78] Broadly, there are two lines of investigation: a search for mutations in oncogenes and tumor suppressor genes, and a search for mutations in transcription factors responsible for negative feedback on TSH.

Thus far, screening of TSHomas has not shown activating mutations of genes encoding for G-protein subunits or for the TRH receptor.[79] The transcription factor Pit-1 is important for differentiation of lacto-, somato- and thyrotropes, but mutations in the Pit-1 gene have not been shown in TSHomas.[80] Neither mutation in proto-oncogenes nor loss of tumor suppressor genes has been found in the few studies conducted. While loss of heterozygosity has been demonstrated on 11q13, where the menin gene is located for other pituitary adenomas and a few TSHomas, sequence analysis in TSHomas has not shown a menin gene mutation. In the setting of MEN1, however, TSHomas have been reported.[80]

In contrast, aberrant alternative splicing of TRβ2 mRNA encoding a TRβ variant lacking the ability to bind T3 has been shown as a mechanism for the impaired negative regulation of TSH in TSHomas.[81] An aggressive phenotype and resistance to somatostatin analog treatment has been associated with LOH and particular polymorphisms at the somatostatin receptor type 5 gene locus.[82] Finally, overexpression of fibroblast growth factor is suggested to play a role in tumor cell proliferation and the development of fibrosis. This is a prominent characteristic of these adenomas, which can be so hard they are called "pituitary stones."[83,84]

Clinical Presentation

The presence of goiter, signs and symptoms of hyperthyroidism, symptoms related to a sellar mass effect, and partial or complete hypopituitarism suggests a TSHoma. The presence of goiter is a prominent feature;

however, thyroid cancer and autonomously functioning nodules are rare occurrences.[85,86]

In general, patients have a long history of hyperthyroidism, with some even carrying a previous incorrect diagnosis of Graves' disease. As a result, approximately 30% have had previous inappropriate thyroid surgery or radio-iodine thyroid ablation.[76,87] In this regard, bilateral exopthalmos is rare, but has been reported in patients subsequently determined to have co-existing autoimmune thyroid disease. Unilateral proptosis has been reported secondary to pituitary tumor invasion of the orbit.[88] It appears that clinical signs and symptoms are milder than in primary hyperthyroidism, and cardiac symptoms, atrial fibrillation and periodic paralysis are rare events.

Because TSH can be co-secreted with other hormones, it is important to consider other hormone over-secretion syndromes. Specifically, acromegaly can be significant and mask symptoms of hyperthyroidism. Disorders of the menstrual cycle can be seen in patients who co-secrete PRL (two-thirds of cases), but are also seen in patients with pure TSHoma (one-third of cases). Further, central hypogonadism and delayed puberty have been reported in males with TSHoma or mixed TSH–FSH adenomas.[80]

Diagnosis

BIOCHEMICAL EVALUATION

Elevated circulating thyroid hormones with a measurable TSH is the biochemical characteristic of central hyperthyroidism. Once disorders in the binding of thyroid hormones or TSH have been excluded, other diagnostic procedures can be carried out. First, the presence of an elevated α-GSU/TSH molar ratio (suggesting hypersecretion of the α subunit) is seen in more than 80% of patients with TSHoma. Some studies suggest that, with improved imaging techniques and ultrasensitive TSH assays, a larger number of tumors are being detected that have normal, not increased, α-GSU levels, and these seem often to be microadenomas.[87,89] Therefore, it is important to perform other tests as noted below. To confirm the effects of elevated thyroid hormone in the peripheral tissues, several parameters are available, including sex-hormone binding globulin (SHBG) and carboxy terminal cross-linked telopeptide of type 1 collagen (ICTP), which are both elevated.[59]

Dynamic testing is also useful in diagnosing TSHoma. Specifically, TSH secretion appears to be independent of T3 levels, and insensitive to stimulation by TRH. Indeed, complete suppression of TSH by T3 (80–100 μg/day × 8–10 days) has not been demonstrated in any case of TSHoma. The T3 suppression test is therefore considered the most useful test in diagnosing TSHoma. In particular, this test is sufficiently sensitive and specific to diagnose TSHoma in those patients with prior thyroidectomy. However, in patients with cardiac disease or in the aged, this test is contraindicated. In this setting, a TRH stimulation test may be employed. In TSHoma, stimulation with TRH does not produce the anticipated increase in either α-GSU or the TSH-β subunit. It is suggested that both of these tests are carried out if possible, as the combination of tests increases the sensitivity and specificity of the work-up. Table 31.1[80] summarizes laboratory parameters that can help to distinguish RTH and TSHoma.

In contrast to the resistance of TSHoma to the anticipated pharmacologic effects of either T3 or TRH, most TSHomas maintain sensitivity to somatostatin or its analogs. In one series, there is a marked decrease of T3 and T4 levels after treatment with long-acting somatostatin analog in most patients with TSHoma; none of the patients with PRTH responded in this manner to somatostatin analog.[90,91] Thus, this potential therapeutic manouver can be used as a test in particularly difficult cases of central hyperthyroidism.

LOCALIZATION

Most TSHomas are detected as pituitary macroadenomas, often with suprasellar invasion. As a result, computed tomography (CT) or magnetic resonance imaging (MRI) of the pituitary are preferred modalities for confirming the presence of a TSHoma. Given the improvements in diagnostic techniques, microadenomas are now reported in 15% of recorded cases.

Ectopic localization of TSHomas is rarely observed; there are two reported cases in the nasopharyngeal region.[92,93] For this purpose, scintigraphy with radiolabeled octreotide (octreoscan) maybe helpful, as many TSHomas express somatostatin receptors. This test should not be a first-line imaging modality because it is not sufficiently specific, given that many different types of pituitary tumors can express somatostatin receptors.[94]

TABLE 31.1 Features to Distinguish Resistance to Thyroid Hormone (RTH) from Thyrotropin Secreting Adenoma (TSHoma)

	RTH	TSHoma
Elevated α-GSU levels	3%	79%
Elevated α-GSU/TSH molar ratio	2%	83%
Serum SHBG (nmol/l)	61 ± 4	117 ± 18
Blunted TSH response to TRH	2%	94%

Adapted from Beck-Peccoz and Persani (2009).[80]

Treatment

For curative treatment of TSHoma, the first step is removal of the tumor by either transphenoidal or subfrontal adenomectomy. The goal is to remove the mass and restore euthyroidism. One study suggests that up to 86% of patients will regain euthyroidism; however, normalization of other parameters such as α-subunit/TSH molar ratio normalized in only about 50%. Furthermore, TSH was normally suppressed by T3 in only 40%. It appeared that an undetectable TSH 1 week after surgery was highly predictive of success.[95]

To reduce the preoperative risk due to hyperthyroidism, methimazole, somatostatin analogs and propranolol may be employed.[80] The restoration of euthyroidism with the above mentioned methods may interfere with the use of TSH as a marker for residual tumor, as non-tumoral thyrotropes will be reactivated prior to surgery and a measurable TSH may be found postoperatively that does not represent secretion from a residual adenoma.

Often, because of the size of the mass, suprasellar extension, or invasion of surrounding structures such as the cavernous sinus or internal carotid artery, surgery may be too dangerous. In such a case, or if surgery is refused, pituitary radiotherapy should be considered.

With surgery or radiation, improvement can be expected in two-thirds of patients. About one-third are cured, while the other one-third have normalization of thyroid hormone levels without complete removal of the adenoma. In those who are cured, postoperative TSH may remain suppressed even up to several weeks, and so the patient may suffer from central hypothyroidism. Further, there is the risk of permanent damage to normal thyrotropes, and indeed the other portions of the pituitary, either because of the mass or because of surgery. Therefore, it is important to test for the function of all axes controlled by the pituitary, and in particular the competency of the hypothalamic–pituitary–adrenal axis must be assured.

In those patients who are not cured, two options remain. First, if the risk of thyroid storm is high, then thyroidectomy can be performed. Otherwise, medical therapy can be used to control the size of the pituitary mass and restore euthyroidism. At this point, medical therapy is based on the use of long-acting somatostatin receptor (SSR) analogs such as Sandostatin and Lanreotide autogel. The rationale for this treatment is based on the *in vitro* finding of SSRs in all TSHomas. The highest binding of somatostatin has been found in mixed TSH/GH adenomas. The somatostatin binding capacity of a tumor seems to correlate with biological response in terms of inhibition of TSH secretion.[90] This treatment is quite effective, with reduction of α-GSU and TSH levels in nearly all cases, and restoration of euthyroidism in 96% of cases (who did not previously have thyroidectomy).[96–98] There is a heterogeneity of response and dose must be tailored to each patient, taking into account the side effects, which include gastrointestinal intolerance and, more seriously, cholelithiasis and glucose intolerance. The GI side effects seem to be transient with long-acting somatostatin analogs. The therapeutic effect can be so profound that hypothyroidism may ensue, and replacement with levothyroxine becomes necessary. Even in patients with prior thyroidectomy, somatostatin analogs should be considered to control tumoral growth. Also, it is notable that there are patients who may have a negative octreoscan but still respond to somatostatin analog treatment. Finally, some have proposed that somatostatin analogs, based on the > 90% treatment efficacy, can be used instead of adenomectomy or radiation. This remains to be further researched.

In TSH–PRL co-secreting tumors the use of dopamine agonists can be considered, though the response to such drugs as bromocriptine and cabergoline has been quite heterogeneous. Also, tumor shrinkage has not been reported with dopamine agonists. Furthermore, there are reports of paradoxical increase of TSH after initiating dopamine agonist therapy. This clinical finding, coupled with reports that dopamine receptors may be found on the TSHoma cells, suggests that dopamine agonists should be used with caution.[80,99]

DISORDERS WITH DECREASED THYROID HORMONE LEVELS AND INAPPROPRIATELY LOW OR NORMAL TSH LEVELS

Central hypothyroidism is defined as pathologically low thyroxine levels due to either decreased secretion of thyroid-stimulating hormone (TSH), or secretion of biologically inactive TSH. Central hypothyroidism (CH) is relatively rare. Indirect estimates based on the prevalence of pituitary tumors indicate that central hypothyroidism occurs in 0.0002% of the population.[100] Other groups suggest that the frequency of central hypothyroidism is 0.005%.[101] This is in contrast to a prevalence of 1–2% for primary hypothyroidism. Also, in newborns, hypothyroidism due to pituitary or hypothalamic dysfunction occurs with an incidence of 1 : 25,000–100,000 births.[102,103] There is an equal sex distribution. Most commonly, central hypothyroidism is due to sellar or suprasellar masses that cause a decreased mass of functional thyrotopes or an interruption of the hypothalamic–pituitary portal circulation. Thus, the diagnosis of central hypothyroidism often occurs in the setting of deficiency or excess of other pituitary hormones.[104] Finally, it is important to note that TSH values may be low, normal or slightly elevated in patients with central hypothyroidism, and thus measurement of TSH value alone cannot be used to either diagnose or monitor patients.

Pathogenesis

The pathogenesis of central hypothyroidism is due to decreased secretion of thyroid-stimulating hormone (TSH), or secretion of biologically inactive TSH. This can occur for the following reasons.

1. Destruction or compression of existing thyrotropes. This is usually due to pituitary masses but may also be caused by vascular, infectious, or inflammatory destruction of thyrotropes cells.
2. Deficient or defective secretion of TRH. This may be due to interruption of the hypophyseal portal blood flow to the anterior pituitary or due to hypothalamic lesions. Theoretically this is due to TRH deficiency alone and thus replacement of TRH could result in normal TSH secretion. One study provided indirect evidence for this notion by correcting hypothyroxinemia with repeated injections of TRH.[105]
3. Congenital defects in thyrotrope development or secretion of biologically active TSH.
4. Functional defects in TSH secretion.

A common finding in central hypothyroidism is disruption of the normal diurnal rhythm of TSH secretion. TSH secretion occurs in a circadian pattern, with a nocturnal surge beginning at about 1900 h and peaking at about 2400 h.[106] In central hypothyroidism, this nocturnal surge is either absent or blunted in nearly all cases (see Fig. 31.3). This occurs due to a decrease in the TSH pulse amplitude at night; pulse frequency is preserved.[101] It appears that the nocturnal surge is important, as the thyroid is particularly stimulated at night. Based on this, some authors have suggested that testing for the absence of the nocturnal TSH surge is a sensitive way to identify cases of central hypothyroidism.

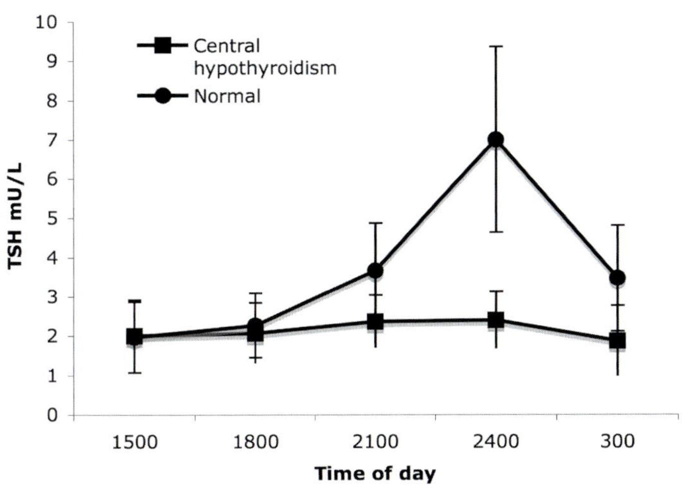

FIGURE 31.3 Normal individuals have a surge in thyrotropin (TSH), peaking at about 2400 h. In contrast, patients with central hypothyroidism have blunted nocturnal surge. *Adapted from Rose (2000).*[106]

Biologically Inactive TSH

The presence of low thyroxine levels in association with normal levels of TSH in patients with known pituitary or hypothalamic disease has been recognized in patients not likely to have primary hypothyroidism because of the absence of auto-immunity and an adequate response to exogenous TSH. Furthermore, exogenous TRH produced an appropriate or even exaggerated rise in TSH, but with an inadequate increase of T3 levels. Assays for TSH biological activity, such as CHO cells transfected[107] with TSH-receptor or adenylate cyclase activity from FRTL-5 cells,[108] suggested that the ratio of biological to immunological TSH was reduced in some patients with central hypothyroidism. Further, it was shown that chronic oral TRH treatment but not acute intravenous treatment could restore TSH biological activity, suggesting a requirement of TRH for full biological activity of TSH.[109] Subsequently, an abnormal glycosylation pattern in a rat model of central hypothyroidism was shown to be reversed by TRH treatment, with a restoration of normal TSH values.[110] It has now been established that TRH ensures the proper glycosylation of TSH, and that there are abnormal glycosylation patterns associated with CH. Investigators have demonstrated abnormal glycosylation patterns, including reduced sialyation of TSH, in central vs primary hypothyroidism.[111–113] Many studies have documented secretion of bioinactive TSH in central hypothyroidism. Furthermore, this may be a common finding in cases of both pituitary and hypothalamic disease.[112]

Etiology

PITUITARY MASS LESIONS

Pituitary mass lesions are the most common cause of central hypothyroidism, accounting for almost 75% of hypopituitarism cases in one series.[114] Most often this is due to non-functioning or functioning adenomas of the pituitary, and is often seen in conjunction with other pituitary hormone deficits. The cause of hypothyroidism is likely from compression of non-tumorous thyrotropes, or interruption of the hypophysial portal circulation due to involvement of the stalk from suprasellar extension.[101] Less commonly, pituitary tumor apoplexy has also been associated with hypopituitarism, and warrants work-up.[101] Interestingly, after surgery for pituitary adenomas, 57% of those who were hypothyroid preoperatively recovered function.[115] A blunted response to TRH preoperatively predicted the inability to recover anterior pituitary function in general.

A similar rate of recovery occurs after surgical decompression to treat pituitary tumor apoplexy.[116] However, surgery for larger pituitary tumors is also associated with the development of pituitary deficits postoperatively, with hypothyroidism occurring in about 10% of cases.[101] Metastatic tumors to the hypothalamic–pituitary region can occur with breast, lung and some other cancers.[104] Usually this is in the setting of advanced disease, and the uncommon finding of hypopituitarism is often associated with diabetes insipidus.

EXTRASELLAR MASSES

Extrasellar masses can cause central hypothyroidism. *Craniopharyngiomas*, especially in younger patients, are frequently associated with central hypothyroidism. TSH deficiency is present in 21–42% of cases.[117] In contrast to pituitary adenomas, surgery rarely resolves the hormone deficiency, and the 10-year probability of hormone deficiency is high. *Meningiomas*, *cysts* and *abscesses* are also reported causes of hypopituitarism.[101]

CONGENITAL DEFECTS

Combined pituitary hormone deficiency (CPHD) is a syndrome of deficient secretion of the anterior pituitary hormones prolactin (PRL), growth hormone (GH) and TSH. Research into the stages of pituitary development has revealed a cascade of transcription factors that control when and how the different anterior pituitary cell lineages develop. Molecular defects have been identified in many of these transcription factors.

1. Pit-1, the pituitary specific transcription factor (or Pou1f1) is expressed relatively late in the pituitary development cascade, and its expression is preserved throughout life. While GH and PRL deficiencies are present early in life, TSH deficiency may present later in childhood. The first described mutations were homozygous non-sense mutations,[118] or acted as dominant negatives.[119]
2. Prop-1 is believed to be required for Pit-1 expression, and does not persist into adulthood. Mutations in Prop-1 are associated with GH, TSH, PRL and gonadotropin deficiencies. Deficiency of GH, TSH and PRL is milder for mutations within Prop-1 than with Pit-1 mutations. Although most patients present with early-onset growth hormone deficiency, TSH deficiency is variable and may not be present at birth. Hypogonadism is also variable in onset and severity. Finally, there is an evolving cortisol deficiency with age, but the cause of this ACTH deficiency is not known.[120,121] In some series of CPHD, mutations in Prop-1 appear to be the most common cause of both sporadic and familial CPHD.[122]
3. HESX-1 is a member of the paired-like class homeobox genes, and is one of the earliest markers of the murine pituitary primordium. It acts as a transcriptional repressor, but its downstream targets are not known. In humans, mutations in HESX-1 were first identified in association with septo-optic dysplasia.[123]
4. PITX2 heterozygous mutations are associated with Rieger syndrome, characterized mainly by defects in the anterior chamber of the eye, and by tooth malformations in humans.[124]
5. Lhx3 and Lhx4 belong to LIM family homeobox genes expressed early in Rathke's pouch and into adulthood. Lhx3 is believed to act synergistically with Pou1f1 to activate TSH-β and PRL promoters. Mutations in humans are identified, and have hormone deficiencies similar to patients with PROP-1 mut (that is GH, TSH PRL in addition to gonadotropin deficiency), but also have a short rigid cervical spine with limited head rotation.[125] There is one report of Lhx4 mutation associated with a poorly formed sella turcica, ectopic posterior pituitary, absent infundibulum and pointed cerebellar tonsils.[126]

Investigation into familial cases of central hypothyroidism has revealed mutations in the TSH-β subunit gene. Mutations near the NH2-terminus have been identified that result in an altered TSH-β subunit unable to bind the α subunit.[127,128] The mutations are in the so-called CAGYC region, which is conserved in the β-subunit glycoproteins. The mutations resulted in an arginine substitution for glycine. Patients were described as homozygous for the mutation. Another group identified three patients in two related Greek families with non-sense mutations in exon 2 giving rise to a truncated peptide including only the first 11 amino acids of the TSH-β peptide.[129] In all of the above cases, TSH was undetectable. Similar other mutations have been reported (E12X, G29R, Q49X, a splice site mutation at exon 2, and 114 X).[130,131] In another report, two related Brazilian families with central hypothyroidism were determined to have a detectable but biologically inactive from of TSH based on both *in vitro* and *in vivo* studies.[132] The mutation was in a critical carboxy-terminal cysteine residue (C105V) in the TSH-β subunit gene. Crystallographic studies of chorionic gonadotropin suggest that cysteine residues C19 and C105 in the TSH-β subunit form the "buckle" of a "seat belt" that surrounds the α subunit and maintains bioactivity of the hormone. Some have suggested that codon 105 represents a "hot spot" for mutation, although a founder effect cannot be ruled out.[130]

Finally, a defective TRH receptor was identified in a single patient with isolated central hypothyroidism. Defects in the TRH receptor were suspected when the patient did not show expected changes in TSH and prolactin after administration of TRH. The patient was found to be compound heterozygous for mutations in the TRH receptor. The mutations were in the 5′-part of the gene, and the mutated receptors did not bind TRH.[36]

INFILTRATIVE DISEASES

Sarcoidosis of the central nervous system occurs in ~5% of sarcoid cases.[133] Of these, hypopituitarism is seen in 10%, usually with profound endocrine dysfunction. TSH deficiency occurs in about 50% of cases.[133] However, sarcoidosis can also present as isolated involvement of the CNS without pituitary dysfunction. On imaging, it may present as a sellar or suprasellar mass, or as a cystic sellar mass. More suggestive findings on MRI include thickening of the central portion of the pituitary stock, and loss of the posterior pituitary bright spot. Langerhans cell histiocytosis (LCH) is a multisystem disorder with a propensity for hypothalamic–pituitary involvement. The most common endocrine disorder is diabetes insipidus, which occurs in about 50% of cases of multisystem LCH. Central hypothyroidism occurs in about 10–15% of cases, and can occur after many years of disease.[134,135] Lymphocytic hypophysitis (LH) is an infiltrative disease of the pituitary of unknown, though likely auto-immune, etiology. Endocrine dysfunction can be seen in approximately 80% of cases, with an interesting predilection for thyrotropes and corticotropes.[136] Hypothyroidism can occur in 50% of cases of LH. Imaging often shows an enlarged pituitary with symmetric shape, homogeneous gadolinium enhancement, relatively low signal on T1-weighted images and high signal on T2-weighted images, serving to distinguish LH from a pituitary adenoma. The frequent co-occurrence of LH with Hashimoto's thyroiditis, type 1 DM, Addison's disease and pernicious anemia has led it to be considered as part of the autoimmune polyglandular syndrome type 1.[137] Hereditary hemochromatosis is due to increased absorption of dietary iron, leading to parenchymal iron overload. Deposition of iron in the anterior pituitary most commonly causes hypogonatropic hypogonadism.[138] TSH deficiency can occur, is quite rare, and can be resolved with iron-depletion treatment.[139] In thalessemia patients who receive multiple transfusions, pituitary iron deposition may occur. Further, thyroid dysfunction is observed; however, it appears that this is not commonly due to central hypothyroidism.[140]

INFECTIONS

Tuberculosis, toxoplasmosis, African trypanosomiasis and syphilis have been associated with central hypothyroidism in association with pituitary disease.[101]

IATROGENIC CAUSES

If the hypothalamic–pituitary axis falls within the field of external radiation, hypopituitarism may ensue. This has long been recognized for patients receiving radiation therapy for primary brain tumors,[141,142] total body irradiation for other illnesses and, importantly, in children undergoing cranial irradiation for prophylaxis in acute lymphoblastic leukemia. Furthermore, the incidence appears to be dose related, with doses of > 30 Gy being associated with a 90% incidence of hypopituitarism at 10 years.[143] While in general it is believed that GH is affected earlier and more commonly, one study suggested that 50% of patients receiving cranial irradiation developed central hypothyroidism and in particular had evidence suggesting increased hypothalamic susceptibility to XRT.[142] In addition, longer-term follow-up studies of survivors of childhood cancers suggest that central hypothyroidism occurs in 65% of patients after brain or nasopharyngeal tumors, in 35% after bone marrow transplantation, and in 15% after leukemia.[144] Rexinoids have been shown to decrease TSH. Specifically, bexarotene, a selective RXR-α agonist, was found to suppress TSH in the majority of patients being treated for cutaneous T-cell lymphoma, and many patients had hypothyroidism.[145] Subsequently, it was shown that bexarotene can suppress TSH even in healthy subjects.[146] It is believed that this may be due to a direct effect of RXR-α on the TSH-β promoter, and is reversible. There is also a report of ribavirin and interferon-α treatment of hepatitis C being associated with central hypothyroidism and hypophysitis that is apparently different from the autoimmune thyroid disease seen with interferon-α treatment.[147]

TRAUMA

Hypopituitarism can occur following head trauma. While gonadotropin deficiency is the most common finding, central hypothyroidism occurs in ~50% of post-traumatic hypopituitarism.[148] Some studies suggest that in cases of "unexplained" central hypothyroidism, a history of head trauma should be sought. Furthermore, the incidence of hypothyroidism following head trauma may be under-appreciated.[149] In addition, patients who had poor outcomes had lower thyroid hormone values in the first days following injury as compared to patients who recovered well.[150]

CLINICAL MANIFESTATIONS ASSOCIATED WITH FUNCTIONAL DEFECTS IN TSH SECRETION

There are several clinical scenarios associated with functional defects in TSH secretion. In most cases this defect need not be treated, as there is no hypothyroidism *per se*.

1. After withdrawal of suppressive T4 therapy for non-toxic goiter or after treatment of hyperthyroidism with radio-iodine, anti-thyroid agents or surgery, the serum TSH may remain low for ~25 days.[151] During

this period, if thyroid hormone replacement is not initiated then T4 values could fall to below normal. Thus, a low TSH and low free T4 may incorrectly suggest central hypothyroidism without knowledge of the patient's history.

2. Similarly, during recovery from the thyrotoxic phase of subacute thyroiditis, TSH levels may remain low even as the T4 values become normal.
3. Non-thyroidal illness (NTI) is characterized by low T4 associated with low TSH.[152] One reason for this may be the induction of type II deiodinase,[153] which is present in hypothalamic and pituitary cells and thus plays an important role in feedback regulation of TSH. Others have noted a defective nocturnal surge in different types of NTI.[154] Increases in IL-6 have also been noted but whether this is part of the pathogenesis of non-thyroidal illness is not known.[155,156] Further, TRH mRNA levels may be decreased in patients with NTI (see Fig. 31.4).[15] During non-thyroidal illness, however, most patients have a normal response to TRH, and biological activity of the TSH is normal or even increased. Specific conditions with thyroid hormone values resembling CH are decompensated diabetes, chronic renal failure, burn, and post-bone marrow transplantation.[101] Interestingly after glycemic control has been restored, nocturnal TSH surge normalizes, but only in patients with residual C-peptide secretion.[157]

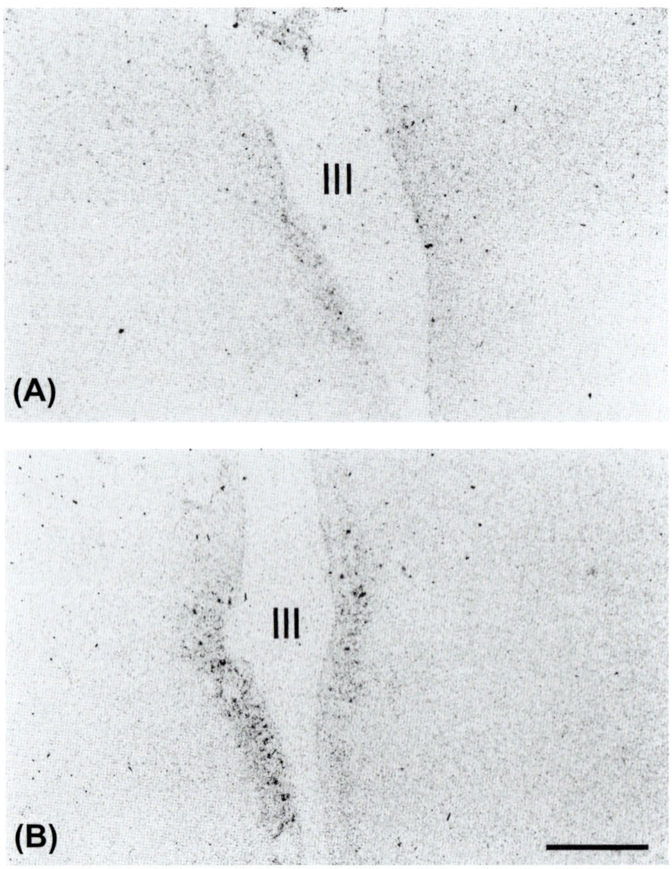

FIGURE 31.4 Panels show *in situ* hybridization for TRH mRNA in PVN of a subject who died with non-thyroidal illness (NTI) (**A**) and a patient who died in an accident (**B**), suggesting reduced TRH mRNA in patients with NTI . *Reproduced from,*[15] *Fliers E* et al. *(1997).*

4. Functional central hypothyroidism can be seen in Cushing's syndrome or prolonged glucocorticoid therapy.[158] In Cushing's syndrome, the nocturnal TSH surge is absent and there may be decreased response of TSH to TRH.[159]
5. Growth hormone (GH) decreases thyroid function by both central and peripheral mechanisms. GH decreases the TSH nocturnal surge and its secretory response to TRH, probably due to stimulation of hypothalamic somatostatin. Furthermore, it increases the metabolism of T4 to T3.[160] There is an interaction between T4 and GH. Both primary and central hypothyroidism can impair endogenous GH secretion, and it is recommended that in children with suspected GH deficiency euthyroidism be verified before provocative testing or treatment with recombinant GH is started. Most studies suggest that treatment of adult growth hormone deficiency patients with recombinant human GH is associated with reduction in T4, increase in T3 and no change in TSH.[161,162] Two studies suggest that 36–47% of patients became hypothyroid within 3–6 months after initiation of GH treatment,[162] and were less likely to show improvement in quality of life.[163] Growth hormone has both central and peripheral effects on thyroid function, as noted above. In cases where there is underlying pituitary pathology, central hypothyroidism may be masked[163] by growth hormone deficiency.
6. Anorexia nervosa is associated with low T3 and normal TSH peak after TRH but a delayed return to baseline.[164]
7. Similarly, depression is also associated with functional disorders of TSH secretion. A blunted response of TSH to TRH may be seen in up to 25% of patients with depression,[165] while others have noted that a decreased nocturnal TSH surge may be more common than absent TRH responses.[166] It is not generally felt that this merits treatment.
8. Aging is associated with a small but progressive decrease in TSH, as well as blunted response of TSH to TRH.[167,168]

Clinical Features

The patient with central hypothyroidism can present in a varied manner depending on the severity and duration of hypothyroidism, the nature of the underlying illness, and the number of other pituitary hormones affected and to what degree. The features related to hypothyroidism, such as cold intolerance, fatigue, slowed speech, weight gain, constipation and loss of mental acuity, are similar to those seen in primary hypothyroidism, although generally they are thought to be less severe. Onset in childhood may be characterized by delay in sexual maturation and/or bone development, in addition to growth failure. Isolated hypothyroidism due to inherited defects presents as dwarfism or cretinism. On physical examination, bradycardia, hypothermia and delay in the relaxation phase of deep tendon reflexes can be seen. Because of the strong likelihood of associated deficiencies in other pituitary hormones, the skin may not be coarse and dry but only pale and cool. Instead of periorbital or peripheral edema, there may be fine, dry wrinkling of the skin. Thinning of hair and the lateral eyebrows, and loss of pubic, axillary or facial hair, is more pronounced than in primary hypothyroidism.

As central hypothyroidism often occurs in the setting of other pituitary hormone deficiencies, these symptoms may predominate. Gonadotropin deficiency and growth hormone deficiency often precede thyrotropin deficiency. In children, this manifests as growth failure and delayed sexual maturation. In adults, there may be no symptoms related to growth hormone deficiency. Gonadotropin deficiency in adults may present with decreased pubic and axillary hair. In men, there is decreased libido, impotence, thinning of the beard and testicular atrophy, while in women there is infertility and amenorrhea. Corticotropin deficiency generally occurs after the development of thyrotropin deficiency, and thus patients may complain of nausea, abdominal discomfort and lassitude. Postural hypotension and de-pigmented areas of the skin may also noted. Hypoglycemia can occur with adrenal insufficiency and decreased insulin requirements in a diabetic patient with growth hormone deficiency. It is important to recognize the possibility of concomitant adrenal insufficiency, as there is a risk of life-threatening adrenal crisis in the setting of infection, trauma or surgery.

If the cause of central hypothyroidism is related to a sellar or suprasellar mass, the patient's presentation may be dominated by symptoms and signs related to mass effects. Thus, headache and visual loss may be the presenting signs of a non-functioning tumor extending out of the sella. Similarly, in the setting of a pituitary tumor associated with hormone over-secretion, the patient is likely to present with symptoms related to this problem rather than hypothyroidism. Thus, female patients may present with amenorrhea and galactorrhea in the setting of a prolactinoma, while men may only note decreased libido and impotence. Growth-hormone and ACTH secreting tumors are less common, but in these cases acromegaly and Cushing's syndrome, respectively, will be the presenting findings. Craniopharyngiomas may also cause diabetes insipidus and growth failure in children, and hypogonadism in adults. Other diseases in which diabetes insipidus is common include sarcoidosis and histiocytosis X because of hypothalamic or posterior pituitary involvement. Another interesting feature of hypothalamic lesions is the possibility of meningeal signs early in the course of disease. Finally, because appetite- and temperature-regulating centers are present in the hypothalamus, obesity and abnormal temperature regulation can occur with hypothalamic lesions.

Other etiologies often have unique clinical scenarios. Thus, severe headache is a prominent feature of pituitary apoplexy, as well as sudden visual loss and involvement of the third and fifth cranial nerves. Post-partum pituitary necrosis often occurs in the setting of a complicated delivery involving hemorrhage and shock, followed by deficient lactation and delay in resumption of menses.

Diagnosis

TSH/FREE T4 MEASUREMENTS

Since the development of ultra-sensitive TSH assays, it has been possible to confirm the diagnosis of central hypothyroidism by the presence of low free thyroxine levels in the presence of low, inappropriately normal or even mildly elevated TSH. Even prior to the availability of sensitive TSH assays, one study showed that 35% of patients had undetectable serum TSH, 41% had normal TSH and 25% had elevated TSH.[169] Despite normal or elevated TSH levels, it is believed that these TSH molecules are bio-inactive.

Measurement of either total or free T3 levels is generally not useful, as they are frequently in the normal range. One prospective study noted that free T3 levels are within normal range in approximately 25% of patients with low free T4, while total T4 and T3 values are normal in an even higher percentage of cases.[170] The TSH values in patients with pituitary and hypothalamic disease exhibit considerable overlap, and thus the TSH value alone cannot distinguish the site of pathology.

LOSS OF NOCTURNAL TSH SURGE

As noted above, TSH secretion occurs in a diurnal rhythm, with a surge in the late evening hours that is under hypothalamic control (see Fig. 31.3). Some investigators suggest that serum TSH samples obtained every 30 minutes between 2300 h and 0200 h can reliably document the absence of this nocturnal surge, providing a sensitive measure of central hypothyroidism.[144]

TRH STIMULATION TESTING

The rationale behind using TRH stimulation testing is to identify the site of pathology. That is, based on the control mechanism for TSH, it might be anticipated that patients with pituitary disease would have an absent or blunted TSH response to TRH, while patients with hypothalamic disease would have a preserved response.

In normal individuals, after i.v. administration of TRH (5 μg/kg), TSH values generally peak in 30 minutes and return to basal levels in 120 minutes.[171] Patients with pituitary or hypothalamic disease exhibit a different pattern. In these cases, after TRH administration there is a delayed rise in TSH, often followed by a sustained increase. Maximal TSH response may be blunted or even absent; this is more often seen in patients with pituitary disease, although patients with hypothalamic disease and no apparent pituitary involvement may also have this finding. Lastly, a normal or exaggerated rise in what is likely bio-inactive TSH occurs occasionally after TRH administration in patients with hypothalamic disease and central hypothyroidism.[172] However, occasionally patients with only pituitary pathology exhibit similar responses. For example, in one study the plasma TSH response to TRH was absent in 13.5%, impaired in 16.8%, normal in 47.2% and exaggerated in 22.5% of the cases, with delayed and/or prolonged pattern of response in 65% of the cases.[173] Thus, TRH testing alone cannot reliably distinguish between hypothalamic or pituitary causes of central hypothyroidism. For this reason, it is important to examine the complete clinical picture along with the thyroid function tests to determine the site of the lesion causing hypothyroidism. Further, while a nocturnal TSH surge is very common in pediatric patients with central hypothyroidism, a TRH test may still be unrevealing.[173,174]

However, in newborns, TRH testing may have a role in distinguishing central hypothyroidism from either mild primary hypothyroidism or other conditions associated with low T4 and normal TSH, such as TBG deficiency, sick euthyroid syndrome, and hypothyroxinemia of prematurity.[175] A recent study characterized the neonatal response to TRH stimulation in a small cohort of patients with persistently low T4 and normal TSH values, identified initially only by newborn screening.[176] In such patients, who can be diagnostically challenging because of the absence of other features suggesting hypopituitarism (such as hypoglycemia or mid-line defects), the TRH stimulation test may be useful. Newborns with confirmed pituitary dysmorphology had a higher peak TSH with persistent TSH elevation – although such a response is expected in primary hypothyroidism as well. A second group had absent or diminished response to TSH, and showed male predominance and a lower incidence of multiple hormone deficiency. Given the importance of instituting proper replacement therapy, this test may be a useful diagnostic tool if further studies confirm its applicability.

IMAGING

Computed tomography (CT) and magnetic resonance imaging (MRI) are sensitive techniques to identify hypothalamic and pituitary lesions. Imaging characteristics may help determine the etiology. For example, calcifications and cysts may suggest craniopharyngioma, while loss of bright spot in the posterior pituitary or thickening of the infundibulum may suggest an infiltrative process such as sarcoidosis. As noted above, lymphocytic hypophysitis has specific features that help distinguish it from a pituitary adenoma, which may be important, given the difference in clinical course.

Treatment

The goal of treatment is euthyroidism. Central hypothyroidism is treated similarly to primary hypothyroidism, with levothyroxine in doses of 1.4–1.6 μg/kg per day. However, in patients who may be at risk from over-treatment, such as the elderly or patients with existing cardiovascular disease, treatment is begun at doses of 0.3–0.7 μg/kg of L-T4 and adjusted every 3–4 weeks. In infants or children the dose of L-T4 required may be higher, as the risk of under-replacement is greater because of increased T4 clearance. As the consequences of under-treatment in this setting are mental retardation and delay of physical growth, it is important to provide adequate treatment from the outset. However, TSH is not useful for monitoring treatment, and should not be measured. Clinical assessment must be combined with measurements of free T4 and free T3. There is evidence to suggest that free T3 values can be a more sensitive measure of over-treatment, while free T4 values may disclose under-treatment.[170]

Patients with hypopituitarism have decreased quality of life, despite replacement hormonal therapy.[114] As TSH is not useful for monitoring, the treatment of central hypothyroidism is more challenging as the "targets" for free T4 values in this setting are not well established. For example, one study of patients with pituitary deficiency found lower free T4 values amongst cases compared to the free T4 values of healthy controls. Some have suggested that free T4 in the middle of the normal range indicates appropriate dosage, while others suggest the upper-third of the normal range to be the correct target. Yet the downsides to over-treatment cannot be overlooked, and include decreased bone mineral density and susceptibility to atrial fibrillation in persons above the age of 60.[177] Finally, growth hormone deficiency is common in patients with central hypothyroidism, and is associated with reduced conversion of T4 to T3. Thus, the free T4 level may not reflect adequate tissue levels of T3. A recent study compared empiric T4 treatment (which resulted in an average dose of 1.1 μg/kg body weight) with a body-weight adapted T4 treatment (where T4 was dosed at 1.6 μg/kg body weight). The latter treatment resulted in improved clinical signs and symptoms of hypothyroidism, and parameters such as lipid levels, muscle enzymes and body weight.[178] Another non-randomized study found that aiming treatment at "normal" free T3 and free T4 concentrations led to subtle hypothyroidism.[170] In addition, patients with central hypothyroidism may need higher doses of levothyroxine than patients with primary hypothyroidism. One study examined clinically euthyroid patients with free T4 in the upper half of the normal range, and found that patients with central hypothyroidism received more exogenous T4 (1.9 μg/kg per day) than did patients with primary hypothyroidism from autoimmune thyroiditis or radio-iodone treatment (1.6 μg/kg per day).[179] In this study the authors noted lower TSH values in central vs primary hypothyroidism, and suggest that even small amounts of TSH may be biologically important.

Finally, it is important to evaluate for adrenal insufficiency prior to initiating treatment with T4. T4 accelerates the metabolism of cortisol, and therefore, if T4 treatment is initiated in the setting of untreated adrenal insufficiency, a life-threatening adrenal crisis may be precipitated.

SUMMARY

In conclusion, disorders of the hypothalamic–pituitary–thyroid axis constitute a diverse range of pathologies and present challenging clinical scenarios from the standpoints of both diagnosis and treatment. The availability of ultra-sensitive assays for TSH has made diagnosis of these disorders simpler; however, the practicing clinician must be alert to the possibilities of these conditions so as to avoid unnecessary tests or incorrect treatments.

With respect to the function of the HPT axis, several issues (such as the precise mechanism by which TH regulates TSH) remain unanswered. Furthermore, the role of epigenetic modifications in control of the HPT axis has not been elucidated. New high-throughput technologies, such as ChIP-seq and the discovery of small-molecule inhibitors of enzymes responsible for epigenetic changes, will surely shed new light on the function of this important axis.

References

1. Nillni EA, Sevarino KA. The biology of pro-thyrotropin-releasing hormone-derived peptides. *Endocr Rev.* 1999;20(5):599–648.
2. Nillni EA, Sevarino KA, Jackson IM. Processing of proTRH to its intermediate products occurs before the packing into secretory granules of transfected AtT20 cells. *Endocrinology.* 1993;132(3): 1271–1277.
3. Schaner P, Todd RB, Seidah NG, Nillni EA. Processing of prothyrotropin-releasing hormone by the family of prohormone convertases. *J Biol Chem.* 1997;272(32):19958–19968.
4. Ishikawa K, Taniguchi Y, Kurosumi K, Suzuki M. Origin of septal thyrotropin-releasing hormone in the rat. *Neuroendocrinology.* 1986;44(1):54–58.
5. Hollenberg AN. Role of thyrotropin-releasing hormone in the regulation of the thyroid axis. In: Wondisford FE, Radovick S, eds. *Clinical Management of Thyroid Disease.* Philadelphia: Saunders Elsevier; 2009:71–80.
6. Segerson TP, Hoefler H, Childers H, et al. Localization of thyrotropin-releasing hormone prohormone messenger ribonucleic acid in rat brain in situ hybridization. *Endocrinology.* 1987;121(1):98–107.
7. Dyess EM, Segerson TP, Liposits Z, et al. Triiodothyronine exerts direct cell-specific regulation of thyrotropin-releasing hormone gene expression in the hypothalamic paraventricular nucleus. *Endocrinology.* 1988;123(5):2291–2297.
8. Perello M, Friedman T, Paez-Espinosa V, Shen X, Stuart RC, Nillni EA. Thyroid hormones selectively regulate the post-translational processing of prothyrotropin-releasing hormone in the paraventricular nucleus of the hypothalamus. *Endocrinology.* 2006;147(6):2705–2716.
9. Segerson TP, Kauer J, Wolfe HC, et al. Thyroid hormone regulates TRH biosynthesis in the paraventricular nucleus of the rat hypothalamus. *Science.* 1987;238(4823):78–80.
10. Abel ED, Ahima RS, Boers ME, Elmquist JK, Wondisford FE. Critical role for thyroid hormone receptor beta2 in the regulation of paraventricular thyrotropin-releasing hormone neurons. *J Clin Invest.* 2001;107(8):1017–1023.
11. Hollenberg AN. The role of the thyrotropin-releasing hormone (TRH) neuron as a metabolic sensor. *Thyroid.* 2008; 18(2):131–139.
12. Ahima RS, Prabakaran D, Mantzoros C, et al. Role of leptin in the neuroendocrine response to fasting. *Nature.* 1996;382(6588): 250–252.
13. Harris M, Aschkenasi C, Elias CF, et al. Transcriptional regulation of the thyrotropin-releasing hormone gene by leptin and melanocortin signaling. *J Clin Invest.* 2001;107(1): 111–120.
14. De Groot LJ. Dangerous dogmas in medicine: the nonthyroidal illness syndrome. *J Clin Endocrinol Metab.* 1999; 84(1):151–164.
15. Fliers E, Guldenaar SE, Wiersinga WM, Swaab DF. Decreased hypothalamic thyrotropin-releasing hormone gene expression in patients with nonthyroidal illness. *J Clin Endocrinol Metab.* 1997;82(12):4032–4036.
16. Kakucska I, Romero LI, Clark BD, et al. Suppression of thyrotropin-releasing hormone gene expression by interleukin-1-beta in the rat: implications for nonthyroidal illness. *Neuroendocrinology.* 1994;59(2):129–137.
17. Kondo K, Harbuz MS, Levy A, Lightman SL. Inhibition of the hypothalamic-pituitary-thyroid axis in response to lipopolysaccharide is independent of changes in circulating corticosteroids. *Neuroimmunomodulation.* 1997;4(4):188–194.
18. Pierce JG, Parsons TF. Glycoprotein hormones: structure and function. *Annu Rev Biochem.* 1981;50:465–495.

19. Pierce JG, Parsons TF. Glycoprotein hormones: similar molecules with different functions. *UCLA Forum Med Sci.* 1979;(21):99–117.
20. Hershman JM, Starnes WR. Extraction and characterization of a thyrotropic material from the human placenta. *J Clin Invest.* 1969;48(5):923–929.
21. Farid NR, Szkudlinski MW. Minireview: structural and functional evolution of the thyrotropin receptor. *Endocrinology.* 2004;145(9):4048–4057.
22. Gaillard S, Wondisford FEW. Thyroid-stimulating hormone and thyroid-stimulating hormone receptor. In: Wondisford FE, Radovick S, eds. *Clinical Management of Thyroid Disease.* Philadelphia: Saunders Elsevier; 2009:81–104.
23. Steinfelder HJ, Radovick S, Wondisford FE. Hormonal regulation of the thyrotropin beta-subunit gene by phosphorylation of the pituitary-specific transcription factor Pit-1. *Proc Natl Acad Sci USA.* 1992;89(13):5942–5945.
24. Hashimoto K, Zanger K, Hollenberg AN, Cohen LE, Radovick S, Wondisford FE. cAMP response element-binding protein-binding protein mediates thyrotropin-releasing hormone signaling on thyrotropin subunit genes. *J Biol Chem.* 2000;275(43):33365–33372.
25. Collu R, Tang J, Castagne J, et al. A novel mechanism for isolated central hypothyroidism: inactivating mutations in the thyrotropin-releasing hormone receptor gene. *J Clin Endocrinol Metab.* 1997;82(5):1561–1565.
26. Yamada M, Saga Y, Shibusawa N, et al. Tertiary hypothyroidism and hyperglycemia in mice with targeted disruption of the thyrotropin-releasing hormone gene. *Proc Natl Acad Sci USA.* 1997;94(20):10862–10867.
27. Nikrodhanond AA, Ortiga-Carvalho TM, Shibusawa N, et al. Dominant role of thyrotropin-releasing hormone in the hypothalamic–pituitary–thyroid axis. *J Biol Chem.* 2006;281(8):5000–5007.
28. Stannard BS, Gesundheit N, Ronin C, Burnside J, Weintraub BD. Differential carbohydrate processing and secretion of thyrotropin and free alpha subunit. Effects of 1-deoxynojirimycin. *J Biol Chem.* 1988;263(17):8309–8317.
29. Stannard BS, Gesundheit N, Thotakura NR, Gyves PW, Ronin C, Weintraub BD. Differential effect of inhibitors of oligosaccharide processing on the secretion of thyrotropin from dispersed rodent pituitary cells. *Biochem Biophys Res Commun.* 1989;165(2):788–794.
30. Weintraub BD, Stannard BS, Meyers L. Glycosylation of thyroid-stimulating hormone in pituitary tumor cells: influence of high mannose oligosaccharide units on subunit aggregation, combination, and intracellular degradation. *Endocrinology.* 1983;112(4):1331–1345.
31. Szkudlinski MW, Thotakura NR, Tropea JE, Grossmann M, Weintraub BD. Asparagine-linked oligosaccharide structures determine clearance and organ distribution of pituitary and recombinant thyrotropin. *Endocrinology.* 1995;136(8):3325–3330.
32. Szkudlinski MW, Thotakura NR, Weintraub BD. Subunit-specific functions of N-linked oligosaccharides in human thyrotropin: role of terminal residues of alpha- and beta-subunit oligosaccharides in metabolic clearance and bioactivity. *Proc Natl Acad Sci USA.* 1995;92(20):9062–9066.
33. Joshi L, Murata Y, Wondisford FE, Szkudlinski MW, Desai R, Weintraub BD. Recombinant thyrotropin containing a beta-subunit chimera with the human chorionic gonadotropin-beta carboxy-terminus is biologically active, with a prolonged plasma half-life: role of carbohydrate in bioactivity and metabolic clearance. *Endocrinology.* 1995;136(9):3839–3848.
34. Larsen PR, Davies TF, Schlumberger M-F, Hay ID. Thyroid physiology and diagnostic evaluation of patients with thyroid disorders. In: Reed Larsen P, Kronenberg HM, Melmed S, Polonsky KS, eds. *Williams Textbook of Endocrinology.* 10th edn. Philadelphia: Saunders Elsevier; 2002:331–373.
35. Lazar MA. Thyroid hormone receptors: multiple forms, multiple possibilities. *Endocr Rev.* 1993;14(2):184–193.
36. Jones I, Srinivas M, Ng L, Forrest D. The thyroid hormone receptor beta gene: structure and functions in the brain and sensory systems. *Thyroid.* 2003;13(11):1057–1068.
37. Abel ED, Moura EG, Ahima RS, et al. Dominant inhibition of thyroid hormone action selectively in the pituitary of thyroid hormone receptor-beta null mice abolishes the regulation of thyrotropin by thyroid hormone. *Mol Endocrinol.* 2003;17(9):1767–1776.
38. Bradley DJ, Towle HC, Young 3rd WS. Spatial and temporal expression of alpha- and beta-thyroid hormone receptor mRNAs, including the beta 2-subtype, in the developing mammalian nervous system. *J Neurosci.* 1992;12(6):2288–2302.
39. Bradley DJ, Towle HC, Young 3rd WS. Alpha and beta thyroid hormone receptor (TR) gene expression during auditory neurogenesis: evidence for TR isoform-specific transcriptional regulation. *in vivo. Proc Natl Acad Sci USA.* 1994;91(2):439–443.
40. Jones I, Ng L, Liu H, Forrest D. An intron control region differentially regulates expression of thyroid hormone receptor beta2 in the cochlea, pituitary, and cone photoreceptors. *Mol Endocrinol.* May 2007;21(5):1108–1119.
41. Gothe S, Wang Z, Ng L, et al. Mice devoid of all known thyroid hormone receptors are viable but exhibit disorders of the pituitary-thyroid axis, growth, and bone maturation. *Genes Dev.* May 15, 1999;13(10):1329–1341.
42. Shibusawa N, Hashimoto K, Nikrodhanond AA, et al. Thyroid hormone action in the absence of thyroid hormone receptor DNA-binding *in vivo*. *J Clin Invest.* 2003;112(4):588–597.
43. Shibusawa N, Hollenberg AN, Wondisford FE. Thyroid hormone receptor DNA binding is required for both positive and negative gene regulation. *J Biol Chem.* 2003;278(2):732–738.
44. Refetoff S, DeWind LT, DeGroot LJ. Familial syndrome combining deaf-mutism, stuppled epiphyses, goiter and abnormally high PBI: possible target organ refractoriness to thyroid hormone. *J Clin Endocrinol Metab.* 1967;27(2):279–294.
45. Refetoff S, Weiss RE, Usala SJ. The syndromes of resistance to thyroid hormone. *Endocr Rev.* 1993;14(3):348–399.
46. Usala SJ, Tennyson GE, Bale AE, et al. A base mutation of the C-erbA beta thyroid hormone receptor in a kindred with generalized thyroid hormone resistance. Molecular heterogeneity in two other kindreds. *J Clin Invest.* 1990;85(1):93–100.
47. Sakurai A, Takeda K, Ain K, et al. Generalized resistance to thyroid hormone associated with a mutation in the ligand-binding domain of the human thyroid hormone receptor beta. *Proc Natl Acad Sci USA.* 1989;86(22):8977–8981.
48. Safer JD, Cohen RN, Hollenberg AN, Wondisford FE. Defective release of corepressor by hinge mutants of the thyroid hormone receptor found in patients with resistance to thyroid hormone. *J Biol Chem.* 1998;273(46):30175–30182.
49. Weiss RE, Weinberg M, Refetoff S. Identical mutations in unrelated families with generalized resistance to thyroid hormone occur in cytosine-guanine-rich areas of the thyroid hormone receptor beta gene. Analysis of 15 families. *J Clin Invest.* 1993;91(6):2408–2415.
50. Collingwood TN, Rajanayagam O, Adams M, et al. A natural transactivation mutation in the thyroid hormone beta receptor: impaired interaction with putative transcriptional mediators. *Proc Natl Acad Sci USA.* 1997;94(1):248–253.

51. Collingwood TN, Wagner R, Matthews CH, et al. A role for helix 3 of the TRbeta ligand-binding domain in coactivator recruitment identified by characterization of a third cluster of mutations in resistance to thyroid hormone. *EMBO J.* 1998;17(16):4760–4770.
52. Adams M, Matthews C, Collingwood TN, Tone Y, Beck-Peccoz P, Chatterjee KK. Genetic analysis of 29 kindreds with generalized and pituitary resistance to thyroid hormone. Identification of thirteen novel mutations in the thyroid hormone receptor beta gene. *J Clin Invest.* 1994;94(2):506–515.
53. Yoh SM, Chatterjee VK, Privalsky ML. Thyroid hormone resistance syndrome manifests as an aberrant interaction between mutant T3 receptors and transcriptional corepressors. *Mol Endocrinol.* 1997;11(4):470–480.
54. Takeda K, Sakurai A, DeGroot LJ, Refetoff S. Recessive inheritance of thyroid hormone resistance caused by complete deletion of the protein-coding region of the thyroid hormone receptor-beta gene. *J Clin Endocrinol Metab.* 1992;74(1):49–55.
55. Lafranchi SH, Snyder DB, Sesser DE, et al. Follow-up of newborns with elevated screening T4 concentrations. *J Pediatr.* 2003;143(3):296–301.
56. Beck-Peccoz P, Chatterjee VK. The variable clinical phenotype in thyroid hormone resistance syndrome. *Thyroid.* Summer 1994; 4(2):225–232.
57. Brucker-Davis F, Skarulis MC, Grace MB, et al. Genetic and clinical features of 42 kindreds with resistance to thyroid hormone. The National Institutes of Health Prospective Study. *Ann Intern Med.* 1995;123(8):572–583.
58. Weiss RE, Stein MA, Trommer B, Refetoff S. Attention-deficit hyperactivity disorder and thyroid function. *J Pediatr.* 1993; 123(4):539–545.
59. Persani L, Preziati D, Matthews CH, Sartorio A, Chatterjee VK, Beck-Peccoz P. Serum levels of carboxyterminal cross-linked telopeptide of type I collagen (ICTP) in the differential diagnosis of the syndromes of inappropriate secretion of TSH. *Clin Endocrinol (Oxf).* 1997;47(2):207–214.
60. Anselmo J, Cao D, Karrison T, Weiss RE, Refetoff S. Fetal loss associated with excess thyroid hormone exposure. *J Am Med Assoc.* 2004;292(6):691–695.
61. Anselmo J, Cesar R. Resistance to thyroid hormone: report of 2 kindreds with 35 patients. *Endocr Pract.* 1998;4(6):368–374.
62. Weiss RE, Dumitrescu A, Refetoff S. Approach to the patient with resistance to thyroid hormone and pregnancy. *J Clin Endocrinol Metab.* 2010;95(7):3094–3102.
63. Weiss RE, Refetoff S. Resistance to thyroid hormone. *Rev Endocr Metab Disord.* 2000;1(1–2):97–108.
64. Weiss RE, Refetoff S. Syndromes of resistance to thyroid hormone. In: Wondisford FE, Radovick S, eds. *Clinical Management of Thyroid Disease.* Philadelphia: Saunders Elsevier; 2009: 299–316.
65. Bogazzi F, Manetti L, Tomisti L, et al. Thyroid color flow Doppler sonography: an adjunctive tool for differentiating patients with inappropriate thyrotropin (TSH) secretion due to TSH-secreting pituitary adenoma or resistance to thyroid hormone. *Thyroid.* 2006;16(10):989–995.
66. Safer JD, Colan SD, Fraser LM, Wondisford FE. A pituitary tumor in a patient with thyroid hormone resistance: a diagnostic dilemma. *Thyroid.* 2001;11(3):281–291.
67. Dulgeroff AJ, Geffner ME, Koyal SN, Wong M, Hershman JM. Bromocriptine and triac therapy for hyperthyroidism due to pituitary resistance to thyroid hormone. *J Clin Endocrinol Metab.* 1992;75(4):1071–1075.
68. Koerner D, Surks MI, Oppenheimer JH. *In vitro* demonstration of specific triiodothyronine binding sites in rat liver nuclei. *J Clin Endocrinol Metab.* 1974;38(4):706–709.
69. Takeda T, Suzuki S, Liu RT, DeGroot LJ. Triiodothyroacetic acid has unique potential for therapy of resistance to thyroid hormone. *J Clin Endocrinol Metab.* 1995;80(7):2033–2040.
70. Beck-Peccoz P, Piscitelli G, Cattaneo MG, Faglia G. Successful treatment of hyperthyroidism due to nonneoplastic pituitary TSH hypersecretion with 3,5,3′-triiodothyroacetic acid (TRIAC). *J Endocrinol Invest.* 1983;6(3):217–223.
71. Darendeliler F, Bas F. Successful therapy with 3,5,3′-triiodothyroacetic acid (TRIAC) in pituitary resistance to thyroid hormone. *J Pediatr Endocrinol Metab.* Sep-Oct 1997; 10(5):535–538.
72. Radetti G, Persani L, Molinaro G, et al. Clinical and hormonal outcome after two years of triiodothyroacetic acid treatment in a child with thyroid hormone resistance. *Thyroid.* 1997; 7(5):775–778.
73. Dorey F, Strauch G, Gayno JP. Thyrotoxicosis due to pituitary resistance to thyroid hormones. successful control with D thyroxine: a study in three patients. *Clin Endocrinol (Oxf).* 1990;32(2):221–228.
74. Hamon P, Bovier-Lapierre M, Robert M, Peynaud D, Pugeat M, Orgiazzi J. Hyperthyroidism due to selective pituitary resistance to thyroid hormones in a 15-month-old boy: efficacy of D-thyroxine therapy. *J Clin Endocrinol Metab.* 1988;67(5): 1089–1093.
75. Anselmo J, Refetoff S. Regression of a large goiter in a patient with resistance to thyroid hormone by every other day treatment with triiodothyronine. *Thyroid.* 2004;14(1):71–74.
76. Beck-Peccoz P, Brucker-Davis F, Persani L, Smallridge RC, Weintraub BD. Thyrotropin-secreting pituitary tumors. *Endocr Rev.* 1996;17(6):610–638.
77. Mixson AJ, Friedman TC, Katz DA, et al. Thyrotropin-secreting pituitary carcinoma. *J Clin Endocrinol Metab.* 1993; 76(2):529–533.
78. Ma W, Ikeda H, Watabe N, Kanno M, Yoshimoto T. A plurihormonal TSH-producing pituitary tumor of monoclonal origin in a patient with hypothyroidism. *Horm Res.* 2003;59(5): 257–261.
79. Dong Q, Brucker-Davis F, Weintraub BD, et al. Screening of candidate oncogenes in human thyrotroph tumors: absence of activating mutations of the G alpha q, G alpha 11, G alpha s, or thyrotropin-releasing hormone receptor genes. *J Clin Endocrinol Metab.* 1996;81(3):1134–1140.
80. Beck Peccoz P, Persani L. Thyroid stimulating hormone-induced hyperfunction. In: Wondisford FE, Radovick S, eds. *Clinical Management of Thyroid Disease.* Philadelphia: Saunders Elsevier; 2009:221–228.
81. Ando S, Sarlis NJ, Krishnan J, et al. Aberrant alternative splicing of thyroid hormone receptor in a TSH-secreting pituitary tumor is a mechanism for hormone resistance. *Mol Endocrinol.* 2001;15(9):1529–1538.
82. Filopanti M, Ballare E, Lania AG, et al. Loss of heterozygosity at the SS receptor type 5 locus in human GH- and TSH-secreting pituitary adenomas. *J Endocrinol Invest.* 2004;27(10): 937–942.
83. Ezzat S, Horvath E, Kovacs K, Smyth HS, Singer W, Asa SL. Basic fibroblast growth factor expression by two prolactin and thyrotropin-producing pituitary adenomas. *Endocr Pathol.* Summer 1995;6(2):125–134.
84. Webster J, Peters JR, John R, et al. Pituitary stone: two cases of densely calcified thyrotrophin-secreting pituitary adenomas. *Clin Endocrinol (Oxf).* 1994;40(1):137–143.
85. Abs R, Stevenaert A, Beckers A. Autonomously functioning thyroid nodules in a patient with a thyrotropin-secreting pituitary adenoma: possible cause–effect relationship. *Eur J Endocrinol.* 1994;131(4):355–358.

86. Gasparoni P, Rubello D, Persani L, Beck-Peccoz P. Unusual association between a thyrotropin-secreting pituitary adenoma and a papillary thyroid carcinoma. *Thyroid.* 1998; 8(2):181–183.
87. Brucker-Davis F, Oldfield EH, Skarulis MC, Doppman JL, Weintraub BD. Thyrotropin-secreting pituitary tumors: diagnostic criteria, thyroid hormone sensitivity, and treatment outcome in 25 patients followed at the National Institutes of Health. *J Clin Endocrinol Metab.* 1999;84(2):476–486.
88. Kourides IA, Pekonen F, Weintraub BD. Absence of thyroid-binding immunoglobulins in patients with thyrotropin-mediated hyperthyroidism. *J Clin Endocrinol Metab.* 1980;51(2): 272–274.
89. Socin HV, Chanson P, Delemer B, et al. The changing spectrum of TSH-secreting pituitary adenomas: diagnosis and management in 43 patients. *Eur J Endocrinol.* 2003;148(4):433–442.
90. Bertherat J, Brue T, Enjalbert A, et al. Somatostatin receptors on thyrotropin-secreting pituitary adenomas: comparison with the inhibitory effects of octreotide upon *in vivo* and *in vitro* hormonal secretions. *J Clin Endocrinol Metab.* 1992;75(2): 540–546.
91. Mannavola D, Persani L, Vannucchi G, et al. Different responses to chronic somatostatin analogues in patients with central hyperthyroidism. *Clin Endocrinol (Oxf).* 2005;62(2):176–181.
92. Pasquini E, Faustini-Fustini M, Sciarretta V, et al. Ectopic TSH-secreting pituitary adenoma of the vomerosphenoidal junction. *Eur J Endocrinol.* 2003;148(2):253–257.
93. Cooper DS, Wenig BM. Hyperthyroidism caused by an ectopic TSH-secreting pituitary tumor. *Thyroid.* 1996;6(4):337–343.
94. Losa M, Magnani P, Mortini P, et al. Indium-111 pentetreotide single-photon emission tomography in patients with TSH-secreting pituitary adenomas: correlation with the effect of a single administration of octreotide on serum TSH levels. *Eur J Nucl Med.* 1997;24(7):728–731.
95. Losa M, Giovanelli M, Persani L, Mortini P, Faglia G, Beck-Peccoz P. Criteria of cure and follow-up of central hyperthyroidism due to thyrotropin-secreting pituitary adenomas. *J Clin Endocrinol Metab.* 1996;81(8):3084–3090.
96. Chanson P, Weintraub BD, Harris AG. Octreotide therapy for thyroid-stimulating hormone-secreting pituitary adenomas. A follow-up of 52 patients. *Ann Intern Med.* 1993;119(3):236–240.
97. Gancel A, Vuillermet P, Legrand A, Catus F, Thomas F, Kuhn JM. Effects of a slow-release formulation of the new somatostatin analogue lanreotide in TSH-secreting pituitary adenomas. *Clin Endocrinol (Oxf).* 1994;40(3):421–428.
98. Kuhn JM, Arlot S, Lefebvre H, et al. Evaluation of the treatment of thyrotropin-secreting pituitary adenomas with a slow release formulation of the somatostatin analog lanreotide. *J Clin Endocrinol Metab.* 2000;85(4):1487–1491.
99. Kienitz T, Quinkler M, Strasburger CJ, Ventz M. Long-term management in five cases of TSH-secreting pituitary adenomas: a single center study and review of the literature. *Eur J Endocrinol.* 2007;157(1):39–46.
100. Hershman J. Hypothalamic and pituitary hypothyroidism. In: Bastenie TA, Bonnyns M, Vanhaelst L, eds. *Recent Progress in Diagnosis and Treatment of Hypothyroidism Conditions.* Amsterdam: Excerpta Medica; 1980:40.
101. Martino E, Bartalena L, Pinchera A. Central hypothyroidism. In: Braverman LE, Utiger RD, eds. *Werner & Ingbar's The Thyroid: A Fundamental and Clinical Text.* 8th edn. Philadelphia, PA: Lippincott, Williams & Wilkins; 2000:462–473.
102. Hanna CE, Krainz PL, Skeels MR, Miyahira RS, Sesser DE, LaFranchi SH. Detection of congenital hypopituitary hypothyroidism: ten-year experience in the Northwest Regional Screening Program. *J Pediatr.* 1986;109(6):959–964.
103. Fisher DA, Klein AH. Thyroid development and disorders of thyroid function in the newborn. *N Engl J Med. 19* 1981;304(12): 702–712.
104. Samuels MH, Ridgway EC. Central hypothyroidism. *Endocrinol Metab Clin North Am/* 1992;21(4):903–919.
105. Van den Berghe G, de Zegher F, Baxter RC, et al. Neuroendocrinology of prolonged critical illness: effects of exogenous thyrotropin-releasing hormone and its combination with growth hormone secretagogues. *J Clin Endocrinol Metab.* 1998;83(2):309–319.
106. Rose SR. Disorders of thyrotropin synthesis, secretion, and function. *Curr Opin Pediatr.* 2000;12(4):375–381.
107. Persani L, Tonacchera M, Beck-Peccoz P, et al. Measurement of cAMP accumulation in Chinese hamster ovary cells transfected with the recombinant human TSH receptor (CHO-R): a new bioassay for human thyrotropin. *J Endocrinol Invest.* 1993;16(7):511–519.
108. Beck-Peccoz P, Persani L. Variable biological activity of thyroid-stimulating hormone. *Eur J Endocrinol.* 1994;131(4):331–340.
109. Beck-Peccoz P, Amr S, Menezes-Ferreira MM, Faglia G, Weintraub BD. Decreased receptor binding of biologically inactive thyrotropin in central hypothyroidism. Effect of treatment with thyrotropin-releasing hormone. *N Engl J Med.* 1985;312(17):1085–1090.
110. Taylor T, Weintraub BD. Altered thyrotropin (TSH) carbohydrate structures in hypothalamic hypothyroidism created by paraventricular nuclear lesions are corrected by *in vivo* TSH-releasing hormone administration. *Endocrinology.* 1989;125(4):2198–2203.
111. Miura Y, Perkel VS, Papenberg KA, Johnson MJ, Magner JA. Concanavalin-A, lentil, and ricin lectin affinity binding characteristics of human thyrotropin: differences in the sialylation of thyrotropin in sera of euthyroid, primary, and central hypothyroid patients. *J Clin Endocrinol Metab.* 1989;69(5): 985–995.
112. Persani L, Ferretti E, Borgato S, Faglia G, Beck-Peccoz P. Circulating thyrotropin bioactivity in sporadic central hypothyroidism. *J Clin Endocrinol Metab.* 2000;85(10):3631–3635.
113. Oliveira JH, Persani L, Beck-Peccoz P, Abucham J. Investigating the paradox of hypothyroidism and increased serum thyrotropin (TSH) levels in Sheehan's syndrome: characterization of TSH carbohydrate content and bioactivity. *J Clin Endocrinol Metab.* 2001;86(4):1694–1699.
114. Bates AS, Van't Hoff W, Jones PJ, Clayton RN. The effect of hypopituitarism on life expectancy. *J Clin Endocrinol Metab.* 1996;81(3):1169–1172.
115. Arafah BM. Reversible hypopituitarism in patients with large nonfunctioning pituitary adenomas. *J Clin Endocrinol Metab.* 1986;62(6):1173–1179.
116. Arafah BM, Harrington JF, Madhoun ZT, Selman WR. Improvement of pituitary function after surgical decompression for pituitary tumor apoplexy. *J Clin Endocrinol Metab.* 1990;71(2):323–328.
117. Karavitaki N, Cudlip S, Adams CB, Wass JA. *Craniopharyngiomas. Endocr Rev.* 2006;27(4):371–397.
118. Tatsumi K, Miyai K, Notomi T, et al. Cretinism with combined hormone deficiency caused by a mutation in the PIT1 gene. *Nat Genet.* 1992;1(1):56–58.
119. Radovick S, Nations M, Du Y, Berg LA, Weintraub BD, Wondisford FE. A mutation in the POU-homeodomain of Pit-1 responsible for combined pituitary hormone deficiency. *Science.* 1992;257(5073):1115–1118.
120. Agarwal G, Bhatia V, Cook S, Thomas PQ. Adrenocorticotropin deficiency in combined pituitary hormone deficiency patients homozygous for a novel PROP1 deletion. *J Clin Endocrinol Metab.* 2000;85(12):4556–4561.

121. Mendonca BB, Osorio MG, Latronico AC, Estefan V, Lo LS, Arnhold IJ. Longitudinal hormonal and pituitary imaging changes in two females with combined pituitary hormone deficiency due to deletion of A301, G302 in the PROP1 gene. *J Clin Endocrinol Metab.* 1999;84(3):942–945.
122. Deladoey J, Fluck C, Buyukgebiz A, et al. "Hot spot" in the PROP1 gene responsible for combined pituitary hormone deficiency. *J Clin Endocrinol Metab.* 1999;84(5):1645–1650.
123. Dattani MT, Martinez-Barbera JP, Thomas PQ, et al. Mutations in the homeobox gene HESX1/Hesx1 associated with septo-optic dysplasia in human and mouse. *Nat Genet.* 1998;19(2):125–133.
124. Dattani MT. Growth hormone deficiency and combined pituitary hormone deficiency: does the genotype matter? *Clin Endocrinol (Oxf).* 2005;63(2):121–130.
125. Netchine I, Sobrier ML, Krude H, et al. Mutations in LHX3 result in a new syndrome revealed by combined pituitary hormone deficiency. *Nat Genet.* 2000;25(2):182–186.
126. Machinis K, Pantel J, Netchine I, et al. Syndromic short stature in patients with a germline mutation in the LIM homeobox LHX4. *Am J Hum Genet.* 2001;69(5):961–968.
127. Hayashizaki Y, Hiraoka Y, Endo Y, Miyai K, Matsubara K. Thyroid-stimulating hormone (TSH) deficiency caused by a single base substitution in the CAGYC region of the beta-subunit. *EMBO J.* 1989;8(8):2291–2296.
128. Hayashizaki Y, Hiraoka Y, Tatsumi K, et al. Deoxyribonucleic acid analyses of five families with familial inherited thyroid stimulating hormone deficiency. *J Clin Endocrinol Metab.* 1990;71(4):792–796.
129. Dacou-Voutetakis C, Feltquate DM, Drakopoulou M, Kourides IA, Dracopoli NC. Familial hypothyroidism caused by a nonsense mutation in the thyroid-stimulating hormone beta-subunit gene. *Am J Hum Genet.* 1990;46(5):988–993.
130. Deladoey J, Vuissoz JM, Domene HM, et al. Congenital secondary hypothyroidism due to a mutation C105Vfs114X thyrotropin-beta mutation: genetic study of five unrelated families from Switzerland and Argentina. *Thyroid.* 2003; 13(6):553–559.
131. Borck G, Topaloglu AK, Korsch E, et al. Four new cases of congenital secondary hypothyroidism due to a splice site mutation in the thyrotropin-beta gene: phenotypic variability and founder effect. *J Clin Endocrinol Metab.* 2004; 89(8):4136–4141.
132. Medeiros-Neto G, Herodotou DT, Rajan S, et al. A circulating, biologically inactive thyrotropin caused by a mutation in the beta subunit gene. *J Clin Invest.* 1996;97(5):1250–1256.
133. Freda PU, Silverberg SJ, Post KD, Wardlaw SL. Hypothalamic–pituitary sarcoidosis. *Trends Endocrinol Metab.* 1992; 3(9):321–325.
134. Nanduri VR, Bareille P, Pritchard J, Stanhope R. Growth and endocrine disorders in multisystem Langerhans' cell histiocytosis. *Clin Endocrinol (Oxf).* 2000;53(4):509–515.
135. Kaltsas GA, Powles TB, Evanson J, et al. Hypothalamo-pituitary abnormalities in adult patients with langerhans cell histiocytosis: clinical, endocrinological, and radiological features and response to treatment. *J Clin Endocrinol Metab.* 2000; 85(4):1370–1376.
136. Powrie JK, Powell M, Ayers AB, Lowy C, Sonksen PH. Lymphocytic adenohypophysitis: magnetic resonance imaging features of two new cases and a review of the literature. *Clin Endocrinol (Oxf).* 1995;42(3):315–322.
137. Molitch ME, Gillam MP. Lymphocytic hypophysitis. *Horm Res.* 2007;68(Suppl 5):145–150.
138. McDermott JH, Walsh CH. Hypogonadism in hereditary hemochromatosis. *J Clin Endocrinol Metab.* 2005;90(4): 2451–2455.
139. Hudec M, Grigerova M, Walsh CH. Secondary hypothyroidism in hereditary hemochromatosis: recovery after iron depletion. *Thyroid.* 2008;18(2):255–257.
140. Zervas A, Katopodi A, Protonotariou A, et al. Assessment of thyroid function in two hundred patients with beta-thalassemia major. *Thyroid.* 2002;12(2):151–154.
141. Harrop JS, Davies TJ, Capra LG, Marks V. Hypothalamic-pituitary function following successful treatment of intracranial tumours. *Clin Endocrinol (Oxf).* 1976;5(4):313–321.
142. Constine LS, Woolf PD, Cann D, et al. Hypothalamic-pituitary dysfunction after radiation for brain tumors. *N Engl J Med.* 1993;328(2):87–94.
143. Littley MD, Shalet SM, Beardwell CG, Robinson EL, Sutton ML. Radiation-induced hypopituitarism is dose-dependent. *Clin Endocrinol (Oxf).* 1989;31(3):363–373.
144. Rose SR, Lustig RH, Pitukcheewanont P, et al. Diagnosis of hidden central hypothyroidism in survivors of childhood cancer. *J Clin Endocrinol Metab.* 1999;84(12):4472–4479.
145. Sherman SI, Gopal J, Haugen BR, et al. Central hypothyroidism associated with retinoid X receptor-selective ligands. *N Engl J Med. 8.* 1999;340(14):1075–1079.
146. Golden WM, Weber KB, Hernandez TL, Sherman SI, Woodmansee WW, Haugen BR. Single-dose rexinoid rapidly and specifically suppresses serum thyrotropin in normal subjects. *J Clin Endocrinol Metab.* 2007;92(1):124–130.
147. Ridruejo E, Christensen AF, Mando OG. Central hypothyroidism and hypophysitis during treatment of chronic hepatitis C with pegylated interferon alpha and ribavirin. *Eur J Gastroenterol Hepatol.* 2006;18(6):693–694.
148. Benvenga S, Campenni A, Ruggeri RM, Trimarchi F. Clinical review 113: Hypopituitarism secondary to head trauma. *J Clin Endocrinol Metab.* 2000;85(4):1353–1361.
149. Benvenga S, Vigo T, Ruggeri RM, et al. Severe head trauma in patients with unexplained central hypothyroidism. *Am J Med.* 2004;116(11):767–771. 1.
150. Woolf PD, Lee LA, Hamill RW, McDonald JV. Thyroid test abnormalities in traumatic brain injury: correlation with neurologic impairment and sympathetic nervous system activation. *Am J Med.* 1988;84(2):201–208.
151. Uy HL, Reasner CA, Samuels MH. Pattern of recovery of the hypothalamic-pituitary-thyroid axis following radioactive iodine therapy in patients with Graves' disease. *Am J Med.* 1995;99(2):173–179.
152. Wehmann RE, Gregerman RI, Burns WH, Saral R, Santos GW. Suppression of thyrotropin in the low-thyroxine state of severe nonthyroidal illness. *N Engl J Med.* 1985;312(9):546–552. 28.
153. Fekete C, Gereben B, Doleschall M, et al. Lipopolysaccharide induces type 2 iodothyronine deiodinase in the mediobasal hypothalamus: implications for the nonthyroidal illness syndrome. *Endocrinology.* 2004;145(4):1649–1655.
154. Adriaanse R, Romijn JA, Brabant G, Endert E, Wiersinga WM. Pulsatile thyrotropin secretion in nonthyroidal illness. *J Clin Endocrinol Metab.* 1993;77(5):1313–1317.
155. Davies PH, Black EG, Sheppard MC, Franklyn JA. Relation between serum interleukin-6 and thyroid hormone concentrations in 270 hospital in-patients with non-thyroidal illness. *Clin Endocrinol (Oxf).* 1996;44(2):199–205.
156. Bartalena L, Grasso L, Brogioni S, Martino E. Interleukin 6 effects on the pituitary-thyroid axis in the rat. *Eur J Endocrinol.* 1994;131(3):302–306.
157. Coiro V, Volpi R, Marchesi C, et al. Influence of residual C-peptide secretion on nocturnal serum TSH peak in well-controlled diabetic patients. *Clin Endocrinol (Oxf).* 1997; 47(3):305–310.

158. Samuels MH, Luther M, Henry P, Ridgway EC. Effects of hydrocortisone on pulsatile pituitary glycoprotein secretion. *J Clin Endocrinol Metab.* 1994;78(1):211–215.
159. Duick DS, Wahner HW. Thyroid axis in patients with Cushing's syndrome. *Arch Intern Med.* 1979;139(7):767–772.
160. Laron Z. Interactions between the thyroid hormones and the hormones of the growth hormone axis. *Pediatr Endocrinol Rev.* 2003;1(Suppl 2):244–249. discussion 250.
161. Jorgensen JO, Moller J, Laursen T, Orskov H, Christiansen JS, Weeke J. Growth hormone administration stimulates energy expenditure and extrathyroidal conversion of thyroxine to triiodothyronine in a dose-dependent manner and suppresses circadian thyrotrophin levels: studies in GH-deficient adults. *Clin Endocrinol (Oxf).* 1994;41(5):609–614.
162. Porretti S, Giavoli C, Ronchi C, et al. Recombinant human GH replacement therapy and thyroid function in a large group of adult GH-deficient patients: when does L-T(4) therapy become mandatory? *J Clin Endocrinol Metab.* 2002;87(5): 2042–2045.
163. Agha A, Walker D, Perry L, et al. Unmasking of central hypothyroidism following growth hormone replacement in adult hypopituitary patients. *Clin Endocrinol (Oxf).* 2007; 66(1):72–77.
164. Croxson MS, Ibbertson HK. Low serum triiodothyronine (T3) and hypothyroidism in anorexia nervosa. *J Clin Endocrinol Metab.* 1977;44(1):167–174.
165. Hein MD, Jackson IM. Review: thyroid function in psychiatric illness. *Gen Hosp Psychiatry.* 1990;12(4):232–244.
166. Bartalena L, Placidi GF, Martino E, et al. Nocturnal serum thyrotropin (TSH) surge and the TSH response to TSH-releasing hormone: dissociated behavior in untreated depressives. *J Clin Endocrinol Metab.* 1990;71(3):650–655.
167. Sell MA, Schott M, Tharandt L, Cissewski K, Scherbaum WA, Willenberg HS. Functional central hypothyroidism in the elderly. *Aging Clin Exp Res.* 2008;20(3):207–210.
168. Mariotti S, Barbesino G, Caturegli P, et al. Complex alteration of thyroid function in healthy centenarians. *J Clin Endocrinol Metab.* 1993;77(5):1130–1134.
169. Faglia G, Bitensky L, Pinchera A, et al. Thyrotropin secretion in patients with central hypothyroidism: evidence for reduced biological activity of immunoreactive thyrotropin. *J Clin Endocrinol Metab.* 1979;48(6):989–998.
170. Ferretti E, Persani L, Jaffrain-Rea ML, Giambona S, Tamburrano G, Beck-Peccoz P. Evaluation of the adequacy of levothyroxine replacement therapy in patients with central hypothyroidism. *J Clin Endocrinol Metab.* 1999;84(3):924–929.
171. Snyder PJ, Utiger RD. Response to thyrotropin releasing hormone (TRH) in normal man. *J Clin Endocrinol Metab.* 1972;34(2):380–385.
172. Faglia G, Beck-Peccoz P, Ferrari C, et al. Plasma thyrotropin response to thyrotropin-releasing hormone in patients with pituitary and hypothalamic disorders. *J Clin Endocrinol Metab.* 1973;37(4):595–601.
173. Caron PJ, Nieman LK, Rose SR, Nisula BC. Deficient nocturnal surge of thyrotropin in central hypothyroidism. *J Clin Endocrinol Metab.* 1986;62(5):960–964.
174. Rose SR, Manasco PK, Pearce S, Nisula BC. Hypothyroidism and deficiency of the nocturnal thyrotropin surge in children with hypothalamic-pituitary disorders. *J Clin Endocrinol Metab.* 1990;70(6):1750–1755.
175. Divall SA, Wondisford FE. TRH testing in its infancy. *J Clin Endocrinol Metab.* 2008;93(2):378–379.
176. van Tijn DA, de Vijlder JJ, Vulsma T. Role of the thyrotropin-releasing hormone stimulation test in diagnosis of congenital central hypothyroidism in infants. *J Clin Endocrinol Metab.* 2008;93(2):410–419.
177. Sawin CT, Geller A, Wolf PA, et al. Low serum thyrotropin concentrations as a risk factor for atrial fibrillation in older persons. *N Engl J Med.* 1994;331(19):1249–1252.
178. Slawik M, Klawitter B, Meiser E, et al. Thyroid hormone replacement for central hypothyroidism: a randomized controlled trial comparing two doses of thyroxine (T4) with a combination of T4 and triiodothyronine. *J Clin Endocrinol Metab.* 2007;92(11):4115–4122.
179. Gordon MB, Gordon MS. Variations in adequate levothyroxine replacement therapy in patients with different causes of hypothyroidism. *Endocr Pract.* Sep-Oct 1999;5(5):233–238.

CHAPTER

32

Neuroendocrine Growth Disorders – Dwarfism, Gigantism

Ali Mohamadi, Roberto Salvatori

Division of Endocrinology, Johns Hopkins School of Medicine, Baltimore, MD, USA

OUTLINE

Summary

The growth hormone–IGF-1 axis plays a paramount role in controlling the rate and progression of linear growth. Alterations in this system may reflect lesions in the hypothalamus, pituitary, or target cells, and may cause growth deficit or excess. Dwarfism refers to the arrest of linear growth resulting in compromised final adult height, and may result from growth hormone deficiency or resistance. Conversely, gigantism is most commonly associated with growth hormone excess, and is characterized by excessive longitudinal growth beyond the patient's genetic target. This chapter reviews the presentations and most common causes of these neuroendocrine growth disorders, highlighting the various gene mutations, endocrine syndromes, and acquired defects that result in states of growth hormone deficiency or excess.

INTRODUCTION

Postnatal somatic growth requires the integrity of several systems, and for this reason it is not solely dependent upon the neuroendocrine apparatus. Nevertheless, growth hormone (GH) and its main effector (insulin-like growth factor-1, IGF-1), play a paramount role in controlling growth and orchestrating its progression. Although growth abnormalities can occur in the setting of a normal GH–IGF axis, there is no doubt that alteration in this system may cause growth deficit or excess (Fig. 32.1).

GH is secreted by the somatotrope cells of the anterior pituitary gland. Although its main function is to promote postnatal longitudinal growth, GH exerts numerous additional functions, such as regulation of glucose and lipid metabolism, bone apposition, skeletal and cardiac muscle mass growth, and arterial pressure. GH secretion is under is dual hypothalamic control, stimulated by GH-releasing hormone (GHRH) and inhibited by somatostatin (SRIF). These two factors act

FIGURE 32.1 Brothers with acromegalic gigantism (230 cm in stature) side-by-side with a pituitary dwarf (69 cm) from the late 1800s. *Reproduced from Enderle A. Dwarfism and gigantism in historical picture postcards.* J R Soc Med. *1998;91:273–278.*

on the somatotrope cells by binding to specific cell surface G-protein coupled receptors (GPCR), which lead to signal transduction that increases (GHRH) or decreases (SRIF) synthesis of intracellular cyclic AMP (cAMP). A third peptide, ghrelin, stimulates GH secretion by activating a distinct receptor (GH secretagog receptor, GHS-R), expressed both in the pituitary and the hypothalamus. The physiological role of ghrelin (mostly produced in the stomach) in regulation GH secretion is still debated. Upon binding of GH to its receptor (GHR), IGF-1 is produced both in the liver, from where it is released into the general circulation, and locally in the peripheral tissues, such as bone, cartilage and muscle, where it acts in a paracrine fashion. In this chapter, we will discuss some of the most common etiologies of alterations in the GH–IGF axis leading to dwarfism and gigantism, and provide insight into the diagnosis of these rare but important disorders.

DWARFISM

The term "dwarfism" refers to any one of a number of defects, acquired or congenital, that results in an arrest linear growth and profound short stature that, if untreated, results in compromised final adult height. Contrary to many forms of dwarfism (such as achondroplasia or other bone dysplasias), GH deficiency (GHD) (commonly referred to as "pituitary dwarfism") and GH resistance are characterized by normal body proportions. GHD may be due to congenital or acquired defects resulting in isolated GHD (IGHD) or combined pituitary hormone deficiency (CPHD).

Clinical Features of Pituitary Dwarfism

Although the definition of short stature, ultimately, is arbitrary, there can be no mistaking the effect of GHD on linear growth in the developing child. Height attenuation is the most common presenting complaint for a patient with true GHD, and there are a number of factors a pediatrician must consider when deciding whether to pursue a work-up for GHD. Foremost is assessing whether the case is, indeed, one of short stature. A consensus statement of the Growth Hormone Research Society defined short stature as 2 SD ($z = 2$, or 2.3 percentile) less than the population mean, and suggested a cutoff of 3 SD ($z = -3$, or 0.13 percentile) as the definition of "severe short stature" that is sufficient by itself to warrant diagnostic investigation.[1]

Consideration must also be taken of the child's predicted adult height, as determined from the heights of his parents. A patient who at first glance may appear to warrant a work-up for GHD based on a height greater than 2 SD below the mean may have a normal growth based on his mid-parental height. For this reason, stock should be placed on the child's height velocity, which is a strong predictor of GHD. Patients who demonstrate a marked and sustained fall in growth velocity (with significant downward deflection across percentiles on their growth chart; Fig. 32.2) deserve a work-up for GHD, as those with height increases deserve a work-up for GH excess.

Numerous other conditions, such as renal insufficiency, malnourishment, malabsorption or excessive exposure to glucocorticoid may result in growth failure despite a normal GH–IGF axis. Although most children born small for gestational age (SGA) experience catch-up growth by 2 years of age (differentiating it from postnatal growth failure), those who fail to reach their target height are easily confused with patients with true GHD when reviewing their growth chart alone.[2] Perhaps most confounding are those children with short stature who lack an apparent etiology, classified as idiopathic short stature (ISS). Within this category are children with

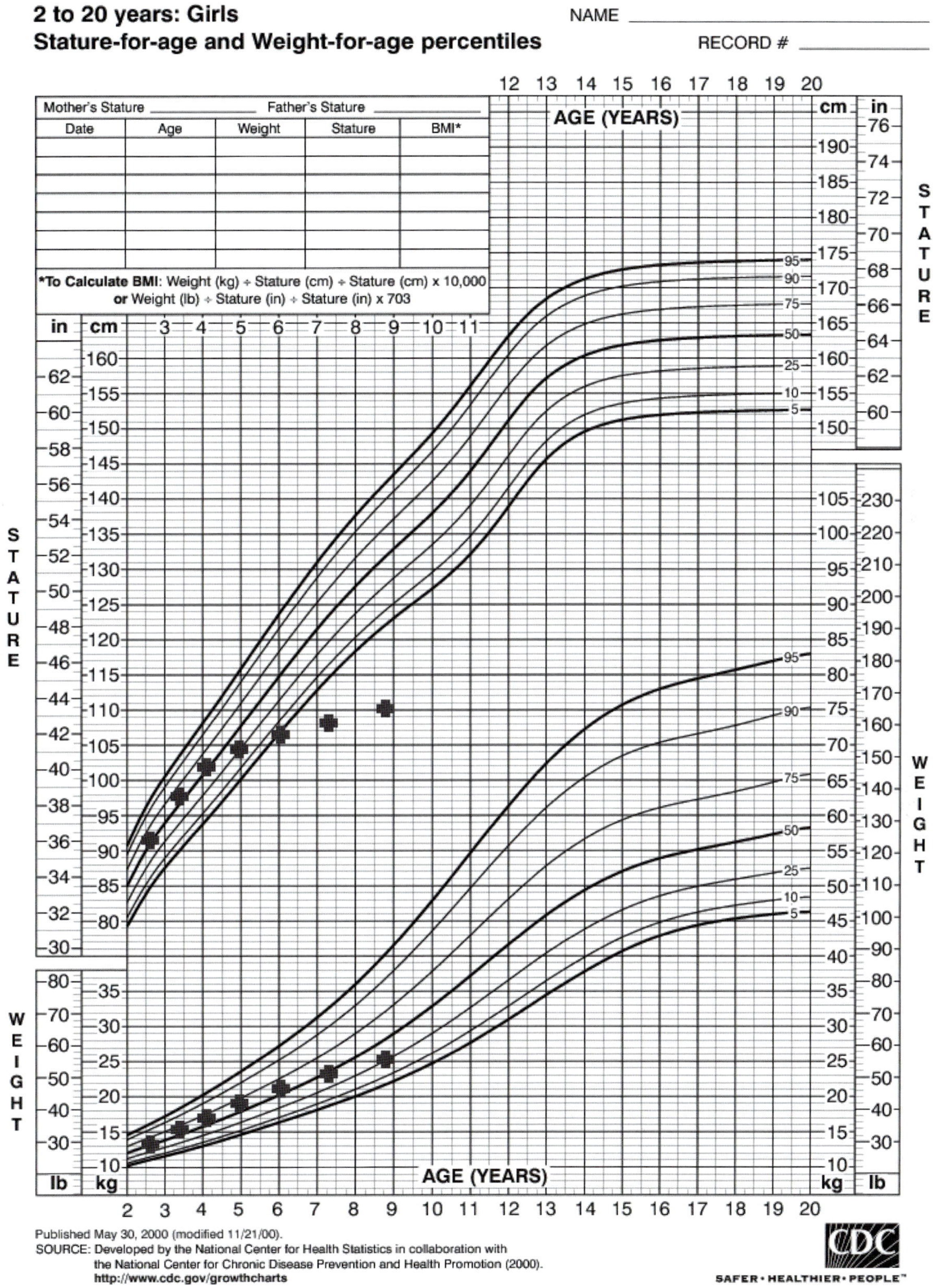

FIGURE 32.2 Arrest in linear growth ("downward deflection" on growth chart) witnessed in a girl with GHD due to craniopharyngioma diagnosed at age 8 + 10/12. As shown on the stature-for-age curve (top), the child's linear growth velocity was steady until age 4 years, at which time she experienced a pronounced decrease in stature from the 50th percentile to below the 5th percentile by age 6. The weight-for-age curve (bottom) demonstrates a slowing of weight gain that begins at age 6, lagging slightly behind the linear growth attenuation.

constitutional delay of growth and puberty and familial short stature, who can be differentiated from those with GHD based on a normal height velocity.

Ultimately, if the result of clinical presentation and a delayed bone age X-ray suggest the presence of GHD, the endocrinologist must pursue formal biochemical assessment of the GH–IGF axis. Exactly how to proceed with this assessment is subject to some controversy. Serum IGF-1 levels are faulted by poor sensitivity and specificity. While GH stimulation testing with provocative agents is an important diagnostic tool for unequivocal GHD, sensitivity, specificity and reproducibility for less severe ("partial") GHD is poor. With all of these factors in mind, the physician who diagnoses a patient with GHD must consider the myriad causes of this disorder, both congenital and acquired, in order to predict how the patient will respond to therapy.

Congenital Defects

Inherited diseases may result in IGHD or CPHD. The consequences are different: while the lack of GH principally causes growth failure, the lack of gonadotropins, TSH and ACTH may have a multitude of consequences, including lack of development of secondary sexual characteristics and infertility (FSH and LH), neurological developmental abnormalities (TSH) and, potentially, death due to adrenal failure (ACTH). In this section we will concentrate on those defects that affect the GH–IGF axis and result in dwarfism as their main outcome.

Isolated GH Deficiency (IGHD)

The incidence of IGHD is estimated to be 1 in 3480 live births.[3,4] While most cases are sporadic, anatomic abnormalities in the pituitary or hypothalamus are present in only a minority of patients,[5] suggesting that genetic or functional defects may account for a significant proportion of cases.[6] Accordingly, between 3% and 30% of IGHD patients have an affected first-degree relative, supporting the hypothesis that genetic defects often cause IGHD. Clinical and genetic studies have led to the description of four distinct forms of familial IGHD (Table 32.1), depending on inheritance mode and development of antibodies.

Although mutations in the *SHOX* (Short Stature Homeobox) gene, responsible for short stature in Turner syndrome or Leri-Weill dyschondrosteosis, have been reported in 2.4% children with idiopathic short stature, they have not been found in IGHD.[7]

Mutations in the GHRHR Gene

The native GHRHR contains 423 amino acids, but a post-translational cleavage removes the first 22 amino acids of the extracellular domain, necessary for intracellular trafficking.[8] The GHRHR gene (*GHRHR*) consists of 13 exons. Its expression is restricted to the pituitary gland, the renal medulla and the placenta.

The first human *GHRHR* mutation was described in 2002, in three families from the Indian subcontinent,[9] including a Pakistani kindred that contains 18 affected individuals, referred to as the "Dwarfs of Sindh".[10] All the affected patients in all these kindreds are homozygous for a nonsense mutation at codon 72 (E72X) that creates an early translation stop signal. This mutation was later shown to be highly prevalent in the Indian subcontinent, where it spread from a single founder for unclear reasons, possibly providing a survival advantage.[11] After these initial reports, additional *GHRHR* mutations were identified, including one additional nonsense mutation (Q43X); eight missense mutations (V10G, A137L, L144H, A176V, A222E, F242C, K329E, R357C); four splice donor mutations (IVS1 + 1G → A, IVS1 + 2T → G, IVS3 + G → A, and IVS7 + 1G → C); two small deletions (Del1140–1144 and Del1121–1124); a mutation in the signal peptide and a promoter mutation (−124A → C).[12–20]

The IVS1 + 1G → A was identified in the largest IGHD kindred described to date, with more than 100 affected individuals in 7 generations residing in Itabaianinha, a rural county in the northeastern Brazilian state of Sergipe.[12] A founder effect has been identified, and its spreading has been facilitated by a high incidence of consanguineous marriages, and geographical isolation.

Until very recently, all *GHRHR* mutations were thought to cause IGHD in an autosomal recessive manner. However, one of the missense mutations (V10G) causes autosomal dominant IGHD by altering the cleavage site of the signal peptide, and causing problems in intracellular trafficking of the receptor and preventing expression of the product of the wild-type allele.[20] All the subjects with bi-allelic *GHRH-R* mutations show early growth failure that begins during the first year of life. Untreated individuals achieve adult stature that ranges from 4 to 9 SDS below average, and have a characteristically high-pitched voice, mild frontal bossing, increased abdominal adiposity, hypercholesterolemia and decreased cardiac size despite increased systolic blood pressure.[21] Neither micropenis nor neonatal hypoglycemia has been reported, possibly due to mild residual GH secretion. Both male and female affected subjects have normal fertility (although puberty is often delayed in untreated subjects), and females are able to breast-feed. Affected subjects have frankly low serum levels of IGF-1. Serum GH levels are low but detectable, and fail to increase to > 3 ng/ml in response to provocative stimuli. Most (but not all) patients with *GHRHR* mutations have MRI evidence of anterior pituitary hypoplasia, likely due to reduced mass of

TABLE 32.1 Forms of Familial Isolated Growth Hormone Deficiency

Type	Serum GH	Transmission	Genes involved	Antibodies to exogenous GH
IA	Undetectable	AR	*GH1* (deletion)	Yes
IB	Low	AR	*GHRHR, GH1*	No
II	Low	AD	*GH1* (splicing alteration), *GHRHR* (signal peptide mutation)	No
III	Low	X-Linked	*BTK* (when associated with X-linked agammaglobulinemia)	No

somatotrope cells, although pathological studies in humans have not yet been performed.[13,22–24] The possibility of a "partial phenotype" in heterozygous carriers has been hypothesized but not yet fully proven, although heterozygous subjects have subtle changes in body composition.[23]

GHRHR mutations account for approximately 10% of IGHD IB.[14] All the mutations have been described in kindreds that are related, with the exception of the L144H, occurring in three families from different continents and with two different haplotypes.[25] The presence of families in which affected individuals are compound heterozygotes for distinct *GHRHR* mutations suggests that defective *GHRHR* alleles may be prevalent in the general population.

Mutations in the GH Gene

The GH protein is composed of a 4-α helical bundle with two disulfide bridges.[26] The GH gene (*GH1*) consists of 5 exons encoding for a 217-amino acid protein precursor of the mature molecule. Approximately 85% of circulating GH consists of a 191-amino acid protein with molecular mass of 22 kDa; 5–10% is a smaller form of 20-kDa protein resulting from alternative splicing of part of exon 3. The remaining 5% of circulating GH consist of smaller proteins that lack the entire exon 3 or exons 3 and 4. Although *GH1* mutations are present in approximately two-thirds of IGHD IA, they are rather infrequent in type IB (1.7%), while overall their prevalence in familial IGHD is estimated to be only 18.5%.[27]

IGHD IA

The first *GH1* mutations discovered were large deletions[28] that cause complete absence of GH. The most frequent deletion results in loss of 6.7 kb of DNA. Others result in loss of 7.0, 7.6 or 45 kb of DNA. Patients with the 6.7-kb deletion who are from different ethnic and geographical origin do not share the same haplotypes, indicating that this deletion has likely arisen multiple times.[29] Two nonsense mutations (W7X and E4X) and two frameshift mutations (Del51C and Del243–244) insert premature stop codons.[30]

IGHD IB

A two-base deletion (Del 5938–5939), and three distinct intron 4 splice mutations (IVS4 + 1G→C, IVS4 + 1G→T, and IVS4 + 5G→C) in *GH1* have been identified in IGH IB, with the same phenotype of patients with *GHRHR* mutations.

IGHD II

This form is caused mostly by *GH1* mutations. Two missense mutations (V110F and R183H) generate a protein that inhibits the release of the secretory granules containing the normal GH protein. Most of the cases are caused by mutations that impair proper splicing of exon 3, thereby generating a protein missing amino acids 32–71 that acts as dominant inhibitor by decreasing the stability of the normal GH protein, and disrupting secretory vesicles.[31,32] The IVS3 + 1G→A mutation has been reported in several families with different ethnic backgrounds, and haplotype analysis has demonstrated that the mutation has arisen independently.[33] While the only mutations known to cause IGHD II involve *GH1*, they could not be detected in approximately 16% of the cases, suggesting that other genes may cause this phenotype.[34] Accordingly, recently a mutation in *GHRHR* has been shown to cause autosomal dominant IGHD.[20]

The degree of GH deficiency ranges from moderate to severe, but generally IGHD II patients tend to have less severe growth retardation than patients with IA and IB. Facial dysmorphism is often present, and neonatal hypoglycemia and micropenis can occur. The size of the anterior pituitary in IGHD IA and II is generally normal or only mildly decreased.[35,36] Interestingly, in a subset of patients with these *GH1* splice mutations, other pituitary cells can be damaged and patients can develop additional pituitary hormone deficits.[37]

Mutations in the GHS Receptor

Recently a GHS-R missense mutation was reported in two unrelated families with IGHD. This mutation, which results in decreased cell-surface expression of the receptor, selectively impairs the constitutive activity of the GHSR, which is, however, still able to respond to ghrelin.[38] If confirmed, this would be a new pathogenic mechanism of growth failure in humans, whose prevalence is presently being investigated.

Abnormal GH Molecules ("Bio-inactive GH")

The original report of two children with normal serum GH but low serum IGF-1, who responded to exogenous GH (proving normally functioning GHR), introduced the concept of "bio-inactive GH."[39] The genetic cause of the phenotype of those patients has not been elucidated. Subsequently, two heterozygous *GH1* missense mutations (R77C and D112G) were described in a similar clinical scenario.[40,41] The R77C mutant GH acts as a receptor antagonist, while the D112G prevents receptor dimerization.

GH Insensitivity Syndromes (GHIS)

MUTATIONS IN THE GHR (LARON'S DWARFISM)

The GHR is a membrane-bound 626-amino acid glycoprotein which includes an extracellular domain, a transmembrane domain, and an intracellular domain that couples with the JAK-Stat, PI3kinase and MAP kinase pathways. The GHR exists in two forms: as a membrane-bound receptor, and as soluble GH binding protein (GHBP) corresponding to the extracellular domain of the receptor.[42] Binding of GH to the membrane-bound receptor leads to receptor dimerization, which is necessary for signaling.

The GHR gene (*GHR*) consists of 10 exons, and the coding region spans exons 2–9.[43] After translation, an 18-amino acid signal sequence is removed. Mutations of the *GHR* cause GH insensitivity syndrome (GHIS), also called Laron's dwarfism, after his initial description of this syndrome in 1966.[44] To date, approximately 50 mutations in the *GHR* have been reported, including large and small deletions, nonsense mutations, frame shifts, splice mutations and missense mutations. Most are located in the extracellular domain, and are therefore associated with low serum GHBP. However, about 20–25% of patients have normal or even high serum GHBP levels. This occurs in missense mutations that alter proper expression of the GHR on the cell surface,[45] and in a splice mutation that alters a region of the receptor involved in homodimerization.[46] Most mutations cause autosomal recessive GHIS (compound heterozygosity has been also reported), apart from two reports on dominant-negative defects of the intracellular domain of the GHR.[47]

Patients with GHIS physically resemble those with IGHD, although some of the facial dysmorphic features (sparse hair, protruding forehead, saddle nose and small chin) tend to be more severe.[48] Contrary to IGHD, reduced psychological performance has also been reported. In contrast to IGHD, growth failure is already present at birth, indicating that IGF-1 is required for intrauterine growth.[49] In addition, serum GH levels are high rather than low.

There does not seem to be a direct relationship between the type of mutation and the degree of growth retardation.[50] The adult height of untreated individuals ranges between −2.2 and −12 SDs. As stature is a multigenic trait, and also influenced by environmental factors, it is not surprising that, even within a population of GHIS patients that is genetically homogeneous, adult height ranges widely.[51] The size of the anterior pituitary is normal.[52] Interestingly, sleep disorders are a common feature of adult patients with LS, although this has never been reported in IGHD.[48] As one would expect, GHIS patients do not respond to GH therapy. Even when treated with recombinant IGF-1, the increase in growth velocity is not as marked as that observed with GH treatment of IGHD patients, possibly due to the lack of direct effect of GH, and normal adult stature is never reached. Although it has been hypothesized that heterozygous *GHR* mutations may cause growth failure in a subgroup of patients with idiopathic short stature,[53] this possibility seems unlikely in view of the fact that heterozygous relatives of patients with GHIS have normal stature.[54]

GHR POLYMORPHISM

In 2004 it was reported that a rather common *GHR* polymorphism lacking exon 3 (d3-GHR) is associated with increased responsiveness to GH therapy in GHD, SGA and Turner's syndrome.[55–57] However, studies looking at the d3-GHR in subjects with IGHD preliminarily indicate that, although there is an increase in linear growth velocity compared to patients with two WT alleles in the first 2 years of treatment, final height is not affected.[58,59] In combination, these results seem to indicate that d3-GHR polymorphism may contribute to the variability of response to GH therapy, but there is no evidence that this influences adult stature in non-treated subjects.

MUTATION IN THE GHR SIGNAL TRANSDUCTION APPARATUS

In 2003 a girl was reported with GHIS and compromised immunological function, with a homozygous missense mutation in the STAT5b gene that alters the intracellular mechanisms involved in transmitting the message from the GHR to the nucleus.[60] Since then, other mutations in this gene have been reported, causing an association of growth failure and immunodeficiency, likely due to malfunction of T cells that also require an intact Stat5b signaling.[61]

Mutations in the IGF-1 Gene and its Receptor

In 1996 Woods and colleagues described a boy with intrauterine growth retardation, sensorineural deafness

BOX 32.1

HOW DO WE KNOW ABOUT d3GH RECEPTOR POLYMORPHISM AND RESPONSIVENESS TO GH THERAPY?

The growth response to GH in children is known to vary depending upon the age and size of the patient, as well as to the dose given. In 2004, a seminal study by Dos Santos and colleagues demonstrated that a common polymorphism in the growth hormone receptor (GHR) may also play a critical role in determining how well a child will grow in response to GH treatment.[1]

The human GHR gene consists of two isoforms, which are based upon whether the third exon, encoding for the extracellular domain, is retained or excluded during splicing. A common polymorphism causes the exon to be "skipped" during transcription, and therefore results in a gene product in which exon 3 is spliced (d3-GHR). Previous *in vitro* studies had demonstrated comparable binding kinetics between the full-length and spliced isoforms,[2,3] but since the region is highly conserved among mammalian species, Dos Santos *et al.* hypothesized that although the polymorphism did not affect binding characteristicis, it may affect GH signaling.

The group sought to test this hypothesis by studying the affect of GH treatment in a population of short children both with and without the polymorphism. Children born small for gestational age (SGA) and with idiopathic short stature (ISS) were genotyped for the polymorphism and then followed for the first 2 years of GH therapy. Among a group of 76 short children, response to growth hormone was greater in children carrying at least one allele encoding the d3-GHR isoform ($P < 10{-}5$ and $P < 0.001$ for the first and second years, respectively). Dos Santos *et al.* next performed an *in vitro* study, co-transfecting cells with vectors expressing full-length GHR, d3-GHR, or both. When cells were exposed to various growth hormone concentrations, d3-GHR induced a higher transcriptional activity than full-length GHR.

A number of studies have followed this seminal work, looking at the effect of the polymorphism in various populations with GHD.[4–7] Not all of these reports have confirmed that the presence of the d3-GHR predicts responsiveness to GH in GH-deficient children, particularly in terms of final stature. Nonetheless, in certain populations, the findings of Dos Santos *et al.* have encouraged physicians to rethink their dosing regimen for patients with GHD, allowing them to tailor their dose based on an important genetic polymorphism that may just be the first of many to be discovered.

References

1. Dos Santos C, Essioux L, Teinturier C, Tauber M, Goffin V, Bougneres P. A common polymorphism of the growth hormone receptor is associated with increased responsiveness to growth hormone. *Nat Genet.* 2004;36:720–724.
2. Hintz RL, Attie KM, Baptista J, Roche A. Effect of growth hormone treatment on adult height of children with idiopathic short stature. Genentech Collaborative Group. *N Engl J Med.* 1999;340:502–507.
3. Bryant J, Cave C, Milne R. Recombinant growth hormone for idiopathic short stature in children and adolescents. *Cochrane Database Syst Rev.* 2003. CD004440.
4. Binder G, Baur F, Schweizer R, Ranke MB. The d3-growth hormone (GH) receptor polymorphism is associated with increased responsiveness to GH in Turner syndrome and short small-for-gestational-age children. *J Clin Endocrinol Metab.* 2006;91:659–664.
5. Ko JM, Park JY, Yoo HW. Common exon 3 polymorphism of the GH receptor (*GHR*) gene and effect of GH therapy on growth in Korean children with idiopathic short stature (ISS). *Clinl Endocrinol.* 2009;70:82–87.
6. Pilotta A, Mella P, Filisetti M, Felappi B, Prandi E, Parrinello G, Notarangelo LD, Buzi F. Common polymorphisms of the growth hormone (GH) receptor do not correlate with the growth response to exogenous recombinant humanGHin GH-deficient children. *J Clin Endocrinol Metab.* 2006;91:1178–1180.
7. Räz B, Janner M, Petkovic V, Lochmatter D, Eblé A, Dattani MT, Hindmarsh PC, Flück CE, Mullis PE. Impact influence of growth hormone (GH) receptor deletion of exon 3 and full-length isoforms on GH response and final height in patients with severe GH deficiency. *J Clin Endocrinol Metab.* 2008;93:974–980.

and mental retardation, who had a homozygous deletion in the IGF-1 gene.[62,63] Interestingly, his parents had short stature and borderline low serum IGF-1 concentrations, suggesting that heterozygosity could be prevalent in the population of children with idiopathic short stature. However, follow up studies have failed to show a precise link between this gene and short stature or IUG.[64]

In 2003 Abuzzahab *et al.* reported two children with mutations in the IGF1-receptor (IGFR).[65] One was compound heterozygous for two missense mutations in exon 2, resulting in reduced ligand binding; the second was heterozygous for a nonsense mutation in exon 2, resulting in reduced expression of IGF1R. Both children had severe intrauterine growth retardation and postnatal growth failure. Since that initial report, a heterozygous mutation in the cleavage site of the proreceptor, a missense mutation in the intracellular kinase domain, and 19-nucleotide duplication resulting in haploinsufficiency of IGF1R protein have been reported.[66–68] The prevalence of these mutations is presently unknown.

Mutation in IGF-1 Transport Protein

Approximately 80–85% of serum IGF-1 circulates as part of a 150-kDa ternary complex, which includes one molecule each of IGF-1 IGF binding protein (IGFBP)-3 or -5, and a third protein named "acid labile subunit" (ALS). In 2004, Domené *et al.* reported the case of a boy with slow pubertal progress and minimal slowing of linear growth. Hormonal studies showed normal stimulated GH levels and increased nocturnal GH secretion, with marked reductions in both IGF-1 and IGF-BP3, and undetectable ALS. The boy was homozygous for a deletion in the ALS gene.[69] A second similar case was later reported, with moderate growth failure but normal onset and progression of puberty, showing a modest role for the ternary complex in the regulation of stature.[70]

Combined Pituitary Hormone Deficiency (CPHD) Syndromes

GHD may be the first or most obvious manifestation of CPHD). Several pituitary-specific transcription factors play a role in the determination of the pituitary cell lineages, and patients with hypopituitarism (including GHD) have been found to have mutations in these genes.

Rpx, also known as Hesx1, is one of the earliest known specific markers for the pituitary primordium, suggesting that it has a role in early determination or differentiation of the pituitary.[71] Mice lacking Rpx have abnormalities in the corpus callosum, anterior and hippocampal commisures, and septum pellucidum similar to the defects seen in septo-optic dysplasia (SOD) in humans.[72] To date, five recessive and five dominant mutations have been described in the gene encoding the transcriptional repressor *HESX1* in patients with SOD.[73] Screening of a large cohort of patients with a wide spectrum of congenital hypopituitarism identified three missense mutations of *HESX1* resulting in GHD, with or without other pituitary hormone deficiencies.[74]

Ptx2 is a paired-like homeodomain transcription factor closely related to the mammalian Otx genes expressed in the rostral brain during development.[75] It is present in the fetal pituitary and is expressed in the adult pituitary gland in the thyrotropes, gonadotropes, somatotropes and lactotropes, but not in the corticotropes.[76] RIEG is the human homolog of Ptx2, and mutations of this gene result in Rieger syndrome, an autosomal dominant condition in which a subset of affected individuals demonstrates GH insufficiency.[77]

Lhx3 is a transcription factor expressed in the anterior and intermediate lobes of the pituitary gland during development. It is expressed in somatotropes, as well as lactotropes, gonadotropes and thyrotropes.[78] Patients with complete deficits of GH, PRL, TSH and gonadotropins, as well as a rigid cervical spine leading to limited head rotation, have been found to have mutations in the Lhx3 gene.

Prop-1 is a paired-like homeodomain transcription factor whose expression is limited to the anterior pituitary during development.[79] Several human mutations of Prop-1 resulting in autosomal recessive deficit of GH, PRL and TSH have been described. Some subjects do not produce LH and FSH at a sufficient level to enter puberty spontaneously[80] while others do, suggesting that Prop-1 is not needed for gonadotrope determination but may have a role in gonadotrope differentiation.[79] Mutations in this gene are the most frequent cause of inherited CPHD.

POUF1 (formerly Pit-1) is a transcription factor whose expression is restricted to the anterior pituitary lobe,[81] and it is essential for the development of somatotropes, lactotropes and thyrotropes.[82] A number of patients with CPHD of GH, PRL and TSH have been found to have POUF1 gene mutations. These mutations are transmitted as either an autosomal recessive or an autosomal dominant (via dominant negative effect) trait. The most common mutation is a dominant one (arginine to tryptophan mutation at codon 271) that has been described in several unrelated patients of different ethnic backgrounds.

Acquired Defects

Among acquired defects resulting in GHD in children, intracranial tumors remain the most prevalent. Most acquired defects run a high risk of involving multiple pituitary hormones, warranting a thorough and complete work-up of each hypothalamic–pituitary axis at the time of diagnosis.

Perinatal Pathology

GHD associated with congenital rubella, toxoplasmosis and cytomegalovirus infections have been described.[83] Perinatal trauma, especially associated with forceps delivery, vaginal bleeding and breech presentations, has been reported to cause CPHD or IGHD.[84]

CNS Tumors

Craniopharyngioma is the most common tumor in the hypothalamic–pituitary region to cause GHD in childhood.[85] The tumor usually arises from remnants of Rathke's pouch, an invagination of the epithelium within the third pharyngeal pouch from which the anterior pituitary evolves. Although histologically a benign tumor, it may be locally invasive, involving adjacent structures – especially the optic tracts and base of the third ventricle. It usually has a solid and cystic component that may contain a cholesterol-rich fluid. The clinical presentation is usually characterized with signs and symptoms of increased intracranial pressure and visual disturbances due to the proximity of the optic chiasm. GH deficiency (72%) is the most common endocrine abnormality at clinical presentation, whereas ACTH, TSH and ADH deficiencies were found in approximately 25% of cases.[85] The management of craniopharyngioma is complex and still controversial, and morbidity remains high. It is important to mention that many children who have been surgically treated for craniopharyngioma may continue to grow with a normal growth velocity despite having clearly documented low GH and low IGF-1. Hyperinsulinism associated with hyperphagia and the marked weight gain observed in these children may explain their normal growth velocity.[86]

Germinomas and optic nerve gliomas usually involve the hypothalamic–pituitary axis. In addition to growth failure, germinomas may present with anorexia and weight loss in older boys, and with diabetes insipidus (DI) alone. DI may precede, sometimes by many years, the detection of the tumor itself by imaging studies.[87] Therefore, idiopathic DI must be investigated with repeated CNS imaging.[88] Elevation of serum and possibly cerebrospinal fluid beta-human chorionic gonadotropin (hCG) levels can be used as a tumor marker, but the sensitivity is low. The high immunogenity of germinomas can cause a marked inflammatory reponse that – by pathology – can be confised with lymphocytic hypophysitis.[89] Optic nerve glioma, which occurs more commonly in patients with neurofibromatosis, may also be associated with pituitary deficiency.[90] These tumors can be treated with targeted radiotherapy, which may also cause pituitary deficiency.[91]

The infiltritative lesion of histiocytosis typically involves the hypothalamus and causes diabetes insipidus. Tumors are usually seen in the pituitary stalk, and these lesions may resolve with chemotherapy. In approximately 30% of cases this will be associated with anterior pituitary deficiencies, including GHD.[92]

Intracranial Irradiation

All children who have received CNS irradiation, whether for prophylaxis for leukemia, for tumors distant from or adjacent to the hypothalamic–pituitary region, or during total body irradiation, are at some risk for the development of GHD.[93–95] The sensitivity of the hypothalamo-pituitary axis to irradiation is dependent on the total dose, tissue localization, and the age of the patient. Younger children given single-exposure irradiation tend to have a higher incidence of GHD compared to older ones with fractionated irradiation.[96] These patients require close endocrine monitoring and long-term follow-up. Within 5 years of receiving doses greater than 30 Gy to the hypothalamic region, more than 85% of children will have documented GHD.[93–95]

Traumatic Brain Injury

Recent studies indicate that hypothalamic–pituitary dysfunction is more common in the pediatric population following traumatic brain injury (TBI) than previously thought. GHD is the most prevalent deficit observed post-TBI, with prevalence ranging from 9% to 28%.[97–101] The lateral location of the somatotropes and their predisposition to vascular insufficiency, stalk injury or anoxia may in part explain the relatively high frequency of post-traumatic GHD, but there is no clear pattern linking the mechanism of TBI to the cause of hypopituitarism.[102] These have also been documented after relatively minor TBI without loss of consciousness.[103–105]

GIGANTISM

Gigantism refers to a condition characterized by excessive longitudinal growth beyond the patient's genetic target. By definition, this originates during childhood while epiphyseal growth plates are still open. Although often used to specifically denote GH excess, the term "gigantism" is also applied to a number of non-hormonally mediated overgrowth conditions in children (not covered in this chapter). All forms of gigantism are extremely rare, and have in common a complex pathophysiologic origin and extraordinary clinical manifestations. Similar to GHD, when evaluating a child with tall stature, mid-parental height and growth velocity must be considered to avoid studying children with familial tall stature.

The association between gigantism and GH excess was recognized as early as the 1800s, when it was noted that pituitary giants invariably developed features of acromegaly, which refers to progressive enlargement of the head, face, hands and feet.[106] While acromegaly is uncommon, occurring at an estimated annual rate of 2.8–4 cases per million,[107] GH excess during childhood and adolescence is extremely rare, with the total number of reported cases thus far only in the hundreds.

GH-secreting adenomas may occur sporadically, or may exist within a constellation of abnormalities in the setting of several familial syndromes. In general, the prevalence of germline mutations is higher in patients with earlier tumor onset. We will concentrate this section on the rare syndromic or familial causes of GH excess that result in gigantism.

Multiple Endocrine Neoplasia, Type 1

Multiple Endocrine Neoplasia Type 1 (MEN-1) is a genetic syndrome characterized by primary endocrine hyperactivity involving the pituitary, parathyroid and pancreas. It is an autosomal dominant syndrome with an estimated prevalence of between 1/10,000 and 1/25,000. The disorder is caused by inactivating mutations in the *menin* gene, a tumor suppressor gene localized to chromosome 11q13. The local loss of the normal allele ("second hit") is necessary for the development of tumors, and explains the incomplete penetrance of this syndrome. The lack of menin affects transcriptional regulation, genome stability, cell division and proliferation.[108–111] While hyperparathyroidism is the most prevalent feature of this syndrome, approximately 30–40% of MEN-1 patients develop pituitary adenomas.[112]

Pituitary adenomas in patients with MEN-1 can secrete different pituitary hormones, or may be non-functioning. Prolactin is the most common (~60%) secreted hormone, followed by GH (~20%) and ACTH.[108,113] Although no genotype–phenotype correlation has been noted in *menin* mutation carriers, in familial MEN-1 the frequency of pituitary disease is significantly higher than in sporadic MEN-1 cases.[108] In addition, in MEN-1 patients with pituitary adenoma and acromegaly, an increased female-to-male ratio has been reported for both familial and sporadic cases.[112]

Although MEN-1 can present early enough to produce GH-secreting tumors capable of causing gigantism, because of the low risk of tumorigenesis before the second decade, this is a rare occurrence and most patients who develop GH-secreting adenomas present with acromegaly rather than gigantism.[114] Interestingly, in some MEN-1 kindreds somatotropinomas are completely absent.[115]

Carney Complex

The fact that approximately 40% of GH-secreting adenomas carry somatic activating mutations in the Gs α gene (*GNAS*) shows that mutations leading to over-activation of the cyclic AMP-dependent pathways may cause the development of GH-secreting adenomas.[116] One of them is the Carney Complex (CNC), a rare autosomal dominant disorder that includes a complex of primary pigmented nodular adrenocortical disease, myxomas, lentigines, endocrine over-activity, and a variety of other tumors such as schwannomas and pituitary adenomas. In approximately 60% of the patients who meet diagnostic criteria, an inactivating mutation in the protein kinase A regulatory subunit type 1A (PRKAR1A) gene has been identified.[117] Similarly to MEN-1, acromegaly is more likely presentation than gigantism.[118] Most CNC patients will have some abnormality of GH secretion due to the underlying pituitary hyperplasia.[119,120] For CNC patients with normal IGF-1 levels and normal pituitary imaging, evaluations should be performed annually to assess for changes that may require treatment.

AIP Mutation

Familial isolated pituitary tumors (FIPA) have also been described in the absence of the mutations resulting in CNC or MEN-1. In some of these families, heterozygous germline inactivating mutations have been discovered in the gene that encodes the aryl hydrocarbon receptor-interacting protein (AIP) gene (*AIP*) located on chromosome 11q13.3. This gene had no previously known function in pituitary physiology, and it was identified by combining chip-based technologies with genealogy data. Similar to the menin gene, this is likely a tumor suppression gene that requires a second hit for tumor development. Patients with this mutation present with combinations of somatotropinomas (in the majority of cases), mixed GH- and PRL-secreting adenomas and prolactinomas. Heterozygous *AIP* mutations have subsequently been discovered in approximately 10% of a Scandanavian cohort with sporadic somatotrope adenomas. In fact, most patients lack a strong family history of pituitary tumor, lending to the notion that AIP could be a low-penetrance susceptibility gene to GH-secreting pituitary adenomas.[121] Typically, tumors in patients with *AIP* mutations are larger and diagnosed at a younger age than in patients with sporadic tumors; therefore, this syndrome must be suspected in all cases of gigantism.[122] Furthermore, these tumors tend to secrete higher levels of GH and have more frequent prolactin co-secretion than matched acromegalic patients without *AIP* mutations.[123] More than half are invasive at diagnosis, often posing difficult treatment challenges. Somatotropinomas due to *AIP* mutations require repeat surgery significantly more often than controls, whereas hormonal and tumor responses to somatostatin analogs are significantly lower than controls. *AIP* mutation-related prolactinomas also appear to have aggressive and difficult-to-treat clinical characteristics.[123]

McCune-Albright Syndrome

McCune-Albright Syndrome (MAS) is characterized by endocrine abnormalities (precocious puberty, thyrotoxicosis, pituitary gigantism and Cushing's syndrome), café-au-lait pigmented lesions and polyostotic fibrous dysplasia and, rarely, by other tumors. Its inheritance is sporadic, and it is most frequently diagnosed in females.[124]

MAS is caused by congenital (but not inherited) heterozygous somatic mutation of the *GNAS*. The clinical phenotype is more marked the earlier the appearance of the mutation during embryogenesis. The clinical findings result from the autonomous hyperactivity of tissues that are regulated by intracellular accumulation of cyclic AMP.

GH excess in MAS can be associated with gigantism and significant morbidity due to exacerbation of polyostotic fibrous dysplasia and sarcomatous transformation of bone tumors.[125] When a discrete adenoma is visible, surgical tumor excision remains the mainstay of treatment of MAS-related somatotropinomas.[126,127] In other cases (or when surgery is contraindicated), treatment with somatostatin receptor ligand is necessary.

CONCLUSION

States of both growth hormone deficiency or resistance and excess provide visible testaments to the role of this hormone in normal physiology. Such disorders can reflect lesions in the hypothalamus, in the pituitary, or in target cells. A deficiency state can result not only from underproduction of the hormone, but also from the target cell's unresponsiveness to the hormone. Clinically, growth hormone deficiency or defects in its binding to receptor are seen as growth retardation or dwarfism. Conversely, excess secretion of growth hormone prior to fusion of the epiphysial growth plates results in gigantism. Recent studies have expanded the understanding of some of the genetic mechanisms underlying these processes, which in turn should provide new guidelines for screening and treatment of these rare but important disease processes.

References

1. GH Research Society. Consensus guidelines for the diagnosis and treatment of growth hormone (GH) deficiency in childhood and adolescence: summary statement of the GH Research Society. *J Clin Endocrinol Metab.* 2000;85:3990–3993.
2. Jung H, Rosilio M, Blum WF, Drop SL. Growth hormone treatment for short stature in children born small for gestational age. *Adv Ther.* 2008;25:951–978.
3. Kojima M, Hosoda H, Date Y, Kakazato M, Matsuo H, Kangawa K. Ghrelin is a growth hormone-releasing acylated peptide from stomach. *Nature.* 1999;402:656–660.
4. Sum Y, Ahmed S, Smith RG. Generation and characterization of ghrelin knock out mice. In: *Proceedings of the 84th Annual Meeting of The Endocrine Society,* San Francisco, CA June 19–22, 2002 Abs P1–174.
5. Lindsay R, Feldkamp M, Harris D, Robertson J, Rallison M. Utah growth study: Growth standards and the prevalence of growth hormone deficiency. *J Pediatr.* 1994;125:29–35.
6. Cacciari E, Zucchini S, Carla G, Pirazzoli P, Cicognani A, Mandini M, Busacca M, Trevisan C. Endocrine function and morphological findings in patients with disorders of the hypothalamo-pituitary area: A study with magnetic resonance. *Arch Dis Child.* 1990;65:1199–1202.
7. Cogan JD, Phillips III JA. Growth disorders caused by genetic defects in the growth hormone pathway GH deficiency. In: Barness LA, Morron III G, Rudolph AM, De Vivo DC, Kabace MM, Tunnessen Jr WW, editors. *Advances in Pediatrics.* St Louis: Mosby; 1998:337–361.
8. Perez Jurado LA, Phillips III JA, Francke U. Exclusion of growth hormone-releasing hormone gene mutations in familial isolated growth hormone deficiency by linkage and single strand conformation analysis. *J Clin Endocrinol Metab.* 1994;78:622–628.
9. Rappold GA, Fukami M, Niesler B, Schiller S, Zumkeller W, Bettendorf M, Heinrich U, Vlachopapadoupoulou E, Reinehr T, Onigata K, Ogata T. Deletions of the homeobox gene SHOX (short stature homeobox) are an important cause of growth failure in children with short stature. *J Clin Endocrinol Metab.* 2002;87:1402–1406.
10. DeAlmeida VI, Mayo KE. Identification of binding domains of the growth hormone-releasing hormone receptor by analysis of mutant and chimeric receptor proteins. *Mol Endocrinol.* 1998;12:750–765.
11. Lin SC, Lin CR, Gukovsky I, Lusis AJ, Sawchenko PE, Rosenfeld MG. Molecular basis of the little mouse phenotype and implications for cell-type specific growth. *Nature.* 1993;364:208–213.
12. Salvatori R, Hayashida CY, Aguilar-Oliveira MH, Phillips III JA, Souza AH, Gondo RG, Toledo SP, Conceicão MM, Prince M, Maheshwari HG, Baumann G, Levine MA. Familial dwarfism due to a novel mutation in the growth hormone-releasing hormone receptor. *J Clin Endocrinol Metab.* 1999;84:917–923.
13. Salvatori R, Fan X, Phillips III JA, Espigares-Martin R, Martin de Lara I, Freeman K, Plotnick L, Al-Ashwal A, Levine MA. Three new mutations in the gene for the growth hormone (GH)-releasing hormone receptor in familial isolated GH deficiency type IB. *J Clin Endocrinol Metab.* 2001;86:273–279.
14. Salvatori R, Fan X, Phillips III JA, Prince M, Levine MA. Isolated growth hormone (GH) deficiency due to compound heterozygosity for 2 new mutations in the GH releasing hormone receptor gene. *Clin Endocrinol (Oxf).* 2001;54:681–688.
15. Salvatori R, Fan X, Mullis PE, Haile A, Levine MA. Decreased expression of the growth hormone-releasing hormone receptor gene due to a mutation in a Pit-1 binding site. *Mol Endocrinol.* 2002;16:450–458.
16. Salvatori R, Fan X, Veldhuis J, Couch R. Serum GH response to pharmacological stimuli and physical exercise in two siblings with two new inactivating mutations in the GHReleasing hormone receptor gene. *Eur J Endocrinol.* 2002;147:591–596.
17. Carakushansky M, Whatmore AJ, Clayton PE, Shalet SN, Gleeson HK, Price DA, Levine MA, Salvatori R. A new missense mutation in the growth hormone releasing hormone receptor gene in familial isolated GH deficiency. *Eur J Endocrinol.* 2003;148:25–30.
18. Roelfsema F, Biermasz NR, Veldman RG, Veldhuis JD, Frolich M, Stokvis-Brantsma WH, Wit JM. Growth hormone (GH) secretion in patients with an inactivating defect of the GH Releasing hormone (GHRH) receptor is pulsatile: Evidence for a role for non-GHRH inputs into the generation of GH pulses. *J Clin Endocrinol Metab.* 2001;86:2459–2464.
19. Horikawa R. Isolated GH deficiency due to inactivating mutation of GHRH receptor. *Nippon Rinsho.* 2002;60:297–305.
20. Godi M, Mellone S, Petri A, Arrigo T, Bardelli C, Corrado L, Bellone S, Prodam F, Momigliano-Richiardi P, Bona G, Giordano M. A recurrent signal peptide mutation in the growth

hormone releasing hormone receptor with defective translocation to the cell surface and isolated growth hormone deficiency. *J Clin Endocrinol Metab.* 2009;94:3939–3947.
21. Barreto-Filho JA, Alcântara MRS, Salvatori R, Azevedo M, Barreto M, Sousa AC, Bastos VPC, Souza AHO, Pereira RMC, Clayton PE, Gill MS, Aguiar-Oliveira MH. Familial isolated growth hormone deficiency is associated with increased systolic blood pressure, dyslipidemia and central obesity. *J Clin Endocrinol Metab.* 2002;87:2018–2023.
22. Netchine I, Talon P, Dastot F, Vitaux F, Goosens M, Amselem S. Extensive phenotypic analysis of a family with growth hormone (GH) deficiency caused by a mutation in the GHReleasing hormone receptor gene. *J Clin Endocrinol Metab.* 1998; 83:432–436.
23. Oliveira HA, Salvatori R, Krauss MPO, Oliveira CRP, Silva PRC, Aguiar-Oliveira MH. Magnetic resonance imaging study of pituitary morphology in subjects homozygous and heterozygous for a null mutation of the growth hormone releasing hormone receptor gene. *Eur J Endocrinol.* 2003;148:427–432.
24. Murray RA, Maheshwari HG, Russell EJ, Baumann G. Pituitary hypoplasia in patients with a mutation in the growth hormone-releasing hormone receptor gene. *Am J Neuroradiol.* 2000; 21:685–689.
25. Salvatori R, Aguiar-Oliveira MH, Monte LVB, Hedges L, Santos NL, Pereira RMC, Phillips III JA. Detection of a recurring mutation in the human GH releasing hormone receptor gene. *Clin Endocrinol (Oxf).* 2002;57:77–80.
26. Zhu T, Goh EL, Graichen R, Ling L, Lobie PE. Signal transduction via the growth hormone receptor. *Cell Signal.* 2001;13:599–616.
27. Wagner JK, Eble A, Hindmarsh PC, Mullis PE. Prevalence of human GH-1 alterations in patients with isolated growth hormone deficiency. *Pediatr Res.* 1998;43:105–110.
28. Phillips III JA, Hjelle BL, Seeburg PH, Zachmann M. Molecular basis for familial isolated growth hormone deficiency. *Proc Natl Acad Sci USA.* 1981;78:6372–6375.
29. Vnencak-Jones CL, Phillips III JA. Hot spots for growth hormone gene deletions in homologous regions outside of Alu repeats. *Science.* 1990;250:1745–1748.
30. Thomas PQ, Dattani MT, Brickman JM, McNay D, Warne G, Zacharin M, Cameron F, Hurst J, Woods K, Dunger D, Stanhope R, Forrest S, Robinson IC, Beddington RS. Heterozygous HESX1 mutations associated with isolated congenital pituitary hypoplasia and septo-optic dysplasia. *Hum Mol Genet.* 2001;10:39–45.
31. Lee MS, Wajnrajch MP, Kim SS, Plotnick LP, Wang J, Gertner JM, Leibel RL, Dannies PS. Autosomal dominant growth hormone (GH) deficiency type II: The Del32-71-GH deletion mutant suppresses secretion of wild-type GH. *Endocrinology.* 2000;141:883–890.
32. McGuinness L, Magoulas C, Sesay AK, Mathers K, Carmignac D, Manneville JB, Christian H, Phillips III JA, Robinson IC. Autosomal dominant growth hormone deficiency disrupts secretory vesicles *in vitro* and *in vivo* in transgenic mice. *Endocrinology.* 2003;144:720–731.
33. Moseley CT, Mullis PE, Prince MA, Phillips III JA. An exon splice enhancer mutation causes autosomal dominantGHdeficiency. *J Clin Endocrinol Metab.* 2002;87:847–852.
34. Fintini D, Salvatori R, Salemi S, Otten B, Ubertini G, Cambiaso P, Mullis PE. Autosomal dominant isolated growth hormone deficiency (IGHD type II) with normal *GH1* gene. *Horm Res.* 2006;65:76–82.
35. Zucchini S, Ambrosetto P, Baroncini C, Cacciari E. Normal pituitary size in two patients with growth hormone gene deletion. *J Ped Endocrinol Metab.* 1996;9:545–548.
36. Binder G, Nagel BH, Ranke MB, Mullis PE. Isolated GH deficiency (IGHD) type II: Imaging of the pituitary gland by magnetic resonance reveals characteristic differences in comparison with severe IGHD of unknown origin. *Eur J Endocrinol.* 2002;147:755–760.
37. Mullis PE, Robinson CAF, Salemi S, Eblé A, Besson A, Vuissoz J-M, Deladoey J, Simon D, Czernichow P, Binder G. Isolated autosomal dominant growth hormone deficiency: an evolving pituitary deficit? A multicenter follow-up study. *J Clin Endocrinol Metab.* 2005;90:2089–2096.
38. Pantel J, Legendre M, Cabrol S, Hilal L, Hajaji Y, Morisset S, Nivot S, Vie-Luton MP, Grouselle D, de Kerdanet M, Kadiri A, Epelbaum J, Le Bouc Y, Amselem S. Loss of constitutive activity of the growth hormone secretagogue receptor in familial short stature. *J Clin Invest.* 2006;116:760–768.
39. Kowarski AA, Schneider J, Ben-Galim E, Weldon VV, Daughaday WH. Growth failure with normal serum RIA-GH and low somatomedin activity: Somatomedin restoration and growth acceleration after exogenous GH. *J Clin Endocrinol Metab.* 1978;47:461–464.
40. Takahashi Y, Shirono H, Arisaka O, Takahashi K, Yagi T, Koga J, Kaji H, Okimura Y, Abe H, Tanaka T, Chihara K. Biologically inactive growth hormone caused by an amino acid substitution. *J Clin Invest.* 1997;100:1159–1165.
41. Takahashi Y, Kaji H, Okimura Y, Goji K, Abe H, Chihara K. Brief report: Short stature caused by a mutant growth hormone. *N Engl J Med.* 1996;334:432–436.
42. Leung DW, Spencer SA, Cachianes G, Hammonds RG, Collins C, Henzel WJ, Barnard R, Waters MJ, Wood WI. Growth hormone receptor and serum binding protein: purification, cloning and expression. *Nature.* 1987;330:537–543.
43. Godowski PJ, Leung DW, Meacham LR, Galgani JP, Hellmiss R, Keret R, Rotwein PS, Parks JS, Laron Z, Wood WI. Characterization of the human growth hormone receptor gene and demonstration of a partial gene deletion in two patients with Laron-type dwarfism. *Proc Natl Acad Sci USA.* 1989;86:8083–8087.
44. Laron Z, Pertzelan A, Mannheimer S. Genetic pituitary dwarfism with high serum concentration of growth hormone – A new inborn error of metabolism? *Isr J Med Sci.* 1966;2:152–155.
45. Wojcik J, Berg MA, Esposito N, Geffner ME, Sakati N, Reiter EO, Dower S, Francke U, Postel-Vinay MC, Finidori J. Four contiguous amino acid substitutions, identified in patients with Laron syndrome, differently affect the binding affinity and intracellular trafficking of the growth hormone receptor. *J Clin Endocrinol Metab.* 1998;83:4481–4489.
46. Metherell LA, Akker SA, Munroe PB, Rose SJ, Caulfield M, Savage MO, Chew SL, Clark AJ. Pseudoexon activation is a novel mechanism for disease resulting in atypical growth-hormone insensitivity. *Am J Hum Genet.* 2001;69:641–646.
47. Ayling RM, Ross R, Towner P, Von Laue S, Finidori J, Moutoussamy S, Buchanan CR, Clayton PE, Norman MR. A dominant negative mutation of the growth hormone receptor causes familial short stature. *Nat Genet.* 1997;16:13–14.
48. Laron Z. Laron Syndrome (primary growth hormone resistance or insensitivity): The personal experience 1958–2003. *J Clin Endocrinol Metab.* 2004;89:1031–1044.
49. Laron Z. The essential role of IGF-I: Lessons from the long-term study and treatment of children and adults with Laron syndrome. *J Clin Endocrinol Metab.* 1999;84:4397–4404.
50. Woods KA, Dastot F, Preece MA, Clark AJ, Postel-Vinay MC, Chatelain PG, Ranke MB, Rosenfeld RG, Amselem S, Savage MO. Phenotype:genotype relationships in growth hormone insensitivity syndrome. *J Clin Endocrinol Metab.* 1997;82:3529–3535.

51. Rosenfeld RG, Rosenbloom AL, Guevara-Aguirre J. Growth hormone (GH) insensitivity to primary GH receptor deficiency. *Endocrine Rev.* 1995;15:369–390.
52. Kornreich L, Horev G, Schwarz M, Karmazyn B, Laron Z. Pituitary size in patients with Laron syndrome (primary GH insensitivity). *Eur J Endocrinol.* 2003;148:339–341.
53. Salerno M, Balestrieri B, Matrecano E, Officioso A, Rosenfeld RG, Di Maio S, Fimiani G, Ursini MV, Pignata C. Abnormal GHreceptor signaling in children with idiopathic short stature. *J Clin Endocrinol Metab.* 2001;86:3882–3888.
54. Rosenbloom AL, Guevara-Aguirre J, Berg MA, Francke U. Stature in Ecuadorians heterozygous for growth hormone receptor gene E180 splice mutation does not differ from that of homozygous normal relatives. *J Clin Endocrinol Metab.* 1998;83:2373–2375.
55. Dos Santos C, Essioux L, Teinturier C, Tauber M, Goffin V, Bougneres P. A common polymorphism of the growth hormone receptor is associated with increased responsiveness to growth hormone. *Nat Genet.* 2004;36:720–724.
56. Binder G, Baur F, Schweizer R, Ranke MB. The d3-growth hormone (GH) receptor polymorphism is associated with increased responsiveness to GH in Turner syndrome and short small-for-gestational-age children. *J Clin Endocrinol Metab.* 2006;91:659–664.
57. Ko JM, Park JY, Yoo HW. Common exon 3 polymorphism of the GH receptor (*GHR*) gene and effect of GH therapy on growth in Korean children with idiopathic short stature (ISS). *Clin Endocrinol.* 2009;70:82–87.
58. Pilotta A, Mella P, Filisetti M, Felappi B, Prandi E, Parrinello G, Notarangelo LD, Buzi F. Common polymorphisms of the growth hormone (GH) receptor do not correlate with the growth response to exogenous recombinant humanGHin GH-deficient children. *J Clin Endocrinol Metab.* 2006;91:1178–1180.
59. Raz B, Janner M, Petkoric V, Lochmatter D, Eble A, Dattani MT, Hindmarch PC, Fluck CE, Mullis PE. Raz impact influence of growth hormone (GH) receptor deletion of exon 3 and full-length isoforms on GH response and final height in patients with severe GH deficiency. *J Clin Endocrinol Metab.* 2008;93:974–980.
60. Kofoed EM, Hwa V, Little B, Woods KA, Buckway CK, Tsubaki J, Pratt KL, Bezronik L, Jasper H, Tepper A, Heinrich JJ, Rosenfeld RG. Growth hormone insensitivity associated with *STAT5b* mutation. *N Engl J Med.* 2003;349:1139–1147.
61. Bernasconi A, Marino R, Ribas A, Rossi J, Ciaccio M, Oleastro M, Ornani A, Paz R, Rivarola MA, Zelazko M, Belgorosky A. Characterization of immunodeficiency in a patient with growth hormone insensitivity secondary to a novel STAT5b gene mutation. *Pediatrics.* 2006;118:e1584–e1592.
62. Woods KA, Camacho-Hubner C, Savage MO, Clark AJ. Intrauterine growth retardation and postnatal growth failure associated with deletion of the insulin-like growth factor I gene. *N Engl J Med.* 1996;335:1363–1367.
63. Camacho-Hubner C, Woods KA, Miraki-Moud F, Hindmarsh PC, Clark AJ, Hansson Y, Johnston A, Baxter RC, Savage MO. Effects *GH and IGF-1 Genetic Defects* **23** of recombinant human insulin-like growth factor I (IGF-I) therapy on the growth hormone–IGF system of a patient with a partial IGF-I gene deletion. *J Clin Endocrinol Metab.* 1999;84:1611–1616.
64. Arends N, Johnston L, Hokken-Koelega A, van Duijn C, de Ridder M, Savage M, Clark A. Polymorphism in the IGF-I gene: clinical relevance for short children born small for gestational age (SGA). *J Clin Endocrinol Metab.* 2002;87:2720.
65. Abuzzahab MJ, Schneider A, Goddard A, Grigorescu F, Lautier C, Keller E, Kiess W, Klammt J, Kratzsch J, Osgood D, Pfäffle R, Raile K, Seidel B, Smith RJ, Chernausek SD. Intrauterine Growth Retardation (IUGR) Study Group. IGF-I receptor mutations resulting in intrauterine and postnatal growth retardation. *N Engl J Med.* 2003;349:2211–2222.
66. Kawashima Y, Kanzaki S, Yang F, Kinoshita T, Hanaki K, Nagaishi Ji Ohtsuka Y, Hisatome I, Ninomoya H, Nanba E, Fukushima T, Takahashi SI. Mutation at cleavage site of insulin-like growth factor receptor in a short-stature child born with intrauterine growth retardation. *J Clin Endocrinol Metab.* 2005;90:4679–4687.
67. Walenkamp MJ, van der Kamp HJ, Pereira AM, Kant SG, van Duyvenvoorde HA, Kruithof MF, Breuning MH, Romijn JA, Karperien M, Wit JM. A variable degree of intrauterine and postnatal growth retardation in a family with a missense mutation in the insulin-like growth factor I receptor. *J Clin Endocrinol Metab.* 2006;91:3062–3070.
68. Fang P, Schwartz ID, Johnson BD, Derr MA, Roberts Jr CT, Hwa V, Rosenfeld RG. Familial short stature caused by haploinsufficiency of the insulin-like growth factor-i receptor due to nonsense-mediated messenger ribonucleic acid decay. *J Clin Endocrinol Metab.* 2009;94:1740–1747.
69. Domene HM, Bengolea SV, Martinez AS, Ropelato MG, Pennisi P, Scaglia P, Heinrich JJ, Jasper HG. Deficiency of the circulating insulin-like growth factor system associated with inactivation of the acid-labile subunit gene. *N Engl J Med.* 2004;350:570–577.
70. Hwa V, Haeusler G, Pratt KL, Little BM, Frisch H, Koller D, Rosenfeld RG. Total absence of functional acid labile subunit, resulting in severe insulin-like growth factor deficiency and moderate growth failure. *J Clin Endocrinol Metab.* 2006;91:1826–1831.
71. Hermesz E, Mackem S. Rpx: a novel anterior-restricted homeobox gene progressively activated in the prechordal plage, anterior neural plate and Rathke's pouch of the mouse embryo. *Development.* 1996;122:41–52.
72. Dattani M, Martinez-Barbera J-P, Thomas PQ, Brickman JM, Gupta R, Mårtensson IL, Toresson H, Fox M, Wales JK, Hindmarsh PC, Krauss S, Beddington RS, Robinson IC. Mutations in the homeobox gene HESX1/Hesx1 associated with septo-optic dysplasia in human and mouse. *Nat Genet.* 1998;19:125–133.
73. McNay DE, Turton JP, Kelberman D, Woods KS, Brauner R, Papadimitriou A, Keller E, Keller A, Haufs N, Krude H, Shalet SM, Dattani MT. HESX1 mutations are an uncommon cause of septooptic dysplasia and hypopituitarism. *J Clin Endocrinol Metab.* 2007;92:691–697.
74. Thomas PQ, Dattani MT, Brickman JM, McNay D, Warne G, Zacharin M, Cameron F, Hurst J, Woods K, Dunger D, Stanhope R, Forrest S, Robinson IC, Beddington RS. Heterozygous HESX1 mutatns assoicated with isolated congenital pituitary hypoplasia and septo-optic dysplasia. *Hum Mol Genet.* 2001;10: 39–15.
75. Lamonerie T, Tremblay JJ, Lanctôt C, Therrien M, Gauthier Y, Drouin J. Ptx1, a bicoid-related homeo box transcription factor involved in transcription of the pro-opiomelanocortin gene. *Genes Dev.* 1996;10:1284–1295.
76. Drouin J, Lamolet B, Lamonerie T, Lanctôt C, Tremblay JJ. The PTX family of homeodomain transcription factors during pituitary development. *Mol and Cell Endocrinol.* 1998;140:31–36.
77. Semina EV, Reiter R, Leysens NJ, Alward WL, Small KW, Datson NA, Siegel-Bartelt J, Bierke-Nelson D, Bitoun P, Zabel BU, Carey JC, Murray JC. Cloning and characterization of a novel bicoid-related homeobox transcription factor gene, RIEG, involved in Rieger Syndrome. *Nat Genet.* 1996;14: 392–399.
78. Zhadanov AB, Bertuzzi S, Taira M, Dawid IB, Westphal H. Expression pattern of the murine LIM class homeobox gene

Lhx3 in subsets of neural and neuroendocrine tissues. *Dev Dyn.* 1995;202:354–364.
79. Fluck C, Deladoey J, Rutishauser K, Eblé A, Marti U, Wu W, Mullis PE. Phenotypic variability in familial combined pituitary hormone deficiency caused by a PROP1 gene mutant resulting in the substitution of Arg → Cys at codon 120 (R120C). *J Clin Endocrinol Metab.* 1998;83:3727–3734.
80. Wu W, Cogan JD, Pfaffle RW, Frisch H, O'Connell SM, Flynn SE, Brown MR, Mullis PE, Parks JS, Phillips III JA, Rosenfeld MG. Mutations in PROP1 cause familial combined pituitary hormone deficiency. *Nat Genet.* 1998;18:147–149.
81. Bodner M, Castrillo JL, Theill LE, Deerinck T, Ellisman M, Karin M. The pituitary-specific transcription factor GHF-1 is a homeobox-containing protein. *Cell.* 1988;55:505–518.
82. Simmons DM, Voss JW, Ingraham HA, Holloway JM, Broide RS, Rosenfeld MG, Swanson LW. Pituitary cell phenotypes involve cell-specific Pit-1 mRNA translation and synergistic interactions with other classes of transcription factors. *Genes Dev.* 1990;4: 695–711.
83. Preece MA, Kearney PJ, Marshall WC. Growth-hormone deficiency in congenital rubella. *Lancet.* 1977;2:842–844.
84. Fujita K, Matsuo N, Mori O, Koda N, Mukai E, Okabe Y, Shirakawa N, Tamai S, Itagane Y, Hibi I. The association of hypopituitarism and small pituitary, invisible pituitary stalk, tye 1 Arnold-Chiari malformation, and syringomyelia in seven patients born in breech position: a further proof of birth injury theory on the pathogenesis of idiopathic hypopituitarism. *Eur J Pediatr.* 1992;151:266–270.
85. Thomsett MJ, Conte FA, Kaplan SL, Grumbach MD. Endocrine and neurologic outcome in childhood craniopharyngioma: review of effect of treatment in 42 patients. *J Pediatr.* 1980;97:728–735.
86. Sorva R. Children with craniopharyngioma: early growth failure and rapid postoperative weight gain. *Acta Paediatr Scand Suppl.* 1988;77:587–592.
87. Sklar C, Grumbach MM, Kaplan Sl, Conte FA. Hormonal and metabolic abnormalities associated with central nervous system germinoma in children and adolescents and the effect of therapy: report of 10 patients. *J Clin Endocrinol Metab.* 1981;52:9–16.
88. Bettendorf M, Fehn M, Grulich-Henn J, Selle B, Darge K, Lüdecke DK, Heinrich UE, Saeger W. Lymphocytic hypophysitis with central diabetes insipidus and consequent panhypopituitarism preceding a multifocal, intracranial germinoma in a prepubertal girl. *Eur J Pediatr.* 1999;158:288–292.
89. Gutenberg A, Bell JJ, Lupi I, Tzou SC, Landek-Salgado MA, Kimura H, Su J, Karaviti LP, Salvatori R, Caturegli P. Pituitary and systemic autoimmunity in a case of intrasellar germinoma. *Pituitary.* 2009 May 26 [Epub ahead of print].
90. Brauner R, Malandry F, Rappaport R, Zucker JM, Kalifa C, Pierre-Kahn A, Bataini P, Dufier JL. Growth and endocrine disorders in optic glioma. *Eur J Pediatr.* 1990;149:825–828.
91. Darendelier F, Livesey EA, Hindmarsh PC, Brook CG. Growth and growth hormone secretion in children following treatment of brain tumours with radiotherapy. *Acta Paediatr Scand Suppl.* 1990;79:950–956.
92. Braunstein GD, Kohler PO. Endocrine manifestations of histiocytosis. *Am J Ped Hemat Oncol.* 1981;3:67–75.
93. Rappaport R, Brauner G. Growth and endocrine disorders secondary to cranial irradiation. *Pediatr Res.* 1989;25:561–567.
94. Shalet SM, Crowne EC, Didi MA, Ogilvy-Stuart AL, Wallace WH. Irradiation induced growth failure. *Ballières Clin Endocrinol Metab.* 1992;6:513–526.
95. Ryalls M, Spoudeas HA, Hindmarsh PC, Matthews DR, Tait DM, Meller ST, Brook CG. Short-term endocrine consequences of total body irradiation and bone marrow transplantation in children treated for leukaemia. *J Endocrinol.* 1993;136:331–338.
96. Sanders JE, Guthrie KA, Hoffmeister PA, Woolfrey AE, Carpenter PA, Appelbaum FR. Final adult heights of patients who received hematopoietic cell transplantation in childhood. *Blood.* 2005;105:1348–1354.
97. Acerini CL, Tasker RC, Bellone S, Bona G, Thompson CJ, Savage MO. Hypopituitarism in childhood and adolescence following traumatic brain injury: the case for prospective endocrine investigation. *Eur J Endocrinol.* 2006;155:663–669.
98. Bondanelli M, De Marinis L, Ambrosio MR, Monesi M, Valle D, Zatelli MC, Fusco A, Bianchi A, Farneti M, degli Uberti EC. Occurrence of pituitary dysfunction following traumatic brain injury. *J Neurotrauma.* 2004;21:685–696.
99. Lieberman SA, Oberoi AL, Gilkison CR, Masel BE, Urban RJ. Prevalence of neuroendocrine dysfunction in patients recovering from traumatic brain injury. *J Clin Endocrinol Metab.* 2001;86:2752–2756.
100. Dimopoulou I, Tsagarakis S, Theodorakopoulou M, Douka E, Zervou M, Kouyialis AT, Thalassinos N, Roussos C. Endocrine abnormalities in critical care patients with moderate-to-severe head trauma: incidence, pattern and predisposing factors. *Intensive Care Med.* 2004;30:1051–1057.
101. Aimaretti G, Ambrosio MR, Di Somma C, Fusco A, Cannavò S, Gasperi M, Scaroni C, De Marinis L, Benvenga S, degli Uberti EC, Lombardi G, Mantero F, Martino E, Giordano G, Ghigo E. Traumatic brain injury and subarachnoid haemorrhage are conditions at high risk for hypopituitarism: screening study at 3 months after the brain injury. *Clin Endocrinol (Oxf).* 2004;61:320–326.
102. Popovic V. GH deficiency as the most common pituitary defect after TBI: clinical implications. *Pituitary.* 2005;(3–4):239–243 (Review).
103. Yamanaka C, Momoi T, Fujisawa I, Kikuchi K, Kaji M, Sasaki H, Yorifuji T, Mikawa H. Acquired growth hormone deficiency due to pituitary stalk transection after head trauma in childhood. *Eur J Pediatr.* 1993;152:99–101.
104. Mariani R, Bortoluzzi MN, Richelme C, El Barbary M, Coussement A. Post-traumatic hypopituitarism after skull injury: apropos of 3 cases. *Arch Pediatr.* 1996;3:796–801.
105. Benvenga S, Vigo T, Ruggeri RM, Lapa D, Almoto B, LoGiudice F, Longo M, Blandino A, Campenni A, Cannavo S, Trimarchi F. Severe head trauma in patients with unexplained central hypothyroidism. *Am J Med.* 2004;116:767–771.
106. Daughaday WH. Pituitary gigantism. *Endocrinol Metab Clin North Am.* 1992;1:633–647.
107. Mestron A, Webb SM, Astorga R Benito P, Catala M, Gaztambide S, Gomez JM, Halperin I, Lucas-Morante T, Moreno B, Obiols G, de Pablos P, Paramo C, Pico A, Torres E, Varela C, Vazquez JA, Zamora J, Albareda M, Gilabert M. Epidemiology, clinical characteristics, outcome, morbidity and mortality in acromegaly based on the Spanish Acromegaly Registry (Registro Espanol de Acromegalia, REA). *Eur J Endocrinol.* 2004;151:439–446.
108. Marx SJ, Agarwal SK, Kester MB, Heppner C, Kim YS, Skarulis MC, James LA, Goldsmith PK, Saggar SK, Park SY, Spiegel AM, Burns AL, Debelenko LV, Zhuang Z, Lubensky IA, Liotta LA, Emmert-Buck MR, Guru SC, Manickam P, Crabtree J, Erdos MR, Collins FS, Chandrasekharappa SC. Multiple endocrine neoplasia type 1: clinical and genetic features of the hereditary endocrine neoplasias. *Recent Prog Horm Res.* 1999;54:397–438.
109. Vergès B, Boureille F, Goudet P, Murat A, Beckers A, Sassolas G, Cougard P, Chambe B, Montvernay C, Calender A. Pituitary disease in MEN type 1 (MEN1): data from the France–Belgium

MEN1 multicenter study. *J Clin Endocrinol Metab.* 2002; 87:457–465.

110. Thakker RV, Bouloux P, Wooding C, Chotai K, Broad PM, Spurr NK, Besser GM, O'Riordan JL. Association of parathyroid tumors in multiple endocrine neoplasia type 1 with loss of alleles on chromosome 11. *N Engl J Med.* 1989;321:218–224.
111. Spada A. Genetic aspects of pituitary tumors. *J Pediatr Endocrinol Metab.* 2001;14(Suppl. 5):1213–1216. discussion 1261–1212.
112. Asa SL, Ezzat S. The pathogenesis of pituitary tumours. *Nat Rev Cancer.* 2000;2:836–849.
113. Jagannathan J, Kanter AS, Sheehan JP, Jane Jr JA, Laws Jr ER. Benign brain tumors: sellar/parasellar tumors. *Neurol Clin.* 2007;25:1231–1249.
114. Vergès B, Boureille F, Goudet P, Murat A, Beckers A, Sassolas G, Cougard P, Chambe B, Montvernay C, Calender A. Pituitary disease in MEN type 1 (MEN1): data from the France–Belgium MEN1 multicenter study. *J Clin Endocrinol Metab.* 2002;8:457–465.
115. Burgess JR, Hoffman L, Greenaway TM. Somatotrophinomas in multiple endocrine neoplasia type 1: a review of clinical phenotype and insulin-like growth factor-1 levels in a large multiple endocrine neoplasia type 1 kindred. *Am J Med.* 1996;100:544–547.
116. Lania A, Spada A. G-protein and signalling in pituitary tumours. *Horm Res.* 2009;71(Suppl 2):95–100.
117. Johnston LB, Chew SL, Trainer PJ, Reznek R, Grossman AB, Besser GM, Monson JP, Savage MO. Screening children at risk of developing inherited endocrine neoplasia syndromes. *Clin Endocrinol.* 2000;52:127–136.
118. Watson JC. Koch CA, Kirschner LS, Nguyen T, Carney JA, Oldfield EH. Neurosurgical implications of Carney complex. *J Neurosurg.* 2000;92:413–418.
119. Boikos SA, Stratakis CA. Carney complex: pathology and molecular genetics. *Neuroendocrinology.* 2006;83:189–199.
120. Pack SD, Kirschner LS, Pak E, Zhuang Z, Carney JA, Stratakis CA. Genetic and histologic studies of somatomammotropic pituitary tumors in patients with the "complex of spotty skin pigmentation, myxomas, endocrine overactivity and schwannomas" (Carney complex). *J Clin Endocrinol Metab.* 2000;85:3860–3865.
121. Vierimaa O, Georgitsi M, Lehtonen R, Vahteristo P, Kokko A, Raitila A, Tuppurainen K, Ebeling TM, Salmela PI, Paschke R, Gundogdu S, De Menis E, Makinen MJ, Launonen V, Karhu A, Aaltonen LA. Pituitary adenoma predisposition caused by germline mutations in the AIP gene. *Science.* 2006;312:1228–1230.
122. Daly AF, Vanbellinghen JF, Khoo SK, Jaffrain-Rea ML, Naves LA, Guitelman MA, Murat A, Emy P, Gimenez-Roqueplo AP, Tamburrano G, Raverot G, Barlier A, De Herder W, Penfornis A, Ciccarelli E, Estour B, Lecomte P, Gatta B, Chabre O, Sabaté MI, Bertagna X, Garcia Basavilbaso N, Stalldecker G, Colao A, Ferolla P, Wémeau JL, Caron P, Sadoul JL, Oneto A, Archambeaud F, Calender A, Sinilnikova O, Montañana CF, Cavagnini F, Hana V, Solano A, Delettieres D, Luccio-Camelo DC, Basso A, Rohmer V, Brue T, Bours V, Teh BT, Beckers A. Aryl hydrocarbon receptor-interacting protein gene mutations in familial isolated pituitary adenomas: analysis in 73 families. *J Clin Endocrinol Metab.* 2007;92:1891–1896.
123. Daly AF, Tichomirowa MA, Petrossians P, Heliövaara E, Jaffrain-Rea ML, Barlier A, Naves LA, Ebeling T, Karhu A, Raappana A, Cazabat L, De Menis E, Montañana CF, Raverot G, Weil RJ, Sane T, Maiter D, Neggers S, Yaneva M, Tabarin A, Verrua E, Eloranta E, Murat A, Vierimaa O, Salmela PI, Emy P, Toledo RA, Sabaté MI, Villa C, Popelier M, Salvatori R, Jennings J, Ferrandez Longás A, Labarta Aizpún JI, Georgitsi M, Paschke R, Ronchi C, Valimaki M, Saloranta C, De Herder W, Cozzi R, Guitelman M, Magri F, Lagonigro MS, Halaby G, Corman V, Hagelstein MT, Vanbellinghen JF, Barra GB, Gimenez-Roqueplo AP, Cameron FJ, Borson-Chazot F, Holdaway I, Toledo SP, Stalla GK, Spada A, Zacharieva S, Bertherat J, Brue T, Bours V, Chanson P, Aaltonen LA, Beckers A. Clinical characteristics and therapeutic responses in patients with germ-line AIP mutations and pituitary adenomas: An International Collaborative Study. *J. Clin. Endocrinol. Metab.* 2007;95:E373–E383.
124. Weinstein LS, Shenker A, Gejman PV, Merino MJ, Friedman E, Spiegel AM. Activating mutations of the stimulatory G protein in the McCune-Albright syndrome. *N Engl J Med.* 1991; 325:1688–1695.
125. Kaushik S, Smoker WR, Frable WJ. Malignant transformation of fibrous dysplasia into chondroblastic osteosarcoma. *Skeletal Radiol.* 2002;31:103–106.
126. Kovacs K, Horvath E, Thorner MO, Rogol AD. Mammosomatotroph hyperplasia associated with acromegaly and hyperprolactinemia in a patient with the McCune-Albright syndrome. A histologic, immunocytologic and ultrastructural study of the surgically-removed adenohypophysis. *Virchows Arch. A. Pathol. Anat. Histopathol.* 1984;403:77–86.
127. Cuttler L, Jackson JA, Saeed uz-Zafar M, Levitsky LL, Mellinger RC, Frohman LA. Hypersecretion of growth hormone and prolactin in McCune-Albright syndrome. *J Clin Endocrinol Metab.* 1989;68:1148–1154.

CHAPTER

33

Neuroendocrine Inherited or Induced Aromatase Enzyme Deficits

Wah Chin Boon[1], *Evan R. Simpson*[2]

[1] Florey Neuroscience Institutes, Centre for Neuroscience, University of Melbourne and Dept of Anatomy and Development, Monash University, Melbourne, Australia, [2] Prince Henry's Institute, Melbourne Australia

OUTLINE

Summary

Aromatase is the enzyme that converts androgen to estrogen. It is expressed in many extra-gonadal sites, including multiple regions in the brain of both sexes. Some of the actions of androgen are in fact mediated by this local conversion to estrogen. Inhibition of aromatase disrupts not only reproductive systems, but also sexual behaviors (especially in the males) and metabolism. In addition, aromatase inhibition is also associated with other disorders such as short-term memory deficit and compulsive behavior.

ESTROGEN

On September 8, 1923, a landmark article[1] published work by Edgar Allen and Edward A. Doisy showing that an extract from hog ovarian follicles could induce the estrous stage of the female reproductive tract of ovarectomized mice in which estrous cycles had stopped. Doisy, in 1929, crystallized pure estrogen in the form of estrone from the urine of pregnant women.[2] He coined the term "estrogen" by adding the suffix -gen, which means "that which produces," onto estrous, thus defining the hormone he had discovered as "that which produces the estrous cycle."[3]

Estrogen – A Female Sex Hormone?

Estrogen refers to a family of steroid hormones found in all studied vertebrates. There are three types of naturally occurring estrogen: estrone (E1), 17β-estradiol (E2) and estriol (E3) (Fig. 33.1). They share a basic structure of 18 carbon atoms arranged in three hexane rings and 1 pentane ring. As indicated above, the major source of estrogen is the ovary; two decades later, it was uncovered that the testis is also a producer of estrogen.[4,5] From the site of synthesis, estrogens are released into the bloodstream, where they are either bound to sex hormone-binding proteins or albumins, gaining rapid and relatively unrestricted access to other parts of the body. Dissociated estrogen diffuses into target tissues and binds to its receptors to exert its effects (for review, see

Handbook of Neuroendocrinology, DOI: 10.1016/B978-0-12-375097-6.10033-2

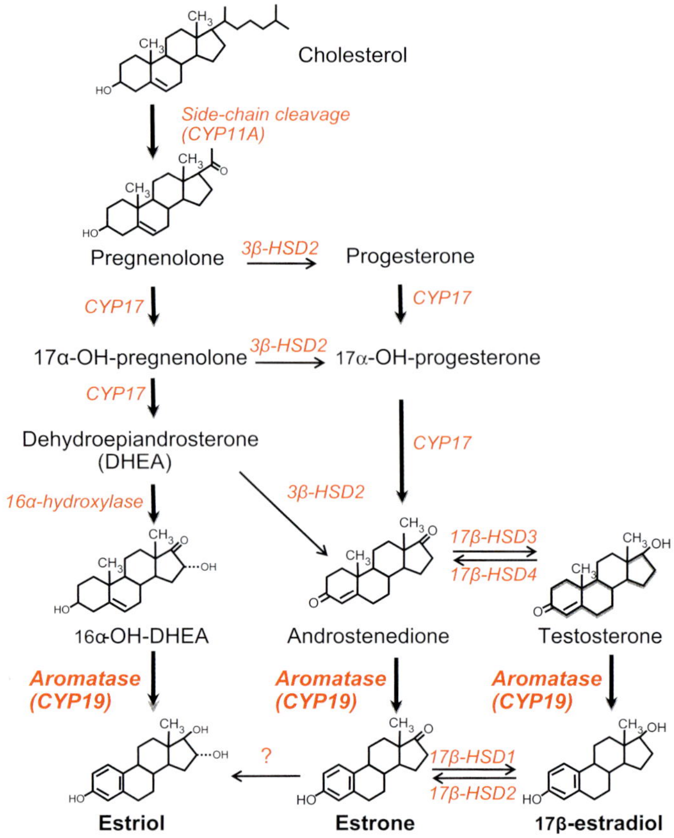

FIGURE 33.1 **The biosynthesis of estrogens and participating steroidogenic enzymes.** *CYP11A*, cholesterol side-chain cleavage; *CYP17*, 17α-hydroxylase; *3β-HSD2*, 3β-hydroxysteroid dehydrogenase type 2; *17β-HSD1*, 17β-hydroxysteroid dehydrogenase type 1; *17β-HSD2*, 17β-hydroxysteroid dehydrogenase type 2; *17β-HSD3*, 17β-hydroxysteroid dehydrogenase type 3; *17β-HSD4*, 17β-hydroxysteroid dehydrogenase type 4; *CYP19*, aromatase. *Reproduced from Boon WC, Chow JD, Simpson ER. The multiple roles of estrogens and the enzyme aromatase.* Prog Brain Res. *2010;181:209–232, with permission.*

Gruber *et al.*[6]), fulfilling its role as an endocrine factor. Estrogen receptors are widespread throughout the body. Major sites of expression include the reproductive system, brain and pituitary gland in males as well as females, leading to the suggestion that this "female" hormone may also have functions in the male.[7] Following the discovery of the estrogen-synthesizing enzyme aromatase, we now know that estrogen is also produced in several extra-gonadal sites in both males and females, where it acts locally in paracrine and intracrine fashions.[8]

By examining humans with aromatase deficiency, and genetically engineered animal models that are estrogen-deficient or -resistant, new roles for estrogen in both females and males have been revealed. In addition to its primary roles for both male and female reproduction,[9] estrogen is essential for proper bone maturation and preventing osteoporosis,[10] plays a role in carbohydrate and lipid metabolism[11] and maintaining healthy vasculature,[12] assumes neuroprotective functions, and regulates behaviors.[13] Male sexual and mating behaviors and masculinization of the brain are estrogen-dependent.[14] Thus, the view that estrogen is solely a female sex hormone has been abandoned.

Aromatase – The Estrogen Synthesizing Enzyme

Aromatase (CYP19) is a member of the cytochrome P450 superfamily. It is found on the endoplasmic reticulum membrane, forming a complex with its co-enzyme/redox partner NADPH-cytochrome P450 reductase. Aromatase is the only known enzyme to catalyze the irreversible and rate-limiting conversion of C19 androgen to form C18 estrogen. This three-step process involves the removal of the 19-methyl group (via two steps of hydroxylation to release formic acid) and aromatization of the steroid A-ring.[15] Each step requires 1 mol of O_2 and 1 mol of NADPH.[16] Unlike other cytochrome P450 enzymes, this heme-protein has unique androgenic specificity[17] (see Box 33.1), and converts androgens such as androstenedione, testosterone and 16α-hydroxytestosterone to estrone, 17β-estradiol and 17β-16α-estriol, respectively (Fig. 33.1).

Aromatase Gene

In humans, aromatase is encoded by a single *CYP19A1* gene spanning 123 kilobases (kb) on chromosome 15 q21.2.[18] The gene consists of nine coding exons (i.e., exons II to exon X, 30 kb) and a complex upstream regulatory region with at least 10 tissue-specific untranslated exons and promoters.[19] Aromatase is expressed in many tissues (including the gonads, placenta, adipose tissues, brain, etc.) under the regulation of tissue-specific promoters, resulting in tissue-specific untranslated first exons in the transcripts. However, the aromatase protein is identical in all tissues, as the untranslated first exons are spliced on to the same acceptor site upstream of the protein-encoding exon II.[20] Table 33.1 summarizes the usage of *CYP19* gene promoter in specific tissues.

Similarly, mouse aromatase is encoded by the *Cyp19A1* gene on chromosome 9 with multiple tissue-specific promoters. On the other hand, the pig has three aromatase isoforms encoded by three *CYP19* genes (*CYP19A1*, *CYP19A2* and *CYP19A3*). These genes are regulated by different promoters and expressed in different tissues.[21] Fish and reptiles have two aromatase isoforms encoded by two separate *Cyp19* genes – the gonadal-specific *Cyp19a1* and the brain-specific *Cyp19b1*.[22]

Central Nervous System

It was first demonstrated in the 1970s that estrogen could be produced in the brain by the local

BOX 33.1

HOW DO WE KNOW ABOUT ANDROGEN SPECIFICITY OF AROMATASE?

Aromatase (CYP19, CYP19A1) is a member of the cytochrome P450 superfamily. Cytochrome P450 (CYP) enzymes are heme-containing enzymes, found throughout nature, which catalyze more bioactivation processes (such as metabolism of sterols, including cholesterol, steroid hormones, etc., and xenobiotic chemicals) than other enzymes.[1] This is because the CYPs collectively have more substrates than any other enzymes, at least among drugs, due to the low specificity of their substrate binding sites.[1] However, the aromatase is highly specific in its substrate (i.e., androgens). To date, aromatase is the only natural mammalian, full-length CYP that has been crystallized.[2] X-ray crystallography of the human placenta aromatase revealed that it has an androgen-specific cleft that "binds the androstenedione molecule snugly with hydrophobic and polar residues exquisitely complement the steroid backbone" (Ghosh *et al.*[2]).

Aromatase catalyzes the conversion of androgens to estrogens in a three-step process, each step requiring 1 mol of O_2, 1 mol of NADPH, and coupling with its redox partner cytochrome P450 reductase.[3] The first two steps are C19-methyl hydroxylation, and the third involves the aromatization of the steroid A-ring, unique to aromatase, with a release of formic acid.[4] The release of a water molecule during the conversion serves as a simple and highly practical way to test for aromatase activity. Basically, this test measures the release of tritiated water (^3HOH) after aromatization of the radioactive substrate [1β-^{3}H]-androstendione or [1β-^{3}H]-testosterone. It was first described in the early 1980s,[5,6] and is still in use with some minor modifications.

References

1. Guengerich FP. Cytochrome P450s and other enzymes in drug metabolism and toxicity. *Am Assoc Pharm Sci J*. 2006;8: E101–E111.
2. Ghosh D, Griswold J, Erman M, Pangborn W. Structural basis for androgen specificity and oestrogen synthesis in human aromatase. *Nature*. 2009;457:219–223.
3. Thompson Jr EA, Siiteri PK. Utilization of oxygen and reduced nicotinamide adenine dinucleotide phosphate by human placental microsomes during aromatization of androstenedione. *J Biol Chem*. 1974;249:5364–5372.
4. Akhtar M, Calder MR, Corina DL, Wright JN. Mechanistic studies on C-19 demethylation in oestrogen biosynthesis. *Biochem J*. 1982;201:569–580.
5. Ackerman GE, Smith ME, Mendelson CR, MacDonald PC, Simpson ER. Aromatization of androstenedione by human adipose tissue stromal cells in monolayer culture. *J Clin Endocrinol Metab*. 1981;53:412–417.
6. Gore-Langton R, McKeracher H, Dorrington J. An alternative method for the study of follicle-stimulating hormone effects on aromatase activity in Sertoli cell cultures. *Endocrinology*. 1980;107:464–471.

aromatization of testosterone.[23] Research in this area has been gaining momentum, and evidence is mounting to show that locally synthesized estrogens from pregnenolone[24] could modulate neuronal functions, in addition to their neuroprotective effects.

Aromatase transcript expression has been detected in rodent and avian brains (see[25,26] for reviews), especially in the limbic system, hypothalamus, preoptic nucleus, sexually dimorphic nucleus, bed nucleus of the striata terminalis (BNST), hippocampus[24] and cerebellum.[27] It has been demonstrated that in rats, the male fetal hypothalamus expresses higher levels of aromatase activity than does the female fetal hypothalamus[28] (see Box 33.1). Male postpubertal porcine hypothalami have been reported to express four-fold higher aromatase activity than female postpubertal hypothalami[21] aromatase transcript levels, as confirmed by immunoblot analysis and reverse transcriptase (RT)-polymerase chain reaction (PCR). Aromatase expression was also detected in rat pituitary by RT-PCR.[29]

In male adult rats, aromatase activity is high in the periventricular preoptic nucleus and medial preoptic nucleus; intermediate in the suprachiasmatic preoptic nucleus, anterior hypothalamus, periventricular anterior hypothalamus and ventromedial nucleus; and low in the arcuate nucleus-median eminence, lateral preoptic nucleus, supraoptic nucleus, dorsomedial nucleus, and lateral hypothalamus.[30]

In immunohistochemistry with a polyclonal antibody raised against the rat aromatase sequence,[31] intense immunostaining was observed in neurons of adult rat amygdaloid structures and supraoptic nucleus, as well as reticular thalamic nucleus, olfactory tract and piriform cortex, while only moderate to light immunoreactivity was observed in the paraventricular and arcuate nuclei and hippocampus. However, contradictory to a previous report,[30] little aromatase immunoreactivity was detected in neurons in the BNST, medial basal hypothalamic and preoptic areas.[31] In contrast, robust aromatase immunoreactivity was detected in the medial

TABLE 33.1 The Expression of Estrogen Receptors and Aromatase in Normal Human Tissues

Human tissue	ERα	ERβ	Aromatase gene promoter
Brain	Hypothalamus and forebrain (Osterlund *et al.* 2000)	Cortex and hippocampus (Gonzalez *et al.* 2007)	Promoter 1f (Honda *et al.* 1994)
Skin	Fibroblasts (Haczynski *et al.* 2002)	Fibroblasts (Haczynski *et al.* 2002)	Promoter 1.4 (Harada 1992)
Cardiovascular	Cardiomyocytes (Mahmoodzadeh *et al.* 2009); vascular endothelium (Cruz *et al.* 2008)	Arteries and vascular stroma (Savolainen *et al.* 2001); smooth muscle cells (Hodges *et al.* 2000)	Promoter 1.7 (endothelial cells) (Sebastian *et al.* 2002)
Bone	Osteoblasts, osteoclasts and osteocytes (Hoyland *et al.* 1997)	Osteoblasts, osteoclasts and osteocytes (Vidal *et al.* 1999)	Promoter 1.6 (Shozu *et al.* 1998)
Adipose	Adipocytes and stromal cells (Price and O'Brien 1993)	(Pedersen *et al.* 2001)	Promoter 1.4/1.3 /II (Mahendroo *et al.* 1991)
Liver	(Grandien 1996)	*Not reported*	N/A (adult) Promoter 1.4 (Fetal hepatocytes) (Zhao *et al.* 1995)
Placenta	Proliferating trophoblasts (Bukovsky *et al.* 2003)	Differentiating trophoblasts (Bukovsky *et al.* 2003)	Promoter 1a/2a/1.2 Promoter 1.8 (Demura *et al.* 2008)
Gonads			
Ovary	Granulosa, theca and epithelium (Saunders *et al.* 2000) *Not detected* (Makinen *et al.* 2001)	Granulosa, theca and epithelium (Saunders *et al.* 2000)	Promoter II (Means *et al.* 1991) Promoter II (Bulun *et al.* 1993)
Testis		Sertoli, Leydig and germ cells (Moore *et al.* 1998; Makinen *et al.* 2001)	
Urogenital tract			
Adrenal glands	*Not detected* (Baquedano *et al.* 2007)	Zona reticularis (Baquedano *et al.* 2007)	Promoter 1.3 / II (Baquedano *et al.* 2007)
Uterus	Proliferative glandular and stromal cells (major) (Mylonas *et al.* 2004)	Proliferative glandular and stromal cells (minor), vascular endothelium (Lecce *et al.* 2001; Mylonas *et al.* 2004)	*Not expressed* (Bulun *et al.* 2005)
Prostate	Stromal nuclei only (Leav *et al.* 2001)	Multiple cell types (Pasquali *et al.* 2001)	Promoter II (Ellem *et al.* 2004)
Fetal tissues	Uterine mesenchyme (Glatstein and Yeh 1995); Ledig cells (Shapiro *et al.* 2005b); Prostate (Shapiro *et al.* 2005a); Neurons (Gonzalez *et al.* 2007)	Gonadal cells (Shapiro *et al.* 2005b); Umbilical vein endothelial cells (Toth *et al.* 2008); prostate (Shapiro *et al.* 2005a); Neurons (Fried *et al.* 2004)	Promoter 1.5 (Toda *et al.* 1994)

Reproduced from Boon WC, Chow JD, Simpson ER. The multiple roles of estrogens and the enzyme aromatase. Prog Brain Res. *2010;181:209–232, with permission.*

preoptic area and hypothalamus of a postnatal day 5 rat,[32] as well as in periventricular regions, ventromedial and arcuate nuclei, the central and medial nuclei of the amygdala, stria terminalis, BNST, lateral septum, medial septum, the diagonal band of Broca, lateral habenula and all areas of the cingulate cortex. This demonstates that aromatase activity can be higher in the brains of neonates than in adults.

In adult male non-human primates (rhesus monkeys and *Cynomolgus macaques*), aromatase activity[33] and messenger RNA (mRNA)[34] have been detected in brain regions such as the hypothalamus and the limbic system. The highest amounts of aromatase activity and its mRNA are found in the BNST, followed by the medial preoptic/anterior hypothalamus, followed by the amygdala. Intermediate amounts can be found in the lateral preoptic area and the ventromedial nucleus of the hypothalamus. Low amounts are found in the lateral septum, lateral hypothalamus, cerebral cortex, hippocampus and cerebellum.[33–37]

Despite various studies having determined aromatase activity and expression in brain regions using animal models,[26,38] studies in human postnatal brain tissue are limited due to the difficulty of obtaining fresh human brain tissue samples. In 1998, Sasano and colleagues[39] examined aromatase expression in various *post mortem* human brain regions using RT-PCR, and demonstrated that aromatase is expressed widely in human brain regions such as the pons, thalamus, hypothalamus and hippocampus. Aromatase expression levels did not differ significantly between temporal lobe tissues from men and women, but aromatase mRNA levels were significantly higher in adults than in children.[40] As an extension to the previous findings, Stoffel-Wagner *et al.*[41] examined the aromatase expression in biopsy samples from 45 female and 54 male epileptic patients using nested competitive RT-PCR. Aromatase transcripts were detected in the hippocampus and temporal and frontal neocortex, with temporal cortex expressing significantly higher levels than frontal neocortex. Again, no expression differences between sexes were observed in any of the brain regions investigated. Recent studies have determined that in the human temporal cortex, aromatase is expressed in a large population of pyramidal neurons and in certain interneurons and astrocytes, suggesting that aromatase serves a significant role in human cerebral cortex.[42]

Aromatase expression, assessed by quantitative RT-PCR and aromatase immunocytochemistry, was detected in normal human pituitary obtained at autopsy (13 males, 6 females, median age 30 years, interquartile ranges 23–63).[43] Although the median relative expression level of pituitary aromatase mRNA of men (median DeltaCt = 42.6, interquartile ranges 7.6–93.9) was higher than in women (median DeltaCt = 3.9, interquartile ranges 0–44.8), the difference was not statistically significant ($P = 0.2$) due to small sample size and large variations within groups.[43] The aromatase levels were also not correlated with the age of the study subjects ($P = 0.42$, $r = -0.21$).[43]

Estrogen exerts its action mainly via the estrogen receptors (ER) α and β (Table 33.1 summarizes the tissue expression sites), which are expressed not only in reproduction-relevant areas such as the hypothalamus and pituitary gland, but also in the hippocampus, cerebral cortex, midbrain, brainstem, caudate putamen and midbrain raphe.[44–46] These regions are associated with mood, cognition, learning and memory, pain sensitivity, attention mechanisms and vulnerability to epilepsy.[47] As both ERα and ERβ are expressed in numerous sites of the brain, estrogens produced locally in the brain could act in a paracrine, intracrine or autocrine manner.

Aromatase and Estrogen Deficiency

Aromatase deficiency results from mutation in the *CYP19* gene. To date, there are only 15 known cases of patients (8 male and 7 female) diagnosed with aromatase deficiency.[48,49] Most of the cases are from consanguineous marriages. Maternal acne during gestation of aromatase-deficient patients was usually reported due to the excess androgens that were not aromatized by the placenta.

Hypothalamus and Pituitary Axis Disruption in Aromatase Deficiency

Gonadotropins such as luteinizing hormone (LH) and follicle-stimulating hormone (FSH) are usually elevated in aromatase-deficient patients. These aromatase-deficient patients confirmed that estrogen exerts a negative feedback on both the pituitary gland and the hypothalamus to modulate the levels of gonadotropin secreted from the anterior pituitary.[50,51] In males, LH stimulates Leydig-cell production of testosterone,[52] whereas in post-pubertal females, an acute rise of LH called the LH surge triggers ovulation and development of the corpus luteum.[53] At the end of the luteal phase, there is a slight rise in FSH that seems to be of importance for the start of the next ovulatory cycle.[53] FSH also stimulates the maturation of germ cells in both males[54] and females.[53] The release of LH and FSH from the anterior pituitary gland is controlled by pulses of gonadotropin-releasing hormone (GnRH).[50] These pulses are subject to the estrogen feedback from the gonads.

Female Aromatase-deficient Patients

As expected, reproductive abnormalities are primary presenting features of aromatase-deficient patients. Female patients are usually diagnosed and treated early in life following signs of pseudohermaphroditism at birth, and pubertal abnormalities such as amenorrhea and delayed maturation of bone development. The phenotype is variable, but the clinical features include mild virilization, cystic ovaries, hypergonadotropinemia, elevated testosterone, low estrogen and enlarged clitoris. The four cases described below will illustrate the variability of the female phenotype in human aromatase deficiency.

One Pakistani proband presented with a homozygous deletion of one of two phenylalanine residues at position 234–235 (TTCTTT to TTT; F234del) of CYP19 in two affected siblings.[55] The first patient with 46,XX karyotype was assigned a male gender identity at birth due to androgenizaton of the external genitalia, was raised as a boy and exhibited male gender behavior. Breast development (Tanner Stage IV, normal adult development) at 14 years of age precipitated severe psychological distress, and, consequently, salpingo-oophorectomy and hysterectomy was performed to reduce hormone production and arrest breast development. Histology revealed cystic ovarian tissue, a well-estrogenized uterus (7 cm in length; mean uterine length for B4, 6.2–0.4 cm) and tortuous fallopian tubes. Gonadotropin concentrations were elevated consistent with prior gonadectomy, although levels of estrogens and androgens were not reported. The patient had a 4-cm phallus. Bone age was delayed by 3 years, bone densitometry showed a z-score of 1.69, and a borderline low high-density lipoprotein (HDL)-cholesterol was detected 18 months after oophorectomy. Mastectomy was performed, and the patient was started on testosterone supplementation with no significant increase in phallic length. His younger sibling, who was raised female, also had significant androgenization at birth, but an infantile uterus was identified on pelvic ultrasound.

A Turkish 46,XX girl who presented with ambiguous genitalia at birth had a homozygous R435C (CGT to TGT) mutation in the CYP19.[55] A history of maternal voice changes was noted during gestation of the patient. At 13.5 years of age, she had Tanner Stage II breast development, Tanner Stage III pubic hair, and androgenized external genitalia. Basal gonadotropin (FSH, LH) and androgen (androstenedione, testosterone) concentrations were elevated, and estradiol was low but detectable. Pelvic ultrasound revealed a 6.3-cm uterus with thin endometrial stripe and bilateral cystic. Bone age was delayed by 2.5 years. At 14.8 years of age, the patient's breast development had not progressed further than Tanner Stage II, and she complained of facial hair. Estrogen replacement fully induced breast development, and cyproterone acetate was given to prevent further facial hair growth.

By contrast, another 46,XX aromatase-deficient patient of Sri Lankan origin had no breast development at 14.5 years of age.[55] Her *CYP19* gene had deletion of a 1600-bp fragment encompassing exon 5. Her clitoral size was within normal limits, and a urogenital sinus with a single opening was found on examination. Her gonadotropins were elevated, serum estradiol was undetectable, and relatively low concentrations of circulating testosterone were measured. She had mild dyslipidemia and a lumbar spine bone mineral density z-score of 1.5, with bone age delay of 4.5 years. Estrogen replacement normalized her lipid profile.

In the last case discussed here,[56] a 46,XX girl inherited compounded heterozygous mutation in the *CYP19* gene from her nonconsanguineous parents – a point mutation (GAA to AAA) at the consensus 5′ splice donor sequence mutation, in exon V (father's allele), and a basepair deletion (A) in Glu 412 (GAA, exon IX), causing a frame shift mutation resulting in a truncated aromatase (mother allele). Similar to the other cases, she presented ambiguous external genitalia at birth and her aromatase deficiency was diagnosed at birth, and therefore her hormonal profile was followed from then onwards. During the first month of postnatal life, serum basal LH, FSH and androgen concentrations were high. From 2 to 6 months of age, serum androgen levels showed a significant increase, along with a significant decrease of serum gonadotropins. At this age, two enlarged follicular cysts were found and removed by laparoscopy. After the first year of life, LH and androgens remained within normal values, with a tendency to increase from age 5.7 years, while FSH levels were elevated above the normal range during prepuberty. A GnRH test in late prepuberty showed higher serum LH peaks than at early prepuberty. The growth pattern was normal until 5 years of age without any treatment. Unexpectedly, at the age of 7.7 years, the patient started spontaneous breast development (Tanner Stage II) and a slight elevation of serum estradiol levels was detected (73.8 pmol/l). A pelvic ultrasound study showed that the size of the uterus had increased slightly, and multiple ovarian cysts were persistently observed. Starting at 7.2 years of age, fasting insulin levels and insulin resistance (IR) increased, and became pronounced after 8.5 years of age. At 9 years of age, glucose intolerance was detected. Estrogen replacement treatment (ERT) was started at 9.6 years of age. During ERT, a pubertal growth spurt and an advance in bone age was observed, as well as adequate progress of breast development and uterine size. ERT was ineffective in suppressing gonadotropin and androgen levels, which remained elevated for chronological age. Ultrasonography showed the presence of ovarian follicular cysts. At 11.3 years of age, type 2 diabetes mellitus and acanthosis nigricans was detected, along with a weight increase above the 97th percentile. GnRH analog (triptorelin acetate 3.75 mg) administration for 1 year resulted in marked decreases in serum gonadotropins (particularly LH) and androgen levels accompanied by a progressive decrease in ovarian volume. Glucose tolerance improved slightly, but IR persisted, even though metformin (drug of choice for treatment of type 2 diabetes) was increased to 1000 mg/day. Lipid profile and liver enzyme levels were normal during the entire follow-up. This case study demonstrated that in girls, aromatase deficiency results in a decrease of estrogen negative feedback regulation of both serum LH and FSH, which can be elevated as early as the second week after birth and persist up to the sixth month of life; during the rest of prepuberty, FSH remains elevated whereas LH returns to normal.

Male Aromatase-deficient Patients

Pubertal development is normal in male aromatase-deficient patients, and so they are usually not diagnosed and treated until adulthood. They have tall stature, delayed bone maturation (unfused epiphyses), osteopenia or osteoporosis, eunuchoid skeletal proportions (i.e., increased arm span, genu valgum, and increased upper-to-lower segment ratio), and are overweight. The effects of estrogen deficiency in the male are diverse, such as infertility, persistent bone growth, and abnormalities in lipid metabolism.[48,49]

The first male patient with aromatase deficiency was described in 1995.[57] Aromatase deficiency in males does not affect reproductive development to the same degree as that in females. Aromatase-deficient men develop progressive infertility in adulthood, with decreased sperm motility and/or low sperm counts as the common causes,[57,58] but sexual activity may also be affected.[59] Metabolic abnormalities (such as slight truncal obesity, hyperinsulinemia, elevated serum triglyceride and LDL cholesterol, and low HDL cholesterol) as well as some liver dysfunction are present in male aromatase-deficient patients (reviewed by Jones *et al.*[48]).

It was not until recent decades that researchers began to focus on the metabolic abnormalities present in male aromatase-deficient patients. Morishima and colleagues[57] first described the metabolic phenotypes of a 27-year-old male aromatase-deficient sibling in 1995. He developed slight truncal obesity, hyperinsulinemia (which was improved with 3 months of oral estrogen replacement therapy), low HDL cholesterol and high LDL and total cholesterol, and elevated serum uric acid, albumin and alanin aminotransferase but normal aspartate aminotransferase levels. The latter tests suggested some liver dysfunction, but no further tests were performed to prove this. Carani *et al.*[58] also found signs of dyslipidemia in another male aromatase-deficient patient, with elevated serum triglyceride and LDL cholesterol, and low HDL cholesterol. Despite normal serum glucose levels, Herrmann and colleagues[60] found the third case of male aromatase-deficient patient with elevated score of homeostasis model assessment of insulin resistance (HOMA-IR), indicating the presence of insulin resistance despite a normal oral glucose tolerance test outcome. The serum lipid profile was normal; however, transdermal estrogen replacement decreased the LDL/HDL cholesterol ratio, which is more favorable in terms of the risk of cardiovascular disease.

The liver phenotype of aromatase-deficient men was first studied by Maffei and colleagues. Hepatic steatosis (fatty liver) was diagnosed by liver biopsy and elevated serum levels of liver enzymes (glutamic oxaloacetic transaminase (GOT), glutamic pyruvic transaminase (GPT), γ-glutamyl-transferase (γ-GT) and alkaline phosphatase), in addition to signs of insulin resistance and dyslipidemia.[61,62] Histological analysis of liver samples showed macro- and microvesicular pattern of steatosis, as well as periportal fibrosis typically observed in non-alcoholic fatty liver disease (NAFLD) and non-alcoholic steatohepatitis (NASH).[61,62] Most importantly, 1 year of transdermal estrogen replacement was able to decrease hepatic lipid accumulation and improve liver function parameters.[61]

It is still uncertain whether untreated female aromatase-deficient patients will also develop fatty liver disease as observed in male patients. Neither cognitive function nor psychological profiles of these patients were reported.

Estrogen-deficient Mouse Models

Initially, castrated male and ovariectomized female animals were used to study estrogen physiology. However, after the discovery that aromatase is expressed in multiple extragonadal tissues, it was realized that gonadectomized models failed to completely eliminate local estrogen synthesis in brain or in peripheral tissues. Interpretation of data can be complicated by these extragonadal sources of estrogen. To overcome this limitation, the Simpson Laboratory[63] generated the aromatase knockout (ArKO) model, which is unable to make endogenous estrogen but can respond to exogenous estrogen. In ArKO mice, exon 9 of the *Cyp19* gene is replaced by a neomycin resistance gene, rendering the final mutated aromatase non-functional.

Two other ArKO models with targeted disruption of the aromatase encoding gene were independently created by Honda *et al.*[64] in Harada's laboratory, and by Toda *et al.*[65] The Harada ArKO model was generated by the disruption of exons 1 and 2 of the *Cyp19* genes, while the Toda model is generated by the same strategy as Simpson's ArKO mice. These ArKO models are important tools to uncover the physiological roles of estrogen. As the distribution of the ArKO mouse progeny genotypes follows the Mendel's law of segregation[63] – i.e., the ratio of wild-type homozygous to heterozygous to knockout homozygous is 1 : 2 : 1 – we can conclude that aromatase deficiency does not result in significant mortality during embryogenesis or fetal development at least for mice.

Neonatal homozygote ArKO mice were phenotypically similar to wild-type (WT) mice. As these mice matured, it became apparent that the male sexual behavioral phenotype was severely affected (see Box 33.2). The absence of estrogen not only affected reproduction, but also led to adiposity with insulin resistance[66] and abnormal bone development.[67] Table 33.2 summarizes the phenotypes of these three aromatase-deficient models. Some of these phenotypes are sexually dimorphic, suggesting that some of the estrogen's physiological roles may be unique to each gender.

ArKO Brain and Behavioral Phenotypes

Interestingly, the ArKO mouse model shows brain and behavioral phenotypes that were not previously observed in gonadectomized animals (for review, see Hill and Boon[68]).

BOX 33.2

HOW DO WE KNOW ABOUT SEXUAL BEHAVIOR IN MALE ARKO MICE?

The male sexual behavior of 12- to 14-week-old wild-type (WT) and ArKO mice were examined by placing a male animal alone with a receptive WT female in a test chamber during the dark phase.[1] When presented with a hormonally primed receptive WT female (which was ovariectomized and injected intraperitoneally with 10 μg 17β-estradiol 48 hours before the test, then with 500 μg progesterone 4–7 hours before the test), WT males approached the female, repeatedly sniffed the anogenital region, and initiated copulation in less than 5 minutes. Within the 20-minute interactive period in the test chamber, male WT mice mounted an average of 11 times. In contrast, the male ArKO mice of the same age demonstrated impaired mounting behavior. When placed with receptive females, ArKO males immediately interacted with the female and appeared to sniff her genital area, similar to the WT males. However, this did not progress to copulation. No attempt to mount was initiated from any of the four male ArKO animals tested. This study clearly demonstrated that aromatase inhibition resulted in severe male sexual behavior impairment.[1]

Reference

1. Robertson KM, Simpson ER, Lacham-Kaplan O, Jones ME. Characterization of the fertility of male aromatase knockout mice. *J Androl.* 2001;22:825–830.

Brain Phenotype

We have reported that neuronal apoptosis occurs in the frontal cortex of aged female ArKO mice[69] and in the hypothalamus of aged male ArKO mice[70] in the absence of external assault (see Box 33.3). This could be a consequence of decreased levels of anti-apoptotic gene expression and increased levels of pro-apoptotic gene expression.[69,71]

The ArKO mouse model developed in Harada's laboratory[64] has been reported to be more susceptible to the neuronal toxins domoic acid[72] or 1-methyl-4-phenyl-1,2,3,6-tetrahydropyridine (MPTP).[73] These male ArKO mice showed significant hilar neuronal loss in the hippocampus after injection of a low dose of domoic acid, which had no effect on the control litter mates. This finding shows that aromatase deficiency increases the vulnerability of neurons to neurotoxic degeneration. These neonatal ArKO animals have decreased Purkinje dendritic growth, spinogenesis and synaptogenesis[74] as compared to the WT animals during the same developmental period, demonstrating that estrogens are important in promoting synapse formation.

Memory, Learning, Compulsion, Depression and Aggression

Our laboratories have demonstrated that both male and female ArKO mice have impaired spatial reference memory,[75] although the female ArKO mice performed as well as the WT in the Morris water maze test[76] (see Box 33.4). Sexually dimorphic behavioral phenotypes have been reported in the ArKO mouse models. Our male but not female ArKO mice developed compulsive behaviors such as excessive grooming and wheel-running activities.[77] The levels of catechol-O-methyl transferase (COMT) enzyme, which degrades catecholamines (dopamine, norepinephrine, epinephrine), were lower in the hypothalamus of the male ArKO mice compared with WT mice. The compulsive behavior of male ArKO mice was ameliorated by estradiol adminstration[77] with concomitant increase in hypothalamus COMT protein concentration to WT levels. On the other hand, depressive-like behavior was observed in the female animals of Harada's ArKO model[78] but not male ArKO mice, which could not be ameliorated by estrogen replacement.

The latter developed infanticide behavior, however, losing their male sexual and aggressive behaviors (for review, see Harada *et al.*[79]). When placed in a cage with an olfactory bulbectomized mouse, chosen because they do not attack other males, adult ArKO males did not show signs of aggression towards the "intruder." Toda and colleagues[80] also reported the loss of aggressive behavior in their male ArKO mice.[81]

Sexual Behavior

Although our female ArKO mice did not show disruption in female sexual behavior when raised on a low phytoestrogen diet (5–15.5 μg isoflavone/g food), animals raised on normal mouse chow containing high levels of phytoestrogens (427–65 μg isoflavone/g food) showed less lordosis when approached by male WT animals.[82] Harada's female ArKO mice, on the other hand, showed less receptivity to male WT mice after

TABLE 33.2 Key Phenotypes of Aromatase Knockout (ArKO)

Phenotypes	ArKO generated in Simpson laboratory *Cyp19A1* exon 9 deletion (Fisher *et al.*, 1998)		ArKO generated in Harada laboratory *Cyp19A1* exon 1 deletion (Honda *et al.*, 1998)		ArKO generated in Shizuta laboratory *Cyp19A1* exon 9 deletion (Nemoto *et al.*, 2000)	
	Male	**Female**	**Male**	**Female**	**Male**	**Female**
CNS/Behavioral	Apoptosis in arcuate nucleus & MPO (Hill *et al.*, 2004; Hill *et al.*, 2007a); ↓sexual activity (Robertson *et al.*, 2001), reduced spatial reference memory (Martin *et al.*, 2003); obsessive-compulsive behavior (Hill *et al.*, 2007b)	Apoptosis in frontal cortex (Hill *et al.*, 2009); reversible middle cerebral artery occlusion resulted in greater total and regional damage in female ArKO mice than ovarectomized WT controls (McCullough *et al.*, 2003); reduced spatial reference memory (Martin *et al.*, 2003) but no change in spatial learning in watermaze test with ↑hippocampus NMDA receptor transcripts (Boon *et al.*, 2005); No change in lordosis when raised on soy-free chow (low phytoestrogen diet) but ↓lordosis when raised on normal mouse chow (high phytoestrogen diet) (Kudwa *et al.*, 2007)	↑Vulneratiblity to neurotoxin (Azcoitia *et al.*, 2001; Morale *et al.*, 2008); prolonged latencies to mount and ↓numbers of mounts in response to receptive stimulus females; deficits in olfactory and visual cues for sexual partner preference (Bakker *et al.*, 2002b); infanticide toward pups (Matsumoto, 2003)	Enhanced response to odor cues (Wesson *et al.*, 2006); depressive-like symptoms (Dalla *et al.*, 2005); ↓lordosis (Bakker *et al.*, 2002a)	Impairment in mounting behavior (Toda *et al.*, 2001a); lack of aggressive behavior (Toda *et al.*, 2001b)	*No info*
Reproductive	Age-progressive infertility and impaired spermatogenesis (Robertson *et al.*, 1999); prostate gland hyperplasia (McPherson *et al.*, 2001)	Infertile with underdeveloped uteri (Fisher *et al.*, 1998); disrupted folliculogenesis; haemorrhagic cystic follicles (Britt *et al.*, 2001)	Infertile; no significant external abnormalities (Honda *et al.*, 1998)	Infertile; underdeveloped uteri; no other significant external abnormalities (Honda *et al.*, 1998)	Infertile (Nemoto *et al.*, 2000)	Infertile (Nemoto *et al.*, 2000); anovulation, depletion of follicles; disorganized interstitial cells; haemorrhages in the ovaries (Toda *et al.*, 2001c)
Metabolic	Hepatic steatosis; ↑serum leptin & insulin; ↓body lean mass; ↓physical activity; ↓calorie intake (Jones *et al.*, 2000; Hewitt *et al.*, 2003; Hewitt *et al.*, 2004)	↑adiposity ↑serum leptin & insulin ↑serum TG, cholesterol & HDL (Misso *et al.*, 2003)	*No info*	*No info*	Hepatic steatosis; Impaired gene expression and hepatic enzyme activities of fatty acid β-oxidation (Nemoto *et al.*, 2000); ↑HDL-cholesterol (Toda *et al.*, 2001b); Insulin resistance (Takeda *et al.*, 2003)	Normal hepatic fatty acid β-oxidation (Toda *et al.*, 2001d)

(*Continued*)

TABLE 33.2 Key Phenotypes of Aromatase Knockout (ArKO)—cont'd

Phenotypes	ArKO generated in Simpson laboratory *Cyp19A1* exon 9 deletion (Fisher *et al.*, 1998)		ArKO generated in Harada laboratory *Cyp19A1* exon 1 deletion (Honda *et al.*, 1998)		ArKO generated in Shizuta laboratory *Cyp19A1* exon 9 deletion (Nemoto *et al.*, 2000)	
	Male	Female	Male	Female	Male	Female
Bone	↓bone formation rate (Oz *et al.*, 2000)	↓bone length & density; ↑bone formation rate; ↑B-cell lymphopioesis (Oz *et al.*, 2000)	*No info*	*No info*	↑bone resorption similar to females, but at 32 weeks bone loss is less compared to females (Miyaura *et al.*, 2001)	Loss of cancellous bone; ↑bone resorption from 9-weeks-old; more severe at 32 weeks (Miyaura *et al.*, 2001)

References:

Azcoitia I, Sierra A, Veiga S, Honda S, Harada N, Garcia-Segura LM. Brain aromatase is neuroprotective. *J Neurobiol.* 2001;47:318–329.

Bakker J, Honda S, Harada N, Balthazart J. The aromatase knock-out mouse provides new evidence that estradiol is required during development in the female for the expression of sociosexual behaviors in adulthood. *J Neurosci.* 2002a;22(20):9104–9112.

Bakker J, Honda S, Harada N, Balthazart J. Sexual partner preference requires a functional aromatase (cyp19) gene in male mice. *Horm Behav.* 2002b;42:158–171.

Boon WC, Diepstraten J, van der Burg J, Jones ME, Simpson ER, van den Buuse M. Hippocampal NMDA receptor subunit expression and watermaze learning in estrogen deficient female mice. *Brain Res Mol Brain Res.* 2005;140:127–132.

Britt KL, Drummond AE, Dyson M, Wreford NG, Jones ME, Simpson ER, Findlay JK. The ovarian phenotype of the aromatase knockout (ArKO) mouse. *J Steroid Biochem Mol Biol.* 2001;79:181–185.

Dalla C, Antoniou K, Papadopoulou-Daifoti Z, Balthazart J, Bakker J. Oestrogen-deficient female aromatase knockout (ArKO) mice exhibit depressive-like symptomatology. *Eur J Neurosci.* 2004 Jul;20(1):217–228.

Fisher CR, Graves KH, Parlow AF, Simpson ER. Characterization of mice deficient in aromatase (ArKO) because of targeted disruption of the cyp19 gene. *Proc Natl Acad Sci USA.* 1998;95:6965–6970.

Hewitt KN, Boon WC, Murata Y, Jones ME, Simpson ER. The aromatase knockout mouse presents with a sexually dimorphic disruption to cholesterol homeostasis. *Endocrinology.* 2003;144:3895–903.

Hewitt KN, Pratis K, Jones ME, Simpson ER. Estrogen replacement reverses the hepatic steatosis phenotype in the male aromatase knockout mouse. *Endocrinology.* 2004;145:1842–1848.

Hill RA, Pompolo S, Jones ME, Simpson ER, Boon WC. Estrogen deficiency leads to apoptosis in dopaminergic neurons in the medial preoptic area and arcuate nucleus of male mice. *Mol Cell Neurosci.* 2004;27:466–476.

Hill RA, Chow J, Fritzemeier K, Simpson ER, Boon WC. Fas/FasL-mediated apoptosis in the arcuate nucleus and medial preoptic area of male ArKO mice is ameliorated by selective estrogen receptor alpha and estrogen receptor beta agonist treatment, respectively. *Mol Cell Neurosci.* 2007a;36:146–157.

Hill RA, McInnes KJ, Gong EC, Jones ME, Simpson ER, Boon WC. Estrogen deficient male mice develop compulsive behavior. *Biol Psychiatry.* 2007b;61:359–366.

Hill RA, Chua HK, Jones ME, Simpson ER, Boon WC. Estrogen deficiency results in apoptosis in the frontal cortex of adult female aromatase knockout mice. *Mol Cell Neurosci.* 2009;41:1–7.

Honda S, Harada N, Ito S, Takagi Y, Maeda S. Disruption of sexual behavior in male aromatase-deficient mice lacking exons 1 and 2 of the cyp19 gene. *Biochem Biophys Res Commun.* 1998;252:445–449.

Jones ME, Thorburn AW, Britt KL, Hewitt KN, Wreford NG, Proietto J, Oz OK, Leury BJ, Robertson KM, Yao S, Simpson ER. Aromatase-deficient (ArKO) mice have a phenotype of increased adiposity. Proc Natl Acad Sci U S A. 2000 Nov 7;97(23):12735–12740.

Kudwa AE, Boon WC, Simpson ER, Handa RJ, Rissman EF. Dietary phytoestrogens dampen female sexual behavior in mice with a disrupted aromatase enzyme gene. *Behav Neurosci.* 2007;121:356–361.

Martin S, Jones M, Simpson E, van den Buuse M. Impaired spatial reference memory in aromatase-deficient (ArKO) mice. *NeuroReport.* 2003;14:1979–1982.

Matsumoto T, Honda S, Harada N. Alteration in sex-specific behaviors in male mice lacking the aromatase gene. *Neuroendocrinology.* 2003;77:416–424.

McCullough LD, Blizzard K, Simpson ER, Oz OK, Hurn PD. Aromatase cytochrome P450 and extragonadal estrogen play a role in ischemic neuroprotection. *J Neurosci.* 2003;23:8701–8705.

McPherson SJ, Wang H, Jones ME, Pedersen J, Iismaa TP, Wreford N, Simpson ER, Risbridger GP. Elevated androgens and prolactin in aromatase-deficient mice cause enlargement, but not malignancy, of the prostate gland. *Endocrinology.* 2001;142:2458–2467.

Misso ML, Murata Y, Boon WC, Jones ME, Britt KL, Simpson ER. Cellular and molecular characterization of the adipose phenotype of the aromatase-deficient mouse. *Endocrinology.* 2003;144:1474–1480.

Miyaura C, Toda K, Inada M, Ohshiba T, Matsumoto C, Okada T, Ito M, Shizuta Y, Ito A. Sex- and age-related response to aromatase deficiency in bone. *Biochem Biophys Res* Commun. 2001;280:1062–1068.

Morale MC, L'Episcopo F, Tirolo C, Giaquinta G, Caniglia S, Testa N, Arcieri P, Serra PA, Lupo G, Alberghina M, Harada N, Honda S, Panzica GC, Marchetti B. Loss of aromatase cytochrome P450 function as a risk factor for Parkinson's disease? *Brain Res Rev.* 2008;57:431–443.

Nemoto Y, Toda K, Ono M, Fujikawa-Adachi K, Saibara T, Onishi S, Enzan H, Okada T, Shizuta Y. Altered expression of fatty acid-metabolizing enzymes in aromatase-deficient mice. *J Clin Invest.* 2000 Jun;105(12):1819–1825.

Oz OK, Zerwekh JE, Fisher C, Graves K, Nanu L, Millsaps R, Simpson ER. Bone has a sexually dimorphic response to aromatase deficiency. *J Bone Miner Res.* 2000;15:507–514.

Robertson KM, O'Donnell L, Jones ME, Meachem SJ, Boon WC, Fisher CR, Graves KH, McLachlan RI, Simpson ER. Impairment of spermatogenesis in mice lacking a functional aromatase (cyp 19) gene. *Proc Natl Acad Sci USA.* 1999;96:7986–7991.

Robertson KM, Simpson ER, Lacham-Kaplan O, Jones ME. Characterization of the fertility of male aromatase knockout mice. *J Androl.* 2001;22:825–830.

Takeda K, Toda K, Saibara T, Nakagawa M, Saika K, Onishi T, Sugiura T, Shizuta Y. Progressive development of insulin resistance phenotype in male mice with complete aromatase (CYP19) deficiency. *J Endocrinol.* 2003;176:237–246.

Toda K, Okada T, Takeda K, Akira S, Saibara T, Shiraishi M, Onishi S, Shizuta Y. Oestrogen at the neonatal stage is critical for the reproductive ability of male mice as revealed by supplementation with 17beta-oestradiol to aromatase gene (Cyp19) knockout mice. *J Endocrinol.* 2001a;168:455–463.

Toda K, Saibara T, Okada T, Onishi S, Shizuta Y. A loss of aggressive behaviour and its reinstatement by oestrogen in mice lacking the aromatase gene (Cyp19). *J Endocrinol.* 2001b;168:217–220.

Toda K, Takeda K, Okada T, Akira S, Saibara T, Kaname T, Yamamura K, Onishi S, Shizuta Y. Targeted disruption of the aromatase P450 gene (Cyp19) in mice and their ovarian and uterine responses to 17beta-oestradiol. *J Endocrinol.* 2001c;170:99–111.

Toda K, Takeda K, Akira S, Saibara T, Okada T, Onishi S, Shizuta Y. Alternations in hepatic expression of fatty-acid metabolizing enzymes in ArKO mice and their reversal by the treatment with 17beta-estradiol or a peroxisome proliferator. *J Steroid Biochem Mol Biol.* 2001d;79:11–17.

Wesson DW, Keller M, Douhard Q, Baum MJ, Bakker J. Enhanced urinary odor discrimination in female aromatase knockout (ArKO) mice. *Horm Behav.* 2006 May;49(5):580–586.

Modified from Boon WC, Chow WD, Simpson ER. The multiple roles of estrogens and the enzyme aromatase. Prog Brain Res. 2010;181:209–232, with permission.

BOX 33.3

HOW DO WE KNOW ABOUT *IN SITU* APOPTOSIS?

During apoptosis, or programmed cell death, cells display several morphological changes, including cellular shrinkage, chromatin condensation, nuclear fragmentation and plasma membrane blebbing, while retaining their plasma membrane integrity (for review, see[1]).

The degradation of nuclear DNA into oligonucleosomal fragments is a hallmark of apoptosis. These DNA fragments are readily detected *in situ* by the TUNEL (terminal deoxynucleotidyl transferase dUTP-mediated nick end labeling) assay, which labels 3′-OH end of DNA breaks with dUTP, which is usually tagged with reporter molecule that can be visualized directly (e.g., a fluorophore such as fluorescein using fluorescence microscopy)[2] or with a secondary detection system (e.g., alkaline phosphatase or horse radish peroxidase) which catalyzes the formation of a color precipitate after addition of a substrate that is detectable by light microscopy.[3]

Apoptosis is initiated and executed by caspases (a family of cysteine-dependent aspartate-directed proteases) and by caspase-activated proteases and nucleases. A member of this family, caspase-3, is a key mediator of apoptosis in neuronal cells (for review, see[4]). Caspase 3 is synthesized as an inactive proenzyme that is processed in cells undergoing apoptosis by self proteolysis and/or cleavage by other upstream proteases (e.g., caspases 8, 9 and 10) at Asp28-Ser29 and Asp175-Ser176). The processed form of caspase 3 consists of large (17-kD) and small (12-kD) subunits that associate to form an active enzyme which in turn proteolytically cleaves and activates other caspases (e.g., caspases 6, 7 and 9), as well as relevant targets in the cells.[4] Therefore, apoptotic cells can also be detected by immunohistochemical staining for active caspase 3.[5]

References

1. He B, Lu N, Zhou Z. Cellular and nuclear degradation during apoptosis. *Curr Opin Cell Biol.* 2009;21:900–912.
2. Hill RA, Pompolo S, Jones MEE, Simpson ER, Boon WC. Estrogen deficiency leads to apoptosis in dopaminergic neurons in the medial pre-optic area and arcuate nucleus of male mice. *Mol Cell Neurosci.* 2004;27:466–476.
3. Robertson KM, O'Donnell L, Jones ME, Meachem SJ, Boon WC, Fisher CR, et al. Impairment of spermatogenesis in mice lacking a functional aromatase (cyp 19) gene. *Proc Natl Acad Sci USA.* 1999;96:7986–7991.
4. D'Amelio M, Cavallucci V, Cecconi F. Neuronal caspase-3 signaling: not only cell death. *Cell Death Differ.* 2010;17: 1104–1114.
5. Hill RA, Chua HK, Jones ME, Simpson ER, Boon WC. Estrogen deficiency results in apoptosis in the frontal cortex of adult female aromatase knockout mice. *Mol Cell Neurosci.* 2009;41:1–7.

BOX 33.4

HOW DO WE KNOW ABOUT LEARNING AND MEMORY IN RODENTS?

Morris Water Maze Test

The Morris water maze test is one of the most commonly used behavioral tests to measure spatial learning in rodents, including rats and normal and genetically modified mice. It was first developed and reported by Richard G. Morris to test the spatial learning behavior of rats,[1] and he showed that a hippocampal lesion would impair spatial learning. The test apparatus consist of a circular water tank (diameter 1–1.8 m, ~60 cm deep) filled with opaque water (usually by adding powdered milk; ~30 cm of depth) and a hidden platform submerged a few centimeters under the water surface in one quadrant of the tank. The tank is surrounded by visual cues. Training of the animals to locate the hidden platform may take a few days (typically 5–7 days). In each training trial, the test animal is allowed to swim in the water for 5 minutes to find the hidden platform. If the animal does not find the platform with in the 5-minute period, it is rescued and placed on the platform. Each animal then spends 5 minutes on the platform before being returned to its homecage. Each animal has five training sessions a day. Given that the platform is hidden, the animal must learn to use the visual cues surrounding the pool to locate the platform. As training progresses, the time used to find the hidden platform (i.e., escape latency) will typically decrease. Such decreased escape latencies will most commonly reflect the adoption of a focal search strategy (i.e., a search strategy centered on the former platform location with little variance) for animals with no impairment in spatial learning. However, reduced escape

(*Continued*)

BOX 33.4 *(cont'd)*

latencies may also reflect the adoption of non-spatial strategies (e.g., mice might learn to swim in concentric circles a fixed distance from the wall). Therefore, to discriminate spatial and non-spatial strategies, mice are usually given a probe test, where the platform is removed from the pool and the mouse is allowed to search for it, typically over a 60-second period. Mice having adopted a spatial strategy will focus their search near the former location of the platform. Tracking software is routinely used to precisely record the position of the mouse throughout the test.

Y-Maze Test

This test is based on the natural drive of rodents to explore novel environments and test the short-term spatial recognition memory.[2] This test requires neither food deprivation (unlike the radial maze) nor electrical foot-shock (unlike the light/dark box), which could modify the motivational and emotional status of the animal. The Y-maze test was successfully applied to study the effect of age, administration of various drugs and food supplements, as well as various stressors, on spatial memory performances.

The Y-maze consists of three identical arms mounted in the shape of a "Y." Each arm is usually 15 inches long and 3.5 inches wide with 3-inch high walls made of opaque material. The procedure consists of two sessions, each about 30 minutes. During the first session, 15 minutes in duration, one arm (the novel arm) is blocked but the mouse can inspect the other two open arms. During the second session, all arms are open. The position of the animal in the maze is usually recorded by a tracking system. Parameters to be analyzed are: first arm entered, number of entries into each arm, and time spent in each arm. The dependent variable is time spent in the novel arm of the Y-maze. During the second session, mice with intact spatial memory will usually make the first turn into the novel (i.e., previously blocked arm) and spend approximately 60%–70% of the time in that arm. Mice with spatial memory deficits do not exhibit a significant difference in arm entries, presumably as a consequence of impaired ability to discriminate between previously seen and novel objects.

In our hands, the female ArKO mice performed worse than the wild-type (WT) mice in the Y-maze test but performed as well if not better than the WT mice in the Morris water maze,[3] thus illustrating that both tests detect different type of cognition functions.

References

1. Morris RG, Garrud P, Rawlins JN, O'Keefe J. Place navigation impaired in rats with hippocampal lesions. *Nature.* 1982;297:681–683.
2. Dellu F, Mayo W, Cherkaoui J, Le Moal M, Simon H. A two-trial memory task with automated recording: study in young and aged rats. *Brain Res.* 1992;588:132–139.
3. Boon WC, Diepstraten J, van der BJ, Jones ME, Simpson ER, van den BM. Hippocampal NMDA receptor subunit expression and watermaze learning in estrogen deficient female mice. *Brain Res Mol Brain Res.* 2005;140:127–132.

hormonal manipulation (estrogen and progesterone administration).[83]

Our male ArKO mice have severe impaired sexual behavior (see Box 33.2),[84] and we have also observed a decrease in synaptophysin immunohistochemical labeling in the male ArKO medial preoptic area of the mice.[85] Synaptophysin is a marker for presynaptic vesicle staining, and we therefore infer that the male behavior deficit of ArKO mice may be a consequence of a neuronal disruption in the medial preoptic area which positively regulates the anticipatory/motivational phase of copulatory behavior.[86]

Severe male sexual behavior deficit was also reported in Harada's ArKO mice. Unlike in their female ArKO mice,[83] estrogen replacement could restore the sexual behavior deficit in male ArKO animals.[87] Furthermore, Bakker *et al.*[88] demonstrated that although 8-week-old male ArKO animals preferred female soiled bedding to male soiled bedding, they failed to show any preference for the volatile body odor of anesthetized male or female stimulus animals. Anesthetized stimulus animals were presented behind opaque doors in a Y-maze. By contrast, male WT mice showed preferences for both female volatile odor and soiled bedding. Estrogen replacement in these adult ArKO mice restored the male sexual behavior without stimulating a volatile female odor preference,[88] indicating that estrogens are required to activate male sexual behavior.

CONCLUSION

In summary, the naturally occurring and genetically induced aromatase deficiency has illustrated that aromatase and hence estrogens play important roles in multiple physiological functions, from the development of genitalia to reproduction (both germ cells develop and copulation), metabolism, brain function and behavior.

The next challenge is to decipher how the local action of estrogen, produced by the aromatase expressed on site, regulates these physiological functions. This could be addressed by generating tissue-specific aromatase knockout or knockin mouse models.

References

1. Allen E, Doisy EA. Landmark article Sep 8, 1923. An ovarian hormone. Preliminary report on its localization, extraction and partial purification, and action in test animals. *J Am Med Assoc.* 1983;250:2681–2683.
2. Doisy EA. Isolation of a crystalline estrogen from urine and the follicular hormone from ovaries. *Am J Obstet Gynecol.* 1972;114: 701–702.
3. Shelfer L. The alchemy of Jargon: etymologies of urologic neologisms. Number 3: the genesis of steroid terminology. *Prostate.* 2009;69:228–230.
4. Berthrong M, Goodwin WE, Scott WW. Estrogen production by the testis. *J Clin Endocrinol Metab.* 1949;9:579–592.
5. Kelch RP, Jenner MR, Weinstein R, Kaplan SL, Grumbach MM. Estradiol and testosterone secretion by human, simian, and canine testes, in males with hypogonadism and in male pseudohermaphrodites with the feminizing testes syndrome. *J Clin Invest.* 1972;51:824–830.
6. Gruber CJ, Tschugguel W, Schneeberger C, Huber JC. Production and actions of estrogens. *N Engl J Med.* 2002;346:340–352.
7. Hess RA, Bunick D, Lee KH, et al. A role for oestrogens in the male reproductive system. *Nature.* 1997;390:509–512.
8. Simpson ER. Aromatase: biologic relevance of tissue-specific expression. *Semin Reprod Med.* 2004;22:11–23.
9. Simpson ER, Misso M, Hewitt KN, et al. Estrogen—the good, the bad, and the unexpected. *Endocr Rev.* 2005;26:322–330.
10. Vandenput L, Ohlsson C. Estrogens as regulators of bone health in men. *Nat Rev Endocrinol.* 2009;5:437–443.
11. Simpson E, Jones M, Misso M, et al. Estrogen, a fundamental player in energy homeostasis. *J Steroid Biochem Mol Biol.* 2005;95: 3–8.
12. Stice JP, Lee JS, Pechenino AS, Knowlton AA. Estrogen, aging and the cardiovascular system. *Future Cardiol.* 2009;5:93–103.
13. Boon WC, Chow JD, Simpson ER. The multiple roles of estrogens and the enzyme aromatase. *Prog Brain Res.* 2010;181:209–232.
14. Sharpe RM. The roles of oestrogen in the male. *Trends Endocrinol Metab.* 1998;9:371–377.
15. Akhtar M, Calder MR, Corina DL, Wright JN. Mechanistic studies on C-19 demethylation in oestrogen biosynthesis. *Biochem J.* 1982;201:569–580.
16. Thompson Jr EA, Siiteri PK. Utilization of oxygen and reduced nicotinamide adenine dinucleotide phosphate by human placental microsomes during aromatization of androstenedione. *J Biol Chem.* 1974;249:5364–5372.
17. Ghosh D, Griswold J, Erman M, Pangborn W. Structural basis for androgen specificity and oestrogen synthesis in human aromatase. *Nature.* 2009;457:219–223.
18. Bulun SE, Sebastian S, Takayama K, Suzuki T, Sasano H, Shozu M. The human CYP19 (aromatase P450) gene: update on physiologic roles and genomic organization of promoters. *J Steroid Biochem Mol Biol.* 2003;86:219–224.
19. Sebastian S, Bulun SE. A highly complex organization of the regulatory region of the human CYP19 (aromatase) gene revealed by the Human Genome Project. *J Clin Endocrinol Metab.* 2001;86: 4600–4602.
20. Simpson ER, Mahendroo MS, Means GD, Kilgore MW, Corbin CJ, Mendelson CR. Tissue-specific promoters regulate aromatase cytochrome P450 expression. *J Steroid Biochem Mol Biol.* 1993;44:321–330.
21. Corbin CJ, Berger T, Ford JJ, et al. Porcine hypothalamic aromatase cytochrome P450: isoform characterization, sex-dependent activity, regional expression, and regulation by enzyme inhibition in neonatal boars. *Biol Reprod.* 2009;81:388–395.
22. Diotel N, Le Page Y, Mouriec K, et al. Aromatase in the brain of teleost fish: expression, regulation and putative functions. *Front Neuroendocrinol.* 2010;31:172–192.
23. Naftolin F, Ryan KJ, Davies IJ, et al. The formation of estrogens by central neuroendocrine tissues. *Recent Prog Horm Res.* 1975;31: 295–319.
24. Hojo Y, Hattori TA, Enami T, et al. Adult male rat hippocampus synthesizes estradiol from pregnenolone by cytochromes P45017alpha and P450 aromatase localized in neurons. *Proc Natl Acad Sci USA.* 2004;101:865–870.
25. Lephart ED, Lund TD, Horvath TL. Brain androgen and progesterone metabolizing enzymes: biosynthesis, distribution and function. *Brain Res Brain Res Rev.* 2001;37:25–37.
26. Naftolin F, Horvath TL, Balthazart J. Estrogen synthetase (aromatase) immunohistochemistry reveals concordance between avian and rodent limbic systems and hypothalami. *Exp Biol Med (Maywood).* 2001;226:717–725.
27. Sakamoto H, Mezaki Y, Shikimi H, Ukena K, Tsutsui K. Dendritic growth and spine formation in response to estrogen in the developing Purkinje cell. *Endocrinology.* 2003;144:4466–4477.
28. Hutchison JB, Beyer C, Hutchison RE, Wozniak A. Sex differences in the regulation of embryonic brain aromatase. *J Steroid Biochem Mol Biol.* 1997;61:315–322.
29. Galmiche G, Corvaisier S, Kottler ML. Aromatase gene expression and regulation in the female rat pituitary. *Ann NY Acad Sci.* 2006;1070:286–292.
30. Roselli CE, Horton LE, Resko JA. Distribution and regulation of aromatase activity in the rat hypothalamus and limbic system. *Endocrinology.* 1985;117:2471–2477.
31. Sanghera MK, Simpson ER, McPhaul MJ, Kozlowski G, Conley AJ, Lephart ED. Immunocytochemical distribution of aromatase cytochrome P450 in the rat brain using peptide-generated polyclonal antibodies. *Endocrinology.* 1991;129:2834–2844.
32. Horvath TL, Roa-Pena L, Jakab RL, Simpson ER, Naftolin F. Aromatase in axonal processes of early postnatal hypothalamic and limbic areas including the cingulate cortex. *J Steroid Biochem Mol Biol.* 1997;61:349–357.
33. Roselli CE, Resko JA. Testosterone regulates aromatase activity in discrete brain areas of male rhesus macaques. *Biol Reprod.* 1989; 40:929–934.
34. Abdelgadir SE, Roselli CE, Choate JV, Resko JA. Distribution of aromatase cytochrome P450 messenger ribonucleic acid in adult rhesus monkey brains. *Biol Reprod.* 1997;57:772–777.
35. Resko JA, Connolly PB, Roselli CE, Abdelgadir SE, Choate JV. Selective activation of androgen receptors in the subcortical brain of male cynomolgus macaques by physiological hormone levels and its relationship to androgen-dependent aromatase activity. *J Clin Endocrinol Metab.* 1993;76:1588–1593.
36. Resko JA, Pereyra-Martinez AC, Stadelman HL, Roselli CE. Region-specific regulation of cytochrome P450 aromatase messenger ribonucleic acid by androgen in brains of male rhesus monkeys. *Biol Reprod.* 2000;62:1818–1822.
37. Roselli CE, Resko JA. Cytochrome P450 aromatase (CYP19) in the non-human primate brain: distribution, regulation, and functional significance. *J Steroid Biochem Mol Biol.* 2001;79:247–253.

38. Naftolin F, MacLusky NJ. Aromatase in the central nervous system. *Cancer Res.* 1982;42:3274s–3276s.
39. Sasano H, Takashashi K, Satoh F, Nagura H, Harada N. Aromatase in the human central nervous system. *Clin Endocrinol (Oxf).* 1998;48:325–329.
40. Stoffel-Wagner B, Watzka M, Steckelbroeck S, et al. Expression of CYP19 (aromatase) mRNA in the human temporal lobe. *Biochem Biophys Res Commun.* 1998;244:768–771.
41. Stoffel-Wagner B, Watzka M, Schramm J, Bidlingmaier F, Klingmuller D. Expression of CYP19 (aromatase) mRNA in different areas of the human brain. *J Steroid Biochem Mol Biol.* 1999;70:237–241.
42. Yague JG, Munoz A, Monasterio-Schrader P, Defelipe J, Garcia-Segura LM, Azcoitia I. Aromatase expression in the human temporal cortex. *Neuroscience.* 2006;138:389–401.
43. Kadioglu P, Oral G, Sayitoglu M, et al. Aromatase cytochrome P450 enzyme expression in human pituitary. *Pituitary.* 2008;11: 29–35.
44. Mitra SW, Hoskin E, Yudkovitz J, et al. Immunolocalization of estrogen receptor beta in the mouse brain: comparison with estrogen receptor alpha. *Endocrinology.* 2003;144:2055–2067.
45. Azcoitia I, Sierra A, Garcia-Segura LM. Localization of estrogen receptor beta-immunoreactivity in astrocytes of the adult rat brain. *Glia.* 1999;26:260–267.
46. Gonzalez M, Cabrera-Socorro A, Perez-Garcia CG, et al. Distribution patterns of estrogen receptor alpha and beta in the human cortex and hippocampus during development and adulthood. *J Comp Neurol.* 2007;503:790–802.
47. McEwen B. Estrogen actions throughout the brain. *Recent Prog Horm Res.* 2002;57:357–384.
48. Jones ME, Boon WC, McInnes K, Maffei L, Carani C, Simpson ER. Recognizing rare disorders: aromatase deficiency. *Nat Clin Pract Endocrinol Metab.* 2007;3:414–421.
49. Lanfranco F, Zirilli L, Baldi M, et al. A novel mutation in the human aromatase gene: insights on the relationship among serum estradiol, longitudinal growth and bone mineral density in an adult man under estrogen replacement treatment. *Bone.* 2008;43:628–635.
50. Finkelstein JS, O'Dea LS, Whitcomb RW, Crowley Jr WF. Sex steroid control of gonadotropin secretion in the human male. II. Effects of estradiol administration in normal and gonadotropin-releasing hormone-deficient men. *J Clin Endocrinol Metab.* 1991; 73:621–628.
51. Hayes FJ, Seminara SB, DeCruz S, Boepple PA, Crowley Jr WF. Aromatase inhibition in the human male reveals a hypothalamic site of estrogen feedback. *J Clin Endocrinol Metab.* 2000;85: 3027–3035.
52. Midzak AS, Chen H, Papadopoulos V, Zirkin BR. Leydig cell aging and the mechanisms of reduced testosterone synthesis. *Mol Cell Endocrinol.* 2009;299:23–31.
53. Christian CA, Moenter SM. The neurobiology of preovulatory and estradiol-induced gonadotropin-releasing hormone surges. *Endocr Rev.* 2010;31:544–577.
54. Cheng CY, Wong EW, Yan HH, Mruk DD. Regulation of spermatogenesis in the microenvironment of the seminiferous epithelium: new insights and advances. *Mol Cell Endocrinol.* 2010; 315:49–56.
55. Lin L, Ercan O, Raza J, et al. Variable phenotypes associated with aromatase (CYP19) insufficiency in humans. *J Clin Endocrinol Metab.* 2007;92:982–990.
56. Belgorosky A, Pepe C, Marino R, et al. Hypothalamic–pituitary–ovarian axis during infancy, early and late prepuberty in an aromatase-deficient girl who is a compound heterocygote for two new point mutations of the CYP19 gene. *J Clin Endocrinol Metab.* 2003;88:5127–5131.
57. Morishima A, Grumbach MM, Simpson ER, Fisher C, Qin K. Aromatase deficiency in male and female siblings caused by a novel mutation and the physiological role of estrogens. *J Clin Endocrinol Metab.* 1995;80:3689–3698.
58. Carani C, Qin K, Simoni M, et al. Effect of testosterone and estradiol in a man with aromatase deficiency. *N Engl J Med.* 1997; 337:91–95.
59. Carani C, Rochira V, Faustini-Fustini M, Balestrieri A, Granata AR. Role of oestrogen in male sexual behaviour: insights from the natural model of aromatase deficiency. *Clin Endocrinol (Oxf).* 1999;51:517–524.
60. Herrmann BL, Saller B, Janssen OE, et al. Impact of estrogen replacement therapy in a male with congenital aromatase deficiency caused by a novel mutation in the CYP19 gene. *J Clin Endocrinol Metab.* 2002;87:5476–5484.
61. Maffei L, Murata Y, Rochira V, et al. Dysmetabolic syndrome in a man with a novel mutation of the aromatase gene: effects of testosterone, alendronate, and estradiol treatment. *J Clin Endocrinol Metab.* 2004;89:61–70.
62. Maffei L, Rochira V, Zirilli L, et al. A novel compound heterozygous mutation of the aromatase gene in an adult man: reinforced evidence on the relationship between congenital oestrogen deficiency, adiposity and the metabolic syndrome. *Clin Endocrinol (Oxf).* 2007;67:218–224.
63. Fisher CR, Graves KH, Parlow AF, Simpson ER. Characterization of mice deficient in aromatase (ArKO) because of targeted disruption of the cyp19 gene. *Proc Natl Acad Sci USA.* 1998;95: 6965–6970.
64. Honda S, Harada N, Ito S, Takagi Y, Maeda S. Disruption of sexual behavior in male aromatase-deficient mice lacking exons 1 and 2 of the cyp19 gene. *Biochem Biophys Res Commun.* 1998;252:445–449.
65. Toda K, Takeda K, Okada T, et al. Targeted disruption of the aromatase P450 gene (Cyp19) in mice and their ovarian and uterine responses to 17beta-oestradiol. *J Endocrinol.* 2001;170:99–111.
66. Jones ME, Thorburn AW, Britt KL, et al. Aromatase-deficient (ArKO) mice have a phenotype of increased adiposity. *Proc Natl Acad Sci USA.* 2000;97:12735–12740.
67. Oz OK, Zerwekh JE, Fisher C, et al. Bone has a sexually dimorphic response to aromatase deficiency. *J Bone Miner Res.* 2000;15: 507–514.
68. Hill RA, Boon WC. Estrogens, brain, and behavior: lessons from knockout mouse models. *Semin Reprod Med.* 2009;27:218–228.
69. Hill RA, Chua HK, Jones ME, Simpson ER, Boon WC. Estrogen deficiency results in apoptosis in the frontal cortex of adult female aromatase knockout mice. *Mol Cell Neurosci.* 2009;41:1–7.
70. Hill RA, Pompolo S, Jones ME, Simpson ER, Boon WC. Estrogen deficiency leads to apoptosis in dopaminergic neurons in the medial preoptic area and arcuate nucleus of male mice. *Mol Cell Neurosci.* 2004;27:466–476.
71. Hill RA, Chow J, Fritzemeier K, Simpson ER, Boon WC. Fas/FasL-mediated apoptosis in the arcuate nucleus and medial preoptic area of male ArKO mice is ameliorated by selective estrogen receptor alpha and estrogen receptor beta agonist treatment, respectively. *Mol Cell Neurosci.* 2007;36:146–157.
72. Azcoitia I, Sierra A, Veiga S, Honda S, Harada N, Garcia-Segura LM. Brain aromatase is neuroprotective. *J Neurobiol.* 2001; 47:318–329.
73. Morale MC, Serra PA, L'episcopo F, et al. Estrogen, neuroinflammation and neuroprotection in Parkinson's disease: glia dictates resistance versus vulnerability to neurodegeneration. *Neuroscience.* 2006;138:869–878.
74. Sasahara K, Shikimi H, Haraguchi S, et al. Mode of action and functional significance of estrogen-inducing dendritic growth, spinogenesis, and synaptogenesis in the developing Purkinje cell. *J Neurosci.* 2007;27:7408–7417.

75. Martin S, Jones M, Simpson E, van den Buuse M. Impaired spatial reference memory in aromatase-deficient (ArKO) mice. *Neuroreport*. 2003;14:1979–1982.
76. Boon WC, Diepstraten J, van der Burg J, Jones ME, Simpson ER, van den Buuse M. Hippocampal NMDA receptor subunit expression and watermaze learning in estrogen deficient female mice. *Brain Res Mol Brain Res*. 2005;140:127–132.
77. Hill RA, McInnes KJ, Gong EC, Jones ME, Simpson ER, Boon WC. Estrogen deficient male mice develop compulsive behavior. *Biol Psychiatry*. 2007;61:359–366.
78. Dalla C, Antoniou K, Papadopoulou-Daifoti Z, Balthazart J, Bakker J. Oestrogen-deficient female aromatase knockout (ArKO) mice exhibit depressive-like symptomatology. *Eur J Neurosci*. 2004;20:217–228.
79. Harada N, Wakatsuki T, Aste N, Yoshimura N, Honda SI. Functional analysis of neurosteroidal oestrogen using gene-disrupted and transgenic mice. *J Neuroendocrinol*. 2009;21:365–369.
80. Toda K, Saibara T, Okada T, Onishi S, Shizuta Y. A loss of aggressive behaviour and its reinstatement by oestrogen in mice lacking the aromatase gene (Cyp19). *J Endocrinol*. 2001;168: 217–220.
81. Nemoto Y, Toda K, Ono M, et al. Altered expression of fatty acid-metabolizing enzymes in aromatase-deficient mice. *J Clin Invest*. 2000;105:1819–1825.
82. Kudwa AE, Boon WC, Simpson ER, Handa RJ, Rissman EF. Dietary phytoestrogens dampen female sexual behavior in mice with a disrupted aromatase enzyme gene. *Behav Neurosci*. 2007; 121:356–361.
83. Bakker J, Honda SI, Harada N, Balthazart J. The aromatase knock-out mouse provides new evidence that estradiol is required during development in the female for the expression of sociosexual behaviors in adulthood. *J Neurosci*. 2002;22: 9104–9112.
84. Robertson KM, Simpson ER, Lacham-Kaplan O, Jones ME. Characterization of the fertility of male aromatase knockout mice. *J Androl*. 2001;22:825–830.
85. Hill RA, Pompolo S, Jones MEE, Simpson ER, Boon WC. Estrogen deficiency leads to apoptosis in dopaminergic neurons in the medial pre-optic area and arcuate nucleus of male mice. *Mol Cell Neurosci*. 2004;27:466–476.
86. Giuliano F, Allard J. Dopamine and male sexual function. *Eur Urol*. 2001;40:601–608.
87. Bakker J, Honda S, Harada N, Balthazart J. Restoration of male sexual behavior by adult exogenous estrogens in male aromatase knockout mice. *Horm Behav*. 2004;46:1–10.
88. Bakker J, Honda S, Harada N, Balthazart J. Sexual partner preference requires a functional aromatase (Cyp19) gene in male mice. *Horm Behav*. 2002;42:158–171.

CHAPTER

34

Pituitary Adenomas

Ines Donangelo, Shlomo Melmed

Pituitary Center, Cedars-Sinai Medical Center, Los Angeles, CA, USA

OUTLINE

Summary

Pituitary tumors are common benign neoplasms that cause symptoms related to excess hormone production, or due to compression of surrounding brain structures. Mechanisms for pituitary tumor formation remain unclear; however, disrupted cell cycle regulation (RB1, p16, p27, PTTG), and signaling pathways (GNAS, CREB, EGFR, FGF) have been implicated in pathogenesis of pituitary adenomas. Changes in pituitary microenvironment, related to humoral and local factors that regulate tissue trophic status may further promote tumor growth. Senescence, a cellular defense protecting against malignant transformation, may underlie the benign nature of some pituitary tumors. Diagnosis of pituitary tumors is based on clinical suspicion of syndromes related to homone excess and/or due symptoms related to tumor mass effect, and should be confirmed with appropriate hormonal and imaging studies. Treatment modalities for pituitary tumors include medical therapy, surgical resection or radiation therapy. The choice of treatment depends on the hormone-secretion subtype and on individual tumor characteristics.

DOI: 10.1016/B978-0-12-375097-6.10034-4

INTRODUCTION

Pituitary adenomas account for about 15% of all intracranial neoplasms. A systematic review that included 18,902 cases noted that unsuspected pituitary adenomas were found at autopsy in 1.5–31% of subjects, with an average frequency of 10.7%.[1] High resolution computed tomography[2] and magnetic resonance imaging (MRI) identifies pituitary lesions ≥ 3 mm in diameter suggestive of adenoma in up to 20% of subjects. The majority of lesions incidentally identified on autopsy or imaging studies are < 10 mm in diameter.[1]

Clinically overt pituitary adenomas are less commonly encountered, with recent studies reporting a population prevalence of ~80 cases per 100,000.[3,4] Pituitary tumors may present with syndromes related to local mass effects or hormonal excess. A large pituitary adenoma may compress the normal pituitary, causing hypopituitarism, visual field loss due to compression of the optic tract, or palsy of cranial nerves due to cavernous sinus invasion. Tumors may arise from any of the differentiated cell types present in the pituitary gland: somatotropes, lactotropes, corticotropes, thyrotropes and gonadotropes. A functional classification of pituitary adenomas based on cell gene product and their relative frequency is depicted in Table 34.1.[5,6] Pituitary tumors secreting adrenocorticotropic hormone (ACTH) result in cortisol excess with Cushing's disease. Overproduction of growth hormone (GH) leads to acral overgrowth and metabolic abnormalities characteristic of acromegaly. Prolactin (PRL) excess results in galactorrhea and hypogonadism. Thyroid-stimulating hormone (TSH) hypersecretion leads to goiter and thyrotoxicosis, and excess secretion of gonadotropin results in gonadal dysfunction. Clinically non-functional tumors do not efficiently secrete their gene products, and are most commonly derived from gonadotrope cells.[7] Most pituitary tumors are sporadic, but some are components of familial tumor syndromes. Both sporadic and familial pituitary adenomas are invariably benign and slow-growing.

TABLE 34.1 Functional Classification of Pituitary Tumors

	Frequency	Clinical syndrome
Clinically functioning		
PRL-secreting adenoma	40%	Hypogonadism, galactorrhea
GH-secreting adenoma	15%	Acromegaly
GH/PRL co-secreting adenoma	6–8%	Acromegaly, hypogonadism
Mixed bicellular GH/PRL cell adenoma		Galactorrhea
Mammosomatotrope cell adenoma		
ACTH-secreting adenoma	10%	Cushing's disease
TSH-secreting adenoma	1%	Hyperthyroidism
Plurihormonal adenoma	1–4%	Mixed
Clinically non-functioning adenomas		
All	35%	
FSH/LH-positive adenomas		Inactive
Silent corticotrope adenomas		Inactive
Null-cell adenomas (hormone negative)		Inactive

PATHOGENESIS OF PITUITARY TUMORS

Mechanisms for development and progression of pituitary adenomas are not fully understood. Several abnormalities have been indentified in pituitary tumors, including imbalances in cell cycle regulation, transcription factors and signaling pathways that result from inherited or spontaneous oncogene activation or tumor suppressor gene inactivation (Table 34.2).[8] The monoclonal origin of pituitary tumors supported by X-chromosome inactivation patterns in tumors of female subjects[9,10] suggests that these tumors are derived from a single cell with intrinsic defects that provide proliferative advantage. Intra-pituitary trophic signals may enhance or restrain expansion of a monoclonal tumor cell population by regulating the extracellular milieu.

Pituitary Trophic Status

Pituitary gland hormone production and proliferation is regulated by central and peripheral signals. Physiological pituitary growth stimulation is observed during puberty and pregnancy/lactation, and results in gland volume increase. Prolonged endocrine gland failure causes pituitary hyperplasia due to removal of tonic inhibition by peripheral hormones directly on pituitary and indirectly on the hypothalamus (e.g., thyrotrope hyperplasia observed in primary hypothyroidism). However, in humans there is no evidence that pituitary hyperplasia consistently precedes pituitary tumor development. Estrogen excess is a recognized cause of pituitary hyperplasia, and although exogenous estrogen resulted in prolactinoma development in an experimental female rat model,[11] there was no association between oral contraceptive use, menopause hormone replacement therapy, lactation or parity and prevalence of pituitary tumors in a case–control study.[12] Similarly, pituitary hyperplasia caused by primary

TABLE 34.2 Selected Molecular Events Related to Tumorigenesis in Human Pituitary Adenoma Subtypes

	Tumor type	Mechanism of activation/inactivation
ACTIVATING		
Gsp	GH adenomas	Activating mutation
CREB	GH adenomas	Increased Ser-phosphorylated CREB promoted by gsp overexpression
Cyclin B2 (*CCNB2*)	All tumor types examined	Overexpression
Cyclin D1 (*CCND1*)	Non-functioning	Overexpression
EGF/EGFR	Non-functioning	Overexpression
PTTG	All tumor types examined	Overexpression
HMGA2	Non-functioning ACTH adenomas Prolactinomas	Overexpression
FGF-4	Prolactinomas	Overexpression
INACTIVATING		
RB1	Negative pRB in ~25% GH adenomas	Promoter methylation
13q14	Aggressive tumors	13q14 loss of heterozygosity
AIP	20% of FIPA 2% Sporadic adenomas	Inactivating mutation
MEN1	Prolactinomas in familial MEN1	Inactivating mutation
P16INK4a (*CDKN2A*)	All tumor types examined	Promoter methylation
P27KIP1 (*CDKN1B*)	All tumor types examined	Reduced expression
MEG3a	Non-functioning GH adenomas	Promoter methylation
Gadd45-γ	Non-functioning GH adenomas Prolactinomas	Promoter methylation

FIPA, familial isolated pituitary adenomas; Gsp, mutation in Guanine nucleotide-activating α-subunit (*GNAS*) gene; CREB, cAMP response element (CRE) binding protein; EGF, epidermal growth factor; EGFR, epidermal growth factor receptor; PTTG, pituitary tumor transforming gene; HMGA, High Mobility Group A gene; FGF-4, fibroblast growth factor; RB1, retinoblastoma susceptibility gene; AIP, aryl hydrocarbon receptor interacting protein gene; *MEN1*, multiple endocrine neoplasia 1; MEG3, maternally expressed 3 gene; Gadd45-γ, growth arrest and DNA damage-inducible gene 45-γ; CDK, cyclin-dependent kinase; CDKI, cyclin-dependent kinase inhibitor.
Modified from Donangelo and Melmed (2011).[8]

hypothyroidism or ectopic tumor production of GHRH is very rarely associated with pituitary adenoma formation.[13] Pituitary tumors are discrete and not surrounded by pituitary hyperplasia, supporting an intrinsic pituitary cell defect in the process of tumor development.

Hypothalamic and circulating hormones and local growth factors likely contribute with a permissive environment that potentiates cell proliferation and mutation, and subsequent tumor growth. Pituitary trophic status may correlate with predisposition to tumor formation, as exemplified in pituitary tumor transforming gene (*PTTG*) mouse models. *PTTG*, identified as the index mammalian securin, regulates sister chromatid separation during mitosis,[14] as discussed below. Mice with targeted pituitary *PTTG* overexpression develop pituitary hyperplasia and microadenomas,[15] while *pttg* inactivation results in pituitary hypotrophy.[16] Heterozygous inactivation of retinoblastoma gene (Rb+/−) in mice results in pituitary tumor formation with high penetrance. Combined pituitary PTTG overexpression and Rb+/− further enhances pituitary hyperplasia and tumor prevalence.[17] Pituitaries of αGSU.*PTTG1* and particularly bitransgenic αGSU.*PTTG1*;*Rb*+/− mice (αGSU.*PTTG1* mice crossbred with *Rb*+/− mice) revealed nuclear enlargement, chromatin redistribution (Fig. 34.1), and enlarged gonadotropes with prominent Golgi complexes and secretory granules. These cell morphology changes are indicative of functionally active

BOX 34.1

HOW DO WE KNOW THERE MAY BE AN INTRINSIC CELL DEFECT IN THE ORIGIN OF PITUITARY ADENOMAS?

1. Pituitary adenomas are monoclonal. Inactivation of the maternal or paternal X chromosome is a random process that takes place early in embryogenesis. As progeny cells retain the X-inactivation pattern of progenitor cells, in females with heterozygosity for X-chromosome-linked polymorphic loci, identification of a single active allele in the tumor indicates monoclonality. X chromosome inactivation analysis of pituitary adenomas indicates almost invariably finding of monoclonality.[1,2]
2. There is absence of pituitary hyperplasia in tissue surrounding pituitary adenomas. Pituitary adenomas are generally discrete, and surrounded by normal (non-hyperplastic) pituitary tissue. Generalized pituitary hyperplasia, such as caused by ectopic GHRH production, is very rarely a cause for pituitary adenoma formation.[3] The simultaneous occurrence of corticotrope hyperplasia and adenoma has only rarely been documented.[4]
3. Surgical resection of well-circumscribed pituitary adenomas controls hormone hypersecretion in > 75% of patients).[5] Reversion to the hormonal pulsatility pattern often occurs after adenoma resection.

References

1. Herman V, Fagin J, Gonsky R, Kovacs K, Melmed S. Clonal origin of pituitary adenomas. *J Clin Endocrinol Metab.* 1990;71:1427–1733.
2. Alexander JM, Biller BM, Bikkal H, Zervas NT, Arnold A, Klibanski A. Clinically nonfunctioning pituitary tumors are monoclonal in origin. *J Clin Invest.* 1990;86:336–340.
3. Thorner MO, Perryman RL, Cronin MJ, Rogol AD, Draznin M, Johanson A, et al. Somatotroph hyperplasia. Successful treatment of acromegaly by removal of a pancreatic islet tumor secreting a growth hormone-releasing factor. *J Clin Invest.* 1982;70:965–977.
4. Haap M, Gallwitz B, Meyermann R, Mittelbronn M. Cushing's disease associated with both pituitary microadenoma and corticotroph hyperplasia. *Exp Clin Endocrinol Diabetes.* 2009;117:289–293.
5. Mortini P, Losa M, Barzaghi R, Boari N, Giovanelli M. Results of transsphenoidal surgery in a large series of patients with pituitary adenoma. *Neurosurgery.* 2005;56:1222–1233. discussion 33.

cells, consistent with hyperplasia. The incidence of anterior lobe tumors increased 3.5-fold in bitransgenic αGSU.*PTTG1*;*Rb*+/− mice compared to Rb+/− mice, suggesting that pituitary *PTTG* overexpression with resulting hyperplasia facilitates tumor formation. Conversely, *pttg* inactivation leads to pituitary hypoplasia, which is protective for tumor formation.[16]

Disruption in Cell Cycle Regulation

Cell proliferation is dependent on a highly organized series of kinases, which rise and fall with progression of the cell cycle (Fig. 34.2). During G1 and G2 cell cycle phases, cells are prepared for completion of the S and M phases, respectively. Disturbance of cell cycle regulation is an important pathway in neoplastic development.

Retinoblastoma Susceptibility Gene (RB1)

The *RB1* gene encodes the pRB protein, a negative regulator of the cell cycle, and tumor suppressor. In its active, hypophosphorylated form pRB binds E2F transcription factors, restraining cell cycle progression from G1 to S phase. Mice with heterozygous *Rb1* inactivation develop pituitary tumors with high penetrance. Loss of heterozygocity of the *RB1* locus in chromosome 13q14 is encountered in aggressive pituitary tumors although pRb remains expressed, suggesting that another suppressor gene may be located adjacent to the *RB1* locus.[2] Nevertheless, approximately 25% of GH-secreting adenomas exhibit loss of pRB expression, and at least in some cases this is associated with promoter hypermethylation; however, this finding did not correlate with tumor behavior.[18,19]

INK4 Family (p16INK4a, p15INK4b and p18INK4c)

The INKA4 family of CDK inhibitors restrains G1/S progression by binding CDK4 and CDK6. CDK inhibitor p16 is encoded by *CDKN2A*, and inhibits CDK4 interaction with cyclin D1, therefore maintaining pRB unphosphorylated (active). P16 expression levels are low in non-functioning pituitary adenomas, especially null cell types, compared to gonadotrope non-functioning tumors,[20] and in a smaller subset of other pituitary tumor subtypes. The mechanism for underlying gene silencing is promoter hypermethylation of *CDKN2A*.[19,21] Mice with a targeted mutation of the *cdkn2c* gene that encodes p18 develop gigantism and organomegalia, and POMC intermediate lobe pituitary hyperplasia and tumors.[22] P18 expression is decreased in

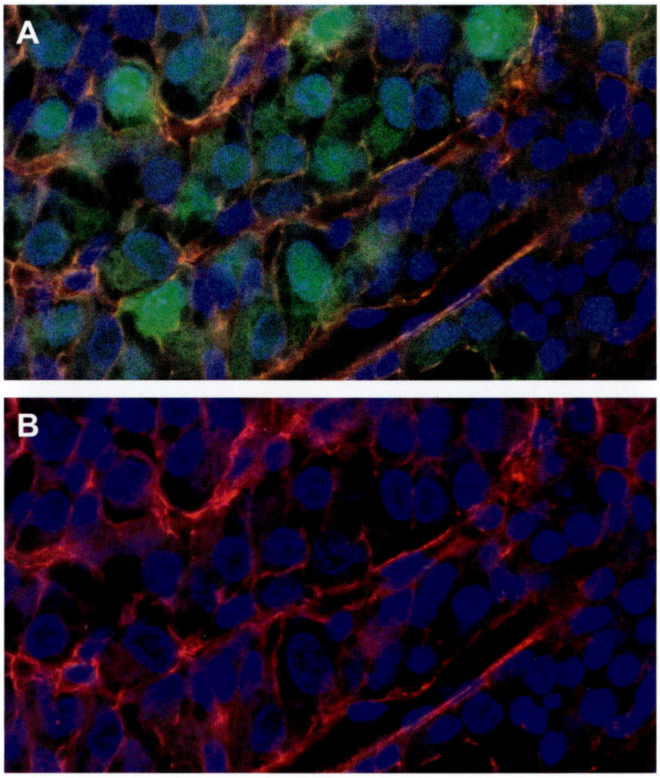

FIGURE 34.1 Targeted PTTG overexpression to anterior lobe pituitary cells results in cell hyperplasia and increased tumor formation. (**A**) and (**B**) are duplicates of the same image, overview of pituitary cells expressing αGSU.PTTG1.IRESeGFP transgene; (**A**) is the untouched image, and in (**B**) the green layer (eGFP) has been hidden for better visualization of nuclear morphology. Contrast between eGFP positive (overexpressing PTTG) and eGFP negative (normal PTTG content) can be appreciated, notably presence of macronuclei and reorganization of chromatin suggestive of hyperplastic cells. *Reproduced from Donangelo I, Gutman S, Horvath E, Kovacs K, Wawrowsky K, Mount M, et al. Pituitary tumor transforming gene overexpression facilitates pituitary tumor development.* Endocrinology *2006;147:4781–4791, with permission.*

a proportion of human non-functioning and ACTH-, GH- and PRL-secreting pituitary adenomas.[22,23] Promoter hypermethylation of the *CDKN2B* gene encoding p15 was also observed in a subset of pituitary tumors.[24]

Cip/Kip Family (p21Cip1, p27Kip1 and p57kip2)

The Cip/Kip family of CDK inhibitors restrains G1/S progression by associating with CDK1 and CDK2 complexes. Mice with a disrupted *Cdkn1b* gene that encodes p27Kip1 develop overall increased body weight, multiorgan hyperplasia, female infertility, POMC intermediate lobe pituitary tumors, and hyperplasia of hematopoietic organs.[25] Homozygous deletion of *Cdkn1b* in rats is associated with a novel MEN-like syndrome, MENX, an autosomal recessive syndrome with overlapping features of human MEN1 and MEN2. These animals develop pheochromocytomas or paragangliomas, parathyroid adenomas, pituitary tumors and thyroid neoplasia.[26] P27 expression is significantly decreased in human corticotrope adenomas and pituitary carcinomas,[27] and other pituitary adenoma types.[28] P21-deficient mice develop spontaneous hematopoietic, endothelial and epithelial tumors later in life, but pituitary tumors are rare.[29]

Cyclins

Cyclins D and E regulate the G1 to S phase of cell cycle progression by forming complexes with CDKs that induce phosphorylation (inactivation) of pRB. Allelic imbalance at the *CCND1* locus occurs more commonly among invasive than non-invasive tumors, and Cyclin D overexpression was observed more frequently in non-functioning tumors than in GH-secreting adenomas.[30] Cyclins A, B and E are also abundantly expressed in larger, more aggressive pituitary adenomas.

Pituitary Tumor Transforming Gene (PTTG)

PTTG was isolated from rat GH-secreting pituitary tumor cells, and *PTTG* overexpression induces cellular transformation *in vitro* and tumor formation in nude mice. PTTG is a mammalian securin, a key regulator of metaphase to anaphase transition during mitosis, and overexpression or suppression of *PTTG* results in aneuploidy, and also plays a role in DNA break repair.[31]

Transgenic mice with human *PTTG* targeted to the pituitary under the α-subunit of glycoprotein hormone (αGSU) promoter exhibited gonadotrope, thyrotrope and somatotrope focal hyperplasia and small adenomas, with elevated serum LH, testosterone, GH and/or IGF-1 levels, and prostate and seminal vesicle hypertrophy.[15] *PTTG* is abundantly expressed in most functioning and non-functioning pituitary tumors, and expression correlates with clinical tumor invasiveness.[32]

Signaling Pathways

Guanine Nucleotide-Activating α-Subunit (GNAS)

GNAS encodes $G_{s\alpha}$, the stimulatory subunit of the G protein that activates adenylyl cyclase to convert ATP to cAMP, regulating cell functions related to protein phosphorylation through protein kinase A (PKA) activation. A mutation in GNAS, termed *gsp*, induces constitutive adenylate cyclase activation. When this mutation occurs as an early postzygotic event it causes McCune-Albright syndrome, characterized by defects in bony skeleton and skin, precocious puberty, thyrotoxicosis, acromegaly, gigantism, or Cushing's syndrome. *Gsp* mutations are found in 30–0% of sporadic GH-secreting adenomas,[33] and *gsp* activates cell proliferation and hormone secretion without ligand binding to the

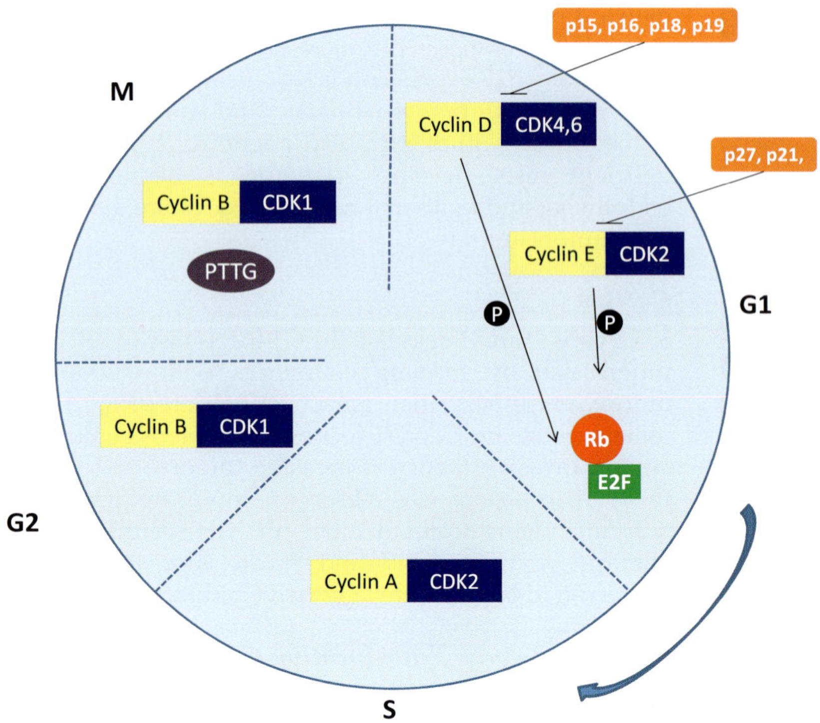

FIGURE 34.2 **Targets for pituitary tumor development in the G1/S cell cycle checkpoint.** Active (non-phosphorylated) pRB blocks progression of cell cycle from G1 to S phase by binding E2F. G1 progression and G1/S transition depends on sequential activation of cyclinD-CDK4 or -CDK6 and cyclinE-CDK2 complexes, leading to inactivation (phosphorylation) of pRB and release of E2F. CDK inhibitors (e.g., p16INK4a, p27Kip1, p21cip1) impede activation cyclin – CDK complexes, and are recognized as important tumor suppressor genes. Imbalances in pRB, p16, p27 and cyclins are present in pituitary adenomas, and the disruption of G1/S checkpoint barrier favors unrestrained cell cycle progression and proliferation. CDK, cyclin-dependent kinase.

GHRH receptor. Expression of $G_{s\alpha}$ is almost exclusively derived from the maternal allele or from the non-mutated paternal allele.[34] In GH-secreting adenomas, activating *gsp* mutation generally occurs in maternal alleles; however, most tumors negative for *gsp* oncogene showed biallelic expression of $G_{s\alpha}$ transcripts.

Activated cAMP-Response Element Binding Proteins (CREB)

cAMP response element (CRE) binding protein (CREB) is a direct target of PKA. Transgenic mice over-expressing a phosphorylation-deficient and transcriptionally inactive CREB mutant in the anterior pituitary exhibit dwarfism and somatotrope hypoplasia, indicating that phosphorylated CREB is required for the somatotrope proliferative response. Higher amounts of phosphorylated (activated) CREB were noted in GH-secreting pituitary tumors, suggesting that constitutively activated CREB may participate in somatotrope transformation.[35]

Growth Factors

Although most pituitary tumors are less vascularized than the normal pituitary gland, microvascular density as a measurement of angiogenesis correlates positively with tumor size and aggressive behavior, especially in prolactinomas. PTTG appears to promote angiogenesis, fibroblast growth factor 2 (FGF-2) and vascular endothelial growth factor (VEGF) expression in an estrogen-induced pituitary tumor model.[36] VEGF increases proliferation and migration of endothelial cells, and induces endothelial permeability and fenestrations. Mean vessel density is higher in macroprolactinomas compared to microprolactionomas.[37] However, VEGF expression does not correlate with aggressive pituitary tumor behavior.[38]

Basic FGF (bFGF or FGF-2) is highly expressed in the pituitary gland.[39] Folliculostellate cells are considered the primary cellular source of pituitary FGF-2, and stimulate lactotrope and folliculostellate cell proliferation, suggesting a role for locally secreted FGF-2 in pituitary trophic cell growth regulation.[40,41] In murine folliculostellate cells, FGF-2 induced autofeedback production of FGF-2, and pttg expression.[40] FGF-4 is encoded by heparin-binding secretory transforming (*hst*) gene, and is expressed in about one-third of prolactinomas, but rarely in non-functioning, GH- and ACTH-secreting pituitary tumors. FGF-4 expression in prolactinomas correlates with tumor invasiveness.[42]

ErbB receptors comprise of four subtypes: epidermal growth factor receptor (EGFR), ErbB2, ErbB3 and ErbB4. EGFR is overexpressed in pituitary adenomas, especially in non-functioning adenomas.[43,44] ErbB2 is expressed in most prolactinomas and GH-secreting adenomas,[45] and ErbB2 and ErbB3 are expressed in most prolactinomas.[46] EGF increases PRL and GH secretion in lacto-somatotrope rat GH3 cells, and inhibition of EGFR signaling results in decreased GH3 proliferation and PRL secretion through inhibition of the EGFR/ERK pathway.[47]

Miscellaneous

Growth Arrest and DNA Damage-Inducible Gene 45γ (GADD45-γ)

GADD45-γ, identified as a pituitary-derived tumor suppressor gene, is expressed in the normal pituitary gland but is absent in most non-functioning, GH-, and PRL-secreting adenomas. *GADD45-γ* is p53-responsive and induced by DNA damage, and participates in growth suppression and apoptosis. Transfection of *GADD45-γ* into immortalized rat pituitary tumor cells decreases cell growth and colony formation. Silencing of *GADD45-γ* in pituitary tumors likely occurs by methylation of CpG islands in the gene promoter.[48]

Maternally Expressed 3 Gene (MEG3)

MEG3 is expressed in human pituitary, brain and other tissues, but is diminished or absent in pituitary tumors and human cancer cell lines.[49] MEG3a is lost in both non-functioning and GH-secreting pituitary adenomas. Expression of *MEG3a* in tumor cell lines suppresses growth and decreases anchorage-dependent colony formation. The MEG3 gene encodes a non-coding RNA that functions to suppress tumor cell growth and increase protein expression of the tumor suppressor p53, and selectively activates p53 target genes.[50] *MEG3* expression is silenced in non-functioning pituitary adenomas, likely by promoter hypermethylation.[51]

High Mobility Group A (HMGA)

The HMGA family includes related HMGA1 and HMGA2 non-histone chromosomal proteins that regulate transcription by altering chromatin structure. Transgenic *HMGA1* and *HMGA2* overexpression in mice causes GH-secreting adenomas and prolactinomas, while trisomy of chromosome 12, the locus for *HMGA2*, is frequently encountered in PRL-secreting adenomas. *HMGA2* is expressed in pituitary adenomas, especially FSH/LH cell adenomas, and in prolactinomas,[52] and *HMGA2* levels correlate with tumor size, invasiveness, and cell proliferation markers. *HMGA2* induces cell proliferation by stimulating cyclin B2 expression through binding *CCNB2* promoter[53] and activating the E2F pathway.[54] MicroRNA *Let-7* suppresses *HMGA2*, and *HMGA2* and *Let-7* expression correlate inversely in human pituitary adenoma.[52]

Hereditary Pituitary Tumors

Pituitary adenomas occur in a familial setting in about 5% of all cases. Although familial pituitary syndromes are quite rarely encountered, molecular defects are better characterized than in sporadic pituitary adenomas. These include multiple endocrine neoplasia 1 (MEN1), familial isolated pituitary adenomas, and Carney complex.

Multiple Endocrine Neoplasia 1 (MEN1)

Over half of familial pituitary adenomas are due to Multiple Endocrine Neoplasia Type 1 (MEN1). MEN1 is an autosomal dominant syndrome with high penetrance, characterized by parathyroid adenoma, pancreatic endocrine tumors and pituitary adenomas. Other tumors encountered in this syndrome include foregut carcinoids and adrenal cortical adenomas. Pituitary adenomas occur in approximately 30–40% of patients, and these may be larger and more aggressive than sporadic counterparts.[55] Most secrete PRL (> 50%), with or without secretion of excess GH, and the remaining are non-functioning, GH-secreting only, ACTH-secreting or, rarely, TSH-secreting adenomas. MEN1 is caused by inactivating mutations of the tumor suppressor gene *MEN1*, which encodes Menin.[20] Menin is a nuclear protein that participates in the regulation of gene transcription, cell proliferation, apoptosis and genome stability.[56] Multiple inactivating *MEN1* mutations have been identified. It remains unclear why *MEN1* mutations cause selected endocrine neoplasms while Menin is ubiquitously expressed.

Approximately 10–20% of patients harboring a clinical diagnosis of MEN1 do not exhibit identifiable *MEN1* mutations. Mutations of other genes may also confer a MEN1-like syndrome. Rarely, inactivation of p27Kip1/*CDKN1B* or, less commonly, other CKI genes (p15, p18, p21) has been identified in families with clinical MEN1, but with no germline *MEN1* mutations.[26,57,58] MEN1 phenotype with p27Kip1/*CDKN1B* inactivation is now termed MEN4.

Familial Isolated Pituitary Adenomas (FIPA)

Familial isolated pituitary adenoma syndrome is defined as two or more members in a family harboring anterior pituitary tumors without evidence of MEN1 or Carney complex. Most of these pituitary tumors are PRL-secreting, GH-secreting, or mixed GH-secreting and PRL-secreting adenomas, but other combinations involving non-functioning pituitary adenomas and ACTH-secreting adenomas have also been reported.[59] Patients with familial disease are on average 4 years younger at diagnosis than those harboring sporadic tumors, and they have significantly larger adenomas.[60]

Analysis of families with pituitary adenomas provided evidence for linkage to chromosome 11q12-11q13, a region previously implicated in isolated familial somatotropinomas (IFS) and MEN1.[61] Mapping of this region identified *AIP*, or aryl hydrocarbon receptor interacting protein gene, and germline *AIP* mutations were detected in some FIPA patients. In the normal pituitary, AIP is expressed in GH- and PRL-secreting cells;[62] however, the exact role of AIP in pituitary function and mechanisms for AIP tumor suppressor function are not clear.

Of the 167 FIPA families for which *AIP* has been sequenced, 38 (22%) have AIP mutations. When only IFS families were analyzed, 30 of 75 (40%) were found to harbor an AIP mutation. Over 1000 patients with sporadic pituitary adenomas have now been assessed and germline mutations in *AIP* have been identified in less than 2%;[59] however, it is unclear whether any of these patients had an incomplete family history.

Carney Complex

Carney complex is a rare autosomal dominant syndrome characterized by cardiac or cutaneous myxoma, cutaneous hyperpigmentation and endocrine overactivity (pituitary tumors, Sertoli cell tumors, pigmented nodular adrenocortical disease). GH-secreting adenomas are the most common pituitary tumors encountered in these patients, and 10% of patients exhibit clinical acromegaly.[63] Approximately 50% of kindreds have an inactivating germline mutation in the *PRKAR1A* gene in chromosome 17q22-24, and an additional locus at 2p16 is associated with disease in some kindreds.[64] Inactivating *PRKAR1A* mutations results in constitutive activation of the PKA catalytic subunit.

Oncogene-Induced Cell Senescence

Cellular senescence denotes the phenomenon of proliferative arrest that occurs in response to diverse mechanisms, including age-linked telomere shortening, DNA damage, oxidative stress, chemotherapy, and oncogene activation. Oncogene-induced senescence (OIS) is a largely irreversible antiproliferative cell response to oncogenic signals that acts as a protective mechanism restraining neoplastic growth. OIS is mediated through activated cell cycle inhibitory pathways.

Cellular senescent markers are elevated in benign tumors but not in malignant carcinomas, and OIS may be protective for malignant transformation. Pituitary tumors are invariably benign, and p21 induction and senescence-associated β-galactosidase activity were found in 70% of human GH-secreting adenomas (Fig. 34.3), but not in pituitary or breast carcinomas.[65] Activation of the pituitary senescence pathway may constrain pituitary tumor progression, and explain the generally indolent nature of pituitary adenomas when compared to cancer.

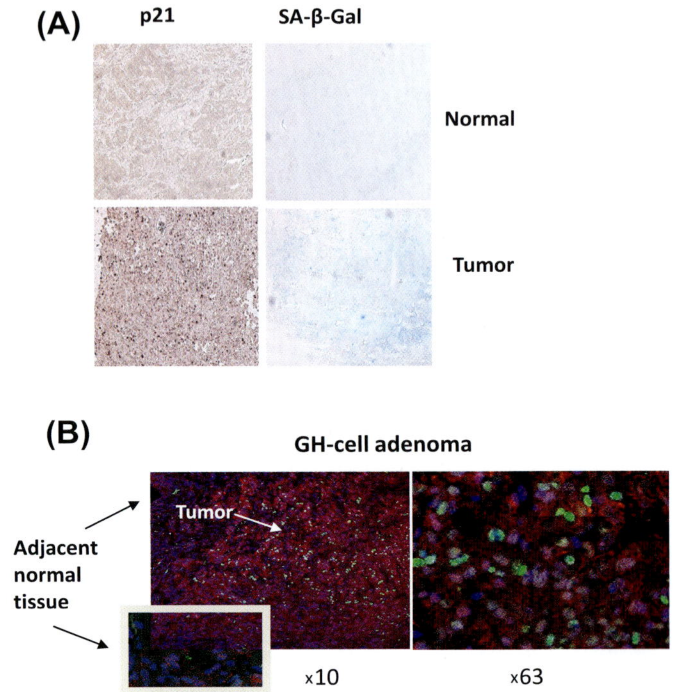

FIGURE 34.3 **Senescence markers in human GH-producing pituitary adenomas.** (**A**) Immunohistochemistry of the same GH-secreting human adenoma sections stained for p21 (brown) and SA-β-gal activity (blue). (**B**) Confocal image of double fluorescence immunohistochemistry of p21 (green) and β-galactosidase (red) proteins coexpression in human pituitary adenoma but not in normal adjacent tissue (left). High resolution (×63) image of the same slide (right). *Reproduced from Chesnokova V, Zonis S, Kovacs K, Ben-Shlomo A, Wawrowsky K, Bannykh S, et al. p21(Cip1) restrains pituitary tumor growth.* Proc Natl Acad Sci USA. *2008;105:17498–17503, with permission.*

TABLE 34.3 Local Effects of Expanding Pituitary Mass

Impacted structure	Clinical effect
Normal pituitary	Growth hormone deficiency – growth failure if before puberty, or adult GH deficiency
	LH/FSH deficiency – secondary hypogonadism
	TSH deficiency – secondary hypothyroidism
	ACTH deficiency – secondary adrenal insufficiency
Optic tract	Superior or bitemporal field defect, scotoma, blindness, homonymous hemianopia
Cavernous sinus	Cranial nerves III, IV, V1, V2, VI – diplopia, ptosis, ophthalmoplegia, and decreased facial sensation
Hypothalamus	Temperature dysregulation; dysfunction of thirst, sleep, appetite, behavioral and autonomic nervous system; diabetes insipidus
Temporal lobe	Seizures
Frontal lobe	Personality disorder, anosmia
Central	Headache, hydrocephalus, psychosis, dementia

DIAGNOSIS AND MANAGEMENT OF PITUITARY TUMORS

Pituitary masses can cause symptoms by secreting hormones and/or by impinging on adjacent normal pituitary tissue and brain structures (Table 34.3). Decompression of a sellar mass may lead to recovery of compromised anterior pituitary function. Pituitary tumors are typically slow–growing over years to decades. Although pituitary carcinoma is very rare, local aggressive behavior of pituitary adenomas with sellar erosion and parasellar extension is not infrequent. Expanding intrasellar lesions can exert significant compressive effects on adjacent normal pituitary tissue, and on surrounding vascular and neurologic structures. The sellar roof presents the least resistance to extension of an expanding tumor growing within the sella turcica, accounting for the vulnerability of the optic chiasm to pituitary mass expansion (Fig. 34.4). Optic chiasm compression affects temporal visual fields, initially superiorly, followed by progression to the entire temporal field. Worsening pressure on the optic nerve could potentially result in blindness, and pituitary-stalk compression of portal vessels in hyperprolactinemia and anterior pituitary dysfunction. Lateral extension of a pituitary mass may invade the cavernous sinus, and may cause palsies of the third, fourth and sixth cranial nerves, as well as the ophthalmic and maxillary branches of the fifth cranial nerve, especially in sudden compression by hemorrhage or tumor infarction. Inferior sphenoid sinus extension through the bony sellar floor can, rarely, cause cerebrospinal fluid leakage.

In a surgical series, more than 90% of sellar tumors are pituitary adenomas.[66] Differential diagnosis of sellar and parasellar masses includes craniopharyngioma, Rathke's cleft cyst, germinoma, optic nerve glioma and meningioma, among others.

Prolactin-Secreting Adenomas

Prolactin secretion is mainly regulated by the inhibitory effect of centrally derived dopamine. Variations in dopamine availability due to physiologic, pharmacologic or pathologic mechanisms interfere with PRL secretion. Disruption of the hypophysial-portal system by pituitary compression or pituitary stalk damage compromises the flow of hypothalamic dopamine to the anterior pituitary, resulting in loss of lactotrope inhibition with ensuing PRL hypersecretion.

Prolactinomas are the most frequently encountered secreting pituitary tumor. The proportion of females with microprolactonomas is higher than males (ratio 20 : 1), although the gender ratio is relatively equal for larger macroprolactinomas.[3,67] Microprolactinomas may be diagnosed earlier in women, due to menstrual irregularities. Men with prolactinomas possess higher basal PRL levels, larger tumor size, and higher degree of tumor invasiveness and proliferative markers than women.[68]

Diagnosis

Tumor size in prolactinomas correlates with serum PRL levels, and prolactin levels >200 ng/ml are strongly indicative of PRL-secreting adenoma, although some neuroleptic drugs may result in marked hyperprolactinemia. Prolactin levels <100 ng/ml in the absence of pituitary mass are most commonly not due to a prolactinoma, and may be related to physiologic, systemic disorders, and medication-induced hyperprolactinemia (Table 34.4).[69] The presence of a pituitary macroadenoma with PRL <100 ng/ml indicates that the sellar tumor is likely not producing PRL, and

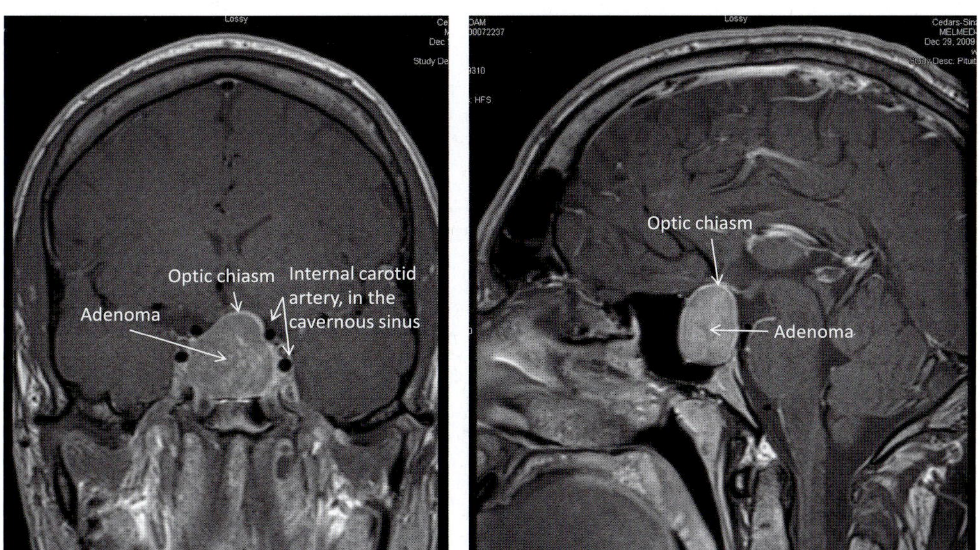

FIGURE 34.4 Pituitary T1-weighted magnetic resonance imaging, coronal (left panel) and sagittal (right panel) demonstrating a sellar mass with compression of the optic chiasm. There is impingement of cavernous sinus without invasion.

a pituitary-stalk mass interrupting the inhibitory effects of hypothalamic dopamine is the cause of hyperprolactinemia. Prolactin levels of 100–200 ng/ml should also be evaluated by considering the causes described in Table 34.4, and are most commonly due to microprolactinomas, sellar mass with stalk compression, or medication-induced hyperprolactinemia. All patients with persistent hyperprolactinemia should undergo pituitary MRI imaging.

Clinical features of prolactinomas vary by sex, tumor size and invasiveness. Elevated PRL causes sexual dysfunction via a feedback effect on gonadotropin pulsatility, likely inhibiting gonadotropin-releasing hormone (GnRH), LH pulse frequency and amplitude.[70] In men with hyperprolactinemia, decreased LH secretion causes low testosterone levels, resulting in diminished libido, erectile dysfunction, oligospermia, lower ejaculate volume and, infrequently, galactorrhea. Women with elevated PRL levels lose pulsatile LH secretion, with resulting low estrogen levels, oligo- or amenorrhea, galactorrhea, and anovulation with infertility. Low estrogen levels cause vaginal dryness, dyspareunia

TABLE 34.4 Etiology of Hyperprolactinemia

1. PHYSIOLOGIC			
Pregnancy			
Lactation			
Stress			
Sleep			
Coitus			
Exercise			
2. PATHOLOGIC			
Hypothalamic-pituitary stalk damage		**Pituitary**	**Systemic disorders**
Tumoral	**CNS disorders**		
Craniopharyngioma	Granulomas	Prolactinoma	Chronic renal failure
Meningioma	Infiltrations	Acromegaly	Polycystic ovarian disease
Dysgerminoma	Rathke's Cyst	Macroadenoma (stalk compression)	Cirrhosis
Hypothalamic metastases	Irradiation	Plurihormonal adenoma	Pseudocyesis
	Pituitary stalk section	Hypophyshitis	Epileptic seizures
	Sellar surgery		Cranial radiation
			Neurogenic chest wall trauma
			Chest wall herpes zoster
			Severe hypothyroidism
3. PHARMACOLOGIC			
Antipsychotics	Haloperidol, Chlorpromazine, Risperidone		
Antidepressants	Tricyclics (amitriptyline, desiplramine), SSRI (sertraline, fluoxetine), MAO-I (pargyline)		
Estrogen	Oral contraceptives		
Anticonvulsants	Phenytoin		
Prokinetics	Metoclopramide		
Antihypertensive	Alpha-methyldopa, Verapamil		
Opiates	Morphine, heroin, methadone		
H2 antagonists	Cimetidine, Ranitidine		
Others	TRH, buspirone, alprazolam		

CNS, central nervous sytem; TRH, thyrotropin-releasing hormone; SSRI, selective serotonin reuptake inhibitors; MAOI, monoamine oxidase inhibitors.
Modified from Melmed and Ho (2011).[69]

and libido loss. Accelerated bone loss occurs due to hypoestrogenemia, and possibly due to a direct negative effect of PRL on bone turnover.[71] Macroprolactinomas may also cause symptoms related to compressed adjacent structures (Table 34.3).

Treatment

Treatment of prolactinomas is aimed at normalization of PRL levels and related signs and symptoms, and reversal of tumor mass effects by tumor removal or shrinkage. Prolactinomas are treated medically with dopamine agonists or, very rarely, surgically.

Dopamine agonists used in treatment of prolactinomas are bromocriptine and cabergoline. Both compounds are ergot-derivate D2 dopamine receptor agonists that lower PRL levels; however, cabergoline has fewer side effects, and is more effective in normalizing PRL levels compared to bromocriptine. Cabergoline treatment results in normalization of PRL levels in 70–80% of patients with prolactinomas.[72,73] About 25% of patients receiving bromocriptine need to discontinue treatment due to side effects (mainly nausea, vomiting and postural hypotension), and about 20% are resistant to the drug. In one study, cabergoline treatment resulted in PRL normalization in 84% of bromocriptine-intolerant and 70% of bromocriptine-resistant patients.[72] Both drugs cause tumor shrinkage and visual field improvement, although these effects have not been directly compared in blinded randomized trials. Given its serotoninergic receptor (5HT2B) activity, high doses of cabergoline used in patients with Parkinson's disease have been associated with cardiac valvular abnormalities. Whether or not cumulative low-dose cabergoline therapy used in treatment of prolactinomas causes valvular insufficiency is currently unclear.[74] The shorter-acting agonist bromocriptine is the dopamine agonist of choice for women with prolactinomas who wish to become pregnant. When pregnancy is confirmed, bromocriptine should be discontinued. Pregnant patients should be carefully monitored for headaches or visual symptoms, and periodic visual field testing is recommended.

BOX 34.2

HOW DO WE KNOW? DIAGNOSTIC APPROACH TO ELEVATED PROLACTIN

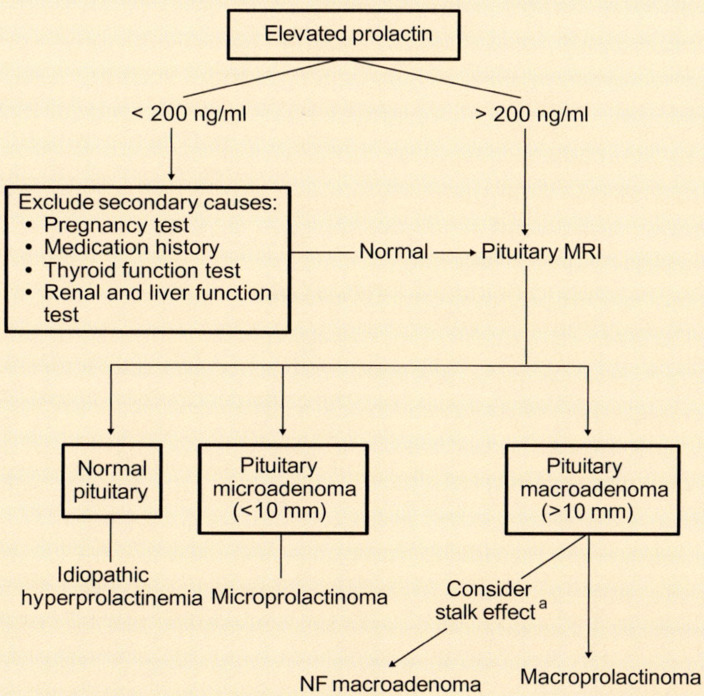

[a]Stalk effect due to non-functioning sellar mass should be considered if prolactin < 100 ng/ml and large pituitary lesion

NF, non-functioning. *Adapted from the Pituitary Center at Cedars Sinai Medical Center institution guidelines.*

Surgical resection of prolactinomas is indicated if there is resistance or intolerance to dopaminergic agonist therapy, or for an invasive macroadenoma causing visual field impairment that fails to rapidly improve with pharmacological therapy. Surgical normalization of serum PRL levels is achieved in ~70% of microprolactinomas but only 30% of macroprolactinomas, and recurrence rates are estimated at about 20% regardless of tumor size. Although radiation therapy controls or reduces prolactinoma size, delayed action (it may take years to achieve full effect), and invariable long-term development of hypopituitarism are disadvantages of this mode of therapy.

Gonadotrope Cell Adenomas (Clinically Non-Functioning Adenomas)

Gonadotrope cell adenomas most frequently present as clinically non-functioning masses. These tumors produce gonadotropin subunits detectable by immunohistochemistry; however, typically no clinically significant serum elevation in gonadotropins is present. Only very rarely do gonadotropin tumors secrete sufficient hormone to cause a clinical syndrome. Most of these tumors immunostain positively for glycoprotein subunits LHβ, FSHβ, and α-subunit.

Diagnosis

Clinically non-functioning pituitary tumors may come to clinical attention due to symptoms related to an expanding pituitary mass (Table 34.3), or may be incidentally detected during brain imaging studies performed for other purposes (incidentaloma). In the latter situation, when a macroadenoma is identified it is often associated with previously unsuspected pituitary hormone deficiency[75] or occasionally with visual field defects when suprasellar tumor expansion impinges the optic chiasm. Evaluation for clinically non-functioning pituitary adenomas includes pituitary MRI, visual field examination and pituitary hormone evaluation for hypopituitarism. In the rarely encountered functioning gonadotrope adenoma, elevated FSH, LH and/or α-subunit may cause pelvic pain due to ovarian hyperstimulation in women, and elevated serum testosterone with acne in men.

Treatment

Treatment is directed towards relieving signs and symptoms related to sellar mass effects. Therefore, large tumors jeopardizing vision, causing headaches, or those associated with hypopituitarism should be considered for surgical resection. Microadenomas and small macroadenomas may be managed with serial imaging studies. Treatment modalities for clinically non-functioning adenomas are surgery and radiation therapy. In a study that evaluated 231 patients with endocrine inactive adenomas,[75] transphenoidal surgery resulted in recovery of at least one pituitary axis in 18% of patients, however new loss of axis function occurred in 7.4% of patients. Improvement of visual disturbances and headache were noted in most patients. Radiation therapy may be considered if the tumor mass re-expands after surgery, endangering vision. Radiation therapy is associated with a high risk for developing new anterior pituitary hormone deficits within several years of follow up, and hormone evaluation should be performed periodically. Medications are not effective in treating non-functioning pituitary adenomas.

Growth Hormone-Secreting Adenomas

GH secretion is regulated by hypothalamic and peripheral factors. Hypothalamic GHRH stimulates GH synthesis and secretion, while somatostatin release-inhibitor factor (SRIF) inhibits GH secretion. SRIF binds to five distinct receptor subtypes (SSTR1 to SSTR5), of which SSTR2 and SSTR5 are expressed in somatotrope cells. Signaling through the SSTR2 and SSTR5 subtypes suppresses secretion of GH. Peripheral GH receptors are most abundant in the liver, and they induce insulin-like growth factor−1 (IGF-1) that mediates GH action on peripheral tissues. IGF-1 inhibits pituitary GH secretion and stimulates hypothalamic SRIF release via feedback mechanisms.

Acromegaly results from excess GH, and is caused by pituitary tumors secreting GH or very rarely by extrapituitary disorders (Table 34.5).[76] Rarely, GH may be secreted by pancreatic tumors or lymphoma. GHRH may be produced by hypothalamic tumors or carcinoid tumors, causing acromegaly through chronic GH-cell overstimulation.[77]

Diagnosis

Manifestations of acromegaly are caused by peripheral actions of GH and IGF-1, and by mass effects due to the pituitary tumor (Table 34.3). More than 70% of GH-secreting tumors are macroadenomas at diagnosis,[77,78] while about one-third of patients with acromegaly present with headache or visual field deficit.[78] Manifestations of excess GH and IGF-1 occur insidiously over several years, with a diagnostic delay of 7−10 years.[78] Effects of excess GH and IGF-1 on bone and soft tissue result in frontal bossing, increased hand and foot size, and mandibular enlargement with prognathism, widening of tooth incisor spaces, coarse facial features, and increased heel pad thickness. GH hypersecretion that occurs before epiphyseal long-bone closure causes pituitary gigantism. Patients with acromegaly exhibit changes in multiple systems, including skin

TABLE 34.5 Causes of Acromegaly

	Prevalence	Hormone product	Clinical features
Pituitary GH excess			
Pituitary adenoma	98%		
Densely granulated GH cell adenoma	30%	GH	Slow-growing
Sparsely granulated adenoma	30%	GH	Rapidly-growing, often invasive
Mixed GH-cell and PRL-cell adenoma	25%	GH and PRL	Variable
Mammosomatotrope cell adenoma	10%	GH and PRL	Common in children
Acidophil stem cell adenoma		PRL and GH	Rapidly-growing, invasive
Plurihormonal adenoma		GH (and PRL, αGSU, FSH, LH, TSH or ACTH)	Often clinically silent
Silent somatotropeinoma		GH	Often clinically silent
Pituitary carcinoma	< 1%	GH	Aggressive, requires documented metastasis
Hereditary syndromes			
Multiple endocrine neoplasia-1 (pituitary adenoma)		GH	Associated with parathyroid hyperplasia and/or pancreatic neuroendocrine tumor
McCune-Albright syndrome (pituitary hyperplasia)		GH, PRL	Associated with polyostotic fibrous dysplasia, café-au-lait macules, precocious puberty, and endocrine hyperfunction
Familial acromegaly (pituitary adenoma)		GH	≥ 2 members in a family with pituitary tumors without evidence of MEN1 or Carney complex
Carney Syndrome (pituitary hyperplasia → adenoma)		GH	Associated with myxomas of the heart, skin hyperpigmentation
Extrapituitary GH excess	< 1%	GH	Systemic features
Pancreatic islet-cell tumor			
Lymphoma			
Iatrogenic			
GHRH excess	< 1%	GHRH	Acromegaly (GH excess) due to somatotrope hyperplasia
Bronchial carcinoid			
Pancreatic islet-cell tumor			
Small-cell lung cancer			
Adrenal adenoma			
Medullary thyroid carcinoma			
Pheochromocytoma			
Hypothalamic tumor			

Modified from Melmed and Kleinberg (2008)[76] and Melmed (2006).[77]

(hyperhidrosis, oily skin), musculoskeletal (jaw malocclusion, arthropathy, carpal tunnel syndrome), cardiovascular (hypertension, left-ventricular hypertrophy, cardiomyopathy with arrhythmias), pulmonary (sleep apnea) and carbohydrate metabolism (insulin resistance, impaired glucose tolerance, diabetes mellitus), and visceromegaly (enlarged tongue, thyroid, liver, thyroid). One-third of patients with acromegaly have colon polyps at diagnosis, and periodic colonoscopy is indicated to evaluate for colon cancer. Overall mortality is elevated, likely due to increased prevalence of cardiovascular and pulmonary disease, and diabetes, with a standardized mortality rate of 1.7.[79] One study identified a history of pituitary radiotherapy, ACTH deficiency, and replacement with higher hydrocortisone doses as independent predictors of mortality.[79]

Biochemical diagnosis of acromegaly should include measurement of serum IGF-1 level, which, if elevated, should be followed by measurement of GH 2 hours after a standard 75-g oral glucose load. Given the episodic nature of GH secretion, a single random GH measurement is not useful for diagnosis. Lack of GH suppression after glucose load, and elevated age- and gender-matched IGF-1 levels, confirm the diagnosis of acromegaly.

Treatment

Control of acromegaly may be achieved by multiple therapeutical modalities – surgery, medications, and radiation therapy. The decision on therapeutic strategy

BOX 34.3

HOW DO WE KNOW? DIAGNOSTIC AND THERAPEUTIC APPROACHES TO ACROMEGALY

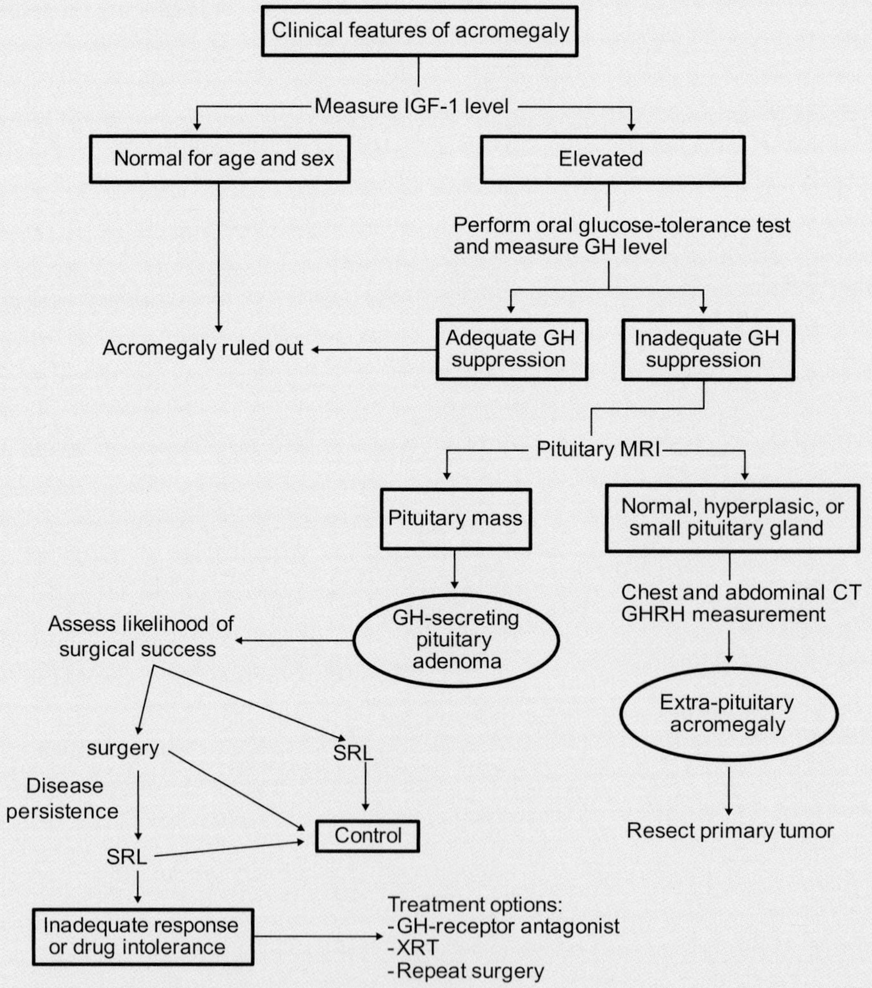

IGF-I, insulin-like growth factor I; MRI, magnetic resonance imaging; CT, computed tomography; SRL, somatostatin receptor ligand; XRT, external beam radiation therapy. *Modified from Melmed (2006).*[1]

Reference

1. Melmed S. Medical progress: Acromegaly. *N Engl J Med.* 2006;355:2558–2573. with permission.

should be tailored towards achieving cure while minimizing side effects. Surgical resection is indicated for GH-secreting pituitary microadenomas, where disease control is achieved in about 70% of patients, and for mass decompression caused by a macroadenoma. Surgical cure occurs in less than 50% of GH-secreting macroadenomas, and almost invariably medical therapy will be required after surgery for control of hormone excess. The incidence of tumor recurrence and post-surgical hypopituitarism is about 10%. Surgery success and overall complication rates highly correlate with the neurosurgeon's experience.

Medical therapy for acromegaly includes somatostatin receptor ligands, and GH receptor antagonists (Fig. 34.5). Somatostatin analogs octreotide and lanreotide are selective for SSTR2 and SSTR5 expressed in normal and adenomatous GH-secreting cells, causing biochemical control of IGF-1 and GH levels in up to 70% of patients.[80] A systematic review of 14 studies found that 37% of patients with acromegaly primarily treated with somatostatin analogs experienced significant tumor shrinkage, defined by decrease in pituitary mass $\geq$ 45%.[81] Somatostatin analogs are generally well tolerated; however, up to 30% of patients develop nausea, diarrhea and flatulence, or asymptomatic gallbladder sludge or stones. An additive effect of the combination of cabergoline and octreotide has been described.[82] Pegvisomant is a GH antagonist that antagonizes GH action by blocking peripheral GH receptor binding, thus lowering serum IGF-1 levels. GH levels may remain elevated, but its biological effect is blocked. Pituitary adenoma size should be monitored periodically. The drug is particularly useful in those patients with persistently elevated IGF-1 levels and controlled GH levels, and those with diabetes. Combined SRIF analog and pegvisomant treatment may be considered for resistant patients.[83]

Radiotherapy should only be considered in tumors that recur or persist after surgery in patients with resistance or intolerance to medical therapy. Alternatively, effects of radiotherapy on GH suppression may take ~9 years, so interim medical therapy is usually required. Most patients develop hypopituitarism years after radiotherapy, and visual deficits, brain necrosis and new tumor formation are rarely described, as is an association between radiotherapy and increased mortality.[79]

ACTH-Secreting Adenomas (Cushing's Disease)

Corticotrope cells express products of proopiomelanocortin gene (POMC), which include ACTH, β-lipotropin and endorphins. ACTH production is regulated by hypothalamic CRH. Other factors, including vasopressin, cytokines, physical stress, exercise and hypoglycemia, also stimulate ACTH to a lesser degree. Glucocorticoids suppress CRH and ACTH. ACTH is secreted in a pulsatile fashion, and is highest in the morning with a nadir late in the evening.

ACTH-secreting adenomas (Cushing's disease) are usually microadenomas. Excess glucocorticoid (Cushing's syndrome) may also be caused by adrenal neoplasms or due to ectopic tumor ACTH production, including small-cell lung cancer, and carcinoid tumors. Iatrogenic glucocorticoid administration is the most common cause of Cushing's syndrome, and should be excluded by history.

Diagnosis

Cushing's syndrome is suspected in patients with clinical features suggestive of glucocorticoid excess, including central obesity, thin skin with easy bruising, hypertension, plethoric moon face, purple striae, glucose intolerance or diabetes mellitus, osteoporosis, and proximal muscle weakness. Cushing's syndrome is confirmed by failure to suppress cortisol levels after a dose of the glucocorticoid dexamethasone, by elevated free cortisol levels in 24-hour urine collections, or by lack of physiological decline of cortisol levels late in the evening. Normal or elevated ACTH levels indicate ACTH-dependent etiologies for Cushing's syndrome, while low ACTH levels point towards an adrenal source of glucocorticoid excess. In general, ACTH-secreting pituitary adenomas retain feedback responsiveness to circulating glucocorticoids, and ACTH levels tend to be higher in ectopic ACTH secretion; however, considerable overlap in ectopic and pituitary ACTH secretion precludes accurate distinction between these disorders. ACTH-dependent Cushing's syndrome (non-suppressed ACTH levels) should be evaluated with pituitary imaging. If a clear pituitary mass > 5–6 mm is visible on MRI scanning, it is appropriate to proceed with transphenoidal resection of the pituitary adenoma. However, most ACTH-secreting adenomas are less than 5 mm in diameter, and may not be visible on MRI scan. In most cases, the distinction between pituitary- vs ectopic-derived ACTH hypersecretion needs to be assessed by bilateral inferior petrosal sinus and peripheral ACTH sampling before and after CRH administration. If this test indicates an ectopic ACTH source, imaging of the chest and abdomen are indicated, as these are the areas where ectopic ACTH-producing tumors are most commonly located.

Treatment

Surgical resection is the preferred treatment modality. ACTH-secreting tumors are usually small, and may be difficult to localize by pituitary MRI or bilateral inferior petrosal sinus sampling for ACTH gradients. Remission rate after surgery varies considerably (22–98%), depending on tumor size, degree of invasiveness, and

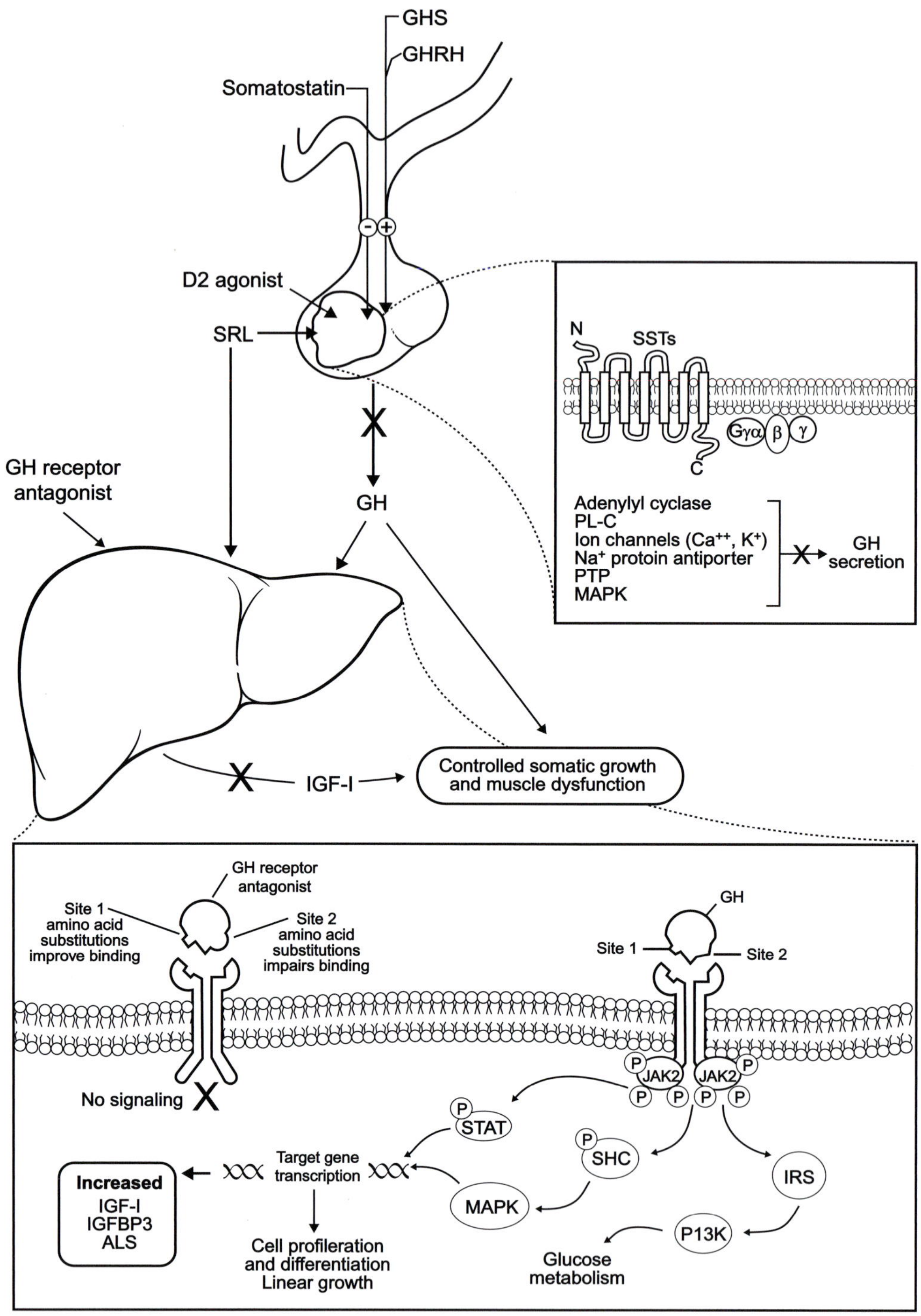

FIGURE 34.5 **Targets for the treatment of acromegaly.** Pituitary somatostatin receptor subtypes and D2 receptors and peripheral growth hormone (GH) receptors are targets for therapeutic ligands. Somatostatin analogs (or somatostatin receptor ligand, SRL) suppress levels of both GH and IGF-I, constrain tumor growth, and inhibit hepatic GH-receptor binding and action. GH-receptor antagonists prevent GH-receptor signaling, which attenuates peripheral IGF-I levels. SST, somatostatin receptor subtype; PL-C, phospholipase C; PTP; protein tyrosine phosphatase; MAPK, mitogenactivated protein kinase; IGFBP3; insulin-like growth factor-binding protein 3; ALS, acid-labile subunit; STAT, signal transducers and activators of transcription; JAK2, Janus kinase 2; PI3K, phosphoinositide 3 kinase; IRS, insulin receptor substrate. *Reproduced from Melmed S. Medical progress: Acromegaly.* N Engl J Med. *2006;355:2558–2573, with permission.*

BOX 34.4

HOW DO WE KNOW? DIAGNOSTIC APPROACH TO CUSHING'S SYNDROME

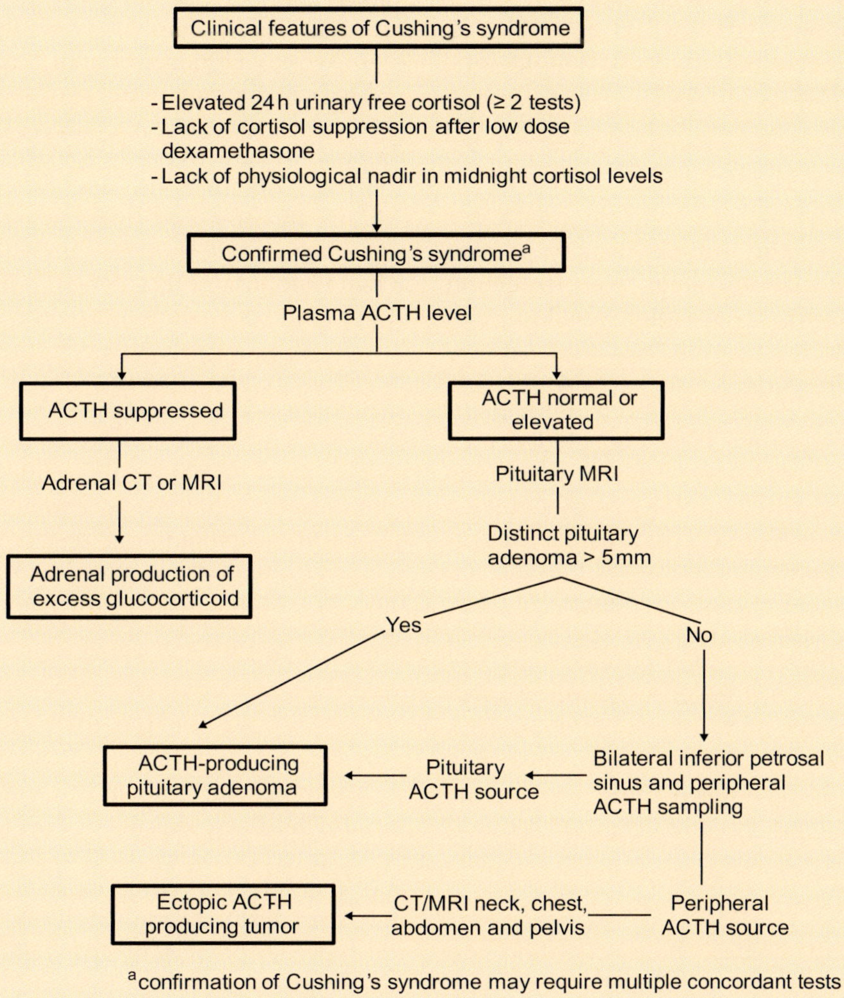

Modified from the Pituitary Center at Cedars Sinai Medical Center institution guidelines.

experience of the surgeon.[84] Overall remission rates are about 80% for microadenomas, but less than 50% for the less common ACTH-secreting macroadenomas. In one study,[85] surgical resection of 167 microadenomas identifiable on MRI resulted in 89% remission. After glucocorticoid excess is corrected by excision of an ACTH-producing tumoral source, there is a lag in recuperation of normal pituitary ACTH-secreting function for up to 1 year. Thus, subnormal cortisol levels in the first day or two after adenomectomy indicate successful removal of an ACTH-secreting adenoma and are associated with long-term remission. Patients with an ACTH-secreting pituitary adenoma not controlled by surgery require pituitary irradiation. Adrenal-directed cortisol-lowering agents, such as ketoconazole, metyrapone, mitotane or aminoglutethimide, may be administered after irradiation to achieve biochemical remission pending the therapeutic impact of radiotherapy. Bilateral adrenalectomy may be required if all these measures fail.

TSH-Secreting Adenomas

TSH synthesis and secretion is mainly stimulated by hypothalamic TRH, while thyroid hormones, dopamine, SRIF and glucocorticoids suppress TSH. Prolonged decrease in thyroid hormone production related to thyroid gland damage caused by thyroidectomy, antithyroid medication, radiation or chronic thyroiditis results in elevation of hypothalamic TRH, causing reversible thyrotrope hyperplasia and pituitary gland enlargement.

TSH-producing pituitary adenomas are very rare; however, they are most common macroadenomas and exhibit invasive behavior. In about one-fourth of the cases TSH-secreting adenomas may co-secrete other anterior pituitary hormones, mainly GH and PRL, as these hormones share common transcription factors, such as PROP-1 and POU1F1 (Pit-1).[86] These tumors express SSTR2 mRNA, and in some cases SSTR3 and SSTR5 mRNA.[87]

Diagnosis

Patients with TSH-secreting tumors usually present with symptoms of thyroid hormone overproduction and goiter or local compressive symptoms (Table 34.3). Symptoms of hyperthyroidism include palpitations, tremor, nervousness and weight loss. The diagnosis is made by demonstrating elevated T4 levels, with inappropriately normal or elevated TSH, and evidence of pituitary adenoma on MRI. These tumors co-secrete α-subunit that is elevated in the majority of patients. Thyroid hormone resistance can produce similar abnormalities in TSH and T4 levels, and this condition may be differentiated from TSH-secreting adenomas by TRH stimulation and T3 suppression tests. In TSH-secreting tumors, the TSH response is blunted to both TRH and T3 administration, while in thyroid hormone insensitivity TSH rises with TRH, and TSH is inhibited with T3 administration. In mixed tumor types, signs or symptoms of acromegaly or hyperprolactinemia may also be presenting complaints, and IGF-I and PRL levels should be determined.

Treatment

Surgical resection and medical therapy are the two main treatment modalities. The main goals of therapy are to decompress tumoral mass effect and restore euthyroidism. Transphenoidal surgery with tumor debulking results in cure – i.e., normalization of circulating thyroid hormone levels and apparent complete removal of tumor mass – in about one-third of patients.[88–90] Limited surgical cure rates are likely related to larger tumor sizes and invasive behavior. Undetectable circulating TSH in the postoperative period is a good predictor of surgical cure, given suppression of normal thyrotrope function by excess tumoral TSH,[91] and thyroid hormone replacement therapy may be necessary for weeks to months.

Medical therapy with propranolol and anti-thyroid medication may be considered as pre-operative adjunct therapy when hyperthyroidism requires immediate control. Somatostatin analogs can be also used as adjunct therapy, and there is growing evidence for its use in primary therapy of TSH-secreting adenomas when vision is not threatened.[88] Treatment with somatostatin analogs leads to reduction of TSH and α-subunit in almost all cases, with restoration of euthyroid state in 95% of the patients.[92,93] Pituitary tumor mass shrinkage occurs in about 50% of the cases.[94] SSTR5 tumor expression appears to predict therapeutic response to somatostatin analogs.[87,95] Therapy with somatostatin analogs is usually well tolerated, and a minority of patients are resistant to this therapy.[94]

Radiation therapy may be employed as adjunctive therapy when surgery and medical therapy are not curative.

FUTURE DIRECTIONS

No molecular imbalance common to all pituitary tumors has been identified. This could be due to limitations in technologies used to identify candidate genes, and the intrinsic complexity of pituitary tumors, as they are derived from at least five distinct pituitary cell types. Advances in investigative resources, such as transcriptomics and proteomics, has enabled identification of new genes and pathways (see Box 34.5). For instance, analysis of human pituitary adenoma proteomic data has reiterated the role of ERK/MAPK signaling and cell cycle regulation abnormalities, and has

BOX 34.5

HOW DO WE KNOW NOVEL GENES ARE INVOLVED IN PITUITARY PATHOGENESIS?

Initial studies on pituitary tumorigenesis used the candidate gene approach, exemplified by identification of disruption of genes involved in cell cycle regulation. However, this methodology has not clarified the proximal cause for most sporadic pituitary tumors, possibly due to involvement of multiple genes. More recently, whole genome analysis resources have yielded identification of novel genes and pathways for pathogenesis of pituitary tumors. Differential display methodology and microarray analysis are some of the most commonly employed resources.

BOX 34.5 *(cont'd)*

Differential Display Methodology

Differential display systematically amplifies mRNA by reverse transcription polymerase chain reaction with oligo-dT primers in combination with a set of random primers, and therefore does not require previous knowledge of mRNA or gene sequences. Resulting subsets of cDNA derived from relevant RNA samples are analyzed in denaturating polyacrylamide gel electrophoresis, indicating differentially expressed cDNA bands that may be retrieved and sequenced.[1] This technique enabled the identification of PTTG1 overexpression from rat pituitary tumor cells,[2] and downregulation of the bone morphogenetic protein 4 (BMP-4) inhibitor noggin from prolactinoma of dopamine receptor-null mice.[3] Similarly, GADD45γ and MEG3 were identified in human pituitary tumors using cDNA representational differential display, a similar whole genome screening resource that utilizes subtractive hybridization technology.[4,5]

Microarray Analysis

Microarray technology is based on a differential hybridization strategy, in which all mRNA that is expressed in the cell is copied into cDNA using reverse transcriptase and then used to hybridize to known sequences of DNA oligonucleotides in microarray platforms. Probe-target hybridization is usually identified by detection of labeled targets.[1]

This technology can be used to identify global changes in transcript expression of thousands of genes in a single experiment. However, noted gene expression imbalances between samples should be confirmed with more specific techniques, such as quantitative real time polymerase chain reaction.

To date, only five studies employing microarray technology for evaluation of human pituitary tumors have been published (reviewed in refs 6–11). The comparison of pituitary adenomas to normal pituitary corroborated earlier described genes with roles in cell proliferation or tumorigenesis, and has identified new candidate genes. Dissimilarity in gene expression profile between the different tumor subtypes was noted, and only six genes with proven roles in cell proliferation or tumorigenesis were differentially expressed in all pituitary tumor subtypes compared to normal pituitary: *GADD45B1, SAT1, ID1, VIM, IGBP5* and *ZFP36L1*.[11] These findings suggest that genes involved in pituitary tumor formation are largely cell-subtype specific. One study[10] compared prolactinomas according to growth behavior, and identified PTTG, CCB1, AURKB, ASK and CENPE as genes involved in tumor proliferation, invasion and aggressiveness. Therefore, elucidation of global gene expression imbalances may help identify new mechanisms for pituitary tumors, with implications for the recognition and identification of possible new therapeutic targets.

References

1. Liang P, Pardee AB. Analysing differential gene expression in cancer. *Nat Rev Cancer.* 2003;3:869–876.
2. Pei L, Melmed S. Isolation and characterization of a pituitary tumor-transforming gene (PTTG). *Mol Endocrinol.* 1997;11:433–441.
3. Paez-Pereda M, Giacomini D, Refojo D, Nagashima AC, Hopfner U, Grubler Y, et al. Involvement of bone morphogenetic protein 4 (BMP-4) in pituitary prolactinoma pathogenesis through a Smad/estrogen receptor crosstalk. *Proc Natl Acad Sci USA.* 2003;100:1034–1039.
4. Zhang X, Sun H, Danila DC, Johnson SR, Zhou Y, Swearingen B, et al. Loss of expression of GADD45 gamma, a growth inhibitory gene, in human pituitary adenomas: implications for tumorigenesis. *J Clin Endocrinol Metab.* 2002;87:1262–1267.
5. Zhang X, Zhou Y, Mehta KR, Danila DC, Scolavino S, Johnson SR, et al. A pituitary-derived MEG3 isoform functions as a growth suppressor in tumor cells. *J Clin Endocrinol Metab.* 2003;88:5119–5126.
6. Evans CO, Young AN, Brown MR, Brat DJ, Parks JS, Neish AS, et al. Novel patterns of gene expression in pituitary adenomas identified by complementary deoxyribonucleic acid microarrays and quantitative reverse transcription-polymerase chain reaction. *J Clin Endocrinol Metab.* 2001;86:3097–3107.
7. Evans CO, Moreno CS, Zhan X, McCabe MT, Vertino PM, Desiderio DM, et al. Molecular pathogenesis of human prolactinomas identified by gene expression profiling, RT-qPCR, and proteomic analyses. *Pituitary.* 2008; 11:231–245.
8. Morris DG, Musat M, Czirjak S, Hanzely Z, Lillington DM, Korbonits M, et al. Differential gene expression in pituitary adenomas by oligonucleotide array analysis. *Eur J Endocrinol.* 2005;153:143–151.
9. Moreno CS, Evans CO, Zhan X, Okor M, Desiderio DM, Oyesiku NM. Novel molecular signaling and classification of human clinically nonfunctional pituitary adenomas identified by gene expression profiling and proteomic analyses. *Cancer Res.* 2005;65:10214–10222.
10. Wierinckx A, Auger C, Devauchelle P, Reynaud A, Chevallier P, Jan M, et al. A diagnostic marker set for invasion, proliferation, and aggressiveness of prolactin pituitary tumors. *Endocr Relat Cancer.* 2007;14:887–900.
11. Wierinckx A, Raverot G, Nazaret N, Jouanneau E, Auger C, Lachuer J, et al. Proliferation markers of human pituitary tumors: contribution of a genome-wide transcriptome approach. *Mol Cell Endocrinol.* 2010;326:30–39.

suggested new pathways for pituitary tumor formation, including mitochondrial dysfunction and oxidative stress deregulation.[96] Results derived from these technologies may provide new insights into pituitary pathogenesis; however, whether or not this new information will be applicable for management of individual pituitary adenoma cases remains unproven.

References

1. Molitch ME. Pituitary tumours: pituitary incidentalomas. *Best Pract Res Clin Endocrinol Metab.* 2009;23:667–675.
2. Pei L, Melmed S, Scheithauer B, Kovacs K, Benedict WF, Prager D. Frequent loss of heterozygosity at the retinoblastoma susceptibility gene (RB) locus in aggressive pituitary tumors: evidence for a chromosome 13 tumor suppressor gene other than RB. *Cancer Res.* 1995;55:1613–1616.
3. Daly AF, Rixhon M, Adam C, Dempegioti A, Tichomirowa MA, Beckers A. High prevalence of pituitary adenomas: a cross-sectional study in the province of Liege. *Belgium. J Clin Endocrinol Metab.* 2006;91:4769–4775.
4. Fernandez A, Karavitaki N, Wass JA. Prevalence of pituitary adenomas: a community-based, cross-sectional study in Banbury (Oxfordshire, UK). *Clin Endocrinol (Oxf).* 2010;72:377–382.
5. Saeger W, Ludecke DK, Buchfelder M, Fahlbusch R, Quabbe HJ, Petersenn S. Pathohistological classification of pituitary tumors: 10 years of experience with the German Pituitary Tumor Registry. *Eur J Endocrinol.* 2007;156:203–216.
6. Clayton RN. Sporadic pituitary tumours: from epidemiology to use of databases. *Baillieres Best Pract Res Clin Endocrinol Metab.* 1999;13:451–460.
7. Melmed S. Mechanisms for pituitary tumorigenesis: the plastic pituitary. *J Clin Invest.* 2003;112:1603–1618.
8. Donangelo I, Melmed S. Molecular pathogenesis of pituitary tumors. In: *Clinical Oxford Textbook of Endocrinology and Metabolism.* Oxford: Oxford University Press; 2011 (in press).
9. Herman V, Fagin J, Gonsky R, Kovacs K, Melmed S. Clonal origin of pituitary adenomas. *J Clin Endocrinol Metab.* 1990;71: 1427–1433.
10. Alexander JM, Biller BM, Bikkal H, Zervas NT, Arnold A, Klibanski A. Clinically nonfunctioning pituitary tumors are monoclonal in origin. *J Clin Invest.* 1990;86:336–340.
11. Sarkar DK. Genesis of prolactinomas: studies using estrogen-treated animals. *Front Horm Res.* 2006;35:32–49.
12. Schoemaker MJ, Swerdlow AJ. Risk factors for pituitary tumors: a case–control study. *Cancer Epidemiol Biomark Prev.* 2009;18: 1492–1500.
13. Thorner MO, Perryman RL, Cronin MJ, Rogol AD, Draznin M, Johanson A, et al. Somatotroph hyperplasia. Successful treatment of acromegaly by removal of a pancreatic islet tumor secreting a growth hormone-releasing factor. *J Clin Invest.* 1982;70:965–977.
14. Zou H, McGarry TJ, Bernal T, Kirschner MW. Identification of a vertebrate sister-chromatid separation inhibitor involved in transformation and tumorigenesis. *Science.* 1999;285:418–422.
15. Abbud RA, Takumi I, Barker EM, Ren SG, Chen DY, Wawrowsky K, et al. Early multipotential pituitary focal hyperplasia in the alpha-subunit of glycoprotein hormone-driven pituitary tumor-transforming gene transgenic mice. *Mol Endocrinol.* 2005;19:1383–1391.
16. Chesnokova V, Kovacs K, Castro AV, Zonis S, Melmed S. Pituitary hypoplasia in Pttg−/− mice is protective for Rb+/− pituitary tumorigenesis. *Mol Endocrinol.* 2005;19:2371–2379.
17. Donangelo I, Gutman S, Horvath E, Kovacs K, Wawrowsky K, Mount M, et al. Pituitary tumor transforming gene overexpression facilitates pituitary tumor development. *Endocrinology.* 2006;147:4781–4791.
18. Simpson DJ, Hibberts NA, McNicol AM, Clayton RN, Farrell WE. Loss of pRb expression in pituitary adenomas is associated with methylation of the RB1 CpG island. *Cancer Res.* 2000;60: 1211–1216.
19. Yoshino A, Katayama Y, Ogino A, Watanabe T, Yachi K, Ohta T, et al. Promoter hypermethylation profile of cell cycle regulator genes in pituitary adenomas. *J Neurooncol.* 2007;83:153–162.
20. Ruebel KH, Jin L, Zhang S, Scheithauer BW, Lloyd RV. Inactivation of the p16 gene in human pituitary nonfunctioning tumors by hypermethylation is more common in null cell adenomas. *Endocr Pathol.* 2001;12:281–289.
21. Simpson DJ, Bicknell JE, McNicol AM, Clayton RN, Farrell WE. Hypermethylation of the p16/CDKN2A/MTSI gene and loss of protein expression is associated with nonfunctional pituitary adenomas but not somatotrophinomas. *Genes Chrom Cancer.* 1999;24:328–336.
22. Kirsch M, Morz M, Pinzer T, Schackert HK, Schackert G. Frequent loss of the CDKN2C (p18INK4c) gene product in pituitary adenomas. *Genes Chrom Cancer.* 2009;48:143–154.
23. Hossain MG, Iwata T, Mizusawa N, Qian ZR, Shima SW, Okutsu T, et al. Expression of p18(INK4C) is down-regulated in human pituitary adenomas. *Endocr Pathol.* 2009;20:114–121.
24. Ogino A, Yoshino A, Katayama Y, Watanabe T, Ota T, Komine C, et al. The p15(INK4b)/p16(INK4a)/RB1 pathway is frequently deregulated in human pituitary adenomas. *J Neuropathol Exp Neurol.* 2005;64:398–403.
25. Kiyokawa H, Kineman RD, Manova-Todorova KO, Soares VC, Hoffman ES, Ono M, et al. Enhanced growth of mice lacking the cyclin-dependent kinase inhibitor function of p27(Kip1). *Cell.* 1996;85:721–732.
26. Pellegata NS, Quintanilla-Martinez L, Siggelkow H, Samson E, Bink K, Hofler H, et al. Germ-line mutations in p27Kip1 cause a multiple endocrine neoplasia syndrome in rats and humans. *Proc Natl Acad Sci USA.* 2006;103:15558–15563.
27. Korbonits M, Chahal HS, Kaltsas G, Jordan S, Urmanova Y, Khalimova Z, et al. Expression of phosphorylated p27(Kip1) protein and Jun activation domain-binding protein 1 in human pituitary tumors. *J Clin Endocrinol Metab.* 2002;87: 2635–2643.
28. Komatsubara K, Tahara S, Umeoka K, Sanno N, Teramoto A, Osamura RY. Immunohistochemical analysis of p27 (Kip1) in human pituitary glands and in various types of pituitary adenomas. *Endocr Pathol.* 2001;12:181–188.
29. Martin-Caballero J, Flores JM, Garcia-Palencia P, Serrano M. Tumor susceptibility of p21(Waf1/Cip1)-deficient mice. *Cancer Res.* 2001;61:6234–6238.
30. Hibberts NA, Simpson DJ, Bicknell JE, Broome JC, Hoban PR, Clayton RN, et al. Analysis of cyclin D1 (CCND1) allelic imbalance and overexpression in sporadic human pituitary tumors. *Clin Cancer Res.* 1999;5:2133–2139.
31. Vlotides G, Eigler T, Melmed S. Pituitary tumor-transforming gene: physiology and implications for tumorigenesis. *Endocr Rev.* 2007;28:165–186.
32. Zhang X, Horwitz GA, Heaney AP, Nakashima M, Prezant TR, Bronstein MD, et al. Pituitary tumor transforming gene (PTTG) expression in pituitary adenomas. *J Clin Endocrinol Metab.* 1999;84:761–767.
33. Landis CA, Harsh G, Lyons J, Davis RL, McCormick F, Bourne HR. Clinical characteristics of acromegalic patients whose pituitary tumors contain mutant Gs protein. *J Clin Endocrinol Metab.* 1990;71:1416–1420.

34. Hayward BE, Barlier A, Korbonits M, Grossman AB, Jacquet P, Enjalbert A, et al. Imprinting of the G(s)alpha gene GNAS1 in the pathogenesis of acromegaly. *J Clin Invest.* 2001;107:R31–R36.
35. Bertherat J, Chanson P, Montminy M. The cyclic adenosine 3′,5′-monophosphate-responsive factor CREB is constitutively activated in human somatotroph adenomas. *Mol Endocrinol.* 1995; 9:777–783.
36. Heaney AP, Horwitz GA, Wang Z, Singson R, Melmed S. Early involvement of estrogen-induced pituitary tumor transforming gene and fibroblast growth factor expression in prolactinoma pathogenesis. *Nat Med.* 1999;5:1317–1321.
37. Turner HE, Harris AL, Melmed S, Wass JA. Angiogenesis in endocrine tumors. *Endocr Rev.* 2003;24:600–632.
38. Fukui S, Otani N, Nawashiro H, Yano A, Nomura N, Tokumaru AM, et al. The association of the expression of vascular endothelial growth factor with the cystic component and haemorrhage in pituitary adenoma. *J Clin Neurosci.* 2003;10:320–324.
39. McCabe CJ, Khaira JS, Boelaert K, Heaney AP, Tannahill LA, Hussain S, et al. Expression of pituitary tumour transforming gene (PTTG) and fibroblast growth factor-2 (FGF-2) in human pituitary adenomas: relationships to clinical tumour behaviour. *Clin Endocrinol (Oxf).* 2003;58:141–150.
40. Vlotides G, Chen YH, Eigler T, Ren SG, Melmed S. Fibroblast growth factor-2 autofeedback regulation in pituitary folliculostellate TtT/GF cells. *Endocrinology.* 2009;150:3252–3258.
41. Hentges S, Boyadjieva N, Sarkar DK. Transforming growth factor-beta3 stimulates lactotrope cell growth by increasing basic fibroblast growth factor from folliculo-stellate cells. *Endocrinology.* 2000;141:859–867.
42. Shimon I, Huttner A, Said J, Spirina OM, Melmed S. Heparin-binding secretory transforming gene (hst) facilitates rat lactotrope cell tumorigenesis and induces prolactin gene transcription. *J Clin Invest.* 1996;97:187–195.
43. Chaidarun SS, Eggo MC, Sheppard MC, Stewart PM. Expression of epidermal growth factor (EGF), its receptor, and related oncoprotein (erbB-2) in human pituitary tumors and response to EGF *in vitro*. *Endocrinology.* 1994;135:2012–2021.
44. Onguru O, Scheithauer BW, Kovacs K, Vidal S, Jin L, Zhang S, et al. Analysis of epidermal growth factor receptor and activated epidermal growth factor receptor expression in pituitary adenomas and carcinomas. *Mod Pathol.* 2004;17:772–780.
45. Botelho CH, Magalhaes AV, Mello PA, Schmitt FC, Casulari LA. Expression of p53, Ki-67 and c-erb B2 in growth hormone and/or prolactin-secreting pituitary adenomas. *Arq Neuropsiquiatr.* 2006; 64:60–66.
46. Vlotides G, Cooper O, Chen YH, Ren SG, Greenman Y, Melmed S. Heregulin regulates prolactinoma gene expression. *Cancer Res.* 2009;69:4209–4216.
47. Vlotides G, Siegel E, Donangelo I, Gutman S, Ren SG, Melmed S. Rat prolactinoma cell growth regulation by epidermal growth factor receptor ligands. *Cancer Res.* 2008;68:6377–6386.
48. Zhang X, Sun H, Danila DC, Johnson SR, Zhou Y, Swearingen B, et al. Loss of expression of GADD45 gamma, a growth inhibitory gene, in human pituitary adenomas: implications for tumorigenesis. *J Clin Endocrinol Metab.* 2002;87:1262–1267.
49. Zhang X, Zhou Y, Mehta KR, Danila DC, Scolavino S, Johnson SR, et al. A pituitary-derived MEG3 isoform functions as a growth suppressor in tumor cells. *J Clin Endocrinol Metab.* 2003;88: 5119–5126.
50. Zhang X, Rice K, Wang Y, Chen W, Zhong Y, Nakayama Y, et al. Maternally expressed gene 3 (MEG3) noncoding ribonucleic acid: isoform structure, expression, and functions. *Endocrinology.* 2010; 151:939–947.
51. Zhao J, Dahle D, Zhou Y, Zhang X, Klibanski A. Hypermethylation of the promoter region is associated with the loss of MEG3 gene expression in human pituitary tumors. *J Clin Endocrinol Metab.* 2005;90:2179–2186.
52. Qian ZR, Asa SL, Siomi H, Siomi MC, Yoshimoto K, Yamada S, et al. Overexpression of HMGA2 relates to reduction of the let-7 and its relationship to clinicopathological features in pituitary adenomas. *Mod Pathol.* 2009;22:431–441.
53. De Martino I, Visone R, Wierinckx A, Palmieri D, Ferraro A, Cappabianca P, et al. HMGA proteins up-regulate CCNB2 gene in mouse and human pituitary adenomas. *Cancer Res.* 2009;69:1844–1850.
54. Fedele M, Visone R, De Martino I, Troncone G, Palmieri D, Battista S, et al. HMGA2 induces pituitary tumorigenesis by enhancing E2F1 activity. *Cancer Cell.* 2006;9:459–471.
55. Verges B, Boureille F, Goudet P, Murat A, Beckers A, Sassolas G, et al. Pituitary disease in MEN type 1 (MEN1): data from the France-Belgium MEN1 multicenter study. *J Clin Endocrinol Metab.* 2002;87:457–465.
56. Yang Y, Hua X. In search of tumor suppressing functions of menin. *Mol Cell Endocrinol.* 2007;265-266:34–41.
57. Agarwal SK, Mateo CM, Marx SJ. Rare germline mutations in cyclin-dependent kinase inhibitor genes in multiple endocrine neoplasia type 1 and related states. *J Clin Endocrinol Metab.* 2009;94:1826–1834.
58. Georgitsi M, Raitila A, Karhu A, van der Luijt RB, Aalfs CM, Sane T, et al. Germline CDKN1B/p27Kip1 mutation in multiple endocrine neoplasia. *J Clin Endocrinol Metab.* 2007;92:3321–3325.
59. Chahal HS, Chapple JP, Frohman LA, Grossman AB, Korbonits M. Clinical, genetic and molecular characterization of patients with familial isolated pituitary adenomas (FIPA). *Trends Endocrinol Metab.* 21:419–427.
60. Beckers A, Daly AF. The clinical, pathological, and genetic features of familial isolated pituitary adenomas. *Eur J Endocrinol.* 2007;157:371–382.
61. Vierimaa O, Georgitsi M, Lehtonen R, Vahteristo P, Kokko A, Raitila A, et al. Pituitary adenoma predisposition caused by germline mutations in the AIP gene. *Science.* 2006;312:1228–1230.
62. Leontiou CA, Gueorguiev M, van der Spuy J, Quinton R, Lolli F, Hassan S, et al. The role of the aryl hydrocarbon receptor-interacting protein gene in familial and sporadic pituitary adenomas. *J Clin Endocrinol Metab.* 2008;93:2390–2401.
63. Kirschner LS, Carney JA, Pack SD, Taymans SE, Giatzakis C, Cho YS, et al. Mutations of the gene encoding the protein kinase A type I-alpha regulatory subunit in patients with the Carney complex. *Nat Genet.* 2000;26:89–92.
64. Boikos SA, Stratakis CA. Carney complex: pathology and molecular genetics. *Neuroendocrinology.* 2006;83:189–199.
65. Chesnokova V, Zonis S, Kovacs K, Ben-Shlomo A, Wawrowsky K, Bannykh S, et al. p21(Cip1) restrains pituitary tumor growth. *Proc Natl Acad Sci USA.* 2008;105:17498–17503.
66. Freda PU, Post KD. Differential diagnosis of sellar masses. *Endocrinol Metab Clin North Am.* 1999;28:81–117. vi.
67. Colao A, Di Sarno A, Cappabianca P, Di Somma C, Pivonello R, Lombardi G. Withdrawal of long-term cabergoline therapy for tumoral and nontumoral hyperprolactinemia. *N Engl J Med.* 2003;349:2023–2033.
68. Delgrange E, Trouillas J, Maiter D, Donckier J, Tourniaire J. Sex-related difference in the growth of prolactinomas: a clinical and proliferation marker study. *J Clin Endocrinol Metab.* 1997;82: 2102–2107.
69. Melmed S, Ho K. Pituitary physiology and diagnostic evaluation. In: Kronenberg HM, Melmed S, Polonsky KS, Larsen PR, eds. *Williams Textbook of Endocrinology.* Philadelphia: Saunders Elsevier; 2011.
70. Milenkovic L, D'Angelo G, Kelly PA, Weiner RI. Inhibition of gonadotropin hormone-releasing hormone release by prolactin

from GT1 neuronal cell lines through prolactin receptors. *Proc Natl Acad Sci USA.* 1994;91:1244–1247.

71. Seriwatanachai D, Thongchote K, Charoenphandhu N, Pandaranandaka J, Tudpor K, Teerapornpuntakit J, et al. Prolactin directly enhances bone turnover by raising osteoblast-expressed receptor activator of nuclear factor kappaB ligand/osteoprotegerin ratio. *Bone.* 2008;42:535–546.
72. Verhelst J, Abs R, Maiter D, van den Bruel A, Vandeweghe M, Velkeniers B, et al. Cabergoline in the treatment of hyperprolactinemia: a study in 455 patients. *J Clin Endocrinol Metab.* 1999;84:2518–2522.
73. Webster J, Piscitelli G, Polli A, Ferrari CI, Ismail I, Scanlon MF. A comparison of cabergoline and bromocriptine in the treatment of hyperprolactinemic amenorrhea. Cabergoline Comparative Study Group. *N Engl J Med.* 1994;331:904–909.
74. Valassi E, Klibanski A, Biller BM. Clinical Review: Potential cardiac valve effects of dopamine agonists in hyperprolactinemia. *J Clin Endocrinol Metab.* 2010;95:1025–1033.
75. Fatemi N, Dusick JR, Mattozo C, McArthur DL, Cohan P, Boscardin J, et al. Pituitary hormonal loss and recovery after transsphenoidal adenoma removal. *Neurosurgery.* 2008;63:709–718. discussion 18-19.
76. Melmed S, Kleinberg D. Pituitary masses and tumors. In: Kronenberg HM, Melmed S, Polonsky KS, Larsen PR, eds. *Williams Textbook of Endocrinology.* Philadelphia: Saunders Elsevier; 2008.
77. Melmed S. Medical progress: Acromegaly. *N Engl J Med.* 2006;355:2558–2573.
78. Drange MR, Fram NR, Herman-Bonert V, Melmed S. Pituitary tumor registry: a novel clinical resource. *J Clin Endocrinol Metab.* 2000;85:168–174.
79. Sherlock M, Reulen RC, Alonso AA, Ayuk J, Clayton RN, Sheppard MC, et al. ACTH deficiency, higher doses of hydrocortisone replacement, and radiotherapy are independent predictors of mortality in patients with acromegaly. *J Clin Endocrinol Metab.* 2009;94:4216–4223.
80. Cozzi R, Montini M, Attanasio R, Albizzi M, Lasio G, Lodrini S, et al. Primary treatment of acromegaly with octreotide LAR: a long-term (up to nine years) prospective study of its efficacy in the control of disease activity and tumor shrinkage. *J Clin Endocrinol Metab.* 2006;91:1397–1403.
81. Melmed S, Sternberg R, Cook D, Klibanski A, Chanson P, Bonert V, et al. A critical analysis of pituitary tumor shrinkage during primary medical therapy in acromegaly. *J Clin Endocrinol Metab.* 2005;90:4405–4410.
82. Mattar P, Alves Martins MR, Abucham J. Short- and long-term efficacy of combined cabergoline and octreotide treatment in controlling IGF-I levels in acromegaly. *Neuroendocrinology.* 2010;92:120–127.
83. Neggers SJ, de Herder WW, Janssen JA, Feelders RA, van der Lely AJ. Combined treatment for acromegaly with long-acting somatostatin analogs and pegvisomant: long-term safety for up to 4.5 years (median 2.2 years) of follow-up in 86 patients. *Eur J Endocrinol.* 2009;160:529–533.
84. Kelly DF. Transsphenoidal surgery for Cushing's disease: a review of success rates, remission predictors, management of failed surgery, and Nelson's Syndrome. *Neurosurg Focus.* 2007;23:E5.
85. Prevedello DM, Pouratian N, Sherman J, Jane Jr JA, Vance ML, Lopes MB, et al. Management of Cushing's disease: outcome in patients with microadenoma detected on pituitary magnetic resonance imaging. *J Neurosurg.* 2008;109:751–759.
86. Cohen LE, Radovick S. Molecular basis of combined pituitary hormone deficiencies. *Endocr Rev.* 2002;23:431–442.
87. Yoshihara A, Isozaki O, Hizuka N, Nozoe Y, Harada C, Ono M, et al. Expression of type 5 somatostatin receptor in TSH-secreting pituitary adenomas: a possible marker for predicting long-term response to octreotide therapy. *Endocr J.* 2007;54:133–138.
88. Ness-Abramof R, Ishay A, Harel G, Sylvetzky N, Baron E, Greenman Y, et al. TSH-secreting pituitary adenomas: follow-up of 11 cases and review of the literature. *Pituitary.* 2007;10:307–310.
89. Beckers A, Abs R, Mahler C, Vandalem JL, Pirens G, Hennen G, et al. Thyrotropin-secreting pituitary adenomas: report of seven cases. *J Clin Endocrinol Metab.* 1991;72:477–483.
90. McCutcheon IE, Weintraub BD, Oldfield EH. Surgical treatment of thyrotropin-secreting pituitary adenomas. *J Neurosurg.* 1990;73:674–683.
91. Losa M, Giovanelli M, Persani L, Mortini P, Faglia G, Beck-Peccoz P. Criteria of cure and follow-up of central hyperthyroidism due to thyrotropin-secreting pituitary adenomas. *J Clin Endocrinol Metab.* 1996;81:3084–3090.
92. Caron P, Arlot S, Bauters C, Chanson P, Kuhn JM, Pugeat M, et al. Efficacy of the long-acting octreotide formulation (octreotide-LAR) in patients with thyrotropin-secreting pituitary adenomas. *J Clin Endocrinol Metab.* 2001;86:2849–2853.
93. Chanson P, Weintraub BD, Harris AG. Octreotide therapy for thyroid-stimulating hormone-secreting pituitary adenomas. A follow-up of 52 patients. *Ann Intern Med.* 1993;119:236–240.
94. Colao A, Pivonello R, Di Somma C, Savastano S, Grasso LF, Lombardi G. Medical therapy of pituitary adenomas: effects on tumor shrinkage. *Rev Endocr Metab Disord.* 2009;10:111–123.
95. Horiguchi K, Yamada M, Umezawa R, Satoh T, Hashimoto K, Tosaka M, et al. Somatostatin receptor subtypes mRNA in TSH-secreting pituitary adenomas: a case showing a dramatic reduction in tumor size during short octreotide treatment. *Endocr J.* 2007;54:371–378.
96. Zhan X, Desiderio DM. Signaling pathway networks mined from human pituitary adenoma proteomics data. *BMC Med Genomics.* 2010;3:13.

CHAPTER

35

Neuroendocrine Tumors

Richard A. Feelders [1], *Leo J. Hofland* [1], *Dik J. Kwekkeboom* [2], *Steven W. Lamberts* [1], *Wouter W. de Herder* [1]

[1] Department of Internal Medicine, Section of Endocrinology, Erasmus Medical Center, Rotterdam, The Netherlands,
[2] Department of Nuclear Medicine, Erasmus Medical Center, Rotterdam, The Netherlands

OUTLINE

Summary

Neuroendocrine tumors (NETs), formerly known as "carcinoid tumors," originate from enterochromaffin cells in the bronchial and gastrointestinal tracts. NETs are rare tumors, although their incidence seems to be increasing. Functional NETs can produce hormones and mediators leading to distinct clinical syndromes, such as the carcinoid syndrome, consisting of flushes and diarrhoea caused by serotonin-producing midgut-NET. Non-functional NETs usually present with symptoms related to mass effects, including abdominal pain, jaundice and small bowel obstruction. NETs can occur as sporadic tumors or as a component of a hereditary tumor syndrome, including multiple endocrine neoplasia type I or von Hippel Lindau syndrome. Based on clinicopathological characteristics, NETs are classified

Handbook of Neuroendocrinology, DOI: 10.1016/B978-0-12-375097-6.10035-6

as well-differentiated neuroendocrine tumors with benign behavior (G1), well-differentiated neuroendocrine carcinomas with low-grade malignant behavior (G2), and poorly differentiated neuroendocrine carcinomas (G3). With this grading system, a prognostic stratification can be made. Chromogranin A serves as biochemical tumor marker for NETs. Imaging modalities that are used to localize NETs and their metastases include computerized tomography, magnetic resonance imaging, endoscopic ultrasound, ^{111}In-pentetreotide scintigraphy, and positron emission tomography. Surgery is indicated in patients with NETs for curative resection, tumor debulking, or to treat complications like small bowel obstruction. Liver metastases can be treated with intervention radiological techniques like hepatic artery (chemo-)embolization or radiofrequency ablation. Somatostatin analogs play an important role in medical therapy for NETs, and have both antisecretory and antiproliferative effects. New drugs that are currently under investigation in patients with a NET are pasireotide, vascular endothelial growth factor (VEGF) monoclonal antibodies, inhibitors of the mammalian target of rapamycin (mTOR) and tyrosine kinase inhibitors. Finally, peptide receptor radionuclide therapy shows promising results with respect to tumor stabilization and regression. Considering the variations in clinical presentation and behavior of NETs, diagnostic and therapeutic management should be performed on a tailor-made basis.

INTRODUCTION

Neuroendocrine tumors (NETs) originate from enterochromaffin cells that are part of the neuroendocrine system of the bronchial and gastrointestinal tracts.[1,2] NETs located in the gastrointestinal tract and pancreas are also referred to as gastroenteropancreatic neuroendocrine tumors (GEP NETs). The nomenclature of these tumors has changed over time. At the beginning of the 20th century, Oberndorfer first described NETs and introduced the term "carcinoid," which combined the carcinoma-like characteristics of these tumors on the one hand and the relatively benign disease course on the other.[2,3] Subsequently, the term carcinoid was coupled to tumor localization (for example, bronchial carcinoid, small bowel carcinoid, etc.) and to a clinical syndrome caused by overproduction of vasoactive mediators by midgut NETs – i.e., "carcinoid syndrome." In addition, aggressive, poorly differentiated NETs were referred to as "atypical carcinoids." Although the term "carcinoid" is still used frequently in several contexts, the recurrent WHO nomenclature has displaced its use for NET with a classification of the tumor according to the degree of differentiation and proliferation[4–6] (Table 35.1). Historically, NETs are classified according to their localization in parts of the embryologic gut (i.e., foregut, midgut and hindgut). Foregut-derived tumors include bronchopulmonary, gastric, (proximal) duodenal and pancreatic NETs; midgut NETs are localized from the distal duodenum up to the ascending colon; and hindgut tumors originate from neuroendocrine cells in the transverse and descending colon and rectum.

TABLE 35.1 WHO Classification of Neuroendocrine Tumors

GRADE	*DEFINITION*	*PROLIFERATION – KI-67 INDEX*
1	Well-differentiated neuroendocrine tumor	≤ 2
2	Well-differentiated neuroendocrine carcinoma	3–20
3	Poorly differentiated neuroendocrine carcinoma	> 20

Table 35.2 shows the incidence of NETs according to localization.[7,8]

NETs are considered to be rare tumors, but recent studies indicate that the incidence of NETs is increasing[9] (Fig. 35.1). It is not clear yet whether this a true increase in NET incidence, the result of increased use of diagnostic procedures, or a combination of both. NETs mostly present as sporadic tumors, but can also be part of an hereditary tumor syndrome such as multiple endocrine neoplasia (MEN) syndrome type I or von Hippel Lindau syndrome. NETs can be divided into "functional" and "non-functional" tumors. Functional NETs are able to produce, store and secrete bioactive peptides, and can present with signs and symptoms related to the biological effects of these substances. Alongside production of peptides that are physiologically produced by neuroendocrine cells, functional NETs can also produce hormones that are normally secreted by endocrine glands – i.e., ectopic hormone production. Non-functional NETs can present with mechanical effects, such as bowel obstruction or ischemia, but are also frequently discovered incidentally during diagnostic procedures.[2]

The clinical course of NETs can be highly variable, and includes a spectrum ranging from well-differentiated, slow-growing tumors to aggressive, highly proliferative

TABLE 35.2 Incidence of NETs According to Localization

NETs	*LOCALIZATION*
Foregut	
Thymus	0.4%
Lung	29.8%
Stomach	4.9%
Midgut	
Small intestine	30.4%
Gallbladder, pancreas	1.0%
Hindgut	
Appendix	5.1%
Colon	9.2%
Rectum	14.5%

Adapted from Modlin et al. *(2003)*[7] *and Bertino* et al. *(2009)*.[8]

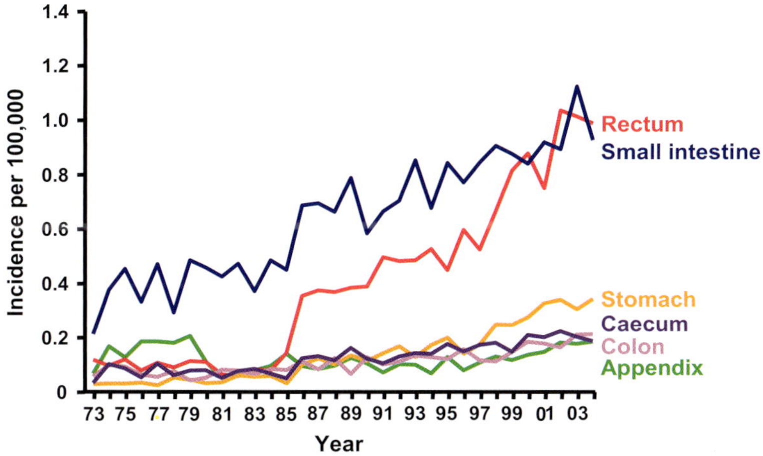

FIGURE 35.1 Annual age-adjusted incidence of NETs in the US population by anatomical location. *Modified from Yao et al. (2008).*[9]

tumors. In recent decades diagnostic procedures have been improved, with new developments in biochemical evaluation and in radiological, scintigraphic and endoscopic imaging techniques. In addition, several treatment modalities have been developed that aim to decrease the morbidity and mortality of patients with NETs. In this chapter, the biological, pathological and genetic aspects of NETs are discussed, as well as their clinical presentation, diagnosis, and treatment.

TUMOR BIOLOGY

Origin and Localization of Neuroendocrine Cells in Fore-, Mid- and Hindgut

Due to the presence of a variety of cells producing different hormones, the gastrointestinal tract may be considered the largest endocrine organ in the human body. Diffuse endocrine cells can be found throughout the gut. The neuroendocrine cell system of the gastrointestinal tract consists of at least 14 distinct cell types, each capable of producing specific peptide hormones or biogenic amines. Regulation of secretions includes the assembly, storage and release of two different secretory vesicles, the large dense core vesicles (LDCVs) being the well-known electron-dense granules of the endocrine cell, as well as synaptic-like microvesicles (SLMVs), which are smaller than the LDCVs and relatively similar to the synaptic vesicles of nerve endings. Endocrine, paracrine and autocrine secretions are all involved in the regulation of various physiological processes in the gut, including secretion, absorption, mucosal cell proliferation, and motility. Typical granular markers associated with LDCVs and SLMCVs are chromogranin and synaptophysin, respectively. Two isoforms of the ATP-dependent vesicular monoamine transporter, VMAT1 and VMAT2, display a differential expression in the LDCVs of gut endocrine cells, with VMAT1 being selectively expressed in serotonin-producing enterochromaffin (EC) cells and VMAT2 in histamine-producing enterochromaffin-like (ECL) cells, as well as in pancreatic islet cells. The endocrine cell compartment in the pancreas and gastrointestinal tract displays a clear compartmentalization. Whereas insulin-producing B cells are exclusively found in the pancreatic islets, and gastric ECL cells in the corpus-fundus of the stomach, several other cell types are restrictively localized in specific gut regions. Therefore, tissue-specific differentiation programs seem to control the "normal" endocrine cell maturation in the gastroenteropancreatic tract.[10] Table 35.3 provides an overview of the 14 identified gastroenteropancreatic cell types, their specific characteristics and their tissue localization. Although gut endocrine cells are highly differentiated cells, incapable of proliferating, endocrine cells in the stomach, pancreas and gut may be capable of changing their number and even phenotype in response to stimuli. Rindi and colleagues hypothesized that "endocrine cell homeostasis is maintained by the entry of new endocrine-committed cells along the differentiation pathway and that such intermediate cells may be sensitive to physiological stimuli as well as transforming agents."[10]

Biology of Neuroendocrine Cells

Depending upon the specific production of peptides and amines produced by the different neuroendocrine cell types, along with their tissue-specific localization patterns, they all influence specific physiological functions. Moreover, the activity of neuroendocrine cells of the gut, like cells of the neuroendocrine cell systems in general, is regulated via a variety of different G-protein coupled as well as growth factor receptors expressed on their cell surface. These receptors are involved in the regulation of secretion and production of the peptides and bioactive amines, as well as in maintaining cellular

TABLE 35.3 Type, Hormonal Content and Main Function, Vesicle Markers, and Tissue Localization of Diffuse Endocrine Cells of Human Pancreas and Gut

	Hormone		Vesicular marker		Localization			
Cell type	**Peptide**	**Amine**	**LDCV**	**SLMV**	**Pancreas**	**Stomach**	**Small intestine**	**Large intestine**
P/D$_1$	Ghrelin		CgA, VMAT2		+	+	+	
EC		5HT	CgA, VMAT1	Syn	+	+	+	+
D	SOM		CgA	Syn	+	+	+	+
L	GLI/PYY		SgII > CgA	Syn			+	+
A	Glucagon		CgA > SgII VMAT2	Syn	+	+		
PP	PP		CgA, SgII VMAT2	Syn	+		+	
B	Insulin		CgA, VMAT2, NESP5	Syn	+			
ECL		Histamine	CgA, VMAT2	Syn		+		
G	Gastrin		CgA	Syn		+	+	
CCK	Cholecystokinin						+	
S	Secretin	5HT	CgA				+	
GIP	GIP/Xenin		CgA				+	
M	Motilin						+	
N	Neurotensin		CgA				+	

LDCV, large dense-core vesicles; SLMV, synaptic-like microvesicles; EC, enterochromaffin cell; 5HT, 5-hydroxytryptamine; SOM, somatostatin; GLI, glucagon-like immunoreactants (glicentin, glucagon-37, glucagon-29, GLP-1, GLP-2); PPY, PP-like peptide with N-terminal tyrosine amide; PP, pancreatic polypeptide; ECL, enterochromaffin-like cell; GIP, gastric inhibitory polypeptide; CgA, chromogranin A; SgII, secretogranin II; VMAT1,2, vesicular monoamine transporter 1,2; NESP5, neuroendocrine secretory protein 55; Syn, synaptophysin.
Adapted from Rindi et al. *(2004).*[10]

homeostasis. The amines and peptides, including growth factors, produced by the diffuse neuroendocrine cells of the gut primarily have a local autocrine and paracrine regulatory role. For pancreatic endocrine cells, auto-, para- and endocrine secretions determine their influence on homeostasis. It is beyond the scope of this chapter to describe in detail the regulation of the function of the individual endocrine cells of the gut and the pancreas.

Pathology of Neuroendocrine Cells

In 1907, Siegfried Oberndorfer first introduced the term "carcinoid" (*karzinoide*, or cancer-like) for small ileal tumors that presented with a relatively benign disease course.[2,3] Tumors arising from diffuse neuroendocrine cells in the gastrointestinal system and pancreas, such as gastroenteropancreatic neuroendocrine tumors (GEP NETs), are relatively rare tumors, with an incidence of approximately 2.5–5 cases per 100,000 population.[7] On the other hand, bronchopulmonary and gastrointestinal NETs represent approximately 0.46% of all malignancies.[7]

NETs of the gastrointestinal system for all sites in which they arise have been classified into three main categories on the basis of histo- and clinicopathological characteristics: G1, well-differentiated neuroendocrine tumors with benign behavior; G1–2, well-differentiated neuroendocrine tumors with uncertain behavior; G2, well-differentiated neuroendocrine carcinomas with low-grade malignant behavior; and G3, poorly differentiated neuroendocrine carcinomas with high-grade malignant behavior.[1,11] This grading system is a useful tool for prognostic stratification, as is shown in Fig. 35.2.[5,6,11] Well-differentiated tumors generally express most of the markers of the neuroendocrine cell lineage, in particular chromogranin A. For this reason, chromogranin A is frequently used as a marker for NETs. Chromogranin A expression is generally lost in poorly differentiated tumors, whereas other well-recognized neuroendocine markers, such as synaptophysin and NSE, are retained. Most NETs are not associated with a clinical syndrome due to hormonal hypersecretion. Only in cases of insulinomas and gastrinomas are overt clinical syndromes present.[1] In addition, in liver-metastasized serotonin-producing carcinoid tumors, an overt and severe clinical syndrome, the "carcinoid syndrome," particularly characterized by flushing and diarrhea, is recognized. Tumors not associated with specific signs and symptoms due to overproduction of

(A)

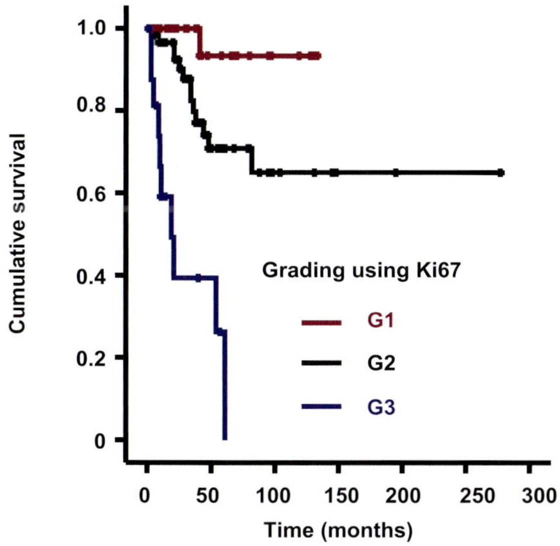

(B)

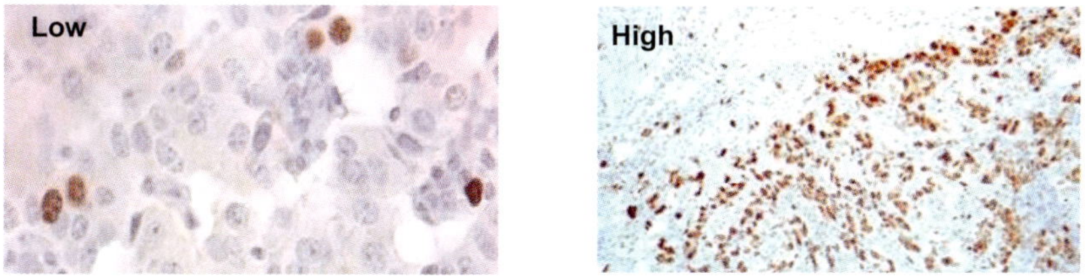

FIGURE 35.2 **(A).** Cumulative survival of patients with foregut neuroendocrine tumors with grading according to Ki67 index. *Modified* from Pape *et al.* (2008).[6] **(B)** Neuroendocrine tumors with low and high Ki67 labeling index.

hormones of bioactive substances are defined as non-functioning NETs.

Genetic Aspects of Neuroendocrine Tumors

The most striking evidence for a genetic background of endocrine tumors has been the identification of the MEN-I syndrome. The MEN-I gene is localized on chromosome 11q13, and acts as a tumor suppressor gene.[1] In patients with MEN-I, the endocrine tumor development is associated with an inherited germ-line mutation of the MEN-I gene, resulting in loss of function of the somatic allele by chromosomal deletion or a point mutation. Although not associated with a particular tumor phenotype, more than 300 different mutations of the MEN-I gene have been identified to date. In pancreatic endocrine tumors, Rindi and colleagues showed allelic losses (loss of heterozygosity; LOH) of 11q13 in 46% of a series of 172 cases.[11] Similar observations were made by other investigators using different techniques.[12,13] Lower abnormality rates were found for MEN-I mutations: 43% in gastrinomas, only 10% in insulinomas, and 15–26% in non-functioning pancreatic endocrine tumors.[11,14] Overall, MEN-I appears to play a role in the pathogenesis of about one-third of pancreatic endocrine tumors.[1] In addition, 9q gains, overexpression of cyclin D1 and chromosome 22q LOH have also been implicated in the pathogenesis of pancreatic endocrine tumors.[1] Genetic imbalances reported to be associated with a more malignant pancreatic endocrine tumor phenotype include deletions of either arm of chromosome 1; allelic loss at 3p25 centromeric to the locus of the von Hippel-Lindau syndrome gene; frequent occurrence of 6q allelic loss for 6p; genetic defects in different oncosuppressor genes; and allelic loss for X-chromosomal markers.[1] P53 mutations in pancreatic endocrine tumors appear to be rare. It should be mentioned, however, that not all these observations have been confirmed by other groups. Most of the studies have used only small series, and confirmation in larger series is needed.[1]

For gastrointestinal tract endocrine tumors, genetic alterations appear to differ between tumors deriving from the primitive foregut (stomach, duodenum and upper jejunum) and those deriving from the midgut and hindgut (ileum, appendix and small intestine). For example, involvement of the MEN-I locus appears to be largely restricted to gastric ECL tumors and duodenal gastrinomas.[1] In well-differentiated midgut endocrine tumors, chromosome 18 seems to be implicated in the pathogenesis. Chromosome 18 imbalance, with a frequent loss of 18q, appears to be a typical feature for this group of neoplasms.[1] Extensive loss of X-chromosomal markers has been implicated in malignant gastric endocrine tumors,[15] whereas such a relationship with a malignant phenotype appears not to exist in midgut endocrine tumors. As in pancreatic endocrine tumors, P53 mutations in well-differentiated tumors of the gut appear rare. On the other hand, both in poorly differentiated pancreatic endocrine tumors and in poorly differentiated carcinomas of the stomach, p53 nuclear hyperexpression/accumulation is observed.[1]

Somatostatin Receptors

Somatostatin (SST) plays an important role in the regulation of the secretions of various organ systems, including the pituitary gland, the pancreas and the gastrointestinal tract. In the brain, SST has a role as a neurotransmitter. The physiological effects of SST are mediated through the activation of seven transmembrane G-protein coupled SST receptors (sst), which are abundantly expressed in the abovementioned organ systems. Five sst subtypes, encoded by different genes, have been characterized. These have been codenamed sst_1, sst_2, sst_3, sst_4 and sst_5. The sst subtypes show a tissue-specific distribution pattern in the body.[16] The main second messenger systems activated following binding of SST to the sst are the lowering of intracellular cAMP concentration though inhibition of adenylyl cyclase, as well as lowering of the intracellular Ca^{2+} concentration through the inhibition of voltage-dependent Ca^{2+} channels.[17] These mechanisms are involved in the potent secretion inhibitory effect of SST. Stimulation of phosphotyrosine phosphatase activity has been shown to be involved in the inhibition of growth-factor stimulated cell proliferation.[17] Both circulating forms of native SST – SST-14 and SST-28, consisting of 14 and 28 amino acids, respectively – bind with high affinity to all five sst subtypes. On the basis of structural and pharmacological properties, two groups of somatostatin receptors can be distinguished. One class, including sst_2, sst_3 and sst_5, binds structural SST analogs as octreotide and lanreotide with high binding affinity, whereas a second class, consisting of sst_1 and sst_4, does not.[18]

NETs are generally characterized by a high expression of somatostatin receptors (sst), although expression levels

BOX 35.1

HOW DO WE KNOW ABOUT SOMATOSTATIN RECEPTORS, KEY TARGETS FOR DIAGNOSIS AND TREATMENT OF NEUROENDOCRINE TUMORS?

Somatostatin receptors (sst) play a key role in the diagnosis and treatment of neuroendocrine tumors (NETs). To date, five somastatin receptors have been identified.[1] The sst subtype profile of NETs is characterized by high expression of sst_2, and variable expression of sst_1, sst_3 and sst_5. ^{111}In-pentetreotide scintigraphy is an important imaging modality in the diagnostic work-up of NETs, and can visualize sst_2 positive primary and metastatic lesions with high sensitivity.[2] The sst_2 preferential somatostatin analogs (SA) octreotide and lanreotide are the mainstays of treatment of NETs, in particular to decrease symptoms by inhibition of hormone release and, to a lesser extent, by inhibition of tumor growth.[3,4] However, some patients do not respond to these SAs, or develop tachyphylaxis. Pasireotide is a new SA with affinity for sst_1 , sst_2, sst_3 and, in particular, sst_5.[5] Recent studies in Cushing's disease (CD) show that pasireotide can suppress adrenocorticotropin (ACTH) secretion by pituitary corticotrope adenomas.[4,7] The sst subtype pattern of corticotrope adenomas is characterized by high sst_5 expression, whereas the expression of sst_2 is relatively low due to selective downregulating effects of high cortisol levels.[8,9] In contrast to corticotrope adenomas, most NETs with ectopic ACTH secretion express sst_2 despite high cortisol levels, possibly because of defects in glucocorticoid receptor signaling.[10] It is not known yet whether combined activation of somatostatin receptors by pasireotide can inhibit hormone secretion or cell proliferation of NETs, but clinical studies are currently performed. *In vitro*, pasireotide reduces cell viability of human bronchial carcinoid cells.[11] Pituitary corticotrope adenomas also express dopamine receptors, and combined treatment with pasireotide and cabergoline has been shown to have synergistic effects in the treatment of patients with CD.[7,8] *In vitro* data indicate that heterodimerization of sst and dopamine receptors can result in enhanced activity.[12] NETs also express dopamine receptors, and the dopamine agonist cabergoline can suppress ectopic ACTH production by NETs.[13] In one patient, combined treatment of ectopic ACTH production with cabergoline and lanreotide was more effective than with either agent alone, and resulted in long-term biochemical control.[14] It is not yet known whether combination therapy of SA and dopamine agonists is effective in other functional NETs. Future studies will reveal whether combined targeting of sst subtypes with pasireotide, and of sst and dopamine receptors with sst_2-preferential SAs or pasireotide and dopamine agonists indeed leads to functional synergism in the treatment of NETs.

BOX 35.1 *(cont'd)*

References

1. De Martino MC, Hofland LJ, Lamberts SW. Somatostatin and somatostatin receptors: from basic concepts to clinical applications. *Prog Brain Res.* 2010;182:255–280.
2. Kwekkeboom DJ, Kam BL, van Essen M, et al. Somatostatin-receptor-based imaging and therapy of gastro-enteropancreatic neuroendocrine tumors. *Endocr Relat Cancer.* 2010;17(1):R53–R73.
3. de Herder WW, Lamberts SW. Somatostatin and somatostatin analogues: diagnostic and therapeutic uses. *Curr Opin Oncol.* 2002;14(1):53–57.
4. de Herder WW, Lamberts SW. Somatostatin analog therapy in treatment of gastrointestinal disorders and tumors. *Endocrine.* 2003;20(3):285–290.
5. Schmid HA. Pasireotide (SOM230): development, mechanism of action and potential applications. *Mol Cell Endocrinol.* 2008;286(1–2):69–74.
6. Boscaro M, Ludlam WH, Atkinson B, et al. Treatment of pituitary-dependent Cushing's disease with the multi-receptor ligand somatostatin analog pasireotide (SOM230): a multicenter, phase II trial. *J Clin Endocrinol Metab.* 2009;94(1):115–122.
7. Feelders RA, de Bruin C, Pereira AM, et al. Pasireotide alone or with cabergoline and ketoconazole in Cushing's disease. *N Engl J Med.* 2010;362(19):1846–1848.
8. de Bruin C, Pereira AM, Feelders RA, et al. Coexpression of dopamine and somatostatin receptor subtypes in corticotroph adenomas. *J Clin Endocrinol Metab.* 2009;94(4):1118–1124.
9. van der Hoek J, van der Lelij AJ, Feelders RA, et al. The somatostatin analogue SOM230, compared with octreotide, induces differential effects in several metabolic pathways in acromegalic patients. *Clin Endocrinol (Oxf).* 2005;63(2):176–184.
10. Ray DW, Littlewood AC, Clark AJ, Davis JR, White A. Human small cell lung cancer cell lines expressing the proopiomelanocortin gene have aberrant glucocorticoid receptor function. *J Clin Invest.* 1994;93(4):1625–1630.
11. Zatelli MC, Minoia M, Martini C, et al. Everolimus as a new potential antiproliferative agent in aggressive human bronchial carcinoids. *Endocr Relat Cancer.* 2010;17(3):719–729.
12. Rocheville M, Lange DC, Kumar U, Patel SC, Patel RC, Patel YC. Receptors for dopamine and somatostatin: formation of hetero-oligomers with enhanced functional activity. *Science.* 2000;288(5463):154–157.
13. Pivonello R, Ferone D, de Herder WW, et al. Dopamine receptor expression and function in corticotroph ectopic tumors. *J Clin Endocrinol Metab.* 2007;92(1):65–69.
14. Pivonello R, Ferone D, Lamberts SW, Colao A. Cabergoline plus lanreotide for ectopic Cushing's syndrome. *N Engl J Med.* 2005;352(23):2457–2458.

may vary considerably among tumors and between different types of tumors. The sst_2 subtype is expressed at the highest level, and explains the successful application of the somatostatin analogs octreotide and lanreotide, which are analogs preferentially binding to sst_2, in the medical treatment of patients with NETs. In addition, the high sst_2 expression in NETs forms the basis for the efficacy of peptide receptor radionuclide therapy (PRRT) using radiolabeled somatostatin analogs in the treatment of NETs.[19,20] The sst_4 is not frequently expressed in NETs, whereas sst_1, sst_3 and sst_5 can be expressed, although with considerable variability among tumors.[21,22] In general, well-differentiated NETs express a high density of somatostatin receptors, whereas lower levels (or even loss of sst expression), is found in poorly differentiated tumors.[23] The clinical utility of the presence of somatostatin receptors on NETs is discussed further later in this chapter.

CLINICAL ASPECTS OF NEUROENDOCRINE TUMORS

Foregut Neuroendocrine Tumors

Bronchopulmonary Tumors

Bronchopulmonary NETs account for approximately 1% of all lung cancers, and comprise about 25% of well-differentiated NETs.[8] Clinical symptoms of bronchopulmonary NETs include cough, hemoptysis, pneumonia and airway obstruction.[24] Carcinoid syndrome – that is, flushing, diarrhea and carcinoid heart disease due to serotonin overproduction – is rare, and occurs in up to 3% of cases.[24] Bronchopulmonary NETs can also be associated with ectopic hormone production leading to Cushing's syndrome (ACTH production), acromegaly (GHRH production) or hypercalcemia (PTH-related proteins).[25–27] Bronchopulmonary NETs are the most frequent cause of Cushing's syndrome due to ectopic ACTH production.[25] Well-differentiated bronchopumonary NETs can be very small and difficult to detect. Imaging techniques used to localize these tumors include CT, somatostatin receptor scintigraphy, positron emission tomography and bronchoscopy.[8] The prognosis of well-differentiated bronchopulmonary NETs is generally good, with a 5-year survival rate above 85%. Poorly differentiated bronchopulmonary NETs, formerly known as "atypical carcinoids," have a less favorable prognosis, with a reported 5-year survival rate of approximately 50%.[8,9,24,28]

Gastric and Duodenal Tumors

Gastric NETs account for approximately 8% of GEP NETs, and are divided into three subtypes.[7,29] Type I gastric NETs, the most frequent type, originate from gastric enterochromaffin-like cells due to sustained growth-promoting effects of elevated gastrin levels in patients with atrophic gastritis with achlorhydria.[29] These tumors are usually benign and multifocal, and are frequently diagnosed during endoscopic procedures of the proximal gastrointestinal tract.[30] Type II gastric NETs develop from enterochromaffin-like cells in response to increased gastrin levels that result from secretion by gastrinomas, and these tumors mostly occur in patients with MEN-I syndrome.[30,31] Type III gastric NETs are sporadic tumors, and usually well-differentiated. These tumors predominantly occur in males, and may be detected during an upper endoscopic procedure, but patients can also present with abdominal pain and weight loss.[29,32] Imaging with endoscopic ultrasound can be useful to assess tumor invasiveness and local lymph node involvement in patients with gastric NETs.

Five types of duodenal NETs can be distinguished: gastrinomas, somatostatinomas, clinically non-functioning NETs, duodenal gangliocytic paragangliomas, and poorly differentiated neuroendocrine carcinomas.[5] Gastrinomas are the most prevalent form of duodenal NETs, and can occur sporadically, or as part of the MEN-I syndrome.[5,33] Next to the duodenum, gastrinomas are located in the pancreas. Gastrinomas overproduce gastrin, resulting in increased gastric acid production, which in turn leads to the Zollinger-Ellison syndrome, characterized by (severe) peptic ulcers, gastroesophageal reflux and diarrhea.[33,34] The widespread use of proton pump inhibitors (PPI) may delay an early diagnosis of gastrinoma.[35] The biochemical diagnosis of gastrinoma is made by measurement of fasting serum gastrin levels. PPI increases serum gastrin levels, and should therefore be temporarily replaced by an H2 blocker, if possible, before measurement of gastrin levels.[33] Localization of small gastrinomas can be difficult; endoscopic ultrasound is useful to detect pancreatic gastrinomas, but is less sensitive in locating duodenal gastrinomas.[36] Progressive growth of gastrinoma occurs in approximately 25% of patients, with the presence of liver metastases being the most important prognostic factor regarding survival.[37]

Pancreatic Tumors

Pancreatic NETs can be divided into functional and non-functional tumors (Table 35.4). Insulinoma is the most prevalent functioning pancreatic NET, with an estimated incidence of 1–3 per million per year.[38] The majority of insulinomas present as small, benign solitary tumors. Malignant behavior is seen in less than 10% of cases. In 5–10% of patients with insulinoma MEN-I syndrome is present, whereas insulinomas occur in approximately 21% of patients with MEN-I syndrome.[39] Typically, the clinical presentation of insulinoma is characterized by Whipple's triad: symptoms of hypoglycemia, a blood glucose level below 40 mg/dl (2.2 mmol/l), and improvement of symptoms after glucose administration. In particular in the fasting condition, insulin overproduction induces hypoglycemia, resulting in neuroglycopenic symptoms and symptoms related to the catecholamine response (Table 35.3).[39] The diagnosis is biochemically established by measurement of inappropriately high insulin, proinsulin and C-peptide levels during hypoglycemia provoked in a 72-hour fasting test.[38,39] The most sensitive imaging modalities to locate insulinomas include three-phase CT, MRI and, in particular, endoscopic ultrasound.[38,40] The sensitivity of somatostatin receptor scintigraphy in detecting insulinomas is less than 50%.[41] In difficult cases, angiography combined with calcium stimulation and transhepatic portal venous sampling can be helpful to determine the pancreatic region of the insulinoma.[42]

TABLE 35.4 Clinical Syndromes of Functioning and Non-functioning Pancreatic Neuroendocrine Tumors

Pancreatic NET type	Clinical syndrome
Insulinoma	Hypoglycemia: symptoms of neuroglycopenia (diplopia, confusion, seizures, coma) and catecholamine response (tachycardia, sweating, tremor, anxiety)
Gastrinoma	Gastroesophageal reflux disease, peptic ulcer disease, diarrhea
VIP-oma	Secretory diarrhea, hypokalemia
Somatostatinoma	Cholelithiasis, diabetes mellitus, diarrhea with steatorrhoea
Glucagonoma	Diabetes mellitus, venous thrombosis, necrolytic migratory erythema, weight loss
NET with ectopic hormone production	Cushing's syndrome (ACTH), acromegaly (GHRH), hypercalcemia (PTH-rp)
Non-functioning NET	Abdominal pain, weight loss, nausea, jaundice

Gastrinomas have been discussed above. Rare functional pancreatic NETs include VIP-omas, somatostatinomas, glucagonomas, and tumors with ectopic hormone secretion leading to Cushing's syndrome (ACTH production), acromegaly (GHRH production) and hypercalcemia (PTH-rp production). The majority of these usually well-differentiated tumors present with malignant disease with liver and regional lymph-node metastases.[5,43–45] These tumors can also be associated with MEN-I syndrome.[46] VIP-omas secrete vasoactive intestinal peptide, which stimulates intestinal secretion, resulting in severe diarrhea, hypokalemia, hypochlorhydria and acidosis, also known as the Verner-Morrison syndrome.[47] Additional symptoms are facial flushing, glycogenolysis and hypercalcemia.[43] Somatostatinomas are pancreatic islet cell tumors that overproduce somatostatin. Somatostatin has many antisecretory effects in the gastrointestinal tract, and the clinical picture of somatostatinomas is classically characterized by diabetes mellitus (inhibition of insulin secretion), cholelithiasis (inhibition of cholecystokinin release) and steatorrhea (inhibition of pancreatic enzyme secretion).[44,48] Somatostatinomas can also arise in the duodenum, but in these patients the aforementioned symptoms are frequently absent for reasons that are not clear.[49] Glucagonomas overproduce glucagon, which can cause diabetes mellitus. Other clinical features of glucagonomas include necrolytic migratory erythema, weight loss, and venous thrombosis.[45,50]

Non-functional pancreatic NETs are divided into: (a) well-differentiated NETs with benign or uncertain behavior, (b) well-differentiated endocrine carcinomas with low-grade malignant behavior, and (c) poorly differentiated carcinomas with high-grade malignant behavior, according to the WHO classification. The majority of these tumors are well-differentiated endocrine carcinomas.[4,5,51] Because non-functional pancreatic NETs are not accompanied by symptoms related to hormone secretion, the diagnosis is often made when the tumor is already at an advanced stage. Clinical symptoms are mostly related to mass effects, and include abdominal pain, weight loss, nausea and jaundice.[52–54] Non-functional pancreatic NETs can occur sporadically, or can be associated with a hereditary tumor syndrome such as MEN-I syndrome, von Hippel-Lindau (VHL) syndrome and tuberous sclerosis.[55–58] Non-functional pancreatic NETs develop in 55% of MEN-I patients, often with multiple localizations, and tend to have a more benign disease course compared to sporadic tumors. Conversely, 8% of patients with non-functional pancreatic NETs have MEN-I syndrome.[55,59] Up to 16% of patients with VHL syndrome develop non-functional pancreatic NETs.[56,57] Imaging modalities that are used for evaluation and follow-up of non-functional pancreatic NETs include CT, MRI, somatostatin receptor scintigraphy and endoscopic ultrasound.[51]

Midgut Neuroendocrine Tumors

Midgut NETs are localized from the distal duodenum up to the ascending colon. NETs of the distal jejunum and ileum, including the appendix, are the most frequent endocrine tumors of the gastrointestinal tract, and account for up to 28% of GEP NETs.[7] Like pancreatic NETs, these tumors can be functional or non-functional. Functional midgut NETs are able to produce, store and secrete mediators like serotonin (5-hydroxytryptamine), histamine, prostaglandins, tachykinins, substance P and dopamine. Serotonin is synthesized from its precursor tryptophan. Physiologically, tryptophan is used for the synthesis of nicotinic acid; however, in patients with a functional midgut NET tryptophan is predominantly converted to serotonin, which may result in nicotinic acid deficiency (pellagra).[60] Continuous or intermittent release of these mediators causes "carcinoid syndrome" in approximately 18% of patients with jejuno-ileal NETs, characterized by flushing, diarrhea, bronchoconstriction and, in the long-term, carcinoid heart disease.[2,60] Carcinoid heart disease is caused by serotonin-induced endocardial fibrosis, and primarily involves the right side of the heart, resulting in tricuspid and pulmonary valve dysfunction. Carcinoid syndrome and carcinoid heart disease usually develop when liver metastases are already present, because mediators produced by the primary tumor will enter the portal circulation and be metabolized by the liver.[2,60] Factors that provoke the release of serotonin and other products include emotion, alcohol ingestion and exercise. Anesthesia, surgery or radiological intervention can cause a so-called carcinoid crisis by inducing massive release of NETs mediators, resulting in severe hypotension, tachycardia and bronchospasm.[60] Serotonin overproduction can be established biochemically by measurement of its degradation product 5-hydroxy-indolacetic acid (5-HIAA) in urine.

Non-functional midgut NETs can be discovered incidentally – for example, during surgical procedures – or when they become symptomatic, mostly presenting with abdominal pain due to local tumor mass effects. Not infrequently, patients have suffered vague abdominal complaints for years, diagnosed as "irritable bowel syndrome." These tumors can also cause mesenteric fibrosis, which can lead to mesenteric ischemia and small bowel obstruction.[2,60] The prognosis of jejuno-ileal NETs depends on the disease stage and differentiation grade at presentation. Overall, the 5-year survival rate of these tumors is 60%, with higher survival rates in patients with regional disease and low proliferative tumors, and lower survival rates in patients with a higher tumor Ki-67 index and distant metastases.[7]

Appendiceal NETs, primarily located in the tip of the appendix, are mostly diagnosed incidentally during appendectomy, but can present with right lower quadrant abdominal pain. NETs of the appendix can produce serotonin; however, because serotonin-positive appendiceal NETs have a very low tendency to metastasize to the liver, the carcinoid syndrome is rarely seen.[5]

Hindgut Neuroendocrine Tumors

Colon NETs are rare, and are often poorly differentiated tumors that have already metastasized to liver, lymph nodes or mesentery at presentation, also due to lack of early symptoms.[5,61] Rectal NETs are more frequent, and comprise 27% of gastrointestinal NETs.[7] Colon and rectal NETs are usually not hormonally active, and are frequently an incidental finding during endoscopy. Rectal NETs generally appear as polypic lesions. Imaging modalities for hindgut NETs include endoscopy, CT/MRI, somatostatin receptor scintigraphy and, for poorly differentiated tumors, FDG-PET. Endoanal ultrasound and/or pelvic MRI are used for staging of rectal NETs and to evaluate resectability.[62] Colon NETs generally have a poor prognosis, due to the advanced stage at diagnosis, with a 5-year survival rate of 40–70%.[7,63] Rectal NETs have a better prognosis, with a 5-year survival rate of approximately 90% in patients with localized disease.[7]

DIAGNOSIS OF NEUROENDOCRINE TUMORS

Biochemical Biomarkers in Neuroendocrine Tumors

Several circulating tumor markers have been evaluated for the diagnosis and follow-up management of neuroendocrine tumors. While these can be very useful for follow-up, isolated elevation of marker levels is generally not sufficient for diagnosis without tissue confirmation.[2,64]

Chromogranin A

The most important marker, chromogranin A (CgA), is a constitutive secretory product of most neuroendocrine tumors. Depending on the extent of the disease, plasma CgA is elevated in > 60% of patients. CgA levels may correlate with tumor volumes, but care should be taken in measuring CgA and interpreting the results. For example, somatostatin analogs are known to affect blood levels of CgA. Spuriously elevated levels of CgA have also been reported in patients using proton pump inhibitors, in patients with renal or liver failure, and in those with chronic gastritis.[65]

24-Hour Urinary Excretion of 5-Hydroxy Indole Acetic Acid (5-HIAA)

Urinary 5-hydroxy indole acetic acid (5-HIAA) (24-hour collection) is a useful laboratory marker for carcinoid tumors. It is a surrogate measure of serotonin metabolism which is tightly linked to the presence of carcinoid syndrome. The specificity of this test has been reported to be 88%. However, certain foods (banana, avocado, plum, eggplant, tomato, plantain, pineapple, kiwi and walnut) and medications (paracetamol, fluorouracil, methysergide, naproxen and caffeine) can increase urinary 5-HIAA levels, and should be avoided during specimen collection.[65]

Neuron-Specific Enolase (NSE)

Neuron-specific enolase (NSE) is present in neurons and neuroendocrine cells. Serum NSE levels are frequently elevated in patients with several neuroendocrine tumors. Like CgA, NSE is a general neuroendocrine marker which cannot differentiate between different subtypes of neuroendocrine tumors. Elevated NSE levels are exclusively associated with poor tumor differentiation.[65]

Insulin

The diagnosis of insulinoma can be established using the following six tight criteria: documented blood glucose levels ≤ 2.2 mmol/l (≤ 40 mg/dl); concomitant insulin levels $\geq 6\,\mu$U/l (≥ 36 pmol/l; $\geq 3\,\mu$U/l using ICMA); C-peptide levels ≥ 200 pmol/l, or proinsulin levels ≥ 5 pmol/l; β-hydroxybutyrate levels ≤ 2.7 mmol/l; and absence of sulfonylurea (metabolites) in the plasma and/or urine.[65,66]

Gastrin

The diagnosis of gastrinoma can be established using determinations of plasma gastrin. Plasma gastrin concentrations in gastrinoma patients are usually strongly elevated ($> 10 \times$ ULN), and are accompanied by a high basal acid output (> 15 mmol/h) and low stomach pH. Serum gastrin also increases after an i.v. injection of 2 U/kg secretin in gastrinoma patients. Because G-cell function is decreased by gastric acid, there is increased circulating gastrin in patients with depressed or absent acid output, as with atrophic gastritis in pernicious anemia. The long-term use of proton pump inhibitors (PPIs) similarly leads to elevated plasma gastrin. The gastric bacterium *Helicobacter pylori* infects the antrum, and is associated with a tendency to increased plasma gastrin concentrations that resolves when the infection is eradicated. Fasting serum gastrin levels are important to differentiate type I and type II neuroendocrine tumors of the stomach from type III.[33,65]

It is recognized that neuroendocrine tumors can sometimes change what (if any) hormones and biomarkers are produced. The general principle of biomarker measurement is to evaluate a large panel of markers at key points of the disease, such as diagnosis or relapse, in order to identify a few biomarkers that are elevated in the particular patient in question, and follow these over time. It is generally not necessary to check every biomarker at every visit.

Imaging of Neuroendocrine Tumors

Imaging studies for neuroendocrine tumors are generally performed for two reasons: for initial evaluation (i.e., identification of primary tumor, staging, assessing the extent of disease and planning therapy), and for follow-up (i.e., surveillance following complete resection or during periods of stability, and evaluation of response following treatment).[2,64]

Initial Evaluation for Extent of Disease in Neuroendocrine Tumors

Imaging studies generally recommended at the time of initial evaluation include somatostatin receptor imaging using ^{111}In-pentetreotide scintigraphy or ^{68}Ga-DOTA-TOC PET, a plain film of the thorax, cross-sectional imaging (CT or MRI) of the (thorax) abdomen and pelvis, and – on an as-needed basis – bone scintigraphy.[2,64]

Follow-up Imaging of Neuroendocrine Tumors

Among patients undergoing surveillance following complete resection, we advocate the use of ^{111}In-labeled pentetreotide scintigraphy or ^{68}Ga-DOTA-TOC PET annually as follow-up for patients without evidence of disease, or on an as-needed basis to resolve difficult issues. For patients with advanced disease, we generally recommend the use of cross-sectional imaging for known sites of disease.

^{111}In-Pentetreotide Scintigraphy

^{111}In-pentetreotide shares the receptor-binding profile of octreotide and lanreotide, rendering it an ideal radiopharmaceutical for imaging of somatostatin receptor subtypes (sst) 2- and 5- (sst_2 and sst_5) positive tumors. The sensitivity of the imaging procedure can be enhanced by the simultaneous use of single positron emission computed tomography (SPECT) imaging. The overall sensitivity of ^{111}In-pentetreotide scintigraphy is approximately 80% to 90%, and it is effective in detecting primary and metastatic lesions not apparent when using conventional radiologic-imaging techniques. Of particular advantage is the fact that a single scan images the entire body, with detection of occult metastatic lesions such as bone metastases (Fig. 35.3). It can also be used to evaluate whether peptide receptor radiotherapy (PRRT) represents a good treatment option[19] (see Fig. 35.3).

Transabdominal Ultrasound

Transabdominal ultrasound can sometimes identify small bowel carcinoids and pancreatic endocrine lesions, and two-thirds of liver metastases, and may also be used to guide percutaneous biopsies of metastases or primary lesions.[67]

CT

Multiphasic CT that includes the arterial phase and the portal venous phase is generally recommended in assessing for metastases. Neuroendocrine tumors are generally vascular tumors that enhance intensely with intravenous contrast during the early arterial phases of imaging, with washout during the delayed portal venous phase. Mass lesions and evidence of calcification and fibrosis define CT scan and MRI findings associated with carcinoid tumors. Radiating strands of fibrosis and spiculation are characteristic hallmarks, especially in conjunction with a mass lesion. There is little difference between the diagnostic efficacy of CT and that of MR.[67] Fig. 35.4 shows an example of a CT scan and ^{111}In-pentetreotide scintigraphy of a NET.

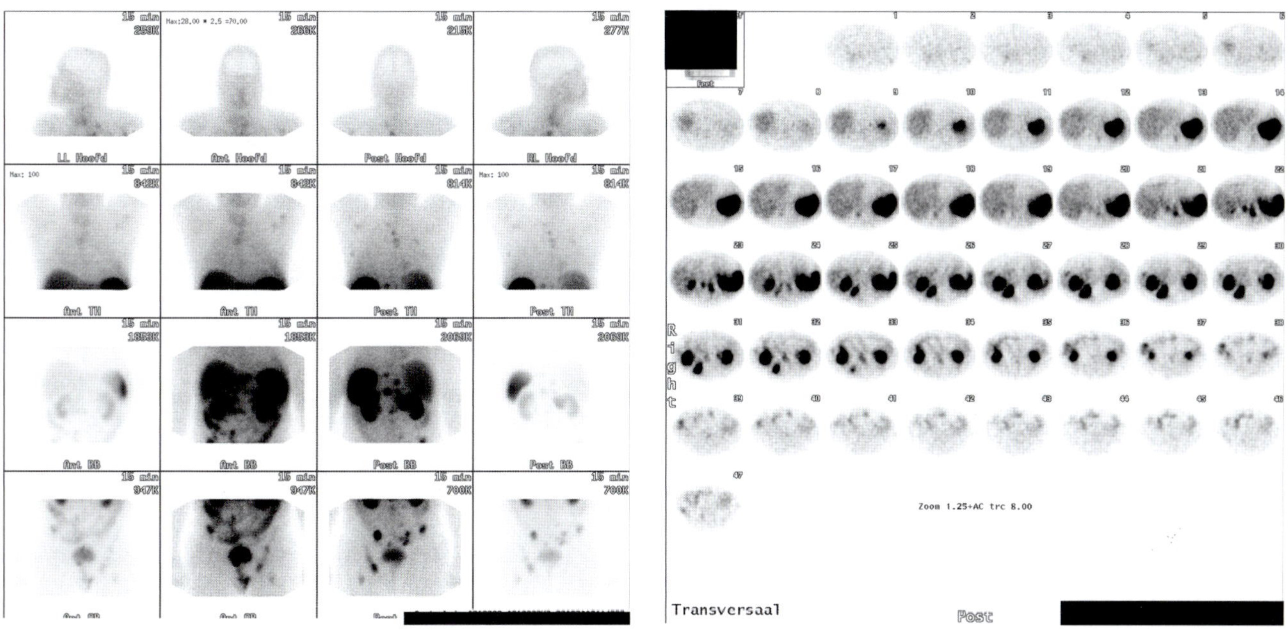

FIGURE 35.3 ^{111}In-pentetreotide scintigraphy detecting bone metastases in a patient with a neuroendocrine tumor. Courtesy RA Feelders.

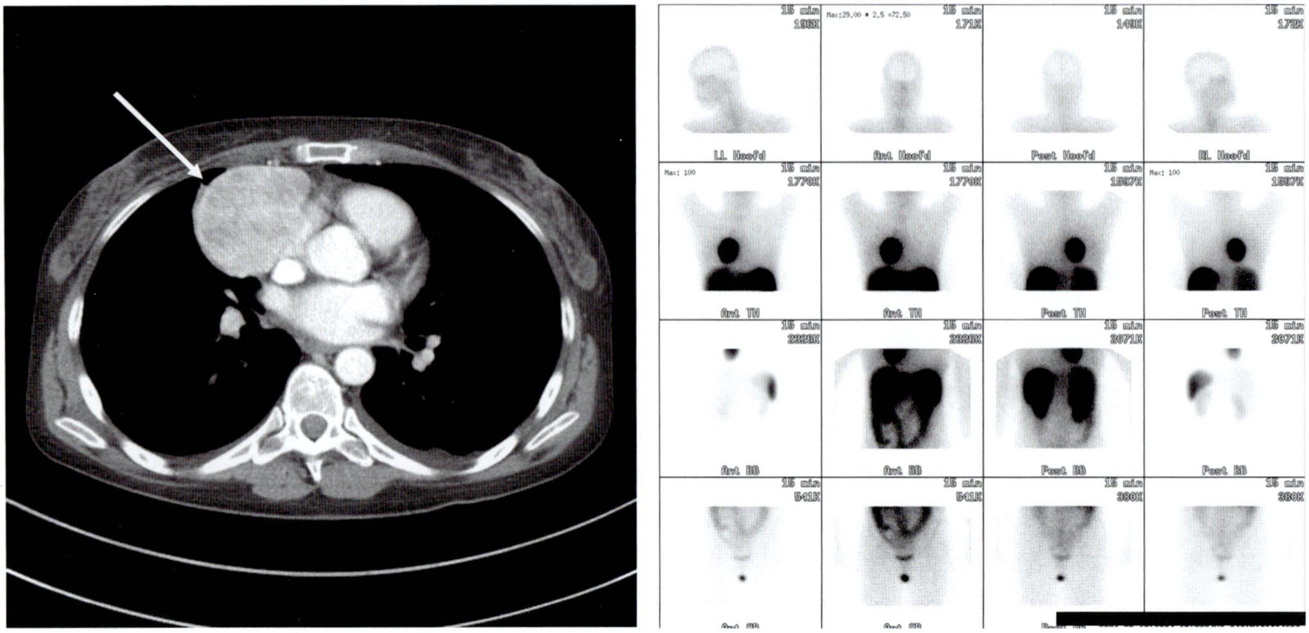

FIGURE 35.4 CT-scan and ^{111}In-pentetreotide scintigraphy of a patient with an ACTH-producing mediastinal neuroendocrine tumor. Courtesy of RA Feelders.

MRI

T1- and T2-weighted imaging and multiphasic (arterial, portal venous and delayed) dynamic MRI for neuroendocrine tumors is generally recommended. Neuroendocrine tumors can have variable appearances on non-contrast MRI. They can be hypo- or iso-intense on T1-weighted images. Metastases to the liver typically show a high signal on T2-weighted images.[67]

^{123}I-MIBG Scintigraphy

Metaiodobenzylguanidine (MIBG) is a guanidine derivative that exploits the specific type 1 amine uptake mechanism at the cell membrane and storage within the intracellular storage vesicles. ^{123}I-MIBG molecular imaging has also been used for neuroendocrine tumors, but has lesser efficacy than in pheochromocytoma, paraganglioma or neuroblastoma. ^{111}In-labeled pentetreotide generally has a higher sensitivity than ^{123}I-MIBG for neuroendocrine tumors of the digestive tract and pancreas.[68]

Positron Emission Tomography (PET)

(PET) is a relatively novel, non-invasive technique that facilitates biochemical and metabolic studies of human tumors. PET [^{18}F]-fluorodeoxyglucose (FDG) imaging, while successful for many solid tumors, has generally not provided additional information about the extent of disease for well-differentiated neuroendocrine tumors because of their generally lower proliferative activity. FDG PET should therefore be used for poorly differentiated neuroendocrine carcinomas, or when ^{111}In-pentetreotide or ^{123}I-MIBG is negative or equivocal.[68]

More recently, ^{68}Ga-DOTA-TOC has been introduced for somatostatin receptor PET imaging, achieving high detection rates and better spatial resolution than ^{111}In-pentetreotide scintigraphy.[69]

Previous studies have shown ^{11}C-5-HTP PET to be a promising imaging modality for the detection of neuroendocrine tumors. ^{11}C-5-HTP-PET proved better than ^{111}In-pentetreotide for tumor visualization. However, it is not widely available.

Other new PET imaging agents for neuroendocrine tumors includes ^{18}F-DOPA.[70]

Bone Scintigraphy

Bone scintigraphy with ^{99m}TcMDP used to be the mainstay for identifying bone metastases associated with neuroendocrine tumors, but this technique might become redundant in the near future when the abovementioned newer PET techniques become widely available.[71]

Endoscopic Imaging of Neuroendocrine Tumors

Endoscopy, Endoscopic Ultrasound and Videocapsule Endoscopy

Upper gastrointestinal endoscopy can identify lesions as far as the ligament of Treitz, and lower and lower gastrointestinal endoscopy can detect some terminal ileal tumors as well as colon and rectal carcinoids.[2]

Endoscopic ultrasound is a highly sensitive method for detecting carcinoid tumors of the stomach and duodenum, as well as for detecting pancreatic neuroendocrine tumors. It is superior to conventional ultrasound, particularly in the detection of small lesions and for determining invasiveness. In pancreatic neuroendocrine tumors it can be useful in the preoperative work-up, as it allows clear visualization of anatomical correlations.[2,72]

(Video)capsule endoscopy has obvious potential for surveillance of the small intestine for carcinoid tumors.[2,73]

TREATMENT OF NEUROENDOCRINE TUMORS

Surgery

Surgery is the primary treatment option for GEP NETs patients with localized disease in order to achieve cure.[2,74] However, at presentation many patients already have metastatic disease with involvement of, in particular, the liver, lymph nodes, mesentery and peritoneum.

Surgery is also performed in advanced stages of disease, for several indications. First, patients with limited hepatic or lymph node involvement may benefit from cytoreductive surgery of primary tumor and metastases, or resection of the primary tumor with subsequent treatment of (liver) metastases with other treatment modalities like embolization or radiofrequency ablation.[2,75,76] In selected cases liver transplantation can be considered, but only when extrahepatic metastases have been excluded.[77] However, to date no randomized controlled trials have been performed that indicate the best strategy to optimize survival. Second, debulking can be performed with intent of symptom palliation.[2,75,76] Third, small bowel obstruction or ischemia due to the primary tumor or associated mesenteral fibrosis is an indication for surgery.[2,75,76] Fourth, surgery is indicated for specific hormonally active tumors like insulinomas and gastrinomas, or NETs with ectopic hormone production. The majority of patients with insulinomas and, to a lesser extent, patients with gastrinomas are disease-free after surgery.[74,78] Finally, patients with carcinoid heart disease and right-sided heart failure can benefit from cardiac valve replacement.[79]

Interventional Radiology

Liver metastases of NETs are usually hypervascular, and occlusion of the hepatic artery by surgery or (chemo-)embolization can induce tumor necrosis and regression.[80,81] Sequential transcatheter arterial chemoembolization (TACE) results in symptom control in 63–100% of patients, and tumor response rates of between 33% and 80%.[81] Whether addition of chemotherapeutic agents to embolization increases efficacy compared to embolization alone has not yet been elucidated. Hepatic artery embolization cannot be performed in the presence of portal vein thrombosis because of dependency of normal hepatic tissue on portal vein blood flow after hepatic artery occlusion. A common complication of embolization procedures is the so-called post-embolization syndrome, consisting of transient pain, fever and elevated liver enzymes. Rare complications include carcinoid crisis, liver and kidney failure, and bleeding peptic ulcer.[80,81] Embolization with yttrium-labeled microspheres can also be effective in inducing tumor regression.[82] Other treatment modalities that aim to destroy liver metastases are percutaneous alcohol injection, radiofrequency ablation (RFA) and cryotherapy. The response to alcohol injection is rather variable, and controlled studies with large patient numbers are lacking.[81] In patients with limited disease and small metastases, RFA is a promising treatment option; however, more studies are needed to evaluate long-term efficacy.[81,83] Cryotherapy (i.e., freezing of liver metastases using a cryoprobe) is applied especially during surgery, and can result in symptom control and tumor response.[81,84] Again, larger studies are needed to establish the place for cryotherapy among other treatment modalities.

Medical Therapy

Somatostatin Analogs

Somatostatin analogs (SAs) have an important role in the medical management of NETs.[2,60] As outlined above, the majority of NETs express somatostatin receptors, predominantly somatostatin receptor subtype (sst) 2. This high sst_2 density not only allows for visualization of NETs with ^{111}In-pentetreotide scintigraphy, but is also a target for treatment with SAs.[85] The currently available SAs, octreotide and lanreotide, preferentially bind with sst_2 and can effectively inhibit NET hormone secretion, resulting in symptom control – i.e., resolution of flushes and diarrhea – in 70–80% of patients with midgut NETs.[86–88] These drugs are generally well tolerated, and a long-acting formulation can be administrated once every 4 weeks. However, in a subset of patients, desensitization to SAs occurs in the long term, with recurrence of symptoms. High-dose intravenously administrated octreotide is indicated in patients with a carcinoid crisis triggered by, for instance, anesthesia, tumor manipulation or radiological intervention.[89]

Apart from symptom control, SAs may also have antiproliferative effects. Although tumor regression occurs in only a small minority of patients, tumor stabilization can

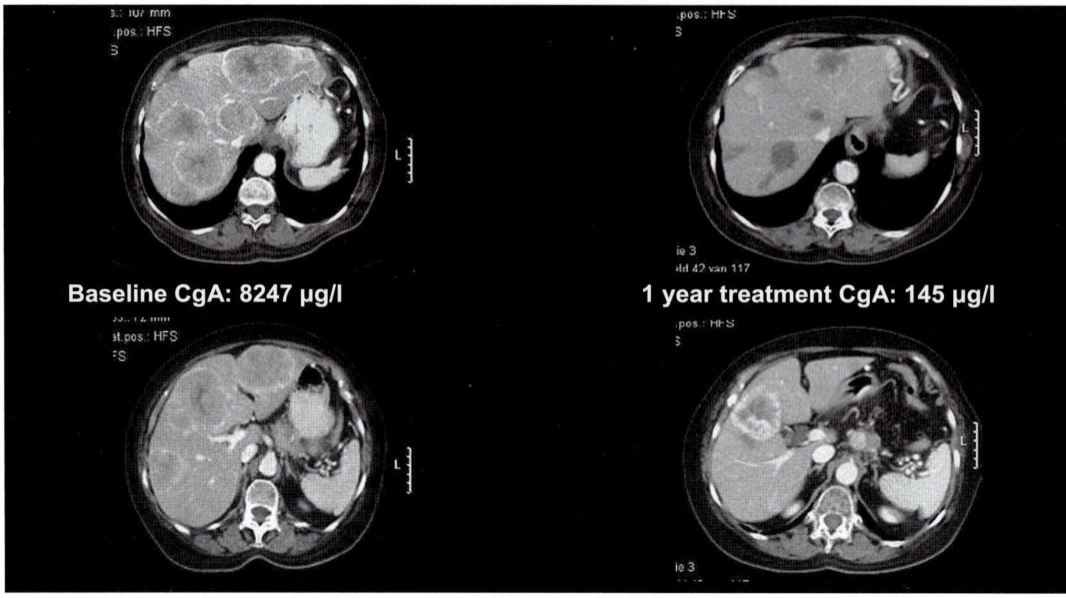

FIGURE 35.5 Effect of one year treatment with Lanreotide Autogel on tumor size in a metastatic clinically nonfunctioning pancreatic NET: tumor regression parelleled by a decrease in chromogranin A (CgA) Levels. Courtesy RA Feelders.

be achieved with long-term SA treatment.[90] The PROMID study showed that, compared to placebo treatment, octreotide LAR significantly prolongs time to tumor progression in patients with a well-differentiated metastatic midgut NET, in particular in those patients with a low hepatic tumor load.[91] This effect was seen in patients with both functional and non-functional midgut NETs. SA may also have antiproliferative effects on pancreatic NETs as shown in Fig. 35.5.However, prospective studies with large patient numbers have not been published yet.

Inhibitory effects of SAs on hormone production and tumor growth may in part explain the increase in survival rates in NET patients since the introduction of SAs in 1987.[9]

Interferon-α and Chemotherapy

Interferon-α can effectively reduce symptoms in patients with NETs and the carcinoid syndrome, and also has antiproliferative effects, but has serious side effects, including fever, fatigue, auto-immune disorders and myelosuppression.[92] Addition of interferon-α to SAs decreases tumor progression more potently compared to SAs alone, although overall survival is not improved.[93,94] Chemotherapy is indicated in patients with poorly differentiated NETs. These highly proliferative NETs are usually treated with etoposide and cisplatin, with a tumor response rate of approximately 50%.[95] Patients with a poorly differentiated or rapidly progressive pancreatic NET can be treated with streptozotocin with fluorouracil or doxorubicin.[95,96] Chemotherapeutic agents like temozolomide and capecitabine may also be useful in patient subsets, but this needs further investigation.

New Developments

Pasireotide is a new SA with affinity not only for sst_2 but also for sst_1, sst_3 and sst_5,[97] and may therefore inhibit hormone secretion and growth of NETs with expression of these somatostatin receptors. The efficacy of pasireotide is currently being investigated in patients with NETs. NETs also express dopamine receptors, and one study demonstrated that the dopamine agonist cabergoline can inhibit ectopic ACTH secretion by NETs.[98]

NETs are well-vascularized tumors that express vascular endothelial growth factor (VEGF). A preliminary study demonstrated that treatment of patients with fore-, mid- and hindgut NETs with the VEGF monoclonal antibody bevacizumab decreased tumor blood flow and had tumor-stabilizing effects.[99]

The mammalian target of rapamycin (mTOR) has been identified as a new target for medical treatment of NETs.[100] The mTOR pathway is activated in neoplastic disease, and plays an important role in the regulation of cell proliferation and angiogenesis.[101] In patients with progressive pancreatic NETs, treatment with the oral mTOR inhibitor everolimus showed promising results with respect to tumor stabilization and progression-free survival.[102,103] The combination of everolimus with octreotide may have synergistic effects, because progression-free survival was 16.7 months in the patients treated with both drugs compared to 9.7 months in patients treated with everolimus alone.[102] The tyrosine-kinase inhibitor sunitinib also seems to be

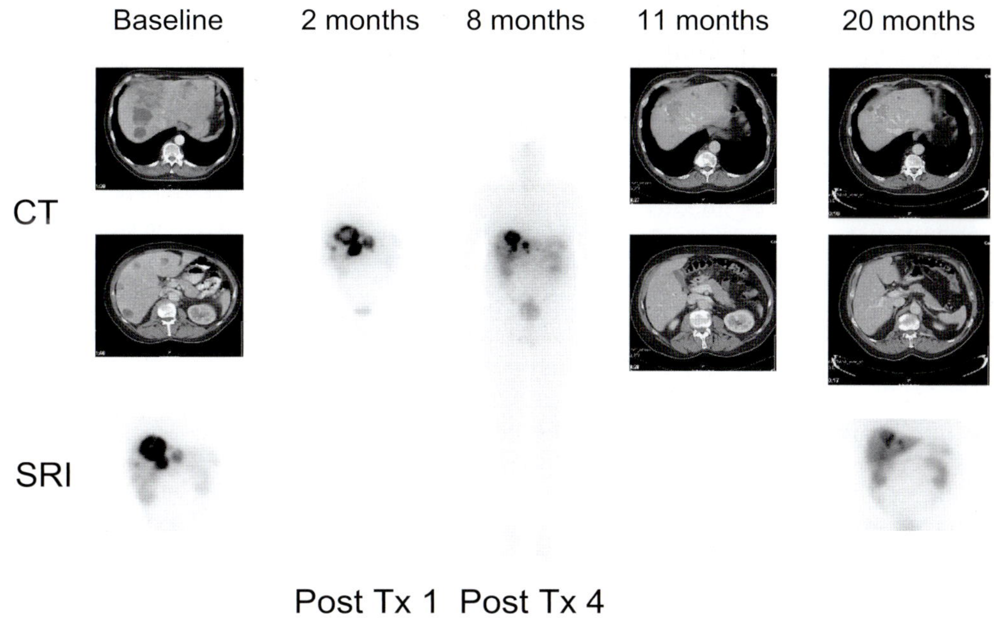

FIGURE 35.6 CT and somatostatin receptor scintigraphy images of a NET patient with liver metastases before and after four treatment cycles with [^{177}Lu-DOTA^0,Tyr^3]octreotate. Post Tx, post-therapy scan. Courtesy D. Kwekkeboom.

effective in patients with advanced well-differentiated pancreatic NETs, as shown in a recent study that demonstrated a median progression-free survival of 11.4 months in sunitinib-treated patients compared to 5.5 months in placebo-treated patients.[104] Current and future studies will evaluate the efficacy and safety of combination therapy with different SAs with bevacizumab, mTOR inhibitors or tyrosine-kinase inhibitors.

Peptide Receptor Radionuclide Therapy

The high sst_2 expression by NETs forms the target for peptide receptor radionuclide therapy (PRRT). The basis of PRRT is that a radionuclide is coupled to a SA. After intravenous administration, this complex is subsequently internalized by the tumor via receptor-mediated endocytosis, resulting in targeted radiation.[19] Radionuclides used for this purpose include ^{111}indium, ^{90}yttrium and ^{177}lutetium. Symptomatic improvement can be achieved with ^{111}In-, ^{90}Y- or ^{177}Lu-labeled SAs, but with respect to tumor regression ^{90}Y- and ^{177}Lu-labeled SAs are the most promising compounds.[19] Treatment of 310 patients with GEP NETs with [^{177}Lu-DOTA^0,-Tyr^3]octreotate resulted in complete remission in 2%, partial remission (decrease in tumor size $>50\%$) in 28%, a minor tumor response (decrease in tumor size between 25 and 50%) in 16%, and stable disease in 35% of patients, with a median time to progression of 40 months.[105] Generally, PRRT is well tolerated; serious adverse events include renal damage and induction of myeloproliferative disorders. Fig. 35.6 shows the effect of [^{177}Lu-DOTA^0,Tyr^3]-octreotate on tumor load in a NET patient with hepatic metastases.

SUMMARY

Neuroendocrine tumors (NETs) are rare tumors originating from enterochromaffin cells of the neuroendocrine system of the bronchial and gastrointestinal tracts. Because NETs have a relatively benign course but also have carcinoma-like characteristics, they were formerly termed "carcinoid" tumors. NETs can be functional or non-functional. Functional NETs are able to produce peptides and mediators that induce distinct clinical symptoms, whereas non-functional NETs usually present with mass effects. The clinicopathological features of NETs represent a broad spectrum ranging from well-differentiated localized tumors to aggressive, poorly differentiated tumors. Based on clinical symptoms and tumor localization, biochemical measurement of specific hormones or their degradation products can reveal hormone overproduction. Imaging of NETs is performed with conventional radiological techniques like computerized tomography and magnetic resonance imaging. In addition, NETs frequently express somatostatin receptors, and can therefore be localized with ^{111}In-pentetreotide scintigraphy. Because of the heterogeneity in clinical symptoms and biological behavior of NETs, diagnostic and therapeutic strategies should be based on an individualized, tailor-made approach. Therapeutic options for NETs include surgery, intervention radiological techniques, medical therapy, and peptide receptor radionuclide therapy. Somatostatin analogs are the mainstay of medical therapy for NETs. Peptide receptor radionuclide therapy also utilizes NETs' somatostatin receptor expression, and targeted radiotherapy with radiolabeled somatostatin analogs shows promising results. New targets for medical therapy include vascular endothelial growth factor (VEGF) and inhibitors of the mammalian target of rapamycin (mTOR). It should be emphasized that, because of the rarity and heterogeneity of NETs, the most efficacious (combination) therapy and optimal order of treatment modalities for NETs (subtypes) are largely unknown. This should be the aim of future studies, in addition to improvement of biochemical monitoring, imaging techniques and prognostic stratification, as well as identification of new molecular targets for medical therapy.

References

1. Rindi G, Bordi C. Highlights of the biology of endocrine tumours of the gut and pancreas. *Endocr Relat Cancer.* 2003;10(4):427–436.
2. Modlin IM, Oberg K, Chung DC, et al. Gastroenteropancreatic neuroendocrine tumours. *Lancet Oncol.* 2008;9(1):61–72.
3. Kloppel G, Dege K, Remmele W, Kapran Y, Tuzlali S, Modlin IM. Siegfried Oberndorfer: a tribute to his work and life between Munich, Kiel, Geneva, and Istanbul. *Virchows Arch.* 2007;451(Suppl. 1):S3–S7.
4. Rindi G, Capella C, Solcia E. Introduction to a revised clinicopathological classification of neuroendocrine tumors of the gastroenteropancreatic tract. *Q J Nucl Med.* 2000;44(1):13–21.
5. Kloppel G, Perren A, Heitz PU. The gastroenteropancreatic neuroendocrine cell system and its tumors: the WHO classification. *Ann NY Acad Sci.* 2004;1014:13–27.
6. Pape UF, Jann H, Muller-Nordhorn J, et al. Prognostic relevance of a novel TNM classification system for upper gastroenteropancreatic neuroendocrine tumors. *Cancer.* 2008;113(2):256–265.
7. Modlin IM, Lye KD, Kidd MA. 5-Decade analysis of 13,715 carcinoid tumors. *Cancer.* 2003;97(4):934–959.
8. Bertino EM, Confer PD, Colonna JE, Ross P, Otterson GA. Pulmonary neuroendocrine/carcinoid tumors: a review article. *Cancer.* 2009;115(19):4434–4441.
9. Yao JC, Hassan M, Phan A, et al. One hundred years after "carcinoid": epidemiology of and prognostic factors for neuroendocrine tumors in 35,825 cases in the United States. *J Clin Oncol.* 2008;26(18):3063–3072.
10. Rindi G, Leiter AB, Kopin AS, Bordi C, Solcia E. The "normal" endocrine cell of the gut: changing concepts and new evidences. *Ann NY Acad Sci.* 2004;1014:1–12.
11. Rindi G, Villanacci V, Ubiali A, Scarpa A. Endocrine tumors of the digestive tract and pancreas: histogenesis, diagnosis and molecular basis. *Expert Rev Mol Diagn.* 2001;1(3):323–333.
12. Speel EJ, Richter J, Moch H, et al. Genetic differences in endocrine pancreatic tumor subtypes detected by comparative genomic hybridization. *Am J Pathol.* 1999;155(6):1787–1794.
13. Rigaud G, Missiaglia E, Moore PS, et al. High resolution allelotype of nonfunctional pancreatic endocrine tumors: identification of two molecular subgroups with clinical implications. *Cancer Res.* 2001;61(1):285–292.
14. Moore PS, Missiaglia E, Antonello D, et al. Role of disease-causing genes in sporadic pancreatic endocrine tumors: MEN1 and VHL. *Genes Chromo Cancer.* 2001;32(2):177–181.
15. D'Adda T, Candidus S, Denk H, Bordi C, Hofler H. Gastric neuroendocrine neoplasms: tumour clonality and malignancy-associated large X-chromosomal deletions. *J Pathol.* 1999;189(3):394–401.
16. De Martino MC, Hofland LJ, Lamberts SW. Somatostatin and somatostatin receptors: from basic concepts to clinical applications. *Prog Brain Res.* 2010;182:255–280.
17. Florio T. Molecular mechanisms of the antiproliferative activity of somatostatin receptors (SSTRs) in neuroendocrine tumors. *Front Biosci.* 2008;13:822–840.
18. Weckbecker G, Lewis I, Albert R, Schmid HA, Hoyer D, Bruns C. Opportunities in somatostatin research: biological, chemical and therapeutic aspects. *Nat Rev Drug Discov.* 2003;2(12):999–1017.
19. Kwekkeboom DJ, Kam BL, van Essen M, et al. Somatostatin-receptor-based imaging and therapy of gastroenteropancreatic neuroendocrine tumors. *Endocr Relat Cancer.* 2010;17(1):R53–R73.
20. Kwekkeboom DJ, de Herder WW, Krenning EP. Somatostatin receptor-targeted radionuclide therapy in patients with gastroenteropancreatic neuroendocrine tumors. *Endocrinol Metab Clin North Am.* 2011;40(1):173–185.
21. Grozinsky-Glasberg S, Grossman AB, Korbonits M. The role of somatostatin analogues in the treatment of neuroendocrine tumours. *Mol Cell Endocrinol.* 2008;286(1-2):238–250.
22. Volante M, Rosas R, Allia E, et al. Somatostatin, cortistatin and their receptors in tumours. *Mol Cell Endocrinol.* 2008;286(1-2):219–229.
23. Lamberts SW, de Herder WW, Hofland LJ. Somatostatin analogs in the diagnosis and treatment of cancer. *Trends Endocrinol Metab.* 2002;13(10):451–457.
24. Fink G, Krelbaum T, Yellin A, et al. Pulmonary carcinoid: presentation, diagnosis, and outcome in 142 cases in Israel and review of 640 cases from the literature. *Chest.* 2001;119(6):1647–1651.
25. Ilias I, Torpy DJ, Pacak K, Mullen N, Wesley RA, Nieman LK. Cushing's syndrome due to ectopic corticotropin secretion: twenty years' experience at the National Institutes of Health. *J Clin Endocrinol Metab.* 2005;90(8):4955–4962.
26. van Hoek M, Hofland LJ, de Rijke YB, et al. Effects of somatostatin analogs on a growth hormone-releasing hormone secreting bronchial carcinoid, *in vivo* and *in vitro* studies. *J Clin Endocrinol Metab.* 2009;94(2):428–433.
27. Docherty HM, Heath DA. Multiple forms of parathyroid hormone-like proteins in a human tumour. *J Mol Endocrinol.* 1989;2(1):11–20.
28. Skuladottir H, Hirsch FR, Hansen HH, Olsen JH. Pulmonary neuroendocrine tumors: incidence and prognosis of histological subtypes. A population-based study in Denmark. *Lung Cancer.* 2002;37(2):127–135.
29. Rindi G, Luinetti O, Cornaggia M, Capella C, Solcia E. Three subtypes of gastric argyrophil carcinoid and the gastric neuroendocrine carcinoma: a clinicopathologic study. *Gastroenterology.* 1993;104(4):994–1006.
30. Ruszniewski P, Delle Fave G, Cadiot G, et al. Well-differentiated gastric tumors/carcinomas. *Neuroendocrinology.* 2006;84(3):158–164.
31. Gibril F, Schumann M, Pace A, Jensen RT. Multiple endocrine neoplasia type 1 and Zollinger-Ellison syndrome: a prospective study of 107 cases and comparison with 1009 cases from the literature. *Medicine (Baltimore).* 2004;83(1):43–83.
32. Borch K, Ahren B, Ahlman H, Falkmer S, Granerus G, Grimelius L. Gastric carcinoids: biologic behavior and prognosis

after differentiated treatment in relation to type. *Ann Surg.* 2005;242(1):64–73.

33. Jensen RT, Niederle B, Mitry E, et al. Gastrinoma (duodenal and pancreatic). *Neuroendocrinology.* 2006;84(3):173–182.
34. Roy PK, Venzon DJ, Shojamanesh H, et al. Zollinger-Ellison syndrome. Clinical presentation in 261 patients. *Medicine (Baltimore).* 2000;79(6):379–411.
35. Corleto VD, Annibale B, Gibril F, et al. Does the widespread use of proton pump inhibitors mask, complicate and/or delay the diagnosis of Zollinger-Ellison syndrome? *Aliment Pharmacol Ther.* 2001;15(10):1555–1561.
36. Anderson MA, Carpenter S, Thompson NW, Nostrant TT, Elta GH, Scheiman JM. Endoscopic ultrasound is highly accurate and directs management in patients with neuroendocrine tumors of the pancreas. *Am J Gastroenterol.* 2000;95(9): 2271–2277.
37. Yu F, Venzon DJ, Serrano J, et al. Prospective study of the clinical course, prognostic factors, causes of death, and survival in patients with long-standing Zollinger-Ellison syndrome. *J Clin Oncol.* 1999;17(2):615–630.
38. de Herder WW, Niederle B, Scoazec JY, et al. Well-differentiated pancreatic tumor/carcinoma: insulinoma. *Neuroendocrinology.* 2006;84(3):183–188.
39. Service FJ. Insulinoma and other islet-cell tumors. *Cancer Treat Res.* 1997;89:335–346.
40. Kann PH, Rothmund M, Zielke A. Endoscopic ultrasound imaging of insulinomas: limitations and clinical relevance. *Exp Clin Endocrinol Diabetes.* 2005;113(8):471–474.
41. Vezzosi D, Bennet A, Rochaix P, et al. Octreotide in insulinoma patients: efficacy on hypoglycemia, relationships with Octreoscan scintigraphy and immunostaining with anti-sst2A and anti-sst5 antibodies. *Eur J Endocrinol.* 2005;152(5):757–767.
42. Jackson JE. Angiography and arterial stimulation venous sampling in the localization of pancreatic neuroendocrine tumours. *Best Pract Res Clin Endocrinol Metab.* 2005;19(2): 229–239.
43. Nikou GC, Toubanakis C, Nikolaou P, et al. VIPomas: an update in diagnosis and management in a series of 11 patients. *Hepatogastroenterology.* 2005;52(64):1259–1265.
44. Soga J, Yakuwa Y. Somatostatinoma/inhibitory syndrome: a statistical evaluation of 173 reported cases as compared to other pancreatic endocrinomas. *J Exp Clin Cancer Res.* 1999;18(1): 13–22.
45. Chastain MA. The glucagonoma syndrome: a review of its features and discussion of new perspectives. *Am J Med Sci.* 2001;321(5):306–320.
46. Levy-Bohbot N, Merle C, Goudet P, et al. Prevalence, characteristics and prognosis of MEN 1-associated glucagonomas, VIPomas, and somatostatinomas: study from the GTE (Groupe des Tumeurs Endocrines) registry. *Gastroenterol Clin Biol.* 2004;28(11):1075–1081.
47. Verner JV, Morrison AB. Endocrine pancreatic islet disease with diarrhea. Report of a case due to diffuse hyperplasia of nonbeta islet tissue with a review of 54 additional cases. *Arch Intern Med.* 1974;133(3):492–499.
48. Garbrecht N, Anlauf M, Schmitt A, et al. Somatostatin-producing neuroendocrine tumors of the duodenum and pancreas: incidence, types, biological behavior, association with inherited syndromes, and functional activity. *Endocr Relat Cancer.* 2008;15(1):229–241.
49. O'Brien TD, Chejfec G, Prinz RA. Clinical features of duodenal somatostatinomas. *Surgery.* 1993;114(6):1144–1147.
50. Bloom SR, Polak JM. Glucagonoma syndrome. *Am J Med.* 1987;82(5B):25–36.
51. Falconi M, Plockinger U, Kwekkeboom DJ, et al. Well-differentiated pancreatic nonfunctioning tumors/carcinoma. *Neuroendocrinology.* 2006;84(3):196–211.
52. Cheslyn-Curtis S, Sitaram V, Williamson RC. Management of non-functioning neuroendocrine tumours of the pancreas. *Br J Surg.* 1993;80(5):625–627.
53. Madura JA, Cummings OW, Wiebke EA, Broadie TA, Goulet Jr RL, Howard TJ. Nonfunctioning islet cell tumors of the pancreas: a difficult diagnosis but one worth the effort. *Am Surg.* 1997;63(7):573–577. discussion 577–8.
54. Matthews BD, Heniford BT, Reardon PR, Brunicardi FC, Greene FL. Surgical experience with nonfunctioning neuroendocrine tumors of the pancreas. *Am Surg.* 2000;66(12): 1116–1122. discussion 1122–3.
55. Thomas-Marques L, Murat A, Delemer B, et al. Prospective endoscopic ultrasonographic evaluation of the frequency of nonfunctioning pancreaticoduodenal endocrine tumors in patients with multiple endocrine neoplasia type 1. *Am J Gastroenterol.* 2006;101(2):266–273.
56. Binkovitz LA, Johnson CD, Stephens DH. Islet cell tumors in von Hippel-Lindau disease: increased prevalence and relationship to the multiple endocrine neoplasias. *Am J Roentgenol.* 1990;155(3):501–505.
57. Hough DM, Stephens DH, Johnson CD, Binkovitz LA. Pancreatic lesions in von Hippel-Lindau disease: prevalence, clinical significance, and CT findings. *Am J Roentgenol.* 1994;162(5): 1091–1094.
58. Verhoef S, van Diemen-Steenvoorde R, Akkersdijk WL, et al. Malignant pancreatic tumour within the spectrum of tuberous sclerosis complex in childhood. *Eur J Pediatr.* 1999;158(4):284–287.
59. Skogseid B, Oberg K, Akerstrom G, et al. Limited tumor involvement found at multiple endocrine neoplasia type I pancreatic exploration: can it be predicted by preoperative tumor localization? *World J Surg.* 1998;22(7):673–677.
60. van der Lely AJ, de Herder WW. Carcinoid syndrome: diagnosis and medical management. *Arq Bras Endocrinol Metabol.* 2005; 49(5):850–860.
61. Grabowski P, Schonfelder J, Ahnert-Hilger G, et al. Expression of neuroendocrine markers: a signature of human undifferentiated carcinoma of the colon and rectum. *Virchows Arch.* 2002;441(3):256–263.
62. Ramage JK, Goretzki PE, Manfredi R, et al. Consensus guidelines for the management of patients with digestive neuroendocrine tumours: well-differentiated colon and rectum tumour/carcinoma. *Neuroendocrinology.* 2008;87(1):31–39.
63. Maggard MA, O'Connell JB, Ko CY. Updated population-based review of carcinoid tumors. *Ann Surg.* 2004;240(1):117–122.
64. Kaltsas GA, Besser GM, Grossman AB. The diagnosis and medical management of advanced neuroendocrine tumors. *Endocr Rev.* 2004;25(3):458–511.
65. de Herder WW. Biochemistry of neuroendocrine tumours. *Best Pract Res Clin Endocrinol Metab.* 2007;21(1):33–41.
66. Service FJ. Hypoglycemic disorders. *N Engl J Med.* 1995; 332(17):1144–1152.
67. Rockall AG, Reznek RH. Imaging of neuroendocrine tumours (CT/MR/US). *Best Pract Res Clin Endocrinol Metab.* 2007; 21(1):43–68.
68. Binderup T, Knigge U, Loft A, et al. Functional imaging of neuroendocrine tumors: a head-to-head comparison of somatostatin receptor scintigraphy, 123I-MIBG scintigraphy, and 18F-FDG PET. *J Nucl Med.* 2010;51(5):704–712.
69. Ruf J, Heuck F, Schiefer J, et al. Impact of Multiphase 68Ga-DOTATOC-PET/CT on therapy management in patients with neuroendocrine tumors. *Neuroendocrinology.* 2010;91(1): 101–109.

70. Koopmans KP, de Vries EG, Kema IP, et al. Staging of carcinoid tumours with 18F-DOPA PET: a prospective, diagnostic accuracy study. *Lancet Oncol.* 2006;7(9):728–734.
71. Putzer D, Gabriel M, Henninger B, et al. Bone metastases in patients with neuroendocrine tumor: 68Ga-DOTA-Tyr3-octreotide PET in comparison to CT and bone scintigraphy. *J Nucl Med.* 2009;50(8):1214–1221.
72. Ruszniewski P, Amouyal P, Amouyal G, et al. Localization of gastrinomas by endoscopic ultrasonography in patients with Zollinger-Ellison syndrome. *Surgery.* 1995;117(6): 629–635.
73. van Tuyl SA, van Noorden JT, Timmer R, Stolk MF, Kuipers EJ, Taal BG. Detection of small-bowel neuroendocrine tumors by video capsule endoscopy. *Gastrointest Endosc.* 2006;64(1): 66–72.
74. Norton JA. Surgery for primary pancreatic neuroendocrine tumors. *J Gastrointest Surg.* 2006;10(3):327–331.
75. de Herder WW, Krenning EP, Van Eijck CH, Lamberts SW. Considerations concerning a tailored, individualized therapeutic management of patients with (neuro)endocrine tumours of the gastrointestinal tract and pancreas. *Endocr Relat Cancer.* 2004;11(1):19–34.
76. Akerstrom G, Hellman P, Hessman O, Osmak L. Management of midgut carcinoids. *J Surg Oncol.* 2005;89(3):161–169.
77. van Vilsteren FG, Baskin-Bey ES, Nagorney DM, et al. Liver transplantation for gastroenteropancreatic neuroendocrine cancers: Defining selection criteria to improve survival. *Liver Transpl.* 2006;12(3):448–456.
78. Norton JA, Jensen RT. Resolved and unresolved controversies in the surgical management of patients with Zollinger-Ellison syndrome. *Ann Surg.* 2004;240(5):757–773.
79. Gustafsson BI, Hauso O, Drozdov I, Kidd M, Modlin IM. Carcinoid heart disease. *Intl J Cardiol.* 2008;129(3):318–324.
80. Ruszniewski P, O'Toole D. Ablative therapies for liver metastases of gastroenteropancreatic endocrine tumors. *Neuroendocrinology.* 2004;80(Suppl 1):74–78.
81. O'Toole D, Ruszniewski P. Chemoembolization and other ablative therapies for liver metastases of gastrointestinal endocrine tumours. *Best Pract Res Clin Gastroenterol.* 2005;19(4): 585–594.
82. Cao CQ, Yan TD, Bester L, Liauw W, Morris DL. Radioembolization with yttrium microspheres for neuroendocrine tumour liver metastases. *Br J Surg.* 2010;97(4):537–543.
83. Hellman P, Ladjevardi S, Skogseid B, Akerstrom G, Elvin A. Radiofrequency tissue ablation using cooled tip for liver metastases of endocrine tumors. *World J Surg.* 2002;26(8): 1052–1056.
84. Seifert JK, Cozzi PJ, Morris DL. Cryotherapy for neuroendocrine liver metastases. *Semin Surg Oncol.* 1998;14(2):175–183.
85. de Herder WW, Lamberts SW. Somatostatin and somatostatin analogues: diagnostic and therapeutic uses. *Curr Opin Oncol.* 2002;14(1):53–57.
86. de Herder WW, Lamberts SW. Somatostatin analog therapy in treatment of gastrointestinal disorders and tumors. *Endocrine.* 2003;20(3):285–290.
87. Rubin J, Ajani J, Schirmer W, et al. Octreotide acetate long-acting formulation versus open-label subcutaneous octreotide acetate in malignant carcinoid syndrome. *J Clin Oncol.* 1999;17(2): 600–606.
88. Wymenga AN, Eriksson B, Salmela PI, et al. Efficacy and safety of prolonged-release lanreotide in patients with gastrointestinal neuroendocrine tumors and hormone-related symptoms. *J Clin Oncol.* 1999;17(4):1111.
89. Kvols LK, Martin JK, Marsh HM, Moertel CG. Rapid reversal of carcinoid crisis with a somatostatin analogue. *N Engl J Med.* 1985;313(19):1229–1230.
90. Shojamanesh H, Gibril F, Louie A, et al. Prospective study of the antitumor efficacy of long-term octreotide treatment in patients with progressive metastatic gastrinoma. *Cancer.* 2002;94(2): 331–343.
91. Rinke A, Muller HH, Schade-Brittinger C, et al. Placebo-controlled, double-blind, prospective, randomized study on the effect of octreotide LAR in the control of tumor growth in patients with metastatic neuroendocrine midgut tumors: a report from the PROMID Study Group. *J Clin Oncol.* 2009; 27(28):4656–4663.
92. Oberg K. Interferon in the management of neuroendocrine GEP-tumors: a review. *Digestion.* 2000;62(Suppl 1):92–97.
93. Kolby L, Persson G, Franzen S, Ahren B. Randomized clinical trial of the effect of interferon alpha on survival in patients with disseminated midgut carcinoid tumours. *Br J Surg.* 2003;90(6): 687–693.
94. Pape UF, Wiedenmann B. Adding interferon-alpha to octreotide slows tumour progression compared with octreotide alone but evidence is lacking for improved survival in people with disseminated midgut carcinoid tumours. *Cancer Treat Rev.* 2003;29(6):565–569.
95. Eriksson B, Annibale B, Bajetta E, et al. ENETS Consensus Guidelines for the Standards of Care in Neuroendocrine Tumors: chemotherapy in patients with neuroendocrine tumors. *Neuroendocrinology.* 2009;90(2):214–219.
96. Kouvaraki MA, Ajani JA, Hoff P, et al. Fluorouracil, doxorubicin, and streptozocin in the treatment of patients with locally advanced and metastatic pancreatic endocrine carcinomas. *J Clin Oncol.* 2004;22(23):4762–4771.
97. Schmid HA. Pasireotide (SOM230): development, mechanism of action and potential applications. *Mol Cell Endocrinol.* 2008; 286(1-2):69–74.
98. Pivonello R, Ferone D, de Herder WW, et al. Dopamine receptor expression and function in corticotroph ectopic tumors. *J Clin Endocrinol Metab.* 2007;92(1):65–69.
99. Yao JC, Phan A, Hoff PM, et al. Targeting vascular endothelial growth factor in advanced carcinoid tumor: a random assignment phase II study of depot octreotide with bevacizumab and pegylated interferon alpha-2b. *J Clin Oncol.* 2008;26(8):1316–1323.
100. Capdevila J, Salazar R, Halperin I, Abad A, Yao JC. Innovations therapy: mammalian target of rapamycin (mTOR) inhibitors for the treatment of neuroendocrine tumors. *Cancer Metastasis Rev.* 2011;30(Suppl 1):27–34.
101. Wullschleger S, Loewith R, Hall MN. TOR signaling in growth and metabolism. *Cell.* 2006;124(3):471–484.
102. Yao JC, Lombard-Bohas C, Baudin E, et al. Daily oral everolimus activity in patients with metastatic pancreatic neuroendocrine tumors after failure of cytotoxic chemotherapy: a phase II trial. *J Clin Oncol.* 2010;28(1):69–76.
103. Yao JC, Shah MH, Ito T, et al. Everolimus for advanced pancreatic neuroendocrine tumors. *N Engl J Med.* 2011;364(6): 514–523.
104. Raymond E, Dahan L, Raoul JL, et al. Sunitinib malate for the treatment of pancreatic neuroendocrine tumors. *N Engl J Med.* 2011;364(6):501–513.
105. Kwekkeboom DJ, de Herder WW, Kam BL, et al. Treatment with the radiolabeled somatostatin analog [177 Lu-DOTA 0, Tyr3] octreotate: toxicity, efficacy, and survival. *J Clin Oncol.* 2008;26(13):2124–2130.

CHAPTER

36

Psychoneuroendocrinology

David R. Rubinow[1], *Virginia L. Crowder*[1], *Peter J. Schmidt*[2], *Katya B. Rubinow*[3]

[1] **Department of Psychiatry, University of North Carolina at Chapel Hill, Chapel Hill, NC, USA,** [2] **Section on Behavioral Endocrinology, NIMH, Bethesda, MD, USA,** [3] **Division of Endocrinology in the Department of Medicine, University of Washington School of Medicine; Seattle, WA, USA**

OUTLINE

Summary

The basic principles of psychoneuroendocrinology can be extracted from the efforts to characterize the complex interactions among the central nervous system, hormones, and behavior. The first set of principles derives from traditional psychoneuroendocrine studies: (1) hormonal disruption can effect behavioral change; (2) exogenously administered hormones can confer psychoactive effects; (3) endocrine dysfunction and psychiatric symptoms may co-occur in both primary endocrinopathies and primary psychiatric disorders. The second set of principles incorporates a more recent, nuanced, and contextual view of physiology in general and the physiological basis of behavioral regulation in particular: (4) altered endocrine function generates behavioral abnormalities in only a subset of individuals; (5) behavioral abnormalities can be etiologically linked to normal endocrine function; and (6) environmental context can modulate this relationship between endocrine function and behavior. The observations giving rise to these latter principles suggest that despite the clear relevance of endocrine function to a variety of affective-behavioral experiences (as implied by the term "psychoneuroendocrinology"), the etiological role of hormones in these experiences is highly context-dependent. Further, identification of the determinants of context dependency is far more likely to inform us about the biological constituents of behavior than is exclusive attention to hormone exposure and level.

INTRODUCTION

The scope of neuroendocrinology has widened considerably beyond its initial exclusive focus on the regulation of hypothalamic and pituitary-mediated hormone secretion. Now, active investigation addresses how the products of both these and hypothalamic-independent neurosecretory pathways in turn can modulate central nervous system (CNS) function. This conceptual expansion is reflected in evolving use of the term "psychoneuroendocrinology." In a restricted and historical sense, psychoneuroendocrinology was principally a descriptive field that sought to identify patterns of

Handbook of Neuroendocrinology, DOI: 10.1016/B978-0-12-375097-6.10036-8

altered hormone secretion associated with psychiatric illness and the psychiatric symptoms of primary hormonal abnormalities. Moreover, disturbances in behavior – a central focus of psychoneuroendocrinology – were regarded as part of the province of literature, morality and psychology, in contradistinction to other, overtly biology-based, fields of medicine. Over recent decades, however, the purview of psychoneuroendocrinology has been transformed as our understanding of the neurobiology of behavior has catapulted behavior from the province of humanistic theory to the very forefront of neuroscience. Indeed, one can question whether the term psychoneuroendocrinology has outlived its usefulness, and whether the "psycho" of psychoneuroendocrinology is biologically and meaningfully distinct from other readouts of CNS function.

In this chapter, we will describe six basic principles that can be extracted from the efforts to characterize the complex interactions among the CNS, hormones and behavior. The first set of principles derives from traditional psychoneuroendocrine studies. The second set of principles incorporates a more recent, nuanced and contextual view of physiology in general, and the physiological basis of behavioral regulation in particular. The observations giving rise to these latter principles suggest that, despite the clear relevance of endocrine function to a variety of affective-behavioral experiences (as implied by the term "psychoneuroendocrinology"), the etiological role of hormones in these experiences is highly context-dependent. Further, identification of the determinants of context dependency is far more likely to inform us about the biological constituents of behavior than is exclusive attention to hormone exposure and level.

HISTORICAL OBSERVATIONS AND PRINCIPLES

Although it is beyond the scope of this chapter to provide a comprehensive history of psychoneuroendocrinology, some historical context is helpful for understanding the evolution of theory in the field. Prior to the 1950s, psychoneuroendocrinology was predicated on three fundamental principles derived from observations of both animals and humans. These preliminary observations provided the foundation for continued psychoneuroendocrine investigations, as well as initial, compelling evidence for a biological basis of behavior.

Principle 1: Hormonal Disruption can Effect Behavioral Change

Long before the term "hormone" was first coined in 1905, speculated links between behavior and the endocrine system (then undefined) were generated through observations of both animals and humans.[1] Many of the earliest observations pertained specifically to reproductive hormones and offered nascent insight into the relationship among hormones, the brain and behavior. In his work the *Historia Animalium* (History of Animals), from 350 BCE, Aristotle noted that castration of male birds "when young" not only prevented the development of physical sex characteristics but also further extinguished sex-specific behaviors such as male singing and sexual interest. Likewise, if castrated when "full grown," the bird would "cease to crow" and lose his prior "desire for sexual intercourse," suggesting that such behaviors were regulated or maintained by some factor removed by castration.[2] Aristotle rendered similar observations about humans in his commentary on eunuchs. As a result of the procedure, Aristotle noted that castrated young men ceased to experience the changes in voice, hair growth and musculature that were observed in the normally developed adult male.[2] Moreover, castration was often performed specifically to effect behavioral outcomes, employed to "eliminate libido" and its attendant distractions and thereby enhance loyalty.[3]

These observations provided compelling evidence that removal of testes could produce physical and behavioral changes in both animals and humans, but centuries elapsed before these changes were ascribed to the withdrawal of "secretions" harbored in the glands. In 1849, Berthold showed that although castration of roosters resulted in behavioral changes, many of the original behaviors could be restored through testicular transplant.[4] These experiments again demonstrated the systemic effects conferred by castration, and further indicated that behavioral changes were not solely consequent to the act of castration itself; rather, testicular transplant could restore a behavioral phenotype in similar, systemic fashion. Berthold reasoned that organs such as the testes must contain secretions, a model that could account for effects observed throughout the "whole organism."[4] He posited that the disruption of these secretions constituted the means by which castration produced behavioral modulation.

Principle 2: Exogenously Administered Hormones can Confer Psychoactive Effects

By the 19th century, these experimental and observational studies culminated in the notion that the administration of testicular or ovarian extract could treat a variety of mood disorders in humans, ranging from depression to the "anergy of senescence."[5] The novel field of organotherapy arose from this premise and became popular after 1889, when Charles

Brown-Séquard proposed to the Society of Biology in Paris that an active ingredient in the testes could treat "debilitating and nervous diseases."[5] Drawing from personal experience with a self-administered dose of orchitic fluid, Brown-Séquard concluded that extracts could produce activating effects on the CNS and promote "increased strength, vigor, and mental activity."[5]

As the appeal of organotherapy grew, physicians began to consider the potential for other organs to harbor therapeutic secretions. Consequently, attempts were made to harvest extractions from nearly every organ in the body, including the thyroid, uterus, ovaries, spleen, thymus, parathyroid and adrenals.[5] Although many of these efforts failed to generate new therapies, extracts from the ovaries, thyroid and adrenals were noted to confer marked psychoactive effects and hence were believed to have utility in the treatment of psychiatric illnesses. For instance, a review of case studies published by Edwards in 1895 described how doses of orchitic tablets successfully rid one woman of her "many fancies," and eliminated the depression and "extreme nervousness" of a perimenopausal woman.[6] Thyroid extracts became a popular treatment for "insanity" as well as (appropriately) for hypothyroidism, then identified as "myxoedema." Extracts from the adrenals were employed for their observed mood-stabilizing effect in manic patients.

Thus, the psychoactive effects of organ extracts led to the use of hormones as early psychotropic therapies. Moreover, these therapies constituted the basis of initial clinical research trials in the study of psychiatric illness. In 1900, for example, Easterbrook published one of the first reviews of the therapeutic effects of common extracts.[7] In a study of 130 patients with various psychiatric disorders treated with the popular thyroid extract, he found clinical improvement in 26–30% of patients, with complete recovery in approximately 12%. He also concluded that thyroid treatment was preferentially efficacious in female patients, particularly for "insanities connected with childbearing."[7] In another study, of 36 patients, Easterbrook determined that ovarian extracts were a useful treatment for female psychiatric patients, particularly those suffering from mania or melancholia.

The allure of organotherapy eventually faded as the cumulative results belied the promised benefits. Nonetheless, the underlying principle re-emerged soon after the isolation of estradiol in 1929, when Werner described the antidepressant efficacy of an estradiol preparation "theelin."[8] This placebo-controlled study established the historical justification and precedent for the use of gonadal steroids, alone or in combination with other agents, for the treatment of depression occurring during periods of reproductive endocrine transition. These reproductive-related depression syndromes included post-partum depression (PPD), premenstrual dysphoria and peri-menopausal depression – disorders recognized long before evolution of the current nomenclature.

In aggregate, then, these initial observations generated a fledgling understanding that disruption of hormonal function can confer behavioral aberrations. Further, it was reasonably inferred that as hormones can modulate behavior, they must exert effects on the CNS. These ideas constitute the requisite foundation for both components of the third principle; namely, endocrinopathies may present with psychiatric symptoms, and, inversely, that primary psychiatric disorders may manifest with abnormal endocrine function.

Principle 3: Behavioral Abnormalities may be Associated with Endocrine Dysfunction

Psychiatric Symptoms Associated with Endocrinopathies

Despite longstanding recognition that endocrine manipulation could alter behavior, it was not until the early 20th century that psychiatric symptoms were identified as clinical features of primary endocrine disorders. Among the first such observations are those found in Harvey Cushing's characterization of the syndrome that bears his name. Central to his description of Cushing's syndrome are a number of associated psychiatric symptoms, including sleep disturbances, irritability and depression.[9] Indeed, the remarkably high prevalence of depression in patients with Cushing's syndrome (50–80%)[10] provided initial evidence for the now well-known link between glucocorticoids and affective disturbance. Most of the psychoneuroendocrine studies of the 1950s and 1960s were devoted to characterizing this link between hypothalamic–pituitary–adrenal (HPA) axis dysregulation and affective illness. During this era glucocorticoid assays first came into widespread use, initially with basal (urine, plasma) measures of glucocorticoids or their metabolites[11,12] and then with dynamic measures of feedback regulation in the form of the dexamethasone suppression test (DST).[13,14]

Similarly, by the 19th century an association had been identified between the presence of a goiter and behavioral anomalies such as hysteria, psychosis and apprehension. Accordingly, Hofrichter commented in 1820 that "women would certainly have long ceased to go about with bare necks, for husbands would have learned to recognize the swelling of this gland as a danger signal of threatening trouble from their better halves."[15] Further, administration of thyroid extract was observed to assuage symptoms of depression associated with myxoedema, whereas excess administration induced psychosis, hallucinations and psychomotor agitation.[7] As these observations antedate the development of thyroid function assays, these psychiatric symptoms were described as elements of clinical syndromes in the absence of a biochemical diagnosis. Nonetheless, the symptoms were understood as phenomena that evolved from a primary disturbance in the thyroid gland and, moreover, responded to treatment with thyroid extract.

Abnormal Endocrine Function Associated with Primary Psychiatric Illness

The presence of psychiatric symptoms in endocrine disorders generated the question as to whether endocrine abnormalities might accompany psychiatric illness. Pursuit of this question was enabled by the isolation and characterization of hormones and subsequent development of biochemical assays that allowed their direct measurement. Blood, urine, saliva, cerebrospinal fluid and post-mortem tissues all were subject to analysis, particularly in early studies that attempted to identify group-specific differences in hormone levels. Many of these initial studies examined the relationship between depression and cortisol and reproduced findings that depressed patients exhibited elevated cortisol levels in plasma, urine and saliva.[16–19] These elevations occurred in the absence of Cushing's syndrome, were affective state-dependent, and were presumed unrelated to stress. In addition to increased basal cortisol levels, abnormalities were identified in cortisol circadian regulation (flattening of the diurnal curve with increased nocturnal secretion)[19] and negative feedback (escape from dexamethasone suppression).[20–22] As additional biochemical assays were developed, further perturbations were identified, including altered central drive (increased CSF CRH, blunted ACTH response to CRH) and an exaggerated response to the CRH-DST test.[23–25] These abnormalities were observed in 40–60% of patients with depression and, therefore, were not inevitable consequences of the depressed state.[26,27] Notably, however, when HPA axis dysregulation was present and sustained despite treatment, it appeared to predict illness recurrence and treatment failure.[28–30]

Although depression was not characterized by abnormal basal levels of thyroid hormone, dynamic disturbances in thyroid function were identified in depressed patients. Specifically, a blunted TSH response to TRH was observed in approximately 25–30% of patients with depression, but the presence of this abnormality did not appear to predict illness course or treatment response.[31,32] Consequently, neither dynamic testing nor adjunctive use of thyroid hormone became mainstays of the evaluation and treatment of depression.

Over the past 20 years, the explosion of available tools for cellular and molecular neurobiology, genetics, and systems neuroscience has dramatically altered the landscape of psychoneuroendocrinology in several critical ways. At the level of the neuron and the neural circuit, we are now able to meaningfully investigate the mechanisms by which abnormal endocrine function can produce behavioral abnormalities, exogenously administered hormones can have psychoactive effects, and behavioral abnormalities can be associated with altered endocrine function. Moreover, the individual variance in the above-described associative principles reveals three novel, fundamental principles that far better characterize the relationship between endocrine function and behavior. These principles emphasize the context-dependency of the effects of hormones on the central nervous system, the differential sensitivity to the mood- and behavior-dysregulating effects of hormones across individuals, and the biological instantiation of susceptibility to the destabilizing effects of hormones on mood and behavior.

Principle 4: Altered Endocrine Function Produces Behavioral Abnormalities in a Subset of Individuals

The associations between endocrinopathies and abnormal behavior modeled the biological "cause" of psychiatric illness, but the lack of uniformity of the observed associations led to conclusions of irrelevance; if all individuals with abnormal endocrine function did not display similar mood or behavioral abnormalities, then mood dysregulation might simply reflect an adverse response to the experience of other endocrinopathy-related symptoms. The strikingly frequent association of depression with Cushing's syndrome argues against this dismissal of a meaningful relationship between endocrine dysfunction and behavior. Nonetheless, the frequency of association alone is inadequate to establish a direct, causal relationship. Are the mood disturbances directly attributable to the abnormal endocrine function, or is the association spurious, temporally coincident but causally indirect?

Evidence in support of a directly causal relationship between altered endocrine function and mood disturbance derives from both case studies and treatment trials. Moreover, the evidence is present comparably in the setting of both endocrinopathies and normal, physiologic alterations in endocrine function. Fig. 36.1 depicts an illustrative example of a young woman who presented with mood and somatic symptoms, including menstrual cycle-related depression.[33] In the course of her evaluation, she was diagnosed with hypothyroidism and initiated on thyroid hormone replacement. As demonstrated in the figure, treatment of her endocrinopathy eliminated the episodic, menstrual cycle phase-entrained affective syndrome with which she had presented. This case study illustrates several points. First, the patient experienced a cyclic mood disorder consequent to a stable endocrinopathy, with symptoms "unmasked" by normal physiologic fluctuations in endogenous gonadal steroid exposure. This suggests that both normal and abnormal changes in endocrine function may influence the susceptibility to mood disorders. Second, the dramatic elimination of the cyclic depression following treatment of her hypothyroidism argues against the attribution of depression to somatic concomitants of the endocrinopathy. Finally, although admittedly an unusual manifestation of hypothyroidism, this case offers compelling evidence for a model of differential sensitivity; though affective symptoms are not universal among patients with hypothyroidism, specific patients are at increased risk of developing an affective syndrome as a direct consequence of an endocrinopathy.

Similarly, the efficacy of estradiol in the treatment of perimenopausal depression[34–36] suggests that either estrogen withdrawal or the hypogonadal levels of estradiol during the perimenopause are causally related to the subsequent onset of depression. Clearly, however, perimenopausal depression is not a universal phenomenon. Interestingly, the response to estradiol is binary; women with perimenopausal depression either respond completely to administration of estradiol or appear not to respond at all. These findings therefore suggest that hormone withdrawal or deficiency might play an etiopathogenic role in some but not all women with perimenopausal depression. This suggestion of differential sensitivity to the behavioral effects of changes in reproductive hormones has been confirmed in a subsequent study. Euthymic women on estradiol replacement therapy were withdrawn from treatment in a blinded,

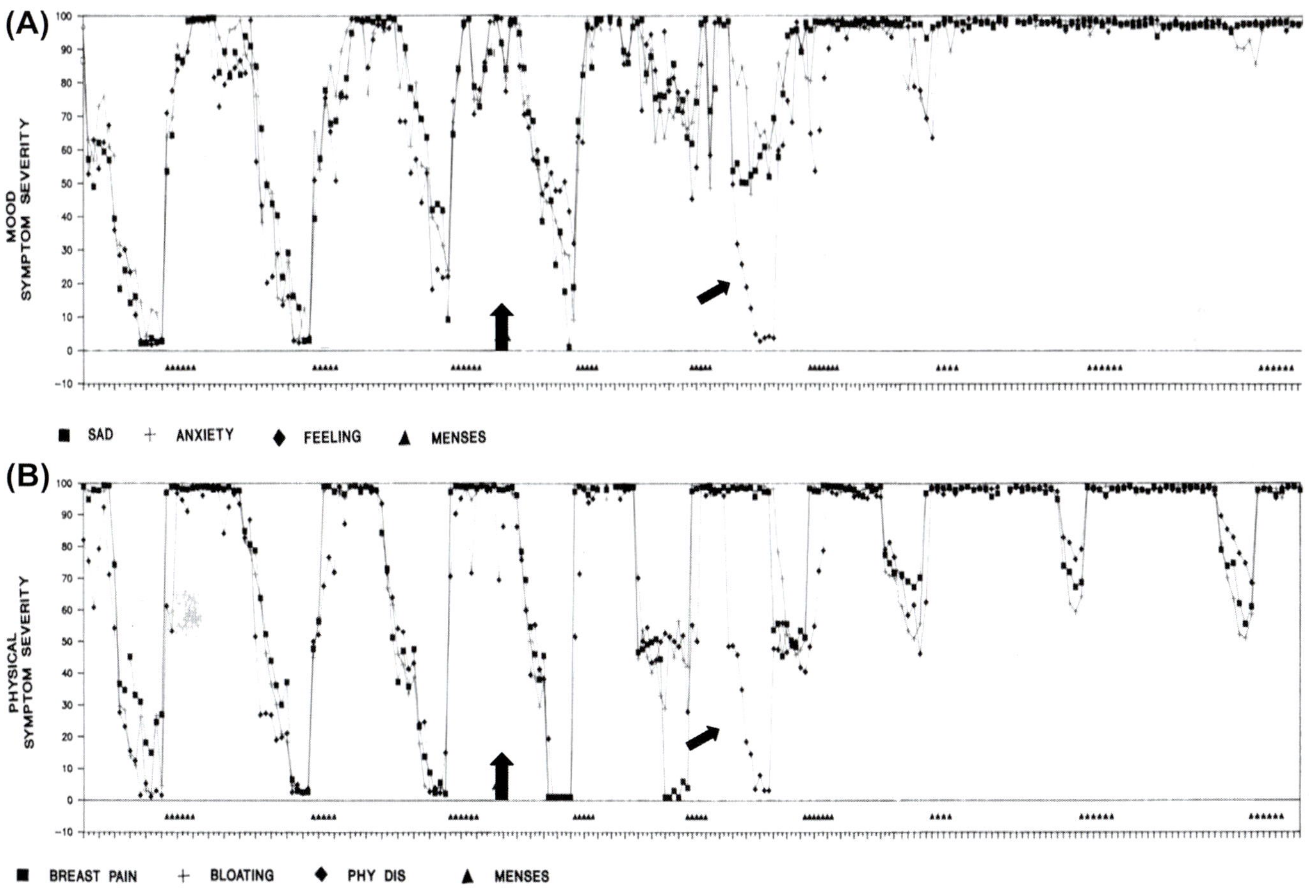

FIGURE 36.1 Daily visual analog self-ratings of sadness, anxiety, and global feelings (**A**) and breast pain, bloating, and physical discomfort (**B**). Y-axis, severity of symptoms (e.g., 0 = "most sad ever," 100 = "most happy ever"); X-axis, days of the calendar month. The first month displayed represents the typical patterns of menstrual related affective and somatic symptoms confirmed during the 3-month screening period. The first arrow indicates when levothyroxine dosage was increased to 0.15 mg/day (after 1 month on 0.10 mg/day). An initial improvement in severity of symptoms was observed after the patient received a dosage of 0.15 mg/day for 6 weeks. The second arrow corresponds to a viral infection that developed, which resulted in a selective decrease in global and somatic ratings during that time. In the ensuing months, all that remained was premenstrual molimina.

BOX 36.1

HOW DO WE KNOW THAT ENDOCRINE FUNCTION AFFECTS BEHAVIORAL AND MOOD DISTURBANCES?

Longitudinal investigation of basal hormone levels in neuroendocrine studies designed to investigate the hormonal etiology of PPD, PMDD, and perimenopausal depression indicated that the behavioral and mood disturbances observed in such disorders occur within the context of normal (although changing) endocrine function. For instance, early studies of women with PMDD and controls revealed no diagnosis-related differences in patterns and levels of progesterone, estradiol, FSH and LH.[1] Likewise, neither studies on perimenopausal[2] nor post-partum depression[3] have found consistent, diagnosis-related differences in the plasma levels of gonadal steroids. As described in Principle 5, hormone manipulation studies have demonstrated that mood disorders can be etiologically linked to normal endocrine function. Thus, normal levels or changes in levels may trigger an altered mood state, but only in a susceptible subgroup.

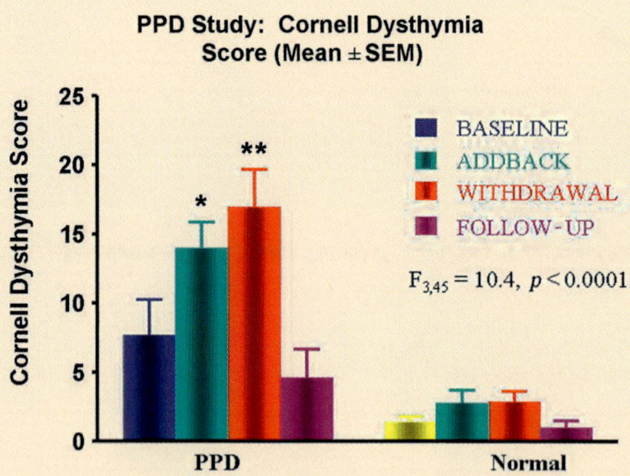

Mean scores on the Cornell Dysthymia Scale before and after estradiol and progesterone replacement in women with a history of post-partum depression ($n = 8$) and normal controls ($n = 8$). [4]study phases: 8-week baseline, when no medications were administered; 12 weeks of leuprolide-induced ovarian suppression, with addback of estradiol (4–10 mg/day) and progesterone (400–900 mg/day) during the past 8 weeks; 4-week early withdrawal, when estradiol and progesterone were withdrawn but leuprolide continued; and 8-week follow-up, when no medications were administered. *Addback vs baseline, $P < 0.05$; **withdrawal vs baseline, $P < 0.01$ (Bonferroni corrected *t*-tests).

For instance, Bloch and colleagues manipulated hormonal conditions to create a scaled-down model of the perinatal period in women with and without a history of PPD.[4] In a blinded fashion, euthymic participants were administered leuprolide and placebo pills for 1 month to stabilize hormone levels, then placebo pills were replaced with high-dose estradiol and progesterone for 2 months, followed by 2 more months of placebo to initiate an abrupt withdrawal phase. Mood symptoms were assessed using the Beck Depression Inventory, the Edinburgh Postnatal Depression Scale and the Cornell Dysthymia Scale. Results showed that women with a history of PPD, but not controls, experienced an emergence of depressive symptoms both during the add-back period and, more prominently, during the early withdrawal period (see figure). Given the achievement of identical hormone conditions in both groups, the hormone-precipitated affective disturbances in those with a history of PPD suggest a differential sensitivity to the potential mood-destabilizing effects of reproductive steroids. These findings parallel those previously demonstrated in women with PMDD.[5] The sources of this susceptibility are unknown, although support for genetic influences has appeared in both animals and humans.[6,7]

References

1. Rubinow DR, Schmidt PJ, Roca CA. Estrogen–serotonin interactions: implications for affective regulation. *Biol Psychiatry.* 1998;44(9):839–850.
2. Schmidt PJ, Murphy JH, Haq N, Danaceau MA, St Clair L. Basal plasma hormone levels in depressed perimenopausal women. *Psychoneuroendocrinology.* 2002;27(8):907–920.
3. Harris B. Biological and hormonal aspects of postpartum depressed mood. *Br J Psychiatry.* 1994;164(3):288–292.
4. Bloch M, Schmidt PJ, Danaceau M, Murphy J, Nieman L, Rubinow DR. Effects of gonadal steroids in women with a history of postpartum depression. *Am J Psychiatry.* 2000; 157(6):924–930.
5. Schmidt PJ, Nieman LK, Danaceau MA, Adams LF, Rubinow DR. Differential behavioral effects of gonadal steroids in women with and in those without premenstrual syndrome. *N Engl J Med.* 1998;338(4):209–216.
6. Huo L, Straub RE, Roca C, Schmidt PJ, Shi K, Vakkalanka R, et al. Risk for premenstrual dysphoric disorder is associated with genetic variation in ESR1, the estrogen receptor alpha gene. *Biol Psychiatry.* 2007;62(8):925–933.
7. Spencer JL, Waters EM, Milner TA, Lee FS, McEwen BS. BDNF variant Val66Met interacts with estrous cycle in the control of hippocampal function. *Proc Natl Acad Sci.* 2010; 107(9):4395–4400.

placebo-controlled fashion; that is, one group was continued on estradiol whereas the second group received placebo. Estradiol withdrawal rapidly precipitated depression in women with a history of estradiol-responsive, perimenopausal depression, but not in those without a history of perimenopausal depression (Schmidt *et al.*, unpublished data).

The evidence presented is in no way comprehensive, but rather representative of the abundant clinical data demonstrating a direct role for changes in endocrine function in the precipitation of affective symptoms. Nonetheless, if we accept the premise that hormonal changes can produce behavioral abnormalities, we must still explain the basis for the apparent differential susceptibility across individuals to these behavior-perturbing effects. Furthermore, we must account for an equally perplexing clinical observation, the evolution of affective disorders in the setting of normal endocrine function, as described below.

Principle 5: Behavioral Abnormalities can be Etiologically Linked to Normal Endocrine Function

Increasingly, the idea that individuals with primary endocrinopathies variably display associated behavioral symptoms reflects shared clinical understanding. Similarly, the remediation of the behavioral symptoms with correction of the abnormal hormone state has been repeatedly observed and indirectly supports the role of the endocrinopathy in the etiopathogenesis of the mood disturbance. In the case of several reproductive endocrine-related mood disorders, however, evidence strongly suggests that *normal* reproductive steroid levels and normal ovarian function can nonetheless trigger mood disorders in a susceptible population.

Women with premenstrual dysphoric disorder (PMDD), for example, exhibit no differences in peripheral hormone levels compared with women lacking the disorder.[37] As symptoms of PMDD are experienced during the luteal phase of the menstrual cycle, a double-blind, placebo-controlled study was designed to examine the effects of mifespristone-induced luteal phase truncation on affective symptoms.[38] Remarkably, women with PMDD experienced typical premenstrual behavioral symptoms in the context of follicular-phase hormonal levels. These observations would appear to suggest the lack of involvement of hormones (particularly luteal-phase hormones) in the precipitation of premenstrual dysphoria. However, in a subsequent double-blind, placebo-controlled study, Schmidt and colleagues employed leuprolide-induced ovarian suppression and demonstrated the complete elimination of cyclic mood symptoms with cessation of gonadal steroid exposure.[39] Moreover, the rapid re-stimulation of mood symptoms followed subsequent blinded administration of either estradiol or progesterone but not placebo. Thus, the ovarian steroids estradiol and progesterone appear to play a direct role in triggering the cyclic mood symptoms seen in women with premenstrual dysphoria. Nonetheless, when identical experimental manipulation was performed in women lacking a history of premenstrual dysphoria, no mood perturbations were observed despite identical hormonal states. Thus, women with a history of premenstrual dysphoria were differentially sensitive to reproductive steroids such that identical hormone exposures induced absolutely no mood effects in normal women but precipitated severe mood perturbations in women with a history of premenstrual dysphoria.

A similar study, by Bloch and colleagues, employed leuprolide in conjunction with high-dose estradiol and progesterone to create a scaled-down model of pregnancy and the puerperium.[40] Using this model, investigators demonstrated that high-dose hormone add-back and subsequent abrupt withdrawal precipitated dramatic depressive symptoms in euthymic women with a history of PPD. Again, no mood symptoms were evident in women lacking that history. This study again demonstrated that despite the absence of consistent evidence of abnormal hormonal levels in women with PPD,[41] reproductive steroids were nonetheless directly involved in triggering the depressive episodes in a susceptible population.

Therefore, in the setting of both normal and aberrant endocrine function, we encounter the challenge of explaining this apparent differential susceptibility to hormone-induced affective disturbance.

Principle 6: Hormonal Effects on Brain and Behavior are Context-dependent

The differential sensitivity to gonadal steroids seen in women with histories of reproductive-related mood disorders emphasizes that the response to a biological signal cannot be inferred without an understanding of the context in which the signal occurs. This context includes current physiological and external environments, prior experience, history of exposure to the hormonal stimulus, and genetic make-up. Thus, hormonal stimuli, whether normal or abnormal, may trigger mood destabilization if they occur in a context of susceptibility. In order to elucidate a model of differential susceptibility, we must first establish the mechanistic potential for hormones to play this "triggering" role.

Neuroscience research has now provided substantial data demonstrating the influence of hormones, particularly of estradiol, on affective regulation. Several animal models demonstrate the antidepressant effects of estradiol and further indicate that these effects are mediated specifically through estrogen receptor-β.[42,43] Estradiol extensively modulates systems implicated in the pathophysiology of depression. It regulates the synthesis, metabolism and receptor concentration/trafficking of the classical neurotransmitters implicated in depression, including serotonin, dopamine and norepinephrine.[44–46] Further, animal studies indicate that estradiol regulates both basal and stress-induced secretion of ACTH and cortisol through effects on the glucocorticoid receptor and CRH secretion.[47–50] Similar to antidepressant therapy, estradiol stimulates brain-derived neurotropic factor (BDNF), a critical growth factor observed to be deficient in depression.[51–53] Notably, stress induces the opposite effect of BDNF suppression.[51,54] In the rat brain, estradiol mimics the molecular effects of mood stabilizers, as it increases CREB activity and trkA and decreases GSK-3β.[53,55,56] Collectively, these functions suggest a neuroprotective role for estradiol. This idea is further supported by studies employing ischemia, glutamate excess and β-amyloid formation as models of neuronal injury.[57,58] Contributing to this neuroprotective effect, estradiol at multiple levels attenuates the inflammatory processes hypothesized to play a pathogenic role in depression.[59]

Finally, estradiol regulates reward circuitry, a system shown to exhibit dysregulation in the setting of depression.[60,61] In human brain imaging studies, estradiol and progesterone modulate cognitive-stimulated activity in the dorsolateral prefrontal cortex (dlPFC) and in BA25, the region implicated in deep brain stimulation treatment of depression.[62] In addition to regulating cerebral blood flow in brain regions (amygdala, dlPFC) implicated in depression, reproductive steroids have been shown to determine the valence of stimuli to which brain regions including the medial PFC react. Protopopescu and colleagues demonstrated that PFC activation was menstrual cycle phase-dependent and, strikingly, found complete, phase-dependent reversal of the affective valance of the stimulus that activated the PFC.[63] Specifically, the PFC responded to positive affective stimuli during the follicular phase of the menstrual cycle but to negative stimuli during the luteal phase.[63] Further, this effect of menstrual cycle phase was seen only in control women and not in those with PMDD.[64] The absence of this finding in women with PMDD could reflect either an effect of the experience of a recurrent mood disorder or a potential source of susceptibility for the disorder.

Studies of thyroid hormone dysregulation additionally lend evidence of the direct role of endocrinopathies in CNS modulation and affective regulation. For example, the severity of the mood and somatic concomitants of hypothyroidism is inversely related to the degree of activation of the anterior cingulate cortex (ACC); further, both symptoms and ACC activation normalize following thyroidal hormone replacement.[65] Gilbert and colleagues found that even modest degrees of thyroid hormone insufficiency in rats during the early postnatal period permanently altered interneuron expression of the calcium binding protein parvalbumin, which may lead to the impairments in synaptic plasticity and cognition characteristic of thyroid hormone insufficiency during development.[66] Furthermore, Hadjab-Lallemend and colleagues, using a TRα1 mutant as a repressive "aporeceptor," demonstrated a critical developmental window during which maternal or early postnatal hypothyroidism would impair GABAergic interneuronal maturation, Cl^- and Ca^{2+} ion homeostasis, and hippocampal excitability. These cellular abnormalities during early development suggest a means by which thyroid hormone deficiency may encode a susceptibility to subsequent functional (and morphological) CNS deficits.[67]

A study by Sheryl Smith and colleagues provides a compelling and illuminating model of hormone-mediated affective regulation.[68] This study demonstrated that the increased stress-induced anxiety seen at puberty in female mice results from the coincident marked decline in levels of the progesterone neurosteroid metabolite allopregnanolone. Allopregnanolone withdrawal results in increased expression of a conformationally altered $GABA_A$ receptor in the hippocampus — a conformation that is inhibited rather than stimulated by allopregnanolone. Consequently, upon exposure to increased allopregnanolone levels during stress, the hippocampal GABA receptors are inhibited. As such, the signaling properties of allopregnanolone are dramatically transformed from hyperpolarizing effects on the pyramidal cells and anxiolysis to depolarizing, anxiogenic effects. Further, allopregnanolone withdrawal-related behavioral effects are revealed in a context-specific fashion — i.e., only under conditions of stress. This model therefore illustrates several key concepts. First, change in hormone level may be as critical a physiological determinant as absolute hormone level. Second, hormones can confer effects that alter the molecular context for subsequent signals, and, as a result of this altered context, the same signal may induce wholly different effects. Finally, as with the stress-dependent display of anxiety behavior, an additional environmental cue may be necessary to unmask the hormone-mediated cellular and molecular changes. Thus, concurrent stimuli from the external environment constitute an important source of variance underlying individual affective response to the same endocrine signal. If such extrinsic differences can alter affective response, the question next arises as to whether intrinsic differences exist that similarly predispose some individuals to mood destabilization.

Genetic context offers a ready explanation for marked differences in response to an endocrine signal across individuals. However, whereas once this appeared an unsubstantiated supposition, neuroendocrine research has established that variations in gene sequence can alter hormonal signaling in a variety of ways. First, changes in the primary structure of hormone receptors can alter the transcriptional effects of the activated receptor, which may then be manifest clinically as an altered sensitivity or susceptibility to hormonal affects. For example, single polymorphic exonic variants in the glucocorticoid receptor confer increased or reduced sensitivity to glucocorticoids.[69–71] Similarly, reductions in the polyglutamine repeat in the androgen receptor confer greater transcriptional efficacy and increased susceptibility to prostate cancer.[72] Second, changes in the sequence of a steroid hormone response element can yield differences in transcriptional activity, even in the absence of altered hormone receptor promoter-binding affinity.[73] Third, even the effects of single nucleotide polymorphisms depend in turn on the greater, intra-genic context; thus, the A645D mutation in the androgen receptor will enhance or reduce the transcriptional effects of testosterone as a function of the relative number of polyglycine and polyglutamine repeats.[74] Finally, sequence variants in genes encoding regulatory enzymes can comparably affect the outcome of hormone signaling. Variants in the thyroid hormone synthetic enzyme 2,3 deiodinase, for example, predict the antidepressant effects of thyroid hormone replacement.[75]

Independent of changes in genetic structure, intracellular signaling is highly context-dependent. The response to a hormonal stimulus is determined by the concentration and type of co-regulators present as well as the presence or absence of other hormone signals. This contextual variance is illustrated by the competing co-integration of the estrogen and glucocorticoid receptors; i.e., CBP/p300 regulation of AP-1 promoter sites is modulated in opposite directions by binding to an estrogen receptor (ER) versus a glucocorticoid receptor (GR).[73]

The most promising explanations, however, for differential sensitivity to the mood destabilizing effects of hormones reside less in genomic sequence variation than in alterations in genetic expression that derive from gene–environment interactions. In a remarkable series of studies, Meaney and colleagues[76] demonstrated that the prenatal licking and grooming behavior of rat dams determined both the stress phenotype (sensitive or resilient) and eventual maternal behavior of female offspring. The investigators further demonstrated that the developmental effects of maternal licking and grooming were the product of enduring, epigenetic modifications of the hippocampal glucocorticoid receptor and pre-optic area estrogen receptor genes. The transduction of environmental events into altered gene expression, via similar epigenetic mechanisms, may prove fundamental to the variance characterizing many long-recognized endocrine–behavioral relationships. Recent clinical data demonstrate that antecedent history of trauma in depressed women, rather than depression itself, predicts hyperactivity of the HPA axis;[77] epigenesis well may prove to underlie these findings, offering a fascinating link between environment, gene expression, and subsequent susceptibility to both affective and endocrine dysregulation.

In aggregate, these studies underscore the role of hormone signals in creating a context for neuronal activity and, ultimately, for overall brain function. An intriguing model for explaining the ability of hormones to regulate affective state is found in the conceptualization of the brain as dynamically assembling and reassembling regions in the service of performing neural computational tasks. These computations then give rise to the emergent states that we call mood or behavior. According to this model, alterations in the activation of a single brain region may bias toward or against the activation of that circuitry, or even generate the "construction" of a new functional neurocircuit.[78] Support for this intriguing notion is found in a recent demonstration that repetitive high-frequency burst spiking of a single rat cortical neuron can trigger a switch between two different global cortical states, i.e., the equivalents of slow-wave and rapid-eye-movement sleep.[79] In conceptual parallel, then, hormone-induced alterations in activity of specific brain regions alter functional networks and effect changes in global brain state.

Notably, all primary endocrinopathies manifest with clinical syndromes that comprise various signs and symptoms, each with its respective rate of prevalence. Interestingly, though none of these signs or symptoms occurs universally, the somatic manifestations are nonetheless readily accepted as a direct consequence of the endocrine dysfunction. Though not all patients with hyperthyroidism experience cardiac symptoms, for example, the variable presence of such symptoms is understood as a reflection of underlying individual differences in cardiovascular physiology. In contrast, this absence of symptom uniformity across patients constitutes the historical premise for questioning the capacity of hormones to regulate affect and behavior. This distinction between somatic and affective symptoms belies the persistent stigma attached to affective disorders and their study, a continued reluctance to recognize a biological basis of behavior with preferential ascription to deficient coping or other personal shortcoming. Long overdue is the recognition that, whether at a physiological level, a systems level or a cellular level, behavior constitutes a CNS readout as relevant an output of endocrine function as any other classical physiologic endpoint. As such, the term psychoneuroendocrinology may well be worth retiring.

References

1. Starling EH. The Croonian lectures on the chemical correlation of the functions of the body. *Lancet.* 1905;166(4275):339–341.
2. Cresswell R. *Aristotle's History of Animals: In Ten Books.* London: George Bell; 1891.
3. Gandelman R. Relative contributions of aggression and reproduction to behavioral endocrinology. *Aggress Behav.* 1984;10(2):123–133.
4. Forbes TR. A.A. Berthold and the first endocrine experiment: Some speculations as to its origin. *Bull Hist Med.* 1949;23: 263–267.
5. Borell M. Brown-Sequard's organotherapy and its appearance in America at the end of the nineteenth century. *Bull Hist Med.* 1976; 50(3):309–320.
6. Edwards HG. Notes on orchitic extract. *Br Med J.* 1895;1(1797): 1261–1262.
7. Easterbrook CC. Organo-therapeutics in mental diseases. *Br Med J.* 1900;2(2073):813–823.
8. Werner AA, Johns GA, Hoctor EF, Ault CC, Kohler LH, Weis MW. Involutional melancholia: Probable etiology and treatment. *J Am Med Assoc.* 1934;103(1):13–16.
9. Cushing H. The basophil adenomas of the pituitary gland and their clinical manifestations (pituitary basophilism). *Bull Johns Hopkins Med Sch.* 1932;50:137–195.
10. Sonino N, Fava GA. Psychiatric disorders associated with Cushing's syndrome. Epidemiology, pathophysiology and treatment. *CNS Drugs.* 2001;15(5):361–373.
11. Lindsay AE, Migeon CJ, Nugent CA, Brown H. The diagnostic value of plasma and urinary 17-hydroxycorticosteroid determinations in Cushing's syndrome. *Am J Med.* 1956;20(1):15–22.
12. Wallace EZ, Christy NP, Jailer JW. Clinical application for the simplified silber-porter method for determining plasma 17-hydroxycorticosteroids. *J Clin Endocrinol Metab.* 1955;15(9): 1073–1088.
13. Liddle GW. Tests of pituitary-adrenal suppressibility in the diagnosis of Cushing's syndrome. *J Clin Endocrinol Metab.* 1960; 20(12):1539–1560.
14. Pavlatos FC, Smilo RP, Forsham PH. A rapid screening test for cushing's syndrome. *J Am Med Assoc.* 1965;193(9):720–723.
15. Kathol RG. Psychiatric abnormalities in endocrine disorders. In: Nemeroff CB, ed. *Neuroendocrinology.* Boca Raton: CRC Press; 1992:396–411.
16. Board F, Wadeson R, Persky H. Depressive affect and endocrine functions; blood levels of adrenal cortex and thyroid hormones in patients suffering from depressive reactions. *AMA Arch Neurol Psychiatry.* 1957;78(6):612–620.
17. Bunney WEJ, Mason JW, Hamburg DA. Correlations between behavioral variables and urinary 17-hydroxycorticosteroids in depressed patients. *Psychosom Med.* 1965;27(4):299–308.
18. Sachar EJ, Hellman L, Fukushima DK, Gallagher TF. Cortisol production in depressive illness. A clinical and biochemical clarification. *Arch Gen Psychiatry.* 1970;23(4):289–298.
19. Sachar EJ, Hellman L, Roffwarg HP, Halpern FS, Fukushima DK, Gallagher TF. Disrupted 24-hour patterns of cortisol secretion in psychotic depression. *Arch Gen Psychiatry.* 1973;28(1):19–24.
20. Carroll BJ, Curtis GC, Mendels J. Neuroendocrine regulation in depression: II. Discrimination of depressed from nondepressed patients. *Arch Gen Psychiatry.* 1976;33(9):1051–1058.
21. Carroll BJ, Martin FIR, Davies B. Resistance to suppression by dexamethasone of plasma 11-O.H.C.S. levels in severe depressive illness. *Br Med J.* 1968;3(5613):285–287.
22. Stokes PE, Pick GR, Stoll PM, Nunn WD. Pituitary-adrenal function in depressed patients: Resistance to dexamethasone suppression. *J Psychiatric Res.* 1975;12(4):271–281.
23. Gold PW, Loriaux DL, Roy A, Kling MA, Calabrese JR, Kellner CH, et al. Responses to corticotropin-releasing hormone in the hypercortisolism of depression and Cushing's disease. Pathophysiologic and diagnostic implications. *N Engl J Med.* 1986;314(21):1329–1335.
24. Holsboer F, von Bardeleben U, Wiedemann K, Müller OA, Stalla GK. Serial assessment of corticotropin-releasing hormone response after dexamethasone in depression Implications for pathophysiology of DST nonsuppression. *Biol Psychiatry.* 1987; 22(2):228–234.
25. Nemeroff CB, Widerlov E, Bissette G, Walleus H, Karlsson I, Eklund K, et al. Elevated concentrations of CSF corticotropin-releasing factor-like immunoreactivity in depressed patients. *Science.* 1984;226(4680):1342–1344.
26. Asnis GM, Sachar EJ, Halbreich U, Nathan RS, Ostrow L, Halpern FS. Cortisol secretion and dexamethasone response in depression. *Am J Psychiatry.* 1981;138(9):1218–1221.
27. Rubin RT, Poland RE, Lesser IM, Winston RA, Blodgett AL. Neuroendocrine aspects of primary endogenous depression. I. Cortisol secretory dynamics in patients and matched controls. *Arch Gen Psychiatry.* 1987;44(4):328–336.
28. Ising M, Horstmann S, Kloiber S, Lucae S, Binder EB, Kern N, et al. Combined dexamethasone/corticotropin releasing hormone test predicts treatment response in major depression-a potential biomarker? *Biol Psychiatry.* 2007;62(1):47–54.
29. Ribeiro SC, Tandon R, Grunhaus L, Greden JF. The DST as a predictor of outcome in depression: a meta-analysis. *Am J Psychiatry.* 1993;150(11):1618–1629.
30. Zobel AW, Nickel T, Sonntag A, Uhr M, Holsboer F, Ising M. Cortisol response in the combined dexamethasone/CRH test as predictor of relapse in patients with remitted depression: a prospective study. *J Psychiatric Res.* 2001;35(2):83–94.
31. Gold MS, Pottash AL, Ryan N, Sweeney DR, Davies RK, Martin DM. TRH-induced TSH response in unipolar, bipolar, and secondary depressions: possible utility in clinical assessment and differential diagnosis. *Psychoneuroendocrinology.* 1980;5(2):147–155.
32. Loosen PT, Prange Jr AJ. Serum thyrotropin response to thyrotropin-releasing hormone in psychiatric patients: a review. *Am J Psychiatry.* 1982;139(4):405–416.
33. Schmidt PJ, Rosenfeld D, Muller KL, Grover GN, Rubinow DR. A case of autoimmune thyroiditis presenting as menstrual related mood disorder. *J Clin Psychiatry.* 1990;51(10):434–436.
34. Schmidt PJ, Nieman L, Danaceau MA, Tobin MB, Roca CA, Murphy JH, et al. Estrogen replacement in perimenopause-related depression: a preliminary report. *Am J Obstet Gynecol.* 2000;183(2):414–420.
35. Soares CN, Poitras JR, Prouty J. Hormone treatment for mood disorders in women. *Expert Rev Neurother.* 2002;2(1):25–34.
36. Zweifel JE, O'Brien WH. A meta-analysis of the effect of hormone replacement therapy upon depressed mood. *Psychoneuroendocrinology.* 1997;22(3):189–212.
37. Rubinow DR, Hoban MC, Grover GN, Galloway DS, Roy-Byrne P, Andersen R, et al. Changes in plasma hormones across the menstrual cycle in patients with menstrually related mood disorder and in control subjects. *Am J Obstet Gynecol.* 1988; 158(1):5–11.
38. Schmidt PJ, Nieman LK, Grover GN, Muller KL, Merriam GR, Rubinow DR. Lack of effect of induced menses on symptoms in women with premenstrual syndrome. *N Engl J Med.* 1991;324(17): 1174–1179.
39. Schmidt PJ, Nieman LK, Danaceau MA, Adams LF, Rubinow DR. Differential behavioral effects of gonadal steroids in women with

and in those without premenstrual syndrome. *N Engl J Med.* 1998;338(4):209–216.
40. Bloch M, Schmidt PJ, Danaceau M, Murphy J, Nieman L, Rubinow DR. Effects of gonadal steroids in women with a history of postpartum depression. *Am J Psychiatry.* 2000;157(6): 924–930.
41. Harris B. Biological and hormonal aspects of postpartum depressed mood. *Br J Psychiatry.* 1994;164(3):288–292.
42. Rocha BA, Fleischer R, Schaeffer JM, Rohrer SP, Hickey GJ. 17 Beta-estradiol-induced antidepressant-like effect in the forced swim test is absent in estrogen receptor-beta knockout (BERKO) mice. *Psychopharmacology (Berl).* 2005;179(3):637–643.
43. Walf AA, Rhodes ME, Frye CA. Antidepressant effects of ERbeta-selective estrogen receptor modulators in the forced swim test. *Pharmacol Biochem Behav.* 2004;78(3):523–529.
44. Fink G, Sumner BE, Rosie R, Grace O, Quinn JP. Estrogen control of central neurotransmission: effect on mood, mental state, and memory. *Cell Mol Neurobiol.* 1996;16(3):325–344.
45. McEwen BS, Alves SE. Estrogen actions in the central nervous system. *Endocr Rev.* 1999;20(3):279–307.
46. Rubinow DR, Schmidt PJ, Roca CA. Estrogen–serotonin interactions: implications for affective regulation. *Biol Psychiatry.* 1998; 44(9):839–850.
47. Burgess LH, Handa RJ. Chronic estrogen-induced alterations in adrenocorticotropin and corticosterone secretion, and glucocorticoid receptor-mediated functions in female rats. *Endocrinology.* 1992;131(3):1261–1269.
48. Kerdelhue B, Jones GS, Gordon K, Seltman H, Lenoir V, Melik Parsadaniantz S, et al. Activation of the hypothalamo-anterior pituitary corticotropin-releasing hormone, adrenocorticotropin hormone and beta-endorphin systems during the estradiol 17 beta-induced plasma LH surge in the ovariectomized monkey. *J Neurosci Res.* 1995;42(2):228–235.
49. Pfeiffer A, Barden N. Glucocorticoid receptor gene expression in rat pituitary gland intermediate lobe following ovariectomy. *Mol Cell Endocrinol.* 1988;55(2-3):115–120.
50. Weiser MJ, Handa RJ. Estrogen impairs glucocorticoid dependent negative feedback on the hypothalamic-pituitary-adrenal axis via estrogen receptor alpha within the hypothalamus. *Neuroscience.* 2009;159(2):883–895.
51. Duman RS. Neuronal damage and protection in the pathophysiology and treatment of psychiatric illness: stress and depression. *Dialogues Clin Neurosci.* 2009;11(3):239–255.
52. Franklin TB, Perrot-Sinal TS. Sex and ovarian steroids modulate brain-derived neurotrophic factor (BDNF) protein levels in rat hippocampus under stressful and non-stressful conditions. *Psychoneuroendocrinology.* 2006;31(1):38–48.
53. Gibbs RB. Levels of trkA and BDNF mRNA, but not NGF mRNA, fluctuate across the estrous cycle and increase in response to acute hormone replacement. *Brain Res.* 1998;787(2):259–268.
54. Allen AL, McCarson KE. Estrogen increases nociception-evoked brain-derived neurotrophic factor gene expression in the female rat. *Neuroendocrinology.* 2005;81(3):193–199.
55. Cardona-Gomez P, Perez M, Avila J, Garcia-Segura LM, Wandosell F. Estradiol inhibits GSK3 and regulates interaction of estrogen receptors, GSK3, and beta-catenin in the hippocampus. *Mol Cell Neurosci.* 2004;25(3):363–373.
56. Carlstrom L, Ke ZJ, Unnerstall JR, Cohen RS, Pandey SC. Estrogen modulation of the cyclic AMP response element-binding protein pathway. Effects of long-term and acute treatments. *Neuroendocrinology.* 2001;74(4):227–243.
57. Green PS, Perez EJ, Calloway T, Simpkins JW. Estradiol attenuation of β-amyloid-induced toxicity: a comparison of MTT and calcein AM assays. *J Neurocytol.* 2000;29(5-6): 419–423.
58. Wise P. Estradiol exerts neuroprotective actions against ischemic brain injury: insights derived from animal models. *Endocrine.* 2003;21(1):11–15.
59. Straub RH. The complex role of estrogens in inflammation. *Endocr Rev.* 2007;28(5):521–574.
60. Dreher JC, Schmidt PJ, Kohn P, Furman D, Rubinow D, Berman KF. Menstrual cycle phase modulates reward-related neural function in women. *Proc Natl Acad Sci USA.* 2007;104(7):2465–2470.
61. Smoski MJ, Felder J, Bizzell J, Green SR, Ernst M, Lynch TR, et al. fMRI of alterations in reward selection, anticipation, and feedback in major depressive disorder. *J Affect Disord.* 2009;118(1-3):69–78.
62. Berman KF, Schmidt PJ, Rubinow DR, Danaceau MA, Van Horn JD, Esposito G, et al. Modulation of cognition-specific cortical activity by gonadal steroids: a positron-emission tomography study in women. *Proc Natl Acad Sci USA.* 1997;94(16):8836–8841.
63. Protopopescu X, Pan H, Altemus M, Tuescher O, Polanecsky M, McEwen B, et al. Orbitofrontal cortex activity related to emotional processing changes across the menstrual cycle. *Proc Natl Acad Sci USA.* 2005;102(44):16060–16065.
64. Protopopescu X, Tuescher O, Pan H, Epstein J, Root J, Chang L, et al. Toward a functional neuroanatomy of premenstrual dysphoric disorder. *J Affect Disord.* 2008;108(1–2):87–94.
65. Bauer M, Silverman DH, Schlagenhauf F, London ED, Geist CL, van Herle K, et al. Brain glucose metabolism in hypothyroidism: a positron emission tomography study before and after thyroid hormone replacement therapy. *J Clin Endocrinol Metab.* 2009;94(8): 2922–2929.
66. Gilbert ME, Sui L, Walker MJ, Anderson W, et al. Thyroid hormone insufficiency during brain development reduces parvalbumin immunoreactivity and inhibitory function in the hippocampus. *Endocrinology.* 2007;148:92–102.
67. Hadjab-Lallemend S, Wallis K, van Hogerlinden M, Dudazy S, Nordstrom K, et al. A mutant thyroid hormone receptor α1 alters hippocampal circuitry and reduces seizure susceptibility in mice. *Neuropharmacology.* 2010;58:1130–1139.
68. Shen H, Gong QH, Aoki C, Yuan M, Ruderman Y, Dattilo M, et al. Reversal of neurosteroid effects at alpha4beta2delta GABAA receptors triggers anxiety at puberty. *Nat Neurosci.* 2007;10(4): 469–477.
69. Huizenga NA, Koper JW, De Lange P, Pols HA, Stolk RP, Burger H, et al. A polymorphism in the glucocorticoid receptor gene may be associated with an increased sensitivity to glucocorticoids *in vivo. J Clin Endocrinol Metab.* 1998;83(1):144–151.
70. Stevens A, Ray DW, Zeggini E, John S, Richards HL, Griffiths CE, et al. Glucocorticoid sensitivity is determined by a specific glucocorticoid receptor haplotype. *J Clin Endocrinol Metab.* 2004; 89(2):892–897.
71. van Rossum EF, Koper JW, Huizenga NA, Uitterlinden AG, Janssen JA, Brinkmann AO, et al. A polymorphism in the glucocorticoid receptor gene, which decreases sensitivity to glucocorticoids *in vivo*, is associated with low insulin and cholesterol levels. *Diabetes.* 2002;51(10):3128–3134.
72. Giovannucci E, Stampfer MJ, Krithivas K, Brown M, Dahl D, Brufsky A, et al. The CAG repeat within the androgen receptor gene and its relationship to prostate cancer. *Proc Natl Acad Sci USA.* 1997;94(7):3320–3323.
73. Uht RM, Anderson CM, Webb P, Kushner PJ. Transcriptional activities of estrogen and glucocorticoid receptors are functionally integrated at the AP-1 response element. *Endocrinology.* 1997; 138(7):2900–2908.
74. Werner R, Holterhus PM, Binder G, Schwarz HP, Morlot M, Struve D, et al. The A645D mutation in the hinge region of the human androgen receptor (AR) gene modulates AR activity, depending on the context of the polymorphic glutamine and glycine repeats. *J Clin Endocrinol Metab.* 2006;91(9):3515–3520.

75. Panicker V, Saravanan P, Vaidya B, Evans J, Hattersley AT, Frayling TM, et al. Common variation in the DIO2 gene predicts baseline psychological well-being and response to combination thyroxine plus triiodothyronine therapy in hypothyroid patients. *J Clin Endocrinol Metab.* 2009;94(5):1623–1629.
76. Meaney MJ, Szyf M, Seckl JR. Epigenetic mechanisms of perinatal programming of hypothalamic-pituitary-adrenal function and health. *Trends Mol Med.* 2007;13(7):269–277.
77. Heim C, Newport DJ, Heit S, Graham YP, Wilcox M, Bonsall R, et al. Pituitary-adrenal and autonomic responses to stress in women after sexual and physical abuse in childhood. *J Am Med Assoc.* 2000;284(5):592–597.
78. Pessoa L. On the relationship between emotion and cognition. *Nat Rev Neurosci.* 2008;9(2):148–158.
79. Li C-YT, Poo M-m, Dan Y. Burst spiking of a single cortical neuron modifies global brain state. *Science.* 2009;324(5927):643–646.

Subject Index

D

E

R

S

T

Author Index